KT-503-857

Biology

sixth edition

Sylvia S. Mader

WCB McGraw-Hill

Boston, Massachusetts Burr Ridge, Illinois Dubuque, Iowa
Madison, Wisconsin New York, New York San Francisco, California St. Louis, Missouri

WCB/McGraw-Hill

A Division of The ***McGraw-Hill*** *Companies*

TITLE: BIOLOGY, 6/E

Copyright © 1998 by The McGraw-Hill Companies, Inc. All rights reserved. Previous editions © 1985, 1987, 1990, 1993, 1996 by Wm. C. Brown Communications, Inc. Printed in the United States of America. Except as permitted under the United States Copyright Act of 1976, no part of this publication may be reproduced or distributed in any form or by any means, or stored in a data base or retrieval system, without the prior written permission of the publisher.

This book is printed on recycled, acid free paper containing 10% postconsumer waste.

2 3 4 5 7 8 9 0 QPD/VNH 9 0 9 8

ISBN 0-697-34080-5

Vice president and editorial director: Kevin T. Kane
Publisher: Michael D. Lange
Sponsoring editor: Patrick Reidy
Developmental editor: Connie Balius-Haakinson
Marketing manager: Julie Joyce Keck
Project manager: Margaret B. Horn
Production supervisor: Sandy Ludovissy
Designer: K. Wayne Harms
Photo research coordinator: Lori Hancock
Art editor: Brenda A. Ernzen
Compositor: Kathleen F. Theis
Typeface: 10/12 Palatino
Printer: Von Hoffmann Press, Inc.
Interior and cover design: Christopher Reese
Cover image: © Tim Davis/Tony Stone Images

Library of Congress Cataloging-in-Publication Data

Mader, Sylvia S.
Biology / Sylvia S. Mader — 6th ed.
p. cm.
Includes bibliographical references and index.
ISBN 0-697-34079-1 (case). — ISBN 0-697-34080-5 (pbk.)
1. Biology. I. Title.
QH308.2.M23 1998
570—dc21 97-29147
CIP

INTERNATIONAL EDITION
Copyright © 1998. Exclusive rights by The McGraw-Hill Companies, Inc. for manufacture and export. This book cannot be re-exported from the country to which it is consigned by McGraw-Hill. The International Edition is not available in North America.

When ordering this title, use ISBN 0-07-115402-7

http://www.mhhe.com

Brief Contents

Table of Contents

* These helpful aids appear in every chapter.

part V
Diversity of Life 490

chapter 43 Digestion and Nutrition 773

chapter 44 Respiration 789

chapter 45 Osmotic Regulation and Excretion 803

chapter 46 Neurons and Nervous Systems 817

chapter 47 Sense Organs 839

chapter 48 Support Systems and Locomotion 855

Preface

Biology is an introductory text that covers the concepts and principles of biology from the structure and function of the cell to the organization of the biosphere. It draws upon the entire world of living things to bring out an evolutionary theme that is introduced from the start.

Scientific Process

Biology emphasizes the scientific process. Notable contributions to biology, including significant experiments, are discussed throughout the text. Chapter 1 explains the scientific method and also illustrates this method by walking students through experiments in the current literature. The text has numerous *Doing Science* readings, many written by contemporary biologists who tell how they go about doing their research and how their findings can be applied to human beings.

Every chapter of *Biology* has been revised and thoroughly updated; however, this edition has not grown significantly in page length. Every chapter has been skillfully revised and rewritten. All illustrations have been carefully correlated to the textual material to ensure that each illustration is on the same or facing page as its textual reference. This direct correlation aids in learning and studying.

Concepts

In this edition the major topics are numbered, and the concepts, which have replaced learning objectives, are grouped according to these topics. This numbering system is used in the textual material, and in the summary, which reviews the concepts according to each major topic. This system allows instructors to assign just certain portions of the chapter and it also allows students to study the chapter in terms of the concepts presented.

New Chapters

Chapter six, "Metabolism: Energy and Enzymes," is new to this edition. This chapter, which explains the laws of thermodynamics, how the cell makes use of ATP, and how enzymes function, lays a foundation for the revised photosynthesis and cellular respiration chapters that follow.

The science of ecology is undergoing fundamental changes, and the ecology chapters have been rewritten to have a modern approach. Four chapters (23–26) are devoted to reviewing the ecological principles that explain how the natural world works. These chapters make frequent reference to how humans impact the environment, and chapter 27 is concerned with this topic alone. Ecological problems of our day are considered.

Ecology is further emphasized in this edition because *Ecology Focus* readings appear in each part, and the part entitled "Behavior and Ecology" has been moved to follow Part III, "Evolution." This sequence is logical because evolution and ecology are intertwined.

Like ecology, systematics is now undergoing changes that promise to revolutionize the science. Cladistics is now challenging the traditional school and has been well received by many. Therefore, students need to be exposed to how cladists go about determining evolutionary relationships. More and more reliance is being placed on molecular data, which suggests there are three domains of life: bacteria, archaea, and eukarya. All of these topics are considered in Part V, including a thorough examination of each kingdom.

New Pedagogy

Consistent with accepted educational methodology, this edition of *Biology* features an introduction and a closing that emphasize the concepts of the chapter. The introduction, which appears on the chapter opening page, sparks student interest and highlights the overall themes for the chapter. *Connecting Concepts,* which closes the text portion of the chapter, stimulates critical thinking and shows how the concepts in the chapter are related to the concepts in other chapters.

Readings

The readings have now been organized into four types. The *Doing Science* readings are often written by contemporary scientists who tell us about their particular type of research and how they became interested in this field of biology. Some of the *Doing Science* readings feature the work of minority researchers such as Barbara McClintock's work regarding "jumping genes" and Susumu Tonegawa's work in antibody diversity. The *Health Focus* readings give practical information concerning some particular topic of interest, such as proper nutrition and how to prevent cancer. The *Ecology Focus* readings draw attention to some particular environmental problem such as the need to preserve tropical rain forests and the relationship between ozone holes and skin cancer. *A Closer Look* readings are designed to expand, in an interesting way, on the core information presented in each chapter.

New Appearance

There are many new illustrations in *Biology*, but special attention was given to Part I in which all art pieces are color consistent with new animal and plant cells in chapter four, "Cell Structure and Function." Location icons are now a part of all organelle illustrations

The appearance of this edition is completely new and improved. Color has been used more effectively, and all text art is now screened in one or two colors. And, the summary statements no longer have a color screen. The end result is a book whose appearance will be pleasing to all.

Technology

Many technology aids are available for use with *Biology*, and each chapter now has its own listing of these. For the student, the Mader Home Page offers exercises to aid learning and resources that expand on the text's content and applications. *Explorations in Human Biology* and *Explorations in Cell Biology and Genetics* are interactive CD-ROMs that bring biology to life. *The Life Science Animations* videotapes include fifty-three additional topics that can be studied in a visually appealing way. *The Dynamic Human* CD-ROM offers three-dimensional visuals that facilitate an understanding of human anatomy and physiology. Other aids are also available, and all of these are listed on the technology page (see page xviii of the preface).

For the Instructor, the *Extended Lecture Outline* makes the contents of the book available in a way that facilitates lecture preparation. The outline is available on the Mader Home Page and on disk by request. *The Visual Resourse Library* on CD-ROM makes the text illustrations available for classroom use. The images and their labels can even be manipulated. To help with the mechanics of teaching there is a computerized version of the *Test Item File* available in Windows and Macintosh formats.

Aids to the Reader

Biology includes a number of aids that have helped students study biology successfully and enjoyably.

▶ New to This Edition

▶ Part IV: Ecology was thoroughly updated for this edition and all chapters have been completely rewritten. *Ecology Focus* readings occur throughout the text.

A new chapter (6) entitled "Metabolism: Energy and Enzymes," the first of three energetics chapters, was rewritten, and the other two chapters (7 and 8) have been revised.

Systematics received special attention and the three domains of life are discussed. Cladistics has been revised and made clearer for the student.

▶ New art appears throughout the text, and in Part I the colors used are consistent with new plant and animal cells. Color is more effectively used throughout the text, giving the book a completely new appearance.

▶ Four types of readings are featured:

Doing Science
Ecology Focus
Health Focus
A Closer Look

▶ The chapter opening page has an integrated outline of the major topics and chapter concepts. The major topics are numbered in the chapter outlines, in the text, and in the Summary. The introduction, which sparks student interest and highlights certain themes, is on the chapter opening page. Each chapter ends with *Connecting Concepts*, which emphasizes how the concepts of the chapter are related to concepts in other chapters.

▶ Technology Aids are correlated to the text. *Explorations in Human Biology* and *Explorations in Cell Biology and Genetics* CD-ROMs offer exciting new ways to understand biological concepts. *The Dynamic Human* on CD-ROM is an interactive three-dimensional visual guide to human anatomy and physiology.

▶ Explore the Mader Home Page for even more information:

http://www.mhhe.com/sciencemath/biology/mader/

History of Biology End Sheets

The inside cover lists major contributions to the field of biology in a concise, chronological manner. Students may refer to these whenever it is appropriate.

Part Introduction

An introduction for each part highlights the central ideas of that part and specifically tells the student how the topics within each part contribute to biological knowledge.

Chapter Concepts

The chapter begins with an integrated outline that numbers the major topics of the chapter and lists the concepts for each topic.

Chapter Introductions

Each chapter has an introduction on the chapter opening page that sparks student interest in the themes for the chapter.

Internal Summary Statements

Internal summaries stress the chapter's key concepts. These appear at the ends of major sections and help students focus their study efforts on the basics.

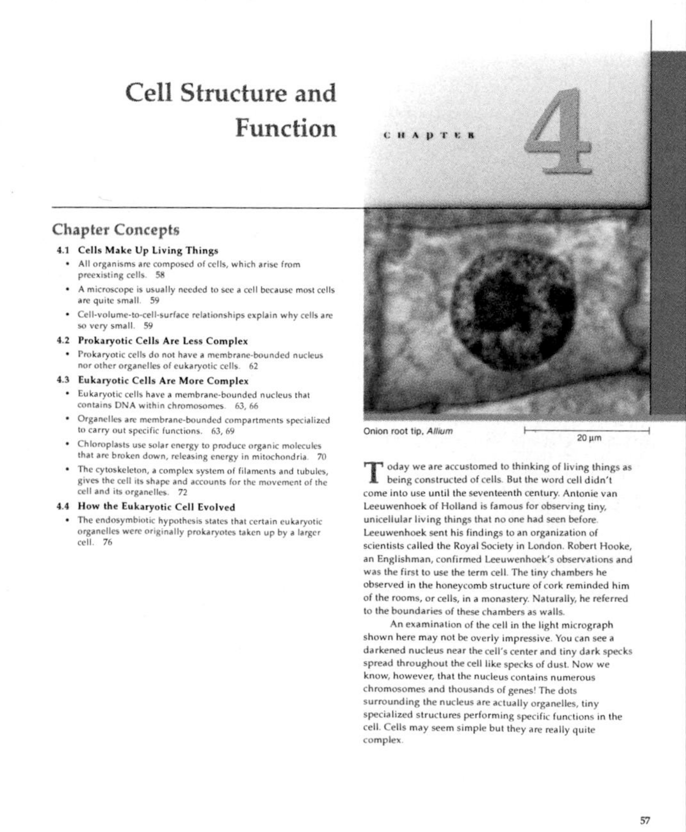

Cell Structure and Function

CHAPTER 4

Chapter Concepts

4.1 **Cells Make Up Living Things**
- All organisms are composed of cells, which arise from preexisting cells. 58
- A microscope is usually needed to see a cell because most cells are quite small. 59
- Cell-volume-to-cell-surface relationships explain why cells are so very small. 59

4.2 **Prokaryotic Cells Are Less Complex**
- Prokaryotic cells do not have a membrane-bounded nucleus nor other organelles of eukaryotic cells. 62

4.3 **Eukaryotic Cells Are More Complex**
- Eukaryotic cells have a membrane-bounded nucleus that contains DNA within chromosomes. 63, 66
- Organelles are membrane-bounded compartments specialized to carry out specific functions. 63, 69
- Chloroplasts use solar energy to produce organic molecules that are broken down, releasing energy in mitochondria. 70
- The cytoskeleton, a complex system of filaments and tubules, gives the cell its shape and accounts for the movement of the cell and its organelles. 72

4.4 **How the Eukaryotic Cell Evolved**
- The endosymbiotic hypothesis states that certain eukaryotic organelles were originally prokaryotes taken up by a larger cell. 76

Onion root tip, *Allium* 20 μm

Today we are accustomed to thinking of living things as being constructed of cells. But the word cell didn't come into use until the seventeenth century. Antonie van Leeuwenhoek of Holland is famous for observing tiny, unicellular living things that no one had seen before. Leeuwenhoek sent his findings to an organization of scientists called the Royal Society in London. Robert Hooke, an Englishman, confirmed Leeuwenhoek's observations and was the first to use the term cell. The tiny chambers he observed in the honeycomb structure of cork reminded him of the rooms, or cells, in a monastery. Naturally, he referred to the boundaries of these chambers as walls.

An examination of the cell in the light micrograph shown here may not be overly impressive. You can see a darkened nucleus near the cell's center and tiny dark specks spread throughout the cell like specks of dust. Now we know, however, that the nucleus contains numerous chromosomes and thousands of genes! The dots surrounding the nucleus are actually organelles, tiny specialized structures performing specific functions in the cell. Cells may seem simple but they are really quite complex.

57

Illustrations and Tables

The illustrations and tables in *Biology* are consistent with multicultural educational goals. Often it is easier to understand a given process by studying a drawing, especially when it is carefully coordinated with the text. Every illustration appears on the same or facing page to its reference.

Readings

Four types of readings are included in the text. *Doing Science* readings invite the reader to share in the excitement of past and current research projects. *Ecology Focus* readings draw attention to some environmental problem. *Health Focus* readings review measures to keep healthy. *A Closer Look* reading is designed to expand in an interesting way on the core information presented in the chapter.

Connecting Concepts

These appear at the close of the text portion of the chapter, and they stimulate critical thinking by showing how the concepts of the chapter are related to others in the text.

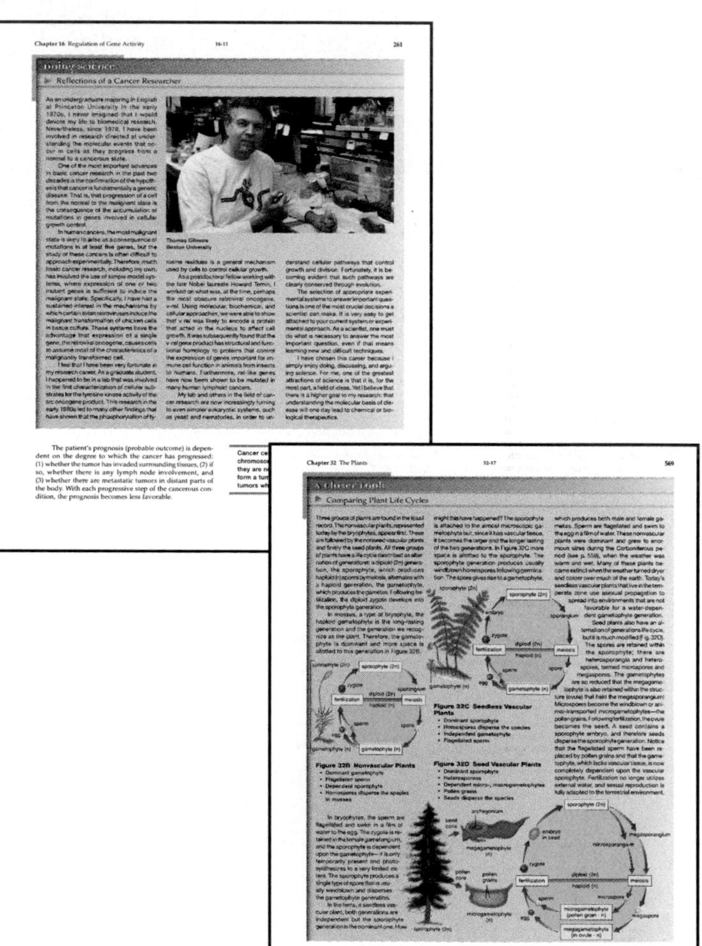

Chapter 16 Regulation of Gene Activity 16-11 261

Doing Science

Reflections of a Cancer Researcher

Chapter 32 The Plants 32-17 569

A Closer Look

Comparing Plant Life Cycles

End of Chapter Pedagogy

The numbered major topics are repeated in the *Summary*, which reviews the concepts for each topic. *Reviewing the Chapter* are a series of study questions that follow the sequence of the chapter. *Testing Yourself* consists of objective questions that allow students to test their ability to answer recall-based questions. Answers to *Testing Yourself* questions are given in Appendix A. *Applying the Concepts* are critical thinking questions based on biological concepts. *Using Technology* lists the available technology, including the Mader Home Page address, for the chapter. *Understanding the Terms* provides a page reference for boldfaced terms in the chapter. A matching exercise allows students to test their knowledge of the terms.

Further Readings

The list of readings at the end of each part suggests references that can be used for further study of the topics covered in the chapters of that part. The references listed in this section were carefully chosen for readability and accessibility. References are followed by a short description and an indication of their level of rigor.

Appendices and Glossary

The appendices contain optional information. *Appendix A* is the answer key to the objective questions found at the end of each chapter; *Appendix B* is an expanded table of chemical elements; *Appendix C* explains the metric system; *Appendix D* is a classification of organisms; and *Appendix E* is a listing of acronyms used in the text along with the complete term. The glossary defines the boldface terms in the text. These terms are the ones most necessary for the successful study of biology. Terms that are difficult to pronounce have a phonetic spelling, and the Greek and Latin derivation is given for selected terms.

402 23-20 Part IV Behavior and Ecology

connecting concepts

Modern ecology began with descriptive studies by nineteenth century naturalists. In fact, an early definition of the field was "scientific natural history." However, modern ecology [illegible] more than a simple descriptive field. Ecology is now very much an experimental, predictive science.

Much of the success in the development of ecology as a predictive science has come from studies of populations and the development of models that examine how populations change over time. The simplest models are based on population growth when there are unlimited resources. This results in exponential growth, a type of population growth that is only rarely seen in nature. Pest species may exhibit exponential growth until they run out of resources. Because so few natural populations exhibit exponential growth, population ecologists realized they must incorporate resource limitation in their models. The simplest models that account for limited resources result in sigmoidal or logistic growth. Populations which exhibit logistic growth will cease growth when they reach the environmental carrying capacity.

Many modern ecological studies are concerned with identifying the factors that place limits on population growth and that set the environmental carrying capacity. A combination of careful descriptive studies, experiments done in nature, and sophisticated models has allowed ecologists to make good predictions about which factors have the greatest influence on population growth. The example of the winter moth is a good case in point.

[illegible] modern ecology has been to try to understand how populations of different species affect each other. This is known as community ecology. Because each population in a community responds to environmental changes in slightly different ways, developing predictive models that explain how communities change has been challenging. However, ecologists are beginning to be able to predict how communities will change through time and to understand what factors influence community properties such as species number, abundance of individuals, and species interactions.

Summary

23.1 Scope of Ecology

Ecology is the study of the interactions of organisms with other organisms and with the physical environment. Ecology encompasses several levels of study: organism, population, community, ecosystem, and finally the biosphere. Ecologists are particularly interested in how interactions affect the distribution and abundance of organisms.

Population density is simply the number of individuals per unit area or volume. Distribution of these individuals can be uniform, random, or clumped. A population's distribution is often determined by limiting factors; that is, abiotic factors like water, temperature, and availability of nutrients.

23.2 Characteristics of Populations

Population size is dependent upon natality (number of births), mortality (number of deaths), immigration, and emigration. The number of births minus the number of deaths results in the net reproductive rate (symbolized as *r*) per capita per unit time.

One model for population growth assumes that the environment offers unlimited resources. In the example given, the members of the population have discrete reproductive events, and therefore the size of next year's population is given by the equation: $N_{t+1} = rN_t$. Under these conditions, exponential growth results in a J-shaped curve.

Most environments restrict growth, and exponential growth cannot continue indefinitely. Under these circumstances an S-shaped or logistic growth curve results. The growth of the population is given by the equation $dN/dt = rN\,(K-N)/K$ for populations in which individuals have repeated reproductive events. The term $(K-N)/K$ represents the unused portion of the carrying capacity (K). When the population reaches carrying capacity, the population stops growing because environmental resistance opposes biotic potential, the maximum net reproductive rate for a population.

Individuals are at different stages of their life span in a population. Mortality (deaths per capita) within a population can be recorded in a life table and illustrated by a survivorship curve. The pattern of population growth is reflected in the age distribution of a population, which consists of prereproductive, reproductive, and postreproductive segments. Populations that are growing exponentially have a pyramid-shaped age distribution pattern.

23.3 Regulation of Population Size

Population growth is limited by density-independent (e.g., weather) and density-dependent factors (predation, competition, and resource availability). Do some populations have an intrinsic means of regulating population growth as opposed to density-independent and density-dependent factors, which are extrinsic means? Territoriality is given as an example of a possible intrinsic means of regulation.

23.4 Life History Patterns

The logistic growth model has been used to suggest that the environment promotes either *r*-selection or *K*-selection. So-called *r*-selection occurs in unpredictable environments where density-independent factors affect population size. Energy is allocated to producing as many small offspring as possible. Adults remain small and do not invest in parental care of offspring. *K*-selection occurs in environments that remain relatively stable, where density-dependent factors affect population size. Energy is allocated to survival and repeated reproductive events. The adults are large and invest in parental care of offspring. Actual life histories contain trade-offs between these two patterns.

23.5 Human Population Growth

The human population is expanding exponentially, and it is unknown when the population size will level off. Most of the expected increase will occur in certain LDCs (less-developed countries) of Africa, Asia, and Latin America. Support for family planning, human development, and delayed child bearing could help prevent an expected increase.

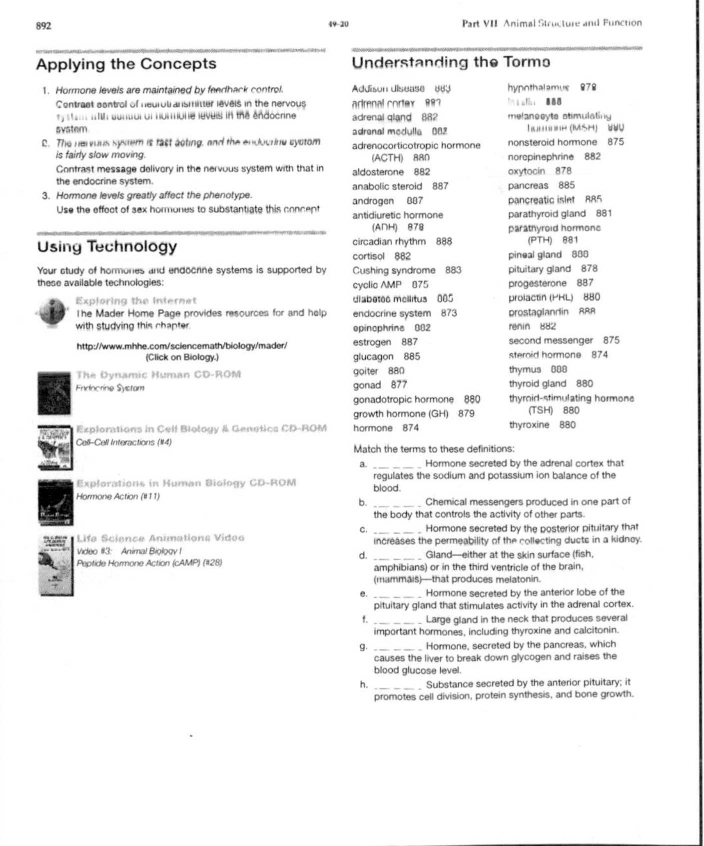

892 49-20 Part VII Animal Structure and Function

Applying the Concepts

1. *Hormone levels are maintained by feedback control.*
   Contrast control of neurotransmitter levels in the nervous system with control of hormone levels in the endocrine system.
2. *The nervous system is fast acting, and the endocrine system is fairly slow moving.*
   Contrast message delivery in the nervous system with that in the endocrine system.
3. *Hormone levels greatly affect the phenotype.*
   Use the effect of sex hormones to substantiate this concept.

Using Technology

Your study of hormones and endocrine systems is supported by these available technologies:

Exploring the Internet
The Mader Home Page provides resources for and help with studying this chapter.

http://www.mhhe.com/sciencemath/biology/mader/
(Click on Biology.)

The Dynamic Human CD-ROM
Endocrine System

Explorations in Cell Biology & Genetics CD-ROM
Cell–Cell Interactions (#4)

Explorations in Human Biology CD-ROM
Hormone Action (#11)

Life Science Animations Video
Video #3: Animal Biology I
Peptide Hormone Action (cAMP) (#28)

Understanding the Terms

Addison disease 883
adrenal cortex 882
adrenal gland 882
adrenal medulla 882
adrenocorticotropic hormone (ACTH) 880
aldosterone 882
anabolic steroid 887
androgen 887
antidiuretic hormone (ADH) 878
circadian rhythm 888
cortisol 882
Cushing syndrome 883
cyclic AMP 875
diabetes mellitus 885
endocrine system 873
epinephrine 882
estrogen 887
glucagon 885
goiter 880
gonad 877
gonadotropic hormone 880
growth hormone (GH) 879
hormone 874
hypothalamus 878
insulin 885
melanocyte-stimulating hormone (MSH) 880
nonsteroid hormone 875
norepinephrine 882
oxytocin 878
pancreas 885
pancreatic islet 885
parathyroid gland 881
parathyroid hormone (PTH) 881
pineal gland 888
pituitary gland 878
progesterone 887
prolactin (PRL) 880
prostaglandin 888
renin 882
second messenger 875
steroid hormone 874
thymus 888
thyroid gland 880
thyroid-stimulating hormone (TSH) 880
thyroxine 880

Match the terms to these definitions:

a. ________ Hormone secreted by the adrenal cortex that regulates the sodium and potassium ion balance of the blood.
b. ________ Chemical messengers produced in one part of the body that controls the activity of other parts.
c. ________ Hormone secreted by the posterior pituitary that increases the permeability of the collecting ducts in a kidney.
d. ________ Gland—either at the skin surface (fish, amphibians) or in the third ventricle of the brain, (mammals)—that produces melatonin.
e. ________ Hormone secreted by the anterior lobe of the pituitary gland that stimulates activity in the adrenal cortex.
f. ________ Large gland in the neck that produces several important hormones, including thyroxine and calcitonin.
g. ________ Hormone, secreted by the pancreas, which causes the liver to break down glycogen and raises the blood glucose level.
h. ________ Substance secreted by the anterior pituitary; it promotes cell division, protein synthesis, and bone growth.

▶ Technology

Several state-of-the-art technology products are available that are correlated to this textbook. These useful and enticing supplements can assist you in teaching and can improve student learning.

Exploring the Internet

http://www.mhhe.com/sciencemath/biology/mader/

The Mader Home Page allows students and teachers from all over the world to communicate. *Biology* has a complete text-specific site developed exclusively for users of the sixth edition. By visiting this site, students can access additional study aids, explore links to other relevant biology sites, catch up on current information, and pursue other activities.

The Internet Primer

by Fritz J. Erickson & John A. Vonk

This short, concise primer shows students and instructors how to access and use the Internet. The guide provides enough information to get started by describing the most critical elements of using the Internet.

The Dynamic Human CD-ROM

This guide to anatomy and physiology interactively illustrates the complex relationships between anatomical structures and their functions in the human body. Realistic, three-dimensional visuals are the premier feature of this exciting learning tool. The program covers each body system, demonstrating to the viewer the anatomy, physiology, histology, and clinical applications of each system. *The Dynamic Human* is listed in the Using Technology section at the end of each systems chapter.

Explorations in Human Biology CD-ROM; Explorations in Cell Biology and Genetics CD-ROM

These interactive CDs, by Dr. George B. Johnson, feature fascinating topics in biology. *Explorations in Human Biology* and *Explorations in Cell Biology and Genetics* have 33 different modules that allow students to study a high-interest biological topic in an interactive way. In this edition of *Biology,* the Explorations that correlate to the chapter are listed in the Using Technology section, which appears at the end of each chapter. The modules are briefly described in the Instructor's Manual.

Life Science Animations Videotapes

Fifty-three animations of key physiological processes are available on videotapes. The animations bring visual movement to biological processes that are difficult to understand on the text page. In this edition of *Biology,* the Explorations that correlate to the chapter are listed in the Using Technology section, which appears at the end of each chapter.

BioSource Videodisc

BioSource Videodisc, by WCB/McGraw-Hill and Sandpiper Multimedia, Inc., features 20 minutes of animations and nearly 10,000 full-color illustrations and photos, many from leading WCB/McGraw-Hill biology textbooks.

Bioethics Forums Videodisc

Bioethics Forums is an interactive program that explores societal dilemmas arising from recent breakthroughs in biology, genetics, and biomedical technology. The scenarios are fictional, but the underlying science and social issues are real. *Bioethics Forums* encourages students to explore the science behind decisions as well as the processes of ethical reasoning and decision-making.

Visual Resource Library

Our electronic art image bank is a CD-ROM that contains hundreds of biological images from *Biology,* Sixth Edition. The CD-ROM contains an easy-to-use program that enables you quickly to view images, and you may easily import the images into PowerPoint to create your own multimedia presentations or use the already prepared PowerPoint presentations. The CD-ROM also includes several video clips featuring key animated biological processes.

Virtual Biology Laboratory CD-ROM

by John T. Beneski and Jack Waber, West Chester University

This CD-ROM is designed primarily for nonscience major students. The exercises are designed to expose students to the types of tools used by biologists, allow students to perform experiments without the use of wet lab setups, and support and illustrate topics and concepts from a traditional biology course.

Virtual Physiology Laboratory CD-ROM

This CD-ROM features ten simulations of the most common and important animal-based experiments ordinarily performed in the physiology component of your laboratory. This revolutionary program allows students to repeat laboratory experiments until they adeptly master the principles involved. The program contains video, audio, and text to clarify complex physiological functions.

The Secret of Life Video Modules

WGBH, Boston and BBC-TV

WGBH has produced eight 15-minute video modules that illuminate the biological universe with unique stories and animation. Each module concludes with a series of stimulating questions for class discussion.

The Secret of Life Videodisc

WGBH, Boston

A two-sided videodisc is available as a companion to *Biology,* Sixth Edition. Topic coverage includes biotechnology, human reproduction, portraits of modern science and research, and human genetics.

Our CD-ROM products may be packaged with the text at a cost savings. Contact your WCB/McGraw-Hill sales representative for details.

Technology Correlations

The Sixth Edition of *Biology* has three technology learning tools that are correlated to the chapters. The Using Technology section at the end of the chapter lists those that are appropriate to that chapter.[1]

The Dynamic Human is an interactive CD-ROM with three-dimensional visuals demonstrating the anatomy, physiology, and histology, along with clinical applications, of each body system.

Explorations in Human Biology and *Explorations in Cell Biology and Genetics* are interactive CD-ROMs consisting of 33 different modules that cover key topics in biology.

Life Science Animations is a set of five videotapes containing 53 animations of processes integral to the study of biology.

Chapter 1 A View of Life
Life Science Animations 52 (Tape 5)

Chapter 2 Basic Chemistry
Life Science Animations 1 (Tape 1)

Chapter 4 Cell Structure and Function
Life Science Animations 2, 3, 4 (Tape 1)

Chapter 5 Membrane Structure and Function
Explorations in Cell Biology 2, 3
Life Science Animations 2, 3 (Tape 1)

Chapter 6 Metabolism: Energy and Enzymes
Explorations in Cell Biology 6, 8
Life Science Animations 7, 11 (Tape 2)

Chapter 7 Photosynthesis
Explorations in Cell Biology 9
Life Science Animations 8, 9, 10 (Tape 1)

Chapter 8 Cellular Respiration
Explorations in Cell Biology 8
Life Science Animations 5, 6, 7, 11 (Tape 1)

[1]Technology aids are described per chapter in the *Instructor's Manual.*

Chapter 9 Cell Division
Explorations in Cell Biology 5
Life Science Animations 12, 50 (Tapes 1 and 5)

Chapter 10 Meiosis and Sexual Reproduction
Explorations in Cell Biology 10
Life Science Animations 13, 14, 19, 20 (Tape 2)

Chapter 12 Chromosomes and Genes
Explorations in Cell Biology 11

Chapter 13 Human Genetics
Explorations in Cell Biology 1, 12, 13
Explorations in Human Biology 1

Chapter 14 DNA: The Genetic Material
Life Science Animations 15 (Tape 2)

Chapter 15 Gene Activity
Explorations in Cell Biology 1, 15
Life Science Animations 16, 17 (Tape 2)

Chapter 16 Regulation of Gene Activity
Explorations in Cell Biology 15, 16
Life Science Animations 18 (Tape 2)

Chapter 17 Recombinant DNA and Biotechnology
Explorations in Cell Biology 14, 17

Chapter 20 Origin and History of Life
Life Science Animations 53 (Tape 5)

Chapter 25 Ecosystems
Life Science Animations 51, 52 (Tape 5)

Chapter 27 Human Impact on the Global Environment
Explorations in Human Biology 16

Chapter 30 The Protists
Life Science Animations 45 (Tape 4)

Chapter 32 The Plants
Life Science Animations 46, 47

Chapter 36 Plant Structure
Life Science Animations 46 (Tape 5)

Chapter 37 Nutrition and Transport in Plants
Life Science Animations 47, 48 (Tape 5)

Chapter 38 Growth and Development in Plants
Life Science Animations 49 (Tape 5)

Chapter 40 Animal Organization and Homeostasis
Dynamic Human, Anatomical Orientation

Chapter 41 Circulation
Explorations in Human Biology 5
Life Science Animations 37, 38, 39, 40 (Tape 4)
Dynamic Human, Cardiovascular System

Chapter 42 Lymph Transport and Immunity
Explorations in Human Biology 12, 13
Life Science Animations 41, 42, 43, 44 (Tape 4)
Dynamic Human, Lymphatic System

Chapter 43 Digestion and Nutrition
Explorations in Human Biology 7
Life Science Animations 33 (Tape 4)
Dynamic Human, Digestive System

Chapter 44 Respiration
Explorations in Human Biology 3, 6
Dynamic Human, Respiratory System

Chapter 45 Osmotic Regulation and Excretion
Dynamic Human, Urinary System

Chapter 46 Neurons and Nervous Systems
Explorations in Cell Biology 8, 9
Explorations in Human Biology 10
Life Science Animations 22, 23, 24, 25 (Tape 3)
Dynamic Human, Nervous System

Chapter 47 Sense Organs
Life Science Animations 26, 27 (Tape 3)

Chapter 48 Support Systems and Locomotion
Explorations in Human Biology 4, 9
Life Science Animations 30, 31 (Tape 3)
Dynamic Human, Muscular System, Skeletal System

Chapter 49 Hormones and Endocrine Systems
Explorations in Cell Biology 4
Explorations in Human Biology 11
Life Science Animations 28 (Tape 3)
Dynamic Human, Endocrine System

Chapter 50 Reproduction
Explorations in Human Biology 13
Dynamic Human, Reproductive System

Chapter 51 Development
Life Science Animations 21, (Tape 2)

More Teaching and Learning Aids

Instructor's Manual/Test Item File

The Instructor's Manual/Test Item File, prepared by Dr. John Richard Schrock, Emporia State University, is designed to assist instructors as they plan and prepare for classes using *Biology*. The first part of the Instructor's Manual pertains to the text chapters and the second part is the Test Item File.

The Instructor's Manual contains both an extended lecture outline and lecture enrichment ideas, which together review in detail the contents of the text chapter. The technology section lists videos and computer software items that are available from outside sources and also those that are available from WCB/McGraw-Hill. Answers to the Applying the Concepts questions appear in the Instructor's Manual.

The Test Item File for each chapter contains approximately 60 objective test questions and several essay questions. These same questions are found in the computerized version of the test item file.

Study Guide

To ensure close coordination with the text, Sylvia Mader has written the *Student Study Guide* that accompanies the text. Each text chapter has a corresponding study guide chapter that includes a listing of objectives, study questions, and a chapter test. Answers to the study questions and the chapter tests are provided to give students immediate feedback.

The concepts in the study guide are the same as those in the text, and the study questions in the study guide are sequenced according to these concepts. Instructors who make their choice of concepts known to the students can thereby direct student learning in an efficient manner. Instructors and students who make use of the *Student Study Guide* should find that student performance increases dramatically.

Laboratory Manual

Sylvia Mader has also written the *Laboratory Manual* to accompany *Biology*. With few exceptions, each chapter in the text has an accompanying laboratory exercise in the manual (some chapters have more than one accompanying exercise). In this way, instructors are better able to emphasize particular portions of the curriculum, if they wish. The 35 laboratory sessions in the manual are designed to further help students appreciate the scientific method and to learn the fundamental concepts of biology and the specific content of each chapter. All exercises have been tested for student interest, preparation time, and feasibility.

Laboratory Resource Guide

More extensive information regarding preparation is found in the *Laboratory Resource Guide*. The guide includes suggested sources for materials and supplies, directions for making up solutions and setting up the laboratory, expected results for the exercises, and suggested answers to questions in the laboratory manual. It is free to all adopters of the laboratory manual.

Transparencies

A set of 300 full color transparency acetates accompany the text. These acetates contain key illustrations from the text. This set of transparencies is also available as slides.

Visuals Testbank

This testbank contains black-and-white versions of 250 of the illustrations available as transparencies. The labels are deleted and copies can be run off for student quizzing or practice.

Micrograph Slides

This ancillary provides a boxed set of 100 color slides of photomicrographs and electron micrographs in the text.

From WCB/McGraw-Hill

How to Study Science, 2nd Edition

by Fred Drewes, Suffolk County Community College

This excellent workbook offers students helpful suggestions for meeting the considerable challenges of a college science course. It offers tips on how to take notes, how to get the most out of laboratories, and how to overcome science anxiety. The book's unique design helps students develop critical thinking skills while facilitating careful notetaking. (ISBN 0–697–15905-1)

A Life Science Living Lexicon

by William N. Marchuk, Red Deer College

This portable, inexpensive reference helps introductory-level students quickly master the vocabulary of the life sciences. Not a dictionary, it carefully explains the rules of word construction and derivation, in addition to giving complete definitions of all important terms. (ISBN 0–697–12133–X)

Biology Study Cards

by Kent Van De Graaff, R. Ward Rhees, and Christopher H. Creek, Brigham Young University

This boxed set of 300 two-sided study cards provides a quick yet thorough visual synopsis of all key biological terms and concepts in the general biology curriculum. Each card features a masterful illustration, pronunciation guide, definition, and description in context. (ISBN 0–697–03069–5)

Critical Thinking Case Study Workbook

by Robert Allen

This ancillary includes 34 critical thinking case studies which are designed to immerse students in the "process of science" and challenge them to solve problems in the same way biologists do. The case studies are divided into three levels of difficulty (introductory, intermediate, and advanced) to afford instructors greater choice and flexibility. An answer key accompanies this workbook. (ISBN 0-697-34250-6)

The AIDS Booklet

by Frank D. Cox

This booklet describes how AIDS and related diseases are commonly spread so that readers can protect themselves and their friends against this debilitating and deadly disease. This booklet is updated quarterly to give readers the most current information. Also visit the Mader Home Page for additional AIDS material. (ISBN 0-697-26261-8)

Chemistry for Biology

by Carolyn Chapman

This workbook is a self-paced introduction or review of the basic principals of chemistry that are most useful in other areas of science. (ISBN 0-697-24121-1)

Biology Startup

by Myles Robinson and Kathleen Pace, Grays Harbor College

Biology Startup is a five-disk Macintosh tutorial that helps nonmajors master challenging biological concepts such as basic chemistry, photosynthesis, and cellular respiration. This program can be a valuable addition to a resource center and is especially helpful as a refresher or for students who need additional assistance to succeed in an introductory biology course. (ISBN 0–697–27227–3)

HealthQuest CD-ROM

HealthQuest CD-ROM is an interactive CD-ROM designed to help students address the behavioral aspects of personal health and wellness. *HealthQuest* allows users to assess their current health and wellness status, determine their health risks and relative life expectancy, explore options, and make decisions to improve the behaviors that impact their health.

Life Science Living Lexicon CD-ROM

by William N. Marchuk, Red Deer College

A Life Science Living Lexicon CD-Rom contains a comprehensive collection of life science terms, including definitions of their roots, prefixes, and suffixes as well as audio pronunciations and illustrations. The Lexicon is student-interactive, providing quizzing and notetaking capabilities. It contains 4,500 terms, which can be broken down for study into the following categories: anatomy and physiology, botany, cell and molecular biology, genetics, ecology and evolution, and zoology. (ISBN 0–697–37993–0)

Acknowledgments

The personnel at WCB/McGraw-Hill have always lent their talents to the success of *Biology*. My editor, Michael Lange, directed the efforts of all. Connie Haakinson, my developmental editor, served as a liaison between the editor, me, and many other people. She met each new challenge in a prompt and most professional way.

The production team worked diligently toward the success of this edition. Margaret Horn was the project manager; Brenda Ernzen, the art editor; Lori Hancock, the photo research coordinator, and K. Wayne Harms was the design coordinator. My thanks to each of them for a job well done!

The Reviewers

Many instructors have contributed not only to this edition of *Biology* but also to previous editions. I am extremely thankful to each one, for they have all worked diligently to remain true to our calling and provide a product that will be the most useful to our students.

In particular, it is appropriate to acknowledge the help of the following individuals. For the Sixth Edition:

Edward Cawley *Loras College*
Diane M. Morris *Bemidji State University*
Donald Kirk *Shasta College*
Michael Mack *Newberry College*
Linda E. Wooten *Bishop State Community College*
William Langley *Butler County Community College*
Dorcas Barnes Noble *Bishop State Community College*
Sue Trammell *Rend Lake College*
Edward J. Greding, Jr. *Del Mar College*
Harland D. Guillory *LA State University at Eunice*
Karen S. Pirc *Moraine Valley Community College*
Charles Paulson *University of Arizona*
Sandra Gibbons *Moraine Valley Community College*
Les M. Brown *Gardner–Webb University*
Anne Bartosh *Howard County Junior College, San Angelo*
Karen J. Dalton *Catonsville Community College*
Randy Lankford *Galveston College*
Jun Tsuji *Siena Heights College*
J. Wesley Bahorik *Kutztown University of Pennsylvania*
Andrew H. Lapinski *Reading Area Community College*
Carolyn Chapman *Suffolk Community College*
Bobby R. Baldridge *Asbury College*
Robert H. Tamarin *University of Massachusetts, Lowell*
Roberta S. Brice *Villa Julie College*
Marcella Piasecki *Lynn University*
E. Juterbock *Ohio State University*
Daniel Pena *William Tyndale College*
William L. Trotter *Des Moines Area Community College*
Charles R. Wert *Linn–Benton Community College*
Charles Smead *KanKaKee Community College*
Mariette S. Cole *Concordia College—St. Paul*
Jeffrey Kassner *Suffolk Community College*
Linda Butler *University of Texas at Austin*
Iona Baldridge *Lubbock Christian University*
Stephen L. Fabritius *Southwestern University*
Garry M. Wallace *Northwest College*
Kirit D. Chapatwala *Selma University*
Jeanette Oliver *Flathead Valley Community College*
Gregg M. Orloff *Emory University*
Ellison Robinson *Midlands Technical College*
Fred Drewes *Suffolk Community College—Selden*
John Sternfeld *SUNY–Cortland*
Susan C. Barber *Oklahoma City University*
Dick T. Stalling *Northwestern State University*
Kathryn Springsteen *Colby–Sawyer College*
Donna S. Emmeluth *Fulton–Montgomery Community College*
Thomas J. Sernka *Westminster College of Salt Lake City*
Mary Sue Gamroth *Joliet Junior College*
Laura H. Ritt *Burlington County College*

Robert W. Yost *Indiana University–Purdue University at Indianapolis*
Gerald DeMoss *Morehead State University*
Carolyn McCracken *Northeast State Technical Community College*
Elizabeth A. Desy *Southwest State University*
John C. Mertz *Delaware Valley College*
Jonas E. Okeagu *Fayetteville State University*
Evert Brown *Casper College*
Timothy J. Bell *Chicago State University*
Bianca L. Graves *Livingstone College*
John Cruzan *Geneva College*
Lawrence C. Parks *University of Louisville*
Kerry S. Kilburn *Old Dominion University*
Katherine J. Denniston *Towson State University*
Wayne Becker *University of Wisconsin–Madison*
Peter I. Ekechukwu *Horry–Georgetown Technical College*
Carol A. Paul *Suffolk Community College*
Nina T. Parker *Shenandoah University*
Tom Arsuffi *Southwest Texas State University*
Kiran Misra *Edinboro University of Pennsylvania*
Linda K. Dion *University of Delaware*
Patrick Guilfoile *Bemidji State University*
Allan Landwer *Hardin–Simmons University*
Barbara C. Liang *Fox Valley Technical College*
Ronald L. Jenkins *Samford University*
Hildagarde Kahn Sanders *Villa Julie College*
Danny Ingold *Muskingum College*
Scott R. Smedley *Cornell University*
Kenneth J. Curry *University of Southern Mississippi*
Lynne Lohmeier *Mississippi Gulf Coast Community College*
Teresa DeGolier *Bethal College*
Felix O. Akojie *UK, Paducah Community College*
William H. Gilbert III *Simpson College*
Joe Coelho *Western Illinois University*
John P. Harley *Eastern Kentucky University*
David J. Hicks *Manchester College*
Jon R. Maki *Eastern Kentucky University*
Terry R. Martin *Kishwaukee College*
Jennifer Carr Burtwistle *Northeast Community College*
Neal D. Mundahl *Winona State University*
Thomas A. Davis *Loras College*
Floyd Sandford *Coe College*

Learn by doing.

You can *do* just that, with these two great CD-ROMs.

Use these CD-ROMs in conjunction with *Biology*. It's easy—and affordable.

See page xix to find out exactly which CD-ROM to use to make the chapter you're reading come to life with interactive CD-ROM technology.

See how biology relates to real life with the "Explorations" CD-ROMs.

Two interactive CD-ROMs by George B. Johnson

Explorations in Human Biology CD-ROM

Macintosh ISBN 0-697-37907-8
Windows ISBN 0-697-37906-X

Comes with Student Workbook (in both English and Spanish versions)

- Observe colorful graphics and interactive animations of vital life processes.
- Manipulate variables, and see how your actions impact the animations.
- Watch subject matter become more lively, more clear—with the help of topic information, glossary (in English and Spanish), and narration (in English and Spanish).
- See how biology applies to life with 16 modules—all on one CD-ROM—exploring topics like Diet and Weight Loss, Smoking and Cancer, Drug Addiction, and Pollution in a Freshwater Lake.

New! Explorations in Cell Biology & Genetics CD-ROM

Hybrid CD-ROM: ISBN 0-697-37908-6

- Set and reset these interactive animations' variables and gauge the results.
- "Pause" an animation for an explanation—with labels—of the process you're seeing.
- Test how well you know your material with quizzes and critical-thinking questions.
- Jump between the glossary and related "Explorations" visuals with the help of hyperlinks.
- Watch film clips and animations of related topics, and hear sound effects.
- Try all 17 modules contained on this cutting-edge CD-ROM, for hot topics like DNA Fingerprinting, Reading DNA, and Heredity in Families.

Learn by doing with these two *affordable* CD-ROMs.
Call toll-free now: 1–800–338–3987.

Look . . .

. . . at the great biology study tools WCB/McGraw-Hill has to offer!

WCB Life Science Animations Videotapes

Series of five videotapes containing animations of complex physiological processes. These animations make challenging concepts easier to understand.

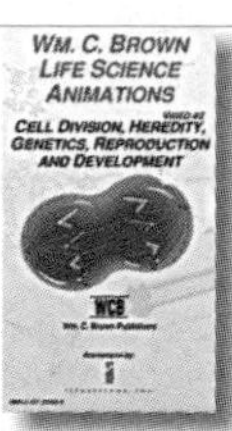

Tape 1 Chemistry, The Cell, and Energetics
ISBN: 0-697-25068-7

Tape 2 Cell Division, Heredity, Genetics, Reproduction, and Development
ISBN: 0-697-25069-5

Tape 3 Animal Biology #1
ISBN: 0-697-25070-9

Tape 4 Animal Biology #2
ISBN: 0-697-25071-7

Tape 5 Plant Biology, Evolution, and Ecology
ISBN: 0-697-26600-1

Life Science Living Lexicon CD-ROM

by William Marchuk

ISBN: 0-697-37993-0

This interactive CD-ROM contains a complete lexicon of life science terminology. Conveniently assembled on an easy-to-use CD-ROM are components such as a glossary of common biological roots, prefixes, and suffixes; a categorized glossary of common biological terms; and a section describing the classification system.

To order any of these products,
contact your bookstore manager,
or call-toll free: 1–800–338–3987.

A View of Life

CHAPTER 1

Chapter Concepts

Marine iguana, *Amblyrhynchus cristatus*

You intuitively know that the marine iguanas and not the rock in the photograph are alive. An iguana is adapted to take nourishment from the environment in order to maintain itself and reproduce. Marine iguanas are members of an ecosystem along the shores of the Galápagos Islands. They have evolved no other place on earth. Unlike their terrestrial relatives they are aquatic and swim beneath the waters in order to feed on leaflike algae that cover the submerged rocks.

Marine iguanas and all living things have a common ancestry as witnessed by shared characteristics such as their cellular organization, metabolic needs, and ability to reproduce. But each organism is adapted differently and interacts with other members and the physical environment within a particular ecosystem. This chapter features the coral reef and the tropical rain forest, productive and diverse ecosystems that are threatened by human activities.

Biology is a science that studies living things and has given us data that substantiates the unity of life and a mechanism by which it diversifies. Science progresses by observation and experimentation and develops concepts that allow us to understand the material world. Biology provides an understanding of life that will hopefully help us preserve living things including humans themselves.

1.1 How to Define Life

Living things are organized, take materials and energy from the environment, respond to stimuli, reproduce and develop, and adapt to the environment.

Living Things Are Organized

The complex organization of living things begins with the cell, the basic unit of life. Cells are made up of molecules that contain atoms, which are the smallest units of matter that can enter into chemical combination. In multicellular organisms, similar cells combine to form a tissue—nerve cells form nerve tissue, for example. Tissues make up organs, as when various tissues combine to form the brain. Organs, in turn, work together in systems; for example, the brain works with the spinal cord and a network of nerves to form the nervous system. A multicellular individual has several organ systems.

There are levels of biological organization that extend beyond the individual organism. All organisms of one type in a particular area belong to a **population.** In a temperate deciduous forest, there is a population of gray squirrels and a population of oak trees. The populations of various animals and plants in the forest make up a community (Fig. 1.1). The populations interact among themselves and with the physical environment (soil, atmosphere, etc.) forming an **ecosystem** [Gk. *oikos*, home, house, and *systema*, ordered arrangement].

Summing the Parts

In the living world, the whole is more than the sum of its parts. Each new level of biological organization has emergent properties that are due to interactions between the parts making up the whole. For example, when cells are broken down into bits of membrane and oozing liquids, these parts themselves cannot carry out the business of living. Slice a frog and arrange the slices, and the frog cannot flick out its tongue and catch flies.

Living things have levels of organization from cells to ecosystems. Each level of organization has emergent properties that cannot be accounted for by a sum of the parts.

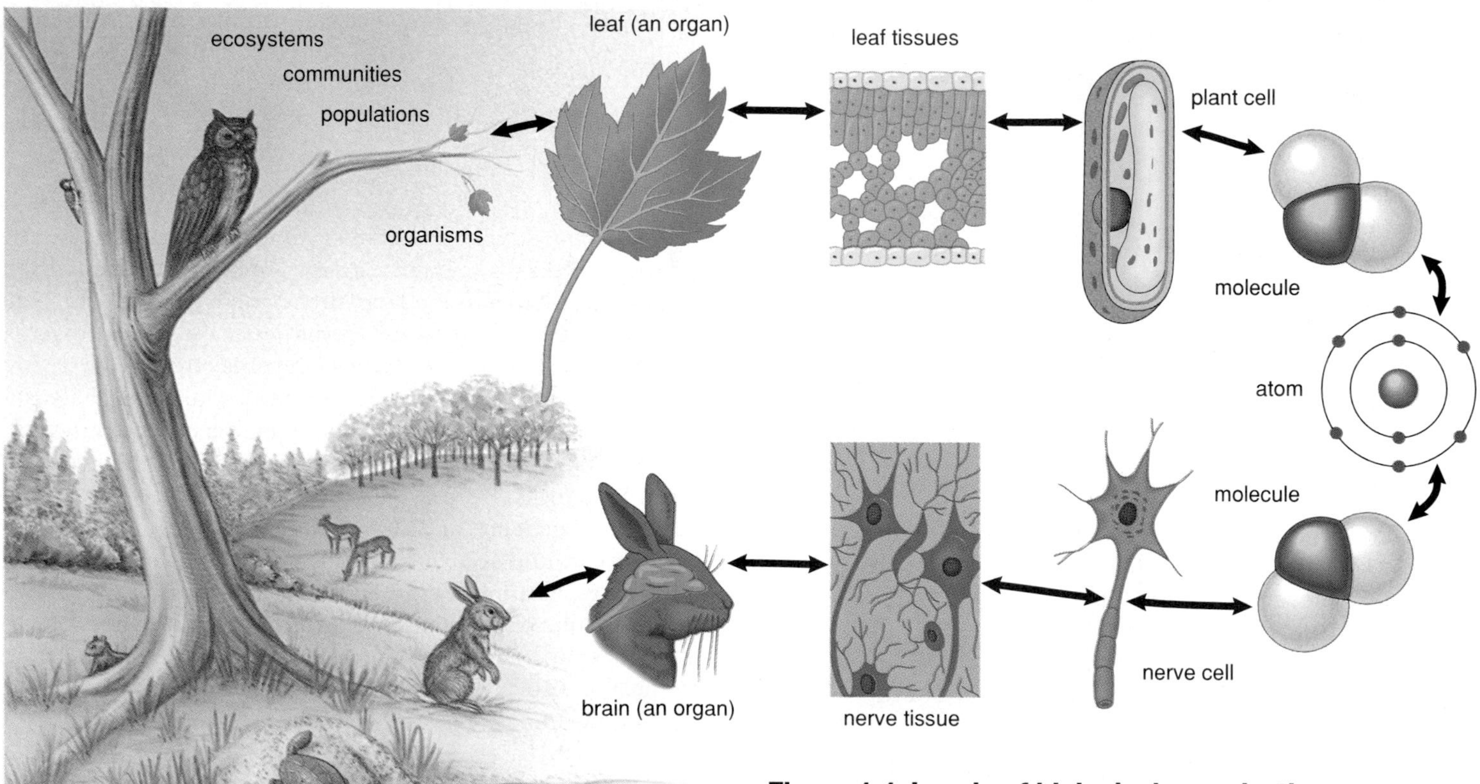

Figure 1.1 Levels of biological organization. An ecosystem is composed of a community of populations along with the physical environment of a particular area (e.g., a forest). A population contains similar types of individuals. The individual organism is composed of organ systems. Each organ is made up of various tissues, which consist of cells. Cells are composed of molecules, which are formed from atoms.

Figure 1.2 Acquiring nutrient materials and energy. Only plants are able to photosynthesize and produce organic nutrients; animals feed directly on plants or other animals. **a.** A lioness charges her prey, a group of zebras. **b.** One zebra of the herd is taken and becomes a meal for hungry lionesses.

Living Things Acquire Materials and Energy

Living things cannot maintain their organization or carry on life's activities without an outside source of materials and energy (Fig. 1.2). Food provides nutrient molecules, which are used as building blocks or for energy. **Energy** is the capacity to do work, and it takes work to maintain the organization of the cell and the organism. When cells use nutrient molecules to make their parts and products, they carry out a sequence of chemical reactions. **Metabolism** [Gk. *meta,* implying change] is all the chemical reactions that occur in a cell.

The ultimate source of energy for nearly all life on earth is the sun. Plants and plantlike organisms are able to capture solar energy and carry on **photosynthesis,** a process that transforms solar energy into chemical energy in the bonds of organic molecules. Animals and plants get energy by metabolizing the organic molecules made by photosynthesis.

An intake of materials and energy is needed if an organism's organization is to be maintained. The ultimate source of energy for life on earth is the sun.

Remaining Homeostatic

For metabolic processes to continue, living things need to keep themselves stable in temperature, moisture level, acidity, and other physiological factors. This is **homeostasis** [Gk. *homoios,* like, resembling, and *stasis,* standing]—the maintenance of internal conditions within certain boundaries.

Many organisms depend on behavior to regulate their internal environment. A chilly lizard may raise its internal temperature by basking in the sun on a hot rock. When it starts to overheat, it scurries for cool shade. Other organisms have control mechanisms that do not require any conscious activity. When a student is so engrossed in her textbook that she forgets to eat lunch, her liver releases stored sugar to keep the blood sugar level within normal limits. Hormones regulate sugar storage and release, but in other instances the nervous system is involved in maintaining homeostasis.

Living Things Respond

Living things find energy and nutrients by interacting with their surroundings. Even unicellular organisms can respond to their environment. In some, the beating of microscopic hairs, and in others, the snapping of whiplike tails, moves them toward or away from light or chemicals. Multicellular organisms can manage more complex responses. A vulture can smell meat a mile away and soar toward dinner. A monarch butterfly can sense the approach of fall and begin its flight south where resources are still abundant.

The ability to respond often results in movement: the leaves of a plant turn toward the sun and animals dart toward safety. The ability to respond helps ensure survival of the organism and allows it to carry on its daily activities. All together, we call these activities the behavior of the organism.

Figure 1.3 Adaptations of rockhopper penguins, *Eudyptes*.
Male and female rockhoppers with their offspring. The stubby forelimbs of penguins are modified as flippers for fast swimming. Only a little over a half meter tall, rockhopper penguins are named for their skill in leaping from rock to rock.

Living Things Have Adaptations

Adaptations [L. *ad*, toward, and *aptus*, fit, suitable] are modifications that make an organism suited to its way of life. For example, penguins are adapted to an aquatic existence in the Antarctic (Fig. 1.3). Most birds have forelimbs proportioned for flying, but a penguin has stubby, flattened wings suitable for swimming. Their feet and tails serve as rudders in the water, but the flat feet also allow them to walk on land. Rockhopper penguins have a bill adapted to eating small shellfish. Their eggs—one, or at most two—are carried on their feet, where they are protected by a pouch of skin. This allows the birds to huddle together for warmth while standing erect and incubating eggs.

The process by which organisms become modified over time is called *natural selection.* Certain members of a **species** [L. *species*, model, kind], defined as a group of interbreeding individuals, may inherit a genetic change that causes them to be better suited to a particular environment. These members can be expected to produce more surviving offspring who also have the favorable characteristic. In this way, the attributes of the species' members change over time.

Living Things Reproduce and Develop

Life comes only from life. Every type of living thing can **reproduce,** or make another organism like itself (Fig. 1.3). Bacteria, protozoa, and other unicellular organisms simply split in two. In most multicellular organisms, the reproductive process begins with the pairing of a sperm from one partner and an egg from the other partner. The union of sperm and egg, followed by many cell divisions, results in an immature individual, which grows and develops through various stages to become the adult.

An embryo develops into a sperm whale or a yellow daffodil because of a blueprint inherited from its parents. The instructions for their organization and metabolism are encoded in the genes. The **genes,** which contain specific information for how the organism is to be ordered, are made of long molecules of DNA (deoxyribonucleic acid). All cells have a copy of the hereditary material, DNA, whose shape resembles a spiral staircase with millions of steps.

Descent with Modification

All living things share the same basic characteristics discussed in this chapter. They are all composed of cells organized in a similar manner. Their genes are composed of DNA, and they carry out the same metabolic reactions to acquire energy and maintain their organization. This unity suggests that all living things are descended from a common ancestor—the first cell or cells. However, **evolution** [L. *evolutio*, an unrolling] is descent with modification. One species can give rise to several species, each adapted to a particular set of environmental conditions. Specific adaptations allow species to play particular roles in an ecosystem. The diversity of organisms is best understood in terms of the many different ways in which organisms carry on their life functions within an ecosystem where they live, acquire energy, and reproduce.

Descent from a common ancestor explains the unity of life. Adaptations to different ways of life account for the great diversity of life-forms.

A closer Look

Evolution: The GUT of Biology

From bacteria to bats, toadstools to trees, whippoorwills to whales—the diversity of the living world boggles the mind! Yet living things are not nearly as different as you might think. All are basically alike. They are all made of cells containing to a large extent the same six elements: carbon, hydrogen, oxygen, nitrogen, phosphorus, and sulfur. Of these six elements, the first four—known as CHON—are the most common in cells. Why CHON? Why not a more exotic mixture, such as europium, gold, krypton, and thulium? The answer is simple. Life is composed of CHON because these elements were abundant when life first began. CHON is the stuff of which stars and galaxies are made—four little elements that are among the five most abundant in the entire universe. Moreover, all are able to combine with one another to form small, sturdy molecules (carbon dioxide, CO_2; water, H_2O; methane, CH_4; ammonia, NH_3; and many others). Some can even pair with atoms of their same type to make up common gases like oxygen, O_2, and nitrogen, N_2. Because all of these simple compounds dissolve in water, all can play an active role in the chemistry of life.

Thus, we and everything else alive are made of the same simple "starstuff." But don't things get terribly complicated when we examine how living things interact to form the ecosystem of the living world? Not at all. In fact, though there are nearly two million living species known to science (and perhaps three to five times that many as yet undiscovered); all live by exactly the same rules. To stay alive, all of these different organisms need to satisfy just two basic necessities. Because all are made of CHON, all need a source of these building blocks of life. And in order to make use of CHON, all need a source of energy. Moreover, for all life, from microbes to man, there are just two strategies to meet these needs: the animal-like strategy of feeding on others (known as *heterotrophy*), and the self-feeding strategy of plants, algae, and plantlike microbes (*autotrophy*). The ecosystem of the entire world is just that simple. There are only two necessities—CHON and energy. And there are only two strategies—heterotrophy and autotrophy. This earth of ours is a planet of the eaters and the eatees!

Although you would not expect it, birds and humans share many characteristics in common. Descent from a common ancestor explains how this came about.

Why is biology this thrifty? Once again, the answer is simple—evolution! Evolution is the *GUT* of biology, the *GRAND UNIFYING THEORY* that links all of life. All organisms, over all of time, are united by a common bond. Just as you are descended from your parents, grandparents, and so forth, going back for many generations, *all* forms of life that have ever lived are tied together by an unbroken evolutionary thread that can be traced back through geologic time to the infancy of our planet.

This linkage of all of life, surmised even by the ancient Greeks, can be seen by comparing the bone structure of a bird and a human. Why should a bird be like a human? After all, birds fly. Rather than having coarse, brittle bones like ours, maybe they would be better served by some strong, lightweight alloy such as titanium. And why do birds have ribs, knees, backbones, and eyeballs in the front of their heads? Are all those necessary for flight? Some, perhaps, but most, probably not. The answer is evolution. Today's life is a product of an unimaginably long evolutionary history, the result of life's development over hundreds and thousands of millions of years. Birds, just like us mammals, are descended from reptiles (lizards, dinosaurs, and the like). Reptiles evolved from amphibians (frogs and salamanders), and amphibians are descendants of fish. As it happened, early evolving fish, more than half a billion years ago, had bones made of bonestuff (technically, hydroxyapatite). Titanium bones for birds (or even for us) might be great, but because of evolutionary ties to earth's earliest living things, it was hydroxyapatite bones or no bones at all. And so it was also for CHON, cells, genes, and even the structure of the world's ecosystem. Our modern world is just a scaled-up version of a microbial menagerie that originated literally billions of years ago!

Because evolution is the *GUT* of biology, the familiar biologic present is firmly rooted in the remote biologic past. But amazingly, recent advances in the life sciences have placed the biologic future in human hands. What evolution accomplished over hundreds of millions—even billions—of years, can now be manipulated and modified in laboratory test tubes within months or even weeks! Heady stuff. We are in the driver's seat, but which of many roads will we take? The future of life on planet earth is in our hands.

J. William Schopf

Director, UCLA Center for the Study of Evolution and the Origin of Life

1.2 Ecosystems Contain Populations

Individual organisms belong to a *population*, all the members of a species within a *community*. All communities taken together make up the **biosphere** [Gk. *bios*, life, and L. *sphaera*, ball], a thin layer of life that encircles earth. The populations within a community interact among themselves and with the physical environment (soil, atmosphere, etc.), thereby forming an *ecosystem*. Two ecosystems of particular interest are coral reefs and tropical rain forests, which are described in this section.

Although an ecosystem like a forest or reef changes—trees fall, fishes come and go, seeds sprout—an ecosystem remains recognizable year after year. We say it is in dynamic balance. In many cases, even the extinction of species (and their replacement by new species through evolution) still allows the dynamic balance of the system to be maintained.

A major feature of the interactions between populations pertains to who eats whom. Plants produce food, and animals that eat plants are food for other animals. Such a sequence of organisms is called a food chain (Fig. 1.4). Both plants and animals interact with the physical environment, as when they exchange gases with the atmosphere.

Nutrients cycle within and between ecosystems. Plants take in inorganic nutrients, like carbon dioxide and water, and produce organic nutrients, such as carbohydrates, that are used by themselves and various levels of animal consumers. When these organisms die and decay, inorganic nutrients are made available to plants once more. The blue arrows in Figure 1.4 show how chemicals cycle through the various populations of an ecosystem.

In contrast, the yellow arrows in Figure 1.4 show how energy flows through an ecosystem: solar energy used by plants to produce organic food is eventually converted to heat when organisms, including plants, use organic food as an energy source. Therefore, a constant supply of solar energy is required for an ecosystem and for life to exist.

In ecosystems, the same nutrients keep cycling through populations, but energy flows because it is eventually converted to heat.

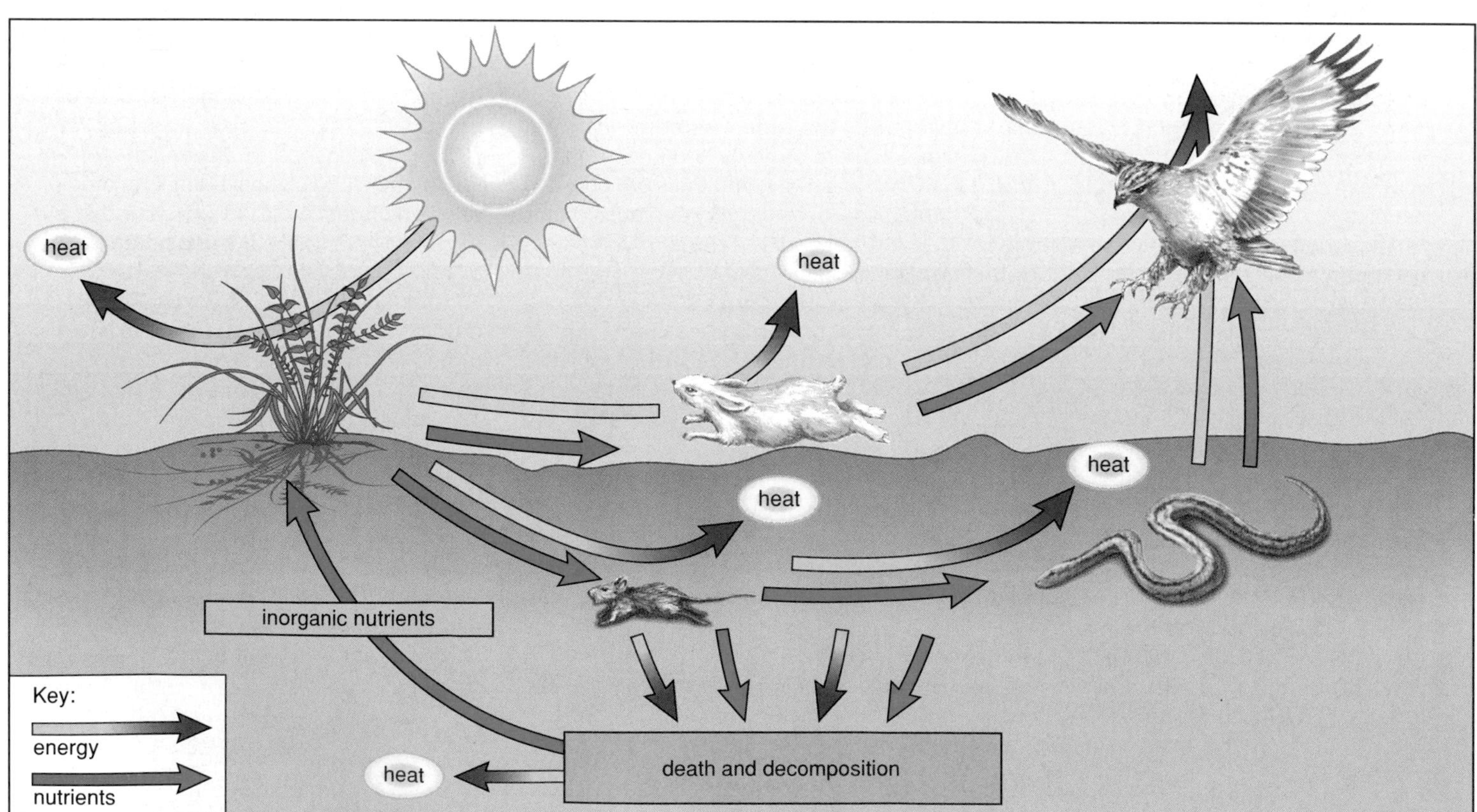

Figure 1.4 Ecosystem organization.
Within an ecosystem, nutrients cycle (see blue arrows): from plants, which use solar energy and inorganic nutrients to produce organic nutrients, to decomposers, which make inorganic nutrients available to plants once more. Energy flows (see yellow to red arrows) from the sun through all populations and eventually is converted to heat.

Coral Reef, a Marine Ecosystem

Coral reefs are found in clear, shallow tropical waters. Nowhere else in the sea is there such an abundance of living things (Fig. 1.5). The reef consists of the skeletons of stony corals, colonial animals that form deposits of calcium carbonate. Only the outer layer of the reef is alive; the rest is an inert structure full of nooks and crannies, where fish can hide from their predators. Some types of fish venture out of hiding only at night, when they feed on plankton, the microscopic organisms that drift through the ocean. The coral animals themselves often feed at night also, taking in whatever comes within reach of their extended tentacles. Other fish are active during the day. These hover near the surface of the coral, grazing on algae and worms or on shrimp and crabs if their jaws, like those of lionfishes, are able to crunch through shells. Parrot fish even grind the stony coral skeletons themselves. All the smaller fishes are prey to large carnivores like groupers, moray eels, and barracudas.

Figure 1.5 Coral reef organization, featuring various animals of a Palau coral reef.

Tropical Rain Forest, a Terrestrial Ecosystem

Tropical rain forests are the most complex ecosystems in the world. They are found at low altitudes near the equator where there is plentiful sun and rainfall the entire year. Major rain forests are located in South America (Fig. 1.6), central and west Africa, and Southeast Asia. Rain forests can be divided into several layers. In the top layer the highest trees rise above the canopy, which is a continuous layer of evergreen trees with broad leaves. If light penetrates through the canopy, there is an understory that consists of shrubs and an undergrowth of ferns and herbs. Most animals live in the canopy where brightly colored birds, such as toucans and macaws, fly around eating fruit, buds, and pollen. Other birds, such as hummingbirds with long bills, feed from nectar often taken from the epiphytes (plants that grow independently on trees). Mammals such as tree sloths and spider monkeys, which also live in the canopy, are preyed upon by jaguars. Other canopy animals include butterflies, tree frogs, and dart-poison frogs. Snakes, spiders, and ants live on the ground. Many animals, such as bats, are active only at night.

Figure 1.6 Tropical rain forest organization, featuring plants and animals of the Amazon basin.

Ecology Focus

Tropical Rain Forests: Can We Live Without Them?

So far, nearly 2 million species of organisms have been discovered and named. Two-thirds of the plant species, 90% of the nonhuman primates, 40% of birds of prey, and 90% of the insects live in the tropics. Many more species of organisms (perhaps as many as 30 million) are estimated to live in the tropical rain forests but have not yet been discovered.

Tropical forests span the planet on both sides of the equator and cover 6–7% of the total land surface of the earth—an area roughly equivalent to our contiguous forty-eight states. Every year humans destroy an area of forest equivalent to the size of Oklahoma. At this rate, these forests and the species they contain will disappear completely in just a few more decades. Even if only the forest areas now legally protected survive, 56–72% of all tropical forest species would still be lost.

The loss of tropical rain forests results from an interplay of social, economic, and political pressures. Many people already live in the forest, and as their numbers increase, more of the land is cleared for farming. People move to the forests because internationally financed projects build roads and open the forests up for exploitation. Small-scale farming accounts for about 60% of tropical deforestation, and this is followed by commercial logging, cattle ranching, and mining. International demand for timber promotes destructive logging of rain forests in Southeast Asia and South America. A market for low-grade beef encourages the conversion of tropical rain forests to pastures for cattle. The lure of gold draws miners to rain forests in Costa Rica and Brazil.

The destruction of tropical rain forests produces only short-term benefits but is expected to cause long-term problems. The forests soak up rainfall during the wet season and release it during the dry season. Without them, a regional yearly regime of flooding followed by drought is expected to destroy property and reduce agricultural harvests. Worldwide, there could be changes in climate that would affect the entire human race. On the other hand, the preservation of tropical rain forests offers benefits. For example, the rich diversity of plants and animals would continue to exist for scientific and pharmacological study. One-fourth of the medicines we currently use come from tropical rain forests. The rosy periwinkle from Madagascar has produced two potent drugs for use against Hodgkin disease, leukemia, and other blood cancers. It is hoped that many of the still-unknown plants will provide medicines for other human ills.

Studies show that if the forests were used as a sustainable source of nonwood products, such as nuts, fruits, and latex rubber, they would generate as much or more revenue while continuing to perform their various ecological functions. And biodiversity could still be preserved. Brazil is exploring the concept of "extractive reserves," in which plant and animal products are harvested but the forest itself is not cleared. Ecologists have also proposed "forest farming" systems, which mimic the natural forest as much as possible while providing abundant yields. But for such plans to work maximally, the human population size and the resource consumption per person must be stabilized.

Preserving tropical rain forests is a wise investment. Such action promotes the survival of most of the world's species—indeed, the human species, too.

The Human Population

The human population tends to modify existing ecosystems for its own purposes. As more and more of the biosphere is converted to towns and cities, fewer of the natural cycles are able to function adequately to sustain the human population. It is important to do all we can to preserve the biosphere, because only then can we be assured that we will continue to exist. The recognition that the workings of the biosphere need to be preserved is one of the most important developments of our new ecological awareness.

Presently, there is great concern about preserving the world's tropical rain forests, as discussed in the reading on this page. The tropical rain forests perform many services for us. For example, they act like a giant sponge and absorb carbon dioxide, a pollutant that pours into the atmosphere from the burning of fossil fuels such as oil and coal. If the rain forests continue to be depleted as they are now, an increased amount of carbon dioxide in the atmosphere is expected to cause an increase in the average daily temperature. Problems with acid rain are also expected to increase, since carbon dioxide combines with water to form carbonic acid, a component of acid rain.

The present *biodiversity* (number and size of populations in a community) of our planet is being threatened. It has been estimated that the number of species in the biosphere may be as high as 80 million species, but thus far fewer than 2 million have been identified and named. Even so we may be presently losing from 24 to even 100 species a day due to human activities. The existence of the species featured in Figure 1.6 is threatened because tropical rain forests are being reduced in size. Most biologists are alarmed over the present rate of extinction and believe the rate may eventually rival the mass extinctions that have occurred during our planet's history.

The human population tends to modify existing ecosystems and to reduce biodiversity. Because all living things are dependent upon the normal functioning of the biosphere, ecosystems should be preserved.

1.3 How Living Things Are Classified

Since life is so diverse (there are over 900,000 known species of insects alone!) it is helpful to have a classification system to group organisms according to their similarities (see Appendix D, Classification of Organisms). **Taxonomy** [Gk. *tasso,* arrange, classify, and *nomos,* usage, law] is the discipline of identifying and classifying organisms according to certain rules. In keeping with a practice started by the Swedish taxonomist Linnaeus, biologists give each living thing a binomial [L. *bis,* two, and *nomen,* name] or two-part name. For example, the scientific name for the garden pea is *Pisum sativum*. The first word is the genus and the second word is the specific epithet of a species within that genus. The members of a species have similar characteristics and reproduce with one another. Similar species are placed in the same genus, then similar genera go into families, families into orders, and so on into five kingdoms (Fig. 1.7). The organizational levels of the taxonomic system are as listed next:

Levels of Classification	Human	Corn
Kingdom	Animalia	Plantae
Phylum*	Chordata	Magnoliophyta
Class	Mammalia	Liliopsida
Order	Primates	Commelinales
Family	Hominidae	Poaceae
Genus	*Homo*	*Zea*
Species	*H. sapiens*	*Z. mays*

* The term division instead of phylum is used in kingdoms Plantae and Fungi

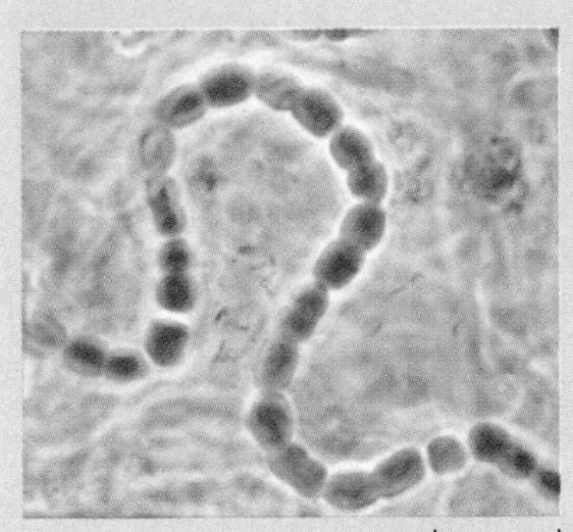

Kingdom Monera:
Nostoc, a cyanobacterium

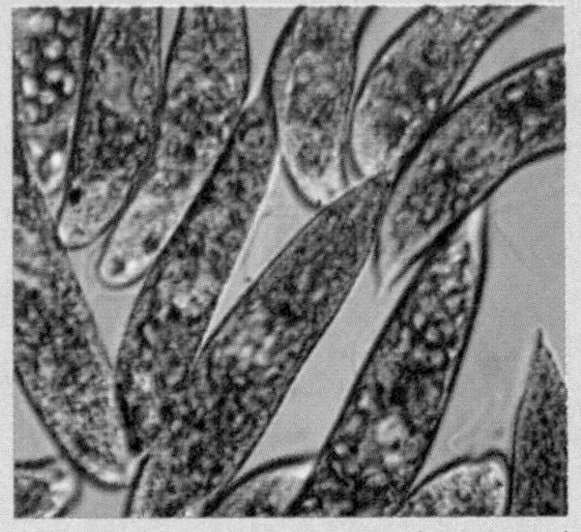

Kingdom Protista:
Euglena, a unicellular organism

Kingdom Fungi:
Coprinus, a shaggy mane mushroom

Kingdom Plantae:
Rosa, a flowering plant

Kingdom Animalia:
Felix, an African lynx

a.

Five Kingdom System of Classification

Kingdom	Organization	Type of Nutrition	Representative Organisms
Monera	Small, simple single cell (sometimes in chains or mats)	Absorb food (some photosynthetic)	Bacteria, including cyanobacteria
Protista (protists)	Complex single cell (sometimes filaments, colonies, or even multicellular)	Absorb, photosynthesize, or ingest food	Protozoa and algae of various types
Fungi	Mostly multicellular and filamentous with specialized, complex cells	Absorb food	Molds and mushrooms
Plantae (plants)	Multicellular with specialized, complex cells	Photosynthesize food	Mosses, ferns, and flowering plants (both woody and nonwoody)
Animalia (animals)	Multicellular with specialized, complex cells	Ingest food	Sponges, worms, insects, fish, amphibians, reptiles, birds, and mammals

b.

Figure 1.7 Five kingdom system of classification.
a. Representatives of the five kingdoms.
b. Brief descriptions of the five kingdoms.

1.4 The Process of Science

Science helps human beings understand the natural world and is concerned solely with information gained by observing and testing that world. Scientists, therefore, ask questions only about events in the natural world and expect that the natural world in turn will provide all the information needed to understand these events. It is the aim of science to be objective rather than subjective, though it is very difficult to make objective observations and to come to objective conclusions because human beings are often influenced by their particular prejudices. Still, anything less than a completely objective observation or conclusion is not considered scientific. Finally, the conclusions of science are subject to change. Quite often in science, new studies, which might utilize new techniques and equipment, reveal that previous conclusions need to be modified or changed entirely.

The ultimate goal of science is to understand the natural world in terms of **theories,** concepts based on the conclusions of experiments and observations (Fig. 1.8). A detective might have a theory about a crime, or a baseball fan might have a theory about the win-loss record of the home team, but in science, the word *theory* is reserved for a conceptual scheme supported by much research and not yet found lacking. Some of the unifying theories of biology are:

Name of Theory	Explanation
Cell	All organisms are composed of cells
Biogenesis	Life comes only from life.
Evolution	All living things have a common ancestor and are adapted to a particular way of life.
Gene	Organisms contain coded information that dictates their form, function, and behavior.

In general, biological theories pertain to various aspects of life. The theory of evolution enables scientists to understand the history of life, the variety of living things, and the anatomy, physiology, and development of organisms—even their behavior. Because the theory of evolution has been supported by so much research for over a hundred years, some biologists refer to the *principle* of evolution. They believe this is the appropriate terminology for any theory that is generally accepted as valid by an overwhelming amount of scientific evidence.

Biologists ask questions and carry on investigations that pertain to the natural world. The conclusions of these investigations may eventually enable biologists to arrive at a theory that is generally accepted by all.

Scientists Have a Method

Scientists, including biologists, employ an approach for gathering information that is known as the **scientific method.** The approach of individual scientists to their work is as varied as they themselves are; still, for the sake of discussion it is possible to speak of the scientific method as consisting of certain steps. Figure 1.8 outlines the essential steps in the scientific method. On the basis of **data** (factual information), which may have been collected by someone previously, a scientist formulates a tentative statement, called a **hypothesis** [Gk. *hypothesis,* assumption]. This is used to guide his or her observations and experimentation, which produce new data.

The new data help a scientist come to a conclusion that either supports or does not support the hypothesis. Because hypotheses are always subject to modification, they can never actually be proven true; however, they can be proven false—that is, hypotheses are falsifiable. When data do not support the hypothesis, it must be rejected; therefore, some think of the body of science as what is left after alternative hypotheses have been rejected.

Scientists working in the same area, such as cell biology, evolution, genetics, or any other area, may eventually suggest a theory, a biological concept that helps biologists understand the natural world.

The scientific method consists of forming a hypothesis, testing it, and coming to a conclusion.

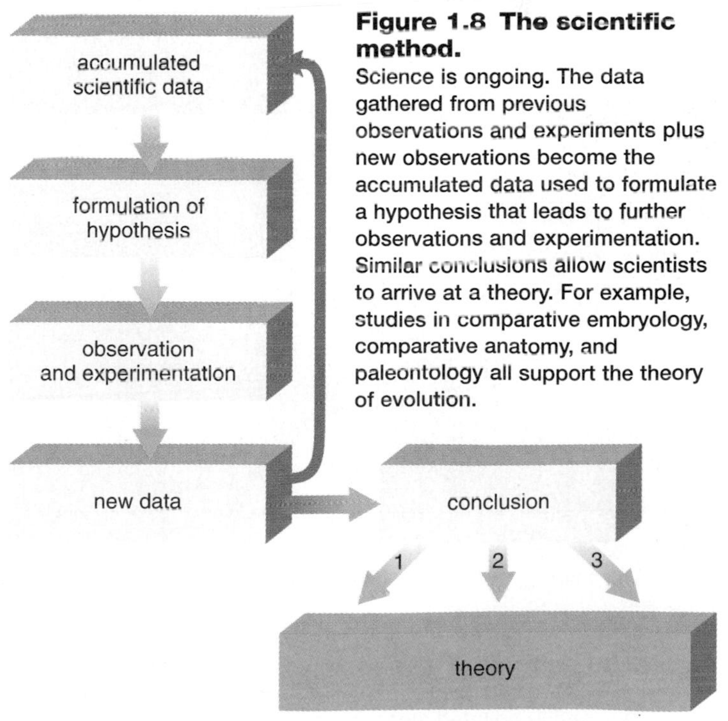

Figure 1.8 The scientific method.
Science is ongoing. The data gathered from previous observations and experiments plus new observations become the accumulated data used to formulate a hypothesis that leads to further observations and experimentation. Similar conclusions allow scientists to arrive at a theory. For example, studies in comparative embryology, comparative anatomy, and paleontology all support the theory of evolution.

An Experiment in Industrial Melanism

In order to examine the scientific method in more detail, we will consider research performed by the British scientist H.B.D. Kettlewell:

Accumulated Scientific Data

Scientists had observed the prevalence of dark-colored peppered moths on trees in polluted areas (where tree trunks are darker) and the prevalence of light-colored peppered moths on trees in nonpolluted areas (where tree trunks are lighter).

Formulating the Hypothesis

Formulating the hypothesis involves **inductive reasoning;** that is, scientists often use isolated facts to arrive at a possible explanation of the observed phenomenon. In this case the hypothesis was that predatory birds are responsible for the unequal distribution of moths because they feed on moths they can see.

Observation and Experimentation

Once the hypothesis has been stated, deductive reasoning comes into play. **Deductive reasoning** begins with a general statement that infers a specific conclusion. It often takes the form of an "if . . . then" statement: *If* predatory birds are responsible for the unequal distribution of dark- and light-colored moths, *then* we should see birds feeding primarily on light-colored moths in polluted areas and on dark-colored moths in nonpolluted areas.

To see if this deduction was correct, Kettlewell performed an experiment. He released equal numbers of the two types of moths in the two different areas. He observed that birds captured more dark-colored moths in nonpolluted areas and more light-colored moths in polluted areas.

New Data

From the unpolluted area, Kettlewell recaptured 13.7% of the light-colored moths and only 4.7% of the dark form. From the polluted area, he recaptured 27.5% of the dark form and only 13% of the light-colored moths (Fig. 1.9). Mathematical data like these are preferred because they are objective and cannot be influenced by the scientist's subjective feelings.

Conclusion

In this instance, the data supported the hypothesis that predatory birds are responsible for the unequal distribution of light- and dark-colored moths.

Reporting the Findings

It is customary to report findings in a scientific journal so that the design and the results of the experiment are available to all. It is necessary to give other researchers details on how experiments were conducted because results must be repeatable; that is, other scientists using the same procedures must get the same results. Otherwise, the hypothesis is no longer supported.

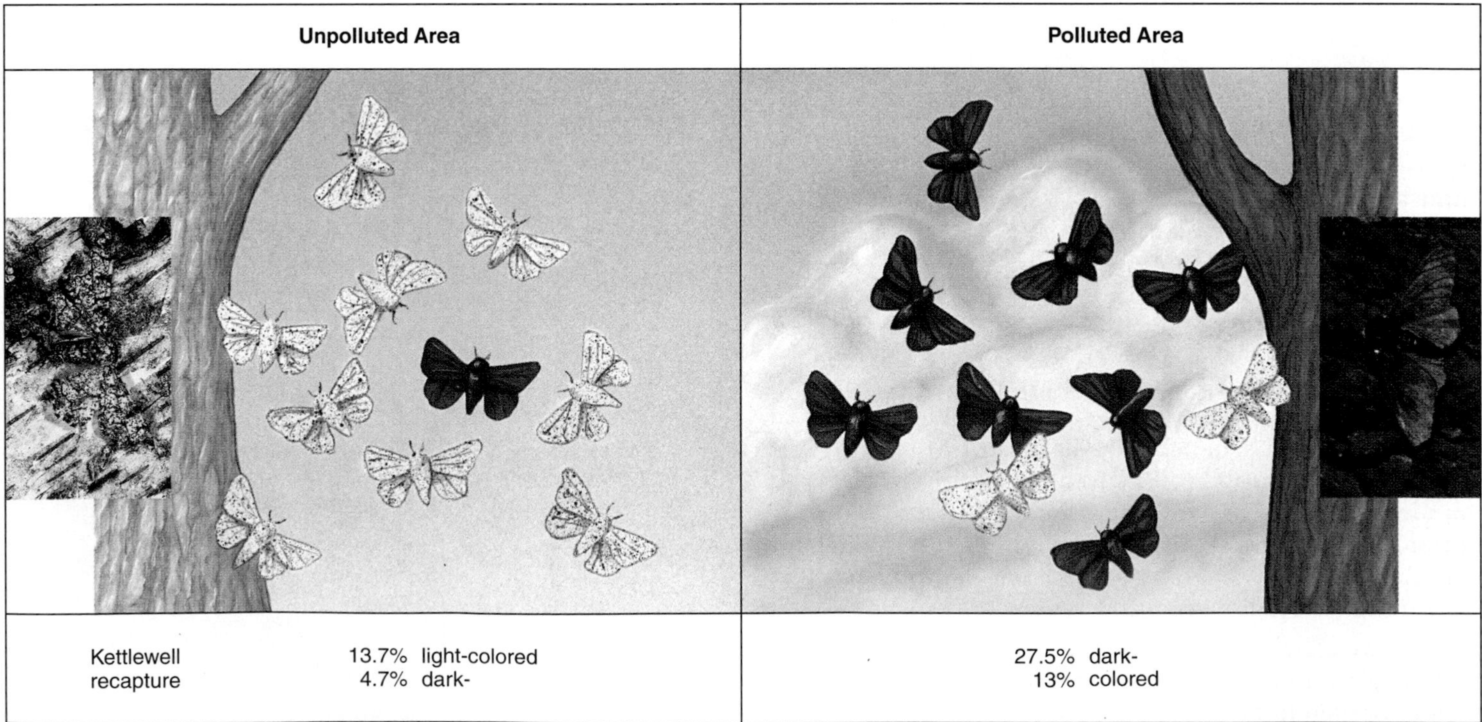

Figure 1.9 Industrial melanism experiment.
Kettlewell,working with *Biston betularia,* hypothesized that in nonpolluted areas, predatory birds fed on dark-colored moths resting on light tree trunks, and in polluted areas, predatory birds fed on light-colored moths resting on dark tree trunks. After releasing the same number of each type of moth in the two areas, he was able to recapture the percentages of each type noted.

Some Investigations Are Controlled Experiments

Most experiments have a **control group** that goes through all the steps of an experiment except the one being tested. The component in an experiment being tested is called the **experimental variable;** in other words, the investigator deliberately manipulates this step of the experiment. Then the investigator observes the effects of the experiment; in other words, he or she observes the **dependent variable:**

Experimental Variable	Dependent Variable
Component of the experiment being tested	Result or change that occurs due to the experimental variable

Example of a Controlled Experiment

As an example of a controlled experiment, suppose physiologists hypothesized that sweetener S is a safe food additive. They then design an experiment in which there are two groups of mice:

Experimental Group	Control Group
sweetener S in diet	no sweetener S in diet

To help ensure that the two groups are identical, inbred (genetically identical) mice are randomly placed into the two groups—say, ten mice per group. It is hoped that if any of the mice are different from the others, random placement will distribute them evenly between the groups. The experimenters also make sure that all environmental conditions—such as availability of water, cage conditions, and temperature of surroundings—are the same for all groups. The food for each group is exactly the same, except for the amount of sweetener S.

At the end of the experiment, both groups of mice are examined for bladder cancer. Let's suppose that 50% of the mice in the experimental group are found to have bladder cancer, while none in the control group have bladder cancer. The results of this experiment do not support the hypothesis that sweetener S is a safe food additive.

Use of a control group gives greater validity to the results of the experiment. In the described experiment, the control group did not show cancer, but the experimental group did. We can conclude that sweetener S in the diet must have brought about the cancer because this was the only difference between the two groups. Suppose, however, that both the experimental and control groups showed equal instances of bladder cancer. If so, we could not conclude that sweetener S in the diet produced the effect (bladder cancer). There must be some other components of the experiment causing the cancer.

At this point, the physiologists might decide to do more experiments. They might hypothesize that sweetener S is safe if the diet contains less than 50% sweetener. They might decide to feed sweetener S to groups of mice at ever-greater percentages of the total intake of food:

Group 1 : no sweetener S in food (control group)
Group 2 : sweetener S in 5% of food
Group 3 : sweetener S in 10% of food
↓
Group 11: sweetener S in 50% of food

Usually the data from experiments such as these are presented in the form of a table or graph (Fig. 1.10). A statistical test may be run to determine if the difference in the number of cases of bladder cancer among the various groups is significant. After all, if a significant number of mice in the control group develop cancer, the results may be invalid. On the basis of the results, the experimenters try to develop a recommendation concerning the safety of adding sweetener S to the food of humans. They may determine, for example, that ever-larger amounts of the sweetener—over 10% of food intake—are expected to cause a progressively increased incidence of bladder cancer.

Controlled experiments have a control group, which is not exposed to the experimental variable.

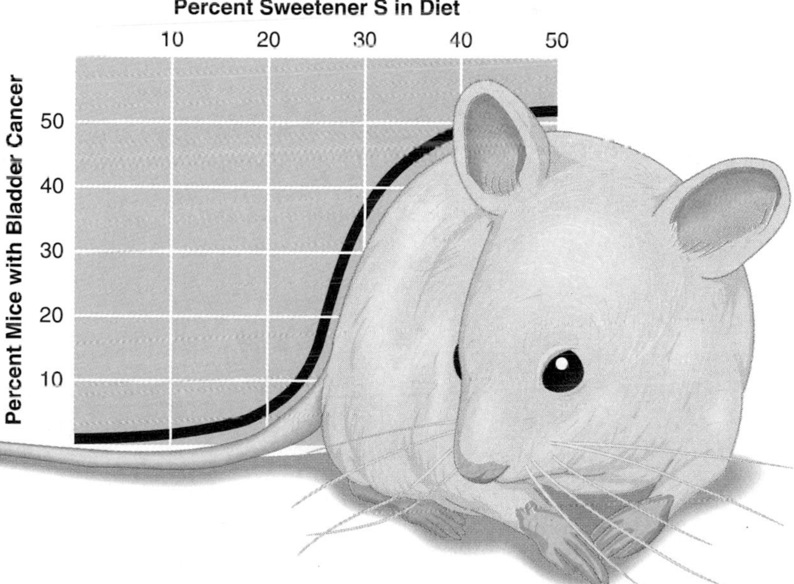

Figure 1.10 Hypothetical sweetener S study. The graph shows a positive correlation between increased amounts of sweetener S and increased incidence of bladder cancer.

Some Investigations Are Observational

Scientists don't always gather data by experimenting. Much of the data they gather is purely observational, but even so, the steps described for the scientific method are still applicable: observations are made, a hypothesis is formulated, predictions are made on the basis of the hypothesis, and data are collected that support or disprove the hypothesis.

Example of Observational Data

Brian C. R. Bertram investigated the reproductive behavior of ostriches, *Struthio camelus*, in East Africa, where as many as seven female ostriches share the same nest. A resident male builds a nest that consists of only a scrape in the ground. He then mates with a female, called the major hen, and she starts to lay eggs. After a few days, other females that may not have mated with the resident male (called minor hens) add their eggs to the nest, and the major hen allows this. However, only the major hen incubates and guards the nest, which eventually can contain up to forty eggs—far too many for her to incubate. She keeps about twenty eggs in the center of the nest, and she pushes the others out to form an outer ring (Fig. 1.11).

Bertram knew of two hypotheses attempting to explain how the eggs in the outer ring could protect the eggs in the center of the nest. If there is a benefit, this would explain the permissive attitude of the major hen toward the minor hens. Local natives thought that, in the case of a savannah fire, the eggs surrounding the nest would rupture and the liquid would prevent the fire from damaging the eggs in the nest. In contrast, two previous investigators thought that these eggs may help to regulate egg temperature.

Bertram carefully studied three ostriches' nests. Every egg was numbered, weighed, measured, photographed, and examined for surface characteristics. He was able to determine which eggs were major hens' eggs by observing the birds laying, using time-lapse photography, and depending on knowledge of egg characteristics. He found that the major hen is able to recognize her own eggs and that she usually pushes only minor hens' eggs into the outer ring (Fig. 1.11). Bertram then observed that predators such as jackals and vultures normally eat only a few eggs during an attack, and that they tend to choose eggs in the outer ring rather than in the center. Therefore, the major hen is protecting her own eggs by pushing minor hens' eggs out to form a ring!

Neither of the original hypotheses was supported, but Bertram used his observations to formulate and support a hypothesis of his own. So far no one has hypothesized why major hens allow some minor hens' eggs to remain in the center of the nest, where they will hatch. Can you formulate a possible hypothesis based on Bertram's data in Figure 1.11? What further observations need to be made?

Much scientific information is based on purely observational (descriptive) data, but the steps previously listed for the scientific method still apply.

Position of Eggs in Three Incubating Ostrich Nests

Eggs Laid by	No. of Eggs in Center (being incubated)	No. of (doomed) Eggs in Outer Ring
Nest A		
Major hen	9	0
Four other hens	10	8
Nest B		
Major hen	13	0
Two other hens	6	5
Nest C		
Major hen	9	1
Four other hens	9	10

Bertram, B.C.R. 1979. Ostriches recognize their own eggs and discard others. *Nature* 279:233.

Figure 1.11 Nesting behavior of ostrich, *Struthio camelus*.
Major hen ostrich rolls eggs out of a communal nest because they belong to other hens. The table shows that her eggs are kept in the center of the nest, where they are protected from predators who choose eggs in the outer ring. This is a benefit to the major hen because it increases her chances of having more living offspring.

connecting concepts

We have learned that there are levels of biological organization and that the cell, the simplest of living things, is composed of nonliving molecules. To understand life, therefore, we must begin with a study of cellular chemistry. We know that all life-forms take energy and materials from the environment and it will be our task in the next few chapters to discover how the cell makes use of energy and materials to maintain itself and reproduce.

Unicellular organisms, like all living things, respond to stimuli, reproduce themselves, and adapt to particular environments, but so do multicellular forms in which like cells form specialized tissues, tissues form organs, and organs form organ systems. Organisms in ecosystems like coral reefs and tropical rain forests have adaptations in structure and function that are immediately observable and emphasize the biodiversity of life.

What we know about biology and what we'll learn in the future results from objective observation and testing of the natural world. The ultimate goal of science is to understand the natural world in terms of theories—conceptual schemes supported by abundant research and not yet found lacking. Evolution is a theory that accounts for the differences that divide and the unity that joins all living things. All living things have the same levels of organization and function similarly because they are related—even back to the first living cells on earth.

Scientific creationism, which states that God created all species as they are today, cannot be considered science because creationism upholds a supernatural cause rather than a natural cause for events. When faith is involved, a hypothesis cannot be tested in a purely objective way. Just as science does not test religious beliefs, it does not make ethical or moral decisions. The general public may want scientists to label certain research as "good" or "bad" and to predict whether any resulting technology will primarily benefit or harm. Yet science, by its very nature, is impartial and simply attempts to study natural phenomena.

Summary

1.1 How to Define Life

Although living things are diverse, they share certain characteristics in common. Living things (a) are organized, and there are levels of organization from the cell to ecosystems, (b) need an outside source of materials and energy, (c) respond to external stimuli, (d) reproduce, passing on genes to their offspring, and (e) have adaptations suitable to their way of life in a particular environment.

The process of evolution explains both the unity and the diversity of life. Descent from a common ancestor explains why all organisms share the same characteristics, and adaptation to various ways of life explains the diversity of life-forms.

1.2 Ecosystems Contain Populations

Within an ecosystem, populations interact with one another and with the physical environment. Nutrients cycle within and between ecosystems, but energy flows unidirectionally, and eventually becomes heat.

Two examples of ecosystems, the coral reef and the tropical rain forest, illustrate that the adaptations of organisms allow them to play particular roles within an ecosystem.

1.3 How Living Things Are Classified

Living things are classified into groups from species to genus, family, order, class, phylum, and kingdom. In the five kingdom system of classification, monera are the unicellular bacteria; protists include protozoa and algae of varying complexity; fungi are multicellular organisms that absorb food; plants are multicellular photosynthesizers; and animals are multicellular organisms that ingest their food.

1.4 The Process of Science

Science seeks to understand the natural world and present this understanding in terms of theories.

The scientific method consists of: using accumulated data and inductive reasoning to formulate a hypothesis; using deductive reasoning to decide how to test by observation and experimentation the hypothesis; using the new data to come to a conclusion. Several conclusions supporting the same general hypothesis may result in a theory.

Scientists often do controlled experiments. In a controlled experiment, the experimental variable is that portion of the experiment being manipulated and the dependent variable is the change due to the experimental variable. The control group is not exposed to the experimental variable.

Reviewing the Chapter

1. What evidence can you cite to show that living things are organized? 2
2. What are the common characteristics of life listed in the text? 3–4
3. Why do living things require an outside source of materials and energy? 3
4. What is passed from generation to generation when organisms reproduce? What has to happen to the hereditary material DNA in order for evolution to occur? 4
5. How does evolution explain both the unity and the diversity of life? 4
6. What is an ecosystem, and why should human beings preserve ecosystems? 6
7. Choose an organism from a coral reef or a tropical rain forest and explain how it is adapted to its environment. 7–8
8. What kingdoms are used in the five kingdom system of classification? What types of organisms are found in each kingdom? 10
9. What is the ultimate goal of science? Give an example that supports your answer. 11
10. Describe the series of steps involved in the scientific method. Which of the steps requires the use of inductive reasoning? deductive reasoning? 12
11. Give an example of a controlled experiment. Name the experimental variable and the dependent variable. 13
12. Explain why you would expect that moral or ethical decisions cannot be made solely on scientific grounds. 15

Testing Yourself

Choose the best answer for each question. For questions 1–4, match the statements in the key with the sentences below.

Key:

a. Living things are organized.
b. Living things metabolize.
c. Living things respond.
d. Living things reproduce.
e. Living things evolve.

1. Genes made up of DNA are passed from parent to child.
2. Zebras run away from approaching lions.
3. Cells use materials and energy for growth and repair.
4. There are many different kinds of living things.
5. Evolution from the first cell(s) best explains why
   a. ecosystems have populations of organisms.
   b. photosynthesizers produce food.
   c. diverse organisms share common characteristics.
   d. All of these are correct.
6. Adaptation to a way of life best explains why living things
   a. display homeostasis.
   b. are diverse.
   c. began as single cells.
   d. are classified into five kingdoms.
7. Into which kingdom would you place a multicellular land organism that carries on photosynthesis?
   a. Protoctista
   b. Fungi
   c. Plantae
   d. Animalia
8. Which is the experimental variable in the experiment concerning sweetener S?
   a. Conditions like temperature and housing are the same for all groups.
   b. The amount of sweetener S in food.
   c. Two percent of the group fed food that was 10% sweetener S got bladder cancer, and 90% of the group fed food that was 50% sweetener S got bladder cancer.
   d. The data were presented as a graph.
9. Which is the control group in this same experiment?
   a. All mice in this group died because a lab assistant forgot to give them water.
   b. All mice in this group received food that was 10% sweetener S.
   c. All mice in this group received no sweetener S in food.
   d. Some mice in all groups got bladder cancer; therefore, there was no control group.
10. Which is an example of an observational investigation?
    a. Jones put broken eggshells in the nest of some gulls and watched the gulls' behavior.
    b. Smith measured the length of the twigs of all trees in the designated area.
    c. Green put pesticide into one jar of amoebas but not into the other jar.
    d. Kettlewell counted how many moths the birds did not eat.
11. An investigator spills dye on a culture plate and then notices that the bacteria live despite exposure to sunlight. He hypothesizes that the dye protects bacteria against death by ultraviolet (UV) light. To test this hypothesis, he decides to expose two culture plates to UV light. One plate contains bacteria and dye; the other plate contains only bacteria. Result: after exposure to UV light, the bacteria on both plates die. Fill in the right-hand portion of this diagram.

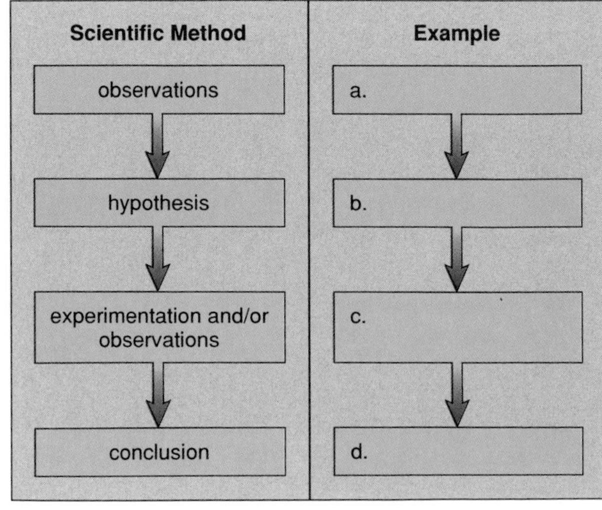

Applying the Concepts

1. *All living things evolved from a common ancestor.*
   How does this concept explain the unity of life-forms?
2. *Organisms are adapted to particular ways of life.*
   How does this concept explain the diversity of life?
3. *Scientists use the scientific method to gather information about the natural world.*
   When utilizing the scientific method, hypotheses can be proven false but not true. Explain.

Using Technology

Your study of biology is supported by these available technologies:

Exploring the Internet

The Mader Home Page provides resources for and help with studying this chapter.

http://www.mhhe.com/sciencemath/biology/mader/
(Click on Biology.)

Life Science Animations Video

Video #5: Plant Biology/Evolution/Ecology
Energy Flow through an Ecosystem (#52)

Understanding the Terms

adaptation 4
biosphere 6
control group 13
datum (pl. data) 11
deductive reasoning 12
dependent variable 13
ecosystem 2
energy 3
evolution 4
experimental variable 13
gene 4
homeostasis 3
hypothesis 11
inductive reasoning 12
metabolism 3
photosynthesis 3
population 2
reproduce 4
scientific method 11
species 4
taxonomy 10
theory 11

Match the terms to these definitions:

a. __________ All of the chemical reactions that occur in a cell during growth and repair.
b. __________ Changes that occur in populations of organisms with the passage of time, often resulting in increased adaptation of organisms to the prevailing environment.
c. __________ Component that is tested in an experiment by manipulating it and observing the results.
d. __________ Process by which plants utilize solar energy to make their own organic food.
e. __________ Sample that goes through all the steps of an experiment except the one being tested; a standard against which results of an experiment are checked.

Further Readings for Chapter One

Barnard, C. et al. 1993. *Asking questions in biology.* Essex: Longman Scientific & Technical. First-year life science students are introduced to the skills of scientific observation and inquiry.

Carey, S. 1994. *A beginner's guide to scientific method.* Belmont, CA: Wadsworth Publishing. The basics of the scientific method are explained.

Cranbrook, E., and Edwards, D. S. 1994. *Belalong: A tropical rain forest.* London: The Royal Geographic Society, and Singapore: Sun Tree Publishing. Provides a very readable, well-illustrated account of biodiversity in a Brunei rain forest.

Drewes, F. 1997. *How to study science.* 2d ed. Dubuque, Iowa: Wm. C. Brown Publishers. Supplements any introductory science text; shows students how to study and take notes and how to interpret text figures.

Johnson, G. B. 1996. *How scientists think.* Dubuque, Iowa: Wm. C. Brown Publishers. Presents the rationale behind 21 important experiments in genetics and molecular biology that became the foundation for today's research.

Marchuk, W. N. 1992. *A life science lexicon.* Dubuque, Iowa: Wm. C. Brown Publishers. Helps students master life sciences terminology.

Margulis, L. et al. 1994. *The illustrated five kingdoms: A guide to the diversity of life on Earth.* New York: HarperCollins College Publishers. Introduces the kingdoms of organisms.

Serafini, A. 1993. *The epic history of biology.* New York: Plenum Press. This is a history of biology from ancient Egyptian medicine to present-day biotechnology.

The Cell

A cell is the basic unit of life, and all living things are composed of cells. Therefore, our knowledge of the structure and function of a cell can be applied directly to the organism, whose organization depends on an ongoing input of matter and energy. Metabolic pathways carry out the transformations needed to change this input into the structure of the cell. In most ecosystems, the energy of the sun maintains life. Plant cells capture solar energy and store it in molecules that later are utilized by all living cells. Knowledge of chemistry, energy transformations, and metabolic pathways increases our understanding of the essence of life.

Basic Chemistry

CHAPTER 2

Chapter Concepts

Cheetah, *Acinonyx jubatus*

A hundred years ago, scientists believed that only nonliving things, like rocks and metals, consisted of chemicals. They thought that living things like sunflowers and cheetahs had a special force, called a vital force, which was necessary for life. Scientific investigation, however, has repeatedly shown that both nonliving and living things have the same physical and chemical bases. Although living things contain molecules not found in inanimate objects, such molecules must still be understood by studying basic chemical properties.

Suppose you have a special interest in cheetahs, and you read books about them and even go to Africa to watch cheetahs in the wild. Still, it would be necessary for you to study chemistry in order to fully understand cheetahs. We perceive the world in terms of whole objects like cheetahs, sunflowers, mushrooms, and humans, but in this chapter you will discover that all organisms consist of atoms and molecules linked together in specific ways that give them properties different from nonliving things. In order to understand how cheetahs run so quickly, the cheetah expert must study the physical and chemical nature of the cat's muscular system.

2.1 Matter Is Composed of Elements

Matter refers to anything that takes up space and has mass. It is helpful to remember that matter can exist as a solid, a liquid, or a gas. Then we can realize that not only are we matter but so too are the water we drink and the air we breathe.

All matter, both nonliving and living, is composed of certain basic substances called **elements.** It is quite remarkable that there are only 92 naturally occurring elements (see Appendix B). We know these are elements because they cannot be broken down to substances with different properties (a property is a chemical or physical characteristic, such as density, solubility, melting point, and reactivity).

Both the earth's crust and organisms are made up of elements, but they differ as to which ones are predominant (Fig. 2.1). Only six elements—carbon, hydrogen, nitrogen, oxygen, phosphorus, and sulfur—make up most (about 98%) of the body weight of most organisms. The acronym CHNOPS helps us remember these six elements. The properties of these elements are essential to the uniqueness of living things from cells to organisms.

All living and nonliving things are matter composed of elements. Six elements in particular are commonly found in living things.

Figure 2.1 Elements of earth's crust and organisms. The earth's crust primarily contains the elements oxygen, silicon, and aluminum. Organisms primarily contain the elements hydrogen, oxygen, carbon, and nitrogen. Along with phosphorus and sulfur, these elements make up most biological molecules.

Elements Contain Atoms

In the early 1800s, the English scientist John Dalton proposed that elements actually contain tiny particles called **atoms** [Gk. *atomos*, uncut, indivisible]. He also deduced that there is only one type of atom in each type of element. You can see why, then, the name assigned to each element is the same as the name assigned to the type of atom it contains. Some of the names we use for the elements (atoms) are derived from English and some are derived from Latin. One or two letters create the *atomic symbol,* which stands for this name. For example, the symbol H stands for a hydrogen atom, and the symbol Na (for *natrium* in Latin) stands for a sodium atom. Table 2.1 gives the atomic symbols for the other elements (atoms) commonly found in living things.

From our discussion of elements, we would expect each atom to have a certain weight. The weight of an atom is in turn dependent upon the presence of certain subatomic particles. Although physicists have identified a number of subatomic particles, we will consider only the most stable of these: **protons, neutrons,** and **electrons** [Gk. *elektron*, amber, electricity]. Protons and neutrons are located within the nucleus of an atom, and electrons move about the nucleus. Figure 2.2 shows the arrangement of the subatomic particles in helium, an atom that has only two electrons. In Figure 2.2*a* the stippling shows the probable location of electrons, and in Figure 2.2*b* the circle shows the average location of electrons.

Table 2.1

Common Elements in Living Things

Element*	Atomic Symbol	Atomic Number	Atomic Weight**
Hydrogen	H	1	1
Carbon	C	6	12
Nitrogen	N	7	14
Oxygen	O	8	16
Sodium	Na	11	23
Magnesium	Mg	12	24
Phosphorus	P	15	31
Sulfur	S	16	32
Chlorine	Cl	17	35
Potassium	K	19	39
Calcium	Ca	20	40

**The Periodic Table of the Elements appears in Appendix B. Note that here and in the full table, elements are arranged in order of ascending atomic number and weight.*

***Average of most common isotopes.*

Note: The atomic number gives the number of protons (and electrons in electrically neutral atoms). The number of neutrons is equal to the atomic weight minus the atomic number.

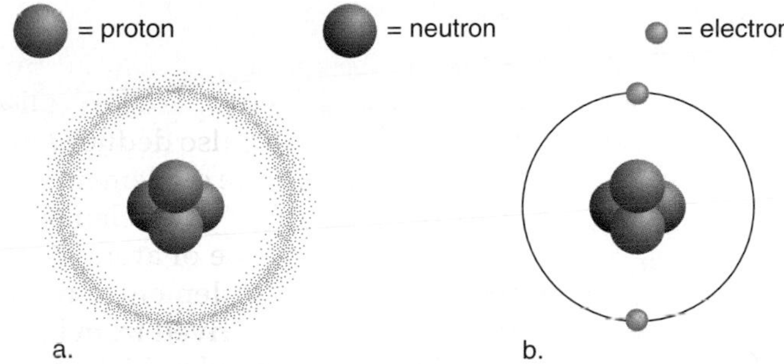

Figure 2.2 Model of helium (He).
Atoms contain subatomic particles, which are located as shown. Protons and neutrons are found within the nucleus, and electrons are outside the nucleus. **a.** The stippling shows the probable location of the electrons in the helium atom. **b.** The average location of electrons is sometimes represented by a circle.

Our concept of an atom has changed greatly since Dalton's day. If we could draw an atom the size of a football field, the nucleus would be like a gumball in the center of the field and the electrons would be tiny specks whirling about in the upper stands. Most of an atom is empty space. We should also realize that we can only indicate where the electrons are expected to be most of the time. In our analogy, the electrons may even stray outside the stadium at times.

The subatomic particles are so tiny that their weight is indicated by special units called atomic mass units (Table 2.2). Protons and neutrons each have about one atomic unit of weight. In comparison, electrons are much lighter; an electron weighs about $^1/_{1,800}$ that of a proton or neutron. Therefore, it is customary to disregard the combined weight of the electrons when calculating the total weight, called the **atomic weight,** of an atom.

All atoms of an element have the same number of protons. This is called the atom's **atomic number.** In Table 2.1, atoms are listed according to increasing atomic number, as they are in the periodic table of the elements found in Appendix B. This indicates that it is the number of protons (i.e., the atomic number) that makes an atom unique. The atomic number is often written as a subscript to the lower left of the atomic symbol. The atomic weight is often written as a superscript to the upper left of the atomic symbol. For example, the carbon atom can be noted in this way:

atomic weight / atomic number — $^{12}_{6}C$ — atomic symbol

Atoms have an atomic symbol, weight, and number. The subatomic particles (protons, neutrons, and electrons) determine the characteristics of atoms.

Table 2.2
Subatomic Particles

Name	Charge	Weight
Electron	One negative unit	Very, very little
Proton	One positive unit	One atomic mass unit
Neutron	No charge	A little more than one atomic mass unit

Isotopes Have Many Uses

The atomic weights given in Table 2.1 are the average weight of each type of atom. This is because atoms of the same type may differ in the number of neutrons. Atoms that have the same atomic number and differ only in the number of neutrons are called **isotopes** [Gk. *isos*, equal, and *topos*, place]. Three isotopes of carbon can be written in the following manner:

$$^{12}_{6}C \quad ^{13}_{6}C \quad ^{14}_{6}C$$

Carbon-12 has six neutrons, carbon-13 has seven neutrons, and carbon-14 has eight neutrons. Unlike the other two isotopes, carbon-14 is unstable; it breaks down into elements with lower atomic numbers. When it decays, it emits radiation in the form of radioactive particles or radiant energy. Therefore carbon-14 (^{14}C) is called a *radioactive isotope.*

Isotopes have many uses. Because proportions of isotopes in various food sources are known, biologists can now determine the proportion of isotopes in mummified or fossilized human tissues to know what ancient peoples ate. Radioactive isotopes are used as tracers in biochemical experiments. For example, ^{14}C was used to detect the sequential biochemical steps that occur during photosynthesis. And because ^{14}C decays at a known rate, the amount of ^{14}C remaining is often used to determine the age of fossils.

Radioactive isotopes are also used in medicine. If a patient is injected with radioactive iodine, the thyroid gland will take it up, and a scan of the thyroid will indicate if any abnormality is present. Glucose labeled with a radioactive isotope can be injected into the body and will be taken up by metabolically active tissues. In PET (positron emission tomography), the radiation given off is used by a computer to generate cross-sectional images that indicate the metabolic activity of various tissues.

Atoms that have the same number of protons but a different number of neutrons and a different weight are called isotopes.

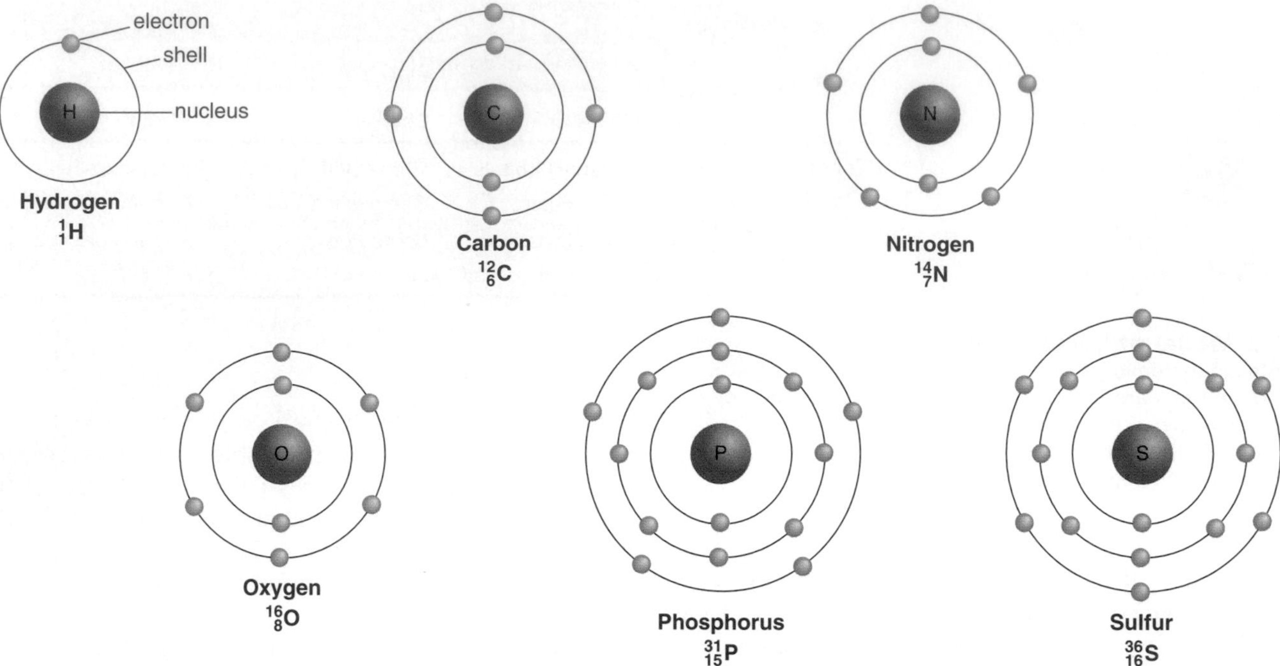

Figure 2.3 Bohr models of atoms.
Electrons are placed in energy levels (electron shells) according to certain rules: the first shell can contain up to two electrons, and each shell thereafter can contain up to eight electrons as long as we consider only atoms with an atomic number of 20 or below. Each shell is to be filled before electrons are placed in the next shell.

Atoms Have Chemical Properties

Protons and electrons carry a charge; protons have a positive (+) electrical charge, and electrons have a negative (–) electrical charge. When an atom is electrically neutral, the number of protons equals the number of electrons. Therefore, in electrically neutral atoms, the atomic number tells you the number of protons and the number of electrons. For example, a carbon atom has six protons, and when electrically neutral, it also has six electrons.

In 1913, the Danish physicist Niels Bohr proposed that electrons orbit in concentric energy levels (called *electron shells*) about the nucleus. He based his model on a previous discovery that although electrons have the same weight and charge, they vary in energy content. **Energy** is defined as the ability to do work. Electrons differ in the amount of their *potential energy,* that is, stored energy ready to do work. The electron shells indicate the relative amounts of stored energy electrons have. Electrons with the least amount of potential energy are located in the shell closest to the nucleus, called the *K* shell. Electrons in the next higher shell, called the *L* shell, have more energy, and so forth, as we proceed from shell to shell outside the nucleus.

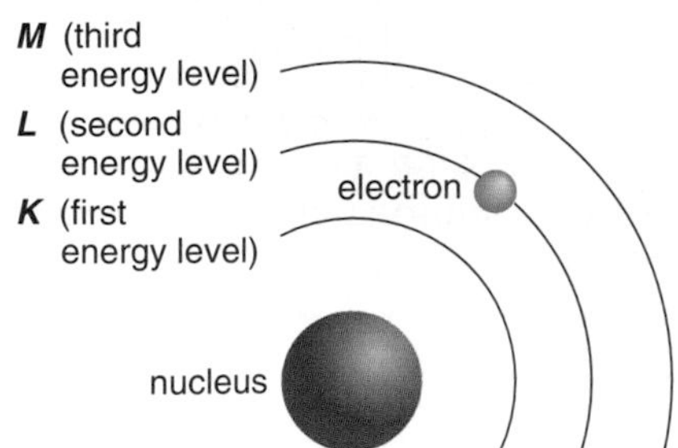

An analogy may help you appreciate that the farther electrons are from the nucleus, the more potential energy they possess. Falling water has energy, as witnessed by how it can turn a waterwheel connected to a shaft that transfers the energy to machinery for grinding grain, cutting wood, or weaving cloth. The higher the waterfall, the greater the amount of energy released per unit amount of water. The potential energy possessed by electrons is not due to gravity; it is due to the attraction between the positively charged protons and the negatively charged electrons. It takes energy to keep an electron farther away from the nucleus as opposed to closer to the nucleus. You have often heard that sunlight provides the energy for photosynthesis in green plants, but it may come as a surprise to learn that when a pigment such as chlorophyll absorbs the energy of the sun, electrons move to higher energy levels about the nuclei.

Atomic Configurations

Figure 2.3 shows you how the Bohr model helps you determine how many electrons are in the outer shell of an atom. For atoms up to calcium, which has an atomic number of 20, the first shell can contain up to two electrons; thereafter, each additional shell can contain eight electrons. For these atoms, each lower shell is filled with electrons before the next higher shell contains any electrons. The sulfur atom, with an atomic number of 16, has three shells (two electrons in the first shell, eight electrons in the second shell, and six electrons in the third, or outer, shell).

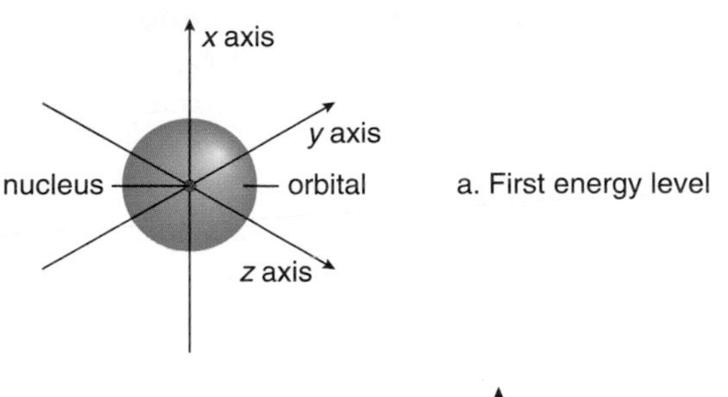

a. First energy level

Figure 2.4 Electron orbitals.
Each electron energy level (see Fig. 2.3) has one or more orbitals, a volume of space in which the rapidly moving electrons are most likely found. The nucleus is at the intersection of axes *x*, *y*, and *z*. **a.** The first energy level (electron shell) has only one orbital, which has a spherical shape. Two electrons can occupy this orbital. **b.** The second energy level has one spherical-shaped orbital and three dumbbell-shaped orbitals at right angles to each other. Since two electrons can occupy each orbital, there are a total of eight electrons in the second electron shell. Each orbital is drawn separately here, but actually the second spherical orbital surrounds the first spherical orbital, and the dumbbell-shaped orbitals pass through the spherical ones.

b. Second energy level

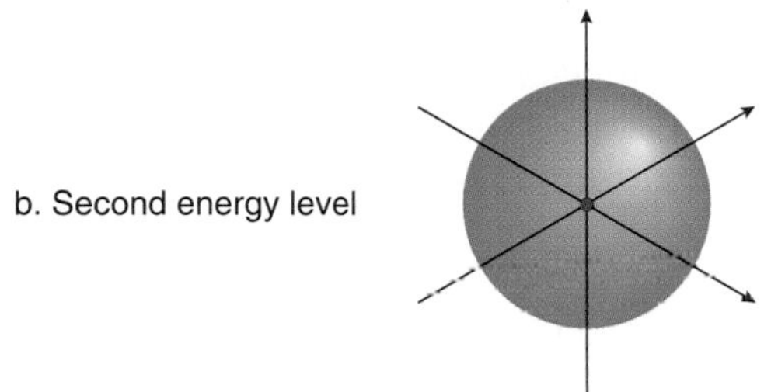
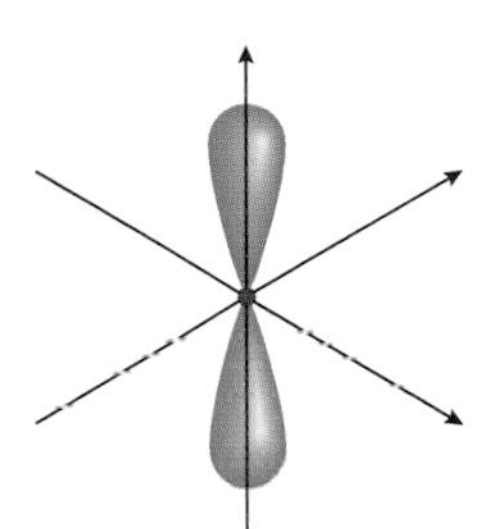
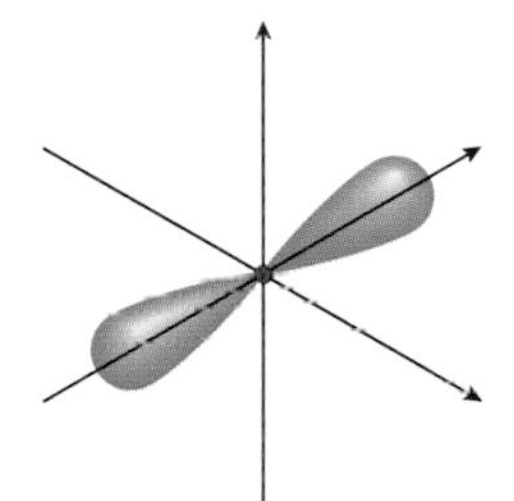
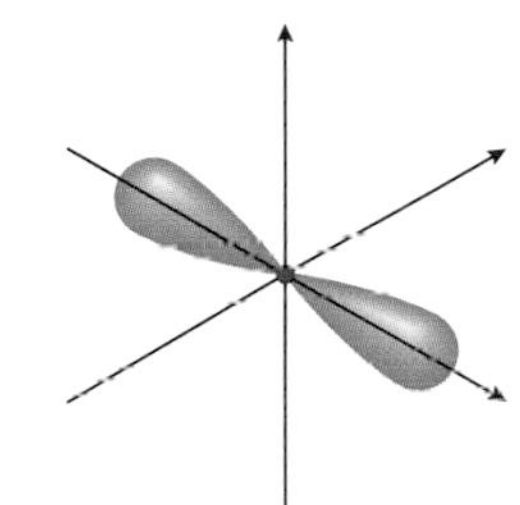

In the periodic table (see Appendix B), elements are arranged in rows according to the number of electrons in the outer shell. If an atom has only one shell, the outer shell is complete when it has two electrons. Otherwise, the *octet rule*, which states that the outer shell is most stable when it has eight electrons, holds. Atoms with eight electrons in the outer shell do not ordinarily react; they are said to be inert. Atoms with fewer than eight electrons in the outer shell react with other atoms in such a way that after the reaction, each has a stable outer shell. Atoms can give up, accept, or share electrons in order to have a stable outer shell.

Electrons Occupy Orbitals

The Bohr model was later modified because it is not actually possible to pinpoint the location of an electron. Rather than an orbit, an electron occupies an **orbital,** which is a volume of space where a rapidly moving electron is statistically predicted to be found (Fig. 2.4). An orbital has a characteristic energy state and a characteristic shape. At the first energy level, there is only a single spherical orbital, where at most two electrons are found about the nucleus. The space is spherical shaped because the most likely location for each electron is a fixed distance in all directions from the nucleus. At the second energy level, there are four orbitals; one of these is spherical shaped but the other three are dumbbell shaped. This is the shape that allows the electrons to be most distant from one another. Since each orbital can hold two electrons, there are a maximum of eight electrons in the *L* shell. Higher shells can be more complex and contain more orbitals if they are an inside shell. If such a shell is the outer shell, it too has only four orbitals and a maximum number of eight electrons.

The number of electrons in the outer shell determines the manner in which atoms react with one another.

Writing Chemical Formulas and Equations

When writing chemical formulas, atomic symbols are used to represent the atoms, and subscripts are used to indicate how many atoms of each type there are in a substance. For example, the chemical formula H_2O (read as H-two-O) indicates that a water molecule contains two hydrogen atoms and one oxygen atom. The chemical formula for glucose contains many atoms:

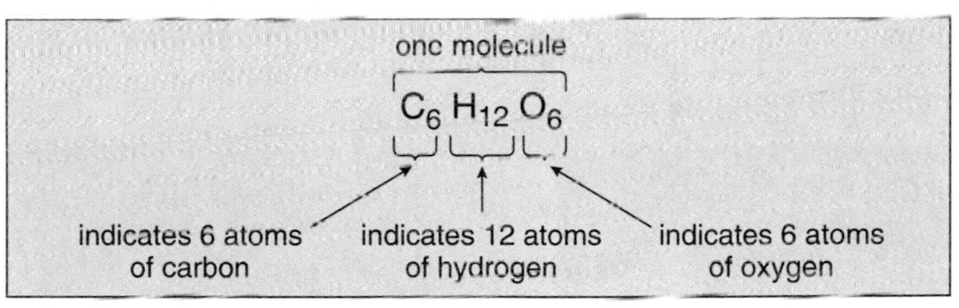

Formulas are used in chemical equations to represent chemical reactions that occur between atoms and molecules:

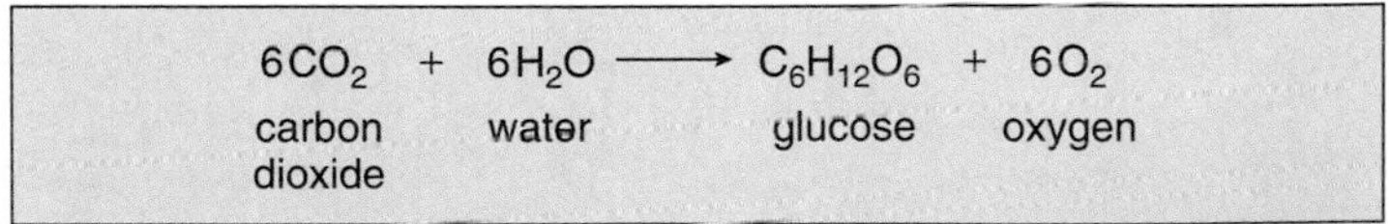

In this equation, which is often used to represent photosynthesis, six molecules of carbon dioxide react with six molecules of water to yield one glucose molecule and six molecules of oxygen. The reactants (molecules that participate in the reaction) are shown on the left of the arrow, and the products (molecules formed by the reaction) are shown on the right. Notice that the equation is "balanced," that is, there are the same number of each type of atom on both sides of the arrow.

2.2 Atoms Form Compounds and Molecules

Molecules can form when two or more atoms of the same element react with one another. Atmospheric oxygen does not exist as a single atom, O; instead, two oxygen atoms are joined as O_2. When atoms of two or more different elements react or bond together, a **compound** results. Water (H_2O) is a compound that contains the elements hydrogen and oxygen. We can also speak of H_2O as a **molecule** [L. *moles,* mass], because a molecule is the smallest part of a compound that still has the properties of that compound.

Electrons possess energy; and the bonds that exist between atoms contain energy. Organisms are dependent upon chemical-bond energy to maintain their organization. When a chemical reaction occurs, electrons shift in their relationship to one another and energy may be given off. This is the energy that we make use of to carry on our daily lives.

Opposite Charges in Ionic Bonding

Ionic bonds form when electrons are transferred from one atom to another. For example, sodium (Na), with only one electron in its third shell, tends to be an electron *donor* (Fig. 2.5*a*). Once it gives up this electron, the second shell, with eight electrons, becomes its outer shell. Chlorine (Cl), on the other hand, tends to be an electron *acceptor.* Its outer shell has seven electrons, so it needs only one more electron to have a completed outer shell. When a sodium atom and a chlorine atom come together, an electron is transferred from the sodium atom to the chlorine atom. Now both atoms have eight electrons in their outer shells.

This electron transfer, however, causes a charge imbalance in each atom. The sodium atom has one more proton than it has electrons; therefore, it has a net charge of +1 (symbolized by Na^+). The chlorine atom has one more electron than it has protons; therefore, it has a net charge of −1 (symbolized by Cl^-). Such charged particles are called **ions.** Sodium (Na^+) and chlorine (Cl^-) are not the only biologically important ions. Some, such as potassium (K^+), are formed by the transfer of a single electron to another atom; others, such as calcium (Ca^{2+}) and magnesium (Mg^{2+}), are formed by the transfer of two electrons.

Ionic compounds are held together by an attraction between the charged ions called an **ionic bond.** When sodium reacts with chlorine, an ionic compound called sodium chloride (NaCl) results, and the reaction is called an ionic reaction. Sodium chloride is a salt commonly known as table salt because it is used to season our food (Fig. 2.5*b*). Salts can exist as dry solids, but when such a compound is placed in water, the ions separate as the salt dissolves, as when NaCl separates into Na^+ and Cl^-. Ionic compounds are most commonly found in this dissociated (ionized) form in biological systems because these systems are 70–90% water.

The transfer of electron(s) between atoms results in ions that are held together by an ionic bond, the attraction of negative and positive charges.

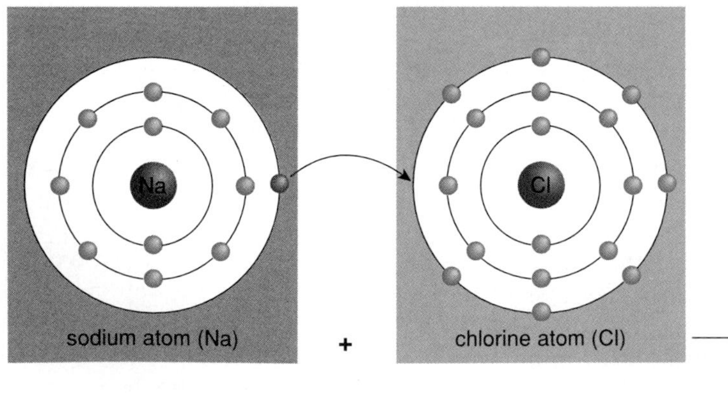

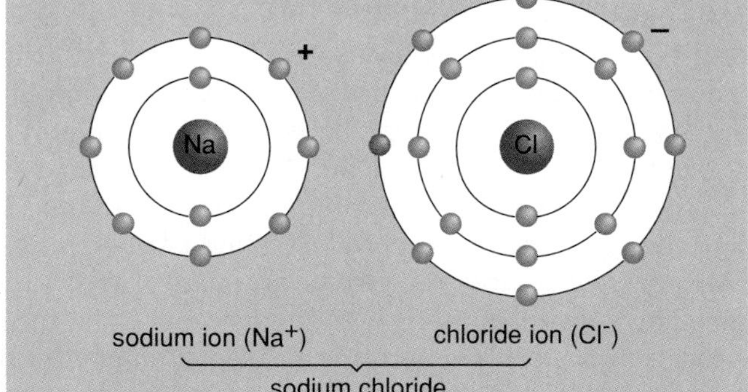

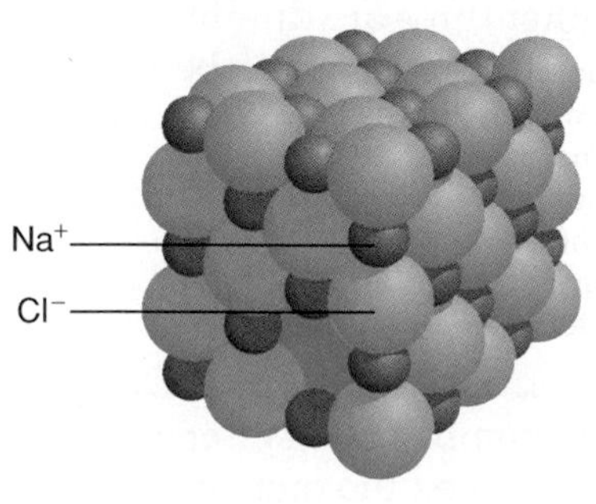

Figure 2.5 Ionic reaction.
a. During the formation of sodium chloride, an electron is transferred from the sodium atom to the chlorine atom. At the completion of the reaction, each atom has eight electrons in the outer shell, but each also carries a charge as shown. **b.** In a sodium chloride crystal, ionic bonding between Na^+ and Cl^- causes the atoms to assume a three-dimensional lattice in which each sodium ion is surrounded by six chlorine ions, and each chlorine ion is surrounded by six sodium ions.

Sharing in Covalent Bonding

A **covalent bond** [L. *co*, together, with, and *valens*, strength] results when two atoms *share* electrons in such a way that each atom has an octet of electrons in the outer shell (or two electrons, in the case of hydrogen). In a hydrogen atom the outer shell is complete when it contains two electrons. If hydrogen is in the presence of a strong electron acceptor, it gives up its electron to become a hydrogen ion (H^+). But if this is not possible, hydrogen can share with another atom and thereby have a completed outer shell. For example, a hydrogen atom can share with another hydrogen atom. In this case, the two orbitals overlap and the electrons are shared between them (Fig. 2.6*a*). Because they share the electron pair, each atom has a completed outer shell. When a reaction results in a covalent molecule, it is called a covalent reaction.

A more common way to symbolize that atoms are sharing electrons is to draw a line between the two atoms as in the structural formula H–H. In a molecular formula, the line is omitted and the molecule is simply written as H_2.

Like a single bond between two hydrogen atoms, a double bond can also allow two atoms to complete their octets. In a double covalent bond, two atoms share two pairs of electrons (Fig. 2.6*b*). In order to show that oxygen gas (O_2) contains a double bond, the molecule can be written as O=O.

It is even possible for atoms to form triple covalent bonds as in nitrogen gas (N_2), which can be written as N≡N. Single covalent bonds between atoms are quite strong, but double and triple bonds are even stronger.

Electron Model	Structural Formula	Molecular Formula
a.	H–H	H_2
b.	O=O	O_2
c.	H–C–H (with H above and below C)	CH_4

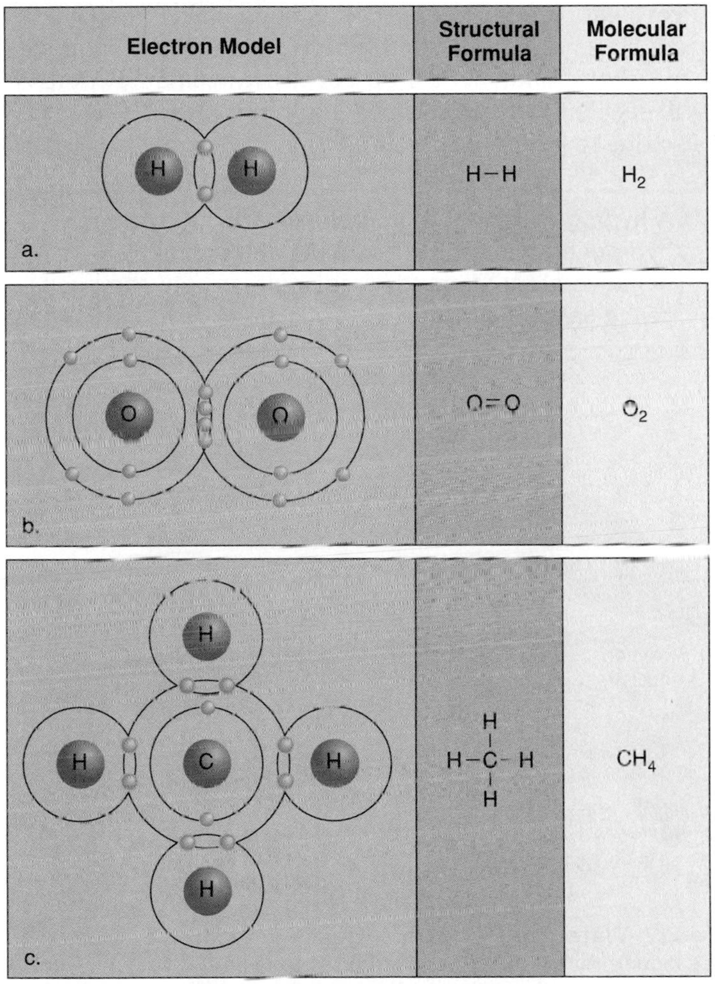

Figure 2.6 Covalently bonded molecules.
In a covalent bond, atoms share electrons so that each atom has a completed outer shell. **a.** A molecule of hydrogen (H_2) contains two hydrogen atoms sharing a pair of electrons. This single covalent bond can be represented in any of the three ways shown. **b.** A molecule of oxygen (O_2) contains two oxygen atoms sharing two pairs of electrons. This results in a double covalent bond. **c.** A molecule of methane (CH_4) contains one carbon atom bonded to four hydrogen atoms. By sharing pairs of electrons, each has a completed outer shell.

In a covalent molecule, atoms share electrons; not only single bonds but also double and even triple bonds are possible.

Oxidation Is the Opposite of Reduction

Oxidation-reduction reactions are an important type of reaction in cells, but the terminology was derived from studying reactions outside of cells. When oxygen combines with a metal, oxygen receives electrons and becomes negatively charged and the metal loses electrons and becomes positively charged.

Today, the terms **oxidation** and **reduction** are applied to many ionic reactions, whether or not oxygen is involved. Very simply, *oxidation refers to the loss of electrons, and reduction refers to the gain of electrons.* In our previous ionic reaction, $Na + Cl \rightarrow NaCl$, the sodium has been oxidized (loss of electron) and the chlorine has been reduced (gain of electron).

The terms oxidation and reduction are also applied to certain covalent reactions. In this case, however, oxidation is the loss of hydrogen atoms, and reduction is the gain of hydrogen atoms. A hydrogen atom contains one proton (symbolized as H^+) and one electron (symbolized as e^-). Therefore, when a molecule loses a hydrogen atom, it has lost an electron, and when a molecule gains a hydrogen atom, it has gained an electron.

When oxidation occurs, an atom is oxidized (loses electrons). When reduction occurs, an atom is reduced (gains electrons). These two processes occur concurrently in oxidation-reduction reactions.

Some Covalent Bonds Are Polar

Normally, the sharing of electrons between two atoms is fairly equal, and the covalent bond is nonpolar. All the molecules in Figure 2.6, including methane (CH_4), are nonpolar. In the case of water (H_2O), however, the sharing of electrons between oxygen and each hydrogen is not completely equal. The larger oxygen atom, with the greater number of protons, dominates the H_2O association. The attraction of an atom for the electrons of a covalent bond is called electronegativity. The oxygen atom is more electronegative than the hydrogen atom, and it can attract the electron pair to a greater extent. In a water molecule, this causes the oxygen atom to assume a slightly negative charge (δ^-), and it causes the hydrogen atoms to assume a slightly positive charge (δ^+). The unequal sharing of electrons in a covalent bond creates a **polar covalent bond,** and, in the case of water, the molecule itself is a polar molecule (Fig. 2.7).

The water molecule is a polar molecule and has a symmetric distribution of charge: one end of the molecule (the oxygen atom) carries a slightly negative charge, and the other ends of the molecule (the hydrogen atoms) carry slightly positive charges.

Hydrogen Bonding

Polarity within a water molecule causes the hydrogen atoms in one molecule to be attracted to the oxygen atoms in other molecules (Fig. 2.7*b*). This attractive force creates a weak bond called a **hydrogen bond.** This bond is often represented by a dotted line because a hydrogen bond is easily broken. Hydrogen bonding is not unique to water. A biological molecule can contain many polar covalent bonds involving hydrogen and usually oxygen or nitrogen. The electropositive hydrogen atom of one molecule is attracted to the electronegative oxygen or nitrogen atom of another molecule within the same or different molecules.

Although a hydrogen bond is more easily broken than a covalent bond, many hydrogen bonds taken together are quite strong. Hydrogen bonds between parts of cellular molecules help maintain their proper structure and function. We will see that some of the important properties of water are also due to hydrogen bonding.

A hydrogen bond occurs between a slightly positive hydrogen atom of one molecule and a slightly negative atom of another molecule or between parts of the same molecule.

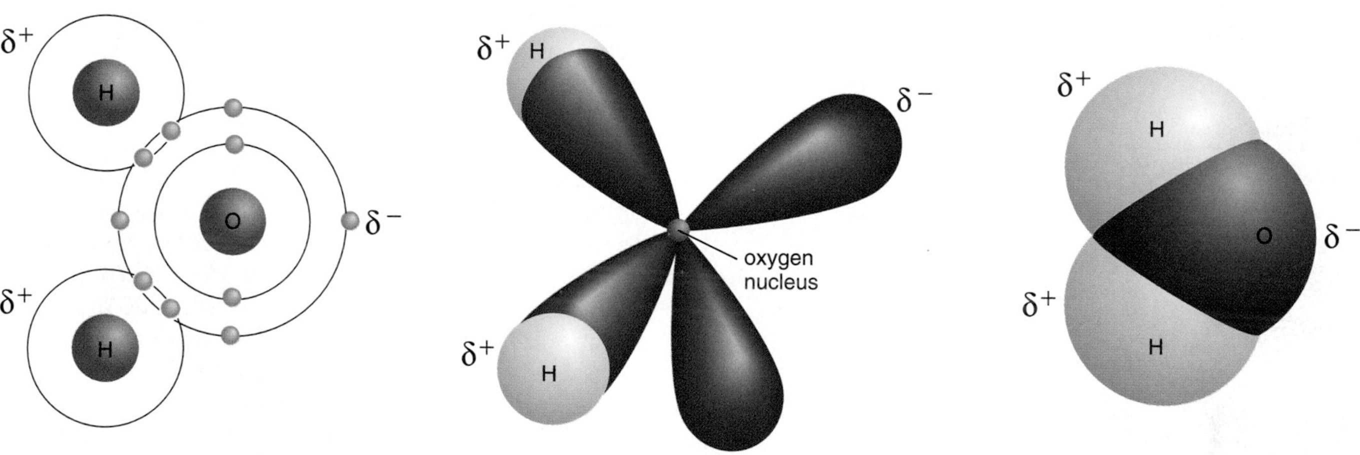

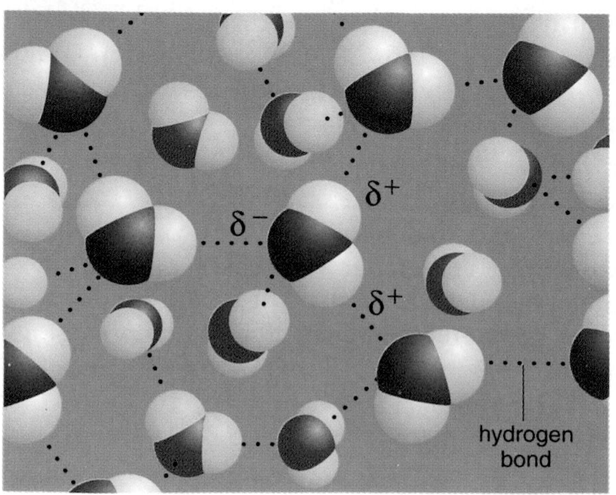

Figure 2.7 Water molecule.
a. Three models for the structure of water. The electron model does not indicate the shape of the molecule. The orbital model shows that the four orbitals of the outer shell (see Fig. 2.4*b*) have rearranged to produce tear-shaped orbitals that point toward the corners of a tetrahedron (a polyhedron having four sides). In water only two of these orbitals are utilized in covalent bonding. The space-filling model shows the V shape of a water molecule. Water is a polar molecule; the oxygen attracts the electrons more strongly than do the hydrogens, and there is a slightly positive charge on each hydrogen and a slightly negative charge on the oxygen. **b.** Hydrogen bonding between water molecules. A hydrogen bond is the attraction of a slightly positive hydrogen to a slightly negative atom in the vicinity. Each water molecule can bond to four other molecules in this manner. When water is in its liquid state, some hydrogen bonds are forming and others are breaking at all times.

2.3 Water Is Essential to Life

The first cell(s) evolved in water, and all living things are 70–90% water. What are the unique properties of water that make water essential to the continuance of life? Water is a polar molecule and water molecules are hydrogen bonded to one another (Fig. 2.7). A hydrogen bond is much weaker than a covalent bond within a water molecule, but taken together, hydrogen bonds cause water molecules to cling together. Without hydrogen bonding between molecules, water would boil at −80 °C and freeze at −100 °C, making life as we know it impossible. But because of hydrogen bonding, water is a liquid at temperatures suitable for life. It boils at 100 °C and freezes at 0 °C.

Water Has Unique Properties

The temperature of liquid water rises and falls more slowly than that of most other liquids. A calorie is the amount of heat energy needed to raise the temperature of one gram of water 1 °C. In comparison, other covalently bonded liquids require only about half this amount of energy to rise in temperature 1 °C. The many hydrogen bonds that link water molecules help water absorb heat without a great change in temperature.

When water cools down, heat is released. Converting one gram of the coldest liquid water to ice requires the loss of 80 calories of heat energy (Fig. 2.8). However, water holds heat, and its temperature falls more slowly than other liquids. This property of water is important not only for aquatic organisms but also for all living things. Water protects organisms from rapid temperature changes and helps them maintain their normal internal temperatures.

Water has a high heat of vaporization. Converting one gram of the hottest water to steam requires an input of 540 calories of heat energy (Fig. 2.8). This high heat of vaporization means that water has a high boiling point. Because water boils at 100 °C, it is in a liquid state at temperatures suitable to living things.

Hydrogen bonds must be broken to change water to steam; this accounts for the very large amount of heat needed for evaporation. This property of water helps moderate the earth's temperature so that it permits the continuance of life. It also gives animals in a hot environment an efficient way to release excess body heat. When an animal sweats, body heat is used to vaporize the sweat, thus cooling the animal.

a.

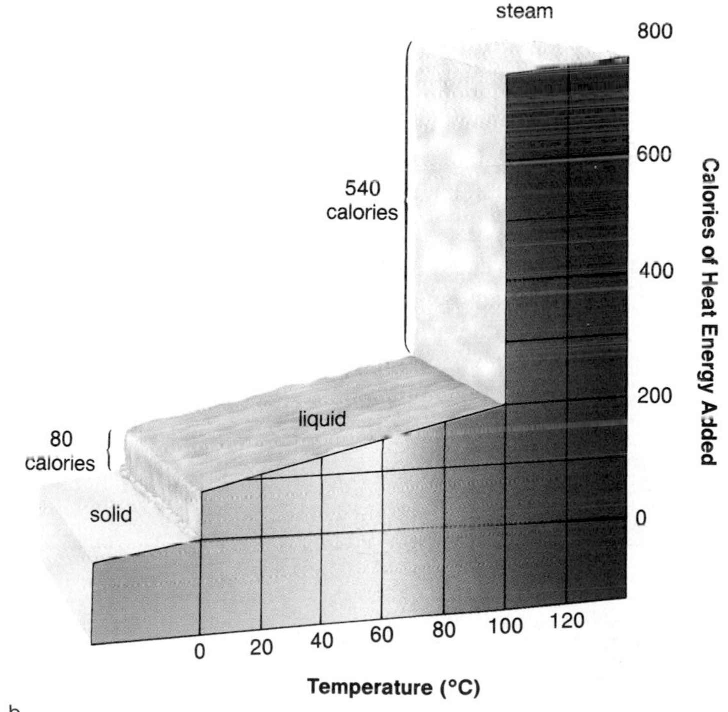

b.

Figure 2.8 Temperature and water.
a. Water has a high heat of vaporization; therefore, splashing water on the body when temperatures rise keeps animals cool. **b.** Water is the only common molecule that can be a solid, a liquid, or a gas according to naturally occurring environmental temperatures. At ordinary temperatures and pressure, water is a liquid, and it takes a large input of heat to change it to steam. In contrast, water gives off heat when it freezes, and this heat will keep the environmental temperature higher than expected. Can you see why there are less severe changes in temperature along the coasts?

Water is the universal solvent and facilitates chemical reactions both outside of and within living systems. Water is the universal solvent in that it dissolves a great number of solutes. When a salt, such as sodium chloride (NaCl), is put into water, the negative ends of the water molecules are attracted to the sodium ions, and the positive ends of the water molecules are attracted to the chloride ions. This causes the sodium ions and the chloride ions to separate and to dissociate in water:

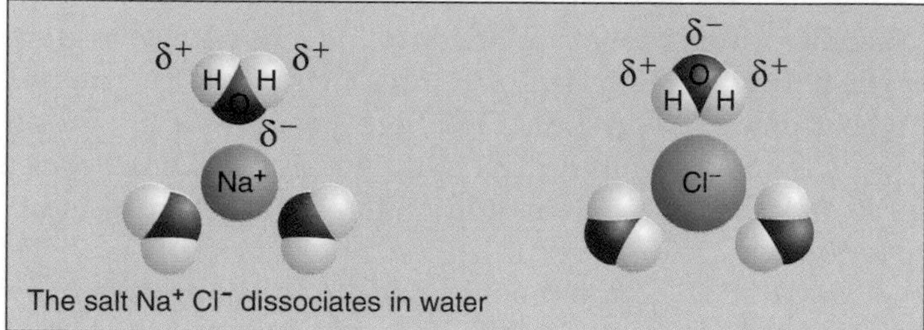

The salt $Na^+ Cl^-$ dissociates in water

Water is also a solvent for larger molecules that contain ionized atoms or are polar molecules:

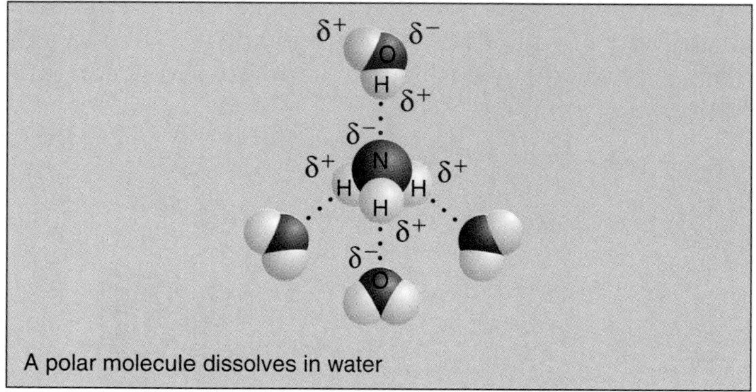

A polar molecule dissolves in water

When ions and molecules disperse in water, they move about and collide, allowing reactions to occur. Those molecules that can attract water are said to be **hydrophilic** [Gk. *hydrias*, of water, and *phileo*, love]. Nonionized and nonpolar molecules that cannot attract water are said to be **hydrophobic** [Gk. *hydrias*, of water, and *phobos*, fear].

Water molecules are cohesive and adhesive. Cohesion is apparent because water flows freely, yet water molecules do not break apart. They cling together because of hydrogen bonding. Because water molecules have a positive and negative pole, they adhere to surfaces, particularly polar surfaces; therefore, water exhibits adhesion. Because water can fill a tubular vessel and still flow, dissolved and suspended molecules are evenly distributed throughout a system. For these reasons, water is an excellent transport system both outside of and within living organisms. One-celled organisms rely on external water to transport nutrient and waste molecules, but multicellular organisms often contain internal vessels in which water serves to transport nutrients and wastes. For example, the liquid portion of our blood is 90% water that contains dissolved and suspended substances.

Cohesion and adhesion both contribute to the transport of water in plants. Plants have their roots anchored in the soil where they absorb water, but the leaves are uplifted and exposed to solar energy. How is it possible for water to rise to the top of even very tall trees (Fig. 2.9)? The properties of water account for the transport of water in plants. A plant contains a system of vessels that reaches from the roots to the leaves. Water evaporating from the leaves is immediately replaced with water molecules from the vessels. Because water molecules are cohesive, a tension is created that pulls water up from the roots. Adhesion of water to the walls of the vessels also helps prevent the water column from breaking apart.

Water has a high surface tension. It's possible to skip rocks on water because it has a high surface tension. Surface tension is measured by determining how difficult it is to break the surface of a liquid. As with cohesion, hydrogen bonding causes water to have a high surface tension. A water strider can even walk on the surface of a pond without breaking the water's surface. Table 2.3 summarizes the properties of water.

Table 2.3

Water

Properties	Chemical Reason	Effect
Resists change of state (from liquid to ice and from liquid to steam)	Hydrogen bonding	Moderates earth's temperature
Resists changes in temperature	Hydrogen bonding	Helps keep body temperatures constant
Universal solvent	Polarity	Facilitates chemical reactions
Is cohesive and adhesive	Hydrogen bonding; polarity	Serves as transport medium
Has a high surface tension	Hydrogen bonding	Difficult to break surface
Less dense as ice than as liquid water	Hydrogen bonding changes	Ice floats on water

Unlike most substances, frozen water is less dense than liquid water. As water cools, the molecules come closer together. They are densest at 4°C, but they are still moving about (Fig. 2.10). At temperatures below 4°C, there is only vibrational movement, and hydrogen bonding becomes more rigid but also more open. This means that water expands as it freezes, which is why cans of Coke burst when placed in a freezer or frost heaves make northern roads bumpy in the winter. It also means that ice is less dense than liquid water, and therefore ice floats on liquid water. If ice did not float on water, ice would sink, and once it began to accumulate at the bottom of ponds, lakes, and perhaps even the ocean, they would freeze solid and never melt, making life impossible in the water and also on land.

Instead, bodies of water always freeze from the top down. When a body of water freezes on the surface, the ice acts as an insulator to prevent the water below it from freezing. This protects many aquatic organisms so that they can survive the winter. As ice melts in the spring, it draws heat from the environment, helping to prevent a sudden change in temperature that might be harmful to life.

Water has unique properties that allow cellular activities to occur and that make life on earth possible.

Figure 2.9 Water as a transport medium.
How does water rise to the top of tall trees? In vessels, the individual vessel elements (members) form a pipeline that is water-filled from the roots to the leaves. When water evaporates from the leaves, this water column is pulled upward due to the cohesion of water molecules with one another and the adhesion of water molecules to the sides of the vessel elements.

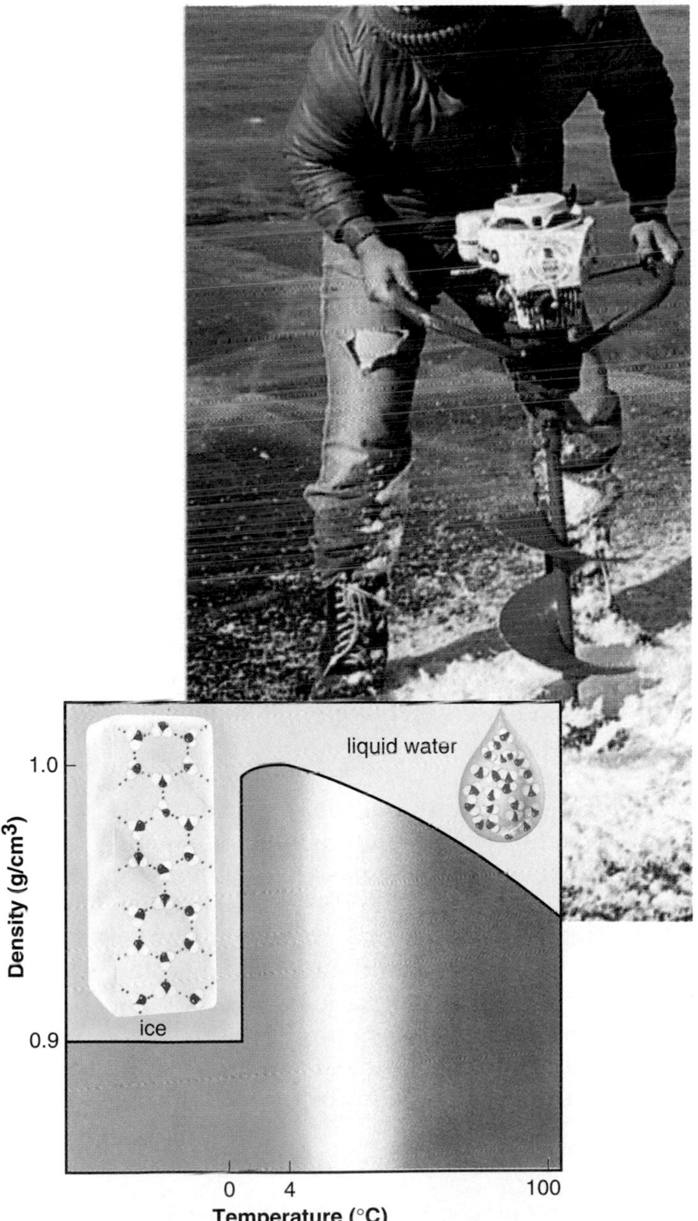

Figure 2.10 Water as ice.
Most substances contract when they solidify, but water expands because the water molecules in ice form a lattice in which the hydrogen bonds are farther apart than in liquid water. Water is more dense at 4°C than at 0°C; therefore, water freezes from the top down, making it necessary to drill a hole in order to go icefishing.

Water and Acids and Bases

When water ionizes, it releases an equal number of **hydrogen ions (H^+)** and **hydroxide ions (OH^-):**

H–O–H ⇌	H^+	+	OH^-
water	hydrogen ion		hydroxide ion

Only a few water molecules at a time are dissociated, and the actual number of these ions is very small (10^{-7} moles[1]/liter).

Acids (High H^+ Concentration)

Lemon juice, vinegar, tomatoes, and coffee are all familiar acids. What do they have in common? **Acids** are molecules that dissociate in water, releasing hydrogen ions (H^+).[2] For example, an important inorganic acid is hydrochloric acid (HCl), which dissociates in this manner:

$$HCl \rightarrow H^+ + Cl^-$$

Dissociation is almost complete; therefore, this is called a strong acid. If hydrochloric acid is added to a beaker of water, the number of hydrogen ions (H^+) increases greatly.

Bases (Low H^+ Concentration)

Milk of magnesia and ammonia are common bases that most people have heard of. **Bases** are molecules that either take up hydrogen ions (H^+) or release hydroxide ions (OH^-). For example, an important inorganic base is sodium hydroxide (NaOH), which dissociates in this manner:

$$NaOH \rightarrow Na^+ + OH^-$$

Dissociation is almost complete; therefore, sodium hydroxide is called a strong base. If sodium hydroxide is added to a beaker of water, the number of hydroxide ions increases.

pH Scale

The **pH scale**[3] is used to indicate the acidity and basicity (alkalinity) of a solution. A pH of exactly 7 is neutral pH. Pure water has an equal number of hydrogen ions (H^+) and hydroxide ions (OH^-), and therefore, one of each is released when water dissociates. One mole of pure water contains only 10^{-7} moles/liter of hydrogen ions, which is the source of the pH value for neutral solutions.

The pH scale was devised to simplify discussion of the hydrogen ion concentration [H^+] and consequently of the hydroxide ion concentration [OH^-]; it eliminates the use of cumbersome numbers. For example,

[H^+] (moles per liter)	pH
1×10^{-6}	6
1×10^{-7}	7
1×10^{-8}	8

Each lower pH unit has 10 times the amount of hydrogen ions (H^+) as the next higher unit.

In order to understand the relationship between hydrogen ion concentration and pH, consider the following question. Of the two other values listed above, which indicates a higher hydrogen ion concentration than pH 7 (neutral pH) and therefore refers to an acidic solution? A number with a smaller negative exponent indicates a greater quantity of hydrogen ions (H^+) than one with a larger negative exponent. Therefore, pH 6 is an acidic solution.

Bases add hydroxide ions (OH^-) to solutions and increase the hydroxide ion concentration [OH^-] of water. Basic (also called alkaline) solutions, then, have fewer hydrogen ions (H^+) compared to hydroxide ions. The pH 8 refers to a basic solution because it indicates a lower hydrogen

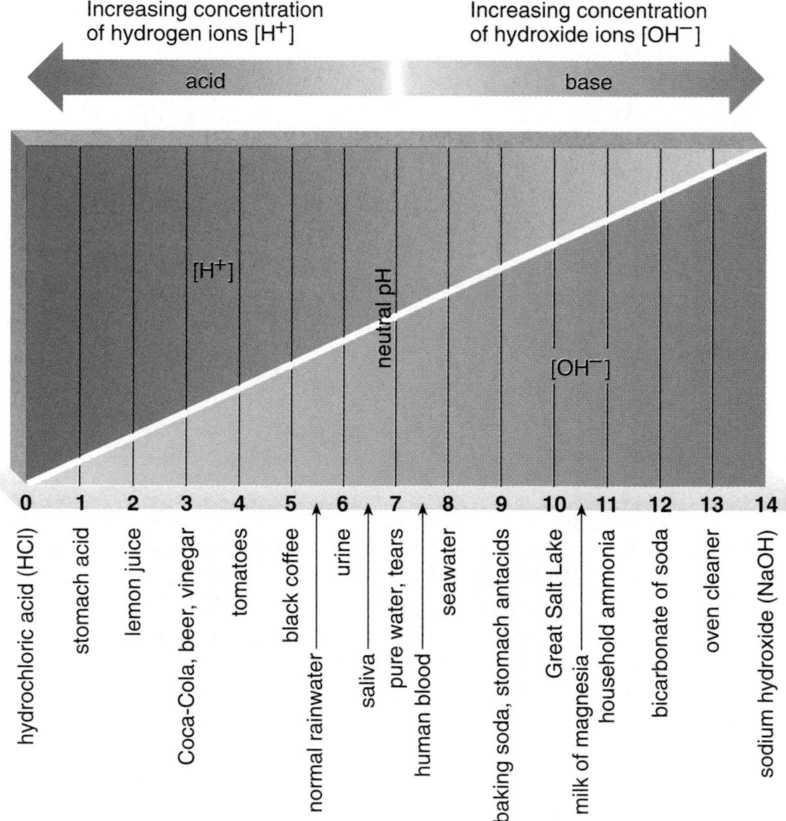

Figure 2.11 The pH scale.
The proportionate amount of hydrogen ions to hydroxide ions is indicated by the diagonal line. Any pH above 7 is basic, while any pH below 7 is acidic.

---

[1] In chemistry, a mole is defined as the amount of matter that contains as many objects (atoms, molecules, ions) as the number of atoms in exactly 12 grams of ^{12}C.

[2] A hydrogen atom contains one electron and one proton. A hydrogen ion has only one proton, so is often called a proton.

[3] pH is defined as the negative log of the hydrogen ion concentration [H^+]. A log is the power to which 10 must be raised to produce a given number.

ion concentration $[H^+]$ (greater hydroxide ion concentration) than pH 7.

The pH scale (Fig. 2.11) ranges from 0 to 14. It uses whole numbers, instead of negative exponents of the number 10, to indicate the hydrogen ion concentration $[H^+]$. As we move down the pH scale, each unit has 10 times the acidity of the previous unit, and as we move up the scale, each unit has 10 times the basicity of the previous unit. A pH of 7 has an equal concentration of hydrogen ions (H^+) and hydroxide ions (OH^-). Above pH 7 there are more hydroxide ions than hydrogen ions, and below pH 7 there are more hydrogen ions than hydroxide ions.

In living things, pH needs to be maintained within a narrow range or there are health consequences.

Buffers Keep pH Steady

The pH of our blood when we are healthy is always about 7.4, that is, just slightly basic (alkaline). Normally, pH stability is possible because the body has built-in mechanisms to prevent pH changes. **Buffers** are the most important of these mechanisms. Buffers help keep the pH within normal limits because they are chemicals or combinations of chemicals that take up excess hydrogen ions (H^+) or hydroxide ions (OH^-). For example, carbonic acid (H_2CO_3) is a weak acid that minimally dissociates and then reforms in the following manner:

$$\underset{\text{carbonic acid}}{H_2CO_3} \underset{\text{re-forms}}{\overset{\text{dissociates}}{\rightleftharpoons}} \underset{\text{bicarbonate ion}}{H^+ + HCO_3^-}$$

Blood always contains some carbonic acid and some bicarbonate ions. When hydrogen ions (H^+) are added to blood, the following reaction occurs:

$$H^+ + HCO_3^- \rightarrow H_2CO_3$$

When hydroxide ions (OH^-) are added to blood, this reaction occurs:

$$OH^- + H_2CO_3 \rightarrow HCO_3^- + H_2O$$

These reactions prevent any significant change in blood pH.

Acids have a pH that is less than 7, and bases have a pH that is greater than 7. Buffers, which can combine with both hydrogen ions and hydroxide ions, help to keep the pH of internal body fluids near pH 7 (neutral).

ecology focus

The Harm Done by Acid Deposition

Normally, rainwater has a pH of about 5.6 because the carbon dioxide in the air combines with water to give a weak solution of carbonic acid. Rain falling in northeastern United States and southeastern Canada now has a pH between 5.0 and 4.0. We have to remember that a pH of 4 is ten times more acidic than a pH of 5 to comprehend the increase in acidity this represents.

There is very strong evidence that this observed increase in rainwater acidity is a result of the burning of fossil fuels, like coal and oil, as well as gasoline derived from oil. When fossil fuels are burned, sulfur dioxide and nitrogen oxides are produced, and they combine with water vapor in the atmosphere to form acids. These acids return to earth contained in rain or snow, a process properly called wet deposition, but more often called acid rain. Dry particles of sulfate and nitrate salts descend from the atmosphere during dry deposition.

Acid deposition adversely affects lakes, particularly in areas where the soil is thin and lacks limestone (calcium carbonate, $CaCO_3$), a buffer to acid deposition. Acid leaches aluminum from the soil, carries aluminum into the lakes, and converts mercury deposits in lake bottom sediments to soluble and toxic methyl mercury. Lakes not only become more acidic, but they also show accumulation of toxic substances. Most fish die at pH 4.5, and in Canada, some 14,000 lakes are almost fishless and an additional 150,000 are in peril because of excess acidity. In the United States, about 9,000 lakes (mostly in the Northeast and Upper Midwest) are threatened, one-third of them seriously.

In forests, acid deposition weakens trees because it leaches away nutrients and releases aluminum. By 1988, most spruce, fir, and other conifers atop North Carolina's Mt. Mitchell were dead from being bathed in ozone and acid fog for years. The soil was so acidic, new seedlings could not survive. More than one-fifth of Europe's forests are now damaged.

These aren't the only effects of acid deposition. Reduction of agricultural yields, damage to marble and limestone monuments and buildings, and even illnesses in humans have been reported. Acid deposition has been implicated in the increased incidence of lung cancer and possibly colon cancer in residents of the East Coast. Tom McMillan, Canadian Minister of the Environment, says that acid rain is "destroying our lakes, killing our fish, undermining our tourism, retarding our forests, harming our agriculture, devastating our heritage, and threatening our health."

connecting concepts

Only 92 elements make up everything on earth, living and nonliving. Even more amazing, living things consist primarily of just six of these elements—carbon, hydrogen, nitrogen, oxygen, phosphorus, and sulphur (CHNOPS for short). These elements combine to form the unique types of molecules found in living cells. Cells consist largely of water, a molecule that contains only hydrogen and oxygen. Covalent bonding between the atoms and hydrogen bonding between the molecules give water properties that sustain life. No other planet we know of has liquid water.

In the next chapter, we will learn that carbon combines covalently with the other five elements (HNOPS) to form the organic molecules of cells. It is these unique molecules that set living forms apart from nonliving objects. Organic molecules can be modified in numerous ways and this accounts for the diversity of life. In a cheetah, for example, organic molecules are modified to become its black and beige fur. An organic molecule called chlorophyll, which is necessary for photosynthesis, gives plants their green color.

It is difficult for us to visualize that a cheetah or a pine tree is a combination of molecules, both organic and inorganic, but later in this book we will learn that even our thoughts of the cat and the tree are simply the result of molecules flowing from one brain cell to another.

Summary

2.1 Matter Is Composed of Elements

Both living and nonliving things are composed of matter consisting of elements. Each element contains atoms of just one type. The acronym CHNOPS tells us the most common elements (atoms) found in living things: carbon, hydrogen, nitrogen, oxygen, phosphorus, and sulphur. Atoms contain subatomic particles. Protons and neutrons in the nucleus determine the weight of an atom. The atomic number indicates the number of protons and the number of electrons in electrically neutral atoms. Protons have a positive charge and electrons have a negative charge. Isotopes are atoms of the same type that differ in their number of neutrons. Radioactive isotopes are used as tracers in biological experiments and medical procedures.

Electrons occupy energy levels (electron shells) at discrete distances from the nucleus. Electron shells have orbitals. The first shell has a single spherical-shaped orbital. The second shell has four orbitals: the first is spherical shaped and the others are dumbbell shaped. The number of electrons in the outer shell determines the reactivity of an atom. The first shell is complete when it has two electrons. In atoms up through calcium, number 20, every shell thereafter is complete with eight electrons. The octet rule states that atoms react with one another in order to have a completed outer shell that contains eight electrons. Most atoms, including those common to living things, do not have completed outer shells. This causes them to react with one another to form compounds and/or molecules. Following the reaction, the atoms have completed outer shells.

2.2 Atoms Form Compounds and Molecules

When an ionic reaction occurs, one or more electrons are transferred from one atom to another. The ionic bond is an attraction between the resulting ions. When a covalent reaction occurs, atoms share electrons. A covalent bond is the sharing of these electrons. There are single, double, and triple covalent bonds. In oxidation-reduction reactions, an atom that has lost electrons (or hydrogen atoms) has been oxidized, and an atom that has gained electrons (or hydrogen atoms) has been reduced.

In polar covalent bonds, the sharing of electrons is not equal; one of the atoms exerts greater attraction for the electrons than the other and a slight charge results on each atom. A hydrogen bond is a weak attraction between a slightly positive hydrogen atom of one molecule and a slightly negative oxygen or nitrogen atom within the same or a different molecule. Hydrogen bonds help maintain the structure and function of cellular molecules.

2.3 Water Is Essential to Life

Water is a polar molecule. The polarity of water molecules causes hydrogen bonding to occur between water molecules. These two features account for the unique properties of water, which are summarized in Table 2.3. These features allow cellular activities to occur and life to exist on Earth.

A small fraction of water dissociates to produce an equal number of hydrogen ions and hydroxide ions. This is termed neutral pH. In acidic solutions, there are more hydrogen ions than hydroxide ions; these solutions have a pH less than 7. In basic solutions, there are more hydroxide ions than hydrogen ions; these solutions have a pH greater than 7. Cells are sensitive to pH changes. Biological systems often contain buffers that help keep the pH within a normal range.

Reviewing the Chapter

1. Name the kinds of subatomic particles studied; list their weight, charge, and location in an atom. Which of these varies in isotopes? 20–21
2. Discuss the Bohr model of atomic structure and the relationship between energy levels and orbitals. 22–23
3. Draw a simplified atomic structure for a carbon atom that has six protons and six neutrons. 23
4. Draw an atomic representation for the molecule $MgCl_2$. Using the octet rule, explain the structure of the compound. 23
5. Explain whether CO_2 (O=C=O) is an ionic or a covalent compound. Why does this arrangement satisfy all atoms involved? 24–25
6. Define oxidation and reduction and note which has been oxidized and which reduced in this equation:

$$4\ Fe + 3\ O_2 \rightarrow 2\ Fe_2\ O_3$$

   What is meant when it is said that an equation is balanced? 23, 25
7. Explain why water is a polar molecule. What is the relationship between the polarity of the molecule and the hydrogen bonding between water molecules? 26
8. Name five properties of water and relate them to the structure of water, including its polarity and hydrogen bonding between molecules. 27–29
9. Define an acid and a base. On the pH scale, which numbers indicate an acid, a base, and a neutral pH? 30
10. What are buffers, and why are they important to life? 31

Testing Yourself

Choose the best answer for each question.

1. An atom that has two electrons in the outer shell would most likely
   a. share to acquire a completed outer shell.
   b. lose these two electrons and become a negatively charged ion.
   c. lose these two electrons and become a positively charged ion.
   d. form hydrogen bonds only.
2. The atomic number tells you the
   a. number of neutrons in the nucleus.
   b. number of protons in the atom.
   c. weight of the atom.
   d. All of these are correct.
3. An orbital
   a. has the same volume and charge as an electron shell.
   b. has the same space and energy content as the energy level.
   c. is the volume of space most likely occupied by an electron.
   d. causes an atom to react with another atom.
4. A covalent bond is indicated by
   a. plus and minus charges attached to atoms.
   b. dotted lines between hydrogen atoms.
   c. concentric circles about a nucleus.
   d. overlapping electron shells or a straight line between atomic symbols.
5. An atom has been oxidized when it
   a. combines with oxygen.
   b. gains an electron.
   c. loses an electron.
   d. Both a and c are correct.
6. In which of these are the electrons shared unequally?
   a. double covalent bond
   b. triple covalent bond
   c. hydrogen bond
   d. polar covalent bond
7. In the molecule

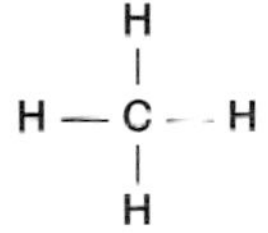

   a. all atoms have eight electrons in the outer shell.
   b. all atoms are sharing electrons.
   c. carbon could accept more hydrogen atoms.
   d. All of these are correct.
8. Which of these properties of water is not due to hydrogen bonding between water molecules?
   a. stabilizes temperature inside and outside cell
   b. molecules are cohesive
   c. is a universal solvent
   d. ice floats on water
9. Acids
   a. release hydrogen ions.
   b. have a pH value above 7.
   c. take up hydroxide ions.
   d. Both a and b are correct.
10. Complete this diagram of a nitrogen atom by placing the correct number of protons and neutrons in the nucleus and electrons in the shells. Explain why the correct formula for ammonia is NH_3 and not NH_4.

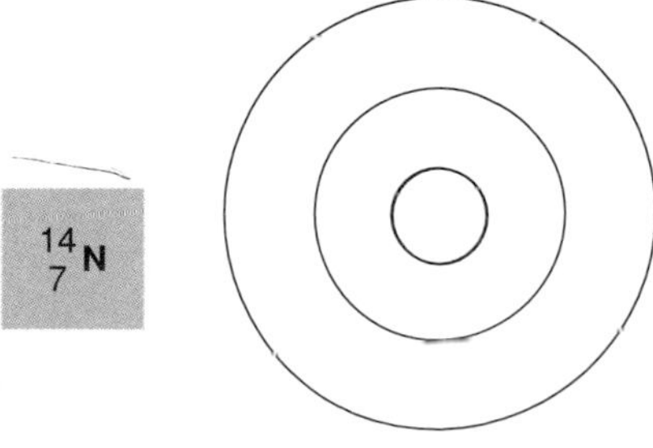

Applying the Concepts

1. *Life has a chemical and physical basis.*
   Give an example from your knowledge of nutrition, medicine, or the environment to show that this concept has an everyday application.
2. *Atomic structure involves electronic energy levels.*
   Show that living things are dependent upon the energy relationships of electrons.
3. *Life is dependent upon the special properties of water.*
   Show that living things are absolutely dependent upon a particular property of water by explaining what would happen if water didn't have this property.

Using Technology

Your study of basic chemistry is supported by these available technologies:

Exploring the Internet
The Mader Home Page provides resources for and help with studying this chapter.

http://www.mhhe.com/sciencemath/biology/mader/
(Click on Biology.)

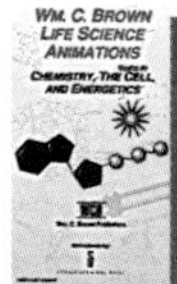

Life Science Animations Video
Video #1: Chemistry, The Cell, and Energetics
Formation of an Ionic Bond (#1)

Understanding the Terms

acid 30
atom 20
atomic number 21
atomic weight 21
base 30
buffer 31
compound 24
covalent bond 25
electron 20
element 20
energy 22
hydrogen bond 26
hydrogen ion 30
hydrophilic 28
hydrophobic 28
hydroxide ion 30
ion 24
ionic bond 24
isotope 21
matter 20
molecule 24
neutron 20
orbital 23
oxidation 25
pH scale 30
polar covalent bond 26
proton 20
reduction 25

Match the terms to these definitions:

a. __________ Bond in which the sharing of electrons between atoms is unequal.
b. __________ Charged particle that carries a negative or positive charge(s).
c. __________ Molecules tending to raise the hydrogen ion concentration in a solution and to lower its pH numerically.
d. __________ Loss of one or more electrons from an atom or molecule; in biological systems, generally the loss of hydrogen atoms.
e. __________ Measurement scale for the hydrogen ion concentration $[H^+]$ of a solution.
f. __________ Union of two or more atoms of the same element; also the smallest part of a compound that retains the properties of the compound.
g. __________ Smallest particle of an element that displays the properties of the element.
h. __________ Substance having two or more different elements united chemically in fixed ratio.
i. __________ Type of molecule that interacts with water by dissolving in water and/or by forming hydrogen bonds with water molecules.
j. __________ Weak bond that arises between a slightly positive hydrogen atom of one molecule and a slightly negative atom of another molecule or between parts of the same molecule.

The Chemistry of Life

CHAPTER 3

Chapter Concepts

Organic molecules first formed in oceans

About 4 billion years ago, the earth's surface was covered with newly formed oceans, but they were devoid of life. The atmosphere consisted of water vapor and a mixture of gases, some containing carbon. Bombarding the planet was an assortment of energies—ultraviolet, volcanic heat, radioactive decay, and lightning—that caused the first carbon-based molecules to form. After that, the first cell(s) came into existence. Laboratory experiments support this possible scenario of how life on earth originated.

Life as we know it is dependent on carbon-based—that is, organic—molecules. Carbon is a versatile atom. It has four electrons in its outermost shell, and this allows it to form covalent bonds with as many as four other atoms. In living things, carbon usually binds to itself, hydrogen, nitrogen, oxygen, and also phosphorus and sulphur. The organic molecules known as carbohydrates, proteins, fats, and nucleic acids make up cells and can be modified in ways that account for the millions of different kinds of organisms on earth.

3.1 Cells Contain Organic Molecules

The most common elements in living things are carbon, hydrogen, nitrogen, and oxygen, which constitute about 95% of your body weight. But we will see that both the sameness and the diversity of life are dependent upon the chemical characteristics of carbon, an atom whose chemistry is essential to living things.

The bonding of hydrogen, oxygen, nitrogen, and other atoms to carbon creates the molecules known as **organic molecules**. Laypeople sometimes use the word organic to mean wholesome. To biologists, organic molecules are those that always contain carbon and hydrogen. It is the organic molecules that characterize the structure and the function of living things like the primrose, the blueshell crab, and the bacterium in Figure 3.1. **Inorganic molecules** constitute nonliving matter, but even so, inorganic molecules like salts (e.g., NaCl) also play important roles in living things. Table 3.1 contrasts inorganic and organic molecules.

Living things are highly organized; therefore, it is easy for us to differentiate between the plant, the animal, and the bacterium in Figure 3.1. Despite their diversity all living things contain the same classes of primary molecules: carbohydrates, proteins, lipids, and nucleic acids. But within these classes there is molecular diversity. For example, the plant, the animal, and the bacterium all utilize a carbohydrate molecule for structural purposes, but the exact carbohydrate is different in each of them.

Table 3.1

Inorganic Versus Organic Molecules

Inorganic Molecules	Organic Molecules
Usually contain positive and negative ions	Always contain carbon and hydrogen
Usually ionic bonding	Always covalent bonding
Always contain a small number of atoms	May be quite large with many atoms
Often associated with nonliving matter	Usually associated with living organisms

Of the elements most common to living things, the chemistry of carbon allows the formation of varied organic molecules. This accounts for both the sameness and the diversity of living things.

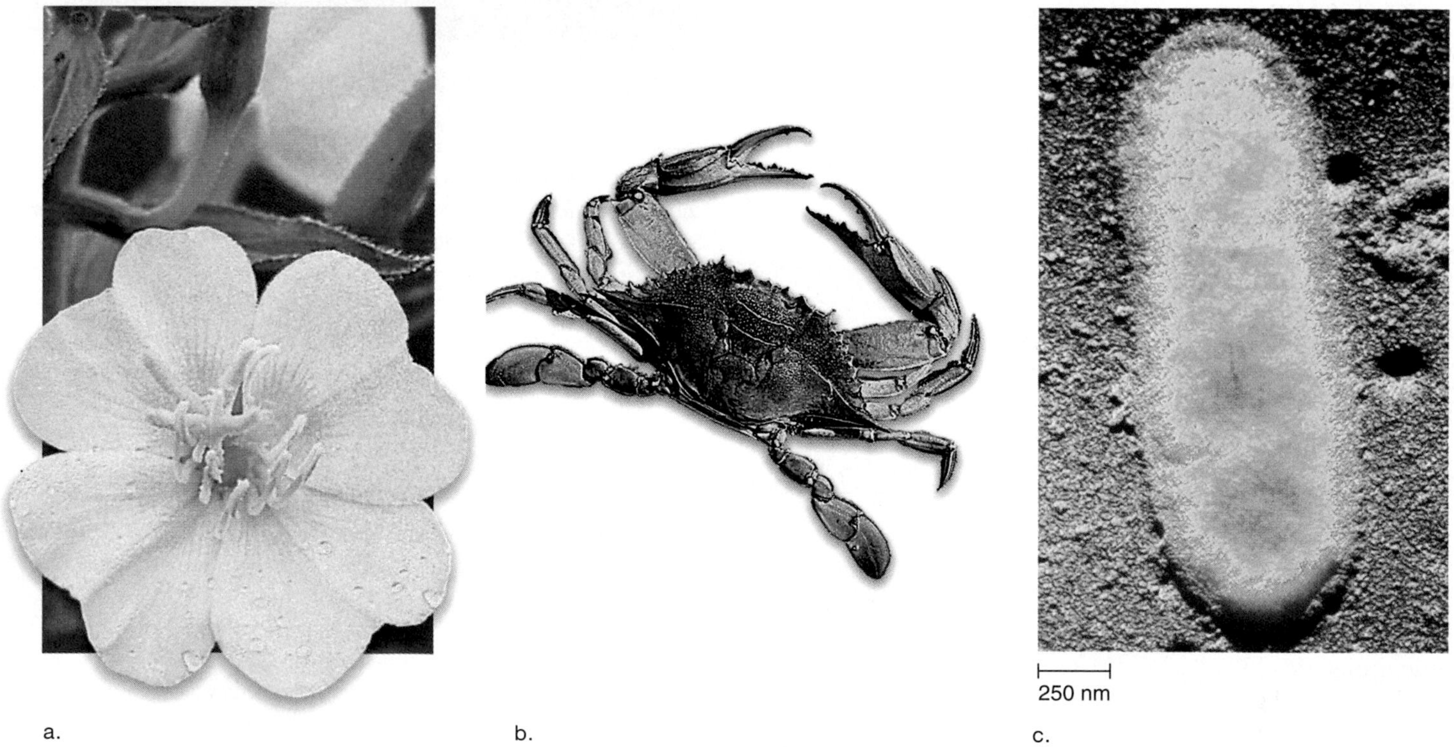

Figure 3.1 Carbohydrates as structural materials.
a. Plants are held erect partly by the incorporation of the polysaccharide cellulose into the cell wall, which surrounds each cell. **b.** The shell of crabs contains chitin, a different type of polysaccharide. **c.** Most bacterial cells, like plant cells, are enclosed by a cell wall, but in this case the wall is strengthened by another type of polysaccharide known as peptidoglycan.

Small Molecules Have Functional Groups

Carbon has four electrons in its outer shell, and this allows it to bond with as many as four other atoms. Moreover, because these bonds are covalent, they are quite strong. Usually carbon bonds to hydrogen, oxygen, nitrogen, or another carbon atom. The ability of carbon to bond to itself makes carbon chains of various lengths and shapes possible. Long chains containing 50 or more carbon atoms are not unusual in living systems. Carbon can also share two pairs of electrons with another atom, giving a double covalent bond. Carbon-to-carbon bonding sometimes results in ring compounds of biological significance.

Carbon chains make up the skeleton or backbone of organic molecules. *Functional groups,* which are clusters of certain atoms that always behave in a certain way, can be attached to the carbon chain. Functional groups also help determine the characteristics of organic molecules that are used in industrial synthetic compounds. Unfortunately, these synthetic compounds can become pollutants, as discussed in the reading on page 45. The small organic molecules in living things—sugars, fatty acids, amino acids, and nucleotides—all have a carbon backbone, and in addition, they often have one or more of the functional groups listed in Figure 3.2.

Hydrocarbon chains, which are composed only of carbon and hydrogen, are **hydrophobic** (not attracted to water). A functional group that ionizes can make an organic molecule **hydrophilic** (attracted to water). For example, the carboxyl (acid) functional group, —COOH, can give up a hydrogen ion (H^+) and become ionized to —COO^-. Then the molecule becomes hydrophilic because the polar (charged) group interacts with water, which is a polar molecule.

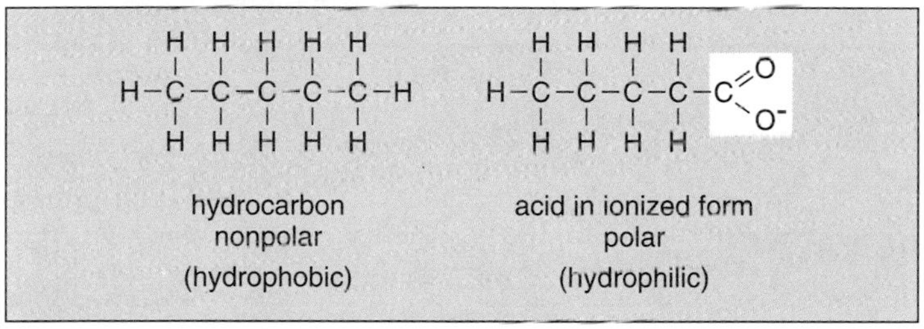

$H-CH_2-CH_2-CH_2-CH_2-CH_2-H$

hydrocarbon
nonpolar
(hydrophobic)

$H-CH_2-CH_2-CH_2-CH_2-C(=O)O^-$

acid in ionized form
polar
(hydrophilic)

Other functional groups (—OH, —CO, and —NH_2) are considered to be polar even though they do not ionize completely. Since cells are 70–90% water, the ability or inability to interact with water profoundly affects the function of organic molecules in cells.

Functional groups add diversity to organic molecules. Isomers also contribute to the diversity of organic molecules. **Isomers** [Gk. *isos,* equal, and *meros,* part, portion] are molecules that have identical molecular formulas, but they are different molecules because the atoms in each are arranged differently. For example, compounds with the molecular formula $C_3H_6O_3$ are assigned different chemical names because the molecules differ structurally (Fig. 3.3).

Functional Groups		
Name	**Structure**	**Found in**
hydroxyl (alcohol)	$R-OH$	sugars
carboxyl (acid)	$R-C(=O)-OH$	fats amino acids
ketone	$R-C(=O)-R$	some sugars
aldehyde	$R-C(=O)-H$	some sugars
amino (amino)	$R-NH_2$	amino acids proteins
sulfhydryl	$R-SH$	some amino acids proteins
phosphate	$R-O-P(=O)(OH)-OH$	phospholipids nucleotides nucleic acids
R = remainder of molecule		

Figure 3.2 Functional groups.
Molecules with the same type of backbone can still differ according to the type of functional group attached to the backbone. Many of these functional groups are polar, helping to make the molecule soluble in water. In this illustration, the remainder of the molecule (aside from the functional groups) is represented by an *R*.

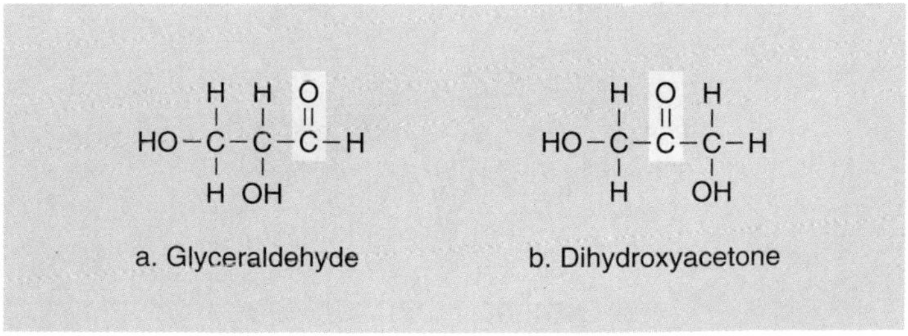

$HO-CH_2-CH(OH)-C(=O)-H$

a. Glyceraldehyde

$HO-CH_2-C(=O)-CH_2-OH$

b. Dihydroxyacetone

Figure 3.3 Isomers.
Isomers have the same molecular formula but different configurations. Both of these compounds have the formula $C_3H_6O_3$.
a. In glyceraldehyde, oxygen is double-bonded to an end carbon.
b. In dihydroxyacetone, oxygen is double-bonded to the middle carbon.

Large Organic Molecules Have Monomers

Each of the small organic molecules already mentioned can be a unit of a large organic molecule, often called a *macromolecule*. A unit is called a **monomer,** and the macromolecule is called a **polymer** [Gk. *polys*, many, and *meros*, part, portion]. Simple sugars (monosaccharides) are the monomers within polysaccharides; fatty acids and glycerol are found in fat, a lipid; amino acids join to form proteins; and nucleotides are the subunits of nucleic acids:

Polymer	Monomer
polysaccharide	monosaccharide
lipid (e.g., fat)	glycerol and fatty acid
protein	amino acid
nucleic acid	nucleotide

Cells contain only these four classes of macromolecules, and each has just one type monomer. Yet the macromolecules of each class can be quite varied. This is because the same type of monomer occurs in different varieties: there are different types of monosaccharides, fatty acids, amino acids, and nucleotides. Also, the same type of monomer can be joined in a way characteristic of a particular macromolecule, or it can have different functional groups. This is why macromolecular carbohydrates (i.e., polysaccharides) can have different characteristics. In the pages that follow, note the names of the various macromolecules, the monomers they contain, and how the monomers are joined. You will also want to know the functions of each type of macromolecule discussed.

There are only a few classes of macromolecules in cells, but these molecules still have great variety and therefore play different roles in cells.

Condensation Is the Reverse of Hydration

Macromolecules are so named because they are very large. A protein can contain hundreds of amino acids, and a nucleic acid can contain hundreds of nucleotides. How do macromolecules get so large? Cells use the modular approach when constructing macromolecules—the size of a macromolecule is dependent on the number of monomers it contains. Cellulose and starch are both polysaccharides containing many glucose monomers. In cellulose, the type of bond between the glucose molecules differs from the type of bond joining glucose molecules in starch. Regardless of the particular macromolecule and the type of bond utilized, monomers are joined by an identical mechanism, which is described in Figure 3.4.

During **condensation,** when two monomers join, a hydroxyl (—OH) group is removed from one monomer and a hydrogen (—H) is removed from the other. Notice in Figure 3.4*a* that water is given off during a condensation reaction. Condensation involves a *dehydration synthesis* because water is removed (dehydration) and a bond is made (synthesis). Condensation does not take place unless the proper enzyme (a molecule that speeds up a chemical reaction in cells) is present and the monomers are in an activated energy-rich form.

Polymers are broken down by **hydrolysis,** which is essentially the reverse of condensation. Notice in Figure 3.4*b* that water is added during hydration: an —OH group from water attaches to one monomer and an —H from water attaches to the other monomer. Hydration involves a hydrolysis reaction because water is used to break a bond. Again, the proper enzyme is required.

Macromolecules are routinely built up in cells by condensation and broken down in cells by hydration.

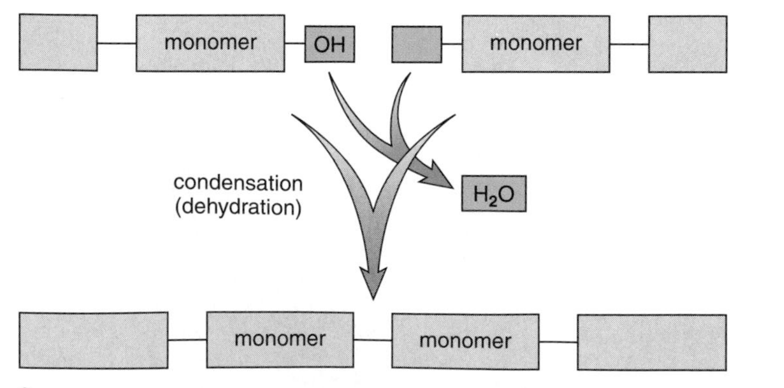

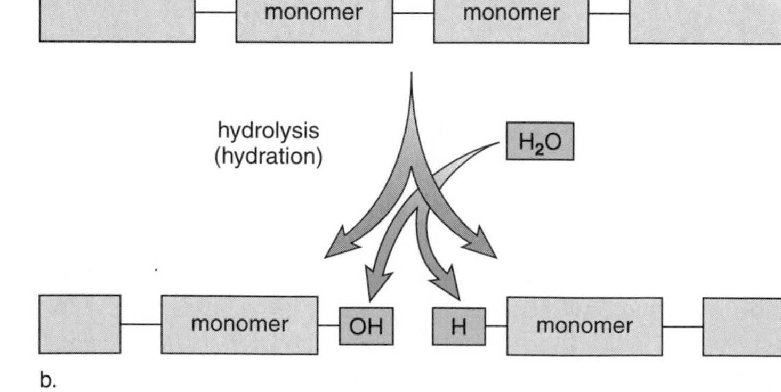

Figure 3.4 Condensation and hydrolysis of polymers.
a. In cells, synthesis often occurs when monomers are joined by condensation (removal of H_2O). **b.** Hydrolysis occurs when the monomers in a polymer separate after the addition of H_2O.

3.2 Carbohydrates

Sugars and polysaccharides belong to a class of compounds called **carbohydrates** [L. *carbo*, charcoal, and Gk. *hydatos*, water]. The formula CH_2O is sometimes used to represent carbohydrates, molecules that bear many hydroxyl groups (Fig. 3.5). Carbohydrates include monosaccharides (one sugar), disaccharides (two sugars joined together), and polysaccharides (many sugars joined together).

Monosaccharides are simple sugars with a carbon backbone of three to seven carbon atoms. The best known sugars are those that have six carbons (*hexoses*). **Glucose** [Gk. *glykys*, sweet, and *osus*, full of] is in the blood of animals, and *fructose* is frequently found in fruits. These sugars are isomers of each other. They both have the molecular formula $C_6H_{12}O_6$, but they differ in structure (Fig. 3.5). Structural differences cause molecules to vary in shape, which is best seen by using space-filling models. Shape is very important in determining how molecules interact with one another.

Ribose and **deoxyribose** are two five-carbon sugars (*pentoses*) of significance because they are found respectively in the nucleic acids RNA and DNA. RNA and DNA are discussed later in the chapter.

A **disaccharide** contains two monosaccharides that have joined by condensation. *Lactose* is a disaccharide that contains galactose and glucose and is found in milk. *Maltose* (composed of two glucose molecules) is a disaccharide of interest because it is found in our digestive tract as a result of starch digestion.

Sucrose is a disaccharide that contains glucose and fructose (Fig. 3.6). Sugar is transported within the body of a plant in the form of sucrose, and this is the sugar we use at the table to sweeten our food. We acquire this sugar from plants, such as sugarcane and sugar beets.

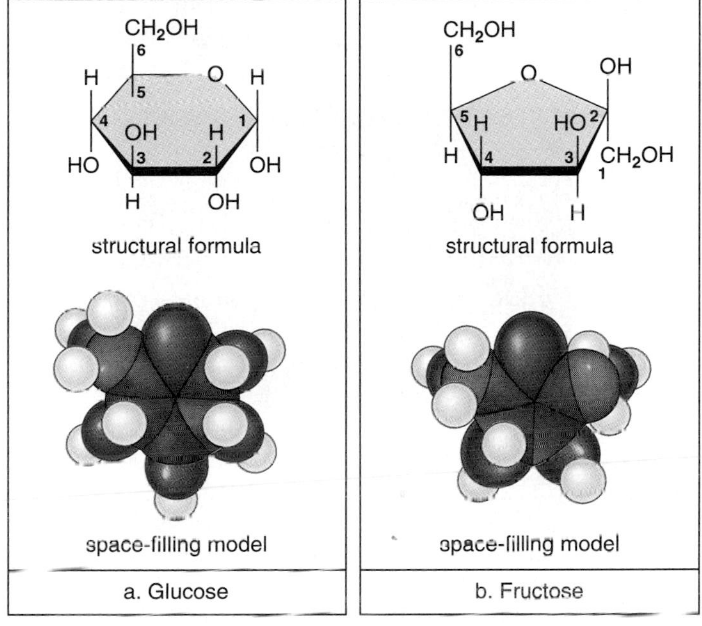

Figure 3.5 Two common six-carbon sugars.
a. Glucose is a six-carbon sugar that usually exists as a ring compound in cells. The shape of glucose is indicated in the space-filling model. b. Fructose is an isomer of glucose. It too has the molecular formula $C_6H_{12}O_6$. Notice, though, that the atoms are bonded in a slightly different manner. The carbon atoms in glucose and fructose are numbered.

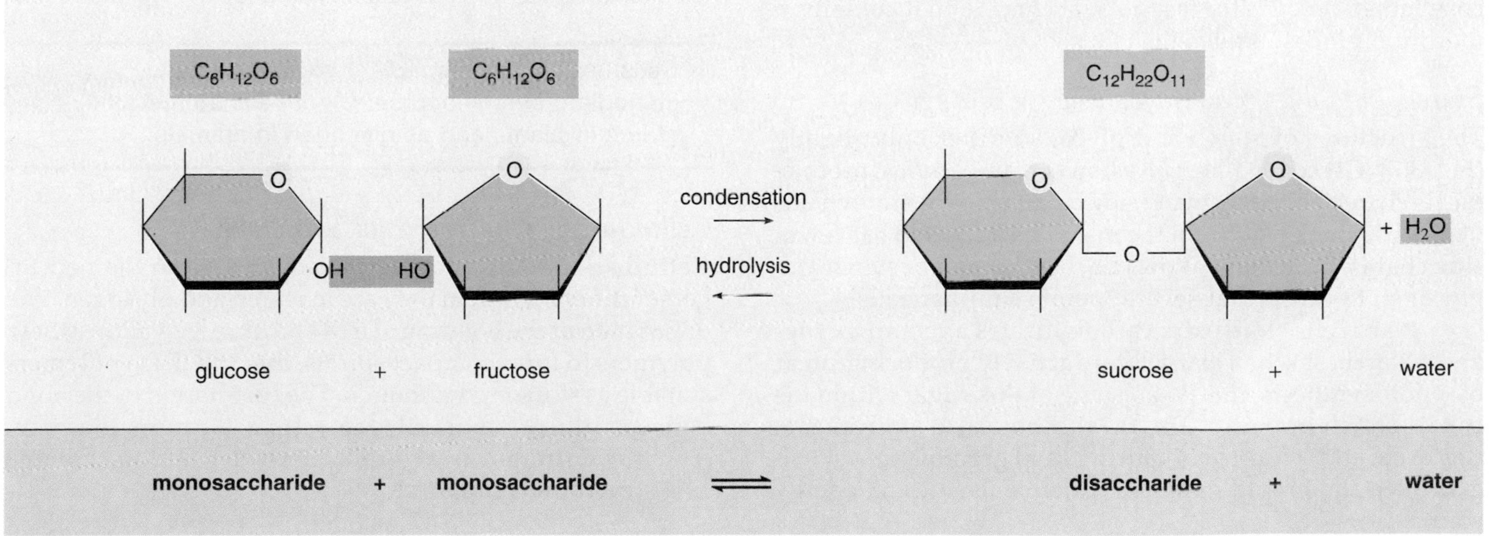

Figure 3.6 Condensation and hydrolysis of sucrose.
During condensation, a bond forms between the glucose and fructose molecules as the components of water are removed. During hydrolysis, the components of water are added to the bond as glucose and fructose re-form.

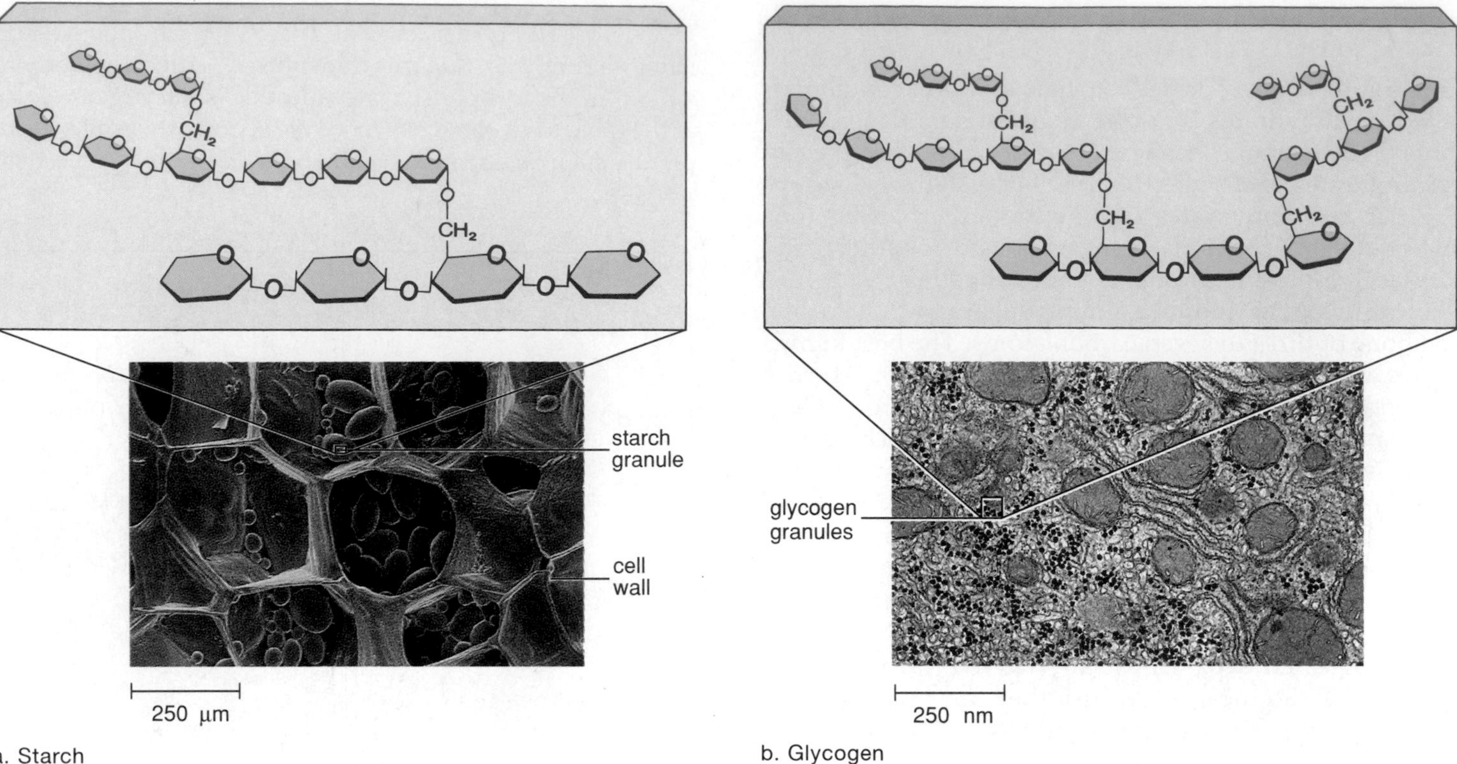

Figure 3.7 Starch and glycogen structure and function.
a. Starch is a chain of glucose molecules that branches as shown. The electron micrograph shows starch granules in plant cells. Glucose is stored in plants as starch. **b.** Glycogen is a highly-branched polymer of glucose molecules. The electron micrograph shows glycogen granules in a portion of a liver cell. Glucose is stored in animals as glycogen.

Polysaccharides Are Varied in Structure and Function

The most common **polysaccharides** in living things are starch, glycogen, and cellulose, which are chains of glucose molecules. Chitin is a polysaccharide that contains a modified glucose molecule.

Starch and Glycogen Are for Energy Storage

The structures of starch and glycogen differ only slightly (Fig. 3.7). **Glycogen** [Gk. *glykys*, sweet, and *genitus*, producing] is characterized by many side branches, which are chains of glucose that go off from the main chain. **Starch** has fewer side chains. Branching allows the breakdown of starch and glycogen to proceed at several points simultaneously.

Plant cells store extra carbohydrates as complex sugars or starches. When leaf cells are actively producing sugar by photosynthesis, they keep some of this sugar within the cell as starch granules (Fig. 3.7*a*). Roots also store sugars as starch, and so do seeds. During seed germination, starch is broken down into maltose, which is used as an energy source.

Animal cells store extra carbohydrates as glycogen, sometimes called "animal starch." After a human eats, the muscles and liver convert glucose molecules to glycogen (Fig. 3.7*b*). Between meals, the liver releases glucose to keep the blood concentration of glucose near the normal 0.1%.

Cells use sugars, especially glucose, as an immediate energy source. Glucose is stored as starch in plants and as glycogen in animals.

Cellulose and Chitin Are for Structure

Cellulose contains glucose molecules that are joined together differently than they are in starch and glycogen. The orientation of the bonds in starch and glycogen allows these polymers to form compact spirals, making these polymers suitable as storage compounds. The orientation of the bond between glucose molecules in cellulose causes the polymers to be straight and fibrous, making cellulose suitable as a structural compound.

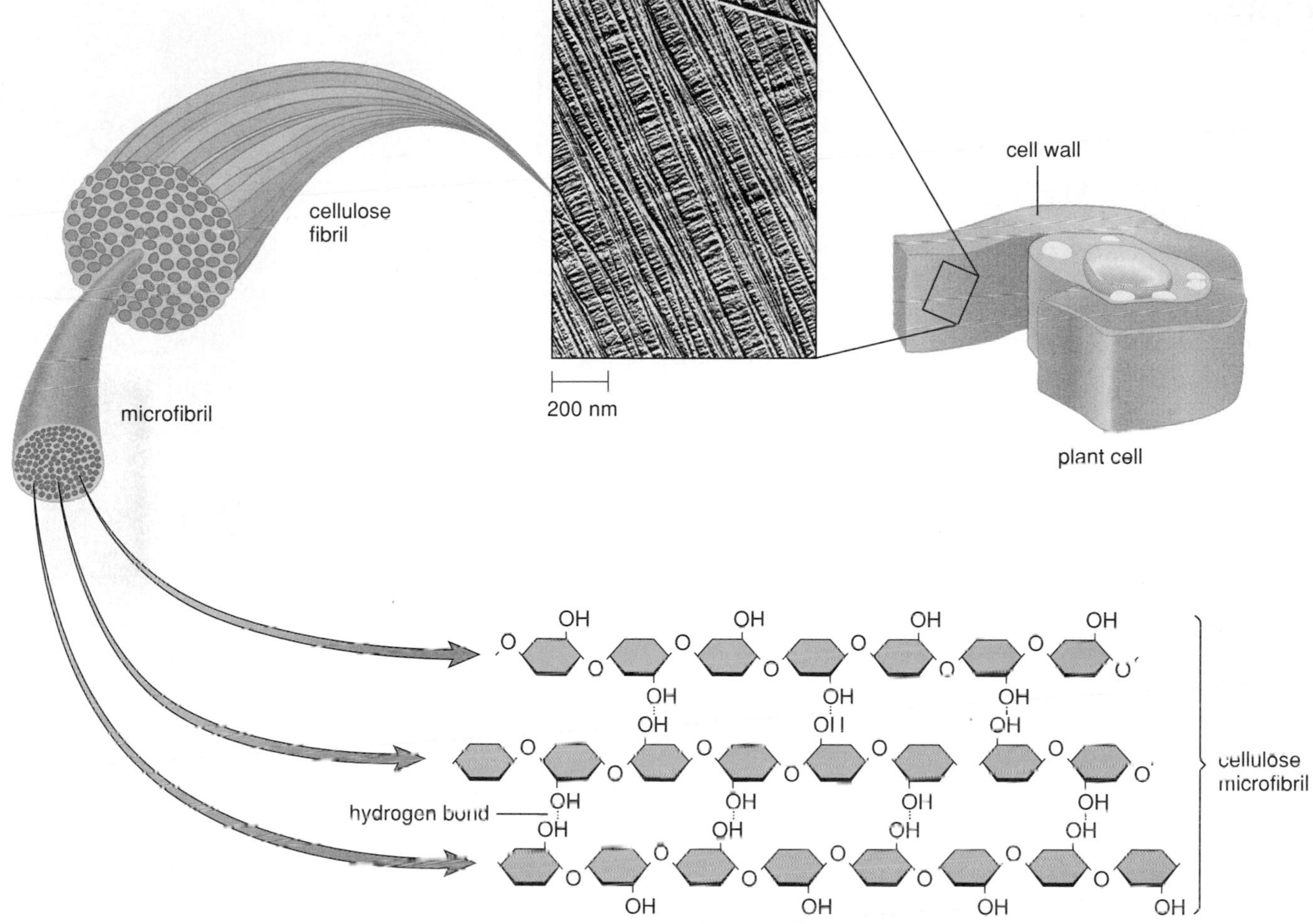

Figure 3.8 Cellulose fibrils.
In plant cell walls, each cellulose fibril contains several microfibrils. Each microfibril contains many polymers of glucose hydrogen-bonded together. Three such polymers are shown.

As shown in Figure 3.8, the long, unbranched polymers of glucose in cellulose are held together by hydrogen bonding within microfibrils. Several microfibrils, in turn, make up a fibril. Layers of cellulose fibrils occur in plant cell walls. Parallel cellulose fibrils occur in layers, and when the layers lie at angles to one another, plant cell walls are stronger.

Humans have found many uses for cellulose. Cotton fibers are almost pure cellulose, and we often wear cotton clothing. Furniture and buildings are made from wood, which contains a high percentage of cellulose. Most animals, including humans, digest little cellulose because the enzymes that digest starch are unable to break the linkage between the glucose molecules in cellulose. Even so, cellulose is a recommended part of our diet because it provides bulk (also called fiber or roughage) that helps the body maintain regularity of elimination. For humans, plants are a source of starch and also fiber.

Cattle and sheep receive nutrients from grass because they have a special stomach chamber, the rumen, where bacteria live that can digest cellulose. Unfortunately, cattle are often kept in feedlots, where they are fed grains instead of grass. This practice wastes fossil fuel energy because it takes energy to grow the grain, to process it, to transport it, and finally to feed it to the cattle. Also, range-fed cattle move about and produce a leaner meat than do cattle raised in a feedlot. There is growing evidence that lean meat is healthier for humans than fatty meat.

Chitin [Gk. *chiton*, tunic], which is found in the exoskeleton of crabs and related animals like lobsters and insects, is also a polymer of glucose. Each glucose unit, however, has an amino group attached to it. The linkage between the glucose molecules is like that found in cellulose; therefore, chitin is not digestible by humans. Recently, scientists have discovered how to turn chitin into thread that can be used as suture material. They hope to find other uses for treated chitin. If so, all those discarded crab shells that pile up beside crabmeat processing plants will not go to waste.

Plant cell walls contain cellulose. The shells of crabs and related animals contain chitin.

3.3 Lipids

A variety of organic compounds are classified as **lipids** [Gk. *lipos*, fat]. Many of these are insoluble in water because they lack any polar groups. The most familiar lipids are those found in fats and oils. Fat is utilized for both insulation and energy reserves by organisms. Fat below the skin of whales is called blubber; in humans it is given slang expressions such as "spare tire" and "love handles."

Phospholipids and steroids are also important lipids found in living things. For example, they are components of cellular membranes including the plasma membrane, a sheetlike structure that encloses the cell.

Lipids are quite varied in structure, but they tend to be insoluble in water.

Fats and Oils Are Similar

Fats and **oils** contain two types of unit molecules: fatty acids and glycerol. Fats, which are most often of animal origin, are solid at room temperature and oils, which are liquid at room temperature, are of plant origin.

Each **fatty acid** consists of a long hydrocarbon chain with a carboxyl (acid) group at one end. Because the carboxyl group is a polar group, fatty acids are soluble in water. Most of the fatty acids in cells contain 16 to 18 carbon atoms per molecule, although smaller ones are also found. Fatty acids are either saturated or unsaturated (Fig. 3.9). *Saturated* fatty acids have no double bonds between the carbon atoms. The carbon chain is saturated, so to speak, with all the hydrogens that can be held. *Unsaturated* fatty acids have double bonds in the carbon chain wherever the number of hydrogens is less than two per carbon atom:

```
  H  H  H  H          H  H  H  H
  |  |  |  |          |  |  |  |
 -C--C--C--C-        -C--C==C--C-
  |  |  |  |          |        |
  H  H  H  H          H        H
   saturated          unsaturated
```

Fats have mostly saturated fatty acids while oils tend to have unsaturated fatty acids. Diets high in fat have been associated with circulatory disorders. Replacement of fat whenever possible with oils such as peanut oil and sunflower oil has been suggested.

Glycerol is a compound with three hydroxyl groups (Fig. 3.10). Hydroxyl groups are polar; therefore, glycerol is soluble in water. When a fat is formed, the acid portions of three fatty acids react with these hydroxyl groups so that fat and three molecules of water result. Again, the larger fat molecule is formed by condensation, and a fat can be hydrolyzed to its components. Since there are three

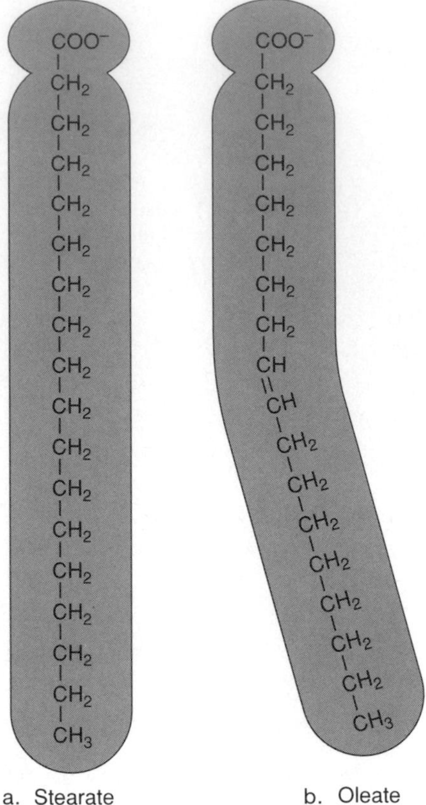

Figure 3.9 Fatty acid structure.
Fatty acids are long hydrocarbon chains ending in a carboxyl (acid) group. a. Stearate is a saturated fatty acid. b. Oleate is an unsaturated fatty acid.

fatty acids per glycerol molecule, fats and oils are sometimes called a **triglyceride.** Triglycerides are often referred to as neutral fats because they, unlike their components, lack polar groups that can interact with water. Therefore, for example, cooking oils do not mix with water even though they are both liquids. Even after shaking, the oil simply separates out.

Triglycerides containing unsaturated hydrocarbon chains melt at a lower temperature than those containing saturated chains. This is because a double bond makes a kink that prevents close packing among the chains (Fig. 3.9). We can reason, then, that butter, which is a solid at room temperature, must contain saturated hydrocarbon chains and that corn oil, which is a liquid even when placed in the refrigerator, must contain unsaturated chains. This difference is useful to living things. For example, the feet of reindeer and penguins contain unsaturated triglycerides, and this helps protect these exposed parts from freezing.

Nearly all animals use fats for long-term energy storage. Fats have mostly C—H bonds, making them a richer

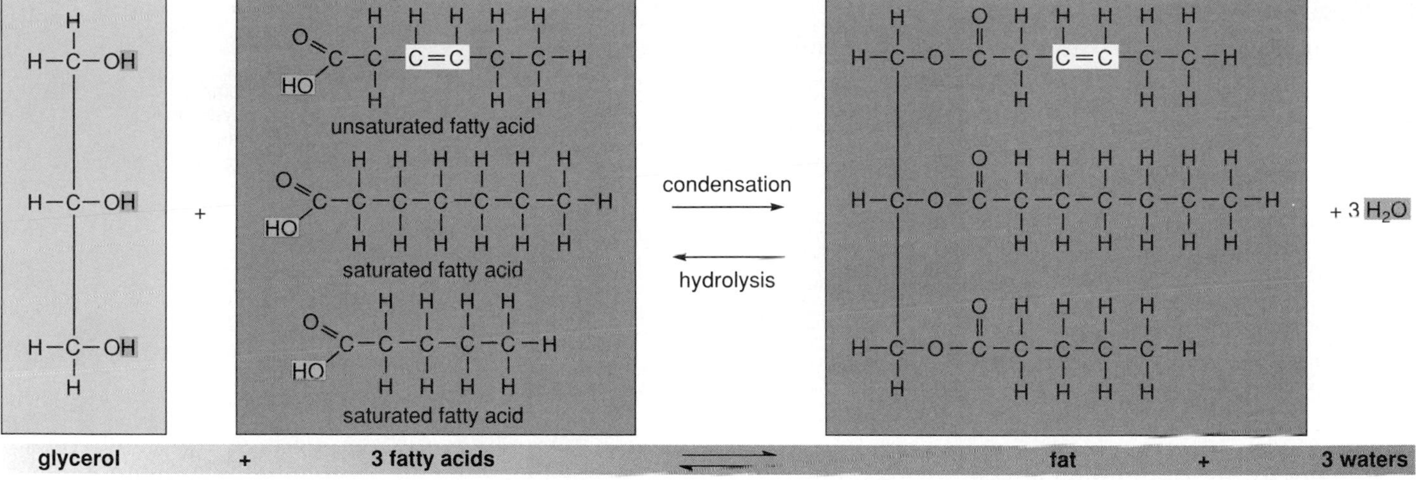

Figure 3.10 Formation of a neutral fat.
Three fatty acids plus glycerol react to produce a fat molecule and three water molecules. A fat molecule plus three water molecules react to produce three fatty acids and glycerol.

supply of chemical energy than carbohydrates, which have many C—OH bonds. Molecule to molecule, animal fat contains over twice as much energy as glycogen; gram to gram, though, fat stores six times as much energy as glycogen. This is because fat droplets, being nonpolar, do not contain water. Small birds, like the broad-tailed hummingbird (Fig. 3.11*a*), store a great deal of fat before they start their long spring and fall migratory flights. About 0.15 gram of fat per gram of body weight is accumulated each day. If the same amount of energy were stored as glycogen, a bird would be so heavy it would not be able to fly.

Fats and oils are triglycerides (one glycerol plus three fatty acids). They are used as long-term energy-storage compounds in plants and animals.

Waxes

In **waxes,** a long-chain fatty acid bonds with a long-chain alcohol. Waxes are solid at normal temperatures because they have a high melting point. Being hydrophobic, they are also waterproof and resistant to degradation. In many plants, waxes form a protective cuticle (covering) that retards the loss of water for all exposed parts. In animals, waxes are involved in skin and fur maintenance. In humans, wax is produced by glands in the outer ear canal. Here its function is to trap dust and dirt particles, preventing them from reaching the eardrum.

A honeybee produces wax in glands on the underside of its abdomen. The wax is used to make the six-sided cells of the comb where honey is stored (Fig. 3.11*b*). Honey contains the sugars fructose and glucose, breakdown products of the sugar sucrose.

a.

b.

Figure 3.11 Use of lipids.
a. Because fat can be concentrated in droplets, it is the preferred way to store energy by broad-tailed hummingbirds, *Selasphorus*, and other birds that migrate long distances. **b.** The comb of a honeybee, *Apis*, is composed of wax, a compound consisting of fatty acids and alcohols. The wax is secreted by the bees' special abdominal glands.

Phospholipids Have a Polar Group

Phospholipids [Gk. *phos*, light, and *lipos*, fat], as implied by their name, contain a phosphate group. A phosphate group is a polar group that can ionize and therefore is hydrophilic:

nonionized phosphate	ionized phosphate
$R{-}P({=}O)({-}OH){-}OH$	$R{-}P({=}O)({-}O^-){-}O^-$

Essentially, phospholipids are constructed like neutral fats, except that in place of the third fatty acid, there is a phosphate group or a grouping that contains both phosphate and nitrogen. This group becomes the polar head of the molecule, while the hydrocarbon chains of the fatty acids become the nonpolar tails (Fig. 3.12). When phospholipid molecules are placed in water, they form a double layer in which the polar heads face outward and the nonpolar tails face each other. This property of phospholipids means that they can form an interface or separation between two solutions, such as the interior and exterior of a cell. The plasma membrane of cells is basically a phospholipid bilayer.

Phospholipids have a polar head and nonpolar tails. In the presence of water they arrange themselves in a double layer, as seen in the plasma membrane of cells, which is surrounded by water.

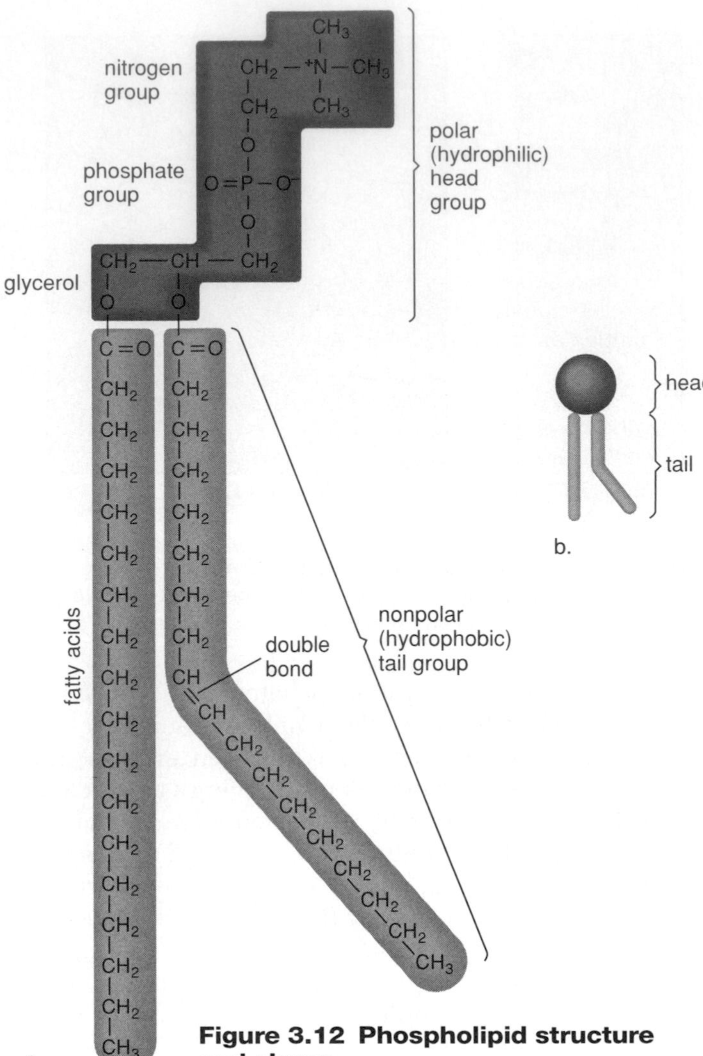

Figure 3.12 Phospholipid structure and shape.
Phospholipids are constructed like fats, except that they contain a phosphate group. **a.** Lecithin, shown here, has a side chain that contains both a phosphate group and a nitrogen-containing group. **b.** The polar (hydrophilic) head group is soluble in water, whereas the two nonpolar (hydrophobic) tail groups are not. This causes the molecule to arrange itself as shown.

Steroids Have Carbon Rings

Steroids are lipids that have an entirely different structure from neutral fats. Each steroid has a backbone of four fused carbon rings and varies from other steroids primarily by the type of functional groups attached to the rings (Fig. 3.13). Cholesterol is the precursor of several other steroids, including the vertebrate hormones such as aldosterone, which helps regulate the sodium content of the blood, and the sex hormones, which help maintain male and female sex characteristics. Their functions vary due primarily to the different attached groups.

As previously mentioned, a diet high in saturated fats and cholesterol can lead to reduced blood flow caused by the deposit of fatty materials on the linings of blood vessels.

Steroids are ring compounds that have a similar backbone but vary according to the attached groups. This causes them to have different functions in the bodies of humans and other animals.

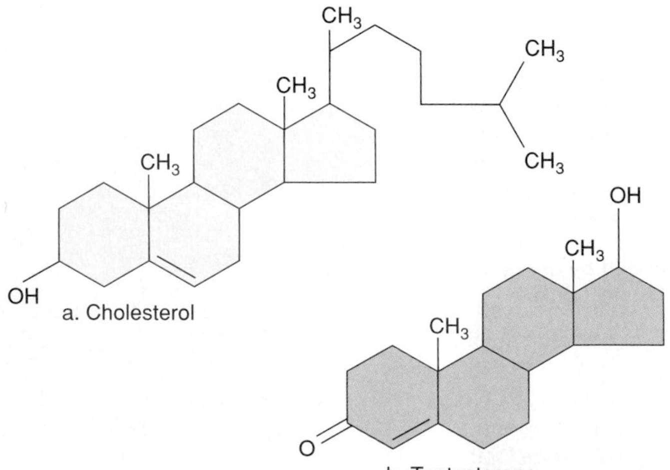

Figure 3.13 Steroid diversity.
a. Cholesterol, like all steroid molecules, has four adjacent rings, but their effects on the body largely depend on the attached groups indicated in red. **b.** Testosterone is the male sex hormone.

Doing science

Environmental Pollution Prevention

As you know, organic compounds include carbohydrates, lipids, proteins, and nucleic acids, which make up our bodies and the bodies of all living things. Other organic substances such as pesticides, herbicides, synthetic rubbers, plastics, and textiles are essential to our modern way of life. Unfortunately these organic substances often become pollutants, chemicals that alter the natural environment and are harmful to living organisms.

Human beings have always polluted their surroundings but in the past it was easier for them to move on and live somewhere else. They knew that given time, the environment would take care of pollution they left behind and they relied on the "out of sight, out of mind" philosophy. Today, an increasing human population, which uses an increasing amount of energy sources, no longer has the luxury to ignore pollution. The human population is expected to reach 7 billion persons by 2000 A.D. and this date is now just around the corner. Our overall energy consumption has gone up by a hundredfold from 2,000 kcal/person/day to 230,000 kcal/person/day in modern industrial nations like the United States. This high energy consumption allows us to mass produce many useful and economically affordable organic products that pollute the atmosphere and groundwater, damage forests and lakes, cause global warming, and even deplete the ozone layer.

Everyone needs and wants clean air to breathe and water to drink. Therefore everyone needs to conserve natural resources and also recycle materials. In 1962, President Richard M. Nixon used his executive powers to create the Environmental Protection Agency (EPA) to manage the country's air, water, and solid waste problems. Since that time Congress has passed a number of laws to regulate pollution. In 1969, the National Environmental Policy Act (NEPA) for the protection, maintenance, and enhancement of our environment became a law. In 1975 Congress enacted the Resource Conservation and Recovery Act (RCRA). This empowers the EPA to identify hazardous wastes, set standards for their management, provides guidelines, and provides financial aid so that each state can have a waste management program. The Comprehensive Environmental Response Compensation and Liability Act (CERCLA), commonly known as Superfund, became law in 1980 and was revised by the Superfund Amendments and Reauthorization Act (SARA) of 1986. The purpose of this act is to regulate the cleanup disposal sites that are leaking hazardous wastes into the environment.

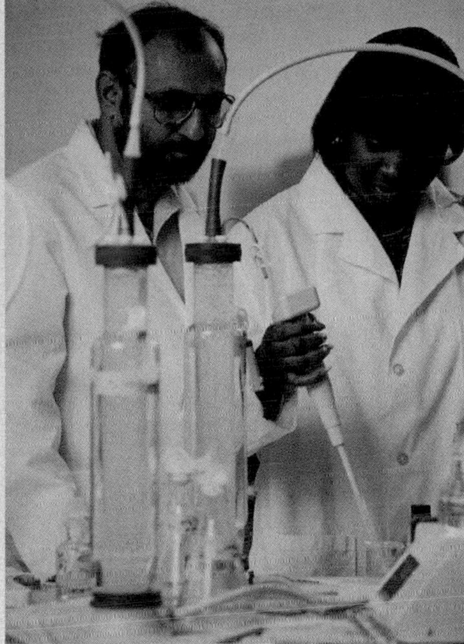

Kirit D. Chapatwala (*left*) with student (*right*). Selma University

As a society, many of our pollution problems arose because we failed to realize that in the environment "Everything is connected to and intermingled with everything else." We need to recognize that air, water, and soil pollution are related, and in general, polluting one of these with toxic chemicals reduces the possibility of natural degradation and/or the recycling capability of the environment. When I began my work as a biochemist, I decided to focus on the possibility of increasing natural degradation of organic toxic substances by isolating bacteria that are capable of breaking down toxic inorganic and organic substances. Bacteria that do this work naturally are more reliable than those that are genetically engineered to do so. Genetically engineered bacteria perform minimally and they are also subject to reverting back to their former state, losing the capability entirely.

My staff, which always includes some students, concentrates on isolating microorganisms capable of degrading certain functional groups. Therefore, these microorganisms can be used to degrade any toxic substance that contains a particular functional group. Cyanide (—CN) is a functional group that is poisonous to a key cellular enzyme and is therefore lethal to aquatic and marine organisms. One of our success stories is the isolation and identification of *Pseudomonas putida,* a bacterium capable of degrading cyanide and nitrites (organic compounds that contain a cyanide group) to carbon dioxide and ammonia, both of which are gases. We place this microorganism, which uses cyanide and nitrites as its sole source of carbon and nitrogen, on alginate (a substance extracted from marine kelp) beads known as biobeads. The use of beads increases degradation capacity and enzymatic stability while decreasing concentration of the compound being degraded.

Currently we have progressed to using a large reactor as opposed to a laboratory flask. Our results with the reactor show that cyanide compounds are degraded to a carbon dioxide and ammonia gas with no loss of microorganisms from the beads. Soon, we hope to find funding for a pilot project. Most of our work is funded by the governmental agencies mentioned above and to a lesser degree by private industries. Our papers, often entitled, "Biodegradation of Inorganic and Organic Compounds Using Immobilized Bead Technology," have been published in various national and international journals, and we have also presented our findings at many local and national meetings.

Once perfected, this technology can easily be adapted for use by various mining industries that release cyanides and industrial processes that create large quantities of nitrites as waste products. Therefore, use of our technology would prevent the release of cyanides and nitrites into the environment as pollutants. And we will have made use of a natural way the environment normally degrades such substances.

3.4 Proteins

Amino acids are the monomers that condense to form **proteins,** which are very large molecules with structural and metabolic functions. Keratin, which makes up hair and nails, and collagen fibers, which support many organs, are examples of structural proteins. The bulk of muscle is composed of the proteins myosin and actin. Many proteins primarily perform metabolic functions. The cellular proteins called **enzymes** are organic catalysts that speed up chemical reactions within cells. Insulin is a hormone that regulates the glucose content of the blood; hemoglobin transports oxygen in the blood. The proteins in the membrane that surrounds each cell have varied functions including enzymatic and transport functions.

In this section, we will see that the role a protein plays in cells is dependent on its molecular properties.

Peptide Bonds Join Amino Acids

All amino acids contain two important functional groups: a carboxyl (acid) group (—COOH) and an amino group (—NH_2), both of which ionize at normal body pH. Therefore, amino acids are hydrophilic:

$H_2N-C(H)(R)-COOH$	$H_3N^+-C(H)(R)-COO^-$
nonionized	ionized

Figure 3.14 shows how two amino acids are joined by a condensation reaction between the carboxyl group of one and the amino group of another. The resulting covalent bond between two amino acids is called a **peptide bond.** The atoms associated with the peptide bond share the electrons unevenly because oxygen is more electronegative than nitrogen. Therefore, the hydrogen attached to the nitrogen has a slightly positive charge, while the oxygen has a slightly negative charge. The polarity of the peptide bond means that hydrogen bonding is possible between parts of a polypeptide.

Amino acids differ in the nature of the *R* group. The *R*, which stands for the remainder of the molecule, ranges in complexity from a single hydrogen to complicated ring compounds. The unique chemical properties of an amino acid depend on those of the *R* group. For example, some *R* groups are polar and some are not. Also, the amino acid cysteine has an *R* group that ends with a sulfhydryl (—SH) group that often serves to connect one chain of amino acids to another by a disulfide bond, —S—S. There are twenty different amino acids commonly found in cells, of which Figure 3.15 gives several examples.

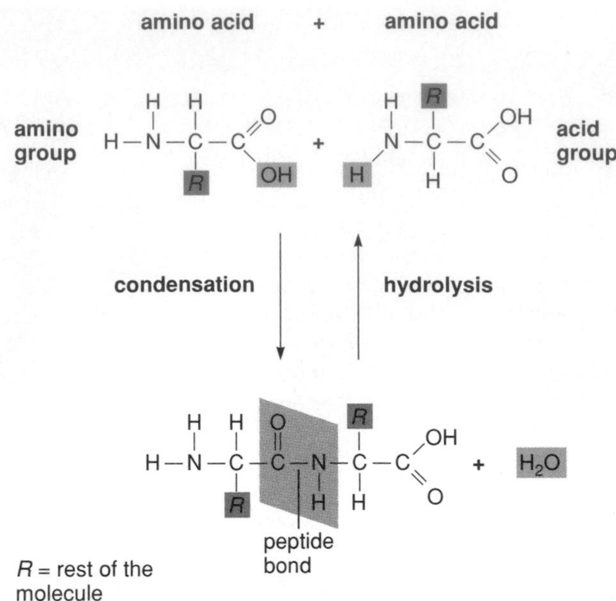

Figure 3.14 Synthesis of a peptide.
The peptide bond forms as the components of water are removed. There is a slightly negative charge on the oxygen and nitrogen and a slightly positive charge on the hydrogen associated with a peptide bond.

A **peptide** is two or more amino acids joined together, and a **polypeptide** is a chain of many amino acids joined by peptide bonds. A protein may contain more than one polypeptide chain; therefore, you can see why a protein could have a very large number of amino acids.

Proteins Can Be Denatured

Both temperature and pH can bring about a change in polypeptide shape. For example, we are all aware that the addition of acid to milk causes curdling; heating causes egg white, which consists mainly of a protein called albumin, to congeal, or coagulate. When a protein loses its normal configuration, it is said to be **denatured.** Denaturation occurs because the normal bonding patterns between parts of a molecule have been disturbed. Once a protein loses its normal shape, it is no longer able to perform its usual function.

If the conditions that caused denaturation were not too severe, and if these are removed, some proteins regain their normal shape and biological activity. This shows that the sequence of amino acids dictates the protein's final shape (Fig. 3.16).

Amino acids are joined by peptide bonds in polypeptides and proteins. The sequence of amino acids determines the shape of the protein, which is important to its function. Some proteins are enzymes that speed up chemical reactions.

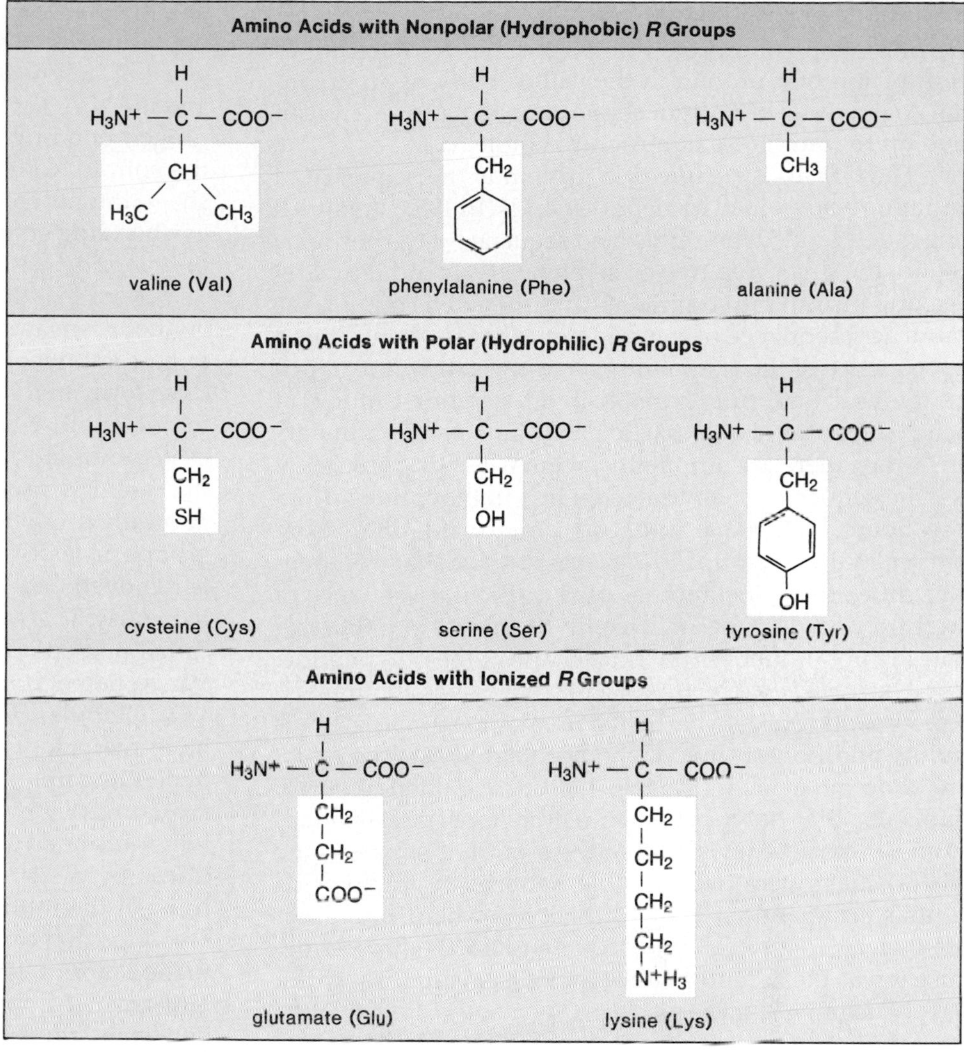

Figure 3.15 Representative amino acids.
The amino acids shown here are in ionized form. The *R* groups are in the contrasting color. Some *R* groups are nonpolar and hydrophobic, some are polar and hydrophilic, and some are ionized and hydrophilic. Proteins contain 20 different kinds of amino acids, of which only eight are shown here.

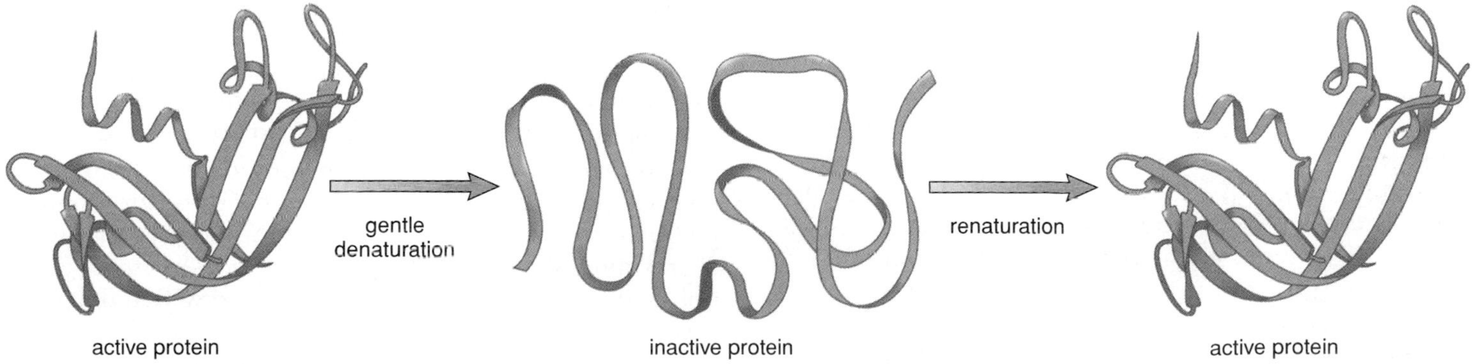

Figure 3.16 Denaturation and renaturation of a protein.
When a protein is denatured, it loses its normal shape and activity. If denaturation is gentle and if the original conditions are restored, some proteins regain their normal shape. This shows that the normal conformation of the molecule is due to the various interactions among a set sequence of amino acids. Each type of protein has a particular sequence of amino acids.

Proteins Have Levels of Structure

The final shape of a protein in large measure determines the function of a protein in the cell or body of an organism. An analysis of protein shape shows that proteins can have up to four levels of structure (Table 3.2).

The *primary structure* of a protein is the sequence of the amino acids joined by peptide bonds. In 1953, Frederick Sanger determined the amino acid sequence of the hormone insulin, the first protein to be sequenced. He did it by first breaking insulin into fragments and then determining the amino acid sequence of the fragments before determining the sequence of the fragments themselves. It was a laborious ten-year task, but it established for the first time that each polypeptide has a particular sequence of amino acids. Today, there are automated sequencers that tell scientists the sequence of amino acids in a polypeptide within a few hours. Notice that since each amino acid differs from another by its *R* group, it is correct to say that polypeptides differ from one another by a particular sequence of the *R* groups. The fact that some of these are polar and some are not influences the final shape of the polypeptide.

The *secondary structure* of a protein comes about when the polypeptide takes a particular orientation in space. Linus Pauling and Robert Corey, who began studying the structure of amino acids in the late 1930s, concluded about 20 years later that polypeptides must have particular orientations in space. They said that the α (alpha) helix and the β (beta) sheet were two possible patterns of amino acids within a polypeptide. They called it the α helix because it was the first pattern they discovered and the β sheet because it was the second pattern they discovered. There is a slightly negative charge on the oxygen and nitrogen and a slightly positive charge on the hydrogen associated with a peptide bond. These charges make it possible for hydrogen bonding to occur between the C=O of one amino acid and the N—H of another amino acid in a polypeptide. Hydrogen bonding between every fourth amino acid holds the spiral shape of an α helix. In the β sheet, or pleated sheet, the polypeptide turns back upon itself and hydrogen bonding occurs between these extended lengths of the polypeptide.

Keratin is the type of protein found in hair, wool, feathers, hooves, claws, beaks, skin, scales, and horns. The keratin in hair and wool is called α keratin because the polypeptides form an α helix. The α helices are covalently bonded to one another by disulfide (S—S) linkages between two cysteine amino acids. When you get a permanent, a reducing agent is used to break the disulfide bonds that are naturally present. (Each —S—S— bond becomes two —SH groups instead.) After the hair is rolled, the reducing agent is removed and new disulfide bonds form that make the hair curly. The keratin in the other structures mentioned is called β keratin because the polypeptides form a β sheet. Silk is another protein in which the polypeptides are β sheets.

Globular proteins, so called because of their shape, have regions of both α helix and β sheet arrangements depending upon the particular amino acids in the primary structure (Fig. 3.17). The polypeptide of a globular protein folds and twists into a *tertiary structure* that is maintained by various types of bonding between the *R* groups. Indeed, the folding and twisting is determined by those *R* groups that can bond with one another, giving stability to the shape of the molecule. Hydrogen bonds, ionic bonds, and covalent bonds are seen. Again, disulfide linkages help maintain the tertiary shape. On the other hand, hydrophobic *R* groups do not bond with other *R* groups and they tend to collect in a common region where they are not exposed to water (Fig. 3.17). These are called hydrophobic interactions. Although hydrophobic interactions are not bonds, they are very important in creating and stabilizing tertiary structure.

Each polypeptide has its own primary, secondary, and tertiary structure. Proteins that consist of only one polypeptide also have these same levels of structure. Some proteins consist of more than one polypeptide, and if so, they go on to have a *quaternary level of structure.* Hemoglobin is a much-studied globular protein that consists of four polypeptides and has a quaternary structure. Each polypeptide has a heme group associated with it that carries oxygen reversibly.

A protein has up to four levels of structure that account for its final three-dimensional shape. The shape of a protein determines its function in cells.

Table 3.2

Levels of Protein Structure

Level of Structure	Description	Type of Bond
Primary	Sequence of amino acids	Covalent (peptide) bond between amino acids
Secondary	Alpha helix and beta sheet	Hydrogen bond between amino acids along the peptide chain
Tertiary	Folding and twisting	Hydrogen, ionic, and covalent (S—S) bonds; hydrophobic interactions between *R* groups
Quaternary	Several polypeptides	Hydrogen, ionic bonds between polypeptide chains

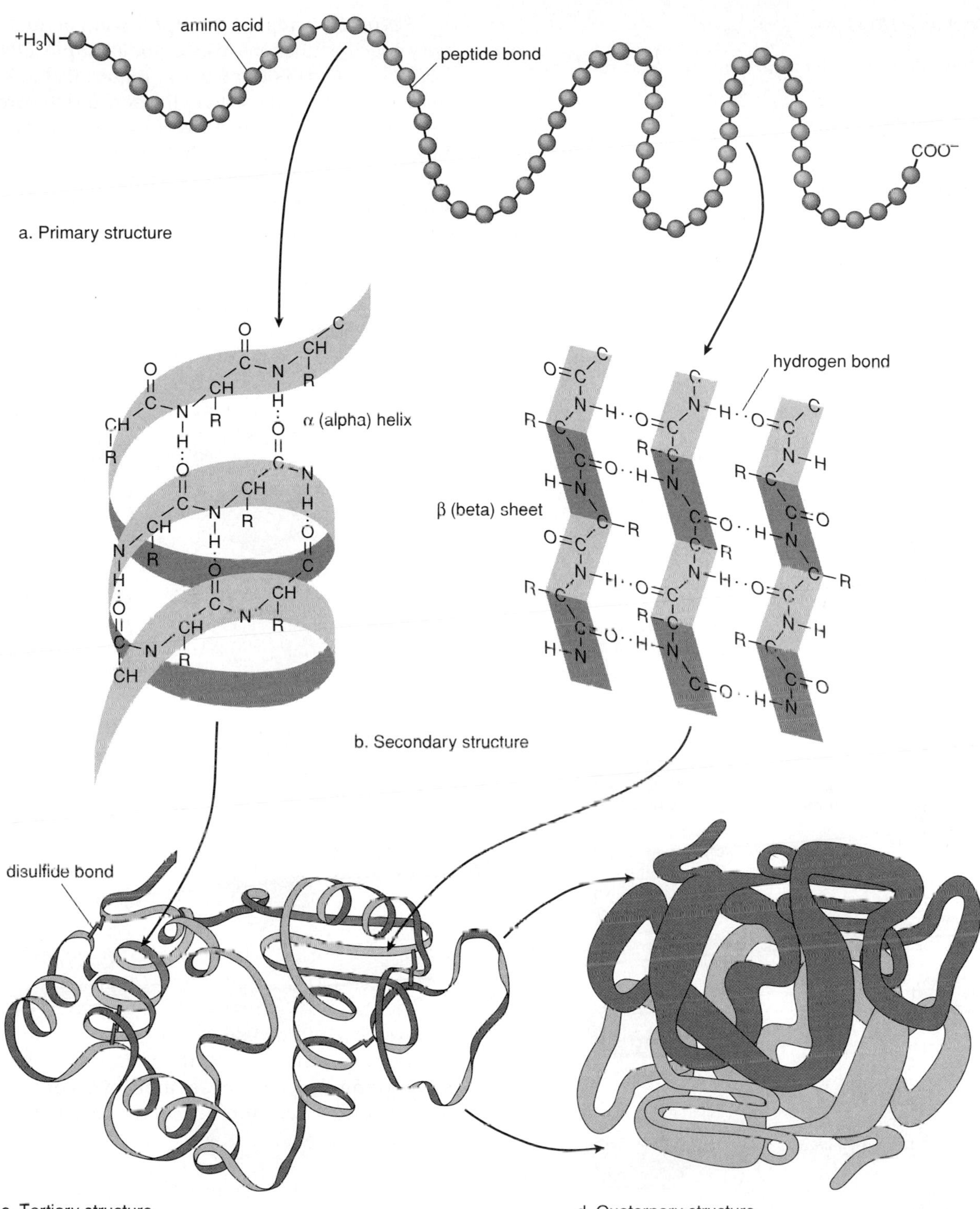

Figure 3.17 Levels of protein structure for a globular protein.
a. The primary structure is the sequence of amino acids. **b.** The secondary structure contains α helix segments and β sheet segments. **c.** The tertiary structure is a twisting and turning of the polypeptide molecule. **d.** The quaternary structure involves more than one polypeptide. Here we show two polypeptides from (c); one is turned 180°.

3.5 Nucleic Acids

Every **nucleotide** is a molecular complex of three types of unit molecules: phosphate (phosphoric acid), a pentose sugar, and a nitrogen-containing base (Fig. 3.18). Nucleotides have metabolic functions in cells. For example, some are components of coenzymes, which facilitate enzymatic reactions. ATP (adenosine triphosphate) is a nucleotide used in cells to supply energy for synthetic reactions and for various other energy-requiring processes. Nucleotides are also monomers in nucleic acids.

Nucleic acids are huge polymers of nucleotides with very specific functions in cells; for example, **DNA** (deoxyribonucleic acid) is the genetic material that stores information regarding its own replication and the order in which amino acids are to be joined to make a protein. Another important nucleic acid, **RNA** (ribonucleic acid) is an intermediary in the process of protein synthesis, conveying information from the DNA regarding the amino acid sequence in a protein.

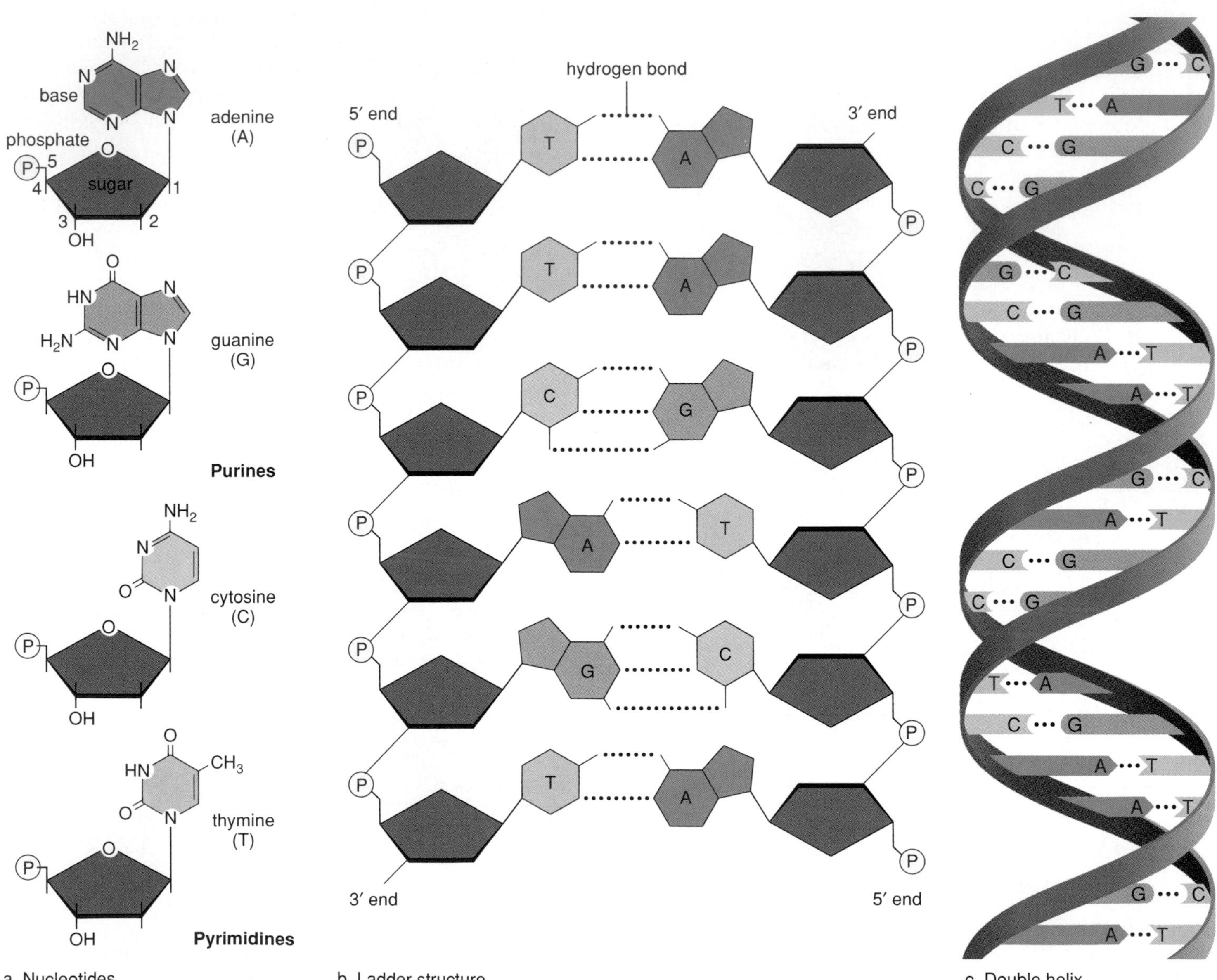

Figure 3.18 DNA structure.
a. There are four different nucleotides in DNA; each contains phosphate, the pentose sugar deoxyribose, and a nitrogen-containing organic base. Two bases are purines: adenine (A) and guanine (G); two bases are pyrimidines: thymine (T) and cytosine (C). **b.** DNA has a ladder structure; the sugar-phosphate molecules make up the sides and the hydrogen-bonded bases make up the rungs. **c.** Actually, DNA is a double helix in which the two strands twist about each other.

In DNA the sugar is deoxyribose, and in RNA the sugar is ribose; the difference accounts for their respective names (Table 3.3). There are four different types of nucleotides in DNA and four in RNA. Figure 3.18 shows the types of nucleotides that are present in DNA. The base can be one of the purines (adenine or guanine), which have a double ring, or one of the pyrimidines (thymine or cytosine), which have a single ring. These structures are called bases because their presence raises the pH of a solution. RNA contains the base uracil instead of thymine.

Nucleotides join in a definite sequence when DNA and RNA form by condensation. The nucleotides form a linear molecule called a strand in which the backbone is made up of phosphate-sugar-phosphate-sugar, with the bases projecting to one side of the backbone. Since the nucleotides occur in a definite order, so do the bases.

RNA is usually single stranded, while DNA is usually double stranded, with the two strands usually twisted about each other in the form of a double helix. The two strands are held together by hydrogen bonds between purine and pyrimidine bases. When unwound, DNA resembles a ladder. The sides of the ladder are made entirely of phosphate and sugar molecules, and the rungs of the ladder are made only of the complementary paired bases. The bases can be in any order within a strand, but between strands, thymine (T) is always paired with adenine (A), and guanine (G) is always paired with cytosine (C). This is called **complementary base pairing.** Therefore, regardless of the order or the quantity of any particular base pair, the number of purine bases (A + G) always equals the number of pyrimidine bases (T + C).

Table 3.3

DNA Structure Compared to RNA Structure

	DNA	RNA
Sugar	Deoxyribose	Ribose
Bases	Adenine, guanine, thymine, cytosine	Adenine, guanine, uracil, cytosine
Strands	Double stranded with base pairing	Single stranded
Helix	Yes	No

DNA has a structure like a twisted ladder: sugar and phosphate molecules make up the sides, and hydrogen-bonded bases make up the rungs of the ladder.

ATP (Adenosine Triphosphate)

ATP is a nucleotide in which *adenosine* is composed of adenine and ribose. Triphosphate stands for the three phosphate groups that are attached to ribose, the pentose sugar (Fig. 3.19). ATP is a high-energy molecule because the last two phosphate bonds are unstable and are easily broken. Usually in cells the terminal phosphate bond is hydrolyzed, leaving the molecule **ADP** (adenosine diphosphate) and a molecule of inorganic phosphate Ⓟ.

The energy that is released by ATP breakdown is coupled to energy-requiring processes in cells, such as the synthesis of macromolecules like carbohydrates and proteins. In muscle cells, the energy is used for muscle contraction, and in nerve cells, it is used for the conduction of nerve impulses. ATP is called the energy currency of cells because when cells carry out an energy-requiring activity or build molecules, they often used ATP as an energy source.

Because energy is released when the last phosphate bonds of ATP are hydrolyzed, it is sometimes called a high-energy bond, symbolized by a wavy line. But this terminology is misleading—the breakdown of ATP releases energy because the products of hydrolysis (ADP and Ⓟ) are more stable than ATP. It is the entire molecule that releases energy and not a particular bond.

ATP is a high-energy molecule. ATP breaks down to ADP + Ⓟ, releasing energy, which is used for all metabolic work done in a cell.

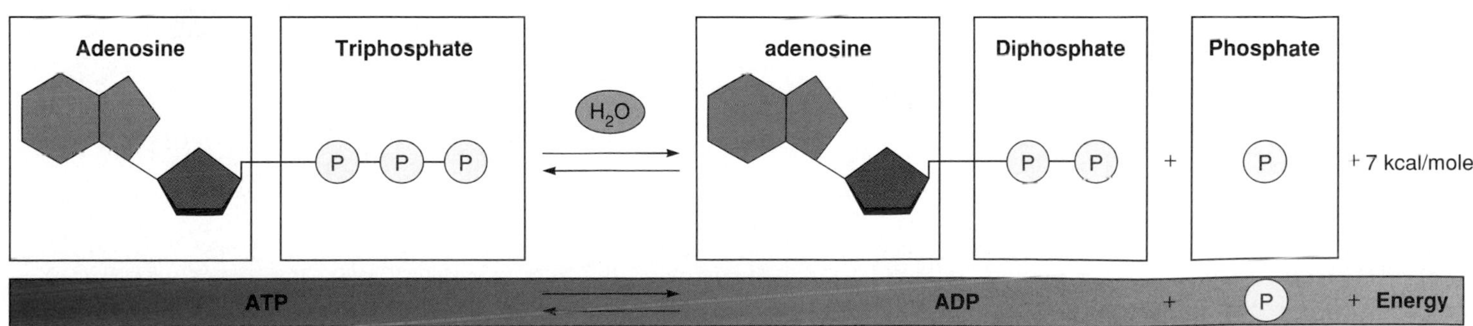

Figure 3.19 ATP reaction.
ATP, the universal energy currency of cells, is composed of adenosine and three phosphate groups. When cells require energy, ATP usually becomes ADP + Ⓟ and energy is released.

Health Focus

Nutrition Labels

As of May 1994, packaged foods must be labeled as depicted in Figure 3A. The nutrition information given here is based on the serving size (i.e., 1 $^1/_4$ cup, 57 grams) of a cereal. A Calorie* is a measurement of energy. One serving of the cereal provides 220 Calories, of which 20 are from fat. At the bottom of the label, the recommended amounts of nutrients are based on a typical diet of 2,000 Calories for women and 2,500 Calories for men. Fats are the nutrient that has the highest energy content: 9 Cal/g compared to 4 Cal/g for carbohydrates and proteins. The body stores fat for later use under the skin and around the organs. It is recommended that a 2,000-Calorie diet contain no more than 65 g (585 Calories) of fat. Dietary fat has been implicated in cancer of the colon, pancreas, ovary, prostate, and breast. Although saturated fat and cholesterol are essential nutrients, dietary consumption of saturated fat and cholesterol should especially be watched and controlled. Cholesterol and saturated fat contribute to the formation of deposits called plaque, which clog arteries and lead to cardiovascular disease, including high blood pressure. Therefore, it is important to know how a serving of the cereal will contribute to the daily maximum recommended amount of fat, saturated fat, and cholesterol for the day. You can find this out by looking at the listing under *% Daily Value:* the total fat in one serving of the cereal provides 3% of the daily recommended amount of fat for the day. How much will a serving of the cereal contribute to the maximum recommended amount of saturated fat? cholesterol?

Carbohydrates (sugars and polysaccharides) are the quickest, most readily available source of energy for the body. Carbohydrates aren't usually associated with health problems, and it is recommended that the largest proportion of the diet be carbohydrates. Complex carbohydrates in breads and cereals are preferable to simple carbohydrates in candy and ice cream because they are likely to contain dietary fiber (nondigestible plant material). Insoluble fiber has a laxative effect and seems to reduce the risk of colon cancer; soluble fiber combines with the cholesterol in food and prevents the cholesterol from entering the body proper.

The body does not store amino acids for the production of proteins, which are found particularly in muscles but also in all cells of the body. Although it is not stated on the label, a woman should have about 44 grams of protein per day and a man should have about 56 grams of protein a day. Red meat is rich in protein, but it is usually also high in saturated fat. Therefore, it is considered best to rely more on protein from plant origins (e.g., whole-grain cereals, dark breads, legumes) more than is customary in the United States.

The amount of dietary sodium (as in table salt) is of concern because excessive sodium intake has been linked to high blood pressure in some people. It is recommended that the intake of sodium be no more than 2,400 mg per day. A serving of this cereal provides what percent of this maximum amount?

Vitamins are essential requirements needed in small amounts in the diet. Each vitamin has a recommended daily intake, and the food label tells what percent of the recommended amount is provided by a serving of this cereal.

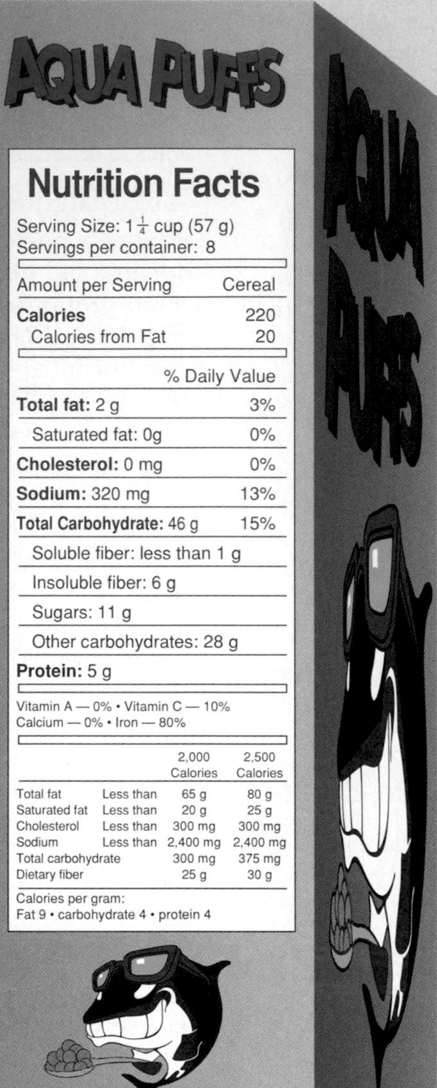

Nutrition Facts

Serving Size: 1 $\frac{1}{4}$ cup (57 g)
Servings per container: 8

Amount per Serving	Cereal
Calories	220
Calories from Fat	20
	% Daily Value
Total fat: 2 g	3%
Saturated fat: 0g	0%
Cholesterol: 0 mg	0%
Sodium: 320 mg	13%
Total Carbohydrate: 46 g	15%
Soluble fiber: less than 1 g	
Insoluble fiber: 6 g	
Sugars: 11 g	
Other carbohydrates: 28 g	
Protein: 5 g	

Vitamin A — 0% • Vitamin C — 10%
Calcium — 0% • Iron — 80%

		2,000 Calories	2,500 Calories
Total fat	Less than	65 g	80 g
Saturated fat	Less than	20 g	25 g
Cholesterol	Less than	300 mg	300 mg
Sodium	Less than	2,400 mg	2,400 mg
Total carbohydrate		300 mg	375 mg
Dietary fiber		25 g	30 g

Calories per gram:
Fat 9 • carbohydrate 4 • protein 4

Figure 3A Nutrition labels.

* A calorie is the amount of heat required to raise the temperature of 1 gram of water 1°C. Food energy is measured in Calories (the capital C means 1,000 calories).

connecting concepts

What does the term organic mean? For some, organic means that food products have been grown without the use of chemicals or have been minimally processed. But you now know that biochemically speaking, organic refers to molecules containing carbon and hydrogen. In biology, organic also refers to living things or anything that has been alive in the past. Therefore, the food we eat and the wood we burn are organic substances. Fossil fuels (coal and oil) formed over 300 million years ago from plant and animal life that, by chance, did not decompose, are also organic. When burned, they release carbon dioxide into the atmosphere just as we do when we respire!

Although living things are very complex, their macromolecules are simply polymers of small organic molecules. Simple sugars are the monomers of complex carbohydrates; amino acids are the monomers of proteins; nucleotides are the monomers of nucleic acids. Fats are composed of fatty acids and glycerol.

This system of forming macromolecules still allows for diversity. Monomers exist in modified forms and can combine in slightly different ways; therefore, a variety of macromolecules can come about. In starch, a plant product, glucose monomers are linked in a slightly different way from glucose monomers in glycogen, an animal product. One protein differs from another by the sequence of the same 20 amino acids.

Clearly, these differences do not negate the fact that the same four types of organic molecules—carbohydrates, proteins, and lipids (fats)—make up the cells of all organisms, both living and nonliving. The plants and animals we consume are composed of the very same organic molecules that make up our bodies. There is no doubt that the chemistry of carbon is the chemistry of life.

Summary

3.1 Cells Contain Organic Molecules

The chemistry of carbon accounts for the diversity of organic molecules found in living things. Carbon can bond with as many as four other atoms. It can also bond with itself to form both chains and rings. Differences in the carbon backbone and attached functional groups cause organic molecules to have different chemical properties. The chemical properties of a molecule determine how it interacts with other molecules and the role the molecule plays in the cell. Some functional groups are hydrophobic and some are hydrophilic.

The large organic molecules in cells are macromolecules, polymers formed by the joining together of monomers. For each bond formed during condensation, a molecule of water is removed, and for each bond broken during hydrolysis, a molecule of water is added.

3.2 Carbohydrates

Monosaccharides, disaccharides, and polysaccharides are all carbohydrates. Therefore, the term carbohydrate includes both the monomers (e.g., glucose) and the polymers (e.g., starch, glycogen, and cellulose). Starch and glycogen are energy-storage compounds, but cellulose and chitin have structural roles in organisms.

3.3 Lipids

Lipids include a wide variety of compounds that are insoluble in water. Fats and oils allow long-term energy storage and contain one glycerol and three fatty acids. Fats tend to contain saturated fatty acids, and oils tend to contain unsaturated fatty acids. In a phospholipid, one of the fatty acids is replaced by a phosphate group. In the presence of water, phospholipids form a double layer because the head of each molecule is polarized and the tails are not. Waxes and steroids are also lipids.

3.4 Proteins

Proteins are polymers of amino acids. Some proteins are enzymes, and other proteins have structural roles in cells and organisms.

A polypeptide is a long chain of amino acids joined by peptide bonds. There are 20 different amino acids in cells that differ only by their *R* groups. Polarity and nonpolarity are important aspects of the *R* groups. The shape of a protein, dependent on its particular sequence of amino acids, influences its biological activity. A protein can be denatured and lose its normal shape and activity, but if renaturation occurs, it assumes both again. This shows that the final shape of a protein is dependent upon the primary sequence of amino acids.

A protein has three levels of structure: the primary level is the sequence of the amino acids; the secondary level contains α (alpha) helices and β (beta) sheets held in place by hydrogen bonding between amino acids along the polypeptide chain; and the tertiary level is the final folding and twisting of the polypeptide that is held in place by bonding and hydrophobic interactions between *R* groups. Proteins that contain more than one polypeptide have a quaternary level of structure, as well.

3.5 Nucleic Acids

Nucleic acids are polymers of nucleotides. Each nucleotide has three components: a sugar, a base, and phosphate (phosphoric acid). DNA, which contains the sugar deoxyribose, is the genetic material that stores information for its own replication and for the order in which amino acids are to be sequenced in proteins. DNA, with the help of RNA, specifies protein synthesis.

ATP, with its unstable phosphate bonds, is the energy currency of cells. Hydrolysis of ATP to ADP + Ⓟ releases energy that is used by the cell to do metabolic work.

Table 3.4 summarizes our coverage of organic molecules in cells.

Table 3.4

Organic Molecules in Cells

	Categories	Elements	Examples	Functions
Carbohydrates	Monosaccharides	C, H, O		
	6-carbon sugar		Glucose	Immediate energy source
	5-carbon sugar		Deoxyribose, ribose	Structure of DNA, RNA
	Disaccharides 12-carbon sugar	C, H, O	Sucrose	Transport sugar in plants
	Polysaccharides polymer of glucose	C, H, O	Starch, glycogen Cellulose	Energy storage in plants, animals Plant cell wall structure
Lipids	Triglycerides 1 glycerol + 3 fatty acids	C, H, O	Fats, oils	Long-term energy storage
	Waxes fatty acid + alcohol	C, H, O	Cuticle Ear wax	Protective covering in plants Protective wax in ears
	Phospholipids like triglyceride except the head group contains phosphate	C, H, O, P	Lecithin	Plasma membrane component
	Steroids backbone of 4 fused rings	C, H, O	Cholesterol Testosterone	Plasma membrane component Male sex hormone
Proteins	Polypeptides polymer of amino acids	C, H, O, N, S	Enzymes Myosin and actin Insulin Hemoglobin Collagen	Speed up cellular reactions Muscle cell components Regulates sugar content of blood Oxygen carrier in blood Fibrous support of body parts
Nucleic Acids	Nucleic acids polymer of nucleotides	C, H, O, N, P	DNA RNA	Genetic material Protein synthesis
	Nucleotides		ATP Coenzymes	Energy carrier Assist enzymes

Reviewing the Chapter

1. How are the chemical characteristics of carbon reflected in the characteristics of organic molecules? 36–37
2. Give examples of functional groups, and discuss the importance of their being hydrophobic or hydrophilic. 37
3. What molecules are monomers of the polymers studied in this chapter? How are monomers joined to produce polymers, and how are polymers broken down to monomers? 38
4. Name several monosaccharides, disaccharides, and polysaccharides, and give a function of each. How are these molecules structurally distinguishable? 39–40
5. Name the different types of lipids, and give a function for each type. What is the difference between a saturated and unsaturated fatty acid? Explain the structure of a fat molecule by stating its components and how they are joined together. 42–43
6. How does the structure of a phospholipid differ from that of a fat? How do phospholipids form a double layer in the presence of water? 44
7. Draw the structure of an amino acid and a dipeptide, pointing out the peptide bond. 46
8. How is the tertiary structure of a polypeptide related to its primary structure? Mention denaturation as evidence of this relationship. 46–48
9. Discuss the four levels of structure of a protein and relate each level to particular bonding patterns. 48
10. How are nucleotides joined to form nucleic acids? Discuss the structure of DNA, and name several differences between the structure of DNA and that of RNA. 50–51
11. Discuss the structure and function of ATP. 51

Testing Yourself

Choose the best answer for each question.

1. Which of these is not a characteristic of carbon?
   a. forms four covalent bonds
   b. bonds with itself
   c. is sometimes ionic
   d. forms long chains
2. The functional group —COOH is
   a. acidic.
   b. basic.
   c. never ionized.
   d. All of these are correct.
3. A hydrophilic group is
   a. attracted to water.
   b. a polar or ionized group.
   c. found in fatty acids.
   d. All of these are correct.

4. Which of these is an example of hydrolysis?
   a. amino acid + amino acid → dipeptide + H_2O
   b. dipeptide + H_2O → amino acid + amino acid
   c. Both of these are correct.
   d. Neither of these is correct.
5. Which of these makes cellulose nondigestible?
   a. a polymer of glucose subunits
   b. a fibrous protein
   c. the linkage between the glucose molecules
   d. the peptide linkage between the amino acid molecules
6. A fatty acid is unsaturated if it
   a. contains hydrogen.
   b. contains double bonds.
   c. contains an acidic group.
   d. bonds to glycogen.
7. Which of these is not a lipid?
   a. steroid
   b. fat
   c. polysaccharide
   d. wax
8. The difference between one amino acid and another is found in the
   a. amino group.
   b. carboxyl group.
   c. *R* group.
   d. peptide bond.
9. The shape of a polypeptide is
   a. maintained by bonding between parts of the polypeptide.
   b. important to its function.
   c. ultimately dependent upon the primary structure.
   d. All of these are correct.
10. Which of these is the peptide bond?

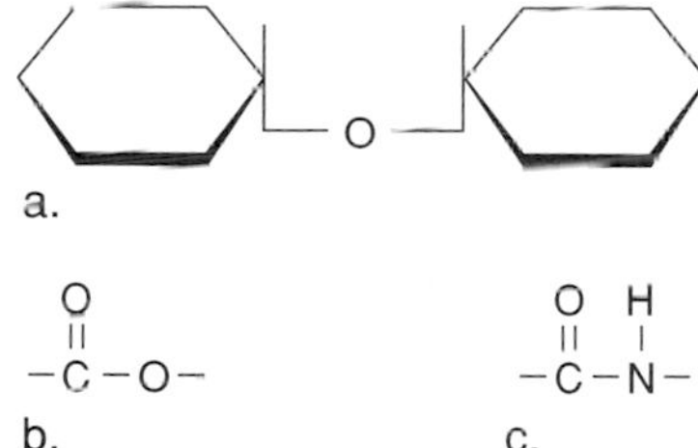

b. −C(=O)−O−

c. −C(=O)−N(H)−

11. Nucleotides
    a. contain a sugar, a nitrogen-containing base, and a phosphate molecule.
    b. are the monomers for fats and polysaccharides.
    c. join together by covalent bonding between the bases.
    d. All of these are correct.
12. Label the following diagram using the terms monomer, hydrolysis, condensation, and polymer, and explain the diagram:

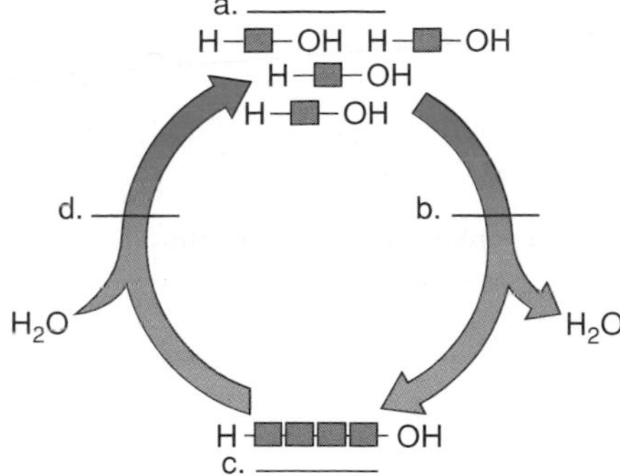

13. Label the levels of protein structure in this diagram of hemoglobin and explain:

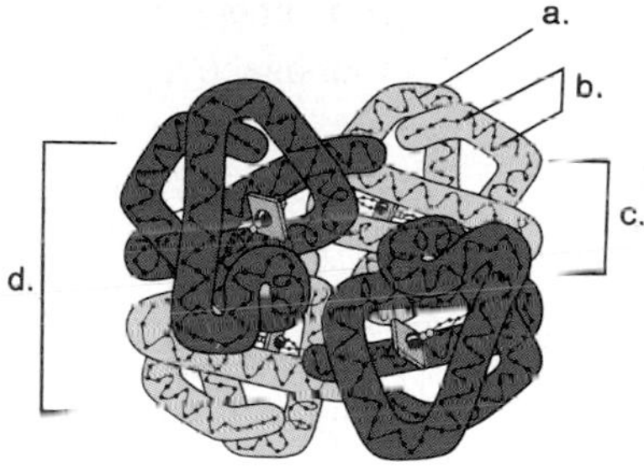

14. ATP
    a. is an amino acid.
    b. has a helical structure.
    c. is a high energy molecule that can react with water to release ADP and phosphate.
    d. provides enzymes for metabolism.

Applying the Concepts

1. *The unity and the diversity of life begin at the molecular level of organization.*
   How do the organisms in Figure 3.1 demonstrate this biological concept?
2. *The atoms and bonds within a molecule determine its chemical and physical properties.*
   Compare fats that contain mostly saturated fatty acids with oils that contain mostly unsaturated fatty acids to demonstrate this concept.
3. *The properties of a molecule determine the role that the molecule plays in the cells or the body of an organism.*
   Choose either a phospholipid or the protein keratin to demonstrate this biological concept.

Using Technology

Your study of the chemistry of life is supported by these available technologies:

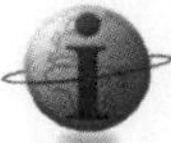

Exploring the Internet

The Mader Home Page provides resources for and help with studying this chapter.

http://www.mhhe.com/sciencemath/biology/mader/
(Click on Biology.)

Understanding the Terms

Match the terms to these definitions:

a. __________ Class of organic compounds consisting of carbon, hydrogen, and oxygen atoms; includes monosaccharides, disaccharides, and polysaccharides.

b. __________ Class of organic compounds that tend to be soluble in nonpolar solvents such as alcohol; includes fats and oils.

c. __________ Macromolecule consisting of covalently bonded monomers.

d. __________ Molecules with the same molecular formula but different structure and, therefore, shape.

e. __________ Organic catalyst, usually a protein molecule, that speeds chemical reactions in living systems.

f. __________ Organic molecule having an amino group and an acid group, that covalently bonds to produce protein molecules.

g. __________ Polymer having, as its primary structure, a sequence of amino acids united through covalent bonding.

h. __________ Polymer of nucleotides—both DNA and RNA.

i. __________ Two or more amino acids joined together by covalent bonding.

Cell Structure and Function

CHAPTER

Chapter Concepts

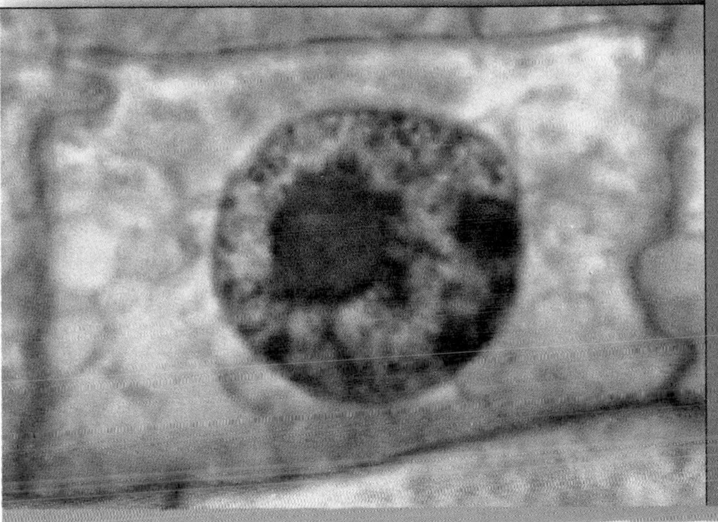

Onion root tip, *Allium* 20 μm

Today we are accustomed to thinking of living things as being constructed of cells. But the word cell didn't come into use until the seventeenth century. Antonie van Leeuwenhoek of Holland is famous for observing tiny, unicellular living things that no one had seen before. Leeuwenhoek sent his findings to an organization of scientists called the Royal Society in London. Robert Hooke, an Englishman, confirmed Leeuwenhoek's observations and was the first to use the term cell. The tiny chambers he observed in the honeycomb structure of cork reminded him of the rooms, or cells, in a monastery. Naturally, he referred to the boundaries of these chambers as walls.

An examination of the cell in the light micrograph shown here may not be overly impressive. You can see a darkened nucleus near the cell's center and tiny dark specks spread throughout the cell like specks of dust. Now we know, however, that the nucleus contains numerous chromosomes and thousands of genes! The dots surrounding the nucleus are actually organelles, tiny specialized structures performing specific functions in the cell. Cells may seem simple but they are really quite complex.

4.1 Cells Make Up Living Things

In the 1830s, Matthias Schleiden stated that all plants are composed of cells and Theodor Schwann stated that all animals are composed of cells. These Germans based their ideas not only on their own work but on the work of all who had studied tissues under microscopes. Today we recognize that virtually all the organisms we see about us are made up of cells. Figure 4.1 illustrates that in our daily lives we observe the whole organism, but if it were possible to view them internally with a microscope, we would then see their cellular nature. A **cell** is the smallest unit of living matter.

There are unicellular organisms but most, including ourselves, are multicellular. A cell is not only the structural unit, it is also the functional unit of organs and, therefore, organisms. This is very evident when you consider certain illnesses of the human body such as diabetes or prostate cancer. It is the cells of the pancreas or the prostate that are malfunctioning, rather than the organ itself.

All organisms are made up of cells, and a cell is the structural and functional unit of organs and, ultimately, organisms.

Cells reproduce. Once a growing cell gets to a certain size, it divides. Unicellular organisms reproduce themselves when they divide. Multicellular organisms grow when their cells divide. Cells are also involved in the reproduction of multicellular organisms. This shows that there is a continuity of cells from generation to generation. Soon after Schleiden and Schwann had presented their findings, another German scientist, Rudolf Virchow, used a microscope to study the life of cells. He came to the conclusion that cells do not arise on their own accord; rather, "every cell comes from a preexisting cell." By the middle of the nineteenth century, biologists clearly recognized that all organisms are composed of self-reproducing cells.

Cells are capable of self-reproduction, and cells come only from preexisting cells.

The previous two highlighted statements are often called the **cell theory.** The cell theory is one of the unifying concepts of biology. There are extensive data to support its two basic principles and it is universally accepted by biologists.

a. Corn, *Zea mays*

c. Rabbit, *Oryctolagus* sp.

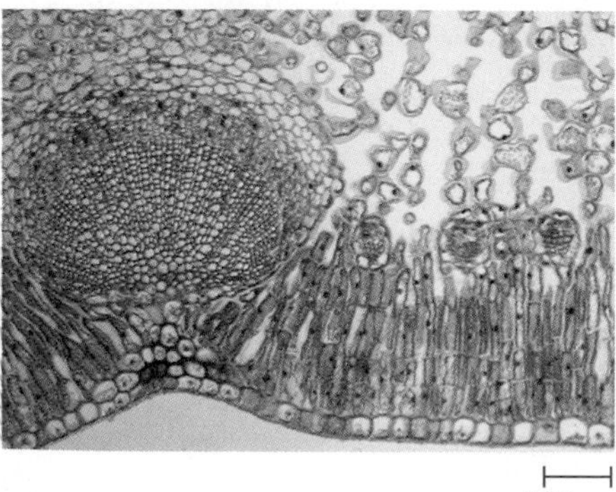

b.

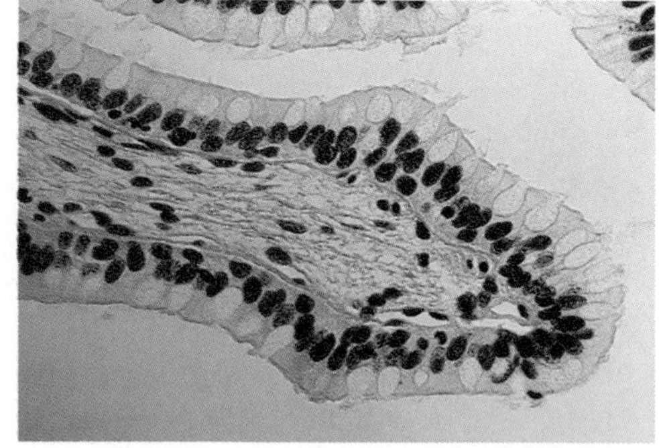

d.

Figure 4.1 Organisms and cells. All organisms, whether plants or animals, are composed of cells. This is not readily apparent because a microscope is usually needed to see the cells. **a.** Corn. **b.** Light micrograph of corn leaf showing many individual cells. **c.** Rabbit. **d.** Light micrograph of a rabbit's intestinal lining showing that it, too, is composed of cells. The dark-staining bodies are nuclei.

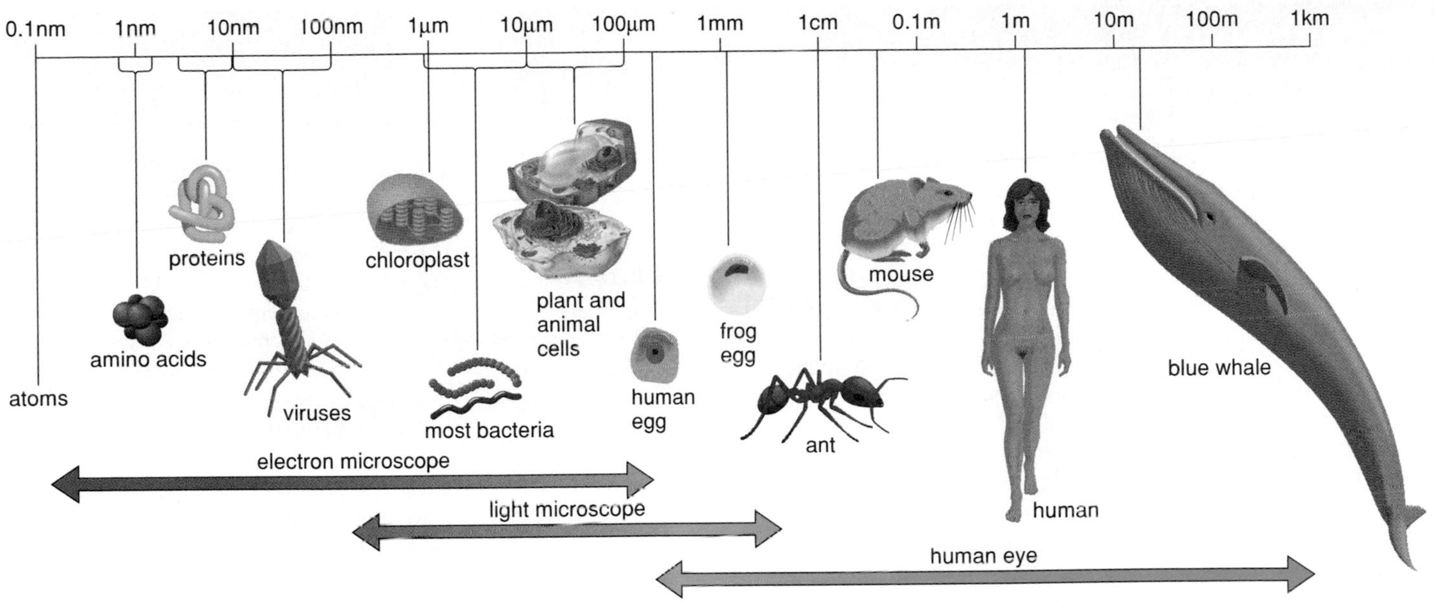

Figure 4.2 The sizes of living things and their components.
Whales', humans', mice's, and even frogs' eggs can be seen by the human eye. It takes a microscope to see most cells and lower levels of biological organization. Cells are visible with the light microscope, but not in much detail. It takes an electron microscope to see organelles in detail and to make out viruses and molecules. Notice that in this illustration each higher unit is 10 times greater than the lower unit. (In the metric system, 1 meter = 10^2 cm = 10^3 mm = 10^6 µm = 10^9 nm—see Appendix C.)

Cells Are Small

Cells are quite small. A frog's egg, at about one millimeter (mm) in diameter, is large enough to be seen by the human eye. Most cells are far smaller than one millimeter; some are even as small as one micrometer (µm)—one thousandth of a millimeter. Cell inclusions and macromolecules are even smaller still than a micrometer and are measured in terms of nanometers (nm).

Why are cells so small? To answer this question consider that a cell needs a surface area large enough to allow adequate nutrients to enter and to rid itself of wastes. A small cube that is 1 mm tall has a surface area of 6 mm^2 and a volume of 1 mm^3. This is a ratio of surface area to volume of 6:1. But a cube that is 2 mm tall has a surface area of 24 mm^2 and a volume of 8 mm^3. This is a ratio of only 3:1. Therefore a small cell has more surface area per volume than does a large cell:

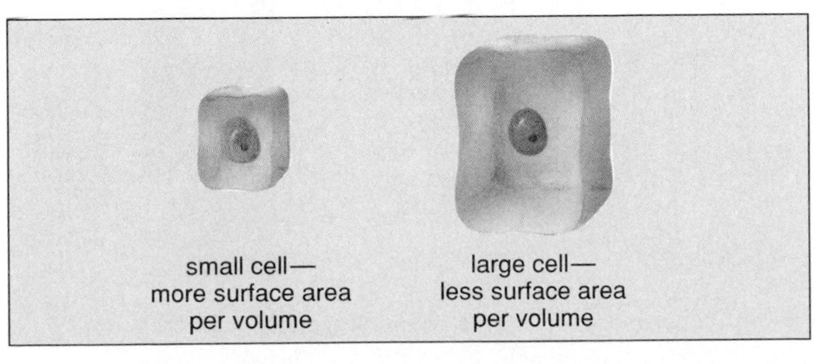

Small cells, not large cells, are likely to have an adequate surface area for exchange of wastes for nutrients. We would expect, then, a size limitation for an actively metabolizing cell. A chicken's egg is several centimeters in diameter, but the egg is not actively metabolizing. Once the egg is incubated and metabolic activity begins, the egg divides repeatedly without growth. Cell division restores the amount of surface area needed for adequate exchange of materials. Further, cells that specialize in absorption have modifications that greatly increase the surface area per volume of the cell. The columnar cells along the surface of the intestinal wall have surface foldings called microvilli (sing., microvillus), which increase their surface area.

A cell needs a surface area that can adequately exchange materials with the environment. Surface-area-to-volume considerations require that cells stay small.

Figure 4.2 outlines the visual range of the eye, light microscope, and electron microscope. The discussion of microscopy in the reading on pages 60–61 explains why the electron microscope allows us to "see" so much more detail than the light microscope does.

Doing Science

Microscopy of Today

Cells were not discovered until the invention of the microscope in the seventeenth century. Since that time, various types of microscopes have been developed for the study of cells and their components.

In the *bright-field light microscope,* light rays passing through a specimen are brought into focus by a set of glass lenses, and the resulting image is then viewed by the human eye. In the *transmission electron microscope,* electrons passing through a specimen are brought into focus by a set of magnetic lenses, and the resulting image is projected onto a fluorescent screen or photographic film. In the *scanning electron microscope* (SEM), a narrow beam of electrons is scanned over the surface of the specimen, which is coated with a thin metal layer. The metal gives off secondary electrons that are collected by a detector to produce an image on a television screen. The SEM permits the development of three-dimensional images (Fig. 4A).

Magnification, Resolution, and Contrast

Almost everyone knows that the magnifying capability of a transmission electron microscope is greater than that of a light microscope. A light microscope can magnify objects a few thousand times, but an electron microscope can magnify them hundreds of thousands of times. The difference lies in the means of illumination. The path of light rays and electrons moving through space is wavelike, but the wavelength of electrons is much shorter than the wavelength of light. This difference in wavelength accounts for the electron microscope's greater magnifying capability and its greater resolving power. The greater the resolving power, the greater the detail eventually seen. Resolution is the minimum distance between two objects at which they can still be seen, or resolved, as two separate objects. If oil is placed between the sample and the objective lens of the light microscope, the resolving power is increased, and if ultraviolet light is used instead of visible light, it is also increased. But typically, a light microscope can resolve down to 0.2 μm, while the electron microscope can resolve down to 0.0002 μm. If the resolving power of the average human eye is set at one, then that of the typical light microscope is about 500, and that of the electron microscope is 100,000. This means that the electron microscope distinguishes much greater detail (Fig. 4A).

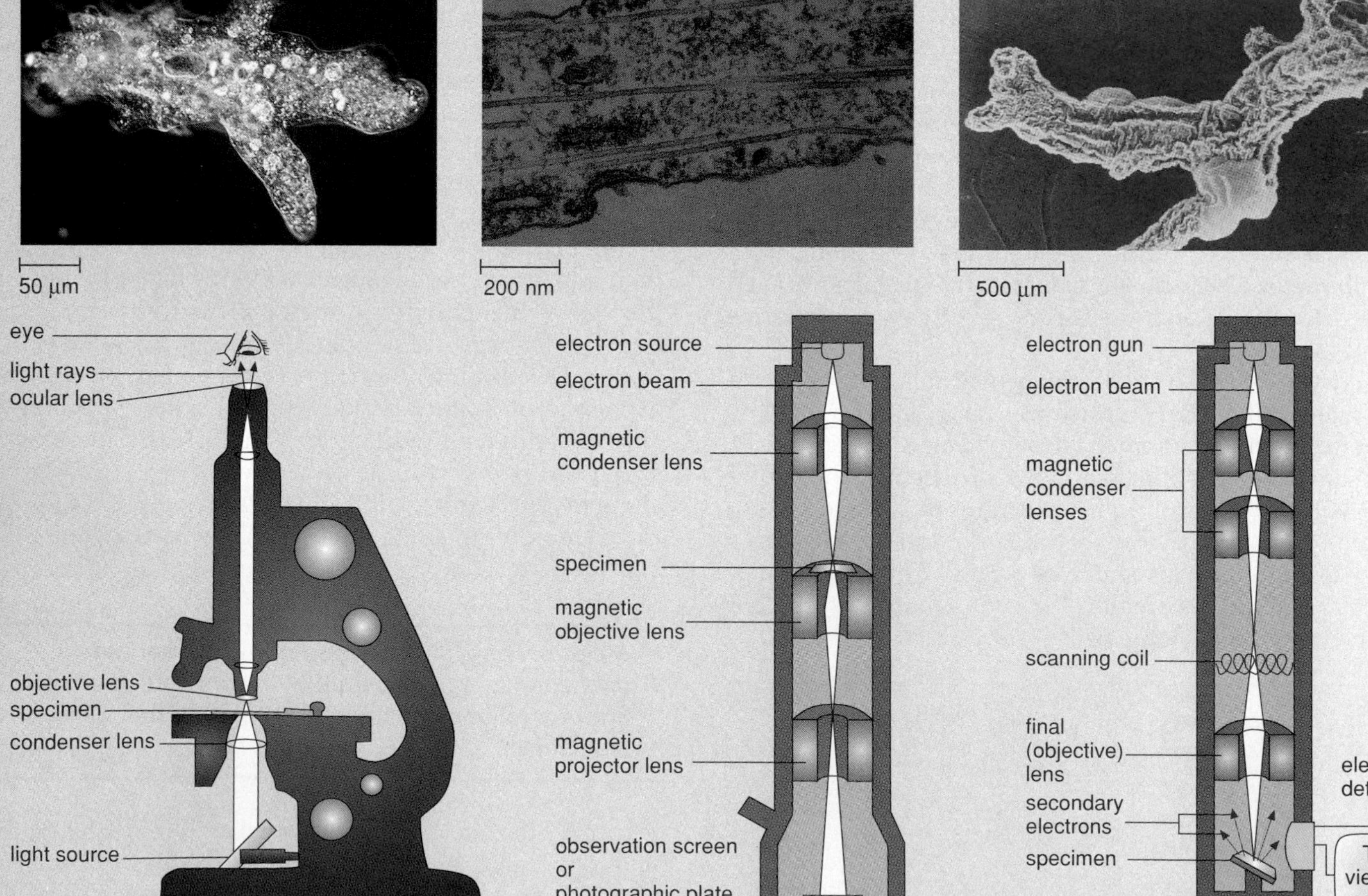

a. Compound light microscope b. Transmission electron microscope c. Scanning electron microscope

Figure 4A Diagram of microscopes with accompanying micrographs of *Amoeba proteus*.

Some microscopes view living specimens, but often specimens are treated prior to observation. Cells are killed, fixed so they do not decompose, and embedded into a matrix. The matrix strengthens the specimen so that it can be thinly sliced. These sections are often stained with colored dyes (light microscopy) or with electron-dense metals (electron microscopy) to provide contrast. Another way to increase contrast is to use optical methods such as phase contrast and differential interference contrast (Fig. 4B). In addition to optical and electronic methods for contrasting transparent cells, a third very prominent research tool is called *immunofluorescence microscopy,* because it uses fluorescent antibodies to reveal the location of a protein in the cell (see Fig. 4.12). The importance of this method is that the cellular distribution of a single type of protein can be examined.

Illumination, Viewing, and Recording

Light rays can be bent (refracted) and brought to a focus as they pass through glass lenses, but electrons do not pass through glass. Electrons have a charge that allows them to be brought into a focus by magnetic lenses. The human eye utilizes light to see an object but cannot utilize electrons for the same purpose. Therefore, electrons leaving the specimen in the electron microscope are directed toward a screen or a photograph plate that is sensitive to their presence. Humans can view the image on the screen or photograph.

A major advancement in illumination has been the introduction of *confocal microscopy,* which uses a laser beam scanned across the specimen to focus on a single shallow plane within the cell. The microscopist can "optically section" the specimen by focusing up and down, and a series of optical sections can be combined in a computer to create a three-dimensional image, which can be displayed and rotated on the computer screen.

An image from a microscope may be recorded by replacing the human eye with a television camera. The television camera converts the light image into an electronic image, which can be entered into a computer. In *video-enhanced contrast microscopy,* the computer makes the darkest areas of the original image much darker and the lightest areas of the original much lighter. The result is a high-contrast image with deep blacks and bright whites. Even more contrast can be introduced by the computer if shades of gray are replaced by colors.

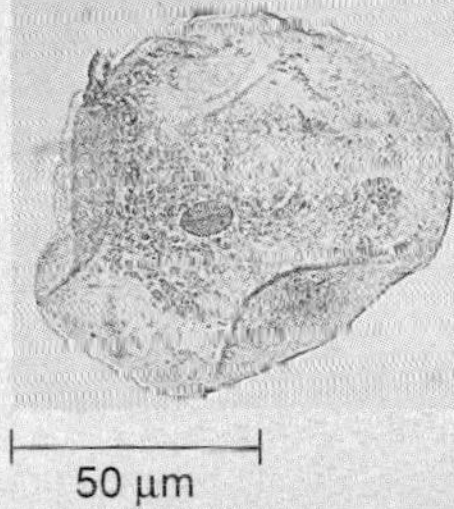

50 μm

Bright-field. Light passing through the specimen is brought directly into focus. Usually, the low level of contrast within the specimen interferes with viewing all but its largest components.

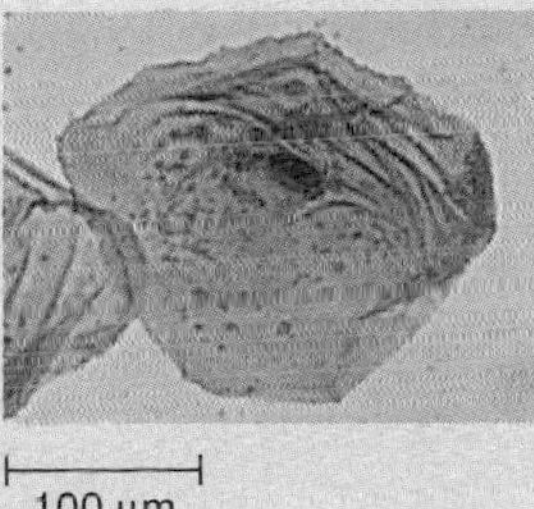

100 μm

Bright-field (stained). Dyes are used to stain the specimen. Certain components take up the dye more than other components, and therefore contrast is enhanced.

100 μm

Differential interference contrast. Optical methods are used to enhance density differences within the specimen so that certain regions appear brighter than others. This technique is used to view living cells, chromosomes, and organelle masses.

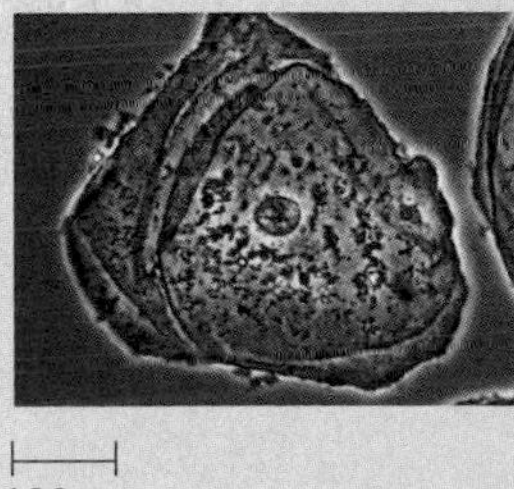

100 μm

Phase contrast. Density differences in the specimen cause light rays to come out of "phase." The microscope enhances these phase differences so that some regions of the specimen appear brighter or darker than others. This technique is widely used to observe living cells and organelles.

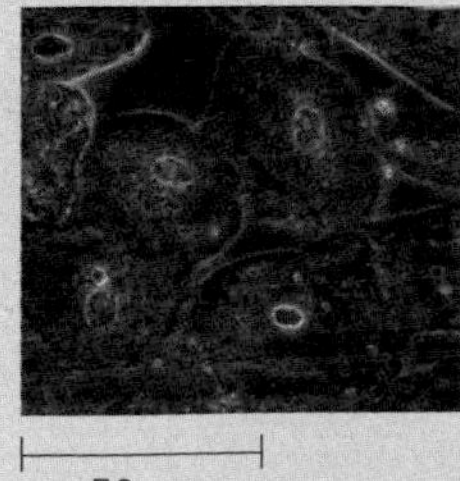

50 μm

Darkfield. Light is passed through the specimen at an oblique angle so that the objective lens receives only light diffracted and scattered by the object. This technique is used to view organelles, which appear quite bright against a dark field.

Figure 4B Photomicrographs of cheek cells, illustrating different types of light microscopy.

4.2 Prokaryotic Cells Are Less Complex

Bacteria [Gk. *bacterion,* rod] are **prokaryotic cells** [Gk. *pro,* before, and *karyon,* kernel, nucleus] in the kingdom Monera. Most bacteria are between 1–10 μm in size; therefore, they are just visible with the light microscope.

Figure 4.3 illustrates the main features of bacterial anatomy. The **cell wall** contains peptidoglycan, a complex molecule with chains of a unique amino disaccharide joined by peptide chains. In some bacteria, the cell wall is further surrounded by a **capsule** and/or gelatinous sheath called a **slime layer.** Motile bacteria usually have long, very thin appendages called flagella (sing., **flagellum**) that are composed of subunits of the protein called flagellin. The flagella, which rotate like propellers, rapidly move the bacterium in a fluid medium. Some bacteria also have *fimbriae,* which are short appendages that help them attach to an appropriate surface.

A membrane called the **plasma membrane** regulates the movement of molecules into and out of the cytoplasm, the interior of the cell. **Cytoplasm** in a prokaryotic cell consists of **cytosol,** a semifluid medium, and thousands of granular inclusions called **ribosomes** that coordinate the synthesis of proteins. In prokaryotes, most genes are found within a single chromosome (loop of DNA, or deoxyribonucleic acid) located within the **nucleoid region** [L. *nucleus,* nucleus, kernel and Gk. *-eides,* like], but they may also have small accessory rings of DNA called *plasmids.* In addition, the photosynthetic cyanobacteria have light-sensitive pigments, usually within the membranes of flattened disks called **thylakoids.**

Although bacteria seem fairly simple, they are actually metabolically diverse. Bacteria are adapted to living in almost any kind of environment and are diversified to the extent that almost any type of organic matter can be used as a nutrient for some particular bacterium. Given an energy source, most bacteria are able to synthesize any kind of molecule they may need. Therefore, the cytoplasm is the site of thousands of chemical reactions and bacteria are more metabolically competent than are human beings. Indeed, the metabolic capability of bacteria is exploited by humans who use them to produce a wide variety of chemicals and products for human use.

Bacteria are prokaryotic cells with these constant features.

Outer boundary:	**cell wall**
	plasma membrane
Cytoplasm:	**ribosomes**
	thylakoids (cyanobacteria)
	innumerable enzymes
Nucleoid:	**chromosome (DNA only)**

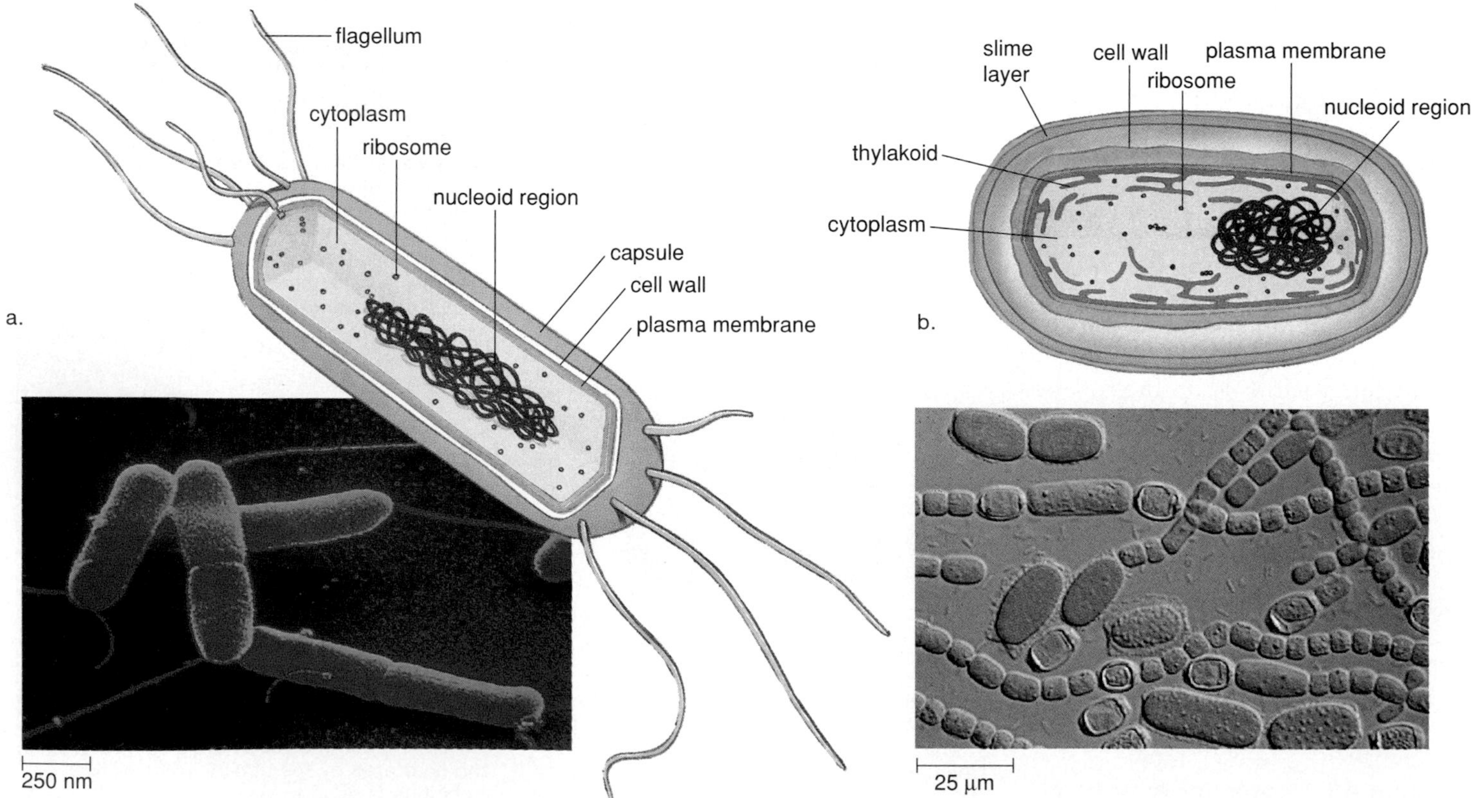

Figure 4.3 Prokaryotic cells.
a. Nonphotosynthetic bacterium. b. Cyanobacterium, a photosynthetic bacterium, formerly called a blue-green alga.

4.3 Eukaryotic Cells Are More Complex

In general, eukaryotic cells are larger than prokaryotic cells. In contrast to prokaryotic cells, **eukaryotic cells** [Gk. *eu*, true, and *karyon*, kernel, nucleus] have a true nucleus. A nucleus is a membrane-bounded structure where DNA is housed within threadlike structures called chromatin.

Eukaryotic cells have a membrane-bounded nucleus, and prokaryotic cells lack a nucleus.

A membrane is a phospholipid bilayer with embedded proteins.

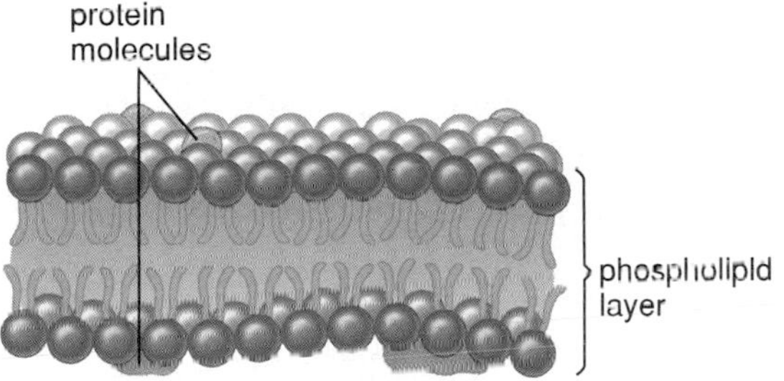

In eukaryotic cells, **organelles** are small membranous bodies, each with a specific structure and function (Table 4.1). The cytosol, which is a semifluid medium outside the nucleus, is divided up and compartmentalized by the organelles. Compartmentalization keeps the cell organized and keeps its various functions separate from one another. Just as prokaryotic cells lack a nucleus, so they also lack organelles. The cytosol in eukaryotic cells has an organized lattice of protein filaments called the *cytoskeleton*.

Eukaryotic cells, like prokaryotic cells, have a plasma membrane that separates the contents of the cell from the environment and regulates the passage of molecules into and out of the cell. Some eukaryotic cells, notably plant cells, also have an outer boundary called a **cell wall.** A plant cell wall contains cellulose fibrils and therefore has a different composition than the cell wall of prokaryotic cells. A cell wall supports and protects the cell but does not interfere with the movement of molecules across the plasma membrane.

Both animal (Fig. 4.4) and plant (Fig. 4.5) cells contain mitochondria while only plant cells have chloroplasts. Only animal cells have centrioles. The color chosen to represent each structure in the plant and animal cell is used for that structure throughout the chapters of this part.

Table 4.1

Eukaryotic Structures in Animal Cells and Plant Cells

Name	Composition	Function
Cell wall*	Contains cellulose fibrils	Support and protection
Plasma membrane	Phospholipid bilayer with embedded proteins	Define cell boundary; regulation of molecule passage into and out of cell
Nucleus	Nuclear envelope surrounding nucleoplasm, chromatin, and nucleoli	Storage of genetic information; synthesis of DNA and RNA
Nucleolus	Concentrated area of chromatin, RNA, and proteins	Ribosomal formation
Ribosome	Protein and RNA in two subunits	Protein synthesis
Endoplasmic reticulum (ER)	Membranous flattened channels and tubular canals	Synthesis and/or modification of proteins and other substances, and transport by vesicle formation
Rough ER	Studded with ribosomes	Protein synthesis
Smooth ER	Having no ribosomes	Various; lipid synthesis in some cells
Golgi apparatus	Stack of membranous saccules	Processing, packaging, and distribution of proteins and lipids
Vacuole and vesicle	Membranous sacs	Storage of substances
Lysosome	Membranous vesicle containing digestive enzymes	Intracellular digestion
Microbody	Membranous vesicle containing specific enzymes	Various metabolic tasks
Mitochondrion	Membranous cristae bounded by an outer membrane	Cellular respiration
Chloroplast*	Membranous grana bounded by two membranes	Photosynthesis
Cytoskeleton	Microtubules, intermediate filaments, actin filaments	Shape of cell and movement of its parts
Cilia and flagella	9 + 2 pattern of microtubules	Movement of cell
Centriole**	9 + 0 pattern of microtubules	Formation of basal bodies

*Plant cells only
**Animal cells only

Figure 4.4 Animal cell anatomy.
a. Generalized drawing. b. Transmission electron micrograph. See Table 4.1 for a description of these structures, along with a listing of their functions.

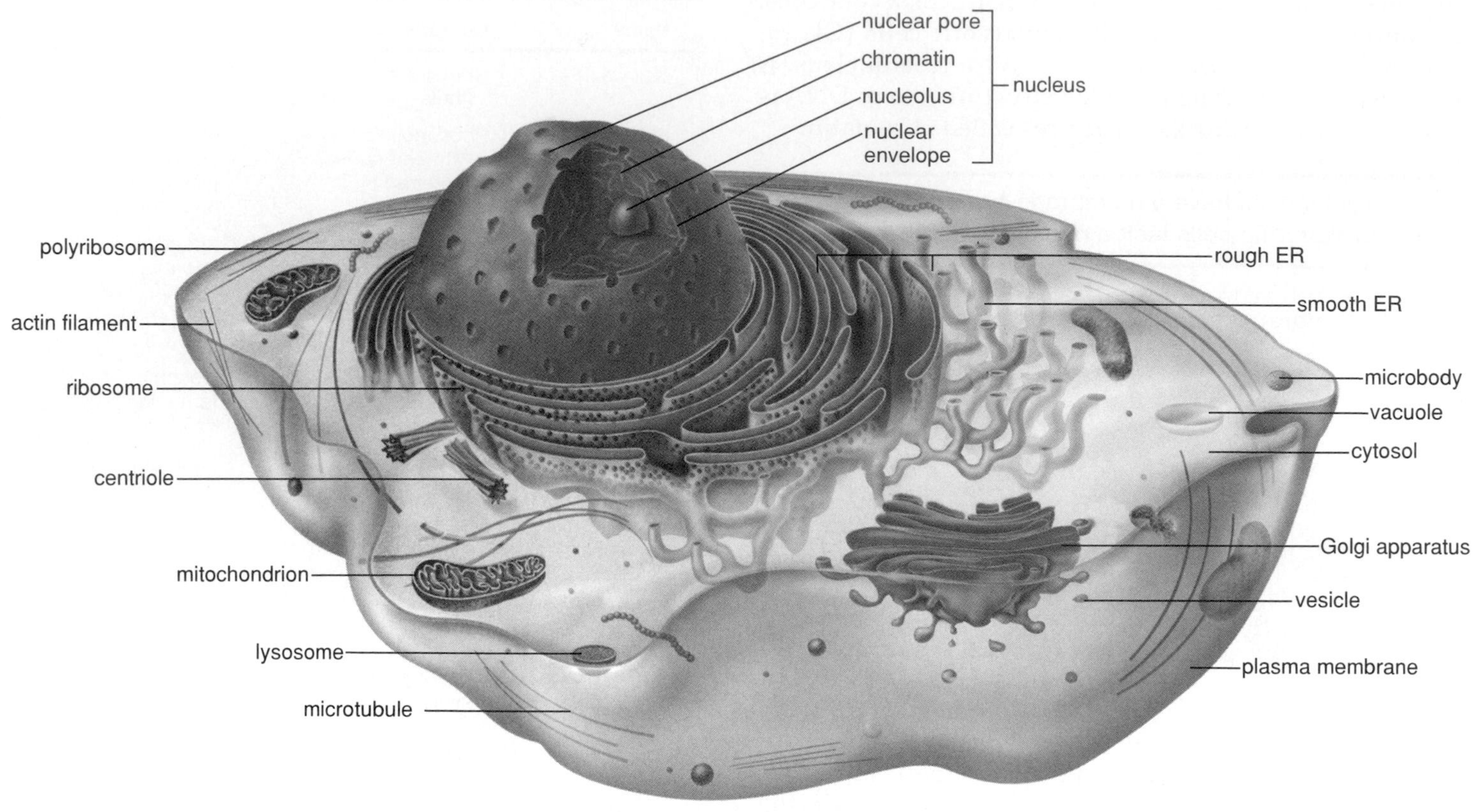

a.

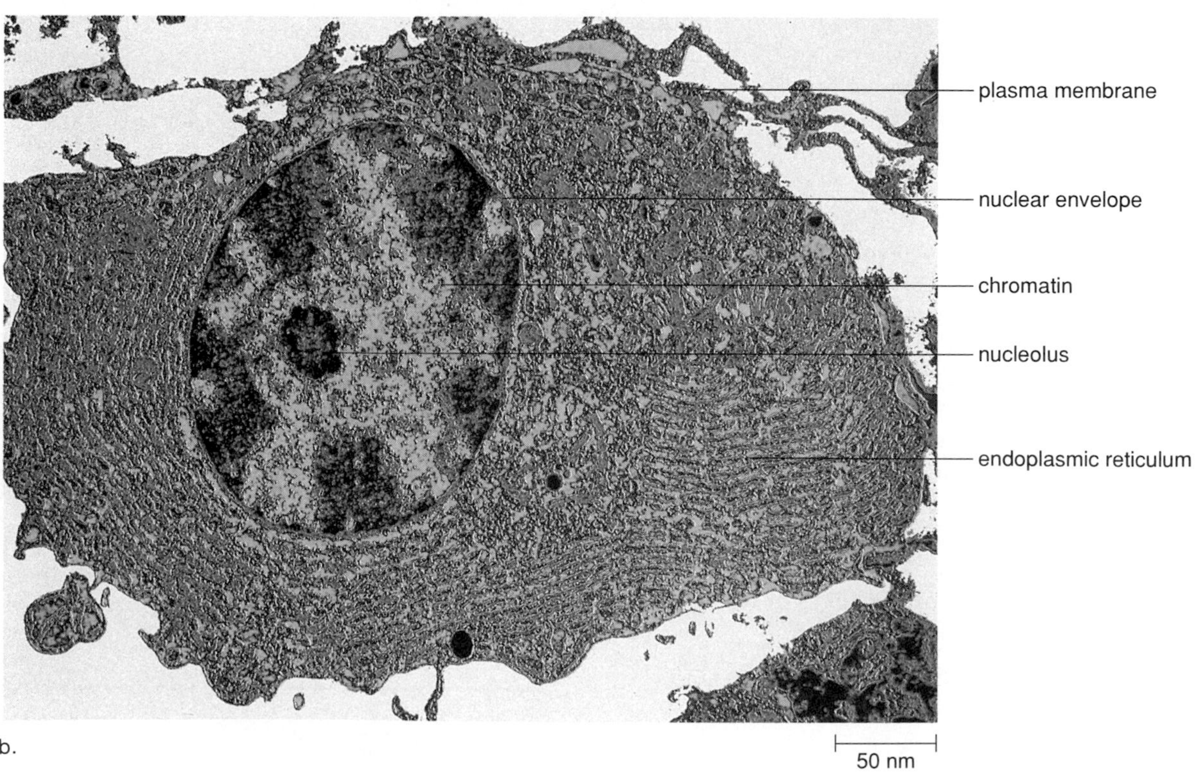

b.

Figure 4.5 Plant cell anatomy.
a. Generalized drawing. b. Transmission electron micrograph of young leaf cell. See Table 4.1 for a description of these structures, along with a listing of their functions.

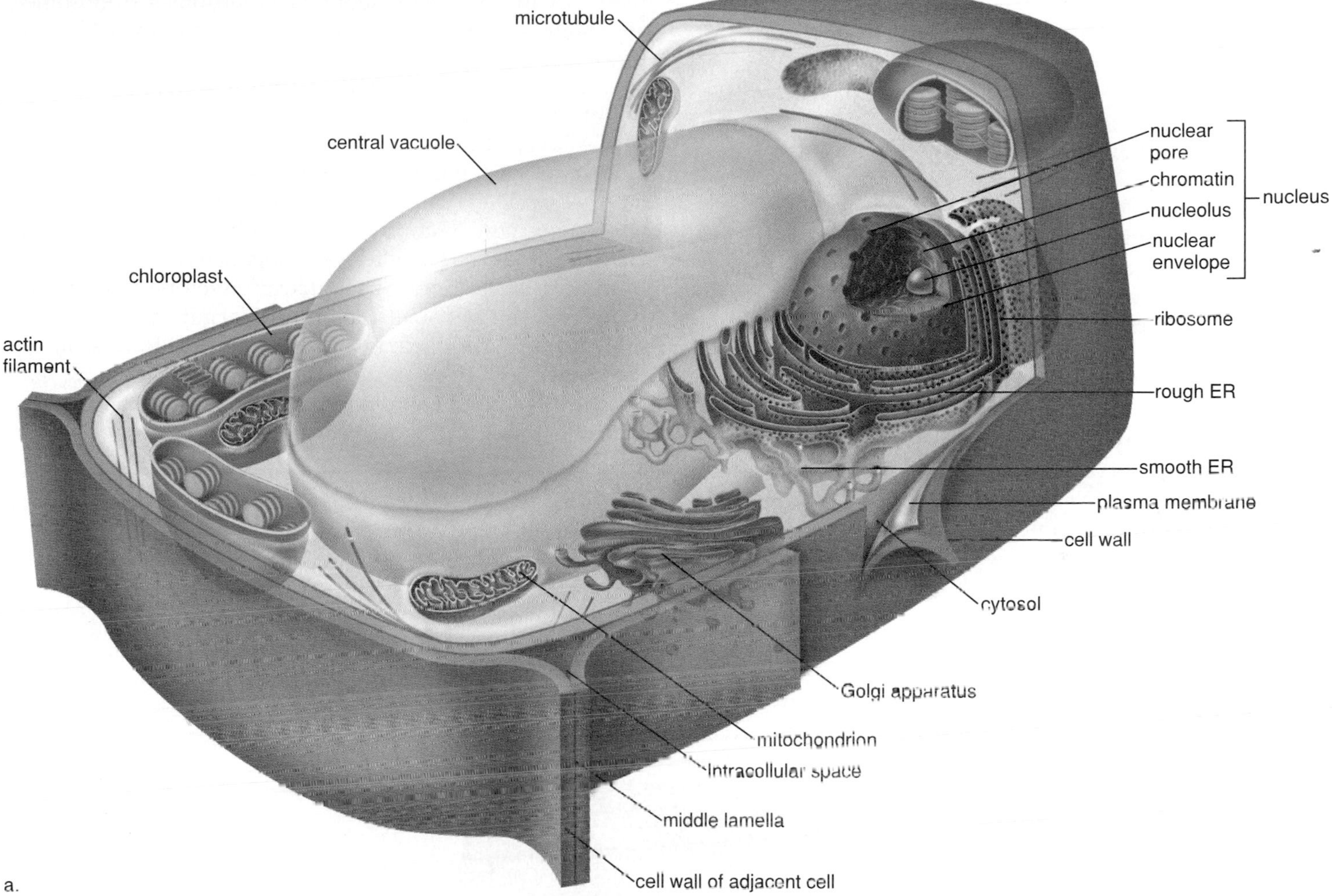

microtubule
central vacuole
chloroplast
actin filament
nuclear pore
chromatin
nucleolus
nuclear envelope
nucleus
ribosome
rough ER
smooth ER
plasma membrane
cell wall
cytosol
Golgi apparatus
mitochondrion
intercellular space
middle lamella
cell wall of adjacent cell

a.

peroxisome
mitochondrion
central vacuole
chloroplast
nucleus
ribosomes
plasma membrane
cell wall
1 μm

b.

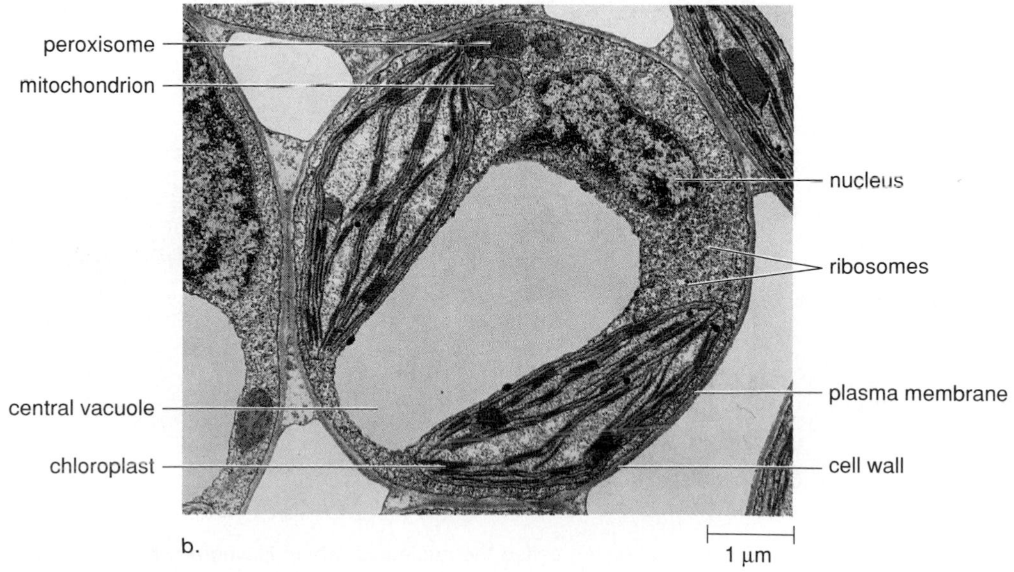

Nucleus Stores Genetic Information

The **nucleus** [L. *nucleus*, kernel], which has a diameter of about 5 μm, is a prominent structure in the eukaryotic cell. The nucleus is of primary importance because it stores genetic information that determines the characteristics of the body's cells and their metabolic functioning. Every cell contains a complex copy of genetic information, but each cell type has certain genes, or segments of DNA, turned on, and others turned off. Which particular genes are active in a cell is controlled at least in part by cytoplasmic molecules capable of entering the nucleus. Activated DNA, with RNA acting as an intermediary, specifies the sequencing of amino acids during protein synthesis. The proteins of a cell determine its structure and the functions it can perform.

When you look at the nucleus, even in an electron micrograph, you cannot see DNA molecules but you can see chromatin. **Chromatin** [Gk. *chroma*, color, and *teino*, stretch] looks grainy, but actually it is a threadlike material that undergoes coiling into rodlike structures called **chromosomes** [Gk. *chroma*, color, and *soma*, body], just before the cell divides. Chemical analysis shows that chromatin, and therefore chromosomes, contains DNA and much protein, and some RNA. Chromatin is immersed in a semifluid medium called the **nucleoplasm.** A difference in pH between the nucleoplasm and cytosol suggests that the nucleoplasm has a different composition.

Most likely, too, when you look at an electron micrograph of a nucleus, you will see one or more regions that look darker than the rest of the chromatin. These are nucleoli (sing., **nucleolus**) where another type of RNA, called ribosomal RNA (rRNA), is produced and where rRNA joins with proteins to form the subunits of ribosomes. (Ribosomes are small bodies in the cytoplasm that contain rRNA and proteins.)

The nucleus is separated from the cytoplasm by a double membrane known as the **nuclear envelope** (Fig. 4.6). A layer of protein fibers called the nuclear lamina is associated with the inner membrane of the nuclear envelope. The nuclear lamina helps maintain the shape of the nucleus, organizes chromatin by providing chromatin attachment sites, and may funnel substances toward or away from the nuclear pores. The nuclear envelope has **nuclear pores** of sufficient size (100 nm) to permit the passage of proteins into the nucleus and ribosomal subunits out of the nucleus. High-power electron micrographs show that the pores have nonmembranous components associated with them that form a nuclear pore complex.

The structural features of the nucleus include the following.

Chromatin: (chromosomes)	**DNA and proteins**
Nucleolus:	**chromatin ribosomal subunits**
Nuclear envelope:	**double membrane with pores**

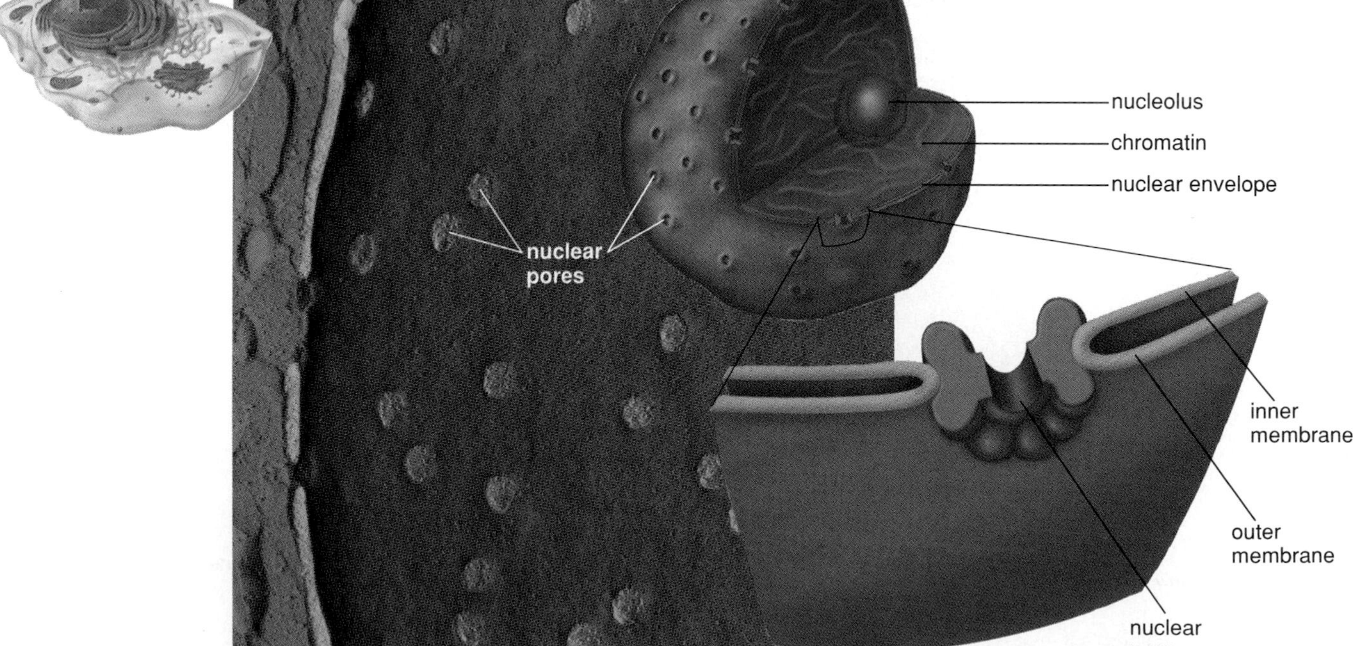

Figure 4.6 Anatomy of the nucleus.
The nucleoplasm contains chromatin. Chromatin has a special region called the nucleolus, which is where rRNA is produced and ribosomal subunits are assembled. The nuclear envelope contains pores, as is shown in this micrograph of a freeze-fractured nuclear envelope. Each pore is lined by a complex of eight proteins.

Ribosomes Are Sites of Protein Synthesis

Ribosomes, as mentioned, are small particles composed of rRNA and proteins. Unlike many of the organelles discussed in this chapter, ribosomes are found in both prokaryotes and eukaryotes. In eukaryotes, ribosomes are 20 nm by 30 nm, and in prokaryotes they are slightly smaller. In both types of cells, ribosomes are composed of two subunits, one large and one small (Fig. 4.7). Each subunit has its own mix of proteins and rRNA. Protein synthesis occurs on the ribosomes.

Ribosomes occur free within the cytosol either singly or in groups called **polyribosomes.** Ribosomes are often attached to the endoplasmic reticulum, a membranous system of saccules and channels discussed in the next section. Proteins synthesized by cytoplasmic ribosomes are used in the cell, such as in the mitochondria and chloroplasts.

Ribosomes are small organelles where protein synthesis occurs. Ribosomes occur in the cytosol, both singly and in groups (i.e., polyribosomes). Numerous ribosomes are attached to the endoplasmic reticulum.

Endomembrane System Is Elaborate

The endomembrane system consists of the nuclear envelope, the endoplasmic reticulum, the Golgi apparatus and several vesicles. This system compartmentalizes the cell so that particular enzymatic reactions are restricted to specific regions. Membranes that make up the endomembrane system are connected by direct physical contact and/or by the transfer of vesicles (tiny membranous sacs) from one part to the other.

Endoplasmic Reticulum

The **endoplasmic reticulum** (ER) [Gk. *endon,* within, *plasma,* something molded, and L. *reticulum,* net], a complicated system of membranous channels and saccules (flattened vesicles), is physically continuous with the outer membrane of the nuclear envelope. Rough ER is studded with ribosomes on the side of the membrane that faces the cytoplasm (Fig. 4.7). Here proteins are synthesized and enter the ER interior where processing and modification begin. Smooth ER, which is continuous with rough ER, does not have attached ribosomes. Smooth ER synthesizes the phospholipids that occur in membranes and has various other functions depending on the particular cell. In the testes, it produces testosterone and, in the liver, it helps detoxify drugs. Regardless of any specialized function, smooth ER also forms *vesicles* in which large molecules are transported to other parts of the cell. Often these vesicles are on their way to the plasma membrane or the Golgi apparatus.

ER is involved in protein synthesis (rough ER) and various other processes such as lipid synthesis (smooth ER). Molecules that are produced or modified in the ER are eventually enclosed in vesicles that often transport them to the Golgi apparatus.

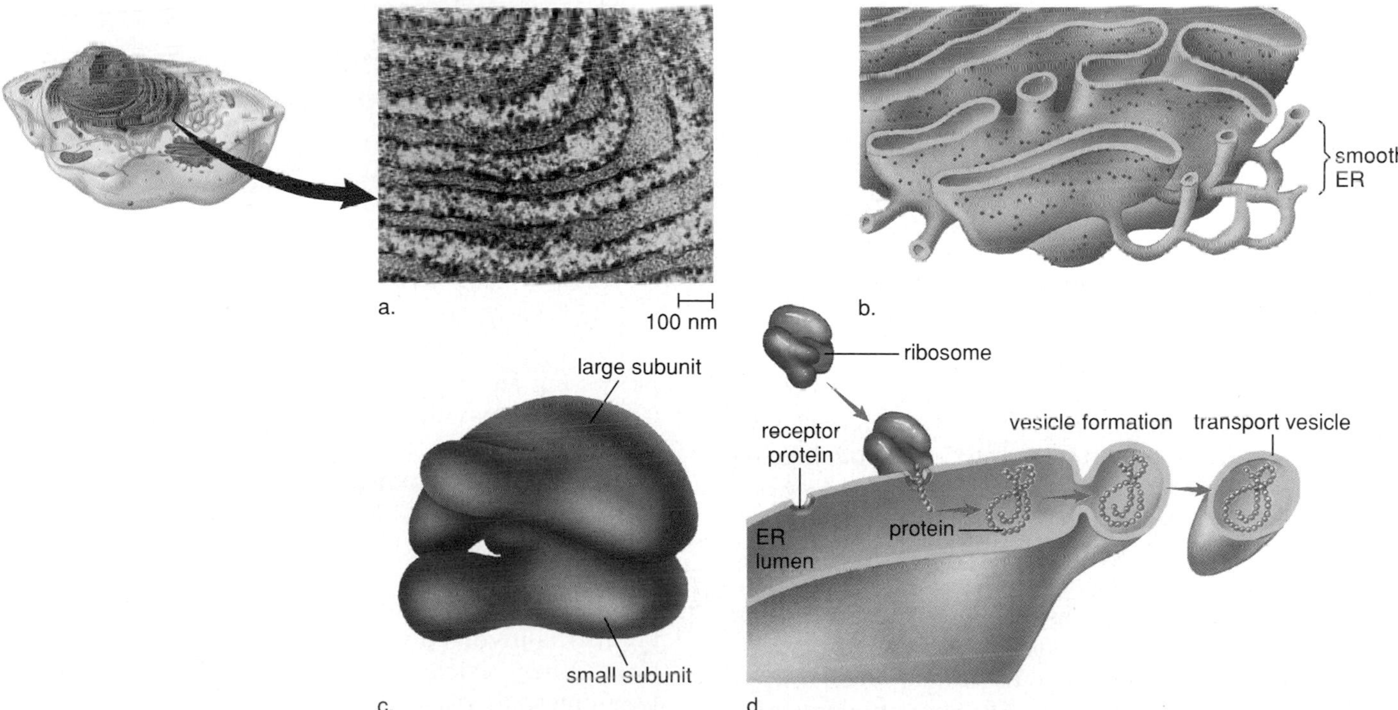

Figure 4.7 Rough endoplasmic reticulum (ER).
a. Electron micrograph of a mouse hepatocyte (liver cell) shows a cross section of many flattened vesicles with ribosomes attached to the side that faces the cytosol. **b.** The three dimensions of the organelle. **c.** Model of a single ribosome illustrates that each one is actually composed of two subunits. **d.** Method by which the ER acts as a transport system.

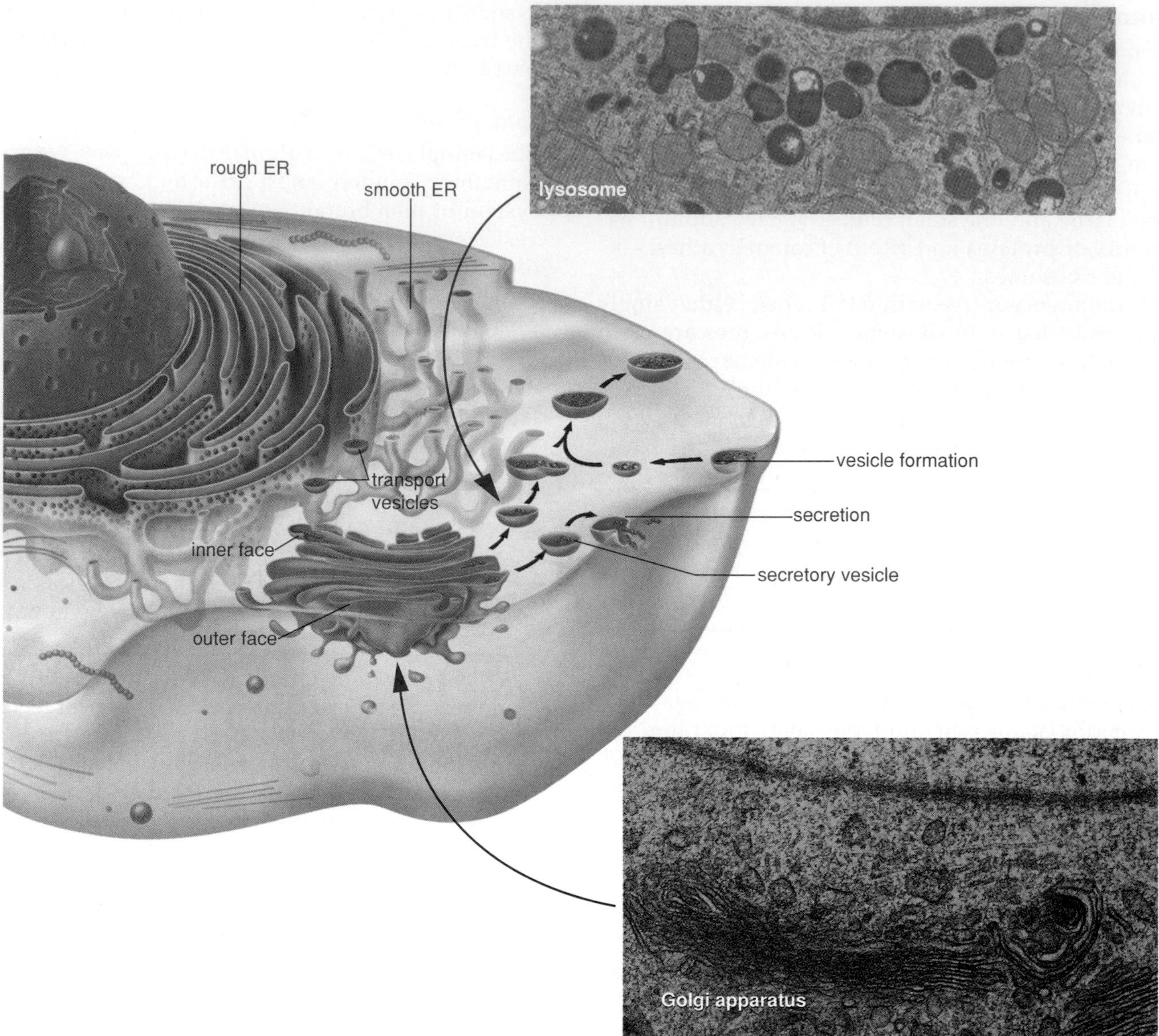

Figure 4.8 Golgi apparatus.
The Golgi apparatus processes proteins that are packaged either in vesicles for secretion from the cell or in lysosomes. Lysosomes function as digestive vesicles.

Golgi Apparatus

The **Golgi apparatus** is named for Camillo Golgi, who discovered its presence in cells in 1898. The Golgi apparatus consists of a stack of three to twenty slightly curved saccules whose appearance can be compared to a stack of pancakes (Fig. 4.8). In animal cells, one side of the stack (the inner face) is directed toward the ER, and the other side of the stack (the outer face) is directed toward the plasma membrane. Vesicles can frequently be seen at the edges of the saccules.

The Golgi apparatus receives protein and/or lipid-filled vesicles that bud from the ER. Some biologists believe that these fuse to form a saccule at the inner face and that this saccule remains as a part of the Golgi apparatus until the molecules are repackaged in new vesicles at the outer face. Others believe that the vesicles from the ER proceed directly to the outer face of the Golgi apparatus, where processing and packaging occurs within its saccules. The Golgi apparatus contains enzymes that modify proteins and lipids; for example, it can add a chain of sugars to them, thereby making them glycoproteins and glycolipids.

The vesicles that leave the Golgi apparatus move to different locations in the cell. Some vesicles proceed to the plasma membrane, where they discharge their contents. Because this is *secretion*, it is often said that the Golgi apparatus is involved in processing, packaging, and secretion. Other vesicles that leave the Golgi apparatus are lysosomes.

The Golgi apparatus processes, packages, and distributes molecules about or from the cell. It is also said to be involved in secretion.

Lysosomes

Lysosomes [Gk. *lyo,* loose, and *soma,* body] are membrane-bounded vesicles produced by the Golgi apparatus in animal cells and plant cells. Lysosomes contain hydrolytic digestive enzymes.

Sometimes macromolecules are brought into a cell by vesicle formation at the plasma membrane (Fig. 4.8). When a lysosome fuses with such a vesicle, its contents are digested by lysosomal enzymes into simpler subunits that then enter the cytoplasm. Some white blood cells defend the body by engulfing bacteria that are then enclosed within vesicles. When lysosomes fuse with these vesicles, the bacteria are digested. It should come as no surprise, then, that even parts of a cell are digested by its own lysosomes (called autodigestion). Normal cell rejuvenation most likely takes place in this matter, but autodigestion is also important during development. For example, when a tadpole becomes a frog, lysosomes digest away the cells of the tail. The fingers of a human embryo are at first webbed, but they are freed from one another as a result of lysosomal action.

Occasionally, a child is born with a metabolic disorder involving a missing or inactive lysosomal enzyme. In these cases, the lysosomes fill to capacity with macromolecules that cannot be broken down. The cells become so full of these lysosomes that the child dies. Someday soon it may be possible to provide the missing enzyme for these children.

Lysosomes are produced by a Golgi apparatus, and their hydrolytic enzymes digest macromolecules from various sources.

Microbodies [Gk. *mikros,* small, little], similar to lysosomes, are membrane-bounded vesicles that contain specific enzymes imported from the cytosol (Fig. 4.9). *Peroxisomes* are microbodies that have enzymes for oxidizing certain organic substances with the formation of hydrogen peroxide (H_2O_2):

$$RH_2 + O_2 \rightarrow R + H_2O_2$$

Hydrogen peroxide, a toxic molecule, is immediately broken down to water and oxygen by another peroxisomal enzyme called catalase. Peroxisomes are abundant in cells that metabolize lipids and in liver cells that metabolize alcohol. They help detoxify alcohol.

Peroxisomes have additional roles in plants. In germinating seeds, peroxisomes, called glyoxysomes, oxidize fatty acids into molecules that can be converted to sugars needed by the growing plant. Peroxisomes also carry out a reaction in leaves that uses up oxygen and releases carbon dioxide that can be used for photosynthesis.

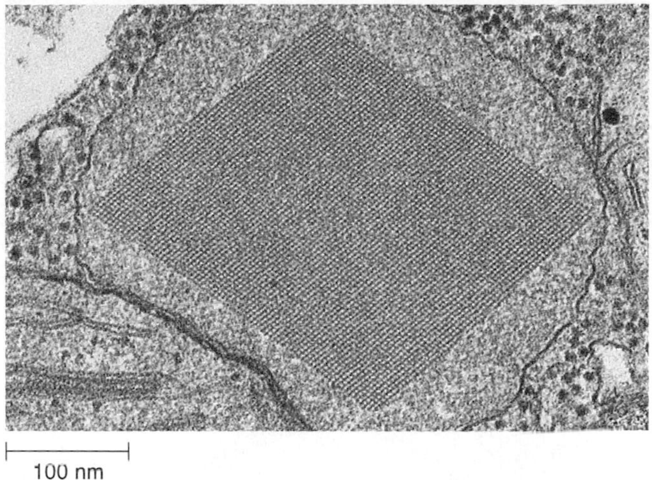

Figure 4.9 Peroxisome in a tobacco leaf.
The crystalline, squarelike core of this microbody cross section is believed to contain the enzyme catalase, which breaks down hydrogen peroxide to form water.

Vacuoles

A **vacuole** is a large membranous sac. A **vesicle** is smaller than a vacuole. Animal cells have vacuoles, but they are much more prominent in plant cells. Typically, plant cells have a large central vacuole so filled with a watery fluid that it gives added support to the cell (see Fig. 4.5).

Vacuoles store substances. Plant vacuoles contain not only water, sugars, and salts but also pigments and toxic molecules. The pigments are responsible for many of the red, blue, or purple colors of flowers and some leaves. The toxic substances help protect a plant from herbivorous animals. The vacuoles present in protozoans are quite specialized, and they include contractile vacuoles for ridding the cell of excess water and digestive vacuoles for breaking down nutrients.

The organelles of the endomembrane system are as follows.

Endoplasmic reticulum (ER): synthesis and modification and transport of proteins and other substances
- Rough ER: protein synthesis
- Smooth ER: lipid synthesis, in particular

Golgi apparatus: processing, packaging, and distribution of protein molecules

Lysosomes: intracellular digestion

Microbodies: various metabolic tasks

Vacuoles: storage areas

Energy-Related Organelles

Life is possible only because of a constant input of energy used to maintain the structure of cells. Chloroplasts and mitochondria are the two eukaryotic membranous organelles that specialize in converting energy to a form that can be used by the cell. **Chloroplasts** use solar energy to synthesize carbohydrates, and carbohydrate-derived products are broken down in mitochondria (sing., **mitochondrion**) to produce ATP molecules.

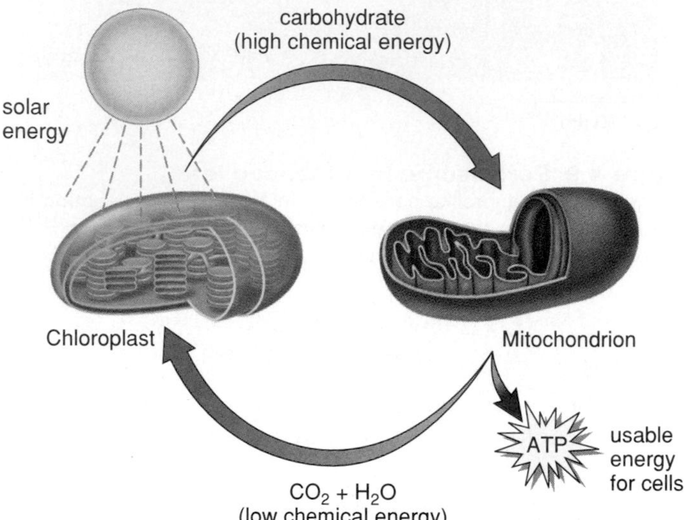

Photosynthesis, which occurs in chloroplasts [Gk. *chloros*, green, and *plastos*, formed, molded], is the process by which solar energy is converted to chemical energy within carbohydrates. Photosynthesis can be represented by this equation:

light energy + carbon dioxide + water ⟶ carbohydrate + oxygen

Here the word *energy* stands for solar energy, the ultimate source of energy for cellular organization. Only plants, algae, and cyanobacteria are capable of carrying on photosynthesis in this manner.

Cellular respiration is the process by which the chemical energy of carbohydrates is converted to that of ATP (adenosine triphosphate), the common carrier of chemical energy in cells. Aerobic cellular respiration can be represented by this equation:

carbohydrate + oxygen ⟶ carbon dioxide + water + energy

Here the word *energy* stands for ATP molecules. When a cell needs energy, ATP supplies it. The energy of ATP is used for synthetic reactions, active transport, and all energy-requiring processes in cells. All organisms carry on cellular respiration, and all organisms except bacteria complete the process of aerobic cellular respiration in mitochondria.

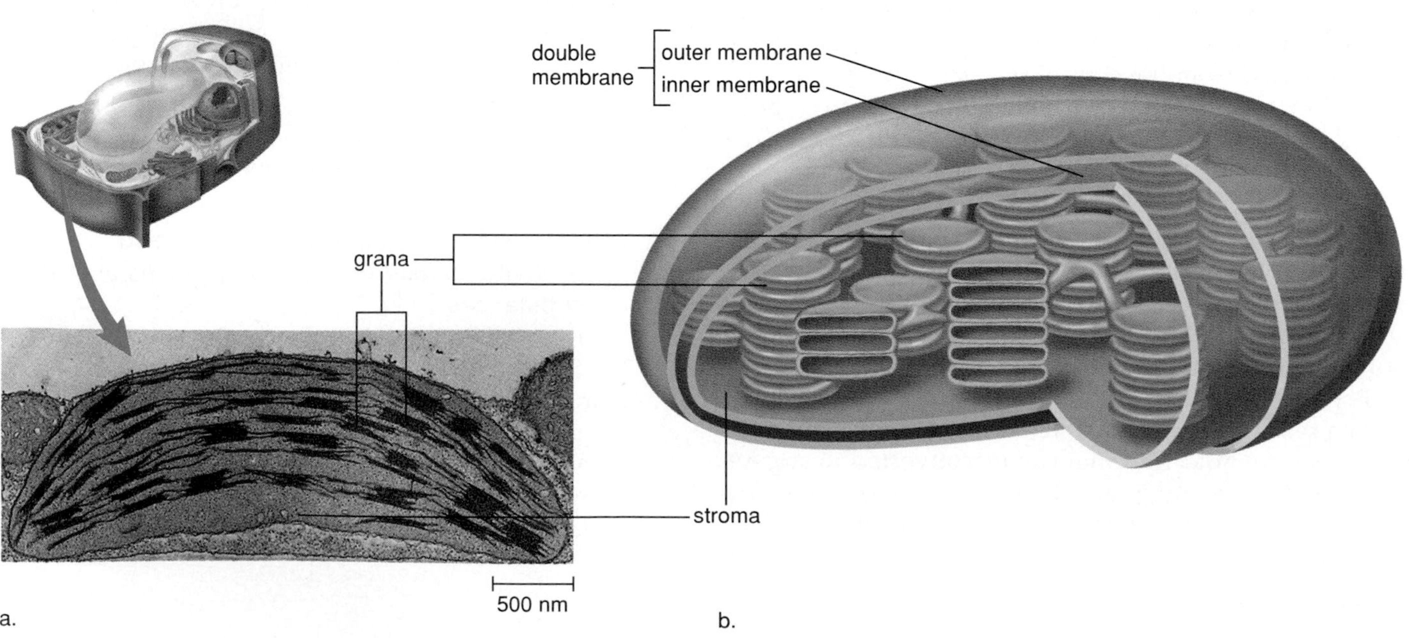

Figure 4.10 Chloroplast structure.
a. Electron micrograph. b. Generalized drawing in which the outer and inner membrane has been cut away to reveal the grana.

Chloroplasts

The photosynthetic cells of algae and plants contain chloroplasts. Chloroplasts are about 4–6 μm in diameter and 1–5 μm in length; they belong to a group of organelles known as plastids. Among the plastids are also the *amyloplasts,* which store starch, and the *chromoplasts,* which contain red and orange pigments.

A chloroplast is bounded by two membranes that enclose a fluid-filled space called the **stroma.** The stroma contains DNA, ribosomes, and enzymes that synthesize carbohydrates. Chloroplasts are able to make some of their own proteins; others are imported into the organelle from the cytosol.

A membrane system within the stroma is organized into interconnected flattened sacs called **thylakoids** [Gk. *thylakos*, sack, and *eides*, like, resembling]. In certain regions, the thylakoids are stacked up in structures called grana (sing., **granum**). There can be hundreds of grana within a single chloroplast (Fig. 4.10). Chlorophyll, a green pigment that is located within the thylakoid membranes of grana, captures solar energy.

There are no chloroplasts in cyanobacteria, which are prokaryotic. Instead, chlorophyll is bound to cytoplasmic thylakoids.

Mitochondria

All eukaryotic cells, whether protozoa, fungi, plant or animal cells, contain mitochondria. This means that algal and plant cells contain both chloroplasts and mitochondria. Most mitochondria are usually 0.5–1.0 μm in diameter and 2–5 μm in length.

Mitochondria, like chloroplasts, are bounded by a double membrane. The inner of these two membranes invaginates to form **cristae.** Cristae provide a much greater surface area to accommodate the protein complexes and other participants in aerobic respiration. The cristae project into the **matrix,** an inner space filled with semifluid medium that contains DNA, ribosomes, and enzymes. These enzymes break down carbohydrate products, releasing energy that is used for ATP production on the cristae (Fig. 4.11).

Mitochondria are able to make some of their own proteins, but others are imported from the cytosol. Mitochondria divide, and in this way their number doubles before cell division occurs.

Chloroplasts and mitochondria are membranous organelles whose structure lends itself to the processes that occur within them.

Figure 4.11 Mitochondrion structure.
a. Electron micrograph. b. Generalized drawing in which the outer membrane and portions of the inner membrane have been cut away to reveal the cristae.

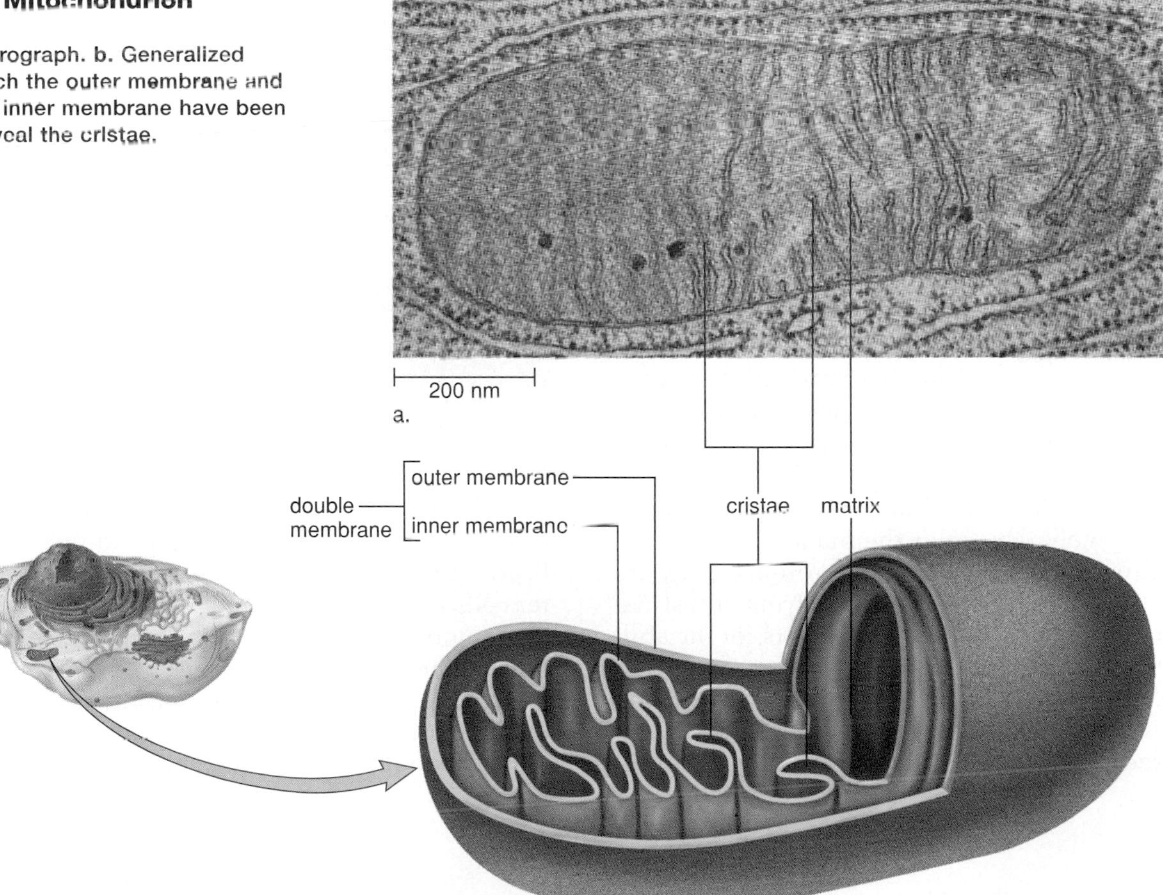

Cytoskeleton Contains Filaments and Microtubules

The **cytoskeleton** [Gk. *kytos*, cell, and *skeleton*, dried body] is a network of interconnected filaments and tubules that extends from the nucleus to the plasma membrane in eukaryotic cells. Prior to the 1970s, it was believed that the cytosol was an unorganized mixture of biomolecules. Then, high-voltage electron microscopes, which can penetrate thicker specimens, showed that the cytosol was instead highly organized. The technique of immunofluorescence microscopy identified the makeup of specific protein fibers within the cytoskeletal network (Fig. 4.12).

The name *cytoskeleton* is convenient in that it allows us to compare the cytoskeleton to the bones and muscles of an animal. Bones and muscles give an animal structure and produce movement. Similarly, we will see that the elements of the cytoskeleton maintain cell shape and cause the cell and its organelles to move. The cytoskeleton is dynamic; elements undergo rapid assembly and disassembly by monomers continuously entering or leaving the polymer. These changes occur at rates that are measured in seconds and minutes. The entire cytoskeletal network can even disappear and reappear at various times in the life of a cell. Before a cell divides, for instance, the elements disassemble and then reassemble into a structure called a spindle that distributes chromosomes in an orderly manner. At the end of cell division, the spindle disassembles and the elements reassemble once again into their former array.

The cytoskeleton contains three types of elements: actin filaments, intermediate filaments, and microtubules, which are responsible for cell shape and movement.

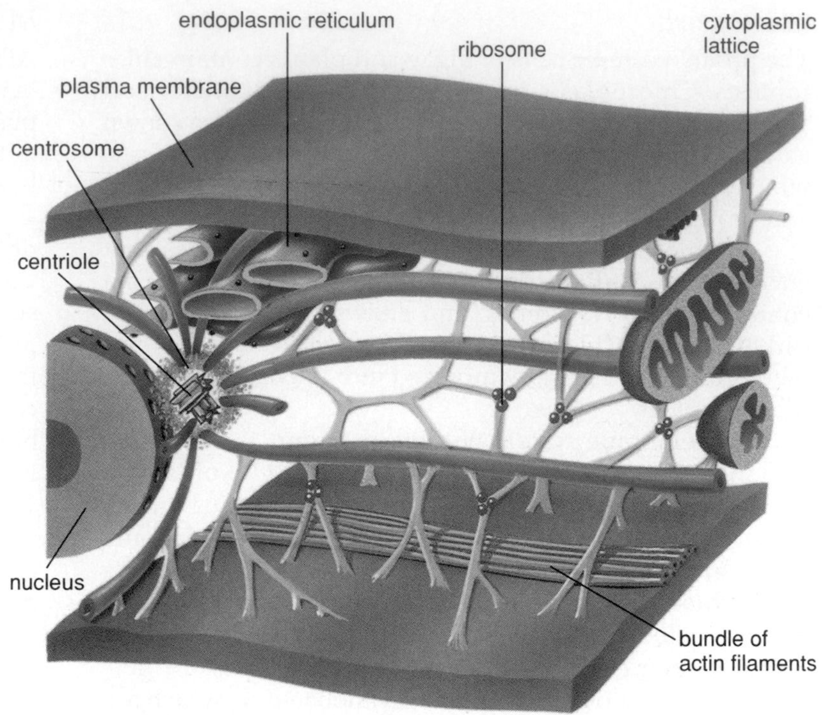

a. Cytoskeleton

Actin Filaments for Structure and Movement

Actin filaments (formerly called microfilaments) are long, extremely thin fibers (about 7 nm in diameter) that occur in bundles or meshlike networks. The actin filament contains two chains of globular actin monomers twisted about one another in a helical manner.

Actin filaments play a structural role when they form a dense complex web just under the plasma membrane, to which they are anchored by special proteins. They are also seen in the microvilli that project from intestinal cells, and their presence most likely accounts for the ability of microvilli to alternately shorten and extend into the intestine. In plant cells, they apparently form the tracks along which chloroplasts circulate or stream in a particular direction.

How are actin filaments involved in the movement of the cell and its organelles? It is well known that myosin interacts with actin filaments in muscle cells to bring about contraction. Actin filaments move because they interact with myosin. Myosin consists of a head plus a tail. After the head combines with and splits ATP, it binds to actin and undergoes a change in configuration that pulls the actin filament forward:

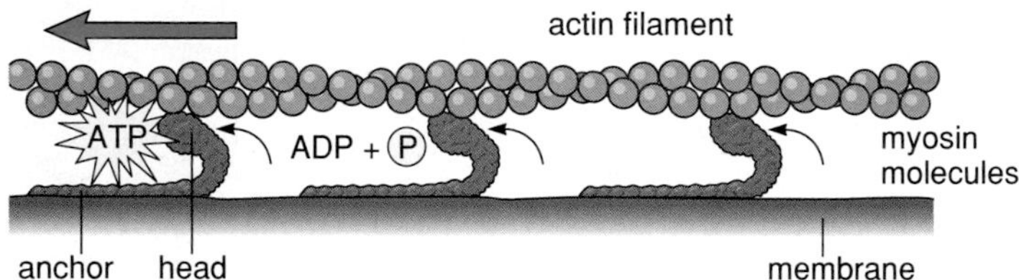

In muscle cells, the tails of several muscle myosin molecules are joined to form a thick filament. In nonmuscle cells, cytoplasmic myosin tails are bound to membranes but the heads still interact with actin. For instance, when daughter cells form during animal cell division, actin filaments (in conjunction with myosin) pinch off the two cells from one another. Also, the presence of a network of actin filaments lying beneath the plasma membrane accounts for the formation of pseudopods, extensions that allow certain cells to move in an amoeboid fashion.

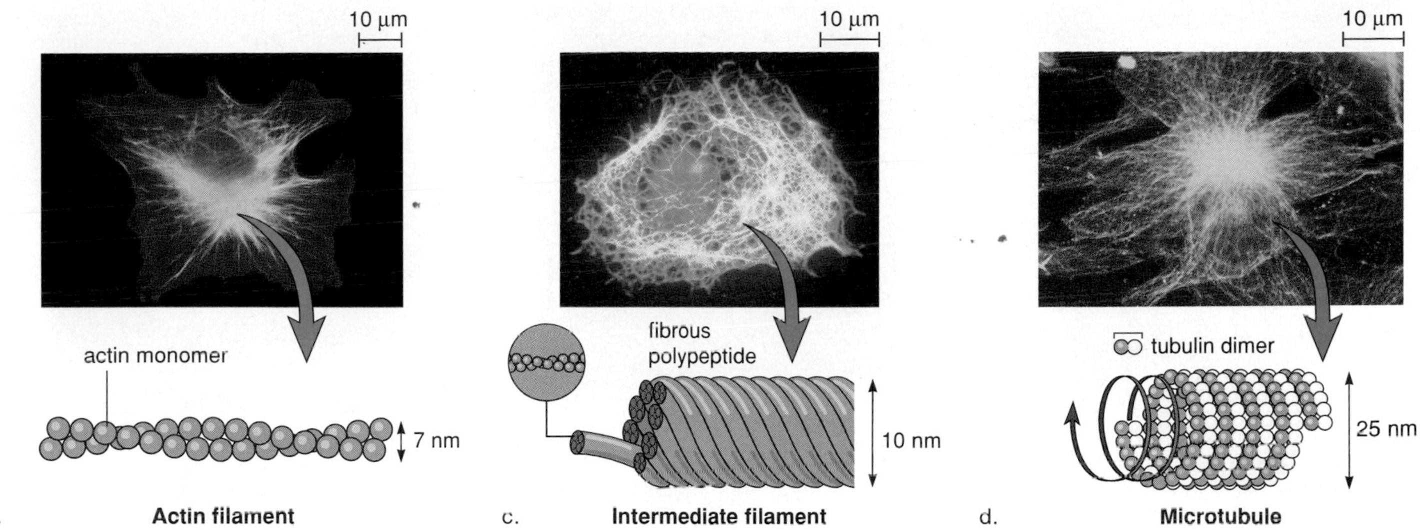

Figure 4.12 The cytoskeleton.
a. The cytoskeleton gives the cell shape, anchors the organelles, and allows them to move. A network of fibers called the cytoplasmic lattice may be a part of the cytoskeleton. Immunofluorescence (a technique based on the binding of fluorescent antibodies to specific proteins) is used to detect the location of (b) actin filaments, (c) intermediate filaments, and (d) microtubules in the cell.

Intermediate Filaments Are Diverse

Intermediate filaments (8–11 nm in diameter) are intermediate in size between actin filaments and microtubules. They are a ropelike assembly of fibrous polypeptides, but the specific type varies according to the tissue. Some intermediate filaments support the nuclear envelope (called nuclear lamina), whereas others support the plasma membrane and take part in the formation of cell-to-cell junctions. In the skin, the filaments, which are made of the protein keratin, give great mechanical strength to skin cells. Recent work has shown intermediate filaments to be highly dynamic. They also assemble and disassemble but need to have phosphate added first by soluble enzymes.

Microtubules Have Tubulin Subunits

Microtubules [Gk. *mikros*, small, little, and L. *tubus*, pipe] are small hollow cylinders about 25 nm in diameter and from 0.2–25 μm in length.

Microtubules are made of a globular protein called tubulin, which occurs as α tubulin and β tubulin. When assembly occurs, these tubulin molecules come together as dimers and the dimers arrange themselves in rows so that an α tubulin is always adjacent to a β tubulin. Microtubules have 13 rows of tubulin dimers surrounding what appears in electron micrographs to be an empty central core.

In many cells the regulation of microtubule assembly is under the control of a microtubule organizing center, called the **centrosome** [Gk. *centrum*, center, and *soma*, body], which lies near the nucleus. Microtubules radiate from the centrosome, helping to maintain the shape of the cell and acting as tracks along which organelles can move. Whereas the *motor molecule* myosin is associated with actin filaments, the motor molecules kinesin and dynein are associated with microtubules. One type of kinesin is responsible for moving vesicles along microtubules, including those that arise from the ER.

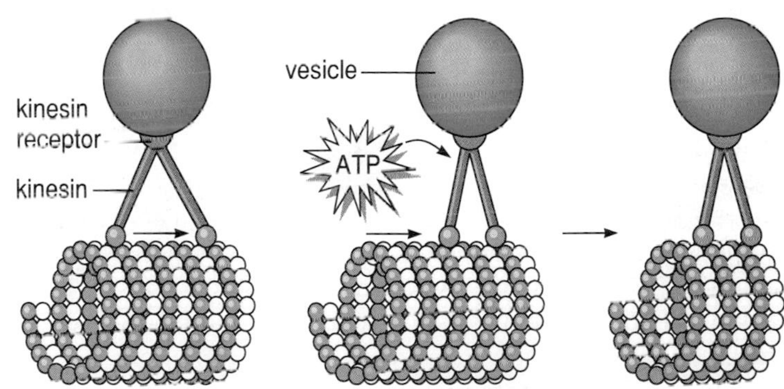

There are different types of kinesin proteins, each specialized to move one kind of vesicle or cellular organelle. A second type of cytoplasmic motor molecule is called cytoplasmic dynein because it is closely related to the molecule dynein found in flagella.

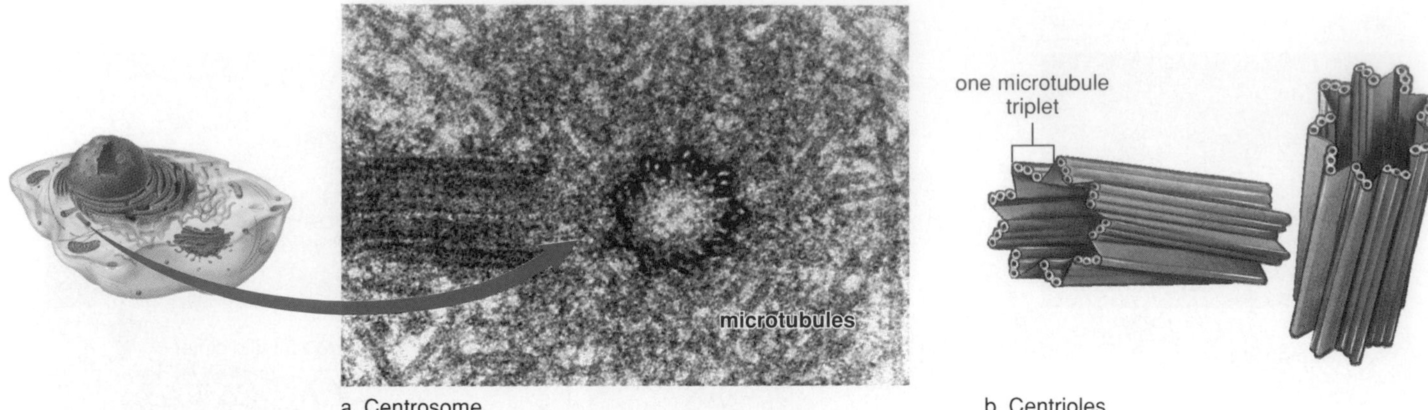

Figure 4.13 Centrioles.
a. A pair of centrioles is found in the centrosome where they lie at right angles to each other. Also, note the large number of free microtubules that radiate out of the centrosome, a microtubule organizing center. **b.** Drawing of centrioles showing their 9 + 0 arrangement of microtubule triplets. (The 0 in this formula means that there are no microtubules in the center of the organelle.)

Centrioles

Centrioles [Gk. *centrum,* center] are short cylinders with a 9 + 0 pattern of microtubule triplets—that is, a ring having nine sets of triplets with none in the middle (Fig. 4.13). In animal cells and most protists, a centrosome contains two centrioles lying at right angles to each other. A centrosome, you will recall, is the major microtubule organizing center for the cell. Therefore, it is possible that centrioles are also involved in the process by which microtubules assemble and disassemble.

Before an animal cell divides, the centrioles replicate and the members of each pair are at right angles to one another. Then each pair becomes part of a separate centrosome. During cell division the centrosomes move apart and may function to organize the mitotic spindle. In any case, each new cell has its own centrosome. Plant and fungal cells have the equivalent of a centrosome but it does not contain centrioles, suggesting that centrioles are not necessary to the assembly of cytoplasmic microtubules.

In cells with cilia and flagella, centrioles are believed to give rise to basal bodies that direct the organization of microtubules within these structures. In other words, a basal body may do for a cilium (or flagellum) what the centrosome does for the cell.

Centrioles, which are short cylinders with a 9 + 0 pattern of microtubule triplets, may be involved in microtubule organization and in the formation of cilia and flagella.

Cilia and Flagella

Cilia [L. *cilium,* eyelash, hair] and **flagella** [L. *flagello,* whip] are hairlike projections that can move either in an undulating fashion, like a whip, or stiffly, like an oar. Cells that have these organelles are capable of movement. For example, unicellular paramecia move by means of cilia, whereas sperm cells move by means of flagella. The cells that line our upper respiratory tract have cilia that sweep debris trapped within mucus back up into the throat, where it can be swallowed. This action helps keep the lungs clean.

In eukaryotic cells, cilia are much shorter than flagella, but they have a similar construction. Both are membrane-bounded cylinders enclosing a matrix area. In the matrix are nine microtubule doublets arranged in a circle around two central microtubules. This is called the 9 + 2 pattern of microtubules. Cilia and flagella move when the microtubule doublets slide past one another (Fig. 4.14).

As mentioned, each cilium and flagellum has a basal body lying in the cytoplasm at its base. Basal bodies have the same circular arrangement of microtubule triplets as centrioles and are believed to be derived from them. It is possible that basal bodies organize the microtubules within cilia and flagella, but this is not supported by the observation that cilia and flagella grow by the addition of tubulin dimers to their tips.

Cilia and flagella, which have a 9 + 2 pattern of microtubules, are involved in the movement of cells.

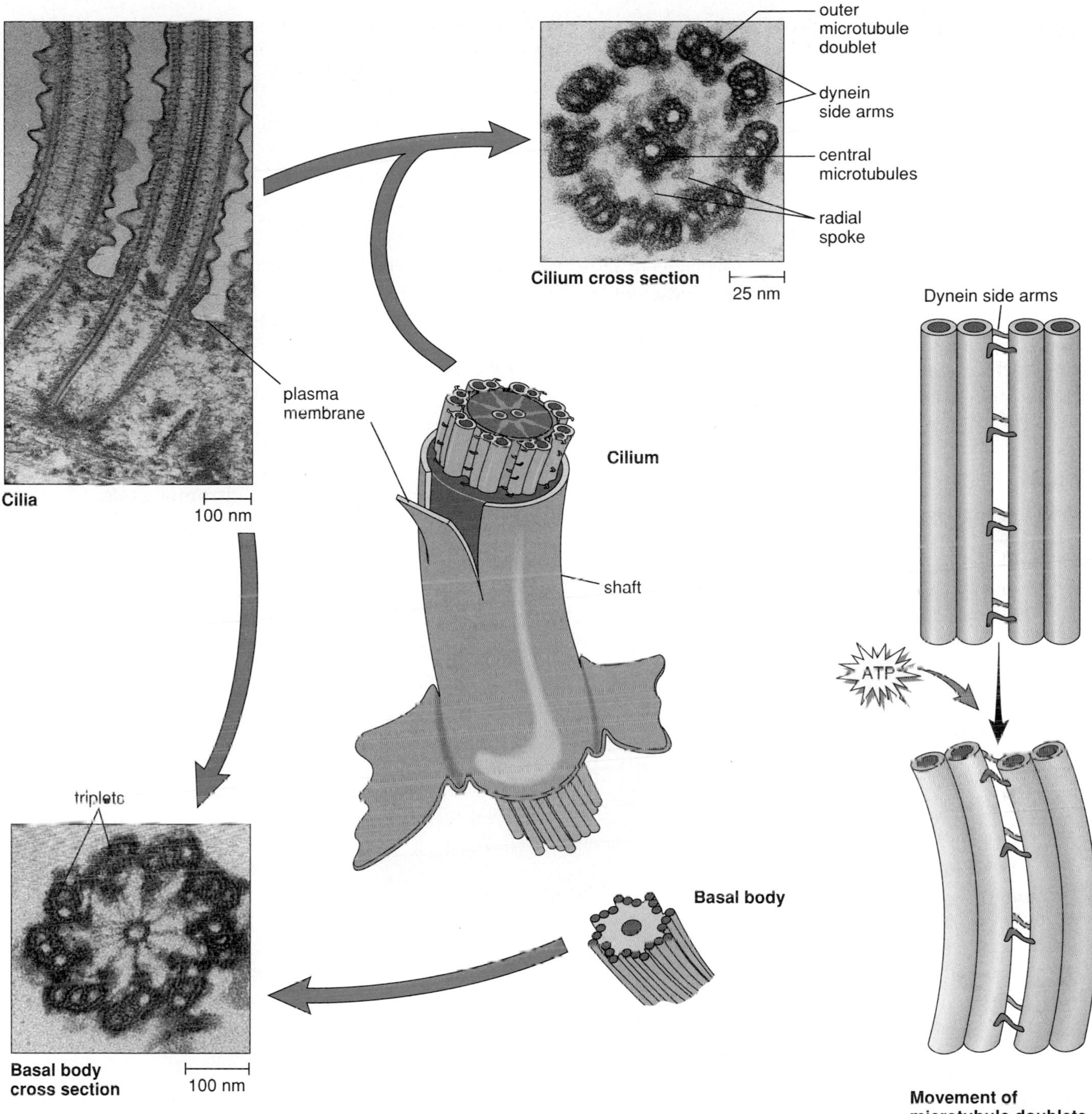

Figure 4.14 Structure of cilium or flagellum.
A cilium has a basal body with a 9 + 0 pattern of microtubule triplets. (Notice the ring of nine triplets, with no central microtubules.) The shaft of the cilium has a 9 + 2 pattern (a ring of nine microtubule doublets surrounds a central pair of single microtubules). Compare the cross section of the basal body to the cross section of the cilium shaft and note that in place of the third microtubule, the outer doublets have side arms of dynein, a motor molecule. In the presence of ATP, the dynein side arms reach out and attempt to move along their neighboring doublet. Because of the radial spokes connecting the doublets to the central microtubules, bending occurs.

4.4 How the Eukaryotic Cell Evolved

Invagination of the plasma membrane might explain the origination of the nuclear envelope and organelles such as the endoplasmic reticulum and the Golgi apparatus. Some believe that the other organelles could also have arisen in this manner. But another hypothesis has been put forth. It has been observed that in the laboratory an amoeba infected with bacteria can become dependent upon them. Some investigators, especially Lynn Margulis, believe that mitochondria and chloroplasts are derived from prokaryotes that were taken up by a much larger cell (Fig. 4.15). Perhaps mitochondria were originally aerobic heterotrophic bacteria and chloroplasts were originally cyanobacteria. The host cell would have benefited from an ability to utilize oxygen or synthesize organic food when by chance the prokaryote was not destroyed. Therefore, after these prokaryotes entered by *endocytosis,* a *symbiotic* relationship was established. Some of the evidence for the endosymbiotic hypothesis is as follows:

1. Mitochondria and chloroplasts are similar to bacteria in size and in structure.
2. Both organelles are bounded by a double membrane—the outer membrane may be derived from the engulfing vesicle, and the inner one may be derived from the plasma membrane of the original prokaryote.
3. Mitochondria and chloroplasts contain a limited amount of genetic material and divide by splitting. Their DNA (deoxyribonucleic acid) is a circular loop like that of bacteria.
4. Although most of the proteins within mitochondria and chloroplasts are now produced by the eukaryotic host, they do have their own ribosomes and they do produce some proteins. Their ribosomes resemble those of bacteria.
5. The RNA (ribonucleic acid) base sequence of their ribosomes suggests a eubacterial origin for chloroplasts and mitochondria.

Margulis even suggests that the flagella of eukaryotes are derived from a spirochete prokaryote that became attached to a host cell (Fig. 4.15). However, it is important to remember that the flagella of eukaryotes but not prokaryotes have the 9 + 2 patterns of microtubules. In any case, the acquisition of basal bodies, which could have become centrioles, may have led to the ability to form a spindle during mitosis and meiosis. The process of meiosis, of course, is associated with sexual reproduction of eukaryotes. Genetic recombination due to crossing-over, independent assortment of chromosomes, and random recombination during fertilization contributes to variation among members of a population and the evolution of new species.

According to the endosymbiotic hypothesis, heterotrophic bacteria became mitochondria and cyanobacteria became chloroplasts after being taken up by precursors to modern-day eukaryotic cells.

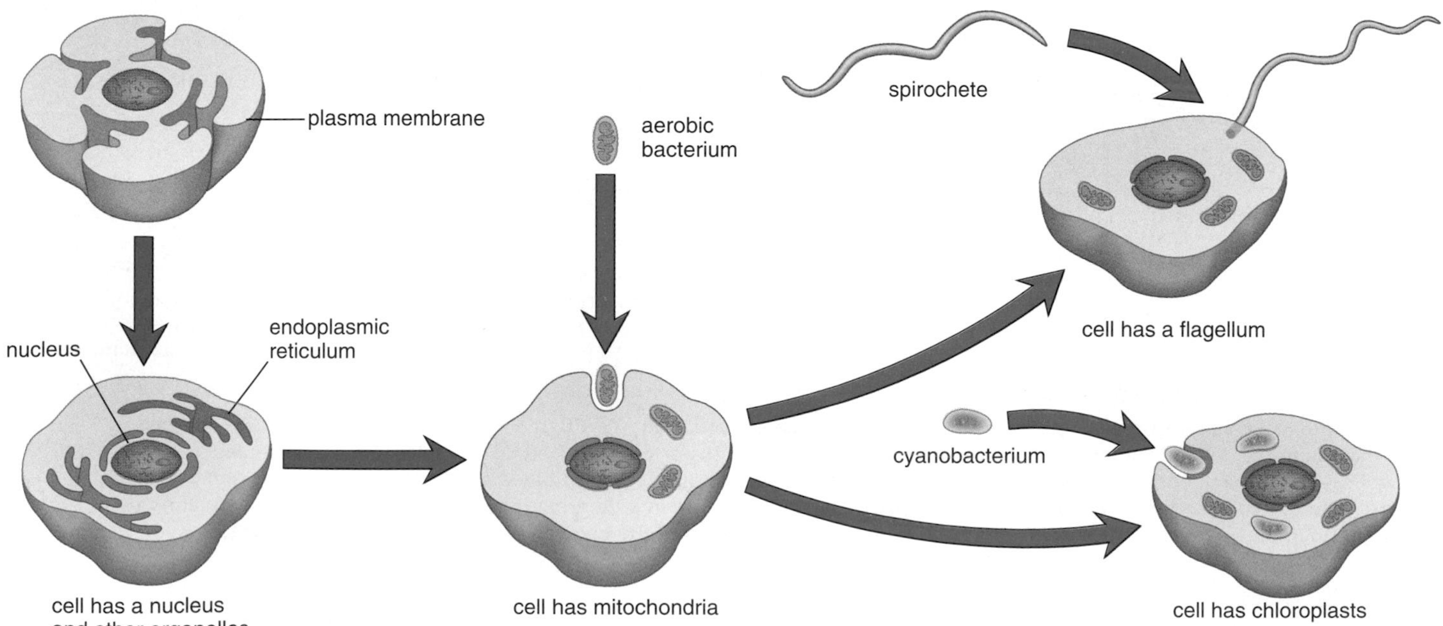

Figure 4.15 Evolution of the eukaryotic cell.
Invagination of the plasma membrane could account for the formation of the nucleus and certain other organelles. The endosymbiotic hypothesis suggests that mitochondria, chloroplasts, and flagella are derived from prokaryotes that were taken up by a much larger eukaryotic cell.

connecting concepts

When the Dutchman Antoine van Leeuwenhoek observed one-celled creatures under his homemade microscope in the 1600s, he probably had no idea the unicellular organisms he found so fascinating had the same basic structure as the cells in his own body. Eukaryotic cells are compartmentalized and therefore can be likened to a miniature factory, where each department has a specific function. The nucleus, which contains the chromosomes, determines the characteristics of the cell. Mitochondria produce high-energy ATP molecules from carbohydrates and oxygen. Proteins are made on the ribosomes and the Golgi apparatus prepares materials to be secreted by the cell.

Unicellular organisms are generalists, while eukaryotic cells are specialized for specific functions. Muscle cells, which depend on a ready supply of ATP, are full of mitochondria; pancreatic cells, which produce insulin, are full of rough ER, and so forth. "Form follows function," a concept developed by architects from designing buildings, applies equally well to cells and even to their organelles. Chloroplasts and mitochondria are organized similarly. In chloroplasts, the pigment chlorophyll is embedded in thylakoid membranes; enzymes within the stroma produce carbohydrates. In mitochondria, protein enzymes within the matrix break down carbohydrate products, and protein complexes within cristae produce ATP.

Prokaryotic cells lack the compartmentalization seen in eukaryotic cells, but they carry out all the functions of eukaryotic cells. They have a plasma membrane, but their single circle of DNA is located within a nucleoid region. They lack a true membrane-bounded nucleus. Like eukaryotes, prokaryotes have ribosomes where proteins are synthesized. Despite their simple organization, bacteria are among the most adaptable and successful forms on earth. This is largely due to their diverse metabolic capabilities; most any organic molecule can be broken down by some type of bacterium. All bacteria produce the many enzymes they need for daily existence.

Summary

4.1 Cells Make Up Living Things

All organisms are composed of cells, the smallest units of living matter. Cells are capable of self-reproduction, and existing cells come only from preexisting cells. Cells are very small and are measured in micrometers. The plasma membrane regulates exchange of materials between the cell and the external environment. Cells must remain small in order to have an adequate amount of surface area per cell volume.

4.2 Prokaryotic Cells Are Less Complex

There are two major groups of cells: prokaryotic and eukaryotic. Both types have a plasma membrane and cytoplasm. Eukaryotic cells also have a nucleus and various organelles. Prokaryotic cells have a nucleoid that is not bounded by a nuclear envelope. They also lack most of the other organelles that compartmentalize eukaryotic cells. Table 4.2 summarizes the similarities and differences between prokaryotic and eukaryotic cells.

4.3 Eukaryotic Cells Are More Complex

The nucleus of eukaryotic cells, represented by animal and plant cells, is bounded by a nuclear envelope containing pores. These pores serve as passageways between the cytoplasm and the nucleoplasm. Within the nucleus, the chromatin undergoes coiling into chromosomes at the time of cell division. The nucleolus is a special region of the chromatin where rRNA is produced and where proteins from the cytoplasm gather to form ribosomal subunits. These subunits are joined in the cytoplasm.

Ribosomes are organelles that function in protein synthesis. They can be bound to ER or exist within the cytosol singly or in groups called polyribosomes.

The endomembrane system includes the ER (both rough and smooth), the Golgi apparatus, the lysosomes, and other types of vesicles and vacuoles. The endomembrane system serves to compartmentalize the cell and keep the various biochemical reactions separate from one another. Newly produced proteins enter the ER lumen, where they may be modified before proceeding to the interior of the smooth ER. The smooth ER has various metabolic functions depending on the cell type, but it also forms vesicles that carry proteins and lipids to different locations, particularly to the Golgi apparatus. The Golgi apparatus processes proteins and repackages them into lysosomes, which carry out intracellular digestion, or into vesicles that fuse with the plasma membrane. Following fusion, secretion occurs. The endomembrane system also includes microbodies that have special enzymatic functions. One significant microbody, the peroxisome, contains enzymes that oxidize molecules by producing hydrogen peroxide that is subsequently broken down. The large single plant cell vacuole not only stores substances but lends support to the plant cell.

Cells require a constant input of energy to maintain their structure. Chloroplasts capture the energy of the sun and carry on photosynthesis, which produces carbohydrate. Carbohydrate-derived products are broken down in mitochondria as ATP is produced. This is an oxygen-requiring process called aerobic respiration.

The cytoskeleton contains actin filaments, intermediate filaments, and microtubules. These maintain cell shape and allow it and the organelles to move. Actin filaments, the thinnest filaments, interact with the motor molecule myosin in muscle cells to bring about contraction; in other cells, they pinch off daughter cells and have other dynamic functions. Intermediate filaments support the nuclear envelope and the plasma membrane and probably participate in cell-to-cell junctions. Microtubules radiate out from the centrosome and are present in centrioles, cilia, and flagella. They serve as tracks, along which vesicles and other organelles move, due to the action of specific motor molecules.

4.4 How the Eukaryotic Cell Evolved

The nuclear envelope most likely evolved through invagination of the plama membrane, but mitochondria and chloroplasts may have arisen through endosymbiotic events.

Table 4.2

Comparison of Prokaryotic Cells and Eukaryotic Cells

	Prokaryotic Cells	Eukaryotic Cells	
		Animal	*Plant*
Size	*Smaller (1–10 μm in diameter)*	*Larger (10–100 μm in diameter)*	
Plasma membrane	Yes	Yes	Yes
Cell wall	Usually (peptidoglycan)	No	Yes (cellulose)
Nuclear envelope	No	Yes	Yes
Nucleolus	No	Yes	Yes
DNA	Yes (single loop)	Yes (chromosomes)	Yes (chromosomes)
Mitochondria	No	Yes	Yes
Chloroplasts	No	No	Yes
Endoplasmic reticulum	No	Yes	Yes
Ribosomes	Yes (smaller)	Yes	Yes
Vacuoles	No	Yes (small)	Yes (usually large, single vacuole)
Golgi apparatus	No	Yes	Yes
Lysosomes	No	Always	Yes
Microbodies	No	Usually	Usually
Cytoskeleton	No	Yes	Yes
Centrioles	No	Yes	No
9 + 2 cilia or flagella	No	Often	No (in flowering plants) Yes (in ferns, cycads, and bryophytes)

Reviewing the Chapter

1. What are the two basic tenets of the cell theory? 58
2. Why is it advantageous for cells to be small? 59
3. What are the contrasting advantages of light microscopy and electron microscopy? 60–61
4. What similar features do prokaryotic cells and eukaryotic cells have? What is their major difference? 62–63
5. Roughly sketch a prokaryotic cell, label its parts, and state a function for each of these. 62
6. Distinguish between the nucleolus, rRNA, and ribosomes. 66–67
7. Describe the structure and the function of the nuclear envelope and the nuclear pores. 66
8. Trace the path of a protein from rough ER to the plasma membrane. 67
9. Give the overall equations for photosynthesis and cellular respiration, contrast the two, and tell how they are related. 70
10. What are the three components of the cytoskeleton? What are their structures and functions? 72–73

Testing Yourself

Choose the best answer for each question.

1. The small size of cells is best correlated with
   a. the fact they are self-reproducing.
   b. their prokaryotic versus eukaryotic nature.
   c. an adequate surface area for exchange of materials.
   d. All of these are correct.
2. Which of these is not a true comparison of the light microscope and the transmission electron microscope?

	Light	**Electron**
a.	uses light to "view" object	uses electrons to "view" object
b.	uses glass lenses for focusing	uses magnetic lenses for focusing
c.	specimen must be killed and stained	specimen may be alive and nonstained
d.	magnification is not as great	magnification is greater

3. Which of these best distinguishes a prokaryotic cell from a eukaryotic cell?
   a. Prokaryotic cells have a cell wall, but eukaryotic cells never do.
   b. Prokaryotic cells are much larger than eukaryotic cells.
   c. Prokaryotic cells have flagella, but eukaryotic cells do not.
   d. Prokaryotic cells do not have a membrane-bounded nucleus, but eukaryotic cells do have such a nucleus.
4. Which of these is not found in the nucleus?
   a. functioning ribosomes
   b. chromatin that condenses to chromosomes
   c. nucleolus that produces rRNA
   d. nucleoplasm instead of cytoplasm
5. Vesicles from the smooth ER most likely are on their way to the
   a. rough ER.
   b. lysosomes.
   c. Golgi apparatus.
   d. plant cell vacuole only.
6. Lysosomes function in
   a. protein synthesis.
   b. processing and packaging.
   c. intracellular digestion.
   d. lipid synthesis.
7. Mitochondria
   a. are involved in cellular respiration.
   b. break down ATP to release energy for cells.
   c. contain grana and cristae.
   d. All of these are correct.
8. Which organelle releases oxygen?
   a. ribosome
   b. Golgi apparatus
   c. mitochondrion
   d. chloroplast
9. Which of these is not true?
   a. Actin filaments are found in muscle cells.
   b. Microtubules radiate out from the ER.
   c. Intermediate filaments sometimes contain keratin.
   d. Motor molecules use microtubules as tracts.
10. Cilia and flagella
    a. bend when microtubules try to slide past one another.
    b. contain myosin that pulls on actin filaments.
    c. are organized by basal bodies derived from centrioles.
    d. Both a and c are correct.
11. Which organelle would not have originated by endosymbiosis?
    a. mitochondria
    b. flagella
    c. nucleus
    d. chloroplasts
12. Study the example given in (a) below. Then for each other organelle listed, state another that is structurally and functionally related. Tell why you paired these two organelles.
    a. The nucleus can be paired with nucleoli because nucleoli are found in the nucleus. Nucleoli occur where chromatin is producing rRNA.
    b. mitochondria
    c. centrioles
    d. ER
13. Label these parts of the cell that are involved in protein synthesis and modification. Give a function for each structure.

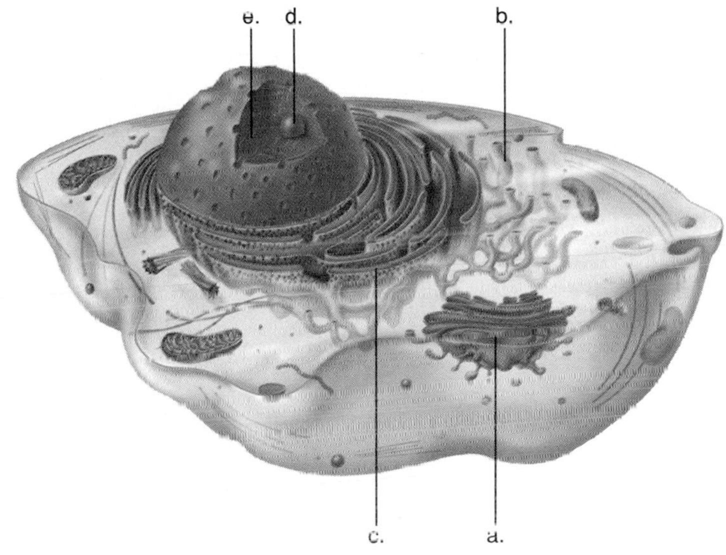

Applying the Concepts

1. *All organisms are made up of cells.*
   What data would you use to convince someone that all organisms are composed of cells? What data would you have to find to nullify the cell theory?
2. *Cells are compartmentalized and highly organized.*
   Substantiate this concept by referring to Table 4.1.
3. *Life begins at the cellular level of organization.*
   Which organelles contribute to the ability of the cell to maintain its structure and grow? What are the functions of these organelles?

Using Technology

Your study of cell structure and function is supported by these available technologies:

Exploring the Internet

The Mader Home Page provides resources for and help with studying this chapter.

http://www.mhhe.com/sciencemath/biology/mader/
(Click on Biology.)

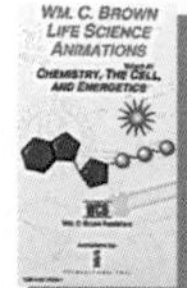

Life Science Animations Video

Video #1: Chemistry, the Cell, and Energetics
Journey into a Cell (#2)
Endocytosis (#3)
Cellular Secretion (#4)

Understanding the Terms

Match the terms to these definitions:

a. _________ Area in prokaryotic cell where DNA is found.
b. _________ Network of fibrils consisting of DNA and associated proteins observed within a nucleus that is not dividing.
c. _________ Dark-staining, spherical body in the cell nucleus that produces ribosomal subunits.
d. _________ Internal framework of the cell, consisting of microtubules, actin filaments, and intermediate filaments.
e. _________ Membrane-bounded vesicle that contains hydrolytic enzymes for digesting macromolecules.
f. _________ Organelle consisting of saccules and vesicles that processes, packages, and distributes molecules about or from the cell.
g. _________ Short, hairlike projection from the plasma membrane, occurring usually in larger numbers.
h. _________ Slender, long extension used for locomotion by some bacteria, protozoa, and sperm.
i. _________ Smallest unit that displays the properties of life; composed of cytoplasm surrounded by a plasma membrane.
j. _________ System of membranous saccules and channels in the cytoplasm, often with attached ribosomes.

Membrane Structure and Function

CHAPTER 5

Chapter Concepts

Plasma membrane of red blood cell

20 nm

A plasma membrane encloses every cell, whether the cell be a unicellular amoeba or one of many from the body of a cockroach, peony, mushroom, or human. Universally, a plasma membrane protects a cell by acting as a barrier between its living contents and the surrounding environment. It regulates what goes into and out of the cell and marks the cell as being unique to the organism. In multicellular organisms, cell junctions requiring specialized features of the plasma membrane connect cells together in specific ways and pass on information to neighboring cells so that the activities of tissues and organs are coordinated.

At first glance, a short pygmy, an overweight diabetic, and a young child with a high cholesterol level seem to have little in common. In reality, however, each suffers from a defect of their cells' plasma membrane. The pygmy's cells won't allow growth hormone to enter; the diabetic's cells stop the entrance of insulin, and the young child's cells prevent the entrance of a lipoprotein. A plasma membrane was essential to the origin of the first cell(s) and its proper functioning is essential to our good health today.

5.1 Membrane Models Have Changed

At the turn of the century, investigators noted that lipid-soluble molecules entered cells more rapidly than water-soluble molecules. This prompted them to suggest that lipids are a component of the plasma membrane. Later, chemical analysis disclosed that the plasma membrane contains phospholipids. In 1925, E. Gorter and G. Grendel measured the amount of phospholipid extracted from red blood cells and determined that there is just enough to form a bilayer around the cells (Fig. 5.1*a*). They further suggested that the nonpolar (hydrophobic) tails are directed inward and the polar (hydrophilic) heads are directed outward, forming a phospholipid bilayer:

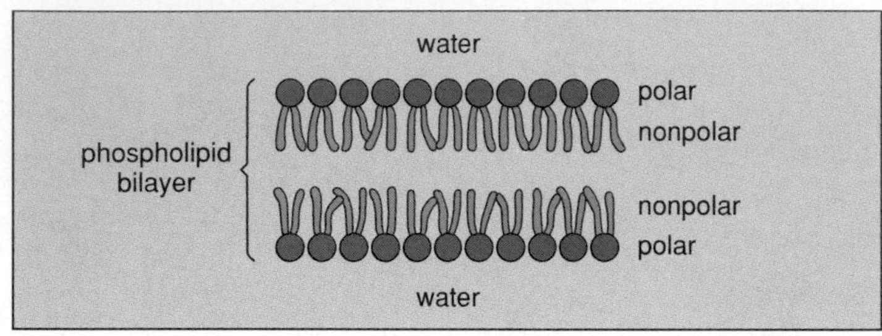

The presence of lipids cannot account for all the properties of the plasma membrane, such as its permeability to certain nonlipid substances. Such observations prompted J. Danielli and H. Davson to suggest in the 1940s that proteins are also a part of the membrane. They proposed a *sandwich model*, later rejected, in which the phospholipid bilayer is a filling between two layers of proteins.

By the late 1950s, electron microscopy had advanced to allow viewing of the plasma membrane and other membranes in the cell. Since the membrane has a sandwichlike appearance, J. D. Robertson assumed that the outer dark layer (stained with heavy metals) contained protein plus the hydrophilic heads of the phospholipids. The interior was assumed to be the hydrophobic tails of these molecules. Robertson went on to suggest that all membranes in various cells have basically the same composition. This proposal was called the *unit membrane model* (Fig. 5.1*b*).

This model of membrane structure was accepted for at least ten years, even though investigators began to doubt its accuracy. For example, not all membranes have the same appearance in electron micrographs, and they certainly do not have the same function. The inner membrane of a mitochondrion, which is coated with rows of particles, functions in cellular respiration and it has a far different appearance from the plasma membrane. Finally, in 1972, S. Singer and G. Nicolson introduced the **fluid-mosaic model** of membrane structure, which proposes in part that the membrane is a fluid phospholipid bilayer in which protein molecules are either partially or wholly embedded. The proteins are scattered throughout the membrane in an irregular pattern that can vary from membrane to membrane. The mosaic distribution of proteins is supported especially by electron micrographs of freeze-fractured membranes (Fig. 5.1*c* and *d*).

a. Electron micrograph of red blood cell plasma membrane

plasma membrane

20 nm

b. Two possible models

Robertson unit membrane

Singer and Nicolson fluid-mosaic model

c. Freeze-fracture of membrane

protein

knife

d. Electron micrograph of freeze-fractured membrane shows presence of particles

Figure 5.1 Membrane structure.
a. The red blood cell plasma membrane typically has a three-layered appearance in electron micrographs. **b.** Robertson's unit membrane model proposed that the outer dark layers in electron micrographs were made up of protein and polar heads of phospholipid molecules, while the inner light layer was composed of the nonpolar tails. The Singer and Nicolson fluid-mosaic model put protein molecules within the lipid bilayer. **c.** A technique called freeze-fracture allows an investigator to view the interior of the membrane. Cells are rapidly frozen in liquid nitrogen and then fractured with a special knife. Platinum and carbon are applied to the fractured surface to produce a faithful replica that is observed by electron microscopy. **d.** A fracture in the middle of the bilayer shows the presence of particles, consistent with the fluid-mosaic model.

The fluid-mosaic model of membrane structure is widely accepted at this time.

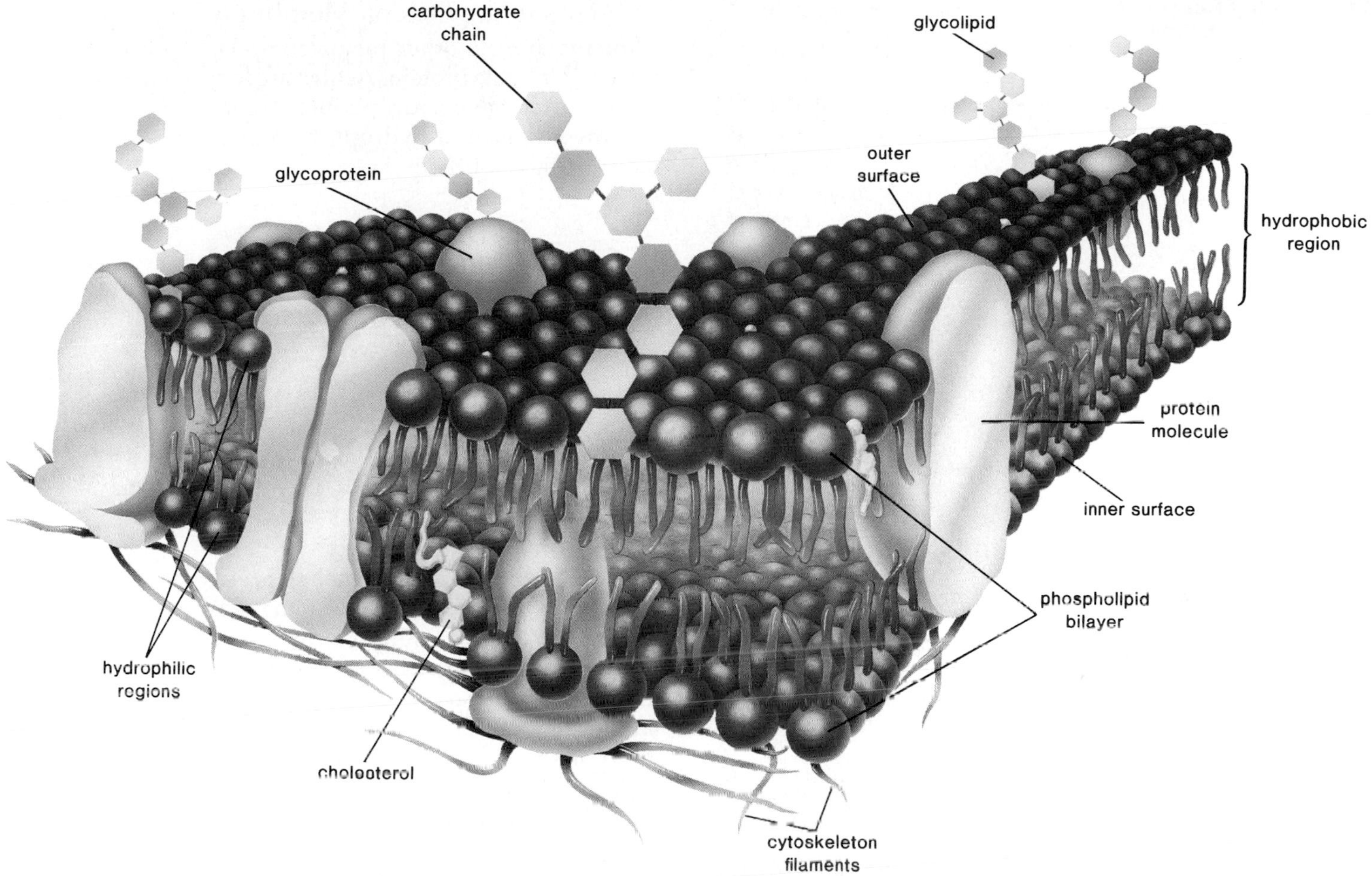

Figure 5.2 Fluid-mosaic model of an animal cell plasma membrane.
The plasma membrane is composed of a phospholipid bilayer with embedded proteins. The hydrophilic heads of the phospholipids are at the surfaces of the membrane, and the hydrophobic tails make up the interior of the membrane. Note the asymmetry of the membrane; for example, carbohydrate chains project externally and cytoskeleton filaments attach to proteins on the cytoplasmic side of the plasma membrane.

5.2 The Plasma Membrane Is Complex

The *fluid-mosaic model* of membrane structure has two components, lipids and proteins. Most of the lipids in the plasma membrane are **phospholipids** [Gk. *phos*, light, and *lipos*, fat], which are known to arrange themselves spontaneously into a bilayer. In a membrane, the hydrophilic (polar) heads of the phospholipid molecules face the intracellular and extracellular fluids. The hydrophobic (nonpolar) tails face each other in the membrane interior (Fig. 5.2).

In addition to phospholipids, there are two other types of lipids in the plasma membrane. **Glycolipids** have a structure similar to phospholipids except that the hydrophilic head is a variety of sugars joined to form a straight or branching carbohydrate chain. Glycolipids have a protective function and also have various other functions to be discussed in this chapter.

Cholesterol is a lipid that is found in animal plasma membranes; related steroids are found in the plasma membrane of plants. Cholesterol reduces the permeability of the membrane to most biological molecules.

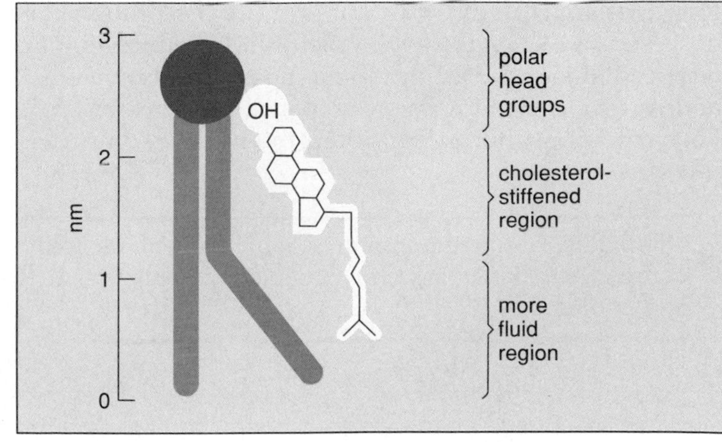

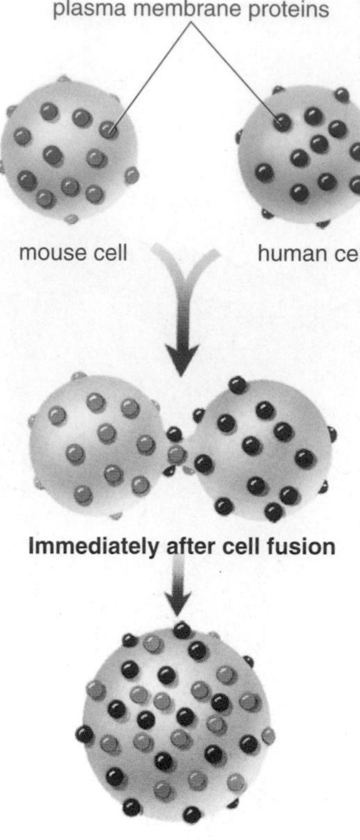

Figure 5.3 Fluidity of plasma membrane. After human and mouse cells were fused, the plasma membrane proteins intermixed within a short time. This experiment illustrates the lateral drifting of proteins due to the fluidity of the membrane.

The Membrane Is Fluid

At body temperature, the phospholipid bilayer of the plasma membrane has the consistency of olive oil. The greater the concentration of unsaturated fatty acid residues, the more fluid the bilayer. In each monolayer, the hydrocarbon tails wiggle, and the entire phospholipid molecule can move sideways at a rate averaging about 2 µm—the length of a prokaryotic cell—per second. Phospholipid molecules rarely flip-flop from one layer to the other, because this would require the hydrophilic head to move through the hydrophobic center of the membrane. The fluidity of a phospholipid bilayer means that cells are pliable. Imagine if they were not—the nerve cells in your neck would crack whenever you nodded your head!

Although some proteins are held in place by cytoskeleton filaments, in general most membrane proteins are observed to drift sideways in the fluid lipid bilayer. This has been demonstrated by fusing mouse and human cells and watching the movement of tagged proteins (Fig. 5.3). Forty minutes after fusion, the proteins are completely intermixed.

The fluidity of the membrane is dependent on its lipid components, including phospholipids, glycolipids, and cholesterol.

Proteins in the Plasma Membrane

Proteins are the other major component of membranes. Transmembrane proteins, which are found within the membrane, often have hydrophobic regions embedded within the membrane and hydrophilic regions that project from both surfaces of the bilayer:

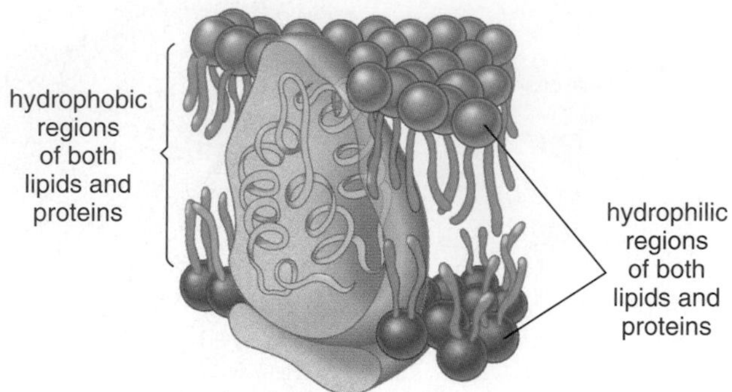

Many plasma membrane proteins are glycoproteins, which have an attached carbohydrate chain. As with glycolipids, the carbohydrate chain projects externally.

Other proteins occur either on the cytoplasmic side or the outer surface side of the membrane. Some of these are anchored to the membrane by a covalently attached lipid or are covalently bonded to the carbohydrate chain of a glycolipid. Still others are held in place by noncovalent interactions, which can be disrupted by gentle shaking or by a change in the pH.

The plasma membrane is asymmetrical; the two halves are not identical. The lipid and protein composition of the inside half differs from the outside half. The carbohydrate chains of the glycolipids and glycoproteins form a carbohydrate coat that envelops the outer surface of the plasma membrane. On the inside some proteins serve as links to the cytoskeletal filaments, and on the outside some serve as links to an extracellular matrix. (The extracellular matrix is discussed on page 94.)

Plasma membrane proteins span the lipid bilayer and often have attached carbohydrate chains. Proteins are also found on the cytoplasmic side or the outer surface side of the membrane.

Cell-Cell Recognition

The carbohydrate chains of glycolipids and glycoproteins serve as the "fingerprints" of the cell. The possible diversity of the chain is enormous; it can vary by the number of sugars (15 is usual, but there can be several hundred), by whether the chain is branched, and by the sequence of the particular sugars. Recalling that glucose has many isomers will help you appreciate how one carbohydrate chain can vary from another.

Glycolipids and glycoproteins vary from species to species, from individual to individual of the same species, and even from cell to cell in the same individual. Therefore, they make cell-cell recognition possible. Researchers working with mouse embryos have shown that as development proceeds, the different type cells of the embryo develop their own carbohydrate chains and that these chains allow the tissues and cells of the embryo to sort themselves out.

As you probably know, transplanted tissues are often rejected by the body. This is because the immune system is able to recognize that the foreign tissue's cells do not have the same glycolipids and glycoproteins as the rest of the body's cells. We now know that a person's particular blood type is due to the presence of particular glycoproteins in the membrane of red blood cells.

Glycoproteins and glycolipids are involved in marking the cell as belonging to a particular individual and tissue.

The Membrane Is a Mosaic

The plasma membrane and the membranes of the various organelles each have their own unique collections of proteins. The proteins form different patterns according to the particular membrane and also within the same membrane at different times. When you consider that a red blood cell plasma membrane contains over 50 different types of proteins, you can see how this would be possible and why the membrane is said to be a mosaic.

The proteins within a membrane determine most of its specific functions. As we will discuss in more detail later, certain plasma membrane proteins are involved in the passage of molecules through the membrane. Some of these have a *channel* through which a substance simply can move across the membrane; others are *carriers* that combine with a substance and help it to move across the membrane. Still other proteins are *receptors;* each type of receptor has a shape that allows a specific molecule to bind to it. The binding of a molecule, such as a hormone (or other signal molecule), can cause the protein to change its shape and bring about an intracellular response. Some plasma membrane proteins have an *enzymatic function* and carry out metabolic reactions directly. Exterior proteins associated with the membrane often have a structural role in that they help stabilize and shape the plasma membrane.

Figure 5.4 depicts the functions of various membrane proteins.

The mosaic pattern of membrane is dependent on proteins, which vary in structure and function.

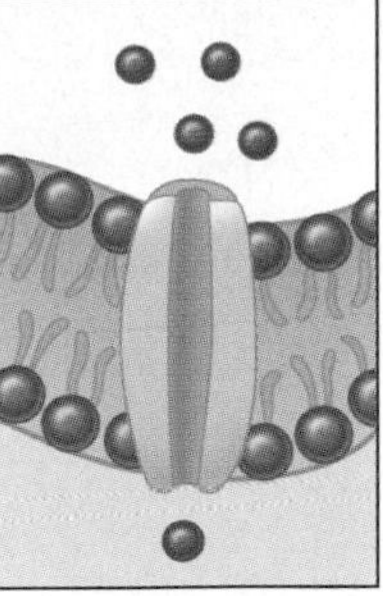

Channel Protein
Allows a particular molecule or ion to cross the plasma membrane freely. Cystic fibrosis, an inherited disorder, is caused by faulty chloride (Cl^-) channel; a thick mucus collects in airways and in pancreatic and liver ducts.

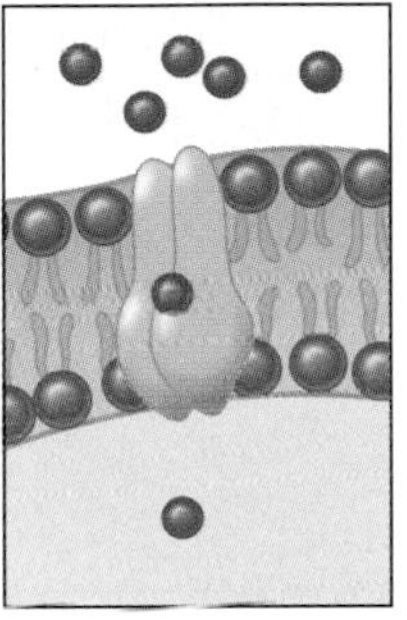

Carrier Protein
Selectively interacts with a specific molecule or ion so that it can cross the plasma membrane. The inability of some persons to use energy for sodium – potassium (Na^+ – K^+) transport has been suggested as the cause of their obesity.

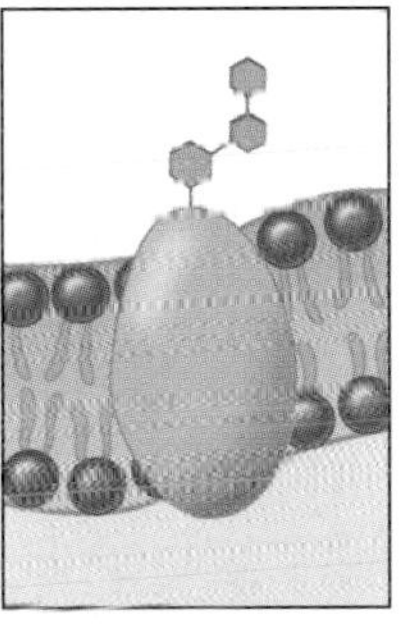

Cell Recognition Protein
The MHC (major histocompatibility complex) glycoproteins are different for each person, so organ transplants are difficult to achieve. Cells with foreign MHC glycoproteins are attacked by blood cells responsible for immunity.

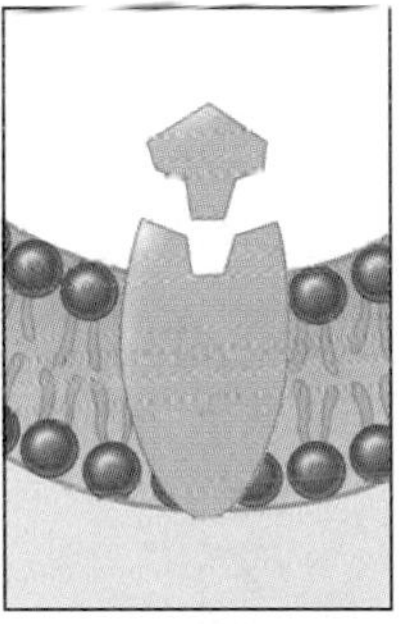

Receptor Protein
Is shaped in such a way that a specific molecule can bind to it. Pygmies are short, not because they do not produce enough growth hormone, but because their plasma membrane growth hormone receptors are faulty and cannot interact with growth hormone.

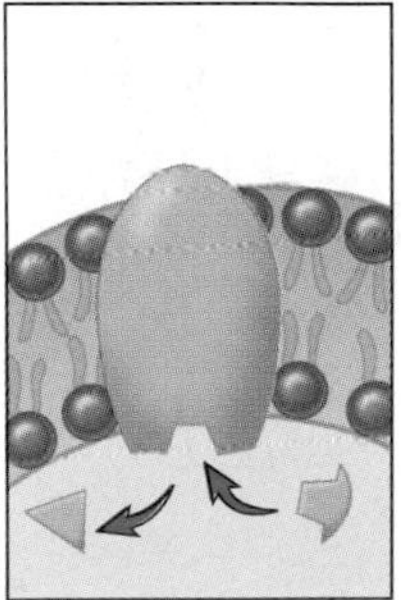

Enzymatic Protein
Catalyzes a specific reaction. The membrane protein, adenylate cyclase, is involved in ATP metabolism. Cholera bacteria release a toxin that interferes with the proper functioning of adenylate cyclase; sodium ions and water leave intestinal cells and the individual dies from severe diarrhea.

Figure 5.4 Membrane protein diversity.
These are some of the functions performed by proteins found in the plasma membrane.

Table 5.1

Passage of Molecules into and out of Cells

	Name	Direction	Requirements	Examples
Passive Transport	Diffusion	Toward lower concentration	Concentration gradient	Lipid-soluble molecules, water, and gases
	Facilitated transport	Toward lower concentration	Carrier and concentration gradient	Sugars and amino acids
Active Transport	Active transport	Toward greater concentration	Carrier plus energy	Sugars, amino acids, and ions
	Exocytosis	Toward outside	Vesicle fuses with plasma membrane	Macromolecules
	Endocytosis Phagocytosis	Toward inside	Vacuole formation	Cells and subcellular material
	Pinocytosis (includes receptor-mediated endocytosis)	Toward inside	Vesicle formation	Macromolecules

5.3 How Molecules Cross the Plasma Membrane

The plasma membrane is semipermeable. A permeable membrane allows all molecules to pass through; an impermeable membrane allows no molecules to pass through; and a semipermeable membrane allows some molecules to pass through. The structure of the plasma membrane affects which types of molecules can freely pass through it. Small, noncharged, lipid-soluble molecules have no difficulty crossing the membrane. Macromolecules cannot freely cross a plasma membrane, and charged ions and molecules have difficulty.

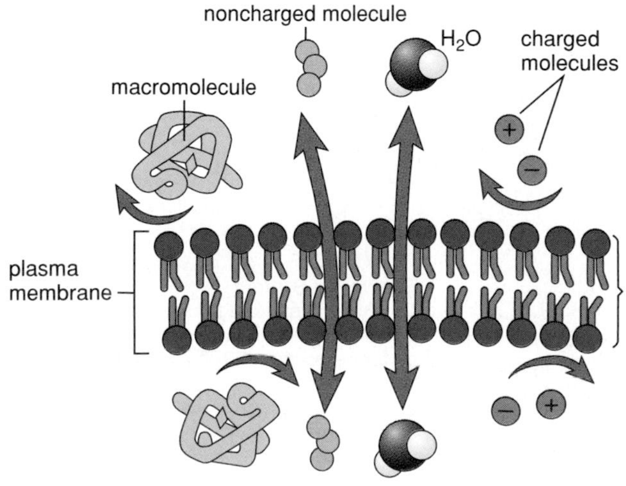

Certain small molecules can cross a plasma membrane, while large molecules cannot. However, some small molecules pass through the plasma membrane quickly, while others have difficulty in passing through or fail to pass through at all. Therefore, a plasma membrane is often regarded as **differentially permeable** (or selectively permeable) as well.

Molecules cross the plasma membrane in both passive and active ways. The active ways use energy (ATP, or adenosine triphosphate, molecules), while the passive ways do not. The *passive ways* involve diffusion and facilitated transport. The *active ways* involve active transport, endocytosis, and exocytosis.

Table 5.1 summarizes the various ways that molecules pass into and out of cells. Notice that facilitated transport and active transport require a carrier molecule. A carrier is a membrane protein that assists the transport of a specific kind of ion or molecules across the plasma membrane.

The plasma membrane is both semipermeable and differentially permeable. Only certain molecules can pass through freely; the others must be assisted across.

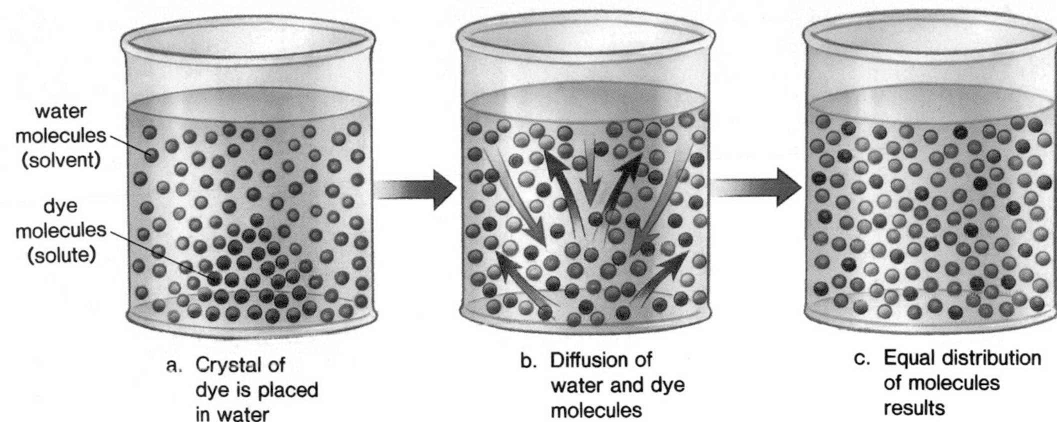

Figure 5.5 Process of diffusion.
Diffusion is spontaneous; no energy is required. **a.** When dye crystals are placed in water, they are concentrated in one area. **b.** The dye dissolves in the water, and a net movement of dye molecules from higher to lower concentration occurs. A net movement of water molecules occurs in the opposite direction. **c.** Eventually, the water and the dye molecules are equally distributed throughout the container.

Use of Diffusion and Osmosis

Diffusion is a physical process that can be observed with any type of molecule. During **diffusion,** molecules move from higher to lower concentration—that is, down their **concentration gradient**—until they are distributed equally. For example, when a few crystals of dye are placed in water (Fig. 5.5), the dye and water molecules move in various directions, but their net movement is toward the region of lower concentration. Eventually, the dye is dissolved in the water, resulting in a colored solution. A solution contains both a solute, usually a solid, and a solvent, usually a liquid. In this case, the **solute** is the dye molecules and the **solvent** is the water molecules. Once the solute and solvent are evenly distributed, they continue to move about, but there is no net movement of either one in any direction.

Diffusion

The chemical and physical properties of the plasma membrane allow just a few types of molecules to enter and exit by diffusion. Lipid-soluble molecules, such as alcohols, can diffuse through the membrane because lipids are the membrane's main structural components.

Gases can also diffuse through the lipid bilayer; this is the mechanism by which oxygen enters cells and carbon dioxide exits cells. As an example, consider the movement of oxygen from the air sacs (alveoli) of the lungs to blood in the lung capillaries (Fig. 5.6). After inhalation (breathing in), the concentration of oxygen in the alveoli is higher than that in the blood; therefore, oxygen diffuses into blood. The principle of diffusion can be, and has been, employed in the treatment of certain human disorders caused by the lack of a particular substance. Cells that produce a substance, like dopamine needed to treat Parkinson disease, are placed in a plastic capsule, and the capsule is implanted in the patient's body. The substance simply diffuses out of the capsule into the body.

Water passes into and out of cells with relative ease. It probably moves through *channels* (see Fig. 5.4), with a pore size large enough to allow the passage of water and prevent the passage of other molecules. The fact that water can penetrate a plasma membrane has important biological consequences, as described in the discussion that follows.

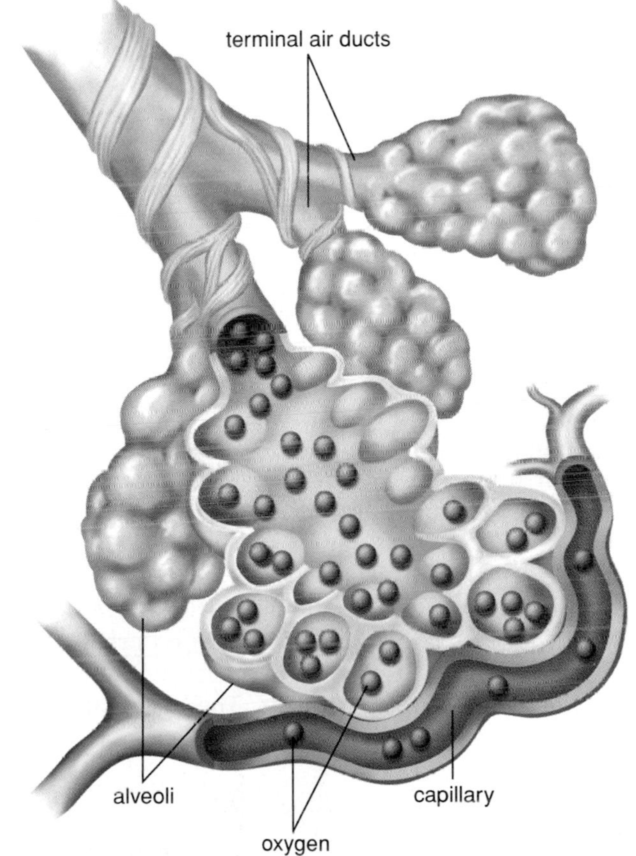

Figure 5.6 Gas exchange in lungs.
Oxygen (O_2) diffuses into the capillaries of the lungs because there is a higher concentration of oxygen in the alveoli (air sacs) than in the capillaries.

Molecules diffuse down their concentration gradients. A few types of small molecules can simply diffuse through the plasma membrane.

Figure 5.7 Osmosis demonstration.
a. A thistle tube, covered at the broad end by a differentially permeable membrane, contains a 10% sugar solution. The beaker contains a 5% sugar solution. **b.** The solute (green circles) is unable to pass through the membrane, but the water passes freely through in both directions. A net movement of water toward the inside of the thistle tube occurs because the thistle tube has a higher percentage of solute. **c.** In the end, the level of the solution rises in the thistle tube until hydrostatic pressure increases to the level of osmotic pressure.

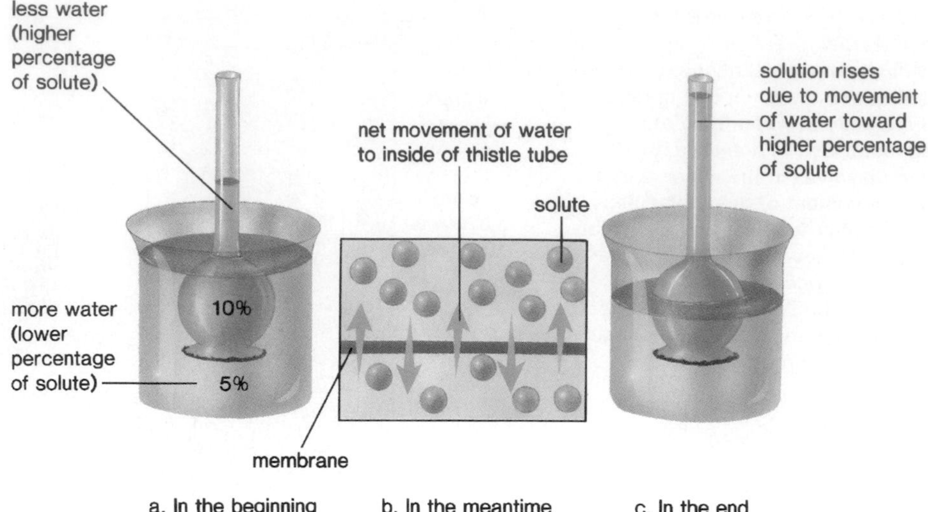

Osmosis

The diffusion of water across a differentially permeable membrane has been given a special name; it is called **osmosis** [Gk. *osmos*, a pushing]. To illustrate osmosis, a thistle tube containing a 10% sugar solution[1] is covered at one end by a differentially permeable membrane and is then placed in a beaker containing a 5% sugar solution (Fig. 5.7). The beaker contains more water molecules (lower percentage of solute) per volume, and the thistle tube contains fewer water molecules (higher percentage of solute) per volume. Under these conditions, there is a net movement of water across the membrane from the beaker to the inside of the thistle tube. The solute is unable to pass through the membrane; therefore, the level of the solution within the thistle tube rises (Fig. 5.7*c*). As water enters the thistle tube, a pressure called hydrostatic pressure builds up and the net movement of water ceases. The hydrostatic pressure is equivalent to the **osmotic pressure** of the solution inside the thistle tube.

Notice the following in this illustration of osmosis:

1. A differentially permeable membrane separates two solutions.
2. The beaker has more water (lower percentage of solute), and the thistle tube has less water (higher percentage of solute).
3. The membrane does not permit passage of the solute.
4. The membrane permits passage of water, and there is a net movement of water from the beaker to the inside of the thistle tube.
5. An osmotic pressure is present: the amount of liquid increases on the side of the membrane with the greater percentage of solute.

[1] Percent solutions are grams of solute per 100 ml of solvent. Therefore, a 10% solution is 10 grams of sugar in 100 ml of water.

These considerations will be important as we discuss osmosis in relation to cells placed in different solutions. The plasma membrane allows such solutes as sugars and salts to pass through, but the difference in permeability between water and these solutes is so great that cells in sugar and salt solutions have to cope with the osmotic movement of water.

Osmosis is the diffusion of water across a differentially permeable membrane. Osmotic pressure develops on the side of the membrane that has the higher solute concentration.

Osmosis occurs constantly in living organisms. For example, due to osmosis, water is absorbed from the human large intestine, is retained by the kidneys, and is taken up by blood. Since living things contain a very high percentage of water, osmosis is a very important physical process that can affect health.

Tonicity

Tonicity refers to the strength of a solution in relationship to osmosis. Cells can be placed in solutions that have the same percentage of solute, a higher percentage of solute, or a lower percentage of solute than the cell. These solutions are called isotonic, hypertonic, and hypotonic, respectively. Figure 5.8 depicts and describes the effects of these solutions on cells.

In the laboratory, cells are normally placed in solutions that cause them neither to gain nor to lose water. Such a solution is said to be an **isotonic solution;** that is, the solute concentration is the same on both sides of the membrane, and therefore there is no net gain or loss of water (Fig. 5.8*a, d*). The prefix *iso* means *the same as* and the term *tonicity* refers to the strength of the solution. A 0.9% solu-

Figure 5.8 Osmosis in animal and plant cells.

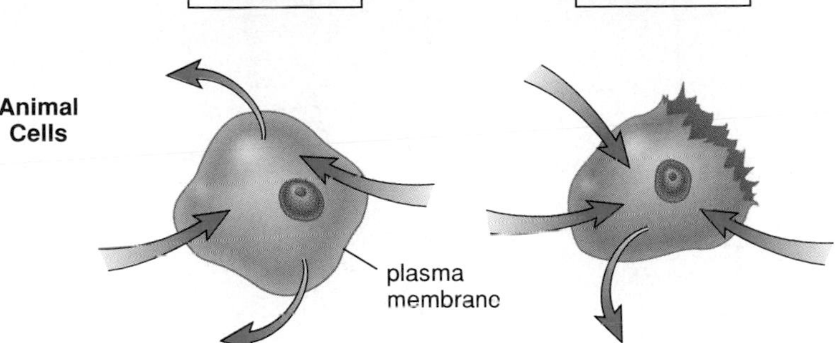

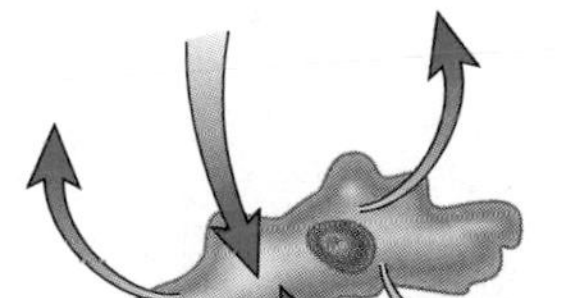

a. No net movement of water into and out of the cell.

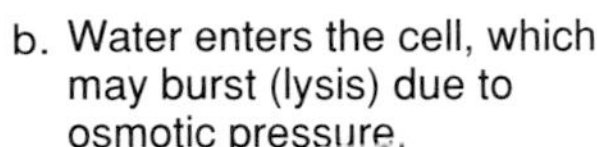

b. Water enters the cell, which may burst (lysis) due to osmotic pressure.

c. Water exits the cell, which shrivels (crenation).

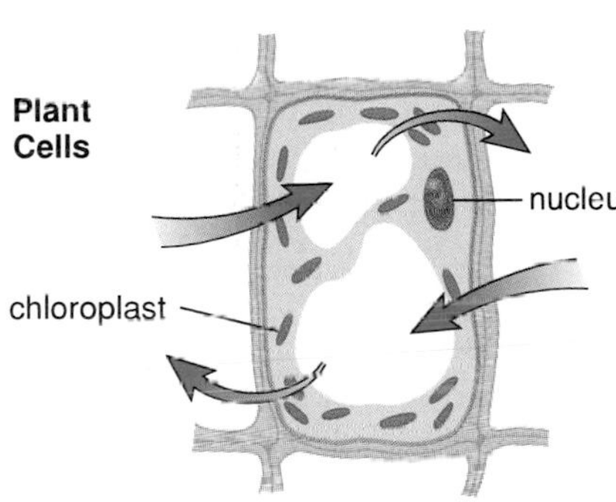

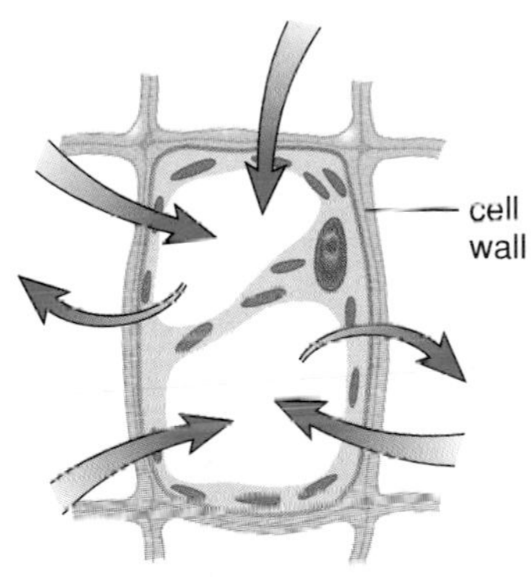

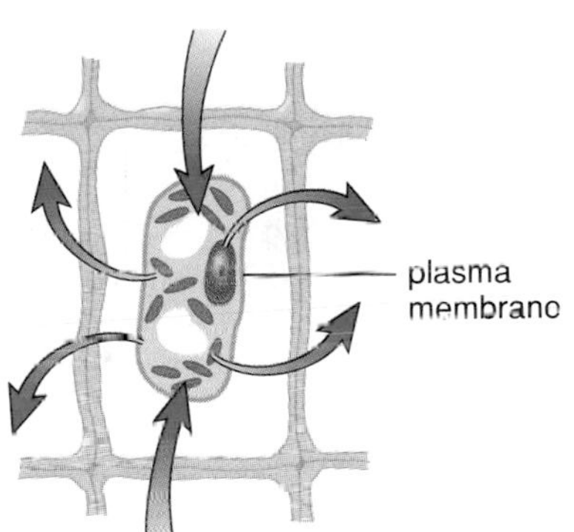

d. No net movement of water into and out of the cell.

e. Vacuoles fill the water, turgor pressure develops, and chloroplasts are seen next to the cell wall.

f. Vacuoles lose water, the cytosol shrinks (plasmolysis), and chloroplasts are seen in the center of the cell.

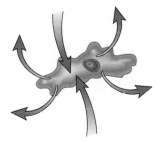

tion of the salt sodium chloride (NaCl) is known to be isotonic to red blood cells because the cells neither swell nor shrink when placed in this solution (Fig. 5.8*a*).

Solutions that cause cells to swell, or even to burst, due to an intake of water are said to be **hypotonic solutions.** The prefix *hypo* means *less than* and refers to a solution with a lower percentage of solute (more water) than the cell. If a cell is placed in a hypotonic solution, water enters the cell; the net movement of water is from the outside to the inside of the cell.

Any concentration of salt solution lower than 0.9% is hypotonic to red blood cells. Red blood cells placed in such a solution expand and sometimes burst due to the buildup of pressure caused by the inward movement of water (Fig. 5.8*b*). The term *lysis* is used to refer to disrupted cells; hemolysis, then, is the process of disrupting red blood cells.

The swelling of a cell in hypotonic solution creates **turgor pressure** [L. *turgor*, swelling]. When a plant cell is placed in a hypotonic solution, we observe expansion of the cytoplasm because the large central vacuole gains water and the plasma membrane pushes against the rigid cell wall (Fig. 5.8*e*). The plant cell does not burst because the cell wall does not give way. Turgor pressure in plant cells is extremely important to the maintenance of the plant's erect position.

Solutions that cause cells to shrink or to shrivel due to a loss of water are said to be **hypertonic solutions.** The prefix *hyper* means *more than* and refers to a solution with a higher percentage of solute (less water) than the cell. If a cell is placed in a hypertonic solution, water leaves the cell; the net movement of water is from the inside to the outside of the cell.

A 10% solution of NaCl is hypertonic to red blood cells. In fact, any solution with a concentration higher than 0.9% sodium chloride is hypertonic to red blood cells. If red blood cells are placed in this solution, they shrink (Fig. 5.8*c*). The term *crenation* [L. *crenatus*, notched, wrinkled] refers to red blood cells in this condition.

When a plant cell is placed in a hypertonic solution, the plasma membrane pulls away from the cell wall as the large central vacuole loses water. This is an example of plasmolysis, a shrinking of the cytosol due to osmosis (Fig. 5.8*f*).

In an isotonic solution, a cell neither gains nor loses water. In a hypotonic solution, a cell gains water and may burst. In a hypertonic solution, a cell loses water and the cytoplasm shrinks.

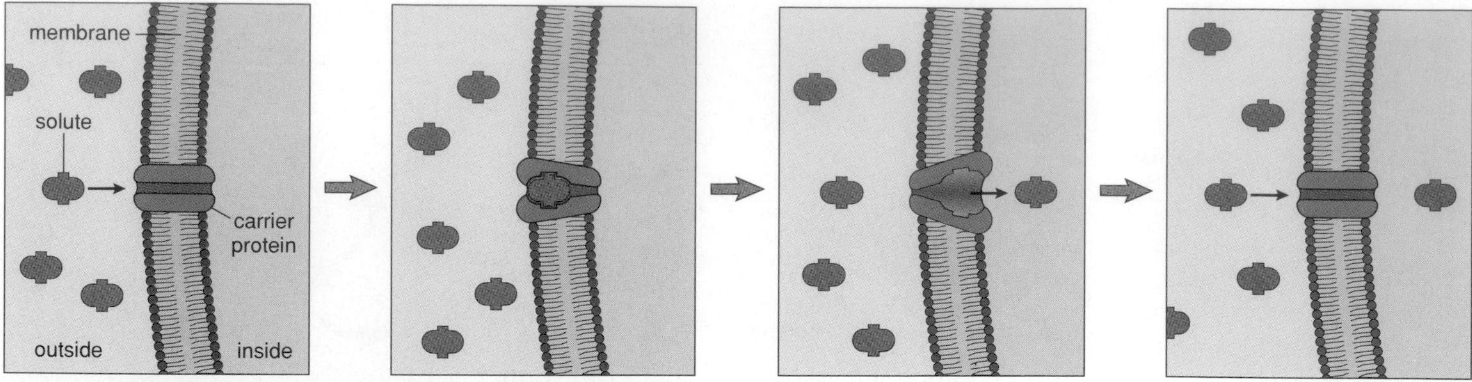

Figure 5.9 Facilitated transport.
During facilitated transport, a carrier protein speeds the rate at which the solute crosses the plasma membrane in the direction of decreasing concentration. Note that the carrier protein undergoes a change in shape (called a conformational change) as it moves a solute across the membrane.

Transport by Carrier Proteins

The plasma membrane impedes the passage of all but a few substances. Yet, biologically useful molecules are able to enter and exit the cell at a rapid rate because there are carrier proteins in the membrane. **Carrier proteins** are specific; each can combine with only a certain type of molecule or ion, which is then transported through the membrane. It is not completely understood how carrier proteins function, but after a carrier combines with a molecule, the carrier is believed to undergo a change in shape that moves the molecule across the membrane. Carrier proteins are required for facilitated transport and active transport (see Table 5.1).

Some of the proteins in the plasma membrane are carriers. They transport biologically useful molecules into and out of the cell.

Facilitated Transport

Facilitated transport explains the passage of such molecules as glucose and amino acids across the plasma membrane even though they are not lipid-soluble. The passage of glucose and amino acids is facilitated by their reversible combination with carrier proteins, which in some manner transport them through the plasma membrane. These carrier proteins are specific. For example, various sugar molecules of identical size might be present inside or outside the cell, but glucose can cross the membrane hundreds of times faster than the other sugars. As stated earlier, this is the reason that the membrane can be called differentially permeable.

A model for facilitated transport (Fig. 5.9) shows that after a carrier has assisted the movement of a molecule to the other side of the membrane, it is free to assist the passage of other similar molecules. Neither diffusion, explained previously, nor facilitated transport requires an expenditure of energy because the molecules are moving down their concentration gradient in the same direction they tend to move anyway.

Active Transport

During **active transport,** molecules or ions move through the plasma membrane, accumulating either inside or outside the cell. For example, iodine collects in the cells of the thyroid gland; glucose is completely absorbed from the gut by the cells lining the digestive tract; and sodium can be almost completely withdrawn from urine by cells lining the kidney tubules. In these instances, molecules have moved to the region of higher concentration, exactly opposite to the process of diffusion.

Both carrier proteins and an expenditure of energy are needed to transport molecules against their concentration gradient. In this case, energy (ATP molecules usually) is required for the carrier to combine with the substance to be transported. Therefore, it is not surprising that cells involved primarily in active transport, such as kidney cells, have a large number of mitochondria near a membrane where active transport is occurring.

Proteins involved in active transport often are called *pumps* because, just as a water pump uses energy to move water against the force of gravity, proteins use energy to move a substance against its concentration gradient. One type of pump that is active in all animal cells, but is especially associated with nerve and muscle cells, moves sodium ions (Na^+) to the outside of the cell and potassium ions (K^+) to the inside of the cell. These two events are linked, and the carrier protein is called a **sodium-potassium pump.** A change in carrier shape after the attachment and again after the detachment of a phosphate group allows it to combine alternately with sodium ions and potassium ions (Fig. 5.10). The phosphate group is donated by ATP when it is broken

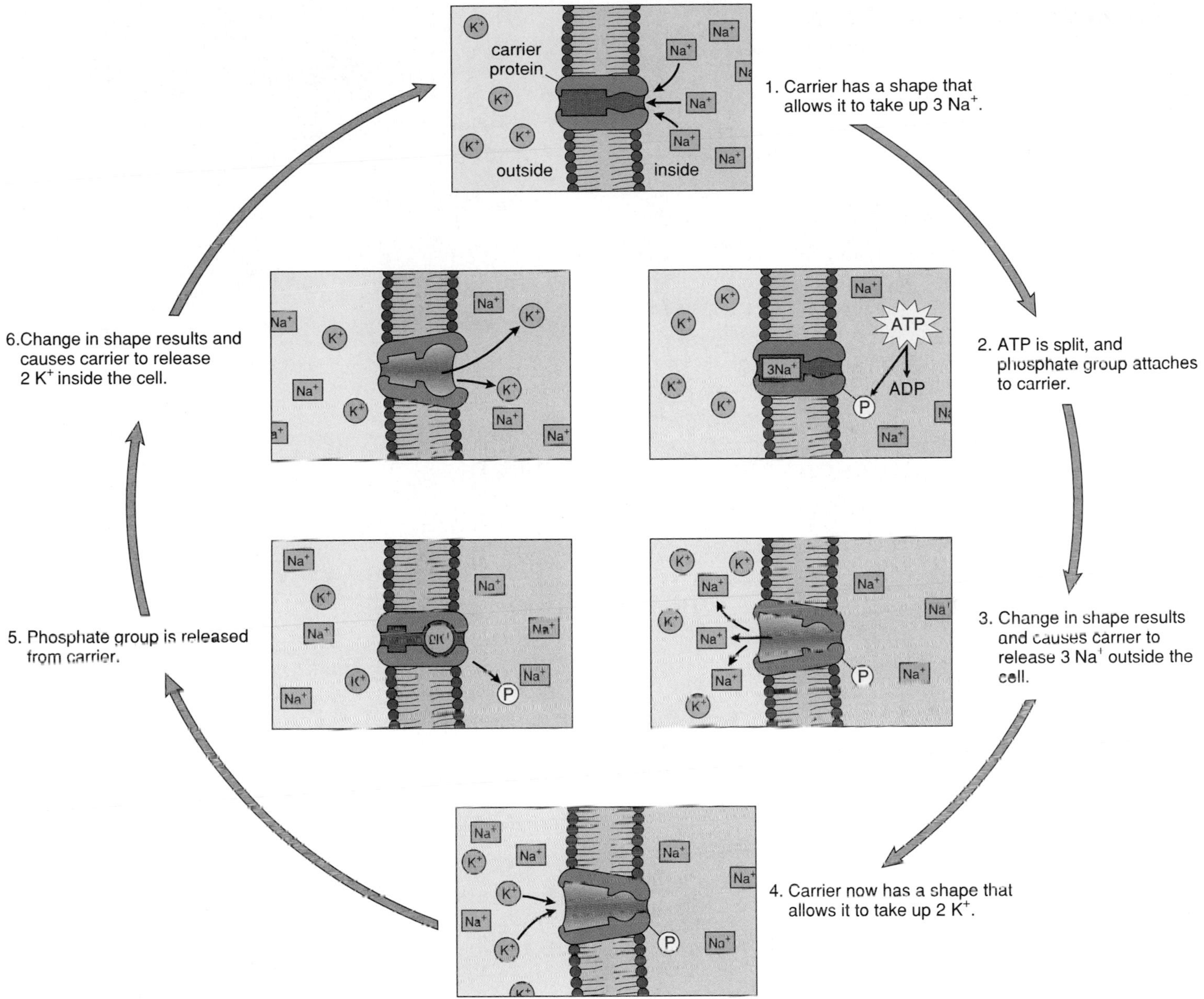

Figure 5.10 The sodium-potassium pump.
The same carrier protein transports sodium ions (Na^+) to the outside of the cell and potassium ions (K^+) to the inside of the cell because it undergoes an ATP-dependent conformational change. Three sodium ions are carried outward for every two potassium ions carried inward; therefore, the inside of the cell is negatively charged compared to the outside.

down enzymatically by the carrier. The sodium-potassium pump results in both a concentration gradient and an electrical gradient for these ions across the plasma membrane.

The passage of salt (NaCl) across a plasma membrane is of primary importance in cells. The chloride ion (Cl^-) usually crosses the plasma membrane because it is attracted by positively charged sodium ions (Na^+). First sodium ions are pumped across a membrane and then chloride ions simply diffuse through channels that allow their passage. As noted in Figure 5.4, the chloride ion channels malfunction in persons with cystic fibrosis, leading to the symptoms of this inherited (genetic) disorder.

During facilitated transport, small molecules follow their concentration gradient. During active transport, small molecules and ions move against their concentration gradient.

Use of Membrane-Assisted Transport

What about the transport of macromolecules such as polypeptides, polysaccharides, or polynucleotides, which are too large to be transported by carrier proteins? They are transported in or out of the cell by vesicle formation, thereby keeping the macromolecules contained so that they do not mix with those in the cytosol.

Exocytosis

During **exocytosis** [Gk. *ex*, out of, and *kytos*, cell], vesicles, often formed by the Golgi apparatus and carrying a specific molecule, fuse with the plasma membrane as secretion occurs. This is the way that insulin leaves insulin-secreting cells, for instance.

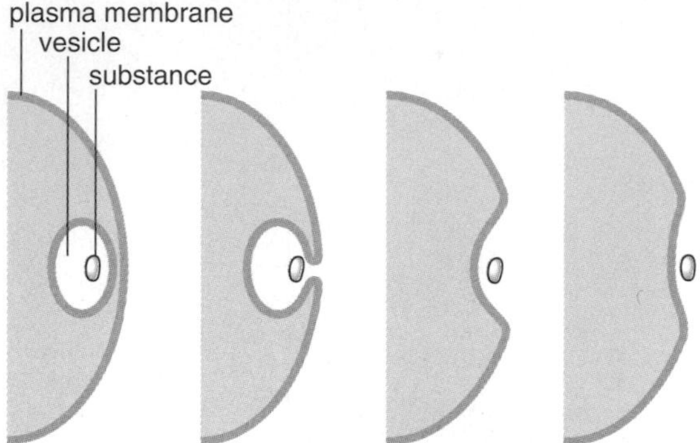

Notice that the membrane of the vesicle becomes a part of the plasma membrane. During cell growth, exocytosis is probably used as a means to enlarge the plasma membrane, whether or not secretion is also taking place.

Endocytosis

During **endocytosis** [Gk. *endon*, within, and *kytos*, cell], cells take in substances by vesicle formation (Fig. 5.11). A portion of the plasma membrane invaginates to envelop the substance, and then the membrane pinches off to form an intracellular vesicle.

When the material taken in by endocytosis is large, such as a food particle or another cell, the process is called **phagocytosis** [Gk. *phagein*, to eat, and *kytos*, cell]. Phagocytosis is common in unicellular organisms like amoebas and in ameboid-type cells like macrophages, which are large cells that engulf bacteria and worn-out red blood cells in mammals. When the endocytic vesicle fuses with a lysosome, digestion occurs.

Pinocytosis [Gk. *pino*, drink, and *kytos*, cell] occurs when vesicles form around a liquid or very small particles. Blood cells, cells that line the kidney tubules or intestinal wall, and plant root cells all use this method of ingesting substances. Whereas phagocytosis can be seen with the light microscope, the electron microscope must be used to observe pinocytic vesicles, which are no larger than 1–2 μm.

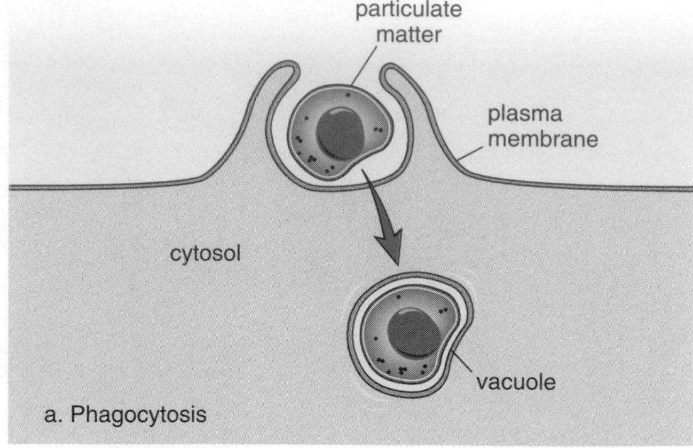

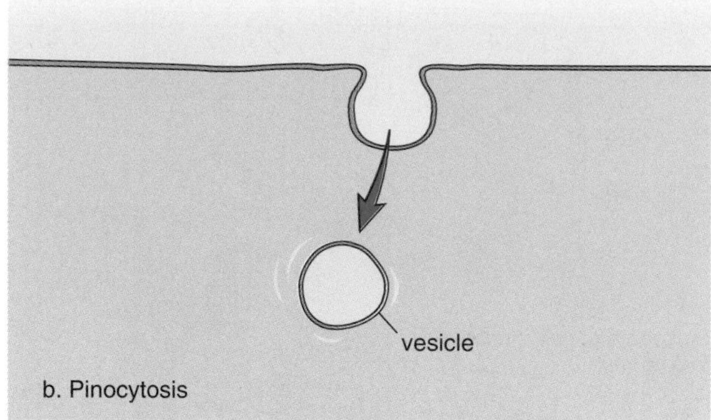

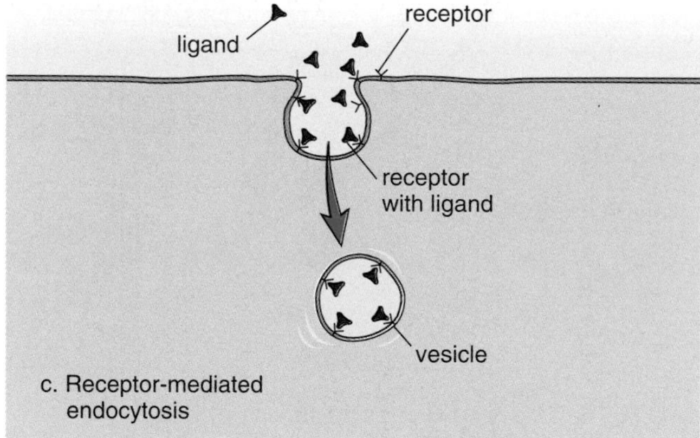

Figure 5.11 Three methods of endocytosis.
a. Phagocytosis occurs when the substance to be transported into the cell is large: white blood cells can engulf bacteria by phagocytosis. Digestion occurs when the resulting vacuole fuses with a lysosome. **b.** Pinocytosis occurs when a macromolecule such as a polypeptide is to be transported into the cell. The result is a small vacuole or vesicle. **c.** Receptor-mediated endocytosis is a form of pinocytosis. The substance to be taken in (a ligand) first binds to a specific receptor protein which migrates to a pit or is already in a pit. The vesicle that forms contains the ligand and its receptor. Sometimes the receptor is recycled, as shown in Figure 5.12.

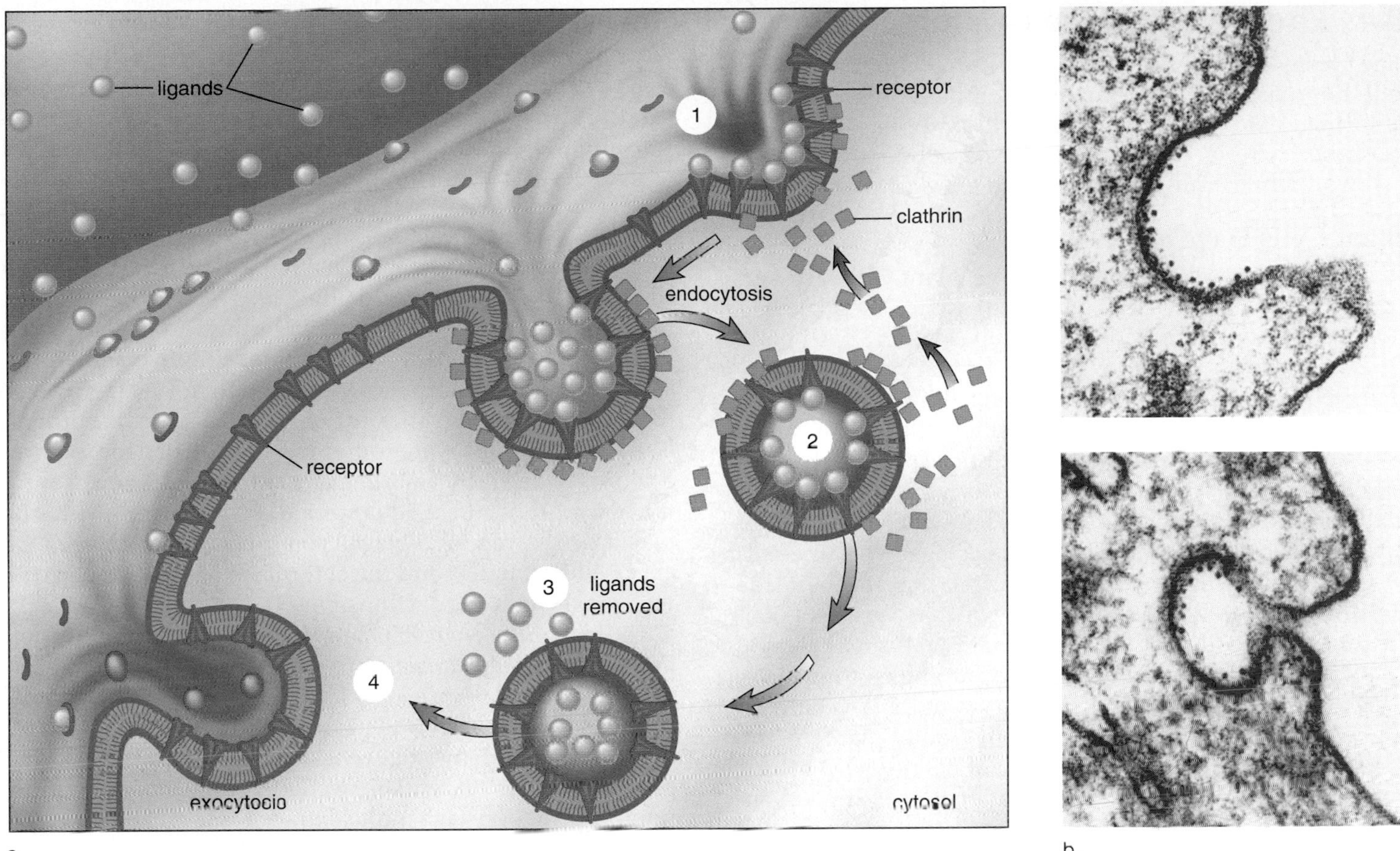

Figure 5.12 Receptor-mediated endocytosis.
a. (1) The receptors in the coated pits combine only with a specific substance, called a ligand. (2) The vesicle that forms is at first coated with the structural protein clathrin, but soon the vesicle loses its coat. (3) Ligands leave the vesicle. (4) When exocytosis occurs, membrane and therefore receptors are returned to the plasma membrane. **b.** Electron micrographs of a coated pit in the process of forming a vesicle.

Receptor-mediated endocytosis is a form of pinocytosis that is quite specific because it involves the use of a **receptor protein** shaped in such a way that a specific molecule can bind to it. A macromolecule that binds to a plasma membrane receptor is called a *ligand* [L. *ligo*, bind, tie]. The binding of ligands to their receptors causes the receptors to gather at one location. This location is called a *coated pit* because there is a layer of fibrous protein, called clathrin, on the cytoplasmic side (see step 1, Fig. 5.12). *Clathrin* is a protein that also coats at least some, if not all, of the vesicles that leave the Golgi apparatus. Clathrin seems to somehow facilitate the formation of a vesicle. Once the vesicle is formed, the clathrin coat is released and the vesicle appears uncoated (see step 2). The fate of the vesicle and its contents depends on the kind of ligand it contains. After a hormone has been received and has acted on a cell, the hormone and receptors are digested after the vesicle fuses with a lysosome. On the other hand, when cholesterol enters a cell (as in step 3) the membrane of the vesicle and therefore the receptors are returned to the plasma membrane (step 4), or the vesicle can go to other membranous locations.

Aside from simply allowing substances to enter cells selectively from an extracellular fluid, coated pits are also involved in the transfer and exchange of substances between cells. Such exchanges take place when substances move from maternal blood into fetal blood at the placenta, for example.

The importance of receptor-mediated endocytosis is demonstrated by a genetic disorder called familial hypercholesterolemia. Cholesterol is transported in blood by a complex of lipids and proteins called low-density lipoprotein (LDL). These individuals have inherited a gene that causes them to have a reduced number and/or defective receptors for LDL in their plasma membranes. Instead of cholesterol entering cells, it accumulates in the walls of arterial blood vessels leading to high blood pressure, occluded (blocked) arteries, and heart attacks.

Receptor-mediated endocytosis allows cells to take up specific kinds of molecules and then sort them within the cell.

5.4 The Cell Surface Is Modified

The plasma membrane is the outer living boundary of the cell, but many cells have an extracellular component that is formed exterior to the membrane. In plants, fungi, algae, and bacteria, the extracellular component is called a cell wall while animal cells have an extracellular matrix.

Plant Cells Have a Cell Wall

In addition to a plasma membrane, plant cells are surrounded by a porous **cell wall** that varies in thickness, depending on the function of the cell (Fig. 5.13). All plant cells have a primary cell wall. The primary cell wall contains cellulose fibrils in which microfibrils are held together by noncellulose substances. Pectins allow the wall to stretch when the cell is growing, and noncellulose polysaccharides harden the wall when the cell is mature. Pectins are especially abundant in the middle lamella, which is a layer of adhesive substances that holds the cells together. Some cells in woody plants have a secondary wall that forms inside the primary cell wall. The secondary wall has a greater quantity of cellulose fibrils than the primary wall, and layers of cellulose fibrils are laid down at right angles to one another. Lignin, a substance that adds strength, is a common ingredient of secondary cell walls in woody plants.

In plant tissues, the cytoplasm of neighboring cells is sometimes connected by plasmodesmata (sing., plasmodesma), numerous narrow channels that pass through the cell wall. Cytoplasmic strands within these channels allow direct exchange of materials between neighboring plant cells. Coordination of cellular activities within a tissue occurs because of these linkages.

Animal Cells Have an Extracellular Matrix

An extracellular matrix is a meshwork of insoluble proteins with carbohydrate chains (glycoproteins) that are produced and secreted by animal cells (Fig. 5.14). The extracellular matrix fills the spaces between animal cells and helps support them. The extracellular matrix influences the development, migration, shape, and function of cells.

Collagen and *elastin fibers* are two well-known structural components of the extracellular matrix. Collagen gives the matrix strength and elastin gives it resilience. *Fibronectins* and *laminins* are two adhesive proteins that seem to play a dynamic role in influencing the behavior of cells. For example, fibronectin and laminin form "highways" that direct the migration of cells during development. Recently, laminins were found to be necessary for the production of milk by mammary gland cells taken from a mouse. Fibronectins and laminins bind to receptors in the plasma membrane and permit communication between the extracellular matrix and the cytoplasm of the cell, perhaps via cytoskeletal connections.

Proteoglycans are glycoproteins that are largely composed of carbohydrate chains containing amino sugars. Proteoglycans provide a packing gel that joins the various proteins in the matrix and most likely regulate the activity of signaling proteins that bind to receptors in the plasma protein. More work will be needed to determine the functions of proteoglycans in the extracellular matrix and how they influence cellular metabolism via signaling proteins like hormones.

Cells have extracellular structures. Cellulose and non-cellulose substances occur in plant cell walls; the extracellular matrix of animal cells is rich in glycoproteins. Certain of these have functions that affect cell behavior.

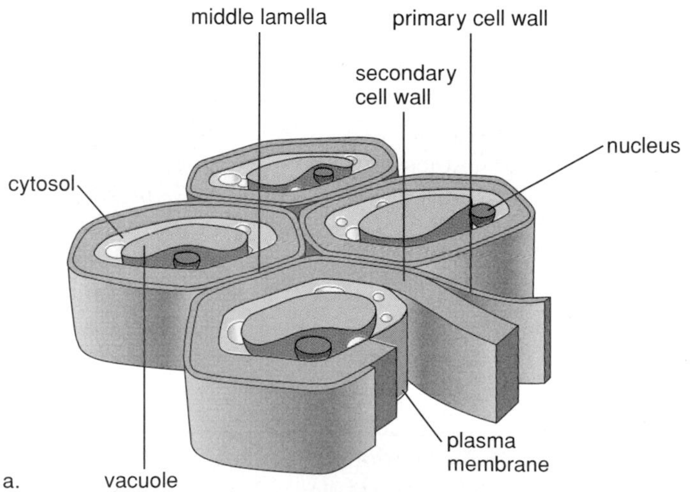

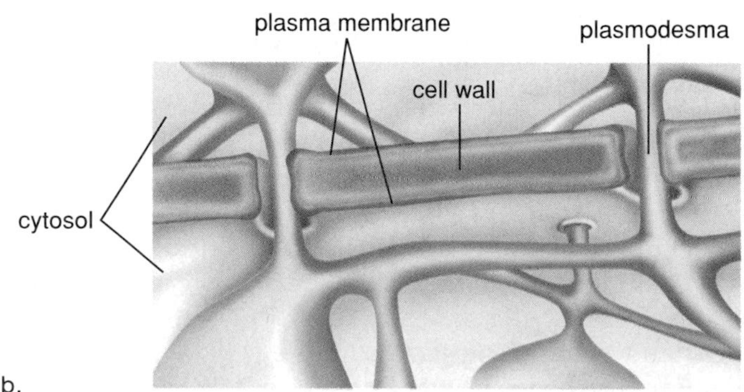

Figure 5.13 Plant cell wall and plasmodesmata.
a. All plant cells have a primary cell wall and some have a secondary cell wall. The cell wall, which lends support to the cell, is freely permeable. **b.** Plasmodesmata are strands of cytoplasm that run in channels connecting certain plant cells.

Animal Cells Have Junctions

For the cells of a tissue to act in a coordinated manner, it is beneficial for the plasma membranes of adjoining cells to interact. The plasmodesmata that link adjacent plant cells and the junctions that occur between animal cells are examples of such cellular interactions.

Junctions in Animal Cells

Three types of junctions are seen between animal cells: adhesion junctions (desmosomes), tight junctions, and gap junctions (Fig. 5.15).

In *adhesion junctions*, internal cytoplasmic plaques, firmly attached to the cytoskeleton within each cell, are joined by intercellular filaments. In some organs—like the heart, stomach, and bladder, where tissues get stretched—adhesion junctions hold the cells together.

Adjacent cells are even more closely joined by *tight junctions*, in which plasma membrane proteins actually attach to each other, producing a zipperlike fastening. The cells of tissues that serve as barriers are held together by tight junctions; in the intestine the digestive juices stay out of the body, and in the kidneys the urine stays within kidney tubules, because the cells are joined by tight junctions.

A *gap junction* allows cells to communicate. A gap junction is formed when two identical plasma membrane channels join. The channel of each cell is lined by six plasma membrane proteins. A gap junction lends strength to the cells, but it also allows small molecules and ions to pass between them. Gap junctions are important in heart muscle and smooth muscle because they permit a flow of ions that is required for the cells to contract.

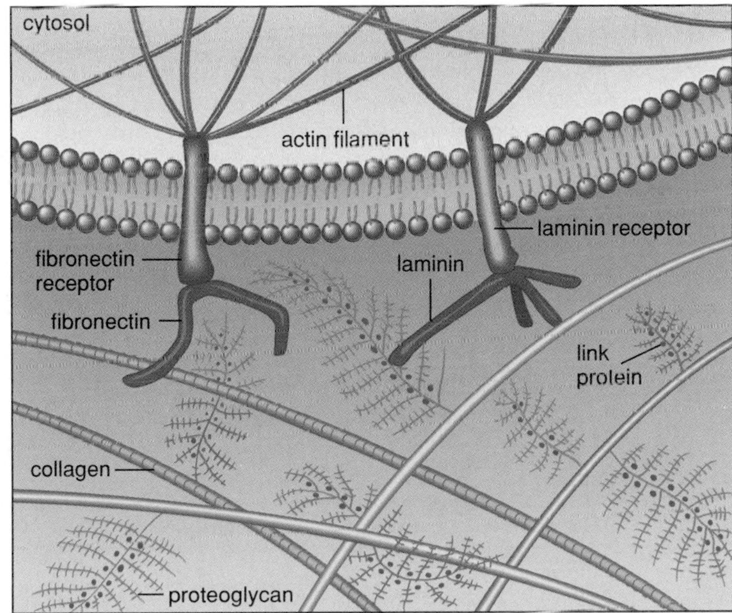

Figure 5.14 Animal cell extracellular matrix.
The glycoproteins in the extracellular matrix support an animal cell and also affect its behavior. Cellulose and elastin also have a support function, while fibronectins and laminins bind to receptors in the plasma membrane and most likely assist cell communication processes.

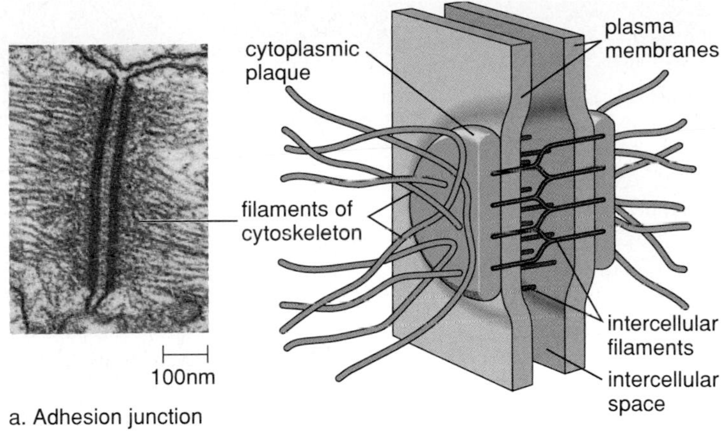

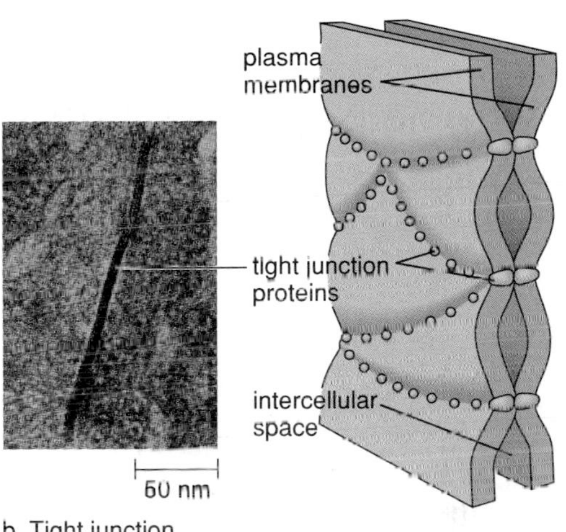

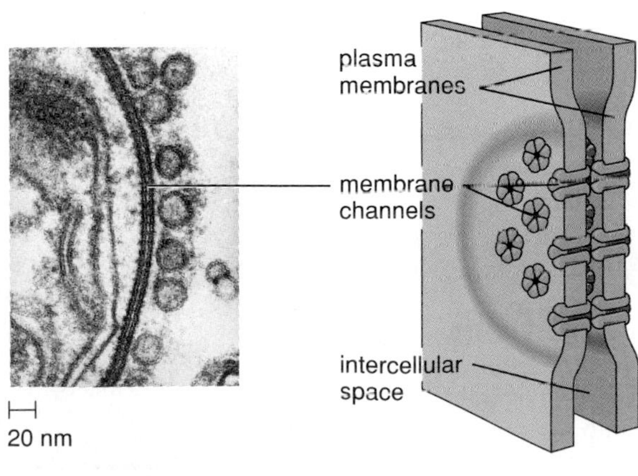

Figure 5.15 Junctions between cells of the intestinal wall.
a. In adhesion junctions (desmosomes), intracellular filaments run between two cells. b. Tight junctions between cells form an impermeable barrier because their adjacent plasma membranes are joined. c. Gap junctions allow communication between two cells because adjacent plasma membrane channels are joined.

connecting concepts

The plasma membrane has quite appropriately been called the gatekeeper of the cell because it maintains the integrity of the cell and stands guard over what enters and leaves the cell. But we have seen that the plasma membrane does so much more than this. Its glycoproteins and glycolipids mark the cell as belonging to the organism. Its numerous proteins allow communication between cells and allow tissues to function as a whole. Now it appears that the extracellular material secreted by cells assists the plasma membrane in its numerous functions.

The progression in our knowledge about the plasma membrane illustrates how science works. Science evolves and changes, and the knowledge we have today is amended and expanded by new investigative work. Basic science has applications that promote the well-being and health of human beings. To know that the plasma membrane is misfunctioning in a pygmy, a diabetic, and a person with a high cholesterol count is a first step toward curing these conditions. Even cancer is sometimes due to receptor proteins that signal the cell to divide even when no growth factor is present.

Our ability to understand the functioning of the plasma membrane is dependent on a thorough understanding of the organic molecules that make up the cell. Today, it is impossible to negate the premise that biology and medicine have a biochemical basis.

Summary

5.1 Membrane Models Have Changed

The fluid-mosaic model of membrane structure developed by Singer and Nicolson was preceded by several other models. Electron micrographs of freeze-fractured membranes support the fluid-mosaic model, and not Robertson's unit membrane concept based on the Danielli and Davson sandwich model.

5.2 The Plasma Membrane Is Complex

There are two components of the plasma membrane, lipids and proteins. In the lipid bilayer, phospholipids are arranged with their hydrophilic (polar) heads at the surfaces and their hydrophobic (nonpolar) tails in the interior. The lipid bilayer has the consistency of oil but acts as a barrier to the entrance and exit of most biological molecules. Membrane glycolipids and glycoproteins are involved in marking the cell as belonging to a particular individual and tissue.

The hydrophobic portion of a transmembrane protein lies in the lipid bilayer of the plasma membrane, and the hydrophilic portion lies at the surfaces. Proteins act as receptors, carry on enzymatic reactions, join cells together, form channels, or act as carriers to move substances across the membrane.

5.3 How Molecules Cross the Plasma Membrane

Some molecules (lipid-soluble compounds, water, and gases) simply diffuse across the membrane from the area of higher concentration to the area of lower concentration. No metabolic energy is required for diffusion to occur.

The diffusion of water across a differentially permeable membrane is called osmosis. Water moves across the membrane into the area of higher solute (less water) content. When cells are in an isotonic solution, they neither gain nor lose water. When cells are in a hypotonic solution, they gain water, and when they are in a hypertonic solution, they lose water (Table 5.2).

Other molecules are transported across the membrane by carrier proteins that span the membrane. During facilitated transport, a carrier protein assists the movement of a molecule down its concentration gradient. No energy is required.

During active transport, a carrier protein acts as a pump that causes a substance to move against its concentration gradient. The sodium-potassium pump carries Na^+ to the outside of the cell and K^+ to the inside of the cell. Energy in the form of ATP molecules is required for active transport to occur.

Larger substances can enter and exit a membrane by exocytosis and endocytosis. Exocytosis involves secretion. Endocytosis includes phagocytosis, pinocytosis, and receptor-mediated endocytosis. Receptor-mediated endocytosis makes use of receptor molecules in the plasma membrane. Once specific substances (e.g., ligands) bind to their receptors, the coated pit becomes a coated vesicle. After losing the coat, the vesicle can join with the lysosome, or after freeing the ligand, the receptor-containing vesicle can fuse with the plasma membrane.

5.4 The Cell Surface Is Modified

Plant cells have a freely permeable cell wall, with cellulose as its main component. Plant cells are joined by small channels called plasmodesmata that span the cell wall and contain strands of cytoplasm that allow materials to pass from one cell to another.

Animal cells have an extracellular matrix that determines their shape and influences their behavior. Junctions between animal cells include adhesion junctions and tight junctions, which help to hold cells together, and gap junctions, which allow passage of small molecules between cells.

Table 5.2

Effect of Osmosis on a Cell

Tonicity of Solution	Concentrations: Solute	Concentrations: Water	Net Movement of Water	Effect on Cell
Isotonic	Same as cell	Same as cell	None	None
Hypotonic	Less than cell	More than cell	Cell gains water	Swells, turgor pressure
Hypertonic	More than cell	Less than cell	Cell loses water	Shrinks, plasmolysis

Reviewing the Chapter

1. Describe the fluid-mosaic model of membrane structure as well as the models that preceded it. Cite the evidence that either disproves or supports these models. 82
2. Tell how the phospholipids are arranged in the plasma membrane. What other lipids are present in the membrane, and what functions do they serve? 83–84
3. Describe how proteins are arranged in the plasma membrane. What are their functions? Describe an experiment indicating that proteins can laterally drift in the membrane. 84
4. What is diffusion, and what substances can diffuse through a differentially permeable membrane? 87
5. Describe an experiment that measures osmotic pressure. 88
6. Tell what happens to an animal cell and a plant cell when placed in isotonic, hypotonic, and hypertonic solutions. 88–89
7. Why do most substances have to be assisted through the plasma membrane? Contrast movement by facilitated transport with movement by active transport. 90
8. Draw and explain a diagram that explains how the sodium-potassium pump words. 90–91
9. Describe and contrast three methods of endocytosis. 92–93
10. Give examples to show that cell surface modifications help plant and animal cells communicate. 94

Testing Yourself

Choose the best answer for each question.

1. Electron micrographs following freeze-fracture of the plasma membrane indicate that
   a. the membrane is a phospholipid bilayer.
   b. some proteins span the membrane.
   c. protein is found only on the surfaces of the membrane.
   d. glycolipids and glycoproteins are antigenic.
2. A phospholipid molecule has a head and two tails. The tails are found
   a. at the surfaces of the membrane.
   b. in the interior of the membrane.
   c. spanning the membrane.
   d. Both a and b are correct.
3. Energy is required for
   a. active transport.
   b. diffusion.
   c. facilitated transport.
   d. All of these are correct.
4. When a cell is placed in a hypotonic solution,
   a. solute exits the cell to equalize the concentration on both sides of the membrane.
   b. water exits the cell toward the area of lower solute concentration.
   c. water enters the cell toward the area of higher solute concentration.
   d. solute exits and water enters the cell.
5. When a cell is placed in a hypertonic solution,
   a. solute exits the cell to equalize the concentration on both sides of the membrane.
   b. water exits the cell toward the area of lower solute concentration.
   c. water exits the cell toward the area of higher solute concentration.
   d. solute exits and water enters the cell.
6. Active transport
   a. requires a carrier protein.
   b. moves a molecule against its concentration gradient.
   c. requires a supply of energy.
   d. All of these are correct.
7. The sodium-potassium pump
   a. helps establish an electrochemical gradient across the membrane.
   b. concentrates sodium on the outside of the membrane.
   c. utilizes a carrier protein and energy.
   d. All of these are correct.
8. Receptor-mediated endocytosis
   a. is no different from phagocytosis.
   b. brings specific substances into the cell.
   c. helps to concentrate proteins in vesicles.
   d. All of these are correct.
9. Plant cells
   a. always have a secondary cell wall, even though the primary one may disappear.
   b. have channels between cells that allow strands of cytoplasm to pass from cell to cell.
   c. develop turgor pressure when water enters the nucleus.
   d. do not have cell-to-cell junctions like animal cells.
10. Write hypotonic solution or hypertonic solution beneath each cell. Justify your conclusions.

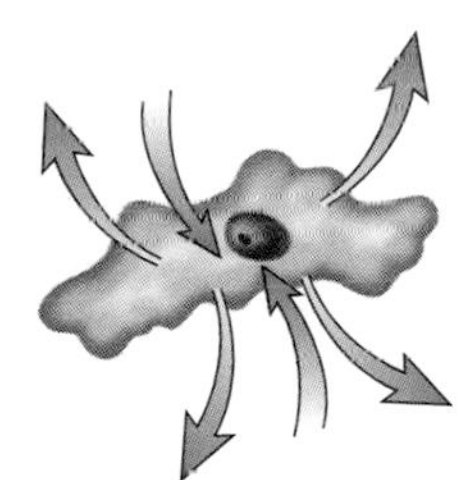

a. ______________________

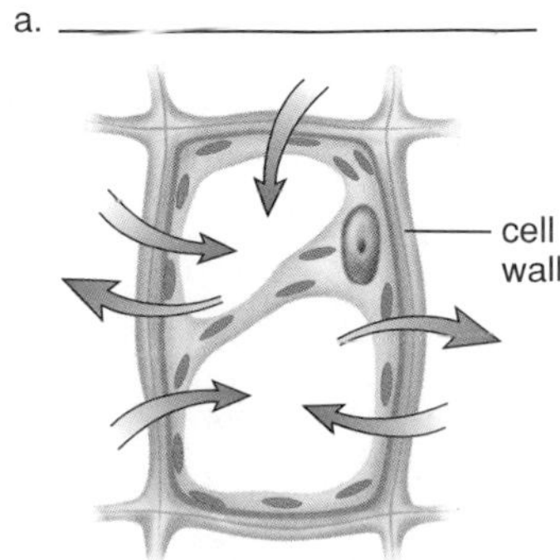

b. ______________________

Applying the Concepts

1. *All cells have a plasma membrane that separates their contents from the extracellular environment and serves to maintain their integrity.*

   In a structural sense, how does the plasma membrane maintain the integrity of the cell? In a functional sense, how does the plasma membrane maintain the integrity of the cell?
2. *To remain alive, cells must have an extracellular environment that is compatible with their continued existence.*

   How does the phenomenon of osmosis demonstrate this concept?
3. *Cells of multicellular organisms function as a unit.*

   For cells to function as a unit, communication is necessary. How does the plasma membrane help distantly located cells communicate with each other? How does the plasma membrane help adjacent cells communicate with each other?

Using Technology

Your study of membrane structure and function is supported by these available technologies:

Exploring the Internet

The Mader Home Page provides resources for and help with studying this chapter.

http://www.mhhe.com/sciencemath/biology/mader/
(Click on Biology.)

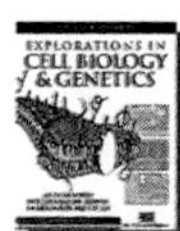

Explorations in Cell Biology & Genetics CD-ROM

Cell Size (#2)
Active Transport (#3)

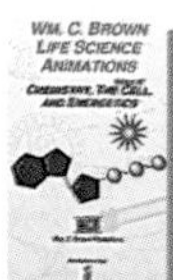

Life Science Animations Video

Video #1: Chemistry, the Cell, and Energetics
Journey into a Cell (#2)
Endocytosis (#3)
Cellular Secretion (#4)

Understanding the Terms

active transport 90
carrier protein 90
cell wall 94
cholesterol 83
concentration gradient 87
differentially permeable 86
diffusion 87
endocytosis 92
exocytosis 92
facilitated transport 90
fluid-mosaic model 82
glycolipid 83
hypertonic solution 89
hypotonic solution 89
isotonic solution 88
osmosis 88
osmotic pressure 88
phagocytosis 92
phospholipid 83
pinocytosis 92
receptor-mediated endocytosis 93
receptor protein 93
sodium-potassium pump 90
solute 87
solvent 87
tonicity 88
turgor pressure 89

Match the terms to these definitions:

a. __________ Ability of plasma membranes to regulate the passage of substances into and out of the cell, allowing some to pass through and preventing the passage of others.

b. __________ Diffusion of water through a differentially permeable membrane.

c. __________ Higher solute concentration (less water) than the cytosol of a cell; causes cell to lose water by osmosis.

d. __________ Measure of the tendency of water to move across a differentially permeable membrane; visible as an increase in liquid on the side of the membrane with higher solute concentration.

e. __________ One of the major lipids found in animal plasma membranes; makes the membrane impermeable to many molecules.

f. __________ Pressure of the cell contents against the cell wall, in plant cells, determined by the water content of the vacuole; gives internal support to the plant cell.

g. __________ Process by which vesicles form around and bring macromolecules into the cell.

h. __________ Protein that combines with and transports a molecule across the plasma membrane.

i. __________ Solution that is equal in solute concentration to that of the cytosol of a cell; causes cell to neither lose nor gain water by osmosis.

j. __________ Carrier protein in the plasma membrane that moves sodium ions out of and potassium ions into animal cells; important in nerve and muscle cells.

Metabolism:
Energy and Enzymes

CHAPTER 6

Chapter Concepts

Painted lady butterfly, *Vanessa cardui*

Living things cannot maintain their organization nor carry on life's other activities without a source of organic food. Green plants utilize solar energy, carbon dioxide, and water to make organic food for themselves and all living things. Animals, like butterflies and human beings, actively seek and eat food.

Food provides nutrient molecules, which are used as a source of building blocks or for energy. Energy is the capacity to do work, and it takes work to maintain the organization of a cell and the organism, including the beautiful wings of this butterfly. When nutrient molecules are broken down completely, they provide the necessary energy to make ATP (adenosine triphosphate), a molecule that fuels the chemical reactions in cells. The use of ATP in all cells is substantial evidence of the relatedness of all life-forms.

Other nutrient molecules become the building blocks that are used by cells to make their parts and products. Metabolism is all the chemical reactions that occur in a cell. Enzymes are protein molecules that speed metabolic reactions in cells at a relatively low temperature. This chapter deals with energy and enzymes, two essential requirements for cellular metabolism.

6.1 Energy

Living things can't grow, reproduce, or exhibit any of the characteristics of life without a ready supply of energy. **Energy,** which is the capacity to do work, occurs in many forms: light energy comes from the sun; electrical energy powers kitchen appliances; and heat energy warms our houses. **Kinetic energy** is the energy of motion. All moving objects have kinetic energy. Thrown baseballs, falling water, and contracting muscles have kinetic energy. **Potential energy** is stored energy. Water behind a dam, or a rock at the top of a hill, or ATP, has potential energy that can be converted to kinetic energy. The energy of a chemical is in the interactions of atoms, one to the other, in the molecule.

Not only do chemicals have potential energy, they also have varying amounts of potential energy. Glucose has much more energy than its breakdown products, carbon dioxide and water.

Two Laws of Thermodynamics

Early researchers who first studied energy and its relationships and exchanges formulated two laws of thermodynamics. *The first law, also called the "law of conservation of energy," says that energy cannot be created or destroyed but can only be changed from one form to another.* Think of the conversions that occur when coal is used to power a locomotive. First, the chemical energy of coal is converted to heat energy and then heat energy is converted to kinetic energy in a steam engine. Similarly, the potential energy of coal or gas is converted to electrical energy by power plants. Do energy transformations occur in the human body? As an example, consider that the chemical energy in the food we eat is changed to the chemical energy of ATP, and then this form of potential energy is converted to the mechanical energy of muscle contraction (Fig. 6.1).

The second law of thermodynamics says that energy cannot be changed from one form to another without a loss of usable energy. Only about 25% of the chemical energy of gasoline is converted to the motion of a car; the rest is lost as heat. Heat, of course, is a form of energy, but heat is the most random form of energy and quickly dissipates into the environment. Or as we have already mentioned, when muscles convert the chemical energy within ATP to the mechanical energy of contraction, some of this energy becomes heat right away. With conversion upon conversion, eventually all usable forms of energy become heat that is lost to the environment. And because heat dissipates, it can never be converted back to a form of potential energy.

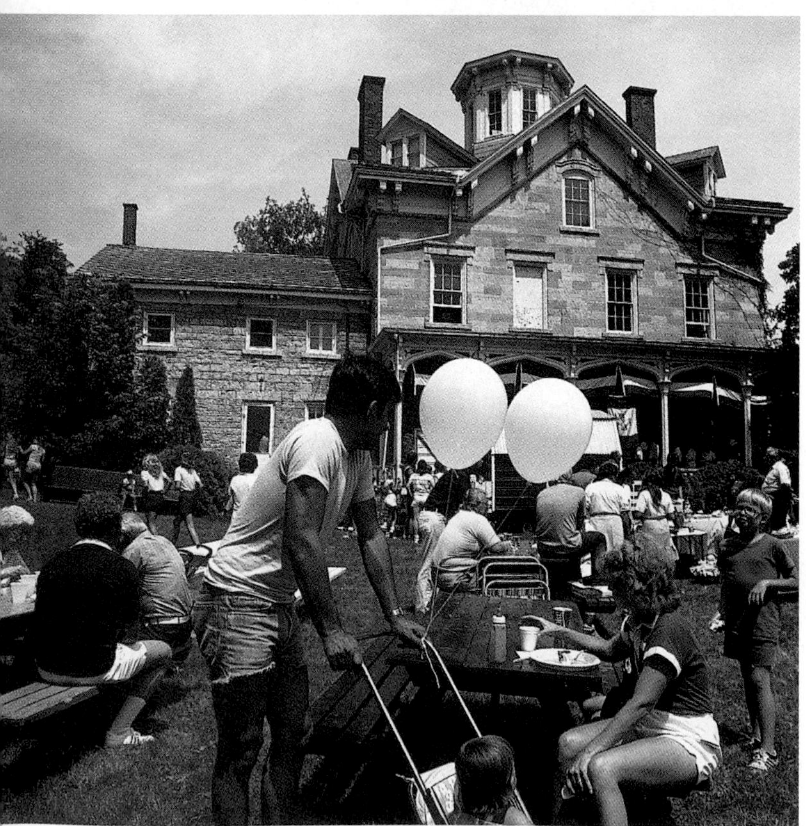

Figure 6.1 Picnickers.
After humans consume food, they often convert its chemical energy into the mechanical energy of muscle contraction.

Entropy

Entropy is a measure of randomness or disorder. An organized, usable form of energy has a low entropy, whereas an unorganized, less stable form of energy such as heat has a high entropy. A neat room has a much lower entropy than a messy room. We know that a neat room always tends toward messiness. In the same way, energy conversions eventually result in heat, and therefore the entropy of the universe is always increasing.

How does an ordered system such as a neat room or an organism come about? You know very well that it takes an input of usable energy to keep your room neat. In the same way, it takes a constant input of usable energy from the food you eat to keep you organized. This input of energy goes through many energy conversions, and the output is finally heat, which increases the entropy of the universe.

A civilized society such as ours uses a great deal of low entropy energy. Fossil fuel energy, such as coal and oil, is used to grow and process the food you need and is used to keep your environment ordered. Building houses and schools and roads all require an input of usable energy, which is finally converted to heat. Our civilized society is increasing the entropy of the universe at a much higher rate than any society in the past.

The laws of thermodynamics explain why the entropy of the universe spontaneously increases and why organisms need a constant input of usable energy to maintain their organization.

6.2 Metabolic Reactions and Energy Transformations

Metabolism is the sum of all the biochemical pathways of the cell. In the reaction A + B → C + D, A and B are the reactants while C and D are the products. How would you know that this reaction will occur spontaneously; that is, without an input of energy? Using the concept of entropy, it is possible to state that a reaction will occur spontaneously if it increases the entropy of the universe. But this is not very helpful in cell biology because we don't wish to consider the entire universe. We simply want to consider this reaction. In such instances, cell biologists use the concept of free energy. **Free energy** is the amount of energy available; that is, energy that is still "free" to do work after a chemical reaction has occurred. Josiah Gibbs, an American physicist, developed the concept of free energy in the 1800s. Therefore, free energy is denoted by the symbol G and is measured as ΔG (delta G). A negative ΔG means that the products have less free energy than the reactants and the reaction will occur spontaneously. In our reaction, if C and D have less free energy than A and B, then the reaction will "go."

Exergonic reactions are ones in which ΔG is negative and energy is released, while endergonic reactions are ones in which the products have more free energy than the reactants. Endergonic reactions can only occur if there is an input of energy.

If the free energy difference in both directions is just about zero, the reaction is reversible and the reaction is at equilibrium. How could you make a reversible reaction "go" in one direction or the other? Very often in cells, as soon as a product is formed, the product is used as a reactant in another reaction. Such occurrences pull the reaction in one direction as opposed to the other direction.

Coupling Reactions

Can the energy released by an exergonic reaction be used to "drive" an endergonic reaction? In the body many reactions such as protein synthesis, nerve conduction, or muscle contraction are endergonic: they require an input of energy. On the other hand, the breakdown of ATP to ADP + Ⓟ is exergonic and energy is released:

ATP → ADP + Ⓟ + energy

Coupling occurs when the energy released by an exergonic reaction is used to drive an endergonic reaction. ATP breakdown is often coupled to cellular reactions that require an input of energy. Coupling, which requires that the exergonic reaction and the endergonic reaction be closely tied, can be symbolized like this:

ATP → ADP + Ⓟ

C + D → A + B

Coupling

How is a cell assured of a supply of ATP? Recall that glucose breakdown during aerobic respiration provides the energy for the buildup of ATP in mitochondria. Only 39% of the free energy of glucose is transformed to ATP; the rest is lost as heat. When ATP breaks down to drive the reactions mentioned, some energy is lost as heat and the overall reaction becomes exergonic (Fig. 6.2).

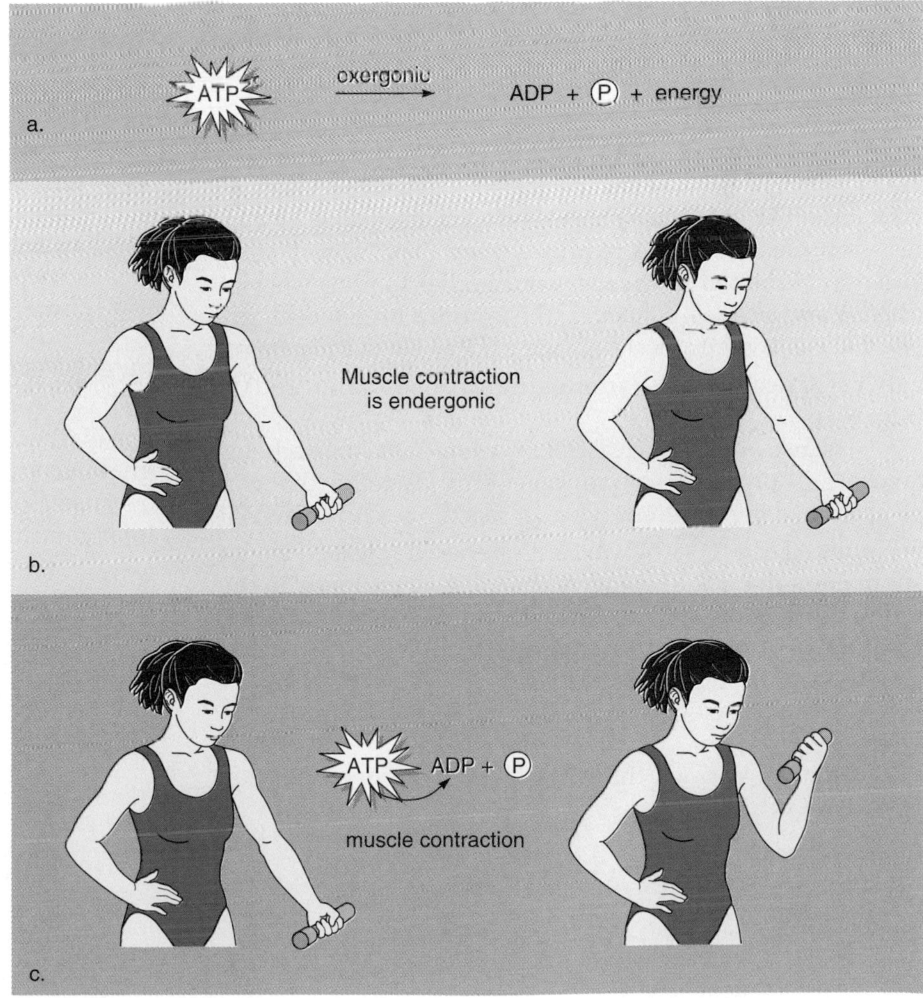

Figure 6.2 Coupled reactions.
a. The breakdown of ATP is exergonic. **b.** Muscle contraction is endergonic and therefore cannot occur without an input of energy. **c.** Muscle contraction is coupled to ATP breakdown, making the overall process exergonic. Now muscle contraction can occur.

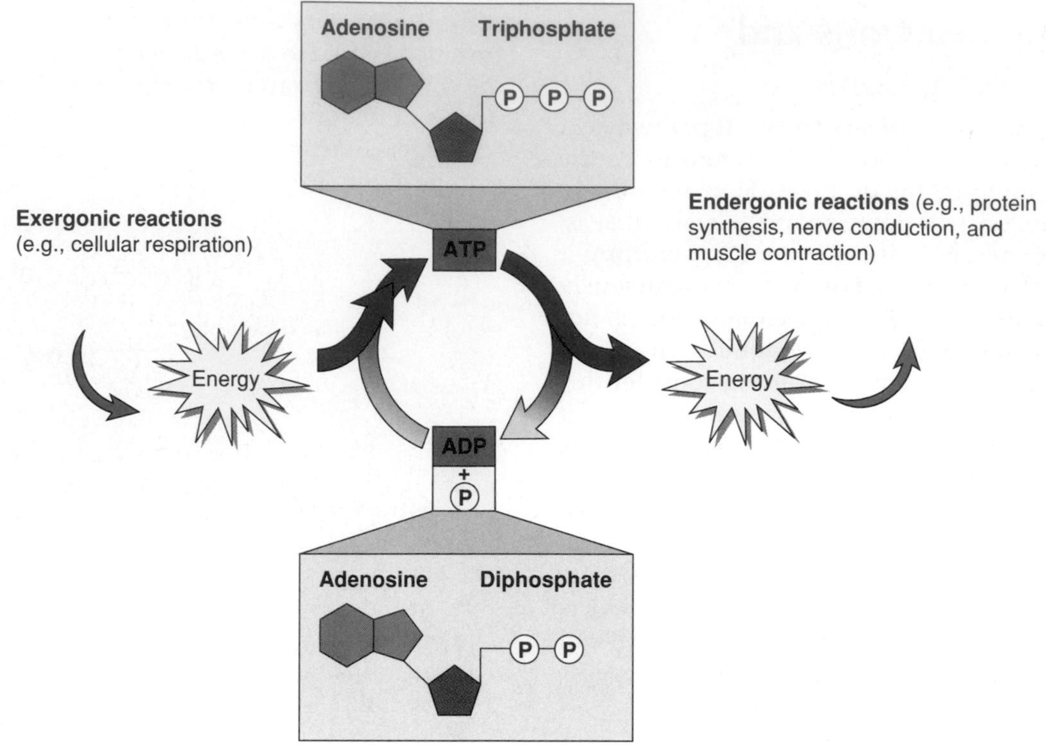

Figure 6.3 The ATP cycle.
In cells, the exergonic breakdown of glucose is coupled to the buildup of ATP, and then the exergonic breakdown of ATP is coupled to endergonic reactions in cells. ATP releases the appropriate amount of energy for most metabolic reactions when a phosphate group is removed by hydrolysis. The high-energy content of ATP comes from the complex interaction of the atoms within the molecule.

ATP: Energy for Cells

ATP (adenosine triphosphate) is the common energy currency of cells: when cells require energy, they "spend" ATP. You may think that this causes our bodies to produce a lot of ATP, and it does; however, the amount on hand at any one moment is minimal because ATP is constantly being generated from **ADP (adenosine diphosphate)** and Ⓟ (Fig. 6.3).

The use of ATP as a carrier of energy has some advantages: (1) It provides a common energy currency that can be used in many different types of reactions. (2) When ATP becomes ADP + Ⓟ, the amount of energy released is just about enough for the biological purposes mentioned in the following section, and so little energy is wasted. (3) ATP breakdown is coupled to endergonic reactions in such a way that it minimizes energy loss.

Function of ATP

Recall that at various times we have mentioned at least three uses for ATP.

Chemical work. Supplies the energy needed to synthesize macromolecules that make up the cell.

Transport work. Supplies the energy needed to pump substances across the plasma membrane.

Mechanical work. Supplies the energy needed to permit muscles to contract, cilia and flagella to beat, chromosomes to move, and so forth.

Structure of ATP

ATP is a nucleotide composed of the base adenine and the sugar ribose (together called adenosine) and three phosphate groups. ATP is called a "high-energy" compound because a phosphate group is easily removed. Under cellular conditions, the amount of energy released when ATP is hydrolyzed to ADP + Ⓟ is about 7.3 kcal per mole.[1]

ATP is a carrier of energy in cells. It is the common energy currency because it supplies energy for many different types of reactions.

[1] A mole is the number of molecules present in the molecular weight of a substance (in grams).

6.3 Metabolic Pathways and Enzymes

Reactions do not occur haphazardly in cells; they are usually a part of a *metabolic pathway,* a series of linked reactions. Metabolic pathways begin with a particular reactant and terminate with an end product. While it is possible to write an overall equation for a pathway as if the beginning reactant went to the end product in one step, there are actually many specific steps in between. In the pathway, one reaction leads to the next reaction, which leads to the next reaction, and so forth in an organized, highly structured manner. This arrangement makes it possible for one pathway to lead to several others, because various pathways have several molecules in common. Also, metabolic energy is captured and utilized more easily if it is released in small increments rather than all at once.

A metabolic pathway can be represented by the following diagram:

$$A \xrightarrow{E_1} B \xrightarrow{E_2} C \xrightarrow{E_3} D \xrightarrow{E_4} E \xrightarrow{E_5} F \xrightarrow{E_6} G$$

In this diagram, the letters A–F are reactants and letters B–G are products in the various reactions. A *reactant* is a substance that participates in a reaction. A *product* is a substance that is formed as a result of a chemical reaction. In the diagram, the letters E_1–E_6 are enzymes.

In most instances, an **enzyme** is a protein molecule[2] that functions as an organic catalyst to speed a chemical reaction. In a crowded ballroom, a mutual friend can cause particular people to interact. In the cell, an enzyme brings together particular molecules and causes them to react with one another.

The reactants in an enzymatic reaction are called the **substrates** for that enzyme. In the first reaction, A is the substrate for E_1 and B is the product. Now B becomes the substrate for E_2, and C is the product. This process continues until the final product G forms.

Any one of the molecules (A–G) in this linear pathway could also be a substrate for an enzyme in another pathway. A diagram showing all the possibilities would be highly branched.

Enzymes Lower the Energy of Activation

Molecules frequently do not react with one another unless they are activated in some way. In the absence of an enzyme, activation is very often achieved by heating the reaction flask to increase the number of effective collisions between molecules. The energy that must be added to cause molecules to react with one another is called the *energy of activation* (E_a). Figure 6.4 compares E_a when an enzyme is not present to when an enzyme is present, illustrating that enzymes lower the amount of energy required for activation to occur.

In baseball, a home-run hitter not only must hit the ball to the fence, the ball must also go over the fence. When enzymes lower the energy of activation, it is like removing the fence; then it is possible to get a home run by simply hitting the ball as far as the fence was.

[2] The discovery of catalytic RNA molecules means that not all enzymes are proteins.

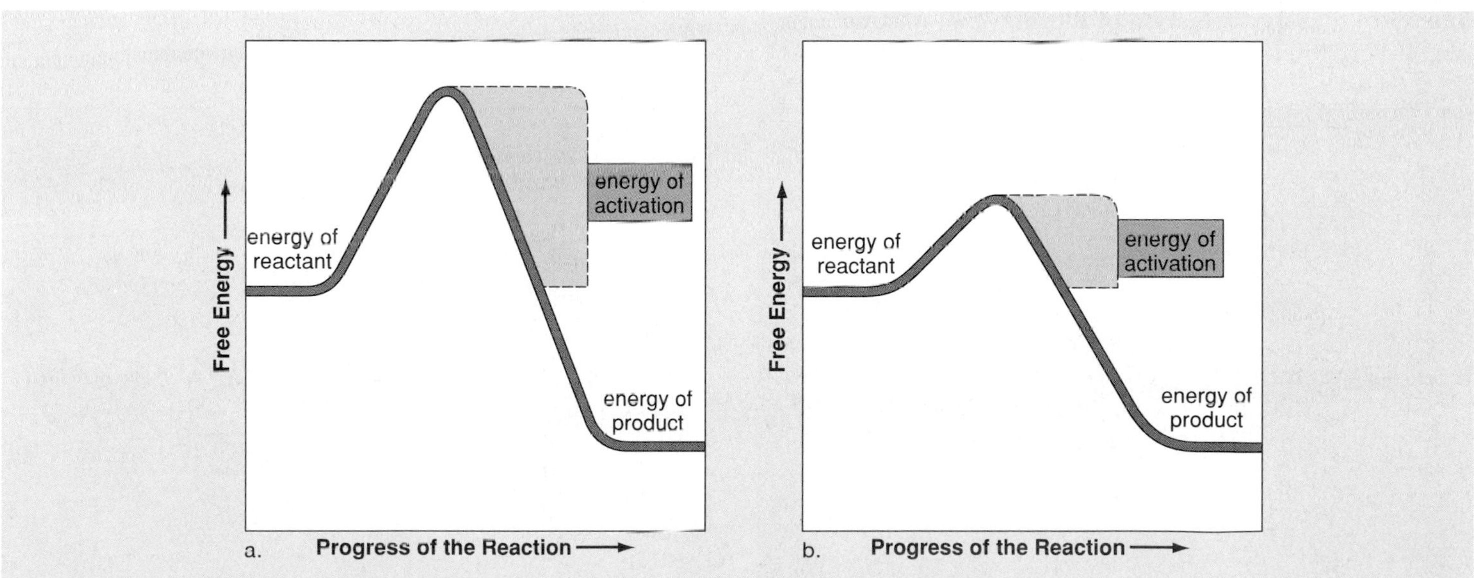

Figure 6.4 Energy of activation (E_a).
Enzymes speed the rate of chemical reactions because they lower the amount of energy required to activate the reactants. **a.** Energy of activation when an enzyme is not present. **b.** Energy of activation when an enzyme is present.

Enzyme-Substrate Complexes

The following equation, which is pictorially shown in Figure 6.5, is often used to indicate that an enzyme forms a complex with its substrate:

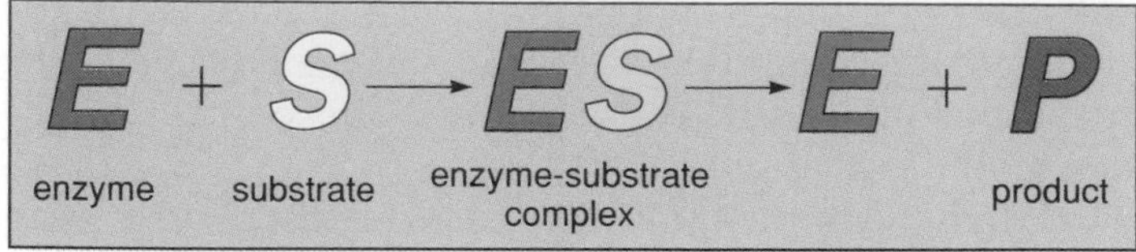

In most instances only one small part of the enzyme, called the **active site,** complexes with the substrate(s). It is here that the enzyme and substrate fit together, seemingly like a key fits a lock; however, it is now known that the active site undergoes a slight change in shape in order to accommodate the substrate(s). This is called the *induced fit model* because the enzyme is induced to undergo a slight alteration to achieve optimum fit.

The change in shape of the active site facilitates the reaction that now occurs. After the reaction has been completed, the product(s) is released, and the active site returns to its original state, ready to bind to another substrate molecule. Only a small amount of enzyme is actually needed in a cell because enzymes are not used up by the reaction.

Some enzymes do more than simply complex with their substrate(s); they actually participate in the reaction. Trypsin digests protein by breaking peptide bonds. The active site of trypsin contains three amino acids with *R* groups that actually interact with members of the peptide bond—first to break the bond and then to introduce the components of water. This illustrates that the formation of the enzyme-substrate complex is very important in speeding up the reaction.

Sometimes it is possible for a particular reactant(s) to produce more than one type of product(s). The presence or absence of an enzyme determines which reaction takes place. If a substance can react to form more than one product, then the enzyme that is present and active determines which product is produced.

Every reaction in a cell requires its specific enzyme. Because enzymes only complex with their substrates, they are named for their substrates, as in the following examples:

Substrate	Enzyme
Lipid	Lipase
Urea	Urease
Maltose	Maltase
Ribonucleic acid	Ribonuclease
Lactose	Lactase

Most enzymes are protein molecules. Enzymes speed chemical reactions by lowering the energy of activation. They do this by forming an enzyme-substrate complex.

Factors That Affect Enzymatic Speed

Enzymatic reactions proceed quite rapidly. Consider, for example, the breakdown of hydrogen peroxide (H_2O_2) as catalyzed by the enzyme catalase: $2\ H_2O_2 \rightarrow 2\ H_2O + O_2$. The breakdown of hydrogen peroxide can occur 600,000 times a second when catalase is present. To achieve maximum product per unit time, there should be enough substrate to fill active sites most of the time. Temperature and optimal pH also increase the rate of an enzymatic reaction.

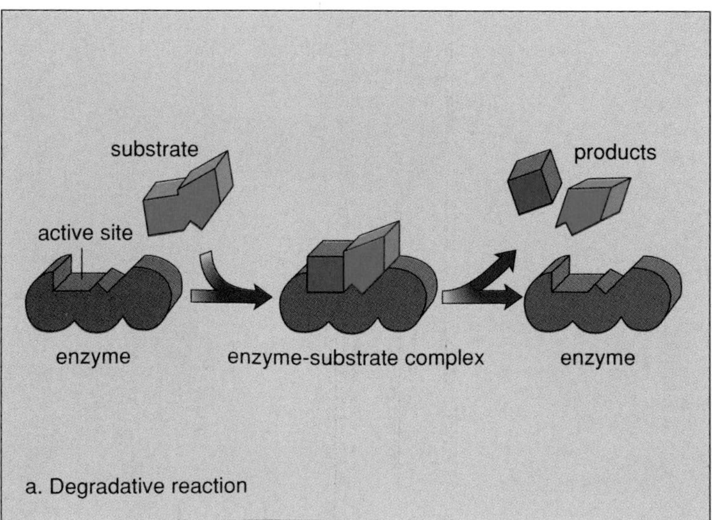

a. Degradative reaction

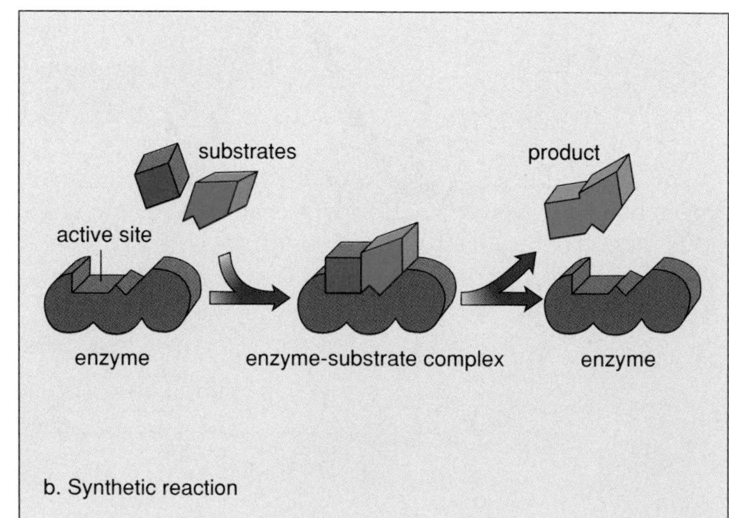

b. Synthetic reaction

Figure 6.5 Enzymatic action.
An enzyme has an active site, which is where the substrates and enzyme fit together in such a way that the substrates are oriented to react. Following the reaction, the products are released and the enzyme is free to act again. **a.** Some enzymes carry out degradative reactions in which the substrate is broken down to smaller molecules. **b.** Other enzymes carry out synthetic reactions in which the substrates are joined to form a larger molecule.

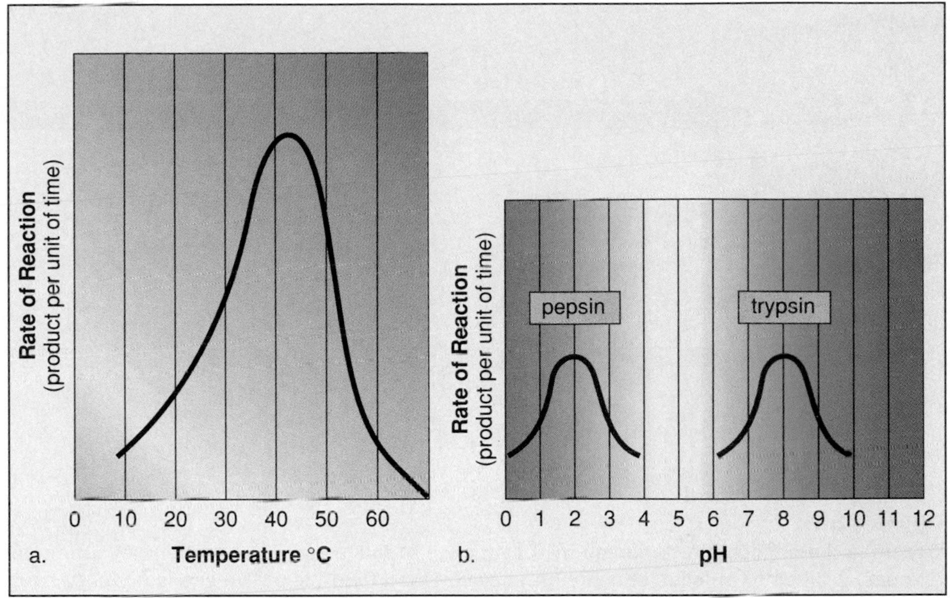

Figure 6.6 Rate of an enzymatic reaction as a function of temperature and pH.
a. At first, as with most chemical reactions, the rate of an enzymatic reaction doubles with every 10°C rise in temperature. In this graph, the rate of reaction is maximum at about 40°C; then it decreases until the reaction stops altogether, because the enzyme has become denatured. **b.** Pepsin, an enzyme found in the stomach, acts best at a pH of about 2, while trypsin, an enzyme found in the small intestine, performs optimally at a pH of about 8. The shape that enables these proteins to bind with their substrates is not properly maintained at other pHs.

Moderate Temperature and Optimal pH Is Best

As the temperature rises, enzyme activity increases (Fig 6.6*a*). This occurs because as the temperature rises there are more effective collisions between enzyme and substrate. However, if the temperature rises beyond a certain point, enzyme activity eventually levels out and then declines rapidly because the enzyme is **denatured.** An enzyme's shape changes during denaturation, and then it can no longer bind its substrate(s) efficiently.

Each enzyme also has an optimal pH at which the rate of the reaction is highest. Figure 6.6*b* shows the optimal pH for the enzymes pepsin and trypsin. At this pH value, these enzymes have their normal configurations. The globular structure of an enzyme is dependent on interactions, such as hydrogen bonding, between *R* groups. A change in pH can alter the ionization of these side chains and disrupt normal interactions, and under extreme conditions of pH, denaturation eventually occurs. Again, the enzyme has an altered shape and is then unable to combine efficiently with its substrate.

Amount of Active Enzyme Affects Speed

Since enzymes are specific, a cell regulates which enzymes are present and/or active at any one time. Otherwise enzymes may be present that are not needed, or one pathway may negate the work of another pathway.

Genes must be turned on to increase the concentration of an enzyme and must be turned off to decrease the concentration of an enzyme.

Another way to control enzyme activity is to activate or deactivate the enzyme. Phosphorylation is one way to activate an enzyme. Molecules received by membrane receptors often turn on kinases, which then activate enzymes by phosphorylating them:

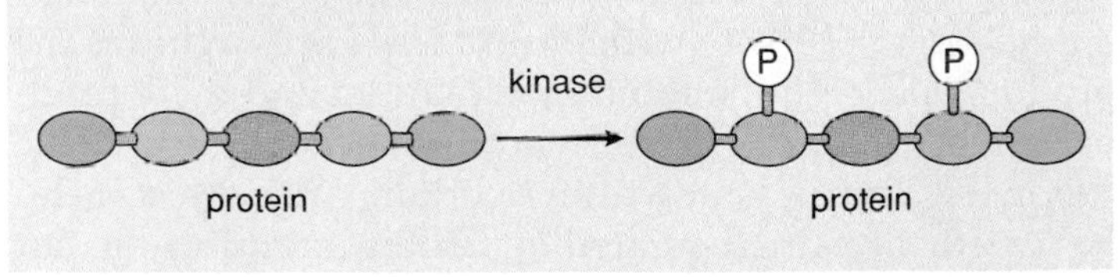

Other enzymes called phosphatases remove phosphate groups from enzymes. Processes controlled by phosphorylation and dephosphorylation include protein synthesis, cell division, activity of motor molecules, nuclear envelope breakdown, and development. Mutated genes that convert normal cells to cancerous cells often code for kinases necessary to cell division. On the other hand, the toxin produced by dinoflagellates that causes the often fatal paralytic seafood poisoning is a potent phosphatase inhibitor.

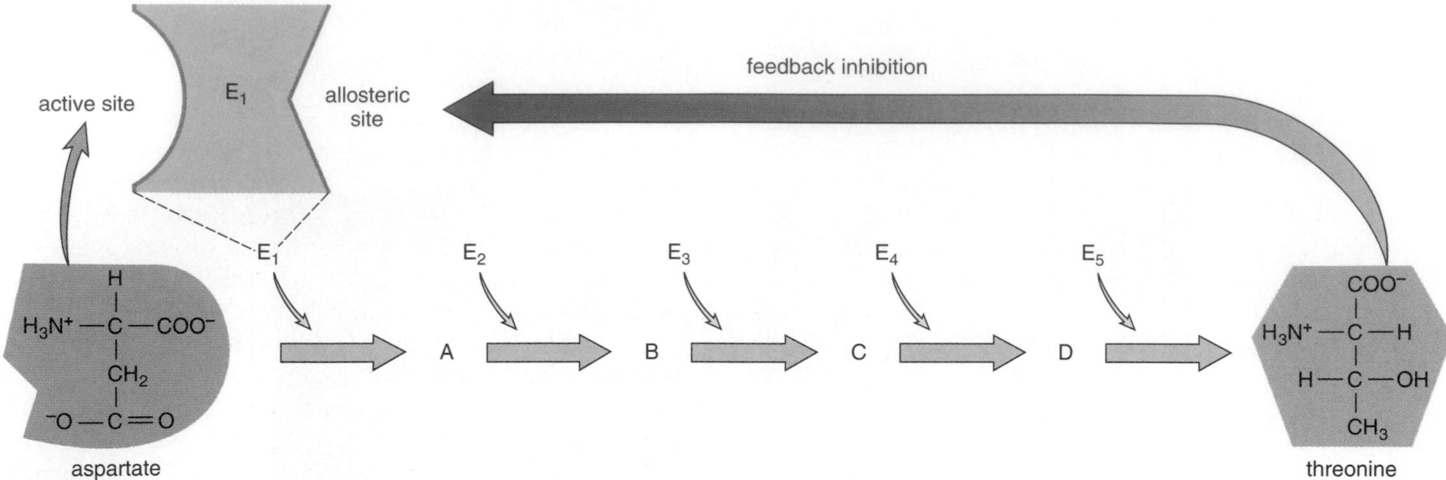

Figure 6.7 Feedback inhibition.
The amino acid aspartate becomes the amino acid threonine by a sequence of five enzymatic reactions. When threonine, the end product of this pathway, is present in excess, it binds to an allosteric site on enzyme 1 and then the active site is no longer able to bind aspartate.

Enzymes Can Be Inhibited

Actually, *inhibition* is a common means by which cells regulate enzyme activity. In *competitive inhibition*, another molecule is so close in shape to the enzyme's substrate that it can compete with the true substrate for the enzyme's active site. This molecule inhibits the reaction because only the binding of the true substrate results in a product. In *noncompetitive inhibition*, a molecule binds to an enzyme, but not at the active site. The other binding site is called the *allosteric site* (*allo*—other; *steric*—space or structure). In this instance, inhibition occurs when binding of a molecule causes a shift in the three-dimensional structure so that the substrate cannot bind to the active site.

The activity of almost every enzyme in a cell can be regulated by its product. When a product is in abundance, it binds competitively with its enzyme's active site; as the product is used up, inhibition is reduced and more product can be produced. In this way, the concentration of the product is always kept within a certain range. Most metabolic pathways are regulated by **feedback inhibition,** but the end product of the pathway binds at an allosteric site on the first enzyme of the pathway (Fig. 6.7). This binding shuts down the pathway, and no more product is produced.

In inhibition, a product binds to the active site or binds to an allosteric site on an enzyme.

Poisons are often enzyme inhibitors. Cyanide is an inhibitor for an essential enzyme (cytochrome oxidase) in all cells, which accounts for its lethal effect on humans. Penicillin blocks the active site of an enzyme unique to bacteria. When penicillin is taken, bacteria die but humans are unaffected.

Cofactors Help Enzymes

Many enzymes require an inorganic ion or organic but nonprotein molecule to function properly; these necessary ions or molecules are called **cofactors.** The inorganic ions are metals such as copper, zinc, or iron. The organic, nonprotein molecules are called **coenzymes.** These cofactors assist the enzyme and may even accept or contribute atoms to the reactions.

It is interesting that vitamins are often components of coenzymes. **Vitamins** are relatively small organic molecules that are required in trace amounts in our diet and in the diet of other animals for synthesis of coenzymes that affect health and physical fitness. The vitamin becomes a part of the coenzyme's molecular structure. These vitamins are necessary to formation of the coenzymes listed:

Vitamin	***Coenzyme***
Niacin	NAD^+
B_2 (riboflavin)	FAD
B_1 (thiamine)	Thiamine pyrophosphate
Pantothenic acid	Coenzyme A (CoA)
B_{12} (cobalamin)	B_{12} coenzymes

A deficiency of any one of these vitamins results in a lack of the coenzyme listed and therefore a lack of certain enzymatic actions. In humans, this eventually results in vitamin-deficiency symptoms: niacin deficiency results in a skin disease called pellagra, and riboflavin deficiency results in cracks at the corners of the mouth.

Coenzymes are nonprotein molecules that assist enzymes in performing their reactions.

6.4 Metabolic Pathways and Living Things

A flow of energy occurs through organisms and through the world of living things (see Fig. 6.11). In eukaryotes, the metabolic pathways of photosynthesis and aerobic respiration transform one form of energy into another. When chloroplasts carry on photosynthesis, solar energy is converted to the energy of carbohydrates; and when mitochondria complete aerobic respiration, the energy stored in carbohydrates is converted to energy temporarily held by ATP. The energy released by ATP breakdown is used by the cell to do various types of work, and eventually it becomes nonusable heat. It cannot be used to re-form ATP. Eventually, all solar energy captured by chloroplasts is converted to heat. Without a constant input of solar energy, life could not exist.

The overall equations for photosynthesis and aerobic respiration are often written in the following manner in order to show they are opposite processes.

$$6\,CO_2 + 6\,H_2O \underset{\text{aerobic respiration}}{\overset{\text{photosynthesis}}{\longleftrightarrow}} C_6H_{12}O_6 + 6\,O_2$$

Glucose in this reaction stands for carbohydrate; photosynthesis is the building up of glucose, while aerobic respiration is the breaking down of glucose.

Electron Transport System

Both photosynthesis and aerobic respiration make use of an electron transport system, which is a series of membrane-bound carriers that pass electrons from one carrier to another. High-energy electrons are delivered to the system and low-energy electrons leave it. Every time electrons are transferred to a new carrier, energy is released; this energy is ultimately used to produce ATP molecules (Fig. 6.8).

Oxidation, you will recall, is the loss of electrons, and reduction is the gain of electrons. Reduction/oxidation (redox) reactions are a major way in which energy transformations occur in cells. In certain redox reactions, the overall result is release of energy, and in others, energy is required. In an electron transport system, each carrier is reduced and then oxidized in turn. The overall effect of oxidation/reduction as electrons are passed from carrier to carrier of the electron transport system is the release of energy for ATP production.

$NADP^+$ and NAD^+

The molecules **$NADP^+$** (nicotinamide adenine dinucleotide phosphate) and **NAD^+** (nicotinamide adenine dinucleotide) are coenzymes of oxidation/reduction. (Notice that $NADP^+$ contains a phosphate group that is lacking in NAD^+.) During photosynthesis, NADPH donates hydrogen atoms ($H^+ + e^-$) to a substrate and it becomes reduced. During aerobic respiration, NAD^+ accepts electrons from a substrate and it becomes oxidized. As NAD^+ accepts electrons, it also takes on a hydrogen ion and becomes NADH.

In chloroplasts, the electrons that enter the electron transport system are taken from water and then energized by solar energy. Following electron transport, the coenzyme $NADP^+$ accepts electrons and an H^+ to become NADPH. NADPH supplies the necessary electrons and ATP supplies the necessary energy to bring about the reduction of carbon dioxide to carbohydrate as the metabolic reactions of photosynthesis proceed.

In mitochondria, the high-energy electrons that enter the electron transport system have been removed from carbohydrate substrates by the coenzyme called NAD^+. The end result of aerobic respiration is the oxidation of glucose to carbon dioxide and water with the concomitant buildup of ATP molecules by way of the electron transport system.

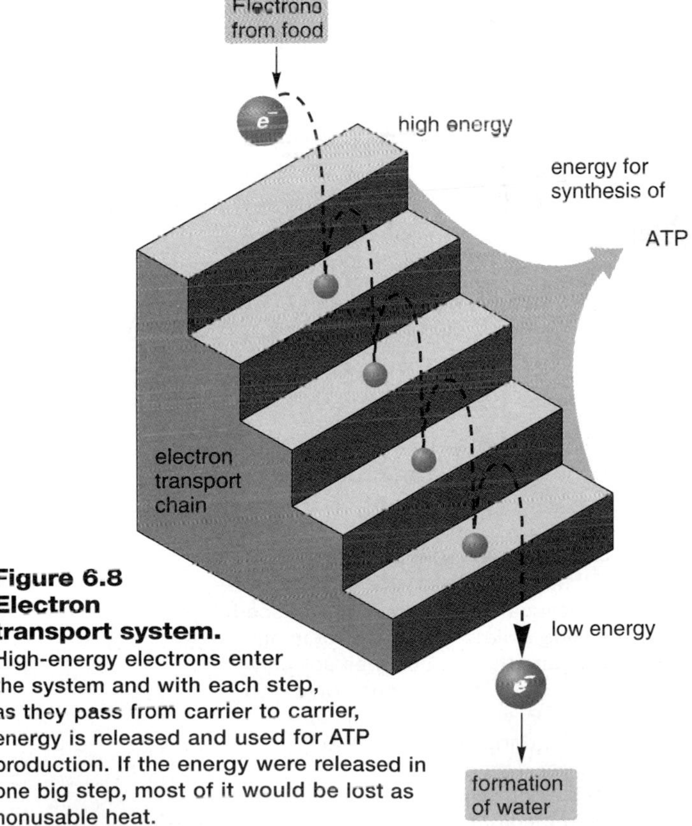

Figure 6.8 Electron transport system. High-energy electrons enter the system and with each step, as they pass from carrier to carrier, energy is released and used for ATP production. If the energy were released in one big step, most of it would be lost as nonusable heat.

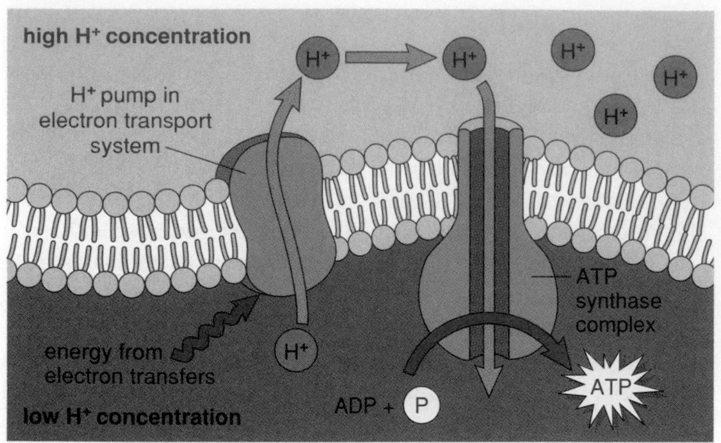

Figure 6.9 Chemiosmosis.
Carriers in the electron transport system pump hydrogen ions (H^+) across a membrane. When the hydrogen ions flow back across the membrane through a protein complex, ATP is synthesized by an enzyme called ATP synthase. Chemiosmosis occurs in mitochondria and chloroplasts.

ATP Production

For many years, it was known that ATP synthesis was somehow coupled to the electron transport system, but the exact mechanism could not be determined. Peter Mitchell, a British biochemist, received a Nobel Prize in 1978 for his *chemiosmotic theory* of ATP production in both mitochondria and chloroplasts.

In mitochondria and chloroplasts, the carriers of the electron transport system are located within a membrane. Hydrogen ions (H^+), which are often referred to as protons in this context, tend to collect on one side of the membrane because they are pumped there by certain carriers. This establishes an electrochemical gradient across the membrane that can be used to provide energy for ATP production. Particles, called *ATP synthase complexes,* span the membrane. Each complex contains a channel that allows hydrogen ions to flow down their electrochemical gradient. The flow of hydrogen ions through the channel provides the energy for the ATP synthase enzyme to produce ATP from ADP + Ⓟ (Fig. 6.9). **Chemiosmosis** [Gk. *osmos,* push] is the production of ATP due to a hydrogen ion gradient across a membrane.

The chemiosmotic theory of ATP production was supported by a now-famous experiment performed by André Jagendorf of Cornell University utilizing chloroplasts (Fig. 6.10). The experiment shows that ATP production is indeed tied only to a hydrogen ion gradient and not to a reaction that is directly light dependent.

Consider this analogy to understand chemiosmosis. The sun's rays evaporate water from the seas and help create the winds that blow clouds to the mountains, where water falls in the form of rain and snow. The water in a mountain reservoir has a higher potential energy than that of water in the ocean. The potential energy is converted to electrical energy when water is released and used to turn turbines in an electrochemical dam before it makes its way to the ocean. The continual release of water results in a continual production of electricity.

Similarly, the sun's energy collected by chloroplasts continually leads to ATP production. ATP is produced because the inner mitochondria membrane acts like a dam to maintain an energy gradient in the form of a concentration gradient of hydrogen ions. The hydrogen ions flow through membrane channels that couple the flow of hydrogen ions to the formation of ATP, like the turbines in a hydroelectric dam system couple the flow of water to the formation of electricity.

The electron transport system deposits hydrogen ions (H^+) on one side of a membrane. When the ions flow down an electrochemical gradient through an ATPase complex, an enzyme utilizes the release of energy to make ATP from ADP and Ⓟ.

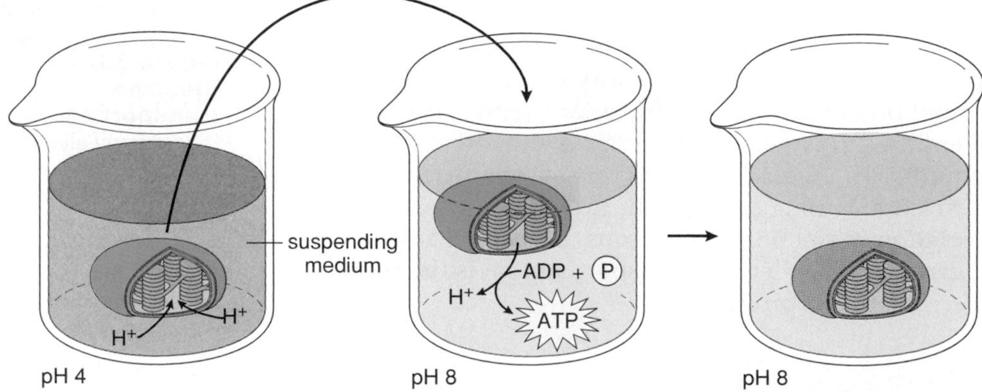

Figure 6.10 An experiment to support the chemiosmotic theory.
An electrochemical gradient was established by first soaking isolated chloroplasts in an acid medium and then the pH was abruptly changed to basic. ATP production occurred after ADP and Ⓟ were added. This all occurred in the dark, proving that ATP synthesis is not directly linked to the chloroplast's electron transport system, which works only in the light.

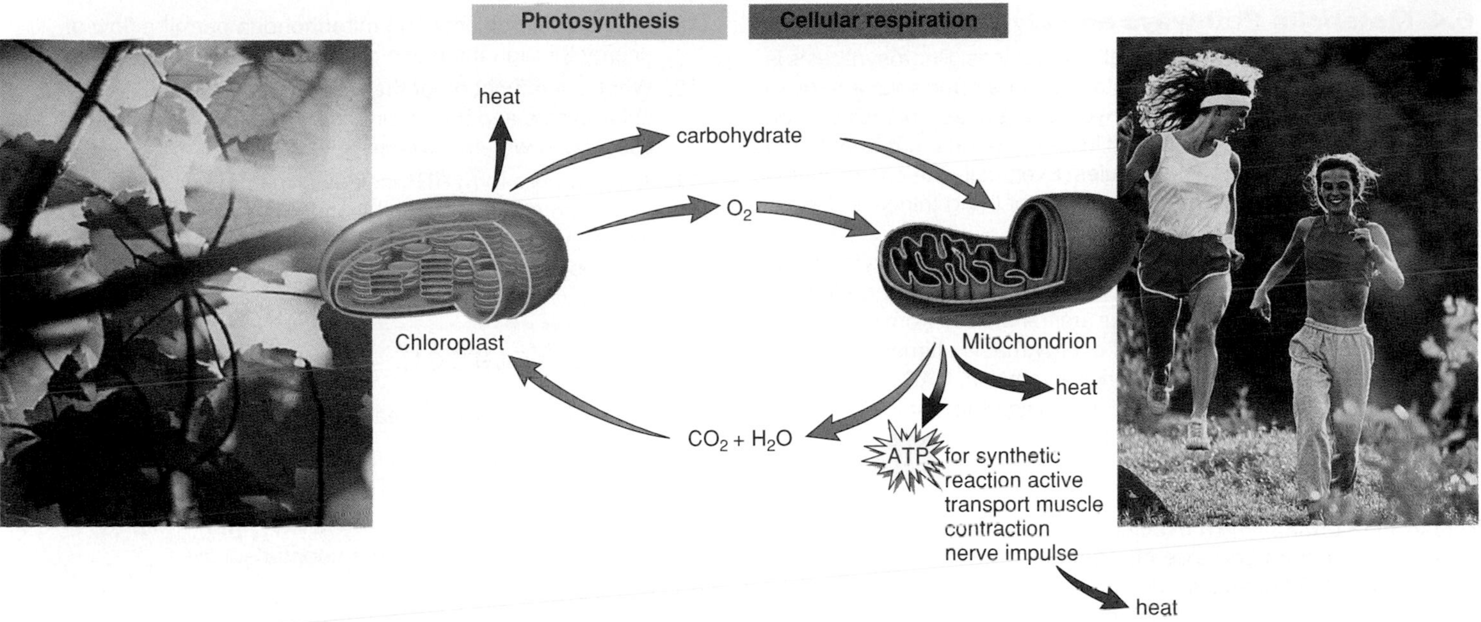

Figure 6.11 Relationship of chloroplasts to mitochondria.
Chloroplasts produce energy-rich carbohydrates. These carbohydrates are broken down in mitochondria, and the energy released is used for the buildup of ATP. There is a loss of usable energy due to the energy conversions of photosynthesis and aerobic respiration; and then, when ATP is used as an energy source, all usable energy is converted to heat.

connecting concepts

All cells use energy. Energy is the ability to do work, to bring about change, to make things happen, whether it's a leaf growing or a human running (Fig. 6.11). The metabolic pathways inside cells utilize the chemical bond energy of ATP to synthesize molecules that cause muscle contraction and even allows us to read these words.

ATP is called the universal energy "currency" of life. This is an apt analogy—before we can spend currency (i.e., money), we must first make some money. Similarly, before we can spend ATP molecules, we must make them. Aerobic respiration in mitochondria transforms the chemical bond energy of carbohydrates to that of ATP molecules. An ATP is spent when it is hydrolyzed and the resulting energy is coupled to an endergonic metabolic reaction. All cells are continually making and breaking down ATPs. If ATP is lacking, the organism dies.

What is the ultimate source of energy for ATP production? Except for a few deep ocean vents and also certain cave communities, the answer is the sun. Photosynthesis inside chloroplasts transforms solar energy into the chemical bond energy of carbohydrates. And then carbohydrate products are broken down in mitochondria as ATP is built up. Chloroplasts and mitochondria are the cellular organelles that permit a flow of energy from the sun through all living things.

Summary

6.1 Energy

There are two energy laws that are basic to understanding energy-use patterns in organisms at the cellular level. The first law states that energy cannot be created or destroyed, but can only be transferred or transformed. The second law states that one usable form of energy cannot be completely converted into another usable form. As a result of these laws, we know that the entropy of the universe is increasing and that only a constant input of energy maintains the organization of living things.

6.2 Metabolic Reactions and Energy Transformations

Metabolism is a term that encompasses all the chemical reactions occurring in a cell. Considering individual reactions, only those that result in a negative free energy difference—that is, the products have less usable energy than the reactants—occur spontaneously. Such reactions, called exergonic reactions, release energy. Endergonic reactions, which require an input of energy, occur only in cells because it is possible to couple an exergonic process with an endergonic process. For example, glucose breakdown is an exergonic metabolic pathway that drives the buildup of many ATP molecules. These ATP molecules then supply energy for cellular work. Thus, ATP goes through a cycle in which it is constantly being built up from, and then broken down to, ADP + Ⓟ.

6.3 Metabolic Pathways and Enzymes

A metabolic pathway is a series of reactions that proceed in an orderly, step-by-step manner. Each reaction requires a specific enzyme. Reaction rates increase when enzymes form a complex with their substrates. Any environmental factor that affects the shape of a protein also affects the ability of an enzyme to do its job. Cellular mechanisms regulate enzyme quantity and activity. The activity of most metabolic pathways is regulated by feedback inhibition. Many enzymes have cofactors or coenzymes that help them carry out a reaction.

6.4 Metabolic Pathways and Living Things

There is a flow of energy through all living things. Photosynthesis is a metabolic pathway in chloroplasts that transforms solar energy to the chemical energy within carbohydrates, and aerobic respiration is a metabolic pathway completed in mitochondria that transforms this energy into that of ATP molecules. Eventually the energy within ATP molecules becomes heat. The world of living things is dependent on a constant input of solar energy.

The overall equation for photosynthesis is the opposite of that for aerobic respiration. Both processes make use of an electron transport system in which electrons are transferred from one carrier to the next one with the release of energy that is ultimately used to produce ATP molecules. During photosynthesis, $NADP^+$ is a coenzyme that reduces substrates, and during aerobic respiration, NAD^+ is a coenzyme that oxidizes substrates. Redox reactions are a major way in which energy transformation occurs in cells.

The chemiosmotic hypothesis explains how the electron transport system produces ATP. The carriers of this system deposit hydrogen ions (H^+) on one side of a membrane. When the ions flow down an electrochemical gradient through an ATPase complex, an enzyme utilizes the release of energy to make ATP from ADP and Ⓟ.

Reviewing the Chapter

1. State the first law of thermodynamics and give an example. 100
2. State the second law of thermodynamics and give an example. 100
3. Explain why the entropy of the universe is always increasing and why an organized system like an organism requires a constant input of useful energy. 100
4. What is the difference between exergonic reactions and endergonic reactions? Why can exergonic but not endergonic reactions occur spontaneously? 101
5. Define coupling and write an equation that shows an endergonic reaction being coupled to ATP breakdown. 101
6. Why is ATP called the energy currency of cells? What is the ATP cycle? 101–2
7. Diagram a metabolic pathway. Label the reactants, products, and enzymes. 103
8. Why is less energy needed for a reaction to occur when an enzyme is present? 103
9. Why are enzymes specific, and why can't each one speed up many different reactions? 104
10. Name and explain the manner in which at least three factors can influence the speed of an enzymatic reaction. How do cells regulate the activity of enzymes? 104–6
11. What are cofactors and coenzymes? 106
12. How do chloroplasts and mitochondria permit a flow of energy through the world of living things? 107
13. What is the function of the electron transport system in chloroplasts and mitochondria? What redox coenzymes are associated with this system? 107
14. Tell how cells form ATP, including the process of chemiosmotic phosphorylation. Tell how ATP is used and explain why the energy released from ATP breakdown cannot be used for ATP buildup. 108

Testing Yourself

Choose the best answer for each question.

1. Exergonic reactions
   a. always make use of ATP.
   b. defy the second law of thermodynamics.
   c. release energy and occur spontaneously.
   d. All of these are correct.
2. Enzymes
   a. make it possible for cells to escape the need for energy.
   b. are nonprotein molecules that help coenzymes.
   c. are not affected by a change in pH.
   d. lower the energy of activation.
3. An allosteric site on an enzyme is
   a. the same as the active site.
   b. where ATP attaches and gives up its energy.
   c. often involved in feedback inhibition.
   d. All of these are correct.
4. A high temperature
   a. can affect the shape of an enzyme.
   b. lowers the energy of activation.
   c. makes cells less susceptible to disease.
   d. Both a and c are correct.
5. Electron transport systems
   a. are found in both mitochondria and chloroplasts.
   b. release energy as electrons are transferred.
   c. are involved in the production of ATP.
   d. All of these are correct.
6. The difference between NAD^+ and $NADP^+$ is that
   a. only NAD^+ production requires niacin in the diet.
   b. one contains high-energy phosphate bonds and the other does not.
   c. one carries electrons to the electron transport system and the other carries them to synthetic reactions.
   d. All of these are correct.

7. ATP
   a. is used only in animal cells and not in plant cells.
   b. carries energy between exergonic pathways and endergonic pathways.
   c. is needed for chemical work, mechanical work, and transport work.
   d. Both b and c are correct.
8. Chemiosmosis is dependent upon
   a. the diffusion of water across a differentially permeable membrane.
   b. an outside supply of phosphate and other chemicals.
   c. the establishment of an electrochemical hydrogen ion (H^+) gradient.
   d. the ability of ADP to join with Ⓟ even in the absence of a supply of energy.
9. Use these terms to label this diagram: substrates, enzyme (used twice), active site, product, and enzyme-substrate complex. Explain the importance of an enzyme's shape to its activity.

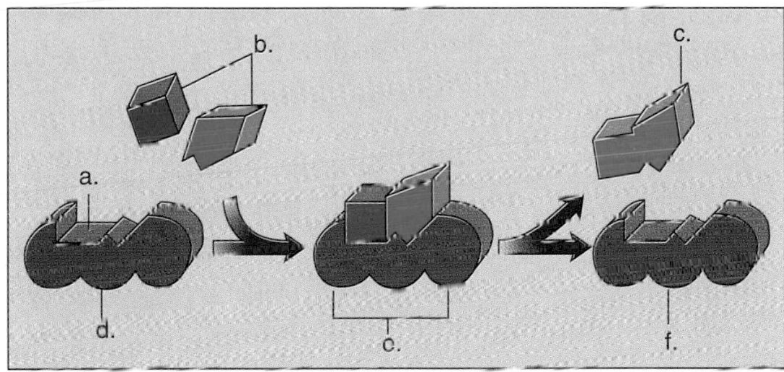

10. Label this diagram describing chemiosmosis.

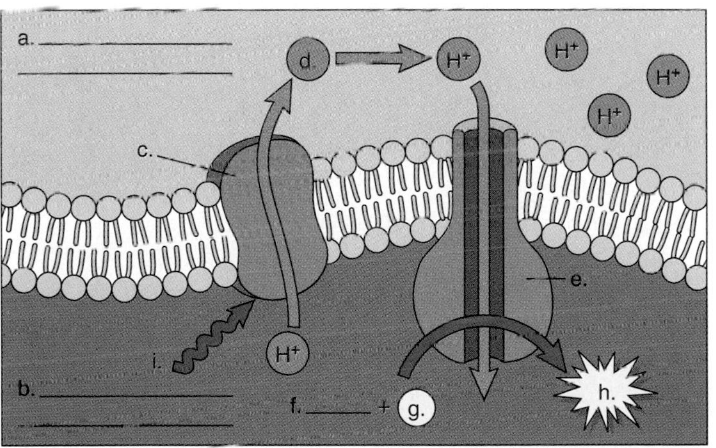

Applying the Concepts

1. *A constant input of energy is required to maintain the organization of the cell.*
   In reference to work performed by the rough ER and the Golgi apparatus, explain specifically why a cell needs energy to maintain its organization.
2. *In keeping with the first and second laws of thermodynamics, a cell takes in useful energy and gives off heat.*
   Explain why you would expect both parts of the ATP cycle to be responsible for the loss of useful energy through heat.
3. *The actions of enzymes allow the cell to maintain itself and grow.*
   Explain why most reactions do not occur in a cell unless a specific enzyme is present.

Using Technology

Your study of cellular energy is supported by these available technologies:

Exploring the Internet

The Mader Home Page provides resources for and help with studying this chapter.

http://www.mhhe.com/sciencemath/biology/mader/
(Click on Biology.)

Explorations in Cell Biology & Genetics CD-ROM

Cell Chemistry: Thermodynamics (#6)
Oxidative Respiration (#8)

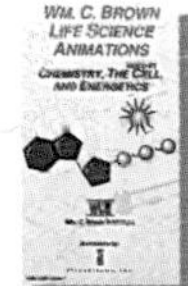

Life Science Animations Video

Video #1: Chemistry, the Cell, and Energetics
The Electron Transport Chain and the Production of ATP (#7)
ATP as an Energy Carrier (#11)

Understanding the Terms

active site 104
ADP (adenosine diphosphate) 102
ATP (adenosine triphosphate) 102
chemiosmosis 108
coenzyme 106
cofactor 106
coupling 101
denatured 105
electron transport system 107
energy 100
entropy 100
enzyme 103
feedback inhibition 106
free energy 101
kinetic energy 100
metabolism 101
NAD^+ 107
$NADP^+$ 107
potential energy 100
substrate 103
vitamin 106

Match the terms to these definitions:

a. _________ All of the chemical reactions that occur in a cell during growth and repair.
b. _________ Nonprotein adjunct required by an enzyme in order to function; many are metal ions, others are coenzymes.
c. _________ Energy associated with motion.
d. _________ Essential requirement in the diet, needed in small amounts. They are often part of coenzymes.
e. _________ Loss of an enzyme's normal shape so that it no longer functions; caused by a less than optimal pH and temperature.
f. _________ Measure of disorder or randomness.
g. _________ Mechanism by which an enzyme is rendered inactive by combining with the product of the reaction or metabolic pathway.
h. _________ Nonprotein organic molecule that aids the action of the enzyme to which it is loosely bound.
i. _________ Nucleotide with two phosphate groups that can accept another phosphate group and become ATP.
j. _________ Organic catalyst, usually a protein, that speeds up a reaction in cells due to its particular shape.

Photosynthesis

CHAPTER

Chapter Concepts

7.1 Sunlight Provides Solar Energy

- Plants make use of solar energy in the visible light range when they carry on photosynthesis. 114

7.2 Photosynthesis Occurs in Chloroplasts

- Photosynthesis takes place in chloroplasts, organelles that contain membranous thylakoids surrounded by a fluid-filled stroma. 116
- Photosynthesis has two sets of reactions: solar energy is captured by the pigments in thylakoids, and carbon dioxide is reduced by enzymes in the stroma. 116

7.3 Solar Energy Is Captured

- Solar energy energizes electrons that enter an electron transport system, releasing energy as electrons pass from one carrier to the next. 118
- Chemiosmosis is dependent on an electrochemical gradient established by the electron transport system. 120

7.4 Carbohydrate Is Synthesized

- Carbon dioxide reduction requires energized electrons and ATP. 121
- Plants use C_3, or C_4, or CAM photosynthesis, which are distinguishable by the manner in which CO_2 is fixed. 123

Koala, *Phascolarctos cinereus*, feeding on *Eucalyptus* leaves

Through photosynthesis, plants and algae produce food for themselves and all other living things. In the oceans, algae, and on land, plants, are the producers at the start of food chains of all types. Animals feed directly on photosynthesizers or on other animals that have fed on photosynthesizers.

When photosynthesis occurs, oxygen is released and carbon dioxide is absorbed. Oxygen is required by all organisms that carry on aerobic respiration. Oxygen also rises high in the atmosphere and forms the ozone shield that protects terrestrial organisms from the damaging effects of the ultraviolet rays of the sun. The absorption of carbon dioxide by photosynthesizers is also critical because excess carbon dioxide in the atmosphere is poisonous to animals and contributes to global warming. Planting trees and saving forests helps purify the air, and it also helps prevent soil erosion because plant roots hold soil in place.

The human population is sustained by the crops we plant. As sources of fabrics, rope, paper, lumber, fuel, and pharmaceuticals, plants make modern society possible. And while we are thanking green plants for human survival, let's not forget the simple beauty of a magnolia bloom or the majesty of an old-growth forest.

7.1 Sunlight Provides Solar Energy

Plants, algae, and a few other types of organisms carry out photosynthesis in this manner:

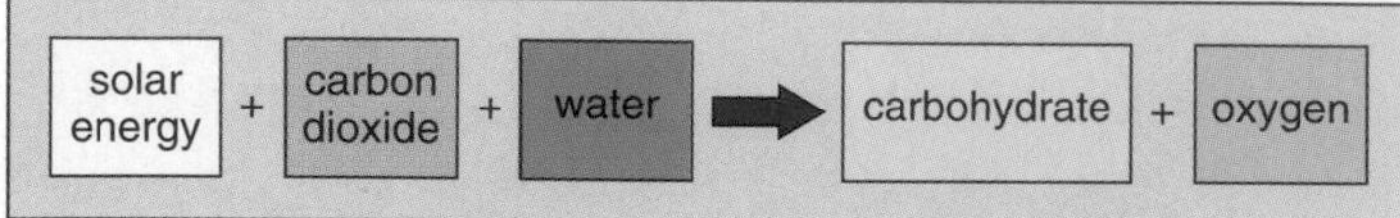

Because **photosynthesis** is an energy transformation in which solar energy in the form of light is converted to the chemical bond energy of carbohydrate molecules, we will begin our discussion of photosynthesis with the energy source—sunlight.

Solar radiation can be described in terms of its energy content and its wavelength. The energy comes in discrete packets called **photons** [Gk. *phos*, light]. So, in other words, you can think of radiation as photons that travel in waves:

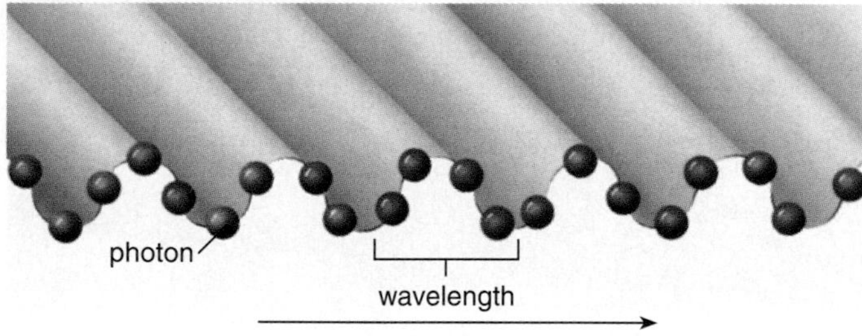

Figure 7.1*a* illustrates that solar radiation, or the **electromagnetic spectrum,** can be divided on the basis of wavelength—gamma rays have the shortest wavelength and radio waves have the longest wavelength. The energy content of photons is inversely proportional to the wavelength of the particular type of radiation; that is, short-wavelength radiation has photons of a higher energy content than long-wavelength radiation. High-energy photons, such as those of short-wavelength ultraviolet radiation, are dangerous to cells because they can break down organic molecules. Low-energy photons, such as those of infrared radiation, do not damage cells because they only increase the vibrational or rotational energy of molecules; they do not break bonds. But photosynthesis utilizes only the portion of the electromagnetic spectrum known as *visible light*. (It is called visible light because it is the part of the spectrum that the eye can see.) Photons of visible light have just the right amount of energy to promote electrons to a higher electron shell in atoms without harming cells. Visible or white light is actually made up of a number of different wavelengths of radiation; when it is passed through a prism (or through raindrops), we see these wavelengths as different colors of light.

Only about 42% of solar radiation passes through the earth's atmosphere and reaches its surface. Most of this radiation is within the visible-light range. Higher energy wavelengths are screened out by the ozone layer in the atmosphere, and lower energy wavelengths are screened out by water vapor and carbon dioxide before they reach the earth's surface. The conclusion is, then, that organic molecules within organisms and processes, like vision and photosynthesis, are chemically adapted to the radiation that is most prevalent in the environment.

The pigments found in photosynthesizing cells are capable of absorbing various portions of visible light. These pigments include chlorophyll *a* and chlorophyll *b* whose absorption spectra are shown in Figure 7.1*b*. Both chlorophyll *a* and chlorophyll *b* absorb violet, blue, and red light better than the light of other colors. Because green light is only minimally absorbed, leaves appear green to us. Other plant pigments such as the carotenoids are shades of yellow and orange and are able to absorb light in the violet-blue-green range. These pigments become noticeable in the

Energy-Balance Sheet

Only 42% of solar energy directed toward earth actually reaches the earth's surface; the rest is absorbed by or reflected into the atmosphere and becomes heat.

Of this usable portion, only about 2% is eventually utilized by plants; the rest becomes heat.

Of this, only 0.1–1.6% is ever incorporated into plant material; the rest becomes heat.*

Of this, only 20% is eaten by herbivores; a large proportion of the remainder becomes heat.*

Of this, only 30% is ever eaten by carnivores; a large proportion becomes heat.*

Conclusion: Most of the available energy is never utilized by living things.

** Plant and animal remains became the fossil fuels that we burn today to provide energy. So eventually this energy becomes heat.*

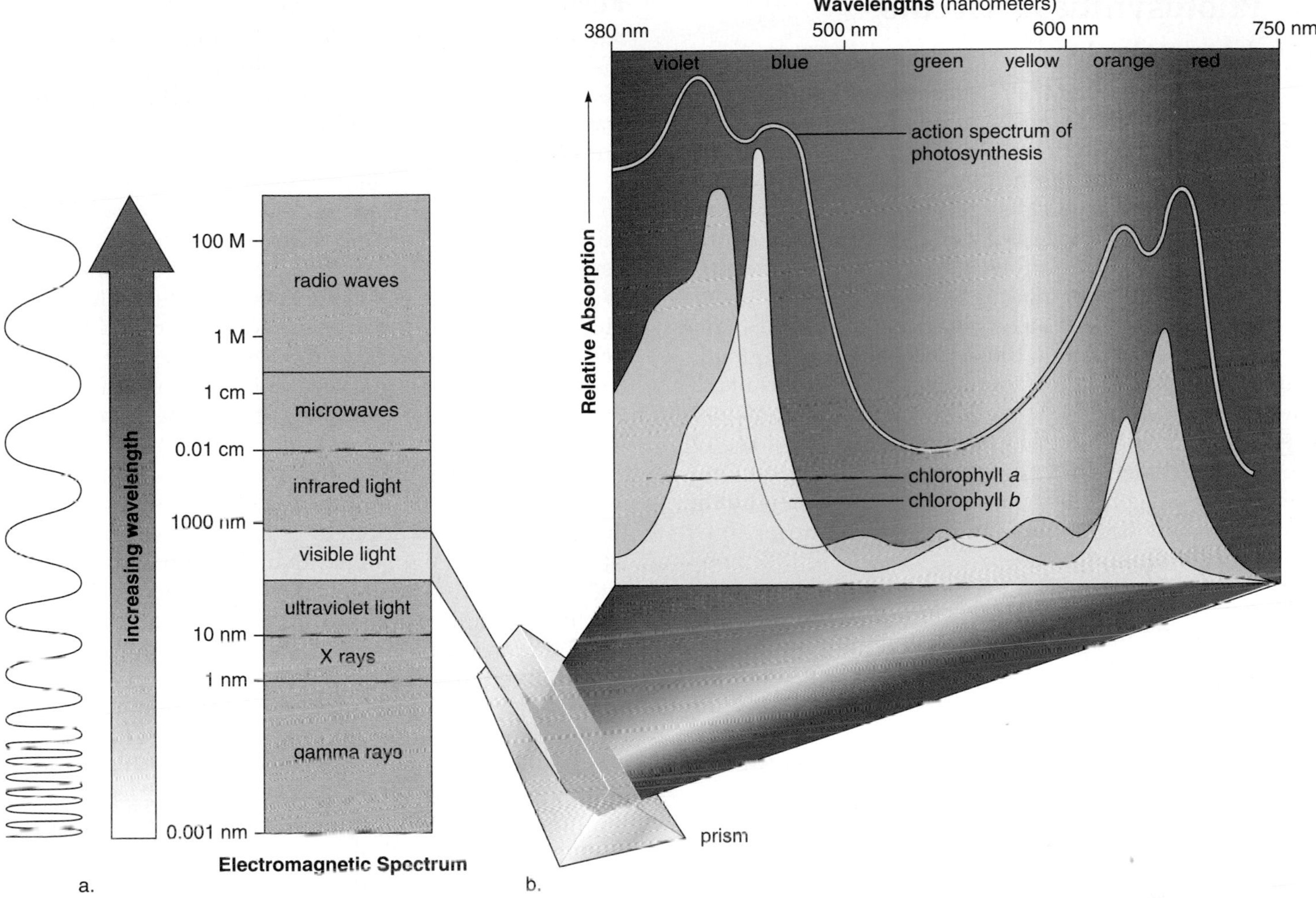

Figure 7.1 The electromagnetic spectrum and chlorophylls *a* and *b*.
a. The electromagnetic spectrum contains forms of energy that differ according to wavelength. Visible light is only a small portion of the electromagnetic spectrum. **b.** Chlorophylls *a* and *b* absorb certain wavelengths within visible light, which accounts for the action spectrum of (wavelengths necessary for) photosynthesis.

fall when chlorophyll breaks down. Photosynthetic pigments capture less than 2% of the solar energy that reaches the earth. Of this only a small percentage ever becomes a part of plant and animal organisms as the box on the previous page shows.

How do you determine the absorption spectrum of the pigments within a chloroplast? To identify the absorption spectrum of a particular pigment, a purified sample is exposed to different wavelengths of light inside an instrument called a spectrophotometer. A spectrophotometer measures the amount of light that passes through the sample, and from this it can be calculated how much was absorbed. The amount of light absorbed at each wavelength is plotted on a graph, and the result is what we call the absorption spectrum (Fig. 7.1*b*). How do we know that the peaks shown in chlorophyll's absorption spectrum indicate the wavelengths used for photosynthesis? Because photosynthesis gives off oxygen, we can use the production rate of oxygen as a means to measure the rate of photosynthesis at each wavelength of light. When such data are plotted, the resulting graph, called the *action spectrum*, is very similar to the absorption spectrum of the chlorophylls. Therefore, we are confident that the light absorbed by the chlorophylls does contribute extensively to photosynthesis.

Photosynthesis utilizes the portion of the electromagnetic spectrum (solar radiation) known as visible light.

7.2 Photosynthesis Occurs in Chloroplasts

It wasn't until the end of the nineteenth century that scientists understood the photosynthetic process and that, in eukaryotes, it occurs in **chloroplasts** [Gk. *chloros,* green, and *plastos,* formed, molded]. The overall equation for photosynthesis given on page 114 suggests that in the presence of sunlight, carbon dioxide and water are combined to produce a high-energy food molecule such as carbohydrate.

In 1930, C. B. van Niel of Stanford University suggested that oxygen given off by photosynthesis comes from water and not from carbon dioxide as had been originally thought. This was proven by two separate experiments. When plants were exposed to carbon dioxide that contained an isotope of oxygen, called heavy oxygen (^{18}O), the O_2 given off by the plant did not contain heavy oxygen. Only when heavy oxygen was a part of water (indicated by the color red in this equation) did this isotope appear in O_2 given off by the plant:

Reduction

$$\text{energy} + CO_2 + 2H_2O \longrightarrow (CH_2O) + H_2O + O_2$$

Oxidation

This equation is preferred by some because it has the advantage of keeping the chemical arithmetic correct.

Chloroplasts Have Two Parts

In a chloroplast, a double membrane surrounds a large central compartment called the **stroma** [Gk. *stroma,* bed, mattress]. The stroma encloses an enzyme-rich solution where CO_2 is first attached to an organic compound and then reduced. A membrane system within the stroma forms flattened sacs called **thylakoids** [Gk. *thylakos,* sack, and *eides,* like, resembling], which in some places are stacked to form grana (sing., **granum**), so called because they looked like piles of seeds to early microscopists. The space within each thylakoid is thought to be connected to the space within every other thylakoid, thereby forming an inner compartment within chloroplasts called the thylakoid space. **Chlorophyll** and other pigments are found within the membranes of the thylakoids. These pigments absorb the solar energy, which will energize the electrons prior to reduction of CO_2 in the stroma (Fig. 7.2).

Chlorophyll within thylakoids absorbs solar energy so that energized electrons are sent to stroma, where CO_2 is reduced.

Photosynthesis Has Two Sets of Reactions

An overall equation tells us the beginning reactants and the end products of a metabolic pathway. What about what goes on in between the first reactants (carbon dioxide and water) and the final products (carbohydrate and water) during photosynthesis?

In 1905, F. F. Blackman suggested that there were probably two sets of reactions involved in photosynthesis because when light is being maximally absorbed, a rise in temperature still increases the rate of photosynthesis. (Enzymatic reactions speed up when temperature is increased.) The first set of photosynthetic reactions is called the **light-dependent reactions** because they cannot take place unless light is present. The second set of reactions is called the **light-independent reactions** because they can take place whether light is present or not.

The light-dependent reactions occur in the thylakoid membrane where the pigments chlorophyll *a* and *b,* plus the carotenoids, are located. These pigments absorb violet, blue, and red light better than the light of other colors. Light can be absorbed by (taken up), reflected by (given off), or transmitted (passed through) a pigment. Chlorophyll *a* appears blue-green and chlorophyll *b* appears yellow-green to us because these are the very colors they do not absorb and are instead reflected to our eyes or transmitted through these pigments (see Fig. 7.1). For the same reason, the carotenoids appear as various shades of yellow and orange to our eyes.

The light-dependent reactions are the *energy-capturing reactions*—low-energy electrons removed from H_2O are energized when thylakoid membrane pigments absorb solar energy. These electrons move from chlorophyll *a* down an electron transport system, which produces ATP from ADP and Ⓟ. Energized electrons are also taken up by $NADP^+$. After $NADP^+$ accepts electrons, it becomes NADPH. This molecule temporarily holds energy, in the form of energized electrons, that will be used to reduce CO_2.

The light-dependent reactions capture solar energy.

The second set of reactions involved in photosynthesis occurs in the stroma of a chloroplast. They are called the light-independent reactions because they can take place in either the light or the dark. The light-independent reactions are the *synthesis reactions,* which use the ATP and NADPH formed in the thylakoids to reduce CO_2.

The light-independent reactions synthesize carbohydrate.

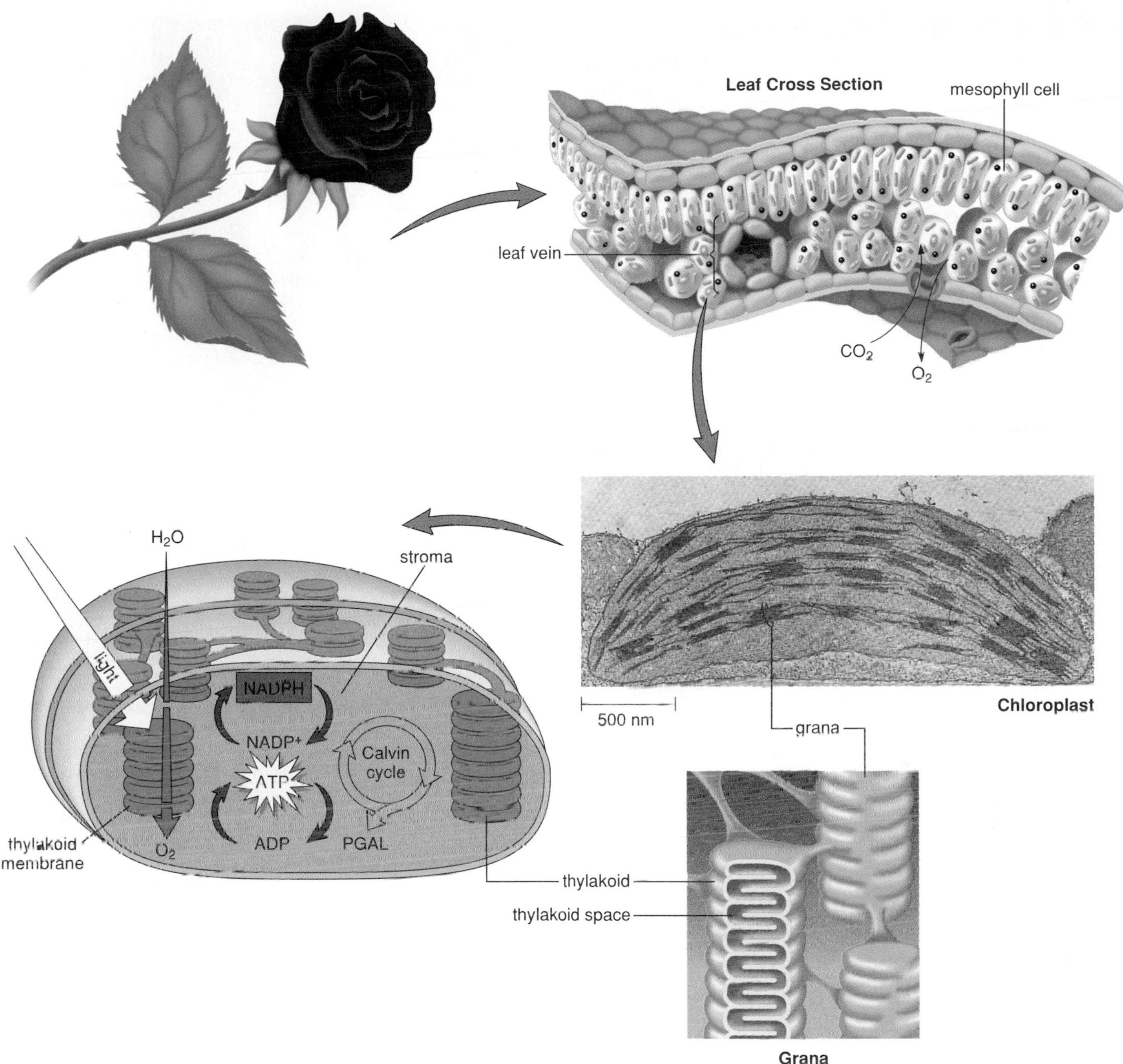

Figure 7.2 Chloroplast structure and function.
A tissue called mesophyll, which forms the bulk of a leaf, contains many chloroplasts. Each chloroplast, bounded by a double membrane, contains a fluid-filled space called the stroma. The membranous thylakoids stacked in grana are interconnected so that there is a single thylakoid space. The light-dependent reactions, which produce ATP and NADPH, occur in the thylakoid membrane. These molecules are used by the light-independent reactions to reduce carbon dioxide during a cyclical series of enzymatic reactions called the Calvin cycle. The Calvin cycle occurs in the stroma.

7.3 Solar Energy Is Captured

The light-dependent reactions that occur in the thylakoid membranes require the participation of two light-gathering units called photosystem I (PS I) and photosystem II (PS II). The photosystems are named for the order in which they were discovered and not for the order in which they occur in the thylakoid membrane. Each **photosystem** has a pigment complex composed of chlorophyll *a* and chlorophyll *b* molecules (primarily green pigments) and accessory pigments, such as carotenoid pigments (primarily orange and yellow pigments). The closely packed pigment molecules in the photosystem serve as an "antenna" for gathering solar energy. Solar energy is passed from one pigment to the other until it is concentrated into one of two particular chlorophyll *a* molecules, the reaction-center chlorophylls. Electrons in the reaction-center chlorophyll *a* molecules become so excited that they escape and move to a nearby electron-acceptor molecule.

In a photosystem, the light-gathering antenna absorbs solar energy and funnels it to a reaction-center chlorophyll *a* molecule, which then sends energized electrons to an electron-acceptor molecule.

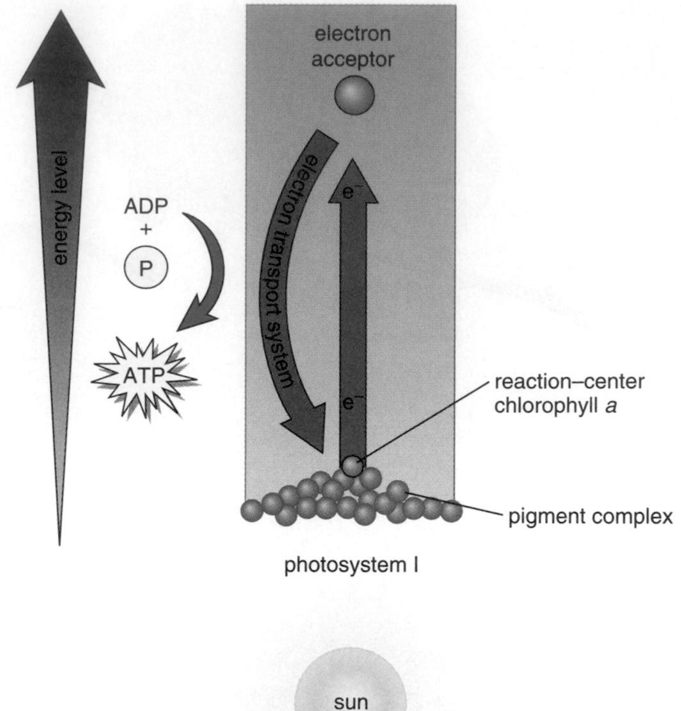

Light–Dependent Reactions

Figure 7.3 The light-dependent reactions: the cyclic electron pathway.
Energized electrons (e^-) leave the photosystem I (PS I) reaction-center chlorophyll *a* and are taken up by an electron acceptor, which passes them down an electron transport system before they return to PS I. Only ATP production results from this pathway.

Electrons Have Two Pathways

Electrons can follow a cyclic electron pathway or a noncyclic electron pathway during the first phase of photosynthesis. The cyclic electron pathway generates only ATP, while the noncyclic pathway results in both NADPH and ATP. ATP is produced through chemiosmotic ATP synthesis in both cases.

ATP production during photosynthesis is sometimes called *photophosphorylation* because light is involved. The production of ATP during the cyclic electron pathway is called *cyclic photophosphorylation* whereas ATP production during the noncyclic electron pathway is called *noncyclic photophosphorylation*.

Cyclic Electron Pathway

The *cyclic electron pathway* (Fig. 7.3) begins after the PS I pigment complex absorbs solar energy. In this pathway, high-energy electrons (e^-) leave the PS I reaction-center chlorophyll *a* molecule but eventually return to it. Before they return, however, the electrons enter an **electron transport system,** a series of carriers that pass electrons from one to the other. As the electrons pass from one carrier to the next, energy that will be used to produce ATP molecules is released and stored in the form of a hydrogen (H^+) gradient. When these hydrogen ions flow down their electrochemical gradient through ATP synthase complexes, ATP production occurs (see page 120).

Some photosynthetic bacteria utilize the cyclic electron pathway only; therefore, this pathway probably evolved early in the history of life. In plants the reactions that occur in the stroma can make use of this extra ATP for the Calvin cycle because this cycle requires a larger number of ATP than NADPH. Also, there are other enzymatic reactions aside from those involving photosynthesis occurring in the stroma that can make use of this ATP. It's possible that the cyclic flow of electrons is utilized alone when carbon dioxide is in such limited supply that carbohydrate is not being produced. At this time there would be no need for NADPH, which is produced only by the noncyclic electron pathway.

The cyclic electron pathway, from PS I back to PS I, has only one effect: production of ATP.

Figure 7.4 The light-dependent reactions: the noncyclic electron pathway.
Electrons, taken from water, move from photosystem II (PS II) to photosystem I (PS I) to NADP⁺. The ATP and NADPH produced will be used by the light-independent reactions to reduce carbon dioxide (CO_2) to a carbohydrate (CH_2O).

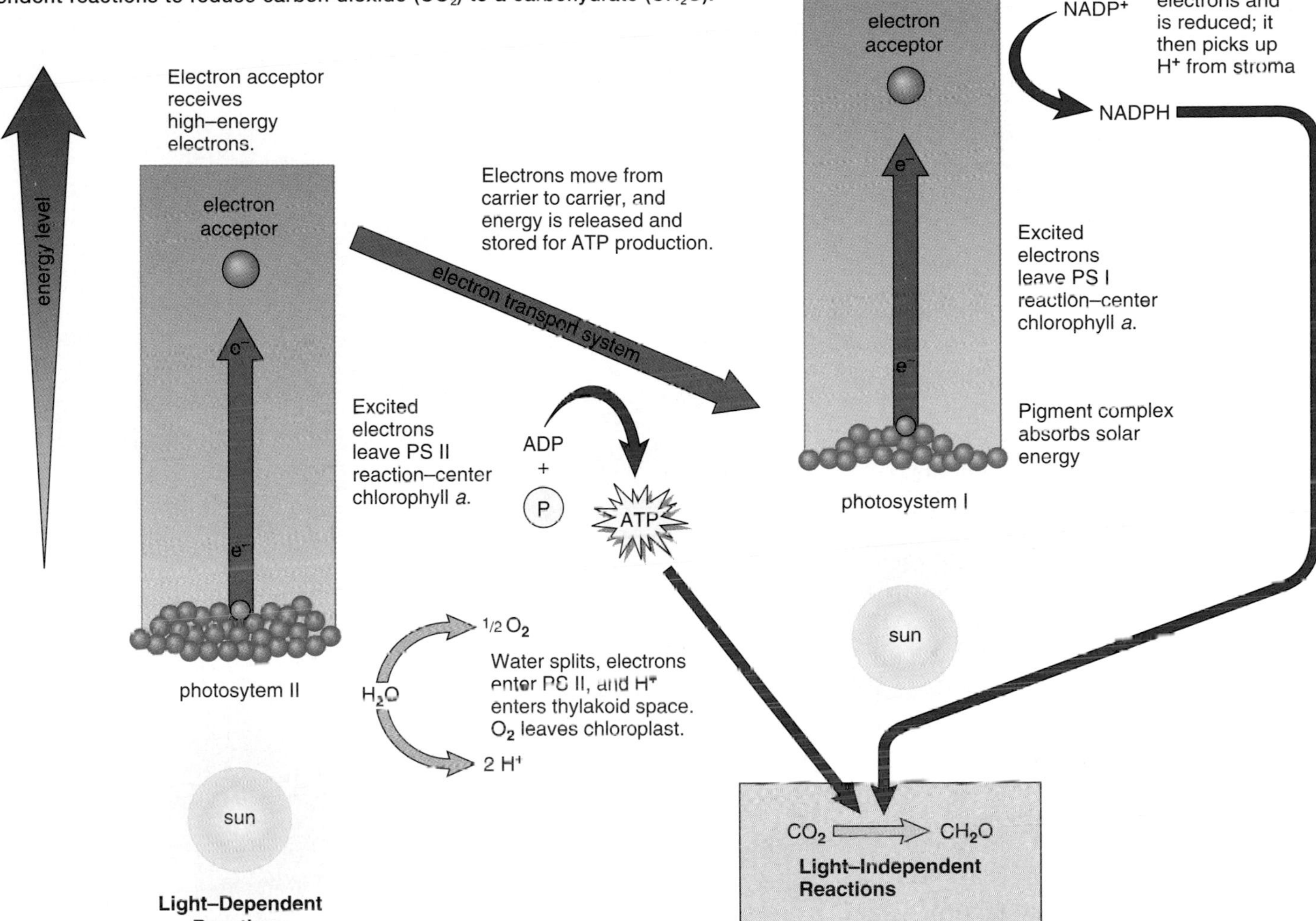

Noncyclic Electron Pathway

During the *noncyclic electron pathway,* electrons move from water (H_2O) through PS II to PS I and then on to $NADP^+$ (Fig. 7.4). This pathway begins when the PS II pigment complex absorbs solar energy and high-energy electrons (e^-) leave the reaction-center chlorophyll *a* molecule. PS II takes replacement electrons from water, which splits, releasing oxygen:

$$H_2O \longrightarrow 2\,H^+ + 2\,e^- + \tfrac{1}{2}\,O_2$$

This oxygen is released from the chloroplast and the plant as oxygen gas. The hydrogen ions (H^+) temporarily stay in the thylakoid space.

The high-energy electrons that leave PS II are captured by an electron acceptor, which sends them to an electron transport system. As the electrons pass from one carrier to the next, energy that will be used to produce ATP molecules is released and stored in the form of a hydrogen (H^+) gradient. When these hydrogen ions flow down their electrochemical gradient through an ATP synthase complex, chemiosmotic ATP synthesis occurs (see page 120).

Low-energy electrons leaving the electron transport system enter PS I. When the PS I pigment complex absorbs solar energy, high-energy electrons leave the reaction-center chlorophyll *a* and are captured by an electron acceptor. This time, the electron acceptor passes the electrons on to $NADP^+$. $NADP^+$ now takes on an H^+ and becomes NADPH:

$$NADP^+ + 2\,e^- + H^+ \longrightarrow NADPH$$

The NADPH and ATP produced by the noncyclic flow of electrons in the thylakoid membrane are used by enzymes in the stroma during the light-independent reactions.

Results of noncyclic electron flow: Water is oxidized (split), yielding H^+, e^-, and O_2; ATP is produced; and $NADP^+$ becomes NADPH.

ATP Production

The thylakoid space acts as a reservoir for hydrogen ions (H^+). First, each time water is oxidized, two H^+ remain in the thylakoid space. Second, as the electrons move from carrier to carrier in the electron transport system, they give up energy, which is used to pump H^+ from the stroma into the thylakoid space. Therefore, there is a large number of H^+ in the thylakoid space compared to the number in the stroma. The flow of H^+ (often referred to as protons in this context) from high to low concentration across the thylakoid membrane provides the energy that allows an *ATP synthase enzyme* to enzymatically produce ATP from ADP + Ⓟ. This method of producing ATP is called *chemiosmosis* because ATP production is tied to an electrochemical gradient.

The Thylakoid Membrane

Both biochemical and structural techniques have been used to determine that there are intact complexes (particles) in the thylakoid membrane (Fig. 7.5):

PS II consists of a protein complex and a light-gathering pigment complex shown to one side. PS II oxidizes water and produces oxygen.

The cytochrome complex acts as the transporter of electrons between PS II and PS I. The pumping of H^+ occurs during electron transport.

PS I consists of a protein complex and a light-gathering pigment complex to one side. Notice that PS I is associated with the enzyme that reduces $NADP^+$ to NADPH.

ATP synthase complex has an H^+ channel and a protruding ATP synthase. As H^+ flows down its concentration gradient through this channel from the thylakoid space into the stroma, ATP is produced from ADP + Ⓟ.

The light-dependent reactions that occur in the thylakoid membrane produce ATP and NADPH and oxidize water so that oxygen is given off.

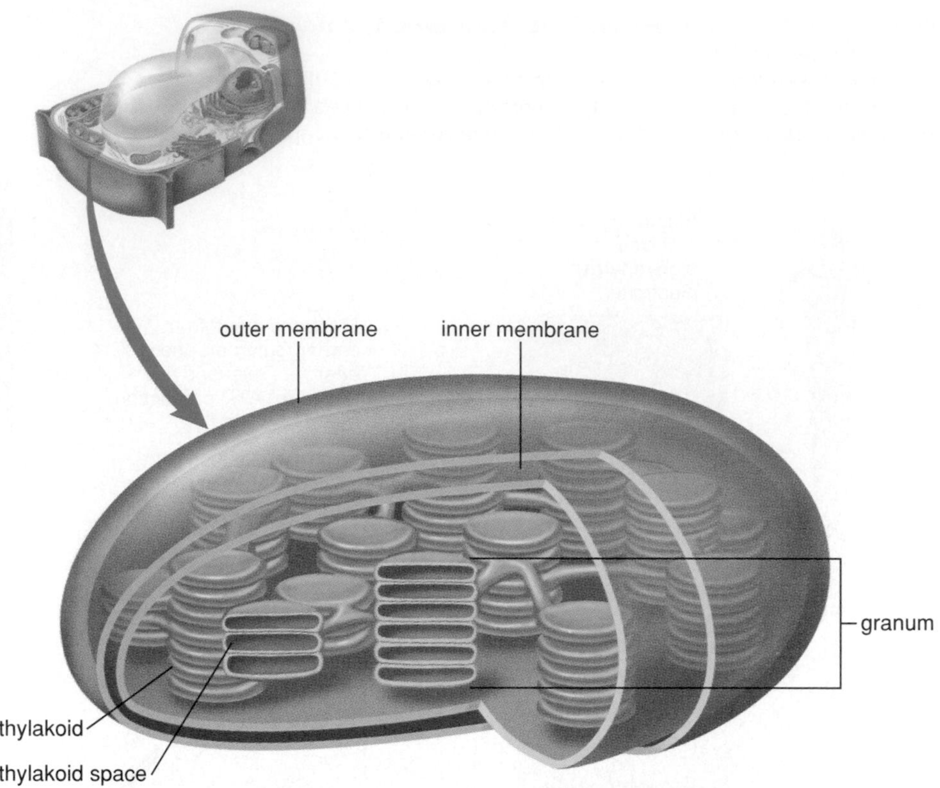

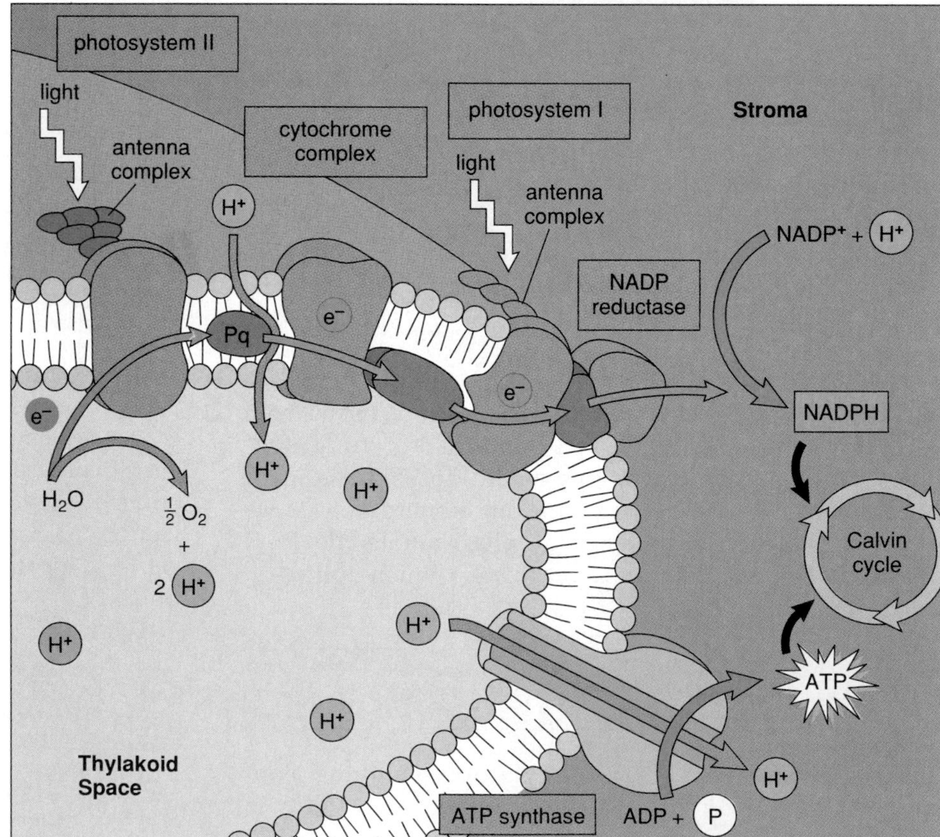

Figure 7.5 Organization of the thylakoid.
Protein complexes within the thylakoid membrane pump hydrogen ions from the stroma into the thylakoid space. (The other functions of the protein complexes are listed to the left. *Pq* is a mobile carrier that transfers hydrogen ions (H^+) from the stroma to the thylakoid space.) When hydrogen ions flow back out of the space into the stroma through the ATP synthase complex, ATP is produced from ADP + Ⓟ.

7.4 Carbohydrate Is Synthesized

The light-independent reactions are the second stage of photosynthesis. They take their name from the fact that light is not directly required for these reactions to proceed. These reactions occur when CO_2 has entered the leaf and ATP and NADPH have been produced during the light-dependent reactions. PS I sets in motion a regulatory mechanism by which the enzymes of the light-independent reactions are turned on.

In this stage of photosynthesis, NADPH and ATP are used to reduce carbon dioxide: CO_2 becomes CH_2O within a carbohydrate molecule. Electrons and energy needed for this reduction synthesis are supplied by NADPH and ATP.

The reduction of carbon dioxide occurs in the stroma of a chloroplast by a series of reactions known as the **Calvin cycle.** In a cycle the final product will also be the first reactant when the cycle of reactions begins again.

The Calvin cycle is named for Melvin Calvin, one of the individuals who was instrumental in identifying the reactions that make up the cycle (Fig. 7.6). Calvin received a Nobel Prize in 1961 for his part in determining these reactions.

The light-independent reactions use the ATP and NADPH from the light-dependent reactions to reduce carbon dioxide.

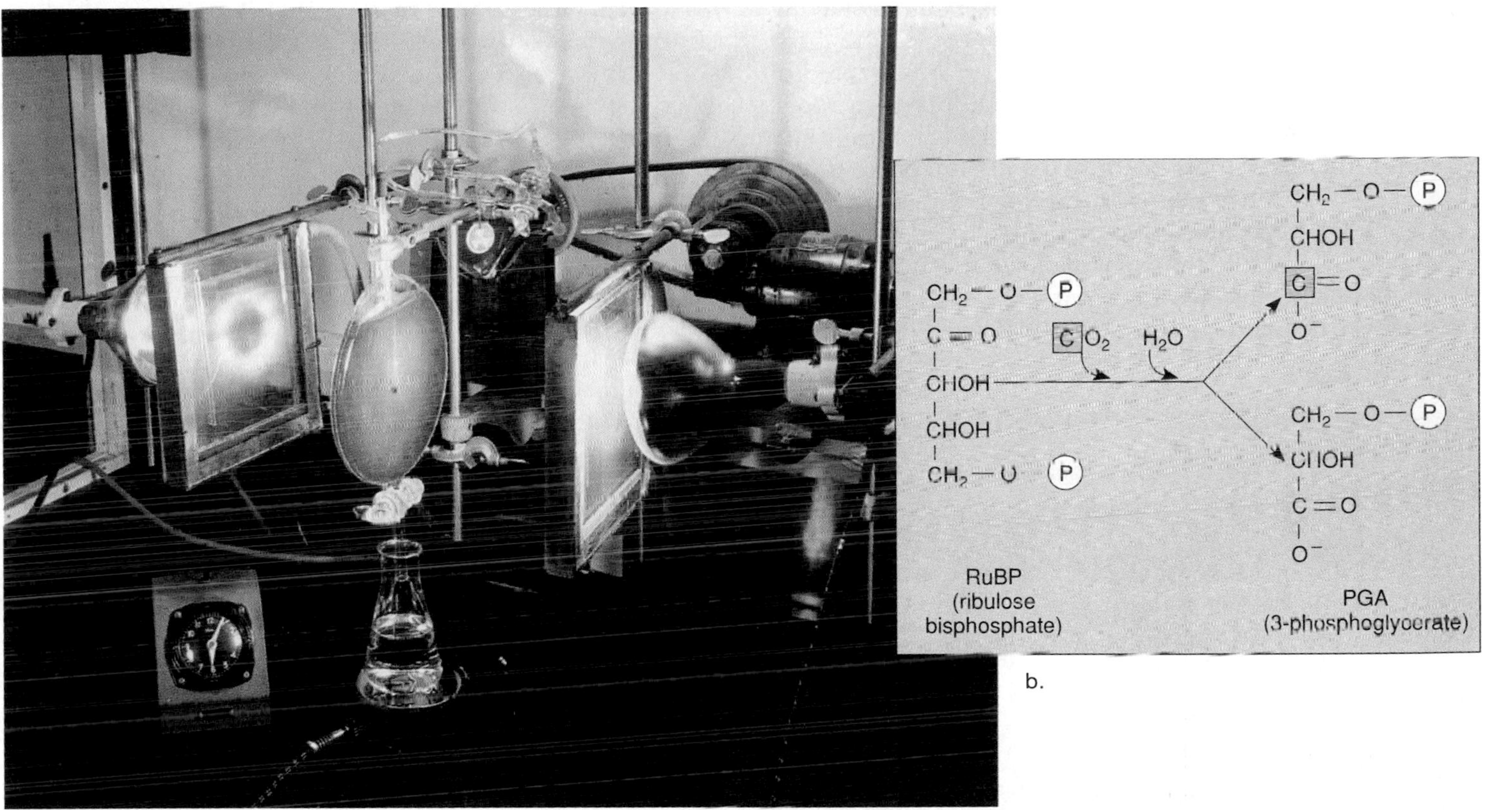

Figure 7.6 Identifying the reactions of the Calvin cycle.
a. Calvin and his colleagues used the apparatus shown; algae *(Chlorella)* were placed in the flat flask, which was illuminated by two lamps. Radioactive carbon dioxide ($^{14}CO_2$) was added, and the algae were killed a short time later (2 seconds, 5 seconds, and so on) by transferring them into boiling alcohol (in the beaker below the flask). This treatment instantly stopped all chemical reactions in the cells. Carbon-based compounds were then extracted from the cells and analyzed. **b.** By tracing the radioactive carbon, Calvin and his colleagues were able to trace the linked reactions, or pathway, by which CO_2 was incorporated. They found that the radioactive carbon (shown in boxes) is first added to RuBP and that the resulting molecule splits to form two molecules of PGA. By gradually increasing the time after initial exposure of the algae to $^{14}CO_2$ before killing, they were able to identify the rest of the molecules in the cycle that is now called the Calvin cycle (see Fig. 7.7).

The Importance of PGAL

PGAL (glyceraldehyde-3-phosphate) is the product of the Calvin cycle that can be converted to all sorts of organic molecules. Compared to animal cells, algae and plants have enormous biochemical capabilities. They use PGAL for the purposes described in Figure 7.7.

Notice that *glucose phosphate* is among the organic molecules that result from PGAL metabolism. This is of interest to us because glucose is the molecule that plants and animals most often metabolize to produce the ATP molecules they require for their energy needs. Glucose is blood sugar in human beings.

Glucose phosphate can be combined with fructose (and the phosphate removed) to form sucrose, the molecule that plants use to transport carbohydrates from one part of the body to the other. Glucose phosphate is also the starting point for the synthesis of starch and cellulose. Starch is the storage form of glucose. Some starch is stored in chloroplasts, but most starch is stored in amyloplasts in roots. Cellulose is a structural component of plant cell walls and becomes fiber in our diet because we are unable to digest it. A plant can utilize the hydrocarbon skeleton of PGAL to form fatty acids and glycerol, which are combined in plant oils. We are all familiar with corn oil, sunflower oil, or olive oil, which we use in cooking. Also, when nitrogen is added to the hydrocarbon skeleton derived from PGAL, amino acids are formed.

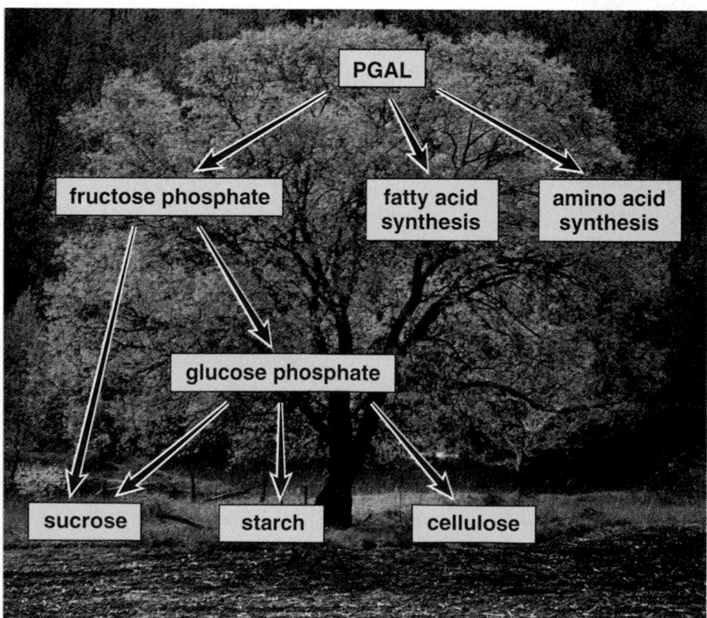

Figure 7.7 Fate of PGAL.
PGAL is the first reactant in a number of plant cell metabolic pathways. Two PGALs are needed to form glucose phosphate: glucose is often considered the end product of photosynthesis. Sucrose is the transport sugar in plants; starch is the storage form of glucose, and cellulose is a major constituent of plant cell walls.

The Calvin Cycle Has Three Stages

The reactions of the Calvin cycle are the light-independent reactions during which carbohydrate is synthesized. The Calvin cycle includes (1) carbon dioxide fixation, (2) carbon dioxide reduction, and (3) regeneration of RuBP (ribulose bisphosphate).

Fixing Carbon Dioxide

Carbon dioxide (CO_2) fixation, the attachment of carbon dioxide to an organic compound, is the first event in the Calvin cycle. **RuBP (ribulose bisphosphate),** a five-carbon molecule, combines with carbon dioxide (Fig. 7.8). The enzyme that speeds this reaction is called RuBP carboxylase, a protein that makes up about 20–50% of the protein content in chloroplasts. The reason for its abundance may be that it is unusually slow (it processes only a few molecules of substrate per second compared to thousands per second for a typical enzyme), and so there has to be a lot of it to keep the Calvin cycle going.

Carbon dioxide fixation occurs when carbon dioxide combines with RuBP.

Reducing PGA

The six-carbon molecule resulting from carbon dioxide fixation immediately breaks down to form two *PGA (3-phosphoglycerate)* three-carbon molecules. Each of the two PGA molecules undergoes reduction to PGAL (glyceraldehyde-3-phosphate) in two steps:

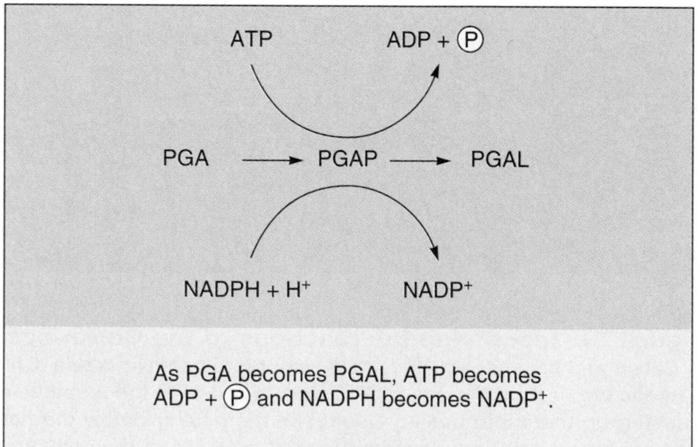

This is the sequence of reactions that uses NADPH and ATP from the light-dependent reactions, and it signifies the reduction of carbon dioxide (CO_2) to a carbohydrate (CH_2O). Electrons and energy are needed for this reduction reaction, and these are supplied by NADPH and ATP, respectively.

Figure 7.8 Light-independent reactions.
The Calvin cycle is divided into three portions: CO_2 fixation, CO_2 reduction, which requires NADPH and ATP, and regeneration of RuBP. Because five PGAL are needed to re-form three RuBP, it takes three turns of the cycle to have a net gain of one PGAL, which can be used to form glucose ($C_6H_{12}O_6$).

Metabolites of the Calvin Cycle	
RuBP	ribulose bisphosphate
PGA	3-phosphoglycerate
PGAP	1,3-bisphosphoglycerate
PGAL	glyceraldehyde-3-phosphate

3 CO_2

(intermediate)

3 C_6

6 PGA C_3

6 ATP

6 ADP

These ATP and NADPH molecules were produced by the light-dependent reactions.

CO_2 Fixation

Calvin cycle

CO_2 Reduction

Regeneration of RuBP

3 RuBP C_5

3 ADP

3 ATP

These ATP molecules were produced by the light-dependent reactions.

6 PGAP C_3

6 NADPH

5 PGAL C_3

6 PGAL C_3

6 $NADP^+$

There is a net gain of one PGAL.

Glucose and other organic compounds

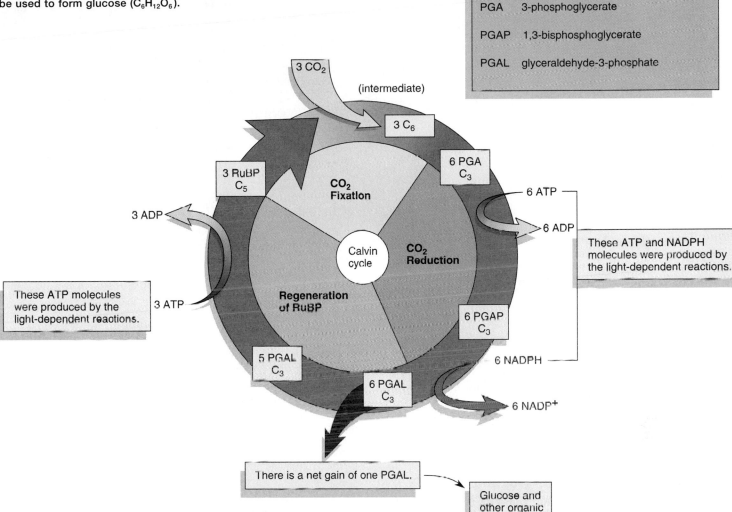

Regenerating RuBP

For every three turns of the Calvin cycle, five molecules of PGAL are used to re-form three molecules of RuBP so that the cycle can continue. Notice that 5 × 3 (carbons in PGAL) = 3 × 5 (carbons in RuBP):

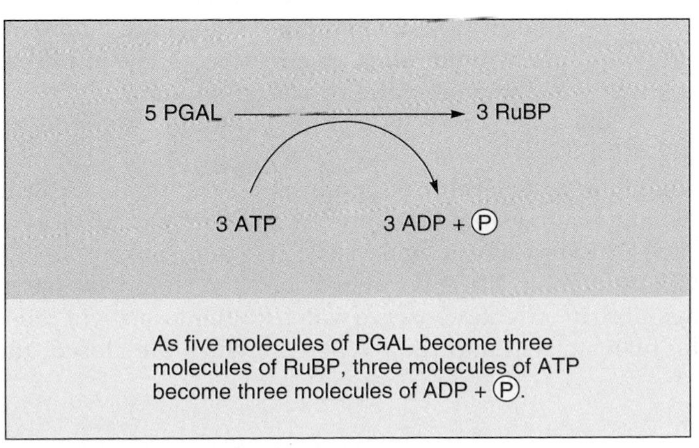

As five molecules of PGAL become three molecules of RuBP, three molecules of ATP become three molecules of ADP + Ⓟ.

The net gain of three turns of the Calvin cycle is one PGAL molecule. This sequence of reactions also utilizes some of the ATP produced by the light-dependent reactions.

Five out of every six PGAL molecules of the Calvin cycle are used to regenerate three RuBP molecules, so the cycle can continue.

The first detectable molecule identified by Melvin Calvin—namely PGA—has three carbons (Fig. 7.8). Therefore the Calvin cycle is also known as the C_3 cycle. Since Melvin Calvin did his research work, it's been discovered that plant species differ in the way they fix carbon dioxide and therefore the first detectable molecule following carbon dioxide fixation is not always a C_3 molecule.

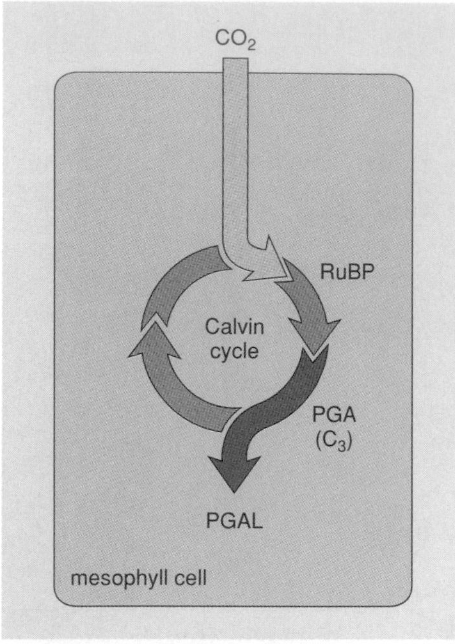

CO_2 fixation in a C_3 plant

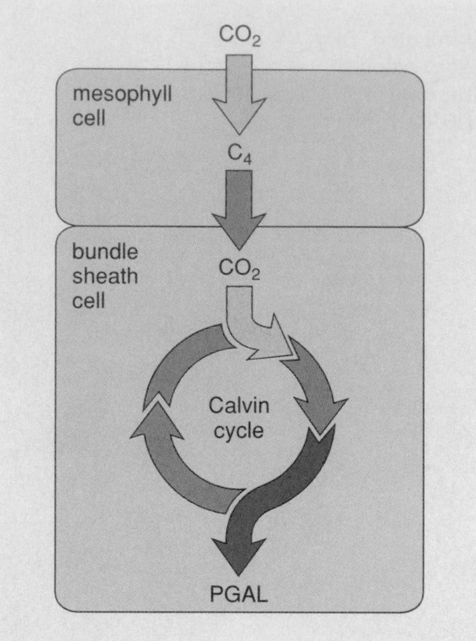

CO_2 fixation in a C_4 plant

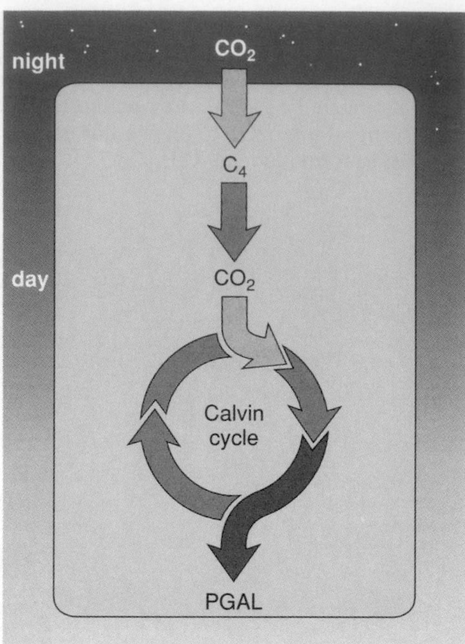

CO_2 fixation in a CAM plant

Figure 7.9 Carbon dioxide fixation.
Plants can be categorized according to the type of carbon dioxide fixation.

Photosynthesis Takes Other Routes

Three modes of photosynthesis are known (Fig. 7.9). In **C_3 plants,** the Calvin cycle fixes carbon dioxide (CO_2) directly, and the first detectable molecule following fixation is PGA, a C_3 molecule. **C_4 plants** fix CO_2 by forming a C_4 molecule prior to the involvement of the Calvin cycle. **CAM plants** fix CO_2 by forming a C_4 molecule at night when stomates can open without much loss of water.

C_4 Plants Flourish When It Is Hot and Dry

The structure of a leaf from a C_3 plant is different from that of a C_4 plant. In a C_3 plant, the *mesophyll cells* contain well-formed chloroplasts and are arranged in parallel layers. In a C_4 leaf, the *bundle sheath cells,* as well as the mesophyll cells, contain chloroplasts. Further, the mesophyll cells are arranged concentrically around the bundle sheath cells:

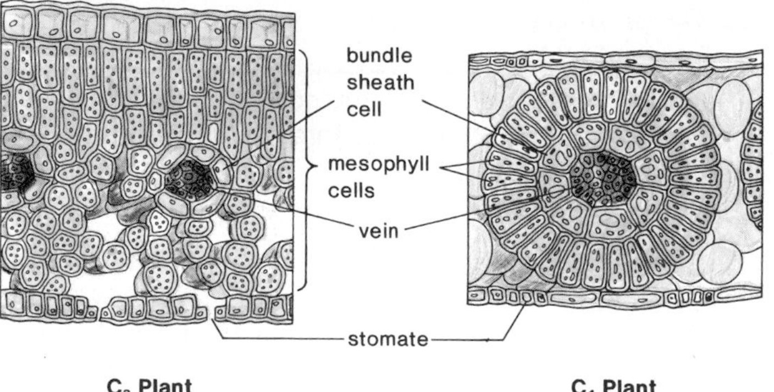

C_3 plants use RuBP carboxylase to fix CO_2 to RuBP in mesophyll cells, and the first detected molecule following fixation is PGA. C_4 plants use the enzyme PEP carboxylase (PEPCase) to fix CO_2 to PEP (phosphoenolpyruvate, a C_3 molecule), and the result is oxaloacetate, a C_4 molecule.

$$\text{RuBP} + CO_2 \xrightarrow{\text{RuBP carboxylase}} 2\ \text{PGA} \qquad (C_3\ \text{plants})$$

$$\text{PEP} + CO_2 \xrightarrow{\text{PEPCase}} \text{oxaloacetate} \qquad (C_4\ \text{plants})$$

In a C_4 plant, CO_2 is taken up in mesophyll cells and then malate, a reduced form of oxaloacetate, is pumped into the bundle sheath cells (Fig. 7.9). Here, CO_2 enters the Calvin cycle. It takes energy to pump molecules, and you would think that the C_4 pathway would be disadvantageous. Yet in hot, dry climates, the net photosynthetic rate of C_4 plants such as sugarcane, corn, and Bermuda grass is about two to three times that of C_3 plants such as wheat, rice, and oats. Why do C_4 plants enjoy such an advantage? The answer is they can avoid photorespiration, discussed next.

Photorespiration

Notice in the adjacent diagrams of leaves, there are little openings called stomates (sing., stomate) in the surfaces of leaves through which water can leave and carbon dioxide (CO_2) can enter. If the weather is hot and dry, these openings close in order to conserve water. (Water loss might cause the plant to wilt and die.) When stomates are closed, the concentration of CO_2 decreases in leaves, while oxygen, a by-product of photosynthesis, increases. RuBP carboxylase is often called rubisco because it can have an oxygenase activity in addition to a carboxylase activity. When oxygen rises in C_3 plants, oxygen competes with CO_2 for the active site of rubisco and the result is only one molecule of PGA:

Photorespiration

$$\text{RuBP} + O_2 \xrightarrow{\text{rubisco}} \text{PGA} + \text{phosphoglycolate} \longrightarrow CO_2$$

This reaction sequence is named **photorespiration** because in the presence of light *(photo)*, oxygen is taken up and CO_2 is produced.

Photorespiration does not occur in C_4 leaves even when stomates are closed because CO_2 is delivered to the Calvin cycle in the bundle sheath cells. When the weather is moderate, C_3 plants have the advantage, but when the weather becomes hot and dry, C_4 plants have the advantage, and we can expect them to predominate. In the early summer, C_3 plants such as Kentucky bluegrass and creeping bent grass predominate in lawns in the cooler parts of the United States, but by midsummer, crabgrass, a C_4 plant, begins to take over.

C_4 plants have an advantage over C_3 plants when the weather is hot and dry because photorespiration does not occur to any extent.

CAM Plants Have an Alternative Way

CAM plants use PEPCase to fix some CO_2 at night, forming malate, a C_4 molecule that is stored in large vacuoles in their mesophyll cells. CAM stands for crassulacean-acid metabolism; the Crassulaceae is a family of flowering succulent (water-containing) plants that live in warm, arid regions of the world. CAM was first discovered in these plants, but now it is known to be prevalent among most succulent plants that grow in desert environments, including the cacti.

Whereas a C_4 plant represents partitioning in space—carbon dioxide fixation occurs in mesophyll cells and the Calvin cycle occurs in bundle sheath cells—CAM is partitioning by the use of time. The C_4 molecules stored in vacuoles during the night release CO_2 during the day to the Calvin cycle within the same cell. During the day NADPH and ATP are available from the light-dependent reactions. The primary reason for this partitioning again has to do with the conservation of water. CAM plants open their stomates only at night, and therefore only at that time is atmospheric CO_2 available. During the day, the stomates close to conserve water and CO_2 cannot enter the plant.

Photosynthesis in a CAM plant is minimal because of the limited amount of CO_2 fixed at night, but it does allow CAM plants to live under stressful conditions.

CAM plants use PEPCase to carry on carbon dioxide fixation at night when the stomates are open. During the day, the stored carbon dioxide can enter the Calvin cycle.

connecting concepts

Through the process of photosynthesis, plants contribute to the global carbon cycle. In the carbon cycle, organisms in both terrestrial and aquatic ecosystems exchange carbon dioxide with the atmosphere. On land, plants take up carbon dioxide from the air, and through photosynthesis, they incorporate carbon into organic nutrients that are used by themselves and others. Because they make their own food, plants are known as autotrophs.

Carbon-based organic nutrients are passed to other organisms when they feed on plants and/or on to other organisms. Organisms, like animals, that take in organic nutrients are called heterotrophs.

Carbon dioxide is returned to the atmosphere when autotrophs and heterotrophs carry on cellular respiration. In this way, the very same carbon atoms cycle from the atmosphere to autotrophs, then to heterotrophs, and then back to autotrophs again.

Living and dead organisms contain organic carbon and serve as a reservoir of carbon in the carbon cycle. Some 300 million years ago, a host of plants died and did not decompose. These plants were compressed to form the coal that we mine and burn today. (Oil has a similar origin, but most likely formed in marine sedimentary rocks that included animal remains.)

The amount of carbon dioxide in the atmosphere is increasing steadily because we humans burn fossil fuels to run our modern industrial society. Many fear that this buildup of carbon dioxide will contribute to global warming, because carbon dioxide is a greenhouse gas. Like the glass of a greenhouse, it allows sunlight to pass through, but then traps the resulting heat. Yet while we are pumping carbon dioxide into the atmosphere due to fossil fuel burning, we are destroying vast tracts of tropical rain forests, which soak up carbon dioxide like a sponge. Clearly, these two actions are not in our self-interest and it would behoove us to reduce the burning of fossil fuels and preserve tropical rain forests.

Summary

7.1 Sunlight Provides Solar Energy

Photosynthesis produces carbohydrates and oxygen, both of which are utilized by the majority of living things. Photosynthesis uses solar energy in the visible-light range; the photons of this range contain the right amount of energy to energize the electrons of chlorophyll molecules. Specifically, chlorophylls *a* and *b* absorb violet, blue, and red wavelengths best. This causes chlorophyll to appear green to us.

7.2 Photosynthesis Occurs in Chloroplasts

A chloroplast is bounded by a double membrane and contains two main components: the liquid stroma and the membranous grana made up of thylakoid sacs. The light-dependent reactions take place in the thylakoids, and the light-independent reactions take place in the stroma.

7.3 Solar Energy Is Captured

In the cyclic electron pathway, electrons energized by the sun leave PS I. But instead of being sent to $NADP^+$, they pass down the electron transport system back to PS I again. It is possible the cyclic pathway occurs only if there is no free $NADP^+$ to receive the electrons; this is a way to prevent an energy overload of PS I. It serves to generate ATP, and in keeping with its name, it probably evolved first and is the only photosynthetic pathway present in certain bacteria today.

The noncyclic electron pathway of the light-dependent reactions begins when solar energy enters PS II. PS II energized electrons are picked up by an electron acceptor. The oxidizing (splitting) of water replaces these electrons in the reaction-center chlorophyll *a* molecule, releasing oxygen to the atmosphere and contributing hydrogen ions (H^+) to the thylakoid space. An acceptor molecule passes electrons to PS I by way of an electron transport (cytochrome) system. When solar energy is absorbed by PS I, energized electrons leave and are ultimately received by $NADP^+$, which also combines with H^+ from the stroma to become NADPH.

The energy made available by the passage of electrons down the electron transport system allows carriers to pump H^+ into the thylakoid space. The buildup of H^+ establishes an electrochemical gradient. When H^+ flows down this gradient through the channel present in ATP synthase complexes, ATP is synthesized from ADP and Ⓟ by ATP synthase.

The thylakoid membrane is highly organized: PS II functions to oxidize (split) water, the cytochrome complex transports electrons and pumps H^+, PS I is associated with an enzyme that reduces $NADP^+$, and ATP synthase produces ATP.

7.4 Carbohydrate Is Synthesized

The energy yield of the light-dependent reactions is stored in ATP and NADPH. These molecules are used by the light-independent reactions to reduce carbon dioxide (CO_2) to carbohydrate, namely PGAL, which is then converted to all the organic molecules a plant needs.

During the first stage of the Calvin cycle, the enzyme RuBP carboxylase fixes CO_2 to RuBP, producing a six-carbon molecule that immediately breaks down to two C_3 molecules. During the second stage, CO_2 is incorporated into an organic molecule and is reduced to carbohydrate (CH_2O). This step requires the NADPH and some of the ATP from the light-dependent reactions. For every three turns of the Calvin cycle, the net gain is one PGAL molecule; the other five PGAL molecules are used to re-form three molecules of RuBP. This step also requires ATP for energy.

In C_4 plants, as opposed to the C_3 plants just described, the enzyme PEPCase fixes carbon dioxide to PEP to form a four-carbon molecule, oxaloacetate, within mesophyll cells. A reduced form of this molecule is pumped into bundle sheath cells where CO_2 is released to the Calvin cycle. C_4 plants avoid photorespiration by a partitioning of pathways in space: carbon dioxide fixation occurs in mesophyll cells and the Calvin cycle occurs in bundle sheath cells.

During CAM photosynthesis, PEPCase fixes CO_2 to PEP at night. The next day, CO_2 is released and enters the Calvin cycle within the same cells. This represents a partitioning of pathways in time: carbon dioxide fixation occurs at night and the Calvin cycle occurs during the day. The plants that carry on CAM are desert plants, in which the stomates only open at night in order to conserve water.

Reviewing the Chapter

1. Why is it proper to say that almost all living things are dependent on solar energy? 114
2. Discuss the electromagnetic spectrum and the absorption spectrum of chlorophyll. Why is chlorophyll *a* green pigment? 114–15
3. Name the two major components of chloroplasts and associate each portion with the two sets of reactions that occur during photosynthesis. How are the two pathways related? 116
4. What roles do PS I and PS II play during the light-dependent reactions? 118
5. Trace the cyclic electron pathway, naming and explaining the main events that occur as the electrons cycle. 118
6. Trace the noncyclic electron pathway, naming and explaining all the events that occur as the electrons move from water to $NADP^+$. 119
7. Explain what is meant by chemiosmotic ATP synthesis, and relate this process to the electron transport system present in the thylakoid membrane. 120
8. How is the thylakoid membrane organized? Name the main complexes in the membrane and give a function for each. 120
9. Describe the three stages of the Calvin cycle. Which stage utilizes the ATP and NADPH from the light-dependent reactions? 122–23
10. Explain C_4 photosynthesis, contrasting the actions of rubisco and PEPCase. 124–25
11. Explain CAM photosynthesis, contrasting it to C_4 photosynthesis in terms of partitioning of a pathway. 125

Testing Yourself

Choose the best answer for each question.

1. The absorption spectrum of chlorophyll
   a. approximates the action spectrum of photosynthesis.
   b. explains why chlorophyll is a green pigment.
   c. shows that some colors of light are absorbed more than others.
   d. All of these are correct.
2. The final acceptor of electrons during the noncyclic electron pathway is
   a. PS I.
   b. PS II.
   c. ATP.
   d. $NADP^+$.
3. A photosystem contains
   a. pigments, a reception center, and an electron acceptor.
   b. ADP, Ⓟ, and hydrogen ions (H^+).
   c. protons, photons, and pigments.
   d. Both b and c are correct.
4. Which of these should not be associated with the electron transport system?
   a. cytochromes
   b. movement of H^+ into the thylakoid space
   c. formation of ATP
   d. absorption of solar energy
5. PEPCase has an advantage compared to rubisco. The advantage is that
   a. PEPCase is present in both mesophyll and bundle sheath cells, but rubisco is not.
   b. rubisco fixes carbon dioxide (CO_2) only in C_4 plants, but PEPCase does it in both C_3 and C_4 plants.
   c. rubisco combines with O_2, but PEPCase does not.
   d. Both b and c are correct.
6. The NADPH and ATP from the light-dependent reactions are used to
   a. cause rubisco to fix CO_2.
   b. reform the photosystems.
   c. cause electrons to move along their pathways.
   d. convert PGA to PGAL.
7. CAM photosynthesis
   a. is the same as C_4 photosynthesis.
   b. is an adaptation to cold environments in the Southern Hemisphere.
   c. is prevalent in desert plants that close their stomates during the day.
   d. occurs in plants that live in marshy areas.
8. Chemiosmotic ATP synthesis depends on
   a. an electrochemical gradient.
   b. a difference in H^+ concentration between the thylakoid space and the stroma.
   c. ATP breaking down to ADP + Ⓟ.
   d. Both a and b are correct.
9. Label this diagram of a chloroplast.

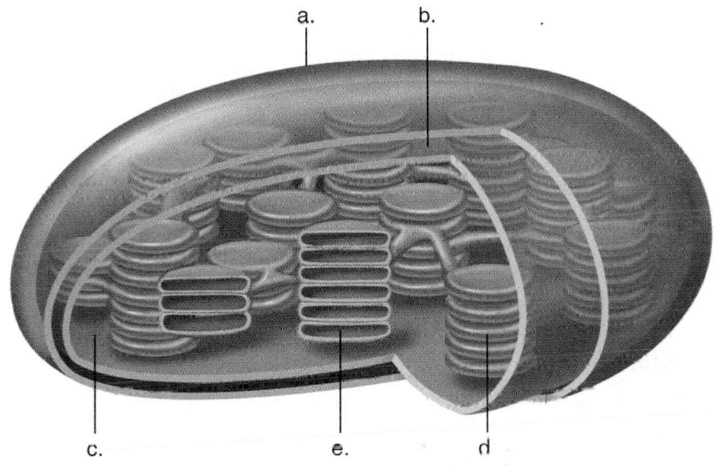

   f. The light-dependent reactions occur in which part of a chloroplast?
   g. The light-independent reactions occur in which part of a chloroplast?
10. Label this diagram using these labels: water, carbohydrate, carbon dioxide, oxygen, ATP, ADP + Ⓟ, NADPH, and $NADP^+$.

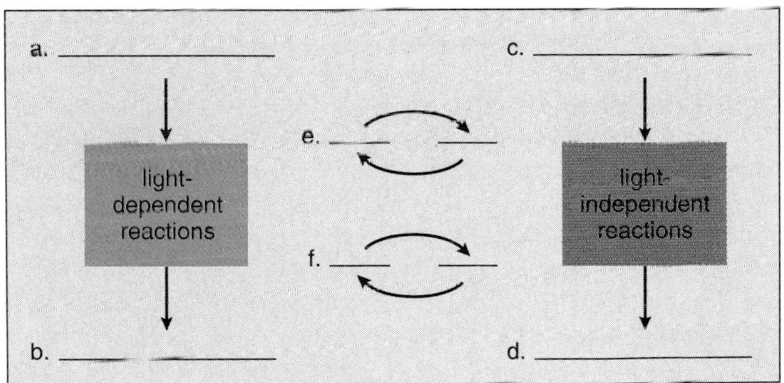

Applying the Concepts

1. *Virtually all living things are dependent on solar energy.* What data would you use to convince a friend that the lives of human beings are dependent on solar energy?
2. *Photosynthesis makes solar energy available to living things.* Explain how photosynthesis makes solar energy available by explaining the roles of chlorophyll, ATP, $NADP^+$, and carbon dioxide (CO_2) in photosynthesis.
3. *The cell and the organelles themselves are compartmentalized, and this assists their proper functioning.* Name two major compartments (separate areas) of a chloroplast, and tell how photosynthesis is dependent on keeping the areas separate.

Using Technology

Your study of photosynthesis is supported by these available technologies:

Exploring the Internet

The Mader Home Page provides resources for and help with studying this chapter.

http://www.mhhe.com/sciencemath/biology/mader/
(Click on Biology.)

Explorations in Cell Biology & Genetics CD-ROM

Photosynthesis (#9)

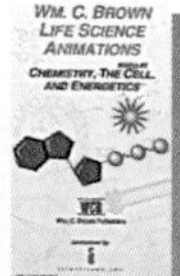

Life Science Animations Video

Video #1: Chemistry, the Cell, and Energetics
The Photosynthetic Electron Transport Chain and Production of ATP (#8)
C_3 Photosynthesis (Calvin Cycle) (#9)
C_4 Photosynthesis (#10)

Understanding the Terms

C_3 plant 124
C_4 plant 124
Calvin cycle 121
CAM plant 124
carbon dioxide (CO_2) fixation 122
chlorophyll 116
chloroplast 116
electromagnetic spectrum 114
electron transport system 118
granum 116
light-dependent reactions 116
light-independent reactions 116
photon 114
photorespiration 125
photosynthesis 114
photosystem 118
RuBP (ribulose bisphosphate) 122
stroma 116
thylakoid 116

Match the terms to these definitions:

a. __________ Discrete packet of solar energy; the amount of energy in a packet is inversely related to the wavelength of the packet.

b. __________ Energy-capturing portion of photosynthesis that takes place in thylakoid membranes of chloroplasts and cannot proceed without solar energy; it produces ATP and NADPH.

c. __________ Green pigment that absorbs solar energy and is important in photosynthesis.

d. __________ Large, central compartment in a chloroplast that is fluid filled and contains enzymes used in photosynthesis.

e. __________ Membrane-bounded organelle with chlorophyll-containing membranous thylakoids; where photosynthesis takes place.

f. __________ Photosynthetic unit where solar energy is absorbed and high-energy electrons are generated; contains a pigment complex and an electron acceptor.

g. __________ Passage of electrons along a series of carrier molecules from a higher to a lower energy level; the energy released is used for the synthesis of ATP.

h. __________ Process usually occurring within chloroplasts whereby chlorophyll traps solar energy and carbon dioxide is reduced to a carbohydrate.

i. __________ Series of photosynthetic reactions in which carbon dioxide is fixed and reduced in the chloroplast.

j. __________ Synthesis portion of photosynthesis that takes place in the stroma of chloroplasts and does not directly require solar energy; it uses the products of the light-dependent reactions to reduce carbon dioxide to a carbohydrate.

Cellular Respiration

CHAPTER 8

Chapter Concepts

Hiking can be strenuous

When you go hiking, take an aerobics class, or just sit around, ATP molecules allow your muscles to contract. ATP molecules are produced during aerobic respiration, a process that requires the participation of mitochondria. There are numerous mitochondria in muscle cells. The glucose and oxygen for aerobic respiration are delivered to cells by the circulatory system, and the end products—water and carbon dioxide—are removed by the circulatory system. Glucose enters blood at the digestive tract, and gas exchange occurs in the lungs. When you breathe hard during exercise, oxygen is entering the blood and carbon dioxide is leaving the blood at a faster rate than usual.

Aerobic respiration consists of many small steps mediated by specific enzymes. High-energy electrons are removed from glucose breakdown products and are passed down an electron transport system located on the cristae of mitochondria. As the electrons move from one carrier to the next, energy is released and captured for the production of ATP molecules. The final acceptor for the energy-spent electrons is oxygen, which is reduced to water. One form of chemical energy (glucose) cannot be transformed completely into another (ATP molecules) without the loss of usable energy in the form of heat. When ATP is produced, heat is given off. Since exercise requires plentiful ATP, your body's internal temperature rises and you begin sweating.

8.1 How Cells Acquire ATP

Cellular respiration includes all the various metabolic pathways which break down carbohydrates and other metabolites with the concomitant buildup of ATP. **Aerobic respiration** [Gk. *aeros*, air], which requires oxygen, involves the complete breakdown of glucose to carbon dioxide (CO_2) and water (H_2O):

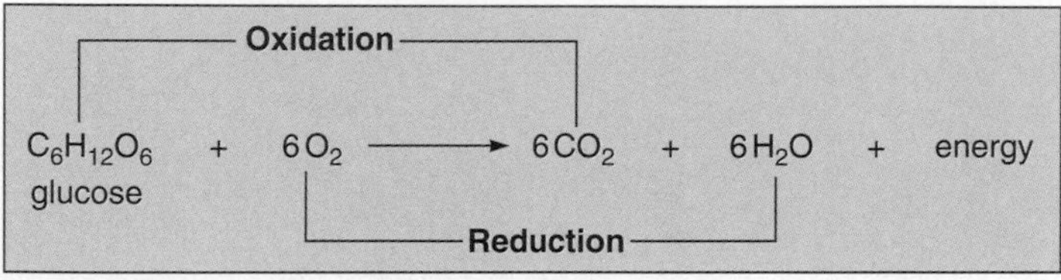

Glucose is a high-energy molecule, and its breakdown products, CO_2 and H_2O, are low-energy molecules. Therefore, we expect the process to be exergonic and release energy. As breakdown occurs, electrons are removed from substrates and eventually are received by oxygen atoms, which then combine with H^+ to become H_2O.

The equation shows changes in regard to hydrogen atom (H) distribution. But remember that a hydrogen atom consists of a hydrogen ion plus an electron ($H^+ + e^-$). When hydrogen atoms are removed from glucose, so are electrons. Since oxidation is the loss of electrons, and reduction is the gain of electrons, glucose breakdown is an oxidation-reduction reaction. Glucose is oxidized and O_2 is reduced.

On the other hand, the buildup of ATP is an endergonic reaction that requires energy.

$$\text{energy} + \text{ADP} + \text{Ⓟ} \longrightarrow \text{ATP}$$

The pathways of aerobic respiration allow the energy within a glucose molecule to be released slowly so that ATP can be produced gradually. Cells would lose a tremendous amount of energy if glucose breakdown occurred all at once—much energy would become nonusable heat. The step-by-step breakdown of glucose to carbon dioxide and water usually realizes a maximum yield of 36 ATP molecules. The energy in 36 ATP molecules is equivalent to about 40% of the energy that was available in glucose.

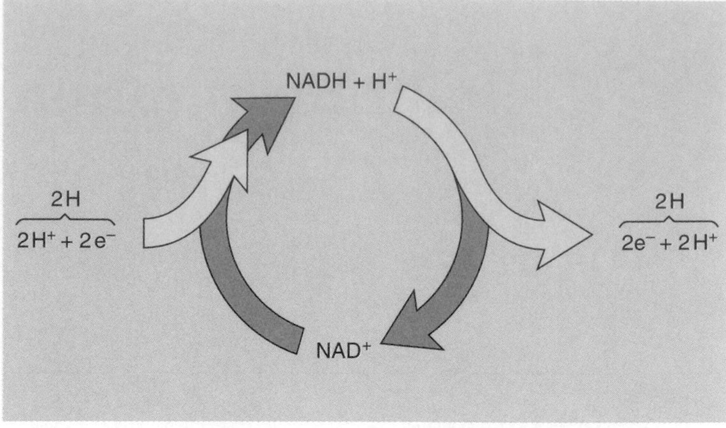

Figure 8.1 The NAD⁺ cycle.
The coenzyme NAD⁺ accepts two electrons (e⁻) plus a hydrogen ion (H⁺) and becomes NADH + H⁺. When NADH passes the electrons to another substrate or carrier and gives off a hydrogen ion, it becomes NAD⁺ again.

NAD⁺ Is a Carrier of Electrons

Cellular respiration involves many individual metabolic reactions, each one catalyzed by its own enzyme. Enzymes of particular significance are the dehydrogenase enzymes that utilize the coenzyme **NAD⁺.** When a metabolite is oxidized, NAD⁺ accepts two electrons plus a hydrogen ion (H^+) and NADH results. The electrons received by NAD⁺ are high-energy electrons that are usually carried to the electron transport system. Figure 8.1 illustrates how NAD⁺ carries electrons.

NAD⁺ is called a coenzyme of oxidation-reduction because it can oxidize a metabolite (accept electrons) and can reduce a metabolite (give up electrons). Only a small amount of NAD⁺ need be present in a cell, because each NAD⁺ molecule is used over and over again. **FAD** is another coenzyme of oxidation-reduction which is sometimes used instead of NAD⁺. FAD accepts two electrons and two hydrogen ions (H^+) to become $FADH_2$.

NAD⁺ and FAD are two coenzymes of oxidation-reduction that are active during cellular respiration.

Metabolic Pathways Are Required

Aerobic respiration involves these metabolic pathways and one individual reaction:

- **Glycolysis** is the breakdown of glucose to two molecules of pyruvate. Enough energy is released for the immediate buildup of two ATP. Glycolysis takes place outside the mitochondria and does not utilize oxygen.
- During the **transition reaction,** pyruvate is oxidized to an acetyl group, and CO_2 is removed.
- The **Krebs cycle** is a cyclical series of oxidation reactions that release CO_2 and produce ATP. Altogether, the Krebs cycle accounts for two immediate ATP molecules per glucose molecule.
- The **electron transport system** is a series of carriers that accept the electrons removed from glucose and pass them along from one carrier to the next until they are finally received by oxygen. As the electrons pass from a higher energy to a lower energy state, energy is released and stored for ATP production.

Pyruvate is a pivotal metabolite in cellular respiration. If oxygen is not available to the cell, fermentation, an anaerobic process, occurs in the cytosol. During fermentation, glucose is incompletely metabolized to lactate or to carbon dioxide and alcohol, depending on the organism. As you can see from Figure 8.2, fermentation results in a net gain of only two ATP molecules per glucose molecule.

The complete breakdown of glucose to carbon dioxide and water is aerobic and requires oxygen, which is utilized in mitochondria. Fermentation is an anaerobic process that takes place in the cytosol and results in an incomplete breakdown of glucose.

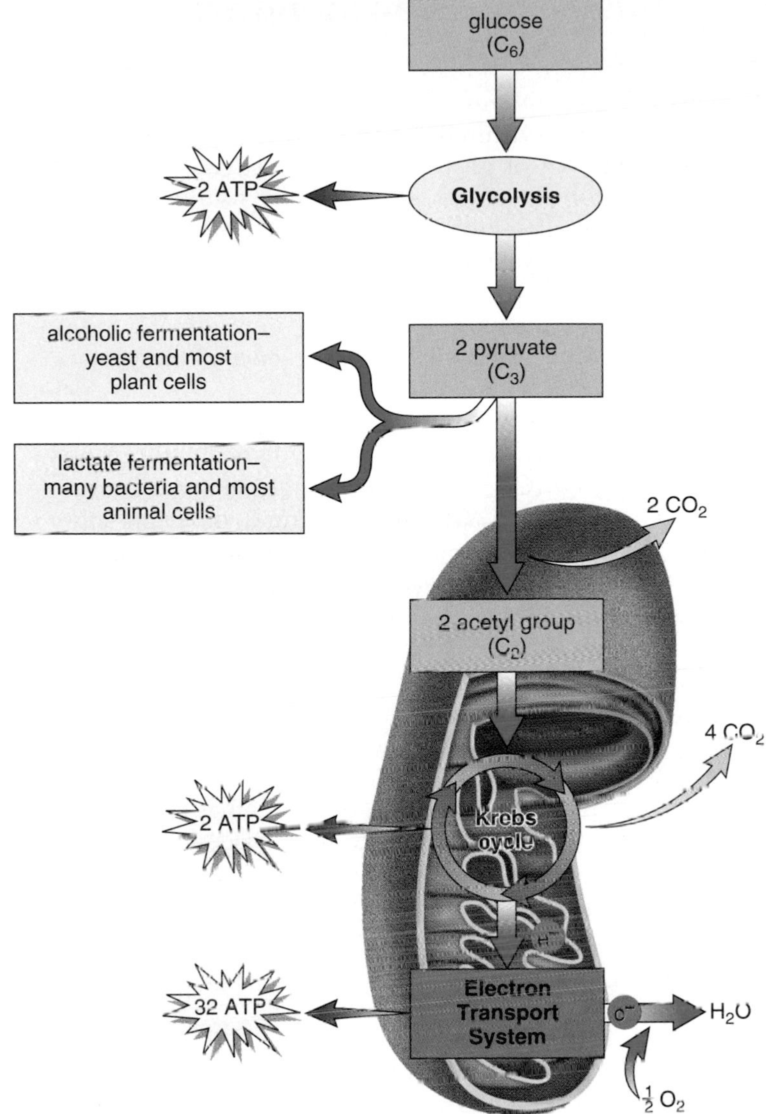

Figure 8.2 Cellular respiration overview.
Glycolysis in the cytosol produces pyruvate, a key intermediary metabolite. If oxygen is not available, fermentation occurs as pyruvate is reduced to either alcohol or lactate, depending on the type of cell. If oxygen is available, aerobic respiration continues in a mitochondrion where pyruvate is metabolized completely to carbon dioxide and water. The net gain of ATP for glycolysis (and fermentation) is two ATP; the net gain of ATP for aerobic respiration is 36 ATP.

8.2 Outside the Mitochondria: Glycolysis

Glycolysis [Gk. *glykeros*, sweet, and *lyo*, loose, dissolve], which takes place within the cytosol outside mitochondria, is the breakdown of glucose to two pyruvate molecules (Fig. 8.3). Since glycolysis is universally found in organisms, it most likely evolved before the Krebs cycle and the electron transport system. This may be why glycolysis occurs in the cytosol and does not require oxygen. Bacteria evolved before other organisms, and there are some bacteria today that are anaerobic; they die in the presence of oxygen.

Energy Investment Steps

As glycolysis begins, the addition of two phosphate groups activates glucose (C_6) to react. This requires two separate reactions and uses two ATP. The resulting C_6 molecule splits into two C_3 molecules, each of which carries a phosphate group. From this point on, each C_3 molecule undergoes the same series of reactions. Because each succeeding reaction occurs twice, it is preceded by a 2× in Figure 8.3.

Energy Harvesting Steps

During glycolysis, oxidation occurs by the removal of electrons which are accepted by NAD. Each NAD accepts two electrons and one hydrogen ion (H^+); altogether two NADH molecules result. Enough energy is released to generate four ATP molecules by **substrate-level phosphorylation** [Gk. *phos*, light, and *phoreus*, carrier], so called because a phosphate is enzymatically transferred from a substrate to ADP. Subtracting the two ATP that were used to get started, there is a net gain of two ATP from glycolysis.

When glycolysis is followed by the breakdown of pyruvate, this molecule enters a mitochondrion, where oxygen is utilized. If oxygen is not available, glycolysis becomes a part of fermentation (see page 141).

Altogether the inputs and outputs of glycolysis are as follows:

Glycolysis	
inputs	outputs
glucose 2 NAD^+ 2 ATP 2 ADP + 2 Ⓟ	2 pyruvate 2 NADH 4 ATP (net 2 ATP)

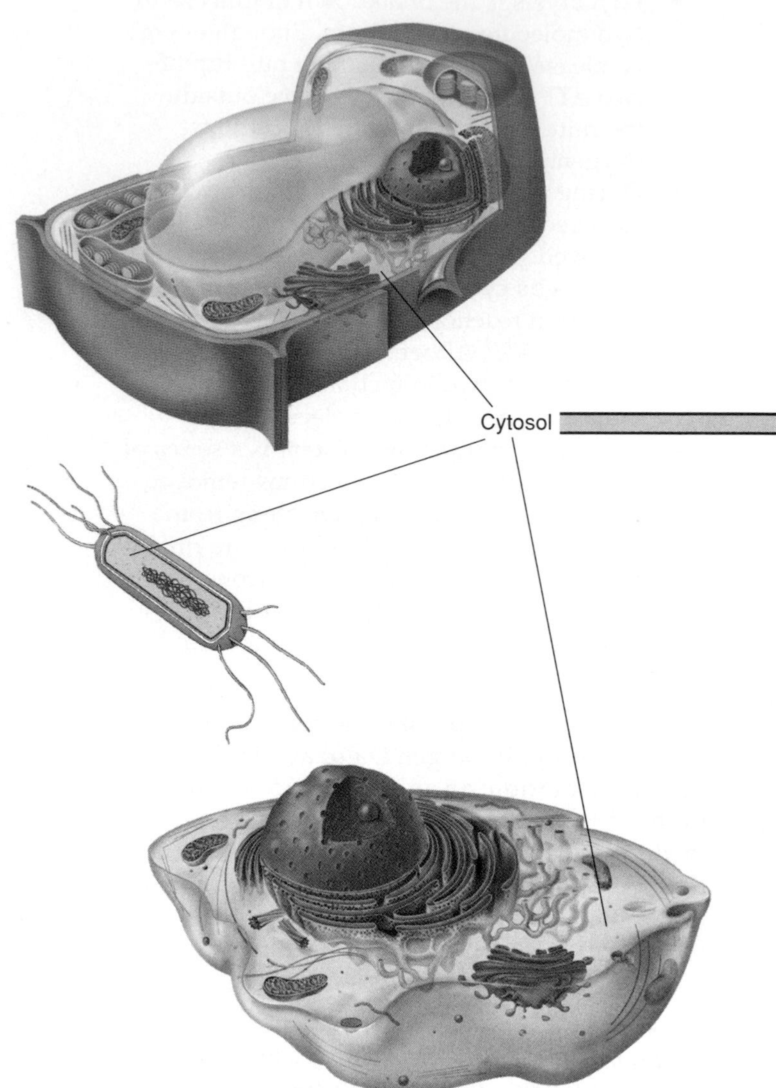

Figure 8.3 Glycolysis.
This metabolic pathway begins with glucose and ends with pyruvate. Net gain of two ATP molecules can be calculated by subtracting those expended from those produced. Text in boxes to the far right explains the reactions.

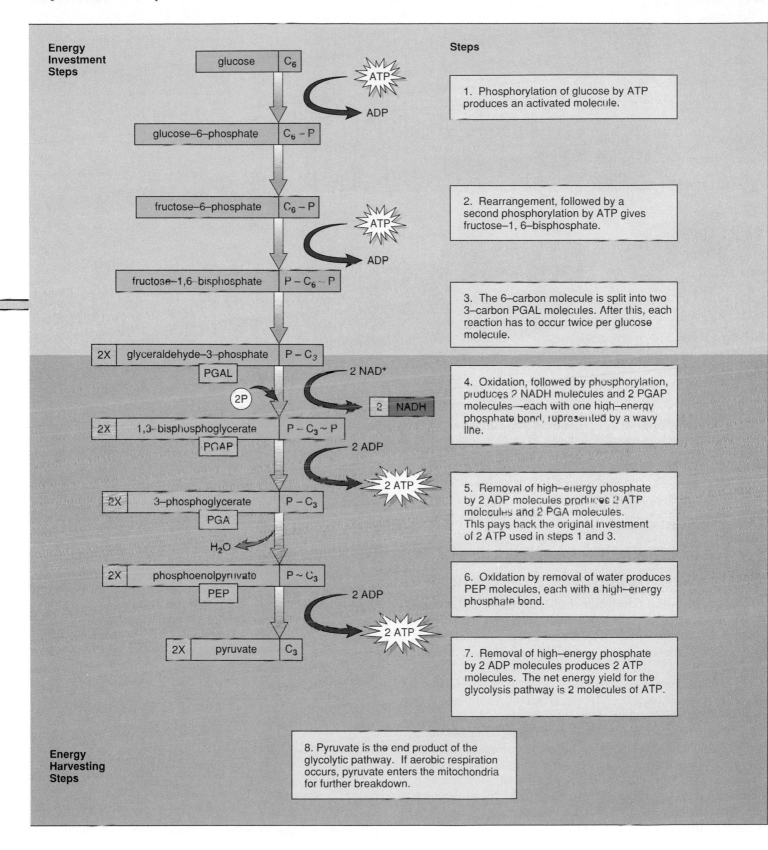

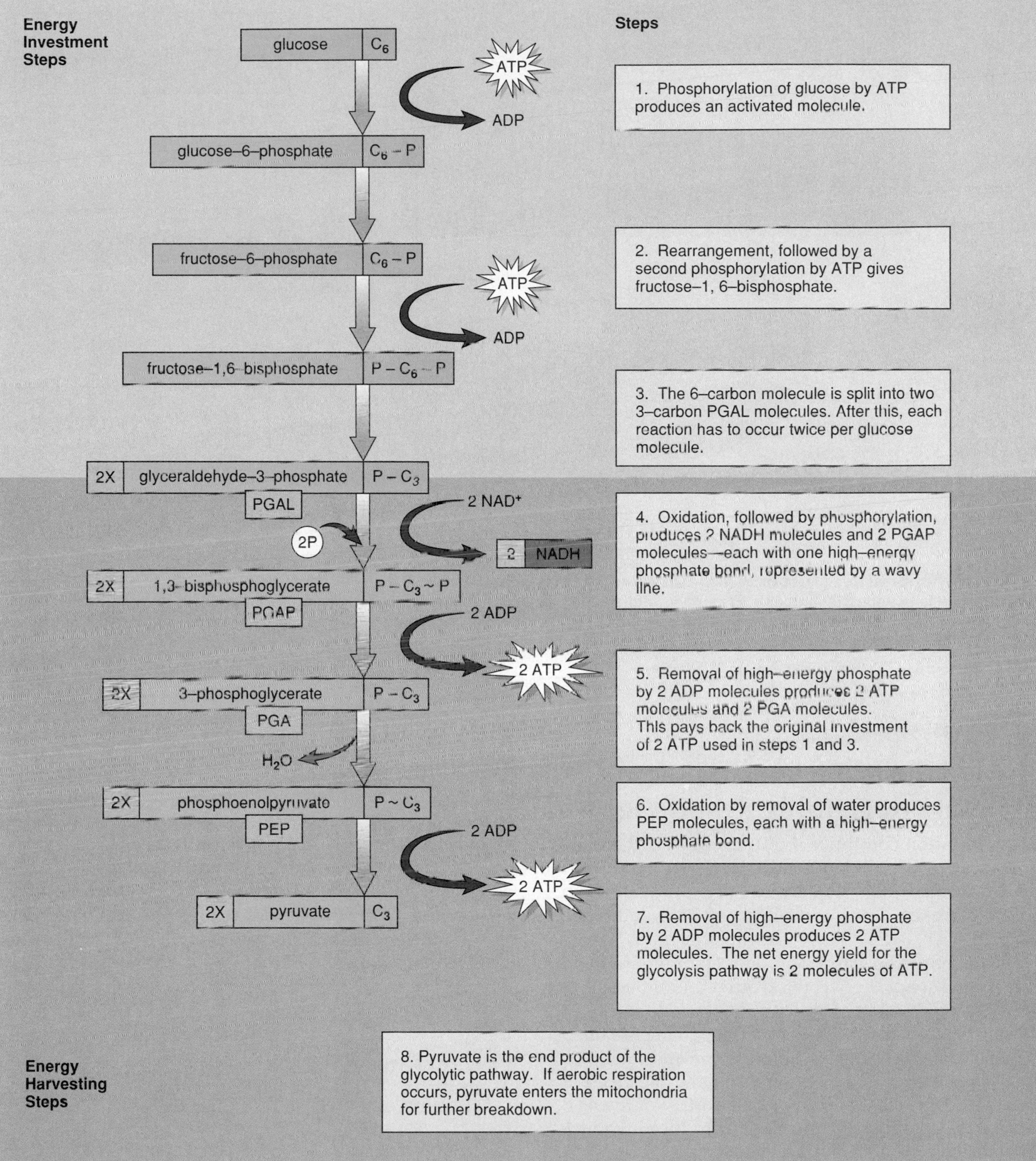
Energy Investment Steps
glucose C_6
ATP
ADP
glucose–6–phosphate $C_6 - P$
fructose–6–phosphate $C_6 - P$
ATP
ADP
fructose–1,6–bisphosphate $P - C_6 - P$
2X glyceraldehyde–3–phosphate $P - C_3$
PGAL
2P
2 NAD^+
2 NADH
2X 1,3–bisphosphoglycerate $P - C_3 \sim P$
PGAP
2 ADP
2 ATP
2X 3–phosphoglycerate $P - C_3$
PGA
H_2O
2X phosphoenolpyruvate $P \sim C_3$
PEP
2 ADP
2 ATP
2X pyruvate C_3
Energy Harvesting Steps
Steps
1. Phosphorylation of glucose by ATP produces an activated molecule.
2. Rearrangement, followed by a second phosphorylation by ATP gives fructose–1, 6–bisphosphate.
3. The 6–carbon molecule is split into two 3–carbon PGAL molecules. After this, each reaction has to occur twice per glucose molecule.
4. Oxidation, followed by phosphorylation, produces 2 NADH molecules and 2 PGAP molecules—each with one high–energy phosphate bond, represented by a wavy line.
5. Removal of high–energy phosphate by 2 ADP molecules produces 2 ATP molecules and 2 PGA molecules. This pays back the original investment of 2 ATP used in steps 1 and 3.
6. Oxidation by removal of water produces PEP molecules, each with a high–energy phosphate bond.
7. Removal of high–energy phosphate by 2 ADP molecules produces 2 ATP molecules. The net energy yield for the glycolysis pathway is 2 molecules of ATP.
8. Pyruvate is the end product of the glycolytic pathway. If aerobic respiration occurs, pyruvate enters the mitochondria for further breakdown.

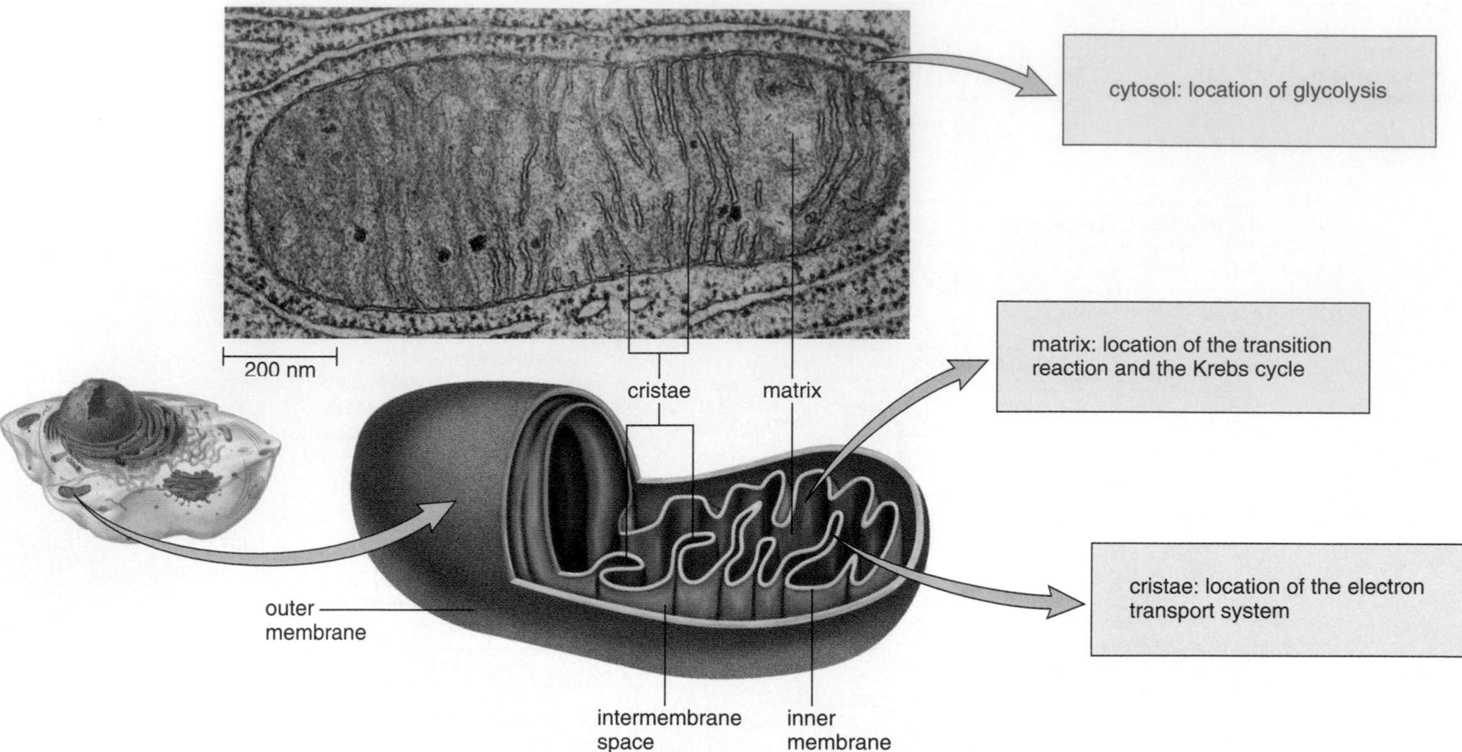

Figure 8.4 Mitochondrion structure and function.
The mitochondrion is bounded by a double membrane with an intermembrane space. The inner membrane invaginates to form the shelflike cristae. Glycolysis takes place in the cytosol outside mitochondria. The Krebs cycle occurs within the matrix of a mitochondrion. The electron transport system is located on the cristae of a mitochondrion.

8.3 Inside the Mitochondria: Completion of Aerobic Respiration

Aerobic respiration continues in mitochondria where the transition reaction, Krebs cycle, and the electron transport system are located (Fig. 8.4). Eventually, pyruvate from glycolysis is broken down completely to carbon dioxide and water.

A **mitochondrion** has a double membrane, with an intermembrane space (between the outer and inner membrane). Cristae are folds of inner membrane that jut out into the matrix, the innermost compartment, which is filled with a gel-like fluid. The transition reaction and the Krebs cycle enzymes are located in the matrix, and the electron transport system is located in the cristae. Most of the ATP produced during cellular respiration is produced in mitochondria; therefore, mitochondria are often called the powerhouses of the cell.

Transition Reaction Releases CO_2

The transition reaction is so called because it connects glycolysis to the Krebs cycle. In this reaction, pyruvate is converted to a two-carbon acetyl group attached to coenzyme A, collectively termed **acetyl-CoA,** and carbon dioxide is given off. This is an oxidation reaction in which electrons are removed from pyruvate by a dehydrogenase that uses NAD^+ as a coenzyme. The reaction occurs twice for each original glucose molecule.

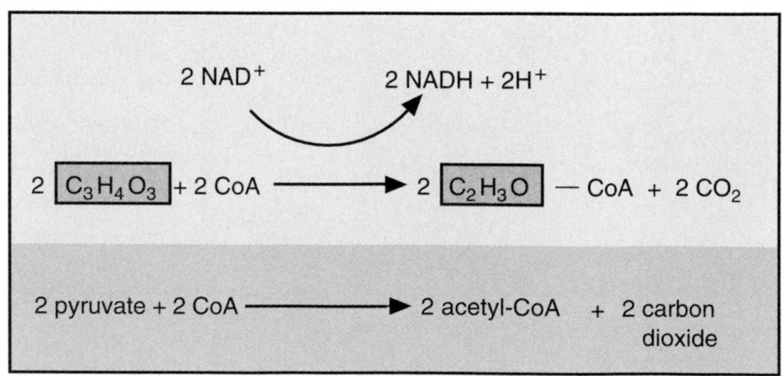

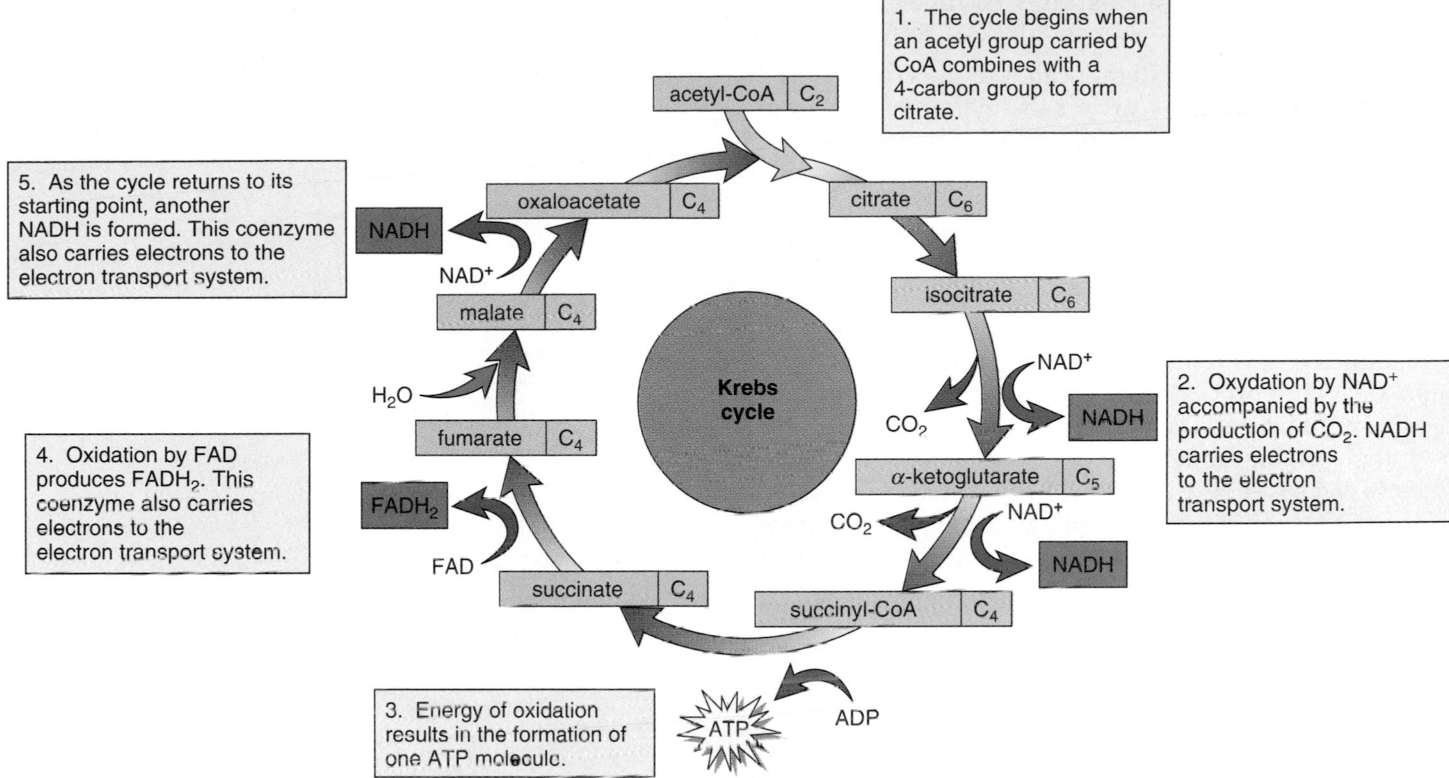

Figure 8.5 Krebs cycle.
The net result of this cycle is the oxidation of an acetyl group to two molecules of carbon dioxide (CO_2). There is a gain of one ATP during the cycle but much energy still resides in the electrons that were removed by three molecules of NAD^+ and one molecule of FAD. NADH and $FADH_2$ then take the electrons to the electron transport system. (Keep in mind that the Krebs cycle turns twice per glucose molecule.)

The Krebs Cycle Finishes Glucose Breakdown

The Krebs cycle is a cyclical metabolic pathway located in the matrix of mitochondria. The Krebs cycle is named for Sir Hans Krebs, a German-born British scientist who received the Nobel Prize for his part in identifying these reactions in the 1930s.

At the start of the Krebs cycle, the C_2 acetyl group produced by the transition reaction joins with a C_4 molecule, forming citrate, a C_6 molecule. For this reason, the cycle is also known as the *citric acid cycle* (Fig. 8.5). As the acetyl group undergoes oxidation by removal of electrons, two molecules of carbon dioxide (CO_2) form. (Carbon dioxide is one of the end products of the complete breakdown of glucose.)

During the oxidation process most of the electrons are accepted by NAD^+ and NADH then forms. But in one instance, electrons are taken by FAD, and $FADH_2$ forms. NADH and $FADH_2$ carry these electrons to the electron transport system. Some of the energy released when oxidation occurs is used immediately to form ATP by substrate-level phosphorylation, as in glycolysis. A high-energy metabolite accepts a phosphate group and subsequently passes it on to ADP so that ATP forms:

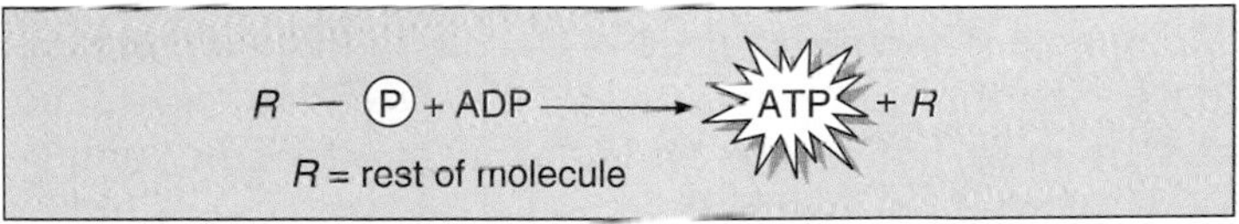

The Krebs cycle turns twice for each original glucose molecule. Therefore, the input and output of the Krebs cycle per glucose molecule are as follows:

Krebs Cycle

inputs	outputs
2 acetyl groups	4 CO_2
2 ADP + 2 (P)	2 ATP
6 NAD^+	6 NADH
2 FAD	2 $FADH_2$

The Electron Transport System Produces Most ATP

The electron transport system located in the cristae of the mitochondria is a series of carriers that pass electrons from one to the other. Notice that some of the protein carriers of the system are cytochrome molecules.

The electrons that enter the electron transport system are carried by NADH and $FADH_2$. Figure 8.6 is arranged to show that high-energy electrons enter the system, and low-energy electrons leave the system. When NADH gives up its electrons, it becomes NAD^+ and the next carrier gains the electrons and is reduced. This oxidation-reduction reaction starts the process, and each of the carriers in turn becomes reduced and then oxidized as the electrons move down the system. As the pair of electrons is passed from carrier to carrier, energy is released and stored to form ATP molecules. Oxygen receives the energy-spent electrons from the last of the carriers. After receiving electrons, oxygen combines with hydrogen ions and water forms:

$$1/2\,O_2 + 2e^- + 2H^+ \longrightarrow H_2O$$

The term **oxidative phosphorylation** refers to the production of ATP as a result of energy released by the electron transport system. Just how ATP production occurs is discussed in the next section.

When NADH delivers electrons to the first carrier of the electron transport system, enough energy is released by the time the electrons are received by O_2 to permit the production of three ATP molecules. When $FADH_2$ delivers electrons to the electron transport system, only two ATP are produced.

The cell needs only a limited supply of the coenzymes NAD^+ and FAD because they are constantly being recycled and reused. Therefore, once NADH has delivered electrons to the electron transport system, it is "free" to return and pick up more hydrogens. In the same manner, the components of ATP are recycled in cells. Energy is required to join ADP + Ⓟ, and then when ATP is used to do cellular work, ADP and Ⓟ are made available once more. The recycling of coenzymes and ADP increases cellular efficiency since it does away with the necessity to synthesize large quantities of NAD^+, FAD, and ADP anew.

As electrons pass down the electron transport system, energy is released and ATP is produced.

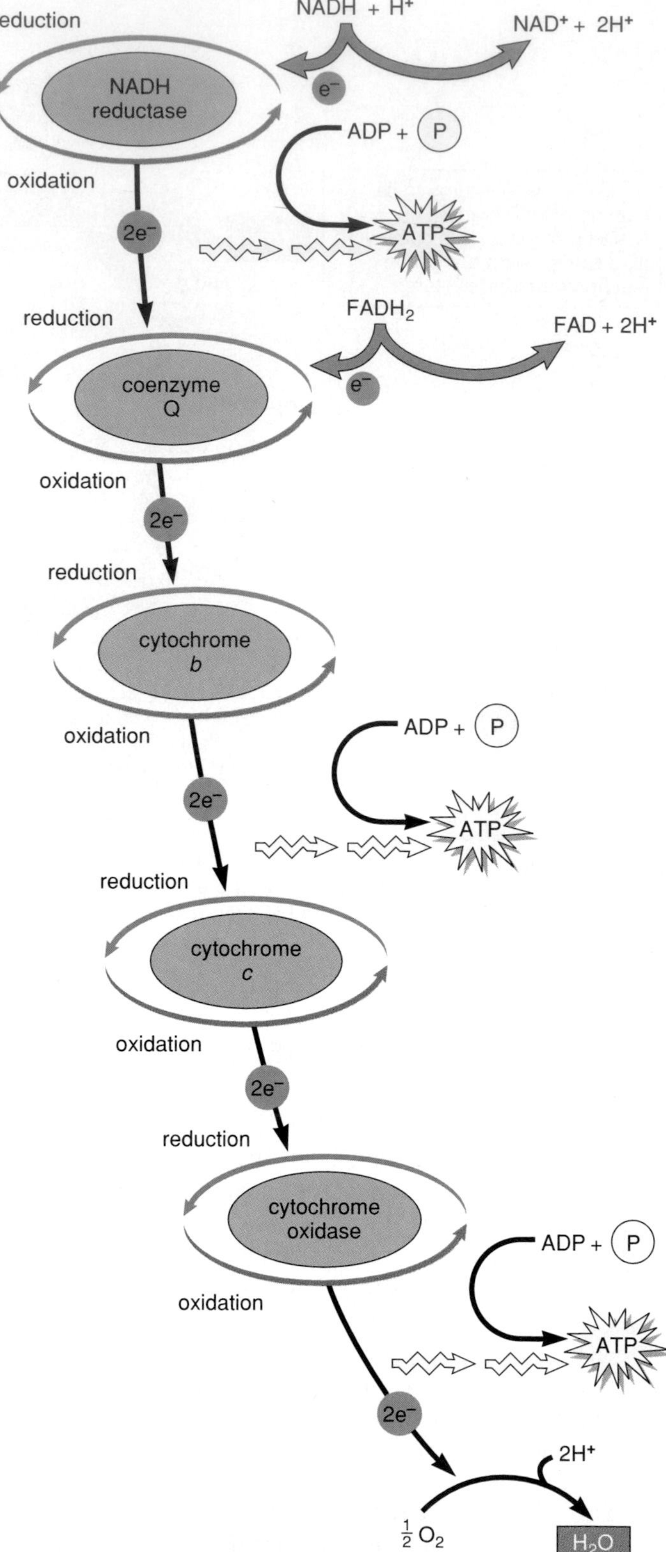

Figure 8.6 The electron transport system.
NADH and $FADH_2$ bring electrons to the electron transport system. As the electrons move down the system, energy is released and used to form ATP. For every pair of electrons that enters by way of NADH, three ATP result. For every pair of electrons that enters by way of $FADH_2$, two ATP result. Oxygen, the final acceptor of the electrons, becomes a part of water.

The Cristae Are Organized

The carriers of the electron transport system and the proteins concerned with ATP synthesis are spatially arranged in a particular manner in the cristae of mitochondria (Fig. 8.7). Essentially the electron transport system consists of three protein complexes and two protein mobile carriers. The mobile carriers transport electrons between the complexes. The complexes are:

NADH dehydrogenase complex, which carries out oxidation of NADH; the electrons enter the electron transport system, and when they are passed on energy becomes available to pump hydrogen ions (H^+) from the matrix into the intermembrane space;

cytochrome b-c *complex,* which receives electrons and also pumps H^+ into the intermembrane space;

cytochrome oxidase complex, which receives electrons and passes them on to oxygen. Once again, energy is made available to pump hydrogen ions (H^+) into the intermembrane space.

Thus far we have been stressing that the carriers accept electrons, which they pass from one to the other. What happens to the hydrogen ions (H^+) carried by NADH and $FADH_2$? The carriers of the electron transport system use the released energy to pump these hydrogen ions from the matrix into the intermembrane space of a mitochondrion. The vertical arrows in Figure 8.7 show that the NADH dehydrogenase complex, the cytochrome *b-c* complex, and the cytochrome oxidase complex all pump H^+ into the intermembrane space. This establishes a strong electrochemical gradient; there are about ten times as many hydrogen ions in the intermembrane space than there are in the matrix.

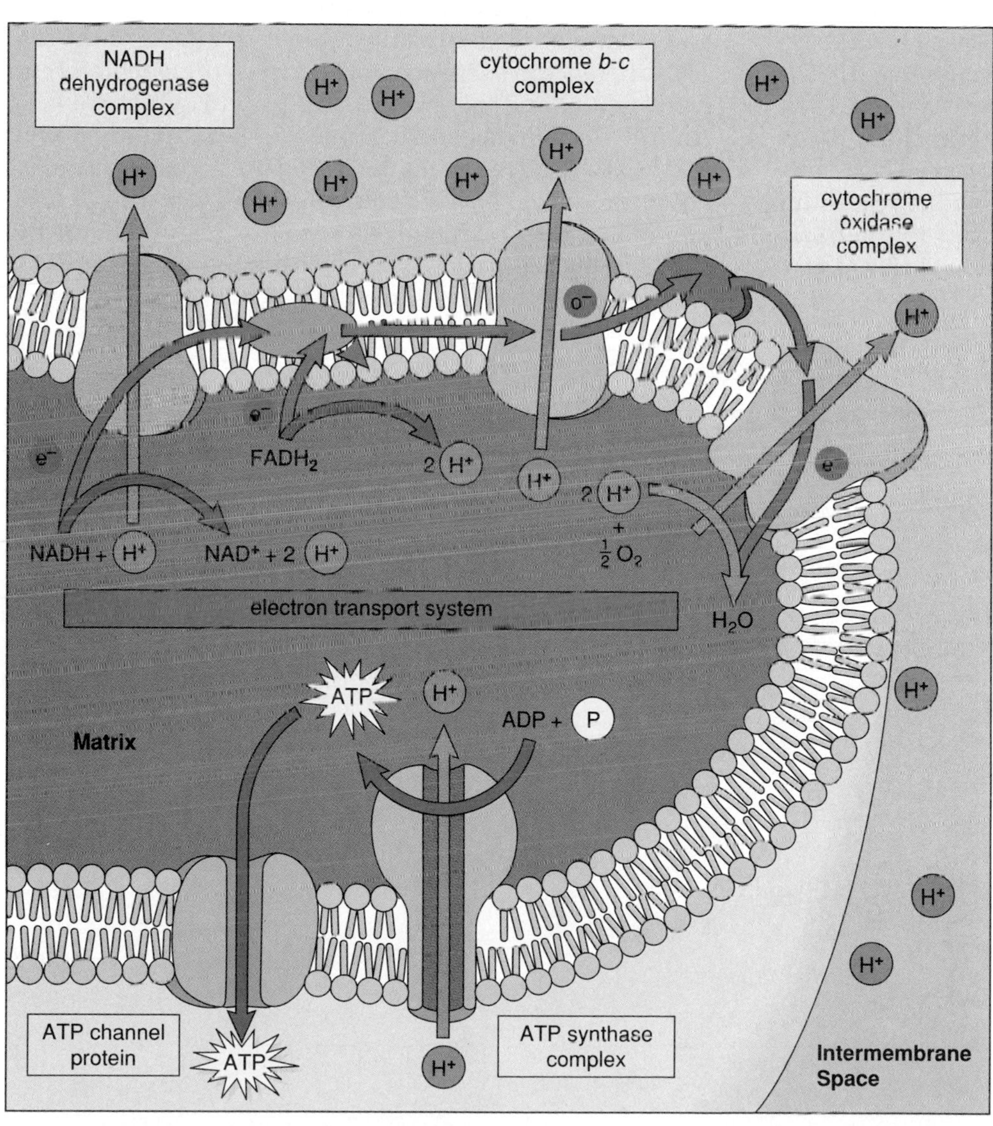

Figure 8.7 Organization of cristae.
The electron transport system is located in the cristae. As electrons move from one complex to the other, hydrogen ions (H^+) are pumped from the matrix into the intermembrane space. As hydrogen ions flow down their concentration gradient from the intermembrane space into the matrix, ATP is synthesized by the enzyme ATP synthase, which is a part of the ATP synthase complex. ATP leaves the matrix by way of a channel protein.

ATP Production The cristae contain an ATP synthase complex through which hydrogen ions flow down their gradient from the intermembrane space into the matrix. As hydrogen ions flow from high to low concentration, the enzyme ATP synthase synthesizes ATP from ADP + Ⓟ. Just how H^+ flow drives ATP synthesis is not known; perhaps the hydrogen ions participate in the reaction, or perhaps they cause a change in the shape of ATP synthase and this brings about ATP synthesis.

The finding of respiratory poisons has lent support to the chemiosmotic theory of ATP synthesis. When one poison inhibits ATP synthesis, the H^+ gradient becomes larger than usual, and when another type makes the membrane leaky to H^+, no ATP is made because of a lack of H^+ gradient.

Once formed, ATP molecules diffuse out of the mitochondrial matrix by way of a channel protein.

Mitochondria synthesize ATP by chemiosmosis. ATP production is dependent upon an electrochemical gradient established by the pumping of H^+ into the intermembrane space.

Calculating the Energy Yield from Glucose Metabolism

Figure 8.8 calculates the ATP yield for the complete breakdown of glucose to CO_2 and H_2O in eukaryotes. The diagram includes the number of ATP produced by substrate phosphorylation and the number that is produced as a result of electrons passing down the electron transport system.

Substrate-Level Phosphorylation: Small Yield

Per glucose molecule, there is a net gain of two ATP from glycolysis in the cytosol and two more from the Krebs cycle in the matrix of mitochondria. Altogether, four ATP are formed outside the electron transport system.

Oxidative Phosphorylation: Great Yield

Most of the ATP produced by aerobic respiration is due to oxidative phosphorylation and energy released by the electron transport system. Per glucose molecule, ten NADH and two $FADH_2$ take electrons to the electron transport system. For each NADH formed *inside* the mitochondria by the Krebs cycle, three ATP result, but for each $FADH_2$, only two ATP are produced. Figure 8.6 explains the reason for this difference: $FADH_2$ delivers its electrons to the transport system after NADH and therefore these electrons cannot account for as much ATP production.

What about the ATP yield of NADH generated *outside* the mitochondria by the glycolytic pathway? NADH cannot cross mitochondrial membranes, but there is a "shuttle" mechanism that allows its electrons to be delivered to the electron transport system inside the mitochondria. The shuttle consists of an organic molecule, which can cross the outer membrane, accept the electrons, and in most cells, deliver them to a FAD molecule in the inner membrane. If FAD is utilized, only two ATP result.

Heart and liver cells, which have high metabolic rates, are exceptions; in these cells cytoplasmic NADH results in the production of three ATP. Also, since prokaryotes (bacteria) do not have mitochondria, each NADH produces three ATP for a total of 38 ATP.

How Efficient Is Aerobic Respiration?

It is interesting to calculate how much of the energy in a glucose molecule eventually becomes available to the cell. The difference in energy content between the reactants (glucose and O_2) and the products (CO_2 and H_2O) is 686 kcal. An ATP phosphate bond has an energy content of 7.3 kcal, and 36 of these are usually produced during glucose breakdown; 36 phosphates are equivalent to a total of 263 kcal. Therefore, 263 kcal/686 kcal × 100, or about 40% of the available energy is usually transferred from glucose to ATP. The rest of the energy is lost in the form of heat.

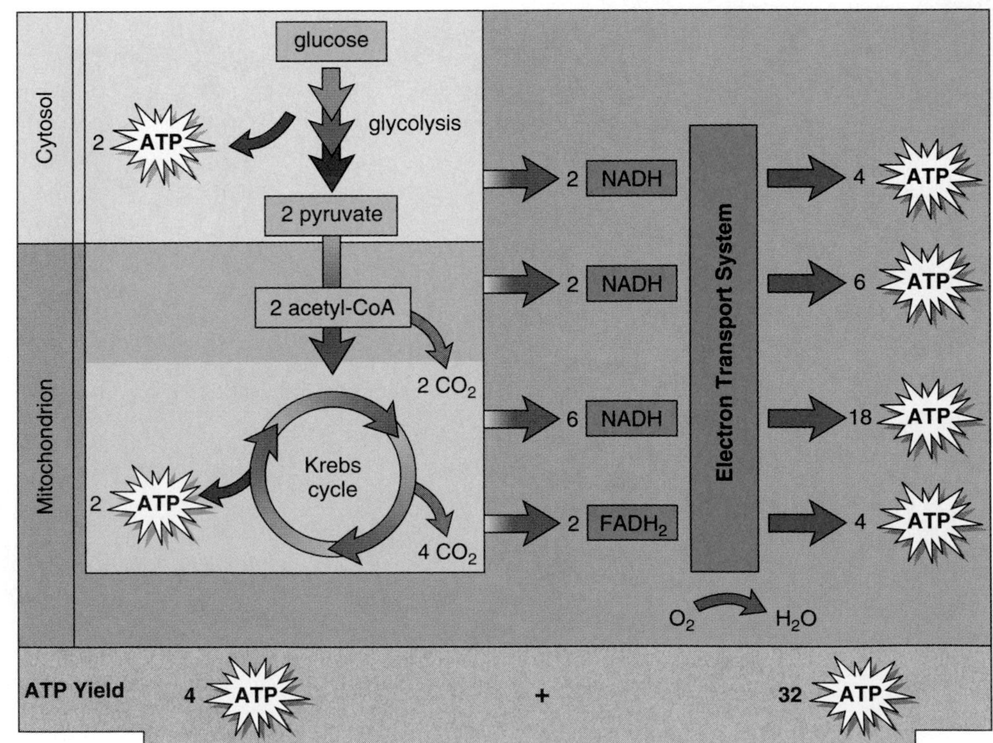

Figure 8.8 Accounting of the energy yield per glucose molecule breakdown in eukaryotes. Substrate level phosphorylation during glycolysis and the Krebs cycle results in the production of four ATP molecules per glucose molecule. Oxidative phosphorylation results in a further 32 ATP, and the grand total of ATP is therefore 36 ATP.

8.4 Metabolic Pool and Biosynthesis

Degradative reactions, which participate in catabolism, break down molecules and tend to be exergonic. Synthetic reactions, which participate in anabolism, tend to be endergonic. Is it correct to say then, that catabolism drives anabolism?

Catabolism: Breaking Down

We already know that glucose is broken down during aerobic cellular respiration. However, other molecules can also undergo catabolism. When a fat is used as an energy source, it breaks down to glycerol and three fatty acids. As Figure 8.9 indicates, glycerol is converted to PGAL, a metabolite in glycolysis. The fatty acids are converted to acetyl-CoA, which enters the Krebs cycle. An 18-carbon fatty acid results in nine acetyl-CoA molecules. In the human body, oxidation of these acetyl-CoA molecules can produce a total of 109 ATP molecules. For this reason, fats are an efficient form of stored energy—there are, after all, three long fatty acid chains per fat molecule.

The carbon skeleton of amino acids can also be broken down. The hydrolysis of proteins results in amino acids whose *R*-group size determines whether the carbon chain is oxidized in glycolysis or the Krebs cycle. The carbon chain is produced in the liver when an amino acid undergoes deamination, or the removal of the amino group. The amino group becomes ammonia (NH_3), which enters the urea cycle and becomes part of urea, the primary excretory product of humans. Just where the carbon skeleton begins degradation is dependent on the length of the *R* group, since this determines the number of carbons left after deamination.

Anabolism: Building Up

We have already mentioned that the ATP produced during catabolism drives anabolism. But there is another way catabolism is related to anabolism. The substrates making up the pathways in Figure 8.9 can be used as starting materials for synthetic reactions. In other words, compounds that enter the pathways are oxidized to substrates that can be used for biosynthesis. This is the cell's **metabolic pool** [Gk. *meta*, implying change] in which one type of molecule can be converted to another. In this way, carbohydrate intake can result in the formation of fat. PGAL can be converted to glycerol, and acetyl groups can be joined to form fatty acids. Fat synthesis follows. This explains why you gain weight from eating too much candy, ice cream, and cake.

Some metabolites of the Krebs cycle can be converted to amino acids through transamination, the transfer of an amino group to an organic acid, forming a different amino acid. Plants are able to synthesize all of the amino acids they need. Animals, however, lack some of the enzymes necessary for synthesis of all amino acids. Adult humans, for example, can synthesize eleven of the common amino acids, but they cannot synthesize the other nine. The amino acids that cannot be synthesized must be supplied by the diet; they are called the *essential amino acids*. The nonessential amino acids can be synthesized. It is quite possible for animals to suffer from protein deficiency if their diet does not contain adequate quantities of all the essential amino acids.

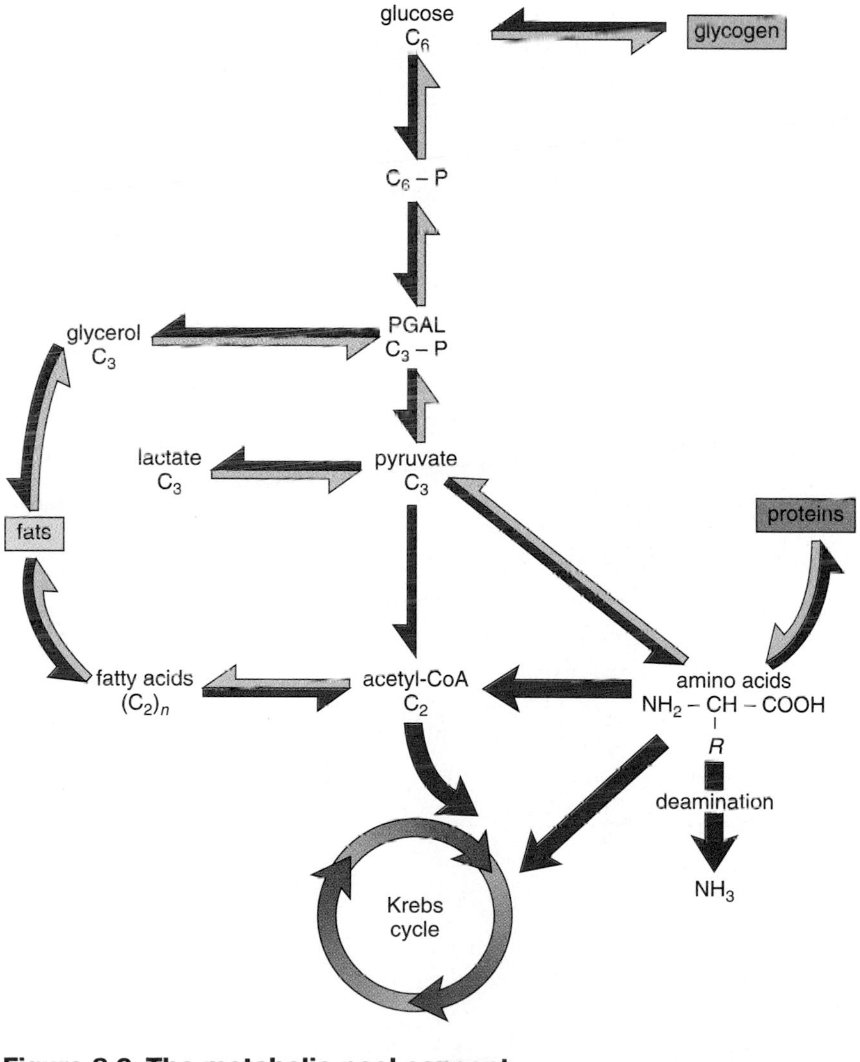

Figure 8.9 The metabolic pool concept.
When they are used as energy sources, carbohydrates, fats, and proteins enter degradative pathways at specific points. Degradation produces metabolites that can also be used for synthesis of other compounds.

All the reactions involved in aerobic cellular respiration are a part of a metabolic pool; the metabolites from the pool can be used for catabolism or for anabolism.

Health Focus

Exercise: A Test of Homeostatic Control

Exercise is a dramatic test of the body's homeostatic control systems—there is a large increase in muscle oxygen (O_2) requirement, and a large amount of carbon dioxide (CO_2) is produced. These changes must be countered by increases in breathing and blood flow to increase oxygen delivery and removal of the metabolically produced carbon dioxide. Also, heavy exercise can produce a large amount of lactic acid due to the utilization of fermentation, an anaerobic process. Both the accumulation of carbon dioxide and lactic acid can lead to an increase in intracellular and extracellular acidity. Further, during heavy exercise, the working muscles produce large amounts of heat that must be removed to prevent overheating. In a strict sense, the body rarely maintains true homeostasis while performing intense exercise or during prolonged exercise in a hot or humid environment. However, a better maintenance of homeostasis is observed in those who have had endurance training.

The number of mitochondria increases in the muscles of persons who train; and, therefore, there is greater reliance on the Krebs cycle and the electron transport system to generate energy. Muscle cells with few mitochondria must have a high ADP concentration to stimulate the limited number of mitochondria to start consuming oxygen. After an endurance training program, the large number of mitochondria start consuming oxygen as soon as the ADP concentration starts rising due to muscle contraction and subsequent breakdown of ATP. Therefore, a steady state of oxygen intake by mitochondria is achieved earlier in the athlete. This faster rise in oxygen uptake at the onset of work means that the oxygen deficit is less, and the formation of lactate due to fermentation is less. Further, any lactate that is produced is removed and processed more quickly.

Training also results in greater reliance on the Krebs cycle and increased fatty acid metabolism, because fatty acids are broken down to acetyl-CoA, which enters the Krebs cycle. This preserves plasma glucose concentration and also helps the body maintain homeostasis.

Source: Scott K. Powers and Edward T. Howley, Exercise Physiology, *2d ed., 1994. McGraw-Hill Companies, Dubuque, IA*

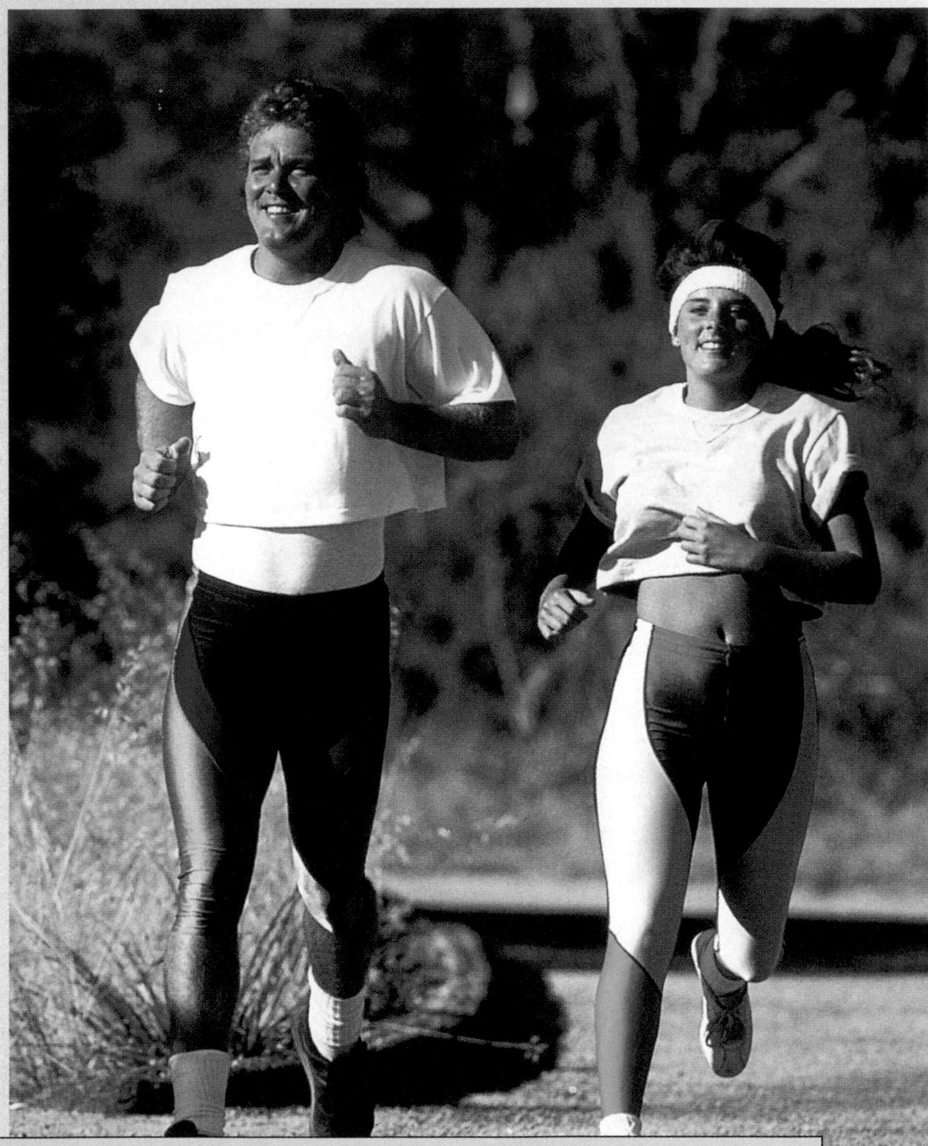

In athletes, there is:

a smaller oxygen deficit due to a more rapid increase in oxygen uptake at the onset of work;

an increase in fat metabolism that spares blood glucose;

a reduction in lactate and hydrogen ion (H^+) formation;

an increase in lactate removal.

8.5 Fermentation

Cellular respiration also includes fermentation. **Fermentation** consists of glycolysis followed by the reduction of pyruvate by NADH to either lactate or alcohol and CO_2 (Fig. 8.10). The pathway operates anaerobically because after NADH transfers its electrons to pyruvate, it is "free" to return and pick up more electrons during the earlier reactions of glycolysis.

Certain anaerobic bacteria such as lactic-acid bacteria, which help us manufacture cheese, consistently produce lactate in this manner. Other bacteria produce chemicals of industrial importance anaerobically: isopropanol, butyric acid, propionic acid, and acetic acid. Yeasts are good examples of organisms that generate alcohol and CO_2. Yeast is used to leaven bread; the CO_2 produced makes bread rise. On the other hand, yeast is used to ferment wine; in that case, it is the ethyl alcohol that is desired. Eventually yeasts are killed by the very alcohol they produce.

Animal, including human, cells are similar to lactic-acid bacteria in that pyruvate, when produced faster than it can be oxidized through the Krebs cycle, is reduced to lactate.

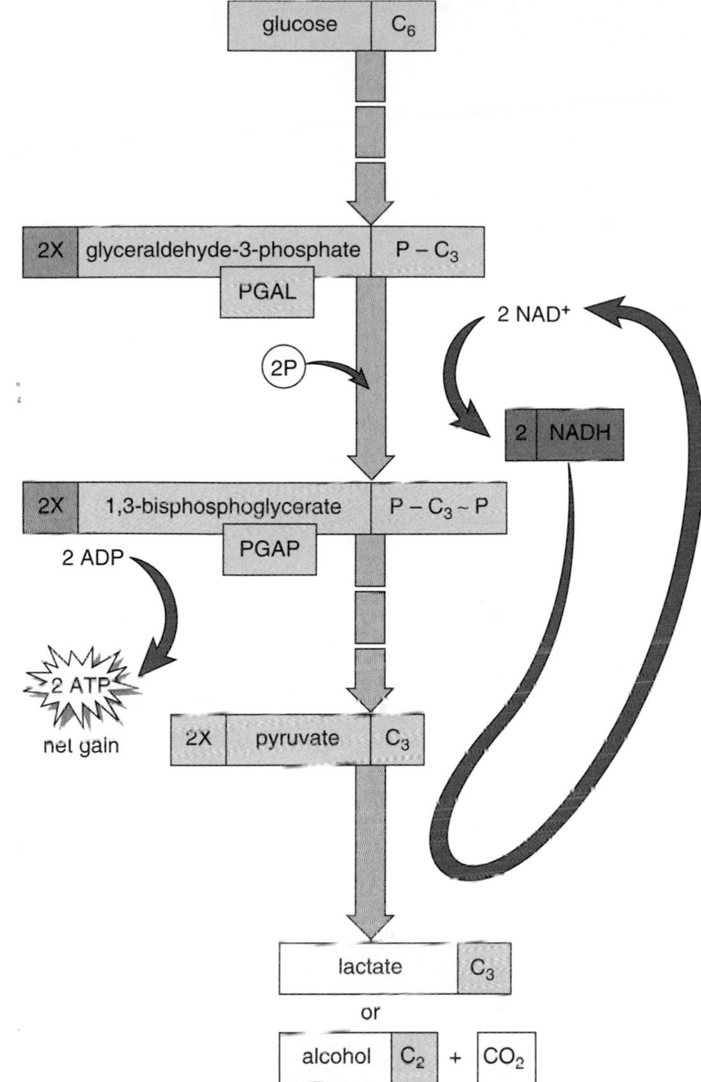

Figure 8.10 Fermentation.
Fermentation consists of glycolysis followed by a reduction of pyruvate. This "frees" NAD^+ and it returns to the glycolytic pathway to pick up more electrons.

Advantages and Disadvantages of Fermentation

Despite its low yield and toxicity, fermentation is essential to humans because it can provide a rapid burst of ATP; muscle cells more than other cells are apt to carry on fermentation. When our muscles are working vigorously over a short period of time, fermentation is a way to produce ATP even though oxygen is temporarily in limited supply.

Lactate, however, is toxic to cells. At first, blood carries away all the lactate formed in muscles. Eventually, however, lactate begins to build up, changing the pH and causing the muscles to fatigue so that they no longer contract. When we stop running, our bodies are in **oxygen debt,** a term that refers to the amount of oxygen needed to restore ATP to its former level and rid the body of lactate. Oxygen debt is signified by the fact that we continue to breathe very heavily for a time. Recovery involves transporting lactate to the liver, where it is converted back to pyruvate. Some of the pyruvate is respired completely, and the rest is converted back to glucose.

How Efficient Is Fermentation?

The two ATP produced per glucose during fermentation is equivalent to 14.6 kcal. Complete glucose breakdown to CO_2 and H_2O represents a possible energy yield of 686 kcal per molecule. Therefore, the efficiency for fermentation is only 14.6 kcal/686 kcal $\times$ 100, or 2.1%. This is much less efficient than the complete breakdown of glucose. The inputs and outputs of fermentation are as shown next.

Fermentation

inputs	outputs
glucose 2 ATP 2 ADP + 2 (P)	2 lactate or 2 alcohol and 2 CO_2 4 ATP (net 2 ATP)

connecting concepts

The organelles of eukaryotic cells have a structure that suits their function. Cells didn't arise until they had a membranous covering, and membrane is also absolutely essential to the organization of chloroplasts and mitochondria. In a chloroplast, membrane forms the grana, which are stacks of interconnected ,flattened membranous sacs called thylakoids. The inner membrane of a mitochondrion invaginates to form the cristae.

The detailed structure of chloroplasts and mitochondria is different, but essentially they operate similarly: an assembly line of particles in the thylakoid membrane and cristae carry out functions necessary to photosynthesis and aerobic respiration, respectively. Pigment complexes containing chlorophyll are present in the thylakoid membrane and absorb solar energy. In both organelles, electron transport system particles pump hydrogen ions into an enclosed space, establishing an electrochemical gradient. When hydrogen ions flow down this gradient through another particle, ATP is produced.

In chloroplasts, the gel-like fluid of the stroma contains the Calvin cycle, which produces carbohydrates, and in mitochondria, the gel-like fluid of the matrix contains the Krebs cycle, which breaks down carbohydrate products. The Calvin cycle takes up carbon dioxide and reduces it to a carbohydrate, utilizing this ATP, while the Krebs cycle releases carbon dioxide and produces some ATP in the process.

According to the endosymbiotic theory, chloroplasts and mitochondria were independent prokaryotic organisms at one time. Indeed, each contains genes not found in the eukaryotic nucleus. Through evolution all organisms are related, and therefore these organelles may be related also. The unity of plan of these organelles suggests that this is the case.

Summary

8.1 How Cells Acquire ATP

The oxidation of glucose to CO_2 and H_2O is an exergonic reaction that drives ATP synthesis, an endergonic reaction. Four events are required: glycolysis, the transition reaction, the Krebs cycle, and passage of electrons along the electron transport system. Oxidation involves the removal of hydrogen atoms ($H^+ + e^-$) from substrate molecules, usually by the coenzyme NAD^+ but in one case by FAD.

8.2 Outside the Mitochondria: Glycolysis

Glycolysis, the breakdown of glucose to two pyruvate, is a series of enzymatic reactions that occur in the cytosol. Breakdown releases enough energy to immediately give a net gain of two ATP by substrate-level phosphorylation. Two NADH are formed.

8.3 Inside the Mitochondria: Completion of Aerobic Respiration

When oxygen is available, pyruvate from glycolysis enters the mitochondria, where the transition reaction takes place. During this reaction, oxidation occurs as CO_2 is removed. NAD^+ is reduced, and CoA receives the C_2 acetyl group that remains. Since the reaction must take place twice per glucose molecule, two NADH result.

The acetyl group enters the Krebs cycle, a cyclical series of reactions located in the mitochondrial matrix. Complete oxidation follows, as two CO_2 molecules, three NADH molecules, and one $FADH_2$ molecule are formed. The cycle also produces one ATP molecule. The entire cycle must turn twice per glucose molecule.

The final stage of glucose breakdown involves the electron transport system located in the cristae of the mitochondria. The electrons received from NADH and $FADH_2$ are passed down a chain of carriers until they are finally received by oxygen, which combines with H^+ to produce water. As the electrons pass down the chain, energy is released and stored for ATP production. The term oxidative phosphorylation is sometimes used for ATP production associated with the electron transport system.

The cristae of the mitochondria contain protein complexes of the electron transport system that pass electrons from one to the other and pump H^+ into the intermembrane space, setting up an electrochemical gradient. When H^+ flows down this gradient through the ATP synthase complex, energy is released and used to form ATP molecules from ADP and Ⓟ. This is ATP synthesis by chemiosmosis.

Of the 36 ATP formed by complete glucose breakdown, four are the result of substrate-level phosphorylation and the rest are produced by oxidative phosphorylation. The energy for the latter comes from the electron transport system. For most NADH molecules that donate electrons to the electron transport system, three ATP molecules are produced. Each NADH formed in the cytosol, however, usually results in only two ATP molecules. This is because the hydrogen atoms must be shuttled across the mitochondrial membrane by a molecule that can cross it. In most cells, these hydrogen atoms are then taken up by FAD. Each molecule of $FADH_2$ results in the formation of only two ATP because its electrons enter the electron transport system at a lower energy level.

8.4 Metabolic Pool and Biosynthesis

Carbohydrate, protein, and fat can be broken down by entering the degradative pathways at different locations. These pathways also provide metabolites needed for the synthesis of various important substances. Degradation and synthesis, therefore, both utilize the same pools of metabolites.

8.5 Fermentation

Fermentation involves glycolysis followed by the reduction of pyruvate by NADH either to lactate or to alcohol and carbon dioxide (CO_2). The reduction process "frees" NAD^+ so that it can accept more hydrogen atoms from glycolysis.

Although fermentation results in only two ATP molecules, it still serves a purpose. In vertebrates, it provides a quick burst of ATP energy for short-term, strenuous muscular activity. The accumulation of lactate puts the individual in oxygen debt because oxygen is needed when lactate is completely metabolized to CO_2 and H_2O.

Reviewing the Chapter

1. What is the overall chemical equation for the complete breakdown of glucose to CO_2 and H_2O? Explain how this is an oxidation-reduction reaction. Why is the reaction able to drive ATP buildup? 130
2. What are NAD^+ and FAD? What are their functions? 130
3. What are the three pathways involved in the complete breakdown of glucose to carbon dioxide (CO_2) and water (H_2O)? What reaction is needed to join two of these pathways? 131
4. Outline the main reactions of glycolysis, emphasizing those that permit ATP formation. 132
5. Give the substrates and products of the transition reaction. Where does it take place? 134
6. What happens to the acetyl group that enters the Krebs cycle? What are the other steps in this cycle? 135
7. What is the electron transport system, and what are its functions? 136
8. Describe the organization of protein complexes within the cristae. Explain how the complexes are involved in ATP production. 137
9. Calculate the energy yield of glycolysis and complete glucose breakdown. Distinguish between substrate-level phosphorylation and oxidative phosphorylation. 138
10. Give examples to support the concept of the metabolic pool. 139
11. What is fermentation and how does it differ from glycolysis? Mention the benefit of pyruvate reduction during fermentation. What types of organisms carry out lactate fermentation, and what types carry out alcoholic fermentation? 141

Testing Yourself

Choose the best answer for each question. For questions 1–8, identify the pathway involved by matching them to the terms in the key.

Key:

a. glycolysis
b. Krebs cycle
c. electron transport system

1. carbon dioxide (CO_2) given off
2. water (H_2O) formed
3. PGAL
4. NADH becomes NAD^+
5. oxidative phosphorylation
6. cytochrome carriers
7. pyruvate
8. FAD becomes $FADH_2$
9. The transition reaction
   a. connects glycolysis to the Krebs cycle.
   b. gives off CO_2.
   c. utilizes NAD^+.
   d. All of these are correct.
10. The greatest contributor of electrons to the electron transport system is
   a. oxygen.
   b. glycolysis.
   c. the Krebs cycle.
   d. the transition reaction.
11. Substrate-level phosphorylation takes place in
   a. glycolysis and the Krebs cycle.
   b. the electron transport system and the transition reaction.
   c. glycolysis and the electron transport system.
   d. the Krebs cycle and the transition reaction.
12. Fatty acids are broken down to
   a. pyruvate molecules, which take electrons to the electron transport system.
   b. acetyl groups, which enter the Krebs cycle.
   c. amino acids, which excrete ammonia.
   d. All of these are correct.
13. Of the 36 ATP molecules that are produced during the complete breakdown of glucose, most are due to the action of
   a. chemiosmosis.
   b. the electron transport system
   c. substrate-level phosphorylation.
   d. Both a and b are correct.
14. Which of these is not true of fermentation?
   a. net gain of only two ATP
   b. occurs in cytosol
   c. NADH donates electrons to electron transport system
   d. begins with glucose

For questions 15–20, match the items below to one of the locations in the key.

Key:

a. matrix of the mitochondrion
b. cristae of the mitochondrion
c. the intermembrane space of mitochondrion
d. in the cytosol

15. electron transport system
16. Krebs cycle
17. glycolysis
18. transition reaction
19. accumulation of hydrogen ions (H^+)
20. ATP synthase complex

21. Label this diagram of a mitochondrion:

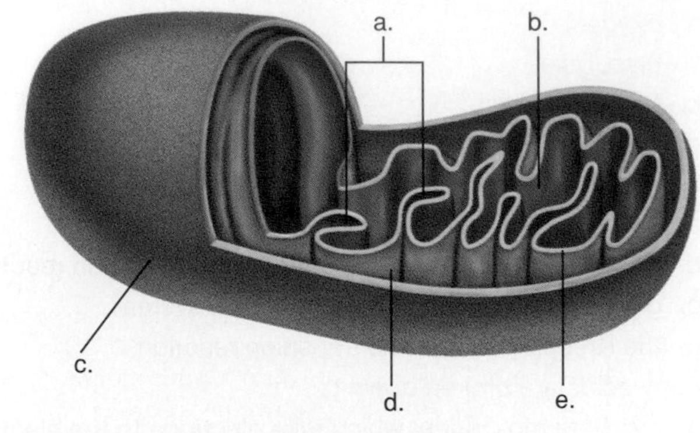

Applying the Concepts

1. *Metabolic pathways consist of a sequence of reactions.*
   What are the advantages to having individual steps in a metabolic pathway?
2. *Solar energy first stored in nutrient molecules is transformed into a form that can be "spent" by the cell.*
   How are nutrient molecules utilized by cells as a source of energy?
3. *Certain metabolic pathways are universal and thereby demonstrate the unity of living things.*
   In what way does glycolysis demonstrate the unity of living things and support the theory of evolution?

Using Technology

Your study of cellular respiration is supported by these available technologies:

Exploring the Internet
The Mader Home Page provides resources for and help with studying this chapter.

http://www.mhhe.com/sciencemath/biology/mader/
(Click on Biology.)

Explorations in Cell Biology & Genetics CD-ROM
Oxidative Respiration (#8)

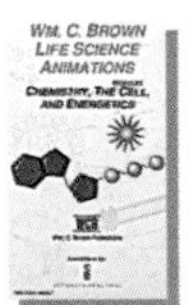

Life Science Animations Video
Video #1: Chemistry, the Cell, and Energetics
Glycolysis (#5)
Oxidative Respiration (including Krebs Cycles) (#6)
The Electron Transport Chain and the Production of ATP (#7)
ATP as an Energy Carrier (#11)

Understanding the Terms

acetyl-CoA 134
aerobic respiration 130
cellular respiration 130
electron transport system 131
FAD 130
fermentation 141
glycolysis 131
Krebs cycle 131
metabolic pool 139
NAD^+ 130
oxidative phosphorylation 136
oxygen debt 141
pyruvate 131
substrate-level phosphorylation 132
transition reaction 131

Match the terms to these definitions:

a. __________ The complete breakdown of glucose to carbon dioxide and water with the resulting buildup of 36 ATP.
b. __________ Anaerobic breakdown of glucose that results in a gain of two ATP and end products such as alcohol and lactate.
c. __________ Anaerobic breakdown of glucose that results in a gain of two ATP and the end product pyrvuate.
d. __________ Cycle of reactions in mitochondria that begins with citric acid; it produces CO_2, ATP, NADH, and $FADH_2$; also called the citric acid cycle.
e. __________ Metabolic reactions that use the energy from carbohydrate or fatty acid or amino acid breakdown to produce ATP molecules.
f. __________ Metabolites that are the products of and/or the substrates for key reactions in cells allowing one type of molecule to be changed into another type, such as the conversion of carbohydrates to fats.
g. __________ Molecule made up of a two-carbon acetyl group attached to coenzyme A. During aerobic cellular respiration, the acetyl group enters the Krebs cycle for further breakdown.
h. __________ The use of oxygen to reconvert lactate, which builds up during anaerobic conditions, to pyruvate.
i. __________ Passage of electrons along a series of membrane-bounded carrier molecules from a higher to lower energy level; the energy released is used for the synthesis of ATP.
j. __________ Coenzyme that delivers electrons from both glycolysis and the Krebs cycle to the electron transport system during aerobic cellular respiration.

Further Readings for Part i

Alberts, B., et al. 1994. *Molecular biology of the cell.* 3d ed. New York: Garland Publishing. An authoritative text concerning modern cell biology and its experimental basis.

Austin, S. M., and Bertsch, G. F. 1995. Halo nuclei. *Scientific American* 272(6):90. New techniques allow scientists to study the nuclei of unstable atoms.

Becker, W. M., and Deamer, D. W. 1996. *The world of the cell.* 3d ed. Redwood City, Calif.: Benjamin/Cummings Publishing. Presents an overview of cell biology.

Carey, S. 1994. *A beginner's guide to the scientific method.* Belmont, Calif.: Wadsworth Publishing. The basics of the scientific method are explained.

Cranbrook, E., and Edwards, D. S. 1994. *Belalong: A tropical rain forest.* London: The Royal Geographic Society, and Singapore: Sun Tree Publishing. Provides a very readable, well-illustrated account of biodiversity in a Brunei rain forest.

Drewes, F. 1997. *How to study science.* 2d ed. Dubuque, Iowa: Wm. C. Brown Publishers. Supplements any introductory science text; shows students how to study and take notes and how to interpret text figures.

Kleinsmith, L. J., and Kish, V. M. 1995. *Principles of cell and molecular biology.* 2d ed. New York: HarperCollins College Publishers. This text introduces students to the fundamental principles that guide cellular organization and function.

Lacy, P. E. July 1995. Treating diabetes with transplanted cells. *Scientific American* 273(1):50. New technology may allow replacement of pancreatic cells in diabetes patients.

Langer, R., and Vacanti, J. P. 1995. Artificial organs. *Scientific American* 273(3):130. Discusses engineering artificial tissue using the body's own cells.

Lanza, R. P., and Chick, W. L. 1995. Encapsulated cell therapy. *Scientific American Science & Medicine* 2(4):16. Implants containing living cells (tissue engineering) within a selectively permeable membrane could provide drug or hormone doses as required for therapy.

Lichtman, J. W. August 1994. Confocal microscopy. *Scientific American* 271(2):40. A confocal microscope focuses at different depths in an organic specimen.

Lodish, H., et al. 1995. *Molecular cell biology.* 3d ed. New York: Scientific American Books. For the more advanced student, this well-illustrated text provides excellent discussions of molecular cell biology.

Marchuk, W. N. 1992. *A life science lexicon.* Dubuque, Iowa: Wm. C. Brown Publishers. Helps students master life sciences terminology.

Margulis, L., et al. 1994. *The illustrated five kingdoms: A guide to the diversity of life on Earth.* New York: HarperCollins College Publishers. Introduces the kingdoms of organisms.

Mathews, C. K., and van Holde, K. E. 1996. *Biochemistry.* 2d ed. Redwood City, Calif.: Benjamin/Cummings Publishing. For advanced students, this text describes the background and concepts behind major biochemical methodologies.

Packer, L. 1994. Vitamin E is nature's master antioxidant. *Science & Medicine* 1(1):54. Vitamin E is proving useful as an antioxidant in reducing oxidative destruction of membrane lipids—a normal process of aging.

Schwartz, A. T., et al. 1994. *Chemistry in context: Applying chemistry to society.* Dubuque, Iowa: Wm. C. Brown Publishers. This introductory text is designed for students in the allied health fields.

Scientific American Science & Medicine. July/August 1995. A is for . . . Nucleosome 2(4):80. Discusses the nucleosome, the basic structural unit of eukaryotic chromosomes.

Stossel, T. P. September 1994. The machinery of cell crawling. *Scientific American* 271(3):54. Discusses how cells move.

Stryer, L. 1995. *Biochemistry*, 5th ed. New York: W.H. Freeman and Company. A well-illustrated, clearly written reference for the advanced student.

Urry, D. W. January 1995. Elastic biomolecular machines. *Scientific American* 272(1):64. Scientists have begun to construct polymer molecules that expand or contract in response to temperature, light, or acidity changes.

Wardlaw, G., et al. 1994. *Contemporary nutrition.* 2d ed. St. Louis: Mosby-Year Book, Inc. This text gives a clear understanding of nutritional information found on product labels.

Zubay, G. L., et al. 1995. *Principles of Biochemistry.* Dubuque, Iowa: Wm. C. Brown Publishers. Presents a focused discussion of basic biochemistry; includes readings on methods of biochemical analyses.

PART II

Genetic Basis of Life

Hereditary information is stored in DNA, molecules that compose the genes located within chromosomes. The chromosomes duplicate prior to cell division and then a complete set is distributed to each and every body cell. A special form of cell division is involved in the production of the sex cells, which contain half the usual number of chromosomes.

Principles of inheritance include those that allow us to predict the chances that an offspring will inherit a particular characteristic from a parent. These have been applied to the breeding of plants and animals and the study of human genetic disorders. But to go further, and to control an organism's characteristics, it is necessary to understand how DNA and RNA function in protein synthesis. The human endeavor known as biotechnology is based on our newfound knowledge of nucleic acid biochemistry.

The principles of inheritance are central to understanding many other topics in biology—from the evolution and diversity of life to the reproduction and development of organisms. There is no topic in biology that stands alone; they are all interrelated!

Cell Division

CHAPTER 9

Chapter Concepts

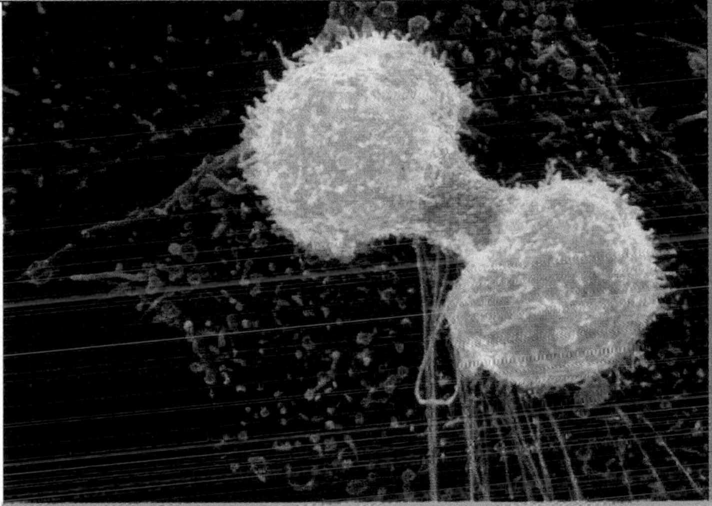

Cell undergoing mitosis

10 μm

Consider the development of a human being. We all begin life as one cell—an egg fertilized by a sperm. Yet in nine short months, we become complex organisms consisting of trillions of cells. How is such a feat possible? Cell division enables a single cell to produce many cells, allowing an organism to grow in size and to replace worn-out tissues.

The instructions for cell division lie in the genes. During the first part of an organism's life, the genes instruct all cells to divide. When adulthood is reached, however, only specific cells—human blood and skin cells, for example—continue to divide. Other tissues, such as nervous tissue, no longer produce new cells.

Why don't all the cells in an adult continue to reproduce? After all, they contain the full complement of genetic material in their nuclei. Although we don't have all the answers to this question yet, cell biologists have recently discovered that specific enzymes regulate the cell cycle, the period that extends from the time a new cell is produced until it completes division. Proper function of these enzymes ensures that cells divide normally. Cancer—abnormal cell division—may result when the enzymes regulating the cell cycle go awry.

9.1 How Prokaryotic Cells Divide

Bacteria are well known prokaryotes, which are unicellular organisms that lack a nucleus and other membranous organelles found in eukaryotic cells.

The Chromosome Is Singular

A **chromosome** is the genetic material of the cell, complexed with proteins. Prokaryotes have a single chromosome that is associated with just a few proteins. A eukaryotic cell has much more protein than a prokaryotic chromosome.

In electron micrographs the bacterial chromosome appears as an electron-dense, irregularly shaped region called the **nucleoid** [L. *nucleus,* nucleus, kernel, and Gk. *-eides,* like], which is not enclosed by a membrane. When stretched out, the chromosome is seen to be a circular loop attached to the inside of the plasma membrane. Its length is up to about 1,000 times the length of the cell, which is why it needs to be folded inside the cell.

The prokaryotic chromosome is largely a single loop of DNA that is tightly folded into a region called the nucleoid.

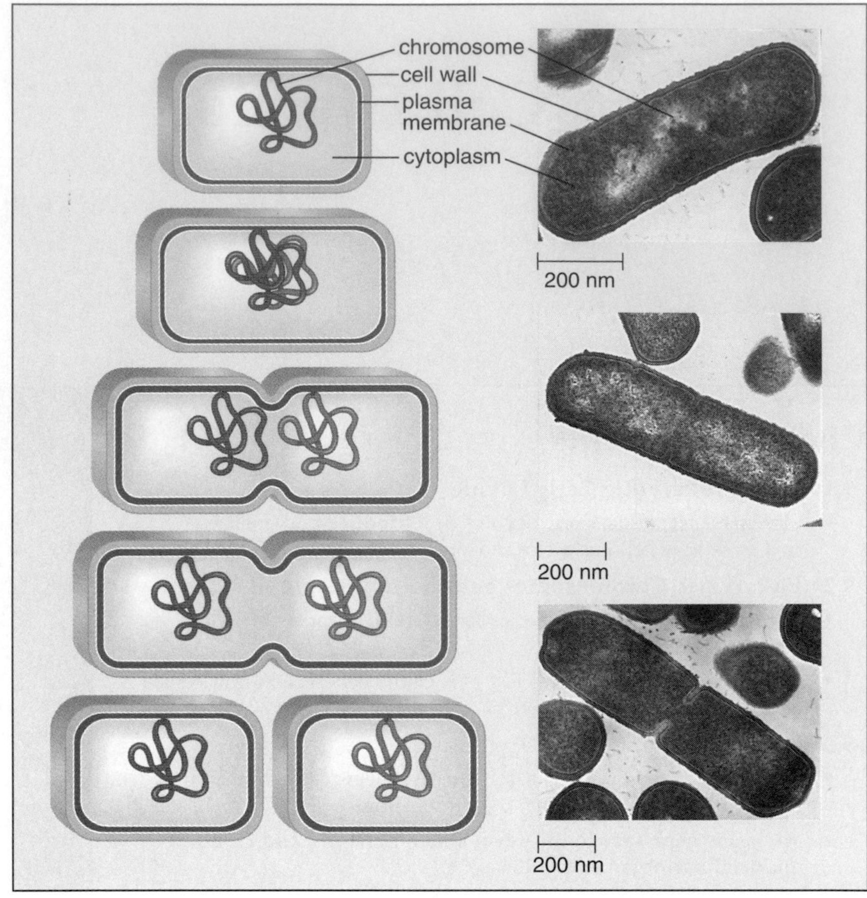

Figure 9.1 Binary fission.
In electron micrographs, it is possible to observe a bacterium dividing to become two bacteria. The diagrams depict chromosomal duplication and distribution. First DNA replicates, and as the plasma membrane lengthens, the two chromosomes separate. Upon fission, each bacterium has its own chromosome.

Division Is by Fission

Asexual reproduction requires a single parent, and the offspring are identical to the parent because they contain the same genes. Prokaryotes reproduce asexually by cell division. The process is termed **binary fission** because division (fission) produces two (binary) daughter cells that are identical to the original parent cell. Before division takes place, DNA is replicated so that there are two chromosomes attached to the inside of the plasma membrane. Following replication, the two chromosomes separate by an elongation of the cell that pulls the chromosomes apart. When the cell is approximately twice its original length, the plasma membrane grows inward and new cell wall forms, dividing the cell into two approximately equal portions (Fig. 9.1).

Escherichia coli, which lives in our intestines, has a generation time (time it takes the cell to divide) of about 20 minutes under favorable conditions. In about seven hours a single cell can increase to over one million cells! Most bacteria, however, have a generation time of from one to three hours and others may require even more than 24 hours.

Asexual reproduction in prokaryotes is by binary fission. DNA replicates and the two resulting chromosomes separate as the cell elongates.

9.2 Eukaryotic Chromosomes and the Cell Cycle

The DNA in the chromosomes of eukaryotes is associated with various proteins including *histone proteins* that are especially involved in organizing chromosomes. When a eukaryotic cell is not undergoing division, the DNA (and associated proteins) within a nucleus is a tangled mass of thin threads called **chromatin** [Gk. *chroma*, color, and *teino*, stretch]. At the time of division, chromatin becomes highly coiled and condensed, and it is easy to see the individual chromosomes.

When the chromosomes are visible it is possible to photograph and count them. Each species has a characteristic chromosome number (Table 9.1); for instance, human cells contain 46 chromosomes, corn has 20 chromosomes, and a crayfish has 200! This is called the full or **diploid (2n) number** [Gk. *diplos*, twofold, and *-eides*, like] of chromosomes that is found in all cells of the body. The diploid number includes two chromosomes of each kind. Half the diploid number, called the **haploid (n) number** [Gk. *haplos*, single, and *-eides*, like] of chromosomes, contains only one of each kind of chromosome. In the life cycle of many animals, only sperm and eggs have the haploid number of chromosomes.

Each type of eukaryote has a characteristic number of chromosomes in the nucleus of each cell.

Cell division in eukaryotes involves nuclear division (karyokinesis) and **cytokinesis**, which is division of the cytoplasm. The nuclei of somatic, or body, cells undergo **mitosis**—nuclear division in which the chromosome number stays constant. A 2n nucleus divides to produce daughter nuclei that are also 2n. Mitosis is the type of nuclear division that is involved in development, growth, and repair of multicellular organisms. Before nuclear division takes place, DNA replicates, duplicating the chromosomes. Each chromosome now has two identical parts called **sister chromatids** (Fig. 9.2). Sister chromatids are genetically identical; that is, they contain exactly the same genes. Sister chromatids are constricted and attached to each other at a region called the **centromere** [Gk. *centrum*, center, and *meros*, part]. During nuclear division the two sister chromatids separate at the centromeres, and in this way each duplicated chromosome gives rise to two daughter chromosomes. These chromosomes, which consist of only one chromatid, are distributed equally to the daughter cells. In this way, each daughter cell gets a copy of each chromosome.

DNA replicates prior to mitosis, so that despite nuclear division the chromosome number stays constant.

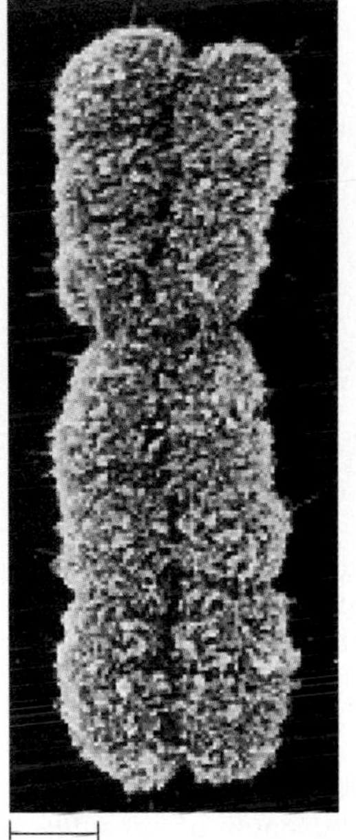

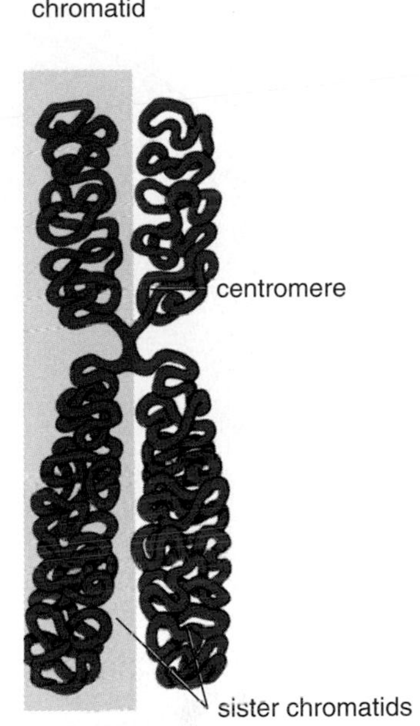

Figure 9.2 Duplicated chromosomes.
A duplicated chromosome contains two sister chromatids, each with copies of the same genes. **a.** Electron micrograph of a highly coiled and compacted chromosome, typical of a nucleus about to divide. **b.** Diagrammatic drawing of a compacted chromosome. One chromatid is screened in blue. The chromatids are held together at a region called the centromere.

Table 9.1
Diploid Chromosome Number of Some Eukaryotes

Type of Organism	Name of Organism	Chromosome Number
Fungi	*Aspergillus nidulans* (mold)	8
	Neurospora crassa (mold)	14
	Saccharomyces cerevisiae (yeast)	34
Plants	*Vicia faba* (broad bean)	12
	Zea mays (corn)	20
	Solanum tuberosum (potato)	48
	Nicotiana tabacum (tobacco)	48
Animals	*Musca domestica* (housefly)	12
	Rana pipiens (frog)	26
	Felis domesticus (cat)	38
	Homo sapiens (human)	46
	Pan troglodytes (chimp)	48
	Equus caballus (horse)	64
	Gallus gallus (chicken)	78
	Canis familiaris (dog)	78

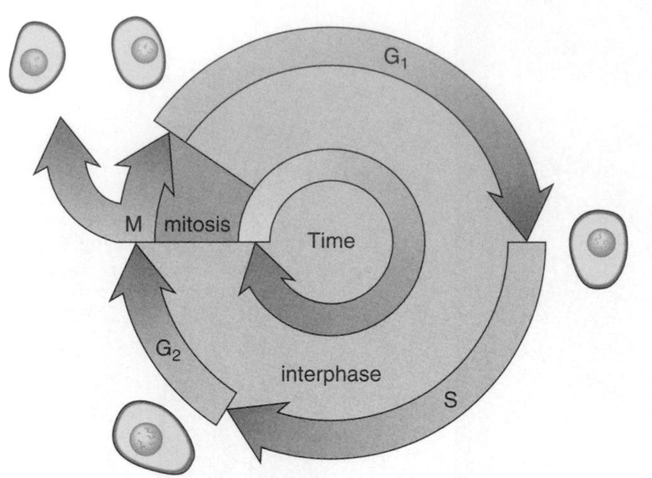

Stage	Main Events	Length of time (hours)	
		Vicia faba	*Homo sapiens* (cultured fibroblasts)
G_1	Organelles begin to double in number	4.9	6.3
S	Replication of DNA	7.5	7.0
G_2	Synthesis of proteins	4.9	2.0
M	Mitosis	2.0	0.7
	Total:	19.3	16.0

(Interphase: G_1, S, G_2)

Figure 9.3 The cell cycle consists of four stages.
(G_1, S for synthesis, G_2, and M for mitosis). The length of the different stages varies both among species and among different cell types in the same individual. The approximate lengths of the phases for the broad bean *(Vicia faba)* and humans *(Homo sapiens)* are given.

How Eukaryotic Cells Cycle

By the 1870s, microscopy could provide detailed and accurate descriptions of chromosomal movements during mitosis, but there was no knowledge of cellular events between divisions. Because there was little visible activity between divisions, this period of time was dismissed as a resting state termed **interphase** [L. *inter,* between, and Gk. *phasis,* appearance]. When it was discovered in the 1950s that DNA replication occurs during interphase, the **cell cycle** concept was proposed.

Cells grow and divide during a cycle that has four stages (Fig. 9.3). The entire cell division stage, including both mitosis and cytokinesis, is termed the *M stage* (M = mitosis). The period of DNA synthesis when replication occurs is termed the *S stage* (S = synthesis) of the cycle. The proteins associated with DNA in eukaryotic chromosomes are also synthesized during this stage. There are two other stages of the cycle. The period of time prior to the S stage is termed the *G_1 stage,* and the period of time prior to the M stage is termed the *G_2 stage.* At first, not much was known about these stages, and they were thought of as G = gap stages. Now we know that during the G_1 stage, the cell grows in size and the cellular organelles increase in number. During the G_2 stage, various metabolic events occur in preparation for mitosis. Some biologists today prefer the designation G = growth for these two G stages. In any case, interphase consists of G_1, S, and G_2 stages.

Cells undergo a cycle that includes the G_1, S, G_2, and M stages.

How the Cycle Is Controlled

Some cells, such as skin cells, divide continuously throughout the life of the organism. Other cells, such as skeletal muscle cells and nerve cells, are arrested in the G_1 stage. If the nucleus from one of these cells is placed in the cytoplasm of an S-stage cell, it finishes the cell cycle. Cardiac muscle cells are arrested in the G_2 stage. If an arrested cell is fused with a cell undergoing mitosis, it too starts to undergo mitosis. It appears, then, that there are stimulatory substances that cause the cell to proceed through two critical checkpoints:

$$G_1 \text{ stage} \longrightarrow S \text{ stage}$$
$$G_2 \text{ stage} \longrightarrow M \text{ stage}$$

Over the past few years biologists have made remarkable progress in identifying the molecules that drive the cell cycle. Many groups, some of whom worked with frog eggs, others with yeast cells, and still others who used various cell cultures as their experimental material, have shown that the activity of enzymes known as *cyclin-dependent kinases (Cdks)* regulates the passage of cells through these checkpoints.

A **kinase** is an enzyme that removes a phosphate group from ATP (the form of chemical energy used by cells) and adds it to a protein. Phosphorylated molecules are a common way for the cell to turn on metabolic pathways. Notice in Figure 9.4 that phosphorylation of a protein precedes the S stage and precedes the M stage of the cell cycle.

The kinases involved in the cell cycle are called cyclin dependent because they are activated when they combine with a protein called a **cyclin.** Cyclins are so named be-

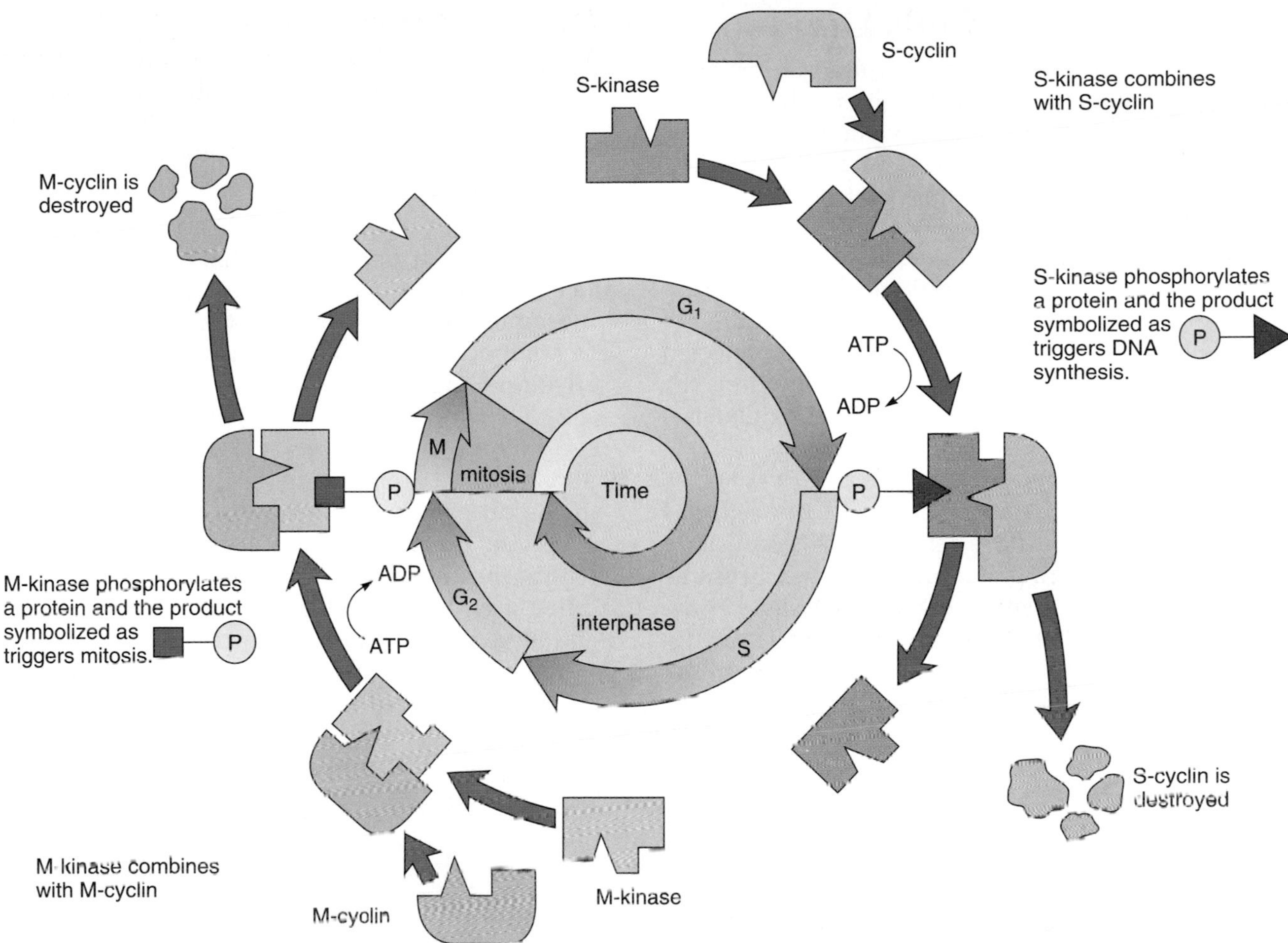

Figure 9.4 Control of the cell cycle.
At two critical checkpoints a kinase combines with a cyclin, and this moves the cell cycle forward. Just before the S stage, S-kinase combines with S-cyclin and synthesis (replication) of DNA takes place; just before the M stage, M-kinase combines with M-cyclin and mitosis occurs. When kinases phosphorylate proteins, the proteins are activated and produce effects appropriate to the particular stage.

cause their quantity is not constant. They increase in amount until they combine with a kinase, but this is a suicidal act because now the kinase not only phosphorylates a protein that drives the cell cycle, it also activates various enzymes, one of which destroys the cyclin.

Figure 9.4 shows how the entire process works. In the diagram, S-kinase is capable of phosphorylating the protein that triggers DNA replication after it has combined with S-cyclin. S-cyclin is now destroyed and S-kinase is no longer active. M-kinase is capable of phosphorylating the protein that turns on mitosis after it has combined with M-cyclin. It is known that this particular phosphorylated protein induces the process of (1) chromosome condensation, (2) nuclear envelope breakdown, and (3) spindle assembly. (The spindle is the structure involved in chromosome movement during mitosis.) Now M-cyclin is destroyed.

Until recently, the mechanics of the cell cycle and the causes of cancer were thought to be distantly related. Now they appear to be intimately related. For example, oncogenes are cancer-causing genes, and it is possible that they code for cyclins that no longer function as they should. Growth factors are molecules that attach to plasma membrane receptors and thereby bring about cell growth. Ordinarily, a cyclin might combine with its kinase only when a growth factor is present. But a cyclin that has gone awry might combine with its kinase even when a growth factor is not present. The result would be uncontrolled cell growth resulting in a tumor. On the other hand, tumor-suppressor genes usually function to prevent cancer from occurring. It has been shown that the product of one major tumor-suppressor gene *(p53)* brings about the production of a protein that can combine with a cyclin-kinase complex and prevent that kinase from becoming active. *p*53 can also help induce apoptosis in cancerous cells. **Apoptosis** [Gk. *apo*, off, and *ptosis*, fall] is a process of programmed cell death involving a cascade of specific cellular events leading to the death and destruction of the cell.

Kinases activated by combination with a cyclin at two critical checkpoints in the cell cycle phosphorylate a protein, and this triggers the beginning of the S stage and the M stage.

9.3 How Eukaryotic Cells Divide

During mitosis, the **spindle** brings about an orderly distribution of chromosomes to the daughter cell nuclei so that the chromosome number stays constant. The spindle contains many fibers, each composed of a bundle of microtubules. Microtubules are hollow cylinders found in the cytoplasm and other structures such as flagella and centrioles. Microtubules, which are made up of the protein tubulin, assemble when tubulin subunits join and disassemble when tubulin subunits become free once more. Microtubules of the cytoskeleton, which is a network of interconnected filaments and tubules, begin to disassemble when the spindle fibers begin forming, probably providing material for spindle formation.

The **centrosome** [Gk. *centrum,* center, and *soma,* body], the main microtubule organizing center of the cell, has divided before mitosis begins. It's believed that centrosomes are responsible for organizing the spindle. Each centrosome contains a pair of barrel-shaped organelles called **centrioles;** however, the fact that plant cells lack centrioles suggests that centrioles are not required for spindle formation.

How Animal Cells Divide

Mitosis, also called *karyokinesis,* is a continuous process that is arbitrarily divided into five phases for convenience of description. These phases are prophase, prometaphase, metaphase, anaphase, and telophase (Fig. 9.5).

Prophase

It is apparent during *prophase* that nuclear division is about to occur because chromatin has condensed and the chromosomes are now visible. As the chromosomes continue to compact, the nucleolus disappears and the nuclear envelope fragments.

The already duplicated chromosomes are composed of two sister chromatids held together at a centromere. Counting the number of centromeres in diagrammatic drawings gives the number of chromosomes for the cell depicted. During prophase, the chromosomes have no apparent orientation within the cell. However, specialized protein complexes called *kinetochores* develop on either side of each centromere, and these are important to future chromosome orientation.

The spindle begins to assemble as pairs of centrosomes migrate away from one another. Short microtubules radi-

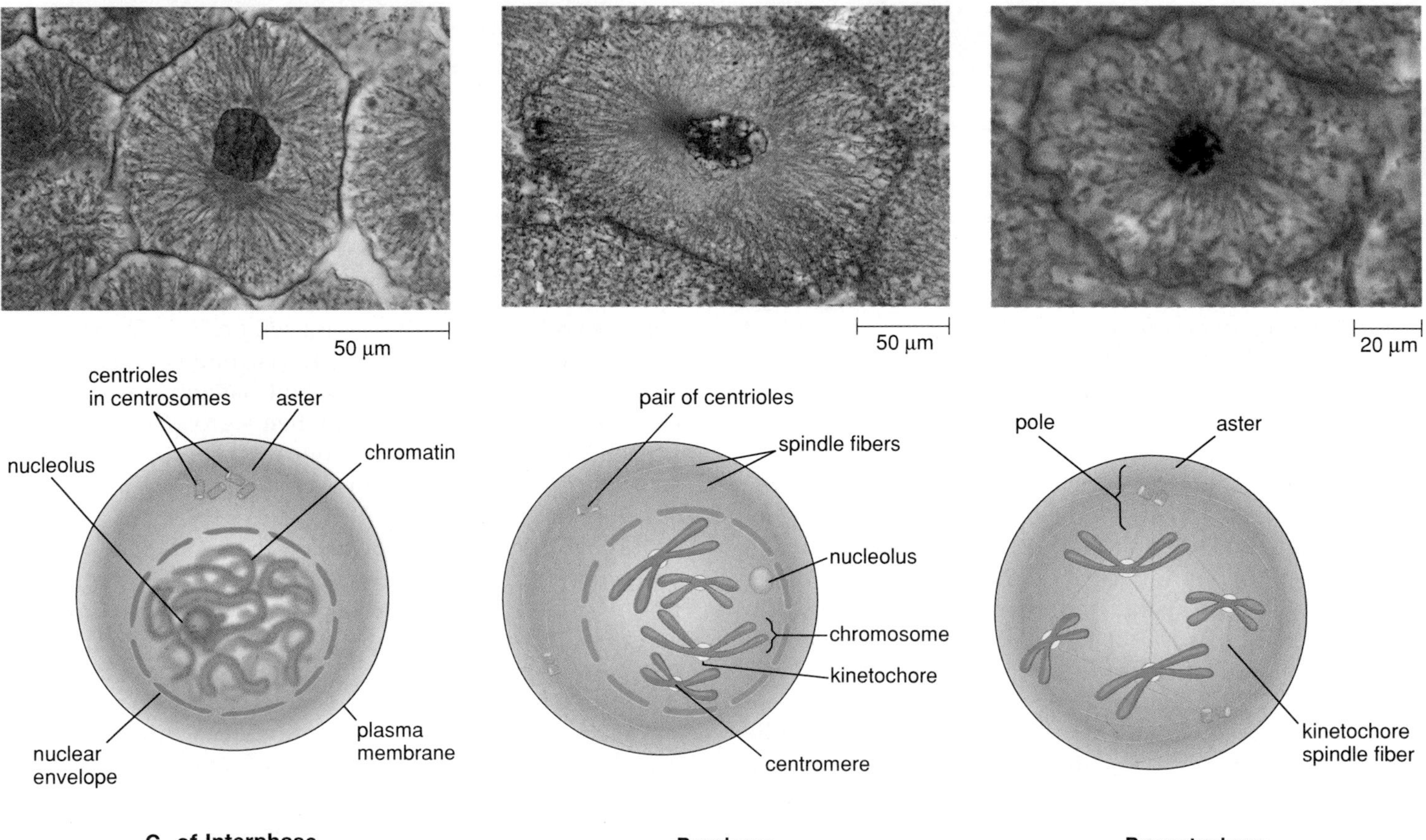

G_2 of Interphase
Chromosomes and centrioles have duplicated in preparation for mitosis.

Prophase
Chromosomes are now distinct; centrosomes begin moving apart and nuclear envelope is fragmenting.

Prometaphase
Spindle is in process of forming and kinetochores of chromosomes are attaching to kinetochore spindle fibers.

Figure 9.5 Phases of animal cell mitosis.

ate out in a starlike **aster** [Gk. *aster,* star] from the pair of centrioles located in each centrosome.

Prometaphase

As *prometaphase* begins, the spindle consists of poles, asters, and fibers, which are bundles of parallel microtubules. An important event during prometaphase is the attachment of the chromosomes to the spindle and their movement as they align at the metaphase plate (equator) of the spindle. The kinetochores of sister chromatids capture spindle fibers coming from opposite poles. Such spindle fibers are called *kinetochore spindle fibers*. In response to attachment by first one kinetochore and then the other, a chromosome moves first toward one pole and then toward the other until the chromosome is aligned at the metaphase plate of the spindle.

Metaphase

During *metaphase,* the chromosomes, attached to kinetochore fibers, are aligned at the metaphase plate. There are many nonattached spindle fibers called *polar spindle fibers,* some of which reach beyond the metaphase plate and overlap.

Anaphase

At the start of *anaphase,* the two sister chromatids of each duplicated chromosome separate at the centromere, giving rise to two daughter chromosomes. Daughter chromosomes, each with a centromere and single chromatid, begin to move toward opposite poles. What accounts for the movement of the daughter chromosomes? First, the polar spindle fibers lengthen as they slide past one another. Second, the kinetochore spindle fibers disassemble at the region of the kinetochores, and this pulls the daughter chromosomes to the poles.

Telophase

During *telophase,* the spindle disappears as new nuclear envelopes form around the daughter chromosomes. Each daughter nucleus contains the same number and kinds of chromosomes as the original parent cell. Remnants of the polar spindle fibers are still visible between the two nuclei.

The chromosomes become more diffuse chromatin once again, and a nucleolus appears in each daughter nucleus. Cytokinesis is nearly complete, and soon there will be two individual daughter cells, each with a nucleus that contains the diploid number of chromosomes.

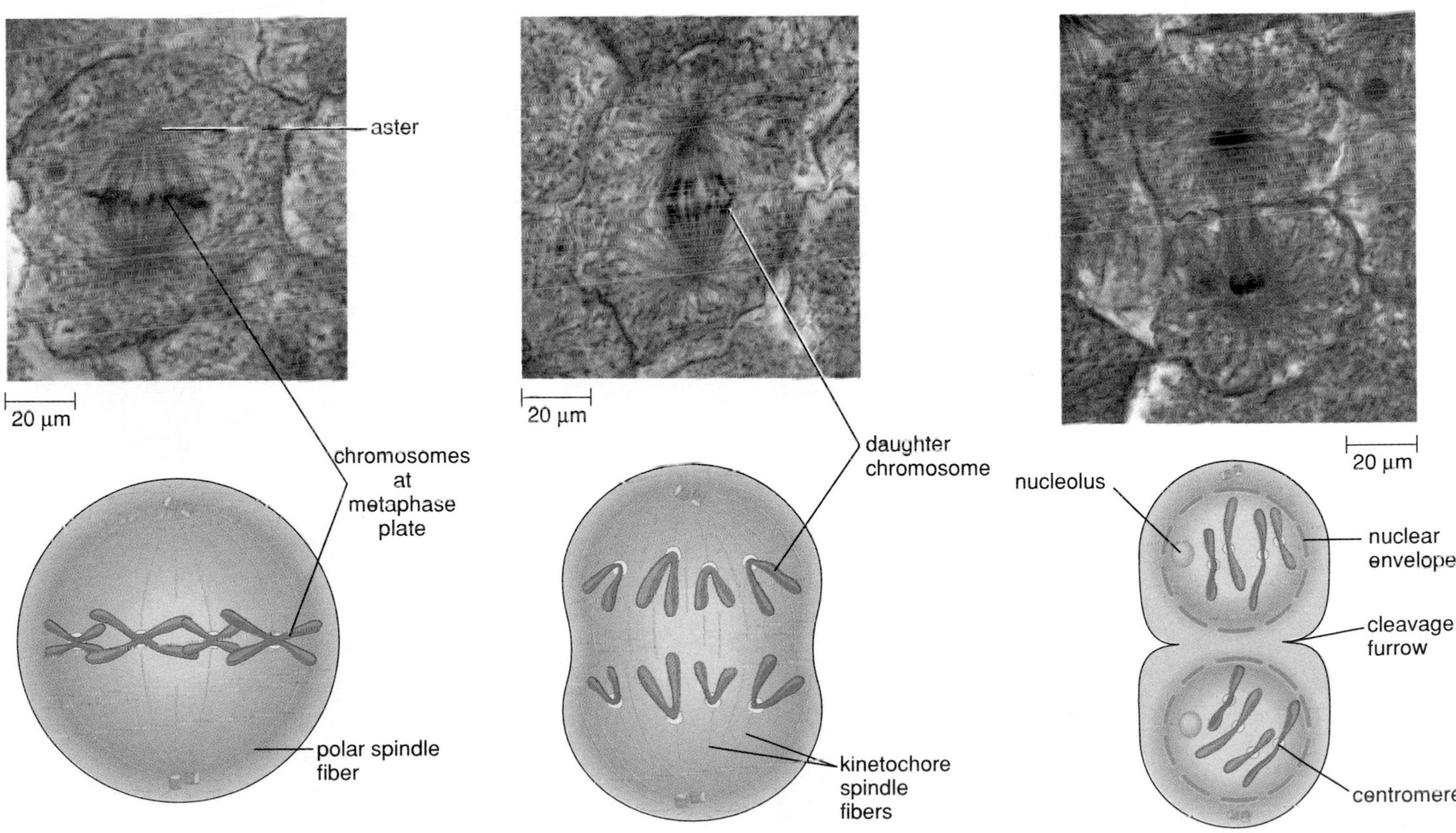

Metaphase

Chromosomes (each consisting of two sister chromatids) are at the metaphase plate (center of fully formed spindle).

Anaphase

Daughter chromosomes (each consisting of one chromatid) are moving to the poles of the spindle.

Telophase

Daughter cells are forming as nuclear envelopes and nucleoli appear. Chromosomes will become indistinct chromatin.

How Plant Cells Divide

As with animal cells, mitosis in plant cells permits growth and repair. Certain plant tissue, called meristematic tissue, retains the ability to divide throughout the life of a plant. Meristematic tissue is found in the root tip and shoot tip of stems. Lateral meristem accounts for the ability of trees to increase their girth each growing season.

Figure 9.6 illustrates mitosis in plant cells. Note that there are exactly the same stages in plant cells as in animal cells. During prophase, the chromatin condenses into scattered previously duplicated chromosomes and the spindle forms; during prometaphase (not illustrated), chromosomes attach to spindle fibers; during metaphase, the chromosomes are at the metaphase plate of the spindle; during anaphase, the daughter chromosomes move to the poles of the spindle; and during telophase, cytokinesis begins. Although plant cells have a centrosome and spindle, there are no centrioles nor asters during cell division.

Mitosis in plant and animal cells ensures the daughter cells receive the same number and kinds of chromosomes as the parent cell.

Cytokinesis in Plant and Animal Cells

Cytokinesis, or cytoplasmic cleavage, usually accompanies mitosis. By the end of mitosis each newly forming cell has received a share of the cytoplasmic organelles which duplicated during interphase. Division of the cytoplasm begins in anaphase, continues in telophase, but does not reach completion until the following interphase begins.

Cytokinesis in Plant Cells

Cytokinesis in plant cells occurs by a process different from that seen in animal cells (Fig. 9.7). The rigid cell wall that surrounds plant cells does not permit cytokinesis by furrowing. Instead, the Golgi apparatus, a membranous organelle in cells, produces membranous sacs called vesicles, which move along microtubules to the midpoint between the two daughter nuclei. These vesicles fuse, forming a **cell plate.** Their membrane completes the plasma membrane for both cells. They also release molecules that signal the formation of plant cell walls, which are strengthened by the addition of cellulose fibrils.

A spindle forms during mitosis in plant cells, but there are no centrioles or asters. Cytokinesis in plant cells involves the formation of a cell plate.

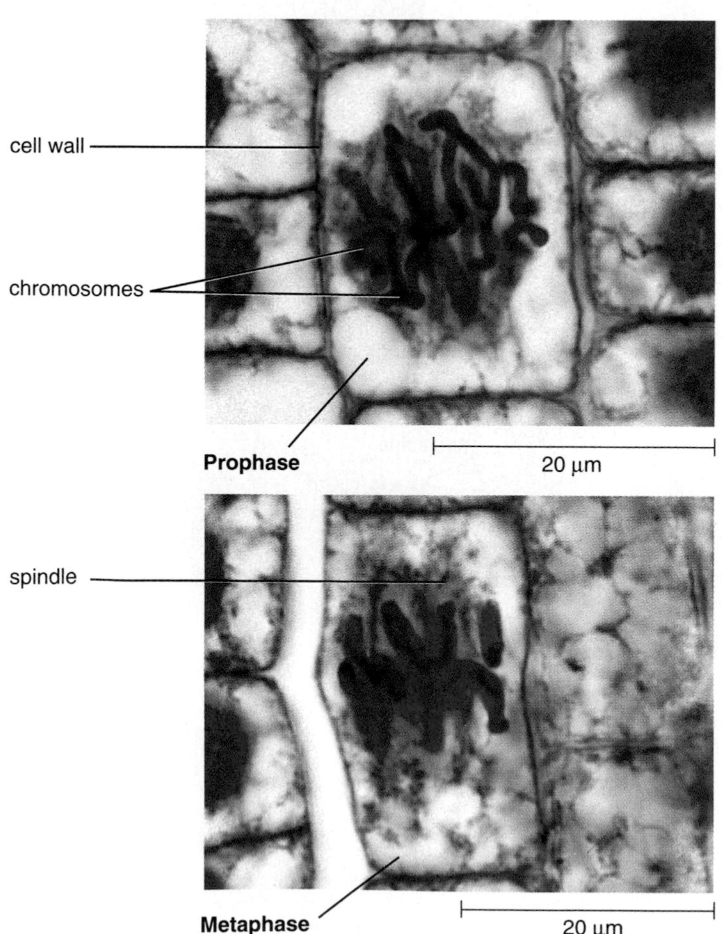

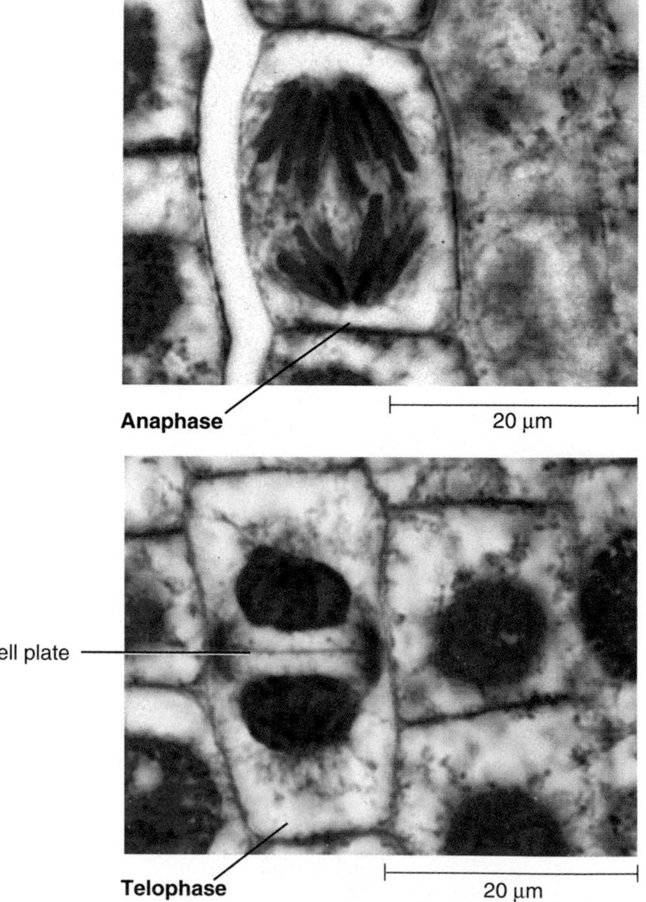

Figure 9.6 Plant cell mitosis.
Note the absence of centrioles and asters and the presence of the cell wall. In telophase, a cell plate develops between the two daughter cells. The cell plate marks the boundary of the new daughter cells, where new plasma membrane and a new cell wall will form for each cell.

Cytokinesis in Animal Cells

In animal cells, a cleavage furrow, which is an indentation of the membrane between the two daughter nuclei, begins as anaphase draws to a close. The cleavage furrow deepens when a band of actin filaments, called the contractile ring, slowly forms a constriction between the two daughter cells. The action of the contractile ring can be likened to pulling a drawstring ever tighter about the middle of a balloon. As the drawstring is pulled tight, the balloon constricts in the middle.

A narrow bridge between the two cells can be seen during telophase, and then the contractile ring continues to separate the cytoplasm until there are two independent daughter cells (Fig. 9.8).

Cytokinesis in animal cells is accomplished by a furrowing process.

Cell Division and Cytokinesis in Other Organisms

Protists and fungi also undergo mitosis and cytokinesis. In fungi and some groups of protists the nuclear envelope does not fragment. It's possible that the first role of microtubules was to support the nuclear envelope during mitosis, and only later did microtubules become attached to the chromosomes themselves. When mitosis is complete in these protists and fungi, the nuclear envelope divides and one nucleus goes to each daughter cell.

In plants and animals, as we have seen, the nuclear envelope fragments and plays no role in mitosis and cytokinesis.

The cells of all organisms divide and new cells only come from preexisting cells. This is a tenet of the cell theory.

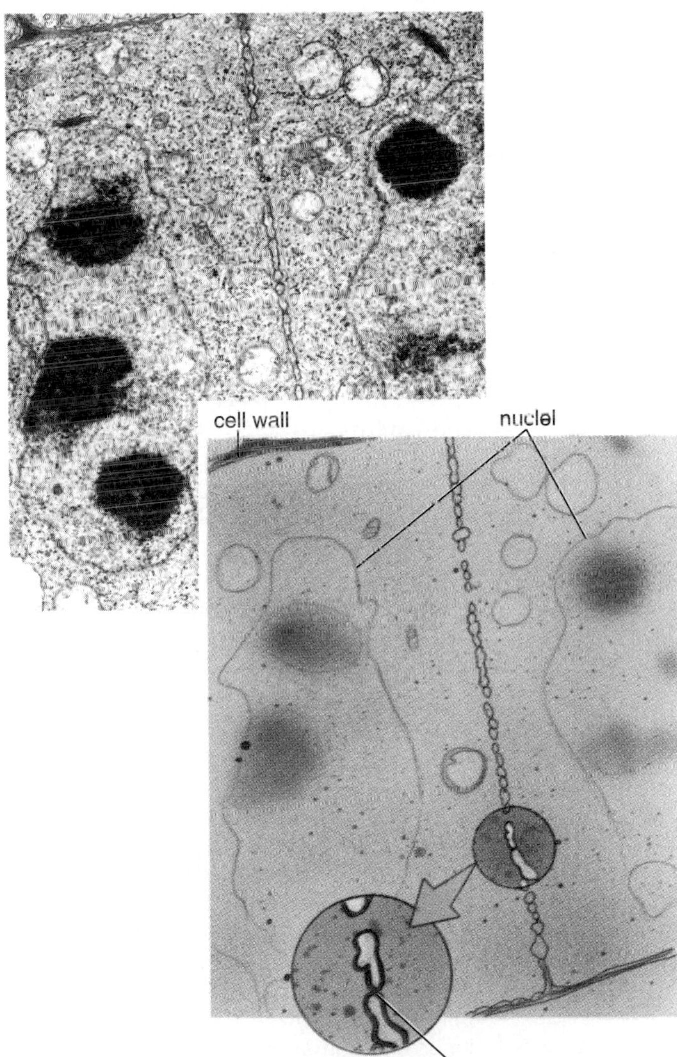

Figure 9.7 Cytokinesis in plant cells.
During cytokinesis in a plant cell, the cell plate forms midway between the two daughter nuclei and extends to the plasma membrane.

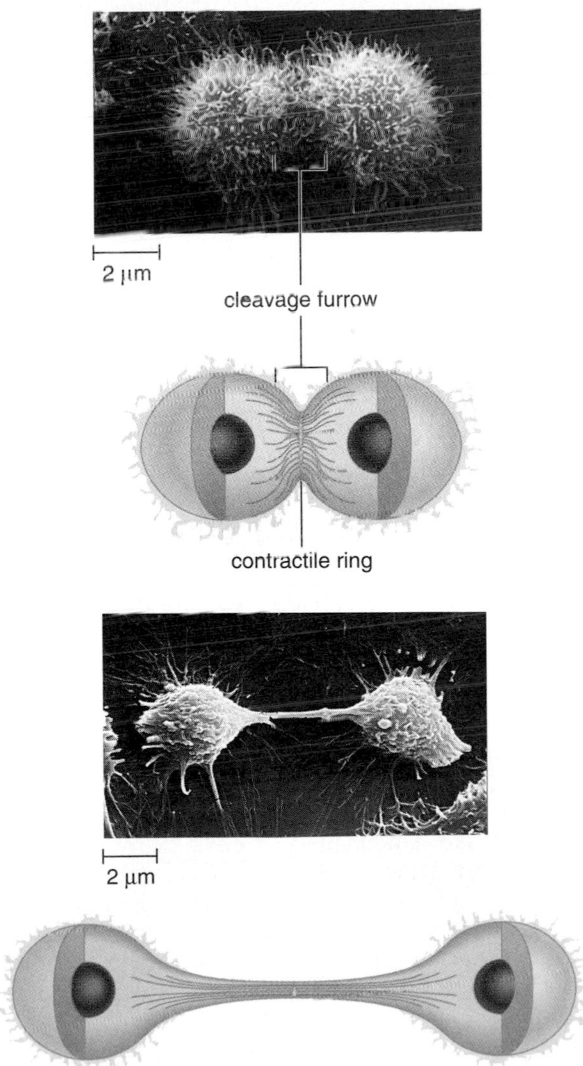

Figure 9.8 Cytokinesis in animal cells.
A single cell becomes two cells by a furrowing process. A contractile ring composed of actin filaments gradually gets smaller, and the cleavage furrow pinches the cell into two cells.

Copyright © R. G. Kessel and C. Y. Shih (1974). *Scanning Electron Microscopy in Biology: A Student's Atlas on Biological Organization* Springer-Verlag, New York.

9.4 Comparing Prokaryotes and Eukaryotes

Binary fission and mitosis ensure that each daughter cell is genetically identical to the parent cell. The genes consist of DNA found in the chromosomes.

Bacteria and protists, such as amoeboids and paramecia, are unicellular. Cell division in unicellular organisms produces two new individuals:

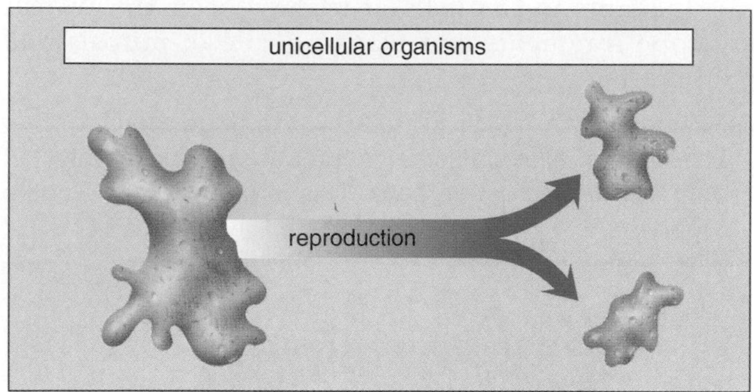

This is a form of asexual reproduction because one parent has produced identical offspring (Table 9.2).

In multicellular forms such as most fungi, plants, and animals, cell division is part of the growth process that produces the multicellular form we recognize as the organism. Cell division is also important in multicellular forms for renewal and repair:

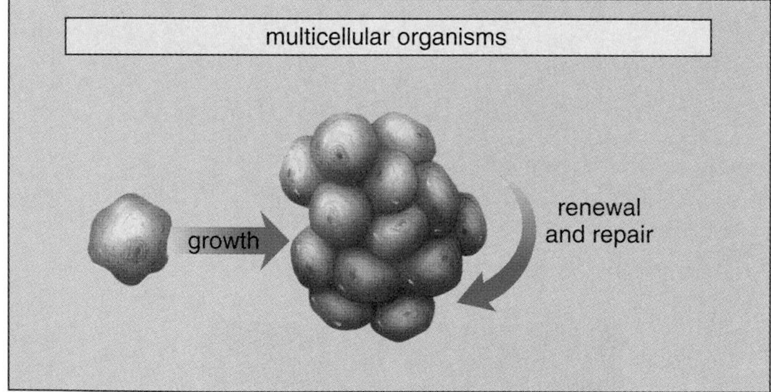

Table 9.2

Functions of Cell Division

Type of Organism	Cell Division	Function
Prokaryotes Bacteria	Binary fission	Asexual reproduction
Eukaryotes Protists Some fungi (yeast)	Mitosis and cytokinesis	Asexual reproduction
Other fungi Plants Animals	Mitosis and cytokinesis	Development, growth, and repair

In prokaryotes, the single chromosome consists largely of DNA with some associated proteins. During binary fission this chromosome duplicates, and each daughter cell receives one copy as the parent cell elongates and a new cell wall forms between the daughter cells. There is no spindle apparatus involved in binary fission.

The chromosomes of eukaryotic cells are composed of DNA and associated proteins. The protein histone organizes a chromosome, allowing it to extend as chromatin during interphase and coil and condense just prior to mitosis. Each species of multicellular eukaryotes has a characteristic number of chromosomes in the nuclei. As a result of mitosis, each daughter cell receives the same number and kinds of chromosomes as the parent cell. The spindle, which appears during mitosis, is involved in distributing the daughter chromosomes to the daughter nuclei. Cytokinesis, either by the formation of a cell plate (plant cells) or by furrowing (animal cells), is division of the cytoplasm.

Due to binary fission and mitosis daughter cells are genetically identical to the parent cell. Cell division allows unicellular organisms to reproduce and is necessary to growth and repair in multicellular organisms.

connecting concepts

The microscopic examination of cells undergoing mitosis may suggest to some that the production of new cells is a simple matter. However, consider what occurs: the chromosomes have duplicated (in the case of humans, this means about a hundred thousand genes replicate); the daughter chromosomes move apart; and finally the cytoplasm divides in order to produce two cells that are usually identical. That this process occurs millions or billions of times in all eukaryotic organisms, without error, is extraordinary.

Still, on occasion, irregularities do occur. Mutations can occur during the process of replication and control of the entire cell cycle can come out of sync. In some people, overproduction of skin cells produces a chronic inflammatory condition known as psoriasis. In many species, uncontrolled cell division of nondifferentiated cells can produce benign tumors or malignant growths. In a rare human genetic condition called progeria, the reproductive capacity of all the body's cells is severely diminished and young people grow old and die at an early age. These abnormalities involve faulty control of the cell cycle and changes in the chromosomes themselves.

The end product of mitosis, regardless of the eukaryotic species, is two new cells, each with the same number and kinds of chromosomes as the parent cell. But what about cell division that produces sex cells, the sperm and egg? As we'll see in the next chapter, a special process produces these cells that contain half the chromosome number.

Summary

9.1 How Prokaryotic Cells Divide

The prokaryotic chromosome is a single long loop of DNA attached to the inside of the plasma membrane. Binary fission involves replication of DNA, followed by an elongation of the cell that pulls the chromosomes apart. Inward growth of the plasma membrane and formation of new cell wall material divide the cell in two.

9.2 Eukaryotic Chromosomes and the Cell Cycle

Between nuclear divisions, the chromosomes are not distinct and are collectively called chromatin. Each eukaryotic species has a characteristic number of chromosomes. The total number is called the diploid number, and half this number is the haploid number.

Among eukaryotes, cell division involves nuclear division (karyokinesis) and division of the cytoplasm (cytokinesis). Mitosis is nuclear division in which the chromosome number stays constant because each chromosome is duplicated and gives rise to two daughter chromosomes.

The cell cycle has four stages. During the G_1 stage the organelles increase in number; during the S stage, DNA replication occurs; during the G_2 stage, various proteins are synthesized; and during the M stage, mitosis occurs.

Biologists have made great strides in understanding how the cell cycle is controlled. Kinases activated when they combine with cyclins phosphorylate proteins that trigger passage from the G_1 to the S stage, and the passage from the G_2 stage to the M stage.

9.3 How Eukaryotic Cells Divide

Mitosis has five phases, which are described here for animal cells.

Prophase—duplicated chromosomes are distinct, the nucleolus is disappearing, the nuclear envelope is fragmenting, and the spindle is forming between centrosomes. Asters radiate from the centrioles within the centrosomes.

Prometaphase—the kinetochores of sister chromatids attach to kinetochore spindle fibers extending from opposite poles. The chromosomes move back and forth until they are aligned at the metaphase plate.

Metaphase—the spindle is fully formed and the duplicated chromosomes are aligned at the metaphase plate. The spindle consists of polar spindle fibers that overlap at the metaphase plate and kinetochore spindle fibers that are attached to chromosomes.

Anaphase—daughter chromosomes move toward the poles. The polar spindle fibers slide past one another and the kinetochore spindle fibers disassemble. Cytokinesis by furrowing begins.

Telophase—nuclear envelopes re-form, chromosomes begin changing back to chromatin, the nucleoli reappear, and the spindle disappears.

Mitosis in plant cells is somewhat different from mitosis in animal cells because plant cells lack centrioles and, therefore, asters. Even so, the mitotic spindle forms and the same five phases are observed.

Cytokinesis in plant cells involves the formation of a cell plate from which the plasma membrane and cell wall are completed. Cytokinesis in animal cells is a furrowing process that divides the cytoplasm.

9.4 Comparing Prokaryotes and Eukaryotes

Binary fission (in prokaryotes) and mitosis (in unicellular eukaryotic protists and fungi) allow organisms to reproduce asexually. Mitosis in multicellular eukaryotes is primarily for the purpose of development, growth, and repair of tissues.

Reviewing the Chapter

1. Describe the prokaryotic chromosome and the process of binary fission. 148
2. Describe the eukaryotic chromosome and the significance of the duplicated chromosome. 149
3. Describe the cell cycle, including a description of interphase. 150
4. How are the two critical checkpoints in the cell cycle controlled? How is this control mechanism apparently related to the development of cancer? 150–51
5. Define the following words: chromosome, chromatin, chromatid, centriole, cytokinesis, centromere, and kinetochore. 148, 149, 152
6. Describe the events that occur during the phases of mitosis. 152–53
7. How does plant cell mitosis differ from animal cell mitosis? 154
8. Contrast cytokinesis in plant cells and animal cells. 154–55
9. Why are binary fission in prokaryotes and mitosis in unicellular eukaryotes forms of asexual reproduction? 156
10. What is the function of mitosis in multicellular eukaryotes? 156

Testing Yourself

Choose the best answer for each question.

1. What feature in prokaryotes substitutes for the spindle action in eukaryotes?
   a. centrioles with asters
   b. fission instead of cytokinesis
   c. elongation of plasma membrane
   d. looped DNA
2. How does a prokaryotic chromosome differ from a eukaryotic chromosome? A prokaryotic chromosome
   a. is shorter and fatter.
   b. is a single loop of DNA.
   c. never replicates.
   d. All of these are correct.
3. The diploid number of chromosomes
   a. is the 2n number.
   b. was in the parent cell and is in the two daughter nuclei following mitosis.
   c. varies according to the particular organism.
   d. All of these are correct.

For questions 4–6, match the descriptions that follow to the terms in the key.

Key:

a. centriole c. chromosome
b. chromatid d. centromere

4. point of attachment for sister chromatids
5. found at a pole in the center of an aster
6. coiled and condensed chromatin
7. If a parent cell has fourteen chromosomes prior to mitosis, how many chromosomes will the daughter cells have?
   a. twenty-eight
   b. fourteen
   c. seven
   d. any number between seven and twenty-eight
8. In which phase of mitosis are the chromosomes moving toward the poles?
   a. prophase c. anaphase
   b. metaphase d. telophase
9. Interphase
   a. is the same as prophase, metaphase, anaphase, and telophase.
   b. includes stages G_1, S, and G_2.
   c. requires the use of polar spindle fibers and kinetochore spindle fibers.
   d. rarely occurs.
10. Cytokinesis
   a. is mitosis in plants.
   b. requires the formation of a cell plate in plant cells.
   c. is the longest part of the cell cycle.
   d. is half a chromosome.
11. Label this diagram of a cell in prophase of mitosis.

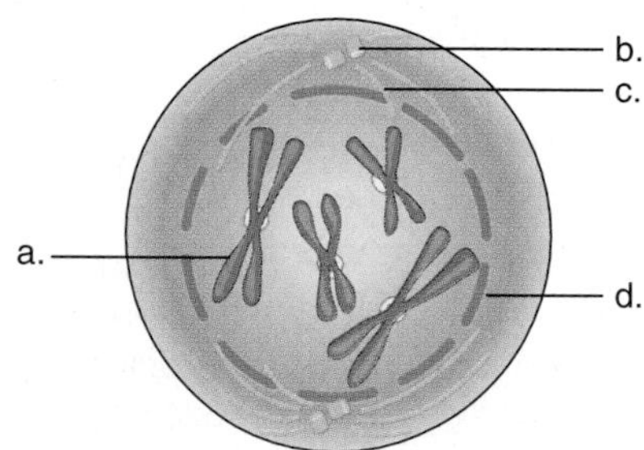

Applying the Concepts

1. *All cells come only from preexisting cells (the principle of biogenesis).*
   Cells can grow and reproduce. How is cell growth related to cell reproduction?
2. *Each body cell contains a full complement of genes.*
   Explain why skin cells arise only from skin cells, and so forth for other tissues.
3. *Most eukaryotic cells divide by means of mitosis.*
   What role does the spindle play in mitosis?

Using Technology

Your study of cell division is supported by these available technologies:

Exploring the Internet

The Mader Home Page provides resources for and help with studying this chapter.

http://www.mhhe.com/sciencemath/biology/mader/
(Click on Biology.)

Explorations in Cell Biology & Genetics CD-ROM

Mitosis: Regulating the Cell Cycle (#5)

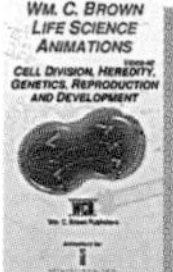

Life Science Animations Video

Video #2: Cell Division/Heredity/Genetics/Reproduction and Development
Mitosis (#12)

Video #5: Plant Biology/Evolution/Ecology
Mitosis and Cell Division in Plants (#50)

Understanding the Terms

Match the terms to these definitions:

a. __________ Full number of chromosomes that occurs in each somatic cell.
b. __________ Central microtubule organizing center of cells consisting of granular material. In animal cells, it contains two centrioles.
c. __________ Constriction where sister chromatids of a chromosome are held together.
d. __________ Division of the cytoplasm following mitosis.
e. __________ Microtubule structure that brings about chromosome movement during nuclear division.
f. __________ One of two genetically identical chromosome units that are the result of DNA replication.
g. __________ Process in which a parent nucleus produces two daughter nuclei, each having the same number and kinds of chromosomes as the parent nucleus.

Meiosis and Sexual Reproduction

CHAPTER

10

Chapter Concepts

10.1 Halving the Chromosome Number

- Due to meiosis, sex cells contain half the total number of chromosomes. 160
- Meiosis occurs at varied times during the life cycle of organisms. 160

10.2 How Meiosis Occurs

- Meiosis requires two cell divisions and results in four daughter cells. 162
- During meiosis nonsister chromatids exchange genetic material. 162

10.3 Meiosis Has Phases

- Meiosis I and meiosis II each have four phases. 164

10.4 Viewing the Human Life Cycle

- In humans meiosis is a part of the production of sperm in males and eggs in females. 166

10.5 Significance of Meiosis

- The process of sexual reproduction which includes meiosis brings about variation and contributes to the evolutionary process. 169

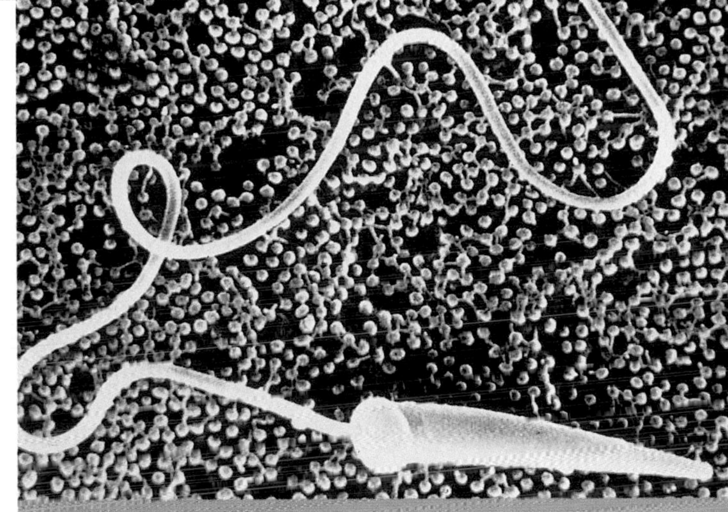

Sea star sperm on egg

1 μm

In asexual reproduction, a single parent passes on a complete set of chromosomes to an offspring. In sexual reproduction, each of two parents give one-half a set of chromosomes to an offspring. Think about what the sex act accomplishes—the male gamete (sperm) fertilizes the female gamete (egg) to produce a diploid zygote. Unless the gametes were haploid, the chromosome number would double with each new generation.

The process of sexual reproduction increases the genetic variability of the next generation in three ways. During the first part of meiosis, the type of nuclear division involved in gamete production, chromosome pairs come together and often swap segments with each other, and this produces different combinations of genes on the chromosomes. Then only one of each chromosome pair ends up in a particular gamete, but which one is up to chance. Finally, the gametes that join during fertilization are usually from two different individuals, and if so, genetic recombination is bound to occur.

As a result of asexual reproduction, the offspring most likely will be quite similar to their single parent. But sexual reproduction usually ensures that the next generation is populated with offspring that are genetically different from their parents and from each other. Genetic variability is the raw material for evolution, and sexual reproduction contributes to the process of evolution, especially if the environment is changing.

10.1 Halving the Chromosome Number

A **gamete** is a haploid sex cell. Gamete formation and then fusion of gametes to form a cell called a *zygote* are integral parts of **sexual reproduction.** Obviously, if the gametes contained the same number of chromosomes as the body cells, the number of chromosomes would double with each new generation. Within a few generations, the cells of the individual would be nothing but chromosomes! The early cytologists (biologists who study cells) realized this, and Pierre-Joseph van Beneden, a Belgian, was gratified to find in 1883 that the sperm and the egg of the worm *Ascaris* each contained only two chromosomes, while the zygote and subsequent embroyonic cells always have four chromosomes.

Life Cycles Vary

The term *life cycle* refers to all the reproductive events that occur from one generation to the next similar generation. A zygote always has the full or **diploid (2n) number** [Gk. *diplos,* twofold, and *-eides,* like] of chromosomes. *Mitosis* is the type of nuclear division that maintains a constant chromosome number. **Meiosis** [Gk. *mio,* less, and *-sis,* act or process of], the topic of this chapter, is the type of nuclear division that reduces the chromosome number from the diploid (2n) number to the haploid (n) number. The **haploid (n) number** [Gk. *haplos,* single, and *-eides,* like] of chromosomes is half the diploid number.

As discussed in the reading on page 161, meiosis occurs at different points during the life cycle of various kinds of organisms. In animals, it occurs during the production of gametes. In plants, meiosis produces spores that divide mitotically to become a haploid generation. The haploid generation produces the gametes. In certain fungi and some algae, meiosis occurs directly after zygote formation and therefore these organisms are haploid.

Notice that all three life cycles depicted in Figure 10A have both a diploid stage and a haploid stage. In animals, the adult is always diploid and the haploid stage consists only of the gametes. In plants, the haploid stage, which produces the gametes, may be larger or smaller than the diploid stage depending on the species. The mosses you see growing on rocks are haploid most of their life cycle while oak trees are diploid most of their life cycle. In fungi (and some algae) only the zygote is diploid, and the organisms we observe in nature are haploid. So the black mold you sometimes see growing on bread and the green coating you see growing on a pond are haploid. It is these haploid individuals that produce gamete nuclei.

Chromosomes Come in Homologous Pairs

In a diploid cell, the chromosomes occur in pairs. The members of each pair are called **homologous chromosomes** or **homologues** [Gk. *homologos,* agreeing, corresponding]. The homologues look alike; they have the same length and centromere position.

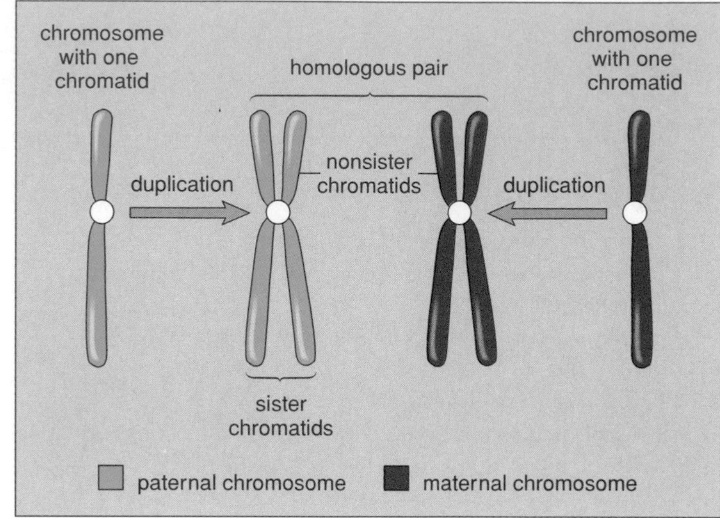

When stained, homologues have a similar banding pattern because they contain the same types of genes. If a gene for length of fingers occurs at a particular locus (location) on one homologue, it also occurs at the same locus on the other homologue. In the one instance, however, the gene might call for short fingers and in the other for long fingers.

The chromosomes in the diagram are duplicated as they would be just before nuclear division. When duplicated, a chromosome is composed of two identical parts called sister chromatids. The sister chromatids are held together at a region called the centromere. Notice that nonsister chromatids do not share the same centromere.

Why does the zygote with the diploid number of chromosomes contain homologous pairs? One member of each homologous pair was inherited from the male parent and the other was inherited from the female parent by way of the gametes. We will see that gametes contain one of each type of chromosome—derived from either the paternal or maternal homologue.

The zygote, which is always diploid, contains homologous chromosomes. Gametes are haploid due to meiosis, which occurs at varied times according to the life cycle.

A closer Look

Life Cycles of Eukaryotes

Sexual reproduction requires two parents, each of which contributes chromosomes (genes) to the offspring by way of sex cells or gametes. The process of meiosis reduces the chromosome number so that the gametes are haploid. After the haploid gametes fuse during fertilization, the zygote will have the diploid number of chromosomes.

Three types of life cycles are typical among eukaryotes. All three life cycles include production of gametes, a haploid phase, fertilization, and a diploid phase before the cycle begins again. The diagrams shown in Figure 10A show the diploid phase above the line and the haploid phase along with gamete production below the line.

Protists and most fungi typically have the haplontic life cycle. In this life cycle, the *zygote undergoes meiosis* and the zygote is the only diploid stage. When the zygote divides, haploid spores form and develop into a haploid adult. *Chlamydomonas* is a unicellular green alga living in freshwater ponds that has the haplontic cycle. *Chlamydomonas* is haploid and when it reproduces sexually, it simply divides producing cells that can fuse and form a diploid zygote. *Chlamydomonas* only forms gametes when environmental conditions are unfavorable—the zygote has a heavy wall that is protective until conditions are favorable again. When the zygote "germinates," it produces haploid zoospores and each one develops into an adult *Chlamydomonas*. Fungi, like black mold that grows on bread, also have a haplontic life cycle. Among fungi the haploid adult that produces the gametes directly is multicellular.

All plants have the alternation of generation life cycle. This life cycle has both a diploid adult stage and a haploid adult stage. A diploid *multicellular adult* (and not the zygote) *produces haploid spores by meiosis*. And then the spores develop into a multicellular haploid adult that produces gametes directly. A pine tree is diploid; that is, all its cells contain the diploid number of chromosomes. Meiosis occurs in the cones, and here the spores become male and female haploid generations that produce sperm and egg. The sperm are carried in pollen to an egg, and the diploid embryo is enclosed within a seed. The seed develops into a new diploid adult.

Animals have the diplontic life cycle. In this life cycle a multicellular *diploid adult produces gametes by a process that involves meiosis.* (There are no spores and the gametes are the only haploid portion of the life cycle.) After the gametes fuse, the zygote develops into the multicellular diploid adult. We are familiar with this life cycle because being animals we also have this life cycle.

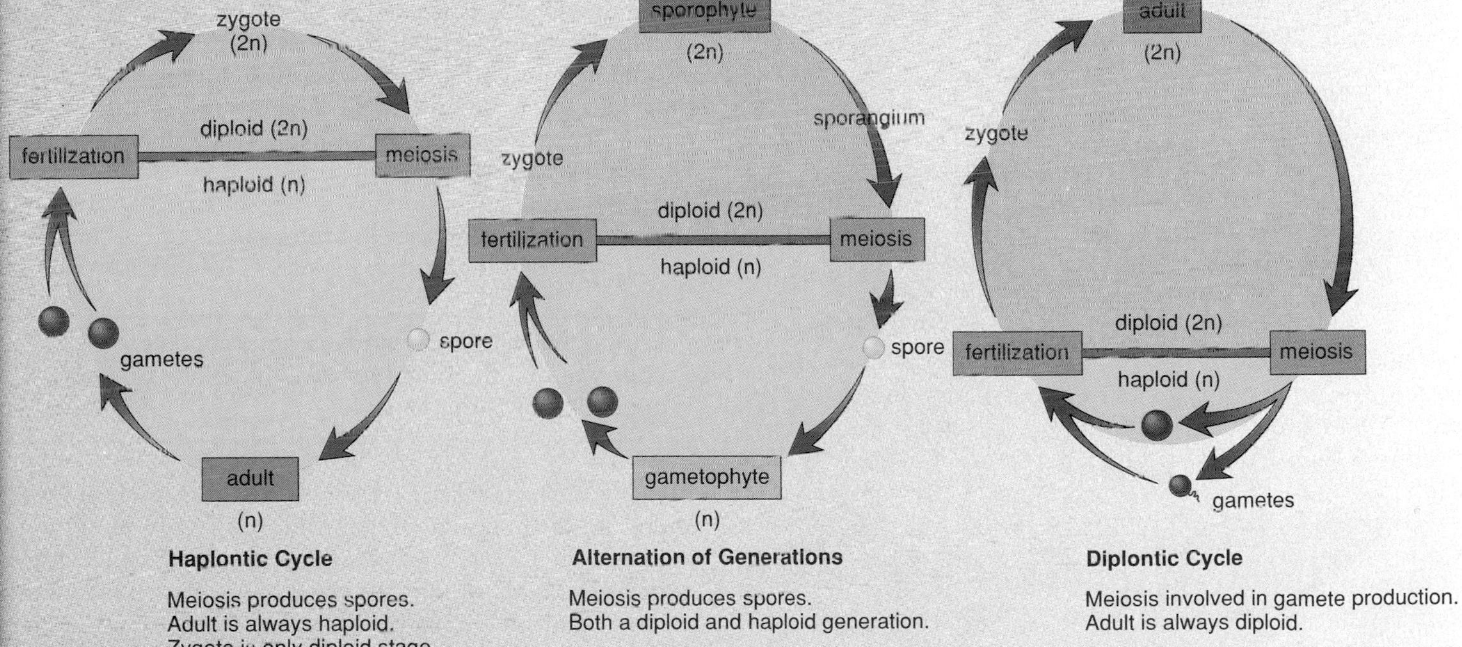

Figure 10A Life cycles of eukaryotes.

10.2 How Meiosis Occurs

Meiosis keeps the chromosome number constant from one generation to the next. It reduces the chromosome number in such a way that the gametes contain only one member of each homologous pair.

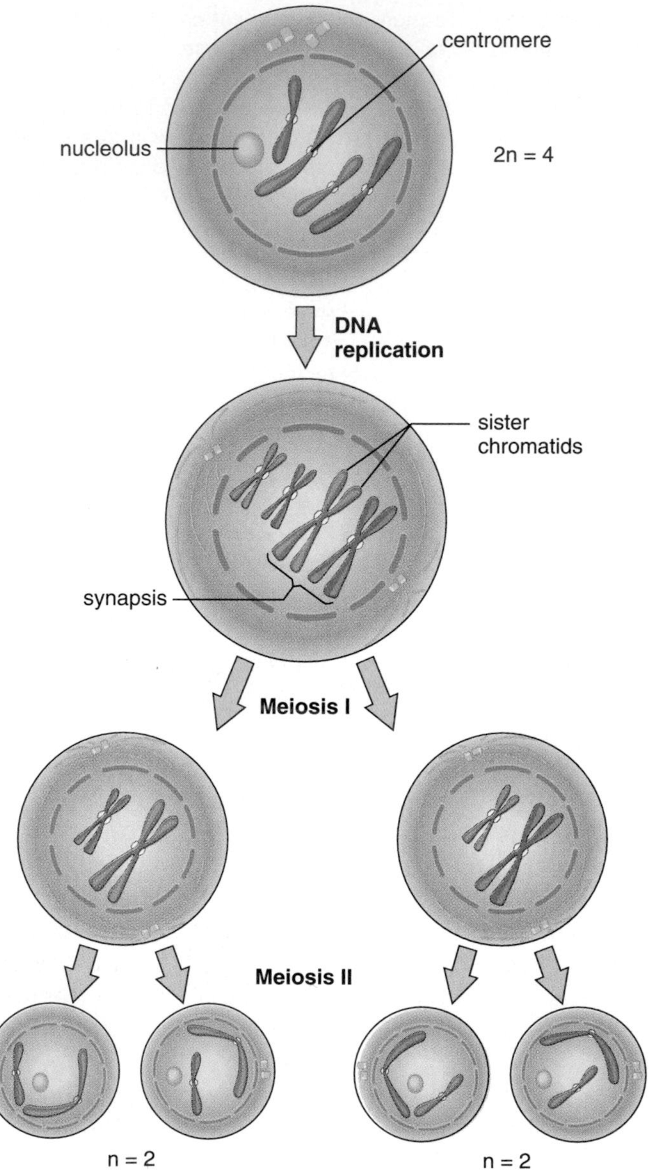

Figure 10.1 Overview of meiosis.
Following DNA replication, each chromosome is duplicated and consists of two chromatids. During meiosis I, the chromosome pairs separate, and during meiosis II, the sister chromatids of each duplicated chromosome separate. At the completion of meiosis, there are four haploid daughter cells. (The blue chromosomes were inherited from one parent, and the red chromosomes were inherited from the other parent.)

Meiosis Has Two Divisions

Meiosis requires two nuclear divisions and produces four haploid daughter cells, each having one of each kind of chromosome and therefore half the total number of chromosomes present in the diploid parent nucleus. The parent cell has the diploid number of chromosomes, while the daughter cells have the haploid number of chromosomes.

Figure 10.1 presents an overview of meiosis, indicating the two cell divisions, meiosis I and meiosis II. Prior to meiosis I, DNA (deoxyribonucleic acid) replication has occurred; therefore, each chromosome has two sister chromatids. During meiosis I, something new happens that does not occur in mitosis. The homologous chromosomes come together and line up side by side due to a means of attraction still unknown. This so-called **synapsis** [Gk. *synaptos*, united, joined together] results in a **bivalent** [L. *bis*, two, and *valens*, strength]—that is, two chromosomes that stay in close association during the first two phases of meiosis I. Sometimes the term tetrad [Gk. *tetra*, four] is used instead of bivalent because, as you can see, a bivalent contains four chromatids. Exchange of genetic material, called crossing-over, may occur between the nonsister chromatids of a tetrad. After crossing-over occurs, the sister chromatids of a chromosome are no longer identical.

Following synapsis, the homologous chromosomes separate. This separation means that only one chromosome from each homologous pair reaches a daughter nucleus. It is important for daughter nuclei to have a member from each pair of homologous chromosomes because only in that way can there be a copy of each kind of chromosome in the daughter nuclei. The members of the homologous pairs separate independently of one another; any particular kind of chromosome can be with any other kind in the daughter nuclei. Therefore, all possible combinations of chromosomes can occur within the gametes that result after meiosis is complete.

During meiosis I, homologous chromosomes separate and the daughter cells have one copy of each kind of chromosome.

No replication of DNA is needed between meiosis I and meiosis II because the chromosomes are already duplicated; they already have two sister chromatids. During meiosis II, the daughter chromosomes derived from sister chromatids move to opposite poles. Therefore, the chromosomes in the four daughter cells have only one chromatid. You can count the number of centromeres to verify that the parent cell has the diploid number of chromosomes and each daughter cell has the haploid number.

Following meiosis II, there are four haploid daughter cells and each chromosome consists of one chromatid.

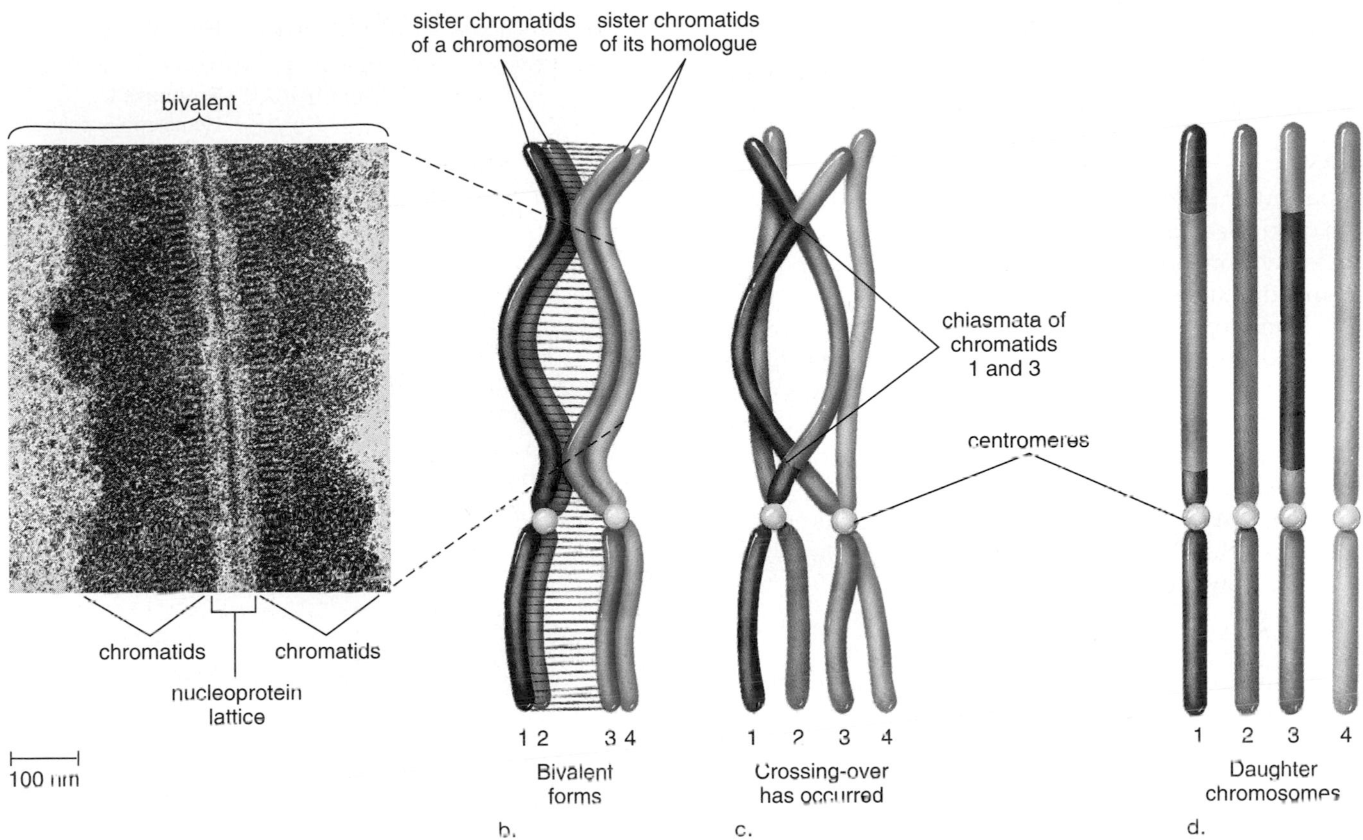

Figure 10.2 Crossing-over occurs during meiosis I.
a. The homologous chromosomes pair up, and a nucleoprotein lattice, called the synaptonemal complex, develops between them. This is an electron micrograph of the complex, which zippers the members of the bivalent together so that corresponding genes are in alignment. b. This diagrammatic representation shows only two places where nonsister chromatids 1 and 3 have come into contact. Actually, the other two nonsister chromatids most likely are also crossing over. c. Chiasmata indicate where crossing-over has occurred. The exchange of color represents the exchange of genetic material. d. Following meiosis II, daughter chromosomes have a new arrangement of genetic material due to crossing-over, which occurred between nonsister chromatids during meiosis I.

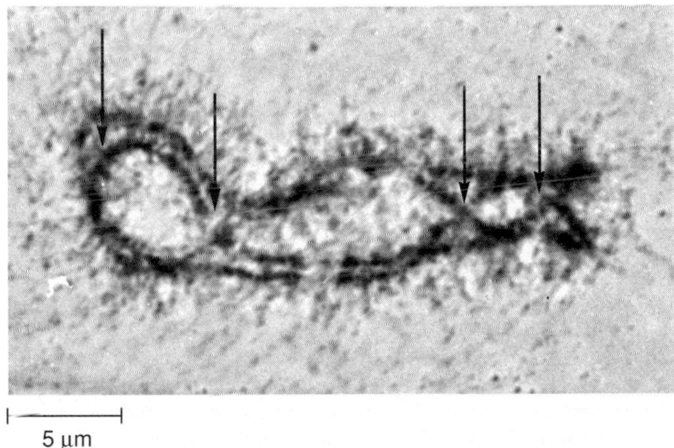

Figure 10.3 Chiasmata (arrows) of a chromosome bivalent, from a testis cell of a grasshopper.
The chiasmata mark the places where crossing-over between nonsister chromosomes of the bivalent has occurred. The chiasmata hold the members of the bivalent together until separation of homologues occurs.

Crossing-Over Introduces Variation

Crossing-over is an exchange of genetic material between nonsister chromatids of a bivalent during meiosis I. At synapsis, homologues line up side by side, and a nucleoprotein lattice (called the synaptonemal complex) appears between them (Fig. 10.2). This lattice holds the bivalent together in such a way that the DNA of the nonsister chromatids is aligned. Now crossing-over occurs. As the lattice breaks down, homologues are temporarily held together by *chiasmata* (sing., chiasma), regions where the nonsister chromatids are attached due to crossing-over (Fig. 10.3). Then homologues separate and are distributed to different daughter cells.

Due to crossing-over of sister chromatids during meiosis I, the daughter chromosomes that move to the poles during meiosis II have a different combination of genes.

Crossing-over is a way to increase genetic variation in the gametes.

10.3 Meiosis Has Phases

Both meiosis I and meiosis II have these phases: prophase, metaphase (preceded by prometaphase), anaphase, and telophase.

Prophase I

It is apparent during prophase I that nuclear division is about to occur because a spindle forms as the centrosomes migrate away from one another. The nuclear envelope fragments, and the nucleolus disappears.

The homologous chromosomes, each having two sister chromatids, undergo synapsis to form bivalents. As depicted in Figure 10.2 by the exchange of color, crossing-over between the nonsister chromatids may occur at this time. After crossing-over, the sister chromatids of a duplicated chromosome are no longer identical.

Throughout prophase I, the chromosomes have been condensing so that by now they have the appearance of metaphase chromosomes.

Metaphase I

During prometaphase I, the bivalents held together by chiasmata (see Fig. 10.3) have moved toward the metaphase plate (equator of the spindle). Metaphase I is characterized by a fully formed spindle and alignment of the bivalents at the metaphase plate. *Kinetochores*, protein complexes just outside the centromeres, are seen, and these are attached to spindle fibers called kinetochore spindle fibers.

Bivalents independently align themselves at the metaphase plate of the spindle. The maternal homologue of each bivalent may be orientated toward either pole, and the paternal homologue of each bivalent may be aligned toward either pole. This means that all possible combinations of chromosomes can occur in the daughter cells.

Anaphase I

During anaphase I, the homologues of each bivalent separate and move to opposite poles. Notice that each chromosome still has two chromatids.

Telophase I

In some species, there is a telophase I stage at the end of meiosis I. If so, the nuclear envelopes re-form and nucleoli appear. This phase may or may not be accompanied by cytokinesis, which is separation of the cytoplasm. Figure 10.4 shows only two of the four possible combinations of haploid chromosomes when the parent cell has two homologous pairs of chromosomes. Can you determine what the other two possible combinations of chromosomes are?

Interkinesis

Interkinesis is similar to interphase between mitotic divisions except that DNA replication does not occur—the chromosomes are already duplicated.

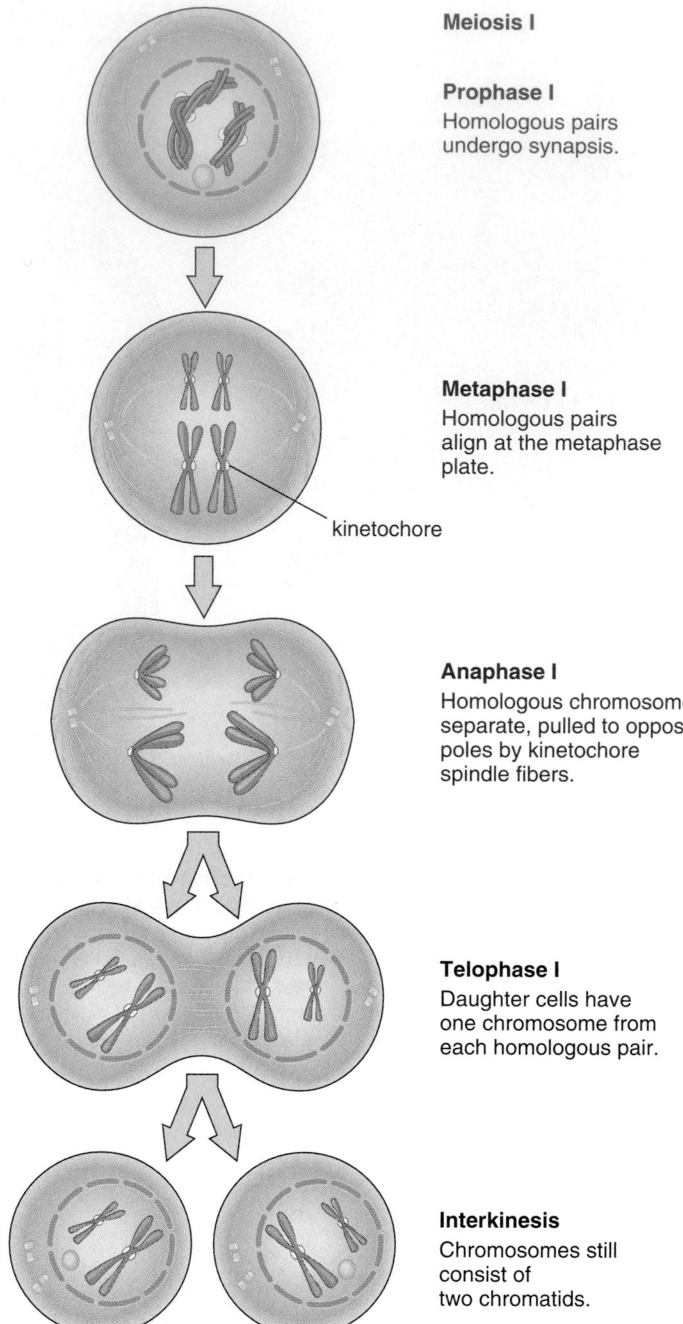
Meiosis I
Prophase I
Homologous pairs undergo synapsis.
Metaphase I
Homologous pairs align at the metaphase plate.
kinetochore
Anaphase I
Homologous chromosomes separate, pulled to opposite poles by kinetochore spindle fibers.
Telophase I
Daughter cells have one chromosome from each homologous pair.
Interkinesis
Chromosomes still consist of two chromatids.

Figure 10.4 Meiosis I.
During meiosis I, homologous chromosomes undergo synapsis and then separate independently—a daughter cell receives one of each kind of chromosome in any of the possible combinations. Following meiosis I, there are two haploid daughter cells and the chromosomes are still duplicated. (The blue chromosomes were inherited from one parent, and the red chromosomes were inherited from the other parent.)

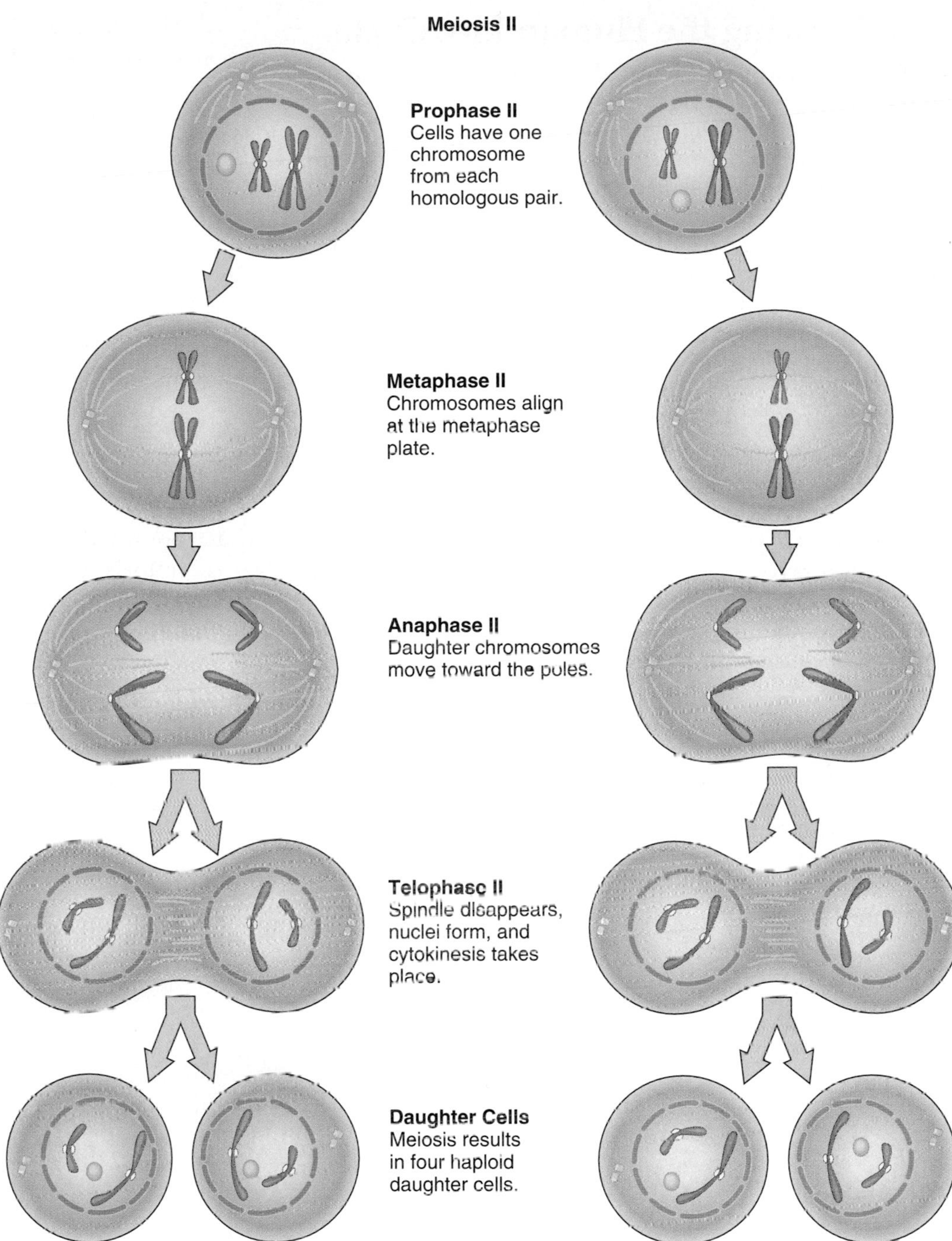

Meiosis II
Prophase II
Cells have one chromosome from each homologous pair.
Metaphase II
Chromosomes align at the metaphase plate.
Anaphase II
Daughter chromosomes move toward the poles.
Telophase II
Spindle disappears, nuclei form, and cytokinesis takes place.
Daughter Cells
Meiosis results in four haploid daughter cells.

Figure 10.5 Meiosis II. During meiosis II, daughter chromosomes consisting of one chromatid each move to the poles. Following meiosis II, there are four haploid daughter cells.

Meiosis II

During metaphase II, the haploid number of chromosomes, which are still duplicated, align at the metaphase plate (Fig. 10.5). During anaphase II, the two sister chromatids separate at the centromere, giving rise to two daughter chromosomes. These daughter chromosomes move toward the poles. At the end of telophase II and cytokinesis, there are four haploid cells. Due to crossing-over of chromatids, each gamete can contain chromosomes with different types of genes.

Following meiosis II, the haploid cells mature and become gametes in animals. In plants, they become spores that divide to produce a haploid adult generation. Plants also have a diploid adult generation (see the reading on page 161). In some fungi and some algae, the zygote that results from gamete fusion immediately undergoes meiosis, and therefore the adult is always haploid.

10.4 Viewing the Human Life Cycle

Mammals, including humans, have a life cycle that requires both meiosis and mitosis (Fig. 10.6). In males, meiosis is a part of **spermatogenesis** [Gk. *sperma*, seed, and L. *genitus*, producing], which occurs in the testes and produces sperm. In females, meiosis is a part of **oogenesis** [Gk. *oon*, egg, and L. *genitus*, producing], which occurs in the ovaries and produces eggs. A sperm and egg join at fertilization and the resulting zygote undergoes mitosis during development of the fetus, which is the stage of development before birth. After birth, mitosis is involved in the continued growth of the child and repair of tissues at any time.

Comparison of Meiosis with Mitosis

The following lists and Figure 10.7 will allow you to compare meiosis to mitosis.

Occurrence

Meiosis occurs only at certain times in the life cycle of sexually reproducing organisms. In humans, meiosis occurs only in the sex organs and produces the gametes. Mitosis is more common because it allows growth and repair of body tissues in multicellular organisms, including humans.

Process

The following are distinctive differences between the processes of meiosis and mitosis.

1. DNA is replicated only once before both meiosis and mitosis; but there are two nuclear divisions during meiosis and only one nuclear division during mitosis.
2. Homologous chromosomes pair and undergo crossing-over during prophase I of meiosis but not during mitosis.
3. Paired homologous chromosomes (bivalents) align at the metaphase plate during metaphase I in meiosis; individual (duplicated) chromosomes align at the metaphase plate during metaphase in mitosis.
4. Homologous chromosomes (with centromeres intact) separate and move to opposite poles during anaphase I in meiosis; and daughter chromosomes move to opposite poles during anaphase in mitosis.
5. The events of meiosis II are just like those of mitosis except that meiosis II nuclei are always haploid.

Daughter Nuclei and Cells

The genetic consequences of meiosis and mitosis are quite different as well.

1. Four daughter cells are produced by meiosis; mitosis results in two daughter cells.
2. The four daughter cells formed by meiosis are haploid; the daughter cells produced by mitosis have the same chromosome number as the parent cell.
3. The daughter cells from meiosis are not genetically identical to each other or to the parent cell. The daughter cells from mitosis are genetically identical to each other and to the parent cell.

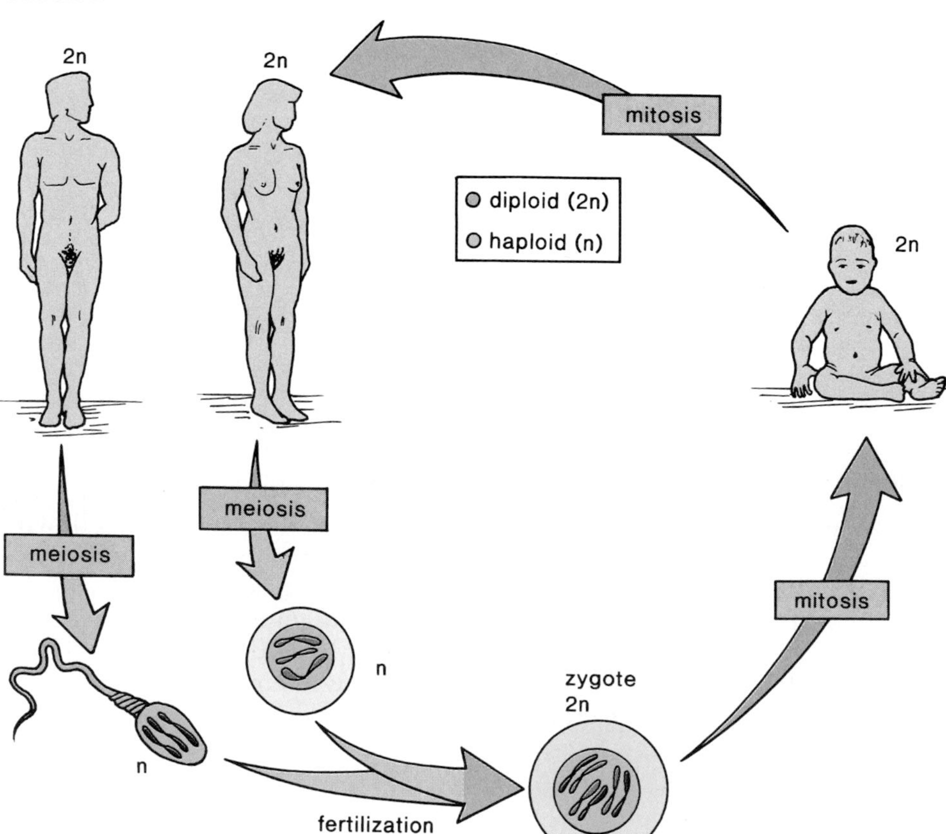

Figure 10.6 Life cycle of humans. Meiosis in males is a part of sperm production, and meiosis in females is a part of egg production. When a haploid sperm fertilizes a haploid egg, the zygote is diploid. The zygote undergoes mitosis as it develops into a newborn child. Mitosis continues after birth until the individual reaches maturity; then the life cycle begins again.

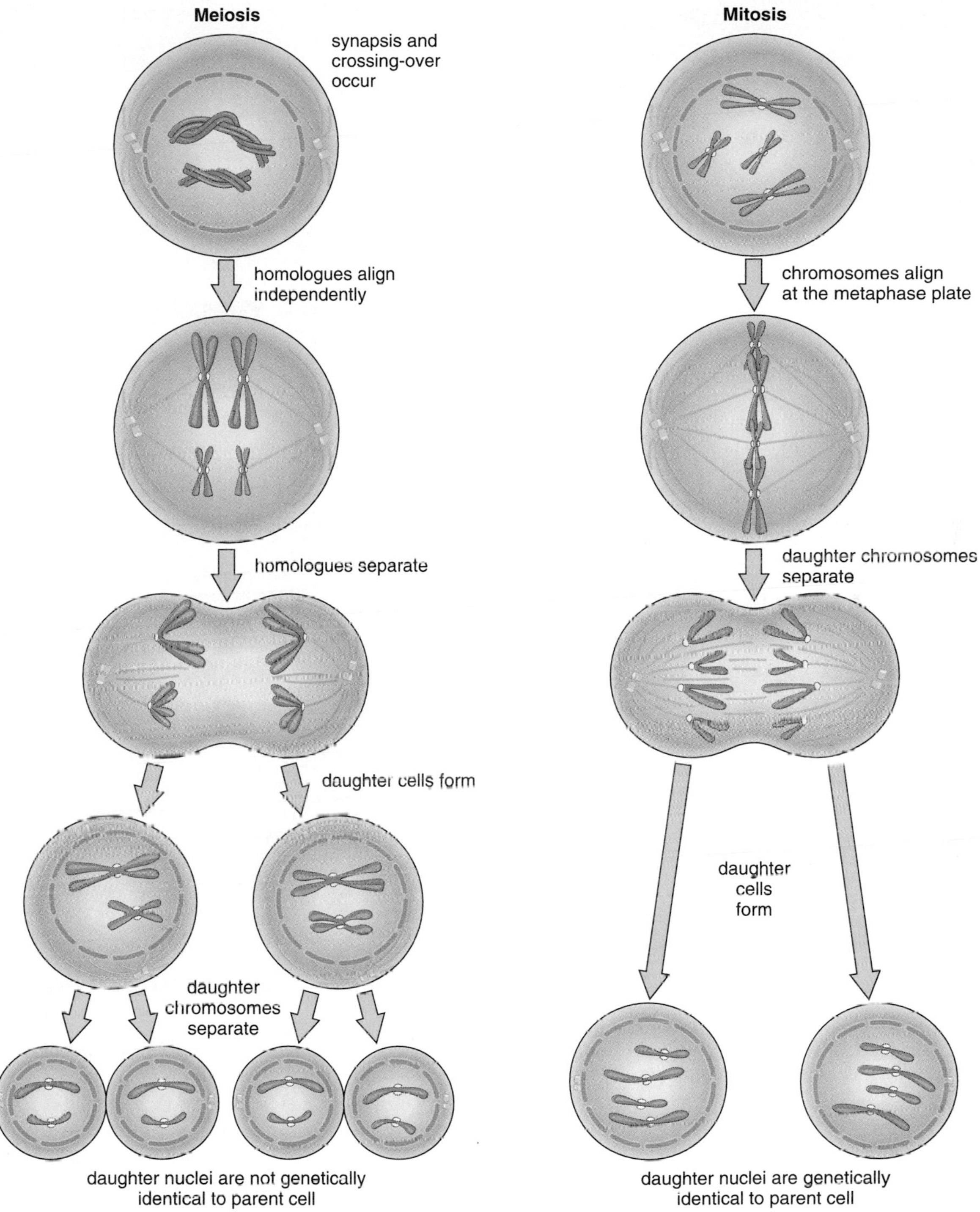
Meiosis
synapsis and crossing-over occur
homologues align independently
homologues separate
daughter cells form
daughter chromosomes separate
daughter nuclei are not genetically identical to parent cell
Mitosis
chromosomes align at the metaphase plate
daughter chromosomes separate
daughter cells form
daughter nuclei are genetically identical to parent cell

Figure 10.7 Comparison of meiosis and mitosis.
(The blue chromosomes were inherited from one parent, and the red chromosomes were inherited from the other parent.)

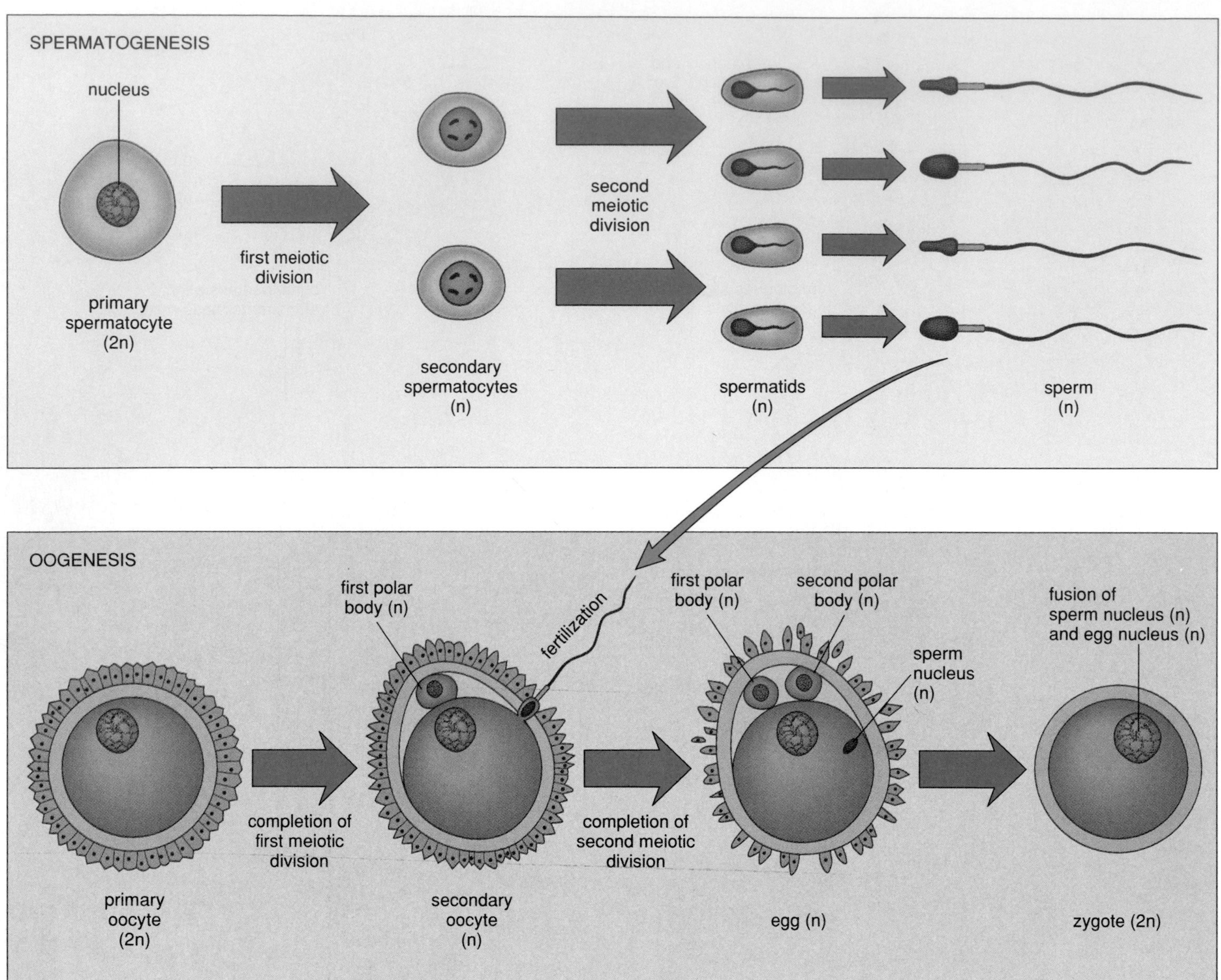

Figure 10.8 Spermatogenesis and oogenesis in mammals.
Spermatogenesis produces four viable sperm, whereas oogenesis produces one egg and at least two polar bodies. In humans, both sperm and egg have 23 chromosomes each; therefore, following fertilization, the zygote has 46 chromosomes.

Spermatogenesis and Oogenesis Produce the Gametes

Figure 10.8 contrasts spermatogenesis with oogenesis, processes that produce the gametes in mammals, including humans. In the testes of males, primary spermatocytes with 46 chromosomes divide to form two secondary spermatocytes, each with 23 duplicated chromosomes. Secondary spermatocytes divide to produce four spermatids, also with 23 daughter chromosomes. Spermatids then differentiate into sperm (spermatozoa). The processs of meiosis in males always results in four cells that become sperm.

In the ovaries of females, a primary oocyte has 46 chromosomes and divides meiotically into two cells, each having 23 chromosomes. One of these cells, termed the **secondary oocyte** [Gk. *oon*, egg, and *kytos*, cell], receives almost all the cytoplasm (Fig. 10.8). The other is a **polar body** that may disintegrate or may divide again. The secondary oocyte begins meiosis II and then stops at metaphase II. Then at ovulation, it leaves the ovary and enters an oviduct where it may be approached by a sperm. If a sperm does enter the oocyte, the oocyte is activated to continue meiosis II to completion. The mature egg has 23 chromosomes. Meiosis in females produces only one egg and possibly three polar bodies. The polar bodies are a way to discard unnecessary chromosomes while retaining much of the cytoplasm in the egg. The cytoplasm serves as a source of nutrients for the developing embryo.

10.5 Significance of Meiosis

In all sexually reproducing organisms, meiosis provides a way to keep the chromosome number constant generation after generation. Without meiosis, the chromosome number of the next generation would continually increase. Not only is the chromosome number halved precisely, each daughter cell receives a copy of each kind of chromosome. This ensures that each daughter cell receives one of each kind of gene.

Meiosis also helps ensure that genetic recombination occurs with each generation. As a result of independent assortment of chromosomes, the chromosomes are distributed to the daughter cells in various combinations. The total number of possible combinations is 2^n, where n is the haploid number of chromosomes. In humans, where n = 23, the number of possible chromosome combinations produced by meiosis is a staggering 2^{23}, or 8,388,608. And this does not even consider the variations that are introduced due to crossing-over. For example, if we assume only one crossover occurs within each bivalent (actually many crossovers usually occur), then in humans crossing-over and independent alignment and separation will generate 4^{23} different possible kinds of gametes, or 70,368,744,000,000.

This is just the variation introduced by meiosis, but if the entire sexual reproductive process is considered, there is another element that introduces variation. Due to *fertilization,* the chromosomes donated by the parents are combined, and in humans, this means that $(2^{23})^2$, or 70,368,744,000,000, chromosomally different zygotes are possible, even assuming no crossing-over. If crossing-over occurs once, then $(4^{23})^2$, or 4,951,760,200,000,000,000,000,000,000, genetically different zygotes are possible for every couple.

In any case, a sexually reproducing population has a tremendous storehouse of genetic recombinations, which may be advantageous for evolution, particularly when the environment is changing. Asexually reproducing organisms, like prokaryotes, depend primarily on mutations to generate variation. This is sufficient because they produce great numbers of offspring within a limited amount of time. Mutation is still the raw material for variation among sexually reproducing organisms, but the shuffling of genetic material due to sexual reproduction may better allow adaptation to a changing environment. If the environment is not changing, the genetic makeup of the parents is most likely still adaptive.

> There are three ways by which genetic recombination comes about due to sexual reproduction:
>
> 1. Independent alignment of bivalents at the metaphase plate means that gametes have different combinations of chromosomes.
> 2. Crossing-over means that the chromosomes in one gamete have a different combination of genes than chromosomes in another gamete.
> 3. Upon fertilization, combining of chromosomes from genetically different gametes occurs.

connecting concepts

Producing a healthy bouncing baby initially depends on producing healthy gametes, gametes that have the correct number of normal chromosomes. This means that meiosis must occur appropriately in both potential parents. In men, meiosis occurs in the testes—it begins at puberty and continues until death. In women, meiosis occurs in the ovaries—it begins at puberty and ends with menopause, when a woman stops producing eggs and menstruating.

Fortunately, meiosis usually proceeds normally, resulting in haploid gametes, each with one copy of each gene. Occasionally, however, a mistake is made. Suppose, for example, that during anaphase I, one pair of homologous chromosomes fails to separate. One gamete will then ultimately receive an extra chromosome, for a total of 24 chromosomes. Suppose this abnormal gamete is an egg that is fertilized by a sperm carrying the normal 23 chromosomes. The result will be a zygote with 47 chromosomes, a major mistake.

Mistakes such as this one sometimes occur spontaneously, that is, for no known reason. Other mistakes seem to be related to exposures to chemicals, drugs, or various energy sources. Ever wonder why a woman's abdomen is shielded with a lead apron during most X rays? It protects her ovaries and therefore her eggs.

Meiosis is an extraordinary process, and the fact that it almost always occurs exactly like it should is even more amazing. Sometimes we take our ability to produce perfect bouncing babies for granted, but when we look more carefully at the process of meiosis, we can appreciate that a precise mechanism is required.

Summary

10.1 Halving the Chromosome Number

Meiosis is involved in any life cycle that involves sexual reproduction. Meiosis ensures that the chromosome number in offspring stays constant generation after generation. In the animal life cycle, only the gametes are haploid; in plants and fungi, there is a multicellular haploid adult that produces the gametes.

The nucleus contains pairs of chromosomes, called homologous chromosomes (homologues). During meiosis, the haploid daughter cells receive one of each kind of chromosome.

10.2 How Meiosis Occurs

Meiosis requires two cell divisions and results in four daughter cells. Replication of DNA takes place before meiosis begins. During meiosis I, the homologues undergo synapsis (resulting in a bivalent) and crossing-over between nonsister chromatids occurs before they independently separate. The daughter cells receive one member of each pair of homologous chromosomes. There is no replication of DNA during interkinesis. During meiosis II, the sister chromatids separate at the centromeres, giving rise to daughter chromosomes that move to opposite poles as they do in mitosis. The four resulting daughter cells are not genetically identical to the parent cell; they are haploid, and due to crossing-over, their chromosomes carry a different combination of genes.

10.3 Meiosis Has Phases

Meiosis I is divided into four phases:

Prophase I—Bivalents form, and crossing-over occurs as chromosomes condense; the nuclear envelope fragments.

Metaphase I—Bivalents independently align at the metaphase plate.

Anaphase I—Homologous chromosomes separate.

Telophase I—Nuclei become haploid, having received one duplicated chromosome from each homologous pair.

Meiosis II is divided into four phases:

Prophase II—Chromosomes condense and the nuclear envelope fragments.

Metaphase II—The haploid number of still duplicated chromosomes align independently at the metaphase plate.

Anaphase II—Daughter chromosomes move to the poles.

Telophase II—Four haploid daughter cells are genetically different from the parent cell.

10.4 Viewing the Human Life Cycle

During the life cycle of humans and many other animals, only the gametes are haploid. Meiosis is involved in spermatogenesis and oogenesis. Fertilization restores the diploid number of chromosomes.

Both mitosis and meiosis occur in the human life cycle. They can be compared in this manner:

Mitosis	**Meiosis I**
Prophase	
No pairing of chromosomes	Pairing of homologous chromosomes
Metaphase	
Duplicated chromosomes at metaphase plate	Bivalents at metaphase plate
Anaphase	
Sister chromatids separate	Homologous chromosomes separate
Telophase	
Daughter nuclei have the parent cell chromosome number	Daughter nuclei are always haploid

Meiosis II is like mitosis except the nuclei are haploid.

Whereas spermatogenesis produces four sperm per meiosis, oogenesis produces one egg and two to three nonfunctional polar bodies. Spermatogenesis occurs in males and oogenesis occurs in females.

10.5 Significance of Meiosis

Mutations are the primary source of genetic variation among asexually reproducing organisms. But among sexually reproducing organisms, meiosis produces variation by independent assortment of homologous chromosomes and crossing-over. Fertilization also contributes to variation. Variation is important to the process of evolution; it promotes the possibility of adaptation to a changing environment.

Reviewing the Chapter

1. Why did early investigators predict that there must be a reduction division in the sexual reproduction process? 160
2. Compare haplontic, alternation of generations, and diplontic life cycles by indicating when meiosis occurs in each. 160
3. Define homologous chromosomes. 160
4. What meiotic events account for the production of four haploid daughter cells with a different genetic makeup than the parent cell? 162
5. Draw and explain a series of diagrams that illustrate synapsis and crossing-over. Indicate in your drawings what holds the homologues together. 162–63
6. Draw and explain a series of diagrams that illustrate the events of meiosis I. 164
7. Draw and explain a series of diagrams that illustrate the events of meiosis II. 165

8. Construct a chart to describe the many differences between meiosis and mitosis. 166
9. What accounts for (1) the genetic similarity between daughter cells and the parent cell following mitosis and (2) the genetic dissimilarity between daughter cells and the parent cell following meiosis? 166
10. Draw a diagram to illustrate the life cycle of humans. Compare spermatogenesis to oogenesis. 166–68
11. List the ways in which sexual reproduction contributes to variation among members of a population. What is the evolutionary significance of this variation? 169

Testing Yourself

Choose the best answer for each question.

1. A bivalent (tetrad) is
   a. a homologous chromosome.
   b. the paired homologous chromosomes.
   c. a duplicated chromosome composed of sister chromatids.
   d. the two daughter cells after meiosis I.
2. If a parent cell has twelve chromosomes, then the daughter cells following meiosis will have
   a. twelve chromosomes.
   b. twenty-four chromosomes.
   c. six chromosomes.
   d. Any one of these could be correct.
3. At the metaphase plate during metaphase I of meiosis, there are
   a. single chromosomes.
   b. unpaired duplicated chromosomes.
   c. bivalents (tetrads).
   d. always twenty-three chromosomes.
4. At the metaphase plate during metaphase II of meiosis, there are
   a. single chromosomes.
   b. unpaired duplicated chromosomes.
   c. bivalents (tetrads).
   d. always twenty-three chromosomes.
5. Gametes contain one of each kind of chromosome because
   a. the homologous chromosomes separate during meiosis.
   b. the chromatids never separate during meiosis.
   c. two replications of DNA occur during meiosis.
   d. crossing-over occurs during prophase I.
6. Crossing-over occurs between
   a. sister chromatids of the same chromosomes.
   b. two different bivalents.
   c. nonsister chromatids of a bivalent.
   d. two daughter nuclei.
7. During which phase of meiosis do homologous chromosomes separate?
   a. prophase II
   b. telophase II
   c. metaphase I
   d. anaphase I
8. Fertilization
   a. is a source of variation during sexual reproduction.
   b. is fusion of the gametes.
   c. occurs in both animal and plant life cycles.
   d. All of these are correct.
9. Which of these is not a difference between spermatogenesis and oogenesis in humans?

	Spermatogenesis	Oogenesis
a.	occurs in males	occurs in females
b.	produces four sperm per meiosis	produces one egg per meiosis
c.	produces haploid cells	produces diploid cells
d.	always goes to completion	does not always go to completion

10. Which of these drawings represents metaphase I? How do you know?

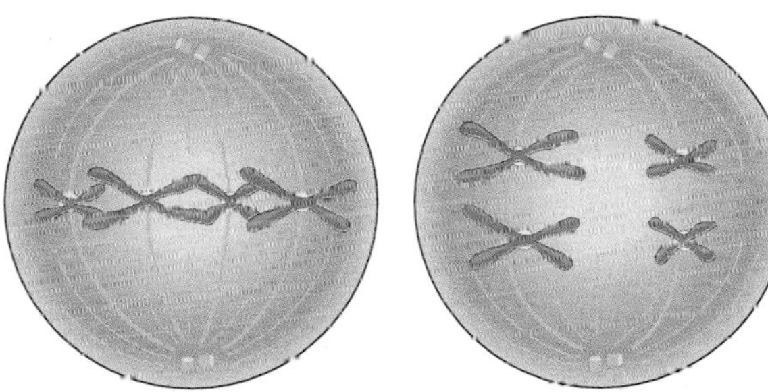

Applying the Concepts

1. *Methods of reproduction support the cell theory (i.e., all organisms are composed of cells).*
   In what way do both asexual and sexual reproduction support the cell theory?
2. *Variation among offspring is greater with sexual reproduction than with asexual reproduction.*
   How does variation occur among organisms that reproduce asexually? How does sexual reproduction introduce variation?
3. *Meiosis is an essential part of the life cycle of eukaryotes.*
   If the zygote undergoes meiosis, the adult is haploid. Haploidy has what advantage and disadvantage?

Using Technology

Your study of meiosis is supported by these available technologies:

Exploring the Internet

The Mader Home Page provides resources for and help with studying this chapter.

http://www.mhhe.com/sciencemath/biology/mader/
(Click on Biology.)

Explorations in Cell Biology & Genetics CD-ROM

Exploring Meiosis (#10)

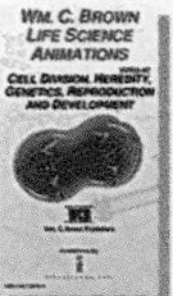

Life Science Animations Video

Video #2: Cell Division/Heredity/Genetics/Reproduction and Development
Meiosis (#13)
Crossing Over (#14)
Spermatogenesis (#19)
Oogenesis (#20)

Understanding the Terms

bivalent 162
crossing-over 163
diploid (2n) number 160
gamete 160
haploid (n) number 160
homologous chromosome 160
homologue 160
meiosis 160
oogenesis 166
polar body 168
secondary oocyte 168
sexual reproduction 160
spermatogenesis 166
synapsis 162

Match the terms to these definitions:

a. __________ Production of sperm in males by the process of meiosis and maturation.
b. __________ Homologous chromosomes at the metaphase plate during meiosis I.
c. __________ In oogenesis, a nonfunctional product; two to three meiotic products are of this type.
d. __________ In oogenesis, the functional product of meiosis I; becomes the egg.
e. __________ Member of a pair of chromosomes that carry genes for the same traits.
f. __________ Pairing of homologous chromosomes during meiosis I.
g. __________ Production of eggs in females by the process of meiosis and maturation.
h. __________ Haploid sex cell.
i. __________ Reproduction involving meiosis, gamete formation, and fertilization; produces offspring with chromosomes inherited from each parent.

Mendelian Patterns of Inheritance

CHAPTER 11

Chapter Concepts

Like begets like

The science of genetics has transformed a general awareness of heredity to precise knowledge of how it is that offspring have a combination of their parents' characteristics. A knowledge of inheritance has been acquired from the study of a varied collection of organisms—peas, fruit flies, bread molds, and bacteria.

Gregor Mendel, the father of genetics, after choosing peas as his experimental material, proposed that each parent donates particulate hereditary factors to the offspring. Mendel worked in the eighteenth century, and his work was largely ignored until the twentieth century, when genetics took a modern turn. It wasn't until then that the term "gene" was coined and it was reasoned that the genes are on the chromosomes.

But what do genes do? Beadle and Tatum X-rayed red bread mold and from their experiments determined that genes in some way control the synthesis of enzymes. It's hard to believe that it wasn't until 1944 that scientists knew that genes are composed of DNA, and 1953 before Watson and Crick deduced the structure of DNA. From then on, work with the bacterium *Escherichia coli* has brought us into the era of modern genetics and the ability to transform the genes of all organisms, including human beings. How is it that genetic manipulations with any type of organism can be applied to ourselves? As you know, all living organisms are related through the process of evolution.

11.1 Introducing Gregor Mendel

Zebras always produce zebras, never bluebirds, and poppies always produce seeds for poppies, never dandelions. Almost everyone who observes such phenomena reasons that parents must pass hereditary information to their offspring. Many also observe, however, that offspring rarely resemble either parent exactly. After all, black-coated mice occasionally produce white-coated mice. The laws of heredity must explain not only the stability but also the variation that is observed between generations of organisms.

Gregor Mendel was an Austrian monk who formulated two fundamental laws of heredity in the early 1860s (Fig. 11.1). Previously, he had studied science and mathematics at the University of Vienna, and at the time of his genetic research he was a substitute natural science teacher at a local technical high school. Various hypotheses about heredity had been proposed before Mendel began his experiments. In particular, investigators were trying to support a blending concept of inheritance at this time.

Blending Concept of Inheritance

When Mendel began his work, most plant and animal breeders acknowledged that both sexes contribute equally to a new individual. They felt that parents of contrasting appearance always produce offspring of intermediate appearance. Therefore, according to this concept, a cross between plants with red flowers and plants with white flowers would yield only plants with pink flowers. When red and white flowers reappeared in future generations, the breeders mistakenly attributed this to an instability in the genetic material.

A blending concept of inheritance offered little help to Charles Darwin, the father of evolution who wanted to give his ideas a genetic basis. If populations contained only intermediate individuals and normally lacked variations, how could diverse forms evolve? Only a particulate theory of inheritance, as proposed by Mendel, can account for the presence of discrete variations (differences) among the members of a population generation after generation. Although Darwin was a contemporary of Mendel, Darwin never learned of Mendel's work because it went unrecognized until 1900. Therefore, Darwin was never able to make use of the particulate theory of inheritance to support his theory of evolution.

At the time Mendel began his study of heredity, the blending concept of inheritance was popular.

Mendel Breaks with the Past

Most likely his background in mathematics prompted Mendel to add a statistical basis to his breeding experiments. He prepared for his experiments carefully and conducted preliminary studies with various animals and plants. He then chose to work with the garden pea, *Pisum sativum* (Fig. 11.2*a*).

Figure 11.1 Mendel working in his garden. Mendel grew and tended the pea plants he used for his experiments. For each experiment, he observed as many offspring as possible. For a cross that required him to count the number of round seeds to wrinkled seeds, he observed and counted a total of 7,324 peas!

The garden pea was a good choice. The plants were easy to cultivate and had a short generation time. And although peas normally self-pollinate (pollen only goes to the same flower), they could be cross-pollinated for the sake of the experiment. Many varieties of peas were available, and Mendel chose twenty-two for his experiments. When these varieties self-pollinated, they were *true-breeding*—the offspring were like the parent plants and like each other. In contrast to his predecessors, Mendel studied the inheritance of relatively simple and distinguishable traits—seed shape, seed color, and flower color (Fig. 11.2*b*).

As Mendel followed the inheritance of individual traits, he kept careful records of the numbers of offspring that expressed each characteristic. And he used his understanding of the mathematical laws of probability to interpret the results. In other words, Mendel simply wanted to observe facts objectively, and if he did have personal beliefs, he set them aside for the sake of the experiment. This is one of the qualities that make his experiments as applicable today as they were in 1860.

Mendel carefully designed his experiments and gathered mathematical data.

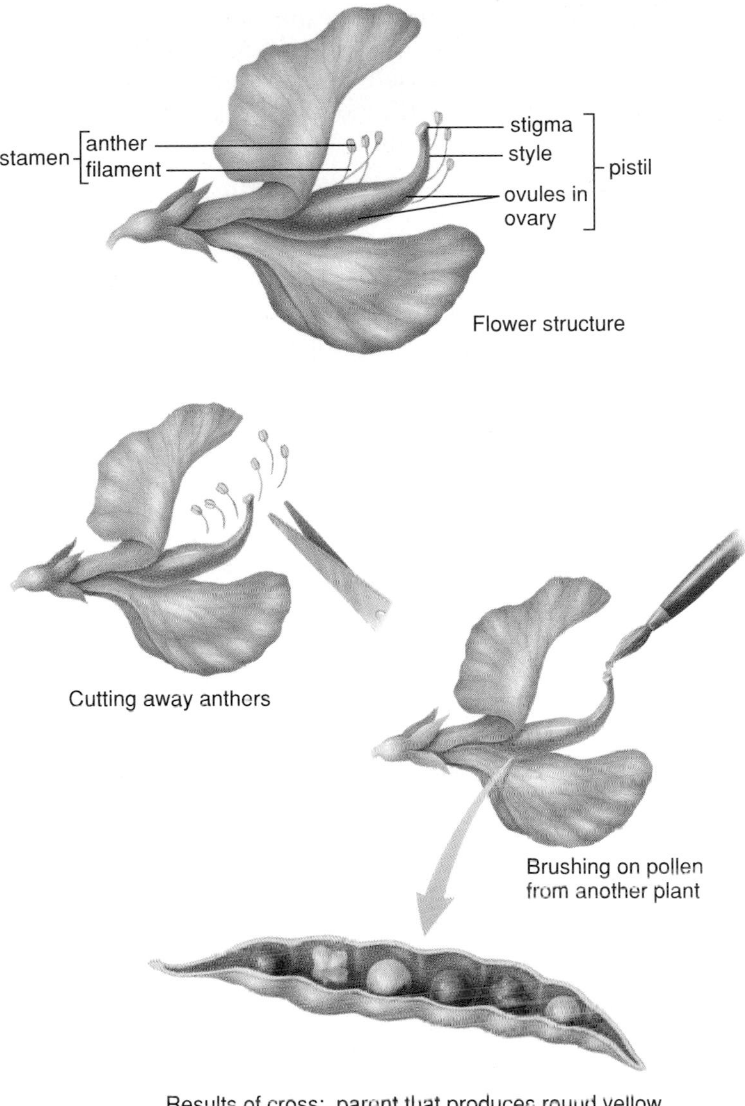

Trait	Characteristics		F_2 Results*	
			Dominant	Recessive
Stem length	Tall	Short	787	277
Pod shape	Inflated	Constricted	882	299
Seed shape	Round	Wrinkled	5,474	1,850
Seed color	Yellow	Green	7,022	2,001
Flower position	Axial	Terminal	651	207
Flower color	Purple	White	705	224
Pod color	Green	Yellow	428	152

*All of these produce approximately a 3:1 ratio. For example,

$$\frac{787}{277} \cong \frac{3}{1}$$

b.

Figure 11.2 Garden pea anatomy and traits.
a. In the garden pea, *Pisum sativum,* each flower produces both male and female gametes. The pollen grains, which upon maturity give rise to sperm, are produced within the anther of the stamen. The ovule, within the ovary of the pistil, eventually contains an egg. Following pollination—when pollen is deposited on the stigma (the top portion of the pistil)—the pollen grain develops a tube through which a sperm reaches the egg. Self-pollination is the rule in the garden pea because the sexual structures of the plant are entirely enclosed by petals. Sometimes Mendel allowed the plant to self-pollinate so that the male and female gametes of the same flower produced offspring. At other times, he performed cross-pollination, so that male and female gametes from different flowers produced offspring. He did this by removing the pollen-producing anthers of one plant and brushing the pollen on the stigmas of another plant. Once the ovules had developed into seeds (peas), they could be observed or planted if necessary to observe the results of a cross. The open pod shows the results of a cross between plants with round, yellow seeds and plants with wrinkled, green seeds. **b.** Other crosses done by Mendel.

11.2 Mendel Did a Monohybrid Cross

After ensuring that his pea plants were true-breeding, Mendel was then ready to perform a cross-pollination experiment between two strains. The initial experiments were monohybrid crosses, so called because the offspring are hybrid—they are the product of two different strains that differ in regard to only one trait. If the blending theory of inheritance were correct, then the cross should yield offspring with an intermediate appearance compared to the parents. For example, the offspring of a cross between a tall plant and short plant should be intermediate in height.

Mendel called the original parents the *P generation* and the first-generation offspring the F_1 (for filial) *generation* (Fig. 11.3). He performed *reciprocal crosses:* first he dusted the pollen of tall plants on the stigmas of short plants, and then he dusted the pollen of short plants on the stigmas of tall plants. In both cases, all F_1 offspring resembled the tall parent.

Certainly, these results were contrary to those predicted by the blending theory of inheritance. Rather than being intermediate, the offspring were tall and resembled only one parent. Did these results mean that the other characteristic (i.e., shortness) had disappeared permanently? Apparently not, because when Mendel allowed the F_1 plants to self-pollinate, $\frac{3}{4}$ of the F_2 *generation* were tall and $\frac{1}{4}$ were short, a 3:1 ratio (Fig. 11.3).

Mendel counted many plants. For this particular cross, he counted a total of 1,064 plants, of which 787 were tall and 277 were short. In all crosses that he performed, he found a 3:1 ratio in the F_2 generation. The characteristic that had disappeared in the F_1 generation reappeared in $\frac{1}{4}$ of the F_2 offspring.

His mathematical approach led Mendel to interpret these results differently from previous breeders. He knew that the same ratio was obtained among the F_2 generation time and time again for the same type cross despite the particular trait, and he sought an explanation. A 3:1 ratio among the F_2 offspring was possible if:

1. the F_1 parents contained two separate copies of each hereditary factor, one of these being dominant and one being recessive;
2. the factors separated when the gametes were formed, and each gamete carried only one copy of each factor; and
3. random fusion of all possible gametes occurred upon fertilization.

In this way, Mendel arrived at the first of his laws of inheritance:

Mendel's law of segregation:
Each organism contains two factors for each trait, and the factors segregate during the formation of gametes so that each gamete contains only one factor for each trait.

Mendel's law of segregation is in keeping with a particulate theory of inheritance because many individual factors are passed on from generation to generation. It is the reshuffling of these factors that explains how variations come about and why offspring differ from their parents.

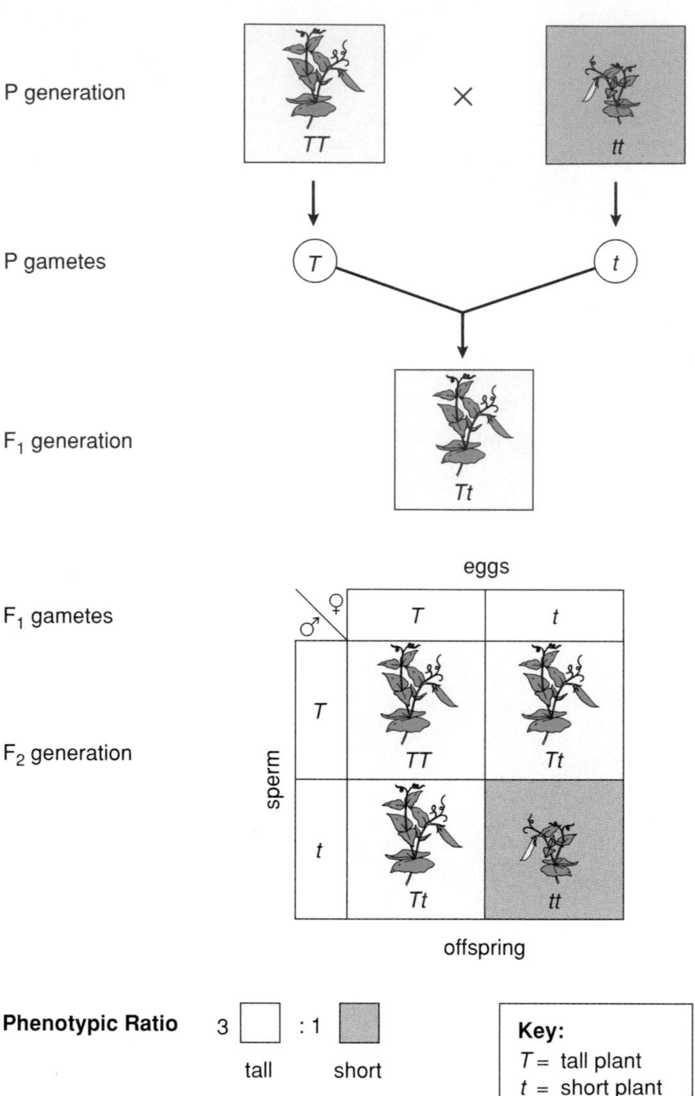

Figure 11.3 Monohybrid cross done by Mendel.
The P generation plants differ in one regard—length of the stem. The F_1 generation are all tall, but the factor for short has not disappeared because $\frac{1}{4}$ of the F_2 generation are short. The 3:1 ratio allowed Mendel to deduce that individuals have two discrete and separate genetic factors for each trait. (♂ = male; ♀ = female)

Modern Genetics Has an Explanation

Figure 11.3 also shows how we now interpret the results of Mendel's experiments on inheritance of stem length in peas. Each trait in a pea plant is controlled by two **alleles** [Gk. *allelon*, reciprocal, parallel], alternate forms of a gene that in this case control the length of the stem. In genetic notation, the alleles are identified by letters, the **dominant allele** (so named because of its ability to mask the expression of the other allele) with an uppercase (capital) letter and the **recessive allele** with the same but lowercase (small) letter. With reference to the cross being discussed, there is an allele for tallness *(T)* and an allele for shortness *(t)*. Alleles occur on a homologous pair of chromosomes at a particular location that is called the **gene locus** (Fig. 11.4).

During meiosis, the type of cell division that reduces the chromosome number, the homologous chromosomes of a bivalent separate during meiosis I. Therefore, the process of meiosis gives an explanation for Mendel's law of segregation and for why there is only one allele for each trait in the gametes.

In Mendel's cross, the original parents (P generation) were true-breeding; therefore, the tall plants had two alleles for tallness *(TT)* and the short plants had two alleles for shortness *(tt)*. When an organism has two identical alleles, as these had, we say it is **homozygous** [Gk. *homo*, same, and *zygos*, balance, yoke]. Because the first parents were homozygous, all gametes produced by the tall plant contained the allele for tallness *(T)*, and all gametes produced by the short plant contained an allele for shortness *(t)*.

After cross-pollination, all the individuals of the resulting F_1 generation had one allele for tallness and one for shortness *(Tt)*. When an organism has two different alleles at a gene locus, we say that it is **heterozygous** [Gk. *hetero*, different, and *zygos*, balance, yoke]. Although the plants of the F_1 generation had one of each type of allele, they were all tall. The allele that is expressed in a heterozygous individual is the dominant allele. The allele that is not expressed in a heterozygote is a recessive allele.

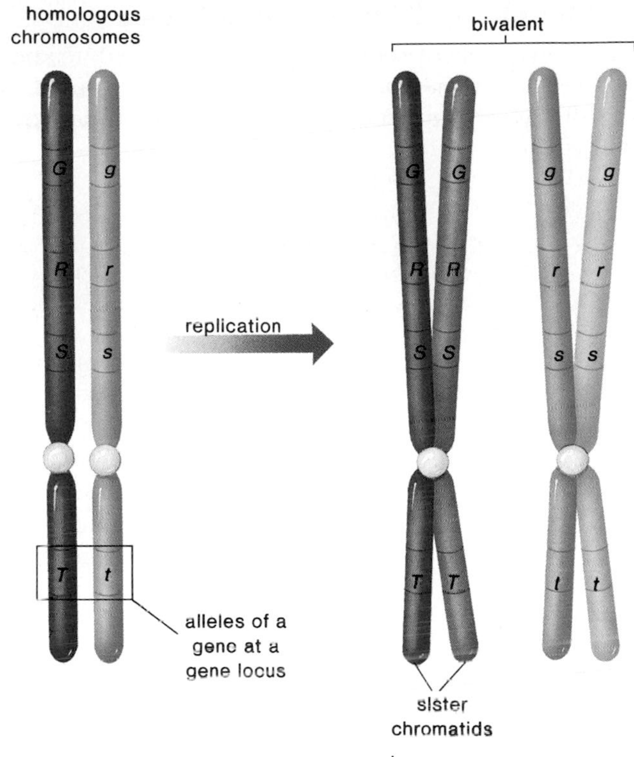

Figure 11.4 Homologous chromosomes.
a. The letters represent alleles; that is, alternate forms of a gene. Each allelic pair, such as *Gg* or *Tt*, is located on homologous chromosomes at a particular gene locus. **b.** Following DNA replication, each sister chromatid carries the same alleles in the same order.

Table 11.1
Genotype versus Phenotype

Genotype	Genotype	Phenotype
TT	Homozygous dominant	Tall plant
Tt	Heterozygous	Tall plant
tt	Homozygous recessive	Short plant

Genotype Versus Phenotype

It is obvious from our discussion that two organisms with different allelic combinations for a trait can have the same outward appearance (*TT* and *Tt* pea plants are both tall). For this reason, it is necessary to distinguish between the alleles present in an organism and the appearance of that organism.

The word **genotype** [Gk. *genos*, birth, origin, race, and *typos*, image, shape] refers to the alleles an individual receives at fertilization. Genotype may be indicated by letters or by short, descriptive phrases. Genotype *TT* is called homozygous dominant, and genotype *tt* is called homozygous recessive. Genotype *Tt* is called heterozygous.

The word **phenotype** [Gk. *phaino*, appear, and *typos*, image, shape] refers to the physical appearance of the individual. The homozygous dominant individual and the heterozygous individual both show the dominant phenotype and are tall, while the homozygous recessive individual shows the recessive phenotype and is short (Table 11.1).

Doing Monohybrid Genetics Problems

When solving genetics problems, it is first necessary to know which characteristic is dominant. For example, the following key indicates that unattached earlobes are dominant over attached earlobes:

Key: *E* = unattached earlobes
e = attached earlobes

If a man, homozygous for unattached earlobes, reproduces with a woman who has attached earlobes, what type of earlobe will the child have? In row 1, P represents the parental generation, and the letters in this row are the genotypes of the parents. Row 2 shows that the gametes of each parent have only one type of allele for earlobes; therefore, the child (F generation) will have a heterozygous genotype and unattached earlobes:

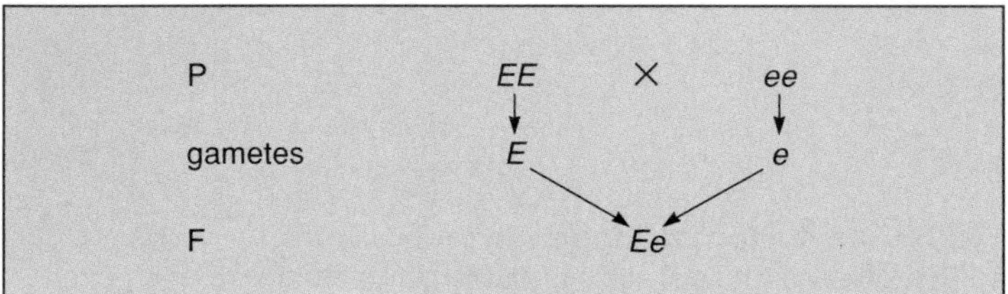

If two heterozygotes reproduce with one another, will the child have unattached or attached earlobes? In the P generation there was only one possible type of gamete for each parent because they were homozygous. Heterozygotes, however, can produce two types of gametes; 1/2 of the gametes will contain an *E,* and 1/2 will contain an *e.*

P	*Ee*	×	*Ee*
gametes	1/2*E*, 1/2*e*		1/2*E*, 1/2*e*

When determining the gametes, it is necessary to keep in mind that although an individual is diploid—that is, has two alleles for each trait—*each gamete is haploid—that is, has only one allele for each trait.* This is true of single-trait crosses as well as multiple-trait crosses.

When doing genetics problems, first decide on the appropriate key, and then determine the genotype of both parents and various types of gametes for both parents.

Practice Problems 1 will help you learn to designate the gametes.

Practice Problems 1*

1. For each of the following genotypes, give all genetically different gametes, noting the proportion of each for the individual.
   a. *WW*
   b. *Ww*
   c. *Tt*
   d. *TT*
2. For each of the following, state whether a genotype (genetic makeup of an organism) or a type of gamete is represented.
   a. *D*
   b. *GG*
   c. *P*

**Answers to Practice Problems appear in Appendix A.*

Figuring the Probable Results

When we calculate the expected results of the cross under consideration, or any genetic cross, we utilize the laws of probability. Imagine flipping a coin; *each time* you flip the coin there is a 50% chance of heads and a 50% chance of tails. In like manner, if the parent has the genotype *Ee*, what is the chance of any child inheriting either an *E* or *e* from that parent?

The chance of $E = 1/2$
The chance of $e = 1/2$

But in the cross $Ee \times Ee$, a child will inherit an allele from both parents. The multiplicative law of probability states that *the chance, or probability, of two or more independent events occurring together is the product (multiplication) of their chance of occurring separately.* Therefore, the probability of receiving these genotypes is as follows:

1. The chance of $EE = 1/2 \times 1/2 = 1/4$
2. The chance of $Ee = 1/2 \times 1/2 = 1/4$
3. The chance of $eE = 1/2 \times 1/2 = 1/4$
4. The chance of $ee = 1/2 \times 1/2 = 1/4$

Now we have to consider the additive law of probability: *the chance of an event that can occur in two or more independent ways is the sum (addition) of the individual chances.* Therefore,

The chance of a child with unattached earlobes (*EE, Ee,* or *eE*) is 3/4 (add **1, 2,** and **3**), or 75%.

The chance of a child with attached earlobes *(ee)* is 1/4 (only **4**), or 25%.

Punnett Square Figures for You

The **Punnett square** was introduced by a prominent poultry geneticist, R. C. Punnett, in the early 1900s as a simple method to figure the probable results of a genetic cross. Figure 11.5 shows how the results of the cross under consideration *(Ee × Ee)* can be determined using a Punnett square. In a Punnett square all possible types of sperm are lined up vertically and all possible types of eggs are lined up horizontally (or vice versa), and every possible combination of alleles is placed within the squares. In our cross each parent has two possible types of gametes (*E* or *e*), so these two types of gametes are lined up vertically and horizontally.

The results of the Punnett square calculations show that the expected genotypes of offspring are 1/4 *EE*, 1/2 *Ee*, and 1/4 *ee*, giving a 1:2:1 genotypic ratio. Since 3/4 have unattached earlobes and 1/4 have attached earlobes, this is a 3:1 phenotypic ratio.

The gametes combine at random, and in practice it is usually necessary to observe a very large number of offspring before a 3:1 ratio can be verified. Only if many offspring are counted can it be assured that all genetically different sperm have had a chance to fertilize all genetically different eggs. If a number of heterozygotes produced 200 offspring, approximately 150 of these would have unattached earlobes and approximately 50 would have attached earlobes (a 3:1 ratio). In terms of genotypes, approximately 50 would be *EE*, about 100 would be *Ee*, and the remaining 50 would be *ee*.

We cannot arrange crosses between humans in order to count a large number of offspring. Therefore, in humans the phenotypic ratio is used to estimate the chances any child has for a particular characteristic. In the cross under consideration, each child has a 3/4 (or 75%) chance of having unattached earlobes and 1/4 (or 25%) chance of having attached earlobes. And we must remember that *chance has no memory:* if two heterozygous parents already have a child with attached earlobes, the next child still has a 25% chance of having attached earlobes.

You will note that the Punnett square makes use of the laws of probability we mentioned in the previous section. First, we recognize there is a 1/2 chance of a child getting the *E* allele or the *e* allele from a parent. In this way we are able to designate the possible gametes. What operation is in keeping with the multiplicative law of probability? The operation of combining the alleles and placing them in the squares. What operation is in keeping with the additive law of probability? The operation of adding the results, and arriving at the phenotypic ratio.

Mendel was quite familiar with the laws of probability and was able to make use of them to see that inheritance of traits depended upon the passage of discrete factors from generation to generation.

The laws of probability allow one to calculate the probable results of one-trait genetic crosses.

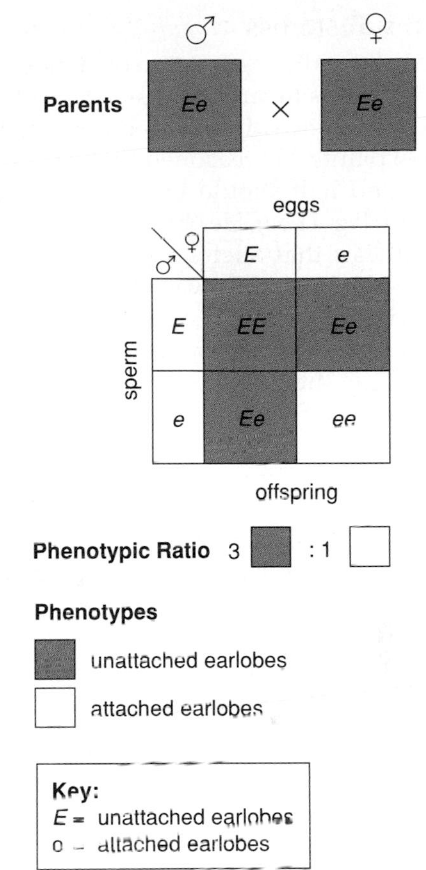

Figure 11.5 Genetic inheritance in humans.
When the parents are heterozygous, each child has a 75% chance of having the dominant phenotype and a 25% chance of having the recessive phenotype. (♂ = male; ♀ = female)

Practice Problems 2*

1. In rabbits, if *B* = dominant black allele and *b* = recessive white allele, which of these genotypes *(Bb, BB, bb)* could a white rabbit have?
2. In pea plants, yellow seed color is dominant over green seed color. When two heterozygote plants are crossed, what percentage of plants would have yellow seeds? green seeds?
3. In humans, freckles is dominant over no freckles. A man with freckles reproduces with a woman having freckles, but the children have no freckles. What chance did each child have for freckles?
4. In horses, trotter *(T)* is dominant over pacer *(t)*. A trotter is mated to a pacer, and the offspring is a pacer. Give the genotype of all horses.

**Answers to Practice Problems appear in Appendix A.*

Mendel Did a Testcross

To test his idea about the segregation of alleles—in modern terms, that the F_1 was heterozygous—Mendel crossed his F_1 generation tall plants with true-breeding, short (homozygous recessive) plants. He reasoned that half the offspring should be tall and half should be short, producing a 1:1 phenotypic ratio (Fig. 11.6*a*). He obtained these results; therefore, his hypothesis that alleles segregate when gametes are formed was supported. It was Mendel's experimental use of simple dominant and recessive traits that allowed him to formulate and to test the law of segregation.

In Figure 11.6, the homozygous recessive parent can produce only one type of gamete—*t*—and so the same results would be obtained if the Punnett square had only one column. This is logical because the *t* means all the gametes that carry a *t*.

Today, a monohybrid **testcross** is used to determine if an individual with the dominant phenotype is homozygous dominant or heterozygous for a particular trait. Since both of these genotypes produce the dominant phenotype, it is not possible to determine the genotype by inspection. Figure 11.6*b* shows that if the F_1 in Mendel's cross had been homozygous dominant, then all the offspring would have been tall.

The results of a testcross indicate whether an individual with the dominant phenotype is heterozygous or homozygous dominant.

Practice Problems 3*

1. In horses, *B* = black coat and *b* = brown coat. What type of cross should be done to best determine whether a black-coated horse is homozygous dominant or heterozygous?
2. In fruit flies, *L* = long wings and *l* = short wings. The offspring exhibit a 1:1 ratio when a long-winged fly is crossed with a short-winged fly. What is the genotype of all flies involved?
3. In the garden pea, round seeds are dominant over wrinkled seeds. An investigator crosses a plant having round seeds with a plant having wrinkled seeds. He counts 400 offspring. How many of the offspring have wrinkled seeds if the plant having round seeds is a heterozygote?

**Answers to Practice Problems appear in Appendix A.*

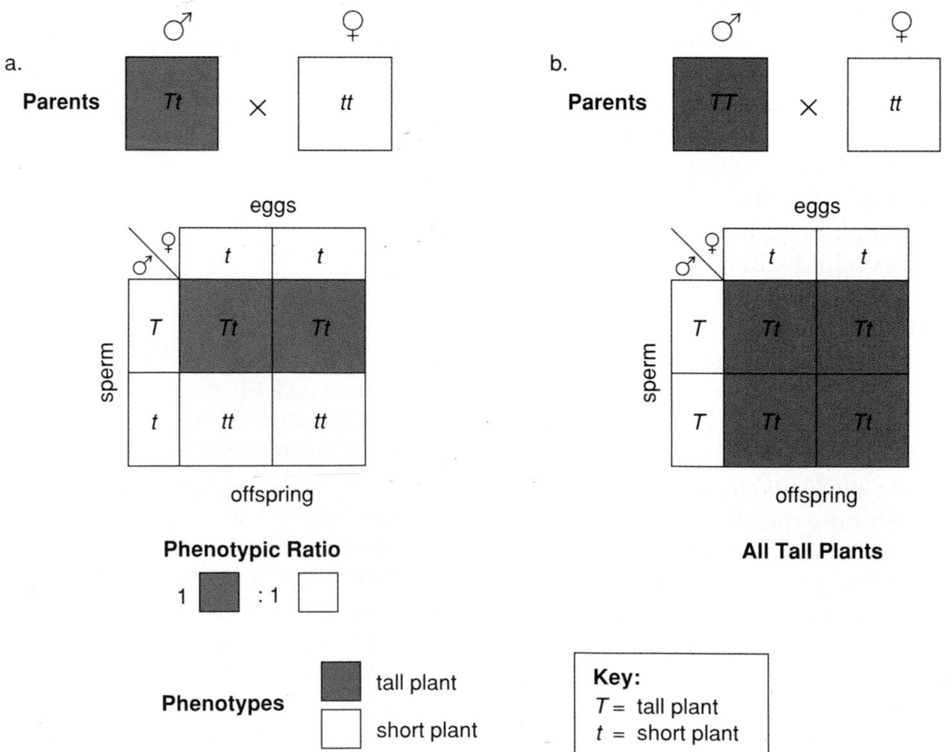

Figure 11.6 Testcross.
Crossing an individual with the dominant phenotype with a recessive individual indicates the genotype. **a.** If a parent with the dominant phenotype is heterozygous, the phenotypic ratio among the offspring is 1:1. **b.** If a parent with the dominant phenotype is homozygous, all offspring have the dominant phenotype. (♂ = male; ♀ = female)

11.3 Mendel Did a Dihybrid Cross

Mendel performed a second series of crosses that are called dihybrid crosses because the offspring are dihybrid—they are the product of two different strains that differ in regard to two traits. For example, he crossed tall plants having green pods with short plants having yellow pods (Fig. 11.7). The F_1 plants showed both dominant characteristics. As before, Mendel then allowed the F_1 plants to self-pollinate. Two possible results could occur in the F_2 generation:

1. If the dominant factors *(TG)* always segregate into the F_1 gametes together, and the recessive factors *(tg)* always stay together, then there would be two phenotypes among the F_2 plants—tall plants with green pods and short plants with yellow pods.
2. If the four factors segregate into the F_1 gametes independently, then there would be four phenotypes among the F_2 plants—tall plants with green pods, tall plants with yellow pods, short plants with green pods, and short plants with yellow pods.

Figure 11.7 shows that Mendel observed four phenotypes among the F_2 plants, supporting the second hypothesis. Therefore, Mendel formulated his second law of heredity:

Mendel's law of independent assortment: Members of one pair of factors separate (assort) independently of members of another pair of factors. Therefore, all possible combinations of factors can occur in the gametes.

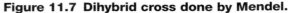

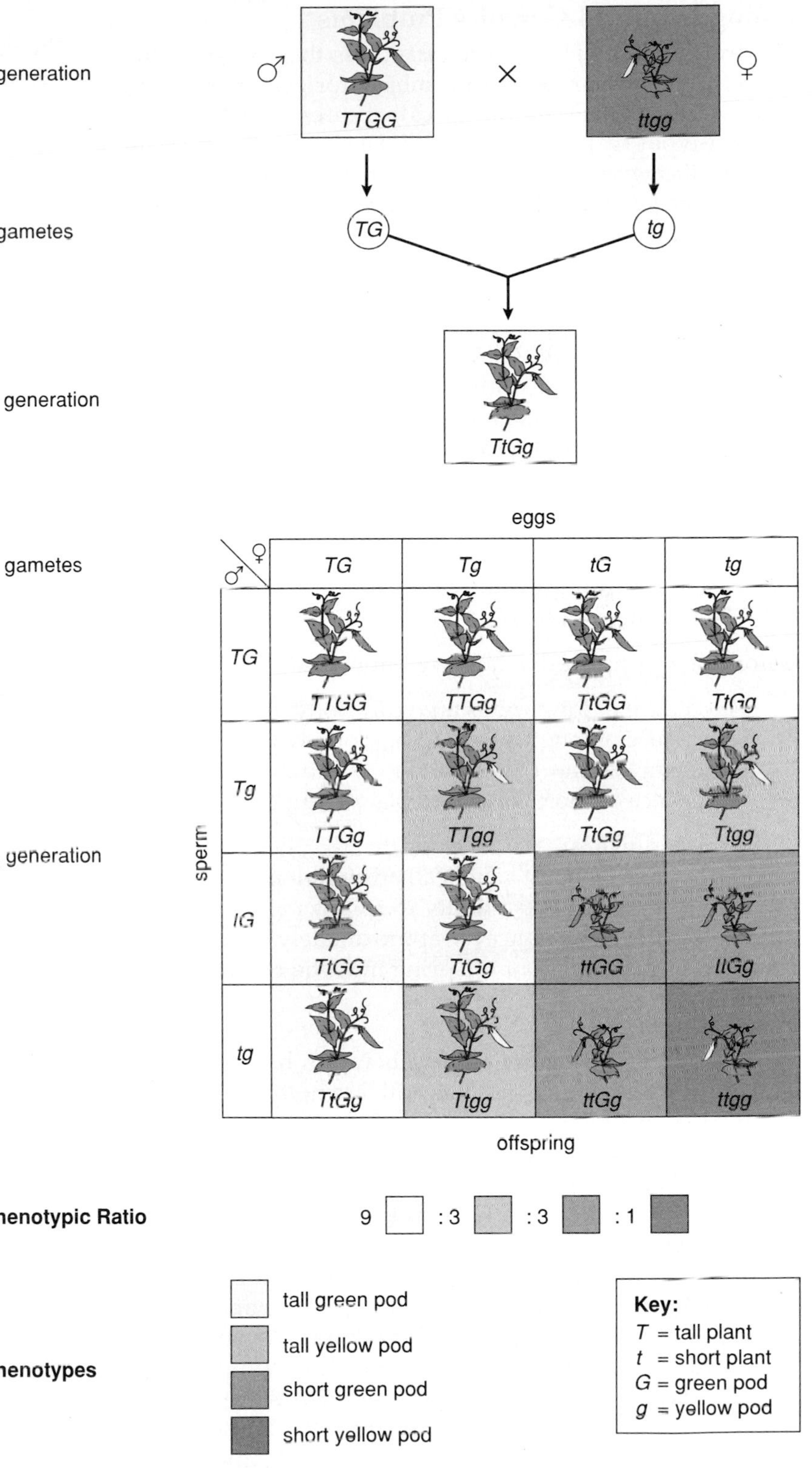

Figure 11.7 Dihybrid cross done by Mendel.
P generation plants differ in two regards—length of the stem and color of the pod. The F_1 generation shows only the dominant traits, but all possible phenotypes appear among the F_2 generation. The 9:3:3:1 ratio allowed Mendel to deduce that factors segregate into gametes independently of other factors.
(♂ = male; ♀ = female)

Practice Problems 4*

1. For each of the following genotypes, give all possible gametes, noting the proportion of each gamete for the individual.
   a. *TtGG*
   b. *TtGg*
   c. *TTGg*
2. For each of the following, state whether a genotype (genetic makeup of an organism) or a type of gamete is represented.
   a. *Tg*
   b. *WwCC*
   c. *TW*

**Answers to Practice Problems appear in Appendix A.*

Doing Dihybrid Genetics Problems

The fruit fly, *Drosophila melanogaster*, less than one-fifth the size of a housefly, is a favorite subject for genetic research because it has mutant characteristics that are easily determined. A "wild-type" fly has long wings and a gray body. There are mutant flies with short (vestigial) wings and black (ebony) bodies. The key for a cross involving these traits is L = long wing, l = short wing, G = gray body, and g = black body.

Figuring Probable Results Again

If two flies heterozygous for both traits are crossed, what are the probable results? Since each characteristic is inherited separately from any other, it is possible to apply again the laws of probability mentioned on page 178. For example, we know the F_2 results for two separate monohybrid crosses are as listed here:

1. The chance of long wings $= 3/4$
   The chance of short wings $= 1/4$
2. The chance of gray body $= 3/4$
   The chance of black body $= 1/4$

Using the multiplicative law, we know that:

The chance of long wings and gray body $= 3/4 \times 3/4 = 9/16$
The chance of long wings and black body $= 3/4 \times 1/4 = 3/16$
The chance of short wings and gray body $= 1/4 \times 3/4 = 3/16$
The chance of short wings and black body $= 1/4 \times 1/4 = 1/16$

Using the additive law, we conclude that the phenotypic ratio is 9:3:3:1. Again, since all genetically different male gametes must have an equal opportunity to fertilize all genetically different female gametes to even approximately achieve these results, a large number of offspring must be counted.

Punnett Square

In Figure 11.8, the flies of the P generation have only one possible type of gamete because both are homozygous. All the F_1 flies are heterozygous (*LlGg*) and have the same phenotype (long wings, gray body).

The Punnett square in Figure 11.8 shows the expected results when the F_1 flies are crossed, assuming that all genetically different sperm have an equal opportunity to fertilize all genetically different eggs. Notice that $9/16$ of the offspring have long wings and a gray body, $9/16$ have long wings and a black body, $3/16$ have short wings and a gray body, and $1/16$ have short wings and a black body. This phenotypic ratio of 9:3:3:1 is expected whenever a heterozygote for two traits is crossed with another heterozygote for two traits and simple dominance is present in both genes.

A Punnett square can also be used to predict the chances of an offspring having a particular phenotype. What are the chances of an offspring with long wings and gray body? The chances are $9/16$. What are the chances of an offspring with short wings and gray body? The chances are $3/16$, and so forth.

Today, we know that these results are obtained and the law of independent assortment holds because of the events of meiosis. The gametes contain one allele for each trait and in all possible combinations because homologues separate independently during meiosis I.

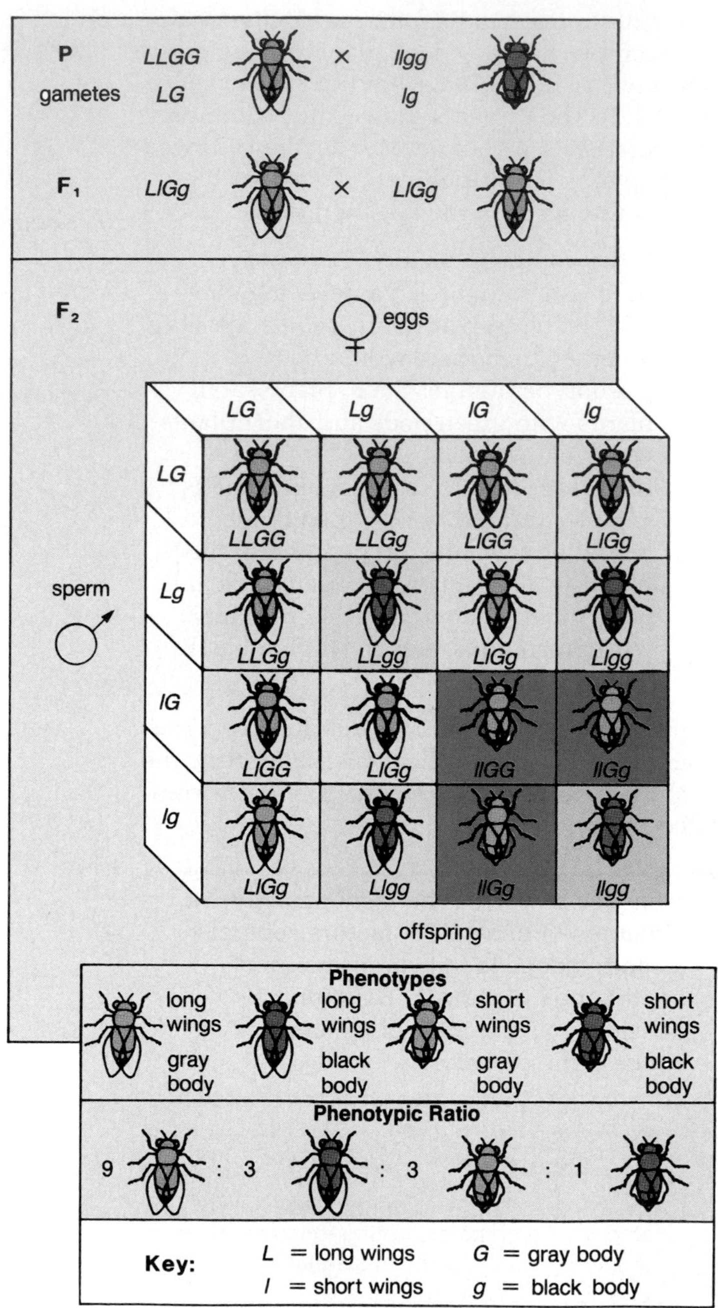

Figure 11.8 Inheritance in fruit flies.
Each F_1 fly (*LlGg*) produces four types of gametes because all possible combinations of alleles can occur in the gametes. Therefore, all possible phenotypes appear among the F_2 offspring. (♂ = male; ♀ = female)

Dihybrids Can Be Tested Also

A dihybrid testcross is used to determine if an individual is homozygous dominant or heterozygous for either of the two traits. Since it is not possible to determine the genotype of a long-winged, gray-bodied fly by inspection, the genotype may be represented as *L*__[1] *G*__.

1 The blank means that a dominant or recessive allele can be present.

When doing a dihybrid testcross, an individual with the dominant phenotype is crossed with an individual with the recessive phenotype. For example, a long-winged, gray-bodied fly is crossed with a short-winged, black-bodied fly. A long-winged, gray-bodied fly heterozygous for both traits will form four different types of gametes. The homozygous fly with short wings and a black body can form only one kind of gamete:

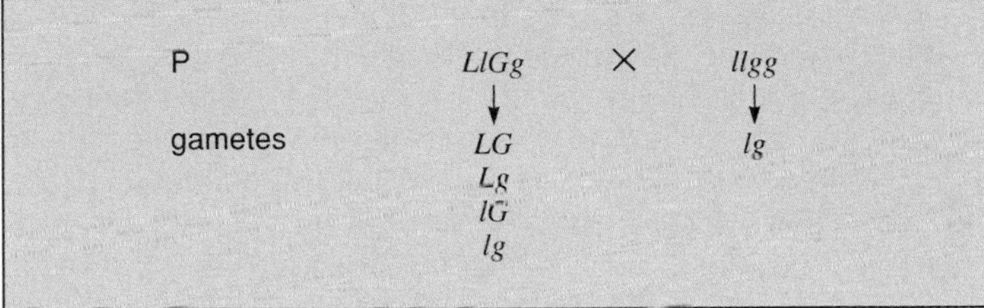

Figure 11.9 shows that ¼ of the expected offspring have long wings and a gray body; ¼ have long wings and a black body; ¼ have short wings and a gray body; and ¼ have short wings and a black body. This is a 1:1:1:1 phenotypic ratio. The presence of offspring with short wings and a black body shows that the *L*__*G*__ fly is heterozygous for both traits and has the genotype *LlGg*.

If the *L*__*G*__ fly is homozygous for both traits, then no offspring will have short wings or a black body. If the *L*__*G*__ fly is heterozygous for one trait but not the other, what is the expected phenotypic ratio among the offspring?

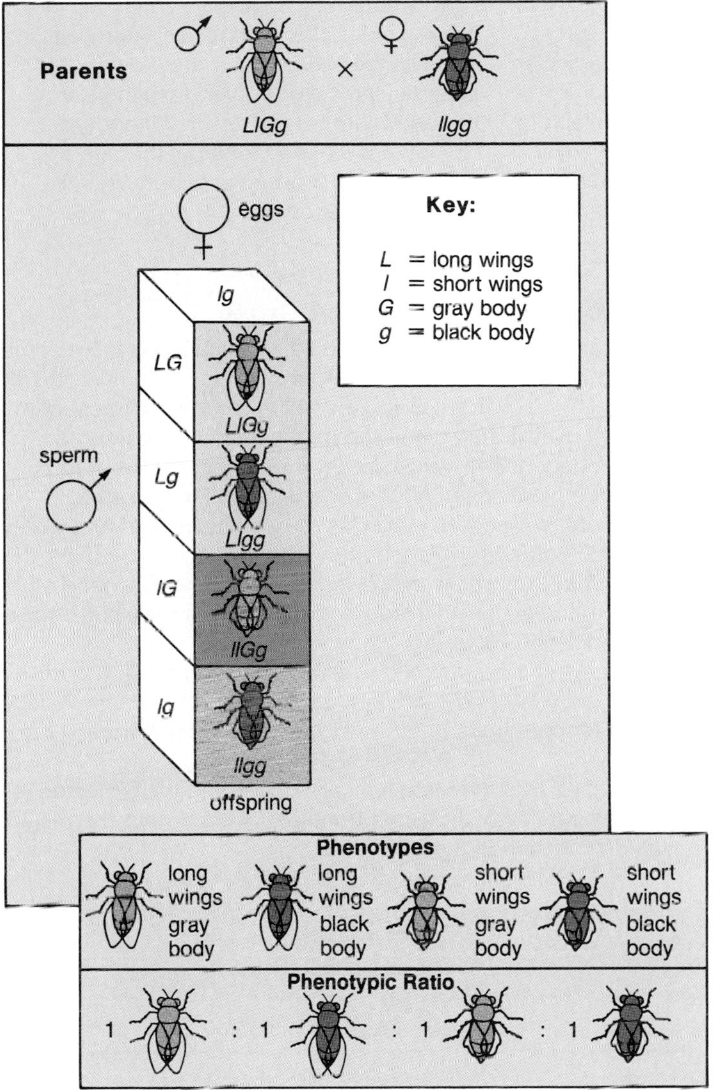

Figure 11.9 Testcross.
Testcross to determine if a fly is heterozygous for both traits. If a fly heterozygous for both traits is crossed with a fly that is recessive for both traits, the expected ratio of phenotypes is 1:1:1:1. What would the result be if the test individual were homozygous dominant for both traits? homozygous dominant for one trait but heterozygous for the other?

Practice Problems 5*

1. In horses, *B* = black coat, *b* = brown coat, *T* = trotter, and *t* = pacer. A black pacer mated to a brown trotter produces a black trotter offspring. Give all possible genotypes for this offspring.
2. In fruit flies, long wings *(L)* is dominant over short wings *(l)*, and gray body *(G)* is dominant over black body *(g)*. In each instance, what are the most likely genotypes of the previous generation if a student gets the following phenotypic results?
   a. 1:1:1:1 (all possible combinations in equal number)
   b. 9:3:3:1 (9 dominant; 3 mixed; 3 mixed; 1 recessive)
3. In humans, short fingers and widow's peak are dominant over long fingers and continuous hairline. A heterozygote in both regards reproduces with a similar heterozygote. What is the chance of any one child having the same phenotype as the parents?

Answers to Practice Problems appear in Appendix A.

connecting concepts

Long before Mendel was born, humans had been breeding plants and animals for desirable traits. Thousands of years ago, Chinese emperors bred birds of the thrush family in order to have birds with the most melodic songs. Farmers throughout history have known that the offspring of certain plants would give a higher yield than others, and selective breeding of livestock could produce sheep with thicker coats and chickens that lay larger eggs.

Mendel's work gave us an explanation for why selective breeding works. His laws of segregation and independent assortment explain why some plants and animals tend to produce more offspring with desired traits than other plants and animals. Do Mendel's laws also apply to humans? Yes, certainly they do. Cystic fibrosis is a lethal genetic disease involving problems with the functioning of mucous membranes in the lungs and pancreatic ducts. Parents who appear to be normal can still produce a child that has cystic fibrosis. Why? Because each parent can have a recessive allele for cystic fibrosis. What are the chances that any particular offspring will have cystic fibrosis if the parents are heterozygous? Each offspring then has a 25% chance of having cystic fibrosis. Is it possible that every child of this couple could have cystic fibrosis? Yes, it is possible because "chance has no memory"—that is, each child has the same chance as any other child.

In other instances, we will see that simple Mendelian inheritance does not apply. For example, how is it possible for mammals, including humans, to have a range of hair colors? As we shall see in the next chapter, some traits are controlled by a number of alleles that can interact to produce the phenotype. In such instances, in particular, it is possible to see the effect of the environment on the phenotype. Sometimes we tend to forget that the environment, as well as the genotype, influences the phenotype.

Summary

11.1 Introducing Gregor Mendel

At the time Mendel began his hybridization experiments, the blending concept of inheritance was popular. This concept stated that whenever the parents were distinctly different, the offspring are intermediate between them. In contrast to preceding plant breeders, Mendel decided to do a statistical study, and he chose to study just seven distinctive traits of the garden pea.

11.2 Mendel Did a Monohybrid Cross

When Mendel did monohybrid crosses, he found that the recessive phenotype reappeared in about $^1/_4$ of the F_2 plants. This allowed Mendel to deduce his law of segregation, which states that the individual has two factors for each trait and the factors segregate into the gametes.

The laws of probability can be used to calculate the expected phenotypic ratio of a cross. In practice, a large number of offspring must be counted in order to observe the expected results, because only in that way can it be ensured that all possible types of sperm have fertilized all possible types of eggs.

Because humans do not produce a large number of offspring, it is best to use a predicted ratio as a means of estimating the chances of an individual inheriting a particular characteristic.

Mendel also crossed his F_1 plants with homozygous recessive plants. The results indicated that the recessive factor was present in the F_1 plants (i.e., that they were heterozygous). Today, we call this a testcross, because it is used to test whether an individual showing the dominant characteristic is homozygous dominant or heterozygous.

11.3 Mendel Did a Dihybrid Cross

Mendel did dihybrid crosses, in which the F_1 individuals were dominant in both traits, but there were four phenotypes among the F_2 offspring. This allowed Mendel to deduce the law of independent assortment, which states that the members of one pair of factors separate independently of those from another pair. Therefore, all possible combinations of factors can occur in the gametes.

In regard to Mendel's dihybrid crosses, the laws of probability show that $^9/_{16}$ of the F_2 offspring have the two dominant traits, $^3/_{16}$ have one dominant trait with one recessive trait, $^3/_{16}$ have the other dominant trait with the other recessive trait, and $^1/_{16}$ have both recessive traits, for a 9:3:3:1 ratio.

The dihybrid testcross allows an investigator to test whether an individual showing two dominant characteristics is homozygous dominant for both traits or for one trait only, or is heterozygous for both traits.

Table 11.2 lists the most typical crosses and the expected results. Learning these saves the trouble of working out the results repeatedly for each of these crosses.

Table 11.2
Common Crosses Involving Simple Dominance

Examples	Phenotypic Ratios
Tt × *Tt*	3:1 (dominant to recessive)
Tt × *tt*	1:1 (dominant to recessive)
TtYy × *TtYy*	9:3:3:1 (9 dominant; 3 mixed; 3 mixed; 1 recessive)
TtYy × *ttyy*	1:1:1:1 (all possible combinations in equal number)

Reviewing the Chapter

1. How did Mendel's procedure differ from that of his predecessors? What mechanism did he use to set aside any personal beliefs he may have had? 174
2. How did the monohybrid crosses performed by Mendel refute the blending concept of inheritance? 174
3. Using Mendel's monohybrid cross as an example, trace his reasoning to arrive at the law of segregation. 176
4. Define these terms: allele, dominant allele, recessive allele, genotype, homozygous, heterozygous, and phenotype. 177
5. Use a Punnett square and the laws of probability to show the results of a cross between two individuals who are heterozygous for one trait. What are the chances of an offspring having the dominant phenotype? the recessive phenotype? 179
6. In what way does a monohybrid testcross support Mendel's law of segregation? 180
7. How is a monohybrid testcross used today? 180
8. Using Mendel's dihybrid cross as an example, trace his reasoning to arrive at the law of independent assortment. 181
9. Use a Punnett square and the laws of probability to show the results of a cross between two individuals who are heterozygous for two traits. What are the chances of an offspring having the dominant phenotype for both traits? having the recessive phenotype for both traits? 182
10. What would the results of a dihybrid testcross be if an individual was heterozygous for two traits? heterozygous for only one trait? homozygous dominant for both traits? 183

Testing Yourself

Choose the best answer for each question.
For questions 1–4, match the cross with the results in the key:

Key:

a. 3:1
b. 9:3:3:1
c. 1:1
d. 1:1:1:1

1. *TtYy* × *TtYy*
2. *Tt* × *Tt*
3. *Tt* × *tt*
4. *TtYy* × *ttyy*
5. Which of these could be a normal gamete?
   a. *GgRr*
   b. *GRr*
   c. *Gr*
   d. None of these are correct.
6. Which of these properly describes a cross between an individual who is homozygous dominant for hairline but heterozygous for finger length and an individual who is recessive for both characteristics? (*W* = widow's peak, *w* = continuous hairline, *S* = short fingers, *s* = long fingers)
   a. *WwSs* × *WwSs*
   b. *WWSs* × *wwSs*
   c. *Ws* × *ws*
   d. *WWSs* × *wwss*
7. In peas, yellow seed (*Y*) is dominant over green seed (*y*). In the F_2 generation of a monohybrid cross that begins when a dominant homozygote is crossed with a recessive homozygote, you would expect
   a. plants that produce three yellow seeds to every green seed.
   b. plants with one yellow seed for every green seed.
   c. only plants with the genotype *YY* or *yy*.
   d. Both a and c are correct.
8. In humans, pointed eyebrows (*B*) are dominant over smooth eyebrows (*b*). Mary's father has pointed eyebrows, but she and her mother have smooth. What is the genotype of the father?
   a. *BB*
   b. *Bb*
   c. *bb*
   d. Any one of these is correct.
9. In guinea pigs, smooth coat (*S*) is dominant over rough coat (*s*) and black coat (*B*) is dominant over white coat (*b*). In the cross *SsBb* × *SsBb*, how many of the offspring will have a smooth black coat on average?
   a. 9 only
   b. about 9/16
   c. 1/16
   d. 6/16
10. In horses, *B* = black coat, *b* = brown coat, *T* = trotter, and *t* = pacer. A black trotter that has a brown pacer offspring is
    a. *BT.*
    b. *BbTt.*
    c. *bbtt.*
    d. *BBtt.*
11. In tomatoes, red fruit (*R*) is dominant over yellow fruit (*r*) and tallness (*T*) is dominant over shortness (*t*). A plant that is *RrTT* is crossed with a plant that is *rrTt*. What are the chances of an offspring being heterozygous for both traits?
    a. none
    b. 1/2
    c. 1/4
    d. 9/16
12. In the cross *RrTt* × *rrtt*,
    a. all the offspring will be tall with red fruit.
    b. 75% (3/4) will be tall with red fruit.
    c. 50% (1/2) will be tall with red fruit.
    d. 25% (1/4) will be tall with red fruit.

Additional Genetics Problems*

1. If a man homozygous for widow's peak (dominant) reproduces with a woman homozygous for continuous hairline (recessive), what are the chances of their children having a widow's peak? a continuous hairline?
2. John has unattached earlobes (recessive) like his father, but his mother has attached earlobes (dominant). What is John's genotype?
3. In humans, the allele for short fingers is dominant over that for long fingers. If a person with short fingers who had one parent with long fingers reproduces with a person having long fingers, what are the chances of each child having short fingers?
4. In a fruit fly experiment (see key on page 182), two gray-bodied fruit flies produce mostly gray-bodied offspring, but some offspring have black bodies. If there are 280 offspring, how many do you predict will have gray bodies and how many will have black bodies? How many of the 280 offspring do you predict will be heterozygous? If you wanted to test whether a particular gray-bodied fly was homozygous dominant or heterozygous, what cross would you do?
5. Using the term 2n in which n = the number of heterozygous gene pairs, determine the number of possible gametes for the genotypes *BBHH, BbHh,* and *BBHh.*
6. In rabbits, black color *(B)* is dominant over brown *(b)* and short hair *(S)* is dominant over long *(s).* In a cross between a homozygous black, long-haired rabbit and a brown, homozygous short-haired one, what would the F_1 generation look like? the F_2 generation? If one of the F_1 rabbits reproduced with a brown, long-haired rabbit, what phenotypes and in what ratio would you expect?
7. In horses, black coat *(B)* is dominant over brown coat *(b)* and being a trotter *(T)* is dominant over being a pacer *(t).* A black pacer is crossed with a brown trotter. The offspring is a brown pacer. Give the genotypes of all these horses.
8. The complete genotype of a long-winged, gray-bodied fruit fly is unknown (see key on page 182). When this fly is crossed with a short-winged, black-bodied fruit fly, the offspring all have a gray body but about half of them have short wings. What is the genotype of the long-winged, gray-bodied fly?
9. In humans, widow's peak hairline is dominant over continuous hairline, and short fingers are dominant over long fingers. If an individual who is heterozygous for both traits reproduces with an individual who is recessive for both traits, what are the chances of their child also being recessive for both traits?

*Answers to Additional Genetics Problems appear in Appendix A.

Applying the Concepts

1. *The laws of heredity explain genetic stability and variation between generations.*
   Explain why you are similar in appearance to your parents but do not resemble either one exactly.
2. *The laws of heredity are the same for all organisms.*
   Why would you expect the same laws of heredity to apply to both plants and animals?
3. *Alleles, carried by way of gametes from generation to generation, are not affected by the phenotype.*
   Explain why an individual, deformed by accident, can have children who have no such deformities.

Using Technology

Your study of genetics is supported by these available technologies:

Exploring the Internet
The Mader Home Page provides resources for and help with studying this chapter.

http://www.mhhe.com/sciencemath/biology/mader/
(Click on Biology.)

Understanding the Terms

allele 177	homozygous 177
dominant allele 177	phenotype 177
gene locus 177	Punnett square 179
genotype 177	recessive allele 177
heterozygous 177	testcross 180

Match the terms to these definitions:

a. __________ Allele that exerts its phenotypic effect only in the homozygote; its expression is masked by a dominant allele.
b. __________ Alternative forms of a gene—which occur at the same locus on homologous chromosomes.
c. __________ Allele that exerts its phenotypic effect in the heterozygote; it masks the expression of the recessive allele.
d. __________ Cross between an individual with the dominant phenotype and an individual with the recessive phenotype to see if the individual with the dominant phenotype is homozygous or heterozygous.
e. __________ Genes of an organism for a particular trait or traits; for example, *BB* or *Aa*.

Chromosomes and Genes

CHAPTER 12

Chapter Concepts

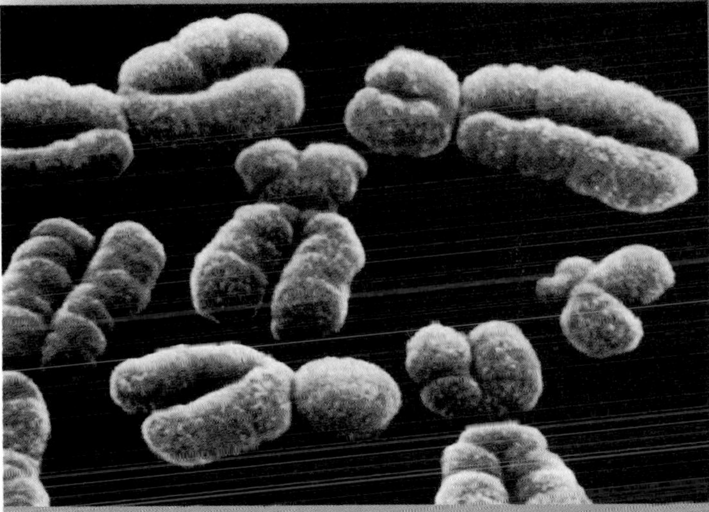

Human chromosomes

2 μm

Considering that Mendel knew nothing about chromosomes, genes, or DNA, it is astounding that his laws about the transfer of genetic information from parent to offspring do apply widely in instances now called "simple Mendelian inheritance." As brilliant and groundbreaking as Mendel's work was, it was only the first step toward a more thorough understanding of how genes function. Today, we know that dominance can occur in degrees, traits can be controlled by more than one gene, one gene may modify the effects of another, and the environment can have a major effect on the phenotype.

The similarity of genes and chromosomal behavior led to the conclusion in the early 1900s that the genes are on the chromosomes, and later it became obvious that Mendel was lucky he happened to do his work with alleles that were located on separate chromosomes. Determining the sequence of genes on the chromosomes became an endeavor that continues even today in genetics laboratories around the globe. Chromosomes can undergo mutations, permanent changes in number and structure, and you can well imagine that such changes have a profound effect on the individual.

12.1 Going Beyond Mendel

Since Mendel's day, patterns of inheritance in addition to simple dominance have been discovered.

Dominance Has Degrees

Mendel always observed simple dominance of one characteristic over another in his experiments. If Mendel had chosen other characteristics, he most likely would have observed **incomplete dominance,** in which neither allele is fully dominant. In a cross between a true-breeding, red-flowered four o'clock strain and a true-breeding, white-flowered strain, the offspring have pink flowers (Fig. 12.1). This is not an example that supports the blending theory of inheritance prevalent when Mendel did his work, however. If these F_1 plants self-pollinate, the F_2 generation has a phenotypic ratio of 1 red-flowered : 2 pink-flowered : 1 white-flowered plant. Because the parental phenotypes reappear in the F_2 generation, it is obvious that we are still dealing with an example of particulate inheritance of the type described by Mendel.

There is a biochemical explanation for incomplete dominance. With simple dominance, the level of a gene-directed protein product is often in between that of the two homozygotes. In other words, the dominant allele is coding for the production of a protein, while the recessive allele is not coding for an effective protein. In our example of incomplete dominance, it can be reasoned that red flowers have double the amount of pigment than pink flowers and that white flowers lack pigment.

There are also examples of **codominance,** in which both alleles are fully expressed. An individual with the blood type AB is exhibiting codominant characteristics (see also page 213). We know this because there are also some individuals who have blood type A and others who have blood type B. The recessive individual has blood type O and exhibits neither dominant characteristic. With codominance, both alleles produce an effective product. In an individual with blood type AB, two different types of glycoproteins appear on the red blood cells.

A Gene That Controls Many Traits

The term **pleiotropy** [Gk. *pleion,* more, and *tropos,* turning] is used to describe a gene that affects more than one characteristic of the individual. Individuals with Marfan syndrome tend to be tall and thin with long legs, arms, and fingers. They are nearsighted, and the wall of the aorta is weak, causing it to enlarge and, eventually, to split. All of these effects are caused by an inability to produce a normal extracellular matrix protein called fibrillin. Fibrillin strengthens elastic fibers and allows extracellular components to adhere to one another.

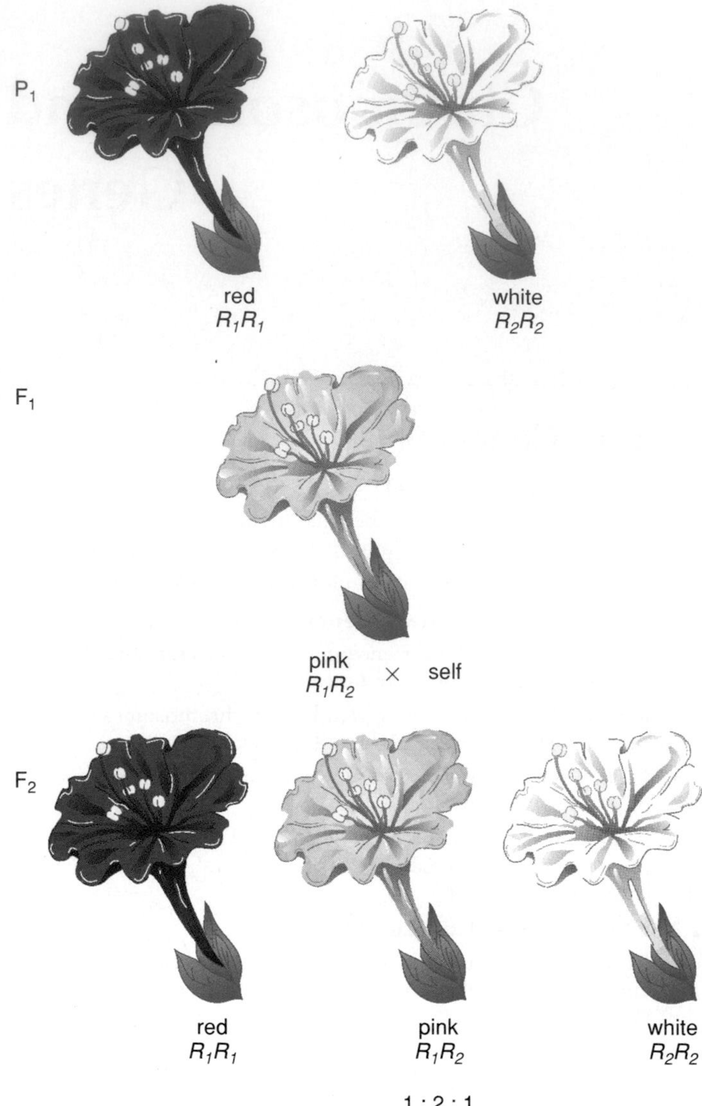

Figure 12.1 Incomplete dominance.
Incomplete dominance is illustrated by a cross between red- and white-flowered four o'clocks. The F_1 plants are pink, a phenotype intermediate between those of the P generation; however, the F_2 results show that neither the red nor the white allele has disappeared. Therefore, this is still an example of particulate inheritance, though the heterozygote R_1R_2 is pink.

There are some who suggest that Abraham Lincoln had Marfan syndrome. Certainly it would account for his lanky frame and the symptoms he described for posterity. If he had not been shot, it's possible he still would have died shortly thereafter from circulatory failure.

In pleiotropy, an allele affects more than one phenotypic characteristic.

Genes That Interact

Many traits are controlled by more than one pair of genes. Often, too, an interaction of these genes affects the phenotype. In these instances, a recessive pair of alleles at one locus might prevent the expression of a dominant allele at another locus. In sweet peas, it would appear that there are two genes affecting pigmentation and that being homozygous recessive in either gene results in a lack of color. If the two varieties of white plants are crossed, the F_1 have purple flowers but among the F_2, $^9/_{16}$ have purple flowers and $^7/_{16}$ have white flowers. Since the ratio is in sixteenths, the F_1 plants must have been dihybrids as shown in Figure 12.2*a*. The 9:7 ratio instead of a 9:3:3:1 ratio can be explained by assuming that both a dominant *A* allele and a dominant *B* allele are required for pigmentation to result. Although the exact details of pigmentation synthesis are known, the metabolic pathway shown in Figure 12.2*b* is hypothesized.

If the *A* allele codes for the first enzyme and the *B* allele codes for the second enzyme, then being homozygous recessive for either gene would result in white instead of purple flowers.

When a recessive pair of alleles at one locus prevents the expression of a dominant allele at another locus, *epistasis*, meaning a "covering-up," has occurred. A similar situation occurs in mammalian animals. If individuals inherit any one of several defects in the metabolic pathway for the synthesis of melanin, the individual is an albino.

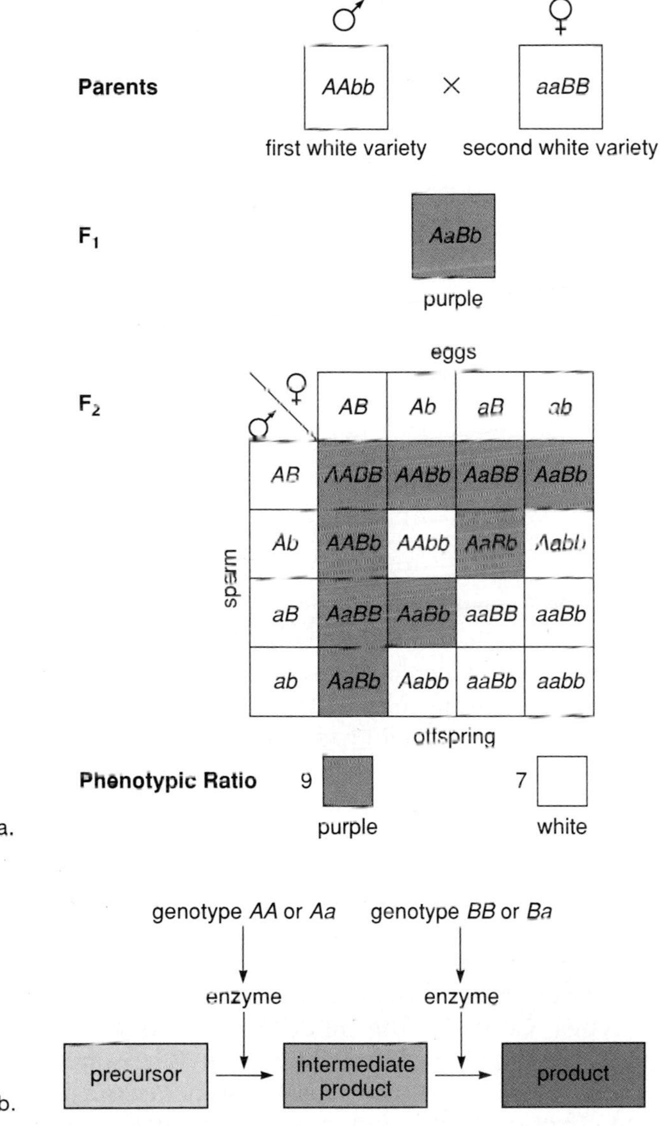

Figure 12.2 Epistasis.
a. Two pairs of genes are believed to be involved in color production in sweet peas. Being homozygous recessive for either gene results in lack of color. b. It is hypothesized that each gene codes for an enzyme in a metabolic pathway that results in pigmentation. If either enzyme is lacking, an absence of color results.

Multiple Alleles

Inheritance can get even more complicated because sometimes there are a series of alleles for a given chromosomal locus, although each individual has only two of these. To take a simple case, three different peppered moth phenotypes are believed to result from the existence of three possible alleles for a single gene. (*m*) is recessive to a second allele, the mottled insularia (*M′*), and the nearly black melanic allele (*M*) is dominant to the mottled insularia (*M′*). Therefore, there are six possible genotypes but only three possible phenotypes (Fig. 12.3).

Inheritance by multiple alleles results in more than two possible phenotypes for a particular trait. Blood type in humans is controlled by three alleles, and there are four possible phenotypes: A, B, AB, and O types of blood (see page 213).

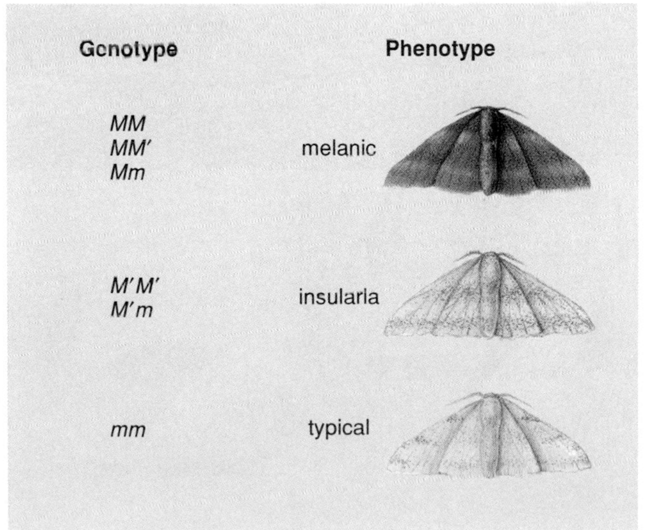

Figure 12.3 Multiple alleles.
Three alleles for one gene locus (in order of dominance *M*, *M′*, *m*) account for the occurrence of three phenotypes of the peppered moth, *Biston betularia*.

Environment Affects the Phenotype

There is no doubt that both the genotype and the environment affect the phenotype. The relative importance of genetic and environmental influences on the phenotype can vary, but in some instances the environment seems to have an extreme effect. In the water buttercup, *Ranunculus peltatus,* the submerged part of the plant has a different appearance from the part above water. Apparently, the presence or absence of an aquatic environment dramatically influences the phenotype.

Temperature can also have a dramatic effect on the phenotypes of plants. Primroses have white flowers when grown above 32°C and red flowers when grown at 24°C. The coats of Siamese cats and Himalayan rabbits are darker in color at the ears, nose, paws, and tail. Himalayan rabbits are known to be homozygous for the allele *ch,* which is involved in production of melanin. Experimental evidence suggests that the enzyme encoded by this gene is active only at a low temperature and that, therefore, black fur only occurs at the extremities where body heat is lost to the environment (Fig. 12.4). When the animal is placed in a warmer environment, new fur on these body parts is light in color.

a.

b.

c.

Figure 12.4 Coat color in Himalayan rabbits.
a. A Himalayan rabbit which has black extremities is usually homozygous recessive for the allele *ch,* which encodes for an enzyme involved in the production of melanin. **b.** The influence of the environment on the phenotype has been demonstrated by plucking out the fur from one area and applying an ice pack. **c.** The new fur which grew in was black instead of white, showing that the enzyme is active only at low temperatures.

Genes That Add Up

Polygenic inheritance [Gk. *polys,* many, and L. *genitus,* producing] occurs when one trait is governed by several genes occupying different loci on the same homologous pair of chromosomes or on different homologous pairs of chromosomes. Each gene has a contributing and noncontributing allele. The contributing allele is represented by a capital letter, and the noncontributing allele is represented by a small letter. Each contributing allele has a quantitative effect on the phenotype, and therefore the allelic effects are additive.

For example, H. Nilsson-Ehle studied the inheritance of seed color in wheat. Genes at three different loci determine seed color. After he crossed plants that produced white and dark red seeds, the F_1 plants were allowed to self-pollinate. Each of the F_2 produced seeds having one of the following seven colors:

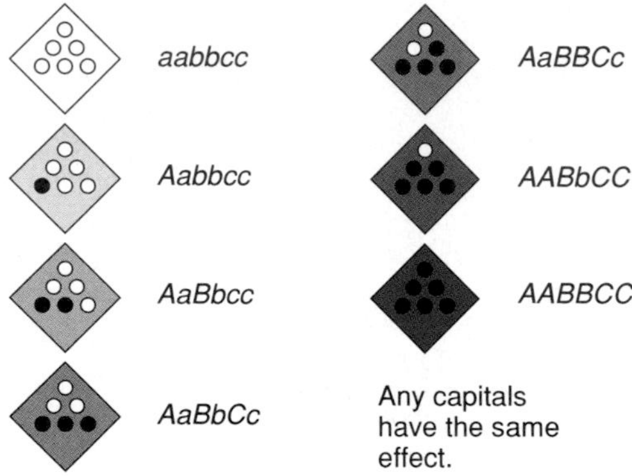

The alleles responsible for the color of the seeds are represented by dots in the diamond shapes. The number of capitals is represented by coloring in the dots. Any capital has the same effect on the color of the seed. Notice that each capital (contributing allele) has a small but equal quantitative effect, and that this accounts for the various degrees of color and the observed range in F_2 phenotypes. The proportions of the phenotypes can be graphed as in Figure 12.5, and these proportions can be connected to form a bell-shaped curve.

If we were observing the phenotypes in nature, we would probably arrive directly at a bell-shaped curve because environmental effects cause many intervening phenotypes. Consider the inheritance of skin color or height in humans, which are believed to be examples of polygenic inheritance. In the first instance, exposure to sun can affect skin color and increase the number of phenotypic variations. In the second instance, nutrition can affect height, resulting in many more phenotypes than those expected.

In polygenic inheritance, each contributing allele adds to the phenotype. A bell-shaped curve of phenotypes is due also to environmental effects.

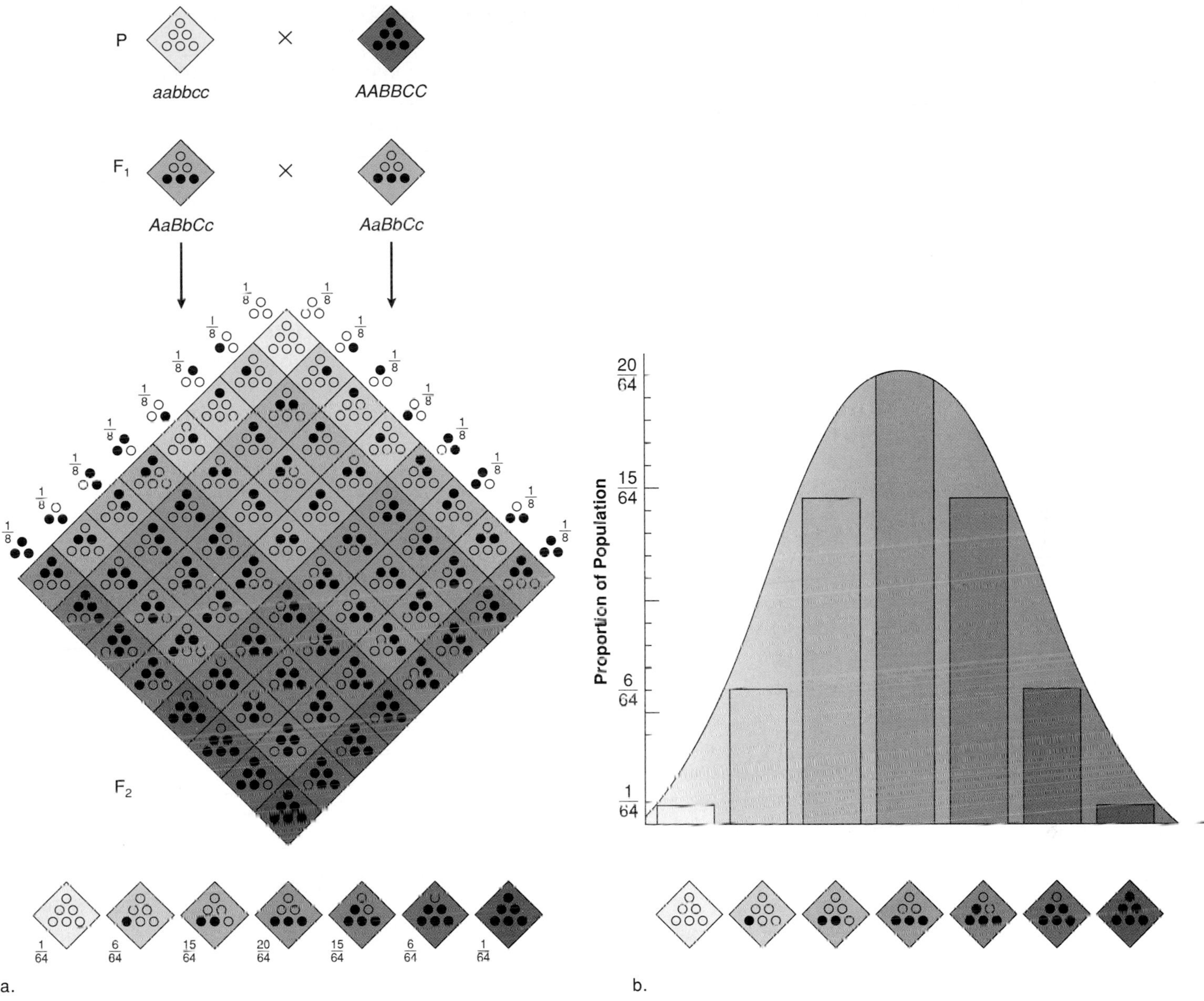

Figure 12.5 Polygenic inheritance.
a. In this example, a trait is controlled by three genes; only the alleles represented by a capital letter are contributing alleles. When two intermediate phenotypes are crossed, seven phenotypes are seen among the F_2 generation. **b.** The phenotype proportions are graphed. If environmental effects are considered, a bell-shaped curve results because of many intervening phenotypes.

Practice Problems 1*

1. What genotypes and phenotypes are possible among the offspring if a moth with the genotype *Mm* is mated to a moth with the genotype *M′m?*
2. What are the results of a testcross involving incomplete dominance when pink-flowered four o'clocks are crossed with white-flowered ones?
3. Breeders of dogs note various colors among offspring that range from white to black. If the coat color is controlled by three pairs of alleles, how many different shades are possible if all dominant genes have the same quantitative effect?
4. Investigators note that albino tigers usually have crossed eyes. What two inheritance patterns can account for a phenotype such as this?

**Answers to Practice Problems appear in Appendix A.*

12.2 Chromosomes Contain Genes

The behavior of chromosomes during mitosis was described by 1875, and for meiosis in the 1890s. By 1902, both Theodor Boveri, a German, and Walter S. Sutton, an American, had independently noted the parallel behavior of genes and chromosomes and had proposed the **chromosomal theory of inheritance,** which states that the genes are located on the chromosomes. This theory is supported by the following observations:

1. Both chromosomes and factors (now called alleles) are paired in diploid cells.
2. Both homologous chromosomes and alleles of each pair separate during meiosis so that the gametes have one-half the total number.
3. Both homologous chromosomes and alleles of each pair separate independently so that the gametes contain all possible combinations.
4. Fertilization restores both the diploid chromosome number and the paired condition for alleles in the zygote.

The genes are on the chromosomes; therefore, they behave similarly during meiosis and fertilization.

Sex Chromosomes Determine Gender

In animal species, the **autosomes,** or nonsex chromosomes, are the same between the sexes. The members of each pair of autosomes are homologous. One special pair of chromosomes is called the **sex chromosomes** because this pair determines the sex of the individual. The sex chromosomes in the human female are XX and those in the male are XY. Because human males can produce two different types of gametes—those that contain an X and those that contain a Y—normally males determine the sex of the new individual.

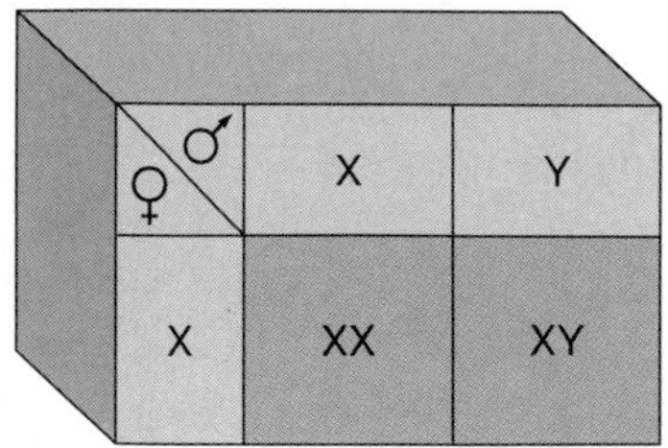

In addition to genes that determine sex, the sex chromosomes carry genes for traits that have nothing to do with the sex of the individual. By tradition, the term sex-linked or **X-linked** is used for genes carried on the X chromosome. The Y chromosome does not carry these genes and indeed carries very few genes.

Genes That Are on the X Chromosomes

The hypothesis that genes are on the chromosomes was substantiated by experiments performed by a Columbia University group of *Drosophila* geneticists, headed by Thomas Hunt Morgan. Fruit flies are even better subjects for genetic studies than garden peas: they can be easily and inexpensively raised in simple laboratory glassware; females mate only once and then lay hundreds of eggs during their lifetimes; and the generation time is short, taking only about ten days when conditions are favorable (Fig. 12.6).

Drosophila flies have the same sex chromosome pattern as humans, and this facilitates our understanding of a cross performed by Morgan. Morgan took a newly discovered mutant male with white eyes and crossed it with a red-eyed female:

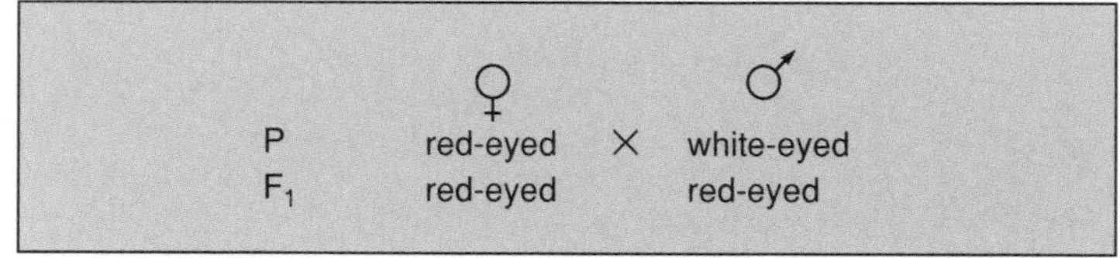

	♀		♂
P	red-eyed	×	white-eyed
F_1	red-eyed		red-eyed

From these results, he knew that red eyes are dominant over white eyes. He then crossed the F_1 flies. In the F_2 generation, there was the expected 3 red-eyed : 1 white-eyed ratio, but it struck him as odd that all of the white-eyed flies were males:

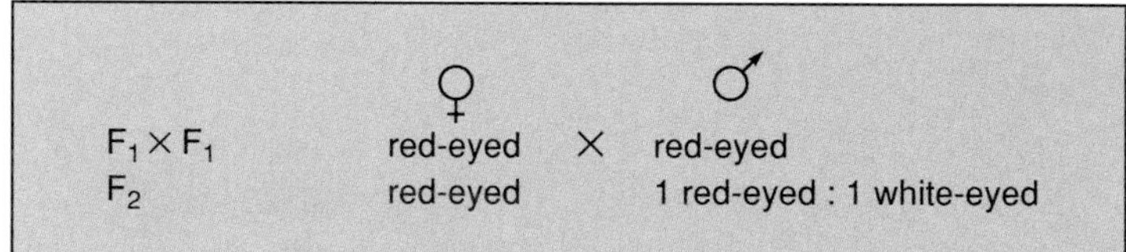

	♀		♂
$F_1 \times F_1$	red-eyed	×	red-eyed
F_2	red-eyed		1 red-eyed : 1 white-eyed

Obviously, a major difference between the male flies and the female flies was their sex chromosomes. Could it be possible that an allele for eye color was on the Y chromosome but not on the X? This idea could be quickly discarded because normal females have red eyes, and they have no Y chromosome. Perhaps an allele for eye color was on the X, but not on the Y, chromosome. Figure 12.6*b* indicates that this explanation would match the results obtained in the experiment. These results support the chromosomal theory of inheritance by showing that the behavior of a specific allele corresponds exactly with that of a specific chromosome—the X chromosome in *Drosophila.*

X-linked alleles are on the X chromosome, and the Y chromosome is blank for these alleles.

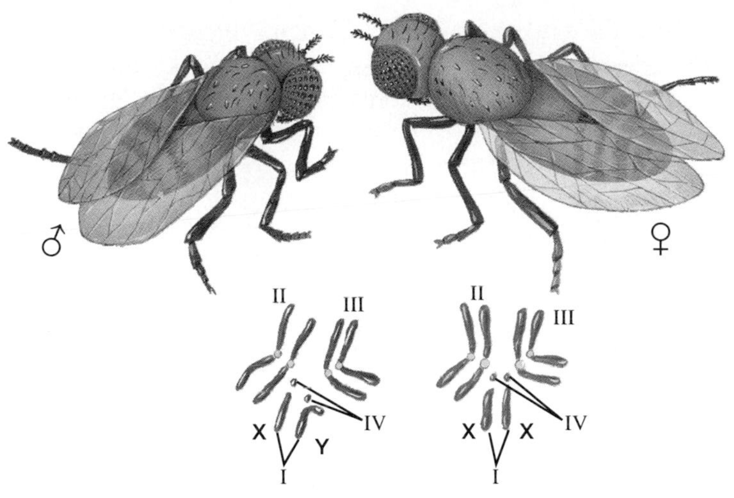

a.

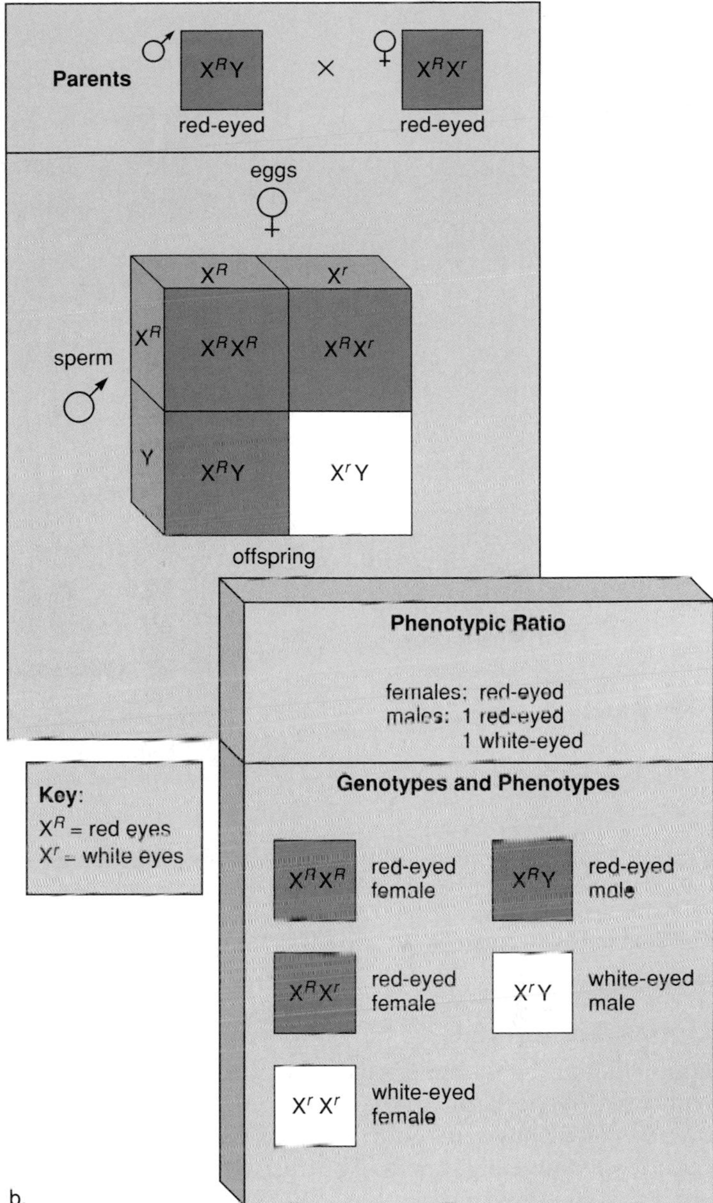

b.

Figure 12.6 X-linked inheritance.
a. Both males and females have eight chromosomes, three pairs of autosomes (II–IV), and one pair of sex chromosomes (I). Males are XY and females are XX. **b.** When solving X-linked genetics problems, males do not have a superscript attached to the Y chromosome because they lack the gene for eye color on that chromosome. Among the offspring in this cross, 50% of the males have white eyes. Because the males receive a Y chromosome from the male parent, they express whichever allele is on the X chromosome inherited from the female parent. In contrast, 50% of the female offspring are red-eyed heterozygotes—when an X^r is received from the female parent, it is masked by an X^R received from the male parent.

Doing X-Linked Problems

Recall that when solving autosomal genetics problems, the key and genotypes are represented as follows:

Key:	**Genotypes:**
L = long wing	*LL, Ll, ll*
l = short wings	

As noted in Figure 12.6, however, the key for an X-linked gene shows an allele attached to the X:

Key:

X^R = red eyes
X^r = white eyes

Notice, too, that there are three possible genotypes for females, but only two for males. Females can be heterozygous X^RX^r, in which case they are carriers. Carriers usually do not show a recessive abnormality, but they are capable of passing on a recessive allele for an abnormality. Males cannot be carriers; if the dominant allele is on the single X chromosome, they have the dominant phenotype, and if the recessive allele is on the single X chromosome, they show the recessive phenotype.

Practice Problems 2*

1. Using the key X^B = bar eyes and X^b = normal eyes, give the three possible genotypes for females and the two possible genotypes for males. What are the possible gametes for these males?
2. Which *Drosophila* cross would produce white-eyed males? In what ratio?
   a. $X^RX^R \times X^rY$
   b. $X^RX^r \times X^RY$
3. A woman is color blind (X-linked recessive). What are the chances of her sons being color blind? If she reproduces with a man having normal vision, what are the chances of her daughters being color blind? Being carriers?

**Answers to Practice Problems appear in Appendix A.*

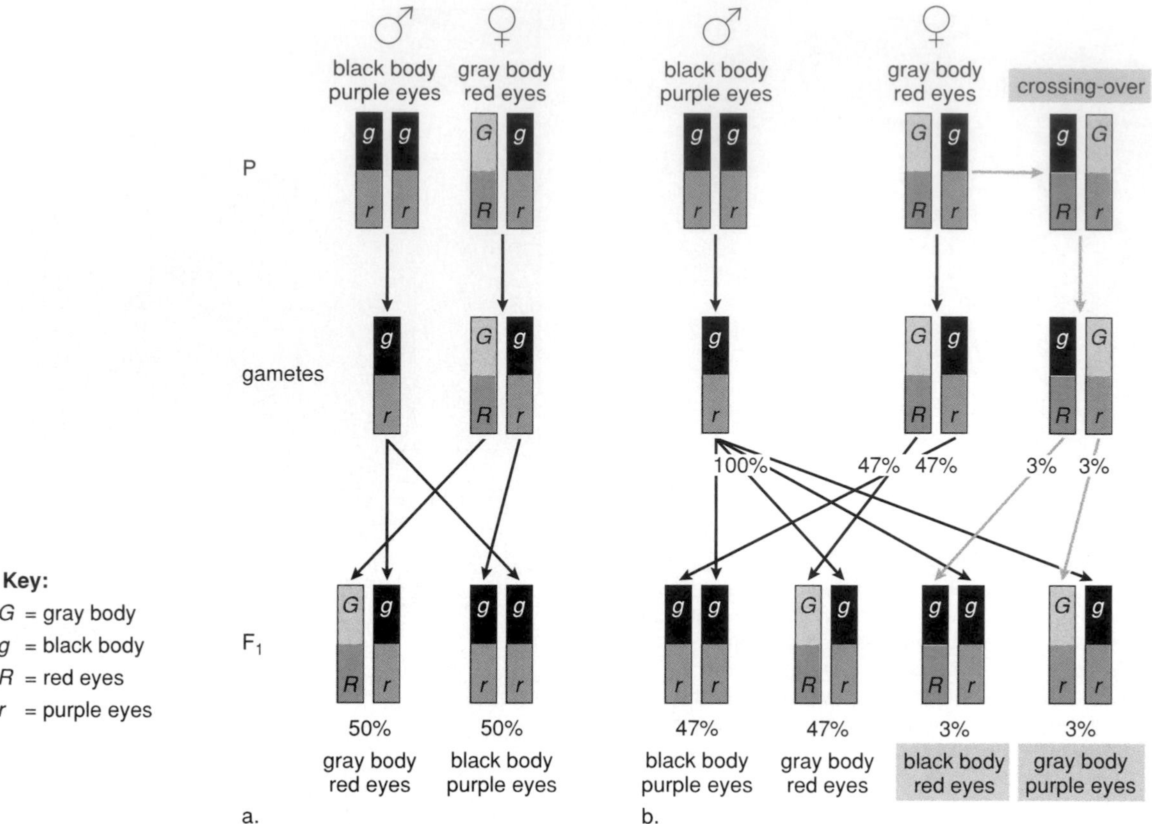

Figure 12.7 Complete linkage versus incomplete linkage.
a. Hypothetical cross in which genes for body and eye color on chromosome 2 of *Drosophila* are completely linked. Instead of the expected 1:1:1:1 ratio among the offspring, there would be a 1:1 ratio. **b.** Crossing-over occurs, and 6% of the offspring show recombinant phenotypes. The percentage of recombinant phenotypes is used to map the chromosomes (1% = 1 map unit).

Genes Are Linked

Drosophila probably has thousands of different genes controlling all aspects of its structure, biochemistry, and behavior. Yet it has only four pairs of chromosomes. This paradox led investigators like Sutton to conclude that each chromosome must carry a large number of genes. For example, it is now known that genes controlling eye color, wing type, body color, leg length, and antennae type are all located on the chromosomes numbered II in Figure 12.6. The alleles for these genes are said to form a **linkage group** because they are found on the same chromosome.

It is easy to predict that crosses involving linked genes will not give the same results as those involving unlinked genes (Fig. 12.7). Suppose you are doing a cross between a gray-bodied, red-eyed heterozygote and a black-bodied, purple-eyed fly. Since the alleles governing these traits are both on chromosome II, you predict that the results will be 1:1 instead of 1:1:1:1, as is the case for unlinked genes. The reason, of course, is that linked alleles tend to stay together and do not separate independently, as predicted by Mendel's laws. Under these circumstances, the heterozygote forms only two types of gametes and produces offspring with only two phenotypes (Fig. 12.7*a*).

When you do the cross, however, you find that a very small number of offspring show recombinant phenotypes (i.e., those that are different from the original parents). Specifically, you find that 47% of the offspring have black bodies and purple eyes, 47% have gray bodies and red eyes, 3% have black bodies and red eyes, and 3% have gray bodies and purple eyes (Fig. 12.7*b*). What happened?

Crossing-over can occur between homologous chromosomes when they are paired during meiosis, and crossing-over contributes to recombinant gametes (Fig. 12.8). In this instance, crossing-over led to a very small number of *recombinant gametes* (Fig. 12.8), and when these were fertilized, *recombinant phenotypes* were observed in the offspring. An examination of chromosome II shows that these two sets of alleles are very close together. Doesn't it stand to reason that the closer together two genes are, the less likely they are to cross over? This is exactly what various crosses have repeatedly shown.

All the genes on one chromosome form a linkage group that tends to stay together, except when crossing-over occurs.

Mapping the Chromosomes

You can use the percentage of recombinant phenotypes to map the chromosomes, because there is a direct relationship between the frequency of crossing-over and the percentage of recombinant phenotypes. In our example (Fig. 12.7), a total of 6% of the offspring are recombinants, and for the sake of mapping the chromosomes, it is assumed that 1% of crossing-over equals one map unit. Therefore, the allele for black body and the allele for purple eyes are six map units apart.

Suppose you want to determine the order of any three genes on the chromosomes. To do so, you can perform crosses that tell you the map distance between all three pairs of alleles. If you know, for instance, that:

1. the distance between the black-body and purple-eye alleles = 6 map units,
2. the distance between the purple-eye and vestigial-wing alleles = 12.5 units, and
3. the distance between the black-body and vestigial-wing alleles = 18.5 units, then the order of the alleles must be as shown here:

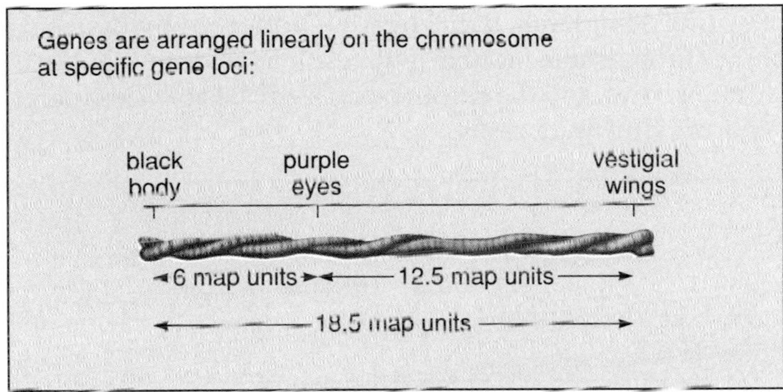

Of the three possible orders (any one of the three genes can be in the middle), only this order gives the proper distance between all three allelic pairs. Because it is possible to map the chromosomes, we conclude that genes are indeed located on chromosomes, they have a definite location, and they occur in a definite order.

The crossing-over frequency is proportional to the recombinant phenotypic frequency, which, in crosses involving linked genes, indicates the distance between genes on the chromosomes.

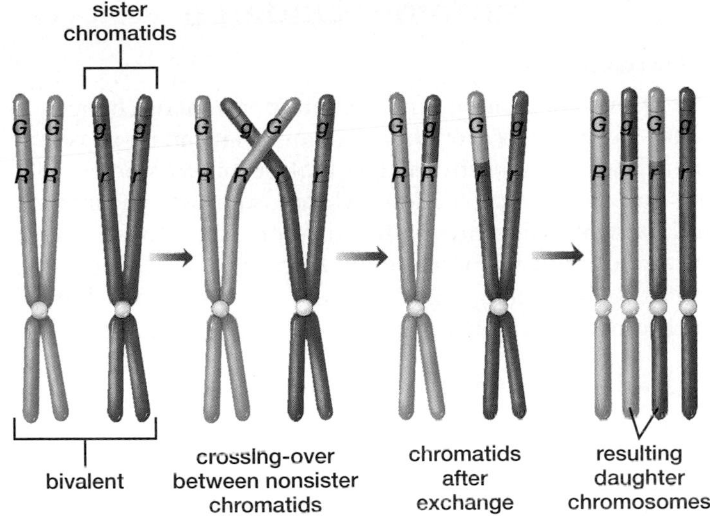

Figure 12.8 Crossing-over.
When homologous chromosomes are in synapsis, the nonsister chromatids exchange genetic material. Following crossing-over, recombinant chromosomes occur. Recombinant chromosomes contribute to recombinant gametes.

Practice Problems 3*

1. When *AaBb* individuals are allowed to self-breed, the phenotypic ratio is just about 3:1. What ratio was expected? What may have caused the observed ratio?
2. In two sweet pea strains, *B* = blue flowers, *b* = red flowers, *L* = long pollen grains, and *l* = round pollen grains. In a cross between a heterozygous plant with blue flowers and long pollen grains and a plant with red flowers and round pollen grains, 44% of offspring are blue, long; 44% are red, round; 6% are blue, round; and 6% are red, long. How many map units separate these two sets of alleles?
3. Investigators performed crosses that indicated bar eye and garnet eye alleles are 13 map units apart, scallop wing and bar eye alleles are 6 units apart, and garnet eye and scallop wing alleles are 7 units apart. What is the order of these alleles on the chromosome?

Answers to Practice Problems appear in Appendix A.

12.3 Chromosomes Undergo Mutations

Mutations [L. *mutatus*, change] are permanent changes in genes (see page 247) or chromosomes that can be passed to offspring if they occur in cells that become gametes. Like crossing-over, recombination of chromosomes during meiosis, and gamete fusion during fertilization, mutations increase the amount of variation among offspring. *Chromosomal mutations* include changes in chromosome number and changes in chromosomal structure.

Changing Chromosome Number

Changes in the chromosome number include monosomies, trisomies, and polyploidy.

Monosomy and Trisomy

Monosomy occurs when an individual has only one of a particular type of chromosome (2n – 1), and *trisomy* occurs when an individual has three of a particular type of chromosome (2n + 1). The usual cause of monosomy and trisomy is nondisjunction during meiosis. *Nondisjunction* can occur during meiosis I if members of a homologous pair fail to separate, and during meiosis II if the daughter chromosomes go into the same daughter cell (see page 203). Although nondisjunction can also occur during mitosis, it is more common during meiosis.

Monosomy and trisomy occur in both plants and animals. In animals, autosomal monosomies and trisomies are generally lethal, although a trisomic individual is more likely to survive than a monosomic one. The survivors are characterized by a distinctive set of physical and mental abnormalities called a syndrome. In humans, Turner syndrome is a monosomy involving the sex chromosomes—the individual inherits a single X chromosome. The most common trisomy among humans is Down syndrome, which involves chromosome 21. Individuals with Turner syndrome are expected to live a full life span, and although those with Down syndrome are subject to various medical conditions, they generally do well also.

Polyploidy

Some mutant eukaryotes have more than two sets of chromosomes. They are called **polyploids** [Gk. *polys*, many, and *plo*, fold]. Polyploid organisms are named according to the number of sets of chromosomes they have. Triploids (3n) have three of each kind of chromosome, tetraploids (4n) have four sets, pentaploids (5n) have five sets, and so on.

Polyploidy is not seen in animals most likely because judging from trisomies, multiple copies of chromosomes would be lethal. Polyploidy is a major evolutionary mechanism in plants. It is estimated that 47% of all flowering plants are polyploids. Among these are many of our most important crops, such as wheat, corn, cotton, sugarcane, and fruits such as watermelons, bananas, and apples (Fig. 12.9). Also many attractive flowers, such as chrysanthemums and daylilies, are polyploids.

Polyploidy generally arises following hybridization. When two different species reproduce, the resulting organism, called a hybrid, may have an odd number of chromosomes. If so, the chromosomes cannot pair evenly during meiosis, and the organism will not be fertile. (Hybridization in animals can also lead to sterility—as when a horse and donkey produce a mule, which is unable to produce offspring.) In plants, hybridization, followed by doubling of the chromosome number, will result in an even number of chromosomes, and the chromosomes will be able to undergo synapsis during meiosis.

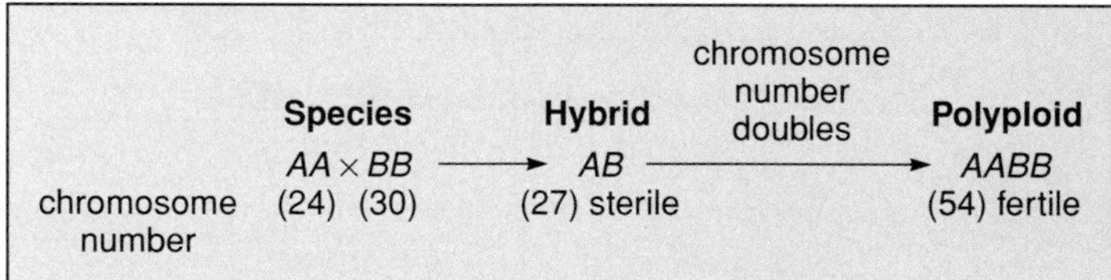

Monosomies (2n – 1) and trisomies (2n + 1) cause abnormalities. In plants, hybridization followed by polyploidy is a positive force for change.

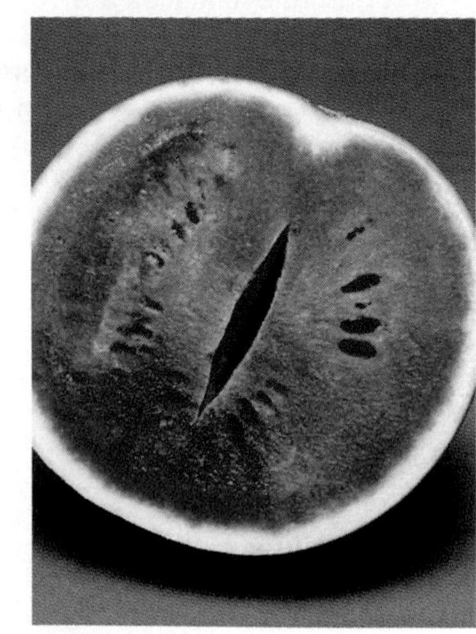
a.

b.

c.

Figure 12.9 Polyploid plants. Many food sources are specially developed polyploid plants. **a, b.** Among these are seedless watermelons and bananas. They are infertile (seeds poorly developed) triploids, but they can be propagated by asexual means. **c.** Polyploidy makes plants and their fruits larger, such as these jumbo McIntosh apples.

Changing Chromosomal Structure

There are various agents in the environment, such as radiation, certain organic chemicals, or even viruses, that can cause chromosomes to break. Sometimes, however, the broken ends of one or more chromosomes do not rejoin in the same pattern as before, and this results in a change in chromosomal structure.

An *inversion* occurs when a segment of a chromosome is turned around 180° (Fig. 12.10). You might think this is not a problem because the same genes are present, but the new position might lead to altered gene activity.

A *translocation* is the movement of a chromosome segment from one chromosome to another, nonhomologous chromosome. Translocation heterozygotes usually have reduced fertility due to production of abnormal gametes.

A *deletion* occurs when an end of a chromosome breaks off or when two simultaneous breaks lead to the loss of an internal segment. Even when only one member of a pair of chromosomes is affected, a deletion often causes abnormalities. An example is cri du chat (cat's cry) syndrome. The affected individual has a small head, is mentally retarded, and has facial abnormalities. Abnormal development of the glottis and larynx results in the most characteristic symptom—the infant's cry resembles that of a cat.

A *duplication* is the doubling of a chromosome segment. There are several ways a duplication can occur. A broken segment from one chromosome can simply attach to its homologue, or unequal crossing-over may occur, leading to a duplication and a deletion:

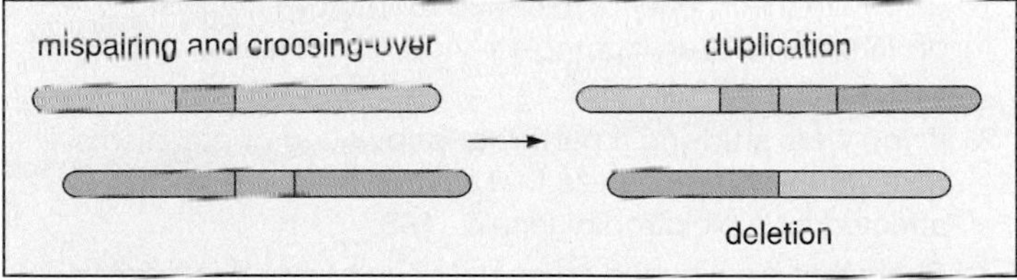

Multiple copies of genes can mutate differently and thereby provide additional genetic variation for the species. For example, there are several closely linked genes for human globin (globin is a part of hemoglobin, which is present in red blood cells and carries oxygen). This may have arisen by a process of duplication followed by different mutations in each gene.

Chromosomal mutations include various changes in structure, which can lead to abnormal gametes and offspring.

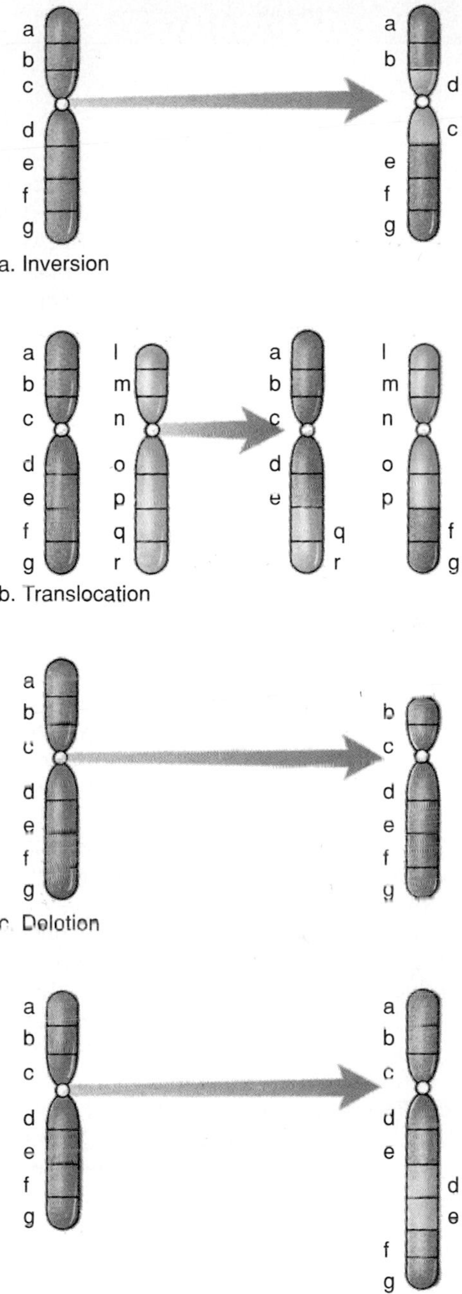

Figure 12.10 Types of chromosomal mutations.
a. Inversion occurs when a piece of chromosome breaks loose and then rejoins in the reversed direction. **b.** Translocation is the exchange of chromosome pieces between nonhomologous pairs. **c.** Deletion is the loss of a chromosome piece. **d.** Duplication occurs when the same piece is repeated within the chromosome.

connecting concepts

Mendel's statistical studies allowed him to formulate patterns of inheritance still in use today to predict phenotypic ratios among offspring. His conclusion that genetic information is transmitted in discrete units (now called genes) is a tribute to Mendel's brilliance. In Mendel's day, little was known about cells, much less genes.

The many other patterns of inheritance that have been discovered since Mendel's time expand on this original contribution. Also, Mendel's laws, developed in the 1860s, helped establish the chromosomal theory of inheritance, which states that the genes are on the chromosomes. The chromosomal theory of inheritance was a very fruitful theory because it not only led to the discovery that DNA is the genetic material, it also set the stage for our ability to manipulate genes. Gene therapy for some human disorders is now a reality. Science progresses. The history of genetics shows that an original discovery can be modified and built upon as more and more research is done in an area.

It is important to realize that human beings are like, not dissimilar from, other organisms. Human genetics is also founded in Mendelian genetics, and as we'll see in the following chapters, human genes and chromosomes must obey the same laws of inheritance that other species obey—another indication that we, too, have an evolutionary history that ties us to other living things.

Summary

12.1 Going Beyond Mendel

There are patterns of inheritance discovered since Mendel's original contribution. For example, different degrees of dominance have been observed. With incomplete dominance, the F_1 individuals are intermediate between the parental types; this does not support the blending theory because the parental phenotypes reappear in F_2. With codominance, the F_1 individuals express both alleles equally. In pleiotropy, a dominant allele causes an effect that drastically alters the phenotype in many regards. Also, genes may interact. A homozygous recessive condition sometimes masks the effect of a dominant allele at another locus (epistasis). Some genes have multiple alleles, although each individual organism has only two alleles. Coat color in rabbits is controlled by multiple alleles, and four different type coats are possible.

The relative influence of the genotype and the environment on the phenotype can vary—there are examples of extreme environmental influence. Polygenic traits are controlled by genes that have an additive effect on the phenotype, resulting in quantitative variations. A bell-shaped curve is seen because environmental influences bring about many intervening phenotypes.

12.2 Chromosomes Contain Genes

The chromosomal theory of inheritance says that the genes are located on the chromosomes, which accounts for the similarity of their behavior during meiosis and fertilization.

Sex determination in animals is dependent upon the chromosomes. Usually females are XX and males are XY. Solid experimental support for the chromosomal theory of inheritance came when Morgan and his group were able to determine that the white-eye allele in *Drosophila* is on the X chromosome.

Alleles on the X chromosome are called X-linked alleles. Therefore, when doing X-linked genetics problems, it is the custom to indicate the sexes by using sex chromosomes and to indicate the alleles by superscripts attached to the X. The Y is blank because it does not carry these genes.

All the alleles on a chromosome form a linkage group. Linked alleles do not obey Mendel's laws because they tend to go into the same gamete together. Crossing-over can cause recombinant gametes and recombinant phenotypes to occur. The percentage of recombinant phenotypes is used to measure the distance between genes and to map the chromosomes.

12.3 Chromosomes Undergo Mutations

Chromosomal mutations fall into two categories: changes in chromosome number and changes in chromosomal structure.

Monosomy occurs when an individual has only one of a particular type of chromosome (2n – 1); trisomy occurs when an individual has three of a particular type of chromosome (2n + 1). Polyploidy occurs when the eukaryotic individual has more than two complete sets of chromosomes.

Changes in chromosomal structure include inversions, translocations, deletions, and duplications.

Reviewing the Chapter

1. Compare the F_2 phenotypic ratio for a cross involving a dominant allele, an incompletely dominant allele, and codominant alleles. 188
2. If you were studying a particular population of organisms, how would you recognize that several traits are being affected by a pleiotropic gene? 188
3. Show that if a dihybrid cross between two heterozygotes involves an epistatic gene, you do not get a ratio of 9:3:3:1. 189
4. Explain inheritance by multiple alleles. List the human blood types and give the possible genotypes for each. 189
5. Give examples to show that the environment can have an extreme effect on the phenotype despite the genotype. 190
6. Explain why traits controlled by polygenes show continuous variation that can be measured quantitatively, have many intermediate forms, and produce a distribution in the F_2 generation that follows a bell-shaped curve. 190
7. What is the chromosomal theory of inheritance? List the ways in which genes and chromosomes behave similarly during meiosis and fertilization. 192
8. How is sex determined in humans? Which sex determines the sex of the offspring? 192
9. How did a *Drosophila* cross involving an X-linked gene help investigators show that certain genes are carried on certain chromosomes? 192

10. Show that a dihybrid cross between a heterozygote and a recessive homozygote involving linked genes does not produce the expected 1:1:1:1 ratio. What is the significance of the small percentage of recombinants that occurs among the offspring? 194
11. What are the two types of chromosomal mutations? What is a monosomy? A trisomy? Why would you expect a monosomy to be more lethal than a trisomy? Why is polyploidy not generally found in animals? In what way may polyploidy have assisted plant evolution? 196
12. What four types of changes in chromosomal structure were discussed? How might it be possible for a duplication and a deletion to occur at the same time? 197

Testing Yourself

Choose the best answer for each question.
For questions 1–4, match the statements that follow to the items in the key.

Key:

a. multiple alleles
b. incomplete dominance
c. polygenes
d. epistatic gene
e. pleiotropic gene

1. A cross between oblong and round squash produced oval squash.
2. Although most people have an IQ of about 100, IQ generally ranges from about 50 to 150.
3. Investigators noted that whenever a particular species of plant had narrow instead of broad leaves, it was also short and yellow, instead of tall and green.
4. In humans, there are three alleles possible at the chromosomal locus that determine blood type.
5. When a white-eyed *Drosophila* female occurs,
   a. both parents could have red eyes.
   b. the female parent could have red eyes, but the male parent has to have white eyes.
   c. both parents must have white eyes.
   d. Both a and b are correct.
6. Investigators found that a cross involving the mutant genes *a* and *b* produced 30% recombinants, a cross involving *a* and *c* produced 5% recombinants, and a cross involving *c* and *b* produced 25% recombinants. Which is the correct order of the genes?
   a. *a, b, c*
   b. *a, c, b*
   c. *b, a, c*
7. A boy is color blind (X-linked recessive) and has a continuous hairline (autosomal recessive). Which could be the genotype of his mother?
   a. *bbww*
   b. X^bYWw
   c. bbX^wX^w
   d. X^BX^bWw
8. Which two of these chromosomal mutations are most likely to occur when homologous chromosomes are undergoing synapsis?
   a. inversion and translocation
   b. deletion and duplication
   c. deletion and inversion
   d. duplication and translocation
9. Investigators do a dihybrid cross between two heterozygotes and get about a 3:1 ratio among the offspring. The reason must be due to
   a. polygenes.
   b. pleiotropic genes.
   c. linked genes.
   d. epistatic genes.
10. In snakes, the recessive genotype *cc* causes the animal to be albino despite the inheritance of the dominant allele (*B* = black). What would be the results of a cross between two snakes with the genotype *CcBB?*
    a. all black snakes
    b. all white snakes
    c. 9:3:3:1
    d. 3 black : 1 albino

Additional Genetics Problems*

1. Chickens that are homozygous for the frizzled trait have feathers that are weak and stringy. When raised at low temperatures, the birds have circulatory, digestive, and hormonal problems. What inheritance pattern explains these results?
2. Chickens having black feathers are crossed with chickens having white feathers, and the result is chickens that appear to have blue feathers. What inheritance pattern explains these results, and what key do you suggest for this cross?
3. In radish plants, the shape of the radish may be long *(LL)*, round *(ll)*, or oval *(Ll)*. If oval is crossed with oval, what proportion of offspring will also be oval?
4. If blood type in cats were controlled by three codominant and multiple alleles, how many genotypes and phenotypes would occur? Could the cross *AC* X *BC* produce a cat with *AB* blood type?
5. In *Drosophila, S* = normal, *s* = sable body; *W* = normal, *w* = miniature wing. In a cross between a heterozygous normal fly and a sable-bodied, miniature-winged fly, the results were 99 normal flies, 99 with a sable body and miniature wings, 11 with a normal body and miniature wings, and 11 with a sable body and normal wings. What inheritance pattern explains these results? How many map units separate the genes for sable body and miniature wing?

*Answers to Additional Genetics Problems appear in Appendix A.

6. In *Drosophila,* the gene that controls red eye color (dominant) versus white eye color is on the X chromosome. What are the expected phenotypic results if a heterozygous female is crossed with a white-eyed male?
7. Bar eyes in *Drosophila* is dominant and X-linked. What phenotypic ratio is expected for reciprocal crosses between pure-breeding flies?
8. In *Drosophila,* a male with bar eyes and miniature wings is crossed with a female who has normal eyes and normal wings. Half the female offspring have bar eyes and miniature wings and half have bar eyes and normal wings. What is the genotype of the female parent?
9. In cats, S = short hair, s = long hair, X^C = black coat, X^c = yellow coat, and $X^C X^c$ = tortoiseshell (calico) coat. If a long-haired yellow male is crossed with a tortoiseshell female homozygous for short hair, what are the expected phenotypic results?

Applying the Concepts

1. *Genes often work together to produce the phenotype.*
   Give examples discussed in this chapter to demonstrate that the phenotype is controlled by the entire genome.
2. *Genes are located on the chromosomes.*
   Give evidence that the genes are on the chromosomes and that the genes are linearly arranged.
3. *Chromosomes undergo mutations.*
   Why would you expect chromosomal mutations to have a profound effect on the phenotype?

Using Technology

Your study of chromosomes and genes is supported by these available technologies:

Exploring the Internet

The Mader Home Page provides resources for and help with studying this chapter.

http://www.mhhe.com/sciencemath/biology/mader/
(Click on Biology.)

Explorations in Cell Biology & Genetics CD-ROM

Three-Point Genetic Cross (#11)

Understanding the Terms

autosome 192
chromosomal theory of inheritance 192
codominance 188
incomplete dominance 188
linkage group 194
multiple allele 189
mutation 196
pleiotropy 188
polygenic inheritance 190
polyploid (polyploidy) 196
sex chromosome 192
X-linked 192

Match the terms to these definitions:

a. __________ Alteration in chromosomal structure or number and also an alteration in a gene due to a change in DNA composition.
b. __________ Any chromosome other than a sex chromosome.
c. __________ Chromosome that determines the sex of an individual; in animals, females have two X chromosomes and males have an X and Y chromosome.
d. __________ Condition in which an organism has more than two complete sets of chromosomes.
e. __________ Gene located on the X chromosomes that does not control a sexual feature of the organism.
f. __________ Pattern of inheritance in which a trait is controlled by several allelic pairs; each dominant allele contributes in an additive and like manner.
g. __________ Pattern of inheritance in which both alleles of a gene are equally expressed.
h. __________ Pattern of inheritance in which the offspring shows characteristics intermediate between two extreme parental characteristics—for example, a red and a white flower producing pink offspring.
i. __________ Condition in which one gene affects many characteristics of the individual.
j. __________ Theory that the genes are on the chromosomes, accounting for their similar behavior.

Human Genetics

CHAPTER 13

Chapter Concepts

A human family

Scarcely a week goes by without a newspaper or television report of another discovery in the exploding field of human genetics. A considerable amount of knowledge about the human genome has already been accumulated. We know, for example, that our somatic cells contain forty-six chromosomes with about a hundred thousand genes. We have genetic markers for many diseases, and we even know the locations of some potential cancer-causing genes.

Genetic counseling is more extensive today because of the availability of so many prenatal tests to detect chromosomal and gene mutations. Amniocentesis followed by karyotyping of fetal or embryonic cells can indicate whether a developing child has Down syndrome or one of the other chromosomal abnormalities. Specific chromosomes can be tested for the presence of a mutant allele for cystic fibrosis, neurofibromatosis, and sickle-cell disease, among others.

Recent studies have also implicated genes in such multifaceted behaviors as sexual orientation, mental illnesses, and addictions. While genes undoubtedly play a role in almost all human behavior, they act in conjunction *with* the environment. Few aspects of human behavior are controlled only by genes or only by environment. The question of nature versus nurture—what percentage of the distribution of a trait is controlled by genes and what percentage is controlled by environment—cannot now nor may ever be answered for most traits; both are important.

13.1 Considering the Chromosomes

It has now been established that somatic (body) cells in humans have 46 chromosomes. To view the chromosomes, a cell can be photographed just prior to division so that a picture of the chromosomes is obtained. The picture may be entered into a computer and the chromosomes electronically arranged by pairs (Fig. 13.1). The members of a pair have the same size, shape, and constriction (location of the kinetochore), and they also have the same characteristic banding pattern upon staining. The resulting display of pairs of chromosomes is called a **karyotype.** Both males and females normally have 23 pairs of chromosomes, but one of these pairs is of unequal length in males. The larger chromosome of this pair is the X chromosome and the smaller is the Y chromosome. These are called the **sex chromosomes** because they contain the genes that determine sex. The other chromosomes, known as **autosomes,** include all the pairs of chromosomes except the X and Y chromosomes. In a karyotype, autosomes are usually ordered by size and numbered from the largest to smallest; the sex chromosomes are identified separately.

A normal human karyotype shows 22 pairs of autosomes and one pair of sex chromosomes. Males have an X and a Y chromosome; females have two X chromosomes.

1. Blood is centrifuged to separate out blood cells.
2. Only white blood cells are transferred and treated to stop cell division.
3. Sample is fixed, stained, and spread on a microscope slide.

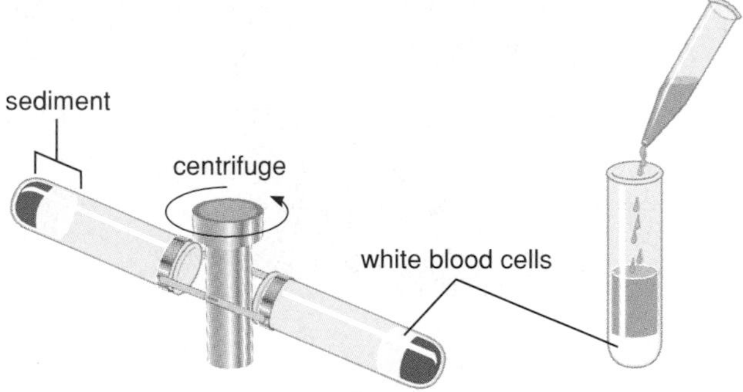

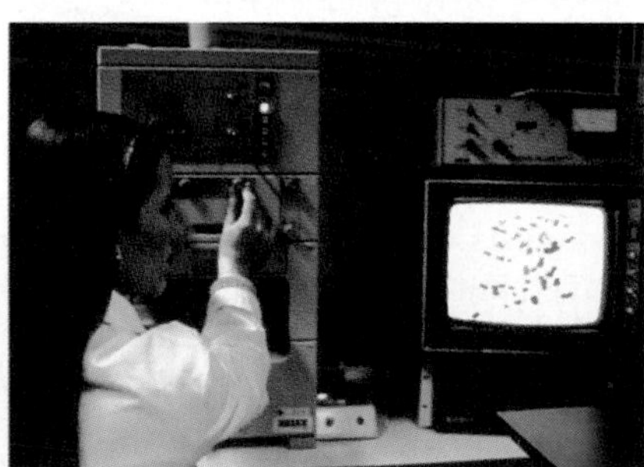

4. Slide is examined microscopically, and the chromosomes are photographed. Computer arranges the chromosomes into pairs.

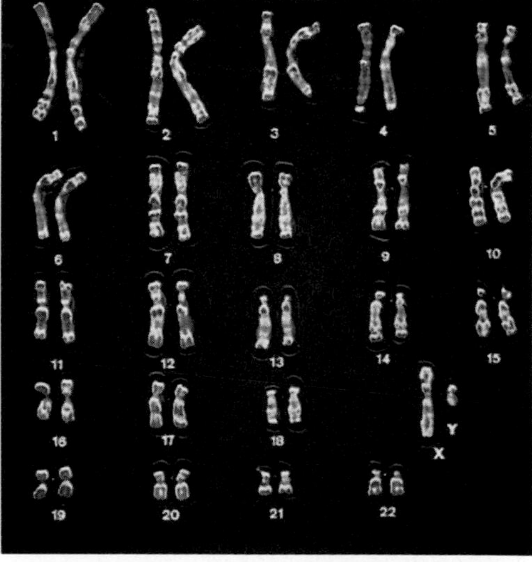

5. Karyotype: Chromosomes are paired by size, centromere location, and banding patterns.

Figure 13.1 Human karyotype preparation.
As illustrated here, the stain used can result in chromosomes with a banded appearance. The bands help researchers identify and analyze the chromosomes.

Nondisjunction Causes Abnormalities

Gamete formation in humans involves meiosis, the type of cell division that reduces the chromosome number by one-half. **Nondisjunction** is the failure of homologous chromosomes to separate during meiosis I or daughter chromosomes to separate during meiosis II (Fig. 13.2). Nondisjunction leads to gametes that have too few (n – 1) or too many (n + 1) chromosomes.

When these abnormal gametes fuse with normal gametes at fertilization, a monosomy (2n – 1) or a trisomy (2n + 1) can result. A study of spontaneous abortions suggests that many trisomies and nearly all monosomies are fatal. The most common autosomal trisomy seen among humans is trisomy 21 (Down syndrome), which occurs in one out of 800 live births (Fig. 13.2*b*). Individuals with trisomy 13 (Patau syndrome) and trisomy 18 (Edward syndrome) have an average life span of less than one year. Heart and nervous system defects prevent normal development. A *syndrome* is a group of symptoms that appear together and tend to indicate the presence of a particular disorder.

Nondisjunction of the sex chromosomes in humans also occurs. Four common, but abnormal, chromosome types are XO, XXX, XXY, and XYY. The symbol O means that a sex chromosome is missing.

Nondisjunction causes an abnormal chromosome number in the gametes. Offspring inherit an extra chromosome (trisomy) or are missing a chromosome (monosomy).

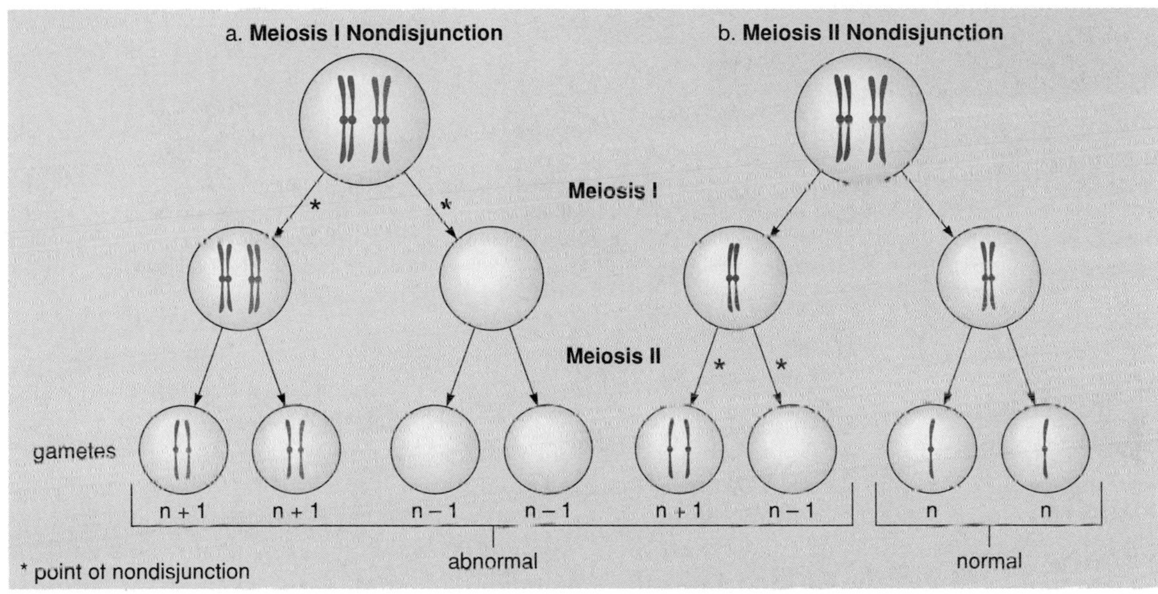

a.

Syndrome	Sex	Chromosomes	Frequency	
			Abortuses	*Births*
Down	M or F	Trisomy 21	1/40	1/800
Patau	M or F	Trisomy 13	1/33	1/15,000
Edward	M or F	Trisomy 18	1/200	1/6,000
Jacob	M	XYY	?	1/1,000
Turner	F	XO	1/18	1/6,000
Klinefelter	M	XXY (or XXXY)	0	1/1,500
Triplo-X	F	XXX (or XXXX)	0	1/1,500

b.

From Robert F. Weaver and Philip W. Hedrick, Genetics, *2d ed. Copyright © 1992 WCB.*

Figure 13.2 Nondisjunction of autosomes during meiosis.
a. Nondisjunction can occur during meiosis I if homologous chromosomes fail to separate, and during meiosis II if the sister chromatids fail to separate completely. In either case, certain abnormal gametes carry an extra chromosome or lack a chromosome. **b.** Frequency of syndromes.

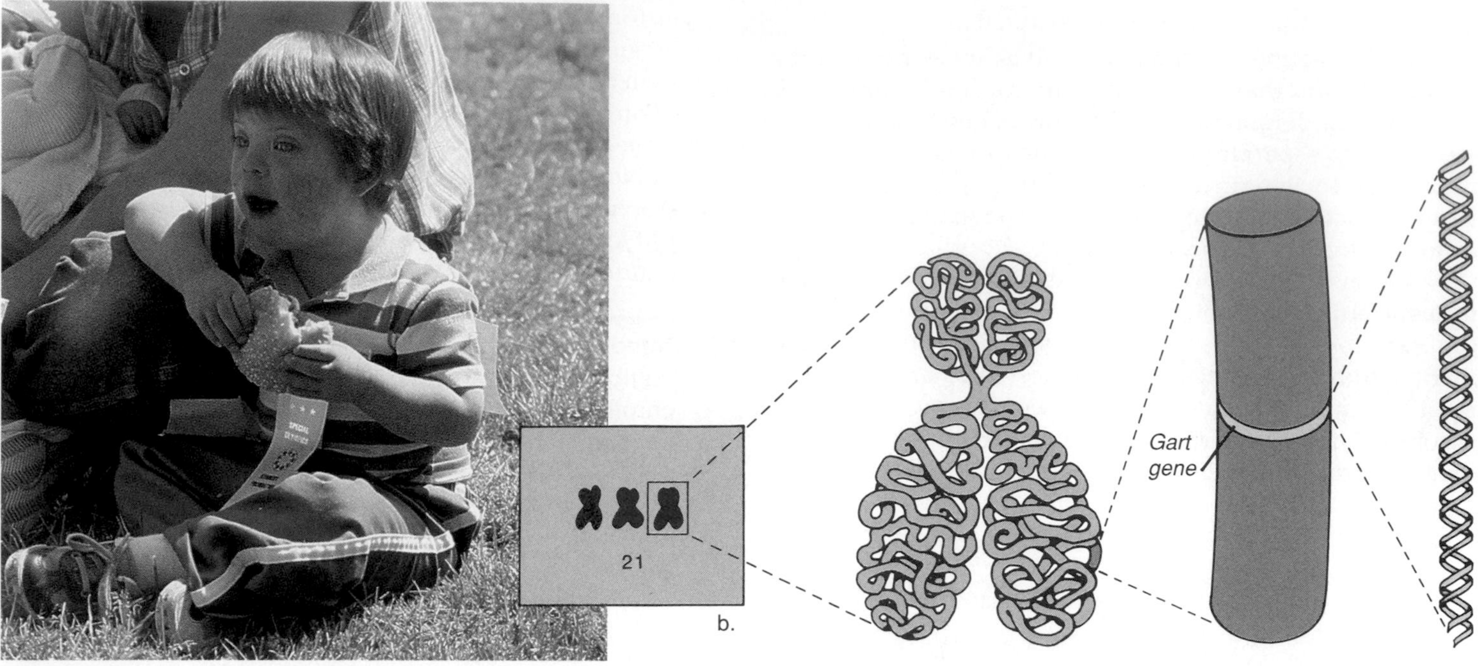

Figure 13.3 Down syndrome.
a. Characteristics of Down syndrome include a wide, rounded face and narrow, slanting eyelids. Mental retardation, along with an enlarged tongue, makes it difficult for a person with Down syndrome to speak distinctly. b. Karyotype of an individual with Down syndrome shows an extra chromosome 21. More sophisticated technologies allow investigators to pinpoint the location of specific genes associated with the syndrome. An extra copy of the *Gart* gene, which leads to a high level of purines, may account for the mental retardation of persons with Down syndrome.

Down Syndrome

Persons with Down syndrome, described in Figure 13.3, usually have three copies of chromosome 21 in their karyotype because the egg had two copies instead of one. The chances of a woman having a Down syndrome child increase rapidly with age, starting at about age 40, and the reasons for this are still being determined. In 23% of the cases of *Down syndrome* studied, the sperm, rather than the egg, had the extra chromosome 21. In about 5% of cases in either the sperm or the egg, the extra chromosome is attached to another chromosome, often chromosome 14. This abnormal chromosome arose because a translocation occurred between chromosomes 14 and 21 in one of the parents or even in a relative who lived generations earlier. Therefore, when Down syndrome is due to a translocation, it is not age related and instead tends to run in the family of either the father or the mother.

Although an older woman is more likely to have a Down syndrome child, most babies with Down syndrome are born to women younger than age 40 because this is the age group having the most babies. As discussed in the reading on page 206, chorionic villi testing and amniocentesis followed by karyotyping can detect a Down syndrome child. However, young women are not encouraged to undergo such procedures because the risk of complications resulting from these tests is greater than the risk of having a Down syndrome child. Fortunately, there is now a test based on substances in the blood that can identify mothers who might be carrying a Down syndrome child, and only these individuals need to undergo further testing.

It is known that the genes that cause Down syndrome are located on the bottom third of chromosome 21 (Fig. 13.3*b*), and extensive investigative work has focused on discovering the specific genes responsible for the characteristics of the syndrome. Thus far, investigators have discovered several genes that may account for various conditions seen in persons with Down syndrome. For example, they have located genes most likely responsible for the increased tendency toward leukemia, cataracts, accelerated rate of aging, and mental retardation. The gene for mental retardation, dubbed the *Gart* gene, causes an increased level of purines in blood, a finding associated with mental retardation. It is hoped that someday it will be possible to control the expression of the *Gart* gene even before birth so that at least this symptom of Down syndrome can be suppressed.

X and Y Numbers Also Change

Sex chromosome abnormalities are often due to nondisjunction of these chromosomes during meiosis. Therefore, the gametes contain an abnormal number of sex chromosomes.

XYY males with Jacob syndrome have two Y chromosomes instead of one. Affected males usually are taller than average, suffer from persistent acne, and tend to have barely normal intelligence. At one time, it was suggested that these men were likely to be criminally aggressive, but the overwhelming majority of these men do not exhibit any behavioral problems.

From birth, an XO individual with *Turner syndrome* has only one sex chromosome, an X; the O signifies the absence of a second sex chromosome. Turner females are short, have a broad chest, and webbed neck. The ovaries, oviducts, and uterus are very small and nonfunctional. Turner females do not undergo puberty or menstruate, and there is a lack of breast development (Fig. 13.4*a*). They are usually of normal intelligence and can lead fairly normal lives, but they are also infertile even if they receive hormone supplements.

A male with *Klinefelter syndrome* has two or more X chromosomes in addition to a Y chromosome and is sterile. The testes and prostate gland are underdeveloped, and there is no facial hair. There may be some breast development (Fig. 13.4*b*). Affected individuals have large hands and feet and very long arms and legs. They are usually slow to learn but not mentally retarded unless they inherit more than two X chromosomes.

A *triplo-X* individual has more than two X chromosomes. It might be supposed that the XXX female is especially feminine, but this is not the case. Although in some cases there is a tendency toward learning disabilities, most triplo-X females have no apparent physical abnormalities except there may be menstrual irregularities, including early onset of menopause.

One sex chromosome monosomy (Turner syndrome) and three trisomies (Klinefelter syndrome, triplo-X, and Jacob syndrome) are known in human beings.

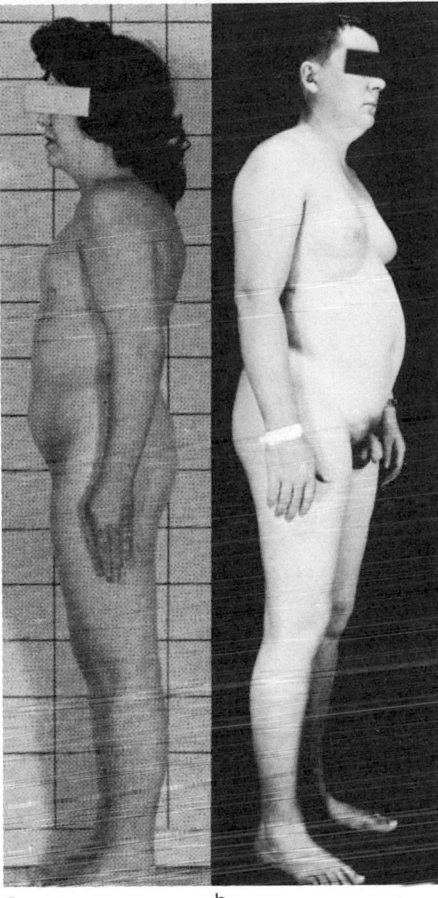

Figure 13.4 Abnormal sex chromosome inheritance. **a.** A female with Turner syndrome (XO) is distinguished by a thick neck, short stature, and immature sexual features. **b.** A male with Klinefelter syndrome (XXY) has immature sex organs and shows breast development.

Fragile X Syndrome

Males outnumber females by about 25% in institutions for the mentally retarded. In some of these males, the X chromosome is nearly broken, leaving the tip hanging by a flimsy thread. These males are said to have fragile X syndrome.

As children, fragile X syndrome individuals may be hyperactive or autistic; their speech is delayed in development and often repetitive in nature. As adults, they have large testes and big, usually protruding ears. They are short in stature, but the jaw is prominent and the face is long and narrow. Stubby hands, lax joints, and a heart defect may also occur. Fragile X chromosomes also occurs in females, but when symptoms do appear in females, they tend to be less severe.

Fragile X syndrome is inherited in an unusual pattern; it passes from symptomless male carrier to a severely affected grandson, as discussed in the reading on page 210. The DNA sequence at the fragile site was isolated and found to have trinucleotide repeats. The base triplet CGG was repeated over and over again. There are about six to 50 copies of this repeat in normal persons but over 230 copies in persons with fragile X syndrome.

Individuals with fragile X syndrome have a large number of triplet repeats at the fragile site.

Health Focus

Detecting Genetic Disorders

Genetic counseling can be more extensive today because of the availability of so many prenatal tests to detect chromosome and gene mutations. Karyotyping of fetal or embryonic cells can indicate whether a developing child has Down syndrome or one of the other chromosomal abnormalities listed in Figure 13.2. Also, tests can be performed to detect the presence of a mutant allele for cystic fibrosis, neurofibromatosis, and sickle-cell disease, among others. For example, neural tube defect occurs early when the nervous system is first forming, and the presence of a chemical called α-fetoprotein indicates that the fetus has this condition. Biochemical tests are available for dozens of enzymes such as hexosaminidase A, the missing enzyme in persons with Tay-Sachs disease.

Amniocentesis is one way to obtain fetal cells for testing of chromosomal abnormalities and genetic disorders (Fig. 13A *a*). The procedure is considered more than 99% safe when done by an experienced physician. However, the procedure does carry a possible risk to the fetus, as well as to the mother. In amniocentesis, a long needle is passed through the abdominal wall, the uterus, and the membranes surrounding the fetus; then a small amount of amniotic fluid, along with fetal cells, is removed. These fetal cells have been sloughed off from the skin or the respiratory or urinary systems. Since there are only a few cells in the fluid and they are nondividing (dividing cells are needed for karyotyping), it may be necessary to culture the cells. If so, it might be four weeks until the cells have grown and multiplied in cell culture in adequate numbers for testing purposes. This is a distinct disadvantage to those who wish to have the option of aborting the pregnancy if an abnormality is found, because it might then be too late for a safe abortion.

There is a new technique, however, that can be used to detect many chromosomal abnormalities just one or two days after cells are obtained by amniocentesis. Fluorescent probes specific for certain regions of chromosomes can be applied to amniotic cells, and they will bind to and light up these regions. For example, a probe for the X chromosome can indicate how many X chromosomes are present in each cell.

Chorionic villi sampling is a way to obtain embryonic cells for testing as early as the fifth week of pregnancy (Fig. 13A *b*). Chorionic villi sampling has been done routinely for ten years, and thus far, the safety of the procedure seems to be comparable to that of amniocentesis. The doctor inserts a long, thin tube through the vagina and into the uterus. With the help of ultrasound imaging, which gives a picture of the uterine contents, the tube is placed between the lining of the uterus and the chorion, a membrane that surrounds the embryo and has projections called chorionic villi. Suction is used to remove a sampling of the chorionic villi cells. Or alternately, a needle is inserted through the abdominal and uterine walls as is done with amniocentesis, except that the needle does not penetrate the fetal membranes. The chorionic villi contain cells with the same chromosomal and genetic makeup as the fetus, and therefore these cells can be tested to obtain information about possible fetal abnormalities. The advantage of chorionic villi sampling is that testing can be done much earlier and an abortion, if desired, can be done during the first trimester when it is considered medically safer for the mother.

Screening eggs for genetic defects is a new technique. Preovulation eggs are removed by aspiration after a telescope with a fiber–optic illuminator, called a laparoscope, is inserted into the abdominal cavity through a small incision in the region of the navel. The prior administration of the appropriate sex hormones ensures that several eggs are available for screening. Only the chromosomes within the first polar body are tested because if the woman is heterozygous for a genetic defect, and the faulty gene is found in the polar body, then the egg must be normal. Normal eggs undergo in vitro (in an artificial environment) fertilization and are placed in the prepared uterus. At present, only one in ten attempts results in a birth, but it is known ahead of time that the child will be normal.

Recently, each year has seen the discovery of more and more genetic mutations associated with disorders ranging from Alzheimer disease to psychological disorders to colon cancer. Expectant couples sometimes ask for genetic tests even when there is little known about the medical risk involved with the potential disorder—and even when the potential disorder is apt to occur in middle age or even later in life. Consequently, many are beginning to question the degree to which prenatal genetic testing should be done, especially when the presence of a mutation may only indicate a *possible* risk of developing the condition. The general public thinks in terms of either-or: either the newborn will develop the disorder or it will not. It is difficult to come to a conclusion when the results indicate a 50% risk of developing a disorder. As a society we have not yet decided how many and what types of genetic tests should be done and how to interpret some of the results once they are done. Therefore, genetic counselors and physicians, too, don't always know how to present the results to patients or if the results should be made known at all. For example, suppose one prenatal twin is XY and the other is XYY—do you think the parents should be informed? Many XYY individuals live a normal life, and revealing the chromosomal abnormality might cause the parents to abort the pregnancy or to treat the XYY child differently. If you were one of the parents, would you want to be told?

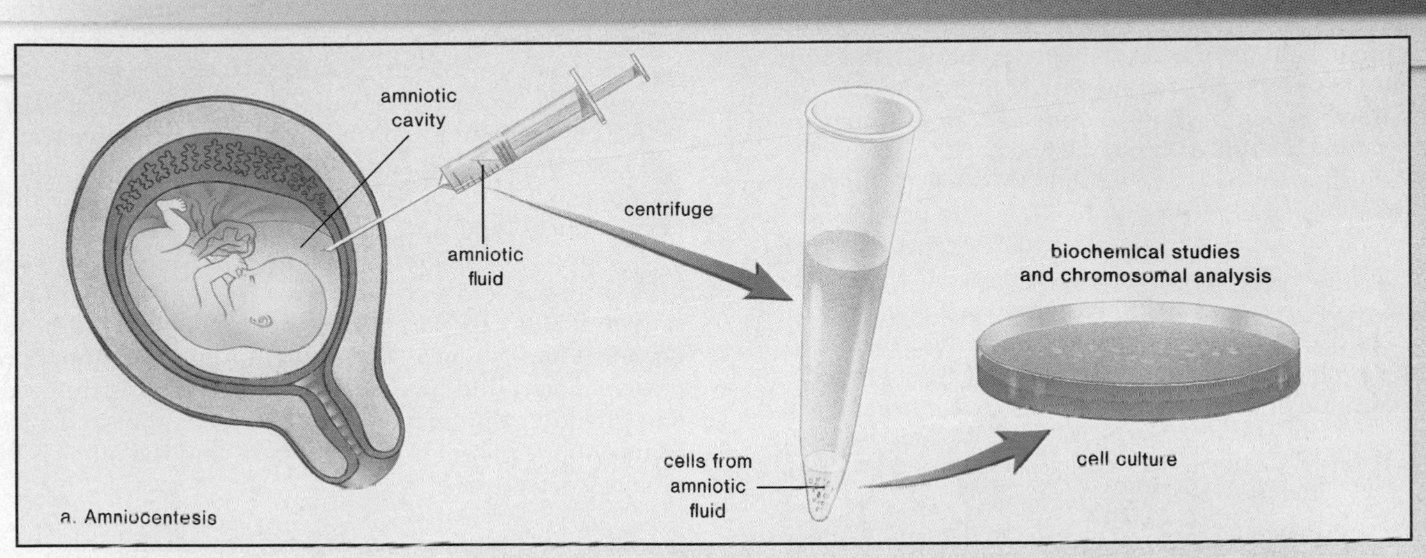

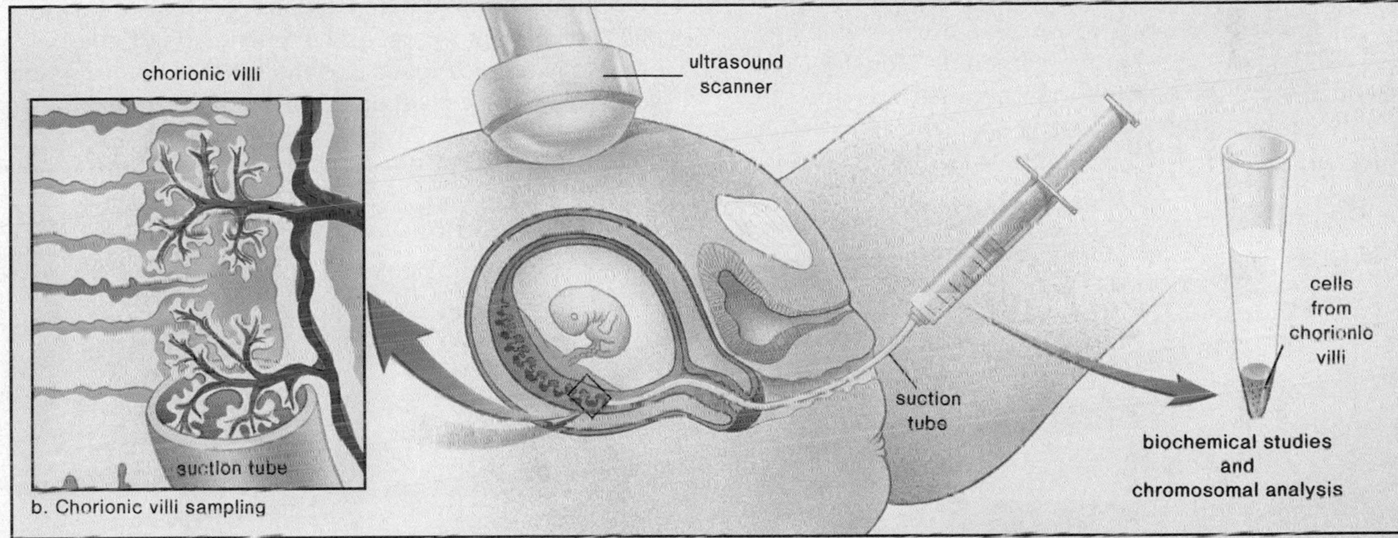

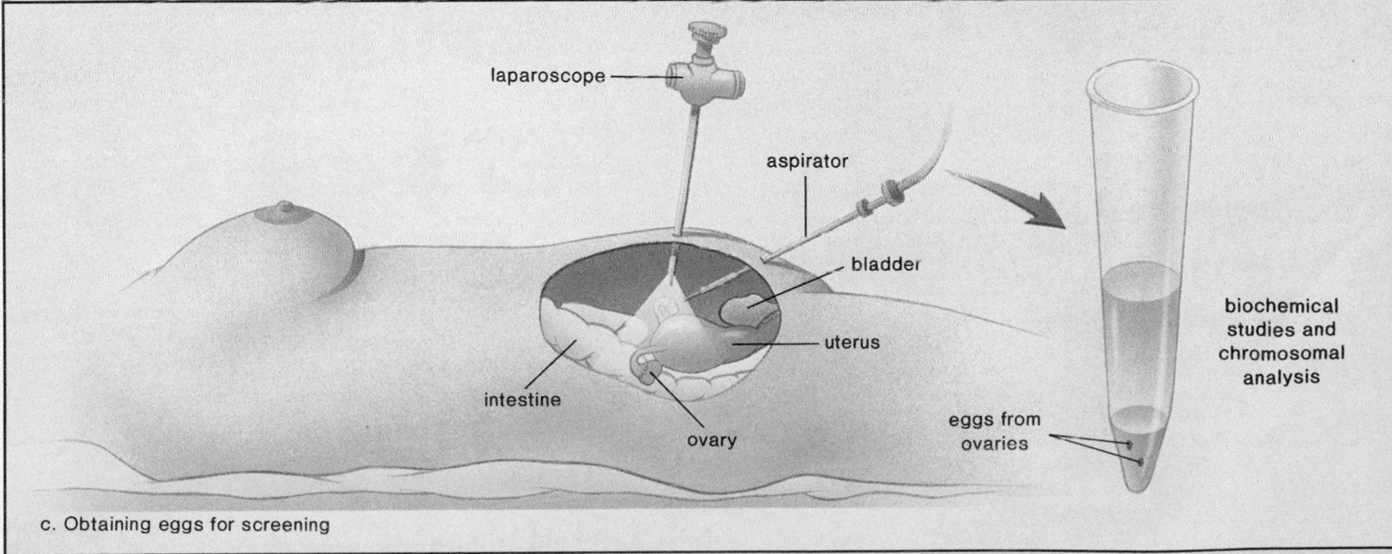

Figure 13A Genetic defect testing before birth.
a. Amniocentesis is used to obtain fetal cells for testing. **b.** Chorionic villi sampling is used earlier in the pregnancy to obtain embryonic cells. **c.** A laparoscope is used to obtain eggs for genetic and biochemical testing prior to pregnancy.

13.2 Considering Autosomal Traits

Many times parents would like to know the chances of an individual child having a certain genotype and, therefore, a certain phenotype. In keeping with Mendel's laws, if one of the parents is homozygous dominant *(EE),* the chances of their having a child with unattached earlobes is 100%, because this parent has only a dominant allele *(E)* to pass on to the offspring. On the other hand, if both parents are homozygous recessive *(ee),* there is a 100% chance that each of their children will have attached earlobes. However, if both parents are heterozygous, then what are the chances that their child will have unattached or attached earlobes? To solve a problem of this type, it is customary *first* to indicate the genotype of the parents and their possible gametes.

Genotypes:	*Ee*	*Ee*
Gametes:	*E* and *e*	*E* and *e*

Second, a **Punnett square** can be used to determine the phenotypic ratio among the offspring when all possible sperm are given an equal chance to fertilize all possible eggs (Fig. 13.5). The possible sperm are lined up along one side of the square, and the possible eggs are lined up along the other side of the square (or vice versa). The ratio among the offspring in this case is 3:1 (three children with unattached earlobes to one with attached earlobes). This means that there is a $3/4$ chance (75%) for each child to have unattached earlobes and a $1/4$ chance (25%) for each child to have attached earlobes.

Another cross of particular interest is that between a heterozygous individual *(Ee)* and a pure recessive *(ee).* In this case, the Punnett square shows that the ratio among the offspring is 1:1, and the chance of the dominant or recessive phenotype is $1/2$, or 50% (Fig. 13.6).

Comparing the two crosses (Fig. 13.5 and Fig. 13.6), each child has a 75% chance of having the dominant phenotype if the two parents are heterozygous and a 50% chance if one parent is heterozygous and the other is recessive. Each child has a 25% chance of having the recessive phenotype if the parents are heterozygous and a 50% chance if one parent is heterozygous and the other is homozygous recessive.

If both parents are heterozygous, each child has a 25% chance of the recessive phenotype. If one parent is heterozygous and the other is homozygous recessive, each child has a 50% chance of exhibiting the recessive phenotype.

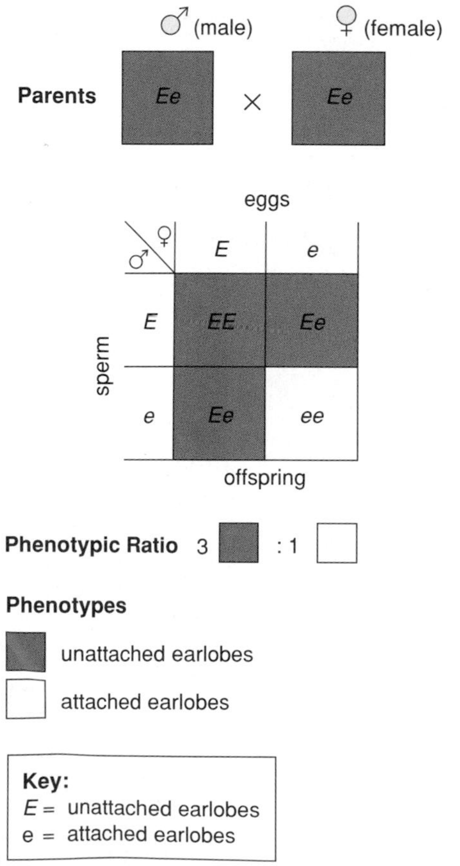

Figure 13.5 Heterozygous by heterozygous cross. When the parents are heterozygous, each child has a 75% chance of having the dominant phenotype and a 25% chance of having the recessive phenotype.

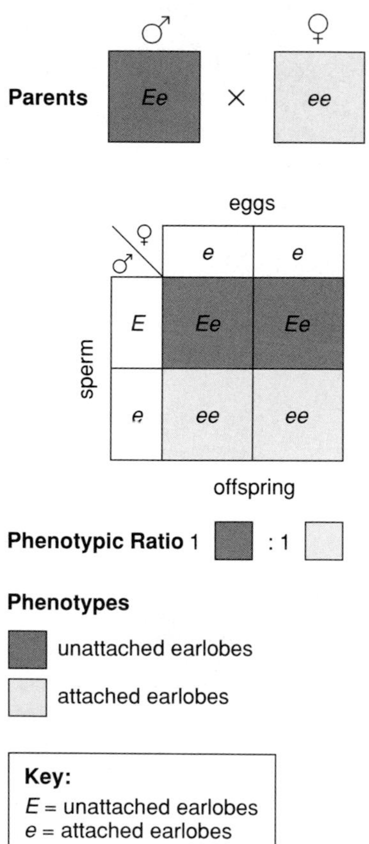

Figure 13.6 Heterozygous by homozygous recessive cross. When one parent is heterozygous and the other recessive, each child has a 50% chance of having the dominant phenotype and a 50% chance of having the recessive phenotype.

Some Disorders Are Dominant

Of the many autosomal dominant disorders, we will discuss only two.

Neurofibromatosis

Neurofibromatosis[1] is one of the most common genetic disorders. It affects roughly one in 3,000 people, including an estimated 100,000 in the United States. It is seen equally in every racial and ethnic group throughout the world.

At birth or later, the affected individual may have six or more large, tan spots (known as cafe-au-lait) on the skin. Such spots may increase in size and number and may get darker. Small benign tumors (lumps) called neurofibromas may occur under the skin or in various organs. Neurofibromas are made up of nerve cells and other cell types.

The expression of neurofibromatosis varies. In most cases, symptoms are mild and patients live a normal life. In some cases, however, the effects are severe. Skeletal deformities, including a large head, are seen, and eye and ear tumors can lead to blindness and hearing loss. Many children with neurofibromatosis have learning disabilities and are hyperactive.

In 1990, researchers isolated the gene for neurofibromatosis, which was known to be on chromosome 17. By analyzing the DNA (deoxyribonucleic acid), they determined that the gene was huge and actually included three smaller genes. This was only the second time that nested genes have been found in humans. The gene for neurofibromatosis is a tumor-suppressor gene active in controlling cell division. When it mutates, a benign tumor develops.

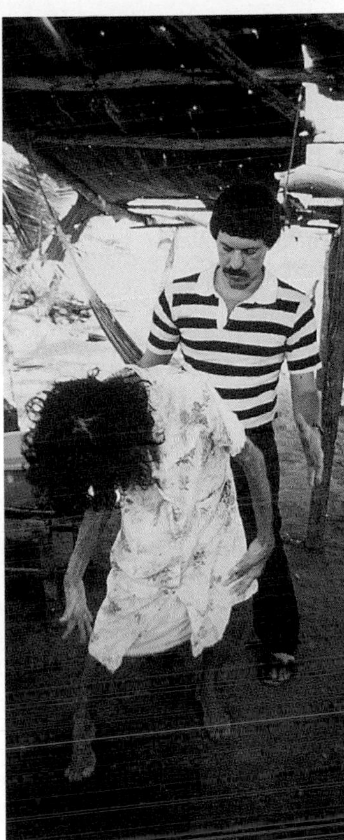

Figure 13.7 Huntington disease. Persons with this condition gradually lose psychomotor control of the body. At first there are only minor disturbances, but the symptoms become worse over time.

Huntington Disease

One in 20,000 persons in the United States has *Huntington disease*, a neurological disorder that leads to progressive degeneration of brain cells, which in turn causes severe muscle spasms and personality disorders (Fig. 13.7). Most people appear normal until they are of middle age and have already had children who might also be stricken. Occasionally, the first signs of the disease are seen in these children when they are teenagers or even younger. There is no effective treatment, and death comes ten to 15 years after the onset of symptoms.

Several years ago, researchers found that the gene for Huntington disease was located on chromosome 4. A test was developed for the presence of the gene, but few want to know if they have inherited the gene because as yet there is no treatment for Huntington disease. After the gene was isolated in 1993, an analysis revealed that it contains many repeats of the base triplet CAG (cytosine, adenine, and guanine). Normal persons have 11 to 34 copies of the triplet, but affected persons tend to have 42 to more than 120 copies. The more repeats present, the earlier the onset of Huntington disease and the more severe the symptoms. It also appears that persons most at risk have inherited the disorder from their fathers. The latter observation is consistent with a new hypothesis called genomic imprinting. The genes are imprinted differently during formation of sperm and egg, and therefore the sex of the parent passing on the disorder becomes important.

It is now known that there are a number of other genetic diseases whose severity and time of onset vary according to the number of triplet repeats present within the gene. Moreover, the genes for these disorders are also subject to genomic imprinting.

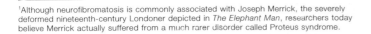

There are many autosomal recessive disorders in humans. Among these are neurofibromatosis and Huntington disease.

[1]Although neurofibromatosis is commonly associated with Joseph Merrick, the severely deformed nineteenth-century Londoner depicted in *The Elephant Man*, researchers today believe Merrick actually suffered from a much rarer disorder called Proteus syndrome.

A Closer Look

Fragile X Syndrome

Fragile X syndrome is one of the most common genetic causes of mental retardation, second only to Down syndrome. It affects about one in 1,500 males and one in 2,500 females and is seen in all ethnic groups. It is called fragile X syndrome because its diagnosis used to be dependent upon observing an X chromosome whose tip is attached to the rest of the chromosome by a thin thread.

The inheritance pattern of fragile X syndrome is not like any other pattern we have studied (Fig. 13B). The chance of being affected increases in successive generations almost as if the pattern of inheritance switches from being a recessive one to a dominant one. Then, too, an unaffected grandfather can have grandchildren with the disorder; in other words, he is a carrier for an X-linked disease. This is contrary to what occurs with other mutant alleles on the X chromosome; in those cases, the male always shows the disorder.

In 1991, the DNA sequence at the fragile site was isolated and found to have trinucleotide repeats. The base triplet, CGG, was repeated over and over again. There are about 6–50 copies of this repeat in normal persons but over 230 copies in persons with fragile X syndrome. Carrier males have what is now termed a premutation; they have between 50 and 230 copies of the repeat and no symptoms. Both daughters and sons receive the premutation but only the daughters pass on the full mutation—that is, over 230 copies of the repeat. Any male, even those with fragile X syndrome and over 230 repeats, passes on at most the premutation number of repeats. It is unknown what causes the difference between males and females.

This type of mutation—called by some a dynamic mutation, because it changes, and by others an expanded trinucleotide repeat, because the number of triplet copies increases—is now known to characterize other conditions, such as certain forms of dystrophy, spinocerebellar ataxia type 1, and Huntington disease. With Huntington disease, the age of onset of the disorder is roughly correlated with the number of repeats, and the disorder is more likely to have been inherited from the father. For autosomal conditions, we would expect the sex of the parent to play no role in inheritance. The present exceptions have led to the genomic imprinting hypothesis—that the sperm and egg carry chromosomes that have been "imprinted" differently. Imprinting is believed to occur during gamete formation, and thereafter, the genes are expressed one way if donated by the father and another way if donated by the mother. Perhaps when we discover why more repeats are passed on by one parent than the other, we will discover the cause of so-called genomic imprinting.

What might cause repeats to occur in the first place? Something must go wrong during DNA replication prior to cell division. The difficulty that causes triplet repeats is not known. But when DNA codes for cellular proteins, the presence of repeats undoubtedly leads to nonfunctioning or malfunctioning proteins.

Scientists have developed a new technique that can identify repeats in DNA, and they expect that this technique will help them find the genes for other human disorders. They expect expanded trinucleotide repeats to be a very common mutation indeed.

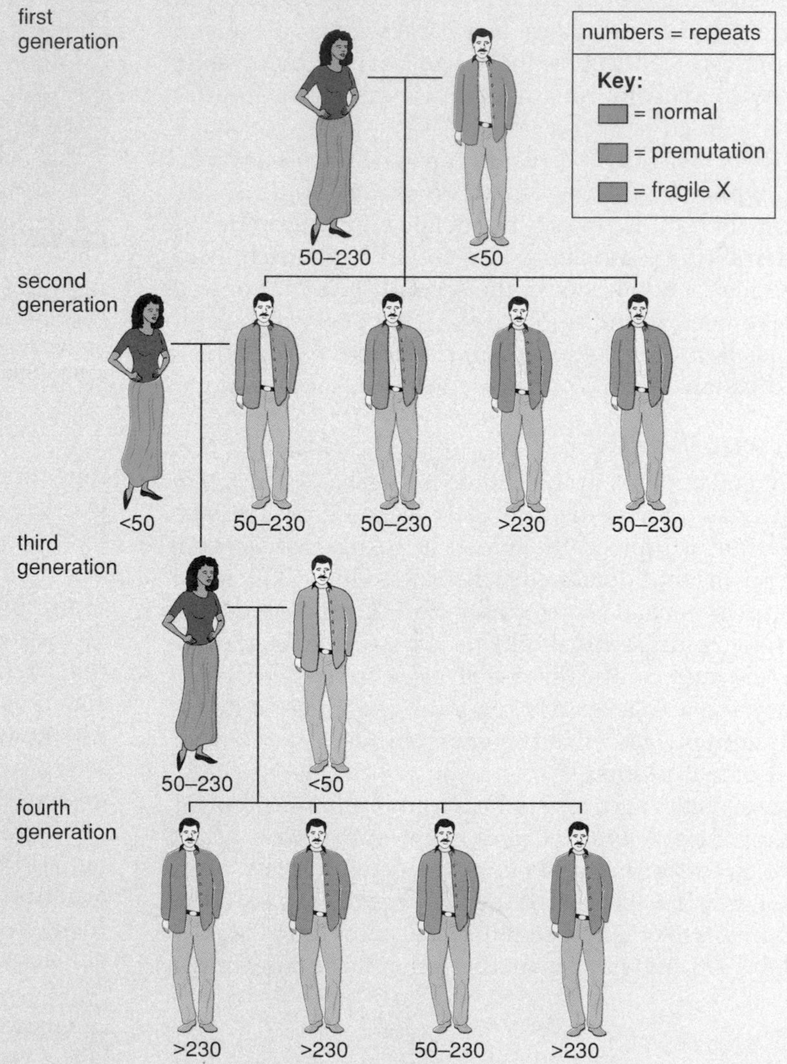

Figure 13B Fragile X syndrome.
Fragile X syndrome is a disorder caused by the presence of triplet repeats at a particular locus. Affected persons have over 230 repeats, a premutation is 50–230 repeats, and normal persons have fewer than 50 repeats. In each successive generation there are more affected individuals, usually males. The mutation is passed on by women with only a premutation number of repeats; men, even if affected, do not pass on the mutation.

Some Disorders Are Recessive

Of the many autosomal recessive disorders, we will discuss only three.

Tay-Sachs Disease

Tay-Sachs disease is a well-known genetic disease that usually occurs among Jewish people in the United States, most of whom are of central and eastern European descent. At first, it is not apparent that a baby has Tay-Sachs disease. However, development begins to slow down between four months and eight months of age, and neurological impairment and psychomotor difficulties then become apparent. The child gradually becomes blind and helpless, develops uncontrollable seizures, and eventually becomes paralyzed. There is no treatment or cure for Tay-Sachs disease, and most affected individuals die by the age of three or four.

Tay-Sachs disease results from a lack of the enzyme hexosaminidase A (Hex A) and the subsequent storage of its substrate, a fatty substance known as glycosphingolipid, in lysosomes. Although more and more lysosomes build up in many body cells, the primary sites of storage are the cells of the brain, which accounts for the onset, and the progressive deterioration of psychomotor functions.

Persons heterozygous for Tay-Sachs have about half the level of Hex A activity found in normal individuals. Prenatal diagnosis of the disease also is possible following either amniocentesis or chorionic villi sampling.

Cystic Fibrosis

Cystic fibrosis is the most common lethal genetic disease among Caucasians in the United States. About one in 20 Caucasians is a carrier, and about one in 2,500 births has the disorder. In these children, the mucus in the bronchial tubes and pancreatic ducts is particularly thick and viscous, interfering with the function of the lungs and pancreas. To ease breathing, the thick mucus in the lungs has to be manually loosened periodically (Fig. 13.8), but still the lungs become infected frequently. The clogged pancreatic ducts prevent digestive enzymes from reaching the small intestine, and to improve digestion patients take digestive enzymes mixed with applesauce before every meal.

In the past few years, much progress has been made in our understanding of cystic fibrosis, and new treatments have raised the average life expectancy to 28 years of age. Research has demonstrated that chloride ions (Cl^-) fail to pass through plasma membrane channel proteins in these patients. Ordinarily, after chloride ions have passed through the membrane, water follows. It is believed that lack of water is the cause of abnormally thick mucus in bronchial tubes and pancreatic ducts. The cystic fibrosis gene, which is located on chromosome 7, has been isolated, and attempts have been made to insert it into nasal epithelium, so far with little success. Genetic testing for the gene in adult carriers and in fetuses is possible; if present, couples have to decide whether to risk having a child with the condition or whether abortion is an option or not.

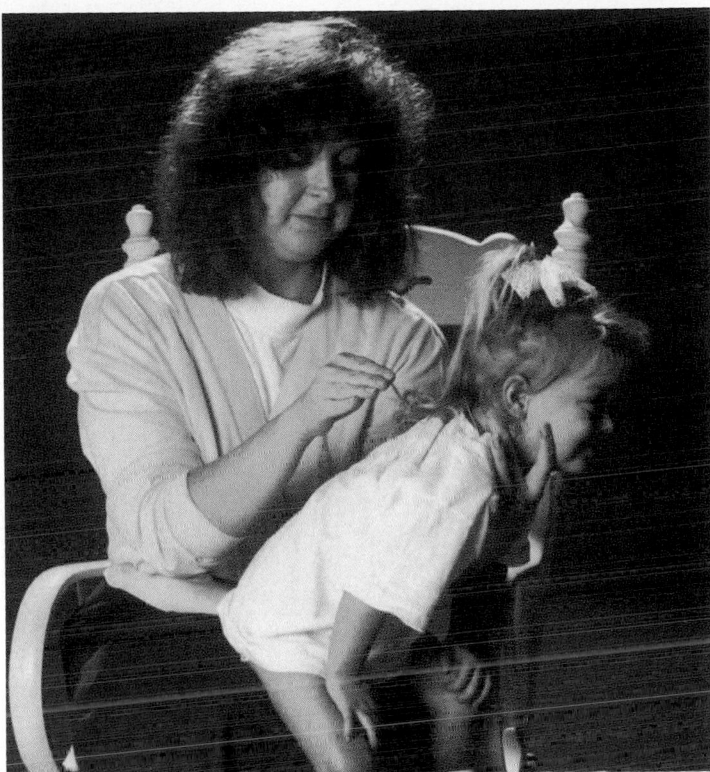

Figure 13.8 Cystic fibrosis.
The mucus in the lungs of a child with cystic fibrosis should be periodically loosened by clapping the back. A new treatment destroys the cells that tend to build up in the lungs. The white blood cells are destroyed by one drug, and another drug does away with the DNA the cells leave behind. Judicious use of antibiotics controls pulmonary infection, and aggressive use of other drugs thins mucous secretions.

Phenylketonuria (PKU)

Phenylketonuria (PKU) occurs once in 5,000 births, so it is not as frequent as the disorders previously discussed. However, it is the most commonly inherited metabolic disorder to affect nervous system development. Close relatives are more apt to have a PKU child.

Affected individuals lack an enzyme that is needed for the normal metabolism of the amino acid phenylalanine, and an abnormal breakdown product, a phenylketone, accumulates in the urine. The PKU gene is located on chromosome 12, and there is a prenatal DNA test for the presence of this allele. Years ago, the urine of newborns was tested at home for phenylketone in order to detect PKU. Presently, newborns are routinely tested in the hospital for elevated levels of phenylalanine in the blood. If necessary, newborns are placed on a diet low in phenylalanine, which must be continued until the brain is fully developed, around age 7, or else severe mental retardation develops.

There are many autosomal recessive disorders in humans. Among these are Tay-Sachs disease, cystic fibrosis, and phenylketonuria (PKU).

Pedigree Charts

When a genetic disorder is a simple autosomal dominant, an individual with the alleles *AA* or *Aa* will have the disorder. When a genetic disorder is a simple autosomal recessive, only individuals with the alleles *aa* will have the disorder. Genetic counselors often construct pedigree charts to determine whether a condition is dominant or recessive. A pedigree chart shows the pattern of inheritance for a particular condition. Consider these two possible patterns of inheritance:

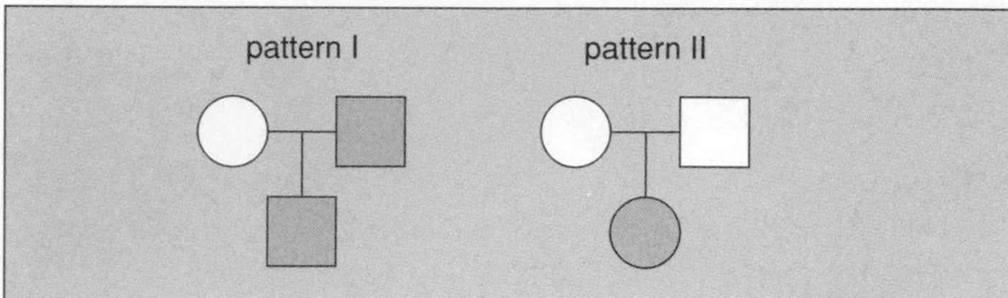

In both patterns, males are designated by squares and females by circles. Shaded circles and squares are affected individuals. A line between a square and a circle represents a union. A vertical line going downward leads, in these patterns, to a single child. (If there are more children, they are placed off a horizontal line.) Which pattern of inheritance do you suppose represents an autosomal dominant characteristic, and which represents an autosomal recessive characteristic?

In pattern I, the child is affected, as is one of the parents. When a disorder is dominant, an affected child usually has at least one affected parent. Of the two patterns, this one shows a dominant pattern of inheritance. Figure 13.9 shows a typical pedigree chart for a dominant disorder. Other ways to recognize an autosomal dominant pattern of inheritance are also given.

In pattern II, the child is affected but neither parent is; this can happen if the condition is recessive and the parents are *Aa*. Notice that the parents are carriers because they appear to be normal but are capable of having a child with a genetic disorder. Figure 13.10 shows a typical pedigree chart for a recessive genetic disorder. Other ways to recognize an autosomal recessive pattern of inheritance are also given in the figure.

It is important to realize that "chance has no memory," therefore, each child born to heterozygous parents has a 25% chance of having the disorder. In other words, it is possible that if a heterozygous couple has four children, each child might have the condition.

Dominant and recessive alleles have different patterns of inheritance.

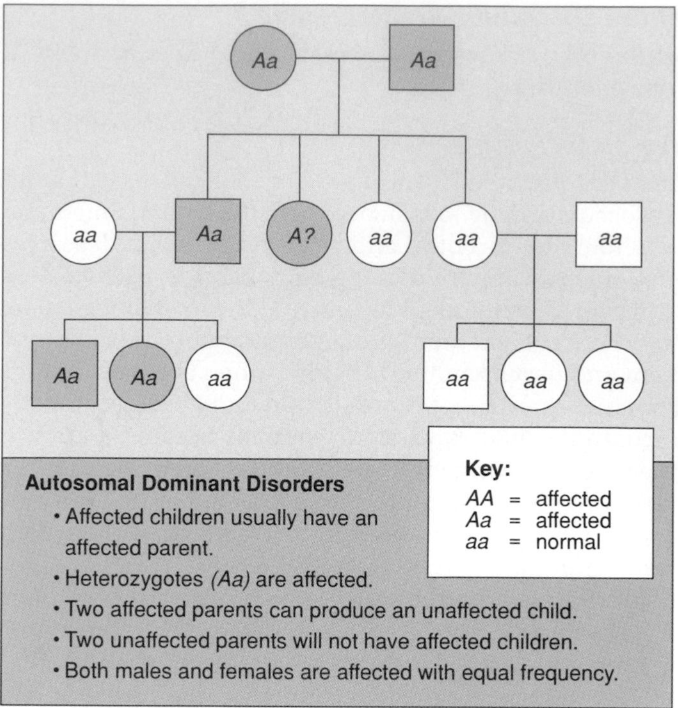

Aa
Aa
aa
Aa
A?
aa
aa
aa
Aa
Aa
aa
aa
aa
aa
Key:
AA = affected
Aa = affected
aa = normal
Autosomal Dominant Disorders
• Affected children usually have an affected parent.
• Heterozygotes (Aa) are affected.
• Two affected parents can produce an unaffected child.
• Two unaffected parents will not have affected children.
• Both males and females are affected with equal frequency.

Figure 13.9 Autosomal dominant pedigree chart.
The list gives ways to recognize an autosomal dominant disorder.

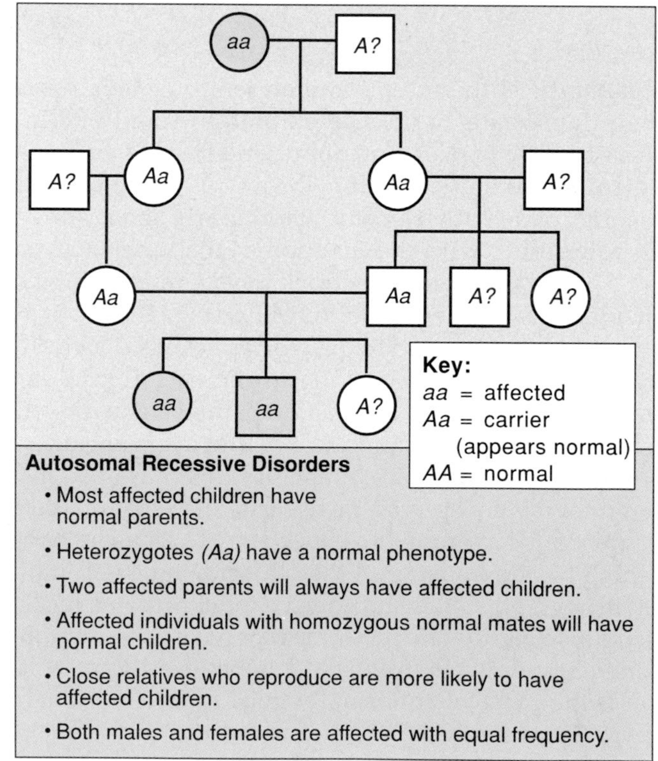

aa
A?
A?
Aa
Aa
A?
Aa
Aa
A?
A?
aa
aa
A?
Key:
aa = affected
Aa = carrier (appears normal)
AA = normal
Autosomal Recessive Disorders
• Most affected children have normal parents.
• Heterozygotes (Aa) have a normal phenotype.
• Two affected parents will always have affected children.
• Affected individuals with homozygous normal mates will have normal children.
• Close relatives who reproduce are more likely to have affected children.
• Both males and females are affected with equal frequency.

Figure 13.10 Autosomal recessive pedigree chart.
Only those affected with the recessive genetic disorder are shaded. The list gives ways to recognize an autosomal recessive disorder.

Other Inheritance Patterns

There are patterns of inheritance other than simple dominant and recessive traits.

Polygenic Traits

Polygenic inheritance occurs when one trait is governed by two or more loci. Each dominant allele has a quantitative effect on the phenotype, and these effects are additive. The result is a continuous variation of phenotypes, resulting in a distribution of these phenotypes that resembles a bell-shaped curve. The more genes involved, the more continuous the variation and the distribution of the phenotypes. Also, environmental effects cause many intervening phenotypes; in the case of height, differences in nutrition assure a bell-shaped curve.

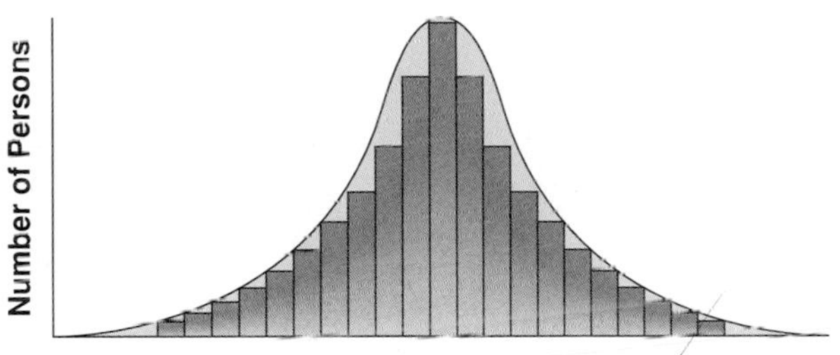

Just how many pairs of alleles control skin color is not known, but a range in colors can be explained on the basis of two pairs. When a very dark person reproduces with a very light person, the children have medium brown skin; and when two people with medium brown skin reproduce with one another, the children range in skin color from very dark to very light. This can be explained by assuming that skin color is controlled by two pairs of alleles and that each capital letter contributes to the color of the skin:

Phenotype	Genotypes
Very dark	*AABB*
Dark	*AABb* or *AaBB*
Medium brown	*AaBb* or *AAbb* or *aaBB*
Light	*Aabb* or *aaBb*
Very light	*aabb*

Notice again that there is a range in phenotypes and that there are several possible phenotypes in between the two extremes. Therefore, the distribution of these phenotypes is expected to follow a bell-shaped curve—few people have the extreme phenotypes, and most people have the phenotype that lies in the middle between the extremes.

Many human disorders, such as cleft lip and/or palate, clubfoot, congenital dislocations of the hip, hypertension, diabetes, schizophrenia, and even allergies and cancers, are most likely controlled by polygenes and subject to environmental influences. Therefore, many investigators are in the process of considering the nature versus nurture question; that is, what percentage of the distribution of the trait is controlled by genes and what percentage is controlled by the environment? Thus far, it has not been possible to come to precise, generally accepted percentages for any particular trait.

In recent years, reports have surfaced that all sorts of behavioral traits such as alcoholism, homosexuality, phobias, and even suicide can be associated with particular genes. No doubt behavioral traits are to a degree controlled by genes, but again, it is impossible at this time to determine to what degree.

Many human traits most likely controlled by polygenes are subject to environmental influences. The frequency of the phenotypes of such traits follows a bell-shaped curve.

Multiple Alleles

Some traits are controlled by **multiple alleles;** that is, more than two alleles. However, each person inherits only two of the total possible number of alleles. Three alleles for the same gene control the inheritance of ABO blood types: *A* = A antigen on red blood cells; *B* = B antigen on red blood cells; *O* = no antigens on red blood cells. Both *A* and *B* are dominant over *O*; therefore, there are two possible genotypes for type A blood and two possible genotypes for type B blood. If a person inherits one of each of these alleles, the blood type will be AB. Type O can only result from the inheritance of two *O* alleles:

Phenotype	Possible Genotype
A	*AA, AO*
B	*BB, BO*
AB	*AB*
O	*OO*

An examination of possible matings between different blood types sometimes produces surprising results. For example, if the cross is *AO* × *BO*, the possible genotypes of children are *AB, OO, AO,* and *BO*.

Blood typing can sometimes aid in paternity suits. A man with type A blood (having genotype *AO*) could possibly be the father of a child with type O blood. On the other hand, a man with type AB blood cannot possibly be the father of a child with type O blood. Therefore, blood tests are legally used only to exclude a man from possible paternity.

The Rh factor is inherited separately from A, B, AB, or O blood types. In each instance, it is possible to be Rh positive (Rh^+) or Rh negative (Rh^-). It can be assumed that the inheritance of this antigen on red blood cells is controlled by a single allelic pair in which simple dominance prevails: the Rh-positive allele is dominant over the Rh-negative allele.

Inheritance by multiple alleles occurs when a gene exists in more than two allelic forms. However, each individual usually inherits only two alleles for these genes.

Dominance Has Degrees

The field of human genetics also has examples of codominance and incomplete dominance. *Codominance* occurs when alleles are equally expressed in a heterozygote. We have already mentioned that the multiple alleles controlling blood type are codominant. An individual with the genotype *AB* has type AB blood.

Sickle-Cell Disease Sickle-cell disease is an example of a human disorder that is controlled by incompletely dominant alleles. Individuals with the Hb^AHb^A genotype are normal, those with the Hb^SHb^S genotype have sickle-cell disease, and those with the Hb^AHb^S genotype have the sickle-cell trait. Two individuals with sickle-cell trait can produce children with all three phenotypes, as indicated in Figure 13.11.

In persons with sickle-cell disease, the red blood cells aren't biconcave disks like normal red blood cells; they are irregular. In fact, many are sickle shaped. The defect is caused by an abnormal hemoglobin that accumulates inside the cells. Because the sickle-shaped cells can't pass along narrow capillary passageways like disk-shaped cells, they clog the vessels and break down. This is why persons with sickle-cell disease suffer from poor circulation, anemia, and poor resistance to infection. Internal hemorrhaging leads to further complications, such as jaundice, episodic pain of the abdomen and joints, and damage to internal organs.

Persons with sickle-cell trait do not usually have any sickle-shaped cells unless they experience dehydration or mild oxygen deprivation. Although a recent study found that army recruits with sickle-cell trait are more likely to die when subjected to extreme exercise, previous studies of athletes do not substantiate these findings. At present, most investigators believe that no restrictions on physical activity are needed for persons with the sickle-cell trait.

Among regions of malaria-infested Africa, infants with sickle-cell disease die, but infants with sickle-cell trait have a better chance of survival than the normal homozygote. When the parasite infects their red blood cells, the cells become sickle shaped and thereafter the cells lose potassium. This causes the parasite to die. The protection afforded by the sickle-cell trait keeps the allele for sickle-cell prevalent in populations exposed to malaria. As many as 60% of the population in malaria-infected regions of Africa have the allele. In the United States, about 10% of the African-American population carries the allele.

In a recent study, 22 children around the world with sickle-cell disease were given bone marrow transplants from healthy siblings and 16 were completely cured of the disease. Optimism over these results is dampened by the knowledge that only 6% of patients met the medical criteria for receiving treatment and the drugs used in the procedure result in infertility. Also, if unsuccessful, health could worsen instead of improve and there is a 10% risk of death from the treatment.

Innovative therapies are still being explored. For example, persons with sickle-cell disease produce normal fetal hemoglobin during development, and drugs that turn on the genes for fetal hemoglobin in adults are being developed. Mice have been genetically engineered to produce sickled red blood cells in order to test new antisickling drugs and various genetic therapies.

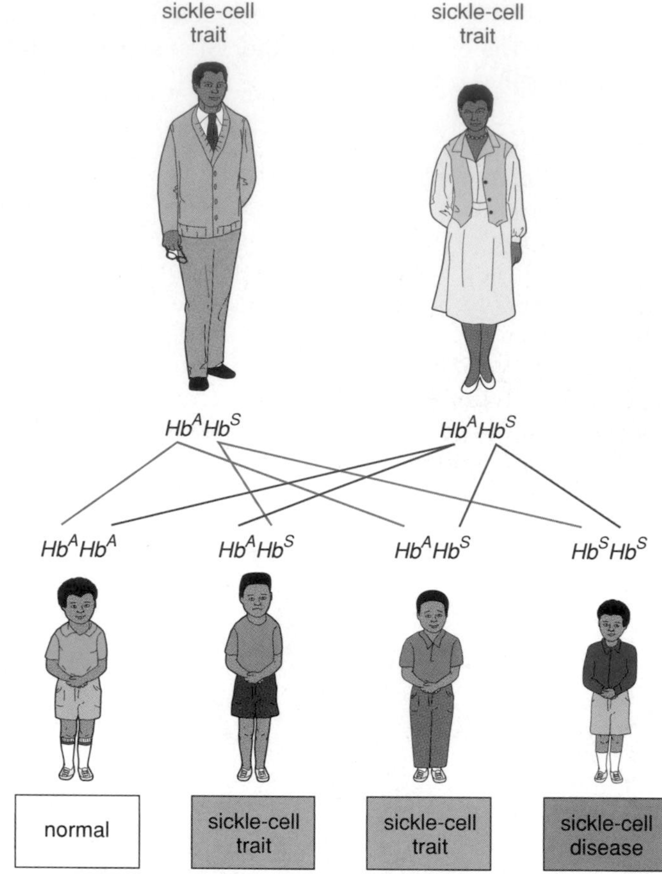

Figure 13.11 Inheritance of sickle-cell disease.
a. In this example, both parents have the sickle-cell trait. Therefore, each child has a 25% chance of having sickle-cell disease or of being perfectly normal and a 50% chance of having the sickle-cell trait. b. Sickled cells. Individuals with sickle-cell disease have sickled red blood cells that tend to clump, as illustrated here.

Sickle-cell disease is an inherited lifelong disorder that is being investigated on many fronts.

13.3 Considering Sex-Linked Traits

The sex chromosomes contain genes just as the autosomal chromosomes do. Some of these genes determine whether the individual is a male or a female. Investigators have found that the Y chromosome has an *SRY* gene (sex-determining region Y gene). When this gene is lacking from the Y chromosome, the individual is a female even though the chromosomal inheritance is XY.

Traits controlled by alleles on the *sex chromosomes* are said to be **sex-linked;** an allele that is only on the X chromosome is **X-linked,** and an allele that is only on the Y chromosome is Y-linked. Most sex-linked alleles are on the X chromosome, and the Y chromosome is blank for these. Very few alleles have been found on the Y chromosome.

The X chromosomes carry many genes unrelated to the sex of the individual, and we will look at a few of these in depth. It would be logical to suppose that a sex-linked trait is passed from father to son or from mother to daughter, but this is not the case. A male always receives a sex-linked condition from his mother, from whom he inherited an X chromosome. The Y chromosome from the father does not carry an allele for the trait. Usually the trait is recessive; therefore a female must receive two alleles, one from each parent, before she has the condition.

X-Linked Alleles

When examining X-linked traits, the allele on the X chromosome appears as a letter attached to the X chromosome (Fig. 13.12). For example, the key for color blindness is as follows:

X^B = normal vision

X^b = color blindness

The possible genotypes in both males and females are as follows:

X^BX^B = female who has normal color vision

X^BX^b = carrier female who has normal color vision

X^bX^b = female who is color blind

X^BY = male who has normal vision

X^bY = male who is color blind

Carriers are individuals that appear normal but can pass on an allele for a genetic disorder. Note that the second genotype is a carrier female because although a female with this genotype appears normal, she is capable of passing on an allele for color blindness. Color-blind females are rare because they must receive the allele from both parents; color-blind males are more common since they need only one recessive allele to be color blind. The allele for color blindness has to be inherited from their mother because it is on the X chromosome; males only inherit the Y chromosome from their father.

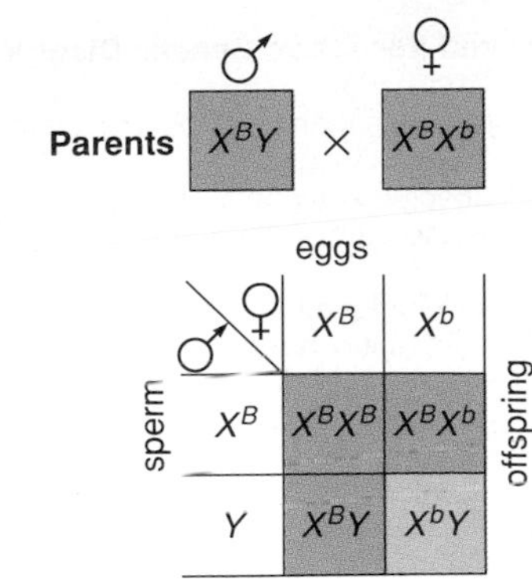

Results: females—all normal
males—1 normal: 1 color blind

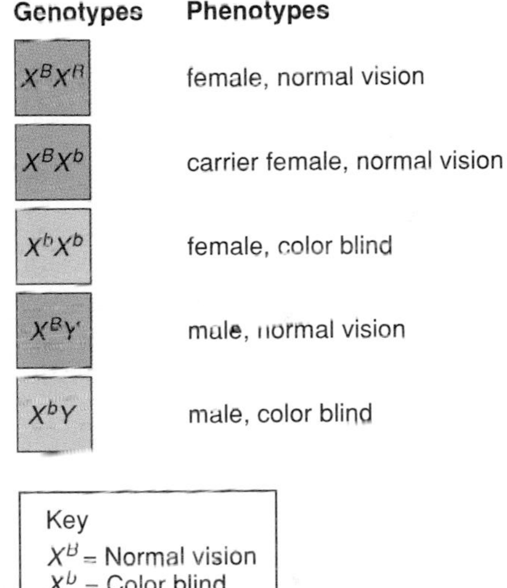

Genotypes
Phenotypes
X^BX^B
female, normal vision
X^BX^b
carrier female, normal vision
X^bX^b
female, color blind
X^BY
male, normal vision
X^bY
male, color blind
Key
X^B = Normal vision
X^b = Color blind

Figure 13.12 Cross involving an X-linked allele. The male parent is normal, but the female parent is a carrier; an allele for color blindness is located on one of her X chromosomes. Therefore, each son stands a 50% chance of being color blind.

Now, let us consider a particular cross. If a heterozygous woman reproduces with a man with normal vision, what are the chances of their having a color-blind daughter? a color-blind son?

Parents: $X^BX^b \times X^BY$

All daughters will have normal color vision because they all receive an X^B from their father. The sons, however, have a 50% chance of being color blind, depending on whether they receive an X^B or an X^b from their mother. The inheritance of a Y chromosome from their father cannot offset the inheritance of an X^b from their mother. Notice in Figure 13.12, the phenotypic results for X-linked problems are given separately for males and females.

The X chromosome carries alleles that are not on the Y chromosome. Therefore, a recessive allele on the X chromosome is expressed in males.

X-linked Recessive Genetic Disorders

- More males than females are affected.
- An affected son can have parents who have the normal phenotype.
- For a female to have the characteristic, her father must also have it. Her mother must have it or be a carrier.
- The characteristic often skips a generation from the grandfather to the grandson.
- If a woman has the characteristic, all of her sons will have it.

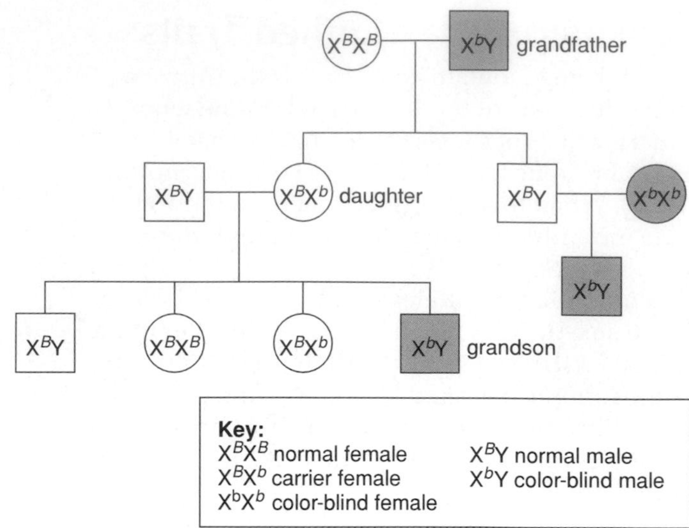

Figure 13.13 X-linked recessive pedigree chart.
The list gives ways of recognizing an X-linked recessive disorder.

Some Disorders Are X-Linked

Figure 13.13 gives a pedigree chart for an X-linked recessive condition. It also lists ways to recognize this pattern of inheritance. X-linked conditions can be dominant or recessive, but most known are recessive. More males than females have the trait because recessive alleles on the X chromosome are always expressed in males since the Y chromosome does not have a corresponding allele. If a male has an X-linked recessive condition, his daughters are often carriers; therefore, the condition passes from grandfather to grandson. Females who have the condition inherited the allele from both their mother and their father; and all the sons of such a female will have the condition. Three well-known X-linked recessive disorders are color blindness, muscular dystrophy, and hemophilia.

Color Blindness

In humans, there are three different classes of cone cells, the receptors for color vision in the retina of the eyes. Only one pigment protein is present in each type of cone cell; there are blue-sensitive, red-sensitive, and green-sensitive cone cells. The gene for the blue-sensitive protein is autosomal, but the genes for the red- and green-sensitive proteins are on the X chromosome. About 8% of Caucasian men have red/green color blindness. Most of these see brighter greens as tans, olive greens as browns, and reds as reddish-browns. A few cannot tell reds from greens at all. They see only yellows, blues, blacks, whites, and grays. Opticians have special charts by which they detect those who are color blind.

Duchenne Muscular Dystrophy

Muscular dystrophy, as the name implies, is characterized by a wasting away of the muscles. The most common form, *Duchenne muscular dystrophy,* is X-linked and occurs in about one out of every 3,600 male births. Symptoms, such as waddling gait, toe walking, frequent falls, and difficulty in rising, may appear as soon as the child starts to walk. Muscle weakness intensifies until the individual is confined to a wheelchair. Death usually occurs by age 20; therefore, affected males are rarely fathers. The recessive allele remains in the population by passage from carrier mother to carrier daughter.

Recently, the gene for muscular dystrophy was isolated, and it was discovered that the absence of a protein now called dystrophin is the cause of the disorder. Much investigative work determined that dystrophin is involved in the release of calcium from the calcium-storage sacs in muscle fibers. The lack of dystrophin causes calcium to leak into the cell, which promotes the action of an enzyme that dissolves muscle fibers. When the body attempts to repair the tissue, fibrous tissue forms, and this cuts off the blood supply so that more and more cells die.

A test is now available to detect carriers for Duchenne muscular dystrophy. Also, various treatments are being attempted. Immature muscle cells can be injected into muscles, and for every 100,000 cells injected, dystrophin production occurs in 30–40% of the patient's muscle fibers. The gene for dystrophin has been inserted into the thigh muscle cells of mice, and about 1% of these cells then produced dystrophin.

Hemophilia

About one in 10,000 males is a hemophiliac. There are two common types of hemophilia: hemophilia A is due to the absence or minimal presence of a clotting factor known as factor IX, and hemophilia B is due to the absence of clotting factor VIII.

Hemophilia is called the bleeder's disease because the affected person's blood does not clot. Although hemophiliacs bleed externally after an injury, they also suffer from internal bleeding, particularly around joints. Hemorrhages can be checked with transfusions of fresh blood (or plasma) or concentrates of the clotting protein. Unfortunately, some hemophiliacs have contracted AIDS after receiving blood or using a blood concentrate, but donors are now screened more closely, and donated blood is now tested for HIV. Also, factor VIII is now available as a genetic engineering product.

At the turn of the century, hemophilia was prevalent among the royal families of Europe, and all of the affected males could trace their ancestry to Queen Victoria of England. Of Queen Victoria's 26 grandchildren, five grandsons had hemophilia and four granddaughters were carriers. Because none of Queen Victoria's forebears or relatives was affected, it seems that the faulty allele she carried arose by mutation either in Victoria or in one of her parents. Her carrier daughters, Alice and Beatrice, introduced the gene into the ruling houses of Europe. Alexis, the last heir to the Russian throne before the Russian Revolution, was a hemophiliac. There are no hemophiliacs in the present British royal family because Victoria's eldest son, King Edward VII, did not receive the gene and therefore could not pass it on to any of his descendants.

Certain traits that have nothing to do with the gender of the individual are controlled by genes on the X chromosomes. Males have only one X chromosome, and therefore X-linked recessive alleles are expressed.

Some Traits Are Sex-Influenced

Some genes not located on the X or Y chromosomes are expressed differently in the two sexes, and therefore the traits they control are referred to as **sex-influenced traits.** Pattern baldness is caused by an autosomal allele that is dominant in males due to the presence of testosterone, the male sex hormone. Heterozygous women with adrenal tumors develop pattern baldness, but hair returns when the tumor is removed. The adrenal glands normally produce some testosterone, even in women. Therefore the hair loss is due to an abnormally large amount of testosterone being produced by the malfunctioning adrenal gland.

connecting concepts

These days, more couples are seeking genetic counseling because they are becoming aware that many illnesses are caused by faulty genes. The counselor examines the background of the couple and tries to determine if any deceased or living family members may have had, or has, a genetic disorder. A pedigree chart may be constructed. Then, the counselor studies the couple itself, and as much as possible, laboratory tests are performed on all persons involved.

Tests are now available for a large number of potential genetic diseases such as Huntington disease and cystic fibrosis. Blood tests can identify carriers of sickle-cell disease. By measuring enzyme levels in blood, skin, or tears, carriers of enzymatic defects can also be identified for certain inborn metabolic errors, such as the fatal Tay-Sachs disease. From this information, the counselor can sometimes predict the chances of a child having a genetic disorder.

Due to our expanding knowledge, potential parents can be better informed when deciding to have children. In addition, some people want to know if they have a faulty gene so they are aware of their particular health risks. For example, it is now sometimes possible to predict which female family members are more likely to get breast cancer. If so, these individuals can be sure to get regular mammograms so that the condition is caught early.

This growing volume of genetic information also raises some tough ethical questions for us all. Who, for example, has the right to see the results of an individual's genetic tests? Can insurance companies deny coverage to individuals who have a gene for a serious illness? Can employers require genetic testing before granting employment? These are hard questions and will require wise decisions on the part of the entire human society.

Summary

13.1 Considering the Chromosomes

It is possible to treat and photograph the chromosomes of a cell so that they can be sorted and arranged in pairs. The resulting karyotype can be used to diagnose chromosomal abnormalities.

Down syndrome (trisomy 21) is the most common autosomal abnormality. The occurrence of this syndrome, which is often related to the mother's age, can be detected by amniocentesis. Most often, Down syndrome is due to nondisjunction during gamete formation, but in a small percentage of cases, there has been a translocation between chromosomes 14 and 21, in which chromosome 21 is attached to chromosome 14.

Turner syndrome (XO) is a monosomy for the X chromosome. There are several trisomies: triplo-X, which is XXX; Klinefelter syndrome, which is XXY; and Jacob syndrome. Fragile X syndrome, which has an unusual pattern of inheritance, is due to the presence of an abnormal X chromosome.

13.2 Considering Autosomal Traits

Certain genetic disorders are inherited in a simple Mendelian manner. Neurofibromatosis is an autosomal dominant disorder that is inherited in this manner. Tay-Sachs disease, cystic fibrosis, and PKU are autosomal recessive disorders that have been studied in detail. (We now know that Huntington disease, long classified as an autosomal dominant disease, is due to expanded trinucleotide repeats, and the greater the number of repeats, the earlier the onset of the disease. Also, the disease is more likely to have been inherited from the father for reasons that are still being explored.)

When studying human genes, biologists often construct pedigree charts to show the pattern of inheritance of a characteristic within a family. The particular pattern indicates the manner in which a characteristic is inherited.

Polygenic traits are controlled by more than one gene loci and the individual inherits an allelic pair for each loci. Skin color illustrates polygenic inheritance. Traits controlled by polygenes are subject to environmental effects and show continuous variations whose frequency distribution forms a bell-shaped curve.

ABO blood type is an example of a human trait controlled by multiple alleles. Sickle-cell disease is a human disorder that is controlled by incompletely dominant alleles.

13.3 Considering Sex-Linked Traits

X-linked genes are on the X chromosome. Since the Y chromosome is blank for X-linked alleles, males are more apt to exhibit the recessive phenotype, which is inherited from their mothers. Color blindness, hemophilia, and Duchenne muscular dystrophy are X-linked recessive disorders.

Some traits, like pattern baldness, are sex-influenced traits controlled by autosomal genes.

Reviewing the Chapter

1. What does the normal human karyotype look like? 202
2. Diagram how nondisjunction can occur during meiosis I and meiosis II. 203
3. What are the characteristics of Down syndrome? What is the most frequent cause of this condition? 203–4
4. What is the only known sex chromosome monosomy in humans? Name and describe three sex chromosome trisomies. 205
5. Describe the symptoms of neurofibromatosis and Huntington disease. How does neurofibromatosis illustrate variable expressivity? For most autosomal dominant disorders, what are the chances of a heterozygote and a normal individual having an affected child? 209
6. Describe the symptoms of Tay-Sachs disease, cystic fibrosis, and PKU. For any one of these, what are the chances of two carriers having an affected child? 211–12
7. How might you distinguish an autosomal dominant trait from an autosomal recessive trait when viewing a pedigree chart? 212
8. What observational data would allow you to hypothesize that a human trait is controlled by polygenes? 213
9. Why is ABO blood type an example of inheritance by multiple alleles? 213
10. Explain how sickle-cell disease is inherited. What are the symptoms of sickle-cell trait and sickle-cell disease? If two persons with the sickle-cell trait reproduce, what are the chances of having a child with the sickle-cell trait? with sickle-cell disease? What race of people is more likely to have this condition? Why? 214
11. Explain how color blindness, hemophilia, and Duchenne muscular dystrophy are inherited. What are the symptoms of these conditions? What are the chances of having an affected male offspring if the mother is a carrier and the father is normal? 216–17
12. What is a sex-influenced trait? Give two examples of such traits. 217

Testing Yourself

Choose the best answer for each question.
For questions 1–3, match the conditions in the key with the descriptions below:

Key:

a. Down syndrome
b. Turner syndrome
c. Klinefelter syndrome
d. XYY

1. male with underdeveloped testes and some breast development
2. trisomy 21
3. XO female
4. Down syndrome
   a. is always caused by nondisjunction of chromosome 21.
   b. shows no overt abnormalities.
   c. is more often seen in children of mothers past the age of thirty-five.
   d. Both a and c are correct.
5. A person has a genetic disorder. Which of these is inconsistent with autosomal recessive inheritance?
   a. Both parents have the disorder.
   b. Both parents do not have the disorder.
   c. All the children (males and females) have the disorder.
   d. All of these are consistent.
   e. All of these are inconsistent.
6. A male has a genetic disorder. Which one of these is inconsistent with X-linked recessive inheritance?
   a. Both parents do not have the disorder.
   b. Only males in a pedigree chart have the disorder.
   c. Only females in previous generations have the disorder.
   d. Both a and c are inconsistent.

For questions 7–10, match the conditions in the key with the following descriptions:

Key:

a. cystic fibrosis
b. Huntington disease
c. hemophilia
d. Tay-Sachs disease

7. autosomal dominant
8. most often seen among Jewish people
9. X-linked recessive
10. thick mucus in lungs and pancreatic ducts
11. Determine if the characteristic possessed by the darkened squares (males) and circles (females) below is an autosomal dominant, an autosomal recessive, or an X-linked recessive disorder.

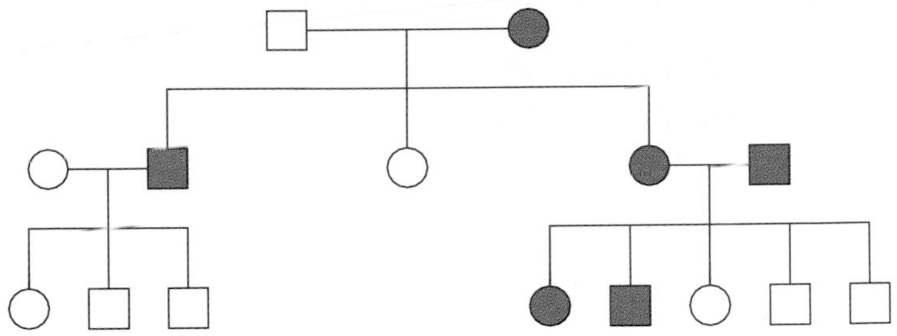

Additional Genetics Problems*

1. A hemophilic (X-linked recessive) man reproduces with a homozygous normal woman. What are the chances that their sons will be hemophiliacs? that their daughters will be hemophiliacs? that their daughters will be carriers?
2. A son with cystic fibrosis (autosomal recessive) is born to a couple who appear to be normal. What are the chances that any child born to this couple will have cystic fibrosis?
3. A man has type AB blood. What is his genotype? Could this man be the father of a child with type B blood? If so, what blood types could the child's mother have?
4. What is the genotype of a man who is color blind (X-linked recessive) and has a continuous hairline (autosomal recessive)? If this man has children by a woman who is homozygous dominant for normal color vision and widow's peak, what will be the genotype and phenotype of the children?
5. Determine if the characteristic possessed by the darkened squares (males) is dominant, recessive, or X-linked recessive. Write in the genotypes for the starred individual.

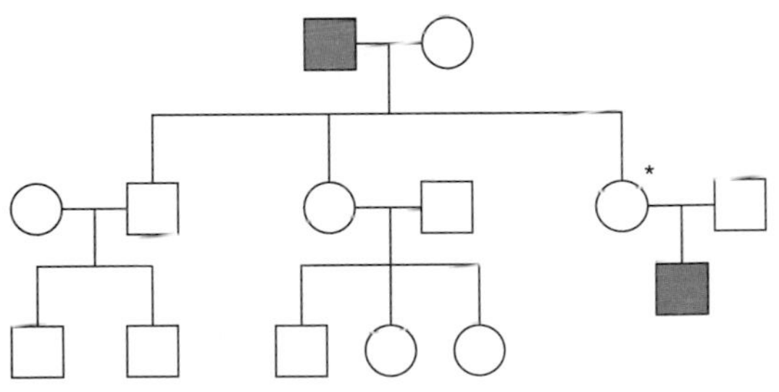

*Answers to Additional Genetics Problems appear in Appendix A.

Applying the Concepts

1. *The principles of genetics are the same for all organisms.*
   What are some similarities and differences between plant genetics and animal genetics? Considering that there are some differences, does this concept still hold?
2. *Inheritance plays a significant role in determining our mental and physical characteristics.*
   Without regard to ethical considerations, design an experiment to solve the nature-nurture question.
3. *Genetic disorders, like other conditions, are sometimes preventable and treatable.*
   What measures, if any, should be taken to prevent genetic diseases in the next generation?

Using Technology

Your study of human genetics is supported by these available technologies:

Exploring the Internet
The Mader Home Page provides resources for and help with studying this chapter.

http://www.mhhe.com/sciencemath/biology/mader/
(Click on Biology.)

Explorations in Human Biology CD-ROM
Cystic Fibrosis (#1)

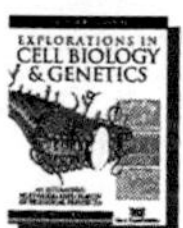

Explorations in Cell Biology & Genetics CD-ROM
Heredity in Families (#12)
Gene Segregation Within Families (#13)

Understanding the Terms

autosome 202	Punnett square 208
carrier 215	sex chromosome 202
karyotype 202	sex-influenced trait 217
multiple allele 213	sex-linked 215
nondisjunction 203	X-linked 215

Match the terms to these definitions:

a. __________ Any chromosome other than a sex chromosome.
b. __________ Autosomal phenotype controlled by an allele that is expressed differently in the two sexes; for example, the possibility of pattern baldness is increased by the presence of testosterone in males.
c. __________ Grid that enables one to calculate the results of simple genetic crosses.
d. __________ Chromosomes arranged by pairs according to their size, shape, and general appearance in mitotic metaphase.
e. __________ Failure of homologous chromosomes or daughter chromosomes to separate during meiosis I and meiosis II, respectively.
f. __________ Heterozygous individual who has no apparent abnormality but can pass on an allele for a recessively inherited genetic disorder.
g. __________ One form of a gene that has more than several alleles, although each individual has only two of these alleles.

DNA: The Genetic Material

CHAPTER 14

Chapter Concepts

The structure of DNA (brown and tan) resembles a spiral staircase.

One of the most exciting periods of scientific activity in history occurred during the thirty short years between the 1930s and 1960s. During this time, researchers in the United States and Europe painstakingly began peeling away the layers of confusion surrounding one of the most elemental questions ever asked by science—*what exactly is the genetic material?* The answer to this fundamental question did not come easily.

Geneticists knew that chromosomes contain protein and DNA (deoxyribonucleic acid). Of these two organic molecules, proteins are seemingly more complicated; they consist of countless sequences of 20 amino acids, which can coil and fold into complex shapes. DNA, on the other hand, contains only four different nucleotides. Surely, the diversity of life-forms on earth must be the result of the endless varieties of proteins.

Due to several elegantly executed experiments, by the mid-1950s researchers realized that DNA, not protein, is the genetic material. But this finding only led to another fundamental question—*what exactly is the structure of DNA?* The biological community at the time knew that whoever determined the structure of DNA would go down in history. Consequently, researchers were racing against time and each other. The story of the discovery of DNA resembles a mystery, with each clue adding to the total picture until the breathtaking design of DNA—a double helix—was finally unraveled.

14.1 Searching for the Genetic Material

Even though previous investigators were able to confirm that the genes are on the chromosomes and were even able to map the *Drosophila* chromosomes, they still didn't know just what genes consisted of. You can well imagine, then, that the search for the genetic material was of utmost importance to biologists at the beginning of the twentieth century. They knew that this material must be:

1. able to *store information* that pertains to both the development and the metabolic activities of the cell or organism;
2. stable so that it *can be replicated* with high fidelity during cell division and be transmitted from generation to generation;
3. able to *undergo rare changes* called **mutations** [L. *muta*, change] that provide the genetic variability required for evolution to occur.

The genetic material must be able to store information, be replicated, and undergo mutations.

Knowledge about the chemistry of DNA was absolutely essential in order to come to the conclusion that DNA is the genetic material. In 1869, the Swiss chemist Friedrich Miescher removed nuclei from pus cells (these cells have little cytoplasm) and found that they contained a chemical he called *nuclein.* Nuclein, he said, was rich in phosphorus and had no sulfur, properties that distinguished it from protein. Later, other chemists did further work with nuclein and said that it contained an acidic substance they called **nucleic acid.** Soon it was realized that there are two types of nucleic acids: **DNA (deoxyribonucleic acid)** and **RNA (ribonucleic acid).**

When it was discovered early in the twentieth century that nucleic acids contain four types of nucleotides, a misguided idea called the *tetranucleotide* (four-nucleotide) *hypothesis* arose. It said that DNA was composed of repeating units, and each unit always had just one of each of the four different nucleotides. In other words, DNA could not vary between genes and therefore could not be the genetic material! Everyone thought that the protein component of chromosomes must be the genetic material because proteins contain 20 different amino acids that can be sequenced in any particular way.

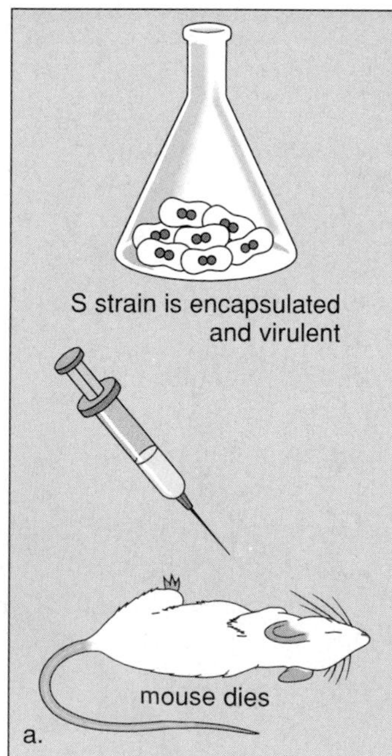
S strain is encapsulated and virulent
mouse dies
a.

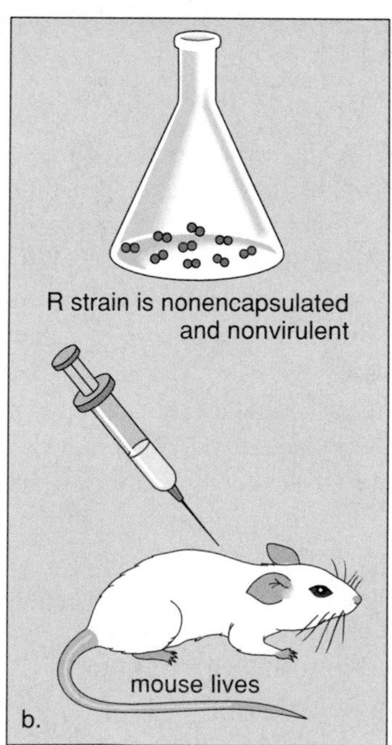
R strain is nonencapsulated and nonvirulent
mouse lives
b.

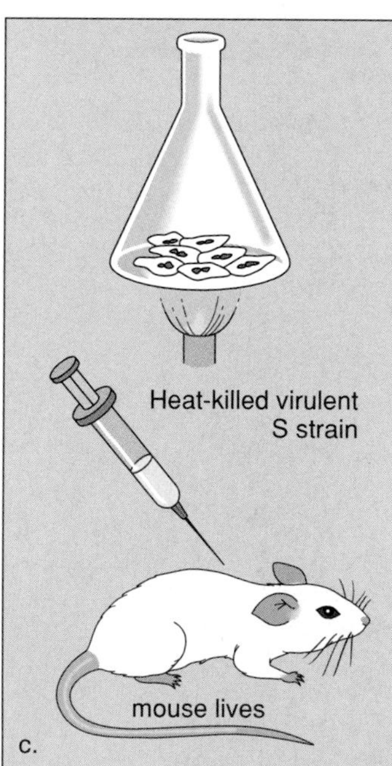
Heat-killed virulent S strain
mouse lives
c.

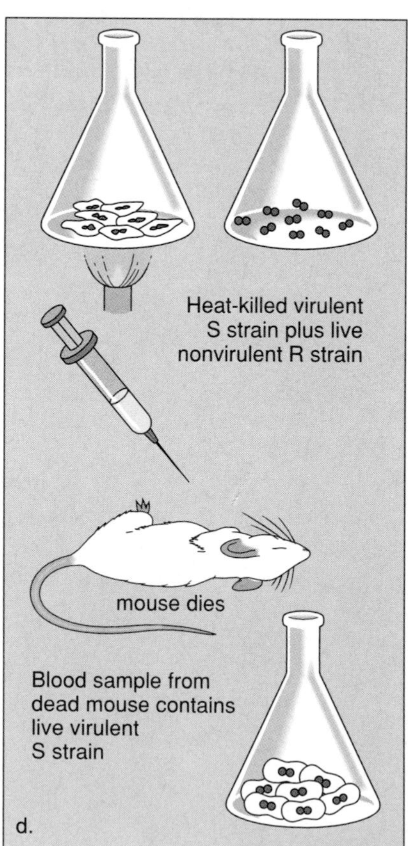
Heat-killed virulent S strain plus live nonvirulent R strain
mouse dies
Blood sample from dead mouse contains live virulent S strain
d.

Figure 14.1 Griffith's transformation experiment.
a. Encapsulated S strain is virulent and kills the mouse. **b.** Nonencapsulated R strain is not virulent and does not kill the mouse. **c.** Heat-killed S strain bacteria do not kill the mouse. **d.** If heat-killed S strain bacteria and R strain bacteria are both injected into a mouse, it dies because the R strain bacteria have been transformed into the virulent S strain.

Bacteria Can Be Transformed

In 1931, the bacteriologist Frederick Griffith performed an experiment with a bacterium (*Streptococcus pneumoniae*, or pneumococcus for short) that causes pneumonia in mammals. He noticed that when these bacteria are grown on culture plates, some, called S strain bacteria, produce shiny, smooth colonies and others, called R strain bacteria, produce colonies that have a rough appearance. Under the microscope, S strain bacteria have a capsule (mucous coat) but R strain bacteria do not. When Griffith injected mice with the S strain of bacteria, the mice died, and when he injected mice with the R strain, the mice did not die (Fig. 14.1). In an effort to determine if the capsule alone was responsible for the virulence (ability to kill) of the S strain bacteria, he injected mice with heat-killed S strain bacteria. The mice did not die.

Finally, Griffith injected the mice with a mixture of heat-killed S strain and live R strain bacteria. Most unexpectedly, the mice died and living S strain bacteria were recovered from the bodies! Griffith concluded that some substance necessary to the synthesis of a capsule and, therefore, virulence must have passed from the dead S strain bacteria to the living R strain bacteria so that the R strain bacteria were *transformed* (Fig. 14.1*d*). This change in the phenotype of the R strain bacteria must be due to a change in their genotype. Indeed, couldn't the transforming substance that passed from S strain to R strain be genetic material? Reasoning such as this prompted investigators at the time to begin looking for the transforming substance to determine the chemical nature of the genetic material.

Finding the Transforming Substance

Obviously, it is not convenient to look for the transforming substance in mice. It is not surprising, then, that the next group of investigators, led by Oswald Avery, isolated the genetic material in vitro (in laboratory glassware). After 16 years of research, this group published a paper demonstrating that the transforming substance is DNA. Their evidence included the following observations:

1. DNA from S strain bacteria causes R strain bacteria to be transformed.
2. Enzymes that degrade proteins cannot prevent transformation, nor did RNase, an enzyme that digests RNA.
3. Enzymatic digestion of the transforming substance with DNase, an enzyme that digests DNA, does prevent transformation.
4. The molecular weight of the transforming substance is so great that it must contain about 1,600 nucleotides! Certainly this is enough for some genetic variability.

These experiments showed not only that DNA is the genetic material, but also that DNA could transform the biosynthetic properties of a cell. While these experiments seem quite convincing, remember that at the time biologists knew proteins were very complex, but they still were not sure about DNA. Some thought that perhaps these results pertained only to this experiment.

Viruses Also Have Genetic Material

In the twentieth century, many geneticists who were interested in determining the chemical nature of the genetic material began to work with **bacteriophages** [Gk. *bacterion*, rod, and *phagein*, to eat], viruses that attack bacteria (Fig. 14.2). The most intensely studied species of bacterium is *Escherichia coli* (*E. coli*), which normally lives within the human gut. Bacteria are unicellular organisms with a very

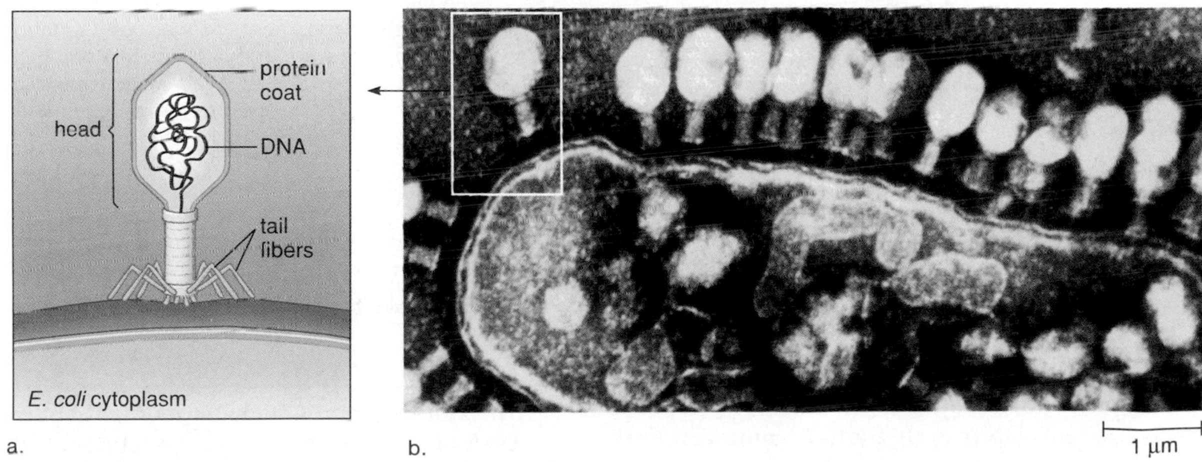

Figure 14.2 Bacteria and bacteriophages.
a. A T virus, which has the structure shown, was the experimental material of choice for determining the physical and chemical characteristics of the genetic material. b. T viruses are pictured attacking an *E. coli* cell.

short generation time. If a few thousand are placed in a test tube with a few salts and an energy source, such as the sugar glucose, there will be as many as 3×10^9 within a few hours. Bacteriophages (or phages) consist only of a protein coat surrounding a nucleic acid core. Enormous populations of a phage can be easily obtained—up to 10^{11} per ml or more in an infected bacterial culture is not unusual.

In 1952, two experimenters, Alfred D. Hershey and Martha Chase, chose a bacteriophage known as T2 for their experimental material. They decided to see which of the bacteriophage components—protein or DNA—entered bacterial cells and directed reproduction of the virus. Whichever did this, they reasoned, was the genetic material.

There is a chemical distinction between DNA and protein. Phosphorus is absent, but *sulfur (S) is present in some amino acids* of a protein. On the other hand, sulfur is absent but *phosphorus (P) is present in DNA* in high amounts. Therefore, it was possible for Hershey and Chase to prepare two batches of phages: one that had the DNA labeled with radioactive ^{32}P, and another that had the protein coat labeled with radioactive ^{35}S. Radioactive phages were allowed to attach briefly to *E. coli* bacterial cells; once the infection process had started, most of the adhering phage coats were sheared from the bacterial cells by agitation in a kitchen blender. Centrifugation then caused the bacterial cells to collect as a pellet at the bottom of the tube. In one experiment (Fig. 14.3*a*), they found that most of the ^{32}P-labeled DNA remained in the pellet:

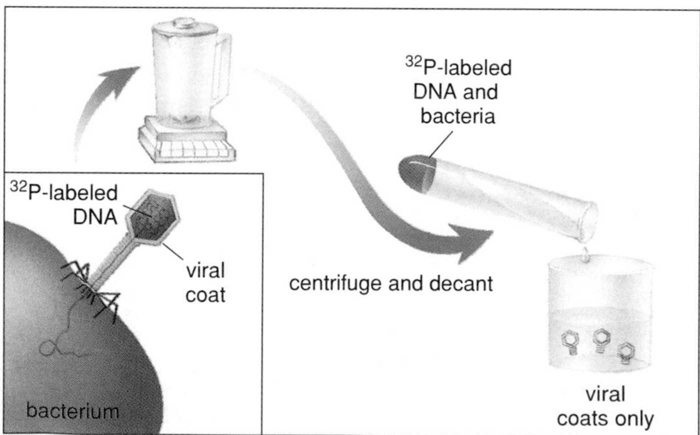

In the other experiment (Fig. 14.3*b*), they found that most of the ^{35}S-labeled protein remained in the phage coats. These results indicated that the DNA of the virus (and not the protein) enters the host, where viral reproduction takes place. Therefore, DNA is the genetic material that contains all necessary information for bacteriophage T2 reproduction.

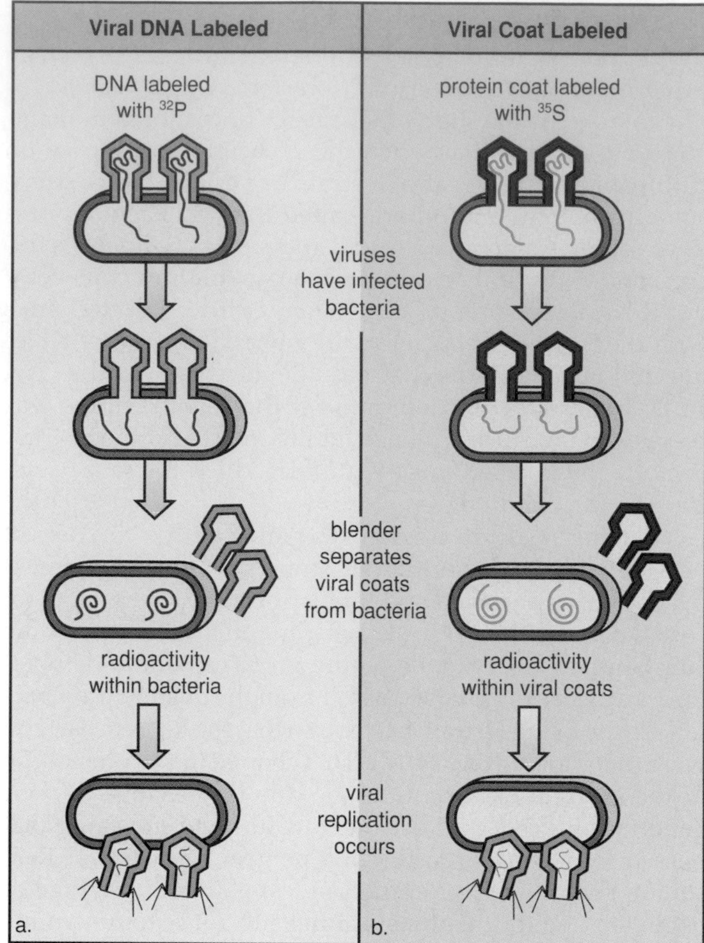

Figure 14.3 Hershey and Chase experiment.
A T virus contains DNA and has a protein coat. It was reasoned that whichever of these enters a bacterium and controls viral reproduction is the genetic material. **a.** In this experiment, ^{32}P was used to label viral DNA. The coats were removed by agitation in a blender, and the radioactively labeled DNA entered the cell. Because replication proceeded normally, DNA is the genetic material. **b.** In this experiment, ^{35}S was used to label the protein coat. When the cells were agitated in a blender, the radioactively labeled protein coats were removed. Because replication proceeded normally, protein is not the genetic material.

Biologists, in general, were persuaded that DNA is the genetic material by this experiment, possibly because recent chemical studies of the composition of DNA had finally shown that the tetranucleotide hypothesis was not true.

The Hershey and Chase experiment showed that DNA and not protein is the genetic material of the T2 bacteriophage.

14.2 Finding the Structure of DNA

During the same period of time that biologists were using viruses to show that DNA is the genetic material, biochemists were busy trying to determine its structure. The structure of DNA shows how DNA stores information, as is required of the genetic material.

Making Use of Nucleotide Data

With the development of new chemical techniques in the 1940s, it was possible for Erwin Chargaff to analyze in detail the base content of DNA. It was known that DNA contains four different types of nucleotides: two with **purine** bases, **adenine (A)** and **guanine (G)**, which have a double ring, and two with **pyrimidine** bases, **thymine (T)** and **cytosine (C)**, which have a single ring (Fig. 14.4*a* and *b*). Did each species contain 25% of each kind of nucleotide, as suggested by the tetranucleotide hypothesis?

A sample of Chargaff's data is seen in Figure 14.4*c*. You can see that while some species—*E. coli* and *Zea mays* (corn), for example—do have approximately 25% of each type of nucleotide, as suggested by the tetranucleotide hypothesis, most do not. Further, the percentage of each type of nucleotide differs from species to species. Therefore, DNA does have the *variability* between species required of the genetic material.

Within each species, however, DNA has the *constancy* required of the genetic material. Further, the percentage of A equals the percentage of T, and the percentage of G equals the percentage of C. The percentage of A + G equals 50%, and the percentage of T + C equals 50%. These relationships are called Chargaff's rules.

Chargaff's rules:

1. **The amount of A, T, G, and C in DNA varies from species to species.**
2. **In each species, the amount of A = T and the amount of G = C.**

The tetranucleotide hypothesis, which said that DNA has repeating units, each with one of the four bases, was not supported by these data. Each species had its own constant base composition.

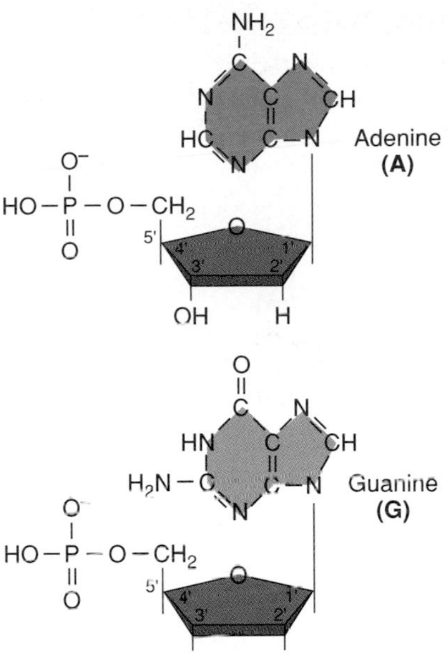

a. **Purine nucleotides**

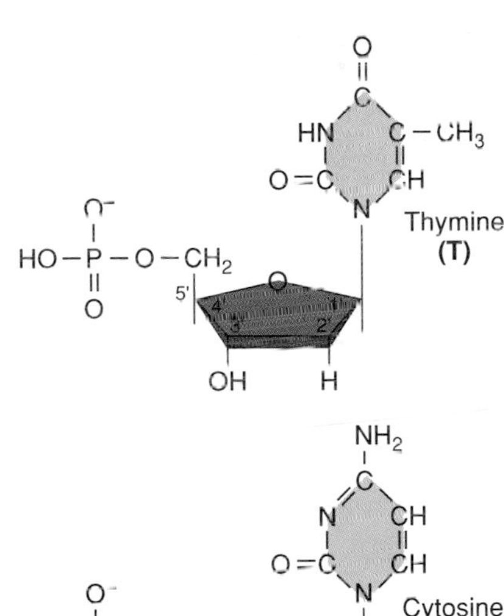

b. **Pyrimidine nucleotides**

Chargaff's DNA Database Composition in Various Species (%)				
Species	**A**	**T**	**G**	**C**
Homo sapiens	31.0	31.5	19.1	18.4
Drosophila melanogaster	27.3	27.6	22.5	22.5
Zea mays	25.6	25.3	24.5	24.6
Neurospora crassa	23.0	23.3	27.1	26.6
Escherichia coli	24.6	24.3	25.5	25.6
Bacillus subtilis	28.4	29.0	21.0	21.6

c.

Figure 14.4 Nucleotide composition of DNA. All nucleotides contain phosphate, a five-carbon sugar, and a nitrogen-containing base. In DNA, the sugar is deoxyribose—there is an absence of oxygen in the 2′ position—and the nitrogen-containing bases are (a) the purines adenine and guanine, which have a double ring, or (b) the pyrimidines thymine and cytosine, which have a single ring. c. Chargaff's data show that the DNA of various species differs. For example, in humans the A, T percentage is about 31%, but in fruit flies the percentage is about 27%.

Base Sequence Varies

Chargaff's data suggest that A is always paired with T and G is always paired with C. The paired bases occur in any order:

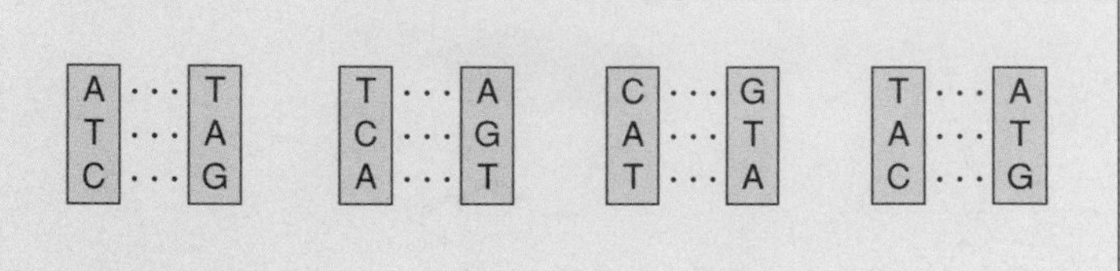

The variability that can be obtained is overwhelming. For example, it has been calculated that the human chromosome contains on the average about 140 million base pairs. Since any of the four possible nucleotides can be present at each nucleotide position, the total number of possible nucleotide sequences is $4^{140 \times 10^6}$ or $4^{140,000,000}$. No wonder each species has its own base percentages!

Making Use of Diffraction Data

Rosalind Franklin, a student of M. H. F. Wilkins at King's College in London, studied the structure of DNA using X rays. She found that if a concentrated, viscous solution of DNA is made, it can be separated into fibers. Under the right conditions, the fibers are enough like a crystal (a solid substance whose atoms are arranged in a definite manner) that an X-ray pattern can be formed on a photographic film. Franklin's picture of DNA showed that DNA is a helix (Fig. 14.5). The helical shape is indicated by the crossed (X) pattern in the center of the photograph. The dark portions at the top and bottom of the photograph indicate that some portion of the helix is repeated.

Watson and Crick Build a Model

James Watson, an American, was on a postdoctoral fellowship at Cavendish Laboratories in Cambridge, England, and while there he began to work with the biophysicist Francis H. C. Crick. Using the data we have just presented, they constructed a model of DNA for which they received a Nobel Prize in 1954 (Fig. 14.6).

Watson and Crick knew, of course, that DNA is a polymer of nucleotides, but they did not know how the nucleotides were arranged within the molecule. This is what they hypothesized.

1. The Watson and Crick model shows that DNA is a double helix with sugar-phosphate backbones on the outside and paired bases on the inside. This arrangement fits the mathematical measurements provided by the X-ray diffraction data for the spacing between the base pairs (0.34 nm) and for a complete turn of the double helix (3.4 nm).
2. Chargaff's rules said that A = T and G = C. The model shows that A is hydrogen-bonded to T and G is hydrogen-bonded to C. This so-called **complementary base pairing** means that a purine is always bonded to a pyrimidine. Only in this way will the molecule have the width (2 nm) dictated by its X-ray diffraction pattern, since two pyrimidines together are too narrow and two purines together are too wide.

The double-helix model of DNA is like a twisted ladder; sugar-phosphate backbones make up the sides, and hydrogen-bonded bases make up the rungs, or steps, of the ladder.

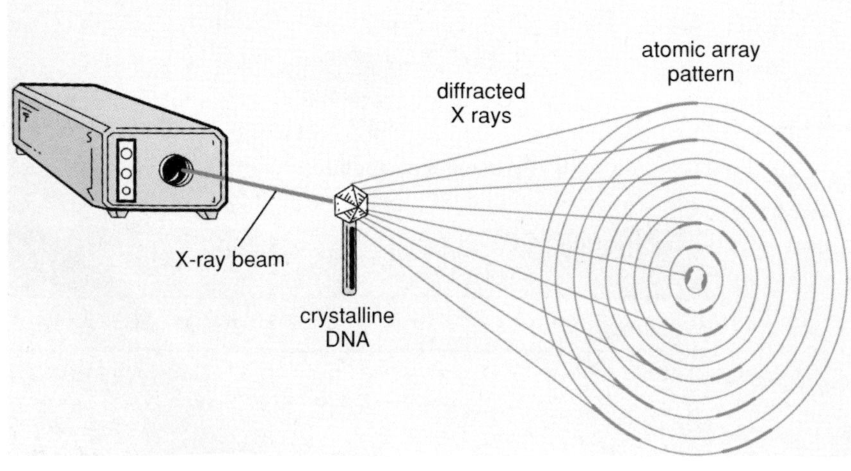

a.

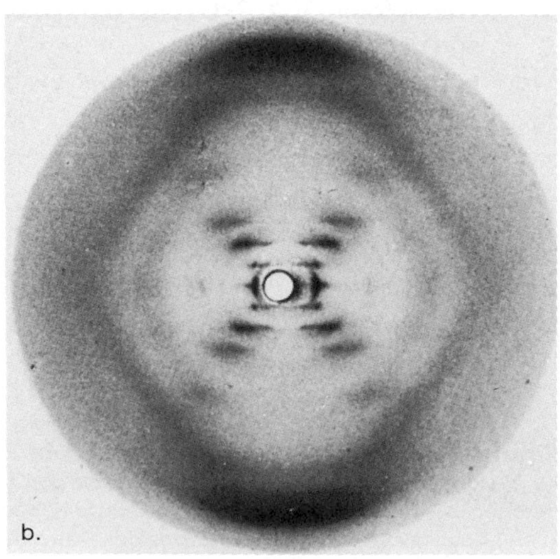

b.

Figure 14.5 X-ray diffraction of DNA.
a. When a crystal is X-rayed, the way in which the beam is diffracted reflects the pattern of the molecules in the crystal. The closer together two repeating structures are in the crystal, the farther from the center the beam is diffracted. **b.** The diffraction pattern of DNA produced by Rosalind Franklin. The crossed (X) pattern in the center told investigators that DNA is a helix, and the dark portions at the top and the bottom told them that some feature is repeated over and over. Watson and Crick determined that this feature was the hydrogen-bonded bases.

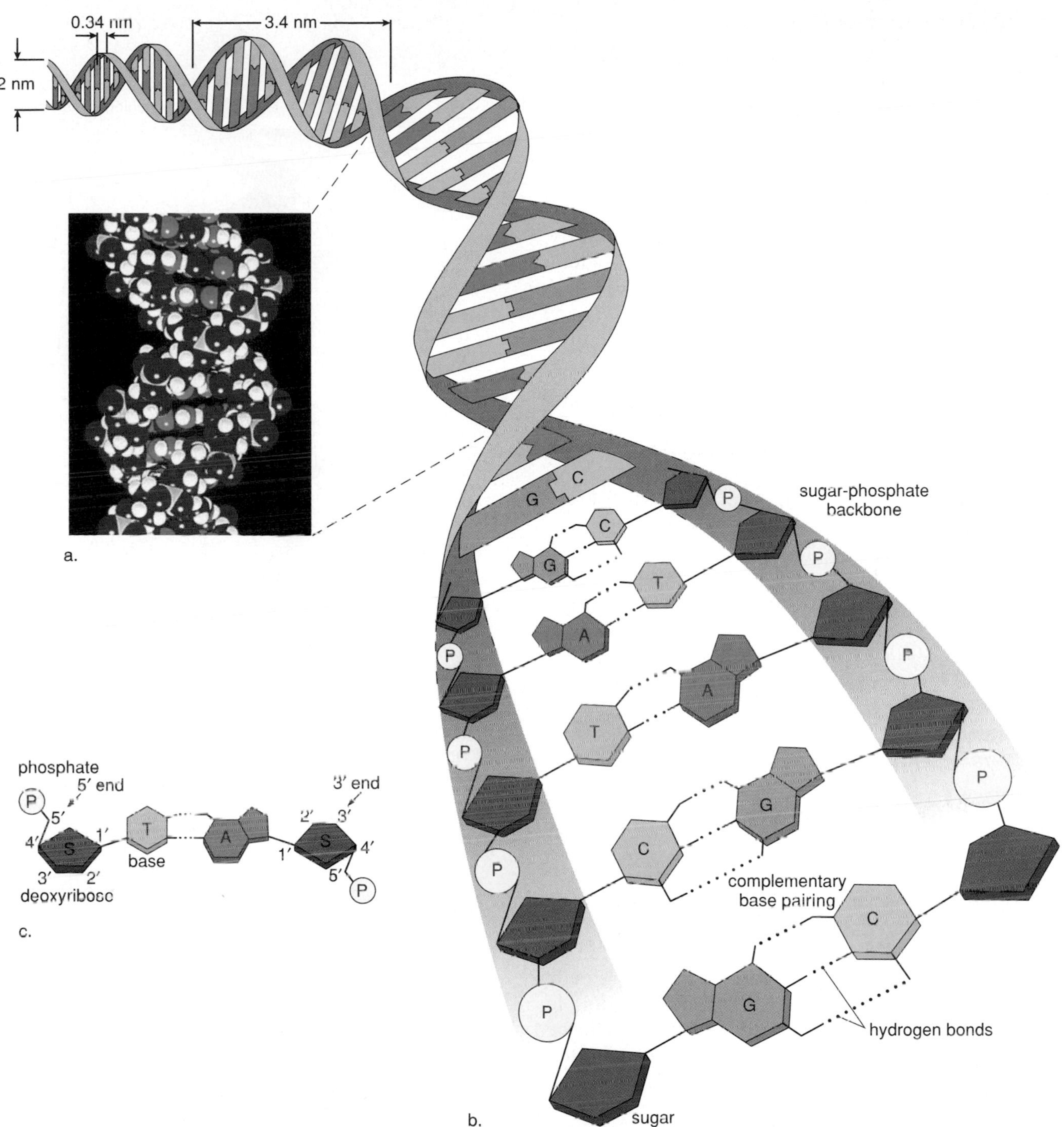

Figure 14.6 Watson and Crick model of DNA.
a. A space-filling model of DNA. Notice the close stacking of the paired bases, as determined by the X-ray diffraction pattern of DNA. (Color code for atoms: yellow = phosphate, dark blue = carbon, red = oxygen, turquoise = nitrogen, and white = hydrogen.) **b.** Diagram of DNA double helix shows that the molecule resembles a twisted ladder. Sugar-phosphate backbones make up the sides of the ladder, and hydrogen-bonded bases make up the rungs of the ladder. Complementary base pairing dictates that A is bonded to T and G is bonded to C. **c.** The two strands of the molecule are antiparallel; that is, the sugar-phosphate groups are oriented in different directions: the 5′ end of one strand is opposite the 3′ end of the other strand.

14.3 DNA Can Be Replicated

Ability to be replicated is one of the key requirements of a genetic material. As soon as Watson and Crick developed their double-helix model, they commented, "It has not escaped our notice that the specific pairing we have postulated immediately suggests a possible copying mechanism for the genetic material."

It has now been confirmed that DNA is replicated by means of complementary base pairing. During replication, each old DNA strand of the parent molecule serves as a template for a new strand in a daughter molecule (Fig. 14.7). A template is most often a mold used to produce a shape complementary to itself. DNA replication is termed **semiconservative replication** because one of the old strands is conserved, or present, in each daughter molecule.

Replication requires the following steps:

1. *Unwinding.* The old strands that make up the parent DNA molecule are unwound and "unzipped" (i.e., the weak hydrogen bonds between the paired bases are broken). There is a special enzyme called helicase that unwinds the molecule.
2. *Complementary base pairing.* New complementary nucleotides, always present in the nucleus, are positioned by the process of complementary base pairing.
3. *Joining.* The complementary nucleotides join to form new strands. Each daughter DNA molecule contains an old strand and a new strand.

Steps 2 and 3 are carried out by an enzyme complex called **DNA polymerase.** DNA polymerase works in the test tube as well as in cells.

In Figure 14.7, the backbones of the parent molecule (original double strand) are bluish, and each base is given a particular color. Following replication, the daughter molecules each have a pinkish backbone (new strand) and a bluish backbone (old strand). A daughter DNA double helix has the same sequence of bases as the parent DNA double helix had originally. Although DNA replication can be explained easily in this manner, it is actually a complicated process. Some of the more precise molecular events are discussed in the reading on page 230.

DNA replication must occur before a cell can divide. Cancer, which is characterized by rapidly dividing cells, is sometimes treated with chemotherapeutic drugs that are analogs (have a similar, but not identical structure) to one of the four nucleotides in DNA. When these are mistakenly used by the cancer cells to synthesize DNA, replication stops and the cells die off.

During DNA replication, the parent DNA molecule unwinds and unzips. Then each old strand serves as a template for a new strand.

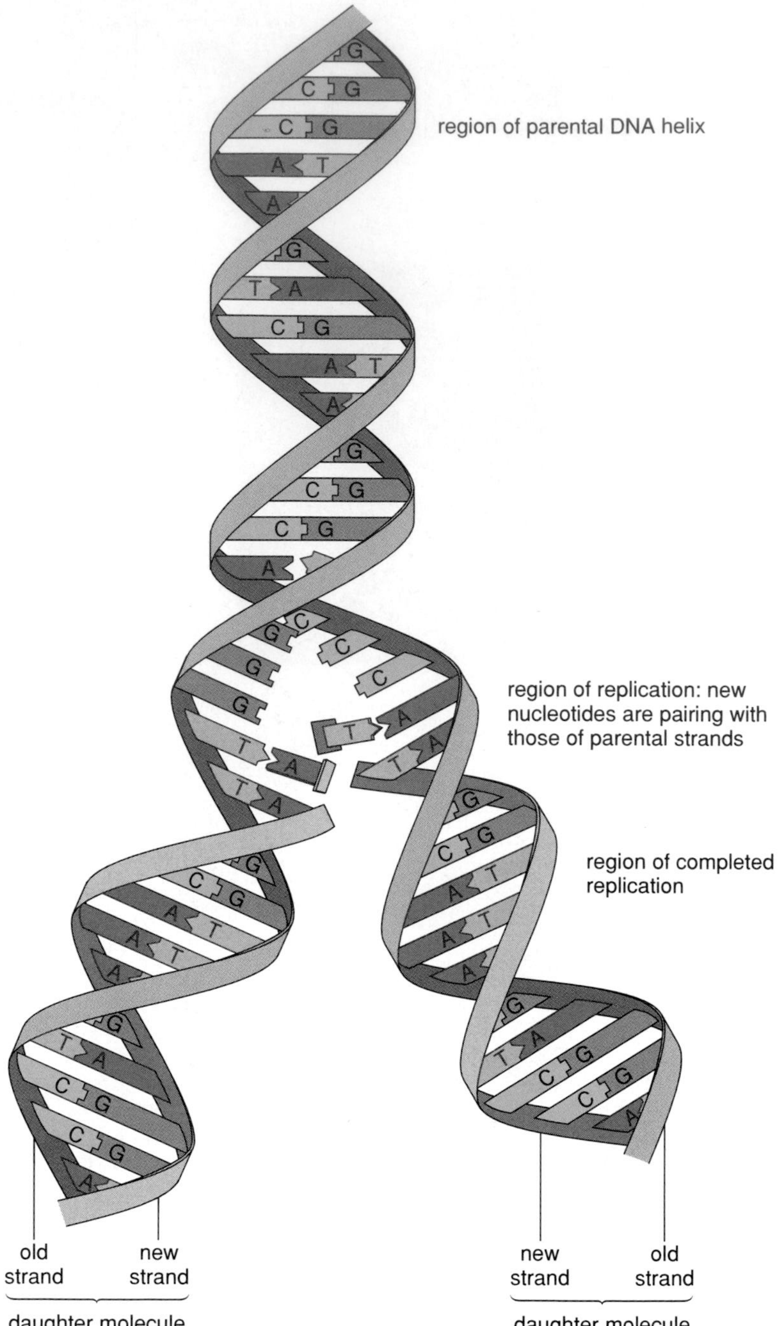

Figure 14.7 Semiconservative replication (simplified). After the DNA molecule unwinds, each old strand serves as a template for the formation of the new strand. Complementary nucleotides available in the cell pair with those of the old strand and then are joined together to form a strand. After replication is complete, there are two daughter DNA molecules. Each is composed of an old strand and a new strand. Each daughter molecule has the same sequence of base pairs as the parent molecule had before unwinding occurred.

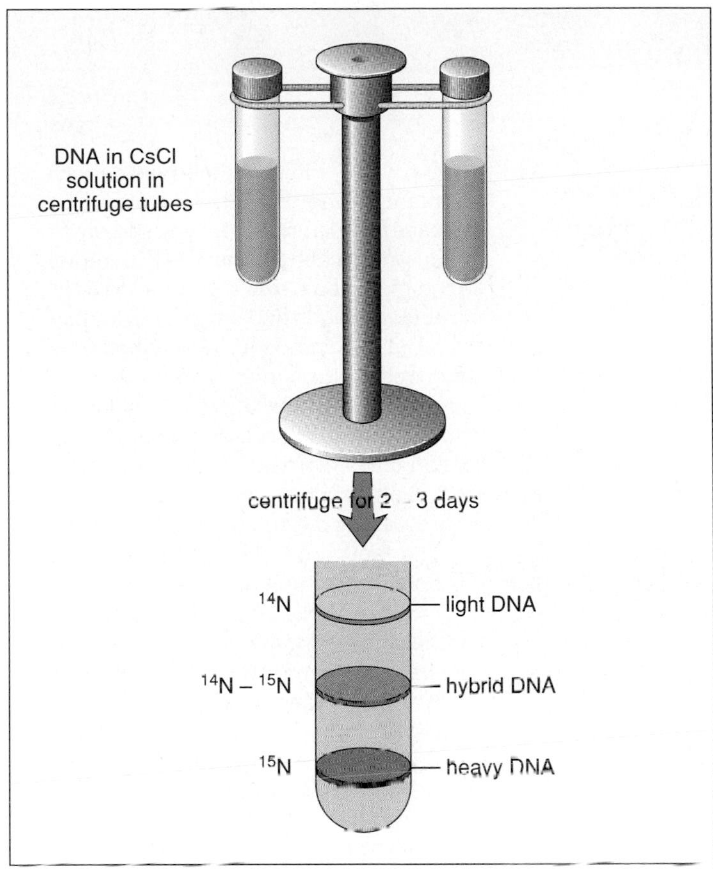

a. Possible results when DNA is centrifuged in CsCl

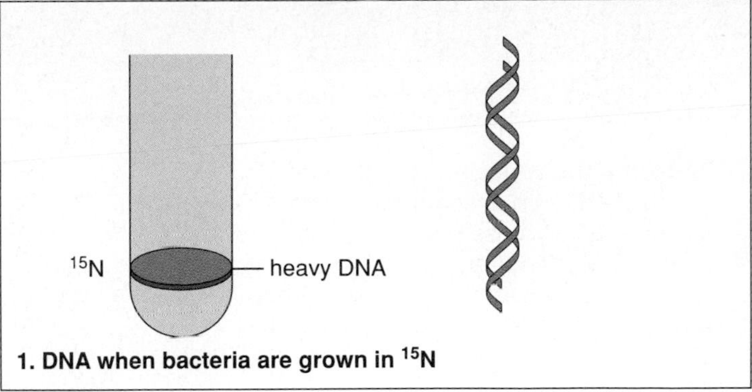

1. DNA when bacteria are grown in ^{15}N

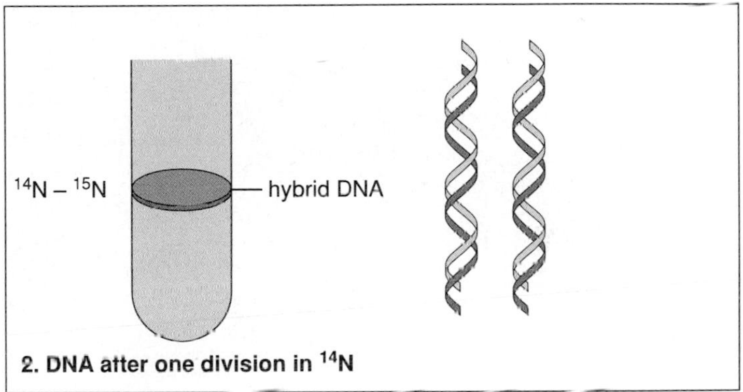

2. DNA after one division in ^{14}N

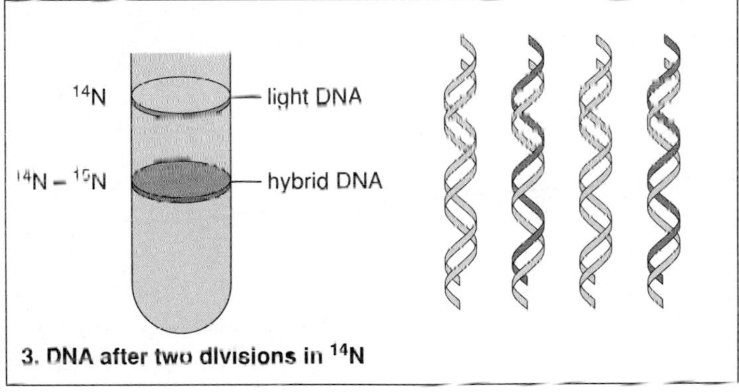

3. DNA after two divisions in ^{14}N

b. Steps in Meselson and Stahl experiment

Figure 14.8 Meselson and Stahl's DNA replication experiment.
a. When DNA molecules are centrifuged in a CsCl density gradient, they separate on the basis of density. b. Cells grown in heavy nitrogen (^{15}N) have dense strands. After one division in light nitrogen (^{14}N), DNA molecules are hybrid and have intermediate density. After two divisions, DNA molecules separate into two bands—one for light DNA and one for hybrid DNA.

Replication Is Semiconservative

DNA replication is termed semiconservative because each daughter double helix contains an old strand and a new strand. Semiconservative replication was experimentally confirmed by Matthew Meselson and Franklin Stahl in 1958.

Centrifuges spin tubes and in that way separate particles from the suspending fluid. Meselson and Stahl knew that it would be possible to centrifuge DNA molecules in a suspending fluid that would separate them on the basis of their different densities (Fig. 14.8*a*). A DNA molecule in which both strands contained heavy nitrogen (^{15}N, with an atomic weight of 15) is most dense. A DNA molecule in which both strands contained light nitrogen (^{14}N, with an atomic weight of 14) is least dense. A hybrid DNA molecule in which one strand is heavy and one is light has an intermediate density.

Meselson and Stahl first grew bacteria in a medium containing ^{15}N so that only heavy DNA molecules were present in the cells. Then they switched the bacteria to a medium containing ^{14}N. After one division, only hybrid DNA molecules were in the cells. After two divisions, half of the DNA molecules were light and half were hybrid. These were exactly the results to be expected if DNA replication is semiconservative.

DNA replication is semiconservative. Each new strand contains an old and a new strand.

A closer Look

Aspects of DNA Replication

Watson and Crick realized that the strands in DNA had to be antiparallel to allow for complementary base pairing. This opposite polarity of the strands introduces complications for DNA replication, as we will now see. First, it is important to take a look at a deoxyribose molecule in which the carbon atoms are numbered (Fig. 14A*a*). Use the structure to see that one of the strands of DNA (Fig. 14A*b*) runs in the $5' \rightarrow 3'$ direction and the other runs in the $3' \rightarrow 5'$ direction. You can tell the 5′ end because that's the last phosphate.

During replication (Fig. 14A*b*), one nucleotide is joined to another. Each new nucleotide already has a phosphate group at the 5′ carbon atom and it is joined to the 3′ carbon atom of a sugar already in place. Therefore it is said that *DNA polymerase synthesizes the replicated strand in the $5' \rightarrow 3'$ direction.* This presents a problem at the replication fork (Fig. 14A*c*), where DNA is unwound and unzipped and where replication is occurring.

At a replication fork, only one of the new (daughter) strands runs in the $5' \rightarrow 3'$ direction (the template for this strand, of course, runs in the $3' \rightarrow 5'$ direction). The new strand running in the $5' \rightarrow 3'$ direction can be synthesized continuously and is called the *leading strand.* But what about the situation when the $5' \rightarrow 3'$ parental strand is serving as the template? Synthesis of the new strand must also be in the $5' \rightarrow 3'$ direction, and therefore synthesis has to begin at the fork. Although synthesis is in the $5' \rightarrow 3'$ direction, this new daughter strand in the end runs from $3' \rightarrow 5'$, opposite to its template. (Do you see why?) Because replication of the $5' \rightarrow 3'$ parental strand must begin repeatedly as the DNA molecule unwinds and unzips, replication is discontinuous; indeed, replication of this strand results in segments called Okazaki fragments, after the Japanese scientist who discovered them. Discontinuous replication takes more time than continuous replication, and therefore the new strand in this case is called the *lagging strand.*

The fact that DNA polymerase can only join a nucleotide to the free 3′ end of another nucleotide presents another problem: *DNA polymerase cannot start the synthesis of a new DNA chain* at the origin of replication. (In the lagging strand there are many origins of replication.) Here, an RNA polymerase lays down a short amount of RNA, called an RNA primer, that is complementary to the DNA strand being replicated. Now DNA polymerase can add DNA nucleotides in the $5' \rightarrow 3'$ direction. Later, while proofreading, DNA polymerase removes the RNA primer and replaces it with complementary DNA nucleotides. Another enzyme, called DNA ligase, joins the 3′ end of each fragment to the 5′ end of another.

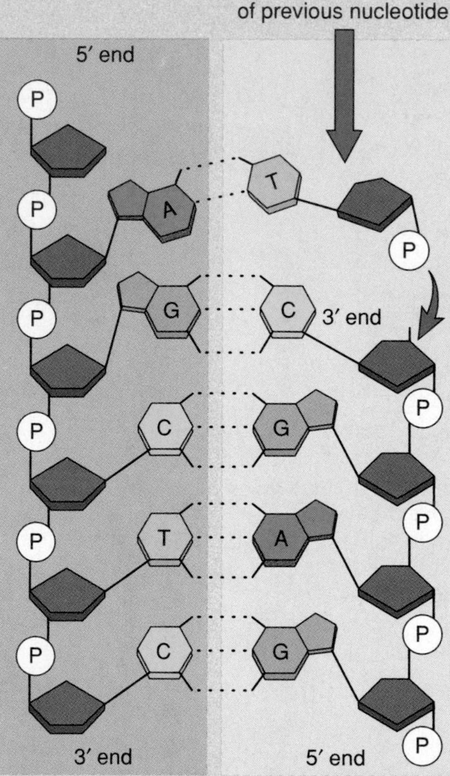

Figure 14A DNA replication (in depth).
a. Structure of deoxyribose, showing where nitrogen base and phosphate groups are attached. **b.** Template strand and replicated strand, the latter of which always grows from the 5′ end toward the 3′ end. **c.** Both parental strands are templates for a daughter strand. Replication is a continuous process for one daughter strand and a discontinuous process for the other daughter strand. The enzymes are much larger than shown.

Replication Errors Do Occur

The ability to mutate is one of the requirements for the genetic material, and base changes during replication are one way mutations—random changes in genes—can occur. During the replication process, DNA polymerase chooses complementary nucleotide triphosphates from the cellular pool. Then the nucleotide triphosphate is converted to a nucleotide monophosphate and aligned with the template nucleotide.[1]

A mismatched nucleotide slips through this selection process only once per 100,000 base pairs at most. The mismatched nucleotide causes a pause in replication, during which time it is excised from the daughter strand and replaced with the correct nucleotide. After this so-called proofreading has occurred, the error rate is only one mistake per one billion base pairs.

The errors that slip through nucleotide selection and proofreading may cause a gene mutation to occur. Actually it is of benefit for mutations to occur occasionally because variation is the raw material for the evolutionary process.

Errors in replication are minimized because DNA polymerase has a "proofreading" function. Mutations occur when a mispairing slips through the proofreading process.

Prokaryotic Versus Eukaryotic Replication

Prokaryotes differ from eukaryotes in a number of ways, including how DNA replication takes place.

DNA Replicates in Prokaryotes

Bacteria have a single circular loop of DNA that must be replicated before the cell divides. In some circular DNA molecules, replication moves around the DNA molecule in one direction only. In others, as shown here, replication starts at the origin but moves in opposite directions. The process always occurs in the 5′ to 3′ direction.

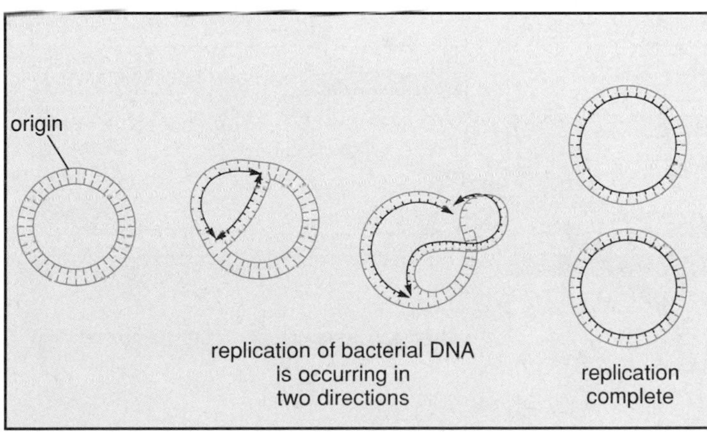

In prokaryotes, the single chromosome is attached to the plasma membrane; after it is replicated the two copies separate as the cell enlarges. Newly formed plasma membrane and cell wall separate the cell into two cells. This process is called binary fission.

Bacterial cells are able to replicate their DNA at a rate of about 10^6 base pairs per minute, and about 40 minutes are required to replicate the complete chromosome. Because bacterial cells are able to divide as often as once every 20 minutes, it is possible for a new round of DNA replication to begin even before the previous round is completed!

DNA Replicates in Eukaryotes

In eukaryotes, DNA replication begins at numerous origins of replication along the length of the chromosome, and the so-called replication bubbles spread bidirectionally until they meet. Notice that there is a V shape wherever DNA is being replicated. This is called a *replication fork.*

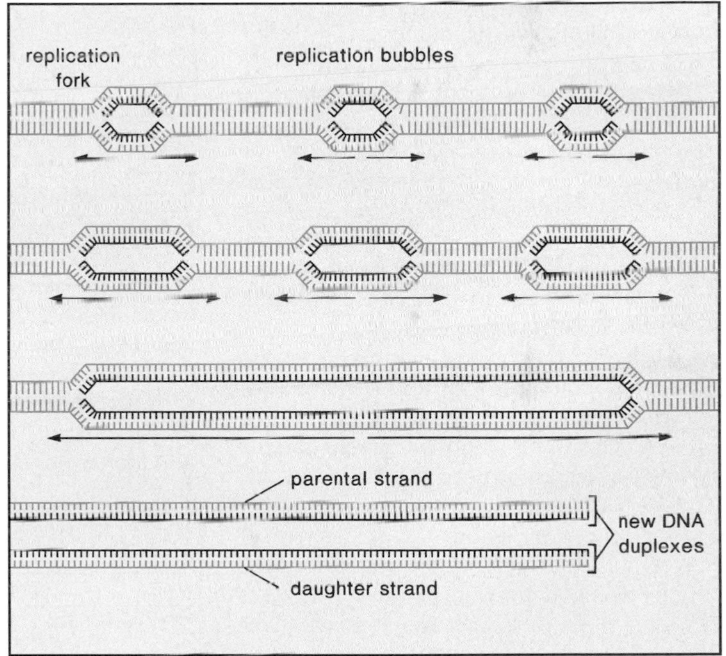

Although eukaryotes replicate their DNA at a slower rate—500 to 5,000 base pairs per minute—there are many individual origins of replication. Therefore, eukaryotic cells complete the replication of the diploid amount of DNA (in humans over 6 billion base pairs) in a matter of hours!

Replication precedes cell division in eukaryotes. The cell cycle contains interphase, during which time replication occurs, and mitosis, during which time the cell divides. A mitotic spindle consisting of microtubules serves to orchestrate the separation of daughter chromosomes into daughter nuclei.

[1] When nucleotide monophosphates are cleaved from nucleotide triphosphates, energy is provided for DNA synthesis.

Doing Science

Barbara McClintock and the Discovery of Jumping Genes

When Barbara McClintock (Fig. 14B) first began studying inheritance in corn (maize) plants, geneticists believed that each gene had a fixed locus on a chromosome. Just as Morgan and his colleagues mapped the chromosomes of *Drosophila*, so McClintock was busy mapping those of corn. In the course of her studies, she came to the conclusion that "controlling elements" could move from one location to another on the chromosome. If a controlling element landed in the middle of a gene, it prevented the expression of that gene. Today, McClintock's controlling elements are called moveable genetic elements, transposons, or (in slang), "jumping genes."

Based on her experiments with maize, McClintock showed that because transposons are capable of suppressing gene expression, they could account for the pigment pattern of the corn strain popularly known as Indian corn. A colorless corn kernel results when cells are unable to produce a purple pigment due to the presence of a transposon within a particular gene needed to synthesize the pigment. While mutations are usually stable, a transposition is very unstable. When the transposon jumps to another chromosomal location, some cells regain the ability to produce the purple pigment and the result is a corn kernel with a speckled pattern (Fig. 14C).

When McClintock first published her results in the 1950s, the scientific community did not know what to make of it. Years later, when molecular genetics was well established, transposons were also discovered in bacteria, yeasts, plants, flies, and humans. Geneticists now believe that transposons:

1. can cause localized mutations; that is, mutations that occur in certain cells and not others.
2. can carry a copy of certain host genes with them when they jump. Therefore, they can be a source of chromosomal mutations such as translocations, deletions, and inversions.
3. can leave copies of themselves and certain host genes before jumping. Therefore, they can be a source of a duplication, another type of chromosomal mutation.
4. in bacteria they can contain one or more genes that make a bacterium resistant to antibiotics.

Considering that transposition has a powerful effect on genotype and phenotype (Fig. 14D), it most likely has played an important role in evolution. For her discovery of transposons, McClintock was, in 1983, finally awarded the Nobel Prize in Physiology or Medicine.

During the years when she received no feedback for her work, McClintock never gave up. She worked steadily and often went through the entire day without an opportunity to speak to anyone. She found it difficult to give speeches to those who were unfamiliar with corn genetics, but her closest colleagues said she liked to discuss her work and answer students' questions. Through it all, McClintock was content to communicate with corn. In her Nobel Prize acceptance speech, the 81-year-old scientist proclaimed that "it might seem unfair to reward a person for having so much pleasure over the years, asking the maize plant to solve specific problems and then watching its responses."

Figure 14B Barbara McClintock (1902–92).

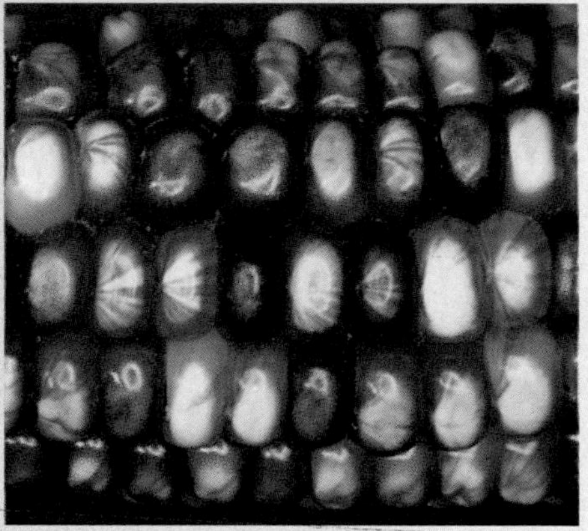

Figure 14C Corn kernels.
Some corn kernels are purple, some are colorless, and some are speckled.

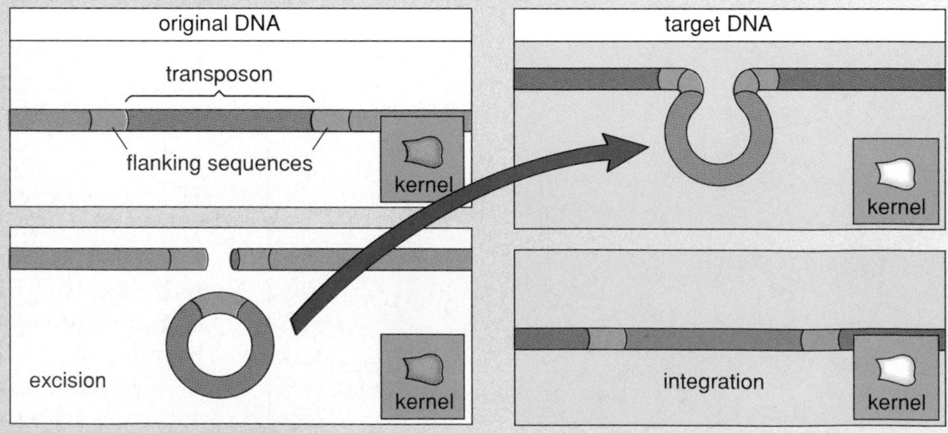

Figure 14D Transposon.
In its original location, a transposon is interrupting a gene (brown) that is not involved in kernel pigmentation. When the transposon moves to another chromosome position, it blocks the action of a gene (blue) required for synthesis of a purple pigment, and the kernel is now colorless.

connecting concepts

Only four nucleotide bases occur in DNA—thymine, cytosine, guanine, and adenine. Erwin Chargaff was the first to determine that the amount of adenine (A) in DNA equals the amount of thymine (T), and the amount of cytosine (C) equals the amount of guanine (G). Now we know that this occurs because A is always paired with T, and C is always paired with G. The simplicity of complementary base pairing enables chromosomes to replicate in a relatively straightforward way. Somewhat like a zipper, the double helix of a chromosome opens up and each old strand is paired with complementary bases. These nucleotides are then joined to form a new strand so each daughter DNA molecule contains an old strand and a new strand.

A human gene can contain over a hundred million base pairs, and these base pairs can occur in any sequence. Because base pairs can occur in any order, the total number of possible sequences in humans runs into billions upon billions. In other words, the possible sequences of base pairs in DNA is virtually unlimited. This accounts for the variations we see among human beings and among all living things.

Every three bases in DNA codes for a particular amino acid, and in this way, a genes specifies the order of amino acids in a particular protein. Ordinarily, the base sequence in DNA causes cells to produce a protein that has the usual structure and function. Occasionally, however, a mutation—a change in base sequence—occurs. Although mutations are often detected and repaired by the cell, some escape detection, and a new variation in DNA base sequence may lead to a change in protein structure and function.

Genetic changes can bring about a modification that gives an individual an advantage over other members of a population, particularly if the environment is changing. If so, this individual will contribute more offspring to the next generation than other members of the population. In this way a population can become adapted to a new and different environment.

Summary

14.1 Searching for the Genetic Material

Early work on the biochemistry of DNA wrongly suggested that DNA lacks the variability necessary for the genetic material.

Griffith injected strains of pneumococcus into mice and observed that smooth (S) strain bacteria are virulent but rough (R) strain bacteria are not. When heat-killed S strain bacteria were injected along with live R strain bacteria, however, virulent S strain bacteria were recovered from the dead mice. Griffith said that the R strain had been transformed by some substance passing from the dead S strain to the live R strain. Twenty years later, Avery and his colleagues reported that the transforming substance is DNA. Hershey and Chase turned to bacteriophage T2 as their experimental material. In two separate experiments, they labeled the protein coat with ^{35}S and the DNA with ^{32}P. They then showed that the radioactive P alone is largely taken up by the bacterial host and that reproduction of viruses proceeds normally. This convinced most researchers that DNA is the genetic material.

14.2 Finding the Structure of DNA

Chargaff did a chemical analysis of DNA and found that A = T and G = C, and that the amount of purine equals the amount of pyrimidine. Franklin prepared an X-ray photograph of DNA that showed it is helical, has repeating structural features, and has certain dimensions. Watson and Crick built a model of DNA in which the sugar-phosphate molecules made up the sides of a twisted ladder and the complementary-paired bases were the rungs of the ladder.

14.3 DNA Can Be Replicated

The Watson and Crick model immediately suggested a method by which DNA could be replicated. The two strands unwind and unzip, and each parental strand acts as a template for a new (daughter) strand. In the end, each new duplex is like the other and like the parental duplex.

Meselson and Stahl demonstrated that replication is semiconservative by the following experiment: bacteria were grown in heavy nitrogen (^{15}N) and then switched to light nitrogen (^{14}N). The density of the DNA following replication was intermediate between these two, as measured by centrifugation of the molecules through a salt gradient.

The enzyme DNA polymerase joins the nucleotides together and proofreads them to make sure the bases have been paired correctly. Incorrect base pairs that survive the process are a mutation. Replication in prokaryotes proceeds from one point of origin until there are two copies of the circular chromosome. Replication in eukaryotes has many points of origin and many bubbles (places where the DNA strands are separating and replication is occurring). Replication occurs at the ends of the bubbles—at replication forks.

Reviewing the Chapter

1. List and discuss the requirements for genetic material. 222
2. What is the tetranucleotide hypothesis, and why did this hypothesis hinder the acceptance of DNA as the genetic material? 222
3. Describe Griffith's experiments with pneumococcus, his surprising results, and his conclusion. 223
4. How did Avery and his colleagues demonstrate that the transforming substance is DNA? 223
5. Describe the experiment of Hershey and Chase, and explain how it shows that DNA is the genetic material. 224
6. What are Chargaff's rules? 225
7. Describe the Watson and Crick model of DNA structure. How did it fit the data provided by Chargaff and the X-ray diffraction pattern? 226–27
8. Explain how DNA replicates semiconservatively. What role does DNA polymerase play? What role does helicase play? 228–29
9. How did Meselson and Stahl demonstrate semiconservative replication? 229
10. Explain how the replication process is a source of mutations. List and discuss differences between prokaryotic and eukaryotic replication of DNA. 231

Testing Yourself

Choose the best answer for each question.
For questions 1–4, match the names in the key to the statements below:

Key:

a. Griffith
b. Chargaff
c. Meselson and Stahl
d. Hershey and Chase

1. A = T and G = C.
2. Only the DNA from T2 enters the bacteria.
3. R strain bacteria became an S strain through transformation.
4. DNA replication is semiconservative.
5. If 30% of an organism's DNA is thymine, then
   a. 70% is purine.
   b. 20% is guanine.
   c. 30% is adenine.
   d. Both b and c are correct.
6. If you grew bacteria in heavy nitrogen and then switched them to light nitrogen, how many generations after switching would you have some light/light DNA?
   a. never, because replication is semiconservative
   b. the first generation
   c. the second generation
   d. only the third generation
7. The double-helix model of DNA resembles a twisted ladder in which the rungs of the ladder are
   a. a purine paired with a pyrimidine.
   b. A paired with G and C paired with T.
   c. sugar-phosphate paired with sugar-phosphate.
   d. Both a and b are correct.
8. Cell division requires that the genetic material be able to
   a. store information.
   b. be replicated.
   c. undergo rare mutations.
   d. All of these are correct.
9. In a DNA molecule, the
   a. bases are covalently bonded to the sugars.
   b. sugars are covalently bonded to the phosphates.
   c. bases are hydrogen-bonded to one another.
   d. All of these are correct.
10. In the following diagram, blue stands for heavy DNA (contains ^{15}N) and red stands for light DNA (does not contain ^{15}N). Label each strand of all three DNA molecules as heavy or light DNA, and explain why the diagram is in keeping with the semiconservative replication of DNA.

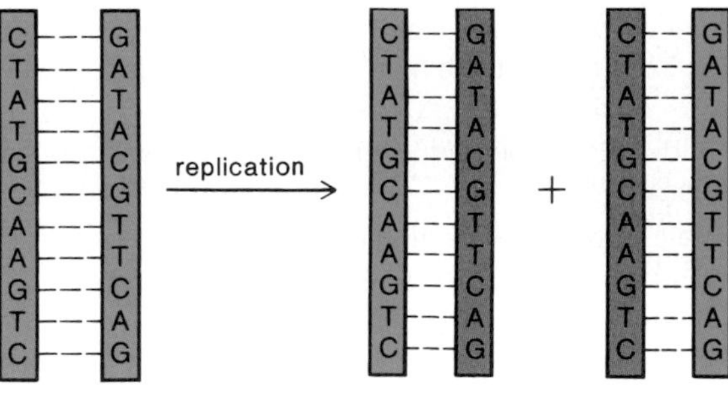

Applying the Concepts

1. *DNA is the genetic material that dictates the form, the function, and the behavior of organisms.*
   Explain the rationale of the Hershey and Chase experiment.
2. *The genetic material has to be capable of replication.*
   To explain the importance of DNA replication, associate the necessity of replication with the cell cycle and the life cycle.
3. *Evolution is dependent on the ability of the genetic material to mutate.*
   Explain the importance of mutation to the evolutionary process.

Using Technology

Your study of DNA is supported by these available technologies:

Exploring the Internet
The Mader Home Page provides resources for and help with studying this chapter.

http://www.mhhe.com/sciencemath/biology/mader/
(Click on Biology.)

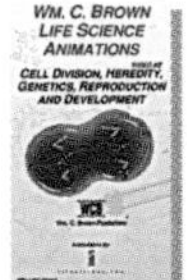

Life Science Animations Video
Video #2: Cell Division/Heredity/Genetics/Reproduction and Development: DNA Replication (#15)

Understanding the Terms

Match the terms to these definitions:

a. __________ Alteration in DNA composition and also an alteration in chromosome structure and number.
b. __________ Bonding between particular purines and pyrimidines in DNA.
c. __________ During replication, an enzyme that joins the nucleotides complementary to a DNA template.
d. __________ Type of nitrogen-containing base, such as adenine and guanine, having a double-ring structure.
e. __________ Virus that infects bacteria.
f. __________ One of four nitrogen-containing bases in nucleotides composing the structure of DNA that is complementary to adenine.

Gene Activity

CHAPTER 15

Chapter Concepts

Garden pea *(Pisum sativum)*

The genotype plus environmental influences determine the phenotype, whether we are speaking of peas or human beings. Yet one gene differs from another only by the sequence of the nucleotide bases in DNA. How does a difference in nucleotide sequence cause the uniqueness of the individual—that is, for example, whether you have blue or brown or hazel eye pigments?

Mendel realized that garden pea plants have varying traits, and today we know why in certain instances. For example, peas are smooth or wrinkled according to the presence or absence of a starch-forming enzyme. It would appear that the allele *S* in peas dictates the presence of the starch-forming enzyme, whereas the allele *s* does not. By studying the activity of genes in cells, geneticists have confirmed that proteins are the link between the genotype and the phenotype.

This holds true for human beings also. You have blue or brown or hazel eye pigments because of the type of enzymes contained within your cells. Through its ability to specify proteins, DNA also brings about the development of one structure over another. When studying protein synthesis in this chapter, keep in mind this flow diagram: DNA's sequence of nucleotides → sequences of amino acids → polypeptides of proteins → enzymes → structures in an organism.

15.1 What Genes Do

In the early 1900s, the English physician Sir Archibald Garrod suggested that there is a relationship between inheritance and metabolic diseases. He introduced the phrase *inborn error of metabolism* to dramatize this relationship. Garrod observed that family members often had the same disorder, and he said this inherited defect could be caused by the lack of a particular enzyme in a metabolic pathway. Since it was known at the time that enzymes are proteins, Garrod was among the first to hypothesize a link between genes and proteins.

Genes Specify Enzymes

Many years later, in 1940, George Beadle and Edward Tatum performed a series of experiments on *Neurospora crassa*, the red bread mold fungus, which reproduces by means of spores. Normally, the spores become mold capable of growing on minimal medium (containing only a sugar, mineral salts, and the vitamin biotin) because mold can produce all the enzymes it needs. In their experiments, Beadle and Tatum used X rays to induce mutations in asexually produced haploid spores. Some of the X-rayed spores could no longer become mold capable of growing on minimal medium; however, growth was possible on medium enriched by certain metabolites. In the example given in Figure 15.1, the mold grows only when supplied with enriched medium that includes all metabolites or C and D alone. Since C and D are a part of this hypothetical pathway

$$A \xrightarrow{1} B \xrightarrow{2} C \xrightarrow{3} D$$

in which the numbers are enzymes and the letters are metabolites, it is concluded that the mold lacks enzyme 2. Beadle and Tatum further found that each of the mutant strains has only one defective gene leading to one defective enzyme and one additional growth requirement. Therefore, they proposed that each gene specifies the synthesis of one enzyme. This is called the *one gene–one enzyme hypothesis.*

Genes Specify a Polypeptide

The one gene-one enzyme hypothesis suggests that a gene mutation causes a change in the structure of a protein. To test this idea, Linus Pauling and Harvey Itano decided to see if the hemoglobin in the red blood cells of persons with sickle-cell disease has a structure different from that of the red blood cells of normal individuals (Fig. 15.2). Recall that proteins are polymers of amino acids, some of which carry a charge. These investigators decided to see if there was a charge difference between normal hemoglobin *(Hb^A)* and sickle-cell hemoglobin *(Hb^S)*. To determine this, they sub-

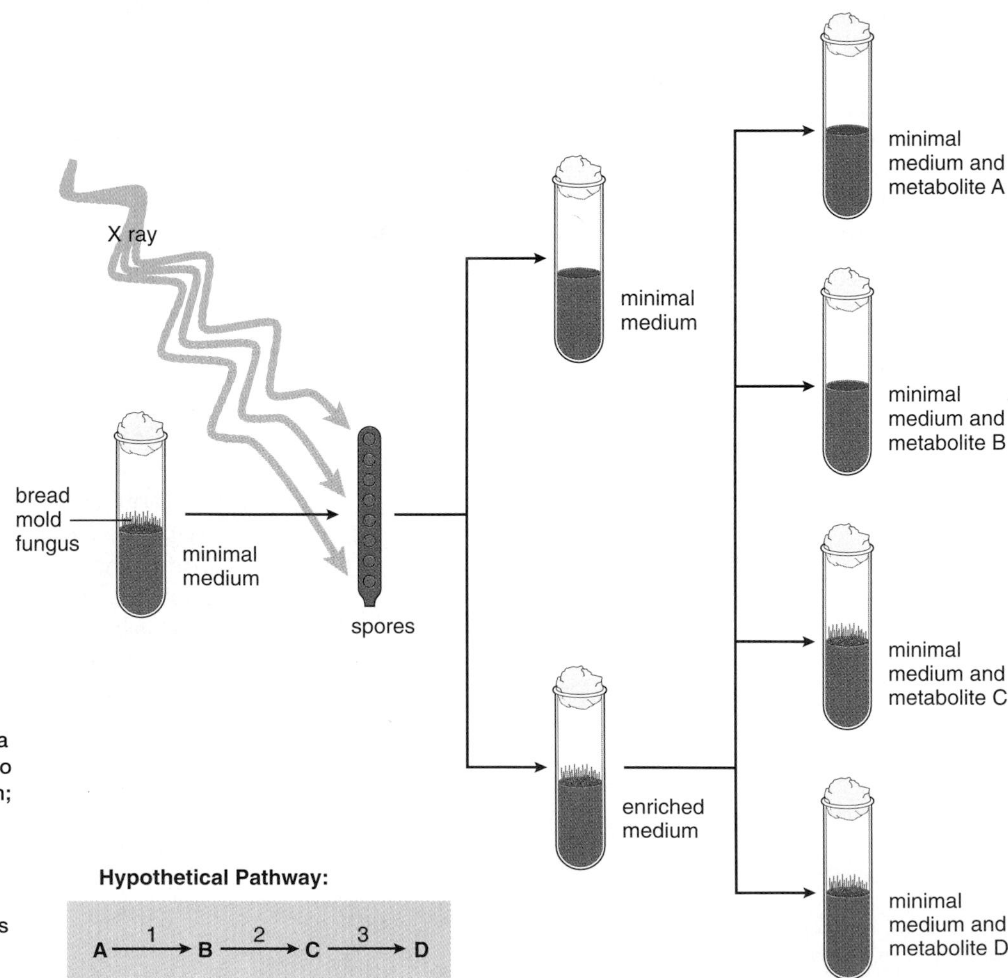

Figure 15.1 Beadle and Tatum experiment.
When haploid spores of *Neurospora crassa*, a bread mold fungus, are X-rayed, some are no longer able to germinate on minimal medium; however, they can germinate on enriched medium. In this example, the mycelia produced do not grow on minimal medium plus metabolite A or B, but they do grow on minimal medium plus metabolite C or D. This shows that enzyme 2 is missing from the hypothetical pathway.

jected hemoglobin collected from normal individuals, sickle-cell trait individuals, and sickle-cell disease individuals to electrophoresis, a procedure that separates molecules according to their size and charge, whether (+) or (–). Here is what they found:

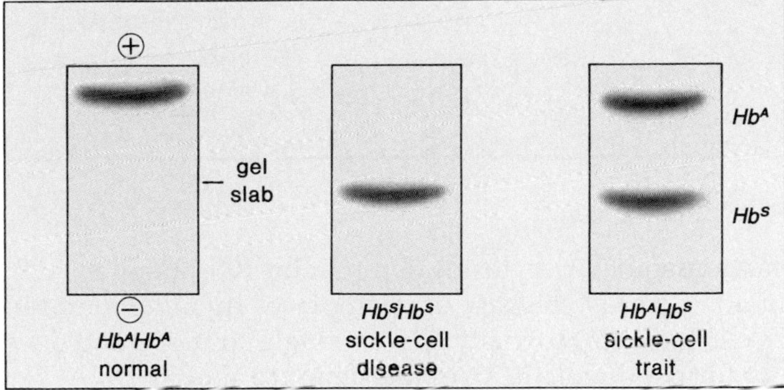

As you can see, there is a difference in migration rate toward the positive pole between normal hemoglobin and sickle-cell hemoglobin. Further, hemoglobin from those with sickle-cell trait separates into two distinct bands, one corresponding to that for Hb^A hemoglobin and the other corresponding to that for Hb^S hemoglobin. Pauling and Itano therefore demonstrated that a mutation leads to a change in the structure of a protein.

Several years later, Vernon Ingram was able to determine the structural difference between Hb^A and Hb^S. Normal hemoglobin Hb^A contains negatively charged glutamate; in sickle-cell hemoglobin the glutamate is replaced by nonpolar valine (Fig. 15.2*c*). This causes Hb^S to be less soluble and to precipitate out of solution, especially when environmental oxygen is low. At these times, the Hb^S molecules stack up into long, semirigid rods that push against the plasma membrane and distort the red blood cell into the sickle shape.

Hemoglobin contains two types of polypeptide chains, designated α (alpha) and β (beta). Only the β chain is affected in persons with sickle-cell trait and sickle-cell disease; therefore, there must be a gene for each type of chain. A refinement of the one gene-one enzyme hypothesis was needed, and it was replaced by the *one gene-one polypeptide hypothesis.*

Each gene specifies one polypeptide of a protein, a molecule that may contain one or more different polypeptides.

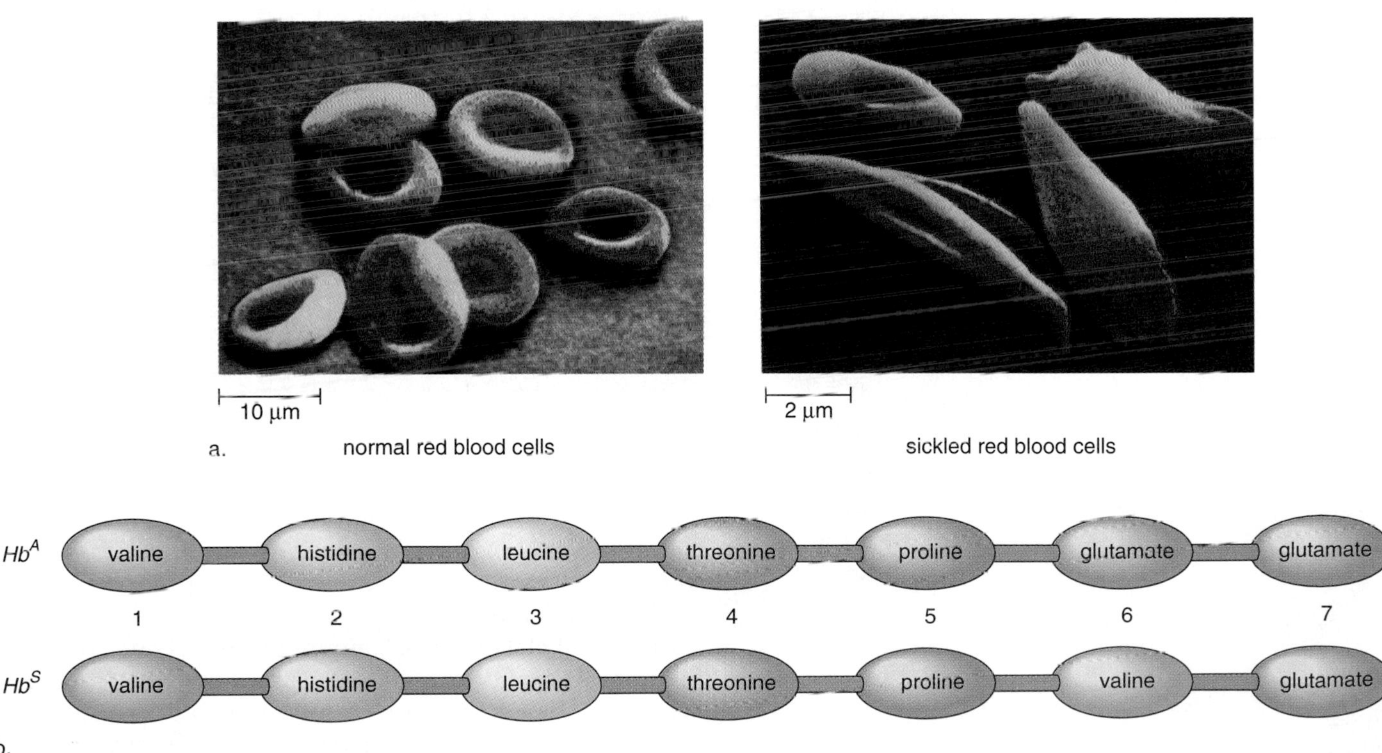

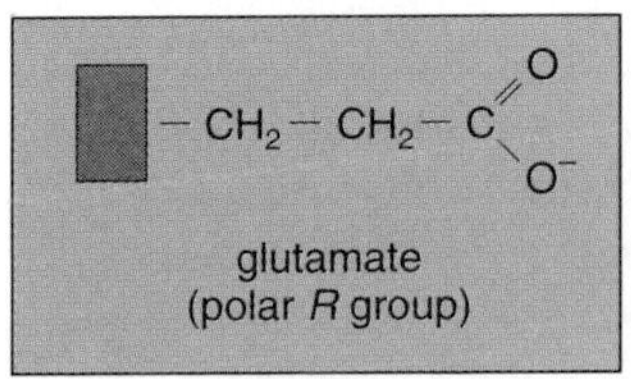

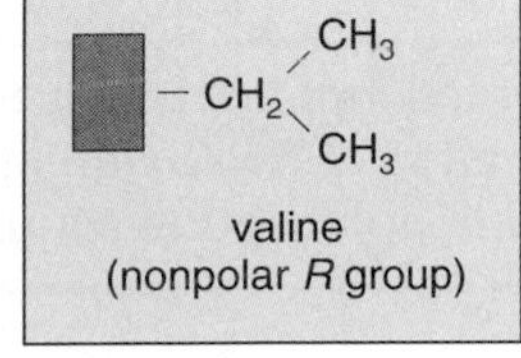

c.

Figure 15.2 Sickle-cell disease in humans.
a. Scanning electron micrograph of normal *(left)* and sickled *(right)* red blood cells. b. Portion of the chain in normal hemoglobin Hb^A and in sickle-cell hemoglobin Hb^S. Although the chain is 146 amino acids long, the one change from glutamate to valine in the sixth position results in sickle-cell disease. c. Glutamate has a polar *R* group, while valine has a nonpolar *R* group, and this causes Hb^S to be less soluble and to precipitate out of solution, distorting the red blood cell into the sickle shape.

15.2 How Genes Are Expressed

Classical geneticists thought of a gene as a particle on a chromosome. To molecular geneticists, however, a gene is a sequence of DNA nucleotide bases that codes for a product. Most often the gene product is a polypeptide. Therefore, a gene does not affect the phenotype directly; rather, the gene product affects the phenotype. As we have seen, the gene for hemoglobin-S protein has a product that causes red blood cells to sickle, and this in turn causes sickle-cell disease.

DNA is a linear polymer of nucleotides, and polypeptides are linear polymers of amino acids. This colinearity suggests that the nucleotide sequence of DNA somehow determines the order of amino acids in proteins. A DNA molecule, however, cannot directly control the sequence of amino acids because DNA is found in the nucleus of eukaryotes (or the nucleoid in prokaryotes), and protein synthesis occurs in the cytoplasm. Therefore, there must be a molecule that acts as a go-between, and the most likely candidate is RNA (ribonucleic acid), a nucleic acid found in both the nucleus and the cytoplasm (Fig. 15.3).

RNA Is Involved

Like DNA, RNA is a polymer of nucleotides. The nucleotides in RNA, however, contain the sugar ribose and the bases adenine (A), cytosine (C), guanine (G), and uracil (U). In other words, the base uracil replaces thymine found in DNA (Table 15.1). Finally, RNA is single stranded and does not form a double helix in the same manner as DNA.

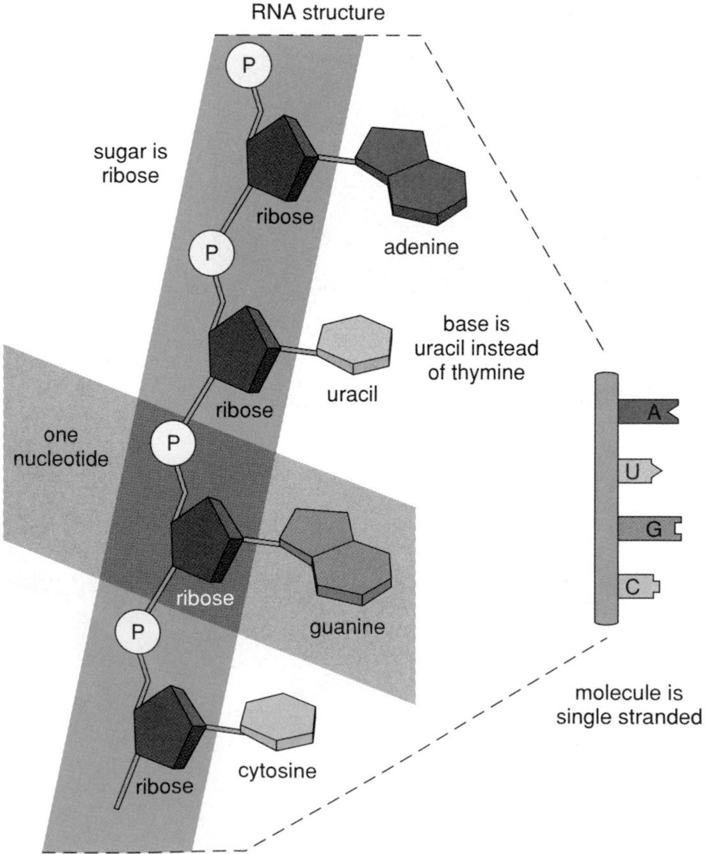

Figure 15.3 Structure of RNA.
Like DNA, RNA is a polymer of nucleotides. RNA, however, is single stranded, the pentose sugar is ribose, and uracil replaces thymine as one of the pyrimidine bases.

Table 15.1
RNA Structure Compared to DNA Structure

	RNA	DNA
Sugar	Ribose	Deoxyribose
Bases	Adenine, guanine, uracil, cytosine	Adenine, guanine, thymine, cytosine
Strands	Single stranded	Double stranded with base pairing
Helix	No	Yes

There are three major classes of RNA, each with specific functions in protein synthesis:

messenger RNA (mRNA): takes a message from DNA in the nucleus to the ribosomes in the cytoplasm.
ribosomal RNA (rRNA): along with proteins, makes up the ribosomes, where proteins are synthesized.
transfer RNA (tRNA): transfers amino acids to the ribosomes.

Several Steps Are Involved

The central dogma of molecular biology explains the manner in which genes are expressed; that is, produce a product:

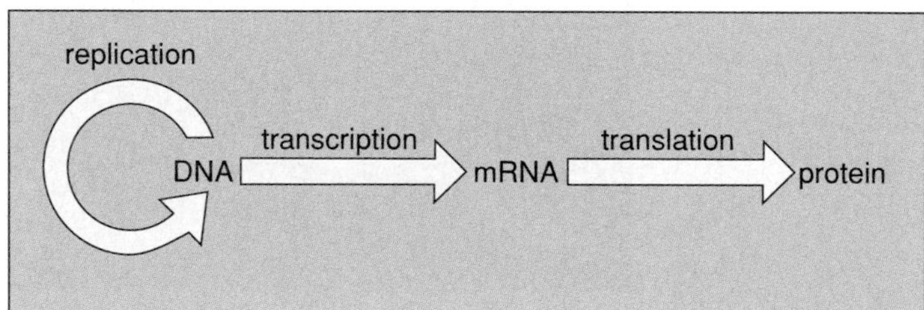

This diagram indicates that DNA not only serves as a template for its own replication, it is also a template for RNA formation. The process by which an RNA copy is made of a portion of DNA is called **transcription** [L. *trans,* across, and *scriptio,* a writing]. Following transcription, RNA molecules move into the cytoplasm. Photographic data show radioactively labeled RNA moving from the nucleus to the cytoplasm, where protein synthesis occurs. The process by which an mRNA transcript directs the sequence of amino acids in a polypeptide is called **translation** [L. *trans,* across, and *latus,* carry, bear].

During transcription, DNA serves as a template for the formation of RNA. During translation, mRNA is involved in polypeptide synthesis.

15.3 How Genes Code for Amino Acids

The central dogma of molecular biology suggests that the sequence of nucleotides in DNA specifies (and a copy of this sequence in RNA directs) the order of amino acids in a polypeptide. It would seem then that there must be a code for each of the 20 amino acids found in proteins. But can four nucleotides provide enough combinations to code for 20 amino acids? If each code word, called a **codon,** were made up of two bases, such as AG, there could be only 16 codons (4 × 4)—not enough to code for 20 amino acids. But if each codon were made up of three bases, such as AGC, there would be 64 codons (4 × 4 × 4)—more than enough to code for 20 different amino acids:

Number of Bases in Genetic Code	Number of Different Amino Acids That Can Be Specified
1	4
2	16
3	64

It is expected, then, that the genetic code is a **triplet code** and that each codon, therefore, consists of three nucleotide bases.

Finding the Genetic Code

In 1961, Marshall Nirenberg and J. Heinrich Matthei performed an experiment that laid the groundwork for cracking the genetic code. First, they found a cellular enzyme could be used to construct a synthetic RNA (one that does not occur in cells), and then they found that the synthetic polymer could be translated in a cell-free system (a test tube that contains "freed" cytoplasmic contents of a cell). Their first synthetic RNA was composed only of uracil, and the protein that resulted was composed only of the amino acid phenylalanine. Therefore, the codon for phenylalanine was known to be UUU. Later, a cell-free system was developed by Nirenberg and Philip Leder in which only three nucleotides at a time were translated; in that way it was possible to assign an amino acid to each of the RNA codons (Fig. 15.4).

A number of important properties of the genetic code can be seen by careful inspection of Figure 15.4:

1. The genetic code is degenerate. This means that most amino acids have more than one codon; leucine, serine, and arginine have six different codons, for example. The degeneracy of the code is protective against potentially harmful effects of mutations.
2. The genetic code is unambiguous. Each triplet codon has only one meaning.
3. The code has start and stop signals. There is only one start signal but three stop signals.

The Code Is Universal

The genetic code in Figure 15.4 is just about universally used by living things. This suggests that the code dates back to the very first organisms on earth and that all living things are related. Once the code was established, changes in it would have been very disruptive, and we know that it has remained largely unchanged over eons.

The genetic code is a triplet code. Each codon consists of three DNA nucleotides. Except for the stop codons, all the codons code for amino acids.

First Base	Second Base				Third Base
	U	C	A	G	
U	UUU phenylalanine	UCU serine	UAU tyrosine	UGU cysteine	U
	UUC phenylalanine	UCC serine	UAC tyrosine	UGC cysteine	C
	UUA leucine	UCA serine	UAA *stop*	UGA *stop*	A
	UUG leucine	UCG serine	UAG *stop*	UGG tryptophan	G
C	CUU leucine	CCU proline	CAU histidine	CGU arginine	U
	CUC leucine	CCC proline	CAC histidine	CGC arginine	C
	CUA leucine	CCA proline	CAA glutamine	CGA arginine	A
	CUG leucine	CCG proline	CAG glutamine	CGG arginine	G
A	AUU isoleucine	ACU threonine	AAU asparagine	AGU serine	U
	AUC isoleucine	ACC threonine	AAC asparagine	AGC serine	C
	AUA isoleucine	ACA threonine	AAA lysine	AGA arginine	A
	AUG (*start*) methionine	ACG threonine	AAG lysine	AGG arginine	G
G	GUU valine	GCU alanine	GAU aspartate	GGU glycine	U
	GUC valine	GCC alanine	GAC aspartate	GGC glycine	C
	GUA valine	GCA alanine	GAA glutamate	GGA glycine	A
	GUG valine	GCG alanine	GAG glutamate	GGG glycine	G

Figure 15.4 Messenger RNA codons.
Notice that in this chart, each of the codons (blue squares) is composed of three letters representing the first base, second base, and third base. For example, find the blue square where C for the first base and A for the second base intersect. You will see that U, C, A, or G can be the third base. The three bases CAU and CAC are codons for histidine; the three bases CAA and CAG are codons for glutamine.

15.4 How Transcription Occurs

Transcription, which takes place in the nucleus of eukaryotic cells, is the first step required for gene expression—the making of a gene product. The formation of mRNA leads to a polypeptide as the gene product. The molecules tRNA and rRNA are also transcribed from DNA templates, and these are products in and of themselves. Enzymes called **RNA polymerases** are involved in transcription.

Messenger RNA Is Formed

During transcription, an mRNA molecule is formed that has a sequence of bases complementary to a portion of one DNA strand; wherever A, T, G, or C is present in the DNA template, U, A, C, or G, respectively, is incorporated into the mRNA molecule (Fig. 15.5). A segment of the DNA helix unwinds and unzips, and complementary RNA nucleotides pair with DNA nucleotides of the strand that is being transcribed. An RNA polymerase joins the nucleotides together in the 5′→3′ direction. In other words, an RNA polymerase only adds a nucleotide to the 3′ end of the polymer under construction.

Transcription begins when RNA polymerase attaches to a region of DNA called a promoter. A **promoter** defines the start of a gene, the direction of transcription, and the strand to be copied. (The presence of promoter tells which DNA strand is to be transcribed.) Elongation of the mRNA molecule occurs as long as transcription proceeds. The RNA-DNA association is not as stable as the DNA helix. Therefore, only the newest portion of an RNA molecule that is associated with RNA polymerase is bound to the DNA, and the rest dangles off to the side. Finally, RNA polymerase comes to a terminator sequence at the other end of the gene being transcribed. The terminator causes RNA polymerase to stop transcribing the DNA and to release the mRNA molecule, now called an RNA transcript.

Many RNA polymerase molecules can be working to produce mRNA transcripts at the same time (Fig. 15.6). This allows the cell to produce many thousands of copies of the same mRNA molecule and eventually many copies of the same protein within a shorter period of time than if the single copy of DNA were used to direct protein synthesis.

As a result of transcription, there will be many mRNA molecules directing protein synthesis. Many more copies of a protein are produced within a given time versus using DNA directly.

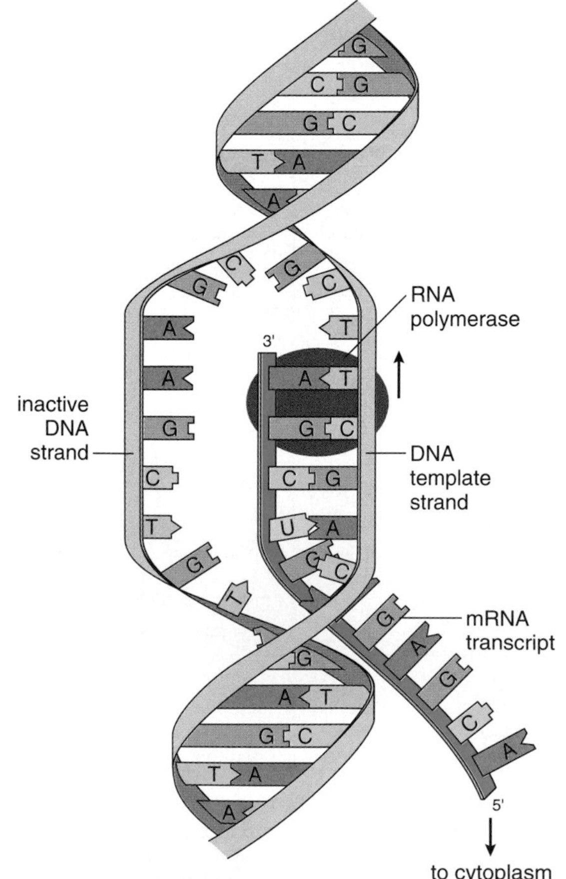

Figure 15.5 Transcription.
During transcription, complementary RNA is made from a DNA template. At the point of attachment of RNA polymerase, the DNA helix unwinds and unzips, and complementary RNA nucleotides are joined together. After RNA polymerase has passed by, the DNA strands rejoin and the RNA transcript dangles to the side.

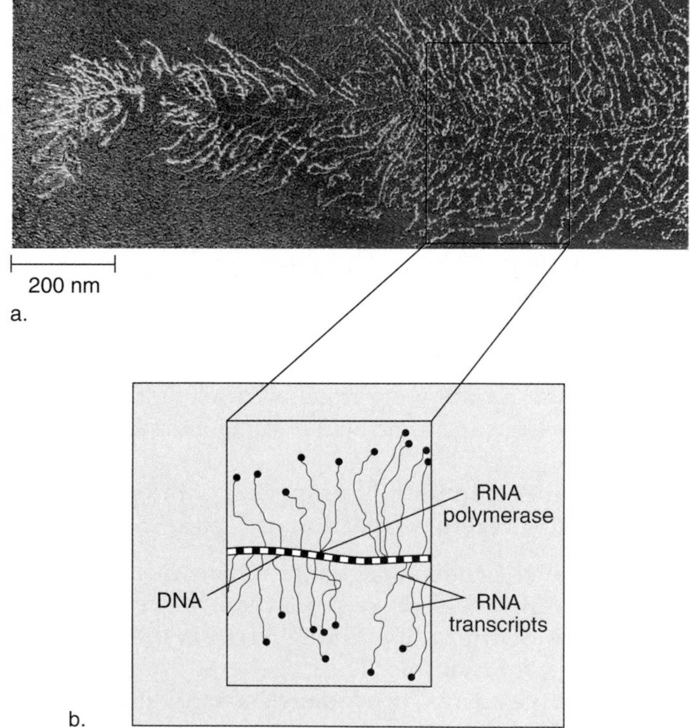

Figure 15.6 RNA polymerase.
a. Numerous RNA transcripts extend from a horizontal gene in an amphibian egg cell. **b.** The strands get progressively longer because transcription begins to the left. The dark dots along the DNA are RNA polymerase molecules. The dots at the end of the strands are spliceosomes involved in RNA processing (Fig. 15.7).

Messenger RNA Is Processed

The newly formed mRNA molecule, called the primary *mRNA* transcript, is modified before it leaves the eukaryotic nucleus. First, the ends of the mRNA molecule are altered: a cap is put onto the 5′ end and a poly-A tail is put onto the 3′ end (Fig. 15.7). The "cap" is a modified guanine (G) nucleotide that helps tell a ribosome where to attach when translation begins. The "poly-A tail" consists of a chain of 150 to 200 adenine (A) nucleotides whose function is not known. Perhaps the tail facilitates the transport of the mRNA out of the nucleus or perhaps it also regulates translation in some way.

Second, portions of the primary mRNA transcript are removed. The portions that remain are called **exons** because they will be expressed. In between the exons are intervening segments called **introns** that are removed during an mRNA processing event called mRNA splicing (Fig. 15.7). The mRNA that has now been processed and leaves the nucleus is called mature mRNA.

How is mRNA processing carried out? And what is the function of introns in the first place? In most cases mRNA splicing is often done by *spliceosomes*, a complex that contains several kinds of ribonucleoproteins. A spliceosome cuts the primary mRNA and then rejoins the adjacent exons. There has been much speculation about the possible role of introns in the eukaryotic genome. It's possible that introns divide a gene into regions that can be joined in different combinations to produce various products in different cells. For instance, it's been shown that the thyroid gland and pituitary gland process the same primary mRNA transcript differently and therefore produce related but different hormones.

Some researchers are trying to determine whether introns exist in all organisms. They have found that the more simple the eukaryote, the less the likelihood of introns in the genes. An intron has been discovered in the gene for a tRNA molecule in *Anabaena*, a cyanobacterium. This particular intron is of interest because it is "self-splicing," that is, it has the capability of splicing itself out of an RNA transcript. RNAs with an enzymatic function, now called **ribozymes,** did away with the belief that only proteins can function as enzymes. Ribozymes, however, are restricted in their function since each one cleaves RNA only at specific locations. Still, the discovery of ribozymes in prokaryotes supports the belief that RNA could have served as both the genetic material and as the first enzymes in the earliest living organisms. This hypothesis does away with the dilemma of what came first, DNA or protein. Perhaps RNA came first.

Particularly in eukaryotes, the primary mRNA transcript is processed before it becomes a mature mRNA transcript.

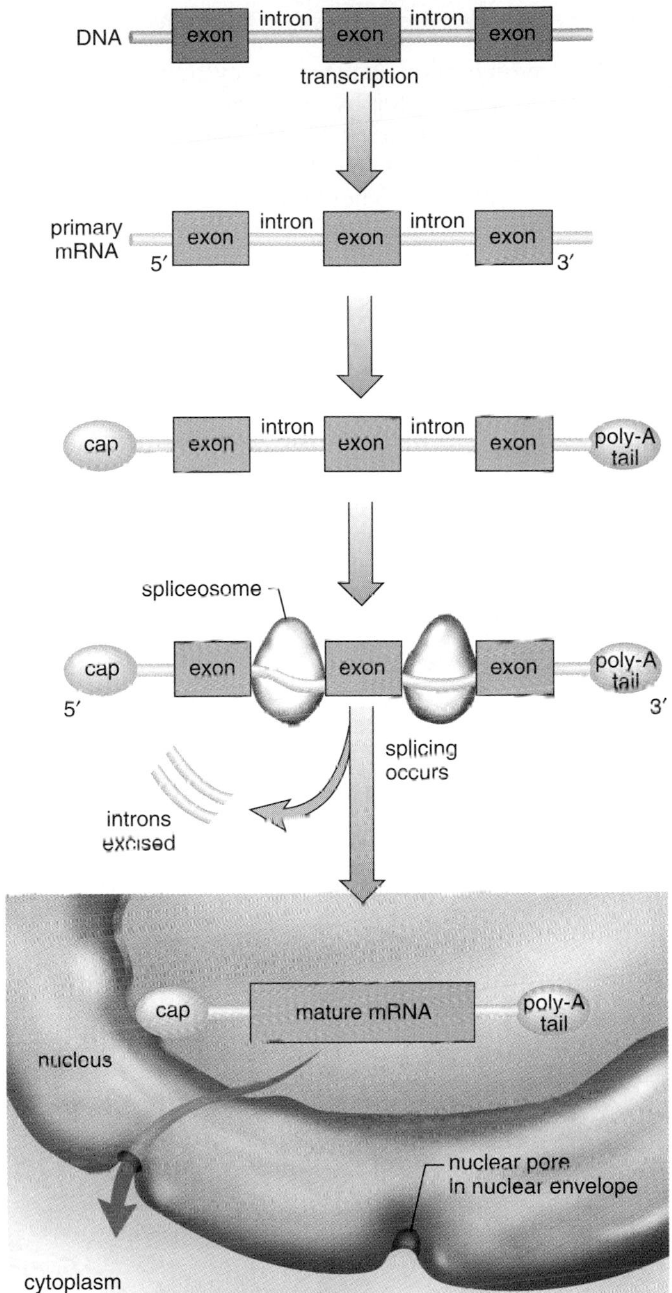

Figure 15.7 Messenger RNA (mRNA) processing in eukaryotes.
DNA contains both exons (coding sequences) and introns (noncoding sequences). Both of these are transcribed and are present in primary mRNA. During processing, a cap and a poly-A tail (a series of adenine nucleotides) are added to the molecule. Also, there is excision of the introns and a splicing together of the exons. This is accomplished by complexes called spliceosomes. Then the mature mRNA molecule is ready to leave the nucleus.

15.5 How Translation Occurs

Translation, which takes place in the cytoplasm of eukaryotic cells, is the second step by which gene expression leads to protein synthesis. During translation, the sequence of codons in mRNA at a ribosome directs the sequence of amino acids in a polypeptide. In other words, one language (nucleic acids) gets translated into another language (protein).

Transfer RNA Transfers Amino Acids

Transfer RNA (tRNA) molecules transfer amino acids to the ribosomes. A tRNA molecule is a single-stranded nucleic acid that doubles back on itself to create regions where complementary bases are hydrogen-bonded to one another. The structure of a tRNA molecule is generally drawn as a flat cloverleaf, but the space-filling model shows the molecule's three-dimensional shape (Fig. 15.8).

There is at least one tRNA molecule for each of the 20 amino acids found in proteins. The amino acid binds to the 3′ end. The opposite end of the molecule contains an **anticodon,** a group of three bases that is complementary to a specific codon of mRNA. For example, a tRNA that has the anticodon GAA binds to the codon CUU and carries the amino acid leucine. (There are fewer tRNAs than there are codons because some tRNAs can pair with more than one codon. If the anticodon contains a U in the third position, it will pair with either an A or G in the third position of a codon. This is called the *wobble effect.*)

How does the correct amino acid become attached to the correct tRNA molecule? This task is carried out by amino acid–activating enzymes, called tRNA synthetases. Just as a key fits a lock, each enzyme has a recognition site for the amino acid to be joined to a particular tRNA. This is an energy-requiring process that utilizes ATP. Once the amino acid–tRNA complex is formed, it travels through the cytoplasm to a ribosome, where protein synthesis is occurring.

Transfer RNA (tRNA) molecules bind with one particular amino acid, and they bear an anticodon that is complementary to a particular codon—the codon for that amino acid.

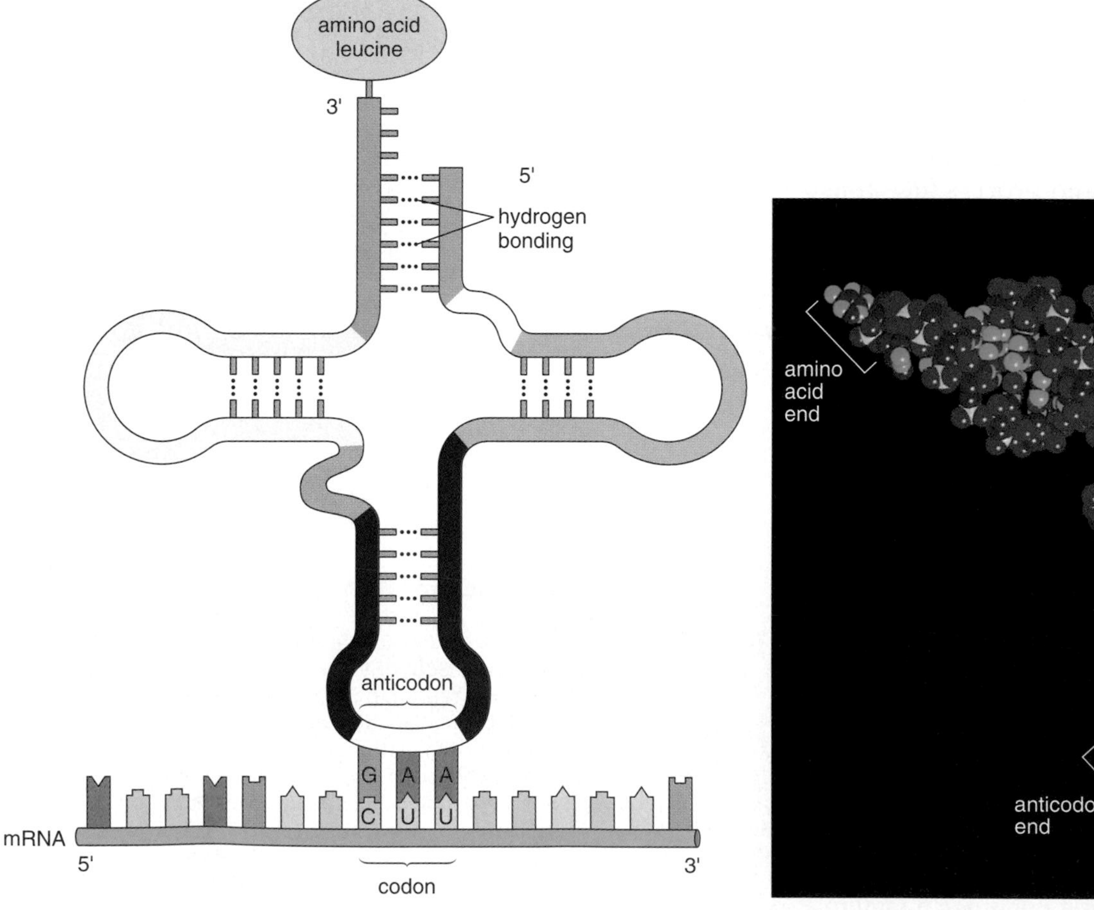

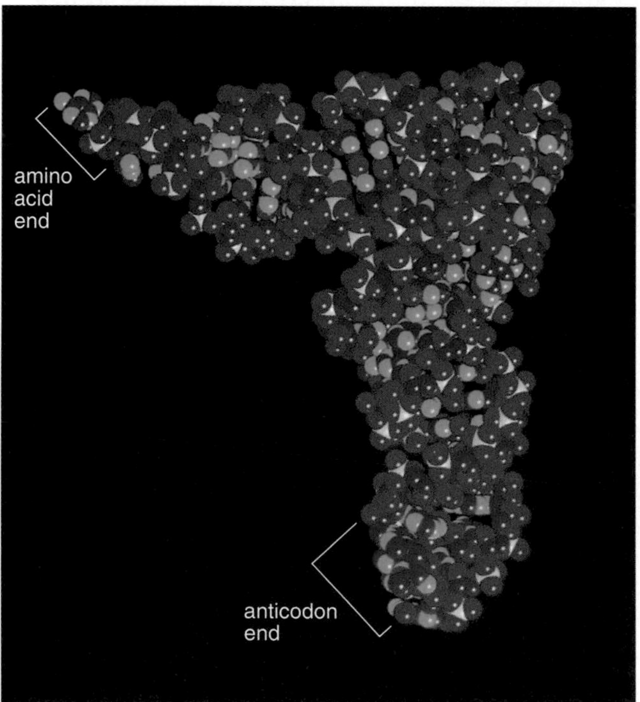

Figure 15.8 Structure of a transfer RNA (tRNA) molecule.
a. Complementary base pairing indicated by hydrogen bonding occurs between nucleotides of the molecule, and this causes it to form its characteristic loops. The anticodon that base-pairs with a particular messenger RNA (mRNA) codon occurs at one end of the folded molecule; the other two loops help hold the molecule at the ribosome. An appropriate amino acid is attached at the 3′ end of the molecule. For this mRNA codon and tRNA anticodon, the appropriate amino acid is leucine. **b.** Space-filling model of tRNA molecule.

Ribosomal RNA Is Structural

Ribosomal RNA (rRNA) is produced off a DNA template in the nucleolus of a nucleus. Then the rRNA is packaged with a variety of proteins into ribosomal subunits, one of which is larger than the other. The subunits move separately through nuclear envelope pores into the cytoplasm where they combine when translation begins. Ribosomes can remain in the cytosol or they can become attached to endoplasmic reticulum.

Prokaryotic cells contain about 10,000 ribosomes, but eukaryotic cells contain many times that number. Ribosomes play a significant role in protein synthesis (Fig. 15.9). Ribosomes have a binding site for mRNA and binding sites for two transfer RNA (tRNA) molecules at a time. These binding sites facilitate complementary base pairing between tRNA anticodons and mRNA codons. Ribosomal RNA (i.e., a ribozyme) is also believed to join amino acids together, in a manner to be described, so that a polypeptide results. As the ribosome moves down the mRNA molecule, new tRNAs arrive; amino acids are joined, and the polypeptide forms and grows longer. Translation terminates once the polypeptide is fully formed; the ribosome dissociates into its two subunits and falls off the mRNA molecule.

As soon as the initial portion of mRNA has been translated by one ribosome, and the ribosome has begun to move down the mRNA, another ribosome attaches to the mRNA. Therefore, several ribosomes are often attached to and translating the same mRNA. The entire complex is called a **polyribosome** (Fig. 15.9).

What determines whether a ribosome remains in the cytosol or binds to endoplasmic reticulum (ER)? All ribosomes are at first free within the cytosol. But once synthesis begins, some polypeptides have a series of amino acids called the signal sequence, which enables a ribosome to bind to rough endoplasmic reticulum (rough ER). Then the polypeptide enters the rough ER lumen and is eventually secreted from the cell. Nearly all polypeptides have a signal sequence that targets them for their final location in the cell.

Ribosomal RNA (rRNA) is found in the ribosomes, structures that facilitate complementary base pairing between mRNA and tRNAs and join amino acids together.

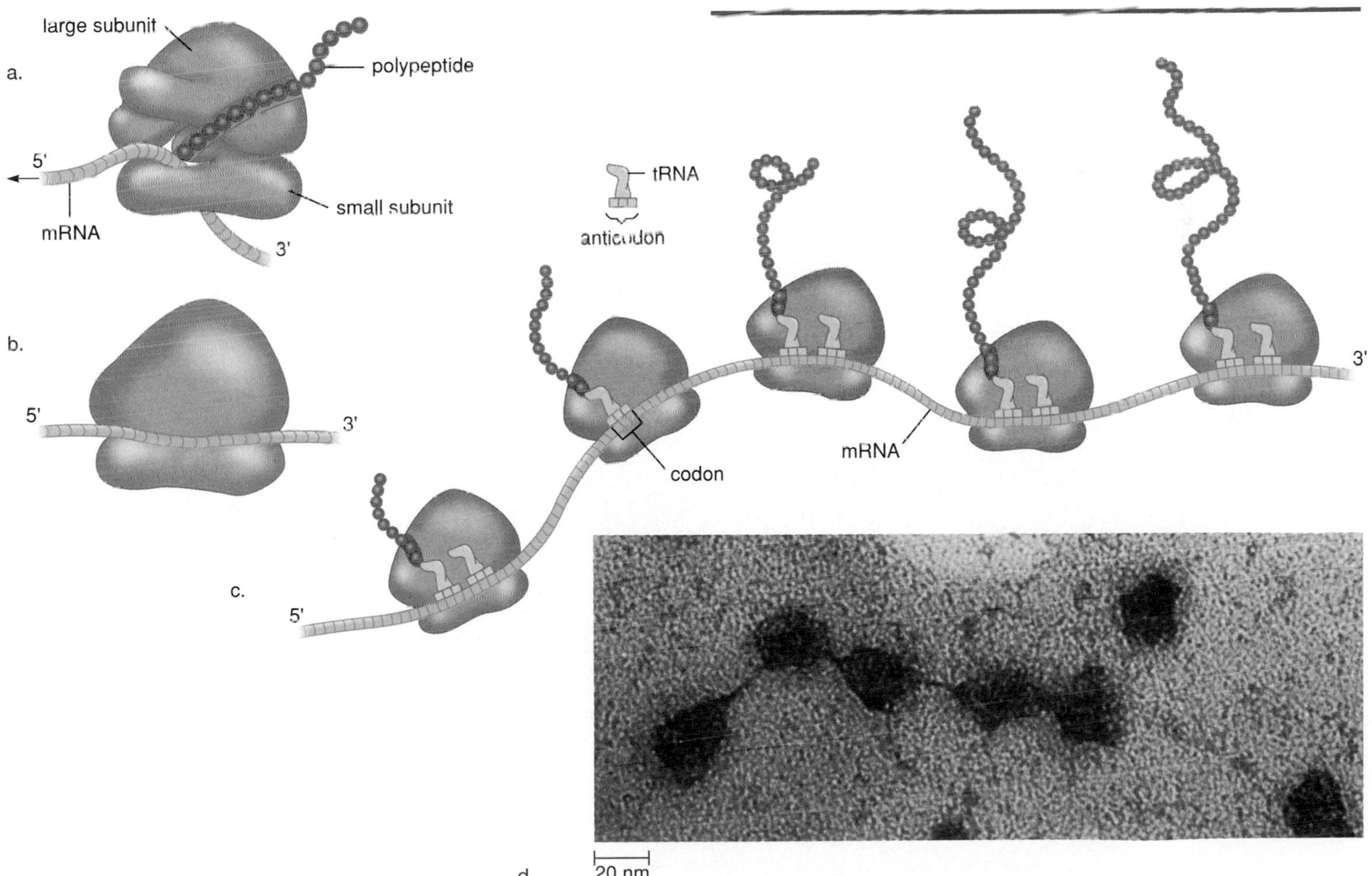

Figure 15.9 Polyribosome structure and function.
a. Side view of ribosome shows positioning of mRNA and growing polypeptide. **b.** An frontal view of a ribosome. **c.** Several ribosomes, collectively called a polyribosome, move along an mRNA at one time. Therefore, several polypeptides can be made at the same time. **d.** Electron micrograph of a polyribosome.

Translation Requires Three Steps

During translation, the codons of an mRNA base-pair with the anticodons of tRNA molecules carrying specific amino acids. The order of the codons determines the order of the tRNA molecules and the sequence of amino acids in a polypeptide. The process of translation must be extremely orderly so that the amino acids of a polypeptide are sequenced correctly.

Protein synthesis involves three steps: chain initiation, chain elongation, and chain termination. It should be kept in mind that enzymes are required for each of the steps to occur, and energy is needed for the first two steps also.

1. Chain initiation: In prokaryotes, a small ribosomal subunit attaches to the mRNA in the vicinity of the *start codon* (AUG). The first or initiator tRNA pairs with this codon. Then a large ribosomal subunit joins to the small subunit (Fig. 15.10*a).*

   Notice that a ribosome has two binding sites for tRNAs. One of these is called the P (for peptide) site, and the other is called the A (for amino acid) site. The initiator tRNA is capable of binding to the P site even though it carries only one amino acid, methionine. The A site is for the next tRNA.

   Proteins called initiation factors are required to bring the necessary translation components (small ribosomal subunit, mRNA, initiator tRNA, and large ribosomal subunit) together. Energy is also expended.

Following chain initiation, mRNA and the first tRNA are now at a ribosome.

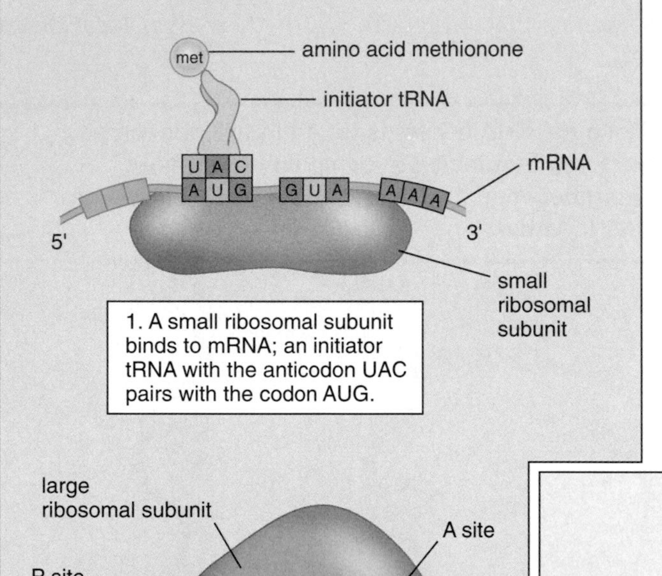

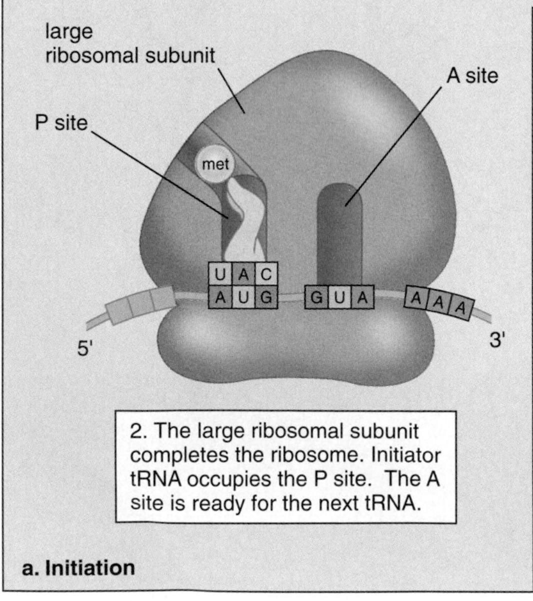

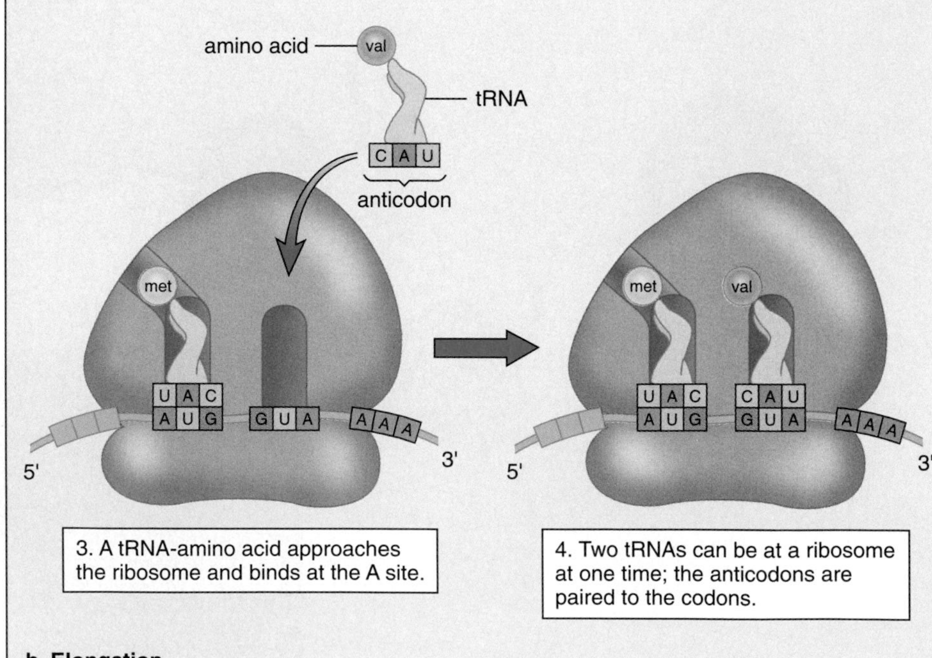

Figure 15.10 Protein synthesis.

2. Chain elongation: During elongation, a tRNA with an attached peptide is already at the P site, and a tRNA carrying its appropriate amino acid is just arriving at the A site. Proteins called elongation factors facilitate complementary base pairing between the tRNA anticodon and the mRNA codon (Fig. 15.10*b*).

   The polypeptide is transferred and attached by peptide bond formation to the newly arrived amino acid. A ribozyme, which is a part of the larger ribosomal subunit, and energy are needed to bring about this transfer. Now the tRNA molecule at the P site leaves.

   Next *translocation* occurs: the mRNA, along with the peptide-bearing tRNA, moves from the A site to the empty P site. Since the ribosome has moved forward three nucleotides, there is a new codon now located at the empty A site.

   The complete cycle—complementary base pairing of new tRNA, transfer of peptide chain, translocation—is repeated at a rapid rate (about 15 times each second in *Escherichia coli*).

During chain elongation, amino acids are added one at a time to the growing polypeptide.

3. Chain termination: Termination of polypeptide synthesis occurs at a stop codon, which does not code for an amino acid (Fig. 15.10*c*). The polypeptide is enzymatically cleaved from the last tRNA by a release factor. The tRNA and polypeptide leave the ribosome, which dissociates into its two subunits (see Fig. 15.9).

During chain termination, the ribosome separates into its two subunits and the polypeptide is released.

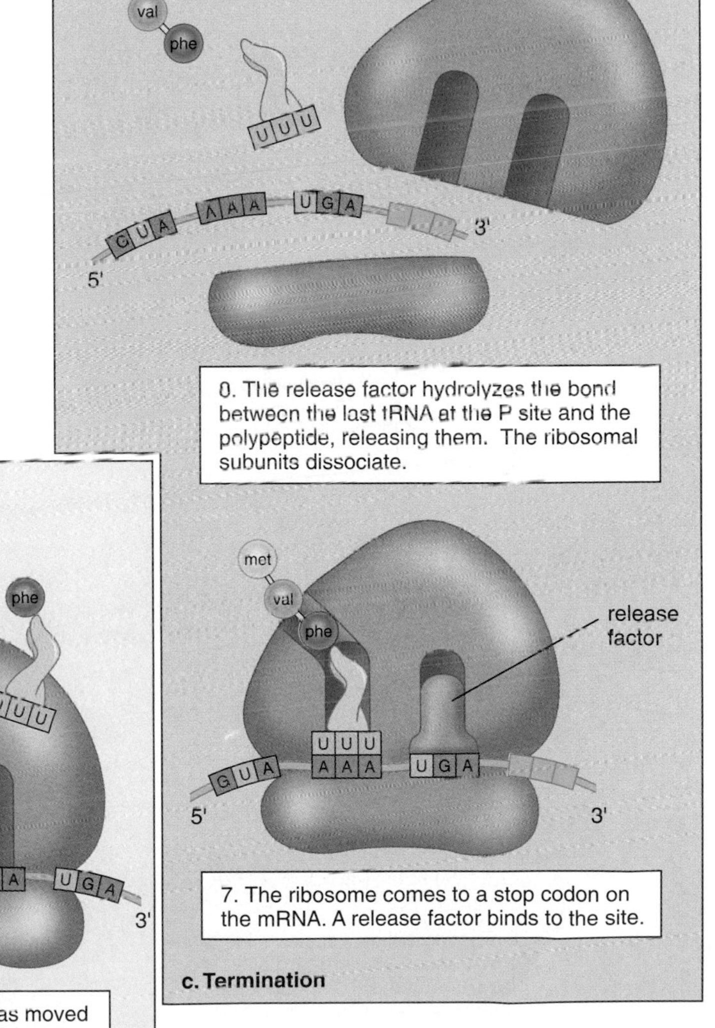
met
val
phe
UUU
GUA
AAA
UGA
5'
3'
8. The release factor hydrolyzes the bond between the last tRNA at the P site and the polypeptide, releasing them. The ribosomal subunits dissociate.
met
val
phe
release factor
UUU
AAA
UGA
GUA
7. The ribosome comes to a stop codon on the mRNA. A release factor binds to the site.

c. Termination

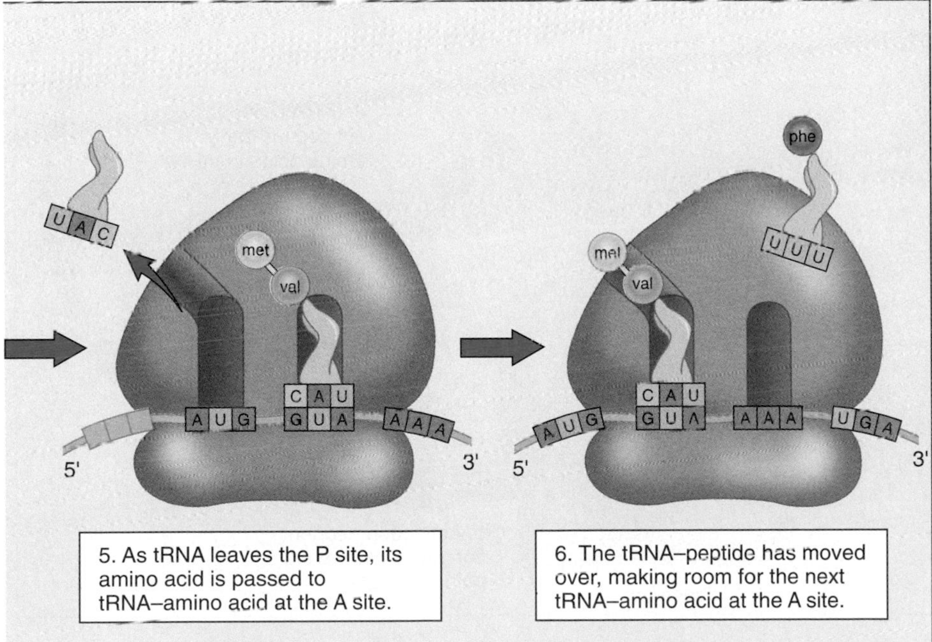
UAC
met
val
CAU
AUG
GUA
AAA
5'
3'
5. As tRNA leaves the P site, its amino acid is passed to tRNA–amino acid at the A site.
phe
met
val
UUU
CAU
AUG
GUA
AAA
UGA
6. The tRNA–peptide has moved over, making room for the next tRNA–amino acid at the A site.

Let's Review Gene Expression

The following list, along with Figure 15.11, provides a brief summary of the events involved in gene expression that results in a protein product.

1. DNA contains genetic information. The sequence of its bases determines the sequence of amino acids in a polypeptide.
2. During transcription, one strand of DNA serves as a template for the formation of mRNA. The bases in mRNA are complementary to those in DNA; every three bases is a codon that codes for an amino acid.
3. Messenger RNA (mRNA) is processed before it leaves the nucleus, during which time the introns are removed.
4. Messenger RNA (mRNA) carries a sequence of codons to the ribosomes, which are composed of rRNA and proteins.
5. Transfer RNA (tRNA) molecules, each of which is bonded to a particular amino acid, have anticodons that pair complementarily to the codons in mRNA.
6. During translation, tRNA molecules and their attached amino acids arrive at the ribosomes, and the linear sequence of codons of the mRNA determines the order in which the amino acids become incorporated into a protein.

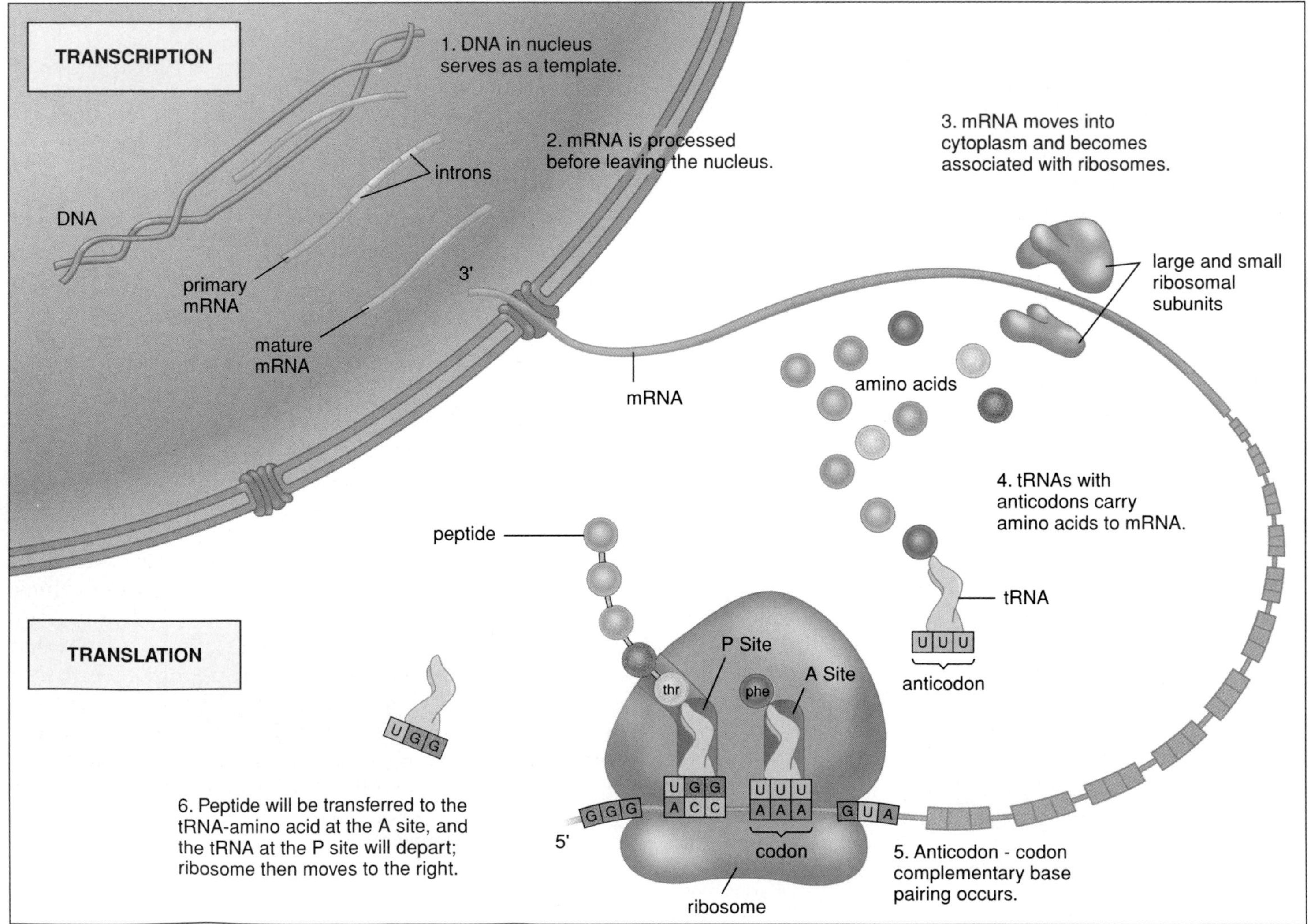

Figure 15.11 Summary of protein synthesis.
Gene expression leads to the formation of a product, most often a protein. Transcription, which occurs in the nucleus, and translation, which occurs in the cytoplasm at the ribosomes, are two steps required for gene expression.

15.6 Mutations Are Base Changes

Early geneticists understood that genes undergo mutations, but they didn't know what causes mutations. It is apparent today that a *gene mutation* is a change in the nucleotide sequence of DNA. If this sequence changes, then the codons change and the sequence of amino acids in a polypeptide changes.

Frameshift Mutations Are Drastic

The term *reading frame* applies to the sequence of codons because they are read from some specific starting point as in this sentence: THE CAT ATE THE RAT. If the letter C is deleted from this sentence and the reading frame is shifted, we read THE ATA TET HER AT—something that doesn't make sense. *Frameshift mutations* occur most often because one or more nucleotides are either inserted or deleted from DNA. The result of a frameshift mutation can be a completely new sequence of codons and a nonfunctional protein.

Point Mutations Can Be Drastic

Point mutations involve a change in a single nucleotide and therefore a change in a specific codon. When one base is substituted for another, the results can be variable. For example, in Figure 15.12, if UAC is changed to UAU, there is no noticeable effect, because both of these codons code for tyrosine. Therefore, this is called a silent mutation. If UAC is changed to UAG, however, the result could very well be a drastic one because UAG is a stop codon. If this substitution occurs early in the gene, the resulting protein may be too short and may be unable to function. Such an effect is called a nonsense mutation. Finally, if UAC is changed to CAC, then histidine is incorporated into the protein instead of tyrosine. This is a missense mutation. A change in one amino acid may not have an effect if the change occurs in a noncritical area or if the two amino acids have the same chemical properties. In this instance, however, the polarities of tyrosine and histidine differ; therefore, this substitution most likely will have a deleterious effect on the functioning of the protein. Recall that the occurrence of valine instead of glutamate in the β chain of hemoglobin results in sickle-cell disease (see Fig. 15.2).

Cause and Repair of Mutations

Mutations due to DNA replication errors are very rare. DNA polymerase, the enzyme that carries out replication, proofreads the new strand against the old strand and detects any mismatched nucleotides, and each is replaced with a correct nucleotide. In the end, there is only one mistake for every one billion nucleotide pairs replicated.

Mutagens [L. *mutatas*, change, and *genitus*, producing], environmental substances that cause random DNA base changes, include radiation (e.g., radioactive elements and X rays) and organic chemicals (e.g., certain pesticides and cigarette smoke). If these mutagens bring about a mutation in the gametes, then the offspring of the individual may be affected. On the other hand, if the mutation occurs in the body cells, then cancer may be the result. Various genes control the cell cycle, and if these mutate the result can be an increased cell division rate and a tumor.

Ultraviolet (UV) radiation is a mutagen that everyone is exposed to; it easily penetrates the skin and breaks the bonds of DNA in underlying tissues. Wherever there are two thymine molecules next to one another, ultraviolet radiation may cause them to bond together, forming thymine dimers. Usually, these dimers are removed from damaged DNA by special enzymes called repair enzymes. Repair enzymes constantly monitor DNA and repair any irregularities. One enzyme excises a portion of DNA that contains the dimer; another makes a new section by using the other strand as a template; and still another seals the new section in place. The importance of repair enzymes is exemplified by individuals with the condition known as xeroderma pigmentosum. They lack some of these repair enzymes, and as a consequence these individuals have a high incidence of skin cancer.

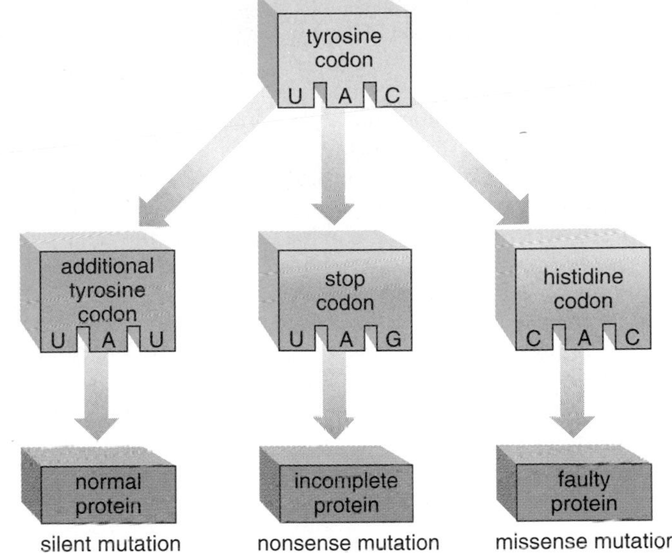

Figure 15.12 Point mutation.
The effect of a base alteration can vary. Starting at the left, if the base change codes for the same amino acid, there is no noticeable effect; if the base change codes for a stop codon, the resulting protein will be incomplete; and if the base change codes for a different amino acid, a faulty protein is possible.

A gene mutation is an alteration in the nucleotide sequence of a gene. The usual rate of mutation is low because DNA repair enzymes constantly monitor and repair any irregularities.

connecting concepts

When geneticists began examining gene activity, one problem immediately became apparent—how can genes dictate the production of certain proteins if the genes remain in the nucleus of eukaryotic cells and protein synthesis occurs in the cytoplasm? A go-between, messenger RNA (mRNA), answers this question.

By having a copy of DNA's nucleotide base sequence, mRNA takes a message into the cytoplasm and to the ribosomes where protein synthesis occurs. Transfer RNAs (tRNAs) bring the amino acids to the ribosomes where they are connected in the order directed by the sequence of codons in mRNA. In this way, the nucleotide sequence in DNA specifies the amino acid sequence in a polypeptide.

Reading about gene activity can sometimes cause you to miss the forest for the trees; it's easy to get caught up in the details and miss the overall point. One cell differs from another by its enzymatic content and, therefore, we might be tempted to say that living things are the products of their enzymes and forget about DNA. But remember that one enzyme differs from another by the sequence of its amino acids, and that the sequence of these amino acids is specified by DNA. Because DNA specifies the order in which amino acids are strung together, it is DNA that controls the structure of the cell, and ultimately the entire phenotype.

So, for example, a brown-eyed person differs from a blue-eyed person because the brown-eyed person has a slightly different mix of enzymes for eye color than does a blue-eyed person. This is due, of course, to the brown-eyed person having slightly different DNA nucleotide base sequences in his or her genes than a blue-eyed person.

Summary

15.1 What Genes Do

Several investigators contributed to our knowledge of what genes do. Garrod is associated with the phrase "inborn error of metabolism" because he suggested that some of his patients had inherited an inability to carry out certain enzymatic reactions.

Beadle and Tatum X-rayed spores of *Neurospora crassa* and found that some of the subsequent cultures lacked a particular enzyme needed for growth on minimal medium. Since they found that the mutation of one gene results in the lack of a single enzyme, they suggested the one gene-one enzyme hypothesis.

Pauling and Itano found that the chemical properties of the β (beta) chain of sickle-cell hemoglobin differ from those of normal hemoglobin, and therefore the one gene-one polypeptide hypothesis was formulated instead. Later, Ingram showed that the biochemical change is due to the substitution of the amino acid valine (nonpolar) for glutamate (polar).

15.2 How Genes Are Expressed

RNA differs from DNA in these ways: (1) the pentose sugar is ribose, not deoxyribose; (2) the base uracil replaces thymine; (3) RNA is single stranded.

The central dogma of molecular biology says that (1) DNA is a template for its own replication and also for RNA formation during transcription, and (2) the complementary sequence of nucleotides in mRNA directs the correct sequence of amino acids of a polypeptide during translation.

15.3 How Genes Code for Amino Acids

The sequence of bases in DNA specifies the proper sequence of amino acids in a polypeptide. The genetic code is a triplet code, and each codon (code word) consists of three bases. The code is degenerate; that is, more than one codon exists for most amino acids. There are also one start and three stop codons. The genetic code is just about universal.

15.4 How Transcription Occurs

Transcription begins when RNA polymerase attaches to a promoter. Elongation occurs until there is a terminator sequence.

Messenger RNA (mRNA) is processed following transcription, a cap is put onto the 5′ end and a poly-A tail is put onto the 3′ end; introns are removed in eukaryotes by spliceosomes.

15.5 How Translation Occurs

Translation requires mRNA, ribosomal RNA (rRNA), and transfer RNA (tRNA). Each tRNA has an anticodon at one end and an amino acid at the other; there are amino acid–activating enzymes that ensure that the correct amino acid is attached to the correct tRNA. When tRNAs bind with their codon at a ribosome, the amino acids are correctly sequenced in a polypeptide according to the order predetermined by DNA.

Many ribosomes move along the same mRNA at a time. Collectively, these are called a polyribosome.

Translation requires these steps. During initiation, mRNA, the first (initiator) tRNA, and the ribosome all come together in the proper orientation at a start codon. During elongation, as the tRNAs bind to their codons, the growing peptide chain is transferred by peptide bonding to the next amino acid in the chain. During termination at a stop codon, the polypeptide is cleaved from the last tRNA. The ribosome now dissociates.

15.6 Mutations Are Base Changes

In molecular terms, a gene is a sequence of DNA nucleotide bases, and a gene mutation is a change in this sequence. Frameshift mutations result when a base is added or deleted, and the result is usually a nonfunctioning protein. Point mutations can range in effect depending on the particular codon change. Gene mutation rates are rather low because DNA polymerase proofreads the new strand during replication and because there are repair enzymes that constantly monitor the DNA.

Reviewing the Chapter

1. What made Garrod think there were inborn errors of metabolism? 236
2. Explain Beadle and Tatum's experimental procedure. 236
3. How did Pauling and Itano know that the chemical properties of Hb^S differed from those of Hb^A? What change does a substitution of valine for glutamate cause in the chemical properties of these molecules? 236–37
4. What are the biochemical differences between RNA and DNA? 238
5. Draw a diagram for the central dogma of molecular biology and tell where the various events occur in a eukaryotic cell. 238
6. How did investigators reason that the code must be a triplet code, and in what manner was the code cracked? Why is it said that the code is degenerate, unambiguous, and almost universal? 239
7. What are the specific steps that occur during transcription of RNA off a DNA template? 240–41
8. How is messenger RNA (mRNA) processed before leaving the eukaryotic nucleus? 241
9. Compare the functions of mRNA, ribosomal RNA (rRNA), and transfer RNA (tRNA) during protein synthesis. What are the specific events of translation? 242–46
10. Why is a frameshift mutation always expected to result in a faulty protein, while a substitution of one base for another may not always result in a faulty protein? 247

Testing Yourself

Choose the best answer for each question.

For questions 1–4, match the investigator to the phrase in the key:

Key:

a. one gene-one enzyme hypothesis
b. inborn error of metabolism
c. one gene-one polypeptide hypothesis

1. Sir Archibald Garrod
2. George Beadle and Edward Tatum
3. Linus Pauling and Harvey Itano
4. Considering the following pathway, if Beadle and Tatum found that *Neurospora* cannot grow if metabolite A is provided, but can grow if B, C, or D is provided, what enzyme would be missing?

$$A \xrightarrow{1} B \xrightarrow{2} C \xrightarrow{3} D$$

   a. enzyme 1
   b. enzyme 2
   c. enzyme 3
   d. All of these are correct.

5. The central dogma of molecular biology
   a. states that DNA is a template for all RNA production.
   b. states that DNA is a template only for DNA replication.
   c. states that translation precedes transcription.
   d. pertains only to prokaryotes because humans are unique.
6. If the sequence of bases in DNA is TAGC, then the sequence of bases in RNA will be
   a. ATCG.
   b. TAGC.
   c. AUCG.
   d. Both a and b are correct.
7. RNA processing is
   a. the same as transcription.
   b. an event that occurs after RNA is transcribed.
   c. the rejection of old, worn-out RNA.
   d. Both b and c are correct.
8. During protein synthesis, an anticodon on transfer RNA (tRNA) pairs with
   a. DNA nucleotide bases.
   b. ribosomal RNA (rRNA) nucleotide bases.
   c. messenger RNA (mRNA) nucleotide bases.
   d. other tRNA nucleotide bases.
9. If the DNA codons are CAT CAT CAT, and a guanine base is added at the beginning, then which would result?
   a. G CAT CAT CAT
   b. GCA TCA TCA T
   c. frameshift mutation
   d. Both b and c are correct.
10. This is a segment of a DNA molecule. (Remember that the template strand only is transcribed.) What are (a) the RNA codons, (b) the tRNA anticodons, and (c) the sequence of amino acids in a protein?

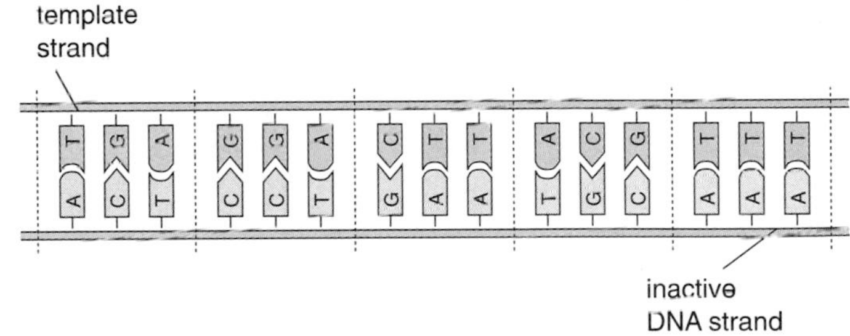

Applying the Concepts

1. *DNA stores genetic information.*
   What is the genetic information stored by DNA, and where is it stored in the molecule?
2. *The genetic code is almost universal.*
   What is the evolutionary significance of a universal genetic code?
3. *There is a flow of information from DNA to RNA to protein.*
   DNA and a protein are colinear molecules. Explain.

Using Technology

Your study of gene activity is supported by these available technologies:

Exploring the Internet

The Mader Home Page provides resources for and help with studying this chapter.

http://www.mhhe.com/sciencemath/biology/mader/
(Click on Biology.)

Explorations in Cell Biology & Genetics CD-ROM

How Proteins Function: Hemoglobin (#1)
Reading DNA (#15)

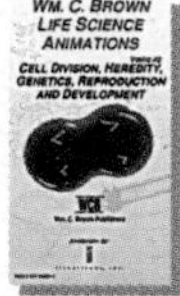

Life Science Animations Video

Video #2: Cell Division/Heredity/Genetics/Reproduction and Development
Transcription of a Gene (#16)
Protein Synthesis (#17)

Understanding the Terms

anticodon 242
codon 239
exon 241
intron 241
messenger RNA (mRNA) 238
mutagen 247
polyribosome 243
promoter 240
ribosomal RNA (rRNA) 238
ribozyme 241
RNA polymerase 240
transcription 238
transfer RNA (tRNA) 238
translation 238
triplet code 239

Match the terms to these definitions:

a. __________ Agent, such as radiation or a chemical, that brings about a mutation.
b. __________ Enzyme that speeds the formation of RNA from a DNA template.
c. __________ In a gene, the portion of the DNA code that is expressed as the result of polypeptide formation.
d. __________ Noncoding segments of DNA that are transcribed but removed before mRNA leaves the nucleus.
e. __________ Process whereby a DNA strand serves as a template for the formation of mRNA.
f. __________ Process whereby the sequence of codons in mRNA determines (is translated into) the sequence of amino acids in a polypeptide.
g. __________ String of ribosomes, simultaneously translating different regions of the same mRNA strand during protein synthesis.
h. __________ Three nucleotides of DNA or mRNA; it codes for a particular amino acid or termination of translation.
i. __________ Three nucleotides on a tRNA molecule attracted to a complementary codon on mRNA.
j. __________ Type of RNA that transfers a particular amino acid to a ribosome during protein synthesis; at one end it binds to the amino acid and at the other end it has an anticodon that binds to an mRNA codon.

Regulation of Gene Activity

CHAPTER 16

Chapter Concepts

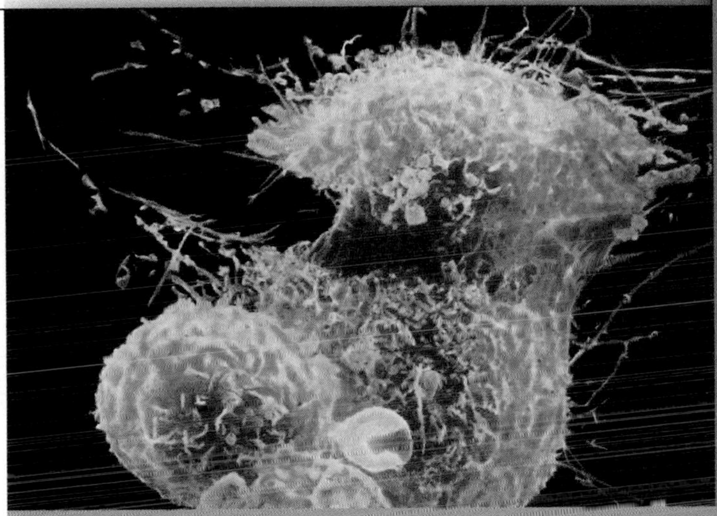

Cancer cells dividing

We all begin life as a one-celled zygote that has 46 chromosomes, and these same chromosomes are passed to all the daughter cells during mitosis. How do the same chromosomes direct certain body cells to become the cardiac muscle of the heart, the lining of the digestive tract, or the nervous tissue of the brain, and all the other types of tissues in the human body? All human cells have genes for digestive enzymes and muscle proteins, but these genes are expressed only in cells that line the digestive tract and in muscle cells, respectively. It must be possible to control gene expression in each particular type of cell.

In the previous chapter, we saw that genes, through transcription and translation, control which enzymes are present in cells. It would seem that a difference in enzyme activity determines what each cell is like, but what regulates whether transcription occurs and whether an enzyme is active in a cell? In this chapter we will see that control of gene expression extends from the nucleus (whether a gene is transcribed) to the cytoplasm (whether an enzyme is active).

This knowledge has far-ranging applications, from an understanding of how development progresses to what causes the occurrence of various disorders, including cancer. Cancer develops when genes that promote cell division are expressed and when genes that suppress cell division are not expressed. It has become increasingly apparent that the control of gene expression is of major importance to the health of all species, including human beings.

16.1 Prokaryotes Utilize Operons

Bacteria don't need the same enzymes (and possibly other proteins) all the time. Suppose, for example, the environment supplies various nutrients—wouldn't it be advantageous to produce just the enzymes needed to break down that nutrient?

In 1961, French microbiologists François Jacob and Jacques Monod showed that *Escherichia coli* is capable of regulating the expression of genes necessary for lactose breakdown. They proposed what is called the operon [L. *opera*, work] model to explain gene regulation in prokaryotes and later received a Nobel Prize for their investigations.

An **operon** consists of a series of genes, called structural genes, plus a portion of DNA called an operator and another portion called a promoter. Each structural gene codes for an enzyme in the same metabolic pathway. A regulatory gene, located outside the operon, codes for a repressor protein that can bind to the operator and switch off the operon.

An operon includes the following elements:

Promoter—a short sequence of DNA where RNA polymerase first attaches when a gene is to be transcribed.

Operator—a short portion of DNA where the repressor protein, coded for by a regulatory gene, can bind. When the repressor protein is bound to the operator, RNA polymerase cannot attach to the promoter and transcription does not occur.

Structural genes—one to several genes coding for enzymes of a metabolic pathway that are transcribed as a unit.

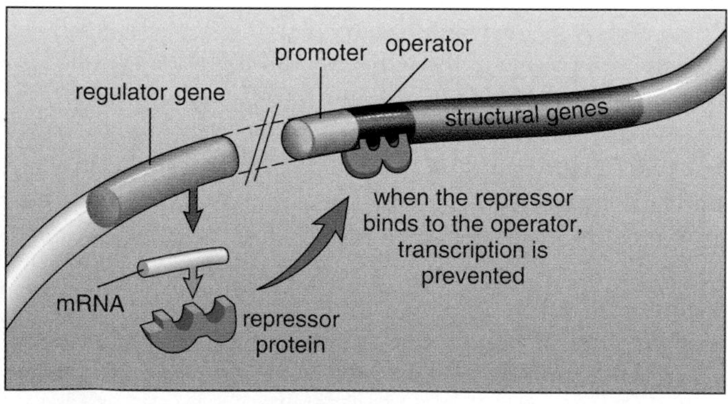

Looking at the *lac* Operon

When *E. coli* is denied glucose and is given the milk sugar lactose instead, it immediately begins to make the three enzymes needed for the metabolism of lactose. These enzymes are encoded by three genes: one gene is for an enzyme called β-galactosidase, which breaks down the disaccharide lactose to glucose and galactose; a second gene codes for a permease that facilitates the entry of lactose into the cell; and a third gene codes for an enzyme called transacetylase, which has an accessory function in lactose metabolism.

The three genes are adjacent to one another on the chromosome and are under the control of a single promoter and a single operator (Fig. 16.1). The **regulator gene,** directly ahead of the promoter, codes for a *lac* operon repressor protein that ordinarily binds to the operator and prevents transcription of the three genes. But when lactose is present, the lactose binds to the **repressor,** and the repressor undergoes a change in shape that prevents it from binding to the operator. Because the repressor is unable to bind to the operator, RNA polymerase is better able to bind to the promoter. When RNA polymerase carries out transcription, the three enzymes for lactose metabolism are produced. Because the presence of lactose brings about production of enzymes, it is called an **inducer** of the *lac* operon: the enzymes are said to be inducible enzymes, and the entire unit is called an **inducible operon.** Inducible operons are usually necessary to metabolic pathways that break down a nutrient.

E. coli preferentially breaks down glucose, and the bacterium has a way to ensure that the lactose operon is maximally turned on only when glucose is absent. When glucose is absent, a molecule called *cyclic AMP (cAMP)* accumulates. (Cyclic AMP, which is derived from ATP, has only one phosphate group, which is attached to ribose at two locations.)

adenine
5′ CH_2 O
P
3′
OH

Cyclic AMP
(cAMP)

Cyclic ATP binds to a molecule called a *catabolite activator protein (CAP)* and the complex attaches to the *lac* promoter, at a binding site. Now RNA polymerase is better able to bind to the promoter.

When glucose is present there is little cAMP in the cell; CAP is inactive and the lactose operon does not function maximally. CAP affects other operons as well and takes its name for inhibiting the catabolism of all metabolites except glucose.

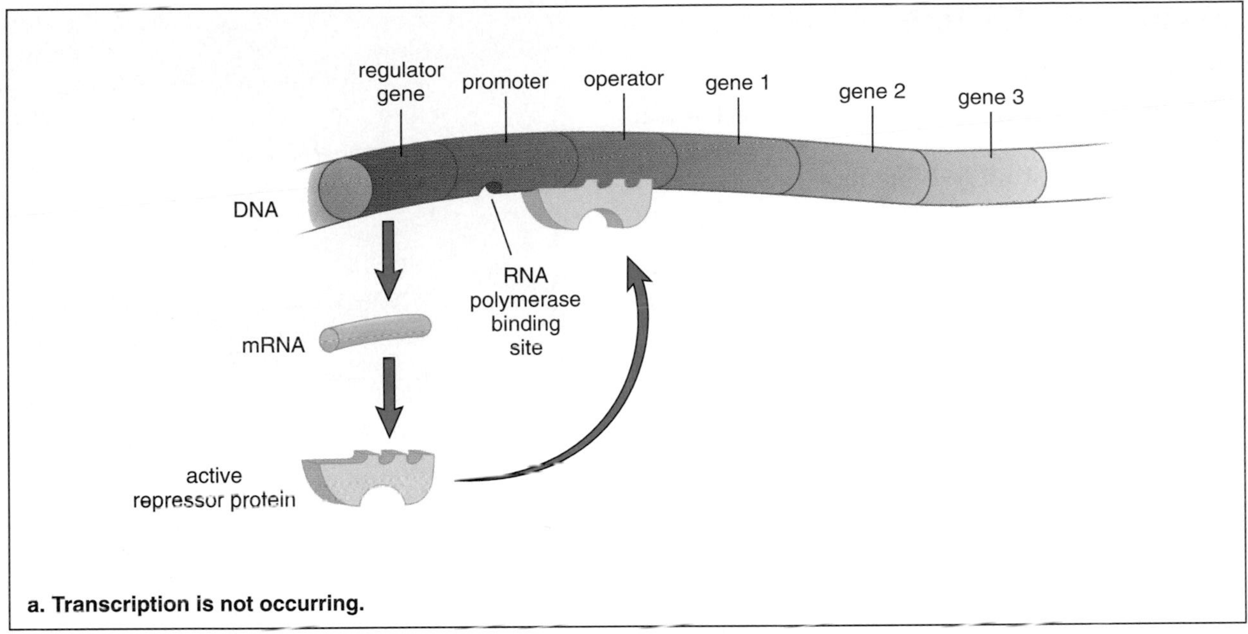

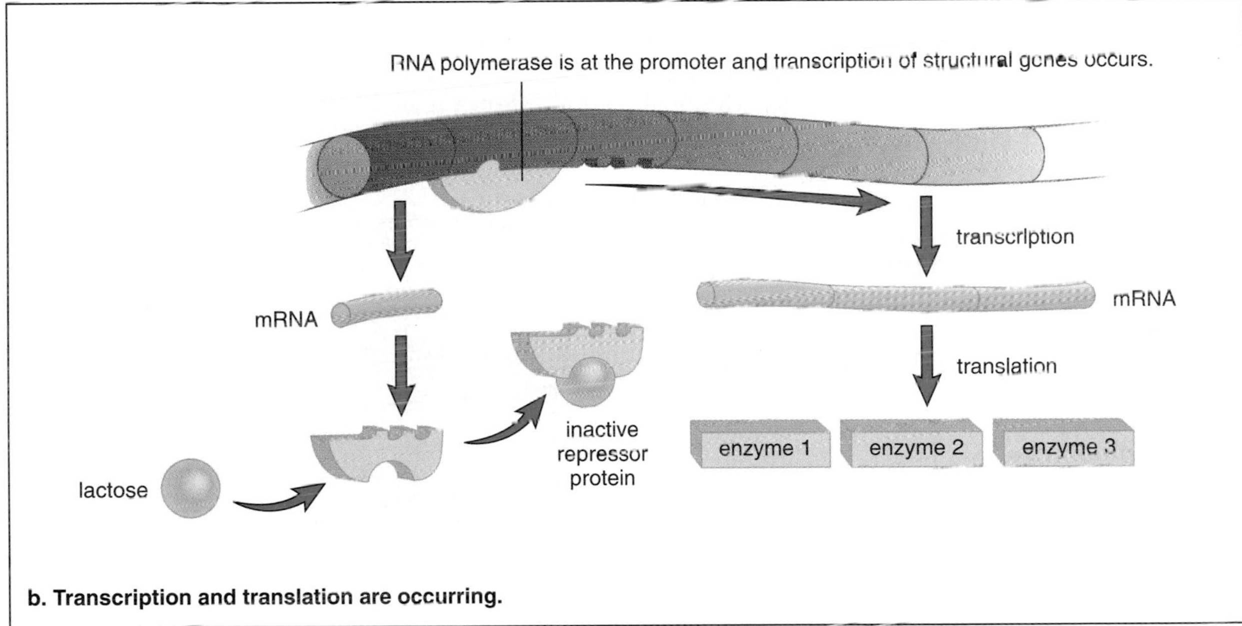

Figure 16.1 The *lac* operon.
a. The regulator gene codes for a repressor protein that is normally active. When active, the repressor protein binds to the operator and prevents RNA polymerase from attaching to the promoter. Therefore, transcription of the three structural genes does not occur. **b.** When lactose (or more correctly, allolactose, an isomer formed from lactose) is present, it binds to the repressor protein, changing its shape so that it can no longer bind to the operator. Now, RNA polymerase binds to the promoter; transcription and translation of the three structural genes follow.

Looking at the *trp* Operon

Jacob and Monod found that other operons in *E. coli* usually exist in the on rather than off condition. For example, the prokaryotic cell ordinarily produces five enzymes that are part of an anabolic pathway for the synthesis of the amino acid tryptophan. If tryptophan is already present in the medium, these enzymes are not needed by the cell. In the *trp* operon, the regulator codes for a repressor that ordinarily is unable to attach to the operator. The repressor has a binding site for tryptophan, and if tryptophan is present it binds to the repressor. Now, a change in shape allows the repressor to bind to the operator. The enzymes are said to be repressible, and the entire unit is called a **repressible operon.** Tryptophan is called the **corepressor.** Repressible operons are usually involved in anabolic pathways that synthesize a substance needed by the cell.

Bacterial DNA contains control regions whose sole purpose is to regulate the transcription of structural genes that code for enzymes or other products.

16.2 Eukaryotes Utilize Various Methods

In eukaryotes, there is no universal regulatory mechanism to control the *expression of genes* that code for a protein product. Many levels of control are possible, and different genes are regulated in various ways. Regulation at the level of transcription is possible, as is regulation at any point in the pathway from gene to functional protein. In eukaryotic cells, there are four primary levels of control of gene activity (Fig. 16.2).

1. **Transcriptional control:** In the nucleus a number of mechanisms serve to control which structural genes are transcribed or the rate at which transcription of the genes occurs. These include the organization of chromatin and the use of transcription factors that initiate transcription, the first step in the process of gene expression.
2. **Posttranscriptional control:** Posttranscriptional control occurs in the nucleus after DNA is transcribed and primary mRNA is formed. Differential processing of preliminary mRNA before it leaves the nucleus, and also the speed with which mature mRNA leaves the nucleus, can affect the ultimate amount of gene product.
3. **Translational control:** Translational control occurs in the cytoplasm after mRNA leaves the nucleus and before there is a protein product. The life expectancy of mRNA molecules (how long they exist in the cytoplasm) can vary, as can their ability to bind ribosomes. It is also possible that some mRNAs may need additional changes before they are translated at all.
4. **Posttranslational control:** Posttranslational control, which also takes place in the cytoplasm, occurs after protein synthesis. The polypeptide product may have to undergo additional changes before it is biologically functional. Also, a functional enzyme is subject to feedback control—the binding of an end product can change the shape of an enzyme so that it is no longer able to carry out its reaction.

Control of gene expression occurs at four levels in eukaryotes. In the nucleus there is transcriptional and posttranscriptional control; in the cytoplasm there is translational and posttranslational control.

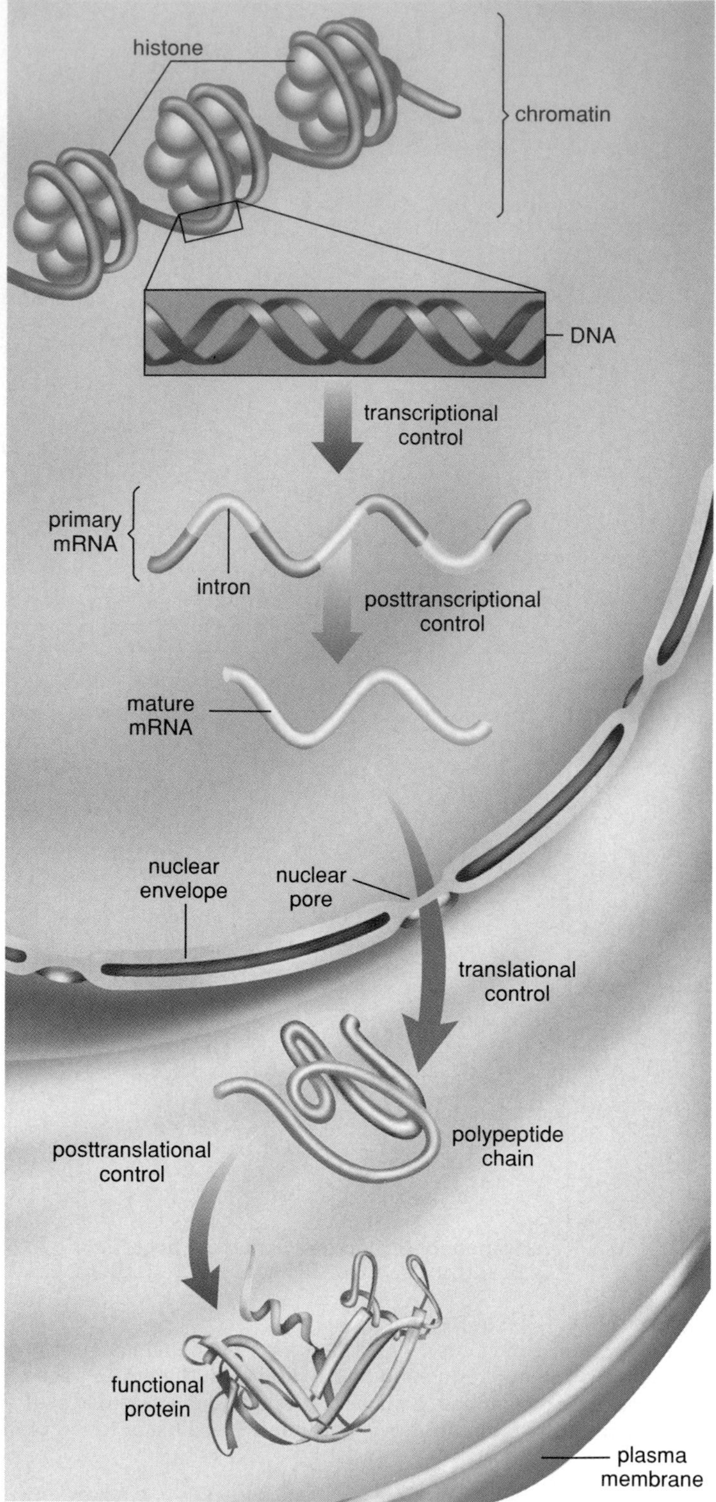

Figure 16.2 Levels of gene expression control in eukaryotic cells.
Transcriptional and posttranscriptional control occur in the nucleus. Translational and posttranslational control occur in the cytoplasm.

A closer look

What's in a Chromosome?

By the mid-1900s, it was known that chromosomes are made up of both DNA and protein, largely histones. Much more is now known about histone proteins.

There are five primary types of histone molecules designated H1, H2A, H2B, H3, and H4. Remarkably, the amino acid sequences of H3 and H4 vary little between organisms. For example, the H4 of peas is only two amino acids different from the H4 of cattle. This similarity suggests that there have been few mutations in the histone proteins during the course of evolution and that the histones therefore have very important functions.

A human cell contains at least 2 meters of DNA. Yet all of this DNA is packed into a nucleus that is about 5 µm in diameter. The histones are responsible for packaging the DNA so that it can fit into such a small space. First the DNA double helix is wound at intervals around a core of eight histone molecules (two copies each of H2A, H2B, H3, and H4), giving the appearance of a string of beads (Fig. 16A*a* and *b*). Each bead is called a nucleosome, and the nucleosomes are said to be joined by "linker" DNA. This string is coiled tightly into a fiber that has six nucleosomes per turn (Fig. 16A*c*). The H1 histone appears to mediate this coiling process. The fiber loops back and forth (Fig. 16A*d* and *e*) and can condense to produce a highly compacted form (Fig. 16A*f*) characteristic of metaphase chromosomes.

An interphase nucleus shows darkly stained chromatin regions called heterochromatin and lightly stained chromatin regions called euchromatin (Fig. 16B). Heterochromatin is genetically inactive (Fig. 16A*e*). The mechanism of inactivation is unknown, but it may involve methylation (adding methyl groups—CH_3—to the bases of DNA). Euchromatin is genetically active (Fig. 16A*c*). It's believed that in this form, the nucleosomes allow access to the DNA, so that a gene can be turned on and be expressed in the cell.

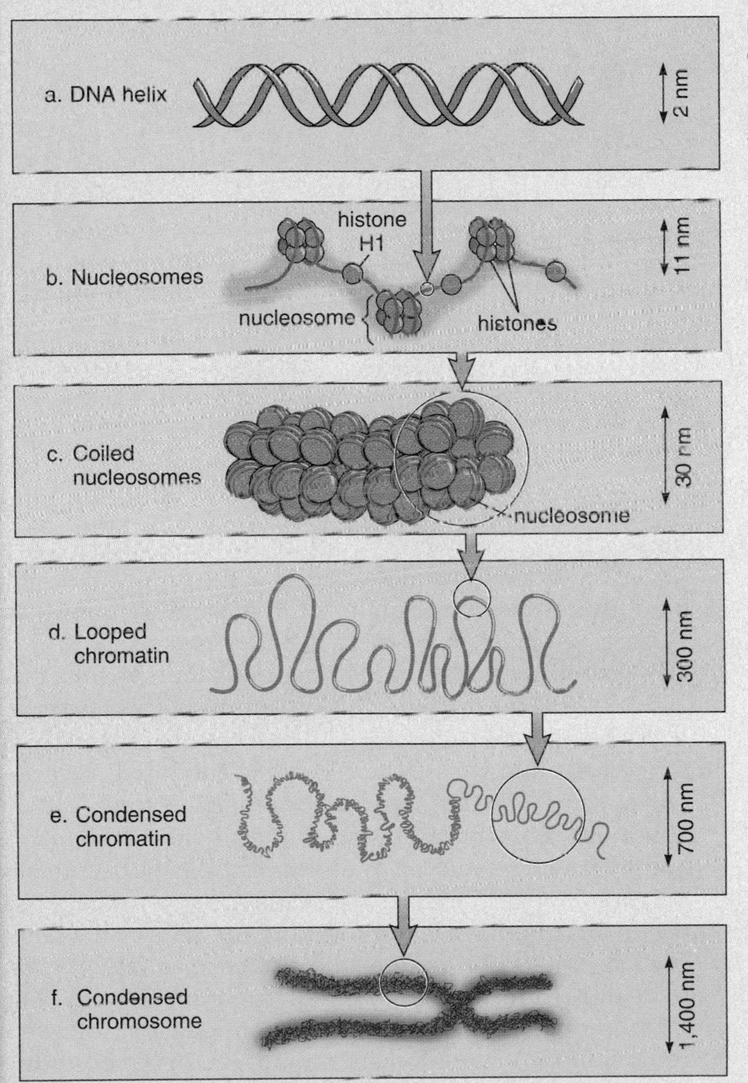

Figure 16A Levels of chromosomal structure.
Each drawing has a scale giving a measurement of length for that drawing. Notice that each measurement represents an ever-increasing length; therefore, it would take a much higher magnification to see the structure in (a) compared to that in (f).

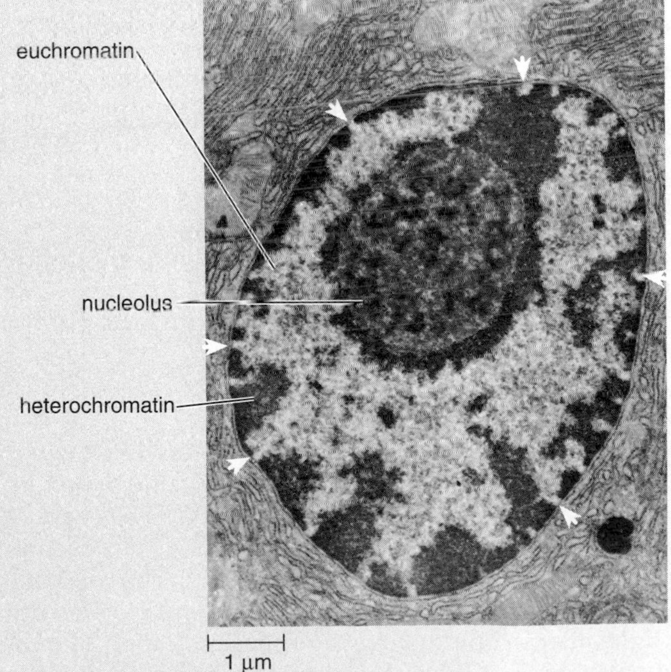

Figure 16B Eukaryotic nucleus.
The nucleus contains chromatin, DNA at two different levels of coiling and condensation. Euchromatin is at the level of coiled nucleosomes (30 nm), and heterochromatin is at the level of condensed chromatin (700 nm) in Figure 16A. Arrows indicate nuclear pores.

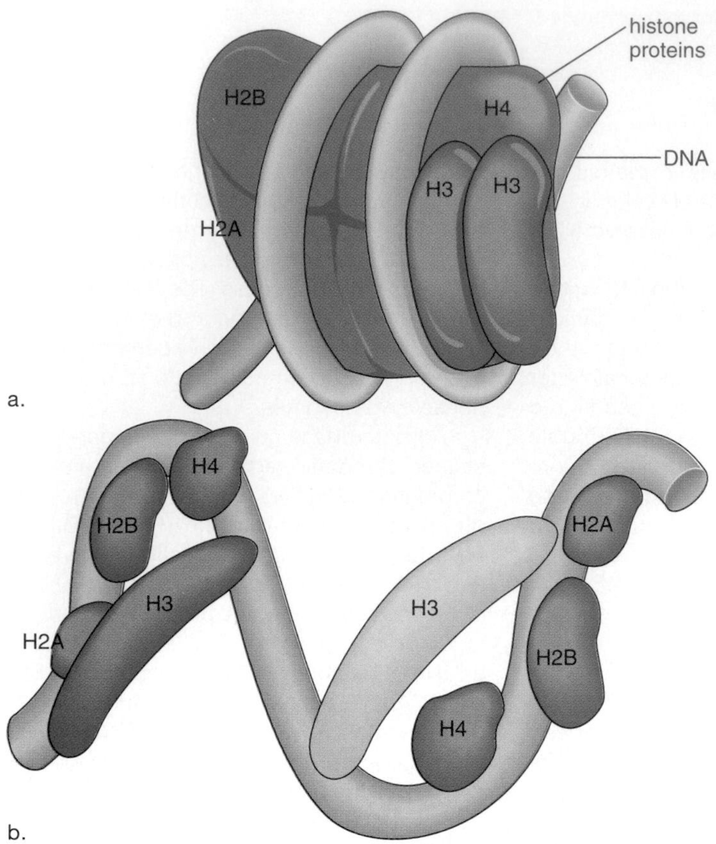

Figure 16.3 Nucleosome structure.
a. In a nucleosome, the DNA double helix winds twice around a group of histone proteins. Such tight packing is characteristic of heterochromatin. **b.** This model shows how it would be possible for the histones to loosen up, allowing transformation from heterochromatin to euchromatin as necessary for transcription to occur.

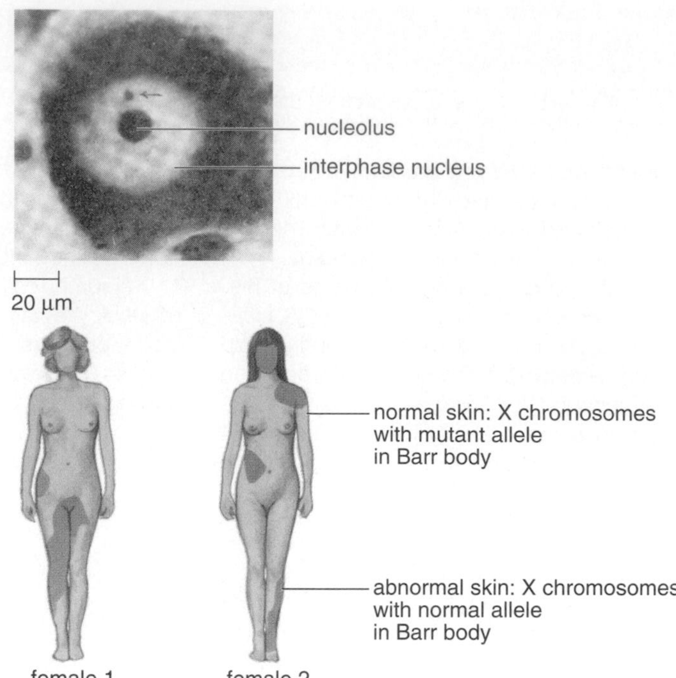

Figure 16.4 Barr body.
Each female nucleus contains a Barr body, which is a condensed, inactive X chromosome. Females are mosaics for alleles carried on the X chromosome because chance alone dictates which X chromosome becomes a Barr body. These two females are heterozygous and have inherited one allele for normal sweat glands and another allele for inactive sweat glands. Where in the body the mutant allele expresses itself (purple color) varies from female to female.

Looking at Transcriptional Control

We might expect transcriptional control in eukaryotes (1) to involve the organization of chromatin and (2) to include regulatory proteins such as those we have just observed in prokaryotes.

Chromatin Plays a Role

In eukaryotes, DNA is associated with proteins. DNA winds around histone proteins to form a spool, and the entire structure is called a nucleosome (Fig. 16.3*a*). The chain of spools coils, loops, and winds to form visible chromosomes during cell division as discussed in the reading on page 255. Chromatin that is highly compacted and visible with the light microscope during interphase is called **heterochromatin.** More diffuse chromatin is called **euchromatin.**

The histone molecules labeled H2–H4 may regulate this change in chromosomal structure (Fig. 16.3*b*). Recent studies seem to suggest that histones can both repress and activate genes. Analysis of the amino acid sequences of histones has established that the histones differ little between species. This suggests that the function of histones is vital to the cell.

Heterochromatin Is Inactive To demonstrate that heterochromatin is genetically inactive, we can refer to the **Barr body,** a highly condensed structure. In mammalian females, one of the X chromosomes is found in the Barr body; chance alone determines which of the X chromosomes is condensed. For a given gene, if a female is heterozygous, 50% of her cells have Barr bodies containing one allele and the other 50% have the other allele. This causes the body of heterozygous females to be mosaic, with patches of genetically different cells (Fig. 16.4). The mosaic is exhibited in various ways: some females show columns of both normal and abnormal dental enamel; some have patches of pigmented and nonpigmented cells at the back of the eye; some have patches of normal muscle tissue and degenerative muscle tissue. Barr bodies are not found in the cells of the female gonads, where the genes of both X chromosomes appear to be needed for development of the immature egg. Likewise, it is clear that the inactivation of one X chromosome in the female zygote lowers the dosage of gene products to that seen in males. Most likely, development has become adjusted to this lower dosage, and a higher dosage would bring about abnormalities. We certainly know that, in regard to other chromosomes, any imbalance causes a greatly altered phenotype.

It appears that transcription is not occurring when chromatin is in the heterochromatin state.

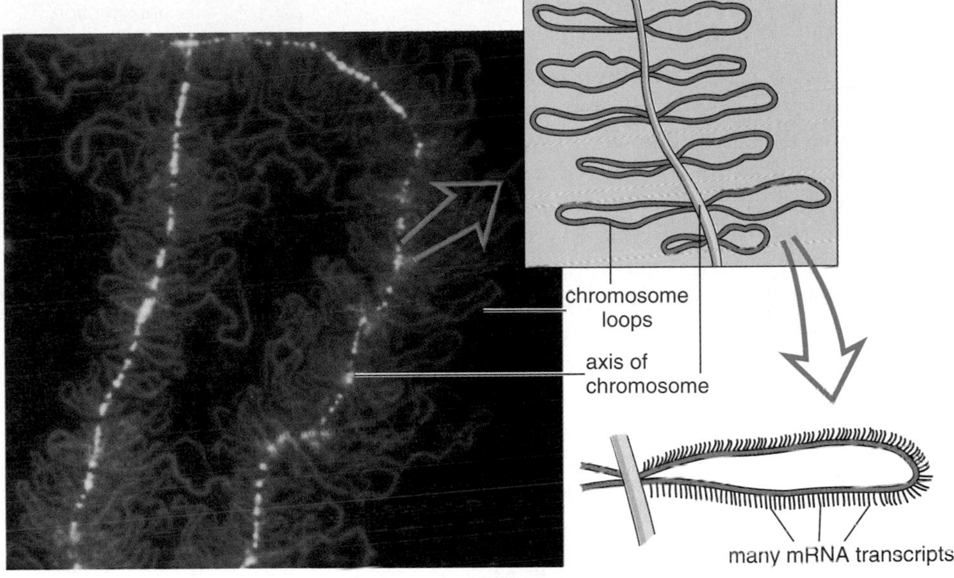

Figure 16.5 Lampbrush chromosomes.
These chromosomes are present in maturing amphibian egg cells and give evidence that when mRNA is being synthesized, chromosomes most likely decondense. Each chromosome has many loops extended from its axis (white). Many mRNA transcripts are being made off of these DNA loops (red).

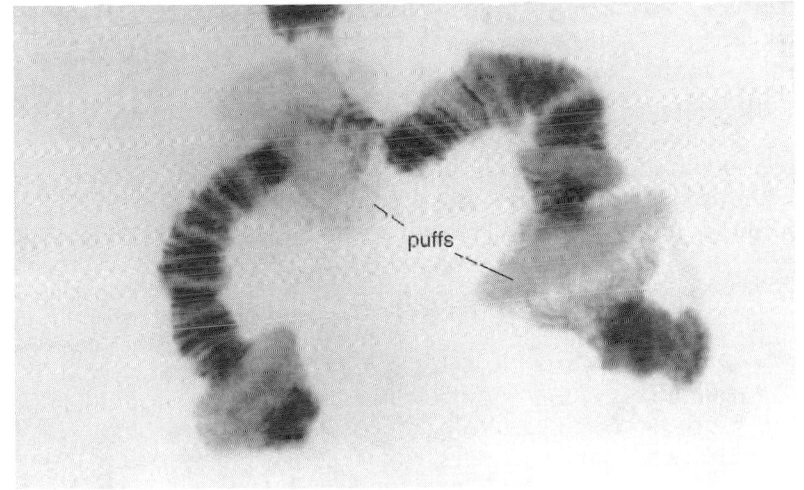

Figure 16.6 Polytene chromosome.
Polytene chromosomes, such as this one from an insect, *Trichosia pubescens,* have about 1,000 parallel chromatids. As development occurs, puffs appear at specific sites, perhaps indicating where DNA is being transcribed. In the laboratory an addition of ecdysone, a hormone that stimulates molting, can cause a pattern of puffs characteristic of natural molting.

Euchromatin Is Active In contrast to heterochromatin, euchromatin is genetically active. The chromosomes within the developing egg cells of many vertebrates are called lampbrush chromosomes because they have many loops that appear to be bristles (Fig. 16.5). Here mRNA is being synthesized in great quantity as development proceeds.

In the salivary glands and other tissues of larval flies, the chromosomes duplicate and reduplicate many times without dividing mitotically. The homologues, each consisting of about 1,000 sister chromatids, synapse together to form giant chromosomes called polytene chromosomes [Gk. *poly,* many, and L. *taenia,* band, ribbon]. It is observed that as a larva develops, first one and then another of the chromosomal regions bulge out, forming chromosomal puffs (Fig. 16.6). The use of radioactive uridine, a label specific for RNA, indicates that DNA is being actively transcribed at these chromosomal puffs. It appears that the chromosome is decondensing at the puffs, allowing RNA polymerase to attach to a section of DNA.

Varying the amount of chromatin as in polytene chromosomes is another form of transcriptional control. When the gene product is either transfer RNA (tRNA) or ribosomal RNA (rRNA), only an increased amount of transcription permits an increase in the amount of gene product. In *Xenopus* (a frog) germ cells that are producing eggs, the number of nucleoli, and therefore copies of RNA genes, actually increases to nearly 1,000. This ensures an immense number of ribosomes in the egg cytoplasm to support protein synthesis during the early stages of development. This form of control is called *gene amplification.*

Transcriptional control in eukaryotes may involve changes in chromatin structure and in the amount of chromatin for certain genes.

Transcription Factors Are Regulatory Proteins

Although no operons like those of prokaryotic cells have been found in eukaryotic cells, investigations suggest that transcription is controlled by DNA-binding proteins called **transcription factors.** Every cell contains many different types of transcription factors, and a different combination is believed to regulate the activity of any particular gene. A group of transcription factors binds to a promoter adjacent to a gene and then the complex attracts and binds RNA polymerase, but transcription may still not begin. Enhancers are also regions where factors that help regulate transcription of the gene can bind. Enhancers can be quite a distance from the promoter, but a hairpin loop in the DNA can bring the factor attached to the enhancer into contact with the transcription factors and polymerase at the promoter. Now transcription begins (Fig. 16.7).

Transcription factors are always present in a cell and most likely they have to be activated in some way before they bind to DNA. The cell has a growth control network that reaches from receptors in the plasma membrane to the nucleus. Kinases, which add a phosphate group to molecules, and phosphatases, which remove a phosphate group, are most likely signaling proteins involved in the growth control network. Kinases are believed to play a significant role in controlling the cell cycle.

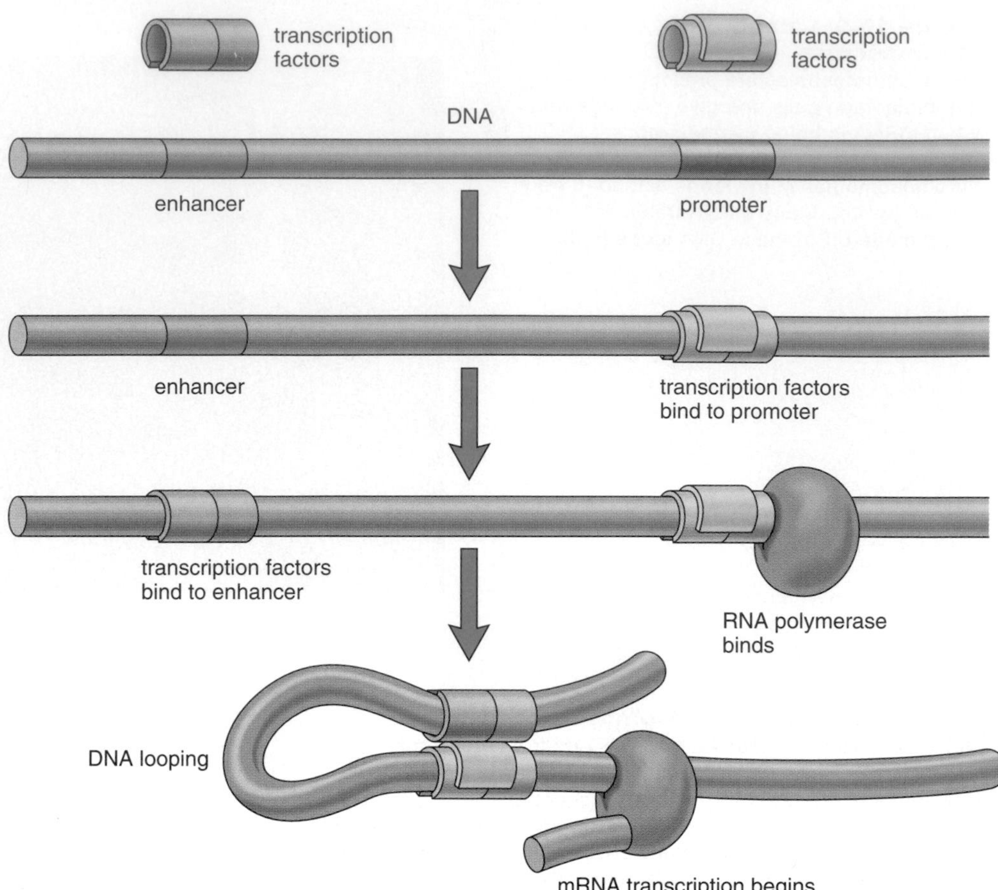

Figure 16.7 Transcription factors.
In this model of how transcription factors in eukaryotic cells work, transcription begins after transcription factors bind to a promoter and to an enhancer. The enhancer is far from the promoter, but physical contact would be possible if the DNA loops bring all factors into contact. Only then does transcription begin.

Transcriptional control in eukaryotes involves organization of chromatin and the use of transcription factors, which are DNA-binding proteins that initiate or regulate transcription.

Looking at Posttranscriptional Control

In eukaryotes, genes have both exons (coding regions) and introns (noncoding regions). Messenger RNA (mRNA) molecules (and other RNAs) are processed before they leave the nucleus and pass into the cytoplasm. Differential excision of introns and splicing of mRNA can vary the type of mRNA that leaves the nucleus. For example, both the hypothalamus and the thyroid gland produce a hormone called calcitonin. The mRNA that leaves the nucleus is not the same in both types of cells, however; radioactive labeling studies show that they vary because of a difference in mRNA splicing (Fig. 16.8). Evidence of different patterns of mRNA splicing is found in other cells, such as those that produce neurotransmitters, muscle regulatory proteins, and antibodies.

The speed of transport of mRNA from the nucleus into the cytoplasm can ultimately affect the amount of gene product realized per unit time following transcription. There is evidence that there is a difference in the length of time it takes various mRNA molecules to pass through a nuclear pore.

Posttranscriptional control in eukaryotes involves differential mRNA processing and factors that affect the length of time it takes mRNA to travel to the cytoplasm.

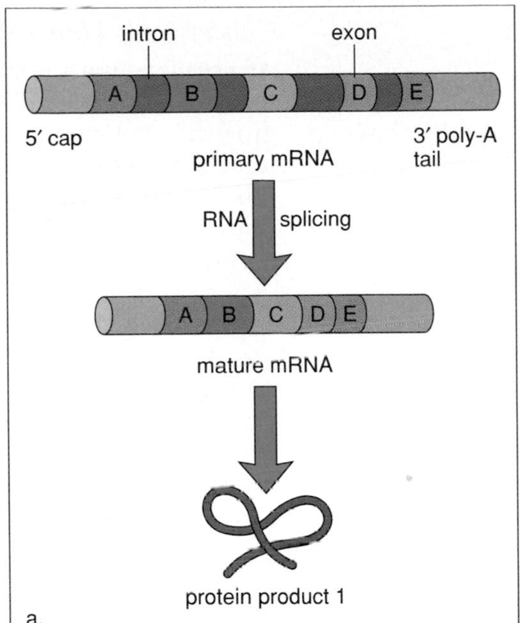

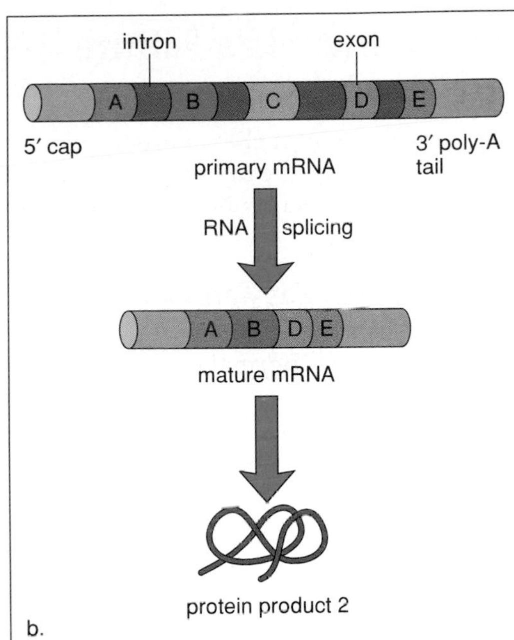

Figure 16.8 Processing of mRNA transcripts.
Because the primary mRNAs are processed differently in these two cells (a and b), distinct proteins result.

Looking at Translational Control

The eggs of frogs and certain other animals contain mRNA molecules that are not translated at all until fertilization occurs. These mRNAs are called *masked messengers* because apparently the cell is unable to recognize them as available for translation. As soon as fertilization occurs, unmasking takes place, and there is a rapid burst of specific gene product synthesis.

Obviously, the longer an active mRNA molecule remains in the cytoplasm, the more product there will eventually be. During maturation, mammalian red blood cells eject their nucleus, and yet they continue to synthesize hemoglobin for several months thereafter. This means that the necessary mRNAs are able to persist all this time. The agents of mRNA destruction are ribonucleases, enzymes that have been found to be associated with ribosomes. Every mature mRNA has a cap consisting of a modified form of guanine nucleotide at the 5′ end and a poly-A tail consisting of 30 to 200 adenine nucleotides at the 3′ end. Differences in these noncoding segments might influence how long the mRNA remains active.

Hormones seem to cause the stabilization of certain mRNA transcripts. For example, the hormone prolactin promotes milk production in mammary glands, but it does so primarily by affecting the length of time the mRNA for casein, a major protein in milk, persists and is translated. Similarly, in estrogen-treated amphibian cells, an mRNA for a phosphoprotein called vitellin persists three weeks, compared to 16 hours in untreated cells. There is evidence to suggest that estrogen binds to and thereby interferes with the action of a ribonuclease in these cells.

Translational control directly affects whether an mRNA is translated and the amount of gene product eventually produced.

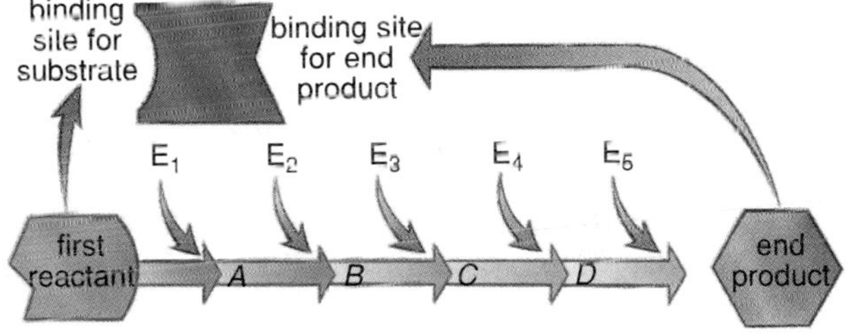

Figure 16.9 Posttranslational control.
The end product of this metabolic pathway can bind to enzyme E_1. When it does, the pathway is shut down because the binding of the end product changes the enzyme's shape so that the substrate cannot bind to the active site.

Looking at Posttranslational Control

Some proteins are not active immediately after translation. For example, bovine proinsulin is at first biologically inactive. After the single long polypeptide folds into a three-dimensional structure, a sequence of about 30 amino acids is enzymatically removed from the middle of the molecule. This leaves two polypeptide chains that are bonded together by disulfide (S—S) bonds and results in an active protein.

Finally, the metabolic activity of proteins is often under feedback control (Fig. 16.9).

Posttranslational control affects the activity of a protein product, whether or not it is functional, and the length of time it is functional.

16.3 Cancer Is a Failure in Genetic Control

Cancer cells exhibit characteristics that indicate they have experienced a severe failure in the regulation of genes that code for products involved in determining the occurrence of cell division.

Characteristics of Cancer Cells

Cancer cells have the following characteristics that distinguish them from normal cells.

Cancer Cells Lack Differentiation

Most cells are specialized; they have a specific form and function that suits them to the role they play in the body. Cancer cells are nonspecialized and do not contribute to the functioning of a body part. A cancer cell does not look like a differentiated muscle, nervous, or connective tissue cell and instead has a shape and form that is distinctly abnormal (Fig. 16.10). Normal cells can enter the cell cycle for about 50 times, and then they die. Cancer cells can enter the cell cycle repeatedly, and in this way they are immortal. In cell tissue culture, they die only because they run out of nutrients or are killed by their own toxic waste products.

Cancer Cells Have Abnormal Nuclei

The nuclei of cancer cells are enlarged, and there may be an abnormal number of chromosomes. The chromosomes have mutated; some parts may be duplicated and some may be deleted. In addition, gene amplification (extra copies of specific genes) is seen much more frequently than in normal cells.

Cancer Cells Form Tumors

Normal cells anchor themselves to a substratum or adhere to their neighbors. They exhibit contact inhibition—when they come in contact with a neighbor, they stop dividing. In culture, normal cells form a single layer that covers the bottom of a petri dish. Cancer cells have lost all restraint; they pile on top of one another and grow in multiple layers. They have a reduced need for growth factors. *Growth factors* are signaling proteins received by receptors in the plasma membrane that are needed by normal cells in order to grow.

In the body, a cancer cell divides to form an abnormal mass of cells called a **tumor** [L. *tumor,* swelling], which invades and destroys neighboring tissue. This new growth, termed *neoplasia,* is made of cells that are disorganized, a condition termed *anaplasia.* A *benign tumor* is a disorganized, usually encapsulated, mass that does not invade adjacent tissue.

Cancer Cells Undergo Angiogenesis and Metastasis

Angiogenesis, the formation of new blood vessels, is required to bring nutrients and oxygen to a cancerous tumor whose growth is not contained within a capsule. Cancer cells release a growth factor that causes neighboring blood vessels to branch into the cancerous tissue. Some modes of cancer treatment are aimed at preventing angiogenesis from occurring.

Cancer in situ is found in its place of origin without any invasion of normal tissue. Malignancy is present when **metastasis** [Gk. *meta,* between, and L. *stasis,* standing, a position] establishes new tumors that are distant from the primary tumor. To accomplish metastasis, cancer cells must first make their way across the extracellular matrix (substances including fibers secreted by the cell) and into a blood vessel or lymphatic vessel. It has been discovered that cancer cells have receptors that allow them to adhere to a component of the extracellular matrix; they also produce proteinase enzymes that degrade the matrix and allow them to invade underlying tissues. Cancer cells tend to be motile, have a disorganized internal cytoskeleton, and lack intact actin filament bundles. After traveling through the blood or lymph, cancer cells may start tumors elsewhere in the body.

Normal Cells

controlled growth

contact inhibition

one organized layer

differentiated cells

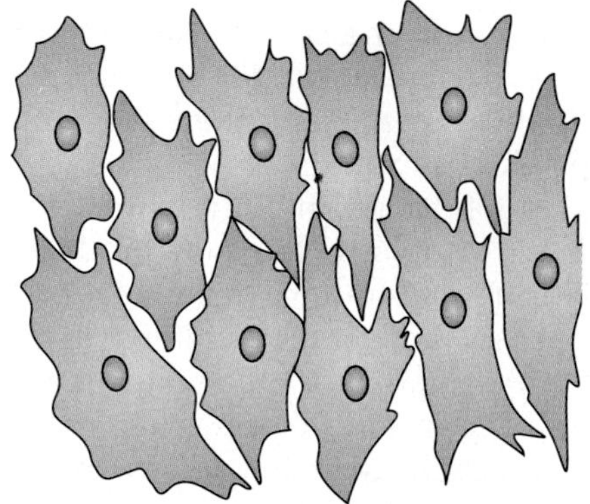

Cancer Cells

uncontrolled growth

no contact inhibition

disorganized, multilayered

nondifferentiated cells

abnormal nuclei

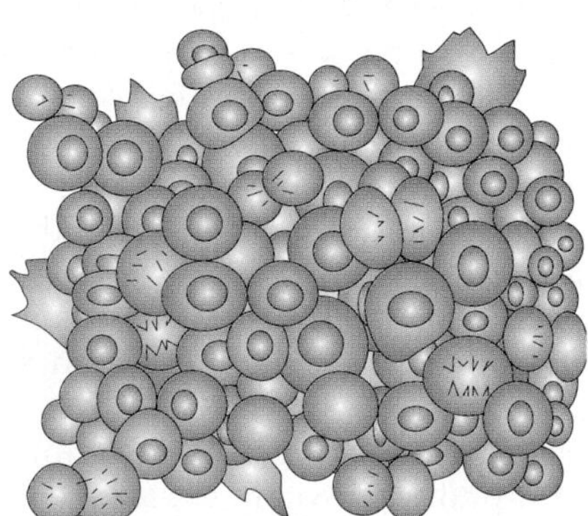

Figure 16.10 Cancer cells.
Cancer cells differ from normal cells in the ways noted.

Doing science

Reflections of a Cancer Researcher

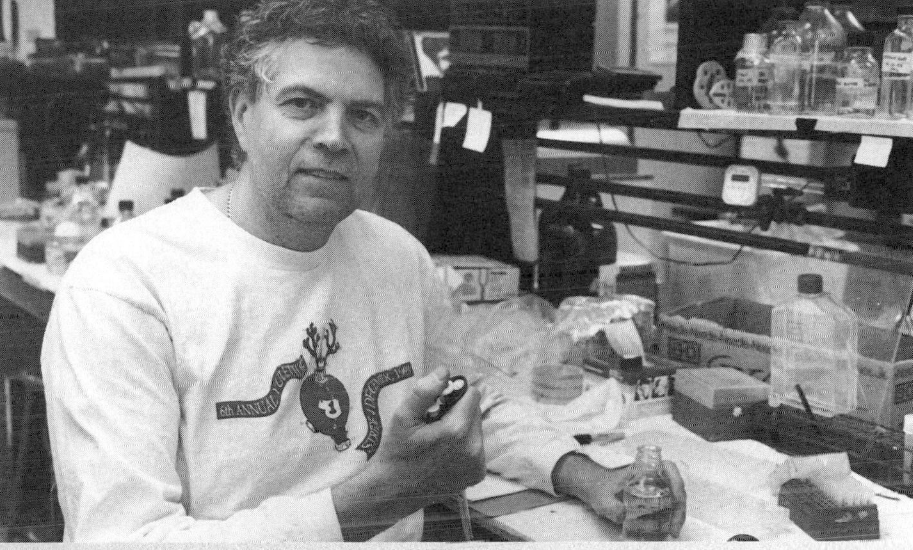
Thomas Gilmore
Boston University

As an undergraduate majoring in English at Princeton University in the early 1970s, I never imagined that I would devote my life to biomedical research. Nevertheless, since 1978, I have been involved in research directed at understanding the molecular events that occur in cells as they progress from a normal to a cancerous state.

One of the most important advances in basic cancer research in the past two decades is the confirmation of the hypothesis that cancer is fundamentally a genetic disease. That is, that progression of a cell from the normal to the malignant state is the consequence of the accumulation of mutations in genes involved in cellular growth control.

In human cancers, the most malignant state is likely to arise as a consequence of mutations in at least five genes, but the study of these cancers is often difficult to approach experimentally. Therefore, much basic cancer research, including my own, has involved the use of simple model systems, where expression of one or two mutant genes is sufficient to induce the malignant state. Specifically, I have had a sustained interest in the mechanisms by which certain avian retroviruses induce the malignant transformation of chicken cells in tissue culture. These systems have the advantage that expression of a single gene, the retroviral oncogene, causes cells to assume most of the characteristics of a malignantly transformed cell.

I feel that I have been very fortunate in my research career. As a graduate student, I happened to be in a lab that was involved in the first characterization of cellular substrates for the tyrosine kinase activity of the *src* oncogene product. This research in the early 1980s led to many other findings that have shown that the phosphorylation of tyrosine residues is a general mechanism used by cells to control cellular growth.

As a postdoctoral fellow working with the late Nobel laureate Howard Temin, I worked on what was, at the time, perhaps the most obscure retroviral oncogene, v-*rel*. Using molecular, biochemical, and cellular approaches, we were able to show that v-*rel* was likely to encode a protein that acted in the nucleus to affect cell growth. It was subsequently found that the v-*rel* gene product has structural and functional homology to proteins that control the expression of genes important for immune cell function in animals from insects to humans. Furthermore, *rel*-like genes have now been shown to be mutated in many human lymphoid cancers.

My lab and others in the field of cancer research are now increasingly turning to even simpler eukaryotic systems, such as yeast and nematodes, in order to understand cellular pathways that control growth and division. Fortunately, it is becoming evident that such pathways are clearly conserved through evolution.

The selection of appropriate experimental systems to answer important questions is one of the most crucial decisions a scientist can make. It is very easy to get attached to your current system or experimental approach. As a scientist, one must do what is necessary to answer the most important question, even if that means learning new and difficult techniques.

I have chosen this career because I simply enjoy doing, discussing, and arguing science. For me, one of the greatest attractions of science is that it is, for the most part, a field of ideas. Yet I believe that there is a higher goal to my research: that understanding the molecular basis of disease will one day lead to chemical or biological therapeutics.

The patient's prognosis (probable outcome) is dependent on the degree to which the cancer has progressed: (1) whether the tumor has invaded surrounding tissues, (2) if so, whether there is any lymph node involvement, and (3) whether there are metastatic tumors in distant parts of the body. With each progressive step of the cancerous condition, the prognosis becomes less favorable.

Cancer cells are nonspecialized, have abnormal chromosomes, and divide uncontrollably. Because they are not constrained by their neighbors, they form a tumor. Then they metastasize, forming new tumors wherever they relocate.

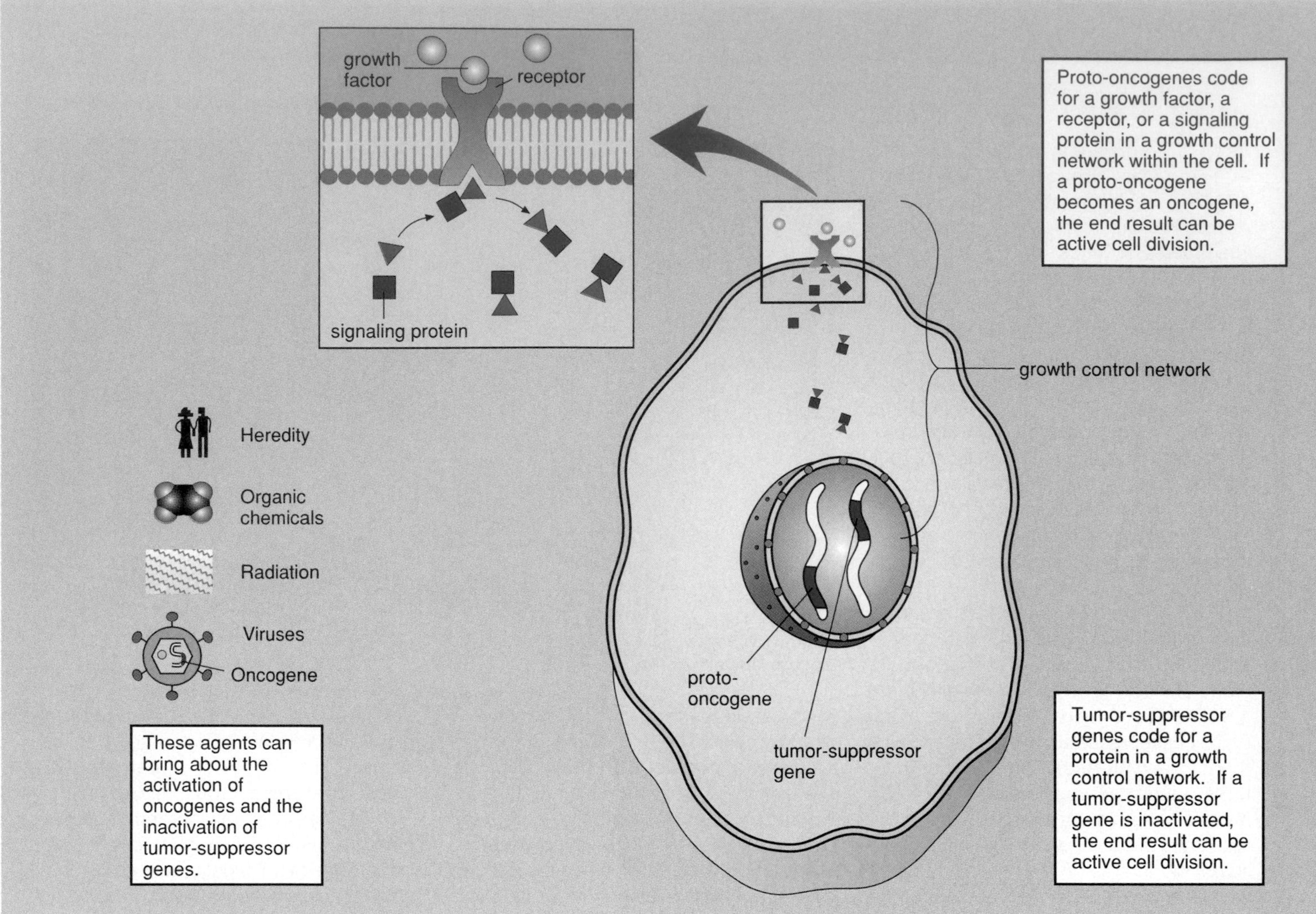
growth factor
receptor
signaling protein
Proto-oncogenes code for a growth factor, a receptor, or a signaling protein in a growth control network within the cell. If a proto-oncogene becomes an oncogene, the end result can be active cell division.
growth control network
Heredity
Organic chemicals
Radiation
Viruses
Oncogene
These agents can bring about the activation of oncogenes and the inactivation of tumor-suppressor genes.
proto-oncogene
tumor-suppressor gene
Tumor-suppressor genes code for a protein in a growth control network. If a tumor-suppressor gene is inactivated, the end result can be active cell division.

Figure 16.11 Growth control network.
The growth control network includes growth factors, their plasma membrane receptors, intracellular reactions, and the genes, notably proto-oncogenes and tumor-suppressor genes. Changes in the growth control network can lead to uncontrolled growth and a tumor. Cancer still does not develop unless the immune system fails to respond and kill these abnormal cells.

Oncogenes and Tumor-Suppressor Genes

A *growth control network* controls cell division in cells. The growth control network influences whether a cell enters or completes the cell cycle, which includes the phases of cell division.

A **growth factor** is a signaling protein that causes cells to divide after it is received by a receptor in the plasma membrane. Cells can secrete growth factors that stimulate themselves or other cells. As Figure 16.11 shows, the growth control network begins with a growth factor, and its receptor in the plasma membrane. It continues with a host of proteins, many of which regulate reactions within cytoplasmic metabolic pathways or regulate the turning on or off of genes within the nucleus. A change occurring anywhere within the growth control network can possibly contribute to the occurrence of abnormal cellular growth.

Two types of genes are known to regulate the growth control network. They are called proto-oncogenes and tumor-suppressor genes.

Proto-Oncogenes and Oncogenes

Proto-oncogenes are normal genes that code for proteins in the growth control network. Proto-oncogenes code for growth factors, or for growth factor receptors, or for regulatory proteins located in the cytoplasm or even in the nucleus.

When a proto-oncogene mutates, the result can be an oncogene [Gk. *onco,* a swelling, and L. *genitus,* producing]. **Oncogenes** are cancer-causing genes, and their products bring about unbridled cell division. For example, an oncogene might cause the cell to produce too much growth factor, or it might code for a receptor that is always active even when no growth factor is present. Or its protein product might promote a certain reaction in the cytoplasm or cause a gene to be turned on that ordinarily is not expressed in a mature cell.

The *ras* oncogenes have been implicated in several types of cancers. An alteration of only a single nucleotide pair is sufficient to convert a normally functioning *ras* proto-oncogene to an oncogene. The *ras*K oncogene is found in

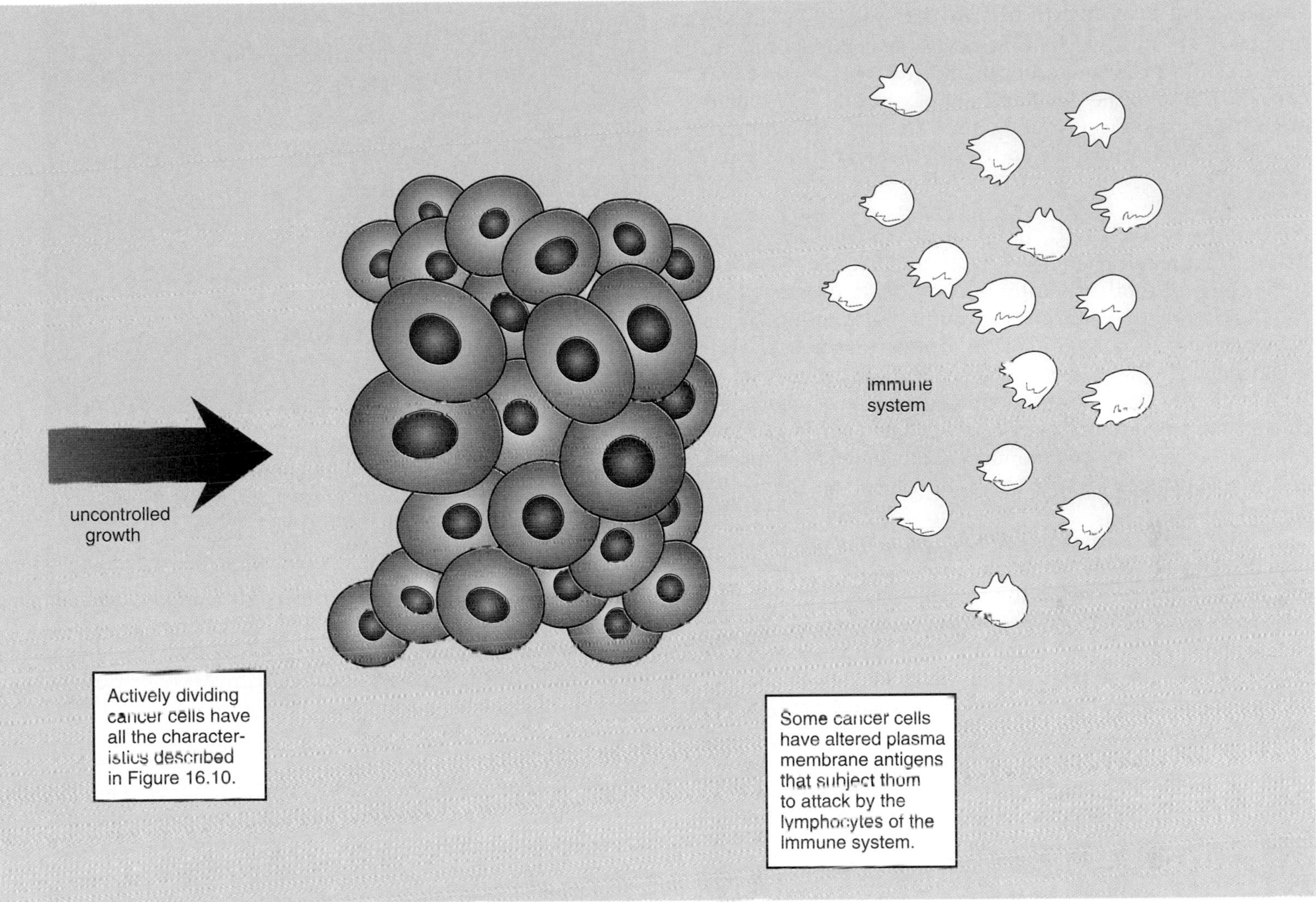
immune
system
uncontrolled
growth
Actively dividing cancer cells have all the characteristics described in Figure 16.10.
Some cancer cells have altered plasma membrane antigens that subject them to attack by the lymphocytes of the immune system.

about 25% of lung cancers, 50% of colon cancers, and 90% of pancreatic cancers. The *ras*N oncogene is associated with leukemias (cancer of blood-forming cells) and lymphomas (cancers of lymphoid tissue), and both *ras* oncogenes are frequently found in thyroid cancers. The proteins encoded by mutant *ras* genes continuously stimulate reactions in the growth control network when no growth factor has been received.

Tumor-Suppressor Genes

Another class of genes plays a major role in triggering cancer. **Tumor-suppressor genes** code for regulatory proteins that can stop reactions from occurring that ordinarily lead to cell division. Tumor-suppressor genes can dampen the effects brought about by oncogenes. Mutated tumor-suppressor genes have lost the ability to prevent reactions that lead to cell division, and they can no longer counteract the presence of an oncogene.

Researchers have identified about a half-dozen or so tumor-suppressor genes. The *RB* tumor-suppressor gene was discovered by studying the inherited condition retinoblastoma. When a child receives only one normal gene, and that gene mutates, eye tumors develop in the retina by age three. *RB* has now been found to malfunction in other cancers such as breast, prostate, and bladder cancers.

We now know that there are signaling proteins that act like negative growth factors. When the substance *TGF* attaches to a plasma membrane receptor, the RB protein is activated. An active RB protein turns off the expression of the proto-oncogene c-*myc*. When the RB protein is not present, the protein product of the c-*myc* gene is thought to cause the expression of other genes whose products lead to abnormal cell division and cancer.

p53 is another tumor-suppressor gene that helps regulate the cell cycle. It also is involved in apoptosis. **Apoptosis** is a process of programmed cell death involving a cascade of specific cellular events leading to the death and destruction of the cell.

Each cell contains a growth control network involving proto-oncogenes and tumor-suppressor genes. When oncogenes are present, and tumor-suppressor genes mutate, uncontrolled growth results.

Causes and Prevention of Cancer

A mutagen is an agent that increases the chances of a mutation, while a **carcinogen** is an environmental agent that can contribute to the development of cancer. Carcinogens are often mutagenic. Carcinogens, heredity, and immunodeficiencies all contribute to the development of cancer.

Carcinogens

Among the best-known mutagenic carcinogens are (1) radiation, (2) organic chemicals (e.g., found in tobacco smoke, some types of foods, pollutants), and (3) viruses.

Ultraviolet radiation in sunlight and tanning lamps is most likely responsible for the dramatic increases seen in skin cancer the past several years. Today, there are at least six cases of skin cancer for every one case of lung cancer. Nonmelanoma skin cancers are usually curable through surgery, but melanoma skin cancer tends to spread and is responsible for 1–2% of total cancer deaths in the United States.

Diagnostic X rays account for most of our exposure to artificial sources of radiation. The benefits of these procedures can far outweigh the possible risk, but it is still wise to avoid any X-ray procedures that are not medically warranted. Despite much publicity, scientists have not been able to show a clear relationship between cancer and radiation from electric power lines, household appliances, and cellular telephones.

Tobacco smoke contains a number of organic chemicals that are known carcinogens, and it is estimated that one-third of all cancer deaths can be attributed to smoking. Lung cancer is the most frequent lethal cancer in the United States, and smoking is also implicated in the development of cancers of the mouth, larynx, bladder, kidney, and pancreas. The greater the number of cigarettes smoked per day, the earlier the habit starts, and the higher the tar content, the greater the possibility of cancer. When smoking is combined with drinking alcohol, the risk of these cancers increases even more (Fig. 16.12).

Passive smoking, or inhalation of someone else's tobacco smoke, is also dangerous and probably causes a few thousand deaths each year.

Statistical studies suggest that a diet rich in saturated fats and low in fiber rivals tobacco in causing colon, rectal, and prostate cancer. As discussed in the reading on the next page, the good news is that eating vegetables and fruits can be protective against the development of cancer.

Industrial chemicals, such as benzene and carbon tetrachloride, and industrial materials, such as vinyl chloride and asbestos fibers, are also associated with the development of cancer. Pesticides and herbicides are dangerous not only to pets and plants but also to our own health because they contain organic chemicals that can cause mutations.

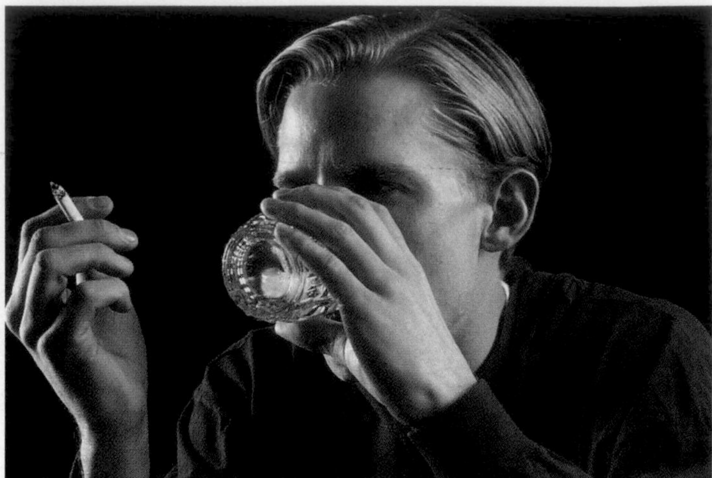

Figure 16.12 Tobacco smoke and alcohol.
Smoking cigarettes, especially when combined with drinking of alcohol, is associated with cancer of the lungs, mouth, larynx, kidney, bladder, pancreas, and many other cancers.

Certain DNA viruses have been linked to human cancers. For example, hepatitis B virus is associated with liver cancer, human papillomavirus with cancer of the cervix, and Epstein-Barr virus with Burkitt's cancer of the lymph nodes and with nasopharyngeal cancer. Another type of virus is the cause of adult T-cell leukemia.

Heredity

Particular types of cancer seem to run in families. The risk of developing breast, lung, and colon cancers increases two- to threefold when first-degree relatives have had these cancers. Investigators have pinpointed the location of a gene called *BRCA1*, that was found by studying families whose female members are prone to breast cancer.

Immunodeficiencies

Cancer is apt to develop in individuals who exhibit an immunodeficiency. For example, cervical cancer develops in women and Kaposi's sarcoma, a cancer of the blood vessels, develops in many persons with AIDS. Transplant patients who are on immunosuppressive drugs are more apt to develop lymphomas and Kaposi's sarcoma.

It appears, then, that an active immune system can help protect us from cancer. Mutated cells may display antigens that normally subject them to attack by T lymphocytes and possibly also antibodies. Cancer is seen more often among the elderly, perhaps because the immune system weakens as we age.

Specific mutations cause cancer. Carcinogens (e.g., tobacco smoke), heredity, and immunodeficiency all play a role in the development of cancer.

Health focus

Prevention of Cancer

There is clear evidence that the risk of certain types of cancer can be reduced by adopting protective behaviors and the right diet.

Protective Behaviors

These behaviors help prevent cancer:

Don't Smoke Cigarette smoking accounts for about 30% of all cancer deaths. Smoking is responsible for 90% of lung cancer cases among men and 79% among women—about 87% altogether. Those who smoke two or more packs of cigarettes a day have lung cancer mortality rates 15 to 25 times greater than nonsmokers. Smokeless tobacco (chewing tobacco or snuff) increases the risk of cancers of the mouth, larynx, throat, and esophagus.

Don't Sunbathe Almost all cases of basal and squamous cell skin cancers are considered to be sun related. Further, sun exposure is a major factor in the development of melanoma, and the incidence of this cancer increases for those living near the equator.

Avoid Alcohol Cancers of the mouth, throat, esophagus, larynx, and liver occur more frequently among heavy drinkers, especially when accompanied by tobacco use (cigarettes or chewing tobacco).

Avoid Radiation Excessive exposure to ionizing radiation can increase cancer risk. Even though most medical and dental X rays are adjusted to deliver the lowest dose possible, unnecessary X rays should be avoided. Excessive radon exposure in homes increases the risk of lung cancer, especially in cigarette smokers. It is best to test your home and take the proper remedial actions.

Be Tested for Cancer Learn the appropriate way to check yourself for breast cancer or testicular cancer. Have other exams done regularly by a physician.

Figure 16C Diet and cancer. Fresh fruits, especially those high in vitamins A and C, and vegetables, especially those in the cabbage family, are believed to reduce the risk of cancer.

Be Aware of Occupational Hazards Exposure to several different industrial agents (nickel, chromate, asbestos, vinyl chloride, etc.) or radiation increases the risk of various cancers. Risk from asbestos is greatly increased when combined with cigarette smoking.

Be Aware of Hormone Therapy Estrogen therapy to control menopausal symptoms increases the risk of endometrial cancer. However, including progesterone in estrogen replacement therapy helps to minimize this risk.

The Right Diet

Statistical studies have suggested that persons who follow certain dietary guidelines are less likely to have cancer. The following dietary guidelines greatly reduce your risk of developing cancer:

1. Avoid obesity. The risk of cancer (especially colon, breast, and uterine cancers) is 55% greater among obese women and 33% greater among obese men, compared to people of normal weight.
2. Lower total fat intake. A high-fat intake has been linked to development of colon, prostate, and possibly breast cancers.
3. Eat plenty of high-fiber foods. These include whole-grain cereals, fruits, and vegetables. Studies have indicated that a high-fiber diet protects against colon cancer, a frequent cause of cancer deaths. It is worth noting that foods high in fiber also tend to be low in fat!
4. Increase consumption of foods that are rich in vitamins A and C. Beta-carotene, a precursor of vitamin A, is found in dark green leafy vegetables, carrots, and various fruits. Vitamin C is present in citrus fruits. These vitamins are called antioxidants because in cells they prevent the formation of free radicals (organic ions that have an unpaired electron) that can possibly damage DNA. Vitamin C also prevents the conversion of nitrates and nitrites into carcinogenic nitrosamines in the digestive tract.
5. Cut down on consumption of salt-cured, smoked, or nitrite-cured foods. Salt-cured or pickled foods may increase the risk of stomach and esophageal cancer. Smoked foods like ham and sausage contain chemical carcinogens similar to those in tobacco smoke. Nitrites are sometimes added to processed meats (e.g., hot dogs and cold cuts) and other foods to protect them from spoilage; as mentioned previously, nitrites are converted to nitrosamines in the digestive tract.
6. Include vegetables from the cabbage family in the diet. The cabbage family includes cabbage, broccoli, Brussels sprouts, kohlrabi, and cauliflower. These vegetables may reduce the risk of gastrointestinal and respiratory tract cancers.
7. Be moderate in the consumption of alcohol. People who drink and smoke are at an unusually high risk for cancers of the mouth, larynx, and esophagus.

connecting concepts

Determining what regulates gene activity is one of the most important questions facing geneticists today. Although gene regulation in prokaryotic cells generally occurs at the transcription level, in eukaryotic cells gene regulation occurs at several different levels.

In the nucleus, transcription factors that bind to DNA control the initiation of transcription. After mRNA is made, it can be chemically altered, and this will affect the specific protein produced by the cell. The amount of protein produced varies according to the length of time it takes mRNA to leave the nucleus and how long mRNA remains active in the cytoplasm. Finally, some proteins are made but remain inactive in the cytoplasm for a period of time.

The study of gene regulation and the study of cancer go hand in hand. Genes direct cells to specialize—to become muscle, nerve, or skin cells, for example. Cancer cells lack such differentiation. While normal cells can only divide a finite number of times, cancer cells have an unlimited ability to divide. Normal cells stop dividing when they touch neighboring cells, cancer cells lack this contact inhibition. Consequently, rapidly growing cancer cells invade and destroy normal healthy tissue.

All of us have known someone fighting cancer. It is an equal opportunity disease—both sexes, all races, all ages and people with very different lifestyles suffer from cancer. Researchers now know that we all carry proto-oncogenes, normal genes that can mutate into oncogenes, cancer-causing genes. We also have tumor-suppressor genes that inhibit cancer growth; if these genes mutate, cancers can flourish.

Because cancer is so widespread, we all need to take preventative measures. Prevention includes common sense behaviors such as not smoking or sunbathing and making changes in the diet such as eating more fruits, vegetables, and high-fiber foods.

Summary

16.1 Prokaryotes Utilize Operons

Regulation in prokaryotes usually occurs at the level of transcription. The operon model developed by Jacob and Monod says that a regulator gene codes for a repressor, which sometimes binds to the operator. When it does, RNA polymerase is unable to bind to the promoter, and transcription of structural genes cannot take place.

The *lac* operon is an example of an inducible operon because when lactose, the inducer, is present it binds to the repressor. The repressor is unable to bind to the operator, and transcription of structural genes takes place.

The *trp* operon is an example of a repressible operon because when tryptophan, the corepressor, is present, it binds to the repressor. The repressor is then able to bind to the operator, and transcription of structural genes does not take place.

16.2 Eukaryotes Utilize Various Methods

The following levels of control of gene expression are possible in eukaryotes: transcriptional control, posttranscriptional control, translational control, and posttranslational control.

Chromatin organization helps regulate transcription. Highly compacted heterochromatin is genetically inactive, as exemplified by Barr bodies. Less compacted euchromatin is genetically active, as exemplified by lampbrush chromosomes in vertebrates and polytene chromosomal puffs in insects. Gene amplification is the replication of a gene such that there are more copies than were originally present in the zygotic nucleus. Regulatory proteins called transcription factors, as well as DNA sequences called enhancers, play a role in controlling transcription in eukaryotes.

Posttranscriptional control refers to variations in messenger RNA (mRNA) processing and the speed with which a particular mRNA molecule leaves the nucleus.

Translational control affects mRNA translation and the length of time it is translated. Posttranslational control affects whether or not an enzyme is active and how long it is active.

16.3 Cancer Is a Failure in Genetic Control

Cancer cells are nondifferentiated, divide repeatedly, have abnormal nuclei, do not require growth factors, and are not constrained by their neighbors. After forming a tumor, cancer cells metastasize and start new tumors elsewhere in the body.

Each cell contains a growth control network that leads to cell division. When proto-oncogenes mutate, becoming oncogenes, and when tumor-suppressor genes mutate, the growth control network no longer functions as it should, and uncontrolled growth results.

Cancer-causing mutations accumulate during the development of cancer. A cancer-causing mutation can arise due to exposure to radiation, such as sunlight, certain organic chemicals, such as tobacco smoke, and a few types of viruses. Also, a mutation leading to cancer can be inherited. Even so, cancer is most apt to occur in the immunodeficient individual. We are beginning to learn how to prevent cancer by adopting certain behaviors.

Reviewing the Chapter

1. Name and state the function of the three components of operons. 252
2. Explain the operation of the *lac* operon, and note why it is considered an inducible operon. 252
3. Explain the operation of the *trp* operon, and note why it is considered a repressible operon. 253
4. With regard to transcriptional control in eukaryotes explain how Barr bodies show that heterochromatin is genetically inactive. 256
5. Explain how lampbrush chromosomes in vertebrates and giant (polytene) chromosomal puffs in insects show that euchromatin is genetically active. 257
6. What do transcription factors do in eukaryotic cells? What are enhancers? 258
7. Give examples of posttranscriptional, translational, and posttranslational control in eukaryotes. 258–59
8. List and discuss four characteristics of cancer cells that distinguish them from the characteristics of normal cells. 260
9. What are oncogenes and tumor-suppressor genes? What role do they play in a regulatory network that controls cell division and involves plasma membrane receptors and signaling proteins? 262–63
10. List four carcinogens to be avoided to prevent the development of cancer. Give the dietary guidelines that can possibly prevent the development of cancer. 264–65

Testing Yourself

Choose the best answer for each question.

1. Which type of prokaryotic cell would be more successful as judged by its growth potential?
   a. One that is able to express all its genes all the time.
   b. One that is unable to express any of its genes any of the time.
   c. One that expresses some of its genes some of the time.
   d. One that divides only when all types of amino acids and sugars are present in the medium.
2. When lactose is present
   a. the repressor is able to bind to the operator.
   b. the repressor is unable to bind to the operator.
   c. transcription of structural genes occurs.
   d. Both b and c are correct.
3. When tryptophan is present
   a. the repressor is able to bind to the operator.
   b. the repressor is unable to bind to the operator.
   c. transcription of structural genes occurs.
   d. Both b and c are correct.
4. Which of these is mismatched?
   a. posttranslational control—nucleus
   b. transcriptional control—nucleus
   c. translational control—cytoplasm
   d. posttranscriptional control—nucleus
5. RNA processing varies in different cells. This is an example of __________ control of gene expression.
   a. transcriptional
   b. posttranscriptional
   c. translational
   d. posttranslational
6. A scientist adds radioactive uridine (label for RNA) to a culture of cells and examines an autoradiograph. Which type of chromatin is apt to be labeled?
   a. heterochromatin
   b. euchromatin
   c. Both a and b are correct.
   d. Neither a nor b is correct.
7. If Barr bodies were genetically active, females heterozygous for an X-linked gene would
   a. be mosaics.
   b. not be mosaics.
   c. die.
   d. Both a and c are correct.
8. Which of these might cause a proto-oncogene to become an oncogene?
   a. exposure of cell to radiation
   b. exposure of cell to certain chemicals
   c. viral infection of the cell
   d. All of these are correct.
9. A cell is cancerous. Where might you find an abnormality?
   a. only in the nucleus
   b. only in the plasma membrane receptors
   c. only in cytoplasmic reactions
   d. in any part of the cell concerned with growth and cell division
10. A tumor-suppressor gene
    a. inhibits cell division.
    b. opposes oncogenes.
    c. prevents cancer.
    d. All of these are correct.
11. Label the diagram of a *lac* operon. Note that transcription is not occurring. What one change is needed in this drawing to have the diagram represent a *trp* operon when no transcription is occurring?

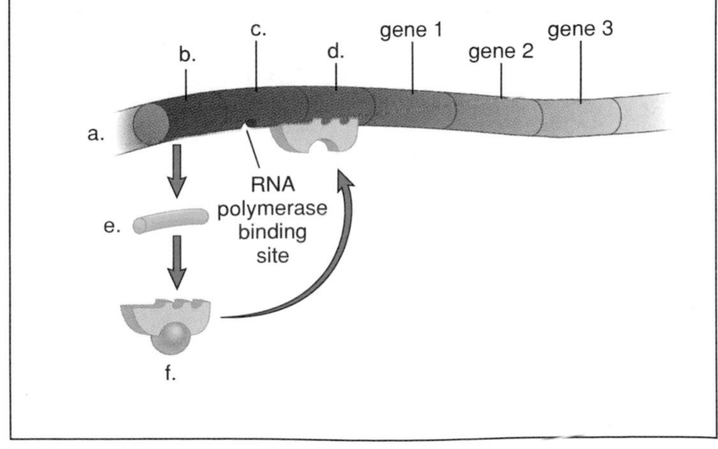

Applying the Concepts

1. *Cells are able to regulate the expression of genes so that only certain ones are expressed at any particular time.*
   What firsthand evidence is there that only certain genes are active in certain cells?
2. *Genes are controlled at various levels that involve both transcription and translation.*
   Define gene expression, and then explain why it is appropriate to speak of posttranslational control of genes.
3. *Faulty regulation of genes that promote cell division and/or genes that inhibit cell division causes the development of cancer.*
   Why would you expect normal cells to have opposing genes and opposing enzymes involved in the regulation of cell division?

Using Technology

Your study of gene activity is supported by these available technologies:

Exploring the Internet

The Mader Home Page provides resources for and help with studying this chapter.

http://www.mhhe.com/sciencemath/biology/mader/
(Click on Biology.)

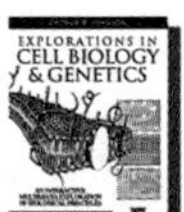

Explorations in Cell Biology & Genetics CD-ROM

Reading DNA (#15)
Gene Regulation (#16)

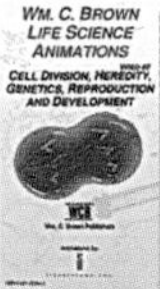

Life Science Animations Video

Video #2: Cell Division/Heredity/Genetics/Reproduction and Development
Regulation of lac *Operon (#18)*

Understanding the Terms

apoptosis 263	operator 252
Barr body 256	operon 252
cancer 260	promoter 252
carcinogen 264	proto-oncogene 262
corepressor 253	regulator gene 252
euchromatin 256	repressible operon 253
growth factor 262	repressor 252
heterochromatin 256	structural gene 252
inducer 252	transcription factor 258
inducible operon 252	tumor 260
metastasis 260	tumor-suppressor gene 263
oncogene 262	

Match the terms to these definitions:

a. __________ Cancer-causing gene.
b. __________ Abnormal growth derived from a single mutated cell that has repeatedly undergone cell division; may be benign or malignant.
c. __________ Dark-staining body in the nuclei of female mammals that contains a condensed, inactive X chromosome.
d. __________ Diffuse chromatin, which is being transcribed.
e. __________ Environmental agent that causes mutations leading to the development of cancer.
f. __________ Gene that codes for a protein that ordinarily suppresses cell division.
g. __________ Gene that codes for an enzyme in a metabolic pathway.
h. __________ Group of structural and regulating genes that functions as a single unit.
i. __________ Highly compacted chromatin that is not being transcribed.
j. __________ In an operon, a sequence of DNA where RNA polymerase begins transcription.

Recombinant DNA and Biotechnology

CHAPTER 17

Chapter Concepts

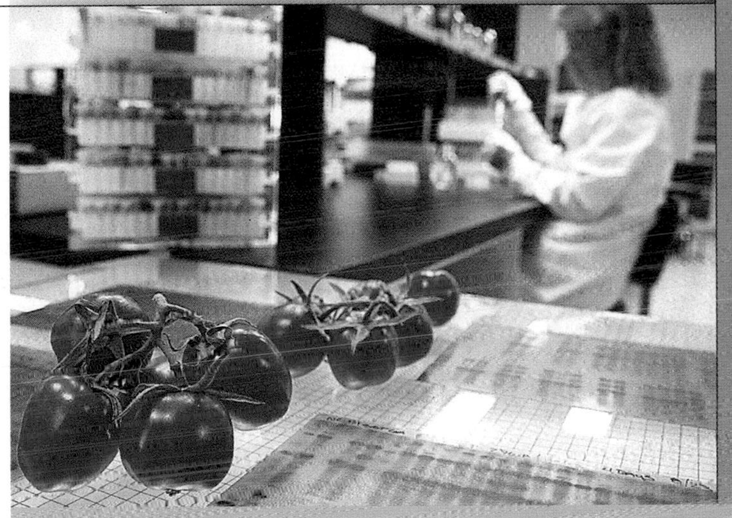

The tomatoes in the foreground were genetically engineered to be insect resistant

Since Mendel's work was rediscovered in 1900, geneticists have made startling advances which have led to a new era of DNA technology. Biotechnology refers to the use of a natural biological system to produce a product or to achieve an end desired by humans. This isn't a new idea; plants and animals have been bred to express particular phenotypes since the dawn of civilization.

Biotechnology based on the manipulation of DNA outside of living cells, however, is a newer and more powerful type of technology. New techniques enable genes to be removed from one organism and inserted into another in order to produce a desired substance, insulin for example. Not very long ago, people with insulin-dependent diabetes mellitus received the insulin they needed for survival from dead animals. Today, they receive human insulin, a product of biotechnology. Since the 1980s, biotechnology has produced drugs and vaccines to curb human illnesses, and nucleic acids for laboratory research.

Genetically engineered bacteria have been used to clean up environmental pollutants, increase the fertility of the soil, and kill insect pests. Biotechnology also extends beyond unicellular organisms; it is now possible to alter the genotype and subsequently the phenotype of plants and animals. Indeed, gene therapy in humans—attempting to repair a faulty gene—is already undergoing clinical trials.

17.1 Cloning of a Gene

Genetic engineering can produce cells that contain recombinant DNA and are capable of producing a new and different protein.

Using Recombinant DNA

Recombinant DNA (rDNA) contains DNA from two different sources. To make recombinant DNA, a technician often begins by selecting a **vector,** the means by which recombinant DNA is introduced into a host cell. One common type of vector is a plasmid. **Plasmids** are small accessory rings of DNA. The ring is not part of the bacterial chromosome and can be replicated independently. Plasmids were discovered by investigators studying the sex life of the intestinal bacterium *Escherichia coli.*

Two enzymes are needed to introduce foreign DNA into vector DNA (Fig. 17.1). The first enzyme, called a **restriction enzyme,** cleaves plasmid DNA and the second, called **DNA ligase,** seals foreign DNA into the opening created by the restriction enzyme.

Restriction enzymes occur naturally in bacteria, where they stop viral reproduction by cutting up viral DNA. They are called restriction enzymes because they *restrict* the growth of viruses. In 1970, Hamilton Smith, at Johns Hopkins University, isolated the first restriction enzyme; now hundreds of different restriction enzymes have been isolated and purified. Each one cuts DNA at a specific cleavage site. For example, the restriction enzyme called *Eco*RI always cuts double-stranded DNA when it has this sequence of bases at the cleavage site:

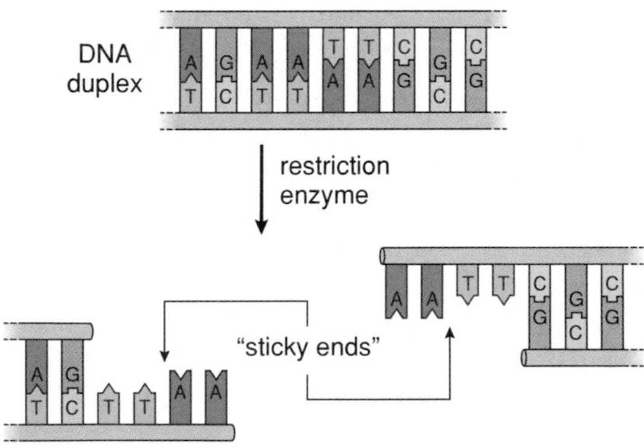

Notice there is now a gap into which a piece of foreign DNA can be placed if it ends in bases complementary to those exposed by the restriction enzyme. To assure this, it is only necessary to cleave the foreign DNA with the same type of restriction enzyme. The single-stranded but complementary ends of the two DNA molecules are called "sticky ends" because they can bind by complementary base pairing. They therefore facilitate the insertion of foreign DNA into vector DNA.

The second enzyme needed for preparation of rDNA, called DNA ligase [L. *ligo,* bind, tie], is a cellular enzyme that seals any breaks in a DNA molecule. Genetic engineers use this enzyme to seal the foreign piece of DNA into the vector. DNA splicing is now complete; an rDNA molecule has been prepared.

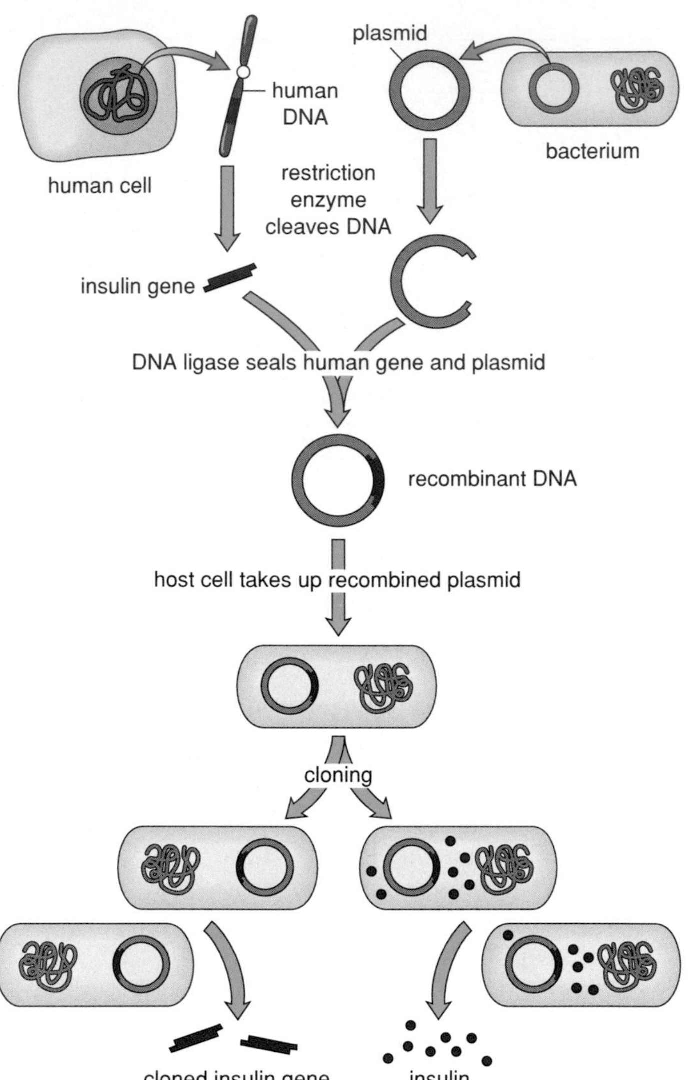

Figure 17.1 Cloning of a gene.
Human DNA and plasmid DNA are cleaved by the same type of restriction enzyme and spliced together by the enzyme DNA ligase. Gene cloning is achieved when a host cell takes up the recombined plasmid and the plasmid reproduces. Multiple copies of the gene are now available to an investigator. If the insulin gene functions normally as expected, the product (insulin) may also be retrieved.

Getting the Product

Bacterial cells take up recombined plasmids, especially if they are treated with calcium chloride to make them more permeable. Thereafter, as the host cell reproduces, a bacterial clone forms. Each of the bacteria contains the foreign gene, which is behaving just as if it were in its original cell. The investigator can recover either the cloned gene, or a protein product from this bacterial clone (Fig. 17.2*a*).

Plasmids can be used as vectors for carrying foreign DNA, and so can viruses, such as the bacteriophage known as lambda (Fig. 17.2*b*). After lambda attaches to a cell, the DNA is released from the virus and enters the bacterium. Here it may direct the reproduction of many more viruses. Each virus in the bacteriophage clone contains a copy of the foreign gene. Notice that a **clone** is a large number of molecules (i.e., cloned genes) or viruses (i.e., cloned bacteriophages) or cells (i.e., cloned bacteria) that are identical to an original specimen.

Using a Genomic Library

A **genome** is the full set of genes of an individual. A *genomic library* is a collection of bacterial or bacteriophage clones; each clone contains a particular segment of the DNA from a foreign cell. When you make a genomic library, an organism's DNA is simply sliced up into pieces, and the pieces are put into vectors (i.e., plasmids or viruses) that are taken up by the host bacteria. The entire collection of bacterial or bacteriophage clones that result therefore contains all the genes of that organism.

In order for mammalian gene expression to occur in a bacterium, the gene has to be accompanied by the proper regulatory regions. Also, the gene should not contain introns because bacterial cells do not have the necessary enzymes to process primary messenger RNA (mRNA; ribonucleic acid). It's possible to make a mammalian genome that lacks introns, however. The enzyme called reverse transcriptase can be used to make a DNA copy of all the mature mRNA molecules from a cell. This DNA molecule, called *complementary DNA (cDNA)*, does not contain introns. Notice that the genomic library made from cDNA contains only the genes currently being expressed in the source cell.

You can also use a particular probe to search a genomic library for a certain gene. A **probe** is a single-stranded nucleotide sequence that will hybridize (pair) with a certain piece of DNA. Location of the probe is possible because the probe is either radioactive or fluorescent. As

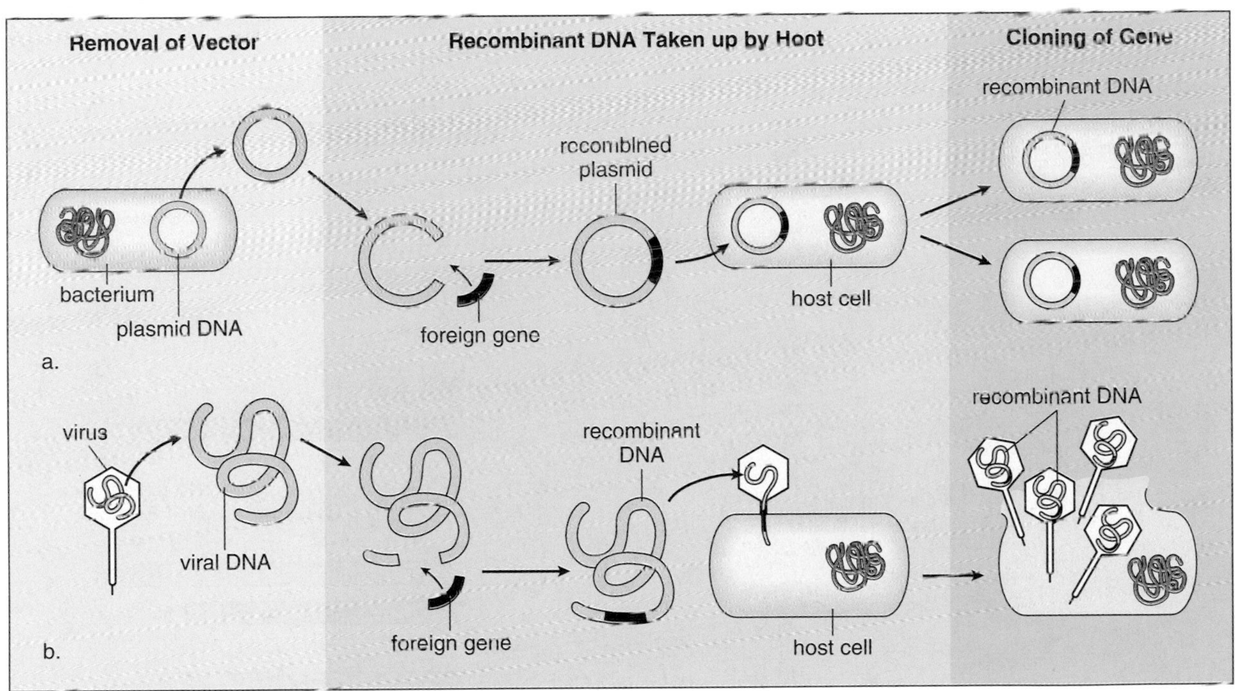

Figure 17.2 Preparation of a genomic library.
Each bacterial or viral clone in a genomic library contains a segment of DNA from a foreign cell. **a.** A plasmid is removed from a bacterium and is used to make recombinant DNA. After the recombined plasmid is taken up by a host cell, replication produces many copies. **b.** Viral DNA is removed from a bacteriophage such as lambda and is used to make recombinant DNA. The virus containing the recombinant DNA infects a host bacterium. Cloning is achieved when the virus reproduces and then leaves the host cell.

illustrated in Figure 17.3, bacterial cells, each carrying a particular DNA fragment, can be plated onto agar in a petri dish. After the probe hybridizes with the gene of interest, the gene can be isolated from the fragment. Now this particular fragment can be cloned further or even analyzed for its particular DNA sequence.

Replicating Small DNA Segments

The polymerase chain reaction (PCR) can create millions of copies of a single gene or any specific piece of DNA in laboratory glassware (like a test tube). PCR is very specific—the targeted DNA sequence can be less than one part in a million of the total DNA sample! This means that a single gene among all the human genes can be amplified (copied) using PCR.

PCR takes its name from DNA polymerase, the enzyme that carries out DNA replication in a cell. It is considered a chain reaction because DNA polymerase will carry out replication over and over again, until there are millions of copies of the targeted DNA. PCR does not replace gene cloning; cloning is still used whenever a large quantity of a protein product is needed, or if the DNA needs to be in its accustomed configuration before being sequenced.

Before carrying out PCR, primers—sequences of about 20 bases that are complementary to the bases on either side of the "target DNA"—must be available. The primers are needed because DNA polymerase does not start the replication process; it only continues or extends the process. After the primers bind by complementary base pairing to the DNA strand, DNA polymerase copies the target DNA (Fig. 17.4).

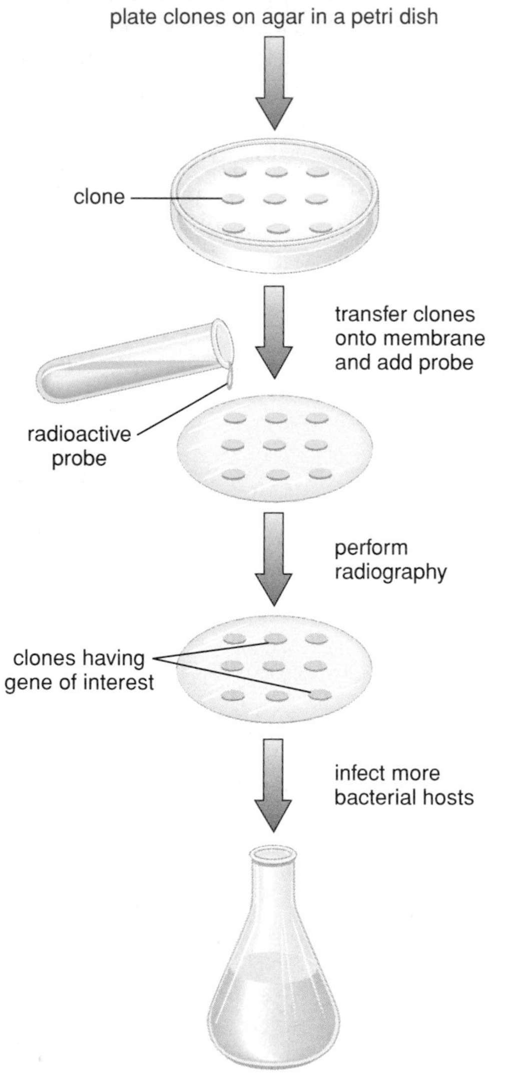

Figure 17.3 Identification of a cloned gene. A probe, which may be single-stranded DNA or mRNA, hybridizes with the cloned fragments containing the gene of interest. To see which clones (fragments) contain the gene, the probe has to be radioactive or fluorescent. Autoradiography is the use of a film that reveals the areas of radioactivity.

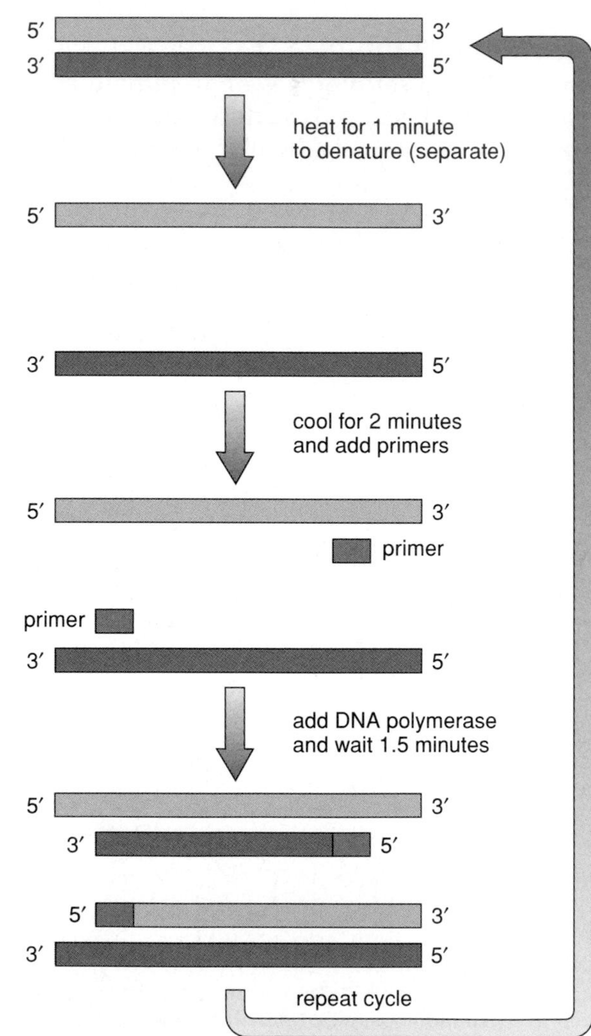

Figure 17.4 Polymerase chain reaction (PCR). PCR is performed in laboratory glassware. Primers (red), which are DNA sequences complementary to the 3′ end of the targeted DNA, are necessary for DNA polymerase to make a copy of the DNA strand.

PCR has been in use for several years, and now almost every laboratory has automated PCR machines to carry out the procedure. Automation became possible after a temperature-insensitive (thermostable) DNA polymerase was extracted from the bacterium *Thermus aquaticus,* which lives in hot springs. This enzyme can withstand the high temperature used to separate double-stranded DNA; therefore, replication need not be interrupted by the need to add more enzyme.

Analyzing DNA

DNA can be subjected to **DNA fingerprinting,** a process described in Figure 17.5. When the DNA of an organism is treated with restriction enzymes, the result is a unique collection of different-sized fragments. Therefore, *restriction fragment length polymorphisms (RFLPs)* exist between individuals. During a process called gel electrophoresis, the fragments can be separated according to their lengths, and the result is a number of bands that are so close together they appear as a smear. However, the use of probes for genetic markers results in a distinctive pattern that can be recorded on X-ray film.

A DNA fingerprint is inherited and therefore resembles that of one's parents. DNA fingerprinting successfully identified the remains of a teenager who had been murdered eight years before because the skeletal DNA was similar to that of the parents' DNA. DNA fingerprinting has also been helpful to evolutionists. For example, it was used to determine that the quagga, an extinct zebralike animal, was a zebra rather than a horse. The only remains of the quagga consisted of dried skin.

Following PCR, DNA segments can be analyzed directly. A probe is not needed because the segments appear as distinctive bands following gel electrophoresis. PCR amplification and analysis has been used (1) to diagnose viral infections, genetic disorders, and cancer; (2) in forensic laboratories to identify criminals; and (3) to determine the evolutionary relationships of various organisms. When the amplified DNA matches that of a virus, mutated gene, or oncogene, then we know that a viral infection, genetic disorder, or cancer is present. And the DNA from a single sperm is enough to identify a suspected rapist. To determine evolutionary relationships, it is often necessary to sequence the DNA. Sequencing mitochondrial DNA segments helped determine the evolutionary history of human populations. It has even been possible to sequence DNA taken from a 76,000-year-old mummified human brain and from a 17- to 20-million-year-old plant fossil following PCR amplification. Nucleotide sequences of DNA segments can be determined quickly with present-day DNA sequencers that make use of computers. The use of PCR, followed by sequencing of DNA, is essential to the Human Genome Project discussed on page 279.

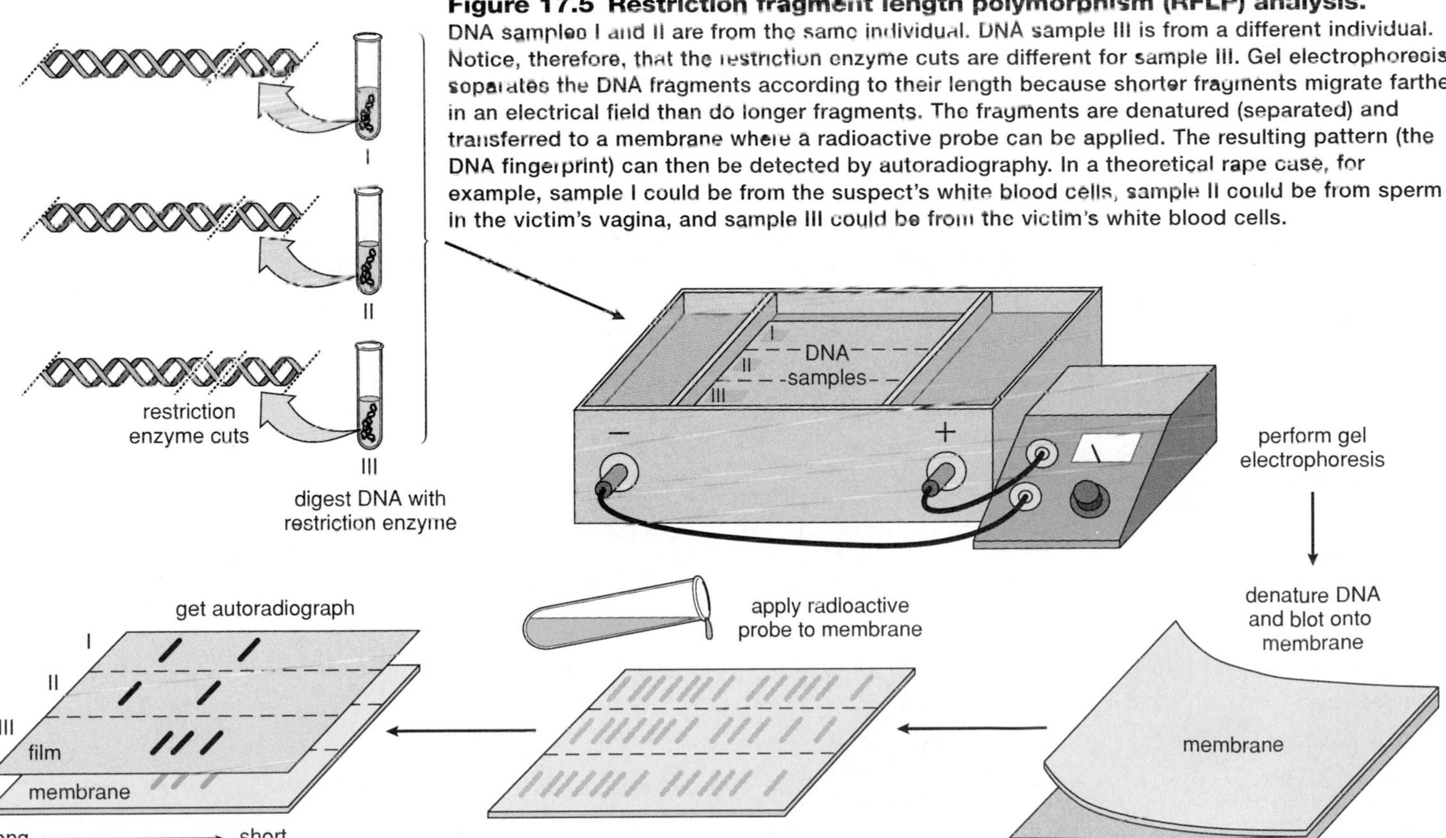

Figure 17.5 Restriction fragment length polymorphism (RFLP) analysis. DNA samples I and II are from the same individual. DNA sample III is from a different individual. Notice, therefore, that the restriction enzyme cuts are different for sample III. Gel electrophoresis separates the DNA fragments according to their length because shorter fragments migrate farther in an electrical field than do longer fragments. The fragments are denatured (separated) and transferred to a membrane where a radioactive probe can be applied. The resulting pattern (the DNA fingerprint) can then be detected by autoradiography. In a theoretical rape case, for example, sample I could be from the suspect's white blood cells, sample II could be from sperm in the victim's vagina, and sample III could be from the victim's white blood cells.

17.2 Biotechnology Products Are Many

Table 17.1 shows at a glance some of the biotechnology products now available. These products include hormones and similar types of proteins or vaccines.

Hormones and Similar Types of Proteins

One impressive advantage of biotechnology is that it allows mass production of proteins that are very difficult to obtain otherwise. The human growth hormone, which is used to treat people who are growing more slowly than normal, was previously extracted from the pituitary glands of cadavers, and it took 50 glands to obtain enough for one dose. Now human growth hormone is produced in quantity by biotechnology. Insulin, to treat diabetics, was previously extracted from the pancreatic glands of slaughtered cattle and pigs; it was expensive and sometimes caused allergic reactions in recipients. And few of us knew of tPA (tissue plasminogen activator), a protein present in the body in minute quantities that activates an enzyme to dissolve blood clots. Now tPA produced by biotechnology is used to treat heart attack victims by dissolving blood clots that are blocking the flow of blood in the coronary arteries of the heart.

The last few entries of Table 17.1 list other troublesome and serious afflictions in humans that may soon be treatable by biotechnology products. Clotting factor VIII treats hemophilia; human lung surfactant treats respiratory distress syndrome in premature infants; atrial natriuretic factor helps control hypertension. And the list will grow, because bacteria (or other cells) can be engineered to produce virtually any protein.

Several hormones produced by biotechnology are for use in animals. Farm animals can now be given growth hormone instead of steroids. Such animals produce a leaner meat that is healthier for humans. When cows are given bovine growth hormone (bGH) they produce 25% more milk than usual, making it possible for dairy farmers to reduce the number of their cows. However, an increased incidence of udder infections can require greater dosages of antibiotic therapy.

Safer Vaccines

Vaccines are used to make people immune to an infectious organism so they do not become ill when exposed to it. In the past, vaccines were made from treated bacteria or viruses, and on occasion they caused the illness they were suppose to prevent.

Vaccines produced through biotechnology do not cause illness. Bacteria and viruses have surface proteins, and a gene for just one of these can be used to genetically engineer bacteria. The copies of the surface protein that result can be used as a vaccine. A vaccine for hepatitis B is now available, and potential vaccines for chlamydia, malaria, and AIDS are in experimental stages (Table 17.2).

Vaccines are available through biotechnology for the inoculation of farm animals. There are vaccines for such illnesses as hoof-and-mouth disease and dysentery. These animal ailments were once a severe drain on the time, energy, and resources of farmers because they caused an untold number of animal illnesses and deaths each year.

Biotechnology products include hormones and similar types of proteins and vaccines. These products are of enormous importance to the fields of medicine and animal husbandry.

Table 17.1

Biotechnology Products: Hormones and Similar Types of Proteins

Treatment of Humans	For
Insulin	Diabetes
Growth hormone	Pituitary dwarfism
tPA (tissue plasminogen activator)	Heart attack
Interferons	Cancer
Erythropoietin	Anemia
Ceredase	Gaucher disease*
Interleukin-2	Cancer
Tumor necrosis factor	Cancer
Clotting factor VIII	Hemophilia
Human lung surfactant	Respiratory distress syndrome
Atrial natriuretic factor	High blood pressure

**A lysosomal storage disorder*

Table 17.2

Biotechnology Products: Potential Vaccines

Prevention in Humans of
AIDS
Herpes (oral and genital)
Hepatitis A, B*, and C
Lyme disease
Whooping cough
Chlamydia

**Now available*

17.3 Making Transgenic Organisms

Free living organisms in the environment that have had a foreign gene inserted into them are called **transgenic organisms.**

Transgenic Bacteria Perform Services

As you know, bacteria are used to clone a gene or to mass-produce a product. They are also genetically engineered to perform other services.

Protection and Enhancement of Plants Genetically engineered bacteria can be used to promote the health of plants. For example, bacteria that normally live on plants and encourage the formation of ice crystals have been changed from frost-plus to frost-minus bacteria. Field tests showed that these genetically engineered bacteria protect the vegetative parts of plants from frost damage. Also, a bacterium that normally colonizes the roots of corn plants has now been endowed with genes (from another bacterium) that code for an insect toxin. The toxin is expected to protect the roots from insects.

Many other genetic engineering applications are thought to be possible in agriculture. For example, *Rhizobium* is a bacterium that lives in nodules on the roots of leguminous plants, such as bean plants. Here, the bacteria fix atmospheric nitrogen into a form that can be used by the plant. It might be possible to genetically engineer bacteria and nonleguminous plants, such as corn, rice, and wheat, so that bacteria can infect them and produce nodules. This would reduce the amount of fertilizer needed on agricultural fields.

Bioremediation Bacteria can be selected for their ability to degrade a particular substance, and then this ability can be enhanced by genetic engineering. For instance, naturally occurring bacteria that eat oil can be genetically engineered to do an even better job of cleaning up beaches after oil spills (Fig. 17.6). Industry has found that bacteria can be used as biofilters to prevent airborne chemical pollutants from being vented into the air. They can also remove sulfur from coal before it is burned and help clean up toxic waste dumps. One such strain was given genes that allowed it to clean up levels of toxins that would have killed other strains. Further, these bacteria were given "suicide" genes that caused them to self-destruct when the job had been accomplished.

Chemical Production Organic chemicals are often synthesized by having catalysts act on precursor molecules or by using bacteria to carry out the synthesis. Today, it is possible to go one step further and to manipulate the genes that code for these enzymes. For instance, biochemists discovered a strain of bacteria that is especially good at producing phenylalanine, an amino acid needed to make aspartame, the dipeptide sweetener better known as NutraSweet. They isolated, altered, and formed a vector for the appropriate genes so that various bacteria could be genetically engineered to produce phenylalanine.

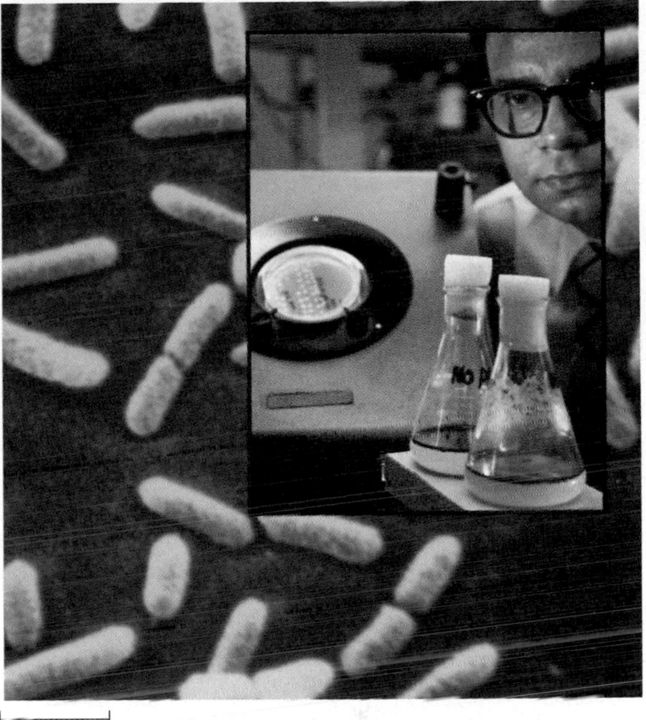

Figure 17.6 Bioremediation.
Bacteria capable of decomposing oil have been engineered and patented by the investigator, Dr. Chakrabarty. In the inset, the flask toward the rear contains oil and no bacteria; the flask toward the front contains the bacteria and is almost clear of oil. Now that engineered organisms (e.g., bacteria and plants) can be patented, there is an even greater impetus to create them.

Mineral Processing Many major mining companies already use bacteria to obtain various metals. Genetic engineering may enhance the ability of bacteria to extract copper, uranium, and gold from low-grade sources. At least two mining companies are testing genetically engineered organisms having improved bioleaching capabilities.

Bacteria are being genetically modified to perform all sorts of tasks not only in the factory but also in the environment. Safety concerns are discussed in the reading on page 277.

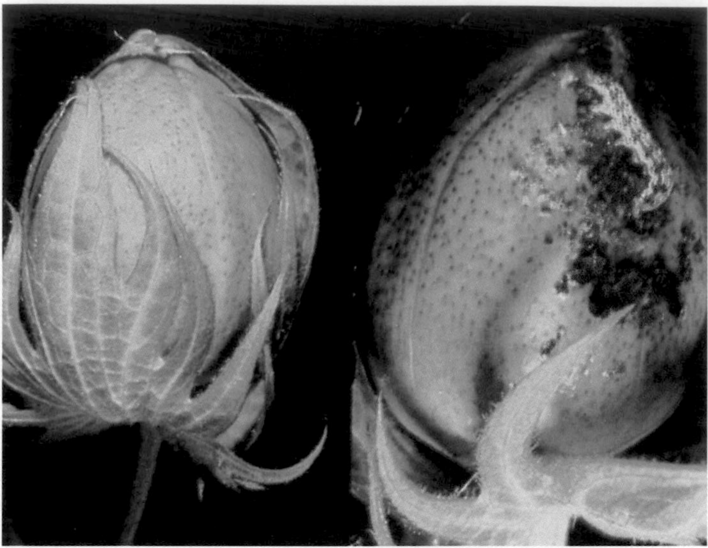

Figure 17.7 Genetically engineered plants. The cotton boll on the right is from a plant that has not been genetically engineered; the boll on the left is from a plant that was genetically engineered to resist cotton bollworm larvae and will go on to produce a normal yield of cotton.

Figure 17.8 Genetically engineered animals. These goats are genetically engineered to produce antithrombin III, which is secreted in their milk.

Transgenic Plants Are Here

The only possible plasmid for genetically engineering plant cells belongs to the bacterium *Agrobacterium,* which will infect many but not all plants. Therefore, other techniques have been developed to introduce foreign DNA into plant cells that have had the cell wall removed and are called protoplasts. It is possible to treat protoplasts with an electric current while they are suspended in a liquid containing foreign DNA. The electric current makes tiny, self-sealing holes in the plasma membrane through which genetic material can enter. Then a protoplast will develop into a complete plant.

Presently, about 50 types of genetically engineered plants that resist insects, viruses, or herbicides have entered small-scale field trials (Fig. 17.7). The major crops that have been improved in this way are soybean, cotton, alfalfa, rice, and potato; however, even genetically engineered corn may reach the marketplace by the year 2000.

It is hoped that one day genetically engineered plants will have one or more of these attributes: (1) they will be heat-, cold-, drought-, or salt-tolerant; (2) they will be more nutritious; (3) they can be stored and transported without fear of damage; (4) they will require less fertilizer; or (5) they will produce chemicals and drugs that are of interest to humans. Plants have been engineered to produce human proteins, such as hormones, in their seeds. A weed called mouse-eared cress has been engineered to produce a biodegradable plastic (polyhydroxybutyrate, or PHB) in cell granules.

Transgenic Animals Are Here

Animals, too, are being genetically engineered. Because animal cells will not take up bacterial plasmids, other methods are used to insert genes into their eggs. It is possible to microinject foreign genes into eggs by hand, but another method uses vortex mixing. The eggs are placed in an agitator with DNA and silicon-carbide needles, and the needles make tiny holes through which the DNA can enter. Using this technique, many types of animal eggs have taken up bovine growth hormone (bGH). The procedure has been used to produce larger fishes, cows, pigs, rabbits, and sheep. Genetically engineered fishes are now being kept in ponds that offer no escape to the wild because there is much concern that they will upset or destroy natural ecosystems.

Gene pharming, the use of transgenic farm animals to produce pharmaceuticals, is being pursued by a number of firms. It is advantageous to use animals because the product is obtainable from the milk of females. Genes that code for therapeutic and diagnostic proteins are incorporated into the animal's DNA, and the proteins appear in the animal's milk. In one instance a bull was genetically engineered to carry a gene for human lactoferrin, a drug for gastrointestinal tract infections, and he passed the gene to many offspring, among them several females. There are also plans to produce drugs for the treatment of cystic fibrosis, cancer, blood diseases, and other disorders. Clinical testing of antithrombin III, for preventing blood clots during surgery, is currently being produced by a herd of goats and clinical trials are to begin soon (Fig. 17.8).

Ecology Focus

Biotechnology: Friend or Foe?

Popular magazines often report biotechnology news such as this example:

> In a feat that could boost wheat production worldwide, plant biologists have for the first time permanently transferred a foreign gene into wheat. The gene makes wheat resistant to the herbicide phosphinothricin, which normally kills any plant it touches, weed or crop. Plant breeders say the herbicide-resistant wheat should enable farmers to spray their fields with the powerful herbicide to eradicate weeds without harming their harvest.*

In a hungry world you would expect such news to be greeted with enthusiasm—this new strain of wheat offers the possibility of a more bountiful harvest and the feeding of many more people. There are some, however, who see dangers lurking in the use of biotechnology to develop new and different strains of plants and animals. And these doomsayers are not just anybody—they are ecologists.

First, we have to consider that herbicide-resistant wheat will allow farmers to use more herbicide than usual in order to kill off weeds. Then, too, suppose this new form of wheat is better able to compete in the wild. Certainly we know of plants that have become pests when transported to a new environment: prickly pear cactus took over many acres of Australia; an ornamental tree, the melaleuca, has invaded and is drying up many of the swamps in Florida. Such plants spread because they are able to overrun the native plants of an area. Perhaps genetically engineered plants will also spread everywhere and be out of control. Or worse, suppose herbicide-resistant wheat were to hybridize with a weed, making the weed also resistant and able to take over other agricultural fields. As more and different herbicides are used to kill off the weed, the environment would be degraded. And similar concerns pertain to any transgenic organism, whether a bacterium, plant, or animal (Fig. 17A).

In the past, humans have been quick to believe that a new advance was the answer to a particular problem. The new pesticide DDT was going to kill off mosquitoes, making malaria a disease of the past. Instead, mosquitoes become resistant, and DDT accumulates in the tissues of humans, possibly contributing to all manner of health problems, from reduced immunity to reproductive infertility. When antibiotics were first introduced, it was hoped a disease like tuberculosis would be eliminated forever. Resistant strains of tuberculosis have now evolved to threaten us all.

More and more transgenic varieties have been developed and are being tested in agricultural fields. A few of these, like Freedom2 Squash, have a weedy relative in the wild with which they could hybridize. Agricultural officials point out, however, that Freedom2 Squash has been growing in fields since 1994, and although hybridization with its weedy relative, a Texas gourd, may have occurred, a "superweed" has not taken over Texas yet. Still, say botanists, it could happen in the future. Laboratory studies in Denmark showed that a hybrid of transgenic oilseed rape and field mustard did resist herbicides and was able to produce highly fertile pollen. Ecologists maintain it may be only a matter of time before a "superweed" does appear in the wild.

Figure 17A Transgenic squash. This squash may look like any other plump yellow squash, but it has been genetically engineered to resist viruses and therefore out produce its cousins which grow in the wild. Could hybridization with one of its cousins create a "superweed" that would overgrow in the wild, choke out native plants, and become an environmental menace?

*Research Notes, "Biology," Science News 141(1992):379.

Ecological Concerns

There are those who are very concerned about the deliberate release of genetically modified microbes into the environment. If these bacteria displaced those that normally reside in an ecosystem, the effects could be deleterious. However, tools are now available to detect, measure, and even stop cell activity in the natural environment. It is hoped that these tools will eventually pave the way for genetically modified microbes to play a significant role in agriculture and in environmental protection.

The same ecological concerns regarding genetically engineered bacteria also pertain to genetically engineered plants and animals. The ecology reading for this chapter uses herbicide-resistant wheat as an example, but the very same arguments can be waged against any genetically engineered organism.

17.4 Gene Therapy Is a Reality

Gene therapy will hopefully give a patient healthy genes to make up for a faulty gene. Gene therapy also includes the use of genes to treat genetic disorders and various other human illnesses. There are ex vivo and in vivo methods of gene therapy.

Some Methods Are Ex Vivo

During ex vivo (outside the living organism) therapy, cells are removed from a patient, treated, and returned to the patient. A retrovirus, which has an RNA genome, is often used as a vector to carry normal genes into the cells of the patient. After recombinant RNA from the retrovirus enters a human cell, such as a bone marrow stem cell, reverse transcription occurs. (During reverse transcription, RNA is used as a template for the formation of cDNA.) It is the resulting DNA that carries the normal gene into the human genome (Fig. 17.9).

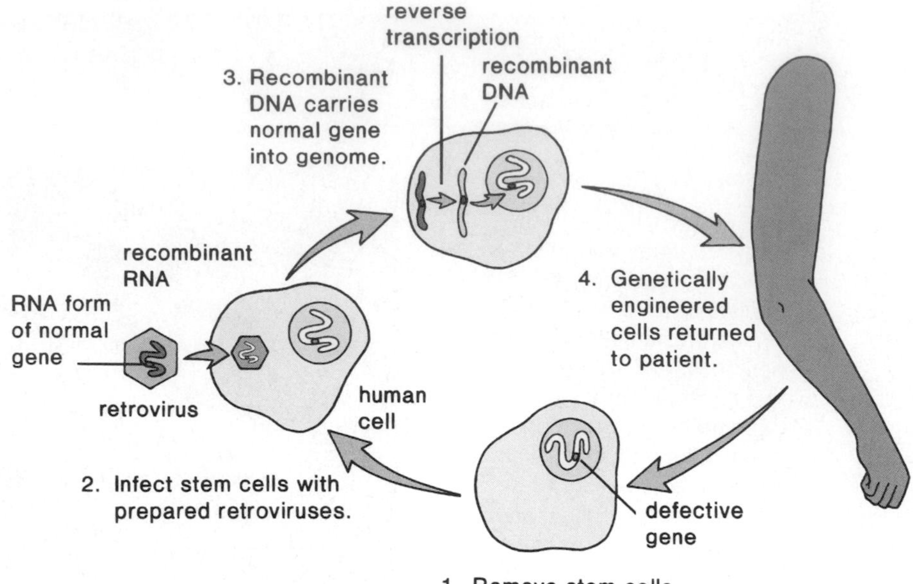

Figure 17.9 Ex vivo gene therapy in humans.
Bone marrow stem cells are withdrawn from the body, a normal gene is inserted into them, and they are then returned to the body.

Two young girls with severe combined immunodeficiency syndrome (SCID)[1] underwent ex vivo gene therapy several years ago. These girls lack an enzyme that is involved in the maturation of T and B cells, and therefore they were subject to life-threatening infections. White blood cells were removed from their blood and infected with a retrovirus that carried a normal gene for the enzyme. Then the cells were returned to the girls. Later, one of these girls and two newborn boys received bone marrow stem cells genetically engineered in the same way. Genetically engineered stem cells are preferred because they are long-lived, and their use may result in a permanent cure.

Among 100-plus gene therapy trials, gene therapy is being used for treatment of familial hypercholesterolemia, a condition that develops when liver cells lack a receptor for removing cholesterol from the blood. The high levels of blood cholesterol make the patient subject to fatal heart attacks at a young age. In a newly developed procedure, a small portion of the liver is surgically excised and infected with a retrovirus containing a normal gene for the receptor. Chemotherapy in cancer patients often kills off healthy cells as well as cancer cells. In clinical trials, researchers have given genes to cancer patients that either make healthy cells more tolerant of chemotherapy or make tumors more vulnerable to it. In one trial, bone marrow stem cells from about 30 women with late-stage ovarian cancer were infected with a virus carrying a gene for multiple-drug resistance.

Some Methods Are In Vivo

Other gene therapy procedures use viruses, laboratory-grown cells, or even synthetic carriers to introduce genes directly into patients. If in vivo (inside the living organism) therapy is used, no cells are removed from the patient. For example, liposomes are microscopic vesicles that spontaneously form when lipoproteins are put in a solution. Liposomes have been coated with healthy cystic fibrosis genes and sprayed into patients' nostrils as a possible treatment for cystic fibrosis. Retroviruses can be used to carry genes for cytokines, soluble hormones of the immune system, directly into the tumors of patients. It has been observed that the presence of cytokines stimulates the immune system to rid the body of cancer cells.

Perhaps it will be used one day for in vivo therapy to cure hemophilia, diabetes, Parkinson disease, or AIDS. To treat hemophilia, patients could get regular doses of cells that contain normal clotting-factor genes. Or such cells could be placed in *organoids*, artificial organs that can be implanted in the abdominal cavity. To cure Parkinson disease, dopamine-producing cells could be grafted directly into the brain. These procedures will use laboratory-grown cells that have been stripped of antigens (to decrease the possibility of an immune system attack).

Gene therapy is now being actively investigated, and researchers are envisioning all sorts of applications aimed at curing human genetic disorders as well as many other types of illnesses.

[1]SCID is often called the "bubble-baby disease" after David, a young person who lived under a plastic dome to prevent infection.

17.5 Mapping the Human Chromosomes

The goal of the *Human Genome Project* is to identify the location of the approximately 100,000 human genes on all the chromosomes (called the *genetic map*) and to determine the sequence of the three billion base pairs (called the *physical map*) in the human genome.

The Genetic Map

Several methods have been used to attempt mapping the human chromosomes (Fig. 17.10). On occasion, the mRNA of a particular gene has been isolated, and reverse transcriptase has been used to produce a cDNA copy of this gene. The cDNA copy can be attached to a fluorescent dye and used as a probe to determine to which chromosome, and indeed, to which band of this chromosome, it belongs. Base pairing between probe and chromosome will occur right on a microscope slide, and the use of fluorescence allows us to see precisely where the gene is located.

Through studying the DNA of related persons, it is sometimes possible to discover a genetic marker; that is, a particular sequence of DNA bases that is closely linked to a disease-causing gene. When a marker is inherited with the gene 98% of the time, it can be used to test for the genetic disorder. The available tests for sickle-cell disease, Huntington disease, and Duchenne muscular dystrophy all make use of a genetic marker. Once a marker has been discovered, it is possible to research DNA libraries until the exact location of the gene is discovered. Widely used are cDNA libraries that contain only the expressed genes of a particular cell type. If you are looking for a gene that causes a muscle defect, you would look for it only in a cDNA library for muscle cells.

The Physical Map

The completed physical map of the human genome will indicate the sequence of all the DNA pairs on the chromosomes. In order to create a physical map, researchers use laboratory procedures such as that briefly described in Figure 17.11. Recently researchers published a map that largely contains expressed sequence tags (ESTs), which are short sequences of DNA from expressed genes. The ESTs are valuable for their ability to speed efforts toward locating disease-causing genes. The applicability of the Human Genome Project to human genetic disease is a primary motivation for its continuance. As genetic testing becomes more frequent, ethical guidelines need to be developed for when and how to use this type of testing.

The Human Genome Project seeks to establish the sequence of all the genes on all the chromosomes and the sequence of all the DNA base pairs on all the chromosomes.

ichthyosis, X linked
hypophosphatemia
ocular albinism
Duchenne muscular dystrophy
retinitis pigmentosa
Lesch-Nyhan syndrome
hemophilia B
fragile X syndrome
hemophilia A
color blindness (several forms)
spastic paraplegia, X linked

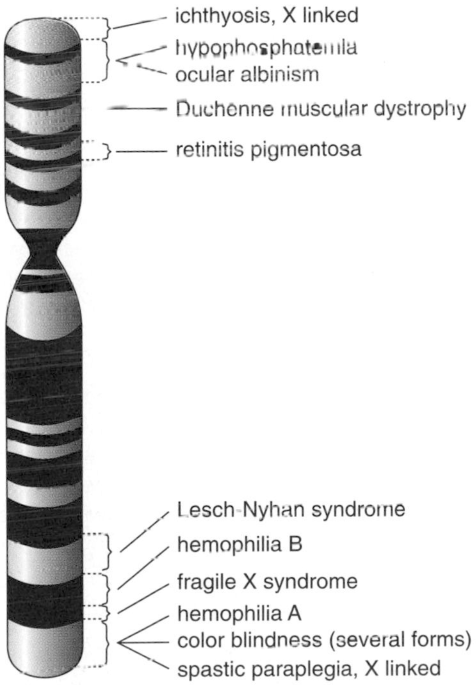

Figure 17.10 Genetic map of the X chromosome.
The human X chromosome has been partially mapped, and this is the order of some of the genes now known to be on this chromosome.

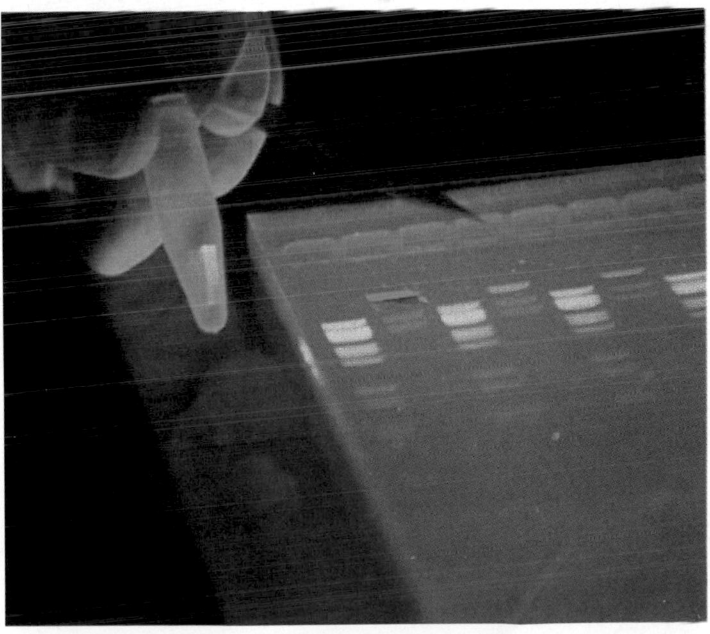

Figure 17.11 Sequencing DNA base pairs.
The DNA is cleaved into fragments that vary in length by only one nucleotide. Each fragment ends with an A, T, C, or G marked by one of four fluorescent dyes. The fragments are subjected to gel electrophoresis, and the investigator simply reads off the sequence of the nucleotides.

connecting concepts

A mere 40 years ago, James Watson and Francis Crick gave the world one of the most significant pieces of information ever unraveled—the structure of DNA. Today, biotechnology is helping us produce drugs, test unborn children for genetic disorders, identify carriers of genetic diseases, and we even hope to replace faulty genes with working copies one day.

This recent explosion in genetic research and biotechnology is not without its critics and possible costs. Some opponents believe humans are tampering with life itself or "playing God." Others fear that new, genetically manufactured organisms may disrupt natural ecosystems or cause as yet unknown diseases. One thing is clear, however—genetic research isn't going away, so we as individuals and as a society have to decide who has a right to the resulting information and how it is to be used. Since the societal ramifications of genetic research have to be considered by us all, each of us needs to know as much as possible about the subject.

A knowledge of genetics will also allow you to have an appreciation of the grand unifying theory (GUT) of biology—evolution. Species differences can be traced to differences in their genes. Why are there many thousands, perhaps millions, of different species on earth and how are they related? What has caused some phenotypes to last for billions of years and others to die out over time?

We will see that general knowledge of genetics and evolution go hand in hand. Today, we have a good idea of how evolution proceeds because we know how genes are transmitted from parent to offspring. As discussed in the next chapter, it is remarkable that two men in the 1800s reached their conclusions about evolutionary process with no knowledge of chromosomes, genes, or DNA.

Summary

17.1 Cloning of a Gene

To clone a gene (assuming that a plasmid is to be the vector), a restriction enzyme is used to cleave plasmid DNA and to cleave foreign DNA. The "sticky ends" produced facilitate the insertion of foreign DNA into vector DNA. The foreign gene is sealed into the vector DNA by DNA ligase.

A genomic library can be used as a source of genes to be cloned. A radioactive or fluorescent probe is used to identify the location of a single gene among the cloned fragments of an organism's DNA. The polymerase chain reaction (PCR) uses the enzyme DNA polymerase to carry out multiple replications of target DNA.

DNA can be subjected to DNA fingerprinting. After the DNA of an organism is treated with restriction enzymes, the fragments are subjected to gel electrophoresis. If the entire genome is used for fingerprinting, the use of probes results in a distinctive pattern that can be recorded on X-ray film. If DNA fingerprinting follows PCR, no probes are needed because of the limited amount of DNA involved. PCR plus analysis, which may even involve sequencing the bases of a DNA segment, has proved to be invaluable.

17.2 Biotechnology Products Are Many

Genetically engineered bacteria produce many products of interest to humans, such as hormones and vaccines.

17.3 Making Transgenic Organisms

Transgenic organisms have also been made. Bacteria have been produced to promote the health of plants, perform bioremediation, extract minerals, and produce chemicals.

Plant cells, genetically engineered while they are protoplasts in tissue culture, grow to be transgenic plants.

Transgenic plants have been produced that have a natural resistance to herbicides and pests. In the future, plants may have an ability to fix atmospheric nitrogen, an increased ability to grow in arid and salty soils, and greater nutritional value.

Genetic engineering of animals has made much progress. Many firms are interested in gene pharming, the use of genetically engineered animals to produce pharmaceuticals in milk.

17.4 Gene Therapy Is a Reality

Human gene therapy is undergoing clinical trials. Ex vivo therapy involves withdrawing cells from the patient, inserting a functioning gene, usually a retrovirus, and then returning the treated cells to the patient. Many investigators are trying to develop in vivo therapy, in which viruses, laboratory-grown cells, or synthetic carriers will be used to carry healthy genes directly into the patient.

17.5 Mapping the Human Chromosomes

The Human Genome Project utilizes base pairing between a probe and a chromosome to find the location of genes. Genetic marker data help assign genes to chromosomes. There are laboratory procedures to develop a physical map; that is, the sequence of DNA base pairs in the human genome. These data will be used to locate disease-causing genes.

Reviewing the Chapter

1. What is the methodology for producing recombinant DNA to be used in gene cloning? 270
2. What is a genomic library, and how do you locate a gene of interest in the library? 271
3. What is the polymerase chain reaction (PCR), and how is it carried out? What are some practical applications for PCR? 272–73
4. Categorize and give examples of the types of biotechnology products available today. 274
5. Bacteria have been genetically engineered to perform what services? What are the ecological concerns regarding their release into the environment? 275–77
6. In what ways have plants been genetically engineered, and what type of genetic engineering is expected in the future? 276
7. In what ways have animals been genetically engineered? 276
8. Explain and give examples of gene therapies in humans. 278
9. Explain the two primary goals of the Human Genome Project. 279

Testing Yourself

Choose the best answer for each question.

1. Which of these is a true statement?
   a. Both plasmids and viruses can serve as vectors.
   b. Plasmids can carry recombinant DNA but viruses cannot.
   c. Vectors carry only the foreign gene into the host cell.
   d. All of these statements are true.
2. Which of these is a benefit to having insulin produced by biotechnology?
   a. It can be mass-produced.
   b. It is nonallergenic.
   c. It is less expensive.
   d. All of these are correct.
3. Restriction fragment length polymorphisms (RFLPs)
   a. identify individuals genetically.
   b. are the basis for DNA fingerprints.
   c. can be subjected to gel electrophoresis.
   d. All of these are correct.
4. Which of these would you not expect to be a biotechnology product?
   a. modified enzyme
   b. DNA probe
   c. protein hormone
   d. steroid hormone
5. What is the benefit of using a retrovirus as a vector in gene therapy?
   a. It is not able to enter cells.
   b. It incorporates the foreign gene into the host chromosome.
   c. It eliminates a lot of unnecessary steps.
   d. Both b and c are correct.
6. Gel electrophoresis
   a. measures the size of plasmids.
   b. tells whether viruses are infectious.
   c. measures the charge and size of proteins and DNA fragments.
   d. All of these are correct.
7. Using this key, put the phrases in the correct order to form a plasmid carrying recombinant DNA.

Key:

1 use restriction enzymes
2 use DNA ligase
3 remove plasmid from parent bacterium
4 introduce plasmid into new host bacterium

   a. 1,2,3,4
   b. 4,3,2,1
   c. 3,1,2,4
   d. 2,3,1,4
8. Which of these is incorrectly matched?
   a. protoplast—plant cell engineering
   b. RFLPs—DNA fingerprinting
   c. DNA polymerase—PCR
   d. DNA ligase—mapping human chromosomes
9. The restriction enzyme called *Eco*RI has cut double-stranded DNA in this manner. The piece of foreign DNA to be inserted ends in what bases (a) to the left? (b) to the right?

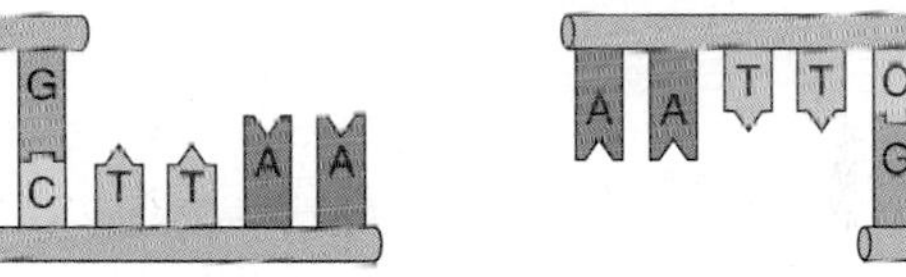

10. Label the following drawings, using these terms: retrovirus, recombinant RNA (twice), defective gene, recombinant DNA, reverse transcription, and human genome.

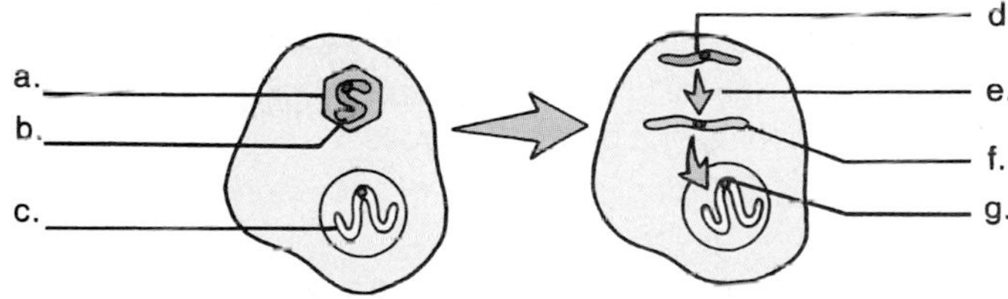

Applying the Concepts

1. *Organisms are chemical and physical machines.*
   Provide examples to show that DNA can be manipulated, and explain why you think some people would find DNA manipulation disturbing.
2. *DNA is the common genetic material of all organisms.*
   How does recombinant DNA technology support this concept?
3. *Biotechnology products can be a boon to humans.*
   Why would you expect meat from cattle genetically engineered to contain extra growth hormone to be safe for human consumption?

Using Technology

Your study of recombinant DNA and biotechnology is supported by these available technologies:

Exploring the Internet

The Mader Home Page provides resources for and help with studying this chapter.

http://www.mhhe.com/sciencemath/biology/mader/
(Click on Biology.)

Explorations in Cell Biology & Genetics CD-ROM

DNA Fingerprinting: You Be the Judge (#14)
Making a Restriction Map (#17)

Understanding the Terms

clone 271	plasmid 270
DNA fingerprinting 273	probe 271
DNA ligase 270	recombinant DNA (rDNA) 270
gene therapy 278	restriction enzyme 270
genetic engineering 270	transgenic organism 275
genome 271	vector 270

Match the terms to these definitions:

a. __________ Bacterial enzyme that stops viral reproduction by cleaving viral DNA; used to cut DNA at specific points during production of recombinant DNA.

b. __________ DNA that contains genes from more than one source.

c. __________ Free-living organisms in the environment that have had a foreign gene inserted into them.

d. __________ In genetic engineering, a means to transfer foreign genetic material into a cell—for example, a plasmid.

e. __________ Known sequence of DNA that is used to find complementary DNA strands; can be used diagnostically to determine the presence of particular genes.

f. __________ Identical copy; in genetic engineering, many identical copies of a gene are made.

g. __________ Self-duplicating ring of accessory DNA in the cytoplasm of bacteria.

h. __________ Use of bioengineered cells or other biotechnology techniques to treat human genetic disorders.

i. __________ Enzyme that links DNA fragments; used during production of recombinant DNA to join foreign DNA to the vector DNA.

Further Readings for Part ii

Anderson, W. F. September 1995. Gene therapy. *Scientific American* 273(3):124. Presents gene therapy now and discusses prospects for the future.

Beardsley, T. March 1996. Vital data. *Scientific American* 274(3):100. DNA tests for a wide array of conditions are becoming available.

Capecchi, M. March 1994. Targeted gene replacement. *Scientific American* 270(3):52. Researchers are deciphering DNA segments that control development and immunity.

Cavanee, W. K., and White, R. L. March 1995. The genetic basis of cancer. *Scientific American* 272(3):72. Medical intervention may be possible during the mutation process of cancerous cells.

Cohen, J. S., and Hogan, M. E. December 1994. The new genetic medicines. *Scientific American* 271(6):76. Artificial strings of nucleic acids can silence genes responsible for many illnesses.

Cummings, M. R. 1994. *Human heredity.* 3d ed. St. Paul, Minn.: West Publishing Co. This introductory text is written in a straightforward, interesting manner, and contains current information in the changing field of genetics.

Davis, D. L., and Bradlow, H. L. April 1995. Can environmental estrogens cause breast cancer? *Scientific American* 273(4):166. Estrogenlike compounds found in the environment may contribute to breast cancer.

Dawkins, R. November 1995. God's utility function. *Scientific American* 273(5):80. The role of genetics in evolution and natural selection is discussed.

Duke, R. C., et al. December 1996. Cell suicide in health and disease. *Scientific American* 275(6):80. Failures in the processes of cellular self-destruction may give rise to cancer, AIDS, Alzheimer disease, and some genetic diseases.

Frederick, R. J., and Egan, M. September 1994. Environmentally compatible applications of biotechnology. *BioScience* 44(8):529. Explains the use of living organisms to restore and safeguard the environment.

Garnick, M. April 1994. The dilemmas of prostate cancer. *Scientific American* 270(4):72. Prostate cancer has been detected with increasing frequency in recent years.

Gibbs, W. W. August 1996. Gaining on fat. *Scientific American* 275(2):88. Some weight problems are genetic or physiological in origin. New treatments might help.

Greenspan, R. J. April 1995. Understanding the genetic construction of behavior. *Scientific American* 272(4):72. Research indicates that even in simple organisms, behavior is influenced by a number of genes.

Greider, C. W., and Blackburn, E. H. February 1996. Telomeres, telomerase, and cancer. *Scientific American* 274(2):92. The enzyme telomerase rebuilds the chromosomes of tumor cells; this enzyme is being researched as a target for anticancer treatments.

Hartl, D. 1994. *Genetics.* 3d ed. Boston: Jones and Bartlett Publishers. Provides a clear, comprehensive, and straightforward introduction to the principles of genetics at the college level.

Haseltine, W. A. March 1997. Discovering genes for new medicines. *Scientific American* 276(3):92. New medical products being developed are a result of recent genetic analyses of the human genome.

Johnson, G. B. 1996. *How scientists think*. Dubuque, Iowa: Wm. C. Brown Publishers. Presents the rationale behind 21 important experiments in genetics and molecular biology that became the foundation for today's research.

Kleinsmith, L. J., and Kish, V. M. 1995. *Principles of cell and molecular biology*. 2d ed. New York: HarperCollins College Publishers. This text introduces students to the fundamental principles that guide cellular organization and function.

Langer, R., and Vacanti, J. P. September 1995. Artificial organs. *Scientific American* 273(3):130. Discusses engineering artificial tissue using the body's own cells.

Lanza, R. P., and Chick, W. L. 1995. Encapsulated cell therapy. *Scientific American Science & Medicine* 2(4):16. Implants containing living cells (tissue engineering) within a selectively permeable membrane could provide drug or hormone doses as required for therapy.

Lasic, D. D. May/June 1996. Liposomes. *Science & Medicine* 3(3):34. Liposomes can be used to deliver drugs or genes for gene therapy.

Leffell, D. J., and Brash, D. E. July 1996. Sunlight and skin cancer. *Scientific American* 275(1):52. Discusses the sequence of changes that may occur in skin cells after exposure to UV rays.

Lyon, J., and Gorner, P. 1995. *Altered fates: Gene therapy and the retooling of human life*. New York: W. W. Norton & Company, Inc. This nonfiction work details the saga of gene therapy, from the scientists to the patients and their families.

Mange, E., and Mange, A. 1994. *Basic human genetics*. Sunderland, Mass.: Sinauer Associates. Presents the general principles of genetics and discusses the implications for individuals and society.

McConkey, E. H. 1993. *Human genetics: The molecular revolution*. Boston: Jones and Bartlett Publishers. This text for professionals and advanced students emphasizes the impact of molecular information while surveying the current status of understanding and treatment of genetic disease.

McGinnis, W., and Kuziora, M. February 1994. The molecular architects of body design. *Scientific American* 270(2):58. The ability to transfer genes between species provides a way to study how genes control development.

Moses, V., and Moses, S. 1995. *Exploiting biotechnology*. Chur, Switzerland: Harwood Academic Publishers. Provides a general understanding of biotechnology and presents its commercial and industrial applications.

Nicolaou, K. C., et al. June 1996. Taxoids: New weapons against cancer. *Scientific American* 274(6):94. Chemists are synthesizing a family of drugs related to taxol for the treatment of cancer.

Noble, E. P. March/April 1996. The gene that rewards alcoholism. *Scientific American Science & Medicine* 3(2):52. A dopamine receptor gene is linked to severe alcoholism.

Perera, F. P. May 1996. Uncovering new clues to cancer risk. *Scientific American* 274(5):54. Molecular epidemiology finds biological markers that explain what makes people susceptible to cancer.

Peters, P. 1994. *Biotechnology: A guide to genetic engineering*. Dubuque, Iowa: Wm. C. Brown Publishers. DNA structure and function are reviewed before biotechnology, including genetic engineering, are explained.

Rennie, J. July 1994. Immortal's enzyme. *Scientific American* 271(1):14. An enzyme that maintains telomeres may make tumor cells immortal.

Rennie, J. June 1994. Grading the gene tests. *Scientific American* 270(6):88. Raises ethical questions about screening embryos for genetic disease before implantation.

Rodgers, G. P., et al. October 1994. Sickle cell anemia. *Scientific American Science & Medicine* 1(4):48. Promising therapies may be forthcoming, despite obstacles to research.

Russell, P. J. 1996. *Genetics*. 4th ed. New York: HarperCollins College Publishers. This easy-to-read text emphasizes an inquiry-based approach to genetics, and explores many of the research experiments that led to important genetic advances.

Scientific American Science & Medicine. January/February 1996. 3(1). Issue contains several articles concerning genetic testing, imaging, and therapy for breast cancer.

Scientific American Science & Medicine. November/December 1995. 2(6). This issue includes articles on targeting proteins for anticancer therapy, and the goals of the Human Genome Project.

Scientific American Special Issue. What you need to know about cancer. September 1996. 275(3). The entire issue is devoted to the causes, prevention, and early detection of cancer, and cancer therapies—conventional and future.

Spicer, D. V., and Pike, M. C. July/August 1995. Hormonal manipulation to prevent breast cancer. *Scientific American Science & Medicine* 2(4):58. This contraceptive regimen could reduce the risk of breast cancer.

Tamarin, R. 1996. *Principles of genetics*. 5th ed. Dubuque, Iowa: Wm. C. Brown Publishers. This text emphasizes the major areas of genetics and genetic research.

Varmus, H., and Weinberg, R. A. 1993. *Genes and the biology of cancer*. New York: *Scientific American* Library. The origins, nature, and mechanisms of carcinogenesis are explored in this book written for the nonexpert.

Velander, W. H., et al. January 1997. Transgenic livestock as drug factories. *Scientific American* 276(1):70. Pigs, cows, sheep, and other farm animals can be bred to produce large amounts of medicinal proteins in their milk.

Weaver, R. F., and Hedrick, P. W. 1997. *Basic genetics*. 3d ed. Dubuque, Iowa: Wm. C. Brown Publishers. Basic concepts are presented in a concise, easy-to-understand manner. There is an emphasis on problem solving in both the text and chapter-end matter.

Welsh, M. J., and Smith, A. E. December 1995. Cystic fibrosis. *Scientific American* 273(6):52. The genetic defects that cause CF hamper or prevent ion transport by affected lung cells.

Wolffe, A. P. November/December 1995. Genetic effects of DNA packaging. *Science & Medicine* 2(6):68. The regulation of DNA coiling in the chromosomes adds to the properties of the genes involved in several genetic diseases.

PART III

Evolution

Evolution refers to descent with modification and adaptation to the environment. Descent from a common ancestor explains the unity of life—living things share a common chemistry and cellular structure because they are all descended from an original source. Each type of living thing has a history that can be traced by way of the fossil record and discerned from a comparative study of other living things. Human evolution can also be understood by the application of these evolutionary principles.

Adaptation to the environment explains the diversity of life as a process that can now be understood in terms of modern genetics. Genetic changes produce the variations that result in the origination of new species (a group that cannot reproduce with other groups no matter how similar in appearance). Each species is adapted to its environment and has unique, evolved solutions to life's problems such as acquiring nutrients, finding a mate, and reproducing. Natural selection is a mechanism that results in adaptation to the environment.

Darwin and Evolution

CHAPTER 18

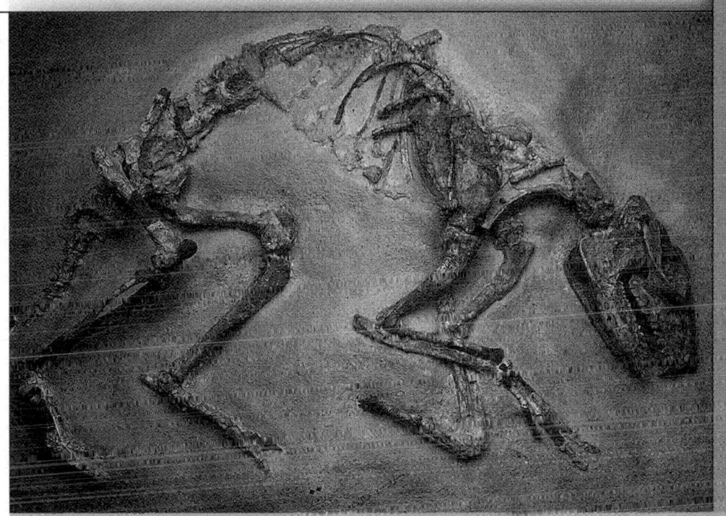

Tertiary fossil, ungulate order Notoungulata

Chapter Concepts

18.1 The World Was Ready

- In the eighteenth century, scientists became especially interested in classifying and understanding the relationship between all the many forms of present and past life. 286
- Gradually in the eighteenth century, scientists began to accept the view that the earth is very old and that life-forms do evolve or change over time. 287
- Evolution has two aspects: descent from a common ancestor and adaptation to the environment. 287

18.2 Darwin Develops a Theory

- Charles Darwin's trip around the Southern Hemisphere aboard the HMS *Beagle* provided him with evidence that the earth is very old and that descent with modification does occur. 289
- Both Darwin and Alfred Wallace proposed natural selection as a mechanism by which adaptation to the environment occurs. This mechanism is consistent with our present-day knowledge of genetics. 292, 295

18.3 Evidence Accumulates

- The fossil record, biogeography, comparative anatomy, and comparative biochemistry support a hypothesis of common descent. 296

Modern geologists believe the earth is over 4 billion years old. They speculate that the mountains, deserts, and oceans of today formed from slow, gradual, but continuous processes of erosion and uplifting. Then about 3.5 billion years ago, life began. From simple, unicellular organisms, new-life forms arose and changed in response to environmental pressures, producing the past and present biodiversity.

However, these are relatively new ideas. Prior to the 1800s, most people thought the earth was only a few thousand years old. They also believed that species were specially created and fixed in time. Remains of living things no longer on earth (i.e., fossils) were explained by the notion of global catastrophes—mass extinctions and repopulations of species—which occurred periodically throughout history.

Into this prevailing climate came Charles Darwin, Alfred Wallace, and other innovative minds who changed the field of biology forever. On a five-year ocean voyage, Darwin observed many diverse life-forms, and he eventually concluded that species evolve in response to their environment. All living things share common characteristics, he believed, because we are descended from a common ancestor. This idea challenged the widely accepted biblical account of creation, and consequently, the concept of evolution required an intellectual revolution of great magnitude.

18.1 The World Was Ready

Charles Darwin was only 22 in 1831 when he accepted the position of naturalist aboard the HMS *Beagle*, a British naval ship about to sail around the world (Fig. 18.1). Darwin's major mission was to expand the navy's knowledge of natural resources (e.g., water and food) in foreign lands. The captain was also hopeful that Darwin would find evidence of the biblical account of creation. The results of Darwin's observations were just the opposite, however, as we shall examine later in the chapter.

Table 18.1 tells us that prior to Darwin, people had an entirely different way of looking at the world. Their mindset was determined by deep-seated beliefs held to be intractable truths. To turn from these beliefs and accept the Darwinian view of the world required an intellectual revolution of great magnitude. This revolution was fostered by changes in both the scientific and the social realms. Here,

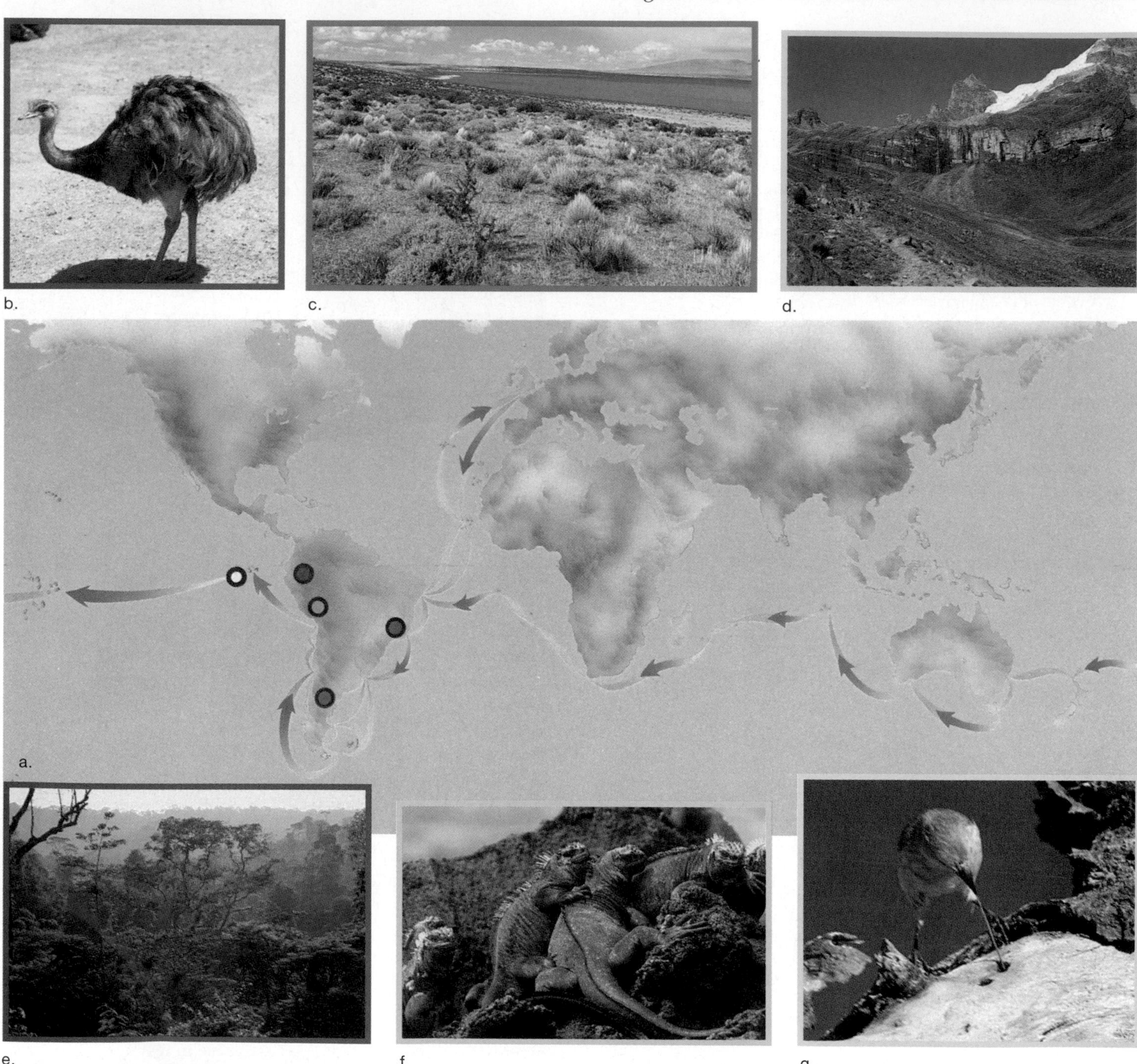

Figure 18.1 Voyage of the HMS *Beagle*.
a. Map shows the journey of the HMS *Beagle* around the world. Notice the encircled colors are keyed to the frames of the photographs, which show us what Charles Darwin may have observed. **b.** As Darwin traveled along the east coast of South America, he noted that a bird called a rhea looked like the African ostrich. **c.** The sparse vegetation of the Patagonian Desert is in the southern part of the continent. **d.** The Andes Mountains of the west coast, with strata containing fossilized animals. **e.** The lush vegetation of a rain forest. **f.** On the Galápagos Islands, marine iguanas have large claws to help them cling to rocks and blunt snouts for eating seaweed. **g.** Galápagos finches are specialized to feed on various foods.

Table 18.1

Contrast of Worldviews

Pre-Darwinian View	Post-Darwinian View
1. Earth is relatively young—age is measured in thousands of years.	1. Earth is relatively old—age is measured in billions of years.
2. Each species is specially created; species don't change, and the number of species remains the same.	2. Species are related by descent—it is possible to piece together a history of life on earth.
3. Adaptation to the environment is the work of a creator, who decided the structure and function of each type of organism. Any variations are imperfections.	3. Adaptation to the environment is the interplay of random variations and environmental conditions.
4. Observations are supposed to substantiate the prevailing worldview.	4. Observation and experimentation are used to test hypotheses, including hypotheses about evolution.

we will only touch on a few of the scientific contributions that helped bring about the new worldview.

Although it is often believed that Darwin forged this change in worldview by himself, biologists during the preceding century had slowly begun to accept the idea of **evolution** [L. *evolutio*, an unrolling]; that is, that species change with time. Evolution explains the unity and diversity of life. Living things share common characteristics because they are descended from a common ancestor. Living things are diverse because each species is adapted to its habitat and way of life. We will see that the history of evolutionary thought is a history of ideas about descent and adaptation. Darwin himself used the expression "descent with modification," by which he meant that as descent occurs through time so does diversification. He saw the process of adaptation as the means by which diversification comes about.

Mid-Eighteenth-Century Contributions

Taxonomy, the science of classifying organisms, was an important endeavor during the mid-eighteenth century. Chief among the taxonomists was Carolus Linnaeus (1707–78), who gave us the binomial system of nomenclature (a two-part name for species, such as *Homo sapiens*) and who developed a system of classification for all known plants. Linnaeus, like other taxonomists of his time, believed in the ideas of *special creation* and *fixity of species* (Fig. 18.2). He thought each species had an "ideal" structure and function and also a place in the *scala naturae*, a sequential ladder of life. The simplest and most material of beings was on the lowest rung of the ladder, and the most complex and most spiritual of beings was on the highest rung. Human beings occupied the last rung of the ladder.

These ideas were Christian dogma, which can be traced to the works of the famous Greek philosophers Plato and Aristotle. Plato said that every object on earth was an imperfect copy of an ideal form, which can be deduced upon reflection and study. To Plato, individual variations were imperfections that only distract the observer. Aristotle saw that organisms were diverse, and some were more complex than others. His belief that all organisms could be arranged in order of increasing complexity became the scala naturae just described.

Figure 18.2 Linnaeus. For the most part, Linnaeus believed that each species was created separately and that classification should reveal God's divine plan. Variations were imperfections of no real consequence.

Linnaeus and other taxonomists wanted to describe the ideal characteristics of each species and also wanted to discover the proper place for each species in the scala naturae. Therefore, Linnaeus did not even consider the possibility of evolutionary change for most of his working life. There is evidence, however, that he did eventually perform hybridization experiments, which made him think that at least species, if not higher categories of classification, might change with time.

Georges-Louis Leclerc, better known by his title, Count Buffon (1707–88), was a French naturalist who devoted many years of his life to writing a 44-volume natural history that described all known plants and animals. He provided evidence of descent with modification, and he even speculated on various mechanisms. Throughout his writings, he mentioned that the following factors could influence evolutionary change: direct influences of the environment, migration, geographical isolation, overcrowding, and the struggle for existence. Buffon seemed to vacillate, however, as to whether or not he believed in evolutionary descent, and often he professed to believe in special creation and the fixity of species.

Erasmus Darwin (1731–1802), Charles Darwin's grandfather, was a physician and a naturalist. His writings on both botany and zoology contained many comments, although they were mostly in footnotes and asides, that suggested the possibility of common descent. He based his conclusions on changes undergone by animals during development, artificial selection by humans, and the presence of vestigial organs (organs that are believed to have been functional in an ancestor but are reduced and nonfunctional in a descendant). Like Buffon, Erasmus Darwin offered no mechanism by which evolutionary descent might occur.

Figure 18.3 A mastodon.
Although mastodons are known only from the fossil record, Cuvier reconstructed the structure of these animals when provided only with a single bone. He said these animals were shorter than modern elephants but had massive, pillarlike legs. The skull was lower and flatter and of generally simpler construction than in modern elephants.

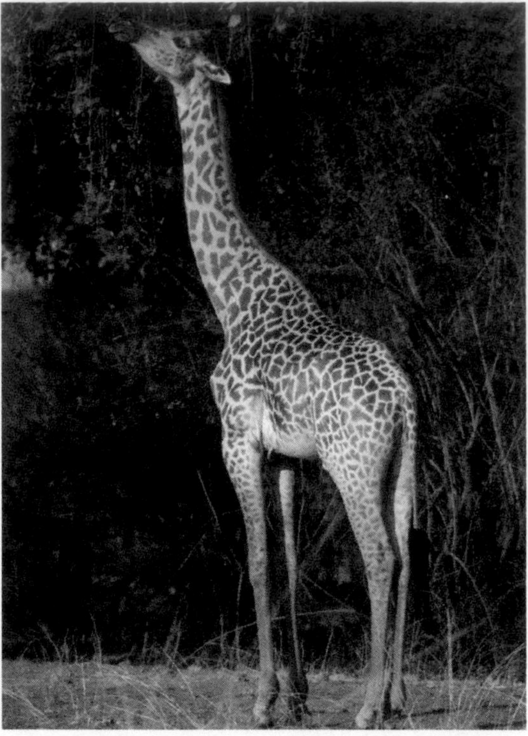

Figure 18.4 Inheritance of acquired characteristics.
Jean-Baptiste de Lamarck is most famous for suggesting that the environment influences heritable traits. The long neck of the giraffe, he said, is due to continual stretching to reach food. Each generation of giraffes stretched a little farther, and this was passed to the next generation. With the advent of modern genetics, it became possible to explain why inheritance of acquired characteristics would not be possible.

Late Eighteenth-Century Contributions

Cuvier and Catastrophism

In addition to taxonomy, comparative anatomy was of interest to biologists prior to Darwin. Explorers and collectors traveled the world and brought back not only currently existing species to be classified but also fossils (remains of once-living organisms) to be studied. Georges Cuvier (1769–1832), a distinguished vertebrate (animals with backbones) zoologist, was the first to use comparative anatomy to develop a system of classifying animals. He also founded the science of **paleontology** [Gk. *palaios,* ancient, old, and *ontos,* having existed; -logy, study of, from *logikos,* rational, sensible], the study of fossils, and suggested that a single fossil bone was all he needed to deduce the entire anatomy of an animal (Fig. 18.3).

Because Cuvier was a staunch advocate of special creation and fixity of species, he faced a real problem when geological evidence of a particular region showed a succession of life-forms in the earth's strata (layers). To explain these observations, he hypothesized that a series of local catastrophes or mass extinctions had occurred whenever a new stratum of that region showed a new mix of fossils. After each catastrophe, the region was repopulated by species from surrounding areas, and this accounted for the appearance of new fossils in the new stratum. The result of all these catastrophes was the appearance of change with time. Some of his followers even suggested that there had been worldwide catastrophes and that after each of these events, God created new sets of species. This explanation of the history of life came to be known as **catastrophism** [Gk. *katastrophe,* calamity, misfortune].

Lamarck's Theory of Evolution

Jean-Baptiste de Lamarck (1744–1829) was the first biologist to believe that evolution does occur and to link diversity with adaptation to the environment. Lamarck's ideas about descent were entirely different from those of Cuvier, perhaps because he was an invertebrate (animals without backbones) zoologist. Lamarck concluded, after studying the succession of life-forms in strata, that more complex organisms are descended from less complex organisms. He mistakenly said, however, that increasing complexity was the result of a natural force—a desire for perfection—that was inherent in all living things.

To explain the process of adaptation to the environment, Lamarck supported the idea of **inheritance of acquired characteristics**—that the environment can bring about inherited change. One example that he gave—and for which he is most famous—is that the long neck of a giraffe developed over time because animals stretched their necks to reach food high in trees and then passed on a long neck to their offspring (Fig. 18.4). The theory of the inheritance of acquired characteristics has never been substantiated by experimentation. The molecular mechanism of inheritance explains why. Phenotypic changes acquired during an organism's lifetime do not result in genetic changes that can be passed to subsequent generations.

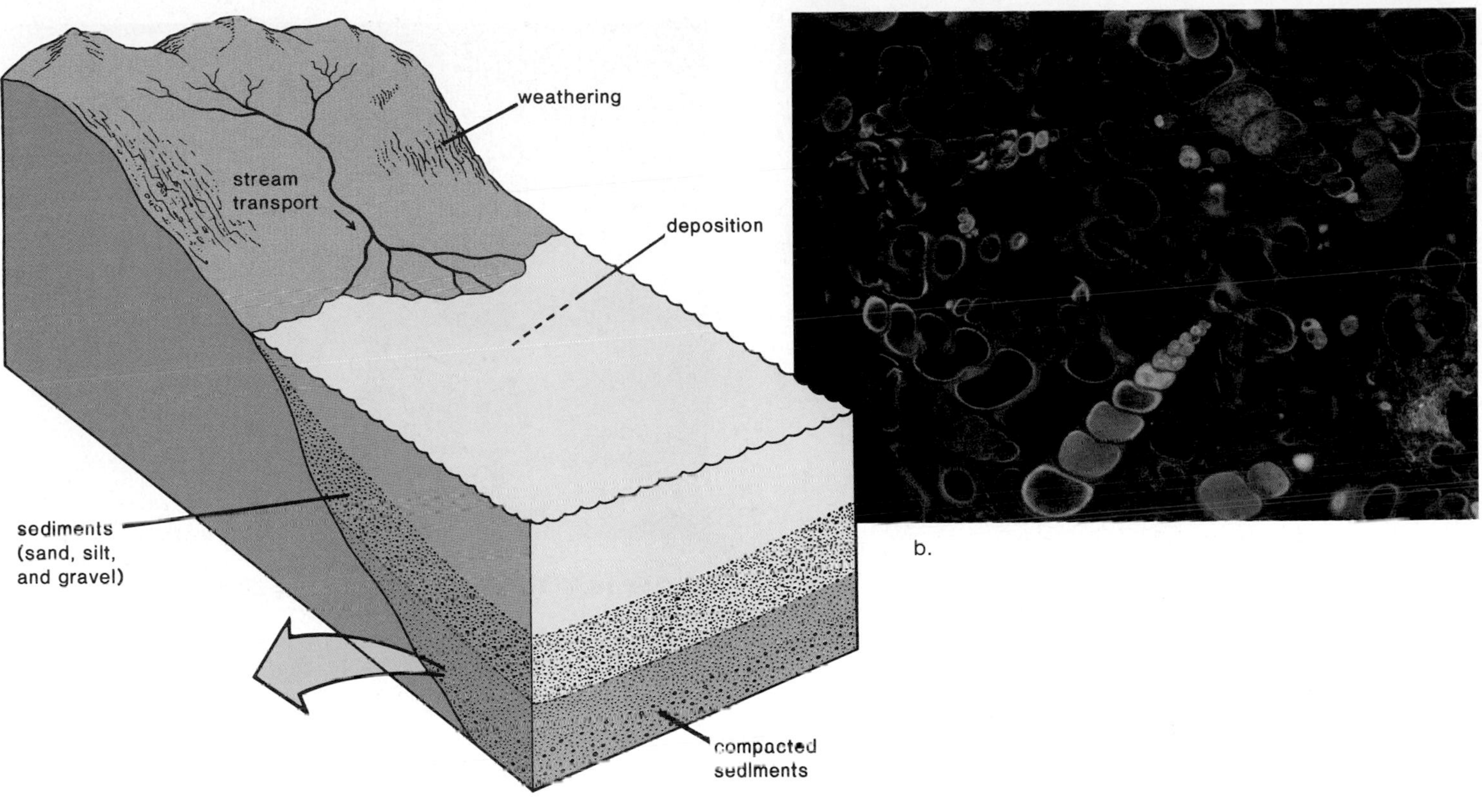

Figure 18.5 Formation of sedimentary rock.
a. This diagram shows how water brings sediments into the sea; the sediments then become compacted to form sedimentary rock. Fossils are often trapped in this rock, and as a result of a later geological upheaval, the rock may be located on land. b. Fossil remains of freshwater snails, *Turritella,* in sedimentary rocks.

18.2 Darwin Develops a Theory

When Darwin signed on as naturalist aboard the HMS *Beagle,* he had a suitable background for the position. He was an ardent student of nature and had long been a collector of insects. His sensitive nature prevented him from studying medicine, and he went to divinity school at Cambridge instead. Even so, he attended many lectures in both biology and geology, and he was also tutored in these subjects by a friend and teacher, the Reverend John Henslow. As a result of arrangements made by Henslow, Darwin spent the summer of 1831 doing fieldwork with Adam Sedgwick, a geologist at Cambridge, and it was Henslow who recommended Darwin for the post aboard the HMS *Beagle.* The trip was to take five years, and the ship was to traverse the Southern Hemisphere (see Fig. 18.1), where we now know that life is most abundant and varied. Along the way, Darwin encountered forms of life very different from those in his native England.

Descent Does Occur

Although it was not his original intent, Darwin began to gather evidence that common descent does occur and that adaptation to various environments results in diversity.

Geology and Fossils Help Darwin Decide

Darwin took Charles Lyell's *Principles of Geology* on the voyage, which presented arguments to support a theory of geological change proposed by James Hutton. In contrast to the catastrophists, Hutton believed the earth was subject to slow but continuous cycles of erosion and uplift. Weather causes erosion; thereafter dirt and rock debris are washed into the rivers and transported to oceans. These loose sediments are deposited in thick layers, which are converted eventually into sedimentary rocks (Fig. 18.5). Then sedimentary rocks, which often contain fossils, are uplifted from below sea level to form land. Hutton concluded that extreme geological changes can be accounted for by slow, natural processes, given enough time. Lyell went on to suggest that these slow changes occurred at a uniform rate. Hutton's general ideas about slow and continual geological change are still accepted today, although modern geologists realize that rates of change have not always been uniform. Darwin was not taken by the idea of uniform change, but he was convinced, as was Lyell, that the earth's massive geological changes are the result of slow processes and that the earth, therefore, must be very old.

Figure 18.6 Glyptodont compared to an armadillo.
a. A giant armadillo-like glyptodont, *Glyptodon,* known only by the study of its fossil remains. Darwin found such fossils and came to the conclusion that this extinct animal must be related to living armadillos. The glyptodont weighed 2,000 kilograms. **b.** A modern armadillo, *Dasypus,* weighs about 4.5 kilograms.

Figure 18.7 The Patagonian hare, *Dolichotis patagonium.*
This animal has the face of a guinea pig and is native to South America, which has no native rabbits. The Patagonian hare has long legs and other adaptations similar to those of rabbits.

On his trip, Darwin observed massive geological changes firsthand. When he explored what is now Argentina, he saw raised beaches for great distances along the coast. When he got to the Andes, he was impressed by their great height. In Chile, he found marine shells inland, well above sea level, and witnessed the effects of an earthquake that caused the land to rise several feet. While Darwin was making geological observations, he also collected fossil specimens. For example, on the east coast of South America, he found the fossil remains of a giant ground sloth and an armadillo-like animal (Fig. 18.6). Darwin accepted the supposition that the earth must be very old, and he began to think that there would have been enough time for descent with modification to occur. Therefore, living forms must be descended from extinct forms known only from the fossil record. It would seem that species were not fixed; instead they changed over time.

Darwin's geological observations were consistent with those of Hutton and Lyell. He began to think that the earth was very old, and there would have been enough time for descent with modification to occur.

Biogeography Also Helps

Biogeography [Gk. *bios,* life, *geo,* earth, and *grapho,* writing] is the study of the geographic distribution of life-forms on earth. Darwin could not help but compare the animals of South America to those with which he was familiar. For example, instead of rabbits he found the Patagonian hare in the grasslands of South America. The Patagonian hare has long legs and ears but the face of a guinea pig, a rodent native to South America (Fig. 18.7). Did the Patagonian hare resemble a rabbit because the two types of animals were adapted to the same type of environment? Both animals ate grass, hid in bushes, and moved rapidly using long hind legs. Did the Patagonian hare have the face of a guinea pig because of common descent with guinea pigs?

As he sailed southward along the eastern coast of the continent of South America, Darwin saw how similar species replaced each other. For example, the greater rhea (an ostrich-like bird) found in the north was replaced by the lesser rhea in the south. Therefore, Darwin reasoned that related species could be modified according to the environment. When he got to the Galápagos Islands, he found further evidence of this. The Galápagos Islands are a small group of volcanic islands off the western coast of South America. The few types of plants and animals found there were slightly different from species Darwin had observed on the mainland, and even more important, they also varied from island to island.

Darwin Observes Tortoises Each of the Galápagos Islands seemed to have its own type of tortoise, and Darwin began to wonder if this could be correlated with a difference in vegetation among islands (Fig. 18.8). Long-necked tortoises seemed to inhabit only dry areas, where food was scarce, most likely because the longer neck was helpful in reaching cacti. In moist regions with relatively abundant ground foliage, short-necked tortoises were found. Had an ancestral tortoise given rise to these different types, each adapted to a different environment?

a.

b.

Figure 18.8 Galápagos tortoises, *Testudo.*
Darwin wondered if all of the tortoises of the various islands were descended from a common ancestor. **a.** The tortoises with dome shells and short necks feed at ground level and are from well-watered islands where grass is available. **b.** Those with shells that flare up in front have long necks and are able to feed on tall treelike cacti. They are from arid islands where prickly pear cactus is the main food source. Only on these islands are the cacti treelike.

a. *Geospiza magnirostris*

b. *Certhidea olivacea*

c. *Cactornis scandens*

Figure 18.9 Galápagos finches.
Each of the present-day thirteen species of finches has a bill adapted to a particular way of life. **a.** For example, the heavy beak of the large ground-dwelling finch is suited to a diet of seeds. **b.** The beak of the warbler-finch is suited to feeding on insects found among ground vegetation or caught in the air. **c.** The long, somewhat decurved beak and the split tongue of the cactus-finch is suited to probing cactus flowers for nectar.

Darwin Observes Finches Although the finches on the Galápagos Islands seemed to Darwin like mainland finches, there are many more types (Fig. 18.9). Today, there are ground-dwelling finches with different-sized beaks, depending on the size of the seeds they feed on, and a cactus-eating finch with a more pointed beak. The beak size of the tree-dwelling finches also varies but according to the size of their insect prey. The most unusual of the finches is a woodpecker-type finch. This bird has a sharp beak to chisel through tree bark but lacks the woodpecker's long tongue to probe for insects. To make up for this, the bird carries a twig or cactus thorn in its beak and uses it to poke into crevices. Once an insect emerges, the finch drops this tool and seizes the insect with its beak.

Later, Darwin speculated whether all the different species of finches he saw could have descended from a type of mainland finch. In other words, he wondered if a mainland finch was the common ancestor to all the types on the Galápagos Islands. Had speciation occurred because the islands allowed isolated populations of birds to evolve independently? Could the present-day species have resulted from accumulated changes occurring within each of these isolated populations?

Biogeography had a powerful influence on Darwin and made him think that adaptation to the environment accounts for diversification; one species can give rise to many species, each adapted differently.

Natural Selection Provides a Mechanism

Once Darwin decided that adaptations develop over time (instead of being the instant work of a creator), he began to think about a mechanism by which adaptations might arise. Both Darwin and Alfred Russel Wallace, who is discussed in the reading on page 293, proposed **natural selection** as a mechanism for evolutionary change. Natural selection is a process in which preconditions (1–3) may result in certain consequences (4–5):

1. The members of a population have heritable variations.
2. In a population, many more individuals are produced each generation than can survive and reproduce.
3. Some individuals have adaptive characteristics that enable them to survive and reproduce better than do other individuals.
4. An increasing proportion of individuals in succeeding generations have the adaptive characteristics.
5. The result of natural selection is a population adapted to its local environment.

Notice that because natural selection utilizes only variations that happen to be provided by genetic changes, it lacks any directedness or anticipation of future needs. Natural selection is an ongoing process because the environment of living things is constantly changing. Extinction (loss of a species) can occur when previous adaptations are no longer suitable to a changed environment.

Figure 18.10 Variation in a population.
For Darwin, variations, such as those seen in a human population, were highly significant and were required for natural selection to result in adaptation to the environment.

Organisms Have Variations

Darwin emphasized that the members of a population vary in their functional, physical, and behavioral characteristics (Fig. 18.10). Before Darwin (see Table 18.1), variations were imperfections that should be ignored since they were not important to the description of a species. For Darwin, variations were essential to the natural selection process. Darwin suspected—but did not have the evidence we have today—that the occurrence of variations is completely random; they arise by accident and for no particular purpose. New variations are just as likely to be harmful as helpful to the organism.

The variations that make adaptation to the environment possible are those that are passed on from generation to generation. The science of genetics was not yet well established, so Darwin was never able to determine the cause of variations or how they are passed on. Today, we realize that genes determine the phenotype of an organism, and that mutations and recombination of alleles during sexual reproduction can cause new variations to arise.

Organisms Struggle to Exist

In Darwin's time, a socioeconomist, Thomas Malthus, stressed the reproductive potential of human beings. He proposed that death and famine were inevitable because the human population tends to increase faster than the supply of food. Darwin applied this concept to all organisms and saw that the available resources were not sufficient for all members of a population to survive. He calculated the reproductive potential of elephants. Assuming a life span of about 100 years and a breeding span of from 30–90 years, a single female probably bears no fewer than six young. If all these young survive and continue to reproduce at the same rate, after only 750 years, the descendants of a single pair of elephants will number about 19 million!

Each generation has the same reproductive potential as the previous generation. Therefore, there is a constant struggle for existence, and only certain members of a population survive and reproduce each generation.

Organisms Differ in Fitness

Fitness is the ability of an organism to survive and reproduce in its local (immediate) environment. The most fit individuals are the ones that survive, that capture a disproportionate amount of resources, and that convert these resources into a larger number of viable offspring. Since organisms vary anatomically and physiologically and the challenges of local environments vary, what determines fitness varies for different populations. For example, among western diamondback rattlesnakes *(Crotalus atrox)* living on lava flows, the most fit are those that are black in color. But among those living on desert soil, the most fit are those with the typical light and dark brown coloring. Background matching helps an animal both capture prey and avoid being captured; therefore, it is expected to lead to survival and increased reproduction.

Doing science

Alfred Russel Wallace

Alfred Russel Wallace (1823–1913) is best known as the English naturalist who independently proposed natural selection as a process to explain the origin of species (Fig. 18A). Like Darwin, Wallace was a collector at home and abroad. Even at age 14, while learning the trade of surveying, he became interested in botany and started collecting plants. While he was a schoolteacher at Leicester in 1844–45, he met Henry Walter Bates, an entomologist, who interested him in insects. Together they went on a collecting trip to the Amazon, which lasted for several years. Wallace's knowledge of the world's extensive flora and fauna was much expanded by a tour he made of the Malay Archipelago from 1854–62. After studying the animals of every important island, he divided the islands into a western group, with animals like those of the Orient, and an eastern group, with animals like those of Australia. The dividing line between the islands of the archipelago is a narrow but deep strait that is now known as the Wallace Line.

Like Darwin, Wallace was a writer of articles and books. As a result of his trip to the Amazon, he wrote two books, entitled *Travels on the Amazon and Rio Negro* and *Palm Trees of the Amazon*. In 1855, during his trip to Malay, he wrote an essay called "On the Law Which Has Regulated the Introduction of New Species." In the essay, he said that "every species has come into existence coincident both in time and space with a preexisting closely allied species." It's clear, then, that by this date he believed in the origin of new species rather than the fixity of species. Later, he said that he had pondered for many years about a mechanism to explain the origin of species. He, too, had read Malthus's treatise on human population increases and, in 1858, while suffering an attack of malaria, the idea of "survival of the fittest" came upon him. He quickly completed an essay discussing a natural selection process, which he chose to send to Darwin for comment. Darwin was stunned upon its receipt. Here before him was the hypothesis he had formulated as early as 1844 but had never dared to publish. In 1856 he had begun to work on a book that would supply copious data to support natural selection as a mechanism for evolutionary change. He told his friend and colleague Charles Lyell that Wallace's ideas were so similar to his own that even Wallace's "terms now stand as heads of my chapters."

Darwin suggested that Wallace's paper be published immediately, even though he as yet had nothing in print. Lyell and others who knew of Darwin's detailed work substantiating the process of natural selection suggested that a joint paper be read to the Linnean society. The title of Wallace's section was "On the Tendency of Varieties to Depart Indefinitely from the Original Type." Darwin presented an abstract of a paper he had written in 1844 and an abstract of his book *On the Origin of Species*, which was published in 1859.

Figure 18A Alfred Russel Wallace.

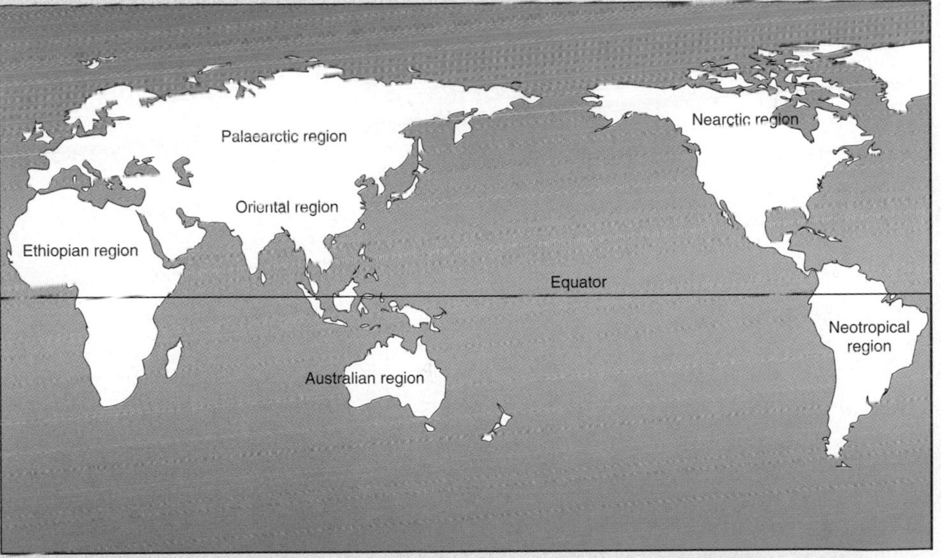

Figure 18B Biogeographical regions.
Aside from presenting a hypothesis that natural selection explains the origin of new species, Alfred Wallace is well known for another contribution. He said that the world can be divided into six biogeographical regions separated by impassable barriers. The deep waters between the Oriental and Australian regions are called the Wallace Line.

Figure 18.11 Artificial selection of animals.
All dogs, *Canis familiaris,* are descended from the wolf, *Canis lupus,* which began to be domesticated about 14,000 years ago. In evolutionary terms, the process of diversification has been exceptionally rapid. Several factors may have contributed: (1) the wolves under domestication were separated from other wolves because human settlements were separate, and (2) humans in each tribe selected for whatever traits appealed to them. Artificial selection of dogs continues even today.

Darwin noted that when humans help carry out *artificial selection,* the process by which a breeder chooses which traits to perpetuate, they select the animals that will reproduce. For example, prehistoric humans probably noted desirable variations among wolves and selected particular animals for breeding. Therefore, the desirable traits increased in frequency in the next generation. This same process was repeated many times over. The result today is that there are many varieties of dogs all descended from the wolf (Fig. 18.11). In a similar way, several varieties of vegetables can be traced to a single ancestor (Fig. 18.12).

In nature, interactions with the environment determine which members of a population reproduce to a greater degree than other members. In contrast to artificial selection, the result of natural selection is not predesired. Natural selection occurs because certain members of a population happen to have a variation that allows them to survive and reproduce. For example, any variation that increases the speed of a hoofed animal helps it escape predators and live longer; a variation that reduces water loss helps a desert plant survive; and one that increases the sense of smell helps a wild dog find its prey. Therefore, we expect organisms with these traits to have increased fitness. *Differential reproduction* occurs when the most fit organisms reproduce and leave more offspring than the less fit.

Organisms Become Adapted

An **adaptation** [L. *ad,* toward, and *aptus,* fit, suitable] is a trait that helps an organism be more suited to its environment. We can especially recognize an adaptation when unrelated organisms, living in a particular environment, display similar characteristics. For example, manatees, penguins, and sea turtles all have flippers, which help them move through the water. Natural selection results in the adaptation of populations to their specific environments. Because of differential reproduction generation after generation, adaptive traits are increasingly represented in each succeeding generation. There are other processes of evolution aside from natural selection, but natural selection is the only process that results in adaptation to the environment.

Darwin Writes a Book

After the HMS *Beagle* returned to England in 1836, Darwin waited over 20 years to publish his ideas. During the intervening years, he used the scientific process to test his hypothesis that today's diverse life-forms arose by descent from a common ancestor and that natural selection is a mechanism by which species can change and even new species can arise. Darwin was prompted to publish his book after reading a similar hypothesis from Alfred Russel Wallace, as discussed in the reading on page 293.

Darwin became convinced that descent with modification explains the history of life. His theory of natural selection proposes a mechanism by which adaptation to the environment occurs.

a.

b.

c.

Figure 18.12 Artificial selection of plants.
All these vegetables are derived from a single species of *Brassica oleracea.* **a.** Chinese cabbage. **b.** Brussels sprouts. **c.** Kohlrabi. Darwin believed that artificial selection provided a model by which to understand natural selection. With natural selection, however, the environment provides the selective force.

18.3 Evidence Accumulates

Many different lines of evidence support the hypothesis of common descent. This is significant, because the more varied the evidence supporting a hypothesis, the more certain it becomes. Darwin cited much of the evidence we will discuss, except he had no knowledge, of course, of the biochemical data that became available after his time.

Fossils Tell a Story

The fossil record is the history of life as recorded by remains from the past. Fossils are at least 10,000 years old and include such items as pieces of bone, impressions of plants pressed into shale, and even insects trapped in tree resin (which we know as amber). Over the last two centuries, paleontologists have studied fossils in the earth's strata (layers) all over the world and have pieced together the story of past life.

The fossil record is rich in information. One of its most striking patterns is the succession of life-forms over time. Catastrophists offered an explanation for the extinction and subsequent replacement of one group of organisms by another group, but they never could explain successive changes that link groups of organisms historically. Particularly interesting are the fossils that serve as transitional links between groups. For example, the famous fossils of *Archaeopteryx* are intermediate between reptiles and birds (Fig. 18.13). The dinosaurlike skeleton of this fossil has reptilian features, including jaws with teeth and a long, jointed tail, but *Archaeopteryx* also had feathers and wings. Other intermediate fossils among the fossil vertebrates include the amphibious fish *Eustheopteron*, the reptile-like amphibian *Seymouria*, and the mammal-like reptiles, or therapsids. These fossils allow us to deduce that fishes preceded amphibians, which preceded reptiles, which preceded both birds and mammals in the history of life.

Sometimes, the fossil record allows us to trace the history of one particular organism, such as the modern-day horse, *Equus*. *Equus* evolved originally from *Hyracotherium*, which was about the size of a dog. This fossil animal had cusped, low-crowned molars, four toes on each front foot, and three toes on each hind foot. When grasslands replaced the forest home of *Hyracotherium*, the ancestors of *Equus* were subject to selective pressure for the development of strength, intelligence, speed, and durable grinding teeth. A larger size provided the strength needed for combat, a larger skull made room for a larger brain, elongated legs ending in hooves provided greater speed to escape enemies, and the durable grinding teeth enabled the animals to feed efficiently on grasses. In all cases, living organisms closely resemble the most recent fossils in their line of descent. Underlying similarities, however, allow us to trace a line of descent over vast amounts of time.

The evidence supports the common descent hypothesis. Fossils can be linked over time because they show a similarity in form, despite observed changes.

The fossil record broadly traces the history of life and more specifically allows us to study the history of particular groups.

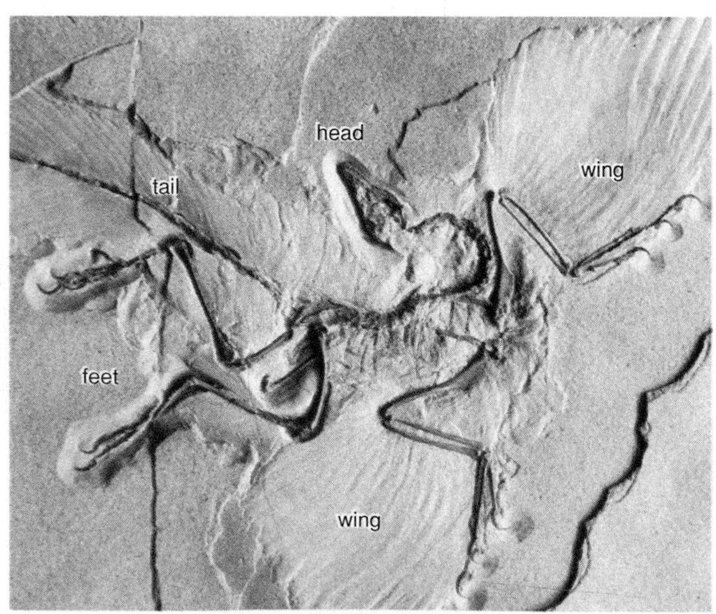

a.

b.

Figure 18.13 Transitional fossils.
a. *Archaeopteryx*, a transitional link between reptiles and birds, had feathers and wing claws. Most likely, it was a poor flier. Perhaps it ran over the ground on strong legs and climbed up into trees with the assistance of these claws. **b.** It also had a feather-covered reptilian-type tail, that shows up well in this artist's representation of the animal.

Biogeographical Separations

Biogeography is the study of the distribution of plants and animals throughout the world. Such distributions are consistent with the hypothesis that related forms evolved in one locale and then spread out into other accessible regions. As previously mentioned, Darwin noted that South America lacked rabbits, even though the environment was quite suitable to them. He concluded there are no rabbits in South America because rabbits originated somewhere else and they had no means to reach South America.

The islands of the world have many unique species of animals and plants found no place else, even when the soil and climate are the same as other places. Why are there so many species of finches on the Galápagos Islands when these same species are not on the mainland? The reasonable explanation is that they are all descended from a common ancestor that came to the islands by chance.

Physical factors, such as the location of continents, often determine where a population can spread. Both cacti and euphorbia are plants adapted similarly to a hot, dry environment—they both are succulent, spiny, flowering plants. Why do cacti grow in North American deserts and euphorbia grow in African deserts when each would do well on the other continent? It seems obvious that they just happened to evolve on their respective continents.

At one time in the history of the earth, South America, Antarctica, and Australia were all connected. Marsupials (pouched mammals) arose at this time and today are found in both South America and Australia. When Australia separated and drifted away, the marsupials diversified into many different forms suited to various environments (Fig. 18.14). They were free to do so because there were no placental mammals in Australia. In other regions such as South America, where there are placental mammals, marsupials are not as diverse.

The distribution of organisms on the earth is explainable by assuming that related forms evolve in one locale. They then diversified as they spread out into other accessible areas.

Coarse-haired wombat, *Vombatus*, nocturnal and living in burrows

Sugar glider, *Petaurista*, a tree dweller

Kangaroo, *Macropus*, a herbivore of plains and forests

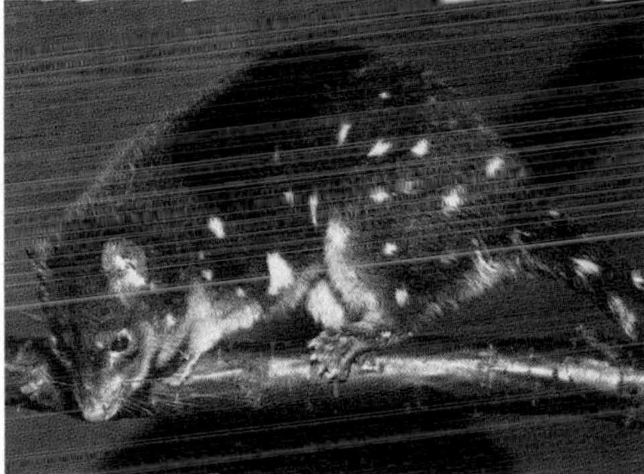

Australian native cat, *Dasyurus*, a carnivore of forests

Tasmanian wolf, *Thylacinus*, a nocturnal carnivore of deserts and plains

Figure 18.14 Biogeography.
Each type of marsupial in Australia is adapted to a different way of life. All of the marsupials in Australia presumably evolved from a common ancestor that entered Australia some 60 million years ago.

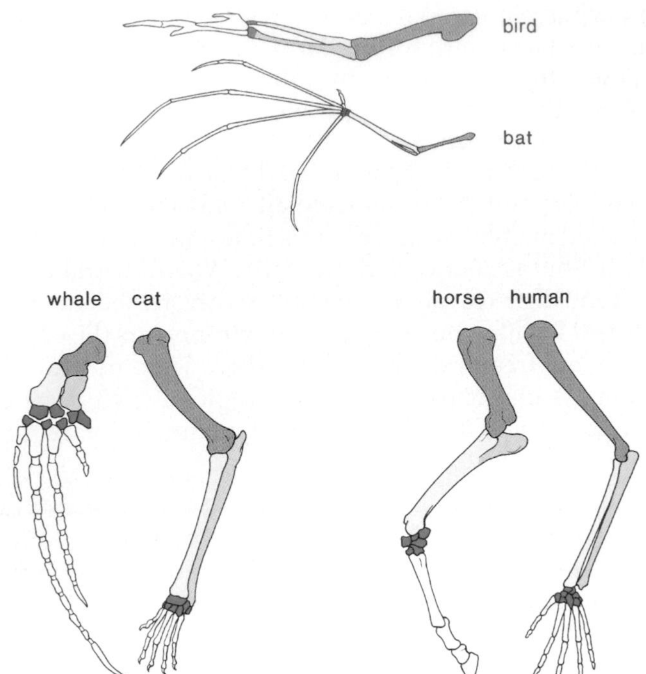

Figure 18.15 Significance of structural similarities. Although the specific design details of vertebrate forelimbs are different, the same bones are present (they are color-coded). This unity of plan is evidence of a common ancestor.

Anatomical Similarities

Darwin was able to show that a common descent hypothesis offers a plausible explanation for anatomical similarities among organisms. Vertebrate forelimbs are used for flight (birds and bats), orientation during swimming (whales and seals), running (horses), climbing (arboreal lizards), or swinging from tree branches (monkeys). Yet all vertebrate forelimbs contain the same sets of bones organized in similar ways, despite their dissimilar functions (Fig. 18.15). The most plausible explanation for this unity is that the basic forelimb plan belonged to a common ancestor, and then the plan was modified in the succeeding groups as each continued along its own evolutionary pathway. Structures that are similar because they were inherited from a common ancestor are called **homologous structures** [Gk. *homologos*, agreeing, corresponding].

Vestigial structures [L. *vestigium*, trace, footprint] are anatomical features that are fully developed in one group of organisms but are reduced and may have no function in similar groups. Most birds, for example, have well-developed wings used for flight. Some bird species (e.g., ostrich), however, have wings that are greatly reduced, and they do not fly. Similarly, snakes have no use for hindlimbs, and yet some have remnants of a pelvic girdle and legs. Humans have a tailbone but no tail. The presence of vestigial structures can be explained by the common descent hypothesis. Vestigial structures occur because organisms inherit their anatomy from their ancestors; they are traces of an organism's evolutionary history.

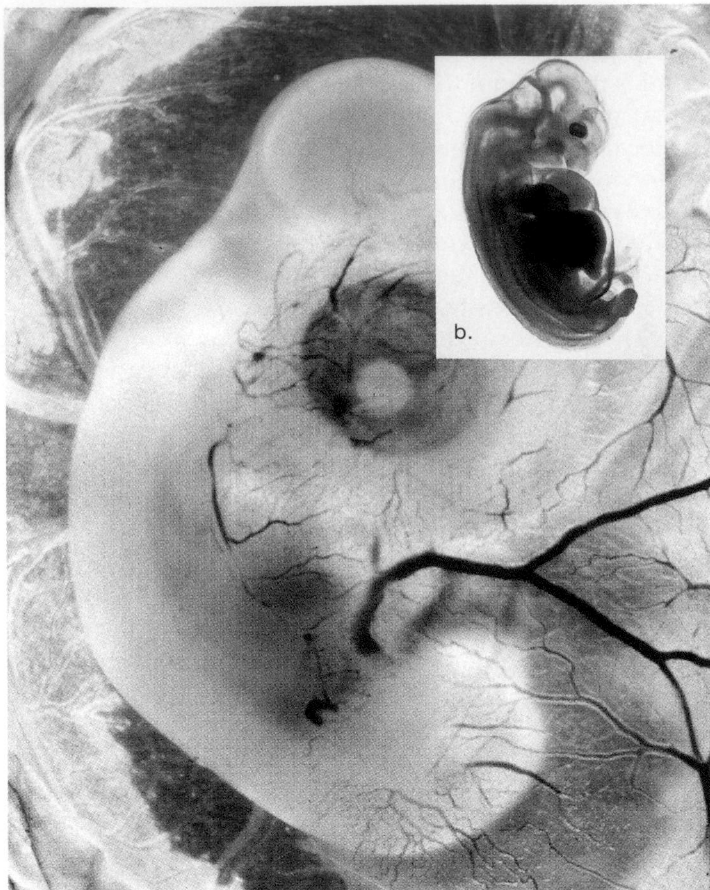

Figure 18.16 Significance of embryological similarities.
a. A chick embryo. **b.** A pig embryo. At these comparable early developmental stages, the two have many features in common, although eventually they are completely different animals. This is evidence that they evolved from a common ancestor.

The unity of plan shared by vertebrates extends to their embryological development (Fig. 18.16). At some time during development, all vertebrates have a supporting dorsal rod, called a notochord, and exhibit paired pharyngeal pouches. In fishes and amphibian larvae, these pouches develop into functioning gills. In humans, the first pair of pouches becomes the cavity of the middle ear and the auditory tube. The second pair becomes the tonsils, while the third and fourth pairs become the thymus and parathyroid glands. Why should terrestrial vertebrates develop and then modify structures like pharyngeal pouches that have lost their original function? The most likely explanation is that fishes are ancestral to other vertebrate groups.

Organisms share a unity of plan when they are closely related because of common descent. This is substantiated by comparative anatomy and embryological development.

Biochemical Differences

Almost all living organisms use the same basic biochemical molecules, including DNA (deoxyribonucleic acid), ATP (adenosine triphosphate), and many identical or nearly identical enzymes. Further, organisms utilize the same DNA triplet code and the same 20 amino acids in their proteins. Organisms even share the same introns and type of repeats. There is obviously no functional reason why these elements need be so similar. But their similarity can be explained by descent from a common ancestor.

When the degree of similarity in DNA base sequences or the degree of similarity in amino acid sequences of proteins is examined, the data are as expected assuming common descent. Cytochrome *c* is a molecule that is used in the electron transport system of all the organisms listed in Figure 18.17. Data regarding differences in the amino acid sequence of the cytochrome *c* show that in a human it differs from that in a monkey by only one amino acid; from that in a duck by 11, and from that in *Candida*, a yeast, by 51 amino acids. These data are consistent with other data regarding the anatomical similarities of these animals and, therefore, their relatedness.

Darwin discovered that many lines of evidence support the hypothesis of common descent. Since his time, it has been found that biochemical evidence also supports the hypothesis. A hypothesis is strengthened when it is supported by many different lines of evidence.

Evolution is no longer considered a hypothesis. It is one of the great unifying theories of biology. In science, the word *theory* is reserved for those conceptual schemes that are supported by a large number of observations and have not yet been found lacking. The theory of evolution has the same status in biology that the germ theory of disease has in medicine.

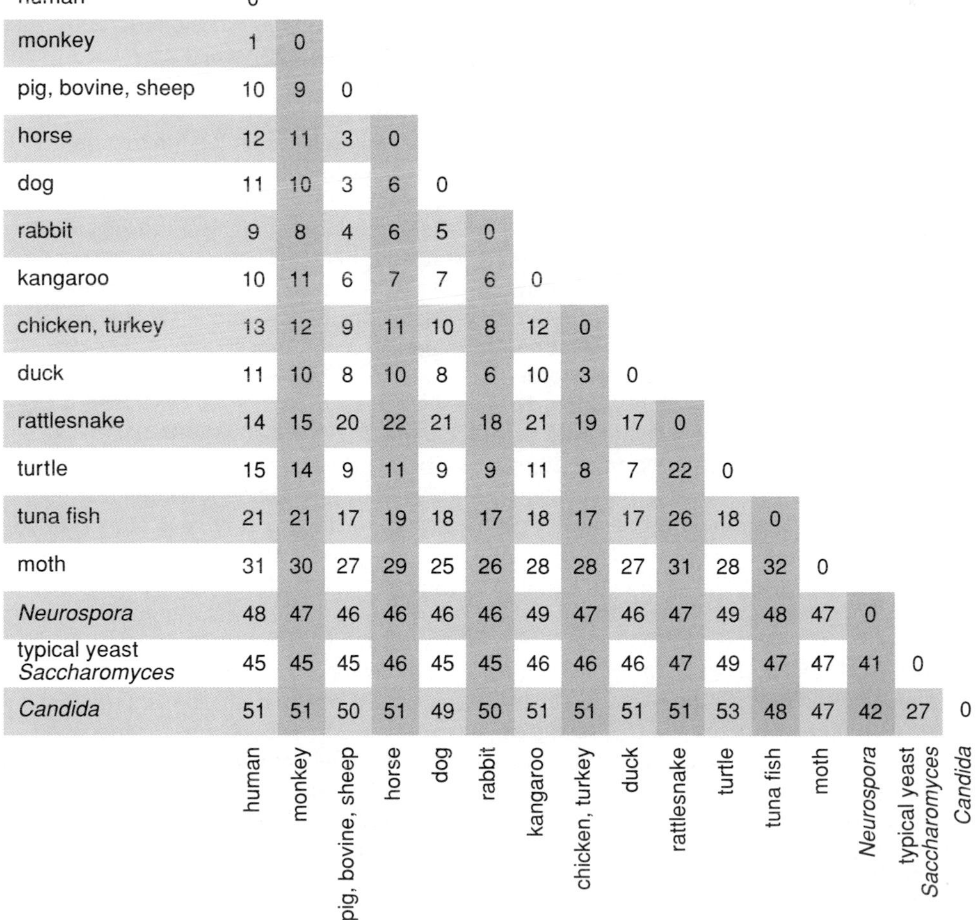

	human	monkey	pig, bovine, sheep	horse	dog	rabbit	kangaroo	chicken, turkey	duck	rattlesnake	turtle	tuna fish	moth	*Neurospora*	typical yeast *Saccharomyces*	*Candida*
human	0															
monkey	1	0														
pig, bovine, sheep	10	9	0													
horse	12	11	3	0												
dog	11	10	3	6	0											
rabbit	9	8	4	6	5	0										
kangaroo	10	11	6	7	7	6	0									
chicken, turkey	13	12	9	11	10	8	12	0								
duck	11	10	8	10	8	6	10	3	0							
rattlesnake	14	15	20	22	21	18	21	19	17	0						
turtle	15	14	9	11	9	9	11	8	7	22	0					
tuna fish	21	21	17	19	18	17	18	17	17	26	18	0				
moth	31	30	27	29	25	26	28	28	27	31	28	32	0			
Neurospora	48	47	46	46	46	46	49	47	46	47	49	48	47	0		
typical yeast *Saccharomyces*	45	45	45	46	45	45	46	46	46	47	49	47	47	41	0	
Candida	51	51	50	51	49	50	51	51	51	51	53	48	47	42	27	0

Figure 18.17 Significance of biochemical differences.
This diagram shows the amino acid differences in cytochrome *c* among the organisms listed. The number where the two organisms intersect (see listing on bottom and sides) indicates the number of amino acid differences. The more closely related the species, the fewer the biochemical differences.

connecting concepts

Before the 1800s, most people believed that the origin and diversity of life on earth was due to the work of a supernatural being who created each species at the beginning of the world, and modern organisms are essentially unchanged descendants of their ancestors. Scientists, however, seek natural, testable hypotheses to explain natural events rather than relying on religious dogma.

At the time Charles Darwin boarded the *Beagle,* he had studied writings of his grandfather Erasmus Darwin, James Hutton, Charles Lyell, Lamarck, Malthus, Linnaeus, and other original thinkers. He had ample time during his five-year voyage to reflect on the ideas of these authors, and collectively, they built a framework that helped support his theory of descent with modifications.

One aspect of scientific genius is the power of astute observation—to see what others miss or fail to appreciate. In this area, Darwin excelled. By the time he reached the Galápagos Islands, Darwin had already begun to hypothesize that species could be modified according to the environment. His observation of the finches on the isolated Galápagos Islands supported his hypothesis. He surmised that the finches on each island varied from each other and from mainland finches because each species had become adapted to a different habitat; therefore, one species had given rise to many. The fact that Alfred Wallace almost simultaneously proposed natural selection as an evolutionary mechanism suggests that the world was ready for a revised view of life on earth.

The theory of evolution has quite rightly been called the grand unifying theory (GUT) of biology. Fossils, comparative anatomy, biogeography, and biochemical data all indicate that living things share common ancestors. Evolutionary principles help us understand why organisms are different and alike, and why some species flourish and others die out. As the earth's habitats changed over millions of years, those individuals with the traits best adapted to new environments survived and reproduced; thus populations changed over time.

Summary

18.1 The World Was Ready

In general, the pre-Darwinian worldview was different from the post-Darwinian worldview (see Table 18.1). The scientific community, however, was ready for Darwin's hypothesis, which received widespread acceptance.

A century before Darwin's trip, classification of organisms had been a main concern of biology. In keeping with the pre-Darwinian worldview, Linnaeus thought that classification should describe the fixed features of species and reveal God's divine plan. Gradually, some naturalists, such as Count Buffon and Erasmus Darwin, began to put forth tentative suggestions that species do change over time.

Georges Cuvier and Jean-Baptiste de Lamarck, contemporaries in the late eighteenth century, differed sharply on evolution. To explain the fossil record of a region, Cuvier proposed that a whole series of catastrophes (extinctions) and repopulations from other regions had occurred. Lamarck said that descent with modification does occur and that organisms do become adapted to their environments; however, he relied on the commonly held but erroneous beliefs (scala naturae and inheritance of acquired characteristics) to substantiate and provide a mechanism for evolutionary change.

18.2 Darwin Develops a Theory

Charles Darwin formulated hypotheses concerning evolution after taking a trip around the world as naturalist aboard the HMS *Beagle* (1831–36). His hypotheses were that common descent does occur and that natural selection results in adaptation to the environment.

Darwin's trip involved two primary types of observations. His study of geology and fossils caused him to concur with Lyell that the observed massive geological changes were caused by slow, continuous changes. Therefore, he concluded that the earth is old enough for descent with modification to occur.

Darwin's study of biogeography, including the animals of the Galápagos Islands, caused him to conclude that adaptation to the environment can cause diversification, including the origin of new species.

Natural selection is the mechanism Darwin proposed for how adaptation comes about. Members of a population exhibit random but inherited variations. (In contrast to the previous worldview, variations are highly significant.) Relying on Malthus's ideas regarding overpopulation, Darwin stressed that there was a struggle for existence. The most fit organisms are those possessing characteristics that allow them to acquire more resources, survive, and reproduce more than the less fit. In this way, natural selection results in adaptation to a local environment.

18.3 Evidence Accumulates

The hypothesis that organisms share a common descent is supported by many lines of evidence. The fossil record, biogeography, comparative anatomy, and comparative biochemistry all support the hypothesis. The fossil record gives us the history of life in general and allows us to trace the descent of a particular group. Biogeography shows that the distribution of organisms on earth is explainable by assuming organisms evolved in one locale. Comparing the anatomy and the development of organisms reveals a unity of plan among those that are closely related. All organisms have certain biochemical molecules in common, and any differences indicate the degree of relatedness. A hypothesis is greatly strengthened when many different lines of evidence support it.

Today, the theory of evolution is one of the great unifying theories of biology because it has been supported by so many different lines of evidence.

Reviewing the Chapter

1. In general, contrast the pre-Darwinian worldview with the post-Darwinian worldview. 287
2. Cite naturalists who made contributions to biology in the mid-eighteenth century, and state their beliefs about evolutionary descent. 287
3. How did Cuvier explain the succession of life-forms in the earth's strata? 288
4. What is meant by the inheritance of acquired characteristics, a hypothesis that Lamarck used to explain adaptation to the environment? 288
5. What reading did Darwin do, and what observations did he make regarding geology? 289
6. What observations did Darwin make regarding biogeography? How did these influence his conclusions about the origin of new species? 290–91
7. What are the essential features of the process of natural selection as proposed by Darwin? 292
8. Distinguish between the concepts of fitness and adaptation to the environment. 292, 295
9. How do data from the fossil record support the concept of common descent? Explain why *Equus* is vastly different from its ancestor *Hyracotherium,* which lived in a forest. 296
10. How do data from biogeography support the concept of common descent? Explain why a diverse assemblage of marsupials evolved in Australia. 297
11. How do data from comparative anatomy support the concept of common descent? Explain why vertebrate forelimbs are similar despite different functions. 298
12. How do data from biochemical studies support the concept of common descent? Explain why the sequence of amino acids in cytochrome *c* differs between two organisms. 299

Testing Yourself

Choose the best answer for each question.

1. According to the theory of acquired inheritance,
   a. if a man loses his hand, then his children will also be missing a hand.
   b. changes in phenotype are passed on by way of the genotype to the next generation.
   c. organisms are able to bring about a change in their phenotype.
   d. All of these are correct.
2. Why was it helpful to Darwin to learn that Lyell thought the earth was very old?
   a. An old earth has more fossils than a new earth.
   b. It meant there was enough time for evolution to have occurred slowly.
   c. Evolution doesn't occur without upheavals.
   d. Darwin said that artificial selection occurs slowly.
3. All the finches on the Galápagos Islands are
   a. completely unrelated.
   b. descended from a common ancestor.
   c. now in competition with one another because they all feed on seeds.
   d. Both a and c are correct.
4. Organisms
   a. compete with other members of their species.
   b. differ in fitness.
   c. are adapted to their environment.
   d. All of these are correct.
5. DNA nucleotide differences between organisms
   a. indicate how closely related organisms are.
   b. indicate that evolution occurs.
   c. explain why there are phenotypic differences.
   d. All of these are correct.
6. If evolution occurs, we would expect different biogeographical regions with similar environments to
   a. all contain the same mix of plants and animals.
   b. each have its own specific mix of plants and animals.
   c. have plants and animals that have similar adaptations.
   d. Both b and c are correct.
7. The fossil record offers direct evidence for common descent because you can
   a. see that the types of fossils change over time.
   b. sometimes find common ancestors.
   c. trace the ancestry of a particular group.
   d. All of these are correct.
8. Organisms adapted to the same way of life will
   a. have structures that show they share a unity of plan.
   b. have similarities that need not indicate a unity of plan.
   c. live in the same biogeographical region.
   d. Both a and c are correct.

For questions 9–12, offer an explanation for each of these observations based on information in the section indicated. Write out your answer.

9. Fossils can be dated according to the strata in which they are located. See Fossils Tell a Story, p. 296.
10. Cacti and euphorbia exist on different continents, but both have spiny, water-storing, leafless stems. See Biogeographical Separations, p. 297.
11. Amphibians, reptiles, birds, and mammals all have pharyngeal pouches at some time during development. See Anatomical Similarities, p. 298.
12. The base sequence of DNA differs from species to species. See Biochemical Differences, p. 299.

Applying the Concepts

1. *All organisms share certain common characteristics because they are descended from a common ancestor.*
   What evidence do you have that organisms as different as bacteria and humans share a common ancestor?
2. *Adaptation to various environments accounts for the diversity of life.*
   The presence of vestigial structures suggests that the process of adaptation is not purposeful. Why?
3. *The scientific method consisting of hypothesis, experimentation, and conclusion has been used to test the unifying theories of biology.*
   Show that the scientific method can be used to test the concept of evolutionary descent.

Using Technology

Your study of Darwin and evolution is supported by these available technologies:

Exploring the Internet
The Mader Home Page provides resources for and help with studying this chapter.

http://www.mhhe.com/sciencemath/biology/mader/
(Click on Biology.)

Understanding the Terms

adaptation 295
biogeography 290
catastrophism 288
evolution 287
fitness 292
homologous structure 298
inheritance of acquired characteristics 288
natural selection 292
paleontology 288
vestigial structure 298

Match the terms to these definitions:

a. __________ Study of the geographical distribution of organisms.
b. __________ Study of fossils that results in knowledge about the history of life.
c. __________ Poorly developed structure that was complete and functional in an ancestor but is no longer functional in a descendant.
d. __________ Organism's modification in structure, function, or behavior suitable to the environment.
e. __________ Lamarckian belief that organisms become adapted to their environment during their lifetime and pass on these adaptations to their offspring.
f. __________ In evolution, a structure that is similar in different organisms because these organisms are derived from a common ancestor.
g. __________ Mechanism of evolution caused by environmental selection of organisms most fit to reproduce, resulting in adaptation.
h. __________ Descent of organisms from common ancestors with the development of genetic and phenotypic changes over time that make them more suited to the environment.
i. __________ Belief espoused by Cuvier that periods of catastrophic extinctions occurred, after which repopulation of surviving species took place, giving the appearance of change through time.
j. __________ Ability of an organism to survive and reproduce in its environment.

Process of Evolution

CHAPTER 19

Chapter Concepts

Crop spraying

When your grandparents were young, infectious diseases such as tuberculosis, pneumonia, and syphilis killed thousands of people every year. Then in the 1940s, penicillin and other antibiotics were developed, and public health officials believed infectious diseases were a thing of the past. Today, however, tuberculosis, pneumonia, and many other ailments are back with a vengeance. What happened? Natural selection occurred.

As Darwin and Wallace emphasized, there is variation among individuals in all populations or groups of organisms. Just as the cats in your neighborhood differ from one another, bacteria that cause disease and pests that eat plants differ from one another. Some of the bacteria and pests just happen to be resistant to drug and pesticides, respectively. The antibiotic (or pesticide) didn't cause this resistance; the organisms were already resistant (preadapted). Only these resistant organisms survive drug treatment or pesticide use to reproduce. Now 50 years later, several strains of "superbugs" that cause various illnesses cannot be killed by antibiotics. One type of bacteria actually feeds on the antibiotic vancomycin!

In this chapter, we will see how individuals with traits best adapted to the present environment are selected to reproduce to a greater extent than other members of a population. Because selection occurs, gene frequencies change in a population; thus, evolution occurs.

19.1 Evolution in a Genetic Context

What Causes Variations?

The members of a population vary from one another. A **population** is all the members of a single species occupying a particular area at the same time. A population could be all the green frogs in a frog pond, all the field mice in a barn, or all the English daisies on a hill. The members of a population reproduce with one another to produce the next generation.

Variations, the raw material for evolutionary change, arise in a population by gene mutations, chromosomal mutations, and recombination.

Gene Mutations

Gene mutations provide new alleles, and therefore they are the ultimate source of variation. A gene mutation is an alteration in the DNA nucleotide base sequence of an allele. We have seen that a change in a single DNA base spells the difference between a normal hemoglobin molecule and sickle-cell hemoglobin. Also, a change in a regulatory gene can increase or decrease the expression of a structural gene.

Mutations occur at random. A mutation is like a shot in the dark; any particular gene can mutate at any particular time. This means that the occurrence of a mutation is not tied to its adaptive value, and the chance of a particular mutation is not expected to increase because the mutation is beneficial to the organism. For example, you might think that it is exposure to a drug that causes bacteria to become resistant. However, drug-resistant mutations can occur in bacteria by chance, before they are exposed to the drug.

Rather than being beneficial, a mutation can be neutral in its effect or even harmful to an organism in its present environment. Many times observed mutations seem harmful, such as those that cause human genetic disease. This may be because members of a population are so adapted to their present environment that only nonbeneficial changes are possible. As discussed later, however, some mutations may remain hidden in a diploid organism to become an important source of variation should the environment change. Or the same mutations may arise again when conditions have changed and they are now favorable mutations.

Chromosomal Mutations

Some chromosomal mutations are simply an alteration in the number of chromosomes inherited. Polyploid plants have more than two sets of chromosomes.

Other chromosomal mutations are an alteration in the arrangement of the alleles on the chromosomes. Inversions (a segment of chromosome is inverted) and translocations (exchange of chromosomal segments between nonhomologous chromosomes) can result in altered allelic activity if the allele thereby comes under the control of a different regulatory gene. Duplication of an allele followed by a change in base sequence of one copy introduces a new allele while still retaining the old allele in the human genome. The best studied example of such an event concerns the globin alleles that occur at different loci but have similar DNA sequences. Adult hemoglobin contains two chains of α (alpha) globin and two chains of β (beta) globin. Besides the ones found in hemoglobin, humans have several other versions of both globin alleles. Most of these versions are expressed only during development, but some are expressed in adults.

Recombination

In sexually reproducing organisms, recombination of alleles and chromosomes is an important source of variation. During meiosis, crossing-over between nonsister chromatids and an independent assortment of chromosomes produce unlike gametes. During fertilization, random gamete union occurs, and this also contributes to a genotype that is unlike those of the parents. Altogether, each population has a storehouse of allelic and chromosomal combinations that potentially exceeds the number of individuals in the population.

The entire genotype, and not individual alleles, is subjected to the natural selection process. A new and different combination of alleles may therefore have great selective value. In a population of snails, stripes and brown color combined might make animals less visible in a woodland habitat. If stripes are controlled by one allele and brown color is controlled by another allele, it is a combination of the two alleles that will be selected for. Recombination may at some time bring the two alleles together so that the combined phenotype can be subjected to natural selection.

Along the same line, consider that many traits having to do with structure, function, and behavior are polygenic. Polygenic traits are controlled by more than one gene, each of which exists at a different locus. Certain combinations of the several alleles might be more adaptive than others in a particular environment. The most favorable combination might not occur until just the right alleles are grouped by recombination. Once the improbable but most favorable phenotype occurs, it can become the prevalent phenotype in the population because of a selection process.

Variations, the raw material for evolutionary change, are controlled by genes. Only gene mutations result in new alleles. Chromosomal mutations and recombination, however, contribute greatly to the production of variant genotypes and phenotypes.

How to Detect Evolution

Not only are variations created, they are also preserved and passed on from one generation to the next. The various alleles at all the gene loci in all individuals make up the **gene pool** of the population. It is customary to describe the gene pool of a population in terms of gene frequencies. Suppose that in a *Drosophila* population, one-fourth of the flies are homozygous dominant for long wings, one-half are heterozygous, and one-fourth are homozygous recessive for short wings. Therefore, in a population of 100 individuals, we have

25 *LL*, 50 *Ll*, and 25 *ll*

What is the number of the allele *L* and the allele *l* in the population?

Number of *L* alleles:			**Number of *l* alleles:**		
LL	$(2\ L \times 25)$	= 50	*LL*	$(0\ l)$	= 0
Ll	$(1\ L \times 50)$	= 50	*Ll*	$(1\ l \times 50)$	= 50
ll	$(0\ L)$	= 0	*ll*	$(2\ l \times 25)$	= 50
		100 *L*			100 *l*

To determine the frequency of each allele, calculate its percentage from the total number of alleles in the population; in each case, 100/200 = 50% = 0.5. The sperm and eggs produced by this population will also contain these alleles in these frequencies. Assuming random mating (all possible gametes have an equal chance to combine with any other), we can calculate the ratio of genotypes in the next generation by using a Punnett square.

There is an important difference between the Punnett square below and one used for a cross between individuals; below the sperm and eggs are those produced by the members of a population—not those produced by a single male and female. As you can see, the results of the Punnett square indicate that the frequency for each allele in the next generation is still 0.5.

		sperm	
		0.5 *L*	0.5 *l*
eggs	0.5 *L*	0.25 *LL*	0.25 *Ll*
	0.5 *l*	0.25 *Ll*	0.25 *ll*

Results:
0.25 *LL* + 0.5 *Ll* + 0.25 *ll*
($\frac{1}{4}$ *LL* + $\frac{1}{2}$ *Ll* + $\frac{1}{4}$ *ll*)

Therefore, sexual reproduction alone cannot bring about a change in allele frequencies. Also, the dominant allele need not increase from one generation to the next. Dominance does not cause an allele to become a common allele. The potential constancy, or equilibrium state, of gene pool frequencies was independently recognized in 1908 by G. H. Hardy, an English mathematician, and W. Weinberg, a German physician. They used the binomial expression $(p^2 + 2pq + q^2)$ to calculate the genotypic and allele frequencies of a population. Figure 19.1 shows you how this is done.

$$p^2 + 2pq + q^2$$

p^2 = % homozygous dominant individuals
p = frequency of dominant allele
q^2 = % homozygous recessive individuals
q = frequency of recessive allele
$2pq$ = % heterozygous individuals

Realize that $p + q = 1$ (there are only 2 alleles)
$p^2 + 2pq + q^2 = 1$ (these are the only genotypes)

Example

An investigator has determined by inspection that 16% of a human population has a continuous hairline (recessive trait). Using this information, we can complete all the genotypic and allele frequencies for the population, provided the conditions for Hardy-Weinberg equilibrium are met.

Given: $q^2 = 16\% = 0.16$ are homozygous recessive individuals

Therefore, $q = \sqrt{0.16} = 0.4$ = frequency of recessive allele
$p = 1.0 - 0.4 = 0.6$ = frequency of dominant allele
$p^2 = (0.6)(0.6) = 0.36 = 36\%$ are homozygous dominant individuals
$2pq = 2(0.6)(0.4) = 0.48 = 48\%$ are heterozygous individuals
or
$= 1.00 - 0.52 = 0.48$

84% have the dominant phenotype

Figure 19.1 Calculating gene pool frequencies using the Hardy-Weinberg equation.

Practice Problems 1*

1. Of the members of a population of pea plants, 9% are short (recessive). What are the frequencies of the recessive allele *t* and the dominant allele *T*? What are the genotypic frequencies in this population?
2. A student places 600 fruit flies with the genotype *Ll* and 400 with the genotype *ll* in a culture bottle. What will the genotypic frequencies be in the next generation and each generation thereafter, assuming a Hardy-Weinberg equilibrium?

**Answers to Practice Problems appear in Appendix A.*

The **Hardy-Weinberg law** states that an equilibrium of allele frequencies in a gene pool, calculated by using the expression $p^2 + 2\,pq + q^2$, will remain in effect in each succeeding generation of a sexually reproducing population as long as five conditions are met.

1. No mutations: allelic changes do not occur, or changes in one direction are balanced by changes in the opposite direction.
2. No gene flow: migration of alleles into or out of the population does not occur.
3. Random mating: individuals pair by chance and not according to their genotypes or phenotypes.
4. No genetic drift: the population is very large, and changes in allele frequencies due to chance alone are insignificant.
5. No selection: no selective force favors one genotype over another.

In real life, these conditions are rarely, if ever, met, and allele frequencies in the gene pool of a population do change from one generation to the next. Therefore, evolution has occurred. The significance of the Hardy-Weinberg law is that it tells us what factors cause evolution—those that violate the conditions listed. Evolution can be detected by noting any deviation from a Hardy-Weinberg equilibrium of allele frequencies in the gene pool of a population.

The accumulation of small changes in the gene pool over a relatively short period of two or more generations is called microevolution. *Microevolution* is involved in the origin of species to be discussed later in this chapter, as well as in the history of life recorded in the fossil record.

A change in allele frequencies results in a change in phenotype frequencies. In Figure 19.2, the original peppered moth population has only 10% dark-colored moths. When dark-colored moths rest on light trunks, they are seen and eaten by predatory birds. With the advent of pollution, the trunks of trees darken and it is the light-colored moths that stand out and are eaten. The birds are acting as a selective agent, and microevolution occurs; the last generation observed has 80% dark-colored moths.

A Hardy-Weinberg equilibrium provides a baseline by which to judge whether evolution has occurred. Any change of allele frequencies in the gene pool of a population signifies that evolution has occurred.

Figure 19.2 Microevolution.
Microevolution has occurred when there is a change in gene pool frequencies—in this case, due to natural selection. On the far left, birds cannot see light moths on light tree trunks and, therefore, light moths are more frequent in the population. On the far right, birds cannot see dark moths on dark tree trunks, and dark moths are more frequent in the population. The percentage of the dark-colored phenotype has increased in the population because predatory birds can see light-colored moths against tree trunks that are now sooty due to pollution.

What Causes Evolution?

The list of conditions for genetic equilibrium implies that the opposite conditions can cause evolutionary change. The conditions that can cause a deviation from the Hardy-Weinberg equilibrium are mutation, gene flow, nonrandom mating, genetic drift, and natural selection. Only natural selection results in adaptation to the environment.

Genes Mutate

As mentioned, mutations provide new alleles which are the raw material for evolutionary change. Mutations underlie all the other mechanisms that provide variation. Due to elaborate DNA replication and DNA repair mechanisms, the mutation rates of individual genes are low (about 10^{-5} per gene per cell cycle), but each organism has many genes, and a population has many individuals. Therefore, per individual and per population, mutations are common, not rare, events, as we can see from the following calculations:

For an individual human: Since there are about 100,000 genes (200,000 alleles) in humans, the average mutation rate per human gamete is $(2 \times 10^5$ alleles$) \times (10^{-5}$ mutation per gene$) = 2$ mutations per human gamete.

For a human population: Since there are about 252 million people in the United States, the mutation rate per generation is $(2.5 \times 10^8$ individuals$) \times (2$ mutations per gamete$) = 5 \times 10^8$ new mutations in each generation.

Therefore, even a modest mutation rate produces a tremendous amount of genetic variation in a population.

In 1966, R. C. Lewontin and J. L. Hubby set out to determine the molecular polymorphism (variation) of 18 different enzymes in natural populations of *Drosophila pseudoobscura.* They extracted various enzymes and subjected them to electrophoresis, a process that separates proteins according to size and charge. They concluded that a fly population is polymorphic at no less than 30% of all its gene loci, and that an individual fly is likely to be heterozygous at about 12% of its loci.

Similar results have been found in studies on many species, demonstrating that high levels of molecular variation are the rule in natural populations. Many of these mutations will not affect the phenotype, and therefore they may not be detected. In a changing environment, even a seemingly harmful mutation can be a source of variation that can help a population become adapted to a new environment. For example, the water flea *Daphnia* ordinarily thrives at temperatures around 20°C, but there is a mutation that requires *Daphnia* to live at temperatures between 25°C and 30°C. The adaptive value of this mutation is entirely dependent on environmental conditions.

Mutations cause a gene pool to have multiple alleles for many genes.

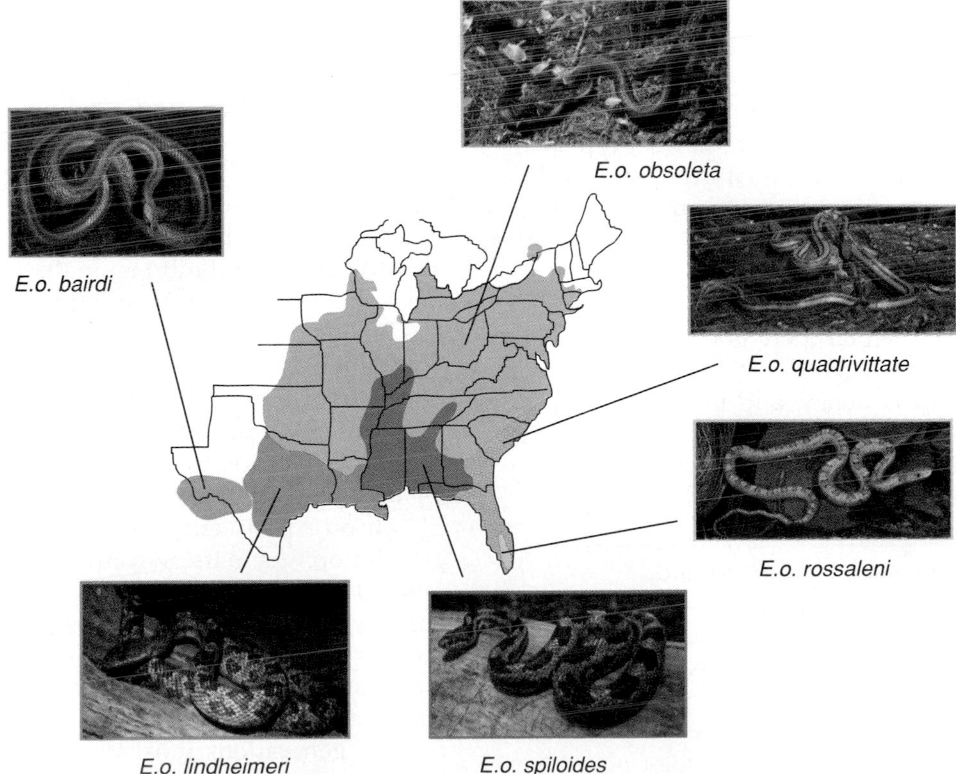

Figure 19.3 Gene flow.
Each rat snake represents a separate population of snakes. Because the populations are adjacent to one another, there is interbreeding and therefore gene flow between the populations. This keeps their gene pools somewhat similar, and each of these populations is a subspecies of the species *Elaphe obsoleta.* Therefore, each has a three-part name.

Gene Flow Brings New Genes

Gene flow, also called gene migration, is the movement of alleles among populations by the migration of breeding individuals. There can be constant gene flow between adjacent animal populations due to the migration of organisms (Fig. 19.3). In order to judge the effect of gene flow, it is necessary to know the difference in gene frequencies between the two populations and the proportion of migrant genes that are introduced.

Gene flow can increase the variation within a population by introducing novel alleles that were produced by mutation in some other population. Continued gene flow among populations makes their gene pools similar and reduces the possibility of allele frequency differences among populations that might be due to natural selection and genetic drift. Indeed, gene flow among populations can prevent speciation from occurring.

Gene flow tends to decrease the diversity among populations, causing their gene pools to become similar.

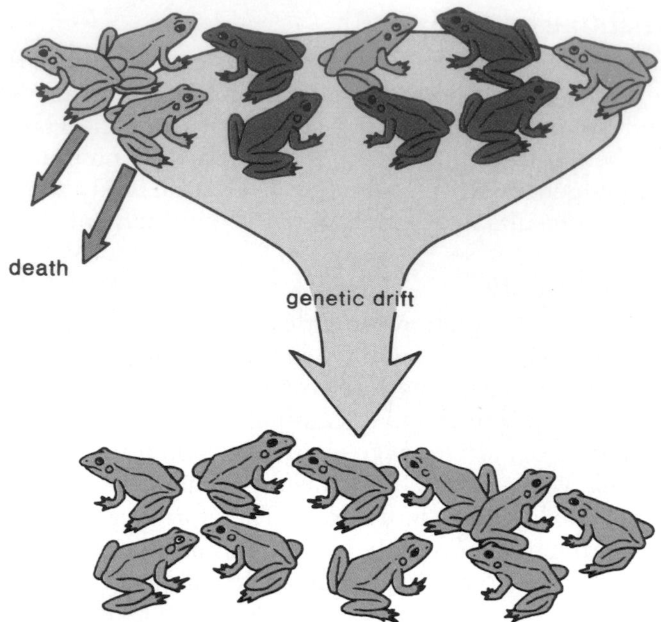

Figure 19.4 Genetic drift.
Genetic drift occurs when by chance only certain members of a population (in this case, green frogs) reproduce and pass on their genes to the next generation. The allele frequencies of the next generation's gene pool may be markedly different from that of the previous generation.

Mating Is Not Always Random

Random mating occurs when individuals pair by chance and not according to their genotypes or phenotypes. Inbreeding, or mating between relatives to a greater extent than by chance, is an example of **nonrandom mating.** This can happen if dispersal is so low that mates are likely to be related. Inbreeding does not change allele frequencies, but it does decrease the proportion of heterozygotes and increase the proportions of both homozygotes at all gene loci. In a human population, inbreeding increases the frequency of recessive abnormalities in the phenotype.

Assortative mating occurs when individuals tend to mate with those that have the same phenotype with respect to some characteristic. For example, in humans tall people seem to prefer to mate with each other. Assortative mating causes the population to subdivide into two phenotypic classes, between which there is reduced gene exchange. Homozygotes for the gene loci that control the trait in question increase in frequency, and heterozygotes for these loci decrease in frequency. Other loci remain in Hardy-Weinberg equilibrium, except for those very closely linked to the loci governing the trait.

Nonrandom mating involves inbreeding and assortative mating. The former results in increased frequency of homozygotes at all loci, and the latter results in increased frequency of homozygotes at only certain loci.

Genetic Drift Promotes Changes

Genetic drift refers to changes in allele frequencies of a gene pool due to chance. Although genetic drift occurs in both large and small populations, a larger population is expected to suffer less of a *sampling error* than a smaller population. Suppose you had a large bag containing 1,000 green balls and 1,000 blue balls, and you randomly drew 10%, or 200, of the balls. Because there is a large number of balls of each color in the bag, you can reasonably expect to draw 100 green balls and 100 blue balls or at least a ratio close to this. But suppose you had a bag containing only 10 green balls and 10 blue balls and you drew 10%, or only 2 balls. The chances of drawing one green ball and one blue ball with a single trial are now considerably less.

When a population is small, there is a greater chance that some rare genotype might not participate at all in the production of the next generation. Suppose there is a small population of frogs in which certain frogs for one reason or another do not pass on their traits. Certainly, the next generation will have a change in allele frequencies. When genetic drift leads to a loss of one or more alleles, other alleles over time become *fixed* in the population (Fig. 19.4).

In an experiment with 107 *Drosophila* populations, each population was in its own culture bottle. Every bottle contained eight heterozygous flies of each sex. There were no homozygous recessive or homozygous dominant flies. From the many offspring, the experimenter chose at random eight males and eight females. These were the parents for the next generation, and so forth for 19 generations. For the first few generations, most populations still contained many heterozygotes. But by the nineteenth generation, 25% of the populations contained only homozygous recessive flies and 25% contained only homozygous dominant flies for a brown-eyed allele.

Genetic drift is a random process, and therefore it is not likely to produce the same results in several populations. The *Drosophila* experiment considered only one allelic pair, but if many pairs had been considered, it is expected that each of the 107 populations would have a different gene pool composition for many allelic pairs after 19 generations. In California, there are a number of cypress groves, each a separate population. The phenotypes within each grove are more similar to one another than they are to the phenotypes in the other groves. Some groves have longitudinally shaped trees, and others have pyramidally shaped trees. The bark is rough in some colonies and smooth in others. The leaves are gray to bright green or bluish, and the cones are small or large. Because the environmental conditions are similar for all groves, and no correlation has been found between phenotype and environment across groves, it is hypothesized that these variations among populations are due to genetic drift.

Founder Effect The **founder effect** is an example of genetic drift in which rare alleles, or combinations of alleles, occur at a higher frequency in a population isolated from the general population. After all, founding individuals contain only a fraction of the total genetic diversity of the original gene pool. Which particular alleles are carried by the founders is dictated by chance alone. The Amish of Lancaster County, Pennsylvania, are an isolated group that was begun by German founders. Today, as many as one in 14 individuals carries a recessive allele that causes an unusual form of dwarfism (affecting only lower arms and legs) and polydactylism (extra fingers) (Fig. 19.5). In the population at large, only one in 1,000 individuals has this allele.

All the Amish with this genetic syndrome can trace their ancestry to a single couple (the Kings) who immigrated to Pennsylvania in 1744. They and their offspring had large families, and therefore the rare allele became concentrated in the population.

Bottleneck Effect Sometimes a population is subjected to near extinction because of a natural disaster (e.g., earthquake or fire) or because of overharvesting and habitat loss. Again, chance alone may determine which individuals survive, and these survivors, in effect, create a genetic bottleneck. The **bottleneck effect** prevents the majority of types of genotypes from participating in the production of the next generation.

The large genetic similarity found in cheetahs is believed to be due to a bottleneck. In a study of 47 different enzymes, each of which can come in several different forms, all the cheetahs had exactly the same form. This demonstrates that genetic drift can cause certain alleles to be lost from a population. Exactly what caused the cheetah bottleneck is not known. It is speculated that perhaps cheetahs were slaughtered by nineteenth-century cattle farmers protecting their herds, or were captured by Egyptians as pets 4,000 years ago, or were decimated by a mass extinction tens of thousands of years ago. Today, cheetahs suffer from relative infertility because of the intense inbreeding that occurred after the bottleneck.

Genetic drift is exaggerated when founder effects and bottleneck effects occur. In these instances, only a few types of genotypes contribute to the gene pool of a population, which is much smaller than that in the original population.

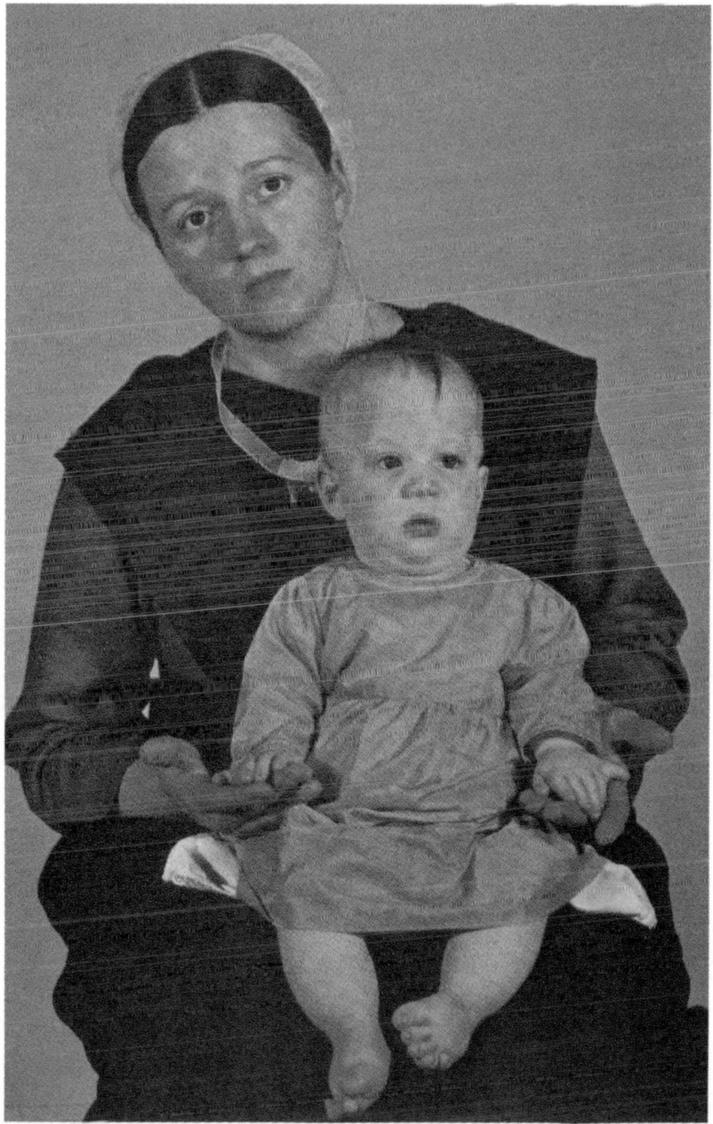

Figure 19.5 Founder effect.
A member of the founding population of Amish in Pennsylvania had a recessive allele for a rare kind of dwarfism. The percentage of the Amish population now carrying this allele is much higher compared to that of the general population.

19.2 Adaptation Occurs Naturally

Natural selection is the process by which populations become adapted to their biotic and abiotic environments. The biotic environment includes organisms that seek resources through competition, predation, and parasitism. The abiotic environment includes weather conditions dependent chiefly upon temperatures and precipitation. In the previous century, Charles Darwin, the father of evolution, became convinced that species evolve (change) with time and suggested that natural selection was the process by which they become adapted to their environment. Here, we restate Darwin's hypothesis of natural selection in the context of modern evolutionary theory.

> Evolution by natural selection requires:
>
> 1. variation. The members of a population differ from one another.
> 2. inheritance. Many of these differences are heritable genetic differences.
> 3. differential adaptedness. Some of these differences affect how well an organism is adapted to its environment.
> 4. differential reproduction. Individuals that are better adapted to their environment are more likely to reproduce, and their fertile offspring will make up a greater proportion of the next generation.

As mentioned previously, random gene mutations are the ultimate source of variation because they provide new alleles. However, in sexually reproducing organisms, recombination of alleles and chromosomes—due to crossing-over during meiosis, independent assortment of chromosomes, and fertilization—contributes greatly to variation. Recombination may at some time produce a more favorable combination of alleles. After all, it is the combined phenotype that is subjected to natural selection.

Fitness

Fitness is the extent to which the traits of an individual allow it to enjoy reproductive success and contributes fertile offspring to the next generation. Selection, which can be described as a composite of the forces that limit reproductive success of an individual, opposes fitness. Fitness is a consequence of adaptation to a particular environment and therefore can vary according to the environment in which an individual lives. That is, the same individual can have a different degree of fitness in different environments. Fitness is relative; it is measured against the reproductive success of other individuals in the same environment. The more fit an individual, the greater is the genetic contribution of that individual to subsequent generations compared to other members of the same population.

Types of Selection

Most of the traits on which natural selection acts are polygenic and controlled by more than one pair of alleles located at different gene loci. Such traits have a range of phenotypes, the frequency distribution of which usually resembles a bell-shaped curve.

Three types of natural selection have been described for any particular trait. They are directional selection, stabilizing selection, and disruptive selection.

Directional Selection

Directional selection occurs when an extreme phenotype is favored and the distribution curve shifts in that direction (Fig. 19.6). Such a shift can occur when a population is adapting to a changing environment. For example, the gradual increase in the size of the modern horse, *Equus*, can be correlated with a change in the environment from forestlike conditions to grassland conditions. Nevertheless, the evolution of the horse should not be viewed as a straight line of descent; we know of many side branches that became extinct.

Industrial Melanism Before the Industrial Revolution in England, collectors of the peppered moth *(Biston betularia)* noted that most moths were light colored, although occasionally a dark-colored (melanistic) moth was captured. Several decades after the Industrial Revolution, however, black moths made up 99% of the moth population in air-polluted areas.

An experiment by H.B.D. Kettlewell of Oxford University showed that when their coloring matches the trees, moths are more likely to avoid being eaten by predatory birds, which act as a selective agent. Moths rest on the trunks of trees during the day (see Fig. 19.2); if they are seen, they are eaten by birds. As long as the trees in the environment are light in color, the light-colored moths live to reproduce. But when the trees turn black from industrial pollution, the dark-colored moths survive and reproduce to a greater extent than the light-colored moths. The dark-colored phenotype, then, becomes more prevalent in the population. Thereafter, when pollution is reduced and the trunks of the trees regain their normal color, the light-colored moths increase in number.

Bacteria and Insects Indiscriminate use of antibiotics and pesticides results in populations of bacteria and insects that are resistant to these chemicals. The mutation that permits resistance is already present before exposure; the chemicals are merely acting as a selective agent. Those organisms that survive exposure have offspring that are resistant, and in this way the population becomes resistant.

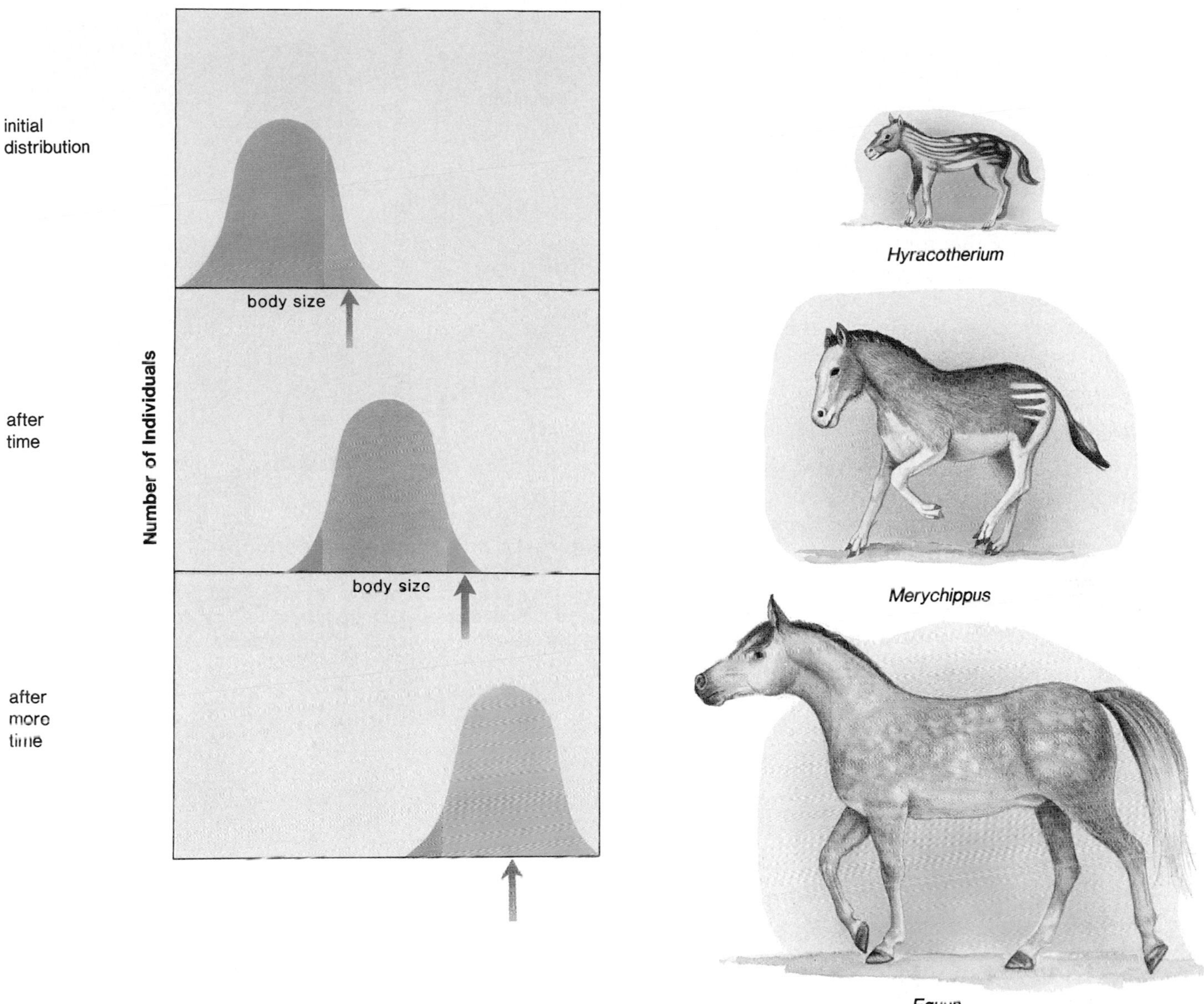

Figure 19.6 Directional selection.
This occurs when natural selection favors one extreme phenotype (see arrows), and there is a shift in the distribution curve. *Equus*, the modern-day horse, evolved from *Hyracotherium*, which was about the size of a dog. This small animal could have hidden among trees and had low-crowned teeth for browsing. When grasslands began to replace forest, the ancestors of *Equus* were subject to selective pressure for the development of strength, intelligence, speed, and durable grinding teeth. A larger size provided the strength needed for combat, a larger skull made room for a larger brain, elongated legs ending in hooves provided greater speed to escape enemies, and the durable grinding teeth enabled the animals to feed efficiently on grasses.

Another example of directional selection is the human struggle against malaria, a disease caused by an infection of the liver and the red blood cells. The *Anopheles* mosquito transmits the disease-causing protozoan *Plasmodium vivax* from person to person. In the early 1960s, international health authorities thought that malaria would soon be eradicated. A new drug, chloroquine, was more effective than the quinine that had been used against *Plasmodium*, and DDT (an insecticide) spraying had reduced the mosquito population. But in the mid-1960s, *Plasmodium* showed signs of chloroquine resistance and, worse yet, mosquitoes were becoming resistant to DDT. A few drug-resistant parasites and a few DDT-resistant mosquitoes had survived and multiplied, making the fight against malaria more difficult than ever. Thus, another avenue has now been taken—a vaccine against malaria is being developed.

Stabilizing Selection

Stabilizing selection occurs when an intermediate phenotype is favored (Fig. 19.7). It can improve adaptation of the population to those aspects of the environment that remain constant. With stabilizing selection, extreme phenotypes are selected against, and individuals near the average are favored. As an example, consider the birth weight of human infants, which ranges from 0.89 to 4.9 kg (2 to 10.8 lb). The death rate is higher for infants who are at these extremes and is lowest for babies who have an intermediate birth weight (about 3.2 kg or 7 lb). Most babies have a birth weight within this range, which gives the best chance of survival. Similar results have been found in other animals, also.

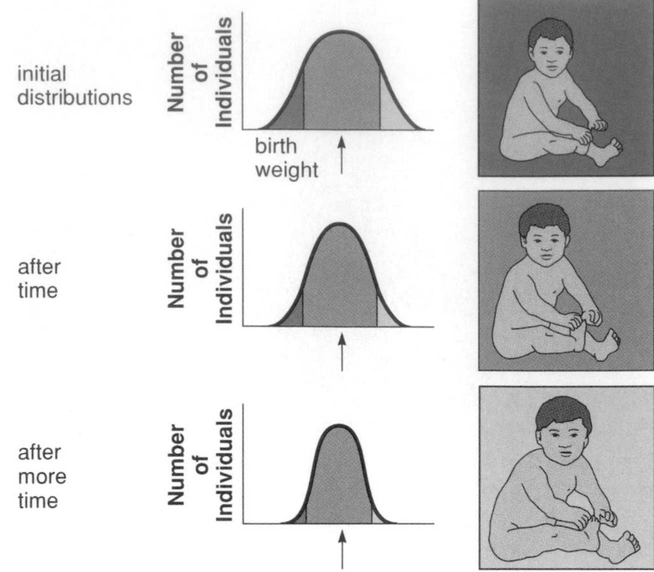

Figure 19.7 Stabilizing selection.
Natural selection favors the intermediate phenotype (see arrows) over the extremes. Today, it is observed that most human babies are of intermediate weight (about 3.2 kg, or 7 lb), and very few babies are either very small or very large.

Disruptive Selection

In **disruptive selection,** two or more extreme phenotypes are favored over any intermediate phenotype (Fig. 19.8). For example, British land snails *(Cepaea nemoralis)* have a wide habitat range that includes low-vegetation areas (grass fields and hedgerows) and forest areas. In low-vegetation areas, thrushes feed mainly on snails with dark shells that lack light bands, and in forest areas, they feed mainly on snails with light-banded shells. Therefore, these two distinctly different phenotypes are found in the population.

Directional selection favors one of the extreme phenotypes; stabilizing selection favors the intermediate phenotype; disruptive selection favors more than one extreme phenotype.

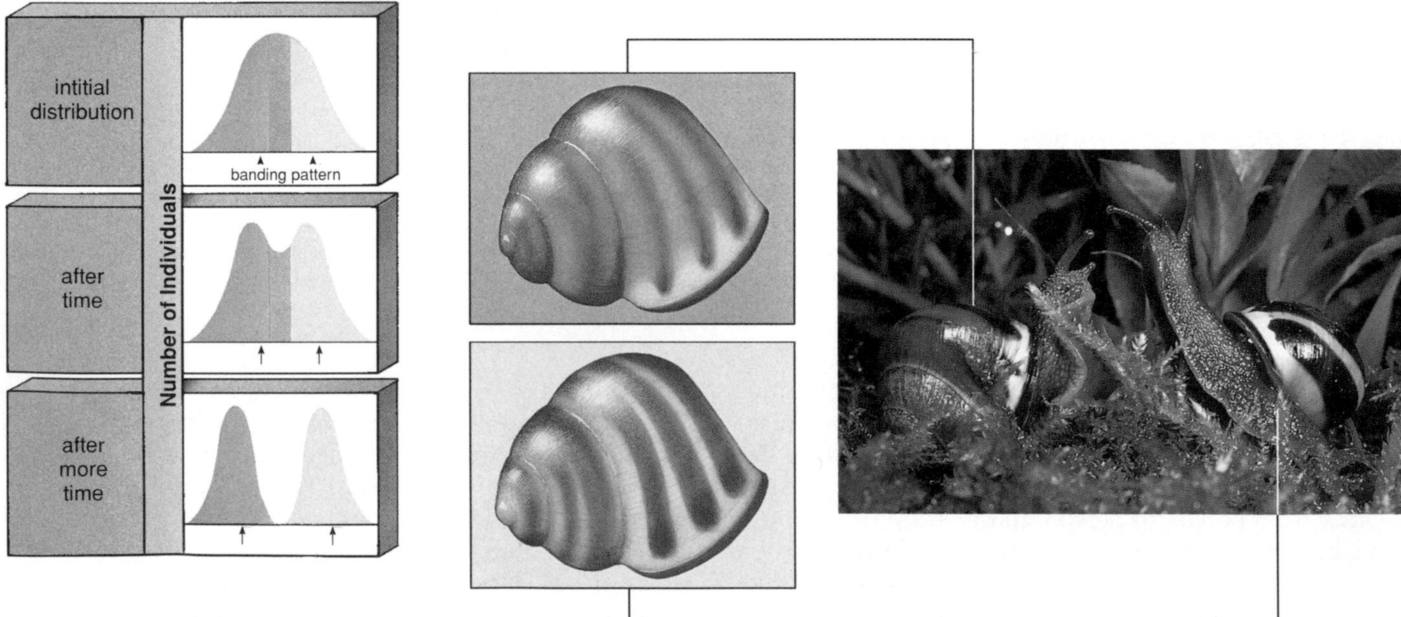

Figure 19.8 Disruptive selection.
Natural selection favors two extreme phenotypes (see arrows). Today, it is observed that British land snails comprise mainly two different phenotypes, each adapted to a particular habitat.

Variations Are Maintained

A population always shows some genotypic variation. The maintenance of variation is beneficial because populations that lack variation may not be able to adapt to new conditions and may become extinct. How can variation be maintained in spite of selection constantly working to reduce it?

First, we must remember that the forces that promote variation are still at work: mutation still creates new alleles, and recombination still recombines these alleles during gametogenesis and fertilization. Second, gene flow might still occur. If the receiving population is small and is mostly homozygous, gene flow can be a significant source of new alleles. Finally, we have seen that natural selection reduces, but does not eliminate, the range of phenotypes. And disruptive selection even promotes polymorphism in a population. There are also other ways variation is maintained.

Diploidy and the Heterozygote

Only alleles that are exposed (cause a phenotypic difference) are subject to natural selection. In diploid organisms, this makes the heterozygote a potential protector of recessive alleles that otherwise would be weeded out of the gene pool. Consider these gene pool frequency distributions:

Frequency of Allele *a* in Gene Pool	Genotypic Frequencies		
	AA	*Aa*	*aa*
0.9	0.01	0.18	0.81
0.1	0.81	0.18	0.01

Notice that even when selection reduces the recessive allele from a frequency of 0.9 to 0.1, the frequency of the heterozygote remains the same. The heterozygote remains a source of the recessive allele for future generations. In a changing environment, the recessive allele may then be favored by natural selection.

Sickle-Cell Disease

In certain regions of Africa, the importance of the heterozygote in maintaining variation is exemplified by sickle-cell disease. The relative fitness of the heterozygote and the two homozygotes will determine their percentages in the population. When the ratio of two or more phenotypes remains the same in each generation, it is called *balanced polymorphism*. The optimum ratio will be the one that correlates with the survival of the most offspring.

Individuals with sickle-cell disease have the genotype Hb^SHb^S and tend to die at an early age due to hemorrhaging and organ destruction. Those who are heterozygous and have sickle-cell trait (Hb^AHb^S) are better off; their red blood cells usually become sickle shaped only when the oxygen content of the environment is low. Geneticists studying the distribution of sickle-cell disease in Africa have found

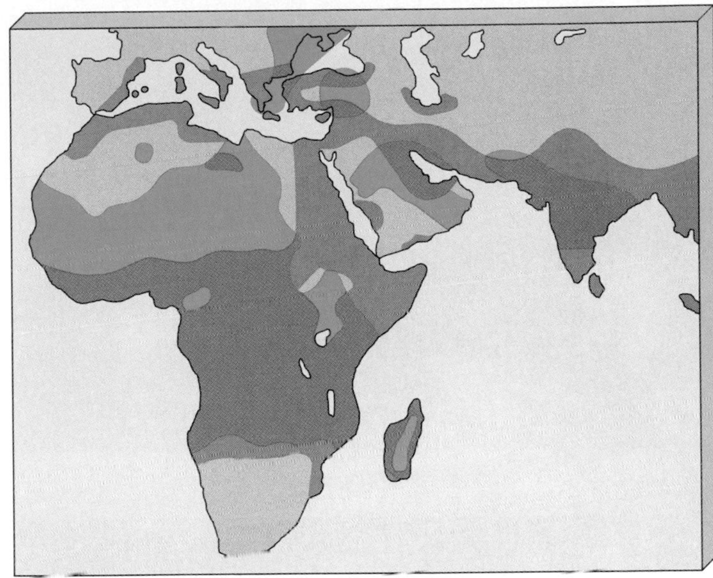

Figure 19.9 Distribution of sickle-cell disease and malaria.
The pink color shows the areas where malaria was prevalent in Africa, the Middle East, and southern Europe in 1920, before eradication programs began; the blue color shows the areas where sickle-cell disease most often occurred. The overlap of these two distributions (purple) suggested that there might be a causal connection.

that the recessive allele (Hb^S) has a higher frequency (0.2 to as high as 0.4 in a few areas) in regions with malaria (Fig. 19.9). Malaria is caused by a parasite that lives in and destroys red blood cells. The frequency of Hb^S is maintained because the heterozygote has an advantage over the normal phenotype in these regions.

Genotype	Phenotype	Result
Hb^AHb^A	Normal	Dies due to malarial infection
Hb^AHb^S	Sickle-cell trait	Lives due to protection from both
Hb^SHb^S	Sickle-cell disease	Dies due to sickle-cell disease

The malarial parasite flourishes in normal homozygotes but dies in heterozygotes because their red blood cells become sickle-shaped when infected with the malaria parasite. Sickle-shaped red blood cells lose potassium and this causes the parasite to die.

Even after populations become adapted to their environments, variation is promoted and maintained by various mechanisms. In diploid organisms, the heterozygote genotype can help maintain variation.

least flycatcher, *Empidonax minimus*

Acadian flycatcher, *Empidonax virescens*

Traill's flycatcher, *Empidonax trailli*

Figure 19.10 Biological definition of a species. Although the three species are nearly identical, we would still expect to find some structural feature that distinguishes them. We know they are separate species because they are reproductively isolated—the members of each species reproduce only with one another. Each species has a characteristic song and has its own particular habitat during the mating season as well.

19.3 Considering Speciation

Speciation is the splitting of one species into two or more species or the transformation of one species into a new species over time. Speciation is the final result of changes in gene pool allele and genotypic frequencies.

What Is a Species?

Sometimes it is very difficult to tell one **species** [L. *species*, a kind] from another. Before we consider the origin of species, then, it is first necessary to define a species. For Linnaeus, the father of taxonomy, one species was separated from another by morphology; that is, their physical traits differed. Darwin saw that similar species, such as the three flycatcher species in Figure 19.10, are related by common descent. The field of population genetics has produced the *biological definition of a species:* the members of one species interbreed and have a shared gene pool, and each species is reproductively isolated from every other species. The flycatchers in Figure 19.10 are members of separate species because they do not interbreed in nature.

Gene flow occurs between the populations of a species but not between populations of different species. For example, the human species has many populations that certainly differ in physical appearance. We know, however, that all humans belong to one species because the members of these populations can produce fertile offspring. On the other hand, the red maple and the sugar maple are separate species. Each species is found over a wide geographical range in the eastern half of the United States and is made up of many populations. The members of each species' populations, however, rarely hybridize in nature. Therefore, these two plants are related but separate species.

With the advent of biochemical genetics, a new way has arisen to distinguish species. The phylogenetic species concept suggests that DNA/DNA comparison techniques can indicate the relatedness of groups of organisms.

Reproductive Isolating Mechanisms

For two species to be separate, they must be reproductively isolated; that is, gene flow must not occur between them. A *reproductive isolating mechanism* is any structural, functional, or behavioral characteristic that prevents successful reproduction from occurring. Table 19.1 lists the mechanisms by which reproductive isolation is maintained.

Premating isolating mechanisms are those that prevent reproduction attempts. Habitat isolation, temporal isolation, behavioral isolation, and mechanical isolation make it highly unlikely that particular genotypes will contribute to the gene pool of a population.

Habitat Isolation. When two species occupy different habitats, even within the same geographic range, they are

Table 19.1

Reproductive Isolating Mechanisms

Isolating Mechanism	Example
Premating	
Habitat isolation	Species at same locale occupy different habitats
Temporal isolation	Species reproduce at different seasons or different times of day
Behavioral isolation	In animals, courtship behavior differs or they respond to different songs, calls, pheromones, or other signals
Mechanical isolation	Genitalia unsuitable for one another
Postmating	
Gamete isolation	Sperm cannot reach or fertilize egg
Zygote mortality	Fertilization occurs, but zygote does not survive
Hybrid sterility	Hybrid survives but is sterile and cannot reproduce
F_2 fitness	Hybrid is fertile but F_2 hybrid has reduced fitness

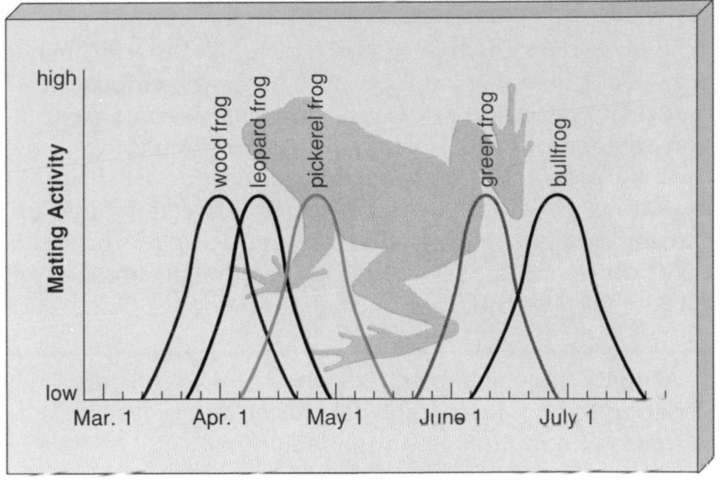

Figure 19.11 Temporal isolation.
Five species of frogs of the genus *Rana* are all found at Ithaca, New York. The species remain separate because the period of most active mating is different for each and because whenever there is an overlap, different breeding sites are used. For example, pickerel frogs are found in streams and ponds on high ground, leopard frogs in lowland swamps, and wood frogs in woodland ponds or shallow water.

less likely to meet and to attempt to reproduce. This is one of the reasons the flycatchers in Figure 19.10 do not mate and the red maple and sugar maple already mentioned do not exchange pollen. In tropical rain forests, many animal species are restricted to a particular level of the forest canopy, and in this way they are isolated from similar species.

Temporal Isolation. Two species can live in the same locale, but if each reproduces at a different time of year, they do not attempt to mate. For example, *Reticulitermes hageni* and *R. virginicus* are two species of termites. The former has mating flights in March through May, whereas the latter mates in the fall and winter months. Similarly, the frogs featured in Figure 19.11 have different periods of most active mating.

Behavioral Isolation. Many animal species have courtship patterns that allow males and females to recognize one another. Male fireflies are recognized by females of their species by the pattern of their flashings; similarly, male crickets are recognized by females of their species by their chirping. Many males recognize females of their species by sensing chemical signals called pheromones. For example, female gypsy moths secrete chemicals from special abdominal glands. These chemicals are detected downwind by receptors on antennae of males.

Mechanical Isolation. When animal genitalia or plant floral structures are incompatible, reproduction cannot occur. Inaccessibility of pollen to certain pollinators can prevent cross-fertilization in plants, and the sexes of many insect species have genitalia that do not match or other characteristics that make mating impossible. For example, male dragonflies have claspers that are suitable for holding only the females of their own species.

Postmating isolating mechanisms prevent hybrid offspring from developing or breeding, even if reproduction attempts have been successful. Gamete isolation, zygote mortality, hybrid sterility, and reduced F_2 fitness all make it unlikely that particular genotypes will contribute to the gene pool of a population.

Gamete Isolation. Even if the gametes of two different species meet, they may not fuse to become a zygote. In animals, the sperm of one species may not be able to survive in the reproductive tract of another species, or the egg may have receptors only for sperm of its species. In plants, the stigma controls which pollen grains can successfully complete pollination.

Zygote Mortality, Hybrid Sterility, and F_2 Fitness. If by chance two of the frog species in Figure 19.11 do form hybrid zygotes, the zygotes fail to complete development or else the offspring are frail. As is well known, a cross between a horse and a donkey produces a mule, which is usually sterile—it cannot reproduce. In some cases, mules are fertile, but the F_2 generation is not. This has also been observed in both evening primrose and cotton plants.

The members of a biological species are able to breed and produce fertile offspring only among themselves. There are several mechanisms that keep species reproductively isolated from one another.

How Do Species Arise?

Whenever reproductive isolation develops, speciation has occurred. Ernst Mayr, at Harvard University, proposed one model of speciation after observing that when a population is geographically isolated from other populations, gene flow stops. Distinguishing variations due to different mutations, genetic drift, and directional selection build up, causing first postmating and then premating reproductive isolation to occur. Mayr called this model **allopatric speciation** [Gk. *allo,* different, and *patri,* fatherland] (Fig. 19.12).

Allopatric speciation occurs when populations are separated by a geographical barrier that prevents them from reproducing with each other.

With **sympatric speciation** [Gk. *sym,* together, and *patri,* fatherland], a population develops into two or more reproductively isolated groups without prior geographic isolation. The best evidence for this type of speciation is found among plants, where it can occur by means of polyploidy. In this case, there would be a multiplication of the chromosome number in certain plants of a single species. Sympatric speciation can also occur due to hybridization between two species, followed by a doubling of the chromosome number. Polyploid plants are reproductively isolated by a postmating mechanism; they can reproduce successfully only with other like polyploids, and backcrosses with diploid parents are sterile.

Sympatric speciation occurs when members of a single population develop a genetic difference that prevents them from reproducing with the parent type.

Figure 19.12 contrasts allopatric speciation with sympatric speciation. Allopatric speciation is by far the more common means of speciation.

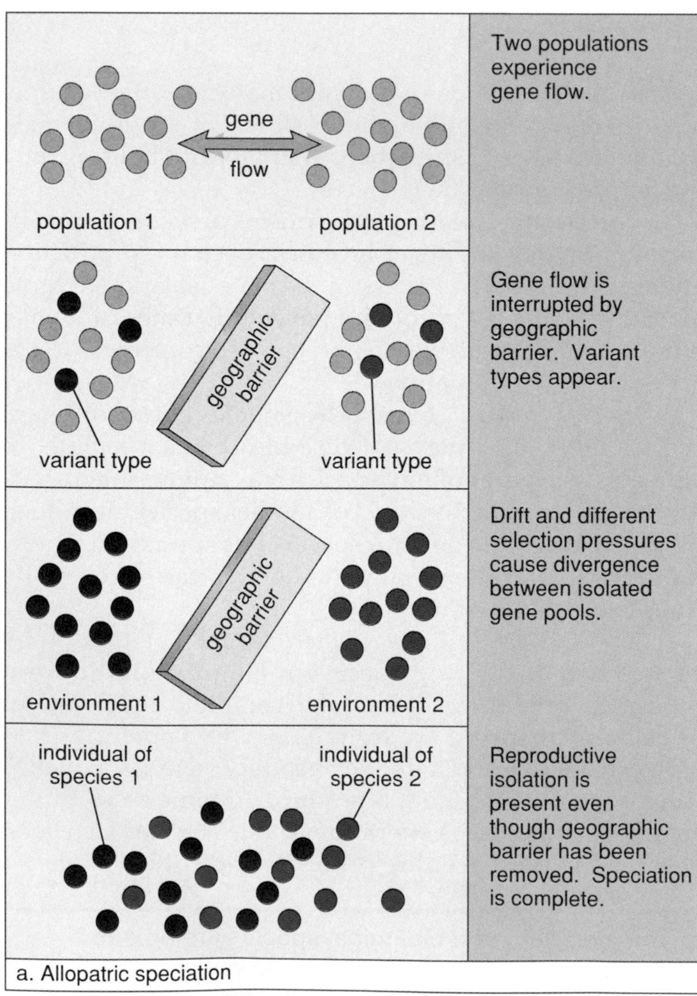

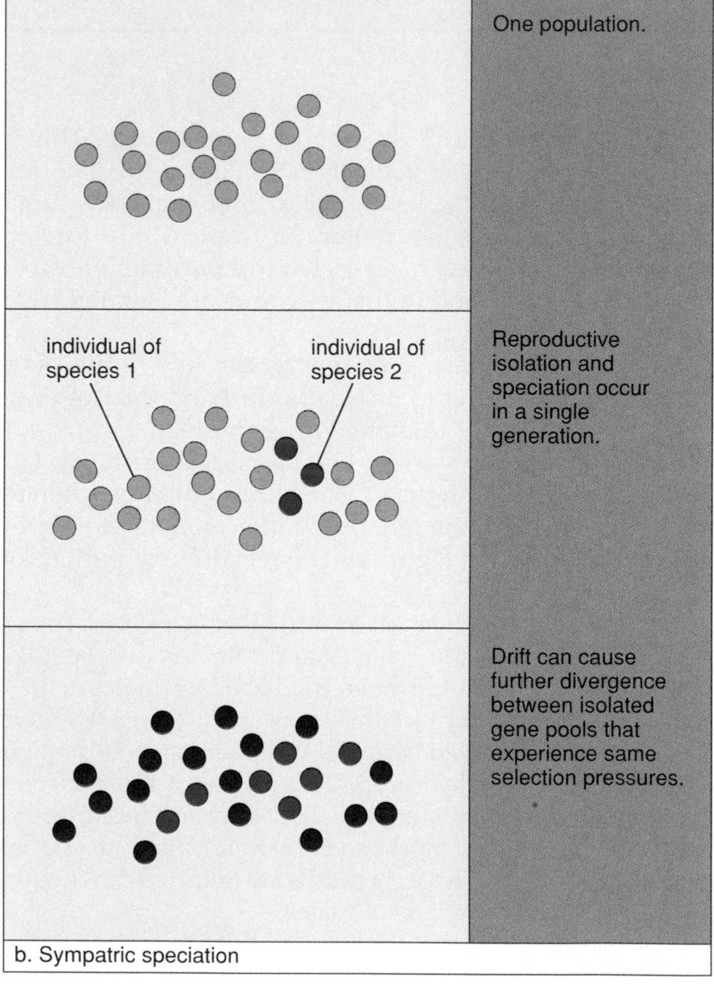

Figure 19.12 Allopatric versus sympatric speciation.
a. Allopatric speciation occurs after a geographic barrier prevents gene flow between populations that originally belonged to a single species.
b. Sympatric speciation occurs when members of a population achieve immediate reproductive isolation without any prior geographic barrier.

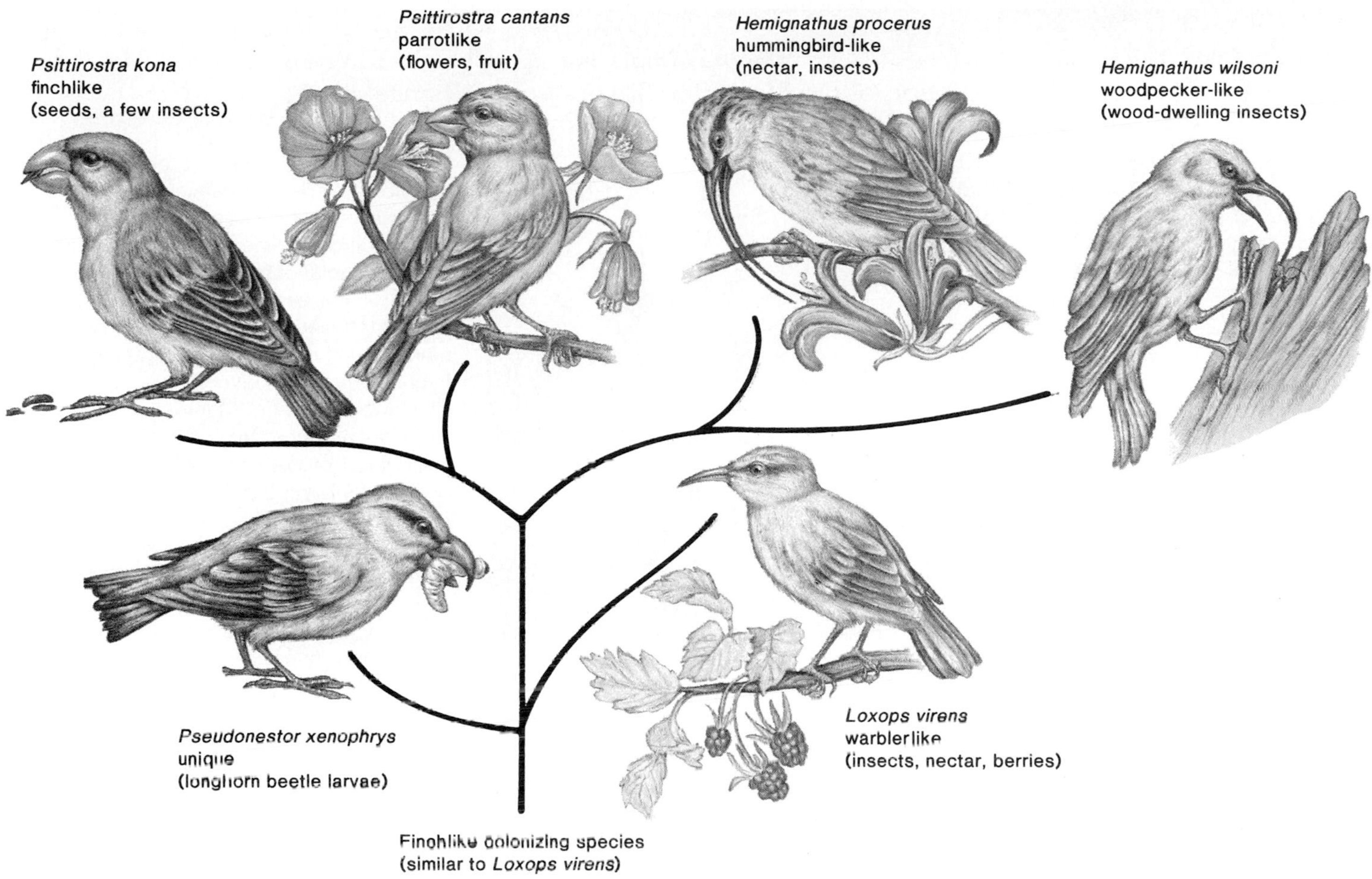

Figure 19.13 Adaptive radiation in Hawaiian honeycreepers.
More than 20 species evolved from a single species of a finchlike bird that colonized the Hawaiian Islands. This illustration shows only six of the more extreme adaptive forms.

Adaptive Radiation Produces Many Species

Adaptive radiation is the rapid development from a single ancestral species of many new species, which have spread out and become adapted to various ways of life. The 13 species of finches that live on the Galápagos Islands are believed to be descended from a type of mainland finch that invaded the islands. As the parent population increased in size, daughter populations were established on the various islands. These daughter populations were subjected to the founder effect and the process of natural selection. Because of natural selection, each population became adapted to a particular habitat on its island. In time, the various populations became so genotypically different that now, when they by chance reside on the same island, they do not interbreed and are therefore separate species. There is evidence that the finches use beak shape to recognize members of the same species during courtship. Rejection of suitors with the wrong type of beak is a behavioral type of premating isolating mechanism.

Similarly, on the Hawaiian Islands, there is a wide variety of honeycreepers that are descended from a common goldfinchlike ancestor that arrived from Asia or North America about 5 million years ago. Today, honeycreepers have a range of beak sizes and shapes for feeding on various food sources, including seeds, fruits, flowers, and insects (Fig. 19.13). Adaptive radiation has also been observed in plants on the Hawaiian Islands, as discussed in the reading on pages 318–19.

After a species undergoes adaptive radiation, there are many new species adapted to living in different habitats.

Doing science

Origin and Adaptive Radiation of the Hawaiian Silversword Alliance

When I was a graduate student at the University of California at Davis, I studied a genus of plants *(Calycadenia)* that belongs to a largely Californian group of plants called tarweeds. I knew that the tarweeds and plants belonging to the silversword alliance (an alliance is an assemblage of closely related species) are in the same large family of plants called Asteraceae (Compositae). But Sherwin Carlquist's anatomical research had suggested that the silversword plants and the tarweeds are even more closely related. The silverswords are found only in Hawaii, so when I took a faculty position at the University of Hawaii, I decided to gather evidence to possibly support Carlquist's interpretation.

There are 28 species of plants in the silversword alliance: five species of *Argyroxiphium,* two species of *Wilkesia*, and 21 species of *Dubautia.* This alliance constitutes one of the most spectacular examples of adaptive radiation among plants. Adaptive radiation characterizes many insular groups that have evolved in isolation away from related and nonrelated organisms. Members of the silversword alliance range in form from matlike subshrubs and rosette shrubs to large trees and climbing vines (lianas). They grow in habitats as diverse as exposed lava, dry scrub, dry woodland, moist forest, wet forest, and bogs. These habitats have a range in elevation from 75 to 3,800 meters (250–12,500 feet), and in annual precipitation from 38 to 1,230 cm (15–485 inches).

Members of the alliance found in moist to wet forest habitats typically exhibit modifications that are adaptive in competition for light, such as increased height, vining habit, and thin leaves with a comparatively large surface area (Fig. 19A). Members of the alliance from open, more arid sites typically show features associated with conservation of water, such as decreased height, thickened leaves with comparatively low surface area, and compact internal tissues (Fig. 19B). Species of *Argyroxiphium* have more or less succulent leaves with compact tissue and channels filled with a water-binding gelatinous matrix of pectin (Fig. 19C). These adaptions help this species survive under the conditions of extreme water stress in its habitats—i.e., largely in oxygen-deficient, acid bogs or dry alpine cinder habitats.

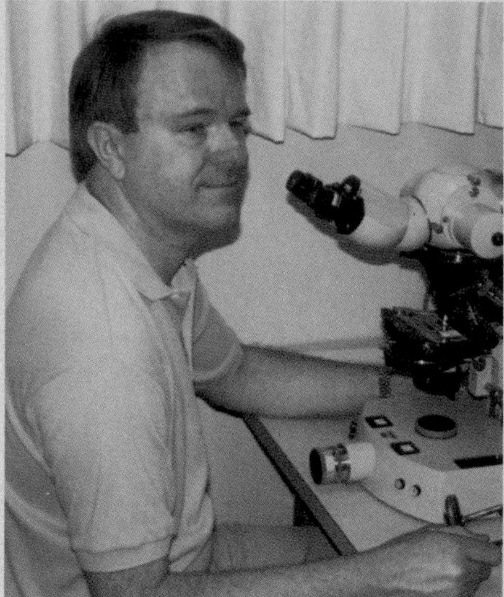

Gerald D. Carr
University of Hawaii at Manoa

I have done cytogenetic analyses of meiotic chromosome pairing in hybrids to determine relationships among the silversword plants and between these plants and the tarweeds. I have sought and found many natural interspecific and intergeneric hybrids among the diverse species of the silversword alliance. Many such hybrids also were produced artificially. Each newly discovered or created hybrid potentially provided information to fill in a missing piece of an intriguing and mysterious natural puzzle of evolution—or sometimes these findings complicated the picture further. Nearly 20 years later, the picture is still not complete. Perhaps the most exciting and personally gratifying result of my involvement with this group is the series of hybrids that have been produced between Hawaiian species and the tarweeds. These hybrids, and also a growing body of molecular data provided by one of my collaborators, Bruce Baldwin, clearly establish the origin of the Hawaiian silversword alliance from the tarweeds.

I like working with others who are innately curious about their natural surroundings and who do research for pure personal satisfaction. My scientific research helps provide insight into the mystery and meaning of the wonderful diversity of nature. An added element of excitement comes in sharing with others a discovery of one of nature's secrets that perhaps no other human has witnessed. Therefore, for me, teaching and research are naturally complementary and rewarding activities.

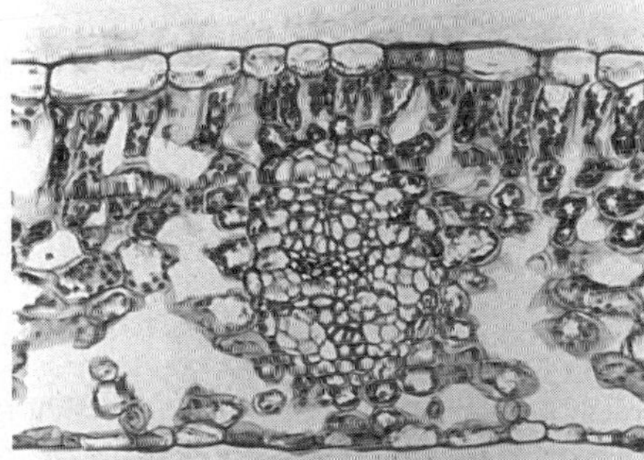

Figure 19A ***Dubautia knudsenii*** **is adapted to living in a moist habitat.**
In the cross section of the leaf, note the loose organization of tissue and thin cuticle on the upper and lower epidermis.

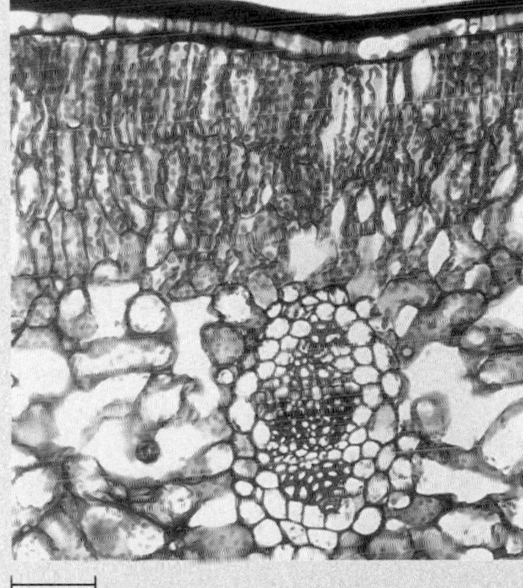

Figure 19B ***Dubautia menziesii*** **is adapted to living in open, more arid sites.**
In cross section, note the thick leaf with compact organization of tissue and very highly developed cuticle on upper and lower epidermis (lower epidermis not shown).

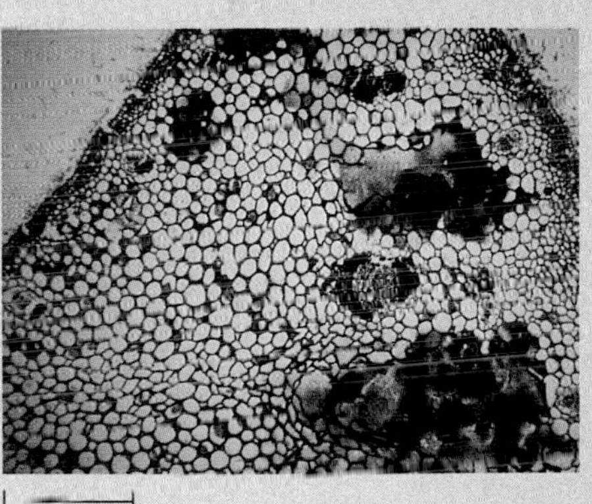

Figure 19C ***Argyroxiphium kauense*** **is adapted to living under conditions of water stress.**
In cross section, note the compact tissue and large channels of water-binding extracellular pectin that alternate with the major vascular bundles.

connecting concepts

We have seen that there are variations among the individuals in any population, whether the population is tuberculosis-causing bacteria in a city, the dandelions on a hill, or the squirrels in your neighborhood. Individuals vary because of the presence of mutations, and in sexually reproducing species, because of the recombination of alleles and chromosomes due to the process of meiosis and fertilization.

The field of population genetics, which utilizes the Hardy-Weinberg law, shows us how the study of evolution can be objective rather than subjective. A change in gene pool allele frequencies defines and signifies that evolution has occurred. There are various agents of evolutionary change, and natural selection is one of these.

Population genetics has also given us the biological definition of a species: the members of one species can successfully reproduce with each other but cannot successfully reproduce with members of another species. This criteria allows us to know when a new species has arisen. Species usually come into being after two populations have been geographically isolated. Suppose the present population of squirrels in your neighborhood was divided by a cavern following an earthquake. Then, there would be two genetically different populations of squirrels that are geographically isolated from each other. Suppose also that the area west of the cavern is slightly drier than that to the east. Over time, each population would become adapted to its particular environment. Eventually, the two populations may become so genetically different that even if members of each population came into contact, they would not be able to produce fertile offspring. Since gene flow between the two populations would not be possible, the squirrels would be considered separate species. By the same process, multiple species can repeatedly arise, as when a common ancestor led to thirteen species of Darwin's finches.

Summary

19.1 Evolution in a Genetic Context

The members of a population vary from one another. Primary sources of variation are gene mutations, which produce new alleles; chromosomal mutations, which can rearrange alleles on a chromosome; and recombination, which can bring certain alleles together into the same genotype. Most traits of evolutionary significance are polygenic.

All the various genes of a population make up its gene pool. The Hardy-Weinberg equilibrium is a constancy of gene pool allele frequencies that remains from generation to generation if certain conditions are met. The conditions are no mutation, no gene flow, random mating, no genetic drift, and no selection. Since these conditions are rarely met, a change in gene pool frequencies is likely. When gene pool frequencies change, evolution has occurred. Deviations from a Hardy-Weinberg equilibrium allow us to determine when evolution has taken place.

The agents of evolutionary change are mutation, nonrandom mating, gene flow, genetic drift, and natural selection. The rate of mutation per gene per generation is low, but since individuals have many genes and a population has many individuals, the mutation rate in a population is an adequate source of new alleles. Genotypic variations may be of evolutionary significance only should the environment change. Nonrandom mating occurs when relatives mate (inbreeding) and assortative mating occurs. Both of these cause an increase in homozygotes. Gene flow occurs when a breeding individual (animals) migrates to another population or when gametes and seeds (plants) are carried into another population. Constant gene flow between two populations causes their gene pools to become similar. Genetic drift occurs when allele frequencies are altered by chance—that is, by sampling error. Genetic drift can cause the gene pools of two isolated populations to become dissimilar as some alleles are lost and others are fixed. Genetic drift is particularly evident when founders start a new population, or after a bottleneck, when severe inbreeding occurs.

19.2 Adaptation Occurs Naturally

The process of natural selection can now be restated in terms of population genetics. A change in gene pool frequencies results in adaptation to the environment.

Three types of selection occur: directional (dark-colored peppered moths are prevalent in polluted areas and light-colored peppered moths are prevalent in nonpolluted areas); stabilizing (the average human birth weight is near the optimum birth weight for survival); and disruptive (*Cepaea* snails vary because a wide geographic range causes selection to vary).

Despite constant natural selection, variation is maintained. Mutation and recombination still occur; gene flow among small populations can introduce new alleles; and natural selection itself sometimes results in variation. In sexually reproducing diploid organisms, the heterozygote acts as a repository for recessive alleles whose frequency is low. In regard to sickle-cell disease, the heterozygote is more fit in areas with malaria and, therefore, the homozygotes are maintained in the population.

19.3 Considering Speciation

The biological definition of a species recognizes that populations of the same species breed only among themselves and are reproductively isolated from other species. Reproductive isolating mechanisms prevent gene flow among species. Premating isolating mechanisms (habitat, temporal, behavioral, and mechanical isolation) prevent mating from being attempted. Postmating isolating mechanisms (gamete isolation, zygote mortality, hybrid sterility, and F_2 fitness) prevent hybrid offspring from surviving and/or reproducing.

Allopatric speciation requires geographic isolation before reproductive isolation occurs. Sympatric speciation does not require geographic isolation for reproductive isolation to develop. The occurrence of polyploidy in plants is an example of this other type of speciation.

Adaptive radiation, as exemplified by the Hawaiian honeycreepers, is a form of allopatric speciation. It occurs because the opportunity exists for new species to adapt to new habitats.

Reviewing the Chapter

1. What are the sources of variation in a population of sexually reproducing diploid organisms? 304
2. What is the Hardy-Weinberg law? 305–6
3. Name and discuss the five agents of evolutionary change. 307–9
4. What is the founder effect, and what is a bottleneck? 309
5. State the steps required for adaptation by natural selection in modern terms. 310
6. Distinguish among directional, stabilizing, and disruptive selection by giving examples. 310–12
7. State ways in which variation is maintained in a population. 313
8. What is the biological definition of a species? 314
9. What is a reproductive isolating mechanism? Give examples of both premating and postmating isolating mechanisms. 314–15
10. How does allopatric speciation occur? sympatric speciation? 316
11. What is adaptive radiation and how is it exemplified by the Galápagos finches and the Hawaiian honeycreepers? 317

Testing Yourself

Choose the best answer for each question.

1. Assuming a Hardy-Weinberg equilibrium, 21% of a population is homozygous dominant, 50% is heterozygous, and 29% is homozygous recessive. What percentage of the next generation is predicted to be homozygous recessive?
   a. 21%
   b. 50%
   c. 29%
   d. 25%
2. A human population has a higher-than-usual percentage of individuals with a genetic disease. The most likely explanation is
   a. gene flow.
   b. natural selection.
   c. genetic drift.
   d. All of these are correct.
3. The offspring of better-adapted individuals are expected to make up a larger proportion of the next generation. The most likely explanation is
   a. mutation.
   b. gene flow.
   c. natural selection.
   d. genetic drift.
4. The continued occurrence of sickle-cell disease in parts of Africa with malaria is due to
   a. continual mutation.
   b. gene flow between populations.
   c. fitness of the heterozygote.
   d. disruptive selection.
5. Which of these is/are necessary to natural selection?
   a. variations
   b. differential reproduction
   c. inheritance of differences
   d. All of these are correct.
6. When a population is small, there is a greater chance of
   a. gene flow.
   b. genetic drift.
   c. natural selection.
   d. mutations occurring.
7. The biological definition of a species is simply the
   a. anatomical and developmental differences between two groups of organisms.
   b. geographic distribution of two groups of organisms.
   c. differences in the adaptations of two groups of organisms.
   d. reproductive isolation of two groups of organisms.
8. Which of these is a premating isolating mechanism?
   a. habitat isolation
   b. temporal isolation
   c. gamete isolation
   d. Both a and b are correct.
9. Male moths recognize females of their species by sensing chemical signals called pheromones. This is an example of
   a. gamete isolation.
   b. habitat isolation.
   c. behavioral isolation.
   d. mechanical isolation.
10. Allopatric but not sympatric speciation requires
    a. reproductive isolation.
    b. geographic isolation.
    c. prior hybridization.
    d. spontaneous differences in males and females.
11. The many species of Galápagos finches were each adapted to eating different foods. This is an example of
    a. gene flow.
    b. adaptive radiation.
    c. sympatric speciation.
    d. All of these are correct.
12. The following diagram represents one species. (a) Use the labels population and gene flow where appropriate. (b) What changes would you make to this diagram in order to symbolize two species?

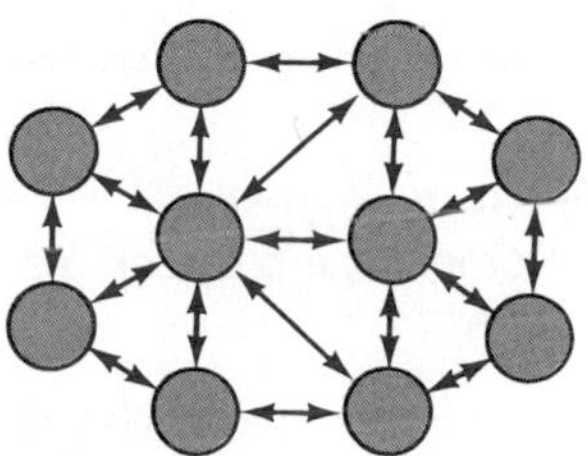

13. The following diagrams represent a distribution of phenotypes in a population. Draw another diagram under (a) to show that disruptive selection has occurred, under (b) to show that stabilizing selection has occurred, and under (c) to show that directional selection has occurred.

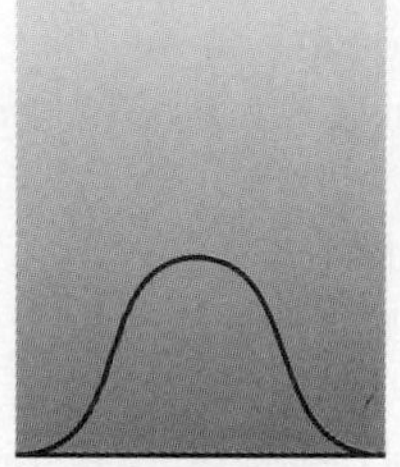
a. Disruptive selection

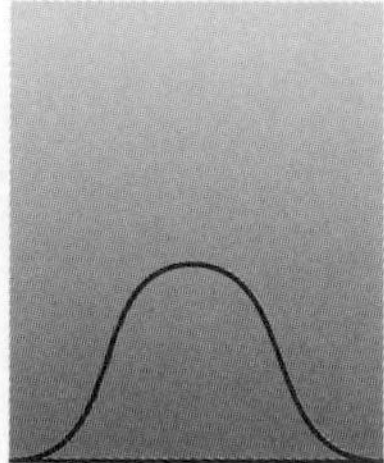
b. Stabilizing selection

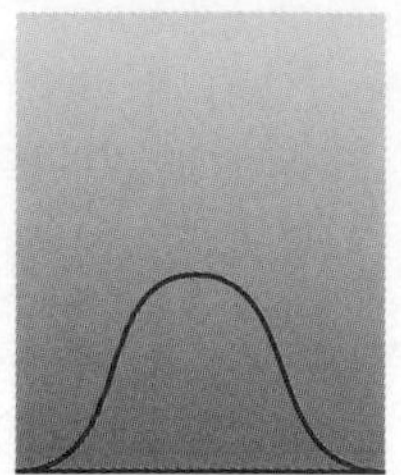
c. Directional selection

Additional Genetics Problems*

1. If $p^2 = 0.36$, what percentage of the population has the recessive phenotype, assuming a Hardy-Weinberg equilibrium?
2. If 1% of a human population has the recessive phenotype, what percentage has the dominant phenotype, assuming a Hardy-Weinberg equilibrium?
3. Four percent of the members of a population of pea plants are short (recessive characteristic). What are the frequencies of both the recessive allele and the dominant allele? What are the genotypic frequencies in this population, assuming a Hardy-Weinberg equilibrium?

Applying the Concepts

1. *The process of evolution has a genetic basis.*
   Why is the process of evolution consistent with a particulate theory of inheritance rather than a blending theory of inheritance?
2. *Selection is the process by which populations accumulate adaptive traits.*
   Reconcile this statement with your knowledge that selection acts at the level of the individual.
3. *The diversity of life is dependent upon the process of speciation.*
   If only new species evolve, then of what use are the higher taxa (classification categories)?

Using Technology

Your study of evolution is supported by these available technologies:

Exploring the Internet

The Mader Home Page provides resources for and help with studying this chapter.

http://www.mhhe.com/sciencemath/biology/mader/
(Click on Biology.)

*Answers to Additional Genetics Problems appear in Appendix A.

Understanding the Terms

adaptive radiation 317
allopatric speciation 316
assortative mating 308
bottleneck effect 309
directional selection 310
disruptive selection 312
fitness 310
founder effect 309
gene flow 308
gene pool 305
genetic drift 308
Hardy-Weinberg law 306
nonrandom mating 308
population 304
postmating isolating mechanism 315
premating isolating mechanism 314
speciation 314
species 314
stabilizing selection 312
sympatric speciation 316

Match the terms to these definitions:

a. __________ Outcome of natural selection in which extreme phenotypes are eliminated and the average phenotype is conserved.

b. __________ Anatomical or physiological difference between two species that prevents successful reproduction after mating has taken place.

c. __________ Change in the genetic makeup of a population due to chance (random) events; important in small populations or when only a few individuals mate.

d. __________ Evolution of a large number of species from a common ancestor.

e. __________ Group of organisms of the same species occupying a certain area and sharing a common gene pool.

f. __________ Taxonomic category whose members are characterized by anatomy and can only breed successfully with each other.

g. __________ Law stating that the allele frequency in a population remains stable under certain assumptions, such as random mating; therefore, no change or evolution occurs.

h. __________ Outcome of natural selection in which extreme phenotypes are favored over the average phenotype, leading to more than one distinct form.

i. __________ Origination of new species due to the evolutionary process of descent with modification.

j. __________ Sharing of genes between two populations through interbreeding.

Origin and History of Life

CHAPTER 20

Chapter Concepts

Deinonychus, carnivorous dinosaurs

One of the most fascinating and frequently asked questions by laypeople and scientists alike is where did life come from? Although various answers have been proposed throughout history, today the most widely accepted hypothesis is that inorganic molecules in Earth's prebiotic oceans combined to produce organic molecules and eventually primitive cells. These earliest cells probably appeared around 3.5 billion years ago. Around two billion years ago, oxygen-releasing photosynthesis began and the atmospheric ozone layer formed. This shield protected the earth's surface from intense ultraviolet radiation, and consequently, life moved from water onto land. Shortly afterward (at least in geologic time), the number of multicellular life-forms increased dramatically.

Millions of species have evolved, changed, and died out since life first began. If we could trace the lineages of all the species ever to have graced our planet, they would resemble a dense bush with countless branches intermingled with others. Some lines of descent are cut off close to the base, indicating that extinction occurred soon after the group arose; the others extend for varying amounts of time. Some that continue to today are groups that have managed to remain relatively unchanged since they first evolved. Other lines have split, producing two or more new groups. Clearly, the history of life on earth has many facets and twists and turns.

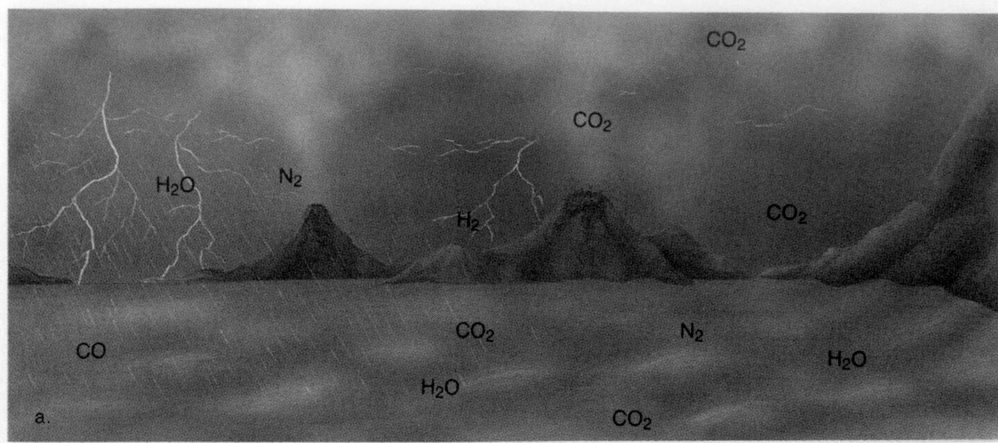

a. The primitive atmosphere contained gases, including water vapor, that escaped from volcanoes; as the water vapor cooled, some gases were washed into the ocean by rain.

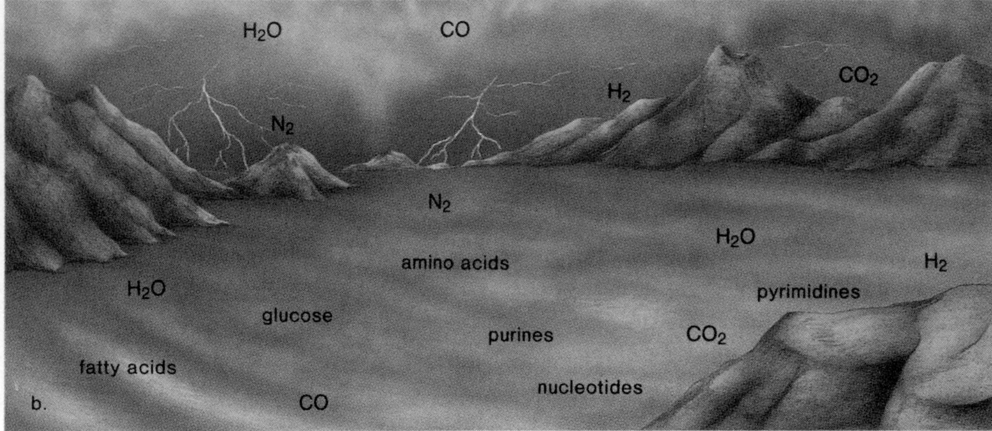

b. The availability of energy from volcanic eruption and lightning allowed gases to form simple organic molecules.

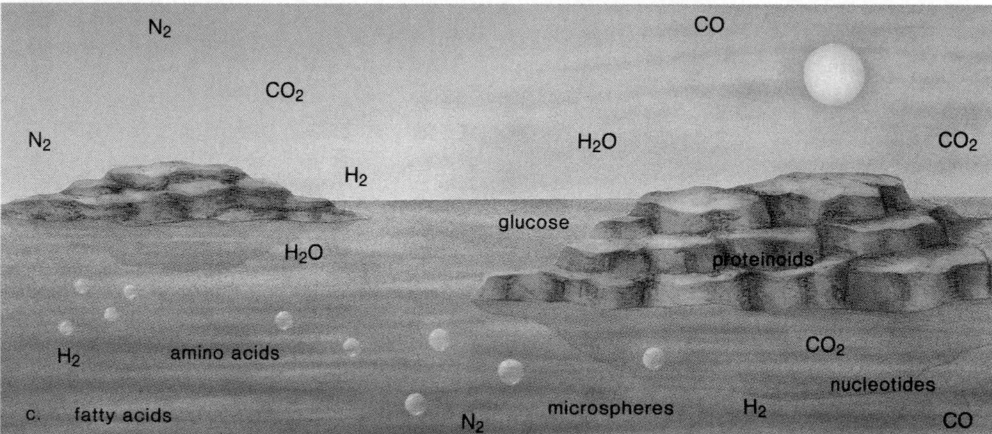

c. Amino acids that splashed up onto rocky coasts could have polymerized into polypeptides (proteinoids) that became microspheres when they reentered the water.

d. Eventually, various types of prokaryotes and then eukaryotes evolved. Some of the prokaryotes were oxygen-producing photosynthesizers.

Figure 20.1 A possible scenario for the origin of life.

20.1 Origin of Life

Today we do not believe that life arises spontaneously from nonlife, and we say that "life comes only from life." But if this is so, how did the first form of life come about? Since it was the very first living thing, it had to come from nonliving chemicals. Could there have been an increase in the complexity of the chemicals—could a **chemical evolution** as depicted in Figure 20.1 have produced the first cell(s) on the primitive earth?

The sun and the planets, including earth, probably formed over a 10-billion-year period from aggregates of dust particles and debris. At 4.6 billion years ago, the solar system was in place. Intense heat produced by gravitational energy and radioactivity caused the earth to become stratified into several layers. Heavier atoms of iron and nickel became the molten liquid core, and dense silicate minerals became the semiliquid mantle. Upwellings of volcanic lava produced the first crust.

The Atmosphere Forms

The size of the earth is such that the gravitational field is strong enough to have an atmosphere. If the earth were smaller and lighter, atmospheric gases would escape into outer space. The earth's *primitive atmosphere* was not the same as today's atmosphere. It is now thought that the primitive atmosphere was produced primarily by outgassing from the interior, particularly by volcanic action. In that case, the primitive atmosphere would have consisted mostly of water vapor (H_2O), nitrogen (N_2), and carbon dioxide (CO_2), with only small amounts of hydrogen (H_2) and carbon monoxide (CO). The primitive atmosphere, with little if any free oxygen, was a *reducing atmosphere* as opposed to the *oxidizing atmosphere* of today. This was fortuitous because oxygen (O_2) attaches to organic molecules, preventing them from joining to form larger molecules.

At first the earth was so hot that water was present only as a vapor that formed dense, thick clouds. Then as the earth cooled, water vapor condensed to liquid water, and rain began to fall (Fig. 20.1*a*). It rained in such enormous quantity over hundreds of millions of years that the oceans of the world were produced. The earth is an appropriate distance from the sun: any closer, water would have evaporated; any farther, water would have frozen.

Small Organic Molecules Evolve

The atmospheric gases, dissolved in rain, were carried down into newly forming oceans. Aleksandr Oparin, a Soviet biochemist, suggested as early as 1938 that organic molecules could have been produced from the gases of the primitive atmosphere in the presence of strong outside *energy sources*. The energy sources on the primitive earth included heat from volcanoes and meteorites, radioactivity from isotopes in the earth's crust, powerful electric discharges in lightning, and solar radiation, especially ultraviolet radiation (Fig. 20.1*b*).

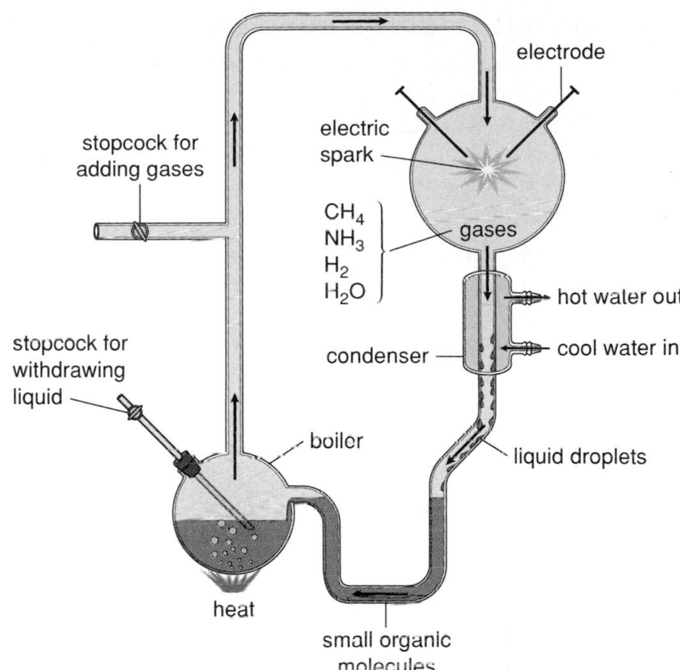

Figure 20.2 Evidence for a chemical evolution. Gases were admitted to the apparatus, circulated past an energy source (electric spark), and cooled to produce a liquid that could be withdrawn. Upon chemical analysis, the liquid was found to contain various small organic molecules.

In 1953, Stanley Miller provided support for Oparin's ideas through an ingenious experiment (Fig. 20.2). Miller placed a mixture resembling a strongly reducing atmosphere—methane (CH_4), ammonia (NH_3), hydrogen (H_2), and water (H_2O)—in a closed system, heated the mixture, and circulated it past an electric spark (simulating lightning). After a week's run, Miller discovered that a variety of amino acids and organic acids had been produced. Since that time, other investigators have achieved similar results by utilizing other, less-reducing combinations of gases dissolved in water.

These experiments support the hypothesis that the primitive gases could have reacted with one another to produce small organic compounds. Neither oxidation (there was no free oxygen) nor decay (there were no bacteria) would have destroyed these molecules, and they would have accumulated in the oceans for hundreds of millions of years. With the accumulation of these small organic compounds, the oceans became a thick, warm organic soup containing a variety of organic molecules.

Atmospheric gases dissolved in the ocean may have reacted with one another to produce simple organic molecules.

Macromolecules Evolve and Interact

The newly formed organic molecules likely polymerized to produce still larger molecules and then macromolecules. There are three primary hypotheses concerning this stage in the origin of life. One is the *RNA-first hypothesis,* which suggests that only the macromolecule RNA (ribonucleic acid) was needed at this time to progress toward formation of the first cell or cells. Thomas Cech and Sidney Altman shared a Nobel Prize in 1989 because they discovered that RNA can be both a substrate and an enzyme. Other types of ribozymes (RNA enzymes) have since been discovered. It would seem, then, that RNA could have carried out the processes of life commonly associated today with DNA (deoxyribonucleic acid, the genetic material) and proteins (enzymes). Some viruses today have RNA genes; therefore, the first genes could have been RNA. And the first enzymes also could have been RNA molecules, since we now know that ribozymes exist. Those who support this hypothesis say that it was an "RNA world" some 4 billion years ago.

Another hypothesis is termed the *protein-first hypothesis.* Sidney Fox has shown that amino acids polymerize abiotically when exposed to dry heat (see Fig. 20.1*c*). He suggests that amino acids collected in shallow puddles along the rocky shore and the heat of the sun caused them to form **proteinoids,** small polypeptides that have some catalytic properties. When proteinoids are returned to water, they form **microspheres** [Gk. *mikros,* small, little, and *sphaera,* ball], structures composed only of protein that have many properties of a cell. It's possible that the first polypeptides had enzymatic properties, and some proved to be more capable than others. Those that led to the first cell or cells had a selective advantage. This hypothesis assumes that DNA genes came after protein enzymes arose. After all, it is protein enzymes that are needed for DNA replication.

The third hypothesis is put forth by Graham Cairns-Smith. He believes that clay was especially helpful in causing polymerization of both proteins and nucleic acids at the same time. Clay attracts small organic molecules and contains iron and zinc, which may have served as inorganic catalysts for polypeptide formation. In addition, clay has a tendency to collect energy from radioactive decay and to discharge it when the temperature and/or humidity changes. This could have been a source of energy for polymerization to take place. Cairns-Smith suggests that RNA nucleotides and amino acids became associated in such a way that polypeptides were ordered by and helped synthesize RNA. It is clear that this hypothesis suggests that both polypeptides and RNA arose at the same time.

Small organic molecules polymerized to produce macromolecules. This could have occurred on heated rocks or in clay.

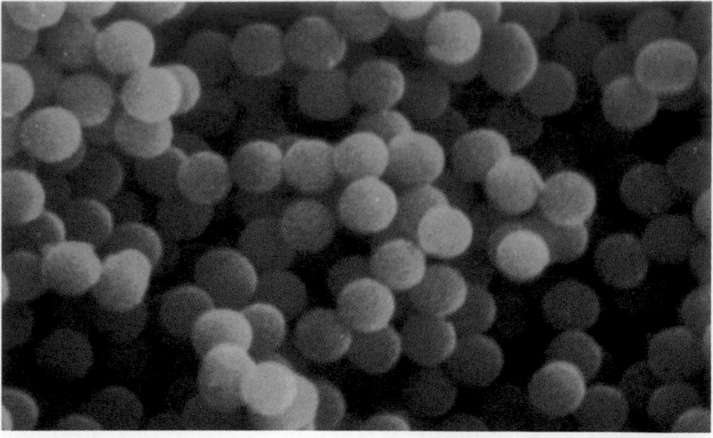

a.

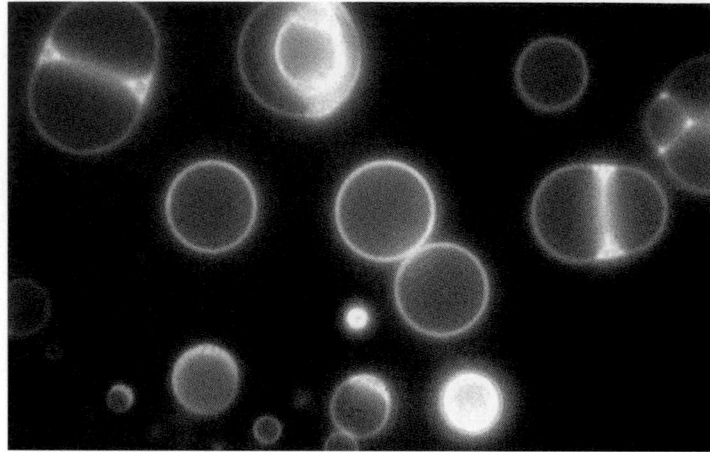

b.

Figure 20.3 Protocell anatomy.
a. Microspheres, which are composed only of protein, have a number of cellular characteristics and could have evolved into the protocell. **b.** Liposomes form automatically when phospholipid molecules are put into water. Plasma membranes may have evolved similarly.

A Protocell Evolves

Before the first true cell arose, there would have been a **protocell** [Gk. *protos,* first], a structure that has a lipid-protein membrane and carries on energy metabolism (Fig. 20.3). Fox has shown that if lipids are made available to microspheres, lipids tend to become associated with microspheres producing a lipid-protein membrane.

Some researchers support the work of Oparin, who was mentioned previously. Oparin showed that under appropriate conditions of temperature, ionic composition, and pH, concentrated mixtures of macromolecules tend to give rise to complex units called *coacervate droplets.* Coacervate droplets have a tendency to absorb and incorporate various substances from the surrounding solution. Eventually, a semipermeable-type boundary may form about the droplet. In a liquid environment, phospholipid molecules automatically form droplets called **liposomes** [Gk. *lipos*, fat, and *soma*, body]. Perhaps the first membrane formed in this manner. In that case, the protocell could have contained only RNA, which functioned as both genetic material and enzymes.

Protocells Were Heterotrophs

The protocell would have had to carry on nutrition so that it could grow. Nutrition was no problem because the protocell existed in the ocean, which at that time contained simple organic molecules that could have served as food. Therefore, the protocell likely was a **heterotroph** [Gk. *hetero*, different, and *trophe*, food], an organism that takes in preformed food. Notice that this suggests that heterotrophs are believed to have preceded **autotrophs** [Gk. *autos*, alone, and *trophe*, food], organisms that make their own food.

At first, the protocell may have used preformed ATP (adenosine triphosphate), but as this supply dwindled, natural selection favored any cells that could extract energy from carbohydrates in order to transform ADP (adenosine diphosphate) to ATP. Glycolysis is a common metabolic pathway in living things, and this testifies to its early evolution in the history of life. Since there was no free oxygen, we can assume that the protocell carried on a form of fermentation.

It seems logical that the protocell at first had limited ability to break down organic molecules and that it took millions of years for glycolysis to evolve completely. It is of interest that Fox has shown that a microsphere from which the protocell may have evolved has some catalytic ability and that Oparin found that coacervates do incorporate enzymes if they are available in the medium.

The protocell is hypothesized to have had a membrane boundary and to have been a heterotrophic fermenter with some degree of enzymatic ability.

A Self-Replication System Evolves

A true cell is a membrane-bounded structure that can carry on protein synthesis needed to produce the enzymes that allow DNA to replicate (Fig. 20.4). The central dogma of genetics states that DNA directs protein synthesis and that there is a flow of information from DNA → RNA → protein. It is possible that this sequence developed in stages.

According to the RNA-first hypothesis, RNA would have been the first to evolve, and the first true cell would have had RNA genes. These genes would have directed and enzymatically carried out protein synthesis. As mentioned, RNA enzymes called ribozymes have been discovered. Also, today we know there are viruses that have RNA genes. These viruses have a protein enzyme called reverse transcriptase that uses RNA as a template to form DNA. Perhaps with time, reverse transcription occurred within the protocell, and this is how DNA genes arose. Once there were DNA genes, then protein synthesis would have been carried out in the manner dictated by the central dogma of genetics.

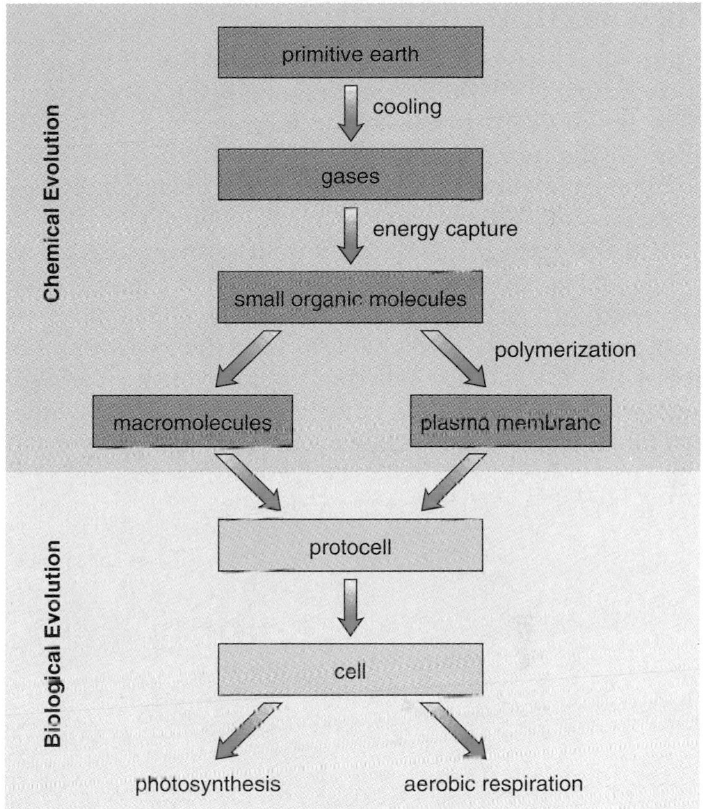

Figure 20.4 Evolution of the protocell.
There was an increase in the complexity of macromolecules, leading to a self-replicating system (DNA → RNA → protein) enclosed by a plasma membrane. The protocell, a heterotrophic fermenter, underwent biological evolution, becoming a true cell, which then diversified.

According to the protein-first hypothesis, proteins, or at least polypeptides, were the first of the three (i.e., DNA, RNA, and protein) to arise. Only after the protocell developed sophisticated enzymes did it have the ability to synthesize DNA and RNA from small molecules provided by the ocean. Researchers point out that a nucleic acid is a very complicated molecule, and the likelihood that RNA arose *de novo* (on its own) is minimal. It seems more likely that enzymes were needed to guide the synthesis of nucleotides and then nucleic acids.

Cairns-Smith proposes that polypeptides and RNA evolved simultaneously. Therefore, the first true cell would have contained RNA genes that could have replicated because of the presence of proteins. This eliminates the baffling chicken-and-egg paradox: which came first, proteins or RNA? But it does mean, however, that two unlikely events would have to happen at the same time.

Once the protocells acquired genes that could replicate, they became cells capable of reproducing, and biological evolution began. The history of life began!

Once the protocell was capable of reproduction, it became a true cell, and biological evolution began.

20.2 History of Life

Figure 20.5 shows the history of the earth as if it had occurred during a 24-hour time span that starts at midnight. (The actual years are shown on an inner ring of the diagram.) This figure illustrates dramatically that only unicellular organisms were present during most (about 80%) of the history of the earth.

If the earth formed at midnight, prokaryotes do not appear until about 5 A.M., eukaryotes are present at approximately 4 P.M., and multicellular forms do not appear until around 8 P.M. Invasion of the land doesn't occur until about 10 P.M., and humans don't appear until 30 seconds before the end of the day. This timetable has been worked out by studying the fossil record.

Fossils Tell a Story

Fossils [L. *fossilis*, dug up] are the remains and traces of past life or any other direct evidence of past life. Traces include trails, footprints, burrows, worm casts, or even preserved droppings. Usually when an organism dies, the soft parts are either consumed by scavengers or undergo bacterial decomposition. Occasionally, the organism is buried quickly and in such a way that decomposition is never completed or is completed so slowly that the soft parts leave an imprint of their structure. Most fossils, however, consist only of hard parts such as shells, bones, or teeth, because these are usually not consumed or destroyed.

The great majority of fossils are found embedded in or recently eroded from sedimentary rock. **Sedimentation**

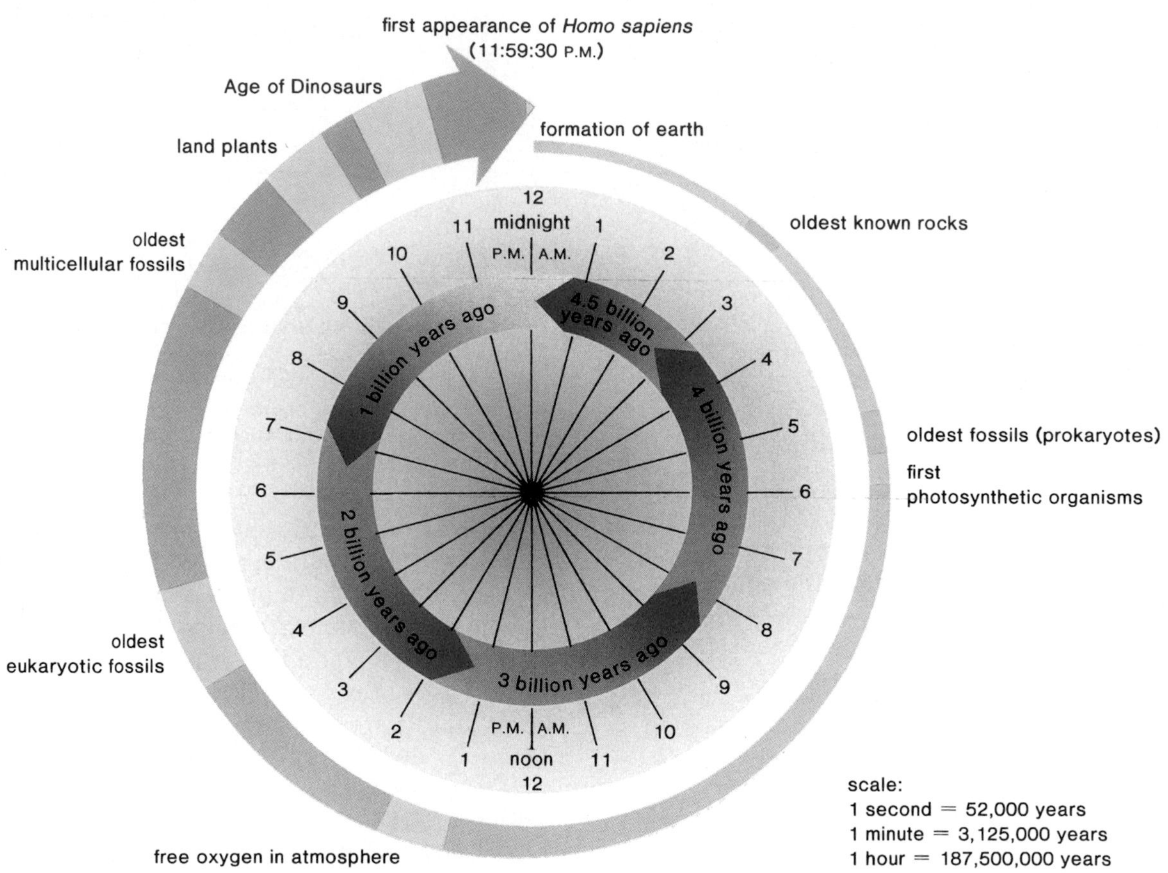

Figure 20.5 History of life according to the fossil record.
The outer ring of this diagram shows the history of life as it would be measured on a 24-hour timescale starting at midnight. (The inner ring shows the actual years starting at 4.5 billion years ago.) The fossil record suggests that a very large portion of life's history was devoted to the evolution of unicellular organisms. The first multicellular organisms do not appear in the fossil record until just before 8 P.M., and humans are not on the scene until less than a minute before midnight.

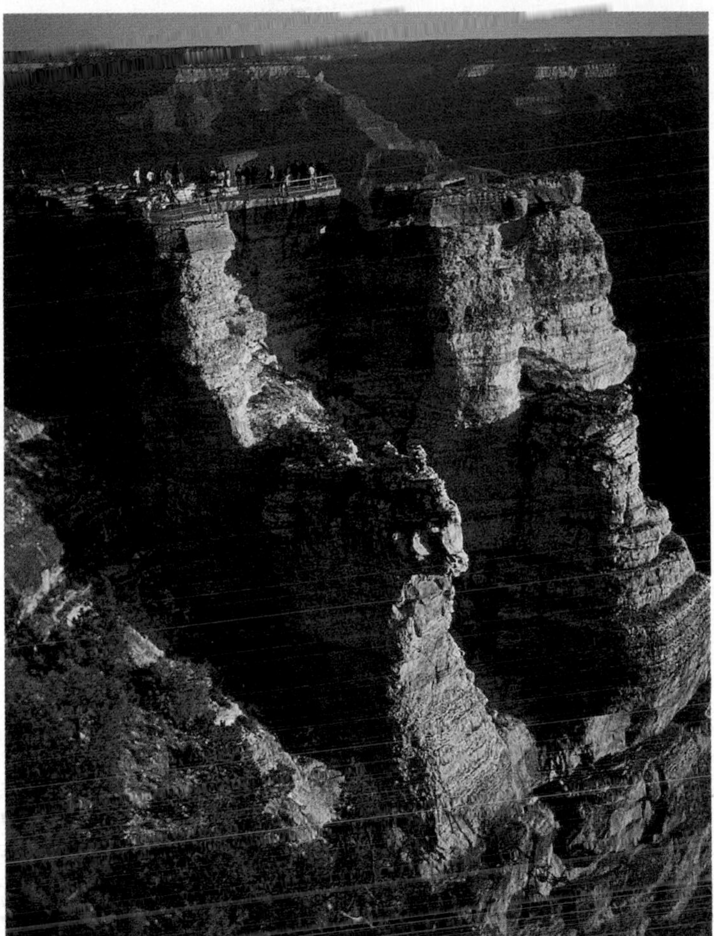

Figure 20.6 The Grand Canyon.
No other place on earth displays such an extensive record of the earth's events for analysis, dating, and study. Even so, it does not have a stratigraphic sequence for the whole of geological time. This was determined by correlating the strata of the canyon to that of other parts of the world.

[L. *sedimentum,* a settling], a process that has been going on since the earth was formed, can take place on land or in bodies of water. Weathering and erosion of rocks produces an accumulation of particles that vary in size and nature and are called sediment. Sediment becomes a *stratum* (pl., strata), a recognizable layer in a stratigraphic sequence (Fig. 20.6). Any given stratum is older than the one above it and younger than the one immediately below it.

The fossils trapped in strata are the fossil record that tells us about the history of life. **Paleontology** [Gk. *palaios,* ancient, old, and *ontos,* having existed; *-logy,* "study of" from *logikos,* rational, sensible] is the science of discovering and studying the fossil record and, from it, making decisions about the history of life. Paleontologists not only want to know the structure and adaptations of an organism, they are also interested in how the organism interacted with others and with the physical environment.

Dating Fossils Relatively

In the early nineteenth century, even before the theory of evolution was formulated, geologists sought to correlate the strata worldwide. The problem was that strata change their character over great distances, and therefore a stratum in England might contain different sediments than one of the same age in Russia. Geologists discovered, however, that a stratum of the same age tend to contain the same fossil, and therefore, fossils could be used for the purpose of *relative dating* of the strata. For example, a particular species of fossil ammonite (an animal related to the chambered nautilus) has been found over a wide range and for a limited time period. Therefore, all strata around the world that contain this fossil must be of the same age.

This approach helped geologists determine the relative dates of the strata despite upheavals, but it was not particularly helpful to biologists who wanted to know the absolute age of fossils in years.

Dating Fossils Absolutely

The absolute dating method that relies on radioactive dating techniques assigns an actual date to a fossil. All radioactive isotopes have a particular half-life, the length of time it takes for half of the radioactive isotope to change into another stable element. If the fossil has organic matter, half of the carbon-14, ^{14}C, will have changed to nitrogen-14, ^{14}N, in 5,730 years. In order to know how much ^{14}C was in the organism to begin with, it is reasoned that organic matter always begins with the same amount of ^{14}C. (In reality, it is known that the ^{14}C levels in the air—and therefore the amount in organisms—can vary from time to time.) Now we need only compare the ^{14}C radioactivity of the fossil to that of a modern sample of organic matter. The amount of radiation left can be converted to the age of the fossil. After 50,000 years, however, the amount of ^{14}C radioactivity is so low it cannot be used to measure the age of a fossil accurately.

^{14}C is the only radioactive isotope contained within organic matter, but it is possible to use others to date rocks and from that infer the age of a fossil contained in the rock. For instance, the ratio of potassium-40 (^{40}K) to argon-40 trapped in rock is often used. If the ratio happens to be 1:1, then half of the ^{40}K has decayed and the rock is 1.3 billion years old. The ratio of isotope uranium-238 to lead-207 can be used only for rocks older than 100 million years. This isotope has such a long half-life that no perceptible decay will have occurred in a shorter length of time.

Fossils, which can be dated relatively according to their location in strata and absolutely according to their content of radioactive isotopes, give us information about the history of life.

How the Story Unfolds

As a result of their study of strata, geologists have divided the history of the earth into eras, then periods and epochs (Table 20.1). We will follow the biologist's tradition of first discussing the Precambrian, a period of time that encompasses the first two eras: the Archean era and the Proterozoic era.

Life Begins in the Precambrian

The Precambrian is a very long period of time—it comprises about 87% of the geologic timescale. It is during this period of time that life arose and the first cells came into existence. The first cells must have been prokaryotes. Prokaryotes do not have a nucleus or any membranous

Table 20.1

The Geological Time Scale: Major Divisions of Geological Time with Some of the Major Evolutionary Events of Each Geological Period

Era	Period	Epoch	Millions of Years Ago	Plant Life	Animal Life
Cenozoic* (from the present to 66.4 million years ago)	Neogene	Holocene	0–0.01	Destruction of tropical rain forests by humans accelerates extinctions	AGE OF HUMAN CIVILIZATION
			Significant Mammalian Extinction		
		Pleistocene	0.01–2	Herbaceous plants spread and diversify	Modern humans appear
		Pliocene	2–6	Herbaceous angiosperms flourish	First hominids appear
		Miocene	6–24	Grasslands spread as forests contract	Apelike mammals and grazing mammals flourish; insects flourish
	Paleogene	Oligocene	24–37	Many modern families of flowering plants evolve	Browsing mammals and monkeylike primates appear
		Eocene	37–58	Subtropical forests with heavy rainfall thrive	All modern orders of mammals are represented
		Paleocene	58–66	Angiosperms diversify	Primitive primates, herbivores, carnivores, and insectivores appear
Mesozoic (from 66.4 to 245 million years ago)		**Mass Extinction: Dinosaurs and Most Reptiles**			
	Cretaceous		66–144	Flowering plants spread; coniferous trees decline	Placental mammals appear; modern insect groups appear
	Jurassic		144–208	Cycads and other gymnosperms flourish	Dinosaurs flourish; birds appear
		Mass Extinction			
	Triassic		208–245	Cycads and ginkgoes appear; forests of gymnosperms and ferns dominate	First mammals appear; first dinosaurs appear; corals and mollusks dominate seas
Paleozoic (from 245 to 570 million years ago)		**Mass Extinction**			
	Permian		245–286	Conifers appear	Reptiles diversify; amphibians decline
	Carboniferous		286–360	Age of great coal-forming forests: club mosses, horsetails, and ferns flourish	Amphibians diversify; first reptiles appear; first great radiation of insects
		Mass Extinction			
	Devonian		360–408	First seed ferns appear	Jawed fishes diversify and dominate the seas; first insects and first amphibians appear
	Silurian		408–438	Low-lying vascular plants appear on land	First jawed fishes appear
		Mass Extinction			
	Ordovician		438–505	Marine algae flourish	Invertebrates spread and diversify; jawless fishes, first vertebrates, appear
	Cambrian		505–570	Marine algae flourish	Invertebrates with skeletons are dominant
Precambrian time (from 570 to 4,600 million years ago)			700 2,100 3,100–3,500 4,600	Multicellular organisms appear First complex (eukaryotic) cells appear First prokaryotic cells in stromatolites appear Earth forms	

Many authorities divide the Cenozoic era into the Tertiary period (contains Paleocene, Eocene, Oligocene, Miocene, and Pliocene) and the Quaternary period (contains Pleistocene and Holocene).

organelles. Of the living prokaryotes today there is a type called archaea that live in the most inhospitable of environments such as hot springs, very salty lakes, and airless swamps—all of which may typify habitats on the primitive earth. The cell wall, plasma membrane, RNA polymerase, and ribosomes of archaea are more like those of eukaryotes than those of other bacteria.

As discussed, the first cell or cells must have been anaerobic heterotrophs (there was no oxygen in the primitive atmosphere). Also, the first photosynthesizers most likely did not give off oxygen. By 2 billion years ago (BYA), however, oxygen-releasing photosynthesis began. Some of the earliest cells, dated about 3.5 BYA, are found in fossilized stromatolites, which are pillarlike structures composed of sedimentary layers containing communities of prokaryotic organisms (Fig. 20.7). Stromatolites containing cyanobacteria, which carry on oxygen-releasing photosynthesis, exist even today in shallow waters off the west coast of Australia.

Due to the action of photosynthesizers, the atmosphere became an oxidizing one instead of a reducing one. Oxygen in the upper atmosphere forms ozone (O_3), which filters out the ultraviolet (UV) rays of the sun. Before the formation of the **ozone shield,** the amount of ultraviolet radiation reaching the earth could have helped create organic molecules, but it would have destroyed any land-dwelling organisms. Once the ozone shield was in place, it meant that living things would be sufficiently protected and would be able to live on land. Life on land is threatened if the ozone shield is reduced. This is why there is such concern today about pollutants that act to break down the ozone shield.

The evolution of photosynthesizing organisms caused oxygen to enter the atmosphere.

The presence of oxygen in the atmosphere meant that most environments were no longer suitable for anaerobic prokaryotes, and they began to decline in importance. Photosynthetic cyanobacteria and aerobic bacteria proliferated as new metabolic pathways evolved.

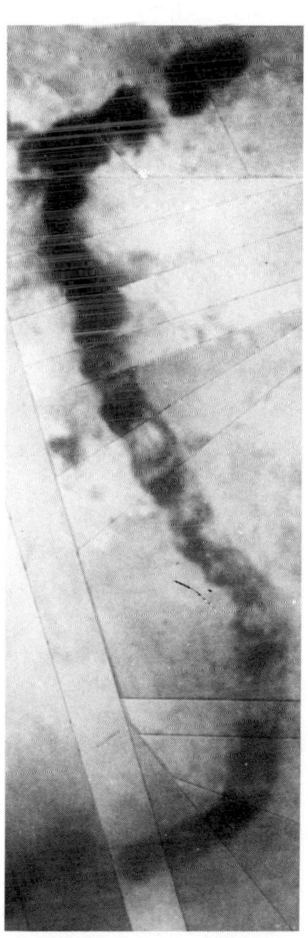

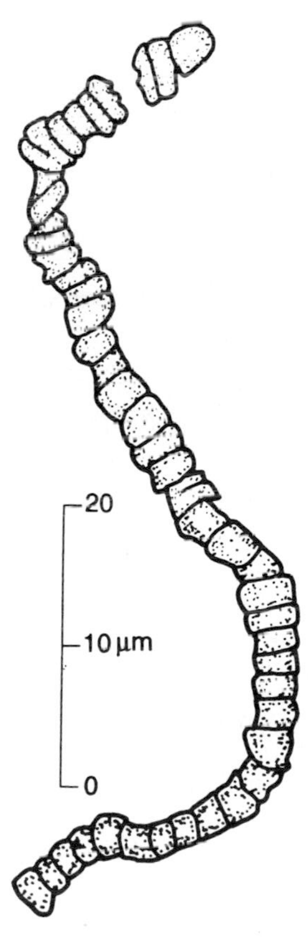

a.

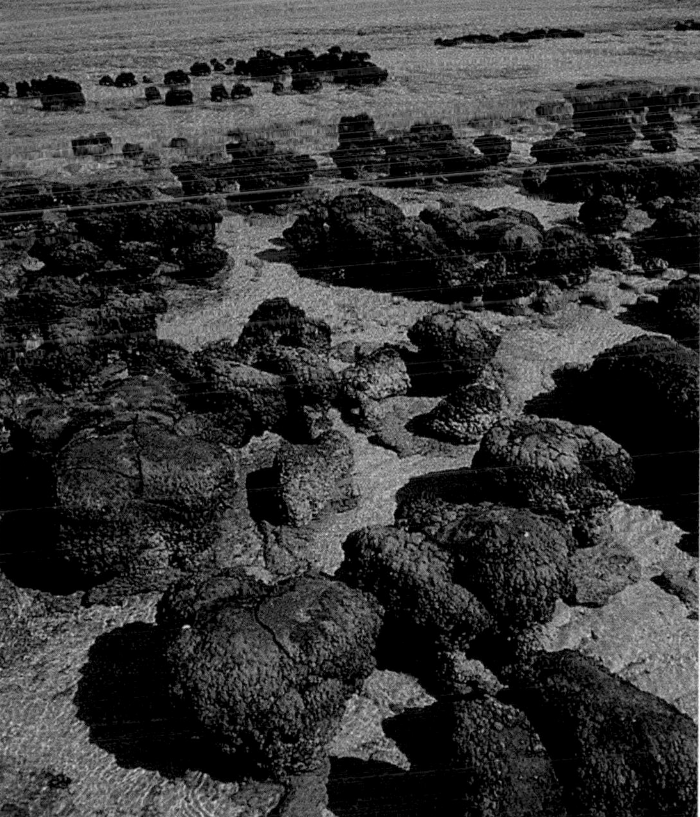

b.

Figure 20.7 Prokaryote fossil of the Precambrian.
a. The prokaryotic microorganism, *Primaevifilum* (with interpretive drawing), was found in a fossilized stromatolite dated about 3.5 billion years ago. **b.** Living stromatolites are located in shallow waters off the shores of western Australia and also in other tropical seas.

Eukaryotic Cells Arise

The eukaryotic cell, which originated about 2.1 BYA, is nearly always aerobic and contains a nucleus as well as other membranous organelles. Most likely the eukaryotic cell acquired its organelles gradually. It may be that the nucleus developed by an invagination of the plasma membrane. The mitochondria of the eukaryotic cell probably were once free-living aerobic prokaryotes, and the chloroplasts probably were free-living photosynthetic prokaryotes. The theory of endosymbiosis says that the nucleated cell engulfed these prokaryotes, which then became organelles. It's been suggested that flagella (and cilia) also arose by endosymbiosis. First, slender undulating prokaryotes could have attached themselves to a host cell in order to take advantage of food leaking from the host's outer membrane. Eventually, these prokaryotes were drawn inside the host cell and became the flagella and cilia we know today. The first eukaryotes were unicellular, as are prokaryotes.

Only unicellular organisms existed on earth during most of its history. During this time many biochemical pathways evolved.

It is not known when multicellularity began, but the very first multicellular forms were most likely microscopic. It's possible that the first multicellular organisms practiced sexual reproduction. Among protists (eukaryotes classified in the kingdom Protista) today, we find colonial forms in which some cells are specialized to produce gametes needed for sexual reproduction. Separation of germ cells, which produce gametes from somatic cells, may have been an important first step toward the development of complex macroscopic animals that appeared about 600 million years ago (MYA). In 1947, fossils of soft-bodied invertebrates of this date were found in the Ediacara Hills in South Australia. Since then, similar fossils have been discovered on a number of other continents. They represent a community of animals that most likely lived on mudflats in shallow marine waters (Fig. 20.8). Many biologists interpret the fossils as being like jellyfish, sea pens (relatives of corals), and segmented worms. Other fossils seem unrelated to the types of animals alive today.

Diverse, complex organisms appear in the fossil record about 600 MYA, perhaps as a result of the evolution of multicellularity in association with sexual reproduction.

a.

b.

Figure 20.8 Sea life of the late Precambrian.
a. Artist's representation of a community of animals, based on Ediacaran fossils. The large frondlike organisms are interpreted here as soft corals, known today as sea pens. Silvery jellyfish swim about, and an elongate, wormlike creature is on the seafloor.
b. *Dickinsonia costata*, a fossil that is interpreted to be a segmented worm.

Complexity Increases in Paleozoic

The Paleozoic era lasted over 300 million years. Many events occurred during this time, which is quite short compared to the length of the Precambrian (see Table 20.1). For one thing, there were three major mass extinctions.

Cambrian Fossils Figure 20.9 shows that the seas teemed with invertebrate life during the Cambrian period. *Invertebrates* are animals without a vertebral column. All of today's groups of animals can trace their ancestry to this time and perhaps earlier according to new molecular clock data (see p. 353 for an explanation of the molecular clock). Fossils, however, are more prevalent during the Cambrian than before. Why are fossils easier to find at this time? Because the animals had protective outer skeletons, and skeletons are capable of surviving the forces that are apt to destroy fossils. For example, Cambrian seafloors were dominated by now-extinct trilobites, which had thick, jointed armor covering them from head to tail. Trilobites are classified as arthropods, a major phylum of animals today. (Some Cambrian species, with most unusual eating and locomotion appendages, have been classified in phyla that no longer exist today.).

Paleontologists have sought an explanation for why animals had skeletons during the Cambrian period but not before. By this time, not only cyanobacteria but also various algae, which are floating photosynthetic organisms, were pumping oxygen into the atmosphere. Perhaps the oxygen supply became great enough to permit aquatic animals to acquire oxygen even though they had outer skeletons. The presence of a skeleton cuts down on possible access to oxygen in seawater. Steven Stanley of Johns Hopkins University suggests that predation may have played a role. Skeletons may have evolved during the Cambrian period because skeletons help protect animals from predators.

The fossil record is rich during the Cambrian period but the animals may have evolved earlier. The richness of the Cambrian period may be due to the evolution of outer skeletons.

Figure 20.9 Sea life of the Cambrian period.
The animals depicted here are found as fossils in the Burgess Shale, a formation of the Rocky Mountains of British Columbia. Some lineages represented by these animals are still evolving today; others have become extinct.

Figure 20.10 Swamp forests of the Carboniferous period.
Vast swamp forests of treelike club mosses and horsetails dominated the land during the Carboniferous period (see Table 20.1). The air contained insects with wide wingspans, such as the dragonfly shown here, and amphibians lumbered from pool to pool.

Later, Land Is Invaded Sometime during the Paleozoic era, algae, which were common in the seas, most likely began to take up residence in bodies of fresh water; from there, they may have invaded damp areas on land. An association of plant roots with fungi called mycorrhizae is credited with allowing plants to live on bare rocks. The fungi are able to absorb minerals, which they pass to the plant, and the plant in turn passes carbohydrates, the product of photosynthesis, to the fungi.

The most prevalent land plants have vascular tissue for water transport, but there are no fossils of vascular plants until the Silurian period. The first vascular plants flourished in the warm swamps of the *Carboniferous period* (Fig. 20.10). Club mosses, horsetails, and seed ferns were the trees of that time, and they grew to enormous size. A wide variety of smaller ferns and fernlike plants formed an underbrush.

The first animals to live on land were scorpions, carnivorous relatives of spiders with a poisonous stinger at the end of their tail and two large pincers on their front legs. Insects enter the fossil record in the Carboniferous period. The outer skeleton and jointed appendages of insects were suitable to a land existence. An outer skeleton prevents drying out, and jointed appendages provide a suitable means of locomotion on land. These traits, plus the evolution of wings, provided advantages that allowed insects to radiate into the most diverse and abundant group of animals today. Flying provides a way to escape enemies and find food.

Vertebrates are animals with a vertebral column. The vertebrate line of descent began in the early Ordovician period with the evolution of fishes. First there were jawless fishes and then fishes with jaws. *Fishes* are ectothermic (cold-blooded), aquatic vertebrates that have gills, scales, and fins. The ray-finned fishes, which include today's most familiar fish, make their appearance in the Devonian period, which is called the Age of Fishes. Amphibians are more directly related to lobed-finned fishes, which may have ventured onto shore to avoid predators. The only vertebrates in the wet Carboniferous forests were *amphibians*, thin-skinned vertebrates that are not fully adapted to life on land, particularly because they must return to water to reproduce. The swamp forests provided the water they needed and amphibians radiated into many different sizes and shapes. Some superficially resembled alligators and were covered with protective scales, others were small and snakelike, and a few were larger plant eaters. The largest measured 6 meters (20 feet) from snout to tail. The Carboniferous period is called the Age of the Amphibians.

There was a change of climate at the end of the Carboniferous period; cold and dry weather brought an end to the Age of the Amphibians and began the process that turned the great Carboniferous forests into the coal we use today to fuel our modern society.

Primitive vascular plants and amphibians were larger and more abundant during the Carboniferous period. Insects appeared and flourished to become the largest animal group today.

Figure 20.11 Dinosaurs of the Mesozoic.
a. The dinosaurs of the Jurassic period included *Apatosaurus*, which fed on cycads and conifers. b. *Tyrannosaurus rex* was a dinosaur of the Cretaceous period, when flowering plants were increasing in dominance.

a.

b.

Dinosaurs Rule in the Mesozoic

Although there was a mass extinction at the end of the Paleozoic era, the evolution of certain types of plants and animals continued into the Triassic, the first period of the Mesozoic era. Nonflowering seed plants (collectively called *gymnosperms*), which had been present in the fossil record of the Permian period, became dominant. Among these largely cone-bearing plants were cycads and conifers. Cycads are short and stout with palmlike leaves; the female plant produces very large cones. Cycads were so prevalent during the Jurassic period that it is sometimes called the Age of the Cycads. By the Cretaceous period of the Mesozoic era, flowering plants (collectively called *angiosperms*) had begun to radiate, and cone-bearing plants declined in importance.

Reptiles, too, can be traced back to the Permian period of the Paleozoic era. Unlike amphibians, *reptiles* can thrive in a dry climate because they have scaly skin and lay a shelled egg that hatches on land. Reptiles underwent an adaptive radiation during the Mesozoic to produce forms that lived in the air, in the sea, and on the land. At the beginning of the Mesozoic era, mammal-like reptiles called therapsids were prevalent. *Therapsids* had vertically positioned limbs that held the body off the ground. During the Jurassic period, large flying reptiles called *pterosaurs* ruled the air and giant marine reptiles with paddlelike limbs ate fishes in the sea, but on land it was the *dinosaurs* (Fig. 20.11) that prevented the evolving mammals from taking center stage.

The dinosaurs were reptiles, some of an enormous size. During the Jurassic period, the gargantuan *Apatosaurus* and the armored tractor sized *Stegosaurus* fed on cycad seeds and conifer trees. During the Cretaceous period, great herds of rhinolike *Triceratops* roamed the plains, as did the infamous *Tyrannosaurus rex*, which was carnivorous and played the same ecological role as lions do today. The size of a dinosaur such as *Apatosaurus* is hard for us to imagine. It was as tall as a four-story office building, and its weight was as much as that of a thousand people! How might dinosaurs have benefited from being so large? One theory is that being ectothermic, the volume-to-surface ratio was favorable for retaining heat. Others present evidence, based largely on bone construction, that at least some of the dinosaurs were endothermic (warm-blooded). Newly found evidence, as discussed in the reading on page 336, suggests that dinosaurs should not be classified as reptiles at all!

During the Jurassic period, one group of dinosaurs called theropods were bipedal and had an elongate, mobile S-shaped neck. The fossil *Archaeopteryx* is a transitional link, which shows that theropods are related to birds. Despite this animal's feathers, its jaw bore teeth, and it had a long lizardlike tail. The fossil record for birds begins at this time and continues on into the next era.

By the end of the Cretaceous period, the dinosaurs were extinct. They were victims of a mass extinction brought about by causes that are still being debated.

During the Mesozoic era, the dinosaurs achieved an enormous size, while mammals remained small and insignificant. Flowering plants replaced cone-bearing plants as the dominant vegetation.

Doing Science

Real Dinosaurs, Stand Up!

Today's paleontologists are setting the record straight about dinosaurs. Because dinosaurs are classified as reptiles, it is assumed that they must have had the characteristics of today's reptiles. They must have been ectothermic, slow moving, and antisocial, right? Wrong!

First of all, not all dinosaurs were great lumbering beasts. Many dinosaurs were less than 1 meter (3 feet) long and their tracks indicate they moved right along. These dinosaurs stood on two legs that were positioned directly under the body. Perhaps they were as agile as ostriches, which are famous for their great speed.

Dinosaurs may have been endothermic. Could they have competed successfully with the preevolving mammals otherwise? They must have been able to hunt prey and escape from predators as well as mammals, which are known to be active because of their high rate of metabolism. Some argue that ectothermic animals have little endurance and cannot keep up. They also believe that the bone structure of dinosaurs also indicates they were endothermic.

Dinosaurs cared for their young much like birds do today. In Montana, paleontologist Jack Horner has studied fossilized nests complete with eggs, embryos, and nestlings (Fig. 20A). The nests are about 7.5 meters (24.6 feet) apart, the space needed for the length of an adult parent. About 20 eggs are laid in neatly arranged circles and may have been covered with decaying vegetation to keep them warm. Many contain the bones of juveniles as much as a meter long. It would seem then that baby dinosaurs remained in the nest to be fed by their parents. They must have obtained this size within a relatively short period of time, again indicating that dinosaurs were endothermic. Ectothermic animals grow slowly and take a long time to reach this size.

Dinosaurs were also social! An enormous herd of dinosaurs found by Horner and colleagues is estimated to have nearly 30 million bones, representing 10,000 animals in one area measuring about 1.6 square miles. Most likely, the herd kept on the move in order to be assured of an adequate food supply, which consisted of flowering plants that could be stripped one season and grow back the next season. The fossilized herd is covered by volcanic ash, suggesting that the dinosaurs died following a volcanic eruption.

Some dinosaurs, such as the duck-billed dinosaurs and horned dinosaurs, have a skull crest. How might it have functioned? Perhaps it was a resonating chamber, used when dinosaurs communicated with one another. Or, as with modern horned animals that live in large groups, the males could have used the skull crest in combat to establish dominance.

If dinosaurs were endothermic, fast-moving, and social animals, should they be classified as reptiles? Some say no!

a.

b.

Figure 20A Behavior of dinosaurs.
a. Nest of fossil dinosaur eggs found in Montana, dating from the Cretaceous period.
b. Bones of a hatchling (about 50 cm [20 inches]) found in the nest. These dinosaurs have been named *Maiosaura,* which means "good mother lizard" in Greek.

Figure 20.12 Mammals of the Oligocene epoch.
The artist's representation of these mammals and their habitat vegetation is based on fossil remains.

Mammals Take Over in the Cenozoic

According to a new system, the Cenozoic era is divided into a Paleogene period and a Neogene period. We are living in the Neogene period.

Mammals Diversify in the Paleogene At the end of the Mesozoic era, mammals began an adaptive radiation into the many habitats now left vacant by the demise of the dinosaurs. Mammals are endothermic and they have hair, which helps keep body heat from escaping. Their name refers to the presence of mammary glands, which produce milk to feed their young. At the start of the Paleocene epoch, mammals were small and resembled a mouse. By the end of the Eocene epoch, mammals had diversified to the point that most of the modern orders were in existence. Bats are mammals that have conquered the air. Whales, dolphins, and other marine mammals live in the sea where vertebrates began their evolution in the first place. Hoofed mammals populate forests and grasslands and are fed upon by diverse carnivores. Many of the types of herbivores and carnivores of the late Paleogene period, however, are extinct today (Fig. 20.12).

Primates Evolve in the Neogene Primates are a type of mammal adapted to living in flowering trees where there is protection from predators and where food in the form of fruit is plentiful. The first primates were small squirrel-like animals, but from them evolved the first monkeys and then apes. During the Miocene epoch, weather changes caused African forests to be replaced by grasslands. It was then that hominids began to walk on two legs and the human line of descent began.

The world's climate was progressively cooler during the Neogene period, so much so that the latter two epochs are known as the Ice Age. During periods of glaciation, snow and ice covered about one-third of the land surface of the earth. The Pleistocene epoch was an age of not only humans, but also giant ground sloths, beavers, wolves, bison, woolly rhinoceroses, mastodons, and mammoths (Fig. 20.13). Humans have survived, but what happened to the oversized mammals just mentioned? Some think that humans became such skilled hunters they are at least partially responsible for the extinction of these awe-inspiring animals.

Figure 20.13 Woolly mammoth of the Pleistocene epoch.
Woolly mammoths, *Mammuthus*, were magnificent animals, which lived along the borders of continental glaciers.

The Cenozoic era is the present era. Only during this time did mammals diversify and human evolution begin.

20.3 Factors That Influence Evolution

It used to be thought that the earth's crust was immobile, that the continents had always been in their present positions, and that the ocean floors were only a catch basin for the debris that washed off the land. In 1920 Alfred Wegener, a German meteorologist, presented data from a number of disciplines to support his hypothesis of **continental drift.** This hypothesis, which was finally confirmed in the 1960s, states that the continents are not fixed; instead, their positions and the position of the oceans have changed over time (Fig. 20.14). About 225 million years ago (MYA), the continents joined to form one supercontinent that Wegener called Pangaea. First, Pangaea divided into two large subcontinents, called Gondwanaland and Laurasia, and then these also split to form the continents of today. Presently, the continents are still drifting in relation to one another.

Continental drift explains why the coastlines of several continents are mirror images of each other—the outline of the west coast of Africa matches that of the east coast of South America. The same geological structures are also found in many of the areas where the continents touched. A single mountain range runs through South America, Antarctica, and Australia. Continental drift also explains the unique distribution patterns of several fossils. Fossils of the same species of seed fern *(Glossopteris)* have been found on all the southern continents. No suitable explanation was possible previously, but now it seems plausible that the plant evolved on one continent and spread to the others when they were joined as one. Similarly, the fossil reptile *Cynognathus* is found in Africa and South America and *Lystrosaurus,* a mammal-like reptile, has now been found in Antarctica, far from Africa and southeast Asia, where it is also found. With mammalian fossils, the situation is different: Australia, South America, and Africa all have their own distinctive mammals because mammals evolved after the continents separated. The mammalian biological diversity of today's world is the result of isolated evolution on separate continents.

The relationship of the continents to one another has affected the biogeography of the earth.

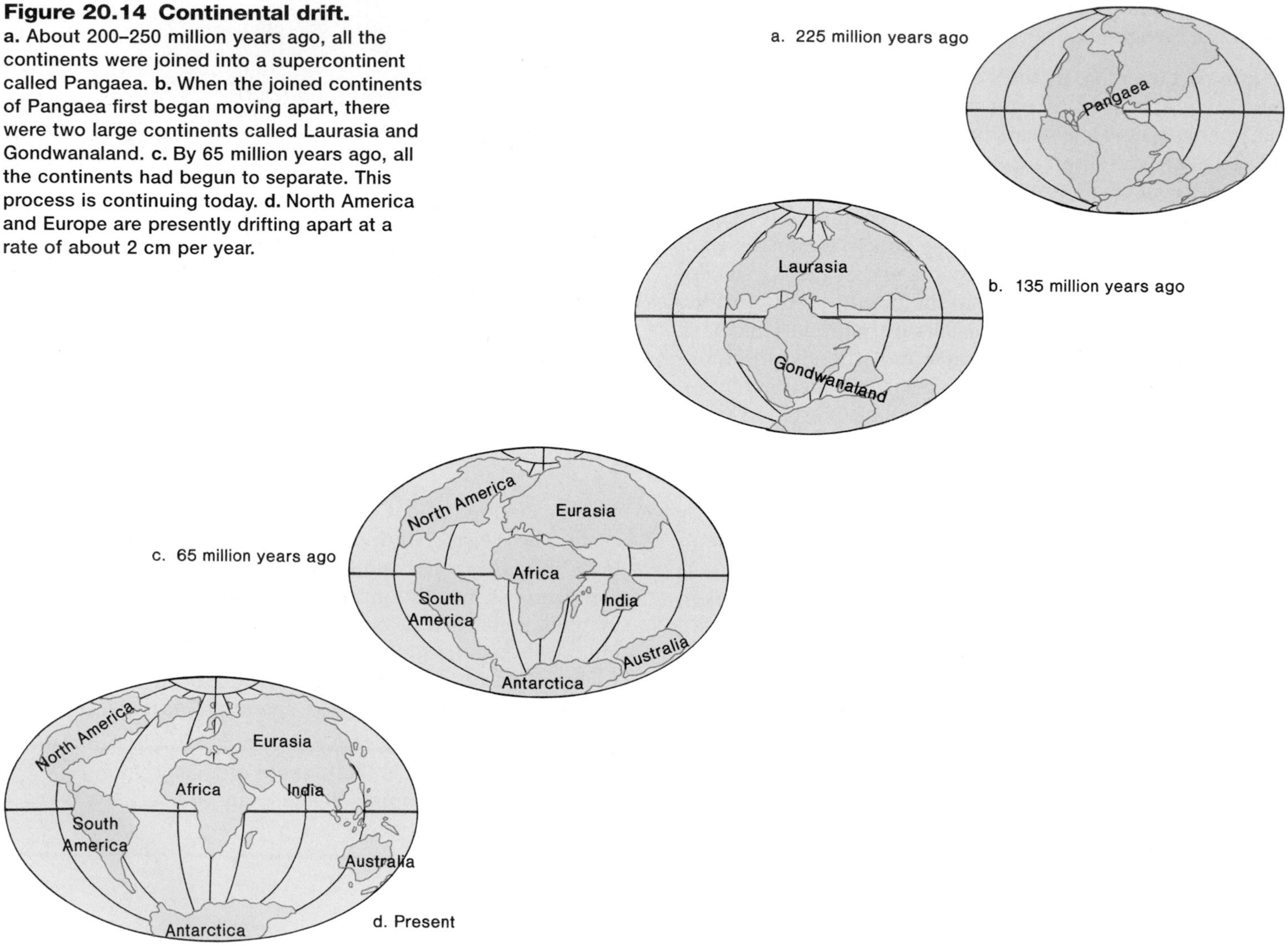

Figure 20.14 Continental drift.
a. About 200–250 million years ago, all the continents were joined into a supercontinent called Pangaea. **b.** When the joined continents of Pangaea first began moving apart, there were two large continents called Laurasia and Gondwanaland. **c.** By 65 million years ago, all the continents had begun to separate. This process is continuing today. **d.** North America and Europe are presently drifting apart at a rate of about 2 cm per year.

a. 225 million years ago

b. 135 million years ago

c. 65 million years ago

d. Present

Why do the continents drift? According to a branch of geology known as **plate tectonics** [Gk. *tektos,* fluid, molten, able to flow] (tectonics refers to movements of the earth's crust), the earth's crust is fragmented into slablike plates that float on a lower hot mantle layer. The continents and the ocean basins are a part of these rigid plates, which move like conveyor belts (Fig. 20.15). At **ocean ridges,** seafloor spreading occurs as molten mantle rock rises and material is added to the ocean floor. Seafloor spreading causes the continents to move a few centimeters a year on the average. At *subduction zones,* the forward edge of a moving plate sinks into the mantle and is destroyed. When an ocean floor is at the leading edge of a plate, a deep trench forms that is bordered by volcanoes or volcanic island chains. When two continents collide, the result is often a mountain range; for example, the Himalayas resulted when India collided with Eurasia. Two plates meet along a *transform boundary* where two plates scrape past one another. The San Andreas *fault* in southern California is at a transform boundary, and the movement of the two plates is responsible for the many earthquakes in that region.

The earth's crust is divided into plates that move because of seafloor spreading at ocean ridges.

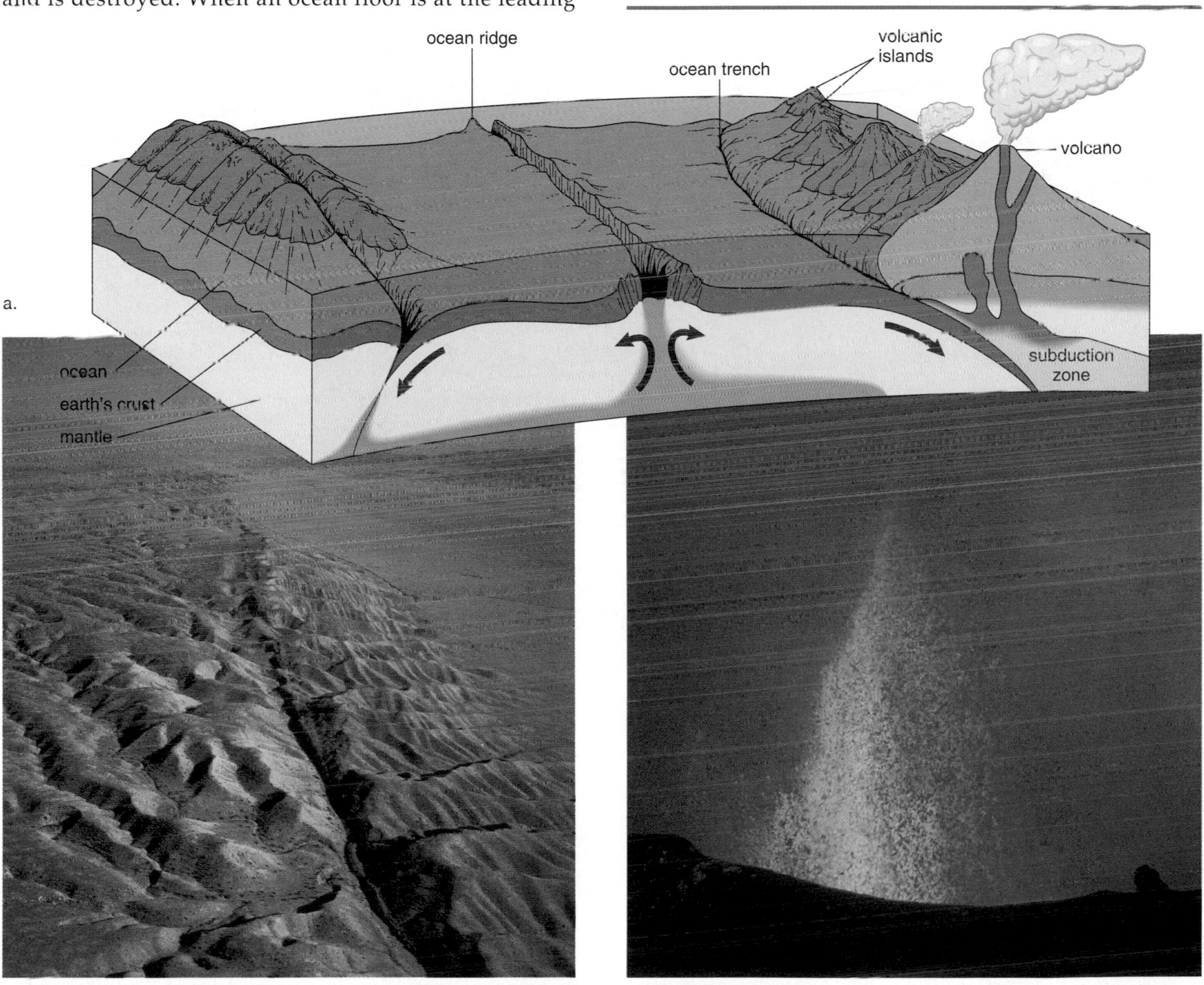

Figure 20.15 Plate tectonics.
a. Plates form and move away from ocean ridges toward subduction zones, where they are carried into the mantle and are destroyed. **b.** A transform boundary occurs where two plates scrape past each other. The San Andreas fault occurs at a transform boundary where earthquakes are apt to occur. **c.** Iceland is one of the few places in the world where an ocean ridge reaches the surface of the sea. The entire island is volcanic in origin.

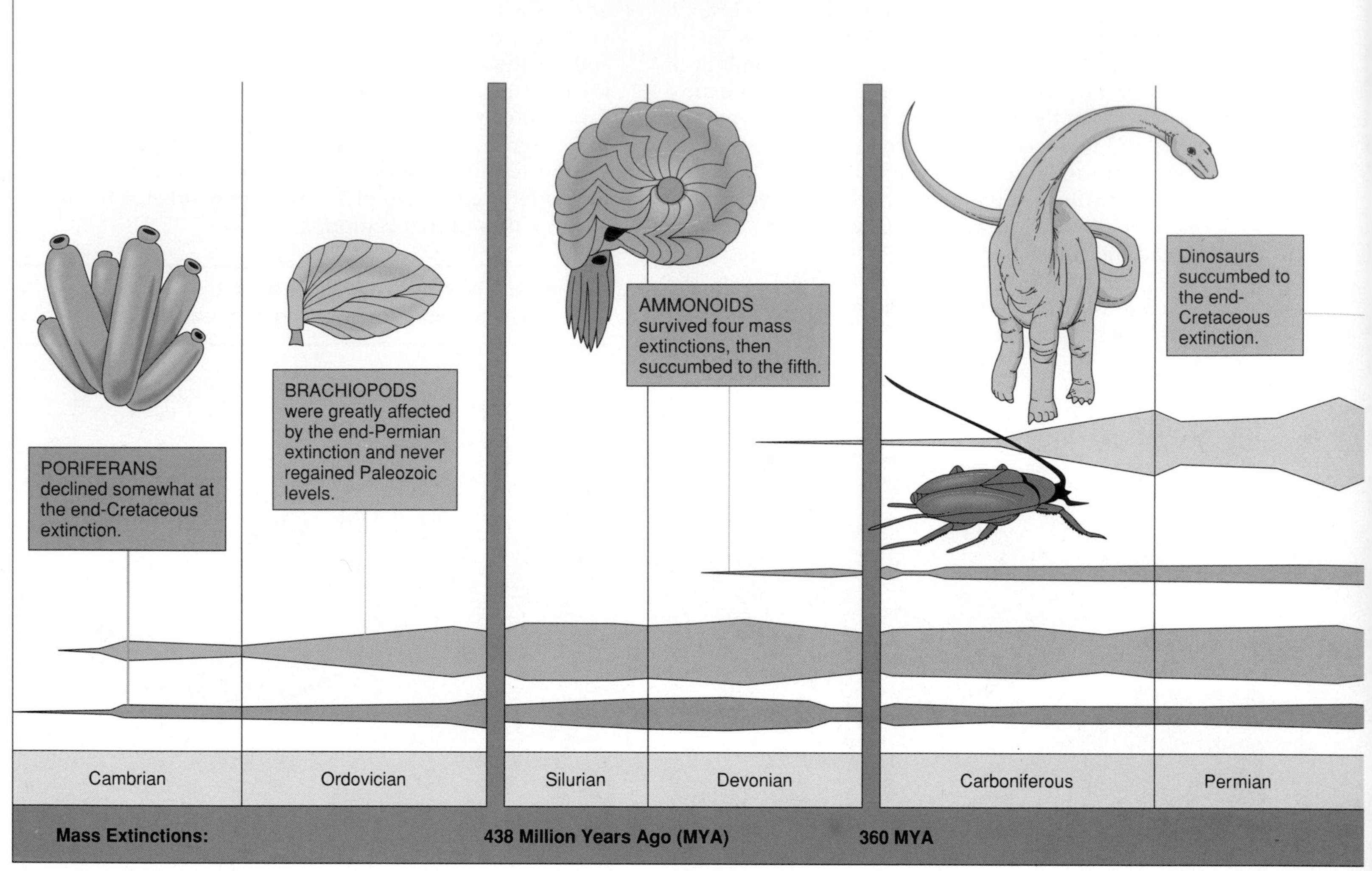

Figure 20.16 Mass extinctions.
Five significant mass extinctions and their effects on the abundance of certain forms of marine and terrestrial life. The width of the horizontal bars indicates the varying abundance of each life-form considered.
Source: Data supplied and illustration reviewed by J. John Sepkoski, Jr., Professor of Paleontology, University of Chicago.

Exploring Mass Extinctions

An **extinction** is the total disappearance of all the members of a species or a higher taxonomic group. A **mass extinction** is the disappearance of a larger number of species or higher taxonomic groups within an interval of just a few million years. There have been at least five mass extinctions throughout history: at the ends of the Ordovician, Devonian, Permian, Triassic, and Cretaceous periods (Fig. 20.16). Mass extinctions are usually followed by at least partial evolutionary recovery in which the remaining groups of organisms undergo adaptive radiations and fill the habitats vacated by those that have become extinct.

Is a mass extinction due to some cataclysmic event, or is it a more gradual process brought on by environmental changes including tectonic, oceanic, and climatic fluctuations? This question was brought to the fore when Walter and Luis Alvarez proposed in 1977 that the Cretaceous extinction was due to a bolide. A *bolide* is an asteroid that explodes, producing meteorites that fall to earth. They found that Cretaceous clay contains an abnormally high level of iridium, an element that is rare in the earth's crust but more common in asteroids (or their fragments, meteorites). The result of a large meteorite striking the earth could have been similar to that from a worldwide atomic bomb explosion: a cloud of dust would have mushroomed into the atmosphere, blocking out the sun and causing plants to freeze and die. Recently, a layer of soot has also been identified in the strata alongside the iridium, and a huge crater that could have been caused by a meteorite was found a few years ago in the Caribbean–Gulf of Mexico region on the Yucatán peninsula.

In 1984, paleontologists David Raup and John Sepkoski suggested that the fossil record of marine animals shows that mass extinctions have occurred every 26 million years and, surprisingly, astronomers can offer an explanation. Our solar system is in a starry galaxy known as the Milky Way. Because of the vertical movement of our sun, our solar system approaches other members of the Milky Way every 26 to 33 million years, producing an unstable situation that could lead to the occurrence of a bolide. Perhaps some mass extinctions are associated with extraterrestrial events.

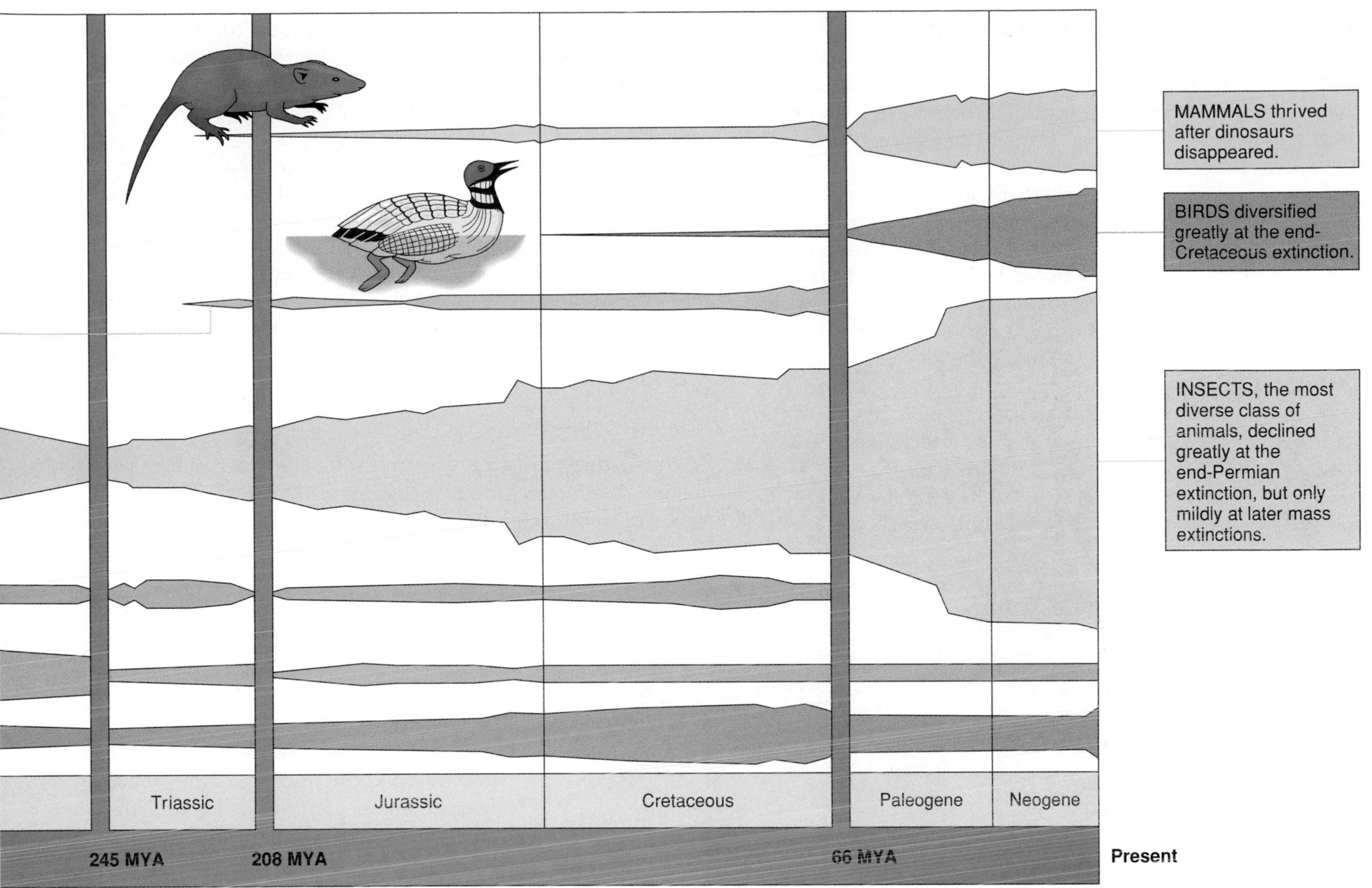
MAMMALS thrived after dinosaurs disappeared.
BIRDS diversified greatly at the end-Cretaceous extinction.
INSECTS, the most diverse class of animals, declined greatly at the end-Permian extinction, but only mildly at later mass extinctions.
Triassic
Jurassic
Cretaceous
Paleogene
Neogene
245 MYA
208 MYA
66 MYA
Present

Certainly continental drift contributed to the Ordovician extinction. This extinction occurred after Gondwanaland arrived at the south pole. Immense glaciers, which drew water from the oceans, chilled even the once-tropical land. Marine invertebrates and coral reefs, which were especially hard hit, didn't recover until Gondwanaland drifted away from the pole and warmth returned. The mass extinction at the end of the Devonian saw an end to 70% of marine invertebrates. Helmont Geldsetzer of Canada's Geological Survey notes that iridium has also been found in Devonian rocks in Australia, suggesting that a bolide event was also involved in the Devonian extinction. Other scientists believe that this mass extinction could have occurred when Gondwanaland headed back over the south pole.

At the end of the Permian period, all land masses joined to form the supercontinent Pangaea. The amount of shallow offshore water shrank dramatically and marine life was greatly affected. The trilobites became extinct and the crinoids (sea lilies) barely hung on. Glaciers originating at either pole poured periodically over the land, and the inland weather grew drier and colder. The swamp forests that had originated in the Carboniferous period and their amphibian inhabitants were particularly affected. The Permian extinction is the worst by far, with nearly 96% of species disappearing.

The extinction at the end of the Triassic period is another that has been attributed to the environmental effects of a meteorite collision with earth. Central Quebec has a crater half the size of Connecticut that some believe is the impact site. The dinosaurs may have benefited from this event because this is the time when the first of the gigantic dinosaurs took charge of the land. The second wave occurred in the Cretaceous period.

Whether a bolide caused or contributed to the Cretaceous extinction is still being investigated. This extinction—which doomed the dinosaurs, pterosaurs, ammonoids, and over 75% of the known species of marine plankton—did not happen overnight. Terrestrial events such as tectonic, oceanic, and climatic fluctuations contributed to sporadic periods of extinction over a period of 0.5 to 5 million years. It could be that an extraterrestrial event dealt the final blow.

Microevolution Versus Macroevolution

As we discussed in the previous chapter, changing gene frequencies in local populations may be considered as *microevolution.* Traditional evolutionists believe that the same processes have been involved in major transformations over geological time, such as those observed in **lineages,** lines of descent from a common ancestor (called *macroevolution*).

The lineage of the horse *Equus* begins with *Hyracotherium,* a dog-sized mammal with a small head and small low-crowned molars with cusps. The feet, with four toes on each forefoot and three toes on each hind foot, were padded. *Hyracotherium* was adapted to the forestlike environment of the Eocene epoch during the Paleogene period. This small animal could have hidden among the trees for protection, and the low-crowned teeth were appropriate for browsing on leaves. In the Miocene epoch, when grasslands began to replace forests, the ancestors of *Equus* were subjected to selective pressure for the development of strength, intelligence, speed, and durable grinding teeth. A large size provided the strength needed for combat, a large skull made room for a larger brain, elongated legs ending in a hoof (derived from a single toe) gave speed to escape enemies, and the durable grinding teeth enabled the animals to feed efficiently on grasses.

Does this history show overall trends such as an increase in size, an increase in the grinding surface of the molar teeth, and a reduction in the number of toes? It does only if we pick and choose among the many fossils available. Closer examination reveals that as *Hyracotherium* evolved into *Equus,* the evolution of every character varied greatly, and there were even times of reversal. If one of the ancestral animals had lived on and *Equus* had become extinct, we no doubt would be discussing a different set of "trends."

Progress Versus Stasis

In the history of life there are a few fossils that can be used to link one major group of organisms to another. But the expression "missing link" points to the fact that few such *transitional links* have been found. *Archaeopteryx,* the most famous of these fossils, links reptiles to birds. Other transitional links include the amphibious fish *Eustheopteron,* the reptile-like amphibian *Seymouria,* and the mammal-like reptiles, or therapsids. To explain the scarcity of transitional links, some paleontologists mention the slim chance of organisms becoming fossils and the subsequent incompleteness of the fossil record.

There are also examples of organisms that are called *living fossils* because they are so similar to an ancestor known from the fossil record. Recently, investigators found exquisitely preserved specimens of cyanobacteria that have the same sizes, shapes, and organization as living forms. These findings suggest that the cyanobacteria of today have not changed at all in over 3 billion years. Among plants, the dawn redwood was thought to be extinct; then a living specimen was discovered in a small area of China. Horseshoe crabs, crocodiles, and coelacanth fish are animals that still resemble their earliest ancestors. Pangolins are also called scaly anteaters. A recently found pangolin fossil from the Eocene epoch shows that these animals have changed minimally in 60 million years (Fig. 20.17). A time of limited evolutionary change in a lineage is called *stasis.*

Figure 20.17 Evolutionary stasis.
An Eocene fossil of a pangolin, *Manis,* found in a mine at Messel in Germany, is the oldest fossil known for this group of animals. It shows that pangolins have changed little since the time they evolved.

Phyletic Gradualism Versus Punctuated Equilibrium

Evolutionists who support **phyletic gradualism** [Gk. *phyle*, tribe] suggest that evolutionary change is rather slow and steady within a lineage after a divergence and is not necessarily dependent upon speciation—that is, the origination of a new species (Fig. 20.18*a*). In other words, fossils of the same species designation can show a trend over time, say from shorter to longer leg length. Indeed the fossil record, even if complete, is unlikely to indicate when speciation has occurred. A species is defined on the basis of reproductive isolation, and reproductive isolation cannot be detected in the fossil record! Since evolution occurs gradually, the expectation is that more transitional links will eventually be found in the fossil record.

Contrary to this evolutionary model, certain paleontologists—Stephen J. Gould, Nile Eldredge, and Steven Stanley in particular—propose that the fossil record demonstrates a model they call **punctuated equilibrium** (Fig. 20.18*b*). Stasis, a period of equilibrium, is due to the failure of a lineage to convert variation into the origination of new species. Living fossils indicate that stasis can persist even for millions of years. In most lineages, however, a period of equilibrium is punctuated by evolutionary change—that is, speciation occurs. With reference to the length of the fossil record (about 3.5 billion years), speciation occurs relatively rapidly. Therefore, transitional links are not likely to become fossils, nor to be found! Indeed, speciation most likely involves only an isolated population at one locale. When a new species evolves and displaces the existing species, then it is likely to show up in the fossil record.

Aside from finding fossils, paleontologists try to determine lineages and develop models of evolutionary change.

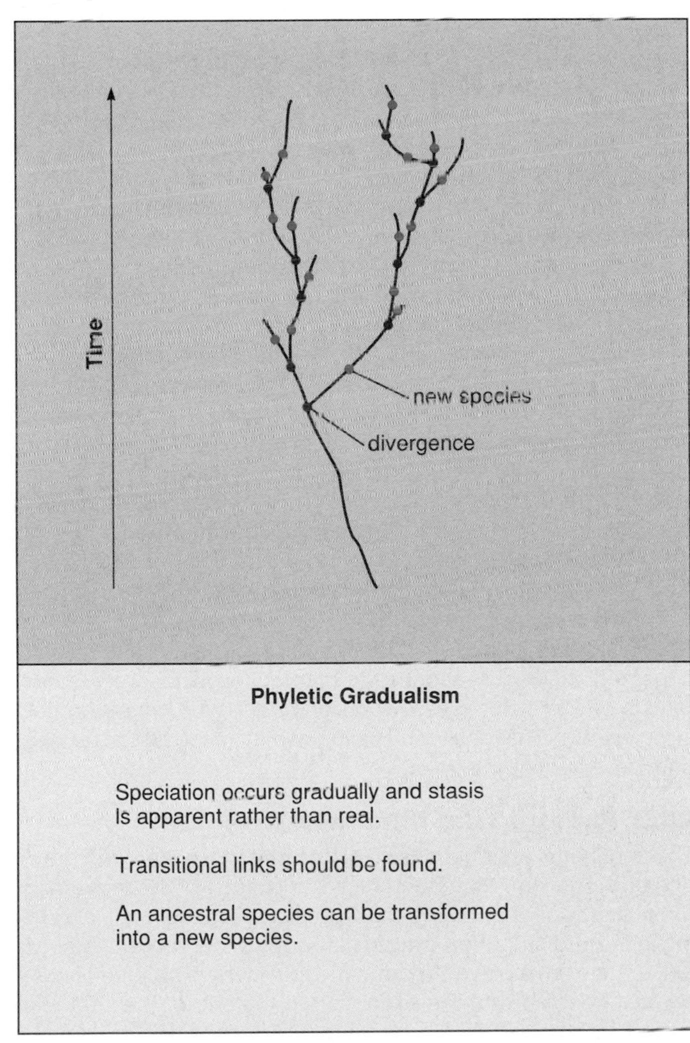
Time
new species
divergence
Phyletic Gradualism
Speciation occurs gradually and stasis
is apparent rather than real.
Transitional links should be found.
An ancestral species can be transformed
into a new species.

a.

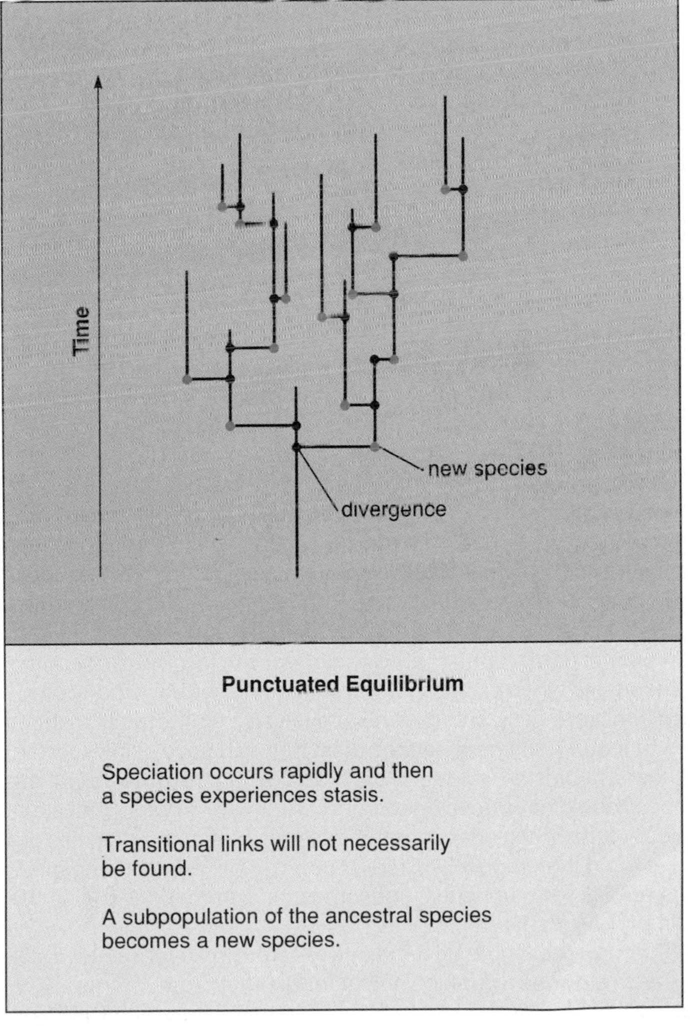
Time
new species
divergence
Punctuated Equilibrium
Speciation occurs rapidly and then
a species experiences stasis.
Transitional links will not necessarily
be found.
A subpopulation of the ancestral species
becomes a new species.

b.

Figure 20.18 Phyletic gradualism versus punctuated equilibrium.
The differences between (a) phyletic gradualism and (b) punctuated equilibrium are reflected in these patterns of time versus speciation.

connecting concepts

Does the process of evolution always have to be the same? The traditional view of the evolutionary process proposes that speciation occurs gradually and steadily over time. A new hypothesis suggests that long periods of little or no evolutionary change are punctuated by periods of relatively rapid speciation. Is it possible that both mechanisms may be at work in different groups of organisms and at different times?

Is it also possible that the history of life could have been different from what it was? The species alive today are the end products of the abiotic and biotic changes which occurred on earth as life evolved. And what if the abiotic and biotic changes had been other than they were? For example, if the continents had not separated 65 million years ago, what types of mammals, if any, would be alive today? Given a different sequence of environments, another mix of plants and animals might very well have resulted than those with which we are familiar.

The history of life on earth, as we know it, is only one possible scenario. If we could rewind the "tape of life" and let history take its course anew, the end result might well be very different, depending on the geologic and biologic events which took place the second time around. As an analogy, consider that if you were born in another time period and in a different country, the end result of "you" might be very different than the "you" of today.

Summary

20.1 Origin of Life

The unique conditions of the primitive earth allowed a chemical evolution to occur. The gases of the primitive atmosphere probably escaped from volcanoes and included water vapor, nitrogen, and carbon dioxide, with small amounts of hydrogen and carbon monoxide. The primitive atmosphere was a reducing atmosphere with no free oxygen gas.

At first the earth was very hot, but when it cooled, the water vapor condensed and the rains began to fall. The rains carried the gases into the newly formed oceans. In the presence of an outside energy source, such as ultraviolet radiation, lightning, or radioactivity, the gases reacted with one another to produce small organic molecules such as amino acids and glucose. This scenario is supported by experiments performed by Stanley Miller and others.

The next step was the formation of macromolecules by polymerization. The RNA-first hypothesis is supported by the discovery of ribozymes, RNA molecules that are present in today's cells and are both substrate and enzyme. The protein-first hypothesis is supported by the observation that amino acids will polymerize in a preferred fashion when heated to drying. Cairns-Smith has suggested that amino acid polymerization and nucleotide polymerization could have occurred in clay, where radioactivity would have supplied the necessary energy.

Polymerized amino acids, called proteinoids, become microspheres when placed in water. Microspheres have many properties similar to cells and may have evolved into protocells, structures that carry on energy metabolism and can have a lipid-protein boundary if lipids are added to the medium. On the other hand, the protocell could have developed from coacervate droplets that automatically form when concentrated mixtures of macromolecules are held at the right temperature, ionic composition, and pH. Phospholipids readily form spherical liposomes, and perhaps this was the origin of the plasma membrane.

The protocell must have been a heterotrophic fermenter living on the preformed organic molecules in the organic soup. The protocell was not a true cell until it contained genes. The RNA-first hypothesis suggests that the first genes and enzymes were RNA molecules. The protein-first hypothesis suggests that protein enzymes were needed before nucleic acids could form. Cairns-Smith suggests that polypeptides and RNA became associated in clay in such a way that the polypeptides catalyzed RNA formation. So the first cell would have contained both genes and protein enzymes at the same time.

20.2 History of Life

The fossil record allows us to trace the history of life. Prokaryotes evolved about 3.5 million years ago during the Precambrian, and they were alone on earth for at least 1.4 million years during which many metabolic pathways evolved. Cyanobacteria were the first to add oxygen to the atmosphere.

A rich animal fossil record starts at the Cambrian period of the Paleozoic era. The occurrence of external skeletons, which seems to explain the increased number of fossils at this time, may have been due to the presence of plentiful oxygen in the atmosphere, or perhaps it was due to predation.

Plants, and then insects, invaded land during the Silurian period. The swamp forests of the Carboniferous period contained primitive vascular plants, insects, and amphibians. This period is sometimes called the Age of Amphibians.

The Mesozoic era was the Age of Cycads and Reptiles. Twice during this era, dinosaurs of enormous size evolved. By the end of the Cretaceous period the dinosaurs were extinct.

The Cenozoic era is divided into the Paleogene period and the Neogene period. The Paleogene is associated with the evolution of mammals and flowering plants that formed vast tropical forests. The Neogene is associated with the evolution of primates; first monkeys appeared, then apes, and then humans. Grasslands were replacing forests and this put pressure on primates, who were adapted to living in trees. The result may have been the evolution of humans—primates who left the trees.

20.3 Factors That Influence Evolution

The continents are on massive plates that move, carrying the continents with them. Plate tectonics is the study of the movement of the plates. Continental drift has affected biogeography and helps explain the distribution pattern of today's land organisms. Mass extinctions have played a dramatic role in the history of life. It has been suggested that the extinction at the end of the Cretaceous period was caused by a large meteorite impact, and evidence has been gathered that other extinctions have a similar cause as well. It has also been suggested that tectonic, oceanic, and climatic fluctuations, particularly due to continental drift, can bring about mass extinctions.

Two models have been proposed to explain evolutionary change in reference to the fossil record. Those who support phyletic gradualism believe that evolution is by slow, gradual change, and that speciation need not necessarily appear in the fossil record. Those who support punctuated equilibrium present evidence that a long period of equilibrium (stasis) is punctuated by speciation that is recorded in the fossil record.

Reviewing the Chapter

Choose the best answer for each question.

1. Describe Stanley Miller's experiment, and discuss its significance. 325
2. Trace the steps by which a chemical evolution may have produced a protocell. 326–27
3. Contrast the RNA-first hypothesis and the protein-first hypothesis. If polymerization occurred in clay, what macromolecules would have been present at the same time? 326
4. What is a protocell? How might a plasma membrane have evolved? 326–27
5. Why is it likely the protocell was a heterotrophic fermenter? 327
6. What is a true cell? How might the first replication system have come about? 327
7. Explain how the fossil record develops and how fossils are dated relatively and absolutely. 328–29
8. Describe the history of life on earth by stating some of the main events of the Precambrian, and the Paleozoic, Mesozoic, and Cenozoic eras. 337
9. What is continental drift, and how is it related to plate tectonics? Give examples to show how biogeography supports the occurrence of continental drift. 338–39
10. When did five significant mass extinctions occur? Note which type of organisms were most affected for each one, and give causes for each of these mass extinctions. 340–41
11. Give an example of an evolutionary lineage. What is a transitional link and a living fossil? How do these pertain to the phyletic gradualism and the punctuated equilibrium models of evolutionary change? 342–43

Testing Yourself

Choose the best answer for each question.

1. Which of these gives a possible sequence of organic chemicals prior to the protocell?
   a. inorganic gases, amino acids, polypeptide, microsphere
   b. inorganic gases, nucleotides, nucleic acids, genes
   c. water, salts, protein, oxygen
   d. Both a and b are correct.
2. Which of these did Stanley Miller place in his experimental system to show that organic molecules could have arisen from inorganic molecules on the primitive earth?
   a. microspheres
   b. purines and pyrimidines
   c. the primitive gases
   d. All of these are correct.
3. Which of these is the chief reason the protocell was probably a fermenter?
   a. It didn't have any enzymes.
   b. The atmosphere didn't have any oxygen.
   c. Fermentation provides the most amount of energy.
   d. All of these are correct.
4. Which of these is a true statement?
   a. The primitive atmosphere was an oxidizing one and today's is a reducing one, making photosynthesis possible.
   b. The primitive atmosphere was 20% oxygen, just like it is today.
   c. The reducing primitive atmosphere contributed to the origin of life, and the oxidizing one of today would hinder it.
   d. It took so long for prokaryote evolution because the primitive atmosphere screened out the ultraviolet radiation from the sun.
5. The significance of liposomes, phospholipid droplets, is that they show that
   a. the first plasma membrane contained protein.
   b. a plasma membrane could have easily evolved.
   c. a biological evolution produced the first cell.
   d. there was water on the primitive earth.
6. Evolution of the DNA→RNA→protein system was a milestone because the protocell
   a. was a heterotrophic fermenter.
   b. could now reproduce.
   c. needed energy to grow.
   d. All of these are correct.
7. Which of these is a true statement?
   a. Eukaryotes evolved before prokaryotes.
   b. Prokaryotes evolved before eukaryotes.
   c. The true cell evolved before the protocell.
   d. Prokaryotes didn't evolve until 1.5 billion years ago.
8. The organisms with the longest evolutionary history are
   a. prokaryotes.
   b. eukaryotes.
   c. photosynthesizers.
   d. plants and animals.
9. Which best describes strata?
   a. sedimentary rock layers that contain fossils
   b. sedimentary rock layers that all date from the same historical time
   c. molten rock that contains radioactive material and is dangerous to the health of humans
   d. All of these are correct.
10. Continental drift helps explain the occurrence of
    a. mass extinctions.
    b. distribution of fossils on earth.
    c. geological upheavals like earthquakes.
    d. All of these are correct.

11. Which of these is mismatched?
    a. Mesozoic—cycads and dinosaurs
    b. Cenozoic—grasses and humans
    c. Paleozoic—prokaryotes and unicellular eukaryotes
    d. Cambrian—marine organisms with external skeletons
12. Place these terms in the diagram in the proper order, starting with (f.) as the first event:
    formation of earth, multicellularity, oxidizing atmosphere, oldest known fossils, origin of eukaryotic cells, Cambrian period begins.

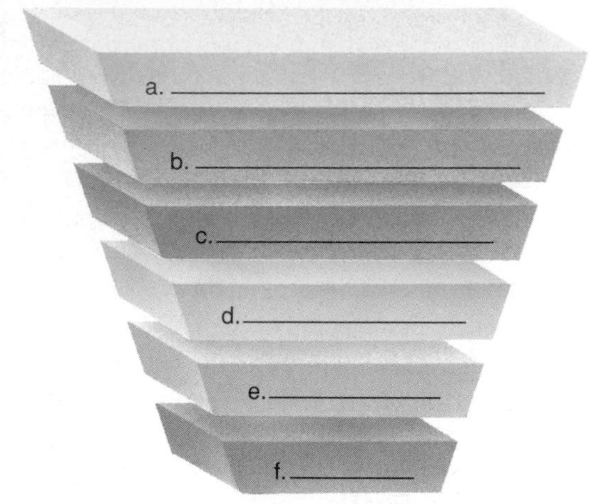

Applying the Concepts

1. *Life is a physical and chemical phenomenon.*
   How do all three hypotheses about the origin of life support this concept?
2. *Cells come only from preexisting cells.*
   Reconcile this concept with the belief that a chemical evolution produced the first cells.
3. *Speciation accounts for the diversity of life.*
   Would this statement be supported by both the phyletic gradualism and the punctuated equilibrium models of evolutionary change? Why or why not?

Using Technology

Your study of the origin and history of life is supported by these available technologies:

Exploring the Internet
The Mader Home Page provides resources for and help with studying this chapter.

http://www.mhhe.com/sciencemath/biology/mader/
(Click on Biology.)

Life Science Animations Video
Video #5: Plant Biology/Evolution/Ecology
History of Continental Drift (#53)

Understanding the Terms

autotroph 327
chemical evolution 325
continental drift 338
extinction 340
fossil 328
heterotroph 327
lineage 342
liposome 327
mass extinction 340
microsphere 326
ocean ridge 339
ozone shield 331
paleontology 329
phyletic gradualism 343
plate tectonics 339
proteinoid 326
protocell 326
punctuated equilibrium 343
sedimentation 328

Match the terms to these definitions:

a. __________ Abiotically polymerized amino acids that are joined in a preferred manner; possible early step in cell evolution.
b. __________ Cell forerunner developed from cell-like microspheres.
c. __________ Droplet of phospholipid molecules formed in a liquid environment.
d. __________ Organism that can make organic molecules from inorganic nutrients.
e. __________ Ridge on the ocean floor where oceanic crust forms and from which it moves laterally in each direction.
f. __________ Evolutionary model that proposes there are periods of rapid change dependent on speciation followed by long periods of stasis.
g. __________ Formed from oxygen in the upper atmosphere, it protects the earth from ultraviolet radiation.
h. __________ Formed from proteinoids exposed to water; has properties similar to today's cells.
i. __________ Increase in the complexity of chemicals that could have led to the first cells.
j. __________ Line of evolutionary descent from a common ancestor.

Human Evolution

CHAPTER 21

Chapter Concepts

21.1 Humans Are Primates

- Humans (*Homo sapiens*) are in the order primates, which are mammals adapted to living in trees. 348
- Primate characteristics include opposable thumbs, an enlarged forebrain, and an emphasis on learned behavior. 349

21.2 Primates Are Diverse

- Prosimians (tarsiers and lemurs) are primates that diverged early from the human line of descent. 349
- Humans are in the suborder Anthropoidea (monkeys, apes, and humans), the superfamily Hominoidea (apes and humans), and the family Hominidea (extinct and modern humans). 351

21.3 Hominids Break Away

- About 4 MYA* *Australopithecus* (the first hominids) evolved in eastern Africa, an area whose climate was no longer suitable to the growth of trees. The australopithecines had a small brain, but they walked erect. 354
- About 2 MYA *Homo habilis*, the first hominid to make tools, most likely was a hunter and may have been able to speak. 357
- About 1.9 to 1 MYA, *Homo erectus* migrated out of Africa and was a big game hunter that possessed fire. *Homo erectus* may have evolved into the so-called archaic *Homo sapiens*. 358
- The out-of-Africa hypothesis says that modern humans evolved in Africa, and after migrating to Europe and Asia (about 100,000 BP*), they replaced the archaic *Homo* species including, perhaps, the Neanderthals. 359
- Cro-Magnon is the name given to modern humans who made sophisticated tools and definitely had culture. 360

*MYA = millions of years ago; BP = before present.

Australopithecus africanus, a hominid

One of the most unfortunate misconceptions concerning human evolution is the belief that Darwin, Wallace, and others have suggested that humans evolved from apes. On the contrary, it is proposed that modern humans and apes evolved from a common apelike ancestor. Today's apes are our cousins, and we couldn't have evolved from our cousins because we are contemporaries— living on earth at the same time. Our relationship to apes is analogous to you and your first cousin being descended from your grandparents.

Some people feel that human evolution is somehow special or different from that of other species. Scientists, on the other hand, study human evolution in the same objective way they approach any other research topic. They have found that while human evolution follows the same patterns of evolutionary descent as other groups, it has more complexity. Various prehuman groups died out, migrated, and interbred with other groups in a short time; nevertheless, recently found fossils have brought knowledge of our evolutionary history closer to an apelike ancestor.

CLASSIFICATION

Humans

Phylum Chordata
- Subphylum Vertebrata
  - Class Mammalia
    - Order Primates
      - Suborder Anthropoidea
        - Superfamily Hominoidea
          - Family Hominidae
            - Genus *Homo*
              - Species *H. sapiens*

21.1 Humans Are Primates

The evolutionary principle of descent with modification applies to every group of organisms, including human beings. It could be argued that the evolution of humans, like that of all living things, begins with the very first cell (or cells). But specifically, as animals our evolutionary history begins in the Cambrian period of the Paleozoic era (Fig. 21.1). Vertebrates (animals with a backbone) arose soon after invertebrates (animals without a backbone) during this era. Fishes, the first vertebrates to evolve, have had a long and successful history. They still dominate the seas today.

The insects (invertebrates) were among the first animals to live on land, and their variety outstrips all living things known. Among vertebrates, the amphibians and then the reptiles held sway on land. **Mammals** [L. *mamma,* breast, teat] (animals that have hair and mammary glands) evolved from the mammal-like reptiles by the Triassic period of the Mesozoic era. Mammals remained small and insignificant while the dinosaurs dominated the land for over 150 million years. Extinction of the dinosaurs marked the end of the Mesozoic era, and then mammals diversified into many groups during the Cenozoic era. Today, there are mammals adapted to living on land, in the water, and in the air.

The dependence of animals upon plants is consistent with the observation that plants invaded land before any group of animals. Angiosperms, which include most trees, evolved at the same time as mammals. Angiosperms are the flowering plants that produce fruits, and **primates** [L. *primus,* first]—mammals adapted to living in trees—often used fruits as a central part of their diet. Many human characteristics are explainable on the basis of adaptation to an arboreal (tree-dwelling) life.

Human beings are related to all living things, especially to other primates, which are mammals adapted to living in trees.

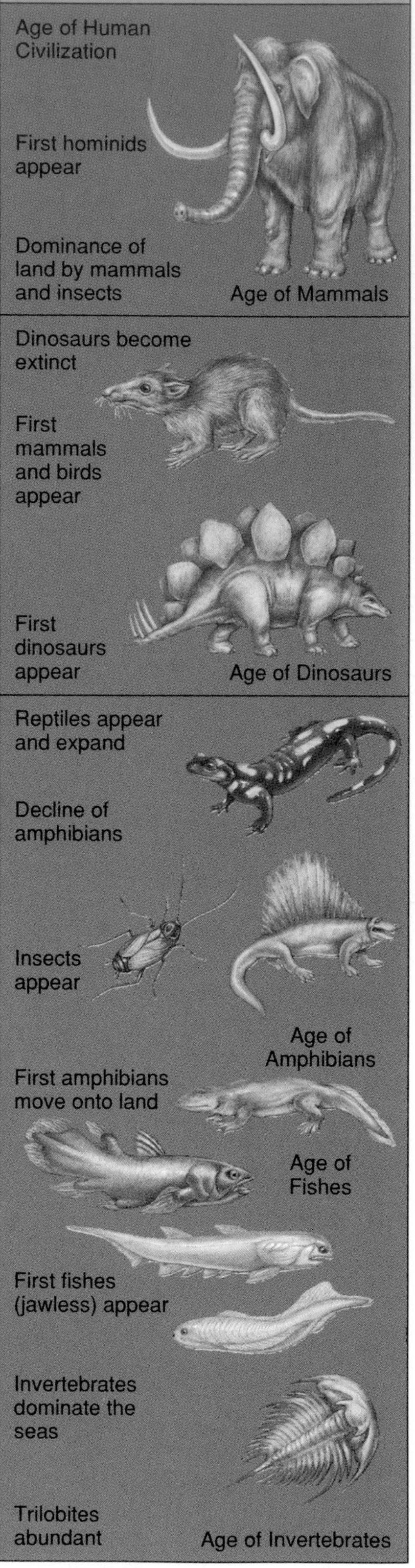

Era	Period	MYA*	Major Biological Events—Animals
Cenozoic	Neogene		Age of Human Civilization
	Paleogene	24	First hominids appear
			Dominance of land by mammals and insects; Age of Mammals
		66	
Mesozoic	Cretaceous		Dinosaurs become extinct
	Jurassic	144	First mammals and birds appear
	Triassic	208	First dinosaurs appear; Age of Dinosaurs
		245	
Paleozoic	Permian		Reptiles appear and expand
		286	Decline of amphibians
	Carboniferous	360	Insects appear
	Devonian		Age of Amphibians; First amphibians move onto land
	Silurian	408	Age of Fishes
	Ordovician	438	First fishes (jawless) appear
		500	Invertebrates dominate the seas
	Cambrian	570	Trilobites abundant; Age of Invertebrates

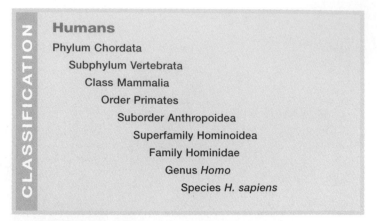

* MYA = millions of years ago.

Figure 21.1 Animal evolution.
An abbreviated geological timescale showing some major evolutionary events of animals.

a. *Tetonius*

b. Mindano tarsier, *Tarsius*

c. Dwarf lemur, *Cheirogaleus*

Figure 21.2 Prosimians.
a. The hypothesized Eocene primate *Tetonius* was small but agile and adapted to eating insects. **b.** Tarsiers are vertical clingers and leapers. The enormous eyes allow the tarsier to judge a safe landing even at night. **c.** There are over 40 species of lemurs on the island of Madagascar. The dwarf lemurs are particularly widespread.

What Are Primate Characteristics?

Primate limbs are mobile, as are the hands, because the thumb (and in nonhuman primates, the big toe as well) is opposable; that is, the thumb can touch each of the other fingers. Therefore, a primate can easily reach out and bring food such as fruit to the mouth. When locomoting, tree limbs can be grasped and released freely because nails have replaced claws.

The sense of smell is of primary importance in animals with a snout. In primates, the snout is shortened considerably, allowing the eyes to move to the front of the head. The stereoscopic vision (or depth perception) that results permits primates to make accurate judgments about the distance and position of adjoining tree limbs.

Gestation is lengthy, allowing time for good forebrain development; the visual portion of the brain is proportionately large, as are those centers responsible for hearing and touch. One birth at a time is the norm in primates; it is difficult to care for several offspring while moving from limb to limb. The juvenile period of dependency is extended, and there is an emphasis on learned behavior and complex social interactions.

These characteristics especially distinguish primates from other mammals:

opposable thumb (and in some cases, big toe)	expanded forebrain
nails (not claws)	emphasis on learned behavior
single birth	extended period of parental care

21.2 Primates Are Diverse

The primate order contains two suborders: prosimians and anthropoids.

Prosimians Came First

Among living primates, the **prosimians** (suborder Prosimii) [L. *pro,* before, and *simia,* ape, monkey] best resemble the first primates (Fig. 21.2). Tarsiers, which are found in the Philippines and East Indies, are curious, mouse-sized creatures with enormous eyes suitable to their nocturnal way of life. Tarsiers are insectivorous, and it's believed that primates may have evolved from mammals that first climbed into the trees to feed on insects. Lemurs, which have a squirrel-like appearance, are confined largely to the island of Madagascar. They feed on plant material, including fruits. Lorises, which are prosimians living in both Africa and Asia, resemble lemurs.

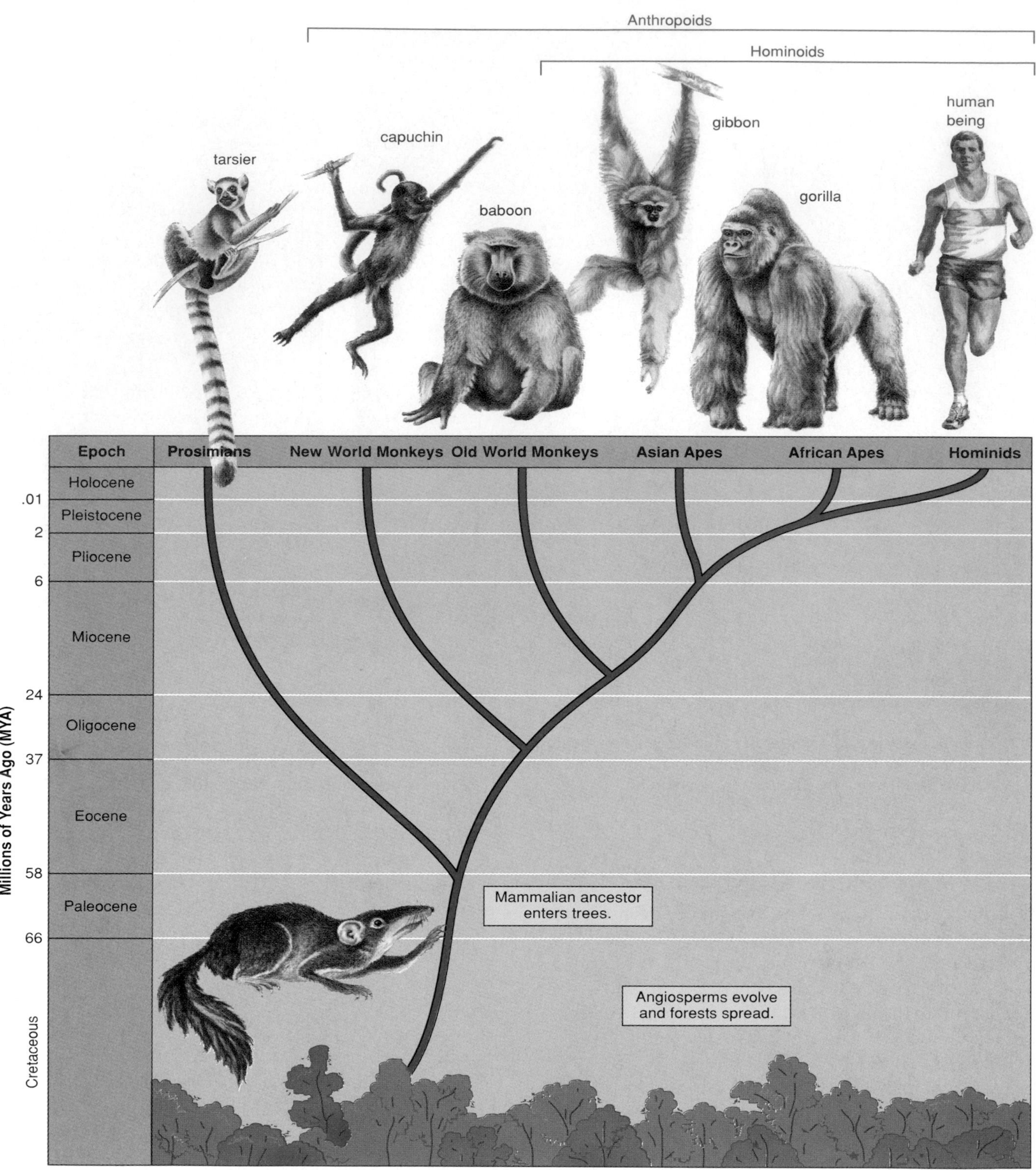

Figure 21.3 Primate evolution.
Evolutionary tree of the primate order began in the late Cretaceous period when a small insect-eating mammal climbed into the spreading angiosperm forests. The descendants of this mammal adapted to a new way of life and developed traits such as a shortened snout and nails instead of claws. Note the epoch in which each type of primate diverged from the main line of descent.

Anthropoids Followed

Figure 21.3 shows the sequence of anthropoid evolution during the Cenozoic era. The surviving **anthropoids** (suborder Anthropoidea) [Gk. *anthropos,* man, and *-eides,* like] are classified into three superfamilies: New World monkeys (Ceboidea), Old World monkeys (Cercopithecoidea), and the hominoids (Hominoidea). The New World monkeys often have long prehensile (grasping) tails and flat noses, and Old World monkeys, which lack such tails, have protruding noses. Two of the well-known New World monkeys are the spider monkey and the capuchin, the "organ grinder's" monkey. Some of the better-known Old World monkeys are now ground dwellers, such as the baboon and the rhesus monkey, which has been used in medical research. The **hominoids** [L. *homo,* man, and Gk. *-eides,* like] include all the apes and humans.

Primates flourished in the forests of northern continents during the Paleocene and Eocene epochs of the Cenozoic era (Fig. 21.3). But as these continents drifted slowly northward, the weather cooled and primates migrated southward. A group of early primates reached South America possibly by crossing the then-narrow Atlantic and evolved there into the New World monkeys. This means that New World monkeys are not more closely related to Old World monkeys than they are to apes.

In Africa, the ancestors of Old World monkeys and of hominoids were most likely four-limbed tree-climbers who fed on fruit. Sometime in the Miocene epoch, the monkeys adapted a more fibrous diet of leaves and diverged from the main line of descent. The hominoids, the more abundant group that fed on fruit, are well represented by apelike forms of that time. The anatomy of ***Proconsul*** is sufficiently primitive to be ancestral to all apes and humans (Fig. 21.4). These hominoids were about the size of a baboon, and the brain was a comparable size at about 165 cc. Their general anatomy seems similar to an Old World monkey also. Although primarily tree-dwellers, *Proconsul* may have also spent time exploring nearby grasslands for food.

At the end of the Miocene epoch, Africarabia (Africa plus the Arabian Peninsula) joined with Asia and the hominoids migrated into Europe and Asia. Two groups can be distinguished: dryomorphs and ramamorphs. While at one time it was believed that ramamorphs were ancestral to the later-appearing hominids, a group that contains humans, ramamorphs are now classified within an ancestral organgutan group.

The apelike *Proconsul,* which was prevalent in Africa during the Miocene epoch, is believed to be ancestral to today's hominoids—apes and humans.

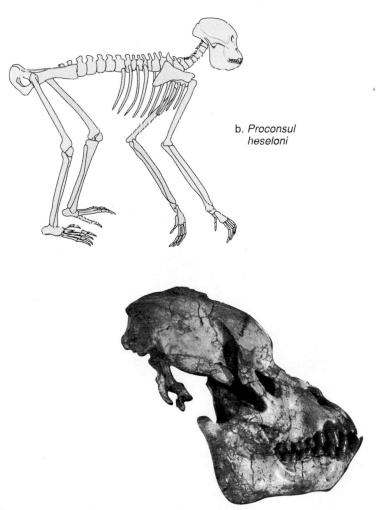

b. *Proconsul heseloni*

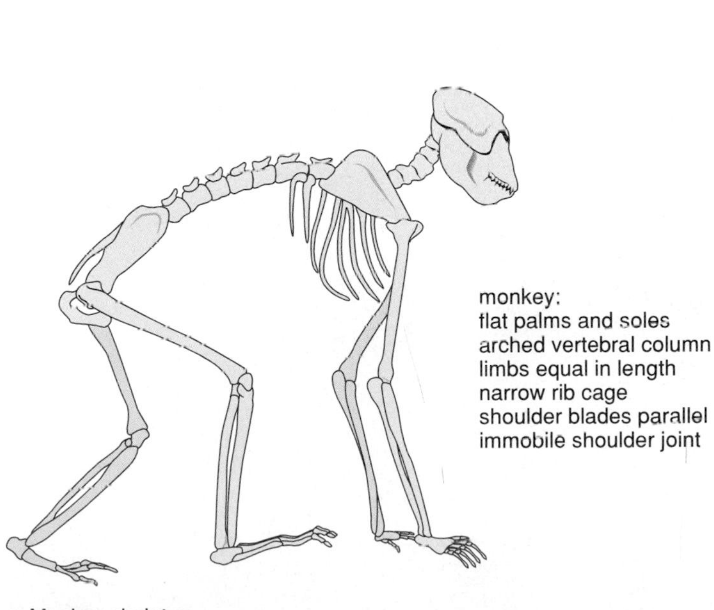

a. Monkey skeleton

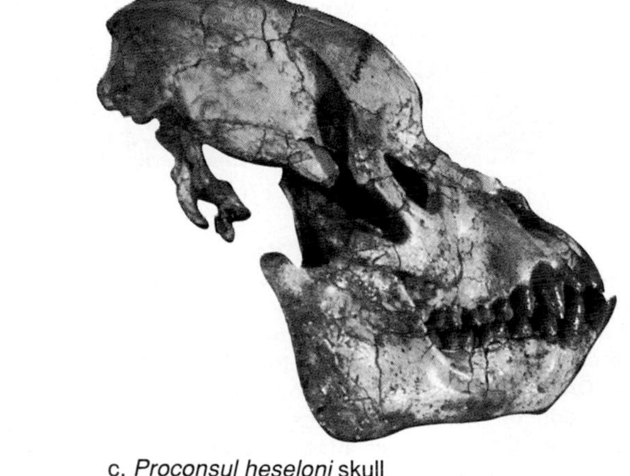

c. *Proconsul heseloni* skull

Figure 21.4 Monkey skeleton compared to *Proconsul* skeleton.
Comparison of monkey skeleton (a) and *Proconsul heseloni* (b), a hominoid of the Miocene epoch. c. Skull of *P. heseloni.*

a. White-handed gibbon, *Hylobates*

b. Orangutan, *Pengo*

c. Mountain gorilla, *Gorilla*

d. Common chimpanzee, *Pan*

Figure 21.5 Ape diversity.
a. Of the apes, gibbons are the most distantly related to humans. They dislike coming down from trees, even at watering holes. They will extend a long arm into the water and then drink collected moisture from the back of the hand. **b.** Orangutans are solitary except when they come together to reproduce. Their name means "forest man"; early Malayans believed that they were intelligent and could speak but did not because they were afraid of being put to work. **c.** Gorillas are terrestrial and live in groups in which a silver-backed male, such as this one, is always dominant. **d.** Of the apes, chimpanzees sometimes seem the most humanlike.

Learning from Living Hominoids

The close relationship between humans and apes is signified by their placement in the same superfamily, Hominoidea. Four types of apes have survived until today: the gibbon and the orangutan are found in Asia; the gorilla and the chimpanzee inhabit Africa (Fig. 21.5). The *gibbon* is the smallest of the apes, with a body weight of about 5.5–11 kilograms. Gibbons have extremely long arms that are specialized for swinging between tree limbs. The *orangutan* is large (75 kilograms) but nevertheless spends a great deal of time in trees. The *gorilla,* the largest of the apes (180 kilograms), spends most of its time on the ground. *Chimpanzees,* which are at home both in trees and on the ground, are the most humanlike of the apes in appearance and are frequently used in psychological experiments.

Molecular biologists have studied our relationship to the apes by comparing the similarity of our proteins and DNA to that of apes. Certain amino acid and DNA base-pair differences are not tied to natural selection and instead occur randomly at a fixed rate. Such differences can be used as a type of **molecular clock** by which we can judge when any two groups of organisms diverged (separated) from one another. The DNA of two species can be compared in the following manner: DNA strands are first separated into single strands. Then a strand from each species is allowed to pair complementarily. The degree of fit is judged by the thermal stability of the hybrid DNA. A molecular study of this sort suggests that monkeys most likely diverged from the primate line of descent about 33 million years ago (MYA), the orangutan diverged 10 MYA, and African apes and humans did not split until around 6 MYA. Therefore, it is believed that the last common ancestor between the African apes and **hominids** (family Hominidae) lived during the Pliocene epoch. Unfortunately, this common ancestor has not yet been found.

The geography of Africa holds a clue as to why the ape and the hominid lines of descent split from each other. A great furrow called the *rift valley* runs north and south in eastern Africa. Tectonic forces that began 12.5 MYA are slowly causing eastern Africa to separate from the rest of the continent. Here, huge mountainous volcanoes and lakes form a geographic barrier that is difficult to cross. Winds that have swept across the African continent for millions of years have deposited rain to the west of the mountains, and the east has become ever more dry. Forests and woodlands have remained in the west but the east is a grassland called a savanna.

The changing environment to the east may have influenced the evolution of hominids, who differ from the apes in the ways noted in Figure 21.6. Anatomical differences of prime importance concern: type of locomotion, which is dependent on skeletal features involving the spine, the pelvis, and the bones of the appendages; jaw shape, which is related to the size and the shape of the teeth; and brain size, which is related to the shape of the head and the brow. The hominid features promoted survival in a grassland as opposed to a forest ecosystem. We can imagine that it was beneficial to be able to stand tall to look over grasses while searching for food or avoiding predators, that jaw and teeth changes were adaptive to a new (perhaps omnivorous), diet, and that an erect posture left the hands free to throw rocks or even manipulate tools as the brain grew larger.

The ape line of descent and the hominid line of descent split about 6 MYA; differences regarding mode of locomotion, shape of jaw, and brain size became distinctive.

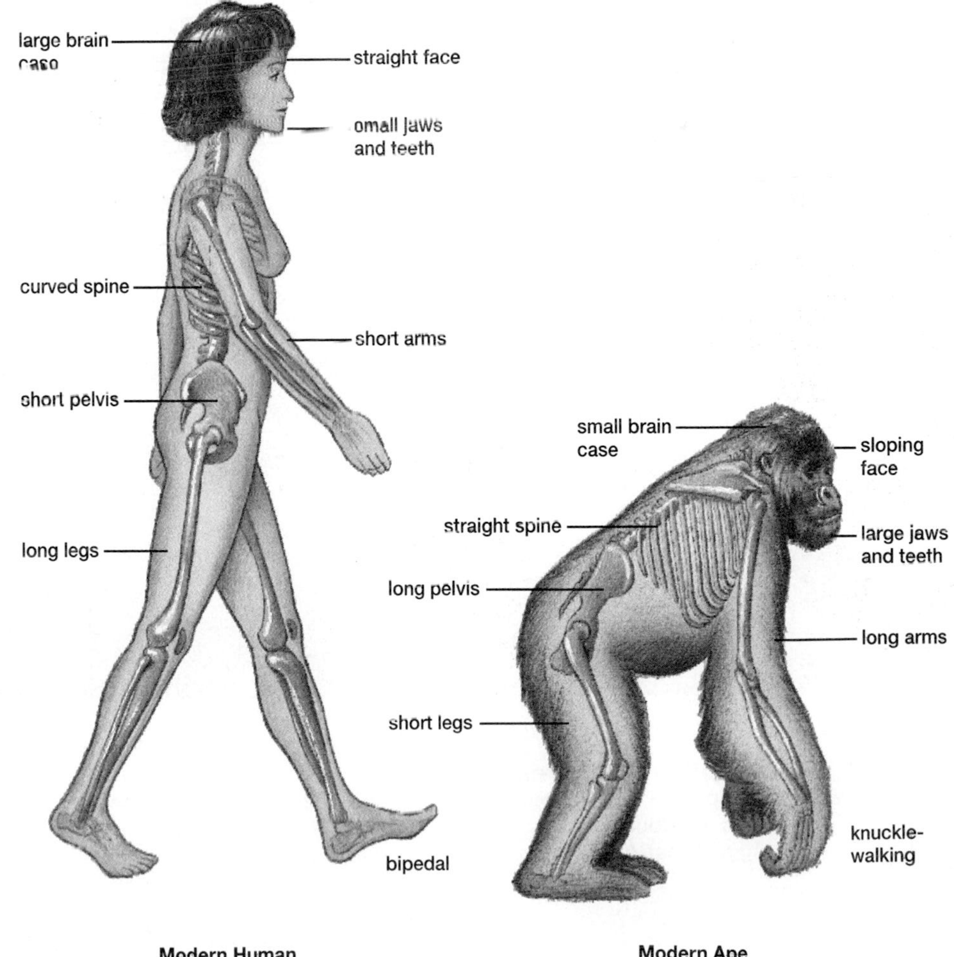

Figure 21.6 Modern human skeletal features and those of a gorilla.

21.3 Hominids Break Away

The hominid line of descent begins with the australopithecines, which evolved and diversified in eastern Africa.

Hominids Walked Erect

It wasn't until the 1960s that paleontologists, following the lead of the renowned couple Louis and Mary Leakey, began to concentrate their efforts in eastern as opposed to southern Africa. It has proven to be worth their while. One 1994 find consisting of skull fragments, teeth, arm bones, and a part of a child's lower jaw has been dated at 4.4 MYA (millions of years ago). Called *Australopithecus ramidus* [L. *australis*, southern, *pithecus*, ape, and *ramus*, branch], it is believed to represent an early stage in human evolution. In comparison to later **australopithecines,** the canine teeth are larger, and the skull is more like that of a chimpanzee. Also, this fossil did not walk erect. However, a 1995 discovery, *Australopithecus anamensis*, dated at just about 4 MYA has jaws like those of an ape but legs like those of humans.

More than 20 years ago, a team led by Donald Johanson unearthed nearly 250 fossils of a hominid called *A. afarensis*. A now-famous female skeleton now dated at 3.18 MYA is known worldwide by its field name, Lucy. (The name derives from the Beatles' song "Lucy in the Sky with Diamonds.") Although her brain was quite small (400 cc), the shapes and relative proportions of her limbs indicate that Lucy did stand upright and walk bipedally (Fig. 21.7). Even better evidence of bipedal locomotion comes from a trail of footprints in Laetoli dated about 3.7 MYA. The larger prints are double—a smaller-sized being was stepping in the footfalls of another—and there are small prints off to the side, within hand-holding distance. Since the australopithecines were apelike above the waist and humanlike below the waist, it seems that human characteristics did not evolve all at one time. The term *mosaic evolution* is applied when different body parts change at different rates and therefore at different times.

The australopithecines were sexually dimorphic. Lucy was about four feet tall and weighed about 30 kilograms. In contrast, the males of the species were five feet tall and weighed up to 45 kilograms. Some have speculated that such size differences may indicate two separate species, but in 1994 new finds, including a more complete skull

a.

b.

Figure 21.7 ***Australopithecus afarensis.***
a. A reconstruction on display at the St. Louis Zoo. **b.** These fossilized footprints occur in ash from a volcanic eruption some 3.7 million years ago. The larger footprints are double—a smaller individual was stepping in the footprints of a larger individual. A youngster was walking to the side. The footprints suggest that *A. afarensis* walked bipedally.

(dubbed the son of Lucy and dated at just about 3 MYA), confirmed the opinion that the fossils belong to one species (Fig. 21.8). Taking into account their smaller body size, the relative brain size is about half of ours and the jaw is heavy. The cheek teeth are enormous, and in males large canine teeth project forward.

It's interesting to speculate that the daily lives of australopithecines may have been comparable to those of baboons. In baboons, as in australopithecines, males are much larger than females. Baboons travel in a troop rather than forming permanent family groups. The troop has a home range, which it constantly travels foraging for food. At night the troop climbs trees to remain safe from predators. Australopithecines had long, curved fingers and toes, and the legs were short in comparison to the mobile arms. This suggests that they, too, still climbed trees on occasion.

A. afarensis had descendants. After a period of stasis that may have lasted a million years (from 3 to 2 MYA), branching speciation occurred. Therefore, instead of thinking about hominid descent in terms of a straight line, it is far better to envision a bush. Some think there may have been as many as ten species of hominids about 2 MYA in Africa, but we will discuss only four of them; three of which are australopithecines. *A. africanus*, which was first named in southern Africa by Raymond Dart in the 1920s, is a gracile (slender) type. *A. boisei* is a robust form from eastern Africa, and *A. robustus* is a similar form from southern Africa. Both gracile and robust forms have a brain size of about 500 cc; their skull differences are essentially due to dental and facial adaptations for different diets. The robust forms have stronger jaws, larger attachments for larger chewing muscles, and bigger grinding teeth because they most likely fed on tougher foods than the gracile form. The robust forms lived in drier habitats where soft fruits and leaves would be harder to come by. In both forms, the pelvis resembles that of Lucy but the hands were more humanlike; possibly they were capable of making tools. Both forms, which are believed to have eaten meat at least occasionally, may have used the tools to process animal carcasses. Of interest is a recent report that the thumb anatomy of *A. robustus* is similar to our own. Modern human hands are adapted to handling tools because of their strong, well-muscled, and opposable thumb.

At one time it was believed that the gracile australopithecine form gave rise to the robust forms. However, in 1985, Alan Walker discovered a robust skull, called the black skull, which was old enough (2.5 MYA) to be ancestral to the robust forms we have been discussing. It has been given the name *A. aethiopicus*. As discussed in the reading on page 356, the earliest *Homo* species also evolved from the *Australopithecus* line of descent.

Several species of australopithecines have been identified. After a period of stasis, during which only *A. afarensis* existed, branching speciation occurred.

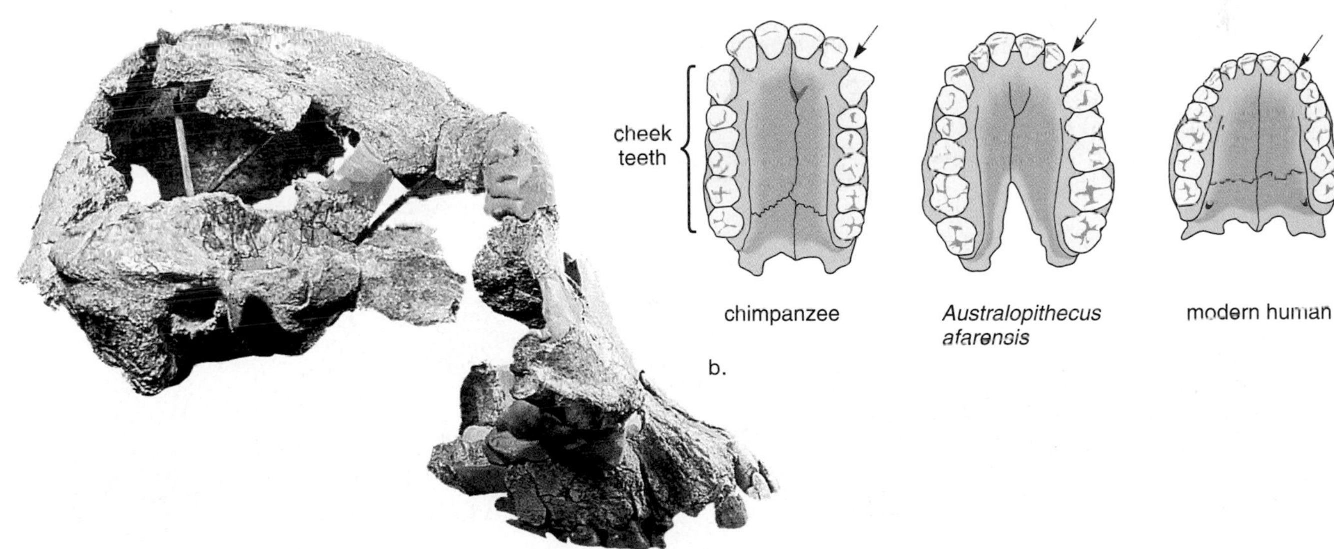

Figure 21.8 *Australopithecus afarensis* skull and dentition.
a. *A. afarensis* had an apelike face with prominent brow ridges and a projecting jaw. **b.** Dentition of an ape, *A. afarensis*, and a modern human. Note the size of the canine teeth, the gap between canine teeth and adjacent incisor (arrow), and the size of the cheek teeth.

Doing Science

Origin of the Genus *Homo*

Fossil evidence shows that the earliest *Homo* species evolved in Africa from the *Australopithecus* line of descent about 2.4 MYA. Remains of australopithecines indicate that they spent part of their time climbing trees and that they retained many apelike traits. Australopithecine arms, like those of an ape, were long compared to the length of the legs. *A. afarensis* also had strong wrists and long, curved fingers and toes. These traits would have served well for climbing, and the australopithecines probably climbed trees for the same reason that chimpanzees do today: to gather fruits and nuts in trees and to sleep above ground at night so as to avoid predatory animals, such as lions and hyenas.

Whereas our brain is about the size of a grapefruit, that of the australopithecines was about the size of an orange. Their brain was only slightly larger than that of a chimpanzee. There is no evidence that the australopithecines manufactured stone tools; presumably, they were not smart enough to do so.

We know that the genus *Homo* evolved from the genus *Australopithecus,* but several years ago I concluded that this could not have happened as long as the australopithecines climbed trees every day. The obstacle relates to the way in which we, members of *Homo,* develop our large brain. Unlike other primates, we retain the high rate of fetal brain growth through the first year after birth. (That is why a one-year-old child has a very large head.) The brain of other primates, including monkeys and apes, grows rapidly before birth, but immediately after birth their brain grows more slowly. An adult human brain is more than three times as large as that of an adult chimpanzee.

A continuation of the high rate of fetal brain growth eventually allowed the genus *Homo* to evolve from the genus *Australopithecus*. But there was a problem in that continued brain growth is linked to underdevelopment of the entire body. Although the human brain becomes much larger, human babies are remarkably weak and uncoordinated. Such helpless infants must be carried about and tended. Human babies are unable to cling to their mothers the way chimpanzee babies can (Fig. 21A).

How, then, did the evolutionary transition from *Australopithecus* to *Homo* take place? Fossilized pollen and other geological evidence reveal that at the time the *Homo* genus evolved, the climate was becoming cooler and drier. This caused forests to shrink and grasslands to expand across broad areas of Africa. The australopithecine line of descent became extinct, just as did many African antelope species that depended on forests for food and shelter. Before extinction occurred, however, humans—who walk erect, rarely climb trees, and have large brains—evolved.

Steven Stanley
The Johns Hopkins University

The origin of the *Homo* genus entailed a great evolutionary compromise. Humans gained a large brain, but they were saddled with the largest interval of infantile helplessness in the entire class Mammalia. The positive value of a large brain must have outweighed the negative aspects of infantile helplessness, however, or natural selection would not have produced the *Homo* genus. Having a larger brain meant that humans were able to outsmart or ward off predators with weapons they were clever enough to manufacture.

Probably very few genetic changes were required to delay the maturation of *Australopithecus* and produce the large brain of *Homo*. The mutation of a regulatory gene that controls one or more other genes most likely could have delayed early maturation. As we learn more about the human genome, we will eventually uncover the particular gene or gene combinations that cause us to have a large brain, and this will be a very exciting discovery.

Figure 21A
A human infant is often cradled and has no means to cling to its mother when she goes about her daily routine.

Homo habilis Made Tools

The oldest fossils to be classified in the genus *Homo* are known as ***Homo habilis*** [L. *homo*, man, and *habilis*, suitable, handy], and his remains dated as early as 2 MYA are often accompanied by stone tools. Why is this hominid classified within our own genus? *H. habilis* was small—about the size of Lucy—but the brain at 700 cc is about 45% larger. In addition, certain portions of the brain thought to be associated with speech areas are enlarged. The cheek teeth are smaller than even those of the gracile australopithecines. Apparently, *H. habilis* had a different way of life than the other hominids of this time.

This hominid lived at a time when the earth was cooling and tropical regions in general were becoming more dry due to the formation of glaciers at the North and South Poles. Many animals, including the robust forms of *Australopithecus*, show dentition that indicates a shift to a more fibrous diet. It is, perhaps, significant that the modern horse appears at this time.

Cut marks on bones that could have been made by stone flakes have been found at many sites throughout eastern Africa that date from 2 MYA. *H. habilis* could have made and used tools in order to strip meat off these bones and, in keeping with the size of the teeth, could have eaten meat to satisfy protein demands. As a scavenger, *H. habilis* may have depended simply on the kills of other animals; or as a predator, may have killed small- to medium-sized prey. This new way of life became available to hominids when they had the ability to make and use tools intelligently.

The stone tools made by *H. habilis* are called Oldowan tools because they were first identified as tools by the Leakeys in Olduvai Gorge (see Fig. 21.14). Oldowan tools are simple and look rather clumsy, but perhaps we are looking at the core that remains after flakes have been removed. The flakes would have been sharp and able to scrape away hide and cut tendons to easily remove meat from a carcass.

H. habilis most likely still ate fruits, berries, seeds, and other plant materials. Perhaps a division of labor arose with certain members of a group serving as hunters and others as gatherers. Speech would have facilitated their cooperative efforts, and later they most likely shared their food and ate together. In this way, society and culture could have begun. *Culture*, which encompasses human behavior and products (such as technology and the arts) is dependent upon the capacity to learn and transmit knowledge through the ability to speak and think abstractly.

Prior to the development of culture, adaptation to the environment necessitated a biological change. The acquisition of culture provided an additional way by which adaptation was possible. And the possession of culture by *H. habilis* may have hastened the extinction of the australopithecines during the early Pleistocene epoch (Fig. 21.9).

H. habilis **warrants classification as a *Homo* because of brain size, posture, and dentition. Circumstantial evidence suggests the use of tools to prepare meat and also the development of culture.**

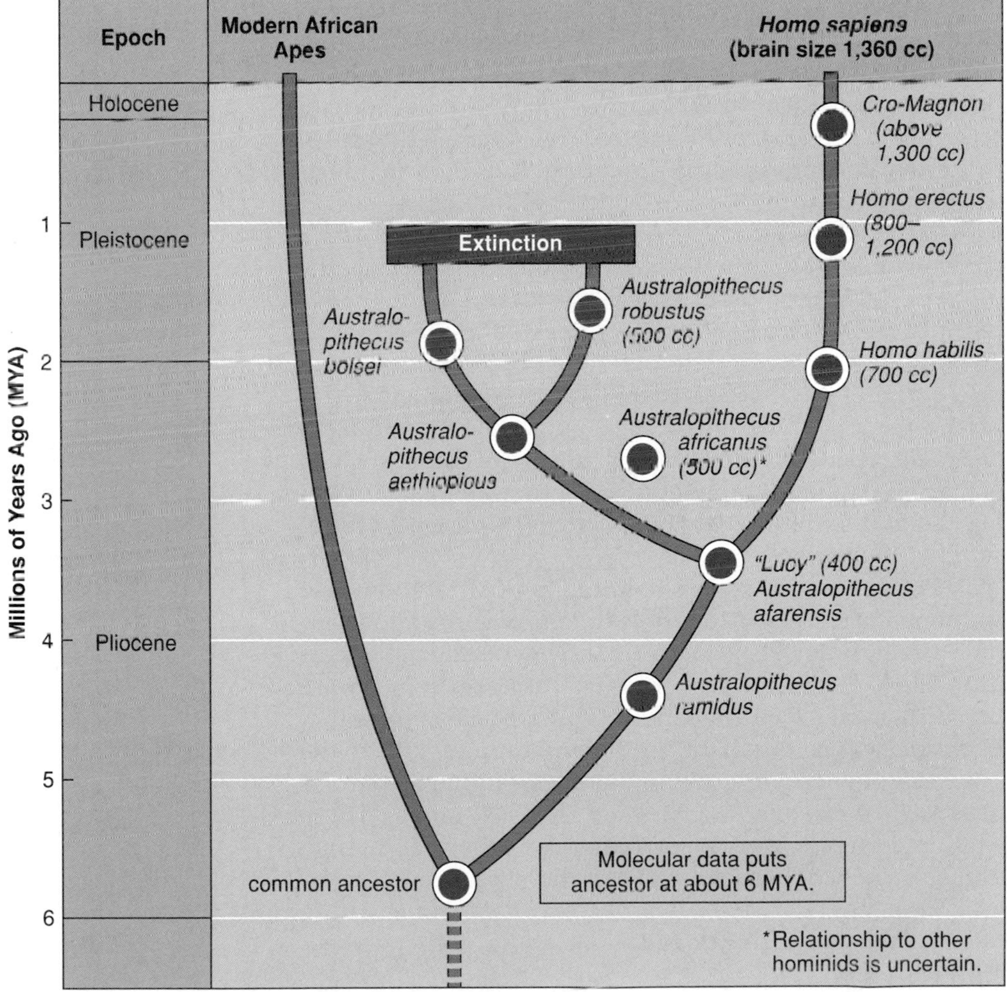

Figure 21.9 Hominoid evolutionary tree.
African apes and hominids split about 6 million years ago (MYA); *Australopithecus ramidus* is the oldest of the hominids; *A. afarensis* lasted about a million years before branching speciation occurred. In particular, there were gracile and robust australopithecine forms and also the first human form, *Homo habilis*.

Homo erectus Traveled

Homo erectus [L. *homo*, man, and *erectus*, upright] is the name assigned to hominid fossils found in Africa, Asia, and Europe and dated between 1.9 and 0.5 MYA. A Dutch anatomist named Eugene Dubois was the first to unearth *H. erectus* bones in Java in 1891, and since that time many other fossils have been found in the same area. Although all fossils assigned the name *H. erectus* are similar in appearance, there is enough discrepancy to suggest that several different species have been included in this group. In particular, some experts believe that the African and Asian forms are two different species.

Compared to *H. habilis*, *H. erectus* had a larger brain (about 1,000 cc), more pronounced brow ridges, a flatter face, and a nose that projects like ours. This type of nose is adaptive for a hot, dry climate because it permits water to be removed before air leaves the body. The recovery of an almost complete skeleton of a ten-year-old boy indicates that *H. erectus* was much taller than the hominids discussed thus far (Fig. 21.10). Males were 1.8 meters (about 6 feet) and females were 1.55 meters (approaching 5 feet) tall. Indeed, these hominids were erect and most likely had a striding gait like ours. But the robust and most likely heavily muscled skeleton still retains some australopithecine features. Even so, the size of the birth canal indicates that infants were born in an immature state that required an extended period of care.

It is believed that *H. erectus* first appeared in Africa and then migrated into Asia and Europe. At one time, the migration was thought to have occurred about 1 MYA, but recently *H. erectus* fossil remains in Java and the Republic of Georgia have been dated at 1.9 and 1.6 MYA, respectively. Therefore, it seems that *H. erectus* either left Africa soon after it evolved, or perhaps an earlier, related form has to be credited with the migration. In any case, such an extensive population movement is a first in the history of humankind and a tribute to the intellectual and physical skills of the species.

H. erectus was the first hominid to use fire, and it also fashioned more advanced tools called Acheulean tools after a site in France (see Fig. 21.14). They used heavy teardrop-shaped axes and cleavers as well as flakes, which were probably used for cutting and scraping. Some believe that *H. erectus* was a systematic hunter and brought kills to the same site over and over again. In one location paleontologists have found over 40,000 bones and 2,647 stones. These sites could have been "home bases" where social interaction occurred and a prolonged childhood allowed time for much learning. Perhaps a language evolved and a culture more like our own developed.

H. erectus, which evolved from *H. habilis*, had a striding gate, made well-fashioned tools (perhaps for hunting), and could control fire. This hominid migrated into Europe and Asia from Africa about 1 MYA.

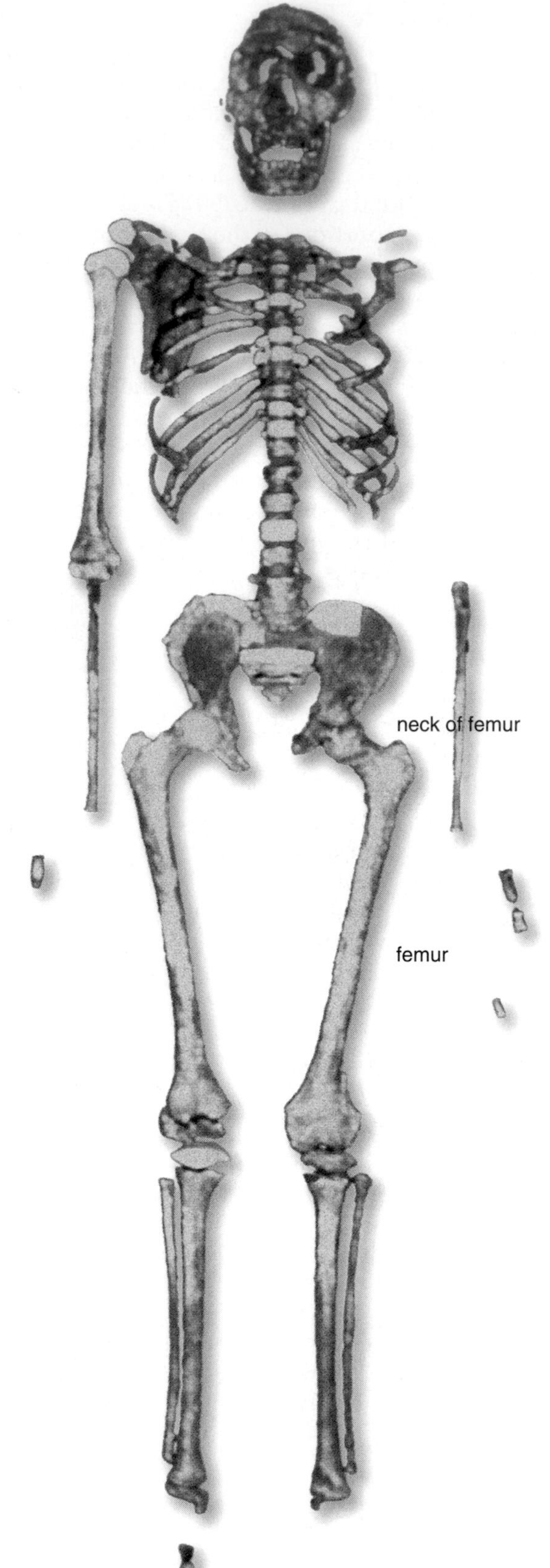

Figure 21.10 ***Homo erectus.***
Skeleton of a ten-year-old boy who lived 1.6 MYA in eastern Africa. The femurs are angled and the necks are much longer than that of a modern human.

Modern Humans Originated How?

It is generally recognized that modern humans originated from *H. erectus*, but where did this occur? About 300,000 years BP (before present), so-called "archaic *Homo sapiens*" were present in Europe, Asia, and Africa. (Neanderthal is a well-known archaic *H. sapiens*.) The hypothesis that each of these individual populations went on to evolve into modern humans is called the *multiregional continuity hypothesis* (Fig. 21.11*a*). This hypothesis, which proposes no migrations, requires that evolution be essentially similar in several different places. Each region would show a continuity of its own anatomical characteristics from about a million years ago, when *H. erectus* first arrived in Eurasia. Opponents argue that it seems highly unlikely that evolution would have produced essentially the same result in these different places. They suggest, instead, the *out-of-Africa hypothesis*, which proposes that archaic *H. sapiens* became fully modern only in Africa, and thereafter they migrated to Europe and Asia about 100,000 years BP. If so, there would be no continuity of characteristics between fossils of 300,000 years BP and fossils of 100,000 years BP. Modern humans may have interbred to a degree with archaic populations but in effect they supplanted them (Fig. 21.11*b*).

According to which hypothesis would modern humans be most genetically alike? With the multiregional hypothesis, human populations have been evolving separately for a long time and therefore genetic differences are expected. With the out-of-Africa hypothesis, we are all descended from a few individuals from about 100,000 years BP. Therefore, the out-of-Africa hypothesis suggests that we are more genetically similar. A few years ago, a study attempted to show that all the people of Europe (and the world for that matter) have essentially the same mitochondrial DNA. Called the mitochondrial Eve hypothesis by the press (note this is a misnomer because no single ancestor is proposed), the statistics that calculated the date of the African migration were found to be flawed. Still, the raw data—which indicate a close genetic relationship among all Europeans—support the out-of-Africa hypothesis.

These opposing hypotheses have sparked many other innovative studies to test them. The final conclusions are still being determined.

Investigators are currently testing two hypotheses: that modern humans (1) evolved separately in Europe, Africa, and Asia, and (2) evolved in Africa and then migrated.

Asia Africa Europe

Present

Homo erectus evolves into modern humans in Asia, Africa, and Europe.

interbreeding

0.5 MYA

interbreeding

migration of *Homo erectus*

migration of *Homo erectus*

1 MYA

a. Multiregional continuity

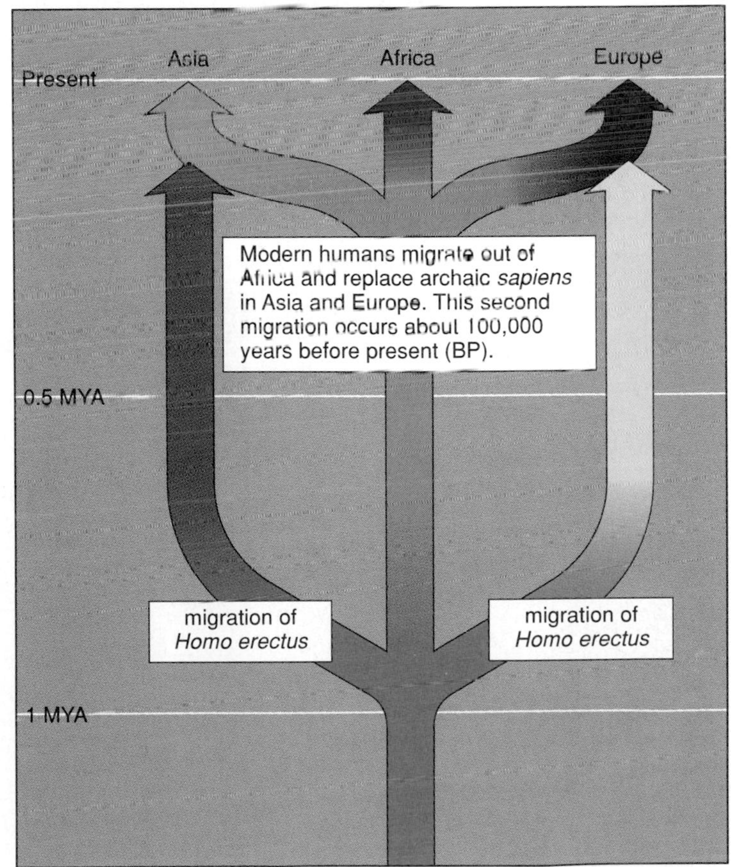

b. Out of Africa

Figure 21.11 Origin of modern humans.
a. The multiregional continuity hypothesis proposes that modern humans evolved separately in at least three different places: Asia, Africa, and Europe. Therefore, continuity of genotypes and phenotypes is expected in these regions. **b.** The out-of-Africa hypothesis proposes that modern humans originated only in Africa; then they migrated and supplanted populations of *Homo* in Asia and Europe about 100,000 years ago.

Neanderthals Were Archaic

A brain capacity larger than 1,000 cc allows a species to be classified as *Homo sapiens*. The **Neanderthals** *(H. sapiens neanderthalensis)* take their name from Germany's Neander Valley, where one of the first Neanderthal skeletons, dated some 200,000 years ago, was discovered. The Neanderthals had massive brow ridges, and the nose, the jaws, and the teeth protruded far forward. The forehead was low and sloping, and the lower jaw sloped back without a chin. New fossils show that the pubic bone is long compared to ours.

Figure 21.12 Neanderthals.
This drawing shows that the nose and the mouth of these people protruded from the face, and their muscles were massive. They made stone tools and were most likely excellent hunters.

At this time, the Neanderthals are thought to be an archaic *H. sapiens* and most likely not in the main line of *Homo* descent. Surprisingly, however, the Neanderthal brain was, on the average, slightly larger than that of modern humans (1,400 cc, compared to 1,360 cc in most modern humans). The Neanderthals were heavily muscled, especially in the shoulders and the neck (Fig. 21.12). The bones of the limbs were shorter and thicker than those of modern humans. It is hypothesized that a larger brain than that of modern humans was required to control the extra musculature. They lived in Eurasia during the last Ice Age, and their sturdy build could have helped conserve heat.

The Neanderthals give evidence of being culturally advanced. Most lived in caves, but those in the open may have built houses. They manufactured a variety of stone tools, including spear points, which could have been used for hunting, and scrapers and knives that would have helped in food preparation. They most likely successfully hunted bears, woolly mammoths, rhinoceroses, reindeer, and other contemporary animals. They used and could control fire, which probably helped in cooking frozen meat and in keeping warm. They even buried their dead with flowers and tools and may have had a religion.

Cro-Magnons Were Modern Humans

The out-of-Africa hypothesis discussed on page 359 is currently receiving much support by both anatomists and molecular biologists. It is increasingly believed that modern humans, by custom called **Cro-Magnons** after a fossil location in France, entered Eurasia 100,000 years ago or even earlier. Cro-Magnons *(H. sapiens sapiens)* had a thoroughly modern appearance (Fig. 21.13). They made advanced stone tools called Aurignacian tools (Fig. 21.14).

Figure 21.13 Cro-Magnon.
Cro-Magnon people are the first to be designated *Homo sapiens sapiens.* Their toolmaking ability and other cultural attributes, such as their artistic talents, are legendary.

Included were compound tools, as when stone flakes were fitted to a wooden handle. They may have been the first to throw spears, enabling them to kill animals from a distance, or to have made knifelike blades. They were such accomplished hunters that some researchers believe they were responsible for the extinction of many larger mammals, such as the giant sloth, the mammoth, the saber-toothed tiger, and the giant ox during the late Pleistocene epoch.

Cro-Magnons hunted cooperatively, and perhaps they were the first to have had a language. They are believed to have lived in small groups, with the men hunting by day, while the women remained at home with the children. It's quite possible that a hunting way of life among prehistoric people influenced our behavior even until today. The Cro-Magnon culture included art. They sculpted small figurines out of reindeer bones and antlers. They also painted beautiful drawings of animals on cave walls in Spain and France (Fig. 21.13).

The main line of hominid descent is now believed to include *A. ramidus, A. afarensis, H. habilis, H. erectus,* and Cro-Magnons.

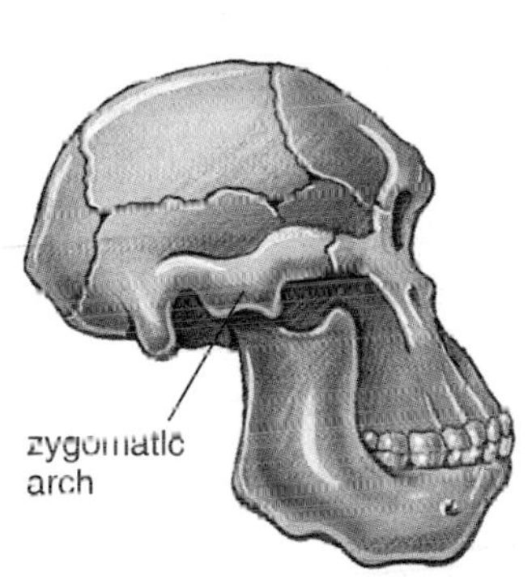

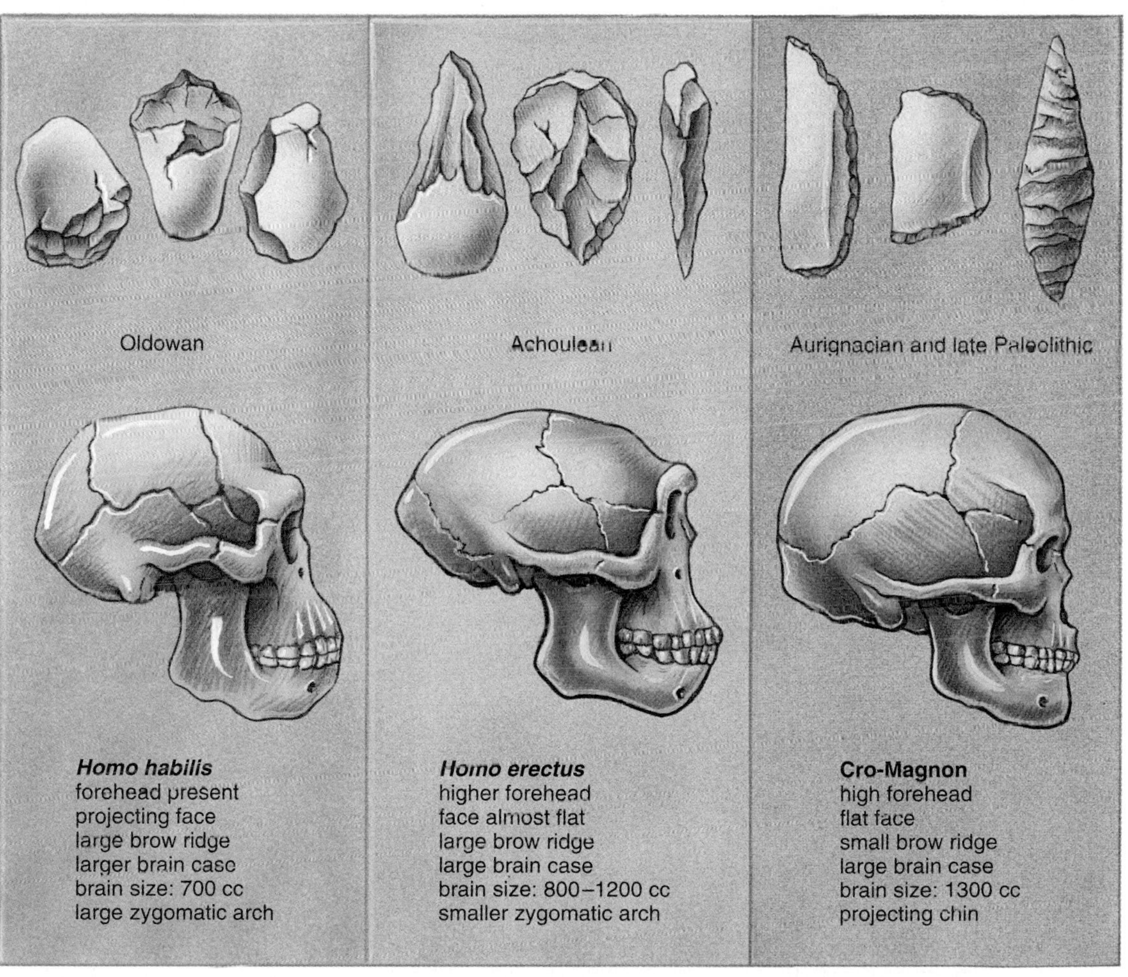

Figure 21.14 Hominid skulls and tools.
The listings compare hominid skulls. The Oldowan tools are crude. Acheulean tools are better made and more varied, and Aurignacian tools are well designed for specific purposes.

connecting concepts

Human beings are diverse, but even so, we are all classified as *H. sapiens sapiens.* This is consistent with the biological definition of species because it is possible for all types of humans to interbreed and to bear fertile offspring. While it may appear that there are various "races," molecular data show that the DNA base sequence varies as much between individuals of the same ethnicity as between individuals of different ethnicity.

It is generally accepted that the phenotype is adapted to the climate of a region. Although it might seem as if dark skin is a protection against the hot rays of the sun, it has been suggested that it is actually a protection against ultraviolet ray absorption. Dark-skinned persons living in southern regions and white-skinned persons living in northern regions absorb the same amount of radiation. (Some absorption is required for vitamin D production.) Other features that correlate with skin color, such as hair type and eye color, may simply be side effects of genes that control skin color.

Differences in body shape represent adaptations to temperature. A squat body with short limbs and a short nose retains more heat than an elongated body with long limbs and a long nose. Also, almond-shaped eyes, a flat nose and forehead, and broad cheeks are believed to be adaptations to the last Ice Age.

While it always has seemed to some that physical differences warrant assigning humans to different "races," this contention is not borne out by the molecular data mentioned previously.

Summary

21.1 Humans Are Primates

The evolution of human beings can be traced from the Cambrian period when animals were prevalent. Among vertebrates, fishes dominated (and still dominate) the seas; on land, amphibians gave way to reptiles as the dominant land group during the Mesozoic era. Mammals evolved from reptiles but did not diversify until the Cenozoic era, after the dinosaurs became extinct. Human beings are primates, animals adapted to living in trees.

Primates have opposable thumbs, and nails replace claws. The forebrain in particular is enlarged, and there is an emphasis on learned behavior during an extended period of parental care.

21.2 Primates Are Diverse

During the evolution of primates, various groups diverged in a particular sequence from the main line of descent. Prosimians (tarsiers and lemurs), which diverged first, are most distantly related to us and most closely related to the original primate.

Anthropoids include New World monkeys, Old World monkeys, and hominoids. New World monkeys took up residence in South America and diverged before Old World monkeys. These two groups are not any more related to each other than to the apes.

The first hominoid ancestor may have been *Proconsul*, whose apelike fossils date from the Miocene epoch. Among living apes, two are Asian (gibbon and orangutan) and two are African (gorilla and chimpanzee). Molecular biologists tell us we are most closely related to the African apes whose ancestry split from ours about 6 MYA (during the Pliocene epoch).

21.3 Hominids Break Away

Hominid evolution continued in eastern Africa, which is isolated from the rest of the continent by the rift valley where tectonic forces are causing eastern Africa to separate from the continent. Eastern Africa became dry and the forests began to disappear.

Hominids classified as australopithecines include *Australopithecus ramidus* dated at 4.4 MYA; *A. afarensis* (Lucy) at 3.4 MYA; and the son of Lucy at just about 3 MYA. Although the brain was small, these hominids walked erect. After a period of stasis, branching speciation occurred. Both gracile australopithecines *(A. africanus)* and robust forms (*A. boisei* and *A. robustus*) existed. Both forms have a brain of about 500 cc and may have made tools and eaten meat. However, the robust forms had a more fibrous grass diet.

Homo habilis, present about 2 MYA, is certain to have made tools. Most likely this species killed small- to medium-sized prey. The larger brain (700 cc) indicates that this hominid may have had speech; if so, culture could have begun. From then on, hunting and culture were inseparable.

H. erectus—with a brain capacity of 1,000 cc and a striding gait—was the first to migrate out of Africa, to have fire, to be a big game hunter, and, possibly, to use home bases.

Two contradicting hypotheses have been suggested about the origination of modern humans. The multiregional continuity hypothesis says that modern humans originated separately in Asia, Europe, and Africa. If so, a distinctive continuity in anatomy and genes is expected in each location. The out-of-Africa hypothesis says that modern humans originated only in Africa and, after migrating into Europe and Asia, they replaced the archaic *Homo* species found there. Many studies are being done to determine the possible validity of these hypotheses.

The Neanderthals may have been an archaic *Homo* species of Europe. Their chinless face, squat frame, and heavy muscles are apparently adaptations to the cold. Cro-Magnon is a name often given to modern humans. Their tools were sophisticated, and they definitely had a culture, as witnessed by the paintings on the walls of caves.

Reviewing the Chapter

1. In general terms, trace the evolution of humans from the time that animals arose. Mention sequential eras and periods and their significant events. 348
2. In what ways are primates adapted to an arboreal life? 349
3. What are the groups of living prosimians, and what is their significance regarding human evolution? 349
4. What are the two groups of living monkeys and how are they related? What is the significance of the fossils classified as *Proconsul?* 351
5. What are the four groups of living apes and how are they related? What is their degree of relatedness to modern humans as determined by molecular biology? 352–53
6. How did geographical changes in Africa affect the evolution of hominids? 353
7. Name the various species of *Australopithecus,* and explain how they are believed to be related to one another and to humans. 354–56
8. Why is *Homo habilis* classified as a *Homo?* If this hominid did make tools, what does this say about its possible way of life? 357
9. What role(s) might *H. erectus* have played in the evolution of modern humans according to the multiregional theory? the out-of-Africa theory? What was this hominid's possible way of life? 358–59
10. Who were the Neanderthals and the Cro-Magnons, and what is their place in the evolution of humans according to the two hypotheses mentioned in question 9? 360–61

Testing Yourself

Choose the best answer for each question.

1. Which of these gives the correct order of divergence from the main line of descent leading to humans?
   a. prosimians, monkeys, gibbons, orangutans, African apes, humans
   b. gibbons, orangutans, prosimians, monkeys, African apes, humans
   c. monkeys, gibbons, prosimians, African apes, orangutans, humans
   d. African apes, gibbons, monkeys, orangutans, prosimians, humans
2. Lucy is a member of what species?
   a. *Homo erectus*
   b. *Australopithecus afarensis*
   c. *H. habilis*
   d. *A. robustus*
3. Hominids did not evolve until
   a. South Africa became hot and tropical.
   b. a rift valley separated eastern Africa.
   c. the climate became drier.
   d. Both b and c are correct.
4. The son of Lucy lived
   a. about 3 MYA.
   b. about 3.5 MYA.
   c. in the same epoch as Lucy.
   d. Both a and c are correct.
5. *H. erectus* could have been the first to
   a. use and control fire.
   b. migrate out of Africa.
   c. make tools.
   d. Both a and b are correct.
6. Which of these characteristics is not consistent with the others?
   a. large brow ridge
   b. large zygomatic arch
   c. high forehead
   d. projecting face
7. Which of these is true? The last common ancestor for apes and hominids
   a. has been found and it resembles a gibbon.
   b. has not been found but it is expected to be from the Pliocene epoch.
   c. has been found and it has been dated at 30 MYA.
   d. is not expected to be found because there was no such common ancestor.
8. Which of these is incorrectly matched?
   a. gibbon—hominoid
   b. *A. africanus*—hominid
   c. tarsier—anthropoid
   d. *H. erectus—Homo*
9. If the multiregional continuity hypothesis is correct, then hominid fossils in China after 100,000 BP
   a. are not expected to resemble earlier fossils.
   b. are expected to resemble earlier fossils.
   c. Either one is consistent with the hypothesis.
10. Which of these is incorrectly matched?
   a. *H. erectus*—made tools
   b. Neanderthal—good hunter
   c. *H. habilis*—controlled fire
   d. Cro-Magnon—good artist
11. Classify humans by filling in the missing lines.

Kingdom Animalia	
Phylum	a. ________
Subphylum	b. ________
c. ________	Mammalia
d. ________	Primates
Suborder	e. ________
Superfamily	f. ________
Family	g. ________
h. ________	*Homo*
Species	i. ________

Applying the Concepts

1. *Humans have an evolutionary history as do all living things.*
   List ways in which you would show that we are closely related to the African apes.
2. *Geographic isolation promotes speciation.*
   Show how this concept pertains to the evolution of humans.
3. *Molecular biology can contribute to our understanding of human* evolution.
   Explain why DNA similarities show relatedness, and give at least two examples of the application of molecular biology to the evolution of humans.

Using Technology

Your study of human evolution is supported by these available technologies:

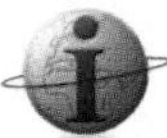

Exploring the Internet
The Mader Home Page provides resources for and help with studying this chapter.

http://www.mhhe.com/sciencemath/biology/mader/
(Click on Biology.)

Understanding the Terms

anthropoid 351
australopithecine 354
Cro-Magnon 360
hominid 353
hominoid 351
Homo erectus 358
Homo habilis 357
mammal 348
molecular clock 353
Neanderthal 360
primate 348
Proconsul 351
prosimian 349

Match the terms to these definitions:

a. _________ Group of primates that includes only monkeys, apes, and humans.
b. _________ Hominid of 2 million years ago; possibly a direct ancestor of modern humans.
c. _________ The common name for the first fossils to be accepted as representative of modern humans.
d. _________ Hominid with a sturdy build who lived during the last Ice Age in Europe and the Middle East; hunted large game, and lived together in a kind of society.
e. _________ Hominid who lived during the Pleistocene epoch; first to have a posture and locomotion similar to modern humans.
f. _________ Member of a class of vertebrates characterized especially by the presence of hair and mammary glands.
g. _________ Member of a superfamily containing humans and the great apes.

Further Readings for Part iii

Coppens, Y. May 1994. East side story: The origin of humankind. *Scientific American* 270(5):88. Explores and provides a theory for why hominids branched off from chimpanzees.

Cowen, R. 1995. *The history of life.* 2d ed. Boston: Blackwell Scientific Publications. From the origin of life through human evolution, this is an introduction to paleontology and paleobiology.

Currie, P. J. May 1996. The great dinosaur egg hunt. *National Geographic* 189(5):96. Fossilized dinosaur eggs, embryos, and nests have been found in China.

Dalziel, I. W. D. January 1995. Earth before Pangaea. *Scientific American* 272(1):58. Geologists search for clues to early continental drift.

Dawkins, R. November 1995. God's utility function. *Scientific American* 273(5):80. The role of genetics in evolution and natural selection is discussed.

deWaal, F. B. M. March 1995. Bonobo sex and society. *Scientific American* 272(3):82. Discusses the Bonobo great ape and its resemblance to humans in behavior, appearance, and intelligence.

Doolittle, R. F., and Bork, P. October 1993. Evolutionary mobile modules in proteins. *Scientific American* 269(4):50. This article discusses the role of proteins in evolution.

Erwin, D. E. July 1996. The mother of mass extinction. *Scientific American* 275(1):72. Global sea level decline and volcanic eruptions may have caused mass extinctions.

Futuyma, D. J. 1995. *Science on trial: The case for evolution.* Sunderland, Mass.: Sinauer Associates. Presents the evidence for evolution and the operation of the evolutionary process versus that proposed by the doctrine of creation.

Gore, R. January 1996. Neanderthals. *National Geographic* 189(1):2. Archeological finds are providing much information on the lives of these early humans.

Horgan, J. October 1995. The new social Darwinists. *Scientific American* 273(4):174. Psychologists and social scientists apply principles of natural selection to mind and behavioral studies.

Johanson, D. C. March 1996. Face-to-face with Lucy's family. *National Geographic* 189(3):96. New fossils from Ethiopia provide more information about human evolution.

Klein, J., et al. December 1993. MHC polymorphism and human origins. *Scientific American* 269(6):78. Explains that human tissue diversity developed long before *Homo sapiens* emerged.

Knecht, H. July 1994. Late Ice Age hunting technology. *Scientific American* 271(1):82. Early humans developed skills for crafting hunting implements.

Leakey, M. September 1995. The dawn of humans. *National Geographic* 188(3):38. An East African species that walked upright 4 million years ago has been discovered.

Levin, H. L. 1994. *The earth through time.* 4th ed. Philadelphia: Saunders College Publishing. Provides an in-depth study of evolutionary and geological eras and events.

Lewin, R. 1993. *The origin of modern humans.* New York: Scientific American Library. This colorfully illustrated book describes discoveries and controversies surrounding human origins.

Lewin, R. 1997. *Patterns in evolution: The new molecular view.* New York: *Scientific American Library.* This easy-to-read book explores how genetic information is providing insight into evolutionary events.

Marshall, L. February 1994. The terror birds of South America. *Scientific American* 270(2):90. These animals were at the top of the food chain 65 million years ago.

Miller, P. December 1995. Jane Goodall. *National Geographic* 188(6):102. This article highlights the research of primatologist, Jane Goodall.

Milton, K. August 1993. Diet and primate evolution. *Scientific American* 269(2):86. Presents the role of diet in primate evolution.

Novacek, M. J., et al. December 1994. Fossils of the flaming cliffs. *Scientific American* 271(6):60. Describes recent discoveries of dinosaur and other fossils in the Gobi desert.

Pääbo, S. November 1993. Ancient DNA. *Scientific American* 269(5):86. Evolutionary change can be observed directly from genetic information in fossil remains.

Royte, E. September 1995. Hawaii's vanishing species. *National Geographic* 188(3):2. Introduced species are threatening Hawaii's native birds, plants, and insects.

Schopf, J. W. 1992. *Major events in the history of life.* Boston: Jones and Bartlett Publishers. Summarizes crucial events that shaped the earth's development and evolution over time.

Scientific American Special Issue October 1994. 271(4). This special issue is devoted to evolution.

Slatkin, M., editor. 1994. *Exploring evolutionary biology.* Readings from *American Scientist.* Sunderland, Mass.: Sinauer Associates. This collection of articles can be used as supplementary material for evolution studies.

Storch, G. February 1992. The mammals of island Europe. *Scientific American* 266(2):64. Describes mammalian fossils found in German shale beds.

Strickberger, M. 1995. *Evolution.* 2d ed. Boston: Jones and Bartlett Publishers. Presents the basics of evolutionary theories.

Webster, D. July 1996. Dinosaurs of the Gobi. *National Geographic* 190(1):70. Well-preserved dinosaur fossils are being unearthed in the Gobi desert.

Wenke, R. 1996. *Patterns in prehistory: Humankind's first three million years.* 4th ed. New York: Oxford University Press. Provides a comprehensive review of world prehistory.

PART IV

Behavior and Ecology

The behavior of an organism is regulated by the nervous and endocrine systems, whose development is under the control of inherited genes but also is influenced by the environment. The genetic basis of behavior suggests that behavior should be subject to natural selection and is adaptive. Organisms do not exist alone; rather, they are part of a population that interacts with both the abiotic (nonliving) and biotic (living) environment. A community is an assemblage of populations in a particular area; but ultimately, ecology considers the distribution and abundance of populations over the surface of the earth: in short, the biosphere. The biosphere performs services for us, such as regulating climate and the quality of the water we drink and the air we breathe. The importance of the biogeochemical cycles that unite all ecosystems has caused humans to place a primary emphasis on the study of ecology.

Animal Behavior

CHAPTER 22

Chapter Concepts

Bowerbirds, *Ptilonorhynchus violaceus*

At the start of the breeding season, male bowerbirds use small sticks and twigs to build elaborate display areas called bowers. They clear the space around the bower and decorate the area with fresh flowers, fruits, pebbles, shells, bits of glass, tinfoil, and any bright baubles they can find. Each species has its own preference in decorations. The satin bowerbird of eastern Australia prefers blue objects, a color that harmonizes with the male's glossy blue-black plumage.

After the bower is complete, a male spends most of his time near his bower, calling to females, renewing his decorations, and guarding his work against possible raids by other males. After inspecting many bowers and their owners, a female approaches one and the male begins a display. He faces her, fluffs up his feathers, and flaps his wings to the beat of a call. The female enters the bower, and if she crouches, the two mate.

Behavior is studied in the same objective manner as any other field of science. A behaviorist would determine how a male bowerbird is organized to perform this behavior and how the behavior helps him secure a mate.

22.1 Behavior Has a Genetic Basis

Two types of questions are central to the study of behavior. *Mechanistic questions* are answered by describing how an animal is biologically organized and equipped to carry out the behavior. *Survival value questions* are answered by describing how the behavior helps an animal exploit resources, avoid predators, or secure a mate. Both types of questions are grounded in the recognition that **behavior,** observable and coordinated responses to environmental stimuli, has at least a partial genetic basis.

Various types of experiments have been performed to determine if behavior has a genetic basis. Birds have the ability to return home, and some migrate great distances spring and fall. Birds that are caged and prevented from migrating exhibit increased activity that is called *migration restlessness.*

Peter Berthold and his colleagues noted that blackcap warblers from Germany migrate to Africa, while those that live in Cape Verde (islands off the west coast of Africa) do not migrate at all. He reasoned that if migration behavior is inherited, then hybrids might show intermediate behavior. Berthold captured birds from Germany and Cape Verde and mated them. He placed the hybrid offspring and Cape Verdean birds in separate cages that were equipped with perches that electronically recorded when birds landed on them. During the next migratory period, the hybrids jumped back and forth between perches every night for many weeks, but the Cape Verdean birds exhibited no migratory restlessness.

Andreas Helbig performed still another experiment with blackcap warblers. German blackcaps fly *southwest* from Germany to Spain and finally to southwest Africa. Austrian blackcaps fly *southeast*, from Austria to Israel and finally to southeast Africa. Helbig created hybrids and then placed the parents and the hybrids in a special funnel cage that allowed the birds to see the stars at night. (Migration studies have shown that birds navigate during migration by using the sun in the day and the stars at night as compasses to determine direction.) The floor and sides of the cages were covered with a special paper that recorded scratch marks as the birds jumped in their preferred direction during the next migratory period. When the choices of hybrid offspring were compared statistically with those of their parents, the hybrids were proved to be intermediate between them (Fig. 22.1). Both Berthold's and Helbig's studies with blackcap warblers support the hypothesis that behavior has at least a partial genetic basis.

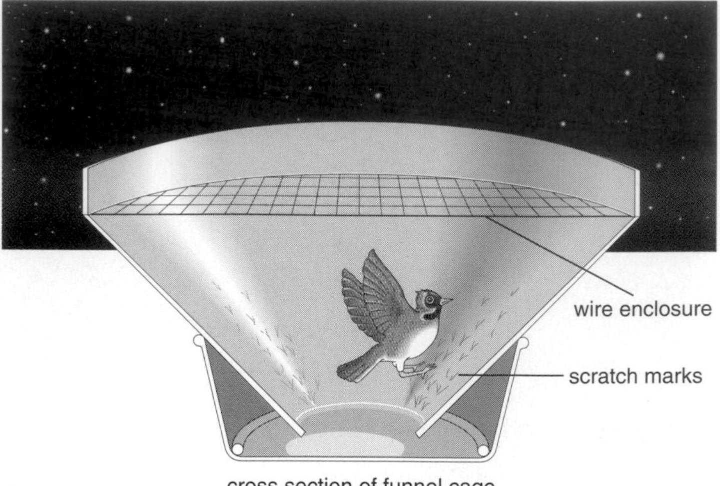

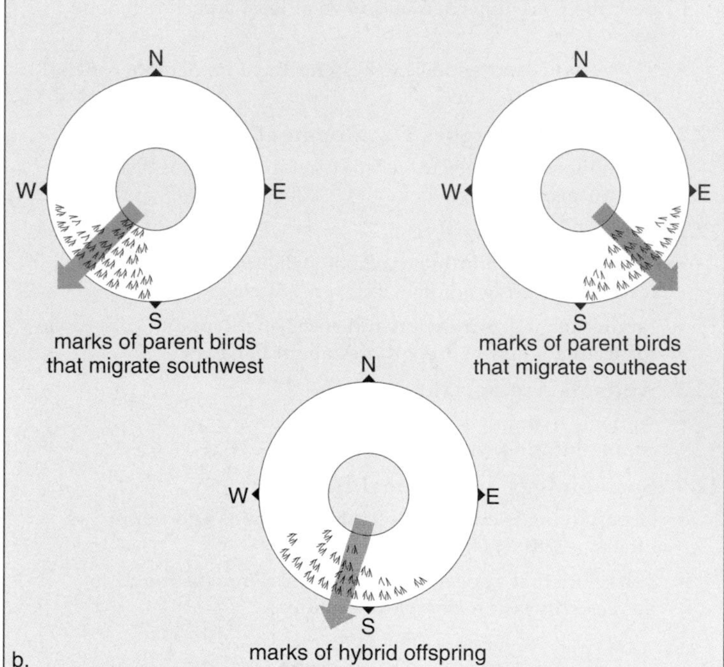

Figure 22.1 Inheritance of migratory behavior in blackcap warblers, *Sylvia.*
a. A funnel-shaped cage allowed birds to see the night sky because birds use the stars as a compass when migrating. b. Experimenters could determine the preferred direction of flight by scratch marks on the floor and sides of the cage. Parents preferred either to fly southeast or southwest; hybrids preferred an intermediate direction.

Steven Arnold performed several experiments with the garter snake *Thamnophis elegans* after he observed a distinct difference between two different types of snake populations in California. Inland populations are aquatic and commonly feed underwater on frogs and fish. Coastal populations are terrestrial and feed mainly on slugs. In the laboratory, inland adult snakes refused to eat slugs, while coastal snakes readily did so. To test for possible genetic differences between the two populations, Arnold arranged matings between inland and coastal individuals and found that isolated newborns show an overall intermediate incidence of slug acceptance.

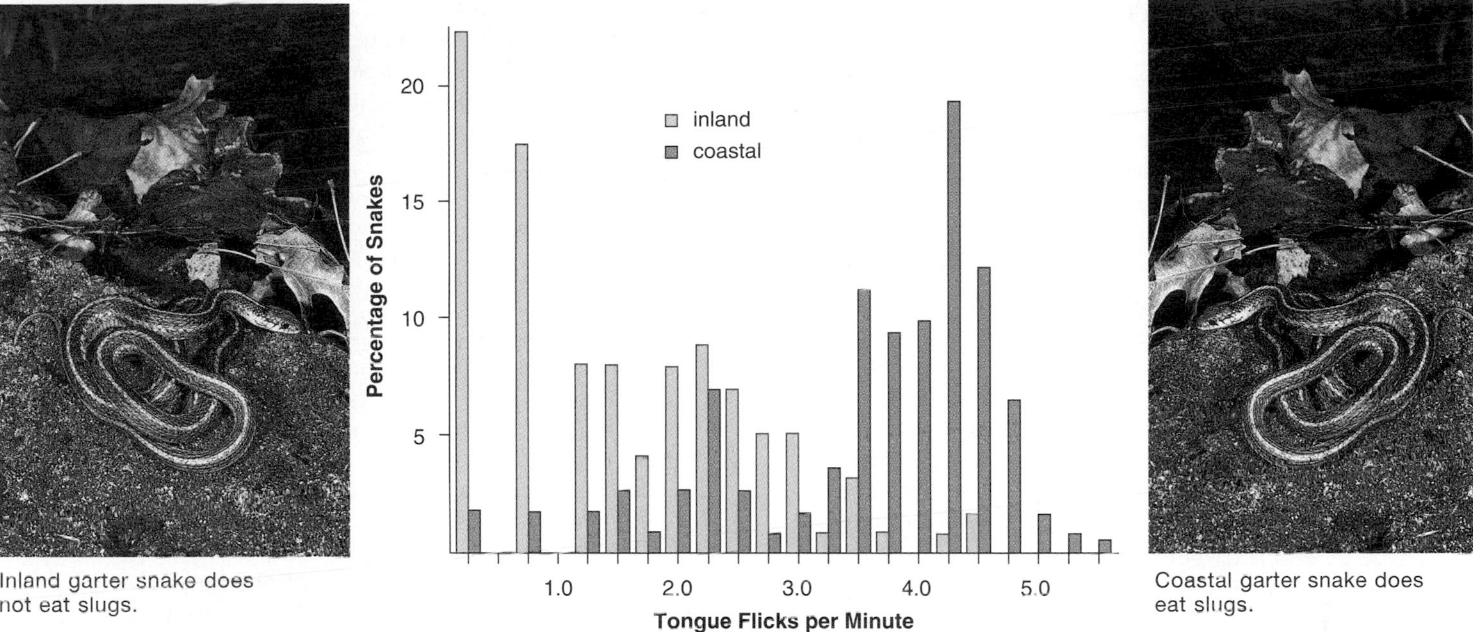

Figure 22.2 Feeding behavior of garter snakes, *Thamnophis elegans.*
The number of tongue flicks by inland and coastal garter snakes as a response to slug extract on cotton swabs. Coastal garter snakes do eat slugs; inland garter snakes do not eat slugs. Coastal snakes tongue-flicked more than inland snakes.

The difference between slug acceptors and slug rejecters appears to be inherited, but what physiological difference have the genes brought about? Arnold devised a clever experiment to answer this causal question. When snakes eat, their tongues carry chemicals to an odor receptor in the roof the mouth. They use tongue flicks to recognize their prey! Even newborns will flick their tongues at cotton swabs dipped in fluids of their prey. Arnold dipped swabs in slug extract and counted the number of tongue flicks for newborn inland and coastal snakes. Coastal snakes had a higher number than inland snakes (Fig. 22.2). Although hybrids showed a great deal of variation in the number of tongue flicks, they were generally intermediate as predicted by the genetic hypothesis. These findings suggest that inland snakes do not eat slugs because they are not sensitive to their smell. It would seem that the genetic differences between the two populations of snakes has resulted in a physiological difference in their nervous systems.

Both nervous and endocrine systems are responsible for the integration of body systems. Is the endocrine system also involved in behavior, and is this system also regulated by the animal's genes? Various studies have been done to show that it is. For example, the egg-laying behavior in the sea slug *Aplysia* involves a set sequence of movements. Following copulation, the animal extrudes long strings of more than a million egg cases. It takes the egg case string in its mouth, covers it with mucus, and waves its head back and forth to wind the string into an irregular mass. This behavior attaches the mass to a solid object, like a rock. Several years ago, scientists isolated and analyzed an egg-laying hormone (ELH) that causes the snail to lay eggs even if it had not mated. ELH was found to be a small protein of 36 amino acids that diffuses into the circulatory system and excites the smooth muscle cells of the reproductive duct, causing them to contract and expel the egg string. Using recombinant DNA techniques, the investigators isolated the ELH gene. The gene's product turned out to be a protein with 271 amino acids. The protein can be cleaved into as many as 11 possible products, and ELH is one of these. ELH alone, or in conjunction with these other products, is thought to control all the components of egg-laying behavior in *Aplysia*.

The results of many types of studies support the hypothesis that behavior has at least a partial genetic basis, and that genes influence the development of neural and hormonal mechanisms that control behavior.

22.2 Behavior Undergoes Development

Given that all behaviors have at least a partial genetic basis, we can go on to ask if environmental experiences after hatching or birth also shape the behavior. Fixed action patterns (FAPs) are stereotyped behaviors—they are always performed the same way each time. FAPs are elicited by a sign stimulus, a cue that sets the behavior in motion. For example, human babies will smile when a flat but face-sized mask with two dark spots for eyes is brought near them. It's possible that some behaviors are FAPs, but increasingly investigators are finding that many behaviors, formerly thought to be FAPs, develop after practice.

One investigator, Jack Hailman, has exhaustively studied laughing gull chicks' begging behavior, which is always performed the same way in response to the parent's red beak. A chick directs a pecking motion toward the parent's beak, grasps it, and strokes it downward (Fig. 22.3). Hailman noticed that sometimes a parent stimulates the begging behavior by swinging its beak gently from side to side. After the chick responds, the parent regurgitates food onto the floor of the nest. If need be, the parent then encourages the chick to eat. This interaction between the chicks and their parents made Hailman think the begging behavior might involve learning. (**Learning** is defined as a durable change in behavior brought about by experience.) To test this hypothesis, Hailman painted diagrammatic pictures of gull heads on small cards and then collected eggs in the field. The eggs were hatched in a dark incubator to eliminate visual stimuli before the test. On the day of hatching, each chick was allowed to make about a dozen pecks at the model. The chicks were returned to the nest and were later retested. The tests showed that on the average, only one-third of the pecks by a newly hatched chick strike the model. But one day after hatching more than half of the pecks are accurate, and two days after hatching the accuracy reaches a level of more than 75%. His treatment and testing of other groups allowed him to conclude that improvement in motor skills, as well as visual experience, strongly effect development of chick begging behavior.

Behavior has a genetic basis, but the development of mechanisms that control behavior is subject to environmental influences, such as practice after birth.

Hailman went on to test how chicks recognize a parent. He found that newly hatched chicks peck equally at any model as long as it has a red beak. Chicks a week old, however, will peck only at models that closely resemble the parent. Hailman speculated that operant conditioning with a reward of food could account for this change in behavior.

Operant conditioning, which is one of many forms of learning, is often defined as the gradual strengthening of stimulus-response connections. In everyday life, most people know that animals can be taught tricks by giving rewards such as food or affection. The trainer presents the stimulus, say a hoop, and then gives a reward (food) for the proper response (jumping through the hoop).

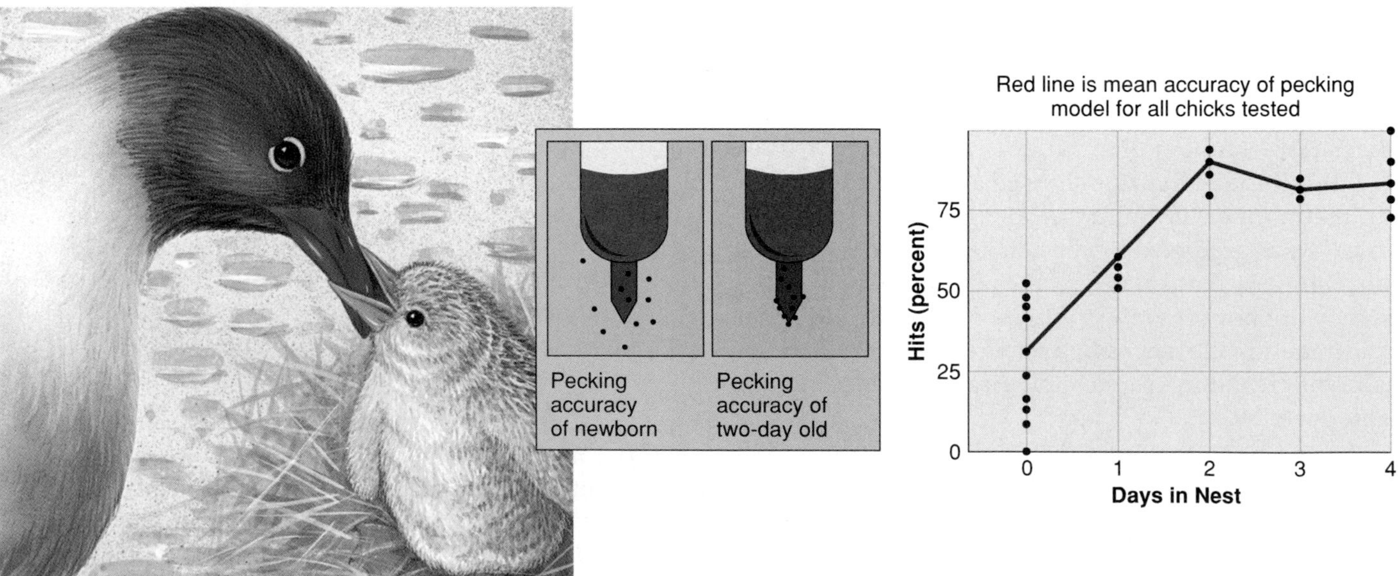

Figure 22.3 Pecking behavior of laughing gull chicks, *Larus atricilla*.
At about three days, laughing gull chicks grasp the beak of a parent, stroking it downward and the parent then regurgitates food.

B. F. Skinner is well known for studying this type of learning in the laboratory. In the simplest type of experiment performed by Skinner, a caged rat happens to press a lever and is rewarded with sugar pellets, which it avidly consumes. Rewarding an animal for a particular behavior *reinforces* the behavior and makes it more likely to happen again. In this way the rat learns to regularly press the lever whenever it is hungry. In more sophisticated experiments, Skinner even taught pigeons to play Ping-Pong by reinforcing desired responses to stimuli.

Imprinting is another form of learning; chicks, ducklings, and goslings will follow the first moving object they see after hatching. This object is ordinarily their mother, but they seemingly can be imprinted on any object—a human or a red ball—if it is the first moving object they see during a sensitive period of two to three days after hatching. The term *sensitive period* means that the behavior only develops during this time. Although the Englishman Douglas Spalding first observed imprinting, the Austrian Konrad Lorenz is well known for investigating it. He found that imprinting not only served the useful purpose of keeping chicks near their mother, it also caused male birds to court a member of the correct species—someone who looks like mother! The goslings who had been imprinted on Lorenz courted human beings later in life.

In-depth studies on imprinting have shown that the process is more complicated than originally thought. Eckhard Hess found that mallard ducklings imprinted on humans in the laboratory would switch to a female mallard that had hatched a clutch of ducklings several hours before. Also, he found that vocalization before and after hatching was normally an important element in the imprinting process. Female mallards cluck during the entire time that imprinting is occurring. Do social interactions influence other forms of learning? Patterns of song learning in birds suggests that they can.

Birds That Learn to Sing

During the past several decades, an increasing number of investigators have studied song learning in birds. Although white-crowned sparrows have a species-specific song, males from different regions have different dialects. This caused Peter Marler to hypothesize that young birds learn how to sing from older birds. To test his hypothesis, Marler housed the birds in cages where he could completely control the songs they heard (Fig. 22.4). A group of birds that *heard no songs at all* sang a song, but it was not fully developed. Birds that *heard tapes of white-crowns singing* sang in that dialect, as long as the tapes were played during a sensitive period from about age ten to fifty days. White-crowned sparrows' dialects (or other species' songs) played before or after this sensitive period had no effect on their song. Apparently, their brain is especially primed to respond to acoustical stimuli during the sensitive period. Neurons that are critical for song production have been located, and they fire when a bird hears a recording of either its own song or its own dialect sung by another bird. Other investigators have shown that birds *given an adult tutor* will sing the song of even a different species— no matter when the tutoring begins! It would appear that social experience has a very strong influence over the development of singing.

Animals have an ability to benefit from experience; learning occurs when a behavior changes with practice.

Isolated bird sings but song is not developed.

Bird sings developed song played during a sensitive period.

Bird sings song of social tutor without regard to sensitive period.

Figure 22.4 Song learning by white-crowned sparrows, *Zonotrichia leucophrys*.
Three different experimental procedures are depicted and the results noted. These results suggest that there is both a genetic basis and an environmental basis for song learning in birds.

22.3 Behavior Is Adaptive

Since genes influence the development of behavior, it is reasonable to assume that behavioral traits (like other traits) can evolve. Our discussion will focus on reproductive behavior—specifically the manner in which animals secure a mate. But we will also touch on the other two survival issues—capturing resources and avoiding predators—because these help an animal survive, and without survival reproduction is impossible. Investigators studying survival value questions seek to test hypotheses that specify how a given trait might improve reproductive success.

Males can father many offspring because they produce sperm in great quantity. We would then expect competition among males to inseminate as many different females as possible. In contrast, females produce few eggs, so the choice of a mate becomes a prevailing consideration. Sexual selection can bring about evolutionary changes in the species. **Sexual selection** is changes in males and females, often due to male competition and female selectivity, leading to reproductive success.

Females Choose

Courtship displays are rituals that serve to prepare the sexes for mating. They help male and female recognize each other so that mating will be successful. They also play a role in a female's choice of a mate.

Gerald Borgia conducted a study of satin bowerbirds (see opening photo, p. 367) to test these two opposing models regarding female choice:

Good genes hypothesis: females choose mates on the basis of traits that are expected to improve the male's chances of survival.

Run-away hypothesis: females choose mates on the basis of traits that make them attractive to females. The term "run away" pertains to the possibility that the trait will be exaggerated in the male until its reproductive unfavorable benefit is checked by the trait's unfavorable survival cost.

In both these models there is no parental investment on the part of males.

Borgia and his assistants watched bowerbirds at feeding stations and also monitored the bowers. They discovered that although males tend to steal blue feathers and/or actively destroy a neighbor's bower, more aggressive and vigorous males were able to keep their bowers in good condition. These were the males usually chosen as mates by females.

Borgia felt that the investigation did not clearly support either hypothesis. It could be that aggressiveness, if inherited, does improve the chances of survival, or it could be that females simply preferred bowers with the most blue feathers.

Bruce Beehler studied the behavior of the birds of paradise in New Guinea (Fig. 22.5). The raggiana bird of paradise is remarkably dimorphic—the males are larger than females and have beautiful orange flank plumes. In contrast, the females are drab. Courting males gather and begin to call. (The gathering of courting males is called a lek.) If a female joins them, the males raise their orange display plumes, shake their wings and hop from side to side, while continuing to call. They then stop calling and lean upside down with the wings projected forward to show off their beautiful feathers.

Female choice can explain why male birds are so much more showy than females, even if it is not known which of Borgia's two hypotheses applies. It's possible that the remarkable plumes of raggiana males do signify health and vigor. Or it's possible that females choose the flamboyant males on the basis that their sons will have an increased chance of being selected by females. Some investigators have hypothesized that extravagant male features could indicate that they are relatively parasite free. Anders Moller has tested this hypothesis using the barn swallow as his experimental material. He artificially shortened and lengthened the tails of male swallows and found that females chose those with the longest tails. Then he showed that males reared in nests sprayed with a mite killer had longer tails than otherwise.

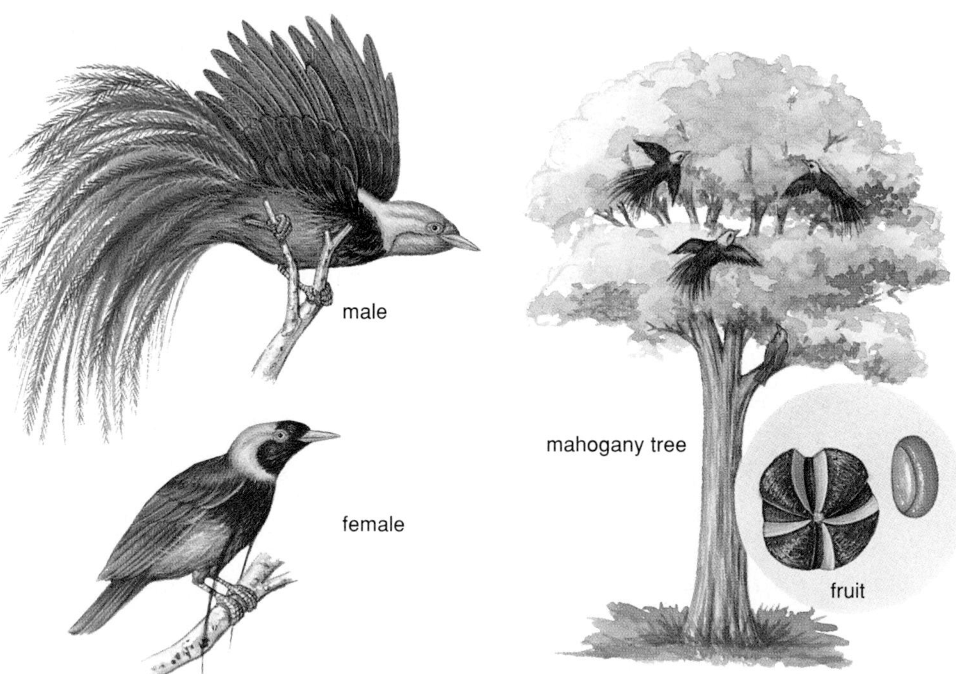

Figure 22.5 Mating behavior in birds of paradise, *Paradisaea raggiana*. Mating behavior in birds of paradise is influenced by species' foraging habits. The raggiana male has resplendent plumage brought about by sexual selection. The females are widely scattered, foraging for complex fruits; the males form leks that females visit to choose a mate.

Figure 22.6 A male olive baboon, *Papio anubis*, displaying full threat.
Males are larger than females and have enlarged canines. Competition between males establishes a dominance hierarchy for the distribution of resources.

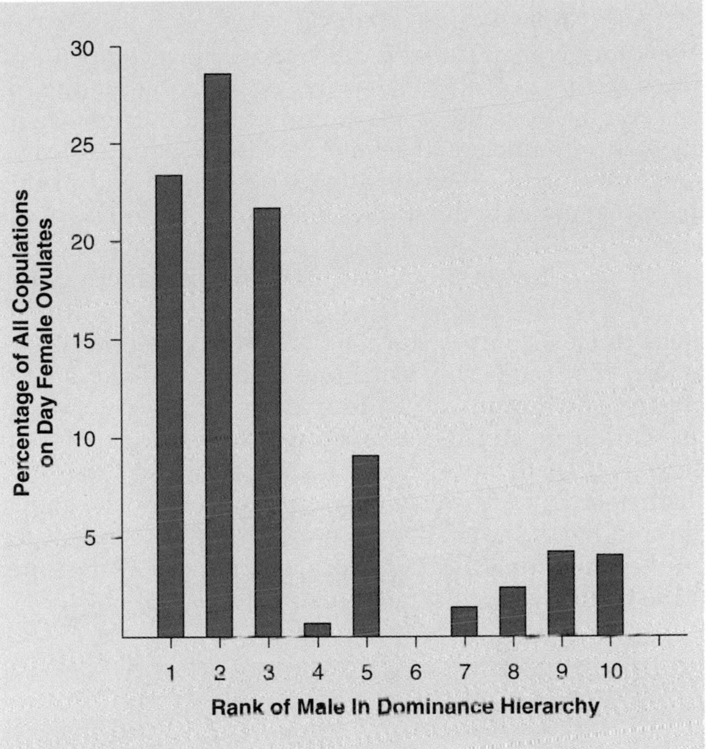

Figure 22.7 Female choice and male dominance among baboons.
Although it may appear that females mate indiscriminately, actually they mate more often with a dominant male when they are most fertile.

Males Compete

A trait that helps a male compete for mates can lead to a positive effect (e.g., more mates), but what are the possible negative effects (e.g., early death)? This type of benefit/cost analysis applies to fitness, traits that favor an individual's reproductive contribution to the subsequent generation. It can be reasoned that when costs are high, it is unlikely that a trait will evolve, but when benefits are high a trait is expected to evolve.

Baboons Have a Dominance Hierarchy

Baboons, a type of Old World monkey, live together in a troop. Males and females have separate **dominance hierarchies** in which a higher ranking animal has greater access to resources than a lower ranking animal. Dominance is decided by confrontations, resulting in one animal giving way to the other.

Baboons are dimorphic; the males are larger than the females and they can threaten others with their long, sharp canines (Fig. 22.6). The baboons travel within a territory, hunting food each day, and sleeping in trees at night. The dominant males decide where and when the troop will move and, if the troop is threatened, they cover the troop as it retreats, and attack when necessary.

Females undergo a period known as estrus, during which they ovulate and are willing to mate. When first receptive during a particular estrus cycle, a female mates with subordinate males and older juveniles. Later a female approaches a dominant male, and the male becomes very interested once ovulation is near. They form a consort pair, which lasts only for an hour or several days. Dominant males are interested in and protective of young baboons, regardless of the true father.

However, the male baboon pays a cost for his dominant position. Being larger means that he needs more food, and being willing and able to fight predators means that he may get hurt, and so forth. Is there a reproductive benefit to his behavior? Glen Hausfater counted copulations of dominant males and found that they do indeed monopolize estrous females when he judged them to be most fertile (Fig. 22.7). Nevertheless, there are other avenues to fathering offspring. Some males act as helpers to particular females and her offspring; the next time she is in estrus she may mate preferentially with him instead of a dominant male. Or subordinate males may form a friendship group that can oppose a dominant male, making him give up a receptive female.

Red Deer Stake Out a Territory

A territory is an area that is defended against competitors. **Territoriality** is protecting an area against other individuals. Vocalization and displays, rather than outright fighting, may be sufficient to defend a territory (Fig. 22.8). Male songbirds signify their territories by singing and in this way other males of the species know when an area is occupied.

T. H. Clutton-Brock has studied reproductive success among red deer (elk) on the Scottish island of Rhum. Stags (males) compete for a harem, a group of hinds (females) that mate only with one stag, called the harem master. The reproductive group occupies a territory that the harem master defends against other stags. Harem masters first attempt to repel challengers by roaring. If the challenger remains, the two lock antlers and push against one another. If the challenger now withdraws, the master pursues him for a short distance, roaring the whole time. If the challenger wins, he becomes the harem master.

After studying the red deer for more than twelve years, Clutton-Brock concluded that a harem master can father two dozen offspring at most, because he is at the peak of his fighting ability for only a short time. And what does it cost him to be able to father offspring? Stags must be large and powerful in order to fight; therefore, they grow faster and have less body fat. During bad times they are more likely to die of starvation and in general have shorter lives. The behavior of harem defense by stags will only persist in the population if its cost (reduction in the potential number of offspring because of a shorter life) is less than its benefit (increased number of offspring due to harem access).

Evolution by sexual selection can occur when females have the opportunity to select among potential mates and/or when males compete among themselves for access to reproductive females.

Figure 22.8 Competition between males among red deer, *Cervus elaphus*.
Male red deer compete for a harem within a particular territory. **a.** Roaring alone may frighten off a challenger. **b.** But outright fighting may be necessary, and the victor is most likely the stronger of the two animals.

A closer Look

How Nature Keeps Time

Certain behaviors in animals reoccur at regular intervals. Behaviors that occur on a daily basis are said to have a circadian (about a day) rhythm. For example, some animals, like humans, are usually active during the day and sleep at night. Others, such as bats, sleep during the day and hunt at night. There are also behaviors that occur on a yearly basis. For example, in the Northern Hemisphere, birds migrate south in the fall, and the young of many animals are born in the spring. Such behaviors have a circannual rhythm.

There are two hypotheses regarding the control of circadian rhythms: either their timing is controlled externally, or there is an internal timing device, often called a biological clock (Fig. 22A). These alternative hypotheses have been tested in crickets, which regularly call every night to attract females. When laboratory crickets are kept in a room under constant conditions with lights continually on or continually off, they continue to call every night, only calling starts as much as 26 hours later than it did on the day before. But exposure to night and day cycles will right the cycles; therefore, it is possible to conclude that there is a biological clock, but it does not keep perfect time and must be reset by environmental stimuli. Similar results have been obtained with all sorts of animals, from fiddler crabs to humans.

Where is the biological clock located? If the optic lobes of crickets are removed from the rest of the brain, crickets will call regardless of the time of day. Therefore, it seems that the clock is located in the optic lobe and that light stimuli received by the eyes reset the clock. Nervous centers for biological clocks have also been located in other animals.

Aside from keeping time, a biological clock must also be able to bring about the change in behavior. In crickets, nerve impulses must be initiated that cause singing to begin. In other instances, it seems that hormones control the behavior in question.

The control of circannual rhythms has also been investigated. In mammals, we know that melatonin is produced at night by the pineal gland, which gets its optical input from the eyes. The amount of melatonin begins to increase in the fall and decrease in the spring. Melatonin is believed to inhibit development of the reproductive organs, which explains why some animals reproduce in the spring. For animals in which the pineal gland is the "third eye," both the clock and the receptor are presumed to be in the pineal gland.

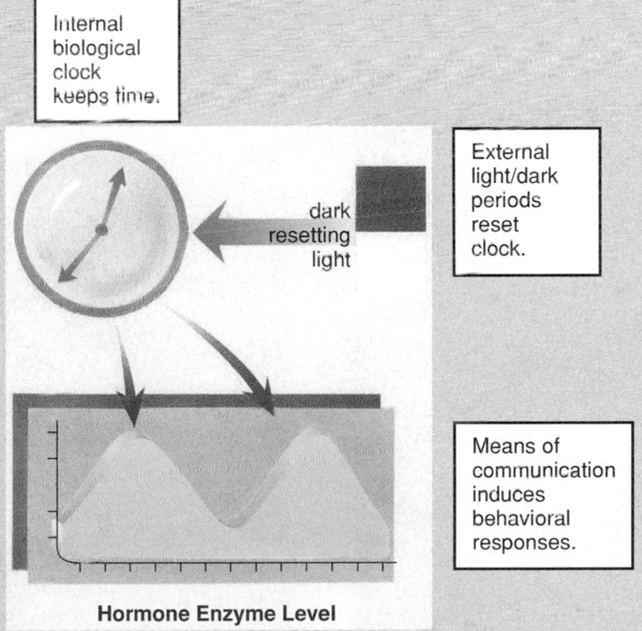

Figure 22A A biological clock system.
A biological clock system has three components, as shown.

22.4 Animals Are Social

Animals exhibit a wide range of degrees of sociality. Some animals are largely solitary and join with a member of the opposite sex only for the purpose of reproduction. Others pair, bond, and cooperate in raising offspring. Still others form a **society** in which members of species are organized in a cooperative manner, extending beyond sexual and parental behavior. We have already had occasion to mention the social groups of baboons and red deer. Social behavior in these and other animals requires that they communicate with one another.

Communication Is Varied

Communication is an action by a sender that influences the behavior of a receiver. The communication can be purposeful but does not have to be purposeful. Bats send out a series of sound pulses and listen for the corresponding echoes in order to find their way through dark caves and locate food at night. Some moths have an ability to hear these sound pulses, and they begin evasive tactics when they sense that a bat is near. Are the bats purposefully communicating with the moths? No, but the bat sounds are a *cue* to the moths that danger is near.

Communication is an action by a sender that affects the behavior of a receiver.

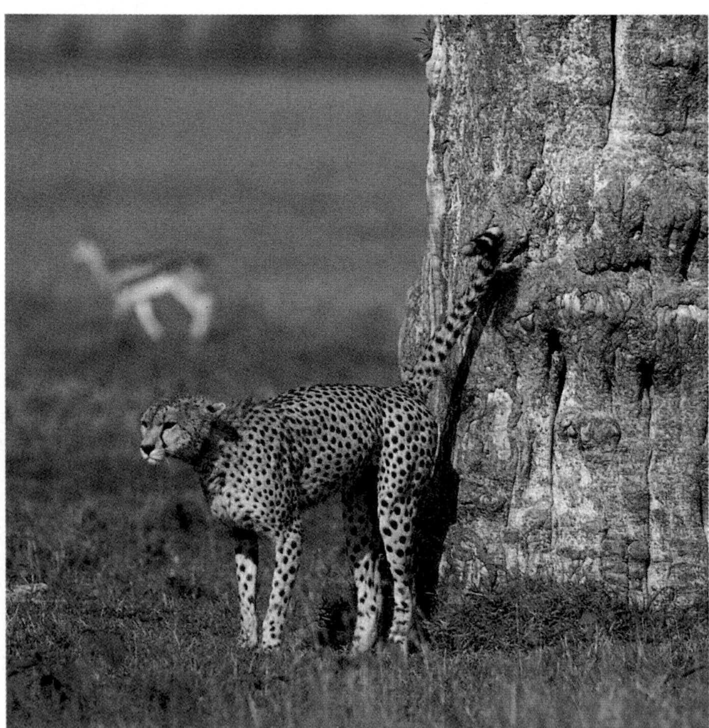

Figure 22.9 Use of pheromone to mark a territory. This male cheetah, *Acinonyx*, is spraying a pheromone onto a tree in order to mark its territory.

Chemical Communication

Chemical signals have the advantage of working both night and day. The term **pheromone** [Gk. *phero*, bear, carry, and *monos*, alone] is used to designate chemical signals that are passed between members of the same species. Female moths secrete chemicals from special abdominal glands, which are detected downwind by receptors on male antennae. The antennae are especially sensitive, and this assures that only male moths of the correct species (and not predators) will be able to detect them.

Cheetahs and other cats mark their territories by depositing urine, feces, and anal gland secretions at the boundaries (Fig. 22.9). Klipspringers (small antelope) use secretions from a gland below the eye to mark twigs and grasses of their territory.

Auditory Communication

Auditory (sound) communication has some advantages over other kinds of communication. It is faster than chemical communication, and it also is effective both night and day. Further, auditory communication can be modified not only by loudness but also by pattern, duration, and repetition. In an experiment with rats, a researcher discovered that an intruder can avoid attack by increasing the frequency with which it makes an appeasement sound.

Male crickets have calls, and male birds have songs for a number of different occasions. For example, birds may have one song for distress, another for courting, and still another for marking territories. Sailors have long heard the songs of humpback whales because they are transmitted through the hull of a ship. But only recently has it been shown

Figure 22.10 A chimpanzee with a researcher. A chimpanzee, *Pan*, with a researcher. Chimpanzees have a limited ability to speak but can learn to use a visual language consisting of symbols. Some believe chimps only mimic their teachers and never understand the cognitive use of a language. Here the experimenter shows Nim the sign for "drink." Nim copies.

that the song has six basic themes, each with its own phrases, that can vary in length and be interspersed with sundry cries and chirps. The purpose of the song is probably sexual and serves to advertise the availability of the singer. Language is the ultimate auditory communication, but only humans have the biological ability to produce a large number of different sounds and to put them together in many different ways. Nonhuman primates have at most only forty different vocalizations, each having a definite meaning, such as the one meaning "baby on the ground," which is uttered by a baboon when a baby baboon falls out of a tree. Although chimpanzees can be taught to use an artificial language, they never progress beyond the capability level of a two-year-old child (Fig. 22.10). It has also been difficult to prove that chimps understand the concept of grammar or can use their language to reason. It still seems as if humans possess a communication ability unparalleled by other animals.

Visual Communication

Visual signals are most often used by species that are active during the day. Contests between males make use of threat postures and possibly prevent outright fighting that might result in reduced fitness. A male baboon displaying full threat is an awesome sight that establishes his dominance and keeps peace within the baboon troop (see Fig. 22.6). Hippopotamuses have territorial displays that include mouth opening.

The plumage of a male raggiana bird of paradise allows him to put on a spectacular courtship dance to attract females and to give them a basis on which to select a suitable mate. Defense and courtship displays are exaggerated and are always performed in the same way so that their meaning is clear.

Tactile Communication

Tactile communication occurs when one animal touches another. For example, gull chicks peck at the parent's beak in order to induce the parent to feed them (see Fig. 22.3). A male leopard nuzzles the female's neck to calm her and to stimulate her willingness to mate. In primates, grooming—one animal cleaning the coat and skin of another—helps cement social bonds within a group.

Honeybees use a combination of tactile and auditory communication to impart information about the environment. Karl von Frisch, another famous behaviorist, did many detailed bee experiments in the 1940s. He discovered that when a foraging bee returns to the hive, it performs a *waggle dance* that indicates the distance and the direction of a food source (Fig. 22.11) As the bee moves between the two loops of a figure 8, it buzzes noisily and shakes its entire body in so-called waggles. Distance to the food source is believed to be indicated by the number of waggles and/or the amount of time taken to complete the straight run. Outside the hive, the dance is done on a horizontal surface and the straight run indicates the direction of the food. Inside the hive, the angle of the straight run to that of the direction of gravity is the same as the angle of the food source to the sun. In other words, a 40° angle to the left of vertical means that food is 40° to the left of the sun. Bees can use the sun as a compass to locate food because their biological clocks, as discussed in the reading on page 375, allow them to compensate for the movement of the sun in the sky.

Animals use a number of different ways to communicate, and communication sometimes facilitates cooperation.

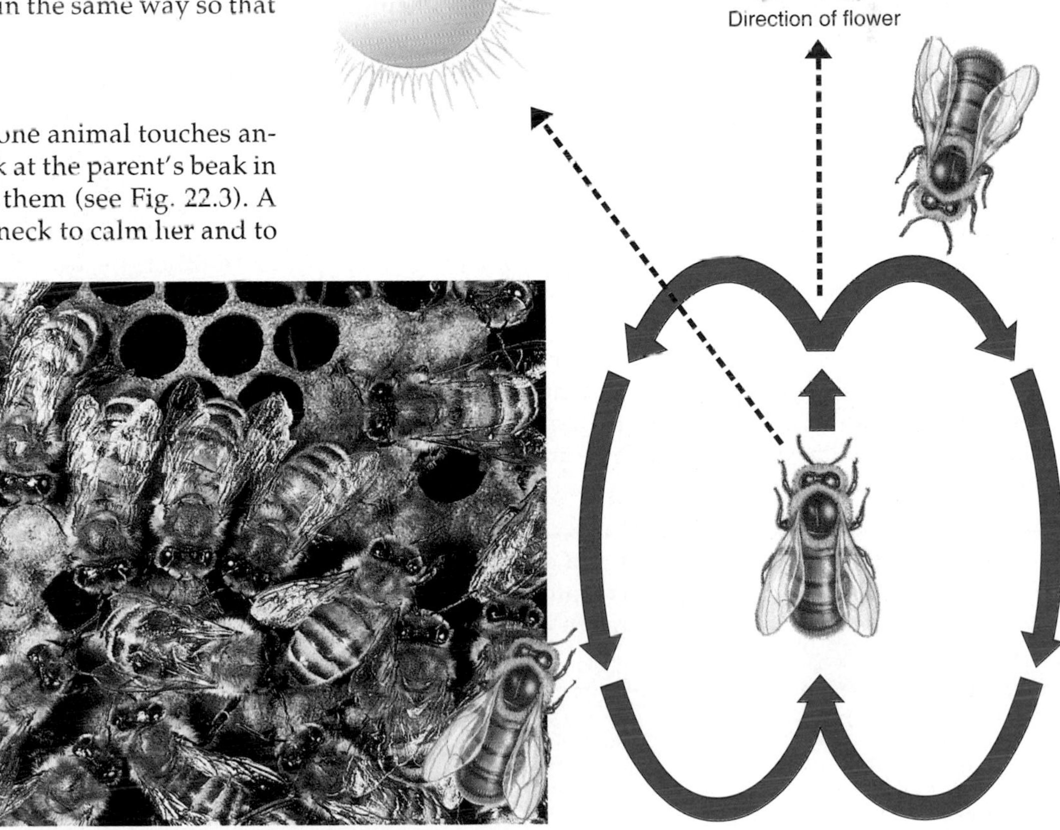

Figure 22.11 Communication among bees.
a. Honeybees, *Apis mellifera*, do a waggle dance to indicate the direction of food. **b.** If the dance is done outside the hive on a horizontal surface, the straight run of the dance will point to the food source.

22.5 Sociobiology and Animal Behavior

Sociobiology applies the principles of evolutionary biology to the study of social behavior in animals. Sociobiologists develop hypotheses about social living based on the assumption that a social individual derives more reproductive benefits than costs from living in a society. Then they may perform a cost-benefit analysis to see if their hypotheses are correct.

Group living does have benefits under certain circumstances. It can help an animal avoid predators, rear offspring, and find food. A group of impalas is more likely to hear an approaching predator than a solitary one. Many fish moving rapidly in many directions might distract a would-be predator.

Whether birds are social can depend on their food source. Raggiana birds feed on nutritious, complex fruits but a related species, the trumpet manucode, *Manucodia keraudrenit,* pair bond, perhaps for life, and both sexes are needed to successfully raise the young. Weaver birds are even more social. They form giant colonies that help protect them from predators, but the birds may also share information about food sources. Primate members of the same troop signal to one another when they have found an especially bountiful fruit tree. Lions working together are able to capture large prey, such as zebra and buffalo.

Group living also has its disadvantages. When animals are crowded together into a small area, disputes can arise over access to the best feeding places and sleeping sites. Dominance hierarchies are one way to apportion resources, but this puts subordinates at a disadvantage. Clutton-Brock found that only large, dominant red deer females can successfully rear sons that tend to result in many offspring. Small subordinate females tend to rear daughters, and daughters tend to result in fewer offspring. But like the subordinate males in a baboon troop, the subordinate red deer females may be more likely to survive and therefore be better off in fitness terms if they stay with a group, despite the cost involved.

Living in close quarters means that illness and parasites can pass from one to the other more rapidly. Baboons and other types of social primates invest much time grooming one another and this most likely helps them remain healthy.

Social living has both advantages and disadvantages. Only if the benefits, in terms of individual reproductive success, outweigh the disadvantages will societies evolve.

Altruism Versus Self-Interest

Altruism [L. *alter,* the other] is behavior that has the potential to decrease the lifetime reproductive success of the altruist while benefiting the reproductive success of another member of the group. In insect societies, especially, reproduction is limited to only one pair, the queen and her mate. For example, among army ants the queen is inseminated only during her nuptial flight, and thereafter she spends her time reproducing. The society has three different sizes of sterile female workers. The smallest workers (3 mm), called the nurses, take care of the queen and larvae, feeding them and keeping them clean. The intermediately sized workers, constituting most of the population, go out on raids to collect food. The soldiers (14 mm), with huge heads and powerful jaws, run along the sides and rear of raiding parties where they can best attack any intruders.

Can the altruistic behavior of sterile workers be explained in terms of reproductive success? In 1964, the English biologist William Hamilton pointed out that a given gene can be passed from one generation to the next in two quite different ways. The first way is direct: a parent can pass the gene directly to an offspring. The second way is indirect: an animal can help a relative reproduce and thereby pass the gene to the next generation via this relative. Direct selection is natural selection that can result in adaptation to the environment when the reproductive success of individuals differs. Indirect selection is natural selection that can result in adaptation to the environment when individuals differ in their effects on the reproductive success of relatives. The **inclusive fitness** of an individual includes both personal reproduction and reproduction of relatives.

Among social bees, social wasps, and the ants, the queen is diploid but her mate is haploid. If the queen has had only one mate, sister workers are more closely related to each other (sharing on average 75% of their genes) than they are to their potential offspring (with which they share on average only 50% of their genes). Therefore, a worker can achieve a greater inclusive fitness benefit by aiding her mother (the queen) to produce additional sisters, than by directly reproducing herself. Under these circumstances, behavior that appears to be altruistic is more likely to evolve.

Indirect selection can also occur among animals whose offspring receive only a half set of genes from both parents. Consider that your brother or sister shares 50% of your genes, your niece or nephew shares 25%, and so forth. This means that the survival of two nieces (or nephews) is worth the survival of one sibling, assuming they both go on to reproduce.

Michael Ghiglieri, when studying chimpanzees in Africa, observed that a female in estrus frequently copulates with several members of the same group, and that the males make no attempt to interfere with each other's matings. How can they be acting in their own self-interest? Genetic relatedness of the males appears to underlie their apparent altruism; they all share genes in common because males never leave the territory in which they are born.

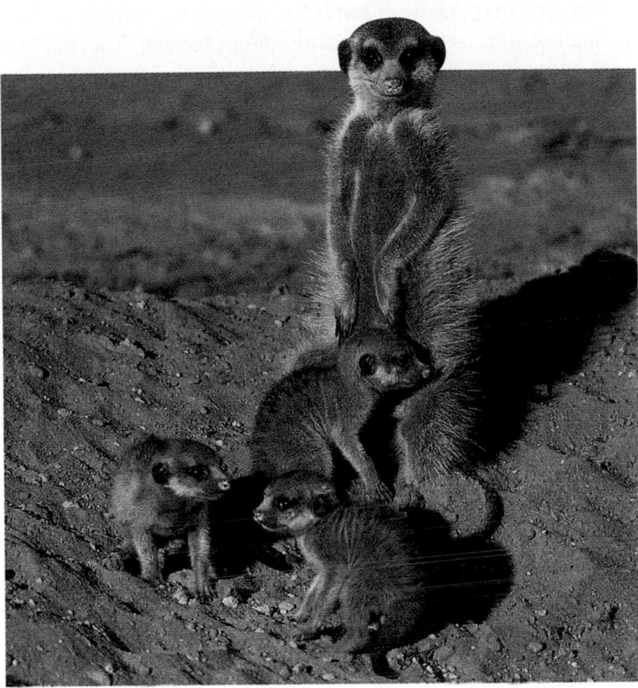

Figure 22.12 Inclusive fitness.
A meerkat, *Suricata*, is acting as a baby-sitter for its young sisters and brothers while their mother is away. Could this helpful behavior contribute to the baby sitter's inclusive fitness?

Helpers at the Nest

In some bird species, offspring from one clutch of eggs may stay at the nest, helping parents rear the next batch of offspring. In a study of Florida scrub jays, the number of fledglings produced by an adult pair doubled when they had helpers. Mammalian offspring are also observed to help their parents (Fig. 22.12). Among jackals in Africa, pairs alone managed to rear an average of 1.4 pups, whereas pairs with helpers reared 3.6 pups.

Is the reproductive success of the helpers increased by their apparent altruistic behavior? It could be if the chance of their reproducing on their own is limited. David and Sandra Ligon studied the breeding behavior of green wood-hoopoes *(Phoeniculus purpurens)*, an insect-eating bird of Africa. A flock may have as many as sixteen members but only one breeding pair; the other sexually mature members help feed and protect the fledglings and protect the home territory from invasion by other green wood-hoopoes. The Ligons found that nest sites are rare, as acacia trees with cavities are relatively rare and the cavities are often occupied by other species. Moreover, predation by snakes within the cavities can be intense; even if a pair of birds acquire an appropriate cavity, they would be unable to protect their offspring by themselves. Therefore, the cost of trying to establish a territory is clearly very high.

What are the benefits of staying behind to help? First, a helper is contributing to the survival of its own kin. Therefore, the helper actually gains a fitness benefit (albeit a smaller benefit than it would achieve were it a breeder). Second, a helper is more likely than a non-helper to inherit a parental territory—including other helpers. Helping, then, involves making a minimal, short-term reproductive sacrifice in order to maximize future reproductive potential. Once again, an apparently altruistic behavior turns out to be an adaptation.

Inclusive fitness is measured by the genes an individual contributes to the next generation, either directly by offspring or indirectly by way of relatives. Many of the behaviors once thought to be altruistic turn out, on closer examination, to be examples of kin selection, and are adaptive.

connecting concepts

Birds build nests, dogs bury bones, cats chase quick-moving objects, and snakes bask in the sun. All animals, including humans, behave—they respond to stimuli, both from the physical environment and from other individuals of the same or different species. Behaviorists are scientists who seek answers to two types of questions: what is the neurophysiological mechanism of the behavior, and how is the behavior beneficial to the organism? Using an evolutionary approach, behaviorists generate hypotheses that can be tested to better understand how the behavior increases individual fitness (i.e., the capacity to produce surviving offspring).

Both types of behavioral studies—those concerned with neurophysiology and those concerned with biological fitness—recognize that behavior has a genetic basis. Inheritance determines, for example, that hawks hunt by using vision rather than smell, and that cats, not dogs, climb trees.

Inheritance, too, produces the behavioral variations that are subject to natural selection, resulting in the most adaptive responses to stimuli. There is survival value, after all, in the ability of male baboons to react aggressively when the troop is under attack.

The study of behavior, however, does not settle the nature-nurture question because we now know that behavior can be modified by experience. Songbirds are born with the ability to sing, but which song or dialect they sing is strongly dependent on the songs they hear from their parents and siblings. A new chimpanzee mother naturally cares for her young, but she is a better mother if she observes other females in her troop raising young.

There is a cost to certain behaviors. Why should subordinate members of a baboon troop or subordinate females in a red deer harem remain in a situation that seemingly leads to reduced fitness? And why should older offspring help younger offspring? The evolutionary answer is that, in the end, there are benefits that outweigh the costs. Otherwise, the behavior would not continue. The evolutionary approach to studying behavior has proved fruitful in helping us understand why birds sing their melodious songs, dolphins frolic in groups, and wondrous birds of paradise display to females.

Summary

22.1 Behavior Has a Genetic Basis

Behaviorists study how animals are organized to perform a particular behavior and how it benefits them to survive and reproduce. Hybrid studies with blackcap warblers produce results consistent with the hypothesis that behavior has at least a partial genetic basis. Garter snake experiments indicate that the nervous system controls behavior. *Aplysia* DNA studies indicate that the endocrine system also controls behavior.

22.2 Behavior Undergoes Development

The environment is influential in the development of behavioral responses, as exemplified by an improvement in laughing gull chick begging behavior and an increased ability of chicks to recognize parents. Modern studies suggest that most behaviors improve with experience. Even behaviors that were formerly thought to be fixed action patterns (FAPs) or otherwise thought to be inflexible sometimes can be modified.

Song learning in birds involves various elements—including the existence of a sensitive period when an animal is primed to learn—and the effect of social interactions.

22.3 Behavior Is Adaptive

Traits that promote reproductive success are expected to be conserved. Males who produce many sperm are expected to compete to inseminate females. Females who produce few eggs are expected to be selective about their mates. Experiments with satin bowerbirds and birds of paradise support these bases for sexual selection.

Other aspects of reproductive behavior are also adaptive. The raggiana bird of paradise males gather in a lek—most likely because females are widely scattered, as is their customary food source. The food source for a related species, the trumpet manucode, is readily available but less nutritious. These birds are monogamous—it takes two parents to rear the young.

A cost-benefit analysis can be applied to competition between males for mates in reference to a dominance hierarchy (e.g., baboons) and territoriality (e.g., red deer). If the behavior is conserved, the benefit most likely outweighs the cost.

22.4 Animals Are Social

Animals that form social groups may communicate with one another. Communications such as chemical, auditory, visual, and tactile signals may foster cooperation that benefits both the sender and the receiver.

There are benefits and costs to living in a social group. If animals live in a social group, it is expected that the advantages (e.g., help to avoid predators, raise young, and find food) will outweigh the disadvantages (e.g., tension between members, spread of illness and parasites, and reduced reproductive potential). This expectation can sometimes be tested.

22.5 Sociobiology and Animal Behavior

In most instances, the individuals of a society act to increase their own reproductive success. Sometimes animals perform apparent altruistic acts, as when individuals help their parents rear siblings. There is a benefit to this behavior when one considers inclusive fitness, which involves both direct selection and indirect selection.

In social insects, apparent altruism is extreme but can be explained on the basis that they are helping a reproducing sibling survive. A study of wood-hoopoes, an African bird, shows that younger siblings may help older siblings who reared them until the younger ones get a chance to reproduce themselves.

Reviewing the Chapter

1. State the two types of questions asked by behaviorists and explain how they differ. 368
2. Describe Helbig's experiment with blackcap warblers, and explain how it shows that behavior has a genetic basis. 368
3. Arnold's experiment with garter snakes showed that what body system is involved in their behavior toward slugs? Explain the experiment and the results. 368–69
4. Studies of *Aplysia* DNA show that the endocrine system is also involved in behavior. Explain. 369
5. Some behaviors require practice before developing completely. How does Hailman's experiment with laughing gull chicks support this statement? 370
6. An argument can be made that social contact is an important element in learning. Explain with reference to imprinting in mallard ducks and song learning in white-crowned sparrows. 371
7. Why would you expect behavior to be subject to natural selection and be adaptive? 372
8. How do Borgia's studies of satin bowerbird behavior and Clutton-Brock's studies with red deer support the belief that reproductive behavior is related to male competition and female selectivity? 372, 374
9. Give examples of the different types of communication among members of a social group. 376–77
10. What is a cost-benefit analysis, and how does it apply to living in a social group? Give examples. 378
11. Give examples of behaviors that appear to be altruistic but actually increase the fitness of an individual. 378–79

Testing Yourself

Choose the best answer for each question.

1. Which of these questions is least likely to interest a behaviorist?
   a. How do genes control the development of the nervous system?
   b. Why do animals living in the tundra have white coats?
   c. Does aggression have a genetic basis?
   d. Why do some animals feed in groups and others feed singly?
2. Female sage grouse are widely scattered throughout the prairie. Which of these would you expect?
   a. A male will maintain a territory large enough to contain at least one female.
   b. Male and female birds will be monogamous, and both will help feed the young.
   c. Males will form a lek where females will choose a mate.
   d. Males will form a dominance hierarchy for the purpose of distributing resources.
3. White-crowned sparrows from two different areas sing with a different dialect. If the behavior is primarily genetic, newly hatched birds from each area will
   a. sing with their own dialect.
   b. need tutors in order to sing in their dialect.
   c. sing only when a female is nearby.
   d. Both a and c are correct.
4. Orangutans are solitary but territorial. This would mean orangutans defend their territory's boundaries against
   a. other male orangutans.
   b. female orangutans.
   c. all types of animals.
   d. Both a and b are correct.
5. The resplendent plumes of a raggiana bird of paradise are due to the fact that birds with the best display
   a. are dominant over other birds.
   b. have the best territories.
   c. are chosen by females as mates.
   d. All of these are correct.
6. Subordinate females in a baboon troop do not produce offspring as often as dominant females. It is clear that
   a. the cost of being in the troop is too high.
   b. the dominant males do not mate with subordinate females.
   c. subordinate females must benefit in some way from being in the troop.
   d. Both a and b are correct.
7. German blackcaps migrate southwest to Africa, and Austrian blackcaps fly southeast to Africa. The fact that hybrids of these two are intermediate shows that
   a. the trait is controlled by the nervous system.
   b. nesting is controlled by hormones.
   c. the behavior is at least partially genetic.
   d. Both a and c are correct.
8. At first laughing gull chicks peck at any model that looks like a red beak; later they will not peck at any model that does not look like a parent. This shows that the behavior
   a. is a fixed action pattern.
   b. undergoes development after birth.
   c. is controlled by the nervous system.
   d. All of these are correct.
9. Which of these could be an answer to a causal question? Males compete because
   a. they have the size and weapons with which to compete.
   b. they produce many sperm for a long time.
   c. the testes produce the hormone testosterone.
   d. Both a and c are correct.
10. Which of these could be an answer to a survival value question? Females are choosy because
    a. they do not have the size and weapons with which to compete.
    b. they invest heavily in the offspring they produce.
    c. the ovaries produce the hormones estrogen and progesterone.
    d. All of these are correct.

Applying the Concepts

1. *All behavior both has a genetic basis and is influenced by lifetime experiences, including learning.*
   Explain why you would expect this statement to apply even to human reasoning.
2. *Behavior increases individual fitness.*
   Why would you expect feeding behavior patterns, territoriality, and reproductive behavior patterns to increase individual fitness?
3. *Individuals live in a society when the benefits to the individual outweigh the costs.*
   Show that apparent altruistic acts may actually increase the fitness of an individual.

Using Technology

Your study of animal behavior is supported by these available technologies:

Exploring the Internet
The Mader Home Page provides resources for and help with studying this chapter.

http://www.mhhe.com/sciencemath/biology/mader/
(Click on Biology.)

Understanding the Terms

Match the terms to these definitions:

a. __________ Changes in males and females, often due to male competition and female selectivity, leading to reproductive success.

b. __________ Chemical released by the body that causes a predictable reaction of another member of the same species.

c. __________ Fitness that results from direct selection and indirect selection.

d. __________ Form of learning that occurs early in the lives of animals; a close association is made that later influences sexual behavior.

e. __________ Form of learning that results from rewarding or reinforcing a particular behavior.

f. __________ Observable, coordinated responses to environmental stimuli.

g. __________ Relatively permanent change in an animal's behavior that results from practice and experience.

h. __________ Signal by a sender that influences the behavior of a receiver.

i. __________ Social interaction that has the potential to decrease the lifetime reproductive success of the member exhibiting the behavior.

j. __________ Social ranking within a group in which a higher ranking individual acquires more resources than a lower ranking individual.

Ecology of Populations

CHAPTER 23

Chapter Concepts

Female units of elephants *(Loxodonta africana)*

Elephants have a large size, are social, live a long time, and produce few offspring. Females live in social family units, and the much larger males visit them only during a breeding season. Females give birth about every five years to a single calf that is well cared for and has a good chance of meeting the challenges of its life style. Normally, an elephant population exists at the carrying capacity of the environment.

A population ecologist studies the distribution and abundance of organisms and relates the hard, cold statistics to the species life history in order to determine what causes population growth or decline. Elephants are threatened because of human population growth, and also because humans tend to slaughter the males. Males don't tend to breed until they have reached their largest size—the size that makes them prized by humans. By now, even a moratorium on killing males will not help much. So few breeding males are left that elephant populations are expected to continue to decline for quite some time.

23.1 Scope of Ecology

In 1866, the German zoologist Ernst Haeckel coined the word ecology from two Greek roots [Gk. *oikos*, home, house, and *-logy*, "study of" from *logikos*, rational, sensible]. He said that **ecology** is the study of the interactions of organisms with other organisms and with the physical environment. And he pointed out that ecology and evolution are intertwined because ecological interactions are selection pressures that result in evolutionary change and evolutionary change affects ecological interactions, and so forth.

Ecology, like so many biological disciplines, is wide-ranging. At one of its lowest levels, ecologists study how the individual organism is adapted to its environment. For example, they study why fishes in a coral reef live only in warm tropical waters and how the fishes feed within that **habitat** (the place where an organism lives) (Fig. 23.1). Most organisms do not exist singly; rather, they are part of a population, a functional unit that interacts with the environment. A **population** is defined as all the organisms within an area belonging to the same species. At this level of study, ecologists are interested in factors that affect the growth and regulation of population size.

A **community** consists of all the various populations interacting at a locale. In a coral reef, there are numerous populations of fishes, crustacea, corals, and so forth. At this level ecologists want to know how interactions like predation and competition affect the organization of a community. An **ecosystem** contains a community of populations and also the abiotic environment. Energy flow and chemical cycling are significant aspects of understanding how an ecosystem functions. The **biosphere** is that portion of the entire earth's surface where living things exist.

Modern ecology is not just descriptive, it is predictive. It analyzes levels of organization and develops models and hypotheses that can be tested. A central goal of modern ecology is to develop models that explain and predict the distribution and abundance of organisms. Ultimately, ecology considers not one particular area, but the distribution and abundance of populations in the biosphere. What factors have brought about the mix of plants and animals in a tropical rain forest at one latitude and in a desert at another? While modern ecology is useful in and of itself, it also has almost unlimited application possibilities, such as the management of wildlife in order to prevent extinction, the maintenance of cultivated food sources, or even the ability to predict the course of an illness such as AIDS.

Ecology is the study of the interactions of organisms with other organisms and with the physical environment. These interactions determine the distribution and abundance of organisms at a particular locale and over the earth's surface.

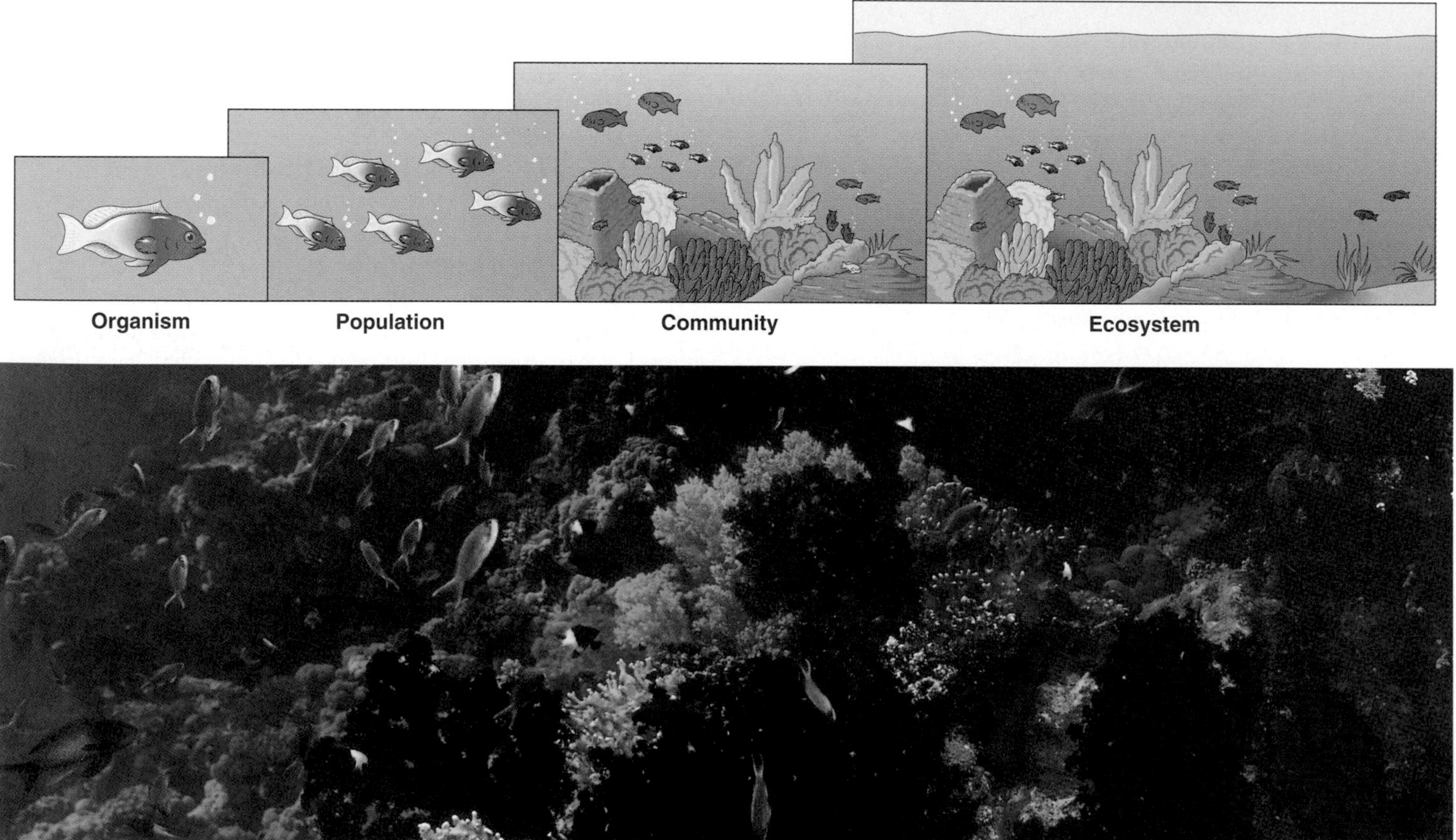

Figure 23.1 Ecological levels.
The study of ecology encompasses various levels from the individual organism to the population, community, and ecosystem.

Density and Distribution of Populations

Population density is the number of individuals per unit area or volume, while **population distribution** is the pattern of dispersal of individuals within the area of interest. In only a few populations are members uniformly distributed, as suggested by considering only population density (Fig. 23.2*a*). In some populations, individuals are randomly distributed as in Figure 23.2*b*, but most are clumped as illustrated in Figure 23.2*c*.

As an example of the relationship between density and distribution, consider that we can calculate the average density of people in the United States, but we know full well that most people live in cities where the number of people per unit area is dramatically higher than in the country. And even within a city more people live in particular neighborhoods than others, and such distributions can change over time. Therefore, basing ecological models solely on population density, as has often been done in the past, can be misleading.

Today ecologists want to analyze and discover what causes the spatial and temporal "patchiness" of organisms. For example, as discussed on page 398, a study of the distribution of hard clams in a bay on the south shore of Long Island, New York, showed that clam abundance is associated with sediment shell content. Indeed, it might be possible to use this information to transform areas that have few clams to high abundance areas.

As with the clams, the distribution of organisms can be due to *abiotic* (nonliving) factors. Physical factors like the type of precipitation, and not just the average temperature, but also the daily and seasonal variations in temperature, and the type of soil can affect where a particular organism lives. In fact, moisture, temperature, or a particular nutrient can be a limiting factor for the distribution of an organism. **Limiting factors** are those factors that particularly determine whether an organism lives in an area. Trout live only in cool mountain streams where there is a high oxygen content, but carp and catfish are found in rivers near the coast because they can tolerate warm waters, which have a low concentration of oxygen. The timberline is the limit of tree growth in mountainous regions or in high latitudes. Trees cannot grow above the high timberline because of low temperature and the fact that water remains frozen most of the year. The distribution of organisms can also be due to a *biotic* (living) factor. In Australia, the red kangaroo does not live outside inland areas because it is adapted to feeding on the arid grasses that grow there.

Ecology as a science includes a study of the distribution of organisms: where and why organisms are located in a particular place at a particular time.

a.

b.

c.

Figure 23.2 Patterns of dispersion within a population. Members of a population may be distributed uniformly, randomly, or usually in clumps. **a.** Golden eagle pair distribution is uniform over a suitable habitat area due to the territoriality of the birds. **b.** The distribution of moose aggregates is random over a suitable habitat. **c.** Cedar trees tend to be clumped near the parent plant because of poor distribution of seeds.

23.2 Characteristics of Populations

At any one point in time, populations have a certain size. **Population size** is the number of individuals contributing to the population's gene pool. Simply counting the number of individuals present in a population is not usually an option; instead it is necessary to estimate the present population size. Methods of estimating population size depend on the kind of species being studied and the validity of the estimate varies according to the method used.

What factors would determine the future size of a population? Just as you might suspect, populations increase in size whenever natality (the number of births) exceeds mortality (the number of deaths) and whenever immigration exceeds emigration.

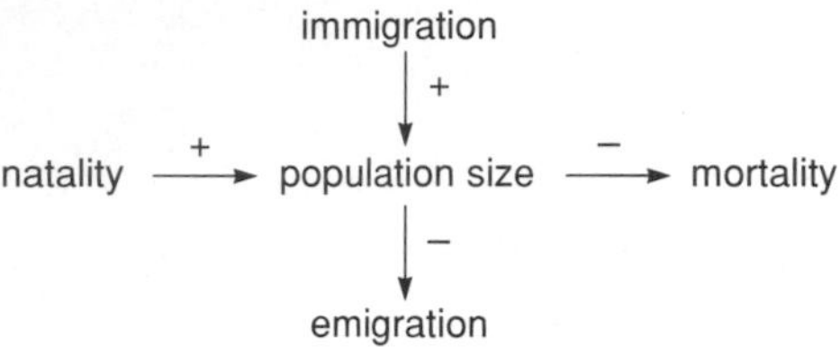

Usually it is possible to assume that immigration and emigration are about equal and therefore it is only necessary to consider the birthrate and the death rate to arrive at the **net reproductive rate,** or simply, ***r*.** Both birthrate and death rate are measured in terms of the individual; that is, *per capita*. For example, suppose a herd of elephants numbers 100. During the year, 10 births and 2 deaths occur; therefore, the birthrate is 5/100 = .05 per elephant per year and the death rate is 2/100 = .02 per elephant per year. We can combine both of these rates to arrive at $r = 0.03$ per capita per year. As we shall see, the net reproductive rate allows us to calculate the growth and size of a population per unit time, such as a year.

Patterns of Population Growth

We shall assume that there are two patterns of population growth. In the first pattern, the members of the population have only a single reproductive event in their lifetime. When this event draws near, the mature adults largely cease to grow, expend all their energy in reproduction, and then they die. Many insects and annual plants reproduce in this manner. They produce offspring or seeds that can survive dryness and/or cold and resume growth the next favorable season. In the other pattern of population increase, members experience many reproductive events throughout their lifetime. During their entire lives, they continue to invest energy in their future survival, increasing their chances of reproducing again. Most vertebrates, bushes, and trees have this pattern of reproduction.

Ecologists have developed mathematical models of population growth based on these two very different patterns of reproduction. But the manner in which organisms reproduce does not always fit either of these two patterns, as exemplified in Figure 23.3. Although the mathematical models we will be using are simplifications, they still are conceptual and help make predictions that can be tested. Testing predictions permit the development of new hypotheses that can be tested.

Ecologists have developed models in the form of mathematical equations that describe population dynamics. These models assume two distinct patterns of reproduction.

Figure 23.3 Patterns of reproduction.
We shall assume that members of populations either have a single suicidal reproductive event or they reproduce repeatedly. But there are complications. **a.** Aphids reproduce repeatedly by asexual reproduction during the summer, and then reproduce sexually but once, right before winter comes on. Therefore, aphids utilize both patterns of reproduction. **b.** The offspring of annual plants can germinate several seasons later. Under these circumstances, population size could fluctuate according to environmental conditions.

Exponential Growth

For the sake of discussion, let's consider a population of insects in which females reproduce only once a year and then they die. Each female produces on the average 2.40 eggs per generation that will survive the winter and become offspring the next year. In the next generation each female will again produce 2.40 eggs; that is, $r = 2.40$ per capita per generation.

Figure 23.4*a* shows how the population would grow year after year for ten years. This growth is equal to the size of the population because all members of the previous generation have died. Figure 23.4*b* shows the growth curve for this population. This growth curve, which has a J shape, depicts exponential growth. With **exponential growth,** the number of individuals added each generation increases as the total number of females increases.

Notice that the curve has these phases:

lag phase: during this phase growth is slow because the population is small.

exponential growth phase: during this phase, growth is accelerating.

Figure 23.4*c* gives the mathematical equation that allows you to calculate growth and size for any population that has discrete (nonoverlapping) generations. In other words, as discussed, all members of the previous generation die off before the new generation appears. In order to use this equation to determine future population size, it is necessary to know *r*, which is the net reproductive rate per capita per generation determined after gathering mathematical data regarding past population increases.

During exponential growth, a population is exhibiting its biotic potential. **Biotic potential** [Gk. *bios,* life] of a population is the maximum population growth that can possibly occur under ideal circumstances. These circumstances include plenty of room for each member of the population, unlimited resources, and no hindrances, such as predators or parasites. Resources include food, water, and a suitable place to exist. For exponential growth to occur, there must be no restrictions placed on growth. It is assumed that there is plenty of room, food, shelter, and any other requirements necessary to sustain unlimited growth. But in reality, exponential growth cannot continue for long because of environmental resistance. **Environmental resistance** is all those environmental conditions such as a limited supply of food, an accumulation of waste products, increased competition between members, or predation (if the population lives in the wild) that prevent populations from achieving their biotic potential.

Exponential growth produces a characteristic J-shaped curve. Because growth accelerates over time, the size of the population can increase dramatically.

Generation	Population size
0	10
1	24
2	57.6
3	138.2
4	331.7
5	796.1
6	1,910.6
7	4,585.4
8	11,005.0
9	26,412.0
10	63,388.8

a.

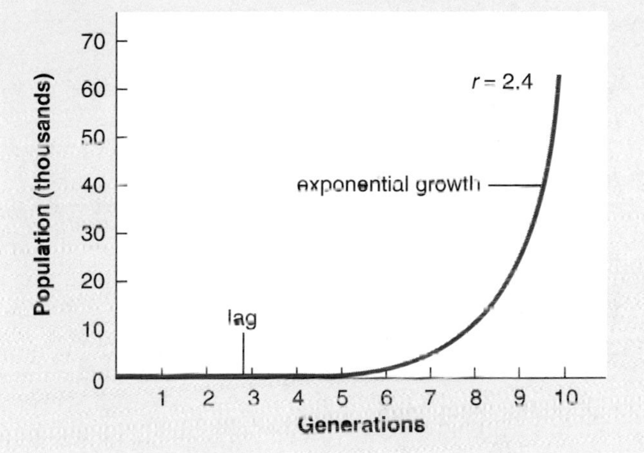

b.

To calculate population size from year to year, use this formula:

$$N_{t+1} = rN_t$$

N_t = number of females already present

r = net reproductive rate

N_{t+1} = population size the following year

c.

Figure 23.4 Model for exponential growth. When the data for discrete reproduction in **(a)** is plotted, the exponential growth curve in **(b)** results. **c.** This formula produces the same results as **(a)** and generates the same graph.

Growth of Yeast Cells in Laboratory Culture

Time (t) (hours)	Number of individuals (N)	Number of individuals added per 2-hour period $\left(\frac{\Delta N}{\Delta t}\right)$
0	9.6	0
2	29.0	19.4
4	71.1	42.1
6	174.6	103.5
8	350.7	176.1
10	513.3	162.6
12	594.4	81.1
14	640.8	46.4
16	655.9	15.1
18	661.8	5.9

a.

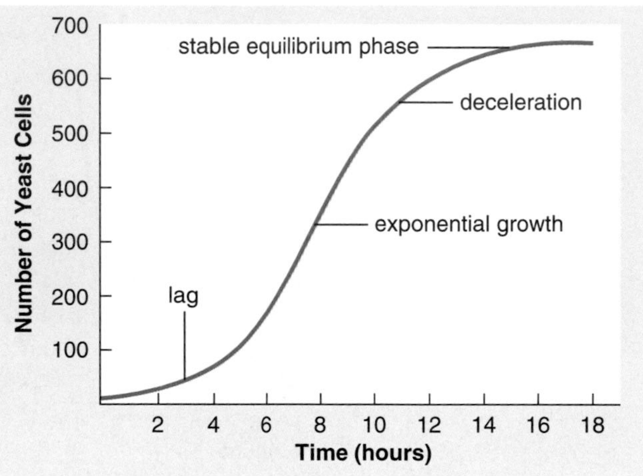

b.

To calculate population growth as time passes, use this formula:

$$\frac{dN}{dt} = r_{max}N\left(\frac{K-N}{N}\right)$$

N = population size

dN/dt = change in population size

r = net reproductive rate

K = carrying capacity

$\frac{K-N}{K}$ = unutilized opportunity for population growth

c.

Figure 23.5 Model for logistic growth.
When the data for repeated reproduction (a) are plotted, the logistic growth curve in (b) results. **c.** This formula produces the same results as (a) and generates the same graph.

Logistic Growth

What type of growth curve results when environmental resistance comes into play? In 1930 Raymond Pearl developed a method for estimating the number of yeast cells present every two hours in a laboratory culture vessel. His data is shown in Figure 23.5*a*. When the data is plotted, the growth curve had the appearance shown in Figure 23.5*b*. This type of growth curve is a sigmoidal (Σ) or S-shaped curve.

Notice **logistic growth** has these phases:

lag phase: during this phase growth is slow because the population is small.
exponential growth phase: during this phase, growth is accelerating.
deceleration phase: during this phase the rate of population growth slows down.
stable equilibrium phase: during this phase, there is little if any growth because births and deaths are about equal.

Figure 23.5*c* gives you the mathematical equation that allows you to calculate logistic growth (so called because the exponential portion of the curve would produce a straight line if the log of N were plotted). The entire equation for logistic growth is:

$$\frac{dt}{dN} = rN\,\frac{(K-N)}{K}$$

but let's consider each portion of the equation separately.

Because the population has repeated reproductive events, we need to consider growth as a function of time:

$$\frac{\Delta N}{\Delta t} = rN$$

If the change in time is very small then we can turn to differential calculus, and the instantaneous population growth is given by:

$$\frac{dN}{dt} = rN$$

This portion of the equation applies to the first two phases of growth—the lag phase and the exponential growth phase. During the exponential growth phase, the population is displaying its biotic potential. What would be the result if exponential growth was experienced by any population for any length of time? Apple trees produce many apples per season, and if each seed became an apple tree, we would soon see nothing but apple trees. Or, to take another example, it has been calculated that if a single female pig had her first litter at nine months and produced two litters a year, each of which contained an average of four females (which, in turn reproduced at the same rate), there would be 2,220 pigs by the end of three years (Fig. 23.6). Because of exponential growth, even species with a low biotic potential would soon cover the face of the earth with their

Figure 23.6 Biotic potential.
The ability of many populations like apple trees or pigs to reproduce exceeds by a wide margin the number necessary to replace those that die.

own kind. Even though elephants require 22 months to produce a single offspring, Charles Darwin calculated that a single pair of elephants could have over 19 million live descendants after 750 years.

Why did the growth curve level off for the yeast population and not for the insect population within the time span of the study? The yeast population, as you know, was grown in a vessel in which food could run short and waste products could accumulate. In other words, the pattern of growth was determined by the environmental resistance. As mentioned previously, environmental resistance opposes exponential growth when a population is displaying its biotic potential. Environmental resistance is all those factors such as a finite amount of resources and an accumulation of waste products that curtail unlimited population growth. Environmental resistance results in the *deceleration phase* and the *stable equilibrium phase* of the logistic growth curve (see Fig. 23.5). Now the population is at the carrying capacity of the environment.

Carrying Capacity

The **carrying capacity** of any environment is the maximum number of individuals of a given species the environment can support. The closer population size gets to the carrying capacity, the greater will be the environmental resistance —as resources become more scarce, the birthrate is expected to decline and the death rate is expected to increase. This will result in a decrease in population growth; eventually, the population stops growing and its size remains stable.

How does our mathematical model for logistic growth take this process into account? To our equation for growth under conditions of exponential growth we add the term

$$\frac{(K-N)}{K}$$

In this expression, K is the carrying capacity of the environment. The easiest way to understand the effects of this term is to consider two extreme possibilities. First, consider a time at which the population size is well below carrying capacity. Resources are relatively unlimited, and we expect rapid, nearly exponential growth to take place. Does the model predict this? Yes, it does. When N is very small relative to K, the term $(K-N)/K$ is very nearly $(K-0)/K$, or approximately 1. Therefore, dN/dt = approximately rN.

Similarly, consider what happens when the population reaches carrying capacity. Here, we predict that growth will stop and the population will stabilize. What happens in the model? When $N= K$, the term $(K-N)/K$ declines from nearly 1 to 0, and the population growth slows to zero.

As mentioned, the model we have developed predicts that exponential growth will occur only when population size is much lower than the carrying capacity. So, as a practical matter, if we are using a fish population as a continuous food source, it would be best to maintain the population size in the exponential phase of the logistic growth curve. Biotic potential is having its full effect and the birthrate is the highest it can be during this phase. If we overfish, the population will sink into the lag phase, and it will be years before exponential growth recurs again. On the other hand, if we are trying to limit the growth of a pest, it is best, if possible, to reduce the carrying capacity rather than reduce the population size. Reducing the population size only encourages exponential growth to begin once again. Farmers can reduce the carrying capacity for a pest by alternating rows of different crops rather than growing one type of crop per the entire field.

Logistic growth curve produces a characteristic S-shaped curve. Population size stabilizes when the carrying capacity of the environment has been reached.

Table 23.1

A Life Table for a *Poa annua* Cohort

Age (months)	Number Observed Alive	Proportion Surviving	Number Dying	Mortality Rate Per Capita	Avg. Number of Seeds/Individual
0–3	843	1,000	121	0.143	0
3–6	722	857	195	0.271	300
6–9	527	625	211	0.400	620
9–12	316	375	172	0.544	430
12–15	144	171	95	0.626	210
15–18	54	64	39	0.722	60
18–21	15	17.8	12	0.800	30
21–24	3	3.6	3	1.000	10
24	0	—	—	—	—

Mortality Patterns

Population growth patterns assume that populations are made up of identical individuals. Actually, the individuals are in different stages of their life span. Investigators studying population dynamics construct life tables that show how many members of an original group of individuals born at the same time, called a **cohort,** are still alive after certain intervals of time. For example, Table 23.1 is a life table for the grass *Poa annua.* The cohort contains 843 individuals. After three months, when 121 individuals have died, the mortality rate is 0.143 per capita. Another way to view this statistic, however, is to consider that 722 individuals are still alive—have survived—after three months. **Survivorship** is the probability of newborn individuals of a cohort surviving to particular ages. If we plot the number surviving, a survivorship curve is produced.

For the sake of discussion, three types of idealized survivorship curves are recognized (Fig. 23.7*a*). The type I curve is characteristic of a population in which most individuals survive well past the midpoint, and death does not come until near the end of the life span. On the other hand, the type III curve is typical of a population in which most individuals die very young. In the type II curve, survivorship decreases at a constant rate throughout the life span.

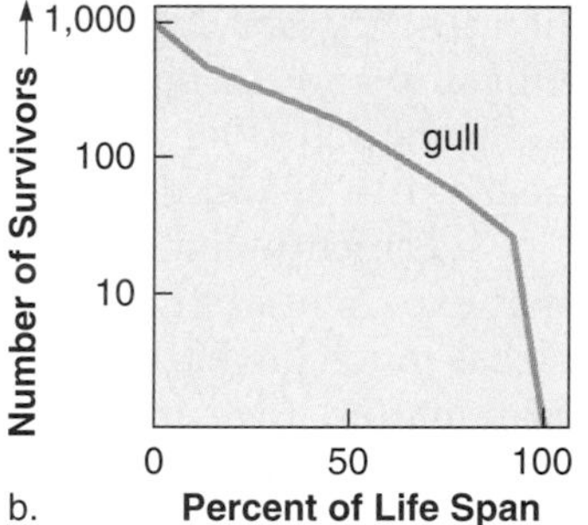

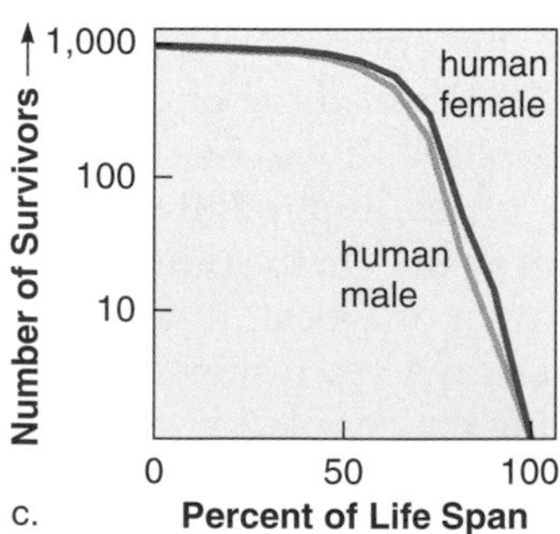

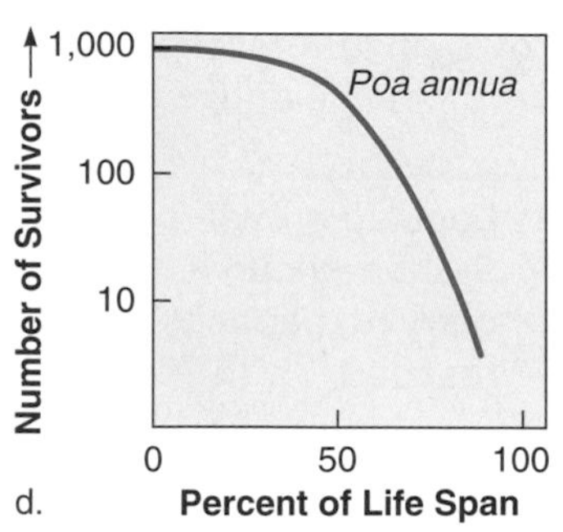

Figure 23.7 Survivorship curves.
Survivorship curves show the number of individuals of a cohort that are still living over time. **a.** Three generalized survivorship curves. **b.** Survivorship curve for gulls fits the type II curve somewhat. **c.** The survivorship curves for human males and females differ slightly but both resemble the type I curve fairly well. **d.** The survivorship curve for *Poa annua* seems to be a combination of the type I and type II curves.

The survivorship curves of natural populations don't fit these three idealized curves exactly. In the *Poa annua* cohort, for example, most individuals survive till six to nine months, and then the chances of survivorship diminish at an increasing rate. Statistics for a gull cohort are close enough to classify the survivorship curve in the type II category, and those for mountain sheep suggest a type I curve. Males and females sometimes have different survivorship curves, as is illustrated for mountain sheep.

There is much that can be learned about the life history of a species by studying its life table and survivorship curve. Would you predict that most or few members of a population with a type III survivorship curve are contributing offspring to the next generation? Obviously since death comes early for most members, only a few are living long enough to reproduce. What about the other two types of survivorship curves? Look again at *Poa annua*'s life table. It tells us that per capita seed production increases as plants mature. How does that compare to the human situation?

Populations have a pattern of mortality/survivorship that becomes apparent from studying the life table and survivorship curve of a cohort.

Age Distribution

When the individuals in a population reproduce repeatedly, several generations may be alive at any given time. From the perspective of population growth, there are three major age groups in a population: prereproductive, reproductive, and postreproductive. Populations differ in what proportion of the population falls in each age group. At least three **age structure diagrams** are possible (Fig. 23.8).

When the prereproductive group is the largest of the three groups, the birthrate is higher than the death rate, and a pyramid-shaped diagram is expected. Under such conditions, even if the growth for that year was matched by the deaths for that year, the population would continue to grow in the following years. Why? Because there are more individuals entering than leaving the reproductive years. Eventually, as the size of the reproductive group equals the size of the prereproductive group, a bell-shaped diagram will result. The postreproductive group will still be the smallest, however, because of mortality. If the birthrate falls below the death rate, the prereproductive group will become smaller than the reproductive group. The age structure diagram will then be urn-shaped because the postreproductive group is now the largest.

The age distribution reflects the past and future history of a population. Because a postwar baby boom occurred in the United States between 1947 and 1960, the reproductive group is now larger than the prereproductive and the postreproductive group. As the baby boomers age, the postreproductive group will be the largest group.

Age distributions contribute to our understanding of the past and future history of a population's growth.

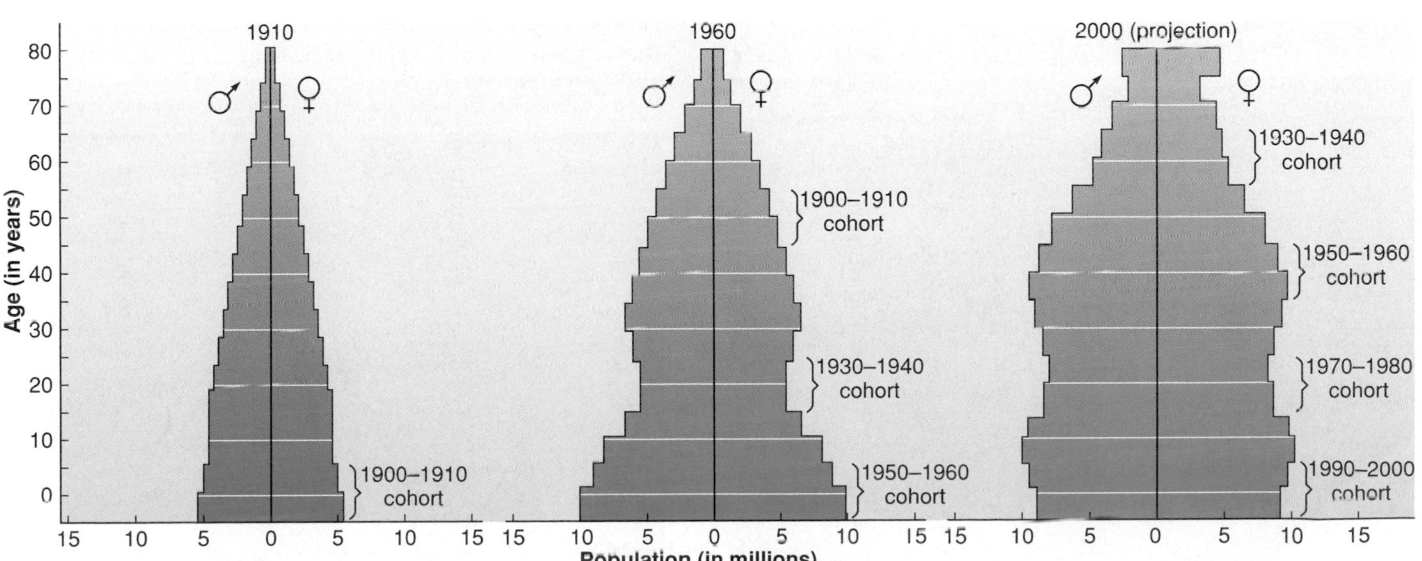

Figure 23.8 U.S. age distributions, 1910, 1960, and 2000 (projected).
In 1910, the distribution was shaped like a pyramid. In 1960, the distribution had shifted toward a bell shape, and in the year 2000, the age distribution is expected to be somewhat urn shaped. The 1950–1960 cohort represents the baby boomers, when an unusually large number of offspring were born following World War II. As the baby boomers grow older, there will be an unusually large number of persons in the postreproductive years.

23.3 Regulation of Population Size

We have developed two models of population growth: exponential growth with its J-shaped curve and logistic growth with its S-shaped curve. In a study of winter moth population dynamics, discussed in the reading on pages 394–95, it was discovered that a large proportion of eggs did not survive the winter and exponential growth never occurred. Perhaps the low number of individuals at the start of each season help prevent the occurrence of exponential growth. This observation raises the question, "How well do the models for exponential and logistic growth predict population growth in natural populations?

Is it possible, for example, that exponential growth may cause population size to rise above the carrying capacity of the environment, and as a consequence a population crash may occur? For example, in 1911, four male and 21 female reindeer *(Rangifer)* were released on St. Paul Island in the Bering Sea off Alaska. St. Paul Island had a completely undisturbed environment—there was little hunting pressure, and there were no predators. The herd grew exponentially to about 2,000 reindeer in 1938, overgrazed the habitat, and then abruptly declined to only eight animals in 1950 (Fig. 23.9).

Even though population growth may not follow our models exactly, we do find that environmental factors do regulate population size in natural environments. When ecologists first considered the question of environmental regulation, they emphasized that the environment contains abiotic (nonliving) and biotic (living) components. It seemed to them that both components could be involved in regulating population size. They suggested that abiotic factors, like weather and natural disasters, were density-independent. By this they meant that the number of organisms present did not influence the effect of the factor. The proportion of organisms killed by accidental fire, for example, is independent of density—fires don't necessarily kill a larger percentage of individuals in dense populations than they do in less dense populations (Fig. 23.10). On the other hand, biotic factors like parasitism, competition, and predation were designated as density-dependent. Predation increases when the prey population gets denser because it is easier for predators to find the prey they are looking for. Consider, for example, a population of crabs in which each crab needs a hole to hide in or else the crab is eaten by shorebirds. If there are only 100 holes, and 102 crabs inhabit the area, two will be without shelter. But since there are only two without shelter they may be hard to find. If neither crab is caught, then the predation rate is 0 captured/2 "available"= 0. However, if 200 crabs inhabit the area, 100 crabs will be without shelter. At this density, they are readily visible to birds and will probably attract a larger number of predators. If half of the exposed crabs are eaten, then the predation rate is

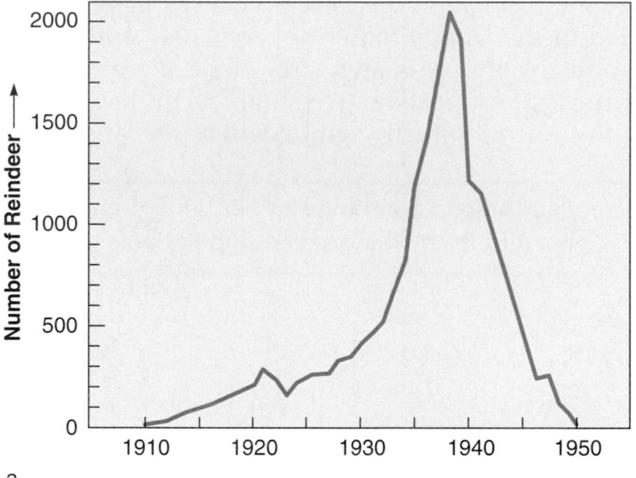

Figure 23.9 Density-dependent effect.
a. On St. Paul Island, Alaska, reindeer grew exponentially for several seasons and then underwent a sharp decline as a result of overgrazing the available range. **b.** A reindeer, *Rangifer.*

Figure 23.10 Density-independent effect.
A fire can start and rage out of control regardless of how many organisms are present.

...er Look

...Winter Moth: A Case Study

...d seem reasonable that many fac... ...ust be involved in regulating popu... ...size. When studies of population ...mics are done, investigators try to ...rmine which are the key factors that ...ermine mortality. For example, George ...rley and George Gradwell studied the ...e cycle of the winter moth *Operophtera ...rumata* and several of its parasites and predators in order to determine which factor correlated best with observable fluctuations in population size.

Life Cycle of a Moth

Like other moths, winter moths have a life cycle that involves these stages: adult moths—larvae—pupae—adult moths. In the fall, adult females emerge from pupae in the soil litter and then climb (because they are wingless) the trunk of the nearest oak tree. After mating with a male (that has wings and can fly), each female lays about 150 eggs on twigs at the top of the tree. The eggs stay there all winter, and then in spring, they hatch when the buds on the tree open. The caterpillars feed on the growing leaves and then, when grown, they spin down to the ground on silk threads, crawl onto the litter, and pupate. In the fall the adults emerge once more (Fig. 23A).

The Parasites

The winter moth was accidently introduced from Europe into Nova Scotia sometime before 1949 and, within a few years, it was causing severe defoliation and beginning to spread westward. Two natural enemies of the moth were imported and became established. The adult female fly *Cyzenis albicans* deposits small resistant eggs on foliage, and these hatch only after they are eaten by the moth. Hatching occurs in the midgut of the moth, and the maggots bore from the gut into the blood cavity and then move to the salivary glands, where they feed slowly until the host pupates. After feeding some more, the maggots eventually form their puparia inside the moth's pupa. The adult flies emerge in the spring from these puparia.

The other parasite is a wasp, *Cratichneumon,* whose females lay an egg in each winter moth pupa they discover. The egg in turn hatches into a larva that consumes the winter moth pupa, and then pupates, later to emerge as an adult.

The Data

Varley and Gradwell counted the size of the population at each stage of the life cycle (Table 23A). They used sticky paper to capture and estimate the number of female moths climbing up the trees. They dissected several female moths to estimate the average number of eggs in each. They sampled the oak leaves to count the average number of caterpillars. They dissected larvae to see how many contained *Cyzenis* maggots. From this they knew that some larvae were killed by other parasites. They counted the number of moth pupae compared to how many *Cratichneumon* wasps emerged. From this they knew that some pupae were killed by predators.

The Conclusion

Varley and Gradwell used a mathematical formula that allowed them to assign a value (called a *k*-value in the table) to each environmental factor that caused moth mortality. Examine column three in Table 23A and note that most mortality was caused by winter disappearance: females laid 658 eggs in the fall, but only 96.4

Table 23A

Life Table for the Winter Moth

	Number Killed (per m²)	Number Alive (per m²)	*k*-value
Adult stage			
Females climbing trees, 1955	—	**4.39**	—
Egg stage			
Females x 150		658.0	—
Larval stage			
Full-grown larvae	551.6	**96.4**	**0.84** = k_1
Attacked by *Cyzenis*	**6.2**	90.2	0.03 = k_2
Attacked by other parasites	**2.6**	87.6	0.01 = k_3
Infected by protozoa	**4.6**	83.0	0.02 = k_4
Pupal stage			
Killed by predators	54.6	28.4	**0.47** = k_5
Killed by *Cratichneumon*	**13.4**	15.0	0.27 = k_6
Adult stage			
Females climbing trees, 1956	—	7.5	—

Note: The figures in bold are those actually measured. The rest of the life table is derived from these. Source: After Varley et al. (1973).

50 captured/100 "available" = 1/2. Thus, increasing the density has increased the proportion of individuals preyed upon.

Is it possible that other types of regulating factors sometimes keep the population size at just about the carrying capacity? Investigators have been collecting data on the population size of the bird called the great tit (*Parus major*) for many years (Fig. 23.11*a*). Yearly counts during the breeding season for a population in Marley Wood, near Oxford, England, are given in Figure 23.11*b*. As you can see, the population size fluctuates above and below the carrying capacity. Clutch size (number of eggs produced per capita) does not seem to fluctuate from year to year, so other hypotheses are needed to explain why these fluctuations occur. Weather, resource abundance (food, shelter, etc.) and the presence and abundance of other animals (such as predators, pathogens, and parasites), are all clearly important—and all are extrinsic to the organism under investigation. Isn't it possible that intrinsic factors—those based on the anatomy, physiology, or behavior of the organisms—might have an effect on population size and growth rates?

Territoriality is apparent when members of a population are spaced out more than would be expected from a random occupation of the area (Fig. 23.11*c*). Twenty to ninety meters exist between nesting great tits. Territoriality in birds and dominance hierarchies in mammals are behaviors that affect population size and growth rates. Recruitment, immigration, and emigration are other social means by which the population size of more complex organisms are regulated by intrinsic means.

Do populations
and growth rates in sp
ing mechanisms? Yes, t
dependent and density-in
populations have an inna
developed models that pre
in even simple systems. For e
of Dungeness crab populations
duce many larvae and then they
not survive, and those that do stay
these circumstances, the model predi
that are now termed *chaos*.

**Density-independent and density-depe
can often explain the population dynami
populations. Both types of factors are ext
organism; perhaps intrinsic factors like terri
birds also play a role.**

a.

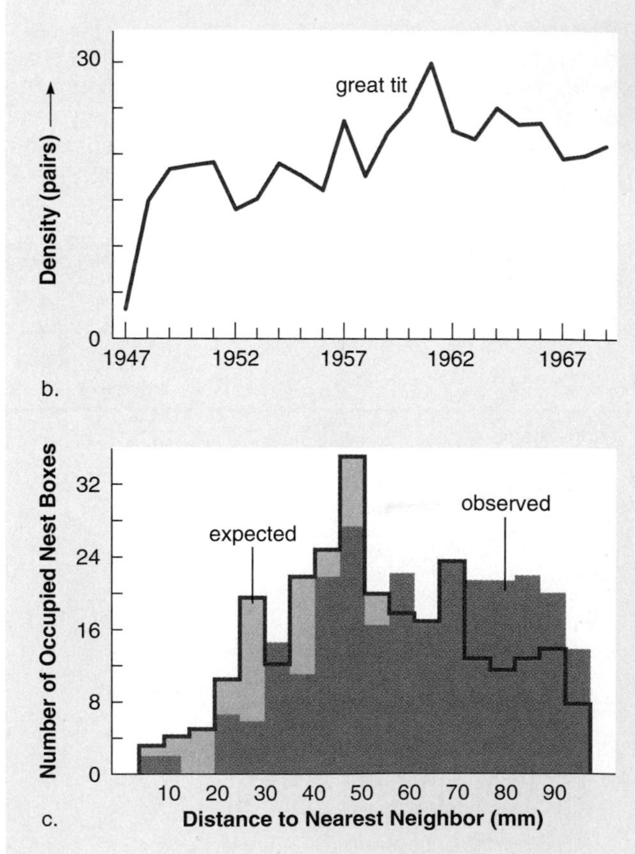

Figure 23.11 Great tit, *Parus major*.
a. Photograph of great tit. **b.** Density dynamics of a breeding population in Marley Wood, near Oxford, England. **c.** Great tits are spaced out more than would be expected if distribution were random.

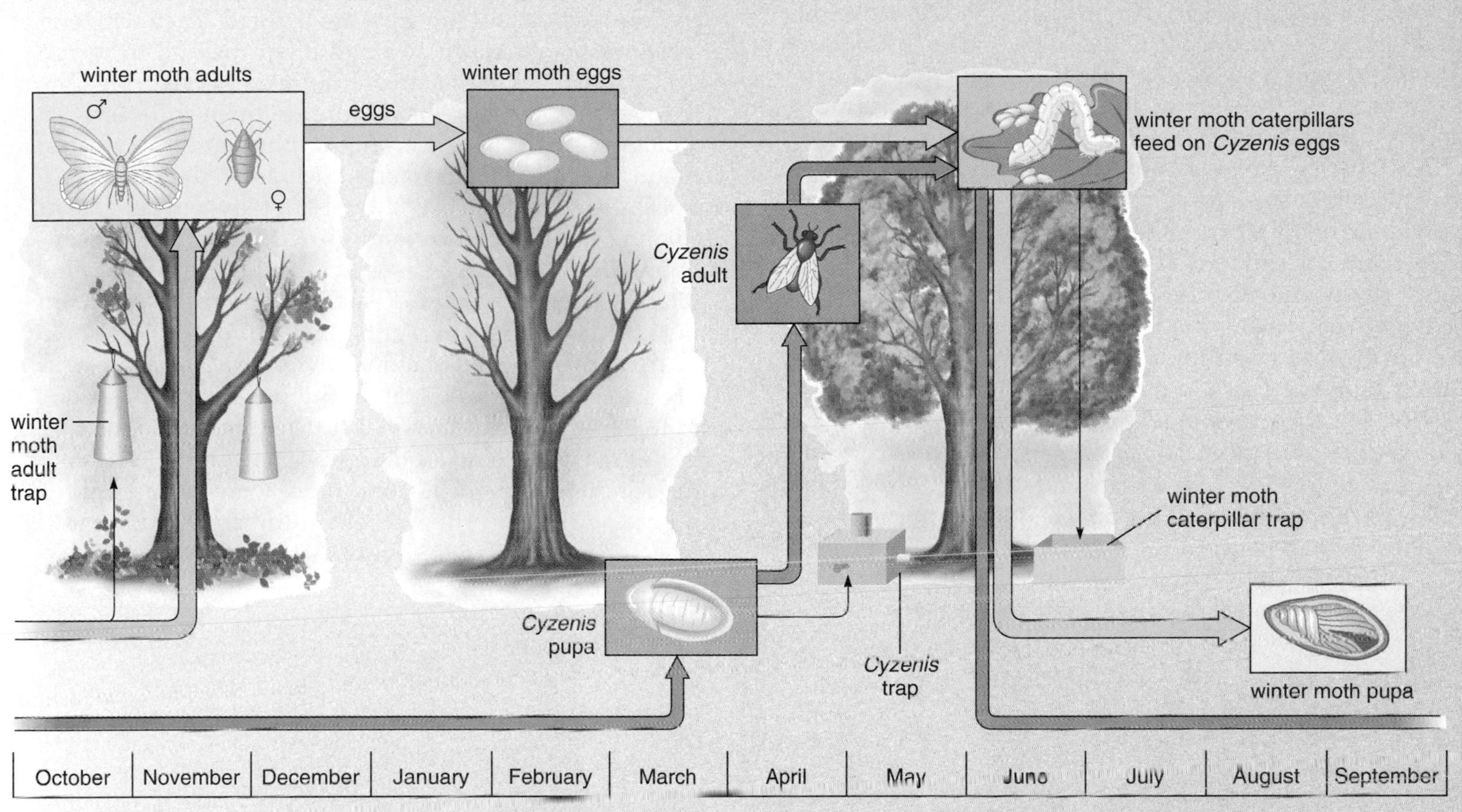

Figure 23A Life history of the winter moth *(Operophtera brumata)* and a parasitic fly *(Cyzenis)*. Winter moth: (one color) 1. In fall, females emerge from pupae and climb trees and mate. 2. Eggs are laid. 3. Caterpillars hatch in spring to eat oak leaves. 4. Caterpillars pupate. *Cyzenis* (second color): a. Flies emerge in spring and lay eggs on leaves, where they are eaten by winter moth caterpillars. b. *Cyzenis* pupates inside of winter moth pupa. The third color in the diagram indicates stages at which collections were made.

caterpillars (larvae) hatched in the spring. The k-value for winter disappearance was 0.84, and the k-values for all other factors are much less. Mortality due to predators such as shrews and beetles has the second largest k-value.

The k values for winter disappearance correlate best with a sum of all mortality k-values as shown in Figure 23B. None of the other k-values, including pupal predation, correlated well. However, this k-value was proportional to the size of the pupae population on which it fed.

The weather cannot be predicted from year to year, but Varley and Gradwell were able to arrive at a mathematical model that predicted moth population size from year to year based on density-dependent factors such as parasitism and predation.

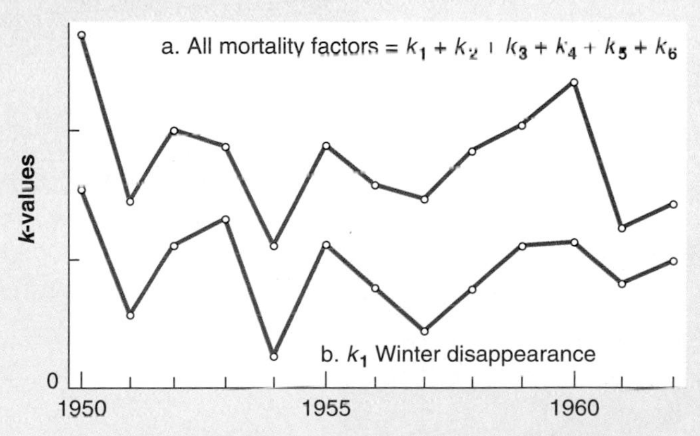

Figure 23B Key factor analysis of the winter moth. Winter moth mortality, expressed as a k-value, due to (a) total mortality because of all factors, correlates well with winter moth mortality due to (b) winter disappearance. Therefore, winter disappearance largely accounts for population fluctuations in the adult population size.

23.4 Life History Patterns

We have already had an opportunity to point out that populations vary on such particulars as the number of births per reproduction, the age of reproduction, life span, the probability of living the entire life span, and so forth. Such particulars are a part of a species' life history. Life histories contain characteristics that can be thought of as trade-offs. Each population is able to capture only so much of the available energy, and how this energy is distributed between its life span (short versus long), reproduction events (few versus many), care of offspring (little versus much), and so forth has evolved over the years. Natural selection shapes the final life history of individual species, and therefore it's not surprising that even related species may have different life history patterns, if they occupy different types of environments (Fig. 23.12).

The logistic growth model has been used to suggest that some individuals are subjected to ***r*-selection** and some are subjected to ***K*-selection.** In fluctuating and/or unpredictable environments, density-independent factors will keep populations in the lag or exponential phase of population growth. Population size is low relative to *K*, and *r*- selection, which favors individuals producing large numbers of offspring relatively early, will be in effect. As a consequence of this pattern of energy allocation, small individuals that mature early and have a short life span are favored. They will tend to produce many relatively small offspring and to forego parental care in favor of a greater number of offspring. The more offspring, the more likely some of them will survive a population crash. Because of low population densities, density-dependent mechanisms such as predation and intraspecific competition are unlikely to play a major role in regulating population size and growth rates. Such organisms are often very good dispersers and colonizers of new habitats. Classic examples of such *opportunistic species* are many insects and annual plants (Fig. 23.13).

In contrast, we can imagine environments that are relatively stable and/or predictable, and in which populations tend to be near *K* with minimal fluctuations in population size. Resources such as food and shelter will be relatively scarce for these individuals, and those who are best able to compete will have the largest number of offspring. *K*-selection favors those individuals that allocate energy to

strawberry arrow-poison frog, *Phyllobates lugubris*

Surinam toad, *Pipa pipa*

wood frog, *Rana sylvatica*

mouth-brooding frog, *Rhinoderma darwinii*

midwife toad, *Alyces obstetricans*

Figure 23.12 Parental care among frogs and toads.
Frogs and toads, which lay their eggs on land, as do wood frogs *(upper right),* exhibit parental care that takes various forms. The midwife toad of Europe *(lower right)* carries strings of eggs entwined around his hind legs and takes them to water when they are ready to hatch. In mouth-brooding frogs of South America *(lower left),* the male carries the larvae in a vocal pouch (brown area), which elongates the full length of his body before the froglets are released. In arrow-poison frogs of Costa Rica *(upper left),* eggs are laid on land and after hatching, the tadpoles wiggle onto the parent's back and are then carried to water. Most amazing, in the Surinam toads of South America *(center),* males fertilize the eggs during a somersaulting bout in which the eggs are placed on the female's back. Each egg develops in a separate pocket, where the tail of the tadpole acts as a placenta to take nourishment from the female's circulatory system.

their own growth and survival and to the growth and survival of their offspring. Therefore they are fairly large, are slow to mature, and have a fairly long life span. Because these organisms, termed *equilibrium species,* are strong competitors, they can become established and exclude opportunistic species. They are specialists rather than colonizers and tend to become extinct when their normal way of life is destroyed. The best possible examples of *K*-strategists are found among birds and mammals, such as bears (Fig. 23.13). The Florida panther is the largest of the animals in the Florida Everglades, requires a very large range, and produces few offspring that must be cared for. Currently, the Florida panther is unable to compensate for a reduction in its range, and is therefore on the verge of extinction.

Nature is actually more complex than these two possible life history patterns. It now appears that our description of *r*-strategist and *K*-strategist populations are at the ends of a continuum, and most populations lie somewhere in between these two extremes. For example, recall that plants have a two-generation life cycle, which includes a sporophyte and gametophyte generation. Ferns, which could be classified as *r*-strategists, distribute many spores and leave the gametophyte to fend for itself, but gymnosperms (e.g., pine trees) and angiosperms (e.g., oak trees), which could be classified as *K*-strategists, retain and protect the gametophyte. They produce seeds that contain the next sporophyte generation plus stored food. The added investment is significant, but these plants still release copious numbers of seeds.

A cod is a rather large fish weighing up to 25 pounds and measuring up to 6 feet in length—but the cod releases gametes in vast numbers, the zygotes form in the sea, and the parents make no further investment in developing offspring. Of the 6 to 7 million eggs released by a single female cod, only a few will become adult fish.

Differences in the environment result in different selection pressures and a range of life history characteristics.

Figure 23.13 Life history strategies. Are dandelions *r*-strategists with the characteristics noted, and are bears *K*-strategists with the characteristics noted? Most often the distinctions between these two possible life strategies are not as clear cut as they may seem.

r-strategists
Small individuals
Short life span
Fast to mature
Many offspring
Little or no care of offspring

K-strategists
Large individuals
Long life span
Slow to mature
Few offspring
Much care of offspring

Doing Science

Distribution of Hard Clams in the Great South Bay

The Great South Bay is a shallow bay located on the south shore of Long Island, New York, whose center is located about 1,010 kilometers east of New York City. The hard clam is a commercially harvested bivalve mollusk that is found throughout the Great South Bay. (It is eaten raw on the half shell, and is the key ingredient in baked clams as well as New England and Manhattan clam chowder.) The Great South Bay has often been referred to as a "hard clam factory" because in the 1970s, over half of the hard clams harvested in the United States came from its waters. For the past 20 years, I have been studying the distribution of hard clam abundance in the eastern third of the bay for the Town of Brookhaven which, by virtue of a 17th century grant from the King of England, owns 6,000 hectares of prime hard clam habitat.

Knowing the distribution and abundance of hard clams, as well as the responsible environmental factors, is important to Brookhaven because its ownership carries with it an obligation for managing a fishing industry for hard clams that has an annual value of $10 million and employs 300 fishermen. Of particular interest to Brookhaven is the possibility of using this information to develop projects, such as supplementing the natural hard clam production using aquaculture technology, that will increase hard clam abundance and hence the hard clam harvest and the economic benefits the hard clam resource provides to Brookhaven.

My study began with the censusing of the hard clam population. For several weeks during the six summers, a barge-mounted crane with a 1 square meter clamshell bucket was used to take bottom grabs at 232 stations located throughout the study area. Each bottom sample was placed in a 1 square meter wire sieve and washed with a high pressure water hose to separate the hard clams from the sediment so that the hard clams could be counted and measured in order to calculate various demographic parameters of the hard clam population. The fieldwork was messy and physically hard, and by the end of the day, my field crew of students and biologists were tired, wet, and covered with mud. The results, however, have proven to be well worth the effort.

Working with Dr. Robert Cerrato of the State University of New York, I was able to draw up a composite census map showing the distribution of hard clam abundance. I was surprised to find that hard clams were not distributed uniformly throughout the study area, but occurred in distinct patches of high and low abundance. I termed the high abundance areas "beds" (a dense assemblage of clams is traditionally referred to as a "clam bed") and six such areas were identified.

When I overlaid the census map on a sediment map that I generated based on my field notes, nearly all of the beds coincided with areas of high shell content sediment associated with formerly productive oyster reefs, or what I call "relict" oyster reefs. This observation is of historical note as well as biological interest because up through the early years of this century, the Great South Bay was a major producer of oysters and was the source of the world famous "Blue Point Oyster." Although oysters are no longer found in the Great South Bay because of environmental changes, they left behind a legacy in the sediment that now supports high abundances of hard clams.

Having found that hard clam abundance was positively associated with relict oyster reefs, I wanted to find out what aspect of the sediment of the relict oyster reefs make them so productive, and conversely why low abundance areas have so few hard clams. This information would be useful to the Town of Brookhaven because it might lead to ways in which the sediment in low abundance areas can be transformed into high abundance areas. I therefore needed a sedimentary "portrait" of high and low abundance areas, and to develop one I borrowed techniques generally associated with other marine science disciplines: from shipwreck hunting, I used a side-scan sonar to map the topography of the bottom; from deep-sea research, an ROV (Remotely Operated Vehicle) to photograph the bottom; from commercial fishing, a fathometer to map the bathymetry; and from pollution studies, a sediment profile camera to photograph the sediment-water interface where hard clams live and feed.

I am now in the process of putting all these different information sources together on a single map (Fig. 23C). Because it will take time and perhaps more studies to develop the portrait, I am concurrently using my finding that hard clam abundance is strongly associated with sediment shell content to explore the feasibility of having shell placed on low abundance areas in order to create new relict oyster reefs. If this strategy works, it would be a tremendous boon to the shellfish industry because nearly three-quarters of the study area is low abundance.

My research project is not only scientifically interesting, but it is also personally rewarding. Over the years, I have become friends with many fishermen and I know that if I am successful, I will be helping them to continue in an occupation that in some cases goes back generations.

Figure 23C
Jeffrey Kassner is preparing a sedimentary "portrait" of high and low clam abundance areas in Great South Bay, Long Island, New York. This information will be used to increase the yield of clams for local fishermen.

23.5 Human Population Growth

The human population has an exponential pattern of growth and a J-shaped growth curve (Fig. 23.14). It is apparent from the position of 1997 on the growth curve in Figure 23.14 that growth is still quite rapid. The equivalent of a medium-sized city (200,000) is added to the world's population every day and 88 million (the equivalent of the combined populations of the United Kingdom, Norway, Ireland, Iceland, Finland, and Denmark) are added every year.

The present situation can be appreciated by considering the doubling time. The **doubling time**—the length of time it takes for the population size to double—is now estimated to be 47 years. Such an increase in population size will put extreme demands on our ability to produce and distribute resources. In 47 years, the world will need double the amount of food, jobs, water, energy, and so on just to maintain the present standard of living.

Many people are gravely concerned that the amount of time needed to add each additional billion persons to the world population has taken less and less time. The first billion didn't occur until 1800; the second billion arrived in 1930; the third billion in 1960, and today there are nearly 6 billion. Only if the net reproductive rate declines can there be zero population growth when the birthrate equals the death rate and population size remains steady. The world's population may level off at 8, 10.5, or 14.2 billion, depending on the speed with which the net reproductive rate declines.

More-Developed Versus Less-Developed Countries

The countries of the world are divided into two groups. The **more-developed countries (MDCs),** typified by countries in North America and Europe, are those in which population growth is low and the people enjoy a good standard of living. The **less-developed countries (LDCs),** such as countries in Latin America, Africa, and Asia, are those in which population growth is expanding rapidly and the majority of people live in poverty. (Sometimes the term *third-world countries* is used to mean the less-developed countries. This term was introduced by those who thought of the United States and Europe as the first world and the former USSR as the second world.)

The more-developed countries (MDCs) doubled their populations between 1850 and 1950. This was largely due to a decline in the death rate, the development of modern medicine, and improved socioeconomic conditions. The decline in the death rate was followed shortly thereafter by a decline in the birthrate, so that populations in the MDCs experienced only modest growth between 1950 and 1975. This sequence of events (i.e., decreased death rate followed by decreased birthrate) is termed a **demographic transition.** Yearly growth of the MDCs as a whole has now stabilized at about 0.1%. The populations of a few of the MDCs—Italy, Denmark, Hungary, Sweden—are not growing or are actually decreasing in size. In contrast, there is no leveling off and no end in sight to U.S. population growth. Although

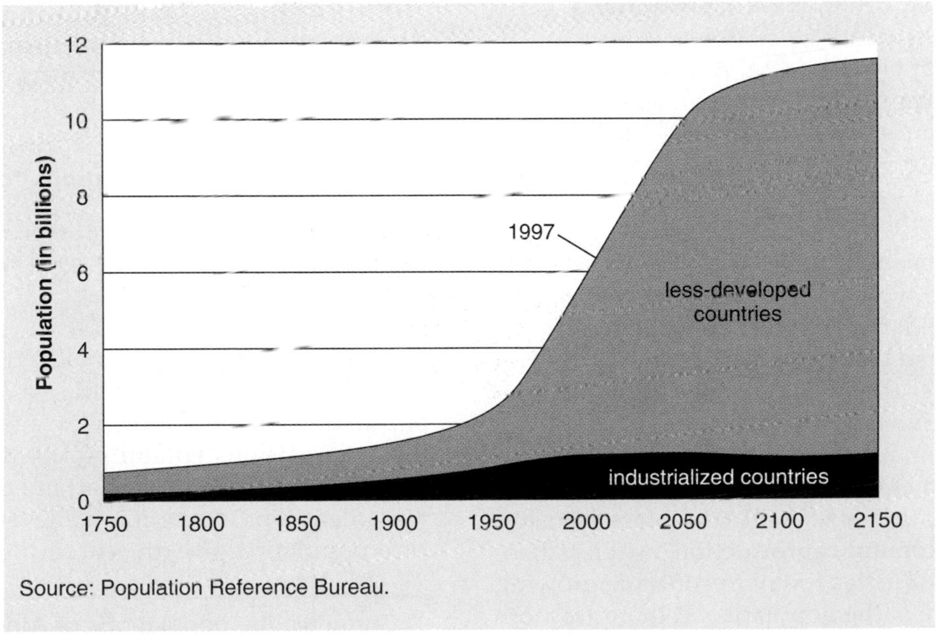

Figure 23.14 World population growth, 1750–2150.
The less-developed countries will contribute most to world population growth.

yearly growth of the United States is only 0.6%, many people immigrate to the United States each year. In addition, a baby boom between 1947 and 1964 means that large number of women are still of reproductive age.

Although the death rate began to decline steeply in the LDCs following World War II with the importation of modern medicine from the MDCs, the birthrate remained high. The yearly growth of the LDCs peaked at 2.5% between 1960 and 1965. Since that time, a demographic transition has begun: the decline in the death rate has slowed and the birthrate has fallen. A yearly growth of 1.8% is expected by the end of the century. Still, because of exponential growth, the population of the LDCs may explode from 4.4 billion today to 10.2 billion in 2100. Most of this growth will occur in Africa, Asia, and Latin America. Ways to greatly reduce the expected increase have been suggested:

1. Establish and/or strengthen family planning programs. A decline in growth is seen in countries with good family planning programs supported by community leaders. Currently, 25% of women in the sub-Saharan Africa say they would like to delay or stop childbearing, yet they are not practicing birth control; likewise, 15% of women in Asia and Latin America have an unmet need of birth control.
2. Use social progress to reduce the desire for large families. Many couples in the LDCs presently desire as many as four to six children. But providing education, raising the status of women, and reducing child mortality are desirable social improvements that could cause them to think differently.
3. Delay the onset of childbearing. A delay in the onset of childbearing and wider spacing of births could cause a temporary decline in the birthrate and reduce the present reproductive rate.

Comparing Age Distributions

The LDCs are experiencing a population momentum because they have more women entering the reproductive years than there are older women leaving them. The age distributions of MDCs and LDCs are compared in Figure 23.15 and as discussed earlier there are three age groups: prereproductive, reproductive, and postreproductive.

Laypeople are sometimes under the impression that if each couple has two children, **zero population growth** (no increase in population size) will take place immediately. However, **replacement reproduction,** as it is called, will still cause most countries today to continue growing due to the age structure of the population. If there are more young women entering the reproductive years than there are older women leaving them, then replacement reproduction will still result in growth of the population.

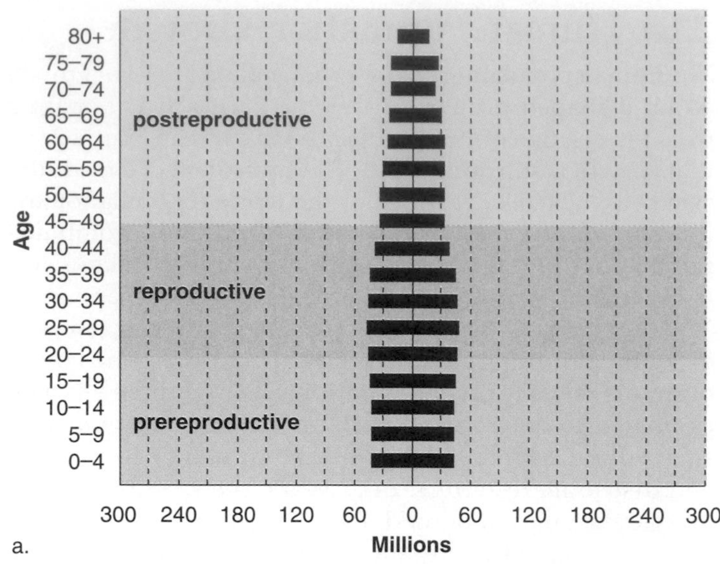

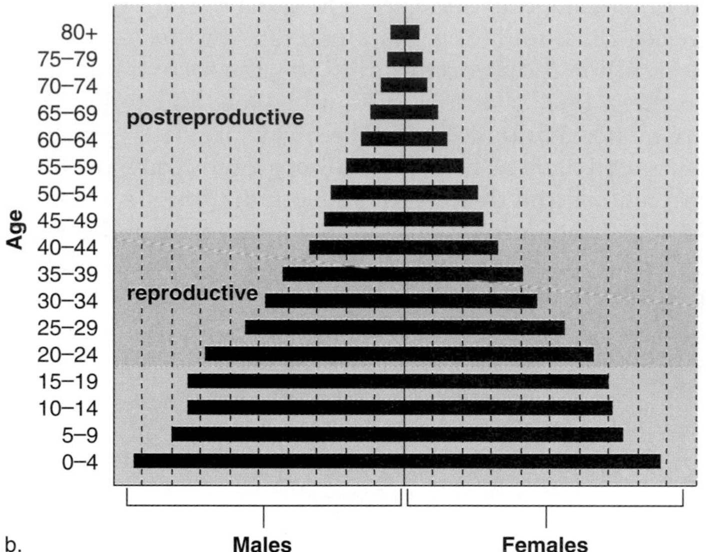

Figure 23.15 Age distributions for MDCs and LDCs (1989).
The diagrams illustrate that (**a**) the MDCs are approaching stabilization, whereas (**b**) the LDCs will expand rapidly due to their age distributions.
Source:Data from *World Population Profiles*, WP-89.

Many MDCs have a stabilized age structure, but most LDCs have a youthful profile—a large proportion of the population is younger than the age of 15. This means that their populations will still expand greatly, even after replacement reproduction is attained. The more quickly replacement reproduction is achieved, however, the sooner zero population growth will result.

Currently, the populations of Africa, Asia, and Latin America are expanding dramatically because of exponential growth.

A Sustainable World

While it may seem like an either-or situation—either preservation of ecosystems or human survival—there is a growing recognition that this is not the case. Natives who harvest rubber from the trees of a tropical rain forest can have a sustainable income from the same trees year after year. One study calculated the market value of rubber and exotic produce, like the aguaje palm fruit, that can be harvested continually from the Amazon rain forest. It concluded that selling these products would yield more than twice the income from either lumbering or cattle ranching, and it would help achieve a sustainable income.

While we are sometimes quick to realize that the growing populations of the LDCs are putting a strain on the environment, we should realize that the excessive resource consumption of the MDCs also stresses the environment. Environmental impact is measured not only in terms of population size, it is also measured in terms of resource consumption and the pollution caused by each person in the population. An average American family, in terms of per capita resource consumption and waste production, is the equivalent of 30 people in India (Fig. 23.16).

Before the establishment of industrialized societies, people felt better connected to the plants and animals on which they depended, and they were then better able to live in a sustainable way. After the industrial revolution, we especially began to think of ourselves as separate from nature and endowed with the right to exploit nature as much as possible. But our industrial society lives on *borrowed carrying capacity*—our cities not only borrow resources from the country, our entire population borrows from the past and future. The forests of the Carboniferous have become the fossil fuels that sustain our way of life today, and the environmental degradation we cause is going to be paid for by our children.

Overpopulation and overconsumption account for increased pollution and also for the mass extinction of wildlife that is going on. We are expected to lose one-third to two-thirds of the earth's species, any one of which could possibly have made a significant contribution to agriculture or medicine. It should never be said, "What use is this organism?" Aside from its contribution to the ecosystem in which it lives, one never knows how a particular organism might someday be useful to humans. Adult sea urchin skeletons are now used as molds for the production of small artificial blood vessels, and armadillos are used in leprosy research.

It is clearly time for a new philosophy. In a **sustainable world,** development will meet economic needs of all peoples while protecting the environment for future generations. Various organizations have singled out communities to serve as models of how to balance ecological and economic goals. For example, in Clinch Valley of southwest Virginia, the Nature Conservancy is helping to revive the traditional method of logging with draft horses. This technique, which allows the selective cutting of trees, preserves the forest and prevents soil erosion, which is so damaging to the environment. The United Nations has an established bioreserve system, a global network of sites that combine preservation with research on sustainable management for human welfare. More than 100 countries are now participants in the program. However, sustainability is more than likely incompatible with the kinds of consumption/waste patterns currently practiced in developed countries.

All peoples can benefit from a sustainable world where economic development and environmental preservation are considered complementary, rather than opposing, processes.

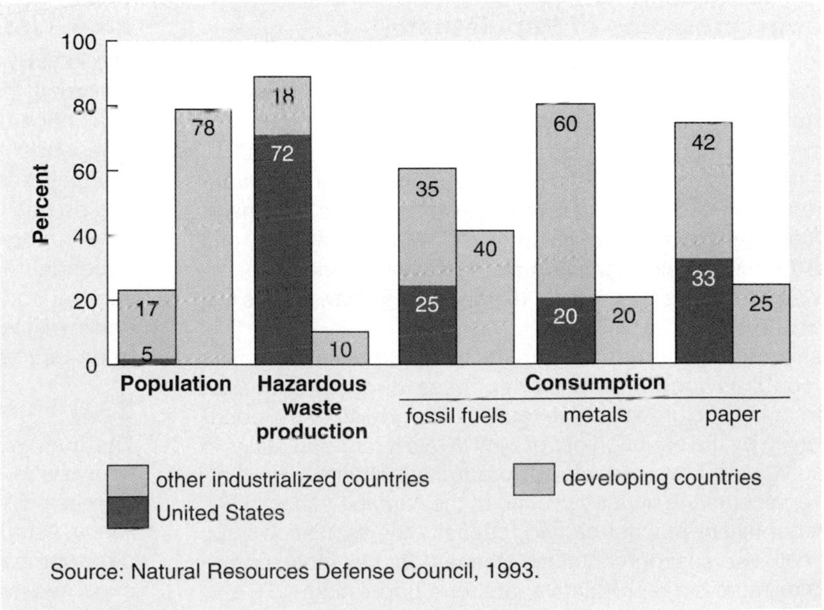

Figure 23.16 Resource consumption for MDCs and LDCs.
The populations of MDCs are smaller than LDCs, but they produce most of the hazardous wastes because their consumption of fossil fuels, metals, and paper, for example, is much greater than the LDCs.

connecting concepts

Modern ecology began with descriptive studies by nineteenth century naturalists. In fact, an early definition of the field was "scientific natural history." However, modern ecology has grown to be much more than a simple descriptive field. Ecology is now very much an experimental, predictive science.

Much of the success in the development of ecology as a predictive science has come from studies of populations and the development of models that examine how populations change over time. The simplest models are based on population growth when there are unlimited resources. This results in exponential growth, a type of population growth that is only rarely seen in nature. Pest species may exhibit exponential growth until they run out of resources. Because so few natural populations exhibit exponential growth, population ecologists realized they must incorporate resource limitation in their models. The simplest models that account for limited resources result in sigmoidal or logistic growth. Populations which exhibit logistic growth will cease growth when they reach the environmental carrying capacity.

Many modern ecological studies are concerned with identifying the factors that place limits on population growth and that set the environmental carrying capacity. A combination of careful descriptive studies, experiments done in nature, and sophisticated models has allowed ecologists to make good predictions about which factors have the greatest influence on population growth. The example of the winter moth is a good case in point.

The next step in the development of modern ecology has been to try to understand how populations of different species affect each other. This is known as community ecology. Because each population in a community responds to environmental changes in slightly different ways, developing predictive models that explain how communities change has been challenging. However, ecologists are beginning to be able to predict how communities will change through time and to understand what factors influence community properties such as species number, abundance of individuals, and species interactions.

Summary

23.1 Scope of Ecology

Ecology is the study of the interactions of organisms with other organisms and with the physical environment. Ecology encompasses several levels of study: organism, population, community, ecosystem, and finally the biosphere. Ecologists are particularly interested in how interactions affect the distribution and abundance of organisms.

Population density is simply the number of individuals per unit area or volume. Distribution of these individuals can be uniform, random, or clumped. A population's distribution is often determined by limiting factors; that is, abiotic factors like water, temperature, and availability of nutrients.

23.2 Characteristics of Populations

Population size is dependent upon natality (number of births), mortality (number of deaths), immigration, and emigration. The number of births minus the number of deaths results in the net reproductive rate (symbolized as r) per capita per unit time.

One model for population growth assumes that the environment offers unlimited resources. In the example given, the members of the population have discrete reproductive events, and therefore the size of next year's population is given by the equation: $N_{t+1} = rN_t$. Under these conditions, exponential growth results in a J-shaped curve.

Most environments restrict growth, and exponential growth cannot continue indefinitely. Under these circumstances an S-shaped or logistic growth curve results. The growth of the population is given by the equation $dN/dt = rN\,(K-N)/K$ for populations in which individuals have repeated reproductive events. The term $(K-N)/K$ represents the unused portion of the carrying capacity (K). When the population reaches carrying capacity, the population stops growing because environmental resistance opposes biotic potential, the maximum net reproductive rate for a population.

Individuals are at different stages of their life span in a population. Mortality (deaths per capita) within a population can be recorded in a life table and illustrated by a survivorship curve. The pattern of population growth is reflected in the age distribution of a population, which consists of prereproductive, reproductive, and postreproductive segments. Populations that are growing exponentially have a pyramid-shaped age distribution pattern.

23.3 Regulation of Population Size

Population growth is limited by density-independent (e.g., weather) and density-dependent factors (predation, competition, and resource availability). Do some populations have an intrinsic means of regulating population growth as opposed to density-independent and density-dependent factors, which are extrinsic means? Territoriality is given as an example of a possible intrinsic means of regulation.

23.4 Life History Patterns

The logistic growth model has been used to suggest that the environment promotes either *r*-selection or *K*-selection. So-called *r*-selection occurs in unpredictable environments where density-independent factors affect population size. Energy is allocated to producing as many small offspring as possible. Adults remain small and do not invest in parental care of offspring. *K*-selection occurs in environments that remain relatively stable, where density-dependent factors affect population size. Energy is allocated to survival and repeated reproductive events. The adults are large and invest in parental care of offspring. Actual life histories contain trade-offs between these two patterns.

23.5 Human Population Growth

The human population is expanding exponentially, and it is unknown when the population size will level off. Most of the expected increase will occur in certain LDCs (less-developed countries) of Africa, Asia, and Latin America. Support for family planning, human development, and delayed child bearing could help prevent an expected increase.

Reviewing the Chapter

1. What are the various levels of ecological study? 384
2. What is density, and why is it sometimes misleading to do studies based on the average number of organisms per unit area? 385
3. What are the four population attributes having to do with population dynamics? 386
4. What type growth curve indicates that exponential growth is occurring? What are the environmental conditions for exponential growth? 387
5. What type growth curve indicates that biotic potential is being opposed by environmental resistance? What environmental conditions are involved in environmental resistance? 388–89
6. What is the carrying capacity of an area? 389
7. What is a life table and the three general types of survivorship curves? 390–91
8. What will the age distribution tell you about the population under study? 391
9. Does population growth of natural populations fit either the exponential growth model or the logistic growth model? Explain why or why not. 392
10. Give support to the belief that intrinsic factors might regulate population size in some populations. 393
11. Why would you expect the life histories of natural populations to vary and contain some characteristics that are so-called *r*-selected and some that are so-called *K*-selected? 396–97
12. What type of growth curve presently describes the population growth of the human populations? In what types of countries is most of this growth occurring, and how might it be curtailed? 399–400

Testing Yourself

Choose the best answer for each question.

1. Which of these levels of ecological study involves both abiotic and biotic components?
   a. organisms
   b. populations
   c. communities
   d. ecosystem
2. When phosphorus is made available to an aquatic community, the algal populations suddenly bloom. This indicates that phosphorus is a
   a. density-dependent regulating factor.
   b. reproductive factor.
   c. limiting factor.
   d. All of these are correct.
3. A J-shaped growth curve should be associated with
   a. exponential growth.
   b. biotic potential.
   c. no environmental resistance.
   d. All of these are correct.
4. An S-shaped growth curve
   a. occurs when there is no environmental resistance.
   b. includes an exponential growth phase.
   c. occurs in natural populations but not laboratory ones.
   d. All of these are correct.
5. If a population has a type I survivorship curve (most live the entire life span), which of these would you also expect?
   a. a single reproductive event per adult
   b. overlapping generations
   c. reproduction occurring near the end of the life span
   d. None of these are correct.
6. A pyramid-shaped age distribution means that the
   a. prereproductive group is the largest group.
   b. population will grow for some time in the future.
   c. country is more likely an LDC rather than an MDC.
   d. All of these are correct.
7. Which of these is a population-independent regulating factor?
   a. competition
   b. predation
   c. weather
   d. resource availability
8. Fluctuations in population growth can correlate to changes in
   a. predation.
   b. weather.
   c. resource availability.
   d. All of these are correct.
9. A species that has repeated reproductive events, lives a long time, but suffers a crash due to the weather is exemplifying
   a. *r*-selection.
   b. *K*-selection.
   c. a mixture of both.
   d. density-dependent and density-independent regulation.
10. The human population
    a. is undergoing exponential growth.
    b. is not subject to environmental resistance.
    c. fluctuates from year to year.
    d. only grows if emigration occurs.

Applying the Concepts

1. *Population sizes are regulated.*
   Explain why the model for logistic growth predicts the regulation of population size.
2. *Populations have discrete or repeated reproductive events.*
   Populations with repeated reproductive events invest energy into survival as well as reproduction. Why might this be a successful strategy?
3. *The human population is subject to environmental resistance.*
   Give examples to show that the human population is not immune to environmental regulations.

Using Technology

Your study of the ecology of populations is supported by these available technologies:

Exploring the Internet
The Mader Home Page provides resources for and help with studying this chapter.

http://www.mhhe.com/sciencemath/biology/mader/
(Click on Biology.)

Understanding the Terms

age structure diagram 391
biosphere 384
biotic potential 387
carrying capacity 389
cohort 390
community 384
demographic transition 399
doubling time 399
ecology 384
ecosystem 384
environmental resistance 387
exponential growth 387
habitat 384
K-selection 396
less-developed country (LDC) 399
limiting factor 385
logistic growth 388
more-developed country (MDC) 399
net reproductive rate 386
population 384
population density 385
population distribution 385
population size 386
replacement reproduction 400
r-selection 396
survivorship 390
sustainable world 401
zero population growth 400

Match the terms to these definitions:

a. __________ Due to industrialization, a decline in the birthrate following a reduction in the death rate so that the population growth rate is lowered.
b. __________ Group of organisms of the same species occupying a certain area and sharing a common gene pool.
c. __________ Growth, particularly of a population, in which the increase occurs in the same manner as compound interest.
d. __________ Largest number of organisms of a particular species that can be maintained indefinitely by a given environment.
e. __________ Maximum population growth rate under ideal conditions.
f. __________ Population in which each person is replaced by only one child.
g. __________ Representation of the number of individuals in each age group in a population.
h. __________ Concept that a high reproductive rate is the most favorable life history strategy for increase in population size under certain conditions.
i. __________ Sum total of factors in the environment that limit the numerical increase of a population in a particular region.
j. __________ Global way of life that can continue indefinitely, because the economic needs of all peoples are met while still protecting the environment.

Community Ecology

CHAPTER 24

Chapter Concepts

24.1 What Is a Biological Community?

- Communities are assemblages of interacting populations, which differ in composition and diversity. 406
- Environmental factors influence community composition and diversity. 407

24.2 Communities Are Organized

- Community organization involves the interactions among species such as competition, predation, parasitism, and mutualism. 409
- The ecological niche is the role an organism plays in its community, including its habitat and its interactions with other organisms. 409
- Competition leads to resource partitioning, which reduces competition between species. 410
- Predation reduces prey population density but can lead to a reduction in predator population density also. 412
- There are a number of different kinds of prey defenses, including mimicry. 414
- Symbiotic relationships include parasitism, commensalism, and mutualism. 416

24.3 Community Structure Changes Over Time

- Ecological succession is a change in species composition and community structure and organization over time. 420

24.4 How to Increase Community Biodiversity

- The intermediate disturbance hypothesis suggests that the presence of patches that contain various species increases biodiversity. 422
- Predation and competition can help maintain biodiversity, and island biogeography suggests how to maintain species richness. 423

Coyotes *(Canis latrans)* and badgers *(Taxidea taxus)* interact within a prairie community

Coyotes and badgers sometimes improve their hunting of prairie dogs by joining forces. When coyotes are around, prairie dogs tend to stay in their burrows where badgers can catch and flush them out to the waiting coyotes. For the coyote, it sure beats wasting time and energy stalking prairie dogs aboveground. Coyotes have even been seen to encourage badgers to move on to more productive hunting grounds by various kinds of antics. This example shows a possible way for two different species living in the same community to interact. (A community is an assemblage of populations in one area.)

There are numerous interactions between species in a community. One species may prey on or parasitize another species or may even assist another species. Often, interactions are not entirely black and white, as when coyotes sometimes kill young badgers. This chapter examines specific interactions and how they determine the distribution and abundance of species in a community. We will see that communities are subject to change, and although an intermediate amount of disturbance seems to increase diversity, extreme disturbances can be damaging to the structure of a community.

24.1 What Is a Biological Community?

Populations do not occur as single entities; they are part of a community. A **community** is an assemblage of populations interacting with one another within the same environment. Communities come in different sizes, and it is sometimes difficult to decide where one community ends and another begins. A fallen log can be considered a community because there are various populations in the log and these populations are interacting with one another. The fungi of decay break down the log and provide food for the various invertebrates living in the log that may feed on one another. Yet it is quite possible that bugs and worms living in the log could be eaten by a passing bird that flies about the entire forest. This interaction means that the log is a part of the forest community. And where does the forest end? Doesn't it gradually fade into the surrounding area? So the demarcation of any community turns out to be somewhat arbitrary.

This chapter examines the various types of community interactions and their importance to the structure of a community. Such interactions illustrate some of the most important selection pressures impinging on individuals. They also help us understand how biodiversity can be preserved.

Community Composition and Diversity

Two characteristics of communities—their composition and diversity—allow us to compare communities. This is a first step toward understanding how a community functions. The *composition* of a community is simply a listing of the various species in the community. The *diversity* includes both species richness (the number of species) and evenness (the relative abundance of individuals of different species).

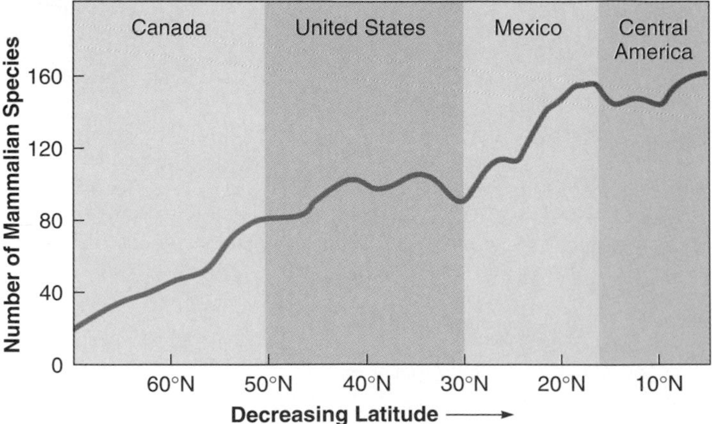

Figure 24.1 Community structure.
Communities differ in their composition, as witnessed by their predominant plants and animals. Diversity of communities is described by the richness of species and their relative abundance. **a.** A coniferous forest. Some mammals found here are listed to the left. **b.** A tropical rain forest. Some mammals found here are listed to the right. **c.** Richness increases with decreasing latitude.

Just glancing at Figure 24.1 makes it easy for us to see that a coniferous forest has a different composition from a tropical rain forest. Pictorially we can see that narrow-leaved evergreen trees are present in a coniferous forest, and broad-leaved evergreen trees are numerous in a tropical rain forest. Mammals also differ between the two communities, as those listed demonstrate.

Diversity of a community goes beyond composition because it includes the number of species and also the abundance of each species. To take an extreme example: a forest in West Virginia has, among other species, 76 yellow poplar trees but only one American elm. If we were simply walking through this forest, we might miss seeing the American elm. If, instead, the forest had 36 poplar trees and 41 American elms, the forest would seem more diverse to us and indeed would be more diverse. The greater the diversity, the greater the number and the more even the distribution of species.

Models Regarding Composition and Diversity

What causes the populations within a community to assemble together in the same place at the same time? Figure 24.1*c* shows that the number of species in a community increases as we move from the northern latitudes to the equator. We know that at the equator the weather is warmer and there is more precipitation. Could it be that community composition is controlled by abiotic factors? A species' range is based on its tolerance for such abiotic factors in the environment as temperature, light, water availability, salinity, and so forth. In order to determine a species' tolerance range, we plot the species' ability to survive and reproduce under a particular environmental condition. The result is a bell-shaped curve that shows the species' range for that abiotic factor (Fig. 24.2*a*). It's possible that species may assemble because their tolerance ranges simply overlap (Fig. 24.2*b*).

This view, advanced by H. L. Gleason in the early 1900s, is called the *individualistic model* because it suggests that each population in a community is there because its own particular abiotic requirements are met by a particular habitat. The individualistic model predicts that species will have independent distributions, and that the boundaries between communities will not be distinct from one another.

For many years, most ecologists supported the *interactive model* of community structure suggested by F. E. Clements in the early 1900s. Clements saw a community as the highest level of organization arising from cell to tissue, to organism, to populations, and finally, to a community. Just as the parts of an organism are dependent on one another, so species are dependent on biotic interactions such as its food source. Further, like an organism, a community remains stable because of homeostatic mechanisms. This theory predicts that the same species will reoccur in communities whose boundaries are distinct from one another.

The data collected by modern ecologists lend support to the individualistic model. For example, F. H. Talbot and colleagues built artificial reefs of a constant size out of cement building blocks and put them in a warm tropical lagoon a few meters from one another. Altogether, 42 species of fish colonized the artificial reefs, but the similarity of species between the reefs was only 32%. Moreover, Talbot found that there was a very high turnover from month to month. From one month to the next, 20–40% of species had been replaced by other species. Therefore, species composition seemed to depend on chance migrations. However, we know that certain animals are more likely to occur where they find their usual food source. Koala bears feed on eucalyptus leaves and are found only in regions where these plants are found. Most likely community structure is dependent on both abiotic and biotic factors.

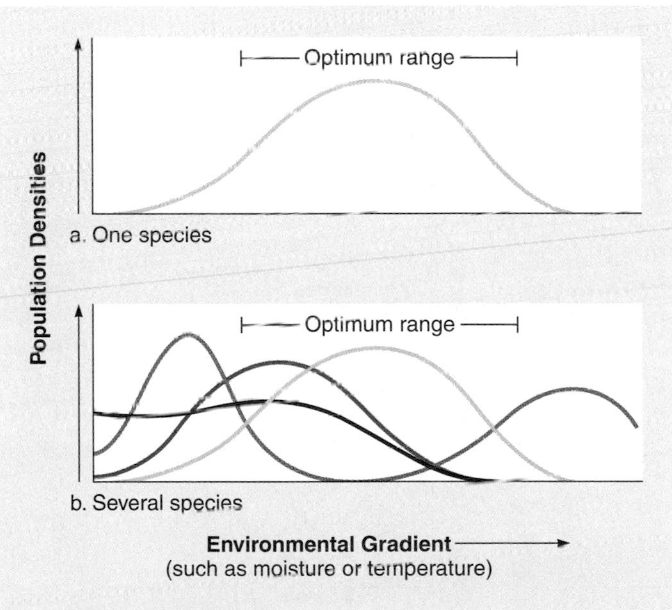

Figure 24.2 Species richness of communities. According to the individualistic hypothesis (**a**) each species is distributed along environmental gradients according to its own tolerance for abiotic factors, and (**b**) a community is believed to be an assemblage of species that happen to occupy the same area because of similar tolerances.

Community composition is probably dependent on both abiotic gradients (e.g. climate, inorganic nutrients) and biotic interactions (e.g., organic food source).

Island Biogeography

Would you expect larger coral reefs to have greater species richness (number of species) than smaller coral reefs? The area (space) occupied by a community can have a profound effect on its diversity. American ecologists Robert MacArthur and E. O. Wilson developed a general theory of *island biogeography* to explain and predict the effects of distance from the mainland and size of an island on community diversity.

Imagine two new islands that as yet contain no species at all. One of these islands is near the mainland, and one is far from the mainland. Which island will receive more immigrants from the mainland? Most likely the near one because it's easier for immigrants to get there (Fig. 24.3*a*).

Similarly, imagine two islands that differ in size. Which island will be able to support a greater number of species? The large one, because a greater amount of resources can support more populations (Fig. 24.3*b*). MacArthur and Wilson studied the diversity on many island chains including the West Indies, and discovered that species richness does correlate positively with island size. Conservationists note that the theory of island biogeography applies to their work because conserved land is like an island surrounded by farms, towns, and cities. The theory tells them that the larger the conserved area, the better the chance of preserving more species.

Is it possible to increase the amount of space without using more area? The *spatial heterogeneity model* holds that if the environment has *patches,* the greater the number of habitats, the greater the diversity. As gardeners, we are urged to create patches in our yards if we wish to attract more butterflies and birds! Stratification is one way to introduce patchiness. Just as a high-rise apartment building allows more human families to live in an area, so stratification within a community provides more and different types of living space for different species. Stratification of its canopy (treetops) is another reason why a tropical rain forest has many more species than a coniferous forest.

The fact that space is limited suggests that there must be an end point beyond which community richness cannot increase. Figure 24.3*c* suggests there will be an equilibrium point for all four types of islands (i.e., near and large, near and small, far and large, far and small). Notice that the equilibrium point is highest for a large island that is near the mainland. An equilibrium is reached when the rate of species immigration matches the rate of species extinction. The equilibrium could be dynamic (new species keep on arriving, and new extinctions keep on occurring), or it could be that the composition of the community will remain steady unless disturbed. This is an idea that we will return to later in the chapter.

The theory of island biogeography says that distance from a population source and size affect species diversity. When immigration and extinction rates are equal an equilibrium in species diversity develops.

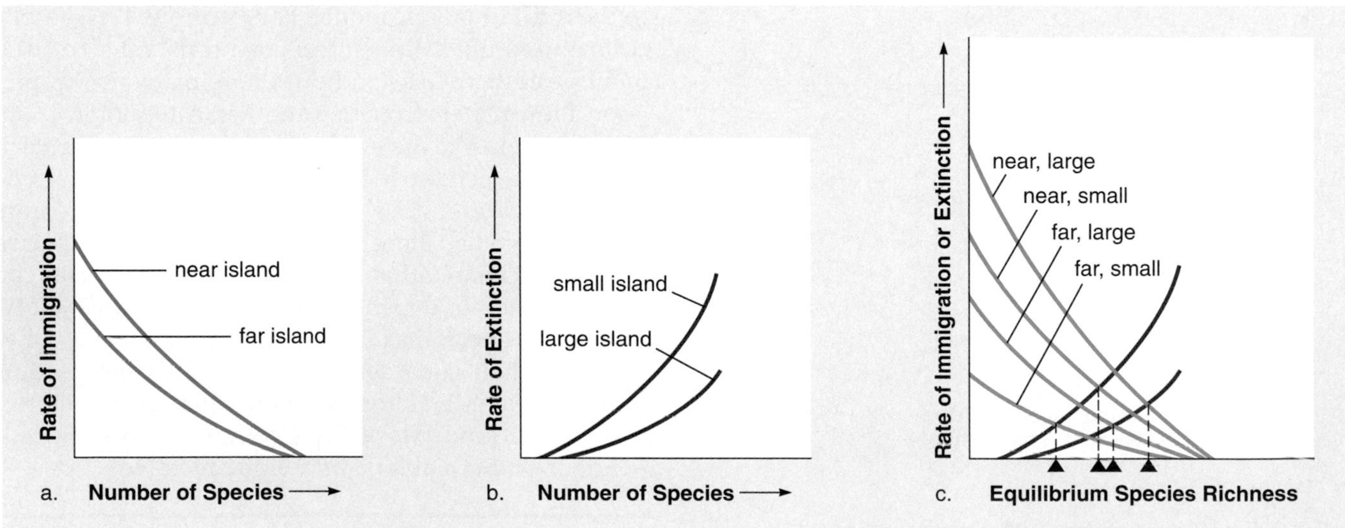

Figure 24.3 Theory of island biogeography.
a. An island that is near the mainland will have a higher immigration rate than an island that is far from the mainland. **b.** A small island will have a higher extinction rate than a large island. **c.** The balance between immigration and extinction for four possible types of islands. A large island that is near the mainland will have a higher equilibrium species richness than the other types of islands.

24.2 Communities Are Organized

Competition for resources, predator-prey, parasite-host, and other types of interactions integrate species into a system of dynamic interacting populations (Table 24.1). Competition for limited resources between two species has a negative effect on the population abundance of both species. In predation, one animal eats another called the prey, and in parasitism one individual obtains nutrients from another, called the host. Predation and parasitism are expected to increase the abundance of the predator population at the expense of the prey and host population, respectively. In commensalism one species is benefited, but the other is not harmed. As an example, commensalism often occurs when one species provides a home or transportation for another. In mutualism, two species help one another and both species are benefited. In Table 24.1 a plus sign (+) is used whenever the fitness of the individual in the interaction is expected to increase, and a minus sign (–) is used whenever the fitness of the individual is expected to decrease. These effects are manifested by a change in the species population abundance as indicated by the table.

Table 24.1

Species Interactions

Interaction	Expected Outcome
Predation (+ –)	Abundance of predator increases and abundance of prey decreases
Symbiotic Relationships	
Parasitism (+ –)	Abundance of both species decreases
Competition (– –)	Abundance of both species decreases
Commensalism (+ 0)	Abundance of one species increases, and the other is not affected
Mutualism (+ +)	Abundance of both species increases

Habitat and Ecological Niche

Each species occupies a particular position in the community, both in a spatial sense (where it lives) and in a functional sense (what role it plays). Its **habitat** is where an organism lives and reproduces in its environment. The habitat of an organism might be the forest floor, a swift stream, or the ocean's edge (Fig. 24.4). The **ecological niche** of an organism is the role it plays in its community, including its habitat and its interactions with other organisms. The niche includes the resources an organism uses to meet its energy, nutrient, and survival demands. For a backswimmer, home is a pond or lake where it eats other insects. The pond must contain vegetation where the backswimmer can hide from its predators such as fish and birds. On the other hand, the water must be clear enough for the backswimmer to see its prey and warm enough for it to be in active pursuit. Since it's difficult to study the total niche of an organism, some observations focus on a certain aspect of an organism's niche, as with the birds featured in Figure 24.4.

Because an organism's niche is affected both by abiotic factors (such as climate and habitat) and biotic factors (such as competitors, parasites, and predators present in the habitat), ecologists distinguish between the fundamental and realized niches. An organism's *fundamental niche* comprises all conditions under which an organism can potentially survive and reproduce; the *realized niche* is the set of conditions under which it exists in nature.

Habitat is where an organism lives, and ecological niche is the role an organism plays in its community, including its habitat and its interactions with other organisms.

Flamingos feed on small mollusks, crustacea and vegetable matter strained from mud pumped through their bills by their powerful tongues.

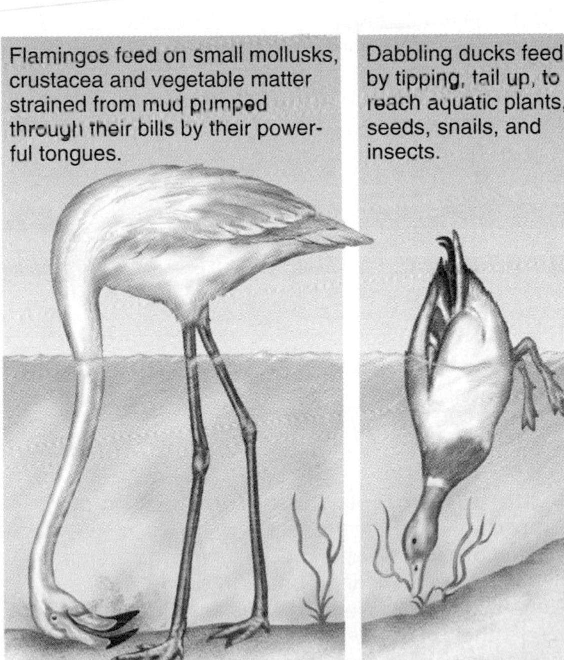

Dabbling ducks feed by tipping, tail up, to reach aquatic plants, seeds, snails, and insects.

Avocets feed on insects, small marine invertebrates, and seeds by sweeping their bills from side to side in shallow water.

Oystercatchers pry open bivalve shells with their knifelike bills and probe sand for worms and crabs.

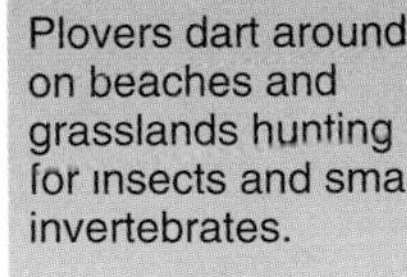

Plovers dart around on beaches and grasslands hunting for insects and small invertebrates.

Figure 24.4 Feeding niches for wading birds. Flamingos feed in deeper water by filter feeding; dabbling ducks feed in shallower areas by upending; avocets feed by sifting. Oystercatchers and plovers are restricted to the shallows by their shorter legs.

Competition Between Populations

Interspecific competition occurs when members of different species try to utilize a resource (like light, space, or nutrients) that is in limited supply. If the resource is not in limited supply, there is no competition. In the 1920s, Lotka and Volterra developed mathematical formulas that predicted competition can lead to the predominance of one species and the virtual elimination of the other. Later, this was demonstrated in the laboratory by G. F. Gause, who grew two species of *Paramecium* in one test tube containing a fixed amount of bacterial food. Although each population survived when grown separately, only one survived when they were grown together (Fig. 24.5). The successful *Paramecium* population had a higher biotic potential than the unsuccessful population. After observing the outcome of similar other experiments in the laboratory, ecologists formulated the **competitive exclusion principle,** which states that no two species can occupy the same niche at the same time.

What does it take to have different ecological niches so that extinction of one species is avoided? In another laboratory experiment, Gause found that two species of *Paramecium* did continue to occupy the same tube when one species fed on bacteria at the bottom of the tube and the other fed on bacteria suspended in solution. Under these circumstances, **resource partitioning** decreased competition between the two species. What could have been one niche became two niches because of a divergence of behavior in this case.

It's possible to observe niche partitioning in nature. When three species of ground finches of the Galápagos Islands occur on separate islands, their bills tend to be the same intermediate size, enabling each to feed on a wider range of seeds. Where they co-occur, selection has favored divergence in beak size because the size of the beak affects the kinds of seeds that can be eaten. In other words, competition has led to resource partitioning. The tendency for characteristics to be more divergent when populations belong to the same community than when they are isolated is termed **character displacement.** Character displacement is often used as evidence that competition and resource partitioning have taken place (Fig. 24.6).

The niche partitioning we observe in nature can be very subtle. Five different species of warblers that occur in North American forests are all about the same size and all feed on bud worms, a type of caterpillar found on spruce trees. In a very famous study, Robert MacArthur recorded the length of time each of five species of warblers spent in different regions of spruce canopies to determine where each species did most of its feeding (Fig. 24.7). He discovered that each type of bird primarily used different parts of the tree canopy and in that way had a different niche.

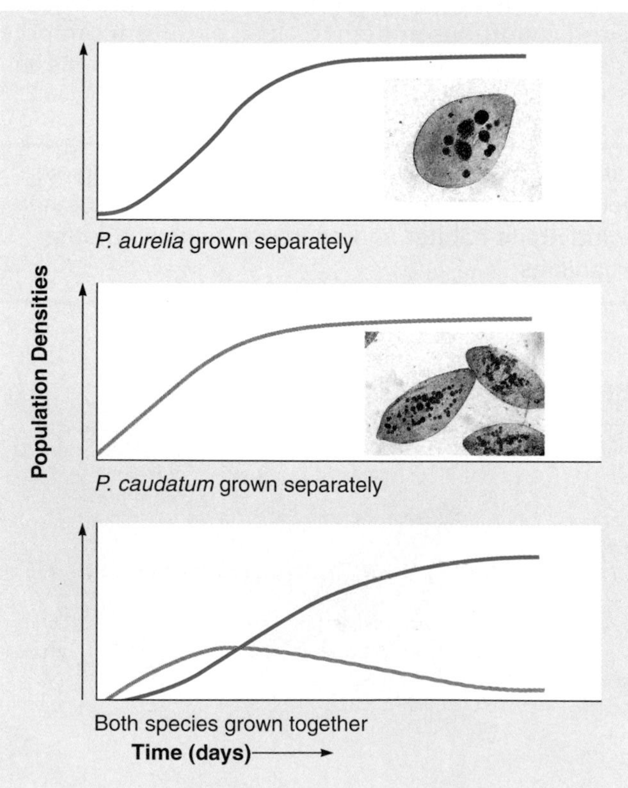

Figure 24.5 Competition between two laboratory populations of *Paramecium*.
When grown alone in pure culture, *Paramecium caudatum* and *Paramecium aurelia* exhibit sigmoidal growth. When the two species are grown together in mixed culture, *P. aurelia* is the better competitor, and *P. caudatum* dies out.

Source: Data from G. F. Gause, *The Struggle for Existence*, 1934, Williams & Wilkins Company, Baltimore, MD.

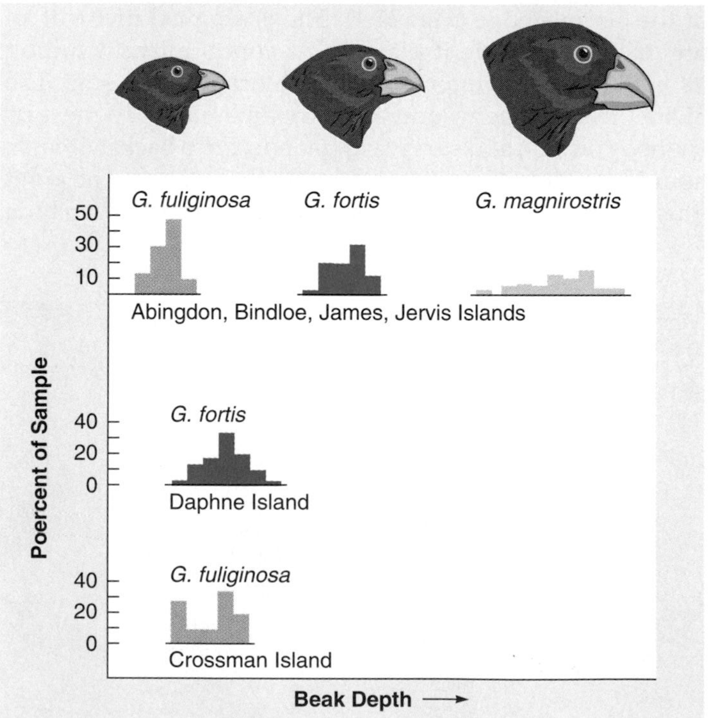

Figure 24.6 Character displacement in finches on Galápagos Islands.
When *Geospiza fulginosa*, *G. fortis,* and *G. magnirostris* are on the same island, their beak sizes are appropriate to eating small-, medium-, and large-size seeds. When *G. fortis* and *G. fuliginosa* are on separate islands, their beaks have the same intermediate size, which allows them to eat seeds that vary in size.

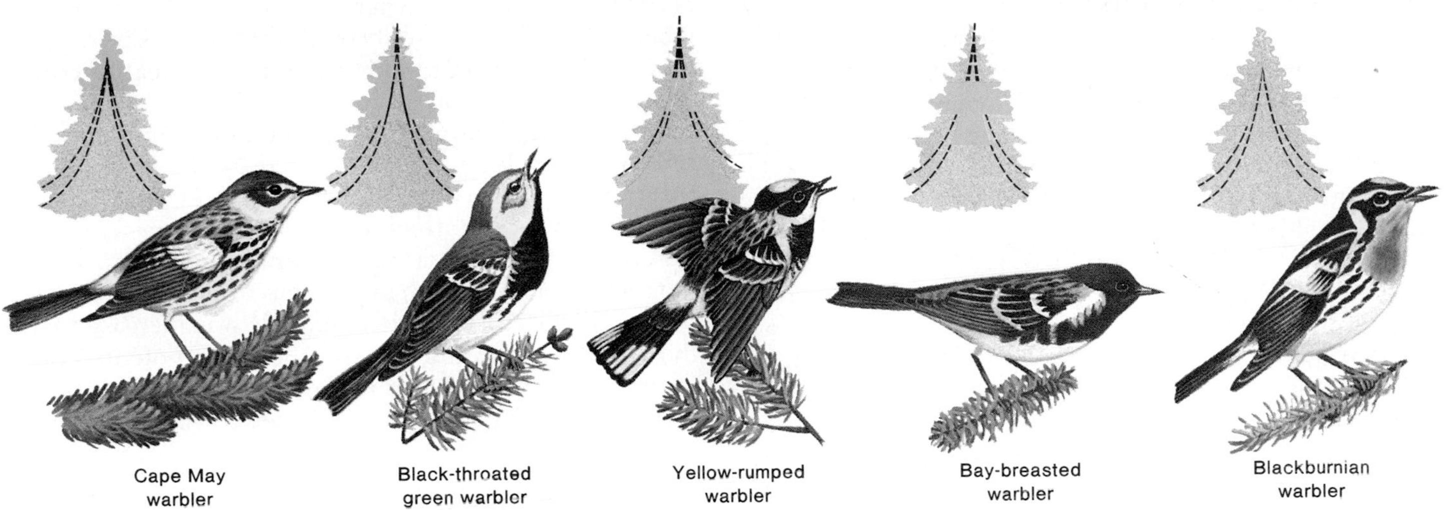

Figure 24.7 Niche partitioning among five species of coexisting warblers.
The diagrams represent spruce trees. The time each species spent in various portions of the trees was determined; each species spent more than half its time in the blue regions.

As another example, consider that swallows, swifts, and martins all eat flying insects and parachuting spiders. These birds even frequently fly in mixed flocks. But each type of bird has different nesting sites and migrate at a slightly different time of year. Therefore they are not competing for the same food source when they are feeding their young.

In all these cases of niche partitioning, we have merely supposed that what we observe today is due to competition in the past. Some ecologists are fond of saying that in doing so we have invoked the "ghosts of competition past." Are there any instances in which competition has actually been observed? Perhaps we have one example. Joseph Connell has studied the distribution of barnacles on the Scottish coast, where a small barnacle (*Chthamalus stellatus*) lives on the high part of the intertidal zone, and a large barnacle (*Balanus balanoides*) lives on the lower part (Fig. 24.8). Free-swimming larvae of both species attach themselves to rocks at any point in the intertidal zone, where they develop into the sessile adult forms. In the lower zone, the large *Balanus* barnacles seem to either force the smaller *Chthamalus* individuals off the rocks or grow over them. To test his observation, Connell removed the larger barnacles and found that the smaller barnacles grew equally well on all parts of the rock. This seems clear evidence that competition is restricting the range of *Chthamalus* on the rocks. *Chthamalus* is more resistant to drying out than is *Balanus*, therefore, it has an advantage that permits it to grow in the upper intertidal zone.

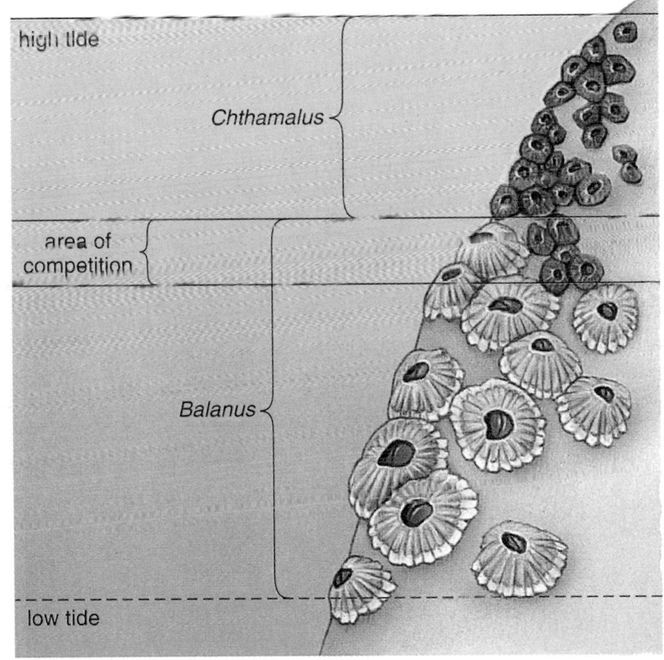

Figure 24.8 Competition between two species of barnacles.
Competition prevents two species of barnacles from occupying as much of the intertidal zone as possible. Both exist in the area of competition between *Chthamalus* and *Balanus*. Above this area only *Chthamalus* survives, and below it only *Balanus* survives.

Interspecific competition may result in niche partitioning. When similar species seem to be occupying the same ecological niche, it is usually possible to find differences that indicate niche partitioning has occurred.

Predator-Prey Interactions

Predation occurs when one living organism, called the **predator,** feeds on another, called the **prey.** In its broadest sense, predaceous consumers include not only animals like lions that kill zebra, but also filter-feeding blue whales that strain krill from ocean waters, parasitic ticks that suck blood from victims, and even herbivorous deer that browse on trees and bushes.

Predator-Prey Population Dynamics

Do predators reduce the population density of prey? On the face of it, you would probably hypothesize that they do, and your hypothesis would certainly be supported by a laboratory study done by Gause. When Gause reared the protozoan *Paramecium caudatum* (prey) and *Didinium nasutum* (predator) together in a culture medium, *Didinium* ate all the *Paramecium* and then died of starvation (Fig. 24.9*a*). In nature, we can find a similar example. When a gardener brought prickly-pear cactus to Australia from South America, the cactus spread out of control until millions of acres were covered with nothing but cacti. The cacti were brought under control when a moth from South America, whose caterpillar feeds only on the cactus, was introduced. Now both cactus and moth are found at greatly reduced densities in Australia.

This raises an interesting point: the population density of the predator can be affected by the prevalence of the prey. In other words, the predator-prey relationship is actually a two-way street. In that context consider that at first the biotic potential (maximum reproductive rate) of the cactus is maximized, but the environmental resistance (factors that oppose biotic potential) was increased when the moth was introduced. And the biotic potential of the moth is maximized when it was first introduced, but the carrying capacity was reduced when its food supply was diminished.

In nature, presence of predators can decrease prey densities and vice-versa.

There are mathematical formulas that predict a cycling of predator and prey populations instead of a steady state:

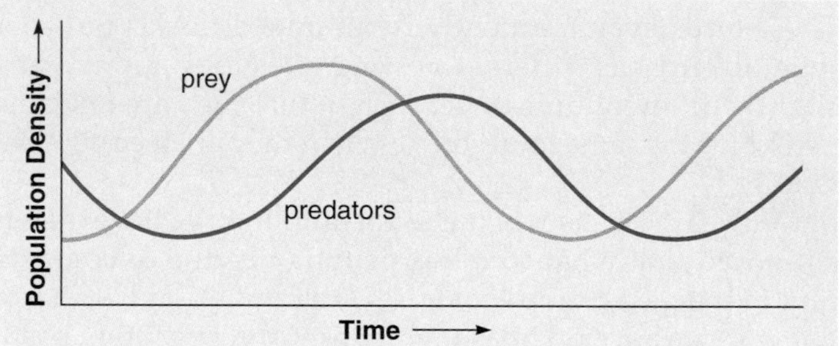

Notice that the predator population is smaller than the prey population, and that it lags behind the prey population. What might cause a cycling like this? We can certainly observe that as the prey population increases, the predator population also increases, most likely because more food has become available. At this point, there are at least two possibilities to account for the reduction in population densities that follow: (1) perhaps the biotic potential (reproductive rate) of the predator is so great, its increased numbers overconsume the prey, and then as the prey population declines, so does the predator population; or (2) perhaps the biotic potential of the prey is unable to keep pace, and the prey population overshoots the carrying capacity and suffers a crash. Now the predator population follows suit because of a lack of food. In either case (1 and 2), the result will be a series of peaks and valleys with the predator population lagging slightly behind the prey.

The famous case of predator/prey cycles occurs between the snowshoe hare and the Canadian lynx, a type of small cat (Fig. 24.10). The snowshoe hare is a common herbivore in the coniferous forests of North America, where it feeds on terminal twigs of various shrubs and small trees.

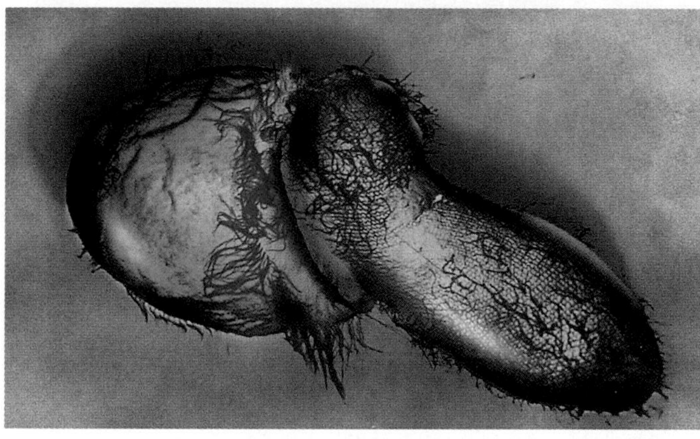

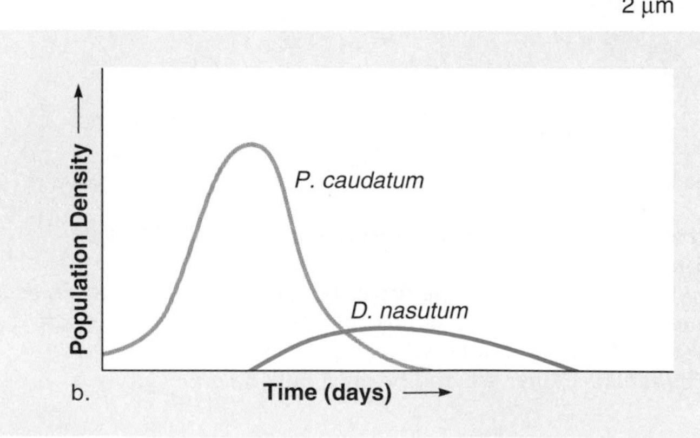

Figure 24.9 Predator-prey interaction between *Paramecium caudatum* and *Didinium nasutum*.
a. Scanning electron micrograph of *Didinium* engulfing *Paramecium*. **b.** In an oat medium, *Didinium* ate all the *Paramecium* and then died out.

Source: Data from G. F. Gause, *The Struggle for Existence*, 1934, Williams & Wilkins Company, Baltimore, MD.

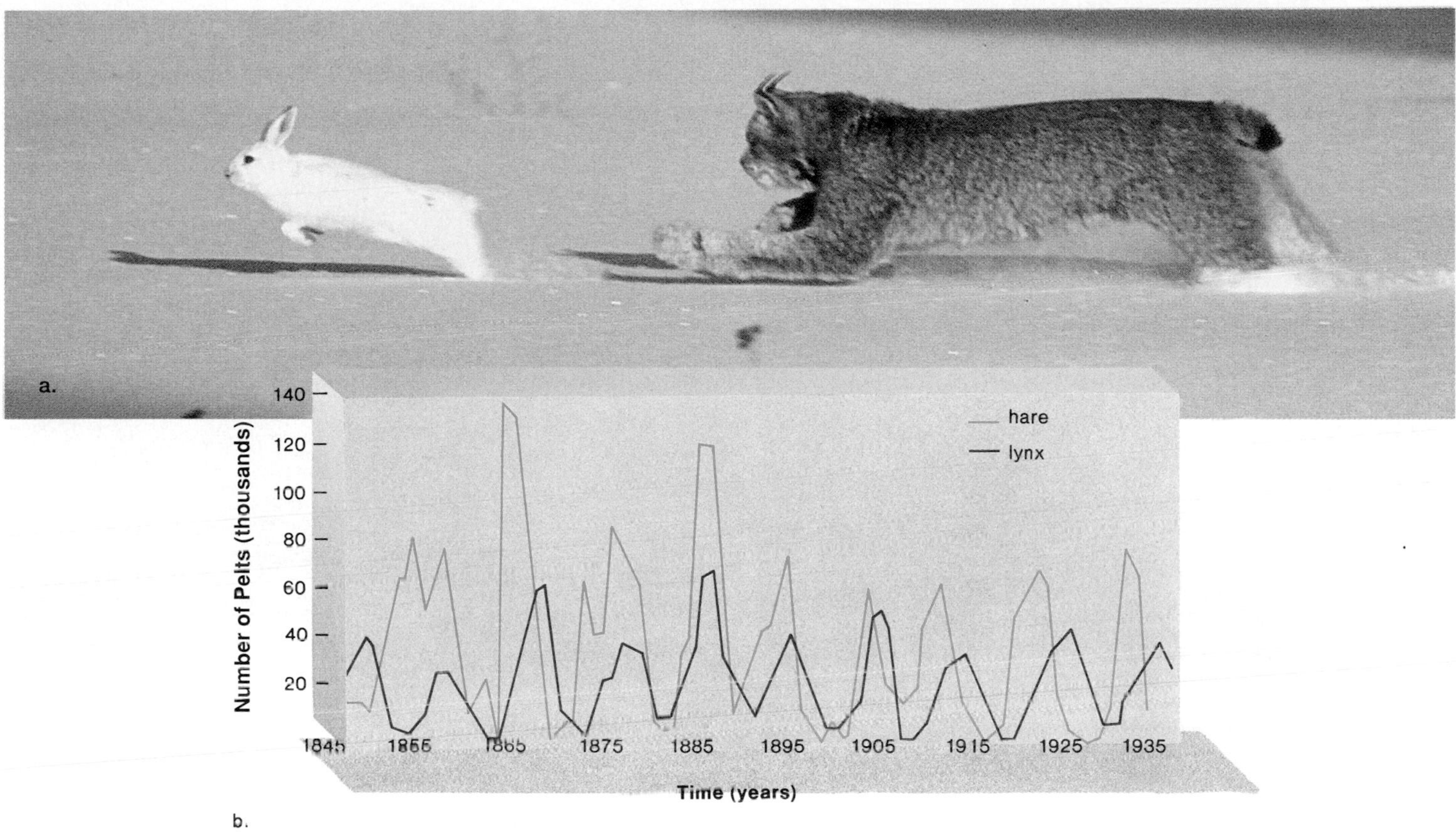

Figure 24.10 Predator-prey interaction between a lynx and a snowshoe hare.
a. A Canadian lynx (*Lynx canadensis*) is a solitary predator. A long, strong forelimb with sharp claws grabs its main prey, the snowshoe hare (*Lepus americanus*). The lynx lives in northern forests. Its brownish-gray coat blends in well against tree trunks, and its long legs enable it to walk through deep snow. b. The number of pelts received yearly by the Hudson Bay Company for almost 100 years shows a pattern of ten-year cycles in population densities. The snowshoe hare population reaches a peak abundance before that of the lynx by a year or more.

The Canadian lynx feeds on snowshoe hares but also on ruffed grouse and spruce grouse, two types of birds. Investigators at first assumed that the first explanation given above on page 412 was sufficient to explain the cycling between the hare and lynx populations. In other words, the lynx had brought about the decline of the hare population. But others noted that the decline in snowshoe hare abundance was accompanied by low growth and reproductive rates that could be signs of a food shortage.

To test whether the predator's (lynx) or the prey's (hare) limited food supply was causing the observed cycling between the two populations, three experiments were done:

1. To do away with both possibilities, a hare population was given a constant supply of food, and predators were excluded. Under these circumstances, the cycling of the hare population ceased.
2. To test the predator effect alone, a hare population was given a constant supply of food, but predators were not excluded. Under these circumstances, the cycling of both populations continued.
3. To test the food effect alone, predators were excluded, but no food was added to the environment of a hare population. Under these circumstances, the cycling of the hare population continued.

These results suggest that both explanations given above apply to the data in Figure 24.10. In other words, both a hare-food cycle and a predator-hare cycle have combined to produce an overall effect, which is observed in Figure 24.10. It's interesting to note that the population densities of the grouse populations also cycle, perhaps because the lynx switches to this food source when the hare population declines. Our discussion points out that predators and prey do not normally exist as simple, two-species systems, and therefore abundance patterns should be viewed with the complete community in mind.

Interactions between predator/prey and between prey and its own food source can interact to produce complex cycles.

a.

b.

Figure 24.11 Camouflage in the anglerfish.
a. A wormlike appendage on the head of an anglerfish (*Lophius americanus*) lures possible prey. **b.** Then it opens its cavernous mouth and devours its prey like this black seabass.

Prey Defenses and Other Interactions

Prey defenses are mechanisms that thwart the possibility of being eaten by a predator. Plants have defenses that discourage herbivores. The sharp spines of the cactus, the pointed leaves of the holly, and the tough, leathery leaves of oak trees make the plant less desirable as a food source. Plants even produce poisonous chemicals, some of which interfere with the normal metabolism of an adult insect, and others that act as hormone analogues that interfere with the development of insect larvae. Some insects can still inhabit trees that produce toxins because the insects have either evolved detoxifying enzymes or a method of holding the toxin within their bodies so it is not harmful to them.

Among animals we find a number of ways that predators fool their prey and prey avoid capture by predators. The anglerfish puts camouflage to work both as a prey defense and predator mechanism. The mottled skin, grotesque jaw fringe, and flat body allow it to blend into the ocean floor where it waits, while a fleshy wormlike appendage atop its head moves slowly back and forth (Fig. 24.11). When a fish draws near, the anglerfish opens its mouth and sucks in its prey. This anglerfish ranges from Newfoundland to North Carolina and appears in seafood stores and menus as "monkfish."

Many examples of protective **camouflage** are known: stick caterpillars look like twigs; katydids look like sprouting green leaves; and some moths resemble the barks of trees. Normally pepper moths are a light color. But in industrial areas where the tree trunks have turned dark due to pollution, the pepper moth populations have mostly dark-colored

a. b. c.

Figure 24.12 Antipredator defenses.
a. The South American lantern fly has a large false head that resembles that of an alligator. This may frighten a predator into thinking it is facing a dangerous animal. **b.** The collared lizard has a display ritual that makes it look larger when confronted by a predator. **c.** The skin secretions of dart-poison frogs are so poisonous that they were used by natives to make their arrows instant lethal weapons. The coloration of these frogs warns others to beware.

members. This change in population statistics is often used as an example of natural selection with predatory birds being the selection agent. The birds capture the more visible light-colored moths on dark tree trunks, and the dark-colored ones remain to contribute a greater number of offspring to the next generation.

Causing fright or harm is another possible prey defense. The South American lantern fly has a large false head that resembles that of an alligator (Fig. 24.12*a*). Other animals have a way to appear larger than they are, such as the cornered collared lizard in Figure 24.12*b*. Many moths have eyelike spots on their underwings that they can flash to startle a bird long enough to escape. The porcupine sends out arrowlike quills with barbs that dig into the predator's flesh and penetrate even deeper as the enemy struggles after being struck. In the meantime, the porcupine runs away.

Association with other prey may sometimes help avoid capture. Flocks of birds, schools of fish, and herds of mammals stick together as protection against predators. Baboons who detect predators visually, and antelopes who detect predators by smell, sometimes forage together, giving double protection against stealthy predators. The gazelle-like springboks of southern Africa jump stiff-legged 8 to 12 feet into the air when alarmed. Such a jumble of shapes and motions might confuse an attacking lion, and the herd can escape.

When you think about it, many prey defenses involve both a structural and behavioral adaptation. Those animals that use camouflage usually stay very still so as not to be detected; those that use fright must spring into action at the right time; and those that use other members as shields must stay together.

Mimicry

Warning coloration is a mechanism used by poisonous animals to prevent attack in the first place. The skin secretions of dart-poison frogs, such as the one shown in Figure 24.12*c*, are so poisonous that South American Indians used these secretions to turn their arrows into instant lethal weapons. **Mimicry** occurs when one species resembles another that possesses an overt antipredator defense. A mimic that lacks the defense of the organism it resembles is called a Batesian mimic (named for Henry Bates, who discovered it). In Figure 24.13, several insects are shown that resemble a wasp, but they lack the ability to sting. Once an animal experiences the defense of the model, it remembers the coloration and avoids all animals that look similar.

There are also examples of species that have the same defense and resemble each other. Among insects, many stinging insects—bees, wasps, and hornets—all have the familiar black and yellow color bands. Mimics that share the same protective defense are called Müllerian mimics after Fritz Müller who suggested that this, too, is a form of mimicry. Each of these Müllerian mimics reinforces the message that insects with this coloration can sting.

Just as with other prey defenses, behavior plays a role in mimicry. Mimicry works better if the mimic acts like the model. For example, beetles that resemble a wasp actively fly from place to place and spend most of their time in the same habitat as the wasp model. Their behavior makes them resemble a wasp to an even greater degree.

Prey escape predation by utilizing camouflage, fright, flocking together, warning coloration, and mimicry.

Figure 24.13 Mimicry among insects with yellow and black stripes.
a. A yellow jacket (*Vespula*) and (b) a bumblebee (*Bombus*) are Müllerian mimics. They have a similar appearance, and they both use stinging as a defense. The other animals—(c) a flower fly (*Chrysotoxum*), (d) a longhorn beetle (*Strophiona*), and (e) a moth (*Aegeria*)—are Batesian mimics because they are incapable of stinging another animal, yet they have the same appearance as the yellow jacket.

a.

b.

c.

d.

e.

Symbiotic Relationships

Symbiosis refers to interactions in which there is a close relationship between members of two populations. Table 24.1 tells how these relationships affect the fitness of the individuals involved.

Parasitism

Parasitism is similar to predation in that an organism called a **parasite** derives nourishment from another called the **host** (just as the predator derives nourishment from its prey). Viruses, such as HIV, that reproduce inside human lymphocytes are always parasitic. Parasites occur in all kingdoms of life also. Bacteria (e.g., strep infection), protists (e.g., malaria), fungi (e.g., rusts and smuts), plants (e.g., mistletoe), and animals (e.g., leeches) all contain parasitic members. While small parasites can be endoparasites, larger ones are more likely to be ectoparasites that remain attached to the exterior of the body by means of specialized organs and appendages. The effects of parasites on the health of the host can range from a slight weakening effect to actually killing them over time. When host populations are at a high density, parasites readily spread from one host to the next, causing intense infestations and a subsequent decline in host density. Parasites that don't kill their host can still play a role in reducing the host's population density because an infected host is less fertile and becomes more susceptible to another cause of death.

In addition to nourishment, host organisms also provide their parasites with a place to live and reproduce, as well as a mechanism for dispersing offspring to new hosts. Many parasites have both a primary and secondary host. The secondary host may be a vector that transmits the parasite to the next primary host. Usually both hosts are required to complete the life cycle. The association between parasite and host is so intimate that parasites are usually specific and require only certain species as hosts.

As an example, we will consider the deer tick called *Ixodes dammini* in the East and *I. ricinus* in the West. Deer ticks are arthropods that go through a number of stages (egg, larva, nymph, adult). They are so named because adults feed and mate on white-tailed deer in the fall. The female lays her eggs on the ground, and when the eggs hatch in the spring they become larvae that feed primarily on white-footed mice. If a mouse is infected with the bacterium *Borrelia burgdorfei*, the larvae become infected also. The fed larvae overwinter and molt the next spring to become nymphs that can, by chance, take a blood meal from a human. It is at this time that the tick may pass the bacterium on to a human who subsequently comes down with Lyme disease characterized by arthritic-like symptoms. The fed nymphs develop into adults and the cycle begins again (Fig. 24.14).

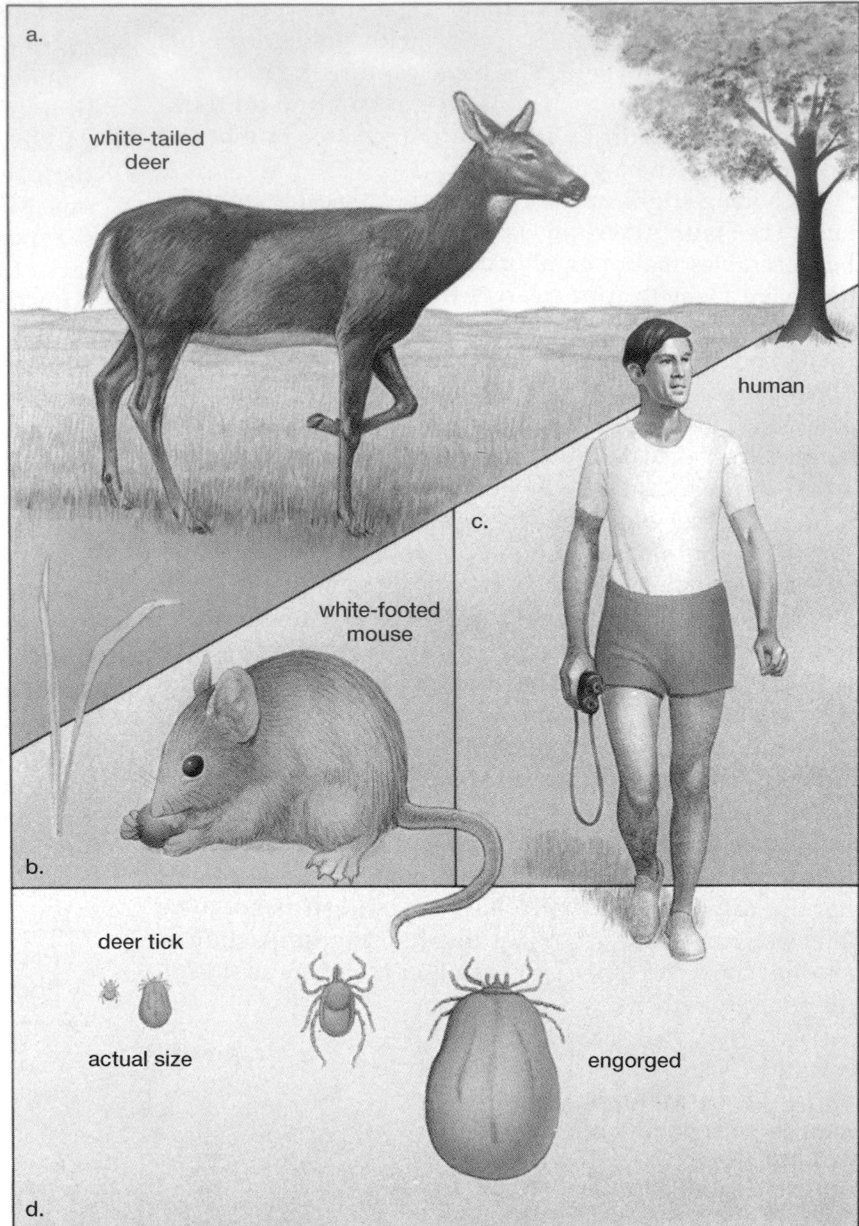

Figure 24.14 The life cycle of a deer tick.
a. Adult ticks feed and mate on white-tailed deer, which accounts for the common name of the ticks. Female adults lay their eggs in soil and then die. **b.** In the spring and summer, larvae feed mainly on white-footed mice. They then overwinter. **c.** During the next summer, nymphs feed on white-footed mice or other animals, including humans. If a nymph is infected with the bacterium *Borrelia burgdorfei*, humans get Lyme disease. **d.** Deer tick before feeding and after feeding. Actual size is shown along with enlarged size.

Parasites are very different from predators. They take nourishment from their host but also use them as a habitat and a way to transmit themselves to the next host. Many complicated associations have evolved.

A closer Look

Interactions and Coevolution

Coevolution is present when two species adapt in response to selective pressure imposed by the other. Symbiosis (close association between two species), which includes parasitism, commensalism, and mutualism is especially prone to the process of coevolution. Flowers that have animal pollinators have features that attract them (see the reading on p. 590). As an example of this type of coevolution, a butterfly-pollinated flower is often a composite, containing many individual flowers; the broad expanse provides room for the butterfly to land, and the butterfly has a proboscis that it inserts into each flower in turn. In this case, coevolution is highly beneficial to the flower because the insect will carry pollen to other flowers of the same type.

Coevolution also occurs between predators and prey. For example, a predator like a cheetah sprints forward to catch its prey, and this behavior might be selective for those gazelles that are able to run away or jump high in the air (like a springbok) to avoid capture. The adaptation of the prey may very well put selective pressure on the predator for an adaptation to the prey's defense mechanism. Hence an "arms race" can develop. The process of coevolution has been studied in the cuckoo, a social parasite that reproduces at the expense of other birds by laying its eggs in their nests. It's a strange sight to see a small bird feeding a cuckoo nestling that is several times its size. How did this strange happening develop? Investigators discovered that in order to "trick" a host bird, the adult cuckoo has to (1) lay an egg that mimics the host's egg, (2) lay its egg very rapidly (only ten seconds are required) in the afternoon while the host is away from the nest, and (3) leave most of the eggs in the nest because hosts will desert a nest that has only one egg in it. (The cuckoo chick hatches first and is adapted to removing any other eggs in the nest [Fig. 24A].) It seems that the host birds may very well now evolve a way to distinguish the cuckoo from their own young.

Coevolution can take many forms. In the case of *Plasmodium,* a cause of malaria, the sexual portion of the life cycle occurs within mosquitoes (the vector), and the asexual portion occurs in humans. The human immune system uses surface proteins to detect pathogens, and *Plasmodium* has numerous genes for surface proteins. Therefore, it is capable of changing its surface proteins repeatedly, and in this way it stays one step ahead of the host's immune system. A similar capability of HIV has added to the difficulty of producing an AIDS vaccine.

The relationship between parasite and host can even include the ability of parasites to seemingly manipulate the behavior of their hosts in self-serving ways. Ants infected with the lance fluke (but not those uninfected) cling to blades of grass with their mouthparts. There the infested ants are eaten by grazing sheep, and the flukes are transmitted to the next host in their life cycle. Similarly, when snails of the genus *Succinea* are parasitized by worms of the genus *Leucochloridium,* they are eaten by birds. As the worms mature, they invade the snail's eyestalks, making them resemble edible caterpillars. Now the birds eat the snails, and the parasites release their eggs, which complete development inside the urinary tracts of birds.

It used to be well accepted that as host and parasite coevolved, each would become more tolerant of the other—if the opposite occurred, the parasite would soon run out of hosts. Parasites could first become commensal, or harmless to the host. Then, given enough time, the parasite and host might even become mutualists. The evolution of the eukaryotic cell by endosymbiosis is even predicated on the supposition that bacteria took up residence inside a larger cell, and then the parasite and cell became mutualists.

This argument is too teleological for some; after all, no organism is capable of "looking ahead" at its evolutionary fate. Rather, if an aggressive parasite can transmit more of itself in less time than a benign one, then aggressiveness would be favored by natural selection. On the other hand, other factors such as the life cycle of the host can determine whether aggressiveness is beneficial or not. For example, a benign parasite of newts will do better than an aggressive one. Why? Because newts take up solitary residence outside ponds in the forest for six years, and parasites have to wait that long before they are likely to meet up with another possible host.

Figure 24A Social parasitism by the Reed warbler-cuckoo.
A cuckoo chick balances the eggs of its hosts on its back one by one, and heaves them from the nest. Its own egg (see inset) mimics and is accepted by the host as its own.

Commensalism

Commensalism is a symbiotic relationship between two species in which one species is benefited and the other is neither benefited nor harmed.

Instances are known in which one species provides a home and/or transportation for the other species. Barnacles that attach themselves to the backs of whales and the shells of horseshoe crabs are provided with both a home and transportation. Remoras are fishes that attach themselves to the bellies of sharks by means of a modified dorsal fin acting as a suction cup. The remoras obtain a free ride and also feed on the remains of the shark's meals. It's possible though that the movement of the host is impeded by the presence of the attached animals, and therefore some are reluctant to use these as instances of commensalism.

Epiphytes grow in the branches of trees, where they receive light, but they take no nourishment from the trees. Instead, their roots obtain nutrients and water from the air. Clownfishes live within the waving mass of tentacles of sea anemones (Fig. 24.15). Because most fishes avoid the poisonous tentacles of the anemones, clownfishes are protected from predators. Perhaps this relationship borders on mutualism, because the clownfishes actually may attract other fishes on which the anemone can feed.

Cases of commensalism may turn out, on closer examination, to be causes of either mutualism or parasitism. Cattle egrets benefit from grazing near cattle because the cattle flush insects and other animals from the vegetation as they graze. Under prey defenses, it was noted that baboons and antelopes normally forage together for added protection. Is this, then, a case of commensalism? To some it seems like wasted effort to try to classify symbiotic relationships into the three categories of parasitism, commensalism, and mutualism. The amount of harm or good two species seem to do one another is dependent on what the investigator chooses to measure.

Figure 24.15 Clownfish among sea anemone's tentacles.
If the clownfish (*Premnas biaculeztus*) performs no service for the sea anemone, this association is a case of commensalism. If the clownfish lures other fish to be eaten by the sea anemone, this is a case of mutualism.

Mutualism

Mutualism is a symbiotic relationship in which both members of the association benefit. Mutualistic relationships often help organisms obtain food or avoid predation.

As with parasitism, it's possible to find examples of mutualism in all kingdoms of life. Bacteria that reside in the human intestinal tract are provided with food, but they also provide us with vitamins, molecules we are unable to synthesize for ourselves. Termites would not even be able to digest wood if it were not for the protozoa that inhabit their intestinal tract. These organisms digest cellulose, which termites cannot. Mycorrhizae are symbiotic associations between the roots of plants and fungal hyphae. It is now known that the roots of most plants form these relationships. The hyphae increase the solubility of minerals in the soil, improve the uptake of nutrients for the plant, protect the plant's roots against pathogens, and produce plant growth hormones. In return the fungus obtains carbohydrates from the plant.

Mutualistic relations need not be equally beneficial to both species. The mutualistic relationship between plants and their animal pollinators has already been mentioned. We can imagine that the relationship began when herbivores feasted on pollen. The provision of nectar may have spared the pollen and at the same time allowed the animal to become an instrument of pollination. Lichens can grow on rocks because their fungal member conserves water and leaches minerals that are provided to the algal partner, which photosynthesizes and provides organic food for both populations. If this were a true mutualistic relationship, each partner would do poorly when grown alone. The algae seem to do fine without the fungus, which suffers when grown alone. For that reason, it's been suggested that the fungus is parasitic at least to a degree on the algae.

Ants form mutualistic relationships with both plants and insects. In tropical America, the bullhorn acacia tree is adapted to provide a home for ants of the species *Pseudomyrmex ferruginea* (Fig. 24.16). Unlike other acacias, this species has swollen thorns with a hollow interior, where ant larvae can grow and develop. In addition to housing the ants, the acacias provide them with food. The ants feed from nectaries at the base of the leaves and eat fat and protein-containing nodules called Beltian bodies, which are found at the tops of some of the leaves. The ants constantly protect the plant from herbivores and other plants that might shade it because, unlike other ants, they are active 24 hours a day. Indeed, when the ants on experimental trees were poisoned, the trees died. Other types of trees, such as those in the genus *Croton*, also have nectaries for ants—the very same ants that form a mutualistic relationship with the butterfly *Thisbe irenea*, whose caterpillars feed on *Croton* saplings. *Thisbe* caterpillars have nectary organs that offer nourishment to ants, keeping them nearby. The caterpillar also exploits the social behavior of ants by releasing chemicals that normally cause ants to defend the ant colony against predators. The end result is that caterpillars are

a.

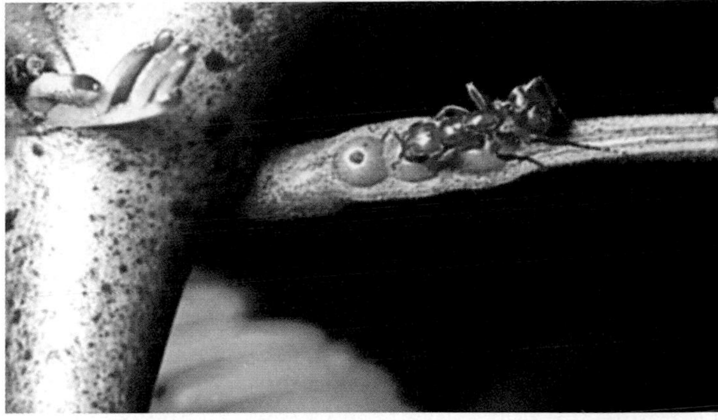

b.

c.

Figure 24.16 Mutualism between the bullhorn acacia tree and ants.
The bullhorn acacia tree (*Acacia*) is adapted to provide nourishment for ants (*Pseudomyrmex ferruginea*). **a.** The thorns are hollow and the ants live inside. **b.** The base of leaves have nectaries (openings) where ants can feed. **c.** Leaves of the bullhorn acacia have bodies at the tips that ants harvest for larval food.

protected while they feast on the trees. In fact, the caterpillars, besides eating the leaves, feed from the ant nectaries on the trees!

Cleaning symbiosis is a symbiotic relationship in which the individual being cleaned is often a vertebrate (Fig. 24.17). Crustacea, fish, and birds act as cleaners and are associated with a variety of vertebrate clients. Large fish in coral reefs line up at cleaning stations and wait their turn to be cleaned by small fish that even enter the mouths of the large fish. Whether cleaning symbiosis is an example of mutualism has been questioned because of the lack of experimental data. If clients respond to tactile stimuli by remaining immobile while cleaners pick at them, then cleaners may be exploiting this response by feeding on host tissues as well as on ectoparasites. Still, it could be that while stimulation may be the immediate cause of client behavior, cleaning could ultimately lead to net gains in client fitness.

Symbiotic relationships do occur between species, but it may be too simplistic to divide them into parasitic, commensalistic, and mutualistic relationships.

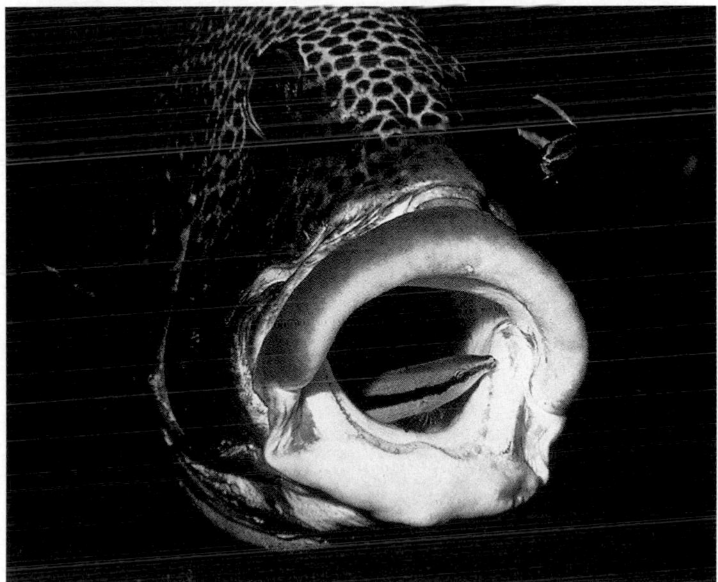

Figure 24.17 Cleaning symbiosis.
A cleaner wrasse (*Labroides dimidiatus*) in the mouth of a spotted sweetlip (*Plectorhincus chaetodontoides*) is feeding off parasites. Does this association improve the health of the sweetlip, or is the sweetlip being exploited? Investigation is under way.

24.3 Community Structure Changes over Time

Each community has a history that can be surveyed over a short time period or even over geological time. We know that the distribution of life has been influenced by dynamic changes occurring during the history of earth. We have previously discussed how continental drifting contributed to various mass extinctions that have occurred in the past. When the continents joined to form the supercontinent Pangaea, many forms of marine life became extinct. Or when the continents drifted toward the poles, immense glaciers drew water from the oceans and even chilled once tropical lands. During an ice age, glaciers moved southward and then, in between ice ages, glaciers retreated, changing the environment and allowing life to colonize the land once again. After time, complex communities came into being.

Many ecologists, however, try to observe changes as they occur during their own lifetime.

Ecological Succession

Communities are subject to disturbances that can range in severity from a storm blowing down a patch of trees to a beaver damming a pond to a volcanic eruption. We know from observation that following these disturbances, we'll see changes in the plant and animal communities over time; often, we'll wind up with the same kind of community with which we started.

Ecological succession is a change within a community following a disturbance. *Primary succession* occurs in areas where there is no soil formation such as following a volcanic eruption or a glacial retreat. Secondary succession begins in areas where soil is present, as when a cultivated field like the cornfield in Figure 24.18 returns to a natural state. Notice that we roughly observe a change from grasses to shrubs to a mixture of shrubs and trees.

a.

b.

c.

d.

e.

Figure 24.18 Secondary succession from a cornfield. **a.** During the first year, only the remains of corn plants are seen. **b.** During the second year, wild grasses have invaded the area. **c.** By the fifth year, the grasses look more mature and sedges have joined them. **d.** During the tenth year, there are goldenrod plants, shrubs (blackberry), and Eastern juniper trees. **e.** After twenty years, the juniper trees are mature and there are also birch and maple trees in addition to the blackberry shrubs.

The first species to begin secondary succession are called **pioneer species,** that is, plants that are invaders of disturbed areas, and then it progresses through a series of stages that are also described in Figure 24.19. Again we observe a series that begins with grasses and proceeds from shrub stages to a mixture of shrubs and trees, until finally there are only trees. Ecologists have tried to determine what the processes and mechanisms are by which the changes described in Figures 24.18 and 24.19 take place. And do these processes always have the same "end point" of community composition and diversity?

Theories About Succession

In 1916, F. E. Clements proposed the *climax-pattern model* of succession, and said that succession in a particular area will always lead to the same type of community, which he called a **climax community.** He believed that climate, in particular, determined whether a desert, a type of grassland, or a particular type of forest results. This is the reason, he said, that there is a coniferous forest in northern latitudes, a deciduous forest in temperate zones, and a tropical rain forest in the tropics. Secondarily, he believed that soil conditions might also affect the results. Shallow, dry soil might produce a grassland where a forest is expected, or the rich soil of a river bank might produce a woodland where a prairie is expected.

Further, Clements believed that each stage facilitated the invasion and replacement by organisms of the next stage. Shrubs can't grow on dunes until dune grass has caused soil to develop. Similarly, in the example given in Figure 24.19, shrubs can't arrive until grasses have made the soil suitable to them. Each successive community prepares the way for the next, so that grass-shrub-forest development occurs in a sequential way.

Aside from the *facilitation model*, there is also an *inhibition model.* The model predicts that colonists hold on to their space and inhibit the growth of other plants until the colonists die or are damaged. Still another model is called a tolerance model. The *tolerance model* predicts that different types of plants can colonize an area at the same time. Sheer chance determines which seeds arrive first, and successional stages may simply reflect the length of time it takes species to mature. This alone could account for the herb-shrub-forest development one often sees (Fig. 24.19). The length of time it takes for trees to develop might simply give the impression that there is a recognizable series of plant communities from the simple to the complex. In reality, the models we have mentioned are not mutually exclusive and succession is probably a multiple complex process.

Although it may not have been apparent to early ecologists, we now recognize that the most outstanding characteristic of natural communities is their dynamic nature. Also, it seems obvious to us now that the most complex communities most likely consist of habitat patches that are at various stages of succession. Each successional stage has its own mix of plants and animals, and if a sample of all stages is present, community diversity would be greatest. Further, we do not know if succession does continue to certain end points, because the process may not be complete anywhere on the face of the earth.

Ecological succession that occurs after a disturbance probably involves complex processes and the end result cannot be foretold ahead of time.

Figure 24.19 Secondary succession in a forest.
In secondary succession in a large conifer plantation in central New York State, certain species are common to particular stages. However, the process of regrowth shows approximately the same stages as secondary succession from a cornfield (see Fig. 24.18).
(After R. L. Smith, 1960.)

24.4 How to Increase Community Biodiversity

Community stability can be recognized in three ways: persistence through time, resistance to change, and recovery once a disturbance has occurred. Persistence occurs when, say, a forest remains just about the same year after year. A forest resists change when its trees are able to regrow their leaves after an insect infestation. And a chaparral community shows resilience when it quickly returns to its normal state after a fire.

The Intermediate Disturbance Hypothesis

The *intermediate disturbance hypothesis* states that a moderate amount of disturbance is required for a high degree of community diversity (Fig. 24.20). Fire, wind, moving water, and severe weather changes are possible abiotic factors that cause a disturbance. For example, fires promote understory plant diversity by preventing longer-lived shrub species from out-competing annuals. In a tropical rain forest, there are a great number of different types of trees, each with their own particular ecological niche requirements. This suggests that the environment is cut up into many small patches, each one slightly different from the other. As long as disturbances are small enough to affect one type of patch and not another, overall stability might still be maintained.

If widespread disturbances occur frequently, diversity is expected to be limited. The community will be dominated by species with a rapid growth rate, a short life span, and a strong ability to colonize disturbed areas. (Refer to Figure 23.13 and note that these attributes characterize *r*-strategists.) If disturbances are less widespread and less frequent, other types of species with slower growth rates and a longer life span will also be able to compete. (Refer to Figure 23.13 and note these are the attributes of *K*-strategists.) Under these circumstances, diversity will be the greatest.

It is possible for a disturbance to be so great, that the community never returns to its original state. Archaeological remains suggest that hundreds of square kilometers were cleared and cultivated by the Maya at Tikal from 300 to 900 A.D. This civilization collapsed for some unknown reason, and then the forest began to regrow. Even today, plant diversity in the community is low, and many of the common tree species are ones known to have been prized by the Maya for wood or for edible fruits or nuts. The animal wildlife is largely restricted to those animals that use these food resources. Even after 1,200 years, the site does not have the expected composition for a tropical rain forest in that area.

The intermediate disturbance hypothesis suggests that environmental heterogeneity increases species richness.

Predation, Competition, and Biodiversity

In certain communities, predation by a particular species reduces competition and increases diversity. Robert Paine was among the first to show this possible effect of predation. He removed the starfish *Pisaster* from test areas along

a.

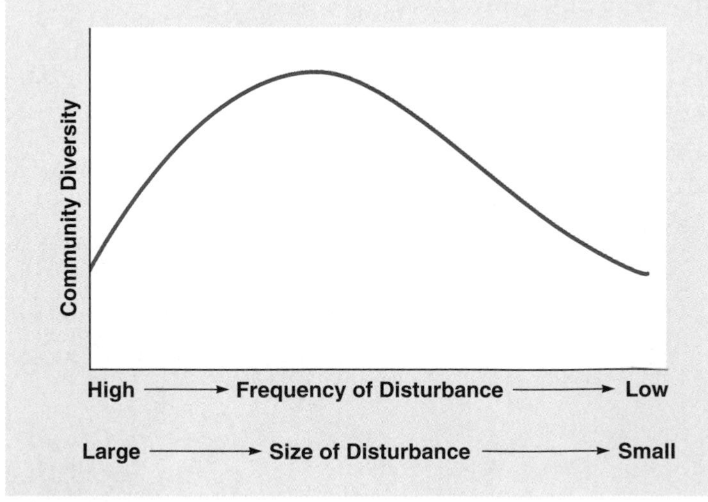

b.

Figure 24.20 The intermediate disturbance hypothesis.
a. Disturbances like the Yellowstone fire in 1988 increase the diversity of a community by providing patches where annual/perennial plants can grow. **b.** This graph suggests that diversity is greatest when disturbances are intermediate in frequency and size.

the rocky intertidal zone on the west coast of North America but not from control areas. In the control areas, there was no change in the number of species, but in the test areas, the mussel *Mytilus* increased in number and excluded other invertebrates and algae from attachment sites on the rocks. The species richness declined drastically (Fig. 24.21). Predators that regulate competition and maintain the diversity of a community are now called **keystone predators.** The elephant is a possible keystone predator in Africa. Elephants feed on shrubs and small trees, causing woodland habitats to become open grassland. This is not beneficial to the elephant, which needs woody species in its diet, but it is beneficial to other ungulates that graze on grasses.

Island Biogeography and Biodiversity

Recall that MacArthur and Wilson have proposed a general theory of island biogeography, which states that a large island close to the mainland (the source of dispersing species) will have greater diversity than a small island that is distant from the mainland (see Fig. 24.3). Of late, humans have created many islands; some of these are not true islands surrounded by water—they are terrestrial patches surrounded by human cities, towns, or farms.

In Panama, Barro Colorado Island (BCI) was created in the 1910s when a river was dammed to form a lake. As predicted by the theory of island biogeography, BCI lost species because it was a small island now cut off from the mainland. Among those species that became extinct were the top predators on the island, namely the jaguar, puma, and ocelot. Thereafter, medium-sized terrestrial mammals such as the coatimundi increased in number. The coatimundi is an avid predator of bird eggs, nestlings, and other small vertebrates. Not surprisingly, there are now fewer bird species on BCI even though the island is large enough to support them. Therefore, it appears that the loss of certain predators from a community can sometimes result in chain reactions that lead to decreased diversity at various levels of community structure.

Such changes as this can also be expected in isolated terrestrial areas. We are finding that preserves need to be very large in order to conserve *K*-strategists, especially top predators.

Exotic Species and Biodiversity

The effects of unbridled competition and predation on community diversity can be extreme, as when humans introduce species into areas where they are not held in check by competitors or predators (see Fig. 27.11*d*). These disturbances are brought on by dispersal of species to new communities where they are not held in check by predators and competitors.

Predation and competition also influence biodiversity. *K*-strategists, which are sometimes top predators, are maintained best on large preserves. Exotic species out-compete and/or prey on native plants and animals.

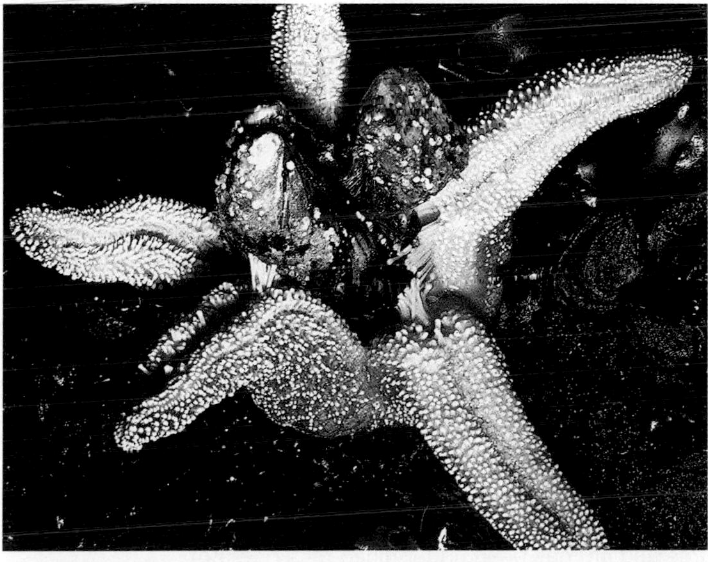

a.

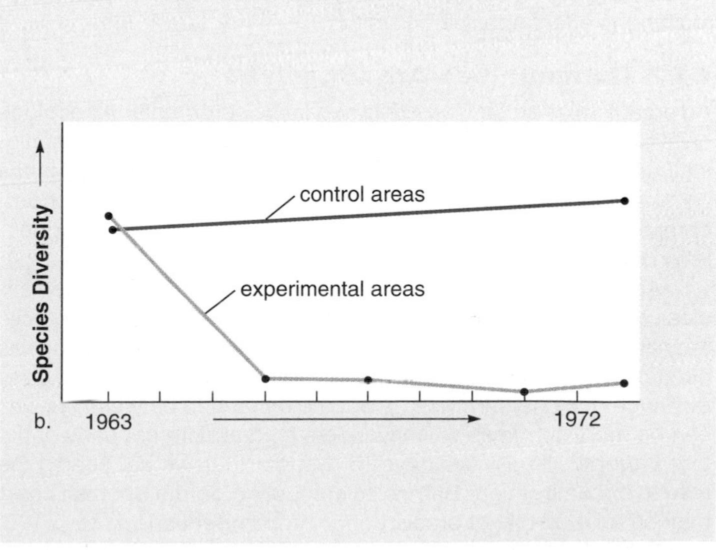

Figure 24.21 Effect of a keystone predator.
a. The sea star *Pisaster ochracheous* is eating a mussel (*Mytilus*), its favorite food. **b.** Along the west coast, *Pisaster* was removed from experimental areas (tidepools) but not from control areas. Diversity decreased in experimental areas because the mussel population increased and crowded out other invertebrates and algae.

connecting concepts

Community ecology is concerned with how populations of different species that occur in a particular place and habitat interact with each other. Historically, there has been considerable debate among ecologists as to how interactions among different populations lead to the kinds of communities we observe in nature. At one extreme, some ecologists felt that communities functioned like "super-organisms", with each species a vital and integral part of the community. At the other extreme, other ecologists felt that each population simply occurred where conditions were best for it and that communities were chance aggregations of species. Although we still do not completely understand how communities are organized, the truth probably lies somewhere in between the two extreme views. We know that a population of any species is distributed where conditions are best for it. However, communities do exhibit emergent properties that result from certain groupings of species co-occurring. For example, tropical rain forests maintain high humidity partially because much of the precipitation that falls in the rain forest is locally recycled due to transpiration (loss of water vapor from plants) by the rain forest's own trees.

The number of individuals in many populations within ecological communities is determined by negative interactions with other species such as interspecific competition, predation, and parasitism. All three of those interactions place limits on the size of the population by limiting survival, reproduction, and growth of populations. Almost all populations are affected by at least one of those negative interactions. Positive interactions such as mutualisms are fairly common in nature (especially for plants), but their effect on population size is not well understood yet.

Perhaps one of the most important recent discoveries about communities is that they are highly dynamic. The number of species, kinds of species, and size of populations within most communities is constantly changing due to disturbances and climatic variability. Species differ in their ability to colonize recently disturbed areas and this also promotes changes in communities over time.

Because the species composition of communities can be so variable, many ecologists have tried to develop methods to study communities that don't rely on species identity. Instead, these ecologists study the movement of energy and nutrients through communities. The physical environment has a large influence on energy and nutrient flow, and thus the nonliving world must be incorporated in our studies of communities. This is the realm of ecosystem ecology. When we study ecosystems we must consider both the living community and the physical environment and how they interact with each other.

Summary

24.1 What Is a Biological Community?

A community is an assemblage of populations interacting with one another within the same environment. Communities differ in their composition (species found there) and their diversity (species richness and relative abundance). Abiotic factors such as latitude and environmental gradients seem to largely control community composition. As suggested by the theory of island biogeography, space limitations affect species diversity.

24.2 Communities Are Organized

An organism's habitat is where it lives in the community. An ecological niche is defined by the role an organism plays in its community, including its habitat and how it interacts with other species in the community. Competition, predator-prey, parasite-host, commensalistic, and mutualistic relationships help organize populations into an intricate dynamic system.

According to the competitive exclusion principle, no two species can occupy the same niche at the same time. When resources are partitioned between two or more species, niche partitioning has occurred. Character displacement is a structural change that gives evidence of niche partitioning. But the difference between species can be more subtle as when warblers feed at different parts of the tree canopy. Usually we have to assume that we are seeing the results of competition. Barnacles competing on the Scottish coast may be an example of present ongoing competition.

Predator-prey interactions between two species are influenced by the biotic potential of the two populations, the carrying capacity of the environment, and the presence of other food sources in the community. Sometimes predation can cause prey populations to decline and remain at relatively low densities, or a cycling of population densities has also been observed.

Prey defenses take many forms: camouflage, use of fright, and warning coloration are three possible mechanisms. Müllerian mimicry occurs when two species with the same warning coloration have the same defenses. Batesian mimicry occurs when one species has the warning coloration but lacks the defense.

We would expect coevolution to occur within a community. The better the predator becomes at catching prey, the better the prey becomes at escaping the predator, for example. Flowers are adapted to attract specific pollinators, and hosts are adapted to cope with potential parasites.

Like predators, prey take nourishment from their host. However, the host usually provides a home and a place where parasites reproduce. Parasites escape detection and manipulate their hosts in various ways. Whether they are aggressive (kill their host) or benign probably depends on which results in the highest fitness, considering the life cycles of both organisms in the relationship.

Perhaps symbiotic relationships are too complicated to try to classify them as commensalistic, parasitic, and mutualistic. Therefore, it is not surprising that certain relationships like clownfishes that live among sea anemone tentacles or lichens that contain fungal and algal members are questionable as to whether they are commensalistic or mutualistic, respectively. *Croton* saplings provide food for ants, but so do the caterpillars that feed on *Croton* saplings! In this case, the possible mutualistic relationship between tree and ant has been undermined by a predator.

24.3 Community Structure Changes Over Time

Directional patterns of community change are called ecological succession. Ecologists have recognized for some time that communities undergo succession both in terms of a geological timescale and present-day timescale. There is now much discussion about the cause of succession and whether it results in so-called climax communities or not. In the past, use of the term climax community has implied that a diverse community is stable in its composition.

24.4 How to Increase Community Biodiversity

It is perfectly obvious now that heterogeneity or the presence of patches in different stages of succession would actually result in the most diversity. The intermediate disturbance hypothesis states that a moderate amount of disturbance is required for a high degree of community diversity.

Predation and competition are examples of interactions that affect biodiversity. Predation by keystone predators increases the diversity of a community. If a preserve is so small that top predators are eliminated, diversity will decrease because other predators are not held in check. Exotic species often out-compete and/or prey on native species so that they are reduced in numbers.

Reviewing the Chapter

1. Contrast the individualistic model of community composition with the interactive model. 407
2. What do experiments with artificial reefs and the island biogeography hypothesis tell us about species richness? 407–8
3. Describe the habitat and ecological niche of some particular organism. 409
4. What is the competitive exclusion principle? How does the principle relate to character displacement and niche partitioning? 410–11
5. What is the special significance of the barnacle experiment off the coast of Scotland? 411
6. Would you expect all predator and prey interactions to produce similar results with regard to population densities? Why or why not? 412–13
7. What is mimicry, and why does it work as a prey defense? 415
8. Give an example of parasitism, commensalism, and mutualism, and show that it is not always easy to distinguish between these symbiotic relationships. 416, 418
9. Describe an example of coevolution between a predator and prey, between a parasite and host, and between mutualists. 417
10. What is ecological succession? What is the present controversy surrounding the concept? 420–21
11. What is the intermediate disturbance hypothesis, and how does it relate to community diversity? 422
12. What interactions affect biodiversity? In what way? 423

Testing Yourself

Choose the best answer for each question.

1. Six species of monkeys are found in a tropical forest. Most likely, they
   a. occupy the same ecological niche.
   b. eat different foods and occupy different ranges.
   c. spend much time fighting each other.
   d. All of these are correct.
2. Leaf cutter ants keep fungal gardens. The ants provide food for the fungus but also feed on the fungus. This is an example of
   a. competition.
   b. predation.
   c. commensalism.
   d. mutualism.
3. Clownfishes live among sea anemone tentacles, where they are protected. If the clownfish provides no service to the anemone, this is an example of
   a. competition.
   b. predation.
   c. commensalism.
   d. mutualism.
4. Two species of barnacles vie for space in the intertidal zone. The one that remains is
   a. the better competitor.
   b. better adapted to the area.
   c. the better predator on the other.
   d. Both a and b are correct.
5. A bullhorn acacia provides a home and nutrients for ants. Which statement is correct?
   a. The plant is under the control of pheromones produced by the ants.
   b. The ants protect the plant.
   c. They have coevolved to occupy different ecological niches.
   d. All of these are correct.
6. The ecological niche of an organism
   a. is the same as its habitat.
   b. includes how it competes and acquires food.
   c. is specific to the organism.
   d. Both b and c are correct.
7. The frilled lizard of Australia suddenly opened its mouth wide and unfurled folds of skin around its neck. Most likely this was a way to
   a. conceal itself.
   b. warn that it was noxious to eat.
   c. scare a predator.
   d. All of these are correct.
8. When one species mimics another species, the mimic sometimes
   a. lacks the defense of the model.
   b. possesses the defense of the model.
   c. is brightly colored.
   d. All of these are correct.

9. Which of these most likely would account for diversity in a coral reef?
   a. Coral reefs can be any size from small to large.
   b. Immigration and emigration are constantly occurring.
   c. Coral reefs are subjected to both exogenous and endogenous disturbances.
   d. All of these are correct.
10. Coevolution probably occurred when
   a. two species were adapted to one another.
   b. both species were adapted to the same environment.
   c. each species affected the population density of the other.
   d. each species had the same tolerance range of an environmental gradient.
11. The presence of patches in an environment may be associated with
   a. spatial and temporal heterogeneity.
   b. a greater degree of diversity.
   c. past disturbances.
   d. All of these are correct.
12. Label this diagram, and tell how it applies to the individualistic concept of community composition.

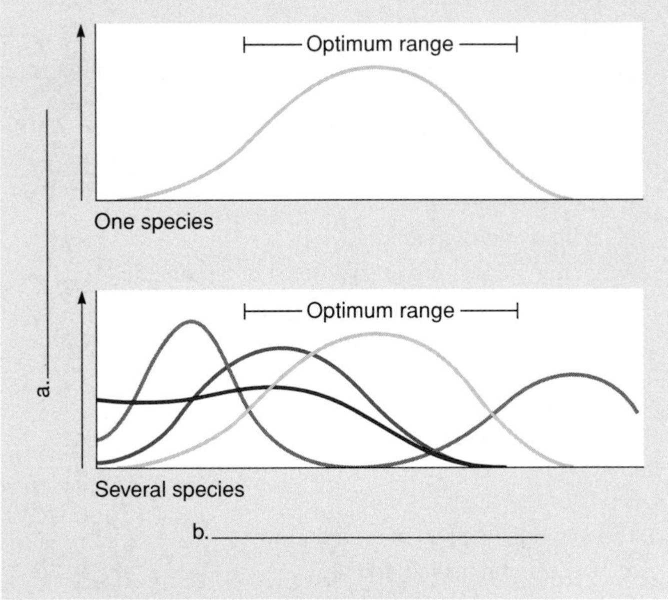

Applying the Concepts

1. *Communities vary in their diversity.*
   How can competition and predation increase community diversity?
2. *Succession has produced present-day communities.*
   How does the use of fire relate to succession and an increase in community diversity?
3. *Symbiotic relationships are important to the fabric of a community.*
   How is parasitism like mutualism, and how is it different?

Using Technology

Your study of community ecology is supported by these available technologies:

Exploring the Internet
The Mader Home Page provides resources for and help with studying this chapter.

http://www.mhhe.com/sciencemath/biology/mader/
(Click on Biology.)

Understanding the Terms

camouflage 414
character displacement 410
climax community 421
coevolution 417
commensalism 418
community 406
community stability 422
competitive exclusion principle 410
ecological niche 409
ecological succession 420
habitat 409
host 416
interspecific competition 410
keystone predator 423
mimicry 415
mutualism 418
ecological niche 409
parasite 416
parasitism 416
pioneer species 421
predation 412
predator 412
prey 412
resource partitioning 410
symbiosis 416

Match the terms to these definitions:

a. _______ Assemblage of populations interacting with one another within the same environment.
b. _______ Place where an organism lives and is able to survive and reproduce.
c. _______ Directional pattern of change in which one community replaces another until a community typical of the area results.
d. _______ Interaction in which one organism uses another, called the prey, as a food source.
e. _______ Joint evolution in which one species exerts selective pressure on the other species.
f. _______ Relationship that occurs when two different species live together in a unique way; it may be beneficial, neutral, or detrimental to one and/or the other species.
g. _______ Role an organism plays in its community, including its habitat and its interactions with other organisms.
h. _______ Symbiotic relationship in which both species benefit, in terms of growth and reproduction.
i. _______ Symbiotic relationship in which one species benefits in terms of growth and reproduction to the harm of the other species.
j. _______ Symbiotic relationship in which one species is benefited, and the other is neither harmed nor benefited.

Ecosystems

CHAPTER 25

Chapter Concepts

Forest and stream ecosystems

An ecosystem, which consists of a community and its physical environment, is characterized by energy flow and chemical cycling. Both of these begin when algae and green plants capture a small percentage of the sun's energy and use it to transform inorganic chemicals like carbon dioxide and water to food sources that are used by themselves and all the other populations in an ecosystem. Eventually when decomposers break down organic matter, the inorganic chemicals are liberated once again, but the energy has dissipated as heat. Ecosystems are not self-contained, they have inputs and outputs to the other ecosystems of the biosphere. Therefore, some ecologists think of the biosphere as a global ecosystem.

Most persons now realize that humans are a part of the biosphere and that our activities affect all of its ecosystems. DDT is an insecticide that was first used during World War II in the Pacific to destroy mosquitoes that carry malaria. Today DDT is found in every body of water on earth and in the tissues of every human being. In the same way, our overuse of resources such as the soil and forests may bring about changes that we find difficult to predict at this time.

25.1 The Nature of Ecosystems

Our planet is unique in many ways. Unlike the other planets in our solar system, earth has water, an atmosphere, and abundant life. In fact, our planet could have been called "water" instead of earth. Water is present in the *hydrosphere,* [Gk. *hydrias,* of water, and L. *sphaera,* ball], which covers over three-quarters of the earth's surface (Fig. 25.1). The warm rays of our sun drive the water cycle. Water evaporates from living things, the ocean, and the rivers, and enters the atmosphere. Thereafter, water condenses and falls as rain. The oceans moderate the temperature of the earth; as surface temperatures rise, the oceans take up a great deal of heat, and then as temperatures cool, they return heat slowly to the atmosphere. This helps keeps the temperature on earth suitable to life.

The *atmosphere* [Gk. *atmos,* vapor, and L. *sphaera,* ball] is concentrated in the lowest 10 kilometers near earth but extends out at least 1,000 kilometers. The atmosphere contains carbon dioxide, nitrogen, and oxygen, gases that are both used and released by living things. Carbon dioxide is necessary for photosynthesis. Oxygen is necessary for cellular respiration, and in the upper atmosphere it becomes ozone, a substance that shields earth from damaging ultraviolet radiation and makes life on land possible.

A rocky substratum called the *lithosphere* [Gk. *lithos,* stone, and L. *sphaera,* ball] extends from Earth's surface to about 100 kilometers deep. The weathering of rocks supplies minerals to plants, which take root in weathered rocks and slowly form soil. Beside minerals, soil contains decaying organic material known as humus. The organisms of decay play a vital role in breaking down organic matter, returning inorganic nutrients to plants so that photosynthesis can continue.

The **biosphere** is that part of the atmosphere, hydrosphere, and lithosphere that contains living things. An ecosystem includes not only a living community but also its physical and chemical environment. The interactions between organisms and their environment help keep the biosphere a suitable place for living things. Human activitivies can alter the interactions between organisms and their environments in a way that reduces the abundance and diversity of life that environments can support. It is important to understand how ecosystems function so that we can repair past damages and predict how human activities might change current conditions.

When organisms interact, they help keep the atmosphere, hydrosphere, and lithosphere suitable for sustaining life.

Biotic Components of an Ecosystem

Ecosystems consist of a group of living organisms and the physical and chemical environment in which they live. The living organisms are the biotic component of an ecosystem, and the physical and chemical environment are a part of the abiotic component of an ecosystem. In ecosystems, living things are classified according to how they get their food: autotrophs make their own food, and heterotrophs feed on other organisms (Fig. 25.2). Heterotrophs include detritivores and decomposers, which feed on organic material in the soil.

Biosphere

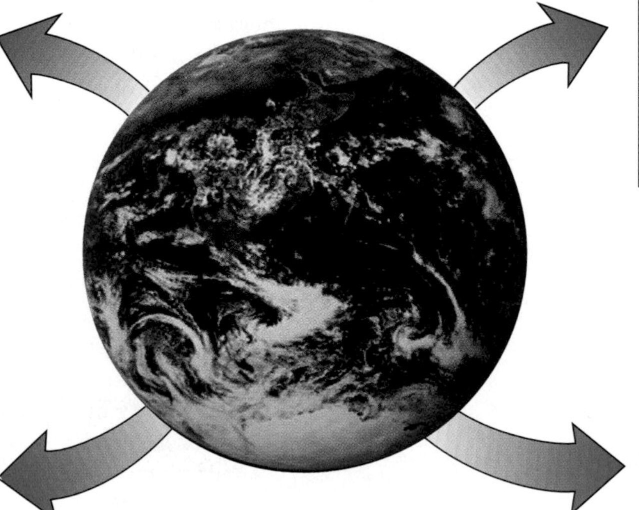

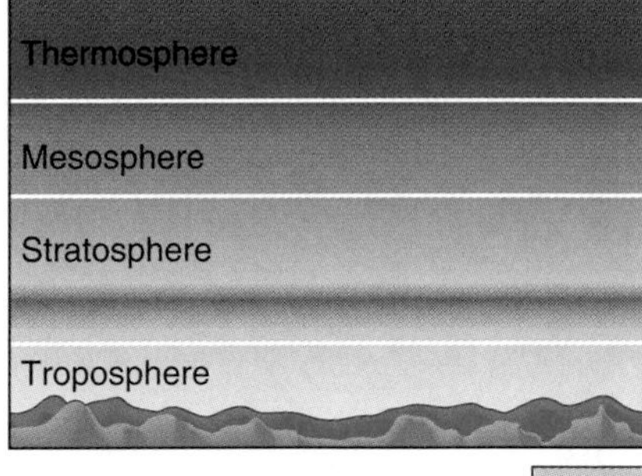

Atmosphere

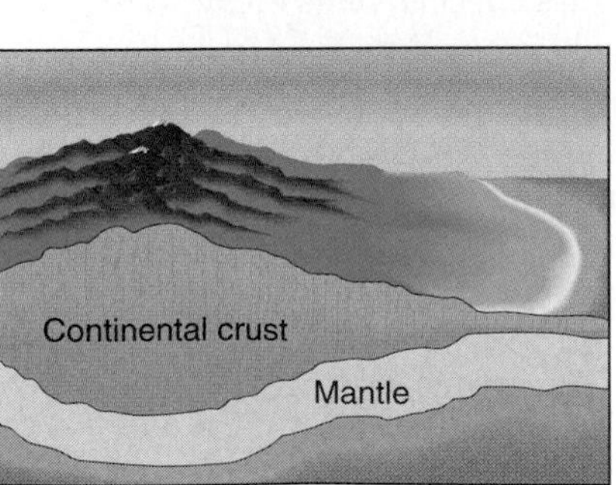

Lithosphere

Hydrosphere

Figure 25.1 Biosphere.
The biosphere, which contains the living things on earth, lies between the lithosphere (crust) and the atmosphere (gases). The hydrosphere contains water.

Autotrophs

Autotrophs produce their own organic nutrients for themselves and other members of the community; therefore, they are called **producers.** Chemoautotrophs are bacteria that obtain energy by oxidizing inorganic compounds such as ammonia, nitrites, and sulfides, and they use this energy to synthesize carbohydrates. Chemoautotrophs have been found to support some cave communities and also deep-sea communities at oceanic ridges where hot minerals spew out of the inner earth. The chemoautotrophs that function in the nitrogen cycle will be discussed in this chapter on page 438.

Photoautotrophs are photosynthesizers that produce most of the organic nutrients for the biosphere. Algae of all types possess chlorophyll and carry on photosynthesis in freshwater habitats and marine habitats. Algae make up the phytoplankton, which are photosynthesizing organisms suspended in water. Green plants are the dominant photosynthesizers on land.

Heterotrophs

Heterotrophs need a preformed source of organic nutrients. They are the **consumers**—they consume food. **Herbivores** are animals that graze directly on plants or algae. In terrestrial habitats, insects are small herbivores, while in aquatic habitats, zooplankton, such as protozoa, play that role. **Carnivores** feed on other animals; birds that feed on insects are carnivores, and so are hawks that feed on birds. This example allows us to mention that there are primary consumers (e.g., insects), secondary consumers (e.g., birds), and tertiary consumers (e.g., hawks). Sometimes tertiary consumers are called top predators. **Omnivores** are animals that feed both on plants and animals. As you most likely know, humans are omnivores.

Detritivores are organisms that feed on **detritus,** which is decomposing particles of organic matter. Fan worms feed off detritus that is floating in marine waters, while clams take it from the substratum. Earthworms and some beetles, termites, and maggots are all terrestrial detritivores. Nonphotosynthetic bacteria and fungi, including mushrooms, are **decomposers;** they carry out **decomposition,** the breakdown of dead organic matter, including animal wastes. Decomposers perform a very valuable service because they release inorganic substances that are taken up by plants once more. Otherwise, plants would be dependent completely on physical processes, such as the release of minerals from rocks, to supply them with inorganic nutrients.

The biotic components of an ecosystem can be classified according to the way they get their food.

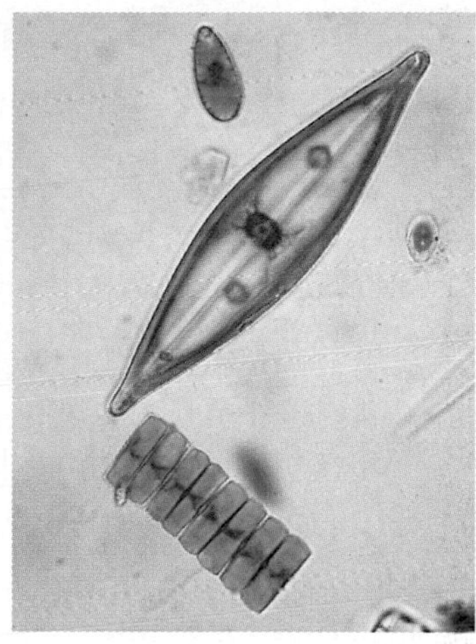

a. Producers

b. Herbivores

c. Carnivores

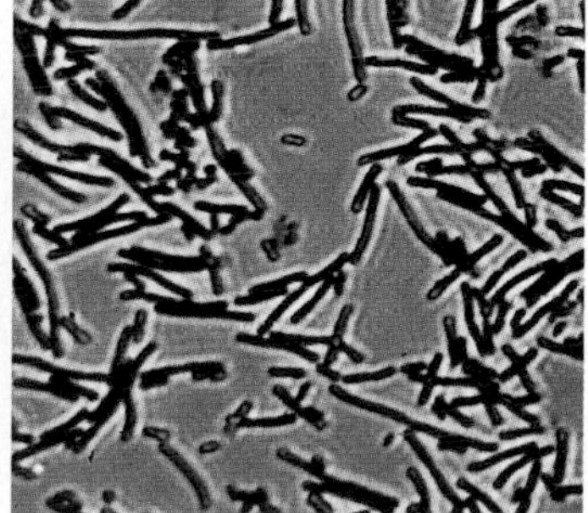

d. Decomposers

Figure 25.2 Biotic components.
Ecosystems include (**a**) autotrophic producers like green plants and algae, such as these diatoms; (**b**) heterotrophic herbivores like giraffes and caterpillars; (**c**) heterotrophic carnivores like the osprey and praying mantis; and (**d**) heterotrophic decomposers like mushrooms and soil-dwelling bacteria.

25.2 Energy Flow and Nutrient Cycling

With few exceptions ecosystems are related by their common dependency on solar energy and shared finite pools of nutrients (Fig. 25.3). Nutrients include elements like CHNOPS (carbon, hydrogen, nitrogen, oxygen, phosphorus, and sulfur), which make up over 98% of the body weight of organisms as well as compounds that contain these and other elements. Carbon dioxide is an inorganic compound that contains carbon, while glucose is an organic compound that contains carbon.

Autotrophs provide organic nutrients for an ecosystem when they photosynthesize and produce organic molecules. These organic molecules are used as a source of energy and organic building blocks by all the heterotrophs. **Primary productivity** is the total amount of energy an ecosystem's producers capture within plant material over a certain length of time. Physical factors like climate (e.g., temperature, rainfall, and humidity) and the nature of the soil can all affect gross primary productivity (Fig. 25.4). Only a portion of this food made by autotrophs is passed on to heterotrophs because plants use organic molecules to fuel their own cellular respiration. Only about 55% of gross primary productivity is available to heterotrophs. This portion of gross primary productivity is called *net primary productivity.*

The laws of thermodynamics are consistent with the observation that energy flows through an ecosystem. The first law states that energy cannot be created (nor destroyed). This explains why ecosystems are dependent on a continual outside source of energy, usually solar energy, which is used by photosynthesizers to produce organic nutrients. The second law states that with every transformation some energy is degraded into a less available form such as heat. Because plants carry on cellular respiration, only about 55% of the original energy absorbed by plants is available to an ecosystem.

Only a small percentage of food taken in by heterotrophs is available to a higher level consumer because of all the processes indicated in this diagram:

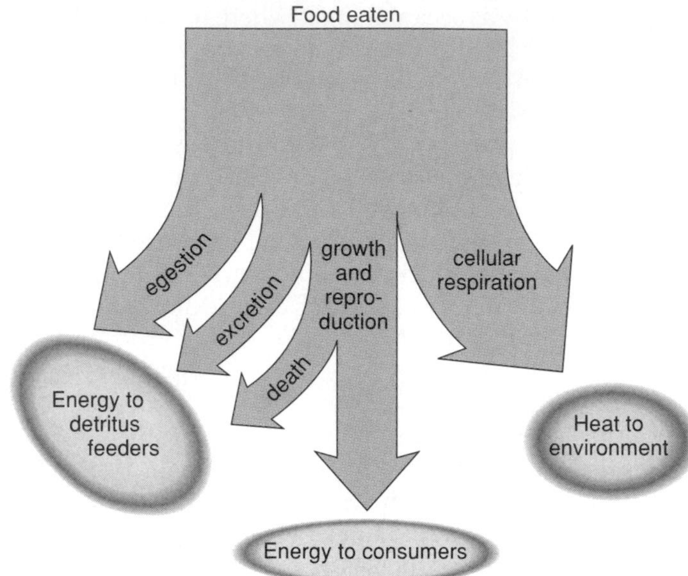

The diagram shows that a certain amount of energy is egested in feces or excreted in urine. Of the assimilated energy, a portion is utilized during cellular respiration and thereafter becomes heat. The remaining portion of energy is converted into increased body weight (or additional offspring). This latter use of energy is called *secondary productivity.*

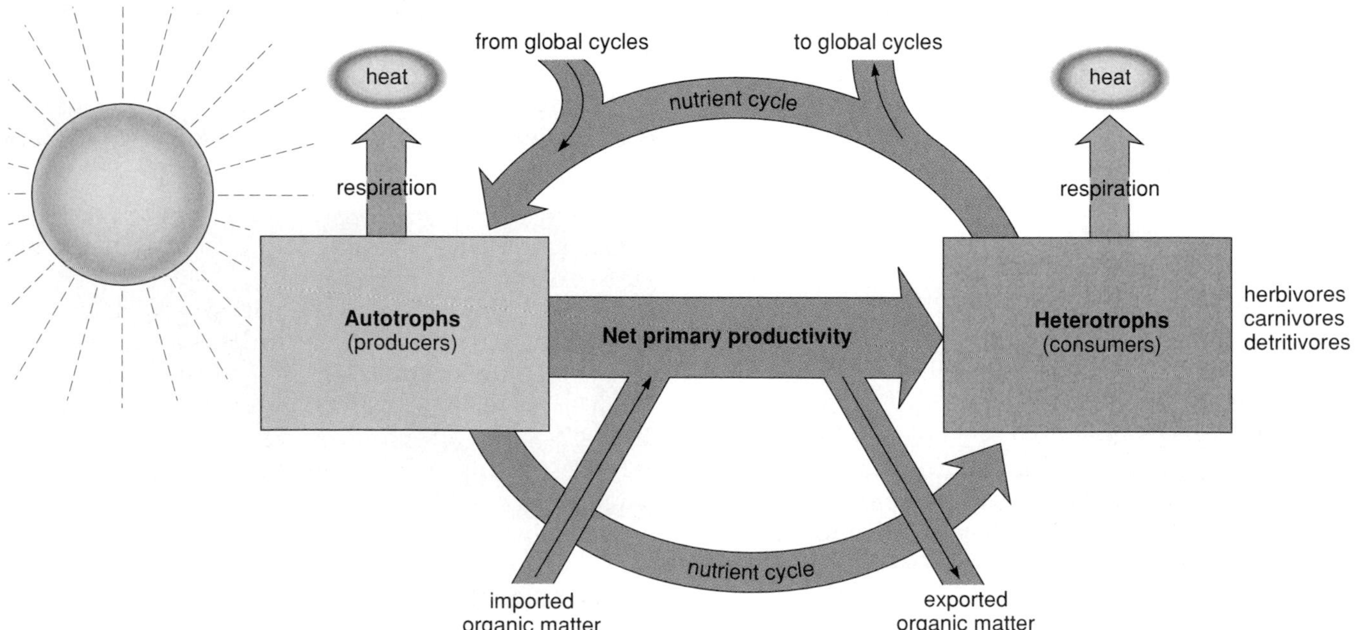

Figure 25.3 Ecosystems.
Relationship between the flow of energy from autotrophs to heterotrophs to the environment and the cycling of nutrients between autotrophs and heterotrophs and in global cycles.

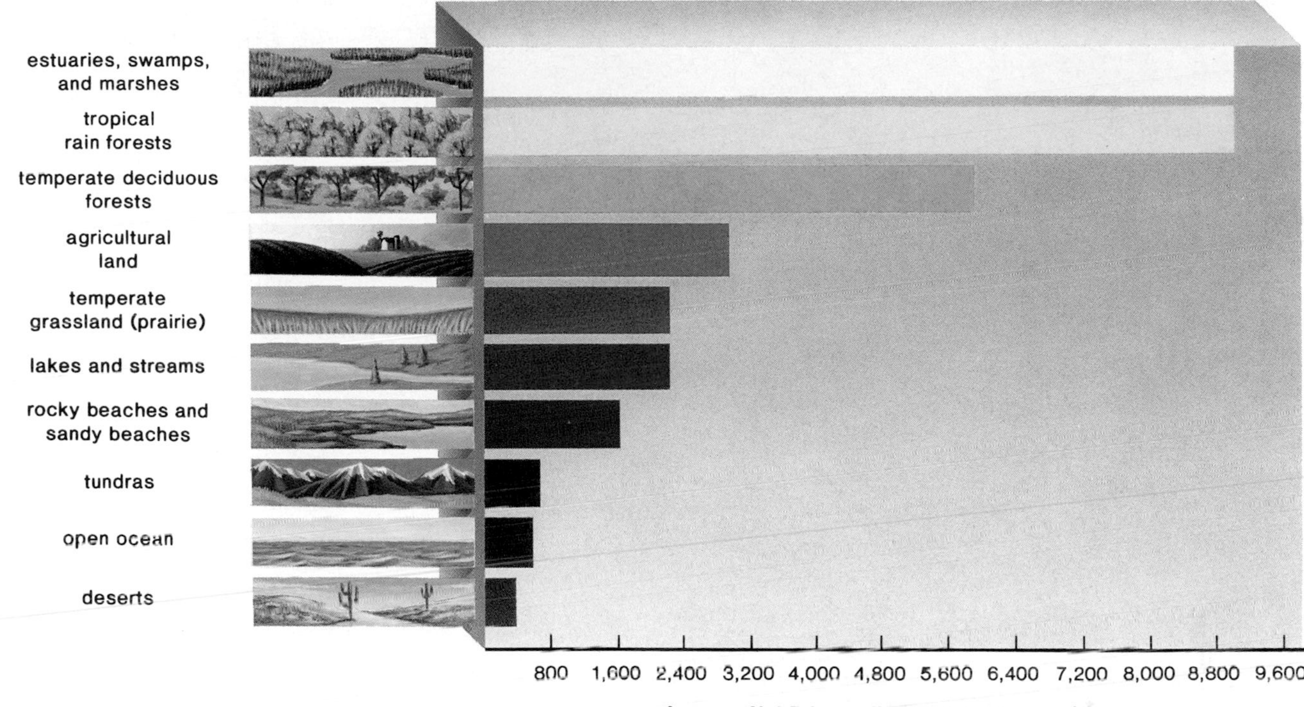

a.

Gross primary productivity (GPP): Total rate of photosynthesis in a specified area, usually expressed as kilocalories of energy produced per square meter per year (kcal/m^2/yr).

Autotrophs (mostly photosynthetic plants and algae) absorb the energy of the sun and produce food for all organisms in a community.

Net primary productivity (NPP): The rat at which plants produce usable food, which contains chemical energy (also expressed as kilocalories per square meter per year).

$$NPP = GPP - RS$$

where RS is the energy used by the autotrophs for respiration.

Since plants respire, they use some of the food they produce (GPP) for respiration, and what is left (NPP) is available for growth and storage and can be used as food for heterotrophs. Usually, between 50% and 90% of GPP remains as NPP.

b.

Figure 25.4 Gross and net primary productivity.
a. Average net primary productivity (NPP) for several communities. NPP is influenced by such factors as temperature and rainfall. Also important is the availability of sunlight and nutrients and the length of the growing season. **b.** The mathematical relationship between gross and net primary productivity.

Perhaps you are thinking that the feces and urine of a heterotroph, and indeed the death of all organisms, does not mean that this amount of energy is lost to the system, and you are correct. This is the energy used by decomposers and other detritivores. Since detritivores are fed upon by other heterotrophs of an ecosystem, the situation can get a bit complicated. Still, we can conceive that all the solar energy that enters an ecosystem eventually becomes heat. And this is consistent with the observation that ecosystems are dependent on a continual supply of solar energy.

Nutrients cycle within and among the earth's ecosystems in global cycles, which we will be discussing. Within an ecosystem, decomposers release nutrients that will become available to producers once more. Decomposers are *saprotrophic;* they send out digestive enzymes into the environment and then retrieve the digestive products.

Figure 25.3 makes it clear that ecosystems are not self-sustaining—they have inputs and outputs. First of all, ecosystems require an input of energy, usually from the sun. (Some ecosystems such as those at deep-sea vents and in certain caves are not directly dependent on solar energy. In these systems, chemosynthetic bacteria are the producers.) Inorganic nutrients enter by way of global cycles, and organic substances enter by rainfall and wind. Animals emigrate and immigrate, bringing nutrients into and taking nutrients out of ecosystems.

Energy flows through an ecosystem, while nutrients cycle within and among ecosystems.

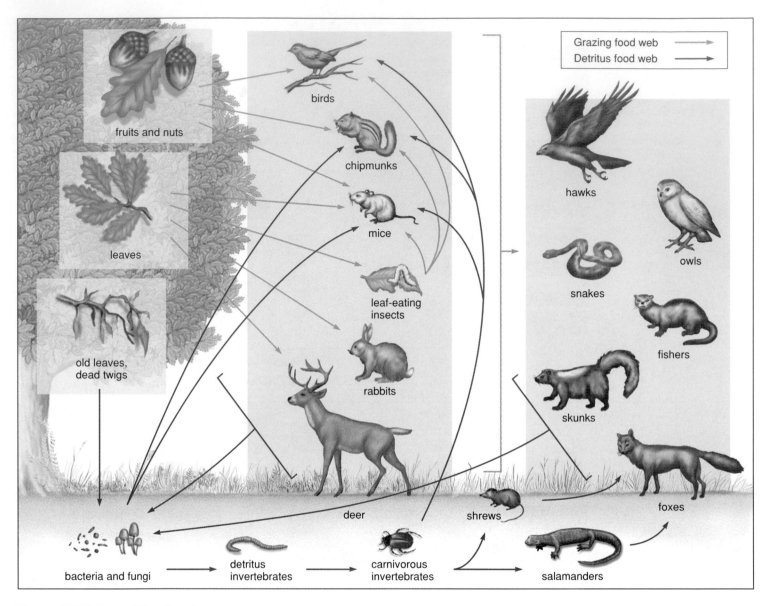

Figure 25.5 Forest food webs.
Two linked food webs are shown for the Hubbard Brook Forest ecosystem: a grazing food web and a detrital food web.

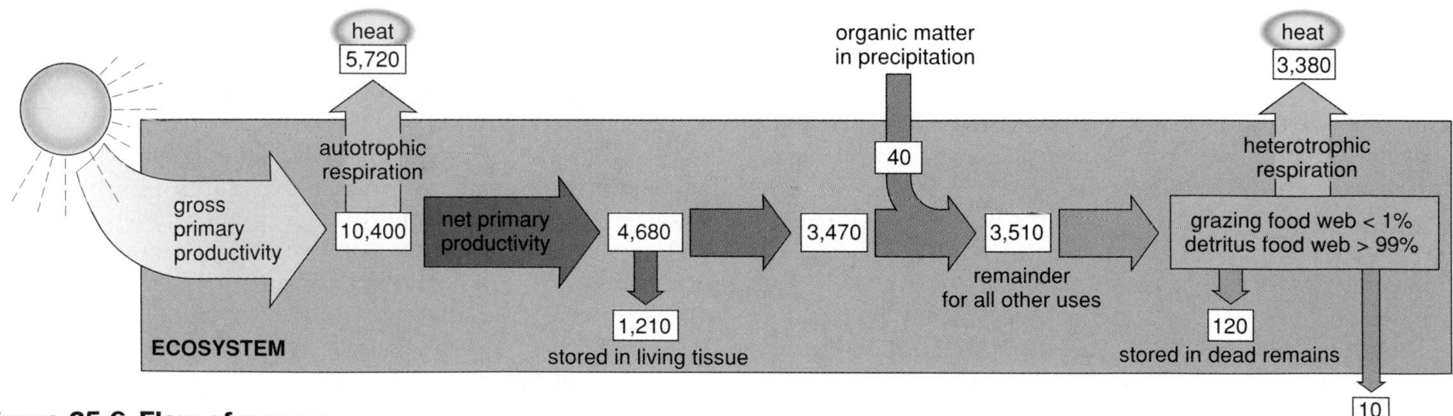

Figure 25.6 Flow of energy.
In this flow of energy diagram for the ecosystem pictured in Figure 25.5, gross primary productivity is 10,400 kcal/m^2/yr, of which 5,720 is used for respiration and becomes heat. Net primary productivity is 4,680 kcal/m^2/yr, of which 3,470 (augmented by an input of 40 kcal/m^2/yr) is passed on to the grazing and detrital food webs. Less than 1% is utilized by the grazing food web, and the rest is funneled through the detrital food web. Of the 3,510 kcal/m^2/yr, 3,380 is used for respiration and becomes heat. A total of 10 kcal/m^2/yr is an output to other ecosystems. Notice that of the original 10,400 kcal/m^2/yr, 1,210 is stored in living plant tissue, and 120 is stored as dead remains in the soil.

Food Webs and Trophic Levels

The principles we have been discussing can now be applied to an actual example—a forest of 132,300 square meters in New Hampshire. In this forest, the producers include sugar maple, beech, and yellow birch trees. The complicated feeding relationships that exist in natural ecosystems are called **food webs**. For example, Figure 25.5 shows that insects in the form of caterpillars feed on leaves, while mice, rabbits, and deer feed on leaf tissue at or near the ground. Birds, chipmunks, and mice feed on fruits and nuts, but they are in fact omnivores because they also feed on caterpillars. These herbivores and omnivores all provide nutrients for a number of different carnivores. This portion of the diagram is called a **grazing food web** because it begins with aboveground plant material.

The lower half of the diagram is devoted to the **detrital food web.** The bacteria and fungi of decay are the decomposers, but they can be food for other detritivores, which feed on organic matter in the soil. Earthworms, shrews and salamanders are examples of detritivores, which become food for aboveground carnivores. In this way, the detritus and the grazing food webs are connected.

We naturally tend to think that aboveground vegetation like trees are the largest storage form of organic matter and energy, but this is not necessarily the case. In this particular forest, the organic matter lying on the forest floor and mixed into the soil contains much more energy than does the leaf matter of living trees. Of this, the largest quantity is in the soil, which contains over twice as much energy as the forest floor. Therefore, more energy may be funneling through the detrital food web than through the grazing food web. Figure 25.6 shows that less than 1% of total available energy moves through the grazing food web; 99% moves through the detrital food web.

Trophic Levels

You can see that Figure 25.5 would allow us to link organisms one to another in a straight-line manner, according to who eats whom. For example, in the grazing food web we could link these organisms:

leaves → caterpillars → mice → hawks

And in the detrital food web we could designate these links:

dead organic matter → soil bacteria
→ earthworms → etc.

Diagrams like this that tell who eats whom are called **food chains.** And a **trophic level** is all the organisms that feed at a particular level in a food chain. In the grazing food web, going from left to right, the trees are primary producers (first trophic level), the first series of animals are primary consumers (second trophic level), and the next group of animals are secondary consumers (third trophic level).

Ecological Pyramids

Sometimes organisms don't fit into one trophic level. For example, chipmunks feed on fruits and nuts, but they also feed on leaf-eating insects. Nevertheless, ecologists portray the energy relationships between trophic levels in the form of **ecological pyramids,** diagrams whose building blocks designate the various trophic levels. A pyramid of numbers simply tells how many organisms there are at each trophic level. It's easy to see that a *pyramid of numbers* could be completely misleading. For example, in Figure 25.5 you would expect each tree to contain numerous caterpillars; therefore there would be more herbivores than autotrophs! The problem, of course, has to do with size. Autotrophs can be tiny, like microscopic algae, or they can be big like beech trees; similarly, herbivores can be small like caterpillars, or they can be large like elephants.

Pyramids of biomass eliminate size as a factor since biomass is the number of organisms multiplied by their weight. You would certainly expect the biomass of producers to be greater than the biomass of the herbivores, and that of the herbivores to be greater than the carnivores, as is shown in Figure 25.7, where the producers are not only algae but also eelgrass. In other aquatic ecosystems such as lakes and open seas, where algae are the only producers, the herbivores may have a greater biomass than

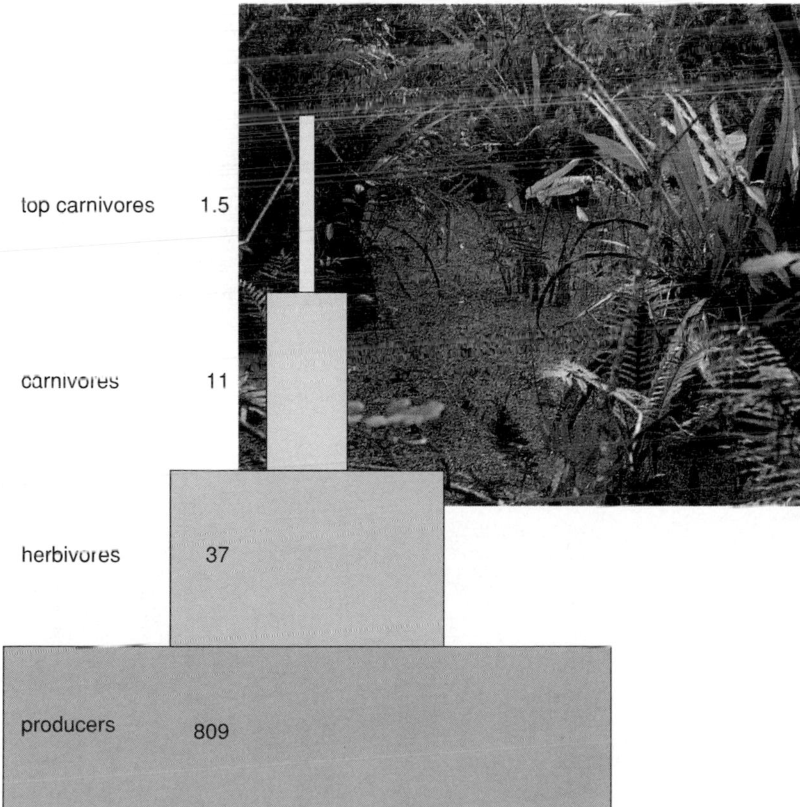

Figure 25.7 Ecological pyramid.
The biomass or dry weight (g/m^2) for trophic levels in a grazing food web in a bog at Silver Springs, Florida. There is a sharp drop in biomass between the producer level and herbivore level, which is consistent with the common knowledge that the detrital food web plays a significant role in bogs.

the producers when you take their measurements. Why? The reason is that over time, the algae reproduce rapidly, but they are also consumed at a high rate. Pyramids like this one, in which there are more herbivores than producers, are called inverted pyramids:

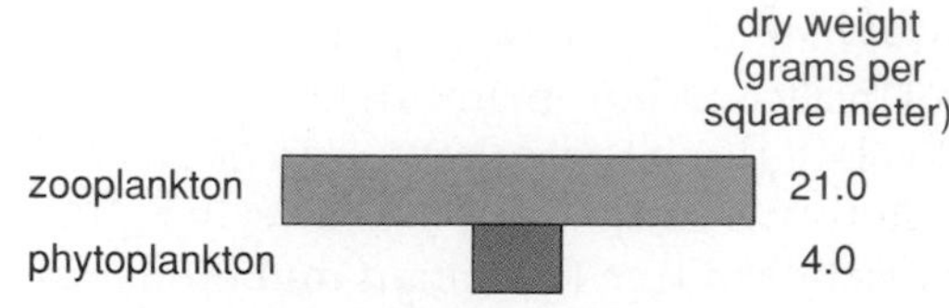

There are ecological *pyramids of energy* also, and they generally have the appearance of Figure 25.7. Ecologists are now beginning to rethink the usefulness of utilizing pyramids to describe energy relationships. One problem is what to do with the detritivores, which are rarely included in pyramids, and yet a large portion of energy becomes detritus in many ecosystems.

There is a rule of 10% with regard to biomass (or energy) pyramids. It says that, in general, the amount of biomass (or energy) from one level to the next is reduced by a magnitude of 10. Thus, if an average of 1,000 (kg or kcal) of plant material is consumed by herbivores, about 100 kg is converted to herbivore tissue, 10 kg (kcal) to first-level carnivore production, and 1 kcal to second-level carnivores. The rule of 10% suggests that few carnivores can be supported in a food web. This is consistent with the observation that each food chain has from three to four links, rarely five.

25.3 Global Biogeochemical Cycles

All organisms require a variety of organic and inorganic nutrients. Carbon dioxide and water are necessary for photosynthesis. Nitrogen is a component of all the structural and functional proteins that sustain living tissues. Phosphorus is essential for ATP and nucleotide production. In contrast to energy, inorganic nutrients are used over and over again by autotrophs.

Since the pahways by which nutrients circulate within and among ecosystems involve both living (biosphere) and nonliving (geological) components, they are known as **biogeochemical cycles.** For each element, the cycling process may involve (1) a reservoir—a source normally unavailable to producers, such as fossilized remains, rocks, and deep-sea sediments; (2) an exchange pool—a source from which organisms do generally take elements, such as the atmosphere, hydrosphere, and lithosphere; and (3) the biotic community—through which chemicals move along food chains (Fig. 25.8).

There are two general categories of biogeochemical cycles. In a *gaseous cycle,* exemplified by the carbon and nitrogen cycles, the element returns to and is withdrawn from the atmosphere as a gas. In the *sedimentary cycle,* exemplified by the phosphorus cycle, the element is absorbed from the soil by plant roots, passed to heterotrophs, and is eventually returned to the soil by decomposers, usually in the same general area.

It would seem that anything put into the environment in one ecosystem could find its way to another ecosystem. The diagrams on the next few pages make it clear that nutrients can flow between terrestrial and aquatic ecosystems. In the nitrogen and phosphorous cycles, these elements run off from a terrestrial to an aquatic ecosystem and in that way enrich aquatic ecosystems. Decaying organic material in aquatic ecosystems can be a source of nutrients for terrestrial inhabitants like fiddler crabs. Seabirds feed on fish but deposit guano (droppings) on land, and in that way phosphorus from the water is deposited on land. Scientists find the soot from urban areas and pesticides from agricultural fields in the snow and animals of the arctic.

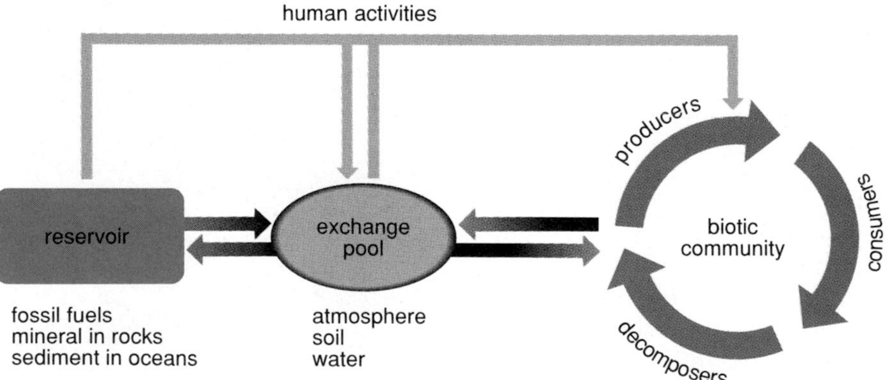

Figure 25.8 Model for nutrient cycling.
Reservoirs, such as fossil fuels, minerals in rocks, and sediments in oceans are normally relatively unavailable sources of nutrients. Pools such as those in the atmosphere, soil, and water are available sources of nutrients for the biotic community. Human activities remove nutrients from reservoirs and pools and make them available to the biotic community, and the result can be pollution.

The Hydrologic (Water) Cycle

The **hydrologic cycle** is described in Figure 25.9. Water cycles through the biosphere in the manner described in Figure 25.9. Fresh water is distilled from salt water. The sun's rays cause fresh water to evaporate from seawater, and the salts are left behind. Vaporized fresh water rises into the atmosphere, cools, and falls as rain over the oceans and the land.

Water evaporates from land and from plants (evaporation from plants is called transpiration). It also evaporates from bodies of fresh water, but since land lies above sea level, gravity eventually returns all fresh water to the sea. In the meantime, water is contained within standing waters (lakes and ponds), flowing water (streams and rivers), and groundwater.

When rain falls, some of the water sinks or percolates into the ground and saturates the earth to a certain level. The top of the saturation zone is called the groundwater table, or simply, the water table. Sometimes groundwater is also located in **aquifers,** rock layers that contain water and will release it in appreciable quantities to wells or springs. Aquifers are recharged when rainfall and melted snow percolate into the soil. In some parts of the country, especially arid areas and southern Florida, withdrawals from aquifers exceed any possibility of recharge. This is called "groundwater mining." In these locations the groundwater is dropping, and residents may run out of groundwater, at least for irrigation purposes, within a few short years. Fresh water, which makes up only about 3% of the world's supply of water, is called a renewable resource because a new supply is always being produced. But it is possible to run out of fresh water when the available supply is not adequate and/or is polluted so that it is not usable.

In the hydrologic cycle, fresh water evaporates from the bodies of water and falls to the earth. Water that falls on land enters the ground, surface waters, or aquifers. Water returns to the ocean—even the quantity that remains in aquifers for some time.

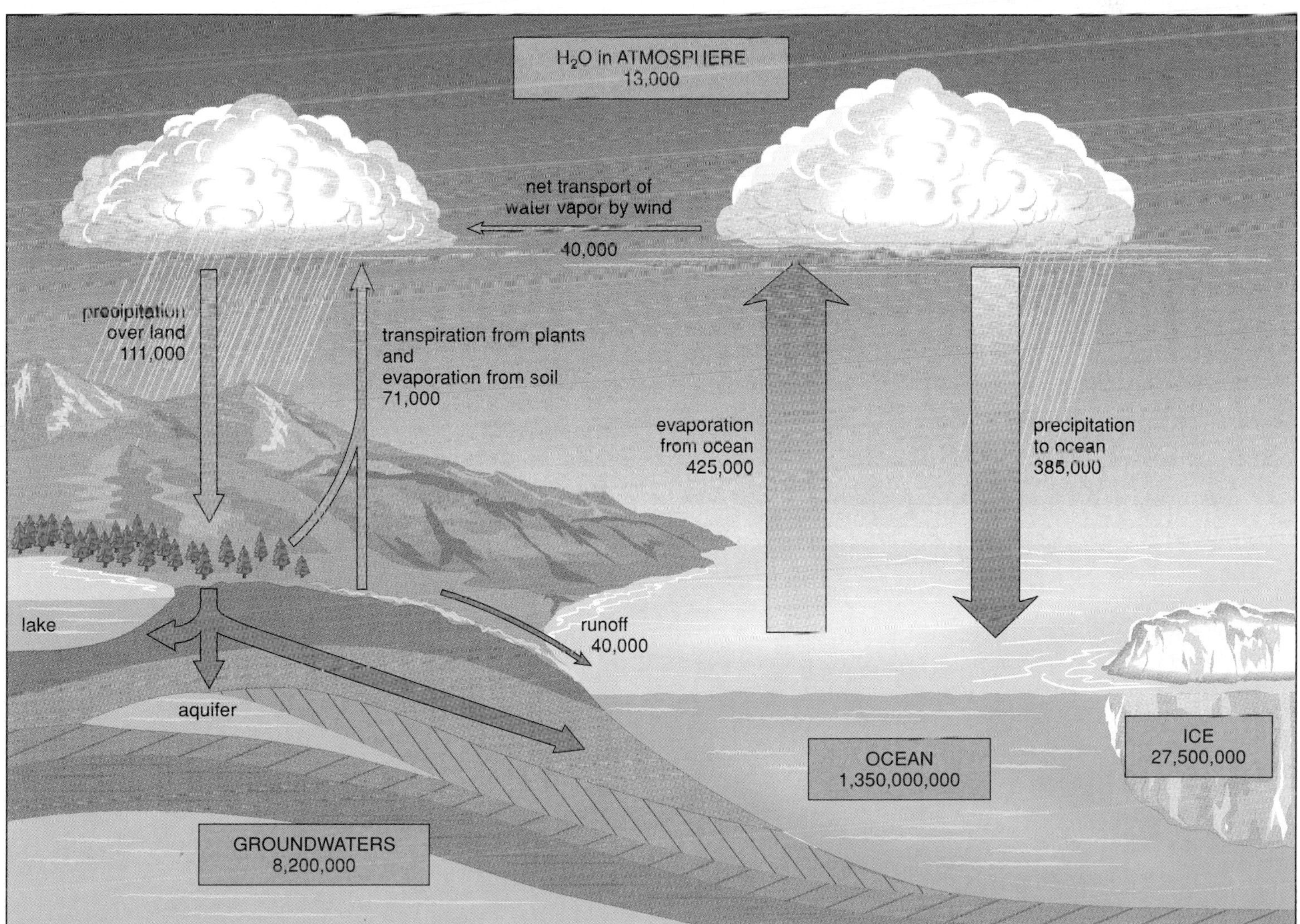

Figure 25.9 The hydrologic (water) cycle.
Evaporation from the ocean exceeds precipitation, so there is a net movement of water vapor onto land where precipitation results in surface water and groundwater that flow back to the sea. On land, transpiration by plants contributes to evaporation. The numbers in this diagram indicate water flow in km^3/yr.

The Carbon Cycle

The relationship between photosynthesis and aerobic cellular respiration should be kept in mind when discussing the **carbon cycle.** Recall that for simplicity's sake, this equation in the forward direction represents aerobic cellular respiration, and in the reverse direction it is used to represent photosynthesis.

$$C_6H_{12}O_6 + 6O_2 \underset{\text{photosynthesis}}{\overset{\text{aerobic cellular respiration}}{\rightleftharpoons}} 6CO_2 + 6H_2O$$

The equation tells us that aerobic cellular respiration releases carbon dioxide, a molecule needed for photosynthesis. However, photosynthesis also releases oxygen, a molecule needed for aerobic respiration. Animals are dependent on green organisms, not only to produce organic food and energy but also to supply the biosphere with oxygen. Since producers both photosynthesize and respire, they can function independently of the animal world. In the carbon cycle, organisms in both terrestrial and aquatic ecosystems exchange carbon dioxide with the atmosphere (Fig. 25.10). On land, plants take up carbon dioxide from the air, and through photosynthesis they incorporate carbon into food that is used by autotrophs and heterotrophs alike. When plants and animals respire, a portion of this carbon is returned to the atmosphere as carbon dioxide.

In aquatic ecosystems, the exchange of carbon dioxide with the atmosphere is indirect. Carbon dioxide from the air combines with water to produce bicarbonate (HCO_3^-), a source of carbon for algae that produce food for themselves and for heterotrophs. Similarly, when aquatic organisms respire, the carbon dioxide they give off becomes bicarbonate. The amount of bicarbonate in the water is in equilibrium with the amount of carbon dioxide in the air.

In addition to the atmosphere, living and dead organisms are available sources for organic carbon. If decomposition fails to occur, dead remains are subject to *carbonification,* physical processes that transform them into coal, oil, and natural gas. We call these reservoirs for carbon the fossil fuels. Most of the fossil fuels were formed during the Carboniferous period, 286 to 360 million years ago, when an exceptionally large amount of organic matter was buried before decomposing. Another reservoir for carbon is calcium carbonate shells. The oceans abound in organisms, some microscopic, with calcium carbonate shells that accumulate in ocean bottom sediments. Limestone is formed from these sediments by geological transformation.

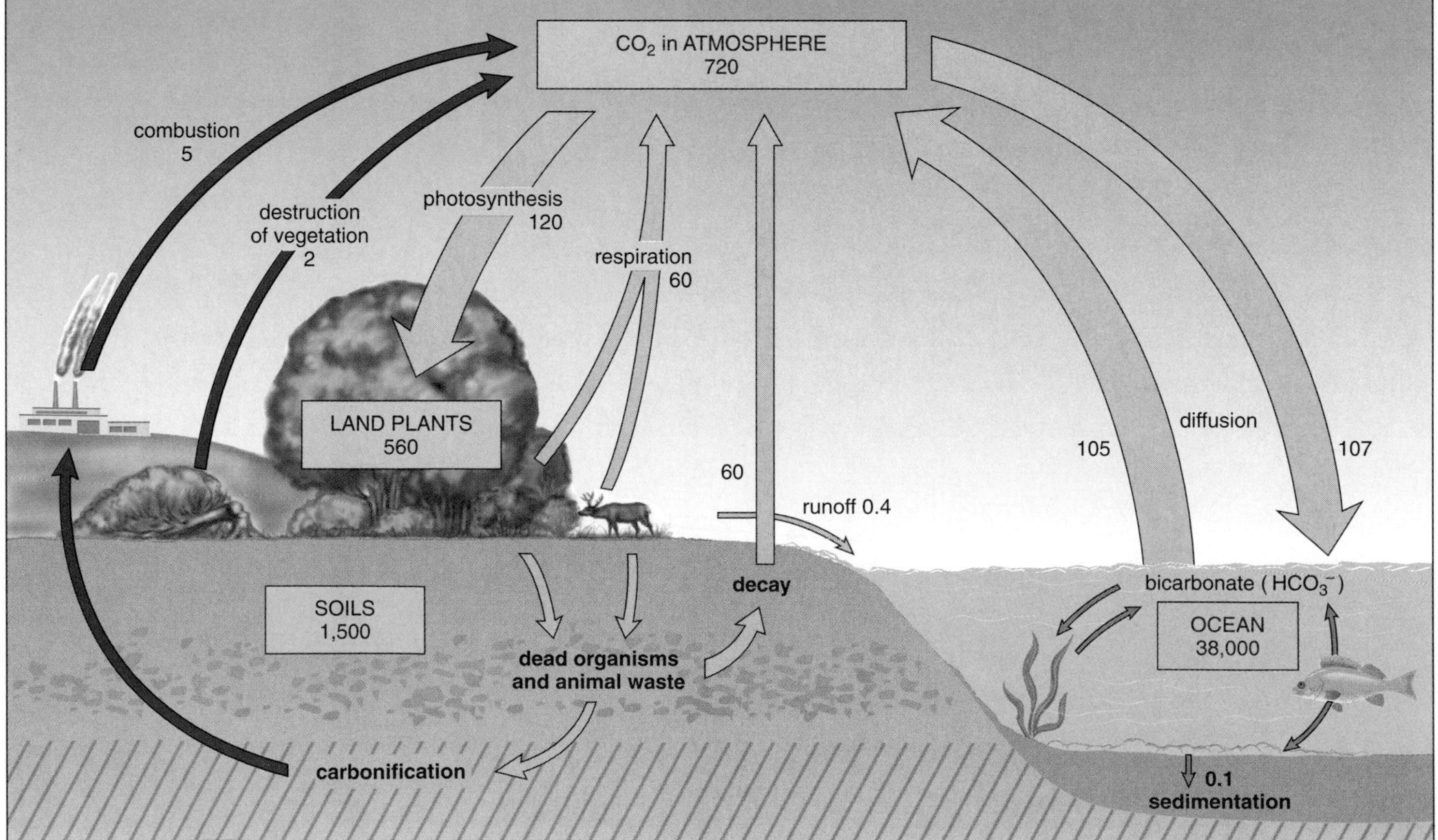

Figure 25.10 The carbon cycle.
The transfer rate of carbon into the atmosphere due to respiration just about matches the rate due to withdrawal by plants for photosynthesis. However, due to the burning of fossil fuels and destruction of vegetation by human activities (dark purple arrows), more carbon dioxide is added to the atmosphere than is withdrawn. The numbers in this diagram indicate 1,016 g C/yr.

Humans Alter Transfer Rates

A *transfer rate* is defined as the amount of a nutrient that moves from one component of the environment to another within a specified period of time. Figure 25.10 shows that the transfer rate of carbon dioxide between land and the atmosphere and between the oceans and the atmosphere, due to photosynthesis and respiration alone, are just about even. However, because humans burn fossil fuels and forests, there is now more carbon dioxide being deposited in the atmosphere than being removed. The activities of human beings increased the amount of carbon dioxide in the atmosphere by at least 6%. It would be even higher except that the oceans are believed to be taking up most of the excess carbon dioxide; the burning of fossil fuels in the last 22 years has probably released 78 billion metric tons of carbon, yet the atmosphere registers an increase of "only" 42 billion metric tons.

Botanists are now in the process of determining if increased photosynthesis will help compensate for higher levels of carbon dioxide in the atmosphere. To observe the results, some have even enclosed patches of agricultural fields or ecosystems under glass and piped in carbon dioxide. They have found that agricultural yields do increase if the application of water, fertilizers, and pesticides keeps pace. However, these inputs result in a greater output of pollutants. Plants in CO_2-rich ecosystems produced tissues that had a lower nitrogen content, and increased decomposition returned carbon dioxide to the atmosphere once again. In one wetland study, the sedges that grew more abundantly were the ones that provided food for methane-producing bacteria, and the range grass that grew best was the type most likely to burn. Based on more than a decade of research, the conclusion is that a carbon dioxide-rich atmosphere will most likely have a negative effect on the composition and operation of ecosystems.

The Nitrogen Cycle

Nitrogen is an abundant element in the atmosphere. Nitrogen (N_2) makes up about 78% of the atmosphere by volume, yet nitrogen deficiency sometimes limits plant growth. Plants cannot incorporate nitrogen into organic compounds and therefore depend on various types of bacteria to make nitrogen available to them in the **nitrogen cycle** (Fig. 25.11).

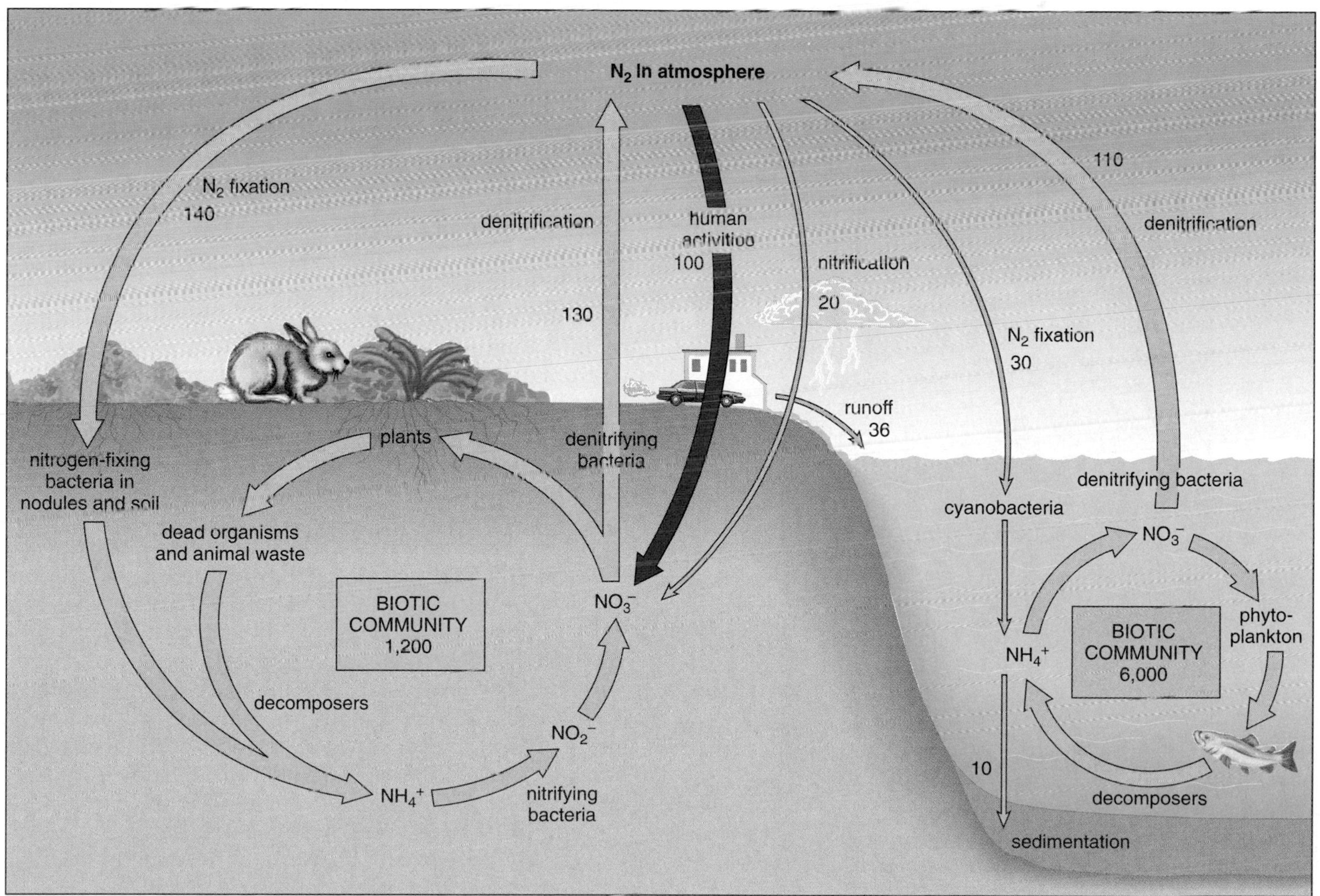

Figure 25.11 The nitrogen cycle.
Nitrogen is made available to biotic communities by internal cycling of the element. Without human activities the amount of nitrogen returned to the atmosphere (denitrification) exceeds withdrawal from the atmosphere (nitrogen fixation and nitrification). Human activities (dark purple arrow) result in an increased amount of NO_3^- in terrestrial communities with resultant runoff to aquatic biotic communities. The numbers in this diagram indicate 1,012 g N/yr.

Nitrogen Gas Becomes Fixed

Nitrogen fixation occurs when nitrogen (N_2) is reduced and added to organic compounds. Some cyanobacteria in aquatic ecosystems and some free-living bacteria in soil are able to reduce nitrogen gas to ammonium (NH_4^+). Other nitrogen-fixing bacteria live in nodules on the roots of legumes (Fig. 25.12). They make reduced nitrogen and organic compounds available to the host plant.

Plants cannot fix atmospheric nitrogen and instead take up both NH_4^+ and nitrate (NO_3^-) from the soil. After plants take up NO_3^- from the soil, it is enzymatically reduced to NH_4^+ and is used to produce amino acids and nucleic acids.

Nitrogen Gas Becomes Nitrates

Nitrification is the production of nitrates. Nitrogen gas (N_2) is converted to nitrate (NO_3^-) in the atmosphere when cosmic radiation, meteor trails, and lightning provide the high energy needed for nitrogen to react with oxygen. Ammonium (NH_4^+) in the soil is converted to nitrate by certain soil bacteria in a two-step process. First, nitrite-producing bacteria convert ammonium to nitrite (NO_2^-), and then nitrate-producing bacteria convert nitrite to nitrate. These two groups of bacteria are called the *nitrifying bacteria*. Notice the subcycle in the nitrogen cycle that involves only ammonium, nitrites, and nitrates. This subcycle need not depend on the presence of nitrogen gas at all (see Fig. 25.11).

Denitrification is the conversion of nitrate to nitrous oxide and nitrogen gas. There are denitrifying bacteria in both aquatic and terrestrial ecosystems. Denitrification counterbalances nitrogen fixation, but not completely. More nitrogen fixation occurs, especially due to fertilizer production.

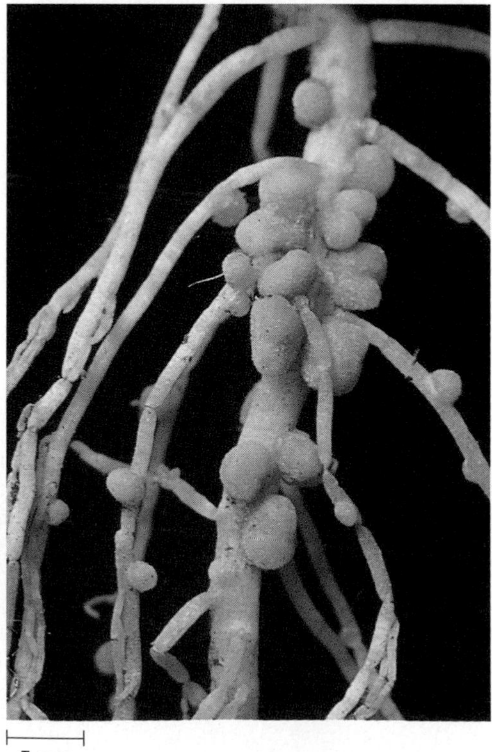

Figure 25.12 Root nodules.
The bacteria that live in these nodules are capable of converting aerial nitrogen to a source that the plant can use.

Humans Alter Transfer Rates

Human activities also are significantly altering a nitrogen cycle transfer rate by producing fertilizers from N_2—in fact, they are nearly doubling the fixation rate. Fertilizer, which also contains phosphate, runs off into lakes and rivers and results in an overgrowth of algae and rooted aquatic plants. The result is *eutrophication* (overenrichment), which can lead to an algal bloom, apparent when green scum floats on the water or when there are excessive mats of filamentous algae. When the algae die off, decomposers use up all available oxygen during cellular respiration. The result is a massive fish kill.

In the nitrogen cycle, nitrogen-fixing bacteria (in nodules and in the soil) reduce nitrogen gas; nitrifying bacteria convert ammonium to nitrate; denitrifying bacteria convert nitrate back to nitrogen gas.

The Phosphorus Cycle

On land, the weathering of rocks makes phosphate ions (PO_4 and $HPO_4^{=}$) available to plants, which take it up from the soil (Fig. 25.13). Some of this phosphate runs off into aquatic ecosystems where algae take phosphate up from the water before it becomes trapped in sediments. Phosphate in sediments only becomes available when a geological upheaval exposes sedimentary rocks to weathering once more. Phosphorus does not enter the atmosphere and, therefore, the **phosphorus cycle** is called a sedimentary cycle.

The phosphate taken up by producers is incorporated into a variety of molecules, including phospholipids and ATP or the nucleotides that become a part of DNA and RNA. Animals eat producers and incorporate some of the phosphate into teeth, bones, and shells that do not decompose for very long periods. Death and decay of all organisms, and also decomposition of animal wastes, do, however, make phosphate ions available to producers once again. Because available phosphate is generally taken up very quickly, it is often a limiting nutrient in most ecosystems. A limiting nutrient is one that regulates the growth of organisms because it is in shorter supply than other nutrients in the environment.

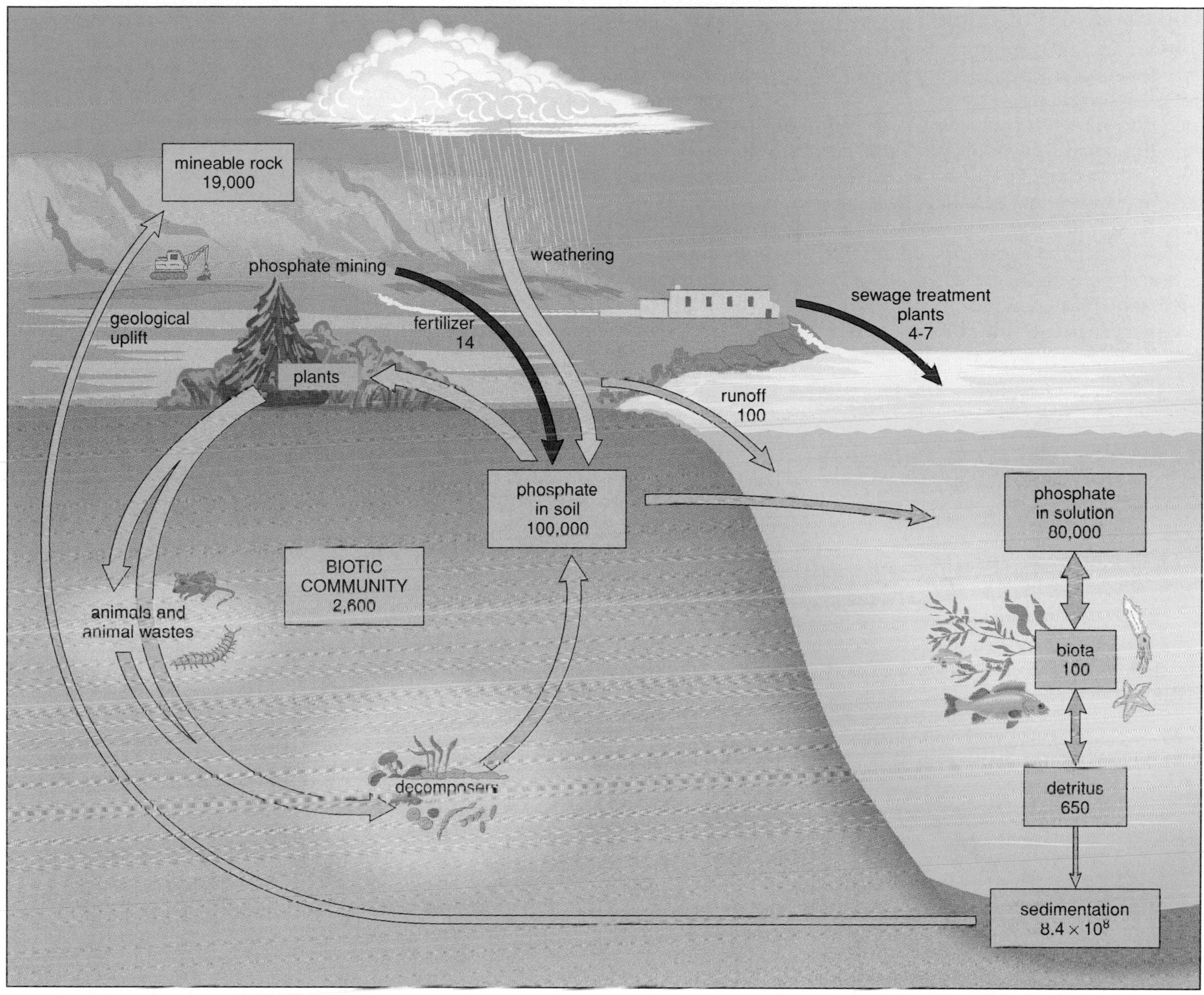

Figure 25.13 The phosphorus cycle.
The weathering of rocks provides phosphorus which cycles locally in both terrestrial and aquatic biota. When phosporus becomes a part of oceanic sediments, it is lost to biotic communities for many years. The dark purple arrows show how humans alter transfer rates. Humans produce fertilizers which add to the amount of phosphorus available to biotic communities—eventually, fertilizers become a part of the runoff which enriches waters. Sewage treatment plants directly add phosphorus to local waters. The numbers in this diagram indicate 1,012 g P/yr.

Humans Alter Transfer Rates

Human beings boost the supply of phosphate by mining phosphate ores for fertilizer production, animal feed supplements, and detergents. One well-known region where phosphorus is strip-mined (so called because earth is removed in strips) lies east of Tampa, Florida. Here, fossilized remains of marine animals were laid down some 10 to 15 million years ago. Phosphate ore is slightly radioactive and, therefore, mining phosphate poses a health threat to all organisms, even those that do the mining! Also, only a portion of this land has been properly reclaimed, and the rest is subject to severe soil erosion.

Runoff of phosphate due to fertilizer use, animal wastes from livestock feedlots, as well as discharge from sewage treatment plants is adding excess phosphate to nearby waters, resulting in eutrophication. Because phosphate is a limiting nutrient in most ecosystems, its potential for causing eutrophication is greater than that of nitrogen.

In the phosphorus cycle, weathering makes phosphate available to producers followed by consumers. Death and decay of all organisms makes phosphate available to consumers once again. Phosphate trapped in sedimentary rock only becomes available following a geological upheaval.

The Cause of Pollution

Human activities are impacting biogeochemical cycles and therefore ecosystems. The pumping of water from aquifers is not normally a part of the water cycle. The burning of fossil fuels and trees is increasing the amount of carbon dioxide in the atmosphere, and the end result may be a global warming. Carbon dioxide allows the sun's rays to pass through, but they absorb and reradiate heat back to the earth, a phenomenon called the *greenhouse effect.* The phosphorus cycle is affected when we produce detergents, and both the nitrogen and phosphate cycles are impacted when we produce fertilizers. Nitrogen and phosphorus runoff in aquatic ecosystems cause eutrophication.

Human activities alter transfer rates because they move an element from one component of an ecosystem to another at a rate that matches or is greater than previous transfer rates. In the carbon cycle, humans increase the transfer rate from fossil fuels (and trees) to the atmosphere and into the ocean. In the nitrogen cycle, humans increase the transfer rate from the atmosphere to the soil and into bodies of water, and in the phosphorous cycle, humans increase the transfer rate from sediments to the soil and into bodies of water. **Pollution** can be defined as a change in transfer rate that can lead directly or indirectly to a degradation of human health or a degradation of plant and animal life.

connecting concepts

Ecosystems consist of the living organisms in an ecological community together with the physical environment which surrounds those organisms. When ecologists study ecosystems they are usually interested in big-scale questions about how ecosystems function and what factors cause particular ecosystems to function differently. Ecosystem functions include such phenomena as energy flow within the ecosystem, storage or loss of energy from the ecosystem, nutrient cycling within an ecosystem, movement of nutrients between ecosystems, decomposition of dead organic matter, storage of water, etc.

One of the most basic ways to study ecosystem function is to study energy flow within ecosystems. By measuring the amount of carbon entering an ecosystem via photosynthesis we can estimate how much solar energy is being fixed by primary producers. We can then determine how much carbon is moving from one trophic level to the next, and finally how much carbon is released through respiration to examine the entire path of energy flow. Different ecosystems differ in how much carbon (energy) is fixed in them and how much travels through their food webs. By determining which factors govern different levels of energy flow (factors such as precipitation, nutrient availability, and seasonality), we can develop good models of basic ecosystem function.

Many ecosystem ecologists feel that while understanding energy flow is key to understanding ecosystem function, it is even more important to understand nutrient cycles and nutrient availability. Nutrient availability does not just influence plant growth, it also determines distributions of animals. Many large grazing mammals preferentially consume plants with high nutrient content and thus tend to occur in areas with high nutrient availability.

Ecosystem studies have become increasingly sophisticated in the last decade. Many early ecosystem studies resulted in the construction of box models (as seen in Fig. 25.3) which allowed good predictions about rates of energy flow and nutrient cycling. However, they did not provide much insight into the mechanisms governing energy flow or nutrient cycling.

Much current research is directed towards "filling in the boxes" so that we are now beginning to learn how particular species influence the function of entire ecosystems. Thus we are now becoming able to predict which species are most important and to determine which species must be preserved in order to have natural ecosystem functioning.

The largest ecosystem of all is the biosphere. In a series of complex feedbacks, living organisms influence the climate of the earth and the climate in turn determines the distribution of biomes (major ecosystem types) within the biosphere. As we shall see, major changes in the different biomes can have large effects on the climate and thus the entire biosphere.

Summary

25.1 The Nature of Ecosystems

The biosphere, which contains living things, includes portions of the atmosphere, lithosphere, and hydrosphere. Ecosystems contain biotic (living) components and abiotic (physical) components where energy flows and nutrients cycle among different global chemical pools. The biotic components of ecosystems are either producers or consumers. Consumers may be herbivores, carnivores, omnivores, detritivores, or decomposers.

25.2 Energy Flow and Nutrient Cycling

Energy flows through an ecosystem. When autotrophs photosynthesize, they transform solar energy into the chemical-bound energy of organic molecules, which become nutrients for themselves and all heterotrophs. As herbivores feed on plants (or algae), and carnivores feed on herbivores, some energy is converted to heat. Feces, urine, and dead bodies become food for decomposers and other detritivores. Eventually all the solar energy that enters an ecosystem is converted to heat, and thus ecosystems require a continual supply of solar energy. Thus ecosystems are open systems.

Nutrients are not lost from the biosphere as is heat. Nutrients recycle within and between ecosystems. Decomposers return some proportion of inorganic nutrients to autotrophs, and other portions are imported or exported between ecosystems in global cycles.

Ecosystems contain food webs, and a diagram of a food web shows how the various organisms are connected by eating relationships. Grazing food webs begin with vegetation that is fed on by herbivores, which become food for carnivores. In detrital food webs, decomposers act on organic material in the soil, and they are fed on by other detritivores, which become food for carnivores in the grazing food web. Thus the two food webs are connected.

It's possible to isolate food chains (straight-line diagrams of who eats whom) from food webs. A trophic level is all the organisms that feed at a particular link in a food chain. Ecological pyramids show trophic levels stacked one on the other like building blocks. Generally they show that biomass and energy content decrease from one trophic level to the next. Most pyramids pertain to grazing food webs and largely ignore the detrital food web portion of an ecosystem.

25.3 Global Biogeochemical Cycles

Nutrients include chemical elements and inorganic and organic compounds that contain these elements, particularly CHNOPS. Reservoirs are components of ecosystems like fossil fuels, sediments, and rocks that contain nutrients available on a limited basis to living things. Pools are components of ecosystems like the atmosphere, soil, and water—which are ready sources of nutrients for living things. Nutrients cycle among the members of the biotic component of an ecosystem. Human activities transfer nutrients from resources and pools to producers at a rate that equals or exceeds the normal transfer rate.

In the water cycle, evaporation over the ocean is not compensated for by precipitation. Evaporation from terrestrial ecosystems includes transpiration from plants. Precipitation over land results in bodies of fresh water plus groundwater, including aquifers. Eventually all water returns to the oceans.

In the carbon cycle, organisms add as much carbon dioxide to the atmosphere as they remove. Human activities such as the burning of fossil fuels and trees are adding carbon dioxide to the atmosphere, and the oceans have been taking it up. Ocean sediments, carbon shells, and limestone are reservoirs for carbon.

In the nitrogen cycle, the biotic community, which includes several types of bacteria, keeps nitrogen recycling back to the producers. A few organisms (cyanobacteria in aquatic habitats and bacteria in root nodules) can fix atmospheric nitrogen, and there are some organisms that return nitrogen to the atmosphere. Human activities convert atmospheric nitrogen to fertilizer, which in substantial quantities runs off into aquatic habitats, causing eutrophication.

In the phosphorus cycle, the biotic community recycles phosphorus back to the producers, and only limited quantities are made available by the weathering of rocks. When phosphorus enters the water, it becomes a part of sediment where it remains until an upheaval occurs. Phosphorus mined by humans becomes a part of fertilizers, which overenrich aquatic habitats.

Pollution is associated with alternations in ecosystem processes that lead to changes in transfer rates within communities and ecosystems.

Reviewing the Chapter

1. Define the biosphere and tell where it is located. 428
2. Distinguish between autotrophs and heterotrophs, and describe four different types of heterotrophs found in natural ecosystems. 429
3. Discuss energy flow in an ecosystem and nutrient cycling within and between ecosystems. Why are ecosystems open systems? 430–31
4. Describe the two types of food webs typically found in ecosystems. Which of these typically moves more energy through an ecosystem? 432–33
5. What is a trophic level? an ecological pyramid? 433–34
6. Give examples of reservoirs and pools in biogeochemical cycles. Which of these is less accessible to living things? 434
7. Draw a diagram to illustrate the water, carbon, nitrogen, and phosphorus biogeochemical cycles and include in your diagram the manner in which humans alter transfer rates. 435–39,
8. Define pollution. 440

Testing Yourself

Choose the best answer for each question.

1. Compare this food chain:

   algae → water fleas → fish → green herons

   to this food chain:

   trees → tent caterpillars → red-eyed vireos → hawks.

   Both water fleas and tent caterpillars are
   a. carnivores.
   b. primary consumers.
   c. detritus feeders.
   d. Both a and b are correct.
2. Consider the components of a food chain:

   producers → herbivores → carnivores → top carnivores.

   Which level contains the most biomass?
   a. producers
   b. herbivores
   c. carnivores
   d. top carnivores
3. Why are ecosystems dependent on a continual supply of solar energy?
   a. Carnivores have a greater biomass than producers.
   b. Decomposers process the greatest amount of energy in an ecosystem.
   c. Energy transformation results in a loss of usable energy to the environment.
   d. Energy cycles within and between ecosystems.
4. Of the total amount of energy that passes from one trophic level to another, about 10% is
   a. respired and becomes heat.
   b. passed out as feces or urine.
   c. stored as body tissue.
   d. All of these are correct.
5. Nutrient cycles always involve
   a. rocks as a reservoir.
   b. movement of nutrients through the biotic community.
   c. the atmosphere as an exchange pool.
   d. loss of the nutrients from the biosphere.
6. Which of the following contribute to the carbon cycle?
   a. respiration
   b. photosynthesis
   c. fossil fuel combustion
   d. All of these are correct.

7. How do plant contribute to the carbon cycle?
   a. When they respire, they release carbon dioxide (CO_2) into the atmosphere.
   b. When they photosynthesize, they consume CO_2 from the atmosphere.
   c. They do not contribute to the carbon cycle.
   d. Both a and b are correct.
8. How do nitrogen-fixing bacteria contribute to the nitrogen cycle?
   a. They return nitrogen (N_2) to the atmosphere.
   b. They change ammonium to nitrate.
   c. They change N_2 to ammonium.
   d. They withdraw nitrate from the soil.
9. In what way are decomposers like producers?
   a. Either one may be the first member of a grazing food chain.
   b. Both produce oxygen for other forms of life.
   c. Both require a source of nutrient molecules and energy.
   d. Both supply organic food for the biosphere.
10. How is a primary consumer like a secondary consumer?
    a. Both pass on less energy to the next trophic level than they received.
    b. Both pass on the same amount of energy to the next trophic level.
    c. Both tend to be herbivores that produce nutrients for plants.
    d. Both are able to convert organic compounds to ATP without the loss of energy.
11. Which statement is true concerning this food chain: grass → rabbits → snakes → hawks?
    a. Each predator population has a greater biomass than its prey population.
    b. Each prey population has a greater biomass than its predator population.
    c. Each population is omnivorous.
    d. Both a and c are correct.
12. Label the trophic levels on the right and tell what the numbers on the left refer to.

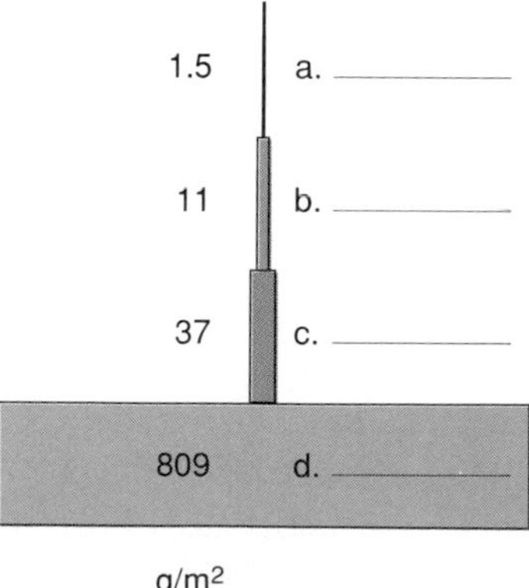

Applying the Concepts

1. *Nutrients cycle within and among ecosystems.*
   How does nutrient cycling involve both abiotic and biotic components of an ecosystem?
2. *Energy flows through an ecosystem.*
   Relate energy flow through an ecosystem to the laws of thermodynamics.
3. *Ecosystems are dynamic.*
   Describe some human activities that lead to pollution, as defined by this chapter?

Using Technology

Your study of ecosystems is supported by these available technologies:

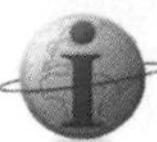

Exploring the Internet

The Mader Home Page provides resources for and help with studying this chapter.

http://www.mhhe.com/sciencemath/biology/mader/
(Click on Biology.)

Life Science Animations Video

Video #5: Plant Biology/Evolution/Ecology
Carbon and Nitrogen Cycles (#51)
Energy Flow through an Ecosystem (#52)

Understanding the Terms

aquifer 435
biogeochemical cycle 434
biosphere 428
carbon cycle 436
carnivore 429
consumer 429
decomposer 429
decomposition 429
denitrification 438
detrital food web 433
detritivore 429
detritus 429
ecological pyramid 433
ecosystem 428
food chain 433
food web 433
grazing food web 433
herbivore 429
hydrologic cycle 435
nitrification 438
nitrogen cycle 437
nitrogen fixation 438
omnivore 429
phosphorus cycle 438
pollution 440
primary productivity 430
producer 429
trophic level 433

Match the terms to these definitions:

a. ________ Biological community together with the associated abiotic environment.
b. ________ Circulating pathway of elements such as carbon and nitrogen from the environment, through biotic communities and also back to the environment.
c. ________ Complex pattern of interlocking and crisscrossing food chains.
d. ________ Feeding level of one or more populations in a food web.
e. ________ Organism at the start of a food chain that makes its own food (e.g., green plants on land and algae in water).
f. ________ Organic matter produced by the decay of a substance such as tissues and animal wastes.
g. ________ Process whereby free atmospheric nitrogen is converted into compounds, such as ammonium and nitrates, usually by soil bacteria.

The Biosphere

CHAPTER 26

Chapter Concepts

26.1 Climate and the Biosphere

- Solar radiation provides the energy that drives climate differences in the biosphere. 444–45
- Global air circulation patterns, ocean currents, and physical features produce the various patterns of temperature and rainfall about the globe. 444–47

26.2 Biomes of the World

- Ecosystems can be divided into a few major classes of terrestrial and aquatic biomes. 449
- Terrestrial and aquatic biomes are linked and interact with one another. 449

26.3 Terrestrial Biomes

- The earth's major terrestrial biomes are forests (broad-leaf and coniferous), shrublands, grasslands (tropical savanna and temperate grasslands), tundra, and deserts. 451

26.4 Aquatic Biomes

- The earth's major aquatic biomes are of two types: fresh water and salt water (usually marine). 459

Montane coniferous forest, Colorado

Terrestrial and aquatic ecosystems are located within the biosphere where living things exist and interact. Portions of the lithosphere, the earth's outer crust, the atmosphere, and the hydrosphere support the biosphere. We already know how ecosystems process solar energy and redistribute materials. Through biogeochemical cycles, driven by solar energy, the terrestrial ecosystems transform the surface of the lithosphere and the composition of the hydrosphere and atmosphere into a life-supporting environment.

In this chapter, we see how ecosystems are distributed over the globe and how climate determines the characteristics of the major biomes, such as forest, deserts, grasslands, and oceans. Each biome has its own mix of plants and animals, which are adapted to living under particular environmental conditions. You can easily learn to associate plants and animals with a particular biome by considering these adaptations.

The farms, towns, and cities of human beings now dominate the biosphere, and most of the biomes of the world have been greatly affected by human activities. We have cleared vast areas of ecosystems after using the plants and killing the animals for our own purposes. It is critical for us to learn to value the services that natural ecosystems perform for us and to work toward preserving the remaining portions of the original biomes you will be reading about in this chapter.

26.1 Climate and the Biosphere

The worldwide currents of air and water affect the abundance and distribution of organisms in the biosphere. The **biosphere** is that portion of our planet that includes the *lithosphere*, the *hydrosphere*, and the atmosphere where living things live. While the biosphere may seem to be huge, it may be likened to the skin of an inflated balloon and is but a thin veneer that covers the surface of the earth.

The massive evergreen forest that stretches along the coast from northern California to British Columbia would not be there if it were not for the winds that blow in off the Pacific Ocean, creating a mild temperature and bringing much rain. The forest in return affects both the local weather and global **climate,** illustrating that the living and nonliving interact in the biosphere.

Global Air and Water Circulation

The global climate and therefore the distribution of biomes in the biosphere is dependent upon (1) variations in reception of solar radiation due to a spherical earth, (2) the tilt of the earth's axis as it rotates about the sun, (3) distribution of landmasses and oceans, and (4) topography (landscape) features. Due to these factors, temperature and rainfall differ throughout the biosphere.

Air Circulation

Because the earth is a sphere, the sun's rays are more direct at the equator and more spread out at polar regions (Fig. 26.1*a*). Therefore, the tropics are warmer than temperate regions. The tilt of the earth as it orbits around the sun causes one pole or the other to be closer to the sun (except at the spring and fall equinoxes), and this accounts for the seasonal changes in climate in all parts of the earth except the equator (Fig. 26.1*b*). When the Northern Hemisphere is having winter, the Southern Hemisphere is having summer and vice versa.

In the atmosphere, heat always passes from warm areas to colder areas. If the earth were standing still, and were a solid, uniform ball, all air movements—which we call winds—would be in two directions. Warm equatorial air would rise and move directly to the poles, creating a zone of lower pressure that would be filled by cold polar air moving equatorward.

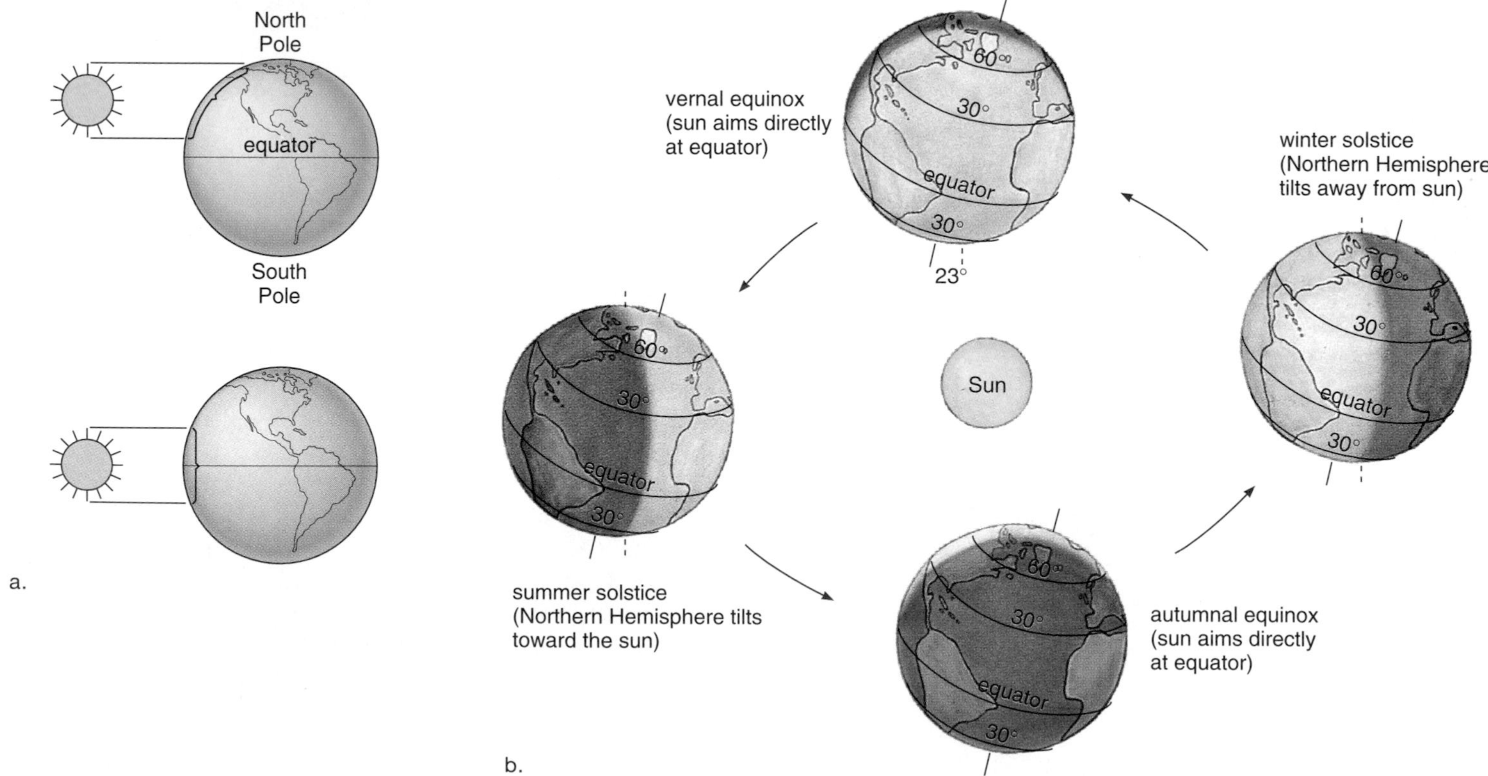

Figure 26.1 Distribution of solar energy.
a. Since the earth is a sphere, beams of solar energy striking the earth near one of the poles is spread over a wider area than similar beams striking the earth at the equator. **b.** The seasons of the Northern and Southern Hemispheres are due to the tilt of the earth on its axis as it rotates about the sun.

Because the earth rotates, and because its surface consists of continents and oceans, the flows of warm and cold air are modified into three large circulation cells in each hemisphere (Fig. 26.2*a*). At the equator, the sun heats the air and evaporates water. The warm moist air rises, cools, and loses most of its moisture as rain. The greatest amounts of rainfall on earth are near the equator. The rising air flows toward the poles, but at about 30° north and south latitude, it sinks toward the earth's surface and reheats. As the air descends and warms, it becomes very dry, creating zones of low rainfall. The great deserts of Africa, Australia, and the Americas occur at these latitudes. At the earth's surface, the air flows both poleward and equatorward. At about 60° north and south latitude, the air rises and cools, producing additional zones of high rainfall. This moisture supports the great forests of Temperate Zone. Part of this rising air flows equatorward, and part continues poleward, descending near the poles, which are zones of low precipitation.

The earth is not standing still; it is rotating on its axis daily. The spinning of the earth affects the winds, so that the major global circulation systems flow toward the east or west rather than directly north or south (Fig. 26.2*b* and *c*). Between about 30° north latitude and 30° south latitude, the winds blow from the east-southeast in the Southern Hemisphere and from the east-northeast in the Northern Hemisphere (the east coasts of continents at these latitudes are wet). These are called trade winds because sailors depended upon them to fill the sails of their trading ships. Between 30° and 60° north and south latitude, strong winds, called the prevailing westerlies, blow from west to east. The west coasts of the continents at these latitudes are wet, as is the Pacific Northwest. The evergreen forest mentioned at the start of the chapter is located here. Weaker winds, called the polar easterlies, blow from east to west at still higher latitudes of their respective hemispheres.

The seasons also affect the winds. In January, cool local winds flow from the cold continental masses of Asia and North America toward the warm, predominantly maritime regions of the equator. In July, the continental masses of Asia and North America have become warmer than the equatorial oceans, so their local air flow is reversed.

The distribution of solar energy caused by a spherical earth, and the rotation and path of the earth about the sun, affect how the winds blow and the amount of rainfall regions of the biosphere receive.

60°
30°
ascending moist air cools and loses moisture
equator
descending dry air warms and retains moisture
30°
60°
a.

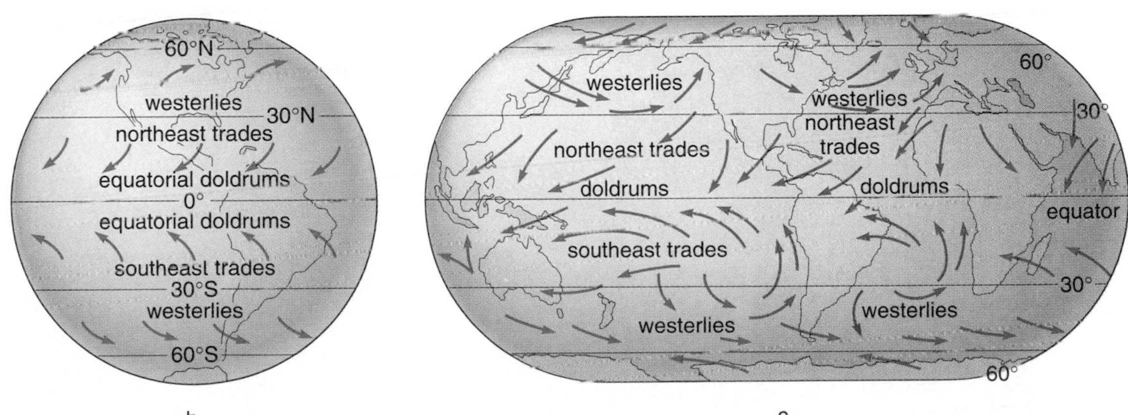

Figure 26.2 Global wind circulation.
a. Warm air rises at the equator and descends at 30°. As air rises it loses moisture, and after descending it gains moisture. Therefore, deserts occur at 30° around the world. Similar patterns occur at the other latitudes as noted. **b.** Because the earth is rotating on its axis, the trade winds move from the northeast to the west in the Northern Hemisphere and from the southeast to the west in the Southern Hemisphere. The westerlies have an easterly direction. **c.** Winds as they are distributed over the entire world.

Ocean Currents

The **hydrosphere,** the portions of the planet Earth that are water, is also warmed by the sun and, as we have discussed, this causes the evaporation of water from the oceans. As moisture evaporates into the air, it carries along the heat of evaporation. When such warm, vapor-laden air continues to rise, the heat remains with the vapor until the air reaches an altitude where cooler temperatures condense the moisture into clouds and precipitation occurs.

Climate is driven by the sun, but the oceans play a major role in redistributing heat in the biosphere. Water tends to be warm at the equator and much cooler at the poles because of the distribution of the sun's rays, as we have discussed before (see Fig. 26.1*a*). Air takes on the temperature of the water below, and warm air moves from the equator to the poles. In other words, the oceans make the winds blow. (The landmasses also play a role, but the oceans hold heat longer and remain cool longer during periods of changing temperature than do solid continents.)

Our story is not finished: when the wind blows strongly and steadily across a great expanse of ocean for a long time, friction from the moving air begins to drag the water along with it. Once the water has been set in motion, its momentum, aided by the wind, keeps it moving in a steady flow we call a current. Because the ocean currents eventually strike land, they move in a circular path—clockwise in the Northern Hemisphere and counterclockwise in the Southern Hemisphere (Fig. 26.3). As the currents flow, they take warm water from the equator to the poles. One such current, called the Gulf Stream, brings tropical Caribbean water to the east coast of North America and the higher latitudes of western Europe. Without the Gulf Stream, Great Britain, which has a relatively warm temperature, would be as cold as Greenland. In the Southern Hemisphere, another major ocean current warms the eastern coast of South America.

Also, in the Southern Hemisphere a current called the Humboldt Current flows toward the equator. The Humboldt Current carries phosphorus-rich cold water northward along the west coast of South America. During a process called **upwelling,** cold offshore winds cause cold nutrient-rich waters to rise and take the place of warm nutrient-poor waters. In South America, the enriched waters cause an abundance of marine life that supports the fisheries of Peru and northern Chile. Birds feeding on these organisms deposit their droppings on land, where it is mined as guano, a commercial source of phosphorus. When the Humboldt Current is not as cool as usual, upwelling does not occur, stagnation results, the fisheries decline, and climate patterns change globally. This phenomenon, which is apt to occur at Christmastime, is called "**El Niño**" (the Christ child).

Major ocean currents move heat from the equator to cooler parts of the biosphere. The Gulf Stream warms the east coast of North America and parts of Europe.

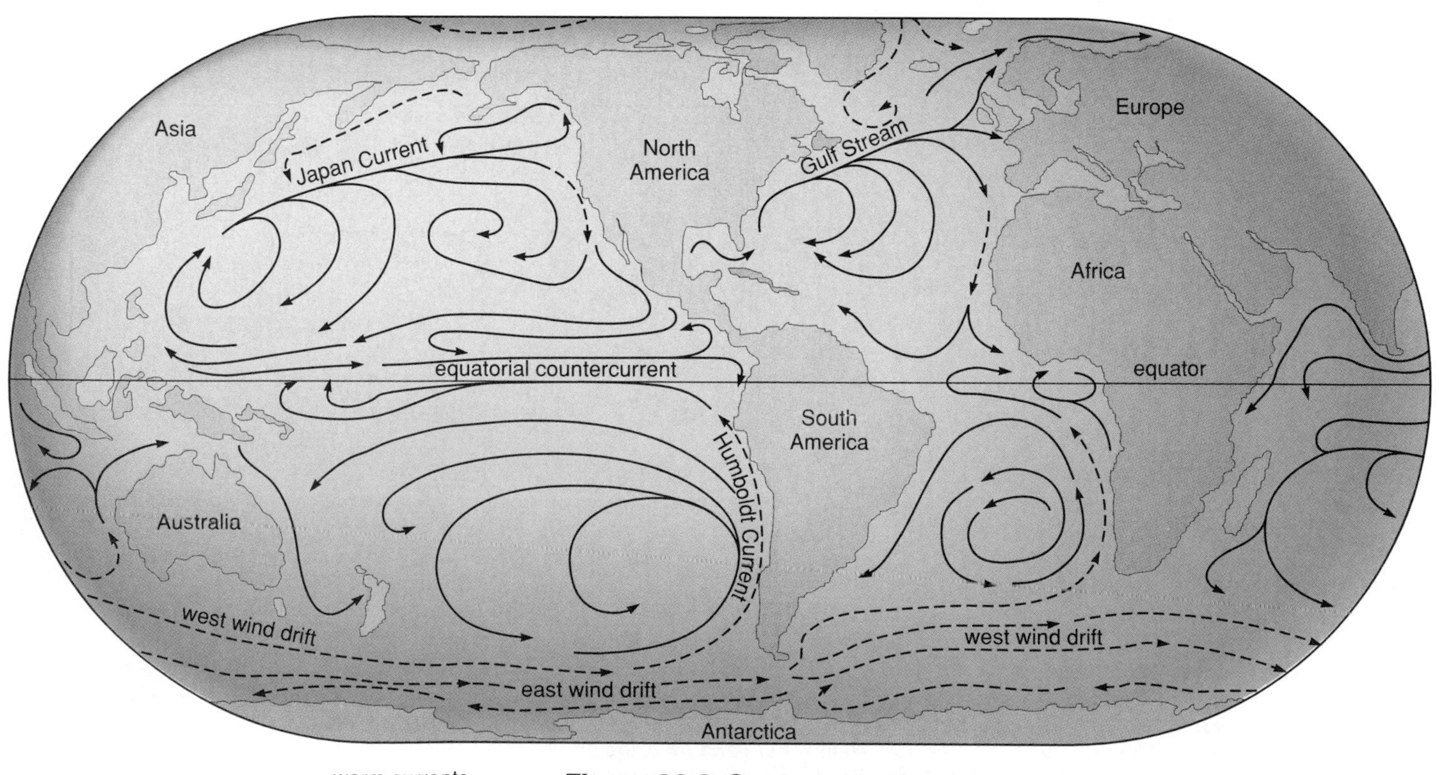

Figure 26.3 Ocean currents.
The arrows on this map indicate the locations and directions of the major ocean currents set in motion by the global wind circulation. By carrying warm water to cool latitudes (e.g., Gulf Stream) and cool water to warm latitudes (e.g., Humboldt Current), these currents have a major effect on the world's climates.

Effects of Topography

Topography means the physical features or "the lay" of the land. One physical feature that affects climate is the presence of mountains. As air blows up and over a mountain range, it rises and cools. This side of the mountain, called the windward side, receives more rainfall than the other side, called the leeward side. On the leeward side, the air descends, picks up moisture and produces clear weather (Fig. 26.4). The difference between the windward side and the leeward side can be quite dramatic. In the Hawaiian Islands, for example, the windward side of the mountains receives more than 750 cm of rain a year, while the leeward side, which is in a **rain shadow,** gets on the average only 50 cm of rain and is generally sunny. In the United States, the western side of the Sierra Nevada Mountains is lush, while the eastern side is a semidesert.

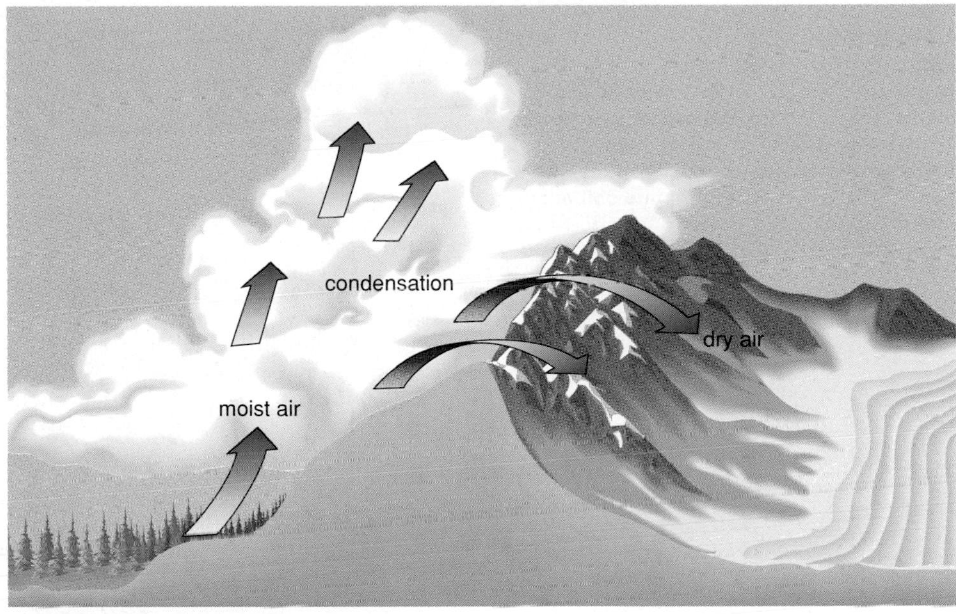

Figure 26.4 Formation of a rain shadow.
When winds from the sea cross a coastal mountain range, they rise and release their moisture as they cool. Therefore, the leeward side of mountains receives relatively little rain and is said to lie in a "rain shadow."

As previously mentioned, the oceans are slower to change temperature—that is, to gain or lose their heat—than landmasses. This causes coasts to have a unique weather pattern that is not seen inland. During the day, the land warms more quickly than the ocean, and the air above the land rises. Then a cool sea breeze blows in from the ocean. At night the reverse happens; the breeze blows from the land to the sea (Fig. 26.5).

India and some other countries in southern Asia have a **monsoon** climate, in which wet ocean winds blow onshore for almost half the year. The land heats more rapidly than the waters of the Indian Ocean during spring. The difference in temperature between the land and ocean causes a gigantic circulation of air: warm air rises over the land, and cooler air comes in off the ocean to replace it. As the warm air rises, it loses its moisture and the *monsoon season* begins. As just discussed, rainfall is particularly heavy on the windward side of hills. Cherrapunji in northern India receives an annual average of 1,090 cm of rain a year because of its high altitude. The weather pattern has reversed by November. The land is now cooler than the ocean; therefore, dry winds blow from the Asian continent across the Indian Ocean. In the winter, the air over the land is dry, the skies cloudless, and temperatures are pleasant. The chief crop of India is rice, which starts to grow when the monsoon rains begin. If the monsoon fails to bring adequate moisture, it can mean hunger or even starvation for the millions of people in the area.

In the United States, people often speak of the "lake effect," meaning that in the winter, Arctic winds blowing over the Great Lakes become warm and moisture laden. When these winds rise and lose their moisture, snow begins to fall. Places such as Buffalo, New York, get heavy snowfalls due to the lake effect, and there is snow on the ground for an average of 90 to 140 days every year. In Great Britain, icy winds from the former Soviet Union blow over the North Sea and cause the eastern coast of England to be overcast and cloudy.

a. Onshore flow during the day

b. Offshore flow during the night

Figure 26.5 Coastal breezes.
a. During the day the land warms faster than the ocean; therefore, a cool sea breeze moves onto land and replaces the warm air that is ascending into the atmosphere. **b.** At night the land cools faster than the ocean; therefore, the wind changes direction and moves out to sea.

Atmospheric circulations between the ocean and landmasses influence regional climate conditions and the distribution of biomes.

Figure 26.6 Pattern of biome distribution.
a. Pattern of world biomes in relation to temperature and moisture. The dashed line encloses a wide range of environments in which either grasses or woody plants can dominate the area, depending on the soil type. b. The same type of biome can occur in different regions of the world, as shown on this global map.

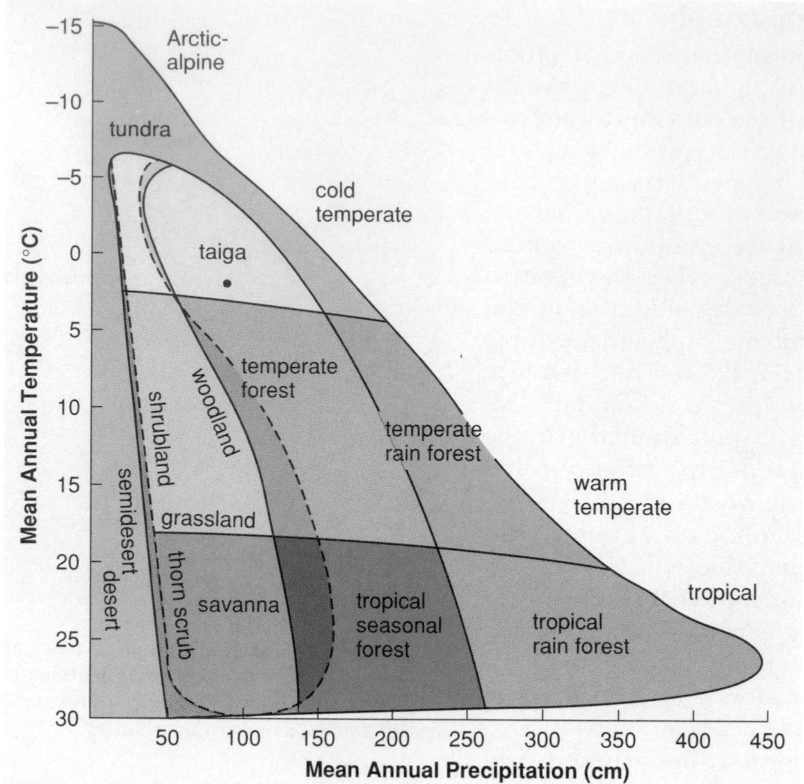

a.

polar ice
tundra
taiga
mountain zone
temperate deciduous forest
temperate evergreen forest
tropical deciduous forest
tropical monsoon forest
tropical rain forest
shrubland
temperate grassland
savanna
semidesert
desert

b.

26.2 Biomes of the World

A **biome** is the largest biogeographical unit of the biosphere. Although the term arose with reference only to terrestrial communities, we will be using it for both terrestrial and aquatic communities. A biome has a particular mix of plants and animals that are adapted to living under certain environmental conditions, of which climate has an overriding influence. For example, when terrestrial biomes are plotted according to their mean annual temperature and mean annual rainfall, a particular pattern results (Fig. 26.6*a*). The distribution of biomes is shown in Figure 26.6*b*. Even though Figure 26.6 shows definite demarcations, the biomes gradually change from one type to the other. Also, although we will be discussing each type of biome separately, we should remember that each biome has inputs from and outputs to all the other terrestrial and aquatic biomes of the biosphere.

The pattern of life on earth is determined principally by climate, which is influenced also by topographical features. The effect of a temperature gradient can be seen not only when we consider latitude but also when we consider altitude. If you travel from the equator to the North Pole, it is possible to observe first a tropical rain forest, followed by a temperate deciduous forest, a coniferous forest, and tundra, in that order, and this sequence is also seen when ascending a mountain (Fig. 26.7). The coniferous forest of a mountain is called a **montane coniferous forest,** and the tundra near the peak of a mountain is called an **alpine tundra.** When going from the equator to the South Pole, you would not reach a region corresponding to a coniferous forest and tundra of the Northern Hemisphere. Why not? Look at the distribution of the landmasses—they are shifted toward the north.

The distribution of biomes is determined by physical factors such as climate (principally temperature and rainfall), which varies according to latitude and altitude.

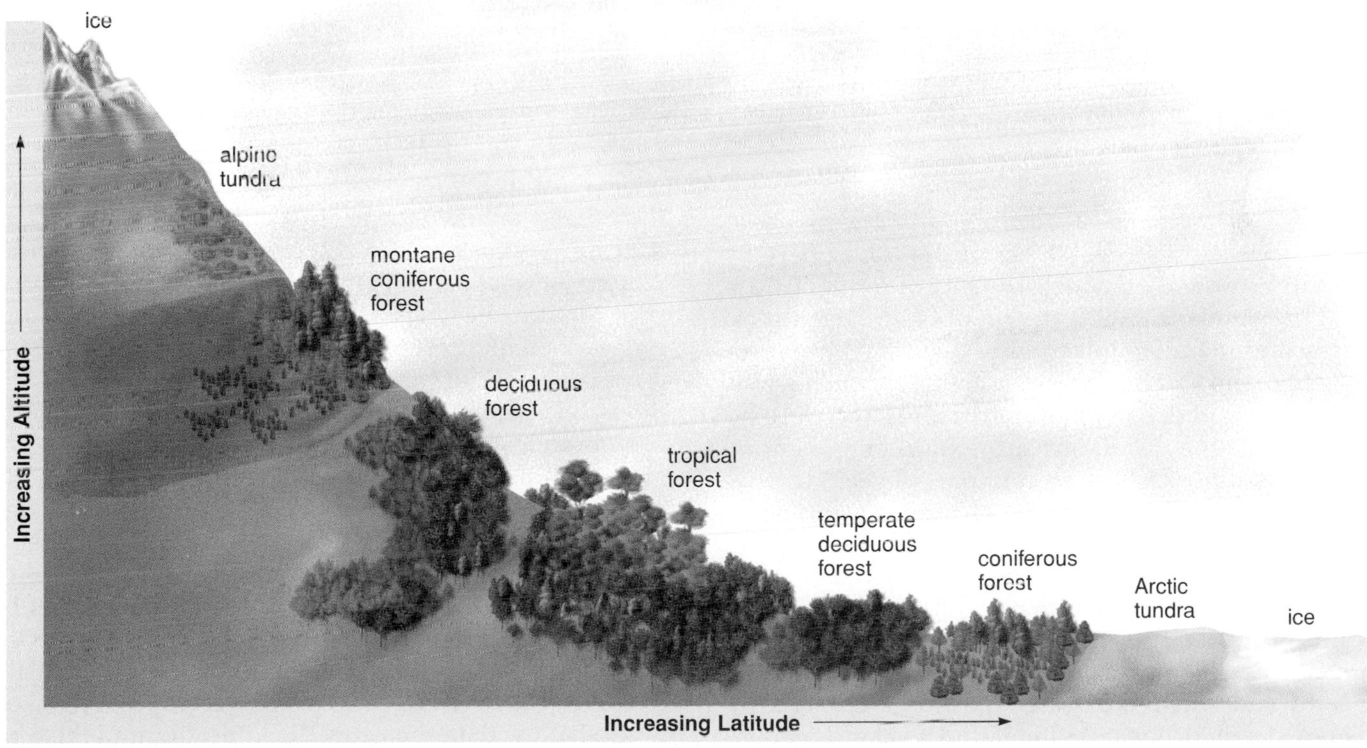

Figure 26.7 Climate and biomes.
Biomes change with altitude just as they do with latitude because vegetation is partly determined by temperature. Rainfall also plays a significant role, which is one reason why grasslands, instead of tropical or deciduous forests, are sometimes found at the base of mountains.

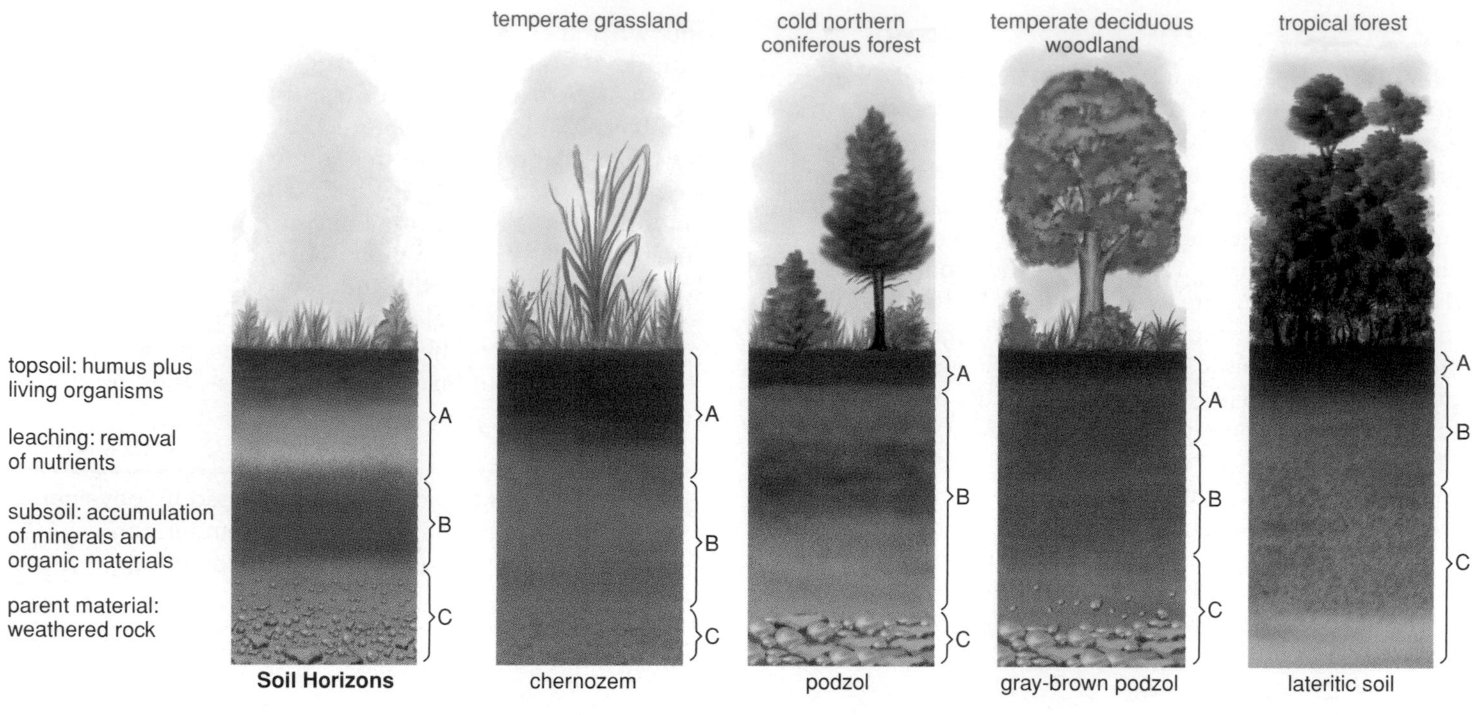

Figure 26.8 Soil types.
a. Soil horizons. The top layer (A horizon) contains most of the organic matter; the next layer (B horizon) accumulates inorganic nutrients as the water carries dissolved minerals downward; and the lowest layer (C horizon) is composed of weathered parent material. b. In temperate grassland soils (called chernozem soils), there is a deep A horizon, and the shallow B horizon does not have sufficient inorganic nutrients to support root growth. In temperate forest soils of the podzol type, the A horizon is shallow, and leaching results in many nutrients in the B horizon. In a tropical forest soil of the lateritic type, there is a reddish residue composed primarily of iron and aluminum oxides.

Soil Types

Soil is the uppermost surface layer of the lithosphere in which plants grow and on which, directly or indirectly, all terrestrial life depends. Soil is the product of the bedrock weathering and the reorganization of this weathered material by percolating water and biotic activities, such as the growth of trees and the work of detritivores and decomposers. *Humus* is the decomposed organic component of soil. It can take thousands, even millions, of years to produce rich topsoil, and therefore soil erosion is a very serious loss of a precious resource for ecosystems and agriculture.

Soil composition plays a major role in determining what kinds of plants, and therefore what kind of animals, are found in a particular region. Soil particles vary from coarse sand and silt to small clay particles. *Loam,* with roughly equal proportions of sand, silt, and clay, together with humus, is among the richest agricultural soils. Soils differ with regard to drainage. The speed with which gravity causes water to drain away correlates with the size of the soil pores, the spaces between the particles of the soil. Sandy soil has the largest pores, and therefore water drains rapidly away. Clay soil, which drains slowly, has the smallest pores because its particles are very tiny. You might think then that plants would grow well in clay, but this is not the case because pores also act as air spaces, and root cells require air in order to carry on cellular respiration.

Mature soil, which may reach to many meters in depth, is described in terms of three soil horizons (Fig. 26.8). The A horizon is the uppermost (or topsoil) layer that contains litter and humus, although most of the soluble chemicals have been leached (washed out). The B horizon has little or no organic matter but does contain the inorganic nutrients leached from A horizon. The C horizon is a layer of weathered and shattered rock. Soils formed in grasslands have a deep A horizon built up from decaying grasses over many years, but because of limited rain, there has been little leaching into the B horizon. In forest soils, both the A and B horizons have enough inorganic nutrients to allow for root growth. In tropical rain forests, the A horizon is more shallow than the generalized profile, and the B horizon is deeper, signifying that leaching is more extensive. Since the topsoil of rain forests lacks nutrients, it can only support crops for several years.

Soil type, determined in part by the amount of rainfall, influences the distribution of biomes.

26.3 Terrestrial Biomes

There are many major terrestrial biomes, and we will consider these: tundra, coniferous forests (taiga, temperate rain forest), temperate deciduous forest, tropical rain forest, savanna, temperate grasslands, shrubland, and desert.

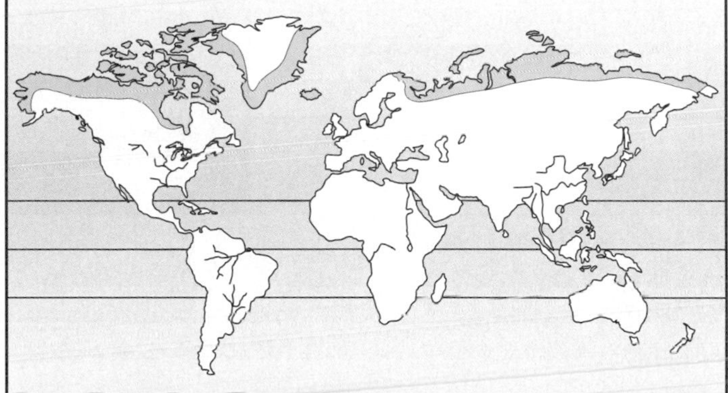

a.

b.

Figure 26.9 The tundra.
a. In this biome, which is nearest the polar regions, the vegetation consists principally of lichens, mosses, grasses, and low-growing shrubs. **b.** Pools of water that do not evaporate nor drain into the permanently frozen ground attract many birds that feed on the plentiful insects in the summer. **c.** Caribou, more plentiful in the summer than the winter, feed on lichens, grasses, and shrubs.

Tundra

The **Arctic tundra** biome, which encircles the earth just south of ice-covered polar seas in the Northern Hemisphere, covers about 20% of the earth's land surface (Fig. 26.9). (A similar community, called the alpine tundra, occurs above the timberline on mountain ranges.) The Arctic tundra is cold and dark much of the year. Because rainfall amounts to only about 20 cm a year, the tundra could possibly be considered a desert, but melting snow creates a landscape of pools and mires in the summer, especially because so little evaporates. Only the topmost layer of earth thaws; the **permafrost** beneath this layer is always frozen, and therefore, drainage is minimal.

Trees are not found in the tundra because the growing season is too short, their roots cannot penetrate the permafrost, and they cannot become anchored in the boggy soil of summer. In the summer, the ground is covered with short grasses and sedges, but there also are numerous patches of lichens and mosses. Dwarf woody shrubs, such as dwarf birch, flower and seed quickly while there is plentiful sun for photosynthesis.

A few animals live in the tundra year-round. For example, the mouselike lemming stays beneath the snow; the ptarmigan, a grouse, burrows in the snow during storms; and the musk ox conserves heat because of its thick coat and short, squat body. In the summer, the tundra is alive with numerous insects and birds, particularly shorebirds and waterfowl that migrate inland. Caribou and reindeer also migrate to and from the tundra, as do the wolves that prey upon them. Polar bears are common near the coast.

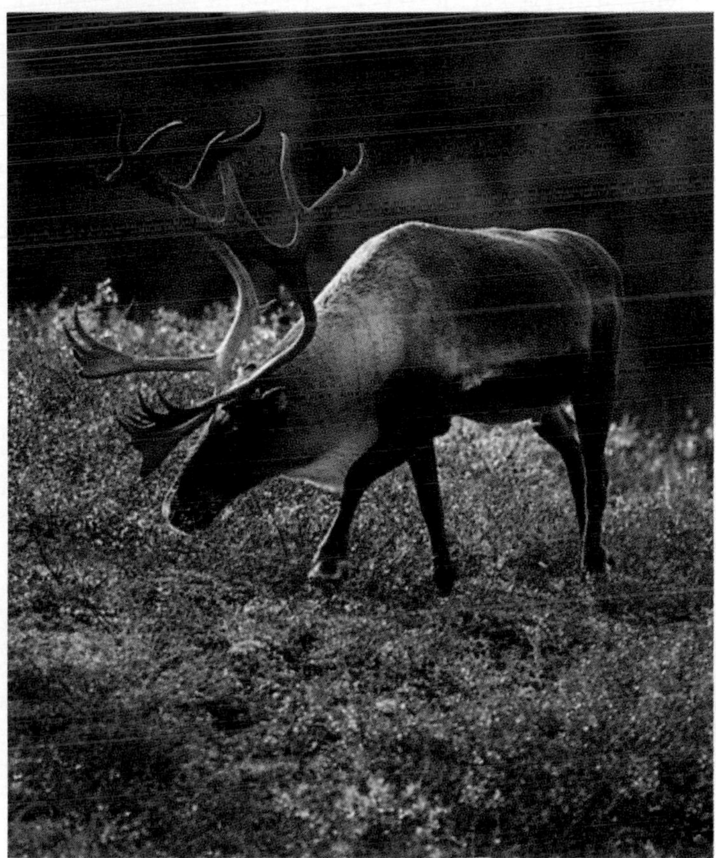

c.

Coniferous Forests

Coniferous forests are found in three locations: in the **taiga,** which extends around the world in the northern part of North America and Eurasia; near mountaintops (where it is called a montane coniferous forest), and also along the Pacific coast of North America, as far south as northern California.

The taiga (Fig. 26.10) typifies the coniferous forest with its cone-bearing trees, such as spruce, fir, and pine. These trees are well adapted to the cold because both the leaves and bark have thick coverings. Also, the needlelike leaves can withstand the weight of heavy snow. There is a limited understory of plants, but the floor is covered by low-lying mosses and lichens beneath the layer of needles. Birds harvest the seeds of the conifers, and bears, deer, moose, beavers, and muskrats live around the cool lakes and along the streams. Wolves prey on these larger mammals. A montane coniferous forest also harbors the wolverine and mountain lion.

The coniferous forest that runs along the west coast of Canada and the United States is sometimes called a **temperate rain forest.** The prevailing winds moving in off the Pacific Ocean lose their moisture when they meet the coastal mountain range. The plentiful rainfall along with a rich soil have produced some of the tallest conifer trees ever in existence, including the coastal redwoods. This forest is also called an old-growth forest because some trees are as old as 800 years. It truly is an evergreen forest because all trees are covered with mosses, ferns, and other plants that grow on their trunks. Whether the limited portion remaining should be preserved has been quite a controversy. Unfortunately, the controversy has centered around the northern spotted owl, which is endemic to this area. The actual concern is conservation of this particular ecosystem.

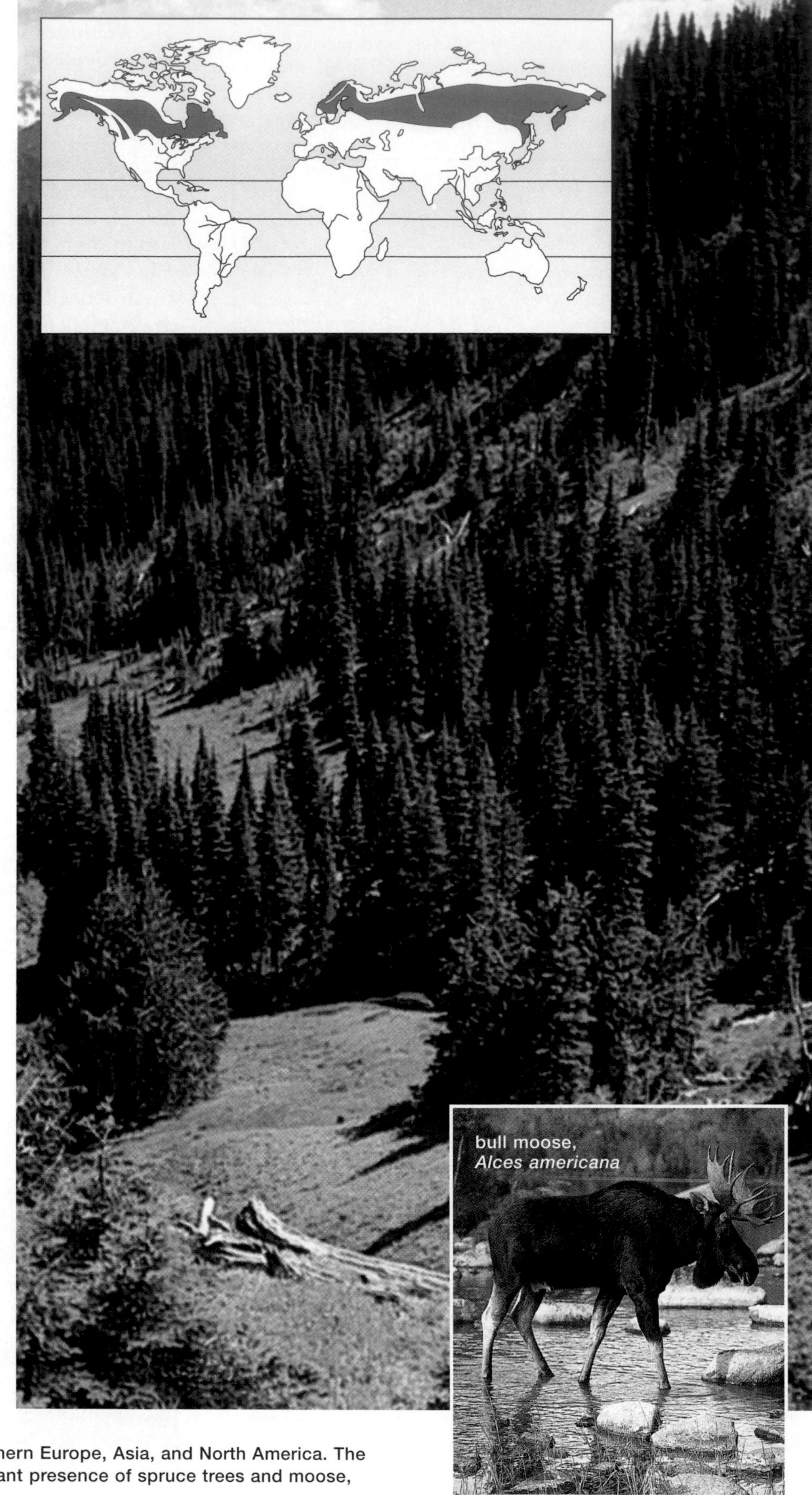

Figure 26.10 The taiga.
The taiga, which means swampland, spans northern Europe, Asia, and North America. The appellation "spruce-moose" refers to the dominant presence of spruce trees and moose, which frequent the ponds.

Temperate Deciduous Forests

Temperate deciduous forests are found south of the taiga in eastern North America (Fig. 26.11), eastern Asia, and much of Europe. The climate in these areas is moderate, with relatively high rainfall (75–150 cm per year). The seasons are well defined, and the growing season ranges between 140 and 300 days. The trees, such as oak, beech, and maple, have broad leaves and are termed deciduous trees; they lose their leaves in the fall and grow them in the spring.

The tallest trees form a canopy, an upper layer of leaves that are the first to receive sunlight. Even so, enough sunlight penetrates to provide energy for another layer of trees called understory trees. Beneath these trees are shrubs that may flower in the spring before the trees have put forth their leaves. Still another layer of plant growth—mosses, lichens, and ferns—resides beneath the shrub layer. This stratification provides a variety of habitats for insects and birds. Ground life is also plentiful. Squirrels, cottontail rabbits, shrews, skunks, woodchucks, and chipmunks are small herbivores. These and ground birds such as turkeys and grouse are preyed on by red foxes. The whitetail deer and black bears, which are herbivores, have increased in number of late. In contrast to the taiga, amphibians and reptiles are found in this biome because the winters are not as cold. Frogs and turtles prefer an aquatic existence, as do the beavers and muskrats, which are mammals.

Autumn fruits, nuts, and berries provide a supply of food for the winter, and the leaves, after turning brilliant colors and falling to the ground, contribute to the rich layer of humus. The minerals within the humus are washed far into the ground by the spring rains, but the deep tree roots capture these and bring them back up into the forest system again.

Figure 26.11 Temperate deciduous forest.
A temperate deciduous forest is home to many and varied plants and animals. Millipedes can be found among leaf litter; chipmunks feed on acorns; and bobcats prey on these and other small mammals.

Tropical Forests

In the **tropical rain forests** of South America, Africa, and the Indo-Malayan region near the equator, the weather is always warm (between 20° and 25°C), and rainfall is plentiful (with a minimum of 190 cm per year). This may be the richest biome, both in terms of number of different kinds of species and their abundance.

A tropical rain forest has a complex structure, with many levels of life (Fig. 26.12). Some of the broad-leaf evergreen trees grow from 15 to 50 meters or more. These tall trees often have trunks buttressed at ground level to prevent their toppling over. Lianas, or woody vines, that encircle the tree as it grows, also help to strengthen the trunk. The diversity of species is enormous—a 10-km^2 area of tropical rain forest may contain 750 species of trees and 1,500 species of flowering plants.

Although there is animal life on the ground (e.g., pacas, agoutis, peccaries, and armadillos), most animals live in the trees (Fig. 26.13). Insect life is so abundant that the majority of species have not been identified yet. Termites play a vital role in the decomposition of woody plant material, and ants are found everywhere, particularly in the trees. The various birds, such as hummingbirds, parakeets, parrots, and toucans, are often beautifully colored. Amphibians and reptiles are well represented by many types of frogs, snakes, and lizards. Lemurs, sloths, and monkeys are well-known primates that feed on the fruits of the trees. The largest carnivores are the big cats—the jaguars in South America and the leopards in Africa and Asia.

Many animals spend their entire life in the canopy, as do some plants. **Epiphytes** are air plants that grow on other plants but have roots of their own that absorb moisture and minerals leached from the canopy; others catch rain and debris in hollows produced by overlapping leaf bases. The most common epiphytes are related to pineapples, orchids, and ferns.

Figure 26.12 Tropical rain forest.
Levels of life in a tropical rain forest. Even the canopy (solid layer of leaves) has levels, and some organisms spend their entire life in one particular level. Long lianas (hanging vines) climb into the canopy, where they produce leaves. Epiphytes are air plants that grow on the trees but do not parasitize them.

While we usually think of tropical forests as being nonseasonal rain forests, there are tropical forests with wet and dry seasons in India, Southeast Asia, West Africa, South and Central America, the West Indies, and northern Australia. Here, there are deciduous trees, with many layers of growth beneath the trees. In addition to the animals just mentioned, certain of these forests also contain elephants, tigers, and hippopotamuses.

Whereas the soil of a temperate deciduous forest biome is rich enough for agricultural purposes, the soil of a tropical rain forest biome is not. Nutrients are cycled directly from the litter to the plants again. Productivity is high because of high temperatures, a yearlong growing season, and the rapid recycling of nutrients from the litter. (In humid tropical forests, iron and aluminum oxides occur at the surface, causing a reddish residue known as laterite. When the trees are cleared, laterite bakes in the hot sun to a bricklike consistency that will not support crops.) Swiden agriculture, often called slash-and-burn agriculture, is a type of agriculture that has been successful, but also destructive, in the tropics. Trees are felled and burned, and the ashes provide enough nutrients for several harvests. Thereafter, the forest must be allowed to regrow, and a new section must be cut and burned.

Figure 26.13 Animals of the tropical rain forest.

Shrublands

It is difficult to define a shrub, but in general shrubs are shorter than trees (4.5–6 meters) with a woody persistent stem, and no central trunk. Shrubs have small but thick evergreen leaves that often are coated with a waxy material that prevents loss of moisture from the leaves. Their thick underground roots survive the dry summers and frequent fires and take deep moisture from the soil. Shrubs are adapted to withstand arid conditions and can also quickly sprout new growth after a fire. As a point of interest, you will recall that a shrub stage is a part of the process of both primary and secondary succession.

Shrublands tend to occur along coasts that have dry summers and receive most of their rainfall in the winter. A shrubland is found along the cape of South Africa, the western coast of North America, and the southwest and southern shores of Australia, around the Mediterranean Sea, and in central Chile. The dense shrubland that occurs in California is known as *chaparral* (Fig. 26.14). This type of shrubland, called the Mediterranean type, lacks an understory and ground litter, and is highly flammable. The seeds of many species require the heat and scarring action of fire to induce germination. Other shrubs sprout from the roots after a fire.

There is also a northern shrub area that lies west of the Rocky Mountains. This area is sometimes classified as a cold desert, but the region is dominated by sagebrush and other hardy plants. Some of the birds found here are dependent upon sagebrush for their existence.

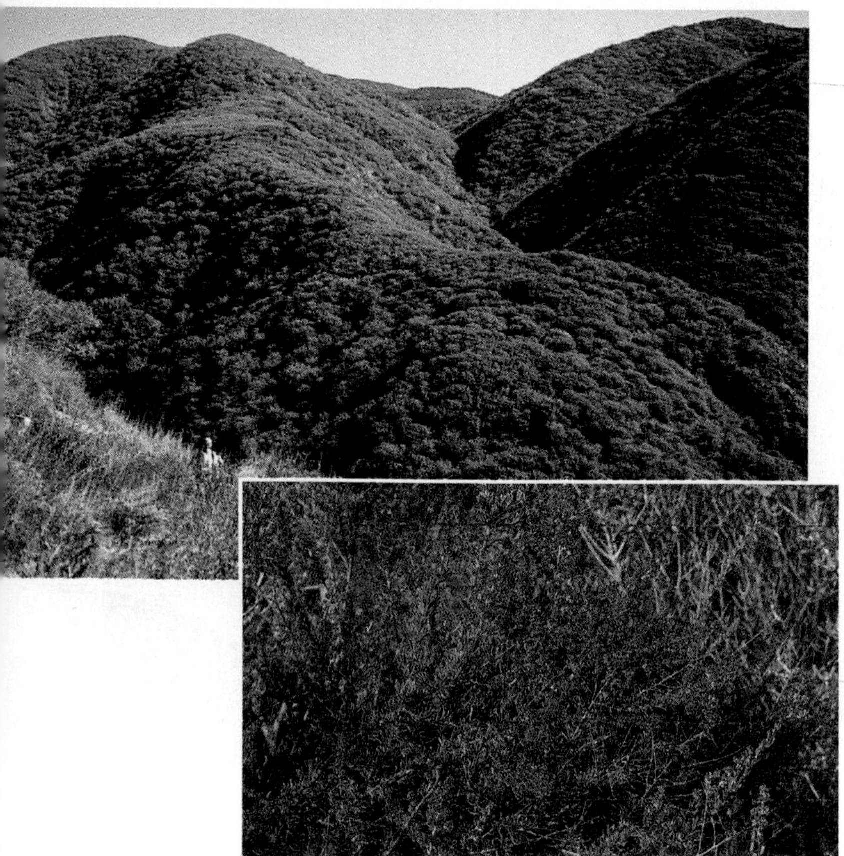

Figure 26.14 Shrubland.
Shrublands, such as chaparral in California, are subject to raging fires, but the shrubs are adapted to quickly regrow.

Grasslands

Grasslands occur where rainfall is greater than 25 cm but is generally insufficient to support trees. Natural grasslands once covered more than 40% of the earth's land surface, but many areas that once were grasslands are now used for the cultivation of crops, such as wheat and corn. In temperate areas, where rainfall is between 10 and 30 inches a year, grasslands occur. Here, it is too dry for forests and too wet for deserts to form.

The grasses are well adapted to a changing environment and can tolerate a high degree of grazing, flooding, drought, and sometimes fire. Where rainfall is high, large tall grasses that reach more than 2 meters in height (e.g., pampas grass) can flourish. In drier areas, shorter grasses between 5 and 10 cm are dominant. Low-growing bunch grasses (e.g., grama grass) grow in the United States near deserts. Grasses also generally grow in different seasons; some grassland animals migrate, and ground squirrels hibernate, when there is little grass for them to eat.

The temperate grasslands include the Russian steppes, the South American pampas, and the North American prairies (Fig. 26.15). When traveling across the United States from east to west, the line between the temperate deciduous forest and a **tall-grass prairie** is roughly along the border between Illinois and Indiana. The tall-grass prairie requires more rainfall than does the **short-grass prairie** that occurs near deserts. Large herds of bison—estimated at hundreds of thousands—once roamed the prairies, as did herds of pronghorn antelope. Now, small mammals, such as mice, prairie dogs, and rabbits, typically live belowground, but usually feed aboveground. Hawks, snakes, badgers, coyotes, and foxes feed on these mammals. Virtually all of these grasslands, however, have been converted to agricultural lands.

Savannas, which are grasslands that contain some trees, occur in regions where a relatively cool dry season is followed by a hot, rainy one (Fig. 26.16). One tree that can survive the severe dry season is the flat-topped acacia, which sheds its leaves during a drought. The African savanna supports the greatest variety and number of large herbivores of all the biomes. Elephants and giraffes are browsers that feed on tree vegetation. Antelopes, zebras, wildebeests, water buffalo, and rhinoceroses are grazers that feed on grasses. Any plant litter that is not consumed by grazers is attacked by a variety of small organisms, among them termites. Termites build towering nests in which they tend fungal gardens, their source of food. The herbivores support a large population of carnivores. Lions and hyenas hunt in packs, cheetahs hunt singly by day, and leopards hunt singly by night.

American bison,
Bison bison

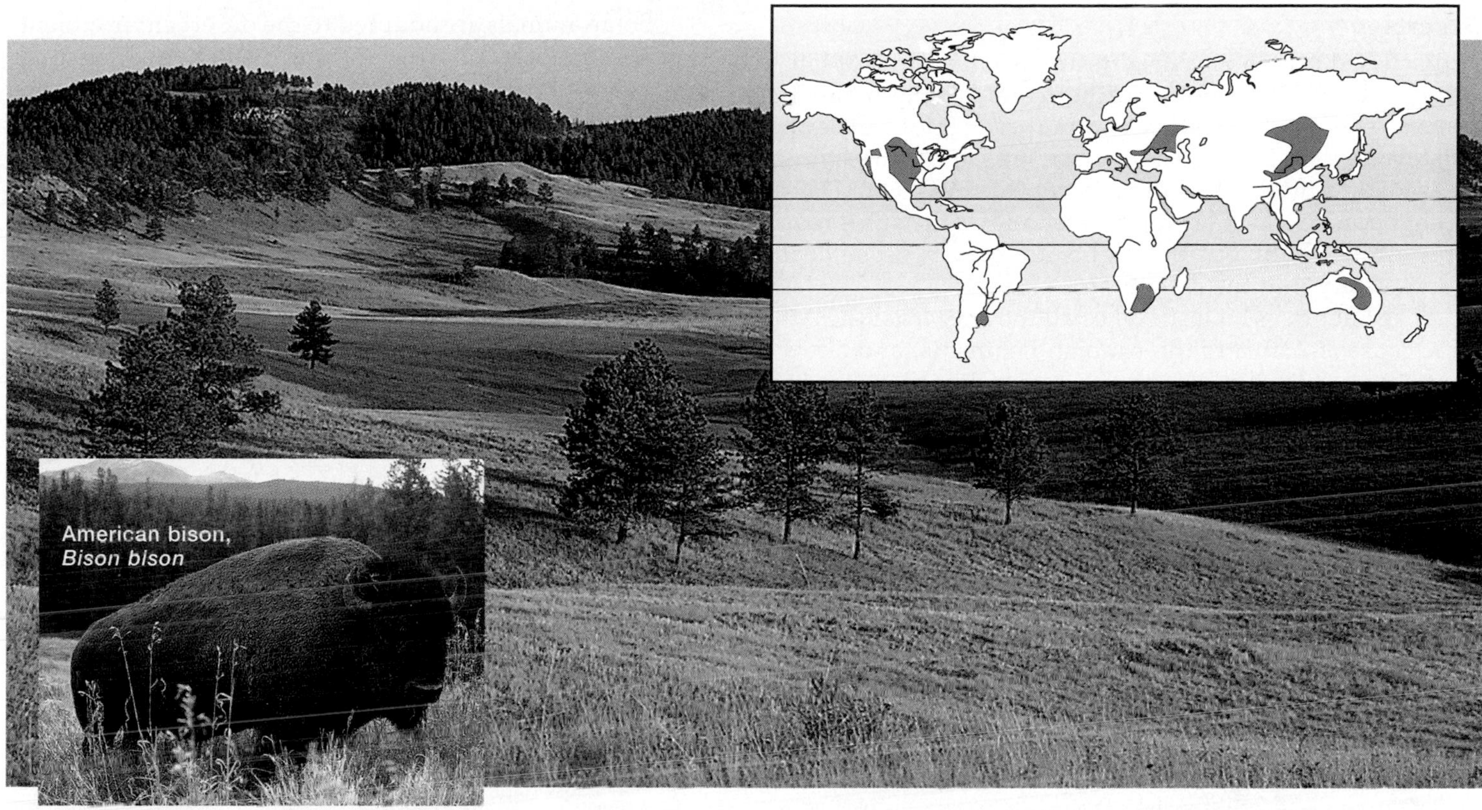

Figure 26.15 The prairie.
Tall-grass prairies are seas of grasses dotted by pines and junipers. Bison, once abundant, are now being reintroduced into certain areas.

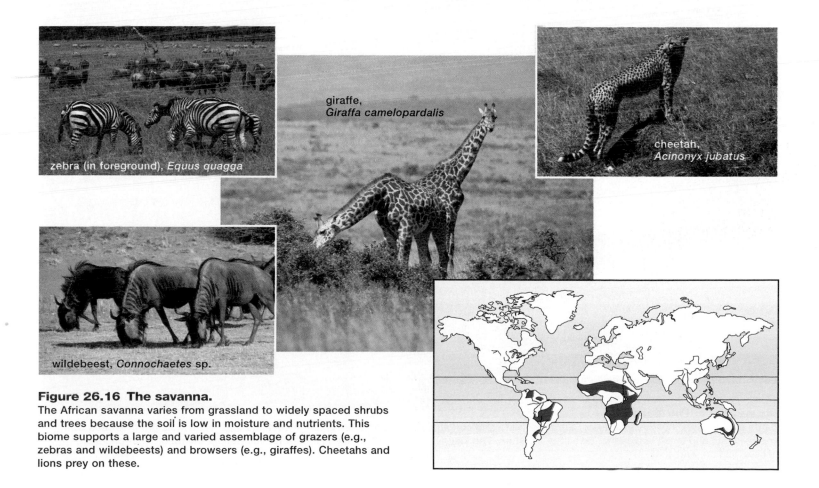

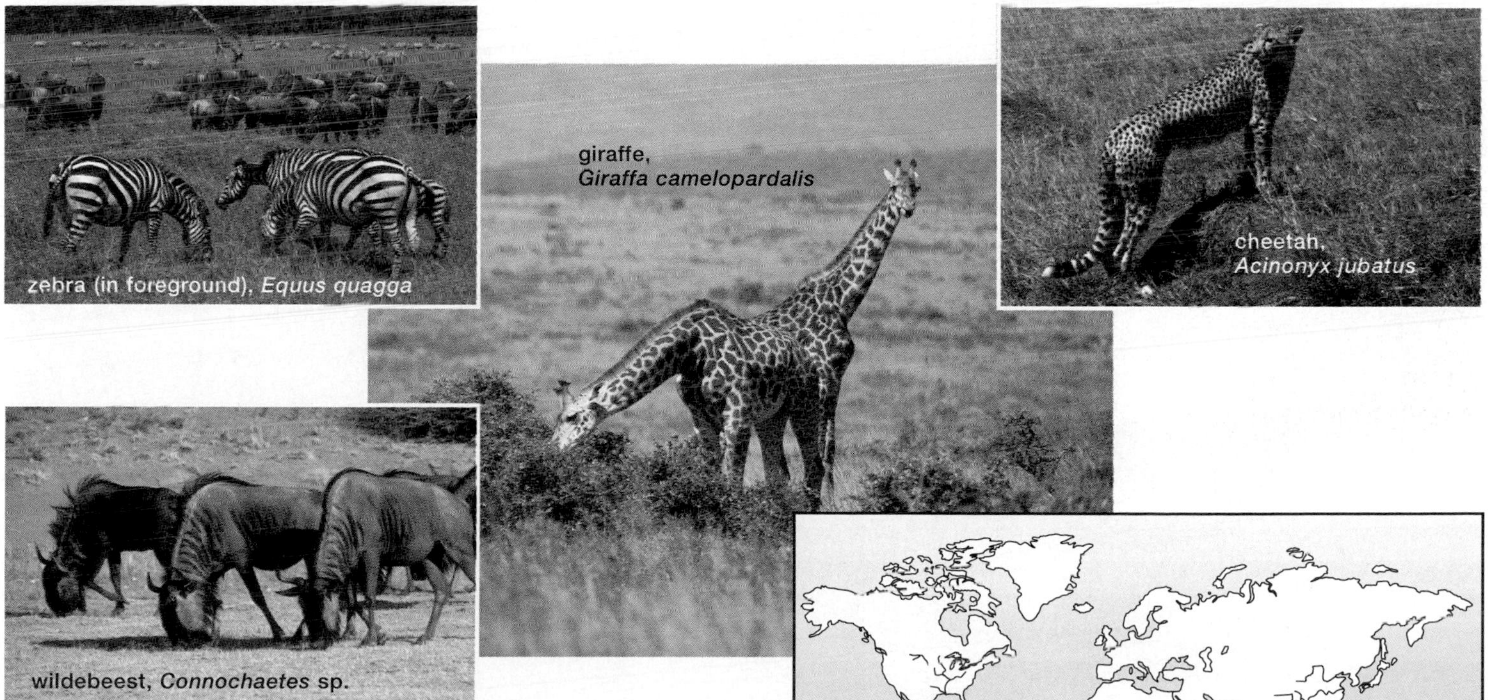

zebra (in foreground), *Equus quagga*
giraffe,
Giraffa camelopardalis
cheetah,
Acinonyx jubatus
wildebeest, *Connochaetes* sp.

Figure 26.16 The savanna.
The African savanna varies from grassland to widely spaced shrubs and trees because the soil is low in moisture and nutrients. This biome supports a large and varied assemblage of grazers (e.g., zebras and wildebeests) and browsers (e.g., giraffes). Cheetahs and lions prey on these.

Deserts

As discussed previously, **deserts** are usually found at latitudes of about 30°, in both Northern and Southern Hemispheres. The winds that descend in these regions pick up moisture before they rise again (see Fig. 26.2). Therefore, the annual rainfall is less than 25 cm (see Fig. 26.6). Days are hot because a lack of cloud cover allows the sun's rays to penetrate easily, but the nights are cold because heat escapes easily into the *atmosphere*.

The Sahara, which stretches all the way from the Atlantic coast of Africa to the Arabian Peninsula, and a few other deserts have little or no vegetation. But most have a variety of plants (Fig. 26.17). The best-known desert perennials in North America are the succulent, spiny-leafed cacti, which have stems that store water and carry on photosynthesis. Also common are nonsucculent shrubs, such as the many-branched sagebrush with silvery gray leaves and the spiny-branched ocotillo that produces leaves during wet periods and sheds them during dry periods.

Some animals are adapted to the desert environment. Reptiles and insects have waterproof outer coverings that conserve water. A desert has numerous insects, which pass through the stages of development from pupa to the next pupa again when there is rain. Reptiles, especially lizards and snakes, are perhaps the most characteristic group of vertebrates found in deserts, but running birds (e.g., the roadrunner) and rodents (e.g., the kangaroo rat) are also well known (Fig. 26.17). Larger mammals, like the coyote, prey on the rodents, as do the hawks.

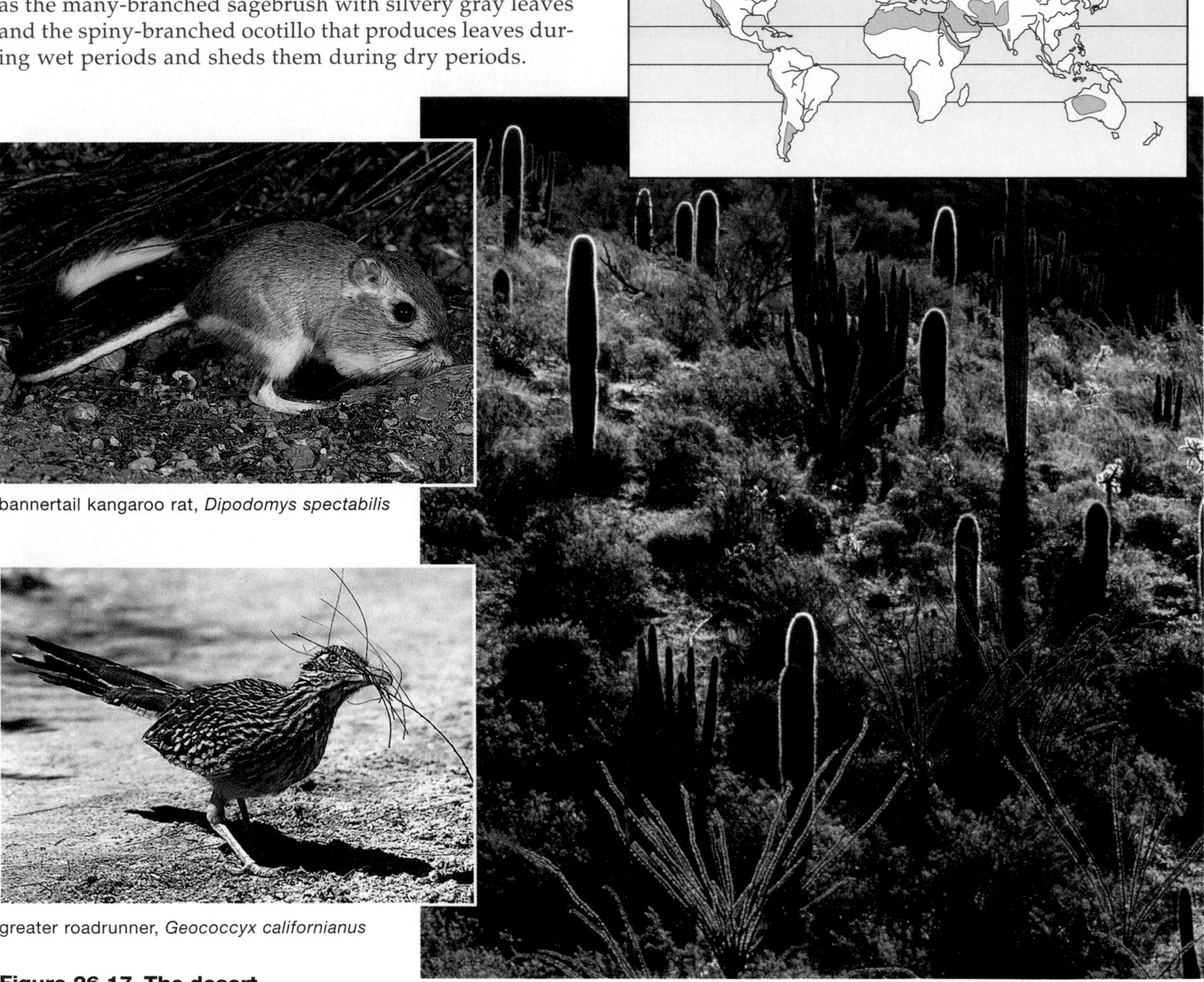

bannertail kangaroo rat, *Dipodomys spectabilis*

greater roadrunner, *Geococcyx californianus*

Figure 26.17 The desert.
Plants and animals that live in a desert are adapted to arid conditions. The plants are either succulents that retain moisture or shrubs with woody stems and small leaves that lose little moisture. The kangaroo rat feeds on seeds and other vegetation; the roadrunner preys on insects, lizards, and snakes.

26.4 Aquatic Biomes

Aquatic biomes are classified as two types: fresh water (inland) or salt water (usually marine). Wetlands that lie near the sea have mixed fresh and salt water, called brackish water. Figure 26.18 shows how these communities are joined physically. They also interact by sharing nutrients and by the biogeochemical cycles. Consideration of the hydrologic cycle (p. 435) shows that fresh water is a distillate of salt water. As the sun's rays cause seawater to evaporate, the salts are left behind. The vaporized fresh water rises into the atmosphere, cools, and falls as rain either over the ocean or over the land. A lesser amount of water also evaporates from and returns to the land. Since land lies above sea level, gravity eventually returns all fresh water to the sea, but in the meantime, it is contained within standing waters (**lakes** and **ponds**), flowing waters (**streams** and **rivers**), and groundwater.

When rain falls, some of the water sinks or percolates into the ground and saturates the earth to a certain level. The top of the saturation zone is called the groundwater table, or simply the water table. Wherever the earth contains basins or channels, water will appear to the level of the water table. The water within basins is called lakes and ponds, and the water within channels is called streams or rivers. Sometimes groundwater is also located in underground rivers called aquifers.

Humans have the habit of channeling aboveground rivers and filling in wetlands (lands that are wet for at least part of the year). These activities degrade ecosystems and eventually cause seasonal flooding. Wetlands provide food and habitats for fish, waterfowl, and other wildlife. They also purify waters by filtering them and by diluting and breaking down toxic wastes and excess nutrients. Wetlands directly absorb storm waters and also absorb overflows from lakes and rivers. In this way they protect farms, cities, and towns from the devastating effects of floods. There are now federal and local laws for the protection of wetlands, but they are not always enforced.

Aquatic biomes can be classified as fresh water or salt water. However, it is important to realize that the two sets of communities interact and are joined by biogeochemical cycles.

stonefly larva, *Plecoptera* sp.

red banded trout, *Salmo gairdneri*

carp, *Cyprinus carpio*

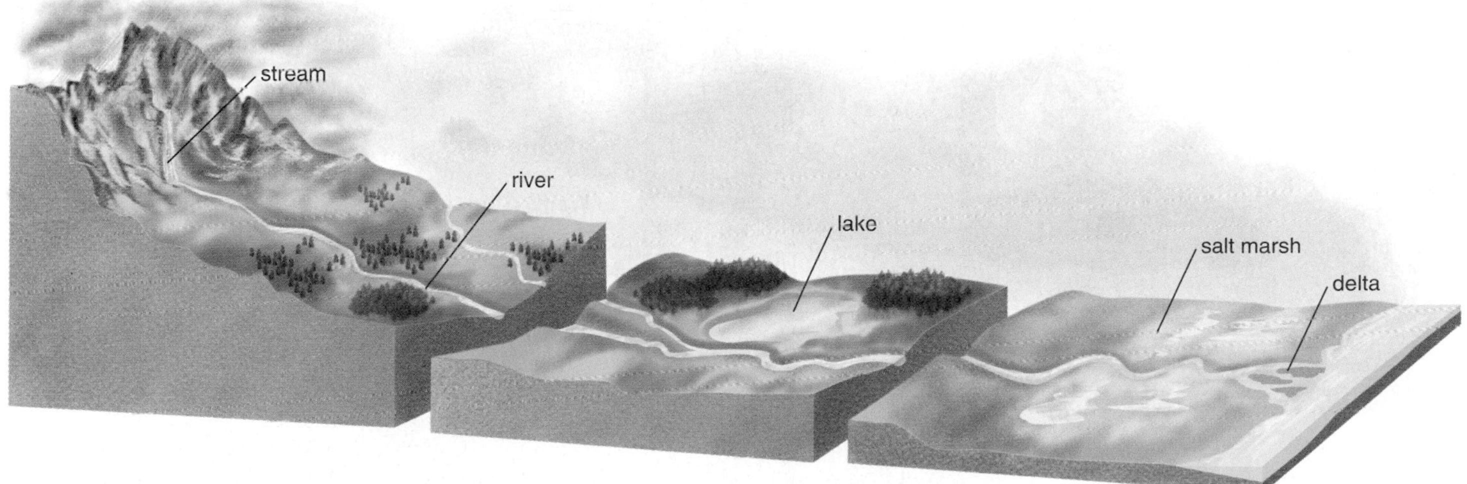

Figure 26.18 Streams and rivers.
Mountain streams have cold, clear water that flows over waterfalls and rapids. The feet of this long-legged stonefly insect larva are clawed, helping it to hold on to stones. Trout are found in occasional pools of the highly oxygenated water. As the streams merge, a river forms that gets increasingly wider and deeper until it meanders across broad, flat valleys. Carp are adapted to water that contains little oxygen and has much sediment. At its mouth, a river may divide into many channels where wetlands and estuaries are located.

Lakes

Lakes are bodies of fresh water often classified by their nutrient status. Oligotrophic (nutrient-poor) lakes are characterized by low organic matter and low productivity. Eutrophic (nutrient-rich) lakes are characterized by high organic matter and high productivity. Such lakes are usually situated in naturally nutrient-rich regions or are enriched by agricultural or urban and suburban runoffs. Oligotrophic lakes can become eutrophic through large inputs of nutrients (Fig. 26.19). This process is called **eutrophication.**

In the temperate zone, deep lakes are stratified in the summer and winter. In summer, lakes in the temperate zone have three layers of water that differ in temperature (Fig. 26.20). The surface layer, the epilimnion, is warm from solar radiation; the middle thermocline experiences an abrupt drop in temperature; and the hypolimnion is cold. These differences in temperature prevent mixing. The warmer, less dense water of the epilimnion "floats" on top of the colder, more dense water of the hypolimnion.

As the season progresses, the epilimnion becomes nutrient-poor, while the hypolimnion begins to be depleted of oxygen. The phytoplankton found in the sunlit epilimnion use up nutrients as they photosynthesize. Photosynthesis releases oxygen, giving this layer a ready supply. Detritus naturally falls by gravity to the bottom of the lake, and here oxygen is used up as decomposition occurs. Decomposition releases nutrients, however.

In the fall, as the epilimnion cools, and in the spring, as it warms, a turnover occurs. In the fall, the upper epilimnion waters become cooler than the hypolimnion waters. This causes the surface water to sink and the deep water to rise. The **fall overturn** continues until the temperature is uniform throughout the lake. At this point, wind aids in the circulation of water so that mixing occurs. Eventually, oxygen and nutrients become evenly distributed.

As winter approaches, the water cools. Ice formation begins at the top, and the ice remains there because ice is less dense than cool water. Ice has an insulating effect, preventing further cooling of the water below. This per-

a.

b.

Figure 26.19 Types of lakes.
Lakes can be classified according to whether they are (**a**) oligotrophic (nutrient-poor) or (**b**) eutrophic (nutrient-rich). Eutrophic lakes tend to have large populations of algae and rooted plants, resulting in a large population of decomposers that use up much of the oxygen, leaving little oxygen for fishes.

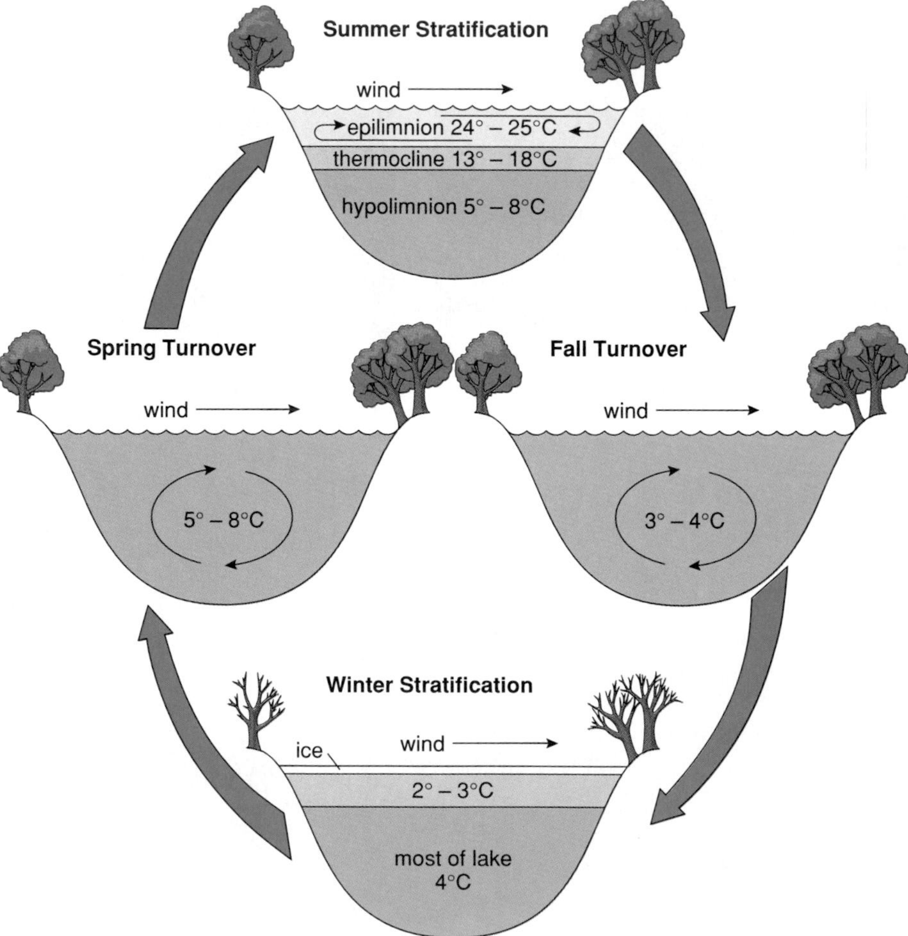

Figure 26.20 Lake stratification.
Temperature profiles of a large oligotrophic lake in a temperate region vary with the season. During spring and fall turnover, the deep waters receive oxygen from surface waters, and surface waters receive nutrients from deep waters.

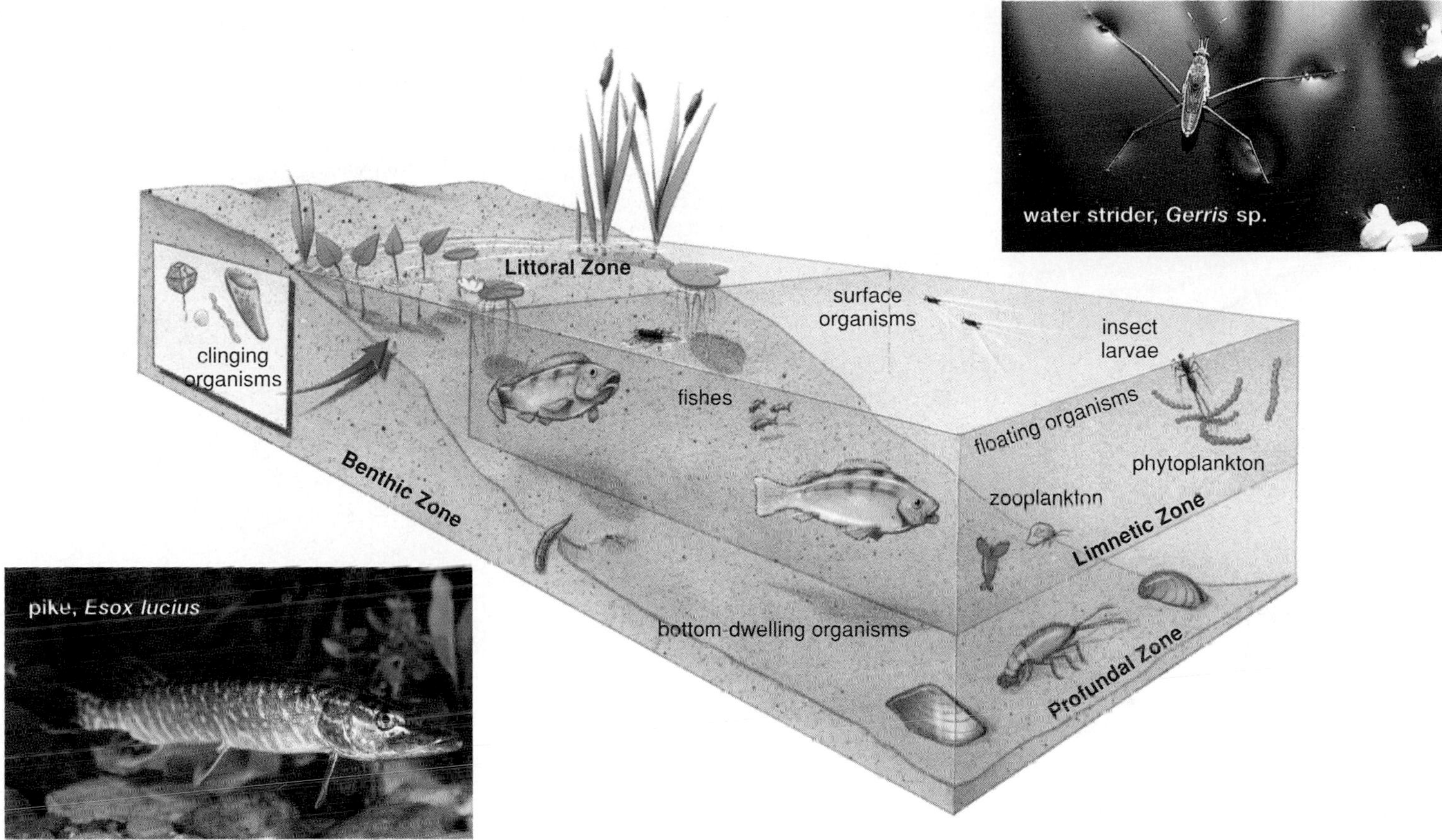

Figure 26.21 Zones of a lake.
Rooted plants and clinging organisms live in the littoral zone. Phytoplankton, zooplankton, and fishes are in the sunlit limnetic zone. Water striders stand on the surface film of water with water-repellent feet. Pike are top carnivores prized by fishermen. Bottom-dwelling organisms, like crayfishes and clams, are in the profundal zone.

mits aquatic organisms to live through the winter in the water beneath the surface of the ice.

In the spring, as the ice melts, the cooler water on top sinks below the warmer water on the bottom. The **spring overturn** continues until the temperature is uniform through the lake. At this point, wind aids in the circulation of water as before. When the surface waters absorb solar radiation, thermal stratification occurs once more.

This vertical stratification and seasonal change of temperatures in a lake basin influence the season distribution of fish and other aquatic life in the lake basin. For example, coldwater fish move to the deeper water in summer and inhabit the upper water in winter. In the fall and spring just after mixing occurs, phytoplankton growth at the surface is most abundant.

Life Zones

In both fresh and salt water, free-drifting microscopic organisms, called **plankton** [Gk. *planktos,* wandering], are important components of the community. **Phytoplankton** [Gk. *phyton,* plant, and *planktos,* wandering] are photosynthesizing plantlike algae that only become noticeable when they reproduce to the extent that a green scum or red tide appears on the water. **Zooplankton** [Gk. *zoon,* animal, *planktos,* wandering] are animals that feed on the phytoplankton. Lakes and ponds can be divided into several life zones. The *littoral zone* is closest to the shore, the *limnetic zone* forms the sunlit body of the lake, and the *profundal zone* is below the level of light penetration (Fig. 26.21). The *benthic zone* is at the soil-water interface. Aquatic plants are rooted in the shallow littoral zone of a lake, and various microscopic organisms cling to these plants and to rocks. Some organisms such as the water strider live at the water-air interface and can literally walk on water. In the limnetic zone, small fishes, such as minnows and killifish feed on plankton and also serve as food for large fishes. In the profundal zone, there are zooplankton and fishes such as whitefish that feed on debris that falls from above. Pike species are "lurking predators." They wait among vegetation around the margins of lakes and surge out to catch passing prey.

A few insect larvae are in the limnetic zone, but they are far more prominent in both the littoral and profundal zones. Midge larvae and ghost worms are common members of the benthos. The benthos are animals that live on the bottom in the benthic zone. In a lake, the benthos include crayfish, snails, clams, and various types of worms and insect larvae.

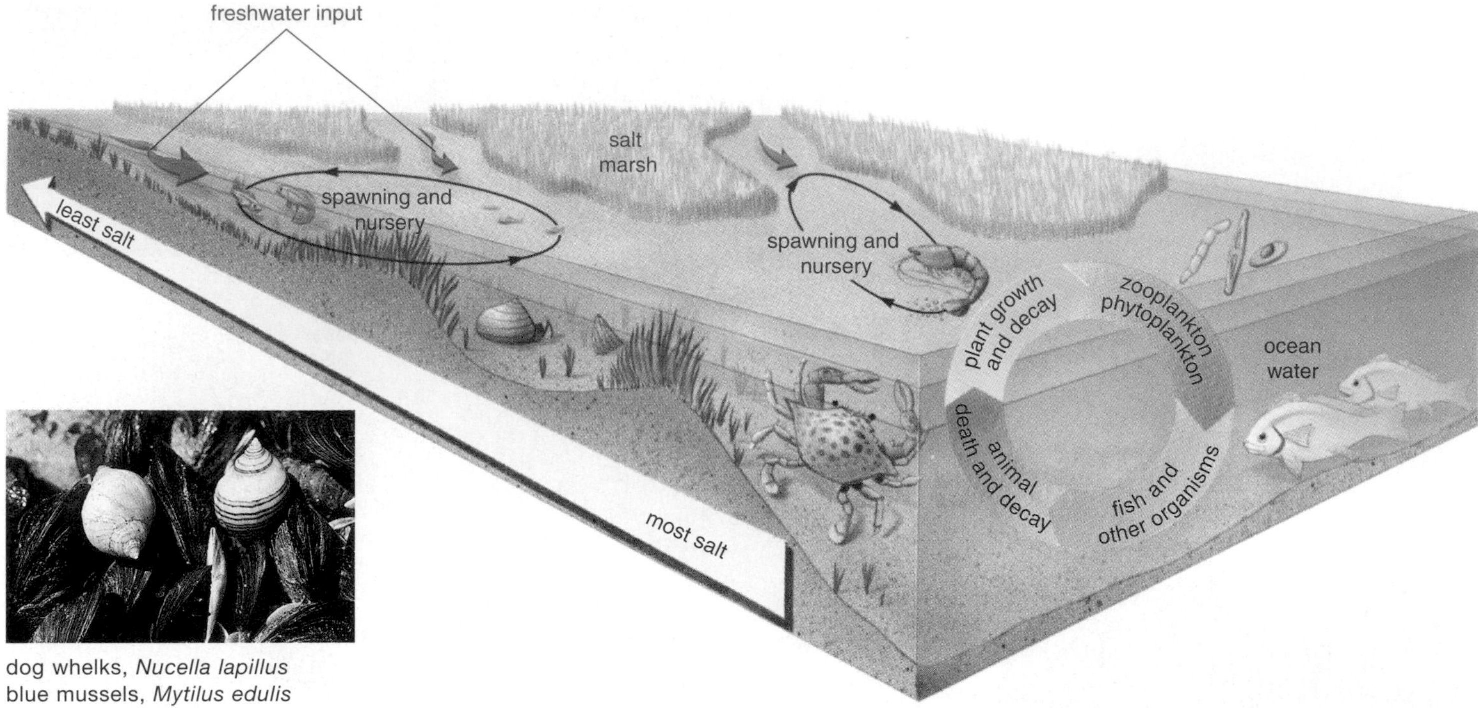

Figure 26.22 Estuary structure and function.
Since an estuary is located where a river flows into the ocean, it receives nutrients from land and also from the sea by way of the tides. Decaying grasses also provide nutrients. Estuaries serve as a nursery for the spawning and rearing of the young for many species of fishes, shrimp, and crustacea, and mollusks such as the blue mussels, and their predator, dog whelks, depicted in the photograph.

Coastal Communities

Near the mouth of a river, a *salt marsh* in the temperate zone and a *mangrove swamp* in the subtropical and tropical zones are likely to develop. Also, the silt carried by a river may form mudflats. It is proper to think of seacoasts and mudflats, salt marshes, and mangrove swamps as belonging to one ecological system.

Estuaries

An **estuary** is a partially enclosed body of water where fresh water and seawater meet and mix (Fig. 26.22). A river brings fresh water into the estuary, and the sea, because of the tides, brings salt water. Coastal bays, tidal marshes, fjords (an inlet of water between high cliffs), some deltas (triangular-shaped areas of land at the mouths of rivers), and lagoons (a body of water separated from the sea by a narrow strip of land) are all examples of estuaries.

Organisms living in an estuary must be able to withstand constant mixing of waters and rapid changes in salinity. Not many organisms are suited to this environment, but for those that are suited, there is an abundance of nutrients. An estuary acts as a nutrient trap because the tides bring nutrients from the sea and at the same time prevent the seaward escape of nutrients brought by the river. Because of this, estuaries produce much organic food.

Although only a few small fish permanently reside in an estuary, many develop there so that there is always an abundance of larval and immature fish. It has been estimated that well over half of all marine fishes develop in the protective environment of an estuary, which explains

a.

b.

Figure 26.23 Types of estuaries.
Many types of regions qualify as estuaries such as the salt marsh depicted in Figure 26.22 and **(a)** mudflats, which are frequented by migrant birds, and **(b)** mangrove swamps skirting the coastlines of many tropical and subtropical lands. The tangled roots of mangrove trees trap sediments and nutrients that sustain many immature forms of sea life.

Figure 26.24 Seacoasts.
a. The littoral zone of a rocky coast, where the tide comes in and out, has different types of shelled and algal organisms at its upper, middle, and lower portions. **b.** Some organisms of a rocky coast live in tidal pools. **c.** A sandy shore looks devoid of life except for the birds that feed there. However, a number of invertebrate species burrow in the sand only during the day or all the time.

why estuaries are called the *nurseries of the sea.* Estuaries are also feeding grounds for many birds, fish, and shellfish because they offer a ready supply of food.

Salt marshes dominated by salt marsh cordgrass are often associated with estuaries. Salt marshes contribute nutrients and detrital material to estuaries in addition to being an important wildlife habitat, a few of which are shown in Figure 26.23.

Seashores

Both rocky and sandy shores are constantly bombarded by the sea as the tides roll in and out (Fig. 26.24). A rocky seashore is divided into zones. The littoral zone is covered and uncovered daily by tides. In the upper portion of the littoral zone, barnacles are glued so tightly to the stone by their own secretions that their calcareous outer plates remain in place even after the enclosed shrimplike animal dies. In the midportion of the littoral zone, brown algae known as rockweed may overlie the barnacles. In the lower portions of the littoral zone, oysters and mussels attach themselves to the rocks by filaments called *byssus threads.* Limpets are snails, as are periwinkles. But periwinkles have a coiled shell and secure themselves by hiding in crevices or under seaweeds, while limpets press their single flattened cone tightly to a rock. Finally, below the littoral zone, macroscopic *seaweeds,* which are the main photosynthesizers, anchor themselves to the rocks by holdfasts.

Organisms cannot attach themselves to shifting, unstable sands on a *sandy beach;* therefore, nearly all the permanent residents dwell underground. They either burrow during the day and surface to feed at night, or they remain permanently within their burrows and tubes. Ghost crabs and sandhoppers (amphipods) burrow themselves above the high tide mark and feed at night when the tide is out. Sandworms and sand (ghost) shrimp remain within their burrows in the **intertidal zone** and feed on detritus whenever possible. Still lower in the beach, clams, cockles, and sand dollars are found.

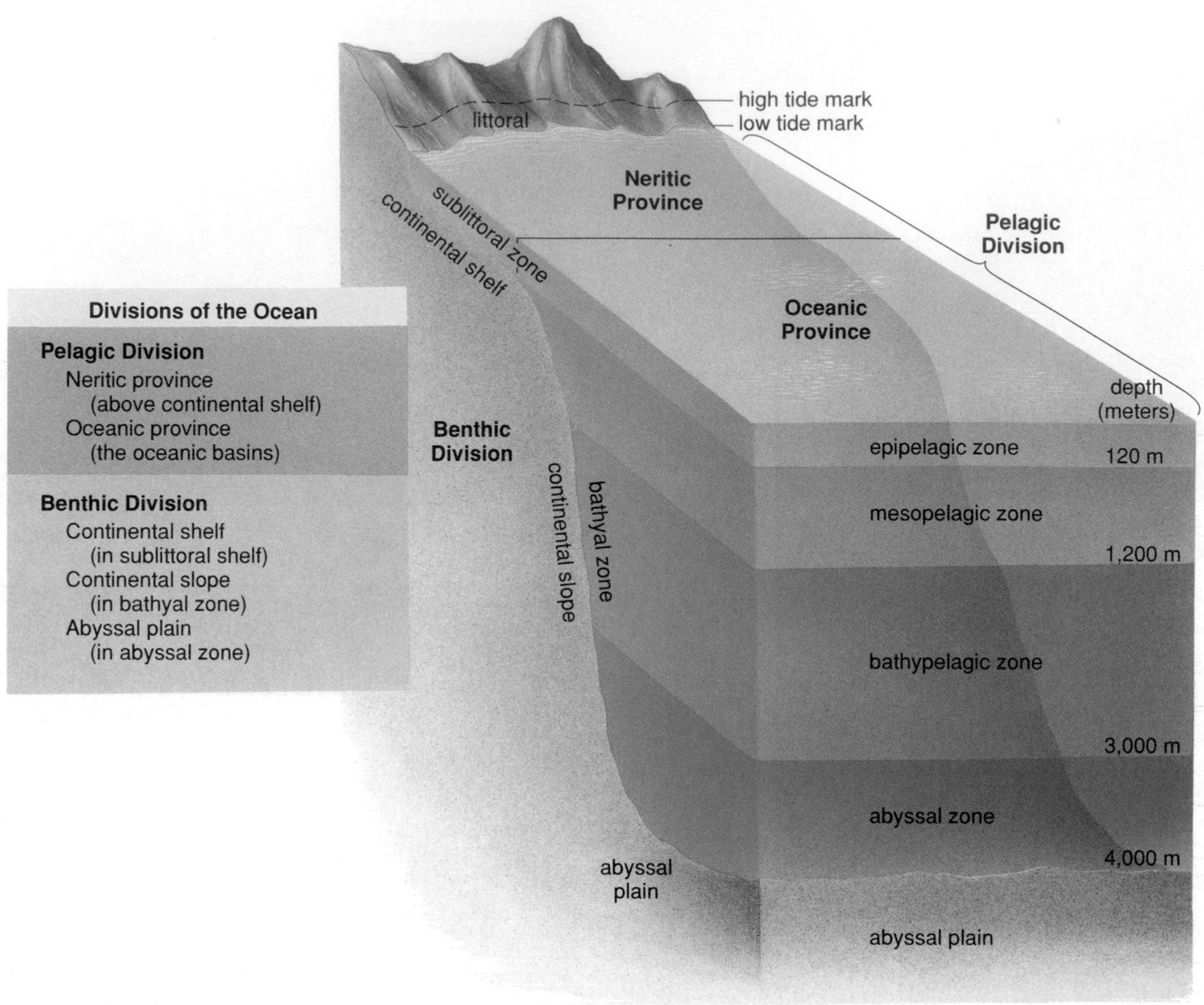

Figure 26.25 Marine environment.
Organisms reside in the pelagic division (blue), where waters are divided as indicated. Organisms also reside in the benthic division (brown), with surfaces divided and in the zones indicated.

Oceans

The oceans cover approximately three-quarters of our planet. The geographic areas and zones of an ocean are shown in Figure 26.25. It is customary to place the organisms of the oceans into either the pelagic division (open waters) or the benthic division (ocean floor).

Pelagic Division

The **pelagic division** [Gk. *pelagos*, the sea] includes the neritic province and the oceanic province. There is a greater concentration of organisms in the neritic province than in the oceanic province because sunlight penetrates the waters of the former, and this is where you find the most nutrients. Phytoplankton consisting of suspended algae is food not only for zooplankton but also for small fishes. These small fishes in turn are food for commercially valuable fishes—herring, cod, and flounder.

In the oceanic province, the epipelagic zone, although it is sunlit, does not have a high concentration of phytoplankton because it lacks nutrients. These photosynthesizers, however, still support a large assembly of zooplankton because the oceans are so large. The zooplankton are food for herrings and bluefishes, which in turn are eaten by larger mackerels, tunas, and sharks. Flying fishes, which glide above the surface, are preyed upon by dolphins (not to be confused with porpoises, which also are present). Whales are other mammals found in the epipelagic zone. Baleen whales strain krill (small crustacea) from the water, and the toothed sperm whales feed primarily on the common squid.

Animals in the mesopelagic zone are carnivores, are adapted to the absence of light, and tend to be translucent, red colored, or even luminescent. There are luminescent shrimps, squids, and fishes, such as lantern and hatchet fishes.

The bathypelagic zone is in complete darkness except for an occasional flash of bioluminescent light. Carnivores and scavengers are found in this zone. Strange-looking fishes with distensible mouths and abdomens and small, tubular eyes feed on infrequent prey.

Benthic Division

The **benthic division** [Gk. *benthos,* depths of the sea] includes organisms that live on or in the soil of the continental shelf, the continental slope, and the abyssal plain (Fig. 26.26). These are the organisms of the sublittoral, bathyal, and abyssal zones.

Seaweed grows in the sublittoral zone, and it can be found in batches on outcroppings as the water gets deeper. Nearly all benthic organisms, however, are dependent on the slow rain of plankton and detritus from the sunlit waters above. There is more diversity of life in the sublittoral and bathyal zones than in the abyssal zone. In these first two zones, clams, worms, and sea urchins are preyed upon by starfishes, lobsters, crabs, and brittle stars.

The abyssal zone is inhabited by animals that live at the soil-water interface on the dark abyssal plain (Fig. 26.26). It once was thought that few animals exist in this zone because of the intense pressure and the extreme cold. Yet many invertebrates live here by feeding on debris floating down from the mesopelagic zone. Sea lilies rise above the seafloor; sea cucumbers and sea urchins crawl around on the sea bottom; and tube worms burrow in the mud.

The flat abyssal plain is interrupted by enormous underwater mountain chains called oceanic ridges. Along the axes of the ridges, crustal plates spread apart and molten magma rises to fill the gap. At **hydrothermal vents,** seawater percolates through cracks and is heated to about 350°C, causing sulfate to react with water and form hydrogen sulfide (H_2S). The water temperature-hydrogen sulfide combination has been found to support communities that contain huge tube worms and clams. Chemosynthetic bacteria that obtain energy from oxidizing hydrogen sulfide live within these organisms. It was a surprise to find communities of organisms living so deep in the ocean, where light never penetrates. Unlike photosynthesis, chemosynthesis does not require light energy.

The neritic province and the epipelagic zone of the oceanic province receive sunlight and contain the organisms with which we are most familiar. The organisms of the benthic division are dependent upon debris that floats down from above.

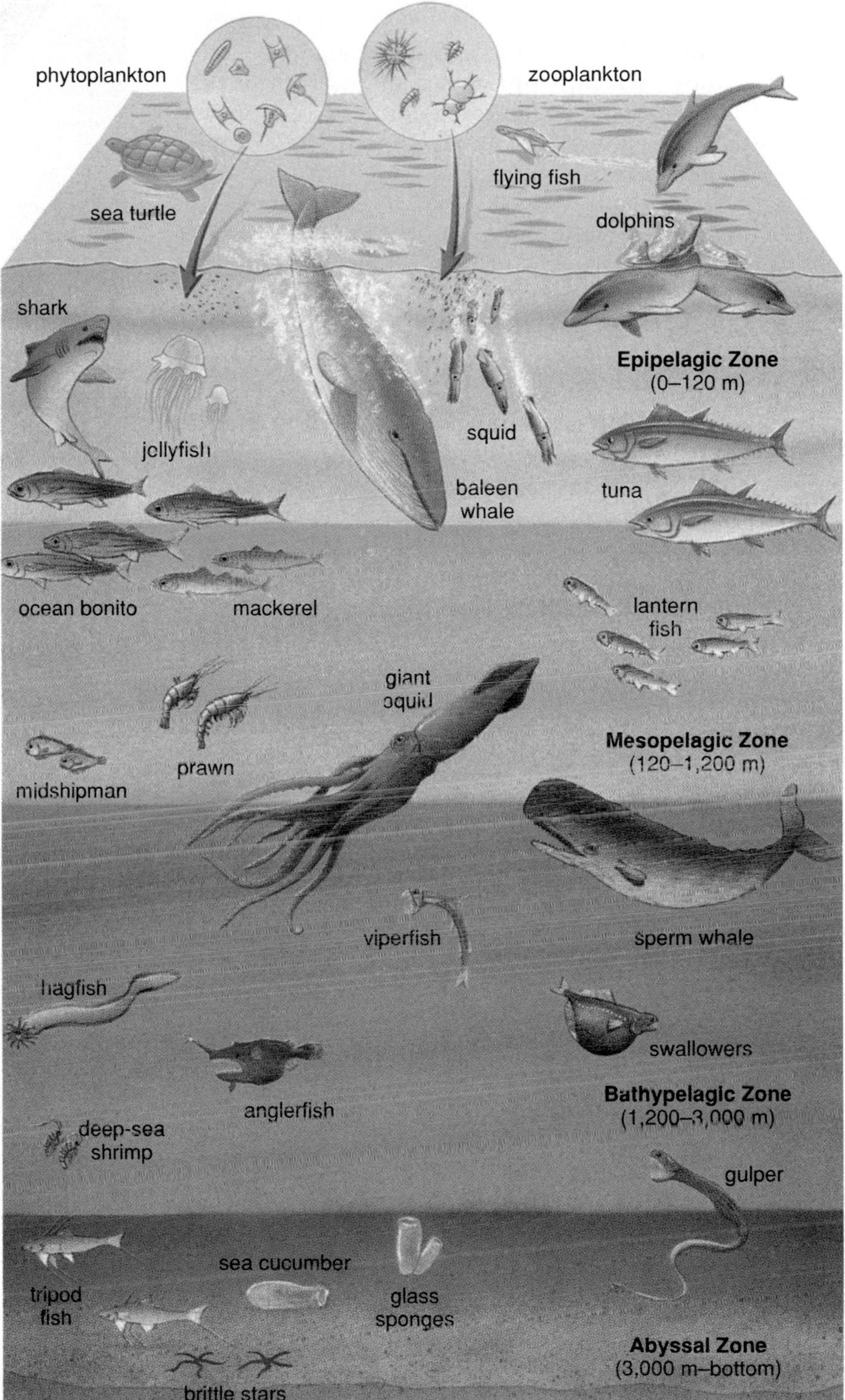

Figure 26.26 Pelagic division.
Organisms of the epipelagic, mesopelagic, and bathypelagic zones are shown. The abyssal zone is a part of the benthic division.

Coral Reefs

Coral reefs (Fig. 26.27) are areas of biological abundance found in shallow, warm, tropical waters just below the surface of the water. Their chief constituents are stony corals, animals that have a calcium carbonate (limestone) exoskeleton, and calcareous red and green algae. Corals do not usually occur individually; rather, they form colonies derived from an individual coral that has reproduced by means of budding. Corals provide a home for a microscopic alga called zooxanthellae. The corals, which feed at night, and the algae, which photosynthesize during the day, are mutualistic and share materials and nutrients. The close relationship of corals and zooxanthellae may be the reason why coral reefs form only in shallow sunlit water—the algae utilize sunlight for photosynthesis. Closely related to corals, sea anemones—the flowers of the sea with tentacles that bear some resemblance to petals—are also plentiful.

A reef is densely populated with life perhaps because reefs are areas of intermediate disturbance (p. 422). The large number of crevices and caves provide shelter for filter feeders (sponges, sea squirts, and fan worms) and for scavengers (crabs and sea urchins). The barracuda, moray eel, and sharks are top predators in coral reefs (Fig. 26.27*b*). There are many types of small beautifully colored fishes. Parrot fishes feed directly on corals, and others feed on plankton or detritus. Small fishes become food for larger fishes like snappers that are caught for human consumption.

Coral reefs are subject the world over to overfishing and oceanic pollution. The crown-of-thorns starfish feeds on coral polyps, and the very existence of the Great Barrier Reef off the coast of Australia is threatened by a plague of these animals. The giant triton is a natural predator, and some believe that the crown-of-thorns is proliferating because humans have killed off the triton for its handsome spiral shell.

Coral reefs, which are areas of biological abundance, occur in tropical seas.

a.

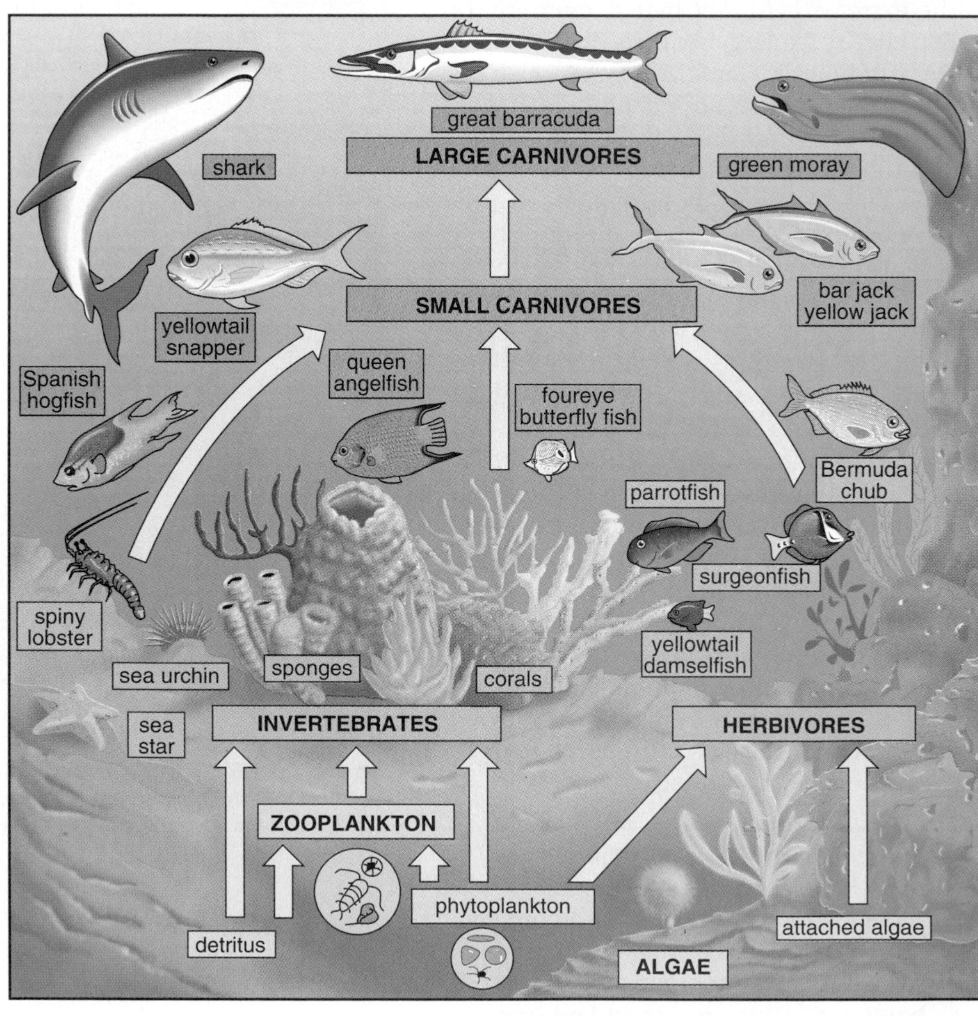

b.

Figure 26.27 Coral reefs.
A coral reef is (a) a unique community of marine organisms. **b.** The abundance of life results in a complex food web—a simplified representation is given here.

connecting concepts

The biosphere is the product of interaction of living organisms with the earth's physical and chemical environment over billions of years. Life has affected the atmosphere, modifying its composition and influencing global climate. Organisms have helped create the chemical and physical conditions of streams, lakes, and the oceans. The soils of terrestrial ecosystems and the sediments of aquatic ecosystems are structured largely by activities of organisms. Reef-building plants and animals have helped build the islands on which millions of humans live. The Earth's diverse biomes also result from interactions of the biota and the abiotic environment.

Over geological time, the biosphere has been changing constantly. The biomes of the age of dinosaurs were strikingly different from those of today. Astrophysical events have triggered some of this change. Changes in the sun's radiation output and in the tilt of the earth's axis have altered the pattern of solar energy reaching the earth's surface. Geological processes have also modified conditions for life. The drifting of continents has changed the arrangement of continents and oceans. Mountain ranges have been thrust up and eroded down. Through these changing conditions, life has evolved, and the structure of the earth's biomes has evolved as well. In the last few million years, humans appeared and learned to exploit the earth's biomes.

Humans have transformed vast areas of many of the terrestrial biomes into farmland, cities, highways, and other developments. Through our resource use and release of pollutants, we have now become an agent of global importance. Yet we still depend on the biodiversity that exists in the earth's biomes, and on the interactions of other organisms within the biosphere. These interactions still influence climate, patterns of nutrient cycling and waste processing, and basic biological productivity. The earth's biotic diversity also provides enjoyment and inspiration to millions of people, who spend billions of dollars to visit coral reefs, deserts, rain forests, and even the Arctic tundra. Unfortunately, as we shall see in the next chapter, many of the earth's biomes may not survive in their present form for our descendants' benefit. Although most of us value earth's diversity, we are causing patterns of global change that are adversely affecting every biome on earth.

Summary

26.1 Climate and the Biosphere

Because the earth is a sphere, the sun's rays are spread out over a larger area than the vertical rays at the equator. The temperature at the surface of the earth therefore decreases from the equator to each pole. The earth is tilted on its axis, and the seasons change as the earth rotates annually about the sun.

Warm air rises near the equator and loses its moisture and then descends at about 30° north and south latitude, and so forth to the poles. When the air descends, it picks up moisture from the land. The great deserts of the world are formed at 30° latitudes. Because the earth rotates on its axis daily, the winds blow in opposite directions above and below the equator. The winds pull ocean waters, setting up currents that move in a circular fashion from the equator. The Gulf Stream is an ocean current that brings warmth to northern Atlantic landmasses.

Topography also plays a role in the distribution of moisture. Air rising over coastal ranges loses its moisture on the windward side, making the leeward side arid. The fact that the oceans hold heat to a greater degree than land causes the monsoon climate in southern Asia.

26.2 Biomes of the World

The term biome is used to refer to terrestrial and aquatic communities. Biomes are distributed according to climate; that is, temperature and rainfall influence the pattern of biomes about the world. The significance of temperature is also seen by observing that the same sequence of biomes is seen when traveling to northern latitudes as traveling up a mountain. Biomes blend one into the other, and there is no actual demarcation between them.

Soil types vary according to the biome. Grasslands have a deep A horizon, which contains organic nutrients, and forests have more of a B horizon, which contains leached inorganic nutrients.

26.3 Terrestrial Biomes

The tundra is the northern-most biome and consists largely of short grasses and sedges and dwarf woody plants. Because of the cold winters and short summers, most of the water in the soil is frozen the year round. This is called the permafrost.

The taiga, a coniferous forest, has less rainfall than other types of forests. The temperate deciduous forest has trees that gain and lose their leaves because of the alternating seasons of summer and winter. Tropical rain forests are the most complex and productive of all biomes.

Shrublands usually occur along coasts that have dry summers and receive most of their rainfall in the winter. Among grasslands, the savanna, a tropical grassland, supports the greatest number of different types of large herbivores. The prairie, found in the United States, has a limited variety of vegetation and animal life.

Deserts are characterized by a lack of water—they are usually found in places with less than 25 cm a year. Some plants, such as cacti, are succulents, and others are shrubs with thick leaves they often lose during dry periods.

26.4 Aquatic Biomes

Aquatic biomes are divided into freshwater communities (lakes, streams, rivers) and marine communities. Rain falls in mountains, and streams develop that join to form a river that runs into the sea. Streams, rivers, and wetlands are different communities.

In deep lakes of the temperate zone, the temperature and the concentration of nutrients and gases in the water vary with depth. In the fall, the water on the surface cools and sinks to the bottom. In the spring, ice on the surface water melts, and this cool water sinks. The entire body of water is therefore cycled twice a year, distributing nutrients from the bottom layers. Lakes and ponds have three life zones. Rooted plants and clinging organisms live in the littoral zone, plankton and fishes are in the sunlit limnetic zone, and bottom-dwelling organisms like crayfishes and mollusks are in the profundal zone.

Estuaries (e.g., salt marshes, mudflats, mangrove forests) are near the mouth of a river. Estuaries are the nurseries of the sea.

Marine communities are divided into coastal communities and the oceans. The coastal communities, especially estuaries, are more productive than the oceans.

An ocean is divided into the pelagic division and the benthic division. The pelagic division (open waters) has three zones. The epipelagic zone receives adequate sunlight and supports the most life. The mesopelagic zone contains organisms adapted to minimum and no light, respectively. The benthic division (ocean floor) includes the organisms that live on the continental shelf in the sublittoral zone, the continental slope in the bathyal zone, and the abyssal plain in the abyssal zone.

Coral reefs are extremely productive communities found in shallow tropical waters. The intermediate disturbance theory is believed to account for the extreme diversity, which results in a complex food web.

Reviewing the Chapter

1. Tell how a spherical earth and the path of the earth about the sun affect climate. 444–45
2. Describe the air circulation about the earth, and tell why deserts are apt to occur at 30° north and south of the equator. Why does the Pacific coast of California get plentiful rainfall? 444–45
3. Describe the ocean currents and how the Gulf Stream accounts for Great Britain having a mild temperature. 446
4. How does a coastal range affect climate? What causes a monsoon climate? 447
5. Name the terrestrial biomes you would expect to find when going from the base to the top of a mountain. 448–49
6. What are the soil horizons, and how do they differ in forests and grasslands? 450
7. Describe the location, the climate, and the populations of the Arctic tundra, coniferous forests (both taiga and temperate rain forest), temperate deciduous forests, tropical forests, shrubland, grasslands (both prairie and savanna), and deserts. 451–56, 458
8. Describe the turnover of a temperate lake and life zones of a lake, and the organisms you would expect to find in each zone. 460–61
9. Describe the coastal communities, and discuss the importance of estuaries to the productivity of the ocean. 462–63
10. Describe the zones of the open ocean and the organisms you would expect to find in each zone. 464–65
11. Describe a coral reef including the varied organisms found there. 466

Testing Yourself

Choose the best answer for each question.

1. The seasons are best explained by
   a. the distribution of temperature and rainfall in biomes.
   b. the tilt of the earth on its axis as it rotates about the sun.
   c. the daily rotation of the earth on its axis.
   d. the fact that the equator is warm and the poles are cold.
2. The mild climate of Great Britain is best explained by
   a. the winds called the westerlies.
   b. the spinning of the earth on its axis.
   c. Great Britain being a mountainous country.
   d. the flow of ocean currents.
3. The location of deserts at 30° is best explained by which statement?
   a. Warm air rises and loses its moisture; cool air descends and becomes warmer and drier.
   b. Ocean currents give up heat as they flow from the equator to the poles.
   c. Cool ocean breezes cool the coast during the day.
   d. All of these are correct.
4. Which of these is mismatched?
   a. tundra—permafrost
   b. savanna—acacia trees
   c. prairie—epiphytes
   d. coniferous forest—evergreen trees
5. All of these phrases describe the tundra except
   a. low-lying vegetation.
   b. northernmost biome.
   c. short growing season.
   d. many different types of species.
6. The forest with a multilevel understory is the
   a. tropical rain forest.
   b. coniferous forest.
   c. tundra.
   d. temperate deciduous forest.
7. All of these phrases describe a tropical rain forest except
   a. nutrient-rich soil.
   b. many arboreal plants and animals.
   c. canopy composed of many layers.
   d. broad-leaved evergreen trees.

8. Which type of animal would you be least likely to find in a grassland biome?
   a. hoofed herbivore
   b. active carnivore
   c. arboreal primate
   d. flying insect
9. Phytoplankton are more likely to be found in which life zone of a lake?
   a. limnetic zone
   b. profundal zone
   c. benthic zone
   d. All of these are correct.
10. An estuary acts as a nutrient trap because of the
    a. action of rivers and tides.
    b. depth at which photosynthesis can occur.
    c. amount of rainfall received.
    d. height of the water table.
11. Which area of an ocean has the greatest concentration of nutrients?
    a. epipelagic zone and benthic zone
    b. epipelagic zone only
    c. benthic zone only
    d. neritic province
12. Use plus signs (4 + is maximum) to indicate the degree of temperature and the amount of rainfall for these biomes.

	Temperature	Rainfall
a. tropical rain forest		
b. savanna		
c. taiga		
d. tundra		

Applying the Concepts

1. *The productivity of communities varies.*
   Use Figure 26.25 to explain why a neritic province is more productive but more polluted than the oceanic province.
2. *The productivity of communities varies.*
   Why is the productivity of the ocean equivalent to that of a desert?
3. *The gross and net primary productivity of communities varies.*
   Explain why you might see this sequence of communities from the base to the top of a mountain: temperate forest, coniferous forest, tundra.

Using Technology

Your study of the biosphere is supported by these available technologies:

Exploring the Internet
The Mader Home Page provides resources for and help with studying this chapter.

http://www.mhhe.com/sciencemath/biology/mader/
(Click on Biology.)

Understanding the Terms

alpine tundra 449
Arctic tundra 451
benthic division 465
biome 449
biosphere 444
climate 444
coral reef 466
desert 458
El Niño 446
epiphyte 454
estuary 462
eutrophication 460
fall overturn 460
hydrosphere 446
hydrothermal vent 465
intertidal zone 463
lake 460
monsoon 447
montane coniferous forest 449
pelagic division 464
permafrost 451
phytoplankton 461
plankton 461
rain shadow 447
savanna 456
short-grass prairie 456
shrubland 456
soil 450
spring overturn 461
stream 459
taiga 452
tall-grass prairie 456
temperate deciduous forest 453
temperate rain forest 452
tropical rain forest 454
upwelling 446
zooplankton 461

Match the terms to these definitions:

a. __________ End of a river where fresh water and salt water mix as they meet.
b. __________ Freshwater and marine organisms that are suspended on or near the surface of the water.
c. __________ Major terrestrial community characterized by certain climatic conditions and dominated by particular types of plants.
d. __________ Ocean floor, which supports a unique set of organisms in contrast to the pelagic division.
e. __________ Open portion of the sea.
f. __________ Terrestrial biome that is a coniferous forest extending in a broad belt across northern Eurasia and North America.
g. __________ Plant that takes its nourishment from the air because its placement in other plants gives it an aerial position.
h. __________ Terrestrial biome that is a grassland in Africa, characterized by few trees and a severe dry season.
i. __________ Zone of air, land, and water at the surface of the earth in which living organisms are found.
j. __________ Mixing process that occurs in spring in stratified lakes whereby the oxygen-rich top waters mix with nutrient-rich bottom waters.

Human Impact on the Global Environment

CHAPTER 27

Chapter Concepts

Diesel fumes from buses in Yosemite National Park

Interactions between the atmosphere, oceans, and land follow natural cycles whose transfer rates have been fairly steady for some time. Now a very large human population is altering these transfer rates and bringing about global environmental changes at a rate and scale that is unmatched in the history of the planet Earth. It could be said that we humans are conducting an experiment whose results are uncertain at this time.

We are dependent on the health of the biosphere because we depend on the abiotic environment and other organisms to provide us with pure water, clean air, food, fiber, medicines, and many other necessities of life. Therefore, we need to consider the possible outcomes of our impact on the biosphere and take steps to prevent or at least minimize the possible degradative effects to all living things. Extinction of nonhuman species is on the increase and has spurred the origination of conservation biology, a new discipline that seeks to address the present biodiversity crisis.

27.1 Human Population and Industrialization

For the first several million years, the number of humans on earth was modest, and their impact on the environment was local in scope. But particularly since the industrial revolution in the eighteenth century, the exponential growth of the human population has been dramatic. Modern mechanized agricultural methods and improved medical care have contributed greatly to a growth rate that now adds 90 million people per year. According to the United Nations, the human population size will reach 10 billion or more by the end of the twenty-first century (Fig. 27.1*a*).

The countries of the world fall into two groups. The so-called more-developed countries (MDCs) became industrialized first. These countries include the United States, Canada, Japan, Russia, Australia, New Zealand, and all the countries of Europe. Industrialization requires resource consumption: energy and minerals are needed to create the lifestyle enjoyed by the peoples of these industrialized countries. The MDCs have 1.2 billion persons, which is only 22% of the world's population, but they account for about 80% of the world's energy and mineral resource consumption. The use of energy to transform raw materials into the goods that make a modern lifestyle possible results in heat and wastes, some of which, like carbon dioxide, alter the transfer rates of the biogeochemical cycles of the biosphere. We don't yet know for certain how this will affect the workings of the biosphere, but certainly change will occur.

Most of the human population increase of the twenty-first century will be generated by the less-developed countries (LDCs) (Fig. 27.1*a*). These countries like those in Latin America, Africa, and Asia are so called because they are not yet fully industrialized. The LDCs contain 78% of the world's population, but they account for only 20% of the world's energy and mineral consumption. Most peoples in the LDCs have a lifestyle that is far below that of the majority of people in the MDCs. An improved standard of living acts as a deterrent to an increase in population size; however, suppose all the people now present in the LDCs strove to have a lifestyle comparable to that of the MDCs? The strain placed on the *carrying capacity* of the environment might be so great that environmental degradation would cause the world population to suffer a crash.

Presently the environmental impact of both the MDCs and the LDCs is about equal as we can realize by examining the follow formula:

Number of people × Number of units of resources used per person × Environmental degradation and pollution per unit of resource used = Environmental impact

However, if the LDCs increased their resource consumption, they would also increase environmental degradation, and the environmental impact of the entire human population would be manyfold more.

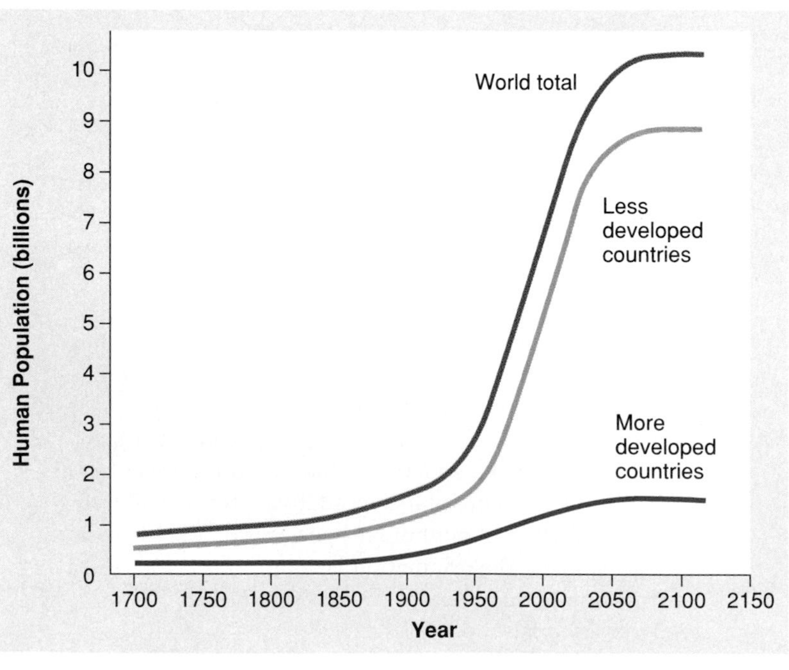

a.

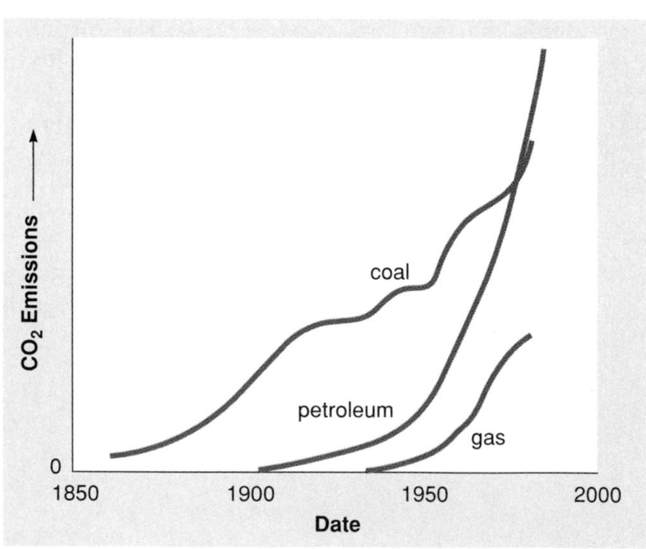

b.

Figure 27.1 Human population and industrialization.
a. Past and projected size of the human population from 1700 to 2120 shows that the less-developed countries will contribute a large proportion of the expected increase. **b.** Use of coal, petroleum, and natural gas from the time of the industrial revolution. Our modern society runs on fossil fuel energy.

By-Products of Industrialization

Industrialization is driven by energy consumption. Coal, petroleum (oil), and natural gas have been and still are the preferred sources of energy for home and industrial use (Fig. 27.1*b*). These are the **fossil fuels,** which were formed by partial decomposition of and pressure of the earth on the remains of plants and animals that lived millions of years ago. Petroleum is used as the starting material for gasoline and for many industrial products, such as plastics, synthetic fibers, and drugs.

In the present context, **pollution** can be defined as any environmental change that adversely affects the lives and health of living things. The burning of fossil fuels results in carbon oxides, hydrocarbons, nitrogen oxides, sulfur oxides, and particulates, which cause air pollution, along with other substances listed in Table 27.1. And some estimate that by 2025 there may be four times as many automobiles in the world as there are today (Fig. 27.2). The release of industrially produced halogen compounds such as the chlorofluorocarbons, and the widespread use of nitrogen fertilizers, are also influencing the chemical composition of the atmosphere.

There are sources of energy that are relatively nonpolluting. Solar energy, which does not add additional heat to the atmosphere, can be collected by solar panels placed on rooftops. Photovoltaic (solar) cells produce electricity directly from sunlight. Falling water is used to produce electricity in hydroelectric plants. Geothermal energy is derived from the heat of the magma in the earth's core. Water, converted to steam by this heat, may be pumped up and used to heat buildings or to generate electricity. Wind power provides enough force to turn vanes, blades, or propellers attached to shafts which, in turn, spin generator motors that produce electricity.

Fossil fuels are **nonrenewable energy sources** because their quantity is limited. After all, we have to mine coal and pump oil and gas out of the ground, and we are ever searching the world for new sources of these fuels. Solar energy and the others mentioned are called **renewable energy sources** because they are replenished by natural physical means.

An increasing human population and the spread of industrialization is threatening not only our well-being, it is also causing a loss of biodiversity. In the 1990s, conservation biology has emerged as a vital new branch of science directed at this environmental crisis. Conservationists want to identify the values of biodiversity and also restore and manage ecosystems for the sustainable benefits to humans. A study of the human impact on global environment necessarily raises questions of ethics which, as we discussed in the first chapter, is not a part of science. However, as a society, we all must be aware of the possible consequences of our activities on the environment and decide what is the appropriate future course to be taken.

Table 27.1

Types of Air Pollutants

Nitrogen oxides	Nitric oxide (NO), nitrogen dioxide (NO_2), and nitrous oxide (N_2O)
Sulfur oxides	Sulfur dioxide (SO_2) and sulfur trioxide (SO_3)
Carbon oxides	Carbon monoxide (CO) and carbon dioxide (CO_2)
Photochemical oxidants	Ozone (O_3) and peroxyacyl nitrates (PANS)
Hydrocarbons	Methane (CH_4) and benzene (C_6H_6)
Organic halogens	Chlorofluorocarbons (CFCs) and halocarbons ($C_xF_xBr_x$)
Aerosols	Suspended particles (dust, smoke, asbestos fibers, metals, etc.) and liquid droplets (sulfuric acid, oils, dioxins, pesticides)

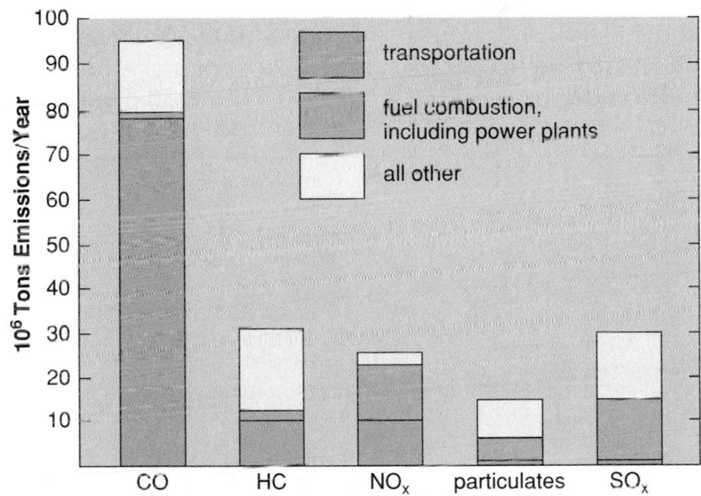

Figure 27.2 Components of air pollution.
The chief cause of carbon monoxide (CO), hydrocarbons (HC), nitrogen oxides (NO_x), particulates (solid matter), and sulfur oxides (SO_x) in the atmosphere is transportation.

The expected increase in the size of the human population and resource consumption may bring about effects that degrade the biosphere and cause a sharp decline in biodiversity.

27.2 Global Climate Change

The earth's climate has fluctuated in the past. We all know that ice ages have occurred in the history of the earth. Presently we are enjoying a moderate temperature that the earth has not seen since about 130,000 years ago. But many are concerned that the global climate will continue to warm and at a rate ten times faster than anytime in the past. Let's examine the reasons why global warming is expected.

Industrialization has significantly altered the transfer rates of the biogeochemical carbon cycle. In 1850, atmospheric carbon dioxide was about 280 parts per million (ppm) and today it is about 350 ppm. This increase is largely due to the burning of fossil fuels and the burning and clearing of forests to make way for farmland and pasture. The oceans are currently taking up about one-half of the carbon dioxide emitted, or else the increase would be much higher than this stated amount. The emission of other gases due to human activities is also taking place. The amount of methane given off by oil and gas wells, rice paddies, and all sorts of organisms including domesticated cows is increasing by about 1% a year. Altogether the following **pollutants** are expected to contribute to global warming:

Gas	*From*
Carbon dioxide (CO_2)	Fossil fuel and wood burning
Nitrous oxide (N_2O)	Fertilizer use and animal wastes
Methane (CH_4)	Biogas (bacterial decomposition, particularly in the guts of animals, in sediments, and in flooded rice paddies)
Chlorofluorocarbons (CFCs)	Freon, a refrigerant
Halons (halocarbons, $C_xF_xBr_x$)	Fire extinguishers
Ozone	Photochemical smog in troposhere

These gases are known as greenhouse gases because just like the panes of a greenhouse they allow solar radiation to pass through but hinder the escape of infrared heat back into space. Figure 27.3*a* shows the earth's radiation and energy balances, and one thing to be learned from this diagram is that water vapor is a greenhouse gas: clouds also reradiate heat back to earth. If the earth's temperature rises due to the **greenhouse effect,** more water will evaporate, forming more clouds, setting up a positive feedback effect that could exacerbate global warming.

Water vapor, carbon dioxide, and methane in the atmosphere played a critical role in the history of the earth because their presence made the earth warm enough to allow living things to evolve. Data based on an analysis of gases trapped in Antartica ice show that indeed the earth's temperature has fluctuated in the past, according to atmospheric levels of carbon dioxide and methane (Fig. 27.3*b*). The greenhouse gases differ in their ability to absorb specific wavelengths of infrared radiation. Atmospheric carbon dioxide may already be absorbing as much infrared radiation as it can. This means that adding a molecule of one of the other greenhouse gases will have more of an effect than adding a molecule of carbon dioxide. The greenhouse effectiveness of methane is now about 25 times that of carbon dioxide. Nitrous oxide is about 200 times more effective than carbon dioxide.

Today data collected around the world show a steady rise in the concentration of the various greenhouse gases. These data are used to generate computer models that predict the earth may warm to temperatures never before experienced by living things. The global climate has already warmed about 0.6°C since the industrial revolution. Computer models are unable to consider all possible variables, but the earth's temperature may rise from 1.5°– 4.5°C by 2060 if greenhouse emissions continue at the current rates (Fig. 27.3*c*).

Predicted Consequences

Global warming will bring about climate changes, which computer models attempt to forecast. It is predicted that as the oceans warm, temperatures in the polar regions will rise to a greater degree than other regions. Glaciers would melt, and sea levels will rise, not only due to this melting but also because water expands as it warms. Water evaporation will increase, and most likely there will be increased precipitation along the coasts and dryer conditions inland. The occurrence of droughts will reduce agricultural yields and also cause trees to die off. Expansion of forests into Arctic areas will most likely not offset the loss of forests in the temperate zones. Coastal agricultural lands such as the deltas of Bangladesh, India, and China would be inundated, and billions will have to be spent to keep coastal cities, like New York, Boston, Miami, and Galveston in the United States, from disappearing into the sea.

Carbon dioxide, methane, and other gases known as the greenhouse gases impede the escape of infrared radiation from the surface of the earth. Ever greater concentrations of these gases are expected to lead to a global warming that will have drastic consequences on the hydrosphere and lithosphere.

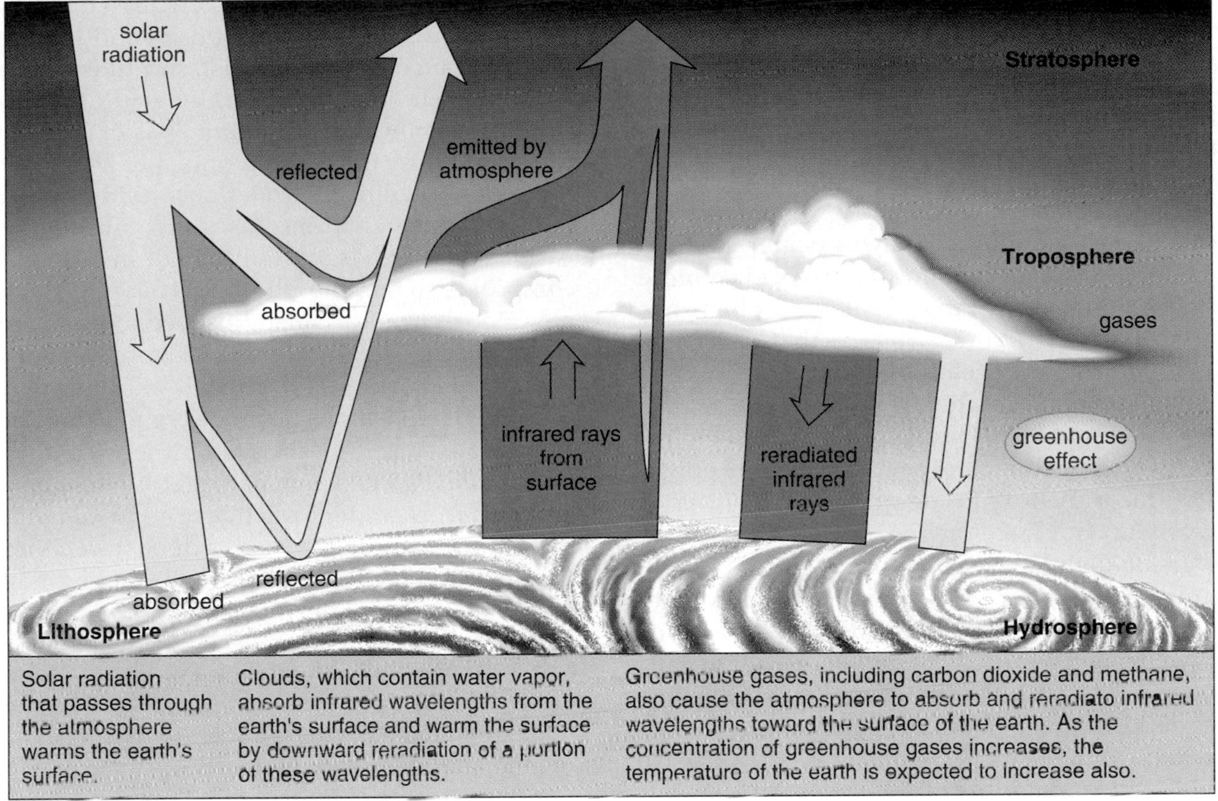

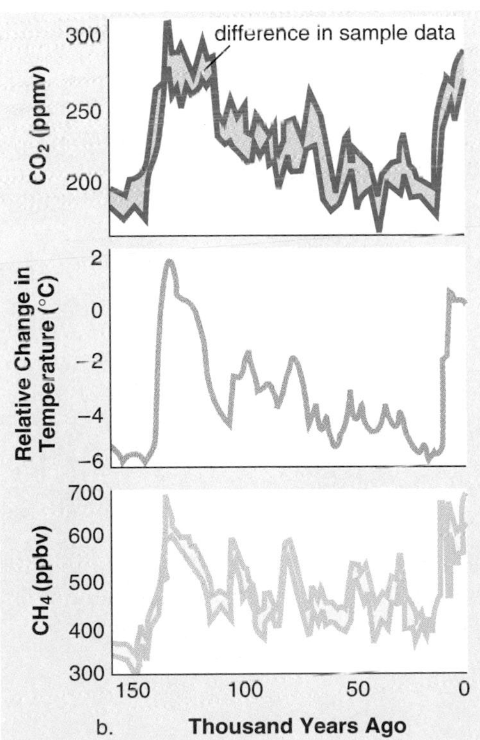

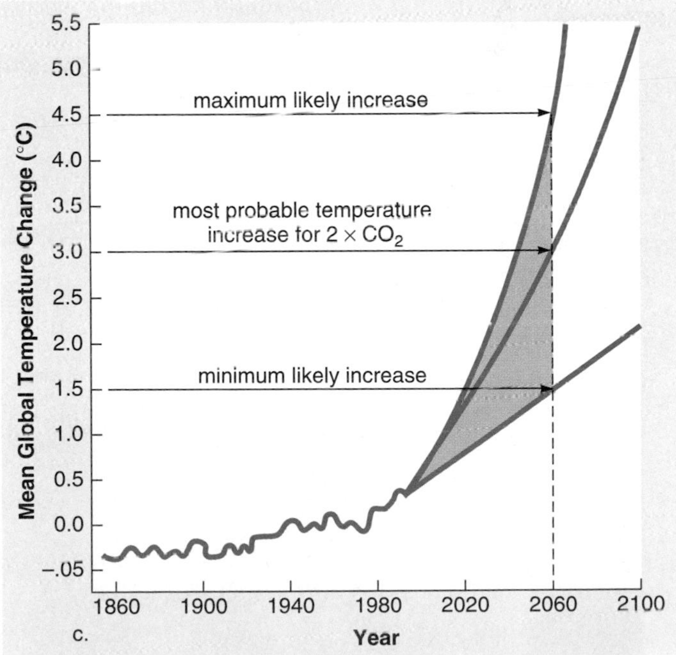

Figure 27.3 Global climate change.
a. Earth's radiation and energy balances showing that the greenhouse gases contribute to global surface warming. **b.** Historical record of atmospheric carbon dioxide, temperature of the earth, and atmospheric methane from Antarctica ice core data. **c.** Mean global temperature change is expected to rise. Because of uncertainties concerning the increase in atmospheric greenhouse gases and in computer modeling, maximum and minimum temperature rises are also given.

27.3 Global Chemical Climate

Pure water has a pH of 7 because dissociation of water releases an equal number of H^+ and OH^- ions. However, the small amount of carbon dioxide in the atmosphere (0.03%) combines with water to produce a weak solution of carbonic acid and an increased number of H^+ ions. Therefore, rain normally has a pH of about 5.6 rather than 7.0. But near urban areas, rain has pH levels nearer 4.0 than 5.6. Cloud and fog droplets are almost always even more acidic than rain. In some fogs, the pH of the droplets has been found to be as low as 1.7—close to that of battery acid. It is no wonder that vegetation and other materials exposed to such fogs deteriorate rapidly.

Our changing global chemical climate is due to human activities. The coal and oil routinely burned by power plants emit sulfur dioxide into the air. Kuwait oil has a high sulfur content, and therefore the oil well fires started during the Persian Gulf War released much sulfur dioxide into the atmosphere. Automobile exhaust routinely puts nitrogen oxides in the air. Both sulfur dioxide and nitrogen oxides are converted to acids when they combine with water vapor in the atmosphere. These acids return to earth as either wet deposition (acid rain or snow) or dry deposition (sulfate and nitrate salts).

The sulfur dioxide and nitrogen oxides are emitted in one locale, but the deposition occurs in another, across state and national boundaries. Increased deposition of acids has drastically affected forests and lakes in northern Europe, Canada, and northeastern United States (Fig. 27.4) because their soils are naturally acidic and their surface waters are only mildly alkaline (basic) to begin with. The forests in these areas are dying, and their waters cannot support normal fish populations. **Acid deposition** causes trees to weaken and increases their susceptibility to disease and insects. It also kills small invertebrates and decomposers so that the entire ecosystem is threatened. The new global chemical climate also reduces agricultural yields and corrodes marble, metal, and stonework, an effect that is noticeable in cities.

Acid deposition causes heavy metals to be present in drinking water. It leaches aluminum from the soils into surrounding waters and also dissolves copper from pipes and lead from solder, which is used to join pipes. A statistical study suggested that acid deposition is implicated in the increased incidence of lung cancer and possibly colon cancer in United States residents of the East Coast.

Acid deposition, which is caused by the emission of sulfur dioxide and nitrogen oxides into the atmosphere, is harmful to ecosystems and human health.

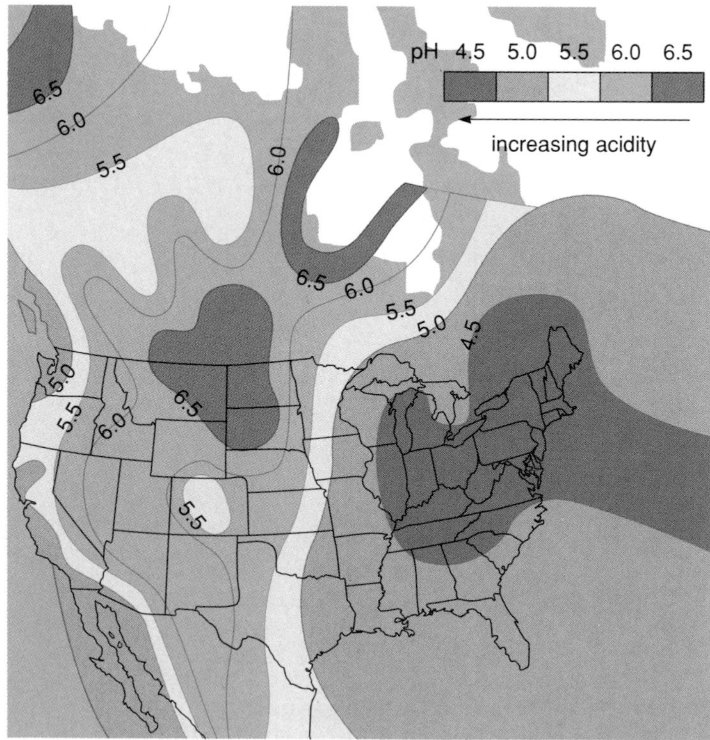

Figure 27.4 Acid deposition.
a. Many forests in higher elevations of northeastern North America and Europe are dying due to acid deposition. b. Air pollution due to emissions from factories and fossil fuel burning is the major cause of acid deposition. c. The numbers in this map of the United States are average pH recordings of deposition. Winds carry acid pollutants toward the northeast from other parts of the country.

Photochemical Smog

Another aspect of the changing chemical climate is increased concentrations of ozone at ground level, resulting from reactions between nitrogen oxides, hydrocarbons, and sunlight. **Photochemical smog** [Gk. *photos,* light, and *chemo,* pertaining to chemicals] contains two air pollutants—nitrogen oxides (NO_x) and hydrocarbons (HC)—that react with one another in the presence of sunlight to produce ozone (O_3) and **PAN (peroxyacetyl nitrate).** Both nitrogen oxides and hydrocarbons come from fossil fuel combustion, but additional hydrocarbons come from various other sources as well, including paint solvents and pesticides.

Ozone and PAN are commonly referred to as oxidants. Breathing ozone affects the respiratory and nervous systems, resulting in respiratory distress, headache, and exhaustion. These symptoms are particularly apt to appear in young people; therefore, in Los Angeles, where ozone levels are often high, schoolchildren must remain inside the school building whenever the ozone level reaches 0.24 ppm (parts per million by weight). Ozone is especially damaging to plants, resulting in leaf mottling and reduced growth.

Carbon monoxide (CO) is another gas that comes from the burning of fossil fuels in the industrial Northern Hemisphere. High levels of carbon monoxide increase the formation of ozone. Carbon monoxide also combines preferentially with hemoglobin and thereby prevents hemoglobin from carrying oxygen. Breathing large quantities of automobile exhaust can even result in death because of this effect. Of late, it has been discovered that the amount of carbon monoxide over the Southern Hemisphere—from the burning of tropical forests—is equal to that over the Northern Hemisphere.

Normally, warm air near the ground is able to escape into the atmosphere. Sometimes, however, air pollutants, including smog and soot, are trapped near the earth due to a thermal inversion. During a **thermal inversion** [Gk. *therme,* heat, and L. *in,* against or over, and *verto,* turn], there is cold air at ground level beneath a layer of warm stagnant air above. This often occurs at sunset, but turbulence usually mixes these layers during the day. Some areas surrounded by hills are particularly susceptible to the effects of a temperature inversion because the air tends to stagnate, and there is little turbulent mixing (Fig. 27.5).

Photochemical smog contains ozone and PAN, which are sometimes trapped near ground level due to a thermal inversion.

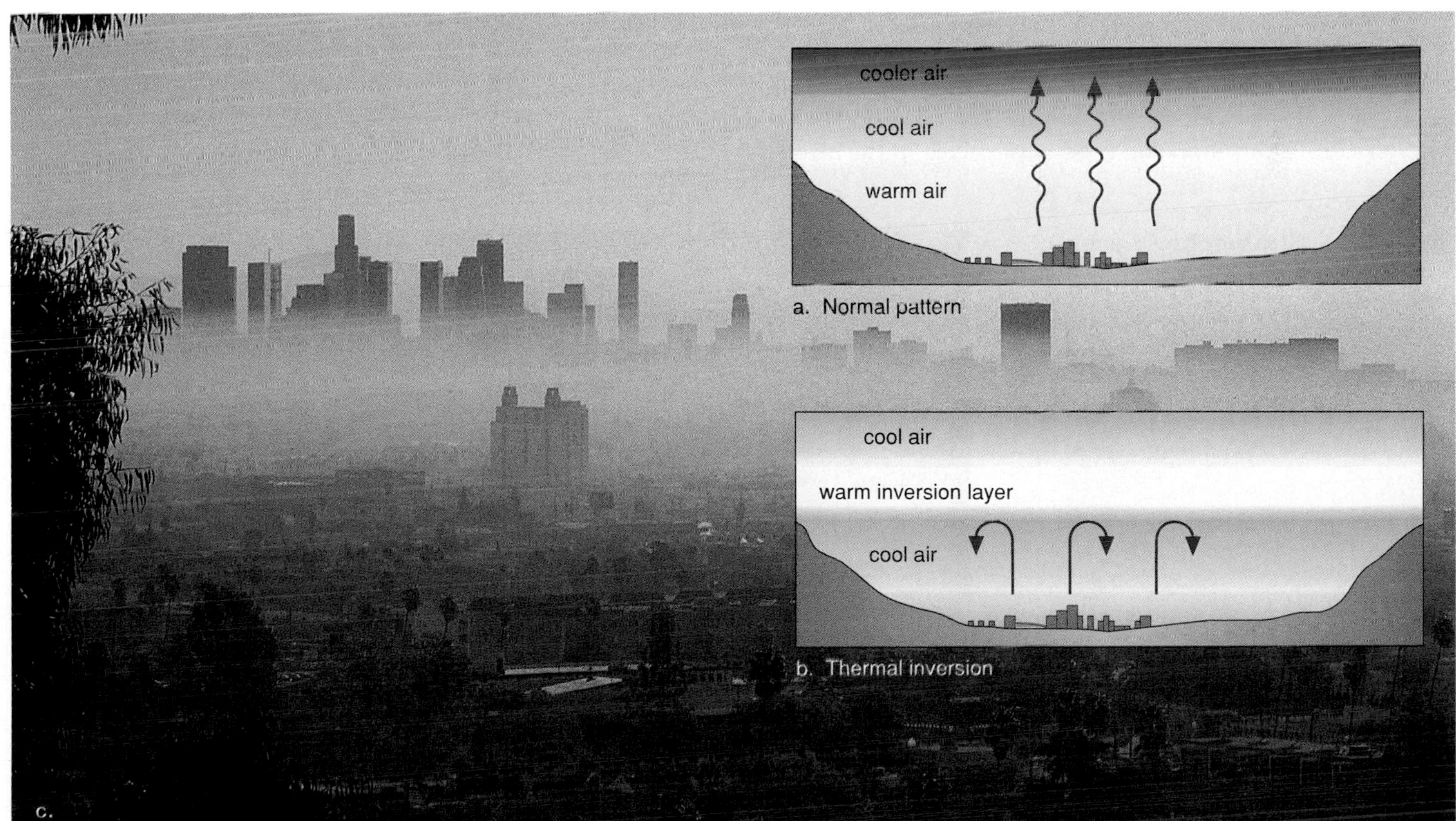

Figure 27.5 Thermal inversion.
a. Normally, pollutants escape into the atmosphere when warm air rises. b. During a thermal inversion, a layer of warm air (warm inversion layer) overlies and traps pollutants in cool air below. c. Los Angeles is particularly susceptible to thermal inversions, and this accounts for why this city is the "air pollution capital" of the United States.

27.4 Stratospheric Ozone Depletion

The earth's atmosphere is divided into layers (see Fig. 27.3). The troposphere envelops us as we go about our day-to-day lives. **Ozone** in the troposphere is a pollutant, but in the stratosphere, some 50 kilometers above the earth, ozone (O_3) forms a layer, called the *ozone shield,* that absorbs most of the wavelengths of harmful ultraviolet (UV) radiation so that they do not strike the earth. In the history of the earth, the development of stratospheric ozone permitted life on land to come into existence because terrestrial organisms are exposed to UV radiation to a greater degree than aquatic organisms. UV radiation impairs crop and tree growth and also kills plankton (microscopic plant and animal life) that sustain oceanic life. Without an adequate ozone shield, food sources and health are threatened. UV radiation causes mutations that can lead to skin cancer and can make the lenses of the eyes develop cataracts. It also is believed to adversely affect the immune system and the ability to resist infectious diseases.

Depletion of the ozone layer within the stratosphere in recent years is, therefore, of serious concern. It became apparent in the 1980s that some worldwide depletion of ozone had occurred, and that by the 1990s there was a severe depletion of some 40–50% above the Antarctic every spring (Fig. 27.6). Severe depletions of the ozone layer are commonly called "**ozone holes.**" Detection devices now tell us that there is an ozone hole above the Arctic as well, and ozone holes could also develop within northern and southern latitudes, where many people live. Whether or not these holes develop depends on prevailing winds, weather conditions, and the type of particles in the atmosphere. A United Nations Environment Program report predicts a 26% rise in cataracts and non-melanoma skin cancers for every 10% drop in the ozone level. A 26% increase translates into 1.75 million additional cases of cataracts and 300,000 more cases of skin cancer every year, worldwide.

The cause of ozone depletion can be traced to chlorine atoms (Cl^-) that are released in the troposphere but rise into the stratosphere. Chlorine atoms combine with ozone and strip away the oxygen atoms one by one. One atom of chlorine can destroy up to 100,000 molecules of ozone before settling to the earth's surface as chloride many years later. These chlorine atoms come from the breakdown of **chlorofluorocarbons (CFCs),** chemicals much in use by humans from 1955 to 1990. The best-known CFC is Freon, a heat transfer agent still found in refrigerators and air conditioners today. CFCs were used as cleaning agents and during the production of Styrofoam found in coffee cups, egg cartons, insulation, and paddings. Their use as a propellent in spray cans has been outlawed in the United States and several other countries but not in western Europe. Although most countries of the world have agreed to stop using CFCs by the year 2000, CFCs already in the atmosphere will be there for over a hundred years before they are spent.

Air pollutants are involved in causing four major environmental effects: global warming, ozone shield destruction, acid deposition, and photochemical smog. Each pollutant may be involved in more than one of these.

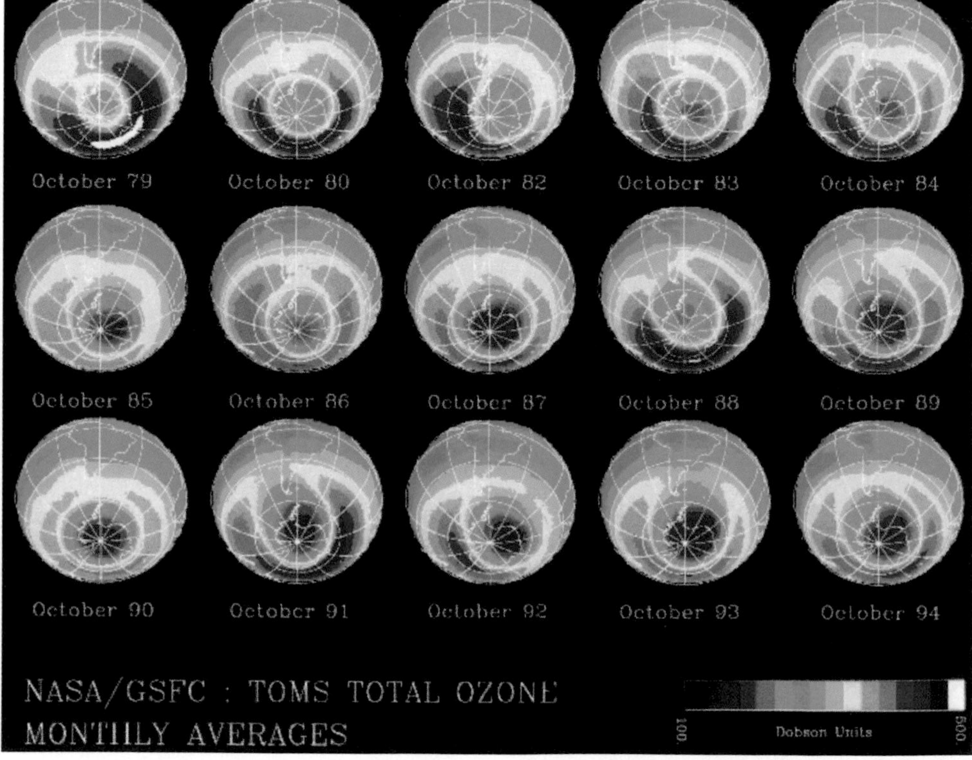

Figure 27.6 Stratospheric ozone. These satellite observations show that the amount of ozone over the Antarctic (the South Pole) between October 1979 and October 1994 fell by more than 50%. Green represents an average amount of ozone, blue less, and purple still less.

27.5 Surface Waters, Aquifers, and Oceans

Fresh water is used for domestic purposes including drinking water, crop irrigation, industrial uses, and energy production. Surface water from rivers, lakes, and underground rivers called **aquifers** are used to meet these needs. The water in aquifers [L. *aqua*, water, and *fero*, to bear, carry] is a vast natural resource, but to ensure a continual supply withdrawals cannot exceed deposits. Countries worldwide are withdrawing water from aquifers. In this country, the farmers of the Midwest withdraw water from aquifers up to 50 times faster than nature replaces it. China, with a population of one billion, is also mining its water to meet the needs of its people, despite the estimate that its aquifers can sustain only 650 million people. In the United States, the government still heavily subsidizes water so that the incentive to use water carefully and efficiently is lacking.

Pollution of Surface Waters

Besides excessive use, pollution of surface water, groundwater, and the oceans is another reason why we are running out of fresh water. Solid wastes include not only household trash but also sewage sludge, agricultural residues, mining refuse, and industrial wastes. Every year, the U.S. population discards billions of tons of solid wastes, some on land and some in fresh and marine waters. *Point sources* are sources of pollution that are easily identifiable and nonpoint sources are those caused by runoff from the land (Fig. 27.7).

The United States spends $9 billion a year on cleanup but only $200 million yearly to prevent pollution. One study showed that it was possible to achieve 70–90% public participation in recycling by spending only thirty cents per household. Likewise, recycling can save industry money as witnessed by the 3M Corporation, which reported savings of $1.2 billion by recycling waste and preventing pollution.

Sewage

Sewage treatment can help degrade organic wastes, which otherwise can cause oxygen depletion in lakes and rivers. As the oxygen level decreases, the diversity of life is greatly reduced. Also, human feces can contain pathogenic microorganisms that cause cholera, typhoid fever, and dysentery. In less-developed countries, where the population is growing and where sewage treatment is practically nonexistent, many children die each year from these diseases. Typically, sewage treatment plants use bacteria to break down organic matter to inorganic nutrients, like nitrates and phosphates, which then enter surface waters. These types of nutrients, which also can enter waters by fertilizer runoff and soil erosion, lead to *cultural eutrophication*, an acceleration of the natural process by which bodies of water fill in and disappear. First, the nutrients cause overgrowth of algae. Then, when the algae die, oxygen is used up by the decomposers, and the water's capacity to support life is reduced. Massive fish kills are sometimes the result of cultural eutrophication.

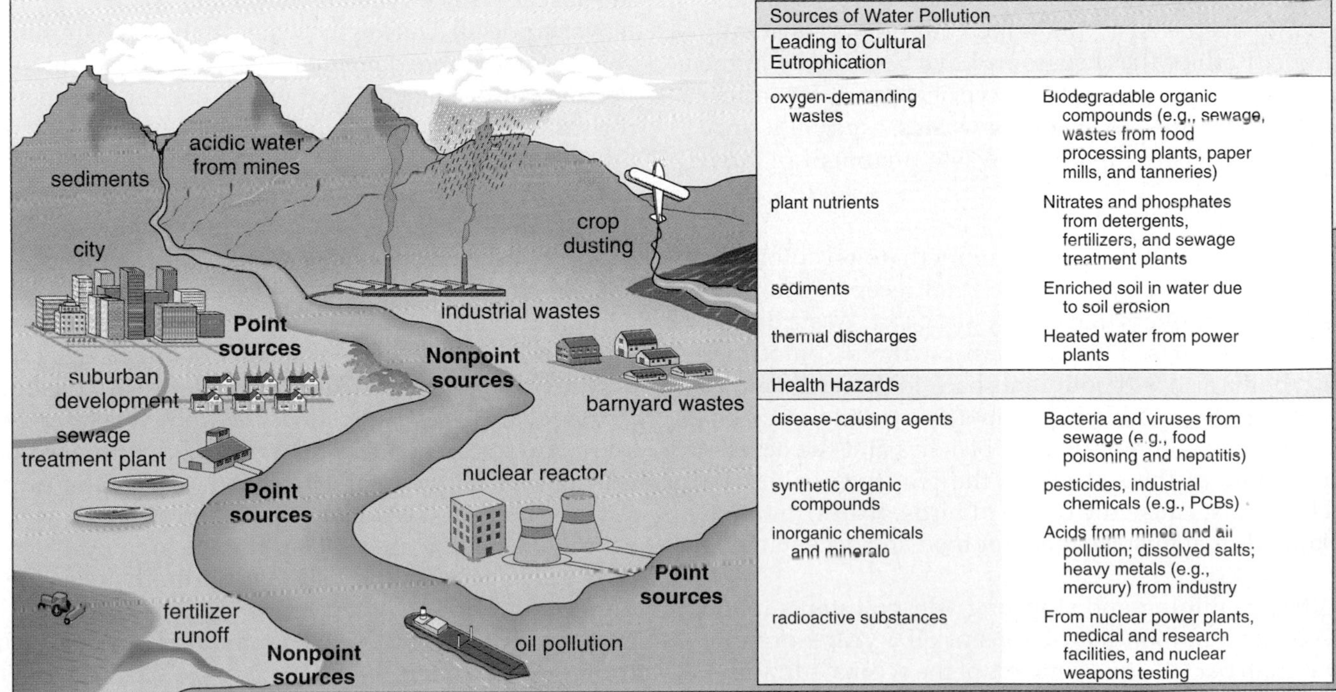

Sources of Water Pollution	
Leading to Cultural Eutrophication	
oxygen-demanding wastes	Biodegradable organic compounds (e.g., sewage, wastes from food processing plants, paper mills, and tanneries)
plant nutrients	Nitrates and phosphates from detergents, fertilizers, and sewage treatment plants
sediments	Enriched soil in water due to soil erosion
thermal discharges	Heated water from power plants
Health Hazards	
disease-causing agents	Bacteria and viruses from sewage (e.g., food poisoning and hepatitis)
synthetic organic compounds	pesticides, industrial chemicals (e.g., PCBs)
inorganic chemicals and minerals	Acids from mines and air pollution; dissolved salts; heavy metals (e.g., mercury) from industry
radioactive substances	From nuclear power plants, medical and research facilities, and nuclear weapons testing

Figure 27.7 Sources of surface water pollution.
Many bodies of water are dying due to the introduction of pollutants from point sources, which are easily identifiable, and nonpoint sources, which cannot be specifically identified.

Agricultural and Industrial Wastes

In areas of intensive animal husbandry, or where there are many septic tanks, ammonium (NH_4^+) released from animal and human waste is converted by soil bacteria to soluble nitrate, which moves down through the soil (percolates) into underground water supplies. Between 5% and 10% of all wells examined in the United States have nitrate levels higher than the recommended maximum.

Industrial wastes can include heavy metals and organochlorides, such as those in some pesticides. These materials are not degraded readily under natural conditions nor in conventional sewage treatment plants. Sometimes, they accumulate in the mud of deltas and estuaries of highly polluted rivers and cause environmental problems if they are disturbed. Industrial pollution is being addressed in many industrialized countries but usually has low priority in less-developed countries. These wastes enter bodies of water and are subject to **biological magnification** (Fig. 27.8). Decomposers are unable to break down these wastes. They enter and remain in the body because they are not excreted. Therefore, they become more concentrated as they pass along a food chain. Notice in Figure 27.8 that the dots representing DDT become more concentrated as they pass from producer to tertiary consumer. Biological magnification is most apt to occur in aquatic food chains, since there are more links in aquatic food chains than there are in terrestrial food chains. Humans are the final consumers in both types of food chains, and in some areas, human milk contains detectable amounts of DDT and PCBs, which are organochlorides.

Industry also pollutes aquifers. Previously, industry ran wastewater into a pit, from which pollutants could seep into the ground. Wastewater and chemical wastes were also injected into deep wells, from which pollutants constantly discharged. Both of these customs have been or are in the process of being phased out. It is very difficult for industry to find other ways to dispose of wastes, especially since citizens do not wish to live near waste treatment plants.

Pollution of Oceans

Coastal regions are not only the immediate receptors for local pollutants, they are also the final receptors for pollutants carried by rivers that empty at a coast. Waste dumping also occurs at sea, and ocean currents sometimes transport both trash and pollutants back to shore. Examples are the nonbiodegradable plastic bottles, pellets, and containers that now commonly litter beaches and the oceans' surfaces. Some of these, such as the plastic that holds a six-pack of cans, cause the death of birds, fishes, and marine mammals that mistake them for food and get entangled in them.

Offshore mining and shipping add pollutants to the oceans. Some 5 million metric tons of oil a year—or more than one gram per 100 square meters of the oceans' surfaces—end up in the oceans. Large oil spills kill plankton, fish fry, and shellfishes, as well as birds and marine mammals. The largest tanker spill in U.S. territorial waters occurred on March 24, 1989, when the tanker *Exxon Valdez* struck a reef in Alaska's Prince William Sound and leaked 44 million liters of crude oil. During the war with Iraq, 120 million liters were released from damaged onshore storage tanks into the Persian Gulf—an event that was called environmental terrorism. Although petroleum is biodegradable, the process takes a long time because the environs do not ordinarily contain a large bacterial population for degrading petroleum. Once the oil washes up onto beaches, it takes many hours of work and millions of dollars to clean it up.

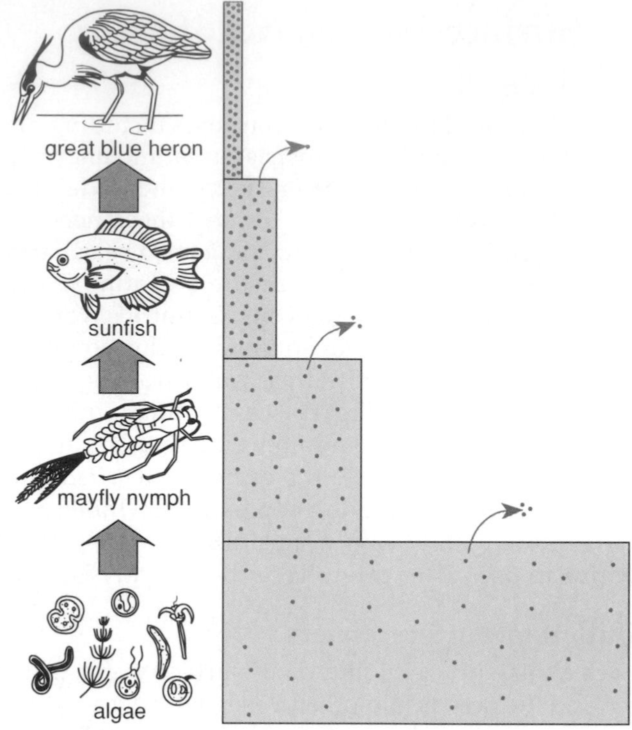

Figure 27.8 Biological magnification.
A poison (dots) such as DDT, which is minimally excreted (arrows) becomes maximally concentrated as it passes along a food chain due to the ever smaller biomass of each higher trophic level.

In the last 50 years, we have polluted the seas and exploited their resources to the point that many species are literally at the brink of extinction. Fisheries once rich and diverse, such as George's Bank off the coast of New England, are in severe decline. Haddock was once the most abundant species in this fishery, and now it accounts for less than 2% of the total catch. Cod and bluefin tuna have suffered a 90% reduction in population size. In warm, tropical regions, many areas of coral reefs are now overgrown with algae because the fish that normally keep the algae under control have been killed off. As discussed in the reading on page 481, sharks may also largely disappear from marine waters.

Ecology Focus

Sharks: The World's Least-Favorite Animal In Decline

Of the many plants and animals threatened by human exploitation, one of the least-loved is the shark. Public perception of these animals is based almost entirely upon news reports of attacks upon humans—which are actually much rarer than one might imagine (Manire and Gruber, 1990)—and gruesome media images, such as the movie *Jaws,* which portray them as merciless, indiscriminate killers. For most people, a shark is little more than a triangular fin, an appetite, and a mouthful of very sharp teeth. The shark's appearance does little to dispel its fearsome reputation. For conservationists concerned with rapidly dwindling shark populations worldwide, the shark's bad reputation is a public relations nightmare.

The single quality that redeems these animals in the public eye is not one that encourages conservation: sharks are a popular item on menus in Chinese restaurants. Shark fishing has become a booming business in the past decade (Waters, 1992). In Asia, sharkfin soup is a delicacy that has created high demand for several species of shark; fins may bring up to $44 per kilogram (Manire and Gruber, 1990; Waters, 1992). The repugnant and wasteful practice called "finning," in which a captured shark is flung back into the water to die after its fins are amputated, has spurred some public sympathy for sharks and has led to calls for banning the practice. A more serious problem, however, is the tendency for sharks to become "by-catches" of commercial fishing done with drift gill nets. More than half of the estimated 200 million annual shark kills are related to accidental gill net catches; sharks caught in this manner are usually simply discarded.

High shark mortality has conservationists concerned for several reasons. Sharks mature very slowly, have long reproductive cycles, and produce only a few young at a time (Manire and Gruber, 1990; Waters, 1992). Fishes such as salmon, which have also been overharvested, can recover rapidly because of the large numbers of offspring they produce annually; sharks do not have this capability. A second problem is that harvesting of sharks by commercial and private fishing concerns is largely unregulated. A few countries, notably Australia and South Africa, have enacted legislation to stem shark losses, but major contributors to commercial shark fishing either see no need for action or are delaying proposed regulations. Recent government-imposed quotas on the total catch of coastal and deep-sea sharks in United States waters are a step in the right direction, but may still be too high to let many vulnerable species recover to their original numbers. Finally, the decimation of shark populations is occurring at a time when very little is known about more than a handful of individual species. Though more than 350 species of sharks exist, management proposals treat all sharks as a single entity because, lacking specific information, management by species is not possible. Species that have been studied, including the lemon shark *(Negaprion brevirostris)*, have demonstrated a precipitous decrease in the number of juveniles observed in the past five years.

The decline of shark populations is a matter for concern in and of itself, but it is also an important factor in a larger problem. Sharks are among the most important predators in marine ecosystems; they feed upon a variety of organisms and are distributed throughout oceans, seas, and lakes worldwide (Manire and Gruber, 1990). Terrestrial ecologists have already observed the benefits of predation for prey populations and the problems that occur when predators are removed from an ecosystem. The decline of sharks could have a significant, and possibly catastrophic, effect upon marine ecosystems. Ironically, sharks have fulfilled their role for some 400 million years, making them one of the longest-lived groups of organisms on the planet; yet their future in the next two decades or so depends upon a change in human attitudes and perceptions. Conservationists have their work cut out for them: they must persuade world governments to look beyond the shark's terrifying aspect and act to preserve an animal vital to the health of the world's oceans.

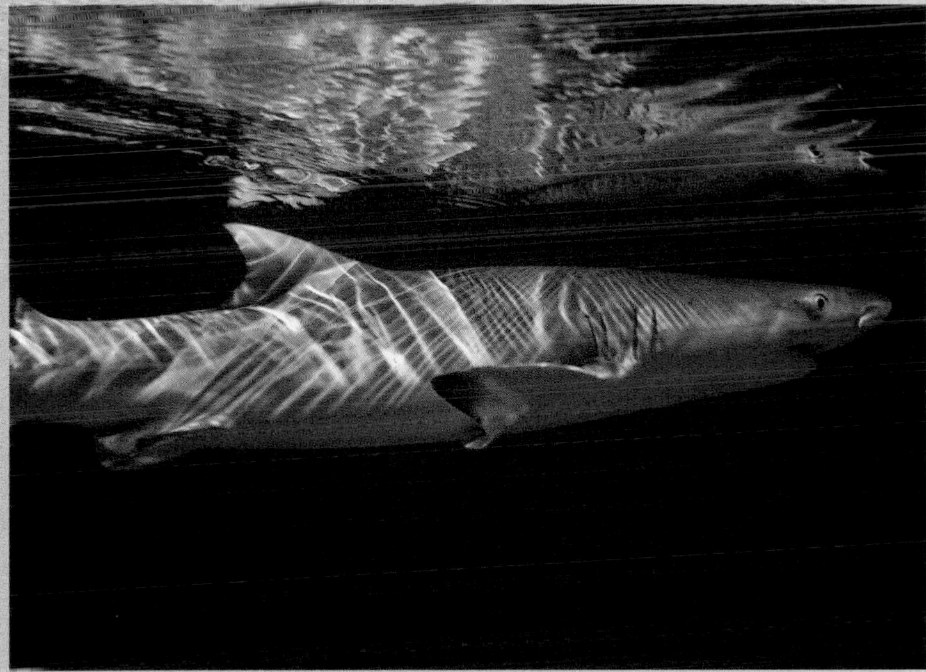

Figure 27A Lemon shark.
An entire population of lemon sharks in the Florida Keys was captured and used as crab bait.

Primack, R. B. 1993. Essentials of conservation biology. *Sunderland, Mass.: Sinauer Associates, Inc., pp. 189–90.*

27.6 Soil Erosion, Desertification, and Deforestation

Whereas 20% of the world's population lived in cities in 1950, it is predicted that 60% will live in cities by the year 2000. Near cities, the development of new housing areas has led to urban sprawl. Such areas tend to take over agricultural land or simply further degrade the land in the area. The land has been degraded in many ways. Here, we will discuss some of the greatest concerns.

Soil Erosion and Desertification

In agricultural areas, wind and rain carry away about 25 billion tons of topsoil yearly, worldwide. If this rate of loss continues, the earth will lose practically all of its topsoil by the middle of the next century. Soil erosion causes a loss of productivity that is compensated by increased use of fertilizers, pesticides, and fossil fuel energy. One answer to the problem of erosion is to adopt soil conservation measures such as are employed by many farmers in the United States. For example, farmers can use strip-cropping and contour farming to control soil erosion.

Desertification is the transformation of marginal lands to desert conditions because of overgrazing and overfarming (Fig. 27.9). Desertification has been particularly evident along the southern edge of the Sahara Desert in Africa, where it is estimated that 240,000 square miles of once-productive grazing land has become desert in the last 50 years. However, desertification also occurs in this country. The U.S. Bureau of Land Management, which opens up federal lands for grazing, reports that much of the rangeland it manages is in poor or bad condition, with much of its topsoil gone and with greatly reduced ability to support forage plants.

Humans use land in various ways. Farms, towns, and cities are located on land. Agricultural land quality is threatened by soil erosion, which can lead to desertification.

Figure 27.9 Desertification.
a. Arid and semiarid areas can be converted to desert by overgrazing. The loss of vegetation increases soil erosion and reduces the amount of water the soil takes up.
b. United Nations Map of World Desertification.

a.

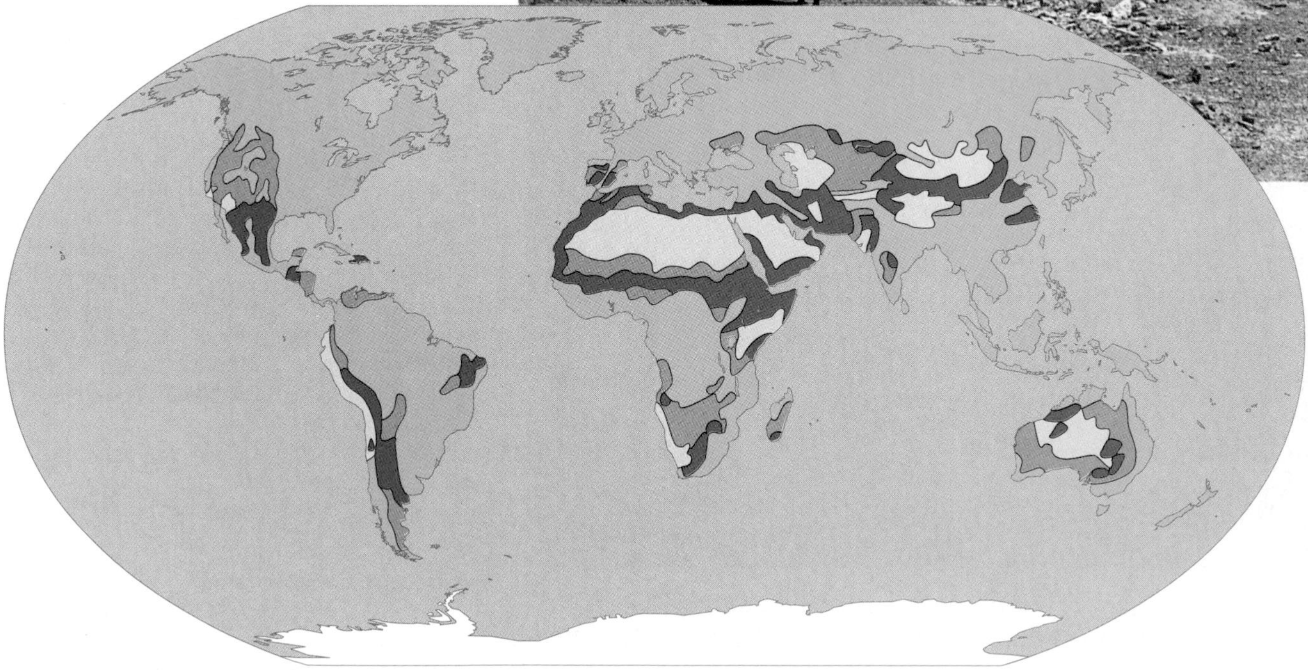

Degree and Risk of Desertification

b. high moderate  existing desert

Deforestation

In Canada, vast stands of trees are scheduled to be felled and turned into paper and wood products like posts and particleboard. Thousands of miles of new logging roads are to be built during the next five to ten years, and this no doubt will bring many visitors to hunt and fish where only indigenous people did so previously. The animals that live there—moose, porcupine, lynx, and snowshoe hare—will also be displaced. Songbirds who migrate there during the summer will no longer find shelter. Although the logging companies are to replant, conservationists wonder if companies have the necessary expertise or if wildlife can sustain themselves in the meantime.

Tropical rain forests are much more biologically diverse than temperate forests. For example, temperate forests across the entire United States contain about 400 tree species. In the rain forest, a typical ten-hectare area holds as many as 750 types of trees. The fresh waters of South America are inhabited by an estimated 5,000 fish species; on the eastern slopes of the Andes, there are 80 or more species of frogs and toads, and in Ecuador, there are more than 1,200 species of birds—roughly twice as many as those inhabiting all of the United States and Canada. Therefore, a very serious side effect of deforestation in tropical countries is the loss of biological diversity.

A National Academy of Sciences study estimated that a million species of plants and animals are in danger of disappearing within 20 years as a result of **deforestation** in tropical countries. Many of these life-forms have never been studied, and yet they may be useful sources of food or medicines.

Logging of tropical forests occurs because industrialized nations prefer furniture made from costly tropical woods and because people want to farm the land (Fig. 27.10). In Brazil, the government allows citizens to own any land they clear in the Amazon forest (along the Amazon River). When they arrive, the people practice slash-and-burn agriculture, in which trees are cut down and burned to provide nutrients and space to raise crops. Unfortunately, the fertility of the land is sufficient to sustain agriculture for only a few years. Once the cleared land is incapable of sustaining crops, the farmer moves on to another part of the rain forest to slash and burn again. In the meantime, cattle ranchers move in. Cattle ranchers are the greatest beneficiaries of deforestation, and increased ranching is therefore another reason for tropical rain forest destruction. A newly begun pig-iron industry in Brazil also indirectly results in further exploitation of the rain forest. The pig iron must be processed before it is exported, and smelting the pig iron requires the use of charcoal (burnt wood).

There is much concern worldwide about the loss of biological diversity due to the destruction of tropical rain forests.

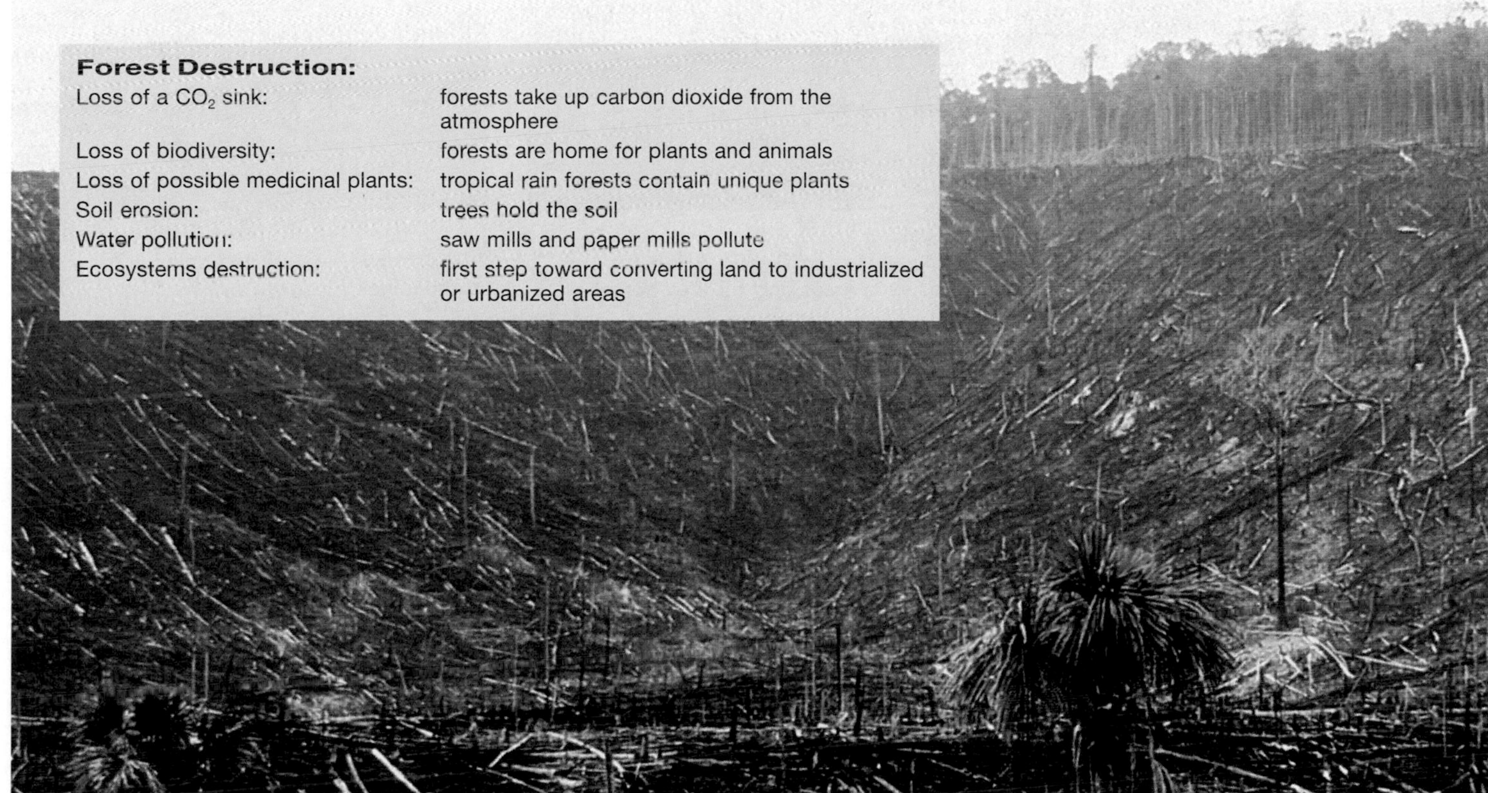

Forest Destruction:
Loss of a CO_2 sink: forests take up carbon dioxide from the atmosphere
Loss of biodiversity: forests are home for plants and animals
Loss of possible medicinal plants: tropical rain forests contain unique plants
Soil erosion: trees hold the soil
Water pollution: saw mills and paper mills pollute
Ecosystems destruction: first step toward converting land to industrialized or urbanized areas

Figure 27.10 Deforestation.
Forest destruction leads to the detrimental effects listed.

27.7 Human Impact on Biodiversity

One level of biodiversity is understood by everyone because it includes all the different species of bacteria, protoctists, fungi, plants, and animals on earth. Genetic variation is a finer level of biodiversity, which helps maintain reproductive vitality and assists adaptation to new environments. And finally different species live in widely different types of communities making up the biosphere. Therefore, biodiversity is considered to have these three levels: genetic diversity, species diversity, and community diversity.

Extinction

There have been several mass extinctions in the history of the earth followed by periods of recovery. Recovery from the last mass extinction resulted in the highest level of biodiversity the earth has known. Human activities that reduce biodiversity may have begun about 30,000 years ago because Cro-Magnon may have killed off many large animals such as the giant sloth, the mammoth, the saber-toothed tiger, and the giant ox. *Hunting* by individuals for food and sport and commercial hunting are collectively one of the major causes for an estimated extinction of 15,000 to 30,000 species a year (Fig. 27.11). The reading on page 481 tells how sharks are disappearing from the oceans due to overexploitation, and we certainly know that many fish stocks are being depleted by overfishing. Similarly, on land humans hunt and collect organisms for both pleasure and profit. Wealthy sportsmen now fly into otherwise unreachable Arctic regions to hunt polar bear from the air. Spotted cats (tigers, cheetahs, leopards, jaguars, etc.) are endangered because customers will pay more for furs as these animals become more and more rare. As with sharks, some species have been hurried toward extinction because only certain parts are desired. Rhino horn powder is used as an aphrodisiac and as a cure for snake bites, kidney disease, and other illnesses. At least a thousand black rhinos a year are killed illegally on preserves in Kenya—only to acquire the horn. Likewise elephants are killed only to acquire their tusks. Cacti are collected from the wild for sale in other countries and, on one expedition, dealers removed all of the native cacti from an island off the Baja California coast. The removal of the cacti threatens the lives of the animals that depend upon them for food and shelter.

Another major cause of extinction is *habitat destruction* outright or by fragmentation into small pieces that cannot support the same species richness as before. By the year 2010, very little undisturbed rain forest will exist outside of national parks and other relatively small protected areas. Animal husbandry, agricultural fields, towns and cities, and also human activities like mining and the building of dams, destroy ecosystems.

Commercial hunting: A trawl net needs to be towed through the water for only a few minutes before it is full.

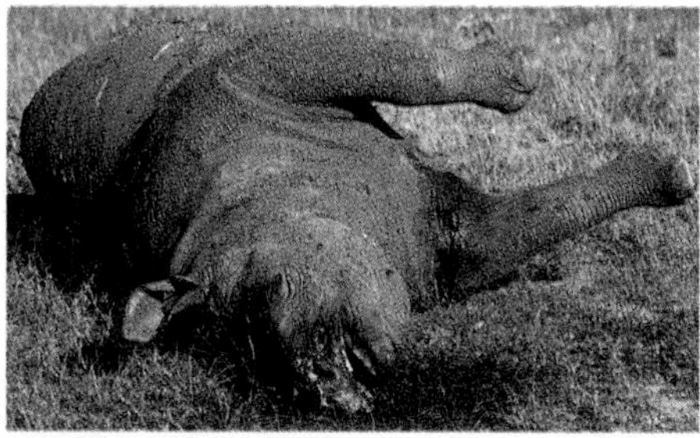
Illegal hunting: In Africa, black rhinoceroses are illegally killed in preserved areas only for their horns.

Destruction of ecosystems: Strip mining for coal destroys habitats and pollutes the surroundings.

Introduction of new species: The brown tree snake *(Boiga irregularis)* has devastated endemic bird populations after being introduced into many Pacific islands.

Pollution: Solid wastes are obvious signs of pollution in ecosystems; pesticides cannot be seen but are deadly to wildlife.

Pollution: Pesticides kill bees, and declining populations of bees threaten the plants that depend on them for pollination.

27.11 Extinction.
Human activities cause extinction in many ways. Some of them are illustrated here.

The accidental or purposeful introduction of *new species* into an ecosystem can cause the extinction of endemic species. The brown tree snake somehow slipped into Guam from southwestern Pacific islands in the late 1940s. Since then, it has wiped out nine of eleven native bird species, leaving the forests eerily quiet. The carp, an Asian fish that can tolerate polluted waters, is now more prevalent than our own native fishes in certain waters.

Pollution also takes it toll. As we have discussed, the global climate may change so rapidly that many species will be unable to adjust their ranges and may become extinct. Pollution can kill species even if biological communities appear to be undisturbed. Biological magnification of pesticides has caused the abundance of predatory birds to decrease, and acid deposition is implicated in a worldwide decline in amphibian populations.

Human activities are on the verge of causing a massive extinction of species and the loss of ecosystems throughout the biosphere. The loss of biodiversity will most likely be detrimental to humans since they depend on the natural environment for raw materials, food, medicines, and other goods and services.

Conservation Biology

As we have discussed, scientists have only begun to formulate theories about the complex interactions within the biosphere and, therefore, they do not yet know how best to counter the rapid and worldwide decline in biodiversity. *Conservation biology* is a relatively new scientific discipline that brings together people and knowledge from many different fields to attempt to solve the biodiversity crisis. Conservation biology wants to understand the effects of human activities on species, communities, and ecosystems, and develop practical approaches to preventing the extinctions of species and the destruction of ecosystems. In the past, ecologists have preferred to study the workings of ecosystems not tainted by human activities, and wildlife managers have been concerned with managing a small number of species for the marketplace and for recreation. Therefore, neither endeavor has addressed the possibility of preserving entire biological communities, although humans are active in the area. Conservation biologists must be able to draw from scientific research and experience in the field to develop a management program that will preserve an ecosystem.

Conservation biology interfaces with many disciplines including ecological ethics. Most conservation biologists believe that biological diversity is a good thing, and that each species has a value all its own, regardless of its direct material value to humans. This viewpoint runs counter to the idea that the value of a species depends on the goods and services it provides or could potentially provide to humans. Still, many conservationists are willing to work with governmental agencies that promote the concept of sustainability: that it is possible for development to meet economic needs while protecting the environment for future generations. Some economists argue that as per capita income increases, environmental degradation first increases, and then begins to decrease as people become affluent enough to begin to protect the environment.

Conservation biology is the scientific study of biodiversity, leading to the preservation of species and the management of ecosystems for sustainable human welfare.

connecting concepts

What will our world be like by the end of this century? One thing is certain: with a human population approaching ten billion, it is going to be a very crowded planet. Moreover, every ecosystem will be adversely affected, even if we could stop population growth now.

As we've seen throughout Part iv, almost all life on earth is dependent on the sun. Using solar energy, carbon dioxide, and water, the photosynthesizers produce glucose and oxygen. The consumers then live off the producers or other consumers. The carbon cycle that helps support the biosphere is now being altered by the burning of fossil fuels and the destruction of rain forests. Both activities add excess carbon dioxide and other gases to our atmosphere, leading to the greenhouse effect and the warming of our atmosphere.

Chemicals produced by human activities create acid rain in many parts of the world. The result is dying forests, unsafe drinking water, and lakes that can't support fish. Air pollution combined with sunlight creates photochemical smog, which causes respiratory problems. Chlorine atoms from CFCs create holes in the earth's ozone layer, leading to skin cancers and cataracts.

Pollution of fresh water and the growing need for clean water to supply earth's expanding population is already causing conflicts over water rights. Perhaps even more frightening, the oceans are also polluted. Historically, people believed that the oceans could withstand any form of human-induced contamination simply because of their massive size. Today, we know that chemical toxins, fossil fuel wastes, and marine garbage damage virtually every marine environment and the organisms that live in them.

Finally, by our sheer numbers, humans are forcing other species from their natural habitats, leading to a decline in biodiversity—the loss of species worldwide. Should we care about declining biodiversity? Absolutely. Consider a house of cards. Every single card has a pivotal position in holding the house together. The removal of one card or even two may not collapse the structure, but eventually the loss of a card will cause the house to come tumbling down. So it is with ecosystems. The loss of a snail darter or spotted owl changes the ecosystem in a way that could be detrimental.

Summary

27.1 Human Population and Industrialization

The human population is increasing, and there may be ten billion people on earth by the end of the twenty-first century. Most persons (78%) live in the less-developed countries, but the 22% of persons in the more-developed countries cause most of the air, water, and land pollution.

27.2 Global Climate Change

Various substances are associated with air pollution, such as sulfur dioxide, hydrocarbons, nitrogen oxides, methane, chlorofluorocarbons (CFCs), carbon dioxide, carbon monoxide, and ozone. These have various sources, but most come from vehicle exhaust and fossil fuel burning.

Like the panes of a greenhouse, carbon dioxide, nitrous oxide, methane, and CFCs allow the sun's rays to pass through but impede the release of infrared wavelengths. It is predicted that a buildup in these "greenhouse gases" will lead to a global warming. The effects of global warming could be a rise in sea level and a change in climate patterns. An effect on agriculture could follow.

27.3 Global Chemical Climate

Sulfur dioxide and nitrogen oxide react with water vapor to form acids that contribute to acid deposition. Acid deposition is killing lakes and forests and also corrodes marble, metal, and stonework.

Hydrocarbons and nitrogen oxides react to form smog, which contains ozone and PAN (peroxyacetyl nitrate). These oxidants are harmful to animal and plant life.

27.4 Stratospheric Ozone Depletion

Ozone shield destruction is particularly associated with CFCs. CFCs rise into the stratosphere and release chlorine. Chlorine causes ozone to break down. Since ozone prevents harmful ultraviolet radiation from reaching the surface of the earth, reduction in the amount of ozone will lead to skin cancer and decreased productivity of the oceans.

27.5 Surface Waters, Aquifers, and Oceans

Water in aquifers is being withdrawn up to 50 times faster than it can be replaced. Pollution of underground water supplies—as well as of lakes and oceans—is also a serious problem.

Solid wastes, including hazardous wastes, are deposited on land. The latter, including metals, organochlorides, and nuclear wastes, may contaminate water supplies. The oceans are the final recipients of wastes deposited in rivers and along the coasts.

27.6 Soil Erosion, Desertification, and Deforestation

Soil erosion reduces the quality of land and leads to desertification.

The Canadian forests and the tropical rain forests in Southeast Asia and Oceania, Central and South America, and Africa are being cut to provide wood for export. Slash-and-burn agriculture also reduces tropical rain forests. The loss of biological diversity due to the destruction of tropical rain forests will be immense. Many of these threatened organisms could possibly be of benefit to humans if we had time to study and domesticate them.

27.7 Human Impact on Biodiversity

Human activities have brought about a biodiversity crisis. Individual and commercial hunting, habitat destruction or fragmentation, and introduction of new species and pollution are all major causes of species extinctions.

Conservation biology is a new discipline that pulls together information from a number of biological fields to determine how best to manage ecosystems for the benefit of all species including human beings.

Reviewing the Chapter

1. Explain why human activities are now impacting the global environment at an ever faster rate and scale. 472–73
2. What substances contribute to air pollution? What are their sources? 473
3. How and why is the global climate expected to change, and what are the predicted consequences of this change? 474
4. What causes acid deposition, and what are its effects? 476
5. How does photochemical smog develop, and what is a thermal inversion? 477
6. Of what benefit is the ozone shield? What pollutant in particular should be associated with stratospheric ozone depletion, and what are the consequences of this depletion? 478
7. What are several ways in which surface waters, aquifers, and oceans can be polluted? What is biological magnification? 479–80
8. Explain how soil erosion and desertification are related. 482
9. What are the primary ecological concerns associated with the destruction of rain forests? 483
10. Explain the primary causes of the biodiversity crisis and the goals of conservation biology. 484–85

Testing Yourself

Choose the best answer for each question.

For questions 1–4, match the terms with those in the key:

Key:

a. sulfur dioxide
b. ozone
c. carbon dioxide
d. chlorofluorocarbons (CFCs)

1. acid deposition
2. ozone shield destruction
3. greenhouse effect
4. photochemical smog

5. Which of these is a true statement?
   a. More-developed countries do not contribute to the destruction of tropical rain forests.
   b. Carbon dioxide from fossil fuel combustion is the primary greenhouse gas.
   c. Water from underground sources is never subject to contamination.
   d. Sulfur dioxide is given off to the same degree by the burning of all fossil fuels.
6. Which of these is a true statement?
   a. Global warming is of no immediate concern.
   b. Global warming is so imminent that nothing can be done.
   c. Reduction in fossil fuel burning will lessen the greenhouse effect.
   d. Since gases not derived from fossil fuel combustion are involved, reduction in fossil fuel burning will not help the greenhouse effect.
7. Tropical rain forest destruction is extremely serious because
   a. it will lead to a severe reduction in biological diversity.
   b. tropical soils cannot support agriculture for long.
   c. large tracts of forest absorb carbon dioxide, reducing the threat of global warming.
   d. All of these are correct.
8. Which of these is mismatched?
   a. fossil fuel burning—carbon dioxide given off
   b. nuclear power—radioactive wastes
   c. solar energy—greenhouse effect
   d. biomass burning—carbon dioxide given off
9. Acid deposition causes
   a. lakes and forests to die.
   b. acid indigestion in humans.
   c. the greenhouse effect to lessen.
   d. All of these are correct.
10. Water is a renewable resource, and
    a. there will always be a plentiful supply.
    b. the oceans can never become polluted.
    c. it is still subject to pollution.
    d. primary sewage treatment plants assure clean drinking water.

Applying the Concepts

1. *Air pollutants have far-reaching effects.*
   Show how air pollutants are involved in all the environmental concerns discussed in this chapter.
2. *Tropical rain forests play a major role in the biosphere.*
   Explain why the tropical rain forests should be saved.
3. *Biodiversity encompasses three levels: genetic diversity, species diversity, and ecosystem diversity.*
   Discuss how these three levels of biodiversity are related to one another.

Using Technology

Your study of human impact on the global environment is supported by these available technologies:

Exploring the Internet
The Mader Home Page provides resources for and help with studying this chapter.

http://www.mhhe.com/sciencemath/biology/mader/
(Click on Biology.)

Explorations in Human Biology CD-ROM
Pollution of a Freshwater Lake (#16)

Understanding the Terms

acid deposition 476
aquifer 479
biological magnification 480
chlorofluorocarbons (CFCs) 478
deforestation 483
desertification 482
fossil fuel 473
greenhouse effect 474
nonrenewable energy source 473
ozone 478
ozone hole 478
PAN (peroxyacetyl nitrate) 477
photochemical smog 477
pollutant 474
pollution 473
renewable energy source 473
sewage treatment 479
thermal inversion 477

Match the terms to these definitions:

a. __________ Air pollution that contains nitrogen oxides and hydrocarbons, which react to produce ozone and PAN (peroxyacetyl nitrate).
b. __________ Formed from oxygen in the upper atmosphere, it protects the earth from ultraviolet radiation.
c. __________ Substance that is added to the environment and leads to undesirable effects for living organisms.
d. __________ Process by which substances become more concentrated in organisms in the higher trophic levels of a food web.
e. __________ Reradiation of solar heat toward the earth, caused by gases in the atmosphere.

Further Readings for Part iv

Allen, J. L., editor. *Environment 96/97.* 1996. Guilford, Conn.: Dushkin Publishing Group/Brown & Benchmark Publishers. This volume is a collection of articles pertaining to the problems and issues of the environment and environmental quality.

Anderson, D. M. August 1994. Red tides. *Scientific American* 271(2):62. The frequency of toxic red tides has been increasing because pollution provides rich nutrients, which encourages algal bloom.

Balick, M. J., and Cox, P. A. 1996. *Plants, people, and culture: The science of ethnobotany.* New York: *Scientific American* Library. This interesting, well-illustrated book discusses the medicinal and cultural uses of plants, and the importance of rain forest conservation.

Begon, M., et al. 1996. *Ecology: Individuals, populations, and communities.* London: Blackwell Science Ltd. This text discusses the distribution and abundance of organisms and their physical and chemical interactions within ecosystems.

Begon, M., et al. 1996. *Population ecology: A unified study of animals and plants.* 3d ed. Oxford: Blackwell Scientific Publications. The present state of population ecology is described in this introductory-level text.

Béland, P. May 1996. The beluga whales of the St. Lawrence River. *Scientific American* 274(5):74. Pollution and hydroelectric projects take their toll on the Beluga whale.

Blaustein, A. R., and Wake, D. B. April 1995. The puzzle of declining amphibian populations. *Scientific American* 272(4):52. Many species are disappearing; this may be a result of changes in the environment, including pollution, disease, and diet.

Borman, F. H., and Likens, G. E. 1994. *Pattern and process in a forested ecosystem.* New York: Springer-Verlag. Disturbance, development, and the steady state based on the Hubbard Brook Ecosystem Study is presented.

Brewer, R. 1994. *The science of ecology.* 2nd ed. Fort Worth: Saunders College Publishing. This text for ecology majors considers the role of humans in nature and in the environment, and uses scientific studies to illustrate ecological principles.

Broecker, W. S. November 1995. Chaotic climate. *Scientific American* 273(5):62. Researchers are beginning to understand past climate patterns; this knowledge may be used to predict future patterns.

Brown, B. E., and Ogden, J. C. 1993. Coral bleaching. *Scientific American* 268(1):64. Coral bleaching usually occurs due to abnormally high seawater temperatures, and this could indicate global warming.

Charlson, R., and Wigley, T. L. February 1994. Sulfate aerosal and climatic change. *Scientific American* 270(2):48. Sulfur aerosals scatter sunlight back into space before it can contribute to global warming.

Collier, M. P., et al. January 1997. Experimental flooding in Grand Canyon. *Scientific American* 276(1):82. Periodic man-made floods may improve the canyon environment.

Cowen, R. 1995. *The history of life.* 2d ed. Boston: Blackwell Scientific Publications. From the origin of life through human evolution, this is an introduction to paleontology and paleobiology.

Cox, G. 1997. *Conservation ecology.* 2d ed. Dubuque, Iowa: Wm. C. Brown Publishers. Discusses the nature of the biosphere, the threats to its integrity, and ecologically sound responses.

Cox, P. A., and Balick, M. J. June 1994. The ethnobotanical approach to drug discovery. *Scientific American* 270(6):82. Many rain forest plants are used by indigenous cultures for medicinal purposes; these flora should be screened for pharmaceutical compounds.

Cranbrook, E., and Edwards, D. S. 1994. *Belalong: A tropical rain forest.* London: The Royal Geographic Society, and Singapore: Sun Tree Publishing. This is a readable, well-illustrated account of biodiversity in a Brunei rain forest.

Cunningham, W. P., and Saigo, B. W. 1997. *Environmental science: A global concern.* Dubuque, Iowa: Wm. C. Brown Publishers. Provides scientific principles plus insights into the social, political, and economic systems impacting the environment.

Dobson, A. P. 1996. *Conservation and biodiversity.* New York: *Scientific American* Library. Discusses the extent and the value of biodiversity, and describes attempts to manage endangered species.

Duellman, W. E. 1992. Reproductive strategies of frogs. *Scientific American* 267(1):80. Frogs have mananged to colonize niches throughout the terrestrial environment.

Eisner, T., and Wilson, E. O., editors. 1994. Readings from Scientific American: *Animal behavior.* San Francisco: W. H. Freeman and Co. A collection of articles on animal behavior.

Frosch, R. A. September 1995. The industrial ecology of the 21st century. *Scientific American* 273(3):178. Taking lessons from the natural ecological system, industries may be able to recycle materials and minimize waste.

Getis, J. 1991. *You can make a difference: Help protect the earth.* Dubuque, Iowa: Wm. C. Brown Publishers. Presents challenges and responses with regard to environmental problems, and interesting alternatives to common household chemicals and cleaners.

Grier, J., and Burk, T. 1992. *Biology of animal behavior.* 2d ed. St. Louis: Mosby-Year Book, Inc. Essays integrate the structure and function of behavior with topics of ethology, comparative psychology, and neurobiology.

Hammond, A. L., et al., editors. 1994. *World resources 1994-95: A report by the World Resources Institute.* New York: Oxford University Press. This report provides accurate information on natural resource management.

Hammond, A. L., et al., editors. 1996. *World resources 1996–97: The urban environment.* New York: Oxford University Press. This report provides accurate information on urban environmental management.

Hedin, L. O., and Likens, G. E. December 1996. Atmospheric dust and acid rain. *Scientific American* 274(6):88. Although there has been a reduction in acidic air pollutants, acid rain continues to be a problem.

Hoagland, W. September 1995. Solar energy. *Scientific American* 273(3):170. Article discusses the use of solar energy to provide fuels.

Holloway, M. April 1994. Nurturing nature. *Scientific American* 270(4):98. Florida's Everglades are serving as a testing ground—and battlefield—for an epic attempt to restore an environment damaged by human activity.

Holloway, M. August 1994. Diversity blues. *Scientific American* 271(2):16. Toxic tides, coastal development, and pollution are increasing; oceanic biodiversity is suffering.

Holloway, M., and Horgan, J. 1991. Soiled shores. *Scientific American* 265(4):102. An evaluation of the effectiveness of cleanup techniques after the *Exxon Valdez* spill, and the implications for Persian Gulf cleanup.

Kirchner, W., and Towne, W. June 1994. The sensory basis of the honeybees' dance language. *Scientific American* 270(6):74. The honeybees' elaborately choreographed dance and its sounds communicate to nestmates where food outside the hive is located.

Klinkenborg, V. December 1995. A farming revolution. *National Geographic* 188(6):60. Farmers are experiencing strong crop yields by practicing farming methods that promote sustainable agriculture.

Kovacs, K. March 1997. Bearded seals. *National Geographic* 191(3):124. This article describes a behavioral study of bearded seals in their natural environment.

Krebs, C. 1994. *Ecology.* 4th ed. New York: Harper-Collins. This book discusses the distribution and abundance of organisms at the population and community levels.

Laman, T. April 1997. Borneo's strangler fig trees. *National Geographic* 191(4):38. Giant parasitic trees in the rain forests of Borneo kill their hosts, but provide food for other animals.

Lents, J. M., and Kelly, W. J. 1993. Clearing the air in Los Angeles. *Scientific American* 269(4):32. During the past two decades, pollution has been cut dramatically, even as population and the number of automobiles soared.

Lohmann, K. 1992. How sea turtles navigate. *Scientific American* 266(1):100. A combination of cues from the earth's magnetic field and the steady seasonal pattern of waves may be the sources of sea turtles' biological maps and compass.

May, R. M. 1992. How many species inhabit the earth? *Scientific American* 267(4):42. Author argues that an accurate census of species is crucial to preserve biological diversity and to manage the earth's resources.

Miller, G. T. 1996. *Living in the environment.* 9th ed. Belmont, Calif.: Wadsworth Publishers. Designed for use in an introductory course on environmental science, this book discusses how the environment is being used and abused, and what individuals can do to protect the environment.

Mitchell, J. G. February 1996. Our polluted runoff. *National Geographic* 189(2):106. 80% of U.S. water pollution is due to land runoff not resulting from municipal or industrial sources.

Morgan, M. et al. 1993. *Environmental science: Managing biological and physical resources.* Dubuque, Iowa: Wm. C. Brown Publishers. Written for the undergraduate, this book explains how various environmental issues are linked.

Murawski, D. A. March 1997. Moths come to light. *National Geographic* 191(3):40. Moths display a variety of disguises for survival.

Nemecek, S. August 1997. Frankly, my dear, I don't want a dam. *Scientific American* 277(2):20. The article discusses how dams affect biodiversity.

Newman, E. 1993. *Applied ecology.* Oxford: Blackwell Scientific Publications. Presents the role of biological science in environmental preservation (a basic knowledge of related fields of science and math is assumed).

Odum, E. 1997. *A bridge between science and society.* 3d ed. Sunderland, Mass.: Sinauer Associates. Introduces the principles of modern ecology as they relate to threats to the biosphere.

Pfenning, D. W., and Sherman, P. W. June 1995. Kin recognition. *Scientific American* 272(6):98. Discusses various methods organisms use to identify siblings.

Pinter, N. and Brandon, M.T. April 1997. How erosion builds mountains. *Scientific American* 276(4):74. The building of mountains depends on the destructive power of water and wind, as well as volcanic eruptions and seismic plate collisions.

Plucknett, D. L., and Winkelmann, D. L. September 1995. Technology for sustainable agriculture. *Scientific American* 273(3):182. The practice of sustainable agriculture, increasing productivity, while protecting the environment, will be difficult to achieve in developing countries.

Pollack, H., and Chapman, D. June 1993. Underground records of changing climate. *Scientific American* 268(6):44. More is being learned about global climate through geological studies.

Primak, R. B. 1995. *A primer of conservation biology.* Sunderland, Mass.: Sinauer Associates. The relatively new discipline of conservation biology addresses the alarming loss of biological diversity throughout the world.

Prugh, T. 1995. *Natural capital and human economic survival.* Solomons, Md: ISEE Press. A concise treatment detailing the necessity of incorporating the dynamics of ecosystems into economic and commercial systems.

Reganold, J. P., et al. 1990. Sustainable agriculture. *Scientific American* 262(6):112. Conservation-minded farming methods combined with modern technology can reduce farmers' dependence on dangerous chemicals.

Rennie, J. 1992. Living together. *Scientific American* 266(1):122. Ecologists are studying intimate associations that develop between host and parasite; these associations may have fundamentally shaped the evolution of all living things.

Rice, R. E., et al. April 1997. Can sustainable management save tropical forests? *Scientific American* 276(4):44. The strategy of replacing trees harvested for lumber in the rain forests often fails.

Robison, B. H. July 1995. Light in the ocean's midwaters. *Scientific American* 273(1):60. New techology allows organisms found in the very deep, dark benthic regions of the ocean to be seen.

Safina, C. November 1995. The world's imperiled fish. *Scientific American* 273(5):46. Article discusses the decline of fish populations due to the commercial fishing industry.

Schmidt, M. J. January 1996. Working elephants. *Scientific American* 274(1):82. In Asia, teams of elephants serve as an alternative to destructive logging equipment.

Smith, R. 1996. Ecology and field biology. 5th ed. New York: Harper & Row. Presents a balanced introduction to ecology—plant and animal, theoretical and applied, physiological and behavioral, and population and ecosystem.

Stiling, P. D. 1996. *Ecology: Theories and applications.* 2nd ed. Upper Saddle River, New Jersey: Prentice Hall. This text for ecology majors stresses the role of evolution and behavior in ecology, and uses scientific studies to illustrate the distribution and abundance of plants and animals.

Strobel, G. A. 1991. Biological control of weeds. *Scientific American* 265(1):72. Presents environmentally compatible alternatives to chemical herbicides.

Toon, O. B., and Turco, R. P. 1991. Polar stratospheric clouds and ozone depletion. *Scientific American* 264(6):68. Clouds and CFCs create the ozone hole in Antarctica.

Tumlinson, J. H., et al. 1993. How parasitic wasps find their hosts. *Scientific American* 266(3):100. Besides recognizing odors from their caterpillar hosts, wasps also learn to identify compounds released by the plant on which the caterpillars feed.

Whitehead, H. November 1995. Sperm whales. *National Geographic* 188(5):56. The social sperm whale forms extended family units.

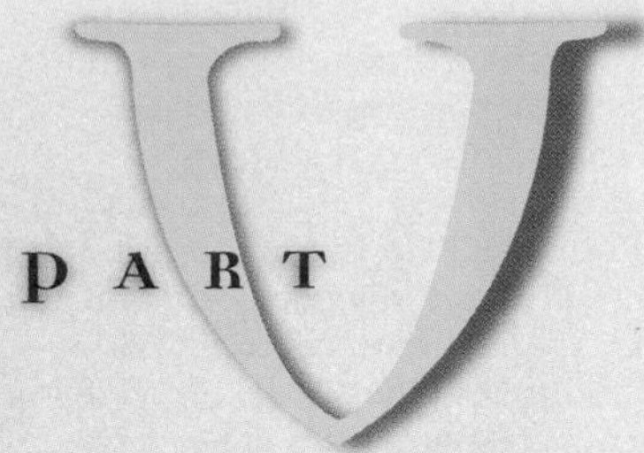

Part V Diversity of Life

Using fossil, molecular, and other types of evidence, biologists have determined the evolutionary relationships between major groups of organisms and know in general the history of life's diversity. Phylogenetic trees help us see these relationships, but it is important to realize that there is no direction to the evolutionary process, and that life does not necessarily progress from the simple to the complex. Each lineage is but a tiny twig on a sprawling bush that has no main trunk.

Adaptation to a particular way of life accounts for life's diversity. Unicellular protozoa and green algae are suited to living in a pond, mushrooms feed on organic matter in your lawn, and insects utilize flight as a way to escape enemies and to seek food. These organisms are just as well adapted to their way of life as are the more complex flowering plants and the vertebrate animals to theirs. The study of biology is the study of the many ways that living organisms have evolved various solutions to the same fundamental problems. All organisms must acquire materials and energy, protect themselves, and reproduce.

Classification of Living Things

CHAPTER 28

Chapter Concepts

28.1 Naming and Classifying Organisms

- Each known species has been given a binomial name consisting of the genus and specific epithet. 492
- Species are distinguished on the basis of structure, metabolism, reproductive isolation, and evolutionary relatedness. This chapter stresses evolutionary relatedness. 494
- Classification involves the assignment of species to a genus, family, order, class, phylum, kingdom, and domain (the largest classification category). 494

28.2 Constructing Phylogenetic Trees

- Systematics encompasses both taxonomy (the naming of organisms) and classification (placing species in the proper categories). 496
- Homology, molecular data, and the fossil record are used to decide the evolutionary relatedness of species and the construction of phylogenetic trees, diagrams that show their relatedness. 497, 499
- There are three main schools of systematics: the traditional, the cladistic, and the numerical phenetic school. 499

28.3 Deciding the Number of Kingdoms

- The five-kingdom system contains: Plantae, Animalia, Fungi, Protista, and Monera. 502
- Recently, three domains have been recognized: Bacteria, Archaea, and Eukarya. 502

Golden toad, *Bufo periglenses*, Costa Rican cloud forest

Faced with the enormous number of living things on earth, scientists realized long ago that we needed a way of classifying and naming individual species. Although the ancient Greek philosopher Aristotle devised a primitive classification system over two thousand years ago, it wasn't until the 1700s that a Swedish biologist, Carolus Linnaeus, developed a systematic method of naming species that is still used today. A species' name consists of two Latin words, as in *Homo sapiens* for humans. No two species have the same scientific name.

An organism is generally classified on the basis of its evolutionary relationship to other species. Suppose, for example, a new toad were found in a rain forest. The animal's reproductive behavior, anatomy, and genetics would all be examined and compared to similar known toads. Once the new toad's relationship to other toads was determined, it would be possible to decide on its name and classification. There are various schools of classification and some of these are quite new. The classification of organisms is not a static field of biology; it changes over time as new discoveries and ideas are developed.

Figure 28.1 Classifying organisms.
How would you classify these organisms? An artificial system would not take into account how they might be related through evolution, as would a natural system.

28.1 Naming and Classifying Organisms

Suppose you were asked to classify the living organisms you know about (Fig. 28.1). Most likely you would begin by making a list, and naturally this would require that you give each one a name. Then you would start assigning the organisms on your list to particular groups. But what criteria would you use—color, size, how the organisms relate to you? Deciding on the number, types, and arrangement of the groups would not be easy, and periodically you might change your mind or even start over. Biologists, too, have not had an easy time deciding how living things should be classified, and changes have been made throughout the history of this field.

Taxonomy [Gk. *tasso*, arrange, classify, and *nomos*, usage, law], the branch of biology concerned with identifying and naming organisms, began with the ancient Greeks and Romans. The scientific names we use today are rendered in Latin. The famous ancient Greek philosopher Aristotle was the first to be interested in taxonomy. Much later, John Ray, a British naturalist of the seventeenth century, also believed that each organism should have a set name. He said, "When men do not know the name and properties of natural objects—they cannot see and record accurately."

Assigning a Two-Part Name

The number of known types of organisms expanded greatly in the mid-eighteenth century due to European travel to distant parts of the world. It was during this time that Carolus Linnaeus (1707–78) developed the **binomial system** of naming species (Fig. 28.2). The name is a binomial because it has two parts. For example, *Lilium buibiferum* and *Lilium canadense* are two different species of lilies. The first word, *Lilium*, is the genus (pl., genera), a classification category that can contain many species. The second word is the specific epithet of the species within that genus. The *specific epithet* sometimes tells us something descriptive about the organism. Notice that the scientific name is in italics; the genus is capitalized while the specific epithet is not. The species is designated by the full

a.

b.

c.

Figure 28.2 Carolus Linnaeus.
a. Linnaeus was the father of taxonomy and gave us the binomial system of classifying organisms. He was particularly interested in classifying plants such as these two lilies. b. *Lilium buiblferum.* c. *Lilium canadense.*

name; in this case either *Lilium buibiferum* or *Lilium canadense.* The specific epithet alone gives no clue as to species—just as the house number alone without the street name gives no clue as to which house is specified. The genus name can be used alone, however, to refer to a group of related species.

Why do organisms need to be given a scientific name in Latin? Why can't we just use common names for organisms? A common name will vary from country to country just because different countries use different languages. Hence the need for a universal language such as Latin, which used to be well known by most scholars. Even those who speak the same language sometimes use different common names for the same organism. The Louisiana heron and the tricolored heron are the same bird found in southern United States. Between countries, the same common name is sometimes given to different organisms. A "robin" in England is very different from a "robin" in the United States, for example. When scientists use the same scientific name, they know they are speaking of the same organism.

The job of identifying and naming the species of the world is a daunting task. Of the estimated 3 to 30 million or even more species now living on earth, we have named a million species of animals and a half million species of plants and microorganisms. We are further along on some groups than others; it's possible we have just about finished the birds, but there may yet be hundreds of thousands of unnamed insects.

The scientific name of an organism consists of its genus and a specific epithet. The complete binomial name indicates the species.

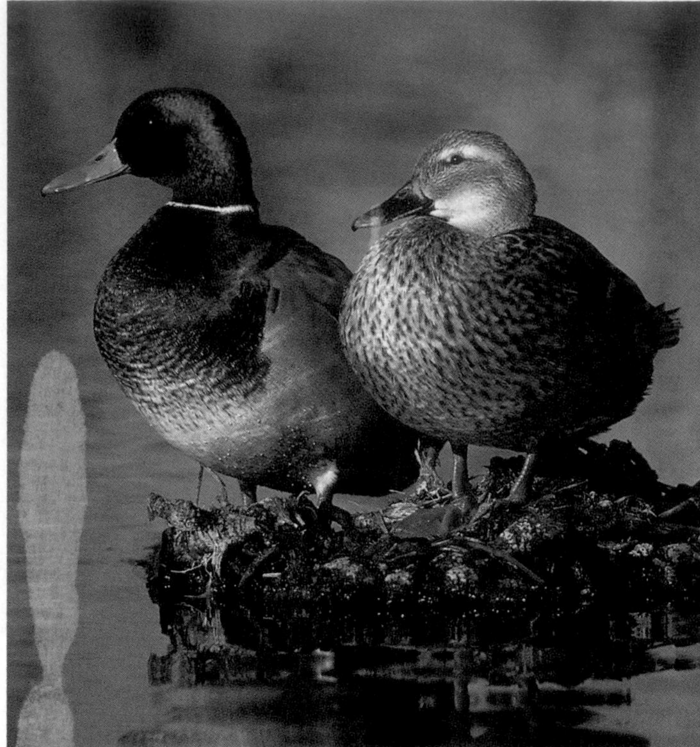

Figure 28.3 Identifying species.
Identifying the members of a species can be difficult—especially when the male and female members do not look alike, as in these mallards, *Anas platyrhynchos.*

Figure 28.4 Hybridization between species.
Zebroids are horse-zebra hybrids. Like mules, zebroids are generally infertile, due to differences in the chromosomes of their parents.

What Is a Species?

There are several ways to distinguish species, and each way has its advantages and disadvantages. For Linnaeus, every species has its own distinctive structural characteristics that are not shared by members of a similar species. In birds, the structural differences can involve the shape, size, and color of the body, feet, bill, or wings. We know very well, however, that variations do occur among members of a species. Differences between males and females or between juveniles and adults may even make it difficult to tell which organism belongs to what species (Fig. 28.3).

The *biological definition* of a species rests on the recognition that distinctive characteristics are passed on from parents to offspring. This definition, which states that members of a species interbreed and share the same gene pool, applies only to sexually reproducing organisms and cannot apply to asexually reproducing organisms. Sexually reproducing organisms are not always as reproductively isolated as we would expect. When a species has a wide geographic range, there may be variant types that tend to interbreed where their populations overlap (see Fig. 19.3). This observation has led to calling these populations subspecies, designated by a three-part name. For example, *Elaphe obsoleta bairdi* and *Elaphe obsoleta obsoleta* are two subspecies within the same snake species *Elaphe obsoleta*. It could be that these subspecies are actually distinct species. Even species that seem to be obviously distinct interbreed on occasion (Fig. 28.4). Therefore, the presence or absence of hybridization may not be informative as to what constitutes a species.

This chapter concerns **classification,** the assignment of organisms to categories on the basis of their **phylogeny**—evolutionary relationship to other organisms. In this context, a species is a taxonomic category below the rank of genus. Species in the same genus share a more recent common ancestor than do species in other taxa. **Taxa** [Gk. *tasso,* arrange, classify] are groups of organisms that fill a particular category of classification; *Rosa* and *Felis* are taxa in the category genus. A **common ancestor** is one that produced at least two lines of descent; there is one ancestor for all the types of roses, for example.

Now Let's Classify

Classification, which begins when an organism is named, includes taxonomy, since genus and species are two classification categories. The individuals we have so far mentioned were taxonomists who contributed to classification. Aristotle divided living things into 14 groups—mammals, birds, fish, and so on. Then he went on to subdivide the groups according to the size of the organisms. Ray used a more natural system, since he grouped animals and plants according to how he thought they were related; but Linnaeus simply used flower part differences to assign plants to these categories, which are still in use today: species, genus, order, and class. His studies were published in a book called *Systema Naturae.*

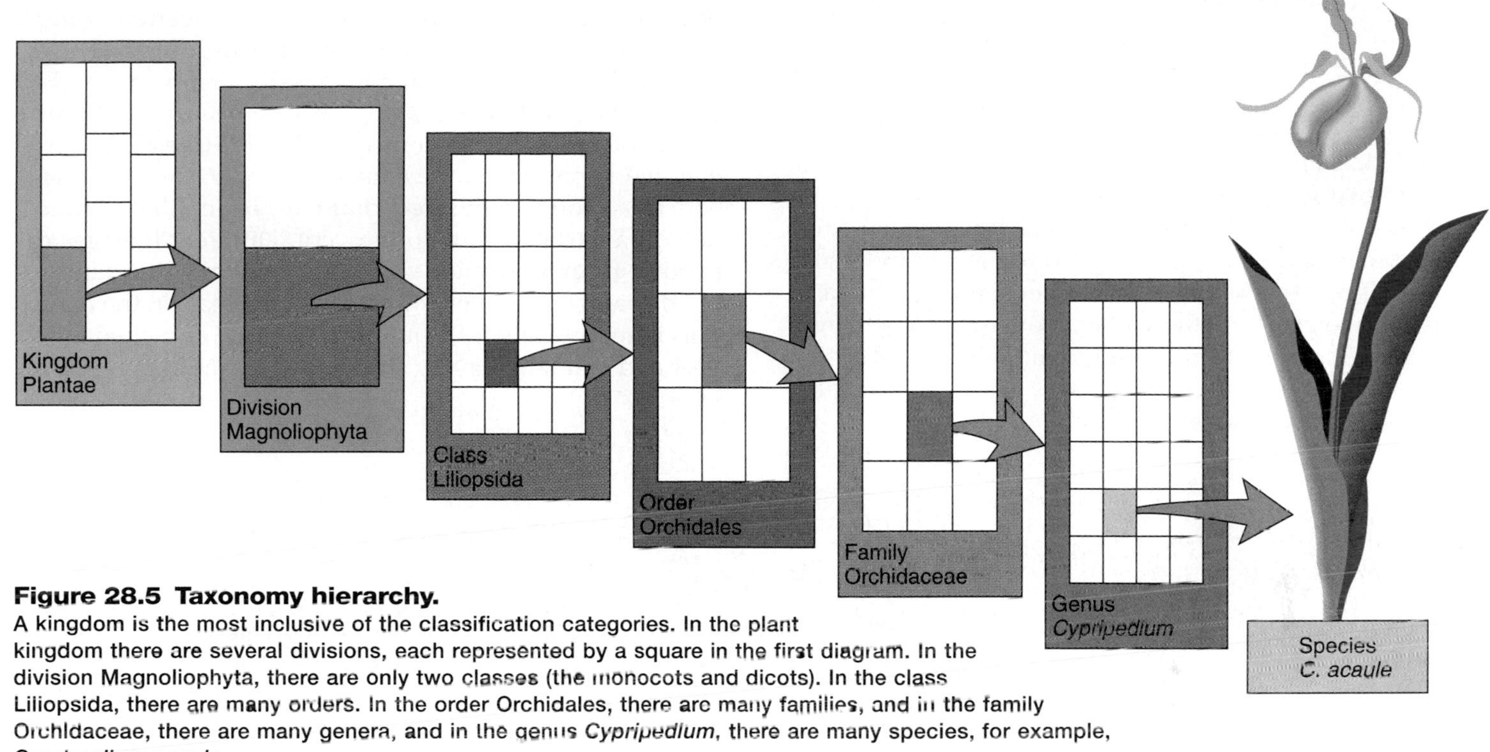

Figure 28.5 Taxonomy hierarchy.
A kingdom is the most inclusive of the classification categories. In the plant kingdom there are several divisions, each represented by a square in the first diagram. In the division Magnoliophyta, there are only two classes (the monocots and dicots). In the class Liliopsida, there are many orders. In the order Orchidales, there are many families, and in the family Orchidaceae, there are many genera, and in the genus *Cypripedium*, there are many species, for example, *Cypripedium acaule.*

Table 28.1

Classification of Humans

Kingdom Animalia	Eukaryotic, usually motile, multicellular organisms, without cell walls or chlorophyll; usually, internal cavity for digestion of nutrients
Phylum Chordata	Organisms that at one time in their life history have a dorsal hollow nerve cord, a notochord, and pharyngeal pouches
Class Mammalia	Warm-blooded vertebrates possessing mammary glands; body more or less covered with hair; well-developed brain
Order Primates	Good brain development, opposable thumb and sometimes big toe; lacking claws, scales, horns, and hoofs
Family Hominidae	Limb anatomy suitable for upright stance and bipedal locomotion
Genus *Homo*	Maximum brain development, especially in regard to particular portions; hand anatomy suitable to the making of tools
Species *sapiens**	Body proportions of modern humans; speech centers of brain well developed

* *To specify an organism, you must use the full name, such as* Homo sapiens.

Today, we make use of at least seven obligatory categories: **species, genus, family, order, class, phylum,** and **kingdom**. (The plant kingdom uses the category *division* instead of *phylum*.) There can be several species within a genus, several genera within a family, and so forth—the higher the category the more inclusive it is (Fig. 28.5). Therefore, there is a *hierarchy* of categories. The organisms that fill a particular classification category are distinguishable from other organisms by sharing a set of characteristics, or simply characters. A *character* is any structural, chromosomal, or molecular feature that distinguishes one group from another. Organisms in the same kingdom have *general* characters in common; those in the same species have quite *specific* characters in common. Table 28.1 lists some of the characters that help classify humans into major categories.

Each of the seven obligatory categories of classification can be subdivided into three additional categories as in superorder, order, suborder, and infraorder. Considering this, there are more than 30 categories of classification.

28.2 Constructing Phylogenetic Trees

Taxonomy and classification are a part of the broader field of **systematics** [Gk. *systema*, an orderly arrangement], which is the study of the diversity of organisms at all levels of organization. One goal of systematics is to determine phylogeny [Gk. *phyle*, tribe, L. *genitus*, producing], or the evolutionary history of a group of organisms. Classification is a part of systematics because it lists the unique characters of each taxon and ideally is designed to reflect phylogeny. A species is most closely related to other species in the same genus, then to genera in the same family, and so forth, from order to class to phylum to kingdom. When we say that two species (or genera, families, etc.) are closely related, we mean they share a recent common ancestor.

Figure 28.6 shows how the classification of groups of organisms allows one to construct a **phylogenetic tree**, a diagram that indicates common ancestors and lines of descent (lineages). In order to classify organisms and to construct a phylogenetic tree, it is necessary to determine the characters of the various taxa. A **primitive character** is one that is present in the common ancestor and all members of a group. A **derived character** is one that is found only in a particular line of descent. Different lineages diverging from a common ancestor may have different derived characters. For example, all the animals in the family Cervidae have antlers, but they are highly branched in red deer and palmate (having the shape of a hand) in reindeer.

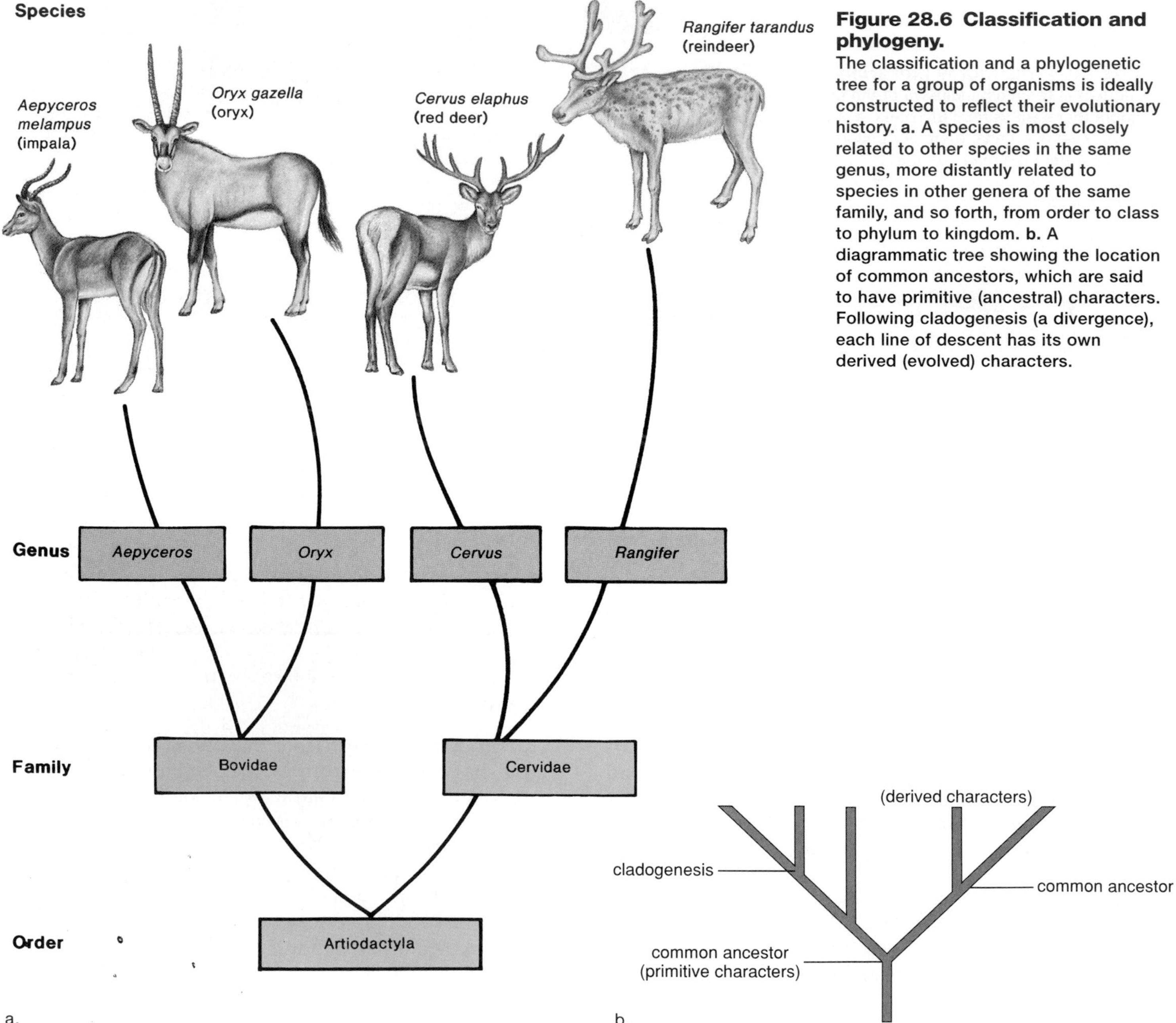

Figure 28.6 Classification and phylogeny.
The classification and a phylogenetic tree for a group of organisms is ideally constructed to reflect their evolutionary history. **a.** A species is most closely related to other species in the same genus, more distantly related to species in other genera of the same family, and so forth, from order to class to phylum to kingdom. **b.** A diagrammatic tree showing the location of common ancestors, which are said to have primitive (ancestral) characters. Following cladogenesis (a divergence), each line of descent has its own derived (evolved) characters.

What Are Our Data?

Systematists gather all sorts of data in order to discover the evolutionary relationship between species. Species placed in one group should have a common ancestry—that is, they should form a *monophyletic* [Gk. *monos*, one, and *phyle*, tribe] group. Systematists rely heavily on homology, molecular data, and the fossil record in order to determine monophyletic groups.

When Parts Are Similar

Homology [Gk. *homologos*, agreeing, corresponding] is character similarity that stems from having a common ancestry. Comparative anatomy, including embryological evidence, provides information regarding homology. The forelimbs of vertebrates are homologous because they contain the same bones organized in the same general way as in a common ancestor. **Homologous structures** are related to each other through common descent, although they may now differ in their structure and function. In contrast, **analogous structures** have the same function in different groups but are not derived from the same organ in a common ancestor. The wings of an insect and the wings of a bat are analogous structures.

Deciphering homology is sometimes difficult because of convergent evolution and parallel evolution. **Convergent evolution** is the acquisition of the same or similar characters in distantly related lines of descent. Convergence occurs when organisms have similar structural and functional traits, not because of a common ancestor but because they are adapted to the same type of environment. Both spurges and cacti are adapted similarly to a hot, dry environment, and they both are succulent, spiny, flowering plants (Fig. 28.7). However, the details of their flower structure indicate that these plants are not closely related. **Parallel evolution** is the acquisition of the same or similar characters in two or more related lineages without it being present in a common ancestor. A similar banding pattern is found in several species of moths, for example. It is sometimes difficult to tell if features are parallel, primitive, or derived.

Prickly pear, *Opuntia*

Spurge, *Euphorbia*

Figure 28.7 Convergent evolution.
Cacti and spurges evolved on different continents, and yet they are both succulent flowering plants with spines. Cacti are adapted to living in American deserts, whereas spurges are adapted to living in tropical habitats of Africa. This is an example of convergent evolution.

When Genes Are Similar

When two lineages first diverge from a common ancestor, the genes and proteins of the lineages are nearly identical. But as time goes by, each lineage accumulates gene changes which lead to RNA and protein changes. Many changes are neutral (not tied to adaptation) and accumulate at a fairly constant rate; such changes can be used as a kind of **molecular clock** to indicate relatedness and evolutionary time.

The genetic difference between two species is determined in a number of ways. Immunological techniques can roughly judge the similarity of plasma membrane proteins. In one procedure, antibodies are produced by transfusing a rabbit with the cells of one species. Cells of the second species are exposed to these antibodies, and the degree of the reaction is observed. The stronger the reaction, the more similar the cells from the two species. Amino acid sequencing of particular proteins, on the other hand, directly indicates when changes have occurred in specific genes.

It is possible to examine and compare the RNA nucleotide sequences of ribosomes to determine relatedness. All cells have ribosomes that are essential for protein synthesis. The genes that code for rRNAs have changed very slowly during evolution in comparison to other genes. Therefore, they provide a reliable indicator of the difference between organisms. This technique has been particularly useful in reclassifying bacteria and archaea as discussed later in the chapter.

It is possible to determine DNA similarities by **DNA-DNA hybridization.** The DNA double helix of each species is separated into single strands. Then strands from both species are allowed to combine. The more closely related the two species, the better the two strands of DNA will stick together. Some long-standing questions in systematics have been resolved by doing DNA-DNA hybridization. The giant panda, which lives in China, was at one time considered to be a bear, but its bones and teeth resemble those of a raccoon. The giant panda eats only bamboo and has a false thumb by which it grasps bamboo stalks. The red panda, which lives in the same area and has the same raccoonlike features, also feeds on bamboo but lacks the false thumb. The results of DNA hybridization studies suggest that after raccoons and bears diverged from a common lineage 50 million years ago, the giant panda diverged from the bear line and the red panda diverged from the raccoon line:

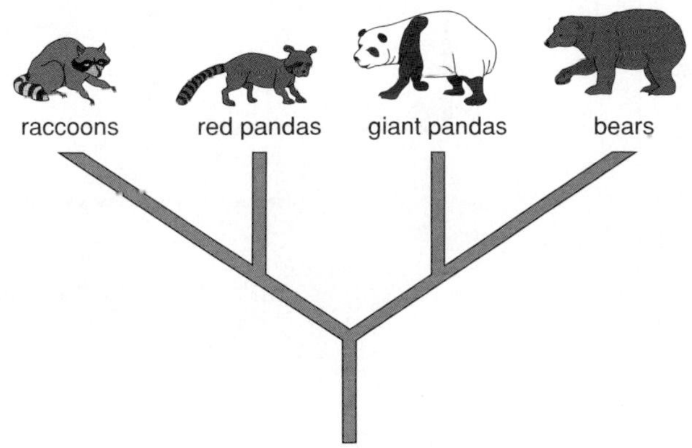

Therefore it can be seen that some of the characters of the giant panda and the red panda are primitive (present in a common ancestor), and some are due to parallel evolution.

Because hybridization studies provide only general information, many researchers prefer to compare data regarding the nucleotide sequence of a particular gene or genes. One study involving DNA differences suggested that chimpanzees are more closely related to humans than they are to other apes (Fig. 28.8). Yet in most classifications, humans and chimpanzees are placed in different families; humans are in the family Hominidae and chimpanzees are in the family Pongidae. In contrast, the rhesus monkey and the green monkey, which have more numerous DNA differences, are placed in the same family (Cerocpithecidae). To be consistent, shouldn't humans and chimpanzees also be in the same family? Some systematists believe that since humans are markedly different from chimpanzees because of adaptation to a different environment, it is justifiable to place humans in a separate family.

When Fossils Are Available

The fossil record shows the history of life in broad terms. However, the fossil record in regard to individual lineages is incomplete for several reasons. First, soft-bodied organisms do not fossilize. The chances of any organism becoming a fossil are not too good; most organisms are eaten or decay before they have a chance to be buried. Second, fossils tend to exist for only harder body parts, such as bones and teeth. Even then, fossils must survive powerful geological processes and end up in a location that allows someone to find them.

Fossils can be dated, and therefore an available fossil can establish the antiquity of a species. In some instances, the fossil record can trace a lineage through time. Even so, the fossil record does not necessarily provide evidence of whether a feature is primitive or derived. The only objective method for distinguishing primitive from derived characters is discussed in the reading on page 501.

Homology, genetic data, and the fossil record help researchers decipher phylogeny and construct phylogenetic trees.

Figure 28.8 Genetic data.
The relationship of certain primate species based on a study of their genomes. The length of the branches indicates the relative number of nucleotide pair differences that were found among groups. With the help of the fossil record, it is possible to suggest a date at which each group diverged from the other.

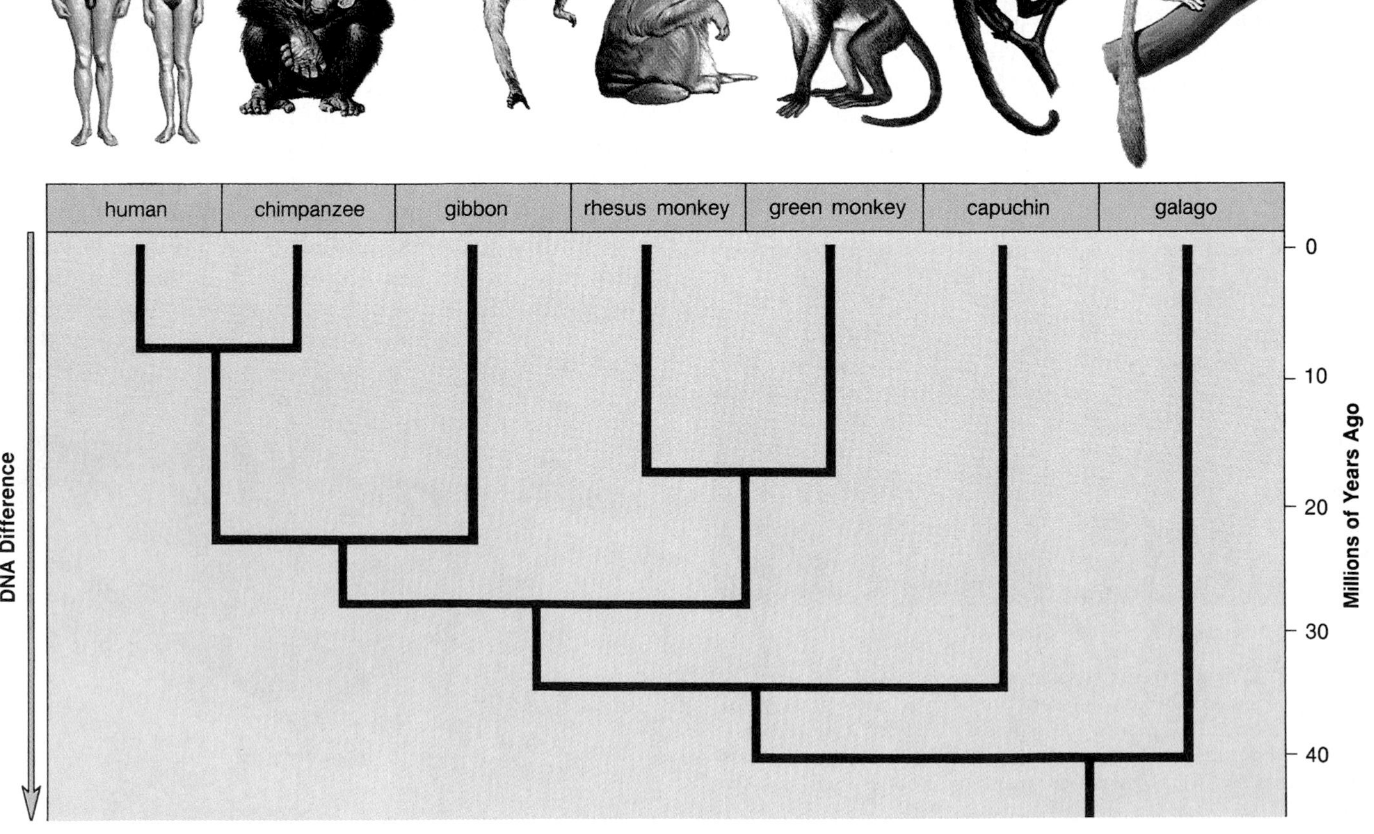

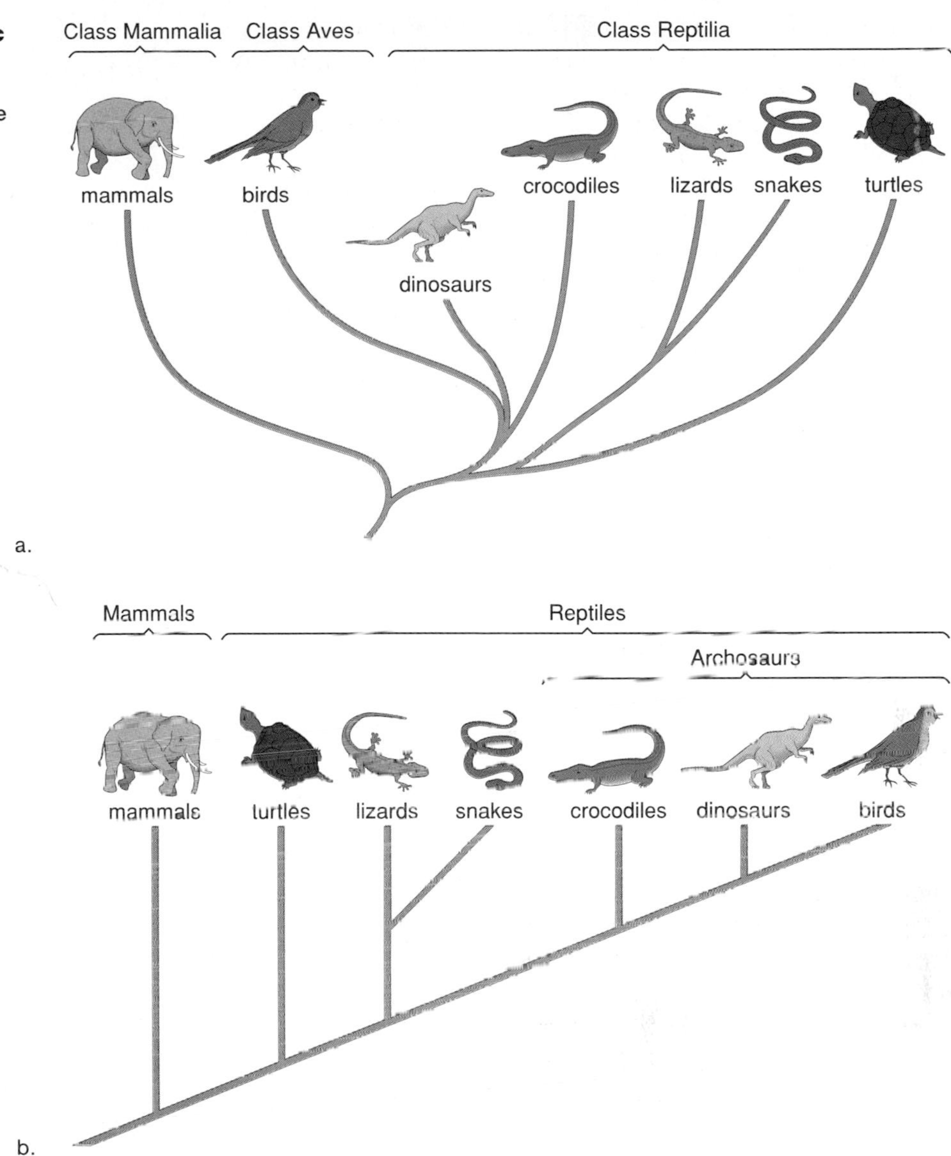

Figure 28.9 Traditional versus cladistic view of reptilian phylogeny.
a. According to traditionalists, mammals and birds are in separate classes. **b.** According to the cladists, crocodiles, dinosaurs, and birds all share a recent common ancestor and should be in the same subclass.

Who Constructs Phylogenetic Trees?

There are three main schools of systematics: traditional, cladistics, and numerical phenetics. The traditional school stresses both common ancestry *and* the degree of structural difference among divergent groups. Therefore, a group that has adapted to a new environment and shows a high degree of evolutionary change is not always classified with the common ancestor from which it evolved. For example, in Figure 28.9*a* birds and mammals are shown as separate lines of descent because it is quite obvious to the most casual observer that mammals (having hair and mammary glands) and birds (having feathers) are quite different in appearance from reptiles (having scaly skin). The traditionalist goes on to say that birds and mammals evolved from reptiles.

Cladists, on the other hand, stress common ancestry and shared derived characters only. They discount adaptations to new environments. In the example cited, cladists are doubtful that "reptiles" should be considered a taxon at all because the only thing dinosaurs, crocodiles, snakes, lizards, and turtles have in common is that they are not birds or mammals.

The cladists prefer the diagram shown in Fig. 28.9*b*. All the animals shown can be placed in one group because they all evolved from a common ancestor, which was an egg-laying creature. All mammals are indeed in one class because they all have hair, mammary glands, and three middle ear bones. Their common ancestor must have had these characteristics also. However, since crocodiles, dinosaurs, and birds all share common derived characteristics, they should all be in the same subclass, called archosaurs. "So does this mean that crocodiles and dinosaurs are birds?" the traditionalist asks. Cladists reply that their method is objective and not subjective like that of the traditionalists.

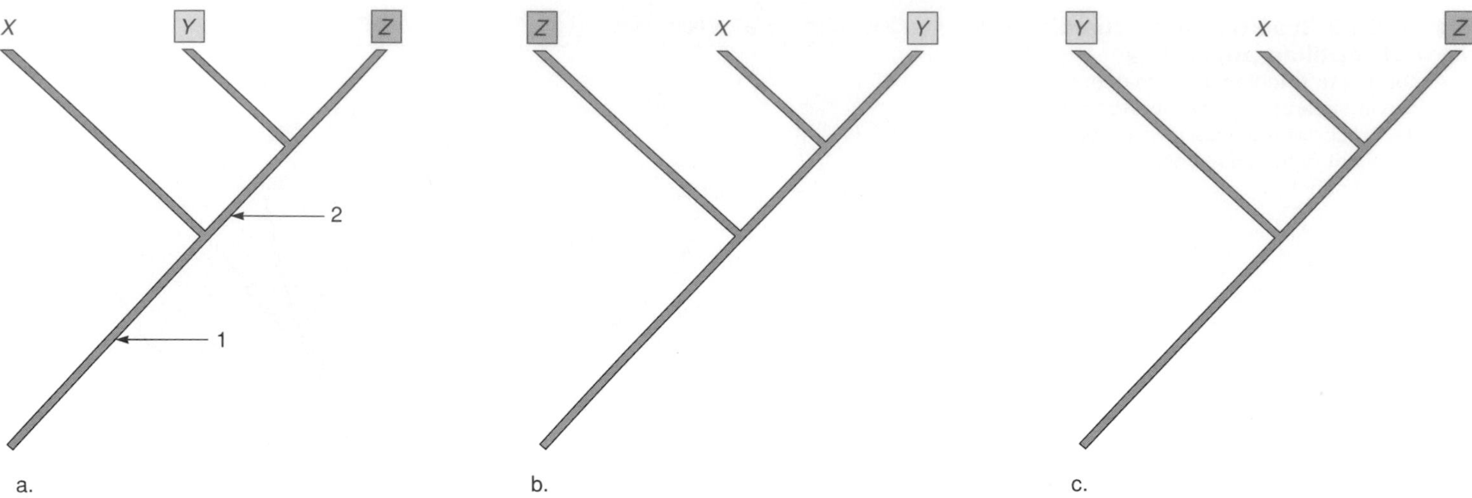

Figure 28.10 Alternate, simplified cladograms.
a. *X, Y, Z* share the same characters, designated by the first arrow, and are judged to form a monophyletic group (share a common ancestor). *Y* and *Z* are grouped together because they share the same derived character, designated by the second arrow and symbolized by the colored box. **b, c.** These cladograms are rejected because in each you would have to assume that the same character (colored boxes) evolved in different groups. Since this seems unlikely, the first branching pattern is chosen as the hypothesis.

Cladistics Is Objective

This school of **cladistics** is based on the work of Willi Hennig, who sought a more objective way of establishing relationships and classifying organisms. He felt that the present traditional system, which was based on general similarities, was subjective and did not produce testable hypotheses. Instead, he suggested that species be grouped according to their *shared derived characters*. As mentioned, all mammals have characters that they share and that distinguish them from other taxa. The different types of mammals also have their own derived characters. For example, all bats have forelimbs modified as wings, and all whales have forelimbs modified as flippers. However, bat wings and whale flippers are homologous, and the presence of homologous structures indicates that taxa share a common ancestry.

Once cladists have assembled their data regarding shared characters, they construct **cladograms** [Gk. *klados*, branch, stem, *gramma*, picture] to show the branching (cladistic) relationships among species in regard to the distribution of shared derived characters. A cladogram should be regarded as a hypothesis that can be tested and either corroborated or refuted on the basis of additional data. A cladogram differs from a traditional phylogenetic tree, particularly because (1) only data (not subjective evaluations) are used in its construction and (2) the data are presented as a part of the cladogram. Figure 28.10 shows a simple cladogram that involves only three species; construction of a more complex cladogram is discussed in the reading on page 501. A cladogram is not considered a phylogenetic tree, but its branching pattern does resemble a true phylogenetic tree.

Figure 28.10 shows a cladogram in which all three species represented by *X, Y,* and *Z* belong to the same monophyletic group, since they all share the derived characters designated by the first arrow. Species *Y* and *Z* are placed in the same subgroup because they share the derived characters designated by the second arrow. How do you know you have done the cladogram correctly, and that the other two patterns shown in Figure 28.10 are not likely? In the other two arrangements, the characters represented by the colored box would have had to evolve twice. Cladists are always guided by the principle of *parsimony*—the minimum number of assumptions is the most logical. That is, they construct the cladogram that minimizes the number of assumed evolutionary changes. However, they must be on the lookout for the possibility that convergent evolution has produced what appears to be common ancestry. Then, too, there is the realization that the reliability of a cladogram is dependent on the knowledge and skill of the particular investigator gathering the data and doing the character analysis.

Phenetics Is Numerical

In numerical phenetics, species are clustered according to the number of their similarities. Systematists of this school believe that you cannot construct a classification that actually reflects phylogeny, and that it is better to rely strictly on a method that does away with personal prejudices. They measure as many traits as possible, count the number the two species share, and then estimate the degree of relatedness. They simply ignore the possibility that some of the shared characters are probably the result of convergence or parallelism, or that some of the characters might depend on one another. For example, a large animal is bound to have larger parts. The results of their analysis are depicted in a phenogram. (Phenograms have been known to vary for the same group of organisms, depending on how the data are collected and handled.) Figure 28.8 is an example of a phenogram that is based solely on the number of DNA differences among the species shown.

A closer Look

How to Construct a Cladogram

Choosing the characters for analysis requires knowledge of the organisms in question. A cladogram (as a scientific hypothesis) is only as reliable as the evidence that produced it. When constructing a cladogram, the first step is to draw up a table that summarizes the characters of the taxa being compared. At least one but preferably several species are studied as outgroups; a taxon that is not one of the study groups. Any character found both in the study groups and the outgroup is considered to be primitive (ancestral) for the study groups. Any character found in all taxa or in only one taxon is excluded from the cladogram. The other characters are considered shared derived characters and are used to construct the cladogram.

Figure 28A is an example of a cladogram. A common ancestor, together with all its descendant species, is a **clade** [Gk. *klados,* branch, stem]; this cladogram has three clades that differ in size because the first includes the other two, and so forth. All the study groups belong to the first clade because they all have vertebrae; newts, snakes, and lizards are in the clade that has lungs and a three-chambered heart, and only snakes and lizards have an amniote egg and internal fertilization. In the event that there is more than one possible cladogram consistent with the evidence, the one that leaves the fewest shared derived characters unexplained or has the fewest evolutionary changes in derived characters is selected.

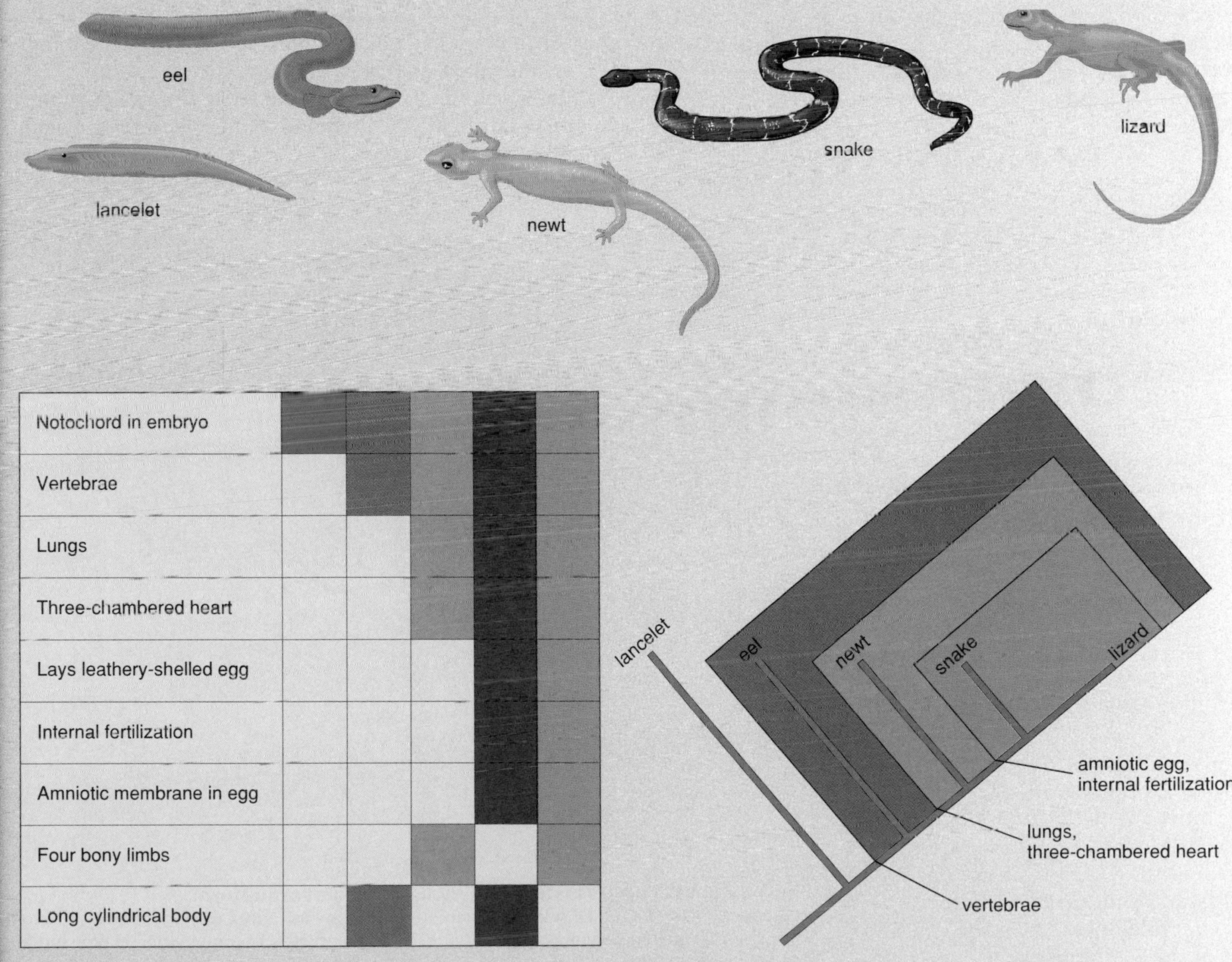

Figure 28A Steps in producing a cladogram.
a. Construction of a table. The table shown here is simplified; a real table would have many more characters. b. Notice that this cladogram does not include the presence of four bony limbs versus a long, cylindrical body. Evidence is needed that snakes have lost their limbs through evolution; otherwise, it appears that eels and snakes share a derived character.

28.3 Deciding the Number of Kingdoms

From Aristotle's time to the middle of the twentieth century, biologists recognized only two kingdoms: kingdom **Plantae** (plants) and kingdom **Animalia** (animals). Plants were literally organisms that were planted and immobile, while animals were animated and moved about. After the light microscope was perfected in the late 1600s, unicellular organisms were revealed that didn't fit neatly into the plant or animal kingdoms. In the 1880s, a German scientist, Ernst Haeckel, proposed adding a third kingdom. He called this **Protista** (protists) in order to separate unicellular microscopic organisms from multicellular ones.

In 1969, R. H. Whittaker expanded the classification system to include five kingdoms: **Plantae, Animalia, Fungi, Protista,** and **Monera**. Organisms were placed into categories based on type of cell (prokaryotic or eukaryotic), organization of cells (unicellular or multicellular), and nutritional requirements. The five-kingdom system of classification depicted in Figure 28.11 became widely accepted at this time.

The five-kindom classification system places the fungi (yeast, mushrooms, and molds) in a separate kingdom. Fungi are eukaryotes that form spores and lack flagella throughout their life cycle. Whittaker argued for a separate kingdom for fungi because they are unicellular or multicellular saprotrophs, organisms that absorb nutrients from decaying organic matter. He pointed out that plants, animals, and fungi are all multicellular eukaryotes, but each has a distinctive nutritional mode: plants are autotrophic by photosynthesis, animals are heterotrophic by ingestion, and fungi are heterotrophic saprotrophs.

The kingdom Protista contains a diverse group of organisms that are hard to classify and define. They are all eukaryotes that are mainly unicellular but may be multicellular filaments, colonies, or sheets. They do not form true tissues. Protists may have ingestive, photosynthetic, or saprotrophic nutrition. In this text, kingdom Protista contains the algae, including multicellular algae, protozoa, water molds, and slime molds. There has been considerable debate over the classification of these organisms.

In the five-kingdom system, the Monera are distinguished by their structure—they are *prokaryotic* (lack a membrane-bounded nucleus)—whereas the organisms in the other kingdoms are *eukaryotic* (have a membrane-bounded nucleus) (Figure 28.12 and Table 28.2). Kingdom Monera originally included all the bacteria. But with the advent of DNA and ribosomal RNA sequencing methods in the 1980s, a new group of microorganisms was identified. These were originally called archaebacteria but are now properly called Archaea (see the next chapter). The archaea have nucleotide sequences that are either close to eukaryotic sequences or are unique and, therefore, not found in either bacteria or eukaryotes.

On the basis of these new data, it has been suggested that classification should include three evolutionary domains: **Bacteria, Archaea,** and **Eukarya.** The bacteria and archaea evolved early in the history of life, while the eukarya evolved later from the archaeal line of descent.

Recently, it has been suggested that there are three evolutionary domains: Bacteria, Archaea, and Eukarya. The five-kingdom system of classification recognizes these kingdoms: Monera, Protista, Fungi, Plantae, and Animalia.

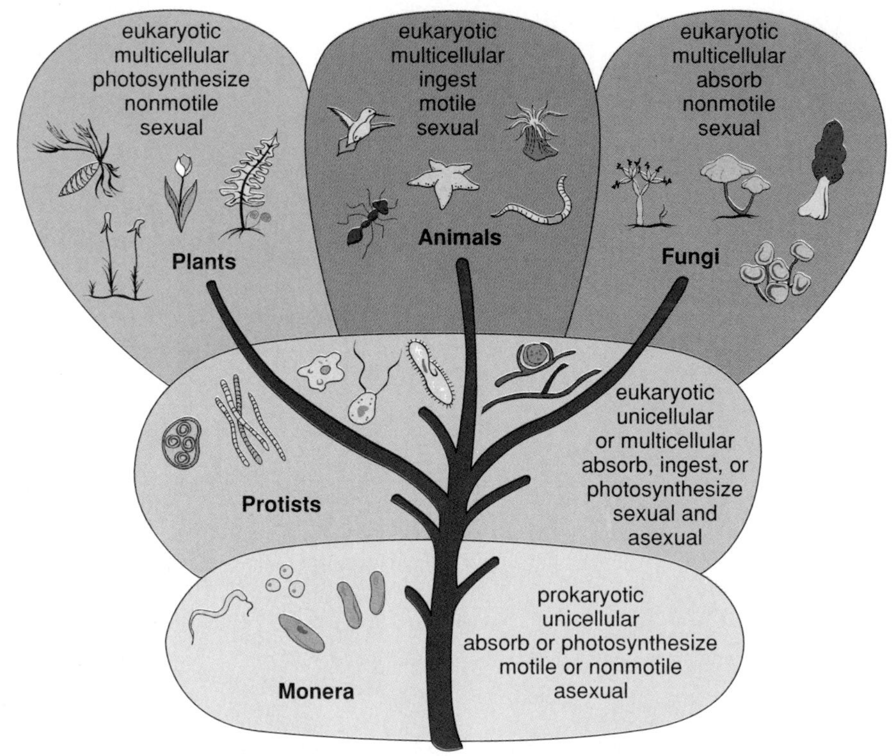

Figure 28.11 The five-kingdom system of classification. Representatives of each kingdom are depicted in the ovals, and a phylogenetic tree roughly indicates the lines of descent. The relationship of eukaryotes to prokaryotes is now under intense investigation.

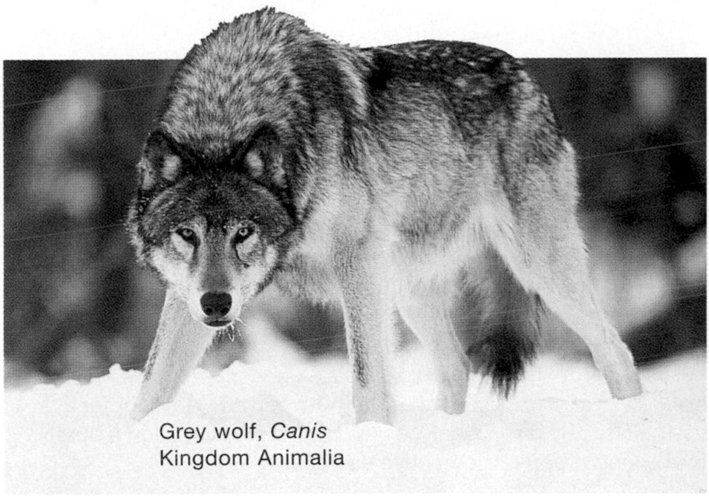
Grey wolf, *Canis*
Kingdom Animalia

Black-eyed Susan, *Rudbeckia*
Kingdom Plantae

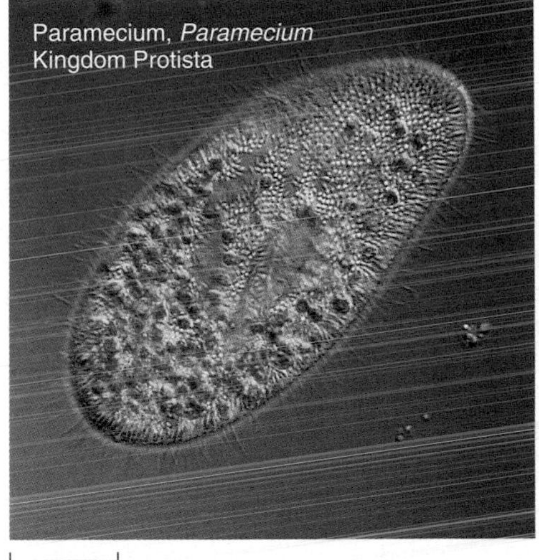
Paramecium, *Paramecium*
Kingdom Protista

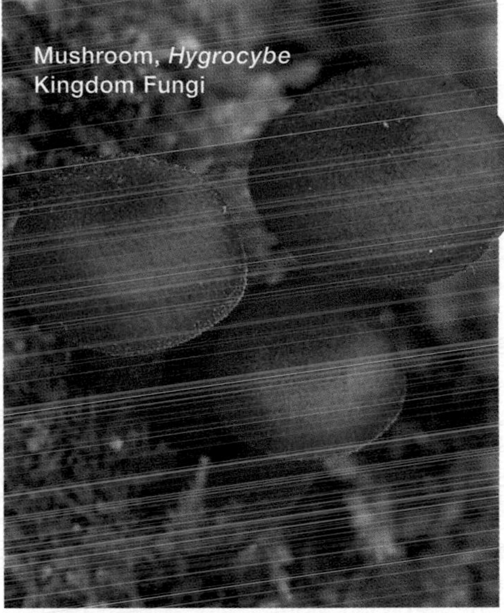
Mushroom, *Hygrocybe*
Kingdom Fungi

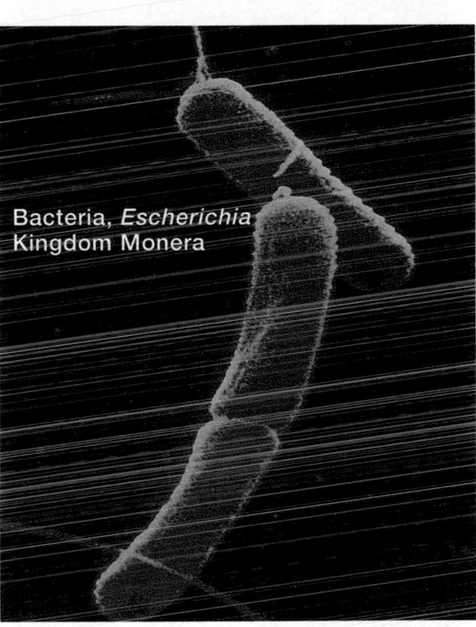
Bacteria, *Escherichia*
Kingdom Monera

Figure 28.12 The five-kingdom system—a pictorial representation.

Table 28.2

Classification Criteria for the Five-Kingdom System

	Monera	Protista	Fungi	Plantae	Animalia
Type of Cell	Prokaryotic	Eukaryotic	Eukaryotic	Eukaryotic	Eukaryotic
Complexity	Unicellular	Unicellular	Unicellular or multicellular	Multicellular	Multicellular
Type of Nutrition	Autotrophic by various or heterotrophic by various	Photosynthetic or heterotrophic by various	Heterotrophic saprotrophs	Photosynthetic	Heterotrophic by ingestion
Motility	Sometimes by flagella	Sometimes by flagella (or cilia)	Nonmotile	Nonmotile	Motile by contractile fibers
Life Cycle*	Asexual usual	Various	Haplontic	Alternation of generations	Diplontic
Internal Protection of Zygote	No	No	No	Yes	Yes
Nervous System	None	Conduction of stimuli in some forms	None	None	Present

*See p. 534.

connecting concepts

We have seen in this chapter that identifying, naming, and classifying living organisms is an ongoing process. Carolus Linnaeus' system of binomial nomenclature is still accepted by virtually all biologists, but many species remain to be found and named (most in the rain forests). For several years now, Whittaker's five-kingdom concept of life on earth has been widely used, but new findings will most likely bring about radical changes from this system. The archaea are creatures structurally similar to bacteria, but it has been found that their ribosomal subunits differ from those of bacteria and are instead similar to those of eukaryotes. Also, some archaeal genes are unique only to the archaea. Hence, the suggestion has been made that there are three domains of life: bacteria, archaea, and eukarya. Most likely, several kingdoms will eventually be recognized among the bacteria and archaea just as there are among the eukarya (protists, fungi, plants, and animals).

Most of today's systematists rely on evolutionary relationships among organisms for classification purposes, but the traditional and cladistic schools differ as to how to determine such relationships. The traditionalist is willing to consider both structural similarities and obvious differences due to adaptations to new environments. The cladists believe that only similarities should be used to classify organisms. In the end, it may be the ability to quickly sequence genes that will do away with the need for any subjective analyses and make classification a purely objective science.

Summary

28.1 Naming and Classifying Organisms

Taxonomy deals with the naming of organisms; each species is given a binomial name consisting of the genus and specific epithet.

Distinguishing species on the basis of structure can be difficult because members of a species can vary in structure. Distinguishing species on the basis of reproductive isolation runs into problems because some species hybridize and because reproductive isolation is very difficult to observe. In this chapter, species is a taxon occurring below the level of genus. Species in the same genus share a more recent common ancestor than species in related genera, etc.

Classification involves the assignment of species to categories. When an organism is named, a species has been assigned to a particular genus. There are seven obligatory categories of classification: species, genus, family, order, class, phylum, and kingdom. Each higher category is more inclusive; members of the same kingdom share general characters, and members of a species share quite specific characters.

28.2 Constructing Phylogenetic Trees

Systematics, a very broad field, encompasses both taxonomy and classification. Classification should reflect phylogeny, and one goal of systematics is to create phylogenetic trees, based on primitive and derived characters.

Homology, molecular data, and the fossil record are used to help decipher phylogenies. Homology helps indicate when species belong to a monophyletic group (share a common ancestor); however, convergent evolution and parallel evolution sometimes make it difficult to distinguish homologous structures from analogous structures. Molecular data are used to indicate relatedness, but DNA base sequence and restriction site data may pertain more directly to phylogenetic characters. Because fossils can be dated, available fossils can establish the antiquity of a species. If the fossil record is complete enough, we can sometimes trace a lineage through time.

Today there are three main schools of systematics: the traditional, the cladistic, and the numerical phenetic schools. The traditional school stresses common ancestry and the degree of structural difference among divergent groups in order to construct phylogenetic trees. The cladistic school analyzes primitive and derived characters and constructs cladograms on the basis of shared derived characters. A clade is a common ancestor, and all the species derived from that common ancestor. Cladograms are diagrams based on objective data. The numerical phenetic school clusters species on the basis of the number of shared similarities regardless of whether they might be convergent, parallel, or depend on one another.

28.3 Deciding the Number of Kingdoms

On the basis of molecular data, it has been suggested that there are three evolutionary domains: Bacteria, Archaea, and Eukarya. The five-kingdom system of classification recognizes these kingdoms: Plantae, Animalia, Fungi, Protista (algae, protozoa, water molds, slime molds), and Monera (the bacteria).

Reviewing the Chapter

1. Explain the binomial system of naming organisms. Why must the species be designated by the complete name? 492–93
2. Why is it necessary to give organisms scientific names? 493
3. Discuss three ways to define a species. Which way relates to classification? 494
4. What are the seven obligatory classification categories? In what way are they a hierarchy? 495
5. How is it that taxonomy and classification are a part of systematics? What three types of data help systematists construct phylogenetic trees? 496–98
6. In what ways do the traditional school, the cladistic school, and the numerical phenetic school of systematics differ? 499
7. Discuss the principles of cladistics, and explain how to construct a cladogram. 500–1
8. Describe the five-kingdom system of classification, and give examples of each kingdom. 502
9. Compare the five kingdoms on the basis of appropriate classification criteria. 502–3
10. What are the three evolutionary domains, and how might they be related? 502

Testing Yourself

Choose the best answer for each question

1. Which is the scientific name of an organism?
   a. *Rosa rugosa*
   b. *Rosa*
   c. *rugosa*
   d. All of these are correct.
2. Which of these best pertains to taxonomy? Species
   a. have three-part names such as *Homo sapiens sapiens.*
   b. are reproductively isolated from other species.
   c. share the most recent common ancestor.
   d. have only primitive characters.
3. The classification category below the level of family is
   a. class.
   b. species.
   c. phylum.
   d. genus.
4. Which kingdom is mismatched?
   a. Monera—fungi
   b. Protista—multicellular algae
   c. Plantae—flowers and mosses
   d. Animalia—arthropods and humans
5. Which kingdom is mismatched?
   a. Fungi—usually saprotrophic
   b. Plantae—usually photosynthetic
   c. Animalia—rarely ingestive
   d. Protista—various modes of nutrition
6. In a phylogenetic tree, which is incorrect?
   a. Dates of divergence are always given.
   b. Common ancestors occur at the notches.
   c. The more recently evolved are at the top of the tree.
   d. Ancestors have primitive characters.
7. Which is mismatched?
   a. homology—character similarity due to a common ancestor
   b. molecular data—DNA strands match
   c. fossil record—bones and teeth
   d. homology—functions always differ
8. One benefit of the fossil record is
   a. that hard parts are more likely to fossilize.
   b. fossils can be dated.
   c. its completeness.
   d. All of these are correct.
9. In the traditional school of systematics, birds are assigned to a different group from reptiles because
   a. they evolved from reptiles.
   b. they are quite different from reptiles.
   c. feathers came from scales.
   d. All of these are correct.
10. In cladistics
    a. a clade must contain the common ancestor plus all its descendants.
    b. derived characters help construct cladograms.
    c. data for the cladogram are presented.
    d. All of these are correct.
11. Answer the following questions about this cladogram.
    a. This cladogram contains how many clades? How are they designated in the diagram?
    b. What character is shared by all the study groups? What characters are shared by only snakes and lizards?
    c. Which groups share a common ancestry? How do you know?

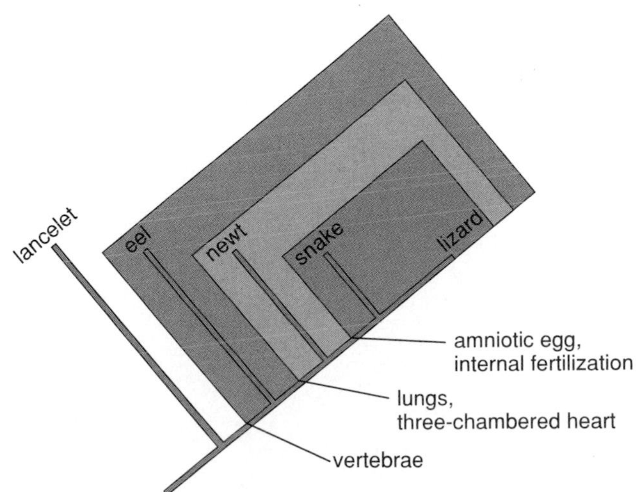

Applying the Concepts

1. *Classification reflects phylogeny.*
   What does the classification of humans tell you about their evolutionary history?
2. *Organisms exist only at the species level.*
   Discuss the difficulties of defining a species.
3. *Classification is based on data.*
   Why do cladists believe that traditionalists are being subjective when they construct phylogenetic trees?

Using Technology

Your study of the classification of living things is supported by these available technologies:

Exploring the Internet
The Mader Home Page provides resources for and help with studying this chapter.

http://www.mhhe.com/sciencemath/biology/mader/
(Click on Biology.)

Understanding the Terms

analogous structure 497
Animalia 502
Archaea 502
Bacteria 502
binomial system 492
clade 501
cladistics 500
cladogram 500
class 495
classification 494
common ancestor 494
convergent evolution 497
derived character 496
DNA-DNA hybridization 497
Eukarya 502
family 495
Fungi 502
genus 495
homologous structure 497
homology 497
kingdom 495
molecular clock 497
Monera 502
order 495
parallel evolution 497
phylogenetic tree 496
phylogeny 494
phylum 495
Plantae 502
primitive character 496
Protista 502
species 495
systematics 496
taxon (pl., taxa) 494
taxonomy 492

Match the terms to these definitions:

a. __________ Branch of biology concerned with identifying, describing, and naming organisms.
b. __________ Diagram that indicates common ancestors and lines of descent.
c. __________ Evolutionary history of a group of organisms.
d. __________ Group of organisms that fills a particular classification category.
e. __________ School of systematics that determines the degree of relatedness by analyzing primitive and derived characters and constructing cladograms.
f. __________ Set of categories to which species are assigned on the basis of their relationship to other species.
g. __________ Similarity in structure due to having a common ancestor.
h. __________ Similarity in structure in related groups that cannot be traced to a common ancestor.
i. __________ Structural, physiological, or behavioral trait that is present in a common ancestor and all members of a group.
j. __________ Study of the diversity of organisms to determine phylogenetic relationships and classify organisms.

Viruses, Bacteria, and Archaea

CHAPTER 29

Chapter Concepts

29.1 Viruses Are Particles

- Viruses are noncellular, while bacteria are fully functioning cellular organisms. 508
- All viruses have an outer capsid composed of protein and an inner core of nucleic acid. Some have an outer membranous envelope. 509
- Viruses are obligate intracellular parasites including bacteriophages (reproduce inside bacteria), and plant and animal viruses. 510

29.2 Bacteria Are Cellular

- Bacteria are prokaryotes; they lack a nucleus and most other cytoplasmic organelles found in eukaryotic cells. 514
- Bacteria reproduce asexually by binary fission; mutations and genetic recombinations by various means introduce variability. 515
- Some bacteria require oxygen; others are obligate anaerobes or facultative anaerobes. 516
- Some bacteria are autotrophs. There are photosynthetic bacteria (cyanobacteria give off oxygen, and purple and green sulfur bacteria do not) and chemosynthetic bacteria. 516
- Most bacteria are aerobic heterotrophs, just as animals are. However, many heterotrophic bacteria are symbiotic, being mutualistic, commensalistic, or parasitic. 516

29.3 How Bacteria Are Classified

- Traditionally, Gram staining is one laboratory method used to differentiate bacteria, which may be rod, round, or spiral shaped. 517

29.4 Archaea Compared to Bacteria

- Three evolutionary domains are now recognized: Bacteria, Archaea, and Eukarya. 519
- The archaea are quite specialized and live in extreme habitats. 519

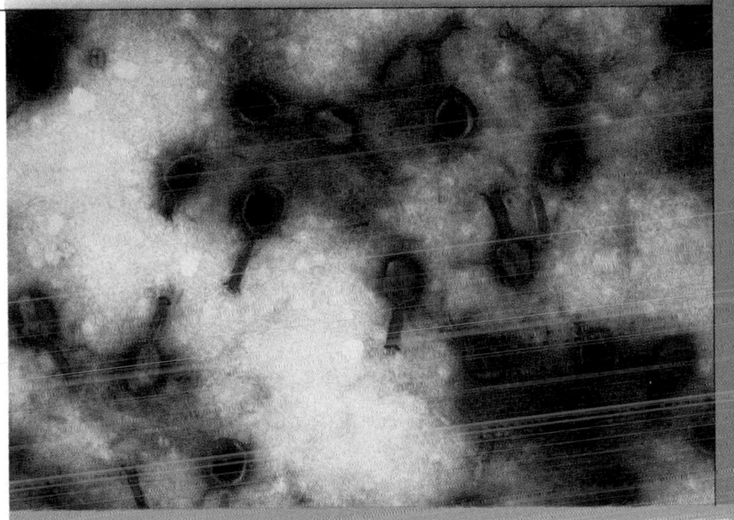

Lysis of *Escherichia coli* liberates many T4 viruses (red)

Although viruses are noncellular and therefore are not in the classification table in Appendix D, we study them in this chapter because of their ability to reproduce in the cells of living organisms. The prokaryotes, on the other hand, are cellular and capable of living in almost every habitat on earth. It has now been suggested on the basis of molecular data that the prokaryotes encompass the bacteria and the archaea, two of the three branches of the tree of life. (The third branch is the eukarya, which are studied in later chapters.)

Lacking a fossil record, most microbiologists had given up the hope of discovering the natural phylogeny of microorganisms, but now they believe that the sequencing of DNA and RNA will allow them to decisively determine the evolutionary relationships among the various types of bacteria and archaea. In the meantime, this chapter primarily reviews the biology of bacteria, whose general structure and function have been well known for many years.

29.1 Viruses Are Particles

Viruses [L. *virus*, poison] are nonliving particles with varied appearance, but even so they share certain common characteristics (Fig. 29.1). They are all infectious. In 1884, French chemist Louis Pasteur suggested that something smaller than a bacterium was the cause of rabies, and it was he who chose the word virus from a Latin word meaning poison. In 1892, Dimitri Ivanowsky, a Russian biologist, was studying a disease of tobacco leaves, called tobacco mosaic disease because of the leaves' mottled appearance. He noticed that an infective extract could be filtered through a fine-pore porcelain filter that retains bacteria, and it still caused disease. This substantiated Pasteur's belief because it meant that the disease-causing agent was smaller than any known bacterium. In the next century electron microscopy was born and viruses were seen for the first time. By the 1950s, virology was an active field of research; the study of viruses has contributed much to our understanding of disease, genetics, and even the characteristics of living things.

Viruses Are Noncellular

The size of a virus is comparable to that of a large protein macromolecule; they are generally smaller than 200 nm in diameter. Many viruses can be purified and crystallized, and the crystals can be stored just as chemicals are stored. Still, viral crystals will become infectious when the viral particles they contain are given the opportunity to invade a host cell.

The following diagram summarizes viral structure:

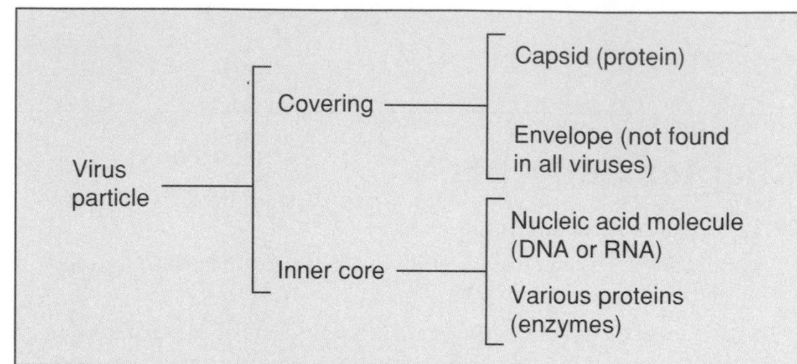

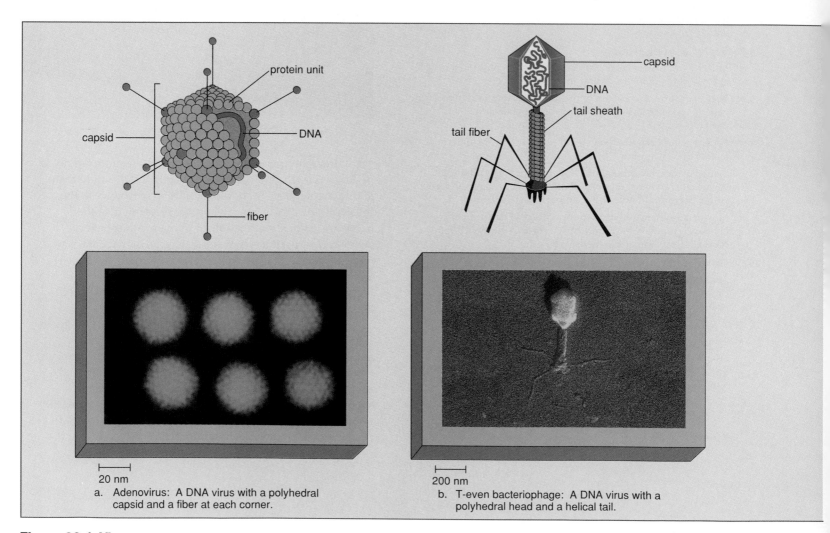

a. Adenovirus: A DNA virus with a polyhedral capsid and a fiber at each corner.

b. T-even bacteriophage: A DNA virus with a polyhedral head and a helical tail.

Figure 29.1 Viruses.
Despite their diversity, all viruses have an outer capsid composed of protein subunits and a nucleic acid core—either DNA or RNA, but not both. Some types of viruses also have a membranous envelope.

Each type of virus always has at least two parts: an outer *capsid* composed of protein subunits and an inner core of nucleic acid—either DNA (deoxyribonucleic acid) or RNA (ribonucleic acid), but not both. The viral genome at most has several hundred genes; a human cell contains thousands of genes. The capsid may be surrounded by an outer membranous envelope; if not, the virus is said to be naked. The envelope is actually a piece of the host's plasma membrane that also contains viral glycoprotein spikes. A viral particle may also contain various proteins, especially enzymes such as the polymerases, needed to produce viral DNA and/or RNA.

The classification of viruses is based on (1) their type of nucleic acid, including whether it is single stranded or double stranded, (2) their size and shape, and (3) the presence or absence of an outer envelope. Viruses differ from bacteria in the ways stated in Table 29.1. From this comparison you can see why viruses are considered nonliving and why they are not in the classification of organisms in Appendix D.

Table 29.1

Comparison of Viruses and Bacteria

Characteristic of Life	Viruses	Bacteria
Consist of cell	No	Yes
Metabolize	No	Yes
Respond to stimuli	No	Yes
Multiply	Yes (always inside living cell)	Yes (usually independently)
Evolve	Yes	Yes

Viruses are noncellular and have at least two parts: an outer capsid composed of protein subunits and an inner core of nucleic acid, either DNA or RNA but not both.

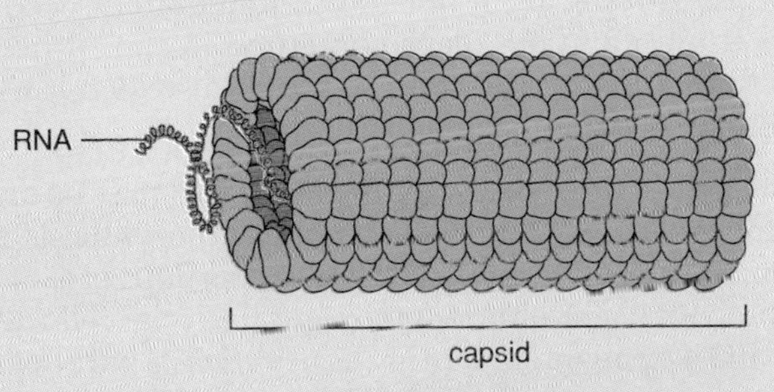

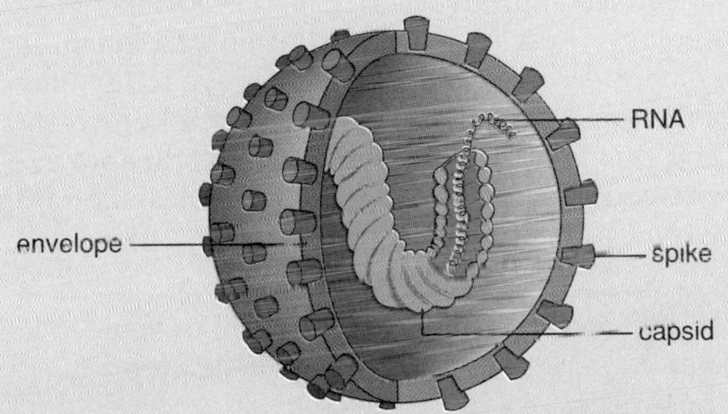

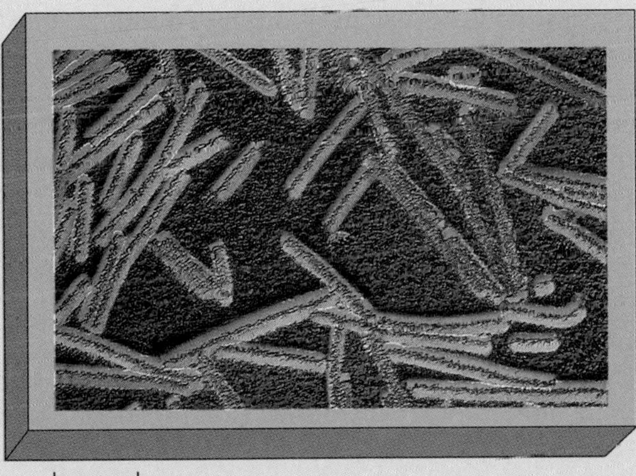

c. Tobacco mosaic virus: An RNA virus with a helical capsid.

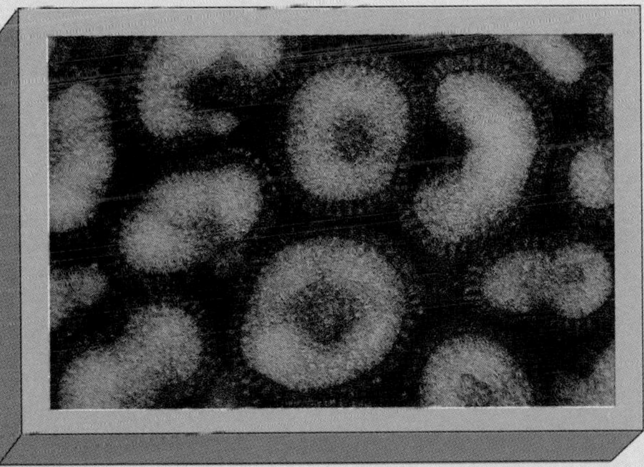

d. Influenza virus: An RNA virus with a helical capsid surrounded by an envelope with spikes.

Viruses Are Parasites

Viruses are *obligate intracellular parasites,* which means they cannot multiply outside a living cell. To maintain animal viruses in the laboratory, they are sometimes injected into live chick embryos (Fig. 29.2). Today, host cells are often maintained in tissue (cell) culture by simply placing a few cells in a glass or plastic container with appropriate medium. The cells can then be infected with the animal virus to be studied. Viruses infect all sorts of cells—from bacterial cells to human cells—but they are very specific. Bacteriophages infect only bacteria, the tobacco mosaic virus infects only plants, and the rabies virus infects only mammals, for example. We know that some human viruses even specialize in a particular tissue. Human immunodeficiency virus (HIV) will enter only certain blood cells, the polio virus reproduces in spinal nerve cells, the hepatitis viruses infect only liver cells. What could cause this remarkable parasite-host cell correlation? It is now believed that viruses are derived from the very cell they infect; the nucleic acid of viruses came from their host cell genomes! Therefore, viruses must have evolved after cells came into existence, and new viruses are probably evolving even now.

Viruses can also mutate; therefore, it is correct to say that they evolve. Those that mutate often can be quite troublesome because a vaccine that is effective today may not be effective tomorrow. Flu viruses are well known for mutating, and this is why you have to have a flu shot every year—antibodies generated from last year's shot are not expected to be effective this year.

Viruses evolve and reproduce but they are obligate intracellular parasites. They only grow inside their specific host cells.

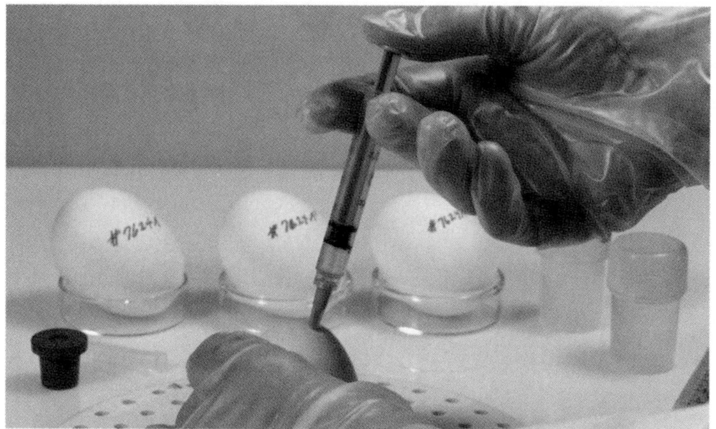

Figure 29.2 Growing viruses.
To grow a virus, you can inoculate live chick eggs with viral particles. A virus reproduces only inside a living cell, not because it uses the cell as nutrients, but rather because it takes over the machinery of the cell.

Viruses Replicate

Viruses gain entry into and are specific to a particular host cell because portions of the capsid (or the spikes of the envelope) adhere in a lock-and-key manner with a receptor on the host cell outer surface. The viral nucleic acid then enters the cell. Once inside, the nucleic acid codes for the protein units in the capsid. In addition, the virus may have genes for a few special enzymes needed for the virus to reproduce and exit from the host cell. In large measure, however, a virus relies on the host's enzymes, ribosomes, transfer RNA (tRNA), and ATP (adenosine triphosphate) for its own reproduction. In other words, a virus *takes over the metabolic machinery of the host cell* when it reproduces.

Bacteriophages Have Two Cycles

Bacteriophages [Gk. *bacterion,* rod, and *phagein,* to eat], or simply phages, are viruses that parasitize bacteria; the bacterium in Figure 29.3 could be *Escherichia coli,* which lives in our intestines, for example. Two types of bacteriophage life cycles, termed the lytic cycle and the lysogenic cycle, have been carefully studied. In the lytic cycle, viral replication occurs and the host cell undergoes *lysis,* a breaking open of the cell to release viral particles. In the lysogenic cycle, viral replication does not immediately occur, but replication may take place sometime in the future. The life cycle of the bacteriophage, lambda, which can carry out either cycle is discussed below.

Lytic Cycle The **lytic cycle** [Gk. *lyo,* loose] may be divided into five stages: attachment, penetration, biosynthesis, maturation, and release. During *attachment,* portions of the capsid combine with a receptor on the rigid bacterial cell wall in a lock-and-key manner. During *penetration,* a viral enzyme digests away part of the cell wall, and viral DNA is injected into the bacterial cell. *Biosynthesis* of viral components begins after the virus brings about inactivation of host genes not necessary to viral replication. The virus takes over the machinery of the cell in order to carry out viral DNA replication and production of multiple copies of the capsid protein subunits. During *maturation,* viral DNA and capsids assemble to produce several hundred viral particles. Lysozyme, an enzyme coded for by a viral gene, is produced; this disrupts the cell wall, and the *release* of phage particles occurs. The bacterial cell dies as a result.

During the lytic cycle, a bacteriophage takes over the machinery of the cell so that viral replication and release occur.

Lysogenic Cycle With the **lysogenic cycle** [Gk. *lyo*, loose, break up, and *genitus*, producing], the infected bacterium does not immediately produce phage but may do so sometime in the future. In the meantime the phage is *latent*—not actively replicating. Following attachment and penetration, viral DNA becomes integrated into bacterial DNA with no destruction of host DNA. While latent, the viral DNA is called a *prophage*. The prophage is replicated along with the host DNA, and all subsequent cells, called lysogenic cells, carry a copy of the prophage. Certain environmental factors, such as ultraviolet radiation, can induce the prophage to enter the lytic stage of biosynthesis, followed by maturation and release.

During the lysogenic cycle, the phage becomes a prophage that is integrated into the host genome. At a later time, the phage may reenter the lytic cycle and replicate itself.

Animal Viruses Also Cycle

Animal viruses replicate in a manner similar to bacteriophages, but there are modifications. If the virus has an envelope, its glycoprotein spikes allow the virus to adhere to plasma membrane receptors. Then the entire virus (not just the nucleic acid) penetrates a host cell by endocytosis. Once inside, the virus is uncoated as the envelope and capsid are removed. The viral genome, either DNA or RNA, is now free of its coverings and biosynthesis proceeds. Another difference among enveloped viruses is that viral release occurs by budding. During budding, the virus picks up its envelope consisting of lipids, proteins, and carbohydrates either from the plasma membrane or the nuclear envelope of the host cell. Other envelope markers, such as the glycoproteins that allow the virus to enter a host cell, are coded for by viral genes. Budding does not necessarily result in the death of the host cell.

Some animal viruses are specific to human cells. Of special concern are those such as the papillomavirus, the

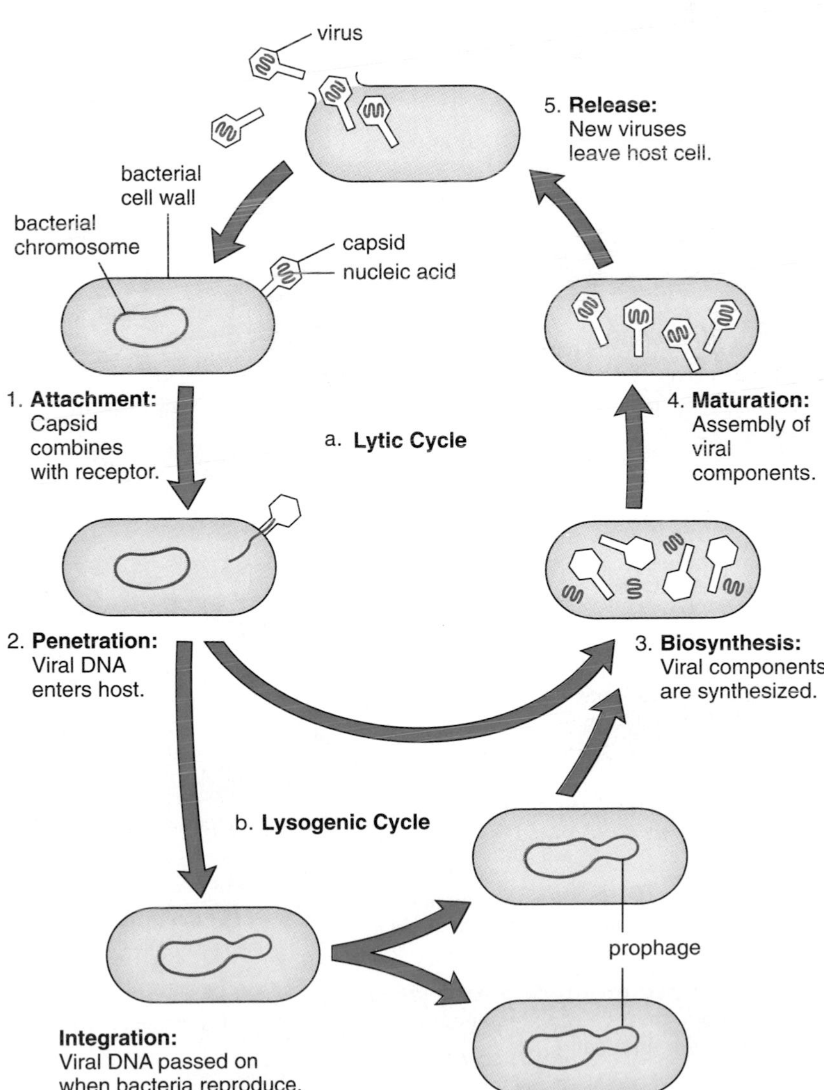
virus
5. Release:
New viruses
leave host cell.
bacterial
cell wall
bacterial
chromosome
capsid
nucleic acid
1. Attachment:
Capsid
combines
with receptor.
a. Lytic Cycle
4. Maturation:
Assembly of
viral
components.
2. Penetration:
Viral DNA
enters host.
3. Biosynthesis:
Viral components
are synthesized.
b. Lysogenic Cycle
prophage
Integration:
Viral DNA passed on
when bacteria reproduce.

Figure 29.3 Lytic and lysogenic cycles in bacteria.
a. In the lytic cycle, viral particles escape when the cell is lysed (broken open). **b.** In the lysogenic cycle, viral DNA is integrated into host DNA. At some time in the future, the lysogenic cycle can be followed by the lytic cycle.

herpes viruses, the hepatitis viruses, and the adenoviruses, which can cause specific types of cancer, in addition to a specific disease. These viruses undergo a period of latency, and their presence so alters the genotype of the cell that the cells become cancerous. Some viruses are cancer-producing because they bring with them *oncogenes,* normal genes that have been transformed and that now can cause the cell to undergo repeated cell division.

After animal viruses enter the host cell, uncoating releases viral DNA or RNA, and replication occurs. If release is by budding, the viral particle acquires a membranous envelope.

Retroviruses Have RNA Genes Retroviruses are RNA animal viruses that have a DNA stage. Figure 29.4 illustrates the reproduction of a retrovirus. A **retrovirus** [L. *retro,* backward, and *virus,* poison] contains a special enzyme called reverse transcriptase, which carries out RNA→cDNA transcription. The DNA is called cDNA because it is a DNA copy of the viral genome. Following replication, the resulting double-stranded DNA is integrated into the host genome. The viral DNA remains in the host genome and is replicated when host DNA is replicated. When and if this DNA is transcribed, new viruses are produced by the steps we have already cited: biosynthesis, maturation, and release—not by destruction of the cell, but by budding.

Retroviruses are of interest because human immunodeficiency viruses (HIV), which cause AIDS, are retroviruses. Retroviruses also cause certain forms of cancer.

Viruses Cause Infections

Viruses are best known for causing infectious diseases in plants and animals, including humans. At least a thousand different viruses cause diseases in plants. Since it is very difficult to distinguish a viral infection from a mineral deficiency and to treat a viral infection in plants, the policy is to propagate plants known to be healthy and to protect them from a viral infection. About a dozen crop diseases have been attributed not to viruses but to **viroids,** which are naked strands of RNA not covered by a capsid. Like viruses, though, viroids direct the cell to produce more viroids.

In humans, viral diseases are also controlled by preventing transmission, administering vaccines, and only recently by the administration of antiviral drugs. Knowing

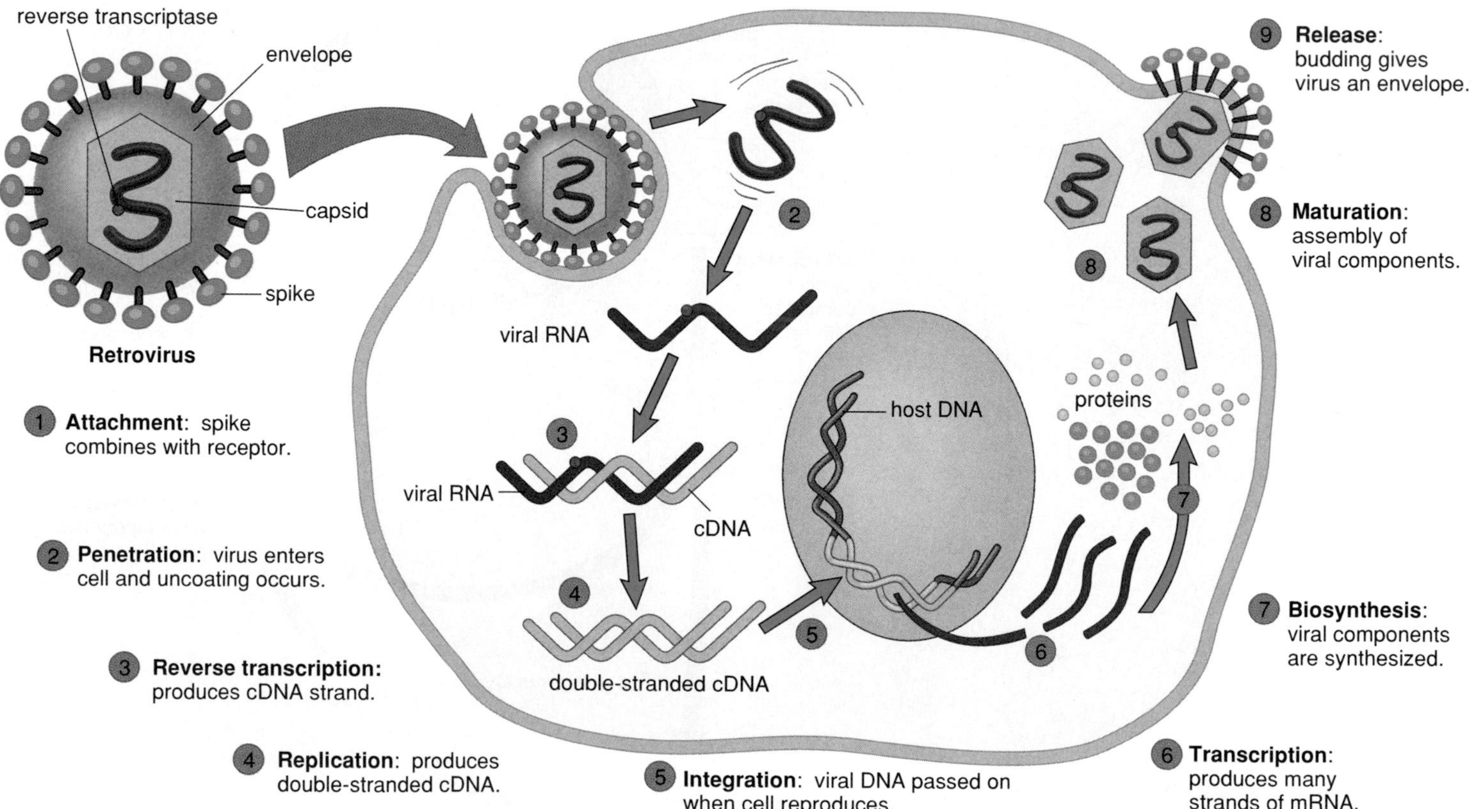

Figure 29.4 Reproduction of a retrovirus.
Notice that the mode of entry requires uncoating, that the virus integrates into the host genome, and that the virus acquires an envelope when it buds from the host cell.

how viral diseases are transmitted can help prevent their spread. Use of a condom will help prevent the transmission of HIV; frequent hand washing during cold season can help prevent a cold. **Vaccines** [L. *vaccinus,* of cows], which are drugs administered to stimulate immunity to a pathogen (a disease-causing agent) are available for some viral diseases such as polio, measles, and mumps. Our study of viral replication should allow you to see why *antibiotics,* which are designed to interfere with bacterial metabolism, have no effect on viral illnesses. Viruses lack most enzymes and instead utilize the enzymes of the host cell. Rarely has it been possible to find a drug that successfully interferes with viral reproduction without also interfering with host metabolism in cells free of the virus. However, the drugs acyclovir for herpes infections and AZT for AIDS do prevent viruses from replicating.

Some diseases in humans have been attributed not to virus and bacteria but to **prions,** which are protein particles that possibly can convert other proteins in the cell to become prions. It seems that prions may have a misshapen tertiary structure and they can cause other similar proteins to convert to this shape also.

29.2 Bacteria Are Cellular

Kingdom Monera includes only **bacteria** [Gk. *bacterion,* rod], which were not discovered until Dutch naturalist Antonie van Leeuwenhoek invented the first microscope in the seventeenth century and examined all sorts of specimens, including scrapings from his own teeth. Leeuwenhoek and others after him believed that the "little animals" he saw could arise spontaneously from inanimate matter. For about 200 years, scientists carried out various experiments to determine the origin of microorganisms, including bacteria. Finally, in about 1850, Louis Pasteur devised an experiment for the French Academy of Sciences that is described in Figure 29.5. It showed that previously sterilized fluid medium cannot become cloudy with bacterial growth unless it is exposed directly to the air. Today we know that bacteria are abundant in air, water, and soil, and on most objects. A single spoonful of earth can contain 10^{10} bacteria. Clearly, the combined number of all bacteria exceeds that of any other type of organism on earth.

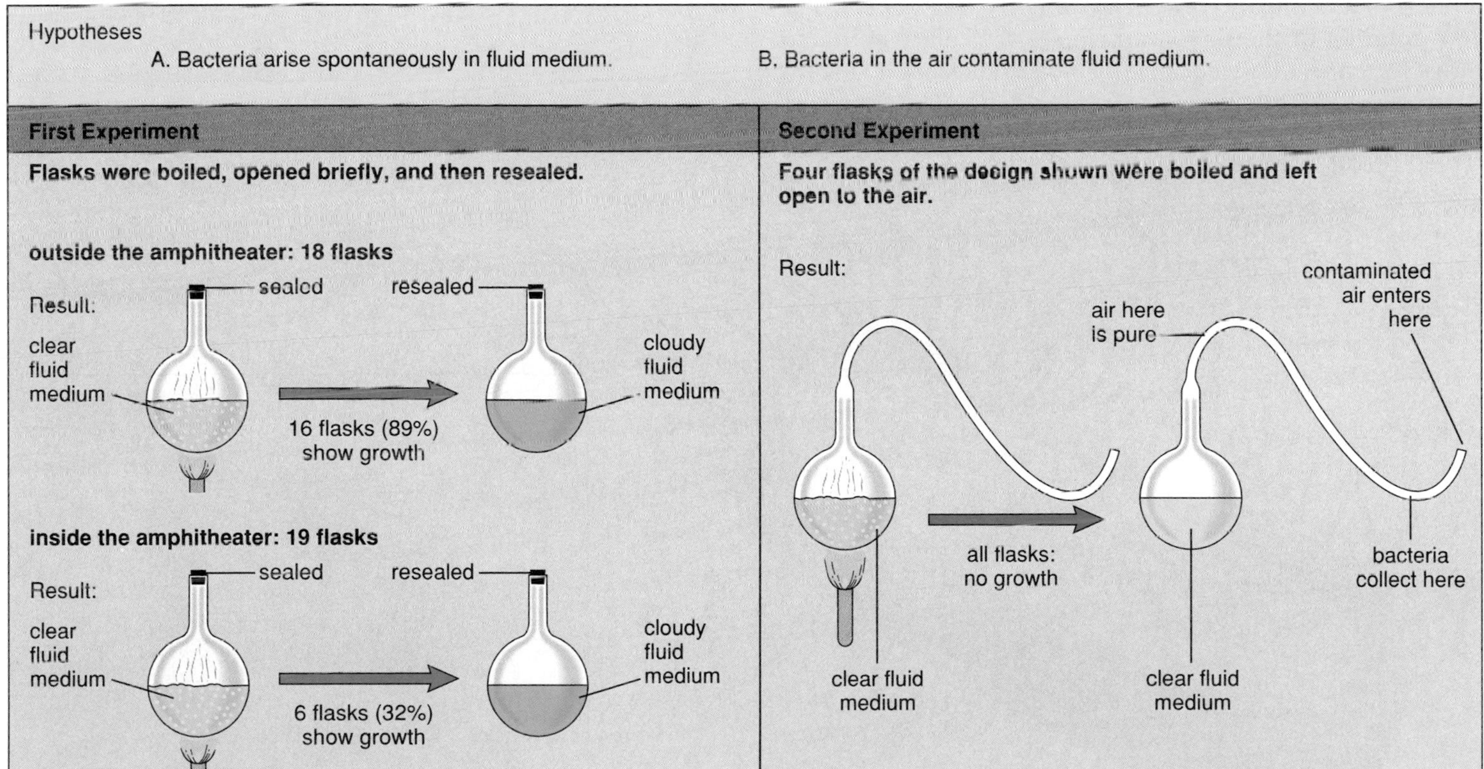

Figure 29.5 Pasteur's experiment.
Pasteur disproved the theory of spontaneous generation of microbes by performing experiments like this one.

Bacterial Structure Is Simple

Bacteria generally range in size from 1–10 μm in length and from 0.7–1.5 μm in width. Bacteria are prokaryotic, a term that means "before a nucleus" and these organisms lack a eukaryotic nucleus (Fig. 29.6). There are prokaryotic fossils dated as long ago as 3.5 billion years, and the fossil record indicates that the prokaryotes were alone on earth for at least 2 billion years. During that length of time, they became extremely diverse, not in structure but in metabolic capabilities. Bacteria are adapted to living in most environments because types differ in the ways they acquire and utilize energy.

Bacteria have an outer cell wall composed of **peptidoglycan,** a molecule that contains chains of a unique amino disaccharide joined by peptide chains. The cell wall, which prevents a bacterium from bursting or collapsing due to osmotic changes, may be surrounded by an attached *capsule* and/or by a loose gelatinous **sheath** called a *slime layer.* In parasitic forms, these outer coverings protect the cell from host defenses.

Some bacteria move by means of **flagella.** The flagellum has a filament composed of three strands of the protein flagellin wound in a helix. The filament is inserted into a hook that is anchored by a basal body (Fig. 29.6*b*). The 360° rotation of the flagellum causes the cell to spin and move forward. Many bacteria adhere to surfaces by means of fimbriae, short hairlike filaments extending from the surface. The fimbriae of *Neisseria gonorrhoeae* allow it to attach to host cells and cause gonorrhea.

A bacterial cell lacks the membranous organelles of a eukaryotic cell, and various metabolic pathways are located on the plasma membrane. Although bacteria do not have a nucleus, they do have a dense area called a **nucleoid** where a single circular strand of DNA, the **bacterial chromosome,** is found. Many bacteria also have accessory rings of DNA called plasmids. Plasmids can be extracted and used as vectors to carry foreign DNA into host bacteria during genetic engineering processes. Protein synthesis in a bacterial cell is carried out by thousands of ribosomes, which are smaller than eukaryotic ribosomes. The following diagram summarizes bacterial cell structure:

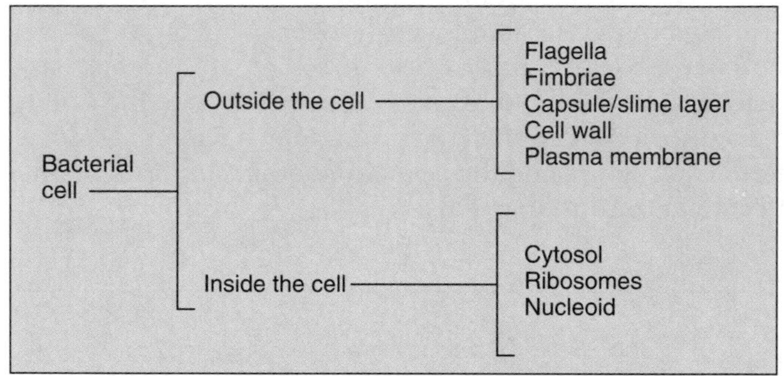

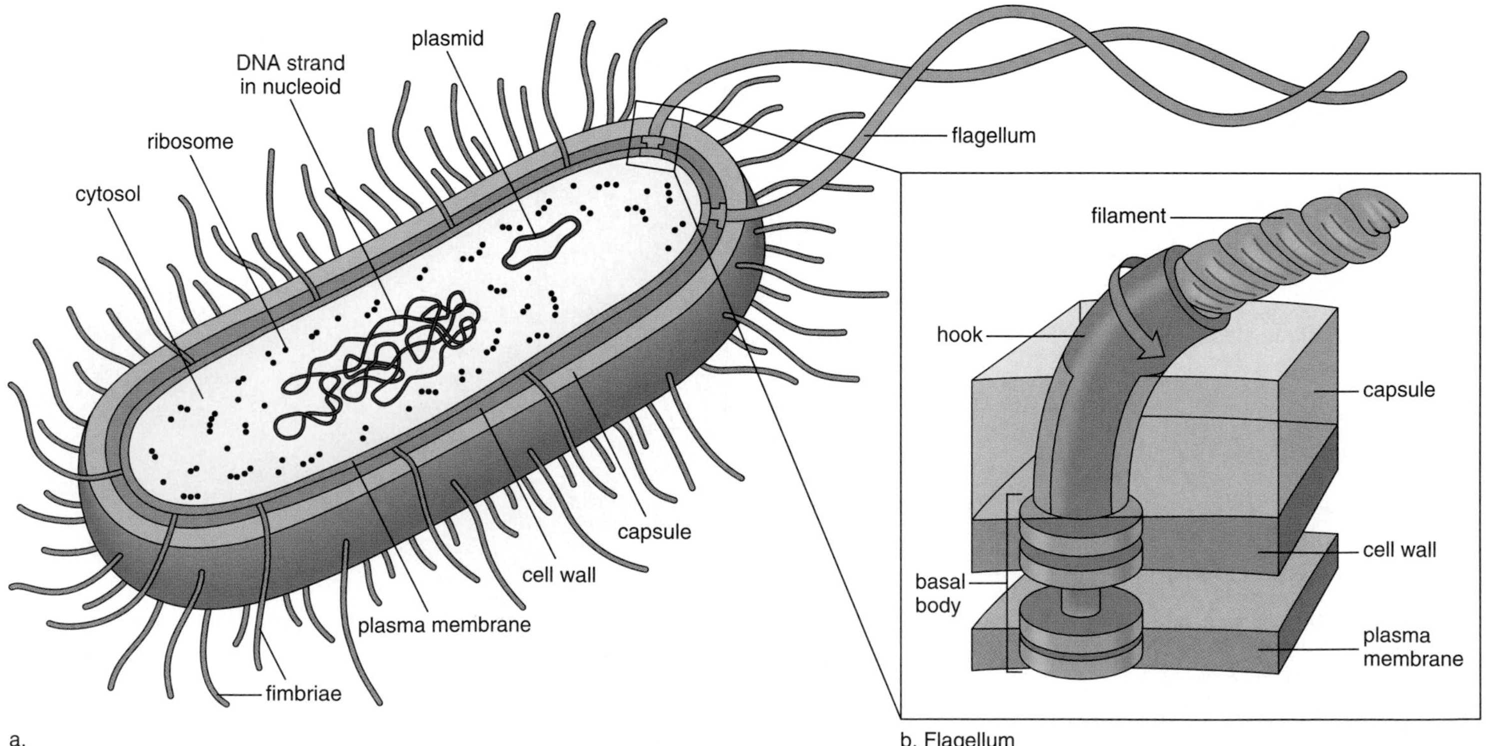

Figure 29.6 Generalized structure of a bacterium.
a. The cell. **b.** The flagellum contains a basal body, a hook, and a filament. The arrow indicates that the hook (and filament) turn 360°.

Bacteria Reproduce Asexually

Bacteria reproduce asexually by means of **binary fission** [L. *binarius*, of two, and *fissura*, cleft, break], a process that can be diagrammed as follows:

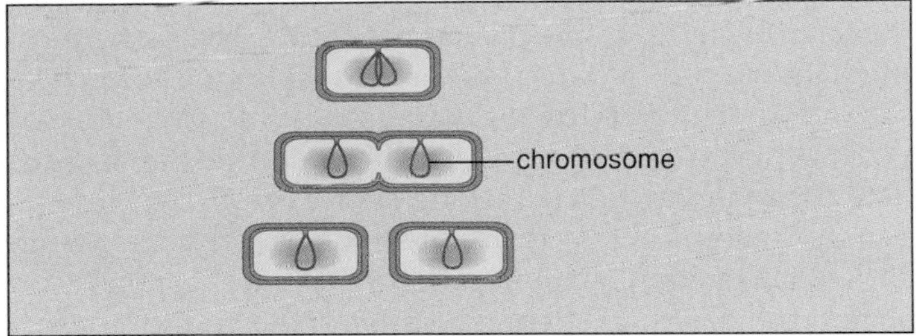

The single circular chromosome replicates, and then two copies separate as the cell enlarges. Newly formed plasma membrane and cell wall separate the cell into two cells. Mitosis, which requires the formation of a spindle apparatus, does not occur in prokaryotes.

In eukaryotes, genetic recombination occurs as a result of sexual reproduction. In bacteria, genetic recombination can occur in three ways. **Conjugation** has been observed between bacteria when the donor cell passes DNA to the recipient cell by way of a sex pilus, which temporarily joins the two cells. Conjugation takes place only between bacteria in the same or closely related species. **Transformation** occurs when a bacterium picks up (from the surroundings) free pieces of DNA secreted by live bacteria or released by dead bacteria. During **transduction,** bacteriophages carry portions of bacterial DNA from one cell to another. Plasmids, which sometimes carry genes for resistance to antibiotics, can be transferred between bacteria by any of these ways.

Because genetic recombination does not occur routinely, mutation becomes a most important avenue for evolutionary change. Because bacteria have a generation time as short as 12 minutes under favorable conditions, mutations are generated and distributed throughout a population more quickly than in eukaryotes. Also, prokaryotes are haploid, and so mutations are immediately subjected to natural selection to assess any possible benefit.

Prokaryotes reproduce asexually by binary fission. Mutations are the chief means of achieving genetic variation.

When faced with unfavorable environmental conditions, some bacteria form **endospores** [Gk. *endon*, within, and *spora*, seed] (Fig. 29.7). A portion of the cytoplasm and a copy of the chromosome dehydrate and are then encased by three heavy, protective spore coats. The rest of the bacterial cell deteriorates and the endospore is released. Spores survive in the harshest of environments—desert heat and dehydration, boiling temperatures, polar ice, and extreme ultraviolet radiation. They also survive for very long periods. When anthrax spores 1,300 years old germinate, they can still cause a severe infection (usually seen in cattle and sheep). To germinate, the endospore absorbs water and grows out of the spore coat. In a few hours' time, it becomes a typical bacterial cell, capable of reproducing once again by binary fission. Humans particularly fear a deadly but uncommon type of food poisoning called botulism that is caused by the germination of endospores inside cans of food. Spore formation is not a means of reproduction, but it does allow survival and dispersal to new places.

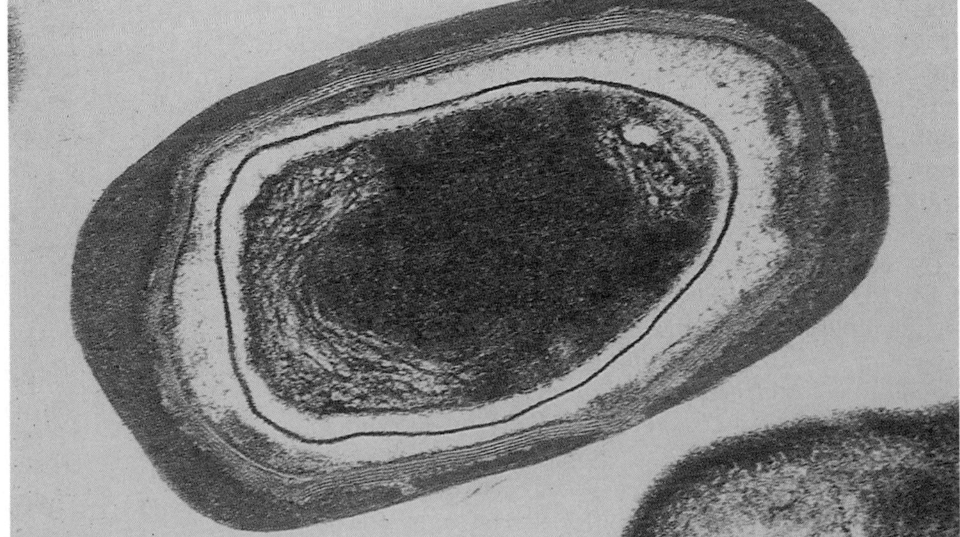

Figure 29.7 The endospore.
An endospore is resistant to extreme environmental conditions. Sterilization, a process that kills all living organisms—even endospores—can be achieved by using an autoclave, a container that maintains steam under pressure. This bacterium, *Bacillus subtilis*, contains an endospore. (For home canning, a pressure cooker is used.)

Bacterial Nutrition Is Diverse

With respect to nutrient requirements, bacteria are not much different from other organisms. One difference, however, concerns the need for oxygen. Some bacteria are *obligate anaerobes* and are unable to grow in the presence of oxygen. A few serious illnesses—such as botulism, gas gangrene, and tetanus—are caused by anaerobic bacteria. Other bacteria, called *facultative anaerobes,* are able to grow in either the presence or the absence of gaseous oxygen. Most bacteria, however, are aerobic and, like animals, require a constant supply of oxygen to carry out cellular respiration.

Autotrophic Bacteria

Autotrophs can utilize carbon dioxide as a carbon source and synthesize all their macromolecules from it. The photosynthetic bacteria are divided into two types—those that evolved first and do not give off oxygen (O_2) and those that evolved later and do give off oxygen.

Do not give off O_2	*Do give off O_2*
Photosystem I only	Photosystems I and II
Unique type of chlorophyll called bacteriochlorophyll	Type of chlorophyll found in plants

The green bacteria and purple bacteria carry on the first type of photosynthesis. These bacteria do not give off oxygen because they do not use water as an electron donor; instead, they use hydrogen sulfide (H_2S):

$$CO_2 + 2\,H_2S \longrightarrow (CH_2O)_n + 2\,S$$

These bacteria usually live in anaerobic conditions such as the muddy bottom of marshes and cannot photosynthesize in the presence of oxygen. In contrast, the cyanobacteria (see Fig. 29.10) contain chlorophyll *a* and carry on photosynthesis in the same manner as algae and plants:

$$CO_2 + H_2O \longrightarrow (CH_2O)_n + O_2$$

Some autotrophic bacteria carry out **chemosynthesis** [Gk. *chemo-,* pertaining to chemicals, *syn,* together, and *thesis,* an arranging]. They oxidize inorganic compounds such as nitrites, hydrogen gas, and hydrogen sulfide to obtain the necessary energy to produce their own organic compounds. *Chemosynthetic* bacteria (also called chemoautotrophs) trap the small amount of energy released from these oxidations to use in the reactions that synthesize carbohydrates.

There are two groups of chemosynthetic bacteria of particular interest. First, at hydrothermal vents 2.5 kilometers below sea level, hot minerals spew out of the earth, providing hydrogen sulfide to chemosynthetic bacteria, which live both freely and within the bodies of giant tube worms. The organic molecules produced by the bacteria support the growth of the worms, as well as clams and crabs. Also, chemosynthetic bacteria carry out reactions that help keep nitrogen cycling within ecosystems. The nitrifying bacteria oxidize ammonia (NH_3) to nitrites (NO_2^-), and nitrites to nitrates (NO_3^-).

Heterotrophic Bacteria

The majority of bacteria are aerobic **saprotrophs** [Gk. *sapros,* rotten, and *trophe,* food]. These so-called **decomposers** can break down such a large variety of molecules that there probably is no natural organic molecule that cannot be digested by at least one bacterial species. The decomposing bacteria play a critical role in recycling matter in ecosystems and in making inorganic molecules available to photosynthesizers. The metabolic capabilities of various heterotrophic bacteria have long been exploited by human beings. Bacteria are used commercially to produce chemicals, such as ethyl alcohol, acetic acid, butyl alcohol, and acetones. Bacterial action is also involved in the production of butter, cheese, sauerkraut, rubber, cotton, silk, coffee, and cocoa. Even antibiotics, discussed later, are produced by some bacteria.

Heterotrophs may be free-living or **symbiotic** [Gk. *sym,* together, and *bios,* life], meaning that they form **mutualistic,** commensalistic, or parasitic relationships. Some mutualistic bacteria are involved in nitrogen cycling. They live in the root nodules of soybean, clover, and alfalfa plants where they reduce atmospheric nitrogen (N_2) to ammonia, a process called **nitrogen fixation** (Fig. 29.8). Plants are unable to fix atmospheric nitrogen, and those without nodules take up nitrate and ammonia from the soil. Other mutualistic bacteria that live in human intestines release vitamins K and B_{12}, which we can use to help produce blood components. In the stomachs of cows and goats, special mutualistic bacteria digest cellulose, enabling these animals to feed on grass.

Commensalism often occurs when one population modifies the environment in such a way that a second population benefits. Obligate anaerobes can live in our intestines only because the bacterium *Escherichia coli* uses up the available oxygen. In a similar manner a primary infection by one bacteria may allow another to take hold.

Figure 29.8 Nodules of a legume.
While there are some free-living bacteria that carry on nitrogen fixation, those of the genus *Rhizobium* invade the roots of legumes with the resultant formation of nodules. Here the bacteria convert atmospheric nitrogen to an organic nitrogen that the plant can use. These are nodules on the roots of a soybean plant, *Glycine.*

29.3 How Bacteria Are Classified

Bacteria are prokaryotes that lack a nucleus and the membranous organelles of the eukaryotic cell. Most bacterial cells are protected by a cell wall that contains a unique molecule, peptidoglycan. Bacteria are commonly differentiated using the Gram stain procedure, which distinguishes different cell wall types. This staining procedure was developed in the late 1880s by Hans Christian Gram, a Danish bacteriologist. Gram-positive bacteria retain a dye-iodine complex and appear purple under the light microscope, while Gram-negative bacteria do not retain the complex and appear pink. This difference is dependent on the construction of the cell wall; namely, the Gram-positive bacteria have a thick layer of peptidoglycan in their cell wall, whereas Gram-negative bacteria have only a thin layer.

Bacteria are found in three basic shapes (Fig. 29.9): rod (bacillus, pl., bacilli); round or spherical (coccus, pl., cocci); and spiral or helical-shaped (spirillum, pl. spirilli). These three basic shapes may be augmented by particular arrangements or shapes of cells. For example, cocci may form clusters (staphylococci, diplococci) or chains (streptococci). Rod-shaped bacteria may appear as very short rods (coccobacilli) or as very long filaments (fusiform).

Historically, bacteria have been subdivided taxonomically into groups based on their cell wall type (Gram-positive or Gram-negative), presence of endospores, metabolism, growth and nutritional characteristics, physiological characteristics, and other criteria. For the past 75 years bacterial taxonomy has been compiled in *Bergey's Manual of Determinative Bacteriology*. The most recent edition of *Bergey's Manual* divided the bacteria into 19 major groups, which were further subdivided into orders, families, genera, and species. The names of the groups—for example, Nonmotile, Gram-negative, curved bacteria or Nonsporeforming Gram-positive rods—reflect the phenotypic bias used to group the bacteria. However, this edition was published just before the recognition of using ribosomal RNA analyses as the basis of phylogenetic grouping.

Since the 1980s, Carl Woese and other researchers have pioneered a new viewpoint of bacterial taxonomy based on the phylogenetic comparisons of bacterial 16S ribosomal RNA sequences. Twelve groups are now recognized. Some of the characteristics and genera of these groups are listed in Table 29.2. Some of the new groups, such as the Spirochetes, are essentially identical to early classification systems. However, other groups contain a diverse assortment of bacteria that appear to be physiologically distant, but nevertheless share common 16S ribosomal RNA sequences. For example, the *Deinococcus* group contains Gram-positive radiation-resistant bacteria and Gram-negative thermophilic bacteria. In spite of their obvious phenotypic differences, these two seemingly diverse bacterial types are genetically related to one another. With the current understanding of how groups of bacteria are related to one another genetically, only further investigations may reveal the relationships of these phenotypic differences.

Cyanobacteria

Cyanobacteria [Gk. *kyanos*, blue, and *bacterion*, rod] are Gram-negative bacteria with a number of unusual traits. They photosynthesize in the same manner as plants and are believed to be responsible for first introducing oxygen into the primitive atmosphere. Formerly, the cyanobacteria were called blue-green algae and were classified with eukaryotic algae, but now we know that they are prokaryotes. They can have other pigments that mask the color of chlorophyll so that they appear, for example, not only blue-green but also red, yellow, brown, or black.

Cyanobacterial cells are rather large and range in size from 1-50 µm in width. They can be unicellular, colonial, or filamentous. Cyanobacteria lack any visible means of locomotion, although some glide when in contact with a

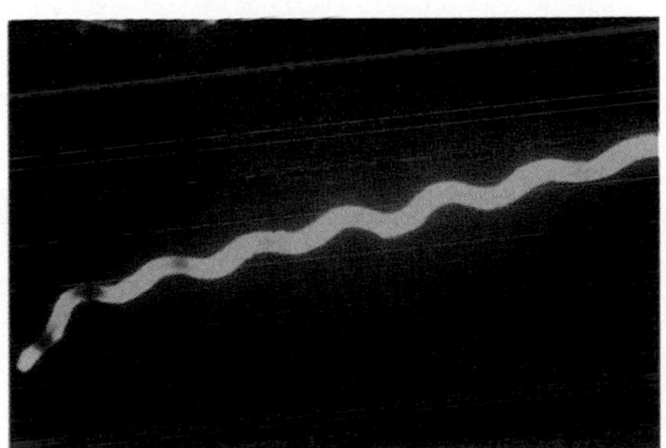

a. A spirillum

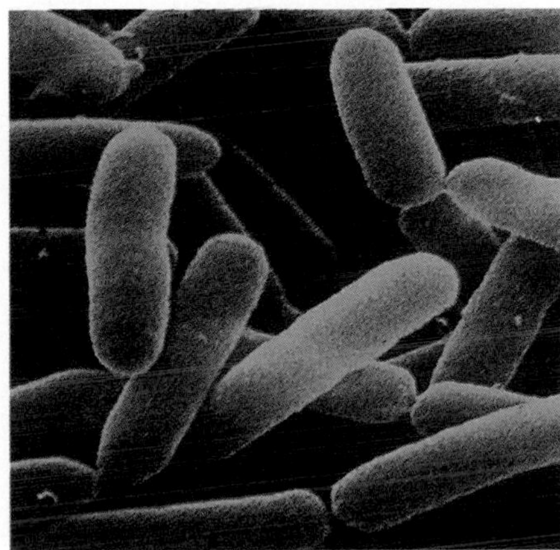

b. Bacilli in pairs

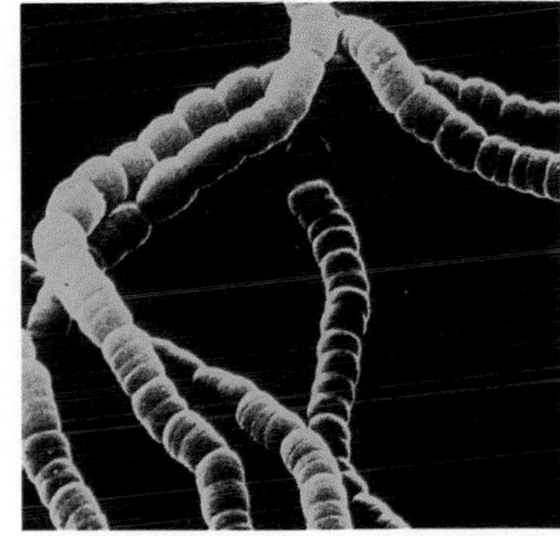

c. Cocci in chains

Figure 29.9 Diversity of bacteria.
a. Spirillum (spiral-shaped) bacterium.
b. Bacillus (rod-shaped) bacterium.
c. Coccus (round) bacterium in chains.

Table 29.2

The Major Phylogenetic Groups of Bacteria

Group	Characteristics of the Group	Representative Genera of the Group
Aquificales	Extremely thermophilic bacteria The oldest branch of the bacterial domain	*Aquifex, Hydrogenobacter*
Thermotogales	Extremely thermophilic bacteria Cells are often surrounded by a toga (sheath)	*Thermotoga, Thermosipho*
Green nonsulfur bacteria	Most bacteria in this group are photosynthetic, some are thermophilic	*Chloroflexus, Herpetosiphon, Thermomicrobium*
Deinococci	Bacteria that have atypical peptidoglycan cell walls (contain ornithine instead of diaminopimelic acid). The *Deinococcus* subgroup are Gram-positive bacteria that are resistant to radiation. The *Thermus* subgroup are Gram-negative thermophilic bacteria.	*Deinococcus, Thermus*
Proteobacteria (purple bacteria)	Includes a phenotypically diverse group of Gram-negative bacteria. Some members of the group are phototrophic but do not produce oxygen during photosynthesis; some are chemotrophic; some members are capable of nitrogen fixation. The group has been further subdivided into alpha, beta, gamma, delta, and epsilon proteobacteria.	*Rhodobacter, Nitrobacter, Beggiatoa, Escherichia, Salmonella, Shigella, Enterobacter, Pseudomonas, Rhizobium, Legionella*
Gram-positive bacteria*	All of the Gram-positive bacteria. This group is subdivided into a low mol% G + C group including endospore-forming bacteria, lactic acid bacteria, anaerobic and aerobic cocci, and mycoplasmas; and a high mol% G + C group including the filamentous actinomycetes and related bacteria.	*Staphylococcus, Clostridium, Bacillus, Mycoplasma, Mycobacterium, Corynebacterium, Streptomyces, Actinomyces*
Cyanobacteria	Phototrophic bacteria that produce oxygen during photosynthesis	*Nostoc, Anabaena, Oscillatoria*
Chlamydiae	Bacteria that lack peptidoglycan and contain a protein cell wall; live as obligate intracellular parasites of animal cells	*Chlamydia*
Planctomycetes	Gram-negative bacteria that divide by budding; many cells produce appendages or stalks	*Planctomyces*
Bacteroides and relatives	Gram-negative, rod-shaped bacteria. This group is subdivided into a *Bacteroides*- subgroup containing fermentative anaerobes, and a *Flavobacterium* subgroup containing respiring aerobes that may be rod- or filamentous-shaped.	*Bacteroides, Fusobacterium, Cytophaga, Flavobacterium*
Green sulfur bacteria	Green phototrophic bacteria that do not produce oxygen during photosynthesis	*Chlorobium*
Spirochetes	Gram-negative helical or spiral-shaped cells with a distinctive corkscrew motility	*Spirochaeta, Borrelia, Treponema, Leptospira*

** G = guanine, C = cytosine*

Lawrence C. Parks, University of Louisville

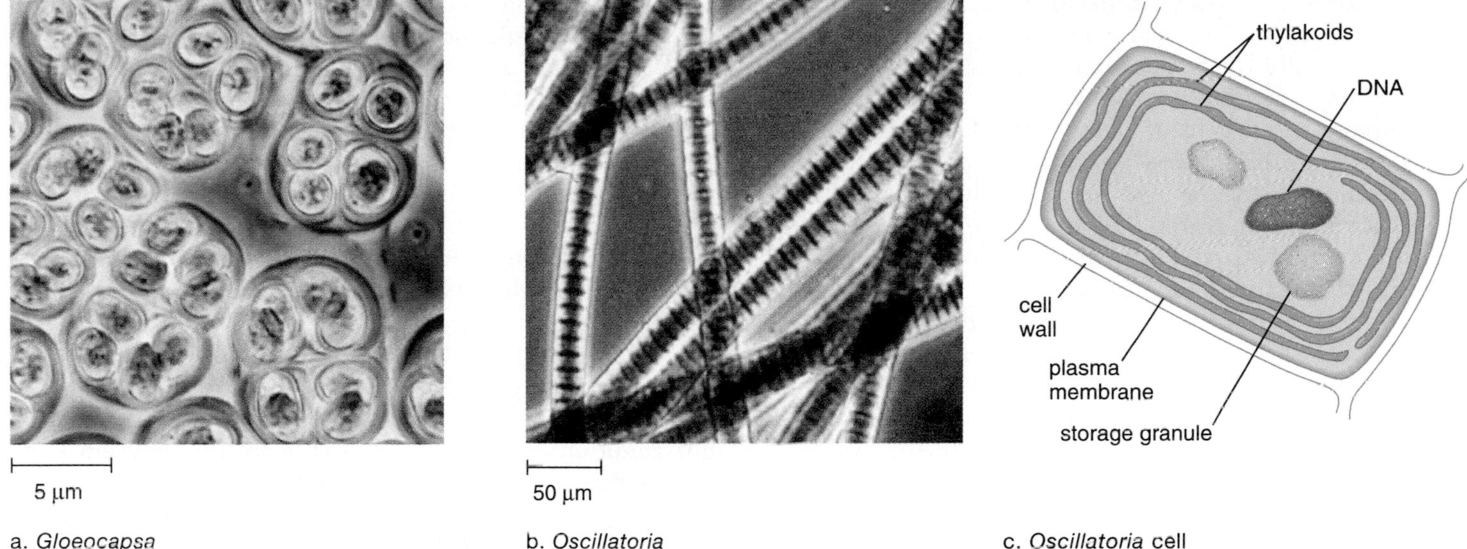

Figure 29.10 Diversity among the cyanobacteria.
a. In *Gloeocapsa*, single cells are grouped in a common gelatinous sheath. b. Filaments of cells occur in *Oscillatoria*. c. One cell of *Oscillatoria* as it appears through the electron microscope.

solid surface and others oscillate (sway back and forth). Some cyanobacteria have a special advantage because they possess heterocysts, which are thick-walled cells without nuclei, where nitrogen fixation occurs. The ability to photosynthesize and also to fix atmospheric nitrogen (N_2) means that their nutritional requirements are minimal. They can serve as food for heterotrophs in ecosystems.

Cyanobacteria (Fig. 29.10) are common in fresh water, in soil, and on moist surfaces, but they are also found in harsh habitats, such as hot springs. They are symbiotic with a number of organisms, such as liverworts, ferns, and even at times invertebrates like corals. In association with fungi, they form **lichens** that can grow on rocks. A lichen is a symbiotic relationship in which the cyanobacterium provides organic nutrients to the fungus, while the fungus possibly protects and furnishes inorganic nutrients to its partner. It is also possible that the fungus is parasitic on the algal component. Lichens help transform rocks into soil; other forms of life then may follow. It is presumed that cyanobacteria were the first colonizers of land during the course of evolution.

Cyanobacteria are ecologically important in still another way. If care is not taken in the disposal of industrial, agricultural, and human wastes, phosphates drain into lakes and ponds, resulting in a "bloom" of these organisms. The surface of the water becomes turbid, and light cannot penetrate to lower levels. When a portion of the cyanobacteria die off, the decomposing bacteria use up the available oxygen, causing fishes to die from lack of oxygen.

Cyanobacteria are photosynthesizers that sometimes can also fix atmospheric nitrogen. In association with fungi, they form lichens, which contribute to soil formation.

29.4 Archaea Compared to Bacteria

One of the important consequences of analyzing bacteria based on 16S ribosomal RNA sequences was the finding that some bacterial groups are as different from most of the bacteria as bacteria are different from eukaryotic cells. These microorganisms were originally called archaebacteria [Gk *archaios*, old]. They are prokaryotic cells that are found in extreme environments thought to be similar to those of the earth. Further investigations have revealed that these unusual microorganisms are different enough from bacteria to be incorporated into their own domain, Archaea.

Archaea are prokaryotes with molecular characteristics that distinguish them from both bacteria and eukaryotes. Analyses of rRNA, tRNA, and DNA suggest that the evolution of living organisms involves three primary domains—Bacteria, Archaea, and Eukarya. Each of these domains has followed its own evolutionary path since they diverged from a common ancestor. Because archaea and some bacteria are both found in extreme environments (hot springs, thermal vents, salt basins), they may have diverged from a common ancestor relatively soon after life began. Then later, the eukarya are believed to have split off from the archaeal line of descent. In other words, the eukarya are believed to be more closely related to the archaea than to the bacteria. Archaea and eukarya share some of the same ribosomal proteins (not found in bacteria), initiate transcription in the same manner, and have similar types of tRNAs.

The plasma membranes of archaea contain unusual lipids that allow them to function at high temperatures. Lipids of archaea contain glycerol linked to branched chain hydrocarbons in contrast to lipids of bacteria that contain glycerol linked to fatty acids. The archaea also evolved diverse cell wall types, which facilitate their survival under extreme conditions. The cell walls of archaea do not contain peptidoglycan. In some archaea, the cell wall is largely composed of polysaccharides, and in others, the wall is pure protein. In a few, there is no cell wall.

Metabolically, the archaea have retained primitive and unique forms of metabolism. *Methanogenesis,* the ability to form methane, is one type of metabolism that is performed only by some archaea. Most archaea are autotrophs and use molecular hydrogen and reduced elemental sulfur, carbon dioxide, and water. There are no photosynthetic archaea, suggesting that chemoautotrophy predated photoautotrophy during evolution.

Similar to separating the bacteria into groups, the archaea are now divided into two or possibly three groups: Chrenarchaeota, Euryarchaeota, and a newly discovered group, Korarchaeota. The euryarchaeota archaea are dominated by the methanogens, but also include the halophiles. *Methanogens* are found in anaerobic environments in swamps, marshes, and the intestinal tracts of animals where they produce methane (CH_4) from hydrogen gas (H_2) and carbon dioxide coupled to the formation of ATP. This methane, which is also called biogas, is released into the atmosphere and contributes to the greenhouse effect and global warming. About 65% of the methane found in our atmosphere is produced by these methanogenic archaea.

The *halophiles* require high salt concentrations (usually 12 to 15%—the ocean is about 3.5% salt) for growth. They have been isolated from highly saline environments such as the Great Salt Lake in Utah, the Dead Sea, solar salt ponds, and hypersaline soils. These archaea have evolved a number of mechanisms to survive in environments that are high in salt. Their proteins have unique chloride pumps that use halorhodopsin (related to the rhodopsin pigment found in our own eyes) to pump chloride inside the cell and bacteriorhodopsin to synthesize ATP in the presence of light.

The third major type of archaea are the *thermoacidophiles,* which are part of the Chrenarchaeota, Euryarchaeota, and Korarchaeota. These archaea are isolated from extremely hot, acidic environments such as hot springs, geysers, submarine thermal vents, and around volcanoes. Most of these archaea survive best at temperatures above 80°C and some can even grow at 105°C (remember that water boils at 100°C)! Many of the thermoacidophiles use sulfate as a source of energy to make ATP. They are, therefore, found in natural environments with high sulfate concentrations and grow best at pH 1 to 2.

connecting concepts

We all know that being diagnosed with a viral infection means there's nothing medical science can do for us because antibiotics are ineffective against viruses. Conversely, any time a medical problem is blamed on bacteria, we feel confident that an appropriate antibiotic will take care of the matter. In recent years, however, many types of bacteria have become resistant to antibiotics. This is a good example of evolution in action—those few bacteria that are resistant to antibiotics survive and reproduce so that in the end, an entire bacterial strain is resistant. Consequently, we are now seeing a re-emergence of diseases once thought to be controlled, such as tuberculosis.

It is erroneous, however, to think of all bacteria as harmful. Most are actually beneficial saprotrophs—they decompose dead organic matter and recycle materials through ecosystems. Mutualistic bacteria fix nitrogen in plant nodules, enable herbivores to digest cellulose, and release certain vitamins in the human intestine.

Cyanobacteria, formerly known as blue-green algae, photosynthesize in the same manner as plants and are believed to have introduced oxygen into the earth's primitive atmosphere. Cyanobacteria may have been the first colonizers of terrestrial environment, and today they join with fungi to form lichens, contributors to soil formation.

Many bacteria and the archaea can live in environments that may represent the kinds of habitats that were available when the earth first formed. We find prokaryotes in such hostile locales as swamps, the Dead Sea, and hot sulfur springs. Clearly, bacteria are fundamental to most ecosystems.

Summary

29.1 Viruses Are Particles

Viruses are noncellular, while bacteria are fully functioning organisms. All viruses have at least two parts: an outer capsid composed of protein subunits and an inner core of nucleic acid, either DNA or RNA but not both. Some also have an outer membranous envelope.

Viruses are obligate intracellular parasites that can be maintained only inside living cells, such as those of a chick egg or those propagated in cell (tissue) culture.

The lytic cycle of a bacteriophage consists of attachment, penetration, biosynthesis, maturation, and release. In the lysogenic cycle of a bacteriophage, viral DNA is integrated into bacterial DNA for an indefinite period of time, but it can undergo the lytic cycle when stimulated.

The reproductive cycle differs for animal viruses with a membranous envelope. Uncoating is needed to free the genome, and budding releases the viral particles from the cell. RNA retroviruses have an enzyme, reverse transcriptase, that carries out reverse transcription. This produces cDNA, which becomes integrated into host DNA. The AIDS virus is a retrovirus.

A viral infection can be controlled by administering vaccines and antiviral drugs. Antiviral drugs prevent DNA replication.

29.2 Bacteria Are Cellular

Kingdom Monera contains the bacteria including cyanobacteria. Bacterial cells lack a nucleus and most other cytoplasmic organelles found in eukaryotic cells. The cell wall contains peptidoglycan. Bacteria reproduce asexually by binary fission; genetic recombination occurs by means of conjugation, transformation, and transduction, but their chief method for achieving genetic variation is mutation. Some bacteria form endospores, which are extremely resistant to destruction; the genetic material can thereby survive unfavorable conditions.

Bacteria differ in their need (and tolerance) for oxygen. There are obligate anaerobes, facultative anaerobes, and aerobic bacteria.

Some bacteria are autotrophic and are either photosynthetic or chemosynthetic. Some photosynthetic bacteria (cyanobacteria) give off oxygen and some (purple and green sulfur bacteria) do not. Chemosynthetic bacteria oxidize inorganic compounds, such as hydrogen sulfide, to acquire energy to make their own food. Surprisingly, these bacteria support communities at hydrothermal vents.

Most bacteria are aerobic heterotrophs and are saprotrophic decomposers that are absolutely essential to the cycling of nutrients in ecosystems. Their metabolic capabilities are so vast that they are used by humans both to dispose of and to produce substances. Many heterotrophic bacteria are symbiotic. The mutualistic nitrogen-fixing bacteria live in nodules on the roots of legumes. Some symbiotes, however, are parasitic and cause plant and animal diseases.

29.3 How Bacteria Are Classified

The classification of bacteria is still being developed. Of primary importance at this time is the shape of the cell and the structure of the cell wall, which affects Gram staining. There are three basic shapes: rod shaped (bacillus), round (coccus), and spiral shaped (spirillum). Of special interest are the cyanobacteria, which were the first organisms to photosynthesize in the same manner as plants. When cyanobacteria are symbionts with fungi, they form lichens.

29.4 Archaea Compared to Bacteria

On the basis of molecular evidence, it is suggested that there are three evolutionary domains: Bacteria, Archaea, and Eukarya. In addition, it appears that the archaea are more closely related to the eukarya than to bacteria. Yet, both archaea and bacteria are prokaryotes. Archaea do not have peptidoglycan in their cell walls, as do the bacteria, and they share genes with the eukarya.

The three types of archaea live under harsh conditions, such an anaerobic marshes (methanogens), salty lakes (halophiles), and hot sulfur springs (thermoacidophiles).

Reviewing the Chapter

1. Contrast viruses with bacteria as per the characteristics of life. 509
2. Describe the general structure of viruses, and tell why they are obligate parasites. 510
3. How are viruses specific, and how can they be cultured in the laboratory? 510
4. Describe both the lytic cycle and the lysogenic cycle of bacteriophages. 510–11
5. How do animal viruses differ in structure and reproductive cycle from bacteriophages? 510–11
6. How do retroviruses differ from other animal viruses? Describe the reproductive cycle of retroviruses in detail. 512
7. How are viral infections controlled, and why don't antibiotics work against viruses? 512–13
8. Explain Pasteur's experiment, which shows that bacteria do not arise spontaneously. 513
9. Describe the general structure of bacteria, and tell how they reproduce. 514–15
10. How does genetic recombination occur in bacteria, and what is the more common source of variation? 515
11. How do some bacteria survive the severest of unfavorable environments? 515
12. How do bacteria differ in their tolerance of and need for oxygen? 516
13. How does photosynthesis differ between the green and purple sulfur bacteria and the cyanobacteria? 516
14. What are chemosynthetic bacteria, and where have they been found to support whole communities? 516
15. Discuss the nutrition of heterotrophic bacteria, note their importance in ecosystems, and give examples of their symbiotic relationships with other types of organisms. 516–17
16. Discuss the importance of cyanobacteria in ecosystems and in the history of the earth. 518–19
17. On the basis of what types of data do microbiologists believe there are three domains of life? What are these domains? 518–20

Testing Yourself

Choose the best answer for each question.

1. Viruses are considered nonliving because
   a. they don't mutate and therefore don't adapt.
   b. they do not locomote.
   c. they cannot reproduce independently.
   d. their nucleic acid does not code for protein.
2. Which of these are found in all viruses?
   a. envelope, nucleic acid, capsid
   b. DNA, RNA, and proteins
   c. proteins and a nucleic acid
   d. proteins, nucleic acids, carbohydrates, and lipids
3. Which step in the lytic cycle follows attachment of virus and release of DNA into the cell?
   a. production of lysozyme
   b. disintegration of host DNA
   c. assemblage
   d. DNA replication
4. Which of these is a true statement?
   a. Viruses carry with them their own ribosomes for protein formation.
   b. New viral ribosomes form after viral DNA enters the cell.
   c. Viruses use the host ribosomes for their own ends.
   d. Viruses do not need ribosomes for protein formation.
5. Which part of an animal virus is not reproduced in multiple copies?
   a. envelope
   b. proteins
   c. capsid
   d. ribosomes
6. RNA retroviruses have a special enzyme that
   a. disintegrates host DNA.
   b. polymerizes host DNA.
   c. transcribes viral RNA to cDNA.
   d. translates host DNA.

7. Which is not true of prokaryotes? They
   a. are living cells.
   b. lack a nucleus.
   c. all are parasitic.
   d. are both archaea and bacteria.
8. Facultative anaerobes
   a. require a constant supply of oxygen.
   b. are killed in an oxygenated environment.
   c. do not always need oxygen.
   d. are photosynthetic.
9. Cyanobacteria, unlike other types of bacteria that photosynthesize, do
   a. not give off oxygen.
   b. give off oxygen.
   c. not have chlorophyll.
   d. not have a cell wall.
10. Chemosynthetic bacteria
    a. are autotrophic.
    b. use the rays of the sun to acquire energy.
    c. oxidize inorganic compounds to acquire energy.
    d. Both a and c are correct.
11. Label this condensed version of bacteriophage reproductive cycles using these terms: bacterial chromosome, penetration, maturation, release, prophage, attachment, and integration.

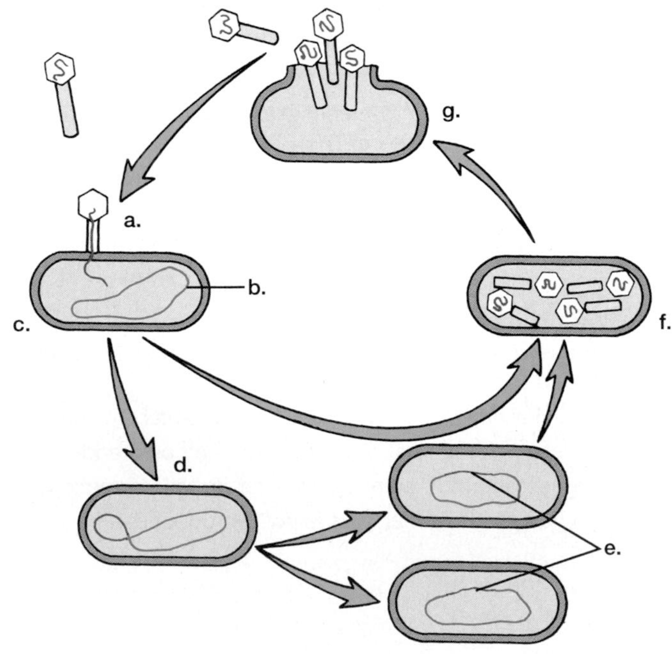

Applying the Concepts

1. *Viruses are obligate parasites.*
   Explain the meaning of this statement.
2. *Prokaryotes are diverse organisms.*
   What evidence do you have that prokaryotes are diverse?
3. *Life evolved from the simple to the complex.*
   How can you reconcile this concept with the former concept?

Using Technology

Your study of viruses, bacteria, and archaea is supported by these available technologies:

Exploring the Internet
The Mader Home Page provides resources for and help with studying this chapter.

http://www.mhhe.com/sciencemath/biology/mader/
(Click on Biology.)

Understanding the Terms

Match the terms to these definitions:

a. __________ Bacteriophage life cycle in which the virus incorporates its DNA into that of the bacterium; only later does it begin a lytic cycle, which ends with the destruction of the bacterium.
b. __________ Nonliving, obligate, intracellular parasite consisting of an outer capsid and an inner core of nucleic acid.
c. __________ One of the bacteriophage life cycles in which the virus takes over the operation of the bacterium immediately upon entering it and subsequently destroys the bacterium.
d. __________ Organism that secretes digestive enzymes and absorbs the resulting nutrients back across the plasma membrane.
e. __________ Photosynthetic bacterium that contains chlorophyll and releases oxygen; formerly called a blue-green alga.
f. __________ Type of prokaryote that is most closely related to eukarya.

The Protists

CHAPTER 30

Chapter Concepts

30.1 Algae Are Plantlike

- Algae are adapted to an aquatic environment and do not protect the gametes and zygote from drying out, as do plants, which are adapted to a terrestrial environment. 524
- Like plants, green algae possess chlorophylls *a* and *b*, have a cell wall of cellulose, and store reserve food as starch inside chloroplasts. 525
- Brown algae (e.g., seaweeds) and golden brown algae (e.g., diatoms) have chlorophylls *a* and *c* plus a brownish carotenoid. 527
- Diatoms along with dinoflagellates are significant producers in marine and freshwater habitats. 528
- Euglenoids are flagellated cells with both animal-like and plantlike characteristics, typifying the difficulty of classifying the protists. 528
- Red algae are filamentous or multicellular seaweeds that are more delicate and grow in warmer waters than brown algae. 529

30.2 Protozoa Are Animal-Like

- While algae are the photoplankton, protozoa are the zooplankton of aquatic habitats. 530
- Protozoa are not considered animals because animals are multicellular, have more than one kind of nonreproductive cell, and undergo embryonic development. 530
- Amoeboids move by pseudopods, ciliates move by cilia, and zooflagellates have flagella. Sporozoa are nonmotile parasites that form spores; *Plasmodium* causes malaria. 530

30.3 Slime Molds and Water Molds Are Funguslike

- Like fungi, slime molds produce nonmotile spores, but unlike fungi they have an amoeboid stage that ingests food. 535
- Like fungi, water molds are filamentous and saprotrophic, but unlike fungi they produce 2n zoospores. 536

Diatom, *Campylodiscus hibernicus*

In the last chapter, we examined the viruses and earth's tiniest life-forms, the prokaryotes. In this chapter, we will be looking at the protists, a group of small eukaryotic organisms: algae, protozoa, and the water and slime molds.

Algae are found in the ocean, fresh water, and on land. Many types are an important part of phytoplankton, floating organisms that photosynthesize and produce food for entire communities. Brown algae constitute seaweed and make up giant underwater kelp forests. The photosynthetic unicellular diatoms occur in enormous numbers, and their remains create diatomaceous earth, which is mined by humans for various purposes.

More animal-like are the protozoa, which are generally heterotrophic unicellular organisms. Protozoa are members of the zooplankton in the oceans and fresh water. The skeletons of marine amoeboids called radiolaria are among the most beautiful forms of nature. The protozoa also cause diseases, ranging from diarrhea to malaria.

The water molds and slime molds each have characteristics comparative to those of fungi. However, their life cycles set them apart from the fungi. Water molds are well known for parasitizing both aquatic and terrestrial forms.

30.1 Algae Are Plantlike

Kingdom Protista contains the **protists,** which are usually unicellular organisms (see classification chart). How the multicellular forms in this kingdom should be classified is a matter of debate, as is classification of the protists in general. Since this kingdom includes groups that are as distantly related as plants are from animals, they probably should be split into several separate kingdoms.

There is no taxonomic category called **algae** [L. *alga,* seaweed]; this term has long been used in biology as a matter of convenience to mean aquatic organisms that carry on photosynthesis. In the ocean and freshwater lakes and ponds, algae are a part of **plankton** [Gk. *planktos,* wandering], organisms that largely drift along with the current or float near the surface. Algae are a part of the **phytoplankton** [Gk. *phyton,* plant, and *planktos,* wandering], which photosynthesize and produce the food that maintains an entire community of organisms. They also produce much of the oxygen in the atmosphere.

Like plants, algae have chloroplasts, and the cells are usually strengthened by the presence of a cell wall. Some algae are colonial; a **colony** is a loose association of independent cells in which there may be cells specialized for reproduction. Some algae are actually multicellular and even have specialized tissues. Why aren't multicellular algae considered plants? In this text, all plants have modifications that protect the gametes and zygote from drying out.

CLASSIFICATION

Kingdom Protista

Eukaryotic; unicellular organisms and their immediate multicellular descendants; sexual reproduction; flagella and cilia with 9 + 2 microtubules

The Algae*
Phylum Chlorophyta: green algae
Phylum Phaeophyta: brown algae
Phylum Chrysophyta: diatoms and allies
Phylum Dinoflagella: dinoflagellates
Phylum Euglenophyta: euglenoids
Phylum Rhodophyta: red algae

The Protozoa*
Phylum Sarcodina: amoebas and allies
Phylum Ciliophora: ciliates
Phylum Zoomastigophora: zooflagellates
Phylum Sporozoa: sporozoa

The Slime Molds*
Phylum Gymnomycota: slime molds

The Water Molds*
Phylum Oomycota: water molds

*Not in the classification of organisms, but added here for clarity.

Obviously these are adaptations to the land environment. Algae, which are adapted to a water environment, do not have these modifications.

For many years it has been customary to classify algae according to their color; therefore, we speak of green, brown, golden brown, and red algae. For the most part, recent biochemical analyses also support these designations. All algae contain at least one type of chlorophyll, but they also contain other types of pigments, and these may mask the color of the chlorophyll. The type of cell wall and the way they store reserve food also help distinguish algae.

Green Algae Are Most Plantlike

Green algae (**phylum Chlorophyta,** 7,000 species) live in the ocean but are more likely found in fresh water and can even be found on land, especially if moisture is available. As discussed on page 540, they associate with fungi in lichens. Some even have modifications that allow them to

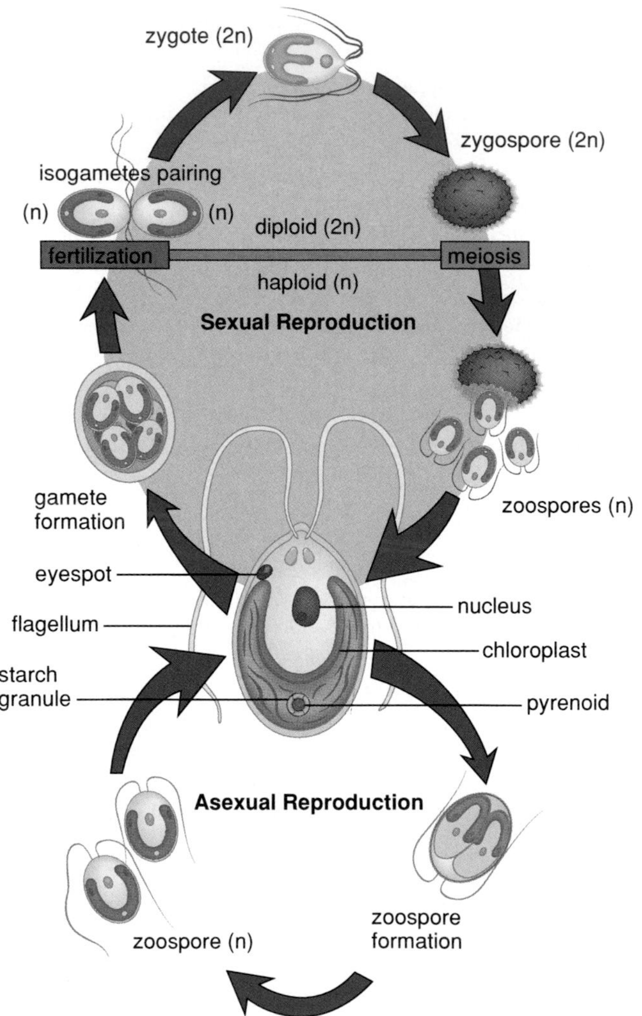

Figure 30.1 ***Chlamydomonas.***
Chlamydomonas is a motile green alga. During asexual reproduction, all structures are haploid; during sexual reproduction, meiosis follows the zygote stage, which is the diploid part of the cycle.

live on tree trunks, even in bright sun. Green algae are believed to be closely related to the first plants because both of these groups (1) have a cell wall that contains cellulose, (2) possess chlorophylls *a* and *b*, and (3) store reserve food as starch inside the chloroplast. (Other types of algae store reserve food outside the chloroplast.) Green algae are not always green because some have pigments that give them an orange, red, or rust color.

Green Algae That Have Flagella

Chlamydomonas is a unicellular green alga usually less than 25 μm long that has been studied in detail using the electron microscope (Fig. 30.1). It has a definite cell wall and a single, large, cup-shaped chloroplast that contains a *pyrenoid,* a dense body where starch is synthesized. The chloroplast also contains a red-pigmented *eyespot* (stigma), which is sensitive to light and helps bring the organism into the light, where photosynthesis can occur. Two long whiplike flagella project from the anterior end of this alga and curl backward to propel the cell freely toward the light.

When growth conditions are favorable, *Chlamydomonas* reproduces asexually. The adult divides, forming *zoospores* (flagellated spores) that resemble the parent cell. A **spore** is a haploid body that develops into a mature adult when conditions are favorable. When growth conditions are unfavorable, *Chlamydomonas* reproduces sexually. Gametes of two different mating types come into contact and join to form a zygote. A heavy wall forms around the zygote, and it becomes a resistant zygospore able to survive until conditions are favorable for germination. When a zygospore germinates, it produces four zoospores by meiosis. In most species, the gametes are identical, a condition known as **isogamy** [Gk. *isos,* equal, and *gamos,* marriage, union]. The gametes are called *isogametes.* In other species, there is a nonmotile egg specialized for storing food, and the motile sperm are specialized for seeking out an egg. This condition is known as **oogamy** [Gk. *oon,* egg, and *gamos,* marriage, union] and the gametes are *heterogametes.*

Doing Science

Behavior in a Unicellular Alga

Susan Dutcher, an investigator at the University of Colorado at Boulder, explains why she has been studying the unicellular, eukaryotic alga Chlamydomonas reinhardtii *for the last 11 years.*

Chlamydomonas, an organism that is only about 10 micrometers (μm) in length, lives in soil and puddles. Most of the cytoplasm is occupied by a large cup-shaped chloroplast that is used for photosynthesis. At the anterior end of the cell are the organelles that attracted me to work on *Chlamydomonas.* These organelles are the two flagella, which are used to propel the cell at velocities of 200 μm per minute. These flagella, like all eukaryotic flagella, have at least 250 different polypeptides assembled into the classic structure of nine doublet microtubules in a circle around two singlet microtubules. The motors of flagella are dyneins, proteins that convert the chemical energy of ATP into the bending action of flagella.

Chlamydomonas uses its flagella for swimming and for several complicated behaviors, including recognition of a mating partner. Generally, a *Chlamydomonas* cell swims with a whiplike motion accompanied by a rolling movement. It can be likened to a swimmer crossing a swimming pool by using the breaststroke but at the same time turning from the stomach, to the side, to the back, and then to the stomach again. Then, too, *Chlamydomonas* can change direction by regulating the strength by which each flagellum beats. When *Chlamydomonas* swims, it orients itself toward a source of light—a behavior known as phototaxis.

Phototaxis most likely helps the organism find light at the correct intensity for photosynthesis. How does *Chlamydomonas* detect light? Like many other algae, *Chlamydomonas* contains a structure known as an eyespot. Its single eyespot appears to utilize a molecule like rhodopsin, which is found in the mammalian eye. The rotation used by *Chlamydomonas* as it swims allows the eyespot to scan the environment. If the light intensity remains the same throughout the rotation, the cell continues to swim straight. If the light intensity is different, the cell adjusts its direction by changing the relative degree of flagellar bending.

Why does one want to study a unicellular green alga instead of a fly, a mouse, or a human? One of the main reasons is the ease of finding mutations that disrupt the biological process being studied. As a geneticist, I dissect a process by removing one piece of the genetic machinery at a time, and then I observe what happens to the structure and function of the cell or organism. Genetic dissection has been used to learn a great deal about the cell cycle, development, and many other processes. Mutations that affect flagella functions have even been identified in humans. There is a class of genetic diseases known as Kartagener syndrome, in which the flagella and cilia are paralyzed. The symptoms include infertility (due to immotile sperm or immotile cilia in the oviduct) and bronchitis (from the failure of immotile cilia in bronchial tubes to clear away mucus). We hope that our studies will help to understand the causes of this particular class of human diseases.

In order to study the role of flagella in phototaxis, my laboratory and others study a variety of mutations that affect the assembly and function of flagella. When *Chlamydomonas* fails to orient toward light, it could be that the organism is unable to detect light or is unable to signal that light is present. Or it could be that a mutation has affected the assembly of flagella so that they cannot respond to the signal. We are interested in understanding this latter class of mutations, as we believe that they will allow us to understand how flagella construction can affect response to a signal. We hope that this type of study in a model genetic organism like *Chlamydomonas* will help to clarify similar processes in other organisms.

Green Algae That Are Filamentous

Filaments [L. *filum,* thread] are end-to-end chains of cells that form after cell division occurs in only one plane. *Spirogyra,* a filamentous green alga, is found in green masses on the surfaces of ponds and streams. It has ribbonlike, spiralled chloroplasts (Fig. 30.2). **Conjugation** [L. *conjugalis,* pertaining to marriage], the temporary union of two individuals during which there is an exchange of genetic material, occurs during sexual reproduction. The two filaments line up parallel to each other, and the cell contents of one filament move into the cells of the other filament, forming diploid zygotes. These zygotes survive the winter, and in the spring they undergo meiosis to produce new haploid filaments.

Oogamy does occur among the filamentous green algae. The genus *Oedogonium* contains filamentous algae in which the cells are cylindrical with netlike chloroplasts; during sexual reproduction there is a definite egg and sperm.

Green Algae That Are Multicellular

Multicellular *Ulva* is commonly called sea lettuce because of its leafy appearance (Fig. 30.3). The thallus (body) is two cells thick and can be a meter long. *Ulva* has an alternation of generations like that of plants except that both generations look exactly alike, the gametes look alike (isogamy), and the spores are flagellated. In plants one generation is typically dominant over (lasts longer than) the other, egg and sperm are produced (oogamy), and the spores are not flagellated (see Fig. 30B).

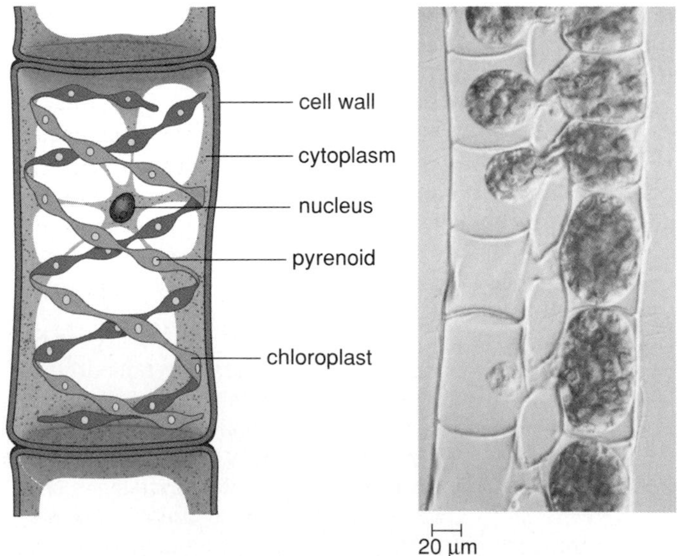

Figure 30.2 ***Spirogyra.***
Spirogyra is a filamentous green alga, in which each cell has a ribbonlike chloroplast. During conjugation the cell contents of one filament enter the cells of another filament. Zygote formation follows.

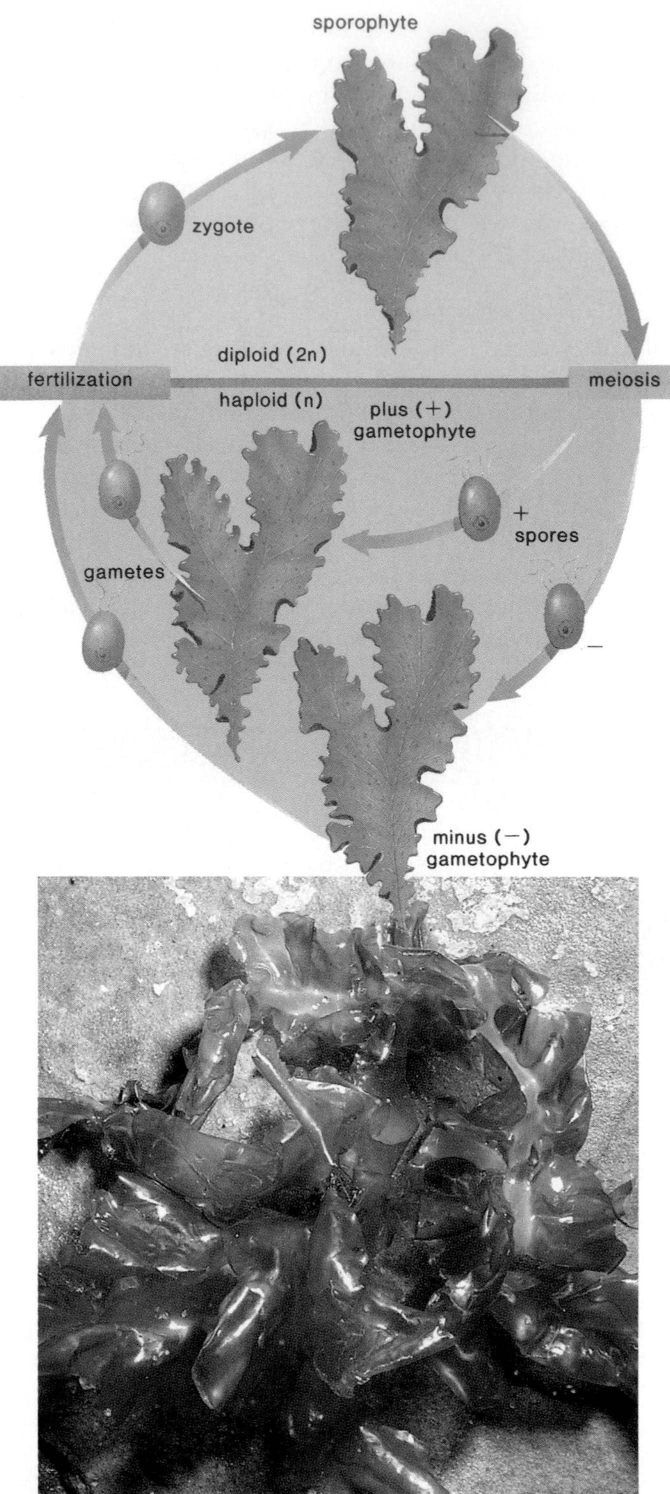

Figure 30.3 ***Ulva.***
Ulva is a multicellular green alga. *Ulva* has an alternation of generations life cycle, as do plants. Notice that the sporophyte and gametophyte have the same appearance and that the gametophyte produces isogametes. These features are unlike plants.

Green Algae That Are Colonial

A number of *colonial* (loose association of cells) forms occur among the flagellated green algae. A *Volvox* colony is a hollow sphere with thousands of cells arranged in a single layer surrounding a watery interior. Each cell of a *Volvox* colony resembles a *Chlamydomonas* cell—perhaps it is derived from daughter cells that fail to separate following zoospore formation. In *Volvox,* the cells cooperate in that the flagella beat in a coordinated fashion. Some cells are specialized for reproduction, and each of these can divide asexually to form a new daughter colony (Fig. 30.4). This daughter colony resides for a time within the parental colony, but then it leaves by releasing an enzyme that dissolves away a portion of the parental colony, allowing it to escape. Sexual reproduction among these algae involves oogamy.

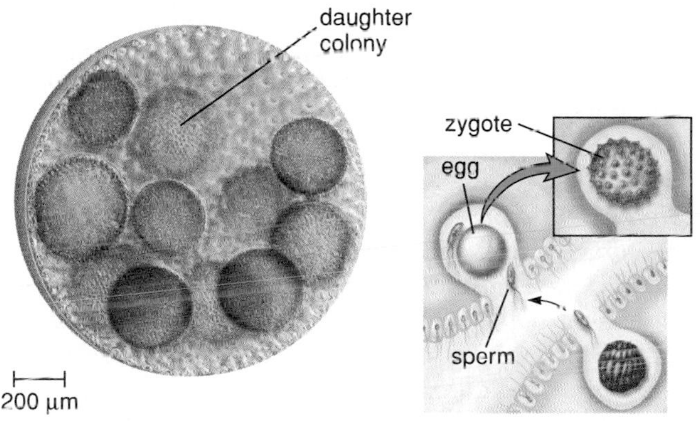

Figure 30.4 ***Volvox.***
Volvox is a colonial green alga. The adult *Volvox* colony often contains daughter colonies, which are asexually produced by special cells. During sexual reproduction, colonies produce a sperm and egg.

Green algae occur in unicellular, colonial, and multicellular forms. During sexual reproduction, the zygote usually undergoes meiosis and the only adult is haploid. *Ulva* has an alternation of generations like plants do.

Brown Algae and Golden Brown Algae Are Biochemically Alike

The brown algae and the golden brown algae have chlorophylls *a* and *c* in their chloroplasts and a type of carotenoid pigment (fucoxanthin) that gives them their color. The reserve food is a carbohydrate called laminarin.

Brown Algae Are Seaweeds

Brown algae (phylum Phaeophyta, 1,500 species) range from small forms with simple filaments to large multicellular forms (50–100 μm long) (Fig. 30.5). The multicellular forms are a type of **seaweed,** a common term for any large, complex alga. Brown algae are often observed along the rocky coasts in the north temperate zone, where they are pounded by waves as the tide comes in and are exposed to dry air as the tide goes out. They do not dry out, however, because their cell walls contain a mucilaginous, water-retaining material.

Both *Laminaria,* commonly called a *kelp,* and *Fucus,* known as rockweed, are examples of brown algae that grow along the shoreline. In deeper waters, the giant kelps (*Nereocystis* and *Macrocystis*) often form spectacular underwater forests. Individuals of the genus *Sargassum* sometimes break off from their holdfasts and form floating masses, where life-forms congregate in the ocean. Brown algae not only provide food and habitat for marine organisms, they are harvested for human food and for fertilizer in several parts of the world. They are also a source of algin, a pectinlike material that is added to ice cream, sherbet, cream cheese, and other products to give them a stable, smooth consistency.

Laminaria is unique among the protists because members of this genus show tissue differentiation—they transport organic nutrients by way of a tissue that resembles phloem in land plants. Most brown algae have the alternation of generations life cycle, but some species of *Fucus* are unique in that meiosis produces gametes and the adult is always diploid, as in animals.

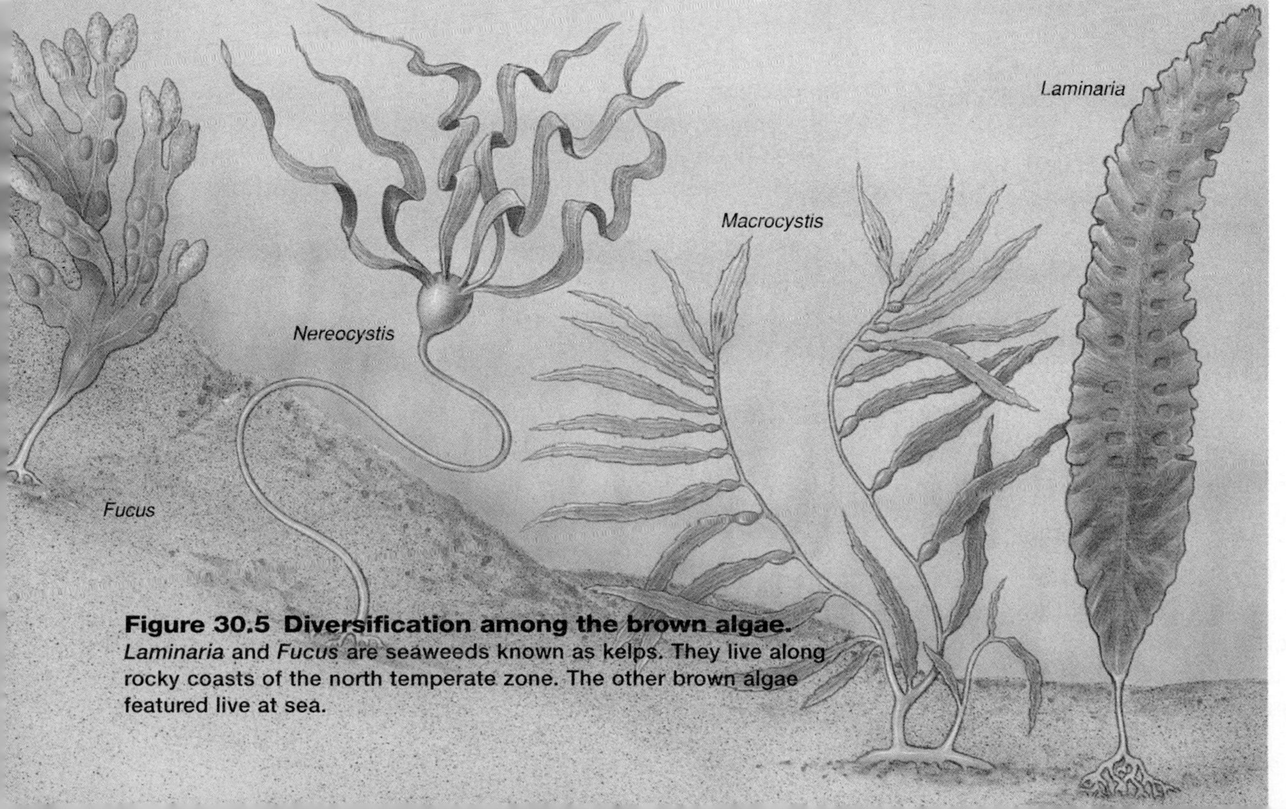

Figure 30.5 Diversification among the brown algae.
Laminaria and *Fucus* are seaweeds known as kelps. They live along rocky coasts of the north temperate zone. The other brown algae featured live at sea.

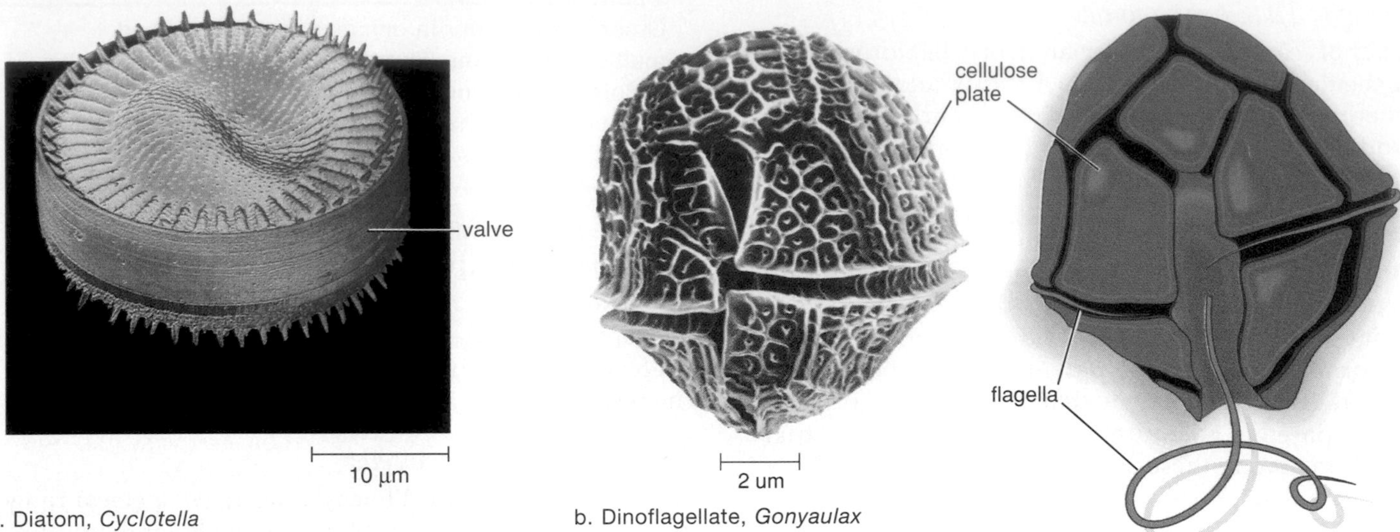

a. Diatom, *Cyclotella*

b. Dinoflagellate, *Gonyaulax*

Figure 30.6 Diatoms and dinoflagellates.
a. Diatoms may be variously colored, but their chloroplasts contain a unique golden brown pigment (fucoxanthin), in addition to chlorophylls *a* and *c*. The beautiful pattern results from markings on the silica-embedded wall. **b.** Dinoflagellates have cellulose plates; these belong to *Gonyaulax*, the dinoflagellate that contains a red pigment and is responsible for occasional "red tides."

Golden Brown Algae Include Diatoms

The **golden brown algae** (**phylum Chrysophyta,** 11,000 species) are distinctive from one another, and some authorities place diatoms in their own phylum called phylum Bacillariophyta. **Diatoms** [Gk. *dia,* through, and *temno,* cut] are the most numerous unicellular algae in the oceans (Fig. 30.6*a*). They are also plentiful in fresh water. Because of their small size, thousands can live in a single millimeter of water. Therefore, they are an important source of food and oxygen for heterotrophs in both freshwater and marine ecosystems.

The structure of a diatom is often compared to a box because the cell wall has two halves or valves, with the larger valve acting as a "lid" for the smaller valve. When diatoms reproduce asexually, each receives one old valve. The new valve fits inside the old one; therefore, the new diatom is smaller than the original one. This continues until diatoms are about 30% of their original size. Then they reproduce sexually. The zygote becomes a structure that grows and then divides mitotically to produce diatoms of normal size.

The cell wall has an outer layer of silica, a common ingredient of glass. The valves are covered with a great variety of striations and markings that form beautiful patterns when observed under the microscope. These are actually depressions or pores through which the organism makes contact with the outside environment. The remains of diatoms, called diatomaceous earth, accumulate on the ocean floor and are mined for use as filtering agents, soundproofing materials, and gentle abrasives.

Dinoflagellates Have Two Flagella

Many **dinoflagellates** [Gk. *dinos,* whirling, and L. *flagello,* whip] (**phylum Dinoflagella,** 1,000 species) are bounded by protective cellulose plates (Fig. 30.6*b*). Most have two flagella; one lies in a longitudinal groove with its distal end free, and the other, which is flat and ribbonlike, lies in a transverse groove that encircles the organism. The longitudinal flagellum acts as a rudder, and the beating of the transverse one causes the cell to spin as it moves forward.

The chloroplasts of a dinoflagellate, which contain chlorophylls *a* and *c* as do those of golden brown algae, are probably derived from an endosymbiotic event (see p. 76). They vary in color from yellow-green to brown. Some species of dinoflagellates are heterotrophic, and it has been suggested that they are really protozoa.

Like the diatoms, dinoflagellates are extremely numerous in the oceans; their density can equal 30,000 in a single millimeter. Under certain conditions, those in the genus *Gymnodinium* and *Gonyaulax* increase in number and cause a "red tide" in the ocean. At these times, they produce a neurotoxin that can kill fish and causes paralytic shellfish poisoning—humans who eat shellfish that have fed on these dinoflagellates suffer paralysis of the respiratory muscles.

Usually the dinoflagellates are an important source of food for small animals in the ocean. They also live as symbionts within the bodies of some invertebrates. For example, corals usually contain large numbers of these organisms, and this allows coral reefs to be one of the most productive ecosystems on earth.

Euglenoids Are Flexible

Euglenoids (**phylum Euglenophyta,** 1,000 species) are small (10–500 µm) freshwater unicellular organisms that typify the problem of classifying protists. One-third of all genera have chloroplasts; the rest do not. Those that lack chloroplasts ingest or absorb their food. In addition, euglenoids grown in the absence of light have been known to lose their chloroplasts and become heterotrophic. This may not be surprising when one knows that their chloroplasts are like those of green algae and are probably derived from a green alga through endosymbiosis. The chloroplasts are surrounded by three rather than two membranes. The pyrenoid is an organ outside the chloroplast that produces an unusual type of carbohydrate polymer (paramylon) not seen in green algae.

Euglenoids have two flagella, one of which typically is much longer than the other and projects out of an anterior vase-shaped invagination (Fig. 30.7). It is called a tinsel flagellum because it has hairs on it. Near the base of this flagellum is an eyespot, which shades a photoreceptor for detecting light. Because euglenoids are bounded by a flexible *pellicle* composed of protein strips lying side by side, they can assume different shapes as the underlying cytoplasm undulates and contracts. As in certain protozoa, there is a contractile vacuole for ridding the body of excess water. Euglenoids reproduce by longitudinal cell division, and sexual reproduction is not known to occur.

Euglenoids have both plantlike and animal-like characteristics. They have chloroplasts but lack a cell wall and swim by flagella.

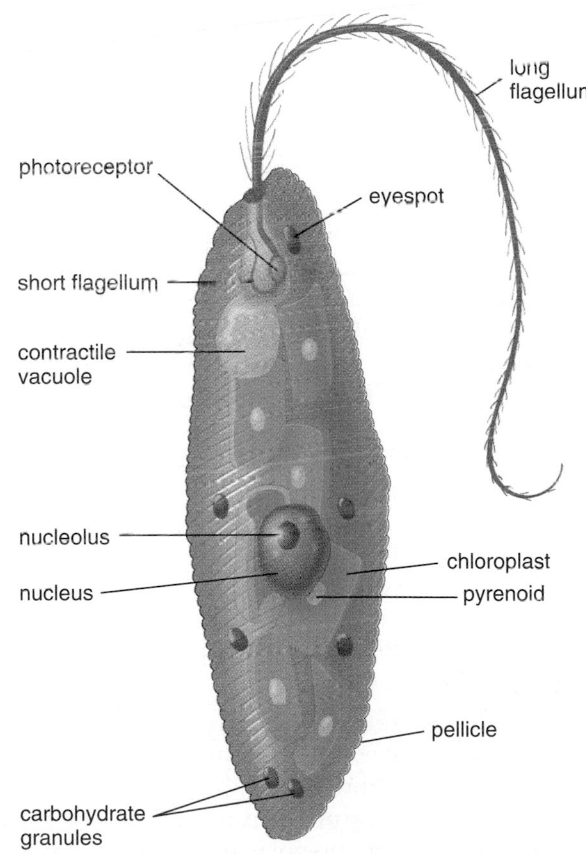

Figure 30.7 ***Euglena*.**
Euglena is a flagellated green alga with both animal-like and plantlike characteristics. A very long flagellum propels the body, which is enveloped by a flexible pellicle. A photoreceptor shaded by an eyespot allows *Euglena* to find light, after which photosynthesis can occur in the numerous chloroplasts. Pyrenoids synthesize a reserve carbohydrate, which is stored in the chloroplasts and also in the cytoplasm.

Red Algae Are Sources of Agar

Like the brown algae, **red algae (phylum Rhodophyta,** 4,000 species) are multicellular, but they live chiefly in warmer seawater, growing in both shallow and deep waters. Red algae are usually much smaller and more delicate than the brown algae, although they can be up to a meter long. Some forms of red algae are simple filaments, but more often they are complexly branched, with the branches having a feathery, flat, or expanded ribbonlike appearance (Fig. 30.8). Coralline algae are red algae that have cell walls impregnated with calcium carbonate. In some instances, they contribute as much to the growth of coral reefs as do coral animals.

Despite their macroscopic size, red algae show little tissue differentiation. Oogamy is seen during sexual reproduction, but the sperm are nonflagellated, an unusual feature in a water environment. Their chloroplasts resemble cyanobacteria in that they contain chlorophyll *a* and a type of pigment called phycobilin. The reserve food resembles glycogen and is called floridean starch.

Like brown algae, red algae are economically important. The mucilaginous material in the cell walls of certain genera of red algae is a source of agar used commercially to make capsules for vitamins and drugs, as a material for making dental impressions, and as a base for cosmetics. In the laboratory, agar is a culture medium for bacteria. When purified, it becomes the gel for electrophoresis, a procedure that separates proteins and nucleotides. Agar is also used in food preparation—as an antidrying agent for baked goods and to make jellies and desserts set rapidly.

Many red algae have filamentous branches or are multicellular.

Figure 30.8 Red alga.
Red algae, represented by *Chondrus crispus*, are smaller and more delicate than brown algae.

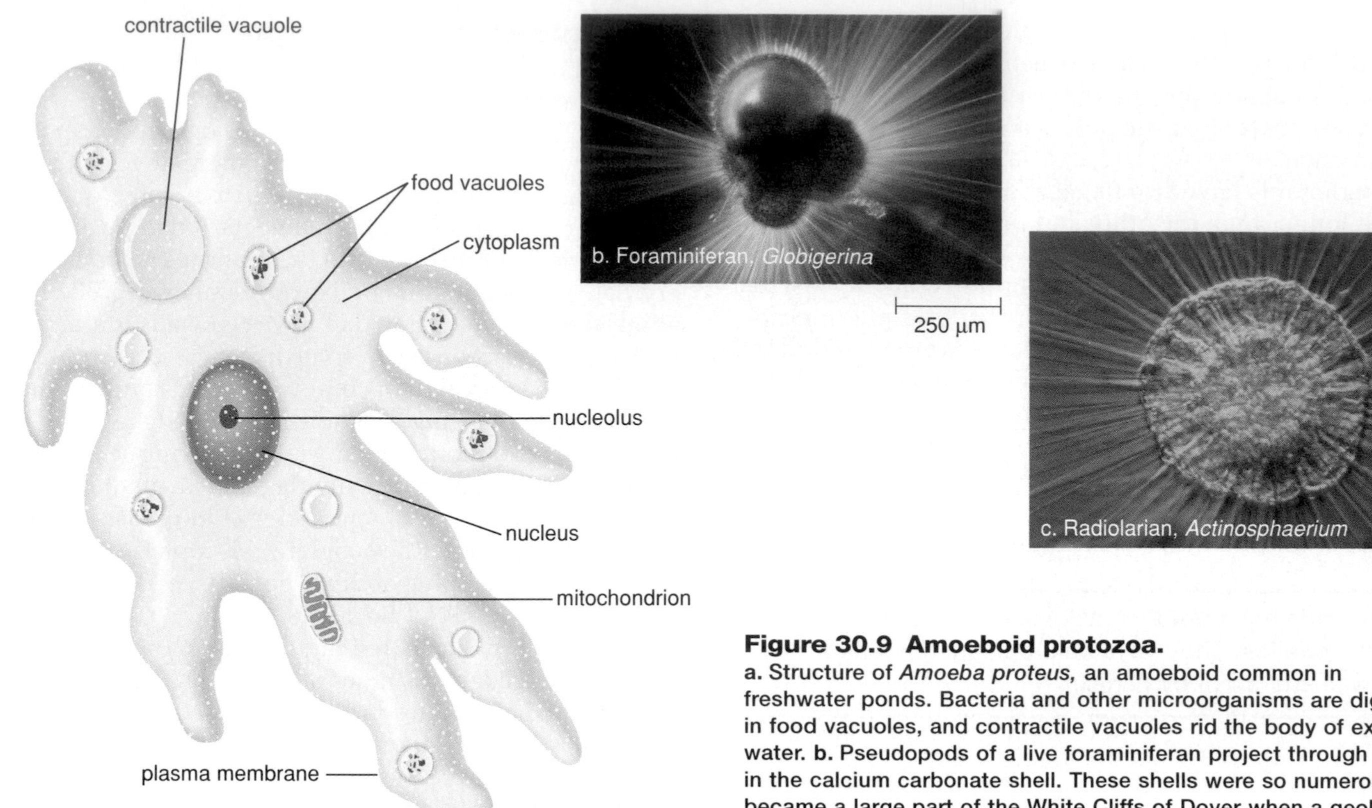

Figure 30.9 Amoeboid protozoa.
a. Structure of *Amoeba proteus*, an amoeboid common in freshwater ponds. Bacteria and other microorganisms are digested in food vacuoles, and contractile vacuoles rid the body of excess water. **b.** Pseudopods of a live foraminiferan project through holes in the calcium carbonate shell. These shells were so numerous they became a large part of the White Cliffs of Dover when a geological upheaval occurred. **c.** Pseudopods of a live radiolarian extend outward through openings in a siliceous shell.

30.2 Protozoa Are Animal-Like

The term protozoa, which is used for convenience, is not a taxonomic category. **Protozoa** [Gk. *protos*, first, and *zoon*, animal] are typically heterotrophic, motile, unicellular organisms of small size (2–1,000 μm). They are not animals because animals in the classification used by this text are multicellular and have more than one kind of nonreproductive cell and undergo embryonic development.

Protozoa usually live in water, but they can also be found in moist soil or inside other organisms. In oceans and freshwater lakes and ponds, they are a part of plankton. Specifically, they are **zooplankton** that feed on phytoplankton. Some protozoa engulf whole food and are termed **holozoic;** others are saprotrophic, and they absorb nutrient molecules across the plasma membrane. Still others are parasitic and are responsible for several significant human infections.

Most protozoa are unicellular, but there are also colonial and even multicellular forms. Even the unicellular ones should not be considered simple organisms. Each cell alone must carry out all the functions performed by specialized tissues and organs in more complex organisms. They have organelles for purposes we have not seen before. Their food is digested inside food vacuoles, and freshwater protozoa have "contractile" vacuoles for the elimination of water.

Although asexual reproduction involving binary fission and mitosis is the rule, many protozoa also reproduce sexually during some part of their life cycle. During times of unfavorable growth, some of them form cysts that have a protective coat and are metabolically inactive. Once favorable growth conditions return, cysts are sites for nuclear reorganization and cell division. Therefore, they are sometimes called reproductive cysts.

The protozoa we will study can be placed in four groups according to their type of locomotor organelle:

Name	Locomotion	Example
Amoeboids	Pseudopods	*Amoeba*
Ciliates	Cilia	*Paramecium*
Zooflagellates	Flagella	*Trypanosoma*
Sporozoa	No locomotion	*Plasmodium*

Amoeboids and Relatives Move by Pseudopods

The **amoeboids** and relatives (**phylum Sarcodina,** 40,000 species) are characterized by pseudopodia. The amoeboids are protists that move and engulf their prey with **pseudopods** [Gk. *pseudes*, false, and *podos*, foot]. Pseudopods form when the cytoplasm streams forward in a particular direction. Many amoeboids have shells, as do the foraminifera and radiolaria.

Amoeba proteus is a commonly studied freshwater member of this group (Fig. 30.9). When amoeboids feed, they **phagocytize** [Gk. *phagein*, eat, and *kytos*, cell]; the pseudopods surround and engulf the prey, which may be algae,

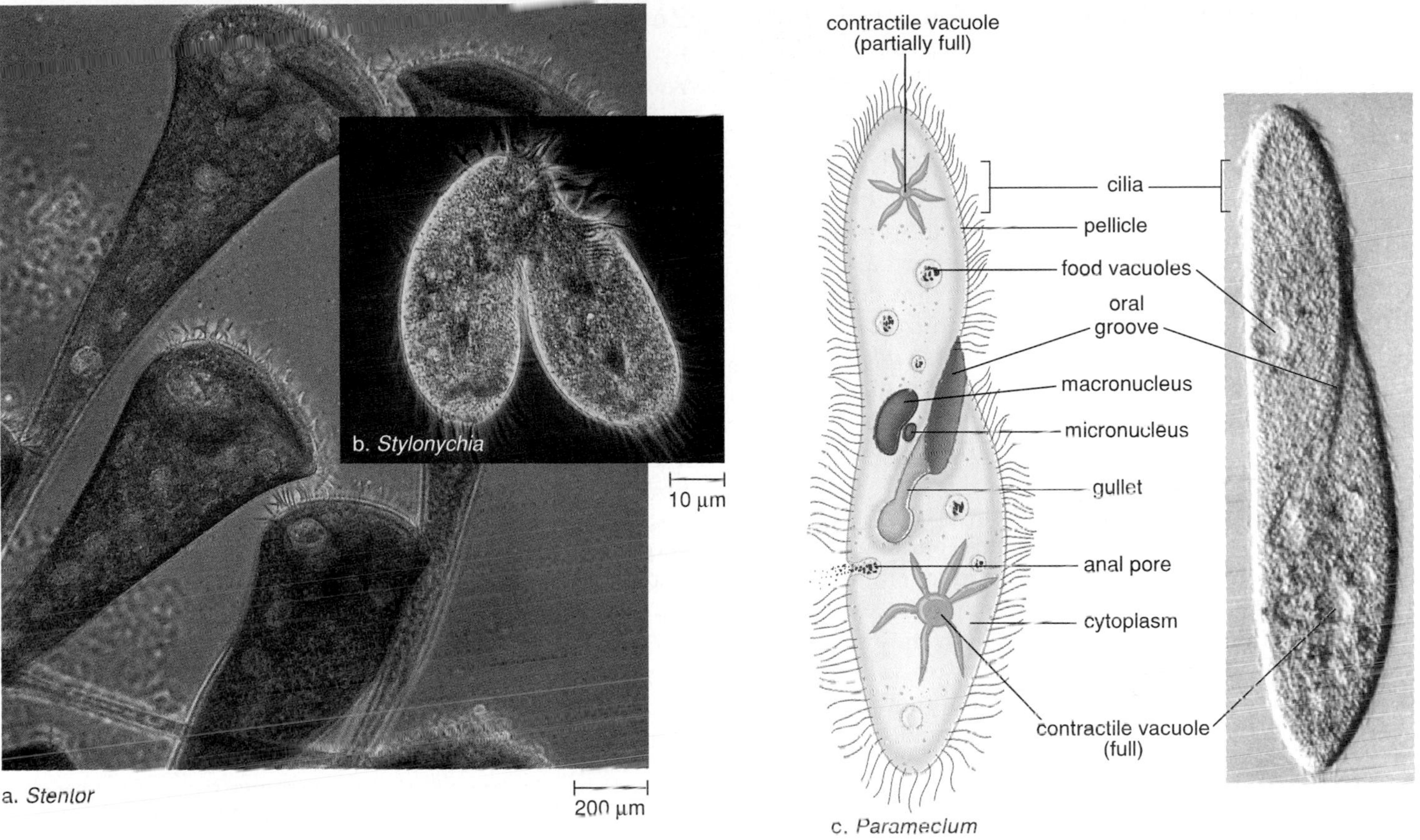

Figure 30.10 Ciliated protozoa.
a. *Stentor,* a large, vase-shaped, freshwater protozoan. b. Two *Stylonychia* conjugating. c. Structure of *Paramecium,* adjacent to an electron micrograph. Ciliates are the most complex of the protozoa. Note the oral groove and the gullet and anal pore.

bacteria, or other protozoa. Digestion then occurs within a food vacuole. Some white blood cells in humans are amoeboid, and they phagocytize debris, parasites, and worn-out cells. Freshwater amoeboids, including *Amoeba proteus,* have contractile vacuoles where excess water from the cytoplasm collects before the vacuole appears to "contract," releasing the water through a temporary opening in the plasma membrane.

Entamoeba histolytica is a parasite that lives in the human intestine and causes amoebic dysentery. Complications arise when this parasite invades the intestinal lining and reproduces there. If the parasites enter the body proper, liver and brain involvement can be fatal.

The *foraminifera,* which are largely marine, have an external calcareous shell (made up of calcium carbonate) with foramina, holes through which long, thin pseudopods extend. The pseudopods branch and join to form a net where the prey is digested. Foraminifera live in the sediment of the ocean floor in incredible numbers—there may be as many as 50,000 shells in a single gram of sediment. Deposits for millions of years, followed by a geological upheaval, formed the White Cliffs of Dover along the southern coast of England. Also, the great Egyptian pyramids are built of foraminiferan limestone.

The *radiolaria,* which have an internal skeleton composed of silica or strontium sulfate, float near the ocean surface. Their exquisite skeletons are found in almost infinite variety. They extend spikelike pseudopods strengthened by microtubules through openings in their shells for feeding purposes.

The amoeboids have pseudopods for locomotion and feeding. *Amoeba proteus* lives in fresh water, but the shelled foraminifera and radiolaria are very abundant in the ocean.

Ciliates Move by Cilia

The **ciliates** (**phylum Ciliophora,** 8,000 species) such as those in the genus *Paramecium* are the most complex of the protozoa (Fig. 30.10). Hundreds of cilia, which beat in a coordinated rhythmic manner, project through tiny holes in a semirigid outer covering, or pellicle. Numerous oval capsules lying in the cytoplasm just beneath the pellicle contain **trichocysts.** Upon mechanical or chemical stimulation, trichocysts discharge long, barbed threads, useful for defense and for capturing prey. *Toxicysts* are similar, but they release a poison that paralyzes prey.

Most ciliates are holozoic. For example, when a paramecium feeds, food is swept down a gullet, below which food vacuoles form. Following digestion, the soluble nutrients are absorbed by the cytoplasm, and the indigestible residue is eliminated at the anal pore.

During asexual reproduction, ciliates divide by transverse binary fission. Ciliates have two types of nuclei: a large *macronucleus* and one or more small *micronuclei*. The macronucleus controls the normal metabolism of the cell, while the micronuclei are concerned with reproduction. Sexual reproduction involves conjugation. The macronucleus disintegrates and, after the micronuclei undergo meiosis, two ciliates exchange a haploid micronucleus. Then the micronuclei give rise to a new macronucleus, which contains copies of only certain housekeeping genes.

The diversity of ciliates is quite remarkable. The barrel-shaped didinia expand to consume paramecia much larger than themselves. Suctoria have an even more dramatic way of getting food. They rest quietly on a stalk until a hapless victim comes along. Then they promptly paralyze it and use their tentacles like straws to suck it dry. *Stentor* may be the prettiest ciliate. It resembles a giant blue vase decorated with stripes (see Fig. 30.10).

Ciliates, which move by cilia, are diverse and very complex. The single cell has specialized regions to carry out various functions.

Zooflagellates Move by Flagella

Protozoa that move by means of flagella are called **zooflagellates (phylum Zoomastigophora)** to distinguish them from unicellular algae that also have flagella. Zooflagellates [Gk. *zoon*, animal, and L. *flagello*, whip] are covered by a pellicle that is often reinforced by underlying microtubules.

Many zooflagellates enter into symbiotic relationships (Fig. 30.11). *Trichonympha collaris* lives in the gut of termites; it contains a bacterium that enzymatically converts the cellulose of wood to soluble carbohydrates that are easily digested by the insect. *Giardia lamblia*, whose cysts are transmitted through contaminated water, attaches to the intestinal wall and causes severe diarrhea. *Trichomonas vaginalis*, a sexually transmitted organism, infects the vagina and urethra of women and the prostate, seminal vesicles, and urethra of men. A **trypanosome,** *Trypanosoma brucei*, transmitted by the bite of the tsetse fly, is the cause of African sleeping sickness. The white blood cells in an infected animal accumulate around the blood vessels leading to the brain and cut off circulation. The lethargy characteristic of the disease is caused by an inadequate supply of oxygen to the brain.

Flagellates usually reproduce by transverse binary fission.

Zooflagellates, which move by flagella, are often symbiotic. The various diseases they cause in humans range from being annoying to being extremely serious.

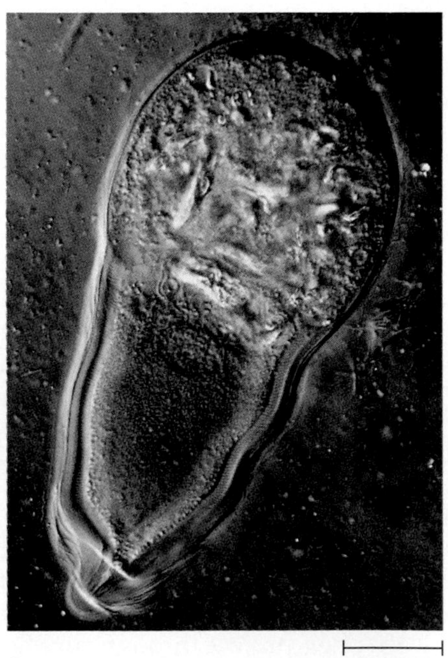

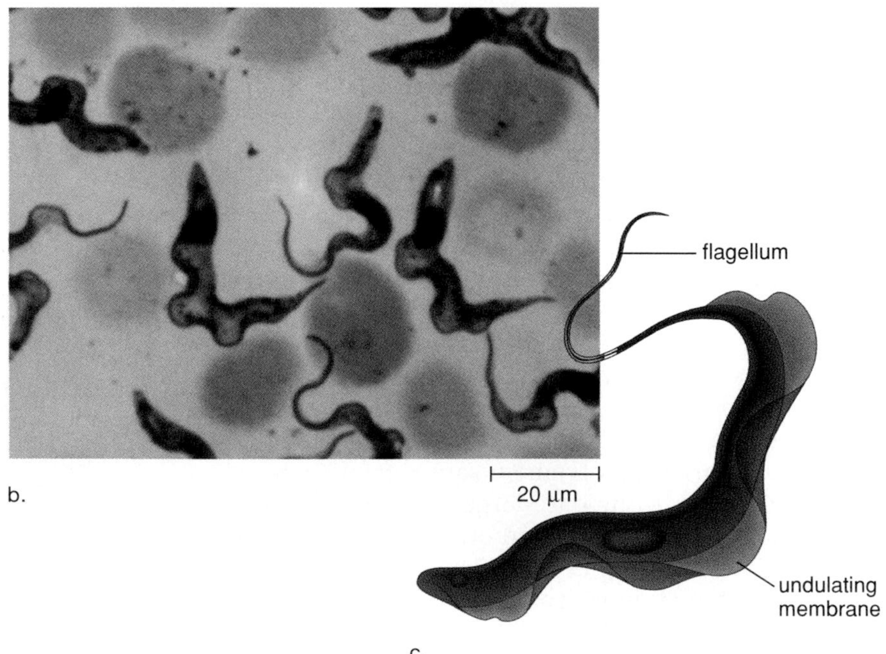

Figure 30.11 Zooflagellates.
a. *Trichonympha collaris* lives in the gut of termites and helps digest cellulose. **b.** Photograph of *Trypanosoma brucei*, the cause of African sleeping sickness, among red blood cells. **c.** The drawing shows its general structure.

Sporozoa Form Spores

Sporozoa (phylum Sporozoa, 3,600 species) are nonmotile parasites. They contain a complex of organelles that may help the organism invade host cells or tissues. Their name recognizes that these organisms form spores at some point in their life cycle. Their complicated life cycle alternates between a sexual and an asexual phase, often with two or more hosts.

Pneumocystis carinii causes the type of pneumonia seen primarily in AIDS patients. During sexual reproduction, thick-walled cysts form in the lining of pulmonary air sacs. The cysts contain spores that successively divide until the cyst bursts and the spores are released. Each spore becomes a new mature organism that can reproduce asexually but may also enter the encysted sexual stage.

The most widespread human parasite is *Plasmodium vivax,* the cause of one type of malaria. When a human is bitten by an infected female *Anopheles* mosquito, the parasite eventually invades the red blood cells. The chills and fever of malaria appear when the infected cells burst and release toxic substances into the blood (Fig. 30.12). Malaria is still a major killer of humans, despite extensive efforts to control it. A resurgence of the disease was caused primarily by the development of insecticide-resistant strains of mosquitoes and by parasites resistant to current antimalarial drugs.

Toxoplasma gondii, another sporozoan, causes toxoplasmosis, particularly in cats but also in people. In pregnant women the parasite can infect the fetus and cause birth defects; in AIDS patients it can infect the brain and cause neurological symptoms.

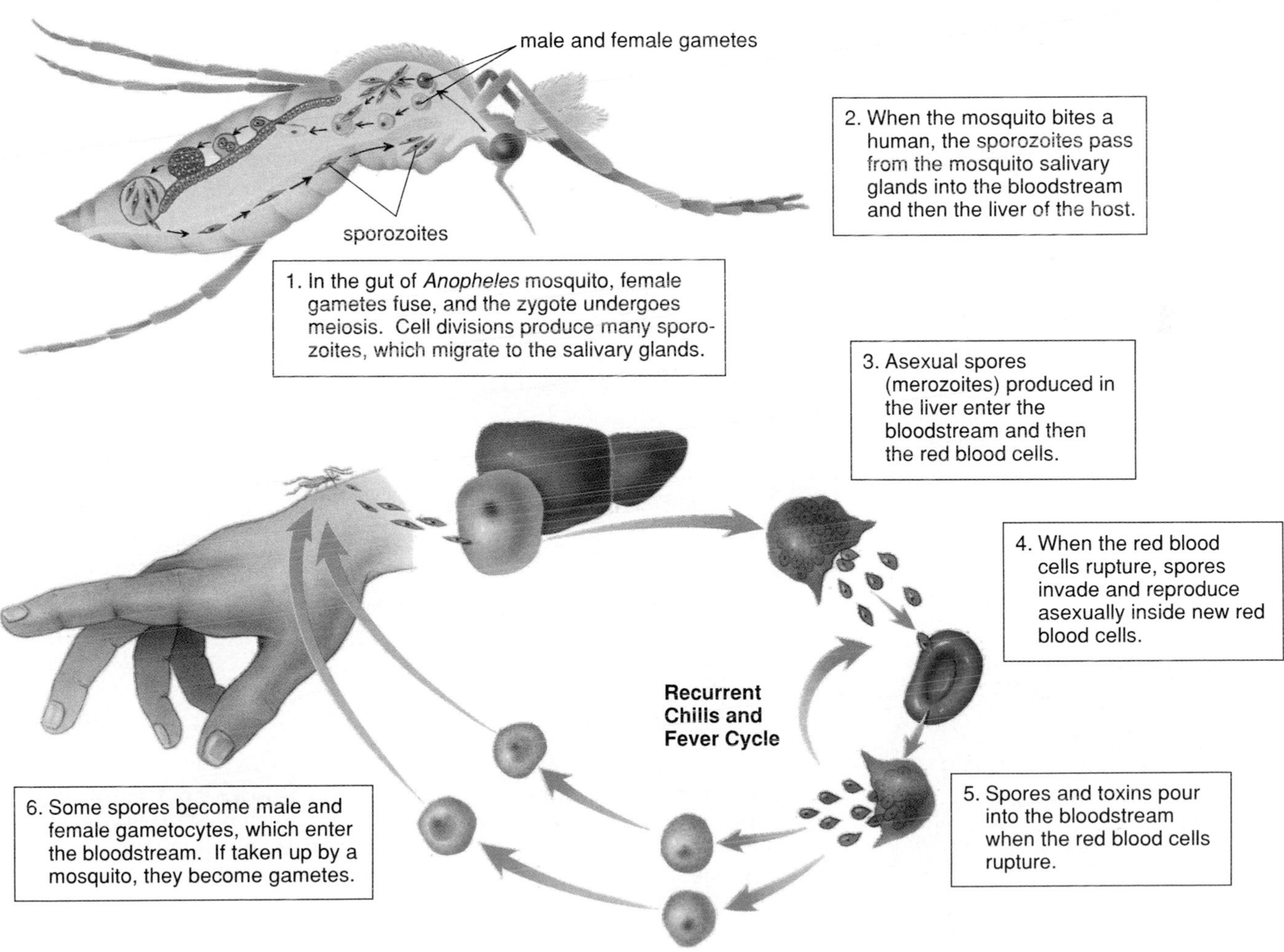

male and female gametes
sporozoites
1. In the gut of *Anopheles* mosquito, female gametes fuse, and the zygote undergoes meiosis. Cell divisions produce many sporozoites, which migrate to the salivary glands.
2. When the mosquito bites a human, the sporozoites pass from the mosquito salivary glands into the bloodstream and then the liver of the host.
3. Asexual spores (merozoites) produced in the liver enter the bloodstream and then the red blood cells.
4. When the red blood cells rupture, spores invade and reproduce asexually inside new red blood cells.
5. Spores and toxins pour into the bloodstream when the red blood cells rupture.
6. Some spores become male and female gametocytes, which enter the bloodstream. If taken up by a mosquito, they become gametes.
Recurrent Chills and Fever Cycle

Figure 30.12 Life cycle of *Plasmodium vivax.*
Asexual reproduction occurs in humans, while sexual reproduction takes place within the *Anopheles* mosquito.

A closer Look

Life Cycles Among the Protists

Organisms reproduce asexually and sexually. Asexual reproduction requires only one parent. The offspring are identical to this parent because the offspring receive a copy of only this parent's genes. There are various modes of asexual reproduction. Single cells such as amoeboids, flagellates, and diatoms simply split in two; *Chlamydomonas* and water molds produce zoospores. In any case, growth alone produces a new adult. Asexual reproduction is the frequent mode of reproduction among protists when the environment is favorable to growth. Individuals identical to the parent are likely to survive and flourish. Sexual reproduction, with its genetic recombination due to fertilization and independent assortment of chromosomes, is more likely to occur among protists when the environment is changing and is unfavorable to growth. Recombination of genes might produce individuals that are more likely to survive extremes in the environment—such as high or low temperatures, acidic or basic pH, or a lack of some particular nutrient.

Sexual reproduction requires two parents, each of which contributes chromosomes (genes) to the offspring by way of gametes. The gametes fuse to produce a diploid zygote. Meiosis occurs during sexual reproduction—just *when* it occurs makes the sexual life cycles diagrammed below differ from one another. In these diagrams, the diploid phase is above and the haploid phase is below the line. In the haplontic cycle (Fig. 30A), the zygote divides by meiosis to form haploid spores that develop into a haploid adult. In aquatic protists, the spores are typically zoospores. The zygote is only the diploid stage in this life cycle, and the haploid adult gives rise to gametes. This sexual cycle is seen in *Chlamydomonas,* a number of algae, protozoa, and fungi.

In alternation of generations (sporic meiosis), the sporophyte (2n) produces haploid spores by meiosis (Fig. 30B). A spore develops into a haploid gametophyte that produces gametes. The gametes fuse to form a diploid zygote, and the zygote develops into the sporophyte. This life cycle is characteristic of some algae (for example, *Ulva* and *Laminaria*) and all plants. In *Ulva,* the haploid and diploid generations have the same appearance. In plants, they are noticeably different from each other.

In the diplontic cycle, a diploid adult produces gametes by meiosis (Fig. 30C). Gametes are the only haploid stage in this cycle. They fuse to form a zygote that develops into the diploid adult. This life cycle is characteristic of a few protists (for example, *Oomycota* and *Fucus*) and all animals. The cycle is isogamous when the gametes look alike—called isogametes—and the cycle is oogamous when the gametes are dissimilar—called heterogametes, usually a small flagellated sperm and a large egg with plentiful cytoplasm.

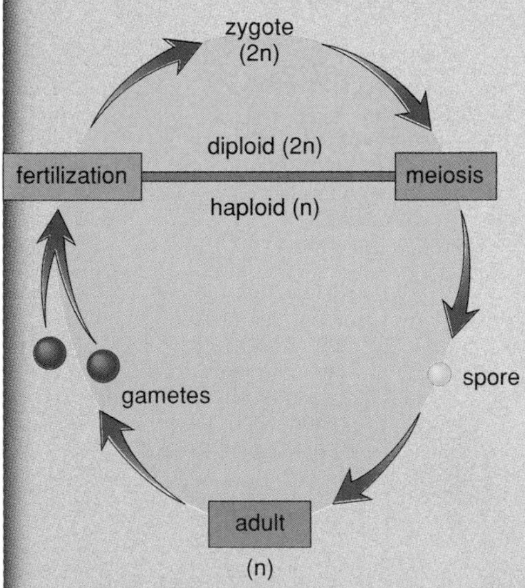

Figure 30A Haplontic cycle.
Zygote is 2n stage.
Meiosis produces spores.
Adult is always haploid.

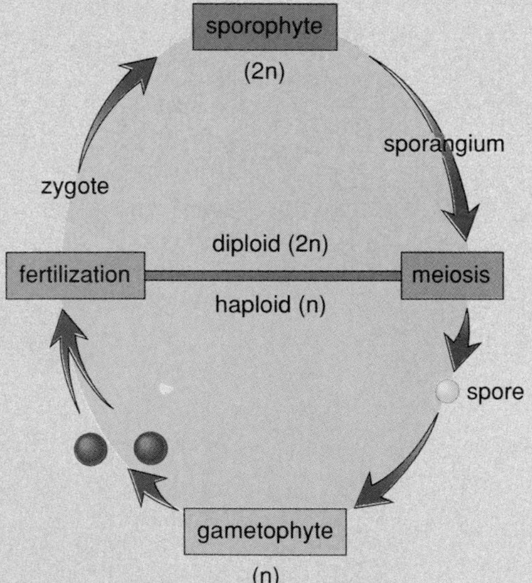

Figure 30B Alternation of generations.
Sporophyte is 2n generation.
Meiosis produces spores.
Gametophyte is haploid generation.

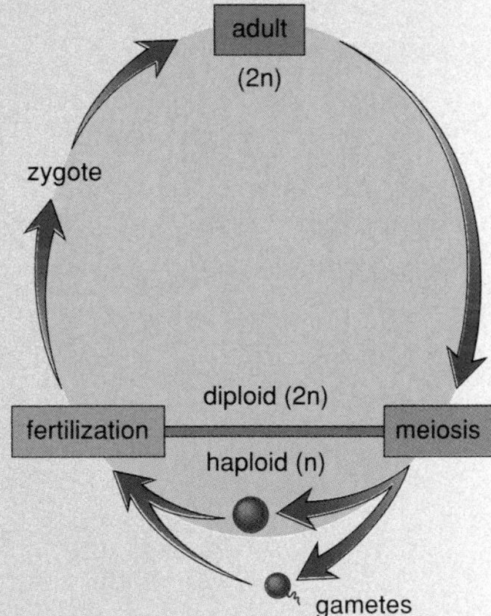

Figure 30C Diplontic cycle.
Adult is always 2n.
Meiosis produces gametes.

30.3 Slime Molds and Water Molds Are Funguslike

Both slime molds and water molds are similar in some respects to fungi. Fungi have a filamentous body and are saprotrophic. A **saprotroph** [Gk. *sapros,* rotten, and *trophe,* food] carries on external digestion and absorbs the resulting nutrients across the plasma membrane. In fungi, meiosis typically follows formation of the zygote, and there are nonmotile spores during both sexual and asexual reproduction. Slime molds and water molds also differ from fungi due to the presence of motility in the life cycle, as discussed below.

Slime Molds Are Amoeboid

Upon a cursory examination, **slime molds (phylum Gymnomycota,** 560 species) might look like fungi but their vegetative state is mobile and amoeboid! At this time they seem more like an amoeboid protozoan than a fungus. Indeed they are heterotrophic by ingestion— they engulf organic material and bacteria. When conditions are unfavorable to growth, they produce and release spores that are resistant to environmental extremes. When the spores germinate, they release cells that begin the life cycle again.

Plasmodial Slime Molds Are a Multinucleated Mass

Usually *plasmodial slime molds* exist as a plasmodium, a diploid multinucleated cytoplasmic mass enveloped by a slime sheath that creeps along, phagocytizing decaying plant material in a forest or agricultural field (Fig. 30.13). The fan-shaped mass contains tubules formed by concentrated cytoplasm in which more liquefied cytoplasm streams. At times unfavorable to growth, such as during a drought, the plasmodium develops many sporangia. A **sporangium** [Gk. *spora,* seed, and *angeion* (dim. of *angos*), vessel] is a reproductive structure that produces spores by meiosis. The spores can survive until moisture is sufficient for them to germinate. In plasmodial slime molds, spores release a haploid flagellated cell or an amoeboid cell. Eventually, two of them fuse to form a zygote that feeds and grows, producing a multinucleated plasmodium once again.

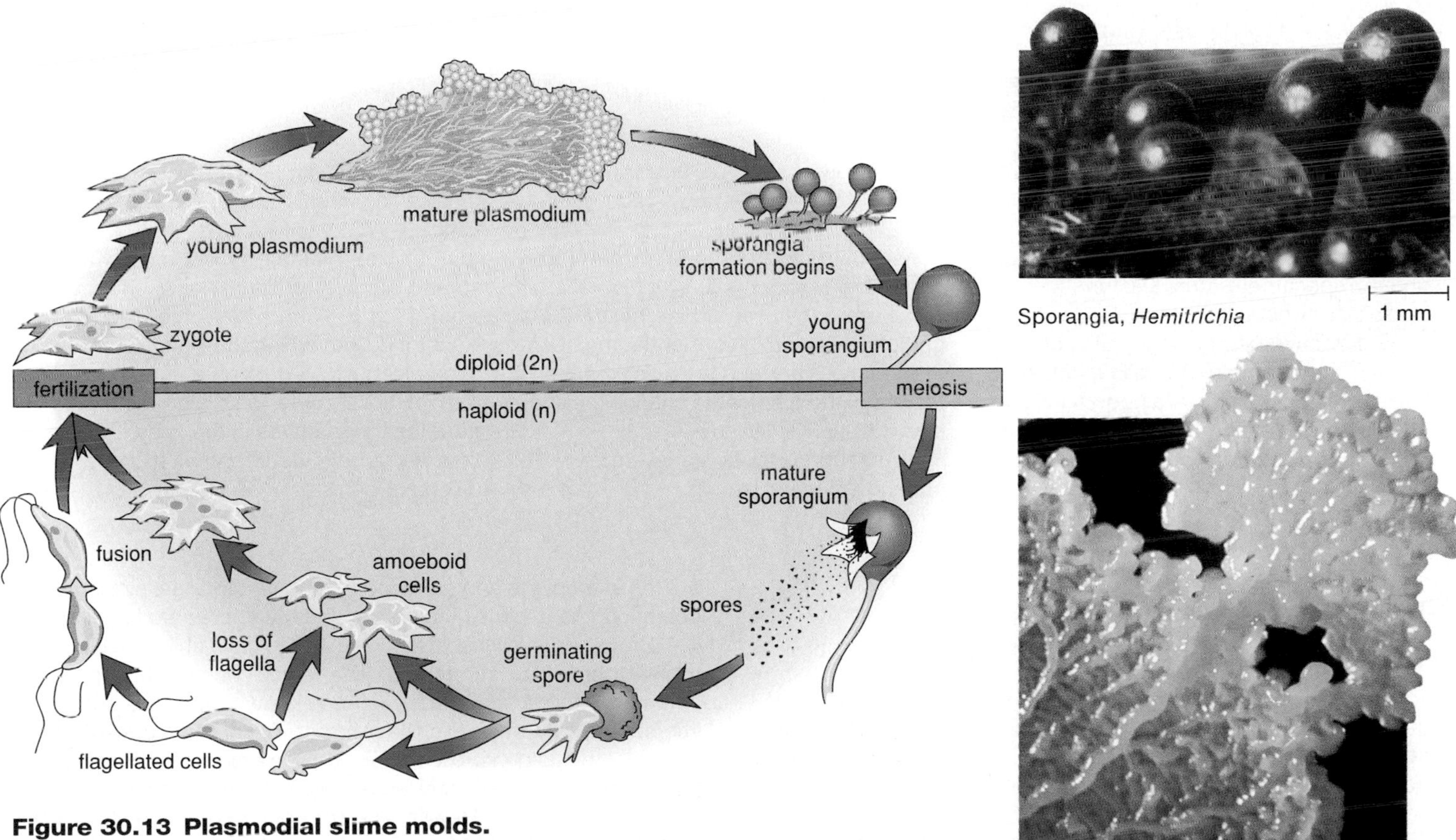

Sporangia, *Hemitrichia*

Plasmodium, *Physarum*

Figure 30.13 Plasmodial slime molds.
During sexual reproduction when conditions are unfavorable to growth, the diploid adult forms sporangia. Haploid spores germinate, releasing haploid amoeboid or flagellated cells that fuse.

Cellular Slime Molds Are Single Cells

Cellular slime molds, as you might expect, exist as individual amoeboid cells. Each lives by phagocytizing bacteria and yeast. As the food supply runs out, the cells release a chemical that causes them to aggregate into a pseudoplasmodium. The pseudoplasmodium stage is temporary and eventually gives rise to a sporangium that produces spores. When favorable conditions return, the spores germinate, releasing haploid amoeboid cells, and this asexual cycle begins again. A sexual cycle is known to occur under very moist conditions.

Slime molds, which produce nonmotile spores, are unlike fungi in that they have an amoeboid stage and are heterotrophic by ingestion.

Water Molds Are Filamentous

The **water molds (phylum Oomycota,** 580 species) usually live in the water, where they parasitize fishes, forming furry growths on their gills. In spite of their common name, others live on land and parasitize insects and plants. A water mold was responsible for the 1840s potato famine in Ireland. Most water molds are saprotrophic and live off dead organic matter, however.

Water molds have a filamentous body as do fungi, but their cell walls are largely composed of cellulose whereas fungi have cell walls of chitin. The life cycle of water molds differs from that of fungi. During asexual reproduction, water molds produce motile spores (2n zoospores), which are flagellated. The adult is diploid (not haploid as in the fungi), and meiosis produces the gametes. Their phylum name refers to the enlarged tips (called oogonia) where eggs are produced.

Water molds have a filamentous body and are saprotrophic. The adult is diploid, and meiosis produces gametes; during asexual reproduction, there are motile zoospores.

---

connecting concepts

Several firsts are seen among the protists. They were the first organisms to be eukaryotes, the first to show signs of multicellularity, and the first to reproduce sexually, using gametes. The eukaryotic cell is larger than the prokaryotic cell and it is compartmentalized. Specific regions have a specific structure and function: cell respiration occurs in the mitochondria, and photosynthesis occurs in the chloroplasts, for example. As cells increase in size, efficiency is maintained as long as there is compartmentalization. As an analogy, consider that as a business grows larger, various departments handle specific tasks.

The sex lives of protists also account for their success. During good times, they reproduce asexually and produce large numbers of individuals. In this case, variations would result only from mutations. When the environment is unfavorable in some way, they reproduce sexually via gametes. Sex ensures that the members of a species vary, and the more variations there are, the greater the chance the species will survive.

Although protists are diverse, we do not know if fungi, plants, and animals are closely related to any of the groups alive today. But it is clear that many of the basic biological features of plants and animals began with the protists. Protists also have medical, economic, and ecological significance. *Plasmodium,* a type of sporozoan that causes malaria, is the most widespread human parasite. The ancient deposits of foraminifera created the limestone from which the Egyptian pyramids are built. Red algae are an integral part of coral reefs, and dinoflagellates produce the red tides in the oceans. Protists are a ubiquitous and vital group of diverse organisms.

Summary

30.1 Algae Are Plantlike

Kingdom Protista contains (1) algae, which carry on photosynthesis in the same manner as plants; (2) protozoa, which, like multicellular animals, are heterotrophic and ingest food; and (3) slime molds and water molds, which have some of the characteristics of fungi.

Algae are a part of phytoplankton, which serves as a source of food in both marine and freshwater ecosystems. They are grouped according to their pigments (color).

Green algae possess chlorophylls *a* and *b,* store reserve food as starch, and have cell walls of cellulose as do plants. Green algae are divided into those that are flagellated (*Chlamydomonas*), filamentous (*Spirogyra* and *Oedogonium*), multicellular (*Ulva*), and colonial (*Volvox*).The life cycle varies among the green algae. In most, the zygote undergoes meiosis and the adult is haploid. Both isogamy and oogamy are seen. Ulva has alternation of generations like plants, but the sporophyte and gametophyte generations are similar in appearance; the gametes look alike, and the spores are flagellated.

Brown algae and golden brown algae have chlorophylls *a* and *c* plus a brownish carotenoid. The large, complex brown algae, commonly called seaweeds, are well known and have economic importance. Diatoms, which have an outer layer of silica, are the best known of the golden brown algae. They are extremely numerous in both marine and freshwater ecosystems. Dinoflagellates have cellulose plates and two flagella, one at a right angle to the other. Their chloroplasts are probably derived from golden brown algae by endosymbiosis. They are extremely numerous in the ocean and on occasion produce a neurotoxin when they form red tides.

Euglenoids are flagellated cells with a pellicle instead of a cell wall. Their chloroplasts are most likely derived from a green alga through endosymbiosis.

Red algae are filamentous or multicellular seaweeds that are more delicate than brown algae, and are usually found in warmer waters. Like brown algae, red algae have economic importance.

30.2 Protozoa Are Animal-Like

Protozoa are part of zooplankton and serve as food for animals in marine and freshwater ecosystems. Protozoa are grouped according to the type of locomotor organelle. Amoeboids move and feed by forming pseudopods. Whereas *Amoeba proteus* is usually studied, the radiolaria and the shelled foraminifera responsible for limestone deposits are more significant. Ciliates, which move by means of cilia, are the most complex protozoa and show great diversity. Zooflagellates, which have flagella, are often symbiotic; *Trichonympha* helps termites digest wood, *Trypanosoma* causes African sleeping sickness in humans. Sporozoa are nonmotile parasites that form spores; *Plasmodium* causes malaria.

30.3 Slime Molds and Water Molds Are Funguslike

Slime molds, which produce nonmotile spores, are unlike fungi in that they have an amoeboid stage and are heterotrophic by ingestion. Water molds, which are filamentous and saprotrophic, are unlike fungi in that they produce 2n zoospores. Meiosis produces the gametes.

Reviewing the Chapter

1. What are the three types of protists, and what are their distinguishing characteristics? 524
2. In general, what is the role of algae in aquatic ecosystems? What primary characteristic is used to group algae? 524
3. Describe the structure of *Chlamydomonas* and *Volvox;* contrast how they reproduce. 524–25, 527
4. Describe the structure of *Spirogyra* and *Oedogonium;* contrast how they reproduce. 526
5. Describe the structure of *Ulva,* and explain how its life cycle differs from that of plants. 526
6. Describe the structure of brown algae and diatoms. Discuss their ecological and economic importance. 527–28
7. Describe the structure of dinoflagellates. What feature do they share with diatoms? What is a red tide? 528
8. Describe the structure of euglenoids. What feature do they share with green algae? 528–29
9. Describe the structure of red algae, and contrast their structure with that of brown algae. Why are both called seaweeds? 529
10. What is the general role of protozoa in aquatic ecosystems, and what primary characteristic is used to group protozoa? Give an example from each group. 530–33
11. Describe the life cycle of *Plasmodium vivax,* the causative agent of malaria. 533
12. What features do slime molds and water molds share with fungi? Describe the life cycle of a plasmodial slime mold. 535–36

Testing Yourself

Choose the best answer for each question.

1. Which of these is not a green alga?
   a. *Volvox*
   b. *Fucus*
   c. *Spirogyra*
   d. *Chlamydomonas*
2. Which is not a characteristic of brown algae?
   a. multicellular
   b. chlorophylls *a* and *b*
   c. lives along rocky coast
   d. harvested for commercial reasons
3. In *Chlamydomonas*
   a. the adult is haploid.
   b. spores survive times of stress.
   c. sexual reproduction occurs.
   d. All of these are correct.
4. Which is found in *Ulva* but not in plants?
   a. alternation of generations
   b. generations look alike
   c. sperm and egg are produced
   d. Both b and c are correct.
5. Which of these algae are not flagellated?
   a. *Volvox*
   b. *Spirogyra*
   c. dinoflagellates
   d. *Chlamydomonas*
6. Which is mismatched?
   a. diatoms—silica shell, boxlike, golden brown
   b. euglenoids—flagella, pellicle, eyespot
   c. *Fucus*—adult is diploid, seaweed, chlorophylls *a* and *c*
   d. *Paramecium*—cilia, calcium carbonate shell, gullet
7. Which is a false statement?
   a. There are flagellated algae and flagellated protozoa.
   b. Among protozoa, both flagellates and sporozoa are symbiotic.
   c. Among the protists, only green algae have a sexual life cycle.
   d. Ciliates exchange genetic material during conjugation.
8. Which is mismatched?
   a. trypanosome—African sleeping sickness
   b. *Plasmodium vivax*—malaria
   c. amoeboid—severe diarrhea
   d. AIDS—*Giardia lamblia*
9. Which is found in slime molds but not fungi?
   a. nonmotile spores
   b. amoeboid adult
   c. zygote formation
   d. photosynthesis

10. Label this diagram of the *Chlamydomonas* life cycle.
11. Name two features that distinguish the asexual from the sexual portion of the cycle shown in the diagram.

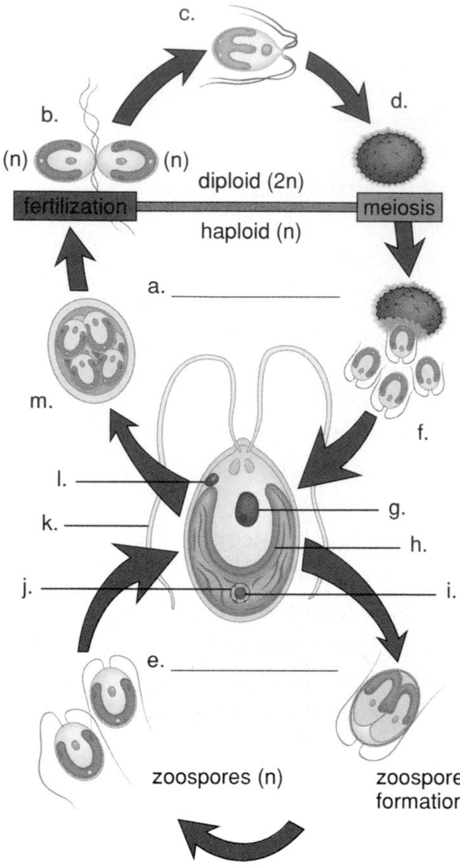

Applying the Concepts

1. *Each organism has one of three types of life cycles.*
   Use the occurrence of meiosis and haploidy to distinguish the various life cycles.
2. *Protists are a diverse group of organisms.*
   The manner in which protists are currently divided is artificial. Explain.
3. *Protozoa are the animal-like protists.*
   Argue that sporozoa should not be considered protozoa on the basis of this concept.

Using Technology

Your study of the protists is supported by these available technologies:

Exploring the Internet

The Mader Home Page provides resources for and help with studying this chapter.

http://www.mhhe.com/sciencemath/biology/mader/
(Click on Biology.)

Life Science Animations Video

Video #4: Animal Biology II
Life Cycle of Malaria (#45)

Understanding the Terms

alga (pl., algae) 524
amoeboid 530
brown alga 527
ciliate 531
colony 524
conjugation 526
diatom 528
dinoflagellate 528
euglenoid 528
filament 526
golden brown alga 528
green alga 524
holozoic 530
isogamy 525
oogamy 525
phagocytize 530
phytoplankton 524
plankton 524
protist 524
protozoan (pl., protozoa) 530
pseudopod 530
red alga 529
saprotroph 535
seaweed 527
slime mold 535
sporangium 535
spore 525
sporozoa 533
trichocyst 531
trypanosome 532
water mold 536
zooflagellate 532
zooplankton 530

Match the terms to these definitions:

a. __________ Aquatic, plantlike organism carrying out photosynthesis and belonging to the kingdom Protista.
b. __________ Asexual reproductive structure that is resistant to unfavorable environmental conditions and develops into a haploid generation.
c. __________ Cytoplasmic extension of amoeboid protists; used for locomotion and engulfing food.
d. __________ Flagellated and flexible freshwater unicellular organism that usually contains chloroplasts and is often characterized as having both animal-like and plantlike characteristics.
e. __________ Freshwater and marine organisms that are suspended on or near the surface of the water.
f. __________ Freshwater or marine unicellular golden brown alga, with a cell wall consisting of two silica-impregnated valves, which is extremely numerous in phytoplankton.
g. __________ Member of a genus of parasitic zooflagellates that cause severe disease in human beings and domestic animals, including a condition called sleeping sickness.
h. __________ Part of plankton containing protozoa and microscopic animals.
i. __________ Protist that moves and engulfs prey with pseudopods.
j. __________ Transfer of genetic material from one cell to another.

The Fungi

CHAPTER 31

Fading scarlet waxy cap mushroom, *Hygrophorus miniatus*

Chapter Concepts

31.1 What Fungi Are Like

- Fungi are saprotrophic decomposers that aid the cycling of inorganic nutrients in ecosystems. 540
- The body of a fungus is multicellular; it is composed of thin filaments called hyphae. 540
- As an adaptation to life on land, fungi produce nonmotile and often windblown spores during asexual and sexual reproduction. 541

31.2 How Fungi Are Classified

- Fungi are classified according to aspects of their sexual life cycle. Zygospore fungi have a dormant stage consisting of a thick-walled zygospore. 542
- During sexual reproduction of sac fungi, saclike cells (asci) produce spores. Asci are located in fruiting bodies. 544
- During sexual reproduction of club fungi, club-shaped structures (basidia) produce spores. Basidia are located in fruiting bodies. 546
- The imperfect fungi always reproduce asexually by conidiospores; sexual reproduction has not yet been observed in these organisms. 549

31.3 Fungi Form Symbiotic Relationships

- Lichens, which may live in stressful environments, are an association between a fungus and a cyanobacterium or a green alga. The fungus may be somewhat parasitic on the alga. 549
- Mycorrhizae is an association between a fungus and the roots of a plant, such that the fungus helps the plant absorb minerals, and the plant supplies the fungus with carbohydrates. 550

What do LSD, athlete's foot, and homemade bread have in common? They all involve species from the kingdom Fungi. Fungi are among earth's obscure life-forms. We find them growing on tree trunks and in the backyard after a rain. They resemble plants, but they lack chloroplasts and do not photosynthesize. Yet, fungi clearly aren't animals, nor do they resemble bacteria or protozoa. Based on their multicellular nature and mode of nutrition, Whittaker placed fungi in their own kingdom. You probably know that mushrooms are fungi, and you might also know that mildew, molds, and morels are fungi, too. From this litany you would think a fungus couldn't be very large. But researchers claim to have found one that spreads over 1,500 acres in Washington State. As fungi spread, they feed off the organic remains of plants and animals. Dead leaves, tree trunks, and even carcasses of animals are their daily fare. They, along with bacteria, are the decomposers that enrich the immediate environs with inorganic nutrients and thereby keep chemicals cycling in ecosystems. Trees grow, grass turns green, and flowers appear—all because fungi, while digesting for themselves, have also digested for others. Perhaps we should think of them as being mutualistic with all of nature.

31.1 What Fungi Are Like

Kingdom Fungi contains the **fungi,** which are mostly multicellular eukaryotes of varied structure that share a common mode of nutrition. Like animals, they are heterotrophic and consume preformed organic matter. Animals, however, are heterotrophic by ingestion; fungi are heterotrophic by absorption. Their cells send out digestive enzymes into the immediate environment and then, when organic matter is broken down, the cells absorb the resulting nutrient molecules.

Most fungi are *saprotrophic decomposers* that break down the waste products and dead remains of plants and animals. Some fungi are parasitic; they live off the tissues of living plants and animals. Fungi can enter leaves through the stomates, and this makes plants especially subject to fungal diseases. Fungal diseases account for millions of dollars in crop losses each year, and they have greatly reduced the numbers of various types of trees. They also cause human diseases such as ringworm, athlete's foot, and yeast infections.

Several types of fungi have a mutualistic relationship with the roots of seed plants; they acquire inorganic nutrients for plants, and in return they are given organic nutrients. Others form an association with a green alga or cyanobacterium within a lichen. Lichens can live on rocks and are important soil formers.

Fungi carry on external digestion and are heterotrophic by absorption. They are saprotrophic or parasitic and form symbiotic relationships.

Most Fungi Are Filamentous

Fungi can be unicellular—yeast are the best known example. But the thallus (body) of most fungi is a multicellular structure known as a mycelium (Fig. 31.1). A **mycelium** [Gk. *mycelium,* fungus filaments] is a network of filaments called **hyphae** [Gk. *hyphe,* web]. Hyphae give the mycelium quite a large surface area per volume of cytoplasm, and this facilitates absorption of nutrients into the body of a fungus. Hyphae grow at their tips, and the mycelium absorbs and then passes nutrients on to the growing tips. When a fungus reproduces, a specific portion of the mycelium becomes a reproductive structure that is then nourished by the rest of the mycelium.

a. Fungal mycelium

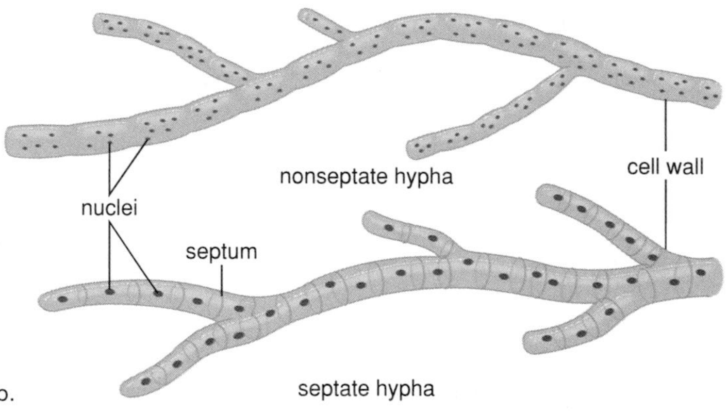

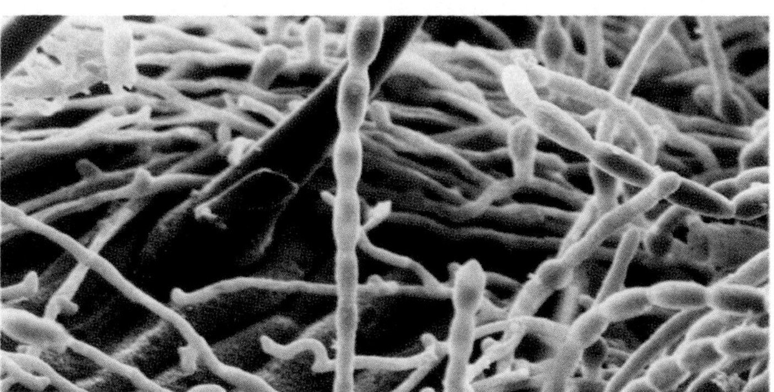

c. Fungal hyphae

10 μm

Figure 31.1 Mycelium of fungi.
a. On a corn tortilla, each mycelium grown from a different spore is quite symmetrical. **b.** Hyphae are either nonseptate (do not have cross walls) or septate (have cross walls). **c.** Scanning electron micrograph showing fungal hyphae.

Fungal cells are quite different from plant cells not only by lacking chloroplasts but also by having a cell wall that contains *chitin* and not cellulose. Chitin, like cellulose, is a polymer of glucose organized into microfibrils. In chitin, however, each glucose molecule has a nitrogen-containing amino group attached to it. Chitin is also found in the external skeleton of insects and all arthropods. The energy reserve of fungi is not starch but glycogen, as in animals. Fungi are nonmotile; they lack basal bodies and do not have flagella at any stage in their life cycle. They move toward a food source by growing toward it. Hyphae can cover as much as a kilometer a day!

Some fungi have no cross walls in their hyphae. These hyphae are called **nonseptate**—septum (L. *septum,* fence, wall) is the term used for cross walls in fungi. Nonseptate fungi are multinucleated; they have many nuclei in the cytoplasm of a hypha. **Septate** fungi have cross walls, but actually the presence of septa makes little difference because pores allow cytoplasm and even sometimes organelles to pass freely from one cell to the other. The septa that separate reproductive cells, however, are complete in all fungal groups.

The nonmotile body of a fungus is made up of hyphae that form a mycelium.

Most Fungi Produce Spores

In general, fungal sexual reproduction involves these stages:

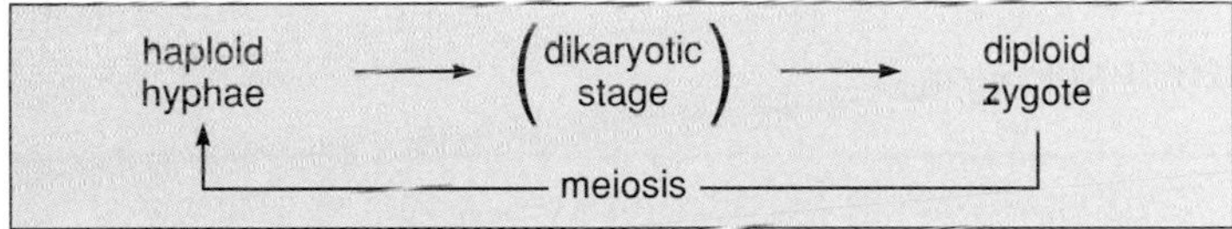

The relative length of time of each phase varies with the species.

During sexual reproduction, hyphae (or a portion thereof) from two different mating types make contact and fuse. You would expect the nuclei from the two mating types to also fuse immediately, and they do in some species. In other species the nuclei pair but do not fuse for days, months, or even years. The nuclei continue to divide in such a way that every cell (in septate hyphae) has at least one of each nucleus. A hypha that contains paired haploid nuclei is said to be n + n or **dikaryotic** [Gk. *dis,* two, and *karyon,* nucleus, kernel]. When the nuclei do eventually fuse, the zygote undergoes meiosis prior to spore formation. Fungal spores germinate directly into haploid hyphae without any noticeable embryological development.

How can a terrestrial and nonmotile organism ensure that the species will be dispersed to new locations? As an adaptation to life on land, fungi produce nonmotile, but usually windblown, spores during both sexual and asexual reproduction (Fig. 31.2). A **spore** is a reproductive cell that can grow directly into a new organism.

Asexual reproduction usually involves the production of spores by a single mycelium. Alternately, asexual reproduction can occur by fragmentation—a portion of a mycelium begins a life of its own. Also, unicellular yeast reproduce asexually by **budding;** a small cell forms and gets pinched off as it grows to full size (see Fig. 31.6).

a.

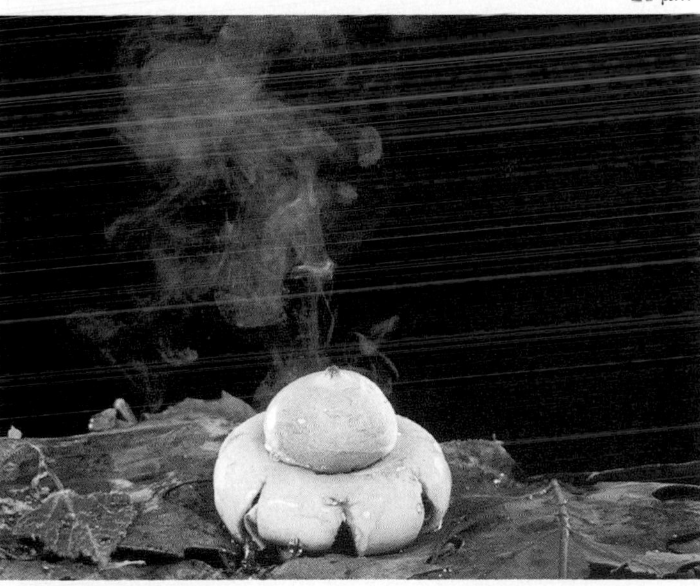

b.

Figure 31.2 Dispersal of spores.
a. Electron micrograph shows the external appearance of spores discharged by **(b)** earth star, *Geastrum*, which releases hordes of spores into the air.

Part a. From C. Y. Shih and R. G. Kessel *Living Images.* Science Books International, Boston, 1982.

31.2 How Fungi Are Classified

CLASSIFICATION

Kingdom Fungi

Multicellular eukaryote; heterotrophic by absorption; lack flagella; nonmotile spores form during both asexual and sexual reproduction

Division Zygomycota: zygospore fungi

Division Ascomycota: sac fungi

Division Basidiomycota: club fungi

Division Deuteromycota: imperfect fungi (i.e., means of sexual reproduction not known)

The ancestry of fungi, which evolved about 570 million years ago, has not been determined. It is possible that the organisms now classified in the **kingdom Fungi** do not share a recent common ancestor. In that case, they probably evolved separately from different protists. It's been suggested, though, that fungi evolved from red algae because both fungi and red algae lack flagella in all stages of the life cycle.

In the past, fungi have been classified in other kingdoms. Originally they were considered a part of the plant kingdom, and then they were placed in the kingdom Protista. But in 1969, R. H. Whittaker argued that their multicellular nature and mode of nutrition (extracellular digestion and absorption of nutrients) meant they should be in their own kingdom. Table 31.1 contrasts the features of fungi with the features of certain other organisms now included in kingdom Protista by this text. In the absence of knowledge of their evolutionary relations, fungal groups are classified according to differences in the life cycle and the type of structure that produces spores. Recently scientists have been using comparative DNA (deoxyribonucleic acid) data and analysis of enzymes and other proteins to decipher evolutionary relationships, but as yet this has not changed the classification of fungi.

Zygospore Fungi Form Zygospores

The **zygospore fungi (division Zygomycota,** 600 species) are mainly saprotrophs living off plant and animal remains in the soil or bakery goods in your own pantry. Some are parasites of small soil protists or worms, and even insects such as the housefly.

The black bread mold, *Rhizopus stolonifer,* is commonly used as an example of this division. The body of this fungus, which is composed of nonseptate hyphae, demonstrates that although there is little cellular differentiation among fungi, the hyphae may be specialized for various purposes. In *Rhizopus, stolons* are horizontal hyphae that exist on the surface of the bread; *rhizoids* grow into the bread, anchor the mycelium and carry out digestion; and *sporangiophores* are stalks that bear sporangia. A **sporangium** is a capsule that produces spores called sporangiospores. During asexual reproduction all structures involved are haploid (Fig. 31.3).

The division name refers to the zygospore, which is seen during sexual reproduction. The hyphae of opposite mating types, termed plus (+) and minus (−), are chemically attracted, and they grow toward each other until they touch. The ends of the hyphae swell as nuclei enter; then cross walls develop a short distance behind each end, forming *gametangia.* The gametangia merge and the result is a large multinucleate cell in which nuclei of the two mating types pair and then fuse. A thick wall develops around the cell, which is now called a **zygospore.** The zygospore undergoes a period of dormancy before meiosis and germination takes place. One or more sporangiophores with sporangia at their tips develop, and many spores are released. The spores, dispersed by air currents, give rise to new haploid mycelia.

Zygospore fungi produce spores within sporangia. During sexual reproduction, a zygospore forms prior to meiosis and production of spores.

Table 31.1

Certain Protists Compared to Fungi

Feature	Fungi	Red Algae	Plasmodial Slime Molds	Water Molds
Body form	Filamentous	Filamentous	Multinucleate plasmodium	Filamentous
Mode of nutrition	Heterotrophic by absorption	Autotrophic by photosynthesis	Heterotrophic by absorption	Heterotrophic by absorption
Basal bodies/flagella	In no stages	In no stages	In one stage	Flagellated zoospores
Cell wall	Contains chitin	Contains cellulose	None	Contains cellulose
Life cycle	Zygotic meiosis (haplontic cycle)	Sporic meiosis (alternation of generations)	Unique	Gametic meiosis (diplontic cycle)

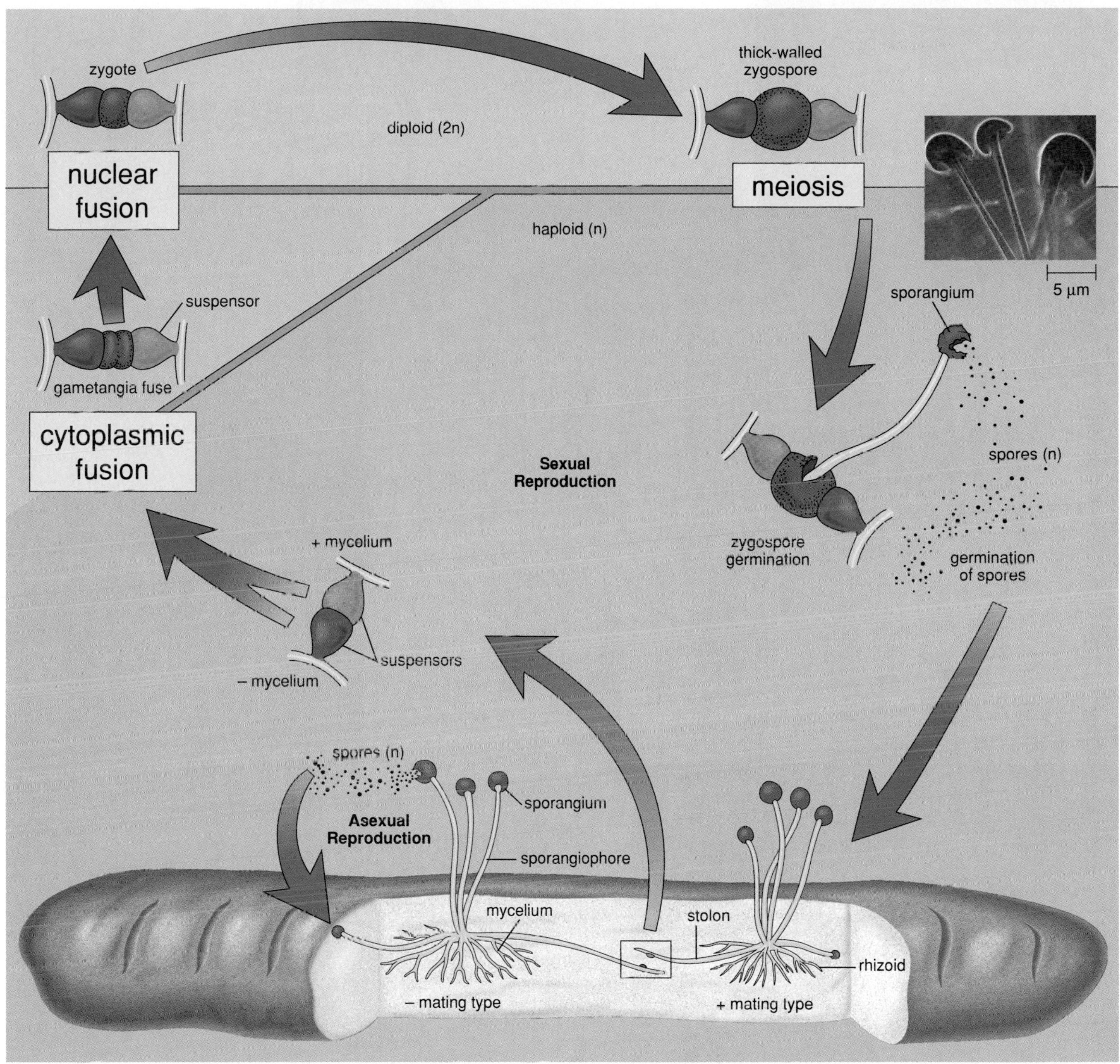

Figure 31.3 Black bread mold *Rhizopus stolonifer.*
Asexual reproduction is the norm. As a result of sexual reproduction, the adult is haploid due to zygomeiosis, which occurs before or as the sporangiospores are produced. After two compatible mating types make contact, first gametangia fuse and then the nuclei fuse. The zygospore is a resting stage that can survive unfavorable growing conditions.

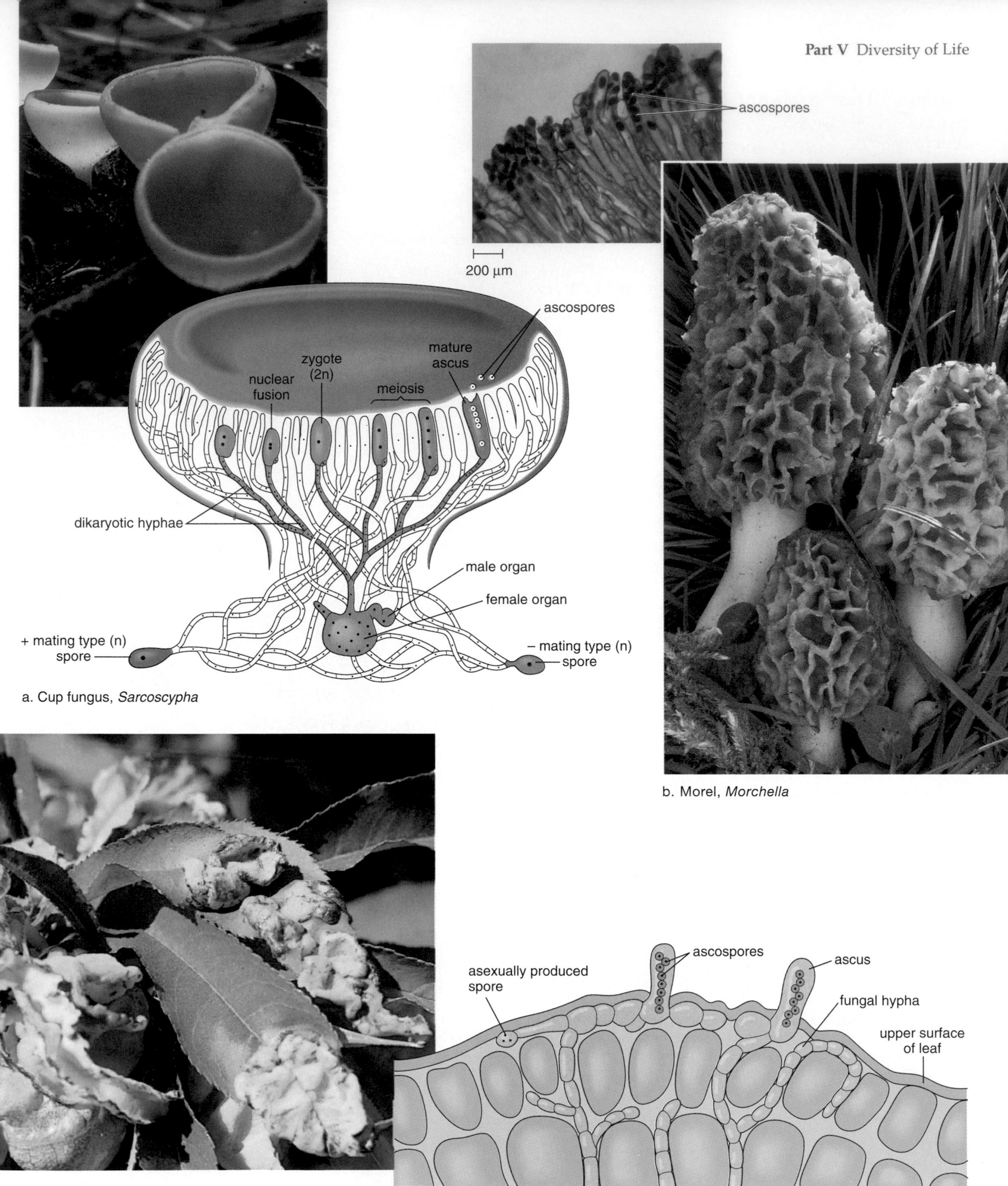

Figure 31.4 Sac fungi.
a. Cup fungi; photo and drawing of ascocarp section. Dikaryotic hyphae terminate in the ascocarp and develop asci where meiosis follows nuclear fusion and spore formation takes place. **b.** Morel; photo and light micrograph, ascocarp section. **c.** Peach leaf curl; photo and drawing.

Sac Fungi Form Ascospores

Most **sac fungi (division Ascomycota,** 30,000 species) are saprotrophs that play an essential ecological role by digesting resistant (not easily decomposed) materials containing cellulose, lignin, or collagen. Red bread molds (e.g., *Neurospora*) are ascomycetes, as are cup fungi, morels, and truffles. Morels and truffles are highly prized as gourmet delicacies. A large number of ascomycetes are parasitic on plants; powdery mildews grow on leaves, as do leaf curl fungi; chestnut blight and Dutch elm disease destroy the trees named. Ergot, a parasitic sac fungus that infects rye and (less commonly) other grains, is discussed in the reading on page 548.

Yeasts are unicellular, but most ascomycetes are composed of septate hyphae. Their division name refers to the **ascus** [Gk. *askos*, bag, sac], a fingerlike sac that develops during sexual reproduction. Ascus-producing hyphae remain dikaryotic (Fig. 31.4) except in the walled-off portion that becomes the ascus where nuclear fusion, meiosis, and ascospore formation take place. Each ascus contains eight haploid nuclei and produces eight ascospores.

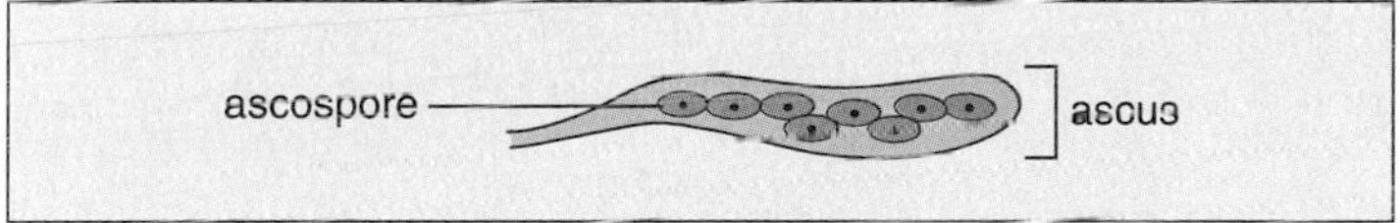

The asci are usually surrounded and protected by sterile hyphae within a fruiting body called an *ascocarp*. A **fruiting body** is a reproductive structure where spores are produced and released. Ascocarps can have different shapes; in cup fungi they are cup shaped, in molds they are flask shaped, and in morels they are stalked and crowned by bell-shaped convoluted tissue that bears the asci. In most ascomycetes, the asci become swollen as they mature and then they burst, expelling the ascospores. If released into the air, the spores are then windblown.

Asexual reproduction, which is the norm among ascomycetes, involves the production of spores called **conidiospores** [Gk. *konis*, dust, and *spora*, seed], or conidia, which vary in size and shape and may be multicellular. There are no sporangia in ascomycetes, and the conidiospores develop directly on the tips of modified aerial hyphae called conidiophores (see Fig. 31.8). When released, they are windblown.

Sac fungi usually produce asexual conidiospores. During sexual reproduction, asci within a fruiting body produce spores.

Yeasts Are Single Cells

Yeasts are unicellular organisms that reproduce asexually either by mitosis or by budding. *Saccharomyces cerevisiae*, brewer's yeast, is representative of budding yeasts: a small cell forms and gets pinched off as it grows to full size (Fig. 31.5). Sexual reproduction, which occurs when the food supply runs out, results in the formation of asci and ascospores. Ascospores from two different mating types can fuse, and the result is a diploid cell that will reproduce asexually before meiosis occurs and ascospores are produced again. The haploid ascospores function directly as new yeast cells.

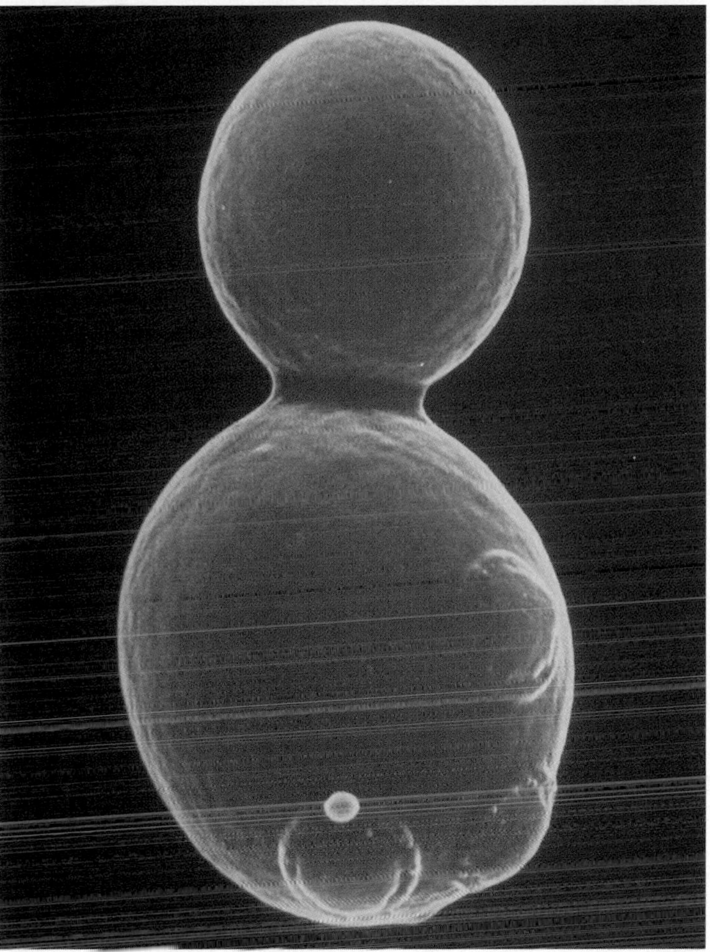

Figure 31.5 Yeast cells.
Saccharomyces cerevisiae is brewer's yeast, which reproduces asexually by budding.

When yeasts ferment, they produce ethanol and carbon dioxide. In the wild, yeasts grow on fruits, and historically the yeasts already present on grapes were used to produce wine. Today selected yeasts are added to relatively sterile grape juice in order to make wine. Also, yeasts are added to prepared grains to make beer. Both the ethanol and the carbon dioxide are retained for beers and sparkling wines; it is released for still wines. In baking, the carbon dioxide given off by yeast is the leavening agent that causes bread to rise.

Yeasts are serviceable to humans in another way. They have become the material of choice in genetic engineering experiments requiring a eukaryote. *Escherichia coli*, the usual experimental material, is a prokaryote and does not function during protein synthesis as a eukaryote would.

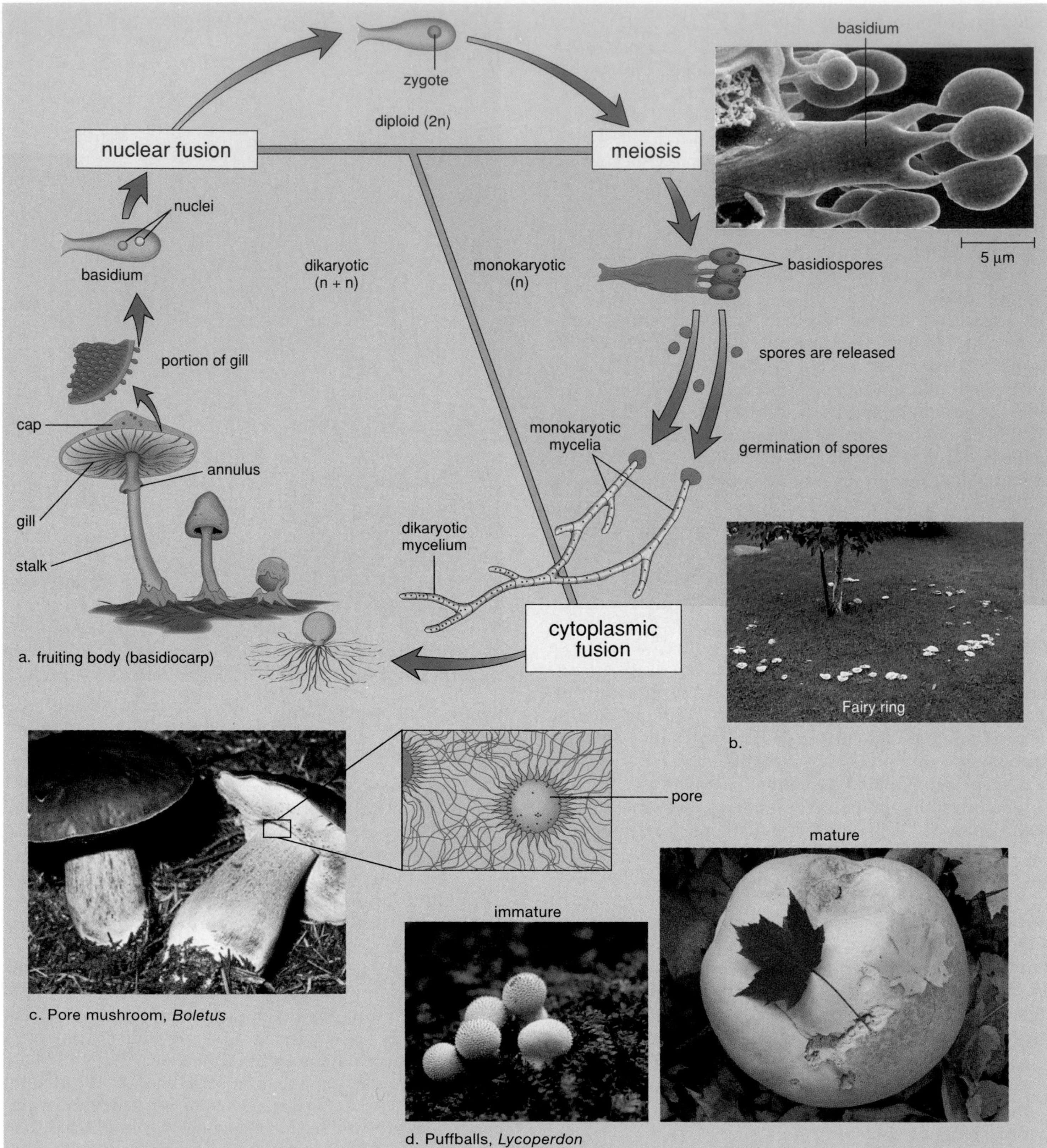

Figure 31.6 Club fungi.
a. Life cycle of a mushroom. Sexual reproduction is the norm. After hyphae from two compatible mating types fuse, the dikaryotic mycelium is long-lasting. Nuclear fusion results in a diploid nucleus within each basidium on the gills of the fruiting body shown. Zygomeiosis and production of basidiospores follow. Germination of a spore results in a haploid mycelium. **b.** Fairy ring. Mushrooms develop in a ring on the outer living fringes of a dikaryotic mycelium. The center has used up its nutrients and is no longer living. **c.** Fruiting bodies of *Boletus.* This mushroom is not gilled; instead it has basidia-lined tubes that open on the undersurface of the cap. **d.** Puffballs grow to be quite large. It is estimated that the mature one shown contains 7 trillion spores.

Club Fungi Have Basidiospores

Club fungi (**division Basidiomycota,** 16,000 species), which have septate hyphae, include the familiar mushrooms growing on lawns and the shelf or bracket fungi found on dead trees. Less well known are puffballs, bird's nest fungi, and stinkhorns. These structures are all fruiting bodies called basidiocarps. Basidiocarps contain the **basidia** [L. *basidium* (dim. originating from Gk. *basis*), pedestal], club-shaped structures that produce basidiospores and from which this division takes its name.

Although club fungi occasionally do produce conidiospores asexually, they usually reproduce sexually. When monokaryotic (n) hyphae of two different mating types meet, they fuse, and a dikaryotic (n + n) mycelium results (Fig. 31.6). The dikaryotic mycelium continues its existence year after year, even hundreds of years on occasion. In mushrooms, the dikaryotic mycelium often radiates out and produces basidiocarps in an ever larger so-called fairy ring. Basidiocarps are composed of nothing but tightly packed hyphae whose walled-off ends become the club-shaped basidia. In the gilled mushrooms, the hyphae terminate in radiating lamellae, and in pore mushrooms and shelf fungi the hyphae terminate in tubes. In any case, the extensive surface area is lined by basidia where nuclear fusion, meiosis, and spore production occur. A basidium has four projections into which cytoplasm and a haploid nucleus enter as the basidiospore forms.

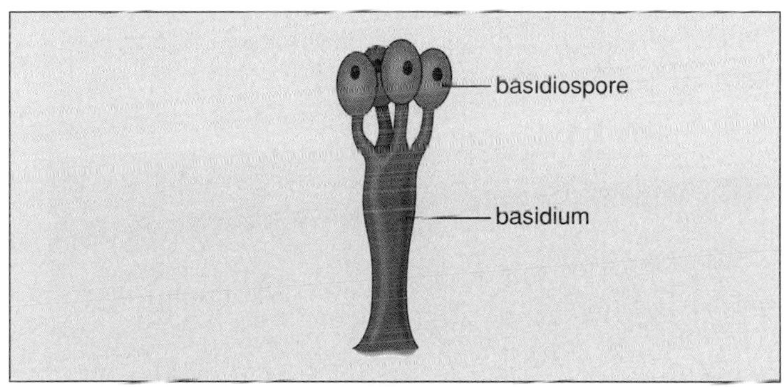

Basidiospores are often windblown; when they germinate, a new haploid mycelium forms.

In puffballs, spores are produced inside parchmentlike membranes, and the spores are released through a pore or when the membrane breaks down. In bird's nest fungi, falling raindrops provide the force that causes the nest's basidiospore-containing "eggs" to fly through the air and land on vegetation that may be eaten by an animal. If so, the spores pass unharmed through the digestive tract. Stinkhorns resemble a mushroom with a spongy stalk and a compact, slimy cap. The long stalk bears the elongated basidiocarp. Stinkhorns emit an incredibly disagreeable odor; flies are attracted by the odor, and when they linger to feed on a sweet jelly, the flies pick up spores that they later distribute.

Club fungi usually reproduce sexually. The dikaryotic stage is prolonged and periodically produces fruiting bodies where spores are produced in basidia.

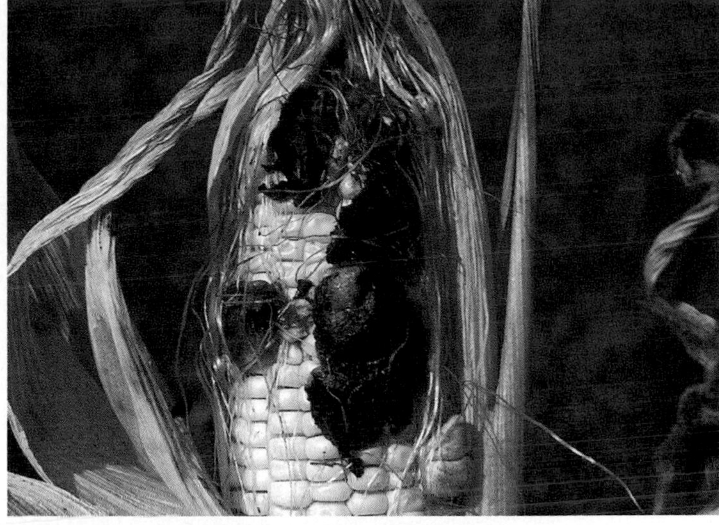

a. Corn smut, *Ustilago*

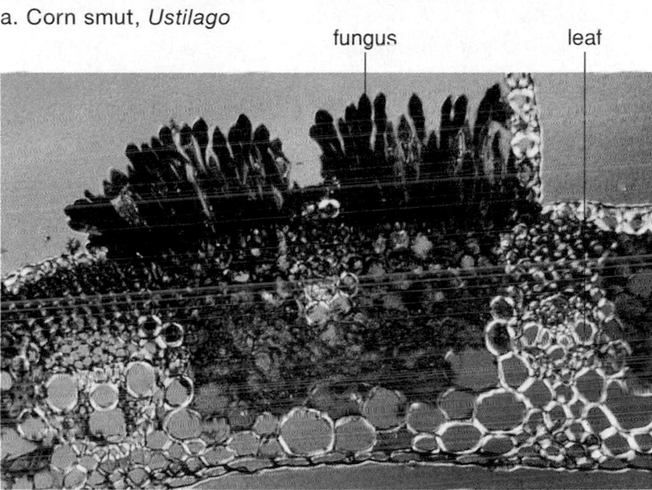

b. Wheat rust, *Puccinia*

Figure 31.7 Smuts and rusts.
a. Corn smut. b. Wheat rust.

Smuts and Rusts Are Parasites

Smuts and *rusts* are club fungi that parasitize cereal crops such as corn, wheat, oats, and rye. They are of great economic importance because of the crop losses they cause every year. Smuts and rusts don't form basidiocarps, and their spores are small and numerous, resembling soot. Some smuts enter seeds and exist inside the plant, becoming visible only near maturity. Other smuts externally infect plants. In corn smut, the mycelium grows between the corn kernels and secretes substances that cause the development of tumors on the ears of corn (Fig. 31.7*a*).

The life cycle of rusts, which may be particularly complex, often requires two different plant host species to complete the cycle. Black stem rust of wheat uses barberry bushes as an alternate host, and blister rust of white pine uses currant and gooseberry bushes. Campaigns to eradicate these bushes in areas where the alternate host grows help keep these rusts in check. Wheat rust (Fig. 31.7*b*) is also controlled by producing new and resistant strains of wheat. The process is continuous, because rust can mutate to cause infection once again.

ecology focus

Deadly Fungi

It is unwise for amateurs to collect mushrooms in the wild because certain mushroom species are poisonous. The red and yellow *Amanitas* are especially dangerous. This species is also known as fly agaric because it was formerly used to kill flies (it was grown and then sprinkled with sugar to attract flies). Its toxins include muscarine and muscaridine, which produce symptoms similar to acute alcoholic intoxication. In one to six hours, the victim staggers, loses consciousness, and becomes delirious, sometimes suffering from hallucinations, manic conditions, and stupor. Luckily it also causes vomiting, which rids the system of the poison, so death occurs in less than 1% of cases. Death Angel (*Amanita phalloides*, Fig. 31A) causes 90% of the fatalities attributed to mushroom poisoning. When this mushroom is eaten, symptoms don't begin until ten to twelve hours later. Abdominal pain, vomiting, delirium, and hallucinations are not the real problem; rather, a poison interferes with RNA (ribonucleic acid) transcription by inhibiting RNA polymerase, and the victim dies from liver and kidney damage.

Some hallucinogenic mushrooms are used in religious ceremonies, particularly among Mexican Indians. *Psilocybe mexicana* contains a chemical called psilocybin that is a structural analogue of LSD and mescaline. It produces a dreamlike state in which visions of colorful patterns and objects seem to fill up space and dance past in endless succession. Other senses are also sharpened to produce a feeling of intense reality.

The only reliable way to tell a nonpoisonous mushroom from a poisonous one is to be able to correctly identify the species. Poisonous mushrooms cannot be identified with simple tests, such as whether they peel easily, have a bad odor, or blacken a silver coin during cooking.

In addition to club fungi, some sac fungi also contain chemicals that can be dangerous to people. *Claviceps purpurea*, the ergot fungus, infects rye and replaces the grain with ergot—hard, purple-black bodies consisting of tightly cemented hyphae containing ascospores (Fig. 31B). When ground with the rye and made into bread, the fungus releases toxic alkaloids that cause the disease ergotism. In humans, vomiting, feelings of intense heat or cold, muscle pain, a yellow face, and hand and feet lesions are accompanied by hysteria and hallucinations. We now know that ergot contains lysergic acid, from which LSD is easily synthesized. From recorded symptoms, historians believe that those who accused their neighbors of witchcraft in Salem, Massachusetts, in the seventeenth century could very well have been suffering from ergotism. As recently as 1951 an epidemic of ergotism occurred in Pont-Saint-Esprit, France. Over 150 persons became hysterical and four died.

Because the alkaloids that cause ergotism stimulate smooth muscle and selectively block the sympathetic nervous system, they can be used in medicine to cause uterine contractions and to treat certain circulatory disorders, including migraine headaches. Although the ergot fungus can be cultured in petri dishes, no one has succeeded in inducing it to form ergot in the laboratory. So far, the only way to obtain ergot, even for medical purposes, is to collect it in an infected field of rye.

Figure 31A Poisonous mushrooms, *Amanita phalloides.*

Figure 31B Ergot infection of rye, caused by *Claviceps purpurea.*

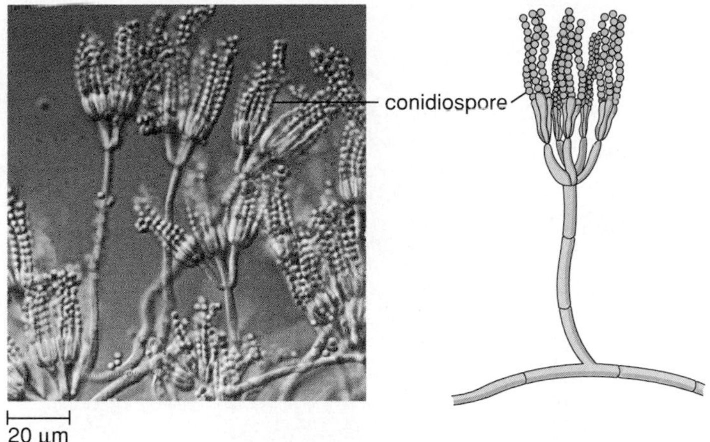

Figure 31.8 Conidiospores.
Sac fungi and imperfect fungi reproduce asexually by producing spores called conidiospores at the ends of certain hyphae. The organism shown here is *Penicillium*, an imperfect fungus.

Imperfect Fungi Reproduce Asexually Only

The **imperfect fungi (division Deuteromycota,** 25,000 species) always reproduce asexually by forming conidiospores (Fig. 31.8). These fungi are "imperfect" in the sense that no sexual stage has yet been observed and may not exist. Without knowing the sexual stage, it is often difficult to classify a fungus as belonging to one of the other divisions. Usually, cellular morphology and biochemistry indicate these fungi are sac fungi that have lost the ability to reproduce sexually. Even so, genetic recombination does occur; hyphae fuse, and crossing-over between chromosomes results in haploid nuclei with varied genetic content.

Several imperfect fungi are serviceable to humans. Some species of the mold *Penicillium* are sources of the antibiotic penicillin, while other species give the characteristic flavor and aroma to cheeses such as Roquefort and Camembert. The bluish streaks in blue cheese are patches of conidiospores. The drug cyclosporine, which is administered to suppress the immune system following an organ transplant operation, is derived from an imperfect fungus found in soil. Another mold, *Aspergillus*, as well as other fungi, are used to produce soy sauce by fermentation of soybeans. In this country *Aspergillus* is used to produce citric and gallic acids, which serve as additives during the manufacture of different products—from inks to chewing gum.

Unfortunately, some imperfect fungi cause disease in humans. Certain dust-borne spores can cause infections of the respiratory tract, while athlete's foot and ringworm are spread by direct contact. *Candida albicans* is a yeastlike organism that causes infections of the vagina, especially in women on the birth-control pill or certain types of antibiotic therapy. This organism also causes thrush, an inflammation of the mouth and throat.

The imperfect fungi cannot be classified into one of the other divisions because their mode of sexual reproduction is unknown.

31.3 Fungi Form Symbiotic Relationships

We have already mentioned several instances in which fungi are parasites of plants and animals. Two other associations are of interest.

Lichens Improve Soil Conditions

Lichens are an association between a fungus and a cyanobacterium or a green alga. The body of a lichen has three layers: the fungus forms a thin, tough upper layer and a loosely packed lower layer that shield the photosynthetic cells in the middle layer (Fig. 31.9). Specialized fungal hyphae, which penetrate or envelop the photosynthetic cells, transfer nutrients directly to the rest of the fungus. Lichens can reproduce asexually by releasing fragments that contain hyphae and an algal cell.

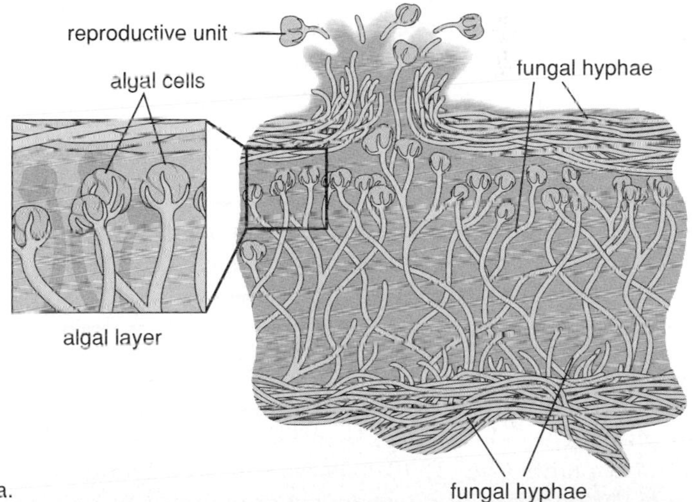

a.

b. Mixture of crustose lichens

c. Fruticose lichen, *Cladonia*

d. Foliose lichen, *Xanthoparmelia*

Figure 31.9 Lichen morphology.
a. A section of a lichen shows the placement of the algal cells and the fungal hyphae, which encircle and penetrate the algal cells. **b.** Crustose lichens are compact. **c.** Fruticose lichens are shrublike. **d.** Foliose lichens are leaflike.

In the past, lichens were assumed to be mutualistic relationships in which the fungus received nutrients from the algal cells, and the algal cells were protected from desiccation by the fungus. Actually, lichens might involve a controlled form of parasitism of the algal cells by the fungus. Algae may not benefit at all from these associations. This is supported by experiments in which the fungal and algal components are removed and grown separately. The algae grow faster when they are alone rather than when they are part of a lichen. On the other hand, it is difficult to cultivate the fungus, which does not naturally grow alone. The different lichen species are identified according to the fungal partner.

Three types of lichens are recognized. Compact crustose lichens are often seen on bare rocks or on tree bark; foliose lichens are leaflike; and fruticose lichens are shrublike (Fig. 31.9). Lichens are efficient at acquiring nutrients and moisture, and therefore they can survive in areas of low moisture and low temperature as well as areas with poor or no soil. They produce and improve the soil, thus making it suitable for plants to invade the area. Unfortunately, lichens also take up pollutants and cannot survive where the air is polluted. Therefore, their presence is an indication that the air is healthy for humans to breathe.

Lichens, an association between fungal hyphae and algal cells, can live in areas of extreme conditions and contribute to the formation of soil.

Mycorrhizae Are "Fungus Roots"

Mycorrhizae [Gk. *mykes*, fungus, and *rhizion*, dim. for root] are mutualistic relationships between soil fungi and the roots of most plants (Fig. 31.10) (also see p. 517). Plants whose roots are invaded by mycorrhizae grow more successfully in poor soils—particularly soils deficient in phosphates—than do plants without mycorrhizae. The fungal partner may enter the cortex of roots but does not enter the cytoplasm of plant cells. Ectomycorrhizae form a mantle that is exterior to the root, and they grow between cell walls. Endomycorrhizae penetrate only into the cell walls. In any case, the presence of the fungus gives the plant a greater absorptive surface for the intake of minerals. The fungus also benefits from the association by receiving carbohydrates from the plant.

Figure 31.10 Plant growth experiment.
Soybean plant, *Glycine*, without mycorrhizae grows poorly compared to two other plants infected with different strains of mycorrhizae.

It is of interest to know that the truffle, a gourmet delight whose ascocarp is somewhat prunelike in appearance, is a mycorrhizal fungus living in association with oak and beech tree roots. It used to be that the French used pigs to sniff out and dig up truffles, but now they have succeeded in cultivating truffles by inoculating the roots of seedlings with the proper mycelium.

Even the earliest fossil plants have mycorrhizae associated with them. It would appear, then, that mycorrhizae helped plants adapt to and flourish on land.

Mycorrhizae (fungus roots), a mutualistic association between a fungus and plant roots, help plants acquire mineral nutrients.

connecting concepts

Fungi clearly deserve their own kingdom. Like animals, they are heterotrophic, but they do not ingest there food—they absorb nutrients. Like plants, they have cell walls, but their cell walls contain chitin instead of cellulose. Fungal cells store energy in the form of glycogen, as do animal cells. At one time fungi were considered a part of the plant kingdom, and then later, they were placed in the kingdom Protista. Whittaker argued that because of their multicellular nature and mode of nutrition, fungi should be in their own kingdom.

The range of species within the kingdom is broad. Some fungi are unicellular (such as yeast), some are multicellular parasites (such as those that cause athlete's foot), and some form mutualistic relationships with other species (such as those found in lichens). Most, however, are saprotrophic decomposers that play a vital role in all ecosystems. The decomposers work with bacteria to break down the waste products and dead remains of plants and animals so organic materials are recycled.

In addition to their various ways of life, fungi have been successful because they have diverse reproductive strategies. Asexual reproduction via spores, fragmentation, and budding is common. During sexual reproduction, the filaments of different mating types typically fuse. Asexual reproduction via spores, fragmentation, and budding is also common.

Fungi have been around for at least 570 million years. What is their evolutionary history? We don't know exactly. It is possible that the species now classified in the kingdom do not share a recent common ancestor, which means they probably evolved from different protists. It is also possible that fungi evolved from red algae because the two groups share similar morphological traits.

Summary

31.1 What Fungi Are Like

Fungi are multicellular eukaryotes that are heterotrophic by absorption. After external digestion, they absorb the resulting nutrient molecules. Most fungi act as saprotrophic decomposers that aid the cycling of chemicals in ecosystems. Some fungi are parasitic, especially on plants, and others are symbiotic with plant roots and algae.

The body of a fungus is composed of thin filaments called hyphae, which collectively are termed a mycelium. The cell wall contains chitin, and the energy reserve is glycogen. Fungi do not have flagella at any stage in their life cycle. Nonseptate hyphae have no cross walls; septate hyphae have cross walls, but there are pores that allow the cytoplasm and even organelles to pass through.

Fungi produce nonmotile and often windblown spores during both asexual and sexual reproduction. During sexual reproduction hyphae tips fuse so that dikaryotic (n + n) hyphae sometimes result, depending on the type of fungus. Following nuclear fusion, zygotic meiosis occurs during the production of the sexual spores.

31.2 How Fungi Are Classified

There are four divisions of fungi: division Zygomycota (zygospore fungi), division Ascomycota (sac fungi), division Basidiomycota (club fungi), and division Deuteromycota (imperfect fungi).

The zygospore fungi are nonseptate, and during sexual reproduction they have a dormant stage consisting of a thick-walled zygospore. When the zygospore germinates, sporangia produce windblown spores. Asexual reproduction occurs when nutrients are plentiful and again sporangia produce spores.

The sac fungi are septate, and during sexual reproduction saclike cells called asci produce spores. Asci are located in fruiting bodies called ascocarps. Asexual reproduction, which is dependent on the production of conidiospores, is more common.

The club fungi are septate, and during sexual reproduction club-shaped structures called basidia produce spores. Basidia are located in fruiting bodies called basidiocarps. Club fungi have a prolonged dikaryotic stage, and asexual reproduction by conidiospores is rare. A dikaryotic mycelium periodically produces fruiting bodies.

The imperfect fungi always reproduce asexually by conidiospores; they have not been observed to reproduce sexually. Therefore, they cannot be placed in one of the other divisions. Several imperfect fungi are of interest; for example, *Penicillium* is the source of penicillin, and *Candida* causes yeast infections.

31.3 Fungi Form Symbiotic Relationships

Lichens are an association between a fungus and a cyanobacterium or a green alga. Traditionally, this association was considered to be mutualistic, but experimentation suggests a controlled parasitism by the fungus on the alga. Lichens may live in extreme environments and on bare rocks; they allow other organisms that will eventually form soil to establish. They are important in primary succession.

The term mycorrhizae refers to an association between a fungus and the roots of a plant. The fungus helps the plant absorb minerals, and the plant supplies the fungus with carbohydrates.

Reviewing the Chapter

1. Which characteristics best define fungi? Describe the body of fungi and how they reproduce. 540–41
2. On what basis are fungi classified? 542
3. Explain the term zygospore fungi. How does black bread mold reproduce asexually? sexually? 542–43
4. Explain the term sac fungi. How do sac fungi reproduce asexually? Describe the structure of an ascocarp. 545
5. Describe the structure of yeasts, and explain how they reproduce. How are yeasts useful to humans? 545
6. Explain the term club fungi. Draw and explain a diagram of the life cycle of a typical mushroom. 546–47
7. What is the economic importance of smuts and rusts? How can their numbers be controlled? 547
8. Explain the term imperfect fungi. Give examples of imperfect fungi that serve humans and examples of those that cause disease. 549
9. Describe the structure of a lichen, and name the three different types. What is the nature of this fungal association? 549–50
10. Describe the association known as mycorrhizae, and explain how each partner benefits. 550

Testing Yourself

Choose the best answer for each question.

1. A decomposer is most likely to utilize which mode of nutrition?
   a. parasitic
   b. saprotrophic
   c. ingestion
   d. Both a and b are correct.
2. Which feature is best associated with hyphae?
   a. strong impermeable walls
   b. rapid growth
   c. large surface area
   d. Both b and c are correct.
3. A spore
   a. contains an embryonic organism.
   b. germinates directly into an organism.
   c. is always windblown.
   d. Both b and c are correct.
4. The taxonomy of fungi is based on
   a. sexual reproductive structures.
   b. shape of the sporocarp.
   c. mode of nutrition.
   d. type of cell wall.
5. In the life cycle of black bread mold, the zygospore
   a. undergoes meiosis and produces zoospores.
   b. produces spores as a part of asexual reproduction.
   c. is a thick-walled dormant stage.
   d. is equivalent to asci and basidia.

6. In an ascocarp
   a. there are fertile and sterile hyphae.
   b. hyphae fuse, forming the dikaryotic stage.
   c. a sperm fertilizes an egg.
   d. hyphae do not have chitinous walls.
7. In which fungus is the dikaryotic stage longer lasting?
   a. zygospore fungi
   b. imperfect fungi
   c. sac fungi
   d. club fungi
8. Conidiospores are formed
   a. by sporangia.
   b. during sexual reproduction.
   c. by all types of fungi.
   d. when nutrients are in short supply.
9. Imperfect fungi are called imperfect because
   a. they have no zygospore.
   b. they cause diseases.
   c. they form conidiospores.
   d. sexual reproduction has not been observed.
10. Lichens
    a. cannot reproduce.
    b. need a nitrogen source to live.
    c. are parasitic on trees.
    d. are able to live in extreme environments.
11. Label this diagram of the life cycle of a mushroom.

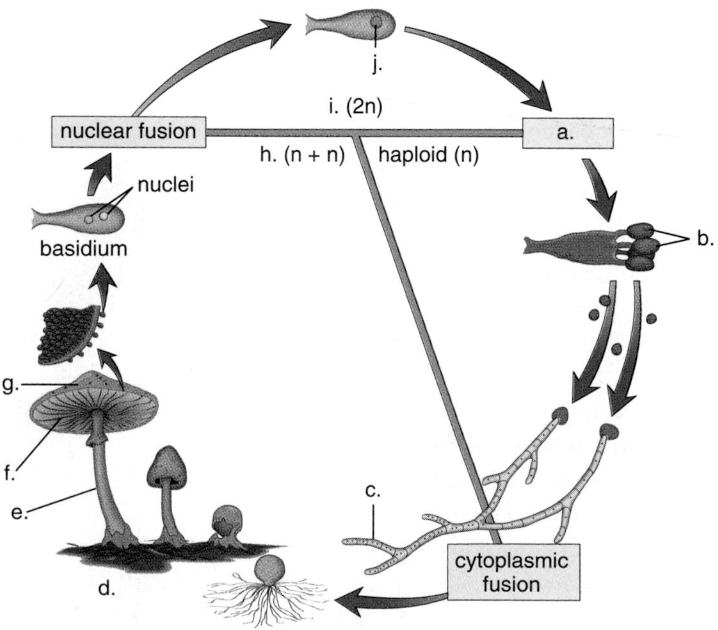

Applying the Concepts

1. *Organisms are adapted to a way of life.*
   How are the structures and reproductive strategies of fungi related to their way of life?
2. *Living things are diverse.*
   All fungi have the same mode of nutrition; in what ways are they diverse?
3. *Classification systems do not always reflect evolutionary relationships.*
   What basic difficulties do you have when trying to draw a phylogenetic tree for fungi?

Using Technology

Your study of the fungi is supported by these available technologies:

Exploring the Internet

The Mader Home Page provides resources for and help with studying this chapter.

http://www.mhhe.com/sciencemath/biology/mader/
(Click on Biology.)

Understanding the Terms

ascus (pl., asci) 545
basidium (pl., basidia) 547
budding 541
conidiospore 545
dikaryotic 541
fruiting body 545
fungus (pl., fungi) 540
hypha (pl., hyphae) 540
lichen 549
mycelium 540
mycorrhiza (pl., mycorrhizae) 550
nonseptate 541
septate 541
sporangium (pl., sporangia) 542
spore 541
zygospore 542

Match the terms to these definitions:

a. __________ Capsule that produces sporangiospores.
b. __________ Clublike structure in which nuclear fusion and meiosis occur during sexual reproduction of club fungi.
c. __________ Filament of the vegetative body of a fungus.
d. __________ Tangled mass of hyphal filaments composing the vegetative body of a fungus.
e. __________ Saprotrophic decomposer; the body is made up of filaments called hyphae that form a mass called a mycelium.
f. __________ Spore produced by sac and club fungi during asexual reproduction.
g. __________ Spore-producing and spore-disseminating structure found in sac and club fungi as a result of sexual reproduction.
h. __________ Symbiotic relationship between certain fungi and algae in which the fungi possibly provide inorganic food or water and the algae provide organic food.
i. __________ Symbiotic relationship between fungal hyphae and roots of vascular plants. The fungus allows the plant to absorb more mineral ions and obtains carbohydrates from the plant.

The Plants

CHAPTER 32

Chapter Concepts

32.1 What Is a Plant?

- Plants are photosynthetic organisms adapted to living on land. Among various adaptations, all plants have an alternation of generations life cycle. 554
- The presence of vascular tissue and reproductive strategy are used to compare plants. 555

32.2 Nonvascular Plants Are Diverse

- The nonvascular plants are the bryophytes (e.g., liverworts and mosses), which are low-growing. In bryophytes, the gametophyte (haploid generation) is dominant, and windblown spores disperse the species. 555

32.3 Vascular Plants Include Seedless and Seed Plants

- In vascular plants, the dominant sporophyte (diploid generation) has transport tissue. 558
- In seedless vascular plants, windblown spores disperse the species, and in seed plants, seeds disperse the species. 558

32.4 Ferns and Allies Are the Seedless Vascular Plants

- The seedless vascular plants were much larger and extremely abundant during the Carboniferous period. 559
- The fern sporophyte is the familiar plant with large fronds that produces windblown spores; the independent gametophyte is a heart-shaped structure that produces flagellated sperm. 562

32.5 Gymnosperms Have Naked Seeds

- There are four divisions of gymnosperms; the familiar conifers and the little-known cycads, ginkgoes, and gnetophytes. 564
- The conifer sporophyte is usually a cone-bearing tree (e.g., pine tree). The pollen cones produce pollen (male gametophyte), which is windblown to the seed cone where windblown seeds are produced. 564

32.6 Angiosperms Have Covered Seeds

- Angiosperms are the very diversified flowering plants, which may depend on an animal pollinator to carry pollen to another flower. The seeds are enclosed within a fruit, which often assists dispersal of seeds. 571

Plants of a bog in Minnesota, including lady's slipper, *Cypripedium reginae*

The biomes of the world are defined by their vegetation: the tropical rain forests, the temperate deciduous forests, and the grasslands, for example. This is appropriate because plants dominate the landscape. They are the major producers on land and were the first organisms to invade the terrestrial environment. Animals cannot live on land without plants to sustain them. Plants store energy as starch and are at the base of most ecological pyramids, including those that sustain the world's human population. They also pull carbon dioxide out of the air and supply the oxygen that organisms require for aerobic respiration.

The plant kingdom includes over 250,000 species, and among them are the earth's largest species—some live oaks and redwoods tower a hundred feet tall and weigh thousands of pounds. The vascular plants have internal tissues that absorb and conduct water, nutrients, and minerals to all parts of their body. The seed plants do not depend on external water for the purpose of reproduction, and the flowering plants have developed unique mutualistic relationships with animals to carry pollen from flower to flower.

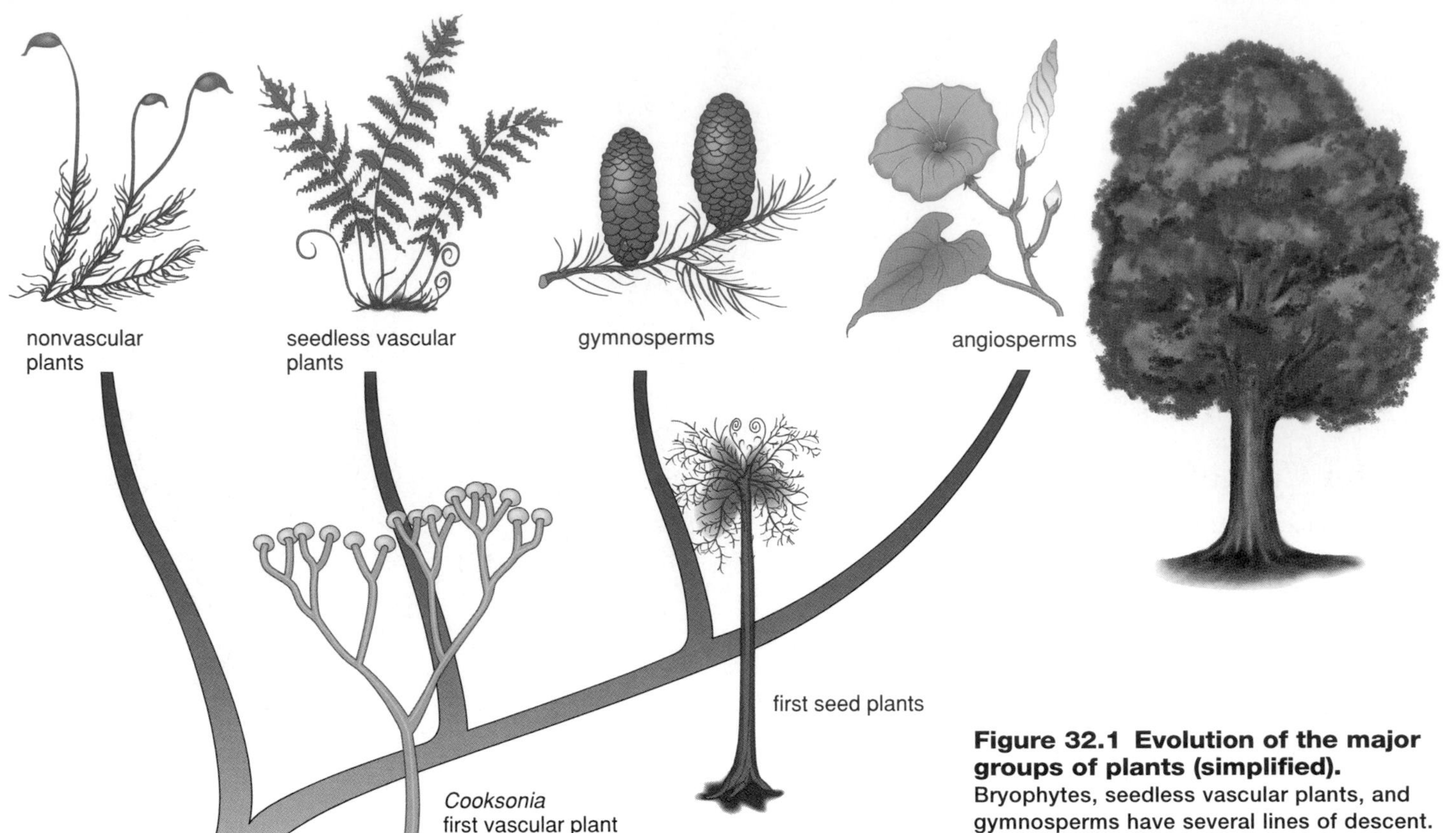

Figure 32.1 Evolution of the major groups of plants (simplified). Bryophytes, seedless vascular plants, and gymnosperms have several lines of descent.

32.1 What Is a Plant?

Plants **(kingdom Plantae)** are multicellular eukaryotes with well-developed tissues (Fig. 32.1). Plants, which are autotrophic by photosynthesis, live in a wide variety of terrestrial environments, from lush forest to dry desert or frozen tundra. A land existence offers some advantages to plants. One advantage is the great availability of light for photosynthesis—water, even if clear, filters light. Also, carbon dioxide and oxygen are present in higher concentrations and diffuse more readily in air than in water.

The land environment, however, requires adaptations to deal with the constant threat of desiccation (drying out). Most plants can obtain water from the substratum by means of roots. And to conserve water, leaves and stems are, at the very least, covered by a waxy **cuticle** that is impervious to water. The surface of leaves is interrupted by pores that can open and close to regulate gas exchange and water loss. Some plants have a vascular system that transports water up to and nutrients down from the leaves. In conjunction with this system, trees have a strong internal skeleton that opposes the force of gravity and lifts the leaves toward the sun. All plants protect the embryo from desiccation, and some protect their entire gametophyte generations. In these plants, sperm-containing pollen grains are transported by wind or motile animal to the egg, and the subsequent embryo becomes a seed.

Plants are adapted to living on land. In general, they tend to have features that allow them to live and reproduce on land.

CLASSIFICATION

Kingdom Plantae

Multicellular; primarily terrestrial eukaryotes with well-developed tissues; autotrophic by photosynthesis; alternation of generations life cycle. Like green algae, plants contain chlorophylls *a* and *b* and carotenoids; store starch in chloroplasts; cell wall contains cellulose.

Nonvascular Plants*

Division Hepatophyta: liverworts

Division Bryophyta: mosses

Division Anthocerotophyta: hornworts

Seedless Vascular Plants*

Division Psilotophyta: whisk ferns

Division Lycopodophyta: club mosses

Division Equisetophyta: horsetails

Division Pteridophyta: ferns

Seed Vascular Plants*

- Gymnosperms*
  - Division Pinophyta: conifers
  - Division Cycadophyta: cycads
  - Division Ginkgophyta: maidenhair tree
  - Division Gnetophyta: gnetophytes
- Angiosperms*
  - Division Magnoliophyta: flowering plants
    - Class Magnoliopsida: dicots
    - Class Liliopsida: monocots

*Not in classification of organisms, but added here for clarity.

Plants are believed to be closely related to green algae because these two groups have several characteristics in common. Both utilize chlorophylls *a* and *b* as well as carotenoid pigments during photosynthesis. The primary food reserve is starch, which is stored inside the chloroplast. Their cell walls contain cellulose, and both types of organisms form a cell plate during cell division.

Why aren't multicellular green algae like *Ulva* considered plants? Plants, but not green algae, have multicellular *gametangia* (sex organs) and sporangia in which reproductive cells are protected by several layers of nonreproductive cells. Later, the zygote and then the embryo are protected within the body of the plant. In algae, fertilization is external and the embryo is not protected from the environment.

Plants and some algae have a two-generation life cycle called alternation of generations that involves sporic meiosis (Fig. 32.2):

1. The **sporophyte** [Gk. *spora*, seed, and *phyton*, plant], the diploid generation, produces haploid spores by meiosis. Spores develop into a haploid generation.
2. The **gametophyte** [Gk. *gamete*, wife, *gametes*, husband, and *phyton*, plant], the haploid generation, produces gametes that unite to form a diploid zygote.

The two generations are dissimilar, and one is dominant over the other. The dominant generation is larger and exists for a longer period of time. In addition, plants are oogamous; the gametes are eggs and sperm. The characteristics of plants are listed in Table 32.1.

Plants, unlike green algae, protect the embryo from drying out. During an alternation of generations life cycle, one generation is dominant over the other.

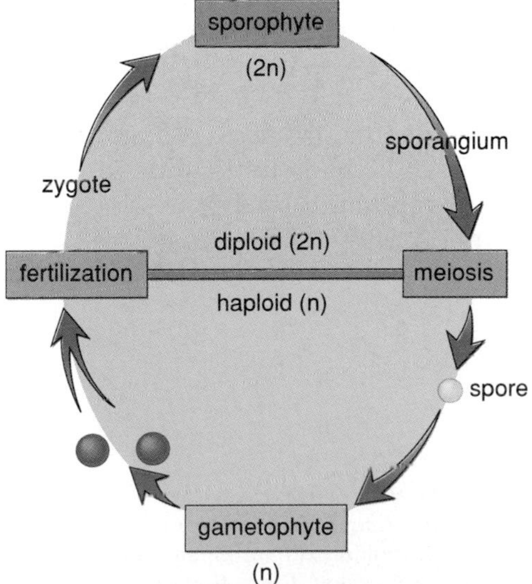

Figure 32.2 Alternation of generations.

32.2 Nonvascular Plants Are Diverse

Plants are currently divided into two main groups: the nonvascular and the vascular plants. The **nonvascular plants** consist of three groups: **hornworts** (division Anthocerotophyta), liverworts (division Hepatophyta), and mosses (division Bryophyta). The nonvascular plants lack vascular plants' specialized means of transporting water and organic nutrients. Although they often have a "leafy" appearance, these plants do not have true roots, stems, and leaves—which by definition must contain true vascular tissue. Therefore, the nonvascular plants are said to have rootlike, stemlike, and leaflike structures.

The gametophyte is the dominant generation in the nonvascular plants—it is the generation we recognize as the plant. Flagellated sperm swim to the vicinity of the egg in a continuous film of water. The sporophyte, when present, is attached to and derives its nourishment from the photosynthetic gametophyte.

The nonvascular plants are quite small; the largest is no more than 20 cm tall. This characteristic is linked to the lack of an efficient means to transport water to any height. And because sexual reproduction involves flagellated sperm, bryophytes are usually found in moist habitats. Nevertheless, mosses compete well in harsh environments because the gametophyte can reproduce asexually, allowing mosses to spread into stressful and even dry habitats. If need be, mosses can dry up and later, when water is available, begin to photosynthesize again.

It is not known how closely related the various nonvascular plants are. The current thinking is that these plants have individual lines of descent, so this text classifies the bryophytes into three separate divisions. Some mosses have a rudimentary form of vascular tissue, and therefore mosses are believed to be more closely related to vascular plants than are liverworts and hornworts. We will be discussing the liverworts and mosses, the more familiar nonvascular plants.

In nonvascular plants, the dominant gametophyte is water dependent because it lacks vascular tissue and produces flagellated sperm.

Table 32.1

Characteristics of Plants

1. Are multicellular and have cells specialized to form tissues and organs.
2. Contain chlorophylls *a* and *b* and carotenoids, store reserve food as starch, and have cellulose cell walls.
3. Have gametangia (sex organs) with an outer layer of nonreproductive cells, which prevents desiccation of developing gametes.
4. Protect the developing embryo from drying out by providing it with water and nutrients within the female reproductive structure.
5. Have a life cycle that is described as alternation of generations.

Liverworts Are Nonvascular

All **liverworts (division Hepatophyta,** 10,000 species) have a flattened thallus but those with a lobed thallus (body) are more familiar than those in which the thallus is divided into "leaves." The name "liverwort" arose in the ninth century when it seemed to some that the plant had lobes like the liver. They thought the plant might therefore be beneficial in treating liver ailments, a hypothesis that has never been supported.

Marchantia is most often used as an example of this group of plants (Fig. 32.3). Each lobe of the thallus is perhaps a centimeter or so in length; the upper surface is smooth, and the lower surface bears numerous **rhizoids** [Gk. *rhizion,* dim. for root] (rootlike hairs) projecting into the soil. *Marchantia* reproduces both asexually and sexually. Gemma cups on the upper surface of the thallus contain *gemmae,* groups of cells that detach from the thallus and can start a new plant. Sexual reproduction depends on disk-headed stalks that bear **antheridia** [Gk. *anthos,* flower, and *-idion,* small], where flagellated sperm are produced, and umbrella-headed stalks that bear **archegonia** [Gk. *arche-,* first, and *gone,* seed], where eggs are produced. Following fertilization a tiny sporophyte no more than a few millimeters in length is composed of a foot, a short stalk, and a capsule. Windblown spores are produced within the capsule.

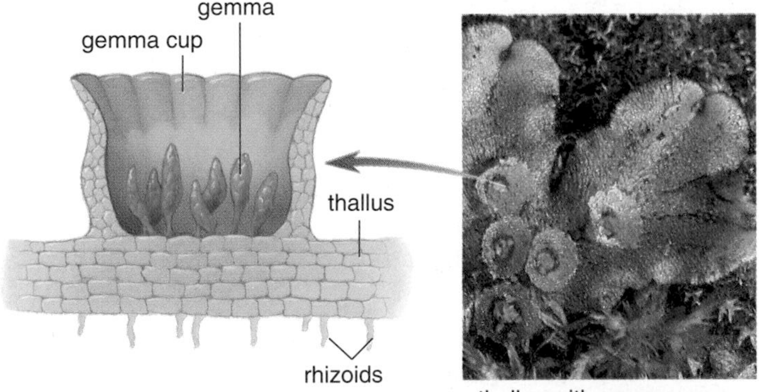

a. Gemma cup

b. Structures that bear antheridia

Structures that bear archegonia

Mosses Are Nonvascular

The body of a **moss (division Bryophyta,** 12,000 species) is usually a leafy shoot although some are secondarily flattened. Mosses can be found from the Arctic through the tropics to parts of the Antarctic. Although most prefer damp, shaded locations in the temperate zone, some survive in deserts and others inhabit bogs and streams. In forests, they frequently form a mat that covers the ground and rotting logs. Mosses can store large quantities of water in their cells, but if a dry spell continues for long, they become dormant. The whole plant shrivels, turns brown, and looks completely dead. As soon as it rains, however, the plant becomes green and resumes metabolic activity.

Some mosses are adapted to unusual conditions. The so-called copper mosses live only in the vicinity of copper and can serve as an indicator plant for copper deposits. Luminous moss, which glows with a golden green light, is found in caves, under the roots of trees, and other dimly lit places. Its cells, which are shaped like tiny lenses, focus what little light there is on the grana of chloroplasts.

The common name of several plants implies they are mosses when they are not. Irish moss is an edible alga that grows in leathery tufts along northern seacoasts. Reindeer moss, a lichen, is the dietary mainstay of reindeer and caribou in northern lands. Club mosses, discussed later in this chapter, are vascular plants; and Spanish moss, which hangs in grayish clusters from trees in the southeastern United States, is a flowering plant of the pineapple family.

Most mosses can reproduce asexually by fragmentation. Just about any part of the plant is able to grow and eventually produce leafy shoots. Figures 32.4 and 32B (see page 569) describe the life cycle of a typical temperate-zone moss. The gametophyte of mosses has two stages. First, there is the algalike *protonema,* a branching filament of cells. Then, after about three days of favorable growing conditions, upright leafy shoots are seen at intervals along the protonema. Rhizoids anchor the shoots, which bear antheridia and archegonia. An antheridium consists of a short stalk, an outer layer of sterile cells, and an inner mass of cells that become the flagellated sperm. An archegonium, which looks like a vase with a long neck, has a single egg located at the base.

The dependent sporophyte consists of a *foot,* which grows down into the gametophyte tissue, a *stalk,* and an upper *capsule,* or **sporangium,** where spores are produced. At first the sporophyte is green and photosynthetic; at maturity it is brown and nonphotosynthetic.

Figure 32.3 Liverwort, *Marchantia.*
a. The thallus can reproduce asexually by forming gemmae, cells that detach from the thallus and are present in gemma cups.
b. During sexual reproduction, the antheridia are present in disk-shaped structures, and the archegonia are present in umbrella-shaped structures.

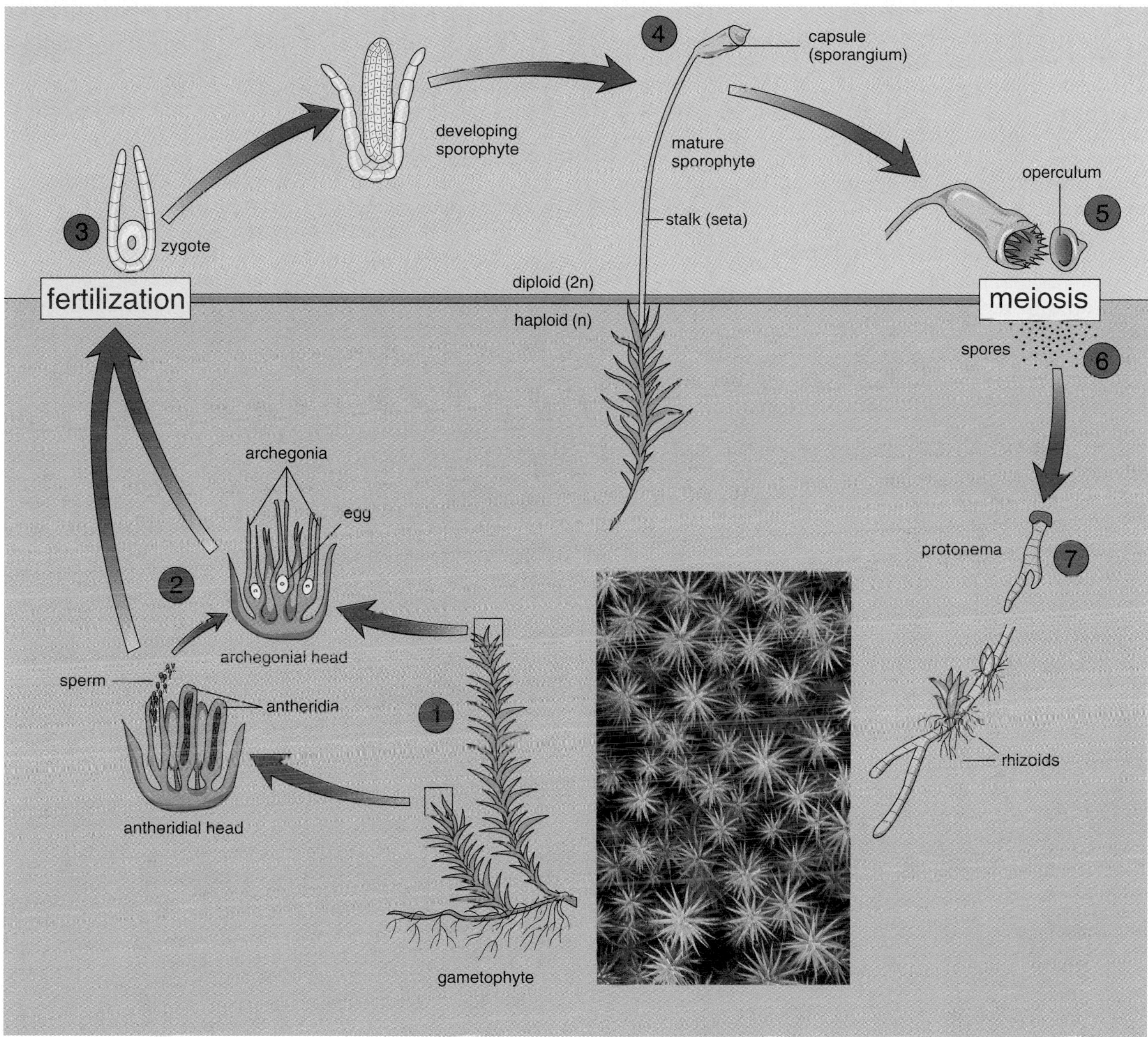

Figure 32.4 Moss life cycle.

The gametophyte is dominant in bryophytes, such as mosses. ① The leafy shoots bear separate male and female gametangia: the antheridia and archegonia. ② Flagellated sperm are produced in antheridia, and these swim in external water to an archegonium that contains a single egg. ③ When the egg is fertilized, the developing embryo is retained within the archegonium. In some species of mosses, a hoodlike covering (calyptra) derived from the archegonium is carried upward by the growing sporophyte. ④ The mature sporophyte growing atop a gametophyte shoot consists of a foot that grows down into the gametophyte tissue, a stalk, and an upper capsule, or sporangium, where meiosis occurs and spores are produced. ⑤ When the covering and capsule lid (operculum) fall off, the spores are mature and are ready to escape. The release of spores is controlled by one or two rings of "teeth" that project inward from the margin of the capsule. The teeth close the opening when the weather is wet but curl up and free the spores when the weather is dry. ⑥ Spores are released at times when they are most likely to be dispersed by air currents. ⑦ When a spore lands on an appropriate site, it germinates into a protonema, the first stage of the gametophyte.

Uses of Nonvascular Plants

Sphagnum, also called bog or peat moss, has commercial importance. The cells of this moss have a tremendous ability to absorb water, which is why peat moss is often used in gardening to improve the water-holding capacity of the soil. In some areas where the ground is wet and acidic, such as bogs, dead mosses, especially sphagnum, accumulate and do not decay. This accumulated moss, called **peat,** can be used as fuel.

Adaptation of Nonvascular Plants

The nonvascular plants are small and simple plants generally found in moist habitats for two reasons: they lack vascular tissue, and their sexual reproduction involves flagellated sperm. Nonvascular plants do not have the complexity of, nor are they found in the variety of habitats occupied by, the vascular plants. Nevertheless, mosses are better than flowering plants at living on stone walls and on fences and even in the shady cracks of hot exposed rocks. They improve newly created soil so that it can be used for the growth of other organisms. For these particular microhabitats, there seems to be a selective advantage to being small and simple (Fig. 32.5).

The nonvascular plants (liverworts and mosses) lack specialized vascular tissue and have a dominant gametophyte. Fertilization requires an outside source of moisture, and they are often found in moist locations.

Figure 32.5 Moss growing on rocks.
The mosses that are adapted to growing on rocks contribute to the formation of soil so that other plants can enter the area.

32.3 Vascular Plants Include Seedless and Seed Plants

Vascular plants [L. *vasculum,* dim. of *vas,* vessel] include ferns and their allies, gymnosperms, and angiosperms. Vascular tissue in these plants consists of **xylem** [Gk. *xylon,* wood], which conducts water and minerals up from the soil, and **phloem** [Gk. *phloios,* bark], which transports organic nutrients from one part of the plant to another. The vascular plants have true roots, stems, and leaves. The **roots** absorb water from the soil and the **stem** conducts water to the leaves. Xylem, with its strong-walled cells, supports the body of the plant against the pull of gravity. The leaves are fully covered by a waxy cuticle except where it is interrupted by **stomates,** little pores whose size can be regulated to control water loss.

The sporophyte generation is dominant in vascular plants. This is the generation that has vascular tissue and the other features just discussed. Another advantage of having a dominant sporophyte relates to its being diploid. If a faulty gene is present, it can be masked by a functional gene. Then, too, the greater the amount of genetic material, the greater the possibility of mutations that will lead to increased variety and complexity. Indeed, the vascular plants are complex, extremely varied, and widely distributed.

The seedless vascular plants (ferns and their allies) disperse the species by producing windblown spores (see Fig. 32C, page 569). When the spores germinate, a relatively large gametophyte is independent of the sporophyte for its nutrition. In these plants, flagellated sperm are released by antheridia and swim in a film of external water to the archegonia, where fertilization occurs. Because the water-dependent gametophyte is independent of the sporophyte, these plants cannot wholly benefit from the adaptations of the sporophyte to a terrestrial environment. As discussed in the reading on the following page, the seedless vascular plants formed the great swamp forest of the Carboniferous period.

In seed plants, there is a separate microgametophyte (male) and megagametophyte (female), as shown in Figure 32D (see page 569). The microgametophyte and megagametophyte are dependent on the sporophyte, which is fully adapted to a dry environment. The mature microgametophyte is the pollen grain. The megagametophyte and subsequent embryo, which is retained within the body of the sporophyte, becomes a seed. The seed disperses the species in seed plants. Seed dispersal occurs by various means. Some seeds are carried by wind and water or by animals to a new location.

The seedless vascular plants are dispersed by windblown spores; those that produce seeds are dispersed by seeds.

Ecology Focus

The First Forests

Our industrial society runs on fossil fuels such as **coal**. The term "fossil fuel" might seem odd at first until one realizes that it refers to the remains of organic material from ancient times. During the Carboniferous period more than 300 million years ago, a great swamp forest (Fig. 32A) encompassed what is now the Appalachian Mountains, northern Europe, and the Ukraine. The weather was warm and humid, and the trees grew very tall. These are not the trees we know today; instead they are related to today's seedless vascular plants: the club mosses, horsetails, and ferns! Club mosses today might stand as high as 30 cm; their ancient relatives were 30.5 meters (over 100 feet) tall and 1 meter wide. The spore-bearing cones were up to 30 cm long and some had leaves more than 1 meter long. Horsetails too—at 15 meters tall—were giants compared to today's specimens. The tree ferns were also taller than today's tree ferns found in the tropics, and there were two other types of trees: seed ferns and ancient gymnosperms. "Seed fern" is a misnomer because it has been shown that these plants, which only resemble ferns, were actually a type of gymnosperm.

The amount of biomass was enormous and occasionally the swampy water rose and the trees fell. Trees under water do not decompose well, and their partially decayed remains became covered by sediment that sometimes changed to sedimentary rock. Sedimentary rock applied pressure and the organic material then became coal, a fossil fuel. This process continued for millions of years, resulting in immense deposits of coal. Geological upheavals raised the deposits to the level they are today, where they can be mined.

With a change of climate, the trees of the Carboniferous period became extinct, and only their herbaceous relatives survived to our time. Without these ancient forests, our life today would be far different than it is because they helped bring about our industrialized society.

Figure 32A Swamp forest of Carboniferous period.

Table 32.2

The Geological Timescale: Some Major Evolutionary Events of Plants

Era	Period	MYA*	Major Biological Events—Plants
Cenozoic	Neogene		
		2	Number of herbaceous plants increases
	Paleogene		
			Land dominated by angiosperms — Age of angiosperms
		66	
Mesozoic	Cretaceous		Angiosperms spread
		144	
	Jurassic		First angiosperms appear
			Age of gymnosperms
		208	
	Triassic		Land dominated by gymnosperms and ferns
		245	
Paleozoic	Permian		Land covered by forests of seedless vascular plants
		286	
	Carboniferous		Age of great coal-forming forests, including club mosses, horsetails, and ferns
		360	Swamp Forest
	Devonian		Expansion of seedless vascular plants on land
		408	
	Silurian		Seedless vascular plants appear on land
		438	
	Ordovician		First plant fossils
		505	
	Cambrian		Unicellular marine algae abundant — Age of algae
		570	

* MYA = millions of years ago.

32.4 Ferns and Allies Are the Seedless Vascular Plants

Because the seedless vascular plants are not closely related, each type is placed in its own division. The seedless vascular plants include whisk ferns (division Psilotophyta), club mosses (division Lycopodophyta), horsetails (division Equisetophyta), and ferns (division Pteridophyta).

Table 32.2 shows the evolutionary history of plants. Among the first vascular plants were the rhyniophytes, which were dominant from the mid-Silurian period until the mid-Devonian period. *Cooksonia* may have been the very first vascular plant, but *Rhynia,* which is also a rhyniophyte, is better known. In *Rhynia,* the erect stems fork repeatedly (are dichotomous). They are attached to a **rhizome,** a horizontal, underground stem that bears rhizoids below. The stem does have vascular tissue, but the plant has no true roots or leaves. Sporangia are at the ends of some of the branches. Ferns and their allies are believed to trace their ancestry back to the rhyniophytes or closely related forms.

Psilotophytes Are Whisk Ferns

The **psilotophytes (division Psilotophyta)** are represented by *Psilotum,* a plant that closely resembles *Rhynia* (Fig. 32.6). **Whisk ferns,** named for their resemblance to whisk brooms, are found in Arizona, Texas, Louisiana, and Florida, as well as Hawaii and Puerto Rico. The whisk ferns have *no leaves or roots.* A branched rhizome has rhizoids, and a mycorrhizal fungus helps gather nutrients. Aerial stems with tiny scales fork repeatedly and carry on photosynthesis. Sporangia are located at the ends of short branches. The independent gametophyte, which is found underground and is penetrated by a mycorrhizal fungus, produces flagellated sperm.

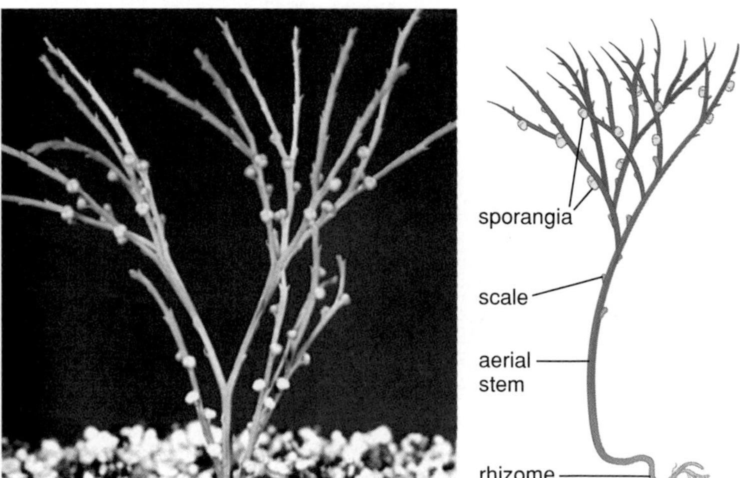

Figure 32.6 Whisk fern, *Psilotum*.
Whisk ferns have no roots or leaves—the branches carry on photosynthesis. The sporangia are yellow.

Figure 32.7 Club moss, *Lycopodium*.
Green photosynthetic stems are covered by scalelike leaves, and sporangia are found on leaves arranged as strobili.

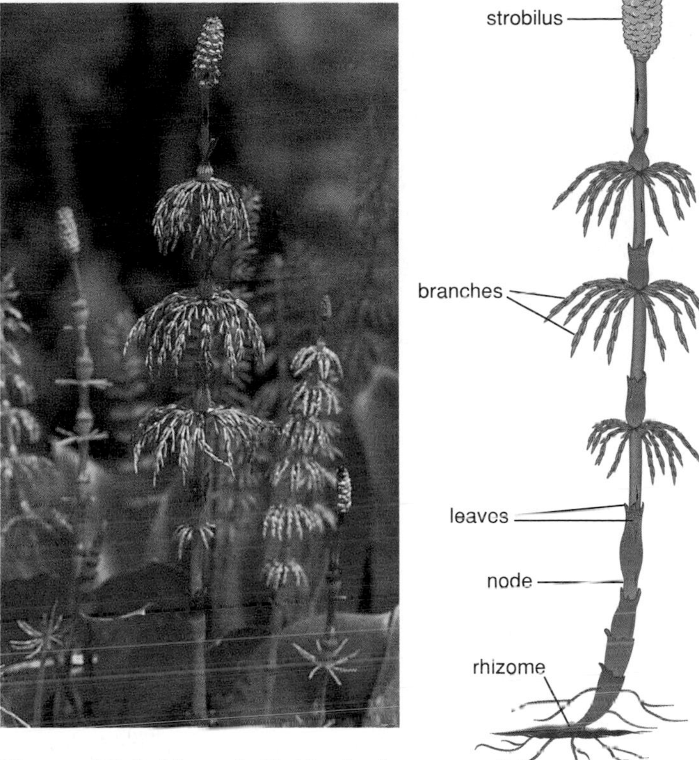

Figure 32.8 Horsetail, *Equisetum*.
There are whorls of branches and tiny leaves at the joints of the stem. The sporangia are borne in strobili.

Club Mosses Have Club-Shaped Strobili

The **club mosses (division Lycopodophyta,** 1,000 species) are common in moist woodlands of the temperate zone where they are known as ground pines. Typically, a branching rhizome sends up aerial stems less than 30 cm tall. Tightly packed, scalelike leaves cover the stem and branches of the plant, which has the appearance shown in Figure 32.7. The leaves are *microphylls,* so called because they have only one strand of vascular tissue. Such a leaf may have evolved from a scale:

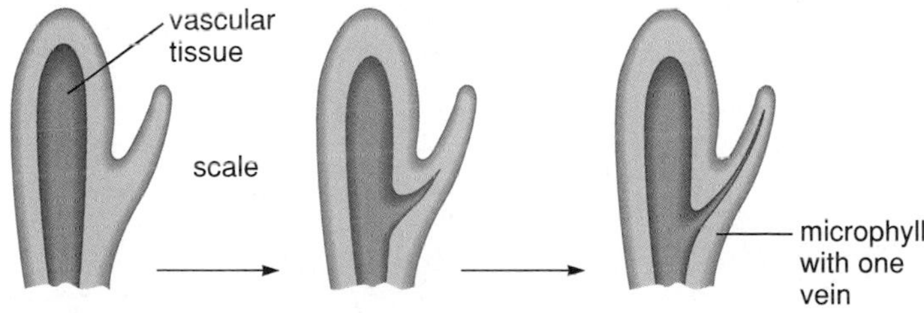

In club mosses, the sporangia are borne on terminal clusters of leaves, called strobili (sing., **strobilus**), which are club shaped. The spores germinate into inconspicuous and independent gametophytes. The majority of club mosses live in the tropics and subtropics where many of them are epiphytes—plants that live on, but are not parasitic on, trees. The closely related spike mosses (*Selaginella*) are extremely varied and include the resurrection plant, which curls up into a tight ball when dry but unfurls as if by magic when moistened.

Club mosses trace their ancestry to the Devonian period (Table 32.2). They are the only living plants to have microphylls, and it would appear that they are indirectly (rather than directly) related to the ferns and other allied groups.

Horsetails Have Jointed Stems

Horsetails (division Equisetophyta, 15 species), which thrive in waste and wet places around the globe, are represented by *Equisetum*, the only genus in existence today (Fig. 32.8). A rhizome produces aerial stems that stand up to 1.3 meters. The whorls of slender green side branches at the joints (nodes) of the stem make the plant bear a fanciful resemblance to a horse's tail. The small, scalelike leaves also form whorls at the nodes. Although the leaves have only a single strand of vascular tissue, they are reduced megaphylls (with many strands of vascular tissue), not microphylls. Many horsetails have strobili at the tips of the stems; others send up special buff-colored stems that bear the strobili. The spores germinate into inconspicuous and independent gametophytes.

The stems are tough and rigid because of silica deposited in cell walls. Early Americans, in particular, used horsetails for scouring pots and called them "scouring rushes." Today they are still used as ingredients in a few abrasive powders.

Ferns Are Leafy

Ferns (division Pteridophyta, 12,000 species) are a widespread group of plants. They are most abundant in warm, moist tropical regions, but they are also found in northern regions and in dry, rocky places. They range in size from those that are low growing and resemble mosses to those that are tall trees. The fronds (leaves) in particular can vary. The royal fern has fronds that stand 6 feet tall; those of the Venus maidenhair fern are branched with broad leaflets. And those of the hart's tongue fern are straplike and leathery. In nearly all ferns, the leaves first appear in a curled-up form called a fiddlehead, which unrolls as it grows. Figure 32.9 shows the usual habitats of ferns and gives examples of their diversity. The life cycle of a typical temperate fern is shown in Figures 32.10 and 32C on page 569.

The fronds are megaphylls secondarily subdivided into leaflets. Megaphylls may have evolved in the following way:

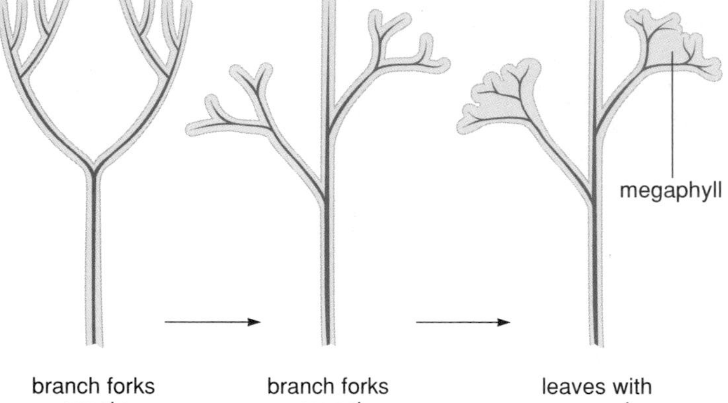

Uses of Ferns

At first it may seem that ferns do not have much economic value, but they are much used by florists in decorative bouquets and as ornamental plants in the home and garden. Wood from tropical tree ferns is often used as a building material because it resists decay, particularly by termites. Ferns have medicinal value; many Indians use ferns as an astringent during childbirth to stop bleeding, and the maidenhair fern is the source of an expectorant.

Adaptation of Ferns

Ferns have true roots, stems, and leaves. The well-developed leaves fan out, capture solar energy, and photosynthesize. The water-dependent gametophyte, which lacks vascular tissue, is separate from the sporophyte. Flagellated sperm require an outside source of water in which to swim to the eggs in the archegonia. Once established, some ferns, like the bracken fern *Pteridium aquilinum,* can spread into drier areas by means of vegetative (asexual) reproduction. Ferns also spread by means of the rhizomes growing horizontally in the soil, producing the fiddleheads that grow up as new fronds.

The seedless vascular plants include ferns and their allies. The sporophyte is dominant, and the species is dispersed by spores. The independent, nonvascular gametophyte produces flagellated sperm.

Maidenhair fern,
Adiantum pedatum

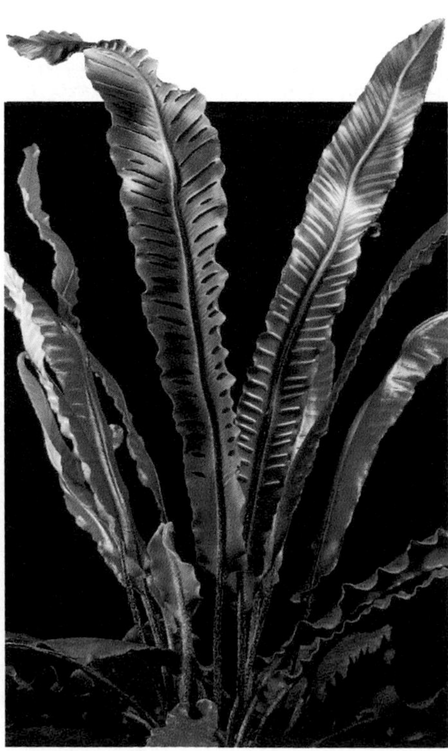

Hart's tongue fern,
Campyloneurum scolopendrium

Royal fern,
Osmunda regalis

Figure 32.9 Diversity of ferns.

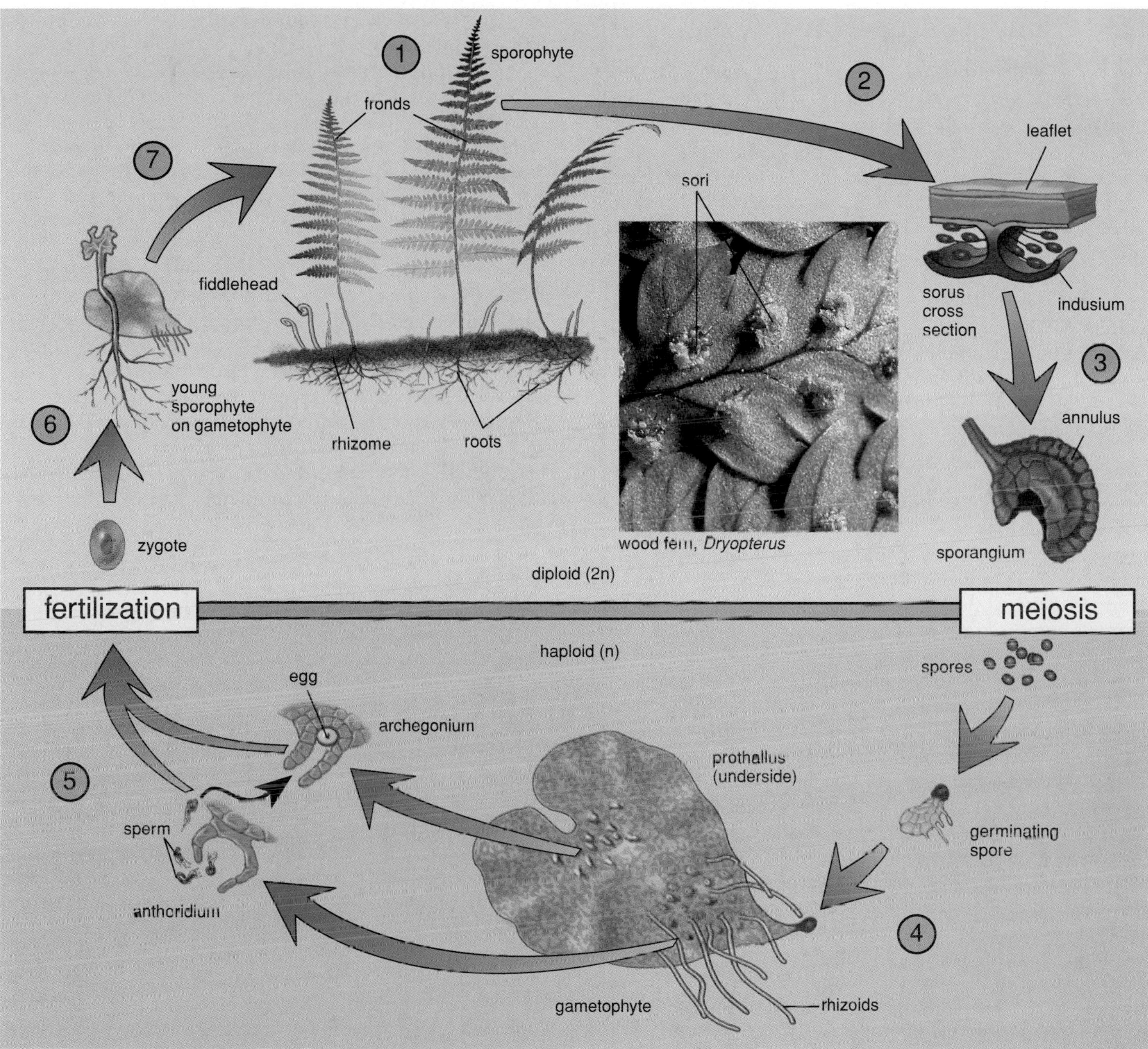

Figure 32.10 Fern life cycle.

① The sporophyte is dominant in ferns. ② In the fern shown here, the sori, protected by an indusium, are on the underside of the leaflets. Within the sporangia, meiosis occurs and spores are produced. ③ As a band of thickened cells on the rim of a sporangium (the annulus) dries out, it moves backward, pulling the sporangium open, and the spores are released. ④ A spore germinates into a prothallus, which bears antheridia and archegonia on the underside. Typically, the archegonia are at the notch and antheridia are toward the tip, between the rhizoids. ⑤ Fertilization takes place when moisture is present, because the flagellated sperm must swim in a film of water from the antheridia to the egg within the archegonium. ⑥ The resulting zygote begins its development inside an archegonium, but the embryo soon outgrows the available space. As a distinctive first leaf appears above the prothallus, and as the roots develop below it, the sporophyte becomes visible. Often the sporophyte tissues and the gametophyte tissues are distinctly different shades of green. ⑦ The young sporophyte develops a root-bearing rhizome from which the fronds project.

32.5 Gymnosperms Have Naked Seeds

A generalized life cycle for seed plants is shown in Figure 32D on page 569. Seed plants, like a few of their predecessors, produce *heterospores* called microspores and megaspores. There are also separate microgametophytes (male) and megagametophytes (female). A **microspore** develops into the immature microgametophyte contained within the **pollen grain** [L. *pollen*, fine dust]. After they are released, pollen grains develop into mature, sperm-bearing microgametophytes. **Pollination** is the transfer of pollen to the vicinity of the megagametophyte. The sperm is delivered to the egg through a pollen tube; therefore, no external water is required for fertilization. Independence from external water when reproducing is a major evolutionary trend in the plant kingdom.

A **megaspore** develops into an egg-bearing megagametophyte while still retained within an ovule. An **ovule** is the sporophyte structure that holds the megasporangium and then the megagametophyte. After fertilization, the zygote becomes an embryonic plant enclosed within the ovule, which then becomes the seed. Notice that megagametophytes and microgametophytes are dependent to a degree upon the diploid sporophyte. This means that the sporophyte can evolve into diverse forms without any corresponding changes in the gametophyte.

In seed plants, seeds disperse the sporophyte, which is the diploid generation. A **seed** is the mature ovule; it contains an embryonic sporophyte and stored food enclosed by a protective seed coat derived from the integuments (coverings) of the ovule. Seeds are resistant to adverse conditions, such as dryness or temperature extremes. They contain an embryo that is already partially developed and a food reserve that supports the emerging seedling until it can exist on its own. The survival value of seeds contributed greatly to the success of seed plants and their present dominance among plants.

Seed plants produce micro- and megagametophytes. This has led to two evolutionary events: replacement of flagellated sperm by pollen grains, and the production of seeds for dispersal of the sporophyte and the species.

Gymnosperm Diversity

There are four groups of **gymnosperms** [Gk. *gymnos*, naked, and *sperma*, seed]: conifers (division Pinophyta), cycads (division Cycadophyta), ginkgo (division Ginkgophyta), and gnetophytes (division Gnetophyta).

In gymnosperms, the seeds are not covered. Instead, they are exposed on the surface of sporophylls, leaves that bear sporangia. (In flowering plants, seeds are enclosed within a fruit.) Reproductive organs are usually borne in **cones** on which the sporophylls are spirally arranged. Other than these features, the four divisions of gymnosperms have little in common.

The gymnosperms didn't begin to flourish until the Mesozoic era. When the supercontinent Pangaea formed during the later part of the former Paleozoic era, mountain ranges arose and deserts appeared on their leeward side. Land that had been swampy became much drier. A mass extinction occurred, and among those that all but vanished were the seedless vascular plants that had made up the great swamp forests of the Carboniferous period. Then the first seed plants, many of which are now extinct (such as the tree ferns), and also modern-day gymnosperms, with their many adaptations to life on land, came into their own.

The gymnosperms didn't begin to flourish until the Mesozoic era, when geological upheavals brought about a change in the climate that fostered their dominance.

Conifers

Conifers [L. *conus*, cone, and *fero*, carry] (**division Pinophyta,** 550 species) are cone-bearing trees such as pines, hemlocks, and spruces. Conifers, which usually have evergreen needle-like leaves, are well adapted to withstand extremes of temperature, humidity, and wind strength. The leaves have a thick cuticle, sunken stomates, and a reduced surface area. Conifers are found in windswept mountaintops, swampy lowlands, and semideserts—from the equator to the frigid far north. The taiga is a coniferous forest extending in a broad belt across northern Eurasia and North America.

Conifers set records for their size and longevity. A giant sequoia named the General Sherman tree, located in California's Sequoia National Park, is 84 meters tall, 10 meters in diameter, and weighs an estimated 1,385 tons. Douglas firs are quite tall and sometimes reach heights of 75 meters, but redwood trees commonly grow more than 90 meters high. Some redwood trees are 2,000 years old, but the gnarled and twisted bristlecone pines in the Nevada mountains are known to be more than 4,500 years old.

The life cycle of pines, outlined in Figure 32.11, demonstrates how fertilization does not require external water and the adaptation of the seed for dispersal by the wind.

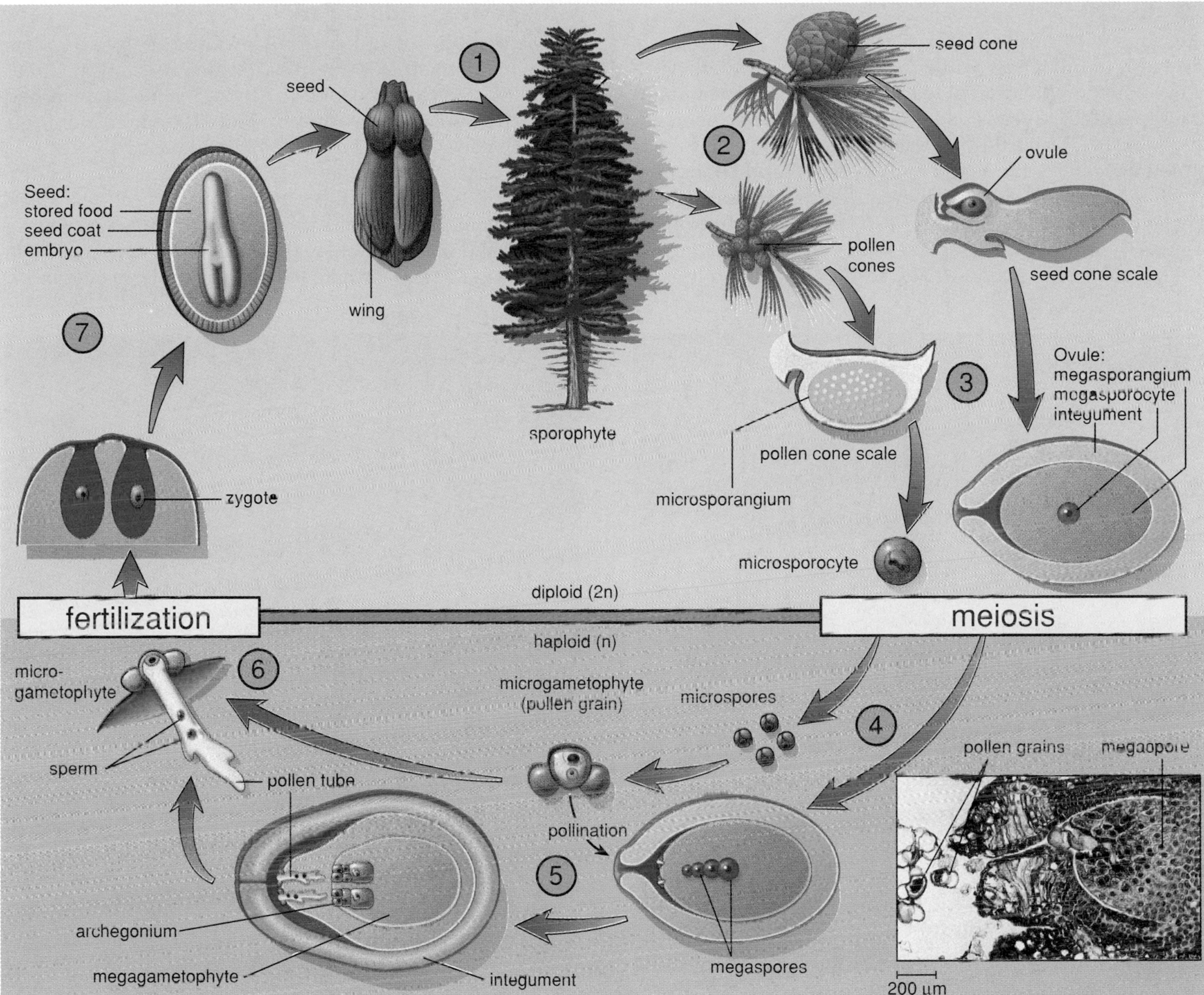

Figure 32.11 Pine life cycle.

① The sporophyte is dominant, and its sporangia are borne in cones. ② There are two types of cones: pollen cones and seed cones. Typically, the pollen cones are quite small and develop near the tips of lower branches. ③ Each scale of a pollen cone has two or more microsporangia on the underside. ④ Within these sporangia, each microsporocyte (microspore mother cell) undergoes meiosis and produces four microspores. ⑤ Each microspore develops into a microgametophyte, which is the pollen grain. The microgametophyte has two wings and is carried by the wind to the seed cone during pollination. The seed cones are larger than the pollen cones and are located near the branch tips. ③ Each scale of the seed cone has two ovules that lie on the upper surface. Each ovule is surrounded by a thick, layered integument, having an opening at one end. ④ The megasporangium is within the ovule, where a megasporocyte (megaspore mother cell) undergoes meiosis, producing four megaspores. ⑤ Only one of these spores develops into a megagametophyte, with two to six archegonia, each containing a single large egg lying near the ovule opening. ⑥ Once a pollen grain is enclosed within the seed cone, it develops a pollen tube that digests its way slowly toward a megagametophyte. The pollen tube discharges two nonflagellated sperm. One of these fertilizes an egg in an archegonium and the other degenerates. Fertilization, which takes place one year after pollination, is an entirely separate event from pollination. ⑦ After fertilization, the ovule matures and becomes the seed, composed of the embryo, the reserve food, and the seed coat. Finally, in the fall of the second season, the seed cone, by now woody and hard, opens to release winged seeds. When a seed germinates, the sporophyte embryo develops into a new pine tree, and the cycle is complete.

Cycads Fed Dinosaurs

The trunk of **cycads (division Cycadophyta,** 100 species) is stout and unbranched, while the large leaves are finely divided. This combination of features gives the plant a palmlike appearance. Cycads flourished during the Mesozoic era (see Table 32.2), and they were so numerous that this era is sometimes referred to as the Age of Cycads and Dinosaurs. Dinosaurs most likely fed on cycad seeds and foliage. Today, cycads are found mainly in tropical and subtropical regions (Fig. 32.12*a*). One species grows as tall as 20 meters and has meter-long cones that weigh almost 90 pounds.

Ginkgo Trees Adorn Parks

Only one species of **ginkgo (division Ginkgophyta),** known as the maidenhair tree (Fig. 32.12*b*), survives today. Growing up to 70 feet, it is covered with forked-veined, fan-shaped leaves that are shed in autumn. Only the microsporangia occur in cones. In modern times, the maidenhair tree was largely restricted to ornamental gardens in China until it was discovered that it does quite well in polluted areas. Because female trees produce seeds with a fleshy covering that fall and make a foul-smelling mess, it is the custom to use only male trees, propagated vegetatively, in city parks.

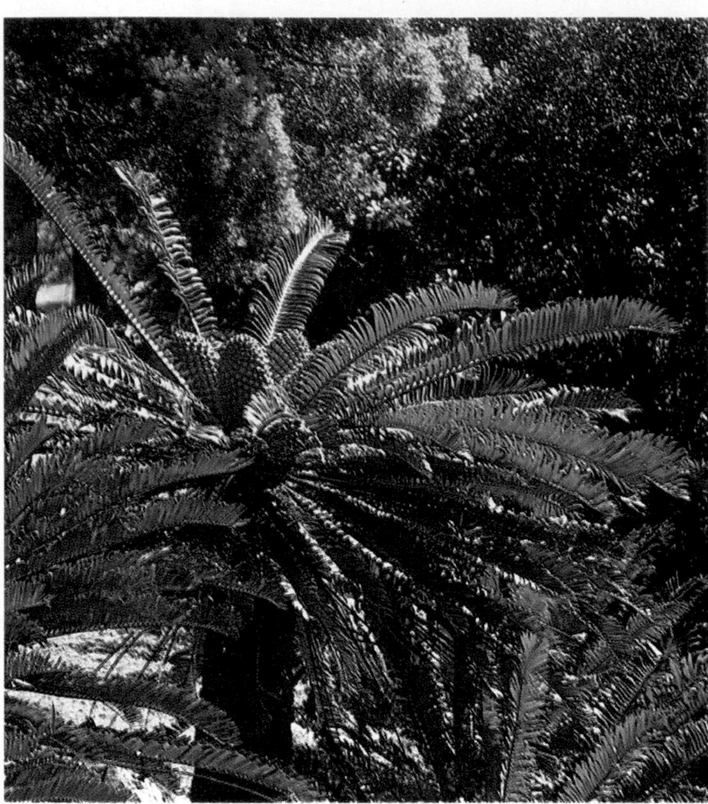

a. Cycad, *Encephalartos humlis*

b. Maidenhair tree, *Ginkgo biloba*

c. Gnetophyte, *Welwitschia mirabilis*

Figure 32.12 Lesser known gymnosperms.
a. Cycads resemble palm trees but are gymnosperms that produce cones. Male plants have pollen cones and female plants have seed cones. **b.** Ginkgoes exist only as a single species, the maidenhair tree. This photograph of a female plant features the seeds. Male plants have pollen cones. **c.** There are three genera of gnetophytes—this is *Welwitschia mirabilis,* a very unusual African plant. The stem, which has two enormous straplike leaves, bears strobili.

Gnetophytes

The **gnetophytes** (**division Gnetophyta,** 70 species), which also live mainly in the tropics, contain only three genera of plants. They look quite different from one another, though they share certain anatomical features (pertaining to their gametophytes and vascular tissue) that make them seem more related to angiosperms than other gymnosperms. Only one genus is found in the United States: *Ephedra* is a many-branched shrub with small, scalelike leaves that is found largely in desert regions. The leaves and stem of this plant are still brewed into Mormon tea. *Welwitschia* belongs to a genus of very unusual plants found in Africa. Most of the plant exists underground. Only a few centimeters of trunk grow aboveground, but the trunk can be over a meter wide! There are only two enormous, straplike leaves, which grow for hundreds of years. These may cover most of the exposed portion of the trunk (Fig. 32.12*c*).

Gymnosperms share a common characteristic: the seeds are naked (not covered). Conifers, with their ability to live under extreme conditions, are still a very prevalent and successful group.

Uses of Gymnosperms

Conifers grow on large areas of the earth's surface and are economically important. They supply much of the wood used for building construction and paper production. They also produce many valuable chemicals, such as those extracted from resin, a viscous liquid substance that protects the conifers from attack by fungi and insects.

Adaptation of Gymnosperms

Gymnosperms have well-developed roots and stems. Many are tall trees that can withstand heat, dryness, and cold. The reproductive pattern of conifers has several important innovations not found in the plants we have considered so far. There is no need of external water for fertilization to occur. Pollen grains are transferred by wind, and the growth of the pollen tube delivers a sperm to an egg. Enclosure of the dependent megagametophyte in an ovule protects it during its development and shelters the developing zygote as well. Finally, the embryo is protected within the seed and is provided with a store of nutrients that supports development for the first period of its growth following germination. All of these factors increase the chance for reproductive success on land.

Conifers are the most prevalent of the gymnosperms. In the conifer life cycle, windblown pollen grains replace flagellated sperm. Following fertilization, the seed develops from the ovule, a structure that is on a cone. The seeds are uncovered and dispersed by the wind.

32.6 Angiosperms Have Covered Seeds

Angiosperms [Gk. *angion,* dim. of *angos,* vessel, and *sperma,* seed] (**division Magnoliophyta**) are the flowering plants; their seeds are enclosed by fruits. With 235,000 species, they are an exceptionally large and successful group of plants. This is six times the number of species of all other plant groups combined. Angiosperms live in all sorts of habitats, from fresh water to desert and from the frigid north to the torrid tropics. They range in size from the tiny, almost microscopic, duckweed to *Eucalyptus* trees that are over 100 meters tall. It would be impossible to exaggerate the importance of angiosperms in our everyday lives. They provide us with clothing, food, medicines, and commercially valuable products.

Origin and Evolution of Angiosperms

Angiosperms evolved from ancient gymnosperms. By the Jurassic period in the middle of the Mesozoic era, many of the gymnosperms had features similar to those we associate with angiosperms. Some had vascular tissue resembling that of angiosperms and some had sporophylls that were beginning to look like parts of flowers. The flowers were visited by beetles, which are among the earliest insects to evolve. Most of these transitional gymnosperms became extinct; only a few lines of descent are in existence today, notably the gnetophytes and the angiosperms. Among today's gymnosperms, the gnetophyte *Ephedra* is believed to be most closely related to angiosperms.

The oldest fossils definitely considered to be angiosperms come from the Cretaceous period (see Table 32.2) of the Mesozoic era, but flowering plants didn't diversify until the Cenozoic era. The continents drifted to their present locations during the Cenozoic era, and the climate grew colder. Perhaps the angiosperms were better able to diversify during these changing times, and this led to their present dominance. Perhaps also helpful was the prevalence of and diversity of insects to assist in pollination.

Rather than being tall trees, the first angiosperms may have been fast-growing woody shrubs that took on the appearance of weeds in open places. Both tall trees and herbaceous plants could have evolved from such an ancestor. Today many angiosperms are nonwoody **herbaceous plants** [L. *herba,* vegetation, plant] rather than being woody shrubs or trees. **Woody plants** have an internal skeleton composed of the xylem of previous seasons. Herbs are able to colonize disturbed places and exist in areas that woody plants cannot tolerate. Angiosperms that are adapted to the temperate, subarctic, and arctic zones are perennials or annuals. *Perennials* are plants that live two or more years; they die back seasonally in herbaceous plants and become dormant in woody plants. *Annuals* are plants that live for only one growing season.

How Flowering Plants Are Classified

Angiosperms are divided into two groups: dicotyledons (class Magnoliopsida) and monocotyledons (class Liliopsida). The **dicotyledons** [Gk, *dis,* two, and *cotyledon,* cup-shaped cavity] (or dicots) are either woody or herbaceous, and they have flower parts usually in fours and fives, net-veined leaves, vascular bundles arranged in a circle within the stem, and two cotyledons, or seed leaves. Dicot families include many familiar plant groups, such as the buttercup, mustard, maple, cactus, pea, and rose families. The rose family includes roses, apples, plums, pears, cherries, peaches, strawberries, raspberries, and a number of other shrubs. The **monocotyledons** (or monocots) are almost always herbaceous and have flower parts in threes, parallel-veined leaves, scattered vascular bundles in the stem, and one cotyledon, or seed leaf. Monocot families include the lily, palm, orchid, iris, and grass families. The grass family includes wheat, rice, corn (maize), and other agriculturally important plants.

The Flower Contains Modified Leaves

The flower consists of several kinds of highly modified leaves that are arranged in concentric rings and attached to a modified stem tip, the receptacle (Fig. 32.13). The sepals, which form the outermost ring, are frequently green and are quite similar to ordinary foliage leaves. They enclose the flower before it opens. Next are the petals, which are often large and colorful. Their color often helps attract pollinators. Within the petals, the stamens form a whorl around the pistil. **Stamens** are the pollen-producing portion of the flower. Each stamen has a slender filament with an anther at the tip. The **anthers** are modified sporophylls containing microsporangia where microspores (pollen grains) are produced. An anther could have evolved by the following stages involving a reduction in its leaflike portion:

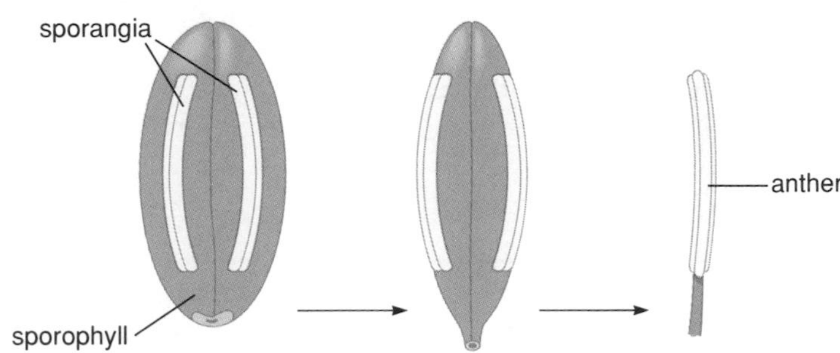

The pollen grains are either blown by the wind or carried by **pollinators** (most often insects) to the **pistil,** the portion of the flower that contains one or more fused carpels. *Carpels* are also modified sporophylls; they contain ovules in which megasporangia are located. A carpel could have evolved by the following stages:

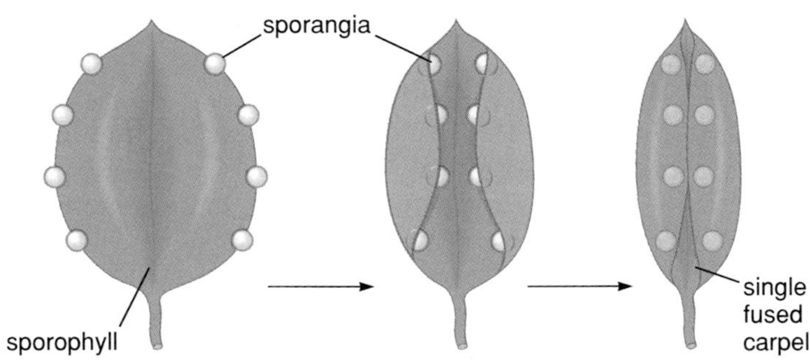

Eventually the pistil came to have its three parts: **stigma, style,** and **ovary.** The ovary contains from one to many ovules, depending on the species of plant. The ovule becomes a seed and the ovary becomes a fruit; therefore, the seeds are enclosed by fruit. A **fruit** is a mature ovary and possible accessory part of a flower that provides a fleshy or dry covering for seeds. Fruits are sometimes specialized to aid in dispersal of seeds. Fleshy fruits may be eaten by animals, which transport the seeds to a new location and then deposit them when they defecate.

The life cycle of flowering plants is given in Figure 32.14.

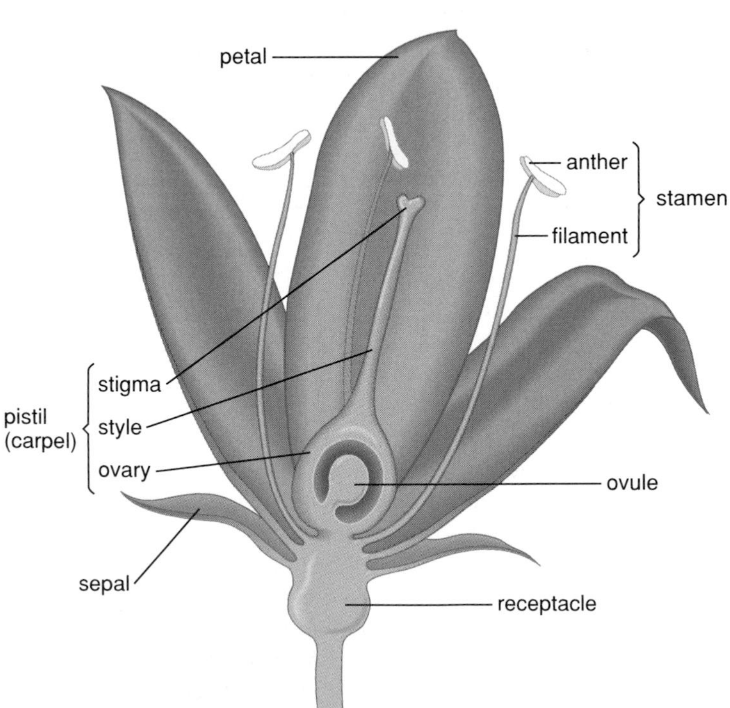

Figure 32.13 Flower structure and diversity of flowering plants.

A closer Look

Comparing Plant Life Cycles

Three groups of plants are found in the fossil record. The nonvascular plants, represented today by the bryophytes, appear first. These are followed by the nonseed vascular plants and finally the seed plants. All three groups of plants have a life cycle described as alternation of generations: a diploid (2n) generation, the sporophyte, which produces haploid (n) spores by meiosis, alternates with a haploid generation, the gametophyte, which produces the gametes. Following fertilization, the diploid zygote develops into the sporophyte generation.

In mosses, a type of bryophyte, the haploid gametophyte is the long-lasting generation and the generation we recognize as the plant. Therefore, the gametophyte is dominant and more space is allotted to this generation in Figure 32B.

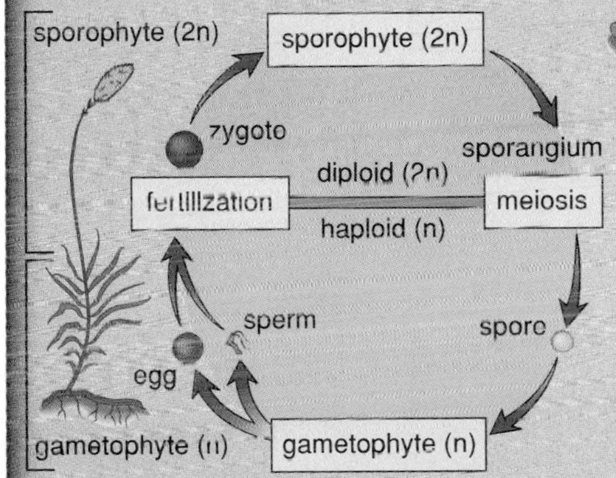

Figure 32B Nonvascular Plants

- Dominant gametophyte
- Flagellated sperm
- Dependent sporophyte
- Homospores disperse the species in mosses

In bryophytes, the sperm are flagellated and swim in a film of water to the egg. The zygote is retained in the female gametangium, and the sporophyte is dependent upon the gametophyte—it is only temporarily present and photosynthesizes to a very limited extent. The sporophyte produces a single type of spore that is usually windblown and disperses the gametophyte generation.

In the ferns, a seedless vascular plant, both generations are independent but the sporophyte generation is the dominant one. How might this have happened? The sporophyte is attached to the almost microscopic gametophyte but, since it has vascular tissue, it becomes the larger and the longer lasting of the two generations. In Figure 32C more space is allotted to the sporophyte. The sporophyte generation produces usually windblown homospores following germination. The spore gives rise to a gametophyte, which produces both male and female gametes. Sperm are flagellated and swim to the egg in a film of water. These nonvascular plants were dominant and grew to enormous sizes during the Carboniferous period (see p. 559), when the weather was warm and wet. Many of these plants became extinct when the weather turned dryer and colder over much of the earth. Today's seedless vascular plants that live in the temperate zone use asexual propagation to spread into environments that are not favorable for a water-dependent gametophyte generation.

Seed plants also have an alternation of generations life cycle, but it is much modified (Fig. 32D). The spores are retained within the sporophyte; there are heterosporangia and heterospores, termed microspores and megaspores. The gametophytes are so reduced that the megagametophyte is also retained within the structure (ovule) that held the megasporangium! Microspores become the windblown or animal-transported microgametophytes—the pollen grains. Following fertilization, the ovule becomes the seed. A seed contains a sporophyte embryo, and therefore seeds disperse the sporophyte generation. Notice that the flagellated sperm have been replaced by pollen grains and that the gametophyte, which lacks vascular tissue, is now completely dependent upon the vascular sporophyte. Fertilization no longer utilizes external water, and sexual reproduction is fully adapted to the terrestrial environment.

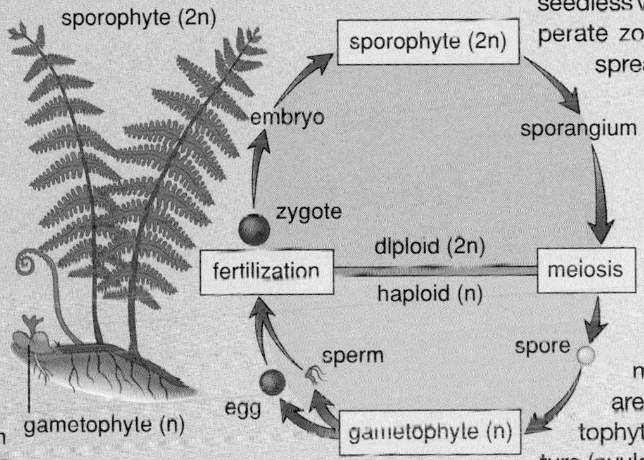

Figure 32C Seedless Vascular Plants

- Dominant sporophyte
- Homospores disperse the species
- Independent gametophyte
- Flagellated sperm

Figure 32D Seed Vascular Plants

- Dominant sporophyte
- Heterosporous
- Dependent micro-, macrogametophytes
- Pollen grains
- Seeds disperse the species

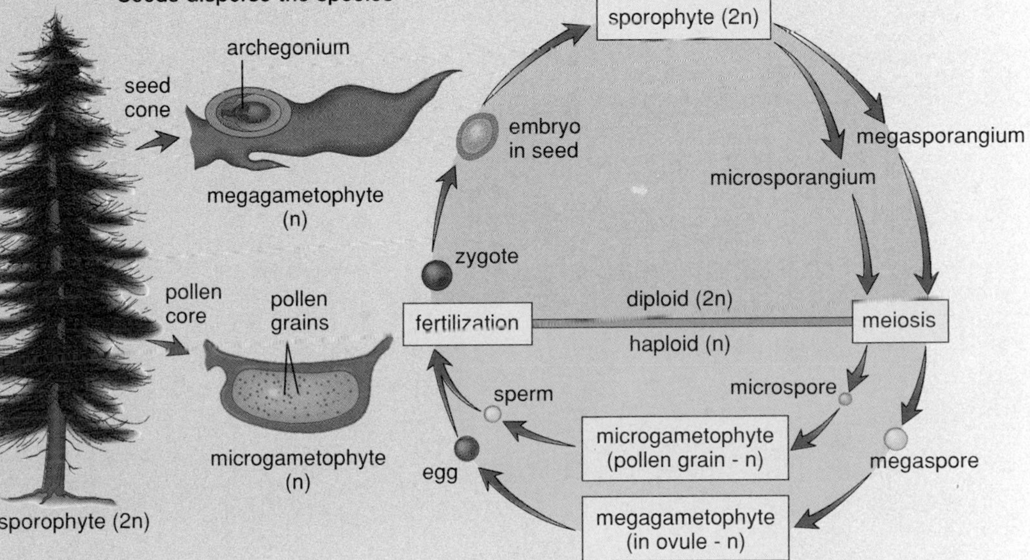

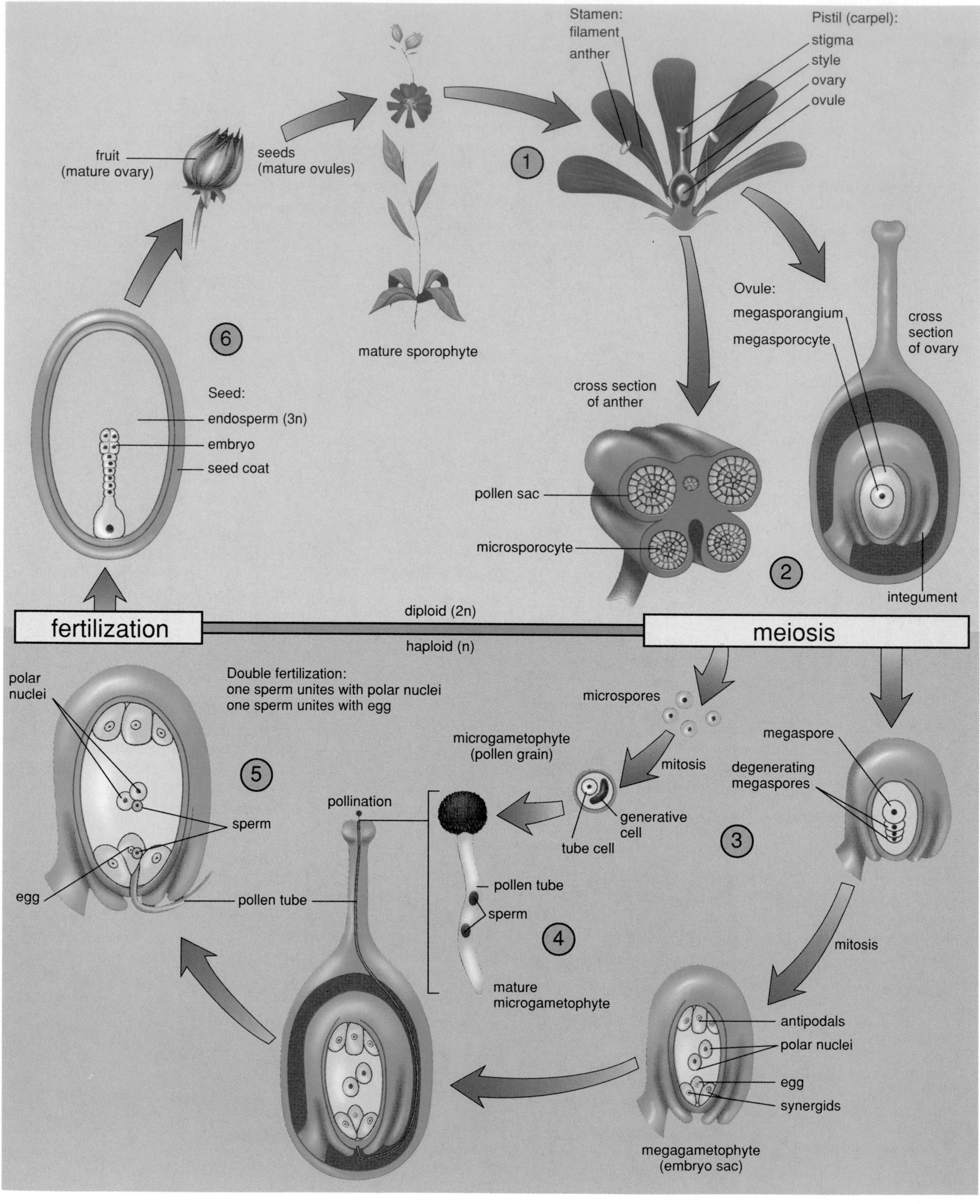

Stamen:
filament
anther
Pistil (carpel):
stigma
style
ovary
ovule
fruit (mature ovary)
seeds (mature ovules)
1
mature sporophyte
Ovule:
megasporangium
megasporocyte
cross section of ovary
cross section of anther
6
Seed:
endosperm (3n)
embryo
seed coat
pollen sac
microsporocyte
2
integument
fertilization
diploid (2n)
haploid (n)
meiosis
polar nuclei
Double fertilization:
one sperm unites with polar nuclei
one sperm unites with egg
microspores
megaspore
microgametophyte (pollen grain)
mitosis
degenerating megaspores
5
pollination
generative cell
tube cell
3
sperm
egg
pollen tube
pollen tube
sperm
4
mitosis
mature microgametophyte
antipodals
polar nuclei
egg
synergids
megagametophyte (embryo sac)

Figure 32.14 Flowering plant life cycle.

① The parts of the flower involved in reproduction are the stamens and the pistil. Reproduction has been divided into development of the megagametophyte, development of the microgametophyte, double fertilization, and the seed. *Development of the megagametophyte.* The ovary at the base of the pistil contains one or more ovules. ② Within an ovule, a megasporocyte (megaspore mother cell) undergoes meiosis to produce four haploid megaspores. ③ Three of these megaspores disintegrate, leaving one functional megaspore, which divides mitotically.④ The result is the megagametophyte, or embryo sac, which typically consists of eight haploid nuclei embedded in a mass of cytoplasm. The cytoplasm differentiates into cells, one of which is an egg and another of which is the endosperm cell with two nuclei (called the polar nuclei). *Development of the microgametophyte.* ① The anther at the top of the stamen has pollen sacs, which contain numerous microsporocytes (microspore mother cells). ② Each microsporocyte undergoes meiosis to produce four haploid cells called microspores. When the microspores separate, each one becomes a microgametophyte or pollen grain. ③ At this point, the young microgametophyte contains two nuclei: the generative cell and the tube cell. Pollination occurs when pollen is windblown or carried by insects, birds, or bats to the stigma of the same type of plant. ④ Only then does a pollen grain germinate and produce a long pollen tube. This pollen tube grows within the style until it reaches an ovule in the ovary. Before fertilization occurs, the generative nucleus divides, producing two sperm, which have no flagella. This germinated pollen grain with its pollen tube and two sperm is the mature microgametophyte. *Double fertilization.* ⑤ On reaching the ovule, the pollen tube discharges the sperm. One of the two sperm migrates to and fertilizes the egg, forming a zygote; the other unites with the two polar nuclei, producing a 3n (triploid) endosperm nucleus. The endosperm nucleus divides to form endosperm, food for the developing plant. This so-called double fertilization is unique to angiosperms. *The seed.* ⑥ The ovule now develops into the seed, which contains an embryo and food enclosed by a protective seed coat. The wall of the ovary and sometimes adjacent parts develop into a fruit that surrounds the seeds. Therefore, angiosperms are said to have *covered seeds.*

Flowers Aid Diversification

We have seen that flowers are involved in the production and development of spores, gametophytes, gametes, and embryos enclosed within seeds. The successful completion of these processes requires the effective dispersal of pollen and then seeds. The various ways in which pollen and seeds can be dispersed has resulted in many different types of flowers.

Although some flowers disperse their pollen by wind, many are adapted to attract specific pollinators such as bees, wasps, flies, butterflies, moths, and even bats, which carry only a particular pollen from flower to flower. For example, bee-pollinated flowers are usually blue or yellow and have ultraviolet shadings that lead the pollinator to the location of nectar at the base of the flower. The mouthparts of bees are fused into a long tube that is able to obtain nectar from this location.

Not only do flowers lend themselves to efficient cross-pollination, they also aid in the dispersal of seeds by producing fruits. There are fruits that utilize wind, gravity, water, and animals for dispersal. Because animals live in particular habitats and/or have particular migration patterns, they are apt to deliver the fruit-enclosed seeds to a suitable location for seed germination (when the embryo begins to grow again) and development of the plant.

Flower diversification may be associated with the numerous means by which flowers are pollinated and fruits are dispersed.

Uses of Angiosperms

Angiosperms provide the food that sustains most animals on land, including humans. Humans also use plant fibers to produce cloth, to provide firewood, and to supply construction materials. Plant oils are not only used in cooking, they are also, for example, used in perfumes and as medicines. Spices and drugs are derived from different parts of various angiosperm plants. The uses of angiosperms are discussed further on pages 567–68.

Adaptation of Angiosperms

Angiosperms have true roots, stems, and leaves. The vascular tissue is well developed, and the leaves are generally broad. Angiosperms are found in all sorts of habitats; some have even returned to the water.

The reproductive organs are in the flowers, which often attract animal pollinators. Food is the reward that pollinators receive in exchange for their services. Some animals eat the pollen while many feed on the nectar secreted by special glands called nectaries that are hidden deep within the flower. Following fertilization the ovules located in ovaries develop into fruits. Therefore, angiosperms produce covered seeds. Fruits often help with dispersal of seeds.

Angiosperms are the flowering plants. The flower both attracts animals (e.g., insects) that aid in pollination and produces seeds enclosed by fruit, which aid dispersal.

connecting concepts

The first plants to invade land didn't have much competition for space and resources. Through the evolutionary process, they adapted to a dry environment and the threat of water loss (desiccation). Today, the nonvascular plants are generally small, and they are usually found in moist habitats. However, mosses can store large amounts of water and even become dormant during dry spells The vascular plants, on the other hand, have specialized tissues to transport water and nutrients from one part of the plant to another. And their leaves are covered by a waxy cuticle interrupted by small pores, which open and close to control water loss.

Reproductive strategies in plants are also adapted to a land environment. Mosses and ferns produce flagellated sperm that require external water, but then windblown spores disperse the species. In seed plants, pollen grains protect the sperm until they fertilize an egg. The sporophyte even retains the egg-producing megagametophyte and the resulting zygote. The seed protects the sporophyte embryo from drying out until it germinates in a new location. Dispersal of plants by spores or seeds reduces competition for resources.

Humans use plants for various purposes. Massive amounts of biomass were submerged by swamps and covered by sediment during the Carboniferous period. Due to extreme pressure, this organic material became the fossil fuel coal, which, along with the other fuels, makes our way of life today possible. Also, we must not forget that plants produce food and oxygen, two resources that keep our biosphere functioning today.

Summary

32.1 What Is a Plant?

Plants are photosynthetic organisms adapted to a land existence. Among the various adaptations, all plants protect the developing embryo from desiccation. All plants have the alternation of generations life cycle. Some types of plants have a dominant gametophyte; others have a dominant sporophyte.

32.2 Nonvascular Plants Are Diverse

The nonvascular plants, which include the liverworts and mosses, are nonvascular plants and therefore lack true roots, stems, and leaves. In the moss life cycle, the gametophyte is dominant. Antheridia produce swimming sperm that need external water to reach the eggs in the archegonia. Following fertilization, the dependent sporophyte consists of a foot, a stalk, and a capsule within which windblown spores are produced by meiosis. Each spore germinates to produce a gametophyte.

32.3 Vascular Plants Include Seedless and Seed Plants

Vascular plants arose during the Silurian period of the Paleozoic era. The extinct rhyniophytes may be ancestral vascular plants. These plants had photosynthetic stems (no leaves or roots) with sporangia at their tips. Most likely, the life cycle was similar to today's ferns. The sporophyte, which is diploid and has vascular tissue, is the dominant generation in ferns. The separate gametophyte produces flagellated sperm. The sporophyte of seed plants produces heterospores which develop into heterogametes. Every aspect of the life cycle is adapted to a dry environment.

32.4 Ferns and Allies Are the Seedless Vascular Plants

The seedless vascular plants include whisk ferns, club mosses, horsetails, and ferns. Lycopods, horsetails, and ferns were also trees during the Carboniferous period, although lycopods and horsetails are limited in diversity and rather small today. Seedless vascular plants have a life cycle like that of the ferns.

In ferns, the separate and water-dependent gametophyte (the heart-shaped prothallus) produces flagellated sperm in antheridia and eggs in archegonia. Following fertilization, the zygote develops into the sporophyte, which has large fronds. On the underside of the fronds are sori (sing., **sorus**), each containing several sporangia. Here meiosis produces windblown spores, each of which develops into a prothallus.

The Mesozoic era saw many geological changes as Pangaea formed and then broke apart. A mass extinction occurred that paved the way for the diversification of the seed plants. Those seed plants that are trees have especially well-developed roots and stems due to secondary growth of vascular tissue. Seed plants produce heterospores, microgametophytes, and megagametophytes. The pollen grain replaces external flagellated sperm, and the megagametophyte is retained within the ovule that develops into the seed.

32.5 Gymnosperms Have Naked Seeds

There are four divisions of gymnosperms (seed plants that bear naked seeds): the familiar conifers and the little-known cycads, ginkgo, and gnetophytes. In conifers, pollen (male) and seed (female) cones are produced by the sporophyte plant. On the underside of a pollen cone scale, there are two microsporangia that produce microspores; each becomes a microgametophyte, or pollen grain. On the upper surface of a seed cone scale, there are two ovules, where meiosis produces one megaspore that develops into the megagametophyte. After windblown pollination, the pollen grain develops a tube through which sperm reach the egg. After fertilization, the ovule matures to be the seed.

32.6 Angiosperms Have Covered Seeds

Angiosperms (seed plants that bear seeds protected by a fruit) are more diverse than the other types of plants. Their success may be associated with climatic changes in the Cenozoic era.

In a flower, the microsporangia develop within the anther portion of a stamen, and the megasporangia develop within ovules located in the ovary of the pistil. Pollination brings the mature microgametophyte (pollen grain) to the pistil, and the pollen tube brings the sperm to the ovule within the ovary. Angiosperms exhibit double fertilization: one sperm fertilizes the egg, and the other unites with the polar nuclei to form the endosperm, which is food for the embryo. The ovule develops into the seed, and the ovary becomes the fruit.

Angiosperms have complex vascular tissue and are found in various habitats. Their reproductive organs are found in flowers. Animal pollination increases the chance of appropriate fertilization, and fruit production often assists the dispersal of seeds.

Reviewing the Chapter

1. What are the characteristics that define plants? 554–55
2. What are the general characteristics of nonvascular plants, and what are the three main types? 555
3. Draw a diagram to describe the life cycle of the moss, pointing out significant features. 557
4. What are the human uses of nonvascular plants, and what are their adaptations? 558
5. What are the general characteristics of vascular plants, and how do features of the seedless plant life cycle differ from that of the seed plant life cycle? 558, 569
6. What is the significance of rhyniophytes in the history of plants? What are the living seedless vascular plants, and describe the period of time in the earth's history when they were larger and more abundant than today. 560–61
7. Draw a diagram to describe the life cycle of the fern, pointing out significant features. What are the human uses of ferns, and what are their adaptations? 562–63
8. List and describe the four divisions of gymnosperms. 564
9. Use a diagram of the pine life cycle to point out significant features, including those that distinguish a seed plant's life cycle from that of a seedless vascular plant. 565
10. What are the human uses of gymnosperms, and what are their adaptations? 567
11. What role may have been played by beetles and insects in the evolution of flowering plants? 567
12. What are the parts of a flower? Use a diagram to explain and point out significant features of the flowering plant life cycle. 568–70
13. Offer an explanation as to why flowering plants are the dominant plants today. What are the human uses of angiosperms, and what are their adaptations? 571

Testing Yourself

Choose the best answer for each question.

1. Which of these are characteristics of plants?
   a. multicellular with specialized tissues and organs
   b. photosynthetic and contain chlorophylls *a* and *b*
   c. protect the developing embryo from desiccation
   d. All of these are correct.
2. In the moss life cycle, the sporophyte
   a. consists of leafy green shoots.
   b. is the heart-shaped prothallus.
   c. consists of a foot, a stalk, and a capsule.
   d. is the dominant generation.
3. The rhyniophytes
   a. are a flourishing group of plants today.
   b. had large leaves like today's ferns.
   c. had sporangia at the tips of their branches.
   d. All of these are correct.
4. You are apt to find ferns in a moist location because they have
   a. a water-dependent sporophyte generation.
   b. flagellated spores.
   c. swimming sperm.
   d. All of these are correct.
5. Which of these is mismatched?
   a. pollen grain—microgametophyte
   b. ovule—megagametophyte
   c. seed—immature sporophyte
   d. pollen tube—spores
6. In the life cycle of the pine tree, the ovules are found on the
   a. needlelike leaves.
   b. seed cones.
   c. pollen cones.
   d. All of these are correct.
7. Which of these is mismatched?
   a. anther—produces microsporangia
   b. pistil—produces pollen
   c. ovule—becomes seed
   d. ovary—becomes fruit
8. Which of these plants contributed the most to our present day supply of coal?
   a. nonvascular plants
   b. seedless vascular plants
   c. conifers
   d. angiosperms
9. Which of these is found in seed plants?
   a. complex vascular tissue
   b. pollen grains replace swimming sperm
   c. retention of megagametophyte within the ovule
   d. All of these are correct.
10. Label the following diagram of alternation of generations.

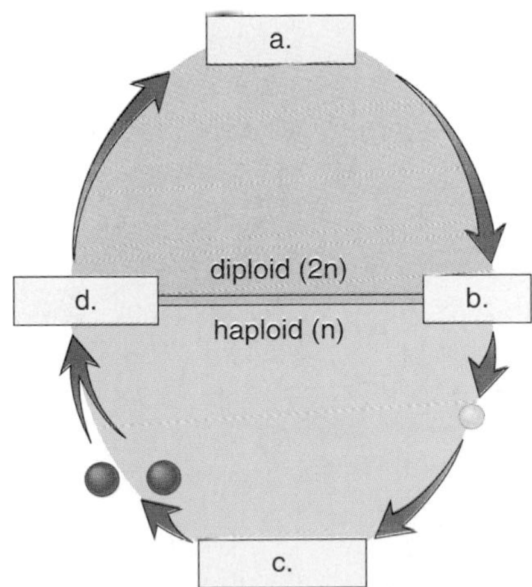

Applying the Concepts

1. *Plants are one of the major groups of organisms alive today.*
   Should all organisms that photosynthesize be placed in the same kingdom? Why or why not?
2. *Plants are adapted to living on land.*
   Vascular plants have roots, stems, and leaves. How does each organ contribute to a plant's adaptation to life on land?
3. *Plants have an alternation of generations life cycle.*
   List several advantages of having a dominant sporophyte, and discuss each one.

Understanding the Terms

angiosperm 567
anther 568
antheridium 556
archegonium 556
club moss 561
coal 559
cone 564
conifer 564
cuticle 554
cycad 566
dicotyledon 568
fern 562
fruit 568
gametangia 555
gametophyte 555
ginkgo 566
gnetophyte 567
gymnosperm 564
herbaceous plant 567
hornwort 555
horsetail 561
liverwort 556
megaspore 564
microspore 564
monocotyledon 568
moss 556
nonvascular plant 555
ovary 568
ovule 564
peat 558
phloem 558
pistil 568
kingdom Plantae 554
pollen grain 564
pollination 564
pollinator 568
psilotophyte 560
rhizoid 556
rhizome 560
root 558
seed 564
sporangium 556
sporophyte 555
stamen 568
stem 558
stigma 568
stomate 558
strobilus 561
style 568
vascular plant 558
whisk fern 560
woody plant 567
xylem 558

Match the terms to these definitions:

a. _________ Diploid generation of the alternation of generations life cycle of a plant; meiosis produces haploid spores that develop into the haploid generation.
b. _________ Flowering plant group; members have one embryonic leaf, parallel-veined leaves, scattered vascular bundles, and other characteristics.
c. _________ Flowering plant group; members have two embryonic leaves, net-veined leaves, cylindrical arrangement of vascular bundles, and other characteristics.
d. _________ Haploid generation of the alternation of generations life cycle of a plant; it produces gametes that unite to form a diploid zygote.
e. _________ Microgametophyte in seed plants.
f. _________ One of the four groups of gymnosperm plants; cone-bearing trees that include pine, cedar, and spruce.
g. _________ Plant that lacks persistent woody tissue.
h. _________ Rootlike hair that anchors a plant and absorbs minerals and water from the soil.
i. _________ Vascular tissue that conducts organic solutes in plants; it contains sieve-tube cells and companion cells.
j. _________ Vascular tissue that transports water and mineral solutes upward through the plant body; it contains vessel elements and tracheids.

Using Technology

Your study of the plant kingdom is supported by these available technologies:

Exploring the Internet
The Mader Home Page provides resources for and help with studying this chapter.

http://www.mhhe.com/sciencemath/biology/mader/
(Click on Biology.)

Life Science Animations Video
Video #5: Plant Biology / Evolution / Ecology
Journey into a Leaf (#46)
How Water Moves Through a Plant (#47)

Animals: Introduction to Invertebrates

CHAPTER 33

Chapter Concepts

Yellow coral polyps, *Parazoanthus garcillis*

Animals are incredibly diverse in structure. They range from worms that can only be seen with a microscope to blue whales, which weigh in at 150 tons. But typically all animals, including humans, are multicellular heterotrophs that ingest their food. Most have tissues and locomote by means of muscle fibers. The adult is always diploid, and during sexual reproduction, the embryo undergoes specific developmental stages. When we think of animals, we tend to imagine birds, dogs, fishes, squirrels, and other vertebrate species. However, most animal species are those that lack a backbone and are commonly known as the invertebrates.

Animals are believed to have evolved from aquatic protozoan-type ancestors some 600 million years ago or earlier. There is no fossil history of early animal evolution because the organisms involved had soft bodies. But biologists are now in the process of using molecular data to trace the evolutionary history of animals. Presently, classification is based on such criteria as body structure and form, type of symmetry, and number of tissue layers. In this chapter, we will examine those invertebrates that lack a true body cavity. A survey of the rest of the animal kingdom follows in the next two chapters.

33.1 How Animals Evolved and Are Classified

Whereas plants are multicellular photosynthetic organisms, animals are multicellular eukaryotes that are heterotrophic by ingesting food (Fig. 33.1). Unlike the fungi that rely on external digestion, animals usually digest their food in a central cavity. Animals produce heterogametes (eggs and sperm), and they follow the diplontic life cycle in which the adult is always diploid. In this cycle, meiosis produces haploid gametes, which join to form a zygote that develops into an adult. Table 33.1 reviews the characteristics of animals.

All animals are believed to have evolved from a protistan ancestor, most likely a protozoan (Fig. 33.2). There are approximately 34 animal phyla, but we will consider in depth only the nine phyla illustrated in the tree. These are the ones recognized as the major animal phyla. All nine phyla contain **invertebrates** [L. *in,* without, and *vertebra,* bones of backbone], which are animals without backbones. The phylum Chordata also contains the vertebrates, which are animals with backbones. Many invertebrates live in the sea, where early animal evolution occurred. All major animal phyla are represented by Cambrian fossils, but it has been very difficult to trace the complete evolutionary history because the fossil record is so much better for hard-shelled animals. In large part, the possible evolutionary relationship of living invertebrates has been worked out by using the anatomical criteria noted in the tree.

One criterion used to classify animals is type of symmetry. **Asymmetry** means that the animal has no particular symmetry. **Radial symmetry** means that the animal is organized circularly and, just as with a wheel, two identical halves are obtained no matter how the animal is sliced longitudinally. **Bilateral symmetry** means that the animal has definite right and left halves; only one longitudinal cut down the center of the animal will produce two equal halves. Radially symmetrical animals tend to be attached to a substrate; that is, they are **sessile.** This type of symmetry is useful to these animals since it allows them to reach out in all directions from one center. Bilaterally symmetrical animals tend to be active and to move forward at an anterior end.

One of the main events during the development of animals is the establishment of *germ layers* from which all other structures are derived. Although a total of three germ layers is seen in most animal embryos, some animals only have two germ layers: ectoderm and endoderm. Such animals have the *tissue level of organization.* Animals with three germ layers—ectoderm, mesoderm, and endoderm—have an *organ level of organization.*

The four animal phyla in the lower half of the phylogenetic tree in Figure 33.2 lack a true **coelom** [Gk. *koiloma,* cavity], an internal body cavity completely lined by mesoderm, where internal organs are found. Some are acoelomates—they have no coelom at all. One phylum contains pseudocoelomates, animals that have a cavity, though it is incompletely lined with mesoderm. There is a layer of

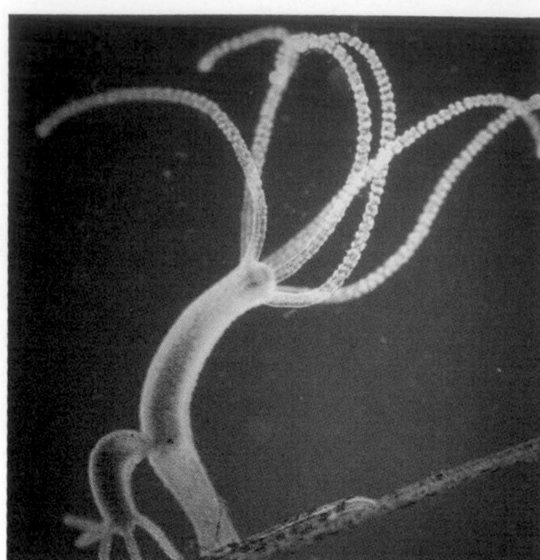

Hydra, *Hydra*

American toad, *Bufo*

Figure 33.1 Animal diversity.
Both a hydra and a toad are multicellular heterotrophic organisms that must take in preformed food.

Table 33.1

Characteristics of Animals

1. Are heterotrophic and usually acquire food by ingestion followed by digestion.
2. Typically have the power of motion or locomotion by means of muscle fibers.
3. Are multicellular, and most have cells specialized to form tissues and organs.
4. Have a life cycle in which the adult is always diploid.
5. Usually practice sexual reproduction and produce an embryo that undergoes specific stages of development.

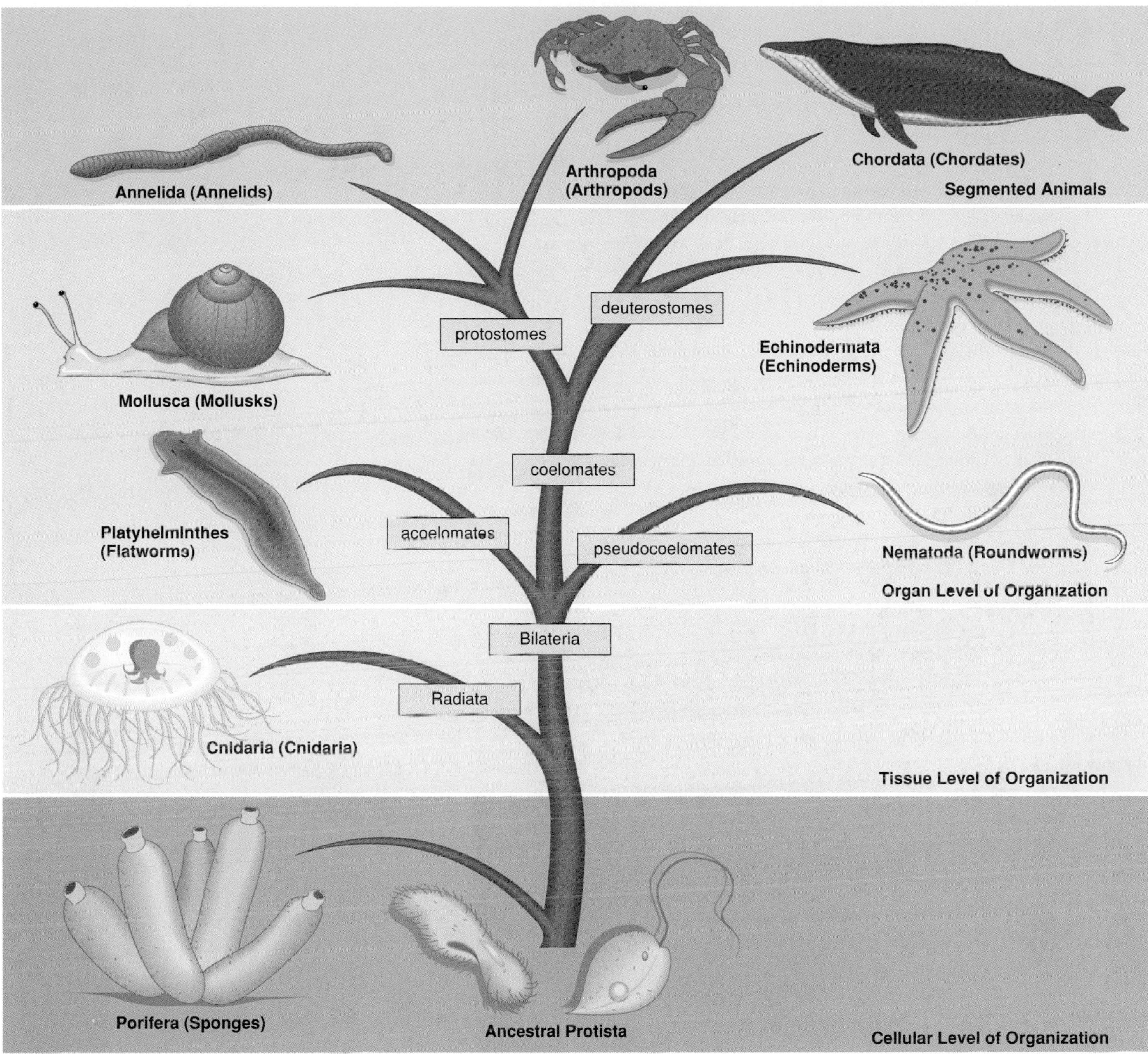

Figure 33.2 Phylogenetic tree of the animal kingdom.
All animals are believed to be descended from protists; however, the sponges may have evolved separately from the rest of the animals.

mesoderm beneath the body wall but not around the gut. All the animal phyla in the two top branches of the tree are true coelomates—they have a coelom that is completely lined with mesoderm. Coelomates are either protostomes or deuterostomes. When the blastopore (the site of invagination of endoderm during development) is associated with the mouth, the animal is a protostome. When the blastopore is associated with the anus and a second opening becomes the mouth, the animal is a deuterostome (see Fig. 34.1).

Coelomates are also divided into those that are nonsegmented and those that are segmented. Segmented animals have repeating units. *Segmentation* leads to specialization of parts because the various segments can become differentiated for specific purposes.

Classification of animals is based on type of symmetry, number of tissue layers, type of coelom, and presence of segmentation.

CLASSIFICATION

Kingdom Animalia

Multicellular organisms with well-developed tissues; usually motile; heterotrophic by ingestion, generally in a digestive cavity; diplontic life cycle. Protostomes include phyla Mollusca, Annelida, and Arthropoda. Deuterostomes include phyla Echinodermata, Hemichordata, and Chordata.

Invertebrates*

- Phylum Porifera: sponges
- Phylum Cnidaria: jellyfishes, sea anemones, corals
- Phylum Ctenophora: comb jellies, sea walnuts
- Phylum Platyhelminthes: flatworms, e.g., planaria, flukes, tapeworms
- Phylum Nemertea: ribbon worms
- Phylum Nematoda: roundworms
- Phylum Rotifera: rotifers
- Phylum Mollusca: chitons, snails, slugs, clams, mussels, squids, octopuses
- Phylum Annelida: segmented worms, e.g., clam worms, earthworms, leeches
- Phylum Onychophora: walking worms
- Phylum Arthropoda: spiders, scorpions, horseshoe crabs, lobsters, crayfish, shrimps, crabs, millipedes, centipedes, insects
- Phylum Echinodermata: sea lilies, sea stars, brittle stars, sea urchins, sand dollars, sea cucumbers, sea daisies
- Phylum Hemichordata: acorn worms
- Phylum Chordata
  - Subphylum Urochordata: tunicates
  - Subphylum Cephalochordata: lancelets

Vertebrates*

- Subphylum Vertebrata
  - Superclass Agnatha: jawless fishes, e.g., lampreys, hagfishes
  - Superclass Gnathostomata: jawed fishes, all tetrapods
    - Class Chondrichthyes: cartilaginous fishes, e.g., sharks, skates, rays
    - Class Osteichthyes: bony fishes, e.g., herring, salmon, cod, eel, flounder
    - Class Amphibia: frogs, toads, salamanders
    - Class Reptilia: snakes, lizards, turtles
    - Class Aves: birds, e.g., sparrows, penguins, ostriches
    - Class Mammalia: mammals, e.g., cats, dogs, horses, rats, humans

* Not in the classification of organisms, but added here for clarity.

33.2 Multicellularity Evolves

All animals are multicellular, but sponges are the only animals to have the *cellular level of organization*. Sponges, which have no symmetry and no tissues, are believed to be out of the mainstream of animal evolution. Most likely they evolved separately from protozoan ancestors and represent a dead-end branch of the evolutionary tree.

Sponges Have Pores

Sponges (phylum Porifera, 5,000 species) are aquatic, largely marine animals, that vary greatly in size, shape, and color. Their saclike bodies are perforated by many pores; the phylum name, Porifera, means pore bearing.

The cellular organization of sponges is demonstrated by experimentation. After a sponge is broken into separate cells, the cells will exist individually until they spontaneously reorganize into a sponge once again. What types of cells are found in a sponge? The outer layer of the wall contains flattened epidermal cells, some of which have contractile fibers; the middle layer is a semifluid matrix with wandering amoeboid cells; and the inner layer is composed of flagellated cells called *collar cells* (or choanocytes) that look like protozoa (Fig. 33.3). There are no nerve cells or other means of coordination between the cells. To some, a sponge can be thought of as a colony of protozoa.

The beating of the flagella of collar cells produces water currents that flow through the pores into the central cav-

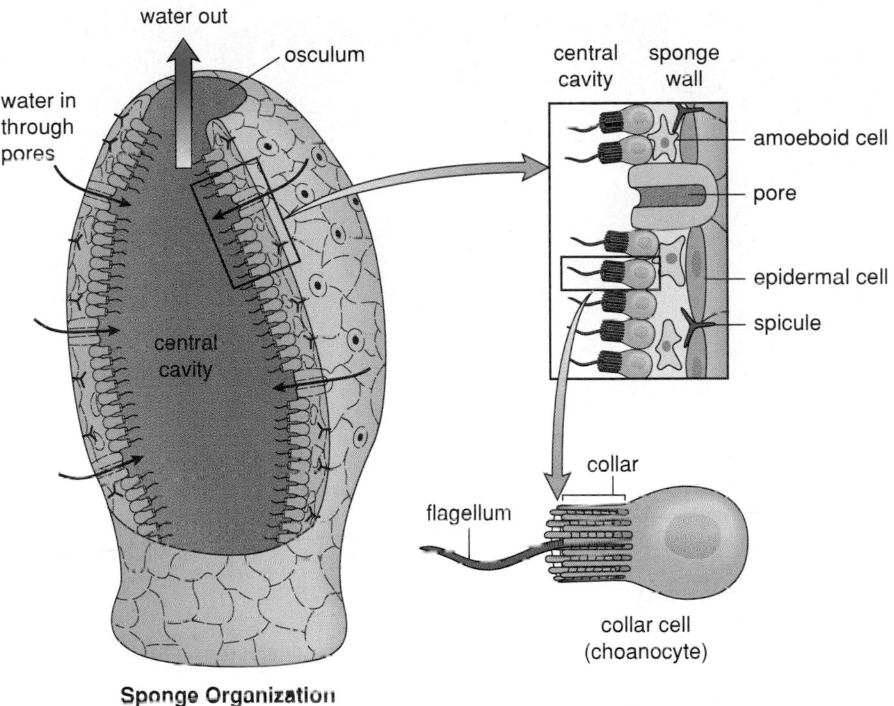

Figure 33.3 Generalized sponge anatomy.
The wall contains two layers of cells: the outer epidermal cells and the inner collar cells. The collar cells (enlarged) have flagella that beat, moving the water through pores as indicated by the arrows. Food particles in the water are trapped by the collar cells and digested within their food vacuoles. Amoeboid cells transport nutrients from cell to cell; spicules compose an internal skeleton of some sponges.

ity and out through the osculum, the upper opening of the body. Although it may seem that sponges can't do much, even a simple one only 10 cm tall is estimated to filter as much as 100 liters of water each day. It takes this much water to supply the needs of the organism. Simple sponges have pores leading directly from the outside into the central cavity. Larger and more complex sponges have canals leading from external to internal pores.

A sponge is a **sessile filter feeder,** an organism that stays in one place and filters its food from the water. Oxygen is also supplied to the sponge, and waste products are taken away by the constant stream of water through the organism. Microscopic food particles brought by the water are engulfed by the collar cells and digested by them in food vacuoles or are passed to the amoeboid cells for digestion. The amoeboid cells also act as a circulatory device to transport nutrients from cell to cell, and they produce the sex cells (the egg and the sperm) and spicules.

Sponges can reproduce asexually by fragmentation or by budding. During budding, a small protuberance appears and gradually increases in size until a complete organism forms. Budding produces colonies of sponges that can become quite large. During sexual reproduction, eggs and sperm are released into the central cavity. Fertilization results in a zygote that develops into a ciliated larva that may swim to a new location. Such a larva assures dispersal of the species for the sessile animal. Like all less specialized organisms, sponges are capable of regeneration, or growth of a whole from a small part. Thus, if a sponge is removed, chopped up, and returned to the water, each piece may grow into a complete sponge.

Sponges are classified on the basis of the type of skeleton they have. Some sponges also have an internal skeleton composed of **spicules** [L. *spicula,* dim. of *spika,* spear], little needle-shaped structures with one to six rays. Chalk sponges have spicules made of calcium carbonate; glass sponges have spicules that contain silica. Most sponges have fibers of spongin, a modified collagen. But some sponges contain only spongin fibers; a bath sponge is the dried spongin skeleton from which all living tissue has been removed. Today, however, commercial "sponges" are usually synthetic.

Sponges have a cellular level of organization and most likely evolved independently from protozoa. They are the only animals in which digestion occurs within cells.

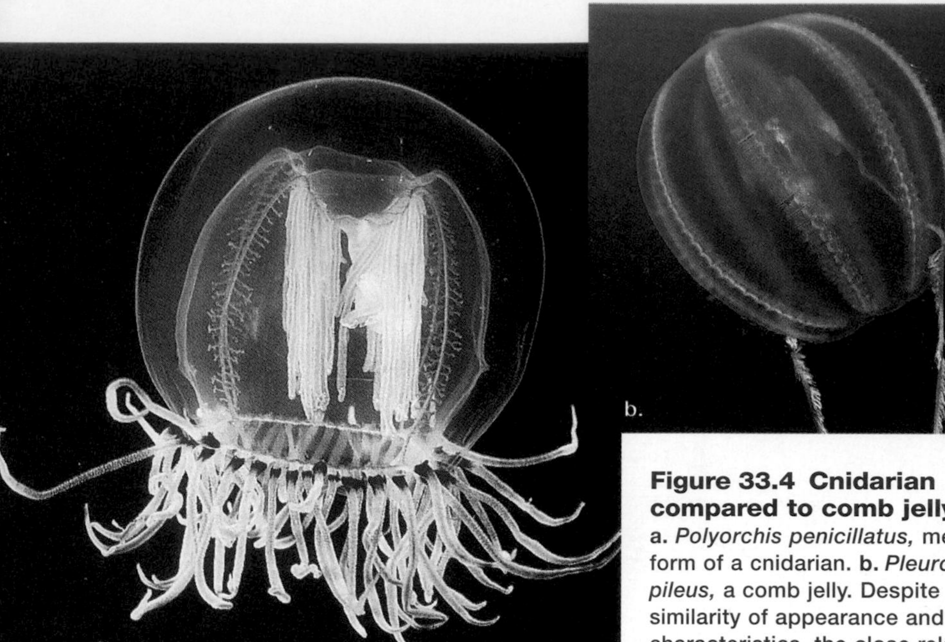

Figure 33.4 Cnidarian compared to comb jelly.
a. *Polyorchis penicillatus*, medusan form of a cnidarian. b. *Pleurobrachia pileus*, a comb jelly. Despite similarity of appearance and shared characteristics, the close relationship of these animals is now in dispute.

33.3 Tissue Layers Evolve

As mentioned, during animal development there are a total of three possible germ layers: ectoderm, endoderm, and mesoderm. Animals in two phyla—comb jellies in **phylum Ctenophora** and cnidaria in phylum Cnidaria—develop only ectoderm and endoderm. Therefore, they are said to be diploblasts (Fig. 33.4). Animals in these phyla are radially *symmetrical*, meaning that any longitudinal cut, such as those shown, produces two identical halves. If an animal is *bilaterally symmetrical*, only the longitudinal cut shown yields two roughly identical halves.

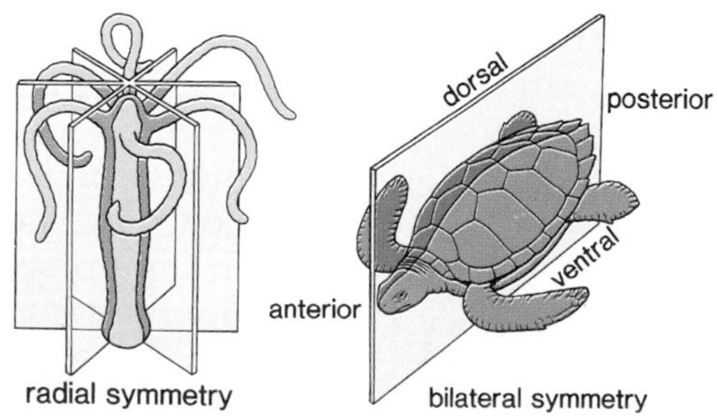

Comb jellies are small (a few cm), transparent, and often luminescent animals that take their name from their eight plates of fused cilia that resemble long combs. Most of their body is a jellylike packing material called **mesoglea** [Gk. *mesos*, middle, and *gloios*, glue]. They are the largest animals to be propelled by the beating of cilia. They capture prey either by means of long tentacles, covered with sticky filaments, or by using the entire body, which is covered with a sticky mucus.

Cnidaria Have Radial Symmetry

Cnidaria (phylum Cnidaria, 9,000 species) are tubular or bell-shaped animals that reside mainly in shallow coastal waters, except for the oceanic jellyfishes. Unique to cnidaria are specialized stinging cells, called cnidocytes, which give the phylum its name. Each cnidocyte has a fluid-filled capsule called a **nematocyst** [Gk. *nema*, thread, and *kystis*, bladder], which contains a long, spirally coiled hollow thread. When the trigger of the cnidocyte is touched, the nematocyst is discharged. Some threads merely trap a prey or predator; others have spines that penetrate and inject paralyzing toxins.

The body of a cnidarian is a two-layered sac. The outer tissue layer is a protective epidermis derived from ectoderm. The inner tissue layer, which is derived from endoderm, secretes digestive juices into the internal cavity, called the **gastrovascular cavity** [Gk. *gastros*, stomach, and L. *vasculum*, dim. of *vas*, vessel] because it serves for digestion of food and circulation of nutrients. The two tissue layers are separated by mesoglea. There are muscle fibers at the base of the epidermal and gastrodermal cells. Nerve cells located below the epidermis near the mesoglea interconnect and form a **nerve net** throughout the body. In contrast to highly organized nervous systems, the nerve net allows transmission of impulses in several directions at once—multiple firings of nematocysts in parts of the body not directly stimulated have been observed. Having both muscle fibers and nerve fibers, these animals are capable of directional movement; the body can contract or extend, and the tentacles that ring the mouth can reach out and grasp prey.

Two basic body forms are seen among cnidaria. The mouth of a *polyp* is directed upward, while the mouth of a jellyfish or *medusa* is directed downward. A medusa has more mesoglea than a polyp, and the tentacles are concentrated on the margin of the bell. At one time, both body forms may have been a part of the life cycle of all cnidaria. When both are present, the sessile polyp stage produces medusae and the motile medusan stage produces egg and sperm, thereby dispersing the species. In some cnidaria, one stage is dominant and the other is reduced; in other species one form is absent altogether.

Cnidaria have a tissue level of organization and are radially symmetrical. They have a sac body plan and exist as polyps or medusae.

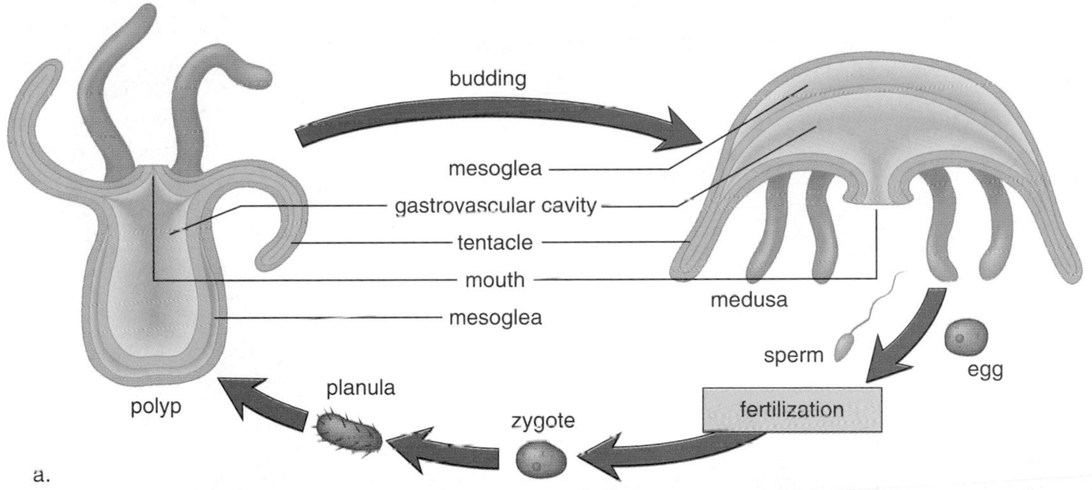

a.

b. Sea anemone, *Apitasia*

c. Cup coral, *Tubastrea*

d. Portuguese man-of-war, *Physalia*

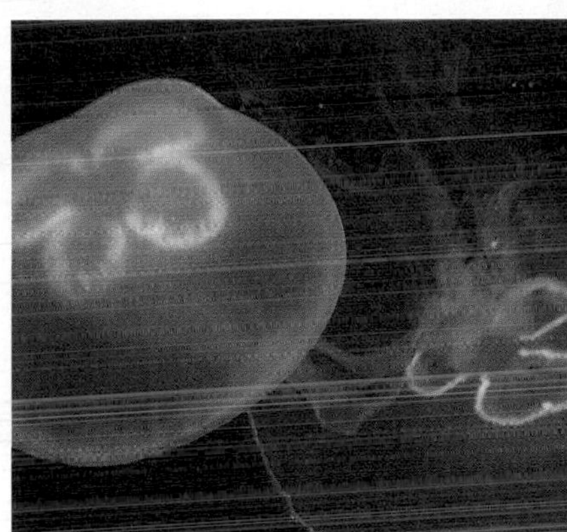

e. Jellyfish, *Aurelia*

Figure 33.5 Cnidarian diversity.
a. The life cycle of a cnidarian. In some cnidaria, there is both a polyp stage and a medusa stage; in others, one stage is dominant, or absent altogether. **b.** The anemone, which is sometimes called the flower of the sea, is a solitary polyp. **c.** Corals are colonial polyps residing in a calcium carbonate or proteinaceous skeleton. **d.** Portuguese man-of-war is a colony of modified polyps and medusae of variously modified polyps. **e.** True jellyfish undergo the complete life cycle; this is the medusa stage.

Cnidarian Diversity

Among the cnidarian classes, class Anthozoa contains the sea anemones and corals (Fig. 33.5*b*, *c*). Sea anemones are solitary polyps that are large enough to be seen with the naked eye. Most sea anemones are 5–100 mm in height and 5–200 mm in diameter; some are much larger. They may be brightly colored and look like beautiful flowers.

The oral disk that bears a mouth of a sea anemone is surrounded by a large number of hollow tentacles. They feed on various invertebrates, and large species can capture fish. Sea anemones live attached to a submerged rock, timber, or shell. A number of species form mutualistic relationships (in which both species benefit) with hermit crabs and live attached to the shell of the crab. The anemone provides protection and camouflage for the crab, and the crab provides locomotion and perhaps some food for the sea anemone.

Corals, which are also anthozoa, resemble sea anemones in appearance. Some corals are solitary, but most are colonial with flat, rounded, or upright and branching colonies. Corals are usually found in shallow waters; this is particularly true of reef-building stony corals whose walls contain mutualistic algae. When threatened, stony corals can retract inside their calcium-carbonate skeletons. The slow accumulation of coral skeletal remains can result in massive structures such as the Great Barrier Reef along the eastern coast of Australia. Several thousand different types of animals interact in coral reefs, areas of biological abundance in tropical seas.

In class Hydrozoa, the polypoid stage is dominant. One of the most unusual hydrozoa is the Portuguese man-of-war, *Physalia*, which looks as if it might be an odd-shaped medusa but actually is a colony of polyps and medusae. (Fig. 33.5*d*). The original polyp becomes a gas-filled float that provides buoyancy—it keeps the colony afloat. Other polyps, which bud from this one, are specialized for feeding or for reproduction. A long single tentacle armed with numerous nematocysts arises from the base of each feeding polyp. Swimmers who accidently come upon a Portuguese man-of-war can receive painful, even serious, injuries from these stinging tentacles.

Class Scyphozoa includes the true jellyfishes, such as *Aurelia* (Fig. 33.5*e*). In jellyfishes, the medusa is the primary stage, and the polyp remains quite small and insignificant. Jellyfishes are a part of the zooplankton of the ocean and as such serve as food for larger animals.

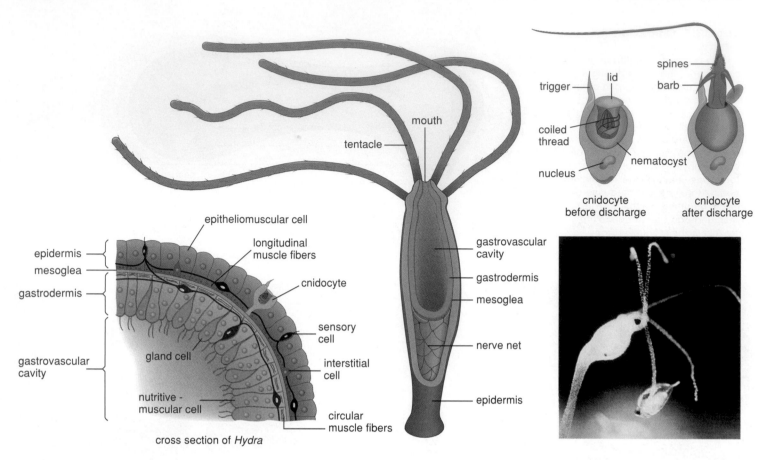

Figure 33.6 Anatomy of *Hydra.*

Hydra and *Obelia* in Depth

Hydra and *Obelia* in class Hydrozoa are two hydrozoa of particular interest. *Hydra* is a solitary polyp and *Obelia* is a colonial form, which has both a polyp and medusa stage in its life cycle.

Hydra *Is a Solitary Polyp*

Hydras [Gk. *hydra,* a many-headed serpent] are freshwater cnidaria. Hydras are likely to be found attached to underwater plants or rocks in most lakes and ponds. The body is a small tubular polyp about one-quarter inch in length. The only opening (the mouth) is in a raised area that is surrounded by four to six tentacles that contain a large number of nematocysts. The central cavity of the animal is the gastrovascular cavity.

Although a hydra usually remains in one location, it can glide along on its base or even move rapidly by means of somersaulting.

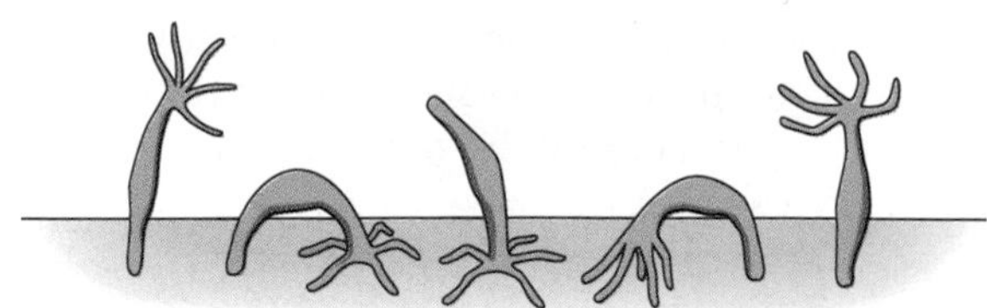

Hydras can respond to stimuli, and if a tentacle is touched with a needle, all the tentacles and the body contract, only to extend later. It is apparent, then, that like other animals capable of locomotion, hydras have both muscular and nerve fibers.

Figure 33.6 shows the microscopic anatomy of *Hydra.* The cells of the epidermis are termed *epitheliomuscular cells* because they contain muscle fibers. Also present in the epidermis are cnidocytes and sensory cells. The latter have long extensions that make contact with the nerve cells within the nerve net. The interstitial, or embryonic, cells also seen in this layer are capable of becoming other types of cells. For example, they can produce an ovary and/or a testis and probably also account for the animal's great regenerative powers. Like the sponges, cnidaria can grow an entire organism from a small piece.

Gland cells of the gastrodermis secrete digestive juices that pour into the gastrovascular cavity. Hydras feed on small prey that are captured when they trigger the release of nematocysts. The tentacles capture and stuff the prey into the gastrovascular cavity, which distends to accommodate the food. The enzymes released by the gland cells begin the digestive process, which is completed within food vacuoles of *nutritive-muscular cells,* the main type of gastrodermal cell. Nutrient molecules are passed by diffusion to the rest of the cells of the body. Nutritive-muscular

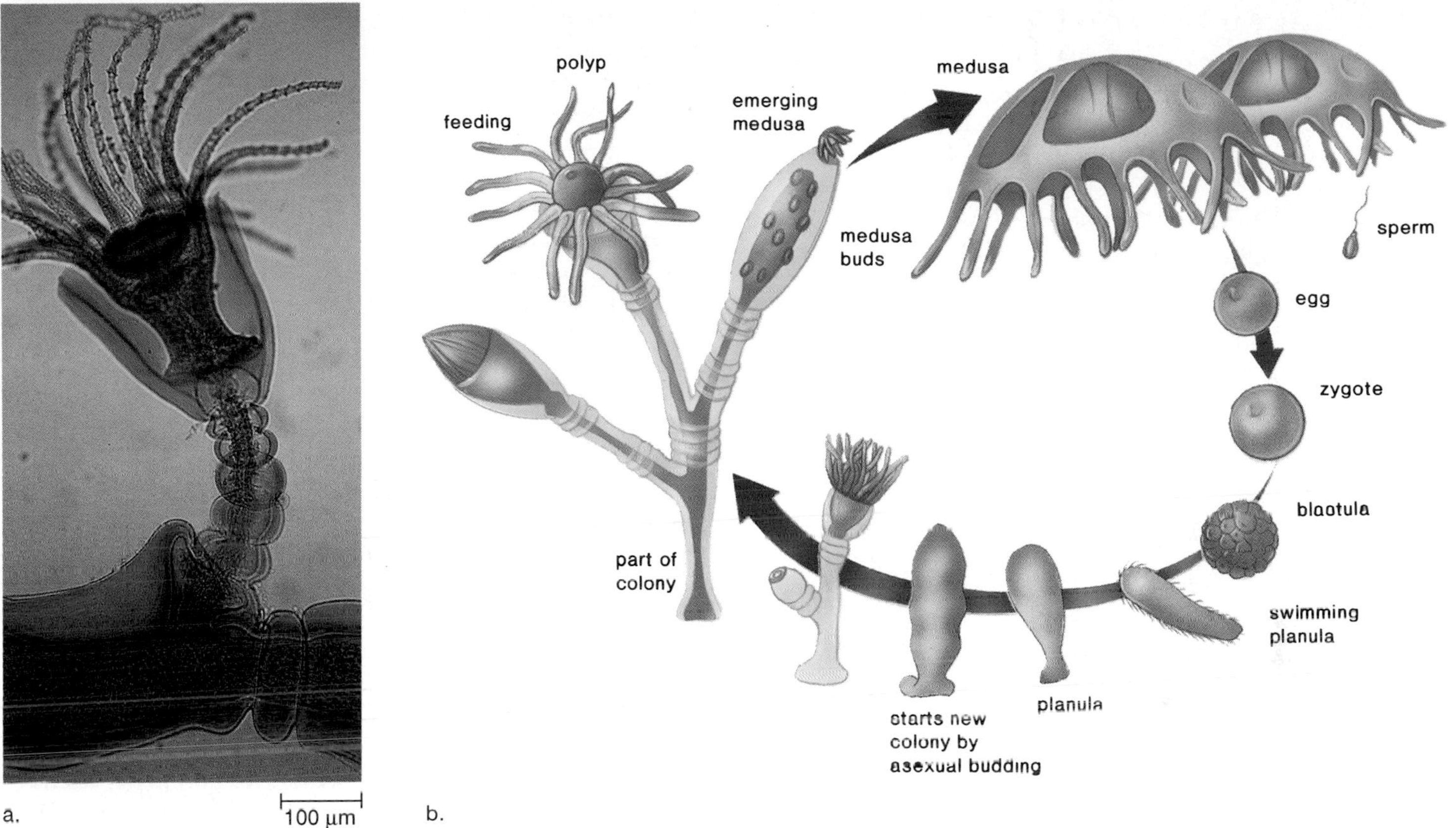

Figure 33.7 *Obelia* structure and life cycle.
Obelia, a colony of feeding polyps and reproductive polyps, undergoes an alternation of generations life cycle. a. Micrograph of feeding polyps. b. Life cycle.

cells also contain contractile fibers that run circularly about the body; when these contract, the animal lengthens.

Hydras can reproduce both asexually and sexually. They reproduce asexually by forming buds, small outgrowths that develop into a complete animal and then detach. When hydras reproduce sexually, sperm from a testis swim to an egg within an ovary. Fertilization and early development occur within the ovary, after which the embryo is encased within a hard, protective shell that allows it to survive until conditions are optimum for it to emerge and develop into a new polyp.

Obelia *Is a Colonial Form*

Obelia (Fig. 33.7) is a colony of polyps that is enclosed by a hard, chitinous covering. There are two types of polyps. The *feeding polyps* extend beyond the covering and can withdraw into it for protection. They have nematocyst-bearing tentacles that can capture and bring prey—such as tiny crustacea, worms, and larvae—into the gastrovascular cavity. The polyps are connected, and the partially digested food is distributed to the rest of the colony.

The colony increases in size asexually by the budding of new polyps. Sexual reproduction involves the production of medusae, which bud from the second type of polyp called *reproductive polyps*. Hydroid *medusae* tend to be smaller than those of the true jellyfishes. The tentacles attached to the bell margin have nematocysts, and they bring food into a gastrovascular cavity that extends even into the tentacles. The nerve net is concentrated into two nerve rings; the bell margin is supplied with sensory cells, such as statocysts, organs of equilibrium, and ocelli, light-sensitive organs.

While some species produce free-swimming medusae, in other species the medusae remain attached to the colony and shed only their gametes. The resulting zygote develops into a ciliated larva called a planula larva. The planula larva settles down and develops into a polyp colony.

Obelia is an example of a colonial hydroid consisting of feeding and reproductive polyps. Aside from this polyp stage, there is also a medusa stage in its life cycle.

The relationship of radially symmetrical cnidaria to the rest of the animal groups, which at some time during their life history are bilaterally symmetrical, has not been easy to determine. However, there are those who believe that a planuloid-type organism could have given rise to both the cnidaria and the flatworms, which are discussed next. The cnidaria have a two-tissue level of organization and radial symmetry. The flatworms have three germ layers and bilateral symmetry.

33.4 Bilateral Symmetry Evolves

All other animals to be studied are bilaterally symmetrical, at least in some stage of development. As embryos, they have three germ layers and are, therefore, called *triploblasts*. As adults, they have the *organ level of organization*. Flatworms in the phylum Platyhelminthes have these features, as do ribbon worms in the phylum Nemertea. Flatworms, however, like cnidaria, have a **sac body plan,** while ribbon worms have a **tube-within-a-tube body plan.** Animals with a sac body plan are said to have an incomplete digestive tract, while those with the tube-within-a-tube plan have a complete digestive tract.

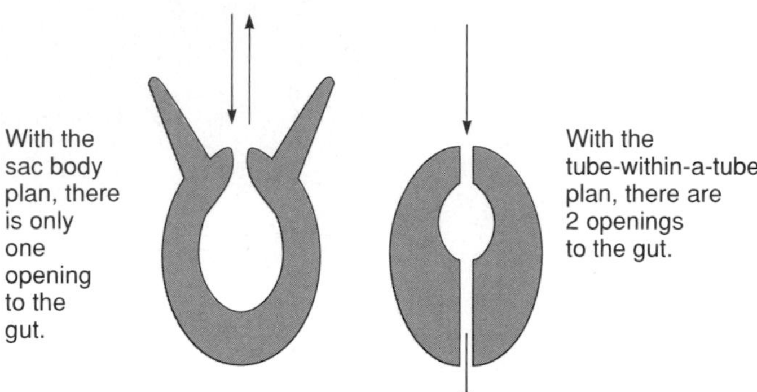

With the sac body plan, there is only one opening to the gut.
With the tube-within-a-tube plan, there are 2 openings to the gut.

In the tube-within-a-tube body plan, there is the possibility of specialization of parts along the length of the tube.

Ribbon worms (**phylum Nemertea,** 650 species), which are mainly marine, have a distinctive proboscis apparatus—a long, hollow tube lying in a cavity called the rhynchocoel (Fig. 33.8). Contraction of the rhynchocoel wall causes the proboscis to evert and shoot outward through a pore located just above the mouth. The proboscis is used primarily for prey capture but also for defense, locomotion, and burrowing. Like flatworms, ribbon worms are acoelomates.

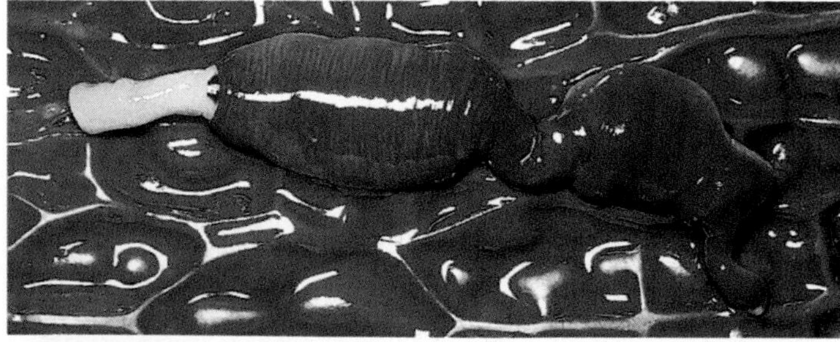

Figure 33.8 Ribbon worm, *Amphiporus*, on giant kelp, *Nacrocystis*.
Ribbon worms, like flatworms, are bilaterally symmetrical, have three germ layers, and demonstrate the organ level of organization. They also have a complete digestive tract. The proboscis is partially extended in this specimen.

Flatworms Are Flat

Flatworms (phylum Platyhelminthes, 13,000 species) can be either free-living or parasitic. Planaria and their relatives are freshwater animals in the class Turbellaria. The majority of flatworms are parasites. The flukes, which are either external or internal parasites, are in the class Trematoda. The tapeworms, which are intestinal parasites of vertebrates, are in the class Cestoda.

Flatworms are complex. In addition to an endodermis derived from endoderm and an epidermis derived from ectoderm, there is a mesoderm layer that gives rise to muscles and reproductive organs. There is no coelom; these animals are, therefore, acoelomates.

Although flatworms have the organ level of organization, there are no specialized circulatory or respiratory structures. How are nutrients distributed about the body? The gastrovascular cavity, which is sometimes highly branched, serves this function. Gas exchange can occur by diffusion because of the animal's flat, thin body. Often, there is an excretory system that functions as an osmotic-regulating system.

Flatworms are *bilaterally symmetrical,* and free-living forms have undergone **cephalization** [Gk. *kaphale,* head]—the development of a head region. There is a *ladder-type nervous system,* so called because the two lateral nerve cords plus the connecting nerves look like a ladder. Paired ganglia (collection of nerve cells) function as a brain; sensory cells are located in the body wall, and the animal is able to respond to various stimuli.

Flatworms have a sac body plan but three germ layers and the organ level of organization. They are bilaterally symmetrical, and cephalization is present.

Planaria Are Free Living

The turbellaria include freshwater planaria such as *Dugesia,* which are small (several mm to several cm) literally flat worms, with brown or black pigmentation (Fig. 33.9). They live in lakes, ponds, and streams, where they feed on small living or dead organisms.

The head is bluntly arrow shaped, with lateral extensions called *auricles* that function as sense organs to detect potential food sources and enemies. There are two light-sensitive eyespots whose pigmentation causes the worm to look cross-eyed. Inside, the brain is connected to a ladder-type nervous system. There are three kinds of muscle layers—an outer circular layer, an inner longitudinal layer, and a diagonal layer—that allow for quite varied movement. In larger forms, locomotion is accomplished by the movement of cilia on the ventral and lat-

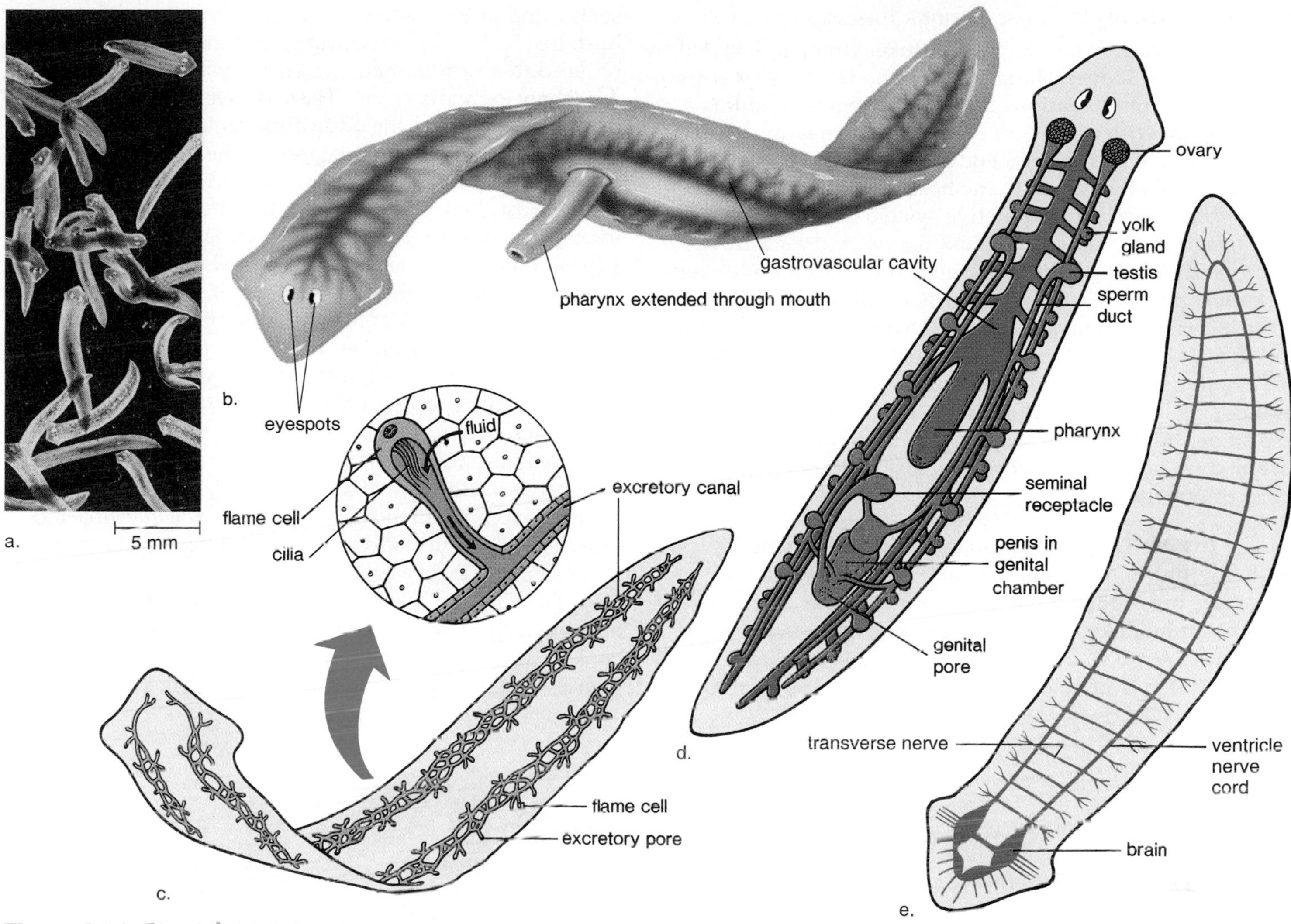

Figure 33.9 Planarian anatomy.
a. The photograph shows that flatworms, *Dugesia*, are bilaterally symmetrical and have a head region with eyespots. **b.** When the pharynx is extended as shown, food is sucked up into a gastrovascular cavity that branches throughout the body. **c.** The excretory system with flame cells is shown in detail. **d.** The reproductive system has both male and female organs, and the digestive system has a single opening. **e.** The nervous system has a ladderlike appearance.

eral surfaces. Numerous gland cells secrete a mucous material upon which the animal moves.

The animal captures food by wrapping itself around the prey, entangling it in slime, and pinning it down. Then a muscular pharynx is extended, and by a sucking motion the food is torn up and swallowed. The pharynx leads into a three-branched gastrovascular cavity in which digestion is both extracellular and intracellular.

Why is it to be expected that *Dugesia,* which live in fresh water would have a well-developed excretory organ system? Their excretory organ functions in osmotic regulation as well as excreting water. The organ consists of a series of interconnecting canals that run the length of the body on each side. Bulblike structures containing cilia are at the ends of the side branches of the canals. The cilia move back and forth, bringing water into the canals that empty at pores. The beating of the cilia reminded some early investigator of the flickering of a flame, and so the excretory organ of the flatworm is called a *flame-cell system.*

Planaria can reproduce asexually. They constrict beneath the pharynx, and each part grows into a whole animal again. Many experiments in development have utilized planaria because of their marked ability to regenerate. Planaria also reproduce sexually. They are **hermaphroditic,** which means that they possess both male and female sex organs. The worms practice cross-fertilization when the penis of one is inserted into the genital pore of the other. The fertilized eggs are enclosed in a cocoon and hatch in two or three weeks as tiny worms.

Free-living planaria best exhibit the bilateral symmetry and organ development—including the nervous system and muscles—of a flatworm.

Parasitic Flatworms Cause Serious Illnesses

Among the *parasitic* flatworms are flukes (trematodes) and tapeworms (cestodes). The structure of both of these worms illustrates the modifications that occur in parasitic animals. Concomitant with the loss of predation, there is an absence of cephalization; the anterior end notably carries hooks and/or suckers for attachment to the host. There is an extensive development of the reproductive system at the expense of the other organs. Well-developed nerves and a gastrovascular cavity are not needed because the animal no longer seeks out and digests prey; instead, it acquires nutrients from its host. Flukes and tapeworms are covered by a *tegument,* a specialized body wall resistant to host digestive juices.

Both flukes and tapeworms utilize a secondary, or intermediate, host to transport the species from primary host to primary host. The primary host is infected with the sexually mature adult; the secondary host contains the larval stage or stages.

Flukes Invade Organs **Trematodes** (class Trematoda) include the flukes, which are usually named for the type of vertebrate organ they inhabit; for example, there are blood, liver, and lung flukes. While the structure may vary slightly, in general the fluke body tends to be oval to elongate. At the anterior end surrounded by sensory papilla, there is an oral sucker and at least one other sucker for attachment to the host. Inside, the digestive system and nervous systems are reduced, and a modified excretory system has a reduced number of excretory canals. There is a well-developed reproductive system, and the adult fluke is usually hermaphroditic, as are planaria. An exception is the blood fluke, which causes **schistosomiasis** and is seen predominantly in the Middle East, Asia, Africa, and South America. About 800,000 infected persons die each year. In this disease, the female flukes deposit their eggs in small blood vessels close to the lumen of the intestine, and the eggs make their way into the digestive tract by a slow migratory process (Fig. 33.10). After the eggs pass out with the feces, they hatch into tiny larvae that swim about in the rice paddies and elsewhere until they enter a particular species of snail. Within the snail asexual reproduction occurs; *sporocysts* that are spore-containing sacs eventually produce new larval forms that leave the snail. If they penetrate the skin of a human, they begin to mature in the liver and implant themselves in the small intestinal blood vessels. Those infected usually die of secondary diseases brought on by their weakened condition.

The Chinese liver fluke requires two hosts: the snail and the fish. Humans become infected when they eat uncooked fish. The adults reside in the liver and deposit their eggs in the bile duct, which carries the eggs to the intestine.

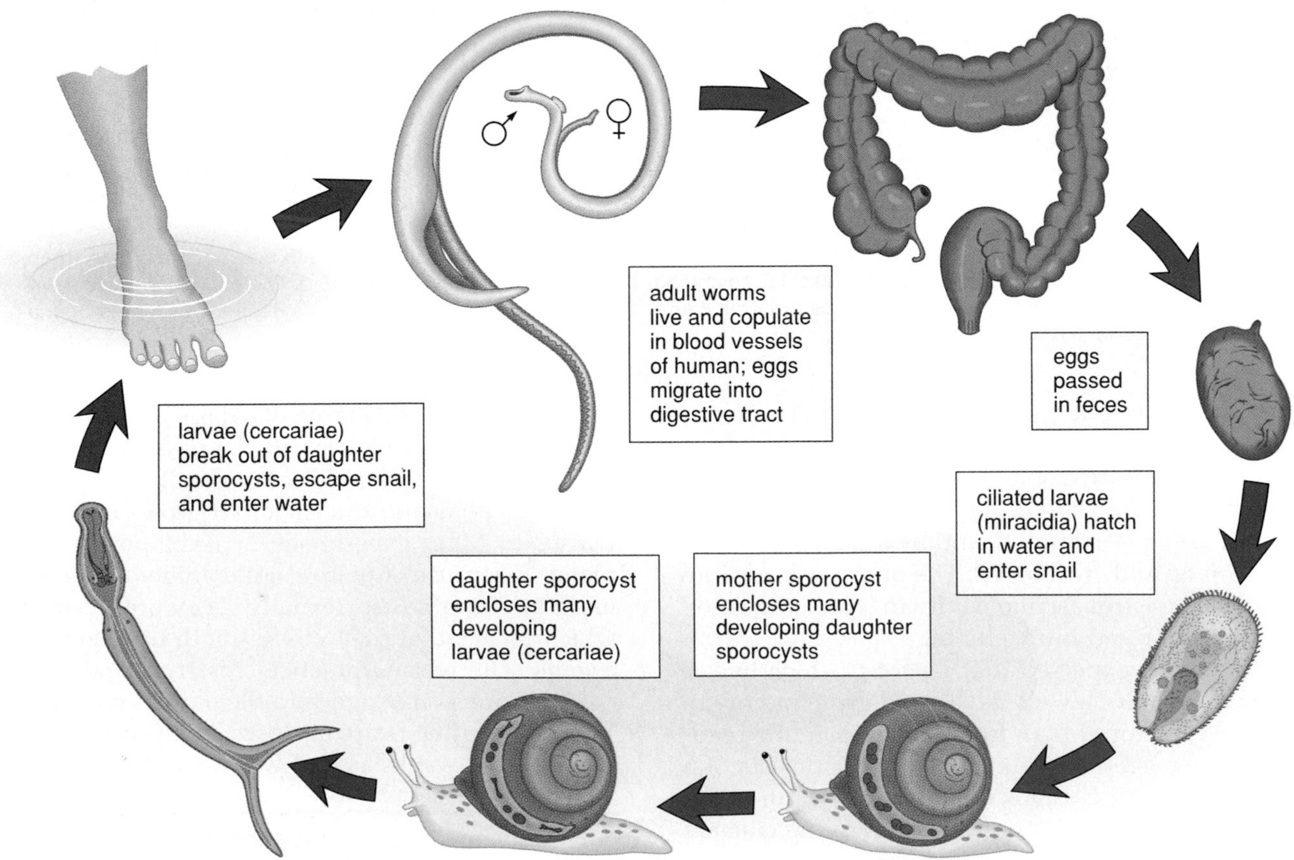
♂
♀
adult worms live and copulate in blood vessels of human; eggs migrate into digestive tract
eggs passed in feces
ciliated larvae (miracidia) hatch in water and enter snail
mother sporocyst encloses many developing daughter sporocysts
daughter sporocyst encloses many developing larvae (cercariae)
larvae (cercariae) break out of daughter sporocysts, escape snail, and enter water

Figure 33.10 Schistosomiasis.
This infection of humans, caused by blood flukes, *Schistosoma*, is an extremely prevalent disease in Egypt—especially since the building of the Aswan High Dam. Standing water in irrigation ditches, combined with unsanitary practices, has created the conditions for widespread infection.

Tapeworms Live in the Gut **Cestodes** (class Cestoda) include the tapeworms. A tapeworm has an anterior region (Fig. 33.11) containing hooks and suckers for attachment to the intestinal wall of the host. Behind the head region, called a **scolex,** there is a short neck and then a long series of proglottids. **Proglottids** are segments, each of which contains a full set of both male and female sex organs and little else. There are excretory canals but no digestive system and only the rudiments of nerves.

After fertilization, the organs disintegrate and the proglottids become nothing but a bag filled with maturing eggs. Gravid proglottids such as these break off and, as they pass out with the feces, the eggs are released.

If eggs are present in feces-contaminated food fed to pigs or cattle, larvae escape when the covering of the eggs is digested away. They burrow through the intestinal wall and travel in the bloodstream to finally lodge and encyst in muscle. Here a **cyst** means a small, hard-walled structure that contains a larva called a *bladder worm.* When humans eat infected meat that has not been thoroughly cooked, the bladder worms break out of the cyst, attach themselves to the intestinal wall, and grow to adulthood. Then the cycle begins again.

Flukes and tapeworms illustrate the modifications that occur when animals take up the parasitic way of life.

Table 33.2 contrasts features of planaria, tapeworms, and flukes to illustrate the anatomical changes to be associated with the parasitic way of life.

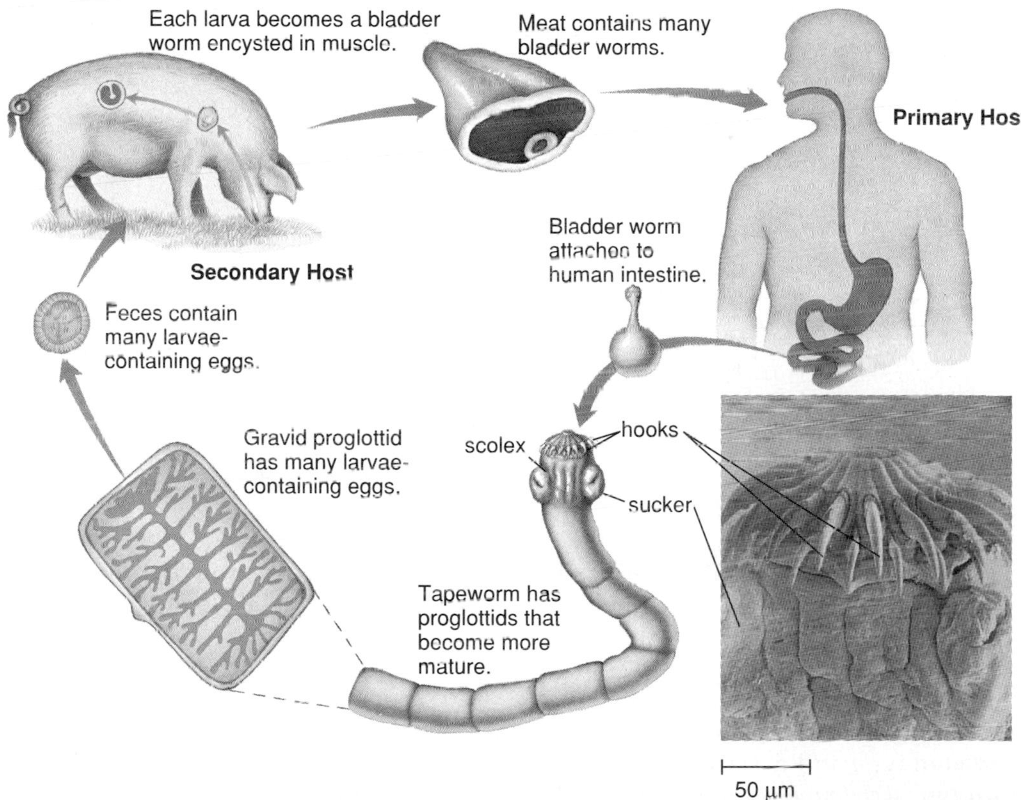
Each larva becomes a bladder worm encysted in muscle.
Meat contains many bladder worms.
Primary Host
Secondary Host
Bladder worm attaches to human intestine.
Feces contain many larvae-containing eggs.
Gravid proglottid has many larvae-containing eggs.
scolex
hooks
sucker
Tapeworm has proglottids that become more mature.
50 μm

Figure 33.11 Life cycle of a tapeworm, *Taenia.*
The life cycle includes a human (primary host) and a pig (secondary host). The adult worm is modified for its parasitic way of life. It consists of a scolex and many proglottids, which become bags of eggs.

Table 33.2

Free-Living Flatworms Versus Parasitic Flatworms

	Free Living	Parasitic	
	Planaria	*Flukes*	*Tapeworms*
Body wall	Ciliated epidermis	Tegument	Glycocalyx covers tegument
Cephalization	Yes—eyespots and auricles	No—oral suckers	No—scolex with hooks and suckers
Nervous system	Nerves and brain	Reduced	Reduced
Digestive organ	Branched	Reduced	Absent
Reproductive organs	Hermaphroditic	Increased in volume	Extensively developed
Larva	Absent	Present	Present

A closer Look

Presence of a Coelom

A coelom is a body cavity surrounding the digestive organ or system. Flatworms are acoelomates (Fig. 33A*a*). They have no body cavity, and the mesoderm is packed solidly between the ectoderm and endoderm. The roundworms are pseudocoelomates. They have a body cavity in which the organs lie loose. Apparently, this is disadvantageous to an increase in size, because most of these worms are small. In pseudocoelomates, a body cavity is incompletely lined by mesoderm. In coelomates, the coelom develops as a cavity within the mesoderm; therefore, it is completely lined with mesoderm. Such a coelom is often called a "true coelom." The organs in a true coelom are held in place by mesenteries, assuring a more stable arrangement with less crowding. Further, the gut is muscular (muscles are derived from mesoderm) and shows specialization of parts not seen in the pseudocoelomates. In the coelomates (all the other animals we will study), the internal organs are more complex.

The coelom also serves other functions. It allows the organs and the body wall to move independently. This means an animal can stretch and bend without putting a strain on the internal organs. The coelom is fluid filled, and this fluid protects and cushions the internal organs. In some animals, the fluid aids in the movement of materials, such as metabolic wastes; in others, this function is taken over by blood vessels. Not only metabolic wastes but also sex cells may be deposited into the coelomic cavity before they are transported away by ducts. The gastrovascular cavity of acoelomates (cnidarians and flatworms) and the fluid-filled coelom of soft-bodied coelomates can act as a hydrostatic skeleton. It offers some resistance to the contraction of muscles and yet permits flexibility, so that the animal can change shape and perform a variety of movements.

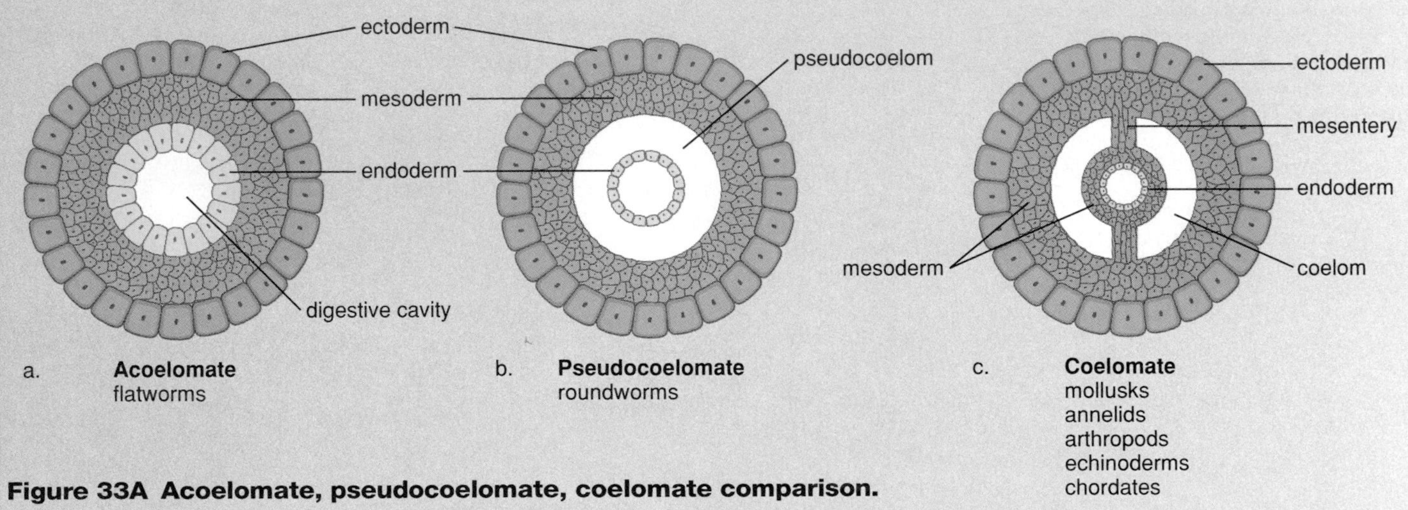

Figure 33A Acoelomate, pseudocoelomate, coelomate comparison.

33.5 A Pseudocoelom Evolves

The accompanying reading compares animals on the basis of coelom type and describes how a coelom can serve as a *hydrostatic skeleton*. A **pseudocoelom** [Gk. *pseudes*, false, and *koiloma*, cavity] is a body cavity that is *incompletely lined by mesoderm*. In other words, mesoderm does not form a complete layer next to the body wall or around the gut. Two animal phyla consist of pseudocoelomates: roundworms (phylum Nematoda) and **rotifers** (**phylum Rotifera,** 2,000 species). These animals also have the tube-within-a-tube body plan. The digestive tract is the inner tube within the rest of the animal, which is the outer tube.

Rotifers, which are abundant in fresh water, are microscopic. Although rotifers are limited in size, they are multicellular, with internal organs. They are named for a crown of cilia (corona) that resembles a rotating wheel and serves both as an organ of locomotion and as an aid in bringing food to the mouth.

Pseudocoelomate animals (e.g., roundworms and rotifers) have a coelom that is incompletely lined by mesoderm. A coelom provides a space for internal organs and can serve as a hydrostatic skeleton.

Roundworms Are Nonsegmented

Roundworms (phylum Nematoda, 500,000 species), as their name implies, have a smooth outside wall, indicating that they are nonsegmented. These worms, which are generally colorless are found almost anywhere—in the sea, in fresh water, and in the soil—in such numbers and size that thousands of them can be found in a small area. Some are predators with teeth and other mouthparts, but many are scavengers or parasites. Among the latter, the pinworms, the hookworms, and *Trichinella* are small parasitic worms that are not easily examined. *Ascaris*, a large parasitic roundworm, is often studied as an example of this phylum.

Ascaris *Infects Humans*

Ascaris males (Fig. 33.12) tend to be smaller (15–31 cm long) than females (20–49 cm long). In males, the posterior end is curved and comes to a point. Both sexes move by means of a characteristic whiplike motion because only longitudinal muscles lie next to the body wall.

The internal organs, including the tubular reproductive organs, lie within the pseudocoelom. Because mating produces eggs that mature in the soil, the parasite is limited to warmer environments. When these eggs are swallowed, larvae escape and burrow through the intestinal wall. Making their way through the organs of the host, they move from the intestine to the liver, the heart, and then the lungs. Within the lungs, molting takes place and, after about ten days, the larvae migrate up the windpipe to the throat where they are swallowed, once again reaching the intestine. Then the mature worms mate and the female deposits eggs that pass out with the feces. In this life cycle, as with that of other roundworms, feces must reach the mouth of the next host; therefore, proper sanitation is the best means to prevent infection with such worms as *Ascaris* and pinworms.

Other Roundworm Parasites

Trichinosis (Fig. 33.12) is a serious infection that humans can contract when they eat rare pork containing encysted larvae. After maturation, the female adult burrows into the wall of the small intestine and produces living offspring that are carried by the bloodstream to the skeletal muscles, where they encyst.

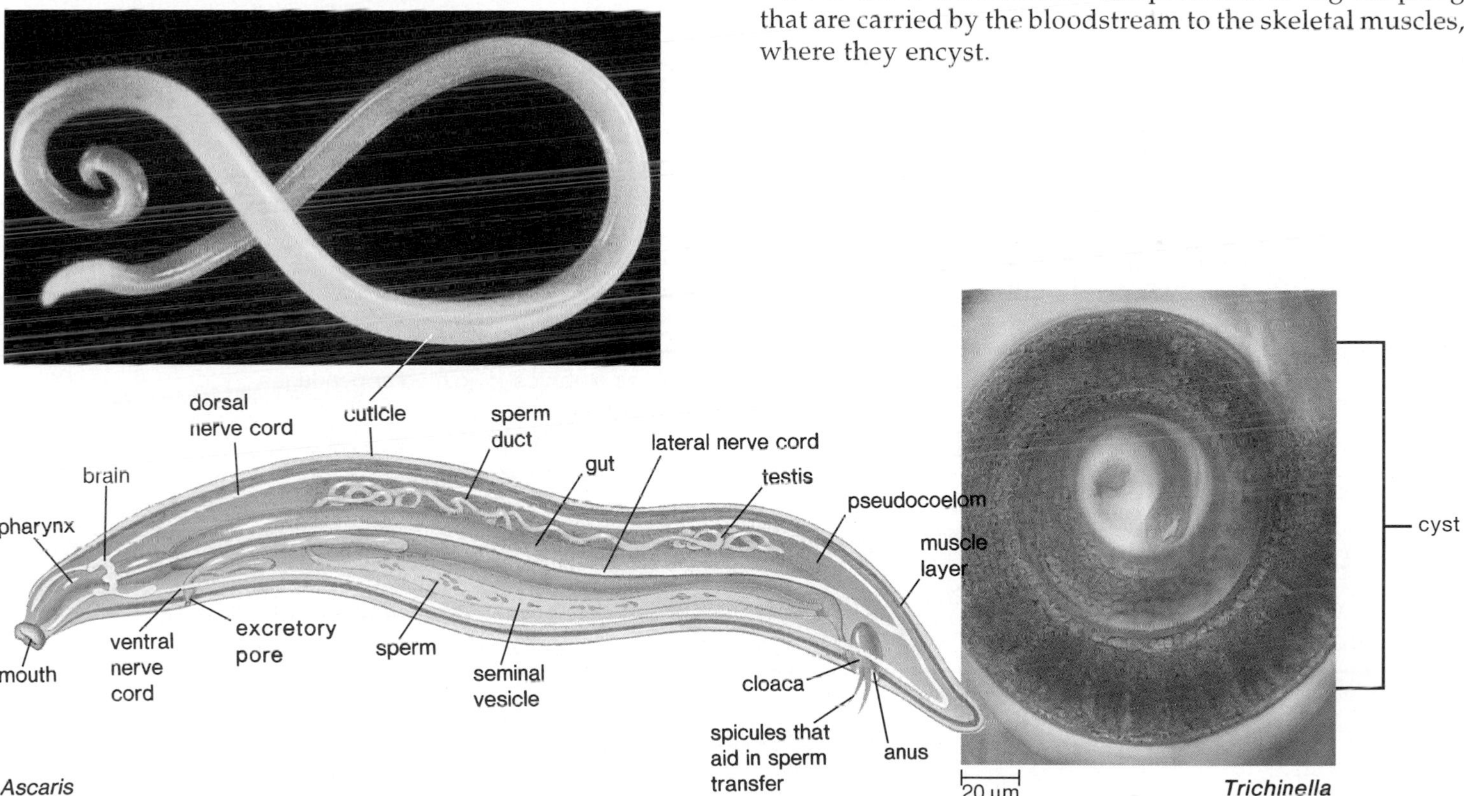

Figure 33.12 Roundworm anatomy.
Note that roundworms such as *Ascaris* have a pseudocoelom and a complete digestive tract with a mouth and an anus. Therefore, roundworms have a tube-within-a-tube body plan. The sexes are separate; this is a male roundworm. The larvae of the roundworm, *Trichinella*, penetrate striated muscle fibers, where they coil in a sheath formed from the muscle fiber.

Filarial worms, a type of roundworm, cause various diseases. *Dirofilaria,* the heartworm of dogs, is a common filarial worm of temperate zones. The cause of **elephantiasis,** a disease restricted to tropical areas of Africa, is caused by a filarial worm that utilizes the mosquito as a secondary host. Because the adult worms reside in lymphatic vessels, collection of fluid is impeded and the limbs of an infected person may swell to a monstrous size (Fig. 33.13). Elephantiasis is treatable in its early stages but not after scar tissue has blocked lymphatic vessels.

Roundworms can be free living or parasitic. Free-living forms live in the soil, or water; parasitic ones cause some common infections such as pinworm and hookworm infections of humans and heartworm infection of dogs.

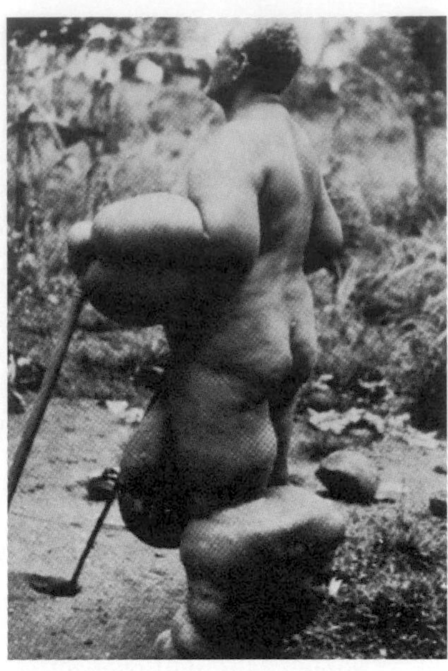

Figure 33.13 Filarial worm.
An infection from a filarial worm, *Wuchereria,* causes elephantiasis, a condition in which the individual experiences extreme swelling in regions where the worms have blocked the lymphatic vessels.

connecting concepts

Animals are a diverse group, but still they have certain features in common. In order to classify animals, zoologists examine and compare their body plan, number of tissue layers derived from germ layers, symmetry, level of organization, and type of body cavity. The sponges, the simplest multicellular animals, are sessile, asymmetric, and have several different kinds of cells but no true tissues. They are significantly different from other phyla, and apparently no other groups evolved from them. For this reason, many biologists consider sponges to be outside of the mainstream of animal evolution. The sea anemones, jellyfish, and related species (cnidaria and ctenophorans) are radially symmetrical and have two germ layers as embryos. Most of these creatures also posses unique stinging cells containing nematocysts, which can put a serious damper on a day at the beach.

Animals more complex than the cnidarian group show bilateral symmetry at some stage in their development. They also have three germ layers as embryos and fairly well-developed organs. The phyla containing the flatworms and the roundworms include both free-living and parasitic species. In fact, members of these two phyla are among the most successful parasites on earth. In the next two chapters we will examine phyla with more morphological complexity. The remaining animal phyla have a body cavity called a coelom. This has led to specialization of parts including a well-developed nervous and musculoskeletal system, Only the arthropods and chordates, however, have jointed appendages.

Table 33.3

Comparison of Animals Without a True Coelom

	Sponges	Cnidaria	Flatworms	Roundworms
Symmetry	Radial or none	Radial	Bilateral	Bilateral
Type of body plan	—	Sac	Sac	Tube-within-a-tube
Tissue layers	—	Two	Three	Three
Level of organization	Cell	Tissue	Organ	Organ
Body cavity	—	—	Acoelomate	Pseudocoelomate

Summary

33.1 How Animals Evolved and Are Classified

Animals are multicellular organisms that are heterotrophic and ingest their food. They have the diplontic life cycle. Typically, they have the power of motion by means of contracting fibers.

It's possible to construct a phylogenetic tree for animals, but this is largely based on a study of today's forms. Type of symmetry, number of tissue layers, type of coelom, and presence or absence of segmentation are criteria that are used in classification. Table 33.3 compares these criteria among the animals studied in this chapter.

33.2 Multicellularity Evolves

Sponges may have evolved separately from other animals, since they have features that set them apart. They have the cellular level of organization and lack tissues and symmetry. Sponges are sessile and depend on a flow of water through the body to acquire food, which is digested in vacuoles within collar cells that line a central cavity.

33.3 Tissue Layers Evolve

Cnidaria and comb jellies have two tissue layers derived from the germ layers ectoderm and endoderm. They are radially symmetrical.

Cnidaria have a sac body plan. They exist as either polyps or medusae, or they can alternate between the two. Hydras and relatives—sea anemones and corals—are polyps; in jellyfishes the medusan stage is dominant. In *Hydra* and other cnidaria, an outer epidermis is separated from an inner gastrodermis by mesoglea. They possess tentacles to capture prey and nematocysts to stun it. A nerve net coordinates movements. Digestion of prey begins in the gastrovascular cavity and is finished within gastrodermal cells.

33.4 Bilateral Symmetry Evolves

Flatworms and ribbon worms have bilateral symmetry and the organ level of organization, including organs derived from mesoderm, a third germ layer, as do all the other phyla of animals to be studied.

Flatworms may be free living or parasitic. Freshwater planaria exemplify the features of flatworms in general and free-living forms in particular. They have muscles and a ladder-type nervous organization, and they show cephalization. They take in food through an extended pharynx leading to a gastrovascular cavity, which extends throughout the body. There is an osmotic-regulating organ that contains flame cells.

Flukes and tapeworms are parasitic. Flukes have two suckers by which they attach to and feed from their hosts. Tapeworms have a scolex with hooks and suckers for attaching to the host intestinal wall. The body of a tapeworm is made up of proglottids, which, when mature, contain thousands of eggs. If these eggs are taken up by pigs or cattle, larvae become encysted in their muscles. If humans eat this meat, they too may become infected with a tapeworm.

33.5 A Pseudocoelom Evolves

Roundworms and rotifers are pseudocoelomates. A coelom provides a space for internal organs and can serve as a hydrostatic skeleton. Roundworms are mostly small and very diverse; they are present almost everywhere in great numbers. The parasite *Ascaris* is representative of the group. Infections can also be caused by *Trichinella,* whose larval stage encysts in the muscles of humans. Elephantiasis is caused by a filarial worm that blocks lymphatic vessels.

Reviewing the Chapter

1. What are the characteristics that separate animals from plants? from fungi? 576
2. What does the phylogenetic tree (see Fig. 33.2) tell you about the evolution of the animals studied in this chapter? 577
3. List the types of cells found in a sponge, and describe their functions. 578–79
4. What features make sponges different from the other organisms placed in the animal kingdom? 579
5. What features do comb jellies and cnidaria have in common? How are they different? 580
6. What are the two body forms found in cnidaria? Explain how they function in the life cycle of various types of cnidaria. 580
7. Describe the anatomy of *Hydra,* pointing out those features that typify cnidaria. 582
8. What features do flatworms and ribbon worms have in common? How are they different? 584
9. Describe the anatomy of a free-living planarian, pointing out those features that typify nonparasitic flatworms. 584–85
10. Describe the parasitic flatworms, and give the life cycle of both the blood fluke that causes schistosomiasis and the pork tapeworm. 586–87
11. What is a pseudocoelom? What are the advantages of having a coelom? What two groups of animals have a pseudocoelom? 588
12. Describe the anatomy of *Ascaris,* pointing out those features that typify roundworms. 589
13. Compare the animals studied in this chapter in terms of symmetry, tissue layers, level of organization, body plan, and type of coelom. 590

Testing Yourself

Choose the best answer for each question.

1. Which of these is not a characteristic of animals?
   a. heterotrophic
   b. diplontic life cycle
   c. have contracting fibers
   d. single cells or colonial
2. The phylogenetic tree of animals shows that
   a. cnidaria evolved directly from sponges.
   b. flatworms evolved directly from roundworms.
   c. both sponges and cnidaria evolved from protista.
   d. All of these are correct.
3. Which of these sponge characteristics is not typical of animals?
   a. They practice sexual reproduction.
   b. They have the cellular level of organization.
   c. They are asymmetrical.
   d. Both b and c are correct.

4. Which of these is mismatched?
   a. sponges—spicules
   b. tapeworms—proglottids
   c. cnidaria—nematocysts
   d. roundworms—cilia
5. Flukes and tapeworms
   a. show cephalization.
   b. have well-developed reproductive systems.
   c. have well-developed nervous systems.
   d. are plant parasites.
6. The presence of mesoderm
   a. restricts the development of a coelom.
   b. is associated with the organ level of organization.
   c. is associated with the development of muscles.
   d. Both b and c are correct.
7. *Ascaris* is a parasitic
   a. roundworm.
   b. flatworm.
   c. hydra.
   d. sponge.
8. Label the following diagram of the cnidarian polyp:

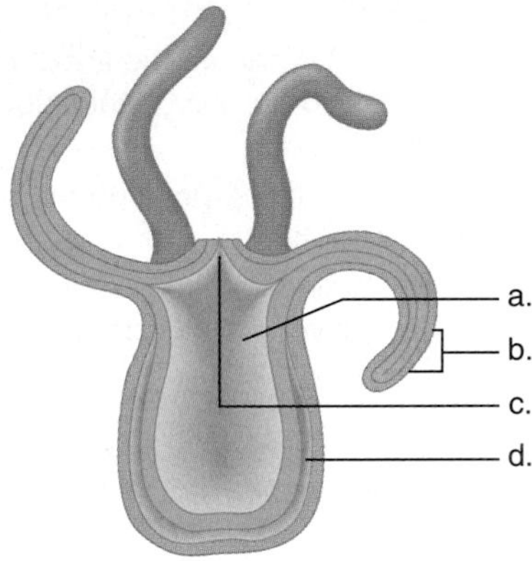

9. Write the correct phylum name beside each of these terms:
   a. radial symmetry
   b. pseudocoelom
   c. tissue level of organization
   d. tube-within-a-tube body plan
10. Write the correct type animal beside each of these terms:
   a. proglottids
   b. eyespots on head
   c. collar cells
   d. cnidocytes

Applying the Concepts

1. *Animals are one of the major groups of organisms alive today.*
   What features might you expect animals to have in common because they are heterotrophic by ingesting food? How might this common trait cause them to be differently adapted?
2. *Animals have a phylogenetic history.*
   A common ancestor for flatworms and roundworms might have had what anatomical features?
3. *Parasites are adapted to their way of life.*
   What types of adaptations are helpful to an internal parasite?

Using Technology

Your study of the first invertebrates is supported by these available technologies:

Exploring the Internet
The Mader Home Page provides resources for and help with studying this chapter.

http://www.mhhe.com/sciencemath/biology/mader/
(Click on Biology.)

Understanding the Terms

asymmetry 576
bilateral symmetry 576
cephalization 584
cestode 587
cnidaria 580
coelom 576
cyst 587
elephantiasis 590
flatworm 584
gastrovascular cavity 580
hermaphroditic 585
hydra 582
invertebrate 576
mesoglea 580
nematocyst 580
nerve net 580
proglottid 587
pseudocoelom 588
radial symmetry 576
ribbon worm 584
rotifer 588
roundworm 589
sac body plan 584
schistosomiasis 586
scolex 587
sessile 576
sessile filter feeder 579
spicule 579
sponge 578
trematode 586
trichinosis 589
tube-within-a-tube body plan 584

Match the terms to these definitions:

a. __________ Blind digestive cavity that also serves a circulatory (transport) function in animals that lack a circulatory system.
b. __________ Body cavity lying between the digestive tract and body wall that is completely lined by mesoderm.
c. __________ Body cavity lying between the digestive tract and body wall that is incompletely lined by mesoderm.
d. __________ Body plan having two corresponding or complementary halves.
e. __________ Body with a digestive tract that has both a mouth and an anus.
f. __________ Characterizes an animal having both male and female sex organs.

Animals: The Protostomes

CHAPTER 34

Chapter Concepts

Garden snail, *Cepaea hortensis*

In the last chapter, we looked at invertebrate animals with which you may not be familiar, but sponges, jellyfish, corals, flatworms, and roundworms have been around for millions of years. It should be remembered that even though they lack the anatomical complexity and the larger size of the animals studied in this chapter, they still have biological and ecological importance.

In this chapter, we examine animals with which you may be more familiar—the mollusks, annelids, and arthropods. The mollusks are a large diversified group with a unique body plan. Most mollusks are aquatic, but snails and slugs are adapted to living on land. Although annelids are a much smaller group, they, too, are quite diversified. Earthworms are well-known annelids with obvious segmentation. The arthropods are also segmented, and they have a jointed exoskeleton that facilitates locomotion on land. The insects, which are arthropods, include more known species than any other group of animals, and most of these groups live on land.

34.1 A Coelom

Protostomes are bilaterally symmetrical, have three germ layers, the organ level of organization, and the tube-within-a-tube body plan. In addition they have a true **coelom** [Gk. *koiloma,* cavity], a body cavity completely lined by mesoderm.

The presence of a coelom has many advantages. Body movements are freer because the outer wall can move independently of the enclosed organs. The ample space of a coelom allows complex organs and organ systems to develop. For example, the digestive tract can coil and provide a greater surface area for absorption of nutrients. And the coelomic cavity can serve as a storage area for eggs and sperm before they are released into the environment. Coelomic fluid protects internal organs against damage and against marked temperature changes. It can also assist in respiration and circulation by providing oxygen and nutrients to nearby cells. Metabolic wastes can accumulate in the cavity prior to being taken away by the excretory system. Also fluid within the cavity protects internal organs and can provide a *hydrostatic skeleton*—muscular contraction pushes against the fluid and allows the animal to move.

When a coelom is present, the digestive system and the body wall can move independently, and internal organs can become more complex. Coelomic fluid can assist respiration, circulation, and excretion. It also serves as a hydrostatic skeleton.

Phyla that have a coelom are divided into the **protostomes** [Gk. *protos,* first, and L. *stoma,* mouth] and **deuterostomes** [Gk. *deuteros,* second, and L. *stoma,* mouth]. There are three major differences in the embryological development of protostomes and deuterostomes (Fig. 34.1). **Cleavage,** the first event of development, is cell division without an increase in size of the cells. In protostomes, **spiral cleavage** occurs, and daughter cells sit in grooves formed by the previous cleavages. It can also be noted that the fate of these cells is fixed and determinate in protostomes; each can contribute to development in only one particular way. In deuterostomes, *radial cleavage* occurs, and the daughter cells sit right on top of the previous cells. The fate of these cells is indeterminate; that is, if they are separated from one another, each cell can go on to become a complete organism.

As development proceeds, a hollow sphere forms, and the indentation that follows produces an opening called the blastopore. In protostomes, the *mouth* appears at or near the blastopore, hence the origin of their name; in deuterostomes, the *anus* appears at or near the blastopore and only later does a new opening form the mouth, hence the origin of their name.

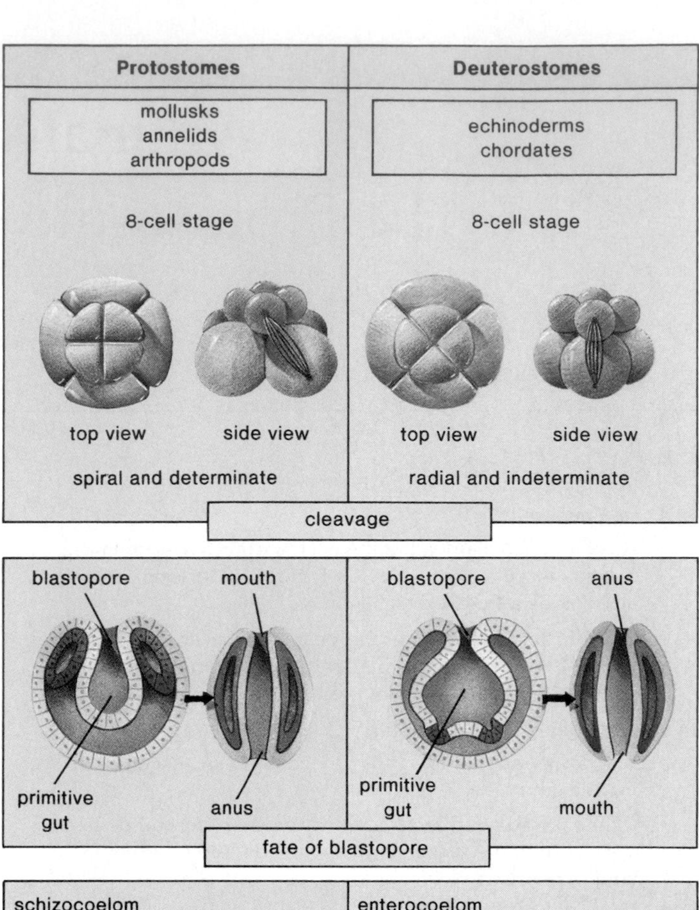

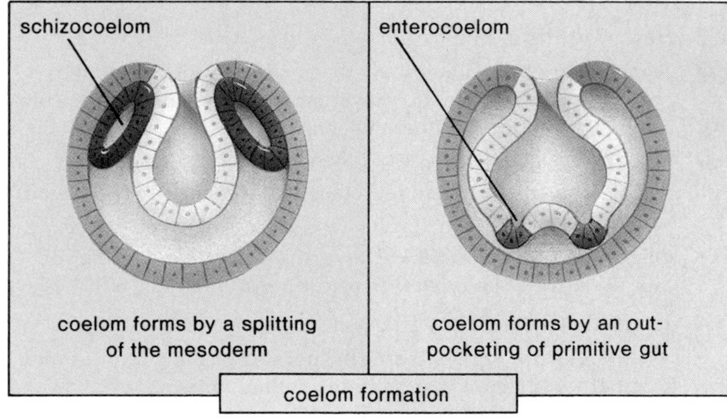

Figure 34.1 Protostomes versus deuterostomes. In the embryo of protostomes (mollusks, annelids, arthropods), cleavage is spiral—new cells are at an angle to old cells—and each cell has limited potential and cannot develop into a complete embryo; the blastopore is associated with the mouth and the coelom is a schizocoelom. In deuterostomes (echinoderms and chordates), cleavage is radial—new cells sit on top of old cells—and each one can develop into a complete embryo; the blastopore is associated with the anus, and the coelom is an enterocoelom.

CLASSIFICATION

Kingdom Animalia

Multicellular organisms with well-developed tissues; usually motile; heterotrophic by ingestion, generally in a digestive cavity; diplontic life cycle. Protostomes include phyla Mollusca, Annelida, and Arthropoda. Deuterostomes include phyla Echinodermata, Hemichordata, and Chordata.

Invertebrates*

- Phylum Porifera: sponges
- Phylum Cnidaria: jellyfishes, sea anemones, corals
- Phylum Ctenophora: comb jellies, sea walnuts
- Phylum Platyhelminthes: flatworms, e.g., planaria, flukes, tapeworms
- Phylum Nemertea: ribbon worms
- Phylum Nematoda: roundworms
- Phylum Rotifera: rotifers
- Phylum Mollusca: chitons, snails, slugs, clams, mussels, squids, octopuses
- Phylum Annelida: segmented worms, e.g., clam worms, earthworms, leeches
- Phylum Onychophora: walking worms
- Phylum Arthropoda: spiders, scorpions, horseshoe crabs, lobsters, crayfish, shrimps, crabs, millipedes, centipedes, insects
- Phylum Echinodermata: sea lilies, sea stars, brittle stars, sea urchins, sand dollars, sea cucumbers, sea daisies
- Phylum Hemichordata: acorn worms
- Phylum Chordata
  - Subphylum Urochordata: tunicates
  - Subphylum Cephalochordata: lancelets

Vertebrates*

- Subphylum Vertebrata
- Superclass Agnatha: jawless fishes, e.g., lampreys, hagfishes
- Superclass Gnathostomata: jawed fishes, all tetrapods
  - Class Chondrichthyes: cartilaginous fishes, e.g., sharks, skates, rays
  - Class Osteichthyes: bony fishes, e.g., herring, salmon, cod, eel, flounder
  - Class Amphibia: frogs, toads, salamanders
  - Class Reptilia: snakes, lizards, turtles
  - Class Aves: birds, e.g., sparrows, penguins, ostriches
  - Class Mammalia: mammals, e.g., cats, dogs, horses, rats, humans

*Not in the classification of organisms, but added here for clarity.

The coelom develops differently in the two groups. In protostomes, the mesoderm arises from cells located near the embryonic blastopore, and a splitting occurs that produces the coelom, called a *schizocoelom*. In deuterostomes, the coelom arises as a pair of mesodermal pouches from the wall of the primitive gut. The pouches enlarge until they meet and fuse, forming an *enterocoelom*.

All complex animals have a true coelom. They are divided into the protostomes (spiral cleavage, blastopore associated with mouth, schizocoelom) and the deuterostomes (radial cleavage, blastopore associated with anus, enterocoelom).

This chapter will discuss the protostomes, and the next chapter will discuss the deuterostomes:

Protostomes	Deuterostomes
mollusks	echinoderms
annelids	chordates
arthropods	

Protostomes, like other animals, evolved in the sea but some ventured onto land and became very successful there. In this chapter, we will have an opportunity to contrast animal adaptations suitable to living in water with adaptations suitable to living on land. Terrestrial existence requires breathing air, preventing desiccation, and having a means of locomotion and reproduction that are not dependent on external water. The excretory system may be modified for the excretion of a solid nitrogenous waste to help conserve water.

34.2 A Three-Part Body Plan

Mollusks have a three-part body plan—visceral mass, mantle, and foot—which distinguishes this phylum from others. The success of this body plan is seen in mollusk diversity, with modifications in methods of feeding and locomotion.

Mollusks Have Soft Bodies

There are over 110,000 living species of **mollusks (phylum Mollusca)**—more than twice the number of vertebrate species! Most are marine, but there are also freshwater and terrestrial mollusks (Fig. 34.2).

The *visceral mass* contains the internal organs, including a highly specialized digestive tract, paired kidneys, and reproductive organs. The phylum name comes from the Latin word *mollusc,* meaning soft, which refers to the visceral mass. The **mantle** is a covering that lies to either side of but does not completely enclose the visceral mass. It may secrete a shell and/or contribute to the development of gills or lungs. The space between the folds of the mantle is called the mantle cavity. The *foot* is a muscular organ that may be adapted for locomotion, attachment, food capture, or a combination of functions. Another feature often present is a *radula,* an organ that bears many rows of teeth and is used to obtain food.

The nervous system of a mollusk consists of several ganglia connected by nerve cords. The *coelom is reduced* and is largely limited to the region around the heart. Most mollusks have an *open circulatory system.* The heart pumps blood, more properly called hemolymph, through vessels into sinuses (cavities) collectively called a hemocoel. Blue hemocyanin, rather than red hemoglobin, is the respiratory pigment.

Some mollusks are slow moving and have no head; many others undergo marked cephalization, having both a head and sense organs, and are active predators. Chitons, class Polyplacophora, have a shell that consists of a row of eight overlapping plates. Their flat foot is used for creeping along or clinging to rocks. The chiton scrapes algae and other plant food from rocks with its well-developed radula.

All mollusks have a visceral mass, mantle, and foot. Many also have a shell and/or radula. A reduced coelom and an open circulatory system are also present.

Figure 34.2 Molluscan diversity.
a. A chiton has a flattened foot and a shell that consists of eight articulating valves. **b.** A scallop, with sensory tentacles extended between the valves. **c.** A chambered nautilus achieves buoyancy by regulating the amount of air in the chambers of its shell. **d.** A nudibranch (sea slug) lacks a shell, gills, and a mantle cavity. Dorsal projections function in gas exchange.

a. Chiton, *Tonicella*

b. Scallop, *Pecten*

c. Chambered nautilus, *Nautilus*

d. Spanish shawl nudibranch, *Flabellina*

Bivalves Have a Double Shell

Clams, oysters, mussels, and scallops are all **bivalves** (class Bivalvia) with a two-part shell that is hinged and closed by powerful muscles. They have no head, no radula, and very little cephalization. Clams use their *hatchet-shaped foot* for burrowing in sandy or muddy soil, and mussels use their foot for production of threads, which attach them to nearby objects. Scallops both burrow and swim; rapid clapping of the valves releases water in spurts and causes the animal to move forward in jerklike fashion for a few feet.

In freshwater clams such as *Anodonta* (Fig. 34.3), the shell, secreted by the mantle, is composed of protein and calcium carbonate with an inner layer of pearl. If a foreign body is placed between the mantle and the shell, pearls form as concentric layers of shell are deposited about the particle. The compressed muscular foot projects ventrally from the shell; by expanding the tip of the foot and pulling the body after it, the clam moves forward.

Within the mantle cavity, the ciliated gills hang down on either side of the visceral mass. The beating of the cilia causes water to enter the mantle cavity by way of the incurrent siphon and to exit by way of the excurrent siphon. The clam is a filter feeder; small particles in this constant stream of water adhere to the gills, and ciliary action sweeps them toward the mouth.

The mouth leads to a stomach and then to an intestine, which coils about in the visceral mass before going right through the heart and ending in an anus. The anus empties at the excurrent siphon. There is also an accessory organ of digestion called a digestive gland. The heart lies just below the hump of the shell within the pericardial cavity, the only remains of the coelom. The circulatory system is open; the heart pumps hemolymph into vessels that open into the hemocoel. The nervous system is composed of three pairs of ganglia (located anteriorly, posteriorly, and in the foot), which are connected by nerves.

There are two excretory kidneys, which lie just below the heart and remove waste from the pericardial cavity for excretion into the mantle cavity. The clam excretes ammonia (NH_3), a toxic substance that requires the concomitant excretion of water.

The sexes are separate. The gonad is located about the coils of the intestine. Certain clams and annelids have the same type of larva, and this indicates a possible evolutionary relationship between mollusks and annelids. The reading on page 607 also offers evidence that these two groups had a common ancestry.

In clams, which are bivalves, the body is protected by a heavy shell. They are filter feeders with a hatchet foot that allows them to burrow slowly in sand and mud.

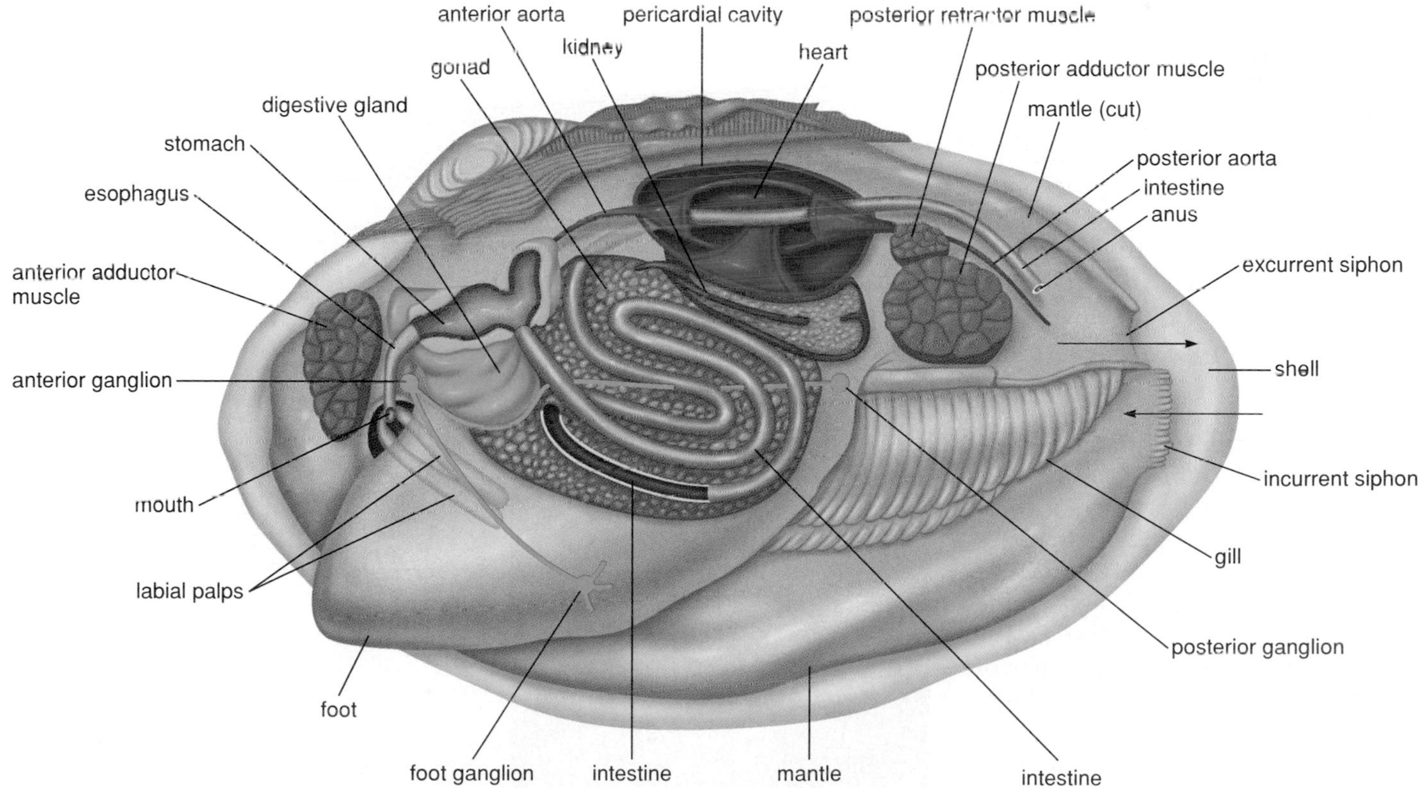

Figure 34.3 Clam, *Anodonta*.
The shell and the mantle have been removed from one side. Trace the path of food from the incurrent siphon to the gills, to the mouth, the stomach, the intestine, the anus, and the excurrent siphon. Locate the three ganglia: anterior, foot, and posterior. The heart lies in the reduced coelom.

Cephalopods Have Heads

Cephalopods [Gk. *kaphale,* head, and *podos,* foot] (class Cephalopoda) include squids, cuttlefish, octopuses, and nautiluses, all of which are fast-swimming predators in the open ocean. Cephalopod means head footed; both squids and octopuses can squeeze their mantle cavity so that water is forced out through a funnel, propelling them by a sort of *jet propulsion.* Also, the tentacles and arms that circle the head capture prey by adhesive secretions or by suckers. A powerful, parrotlike beak is used to tear prey apart. They have well-developed sense organs, including focusing *camera-type eyes* that are very similar to those of **vertebrates.** Cephalopods, particularly octopuses, have well-developed brains and show a remarkable capacity for learning. Nautiluses are enclosed in shells, but squids have a shell that is reduced and internal. Octopuses lack shells entirely. For protection, squids and octopuses possess *ink sacs,* from which they can squirt a cloud of brown or black ink. This action often leaves a potential predator completely confused.

In squids, such as *Loligo* (Fig. 34.4), a tough, muscular mantle contains a vestigial skeleton called the pen that surrounds the visceral mass. Jet propulsion is accomplished in this manner: water freely enters the mantle cavity by way of a space that circles the head. When the cavity is closed off and constricted tightly against the head, water exits the funnel, propelling the squid in the opposite direction. The funnel can be directed anteriorly or posteriorly, resulting in either forward or backward movement. A squid has an effective means of seizing and eating food. For example, *Loligo* darts backward rapidly into a school of young mackerel, seizes a fish with its tentacles, quickly biting the neck and severing the nerve cord with its jaws.

Unlike other mollusks and in keeping with its active life, the squid has a closed circulatory system, meaning that blood is always enclosed within blood vessels or a heart. The squid has three hearts—one of which pumps blood to all the internal organs while the other two pump blood to the gills located in the mantle cavity. This efficient closed system effectively circulates oxygen and nutrients to body parts. The brain is formed from a fusion of the three molluscan ganglia. Nerves leave the brain and supply various parts of the body, including an especially large pair that control the rapid contraction of the mantle. The gonads take up a large part of the visceral mass, and the sexes are separate. Packets called spermatophores contain sperm, which the male passes to the female mantle cavity by means of its specialized tentacle. After the eggs are fertilized, they are attached to the substratum in elongated strings, each containing as many as 100 eggs.

Squids, which are cephalopods, have a closed circulatory system and a well-developed nervous system with cephalization. They are active predators in the deep ocean.

Table 34.1 compares the anatomy and behavior of a clam, which is a filter feeder, and a squid, which is an active predator.

Table 34.1

Comparison of Clam and Squid

	Clam	Squid
Food-intake	Filter feeder	Active predator
Skeleton	Heavy shell for protection	No external skeleton
Circulation	Open	Closed
Cephalization	None	Marked
Locomotion	Hatchet foot	Jet propulsion
Nervous system	Three separate ganglia	Brain and nerves

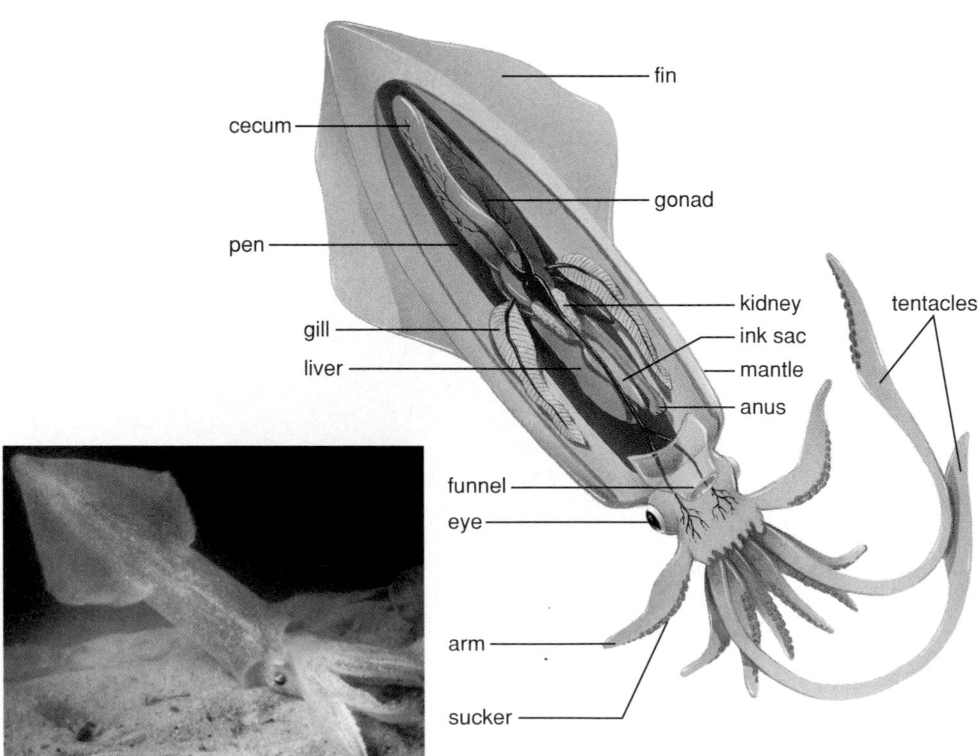

Figure 34.4 Squid, *Loligo.*

Gastropods Undergo Torsion

Gastropods [Gk. *gastros*, stomach, and *podos*, foot] (class Gastropoda)—including snails, whelks, conchs, periwinkles, and sea slugs—are usually found in marine habitats, although they sometimes inhabit freshwater environments. In addition, garden snails and slugs are adapted to terrestrial habitats. Many gastropods are herbivores that use their radula to scrape food from surfaces. Others are carnivores, using their radula to bore through surfaces such as bivalve shells to obtain food.

Gastropods have an elongated, flattened foot. In most, a well-developed head region with eyes and tentacles projects from a coiled shell that protects the visceral mass. Nudibranchs (sea slugs) and terrestrial slugs, however, lack a shell. During development, gastropods undergo a *torsion*, or twisting, that brings the anus and mantle cavity downward, then forward and around to a position above the head. Torsion positions the visceral mass squarely above the foot.

In aquatic gastropods, gills are found in the mantle cavity, but in those adapted to land, the mantle is richly supplied with blood vessels and functions as a lung when air is moved in and out through respiratory pores. As another adaptation to life out of water, terrestrial gastropod development does not include the swimming larval stage found in aquatic species.

Land snails, such as *Helix aspersa*, have three obvious divisions of the body: a head with two pairs of *tentacles*, one pair of which bears *eyes* at the tips; a flat, long, muscular foot; and a visceral mass surrounded by a shell (Fig. 34.5). The shell not only offers protection but also prevents desiccation (drying out). Waves of contraction run from the anterior to the posterior portion of the foot, and a lubricating mucus is secreted to facilitate movement.

Land snails are hermaphroditic; when two snails meet, they shoot calcareous darts into each other's body wall as a part of premating behavior. Then each inserts a *penis* into the vagina of the other to provide sperm for the future fertilization of eggs that are deposited in the soil. Development proceeds directly without the formation of larvae.

The presence of a copulatory organ such as the penis, and even hermaphroditism, are adaptations to life on land. The penis allows easy transfer of sperm from one animal to another; hermaphroditism assures that any two animals can mate. This is especially useful in slow-moving animals that have large ranges.

Table 34.2 compares adaptations of the clam and squid to water with adaptations of the snail to land.

Table 34.2

Comparison of Clam and Squid to Snail

	Clam and Squid	Snail
Skeleton	Protection in clam	Protection and prevention of desiccation
Locomotion	Suitable to beach (clam), water (squid)	Suitable to dry surface
Respiration	Gills that are kept moist by external water	Mantle serves as lungs
Excretion	Ammonia diluted in water	Uric acid as a solid
Reproduction	No penis, separate sexes, larval stage in clam	Penis, hermaphroditism, no larval stage

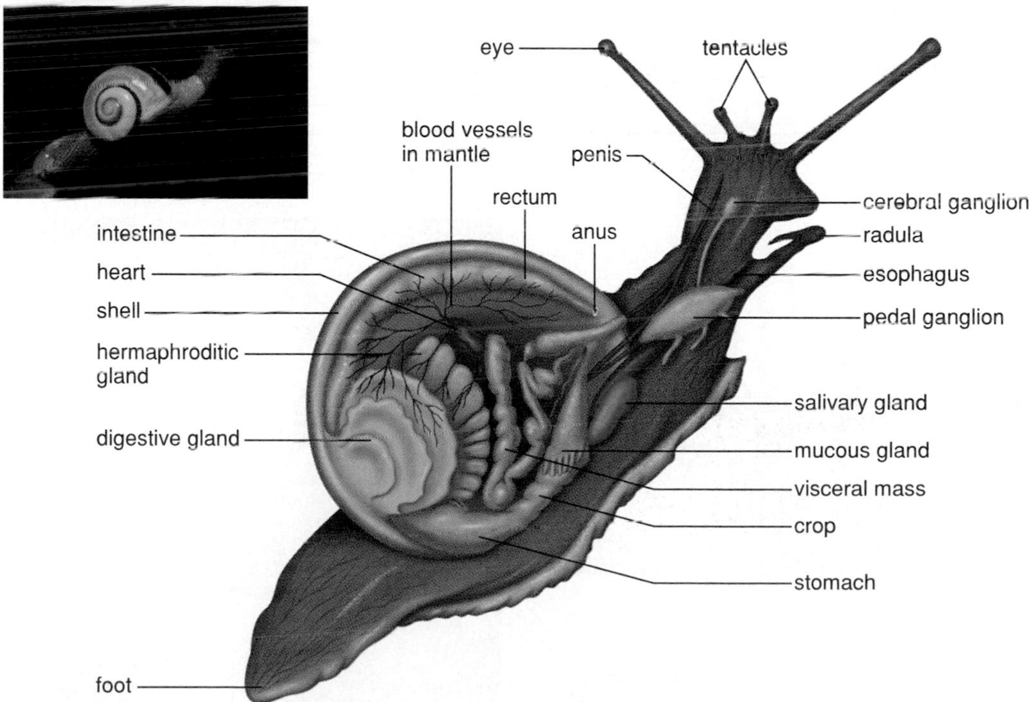

Figure 34.5 Land snail, *Helix*.

Land snails are gastropods, which are adapted to a terrestrial environment. All snails have a coiled shell, a large, flat, muscular foot, and a head region. In addition, the mantle in a garden snail becomes a lung.

34.3 Segmentation Evolves

Segmentation is a subdivision of the body along its length into repeating units, called segments. Segmentation may have evolved in conjunction with a hydrostatic skeleton. When an animal utilizes a hydrostatic skeleton, partitioning of the coelom permits each body segment independence of movement. Now instead of just burrowing in the mud, an animal can crawl on a surface. Once segmentation appeared, it became significant for another reason. Each segment or group of segments can become specialized to perform a particular function.

Annelids Are Segmented Worms

Annelids [L. *annelus,* dim. of *annulus,* ring] (**phylum Annelida,** 12,000 species) are *segmented,* as is externally evidenced by the rings that encircle the body. Internally, *partitions called septa* (sing., septum) *divide the well-developed, fluid-filled coelom,* which acts as a hydrostatic skeleton.

In annelids, the tube-within-a-tube body plan has led to *specialization of the digestive tract.* For example, the digestive system may include a pharynx, a stomach, and accessory glands. They have an extensive *closed circulatory system* with blood vessels that run the length of the body and branch to every segment. The nervous system consists of a *brain connected to a ventral solid nerve cord, with a ganglion in each segment.* The excretory system consists of *paired* **nephridia** [Gk. *nephros,* kidney], which are coiled tubules in each segment that collect waste material from the coelom and excrete it through openings in the body wall.

Annelids are segmented worms. The coelom is divided by septa, and organ systems have repeating parts. There is a closed circulatory system and a ventral solid nerve cord.

Marine Worms Have Parapodia

Most annelids (class Polychaeta) are marine; their class name refers to the presence of many setae. **Setae** [L. *seta,* bristle] are bristles that anchor the worm or help it move. In *polychaetes* [Gk. *polys,* many, and *chaite,* bristle, long hair], the setae are in bundles on *parapodia* [Gk. *para,* beside, and *podos,* foot], which are paddlelike appendages found on most segments. These are used not only in swimming but also as respiratory organs where the expanded surface area allows for exchange of gases. Clam worms (Fig. 34.6*a*) such as *Nereis* are predators. They prey on crustacea and other small animals, which are captured by a pair of strong chitinous jaws that extend with a part of the pharynx when the animal is feeding. Associated with its way of life, *Nereis* undergoes *cephalization* and has a head region with eyes and other sense organs.

Other polychaetes are sedentary (sessile) tube worms, with tentacles that form a funnel-shaped fan. Water currents, created by the action of cilia, trap food particles that are directed toward the mouth. They are *sessile filter feeders;* a sorting mechanism rejects large particles, and only the smaller ones are accepted for consumption.

Polychaetes have breeding seasons, and only during these times do the worms have sex organs. In *Nereis,* many worms concurrently shed a portion of their bodies containing either eggs or sperm, and these float to the surface where fertilization takes place. The zygote rapidly develops into a type of larva that is similar to that in marine clams. The existence of this larva in both the annelids and mollusks shows that these two groups of animals are related.

Polychaetes are marine worms with bundles of setae attached to parapodia.

a. Clam worm, *Nereis*

b. Feather-duster worm, *Sabellastarte*

Figure 34.6 Polychaete diversity.
a. *Nereis* is a predaceous polychaete that undergoes cephalization. Note the parapodia, which are used for swimming and as respiratory organs. **b.** Feather-duster worms (a type of tube worm) are sessile filter feeders whose tentacles form a funnel-shaped fan.

Earthworms Hide Underground

The **oligochaetes** [Gk. *oligos*, few, and *chaite*, bristle, long hair] (class Oligochaeta), which include earthworms, have few setae. Earthworms (e.g., *Lumbricus*) do not have a well-developed head or parapodia (Fig. 34.7). Their setae protrude in clusters directly from the surface of their body. Locomotion, which is accomplished section by section, utilizes muscle contraction and the setae. When longitudinal muscles contract, segments bulge and their setae protrude into the dirt; then when circular muscles contract, the setae are withdrawn and these segments move forward.

Earthworms reside in soil where there is adequate moisture for the body wall to remain moist for gas exchange purposes. They are scavengers that feed on leaves or any other organic matter, living or dead, which can conveniently be taken into the mouth along with dirt. Food drawn into the mouth by the action of the muscular pharynx is stored in a crop and ground up in a thick, muscular gizzard. Digestion and absorption occur in a long intestine whose dorsal surface is expanded by a *typhlosole* that allows additional surface for absorption.

Earthworm segmentation, which is so obvious externally, is also internally evidenced by septa. The long ventral solid nerve cord leading from the brain has ganglionic swellings and lateral nerves in each segment. The paired nephridia, or coiled tubules, in each segment have two openings: one is a ciliated funnel that collects coelomic fluid and the other is an exit in the body wall. Between the two openings is a convoluted region where waste material is removed from the blood vessels about the tubule. Red blood moves anteriorly in a dorsal blood vessel and then is pumped by five pairs of hearts into a ventral vessel. As the ventral vessel takes the blood toward the posterior regions of the worm's body, it gives off branches in every segment. Altogether, segmentation is evidenced by

- body rings
- coelom divided by septa
- setae on each segment
- ganglia and lateral nerves in each segment
- nephridia in most segments
- branch blood vessels in each segment

The worms are *hermaphroditic;* the male organs are the testes, the seminal vesicles, and the sperm ducts, and the female organs are the ovaries, the oviducts, and the seminal receptacles. Two worms lie parallel to each other facing in opposite directions. The fused midbody segment, called a *clitellum,* secretes mucus, protecting the sperm from drying out as they pass between the worms. After the worms

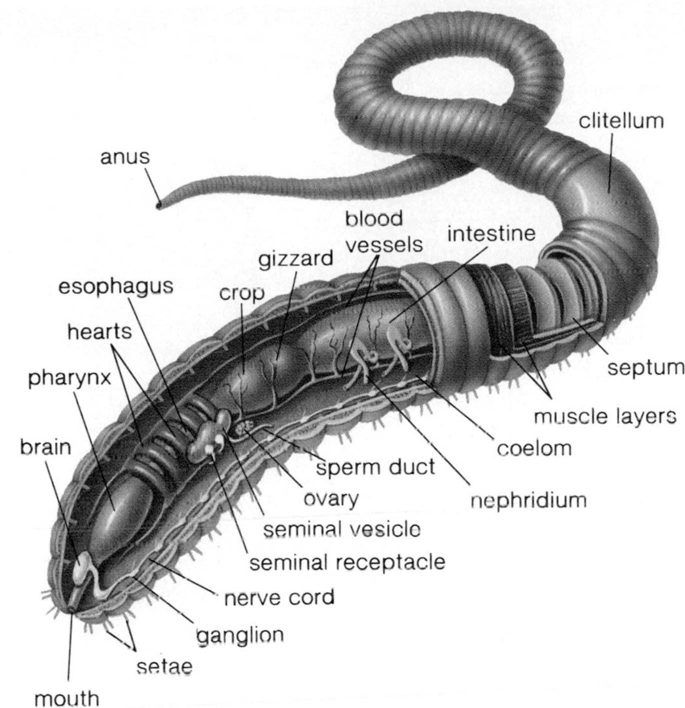

Figure 34.7 Earthworm, *Lumbricus*.
a. Internal anatomy of the anterior part of an earthworm. Notice that each body segment bears a pair of setae and that internal septa divide the coelom into compartments. b. When earthworms mate, they are held in place by a mucus secreted by the clitellum. The worms are hermaphroditic, and sperm pass from the seminal vesicles of each to the seminal receptacles of the other.

Health focus

Bloodsuckers

Parasitic leeches attach themselves to the body of their hosts and suck blood or other body fluids from fishes, turtles, snails, or mammals. The European species of *Hirudo medicinalis* was used in the 1700s and 1800s to "let blood" in feverish patients in order to withdraw any poisons and "excess" blood. This procedure, which is most debilitating for the patient, is no longer practiced, but leeches are still used on occasion to remove blood from bruised skin. The anterior sucker of the medicinal leech has three jaws that saw through the skin, producing a Y-shaped incision through which blood is drawn up by the sucking action of the muscular pharynx (Fig. 34A). The salivary glands secrete an anticoagulant called hirudin that keeps the blood flowing while the leech is feeding. Other salivary ingredients dilate the host's blood vessels and act as an anesthetic. A medicinal leech can take up to five times its body weight in blood because the crop has pouches where the blood can be stored as the animal expands in size. When it has taken its fill, the leech drops off and digests its meal. Complete digestion takes a long time, and it's been suggested that a leech needs to feed only once a year.

Not all leeches suck blood or fluids; some are even predaceous and eat a variety of small invertebrates.

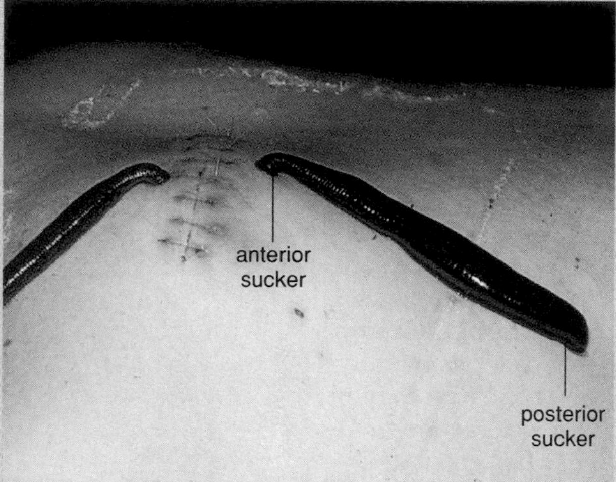

Figure 34A Medicinal leeches, *Hirudo*. They are sometimes used medically to remove blood that has accumulated in damaged tissues.

separate, the clitellum of each produces a slime tube, which is moved along over the head by muscular contractions. As it passes, eggs and the sperm received earlier are deposited and fertilization occurs. The slime tube then forms a cocoon to protect the worms as they develop. There is no larval stage.

It is interesting to compare the anatomy of clam worms with earthworms because it highlights the manner in which earthworms are adapted to life on land. Jaws are not needed by the nonpredatory earthworms that extract organic remains from the soil they eat. The lack of parapodia helps reduce the possibility of water loss and facilitates burrowing in soil. The clam worm makes use of external water while the earthworm provides a mucous secretion to aid fertilization. It is the water form that has the swimming larva and not the land form.

Earthworms, which burrow in the soil, lack cephalization and parapodia. They are hermaphroditic, and there is no larval stage.

Leeches Are Parasites

Leeches (class Hirudinea) are usually found in fresh water, but some are marine or even terrestrial. They have the same body plan as other annelids but they have no setae, and each body ring has several transverse grooves. Most leeches are only 2–6 cm in length but some, including the medicinal leech, are as long as 20 cm.

Among their modifications are two *suckers,* a small oral one around the mouth and a large posterior one. While some leeches are free-living predators, most are fluid feeders that attach themselves to open wounds. Some bloodsuckers, such as the medicinal leech, are able to cut through tissue. Leeches are able to keep blood flowing and prevent clotting by means of a substance in their saliva known as *hirudin,* a powerful anticoagulant. This has added to their potential usefulness in the field of medicine today, as discussed in the accompanying reading.

Leeches are modified in a way that lends itself to the parasitic way of life. Some are external parasites known as bloodsuckers.

34.4 Jointed Appendages Evolve

Arthropods are related to annelids, as discussed in the reading on page 607. Arthropods are segmented, but they also have a rigid, but jointed, **exoskeleton** [Gk. *ex,* out of, and *skeleton,* dried body]. Arthropod literally means "jointed foot," but actually they have freely movable **jointed appendages.** Segmentation and jointed appendages are seen both in arthropods and vertebrates, two groups that are quite successful on land.

Arthropods Have Jointed Appendages

Arthropods [Gk. *arthron,* joint, and *podos,* foot] (**phylum Arthropoda,** over one million species) show such diversity and are adapted to so many different habitats that they are often said to be the most successful of all animals. Some claim the modern era should be called the age of arthropods, not the age of mammals!

What characteristics account for the success of arthropods? First, there is the strong but flexible *exoskeleton.* The exoskeleton of arthropods is composed primarily of **chitin** [Gk. *chiton,* tunic], a strong, flexible, nitrogenous polysaccharide. The *exoskeleton* serves many functions such as protection, attachment for muscles, locomotion, and prevention of desiccation. However, because it is hard and nonexpandable, arthropods must **molt,** or shed, the exoskeleton as they grow larger. Before molting, the body secretes a new, larger exoskeleton, which is soft and wrinkled, underneath the old one. After enzymes partially dissolve and weaken the old exoskeleton, the animal breaks it open and wriggles out. The new exoskeleton then quickly expands and hardens.

Arthropods are segmented but some *segments are fused* into regions, such as a head, a thorax, and an abdomen. In trilobites (subphylum Trilobitomorpha), which flourished during the Cambrian period, there was a pair of appendages on each body segment. In modern arthropods, *appendages are specialized* for such functions as walking, swimming, reproducing, eating, and sensory reception. These modifications account for much of the diversity of arthropods. Several arthropod groups, such as insects, arachnids, centipedes, and millipedes, contain species that are adapted to terrestrial life.

Arthropods have a *well-developed nervous system.* There is a brain and a ventral solid nerve cord. The head bears various types of sense organs, including eyes of two types—compound and simple (Fig. 34.8). The compound eye is composed of many complete visual units, each of which operates independently. The lens of each visual unit focuses an image on the light-sensitive membranes of a small number of photoreceptors within that unit. The simple eye has a single lens that brings the image to focus into many receptors, each of which receives only a portion of the image.

Arthropods have a *variety of respiratory organs.* Marine forms utilize gills, which are vascularized, highly convoluted, thin-walled tissue specialized for gas exchange. Terrestrial forms have book lungs (e.g., spiders) or air tubes called **tracheae** [L. *trachia,* windpipe]. Tracheae serve as a rapid way to transport oxygen directly to the cells.

Finally, the occurrence of metamorphosis has contributed to the success of arthropods. **Metamorphosis** [Gk. *meta,* implying change, and *morphe,* shape, form] is a drastic change in form and physiology that occurs as an immature stage, called a larva, becomes an adult. Among arthropods, the larva eats different food and lives in a different environment than the adult. This reduces competition and allows more members of a species to exist at one time. For example, larval crabs live among and feed on plankton, while adult crabs are bottom dwellers that catch live prey or scavenge dead organic matter. Among insects, such as butterflies, the caterpillar feeds on leafy vegetation while the adult feeds on nectar.

A phylum can be divided into subphyla, and we will consider three arthropod subphyla: the chelicerates (e.g., spiders), the crustacea (e.g., crayfish), and the uniramia (e.g., millipedes, centipedes, and insects). There can also be superclasses, which include more than one class; the millipedes and centipedes are in a superclass and the insects are another superclass.

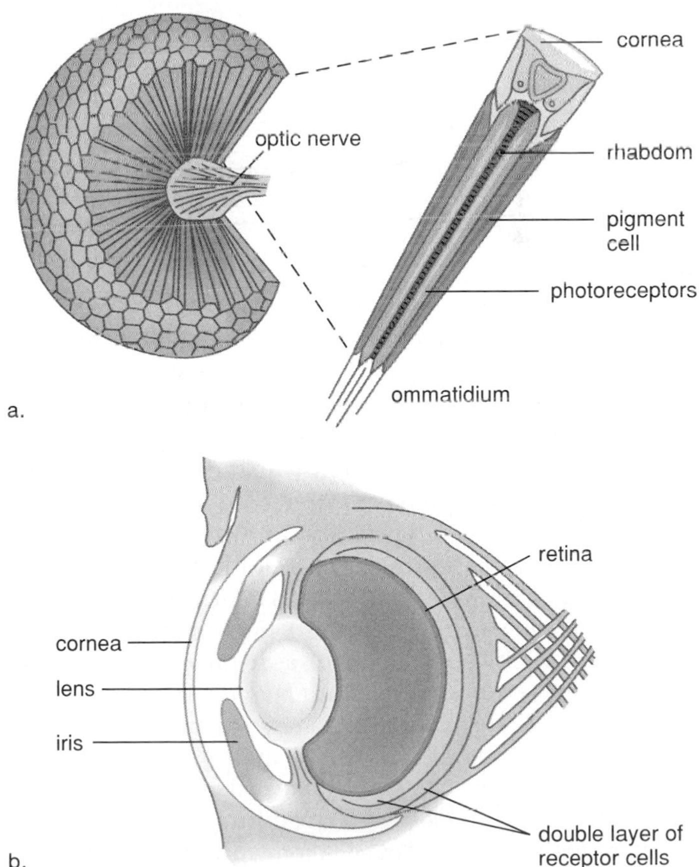

Figure 34.8 Compound versus simple eyes.
a. The compound eye of arthropods contains many individual units, each of which has its own lens. The lens focuses the image onto a rhabdom consisting of the light-sensitive membranes of a small number of photoreceptors. b. The simple eye of cephalopods has a single lens, which focuses the image onto a double layer of receptor cells.

a. Kenyan giant scorpion, *Pandinus*

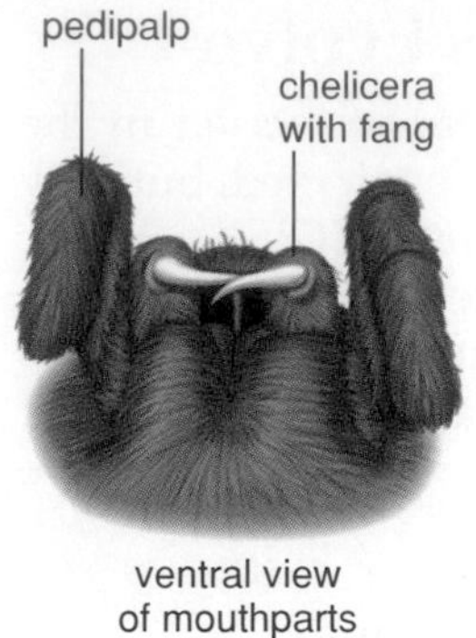

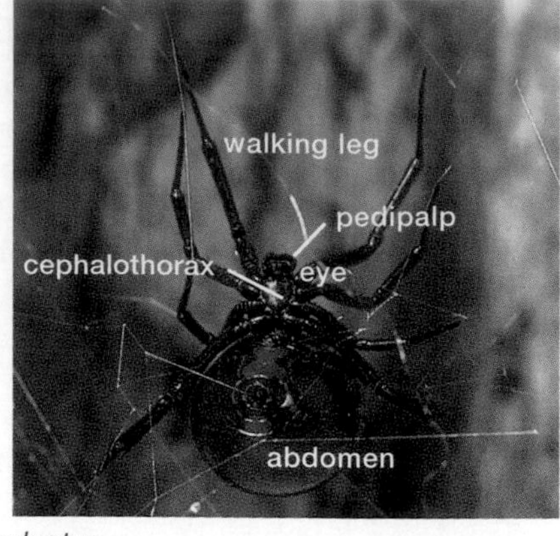

c. Black widow spider, *Latrodectus*

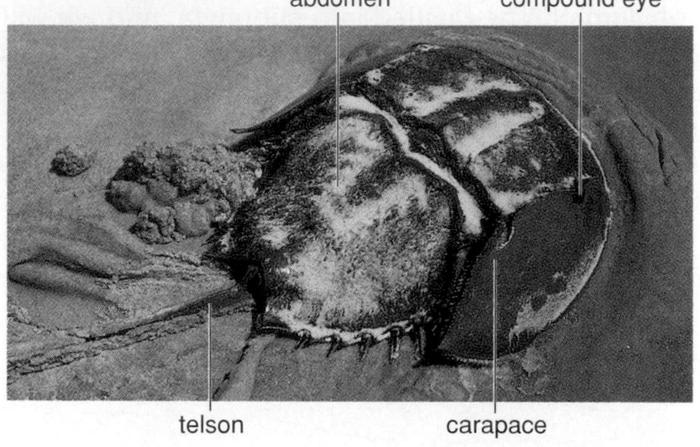

b. Horseshoe crab, *Limulus*

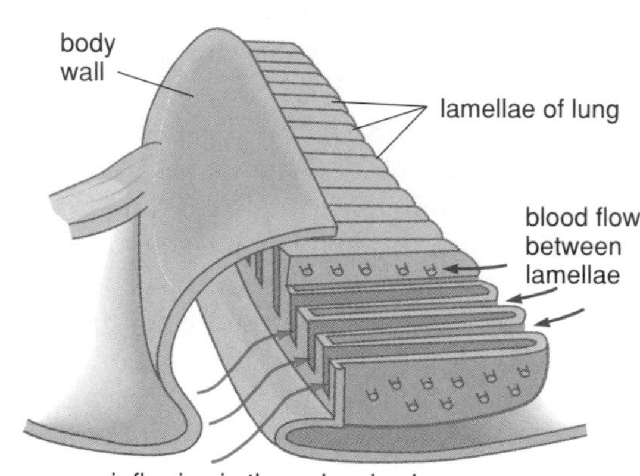

d.

Figure 34.9 Chelicerate diversity.
a. Scorpions are more common in tropical areas. **b.** Horseshoe crabs are common along the east coast. **c.** The black widow spider is a poisonous spider found in the United States. **d.** Terrestrial chelicerates, such as spiders, breathe by means of book lungs.

Chelicerates Have Pincerlike Appendages

The **chelicerates** (**subphylum Chelicerata,** 57,000 species) include terrestrial spiders, scorpions, ticks, mites, and the less familiar marine horseshoe crabs and sea spiders (Fig. 34.9). In this group, the first pair of appendages—the pincerlike *chelicerae*— are feeding organs. The second pair—*pedipalps*—are feeding or sensory in function. The pedipalps are followed by four pairs of walking legs. All of these appendages are attached to a **cephalothorax** [Gk. *kephale,* head, and *thorax,* breastplate] (fused head and thorax), which is followed by an abdomen that contains internal organs. There are no appendages found on the heads of these animals (antennae, mandibles, or maxillae), as in the other subphyla.

Horseshoe crabs of the genus *Limulus* are familiar along the east coast of North America. They scavenge sandy and muddy substrates for annelids, small mollusks, and other invertebrates. The body is covered by exoskeletal shields. The anterior shield is a horseshoe-shaped *carapace,* which bears two prominent compound eyes. A long, unsegmented telson projects to the rear. These marine animals have book gills, named for their resemblance to the pages of a closed book.

Scorpions, which are arachnids, are the oldest terrestrial arthropods. Today, they occur in the tropics, subtropics, and temperate regions of North America. They are nocturnal and spend most of the day hidden under a log or a rock. In these animals, the pedipalps are large pincers, and the long abdomen ends with a stinger that contains venom. Ticks and mites, with 25,000 species, may outnumber all other kinds of arachnids. They are often parasites. Ticks suck the blood of vertebrates and are sometimes transmitters of diseases, such as Rocky Mountain spotted fever or Lyme disease. Chiggers, the larvae of certain mites, feed on the skin of vertebrates.

Spiders, the most familiar arachnids, have a narrow waist that separates the cephalothorax from the abdomen. Spiders don't have compound eyes; instead, they have numerous simple eyes that perform a similar function. The chelicerae are modified as fangs, with ducts from poison glands, and the pedipalps are used to hold, taste, and chew food. The abdomen often contains silk glands, and they spin a web in which to trap their prey. Invaginations of the body wall form lamellae ("pages") of their so-called *book lungs.* Air flowing into the folded lamellae on one side exchanges gases with blood flowing in the opposite direction on the other side.

The chelicerates include horseshoe crabs, spiders, and scorpions. These animals have pincerlike appendages called chelicerae, which are modified as fangs in spiders.

Crustacea Have a Calcified Exoskeleton

Crustacea (subphylum Crustacea, 40,000 species) are successful, largely marine, arthropods. Crustacea are named for their hard shells; the exoskeleton is calcified. Although their anatomies are extremely diverse, the head usually bears a pair of compound eyes and *five pairs of appendages.* The first two pairs, called antennae and antennules, lie in front of the mouth and have sensory functions. The other three pairs (mandibles, first and second maxillae) lie behind the mouth and are usually used in feeding as mouthparts. *Biramous* [Gk. *bis,* two, and *ramus,* a branch] *appendages* on the thorax and abdomen are segmentally arranged; one branch is the gill branch and the other is the leg branch:

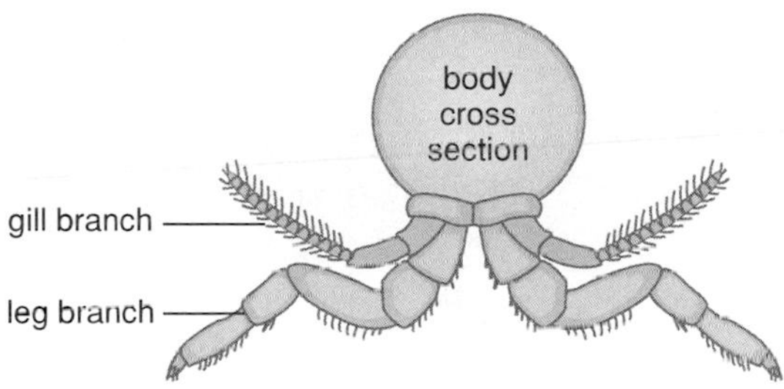

Copepods and krill are small crustacea that live in the water, where they feed on algae. In the marine environment, they serve as food for fishes, sharks, and whales. They are so numerous that despite their small size, some believe that they are harvestable as food. Barnacles are also crustacea but they have a thick, heavy shell as befits their inactive life-style. Stalked (gooseneck) barnacles are attached by a stalk, and stalkless (acorn) barnacles are attached directly by their shells. You see barnacles on wharf pilings, ship hulls, seaside rocks, and even the bodies of whales. They begin life as free-swimming larvae, but they undergo a metamorphosis that transforms their swimming appendages to cirri, feathery structures that are extended and allow them to filter feed when they are submerged.

Decapods are the most familiar and numerous of the crustacea. They include shrimps, lobsters, crayfish, and crabs, in which the thorax bears five pairs of walking legs. The first pair may be modified as claws. Typically, there are gills situated above the walking legs. Figure 34.10*a* gives a view of the external anatomy of the crayfish. The head and thorax are fused into a *cephalothorax,* which is covered on the top and sides by a nonsegmented carapace. The abdominal segments are equipped with *swimmerets,* small paddlelike structures. The first two pairs of swimmerets in the male are quite strong and are used to pass sperm to the female.

The last two segments bear the uropods and the telson, which make up a fan-shaped tail. Ordinarily, a crayfish lies in wait for prey. It faces out from an enclosed spot with the claws extended and the antennae moving about. The claws seize any small animal, either dead ones or live ones that happen by, and carry them to the mouth. When a crayfish moves about, it generally crawls slowly but may swim rapidly by using its heavy abdominal muscles and tail.

The respiratory system consists of gills that lie above the walking legs protected by the carapace. As shown in Figure 34.10*b*, the digestive system includes a stomach, which is divided into two main regions: an anterior portion called the *gastric mill,* equipped with chitinous teeth to grind coarse food, and a posterior region, which acts as a filter to prevent coarse particles from entering the digestive glands where absorption takes place. *Green glands* lying in the head region, anterior to the esophagus, excrete metabolic wastes through a duct that opens externally at the base of the antennae. The *coelom,* which is so well developed in the annelids, *is reduced* in the arthropods and is composed chiefly of the space about the reproductive system. A heart pumps hemolymph containing the respiratory pigment hemocyanin into a **hemocoel** [Gk. *haima,* blood,

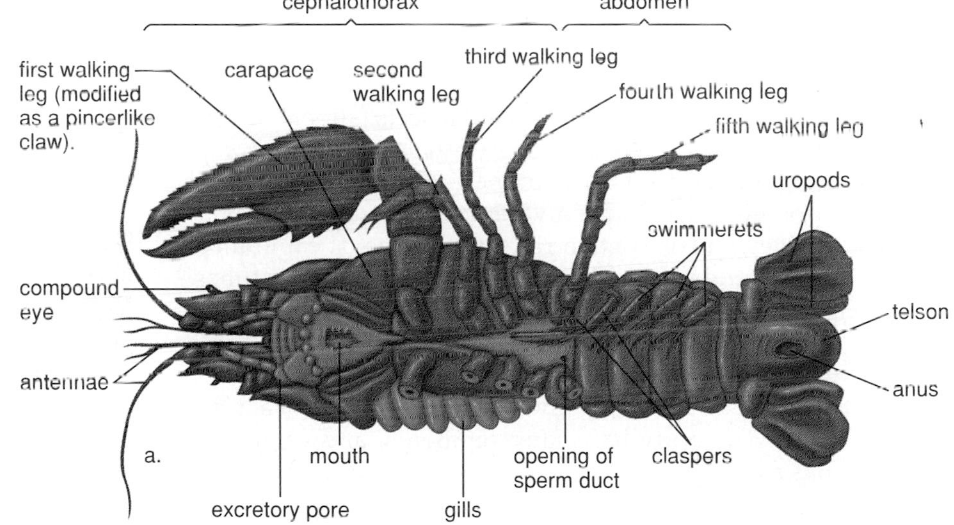

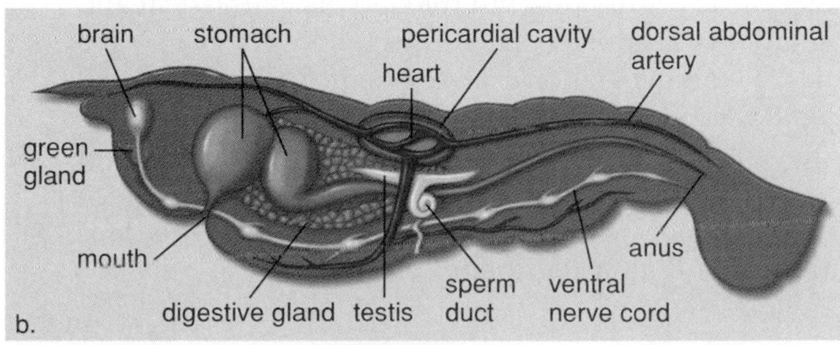

Figure 34.10 Male crayfish, *Cambarus.*
a. Externally, it is possible to observe the jointed appendages, including the swimmerets, the walking legs, and the claws. These appendages, plus a portion of the carapace, have been removed from the right side so that the gills are visible. **b.** Internally, the parts of the digestive system are particularly visible. The circulatory system can also be clearly seen. Note the ventral solid nerve cord.

and *koiloma,* cavity] consisting of sinuses (open spaces) where the hemolymph flows about the organs. (Whereas hemoglobin is a red pigment, hemocyanin is a blue pigment.) This is an open circulatory system because blood is not contained within blood vessels.

The nervous system is quite similar to that of the earthworm. There is a brain, as well as a *ventral nerve cord* that passes posteriorly. Along the length of the nerve cord, periodic ganglia give off lateral nerves.

The sexes are separate in the crayfish, and the gonads are located just ventral to the pericardial cavity. In the male, a coiled sperm duct opens to the outside at the base of the fifth walking leg. Sperm transfer is accomplished by the modified first two swimmerets of the abdomen. In the female, the ovaries open at the bases of the third walking legs. A stiff fold between the bases of the fourth and fifth pairs serves as a seminal receptacle. Following fertilization, the eggs are attached to the swimmerets of the female.

Crustacea are mainly marine arthropods in which the head bears five pairs of appendages, including mandibles and maxillae. Typically, there are biramous appendages.

Uniramia Breathe with Tracheae

Uniramia (subphylum Uniramia) include millipedes, centipedes (superclass Myriapoda), and insects (superclass Insecta) whose *uniramous appendages* attached to the thorax and abdomen have only one branch—the leg branch. The head appendages include only *one pair of antennae,* one pair of mandibles, and one or two pairs of maxillae. The uniramia live on land and breathe by means of a system of air tubes called tracheae.

Centipedes and Millipedes

Centipedes (2,500 species) have a body composed of a head and trunk. The body has many segments, and each segment has a pair of walking legs. They are carnivorous animals, and the head bears antennae and mouthparts with poison fangs.

Millipedes (10,000 species) have the same segmented organization as centipedes, but the body is cylindrical and some segments are fused. Therefore they appear to have two pairs of walking legs on each segment. Millipedes dwell in the soil, feeding on dead organic matter.

The term centipede means one hundred legs and millipede means one thousand legs. Although these arthropods do not have this many legs, a millipede does have more legs than a centipede (Fig. 34.11).

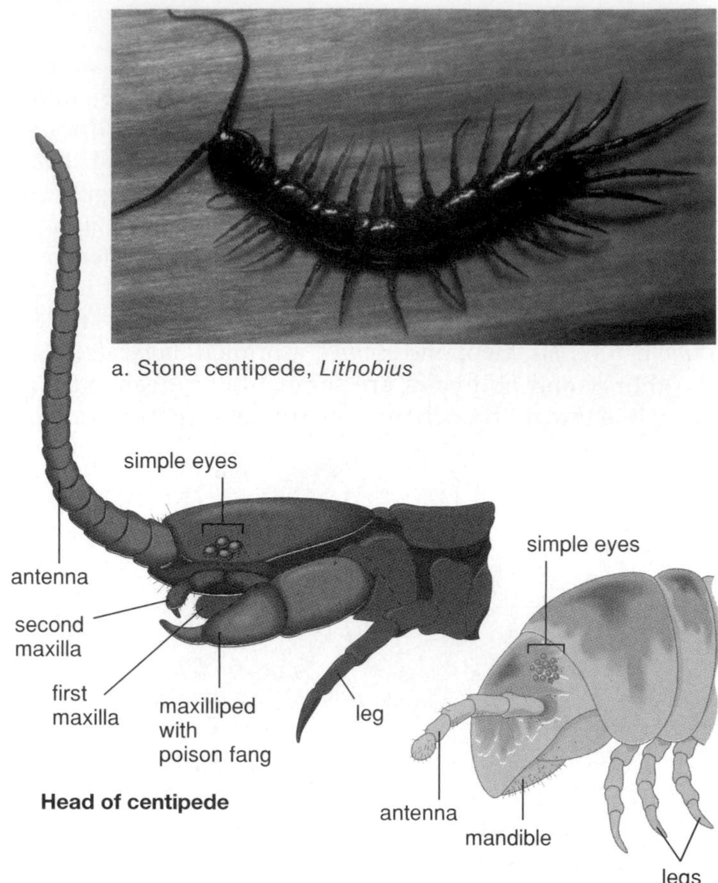

Figure 34.11 Centipede and millipede.
a. A centipede has a pair of appendages on most every segment. **b.** Head of centipede compared to head of millipede. A centipede is a carnivorous animal and kills its prey with a poison fang. Most millipedes are herbivorous and usually feed on decayed plant matter. **c.** A millipede has two pairs of legs on most segments.

A Closer Look

Evolutionary Relationships Among the Protostomes

The protostomes (mollusks, annelids, and arthropods) share embryological features such as spiral and determinate cleavage, the fate of the blastopore, and the manner in which the coelom forms (see Fig. 34.1), but exactly how they are related is not known. However, there is some evidence that the mollusks and annelids share a common ancestor and that annelids and arthropods share a common ancestor.

Mollusks are not segmented as are the annelids except for those in the genus *Neopilina* (Fig. 34B). These unusual mollusks were presumed to be extinct for the past 500 million years, but ten living specimens were dredged up from a depth of more than 3,500 meters in the Pacific Ocean near Costa Rica in 1952. As expected for mollusks, they have a mantle, a single dorsal shell, a ventral muscular foot, and a radula. However, they also have a segmental arrangement of gills, nephridia, and muscles. This evidence of segmentation suggests that *Neoplina* could be similar to a possible common ancestor of both mollusks and annelids. On the other hand, it is possible that the segmentation seen in *Neopilina* arose independently in the molluskan line and that both mollusks and annelids are descended from nonsegmented flatworms.

Similarly, there are animals that show a possible evolutionary relationship between the annelids and arthropods. *Peripatus* in the phylum Onychophora (Fig. 34Ca) are called walking worms because they have 14 to 40 pairs of short, stumpy legs. Although these appendages resemble annelid parapodia, they are arthropod-like in holding the body up off the ground. There is a cuticle of chitin and a well-defined head with a pair of mandibles, both of which are reminiscent of arthropods. However, some internal organs resemble those of annelids and some have arthropod characteristics. The excretory organs and musculature are more like those of annelids, but the circulatory and respiratory systems are more like those of arthropods. The excretory system contains segmentally arranged nephridia that collect fluid from a coelomic sac and open to the outside by nephridiopores. Members of the *Peripatus* genus have an open circulatory system with a dorsal tubular heart. And the circulatory system does not distribute oxygen to the tissues. Instead, there is a tracheal system of air tubes with spiracles that serve this function. Although the appendages of *Peripatus* are not jointed, the fossil record does contain wormlike animals with jointed appendages. It's possible that arthropods and annelids share a common ancestor, or perhaps arthropods are descended directly from annelids.

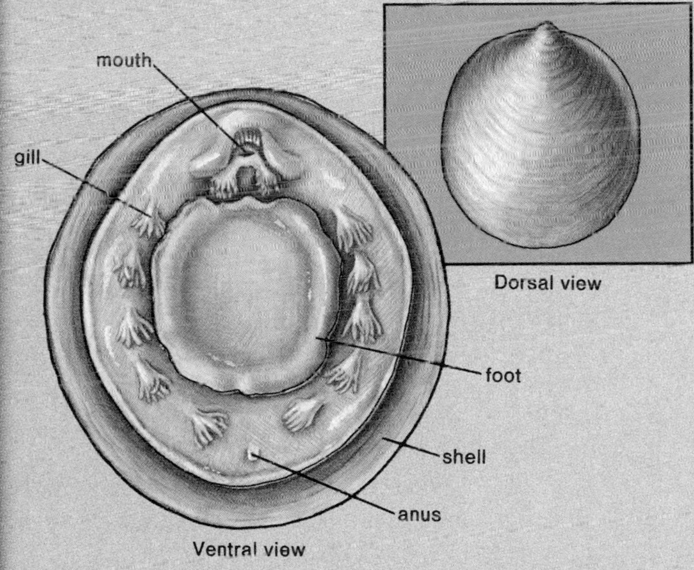

Figure 34B Mollusk, *Neopilina*.

a.

b.

Figure 34C Walking worms, phylum Onychophora. a. General appearance of *Peripatus* sp. b. Anterior region of *Macroperipatus torquatus* viewed from below to show its appendages.

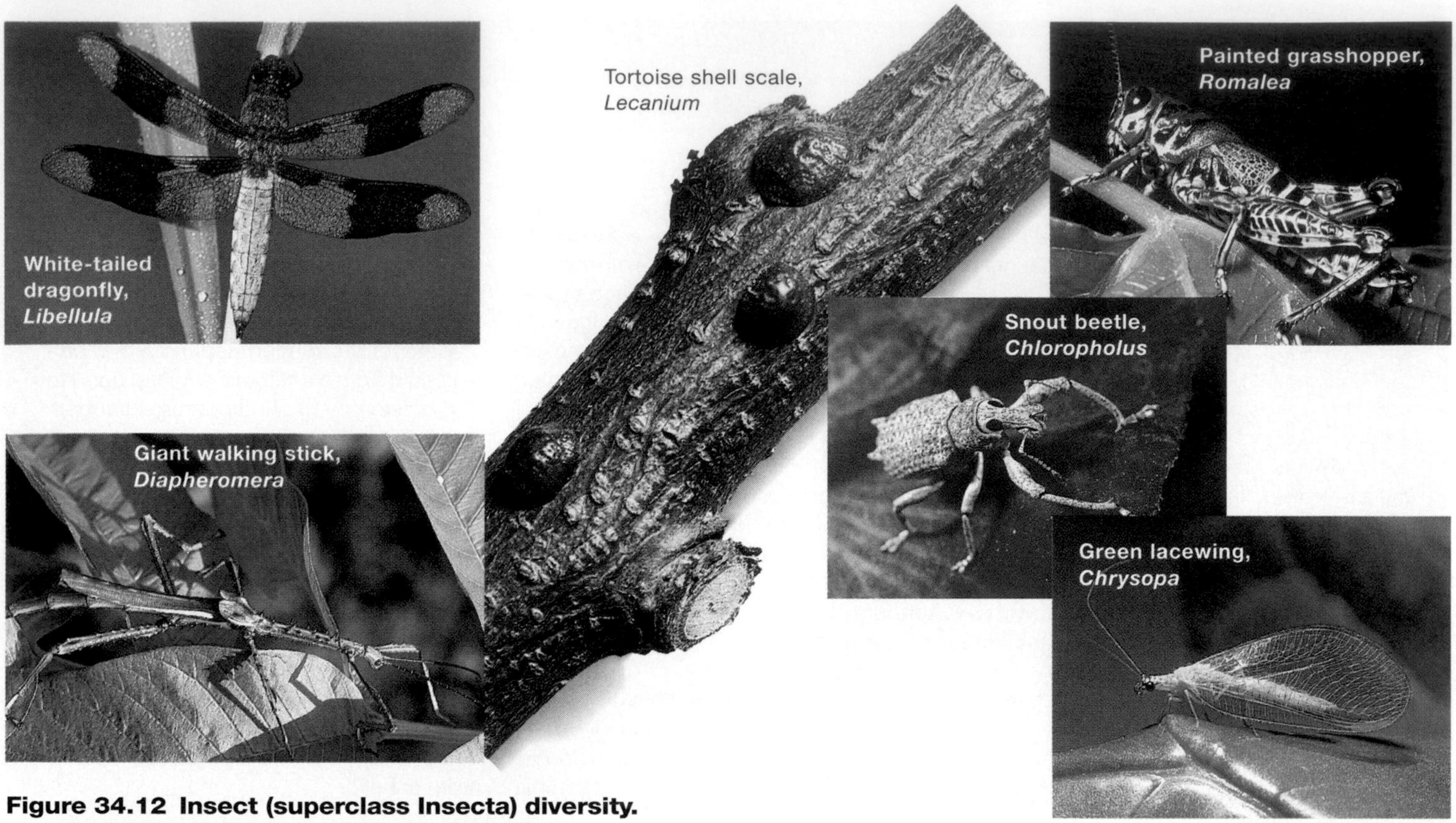

White-tailed dragonfly, *Libellula*
Tortoise shell scale, *Lecanium*
Painted grasshopper, *Romalea*
Giant walking stick, *Diapheromera*
Snout beetle, *Chloropholus*
Green lacewing, *Chrysopa*

Figure 34.12 Insect (superclass Insecta) diversity.

ocelli
right mandible
left mandible
right maxilla with maxillary palp
left maxilla with maxillary palp
labrum
labium with labial palps
Chewing (grasshopper)
maxilla
Siphoning Tube (butterfly)
labrum
Siphoning Tube (housefly)

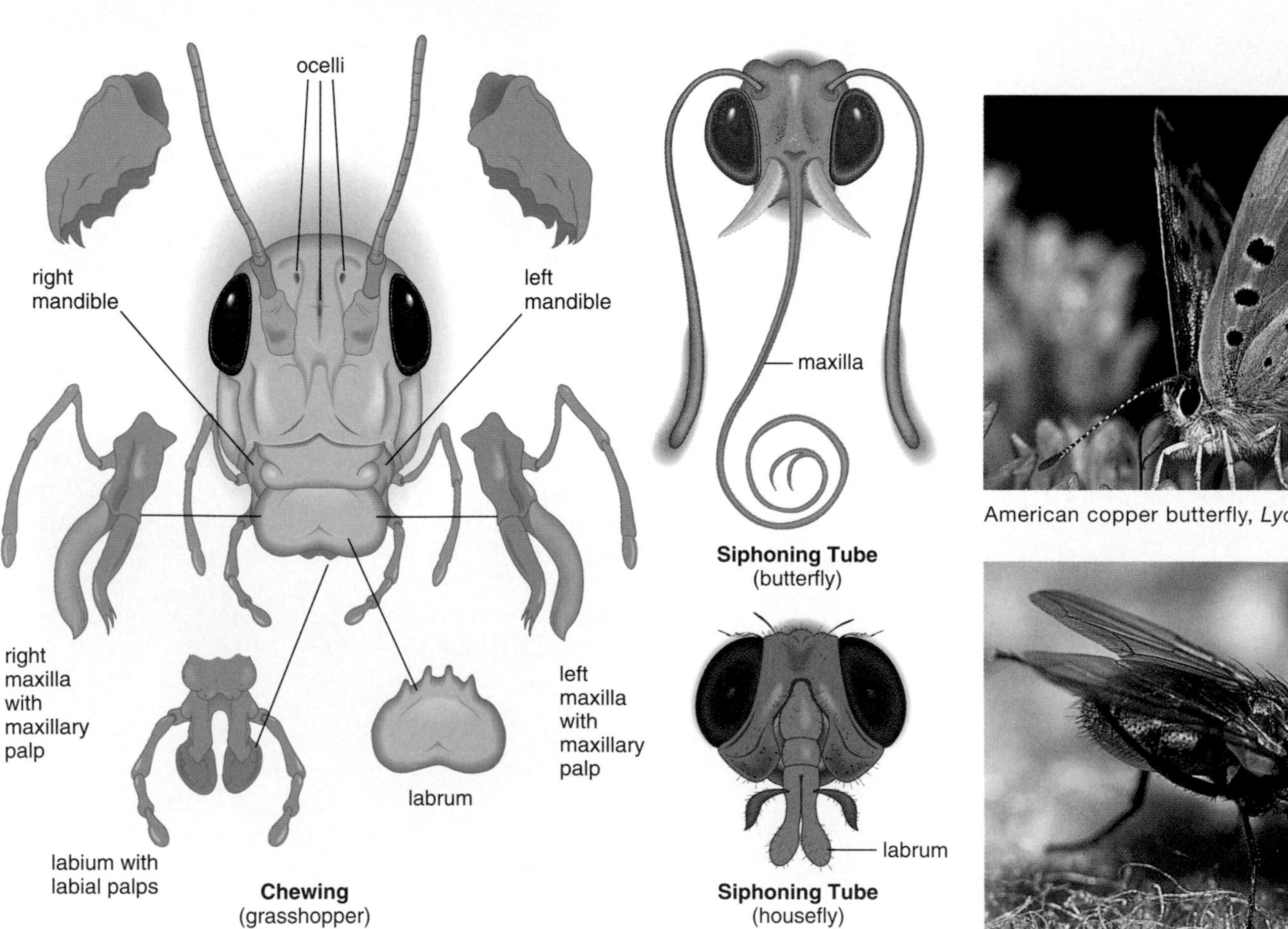

American copper butterfly, *Lycaena*

Housefly, *Musca*

Figure 34.13 Three types of insect mouthparts.

Most Insects Have Wings

Insects (superclass Insecta, 900,000 species) include more known species than all other animal species combined. Only a few types can be represented in Figure 34.12. Insects are adapted for an active life on land, although some have secondarily invaded aquatic habitats. The body of an insect is divided into a head, a thorax, and an abdomen. The head bears the sense organs and mouthparts (Fig. 34.13); the thorax bears three pairs of legs and one or two pairs of wings; and the abdomen contains most of the internal organs. Wings enhance an insect's ability to survive by providing a way of escaping enemies, finding food, facilitating mating, and dispersing the species. The exoskeleton of an insect is lighter and contains less chitin than that of many other arthropods.

In the grasshopper (Fig. 34.14), the third pair of legs is suited to jumping. There are two pairs of wings. The forewings are tough and leathery, and when folded back at rest they protect the broad, thin hindwings. On the lateral surface, the first abdominal segment bears a large tympanum on each side for the reception of sound waves. The posterior region of the exoskeleton in the female has two pairs of projections that form an ovipositor, which is used to dig a hole in which eggs are laid.

The digestive system is suitable for a herbivorous diet. In the mouth, food is broken down mechanically by mouthparts and enzymatically by salivary secretions. Food is temporarily stored in the crop before passing into the gizzard, where it is finely ground. Digestion is completed in the stomach, and nutrients are absorbed into the hemocoel from outpockets called gastric ceca. A *cecum* is a cavity open at one end only. The excretory system consists of **Malpighian tubules,** which extend into a hemocoel and collect nitrogenous wastes that are concentrated and excreted into the digestive tract. The formation of a solid nitrogenous waste, namely uric acid, conserves water.

The respiratory system begins with openings in the exoskeleton called spiracles. From here, the air enters small tubules called tracheae (Fig. 34.14*a*). The tracheae branch and rebranch, finally ending in moist areas where the actual exchange of gases takes place. The movement of air through this complex of tubules is not a passive process; air is pumped through by a series of several bladderlike structures (air sacs), which are attached to the tracheae near the spiracles. Air enters the anterior four spiracles and exits by the posterior six spiracles. Breathing by tracheae may account for the small size of insects (most are less than 60 mm in length) since the tracheae are so tiny and fragile that they would be crushed by any amount of weight.

The circulatory system contains a slender, tubular heart that lies against the dorsal wall of the abdominal exoskeleton and pumps hemolymph into the hemocoel where it circulates before returning to the heart again. The hemolymph is colorless and lacks a respiratory pigment, and the tracheal system transports gases.

Reproduction is adapted to life on land. The male has a penis, which passes sperm to the female. Internal fertilization protects both gametes and zygotes from drying out. The female deposits the fertilized eggs in the ground with her ovipositor.

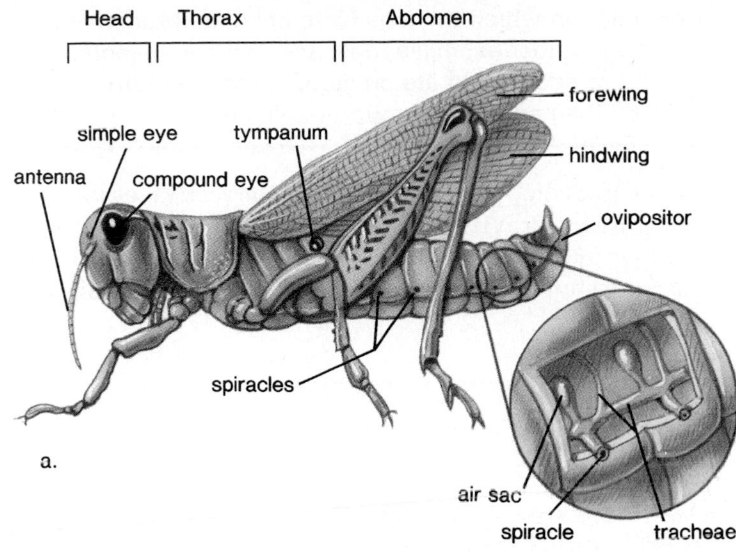

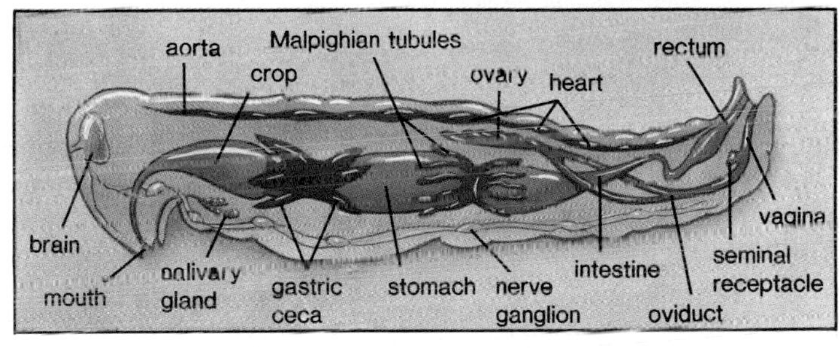

Figure 34.14 Female grasshopper, *Romalea*.
a. Externally, the tympanum uses air waves for sound reception, and the hopping legs and the wings are for locomotion.
b. Internally, the digestive system is specialized. The Malpighian tubules excrete a solid nitrogenous waste (uric acid). A seminal receptacle receives sperm from the male, which has a penis.

Grasshoppers undergo *incomplete metamorphosis,* a gradual change in form as the animal matures. The immature grasshopper, called a nymph, is recognizable as a grasshopper, even though it differs somewhat in shape and form from the adult. Other insects, such as butterflies, undergo *complete metamorphosis,* involving drastic changes in form. At first, the animal is a wormlike larva (caterpillar) with chewing mouthparts. It then forms a case, or cocoon, about itself and becomes a *pupa*. During this stage, the body parts are completely reorganized, the *adult* then emerges from the cocoon. This life cycle allows the larvae and adults to make use of different food sources.

Insects also show remarkable behavior adaptations, exemplified by the social systems of bees, ants, termites, and other colonial insects. Insects are so numerous and so diverse that the study of this one group is a major specialty in biology called entomology.

Uniramia—in which the legs have only one branch—include centipedes, millipedes, and insects. Insects, which are adapted to life on land, comprise more species than any other group of animals.

The grasshopper is adapted to a terrestrial environment while the crayfish is adapted to an aquatic environment. In crayfish, gills take up oxygen from water, while in the grasshopper, tracheae allow oxygen-laden air to enter the body. Appropriately, the crayfish has an oxygen-carrying pigment and a grasshopper has no such pigment in its blood. A liquid waste (ammonia) is excreted by a crayfish, while a solid waste (uric acid) is excreted by a grasshopper. Only in grasshoppers (1) is there a tympanum for the reception of sound waves and (2) do males have a penis and females an ovipositor. To swim, crayfish utilize tail (telson and uropods) and abdominal muscles; a grasshopper has legs for hopping and wings for flying.

connecting concepts

Ninety percent of all animal species have a coelom and are in the phyla Mollusca, Annelida, Arthropoda, and Chordata. (The chordates, which include the vertebrates, will be studied in the following chapter.) This would indicate that there are some adaptive advantages to having a coelom, or perhaps a combination of bilateral symmetry and the complexity made possible by the presence of a coelom. Most animals with a coelom have both a circulatory and respiratory system, which permit them to be active.

Animals with a coelom are divided into the protostomes (mollusks, annelids, and arthropods) and the deuterostomes (echinoderms and chordates) on the basis of embryological development. However, it is of interest that both lines of descent have representatives that are segmented. Segmentation of the annelids and many arthropods is obvious because the body has demarcations that can be seen externally. The repeating vertebrae of the backbone in vertebrates signal that they, too, are segmented. Was a coelom present before the evolution of segmentation? Perhaps it was. Later, a partitioned coelom may have provided a hydrostatic skeleton that enabled a worm to burrow more efficiently in the soil. In other words, there was a selective advantage to having a segmented coelom in organisms that lack limbs. The later evolution of limbs in jointed arthropods (e.g., insects) and vertebrates (e.g., frogs, lizards, horses) was even more adaptive for locomotion on land.

Animals that live on land must also have a means of reproduction that allows fertilization and development to take place without a dependence on external water.

Summary

34.1 A Coelom

Annelids, arthropods, and mollusks are all protostomial coelomates. As adults they all have bilateral symmetry, the tube-within-a-tube body plan, and organs derived from three germ layers.

The presence of a coelom has many advantages. The digestive system and body wall can move independently, and internal organs can become more complex. Coelomic fluids can assist respiration, circulation, and excretion, and can serve as a hydrostatic skeleton.

There are two groups of coelomate animals. In the protostomes, the mouth appears at or near the blastopore, and in the deuterostomes, the anus appears at or near the blastopore. Other differences between the two are spiral cleavage and a schizocoelom in protostomes versus radial cleavage and an enterocoelom in deuterostomes.

34.2 A Three-Part Body Plan

The body of a mollusk typically contains a visceral mass, a mantle, and a foot. Many also have a head and a radula. The nervous system consists of several ganglia connected by nerve cords. There is a reduced coelom and an open circulatory system. Snails (gastropods) are adapted to life on land, squids (cephalopods) to an active life in the sea, and clams (bivalves) to a sedentary coastal life.

Bivalves are filter feeders. Water enters and exits by siphons, and food trapped on the gills is swept toward the mouth. The nervous system has three major ganglia.

Predaceous squids undergo cephalization, move rapidly by jet propulsion, and have a closed circulatory system.

34.3 Segmentation Evolves

Annelids are segmented worms. They have a well-developed coelom divided by septa, a closed circulatory system, a ventral solid nerve cord, and paired nephridia. Polychaetes ("many bristles") are marine worms that have parapodia. They may be predators, with a definite head region, or they may be filter feeders, with ciliated tentacles to filter food from the water. Earthworms are oligochaetes ("few bristles") that use the body wall for gas exchange. Leeches (class Hirudinea) are also part of this phylum.

34.4 Jointed Appendages Evolve

Arthropods are the most varied and numerous of animals. Their success is largely attributable to a flexible exoskeleton, specialization of body regions, and jointed appendages. Also important are a high degree of cephalization, a variety of respiratory organs, and reduced competition through metamorphosis.

The arthropods are classified into four subphyla. Subphylum Trilobitomorpha includes animals that have been extinct for some 200 million years. The subphylum Chelicerata (horseshoe crabs, spiders, scorpions, ticks, and mites) have chelicerae, pedipalps, and four pairs of walking legs attached to a cephalothorax.

Arthropods in the subphylum Crustacea (crayfish, lobsters, shrimps, copepods, krill, and barnacles) have a head that bears compound eyes, antennae, antennules, mandibles, and maxillae. There are biramous (two-branched) appendages on the thorax and abdomen. The crayfish illustrates other features such as an open circulatory system, respiration by gills, and a ventral solid nerve cord.

Arthropods in the subphylum Uniramia (centipedes, millipedes, and insects) have uniramous (one-branched) appendages on thorax and abdomen. The head has one pair of antennae, one pair of mandibles, and one or two pairs of maxillae. These terrestrial animals breathe by means of tracheae.

Insects include butterflies, grasshoppers, bees, and beetles. The anatomy of the grasshopper illustrates insect anatomy and ways they are adapted to life on land. Like other insects, grasshoppers have wings and three pairs of legs attached to the thorax. Grasshoppers have a tympanum for sound reception, a specialized digestive system for a grass diet, excretion of solid, nitrogenous waste by Malpighian tubules, tracheae for respiration, internal fertilization, and incomplete metamorphosis.

Reviewing the Chapter

1. List and discuss several advantages of the coelom. 594
2. Contrast the development of protostomes and deuterostomes in three ways. Tell which phyla belong to each group. 594–95
3. Name at least three members of each phylum studied in this chapter. 595
4. What are the general characteristics of mollusks and the specific features of bivalves, cephalopods, and gastropods? 596–99
5. Contrast the anatomy of the clam, the squid, and the snail, indicating how each is adapted to its way of life. 597–99
6. What are the general characteristics of annelids and the specific features of the three major classes? 600–2
7. Contrast the anatomy of the clam worm and the earthworm, indicating how each is adapted to its way of life. 600–2
8. What are the general characteristics of arthropods and the specific features of the three major subphyla? 603–6
9. Describe the specific features of insects, using the grasshopper as an example. 609–10
10. Contrast the anatomy of the crayfish and the grasshopper, indicating how each is adapted to its way of life. 610

Testing Yourself

Choose the best answer for each question.

1. Which of these does not pertain to a protostome?
   a. spiral cleavage
   b. blastopore is associated with the anus
   c. schizocoelom
   d. annelids, arthropods, and mollusks
2. Which of these best shows that annelids and arthropods are closely related? Both
   a. have a complete digestive tract.
   b. have a ventral solid nerve cord.
   c. are segmented.
   d. All of these are correct.
3. Which of these best shows that snails are not closely related to crayfish?
   a. Snails are terrestrial, and crayfish are aquatic.
   b. Snails have a broad foot, and crayfish have jointed appendages.
   c. Snails are hermaphroditic, and crayfish have separate sexes.
   d. Snails are insects, but crayfish are fishes.
4. Which of these is mismatched?
   a. clam—gills
   b. lobster—gills
   c. grasshopper—book lungs
   d. polychaete—parapodia
5. Which of these is an incorrect statement?
   a. Spiders are carnivores in the phylum Arthropoda.
   b. Clams are filter feeders in the phylum Mollusca.
   c. Earthworms are scavengers in the phylum Arthropoda.
   d. Squids are predators in the phylum Mollusca.
6. A radula is a unique organ for feeding found in
   a. mollusks.
   b. annelids.
   c. arthropods.
   d. All of these are correct.
7. Which of these is mismatched?
   a. crayfish—walking legs
   b. clam—hatchet foot
   c. grasshopper—wings
   d. earthworm—many cilia
8. Which of these is mismatched?
   a. mollusk—schizocoelom
   b. insects—hemocoel
   c. crayfish—coelom divided by septa
   d. clam worm—true coelom
9. Which of these is an adaptation to a terrestrial way of life in the snail?
   a. using the mantle instead of gills for respiration
   b. cephalization with antennae and eyes
   c. hatchet foot for crawling
   d. torsion so that anus is above the head

10. Which of these terms apply to annelids, to arthropods, and/or to mollusks?

Terms:

a. organ system level of organization
b. segmentation
c. true coelom
d. cephalization in some representatives
e. bilateral symmetry
f. complete gut
g. jointed appendages
h. three-part body plan

11. Which of these terms apply to clams and/or to earthworms?

Terms:

a. annelid
b. mollusk
c. three ganglia
d. gills
e. closed circulatory system
f. setae
g. open circulatory system
h. hatchet foot
i. hydrostatic skeleton
j. ventral solid nerve cord

12. Label this diagram.

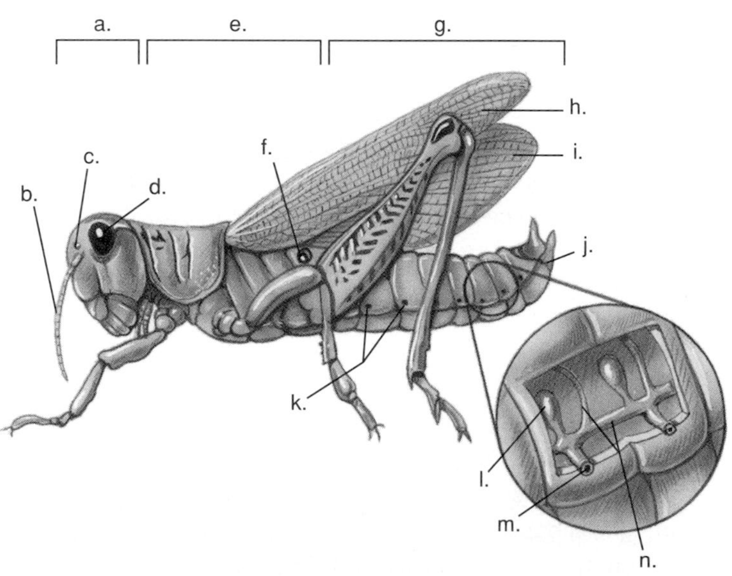

Applying the Concepts

1. *Segmentation leads to specialization of parts.*
   Compare external features of the earthworm to those of the grasshopper to demonstrate this concept.
2. *Animals are adapted to the manner they acquire food.*
   Support this concept in regard to one animal from each of the phyla studied in the chapter.
3. *Insects are very well adapted to a land environment.*
   List and discuss grasshopper features that adapt it to a life on land.

Using Technology

Your study of the protostomes is supported by these available technologies:

Exploring the Internet
The Mader Home Page provides resources for and help with studying this chapter.

http://www.mhhe.com/sciencemath/biology/mader/
(Click on Biology.)

Understanding the Terms

annelid 600
arthropod 603
bivalve 597
centipede 606
cephalopod 598
cephalothorax 604
chelicerate 604
chitin 603
cleavage 594
coelom 594
crustacea 605
decapod 605
deuterostome 594
exoskeleton 603
gastropod 599
hemocoel 605
insect 609
jointed appendage 603
leech 602
Malpighian tubule 609
mantle 596
metamorphosis 603
millipede 606
mollusk 596
molt 603
nephridium 600
oligochaete 601
protostome 594
segmentation 600
seta (pl., setae) 600
spiral cleavage 594
trachea (pl., tracheae) 603
uniramia 606
vertebrate 598

Match the terms to these definitions:

a. __________ Air tube in insects located between the spiracles and the tracheoles.
b. __________ Blind, threadlike excretory tubule near the anterior end of an insect hindgut.
c. __________ Cell division without cytoplasmic addition or enlargement; occurs during the first stage of animal development.
d. __________ Change in shape and form that some animals, such as insects, undergo during development.
e. __________ Freely moveable appendage of arthropods.
f. __________ Periodic shedding of the exoskeleton in arthropods.
g. __________ Protective external skeleton, as in arthropods.
h. __________ Repetition of body units as is seen in the earthworm.
i. __________ Segmentally arranged, paired excretory tubules of many invertebrates, as in the earthworm, where the contents are released through a nephridiopore.
j. __________ Strong but flexible nitrogenous polysaccharide found in the exoskeleton of arthropods.

Animals:
The Deuterostomes

CHAPTER 35

Chapter Concepts

35.1 Radial Symmetry, Again

- In deuterostomes, the blastopore is associated with the anus, and the coelom develops by outpocketing from the primitive gut. 614
- Echinoderms (e.g., sea star) and chordates (e.g., vertebrates) are both related to the hemichordates (e.g., acorn worms). 614, 616
- Echinoderms have radial symmetry and a unique water vascular system for locomotion. 616

35.2 Chordate Characteristics Evolve

- Lancelets are invertebrate chordates with the three chordate characteristics as adults: notochord, a dorsal hollow nerve cord, and pharyngeal pouches. 617

35.3 The Vertebrate Body Plan Evolves

- In vertebrates, the notochord is replaced by the vertebral column. Most vertebrates also have a head region, endoskeleton, and paired appendages. 619
- There are three groups of fishes: jawless (e.g., hagfishes and lampreys), cartilaginous (sharks and rays), and bony fishes (ray-finned and lobe-finned). Most modern-day fishes are ray-finned (e.g., trout, perch, etc.). 619
- Amphibians (e.g., frogs and salamanders), who were more numerous during the Carboniferous period, evolved from lobe-finned fishes. 622

35.4 The Amniote Egg Evolves

- Modern-day reptiles (e.g., snakes, lizards, turtles) are the remnants of an ancient group that evolved from amphibians. 624
- The shelled egg of reptiles, which contains extraembryonic membranes, is an adaptation for reproduction on land. 624

35.5 Wings and Feathers Evolve

- Birds have feathers and skeletal adaptations that enable them to fly. 629

35.6 Homeothermy Pays Off

- Mammals, which evolved from reptiles, were present when the dinosaurs existed. They did not diversify until the dinosaurs became extinct. 629
- Mammals are vertebrates with hair and mammary glands. The former helps them maintain a constant body temperature. 629
- Mammals are classified according to methods of reproduction: monotremes lay eggs, marsupials have a pouch in which newborn mature, and placental mammals retain the offspring in a uterus until birth. 629

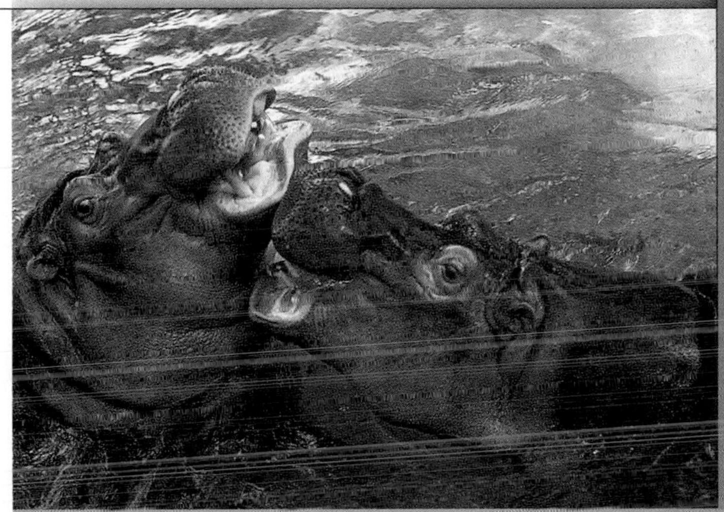

Hippopotami, *Hippopotamus amphibius*, playing

Human beings are vertebrate chordates, a phylum that includes a few invertebrate members. Without evidence to the contrary, who would think the invertebrate chordates are most closely related to the echinoderms—headless, brainless, and unsegmented creatures that spend their lives at sea? Sea stars, sea urchins, brittle stars, and sea lilies are all echinoderms, animals that are radially symmetrical as adults.

The fishes, which are aquatic vertebrates, far outnumber any of the other chordates. Still, there are more types of terrestrial animals among the chordates than in any other phylum. The reptiles were the first vertebrates to successfully reproduce on land by laying a shelled egg, as do birds. But within the egg, the embryo has a watery environment. Despite the live birth of placental mammals, such as ourselves, the embryo is still bathed by water while it develops. Today, birds and placental mammals are the most conspicuous vertebrates on land, because they underwent adaptive radiation after many reptiles, such as the dinosaurs, had become extinct. Hippopotami, pictured above, are hoofed mammals that spend a lot of time in the water because, when the sun is out, their skin loses water at a very high rate.

35.1 Radial Symmetry, Again

Phylum Echinodermata (6,000 species) and **phylum Chordata,** which are discussed in this chapter, contain the **deuterostomes** [Gk. *deuteros,* second, and L. *stoma,* mouth], animals in which the second embryonic opening is associated with the mouth (the first opening is associated with the anus). *Cleavage,* the first series of embryonic cell divisions, is radial in deuterostomes, and the daughter cells sit on top of the previous cells. The cells are *indeterminate;* if separated, each one can develop into a complete organism. Finally, the coelom develops as outpockets from the primitive gut; it is an *enterocoelom* (see Fig. 34.1).

Although the echinoderms are *radially symmetrical as adults,* their larva is a free-swimming planktonic filter feeder with bilateral symmetry. Metamorphosis results in the radially symmetrical adult.

Echinoderms Have a Spiny Skin

Echinoderms [Gk. *echinos,* spiny, and *derma,* skin] have an *endoskeleton* (internal skeleton) consisting of spine-bearing, calcium-rich plates. The *spines,* which stick out through the delicate skin, account for their name. Class Crinoidea (600 species) includes the stalked sea lilies and the motile feather stars. Their branched arms, used for filter feeding, give them a flowerlike appearance. Class Holothuroidea (1,500 species) are the sea cucumbers with a long leathery body that resembles a cucumber, except that there are feeding tentacles about the mouth.

Class Echinoidea (950 species) includes sea urchins and sand dollars, both of which use their spines for locomotion, defense, and burrowing. Sea urchins are well known for their long, blunt spines, and the flattened, somewhat circular skeleton of a sand dollar has a familiar five-part flowerlike pattern. Actually the pattern is due to pores for

CLASSIFICATION

Kingdom Animalia

Multicellular organisms with well-developed tissues; usually motile, heterotrophic by ingestion, generally in a digestive cavity; diplontic life cycle. Protostomes include phyla Mollusca, Annelida, and Arthropoda. Deuterostomes include phyla Echinodermata, Hemichordata, and Chordata.

Invertebrates*

- Phylum Porifera: sponges
- Phylum Cnidaria: hydras, jellyfishes, sea anemones, corals
- Phylum Ctenophora: comb jellies, sea walnuts
- Phylum Platyhelminthes: flatworms, e.g., planaria, flukes, tapeworms
- Phylum Nemertea: ribbon worms
- Phylum Nematoda: roundworms
- Phylum Rotifera: rotifers
- Phylum Mollusca: chitons, snails, slugs, clams, mussels, squids, octopuses
- Phylum Annelida: segmented worms, e.g., clam worms, earthworms, leeches
- Phylum Onychophora: walking worms
- Phylum Arthropoda: spiders, scorpions, horseshoe crabs, lobsters, crayfish, shrimps, crabs, millipedes, centipedes, insects
- Phylum Echinodermata: sea lilies, sea stars, brittle stars, sea urchins, sand dollars, sea cucumbers, sea daisies
- Phylum Hemichordata: acorn worms
- Phylum Chordata
  - Subphylum Urochordata: tunicates
  - Subphylum Cephalochordata: lancelets

Vertebrates*

- Subphylum Vertebrata
- Superclass Agnatha: jawless fishes, e.g., lampreys, hagfishes
- Superclass Gnathostomata: jawed fishes, all tetrapods
  - Class Chondrichthyes: cartilaginous fishes, e.g., sharks, skates, rays
  - Class Osteichthyes: bony fishes, e.g., herring, salmon, cod, eel, flounder
  - Class Amphibia: frogs, toads, salamanders
  - Class Reptilia: snakes, lizards, turtles
  - Class Aves: birds, e.g., sparrows, penguins, ostriches
  - Class Mammalia: mammals, e.g., cats, dogs, horses, rats, humans

*Not in the classification of organisms, but added here for clarity.

skin projections in the living animal. Class Ophiuroidea (2,000 species) contains the brittle stars, which have a central disk from which long, flexible arms radiate. They move quickly by using their arms to push themselves along.

Class Asteroidea (1,500 species) consists of the sea stars, also called the starfishes. The somewhat flattened body of most sea stars has a central disk from which five, or a multiple of five, sturdy arms (rays) extend.

Sea Stars Have Arms

Sea stars are commonly found along rocky coasts where they feed on clams, oysters, and other bivalve mollusks. The *five-rayed body* has an oral, or mouth, side (the underside) and an aboral, or anus, side (the upper side) (Fig. 35.1). Various structures project through the body wall: (1) spines from the endoskeletal plates offer some protection; (2) pincerlike structures called pedicellarie keep the surface free of small particles; and (3) skin gills, tiny fingerlike extensions of the skin, are used for respiration. On the oral surface, each arm has a groove lined by little **tube feet.**

To feed, a sea star positions itself over a bivalve and attaches some of its tube feet to each side of the shell. By working its tube feet in alternation, it pulls the shell open. A very small crack is enough for the sea star to evert its cardiac stomach and push it through the crack, so that it contacts the soft parts of the bivalve. The stomach secretes enzymes, and digestion begins even while the bivalve is attempting to close its shell. Later, partly digested food is taken into the sea star's body, where digestion continues in the pyloric stomach using enzymes from the digestive glands found in each arm. A short intestine opens at the anus on the aboral side.

In each arm the well-developed coelom contains a pair of digestive glands and gonads (either male or female) that

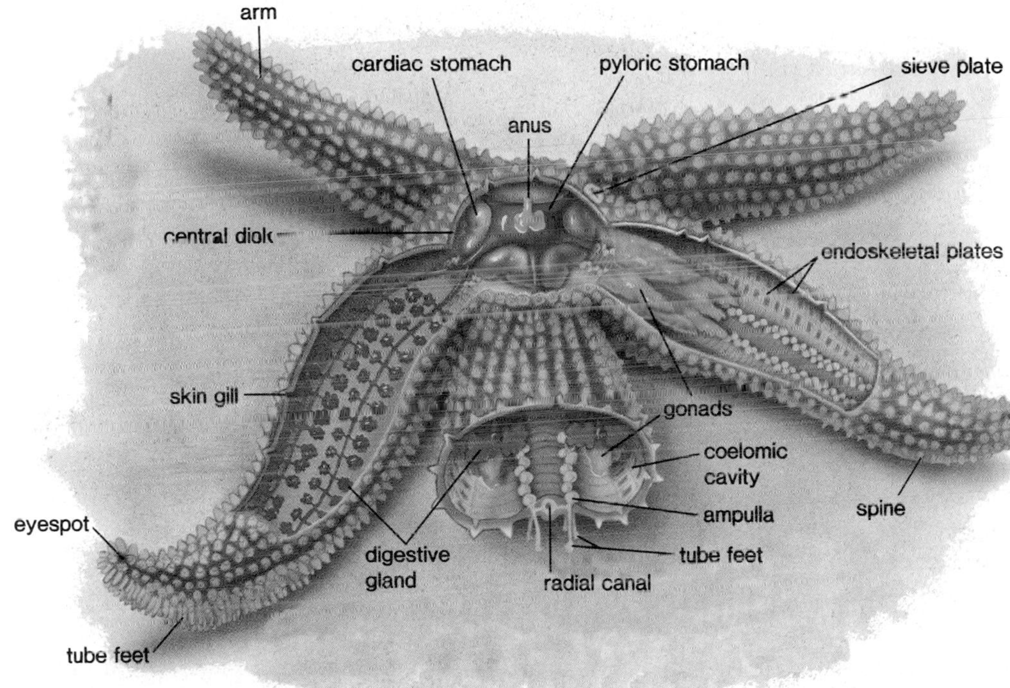

Figure 35.1 Echinoderms.
a. Sea star (starfish) anatomy and habitat. Like other echinoderms, sea stars have a water vascular system shown here in light gold.
b. Sea cucumber. c. Sea urchin.

b. Sea cucumber, *Pseudocolochirus*

a. Red sea star, *Mediastar*

c. Purple sea urchin, *Strongylocentrotus*

open on the aboral surface by very small pores. The nervous system consists of a central nerve ring that gives off radial nerves in each arm. A light-sensitive eyespot is at the tip of each arm. Sea stars are capable of coordinated but slow responses and body movements.

Locomotion depends on the **water vascular system.** Water enters this system through a structure on the aboral side called the sieve plate, or madreporite. From there it passes through a stone canal to a ring canal, which surrounds the mouth, and then to a radial canal in each arm. From the radial canals, many lateral canals extend into the tube feet, each of which has an ampulla. Contraction of an ampulla forces water into the tube foot, expanding it. When the foot touches a surface, the center is withdrawn, giving it suction so that it can adhere to the surface. By alternating the expansion and contraction of the tube feet, a starfish moves slowly along.

Echinoderms don't have a complex respiratory, excretory, or circulatory system. Fluids within the coelomic cavity and the water vascular system carry out many of these functions. For example, gas exchange occurs across the skin gills and the tube feet. Nitrogenous wastes diffuse through the coelomic fluid and the body wall. Cilia on the peritoneum lining the coelom keep the coelomic fluid moving.

Sea stars reproduce asexually and sexually. If the body is fragmented, each fragment can regenerate a whole animal. Fishermen who try to get rid of sea stars by cutting them up and tossing them overboard are merely propagating more sea stars! Sea stars spawn and release either eggs or sperm at the same time. The bilateral larva undergoes a metamorphosis to become the radially symmetrical adult.

Echinoderms have a well-developed coelom and internal organs despite being radially symmetrical. Spines project from their internal skeleton, and there is a unique water vascular system.

Figure 35.2 Hemichordates. The acorn worm, *Glossobalanus*, showing the proboscis, the collar, the pharyngeal regions, and certain internal structures.

35.2 Chordate Characteristics Evolve

To be considered a **chordate** (**phylum Chordata,** 45,000 species), an animal must have the three basic characteristics listed below at some time during its life history:

1. A dorsal supporting rod called a **notochord** [Gk. *notos,* back, and *chorde,* string]. The notochord is located just below the **nerve cord.** Vertebrates have an embryonic notochord that is replaced by the vertebral column during development.
2. A *dorsal hollow nerve cord.* By hollow, it is meant that the cord contains a canal filled with fluid. In vertebrates, the nerve cord, more often called the spinal cord, is protected by the vertebrae.
3. *Pharyngeal pouches.* These are seen only during embryonic development in most vertebrates. In the invertebrate chordates, the fishes, and amphibian larvae, the pharyngeal pouches become functioning **gills.** Water passing into the mouth and the pharynx goes through the **gill slits,** which are supported by gill arches. In terrestrial vertebrates, the pouches are modified for various purposes. In humans, the first pair of pouches become the auditory tubes. The second pair become the tonsils, while the third and fourth pairs become the thymus gland and the parathyroids.

Other features also distinguish chordates. Most have an internal skeleton against which muscles work and—as embryos if not as adults—they have a *post-anal tail,* a tail that extends beyond the anus.

The **hemichordates** (**phylum Hemichordata,** 90 species) are not considered chordates but they have features that resemble those of chordates. Acorn worms, which are hemichordates living in or on tidal mudflats, have a *proboscis,* a *collar,* and a *trunk* (Fig. 35.2). There is a dorsal nerve cord in the collar and trunk that resembles the nerve cord of chordates, and the pharynx just below the collar has gill slits. Most interesting, *the larva of hemichordates resembles that of echinoderms.* Is it possible then that echinoderms and hemichordates share a common ancestor and that the hemichordates and chordates are related by way of a common ancestor? Some think so.

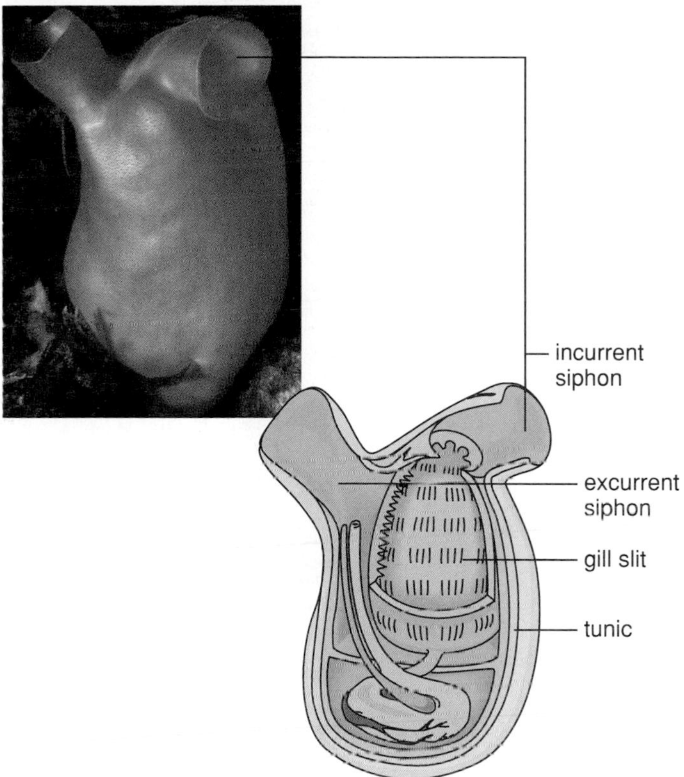

Figure 35.3 Tunicate, *Halocynthia.*
Note that the only chordate characteristic remaining in the adult is gill slits.

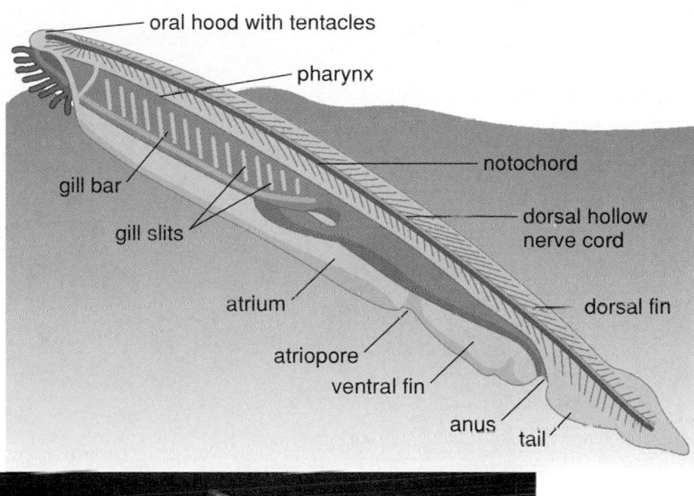

Figure 35.4 Lancelet, *Branchiostoma.*
Lancelets are filter feeders. Water enters the mouth and exits at the atriopore after passing through the gill slits.

Invertebrate Chordates

There are a few *invertebrate chordates* in which the notochord persists and is never replaced by the vertebral column.

Tunicates Have Gill Slits

Tunicates (subphylum Urochordata, 1,250 species) live on the ocean floor and take their name from a tunic that makes them look like thick-walled, squat sacs. They are also called sea squirts because they squirt out water from their excurrent siphon when disturbed. The tunicate larva is bilaterally symmetrical and has the three chordate characteristics. Metamorphosis produces the sessile adult with an incurrent and excurrent siphon (Fig. 35.3).

The pharynx is lined by numerous cilia whose beating creates a current of water that moves into the pharynx and out the numerous *gill slits,* the only chordate characteristic that remains in the adult. Microscopic particles adhere to a mucous secretion and are eaten.

Is it possible that the tunicates are directly related to the vertebrates? It has been suggested that a larva with the three chordate characteristics may have become sexually mature without developing the other adult tunicate characteristics. Then it may have evolved into a fishlike vertebrate. Or perhaps it was a cephalochordate larva (discussed next) that became a vertebrate.

Lancelets Have Three Chordate Characteristics

Lancelets (subphylum Cephalochordata, 23 species) are in the genus *Branchiostoma,* formerly called Amphioxus. These marine chordates, which are only a few centimeters long, are named for their resemblance to a lancet—a small, two-edged surgical knife (Fig. 35.4). Lancelets are found in the shallow water along most coasts where they usually lie partly buried in sandy or muddy substrates with only their anterior mouth and gill apparatus exposed. They feed on microscopic particles filtered out of the constant stream of water that enters the mouth and passes through the gill slits into an atrium that opens at the atriopore.

Lancelets retain the three chordate characteristics as an adult. The notochord extends from the tail to the head and this accounts for their subphylum name, Cephalochordata. In addition, segmentation is present, as witnessed by the fact that the muscles are segmentally arranged and the dorsal hollow nerve cord has periodic branches.

The invertebrate chordates include the tunicates and the lancelets. A lancelet is the best example of a chordate that possesses the three chordate characteristics as an adult.

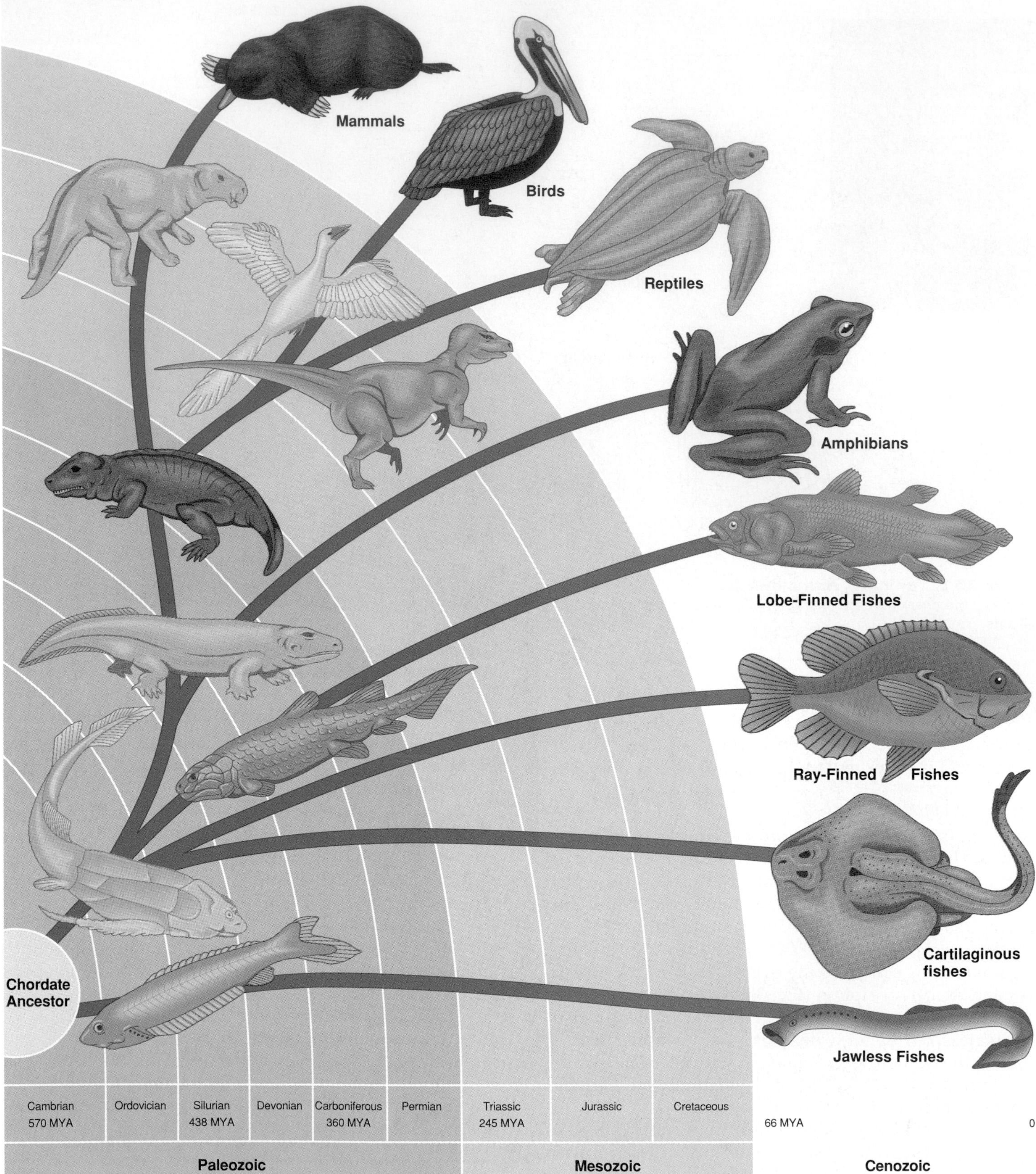

Figure 35.5 Phylogenetic tree of the chordates.
The geological timescale tells when the ancestors to modern-day vertebrates appeared in the fossil record.

35.3 The Vertebrate Body Plan Evolves

At some time in their life history, **vertebrates (subphylum Vertebrata,** 43,700 species) have all three chordate characteristics (notochord, dorsal hollow nerve cord, and pharyngeal pouches). The embryonic notochord, however, is generally replaced by a *vertebral column* composed of individual vertebrae. The vertebral column, which is a part of the flexible but strong endoskeleton, gives evidence that vertebrates are segmented. The vertebrate skeleton (either cartilage or bone) is a living tissue that grows with the animal. It also protects internal organs and serves as a place of attachment for muscles. Together the skeleton and muscles form a system that permits rapid and efficient movement. *Two pairs of appendages* are characteristic. The pectoral and pelvic fins of fish evolved into the jointed appendages that allowed vertebrates to move onto land.

The main axis of the internal jointed skeleton consists of not only the vertebral column, but also a skull that encloses and protects the brain. During vertebrate evolution the brain increased in complexity, and specialized regions developed to carry out specific functions. The high degree of cephalization is accompanied by complex sense organs. The eyes develop as outgrowths of the brain. The ears are primarily equilibrium devices in aquatic vertebrates, but they also function as sound-wave receivers in land vertebrates.

Vertebrates have a complete digestive tract and a large coelom. The circulatory system is closed (the blood is contained entirely within blood vessels). Vertebrates have an efficient means of obtaining oxygen from water or air, as appropriate. The kidneys are important excretory and water-regulating organs that conserve or rid the body of water as necessary. The sexes are generally separate, and reproduction is usually sexual.

Figure 35.5 is a phylogenetic tree that shows the evolution of vertebrates from a chordate ancestor. Vertebrates include fishes, amphibians, reptiles, birds, and mammals.

Vertebrates are distinguished, in particular, by:

- living endoskeleton
- closed circulatory system
- paired appendages
- efficient respiration and excretion
- high degree of cephalization

In short, vertebrates are adapted to an active lifestyle.

Fishes are aquatic, gill-breathing vertebrates that usually have fins and skin covered with **scales.** The first vertebrates were fishlike. The larval form of a modern-day lamprey, which looks like a lancelet, may resemble the first vertebrates. It not only has the three chordate characteristics like the tunicate larva, it also has a two-chambered heart, a three-part brain, and other internal organs that are like those of vertebrates. The small, jawless, and finless **ostracoderms** are the earliest vertebrate fossils. They were filter feeders, but most likely they were capable of moving water through their gills by muscular action. Ostracoderm fossils have been dated from the Cambrian and as late as the Devonian period, but then they apparently became extinct. Although none of the living jawless fishes has any external protection, large defensive head-shields were not uncommon in the early jawless fishes.

Lampreys and Hagfishes

All the jawless fishes are called **agnathans** [Gk. *a*, without, and *gnathos*, jaw] (superclass Agnatha, 63 species). **Lampreys** and **hagfishes** are modern-day jawless fishes that lack a bony skeleton. They are cylindrical, up to a meter long, and have *smooth, nonscaly* skin (Fig. 35.6). The hagfishes are scavengers feeding on soft-bodied invertebrates and dead fishes they suck into their mouths. Many lampreys are filter feeders like their ancestors. Parasitic lampreys have a round, muscular mouth to attach themselves to another fish and suck nutrients from the host's circulatory system. Marine parasitic lampreys gained entrance to the Great Lakes when a canal from the St. Lawrence River was deepened. The lamprey population grew quickly and caused extensive reduction in the trout population of the Great Lakes in the early 1950s.

Figure 35.6 Lamprey, *Petromyzon.*
Note its toothed oral disk attached to aquarium glass. Lampreys, which are members of the vertebrate superclass Agnatha, have an elongated, rounded body and nonscaly skin.

Jaws Evolve

All the other animals to be studied are **gnathostomates** [Gk. *gnathos,* jaw, and L. *stoma,* mouth] (superclass Gnathostomata, 25,000 species); they have **jaws,** tooth-bearing bones of the head. Jaws are believed to have evolved from the first pair of gill arches of agnathans. The second pair of gill arches became support structures for the jaws.

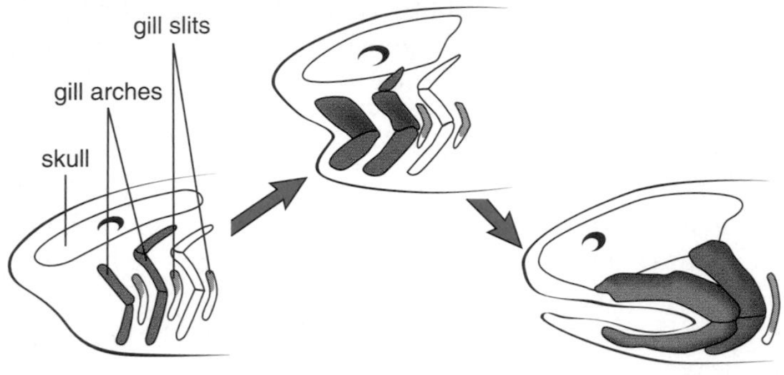

The **placoderms** are well-known extinct jawed fishes of the Devonian period. Placoderms were armored with heavy, bony plates and had strong jaws. Like modern-day fishes, they had paired pectoral and pelvic fins (Fig. 35.7). *Paired fins,* which allow a fish to balance and to maneuver well in the water, are also an asset for predation.

The jawed fishes of today include the cartilaginous fishes and the bony fishes.

Cartilaginous Fishes Include Sharks

Sharks, rays, and skates (class Chondrichthyes, 850 species) are the largely marine **cartilaginous fishes;** they have a skeleton of cartilage instead of bone. There are five to seven gill slits on both sides of the pharynx, and they lack the gill cover of bony fishes. Their body is covered with epidermal *placoid* (toothlike) *scales* that project posteriorly, which is why a shark's skin feels like sandpaper. The menacing teeth of sharks and their relatives are simply larger, specialized versions of these scales. At any one time, a shark such as the great white shark may have up to 3,000 teeth in its mouth, arranged in six to twenty rows. Only the first row or two are actively used for feeding; the other rows are replacement teeth.

Three well-developed senses enable sharks and rays to detect their prey. They have the ability to sense electric currents in water—even those generated by the muscle movements of animals. They have a *lateral line system,* a series of pressure-sensitive cells that lie within canals along both sides of the body, which can sense pressure caused by a fish or other animal swimming nearby. They also have a very keen sense of smell; the part of the brain associated with this sense is very well developed. Sharks can detect about one drop of blood in 115 liters (25 gallons) of water.

The largest sharks are filter feeders, not predators. The basking sharks and whale sharks ingest tons of small crustacea, collectively called krill. Many sharks are fast-swimming predators in the open sea. The great white shark, about 7 meters (23 feet) in length, feeds regularly on dolphins, sea lions, and other seals. Humans are normally not attacked except when mistaken for sharks' usual prey. Tiger sharks, so named because the young have dark bands, reach 6 meters (18 feet) in length and are unquestionably one of the most dangerous sharks. As it swims through the water, a tiger shark will swallow anything it can—including rolls of tar paper, shoes, gasoline cans, paint cans, and even human parts.

In rays and skates (see cartilaginous fishes, Fig. 35.5), which live on the ocean floor, the pectoral fins are greatly enlarged into a pair of large, winglike **fins.** They usually swim slowly along the sea bottom and feed on animals that they dredge up. Stingrays of the genus *Raja* have a tail modified into a defensive lash—the dorsal fin persists as a venomous spine. Members of the electric ray family are slow swimmers that feed on fishes they capture after stunning them with electric shocks. Their large electric organs, located at the bases of their pectoral fins, can discharge over 300 volts. Sawfish rays are named for their large, protruding anterior "saw." Swimming into a school of fishes, they rapidly move their saw back and forth, stunning or killing fish which they later eat.

Bony Fishes Are Most Numerous of Vertebrates

Bony fishes (class Osteichthyes, 20,000 species) have a skeleton of bone. Most bony fishes are ray-finned fishes in which fan-shaped fins are supported by thin, bony rays. The lobe-finned fishes, a very small group of bony fishes, are important because lobe-finned fishes of the Devonian period are ancestral to amphibians.

Ray-Finned Fishes Include Familiar Fishes The **ray-finned fishes** are the most successful and diverse of all the vertebrates. Some, like herrings, are filter feeders; others, like trout, are opportunists; and still others are predaceous carnivores, like the piranhas and barracudas. Often the common names of fishes reflect their appearance. Zebra fish are striped, stonefish resemble stones, sea horses look like tiny upright horses, and porcupine fish (when inflated) are protected by lateral spikes.

Despite their diversity, bony fishes have features in common (Fig. 35.7). The skeleton is of bone and they are covered by *scales* formed *of bone.* The gills do not open separately and instead are covered by an *operculum.* They have a **swim bladder,** a gas-filled sac whose pressure can be altered to change buoyancy and, therefore, their depth in the water. Some fishes (trout, salmon, eels) can move from fresh water to salt water. When in fresh water, their kidneys excrete very dilute urine and their gills absorb salts from the water by active transport.

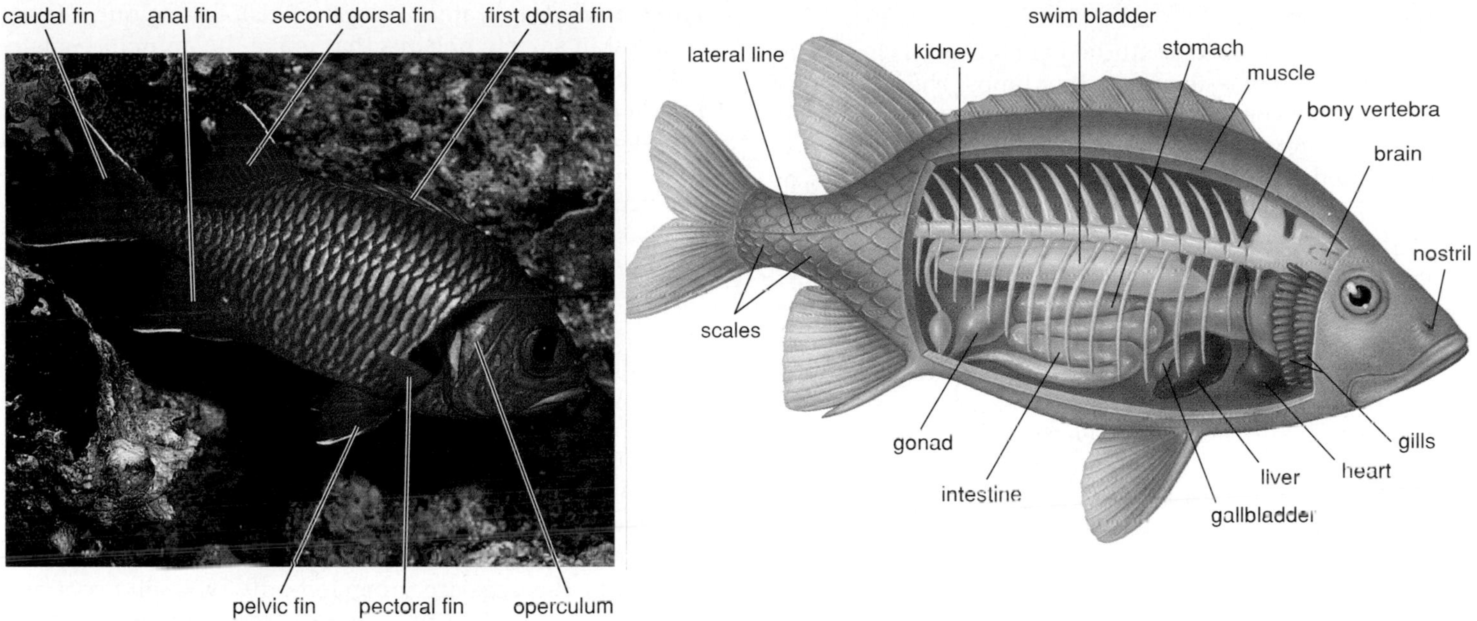

Figure 35.7 Ray-finned fish, *Myripristis*.
External and internal anatomy of a ray-finned fish, commonly called a soldierfish.

Bony fishes depend on color vision to detect both rivals and mates. Usually eggs and sperm are shed in the water or into a nest constructed by the parents. In almost all bony fishes, fertilization and embryo development occur outside the female's body.

Lobe-Finned Fishes In **lobe-finned fishes,** the fleshy fins are supported by central bones. They include six species of lungfishes with lungs and one species of coelacanth with especially noticeable lobes (Fig. 35.8). *Lungfishes* live in Africa, South America, and Australia either in stagnant fresh water or in ponds that dry up annually. *Coelacanths,* in contrast, inhabit deep ocean environments. Because only Mesozoic fossils of these fishes had been found, it was once thought that coelacanths were extinct. Then one was captured off the eastern coast of South Africa in 1938, and now about 200 more have been captured. Comparisons of mitochondrial DNA suggest that lungfishes, and not coelacanths, are the closest living relatives of modern amphibians.

Figure 35.8 Lobe-finned fishes.
The coelacanth, *Latimeria*, is a living fossil—the only survivor of a line once thought extinct.

The cartilaginous and bony fishes are characterized by:

- jaws
- paired appendages
- scales

Limbs Evolve

All the other animals to be studied are **tetrapods** [Gk. *tetra,* four, and *podos,* foot]—they have four limbs. The lobe-finned fishes of the Devonian period are ancestral to the amphibians, the first tetrapods. Figure 35.9 compares the limbs of a lobe-finned fish to those of an early amphibian, and you can see that the same bones are present in both animals. Animals that live on land use limbs to support the body, especially since air is less buoyant than water. Lobe-finned fishes and early amphibians also had lungs and internal nares as a means to respire air.

Two hypotheses have been suggested to account for the evolution of amphibians from lobe-finned fishes. Perhaps lobe-finned fishes, which were capable of using their limbs to move from pond to pond, had an advantage over those who could not do so. Or perhaps the supply of food on land in the form of plants and insects—and the absence of predators—promoted further adaptations to the land environment. In any case, the first amphibians diversified during the Carboniferous period, which is known as the Age of Amphibians.

Amphibians

Aside from presence of limbs and appropriate modifications of the girdles, amphibians have other features not seen in bony fishes. **Amphibians** [Gk. *amphibios,* living on both land and in water] (class Amphibia, 3,900 species) have a tongue for catching prey, eyelids for keeping eyes moist, ears adapted to picking up sound waves, and a voice-producing larynx. The brain is larger than that of a fish and the **cerebral cortex** is more developed.

The *smooth, nonscaly skin* of an amphibian is kept moist by numerous mucous glands. It plays an active role in water balance and respiration and can also help in temperature regulation when on land, through evaporative cooling. A thin, moist skin does mean, however, that most amphibians stay close to water or else they risk drying out. Glands in the skin secrete poisons that make the animal distasteful to eat. Some tropical species with brilliant fluorescent green and red coloration are particularly poisonous. Colombian Indians dip their darts in the deadly secretions of these frogs, aptly called dart-poison frogs.

Amphibians usually have **lungs,** but they are relatively small and respiration is supplemented by exchange of gases across the porous skin. The single-loop circulatory path of the fish has been replaced by a double loop (Fig. 35.10); a *three-chambered heart* pumps blood before and after it has gone to the lungs. Oxygenated blood is partially mixed with deoxygenated blood, however, in the single ventricle (lower chamber of the heart).

The class name, Amphibia, means "on both sides," a reference to the fact that most amphibians return to water for the purpose of reproduction. They shed their eggs and sperm into the water, where external fertilization takes place. Generally, the eggs are protected only by a jelly coat and not by a shell. When the young hatch, they are tadpoles (aquatic larvae with gills) that feed and grow in the water. After they undergo a metamorphosis, amphibians emerge from the water as adults that breathe air (Fig. 35.11). Some amphibians, however, have evolved mechanisms that allow them to reproduce on land.

Amphibians are **ectothermic** [Gk. *ekto,* outer, and *therme,* heat]; they depend on the environment to regulate their body temperature. During winters in the temperate zone, they become inactive and enter *torpor.* The European common frog can survive even if the temperature drops to −6°C.

Among the amphibians are frogs, toads, salamanders, and newts. The salamanders and newts have an elongated body, with a long tail and usually two pairs of legs. Because their legs are set at right angles to the body, they resemble more closely the earliest fossil amphibians. They

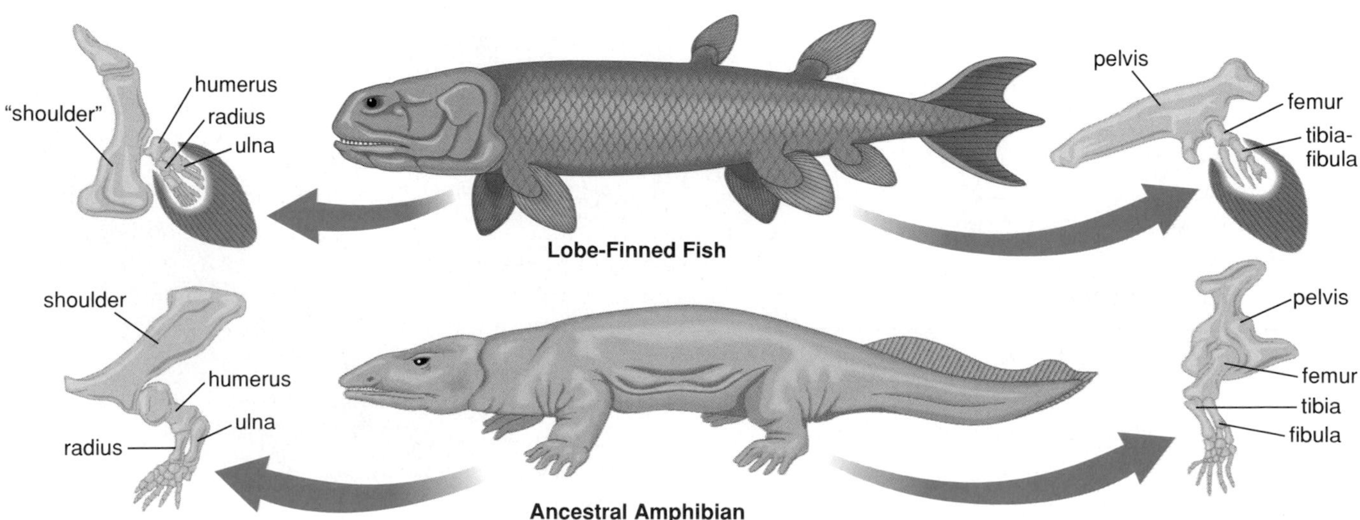

Figure 35.9 Lobe-finned fish versus amphibian.
The drawings show similarities in skeletal structure of the appendages of a lobe-finned fish and an ancestral amphibian.

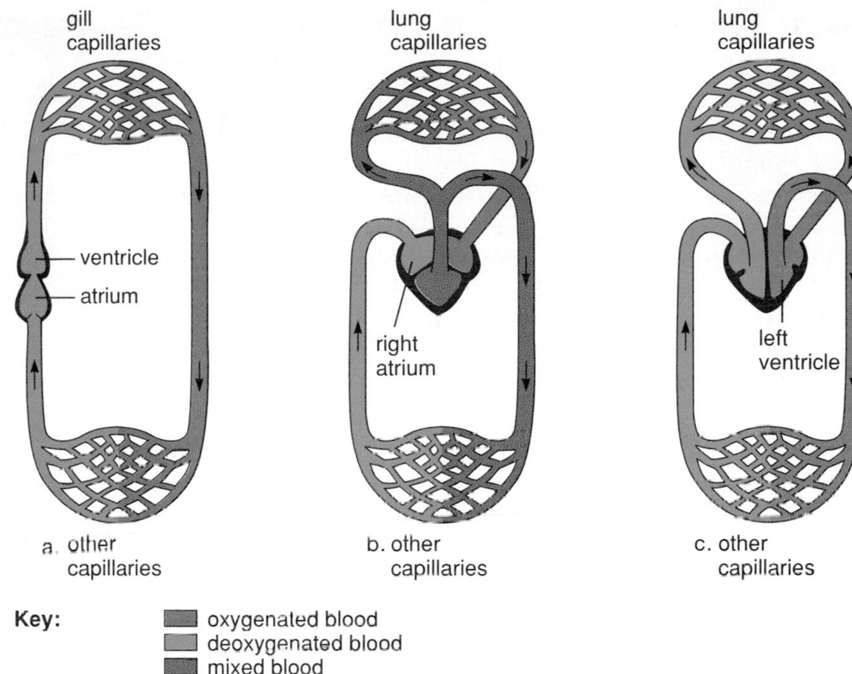

Figure 35.10 Vertebrate circulatory systems. **a.** The single-loop system of fishes utilizes a two-chambered heart. **b.** The double-loop system of amphibians sends blood to the lungs and to the body. There is some mixture of oxygenated and deoxygenated blood in the three-chambered heart. **c.** The four-chambered heart of birds and mammals sends only deoxygenated blood to the lungs and oxygenated blood to the body.

Figure 35.11 Frog metamorphosis.
During metamorphosis the animal changes from an aquatic to a terrestrial organism.

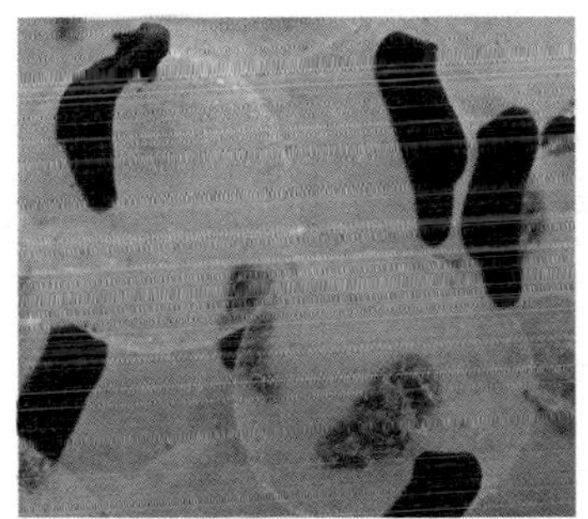

a. Tadpoles hatch

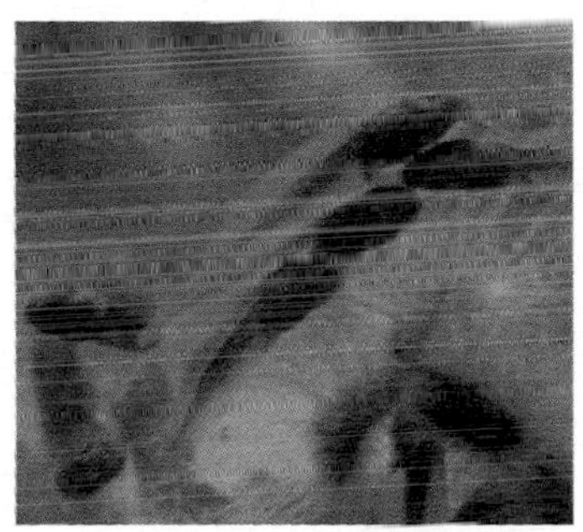

b. Tadpole respires with external gills

c. Front and hind legs are present

d. Frog, *Rana*, respires with lungs; tail resorbed

locomote like a fish, with side-to-side sinusoidal (S-shaped) movements:

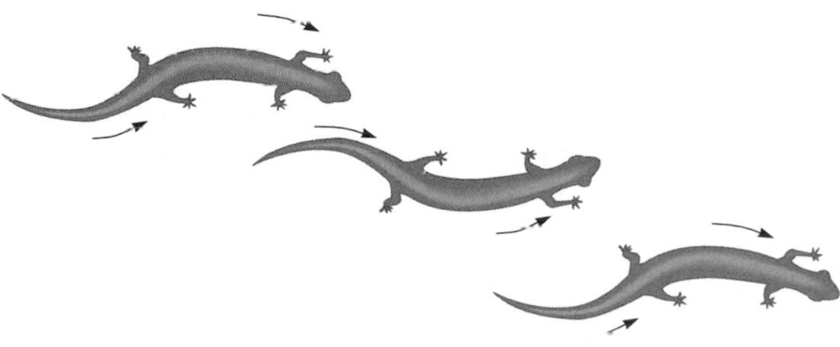

Both salamanders and newts are carnivorous, feeding on small invertebrates such as insects, slugs, snails, and worms. Salamanders practice internal fertilization; in most, males produce a sperm-containing *spermatophore* that females pick up with the **cloaca** (the common receptacle for the urinary, genital, and digestive canals). Then the eggs are laid in water or on land, depending on the species. Some amphibians, such as the mudpuppy of eastern North America, remain in the water and retain the gills of the larva.

Frogs and toads are the tailless amphibians. In these animals, the head and trunk are fused and the hindlimbs are specialized for jumping. Numerous muscles are located in the limbs as well as the trunk. Frogs have smooth skin and long legs, and they live in or near fresh water; toads have stout bodies and warty skin, and they live in dark, damp places away from the water.

These features, in particular, distinguish amphibians:

- usually tetrapods
- usually lungs in adult
- metamorphosis
- smooth and moist skin
- three-chambered heart

35.4 The Amniote Egg Evolves

It is adaptive for land animals to have a means of reproduction that is not dependent on external water. Reptiles practice internal fertilization and lay eggs that are protected by a *leathery shell* (Fig. 35.12). The **amniote egg** contains *extraembryonic membranes,* which protect the embryo, remove nitrogenous wastes, and provide the embryo with oxygen, food, and water. These membranes are not part of the embryo itself and are disposed of after development is complete. One of the membranes, the *amnion,* is a sac that fills with fluid and provides a "private pond," within which the embryo develops.

Reptiles are believed to have evolved from amphibian ancestors sometime during the Carboniferous period when amphibians were already diversifying and living on land (Fig. 35.13). The first reptiles, known as the stem reptiles, gave rise to several other lineages, each adapted to a different way of life. Of interest to us are the pelycosaurs, or sail lizards, so named because of a sail-like web of skin held above the body by slender spines. They are related to the **therapsids,** mammal-like reptiles that came later. Other lines of descent returned to the aquatic environment; one marine reptile (ichthyosaurs) of the Mesozoic era was fishlike while another (plesiosaurs) had a long neck and large rowing paddles for limbs. Reptiles known as thecodonts gave rise to most of the reptiles, living and extinct. The flying reptiles (pterosaurs) of the Mesozoic era had a keel for the attachment of large flight muscles and air spaces in their bones to reduce weight. Their wings were membranous and supported by elongated bones of the fourth finger. Then there were the "ruling reptiles," as the dinosaurs are sometimes called. The dinosaurs were varied in size and behavior but are well remembered for the great size of some. *Brachiosaurus,* a herbivore, was about 23 meters (75 feet) long and about 17 meters (56 feet) tall. *Tyrannosaurus rex,* a carnivore, was 5 meters (16 feet) tall when standing on its hind legs. A bipedal stance frees the forelimbs for seizing prey or fighting off predators. It is also preadaptive for the evolution of wings: birds are descended from dinosaurs—in fact some say birds are actually living dinosaurs.

The dinosaurs and mammal-like reptiles did not sprawl; the limbs supported the body from underneath, providing increased agility and swiftness that otherwise would not be possible. The amniote egg and improved locomotion were both adaptive for surviving on land.

Reptiles dominated the earth for about 170 million years during the Mesozoic era and then most died out. What could have caused the mass extinction that occurred at the end of the Cretaceous period? The answer is not known, but recently a layer of the mineral iridium, which is rare on earth but common in meteorites, has been found in rocks of that age. The impact of a large meteorite could have set off earthquakes and fires, raising enough dust to block out the sun. Death of most plants and animals would have followed. Such a scenario has been proposed by Luis and Walter Alvarez and several others who are still gathering evidence to support their hypothesis.

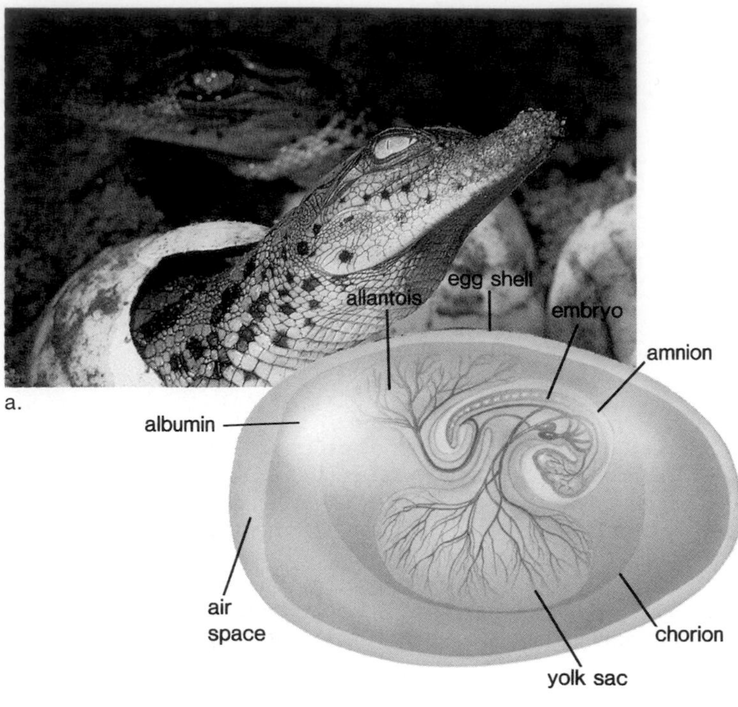

Figure 35.12 Reptilian egg.
a. Baby American crocodile, *Crocodylus,* hatching out of its shell. The shell is leathery and flexible, not brittle like birds' eggs.
b. Inside the egg, the embryo is surrounded by membranes. The chorion aids gas exchange, the yolk sac provides nutrients, the allantois stores waste, and the amnion encloses a fluid that prevents drying out and provides protection.

Reptiles

Most **reptiles** [L. *reptile,* snake] (class Reptilia, 6,000 species) today live in the tropics or subtropics. Among reptiles are lizards and snakes that usually live on or in the soil, and turtles, crocodiles, and alligators that live in the water.

Reptiles have a thick, *scaly skin* that is keratinized and impermeable to water. Keratin is the protein found in hair, fingernails, and feathers. The protective skin prevents water loss but requires several *molts* a year. The reptilian **lung** is more developed, and air rhythmically moves in and out due to the presence of an expandable rib cage, except in turtles. The heart is nearly four chambered in most and is completely four chambered in crocodiles; therefore, deoxygenated blood is completely separate from oxygenated blood. The well-developed kidneys excrete uric acid and, therefore, less water is required to rid the body of nitrogenous waste.

Reptiles, like amphibians, are *ectothermic.* This feature allows them to survive on a fraction of the food per body weight required by birds and mammals. Still, they are adapted behaviorally to maintain a warm body temperature by warming themselves in the sun.

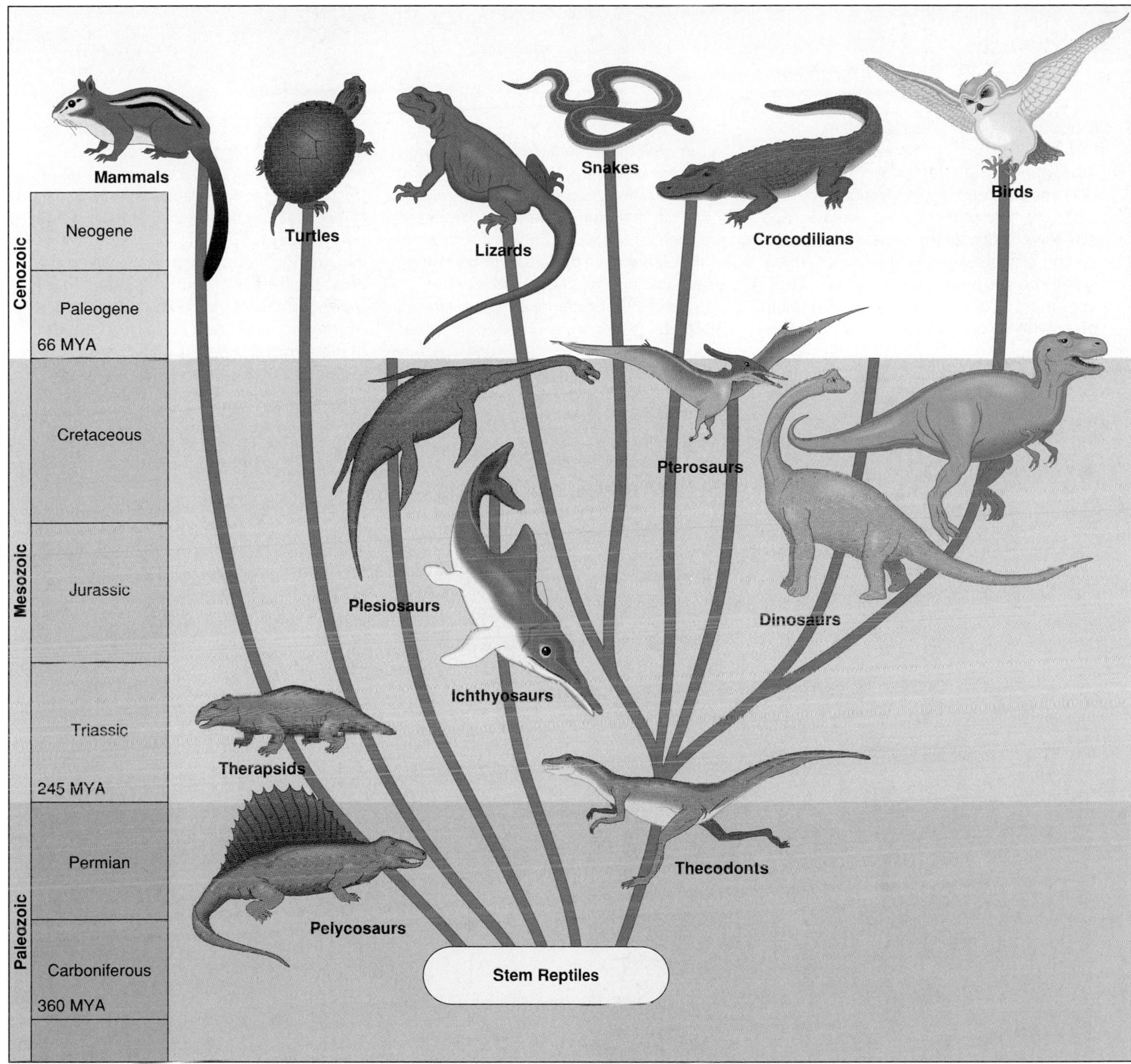

Figure 35.13 Phylogenetic tree.
This phylogenetic tree shows the relationship among reptiles (including the dinosaurs), birds, and mammals.

There are about 6,000 species of snakes and lizards that live mainly in the tropics and deserts. Lizards have four clawed feet and resemble their prehistoric ancestor in appearance. They are carnivorous and feed on insects and small animals, including other lizards. Marine iguanas of the Galápagos Islands are adapted to spend long periods of time at sea where they feed on sea lettuce and other marine vegetation. Chameleons are adapted to live in trees and have long, sticky tongues for catching insects some distance away. They can change color in order to blend in with their background. Other lizards use different means to avoid being eaten. An Australian frilled lizard erects a collar of skin about its neck, which greatly increases its apparent size and makes it look frightening. Worm lizards have adapted a subterranean life and look somewhat like an earthworm, since they lack legs and are blind.

A Closer Look

Venomous Snakes Are Few

Modern snakes and lizards make up 95% of living reptiles. Snakes evolved from lizards during the Cretaceous period and became adapted to burrowing. They lack limbs, so their prey must be subdued and swallowed without the benefit of appendages for manipulating food. Most snakes, like the boas and pythons, are powerful constrictors, suffocating their struggling prey with strong coils. Smaller snakes, such as the familiar garter snakes and water snakes, frequently swallow their food while it is still alive. Still others use toxic saliva to subdue their prey, usually lizards. It is likely that snake venom evolved as a way to obtain food and is used only secondarily in defense.

There are two major groups of venomous snakes. Elapids are represented in the United States by the coral snakes, *Micrurus fulvius* and *Micruroides euryxanthus*, which inhabit the southern states and display bright bands of red, yellow (or white), and black that completely encircle the body (Fig. 35A). In these snakes, the fangs, which are modified teeth, are short and permanently erect. The venom is a powerful neurotoxin that usually paralyzes the nervous system. Actually, coral snakes are responsible for very few bites—probably because of their secretive nature, small size, and relatively mild manner.

Vipers, represented in the United States by pit vipers such as the copperhead and cottonmouth (both *Agkistrodon*) and about 15 species of rattlesnakes (*Sistrurus* and *Crotalus*), make up the remaining venomous snakes of the United States. They have a sophisticated venom delivery system terminating in two large, hollow, needle-like fangs that can be folded against the roof of the mouth when not in use. The venom destroys the victim's red blood cells and causes extensive local tissue damage. These snakes are readily identified by the combination of heat-sensing facial pits, elliptical pupils in the eyes, and a single row of scales on the underside of the tail (Fig. 35B). None of our harmless snakes has any combination of these characteristics.

First aid for snakebite is not advised if medical attention is less than a few hours away; application of a tourniquet, incising the wound to promote bleeding, and other radical treatments often cause more harm than good. The best method of treating snakebite is through the use of prescribed antivenin, a serum containing antibodies to the venom. A hospital stay is required because some people are allergic to the serum.

Most people are not aware that snakes are perhaps the greatest controllers of disease-carrying, crop-destroying rodents because they are well adapted to following such prey into their hiding places. Also, snakes are important food items in the diets of many other carnivores, particularly birds of prey such as hawks and owls. Their presence in an ecosystem demonstrates the overall health of the environment.

Figure 35A Coral snake, *Micrurus fulvius*. This is a front-fanged snake, one of the major groups of venomous snakes.

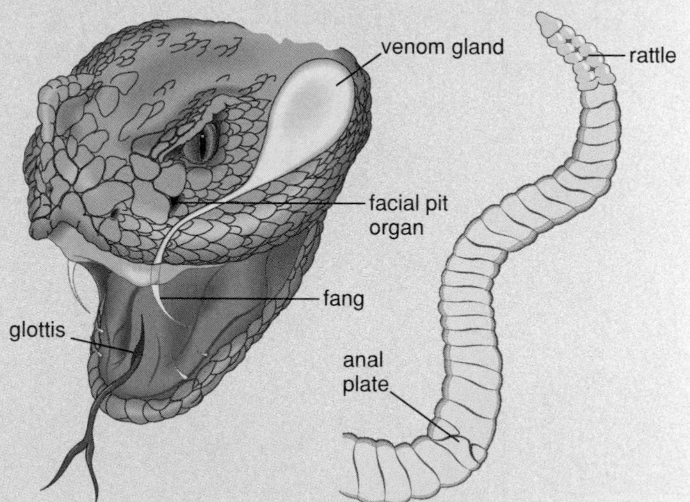

Figure 35B Viper characteristics. Vipers, the other major group of venomous snakes, can be identified by these characteristics.

Snakes evolved from lizards and have also lost their legs as an adaptation to burrowing. They are carnivorous and have a jaw that is only loosely attached to the skull; therefore, they can eat prey, such as a mouse, that is much larger than their head size. When a snake flicks out its tongue, it is collecting airborne molecules and transferring them to a Jacobson's organ for analysis. *Jacobson's organ,* which opens at the roof of the mouth, is an olfactory organ for the analysis of airborne chemicals. Though most snakes are not poisonous, the poisonous rattlesnakes, coral snakes, copperheads, and cobras have fangs for puncturing the skin and injecting venom.

Turtles have a heavy shell to which the ribs and thoracic vertebrae are fused. They lack teeth but have a sharp beak. Most turtles spend some time in water. Sea turtles leave the ocean only to lay their eggs. Their legs are flattened and paddlelike, while tortoises, which are usually terrestrial, have strong legs for walking. Crocodiles and alligators lead a largely aquatic life, feeding on fishes, turtles, and terrestrial animals that venture close enough to be caught (Fig. 35.14).

They have long, powerful jaws with numerous teeth and a muscular tail that serves as both a weapon and a paddle. Although other reptiles are voiceless, male crocodiles and alligators bellow to attract mates. In some species, the male protects the eggs and cares for the young.

These features, in particular, distinguish reptiles:

- **usually tetrapods**
- **lungs with expandable rib cage**
- **shelled egg**
- **dry, scaly skin**

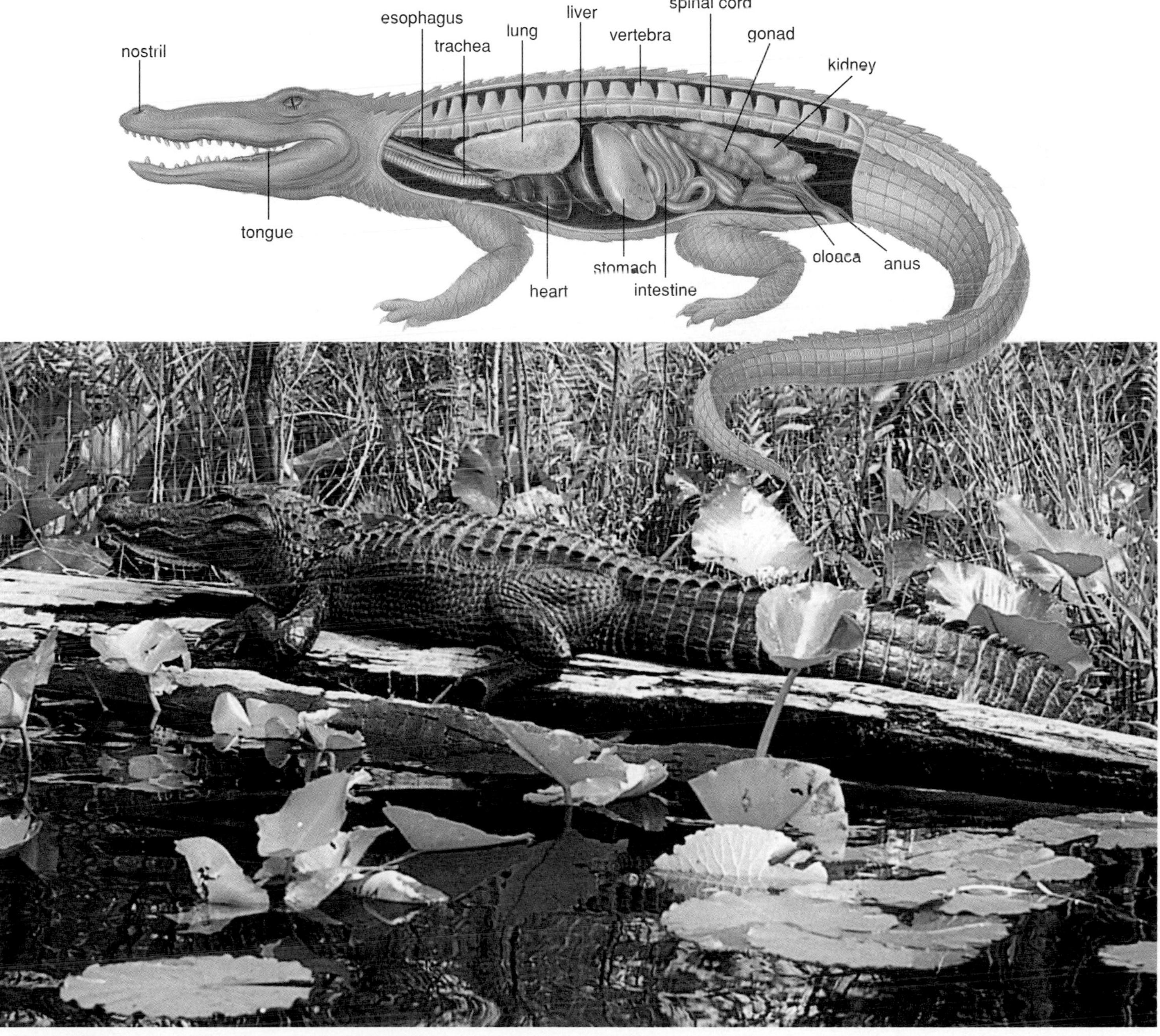

Figure 35.14 American alligator, *Alligator.*

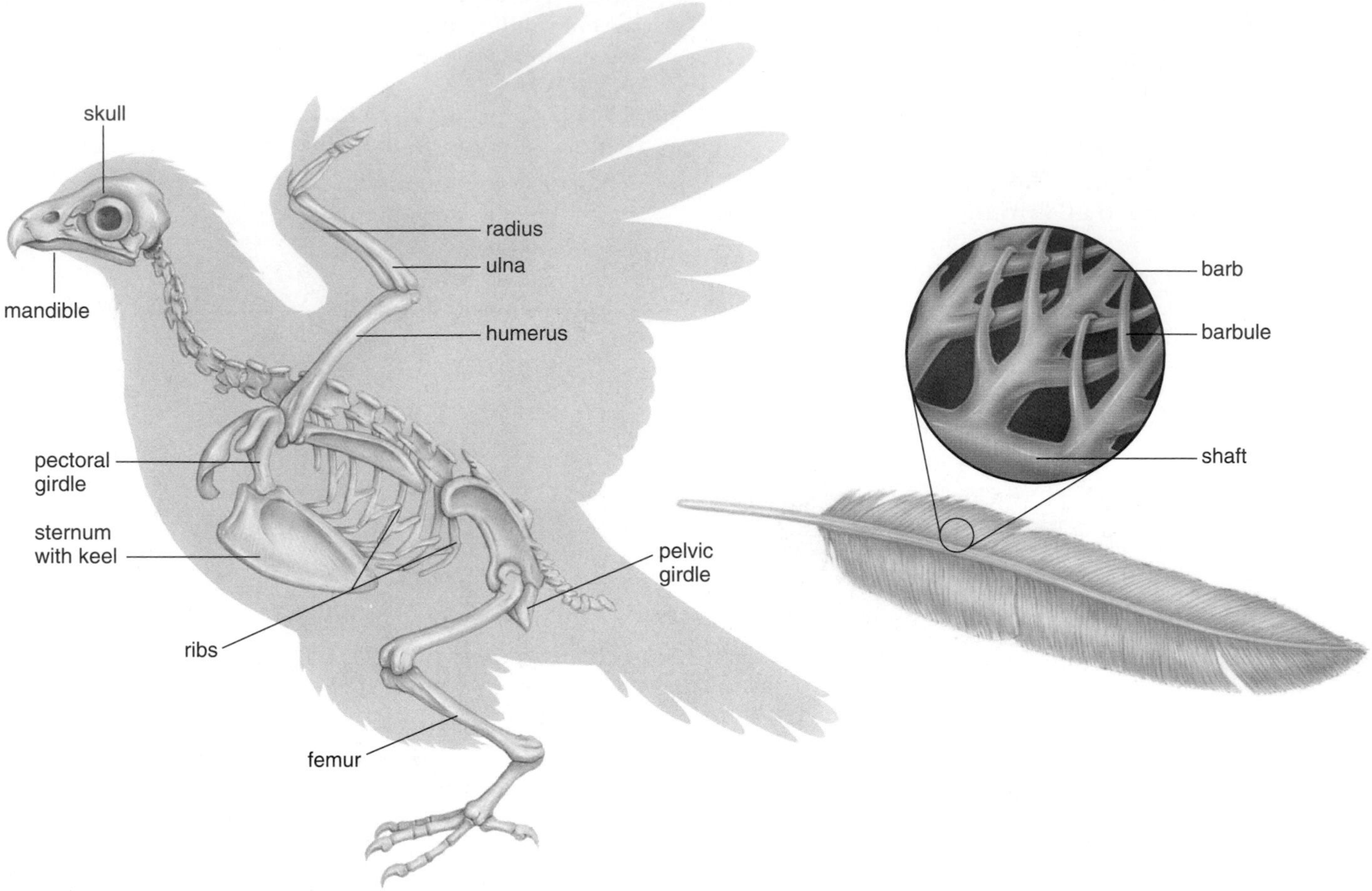

skull
mandible
radius
ulna
humerus
pectoral girdle
sternum with keel
ribs
femur
pelvic girdle
barb
barbule
shaft

a.

upstroke

downstroke

b. Bald eagle, *Haliaetus*

Figure 35.15 Bird anatomy.
a. Birds have a large, keeled breastbone (sternum) to which flight muscles attach. Bird bones are strong but weigh very little because of the air cavities they contain. In feathers, a hollow central shaft gives off barbs and barbules, which interlock in a latticelike array. **b.** Among living vertebrates, only birds and bats rely on true flight—they fly by flapping their wings. Bird flight requires an airstream and a powerful wing downstroke for lift, a force at right angles to the airstream.

35.5 Wings and Feathers Evolve

Today's birds lack teeth and have only a vestigial tail, but they still retain many reptilian features such as the shape of the body, scales on their legs, claws on their toes, and a horny beak. They also lay amniotic eggs, although the *shell is hard* rather than leathery. The exact ancestry of birds is in dispute, but there are those who contend that birds are closely related to bipedal dinosaurs and that they should be classified as such.

Birds Have Feathers

Birds (class Aves, 9,000 species) are the only modern animals to have **feathers,** which are actually modified reptilian scales. Bird feathers are of two types. *Contour feathers* cover the body, and those attached to the wings (called flight feathers) overlap to produce a broad, flat surface beneficial for flight. *Down feathers* provide excellent insulation against body heat loss. This is important because birds are **homeothermic** [Gk. *homoios,* like, and *therme,* heat]; they maintain a constant, relatively high body temperature, which permits them to be continuously active even in cold weather. Perhaps feathers first provided insulation and then became adapted for flight.

Nearly every anatomical feature of birds can be related to their ability to fly (Fig. 35.15). Their forelimbs are modified as *wings.* They have hollow, very light bones laced with air cavities. A horny *beak* has replaced jaws equipped with teeth, and a slender neck connects the head to a rounded, compact torso. The breastbone is enlarged and has a *keel,* to which the strong muscles are attached for flying. Oxygen delivery to these muscles is efficient; there is no dead space in the lungs because air flows one way through the air sacs and the lungs, and the double-loop circulation is improved because the *four-chambered heart* keeps oxygenated blood separated from deoxygenated blood.

Flight requires well-developed sense organs and nervous system. Birds have particularly acute vision and excellent muscle reflexes. Because birds can fly, they migrate and exploit food sources in widely separated habitats. Complex hormonal regulation and behavior responses are involved in bird behavior, which also includes caring for fledglings until they are independent.

Classification of birds is particularly based on beak and foot types, and to some extent on habitat and behavior. The various orders include birds of prey with notched beaks and sharp talons; shorebirds with long, slender, probing bills and long, stiltlike legs; woodpeckers with sharp, chisel-like bills and grasping feet; waterfowl with webbed toes and broad bills; penguins with wings modified as paddles; and songbirds with perching feet.

These features, in particular, distinguish birds:

- feathers
- hard-shelled egg
- four-chambered heart
- usually wings for flying
- air sacs
- homeothermic

35.6 Homeothermy Pays Off

Mammals evolved during the Mesozoic era from therapsids, the mammal-like reptiles. The mammalian skull accommodates a larger brain relative to body size than does the reptilian skull; mammalian cheek teeth are differentiated as premolars and molars; their vertebrae are highly differentiated and the middle region of the backbone is arched, providing more effective movement on land. True mammals appeared during the Jurassic period, about the same time as the first dinosaurs. These first mammals were small, about the size of mice. All the time the dinosaurs flourished (165 million years), mammals were a minor group that changed little. Some of the earliest mammalian groups, represented today by the monotremes and marsupials, are not abundant today. The placental mammals that evolved later went on to occupy the many habitats previously occupied by the dinosaurs.

Mammals Have Hair and Mammary Glands

The chief characteristics of **mammals** [L. *mamma,* breast, teat] (class Mammalia, 4,500 species) are hair and milk-producing mammary glands. Mammals are also homeothermic, as are birds. Many of the adaptations of mammals are related to temperature control. *Hair,* for example, provides insulation against heat loss and allows mammals to be active even in cold weather. Like birds, mammals have efficient respiratory and circulatory systems, which assure a ready oxygen supply to muscles whose contraction produces body heat. Like birds, mammals have a double-loop circulation and a *four-chambered heart.*

Mammary glands enable females to feed (nurse) their young without deserting them to find food. Nursing also creates a bond between mother and offspring that helps ensure parental care while the young are helpless. In most mammals, the young are born alive after a period of development in the uterus, a part of the female reproductive tract. Internal development shelters the young and allows the female to move actively about while the young are maturing. Mammals are classified according to means of reproduction.

Mammals That Lay Eggs

Monotremes [Gk. *monos,* one, and *trema,* hole], mammals that have a cloaca and lay hard-shelled amniote eggs, are represented by the duckbill platypus and the spiny anteater, both of which are found in Australia (Fig. 35.16*a*). The female duckbill platypus lays her eggs in a burrow in the ground. She incubates the eggs and, after hatching, the young lick up milk that seeps from modified sweat glands on the abdomen of both males and females. The spiny anteater has a pouch on the belly side formed by swollen mammary glands and longitudinal muscle. The egg moves from the cloaca to this pouch where hatching takes place, and the young remain for about 53 days. Then they stay in a burrow, where the mother periodically visits and nurses them.

a. Duckbill platypus, *Ornithorhynchus*

b. Koala, *Phascolarctos*

Figure 35.16 Types of mammals.
a. The duckbill platypus is a monotreme, which lays shelled eggs. **b.** The koala is a marsupial, whose young are born immature and complete their development within the mother's pouch. **c.** The white-tailed deer is a placental mammal, whose young develop within a uterus.

Mammals That Have Pouches

The young of **marsupials** [Gk. *marsupium,* pouch] begin their development inside the female's body, but they are born in a very immature condition. Newborns crawl up into a *pouch* on their mother's abdomen. Inside the pouch, they attach to nipples of mammary glands and continue to develop. Frequently, more are born than can be accommodated by the number of nipples, and it's "first come, first served."

Today, marsupial mammals are found mainly in Australia; only a few marsupials, such as the American opossum, are found outside that continent. In Australia, marsupials underwent adaptive radiation for several million years without competition from placental mammals, which arrived there only recently. Among the herbivorous marsupials, koalas are tree-climbing browsers (Fig. 35.16*b*) and kangaroos are grazers. The Tasmanian wolf or tiger, thought to be extinct, was a carnivorous marsupial about the size of a collie dog.

Mammals That Have Placentas

Developing **placental mammals** are dependent on the **placenta,** an organ of exchange between maternal blood and fetal blood. Nutrients are supplied to the growing offspring, and wastes are passed to the mother for excretion. While the fetus is clearly parasitic on the female, in exchange, she is free to move about as she chooses while the fetus develops. The young are born at a relatively advanced stage of development.

Placental mammals lead an active life. The senses are acute and the brain is enlarged due to the expansion of the foremost part—the cerebral hemispheres. These have become convoluted and have expanded to such a degree that they hide many other parts of the brain from view. The brain is not fully developed for some time after birth, and there is a *long period of dependency* on the parents, during which the young learn to take care of themselves.

Placental mammals populate all continents except Antarctica. Most mammals live on land, but some (e.g., whales, dolphins, seals, sea lions, manatees) are secondarily adapted to live in water, and bats are able to fly. Classification is primarily based on mode of locomotion and methods of obtaining food. The following are some of the major orders of mammals:

Order Chiroptera (925 species) includes the nocturnal bats whose wings consist of two layers of skin and connective tissue stretched between the elongated bones of all fingers but the first. Many species use echolocation to locate their usual insect prey. But there are also bird-, fish-, frog-, and plant-eating bats.

Order Primates (180 species) includes lemurs, monkeys, gibbons, chimpanzees, gorillas, and humans. Typically, primates are tree-dwelling fruit eaters although some, like humans, are ground dwellers. They all have a freely movable head, they have five digits with nails (not claws), and the thumb in many (sometimes the large toe) is opposable. Primates, especially humans, are well known for their well-developed brains.

Order Rodentia (1,760 species), the largest order, includes mice, rats, squirrels, beavers, and porcupines. Rodents have incisors that grow continuously. They usually feed on seeds but some are omnivorous and some eat mainly insects.

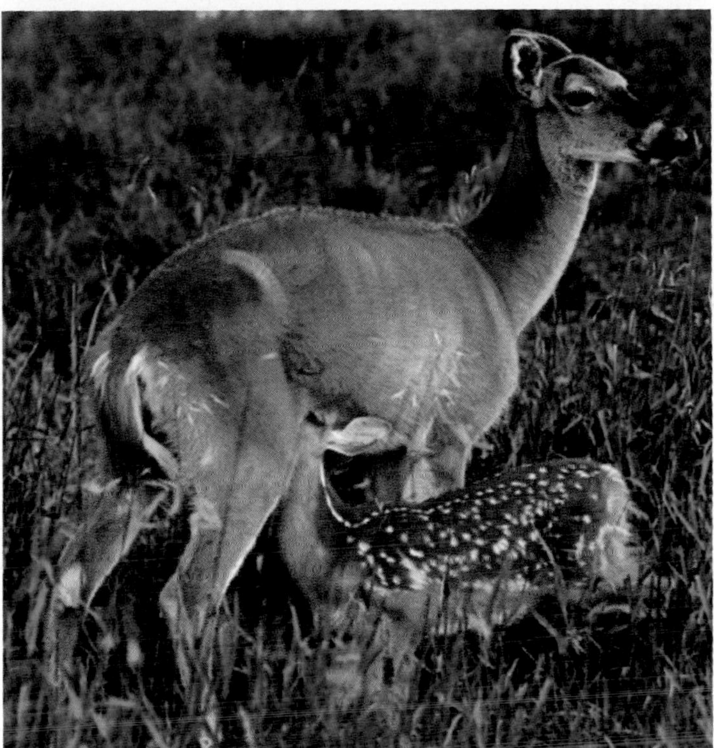

c. White-tailed deer, *Odocoileus*

Order Cetacea (80 species) includes the whales and dolphins, which are mammals despite their lack of hair or fur. Blue whales, the largest animal ever to have lived on this planet, are baleen whales that feed by straining large quantities of water containing plankton. Toothed whales feed mainly on fish and squid.

Order Carnivora (270 species) includes dogs, cats, bears, raccoons, and skunks. All these animals have limbs adapted for running and have a well-developed sense of smell. The canines of meat eaters are large and conical. There are some aquatic carnivores—namely seals, sea lions, and walruses—which must return to land to reproduce.

Order Proboscidea (2 species) includes the elephants, the largest living land mammals. The upper lip and nose have become elongated and muscularized to form a trunk.

The hoofed mammals include the orders Perissodactyla (e.g., horses, zebras, tapirs, rhinoceroses, 17 species) and the Artiodactyla (e.g., pigs, cattle, deer, hippopotami, buffaloes, giraffes, 185 species) whose elongated limbs are adapted for running across open grasslands. Both groups of animals are herbivorous and have large grinding teeth.

Order Lagomorpha (65 species) includes the herbivorous rabbits, hares, and pikas—animals that superficially resemble rodents. They also have two pairs of continually growing incisors, and their hind legs are longer than their front legs.

These features, in particular, distinguish placental mammals:

- body hair
- differentiated teeth
- well-developed brain
- infant dependency
- mammary glands
- homeothermic
- internal development

connecting concepts

How do you measure success? By almost any criterion, vertebrates lose. Vertebrates are eukaryotes that have been assigned to one domain, while the prokaryotes now have two domains. There are more prokaryotes and possibly more types of prokaryotes than any other living form. The unseen world is much larger than the seen world! And prokaryotes are adapted to utilize most any type of energy source and live in most any type of environment.

If it were not for the myriad types of invertebrate insects, there would be more aquatic species than terrestrial ones. The adaptive radiation of mammals has taken place on land, and this might seem impressive to some, but actually the number of mammalian species (4,500) is small compared to, say, the mollusks (110,000 species), which radiated in the sea. On land, vertebrate species are far outnumbered by the number of insect species. Being small has its advantages. To a large animal like a cow, grass is a more or less uniform carpet, but to small insects it is a highly varied habitat. Specific insects feed on the roots, stems, leaves, flowers, pollen, and seeds.

In terms of size, however, the vertebrates might win. A blue whale can weigh as much as 150 tons and have a length of 30 meters. But then again, redwood trees can be over 100 meters tall and there are eucalyptus trees measuring as much as 140 meters. Metabolically speaking, plants are much more capable than animals. They can make their own food and change glucose into all the other organic molecules they require. An ecosystem only needs plants and decomposers to sustain itself.

The size and complexity of the brain is cited as a criterion by which vertebrates are more successful than other living things. However, it can be suggested that this characteristic is associated with others that make an animal prone to extinction. Which type animal is likely to become extinct? As we discussed earlier on page 397, animals that are large, have a long life span, are slow to mature, have few offspring, and expend much energy caring for their offspring tend to become extinct when their normal way of life is destroyed. Vertebrates, in general, are more threatened than other types of organisms by our present ecological crisis—a crisis brought on the activities of the vertebrate with the most complex brain of all—*Homo sapiens*.

Summary

35.1 Radial Symmetry, Again

In the deuterostomes, the blastopore is associated with the anus, radial cleavage produces indeterminate cells, and there is an enterocoelom. Echinoderms and chordates are deuterostomes.

Echinoderms (e.g., sea stars, sea urchins, sea cucumbers, and sea lilies) have radial symmetry as adults (not as larvae) and internal calcium-rich plates with spines. Typical of echinoderms, sea stars have tiny skin gills, a central nerve ring with branches, and a water vascular system for locomotion. Each arm of a sea star contains branches from the nervous, digestive, and reproductive systems.

Hemichordates indicate that the echinoderms are related to the chordates. Their larva resembles the larva of echinoderms, and they have two chordate features (a dorsal hollow nerve cord and gill slits).

35.2 Chordate Characteristics Evolve

Chordates (tunicates, lancelets, and vertebrates) have a notochord, a dorsal hollow nerve cord, and pharyngeal pouches at one time in their life history. Also, there is a post-anal tail.

Tunicates and lancelets are the invertebrate chordates. Tunicates lack chordate characteristics (except gill slits) as adults, but they have a larva that could be ancestral for the vertebrates. Lancelets are the only chordate to have the three characteristics in the adult stage.

35.3 The Vertebrate Body Plan Evolves

Vertebrates are in the phylum Chordata and have the three chordate characteristics as embryos. As adults, the notochord is replaced by the vertebral column. Vertebrates, which undergo cephalization, have an endoskeleton, paired appendages, and well-developed internal organs.

The first vertebrates lacked jaws and paired appendages. They are represented today by the hagfishes and lampreys. Ancestral bony fishes, which had jaws and paired appendages, gave rise during the Devonian period to two groups: today's cartilaginous fishes (skates, rays, and sharks) and the bony fishes, including the ray-finned fishes and the lobe-finned fishes. The ray-finned fishes became the most diverse group among the vertebrates. The lobe-finned fishes, represented today by three species of lungfishes and the coelacanth, gave rise to the amphibians.

Ancestral amphibians were tetrapods that diversified during the Carboniferous period. They are represented primarily today by frogs and salamanders, which usually return to the water to reproduce and then metamorphose into terrestrial adults.

35.4 The Amniote Egg Evolves

Reptiles (today's snakes, lizards, turtles, and crocodiles) lay a shelled egg, which allows them to reproduce on land.

One main group of ancient reptiles, the stem reptiles that presumably evolved from amphibian ancestors, produced a line of descent that evolved into both dinosaurs and birds during the Mesozoic era. A different line of descent from stem reptiles evolved into mammals.

35.5 Wings and Feathers Evolve

Birds are feathered, which helps them maintain a constant body temperature. They are adapted for flight: their bones are hollow, their shape is compact, their breastbone is keeled, and they have well-developed sense organs.

35.6 Homeothermy Pays Off

Mammals remained small and insignificant while the dinosaurs existed, but when the latter became extinct at the end of the Cretaceous period, mammals became dominant land organisms.

Mammals are vertebrates with hair and mammary glands. The former helps them maintain a constant body temperature, and the latter allows them to nurse their young. Monotremes lay eggs; marsupials have a pouch in which the newborn matures; and the placental mammals, which are far more varied and numerous, retain offspring inside the uterus until birth.

Reviewing the Chapter

1. What are the general characteristics of deuterostomes? 614
2. What are the general characteristics of echinoderms? Explain how the water vascular system works in sea stars. 614–16
3. What three characteristics do all chordates have at some time in their life history? How do hemichordates show that echinoderms are related to the chordates? 616
4. Describe the two groups of invertebrate chordates, and explain how the tunicates might be ancestral to vertebrates. 617
5. What is the vertebrate body plan? Discuss the distinguishing characteristics of vertebrates. 619
6. Describe the jawless fishes, including ancient ostracoderms. 619
7. What is the significance of having jaws? Describe the ancient placoderms and today's cartilaginous and bony fishes. The amphibians evolved from what type of fish? 620–22
8. What is the significance of being a tetrapod? Discuss the characteristics of amphibians, stating which ones are especially adaptive to a land existence. Explain how their class name (Amphibia) characterizes these animals. 622–23
9. What is the significance of the amniote egg? What other characteristics make reptiles less dependent on a source of external water? 624
10. Draw a simplified phylogenetic tree for reptiles showing their relationship to birds and mammals. Your tree should also include stem reptiles and dinosaurs. 625
11. What is the significance of wings? In what other ways are birds adapted to flying? 629
12. What is the significance of homeothermy? What are the three subclasses of mammals, and what are their primary characteristics? 629–31

Testing Yourself

Choose the best answer for each question.

1. Which of these does not pertain to a deuterostome?
   a. blastopore is associated with the anus
   b. spiral cleavage
   c. enterocoelom
   d. echinoderms and chordates
2. The tube feet of echinoderms
   a. are their head.
   b. are a part of the water vascular system.
   c. help pass sperm to females during reproduction.
   d. All of these are correct.
3. Which of these is not a chordate characteristic?
   a. dorsal supporting rod, the notochord
   b. dorsal hollow nerve cord
   c. pharyngeal pouches
   d. vertebral column
4. Adult tunicates
   a. do not have all three chordate characteristics.
   b. have a larva that resembles that of hemichordates.
   c. are fishlike in appearance.
   d. All of these are correct.
5. Sharks and bony fishes are different in that only
   a. bony fishes have paired fins.
   b. bony fishes have an operculum.
   c. sharks have a bony skeleton.
   d. sharks are predaceous.
6. Amphibians arose from
   a. cartilaginous fishes.
   b. jawless fishes.
   c. ray-finned fishes.
   d. bony fishes with lungs.
7. Which of these is not a feature of amphibians?
   a. dry skin that resists desiccation
   b. metamorphosis from a swimming form to a land form
   c. small lungs and a supplemental way of gas exchange
   d. reproduction in the water
8. How are frogs different from snakes?
   a. The ancestors of snakes but not salamanders were tetrapods.
   b. Salamanders reproduce on land but snakes do not.
   c. Snakes but not salamanders lay a shelled egg.
   d. Salamanders are carnivores, and snakes are herbivores.
9. Dinosaurs
   a. were dominant during the Mesozoic era.
   b. are closely related to the birds.
   c. lay shelled eggs.
   d. All of these are correct.
10. Which of these is a true statement?
    a. In all mammals, offspring develop within the female.
    b. All mammals have hair and mammary glands.
    c. All mammals are land-dwelling forms.
    d. All of these are true.
11. Complete the following diagram.

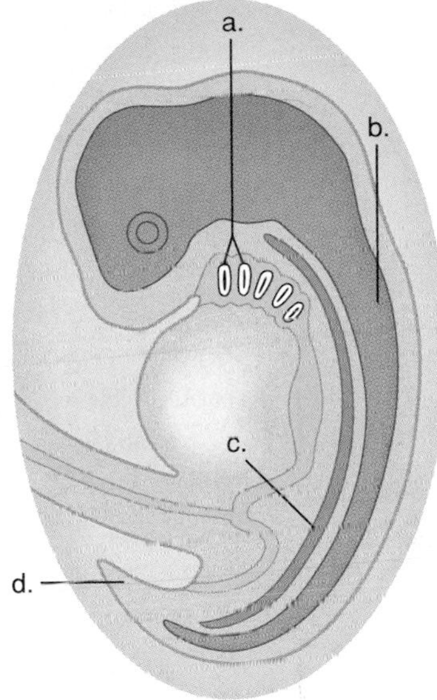

Applying the Concepts

1. *Protostomes and deuterostomes both evolved similar features.*
   Support this statement by referring to common features of arthropods and chordates.
2. *Arthropods and vertebrates are highly successful groups of animals on land.*
   What characteristics shared by arthropods and vertebrates are adaptive to a land existence?
3. *Reptiles are adapted to reproduce on land.*
   Compare the adaptations of reptiles and angiosperms for reproduction on land.

Using Technology

Your study of the deuterostomes is supported by these available technologies:

Exploring the Internet
The Mader Home Page provides resources for and help with studying this chapter.

http://www.mhhe.com/sciencemath/biology/mader/
(Click on Biology.)

Understanding the Terms

agnathan 619
amniote egg 624
amphibian 622
bird 629
bony fish 620
cartilaginous fish 620
cerebral cortex 622
chordate 616
cloaca 623
deuterostome 614
echinoderm 614
ectothermic 622
feather 629
fin 620
fish 619
gill 616
gill slit 616
gnathostomate 620
hagfish 619
hemichordate 616
homeothermic 629
jaw 620
lamprey 619
lancelet 617
lobe-finned fish 621
lung 622
mammal 629
marsupial 630
monotreme 629
nerve cord 616
notochord 616
ostracoderm 619
placenta 630
placental mammal 630
placoderm 620
ray-finned fish 620
reptile 624
scale 619
swim bladder 620
tetrapod 622
therapsid 624
tube foot 615
tunicate 617
vertebrate 619
water vascular system 616

Match the terms to these definitions:

a. __________ Egg that has an amnion, as seen during the development of reptiles, birds, and mammals.
b. __________ Animal (bird or mammal) that maintains a uniform body temperature independent of the environmental temperature.
c. __________ Egg-laying mammal—for example, duckbill platypus and spiny anteater.
d. __________ Mammal bearing immature young nursed in a pouch—for example, kangaroo.
e. __________ Having a body temperature that varies according to the environmental temperature.
f. __________ Member of a class of terrestrial vertebrates that includes frogs, toads, and salamanders; they are still tied to a watery environment for reproduction.
g. __________ Member of a class of terrestrial vertebrates with internal fertilization, scaly skin, and an egg with a leathery shell; includes snakes, lizards, turtles, and crocodiles.
h. __________ Member of a class of vertebrates characterized especially by the presence of hair and mammary glands.
i. __________ Member of a mammalian subclass characterized by a placenta, an organ of exchange for nutrients, gases, and wastes, between maternal and fetal blood.
j. __________ Member of a phylum of marine animals that includes sea stars, sea urchins, and sand dollars; characterized by radial symmetry and a water vascular system.

Further Readings for Part v

Anderson, D. M. August 1994. Red tides. *Scientific American* 271(2):62. Red tide algal blooms release toxins into oceans, killing whales and fish, and causing illness to humans.

Brusca, R. C., and Brusca, G. J. 1990. *Invertebrates*. Sunderland, Mass.: Sinauer Associates. Discusses invertebrate zoology using the *Bauplan* concept of body design, development, evolution, and phylogenetics.

Carmichael, W. January 1994. The toxins of cyanobacteria. *Scientific American* 270(1):78. Cyanobacteria can be hazardous or beneficial, depending on the approach taken.

Castro, P., and Huber, M. 1997. *Marine Biology*. 2d ed. St. Louis: Mosby-Year Book, Inc. This introductory text is designed to provide a stimulating overview of marine biology.

Cowen, R. 1995. *History of life*. 2d ed. Boston: Blackwell Scientific Publications. A readable account for the undergraduate of the history of life based on the fossil record.

Duellman, W. E. July 1992. Reproductive strategies of frogs. *Scientific American* 267(1):80. Frogs have managed to colonize niches throughout the terrestrial environment.

Dusenbery, D. B. 1996. *Life at small scale: The behavior of microbes*. New York: Scientific American Library. This easy to read, well-illustrated text describes how microbes respond to the physical demands of their environment.

Eigen, M. July 1993. Viral quasispecies. *Scientific American* 269(1):42. A more dynamic view of viral populations holds clues to understanding and defeating them.

Goulding, M. March 1993. Flooded forests of the Amazon. *Scientific American* 266(3):114. Unique adaptations allow creatures to thrive in the aquatic ecosystems of the rain forest.

Hickman, C. J., et al. 1993. *Integrated principles of zoology*. 9th ed. St. Louis: Mosby-Year Book, Inc. This introductory text provides the basics for understanding the animal kingdom.

Hotez, P. J., and Pritchard, D. I. June 1995. Hookworm infection. *Scientific American* 272(6):68. Discusses how the biology of parasites offers clues to possible vaccines and also for new treatments for heart disease and immune disorders.

Kantor, F. S. September 1994. Disarming Lyme disease. *Scientific American* 271(3):34. Discusses the progress being made in understanding and treating Lyme disease.

Kardong, K. V. 1995. *Vertebrates: Comparative anatomy, function, evolution*. Dubuque, Iowa: Wm. C. Brown Publishers. An understanding of function and evolution is integrated with discussion of vertebrate morphology.

Kellert, S. R. 1996. *The value of life: Biological diversity and human society*. Washington D.C.: Island Press/Shearwater Books. The importance of biological diversity to the well-being of humanity is explored.

Kent, G. 1992. *Comparative anatomy of the vertebrates*. 7th ed. St. Louis: Mosby-Year Book, Inc. The basics of vertebrate comparative structure and function from an evolutionary perspective.

Le Guenno, B. October 1995. Emerging viruses. *Scientific American* 273(4):56. Article discusses viral disease outbreaks, particularly the deadly hemorrhagic fever viruses.

Levin, H. L. 1994. *The earth through time*. 4th ed. Philadelphia: Saunders College Publishing. Provides an in-depth study of evolutionary and geological eras and events.

Losick, R., and Kaiser, D. February 1997. Why and how bacteria communicate. *Scientific American* 276(2):68. Bacteria can send and receive chemical messages, and can organize into complex structures.

Margulis, L., et al. 1994. *The illustrated five kingdoms: A guide to the diversity of life on earth*. New York: HarperCollins College Publishers. Presents a catalog of the world's living diversity.

Mauseth, J. 1995. *Botany: An introduction to plant biology*. 2d ed. Philadelphia: Saunders College Publishing. The text emphasizes evolution and diversity in botany, as well as general principles of plant physiology and anatomy.

McClanahan, L., et al. March 1994. Frogs and toads in deserts. *Scientific American* 270(3):82. Certain amphibians reveal diverse and unusual adaptations to extreme environments.

Miller, S. A., and Harley, J. P. 1996. *Zoology: The animal kingdom*. Dubuque, Iowa: Wm. C. Brown Publishers. This introductory zoology text emphasizes the interrelationships of life-forms.

Morello, J., et al. 1994. *Microbiology in patient care*. 5th ed. Dubuque, Iowa: Wm. C. Brown Publishers. Provides a basic knowledge of the principles of microbiology and epidemiology.

Murawski, D. A. March 1997. Moths come to light. *National Geographic* 191(3):40. Moths display a variety of disguises and survival techniques.

Norse, E. 1993. *Global marine biological diversity*. Washington D.C.: Island Press. Introduces marine conservation.

Pechenik, J. A. 1991. *Biology of invertebrates*. 2d ed. Dubuque, Iowa: Wm. C. Brown Publishers. Written for the beginning college student, this is a readable, authoritative introduction to the biology of the invertebrates.

Prescott, L., et al. 1993. *Microbiology*. 2d ed. Dubuque, Iowa: Wm. C. Brown Publishers. This introductory text covers all major areas of microbiology.

Raven, P., et al. 1992. *Biology of plants*. 5th ed. New York: Worth Publishers. This book introduces viruses, bacteria, photosynthetic protoctists, fungi, and plants.

Rinderer, T. E., et al. December 1993. Africanized bees in the U. S. *Scientific American* 269(6):84. Africanized honeybees have reached the United States from points south.

Romoser, W. S., and Stoffolano, J. G., Jr. 1994. *The science of entomology*. 3d ed. Dubuque, Iowa: Wm. C. Brown Publishers. Introduces entomology from both a basic and applied perspective.

Schmidt, G. D., and Roberts, L. S. 1996. *Foundations of parasitology*. 5th ed. Dubuque, Iowa: Wm. C. Brown Publishers. Designed for upper-division parasitology classes, this text emphasizes the major parasites of humans and animals.

Schopf, J. W. 1992. *Major events in the history of life*. Boston: Jones and Bartlett Publishers. Summarizes crucial events shaping the earth's development and evolution over time.

Stern, K. 1994. 6th ed. *Introductory plant biology*. Dubuque, Iowa: Wm. C. Brown Publishers. Presents basic botany in a clear, informative manner.

Sumich, J. L. 1996. *An introduction to the biology of marine life*. 6th ed. Dubuque, Iowa: Wm. C. Brown Publishers. This introductory marine biology text covers taxonomy, evolution, ecology, behavior, and physiology of selected groups of marine organisms.

Talaro, K., and Talaro, A. 1993. *Foundations in microbiology*. Dubuque, Iowa: Wm. C. Brown Publishers. Surveys general topics for students entering health careers.

Plant Structure and Function

Flowering plants are terrestrial photosynthesizers. Their leaves, lifted skyward, capture solar energy and have pores that permit gas exchange. Their roots, anchored in the soil, absorb water. A vascular system transports water and minerals up the stem to the leaves and transports the products of photosynthesis to all body parts. Plant cells use sugars as the starting material for the production of all the metabolites they need.

Just like animals, plants are able to respond to environmental stimuli. For example, they bend toward unidirectional light. A complex network of hormones controls plant responses and indeed all aspects of their physiology. Unlike animals, plants grow their entire lives and are ever capable of producing new body parts. In general, trees have a much longer life span than animals: some live for thousands of years!

The adaptations of flowering plants to the land environment is especially evident when one considers the manner in which they carry on reproduction. Gametes, zygotes, and embryos are all protected from desiccation. And though flowering plants are nonmobile, many utilize the mobility of wind and animals to accomplish gamete and seed distribution.

Plant Structure

CHAPTER 36

Chapter Concepts

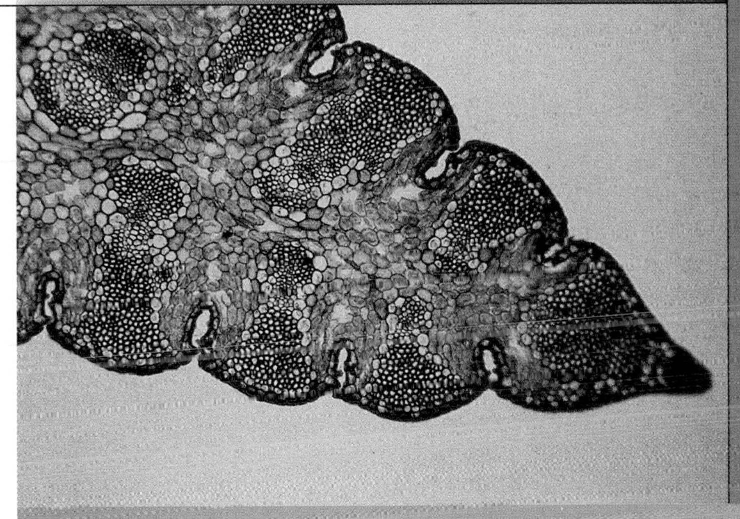

Leaf of Spanish bayonet, *Yucca glauca*

There is a stunning array of plant life covering the earth, and over 80% of all living plants are flowering plants, or angiosperms. Therefore, it is fitting that we set aside one part of the text to examine the structure and the function of flowering plants in particular. The organization of a flowering plant is suitable to photosynthesizing on land. The elevated leaves have a shape that facilitates absorption of solar energy. Strong stems conduct water up to the leaves and organic food down to the roots, which anchor the plant and have many branches for absorption of water. It can be seen, then, that flowering plants are complex organisms whose organs have a specific structure and functions.

A knowledge of plants has many practical applications. Gardening is the number-one hobby in the United States, and people want to know how to keep their plants healthy. Even more importantly, we depend on plants for sustenance, as a source of fibers for making clothes, for construction materials, and for medicines. Our very existence depends on our ability to value, and our understanding of the anatomy and physiology of, plants.

36.1 Plants Have Organs

Flowering plants are extremely diverse because they have adapted to living in varied environments. There are even flowering plants that live in water! Despite their great diversity in size and shape, flowering plants usually have three vegetative organs. An organ contains different types of tissues and performs one or more specific functions. The vegetative organs of a flowering plant—the root, the stem, and the leaf (Fig. 36.1)—allow a plant to live and grow. The flower, which functions during reproduction, contains a number of organs.

Roots Anchor

Although we are accustomed to speaking of the root, it is more appropriate to refer to the root system. The **root system** of a plant is the main root, together with any and all of its **lateral** (side) branches (Fig. 36.2*a*). As a rule of thumb, the root system is at least equivalent in size and extent to the **shoot system** (the part of the plant above ground). An apple tree, then, has a much larger root system than, say, a corn plant. A single corn plant may have roots as deep as 2.5 meters and spread out over 1.5 meters, but a mesquite tree that lives in the desert may have roots that penetrate to a depth of 20 meters. The extensive root system of a plant anchors it in the soil and gives it support.

The root system absorbs water and minerals from the soil, for the entire plant. The cylindrical shape of a root allows it to penetrate the soil as it grows and permits water to be absorbed from all sides. The absorptive capacity of a root is also increased by its many root hairs. Root hairs, which are projections from root hair cells, are especially responsible for the absorption of water. Root hair cells are found in a special zone located near the root tip. Root hairs are so numerous that they increase the absorptive surface of a root tremendously. It's been estimated that a single rye plant has about 14 billion hair cells and, if placed end to end, the root hairs would stretch 10,626 kilometers. Root hair cells are constantly being replaced. So this same rye plant most likely forms about 100 million new root hair cells every day. You are probably familiar with the fact that a plant yanked out of the soil will not fare well when transplanted; this is because the root hairs are torn off. Transplantation is more apt to be successful if you take a part of the surrounding soil along with the plant.

Roots have still other functions. Roots produce hormones that stimulate the growth of stems and coordinate their size with the size of the root. It is more efficient for a plant to have a root and stem size that is appropriate one to the other. Also, **perennial** plants, which die back and then regrow the next season, store the products of photosynthesis in their roots. Carrots and sweet potatoes come from the roots of such plants, for example.

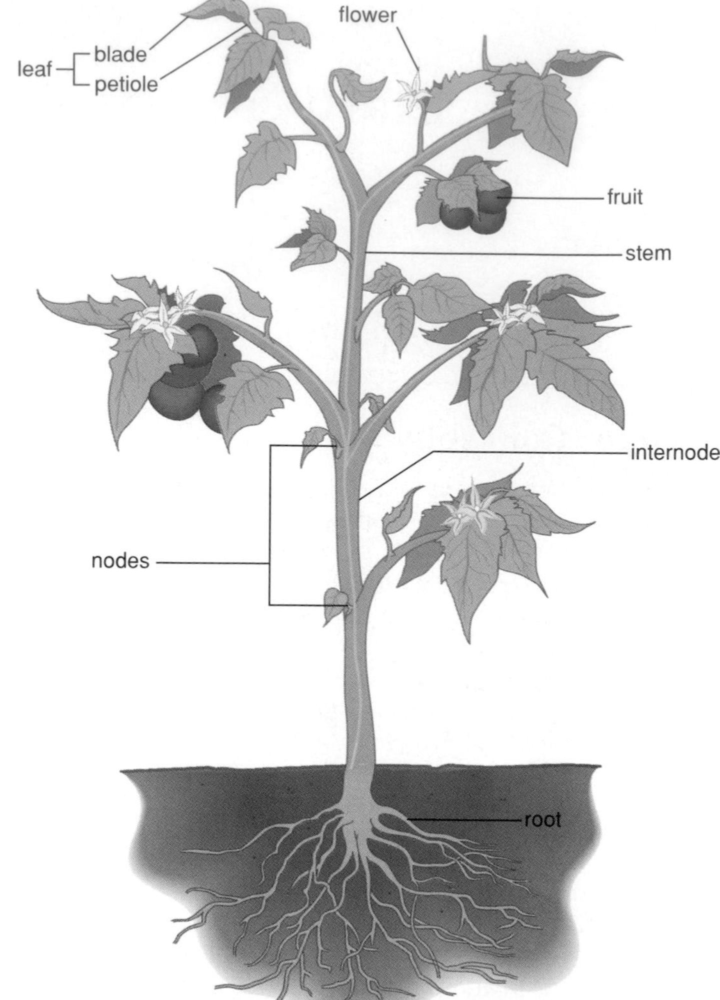

Figure 36.1 Organization of plant body.
Roots, stems, and leaves are vegetative organs. The flower and fruit are reproductive structures. **a.** Drawing of a tomato plant, a dicot plant. **b.** Photo of an onion plant, a monocot plant.

a.

b.

c.

Figure 36.2 Vegetative organs of the tomato, *Lycopersicon.*
a. The root system anchors the plant and absorbs water and minerals. **b.** The stem supports the leaves and transports water and organic nutrients. **c.** The leaf, which is often broad and thin, carries on photosynthesis.

Stems Support

The shoot system of a plant contains both stems and leaves. A **stem** is the main axis of a plant along with its lateral branches (Fig. 36.2*b*). The stem of a flowering plant terminates in tissue that allows the stem to elongate and produce leaves. If upright, as most are, stems support leaves in such a way that each leaf is exposed to as much sunlight as possible. A **node** occurs where leaves are attached to the stem and an **internode** is the region between the nodes. The presence of nodes and internodes is used to identify a stem even if it happens to be an underground stem. In some plants the nodes of horizontal stems asexually produce new plants.

Aside from supporting the leaves, a stem has vascular tissue that transports water and minerals from the roots to the leaves and transports the products of photosynthesis usually in the opposite direction. A cylindrical stem has the best shape to carry on these functions and of necessity the conducting cells are nonliving. Nonliving cells can best function as pipelines for water and mineral transport and for organic nutrient transport. Trees can grow taller every year only because they accumulate nonfunctional woody tissue that adds to the strength of their stems.

Some stems have functions other than transport. Some are specialized for storage. The stem stores water in cactuses and in other plants. Tubers are horizontal stems that store nutrients.

Leaves Photosynthesize

A foliage **leaf** is that part of a plant that usually carries on photosynthesis, a process that requires water, carbon dioxide, and sunlight (Fig. 36.2*c*). Leaves receive water from the root system by way of the stem. In fact, as we shall see in the next chapter, stems and leaves function together to bring about water transport from the roots.

In contrast to the shape of stems, foliage leaves are usually broad and thin. This shape has the maximum surface area for the absorption of carbon dioxide and the collection of solar energy. Also unlike stems, leaves are almost never woody. All their cells are living, and the bulk of a leaf contains tissue specialized to carry on photosynthesis.

The wide portion of a foliage leaf is called the *blade;* the *petiole* is a stalk that attaches the blade to the stem. The upper and acute angle between the petiole and stem is designated the leaf axil, and this is where an **axillary** (lateral) **bud,** which may become a branch or a flower, originates. Not all leaves are foliage leaves. Some are specialized to protect buds, attach to objects (tendrils), and store food (bulbs) or even capture insects.

A flowering plant has three vegetative organs: the root absorbs water and minerals, the stem supports and services leaves, and the leaf carries on photosynthesis.

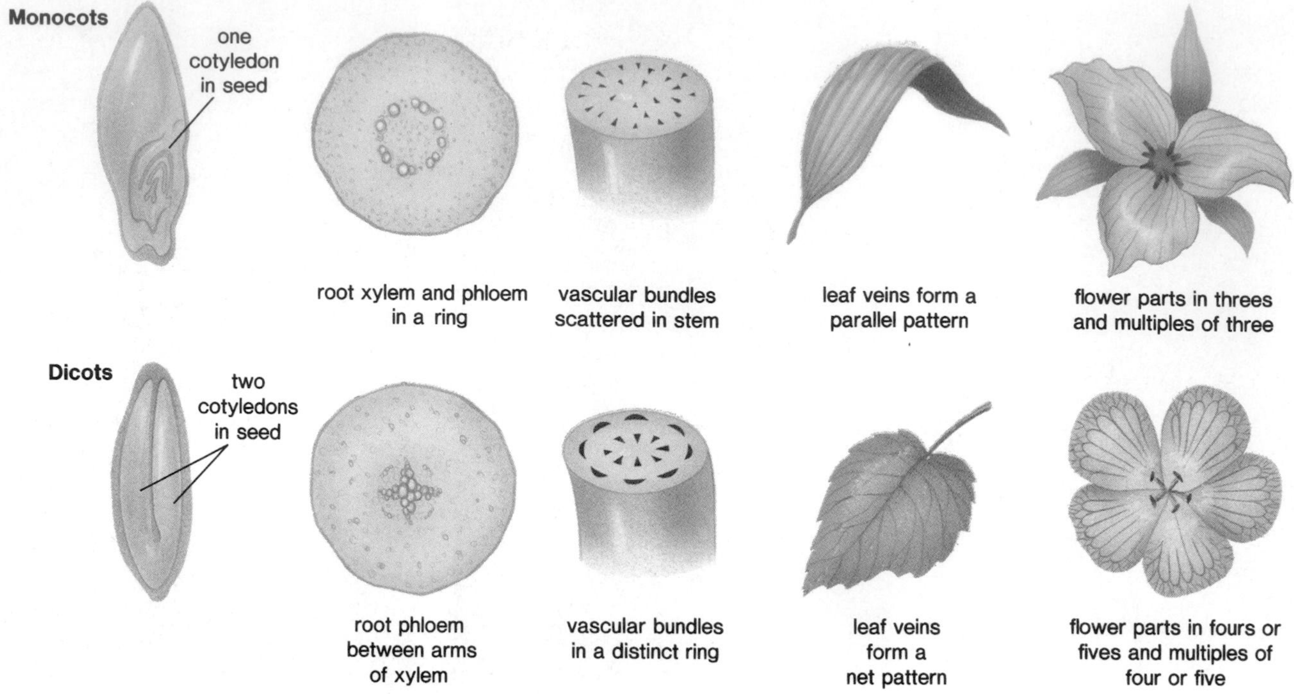

Figure 36.3 Flowering plants are either monocots or dicots.
Five features illustrated here are used to distinguish monocots from dicots: number of cotyledons; the arrangement of vascular tissue in roots, stems, and leaves; and the number of flower parts.

36.2 Monocot Versus Dicot Plants

Flowering plants are divided into the **monocots** and **dicots,** depending on the number of cotyledons or seed leaves in the embryonic plant (Fig. 36.3). **Cotyledons** [Gk. *cotyledon,* cup-shaped cavity] provide nutrient molecules either by storing or absorbing food for growing embryos before the true leaves begin photosynthesizing. Some embryos have one cotyledon, and these plants are known as monocotyledons, or monocots. Other embryos have two cotyledons, and these plants are known as dicotyledons, or dicots.

There is a different arrangement of vascular (transport) tissue in monocot and dicot roots. In the monocot root, *xylem,* the vascular tissue that transports water, occurs in a ring, and in the dicot, xylem has a star shape. **Vascular bundles** contain vascular tissue surrounded by a bundle sheath. In a monocot stem, the vascular bundles are scattered; in a dicot stem, they occur in a ring. Figure 36.3 shows cross sections of stems, but keep in mind that vascular bundles extend lengthwise in the stem. Leaf **veins** are vascular bundles within a leaf. Monocots exhibit parallel venation, and dicots exhibit netted venation, which may be either pinnate or palmate. Pinnate venation means that branch veins originate from points along the centrally placed main vein, and palmate venation means that the branch veins all originate at the point of attachment of the blade to the petiole:

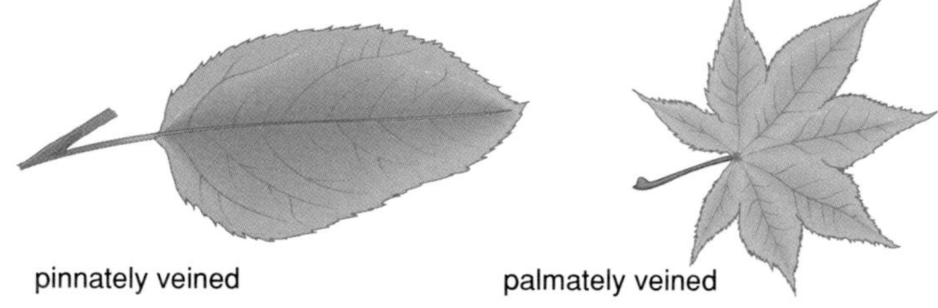

Adult monocots and dicots have other structural differences, such as differences in the usual number of flower parts and the number of apertures (thin areas in the wall) of pollen grains. Dicot pollen grains usually have three apertures, and monocot pollen grains usually have one aperture.

Although the division between monocots and dicots may seem arbitrary, it does in fact represent a separation that most likely dates back to the origin of flowering plants. The dicots are the larger group and include some of our most familiar flowering plants—from dandelions to oak trees. The monocots include grasses, lilies, orchids, and palm trees, among others. Some of our most significant food sources are monocots, including rice, wheat, and corn.

Flowering plants are divided into monocots and dicots on the basis of structural differences.

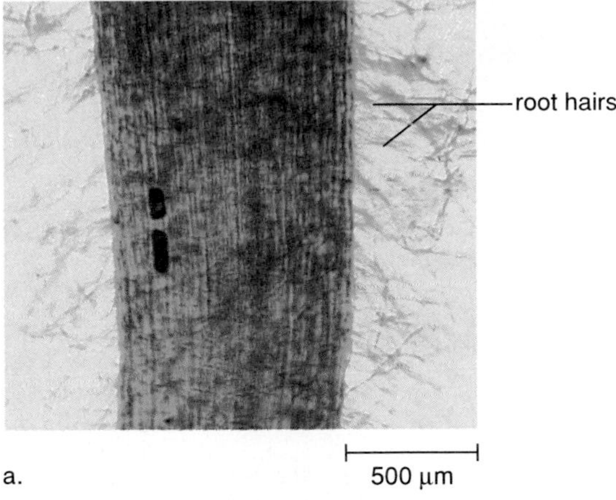

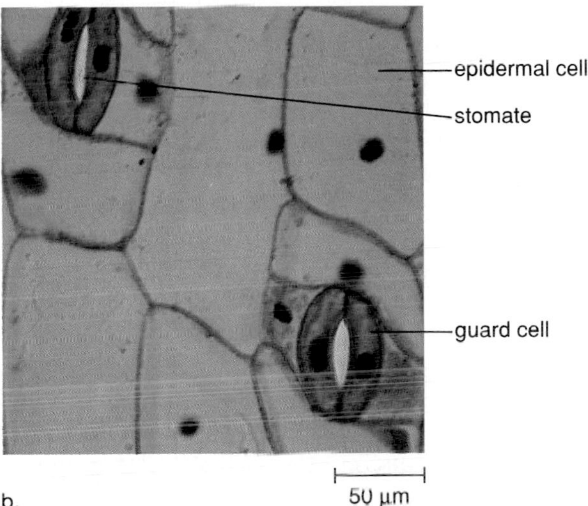

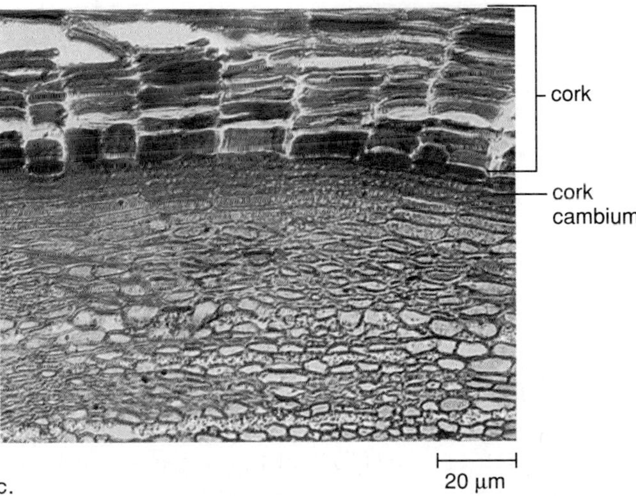

Figure 36.4 Modifications of epidermal tissue.
a. Root epidermis has root hairs to absorb water. **b.** Leaf epidermis contains stomates for gas exchange. **c.** Cork replaces epidermis in older, woody stems.

36.3 Plant Organs Have Tissues

A plant grows its entire life because it has **meristem** (embryonic tissue) located in the stem and root apexes. Three types of meristem continually produce the three types of specialized tissue in the body of a plant: protoderm, the outer most primary meristem, gives rise to epidermis; ground meristem produces ground tissue; and procambium produces vascular tissue.

We will be discussing these three specialized tissues:

1. **Epidermal tissue**—forms the outer protective covering of a plant.
2. **Ground tissue**—fills the interior of a plant.
3. **Vascular tissue**—transports water and nutrients in a plant and provides support.

Epidermal Tissue Protects

The entire bodies of nonwoody (herbaceous) and young woody plants are covered by a layer of **epidermis** [Gk. *epi*, over, and *derma*, skin], which contains closely packed epidermal cells. The walls of epidermal cells that are exposed to air are covered with a waxy **cuticle** to minimize water loss. The cuticle also protects against bacteria and other organisms that might cause disease.

In roots, certain epidermal cells have long, slender projections called **root hairs** (Fig. 36.4*a*). As mentioned, the hairs increase the surface area of the root for absorption of water and minerals; they also help to anchor the plant firmly in place.

Protective hairs of a different nature are produced by epidermal cells of stems and leaves. Epidermal cells may also be modified as glands that secrete protective substances of various types.

In leaves, the lower epidermis in particular contains specialized cells, such as guard cells (Fig. 36.4*b*). The guard cells, which unlike epidermal cells have chloroplasts, surround microscopic pores called **stomates** (also called stomata). When the stomates are open, gas exchange can occur. During the day, carbon dioxide diffuses in and oxygen diffuses out.

In older woody plants, the epidermis of the stem is replaced by cork tissue. **Cork,** the outer covering of the bark of trees, is made up of dead cork cells that may be sloughed off (Fig. 36.4*c*). New cork cells are made by a meristem called cork cambium. As the new cork cells mature, they increase slightly in volume and their walls become encrusted with *suberin*, a lipid material, so that they are waterproof and chemically inert. These nonliving cells protect the plant and make it resistant to attack by fungi, bacteria, and animals.

Epidermal tissue forms the outer protective covering of a herbaceous plant. It is modified in roots, stems, and leaves.

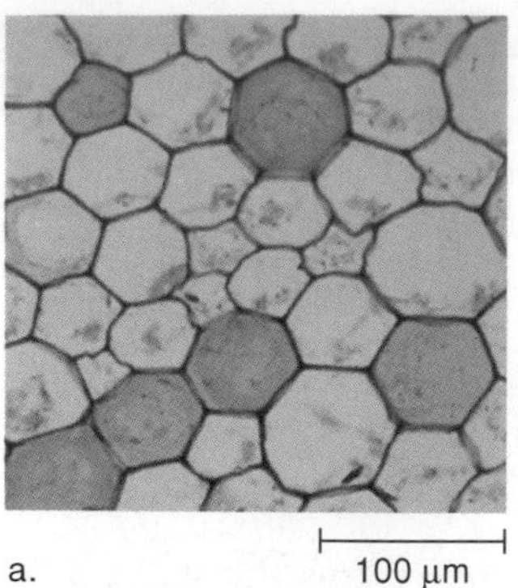

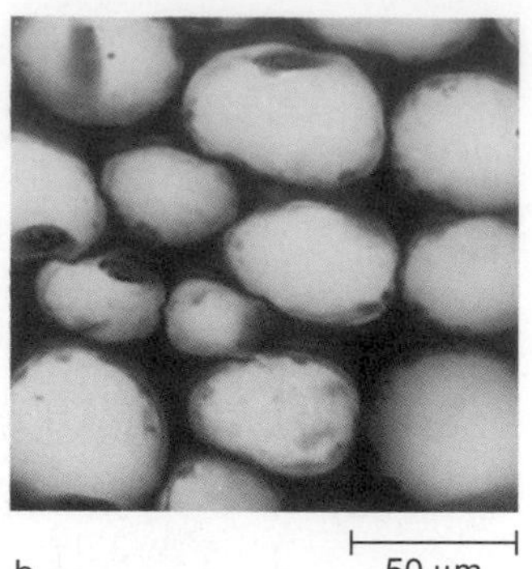

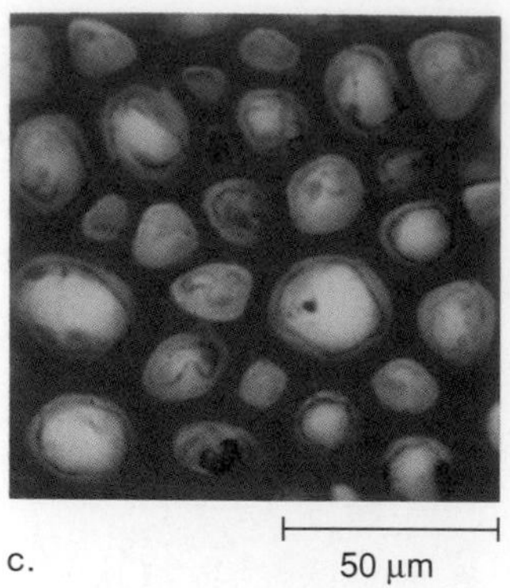

Figure 36.5 Ground tissue cells.
a. Parenchyma cells are the least specialized of the plant cells. **b.** Collenchyma cells. Notice how much thicker the walls are compared to those of parenchyma cells. **c.** Sclerenchyma cells have very thick walls and are nonliving—their only function is to give strong support.

Ground Tissue Fills

Ground tissue forms the bulk of a plant and contains parenchyma, collenchyma, and sclerenchyma cells (Fig. 36.5). **Parenchyma** [Gk. *para,* beside, and *enchyma,* infusion] cells correspond best to the typical plant cell. These are the least specialized of the cell types and are found in all the organs of a plant. They may contain chloroplasts and carry on photosynthesis, or they may contain colorless plastids that store the products of photosynthesis. Parenchyma cells can divide and give rise to more specialized cells, such as when roots develop from stem cuttings placed in water.

Collenchyma cells are like parenchyma cells except they have thicker primary walls. The thickness is uneven and usually involves the corners of the cell. Collenchyma cells often form bundles just beneath the epidermis and give flexible support to immature regions of a plant body. The familiar strands of celery stalks are composed mostly of collenchyma cells.

Sclerenchyma cells have thick secondary cell walls, usually impregnated with *lignin,* which is an organic substance that makes the walls tough and hard. Most sclerenchyma cells are nonliving; their primary function is to support mature regions of a plant. Two types of sclerenchyma cells are fibers and sclereids. Although fibers are occasionally found in ground tissue, most are found in vascular tissue, which is discussed next. Fibers are long and slender and may be found in bundles that are sometimes commercially important. Hemp fibers can be used to make rope, and flax fibers can be woven into linen. Flax fibers, however, are not lignified, which is why linen is soft. Sclereids, which are shorter than fibers and more varied in shape, are found in seed coats and nutshells. They also give pears their characteristic gritty texture.

Vascular Tissue Transports

There are two types of vascular (transport) tissue. **Xylem** transports water and minerals from the roots to the leaves, and **phloem** transports organic nutrients, usually from the leaves to the roots. Xylem contains two types of conducting cells: *tracheids* and *vessel elements* (Fig. 36.6). Both types of conducting cells are hollow and nonliving, but the vessel elements are larger, lack transverse end walls, and are arranged to form a continuous pipeline for water and mineral transport. The elongated tracheids, with tapered ends, form a less obvious means of transport, but water can move across the end walls and sidewalls because there are pits, or depressions, where the secondary wall does not form. In addition to vessel elements and tracheids, xylem contains parenchyma cells that store various substances. Vascular rays, which are flat ribbons or sheets of parenchyma cells located between rows of tracheids, conduct water and minerals laterally. Xylem also contains fibers, sclerenchyma cells that lend support.

The conducting cells of phloem are sieve-tube cells, each of which has a companion cell (Fig. 36.7). *Sieve-tube cells* contain cytoplasm but no nuclei. These cells have channels in their end walls that in cross section make them resemble a sieve. Plasmodesmata (sing., plasmodesma), which are strands of cytoplasm, extend from one cell to another through this so-called *sieve plate.* The smaller *companion cells* are more generalized cells in that they do have a nucleus in addition to cytoplasm. Companion cells are closely connected to sieve-tube cells by numerous plasmodesmata, and the nucleus of the companion cell may control and maintain the life of both cells. The companion cells are also believed to be involved in the transport function of phloem.

It is important to realize that vascular tissue (xylem and phloem) extends from the root to the leaves and vice versa. In the roots, the vascular tissue is located in the **vascular cylinder;** in the stem, it forms vascular bundles; and in the leaves, it is found in leaf veins.

The vascular tissues are xylem and phloem. Xylem transports water and minerals, and phloem transports organic nutrients.

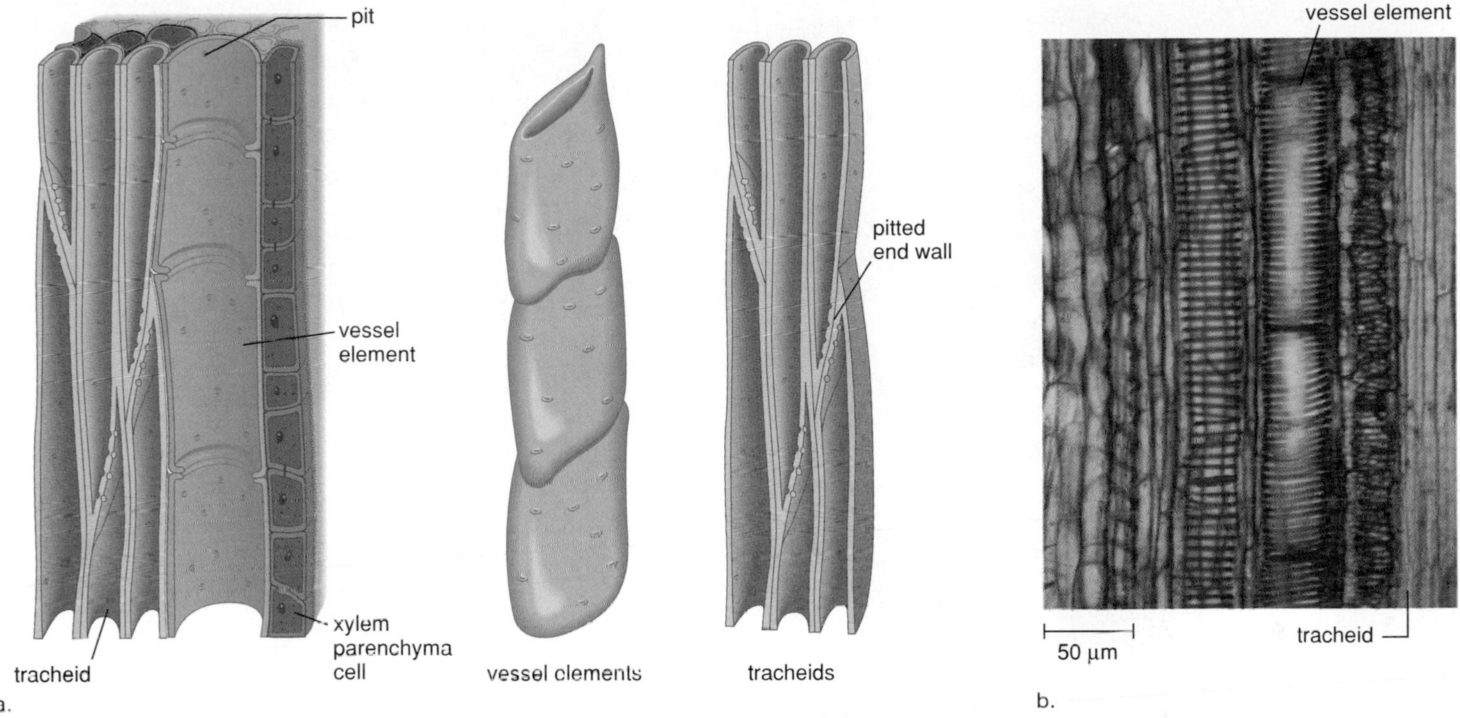

Figure 36.6 Xylem structure.
a. General organization of xylem *(far left)* followed by an external view of vessel elements (stacked on top of each other) and a longitudinal view of several tracheids. **b.** Photomicrograph of xylem vessels.

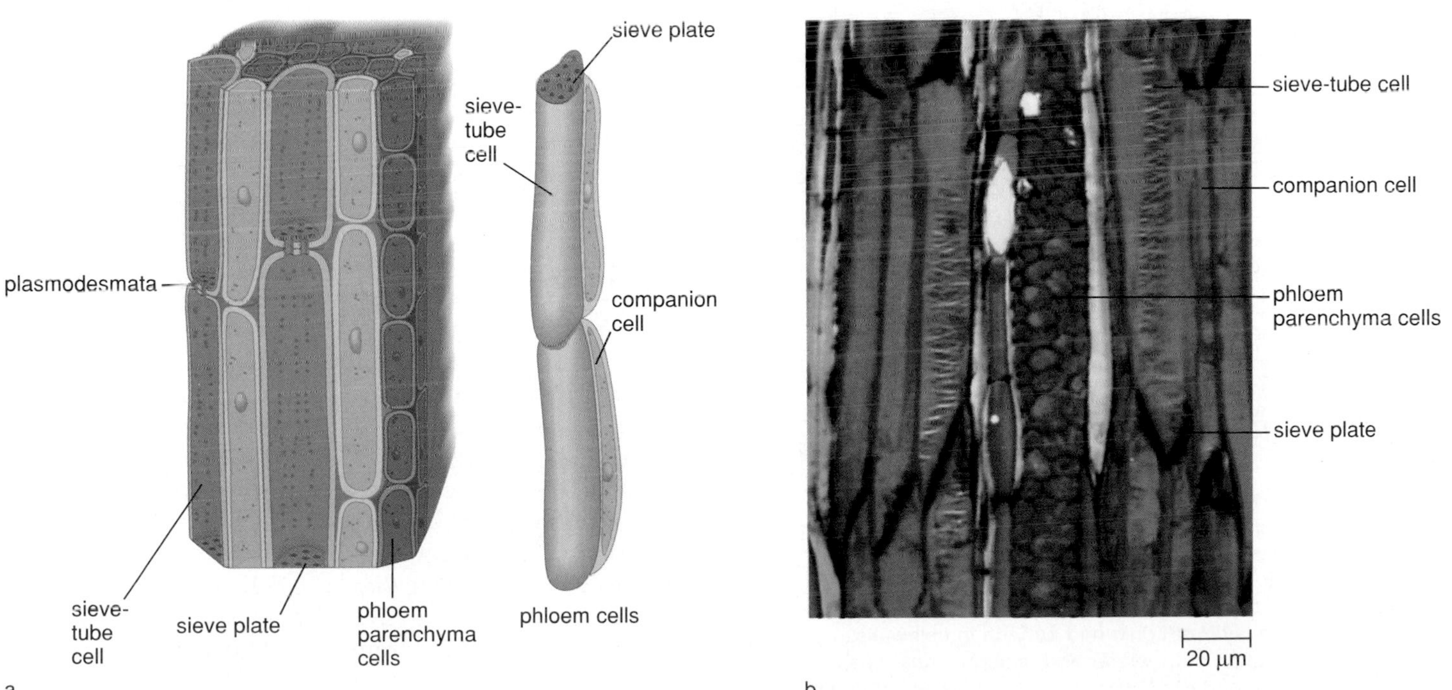

Figure 36.7 Phloem structure.
a. General organization of phloem *(far left)* followed by external view of two sieve-tube cells and their companion cells. **b.** Photomicrograph of phloem sieve tubes.

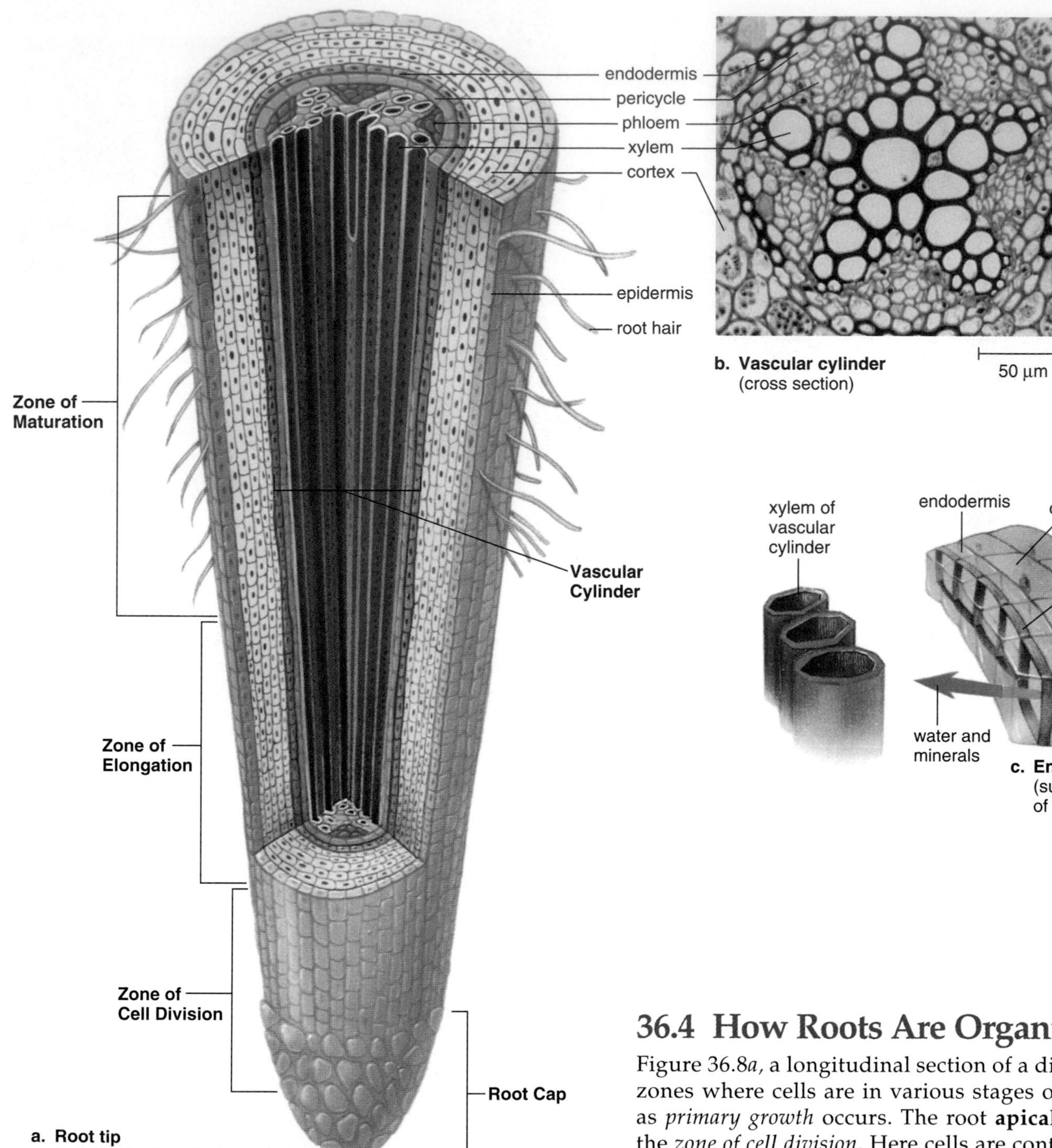

Figure 36.8 Dicot root tip.
a. The root tip is divided into four zones, best seen in a longitudinal section such as this. b. The vascular cylinder of a dicot root contains the vascular tissue. Xylem is typically star shaped, and phloem lies between the points of the star. c. Endodermis showing the Casparian strip, a layer of lignin and suberin in cell walls. Because of the Casparian strip, water and minerals must pass through the cytoplasm of endodermal cells. In this way, endodermal cells regulate the passage of minerals into the vascular cylinder.

36.4 How Roots Are Organized

Figure 36.8*a*, a longitudinal section of a dicot root, reveals zones where cells are in various stages of differentiation as *primary growth* occurs. The root **apical meristem** is in the *zone of cell division*. Here cells are continuously added to the root cap below and the zone of elongation above. The root cap is a protective cover for the root tip. The cells in the root cap have to be replaced constantly, because they are ground off as the root pushes through rough soil particles. In the *zone of elongation*, the cells become longer as they become specialized. In the *zone of maturation*, the cells are mature and fully differentiated. This zone is recognizable even in a whole root because root hairs are borne by many of the epidermal cells.

Tissues of a Dicot Root

Figure 36.8*a* also shows a cross section of a root at the region of maturation. These specialized tissues are identifiable:

Epidermis The epidermis, which forms the outer layer of the root, consists of only a single layer of cells. The majority of epidermal cells are thin walled and rectangular, but in the zone of maturation, many epidermal cells have root hairs. These project as far as 5–8 mm into the soil particles.

Cortex Moving inward, next to the epidermis, large, thin-walled parenchyma cells make up the *cortex* of the root. These irregularly shaped cells are loosely packed, and it is possible for water and minerals to move through the cortex without entering the cells. The cells contain starch granules, and the cortex functions in food storage.

Endodermis The **endodermis** [Gk. *endon*, within, and *derma*, skin] is a single layer of rectangular endodermal cells that forms a boundary between the cortex and the inner vascular cylinder. The endodermal cells fit snugly together and are bordered on four sides (but not the two sides that contact the cortex and the vascular cylinder) by a layer of impermeable lignin and suberin known as the **Casparian strip** (Fig. 36.8*c*). This strip does not permit water and mineral ions to pass between adjacent cell walls. Therefore, the only access to the vascular cylinder is through the endodermal cells themselves, as shown by the arrow in Figure 36.8*c*. It is said that the endodermis regulates the entrance of minerals into the vascular cylinder.

Vascular Tissue The **pericycle,** the first layer of cells within the *vascular cylinder,* has retained its capacity to divide and can start the development of branch or secondary roots (Fig. 36.9). The main portion of the vascular cylinder, though, contains vascular tissue. The xylem appears star shaped in dicots because several arms of tissue radiate from a common center (Fig. 36.8*b*). The phloem is found in separate regions between the arms of the xylem.

How Monocot Roots Are Organized

Monocot roots have the same growth zones as a dicot root, but they do not undergo secondary growth as many dicot roots do. Also, the organization of their tissues is slightly different. In a monocot root's **pith,** which is centrally located, ground tissue is surrounded by a vascular ring composed of alternating xylem and phloem bundles (Fig. 36.10). They also have pericycle, endodermis, cortex, and epidermis.

The root system of a plant absorbs water and minerals, which cross the epidermis and cortex before entering the endodermis, the tissue that regulates the entrance of molecules into the vascular cylinder.

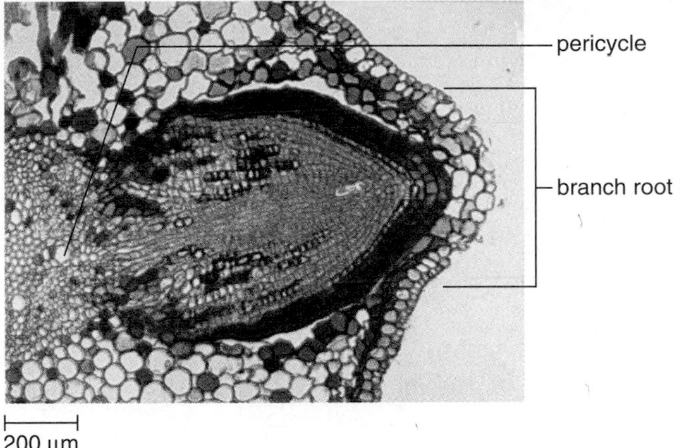

Figure 36.9 Cross section of dicot root.
This cross section of a willow, *Salix*, shows the origination of a branch root from the pericycle.

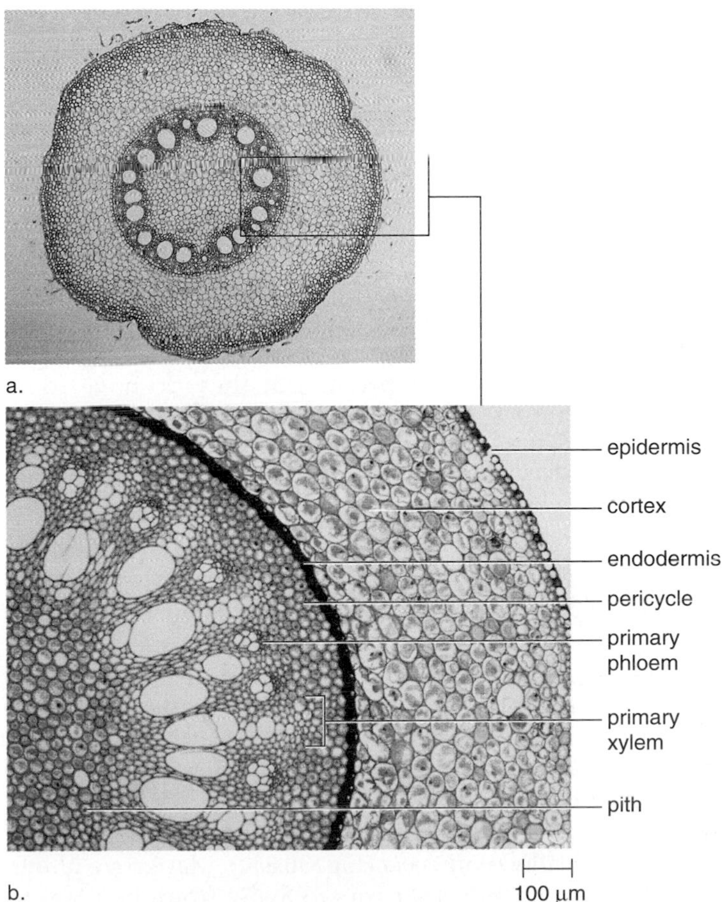

Figure 36.10 Monocot root.
a. In the overall cross section, it is possible to observe that a vascular ring surrounds a central pith. **b.** The enlargement shows the exact placement of various tissues.

a.

b.

c.

Figure 36.11 Three types of roots.
a. A taproot may have secondary roots in addition to a main root. **b.** A fibrous root has many secondary roots with no main root. **c.** Adventitious roots, such as prop roots, extend from the stem.

How Roots Differ

Roots have various adaptations and associations to better perform their functions: anchorage, absorption of water and minerals, and storage of carbohydrates.

In some plants, notably dicots, the first or **primary root** grows straight down and remains the dominant root of the plant. This so-called **taproot** is often fleshy and stores food (Fig. 36.11*a*). Carrots, beets, turnips, and radishes have taproots that we consume as vegetables. Sweet potato plants don't have taproots but they do have roots with special storage areas for starch. We call these storage areas sweet potatoes.

In other plants, notably monocots, there is no single, main root; instead there are a large number of slender roots. These grow from the lower nodes of the stem when the first or primary root dies. These slender roots and their lateral branches make up a **fibrous root system** (Fig. 36.11*b*). Everyone has observed the fibrous root systems of grasses and has noted how these roots can hold the soil.

Since the fibrous roots of monocots develop from organs of the shoot system instead of the root system, they are known as **adventitious roots.** Some of these roots may emerge above the soil line, as they do in corn plants, in which their main function is to help anchor the plant. If so, they are called *prop roots* (Fig. 36.11*c*). Mangrove plants have large prop roots that spread away from the plant to help anchor it in marshy soil, where mangroves are typically found. Black mangroves grow in the water, and their roots have *pneumatophores,* which are root projections that rise above the surface of the water. In this way, the plants acquire oxygen from the air for aerobic cellular respiration. Other examples of adventitious roots are those found on underground stems (rhizomes), or the "holdfast" roots found along the internodes on the aerial shoots of ivy plants.

Some plants such as dodders and broomrapes are parasitic on other plants. Their stems have rootlike projections called *haustoria* that grow into the host plant and make contact with vascular tissue from which they extract water and nutrients.

Mycorrhizae are fungus roots—an association between roots and fungi—which can extract water and minerals from the soil better than roots that lack a fungus partner. This is called a mutualistic relationship because the fungus receives sugars and amino acids from the plant, which receives water and minerals via the fungus. Peas, beans, and other legumes have root nodules where nitrogen-fixing bacteria live.

Plants cannot extract nitrogen from the air, but the bacteria within the nodules can take up and reduce atmospheric nitrogen. This means that the plant is no longer dependent upon a supply of nitrogen (i.e., nitrate or ammonium) in the soil, and indeed these plants are often planted just to bolster the nitrogen supply of the soil.

Roots have various adaptations and associations to enhance their ability to anchor a plant, absorb water and minerals, and store the products of photosynthesis.

doing science

Tree Rings Tell a Story

Each year a tree adds a new growth ring to its trunk. In the spring, growth is fast and the wood is light in color; in the summer, growth is slower and the wood is dark in color. Counting the dark rings tells you the age of the tree, but studying tree rings to determine their shape, thickness, color, and evenness lets you know what a tree has been through (Fig. 36A). Rings can reveal the years in which there were forest fires, smog damage, and leaf damage from caterpillars.

A great deal can also be learned from tree rings about past climatic conditions. For many years, A. E. Douglass, Harold C. Fritts, and others associated with the Laboratory of Tree-Ring Research at the University of Arizona studied ring widths in trees from arid sites. They discovered that very significant statistical relationships exist between growth of trees and climatic data. With the aid of computers and statistical analyses, they developed techniques that take into account subtle climatic and other environmental variables. They were able to reconstruct relatively precise histories of climatic fluctuations and changes dating back hundreds of years. Today, tree ring data are analyzed to try to determine climates dating back to prehistoric times.

Trees do not have to be cut down to examine the rings. A simple instrument called an *increment borer*, which consists primarily of a rigid metal cylinder, is driven into the trunk of a tree, and a core of wood is removed. The hole is then plugged to prevent disease-causing organisms from entering the tree, and the rings revealed by the core are then examined and analyzed.

1914
When the tree was 6 years old, something pushed against it, making it lean. The rings are now wider on the lower side, as the tree builds "reaction wood" to help support it.

1924
The tree is growing straight again. But its neighbors are growing too, and their crowns and root systems take much of the water and sunshine the tree needs.

1927
The surrounding trees are harvested. The larger trees are removed and there is once again ample nourishment and sunlight. The tree can now grow rapidly again.

1930
A fire sweeps through the forest. Fortunately, the tree is only scarred, and year by year more and more of the scar is covered over by newly formed wood.

1942
These narrow rings may have been caused by a prolonged dry spell. One or two dry summers would not have dried the ground enough to slow the tree's growth this much.

1957
Another series of narrow rings may have been caused by an insect like the larva of the sawfly. It eats the leaves and leafbuds of many kinds of coniferous trees.

Figure 36A Life history of a tree.
This tree was planted in 1908 and cut down in 1970. The anatomy of the tree rings correlates to the events listed by date.
Source: Data from St. Regis Paper Company, New York, NY, 1966.

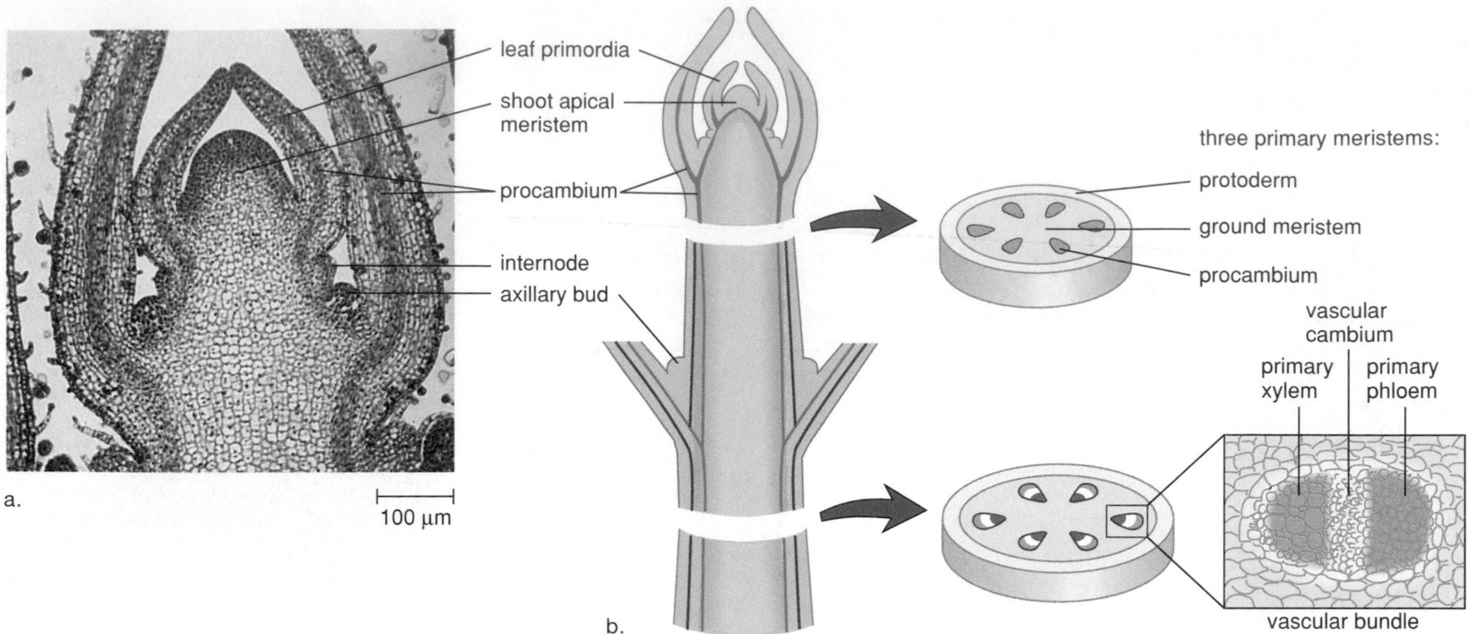

Figure 36.12 Stem tip and primary meristems.
a. The shoot apical meristem within a terminal bud is surrounded by leaf primordia. **b.** The shoot apical meristem produces the primary meristems: protoderm gives rise to epidermis, ground meristem gives rise to pith and cortex, and procambium gives rise to xylem and phloem.

36.5 How Stems Are Organized

Growth of a stem can be compared to the growth of a root. During *primary growth,* apical meristem at the stem tip called **shoot apical meristem** produces new cells that elongate and thereby increase the length of the stem. The shoot apical meristem, however, is protected within a terminal **bud** where *leaf primordia* (immature leaves) envelop it (Fig. 36.12*a*). In the temperate zone, a terminal bud stops growing in the winter and is then protected by bud scales. In the spring, when growth resumes, these scales fall off and leave a scar. You can tell the age of a stem by counting these bud scale scars.

Leaf primordia are produced by the apical meristem at regular intervals called nodes. The portion of a stem between two sequential nodes is called an internode. As a stem grows, the internodes increase in length. Axillary buds, which are usually dormant but may develop into branch shoots or flowers, are seen at the axes of the leaf primordia.

A shoot is a developing structure in which tissues are continually becoming differentiated. In addition to leaf primordia, three specialized types of primary meristem develop from shoot apical meristem (Fig. 36.12*b*). These primary meristems contribute to the length of a shoot. As mentioned, the *protoderm,* the outermost primary meristem, gives rise to epidermis. The *ground meristem* produces two tissues composed of parenchyma cells. The parenchyma tissue in the center of the stem is the pith, and the parenchyma tissue inside the epidermis is the cortex.

The *procambium,* seen as an obvious strand of tissue in Figure 36.12*a,* produces the first xylem cells, called primary xylem, and the first phloem cells, called primary phloem. Differentiation continues as certain cells become the first tracheids or vessel elements of the xylem within a vascular bundle. The first sieve-tube cells of the phloem do not have companion cells and are short lived (some live only a day before being replaced). Mature phloem develops later, when all surrounding cells have stopped expanding and **vascular cambium** [L. *vasculum,* dim. of *vas,* vessel, and *cambio,* exchange] occurs between xylem and phloem.

Herbaceous Stems Are Nonwoody

Mature nonwoody stems, called **herbaceous stems** [L. *herba,* vegetation, plant], exhibit only primary growth. The outermost tissue of herbaceous stems is the epidermis, which is covered by a waxy cuticle to prevent water loss. These stems have distinctive vascular bundles, where xylem and phloem are found. In each bundle, xylem is typically found toward the inside of the stem and phloem is found toward the outside.

In the dicot herbaceous stem, the bundles are arranged in a distinct ring that separates the cortex from the central pith, which stores water and products of photosynthesis (Fig. 36.13). The cortex is sometimes green and carries on photosynthesis, and the pith may function as a storage site. In the monocot stem, the vascular bundles are scattered throughout the stem, and there is no well-defined cortex or well-defined pith (Fig. 36.14).

As a stem grows, the shoot apical meristem produces new leaves and primary meristems. The primary meristems produce the other tissues found in herbaceous stems.

Figure 36.13 Herbaceous dicot stem.
a. A cross section of an alfalfa stem, *Medicago,* shows that the vascular bundles are in a ring. **b.** The drawing of a section of the stem identifies the tissues in the vascular bundle and the stem.

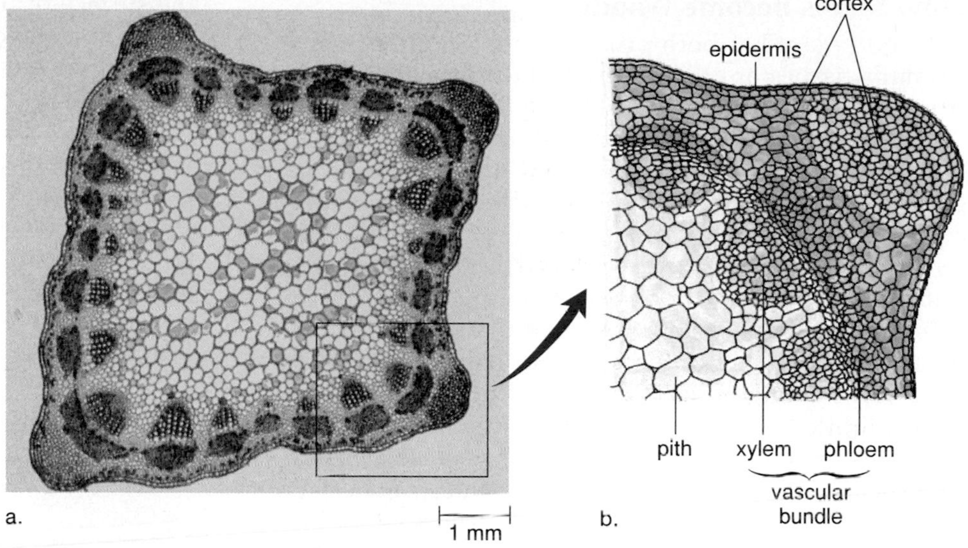

Figure 36.14 Monocot stem.
a. A cross section of corn, *Zea mays,* shows that the vascular bundles are scattered. **b.** An enlargement of the stem shows the vascular bundle in more detail. **c.** An enlargement of one vascular bundle shows the arrangement of tissues in a bundle. Sieve-tube cells and companion cells are part of phloem; vessel elements are part of xylem.

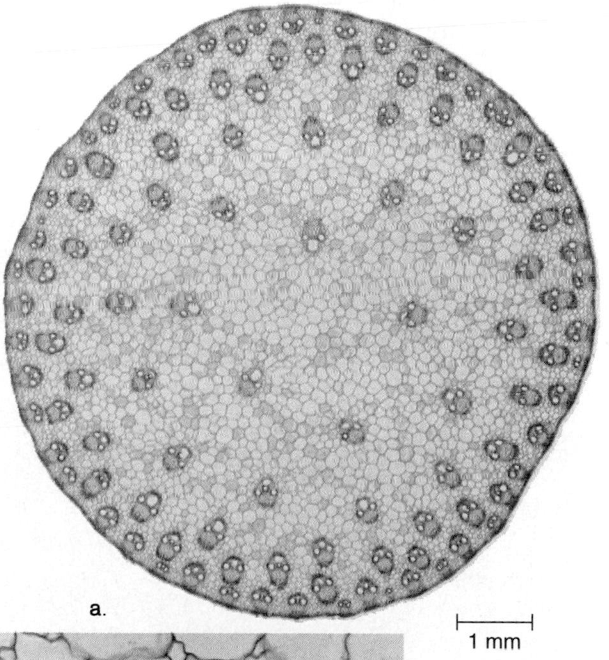

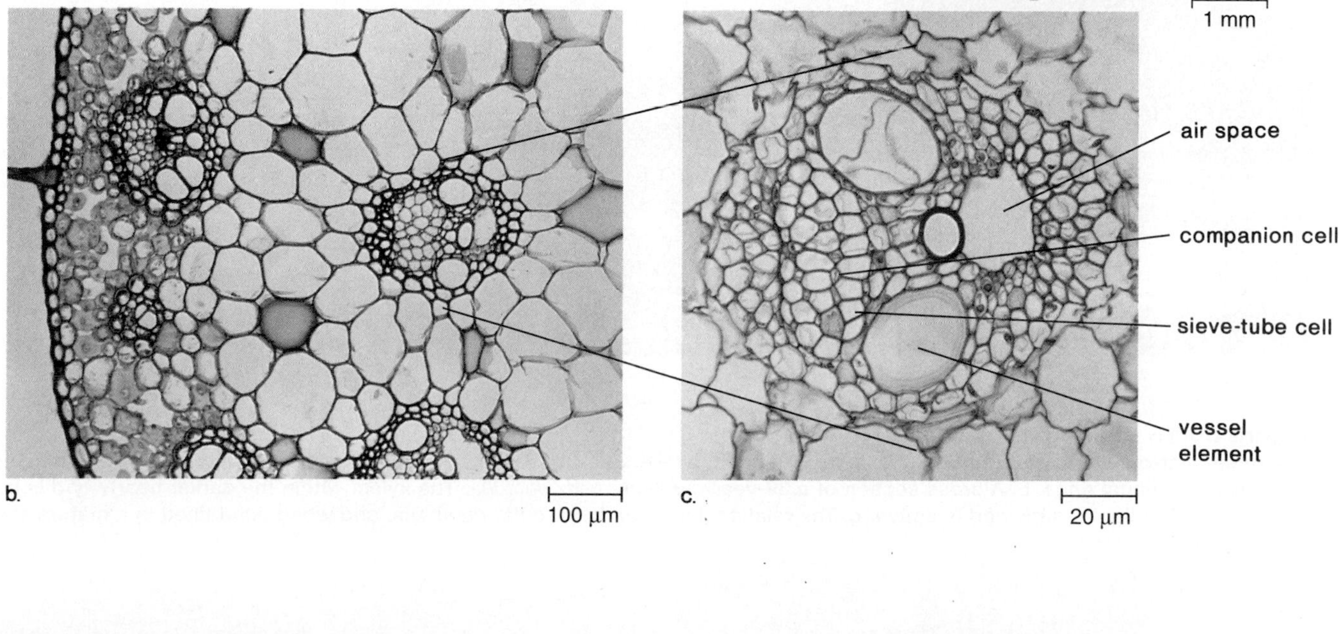

a.
1 mm
air space
companion cell
sieve-tube cell
vessel element
b.
100 μm
c.
20 μm

How Stems Become Woody

A woody plant has both primary and secondary tissues. Primary tissues are those new tissues formed each year from primary meristems right behind an apical meristem. Secondary tissues develop during the second and subsequent years of growth from **lateral meristems.** Primary growth, which occurs in all plants, increases the length of a plant, and *secondary growth,* which occurs only in conifers and some dicots, increases its girth. As a result of secondary growth, a woody dicot stem has an entirely different type of organization than that of a herbaceous dicot stem. After secondary growth has continued for a time, it is no longer possible to make out individual vascular bundles. Instead, a woody stem has three distinct areas: the bark, the wood, and the pith (Fig. 36.15*a*). Rays, such as pith rays, are of living cells that allow materials to move laterally.

Trees undergo secondary growth because of a change in vascular cambium. Recall that vascular cambium begins as a meristem between the xylem and phloem of each vascular bundle. In woody plants, the vascular cambium develops to form a ring of meristem that divides parallel to the surface of the plant. The secondary tissues produced by the vascular cambium, called secondary xylem and secondary phloem, therefore add to the girth of the stem instead of to its length (Fig. 36.16).

In trees that have a growing season, vascular cambium is dormant during the winter. In the spring, when moisture is plentiful and leaves require much water for growth, the xylem [Gk. *xylon,* wood] contains wide vessel elements with thin walls. In this so-called spring wood, wide vessels transport sufficient water to the growing leaves. Later in the season, moisture is scarce and the wood at this time, called summer wood, has a lower proportion of vessels. Strength

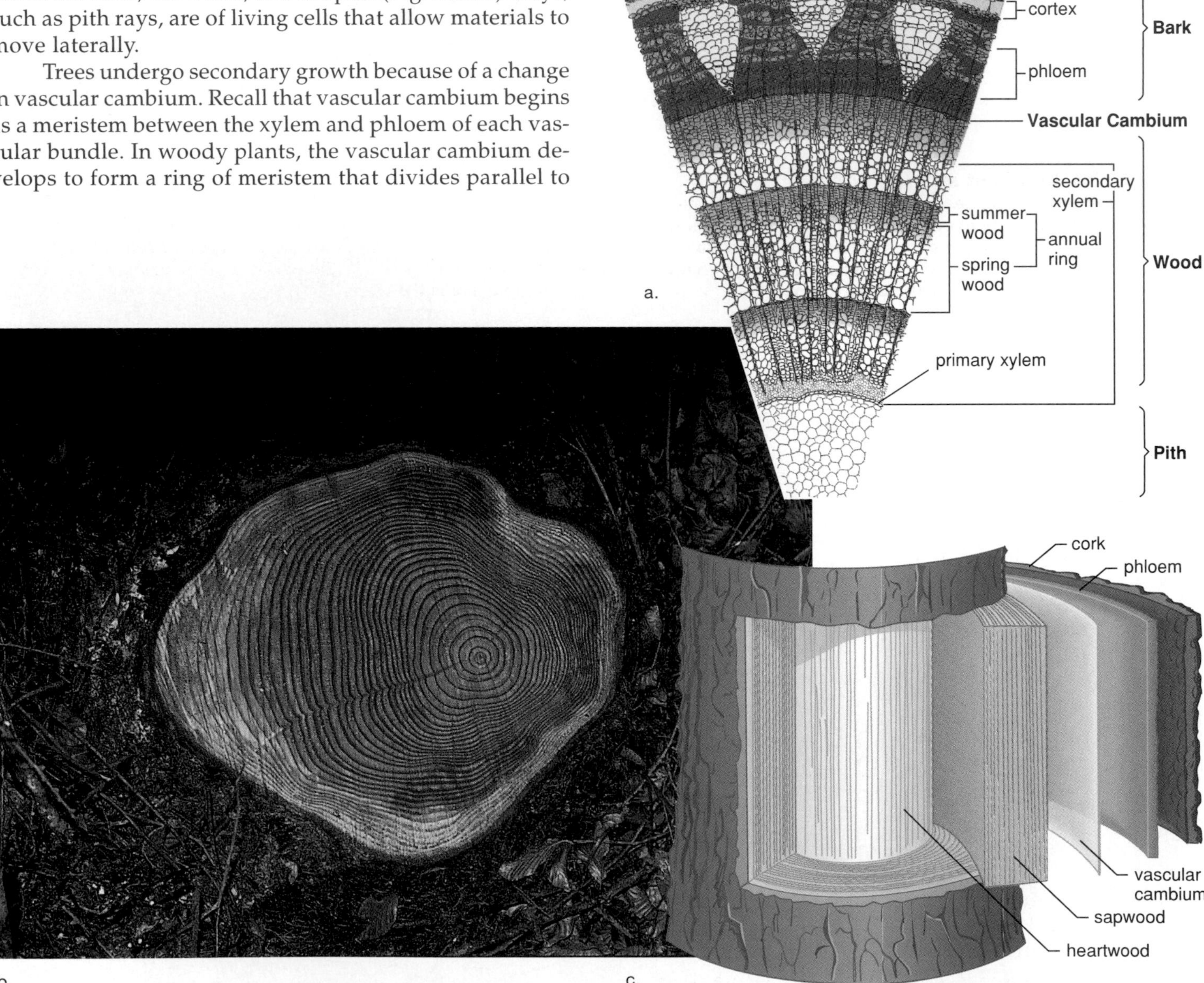

Figure 36.15 Dicot woody stem.
a. A three-year-old stem showing bark, vascular cambium, wood, and pith. The wood consists of secondary xylem, which accumulates and becomes the annual rings. **b.** A cross section of a 39-year-old larch, *Larix decidua.* The xylem within the darker heartwood is now inactive; the xylem within the lighter sapwood is active. **c.** The relationship of bark, vascular cambium, and wood is retained in a mature stem. The pith has disappeared.

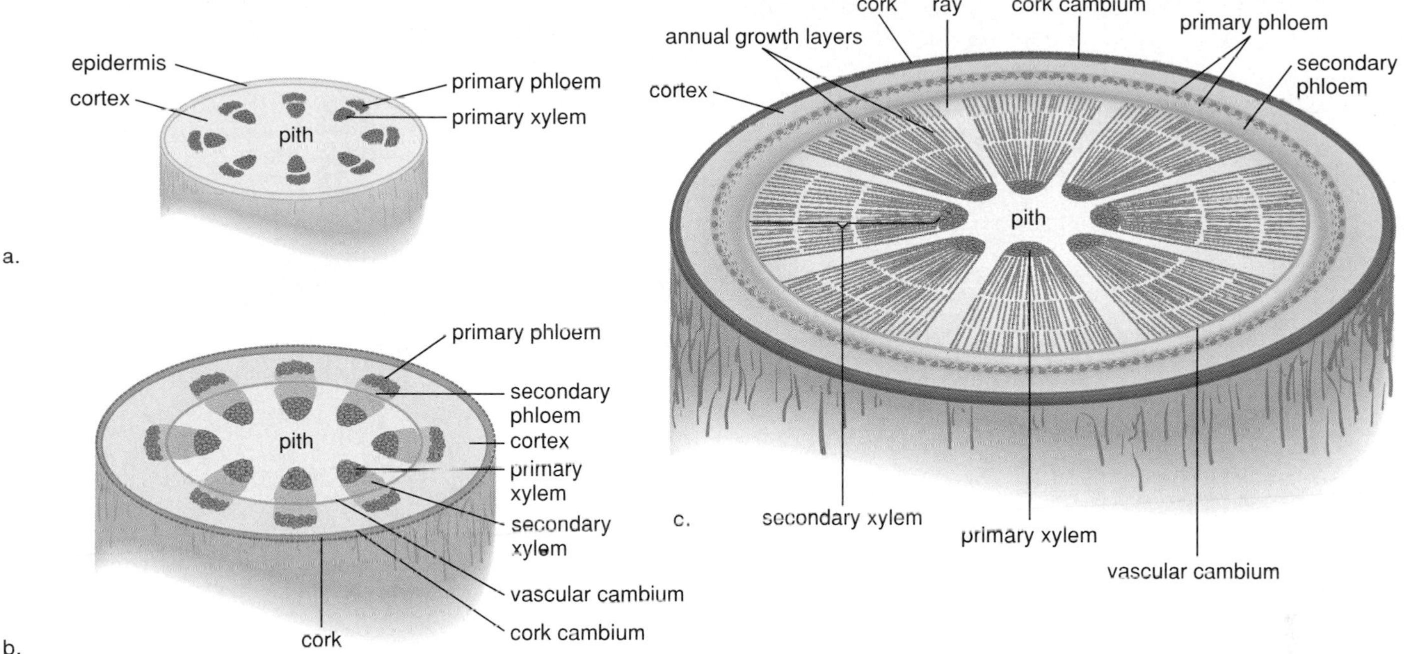

Figure 36.16 Secondary growth of stems.
a. A dicot herbaceous stem before secondary growth begins. b. Secondary growth has begun. Cork has replaced the epidermis. Vascular cambium produces secondary xylem and secondary phloem. c. A three-year-old stem. Cork cambium produces new cork. The primary phloem and cortex will eventually disappear, and only the secondary phloem (within the bark), produced by vascular cambium, will be active that year. The secondary xylem, also produced by vascular cambium, builds up to become the annual rings.

is required because the tree is growing larger and summer wood contains numerous fibers and thick-walled tracheids. At the end of the growing season, just before the cambium becomes dormant again, only heavy fibers with especially thick secondary walls may develop. When the trunk of a tree has spring wood followed by summer wood, the two together make up one year's growth, or **annual ring.** You can tell the age of a tree by counting the annual rings. The outer annual rings, where transport occurs, is called sapwood (Fig. 36.15*b, c*).

In older trees, the inner annual rings, called the heartwood, no longer function in water transport. The cells become plugged with deposits, such as resins, gums, and other substances that inhibit the growth of bacteria and fungi. Heartwood may help support a tree, although some trees stand erect and live for many years after the heartwood has rotted away.

The bark of a tree contains cork, cork cambium, and phloem [Gk. *phloios,* bark]. Although secondary phloem is produced each year by vascular cambium, phloem does not build up for many seasons. The phloem tissue is soft, making it possible to remove the bark of a tree; however, this is very harmful because without phloem there is no conduction of organic nutrients.

Cork cambium [L. *cambio,* exchange] is meristem located beneath the epidermis. When cork cambium begins to divide, it produces tissue that disrupts the epidermis and replaces it with cork cells. Cork cells are impregnated with suberin, a lipid material that makes them waterproof but also causes them to die. This is protective because now there is nothing nutritious for an animal to eat. But an impervious barrier means that gas exchange is impeded except at *lenticels,* which are pockets of loosely arranged cork cells not impregnated with suberin.

The first flowering plants to evolve may have been woody shrubs; herbaceous plants evolved later. Is it advantageous to be woody? If there is adequate rainfall, woody plants can grow taller and have more growth because they have adequate vascular tissue to support and service leaves. However, it takes energy to produce secondary growth and prepare the body for winter if the plant lives in the temperate zone. Also, there is a greater need for defense mechanisms because a long-lasting plant that stays in one spot is likely to be attacked by herbivores and parasites. Then, too, trees don't usually reproduce until they have grown several seasons, by which time they may have succumbed to an accident or disease. Perhaps it is more advantageous for a plant to put most of its energy into simply reproducing rather than being woody.

Woody plants grow in girth due to the presence of vascular cambium and cork cambium. Their bodies have three main parts: bark (which contains cork, cork cambium, and phloem), wood (which contains xylem), and pith.

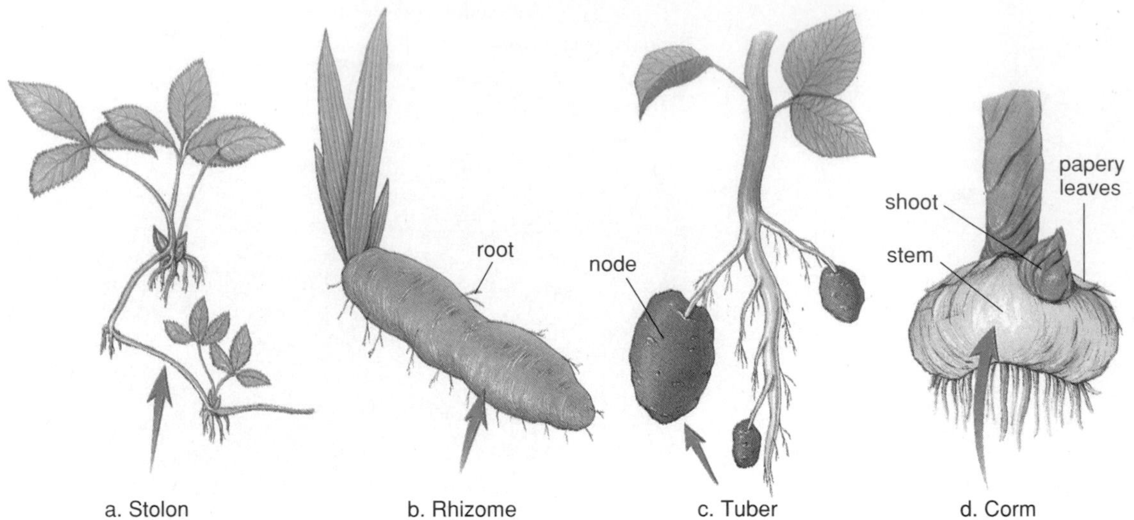

Figure 36.17 Stem modifications.
a. A strawberry plant has aboveground, horizontal stems called stolons. Every other node produces a new shoot system. **b.** The underground horizontal stem of an iris is a fleshy rhizome. **c.** The underground stem of a potato plant has enlargements called tubers. We call the tubers potatoes. **d.** The corm of a gladiolus is a stem covered by papery leaves.

How Stems Differ

Stem modifications are illustrated in Figure 36.17. Aboveground horizontal stems, called **stolons** [L. *stolo,* shoot] or runners, produce new plants where nodes touch the ground. The strawberry plant is a common example of this type of stem.

Underground horizontal stems, **rhizomes** [Gk. *rhiza,* root], may be long and thin, as in sod-forming grasses, or thick and fleshy, as in iris. Rhizomes survive the winter and contribute to asexual reproduction because each node bears a bud. Some rhizomes have enlarged portions called *tubers,* which function in food storage. For example, potatoes are tubers. The eyes of potatoes are buds that mark the nodes.

Corms are bulbous underground stems that lie dormant during the winter, just as rhizomes do. They also produce new plants the next growing season. Gladiolus corms are referred to as bulbs by laypersons, but the botanist reserves the term *bulb* for a structure composed of modified leaves.

Aboveground, vertical stems can also be modified. For example, cacti have succulent stems modified for water storage. The tendrils of grape plants that twine around a support structure are modified stems.

Humans make use of stems in many ways. The stem of the sugarcane plant is a primary source of table sugar. The spice cinnamon and the drug quinine are derived from the bark of different plants. And wood is especially helpful to humans, as discussed next.

Humans Utilize Wood

Humans have used wood for various purposes since antiquity. Much of the wood harvested in the United States is used to produce lumber. There is much waste in lumber production but the waste can be used for pulp, fiberboard, and particleboard. Paper is made from pulp, as is discussed in the reading on the next page.

A veneer is a thin layer or sheet of wood that is uniform in thickness. Many times the last step in furniture-making occurs when an expensive veneer is glued to a piece made from cheaper wood. Plywoods and laminated woods are made by gluing together several layers of veneers in such a way as to give added strength. One advantage of these products is that they can be curved and used for boats, furniture, and other products. Aircraft decking and helicopter blades are made from laminated wood as well.

Particleboard, which is manufactured by gluing particles of wood together, and fiberboard, which is made from fibers, also have many uses. They are used for construction of cabinets, refrigeration cars, toys, concrete forms, and even parts of automobiles.

Poles, posts, and certain mine timbers are wood products in round form. Poles are used to support telephone and electric wires. They are also used as pilings for wharves and building foundations, and are used in fences.

Wood is also important for recreational purposes. Hard maple, which makes fine croquet balls or bowling pins, is also used as the flooring of bowling alleys. White ash is perfect for baseball bats, rackets, and oars. Red spruce is a favorite wood for the making of violins.

A closer Look

Paper Comes From Plants

The word *paper* takes its origin from papyrus, the plant Egyptians used to make the first form of paper. The Egyptians manually made sheets from the treated stems of papyrus grass and then strung them together into scrolls. From that beginning some 5,500 years ago, the production of paper is now a worldwide industry of major importance. The process is fairly simple. Plant material is ground up mechanically, and chemically treated to form a pulp that contains "fibers," which biologists know are the tracheids and vessel elements of a plant. The fibers automatically form a sheet when they are screened from the pulp. Today a revolving wire-screen belt is used to deliver a continuous wet sheet of paper to heavy rollers and heated cylinders which remove most of the remaining moisture and press the paper flat.

There are, of course, different types of paper dependent upon the plant material used and the way it is treated. Among the major plants used to make paper are

Eucalyptus trees. In recent years Brazil has devoted huge areas in the Amazon to the growing of cloned eucalyptus seedlings, specially selected and engineered to be ready for harvest after about seven years.

Temperate hardwood trees. Plantation cultivation in Canada provides birch, beech, chestnut, poplar, and particularly aspen wood for paper making. Tropical hardwoods, usually coming from Southeast Asia, are also used.

Softwood trees. In the United States, several species of pine trees have been genetically improved to have a higher wood density and to be harvestable five years earlier than ordinary pines. Southern Africa, Chile, New Zealand, and Australia also devote thousands of acres to growing pines for paper pulp production.

Bamboo. Several Asian countries, especially India, provide vast quantities of bamboo pulp for the making of paper. Because bamboo is harvested without destroying the roots, and the growing cycle is favorable, this plant, which is actually a grass, is expected to be a significant source of paper pulp despite high processing costs to remove impurities.

Flax and cotton plants. Linen and cotton cloth from textile and garment mills are used to produce *rag paper* whose flexibility and durability are desirable in legal documents, high-grade bond paper, and high-grade stationery.

It has been known for sometime that paper largely consists of the cellulose within plant cell walls. It seems reasonable to suppose, then, that paper could be made from synthetic polymers (e.g., rayon). Indeed, synthetic polymers produce a paper that has qualities superior to those of paper made from natural sources, but the cost thus far is prohibitive. Another consideration, however, is the ecological effects of making paper from trees. Plantations containing stands of uniform trees replace natural ecosystems and when the trees are clear-cut, the land is laid bare. Paper mill wastes, which include caustic chemicals, add significantly to the pollution of rivers and streams.

The use of paper for packaging and to make all sorts of products has increased dramatically in this century. Each person in the United States consumes about 318 kilograms (699 pounds) of paper products per year, and this compares to only 2.3 kilograms (5 pounds) of paper per person in India. It is clear, then, we should take the initiative in recycling paper. When newspaper, office paper, and photocopies are soaked in water, the fibers are released and they can be used to make a new batch of paper. It's estimated that recycling the Sunday newspapers alone would save an estimated number of 500,000 trees each week!

Figure 36B Paper production.
Machine No. 35 at Champion International's Courtland, Alabama, mill produces a 29-foot-wide roll of office paper every 60 minutes.

36.6 How Leaves Are Organized

Leaves are the organs of photosynthesis in vascular plants. As mentioned earlier, a leaf usually consists of a flattened *blade* and a *petiole* connecting the blade to the stem. The blade may be single or composed of several leaflets. Externally, it is possible to see the pattern of the *leaf veins,* which contain vascular tissue. Leaf veins have a net pattern in dicot leaves and a parallel pattern in monocot leaves (see Fig. 36.3).

Figure 36.18 shows a cross section of a typical dicot leaf of a temperate zone plant. At the top and bottom is a layer of epidermal tissue that often bears protective hairs and/or glands that produce irritating substances. These features may prevent the leaf from being eaten by insects. The epidermis characteristically has an outer, waxy cuticle [L. *cutis,* skin] that keeps the leaf from drying out. Unfortunately, the cuticle also prevents gas exchange because it is not gas permeable. However, the epidermis, particularly the lower epidermis, contains stomates that allow gases to move into and out of the leaf. Each stomate has two guard cells that regulate its opening and closing.

The body of a leaf is composed of **mesophyll** [Gk. *mesos,* middle, and *phyllon,* leaf] tissue, which has two distinct regions: **palisade mesophyll,** containing elongated cells, and **spongy mesophyll,** containing irregular cells bounded by air spaces. The parenchyma cells of these layers have many chloroplasts and carry on most of the photosynthesis for the plant. The loosely packed arrangement of the cells in the spongy layer increases the amount of surface area for gas exchange.

How Leaves Differ

The blade of a leaf can be simple or compound, with two or more separate leaflets making up the blade (Fig. 36.19). There are also various vascular arrangements in leaves and innumerable combinations of overall leaf shape, margin, and base modifications.

Leaves are adapted to environmental conditions. Shade plants tend to have broad, wide leaves, and desert plants tend to have reduced leaves with sunken stomates. The leaves of a cactus are the spines attached to the succulent stem (Fig. 36.20*a*). Other succulents have leaves adapted to hold moisture.

An onion *bulb* is made up of leaves surrounding a short stem. In a head of cabbage, large leaves overlap one another. The petiole of a leaf can be thick and fleshy, as in celery and rhubarb. Climbing leaves, such as those of peas and cucumbers, are modified into tendrils that can attach to nearby objects (Fig. 36.20*b*). The leaves of a few plants are specialized for catching insects. The leaves of a sundew have sticky epidermal hairs that trap insects and then secrete digestive enzymes. The Venus's-flytrap has hinged leaves that snap shut and interlock when an insect triggers sensitive hairs (Fig. 36.20*c*). The leaves of a pitcher plant resemble a pitcher and have downward-pointing hairs that lead insects into a pool of digestive enzymes. Insectivorous plants commonly grow in marshy regions, where the supply of soil nitrogen is severely limited. The digested insects provide the plants with a source of organic nitrogen.

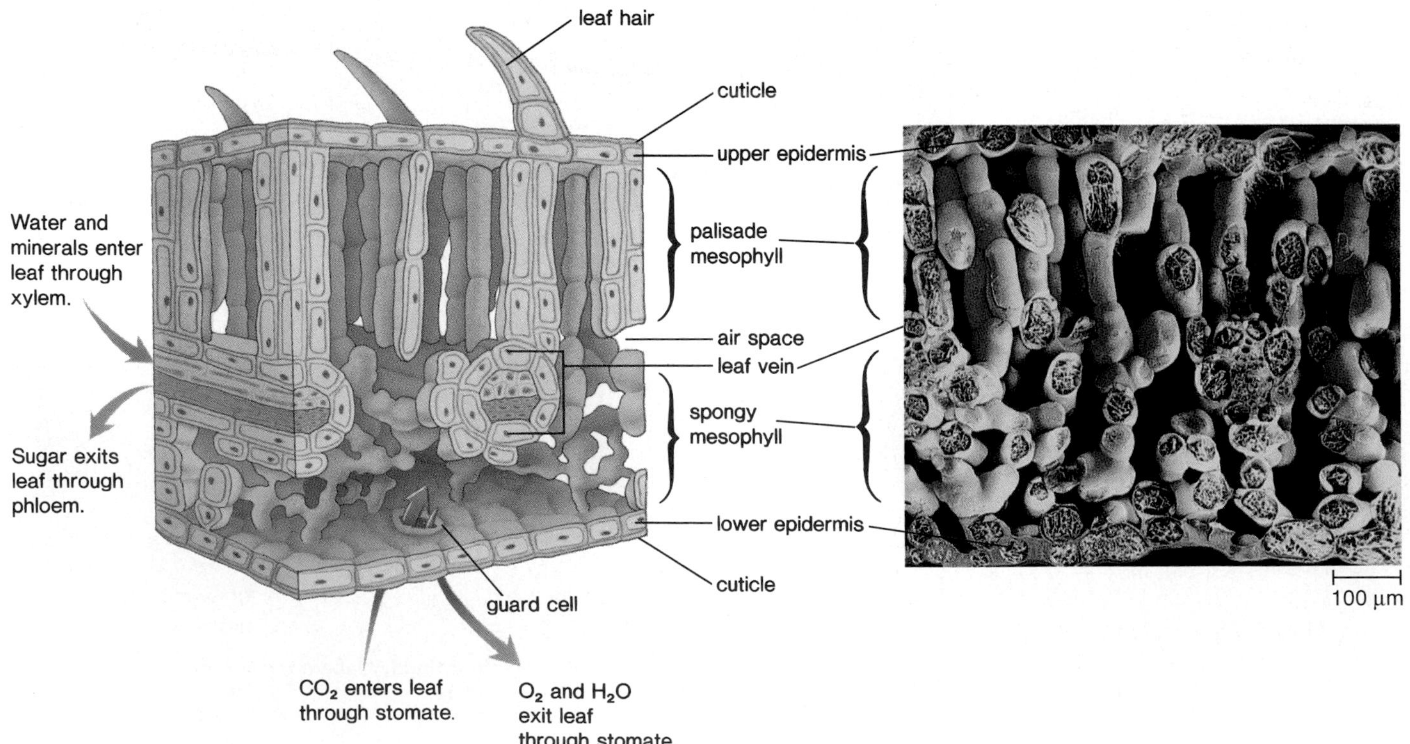

Figure 36.18 Leaf structure.
Photosynthesis takes place in the mesophyll tissue. The veins contain xylem and phloem for the transport of water and solutes. The leaf is enclosed by epidermal cells covered with a waxy layer, the cuticle. The leaf hairs are also protective. A stomate is an opening in the epidermis that permits the exchange of gases.

Figure 36.19 Classification of leaves.
a. The cottonwood tree has a simple leaf. **b.** The shagbark hickory has a pinnately compound leaf. **c.** The honey locust has a twice pinnately compound leaf. **d.** The buckeye has a palmately compound leaf.

a. Simple leaf b. Pinnately compound leaf c. Twice pinnately compound leaf d. Palmately compound leaf

a. Cactus, *Opuntia*

b. Cucumber, *Cucumis*

c. Venus's-flytrap, *Dionaea*

Figure 36.20 Leaf modifications.
a. The spines of a cactus, *Opuntia,* plant are leaves modified to protect the fleshy stem from animal consumption. **b.** The tendrils of a cucumber, *Cucumis,* are leaves modified to attach the plant to a physical support. **c.** The leaves of the Venus's-flytrap, *Dionaea,* are modified to serve as a trap for insect prey. When triggered by an insect, the leaf snaps shut. Once shut, the leaf secretes digestive juices, which break down the soft parts of the insect's body.

connecting concepts

Perhaps a multicellular alga is ancestral to plants because multicellularity would have been adaptive for development of the specialized tissues found in plants—organisms that evolved on land. On land, as opposed to an aquatic environment, there is a danger of drying out. Even humid air is drier than a living cell, and the prevention of water loss is critical for plants. The epidermis and the cuticle it produces help prevent water loss and overheating in sunlight. (It is also protective against invasion by bacteria, fungi, and small insects.) The cork of woody plants is especially protective against water loss. Only the presence of lenticels still allows gas exchange to occur.

In an aquatic environment, water is available to all cells, but on land it is adaptive to have a means of water uptake and transport. In plants, the roots absorb water and have special extensions called root hairs that facilitate the uptake of water. Xylem transports water to all plant parts. The next chapter discusses how the drying effect of air is used by plants to help move water from the roots to the leaves. Roots buried in soil where they absorb water cannot photosynthesize, and the structure of phloem permits the transport of sugars down to the roots.

In an aquatic environment, water buoys up organisms and keeps them afloat, but on land it is adaptive to have a means to oppose the force of gravity. The stems of plants contain strong-walled sclerenchyma cells, tracheids, and vessel elements. The accumulation of annual rings in woody plants offers more support and allows a tree to grow ever taller.

In an aquatic environment, the surrounding water prevents the gametes and zygote from drying out As we shall see in the last of our four chapters in this part, flowering plants do not rely on external water for reproductive purposes and have evolved structures and a means of reproduction that prevent the gametes, zygote, and embryo from drying out.

Table 36.1

Vegetative Organs and Major Tissues

	Roots	Stems	Leaves
Function	Absorb water and minerals	Transport water and nutrients	Carry on photosynthesis
	Anchor plant	Support leaves	
	Store materials	Help store materials	
Tissue			
Epidermis*	Protect inner tissues	Protect inner tissues	Protect inner tissues
	Root hairs absorb water and minerals		Stomates carry on gas exchange
Cortex†	Store water and products of photosynthesis	Carry on photosynthesis, if green Some storage of products of photosynthesis	—
Endodermis†	Regulate passage of minerals in vascular tissue	Regulate passage of minerals in vascular tissue if present	—
Vascular‡	Transport water and nutrients	Transport water and nutrients	Transport water and nutrients
Pith†	Store products of photosynthesis and water	Store products of photosynthesis	—
Mesophyll†	—	—	Primary site of photosynthesis

Note: *Plant tissues belong to one of three tissue systems:*
* *Dermal tissue system*
† *Ground tissue system*
‡ *Vascular tissue system*

Summary

36.1 Plants Have Organs

A flowering plant has three vegetative organs. A root anchors a plant, absorbs water and minerals, and stores the products of photosynthesis (Table 36.1). Stems support leaves, conduct materials to and from roots and leaves, and help store plant products. Leaves carry on photosynthesis.

36.2 Monocot Versus Dicot Plants

Flowering plants are divided into the dicots and monocots according to the number of cotyledons in the seed, the arrangement of vascular tissue in root, stems, and leaves, and the number of flower parts.

36.3 Plant Organs Have Tissues

Three types of meristem continually divide and produce specialized tissues. Protoderm produces epidermis tissue, ground meristem produces ground tissue, and procambium produces primary vascular tissue.

Epidermal tissue contains the epidermis, which is modified in different organs of the plant. In the roots, epidermal cells bear root hairs; in the leaves, the epidermis contains guard cells.

Ground tissue contains parenchyma cells, which are thin walled and capable of photosynthesis when they contain chloroplasts. Collenchyma cells have thicker walls for flexible support. Sclerenchyma cells are hollow, nonliving support cells with secondary walls.

Vascular tissue consists of xylem and phloem. Xylem contains vessel elements and tracheids, which are elongated and tapered with pitted end walls. Xylem transports water and minerals. Phloem contains sieve-tube cells, each of which has a companion cell. Phloem transports organic nutrients.

36.4 How Roots Are Organized

A root tip shows three zones: the zone of cell division (apical meristem) protected by the root cap, the zone of elongation, and the zone of maturation.

A cross section of a herbaceous dicot root reveals the epidermis (protects), the cortex (stores food), the endodermis (regulates the movement of minerals), and the vascular cylinder (vascular tissue). In the vascular cylinder of a dicot, the xylem appears star shaped and the phloem is found in separate regions, between the arms of the xylem. In contrast, a monocot root has a ring of vascular tissue with alternating bundles of xylem and phloem surrounding pith.

Three types of roots are taproots, fibrous roots, and adventitious roots.

36.5 How Stems Are Organized

Primary growth of a stem is due to the activity of the shoot apical meristem, which is protected within a terminal bud. A terminal bud contains leaf primordia at nodes, and internodes between the nodes. When stems grow, the internodes lengthen.

In cross section, a nonwoody dicot has epidermis, cortex tissue, vascular bundles in a ring, and an inner pith. Monocot stems have scattered vascular bundles, and the cortex and pith are not well defined.

Secondary growth of a woody stem is due to vascular cambium, which produces new xylem and phloem every year, and cork cambium, which produces new cork cells when needed. Cork replaces epidermis in woody plants. The cross section of a woody stem shows bark, wood, and pith. The bark contains cork and phloem. Wood contains annual rings of xylem.

Stems are modified in various ways, such as horizontal aboveground and underground stems. Corms and some tendrils are also modified stems.

36.6 How Leaves Are Organized

A cross section of a leaf shows the epidermis, with stomates mostly below. Vascular tissue is present within leaf veins.

Leaves are variously modified. The spines of a cactus are leaves. Other succulents have fleshy leaves. An onion is a bulb with fleshy leaves, and the tendrils of peas are leaves. The Venus's-flytrap has leaves that trap and digest insects.

Reviewing the Chapter

1. Name and discuss the vegetative organs of a plant. 638–40
2. List five differences between monocots and dicots. 640
3. Contrast an epidermal cell with a cork cell. These cells are found in what type of plant tissue? Explain how epidermis is modified in various organs of a plant. 641
4. Contrast the structure and function of parenchyma, collenchyma, and sclerenchyma cells. These cells occur in what type of plant tissue? 642
5. Contrast the structure and function of xylem and phloem. Xylem and phloem occur in what type of plant tissue? 642–43
6. Name and discuss the zones of a root tip. Trace the path of water and minerals from the root hairs to xylem. Be sure to mention the Casparian strip. 644–45
7. Describe three basic types of roots, and give examples of other root modifications. 646
8. Describe the primary and secondary growth of a stem. 648–50
9. Describe the cross section of a monocot, a herbaceous dicot, and a woody stem. 649–51
10. Discuss the adaptation of stems by giving several examples. 652
11. Describe the structure and organization of a typical dicot leaf. 654
12. Note several ways in which leaves are specialized. 654
13. Name and state the function of the main tissues within each plant organ. 656

Testing Yourself

Choose the best answer for each question.

1. Which of these is an incorrect contrast between monocots and dicots?
   monocots—dicots
   a. one cotyledon—two cotyledons
   b. leaf veins parallel—net veined
   c. vascular bundles in a ring—vascular bundles scattered
   d. All of these are incorrect.
2. Which of these types of cells is most likely to divide?
   a. parenchyma
   b. meristem
   c. epidermis
   d. xylem
3. Which of these cells in a plant is apt to be nonliving?
   a. parenchyma
   b. collenchyma
   c. sclerenchyma
   d. epidermal
4. Root hairs are found in the zone of
   a. cell division.
   b. elongation.
   c. maturation.
   d. All of these are correct.
5. Cortex is found in
   a. roots, stems, and leaves.
   b. roots and stems.
   c. roots only.
   d. stems only.
6. Between the bark and the wood in a woody stem, there is a layer of meristem called
   a. cork cambium.
   b. vascular cambium.
   c. apical meristem.
   d. the zone of cell division.
7. Which part of a leaf carries on most of the photosynthesis of a plant?
   a. epidermis
   b. mesophyll
   c. epidermal layer
   d. guard cells
8. Annual rings are the number of
   a. internodes in a stem.
   b. rings of vascular bundles in a monocot stem.
   c. layers of secondary xylem in a stem.
   d. Both b and c are correct.
9. The Casparian strip is found
   a. between all epidermal cells.
   b. between xylem and phloem cells.
   c. on four sides of endodermal cells.
   d. within the secondary wall of parenchyma cells.
10. Which of these is a stem?
   a. taproot of carrots
   b. stolon of strawberry plants
   c. spine of cacti
   d. Both b and c are correct.

11. Label this root using the terms endodermis, phloem, xylem, cortex, and epidermis:

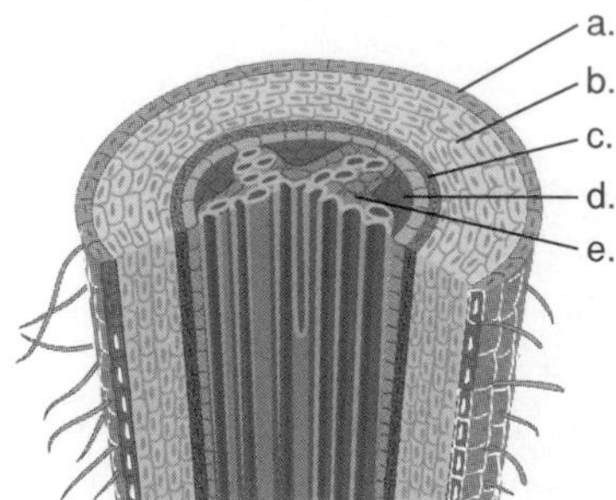

12. Label this leaf using the terms leaf vein, lower epidermis, palisade mesophyll, spongy mesophyll, and upper epidermis:

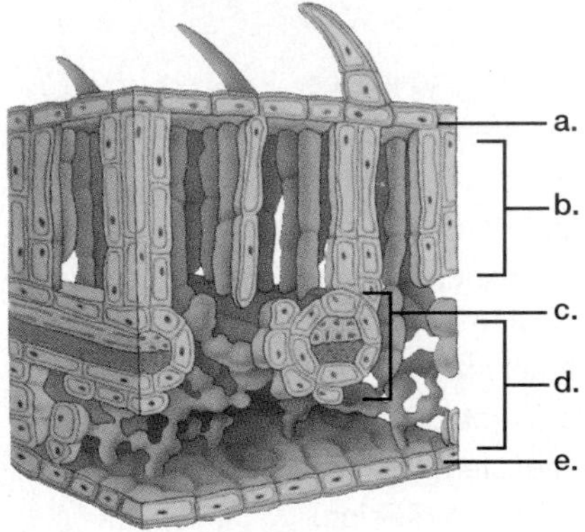

Applying the Concepts

1. *Structure suits the function.*
   How does the structure of a dicot leaf suit its function of carrying on photosynthesis?
2. *Living things are organized.*
   Show that a plant is organized on all levels of its structure.
3. *Organisms are adapted to the environment.*
   In what ways do xylem rings (wood) help a plant live on land?

Using Technology

Your study of plant structure and function is supported by these available technologies:

Exploring the Internet
The Mader Home Page provides resources for and help with studying this chapter.

http://www.mhhe.com/sciencemath/biology/mader/
(Click on Biology.)

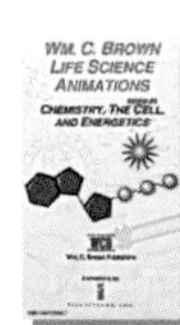

Life Science Animations Video
Tape #5: Plant Biology/Evolution/Ecology
Journey into a Leaf (#46)

Understanding the Terms

adventitious root 646
angiosperm 637
annual ring 651
apical meristem 644
axillary bud 639
bud 648
Casparian strip 645
collenchyma 642
cork 641
cork cambium 651
cortex 645
cotyledon 640
cuticle 641
dicot 640
endodermis 645
epidermal tissue 641
epidermis 641
fibrous root system 646
ground tissue 641
herbaceous stem 648
internode 639
lateral 638
lateral meristem 650
leaf 639
meristem 641
mesophyll 654
monocot 640
node 639
palisade mesophyll 654
parenchyma 642
perennial 638
pericycle 645
phloem 642
pith 645
primary root 646
rhizome 652
root hair 641
root system 638
sclerenchyma 642
shoot apical meristem 648
shoot system 638
spongy mesophyll 654
stem 639
stolon 652
stomate 641
taproot 646
vascular bundle 640
vascular cambium 648
vascular cylinder 642
vascular tissue 641
vein 640
xylem 642

Match the terms to these definitions:

a. __________ Inner, thickest layer of a leaf consisting of palisade and spongy mesophyll; the site of most of photosynthesis.
b. __________ Lateral meristem that produces secondary phloem and secondary xylem.
c. __________ Least specialized of all plant cell or tissue types; contains plastids and is found in all organs of a plant.
d. __________ Outer covering of bark of trees; made of dead cells that may be sloughed off.
e. __________ Rootlike, underground stem.
f. __________ Seed leaf for embryonic plant, provides nutrient molecules for the developing plant before its mature leaves begin to photosynthesize.
g. __________ Stem that grows horizontally along the ground and establishes plantlets periodically when it contacts the soil (e.g., the runners of a strawberry plant).
h. __________ Vascular tissue that conducts organic solutes in plants; contains sieve-tube cells and companion cells.
i. __________ Vascular tissue that transports water and mineral solutes upward through the plant body.
j. __________ Waxy layer covering the epidermis of plants that protects the plant against water loss and disease-causing organisms.

Nutrition and Transport in Plants

CHAPTER 37

Chapter Concepts

The roots of a geranium, *Pelargonium*, are exposed

Carbon (C), oxygen (O), hydrogen (H), nitrogen (N), potassium (K), calcium (Ca), and other elements are required in various amounts by virtually all plants. Gases such as CO_2 and O_2 are taken up by the leaves, but the roots absorb water (H_2O) and minerals in the form of NO_3^-, K^+, and Ca^{2+}. Plants, unlike most animals, have an amazing ability to concentrate minerals—that is, to take them up until they are many times more prevalent in the plant than in the surrounding medium. Some plants, such as those in the legume family, have roots infected with bacteria that make fixed nitrogen (e.g., NH_4^+) available to the plant for the production of proteins. This relationship allows legumes (e.g., beans) to be high in protein.

In humans, blood, which contains nutrients, salts, and other substances, is pumped throughout the body by the heart. In plants, there is no central pumping mechanism, yet materials move from the roots to the leaves and vice versa. How is this accomplished? The unique properties of water account for the movement of water in xylem, while osmosis plays a role in phloem transport.

37.1 Plants Require Inorganic Nutrients

Plants require only inorganic nutrients in order to be able to make their own organic food. **Essential inorganic nutrients** must fulfill these criteria: (1) they must have an identifiable nutritional role, (2) no other element can substitute and fulfill the same role, and (3) a deficiency of these elements causes the plant to die without completing its life cycle.

Carbon, hydrogen, and oxygen make up 96% of a plant's dry weight and are essential inorganic nutrients. Carbon dioxide (CO_2) is the source of carbon for a plant, and water (H_2O) is the source of hydrogen. Oxygen can come from atmospheric oxygen, carbon dioxide, or water. The essential inorganic nutrients listed in Table 37.1 are minerals that come from the soil. They are divided into **macronutrients** and **micronutrients** according to their relative concentration in plant tissue. Nitrogen (N) is a common atmospheric gas, but plants are unable to make use of it. Instead, plants take up either ammonium (NH_4^+) or nitrate (NO_3^-) from the soil. Usually, soil bacteria have converted ammonium to nitrate, and therefore plants take up nitrate. As discussed on pages 662–63, some plants have root nodules containing mutualistic bacteria that fix atmospheric nitrogen and make nitrogen compounds available to their host.

Beneficial nutrients are another category of elements taken up by plants. Beneficial nutrients either are required or enhance the growth of a particular plant. Plants such as *Equisetum* (horsetails) require silicon as a mineral nutrient, and halophytes such as sugar beets show enhanced growth in the presence of sodium. Nickel is a beneficial mineral nutrient in soybeans when root nodules are present.

How Are Requirements Determined?

If you burn a plant, its nitrogen component is given off as ammonia and other gases, but most other mineral elements remain in the ash. The presence of a particular element in the ash, however, does not necessarily mean that the plant normally requires it. The preferred method for determining the mineral requirements of a plant was developed at the end of the nineteenth century. This method is called water culture, or **hydroponics** [Gk. *hydrias*, water, and *ponos*, hard work] (Fig. 37.1). Hydroponics allows plants to grow well if they are supplied with all the mineral nutrients they need. The investigator omits a particular mineral and observes the effect on plant growth. If growth suffers, it can be concluded that the omitted mineral is a required nutrient. This method has been more successful for macronutrients than for micronutrients. For studies involving the latter, the water and the mineral salts used must be absolutely pure; purity is difficult to attain, because even instruments and glassware can introduce micronutrients. Then, too, the element in question may already be present in the seedling used in the experiment. These factors complicate the determination of essential plant micronutrients by means of hydroponics.

What Affects Mineral Availability?

As mentioned previously, the essential minerals that a plant needs are normally taken from the soil. Location of minerals in the soil is critical, because if a root does not come within a few millimeters of an ion, no uptake occurs. The downward movement of minerals through soil is influenced by the quantity and the pH of water. Sometimes water makes mineral ions available to roots, but often it leaches, or removes, mineral ions from the soil zone in which roots grow. The lower the pH, the greater the leaching power of water.

Table 37.1

Some Essential Nutrients in Plants

Elements	Symbol	Ionic Form	Major Functions
Macronutrients			
Nitrogen	N	NO_3^-, NH_4^+	Part of nucleic acids, proteins, chlorophyll, and coenzymes
Potassium	K	K^+	Cofactor for enzymes; involved in water balance and movement of stomates
Calcium	Ca	Ca^{2+}	Regulates responses to stimuli and movement of substances through plasma membrane; involved in formation and stability of cell walls
Phosphorus	P	$H_2PO_4^-$, HPO_4^{2-}	Part of nucleic acids, ATP, and phospholipids
Magnesium	Mg	Mg^{2+}	Part of chlorophyll; activates a number of enzymes
Sulphur	S	SO_4^{2-}	Part of amino acids, some coenzymes
Micronutrients			
Iron	Fe	Fe^{2+}, Fe^{3+}	Part of cytochrome needed for cellular respiration
Manganese	Mn	Mn^{2+}	Activates some enzymes such as those of the Krebs cycle
Boron	B	BO_3^{3-}, $B_4O_7^{2-}$	Participates in nucleic acid synthesis, hormone responses, and membrane function

Figure 37.1 Plant nutrition.
The cause of plant nutrient deficiencies is easily diagnosed when plants are grown in a series of complete nutrient solutions except for the elimination of just one nutrient at a time. Sunflower plants respond rapidly to a deficiency of the following: **a.** Nitrogen is missing in the nutrient solution for plant on the left. **b.** Phosphorus is missing in the nutrient solution for plant on the left. **c.** Calcium is missing in the nutrient solution for plant on the left. In each photograph, healthy plants with a complete nutrient solution are shown on the right.

a.

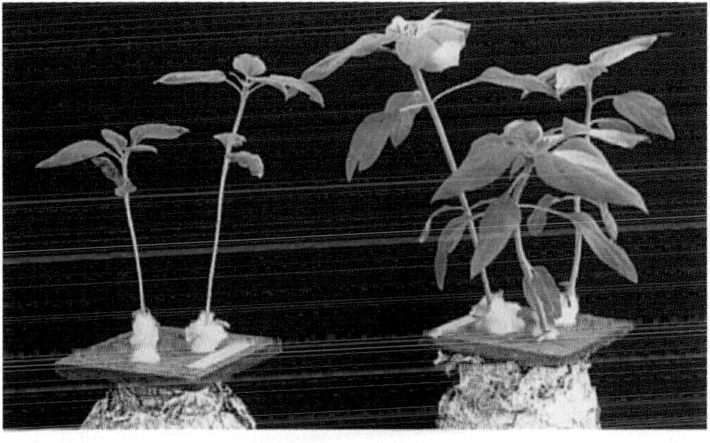

b.

c.

Soil particles, particularly clay and any organic matter in soil, contain negative charges that attract positively charged ions, such as calcium (Ca^{2+}), potassium (K^+), and magnesium (Mg^{2+}). This attraction keeps these ions at a soil level where they are available to plants. In acidic soils, however, hydrogen ions (H^+) replace these other positive ions so that the useful ions float free and are easily leached by water:

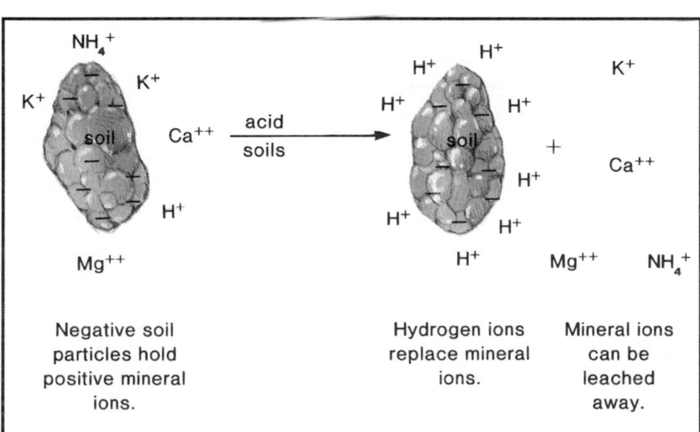

On the other hand, in very acidic soils, aluminum (Al^{3+}) and iron (Fe^{3+}) ions become available to plants. These mineral ions are insoluble above a pH of 4, but they become soluble at a more acidic pH. They displace calcium, potassium, and magnesium ions on soil particles in the same way as hydrogen ions do. High levels of aluminum and iron ions in soils are toxic to plants.

It is clear, then, that acid deposition has two effects that are deleterious to plants: it causes the removal of the mineral ions that plants need for proper nutrition, and it makes available other mineral ions that are toxic at high levels. The death of trees in certain portions of the United States is, in part, the result of acid deposition (from acid rain/snow).

Certain elements are required by plants for good nutrition. Lack of an essential nutrient causes plants to die. Beneficial nutrients serve particular purposes in certain plants.

How Minerals Are Taken in and Distributed

The pathway for mineral transport in a plant is the same as that for water. Like water, minerals can move past the epidermis and through the cortex by way of porous cell walls (see Fig. 37.7). Eventually, however, because of the **Casparian strip,** minerals must enter the cytoplasm of endodermal cells if they are to proceed any farther. In addition, minerals often move directly into the cytoplasm of epidermal cells at the **root hair** region. Plants possess an astonishing ability to concentrate minerals—that is, to take up minerals until they are many times more concentrated in the plant than in the surrounding medium. The concentration of certain minerals in roots is as much as 10,000 times greater than in the surrounding soil.

Following their uptake by root cells, minerals are secreted into xylem and are transported toward the leaves by the upward movement of water. Along the way, minerals can exit xylem and enter those cells that require them. Some eventually reach leaf cells. In any case, minerals must again cross a selectively permeable plasma membrane when they exit xylem and enter living cells. By what mechanism do minerals cross plasma membranes?

Table 37.1 shows that plant cells absorb minerals in the ionic form: nitrogen is absorbed as nitrate (NO_3^-) or ammonium (NH_4^+) ions, phosphorus as phosphate (HPO_4^{2-}), potassium as potassium ions (K^+), and so forth. Since the plasma membrane is charged, ions are likely to have difficulty crossing it. It has long been known that plant cells expend energy to actively take up and concentrate mineral ions. If roots are deprived of oxygen or are poisoned so that cellular respiration cannot occur, mineral ion uptake is diminished. The energy of ATP is required for mineral ion transport, but not directly (Fig. 37.2). A plasma membrane pump, called a proton pump, hydrolyzes ATP and uses the energy released to transport hydrogen ions (H^+) out of the cell. This sets up an electrochemical gradient that drives positively charged ions like K^+ through a channel protein into the cell. Negatively charged mineral ions are transported, along with H^+, by carrier proteins. Since H^+ is moving down its concentration gradient, no energy is required. Notice that this model of mineral ion transport in plant cells is based on chemiosmosis, the establishment of an electrochemical gradient to perform work.

Minerals follow the same path as water. When minerals cross plasma membranes, energy is expended to establish an electrochemical gradient that promotes mineral uptake.

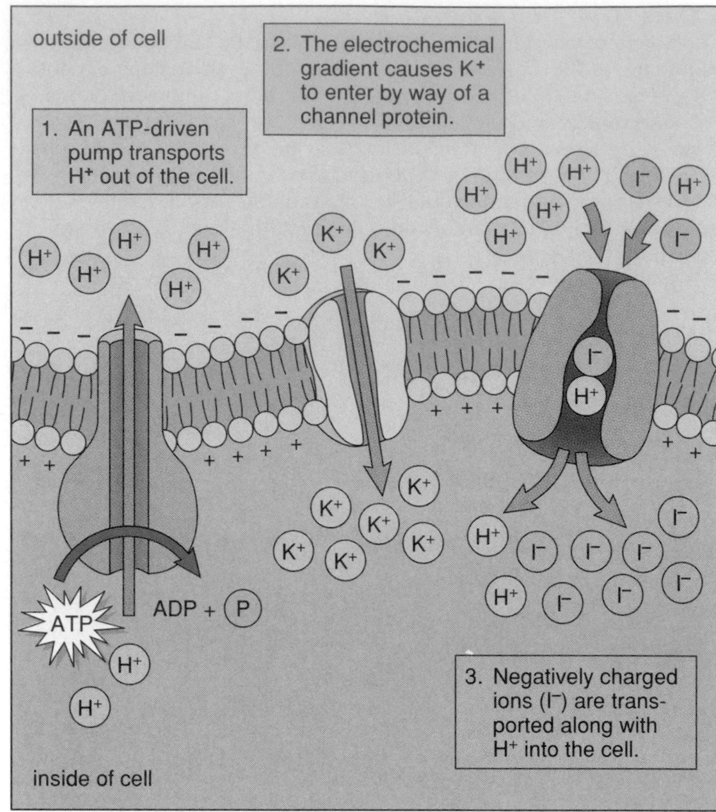

Figure 37.2 Model for mineral transport.
When minerals are transported across the plasma membrane, an ATP-driven pump removes hydrogen ions from the cell. This establishes an electrochemical gradient that allows potassium (K^+) and other positively charged ions to cross the membrane via a channel protein. Negatively charged mineral ions (I^-) can cross the membrane by way of a carrier when they hitch a ride with hydrogen ions (H^+), which are diffusing down their concentration gradient.

How Roots Are Adapted for Uptake

Two symbiotic relationships are known to assist roots in supplying mineral nutrients to a plant. In the first type, plants are unable to make use of atmospheric nitrogen (N_2) because they do not have the cellular enzymes to break the $N \equiv N$ bond. Some plants remove nitrate (NO_3^-) or ammonium (NH_4^+) from the soil. But others—such as legumes, soybeans, and alfalfa—have roots infected by *Rhizobium* bacteria, which can break the $N \equiv N$ bond and reduce nitrogen to NH_4^+ for incorporation into organic compounds. (The reduction of N_2 to NH_4^+ is called **nitrogen fixation.**) The bacteria live in **root nodules** [L. *nodulus,* dim. of *nodus,* knot] and are supplied with carbohydrates by the host plant (Fig. 37.3). The bacteria, in turn, furnish their host with nitrogen compounds. Many other plants are now

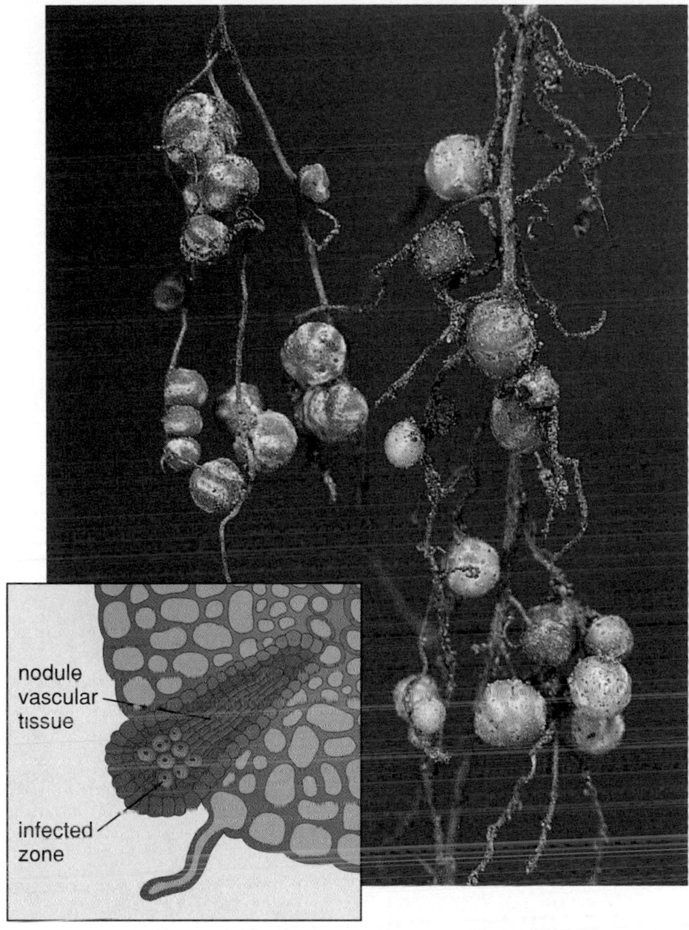

Figure 37.3 Root nodules.
Nitrogen-fixing bacteria infect and live in nodules on the roots of plants, particularly legumes. They make reduced nitrogen available to a plant and the plant passes carbohydrates to the bacteria. The inset shows the inner composition of a nodule.

known to have a semi-symbiotic relationship with free-living, nitrogen-fixing microorganisms in the soil. In some grasses these microorganisms live on the surface of roots and receive carbon compounds exuded by the roots. They use the energy derived from these compounds to fix nitrogen, some of which they release to the root of the plant.

The second type of symbiotic relationship, called a mycorrhizal association, involves fungi and almost all plant roots (Fig. 37.4). Only a small minority of plants do not have **mycorrhizae** [Gk. *mykes,* fungus, and *rhiza,* root], and these plants are most often limited as to the environment in which they can grow. Ectomycorrhizae form a mantle that is exterior to the root, and they grow between cell walls. Ectomycorrhizae can penetrate cell walls. In any case, the fungus increases the surface area available for mineral and

Figure 37.4 Mycorrhizae.
Experimental results show that rough lemon plants with mycorrhizae (MYCO) grow much better than plants without mycorrhizae. Inset shows roots of a lodgepole pine associated with fungus filaments.

water uptake and breaks down organic matter, releasing nutrients that the plant can use. In return, the root furnishes the fungus with sugars and amino acids. Plants are extremely dependent on mycorrhizae. Orchid seeds, which are quite small and contain limited nutrients, do not germinate until a mycorrhizal fungus has invaded their cells. Nonphotosynthetic plants, such as Indian pipe, use their mycorrhizae to extract nutrients from nearby trees.

As a special adaptation, some plants have poorly developed roots or no roots at all because minerals and water are supplied by other mechanisms. Carnivorous plants such as the Venus's-flytrap and sundews obtain some nitrogen and minerals when their leaves capture and digest insects. **Epiphytes** [Gk. *epi,* over, and *phyton,* plant] are "air plants"; they do not grow in soil but on larger plants, which give them support. They do not receive nutrients from their host, however. Some epiphytes have roots that absorb moisture from the atmosphere, and many catch rain and minerals in special pockets at the base of their leaves. Parasitic plants such as dodders, broomrapes, and pinedrops send out rootlike projections called haustoria that grow into the host stem and tap into the xylem and phloem of the host.

Most plants are assisted in acquiring minerals by a symbiotic relationship with microorganisms and/or fungi. Some, such as predaceous or parasitic plants, have special adaptations for mineral acquisition.

37.2 How Water Moves Through a Plant

Vascular plants have evolved a tissue which transports water and minerals (Fig. 37.5). **Xylem,** the vascular tissue that transports water and minerals, contains two types of conduction cells: tracheids and vessel elements. **Tracheids** have end walls that are pitted, but the vessel elements have no end walls at all. The **vessel elements** placed end to end form a completely hollow pipeline from the roots to the leaves. These strong-walled nonliving cells also give trees much-needed internal support.

Similarly, transport tissue is present for organic nutrients, which are produced primarily by leaves but are needed throughout the plant. Roots buried in the soil cannot possibly carry on photosynthesis, but they still require a source of energy in order to carry on cellular metabolism. Vascular plants are able to transport the products of photosynthesis to regions that require them and/or where they will be stored for future use. The conducting cells in **phloem** are **sieve-tube cells,** each of which typically has a companion cell. **Companion cells** can provide energy to sieve-tube cells, which contain cytoplasm but have no nucleus. Their end walls are called sieve plates because they contain numerous pores. The sieve-tube cells are aligned end to end, and strands of cytoplasm called plasmodesmata extend from one cell to the other through the sieve plates. Therefore, there is a continuous pathway for organic nutrient transport throughout the plant.

Vascular plants have transport tissues; xylem transports water and minerals from the roots to the leaves, and phloem transports organic nutrients to various parts, particularly from the leaves to the roots.

Knowing that vascular plants are structured in a way that allows materials to move from one part to another part does not tell us the mechanisms involved. Plant physiologists have performed numerous experiments to determine how water and minerals rise to the top of very tall trees in xylem and how organic nutrients move in the opposite direction in phloem.

We would expect these mechanisms to be mechanical in nature and based on the properties of water because water is a large part of both xylem sap and phloem sap. Keep in mind, therefore, that water molecules diffuse freely across plasma membranes, and that because water molecules are polar, hydrogen bonding occurs between water molecules themselves and between water molecules and other molecules.

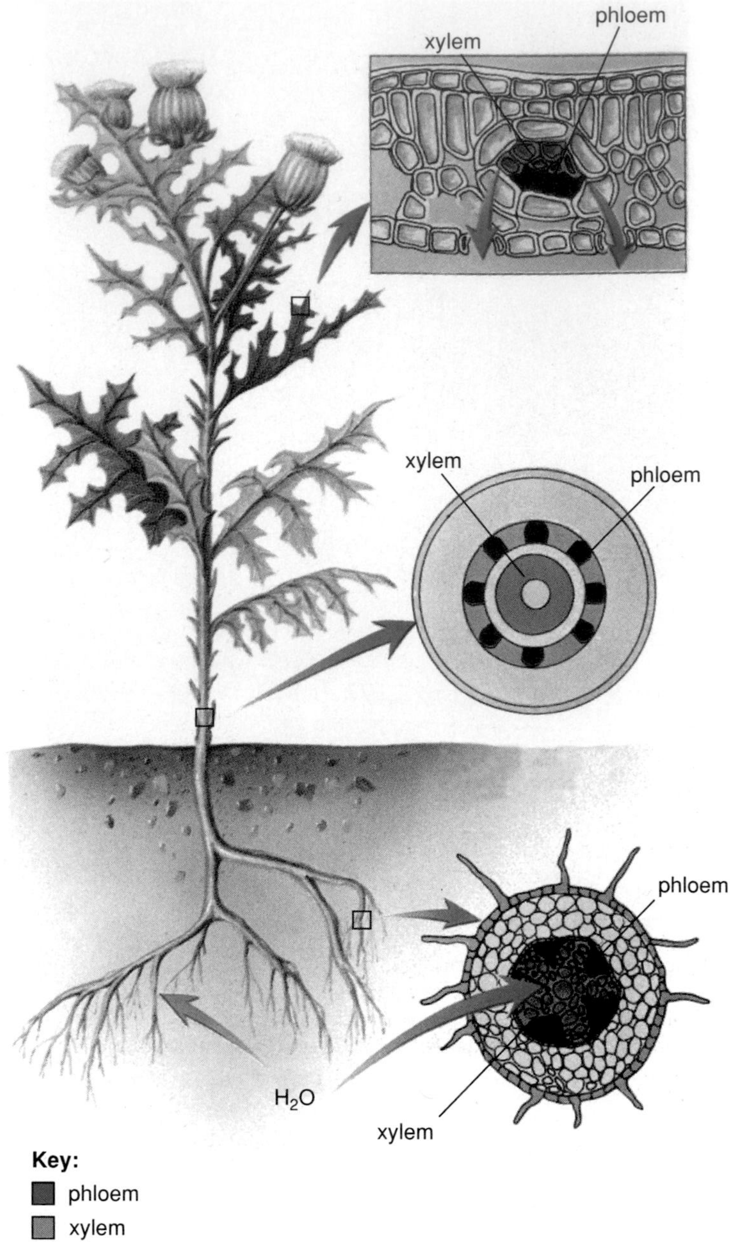

Figure 37.5 Plant transport system.
Vascular tissue in plants includes xylem, which transports water and minerals from the roots to the leaves, and phloem, which transports organic nutrients particularly in the opposite direction. Notice that the xylem and the phloem are in the roots, the stem, and the leaves, which are the vegetative organs of a plant.

In order for water to move from the roots to the leaves, it must cross plasma membranes. Let's now examine a conceptual model developed by plant physiologists to explain how water moves from the roots to the leaves.

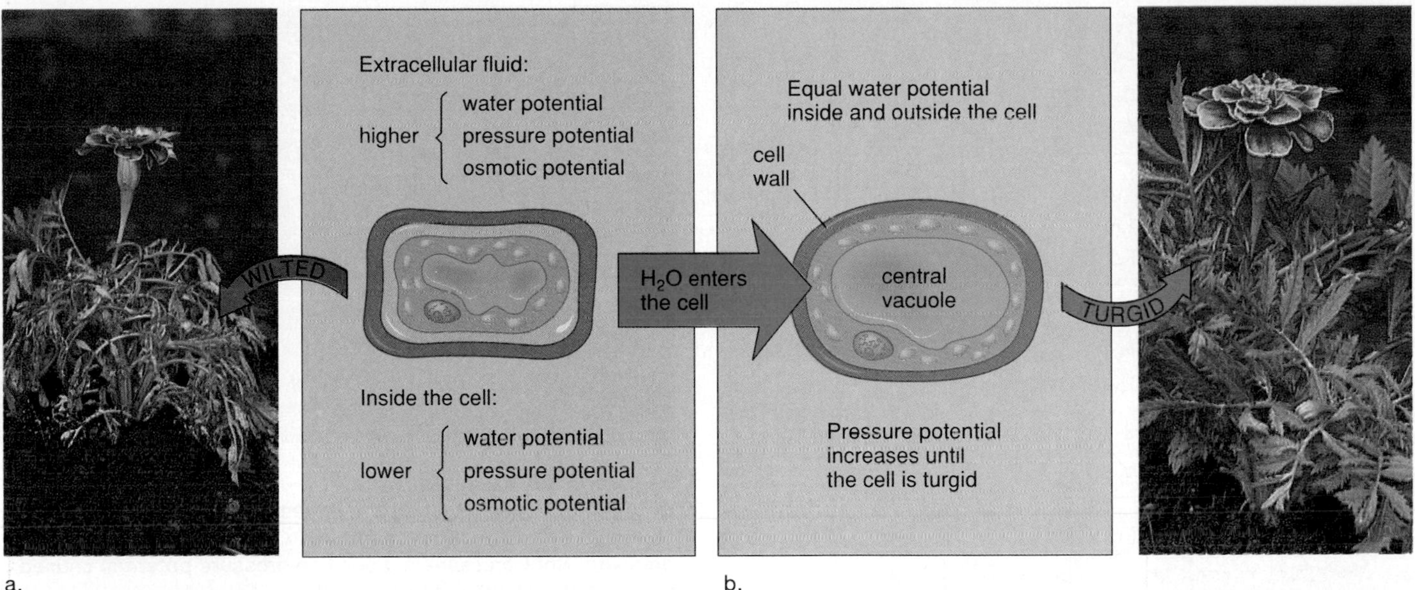

Figure 37.6 Water potential and turgor pressure.
a. The cells of a wilted plant have a lower water potential due to a lower osmotic potential. When available, water therefore enters the cells. b. Equilibrium is achieved when the pressure potential rises in the cells. Cells are now turgid, and the plant is no longer wilted.

Water Potential Is Critical

As you may know, potential energy is stored energy due to the position of an object. A boulder placed at the top of a hill has potential energy. When pushed, the boulder moves down the hill as potential energy is converted into kinetic (motion) energy. The boulder at the bottom of the hill has lost much of its potential energy.

Let us define **water potential** as the potential energy of water. Just like the boulder, water at the top of a waterfall has a higher water potential than water at the bottom of the waterfall. As illustrated by this example, *water moves from a region of higher water potential to a region of lower water potential.*

In terms of cells, two factors usually determine water potential, which in turn determines the direction in which water will move across a plasma membrane (Fig. 37.6). These factors concern differences in:

1. water pressure across a membrane
2. solute concentration across a membrane

Pressure potential is the effect that pressure has on water potential. With regard to pressure, it is obvious to us that water will move across a membrane from the area of higher pressure to the area of lower pressure. The higher the water pressure, the higher the water potential.

Osmotic potential is the effect that solutes have on water potential. The presence of solutes restricts the movement of water because water tends to interact with solutes. Indeed, water tends to move across a membrane from the area of lower solute concentration to the area of higher solute concentration. The lower the concentration of solutes (osmotic potential) the higher the water potential.

Not surprisingly, increasing water pressure will counter the tendency of water to enter a cell because of the presence of solutes. Let us consider a common situation in regard to plant cells. As water enters a plant cell by osmosis, water pressure will increase inside the cell—a plant cell has a strong cell wall that allows water pressure to build up. When will water stop entering the cell? When the pressure potential inside the cell increases and balances the osmotic potential outside the cell.

Pressure potential that increases due to the process of osmosis is often called turgor pressure. Turgor pressure is critical, since plants depend on it to maintain the turgidity of their bodies (Fig. 37.6). The cells of a wilted plant have lost water due to a reversal in the usual water potential difference across plasma membranes.

Water moves across plasma membranes from the area of higher water potential to the area of lower water potential.

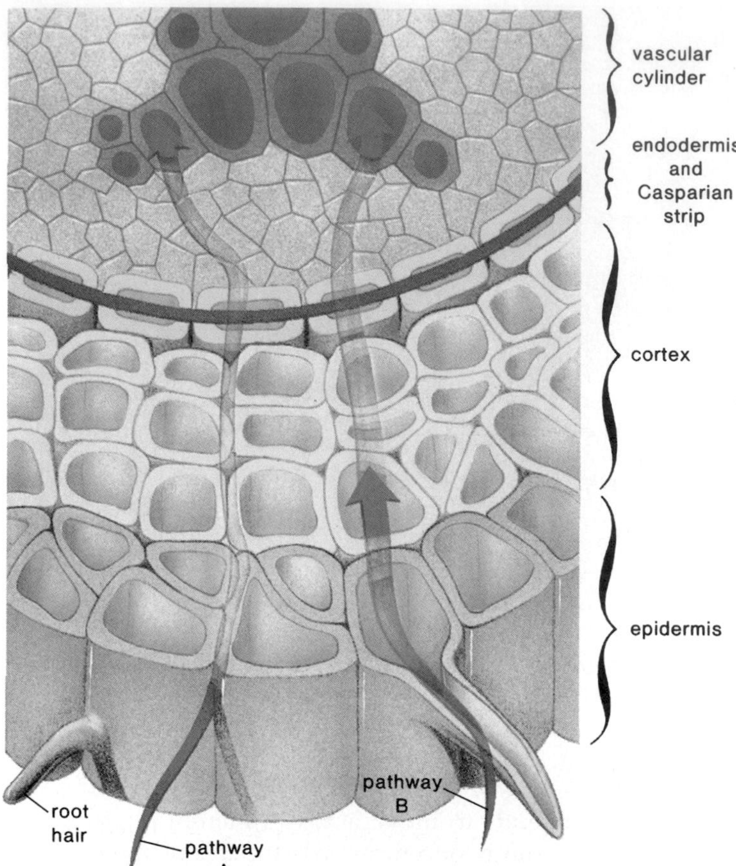

Figure 37.7 Water pathways to xylem in root.
Pathway A: water and minerals travel via porous cell walls and then enter endodermal cells because of the Casparian strip. Pathway B: water and minerals enter a root hair cell and move from cell to cell to finally enter the xylem.

Xylem Transports Water

The movement of water and minerals in a plant involves passage through a root, then in the xylem (particularly the vessels), and finally through the leaves. Special problems dealing with mineral transport are discussed in the next section.

From the Roots

In order for water to reach the xylem, it must pass through the root. As Figure 37.7 shows, water can enter the root of a flowering plant from the soil simply by diffusing *between* the cells. Eventually, however, the *Casparian strip,* a band of suberin and lignin bordering four sides of root endodermal cells, forces water to enter endodermal cells. Alternatively, water can enter root hair cells and then progress from cell to cell across the cortex and **endodermis.** Regardless of the pathway, water enters root cells when they have a lower osmotic potential, and therefore lower water potential, than does soil solution. Water entering root cells creates a positive pressure potential called **root pressure.** Root pressure, which primarily occurs at night, tends to push xylem sap upward. Root pressure is also responsible for **guttation** [L. *gutta,* drops, spots], when drops of water are forced out of vein endings along the edges of leaves (Fig. 37.8). Although root pressure may contribute to the upward movement of water in some instances, it is not believed to be the mechanism by which water can rise to the tops of very tall trees.

Figure 37.8 Guttation.
Drops of guttation water on the edges of a strawberry leaf. Guttation, which occurs at night, is thought to be due to root pressure. Root pressure is a positive pressure potential caused by the entrance of water into root cells.

When Is Water Available?

After a rain or irrigation, gravity causes water in the soil to drain away. The speed with which this happens can be correlated with the permeability of the soil. Sandy soil, which is very permeable, drains rapidly while clay soil, which is much less permeable, drains slowly. The water remaining in the soil after drainage is referred to as its *field capacity.*

Just because water is in the soil does not mean it is available to the plant. The point at which water is no longer available to the roots is called the *permanent wilting point* of the soil. At this point, the soil, especially clay soil, still contains water, but it is held tightly on tiny particles. The soil and root water potentials have become the same or have reversed, and the roots can no longer remove water from the soil.

The ideal soil holds water, has air spaces (roots need oxygen for cellular respiration), and is easily penetrated by roots. The best agricultural soils turn out to be loams that are 10–25% clay, with the rest being equal parts of sand and silt.

Water and minerals must enter root cells before they reach the xylem. Water enters root cells because the water potential within root cells is less than that of the soil solution.

To the Leaves

Once water enters xylem, it must be transported upward the entire height of a tree. This can be a daunting task, given that redwood trees can exceed 110 meters in height.

The **cohesion-tension model** of xylem transport outlined in Figure 37.9 suggests a mechanism for xylem transport. The term *cohesion* refers to the tendency of water molecules to cling together. Because of hydrogen bonding, water molecules interact with one another, and there is a continuous water column in xylem from the leaves to the roots that is not easily broken. Adhesion is a property of water that gives the water column extra strength and prevents it from slipping back. *Adhesion* refers to the ability of water, a polar molecule, to interact with the molecules making up the walls of the vessels in xylem.

Why does the continuous water column move upward? It does so because it is under *tension*—it is pulled up by a *negative pressure potential.* Consider the structure of a leaf. When the sun rises, the stomates (stomata) open and carbon dioxide enters a leaf. Within the leaf the mesophyll cells—particularly the spongy layer—are exposed to the air, which can be quite dry. Water now evaporates from mesophyll cells. Evaporation of water from leaf cells is called **transpiration.** At least 90% of the water taken up by the roots is eventually lost by transpiration. This means that the total amount of water lost by a plant over a long period of time is surprisingly large. A single *Zea mays* (corn) plant loses somewhere between 135 and 200 liters of water through transpiration during a growing season.

As transpiration occurs, the water column is pulled upward—first within the leaf, then from the stem, and finally from the roots. During the day, root cells have a more negative water potential than that of the soil due to transpiration at the leaves! For the negative water potential to reach from the leaves to the root, the water column must be continuous.

What happens if the water column within xylem is broken, as by cutting a stem? The water column "snaps back" down the xylem vessel away from the site of breakage, making it more difficult for conduction to occur. This is why it is best to maximize water conduction by cutting flower stems under water. This effect has also allowed investigators to measure the tension in stems. A device called the pressure bomb measures how much pressure it takes to push the xylem sap back to the cut surface of the stem.

The driving force for the ascent of water in xylem is a negative pressure potential brought about by transpiration at the leaves.

There is an important consequence to the way water is transported in plants; when a plant is under water stress, the

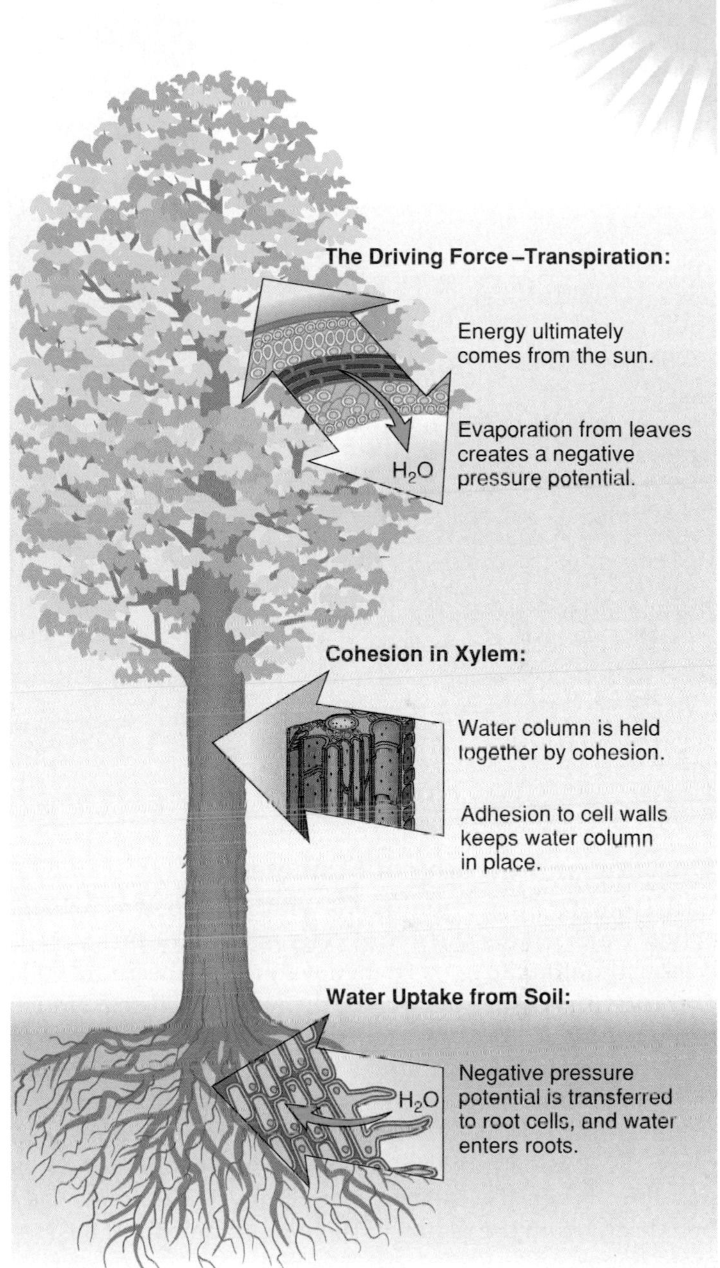

The Driving Force–Transpiration:
Energy ultimately comes from the sun.
Evaporation from leaves creates a negative pressure potential.
H_2O
Cohesion in Xylem:
Water column is held together by cohesion.
Adhesion to cell walls keeps water column in place.
Water Uptake from Soil:
H_2O
Negative pressure potential is transferred to root cells, and water enters roots.

Figure 37.9 Cohesion-tension model of xylem transport.
A negative pressure potential created by evaporation (transpiration) at the leaves pulls water along the length of the plant—even down to the root hairs.

stomates close. Now the plant loses little water because the leaves are protected against water loss by the waxy **cuticle** of the upper and lower epidermis. When the stomates are closed, however, carbon dioxide cannot enter the leaves and plants are unable to photosynthesize. Photosynthesis, therefore, requires an abundant supply of water so that the stomates can remain open and allow carbon dioxide to enter.

Figure 37.10 Opening and closing of stomates.
Stomates open when water enters guard cells and turgor pressure increases. Stomates close when water exits guard cells and there is a loss of turgor pressure.

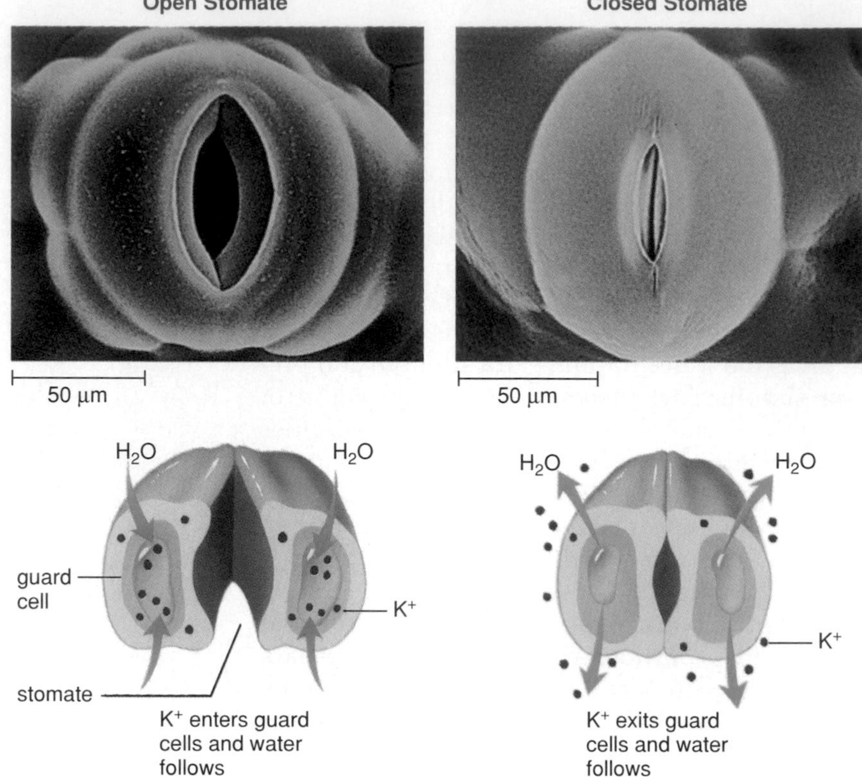

Opening and Closing of Stomates

Each **stomate** has two **guard cells** with a pore between them. When water enters the guard cells and *turgor pressure increases,* the stomate opens; when water exits the guard cells and turgor pressure decreases, the stomate closes. Notice in Figure 37.10 that the guard cells are attached to each other at their ends and that the inner walls are thicker than the outer walls. When water enters, a guard cell's radial expansion is restricted because of cellulose microfibrils in the walls, but lengthwise expansion of the outer walls is possible. When the guard cells expand lengthwise, they buckle out from the region of their attachment and the stomate opens.

Since about 1968 it has been clear that there is an accumulation of potassium ions (K^+) within guard cells when stomates open. In other words, active transport of K^+ results in a water potential that is lower than that of surrounding cells, and this causes water to enter guard cells. Also interesting is the observation that there is an accumulation of hydrogen ions (H^+) outside guard cells as K^+ moves into them. A proton pump run by the breakdown of ATP to ADP + Ⓟ transports H^+ to the outside of the cell. This establishes an electrochemical gradient that allows K^+ to enter by way of a channel protein (see Fig. 37.2).

What regulates the opening and closing of stomates? It appears that the blue-light component of sunlight is a signal that can cause stomates to open. There is evidence to suggest that a flavin pigment absorbs blue light, and then this pigment sets in motion the cytoplasmic response that leads to activation of the proton pump. Similarly, there could be a receptor in the plasma membrane of guard cells that brings about inactivation of the pump when carbon dioxide (CO_2) concentration rises, as might happen when photosynthesis ceases. Abscisic acid (ABA), which is produced by cells in wilting leaves, can also cause stomates to close. Although photosynthesis cannot occur, water is conserved.

If plants are kept in the dark, the stomates open and close on a 24-hour basis just as if they were responding to the presence of sunlight in the daytime and the absence of sunlight at night. This means that there must be some sort of internal *biological clock* that is keeping time. Circadian rhythms (a behavior that occurs every 24 hours) and biological clocks are areas of intense investigation at this time.

When stomates open, first K^+ and then water enters guard cells. Stomates open and close in response to environmental signals, and the exact mechanism is being investigated.

doing science

Competition for Resources and Biodiversity

From my earliest years in school, I have loved both mathematics and biology. In college, the biological issues that intrigued me most were those concerned with the relationship of species to their environment, including the effects of interactions such as competition or predation. It seemed to me that biodiversity might relate to these interactions and, if so, it might be possible to develop a theory, expressed mathematically, to explain biodiversity.

The first step toward this goal is to establish that competition and/or predation do relate to biodiversity.

The Minnesota grasslands in which I work often have more than 100 plant species coexisting in an area the size of a few hectares. Long-term experiments have shown that all these plant species are held in check by (limited by) competition for the same resource, namely soil nitrogen. Mathematical models predict that the number of coexisting species can never be more than the number of resources that limit them. Is there an answer to this seeming paradox? There is if these species are competing with one another on some other basis besides soil nitrogen.

To find out what this basis might be, we did a series of experiments. We found that the factor limiting these species was not insect or mammalian herbivores, light availability, or fire. However, the plant species did differ in their ability to disperse to new sites. We discovered this by planting over 50 different plant species in several plots and finding how well they could germinate, grow, and reproduce there. The best competitors for nitrogen were native bunchgrasses that allocated 85% of their biomass to root but only 0.5% of their biomass to seed. Little bluestem *(Schizachyrium scoparium)* is an example of a bunchgrass (Fig. 37A). The best dispersers allocated 30% of their biomass to seed, but only 40% of their biomass to root. Bent grass *(Agrostis scabra)* is an example of a poor competitor for soil nitrogen but a good disperser (Fig. 37B).

A mathematical model showed that such allocation-based tradeoffs could explain the stable coexistence of a whole range of plant species that differ according to their ability to compete for nitrogen and to disperse to new areas. Coexistence occurs because better competitors for soil nitrogen are poorer dispersers, and therefore they do not occupy all sites. Better dispersers are also better at finding and occupying all sites. Because their ability to compete and disperse vary, a number of species can coexist.

G. David Tilman
University of Minnesota

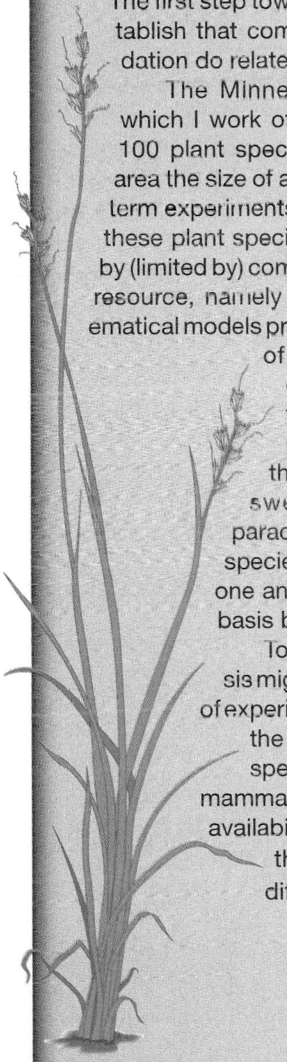

Another issue of interest to me is the relationship between biodiversity and the stability of an ecosystem. My colleagues and I were annually sampling 207 permanent plots when there was a drought in 1987–88. The 1988 drought was the third worst in the past 150 years. We found that plots that contained only one to four species had their productivity (total mass of living plants) fall to between 1/8 and 1/16 of the pre-drought level. But plots that contained 16 to 26 species were able to maintain their productivity at about 1/2 the pre-drought level (Fig. 37C). This suggests that high biodiversity does buffer ecosystems against a disturbance and that it is wise to conserve the biodiversity of ecosystems in all areas—whether in Minnesota, New Jersey, Oregon, or the tropics.

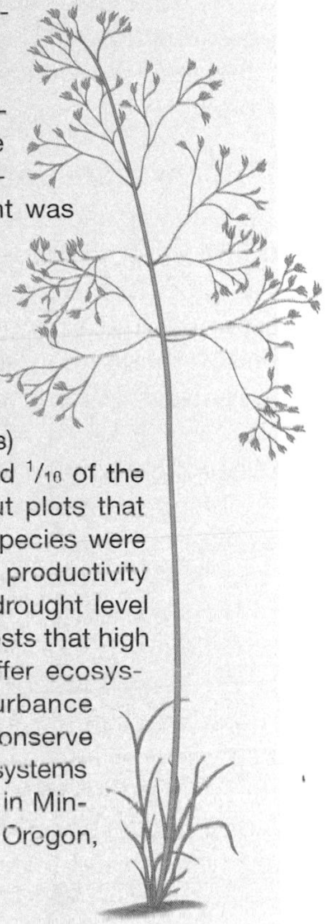

Figure 37A
Little bluestem *(Schizachyrium scoparium)* is a North American bunchgrass that prefers sandy or rocky habitats. It is an excellent competitor for soil nitrogen because of its high allocation of roots.

Figure 37B
Bent grass *(Agrostis scabra)* is a native North American prairie grass that is a poor competitor for soil nitrogen. It disperses rapidly into disturbed areas because of its high allocation of seed.

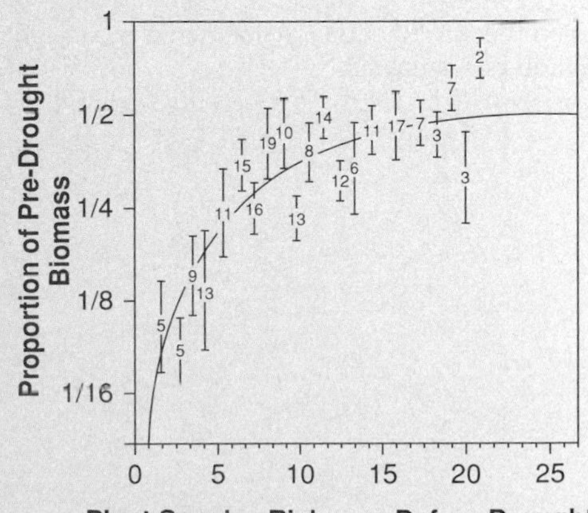

Figure 37C
During drought, plots with higher plant species richness attained plant biomasses that were a larger proportion of their pre-drought plant biomass. This graph compares biomass in 1988, a drought year, with that in 1986, a year before the drought. The numbers placed at the means indicate the number of plots used to calculate that mean. The high and low values for each biomass ratio are also indicated.

37.3 How Organic Nutrients Are Transported

Not only do plants transport water and minerals from the roots to the leaves, they also transport organic nutrients to the parts of plants that have need of them. This includes young leaves that have not yet reached their full photosynthetic potential, flowers that are in the process of making seeds and fruits, and the roots whose location in the soil prohibits them from carrying on photosynthesis.

Phloem Transports Organic Nutrients

As long ago as 1679, Marcello Malpighi suggested that bark is involved in translocating sugars from leaves to roots. He observed the results of removing a strip of bark from around a tree, a procedure called **girdling.** If a tree is girdled below the level of the majority of leaves, the bark swells just above the cut and sugar accumulates in the swollen tissue. We know today that when a tree is girdled, the phloem is removed but the xylem is left intact. Therefore, the results of girdling suggest that phloem is the tissue that transports sugars.

Radioactive tracer studies with ^{14}C have confirmed that phloem transports organic nutrients. When ^{14}C-labeled carbon dioxide (CO_2) is supplied to mature leaves, radioactively labeled sugar is soon found moving down the stem into the roots. This labeled sugar is found mainly in the phloem, not in the xylem. Radioactive tracer studies have also confirmed the role of phloem in transporting other substances, such as amino acids, hormones, and even mineral ions. Hormones are transported from their production sites to target areas, where they exert their regulatory influences. In the autumn, before leaves fall, mineral ions are removed from the leaves and are taken to other locations in the plant.

Chemical analysis of phloem sap has shown that its main component is sucrose, and the concentration of nutrients is usually 10–13% by volume. It is difficult to take samples of sap from the phloem without injuring the phloem, but this problem is solved by using aphids (Fig. 37.11), small insects that are phloem feeders. The aphid drives its stylet, which is a sharp mouthpart that functions like a hypodermic needle, between the epidermal cells and withdraws sap from a sieve-tube cell. If the aphid is anesthetized using ether, its body can be carefully cut away, leaving the stylet. Phloem can then be collected and analyzed by the researcher.

Phloem Transport Uses Positive Pressure

As mentioned, the conducting cells of phloem are sieve-tube cells lined up end to end with their sieve plates abutting. Cytoplasm extends through the sieve plates of adjoining cells to form a continuous sieve-tube system that extends from the roots to the leaves and vice versa. An explanation of phloem transport must account for the movement of fairly large amounts of organic material for long distances in a relatively short period of time. The movement rate of radioactively labeled ^{14}C sugar has been determined by analysis of sap withdrawn from two areas of the stem. Materials appear to move through the phloem at a rate of 60–100 cm per hour and possibly up to 300 cm per hour.

The **pressure-flow model** is a current explanation for the movement of organic materials in phloem (Fig. 37.12). During the growing season, the leaves are photosynthesizing and producing sugar. Sucrose is actively transported into phloem. Again, this is dependent upon an electrochemical gradient established by a proton pump. Sucrose is carried across the membrane in conjunction with

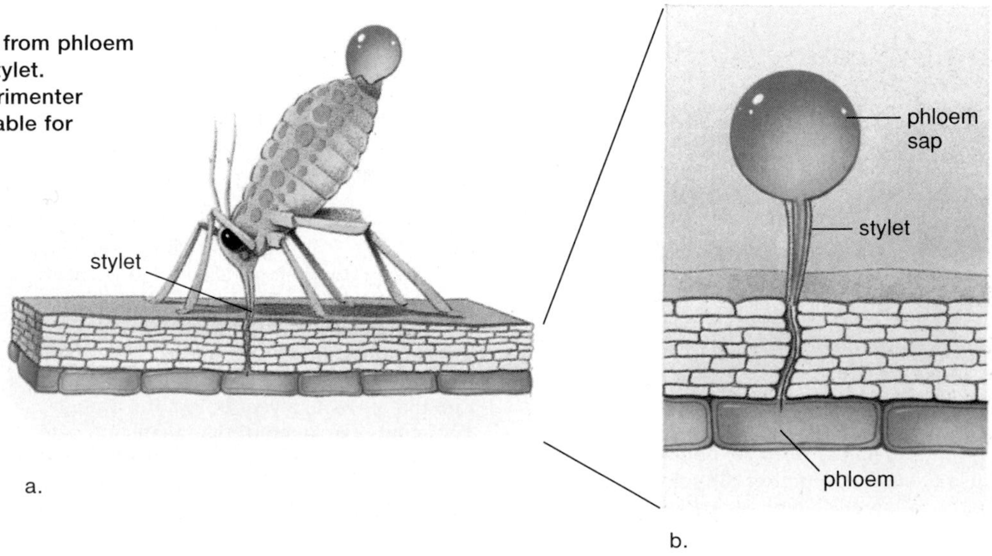

Figure 37.11 Acquiring phloem sap. Aphids are small insects that remove nutrients from phloem by means of a needlelike mouthpart called a stylet. **a.** Aphid with stylet in place. **b.** When the experimenter removes the aphid's body, phloem sap is available for collection and analysis.

hydrogen ions (H^+), which are moving down their concentration gradient (see Fig. 37.2). Water now flows into sieve tubes because of their lower osmotic potential. Transport across sieve-tube membranes is possible because sieve-tube cells have a living plasma membrane. Also, the energy needed for sucrose transport is provided by the companion cells. The buildup of water creates a *positive pressure potential* within the sieve tubes of the leaves compared to the roots.

A positive pressure potential gradient exists from the leaves to the roots because at the roots sucrose is transported out of phloem and water follows. This being the case, the positive pressure potential gradient causes a *flow* of water from the leaves to the roots. As water flows within phloem, it brings sucrose with it.

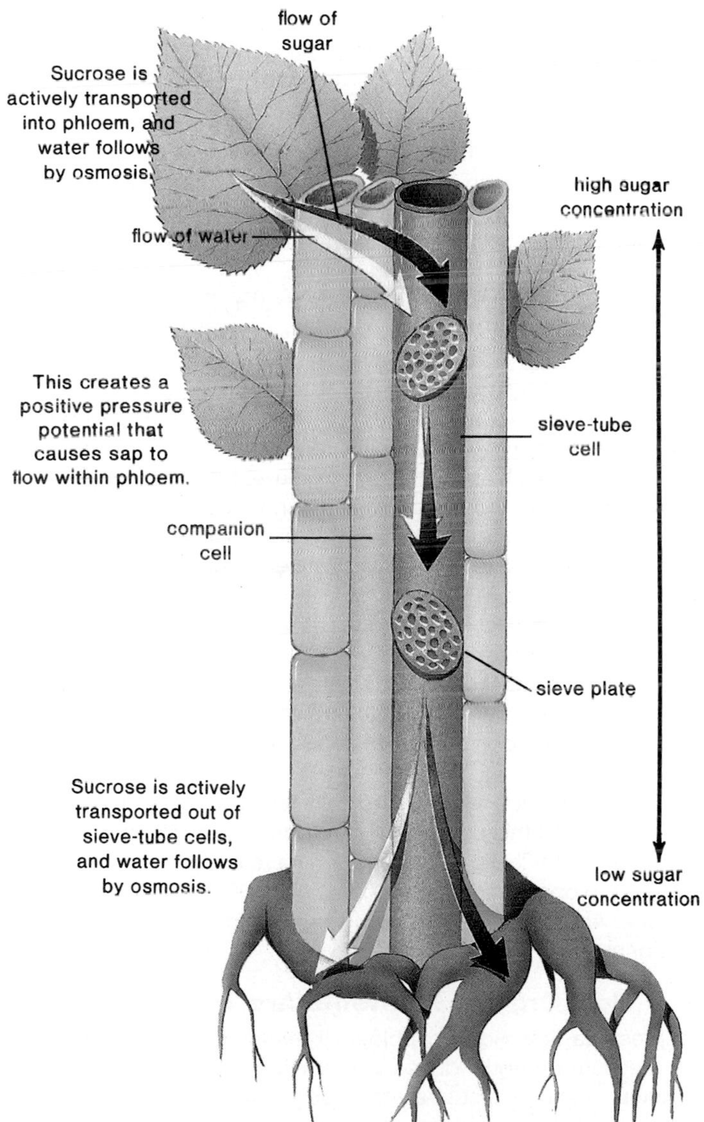

Figure 37.12 Pressure-flow model of phloem transport.
Phloem sap flows from a source (often leaves) to a sink (often roots) because of a positive pressure potential gradient from the leaves to the roots.

The pressure-flow model is supported by the following experiment: two bulbs are connected by a glass tube. The first bulb contains solute at a higher concentration than the second bulb. Each bulb is bounded by a differentially permeable membrane, and the entire apparatus is submerged in distilled water:

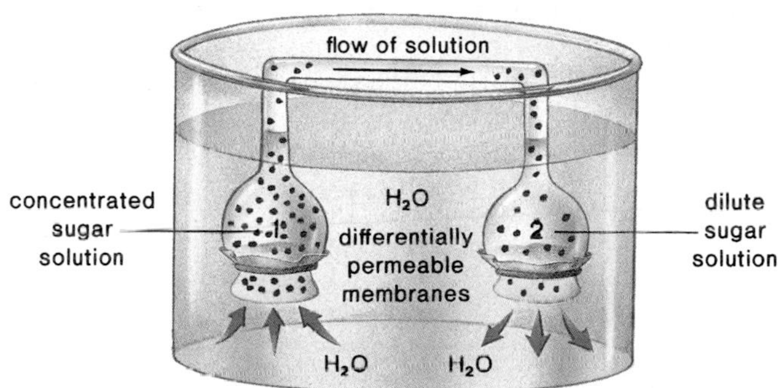

Distilled water flows into the first bulb because it has the lower water potential due to a higher concentration of solute. The entrance of water raises the water potential due to an increase in *pressure potential*, and water *flows* toward the second bulb. This flow not only drives water toward the second bulb, it also provides enough drive to force water out through the membrane of the second bulb—even though the second bulb contains a higher concentration of solute than the distilled water.

The pressure-flow model of phloem transport can account for any direction of flow in sieve tubes if we consider that the direction of flow is always from *source to sink.* In young seedlings, the cotyledons, containing reserve food, are a major source of sucrose and roots are a sink. Therefore, the flow is from the cotyledons to the roots. In older plants, the most recently formed leaves can be a sink and they will receive sucrose from other leaves until they begin to maximally photosynthesize. When a plant is in fruit, phloem flow is monopolized by the fruits, little goes to the rest of the plant, and vegetative growth is slow.

Because phloem sap flows from a source to a sink, situations arise in which there is a bidirectional flow within phloem—not just at different times in the life cycle but even at the same time! It is not known at this time if bidirectional flow is occurring within a single sieve tube or if different sieve tubes are conducting phloem sap in opposite directions.

The pressure-flow model of phloem transport suggests that phloem sap can move either up or down as appropriate for the plant at a particular time in its life cycle.

connecting concepts

The land environment offers many advantages for plants, such as greater availability of light and carbon dioxide for photosynthesis. (Water, even if clear, filters out light, and carbon dioxide concentration and rate of diffusion is less in water.) The evolution of a transport system was critical, however, for plants to make full use of these advantages. Only if a transport system is present can plants elevate the leaves so that they are better exposed to solar energy and carbon dioxide in the air. A transport system brings water, a raw material of photosynthesis, from the roots to the leaves, and also brings the products of photosynthesis down to the roots. Roots lie beneath the soil, and their cells depend on an input of organic food from the leaves in order to remain alive. An efficient transport system allows roots to penetrate deeply into the soil to absorb water and minerals.

The presence of a transport system also allows materials to be distributed to those parts of the body that are growing most rapidly. New leaves and flower buds would grow rather slowly if they had to depend on their own rate of photosynthesis, for example. Height in flowering plants, due to the presence of a transport system, has other benefits aside from elevation of leaves. It is also adaptive to have reproductive structures located where the wind can better distribute pollen and seeds. Once animal pollination came into existence, it was beneficial for flowers to be located where they are more easily seen by animals.

Clearly, plants with a transport system have a competitive edge in the terrestrial environment.

Summary

37.1 Plants Require Inorganic Nutrients

Plants need both essential and beneficial inorganic nutrients. Carbon, hydrogen, and oxygen make up 96% of a plant's dry weight. The other necessary elements are taken up by the roots as mineral ions. Even nitrogen (N), which is present in the atmosphere, is most often taken up as NO_3^-.

You can determine mineral requirements by hydroponics, which is growing plants in a solution. The solution is varied by the omission of one mineral. If the plant grows poorly, then the missing mineral is required for growth.

Roots can only take up the minerals that are available in soil. Acid deposition reduces the availability of some minerals and increases the availability of other minerals. Ions such as calcium (Ca^{2+}), potassium (K^+), and magnesium (Mg^{2+}) are displaced from soil particles by hydrogen ions (H^+) and are leached from the soil. Acid deposition increases the solubility of aluminum (Al^{3+}) and iron (Fe^{3+}) and makes them available in toxic amounts.

Mineral ions cross plasma membranes by a chemiosmotic mechanism. A proton pump transports H^+ out of the cell. This establishes an electrochemical gradient that causes positive ions to flow into the cells. Negative ions are carried across in conjunction with H^+, which is moving along its concentration gradient.

Plants have various adaptations that assist them in acquiring nutrients. Legumes have nodules infected with the bacterium *Rhizobium*, which makes nitrogen compounds available to these plants. Many other plants have mycorrhizae, or fungus roots. The fungus gathers nutrients from the soil, and the root provides the fungus with sugars and amino acids. Some plants have poorly developed roots. Most epiphytes live on, but do not parasitize, trees, whereas mistletoe and some other plants parasitize their host.

37.2 How Water Moves Through a Plant

As an adaptation to life on land, plants have a vascular system that transports water and minerals from the roots to the leaves and must also transport the products of photosynthesis in the opposite direction. Vascular tissue includes xylem and phloem.

Water can enter the root by moving between the cells until it reaches the Casparian strip, after which it passes through an endodermal cell before entering xylem. Water can also enter root hairs, then pass through the cells of the cortex and endodermis to reach xylem.

Water is drawn into root cells because they have the lower water potential. Particularly at night, a pressure potential called root pressure can build in the root. However, this does not contribute significantly to xylem transport.

Much of the water in soil is not available to plants. In sandy soils, it drains quickly, and in clay, it adheres to the tiny particles. Field capacity is the amount of water in a soil after drainage. At the permanent wilting point, no more water is available to the plant.

The cohesion-tension model of xylem transport states that transpiration creates a tension—a negative pressure potential—that pulls water upward in xylem. This means of transport works only because water molecules are cohesive with one another and adhesive with xylem walls.

Most of the water taken in by a plant is lost through stomates by transpiration. Only when there is plenty of water do stomates remain open, allowing carbon dioxide to enter the leaf and photosynthesis to occur.

Stomates open when guard cells take up water. They stretch lengthwise, and this causes them to buckle out. Water enters the guard cells after potassium ions (K^+) have entered. Light signals stomates to open, and high carbon dioxide (CO_2) signals stomates to close. Abscisic acid produced by wilting leaves also signals for closure.

37.3 How Organic Nutrients Are Transported

The pressure-flow model of phloem transport proposes that a positive pressure potential drives phloem sap from the leaves in any direction. Sucrose is actively transported into phloem—also by a chemiosmotic mechanism—at a source, and water follows because of the lower osmotic potential. The resulting increase in pressure potential creates a flow that moves water and sucrose to a sink. A sink can be at the roots or any other part of the plant that requires nutrients.

Reviewing the Chapter

1. Name the elements that make up most of a plant's body. What are essential mineral nutrients and beneficial mineral nutrients? 660
2. Briefly describe two methods used to determine the mineral nutrients of a plant. 660
3. Explain how acid deposition affects availability of minerals to a plant and why this can lead to plant death. 661
4. Describe the chemiosmotic mechanism by which mineral ions cross plasma membranes. 662–63
5. Name two symbiotic relationships that assist plants in taking up minerals and two types of plants that tend not to take up minerals by roots from soil. 662–63
6. Why does a land existence cause plants to need a vascular system? What is their vascular system? 664
7. What is water potential? In regard to cells, what two components determine water potential? When cells are turgid, water potential is equal on both sides of the membrane. Why? 665
8. Give two pathways by which water and minerals can cross the epidermis and cortex of a root. What feature allows endodermal cells to regulate the entrance of molecules into the vascular cylinder? 666
9. What is root pressure, and why can't it account for transport of water in xylem? 666
10. Explain how field capacity and permanent wilting point relate to water availability in soils. Why do soils differ as to water availability? 666
11. Describe and give evidence for the cohesion-tension model of water transport. 667
12. Describe the structure of stomates and explain how they can open and close. By what mechanism do guard cells take up potassium (K^+) ions? 668
13. What data are available to show that phloem transports organic compounds? Explain the pressure-flow model of phloem transport. 670–71

Testing Yourself

Choose the best answer for each question.

1. Which of these is not a nutrient for plants?
   a. water
   b. carbon dioxide gas
   c. mineral ions
   d. nitrogen gas
2. The Casparian strip affects
   a. how water and minerals move into the vascular cylinder.
   b. how water but not minerals move.
   c. how minerals but not water move.
   d. neither the flow of water nor the flow of minerals into a plant.
3. Field capacity is
   a. the same as permanent wilting point.
   b. the amount of water in soil available to plants.
   c. the amount of water in soil after drainage takes place.
   d. higher for sandy soils than clay soils.
4. Stomates are usually open
   a. at night, when the plant requires a supply of oxygen.
   b. during the day, when the plant requires a supply of carbon dioxide.
   c. whenever there is excess water in the soil.
   d. All of these are correct.
5. Which of these is not a mineral ion?
   a. NO_3^-
   b. Mg^+
   c. CO_2
   d. Al^{3+}
6. Water flows from the
   a. higher water potential to the lower water potential.
   b. more positive water potential to the more negative water potential.
   c. more positive osmotic potential to the more negative water potential.
   d. All of these are correct.
7. The pressure-flow model of phloem transport states that
   a. phloem sap always flows from the leaves to the root.
   b. phloem sap always flows from the root to the leaves.
   c. water flow brings sucrose from a source to a sink.
   d. Both a and c are correct.
8. Root hairs do not play a role in
   a. oxygen uptake.
   b. mineral uptake.
   c. water uptake.
   d. carbon dioxide uptake.
9. Explain why this experiment supports the hypothesis that transpiration can cause water to rise to the top of tall trees.

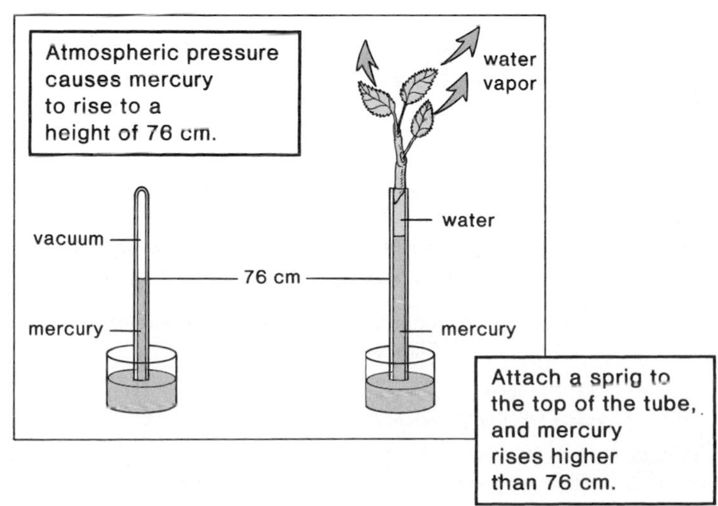

10. Label water (H_2O) and potassium (K^+) ions in these diagrams. What is the role of K^+ in the opening of stomates?

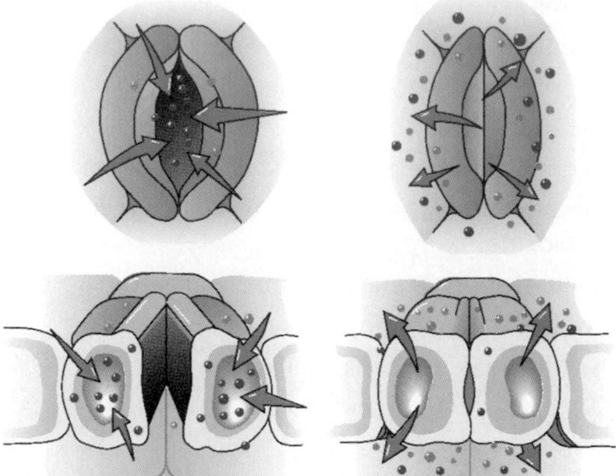

11. Explain why there is a flow of solution from the left bulb to the right bulb.

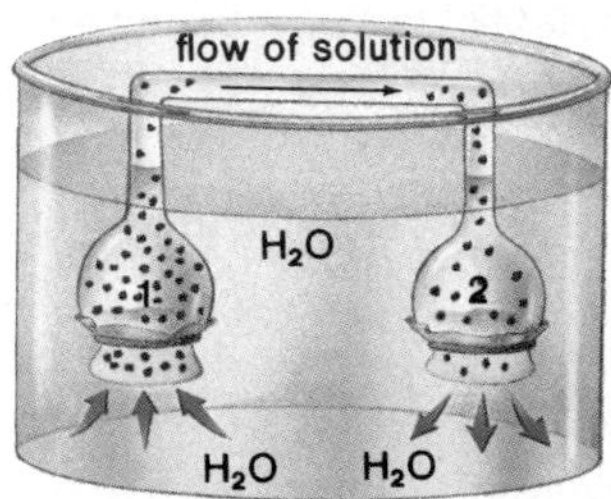

Applying the Concepts

1. *Living things are physical and chemical machines.*
   How does transport of water in xylem support this concept? How does the opening and closing of a stomate support this concept? How does transport of organic substances in phloem support this concept?
2. *Plants, like other living things, are highly organized.*
   How does the structure of xylem contribute to the coordination of plant processes and parts?

Using Technology

Your study of nutrition and transport in plants is supported by these available technologies:

Exploring the Internet
The Mader Home Page provides resources for and help with studying this chapter.

http://www.mhhe.com/sciencemath/biology/mader/
(Click on Biology.)

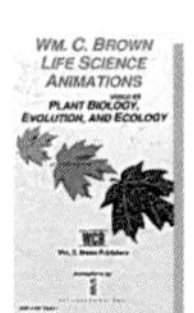

Life Science Animations Video
Video #5: Plant Biology/Evolution/Ecology
How Water Moves Through a Plant (#47)
How Food Moves From a Source to a Sink (#48)

Understanding the Terms

Casparian strip 662
cohesion-tension model 667
companion cell 664
cuticle 667
endodermis 666
epiphyte 663
essential inorganic nutrient 660
girdling 670
guard cell 668
guttation 666
hydroponics 660
macronutrient 660
micronutrient 660
mycorrhiza 663
nitrogen fixation 662
phloem 664
pressure-flow model 670
root hair 662
root nodule 662
root pressure 666
sieve-tube cell 664
stomate 668
tracheid 664
transpiration 667
vessel element 664
water potential 665
xylem 664

Match the terms to these definitions:

a. __________ Force generated by an osmotic gradient that serves to elevate sap through xylem for a short distance.
b. __________ Liberation of water droplets from the edges and tips of leaves.
c. __________ Model explaining transport through sieve tubes of phloem by a positive pressure potential (compared to a sink) due to the active transport of sucrose and the passive transport of water.
d. __________ Plant that takes its nourishment from the air because its attachment to other plants gives it an aerial position.
e. __________ Plant's loss of water to the atmosphere, mainly through evaporation at leaf stomates.
f. __________ Potential energy of water; it is a measure of the capability to release or take up water.
g. __________ Removing a strip of bark from around a tree.
h. __________ Structure on plant roots that contains nitrogen-fixing bacteria.
i. __________ Symbiotic relationship between fungal hyphae and roots of vascular plants. The fungus allows the plant to absorb more mineral ions and obtains carbohydrates from the plant.
j. __________ Type of plant cell that is found in pairs, with one on each side of a leaf stomate; changes in the turgor pressure of these cells regulate the size and passage of gases through the stomate.

Growth and Development in Plants

CHAPTER 38

Chapter Concepts

Germinating coconut of coconut palm, *Cocos nucifera*

All organisms receive stimuli, transform them into meaningful signals, and usually respond in an appropriate way. Plants respond to stimuli like light, gravity, and seasonal changes. In the spring, seeds germinate and growth begins in the presence of sunlight if the soil is warm enough to contain liquid water. In the fall, when temperatures drop, shoot and root apical growth ceases.

Plants do not have eyes, ears, and an elaborate nervous system for quick responses to environmental stimuli. Instead, hormones are involved in the perception of daylength, in the shedding of leaves in the fall, in the breaking of seed dormancy, and a host of other phenomena. Some plant responses are short-term. For example, plants will bend toward the light within a few days because a hormone produced by the growing tip has moved from the sunny side to the shady side of the stem. The bending occurs because cells on the shady side elongate more than the cells facing the light. Similarly, if a potted plant is tipped on its side, the stem will turn upward away from the source of gravity, usually in less than 24 hours.

Even long-term plant responses to environmental conditions involve a change in the pattern of growth and development. The same species exposed to two different sets of stimuli may take on a distinctly different appearance. For example, oak trees exposed to high winds are strong and thickly branched, while those better protected are taller with less branches.

38.1 Plant Responses to Stimuli

Organisms are capable of responding to environmental stimuli, as when you withdraw your hand from a hot stove. It is adaptive for organisms to respond to environmental stimuli because it leads to their longevity and ultimately to the survival of the species. Animals often respond to external stimuli by an appropriate motion. Presented with a nipple, a newborn instinctively begins sucking. Sometimes plants, too, respond rapidly, as when the stomates (also called stomata) open at sunrise. While animals can change their location in response to a stimulus, plants, which are rooted in one place, change their growth pattern. This is why we can determine what a tree has been through—or even the history of the earth's climate—by studying tree rings!

Plants Respond by Growing

Plant growth toward or away from a directional stimulus is called a **tropism** [Gk. *tropos,* turning]. The term directional means that the stimulus is coming from only one direction instead of multiple directions. Growth toward a stimulus is called a positive tropism and growth away from a stimulus is called a negative tropism. Tropisms are due to differential growth—one side of an organ elongates faster than the other, and the result is a curving toward or away from the stimulus (Fig. 38.1). Three well-known tropisms in plants are:

Phototropism: a movement in response to a light stimulus
Gravitropism: a movement in response to gravity
Thigmotropism: a movement in response to touch

What mechanism permits plants to respond to stimuli? When humans respond to light, the stimulus is first received by a pigment in the retina at the back of the eyes, and then nerve impulses are generated that go to the brain. Thereafter, humans respond appropriately. In other words, the first step is *reception* (or perception) of the stimulus. The next step is *transduction,* meaning that the stimulus has been changed into a form that is meaningful to the organism. (In our example, the light stimulus was changed to nerve impulses.) Finally, there is the *response* by the organism. As discussed in the reading on page 683, investigators are beginning to study on a cellular basis how a signal is coupled to a response in plants.

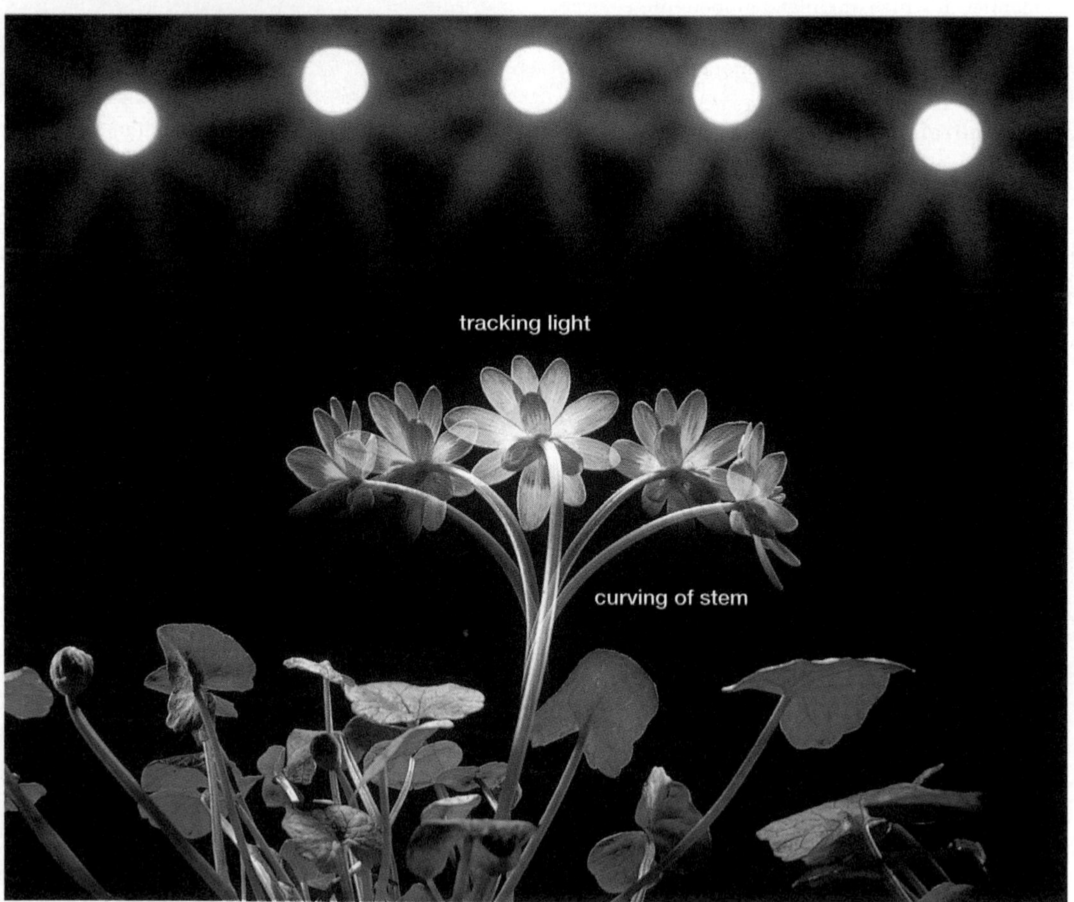

Figure 38.1 Phototropism.
Time-lapse photograph of a buttercup, *Ranunculus ficaria,* curving toward and tracking a source of light.

Plants Respond to Light

Early researchers, including Charles Darwin and his son Francis, observed that plants curve toward the light. The positive **phototropism** [Gk. *photos,* light, and *tropos,* turning] of stems occurs because the cells on the shady side of the stem elongate. Curving away from light is called negative phototropism. Roots, depending on the species examined, are either insensitive to light or they exhibit negative phototropism.

Because blue light in particular causes phototropism to occur, it's believed that a yellow pigment related to the vitamin riboflavin acts as a photoreceptor for light. Following reception, the plant hormone auxin migrates from the bright side to the shady side of a stem. The cells on that side elongate faster than those on the bright side, causing the stem to curve toward the light. It's not yet known how reception of the stimulus (light) is coupled to the production of auxin. That is, it is not known how transduction occurs.

As discussed in the next sections, auxin is also involved in gravitropism, **apical dominance,** root development, and seed development.

Plants Respond to Gravity

When an upright plant is placed on its side, the stem displays negative **gravitropism** [L. *gravis,* heavy, and Gk. *tropos,* turning] because it grows upward, opposite the pull of gravity (Fig. 38.2*a*). Again, Charles Darwin and his son were among the first to say that roots, in contrast to stems, show positive gravitropism (Fig. 38.2*b*). Further, they discovered that if the root cap is removed, roots no longer respond to gravity. Later investigators came up with an explanation. Root cap cells contain sensors called **statoliths,** which are believed by some to be starch grains located within amyloplasts, a type of plastid. (Chloroplasts are a different type of plastid.) Due to gravity, the amyloplasts settle to the lowest part of the cell (Fig. 38.2*c*).

Again, the hormone auxin brings about the positive gravitropism of roots and the negative gravitropism of stems. The two types of tissues respond differently to auxin, which moves to the lower side of both stems and root after gravity has been perceived. Auxin inhibits the growth of root cells; therefore the cells of the upper surface elongate and the root curves downward. Auxin stimulates the growth of stem cells; therefore, the cells of the lower surface elongate and the stem curves upward.

a.

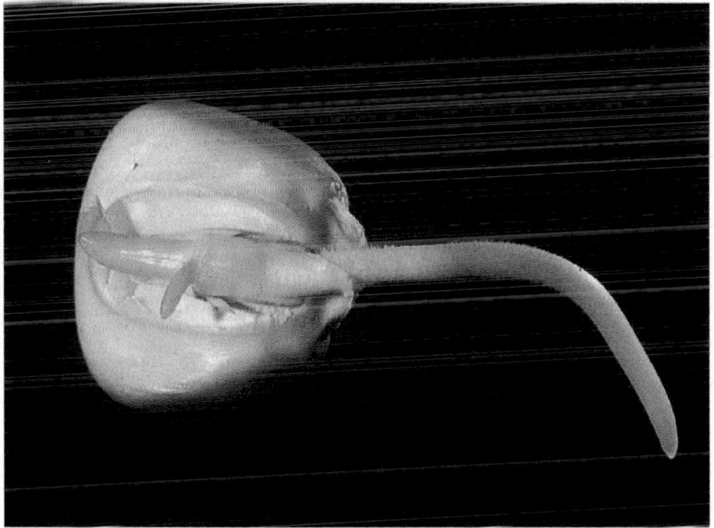

b.

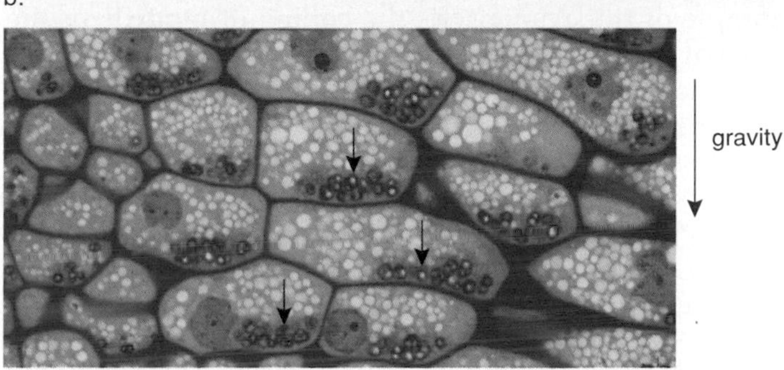

c.

Figure 38.2 Gravitropism.
a. Negative gravitropism of the stem of a *Coleus* plant 24 hours after the plant was placed on its side. **b.** Positive gravitropism of a root emerging from a corn kernel. **c.** Sedimentation of statoliths (see arrows), which are amyloplasts containing starch granules, is thought to explain how roots perceive gravity.

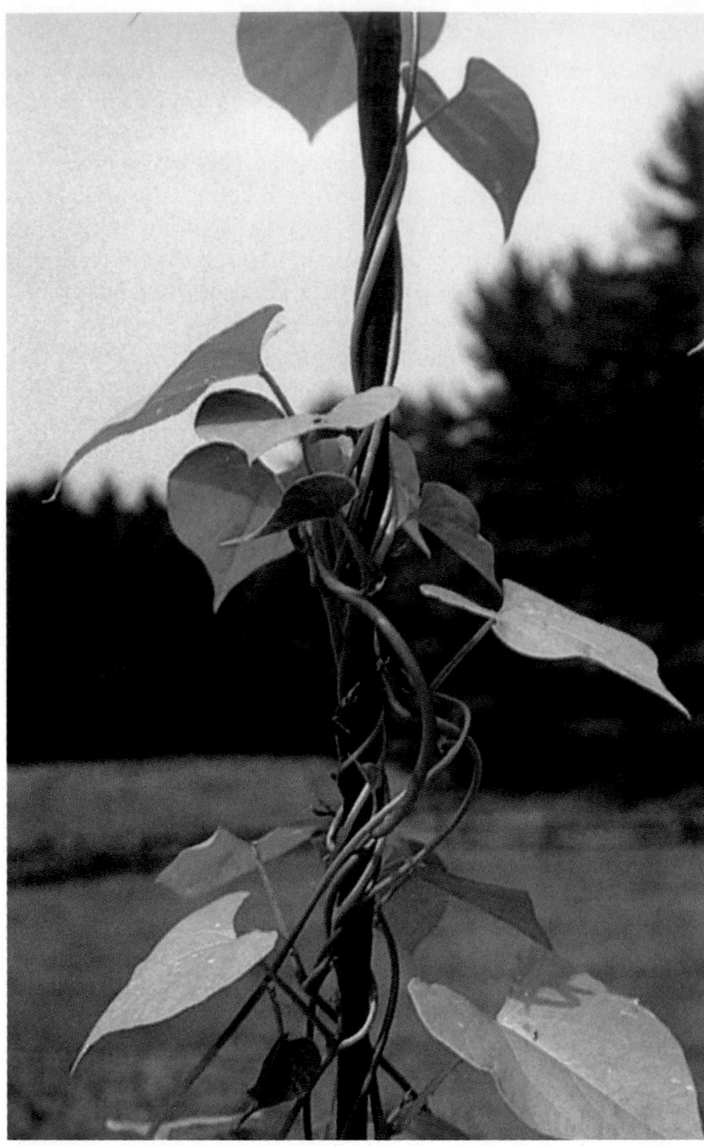

Figure 38.3 Coiling response of a morning glory plant, *Ipomoea*.

Plants Respond to Contact

Unequal growth due to contact with solid objects is called **thigmotropism** [Gk. *thigma,* touch, and *tropos,* turning]. An example of this response is the coiling of **tendrils** or the stems of plants such as morning glory (Fig. 38.3).

The plant grows straight until it touches something. Then the cells in contact with an object, such as a pole, grow less while those on the opposite side elongate. Thigmotropism can be quite rapid; a tendril has been observed to encircle an object within ten minutes. The response endures; a couple of minutes of stroking can bring about a response that lasts for several days. The response can also be delayed; tendrils touched in the dark will respond once they are illuminated. ATP (adenosine triphosphate) rather than light can cause the response; therefore, the need for light may simply be a need for ATP. Also, the hormones auxin and ethylene may be involved since they can induce curvature of tendrils even in the absence of touch.

Thigmomorphogenesis is a touch response related to thigmotropism. In this case, however, the entire plant responds to the presence of environmental stimuli like wind or rain. The same type of tree growing in a windy location often has a shorter, thicker trunk than one growing in a more protected location. Even simple mechanical stimulation like rubbing a plant with a stick can inhibit cellular elongation and produce a sturdier plant with increased amounts of support tissue.

Plants Respond by Turgor Pressure Changes

In contrast to tropisms, *nastic movements* are independent of the direction of the stimulus. *Seismonastic movements* result from touch, shaking, or thermal stimulation. If you touch a *Mimosa pudica* leaf, the leaflets fold because the petiole droops (Fig. 38.4). This response, which takes only a second or two, is due to a loss of turgor pressure within cells located in a thickening, called a pulvinus, at the base of each leaflet. Investigation shows that potassium ions (K^+) move out of the cells and then water follows by osmosis. A single stimulus such as a hot needle is enough to cause all the leaves to respond. There must be some means of communication in order for this to occur, and in fact a nerve impulse–type stimulus has been recorded in these plants!

A Venus's-flytrap has three sensitive hairs at the base of the trap, and if these are touched by an insect, a nerve impulse–type stimulus brings about a closing of the trap. This, too, is due to turgor pressure changes in the leaf cells that form the trap.

Some Plants Have Sleep Movements

A *sleep movement* is a nastic response that occurs daily in response to light and dark changes. One of the most common examples occurs in a houseplant called the prayer plant because at night the leaves fold upward into a shape resembling that of hands at prayer (Fig. 38.5). This movement is also due to changes in the turgor pressure of motor cells in a pulvinus located at the base of each leaf.

An Internal Clock Tells Time Organisms exhibit periodic fluctuations that correspond to environmental changes. For example, your temperature and blood pressure tend to change with the time of day, and you become sleepy at a certain time of night. The prayer plant just mentioned displays rhythmic "sleep" behavior (Fig. 38.5). A biological rhythm with a 24-hour cycle is called a **circadian rhythm** [L. *circum,* about, and *dies,* day].

Figure 38.4 Seismonastic movement.
A leaf of the sensitive plant, *Mimosa pudica*, before and after it is touched.

Figure 38.5 Sleep movement.
Prayer plant, *Maranta leuconeura*, before dark and after dark when the leaves fold up.

Circadian rhythms tend to persist, even if the appropriate environmental cues are no longer present. For example, on a transcontinental flight, you will likely suffer jet lag, and it will take several days to adjust to the time change because your body will still be attuned to the day-night pattern of your previous environment. The internal mechanism by which a biological rhythm is maintained in the absence of appropriate environmental stimuli is termed a **biological clock.** Typically, if organisms are sheltered from environmental stimuli, their circadian rhythms continue, but the cycle extends. In prayer plants, for example, the sleep cycle changes to 26 hours. Therefore, it is believed that biological clocks are synchronized by external stimuli to 24-hour rhythms. The length of daylight compared to the length of darkness, called the photoperiod, sets the clock. Temperature has little or no effect. This is adaptive because the photoperiod indicates seasonal changes better than temperature changes. Spring and fall, in particular, can have both warm and cold days.

There are other examples of circadian rhythms in plants. Stomates and certain flowers usually open in the morning and close at night, and some plants secrete nectar at the same time of the day or night.

38.2 Plant Hormones Coordinate Responses

In order for plants to respond to stimuli, the activities of plant cells and structures have to be coordinated. Almost all communication in a plant is done by **hormones** [Gk. *hormao,* instigate], chemical messengers produced in very low concentrations and active in another part of the organism. A particular response is probably influenced by several hormones and most likely requires a specific ratio of two or more hormones. Hormones are synthesized or stored in one part of the plant, but will travel within phloem or from cell to cell after reception of the appropriate stimulus.

Each naturally occurring hormone has a specific chemical structure. Other chemicals, some of which differ only slightly from the natural hormones, also affect the growth of plants. These and the naturally occurring hormones are sometimes grouped together and called plant growth regulators. The various uses of plant growth regulators are discussed in the reading on page 684.

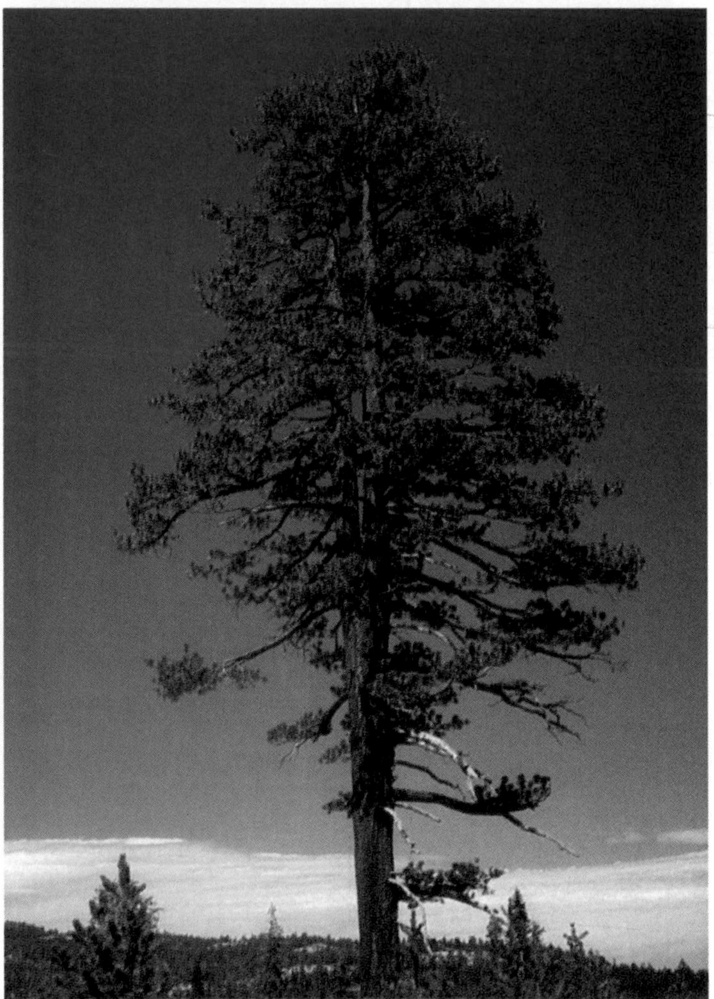

Figure 38.6 Apical dominance.
If the terminal bud is removed, a plant becomes fuller because apical dominance has been removed and axillary buds are free to develop into branches.

Auxin and Its Many Effects

The most common naturally occurring **auxin** [Gk. *auximos,* promoting growth] is indoleacetic acid (IAA). It is produced in shoot apical meristem and is found in young leaves and in flowers and fruits. Therefore, you would expect that auxin would affect many aspects of plant growth and development. Apically produced auxin prevents the growth of axillary buds, a phenomenon called apical dominance. When a terminal bud is removed deliberately or accidentally, the nearest axillary buds begin to grow, and the plant branches. To achieve a fuller look, one generally prunes the top (apical meristem) of the plant. This removes apical dominance and causes more branching of the main body of the plant (Fig. 38.6).

The application of a weak solution of auxin to a woody cutting causes roots to develop. Auxin production by seeds also promotes the growth of fruit. As long as auxin is concentrated in leaves or fruits rather than in the stem, leaves and fruits do not fall off. Therefore, trees can be sprayed with auxin to keep mature fruit from falling to the ground.

As discussed earlier, auxin is also involved in gravitropism and phototropism. After gravity has been perceived, auxin moves to the lower surface of roots and stems. Thereafter, roots curve downward and stems curve upward (see Fig. 38.2). The role of auxin in the positive phototropism of stems has been studied for quite some time. The experimental material of choice has been oat seedlings with coleoptile intact. A **coleoptile** is a protective sheath for the young leaves of the seedling. In 1881, the Darwins found that phototropism will not occur if the tip of the seedling is cut off or is covered by a black cap. They concluded that some influence that causes curvature is transmitted from the coleoptile tip to the rest of the shoot.

In 1926 Frits W. Went cut off the tips of coleoptiles and placed them on agar (a gelatinlike material). Then he placed an agar block to one side of a tipless coleoptile and found that the shoot would curve away from that side. The bending occurred even though the seedlings were not exposed to light (Fig. 38.7). Went concluded that the agar blocks contained a chemical that had been produced by the coleoptile tips. It was this chemical, he decided, that had caused the shoots to curve. He named the chemical substance auxin after the Greek word *aux,* which means to grow.

How Auxin Works

When a plant is exposed to unidirectional light, auxin moves to the shady side where it binds to receptors and activates the ATP-driven proton pump (Fig. 38.8). As hydrogen ions (H^+) are being pumped out of the cell, the cell wall becomes acidic, breaking hydrogen bonds. Cellulose fibrils are weakened, and activated enzymes further degrade the cell wall. The electrochemical gradient established by the proton pump causes solutes to enter the cell and water follows by osmosis. The turgid cell presses against the cell wall, stretching it so that elongation occurs. Auxin-mediated elongation is observed in younger, as opposed to more mature, cells. Perhaps older cells lack auxin receptors.

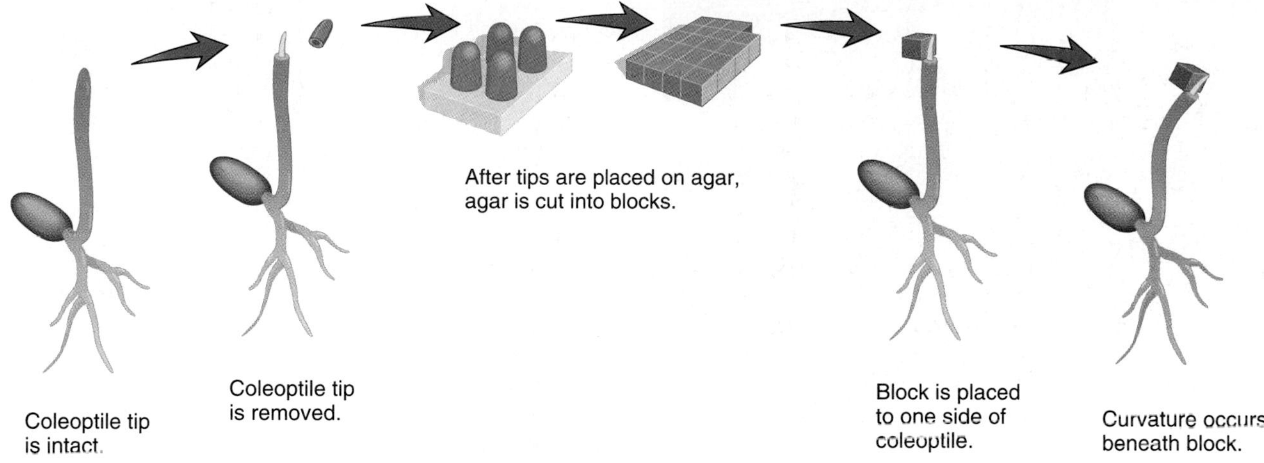
After tips are placed on agar, agar is cut into blocks.
Coleoptile tip is intact.
Coleoptile tip is removed.
Block is placed to one side of coleoptile.
Curvature occurs beneath block.

Figure 38.7 Demonstrating phototropism.
Oat seedlings are protected by a hollow sheath called a coleoptile. After a tip is removed and placed on agar, an agar block placed on one side of the coleoptile can cause it to curve.

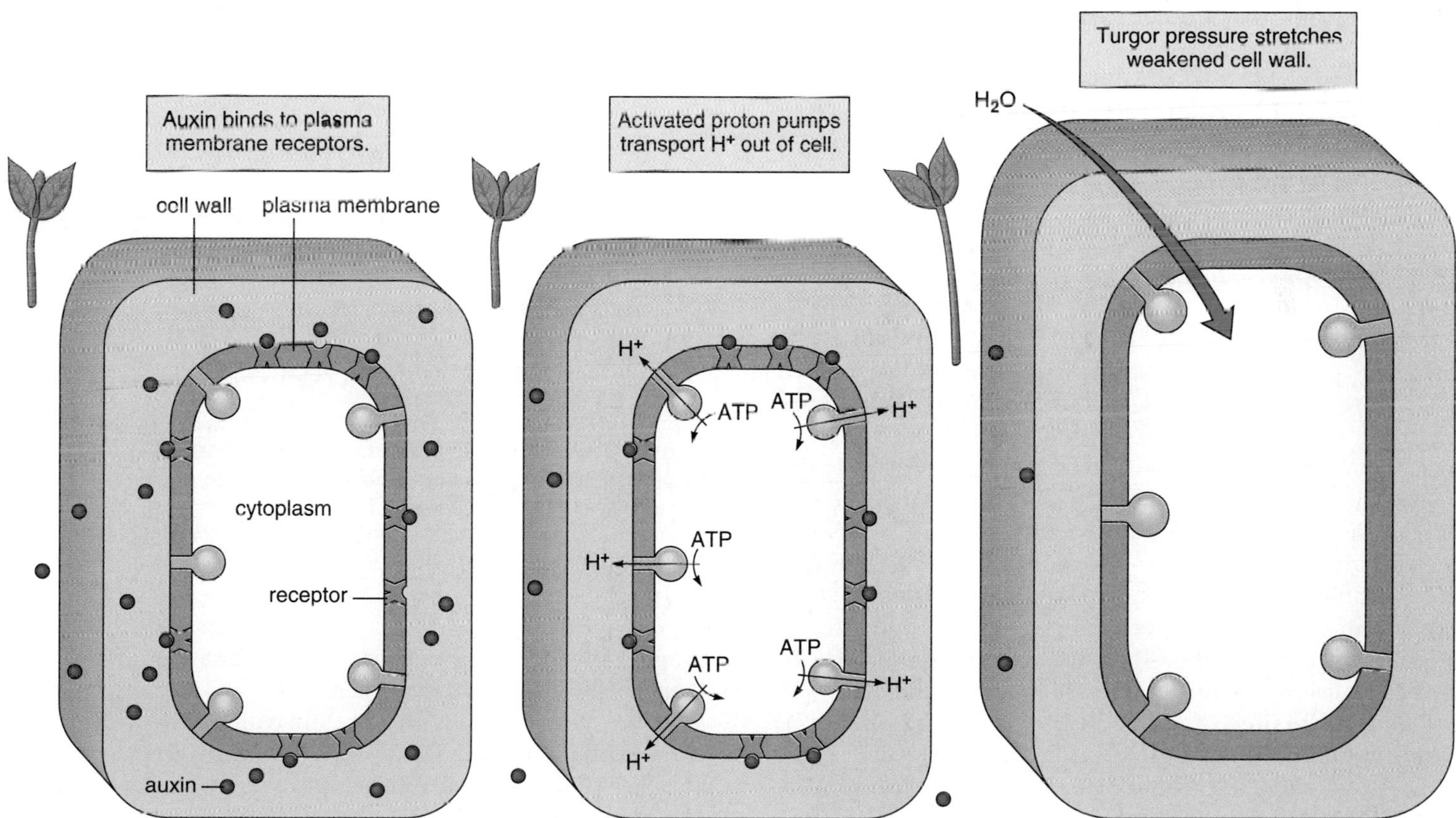
Auxin binds to plasma membrane receptors.
cell wall
plasma membrane
cytoplasm
receptor
auxin
Activated proton pumps transport H⁺ out of cell.
H⁺
ATP
Turgor pressure stretches weakened cell wall.
H₂O

Figure 38.8 Auxin mode of action.
After auxin binds to a receptor, the combination stimulates the proton pump so that hydrogen ions (H^+) are transported out of the cell. The resulting acidity causes the cell wall to weaken, and the electrochemical gradient causes solutes to enter the cell. Water follows by osmosis and the cell elongates.

Figure 38.9 Effect of gibberellins.
The *Cyclamen* plant on the right was treated with gibberellins; the plant on the left was not treated. Gibberellins are often used to promote stem elongation in economically important plants, but the exact mode of action still remains unclear.

Gibberellins and Stem Elongation

We know of about 70 gibberellins that chemically differ only slightly. The most common of these is GA_3 (the subscript designation distinguishes it from other gibberellins). **Gibberellins** [L. = dim. of *gibbus,* bent] are growth promoters that bring about elongation of the resulting cells. When gibberellins are applied externally to plants, the most obvious effect is *stem elongation* (Fig. 38.9). Gibberellins can cause dwarf plants to grow, cabbage plants to become 2 meters tall, and bush beans to become pole beans.

Gibberellins were discovered in 1926, the same year that Went performed his classic experiments with auxin. Ewiti Kurosawa, a Japanese scientist, was investigating a fungal disease of rice plants called "foolish seedling disease." The plants elongated too quickly, causing the stem to weaken and the plant to collapse. Kurosawa found that the fungus infecting the plants produced an excess of a chemical he called gibberellin, named after the fungus *Gibberella fujikuroi.* It wasn't until 1956 that gibberellic acid was isolated from a flowering plant rather than from a fungus. Sources of gibberellin in flowering plant parts are young leaves, roots, embryos, seeds, and fruits.

The **dormancy** of seeds and buds can be broken by applying gibberellins, and research with barley seeds has shown how GA_3 acts as a chemical messenger. Barley seeds have a large, starchy endosperm, which must be broken down into sugars to provide energy for growth. After the embryo produces gibberellins, amylase, an enzyme that breaks down the starch, appears in cells just inside the seed coat. It is hypothesized that GA_3 (the first chemical messenger) attaches to a receptor in the plasma membrane, and then a second messenger inside the cell, namely calcium ions (Ca^{2+}), combines with a protein called calmodulin. This complex is believed to activate the gene that codes for amylase (Fig. 38.10). Amylase then acts on starch to release sugars used as a source of energy by the growing embryo.

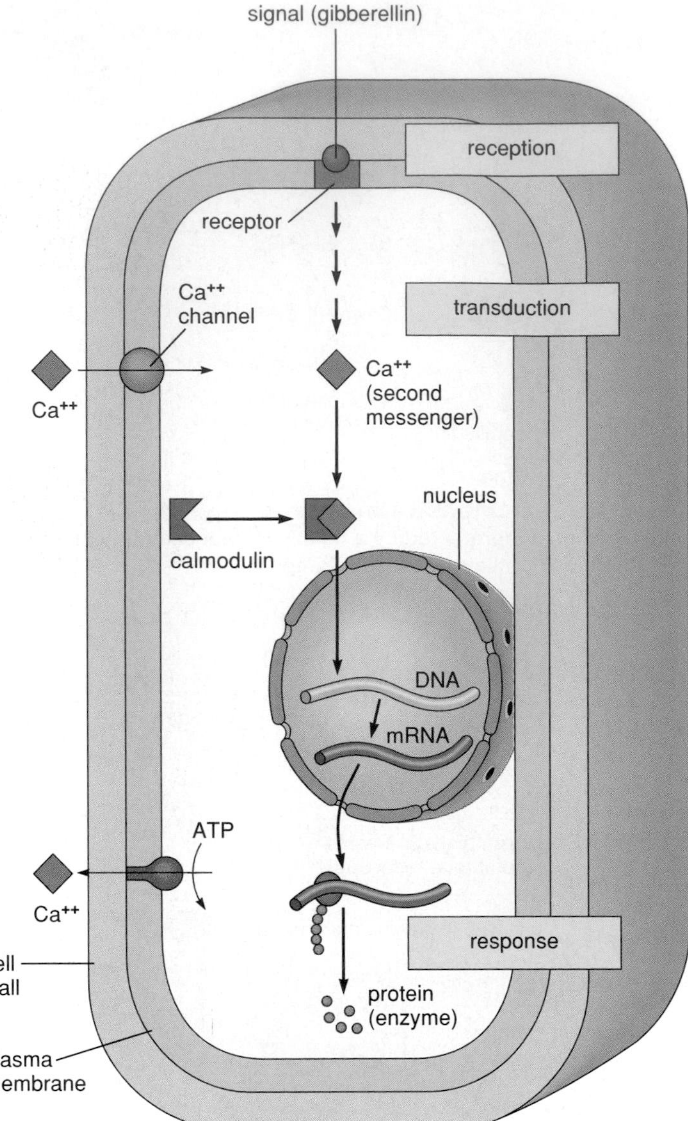

Figure 38.10 Gibberellin mode of action.
It appears that the hormone (gibberellin) binds to a receptor, and a second messenger (Ca^{2+}) inside the cell activates a protein (calmodulin) and the complex binds to DNA. Thereafter, an enzyme is produced.

The action of hormones on the cellular level is another example of the reception-transduction-response pathway described at the beginning of this chapter. The following reading describes the work of two investigators in this area.

Doing Science

Husband and Wife Team Explores Signal Transduction in Plants

Plants perceive and react to a variety of environmental stimuli. Some examples include light intensity and quality, gravity, carbon dioxide levels, pathogen infection, drought, and touch. Plant responses may be short term, such as the rapid localized cell death that occurs at the site of infection by a pathogen. This rapid cell death, termed the hypersensitive response, serves to limit the further progression of infection. Another form of short-term response is the opening or closing of leaf stomates in response to light levels. High light causes the stomates to open, while low light or darkness causes them to close. In this way, the supply of carbon dioxide to the leaves via the stomates is coordinated with light-driven photosynthetic reactions occurring in leaf cells. On the other hand, plant responses to environmental stimuli may be more long term and involve actual plant growth. An example is the plant response to gravity (gravitropism), which results in the downward growth of the root and the upward growth of the stem. In each of these situations, the coupling of a stimulus to an appropriate response represents signal transduction.

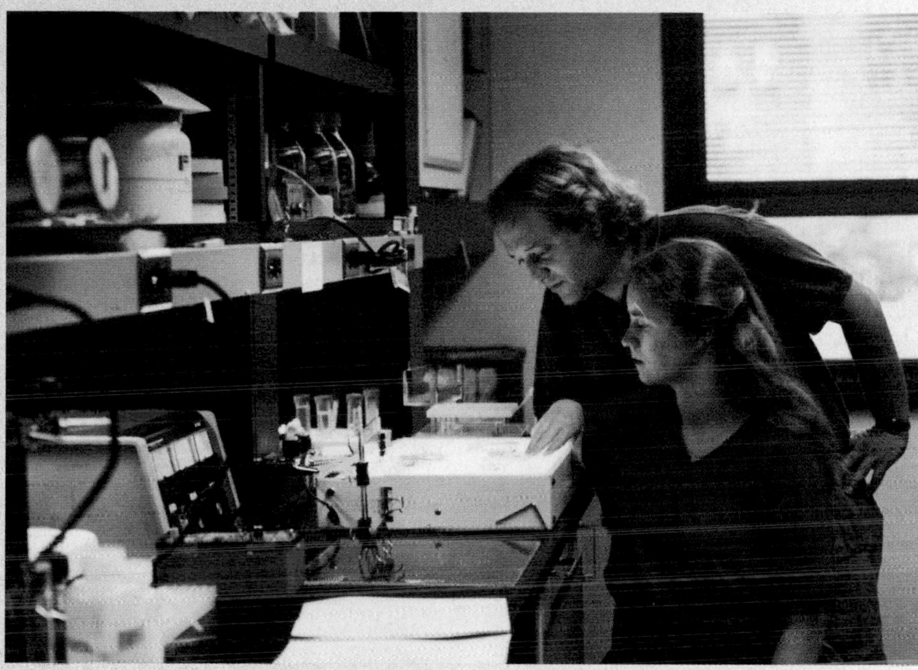

Donald Briskin and Margaret Gawienowski
University of Illinois at Urbana-Champaign

Although we know that plants do not have a nervous system and a brain, some system or process must be present in these organisms to allow the perception of a stimulus and its conversion to an appropriate response. While we are both interested in attempting to understand this process, we approach this complex research area from different perspectives related to our training, expertise, and interests.

As a plant biochemist, Donald focuses on how membrane transport processes are involved in signal transduction. Here, research has shown that membrane transport of calcium ions (Ca^{2+}) can play a fundamental role in the signal transduction process. When a stimulus is perceived by a plant cell, a transient increase in cytoplasmic Ca^{2+} concentration occurs due to the rapid opening of Ca^{2+} channels at plant membranes, which operate not unlike Ca^{2+} channels present in animal membranes. This transient increase in cytoplasmic Ca^{2+} acts as a sort of message linking the stimulus to a response at the biochemical level. In particular, Donald studies how Ca^{2+}-transporting enzymes (called Ca^{2+}-ATPases) and Ca^{2+} channels are coordinated after a stimulus is received.

Just how a single type of message—such as a rapid change in cytoplasmic Ca^{2+}—can elicit so many different types of specific plant responses is Margaret's interest. As a plant molecular biologist, Margaret focuses on how proteins in the plant cytoplasm that bind Ca^{2+} might impart specificity to the response. Of particular interest to her is a Ca^{2+}-binding protein called calmodulin, which controls the activity of a number of other proteins once it binds to Ca^{2+}. Her work has shown that calmodulin in plants is encoded by multiple genes that produce variants of the protein called isoforms. The isoforms are present to a different extent in different plant tissues, have different properties (Ca^{2+} binding, target protein regulation), and may play a role in determining which responses are coupled to transient cytoplasmic Ca^{2+} changes. Margaret is also interested in using a molecular approach to find more calmodulin-binding proteins that may be associated with various plant membranes.

Although our approaches toward understanding plant signal transduction are very different, the work of each of us tends to complement the other's. We view each other's work from a slightly different perspective, and this helps generate ideas for new experiments and strategies. Furthermore, we work together in the same laboratory at the University of Illinois at Urbana-Champaign, and this leads to much discussion and exchange of opinions. For us, research is not only challenging and exciting, but it keeps us wondering what the next day will bring.

A Closer Look

Plant Growth Regulators

Now that the formulas for many plant hormones are known, it is possible to synthesize them, as well as related chemicals, in the laboratory. Collectively, these substances are known as plant growth regulators. Many scientists hope plant growth regulators will bring about an increase in crop yield, just as fertilizers, irrigation, and pesticides have done in the past.

Since auxin was first discovered, researchers have found many agricultural and commercial uses for it. Auxins cause the base of stems to form new roots quickly, so that new plants are easily started from cuttings. When sprayed on apples, pears, and citrus fruits shortly before harvest, auxins prevent fruit from dropping too soon. Because auxins inhibit the growth of lateral buds, potatoes sprayed with an auxin do not sprout and thus have a longer storage life.

In high concentrations, some auxins are used widely in agriculture as **herbicides** to prevent the growth of broad-leaved plants. In addition to their use in weed control, the synthetic auxins known as 2,4D and 2,4,5T were used as defoliants during the Vietnam War. Even though 2,4D has been used for over 35 years, we do not yet know how it works. Apparently, it is structurally different enough from natural auxin that a plant's own enzymes cannot break it down. Its concentration rises until metabolism is disrupted, cellular order is lost, and the cells die.

The other plant hormones studied in this chapter also have agricultural and commercial uses. Gibberellins are used to stimulate seed germination and seedling growth of some grains, beans, and fruits. They also increase the size of some mature plants. Treatment of sugarcane with as little as 2 oz per acre increases the cane yield by more than 5 metric tons. The application of either auxins or gibberellins can cause an ovary and accessory flower parts to develop into fruit, even though pollination and fertilization have not taken place. In this way, it is sometimes possible to produce seedless fruits and vegetables or bigger, more uniform bunches with larger fruit, as shown in Figure 38A.

Figure 38A Effect of gibberellin.
Control Thompson seedless grapes, *Vitis vinifera,* are at left. GA_3 was sprayed at bloom and at fruit set on grapes at right. Almost all grapes sold in stores are now treated with gibberellin.

Because cytokinins retard the aging of leaves and other organs, they are sprayed on vegetables to keep them fresh during shipping and storage. Such treatment of holly, for example, allows it to be harvested many weeks prior to a holiday.

Plant growth regulators are used in tissue culture when seedlings are grown from a few cells in laboratory glassware. New varieties of food crops with particular characteristics—such as tolerance to herbicides and insects—are being developed by genetically engineering the original cells. It may one day be possible to endow most crops with the ability to utilize atmospheric nitrogen and/or make a wider range of proteins.

Several synthetic inhibitors are used to oppose the action of auxins, gibberellins, and cytokinins normally present in plants. Some of these can cause leaf and fruit drop at a time convenient to the farmer. Removing leaves from cotton plants aids in their harvest, and thinning the fruit of young fruit trees results in larger fruit as the trees mature. Retarding the growth of other plants sometimes increases their hardiness. For example, an inhibitor has been used to reduce stem length in wheat plants, so that the plants do not fall over in heavy winds and rain.

The commercial uses of ethylene were greatly increased with the development of ethylene-releasing compounds. Ethylene gas is injected into airtight storage rooms to ripen bananas, honeydew melons, and tomatoes. It will also degreen oranges, lemons, and grapefruit when the rind would otherwise remain green because of a high chlorophyll level. When sprayed on certain fruit and nut crops, ethylene increases the chances that the fruit will detach when the trees are shaken at harvest time.

Today, fields and orchards are often sprayed with synthetic growth regulators.

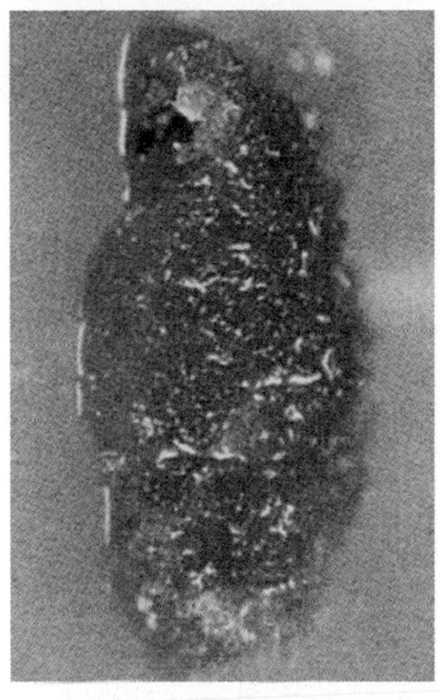
a.

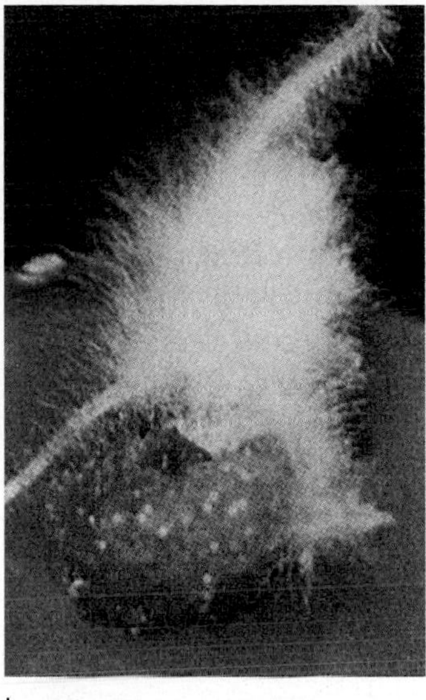
b.

c.

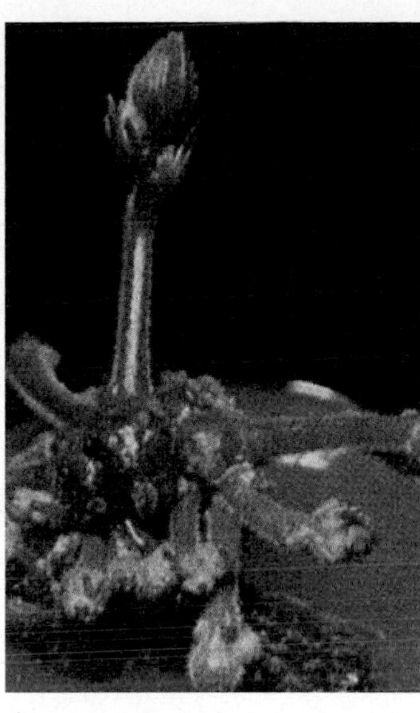
d.

Figure 38.11 Interaction of hormones.
Tissue culture experiments have revealed that auxin and cytokinin interact to affect differentiation during development. **a.** In tissue culture that has the usual amounts of these two hormones, tobacco strips develop into a callus of undifferentiated tissue. **b.** If the ratio of auxin to cytokinin is appropriate, the callus produces roots. **c.** Change the ratio, and vegetative shoots and leaves are produced. **d.** Yet another ratio causes floral shoots. It is now clear that plant hormones rarely act alone; it is the relative concentrations of these hormones that produce an effect. The modern emphasis is to look for an interplay of hormones when a growth response is studied.

Cytokinins and Cell Division

The **cytokinins** [Gk. *kytos*, cell, and *kineo*, move] are a class of plant hormones that promote cell division; *cytokinesis means cell division*. These substances are derivatives of adenine, one of the purine bases in DNA and RNA. A naturally occurring cytokinin called *zeatin* has been extracted from corn kernels. Kinetin, a synthetic cytokinin, also promotes cell division.

The cytokinins were discovered as a result of attempts to grow plant tissue and organs in culture vessels in the 1940s (Fig. 38.11). It was found that cell division occurs when coconut milk (a liquid endosperm) and yeast extract are added to the culture medium. Although the effective agent or agents could not be isolated, they were collectively called cytokinins. Not until 1967 was the naturally occurring cytokinin called zeatin isolated from coconut milk. Cytokinins have been isolated from various seed plants, where they occur in the actively dividing tissues of roots and also in seeds and fruits.

Plant tissue culturing is now common practice, and researchers are well aware that the ratio of auxin to cytokinin and the acidity of the culture medium determine whether the plant tissue forms an undifferentiated mass, called a callus, or differentiates to form roots, vegetative shoots, leaves, or floral shoots (Fig. 38.11). Researchers have reported that chemicals they called oligosaccharins (chemical fragments released from the cell wall) are also effective in directing differentiation. They hypothesize that auxin and cytokinins are a part of a reception-transduction-response pathway (see Fig. 38.10), which leads to the activation of enzymes that release these fragments from the cell wall. Perhaps all plant hormones have a similar effect

Cytokinins Affect Leaves

When a plant organ, such as a leaf, loses its natural color, it is most likely undergoing an aging process called **senescence.** During senescence, large molecules within the leaf are broken down and transported to other parts of the plant. Senescence does not always affect the entire plant at once; for example, as some plants grow taller, they naturally lose their lower leaves. It has been found that senescence of leaves can be prevented by the application of cytokinins. Not only can cytokinins prevent death of leaves, they can also initiate leaf growth. Lateral buds begin to grow despite apical dominance when cytokinin is applied to them.

Cytokinins promote cell division, prevent senescence, and initiate growth. The interaction of hormones is well exemplified by the effect of varying ratios of auxin and cytokinins on differentiation of plant tissues.

Doing science

Arabidopsis thaliana, the Valuable Weed

A powerful new tool is available for the study of plant growth and development. *Arabidopsis thaliana* is a weed that few people knew of 15 years ago. But now it is being studied in hundreds of laboratories all over the world because it lends itself to genetic manipulation (Fig. 38B). Adult plants can be grown from a few cells in tissue culture and, after a short generation time, adult plants will produce 10,000 seeds each in only four to six weeks. Because of its small size, dozens of plants can grow in a single pot. *A. thaliana* has a small genome; there are only five pairs of chromosomes and 100 million nucleotides total. In contrast, tobacco plants, which often were used as experimental material in the past, probably have the same number of genes but 15 times the number of nucleotides. This excess noncoding DNA makes it difficult to find the genes and determine what they do.

There are many natural mutants of *Arabidopsis,* and much can be determined by studying them. In one instance, a dwarf plant was found to be deficient in the amount of gibberellin-producing enzymes. Since the plant showed a lack of internodal elongation but showed normal development of leaves and flowers, it was known that gibberellins were affecting only internodal elongation. Many additional mutants can be produced because these plants are susceptible to infection by *Agrobacterium tumefaciens,* the bacterium with a plasmid that inserts into plant genomes. On occasion the plasmid inserts itself into the middle of a normal gene, disrupting it and preventing it from functioning as it should. Researchers have used this method to determine that there are three classes of genes essential to normal floral pattern formation. Triple mutants that lack all types of genetic activities have flowers that consist entirely of leaves arranged in whorls (compare Fig. 38C and D). These floral organ identity genes appear to be transcriptional regulators that are expressed and required for extended periods.

An international collaborative effort is under way to map the *Arabidopsis* genome and sequence each of its genes. Once the genes of *Arabidopsis* are known, they can be used to identify the same genes in crop plants. It is reasonable to expect that plant cells function similarly; therefore, it is likely we will learn how to improve crops by studying *Arabidopsis.*

Figure 38B Overall appearance of *Arabidopsis thaliana.* Many investigators have turned to this weed as an experimental material to study the actions of genes, including those that control growth and development.

Figure 38C *Arabidopsis thaliana* flower. The structure of the flower is determined in part by three classes of floral organ identity genes.

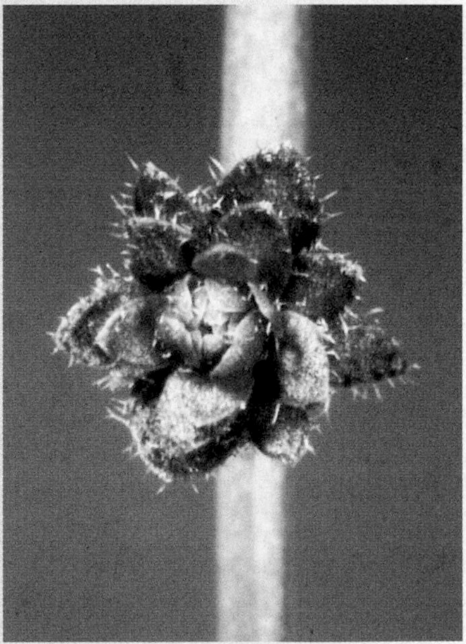

Figure 38D Mutated flower. Mutations affecting all three classes of floral organ identity genes result in flowers that contain leaves arranged in whorls.

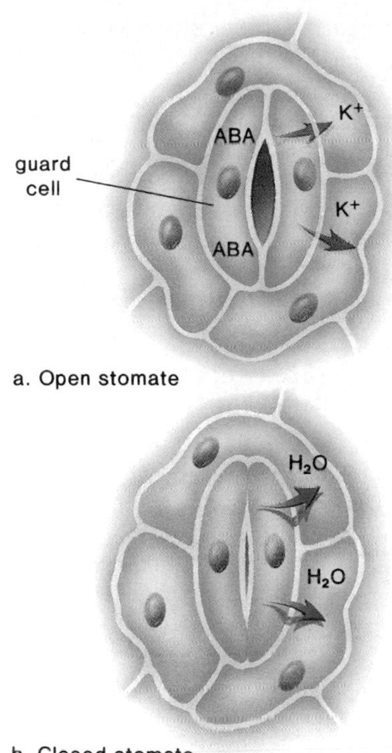

Figure 38.12 Control of stomate opening.
The hormone abscisic acid (ABA) is believed to bring about the closing of the stomates when water stress occurs. First potassium ions (K^+) exit and then water (H_2O) also leaves guard cells.

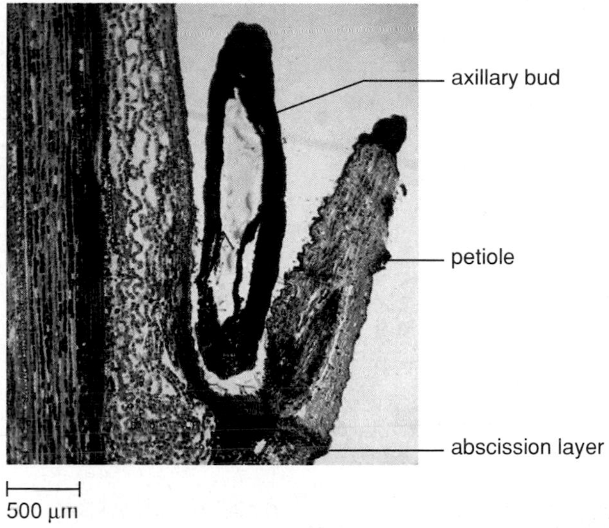

Figure 38.13 Abscission.
Before a leaf falls, a special band of cells called the abscission layer develops at the base of the petiole, the leaf stem. Here the hormone ethylene promotes the breakdown of plant cell walls so that the leaf finally falls. Here also, a layer forms a leaf scar, which protects the plant from possible invasion by microorganisms.

Plant Hormones That Inhibit

In contrast to the hormones discussed so far, the plant hormones abscisic acid and ethylene inhibit growth.

Abscisic Acid Is the Stress Hormone

Abscisic acid (ABA) is sometimes called the stress hormone because it initiates and maintains seed and bud dormancy and brings about the closure of stomates. *Dormancy* occurs when a plant organ readies itself for adverse conditions by stopping growth (even though conditions at the time are favorable for growth). For example, it is believed that abscisic acid moves from leaves to vegetative buds in the fall, and thereafter these buds are converted to winter buds. A winter bud is covered by thick and hardened scales. A reduction in the level of abscisic acid and an increase in the level of gibberellins are believed to break seed and bud dormancy. Then seeds germinate and buds send forth leaves.

Abscisic acid brings about the closing of stomates when a plant is under water stress (Fig. 38.12). In some unknown way, abscisic acid causes potassium ions (K^+) to leave guard cells. Thereafter, the guard cells lose water and the stomates close. Although the external application of abscisic acid promotes abscission, this hormone is no longer believed to function naturally in this process. Instead, the hormone ethylene is believed to bring about abscission (Fig. 38.13). Abscisic acid is produced by any "green tissue" with chloroplasts, monocot endosperm, and roots.

Ethylene Is for Fruit Ripening

In the early 1900s, it was common practice to prepare citrus fruits for market by placing them in a room with a kerosene stove; only later did researchers realize that an incomplete combustion product of kerosene, namely **ethylene,** ripens fruit. It does so by increasing the activity of enzymes that soften fruits. For example, it stimulates the production of *cellulase,* an enzyme that hydrolyzes cellulose in plant cell walls. Because it is a gas, ethylene moves freely through the air—a barrel of ripening apples can induce ripening of a bunch of bananas even some distance away. Ethylene is released at the site of a wound due to physical damage or infection (which is why one rotten apple spoils the whole bunch).

The presence of ethylene in the air inhibits the growth of plants in general; homeowners who use natural gas to heat their homes sometimes report difficulties in growing houseplants. Ethylene is also present in automobile exhaust, and it's possible that plant growth is inhibited by the exhaust that enters the atmosphere. It only takes one part of ethylene per 10 million parts of air to bring about inhibition of plant growth.

Ethylene is also involved in **abscission** [L. *abscissus,* cut off], the dropping of leaves, fruits, and flowers from a plant. Lower levels of auxin and perhaps gibberellin in these areas of the plant (compared to the stem) probably initiate abscission (Fig. 38.13). But once the process of abscission has begun, ethylene stimulates such enzymes as cellulase, which cause leaf, fruit, or flower drop.

38.3 The Photoperiod Controls Seasonal Changes

Many physiological changes in plants are related to a seasonal change in day length. Such changes include seed **germination,** the breaking of bud dormancy, and the onset of senescence. A physiological response prompted by changes in the length of day or night is called **photoperiodism** [Gk. *photos,* light, and *peroidus,* completed course]. In some plants, photoperiodism influences flowering: violets and tulips flower in the spring, and asters and goldenrods flower in the fall.

In the 1920s, when U.S. Department of Agriculture scientists began to study photoperiodism in more detail, they decided to grow plants in a greenhouse, where they could artificially alter the photoperiod. This work led them to conclude that plants can be divided into three groups:

1. **Short-day plants**—flower when the day length is shorter than a critical length (examples are cocklebur, poinsettia, and chrysanthemum).
2. **Long-day plants**—flower when the day length is longer than a critical length (examples are wheat, barley, clover, and spinach).
3. **Day-neutral plants**—flowering is not dependent on day length (examples are tomato and cucumber).

Further, we should note that both a long-day plant and a short-day plant can have the same critical length (Fig. 38.14). Spinach is a long-day plant that has a critical length of 14 hours; ragweed is a short-day plant with the same critical length. Spinach, however, flowers in the summer when the day length increases to 14 hours or more, and ragweed flowers in the fall, when the day length shortens to 14 hours or less. We now know that some plants may require a specific sequence of day lengths in order to flower.

In 1938, K. C. Hammer and J. Bonner began to experiment with artificial lengths of light and dark that did not necessarily correspond to a normal 24-hour day. They discovered that the cocklebur, a short-day plant, flowers as long as the dark period is continuous for 8.5 hours, regardless of the length of the light period. Further, if this dark period is interrupted by a brief flash of white light, the cocklebur does not flower. (Interrupting the light period with darkness has no effect.) Similar results have also been found for long-day plants. They require a dark period that is shorter than a critical length, regardless of the length of the light period. If a slightly longer-than-critical-length night is interrupted by a brief flash of light, however, long-day plants flower. We must conclude, then, that the length of the dark period controls flowering, not the length of the light period. Of course, in nature, short days always go with long nights and vice versa.

> **Short-day plants require a period of darkness that is longer, and long-day plants require a period of darkness that is shorter, than a critical length to flower.**

Phytochrome and Plant Flowering

If flowering is dependent on day and night length, plants must have some way to detect these periods. Many years of research by U.S. Department of Agriculture scientists led to the discovery of a plant pigment called phytochrome. **Phytochrome** [Gk. *phyton,* plant, and *chroma,* color] is a blue-green leaf pigment that alternately exists in two forms. As Figure 38.15 indicates:

P_r (phytochrome red) absorbs red light (of 660 nm wavelength) and is converted to P_{fr}.

P_{fr} (phytochrome far-red) absorbs far-red light (of 730 nm wavelength) and is converted to P_r.

Direct sunlight contains more red light than far-red light; therefore, P_{fr} is apt to be present in plant leaves during the day. In the shade and at sunset, there is more far-red light than red light; therefore, P_{fr} is converted to P_r as

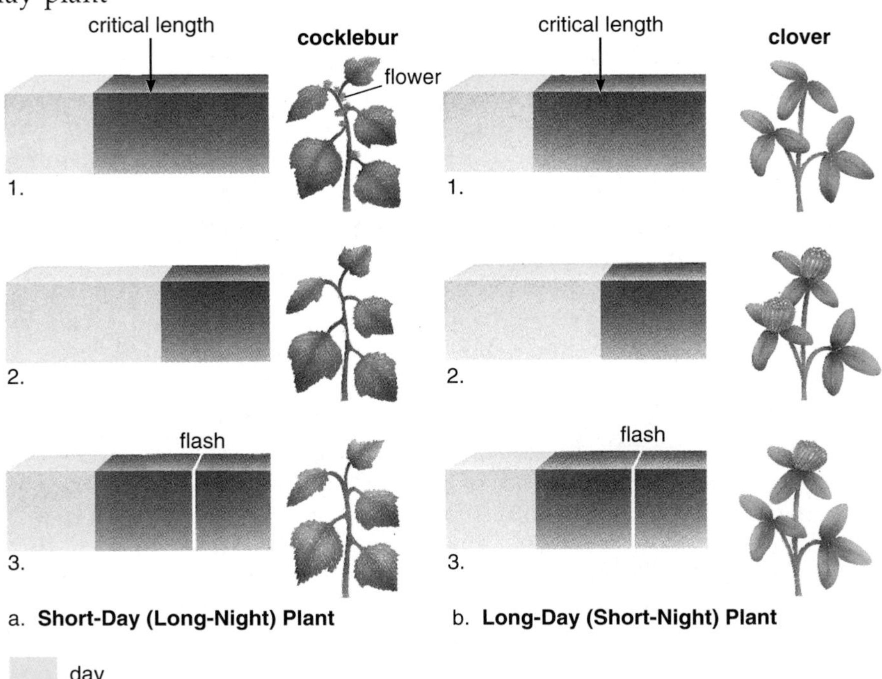

Figure 38.14 Photoperiodism and flowering.
a. Short-day plant. (1) When the day is shorter than a critical length, this type of plant flowers. (2) The plant does not flower when the day is longer than the critical length. (3) It also does not flower if the longer-than-critical-length night is interrupted by a flash of light. **b.** Long-day plant. (1) When the day is shorter than a critical length, this type plant does not flower. (2) The plant flowers when the day is longer than a critical length. (3) It also flowers if the slightly longer-than-critical-length night is interrupted by a flash of light.

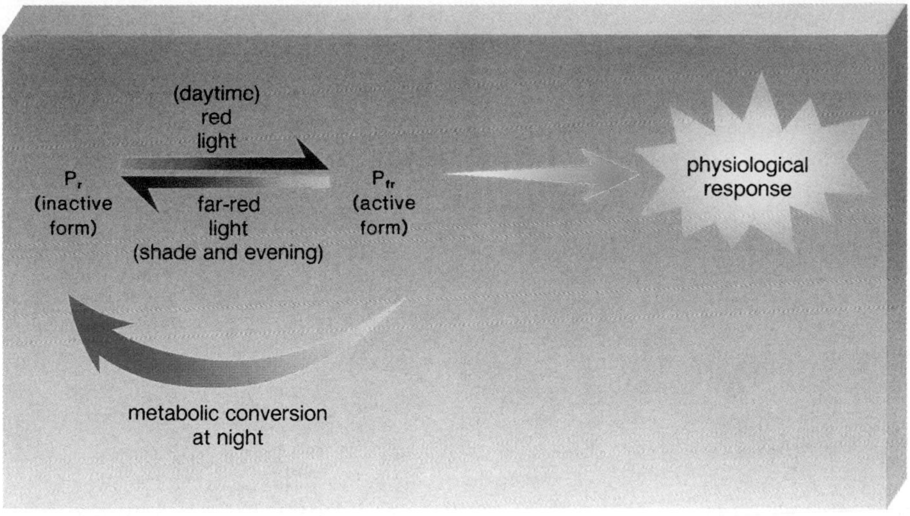

Figure 38.15 Phytochrome conversion cycle.
The inactive form P_r is prevalent during the night. At sunset or in the shade, when there is more far-red light, P_{fr} is converted to P_r. Also during the night, metabolic processes cause P_{fr} to be converted into P_r. P_{fr}, the active form of phytochrome, is prevalent during the day because at that time there is more red light than far-red light.

night approaches. There is also a slow metabolic replacement of P_{fr} by P_r during the night.

It is possible that phytochrome conversion is the first step in a reception-transduction-response pathway that results in flowering. At one time, researchers hypothesized that there was a special flowering hormone called florigen, but such a hormone has never been discovered.

Phytochrome alternates between two forms (P_{fr} during the day and P_r during the night), and this conversion allows a plant to detect photoperiod changes.

Phytochrome Has Other Functions

The $P_r \rightleftharpoons P_{fr}$ conversion cycle is now known to control other growth functions in plants. It promotes seed germination and inhibits stem elongation, for example. The presence of P_{fr} indicates to some seeds that sunlight is present and conditions are favorable for germination. This is why some seeds must be partly covered with soil when planted. Germination of other seeds is inhibited by light, so they must be planted deeper. Following germination, the presence of P_r indicates that stem elongation may be needed to reach sunlight. Seedlings that are grown in the dark etiolate; that is, the stem increases in length and the leaves remain small (Fig. 38.16). Once the seedling is exposed to sunlight and P_r is converted to P_{fr}, the seedling begins to grow normally—the leaves expand and the stem branches. It has now been shown that phytochrome in the P_{fr} form leads to the activation of one or more regulatory proteins in the cytosol. These proteins migrate to the nucleus where they bind to so-called "light-stimulated" genes that code for proteins found in chloroplasts.

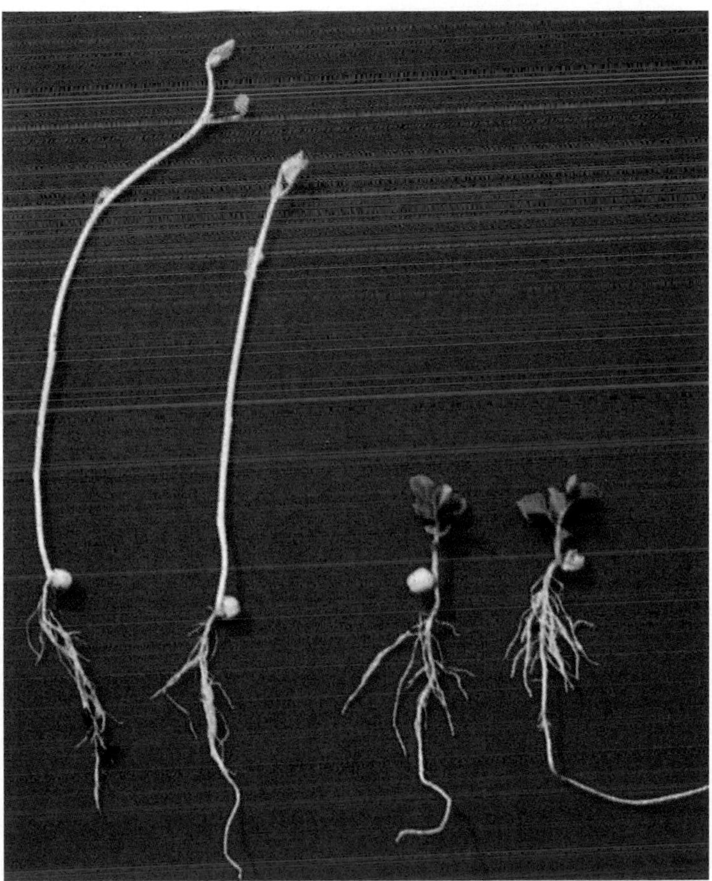

Figure 38.16 Phytochrome control of growth pattern.
If far-red light is prevalent, as it is in the shade, the stem of a seedling elongates but the leaves remain small *(left).* However, if red light is prevalent, as it is in bright sunlight, the stem does not elongate but the leaves expand *(right).* These effects are due to phytochrome.

connecting concepts

Behavior in plants can be understood on three different levels of organization. On the species level, plant responses that promote survival and reproductive success have evolved through the process of natural selection. On the organismal level, hormones coordinate the growth and development of plant parts. And on the cellular level, hormones influence cellular metabolism. Let's take the example, why do plants bend toward the light? To provide an answer on the species level, consider that those plants that bend toward the light will be able to produce more organic food and be the ones to have more offspring. To provide an answer on the organismal level, consider that light causes certain plant parts to produce auxin, and when auxin moves from the lit side of a stem to the shady side, elongation occurs, and thereafter, the plant bends toward the light. To provide an answer on the cellular level, consider that after auxin is received by a plant cell, solutes and water enter the cell, causing its walls to stretch.

The response of both the organism and the cell involve three steps: (1) perception of the stimulus, (2) transduction of the stimulus, and (3) response to the stimulus. Can you designate these steps on the cellular level? Reception of auxin by plasma membrane receptors is the first step, entrance of solutes and water into the plant cell is the second step, and stretching of the cell wall is the third step.

If we were considering animals instead of plants, the same type of biological explanations would apply. Plants and animals and, indeed, all organisms share common ancestors, even back to the very first cell(s). You will recall that evolution explains both the unity and diversity of living things. Organisms are similar because they share common ancestors; they are different because they are adapted to different ways of life.

Summary

38.1 Plant Responses to Stimuli

Like animals, plants utilize a reception (or perception)-transduction-response pathway when they respond to a stimulus.

When plants respond to stimuli, growth and/or movement occurs. Tropisms are growth responses toward or away from unidirectional stimuli. The positive phototropism of stems results in a bending toward light, and the negative gravitropism of stems results in a bending away from the direction of gravity. Roots that bend toward the direction of gravity show positive gravitropism. Thigmotropism occurs when a plant part makes contact with an object, as when tendrils coil about a pole.

Nastic movements are not directional. Due to turgor pressure changes, some plants respond to touch and some perform sleep movements.

Plants exhibit circadian rhythms, which are believed to be controlled by a biological clock. The sleep movements of prayer plants, the closing of stomates, and the daily opening of certain flowers have a 24-hour cycle.

38.2 Plant Hormones Coordinate Responses

Both stimulatory and inhibitory hormones help control certain growth patterns of plants. There are hormones that stimulate growth (auxins, gibberellins, and cytokinins) and hormones that inhibit growth (abscisic acid and ethylene).

Auxin-controlled cell elongation is involved in phototropism and gravitropism. When a plant is exposed to light, auxin moves laterally from the bright to the shady side of a stem. There it activates the proton (H^+) pump, leading to the uptake of solutes and an increase in turgor pressure, and the result is elongation of cells on that side.

Gibberellin research indicates that after a hormone attaches to a plasma membrane receptor, a second chemical messenger inside the cell leads to transcription and protein synthesis.

Cytokinins cause cell division, the effects of which are especially obvious when plant tissues are grown in culture.

Abscisic acid (ABA) and ethylene are two plant growth inhibitors. ABA is well known for causing stomates to close, and ethylene is known for causing fruits to ripen.

38.3 The Photoperiod Controls Seasonal Changes

Photoperiodism is seen in some plants. For example, short-day plants flower only when the days are shorter than a critical length, and long-day plants flower only when the days are longer than a critical length. Actually, research has shown that it is the length of darkness that is critical. Interrupting the dark period with a flash of white light prevents flowering in a short-day plant and induces flowering in a long-day plant.

Phytochrome is a pigment that responds to both red and far-red light and is involved in flowering. Daylight causes phytochrome to exist as P_{fr}, but during the night, it is reconverted to P_r by metabolic processes. Phytochrome in the P_{fr} form leads to activation of regulatory proteins that bind to genes.

In addition to being involved in the flowering process, P_{fr} promotes seed germination, leaf expansion, and stem branching. When P_r dominates, the stem elongates and grows toward sunlight.

Reviewing the Chapter

1. List and describe three tropisms, and indicate the involvement of any hormones in these responses. 676–78
2. What are nastic responses, and how do turgor pressure changes bring about movements of plants? 678
3. What is a biological clock, how does it function, and what is its primary usefulness in plants? 678–79
4. Name three types of hormones that promote growth processes, and give several specific functions for each. 680–85
5. Why does removing a terminal bud cause a plant to get bushier? 680
6. What experiments led to knowledge that a hormone is involved in phototropism? Explain the mechanism by which auxin brings about elongation of cells. 680–81
7. What is the hypothesized mode of hormone action according to gibberellin research? 682
8. Give experimental evidence to suggest that hormones interact when they bring about an effect. 685
9. Name two types of hormones that inhibit growth processes, and give several specific functions for each. 687
10. Define photoperiodism, and discuss its relationship to flowering in certain plants. 688
11. What is the phytochrome conversion cycle, and what are some possible functions of phytochrome in plants? 688–89

Testing Yourself

Choose the best answer for each question.

For questions 1–5, match the items below with the hormones in the key.

Key:

a. auxin
b. gibberellin
c. cytokinin
d. ethylene
e. abscisic acid

1. One rotten apple can spoil the barrel.
2. Cabbage plants bolt (grow tall).
3. Stomates close when a plant is water stressed.
4. Sunflower plants all point toward the sun.
5. Coconut milk causes plant tissues to undergo cell division.
6. After unidirectional light is received, auxin moves to the shady side of a stem, and then the stem bends toward the light. Which step in this sequence represents transduction?
   a. sensitivity of a receptor to light
   b. auxin movement toward the shady side
   c. curving of stem toward light
   d. All of these are correct.
7. Which of these is a correct statement?
   a. Both stems and roots show positive gravitropism.
   b. Both stems and roots show negative gravitropism.
   c. Only stems show positive gravitropism.
   d. Only roots show positive gravitropism.
8. Short-day plants
   a. are the same as long-day plants.
   b. are apt to flower in the fall.
   c. do not have a critical photoperiod.
   d. All of these are correct.
9. A plant requiring a dark period of at least 14 hours will
   a. flower if a 14-hour night is interrupted by a flash of light.
   b. not flower if a 14-hour night is interrupted by a flash of light.
   c. not flower if the days are 14 hours long.
   d. Both b and c are correct
10. Phytochrome
    a. is a plant pigment.
    b. is present as P_{fr} during the day.
    c. activates regulatory proteins.
    d. All of these are correct.
11. Circadian rhythms
    a. require a biological clock.
    b. do not exist in plants.
    c. are involved in the tropisms.
    d. All of these are correct.
12. Label the following diagram. Explain what causes stomates to close when a plant is water stressed.

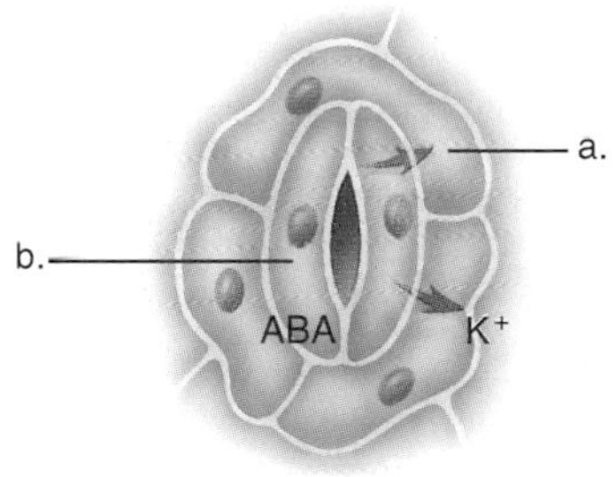

Applying the Concepts

1. *Physiological processes are coordinated in organisms.*
   Give an example to show that plant hormones are involved in coordinating physiological processes.
2. *Biological clocks help organisms maintain adaptive behaviors.*
   (a) Give examples to show that circadian rhythmic behavior in plants is adaptive. (b) How might flowering be controlled by a biological clock system?
3. *Reception-transduction-response pathways operate in all organisms.*
   Relate the elongation of cells brought about by auxin to this pathway, using Figure 38.10 as a guide.

Using Technology

Your study of growth and development in plants is supported by these available technologies:

Exploring the Internet

The Mader Home Page provides resources for and help with studying this chapter.

http://www.mhhe.com/sciencemath/biology/mader/
(Click on Biology.)

Life Science Animations Video

Video #5: Plant Biology/Evolution/Ecology
How Leaves Change Color and Drop in Fall (#49)

Understanding the Terms

Match the terms to these definitions:

a. __________ Biological rhythm with a 24-hour cycle.
b. __________ Directional growth of plants in response to light.
c. __________ Directional growth of plants in response to the earth's gravity.
d. __________ Dropping of leaves, fruits, or flowers from a plant.
e. __________ In plants, a growth response toward or away from a directional stimulus.
f. __________ Plant hormone producing increased stem growth; also involved in flowering and seed germination.
g. __________ Plant hormone regulating growth, particularly cell elongation.
h. __________ Plant hormone that causes ripening of fruit and is also involved in dropping of leaves.
i. __________ Plant hormone that causes stomates to close and that initiates and maintains dormancy.
j. __________ Relative lengths of daylight and darkness that affect the physiology and behavior of an organism.

Reproduction in Plants

CHAPTER 39

Chapter Concepts

A flower of the rosebay willow, *Epilobium*, is approached by a honeybee, *Apis*

Many plants can reproduce both asexually and sexually. Asexual reproduction, which produces a plant like the parent, is advantageous if the environment is stable. Sexual reproduction is a way to produce offspring that are different from the parent, some of whom may be more fit in a different or changing environment. The flowering plants, or angiosperms, are the most diverse and widespread of all the plants, perhaps because they have a means of sexual reproduction that is well adapted to life on land. Pollen protects the sperm nucleus until fertilization takes place, and seeds protect the embryo until dispersal takes place. The structure of the flower allows angiosperms to produce seeds within fruits. The evolution of the flower has no doubt contributed heavily to the enormous success of angiosperms in another way. It permits pollination to take place, not only by wind, but also by animals. Flowering plants that rely on animals for pollination have a mutualistic relationship with them. The flower provides nutrients for the pollinator such as a bee, a fly, a beetle, a bird, or even a bat. The animals in turn inadvertently carry pollen from one flower to another, allowing pollination to occur.

39.1 Flowering Plants Undergo Alternation of Generations

A *life cycle* is the entire sequence of events from the time of fertilization and formation of the zygote to gamete formation once again. In contrast to animals, which have only one type of adult generation in their life cycle, plants have two types—a diploid generation and a haploid generation—that alternate with each other. The diploid generation is called the **sporophyte** because it produces spores. The spores are haploid, and they divide to become the haploid generation, which is called the **gametophyte** because it produces gametes. In flowering plants, the diploid sporophyte is said to be dominant. It is larger, longer lasting, and the generation we generally recognize as the plant. This is the generation that flowers (Fig. 39.1).

A flower produces two types of spores, microspores and megaspores. A **microspore** [Gk. *mikros,* small, little] develops into a microgametophyte; a **megaspore** develops into a megagametophyte. The microgametophyte is the pollen grain, which is either windblown or carried by an animal to the vicinity of the megagametophyte. When a pollen grain matures, it contains nonflagellated sperm cells, which travel by way of a pollen tube to the megagametophyte, which is called an embryo sac. Once a sperm fertilizes an egg in the embryo sac, the life cycle begins again.

Notice that the sporophyte is dominant in flowering plants. It is the generation that contains vascular tissue and has other adaptations suitable to living on land. The life cycle of a flowering plant is also adapted to a land existence. The gametophytes which produce the gametes are microscopic and dependent upon the sporophyte. Since the microgametophyte (pollen grain) is transported to the megagametophyte by wind or an animal, water is not needed for transport. The pollen tube carries a sperm to an egg, and again no outside water is needed. (This may be compared to the manner in which fertilization is accomplished in human beings. At the time of sexual intercourse, the penis deposits sperm into the vagina of the female.) Following fertilization, the diploid zygote develops into an embryo protected within a seed enclosed by a fruit.

Unlike flowering plants, some plants produce only one kind of spore and this develops into a separate, but water-dependent, gametophyte. The evolution of two separate and dependent gametophytes in flowering plants led to the production of pollen and a means to carry on pollination and fertilization without need of external water.

Flowering plants undergo an alternation of generations life cycle that is modified in such a way that a sperm does not require an outside source of water to reach an egg.

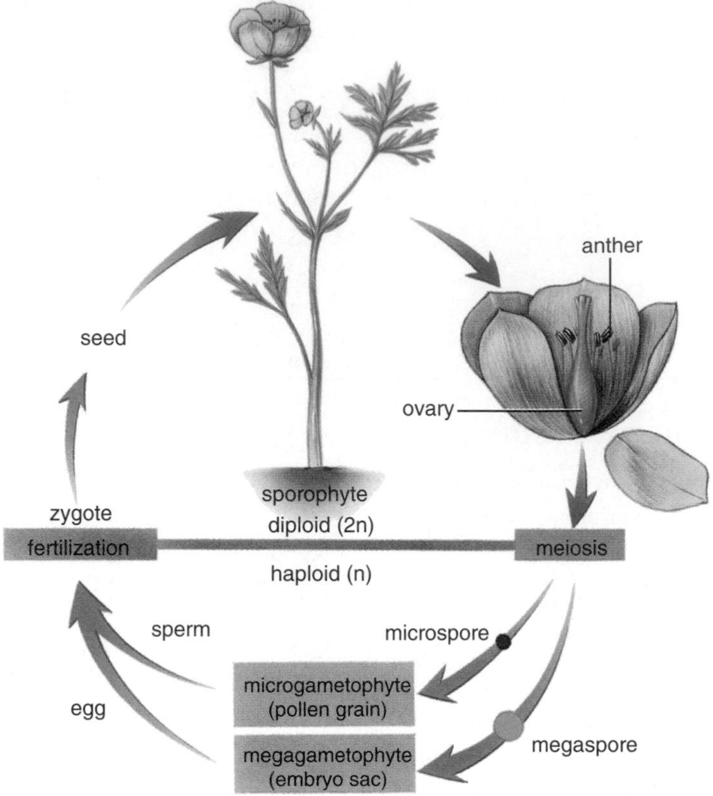

Figure 39.1 Alternation of generations overview. In flowering plants, the diploid zygote develops into an embryo enclosed within a seed. The embryo becomes the sporophyte, which bears flowers. The flower produces microspores and megaspores by meiosis. A megaspore becomes a megagametophyte, which produces an egg, and a microspore becomes a microgametophyte (pollen grain), which produces sperm. Fertilization results in a zygote once more.

A Flower Is a Sporophyte Structure

A **flower,** the reproductive structure of angiosperms, develops within a bud. In many plants, the same shoot apical meristem that previously formed leaves suddenly stops producing leaves and starts producing a flower. In other plants, axillary buds develop directly into flowers. Flower structures are modified leaves attached to a short stem tip called a *receptacle* (Fig. 39.2). In monocots, flower parts occur in threes and multiples of three. In dicots, flower parts are in fours or fives and multiples of four or five. The **sepals,** which are the most leaflike of all the flower parts, are usually green, and they protect the bud as the flower develops within. When the flower opens, there is an outer whorl of sepals and an inner whorl of **petals,** whose color accounts for the attractiveness of many flowers. The size, the shape, and the color of a flower are attractive to a specific pollinator. Wind-pollinated flowers often have no petals at all.

At the very center of a flower is the **pistil** [L. *pistillum,* club-shaped pounder, pestle], which is often vaselike in appearance. A pistil may be simple or compound. A simple pistil contains a single reproductive unit called a **carpel** [Gk. *karpos,* fruit]. A carpel usually has three parts: the **stigma,** an enlarged sticky knob, the **style,** a slender stalk, and the **ovary,** an enlarged base. A compound pistil has multiple carpels, which are often fused:

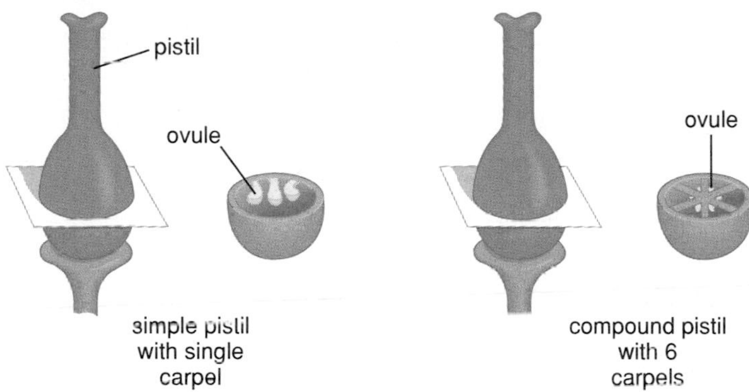

The ovary of a pistil contains a number of **ovules** [L. *ovulum,* dim. of ovum: egg], which play a significant role in the production of megaspores and therefore megagametophytes. Grouped about the pistil are a number of **stamens,** each of which has two parts: the **anther,** a saclike container, and the **filament** [L. *filum,* thread], a slender stalk. Pollen grains develop from microspores in the anther.

Not all flowers have sepals, petals, stamens, and a pistil. Those that do are said to be complete and those that do not are said to be incomplete. Flowers that have both stamens and a pistil are called perfect flowers: those with only stamens are staminate flowers and those with only pistils are pistillate flowers. If staminate flowers and pistillate flowers are on one plant, as in corn, the plant is monoecious [Gk. *monos,* one, and *oikos,* home, house]. If staminate and pistillate flowers are on separate plants, the plant is dioecious. Holly trees are dioecious and, if red berries are a priority, it is necessary to acquire a plant with staminate flowers and another with pistillate flowers.

Sometimes the pistil is called the female part of the flower and the stamens are called the male part of the flower, but this is not strictly correct. The pistil and stamens do not produce gametes; they produce megaspores and microspores, respectively. A microspore matures into a microgametophyte (pollen grain), which produces sperm, and a megaspore matures into a megagametophyte, which produces an egg.

A flower contains four basic parts: sepals, petals, stamens, and a pistil.

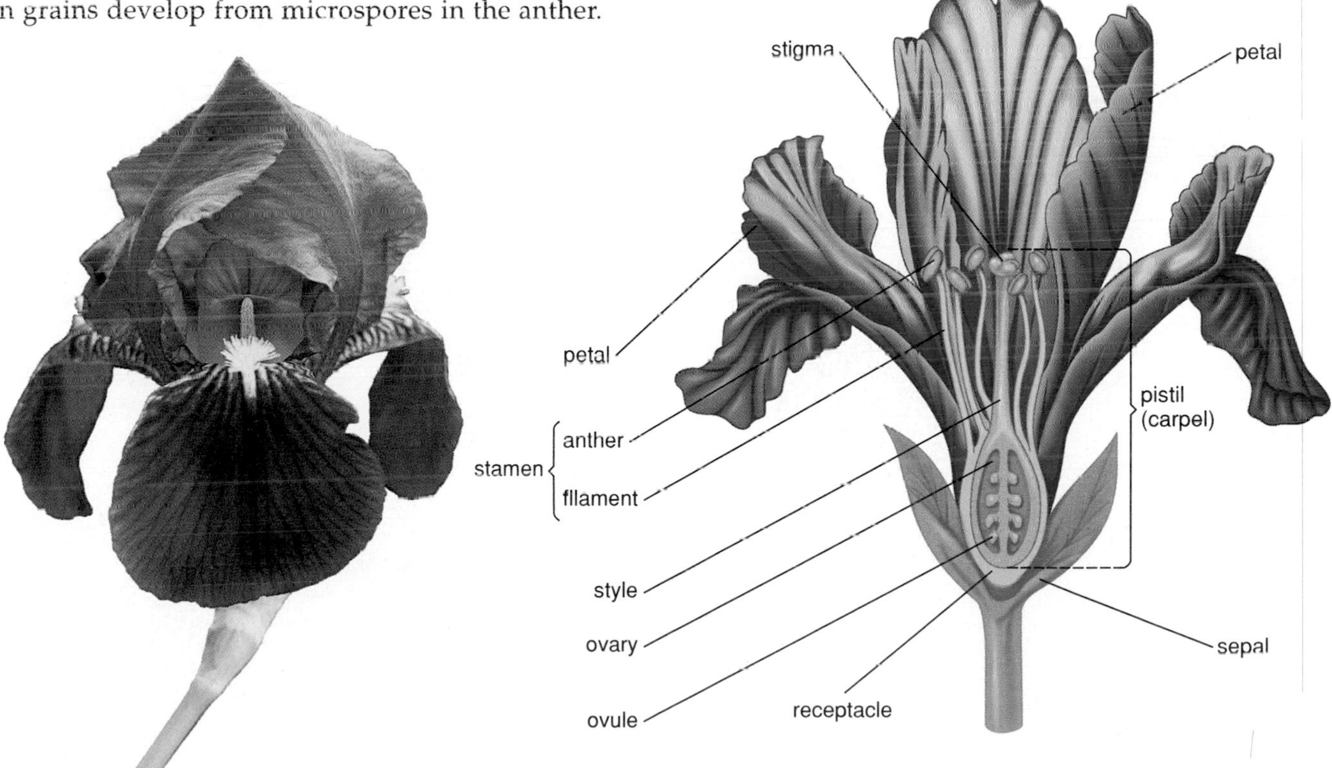

Figure 39.2 Typical flower, *Iris.*
In angiosperms, the sporophyte produces flowers. A megagametophyte develops within an ovule and microgametophytes (pollen grains) develop within an anther.

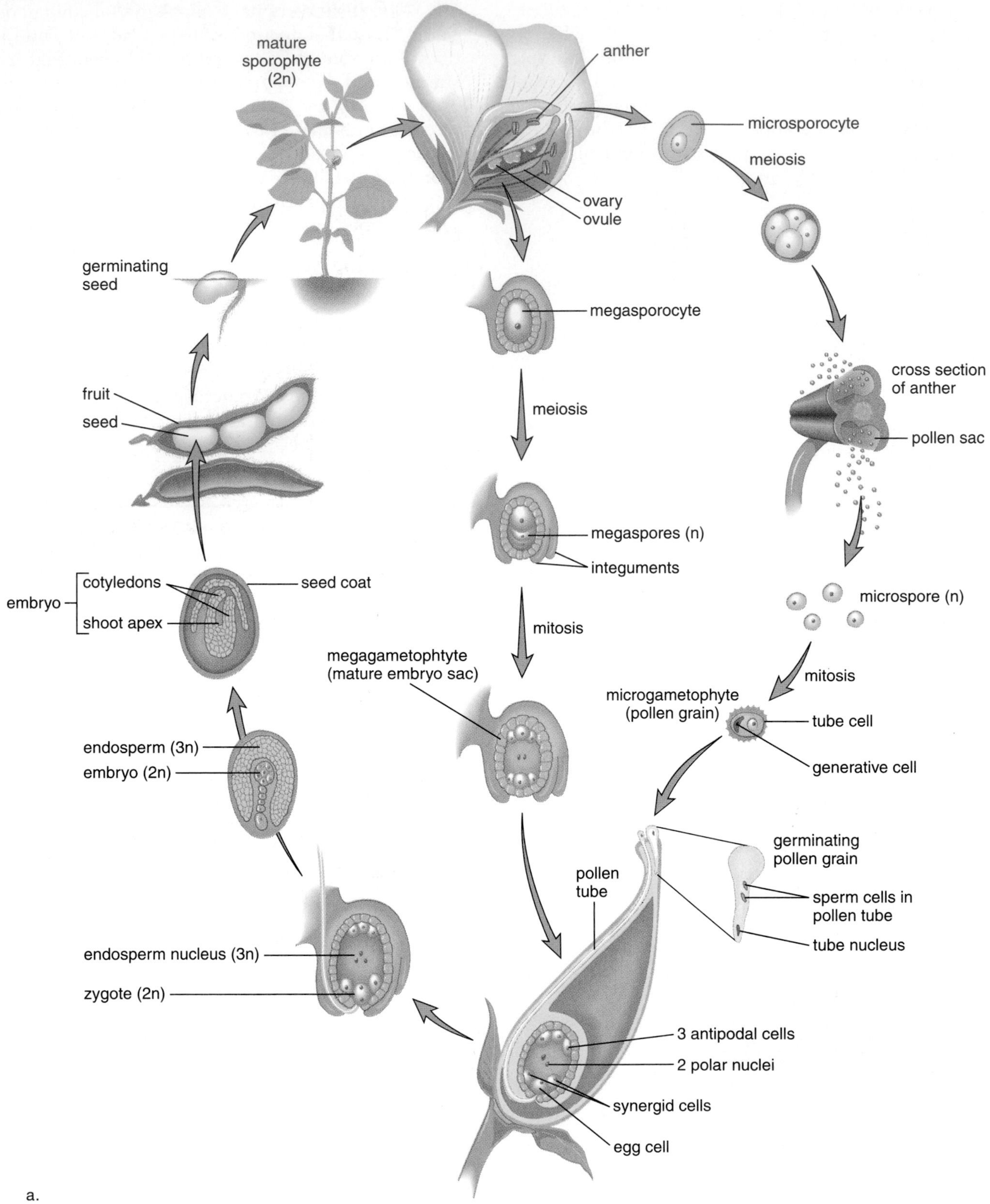
mature
sporophyte
(2n)
anther
microsporocyte
meiosis
ovary
ovule
germinating
seed
megasporocyte
cross section
of anther
fruit
seed
meiosis
pollen sac
megaspores (n)
integuments
embryo
cotyledons
shoot apex
seed coat
microspore (n)
mitosis
megagametophtyte
(mature embryo sac)
mitosis
microgametophyte
(pollen grain)
tube cell
generative cell
endosperm (3n)
embryo (2n)
germinating
pollen grain
pollen
tube
sperm cells in
pollen tube
tube nucleus
endosperm nucleus (3n)
zygote (2n)
3 antipodal cells
2 polar nuclei
synergid cells
egg cell
a.

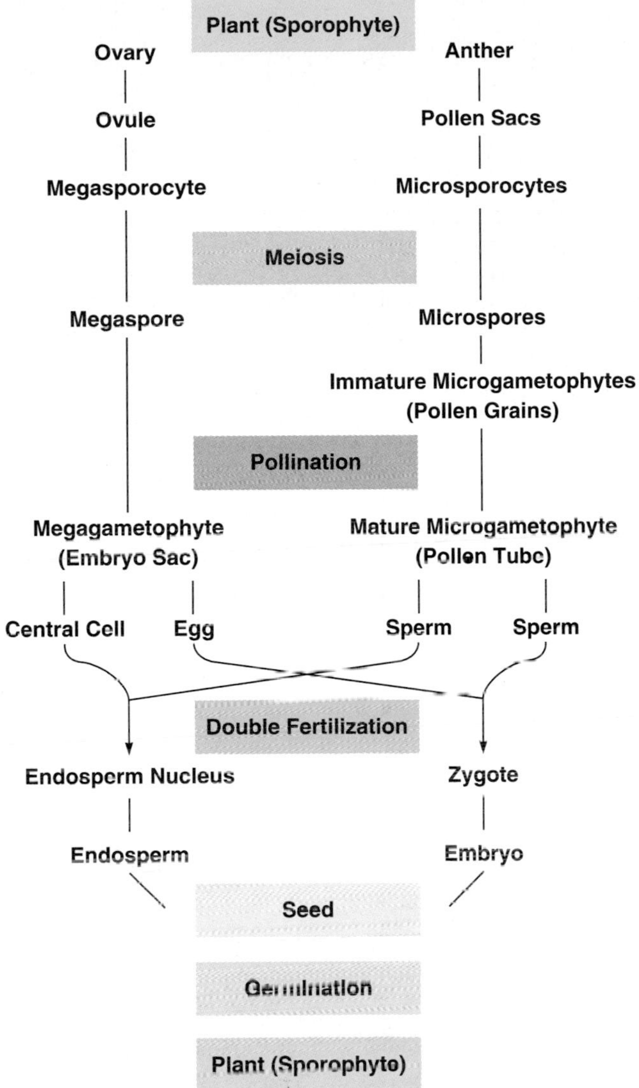

Figure 39.3 Life cycle of a flowering plant.
a. An ovary in the carpel contains an ovule where a megasporocyte produces a megaspore by meiosis. A megaspore develops into an embryo sac containing seven cells, one of which is an egg. A pollen sac in the anther contains microsporocytes, which produce microspores by meiosis. A microspore develops into a pollen grain that contains sperm cells by the time it germinates. The sperm cells travel down a pollen tube; one sperm fertilizes the egg to form a diploid zygote, and the other fuses with the polar nuclei of the central cell to form a triploid endosperm nucleus. A seed contains the developing embryo plus stored food. Germination and growth of a seed results in a new sporophyte plant. **b.** Flow diagram.

The Gametophytes Are Separate

The ovary of a carpel contains one or more ovules (Fig. 39.3*a*, middle). Ovules have a central mass of parenchyma cells almost completely covered by integuments except where there is an opening, the micropyle. One parenchyma cell enlarges to become a **megasporocyte** (megaspore mother cell), which undergoes meiosis, producing four haploid megaspores. Three of these disintegrate, leaving one functional megaspore, whose nucleus divides mitotically until there are eight nuclei of the embryo sac or **megagametophyte** [Gk. *megas*, great, large]. When cell walls form later, there are seven cells, one of which is binucleate. The megagametophyte called the **embryo sac** consists of these seven cells:

one egg cell associated with
two synergid cells
one central cell with two polar nuclei
three antipodal cells

Microgametophytes are produced in the stamens (Fig. 39.3*a*, far right). An anther has four pollen sacs, each containing many **microsporocytes** (microspore mother cells). A microsporocyte undergoes meiosis to produce four haploid microspores. A microspore divides mitotically, forming two cells enclosed by a finely sculptured wall. This structure, called the **pollen grain,** is the immature **microgametophyte** and contains a *tube cell* and a *generative cell*. Either now or later, the generative cell divides mitotically to produce two sperm cells. The walls separating the pollen sacs in the anther break down when the pollen grains are ready to be released.

Look back at Figure 39.1 and realize that the sporophyte, the flowering diploid generation, has now produced a haploid megaspore and haploid microspores. The megaspore has developed into an embryo sac, the egg-producing megagametophyte generation. Each microspore has developed into a pollen grain, the sperm-producing microgametophyte generation. The haploid generation in flowering plants is represented by the megagametophyte and the microgametophyte, which are microscopic. The megagametophyte is retained within the body of the sporophyte. As discussed in the reading on pages 698–99, microgametophytes (pollen grains) are windblown or carried by various kinds of animals to the stigma of a pistil.

In flowering plants, the megagametophyte within the ovule produces an egg. The microgametophyte (pollen grain) produces two sperm.

A closer Look

Plants and Their Pollinators

A plant and its pollinator(s) are adapted to one another. They have a mutualistic relationship in which each benefits—the plant uses its pollinator to ensure that cross-pollination takes place, and the pollinator uses the plant as a source of food. This mutualistic relationship came about through the process of **coevolution;** that is, the codependency of the plant and the pollinator is the result of suitable changes in structure and function of each. The evidence for coevolution is observational. For example, floral coloring and odor are suited to the sense perceptions of the pollinator, the mouthparts of the pollinator are suited to the structure of the flower, the type of food provided is suited to the nutritional needs of the pollinator, and the pollinator forages at the time of day that specific plants are open. The following are examples of such coevolution.

Bee-Pollinated Flowers

There are now 20,000 different species of bees that pollinate flowers. The best-known pollinators are the honeybees (Fig. 39A*a*). Bee eyes see a spectrum of light that is different from the spectrum seen by humans. The bee's visible spectrum is shifted so that they do not see red wavelengths but do see ultraviolet wavelengths. Bee-pollinated flowers are usually brightly colored and are predominantly blue or yellow; they are not entirely red. They may also have ultraviolet shadings called honey guides, which highlight the portion of the flower that contains the reproductive structures. The mouthparts of bees are fused into a long tube that contains a tongue. This tube is an adaptation for sucking up nectar provided by the plant, usually at the base of the flower.

Bee flowers are delicately sweet and fragrant to advertise that nectar is present. The honey guides often point to a narrow floral tube large enough for the bee's feeding apparatus but too small for other insects to reach the nectar. Bees also collect pollen as food for their larvae. Pollen clings to the hairy body of a bee, and the bees also gather it by means of bristles on their legs. They then store the pollen in pollen baskets on the third pair of legs. Bee-pollinated flowers are sturdy and irregular in shape because they often have a landing platform where the bee can alight. The landing platform requires the bee to brush up against the anther and stigma as it moves toward the floral tube to feed. One type of orchid, *Ophrys,* has evolved a unique adaptation. The flower resembles a female bee, and when the male of that species attempts to copulate with the flower, the bee receives pollen.

a.

b.

Figure 39A Pollinators.
a. A bee-pollinated flower is a color other than red (bees cannot detect this color) and has a landing platform where the reproductive structures of the flower brush up against the bee's body. **b.** A butterfly-pollinated flower is often a composite, containing many individual flowers. The broad expanse provides room for the butterfly to land, after which it lowers its proboscis into each flower in turn. **c.** Hummingbird-pollinated flowers are curved back, allowing the bird to insert its beak to reach the rich supply of nectar. While doing this, the bird's forehead and other body parts touch the reproductive structures. **d.** Bat-pollinated flowers are large, sturdy flowers that can take rough treatment. Here the head of the bat is positioned so that its bristly tongue can lap up nectar.

Moth- and Butterfly-Pollinated Flowers

Contrasting moth- and butterfly-pollinated flowers emphasizes the close adaptation between pollinator and flower. Both moths and butterflies have a long, thin, hollow proboscis, but they differ in other characteristics. Moths usually feed at night and have a well-developed sense of smell. The flowers they visit are visible at night because they are lightly shaded (white, pale yellow, or pink), and they have strong, sweet perfume, which helps attract moths. Moths hover when they feed, and their flowers have deep tubes with open margins that allow the hovering moths to reach the nectar with their long proboscis. Butterflies are active in the daytime and have good vision but a weak sense of smell. Their flowers have bright colors—even red because butterflies can see the color red—but the flowers tend to be odorless. Unable to hover, butterflies need a place to land. Flowers that are visited by butterflies often have flat landing platforms (Fig. 39A*b*). The flowers also tend to be composites, with many individual flowers clustered in a head. Each flower has a long, slender floral tube, accessible to the long, thin butterfly proboscis.

Bird- and Bat-Pollinated Flowers

In North America, the most well-known bird pollinators are the hummingbirds. These tiny animals have good eyesight but do not have a well-developed sense of smell. Like moths, they hover when they feed. Typical flowers pollinated by hummingbirds are red, with a slender floral tube and margins that are curved back and out of the way. And although they produce copious amounts of nectar, the flowers have little odor. As a hummingbird feeds on nectar with its long, thin beak, its head comes into contact with the stamens and pistil (Fig. 39A*c*).

Bats are adapted to gathering food in various ways, including feeding on the nectar and pollen of plants. Bats are nocturnal and have an acute sense of smell. Those that are pollinators also have keen vision and a long, extensible, bristly tongue. Typically, bat-pollinated flowers open only at night and are light-colored or white. They have a strong, musty smell similar to the odor that bats produce to attract one another. The flowers are generally large and sturdy and are able to hold up when a bat inserts part of its head to reach the nectar. While the bat is at the flower, its head is dusted with pollen (Fig. 39A*d*).

Coevolution

These examples are evidence of coevolution, but how did coevolution come about? Some 200 million years ago, when seed plants were just beginning to evolve and insects were not as diverse as they are today, wind alone was used to carry pollen. Wind pollination, however, is a hit-or-miss affair. Perhaps beetles feeding on vegetative leaves were the first insects to carry pollen directly from plant to plant by chance. This use of animal motility to achieve cross-fertilization no doubt resulted in the evolution of flowers, which have features, such as the production of nectar, to attract pollinators. Then, if beetles developed the habit of feeding on flowers, other features, such as the protection of ovules within ovaries, may have evolved.

As cross-fertilization continued, more and more flower variations likely developed, and pollinators became increasingly adapted to specific angiosperm species. Today, there are some 235,000 species of flowering plants and over 700,000 species of insects. This diversity suggests that the success of angiosperms has contributed to the success of insects, and vice versa.

c.

d.

Pollination Precedes Fertilization

Pollination and fertilization are two separate events. **Pollination** is simply the transfer of pollen (Fig. 39.4) from the anther to the stigma of a pistil. **Fertilization** [L. *fertilis*, fruitful] is the fusion of nuclei, as when the sperm nucleus and the egg nucleus fuse.

Pollination Transfers Pollen

As mentioned previously, pollination is brought about by the wind or with the assistance of a particular pollinator. Self-pollination occurs if the pollen is from the same plant, and cross-pollination occurs if the pollen is from a different plant. Cross-pollination is preferred because it fosters genetic recombination resulting in new and varied plants.

Fertilization Is Double

When a pollen grain lands on the stigma of the same species, it germinates, forming a pollen tube (Fig. 39.5). The germinated pollen grain, containing a tube cell and two sperm, is the mature microgametophyte. As it grows, the pollen tube passes between the cells of the stigma and the style to reach the micropyle of the ovule. Now **double fertilization** occurs. One sperm nucleus unites with the egg nucleus, forming a 2n zygote, and the other sperm nucleus migrates and unites with the polar nuclei of the central cell, forming a 3n endosperm nucleus. The zygote divides mitotically to become the **embryo,** a young sporophyte, and the endosperm nucleus divides mitotically to become the endosperm. **Endosperm** [Gk. *endon*, within, and *sperma*, seed] is the tissue that will nourish the embryo and seedling as they undergo development.

Flowering plants practice double fertilization. One sperm nucleus unites with the egg nucleus, producing a zygote, and the other unites with the polar nuclei, forming a 3n endosperm cell.

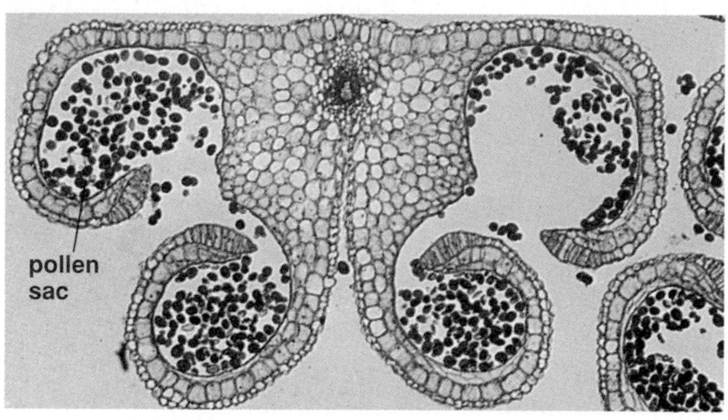

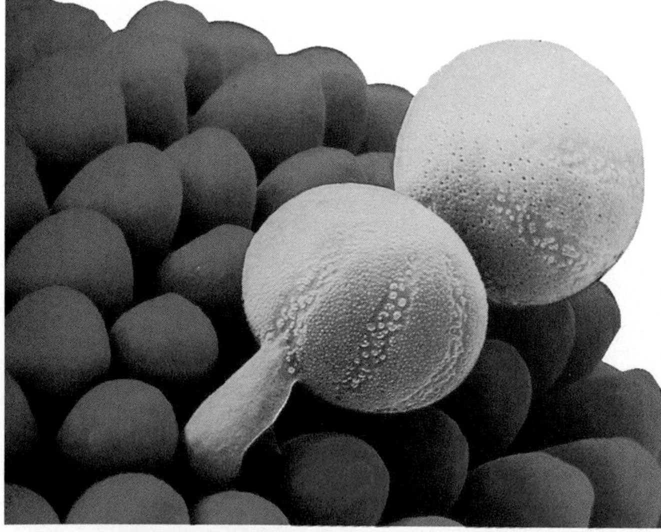

Figure 39.4 Development of pollen grains.
a. A mature anther showing that the walls between the pollen sacs have opened, allowing the pollen grains to be released.
b. Germinating pollen grains of goosegrass, *Galium aparine*, are colored yellow in this false-colored scanning electron micrograph.

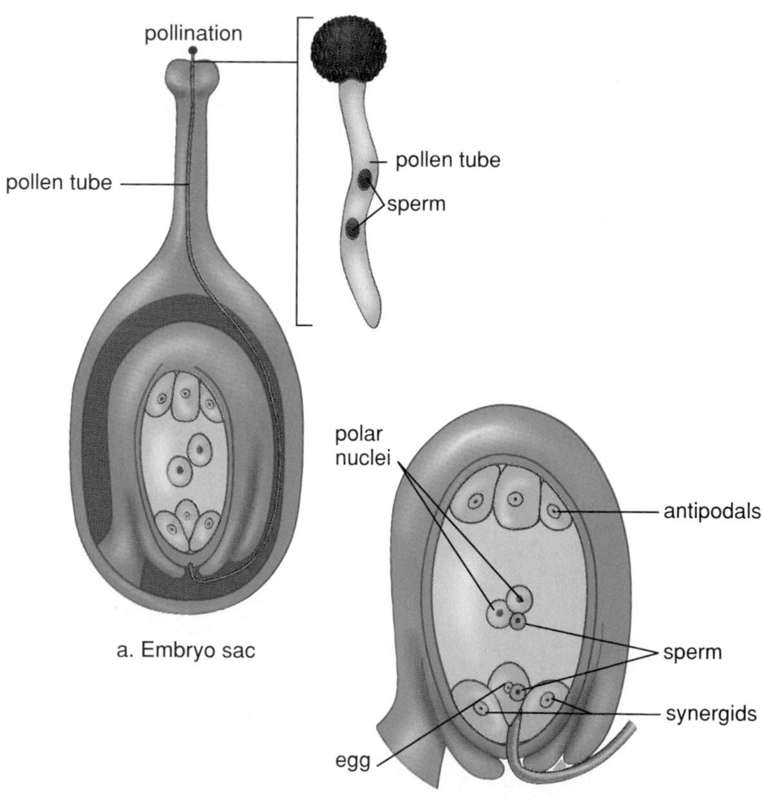

Figure 39.5 Fertilization.
a. The pollen tube grows down the style, carrying with it two sperm cells. The pollen tube grows through the micropyle and releases both sperm. **b.** One of these migrates to the egg; this sperm nucleus fuses with the egg nucleus, forming a zygote. The second sperm migrates into the central cell; this sperm nucleus fuses with the polar nuclei, forming a large triploid endosperm nucleus within a central cell.

39.2 The Embryo Develops in Stages

Stages in the development of a dicot embryo are shown in Figure 39.6. After double fertilization has taken place, the single-celled zygote lies beneath the endosperm nucleus. The endosperm nucleus divides to produce a mass of endosperm tissue surrounding the embryo. The zygote also divides, forming two parts: the upper part is the embryo, and the lower part is the suspensor, which anchors the embryo and transfers nutrients to it from the sporophyte plant. Soon the **cotyledons** [Gk. *cotyledon,* cup-shaped cavity], or seed leaves, can be seen. At this point, the dicot embryo is heart shaped. Later, when it becomes torpedo shaped, it is possible to distinguish the shoot apex and the root apex. These contain apical meristems, the tissues that bring about primary growth in a plant; the shoot apical meristem is responsible for aboveground growth, and the root apical meristem is responsible for underground growth.

Monocots, unlike dicots, have only one cotyledon. Another important difference between monocots and dicots is the manner in which nutrient molecules are stored in the seed. In a monocot, the cotyledon rarely stores food; rather, it absorbs food molecules from the endosperm and passes them to the embryo. During the development of a dicot embryo, the cotyledons usually store the nutrient molecules that the embryo uses. Therefore, in Figure 39.6 we can see that the endosperm seemingly disappears. Actually, it has been taken up by the two cotyledons. In a plant embryo, the epicotyl is above the cotyledon and contributes to shoot development; the hypocotyl is that portion below the cotyledon that contributes to stem development; and the radicle contributes to root development. The embryo plus stored food is now contained within a seed.

The plant embryo (which has gone through a set series of stages), plus its stored food, is contained within a seed.

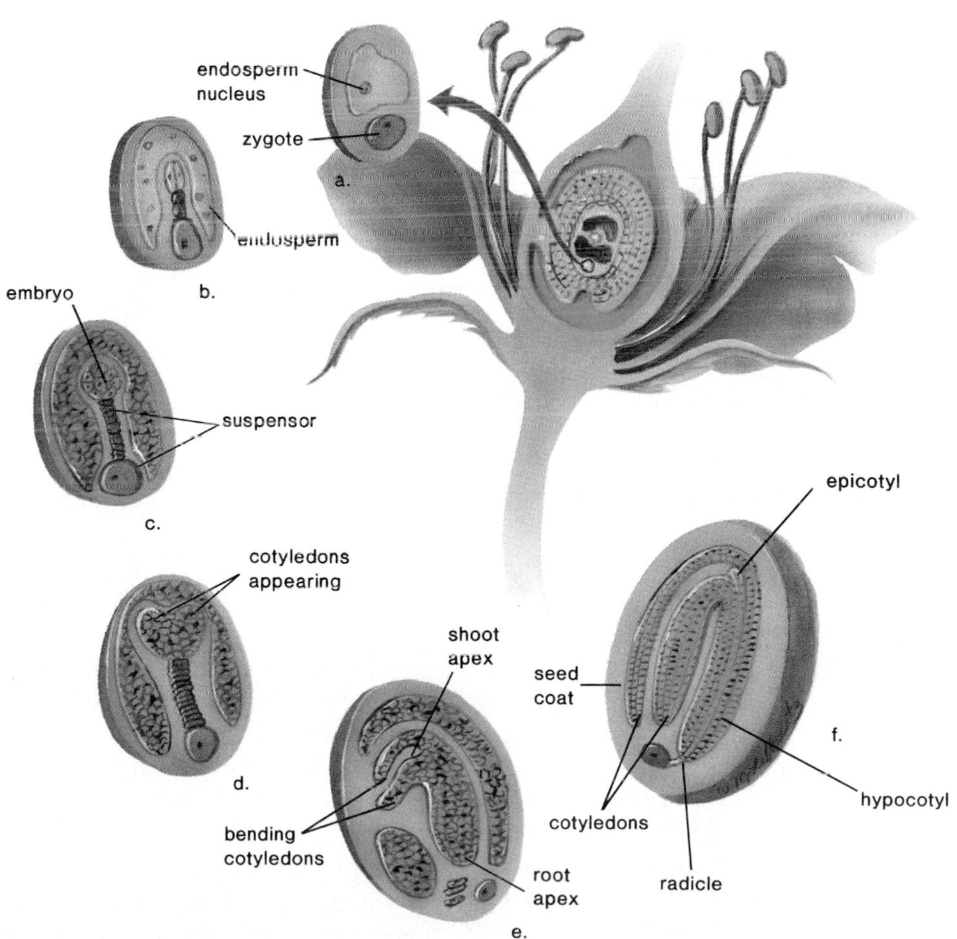

Figure 39.6 Development of a dicot embryo.
a. The single-celled zygote lies beneath the endosperm nucleus. **b, c.** The endosperm is a mass of tissue surrounding the embryo. The embryo is located above the suspensor. **d.** The embryo becomes heart shaped as the cotyledons begin to appear. **e.** There is progressively less endosperm as the embryo differentiates and enlarges. As the cotyledons bend, the embryo takes on a torpedo shape. **f.** The embryo consists of the epicotyl (represented here by the shoot apex), the hypocotyl, and the radicle (the latter of which contains the root apex).

a. Almond, *Prunus*

b. Tomato, *Lycopersicon*

Figure 39.7 Fruit diversity.
a. The almond, *Prunus*, fruit is fleshy, with a single seed enclosed by a hard covering. **b.** The tomato, *Lycopersicon*, is derived from a compound ovary. **c.** Dry fruit of the pea, *Pisum*, plant develops from a simple ovary. **d.** Blackberry, *Rubus*, flowers contain several separate ovaries; each cluster is a berry from a single flower.

39.3 The Seeds Are Enclosed by Fruit

As the zygote develops into an embryo, the integuments of the ovule harden and become the seed coat. A **seed** is a structure formed by the maturation of the ovule; it contains a sporophyte embryo plus stored food. The ovary, and sometimes other floral parts, develops into a fruit. Although peas and beans in their shell, tomatoes, and cucumbers are commonly called vegetables, botanists categorize them as fruits. A **fruit** is a mature ovary that usually contains seeds.

Fruits Are Varied

As fruit develops from an ovary, the ovary wall thickens to become the *pericarp.* Most fruits are *simple fruits;* they are derived from an individual ovary, either simple or compound. In simple fleshy fruits, the pericarp is at least somewhat fleshy. Peaches and plums are good examples of simple fleshy fruits. In almonds, the fleshy part of the pericarp is a husk removed before marketing. We crack the remaining portion of the pericarp to obtain the seed (Fig. 39.7*a*). An apple develops from a compound ovary (one that has several sections), but much of the flesh comes from the receptacle, which grows around the ovary. It's more obvious that a tomato comes from a compound ovary, because in cross section, you can see several seed-filled cavities (Fig. 39.7*b*).

Dry fruits have a dry pericarp (Table 39.1). Legumes, such as peas (Fig. 39.7*c*) and beans, produce a fruit that splits along two sides, or seams. Not all dry fruits split at maturity. The pericarp of a grain is tightly fused to the seed and cannot be separated from it. A corn kernel is a grain, as are the fruits of wheat, rice, and barley plants.

Some fruits develop from several individual ovaries and therefore are compound fruits. A blackberry is an aggregate fruit in which each berry is derived from a separate ovary of a single flower (Fig. 39.7*d*). The strawberry is also an aggregate fruit, but each ovary becomes a one-seeded fruit called an achene. The flesh of a strawberry is from the receptacle. In contrast, a pineapple comes from the fruit of many individual flowers attached to the same fleshy stalk. As the ovaries mature, they fuse to form a large, multiple fruit.

In flowering plants, the seed develops from the ovule and the fruit develops from the ovary.

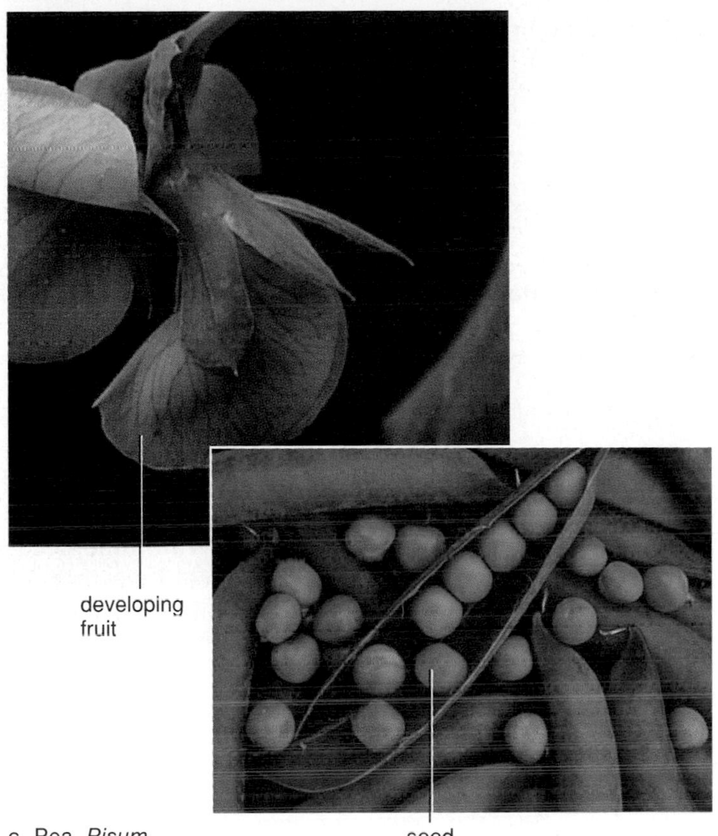

c. Pea, *Pisum*

d. Blackberry, *Rubus*

Table 39.1

Kinds of Fruit

Name	Description	Example
Simple Fruits	*Develop from an individual ovary*	
Fleshy	Pericarp is usually fleshy	
Drupe	From simple ovary with one seed (pit) and soft "skin"	Peach, plum, olive
Berry	From compound ovary with many seeds	Grape, tomato
Pome	From compound ovary; flesh is from accessory flower parts	Apple, pear
Dry	Pericarp is dry	
Follicle	From simple ovary that splits open down one side	Milkweed, peony
Legume	From simple ovary that splits open down both sides	Pea, bean, lentil
Capsule	From compound ovary with capsules that split in various ways	Poppy
Achene	From simple ovary with one-seeded small fruit; pericarp easily removed	Sunflower, dandelion
Nut	From simple ovary with one-seeded fruit; hard pericarp	Acorn, hickory nut, chestnut
Grain	From simple ovary with one-seeded small fruit; pericarp completely united with seed coat	Rice, oat, barley
Compound Fruits	*Develop from a group of individual ovaries*	
Aggregate fruits	Ovaries are from a single flower	Blackberry, raspberry, strawberry
Multiple fruits	Ovaries are from separate flowers clustered together	Pineapple

How Seeds Disperse and Germinate

For plants to be widely distributed, their seeds have to be dispersed—that is, distributed preferably long distances from the parent plant. Following dispersal, the seeds germinate: they begin to grow so that a seedling appears.

Dispersal Requires Help

Plants have various means to ensure that dispersal takes place. The hooks and spines of clover, bur, and cocklebur attach to the fur of animals and the clothing of humans. Birds and mammals sometimes eat fruits, including the seeds, which are then defecated (passed out of the digestive tract with the feces) some distance from the parent plant. Squirrels and other animals gather seeds and fruits, which they bury some distance away.

The fruit of the coconut palm, which can be dispersed by ocean currents, may land many hundreds of kilometers away from the parent plant. Some plants have fruits with trapped air or seeds with inflated sacs that help them float in water. Many seeds are dispersed by wind. Woolly hairs, plumes, and wings are all adaptations for this type of dispersal. The seeds of an orchid are so small and light that they need no special adaptation to carry them far away. The somewhat heavier dandelion fruit uses a tiny "parachute" for dispersal. The winged fruit of a maple tree, which contains two seeds, has been known to travel up to 10 kilometers from its parent. A touch-me-not plant has seed pods that swell as they mature. When the pods finally burst, the ripe seeds are hurled out.

Animals, water, and wind help plants disperse their seeds.

Germination of Seeds

Some seeds do not **germinate** until they have been dormant for a period of time. For seeds, *dormancy* is the time during which no growth occurs, even though conditions may be favorable for growth. In the temperate zone, seeds often have to be exposed to a period of cold weather before dormancy is broken. In deserts, germination does not occur until there is adequate moisture. This requirement helps ensure that seeds do not germinate until the most favorable growing season has arrived. Germination—an

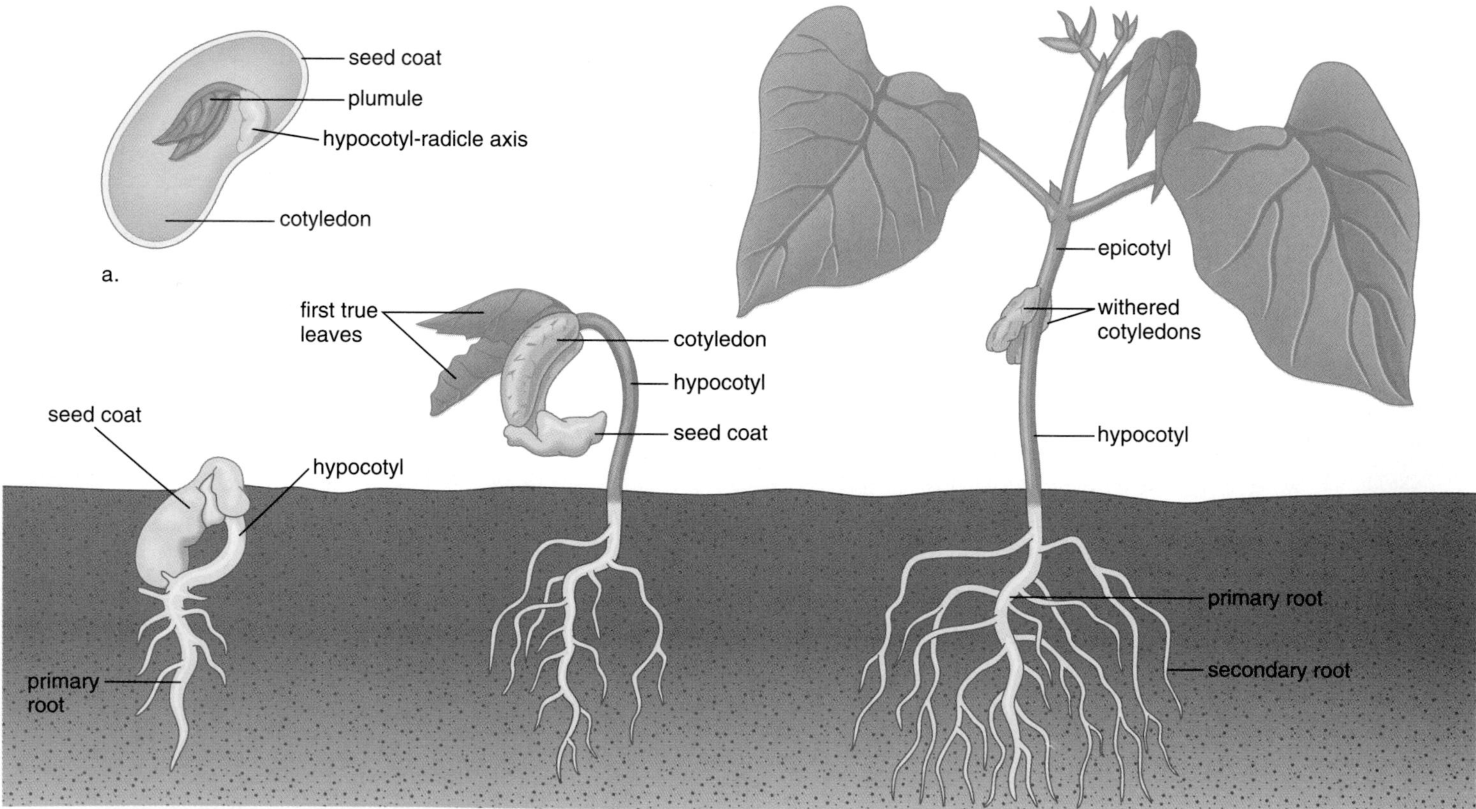

Figure 39.8 Common garden bean structure and germination.
a. Seed structure. b. Germination and development of the seedling.

event that takes place if there is sufficient water, warmth, and oxygen to sustain growth—requires regulation, and both inhibitors and stimulators are known to exist. It is known that fleshy fruits (e.g., apples, pears, oranges, and tomatoes) contain inhibitors so that germination does not occur until the seeds are removed and washed. In contrast, stimulators are present in the seeds of some temperate zone woody plants. Mechanical action may also be required. Water, bacterial action, and even fire can act on the seed coat, allowing it to become permeable to water. The uptake of water causes the seed coat to burst.

Germination in Dicots and Monocots

As mentioned, the embryo of a dicot, such as a bean plant, has two seed leaves, called cotyledons. The cotyledons, which supply nutrients to the embryo and seedling, eventually shrivel and disappear. If the two cotyledons of a bean seed are parted, you can see a rudimentary plant (Fig. 39.8). The epicotyl bears young leaves and is called a **plumule** [L. *plumulla*, dim. of *pluma*: feather]. As the dicot seedling emerges from the soil, the shoot is hook shaped to protect the delicate plumule. The hypocotyl becomes the stem and the radicle develops into the roots. When a seed germinates in darkness, it etiolates—the stem is elongated, the roots and leaves are small, and the plant lacks color and appears spindly. Phytochrome, a pigment that is sensitive to red and far-red light, regulates this response and induces normal growth once proper lighting is available.

A corn plant is a monocot that contains a single cotyledon. Actually, the endosperm is the food-storage tissue in monocots, and the cotyledon does not have a storage role. A corn kernel is a type of fruit called a grain and the outer covering is the pericarp (Fig. 39.9). The plumule and radicle are enclosed in protective sheaths called the coleoptile and the coleorhiza, respectively. The plumule and the radicle burst through these coverings when germination occurs.

Germination is a complex event regulated by many factors. The embryo breaks out of the seed coat and becomes a seedling with leaves, stem, and roots.

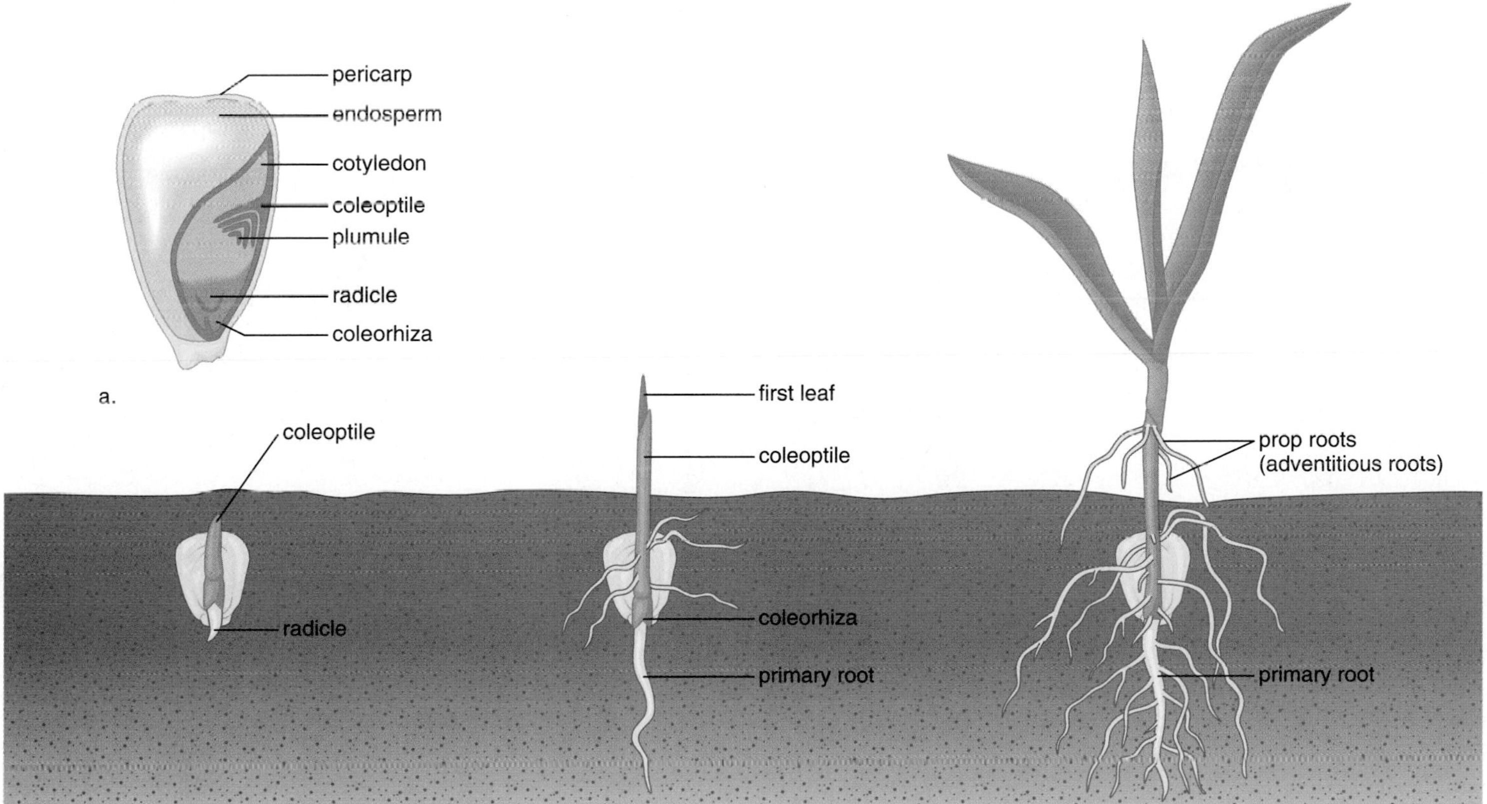

Figure 39.9 Corn kernel structure and germination.
a. Grain structure. b. Germination and development of the seedling.

A closer Look

Economic Importance of Plants

Stop a minute to think about how important plants are in everyday life. Essentially everywhere you look, the many materials you use are directly produced from plants, or their production was modeled after a plant product.

We derive most of our sustenance from three flowering plants: wheat, corn, and rice. All three of these plants are of the grass family and are collectively, along with other species, called grains. Most of the earth's 5.6 billion people live a simple way of life, growing their food on family plots. The continued growth of these plants is essential to human existence. A virus or other disease could hit any one of these three plants and could cause massive loss of life from starvation.

Corn, wheat, and rice originated and were first cultivated in different parts of the globe. Corn, or what is properly called maize, was first cultivated in Central America about 7,000 years ago. Disagreement still exists as to exactly what the wild plant looks like and where it originally grew. A most recent theory is that maize was developed from a plant called teosinte, which grows in the highlands of central Mexico. By the time Europeans were exploring Central America, over 300 varieties were already in existence—growing from Canada to Chile. We now commonly grow six major varieties of corn: sweet, pop, flour, dent, pod, and flint. Rice has its origin in southeastern Asia several thousand years ago, where it grew in swamps. Today we are familiar with white and brown rice, which differ in the extent of processing. The initially harvested rice is thrashed to remove the hulls, creating brown rice. After polishing to remove the seed coat and embryo, the starchy, endosperm, or white rice, is left. Unfortunately, it is the seed coat and embryo that contain the greatest nutritional value. Today rice is grown throughout the tropics and subtropics where water is abundant. Wheat is commonly used in the United States to produce flour and bread. It was first cultivated in the Near East (Iran, Iraq, and neighboring countries) about 8,000 B.C.; hence, it is thought to be one of the earliest cultivated plants. Wheat was brought to North America in 1520 with early settlers; now the United States is one of the world's largest producers of wheat.

Many of us have an "addiction" to sugar. This nifty carbohydrate comes almost exclusively from two plants—sugarcane (grown in South America, Africa, Asia, and the Caribbean) and sugar beets (grown mostly in Europe). Each provides about 50% of the world's sugar.

Numerous other foods are bland or tasteless without spices. In the Middle Ages, wealthy Europeans spared no cost to obtain spices from the Near and Far East. In the fifteenth and sixteenth centuries, major expeditions were launched in an attempt to find better and cheaper routes for spice importation. The queen of Portugal became convinced that Columbus would actually find a shorter route to the Far East by traveling west by ocean rather than east by land. Columbus's idea was sound, but he encountered a little barrier, the New World. This later provided Europe with a wealth of new crops, including corn, potatoes, peppers, and tobacco.

Our most popular drinks—coffee, tea, and cola—also come from plants. Coffee has its origin in Ethiopia, where it was first used (along with animal fat) during long trips for sustenance and to relieve fatigue. Coffee as a drink was not developed until the thirteenth century in Arabia and Turkey, and it did not catch on in Europe until the seventeenth century. Tea is thought to have been developed somewhere in central Asia. Its earlier uses were almost exclusively medicinal, especially among the Chinese. The drink as we now know, was not developed until the fourth century. By the mid-seventeenth century it became popular in Europe. Cola is a common ingredient in tropical drinks and was used around the turn of the century, along with the drug coca (used to make cocaine), in the "original" Coca-Cola.

Until a few decades ago cotton and other natural fibers were our only source of clothing. China is now the largest producer of cotton. Over 30 species of native cotton grow around the world. In the United States, cotton grows as an annual, but in the tropics shrubs over six feet tall

Corn, *Zea*

Wheat, *Triticum*

Rice, *Orza*

Figure 39B Cereal grains. These cereal grains are the principal source of calories and protein for our civilization.

are not uncommon. The cotton fiber itself comes from hair that grows on the seed. In sixteenth-century Europe, cotton was a little understood fiber known only from stories brought back from Asia. Columbus and other explorers were amazed to see the elaborately woven cotton fabrics in the New World. But by 1800 Liverpool was the world's center of cotton trade. Interestingly, when Levi Strauss wanted to make a tough pair of jeans, he needed a stronger fiber than cotton, so he used hemp. Even so, hemp (marijuana) is now known primarily as a hallucinogenic drug.

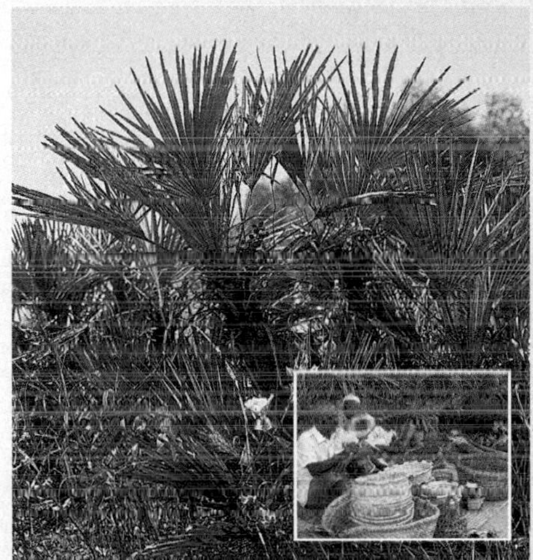

a. Dwarf fan palms, *Chamaerops*

b. Rubber, *Hevea*

Rubber is another plant that has many uses today. The product had its origins in Brazil from the thick, white sap of the rubber tree. Once collected, the sap is placed in a large vat where acid is added to coagulate the latex. When the water is pressed out, the product is formed into sheets or crumbled and placed into bales. Much stronger rubber, such as that in tires, was made by adding sulfur and heating in a process called vulcanization; this produces a flexible material less sensitive to temperature changes. Today though, much rubber is synthetically produced.

Beyond these uses, plants have been used for centuries for a number of important household items, including the house itself. We are most familiar with lumber being used as the major structural portion in buildings. This wood comes mostly from a variety of trees: pine, fir, and spruce, among others. In the tropics, trees and even herbs provide important components for houses. In Central and South America, palm leaves are preferable to tin for roofs, since they last as long as ten years and are quieter during a rainstorm. In the Near East, numerous houses along rivers are made entirely of reeds.

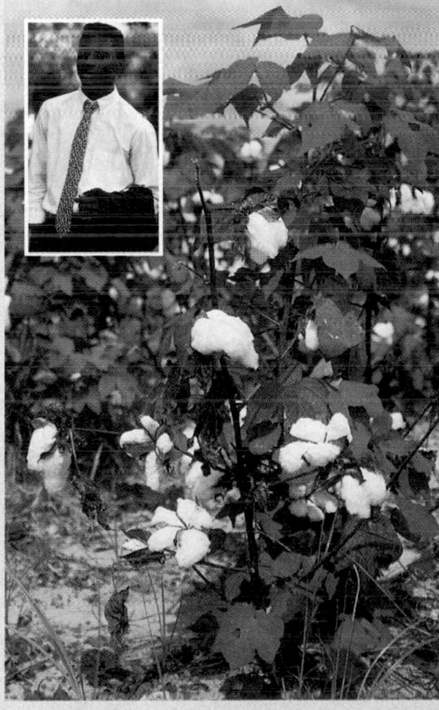

c. Cotton, *Gossypium*

Figure 39C Some uses of plants.
a. Dwarf fan palms can be used to make baskets. **b.** A rubber plant provides latex for making tires. **c.** Cotton becomes clothing worn by all.

An actively researched area of plant use today is that of medicinal plants. Currently about 50% of all pharmaceutical drugs have their origin from plants. The treatment of cancers and AIDS appears to rest in the discovery of miracle plants. Indeed, the National Cancer Institute (NCI) and most pharmaceutical companies have spent millions (or, more likely, billions) of dollars to send botanists out to collect and test plant samples from around the world. Tribal medicine men, or shamen, of South America and Africa have already been of great importance in developing numerous drugs.

One of the most amazing stories is of the development of quinine to treat malaria. Over the centuries, malaria has caused far more human deaths than any other disease. Yet it was not until the 1630s that word of a cure reached Europe. The cure came from the bark of the cinchona tree, common to northeastern South America. In the 1940s a synthetic form of the drug was developed, chloroquine. By the late 1960s it was found that some of the malaria parasites, which live in red blood cells, had become resistant to the synthetically produced drug, especially in Africa. More recently, resistant parasites are showing up in Asia and the Amazon. Today the only 100% effective drug for malaria treatment must come directly from the cinchona tree.

Obviously numerous plant extracts continue to be misused for their hallucinogenic or other effects on the human body: coca for cocaine and crack, opium poppy for morphine, and yam for steroids.

In the end, however, who can neglect the simple beauty of plants in their natural setting? Shade trees are so inviting during the summer, ornamental plants accent an architectural setting, and flowers brighten any yard. We should remember, then, that plants are good for much, much more than producing oxygen for us to breathe.

39.4 Plants Can Reproduce Asexually

Because plants contain nondifferentiated meristem tissue, they routinely reproduce asexually by *vegetative propagation*. In asexual reproduction there is only one parent, instead of two as in sexual reproduction. Complete strawberry plants will grow from the nodes of stolons (aboveground horizontal stems), and violets will grow from the nodes of rhizomes (underground horizontal stems) (Fig. 39.10). White potatoes are actually portions of underground stems, and each eye is a bud that will produce a new potato plant if it is planted with a portion of the swollen tuber. Sweet potatoes are modified roots; they can be propagated by planting sections of the root. You may have noticed that the roots of some fruit trees, such as cherry and apple trees, produce "suckers," small plants that can be used to grow new trees.

In addition to plants already mentioned, sugarcane, pineapple, cassava, and many ornamental plants have been propagated from stem cuttings for some time. In these plants, pieces of stem will automatically produce roots. The discovery that auxin will cause roots to develop has expanded the list of plants that can be propagated from stem cuttings.

Plants Propagate in Tissue Culture

Hydroponics, the growth of plants in aqueous solutions, had begun by the 1860s. This practice, along with the ability of plants to reproduce asexually, led the German botanist Gottleib Haberlandt to speculate in 1902 that entire plants could be produced by tissue culture. **Tissue culture** is the growth of a *tissue* in an artificial liquid *culture* medium. Haberlandt said that plant cells are **totipotent** [L. *totus*, all, whole, and *potens*, powerful]—each cell has the full genetic potential of the organism—and therefore a single cell could become a complete plant. But it wasn't until 1958 that Cornell botanist F. C. Steward grew a complete carrot plant from a tiny piece of phloem (Fig. 39.11). Like former investigators, he provided the cells with sugars, minerals, and vitamins, but he also added coconut milk. (Later, it was discovered that coconut milk contains the hormone cytokinin.) When the cultured cells began dividing, they produced a *callus*, an undifferentiated group of cells. Then the callus differentiated into shoot and roots and developed into complete plants.

Hybridization [L. *hybrida*, mongrel], the crossing of different varieties of plants or even species, is routinely done to produce plants with desirable traits. Hybridization, followed by vegetative propagation of the mature plants, will generate a large number of identical plants with these traits. But tissue culture has led to *micropropagation*, a commercial method of producing thousands, even millions, of identical seedlings in a limited amount of space. One favorite method to accomplish micropropagation is by *meristem culture*. If the correct proportions of auxin and cytokinin are added to a liquid medium, many new shoots will develop from a single shoot apex. When these are removed, more shoots form. Since the shoots are genetically identical, the adult plants that develop from them, called *clonal plants*, all have the same traits. Another advantage to meristem culture is that meristem, unlike other portions of a plant, is virus-free; therefore the plants produced are also virus-free. (The presence of plant viruses weakens plants and makes them less productive.)

Figure 39.10 Asexual reproduction in plants.
Meristem tissue at nodes can generate new plants, as when the "runners" of strawberry plants, *Fragaria*, give rise to new plants.

When flower meristem is used to clone plants, embryolike structures called *somatic embryos* form at the top of the callus. Somatic (asexually produced) embryos that are encapsulated in a protective hydrated gel (and sometimes called artificial seeds) can be shipped anywhere. It's possible to produce millions of somatic embryos at once in large tanks called bioreactors. This is done for certain vegetables like tomato, celery, and asparagus and for ornamental plants like lilies, begonias, and African violets. *Anther culture* is a new technique in which mature anthers are cultured in a medium containing vitamins and growth regulators. The haploid tube cells within the pollen grains divide, producing proembryos consisting of as many as 20 to 40 cells. Finally the pollen grains rupture, releasing haploid embryos. The experimenter can now generate a haploid plant, or chemical agents can be added that encourage chromosomal doubling. After chromosomal doubling, the resulting plants are diploid but homozygous for all their alleles. Anther culture is a direct way to produce plants that express recessive alleles. If the recessive alleles govern desirable traits, the plants have these traits. In contrast to anther culture, it would probably take several generations of hybridized plants to bring out desirable recessive traits.

The culturing of leaf, stem, or root tissues has led to a technique called *cell suspension culture.* Rapidly growing calluses are cut into small pieces and shaken in a liquid nutrient medium so that single cells or small clumps of cells break off and form a suspension. These cells will produce the same chemicals as the entire plant. For example, cell suspension cultures of *Cinchona ledgeriana* produce quinine and those of *Digitalis lanata* produce digitoxin. Scientists envision that it will also be possible to maintain cell suspension cultures in bioreactors for the purpose of producing chemicals used in the production of drugs, cosmetics, and agricultural chemicals. If so, it will no longer be necessary to farm plants simply for the purpose of acquiring the chemicals they produce.

Figure 39.11 Cloning of a plant from tissue cells.
a. Sections of carrot root are cored, and thin slices are placed in a liquid nutrient medium. **b.** After a few days, the cells form a nondifferentiated callus. **c.** After several weeks, the callus begins sprouting cloned carrot plants. **d.** Eventually the carrot plants can be moved from culture to medium to pots.

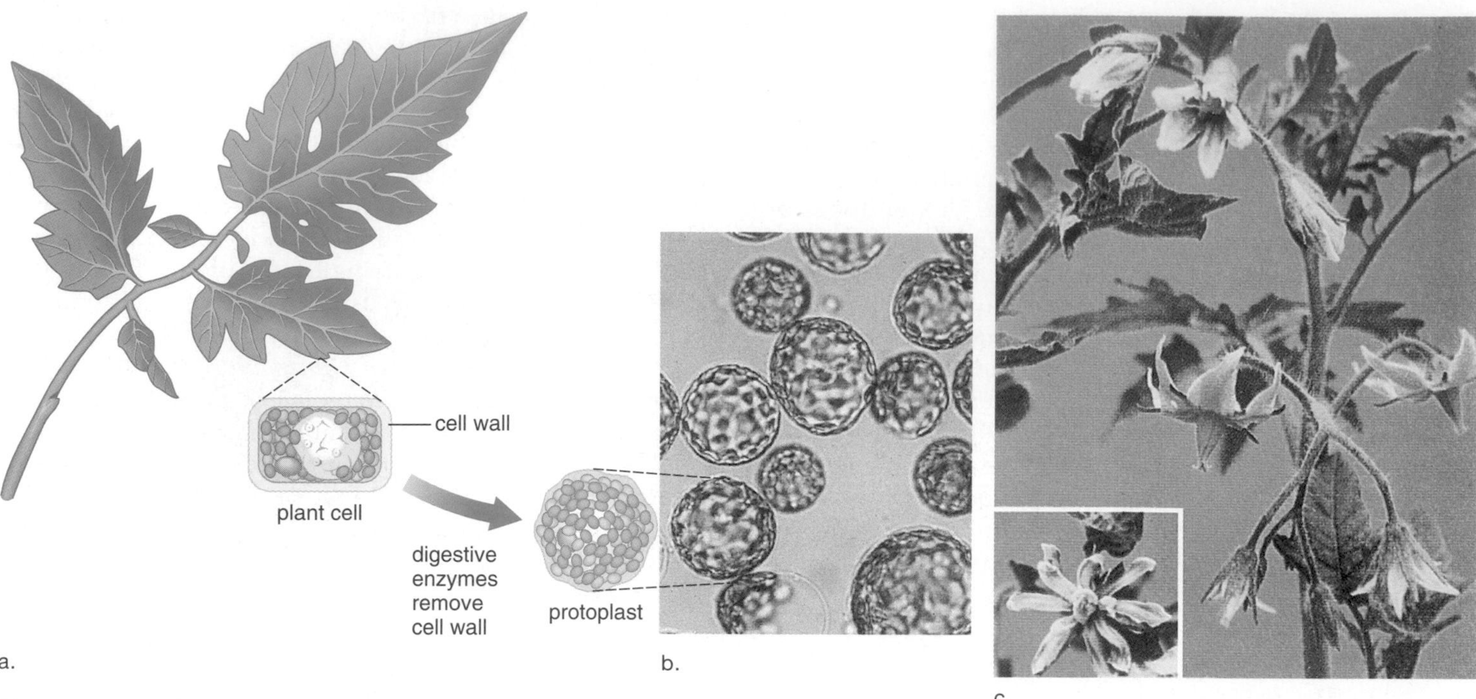

Figure 39.12 Plant protoplasts.
a. When plant cell walls are removed by digestive enzyme action, the result is a naked cell, the protoplast. **b.** Photomicrograph of protoplasts. **c.** Protoplasts from two different species will sometimes fuse to produce hybrid plants. This pomato plant resulted from the fusion of a potato protoplast and a tomato protoplast. The white flowers are like those of a potato plant, and the yellow flowers are like those of a tomato plant.

Because plant cells are totipotent, it should be possible to grow an entire plant from a single cell. This, too, has been done. Enzymes are used to digest the cell walls of a small piece of tissue, usually mesophyll tissue, from a leaf, and the result is naked cells without walls, called **protoplasts** [Gk. *protos*, first, and *plastos*, formed, molded] (Fig. 39.12). The protoplasts regenerate a new cell wall and begin to divide. These clumps of cells can be manipulated to produce somatic embryos or entire plants. Plants generated from the somatic embryos vary somewhat because of mutations that arise during the production process. These so-called *somaclonal variations* are another way to produce new plants with desirable traits.

Genetic Engineering of Plants

It is possible today to alter the genes of organisms so that they have new and different traits. *Transgenic plants* carry a foreign gene that has been introduced into their cells. Since a whole plant will grow from a protoplast, it is necessary only to place the foreign gene into a living protoplast. A foreign gene isolated from any other type of organism is placed in the tissue culture medium. High-voltage electric pulses can then be used to create pores in the plasma membrane so that the DNA enters. For example, after genes for the production of the firefly enzyme luciferase were inserted into tobacco protoplasts, the adult plants glowed when sprayed with the substrate luciferin (Fig. 39.13).

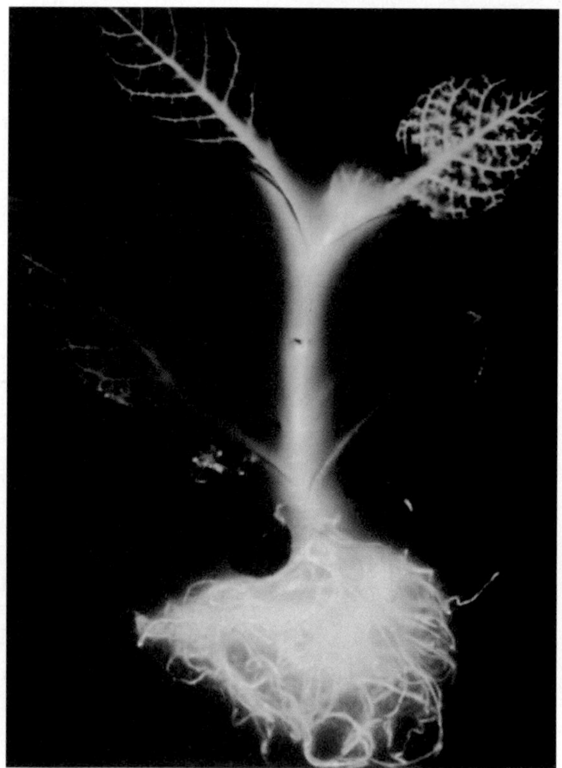

Figure 39.13 Genetically engineered plant.
This transgenic plant glows when sprayed with luciferin because its cells contain the protein luciferase, a firefly enzyme that acts on the chemical luciferin, which emits light.

a.

b.

c.

Some Genetically Engineered Plants

Cereals	Fiber Crops	Food Legumes and Oilseeds	Horticultural Crops		Pastures	Trees
Rice	Cotton	Flax	Carrot	Petunia	Alfalfa	Poplar
Corn		Canola	Cauliflower	Potato	White clover	Apple
Wheat		Soybean	Celery	Sugar beet	Orchard grass	Walnut
Barley		Sunflower	Cucumber	Tobacco		
Rye		Bean	Lettuce	Tomato		
		Pea	Melon			

d.

Figure 39.14 Transgenic crops.
a. Herbicide-tolerant soybeans thrive despite being sprayed with the same herbicide as nontolerant soybeans. **b.** Potato plant that received a gene from *Bacillus thurigiensis* (B.t.) is protected from the Colorado potato beetle, whereas the control plant is not. **c.** Tomatoes that have been engineered to resist the spoilage seen in fruits that have not been so engineered. **d.** A list of crops that have been genetically engineered.

Unfortunately, the regeneration of cereal grains from protoplasts has been difficult. Corn and wheat protoplasts produce infertile plants. As a result, other methods are used to introduce DNA into plant cells that still have their cell walls. In one technique, foreign DNA is inserted into the plasmid of the bacterium *Agrobacterium,* which normally infects plant cells. A plasmid, which is a circular fragment of DNA separated from the bacterial chromosome, can be used to produce recombinant DNA. Recombinant DNA contains genes from different sources, namely those of the plasmid and the foreign genes of interest. When the bacterium infects the plant cells, the plasmid and its DNA are introduced into the plant cells. In 1987, John C. Sanford and Theodore M. Klein of Cornell University developed another method of introducing DNA into intact plant cells. They constructed a device, called a *particle gun,* that bombards plant cells with DNA-coated microscopic metal particles. Many types of plant cells in tissue culture, including corn and wheat, have been genetically engineered using a particle gun. Later, adult plants are generated.

Various crops have been engineered to be resistant to viral infections, insect predation, and herbicides that are judged to be environmentally safe (Fig. 39.14). If crops are resistant to a broad-spectrum herbicide and weeds are not, then the herbicide can be used to kill the weeds. The hope is that in the future it will be possible to produce plants that have a higher protein content and that require less water and fertilizer.

The ability of plants to reproduce asexually has led to the generation of plants in tissue culture. This, in turn, has promoted the genetic engineering of plants.

connecting concepts

Life on earth would not be possible without vascular and specifically flowering plants which now dominate the biosphere. *Homo sapiens* evolved with flowering plants and therefore does not know a world without them. The earliest humans were mostly herbivores—they relied on foods they could gather for survival—fruits, nuts, seeds, tubers, roots and so forth. Plants also provided protection from the environment, offering shelter from heavy rains and noonday sun. Later on, human civilizations could not have begun without the development of agriculture. The majority of the world's population still relies primarily on three plants—corn, wheat, and rice—for the majority of its sustenance. Sugar, coffee, spices of all kinds, cotton, rubber, and tea are plants which have even promoted wars because of their importance to a country's economy. Although we now live in an industrialized society, we are still dependent on plants and have put them to even more uses. They are needed to lubricate the engines of supersonic jets and to make cellulose acetate for films, to take a couple of examples. For millions of urban dwellers, plants are their major contact with the natural world. We grow them not only for food and shelter, but for their simple beauty.

Most people fail to appreciate the importance of plants, but plants may be even more critical to our lives today than they were to our early ancestors on the African plains. Currently, half of all pharmaceutical drugs have their origin in plants. The world's major drug companies are engaged in a frantic rush to collect and test plants in the rain forests for their drug-producing potential. Why the rush? Because the rain forests may be gone before all the possible cures for cancer, AIDS, and other killers have been found. Wild plants not only can help cure human ills, they also serve as a source for genes that can improve the quality of plants that support our way of life.

Summary

39.1 Flowering Plants Undergo Alternation of Generations

Flowering plants produce flowers and often have a mutualistic relationship with an animal pollinator. They also produce seeds that are covered by fruits.

Flowering plants exhibit an alternation of generations life cycle that includes the microgametophyte and megagametophyte. The pollen grain is the microgametophyte. The megagametophyte is called the embryo sac, and it remains within the body of the sporophyte plant.

Typical parts of a flower are: sepals, which are usually green in color and form an outer whorl; petals, often colored, which form an inner whorl; the pistil, which is in the center and contains the carpels, each consisting of stigma, style, and ovary; and the stamens, each having a filament and anther, which are around the base of the carpels. The ovary contains ovules.

Each ovule within the ovary contains a megasporocyte, which divides meiotically to produce four haploid megaspores, only one of which survives. This megaspore divides mitotically to produce the megagametophyte (embryo sac), which usually has eight nuclei. The central cell contains two polar nuclei, and one of the three cells next to the micropyle is an egg cell.

The anthers contain microsporocytes, each of which divides meiotically to produce four haploid microspores. Each of these divides mitotically to produce a two-celled pollen grain. One cell is the tube cell, and the other is the generative cell. The generative cell later divides to produce two sperm cells. The pollen grain is the microgametophyte. After pollination, the pollen grain germinates, and as the pollen tube grows, the sperm cells travel to the embryo sac. Pollination is simply the transfer of pollen from anther to stigma.

Flowering plants practice double fertilization. One sperm nucleus unites with the egg nucleus, forming a 2n zygote, and the other unites with the polar nuclei of the central cell, forming a 3n endosperm cell.

After fertilization, the endosperm cell divides to form the endosperm. The zygote becomes the sporophyte embryo. The ovule matures into the seed (its integuments become the seed coat). The ovary becomes the fruit.

39.2 The Embryo Develops in Stages

As the ovule is becoming a seed, the zygote is becoming an embryo. After the first several divisions, it is possible to discern the embryo and the suspensor. The suspensor attaches the embryo to the ovule and supplies it with nutrients. The dicot embryo becomes first heart shaped and then torpedo shaped. Once you can see the two cotyledons, it is possible to distinguish the shoot apex and the root apex, which contain the apical meristems. In dicot seeds, the cotyledons frequently take up the endosperm.

39.3 The Seeds Are Enclosed by Fruit

The seeds of flowering plants are enclosed by fruits. There are different types of fruits. Simple fruits are derived from a single ovary (which can be simple or compound). Some simple fruits are fleshy, such as a peach or an apple. Others are dry, such as peas, nuts, and grains. Aggregate fruits develop from a number of ovaries of a single flower, and compound fruits develop from multiple ovaries of separate flowers.

Flowering plants have several ways to disperse seeds. Seeds may be blown by the wind, attached to animals that carry them away, eaten by animals that defecate them some distance away, or adapted to water transport.

Prior to germination, you can distinguish a bean (dicot) seed's two cotyledons and plumule, which is the shoot that bears leaves. Also present are the epicotyl, the hypocotyl, and the radicle. In a corn kernel (monocot), the endosperm, the cotyledon, the plumule, and the radicle are visible.

39.4 Plants Can Reproduce Asexually

Many flowering plants reproduce asexually, as when the nodes of stems (either aboveground or underground) give rise to entire plants, or when roots produce new shoots.

The practice of hydroponics—and the recognition that plant cells can be totipotent—led to plant tissue culture, a technique that now has many applications.

Micropropagation, the production of clonal plants as a result of meristem culture in particular, is now a commercial venture. Flower meristem culture results in somatic embryos that can be packaged in gel for worldwide distribution. Anther culture results in homozygous plants that express recessive genes. Leaf, stem, and root culture can result in cell suspensions that may eventually allow the production of plant chemicals in large tanks. Development of adult plants from protoplasts results in somaclonal variations, a new source of plant varieties.

Protoplasts in particular lend themselves to direct genetic engineering in tissue culture. Otherwise, the *Agrobacterium* technique or the particle-gun technique allow foreign genes to be introduced into plant cells, which then develop into adult plants with particular traits.

Reviewing the Chapter

1. Name two unique features regarding the reproduction of flowering plants. 694
2. Draw a diagram of a flower and name the parts. 695
3. Draw a diagram that illustrates the life cycle of flowering plants. Why don't flowering plants require a source of outside water for pollination? 696–97
4. Describe the development of a megagametophyte, from the megasporocyte to the production of an egg. 696–97
5. Describe the development of the microgametophyte, from the microsporocyte to the production of sperm. 696–97
6. What is the difference between pollination and fertilization? 700
7. Describe the sequence of events as a dicot zygote becomes an embryo enclosed within a seed. 701
8. Distinguish between simple fleshy fruits and simple dry fruits. Give an example of each type. What is an aggregate fruit? A multiple fruit? 702–3
9. Name several mechanisms of seed and/or fruit dispersal. Contrast the germination of a bean seed with that of a corn kernel. 704–5
10. In what ways do plants ordinarily reproduce asexually? What is the importance of totipotency in regard to tissue culture? 708
11. What are the benefits of meristem culture to achieve micropropagation; flower meristem culture to produce somatic embryos; anther culture to produce homozygous plants; leaf, stem, or root culture to produce cell suspension cultures; and the maintenance of protoplasts in tissue culture? 708–10
12. How are transgenic plants produced? What types of plants have been produced, and for what purpose have they been genetically engineered? 710–11

Testing Yourself

Choose the best answer for each question.

1. In plants,
   a. gametes become the gametophyte generation.
   b. spores become the sporophyte generation.
   c. sporophyte organisms produce spores.
   d. Both a and b are correct.
2. The flower part that contains ovules is the
   a. carpel.
   b. stamen.
   c. sepal.
   d. petal.
3. The megasporocyte and the microsporocyte
   a. both produce pollen grains.
   b. both divide meiotically.
   c. both divide mitotically.
   d. produce pollen grains and embryo sacs, respectively.
4. A pollen grain is
   a. a haploid structure.
   b. a diploid structure.
   c. first a diploid and then a haploid structure.
   d. first a haploid and then a diploid structure.
5. Which of these is mismatched?
   a. polar nuclei—plumule
   b. egg and sperm—zygote
   c. ovule—seed
   d. ovary—fruit
6. Which of these is not a fruit?
   a. walnut
   b. pea
   c. green bean
   d. peach
7. Animals assist with
   a. both pollination and seed dispersal.
   b. only pollination.
   c. only seed dispersal.
   d. only asexual propagation of plants.
8. A seed contains
   a. a zygote.
   b. an embryo.
   c. stored food.
   d. Both b and c are correct.
9. Which of these is mismatched?
   a. plumule—leaves
   b. cotyledon—seed leaf
   c. epicotyl—root
   d. pericarp—corn kernel
10. Which of these best shows that plant cells are totipotent?
    a. shoot apex culture for the purpose of micropropagation
    b. flower meristem culture for the purpose of somatic embryos
    c. leaf, stem, and root culture for the purpose of cell suspension cultures
    d. protoplast culture for the purpose of genetic engineering of plants

11. Label the following diagram of alternation of generations in flowering plants.

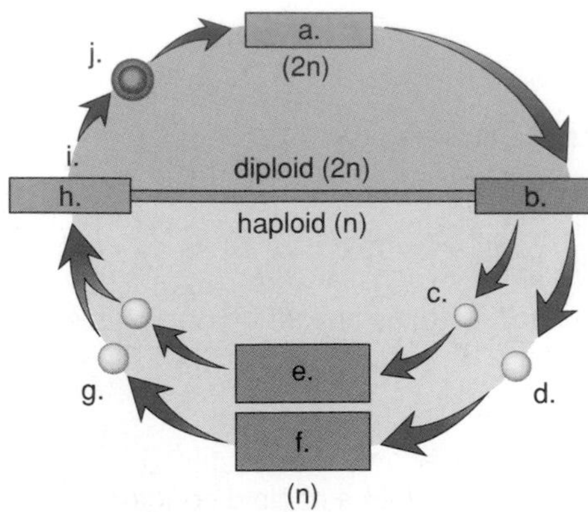

Applying the Concepts

1. *Living things have a reproductive strategy.*
   What is the reproductive strategy of flowering plants?
2. *Living things have life cycles that promote survival of the species.*
   How does the life cycle of flowering plants promote the possibility of gametophyte and sporophyte survival?
3. *Coevolution increases the fitness of both parties.*
   How does coevolution between plants and their pollinators increase the fitness of the pollinators?

Using Technology

Your study of reproduction in plants is supported by these available technologies:

Exploring the Internet
The Mader Home Page provides resources for and help with studying this chapter.

http://www.mhhe.com/sciencemath/biology/mader/
(Click on Biology.)

Understanding the Terms

anther 695
carpel 695
coevolution 698
cotyledon 701
double fertilization 700
embryo 700
embryo sac 697
endosperm 700
fertilization 700
filament 695
flower 694
fruit 702
gametophyte 694
germinate 704
hybridization 708
megagametophyte 697
megaspore 694
megasporocyte 697
microgametophyte 697
microspore 694
microsporocyte 697
ovary 695
ovule 695
petal 694
pistil 695
plumule 705
pollen grain 697
pollination 700
protoplast 710
seed 702
sepal 694
sporophyte 694
stamen 695
stigma 695
style 695
tissue culture 708
totipotent 708

Match the terms to these definitions:

a. __________ Flower structure consisting of an ovary, a style, and a stigma.
b. __________ Flowering plant structure consisting of one or more ripened ovaries that usually contain seeds.
c. __________ In flowering plants, a reproductive unit of a pistil; consisting of three parts—the stigma, the style, and the ovary.
d. __________ In flowering plants, the embryonic plant shoot that bears young leaves.
e. __________ In flowering plants, the transfer of pollen from the anther of the stamen to the stigma of the pistil.
f. __________ In seed plants, the gametophyte that produces an egg; in flowering plants, an embryo sac.
g. __________ Mature ovule that contains an embryo, with stored food enclosed in a protective coat.
h. __________ Microgametophyte in seed plants.
i. __________ Reproductive organ of a flowering plant, consisting of several kinds of modified leaves arranged in concentric rings and attached to a modified stem called the receptacle.
j. __________ Seed leaf for embryo of a flowering plant; provides nutrient molecules for the developing plant before it begins to photosynthesize.

Further Readings for Part vi

Bazzaz, F. A., and Fajer, E. D. January 1992. Plant life in a CO_2-rich world. *Scientific American* 266(1):68. Experiments with plants grown in a CO_2-enriched atmosphere show that the risks to ecosystems outweigh the gains in productivity.

Chrispeels, M., and Sadava, D. 1994. *Plants, genes and agriculture.* Boston: Jones and Bartlett Publishers. Teaches plant biology in an agricultural context.

Cox, P. A. October 1993. Water-pollinated plants. *Scientific American* 269(4):68. Some flowering aquatic species provide evidence for evolutionary convergence toward efficient pollination strategies.

Cox, P. A., and Balick, M. J. June 1994. The ethnobotanical approach to drug discovery. *Scientific American* 270(6):82. Many rain forest plants are used by indigenous cultures for medicinal purposes; these flora should be screened for pharmaceutical compounds.

Govindjee, and Coleman, W. J. February 1990. How plants make oxygen. *Scientific American* 262(2):50. Discusses the mechanism that enables plants and some bacteria to make oxygen.

Lea, P., and Leegood, R. 1993. *Plant biochemistry and molecular biology.* Chichester, England: John Wiley & Sons. Gives an overview of plant metabolism.

Lewington, A. 1990. *Plants for people.* New York: Oxford University Press. This is a full-color account of the ways in which people make use of plant products.

Luoma, J. R. March 1997. The magic of paper. *National Geographic* 191(3):88. The paper-making process is discussed in this article.

Mauseth, J. 1995. *Botany: An introduction to plant biology.* 2d ed. Philadelphia: Saunders College Publishing. Emphasizes evolution and diversity in botany and general principles of plant physiology and anatomy.

Meyerowitz, E. M. Novemher 1994. The genetics of flower development. *Scientific American* 271(5):56. Shows how flower design is determined by genetic signals.

Moore, R., and Clark, W. D. 1995. *Botany, plant form & function.* Dubuque, Iowa: Wm. C. Brown Publishers. This is a comprehensive discussion of general botany that captures the excitement of new developments.

Niklas, K. 1992. *Plant biomechanics.* Chicago: The University of Chicago Press. The book explores how plants function, grow, reproduce, and evolve within limits set by their physical environment.

Northington, D., and Goodin, J. R. 1996. *The botanical world.* 2d ed. St. Louis: Times-Mirror/Mosby College Publishing. This is an account of plant interactions and basic physiology.

Raven, P., et al. 1992. *Biology of plants.* 5th ed. New York: Worth Publishers. Written for the undergraduate, this book covers viruses, bacteria, photosynthetic protists, fungi, and plants.

Redington, C. 1994. *Plants in wetlands.* Dubuque, Iowa: Kendall/Hunt Publishing Company. Explains how specific plants interact within the wetlands ecosystem.

Salisbury, F., and Ross, C. 1992. *Plant physiology.* 4th ed. Belmont, Calif.: Wadsworth Publishers. This text for botany majors emphasizes seed plant physiology.

Seymour, R. S. March 1997. Plants that warm themselves. *Scientific American* 276(3):104. Some plants generate heat to keep blossoms at a constant temperature.

Steeves, T., and Sussex, I. 1990. *Patterns in plant development.* 2d ed. Cambridge: Cambridge University Press. Details plant development, from embryo through secondary growth.

Stern, K. 1994. 6th ed. *Introductory plant biology.* Dubuque, Iowa: Wm. C. Brown Publishers. Presents basic botany in a clear, informative manner.

Stryer, L. 1995. *Biochemistry.* 4th ed. New York: W. H. Freeman and Company. Chapter 22 of this text presents an advanced but understandable treatment of photosynthesis.

Taiz, L., and Zeiger, E. 1991. *Plant physiology.* Redwood City, Calif.. The Benjamin/Cummings Publishing Company, Inc. Presents the dynamic processes of growth, metabolism, and reproduction in plants.

Tedeschi, H. 1993. *Cell physiology.* 2d ed. Dubuque, Iowa: Wm. C. Brown Publishers. This text emphasizes the need to study original articles as the physiology of the cell is studied from the prospective of the cell as an energy converter.

Animal Structure and Function

In contrast to plants, animals are heterotrophic. Their mobility, which is dependent upon nerve fibers and muscle fibers, is essential in finding food. Then the food is digested and the nutrients are distributed to cells. Finally, wastes are expelled.

We will observe a progressive modification from simple to more complex structures to carry out these functions. Genes code for anatomical and physiological solutions to environmental requirements, and evolution results in modifications of these solutions. Complexity has arisen through the evolutionary process as a way to adapt to the environment, which includes both the abiotic and biotic (living) environment.

In this part, we will map out the various lines of animal descent and, in so doing, will trace the evolution of the major types of animals. We will observe the many modifications that allow animals to utilize resources, extract energy, protect themselves, and reproduce.

Animal Organization and Homeostasis

CHAPTER 40

Ring-tailed lemurs, *Lemur catta*

Chapter Concepts

40.1 Animals Are Organized

- Animals have these levels of organization: molecules—cells—tissues—organs—organ systems—organism. 718
- Animal tissues can be categorized into four major types: epithelial, connective, muscular, and nervous tissues. 718
- Epithelial tissues, which line body cavities and cover surfaces, are specialized in structure and function. 718
- Connective tissues, which protect, support, and bind other tissues, include cartilage and bone and also blood, the only liquid tissue. 720
- Muscular tissues, which contract, make body parts move. 722

40.2 Organs Have Structure and Functions

- Organs usually contain several types of tissues. For example, although skin is composed primarily of epithelial tissue and connective tissue, it also contains muscle and nerve fibers. 724
- Organs are grouped into organ systems, each of which has specialized functions. 726
- The coelom, which arises during development is later divided into various cavities where specific organs are located. 726

40.3 Homeostasis Is Necessary

- Homeostasis is the relative constancy of the internal environment. All organ systems contribute to homeostasis in animals. 728

All animals are multicellular, and by exchanging substances with the environment, keep their cells in working order. With a relatively stable internal environment, termed homeostasis, cells can function despite changing conditions outside the body.

The organization of complex animals includes organ systems, such as the digestive and respiratory systems, that service the cells. Organ systems contain organs, which are composed of tissues, and each type of tissue has like cells. This chapter concerns these levels of organization and in particular examines the various types of tissues found in more complex animals, like the ring-tailed lemurs in the photograph. The skin of vertebrates exemplifies an organ, and the functions of the various organ systems is also mentioned.

Homeostasis is achieved through feedback control mechanisms regulated by the nervous and endocrine systems. Homeostasis is our theme for a study of animal structure and function in this part that emphasizes a comparative view of the major types of structural and functional adaptations of animals. Ring-tailed lemurs are adapted to locomotion on land and in trees. They feed primarily on the fruits and leaves of trees.

40.1 Animals Are Organized

All living things are organized, and animals are no exception. Some animals, like sponges, are merely multicellular, and some, like cnidaria, have two tissue layers derived from two germ layers only, but most animals have three germ layers: ectoderm, mesoderm, and endoderm. The germ layers arise during development, and then in vertebrates develop into the structures mentioned:

Embryonic Germ Layer	Vertebrate Adult Structures
Ectoderm (outer layer)	Epidermis of skin; epithelial lining of mouth and rectum; nervous system
Mesoderm (middle layer)	Skeleton; muscular system; dermis of skin; circulatory system; excretory system; reproductive system, including most epithelial linings; outer layers of respiratory and digestive systems
Endoderm (inner layer)	Epithelial lining of digestive tract and respiratory tract; associated glands of these systems; epithelial lining of the urinary bladder

In this chapter we begin at the tissue level of organization, but we also consider the organ and system levels of organization. A tissue is a group of similar cells performing a similar function. Different types of tissues make up organs, and several organs are found within an organ system. The organ systems make up the organism (Fig. 40.1).

The structure and functions of an organ system are dependent upon the structure and functions of the organ, tissue, and cell type contained therein. For instance, the columnar epithelial cells of the intestine, which absorb nutrients, have microvilli, which increase surface area for absorption. Skeletal muscle cells are tubular and contain elements that contract to a shorter length. Nerve cells have long, slender projections that carry nerve impulses to distant body parts.

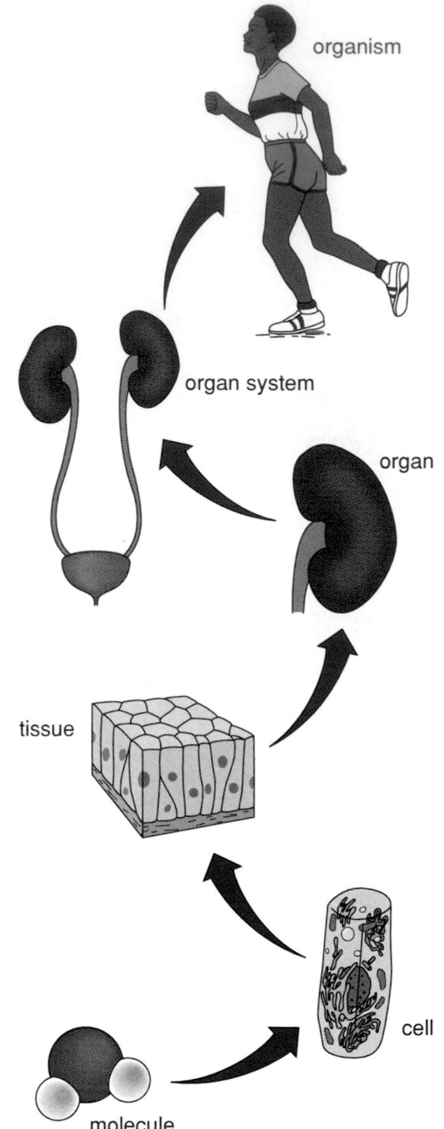

Figure 40.1 Levels of organization.
A cell is composed of molecules, a tissue is made up of cells; an organ is composed of tissues; and the organism contains organ systems.

Epithelial Tissue Covers

There are four major types of **tissue** in vertebrate animals: *epithelial tissue* covers body surfaces and lines body cavities; *connective tissue* binds and supports body parts; *muscular tissue* causes body parts to move; and *nervous tissue* responds to stimuli and transmits impulses from one body part to another.

Epithelial tissue [Gk. *epi,* over, and L. *theca,* case, container], also called epithelium, forms a continuous layer over body surfaces, lines inner cavities, and forms glands. One side of an epithelium is exposed at the skin surface or to a body cavity. The other side is usually attached to a basement membrane, which is a thin mat of specialized extracellular matrix (see Fig. 5.14).

Classifed according to cell shape, there are three types of epithelial tissues. *Squamous epithelium* is composed of flat cells; *cuboidal epithelium* contains cube-shaped cells; and in *columnar epithelium,* the oblong cells resemble pillars or columns. Figure 40.2 describes the structure and function of epithelium (pl., epithelia) in vertebrates. Any epithelium can be simple or stratified. *Simple* means that the tissue has a single layer of cells, and *stratified* means that the tissue has layers piled one on top of the other. One type of epithelium is pseudostratified—it appears to be layered, but actually true layers do not exist because each cell touches the basement membrane.

Epithelial tissues have various functions. Epithelial cells can have hairlike extensions called cilia, which bend and move materials in a particular direction. In the human body, the ciliated epithelium lining the respiratory tract sweeps impurities toward the throat so they do not enter the lungs. An epithelium sometimes secretes a product and is described as glandular. A gland can be a single

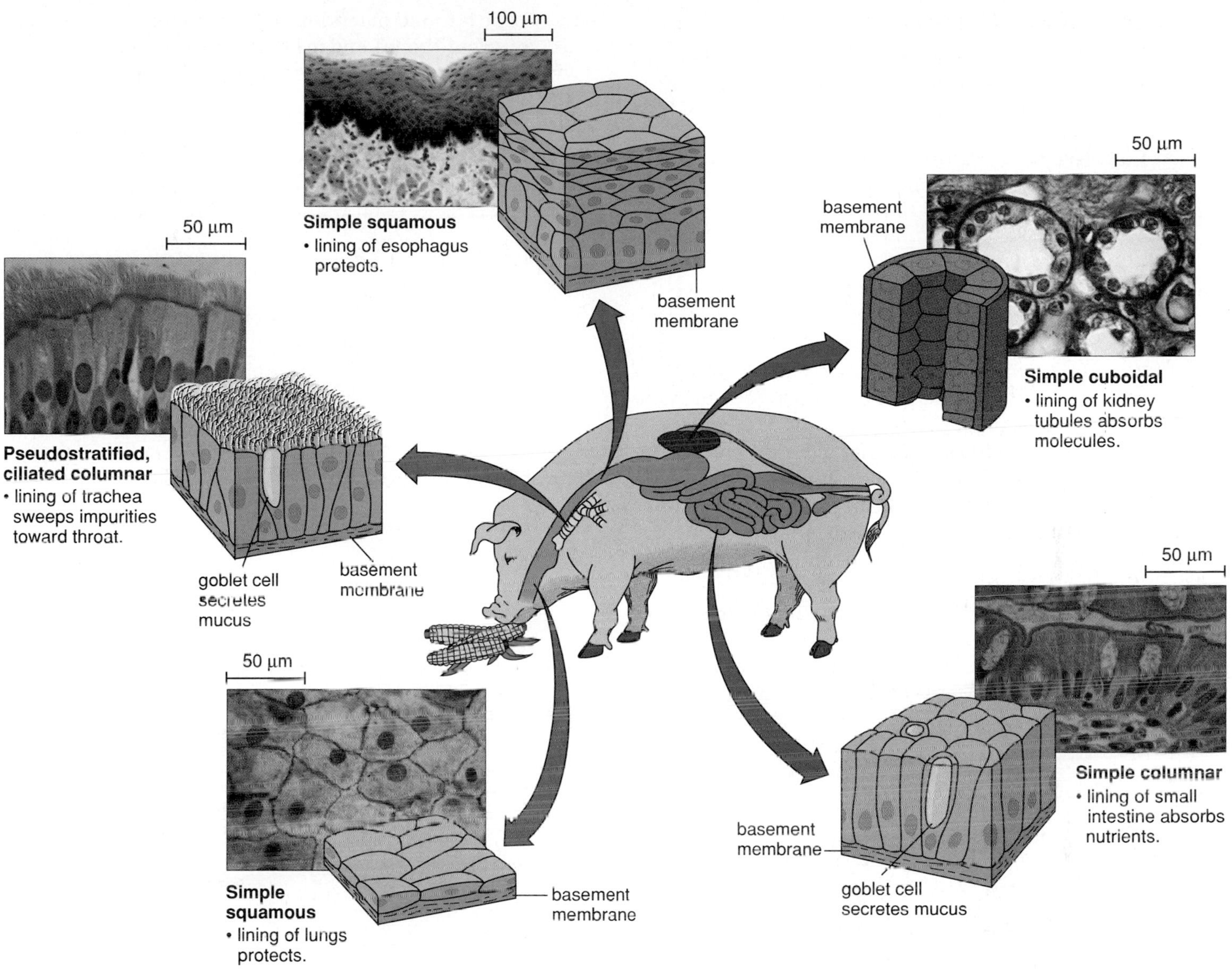

Figure 40.2 Types of epithelial tissues in vertebrates.
Epithelial tissues are classified according to shape of cell, whether they are simple or stratified, and whether they have cilia. Epithelial tissues have functions associated with protection, secretion, and absorption.

epithelial cell, such as the mucus-secreting goblet cells in the lining of the human intestine, or a gland can contain numerous cells. **Exocrine glands** secrete their product into ducts or into cavities, and **endocrine glands** secrete their products directly into the bloodstream.

Epithelium forms the outer layer of skin of most animals. The skin of earthworms and snails is glandular and produces mucus that lubricates the body, helping to ease movement through a dry environment. In roundworms, annelids, and arthropods, an outer nonliving and protective cuticle is produced by epithelium. In terrestrial vertebrates, skin cells contain keratin, a substance that protects the skin from the possible loss of water.

Epithelial tissue cells are packed tightly and joined to one another in one of three ways: tight junctions, adhesion junctions (desmosomes), or gap junctions (see Fig. 5.15). In *tight junctions*, plasma membrane proteins extending between neighboring cells bind them tightly. For example, they prevent digestive juices from passing between the epithelial cells lining the lumen (cavity). In *adhesion junctions*, cytoskeletal elements join internal plaques present in both cells. They allow the skin to withstand considerable stretching and mechanical stress. *Gap junctions* form when two identical plasma membrane channels join. They allow ions and small molecules to pass between the cells.

Epithelial tissue is classified according to cell shape. There can be one or many layers of cells, and the epithelium can be ciliated and/or glandular.

Connective Tissue Connects

Connective tissue binds structures together, provides support and protection, fills spaces, stores fat, and forms blood cells. It provides the source cells for muscle and skeletal cells in animals that can regenerate lost parts.

Connective tissue cells are usually separated widely by a *matrix,* a noncellular material that varies in consistency from solid to semifluid to fluid.

Loose Fibrous and Dense Fibrous

The cells of loose fibrous and dense fibrous connective tissues, called **fibroblasts** [L. *fibra,* thread, and Gk. *blastos,* bud], are located some distance from one another and are separated by a jellylike matrix that contains white collagen fibers and yellow elastic fibers. Collagen fibers provide flexibility and strength; elastic fibers provide elasticity.

Loose fibrous connective tissue supports epithelium and also many internal vertebrate organs (Fig. 40.3*a*). Its presence in lungs, arteries, and the urinary bladder allows these organs to expand. It forms a protective covering encasing many internal organs, such as muscles, blood vessels, and nerves.

Dense fibrous connective tissue contains many collagenous fibers that are packed closely together. This type of tissue has more specific functions in vertebrates than does loose connective tissue. For example, dense fibrous connective tissue is found in **tendons** [L. *tendo,* stretch], which connect muscles to bones, and in **ligaments** [L. *ligamentum,* band], which connect bones to other bones at joints.

Loose fibrous connective tissue and dense fibrous connective tissue contain fibroblasts separated by a matrix, which contains collagen and elastic fibers.

Adipose Tissue and Reticular Connective Tissue

Adipose tissue [L. *adipalis,* fatty], insulates the body and provides padding because the fibroblasts enlarge and store fat (Fig. 40.3*b*). In mammals, adipose tissue is found particularly beneath the skin, around the kidneys, and on the surface of the heart. *Reticular connective tissue* is present in the lymph nodes, the spleen, and the bone marrow. Here, reticular fibers, associated with reticular cells resembling fibroblasts, support many free blood cells.

Cartilage and Bone Are Solid

Cartilage and bone are rigid connective tissues in which structural proteins (cartilage) or calcium salts (bone) are deposited in the intercellular matrix.

In **cartilage** [L. *cartilago,* gristle], the cells lie in small chambers called **lacunae** (sing., lacuna), separated by a matrix that is strong yet flexible (Fig. 40.3*c*). There are various

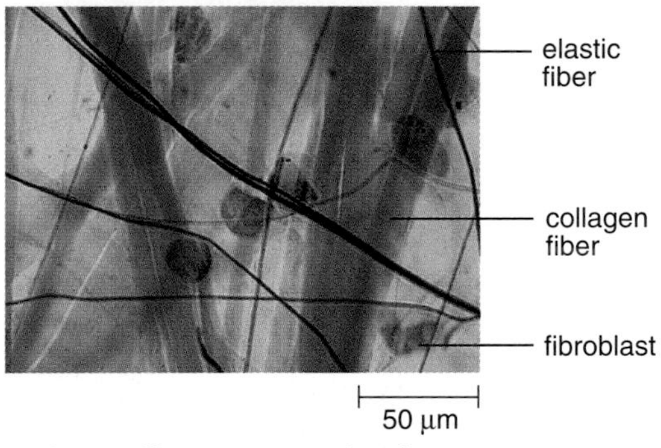

a. **Loose fibrous connective tissue**
- has space between components.
- occurs beneath skin and most epithelial layers.
- functions in support and binds organs.

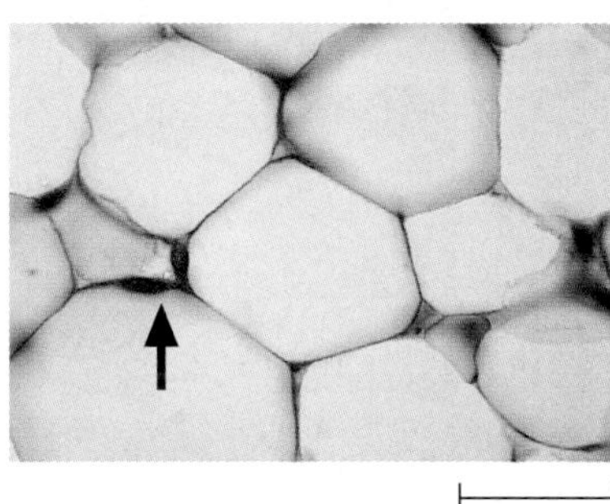

b. **Adipose tissue**
- cells are filled with fat.
- occurs beneath skin, around organs and heart.
- functions in insulation, stores fat.

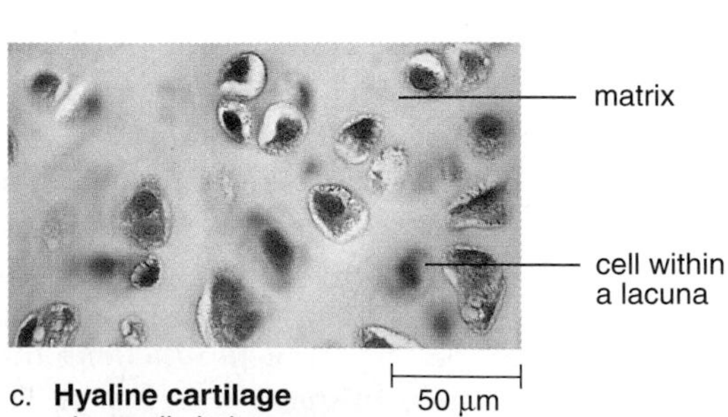

c. **Hyaline cartilage**
- has cells in lacunae.
- occurs in nose and walls of respiratory passages; at ends of bones including ribs.
- functions in support and protection.

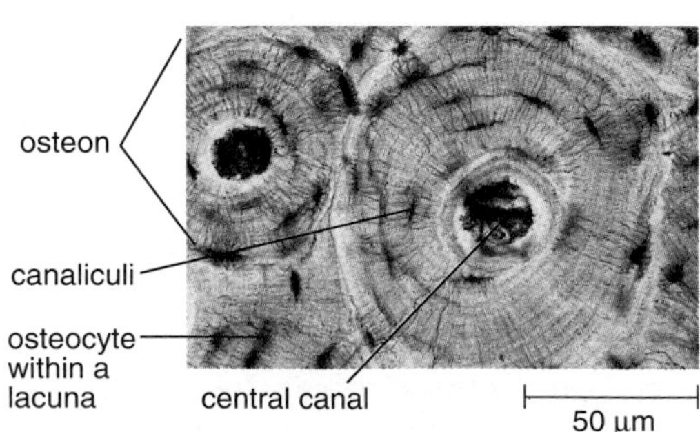

d. **Compact bone**
- has cells in concentric rings.
- occurs in bones of skeleton.
- functions in support and protection.

Figure 40.3 Connective tissue examples.
a. In loose connective tissue, fibroblasts are separated by a matrix that is jellylike but contains fibers. **b.** Adipose tissue cells are filled with fat and the nuclei (arrow) are pushed to one side. **c.** In hyaline cartilage, a flexible matrix has a translucent appearance. **d.** In compact bone, the hard matrix contains concentric rings of osteocytes in elongated cylinders called osteons. The central canal contains blood vessels and nerve fibers.

types of cartilage, which are classified according to type of collagen and elastic fiber found in the matrix. In some vertebrates, notably sharks and rays, the entire skeleton is made of cartilage. In humans, the fetal skeleton is cartilage, but it is later replaced by bone. Cartilage is retained at the ends of long bones, at the end of the nose, in the framework of the ear, in the walls of respiratory ducts, and within intervertebral disks.

In **bone,** the matrix of inorganic, chiefly calcium, salts is deposited around protein fibers, especially collagen fibers. The minerals give bone rigidity, and the protein fibers provide elasticity and strength, much as steel rods do in reinforced concrete.

In *compact bone,* bone cells (e.g., osteocytes) are located in lacunae that are arranged in concentric circles within osteons (Haversian systems) around tiny tubes called central canals (Fig. 40.3*d*). Nerve fibers and blood vessels are in these canals. The latter bring the nutrients that allow bone to renew itself. The nutrients can reach all of the cells because there are minute canals (canaliculi) containing thin processes of the osteocytes that connect them with one another and with the central canals.

The ends of a long bone contain spongy bone, which has an entirely different structure. *Spongy bone* contains numerous bony bars and plates separated by irregular spaces. Although lighter than compact bone, spongy bone still is designed for strength. Just as braces are used for support in buildings, the solid portions of spongy bone follow lines of stress.

Cartilage and bone are support tissues. Cartilage is more flexible than bone because the matrix is rich in protein; bone is rich in calcium salts.

Blood Is a Liquid

Blood has transporting, regulating, and protective functions. It transports nutrients and oxygen to cells and removes carbon dioxide and other wastes. It helps distribute heat and also plays a role in fluid, ion, and pH balance. Blood is a connective tissue in which the cells are separated by a liquid called plasma (Table 40.1). In vertebrates, blood cells are primarily two types: red blood cells (erythrocytes), which carry oxygen, and white blood cells (leukocytes), which aid in fighting infection (Fig. 40.4). Also present in plasma are platelets, which are important to the initiation of blood clotting. Platelets are not complete cells; rather, they are fragments of giant cells found in the bone marrow.

Blood is unlike other types of connective tissue in that the intercellular matrix (i.e., plasma) is not made by the cells. Plasma is a mixture of different types of molecules that enter the blood at various locations.

Blood is a connective tissue in which the matrix is plasma.

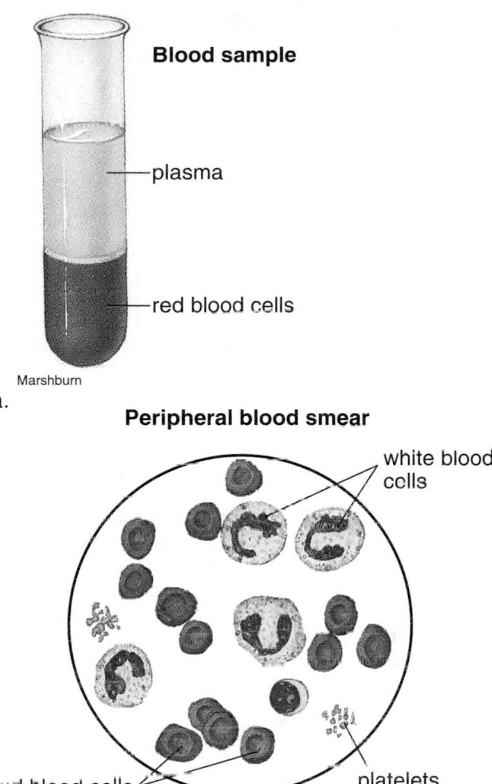

Figure 40.4 Blood, a liquid tissue.
a. Blood is classified as connective tissue because the cells are separated by a matrix—plasma. Plasma, the liquid portion of blood, usually contains several types of cells. This shows the cells in human blood (red blood cells, white blood cells, and platelets, which are actually fragments of a larger cell). **b.** Drawing of the components of blood.

Table 40.1

Blood Plasma

Water (92% of Total)	
Solutes (8% of Total)	
Inorganic ions (salts)	Na^+, Ca^{2+}, K^+, Mg^{2+}, Cl^-, HCO_3^-, HPO_4^{2+}, SO_4^{2+}
Gases	O_2, CO_2
Plasma proteins	Albumin, globulins, fibrinogen
Organic nutrients	Glucose, fats, phospholipids, amino acids, etc.
Nitrogenous waste products	Urea, ammonia, uric acid
Regulatory substances	Hormones, enzymes

Muscular Tissue Contracts

Muscular (contractile) tissue is composed of cells called *muscle fibers.* Muscle fibers contain actin filaments and myosin filaments, whose interactions accounts for the movements we associate with animals. There are three types of vertebrate muscle tissue: skeletal, cardiac, and smooth.

Skeletal muscle also called striated muscle is attached via tendons to the bones of the skeleton; it moves body parts. It is under voluntary control and contracts faster than all the other muscle types. Skeletal muscle fibers are cylindrical and quite long—sometimes they run the length of the muscle. They arise during development when several cells fuse, producing one muscle fiber with multiple nuclei. The nuclei are located at the periphery of the fiber, just inside the plasma membrane. Skeletal muscle fibers are **striated** [L. *stria,* lined]. Light and dark bands run perpendicular to the length of the fiber. These bands are due to the placement of actin filaments and myosin filaments in the cell (Fig. 40.5*a*).

Cardiac muscle, which is found only in the wall of the heart, is responsible for the heartbeat, which pumps blood. Cardiac muscle seems to combine features of both smooth muscle and skeletal muscle (Fig. 40.5*b*). Its fibers have striations like skeletal muscle, but the contraction of the heart is involuntary for the most part. Cardiac muscle fibers also differ from skeletal muscle fibers in that they have a single, centrally placed nucleus. These fibers are branched and seemingly fused one with the other, and the heart appears to be composed of one large interconnecting mass of muscle fibers. Actually, cardiac muscle fibers are separate and individual, but they are bound end to end at intercalated disks, areas where folded plasma membranes between two fibers contain desmosomes and gap junctions. *Intercalated disks* allow impulses to move from cell to cell so that the beat of the heart is coordinated.

Smooth (visceral) muscle is so named because its fibers lack striations. The spindle-shaped fibers form layers in which the thick middle portion of one fiber is opposite the thin ends of adjacent fibers. Consequently, the nuclei form an irregular pattern in the tissue (Fig. 40.5*c*). Smooth muscle is not under voluntary control and therefore is said to be involuntary. Smooth muscle, found in walls of viscera (intestine, stomach, and other internal organs) and blood vessels, contracts more slowly than skeletal muscle but can remain contracted for a longer time. When the smooth muscle of the intestine contracts, it moves the food along, and when the smooth muscle of the blood vessels contracts, it constricts the blood vessels, helping to raise the blood pressure.

All muscle tissue contains actin and myosin filaments; these overlap, forming striations in skeletal and cardiac muscle but not in smooth muscle.

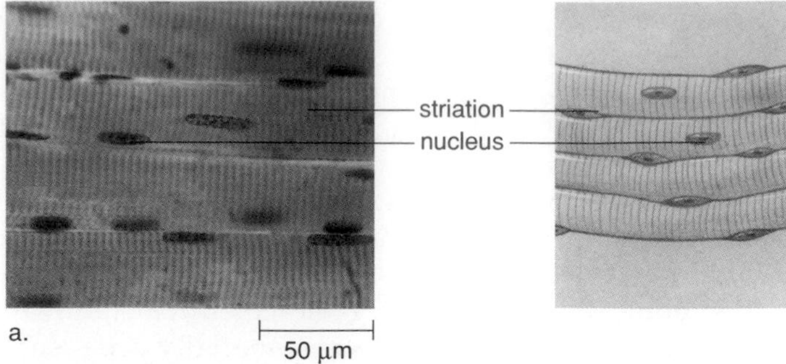

Skeletal muscle
- has striated cells with multiple nuclei.
- usually attached to skeleton.
- functions in voluntary movement.

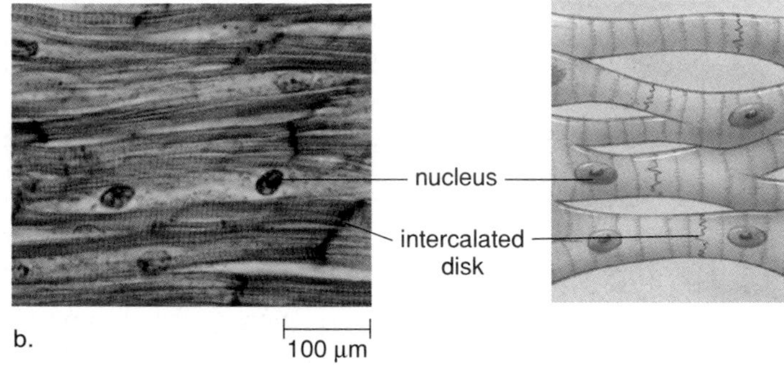

Cardiac muscle
- has branching striated cells, each with a single nucleus.
- occurs in the wall of the heart.
- functions in the pumping of blood.

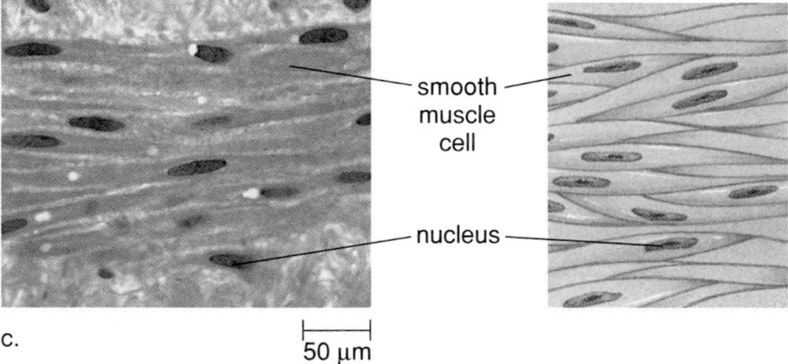

Smooth muscle
- has spindle-shaped cells, each with a single nucleus.
- occurs in walls of internal organs.
- functions in movement of substances in lumens of body.

Figure 40.5 Muscular tissue.
a. Skeletal muscle is voluntary and striated. **b.** Cardiac muscle is involuntary and striated. The plasma membranes of the branching cells fit together at intercalated disks. **c.** Smooth muscle is involuntary and nonstriated.

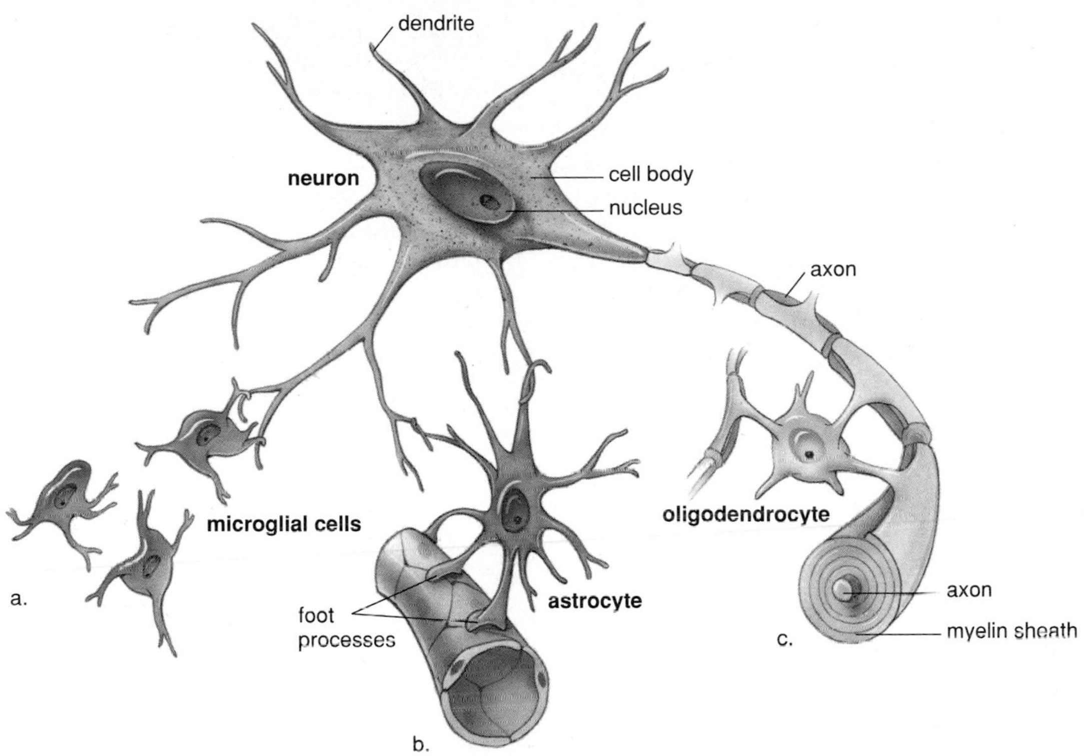

Figure 40.6 Neuroglial cells in the central nervous system.
a. Microglial cells are phagocytes that clean up debris. b. Astrocytes lie between neurons and a capillary; therefore, substances entering neurons from the blood must first pass through astrocytes. c. Oligodendrocytes form the myelin sheaths around fibers in the central nervous system.

Nervous Tissue Conducts

Nervous tissue, which contains nerve cells called **neurons** [Gk. *neuron,* nerve] (Fig. 40.6), is present in the brain and spinal cord. A neuron is a specialized cell that has three parts: (1) dendrites conduct signals to the cell body; (2) the cell body contains the major concentration of the cytoplasm and the nucleus of the neuron; and (3) the axon conducts nerve impulses away from the cell body.

Axons and dendrites are called *neuron* fibers. Fibers can be quite long, and outside the brain and spinal cord, long fibers, bound by connective tissue, form nerves. Nerves conduct impulses from receptors to the spinal cord and the brain, where the phenomenon called sensation occurs. They also conduct nerve impulses away from the spinal cord and brain to the muscles and glands, causing them to contract and secrete, respectively. In this way, a coordinated response to the **stimulus** is achieved. In addition to neurons, nervous tissue contains neuroglial cells.

Neuroglial Cells Are Versatile

There are several different types of neuroglial cells in the CNS (Fig. 40.6), and much research is currently being conducted to determine how much "glial" cells contribute to the functioning of the central nervous system. Neuroglial cells outnumber neurons nine to one and take up more than half the volume of the brain, but until recently, they were thought to merely support and nourish neurons. Oligodendrocytes form myelin, and microglial cells, in addition to supporting neurons, also phagocytize bacterial and cellular debris. Aside from providing nutrients to neurons, astrocytes absorb the neurotransmitter glutamate, and they produce a growth factor known as *glial-derived growth factor (GDGF)* that someday might be used as a cure for Parkinson disease and other diseases caused by neuron degeneration. Neuroglial cells don't have long processes, but even so, researchers are now beginning to gather evidence that they do communicate among themselves and with neurons!

Nerve cells, called neurons, have fibers (processes) called axons and dendrites. Long fibers are found in nerves.

40.2 Organs Have Structure and Functions

We tend to associate particular tissues with particular organs. For example, we associate muscular tissue with muscles and nervous tissue with the brain. In actuality, however, an **organ** is a structure that is composed of two or more types of tissues working together to perform particular functions. An **organ system** contains many different organs that cooperate to carry out a process such as digestion of food. We are going to examine human skin as an example of an organ. Some authorities even call skin the *integumentary system* (especially since it cannot be placed in one of the other systems). They maintain that the hair follicles, the oil and sweat glands, the sensory receptors, and the skin are separate organs, and these organs work together to perform various functions.

Human Skin Is an Organ

Human skin (Fig. 40.7) covers the body, protecting underlying parts from physical trauma, microbial invasion, and water loss. Skin cells manufacture a precursor molecule that is converted to vitamin D in the body after it is exposed to ultraviolet (UV) light. Only a small amount of UV radiation is needed to change the precursor to vitamin D. The skin also helps regulate body temperature, and because

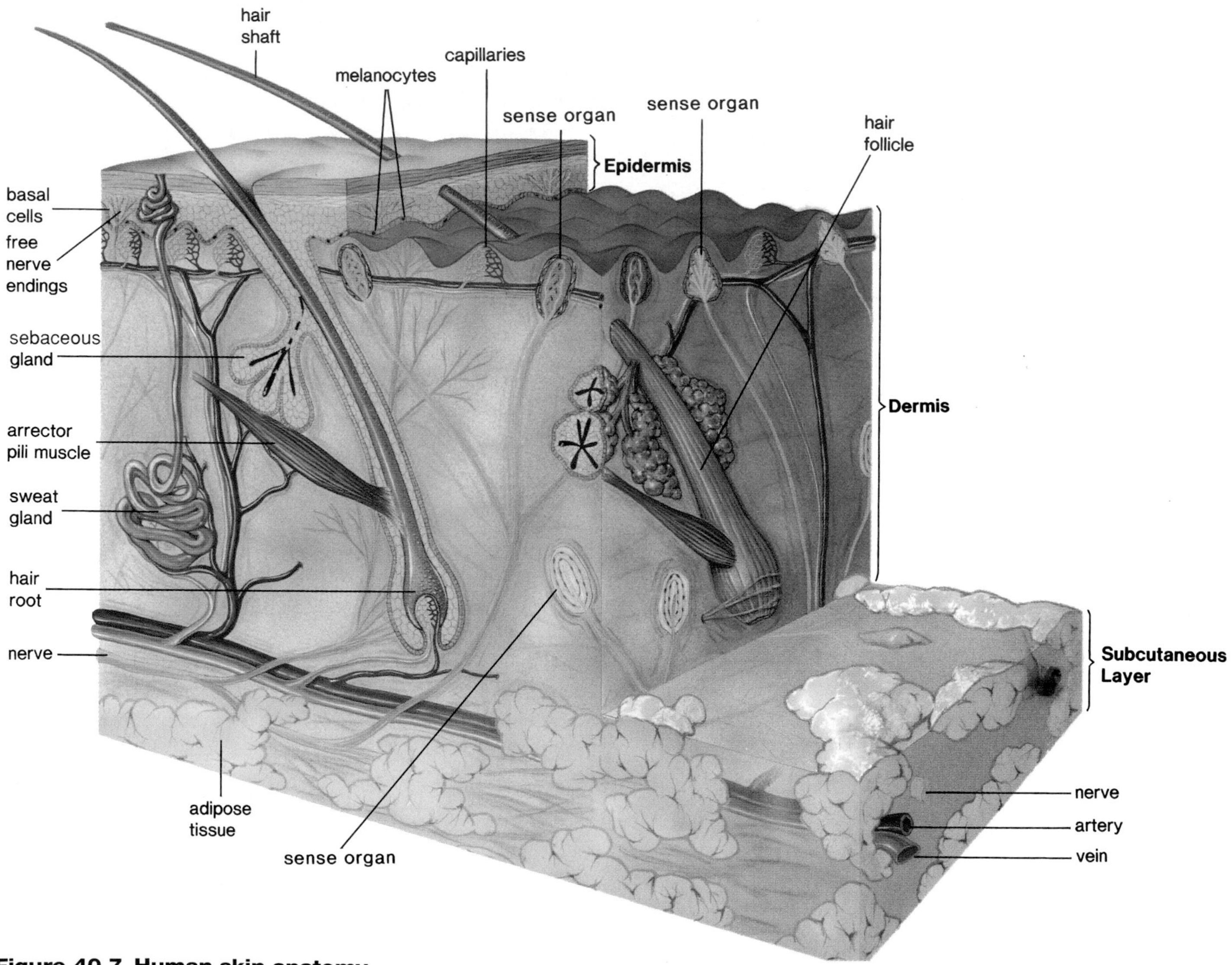

Figure 40.7 Human skin anatomy.
Skin contains three layers: epidermis, dermis, and subcutaneous.

it contains sensory receptors, the skin helps us to be aware of our surroundings and to know when to communicate with others.

Skin has an outer epidermal layer (the epidermis) and an inner layer (the dermis). Beneath the dermis, a subcutaneous layer binds the skin to underlying organs.

Skin Has Layers

The **epidermis** [Gk. *epi*, over, and *derma*, skin] is the outer, thinner layer of the skin. It is a stratified squamous epithelium, and its cells are derived from the basal cells, which undergo continuous cell division. As newly formed cells are pushed to the surface away from their blood supply, they gradually flatten and harden. Eventually, they die and are sloughed off. Hardening is caused by cellular production of a waterproof protein called *keratin*. Over much of the body, keratinization is minimal, but the palm of the hand and the sole of the foot have a particularly thick outer layer of dead keratinized cells.

Specialized cells in the dividing layer of epidermis, called melanocytes, produce melanin, the pigment responsible for skin color in dark-skinned persons. When you sunbathe, the melanocytes become more active, producing melanin which protects the skin from the damaging effects of the ultraviolet (UV) radiation in sunlight.

Nails grow from special epidermal cells at the base of the nail in the region called the nail root. These cells become keratinized as they grow out over the nail bed. The visible portion of the nail is called the nail body. The pink color of nails is due to the vascularized dermal tissue beneath the nail. The whitish color of the half-moon–shaped base results from the thicker germinal layer in this area. Ordinarily, nails grow only about one millimeter a week.

The epidermis of skin is made up of stratified squamous epithelium. In this layer, new cells are pushed outward, become keratinized, die, and are sloughed off.

The **dermis** [Gk. *derma*, skin] is a layer of fibrous connective tissue that is deeper and thicker than the epidermis. It contains elastic fibers and collagen fibers. The collagen fibers form bundles that interlace and run, for the most part, parallel to the skin surface. There are several types of structures in the dermis. A hair, except for the root, is formed of dead, hardened epidermal cells; the root is alive and resides in a hair follicle found in the dermis. Each follicle has one or more oil (sebaceous) glands that secrete sebum, an oily substance that lubricates the hair and the skin. A smooth muscle called the arrector pili muscle is attached to the hair follicle in such a way that when contracted, the muscle causes the hair to stand on end. When you are frightened or cold, goose bumps appear due to a mounding up of the skin from the contraction of these muscles.

Sweat (sudoriferous) glands are present in most regions of the skin. A sweat gland begins as a coiled tubule within the dermis, but then it straightens out near its opening. Some sweat glands open into hair follicles, but most open onto the surface of the skin.

Small receptors are present in the dermis. There are different receptors for pressure, touch, temperature, and pain. Pressure receptors are in onion-shaped sense organs called Pacinian corpuscles that lie deep inside the dermis and around joints and tendons. They are also believed to provide instant information about how and where we move. In cats, Pacinian corpuscles are concentrated on the paws, the leg joints, and the connective tissue of the abdomen. Those close to the ground may provide information about the location of prey. Closely related sensors on the tongues of woodpeckers help them find insects in tree bark.

Touch receptors, which are flat and oval shaped, are concentrated in the fingertips, the palms, the lips, the tongue, the nipples, the penis, and the clitoris. Their prevalence is thought to provide these regions with special sensitivity. Heat and cold sense organs are encapsulated by sheaths of connective tissue and contain lacy networks of nerve fibers. Nerve fibers branch out through all skin, and free nerve endings are believed to be the receptors for pain.

The dermis also contains blood vessels. When blood rushes into these vessels, a person blushes, and when blood volume is reduced in them, a person turns ashen or white.

The dermis, composed of fibrous connective tissue, lies beneath the epidermis. It contains hair follicles, sebaceous glands, and sweat glands. It also contains receptors, blood vessels, and nerve fibers.

The subcutaneous layer, which lies below the dermis, is composed of loose connective tissue, including adipose tissue. Adipose tissue helps to insulate the body by minimizing both heat gain and heat loss. A well-developed subcutaneous layer gives a rounded appearance to the body. Excessive development of this layer accompanies obesity.

Skin Cancer Can Develop

In recent years, there has been a great increase in the number of persons with skin cancer, and physicians believe this is due to sunbathing or even to the use of tanning machines. Protecting your skin from the sun's damaging rays is discussed in the reading on page 727.

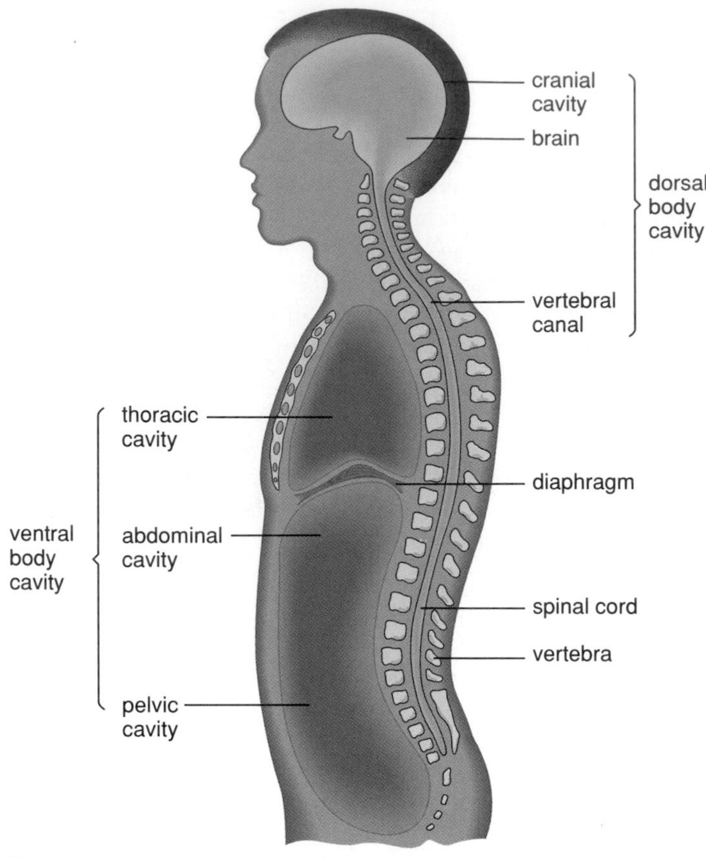

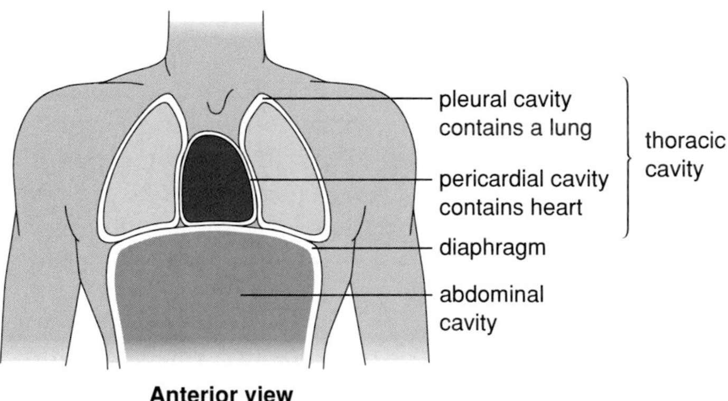

Figure 40.8 Mammalian body cavities.
a. Side view. There is a dorsal cavity, which contains the cranial cavity and the vertebral canal. The brain is in the cranial cavity, and the spinal cord is in the vertebral canal. There is a well-developed ventral cavity, which is divided by the diaphragm into the thoracic cavity and the abdominal cavity. The heart and lungs are in the thoracic cavity, and most other internal organs are in the abdominal cavity. **b.** Anterior view of the thoracic cavity.

Organs Form Systems

In most animals, individual organs function as part of an *organ system,* the next higher level of animal organization (see Fig. 40.1). These same systems are found in all vertebrate animals. The organ systems of vertebrates carry out the life processes that are common to all animals, and indeed to all organisms:

Life Processes	Human Systems
Coordinate body activities	Nervous system Endocrine system
Acquire materials and energy (food)	Skeletal system Muscular system Digestive system
Maintain body shape	Skeletal system Muscular system
Exchange gases	Respiratory system
Transport materials	Cardiovascular system
Excrete wastes	Urinary system
Protect the body from disease	Lymphatic system Immune system
Produce offspring	Reproductive system

Bodies Have Cavities

Each organ system has a particular distribution within the human body. There are two main body cavities: the smaller dorsal body cavity and the larger ventral body cavity (Fig. 40.8). The brain and the spinal cord are in the dorsal body cavity.

During development, the ventral body cavity develops from the *coelom.* In humans and other mammals, the coelom is divided by a muscular diaphragm that assists breathing. The heart (a pump for the closed cardiovascular system) and the lungs are located in the upper (thoracic or chest) cavity. The major portions of the digestive system, the excretory system, and much of the reproductive system are located in the lower (abdominal) cavity. The major organs of the urinary system are the paired kidneys. The digestive system has accessory organs such as the liver and pancreas. Each sex has characteristic sex organs.

The animal body is organized; the organs have a specific structure, function, and location. The performance of the organs of each system are coordinated.

Health Focus

Skin Cancer on the Rise

In Victorian days, Caucasian women carried parasols to keep their skin fair. But early in this century, some fair-skinned people began to prefer the golden-brown look, and they took up sunbathing as a way to achieve a tan. A few hours after exposure to the sun, pain and redness due to dilation of blood vessels occur. Tanning occurs when melanin granules increase in keratinized cells at the surface of the skin as a way to prevent any further damage by ultraviolet (UV) rays. The sun gives off two types of UV rays: UV-A rays and UV-B rays. UV-A rays penetrate the skin deeply, affect connective tissue, and cause the skin to sag and wrinkle. UV-A rays are also believed to increase the effects of the UV-B rays, which are the cancer-causing rays. UV-B rays are more prevalent at midday.

a.

U.S. City	UV Index
Anchorage	100
Seattle	477
Minneapolis	570
Boston	591
Chicago	637
New York City	639
Philadelphia	656
Washington, D.C.	683
Columbus	698
St. Louis	714
San Francisco	715
Boise	715
Los Angeles	824
Dallas	871
Atlanta	875
Las Vegas	876
Phoenix	889
Denver	951
Houston	999
Miami	1028
Honolulu	1147

b.

Figure 40A Sunbathing.
a. Sunbathing is even more dangerous today because of ozone depletion. **b.** The UV index measures the relative amount of solar UV radiation for cities. An ozone hole will increase each city's index by about 20%; as the index rises, so does the incidence of skin cancer.

Skin cancer is categorized as either nonmelanoma or melanoma. Nonmelanoma cancers are of two types. Basal cell carcinoma, the most common type, begins when UV radiation causes epidermal basal cells to form a tumor, while at the same time suppressing the immune system's ability to detect the tumor. The signs of a tumor are varied. They include an open sore that will not heal, a recurring reddish patch, a smooth, circular growth with a raised edge, a shiny bump, or a pale mark. About 95% of patients are easily cured, but reoccurrence is common.

Squamous cell carcinoma begins in the epidermis proper. Squamous cell carcinoma is five times less common than basal cell carcinoma, but it is more likely to spread to nearby organs, and the death rate is about 1% of cases. The signs of squamous cell carcinoma are the same as that for basal cell carcinoma, except that it may also show itself as a wart that bleeds and scabs.

Melanoma that starts in the melanocytes has the appearance of an unusual mole. Unlike a dark mole that is circular and confined, melanoma moles look like spilled ink spots. A variety of shades can be seen in the same mole, and they can itch, hurt, or feel numb. The skin around the mole turns grey, white, or red. Melanoma is most apt to appear in persons who have fair skin, particularly if they have suffered occasional severe burns as children. The chance of melanoma increases with the number of moles a person has. Most moles appear before the age of 14, and their appearance is linked to sun exposure. Melanoma rates have risen since the turn of the century, but the incidence has doubled in the last decade. Now about 32,000 cases of melanoma are diagnosed each year, and one in five persons diagnosed dies within five years.

Since the incidence of skin cancer is related to UV exposure, scientists have developed a UV index to determine how powerful the solar rays are in different U.S. cities (Fig. 40A*b*). The index assigns a baseline level of 100 to Anchorage, Alaska. In general, the more southern the city, the higher the UV index, and the greater the risk of skin cancer.

Looking ahead, we should note that for every 10% decrease in the ozone layer, the UV index per city rises by 13–20%, and the chance of skin cancer rises as well. Even if you live in Anchorage, you should take the following steps to protect yourself from the sun.

To prevent the possible occurrence of skin cancer, observe the following:

- Use a broad-spectrum sunscreen, which protects you from both UV-A and UV-B radiation, with an SPF (sun protection factor) of at least 15. (This means, for example, that if you usually burn after a 20-minute exposure, it will take 15 times that long before you will burn.)
- Wear protective clothing. Choose fabrics with a tight weave and wear a wide-brimmed hat. A baseball cap does not protect the rims of the ears.
- Stay out of the sun altogether between the hours of 10 A.M. and 3 P.M. This will reduce your annual exposure by as much as 60%.
- Wear sunglasses that have been treated to absorb both UV-A and UV-B radiation. Otherwise, sunglasses can expose your eyes to more damage than usual because pupils dilate in the shade.
- Avoid tanning machines. Although most tanning devices use high levels of only UV-A, the deep layers of the skin become more vulnerable to UV-B radiation when you are later exposed to the sun.

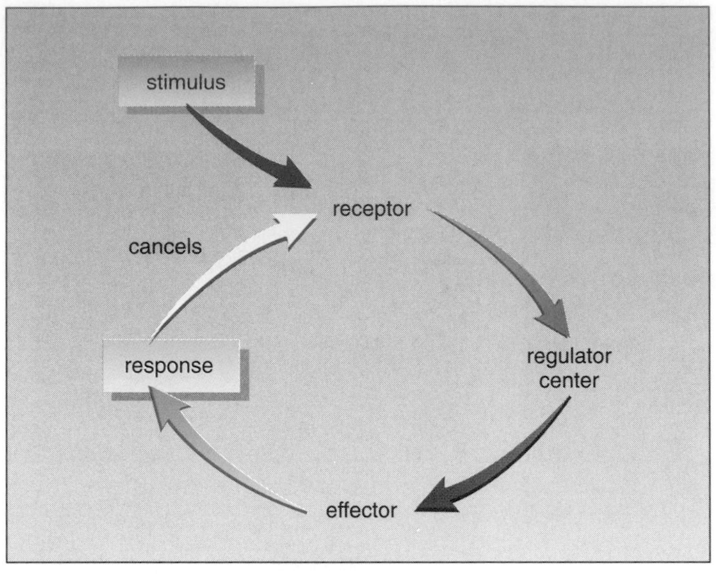

a.

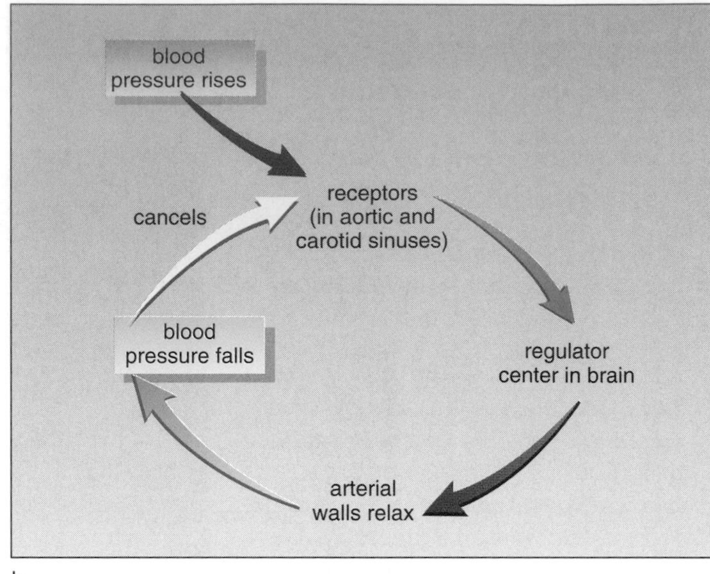

b.

Figure 40.9 Negative feedback control.
a. A stimulus causes a receptor to signal a regulator center in the brain. The regulator center signals effectors to respond, and a response cancels the stimulus. **b.** For example, when blood pressure rises, special receptors in blood vessels signal a particular center in the brain. The brain signals the arteries to relax, and blood pressure falls.

40.3 Homeostasis Is Necessary

Claude Bernard, a famous French physiologist, pointed out in 1859 that while an animal lives in an external environment, the cells of the body lie within an internal environment. The *internal environment* is a fluid, called tissue fluid, that bathes the cells of the body. Bernard said that the relative stability of the internal environment allows animals to live in an external environment that can vary considerably. Later, Walter Cannon, an American physiologist, introduced the term **homeostasis** [Gk. *homoios,* like, resembling, and *stasis,* standing]. He said there is a dynamic interplay between events that tend to change the internal environment and those that act against this possibility. To achieve homeostasis, the composition of blood and tissue fluid, the body temperature, and the blood pressure, for example, must stay within a normal range.

The internal environment of an animal's body consists of tissue fluid, which bathes the cells.

Most organ systems of the human body contribute to homeostasis. The digestive system takes in and digests food, providing nutrient molecules that enter the blood and replace the nutrients that are constantly being used by the body cells. The respiratory system adds oxygen to the blood and removes carbon dioxide. The amount of oxygen taken in and carbon dioxide given off can be increased to meet body needs. The liver and the kidneys contribute greatly to homeostasis. For example, immediately after glucose enters the blood, it can be removed by the liver and stored as glycogen. Later, glycogen is broken down to replace the glucose used by the body cells; in this way, the glucose composition of the blood remains constant. The hormone insulin, secreted by the pancreas, regulates glycogen storage. The kidneys are also under hormonal control as they excrete wastes and salts, substances that can affect the pH level of the blood.

Although homeostasis is, to a degree, controlled by hormones, it is ultimately controlled by the nervous system. In humans, the brain contains centers that regulate such factors as blood pressure. Maintaining blood pressure levels requires a receptor that detects unacceptable levels and signals a regulator center. If a correction is required, the center then directs an adaptive response (Fig. 40.9). Once normalcy is obtained, the receptor is no longer stimulated. This is called control by **negative feedback** because the regulator centers bring about a response that is opposite to present conditions. A useful analogy that helps illustrate how such a negative feedback mechanism works is the control of the heating system of a house by a thermostat. Through negative feedback control, indoor temperature can be maintained within a relatively narrow range with a slight fluctuation above and below a mean. **Positive feedback** also occurs on occasion. In these instances, certain events increase the likelihood of a particular response. For example, once the childbirth process begins, each succeeding event makes it more likely that the process will continue until completion.

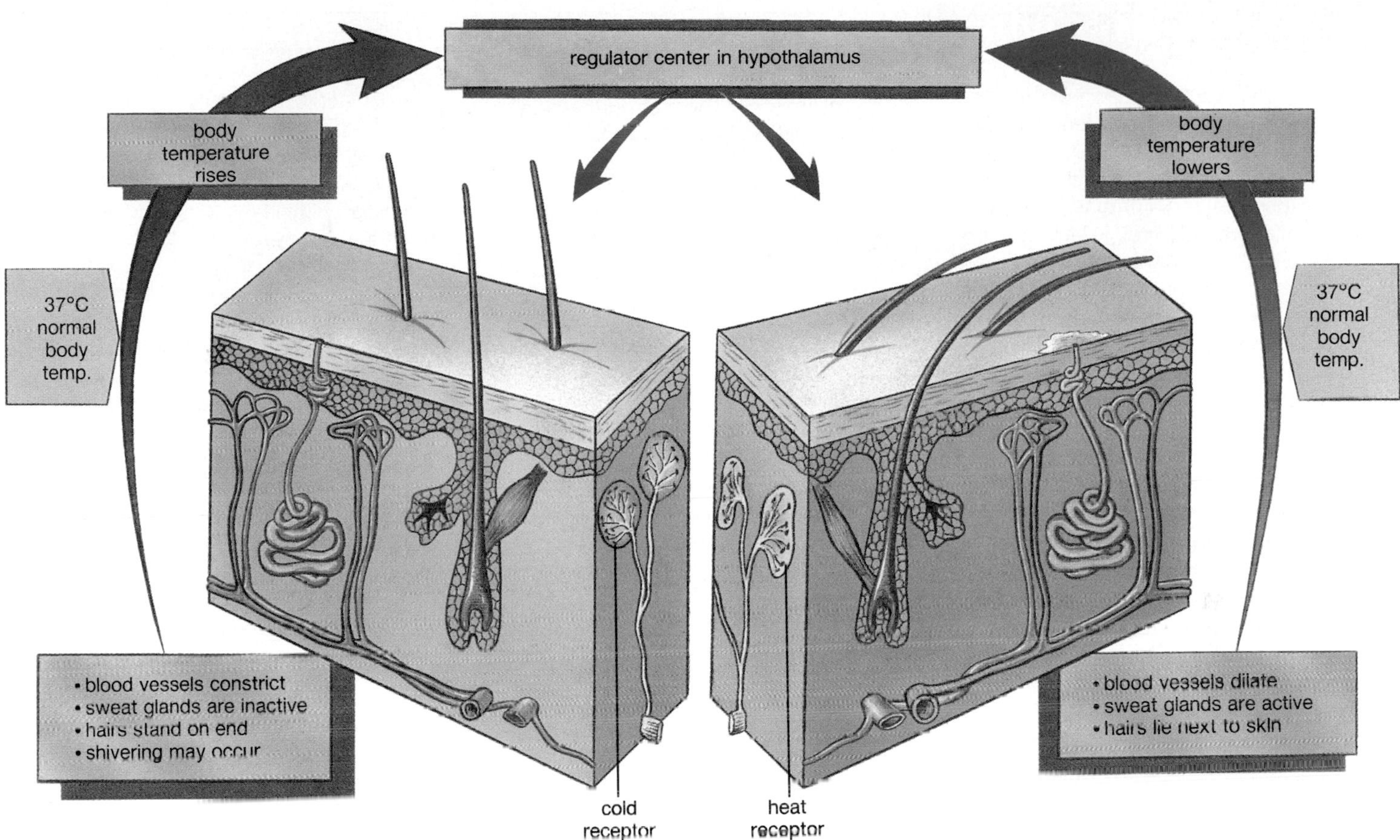
regulator center in hypothalamus
body temperature rises
body temperature lowers
37°C normal body temp.
37°C normal body temp.
• blood vessels constrict
• sweat glands are inactive
• hairs stand on end
• shivering may occur
• blood vessels dilate
• sweat glands are active
• hairs lie next to skin
cold receptor
heat receptor

Figure 40.10 Homeostasis and temperature control.
When the body temperature rises above normal, the regulator center directs the blood vessels to dilate and the sweat glands to be active so that the body temperature returns to 37°C. When the body temperature lowers below normal, the regulator center directs the blood vessels to constrict, causing the hairs to stand on end, and shivering to occur. The body temperature now returns to 37°C.

How Body Temperature Is Regulated

Although the skin senses a change in external temperature, the regulator center for body temperature, located in the hypothalamus, is sensitive only to the temperature of the blood. When the body temperature falls below normal, the regulator center directs (via nerve impulses) the blood vessels of the skin to constrict (Fig. 40.10). This conserves heat. Also, the arrector pili muscles pull hairs erect, and a layer of insulating air is trapped next to the skin. If body temperature falls even lower, the regulator center sends nerve impulses to the skeletal muscles, and shivering occurs. Shivering generates heat, and gradually body temperature rises to 37°C. If the body temperature is normal, the regulator center is not active.

When the body temperature is higher than normal, the regulator center is activated and then directs the blood vessels of the skin to dilate. This allows more blood to flow near the surface of the body, where heat can be lost to the environment. In addition, the regulator center activates the sweat glands, and the evaporation of sweat also helps lower body temperature. Gradually, body temperature decreases to 37°C.

The temperature of the human body is maintained at about 37°C due to the action of a regulator center in the hypothalamus.

---

connecting concepts

This chapter has concentrated on the coelomate animals and particularly vertebrates. But not all animals are coelomates. A sponge is merely multicellular, and there are neither tissues nor organs. The Radiata (e.g., cnidaria and comb jellies) have only two tissue layers derived from the germ layers endoderm and ectoderm. All other animals, termed Bilateria, have three germ layers during development and various types of tissues as adults. Flatworms are acoelomates and do not have organ systems, but all other animals have some sort of coelom in which internal organs, such as those belonging to the digestive, respiratory, and reproductive systems, are found. The degree of organization in animals increases as complexity increases.

Homeostasis in unicellular organisms and thin aceolomate animals occurs only at the cellular level, and each cell must carry out its own exchanges with the external environment to maintain a relative constancy of cytoplasm. In complex animals, there are localized boundaries where materials are exchanged with the external environment. In terrestrial animals, gas exchange usually occurs within lungs, food is digested within a digestive tract, and kidneys collect and excrete metabolic wastes. Exchange boundaries are an effective way to regulate the internal environment if there is a transport system to carry materials from one body part to another. Circulation in invertebrates and vertebrates carries out this function.

Regulating mechanisms occur at all levels of organization. At the cellular level, the actions of enzymes are often controlled by feedback mechanisms. However, in animals with organ systems, the nervous and endocrine systems regulate the actions of organs. In addition, an animal's nervous system gathers and processes information about the external environment. Sense receptors act as specialized boundaries through which external stimuli are received and converted into a form that can be processed by the nervous system. The information received and processed by the nervous system may then influence the organism's behavior in a way that contributes to homeostasis.

Summary

40.1 Animals Are Organized

Tissues arise from the embryonic germ layers: ectoderm, mesoderm, and endoderm. Ectoderm produces nervous tissue; mesoderm produces muscle and certain systems (circulatory, excretory, reproductive); endoderm produces the linings of the digestive and respiratory tracts and associated organs. Epithelial tissue comes from all three layers.

Epithelial tissue covers the body and lines cavities. There is squamous, cuboidal, and columnar epithelium. Each type can be simple or stratified; it can also be glandular, or have modifications like cilia. Epithelial tissue protects, absorbs, secretes, and excretes.

Connective tissue has a matrix between cells. Loose connective tissue and dense fibrous connective tissue contain fibroblasts and fibers (collagen and elastic). Adipose tissue is a type of loose connective tissue; tendons and ligaments are dense fibrous connective tissue.

Cartilage (protein matrix) and bone (calcium matrix) are rigid connective tissues. Blood is connective tissue in which plasma is the matrix.

Muscular (contractile) tissue can be smooth or striated (both skeletal and cardiac). In humans, smooth muscle is in the wall of internal organs, skeletal muscle is attached to bone, and cardiac muscle is in the heart.

Nervous tissue contains neurons having three parts: dendrites, a cell body, and an axon. Neuroglial cells support and protect neurons.

40.2 Organs Have Structure and Functions

Organs contain various tissues. Skin is an organ that has tissue layers. Epidermis (stratified squamous epithelium) overlies the dermis (fibrous connective tissue containing sensory receptors, hair follicles, blood vessels, and nerves). The subcutaneous layer is composed of loose connective tissue.

Organ systems contain several organs. Organ systems of humans carry out the life processes that are common to all organisms and also have specific functions.

The human body contains two main cavities. The dorsal cavity contains the brain and spinal cord. The ventral cavity is divided into the thoracic cavity (heart and lungs) and the abdominal cavity (most other internal organs).

40.3 Homeostasis Is Necessary

Homeostasis is the relative constancy of the internal environment so that blood and tissue constituents and values stay within a normal range. All organ systems contribute to homeostasis but special contributions are made by the liver, which keeps the blood glucose constant, and the kidneys, which regulate the pH. The nervous and hormonal systems regulate the other body systems. Both of these are controlled by negative feedback mechanisms, which result in slight fluctuations above and below desired levels. Body temperature is regulated by a center in the hypothalamus.

Reviewing the Chapter

1. Name the four major types of tissues. 718
2. Describe the structure and the functions of three types of epithelial tissue. 718–19
3. Describe the structure and the functions of six types of connective tissue. 720–21
4. Describe the structure and the functions of three types of muscular tissue. 722
5. Nervous tissue contains what types of cells? 723
6. Describe the structure of skin, and state at least two functions of this organ. 724–25
7. In general terms, describe the location of the human organ systems. 726
8. Tell how the various systems of the body contribute to homeostasis. 728
9. What is the function of receptors, the regulator center, and effectors in a negative feedback mechanism? Why is it called negative feedback? 728

Testing Yourself

Choose the best answer for each question.

1. Which of these is mismatched?
   a. epithelial tissue—protection and absorption
   b. muscular tissue—contraction and conduction
   c. connective tissue—binding and support
   d. nervous tissue—conduction and message sending
2. Which of these is not epithelial tissue?
   a. simple cuboidal and stratified columnar
   b. bone and cartilage
   c. stratified squamous and simple squamous
   d. All of these are epithelial tissue.
3. Which tissue is more apt to line a lumen?
   a. epithelial tissue
   b. connective tissue
   c. nervous tissue
   d. muscular tissue
4. Tendons and ligaments are
   a. connective tissue.
   b. associated with the bones.
   c. found in vertebrates.
   d. All of these are correct.
5. Which tissue has cells in lacunae?
   a. epithelial tissue
   b. cartilage
   c. bone
   d. Both b and c are correct.
6. Cardiac muscle is
   a. striated.
   b. involuntary.
   c. smooth.
   d. Both a and b are correct.
7. Which of these components of blood fights infection?
   a. red blood cells
   b. white blood cells
   c. platelets
   d. All of these are correct.
8. Which of these body systems contribute to homeostasis?
   a. digestive and excretory systems
   b. respiratory and nervous systems
   c. nervous and endocrine systems
   d. All of these are correct.
9. With negative feedback
   a. the output cancels the input.
   b. there is a fluctuation above and below the average.
   c. there is self-regulation.
   d. All of these are correct.
10. When a person is cold, the blood vessels
    a. dilate, and the sweat glands are inactive.
    b. dilate, and the sweat glands are active.
    c. constrict, and the sweat glands are inactive.
    d. constrict, and the sweat glands are active.
11. Identify each of these tissues, and tell whether the tissue is a type of epithelial tissue, muscular tissue, nervous tissue, or connective tissue.

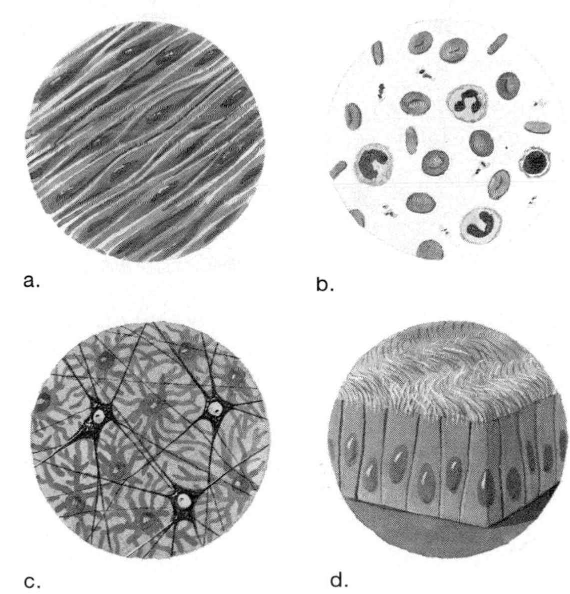

Applying the Concepts

1. *An animal's body has levels of organization.*
   How is each level of organization more than the sum of its parts?
2. *All organisms carry out certain life processes.*
   Review the chart on page 726 and explain the associations listed.
3. *The internal environment of organisms stays relatively constant.*
   How does each of the life processes (except producing offspring) help keep the internal environment relatively constant?

Using Technology

Your study of animal organization and homeostasis is supported by these available technologies:

Exploring the Internet
The Mader Home Page provides resources for and help with studying this chapter.

http://www.mhhe.com/sciencemath/biology/mader/
(Click on Biology.)

The Dynamic Human CD-ROM
Anatomical Orientation

Understanding the Terms

adipose tissue 720
blood 721
bone 721
cardiac muscle 722
cartilage 720
connective tissue 720
dense fibrous connective tissue 720
dermis 725
endocrine gland 719
epidermis 725
epithelial tissue 718
exocrine gland 719
fibroblast 720
homeostasis 728
lacuna 720
ligament 720
loose fibrous connective tissue 720
muscular (contractile) tissue 722
negative feedback 728
nervous tissue 723
neuron 723
organ 724
organ system 724
positive feedback 728
skeletal muscle 722
smooth (visceral) muscle 722
stimulus 723
striated 722
tendon 720
tissue 718

Match the terms to these definitions:

a. __________ Connective tissue in which fat is stored.
b. __________ Deeper, thicker layer of the skin that consists of fibrous connective tissue and contains various structures such as sense organs.
c. __________ Having bands; cardiac and skeletal muscle have alternating light and dark crossbands that are visible in certain muscle fibers and are produced by the distribution of contractile proteins.
d. __________ Maintenance of normal internal conditions in a cell or an organism by means of a self-regulating mechanism.
e. __________ Nerve cell that characteristically has three parts: dendrites, cell body, and axon.
f. __________ Outer, protective layer of the skin.
g. __________ Strap of dense fibrous connective tissue that joins skeletal muscle to bone.
h. __________ Striated, involuntary muscle tissue found only in the heart.
i. __________ Type of animal tissue forming a continuous layer over most body surfaces (i.e., skin) and inner cavities; squamous, cuboidal, and columnar are the three types of this tissue.
j. __________ Type of connective tissue in which cells are separated by a liquid called plasma.

Circulation

CHAPTER

Chapter Concepts

Wild horses, *Equus*, running in the Patagonian desert

All animal cells acquire nutrients and oxygen from the environment and give off carbon dioxide and other wastes to the environment. In small aquatic animals, each cell directly exchanges materials with the external environment by utilizing diffusion and plasma membrane transport mechanisms. These animals have no need of a circulatory system.

Larger, more active, animals have special organs with localized boundaries for exchange of substances with the external environment. These animals have a circulatory system in which the heart pumps a fluid about the body to all organs. In this way, nutrients, gases, and metabolic wastes are transported to or from exchange boundaries to all the cells of the body. In vertebrates, such as the wild horses in the photograph, the circulatory system can be depended upon to supply muscles with the glucose and oxygen they need to run quickly.

Most circulatory systems also have other functions that assist homeostasis, such as when clotting occurs and prevents loss of body fluids following minor injuries. Also, circulatory fluids often contain phagocytic cells that are capable of removing microbes and other substances that are foreign to the body. Some blood cells assist immunity by producing antibodies.

41.1 Transport in Invertebrates

Unicellular animal-like protozoa with a high surface area to volume ratio rely on diffusion for gas nutrients, and waste exchanges with the external environment that surrounds them. Even some small multicellular animals do not have an internal transport system because their cells can be serviced without one (Fig. 41.1). Larger invertebrates usually have a circulatory system. Some of these have an open—and others have a closed—circulatory system.

Invertebrates With a Cavity

Sea anemones, which are cnidaria, and planaria, which are flatworms, have a sac body plan (Fig. 41.1*b, c*). This body plan makes a circulatory system unnecessary. In a sea anemone, cells are either part of an external layer or they line the gastrovascular cavity. In either case, each cell is exposed to water and can independently exchange gases and rid itself of wastes. The cells that line the gastrovascular cavity are specialized to carry out digestion. They pass nutrient molecules to other cells by diffusion. In a planarian, a trilobed gastrovascular cavity ramifies throughout the small and flattened body. No cell is very far from one of the three digestive branches, so nutrient molecules can diffuse from cell to cell. Similarly, diffusion meets the respiratory and excretory needs of the cells.

Pseudocoelomate invertebrates, such as nematodes, use the coelomic fluid of their body cavity for transport purposes. Echinoderms also rely on movement of coelomic fluid within a body cavity as a circulatory system.

With an Open or a Closed System

All other animals have a **circulatory system** in which a pumping heart moves a fluid into blood vessels. There are two types of circulatory fluids: **blood,** which is always contained within blood vessels, and **hemolymph,** which flows into a body cavity called a hemocoel. Hemolymph is a mixture of blood and interstitial fluid.

Figure 41.1 Aquatic organisms without a circulatory system.
a. A paramecium is a unicellular organism that carries on gas exchange across its cell surface. Food particles that flow into a specialized region called a gullet are enclosed within food vacuoles (green), where digestion occurs. Molecules leave these vacuoles as they are distributed about the cell by a movement of the cytoplasm. **b.** In a sea anemone, a cnidarian, digestion takes place inside the gastrovascular cavity, so named because it (like a vascular system) makes digested material available to the cells that line the cavity. These cells can also acquire oxygen from the watery contents of the cavity and discharge their wastes there. **c.** In a planarian, a flatworm, the gastrovascular cavity ramifies throughout the body, bringing nutrients to body cells. Diffusion is sufficient to pass molecules to every cell from either the cavity or the exterior surface.

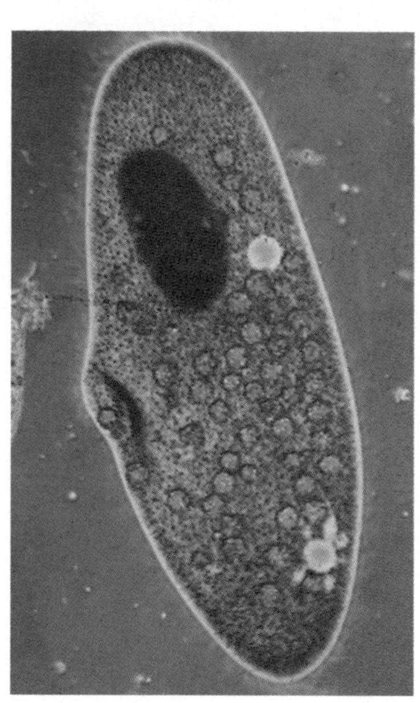

a. Paramecium, *Paramecium* 20 µm

b. Sea anemone, *Apitasia*

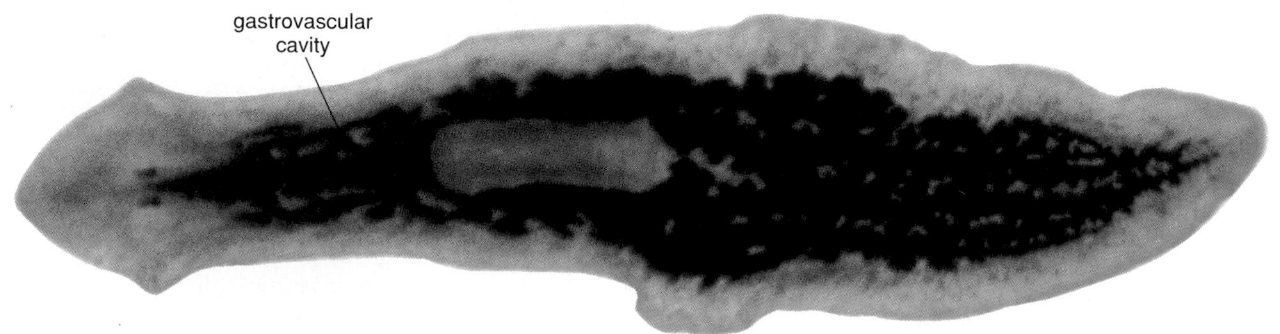

c. Flatworm, *Dugesia* 200 µm

Hemolymph is seen in animals that have an open system. In most mollusks and arthropods, the heart pumps hemolymph via vessels into tissue spaces that are sometimes enlarged into saclike sinuses (Fig. 41.2*a*). Eventually hemolymph drains back to the heart. In the grasshopper, the dorsal heart pumps hemolymph into a dorsal aorta, which empties into the *hemocoel*. When the heart contracts, openings called ostia (sing., ostium) are closed; when the heart relaxes, the hemolymph is sucked back into the heart by way of the ostia. The hemolymph of a grasshopper is colorless because it does not contain hemoglobin or any other respiratory pigment. It carries nutrients but no oxygen. Oxygen is taken to cells and carbon dioxide is removed from them by way of air tubes, called tracheae, which are found throughout the body. The tracheae provide an efficient transport and delivery of respiratory gases while at the same time restricting water loss.

Some invertebrates (for example, earthworms—annelids; squids and octopuses—mollusks), have a *closed circulatory system*. Blood, which usually consists of cells and plasma, is pumped by the heart into a system of blood vessels (Fig. 41.2*b*). There are valves that prevent the backward flow of blood. In the segmented earthworm, five pairs of anterior lateral vessels pump blood into the ventral blood vessel, which has a branch in every segment of the worm's body. Blood moves through these branches into capillaries, where exchanges with tissue fluid take place. Blood then moves into veins that return it to the dorsal blood vessel. This dorsal blood vessel returns blood to the heart for repumping.

The earthworm has red blood that contains the respiratory pigment hemoglobin. Hemoglobin is dissolved in the blood and is not contained within cells. The earthworm has no specialized boundary (e.g., lungs) for gas exchange with the external environment. Gas exchange takes place across the body wall, which must always remain moist for this purpose.

Animals with a gastrovascular cavity use this cavity for transport purposes. Other animals have an open or a closed circulatory system.

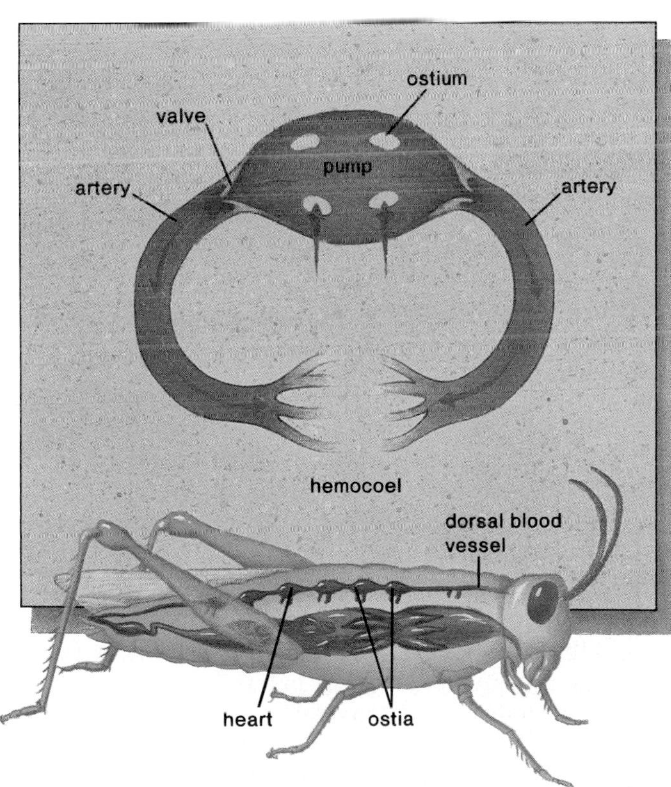

a. Open circulatory system

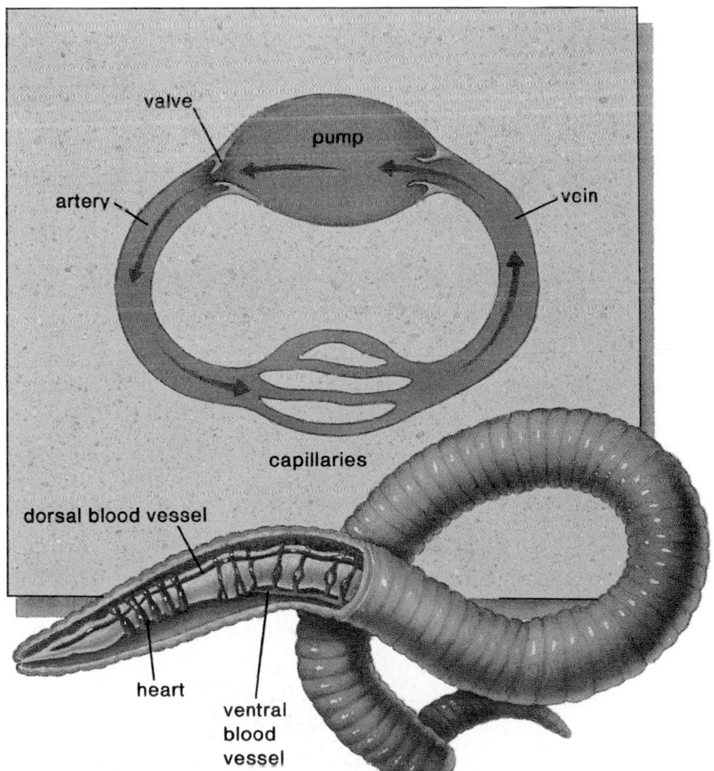

b. Closed circulatory system

Figure 41.2 Open versus closed circulatory system.
a. The grasshopper has an open circulatory system. A hemocoel is a body cavity filled with hemolymph, which freely bathes the internal organs. The heart keeps the hemolymph moving, but this open system probably could not supply oxygen to wing muscles rapidly enough. These muscles receive oxygen directly from tracheae (air tubes). **b.** The earthworm has a closed circulatory system. The dorsal and ventral blood vessels are joined by five pairs of anterior hearts and by branch vessels in the rest of the worm.

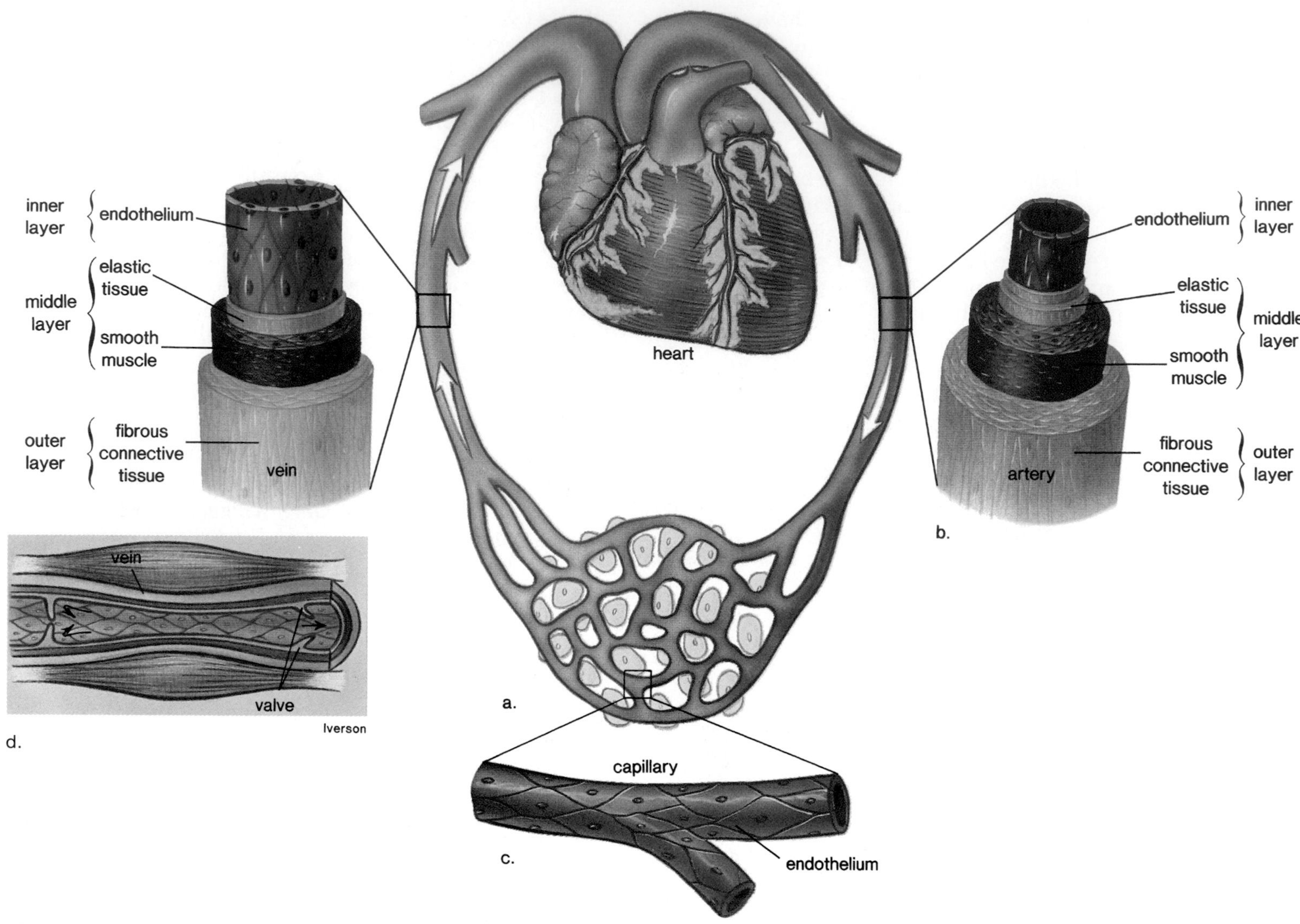

Figure 41.3 Circulation in birds and mammals.
a. Blood leaving the heart moves from an artery to arterioles to capillaries to venules and then returns to the heart by way of a vein. **b.** Arteries have well-developed walls with a thick middle layer of elastic tissue and smooth muscle. **c.** Capillary walls are only one cell thick. **d.** Veins have flabby walls, particularly because the middle layer is not as thick as in arteries. Veins have valves, which point toward the heart.

41.2 Transport in Vertebrates

All vertebrate animals have a closed circulatory system, which is called a **cardiovascular system** [Gk. *kardia,* heart, L. *vascular,* vessel]. It consists of a strong, muscular heart, in which the *atria* (sing., atrium) receive blood and the muscular *ventricles* pump blood out through the blood vessels. There are three kinds of blood vessels: **arteries,** which carry blood away from the heart; **capillaries** [L. *capillus,* hair], which exchange materials with tissue fluid; and **veins** [L. *vena,* blood vessel], which return blood to the heart (Fig. 41.3).

Arteries have thick walls, and those attached to the heart are resilient, meaning that they are able to expand and accommodate the sudden increase in blood volume that results after each heartbeat. **Arterioles** are small arteries whose diameter can be regulated by the nervous system. Arteriole constriction and dilation affect blood pressure in general. The greater the number of vessels dilated, the lower the blood pressure.

Arterioles branch into capillaries, which are extremely narrow, microscopic tubes with a wall composed of only one layer of cells. Capillary beds (many capillaries interconnected) are so prevalent that in humans, all cells are within 60–80 μm of a capillary. But only about 5% of the capillary beds are open at the same time. After an animal has eaten, the capillary beds in the digestive tract are usually open, and during muscular exercise, the capillary beds of the muscles are open. Capillaries, which are usually so narrow that red blood cells pass through in single file, allow exchange of nutrient and waste molecules across their thin walls.

Venules and veins collect blood from the capillary beds and take it to the heart. First the venules drain the blood from the capillaries, and then they join to form a vein. The wall of a vein is much thinner than that of an artery and this may be associated with a lower blood pressure in the veins. Valves within the veins point, or open, toward the heart, preventing a backflow of blood when they close (Fig. 41.3*d*).

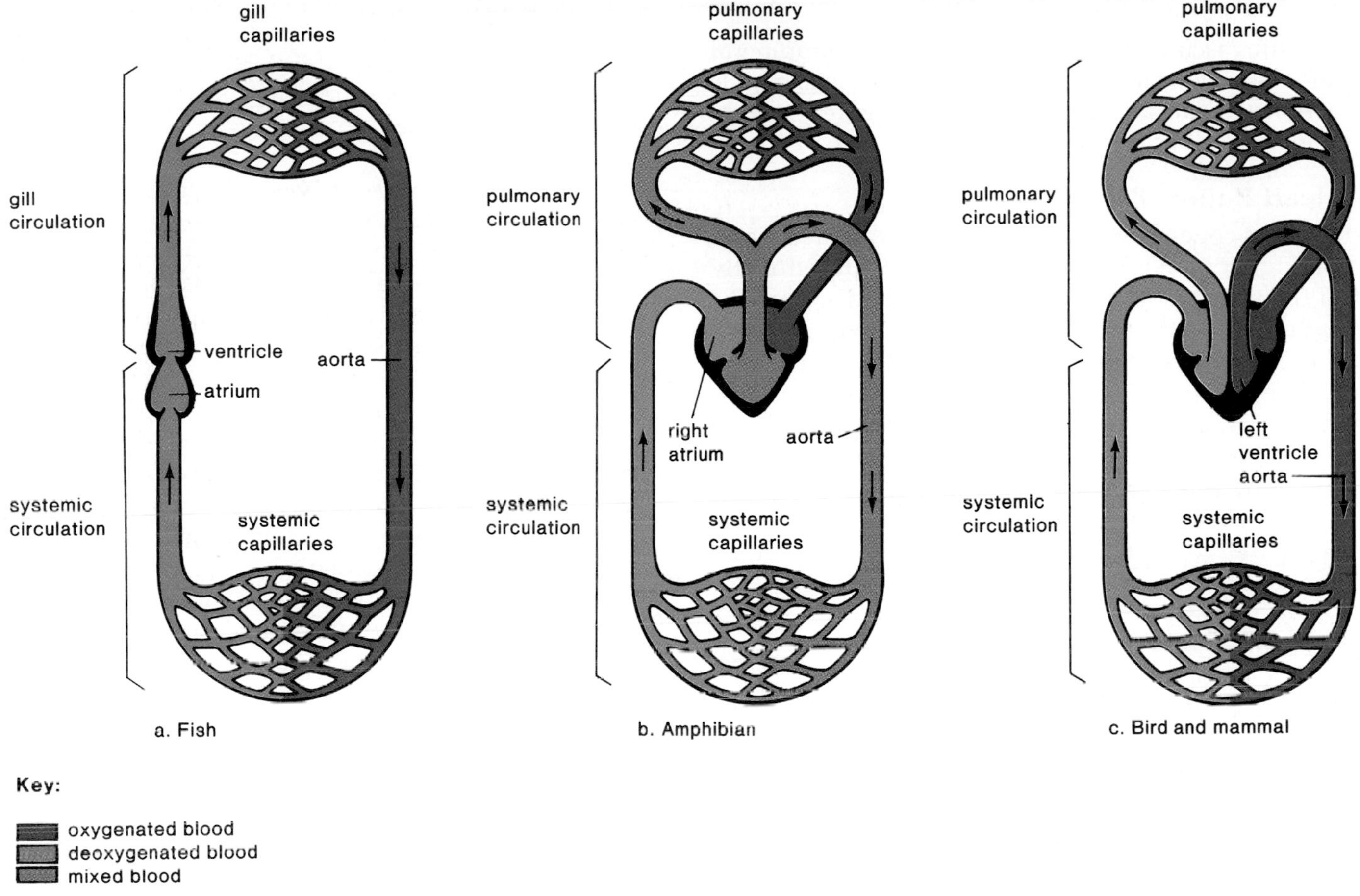

Figure 41.4 Comparison of circulatory loops in vertebrates.
a. In a fish, the blood moves in a single loop. The heart has a single atrium and ventricle and pumps the blood into the gill region, where gas exchange takes place. Blood pressure created by the pumping of the heart is dissipated after the blood passes through the gill capillaries. This is a disadvantage of this single-loop system. **b.** Amphibians have a double-loop system in which the heart pumps blood to both the lungs and the body itself. Although there is a single ventricle, there is little mixing of oxygenated and deoxygenated blood. **c.** The pulmonary and systemic systems are completely separate in birds and mammals, since the heart is divided by a septum into a right and left half. The right side pumps blood to the lungs, and the left side pumps blood to the body proper.

Comparing Circulatory Pathways

Among vertebrate animals, there are three different types of circulatory pathways. In fishes, blood follows a one-circuit (single-loop circulatory) pathway through the body. The heart has a single atrium and a single ventricle (Fig. 41.4*a*). The pumping action of the ventricle sends blood under pressure to the gills, where it is oxygenated. After passing through the gills, blood is under reduced pressure and flow. However, this single circulatory loop has advantages in that the gill capillaries get deoxygenated blood and the systemic capillaries get fully oxygenated blood.

As a result of evolutionary changes, the other vertebrates have a two-circuit (double-loop circulatory) pathway. The heart pumps blood to the tissues, called *systemic circulation,* and also pumps blood to the lungs, called *pulmonary circulation.* This double pumping action is an adaptation to breathing air on land.

In amphibians, the heart has two atria, but there is only a single ventricle (Fig. 41.4*b*). The same holds true for most reptiles, except that the ventricle has a partial septum. The hearts of crocodiles, which are reptiles, and all birds and mammals are divided into right and left halves (Fig. 41.4*c*). The right ventricle pumps blood to the lungs, and the left ventricle, which is larger than the right ventricle, pumps blood to the rest of the body. This arrangement provides adequate blood pressure for both the pulmonary and systemic circulations.

In mammals, birds, and crocodiles, the heart is divided into a right side and a left side; this ensures adequate blood pressure for both the pulmonary and systemic circulations.

41.3 Transport in Humans

In the cardiovascular system of humans, the pumping of the heart keeps blood moving in the arteries. Skeletal muscle contraction pressing against veins is primarily responsible for the movement of blood in the veins.

The Heart Pumps Blood

The **heart** is a cone-shaped, muscular organ about the size of a fist (Fig. 41.5). It is located between the lungs directly behind the sternum (breastbone) and is tilted so that the apex is directed to the left. The major portion of the heart, called the **myocardium** [Gk. *myos*, muscle, and *kardia*, heart], consists of cardiac muscle tissue. The muscle fibers of the myocardium are branched and tightly joined to one another. The heart lies within the pericardium, a thick, membranous sac that contains pericardial fluid, which has a cushioning and lubricating effect. The inner surface of the heart is lined with endocardium, which consists of connective tissue and endothelial tissue.

Internally, a wall called the septum separates the heart into a right side and a left side (Fig. 41.6). The heart has four chambers: two upper, thin-walled **atria** and two lower, thick-walled **ventricles.** The atria are much smaller and weaker than the muscular ventricles, but they hold the same volume of blood.

The heart also has **valves,** which direct the flow of blood and prevent its backward movement. These valves are supported by strong fibrous strings called chordae tendineae. The chordae, which are attached to muscular projections of the ventricular walls, support the valves and prevent them from inverting when the heart contracts. The **atrioventricular valve** on the right side is called the *tricuspid valve* because it has three cusps, or flaps. The valve on the left side is called the *bicuspid* (or the *mitral*) because it has two flaps. There are also semilunar valves between the ventricles and their attached vessels. A *semilunar valve* has three pockets of tissue, and each pocket resembles a half-moon. The pulmonary semilunar valve lies between the right ventricle and the pulmonary trunk. The aortic semilunar valve lies between the left ventricle and the aorta.

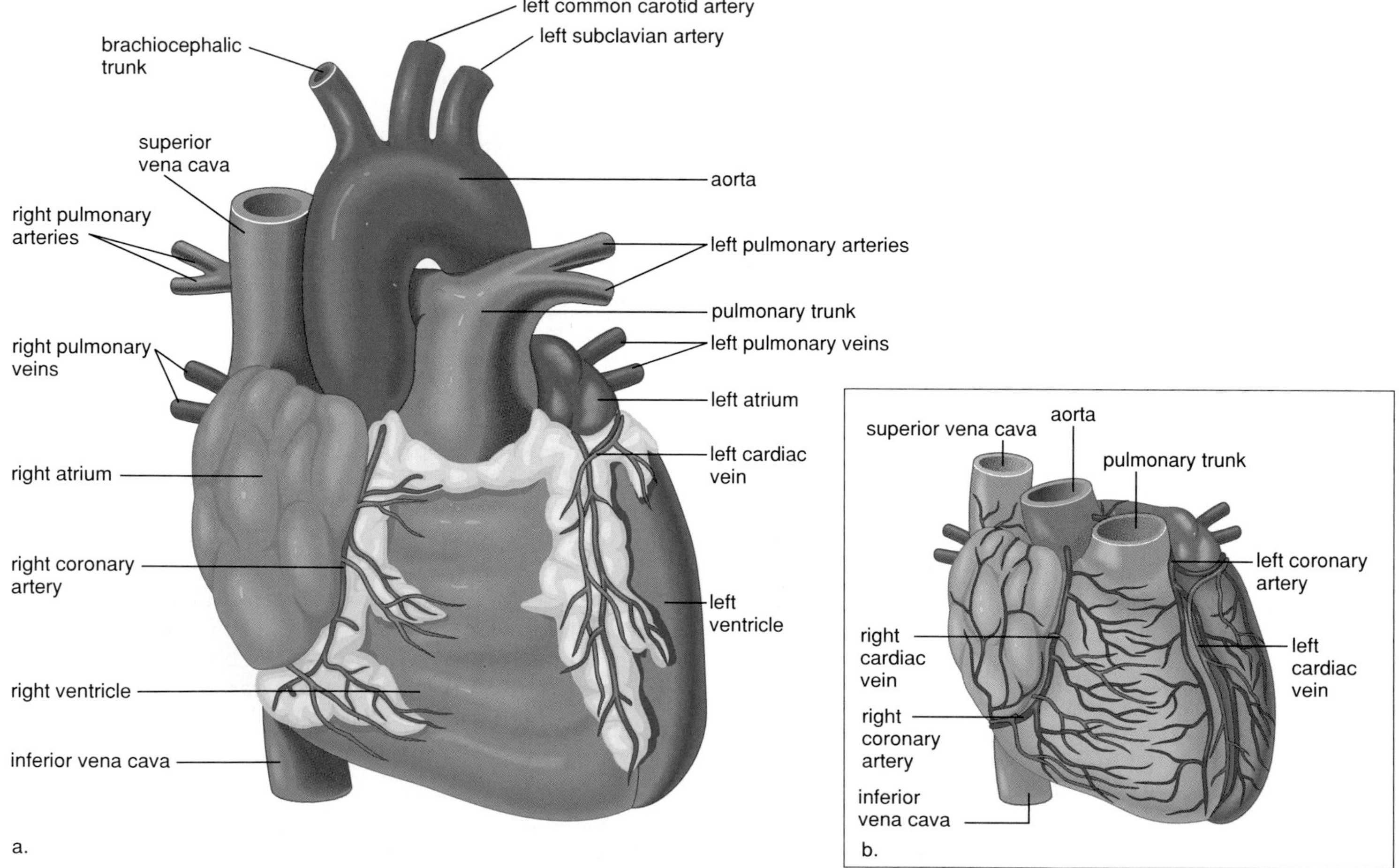

Figure 41.5 External heart anatomy.
a. The venae cavae bring deoxygenated blood to the right side of the heart from the body, and the pulmonary arteries take this blood to the lungs. The pulmonary veins bring oxygenated blood from the lungs to the left side of the heart, and the aorta takes this blood to the body. **b.** The coronary arteries and cardiac veins pervade cardiac muscle. They bring oxygen and nutrients to cardiac cells, which derive no benefit from blood coursing through the heart.

Humans have a four-chambered heart (two atria and two ventricles). A septum separates the right side from the left side.

How the Blood Moves

We can trace the path of blood through the heart (Fig. 41.6) in the following manner:

The superior vena cava and the inferior vena cava, both carrying deoxygenated blood, enter the right atrium.

The right atrium sends blood through an atrioventricular valve (the tricuspid valve) to the right ventricle.

The right ventricle sends blood through the pulmonary semilunar valve into the pulmonary trunk and the pulmonary arteries to the lungs.

The pulmonary veins, carrying oxygenated blood from the lungs, enter the left atrium.

The left atrium sends blood through an atrioventricular valve (the bicuspid or mitral valve) to the left ventricle.

The left ventricle sends blood through the aortic semilunar valve into the aorta and to the body proper.

From this description, you can see that blood must go through the lungs in order to pass from the right side to the left side of the heart. The heart is a double pump because the right side of the heart sends blood through the lungs, and the left side sends blood throughout the body. Since the left ventricle has the harder job of pumping blood to the entire body, its walls are thicker than those of the right ventricle.

The right side of the heart pumps blood to the lungs, and the left side of the heart pumps blood throughout the body.

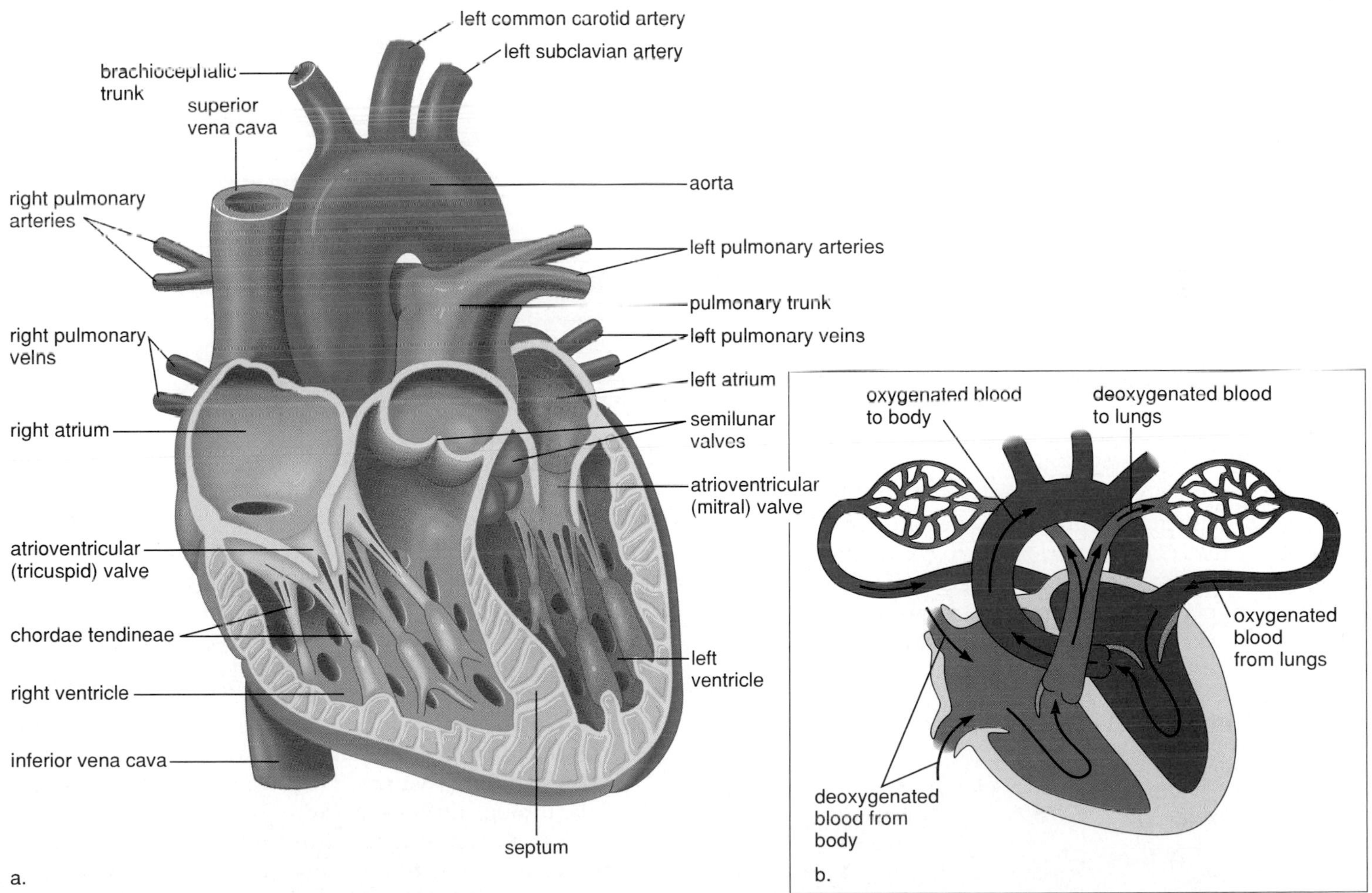

Figure 41.6 Internal view of the heart.
a. The right side of the heart contains deoxygenated blood, and the left side of the heart contains oxygenated blood. **b.** This diagrammatic representation of the heart allows you to trace the path of the blood. On the right side of the heart: venae cavae, right atrium, right ventricle, pulmonary arteries to lungs. On the left side of the heart: pulmonary veins, left atrium, left ventricle, aorta to body.

A Closer Look

William Harvey and Circulation of the Blood

William Harvey (1578–1657) was the first to offer proof that the blood circulates in the body of humans and other animals. Harvey was an English scientist of the seventeenth century, a time of renewed interest in the collection of facts, the use of the hypothesis, the use of experimentation, and the respect for mathematics. This time period is known as the scientific revolution.

After many years of research and study, Harvey hypothesized that the heart is a pump for the entire circulatory system and that blood flows in a circuit. In contrast to former anatomists, Harvey dissected not only dead but also live organisms and observed that when the heart beats, it contracts, forcing blood into the aorta. Had this blood come from the right side of the heart? To do away with the complication of the lungs (pulmonary circulation), Harvey turned to fishes and noticed that the heart first receives and then pumps the blood forward. He observed that blood in the mammalian fetus passes directly from the right side of the heart through the septum to the left side. He felt confident that in mature higher organisms, all blood moves from the right to the left side of the heart by way of the lungs.

Harvey then wanted to show an intimate connection between the arteries and veins in the tissues of the body. Again using live organisms, he demonstrated that if an artery is slit, the whole blood system empties, including arteries and veins. He measured the capacity of the left ventricle in humans and found it to be 2 oz. Since the heart beats 70 times a minute, in one hour the left ventricle forces into the aorta no less than $70 \times 60 \times 2$ = 8,400 oz = 525 lb, or about three times a man's weight! Could so much blood be created and consumed every hour? The same blood must return again and again to the heart.

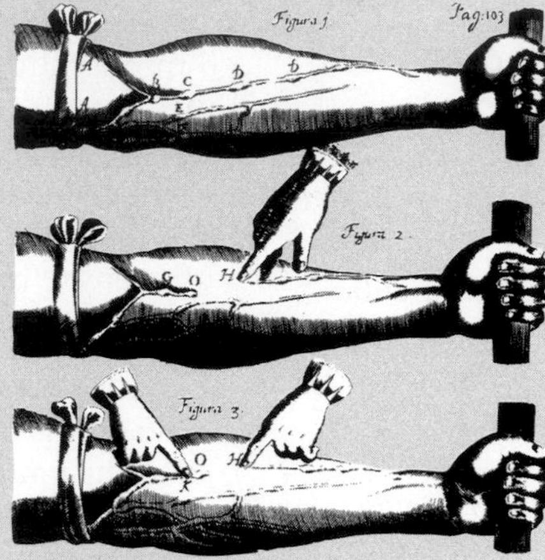

Figure 41A Harvey's drawings showing how he concluded that the veins return blood to the heart.
Harvey tied a ligature (Figure 1) above the elbow to observe the accumulation of blood in the veins. Blood can be forced past a valve from H to O (Figure 2), but not in the opposite direction (Figure 3). Therefore, it can be deduced that blood ordinarily moves toward the heart in the veins.

Harvey also studied the valves in the veins and suggested their true purpose. By the use of ligatures, he demonstrated that a tight ligature on the arm causes the artery to swell on the side of the heart, and a slack ligature causes the vein to swell on the opposite side (Fig. 41A). He said, "This is an obvious indication the blood passes from the arteries into the veins . . . and there is an anastomosis of the two orders of vessels."

Harvey's methods showed how fruitful research might be done. He established that physical and mechanical evidence could provide data for a theory of circulation. He erred, however, when he speculated that the heart heated the blood, and the lung served to cool it or control the degree of heat. His basic method, however, contributed to the scientific revolution and set an example for others to follow.

How the Heartbeat Occurs

The heart contracts, or beats, about seventy times a minute, and each heartbeat lasts about 0.85 second. The term **systole** [Gk. *systole,* contraction] refers to contraction of heart chambers, and the word **diastole** [Gk. *diastole,* dilation, spreading] refers to relaxation of these chambers. Each heartbeat, or *cardiac cycle,* consists of the following phases:

Cardiac Cycle		
Time	**Atria**	**Ventricles**
0.15 sec	Systole	Diastole
0.30 sec	Diastole	Systole
0.40 sec	Diastole	Diastole

First the atria contract (while the ventricles relax), then the ventricles contract (while the atria relax), and then all chambers rest. The short systole of the atria is appropriate since the atria send blood only into the ventricles. It is the muscular ventricles that actually pump blood out into the circulatory system proper. When the word *systole* is used alone, it usually refers to the left ventricular systole.

When the heart beats, the familiar lub-dub sound is heard as the valves of the heart close. The *lub* is caused by vibrations of the heart when the atrioventricular valves close, and the *dub* is heard when vibrations occur due to the closing of the semilunar valves. These valves aid circulation of the blood through the heart because they permit only one-way flow and do not allow a backward flow of blood. The *pulse* is a wave effect that passes down the walls of the arterial blood vessels when the aorta expands and then recoils following ventricle systole. Because there is one arterial pulse per ventricular systole, the arterial pulse rate can be used to determine the heart rate.

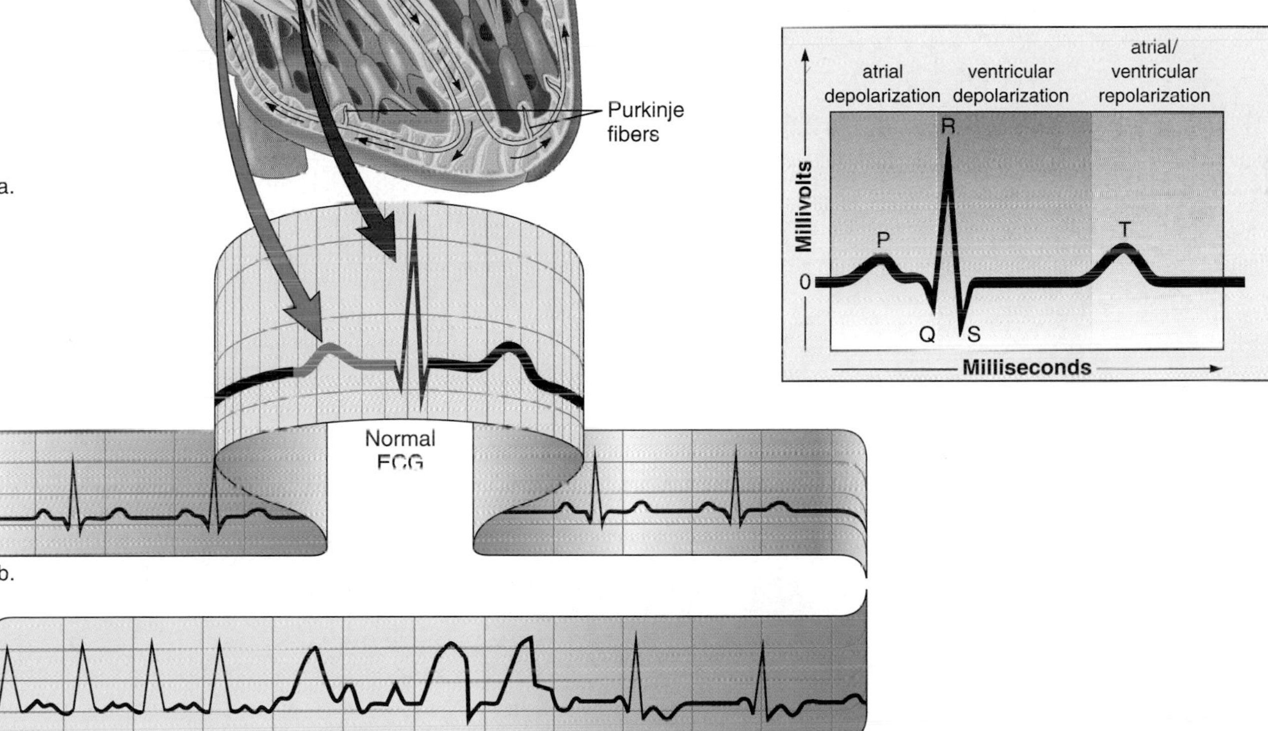

Figure 41.7 Control of the cardiac cycle.
a. The SA (sinoatrial) node sends out a stimulus (black arrows), which causes the right and left atria to contract. When this stimulus (black arrows) reaches the AV (atrioventricular) node, it signals the ventricles to contract by way of the Purkinje fibers. **b.** A normal electrocardiogram (ECG) indicates that the heart is functioning properly. The *P* wave represents excitation and occurs just prior to contraction of the atria. The second wave, or the *QRS* complex, occurs just prior to ventricular contraction. The third, or *T,* wave occurs just before the ventricles relax. **c.** Abnormal ECGS: sinus tachycardia is an abnormally fast heartbeat due to a fast SA node stimulus; ventricular fibrillation is an irregular heartbeat due to irregular stimulation of the ventricles; and mitral stenosis occurs because the bicuspid (mitral) valve is obstructed.

The contraction of the heart is intrinsic, meaning that the heart will beat independent of any nervous stimulation. In fact, it is possible to dissect out a small heart, such as a frog's heart, and watch it undergo contraction in a petri dish. This is due to a unique type of tissue called nodal tissue, with both muscular and nervous characteristics, located in two regions of the heart. Of these regions, the SA (*sinoatrial*) *node,* is found in the upper dorsal wall of the right atrium; the other, the AV (*atrioventricular*) *node,* is found in the base of the right atrium very near the septum (Fig. 41.7*a*). The SA node initiates the heartbeat and every 0.85 second automatically sends out an excitation impulse, which causes the atria to contract. Therefore the SA node is called the *pacemaker* because it usually keeps the heartbeat regular. When the impulse reaches the AV node, this node, by way of specialized fibers, sends out impulses that cause the ventricles to contract. Although the beat of the heart is intrinsic, it is regulated by the nervous system, which can increase or decrease the heartbeat rate.

With the contraction of any muscle, including the myocardium, ionic changes occur; these can be detected with electrical recording devices. The complex graph that results, called an electrocardiogram (ECG or EKG), is associated with the cardiac cycle as indicated (Fig. 41.7*b, c*). An examination of the electrocardiogram indicates whether the heartbeat has a normal or an irregular pattern.

The Vessels Conduct Blood

The human cardiovascular system includes two major circular pathways, the **pulmonary circuit** [L. *pulmonarius*, of the lungs] and the **systemic circuit** (Fig. 41.8).

To the Lungs

In the pulmonary circuit, the path of blood can be traced as follows. Deoxygenated blood from all regions of the body collects in the right atrium and then passes into the right ventricle, which pumps it into the pulmonary trunk. The pulmonary trunk divides into the right and left pulmonary arteries, which carry blood to the lungs. As blood passes through the pulmonary capillaries, carbon dioxide is given off and oxygen is picked up. Blood returns to the left atrium of the heart, through pulmonary venules that join to form pulmonary veins.

To the Body

The **aorta** [L. *aorte*, great artery] and the **venae cavae** (sing., vena cava) [L. *vena*, blood vessel, and *cavus*, hollow] serve as the major pathways for blood in the systemic circuit. To trace the path of blood to any organ in the body, you need only mention the aorta, the proper branch of the aorta, the organ, and the vein returning blood to the vena cava. In the systemic circuit, arteries contain oxygenated blood and have a bright red color, but veins contain deoxygenated blood and appear dull red or, when viewed through the skin, blue.

The *coronary arteries* are extremely important because they serve the heart muscle itself (see Fig. 41.5). (The heart is not nourished by the blood in its chambers.) The coronary arteries arise from the aorta just above the aortic semilunar valve. They lie on the exterior surface of the heart, where they branch into arterioles and then capillaries. The capillary beds enter venules, which join to form the cardiac veins, and these empty into the right atrium.

A **portal system** [L. *porto*, carry, transport] is one that begins and ends in capillaries. One place in the human body where a portal system is found is between the small intestine and the liver. Blood passes from the intestinal villi into venules that join to form the hepatic portal vein, a vessel that connects the intestine with the liver. The hepatic vein leaves the liver and enters the inferior vena cava.

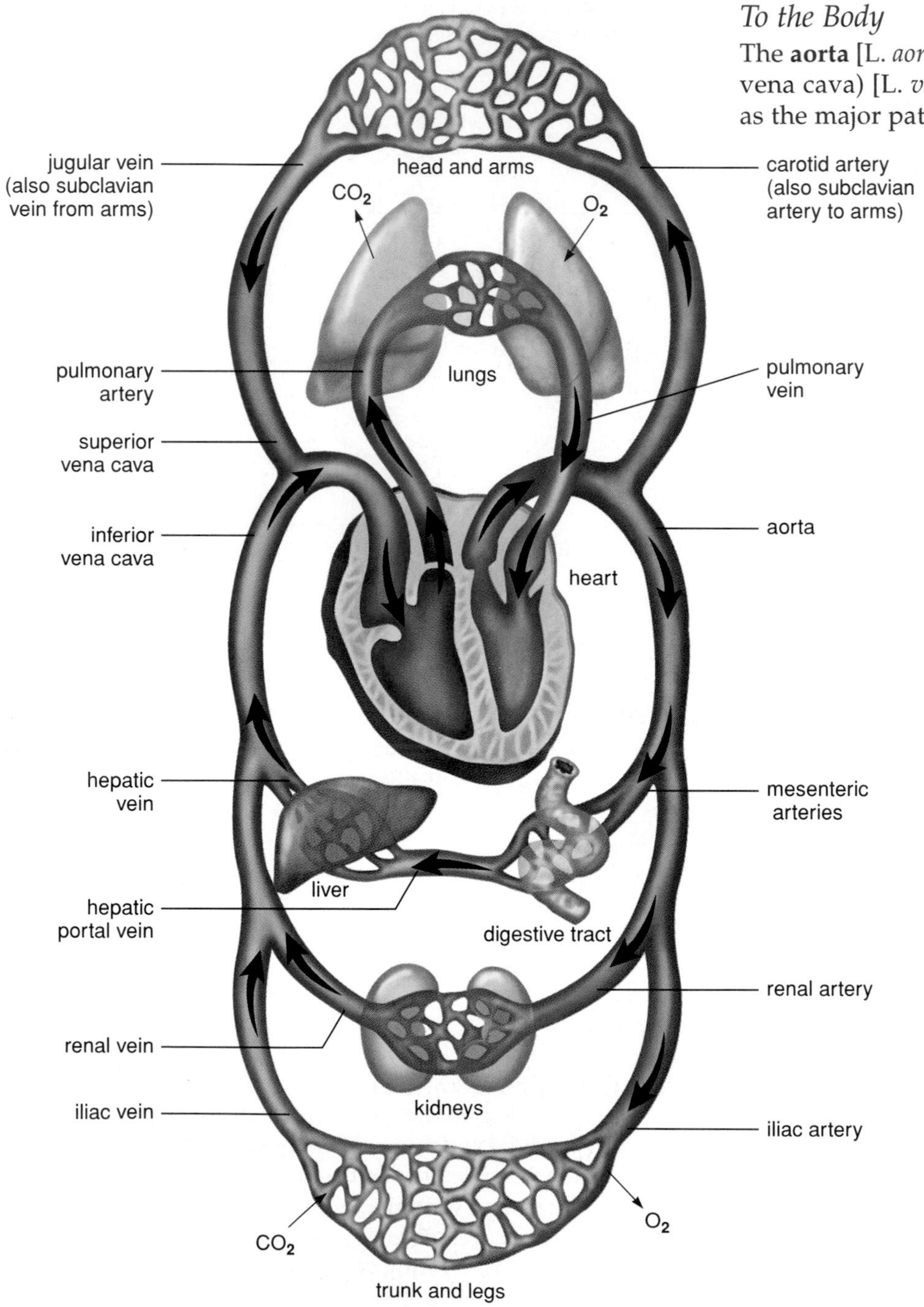

Figure 41.8 Path of blood.
In order to trace blood from the right to the left side of the heart, you must begin at the lungs. In order to trace blood from the capillaries of the digestive tract to the right atrium, you must consider the hepatic portal vein and the hepatic vein. The blue-colored vessels carry deoxygenated blood, and the red-colored vessels carry oxygenated blood; the arrows indicate the flow of blood.

The pulmonary circuit takes deoxygenated blood to the lungs and oxygenated blood to the heart. The systemic circuit takes blood throughout the body from the aorta to the venae cavae.

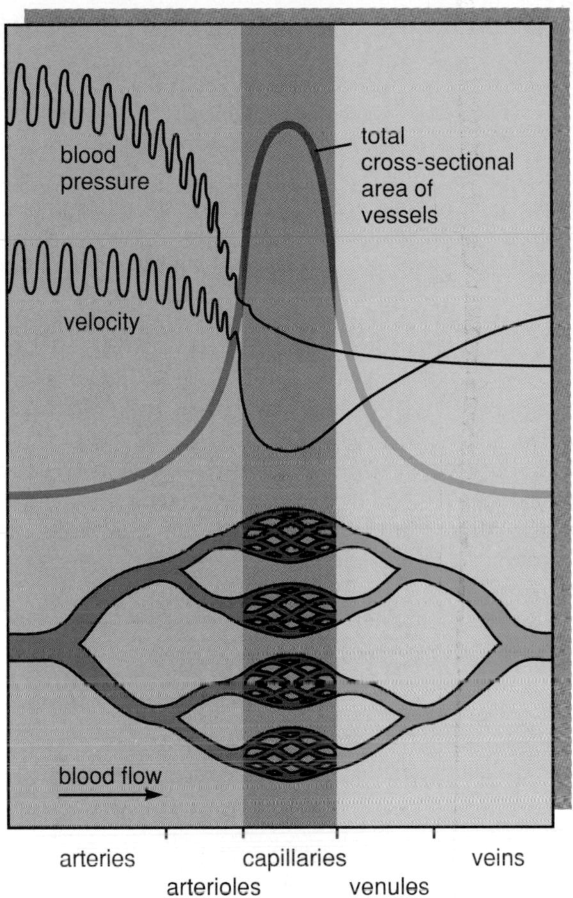

Figure 41.9 Velocity and blood pressure related to vascular cross-sectional area.
Capillaries have the greatest cross-sectional area, and blood is under minimal pressure and has the least velocity. Skeletal muscle contraction, and not blood pressure, accounts for the velocity of blood in the veins.

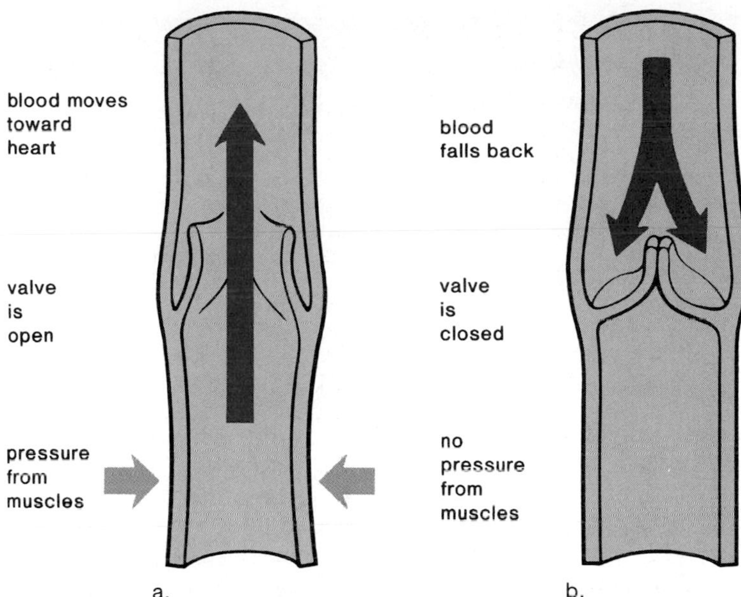

Figure 41.10 Cross section of a valve in a vein.
a. Pressure on the walls of a vein, exerted by skeletal muscles, increases blood pressure within the vein and forces a valve open. b. When external pressure is no longer applied to the vein, blood pressure decreases, and back pressure forces the valve closed. Closure of the valves prevents the blood from flowing in the opposite direction.

Blood Pressure Drops Off

When the left ventricle contracts, blood is forced into the systemic arteries under pressure. *Systolic pressure* results from blood being forced into the arteries during ventricular systole, and *diastolic pressure* is the pressure in the arteries during ventricular diastole. Human **blood pressure** is measured with a sphygmomanometer, which has a pressure cuff that measures the amount of pressure required to stop the flow of blood through an artery. Blood pressure is normally measured on the brachial artery, which is in the upper arm, and is stated in millimeters of mercury (mm Hg). A blood pressure reading consists of two numbers, for example, 120/80—which represents systolic and diastolic pressures, respectively.

As blood flows from the aorta into the various arteries and arterioles, blood pressure falls. Also, the difference between systolic and diastolic pressure gradually diminishes. In the capillaries, there is a slow, fairly even, flow of blood. This may be related to the very high total cross-sectional area of the capillaries (Fig. 41.9). It's even been suggested that if all the blood vessels in a human were connected end to end, the total distance would reach around the earth two times at the equator, or about 60,000 miles. A large portion of this distance would be due to the quantity of capillaries.

Blood pressure in the veins is low and cannot move blood back to the heart, especially from the limbs of the body. When skeletal muscles near veins contract, they put pressure on the collapsible walls of the veins and on blood contained in these vessels. Veins, however, have *valves* that prevent the backward flow of blood, and therefore pressure from muscle contraction is sufficient to move blood through veins toward the heart (Fig. 41.10). *Varicose veins*, abnormal dilations in superficial veins, develop when the valves of the veins become weak and ineffective due to a backward pressure of the blood.

The beat of the heart supplies the pressure that keeps blood moving in the arteries, and skeletal muscle contraction pushes blood in the veins back to the heart.

Health Focus

Prevention of Cardiovascular Disease

All of us can take steps to prevent the occurrence of cardiovascular disease, the most frequent cause of death in the United States. There are genetic factors that predispose an individual to cardiovascular disease, such as family history of heart attack under age 55, male gender, and ethnicity (African Americans are at greater risk). Those with one or more of these risk factors need not despair, however. It means only that they need to pay particular attention to these guidelines for a heart-healthy lifestyle.

The Don'ts

Smoking and Drug Abuse

Hypertension is well recognized as a major contributor to cardiovascular disease. When a person smokes, the drug nicotine, present in cigarette smoke, enters the bloodstream. Nicotine causes the arterioles to constrict and the blood pressure to rise. Restricted blood flow and cold hands are associated with smoking by most people. More serious is the need for the heart to pump harder to propel the blood through the lungs at a time when the oxygen-carrying capacity of the blood is reduced.

Stimulants, such as cocaine and amphetamines, can cause an irregular heartbeat and lead to heart attacks and strokes in people who are using drugs even for the first time. Intravenous drug use may result in a cerebral embolism.

Obesity

Hypertension occurs more often in persons who are obese–those who are more than 20% above recommended weight. More tissues require servicing, and the heart sends the extra blood out under greater pressure in those who are overweight. Since it is very difficult for obese individuals to lose weight, it is recommended that weight control be a lifelong endeavor. Even a slight decrease in weight can bring with it a reduction in hypertension. A 4.5-kilogram weight loss doubles the chance that blood pressure can be normalized without drugs.

The Do's

Healthy Diet

It was once thought that a low-salt diet was protective against cardiovascular disease, and it still may be in certain persons. Theoretically, hypertension occurs because the more salty the blood, the greater the osmotic pressure and the higher the water content. In recent years, the emphasis has switched to a diet low in saturated fats and cholesterol as protective against cardiovascular disease. Cholesterol is ferried in the blood by two types of plasma proteins called LDL (low-density lipoprotein) and HDL (high-density lipoprotein). LDL (called "bad" lipoprotein) takes cholesterol from the liver to the tissues, and HDL (called "good" lipoprotein) transports cholesterol out of the tissues to the liver. When the LDL level in blood is abnormally high or the HDL level is abnormally low, cholesterol accumulates in the cells. When cholesterol-laden cells line the arteries, plaque develops, which interferes with circulation (Fig. 41B).

It is recommended that everyone know his or her blood cholesterol level. Individuals with a high blood cholesterol level (240 mg/100 ml) should be further tested to determine their LDL-cholesterol level. The LDL-cholesterol level together with other risk factors such as age, family history, general health, and whether the patient smokes will determine who needs dietary therapy to lower their LDL. Drugs are to be reserved for high-risk patients.

Evidence is mounting to suggest a role for antioxidant vitamins (A, E, and C) in the prevention of cardiovascular disease. Antioxidants protect the body from free radicals that may damage HDL cholesterol through oxidation or damage the lining of an artery, leading to a blood clot that can block the vessel. Nutritionists believe that the consumption of at least five servings of fruit and vegetables a day may be protective against cardiovascular disease.

41.4 Cardiovascular Disorders in Humans

Cardiovascular disease (CVD) is the leading cause of untimely death in the Western countries. Modern research efforts have resulted in improved diagnosis, treatment, and prevention. This section discusses the range of advances that have been made in these areas. The reading on this page emphasizes how to prevent CVD from developing in the first place.

Hypertension Is Deadly

It is estimated that about 20% of all Americans suffer from *hypertension,* which is high blood pressure. Women of any age are considered to have hypertension if their blood pressure reading is 160/95 or above. For a man under age 45, a reading above 130/90 is hypertensive, and beyond age 45, a reading above 140/95 is considered hypertensive. While both systolic and diastolic pressures are considered important, it is the diastolic pressure that is emphasized when medical treatment is being considered.

Hypertension is sometimes called a silent killer because it may not be detected until a stroke or heart attack occurs. It has long been thought that a certain genetic makeup might account for the development of hypertension. Now researchers have discovered two genes that may be involved in some individuals. One gene codes for angiotensinogen, a plasma protein that is converted to a powerful vasoconstrictor in part by the product of the second gene. Persons with hypertension due to overactivity of these genes might one day be cured by gene therapy.

Atherosclerosis and Fatty Arteries

Hypertension also is seen in individuals who have *atherosclerosis* (formerly called arteriosclerosis), an accumulation of soft masses of fatty materials, particularly cholesterol, beneath the inner linings of arteries (Fig. 41B). Such deposits are called plaque. As deposits occur, plaque

Exercise

Those who exercise are less apt to have cardiovascular disease. One study found that moderately active men who spent an average of 48 minutes a day on a leisure-time activity such as gardening, bowling, or dancing had one-third fewer heart attacks than peers who spent an average of only 16 minutes each day. Exercise helps to keep weight under control, may help minimize stress, and reduces hypertension. The heart beats faster when exercising, but exercise slowly increases its capacity. This means that the heart can beat slower when we are at rest and still do the same amount of work. One physician recommends that his cardiovascular patients walk for one hour, three times a week, and in addition they are to practice meditation and yoga-like stretching and breathing exercises to reduce stress.

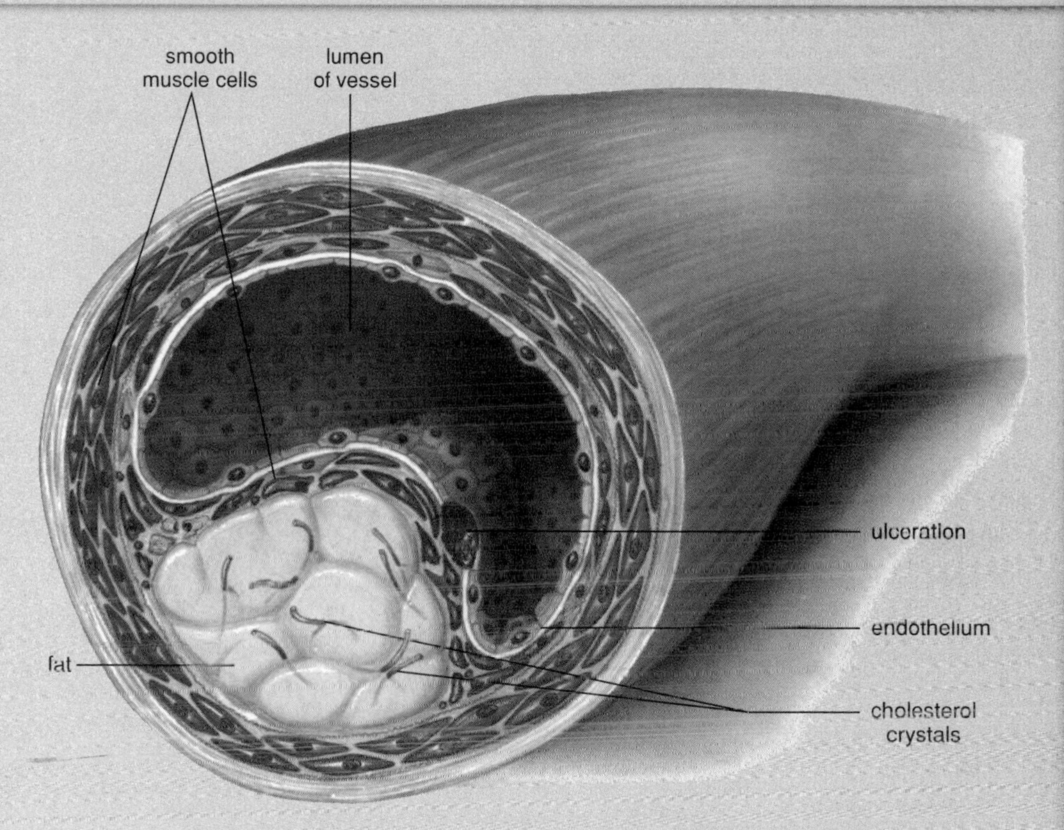

Figure 41B Plaque.
Cross section of plaque shows its composition and indicates how it bulges out into the lumen of an artery, obstructing blood flow. Now a blood clot is more likely to occur and block the artery.

tends to protrude into the lumen of the vessel, interfering with the flow of blood. Atherosclerosis begins in early adulthood and develops progressively through middle age, but symptoms may not appear until an individual is 50 or older. To prevent its onset and development, the American Heart Association and other organizations recommend a diet low in saturated fat and cholesterol and rich in fruits and vegetables.

Plaque can cause a clot to form on the irregular arterial wall. As long as the clot remains stationary, it is called a *thrombus,* but when and if it dislodges and moves along with the blood, it is called an *embolus.* If *thromboembolism* is not treated, complications, which are mentioned in the following section, can arise.

In certain families, atherosclerosis is due to an inherited condition such as familial hypercholesterolemia. The presence of the disease-associated mutation can be detected, and this information is helpful if measures are taken to prevent the occurrence of the disease.

Stroke and Heart Attack

Stroke, heart attack, and aneurysm are associated with hypertension and atherosclerosis. A cardiovascular accident (CVA), also called a *stroke,* often results when a small cranial arteriole bursts or is blocked by an embolus. A lack of oxygen causes a portion of the brain to die, and paralysis or death can result. A person sometimes is forewarned of a stroke by a feeling of numbness in the hands or the face, difficulty in speaking, or temporary blindness in one eye.

A myocardial infarction (MI), also called a *heart attack,* occurs when a portion of the heart muscle dies due to a lack of oxygen. If a coronary artery becomes partially blocked, the individual may then suffer from *angina pectoris,* characterized by chest pains and/or a radiating pain in the left arm. Nitroglycerin or related drugs dilate blood vessels and help relieve the pain. When a coronary artery is completely blocked, perhaps because of thromboembolism, a heart attack occurs.

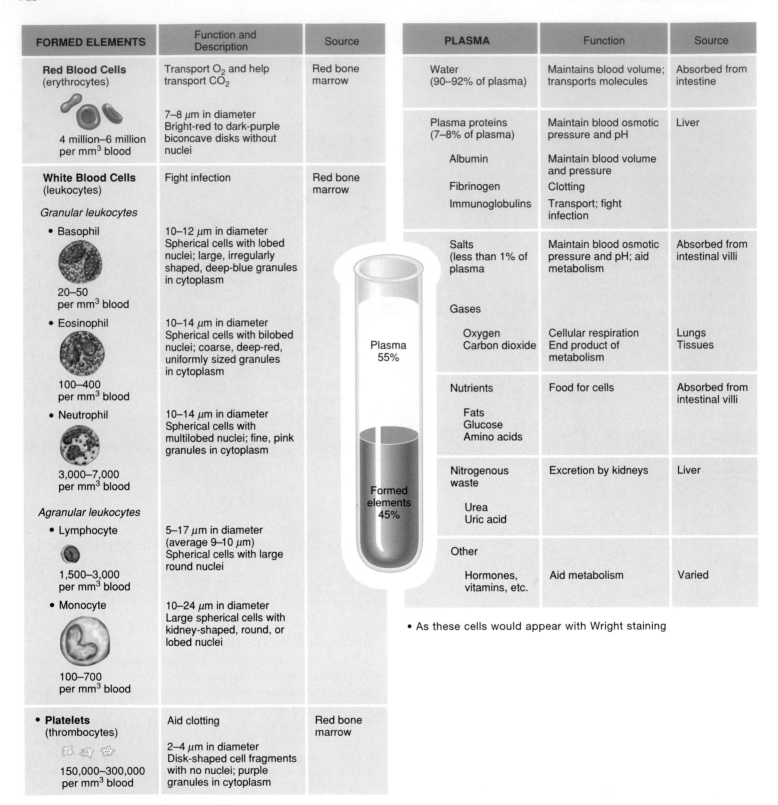

FORMED ELEMENTS	Function and Description	Source
Red Blood Cells (erythrocytes) 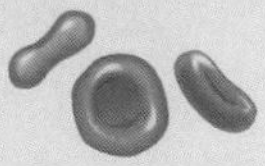 4 million–6 million per mm^3 blood	Transport O_2 and help transport CO_2 7–8 μm in diameter Bright-red to dark-purple biconcave disks without nuclei	Red bone marrow
White Blood Cells (leukocytes)	Fight infection	Red bone marrow
Granular leukocytes		
• Basophil 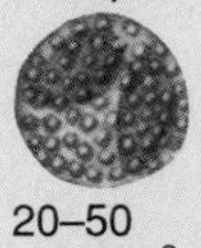 20–50 per mm^3 blood	10–12 μm in diameter Spherical cells with lobed nuclei; large, irregularly shaped, deep-blue granules in cytoplasm	
• Eosinophil 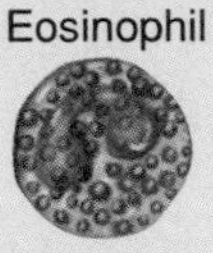 100–400 per mm^3 blood	10–14 μm in diameter Spherical cells with bilobed nuclei; coarse, deep-red, uniformly sized granules in cytoplasm	
• Neutrophil 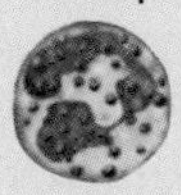 3,000–7,000 per mm^3 blood	10–14 μm in diameter Spherical cells with multilobed nuclei; fine, pink granules in cytoplasm	
Agranular leukocytes		
• Lymphocyte 1,500–3,000 per mm^3 blood	5–17 μm in diameter (average 9–10 μm) Spherical cells with large round nuclei	
• Monocyte 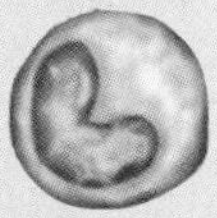 100–700 per mm^3 blood	10–24 μm in diameter Large spherical cells with kidney-shaped, round, or lobed nuclei	
• **Platelets** (thrombocytes)  150,000–300,000 per mm^3 blood	Aid clotting 2–4 μm in diameter Disk-shaped cell fragments with no nuclei; purple granules in cytoplasm	Red bone marrow

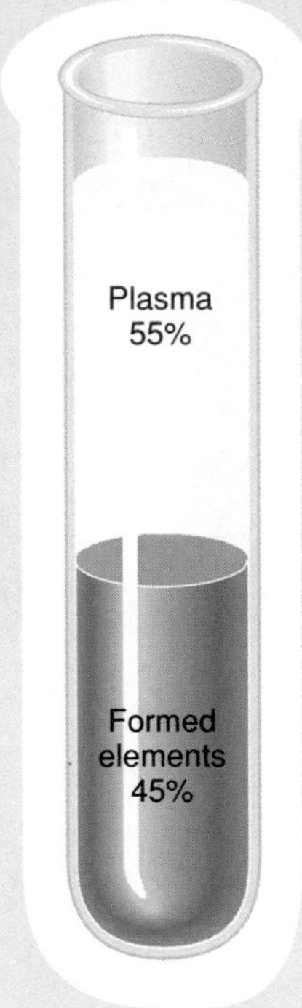

Plasma 55%

Formed elements 45%

PLASMA	Function	Source
Water (90–92% of plasma)	Maintains blood volume; transports molecules	Absorbed from intestine
Plasma proteins (7–8% of plasma)	Maintain blood osmotic pressure and pH	Liver
Albumin	Maintain blood volume and pressure	
Fibrinogen	Clotting	
Immunoglobulins	Transport; fight infection	
Salts (less than 1% of plasma	Maintain blood osmotic pressure and pH; aid metabolism	Absorbed from intestinal villi
Gases		
Oxygen	Cellular respiration	Lungs
Carbon dioxide	End product of metabolism	Tissues
Nutrients	Food for cells	Absorbed from intestinal villi
Fats Glucose Amino acids		
Nitrogenous waste	Excretion by kidneys	Liver
Urea Uric acid		
Other		
Hormones, vitamins, etc.	Aid metabolism	Varied

• As these cells would appear with Wright staining

Figure 41.11 Composition of blood.
When blood is transferred to a test tube and is prevented from clotting, it forms two layers. The transparent yellow top layer is plasma, the liquid portion of blood. The formed elements are in the bottom layer. The tables describe these components in detail.

41.5 Blood Is Transport Medium

The blood of mammals has numerous functions that help maintain homeostasis. Blood (1) transports substances to and from the capillaries, where exchanges with tissue fluid take place; (2) helps guard against invasion by microbes, such as bacteria and viruses; (3) helps regulate body temperature; and (4) clots, preventing a potentially life-threatening loss of blood.

In humans, blood has two main portions: the liquid portion, called plasma, and the formed elements consisting of various cells and platelets (Fig. 41.11). **Plasma** [Gk. *plasma*, something molded] contains many types of molecules, including nutrients, wastes, salts, and proteins. The salts and proteins are involved in buffering the blood, effectively keeping the pH near 7.4. They also maintain blood's osmotic pressure so that water has an automatic tendency to enter blood capillaries. Several plasma proteins are involved in blood clotting and others transport large organic molecules in the blood. Albumin, the most plentiful of the plasma proteins transport bilirubin, a breakdown product of hemoglobin. Globulins have various functions and among them are the lipoproteins that transport cholesterol.

Formed Elements Are Varied

The formed elements are of three types: **red blood cells** (RBCs), or erythrocytes [Gk. *erythros*, red, *kytos*, cell]; **white blood cells** (WBCs), or leukocytes [Gk. *leukos*, white, *kytos*, cell]; and **platelets**, or thrombocytes [Gk. *thrombos*, blood clot, *kytos*, cell].

RBCs Transport Oxygen

Red blood cells are small biconcave disks that at maturity lack a nucleus and contain the respiratory pigment hemoglobin. There are 6 million red blood cells per mm^3 of whole blood, and each one of these cells contains about 250 million hemoglobin molecules. **Hemoglobin** [Gk. *haima*, blood, and L. *globus*, ball] contains four globin protein chains, each associated with heme, an iron-containing group. Iron combines loosely with oxygen, and in this way oxygen is carried in the blood. If there is an insufficient number of red blood cells, or if the cells do not have enough hemoglobin, the individual suffers from anemia and has a tired, rundown feeling.

Red blood cells are manufactured continuously in the red bone marrow of the skull, the ribs, the vertebrae, and the ends of the long bones. The growth factor erythropoietin, which is produced when an enzyme from the kidneys acts on a precursor made by the liver, stimulates the production of red blood cells. Now available as a drug, erythropoietin is helpful to persons with anemia and is also sometimes abused by athletes who want to increase performance.

Before they are released from the bone marrow into blood, red blood cells lose their nucleus and synthesize hemoglobin. After living about 120 days, they are destroyed chiefly in the liver and the spleen, where they are engulfed by large phagocytic cells. When red blood cells are destroyed, hemoglobin is released. The iron is recovered and is returned to the red bone marrow for reuse. The heme portions of the molecules undergo chemical degradation and are excreted by the liver as bile pigments in the bile. The bile pigments are primarily responsible for the color of feces.

WBCs Fight Infection

White blood cells differ from red blood cells in that they are usually larger and have a nucleus, they lack hemoglobin, and, without staining, they appear translucent. With staining, white blood cells appear light blue unless they have granules that bind with certain stains. The following are *granular leukocytes* with a lobed nucleus: neutrophils, which have granules that stain slightly pink; eosinophils, which have granules that take up the red dye eosin; and basophils, which have granules that take up a basic dye, staining them a deep blue. The *agranular leukocytes* with no granules and a circular or indented nucleus are the larger monocytes and the smaller lymphocytes. The newly discovered stem cell growth factor (SGF) can be used to increase the production of all white blood cells, and there are also various specific stimulating factors that can be used to stimulate the production of specific stem cells. These growth factors should be helpful to patients with low immunity, such as AIDS patients.

When microorganisms enter the body due to an injury, the response is called an *inflammatory reaction* because there is swelling and reddening at the injured site. The damaged tissue has released kinins, which cause vasodilation, and histamines, which cause increased capillary permeability. **Neutrophils** [Gk. *neuter*, neither, and *phileo*, love], which are amoeboid, squeeze through the capillary wall and enter the tissue fluid, where they phagocytize foreign material. Monocytes appear and are transformed into **macrophages** [Gk. *makros*, long, and *phagein*, to eat], which are large phagocytizing cells that release white blood cell growth factors. Soon there is an explosive increase in the number of leukocytes. The thick, yellowish fluid called pus contains a larger proportion of dead white blood cells that have fought the infection.

Lymphocytes [L. *lympha*, clear water, and Gk. *kytos*, cell] also play an important role in fighting infection. Certain lymphocytes called T cells attack infected cells that contain viruses. Other lymphocytes called B cells produce antibodies. Each B cell produces just one type of antibody, which is specific for one type of antigen. An **antigen** [Gk. *anti*, against, and L. *genitus*, forming, causing], which is most often a protein but is sometimes a polysaccharide, causes the body to produce an antibody because the antigen doesn't belong to the body. Antigens are found in the outer covering of parasites or are present in their toxins. When **antibodies** [Gk. *anti*, against] combine with antigens, the complex is often phagocytized by a macrophage. An individual is actively immune when a large number of B cells are all producing the specific antibody needed for a particular infection.

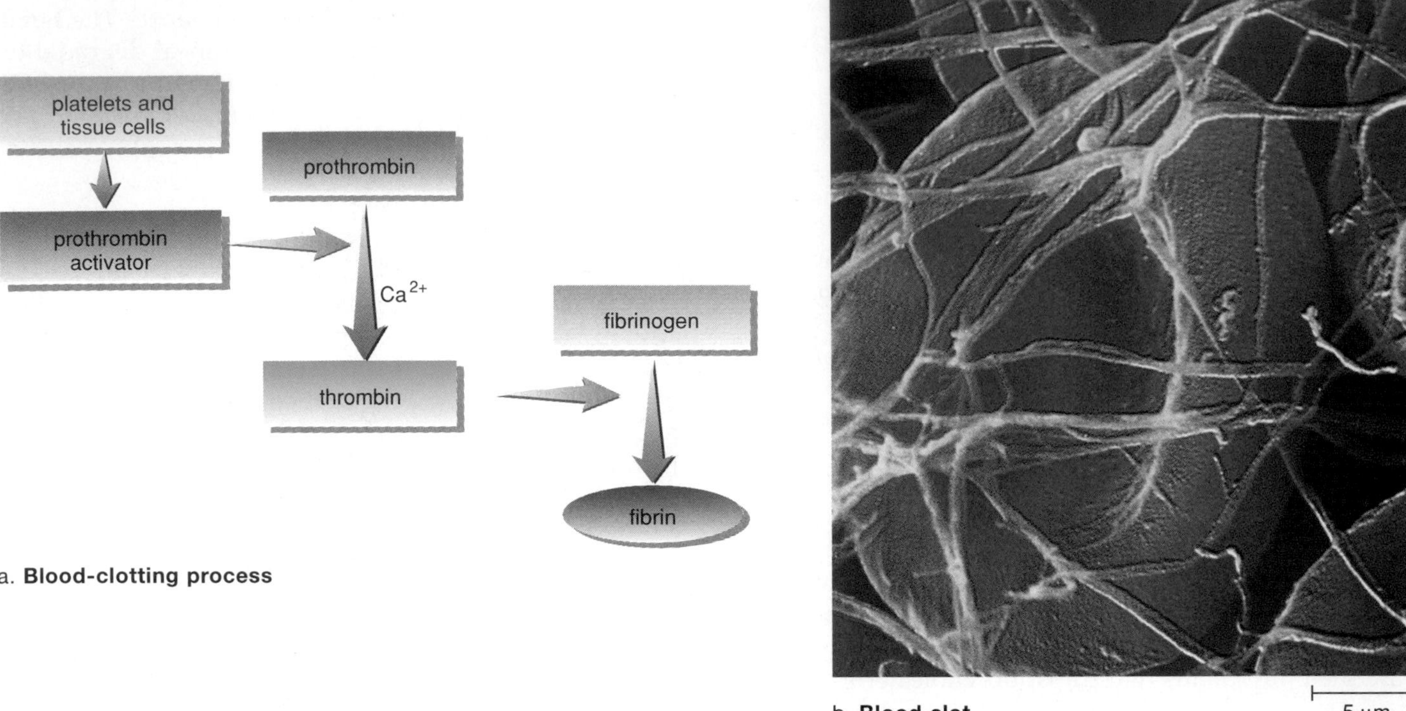

Figure 41.12 Blood clotting.
a. Platelets and damaged tissue cells release prothrombin activator, which acts on prothrombin in the presence of calcium ions (Ca^{2+}) to produce thrombin, which acts on fibrinogen to form fibrin. **b.** A scanning electron micrograph shows a red blood cell caught in the fibrin threads of a clot.

Platelets Assist Blood Clotting

Platelets (thrombocytes) result from fragmentation of certain large cells, called *megakaryocytes,* in the red bone marrow. Platelets are produced at a rate of 200 billion a day, and the blood contains 150,000–300,000 per mm^3. These formed elements are involved in the process of blood clotting, or coagulation.

There are at least 12 clotting factors in the blood that participate in the formation of a blood clot. We will discuss the roles played by platelets, prothrombin, and fibrinogen. Fibrinogen and prothrombin are proteins manufactured and deposited in blood by the liver. Vitamin K, found in green vegetables and also formed by intestinal bacteria, is necessary for the production of prothrombin, and if by chance this vitamin is missing from the diet, hemorrhagic disorders develop.

Blood Clotting Has Steps When a blood vessel in the body is damaged, platelets clump at the site of the puncture and partially seal the leak. They and the injured tissues release a clotting factor called *prothrombin activator* that converts prothrombin to thrombin. This reaction requires calcium ions (Ca^{2+}). **Thrombin,** in turn, acts as an enzyme that severs two short amino acid chains from each fibrinogen molecule. These activated fragments then join end to end, forming long threads of *fibrin.* Fibrin threads wind around the platelet plug in the damaged area of the blood vessel and provide the framework for the clot. Red blood cells also are trapped within the fibrin threads; these cells make a clot appear red (Fig. 41.12). A fibrin clot is present only temporarily. As soon as blood vessel repair is initiated, an enzyme called plasmin destroys the fibrin network and restores the fluidity of plasma.

If blood is allowed to clot in a test tube, a yellowish fluid develops above the clotted material. This fluid is called serum, and it contains all the components of plasma except fibrinogen. Table 41.1 reviews the many different terms we have used to refer to various body fluids related to blood.

A blood clot consists of platelets and red blood cells entangled within fibrin threads.

Table 41.1
Body Fluids

Name	Composition
Blood	Formed elements and plasma
Plasma	Liquid portion of blood
Serum	Plasma minus fibrinogen
Tissue fluid	Plasma minus most proteins
Lymph	Tissue fluid within lymphatic vessels

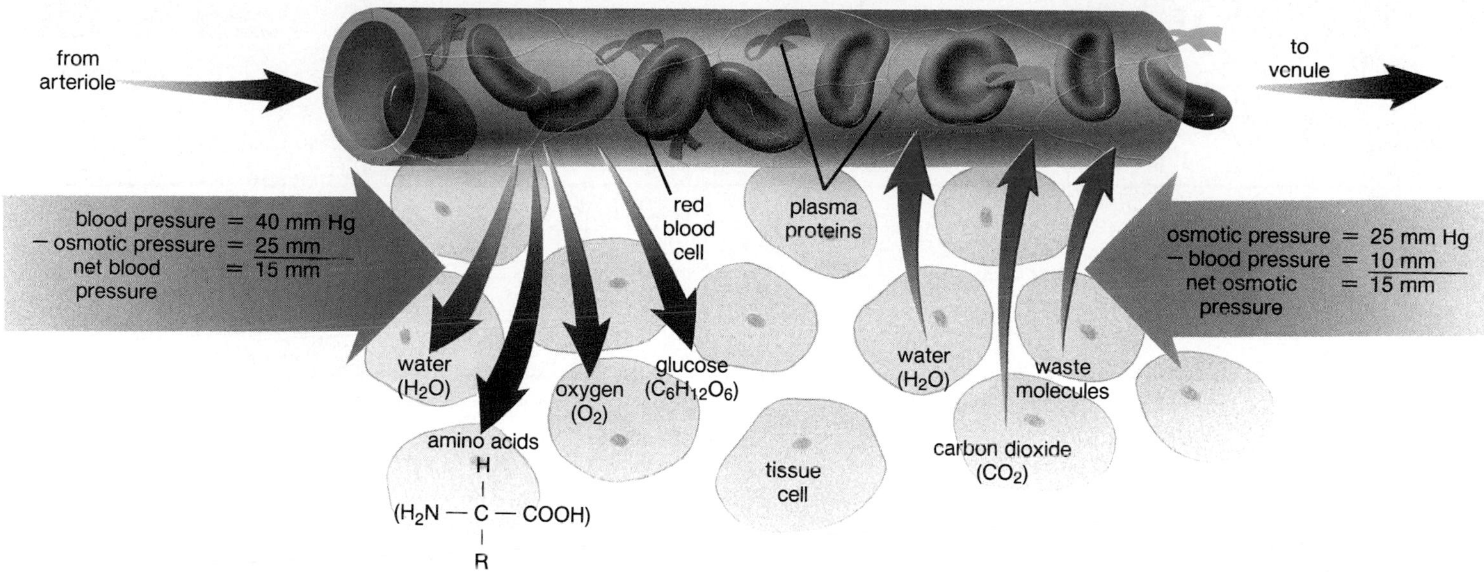

Figure 41.13 Capillary exchange.
A capillary, illustrating the exchanges that take place and the forces that aid the process. At the arterial end of a capillary, the blood pressure is higher than the osmotic pressure; therefore, water (H_2O) and nutrients tend to leave the bloodstream. In the midsection, molecules including oxygen (O_2) follow their concentration gradients. At the venous end of a capillary, the osmotic pressure is higher than the blood pressure; therefore, water and wastes tend to enter the bloodstream. Notice that the red blood cells and the plasma proteins are too large to exit a capillary.

Exchanges Between Blood and Tissue Fluid

Two forces primarily control movement of fluid through the capillary wall: osmotic pressure, which tends to cause water to move from tissue fluid to blood, and blood pressure, which tends to cause water to move in the opposite direction. At the arterial end of a capillary, blood pressure (40 mm Hg) is higher than the osmotic pressure of blood (25 mm Hg) (Fig. 41.13). Osmotic pressure is created by the presence of salts and the plasma proteins. Because blood pressure is higher than osmotic pressure at the arterial end of a capillary, water exits a capillary at this end.

Midway along the capillary, where blood pressure is lower, the two forces essentially cancel each other, and there is no net movement of water. Solutes now diffuse according to their concentration gradient—nutrients (glucose and oxygen) diffuse out of the capillary, and wastes (carbon dioxide) diffuse into the capillary. Red blood cells and almost all plasma proteins remain in the capillaries, but small substances leave. The substances that leave a capillary contribute to **tissue fluid,** the fluid between the body's cells. Since plasma proteins are too large to readily pass out of the capillary, tissue fluid tends to contain all components of plasma except much less protein.

At the venule end of a capillary, where blood pressure has fallen even more, osmotic pressure is greater than blood pressure, and water tends to move into the capillary. Almost the same amount of fluid that left the capillary returns to it, although there is always some excess tissue fluid collected by the lymphatic capillaries (Fig. 41.14). Tissue fluid contained within lymphatic vessels is called **lymph.** Lymph is returned to the systemic venous blood when the major lymphatic vessels enter the subclavian veins in the shoulder region.

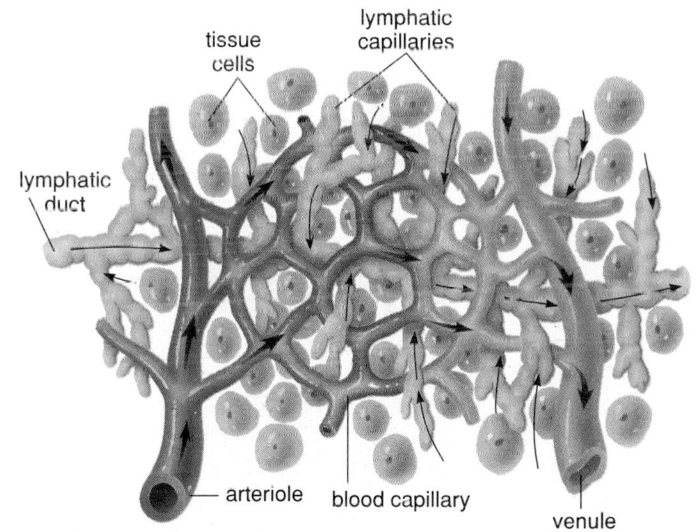

Figure 41.14 Lymphatic capillaries.
Arrows indicate that lymph is formed when lymphatic capillaries take up excess tissue fluid. Lymphatic capillaries lie near blood capillaries.

Oxygen and nutrient substances exit a capillary near the arterial end; carbon dioxide and waste molecules enter a capillary near the venous end.

Table 41.2

The ABO System

Blood Type	Antigen on Red Blood Cells	Antibody in Plasma	% U.S. African American	% U.S. Caucasian	% U.S. Asian	% North American Indians	% Americans of Chinese Descent
A	A	Anti-B	27	41	28	8	25
B	B	Anti-A	20	9	27	1	35
AB	A, B	None	4	3	5	0	10
O	None	Anti-A and anti-B	49	47	40	92	30

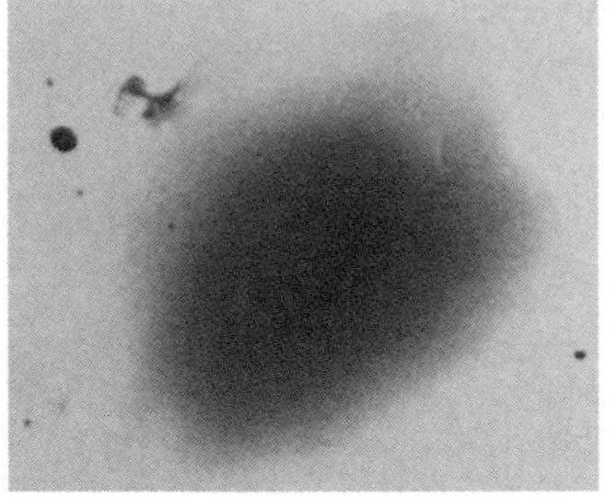

Antigen not present and agglutination does not occur.

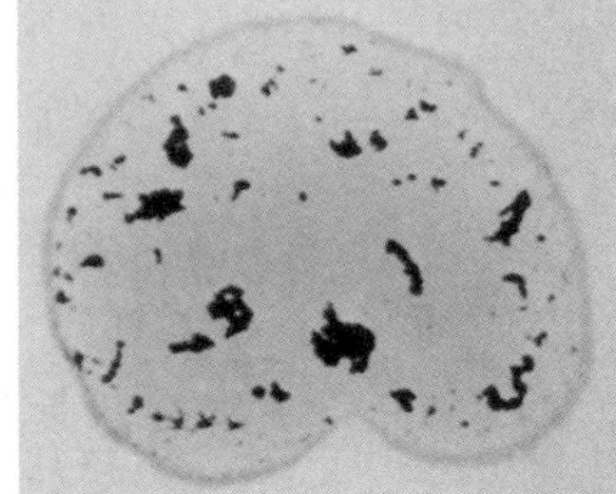

Antigen present and agglutination occurs.

a.

anti-A anti-B anti-Rh type blood

O+

A–

B+

AB–

b.

Figure 41.15 Blood typing.
The standard test to determine ABO and Rh blood type consists of putting a drop of anti-A antibodies, anti-B antibodies, and anti-Rh antibodies on a slide. To each of these, a drop of the person's blood is added. **a.** If agglutination occurs, as seen in the photo on the right, the person has this antigen on red blood cells. **b.** Several possible results. The blood type reveals the antigens that are present on the red blood cells (represented by dots).

Know Your Blood Type

Although there are at least 12 well-known blood type identification systems, the ABO system and the Rh system are most often used to determine blood type.

ABO System Is Common

Before the twentieth century, blood transfusions sometimes resulted in adverse reactions or even death. A concerned Viennese physician, Karl Landsteiner, began to study the matter by mixing different samples of blood and examining the effect under the microscope. In the end, he and his associates determined that there are four major blood groups among humans. They designated the types of blood as A, B, AB, and O (Table 41.2). The types of blood are dependent on whether A antigen and/or B antigen is present on red blood cells. Type O blood has neither the A antigen nor the B antigen on red blood cells; the other types of blood have one or both of the antigen(s) present. A (or B) is not an antigen to an individual with blood type A, but it can be an antigen to a recipient with a different blood type.

Within the plasma of the individual, there are antibodies to the antigens that are not present on the red blood cells. Therefore, for example, type A has an antibody called anti-B in the plasma. Type AB blood has neither anti-A nor anti-B antibodies because both antigens are on the red blood cells. This is reasonable because if these antibodies were present, **agglutination** [L. *ad,* to, *glutinis,* be sticky], or clumping of red blood cells, would occur.

For a recipient to receive blood from a donor, the recipient's plasma must not have an antibody that causes the donor's cells to agglutinate. For this reason it is important to determine each person's blood type. Figure 41.15 demonstrates a way to use the antibodies derived from plasma to determine blood type. If clumping occurs after a sample of blood is exposed to anti-A or anti-B an-

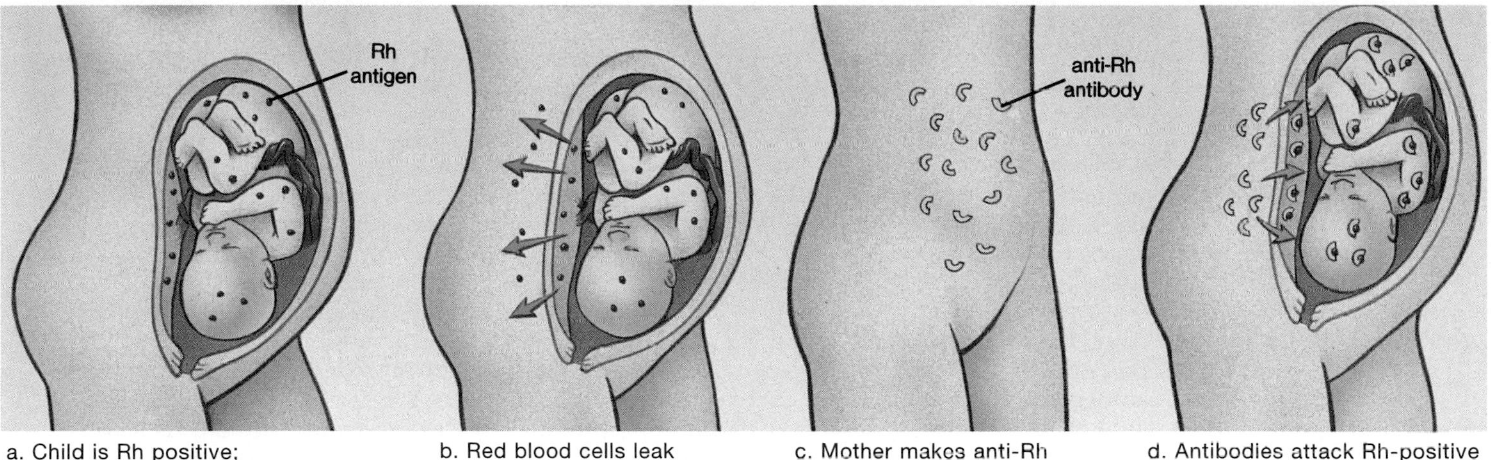

a. Child is Rh positive; mother is Rh negative

b. Red blood cells leak across placenta

c. Mother makes anti-Rh antibodies

d. Antibodies attack Rh-positive red blood cells in child

Figure 41.16 Hemolytic disease of the newborn.
In this series of drawings, the first pregnancy with an Rh-positive child is represented by (a) and (b). The second pregnancy with an Rh-positive child is represented by (d).

tibody, the person has that antigen on the red blood cells. In the first example given, the individual's blood sample does not react to either anti-A or anti-B antibody; therefore, the blood type is O. The + and – are discussed in the next section.

In the ABO system, there are four blood types (A, B, AB, and O) designated by the antigens present on red blood cells.

Rh System Also

Another important antigen in matching blood types is the Rh (for rhesus monkey) factor. About 85% of the U.S. population has the Rh factor on the red blood cells and are Rh positive (Rh^+). About 15% do not have the antigen and are Rh negative (Rh^-). Neither Rh^+ nor Rh^- individuals normally make antibodies to the Rh factor, but Rh^- individuals will make them when exposed to the Rh^+ red blood cells.

To test if an individual is Rh^- or Rh^+, blood is mixed with anti-Rh antibodies. When Rh-positive blood is mixed with anti-Rh antibodies, agglutination occurs (Fig. 41.15). The designation of blood type usually also includes whether the person has or does not have the Rh factor on the red blood cells.

The Rh factor is particularly important during pregnancy. Rh positive is dominant over Rh negative. If the mother is Rh negative and the father is heterozygous, the child has a 50% chance of being Rh positive (Fig. 41.16). The red blood cells of an Rh^+ child will leak across the placental barrier into the mother's circulatory system because placental tissues normally break down before and at birth. The presence of these Rh antigens causes the mother to produce anti-Rh antibodies. In this or a subsequent pregnancy with an Rh-positive baby, anti-Rh antibodies produced by the mother may cross the placenta and destroy this child's red blood cells. This is called *hemolytic disease of the newborn* (HDN), because hemolysis continues after the baby is born. Due to red blood cell destruction followed by heme breakdown, bilirubin rises in the blood. Excess bilirubin can lead to brain damage, mental retardation, or even death.

The Rh problem has been solved by giving Rh-negative women an Rh immunoglobulin injection either midway through the first pregnancy or no later than 72 hours after giving birth to an Rh-positive child. This injection contains anti-Rh antibodies that attack any of the baby's red blood cells in the mother's blood before these cells stimulate her immune system to produce her own antibodies. This injection is not beneficial if the woman has already begun to produce antibodies; therefore, the timing of the injection is most important.

The possibility of hemolytic disease of the newborn exists when the mother is Rh negative and the father is Rh positive.

connecting concepts

Small aquatic animals with no circulatory system may rely on the passage of external water through a gastrovascular cavity to service cells. Roundworms and other pseudocoelomates use a fluid-filled body cavity as a means of transporting substances from and to the digestive tract. A fluid-filled cavity, whether it is a gastrovascular cavity, pseudocoelom, or a coelom as in earthworms, can act as a hydrostatic skeleton (see also page 856). Unlike a hard endo- or exoskeleton, a hydrostatic skeleton doesn't take much energy to carry around. Even animals that have a rigid skeleton may still rely on body fluids for the purpose of locomotion. Bivalves, for example, pump hemolymph into the foot, thereby expanding it for digging into the mud.

Most animals do have a circulatory system that uses either hemolymph or blood as a circulatory fluid for moving gases, nutrients, and wastes from one part of the body to another. The respiratory function of a circulatory fluid is often augmented by the presence of a pigment that combines with and carries oxygen. Body fluids make ideal culture media for the growth of infectious parasites, and these fluids often have some way to ward off an invasion. In mammals, the white blood cells perform this function by phagocytizing microbes and producing antigens.

In birds and mammals, which are homeothermic, blood plays a role in temperature regulation. In animals that live in cold climates, blood vessels are arranged so that heat can pass from the arteries to the veins within exposed limbs. This conserves heat and also keeps the limbs from freezing. In vertebrates that dive under water, most of the blood pumped by the heart flows through the capillaries of the brain and heart. Closure of other possible pathways keeps these organs active despite the cessation of breathing. These extreme examples shows how the circulatory system can be modified as an adaptation to the environment.

Summary

41.1 Transport in Invertebrates

Some invertebrates do not have a transport system. The presence of a gastrovascular cavity allows diffusion alone to supply the needs of cells in cnidaria and flatworms.

Other invertebrates do have a transport system. Insects have an open circulatory system, and earthworms have a closed one.

41.2 Transport in Vertebrates

Vertebrates have a closed system in which arteries carry blood away from the heart to capillaries, where exchange takes place, and veins carry blood to the heart.

Fishes have a single-loop circulatory pathway because the heart, with the single atrium and ventricle, pumps blood only to the gills. The other vertebrates have both pulmonary and systemic circulation. Amphibians have two atria but a single ventricle. Birds and mammals, including humans, have a heart with two atria and two ventricles, in which oxygenated blood is always separate from deoxygenated blood.

41.3 Transport in Humans

The heartbeat in humans begins when the SA node (pacemaker) causes the two atria to contract and blood moves through the atrioventricular valves to the two ventricles. The SA node also stimulates the AV (atrioventricular) node, which in turn causes the two ventricles to contract. Ventricular contraction sends blood through the semilunar valves to the pulmonary trunk and the aorta. Now all chambers rest. The heart sounds, lub-dub, are caused by the closing of the valves.

Blood pressure created by the beat of the heart accounts for the flow of blood in the arteries, but skeletal muscle contraction is largely responsible for the flow of blood in the veins, which have valves preventing a backward flow.

41.4 Cardiovascular Disorders in Humans

Hypertension and atherosclerosis are two circulatory disorders that lead to heart attack and to stroke. A heart-healthy diet, getting regular exercise, maintaining a proper weight, and not smoking cigarettes are protective against the development of these conditions.

41.5 Blood Is Transport Medium

Blood has two main parts: plasma and formed elements. Plasma contains mostly water (90–92%) and proteins (7–8%) but also nutrients and wastes.

The red blood cells contain hemoglobin and function in oxygen transport. Defense against disease depends on the various types of white blood cells. Neutrophils and monocytes are phagocytic and are especially responsible for the inflammatory reaction. Lymphocytes are involved in the development of immunity to disease.

The platelets and two plasma proteins, prothrombin and fibrinogen, function in blood clotting, an enzymatic process that results in fibrin threads. Blood clotting is a complex process that includes three major events: platelets and injured tissue release prothrombin activator, which enzymatically changes prothrombin to thrombin; thrombin is an enzyme that causes fibrinogen to be converted to fibrin threads.

When blood reaches a capillary, water moves out at the arterial end, due to blood pressure. At the venule end, water moves in, due to osmotic pressure. In between, nutrients diffuse out and wastes diffuse in.

In the ABO blood system there are four types of blood: A, B, AB, and O, depending on the type of antigen present on red blood cells. If the red blood cells have a particular antigen, they clump when exposed to the corresponding antibody. Table 41.1 tells which types of antibodies are present in the plasma for the various ABO blood groups.

Reviewing the Chapter

1. Describe transport in those invertebrates that have no circulatory system; in those that have an open circulatory system; and in those that have a closed circulatory system. 734–35
2. Compare the circulatory systems of a fish, an amphibian, and a mammal. 737
3. Trace the path of blood in humans from the right ventricle to the left atrium; from the left ventricle to the kidneys and to the right atrium; from the left ventricle to the small intestine and to the right atrium. 739
4. Describe the mechanism of a heartbeat, mentioning all the factors that account for this repetitive process. Describe how the heartbeat affects blood flow. What other factors are involved in blood flow? 740

5. Define these terms: pulmonary circuit, systemic circuit, and portal system. 742
6. Discuss the life cycle and function of red blood cells. 747
7. How are white blood cells classified? What are the functions of neutrophils, monocytes, and lymphocytes? 747
8. Name the steps that take place when blood clots. Which substances are present in blood at all times, and which appear during the clotting process? 748
9. What forces facilitate exchange of molecules across the capillary wall? 749
10. What are the four ABO blood types? For each, state the antigen(s) on red blood cells and the antibody(ies) in the plasma. 750
11. Problems can arise during childbearing if the mother is which Rh type and the father is which Rh type? Explain why. 751

Testing Yourself

Choose the best answer for each question.

1. Which one of these would you expect to be part of a closed, but not an open, circulatory system?
   a. ostia
   b. capillary beds
   c. hemocoel
   d. heart
2. In a one-circuit circulatory system, blood pressure
   a. is constant throughout the system.
   b. drops significantly after gas exchange has taken place.
   c. is higher at the intestinal capillaries than at the gill capillaries.
   d. cannot be determined.
3. In which of the following animals is the blood entering the aorta incompletely oxygenated?
   a. frog
   b. chicken
   c. monkey
   d. fish
4. Which of these factors has little effect on blood flow in arteries?
   a. heartbeat
   b. blood pressure
   c. total cross-sectional area of vessels
   d. skeletal muscle contraction
5. In humans, blood returning to the heart from the lungs returns to the
   a. right ventricle.
   b. right atrium.
   c. left ventricle.
   d. left atrium.
6. Systole refers to the contraction of the
   a. major arteries.
   b. SA node.
   c. atria and ventricles.
   d. All of these are correct.
7. Which of these associations is incorrect?
   a. white blood cells—infection fighting
   b. red blood cells—blood clotting
   c. plasma—water, nutrients, and wastes
   d. platelets—blood clotting
8. Water enters capillaries on the venule side as a result of
   a. active transport from tissue fluid.
   b. an osmotic pressure gradient.
   c. increased blood pressure on this side.
   d. higher red blood cell concentration on this side.
9. The last step in blood clotting
   a. is the only step that requires calcium ions.
   b. occurs outside the bloodstream.
   c. converts prothrombin to thrombin.
   d. converts fibrinogen to fibrin.
10. During blood typing, agglutination indicates that the
    a. plasma contains certain antibodies.
    b. red blood cells carry certain antigens.
    c. plasma contains certain antigens.
    d. red blood cells carry certain antibodies.
11. A baby born to which of these couples is most likely to suffer from hemolytic disease of the newborn?
    a. Rh^+ mother and Rh^- father
    b. Rh^- mother and Rh^- father
    c. Rh^+ mother and Rh^+ father
    d. Rh^- mother and Rh^+ father
12. Label this diagram of the heart:

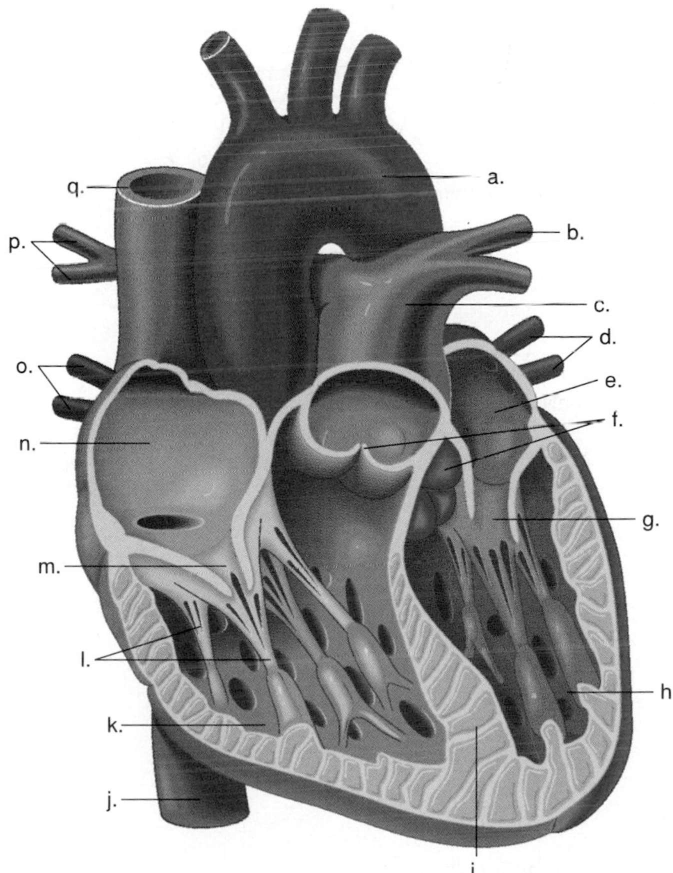

Applying the Concepts

1. *The internal environment of most animals remains relatively constant.*
   How does the composition of tissue fluid stay relatively constant in vertebrates?
2. *A circulatory system is necessary to the life of a complex animal.*
   Why is a circulatory system imperative for complex animals?
3. *Animals are physical and chemical machines; for example, energy must be exerted in order to keep the blood circulating.*
   What type of energy keeps the blood circulating? How do animals acquire the necessary energy?

Using Technology

Your study of the circulation is supported by these available technologies:

Exploring the Internet

The Mader Home Page provides resources for and help with studying this chapter.

http://www.mhhe.com/sciencemath/biology/mader/
(Click on Biology.)

The Dynamic Human CD-ROM

Cardiovascular System

Explorations in Human Biology CD-ROM

Evolution of the Heart (#5)

Life Science Animations Video

Video #4: Animal Biology II
The Cardiac Cycle & Production of Sounds (#32)
Blood Circulation (#37)
Production of Electrocardiogram (#38)
Common Congenital Defects of the Heart (#39)
A, B, O Blood Types (#40)

Understanding the Terms

agglutination 750
antibody 747
antigen 747
aorta 742
arteriole 736
artery 736
atrioventricular valve 738
atrium (pl., atria) 738
blood pressure 743
capillary 736
circulatory (or cardiovascular) system 734
diastole 740
erythrocyte 747
heart 738
hemoglobin 747
hemolymph 735
leukocyte 747
lymph 749
lymphocyte 747
macrophage 747
myocardium 738
neutrophil 747
plasma 747
platelet 748
portal system 742
pulmonary circuit 742
systemic circuit 742
systole 740
thrombin 748
thrombocyte 747
tissue fluid 749
valve 738
vein 736
vena cava 742
ventricle 738
venule 736

Match the terms to these definitions:

a. __________ Blood vessel that transports blood away from the heart.
b. __________ Clumping of red blood cells due to a reaction between antigens on red blood cell plasma membranes and antibodies in the plasma.
c. __________ Force of blood pushing against the inside wall of a blood vessel.
d. __________ Red blood cell that contains hemoglobin and carries oxygen from the lungs or gills to the tissues in vertebrates.
e. __________ Heart valve located between an atrium and a ventricle.
f. __________ Large phagocytic cell derived from a monocyte that ingests microbes and debris.
g. __________ In vertebrates, the liquid portion of blood; contains nutrients, wastes, salts, and proteins.
h. __________ In humans, the major systemic artery that takes blood from the heart to the tissues.
i. __________ Iron-containing respiratory pigment occurring in vertebrate red blood cells and in blood plasma of many invertebrates.
j. __________ Vessel that takes blood from an artery to capillaries.

Lymph Transport and Immunity

CHAPTER

Chapter Concepts

42.1 Lymphatic System Helps Cardiovascular System

- The lymphatic vessels form a one-way system, which transports lymph from the tissues and fat from the lacteals to certain cardiovascular veins. 756

42.2 Some Defenses Are Nonspecific

- Immunity consists of nonspecific and specific defenses to protect the body against disease. 758
- Nonspecific defenses consist of barriers to entry, the inflammatory reaction, and protective proteins. 758

42.3 Other Defenses Are Specific

- Specific defenses require two types of lymphocytes: B lymphocytes and T lymphocytes. 760

42.4 Immunity Can Be Induced

- Induced immunity for medical purposes involves the use of vaccines to achieve long-lasting immunity and the use of antibodies to provide temporary immunity. 767

42.5 Immunity Has Side Effects

- While immunity preserves our existence, it also is responsible for certain undesirable effects, such as tissue rejection, allergies, and autoimmune diseases. 769

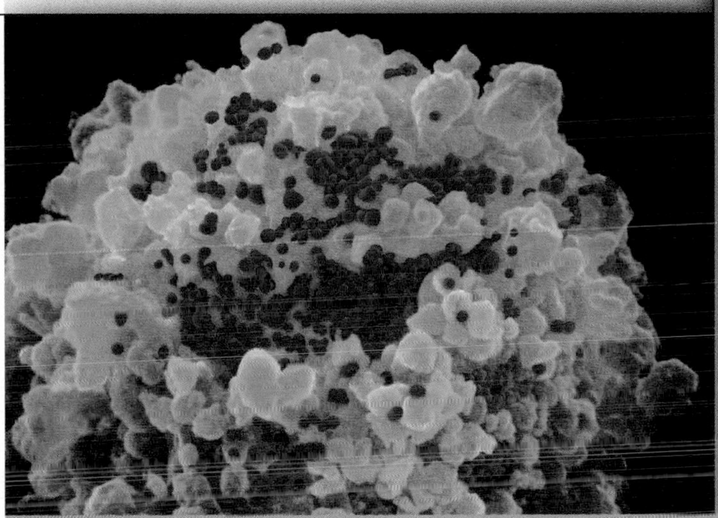

HIV virus (red) buds from a T-lymphocyte blood cell

Since HIV came on the scene in the 1970s, the general public has learned more about the immune system than past generations ever dreamed of knowing. HIV—human immunodeficiency virus—is a group of viruses transmitted by the exchange of internal body fluids. HIV does not kill directly; it kills because it destroys the natural immune system of its host, thereby enabling pathogenic microbes to invade the body. Individuals with an intact immune system can usually defend themselves against fungal infections, bacteria, and even abnormal body cells. HIV sufferers cannot, so they succumb to these opportunistic intruders.

An efficient immune system is a necessity for all species. Homeostasis would be impossible without a way to resist a possible takeover by parasites and/or toxins. Immunity includes all the possible means to defend ourselves, including assistance from the lymphatic system. The mammalian lymphatic system consists of one-way lymph vessels, lymph nodes, lymphatic fluid, and several organs. In addition to its role in defense, the lymphatic system has several other vital functions. It ensures that fluids, which leak into the tissues, are returned back to the bloodstream. It also absorbs fats in the small intestine and transports them to the cardiovascular system.

42.1 Lymphatic System Helps Cardiovascular System

The mammalian **lymphatic system** [L. *lympha,* clear water] consists of lymphatic vessels and the lymphoid organs. This system, which is closely associated with the cardiovascular system, has three main functions: (1) lymphatic vessels take up excess tissue fluid and return it to the bloodstream; (2) lymphatic capillaries absorb fats at the intestinal villi and transport them to the bloodstream; and (3) the lymphatic system helps to defend the body against disease.

Lymphatic Vessels Transport One Way

Lymphatic vessels are quite extensive; most regions of the body are richly supplied with lymphatic capillaries (Fig. 42.1). The construction of the larger lymphatic vessels is similar to that of cardiovascular veins, including the presence of valves. Also, the movement of lymph within these vessels is dependent upon skeletal muscle contraction. When the muscles contract, the lymph is squeezed past a valve that closes, preventing the lymph from flowing backwards.

The lymphatic system is a one-way system that begins with lymphatic capillaries. These capillaries take up fluid that has diffused from and has not been reabsorbed by the blood capillaries. **Edema** [Gk. *oidema,* swelling with fluid] is localized swelling caused by the accumulation of tissue fluid. This can happen if too much tissue fluid is made and/or not enough of it is drained away. Once tissue fluid enters the lymphatic vessels, it is called **lymph** [L. *lympha,* clear water]. The lymphatic capillaries join to form lymphatic vessels that merge before entering one of two ducts: the thoracic duct or the right lymphatic duct. The *thoracic duct* is much larger than the right lymphatic duct. It serves the lower extremities, the abdomen, the left arm, and the left side of both the head and the neck. The *right lymphatic duct* serves the right arm, the right side of both the head and the neck, and the right thoracic area. The lymphatic ducts enter the subclavian veins, which are cardiovascular veins in the thoracic region.

Lymph flows one way from a capillary to ever-larger lymphatic vessels and finally to a lymphatic duct, which enters a subclavian vein.

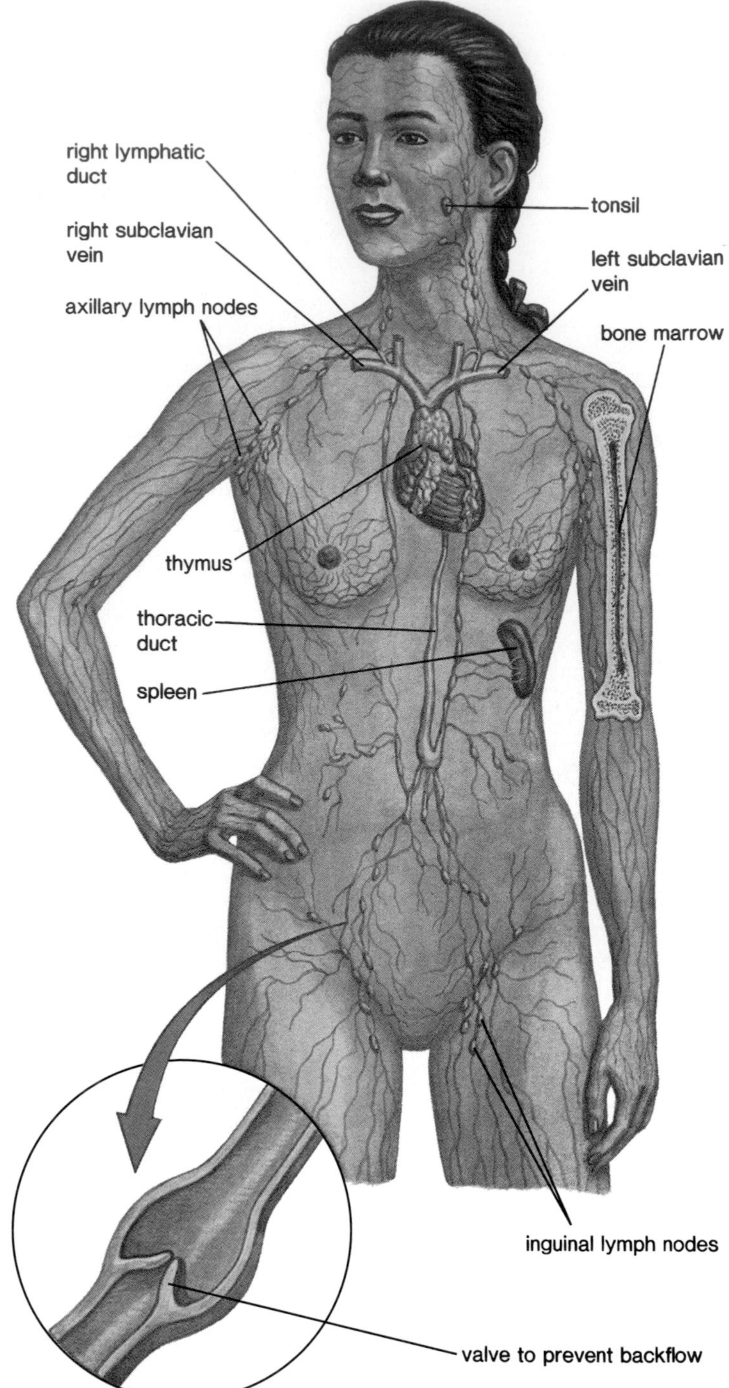

Figure 42.1 Lymphatic system.
The lymphatic vessels drain excess fluid from the tissues and return it to the cardiovascular system. The enlargement shows that lymphatic vessels have valves to prevent backward flow.

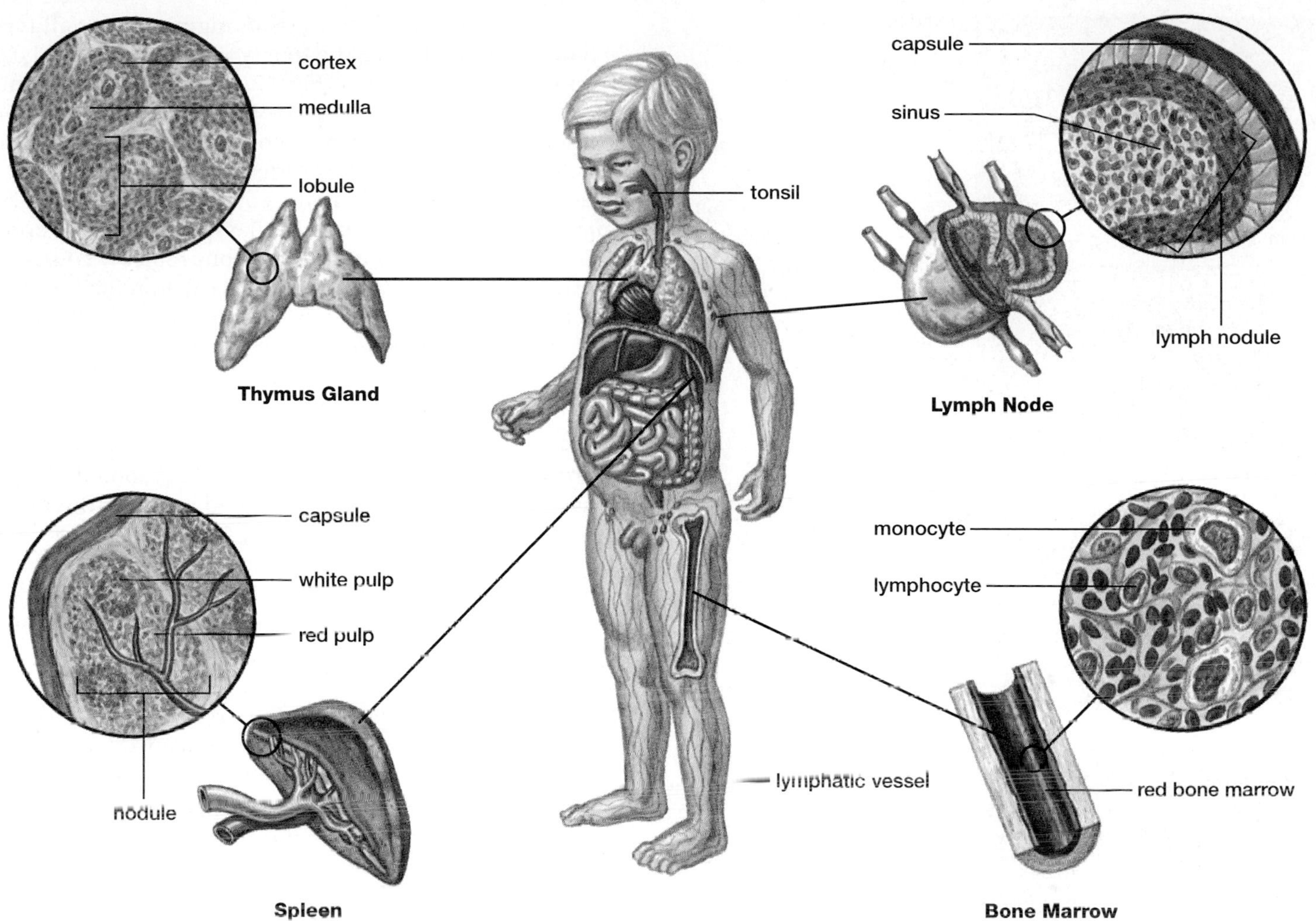

Figure 42.2 The lymphoid organs.
The lymphoid organs include the lymph nodes, the spleen, the thymus gland, and the red bone marrow, which all contain lymphocytes.

Lymphoid Organs Assist Immunity

The lymphoid organs of special interest are the lymph nodes, the spleen, the thymus gland, and the bone marrow (Fig. 42.2).

Lymph nodes, which are small (about 1–25 mm) ovoid or round structures, are found at certain points along lymphatic vessels. A lymph node has a fibrous connective tissue capsule penetrated by incoming and outgoing lymphatic vessels. Connective tissue also divides a node into nodules. Each nodule contains a sinus (open space) filled with many lymphocytes and macrophages. As lymph passes through the sinuses, the macrophages purify it of infectious organisms and any other debris.

Nodules can occur singly or in groups. The *tonsils,* located in back of the mouth on either side of the tongue, and the *adenoids,* located on the posterior wall above the border of the soft palate, are composed of partly encapsulated lymph nodules. Also, nodules called *Peyer's patches* are found within the intestinal wall.

The lymph nodes are found in groups in certain regions of the body. For example, the inguinal nodes are in the groin and the axillary nodes are in the armpits.

The *spleen* is located in the upper left region of the abdominal cavity just beneath the diaphragm. The construction of the spleen is similar to that of a lymph node. Outer connective tissue divides the organ into lobules, which contain sinuses. In the spleen, however, the sinuses are filled with blood instead of lymph. The blood vessels of the spleen can expand, and this enhances the carrying capacity of this organ to serve as a blood reservoir and to make blood available in times of low pressure or when the body needs extra oxygen carrying capacity.

A spleen nodule contains red pulp and white pulp. The white pulp contains mostly lymphocytes. Red pulp contains red blood cells, lymphocytes, and macrophages. The red pulp helps to purify blood that passes through the spleen by removing bacteria and worn-out or damaged red blood cells. If the spleen ruptures due to injury, it can be removed. Although its functions are replaced by other organs, the individual is often slightly more susceptible to infections and may have to receive antibiotic therapy indefinitely.

The *thymus gland* is located along the trachea behind the sternum in the upper thoracic cavity. This gland varies in

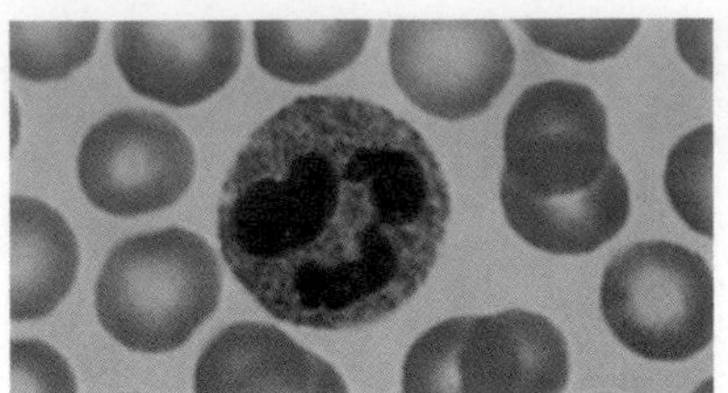

Neutrophil
40–70%
Phagocytizes primarily bacteria

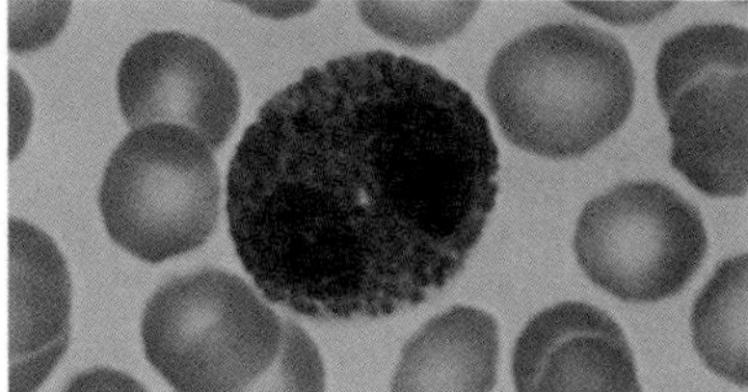

Eosinophil
1–4%
Phagocytizes and destroys antigen-antibody complexes

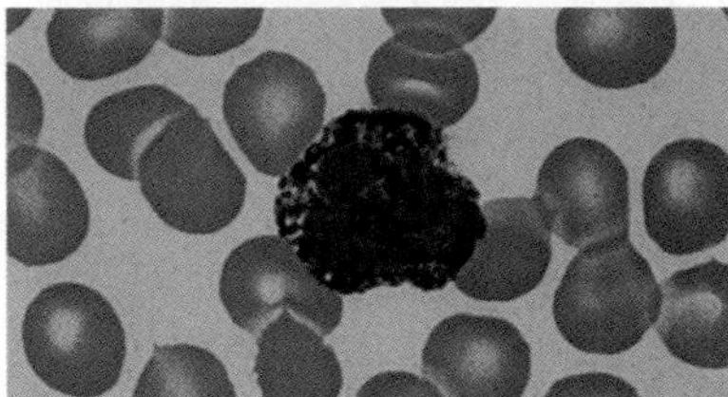

Basophil
0–1%
Releases histamine when stimulated

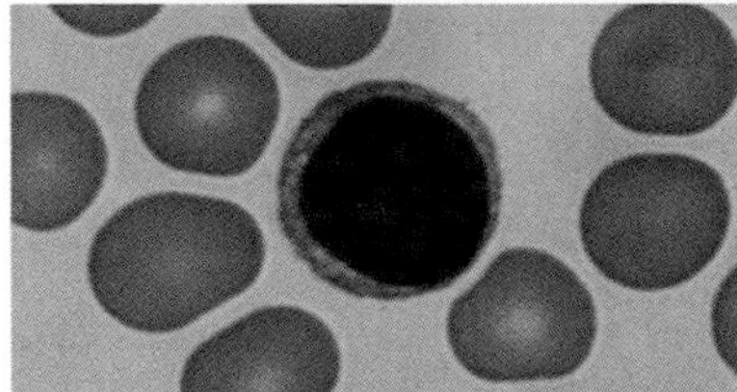

Lymphocyte
20–45%
B type produces antibodies in blood and lymph;
T type kills virus-containing cells

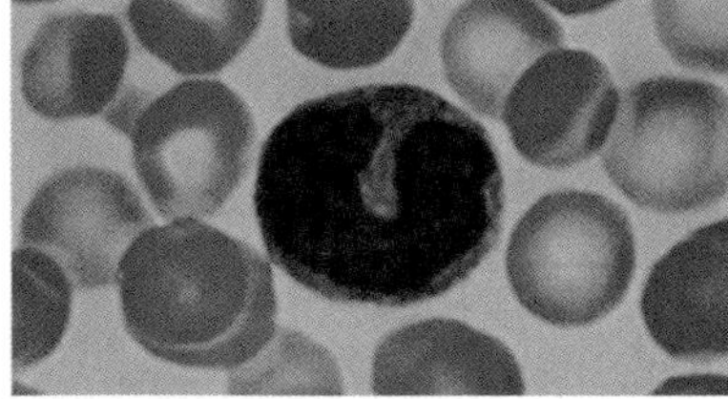

Monocyte
4–8%
Becomes macrophage—phagocytizes bacteria and viruses

Figure 42.3 Leukocytes.
There are five types of white blood cells, which differ according to structure and function. The frequency of each type of cell is given as a percentage of the total white blood cell count.

size, but it is larger in children than in adults and may disappear completely in old age. The thymus is divided into lobules by connective tissue. The T lymphocytes mature in these lobules. The interior (medulla) of the lobule, which consists mostly of epithelial cells, stains lighter. It produces thymic hormones, such as thymosin, that are thought to aid in maturation of T lymphocytes. Thymosin may also have other functions in immunity.

Red bone marrow is the site of origination for all types of blood cells, including the five types of white blood cells that function in immunity (Fig. 42.3). The marrow contains stem cells that are ever capable of dividing and producing cells that go on to differentiate into the various types of blood cells. In a child, most bones have red bone marrow, but in an adult it is present only in the bones of the skull, the sternum (breastbone), the ribs, the clavicle, the pelvic bones, and the vertebral column. The red bone marrow consists of a network of connective tissue fibers, called reticular fibers, which are produced by cells called reticular cells. These and the stem cells and their progeny are packed about thin-walled sinuses filled with venous blood. Differentiated blood cells enter the bloodstream at these sinuses.

The lymphoid organs have specific functions that assist immunity. Lymph is cleansed in lymph nodes; blood is cleansed in the spleen; T lymphocytes mature in the thymus; and white blood cells are made in the bone marrow.

42.2 Some Defenses Are Nonspecific

Immunity [L. *immunis*, exempt, free] is the ability of the body to defend itself against infectious agents, foreign cells, and even abnormal body cells, such as cancer cells. Natural immunity ordinarily keeps us free of diseases, including those caused by *microbes* [Gk. *mikros*, small, little] and also cancer.

Immunity includes nonspecific and specific defenses. The three types of nonspecific defenses—barriers to entry, the inflammatory reaction, and protective proteins—are effective against many types of infectious agents.

Barring Entry

Skin and the mucous membranes lining the respiratory, digestive, and urinary tracts serve as mechanical barriers to entry by microbes. Oil gland secretions contain chemicals that weaken or kill bacteria on skin. The upper respiratory tract is lined by ciliated cells that sweep mucus and trapped particles up into the throat, where they can be swallowed, or expectorated (coughed out). The stomach has an acidic pH, which inhibits the growth of many types of bacteria. The various bacteria that normally reside in the intestine and other areas, such as the vagina, prevent pathogens from taking up residence.

Inflammatory Reaction

Whenever the skin is broken due to a minor injury, a series of events occurs that is known as the **inflammatory reaction** [L. *in*, into, and *flamma*, fire]. The inflamed area has four symptoms: redness, pain, swelling, and heat. Figure 42.4 illustrates the participants in the inflammatory reaction. The mast cells, one type of participant, are derived from basophils (Fig. 42.3), which take up residence in the tissues.

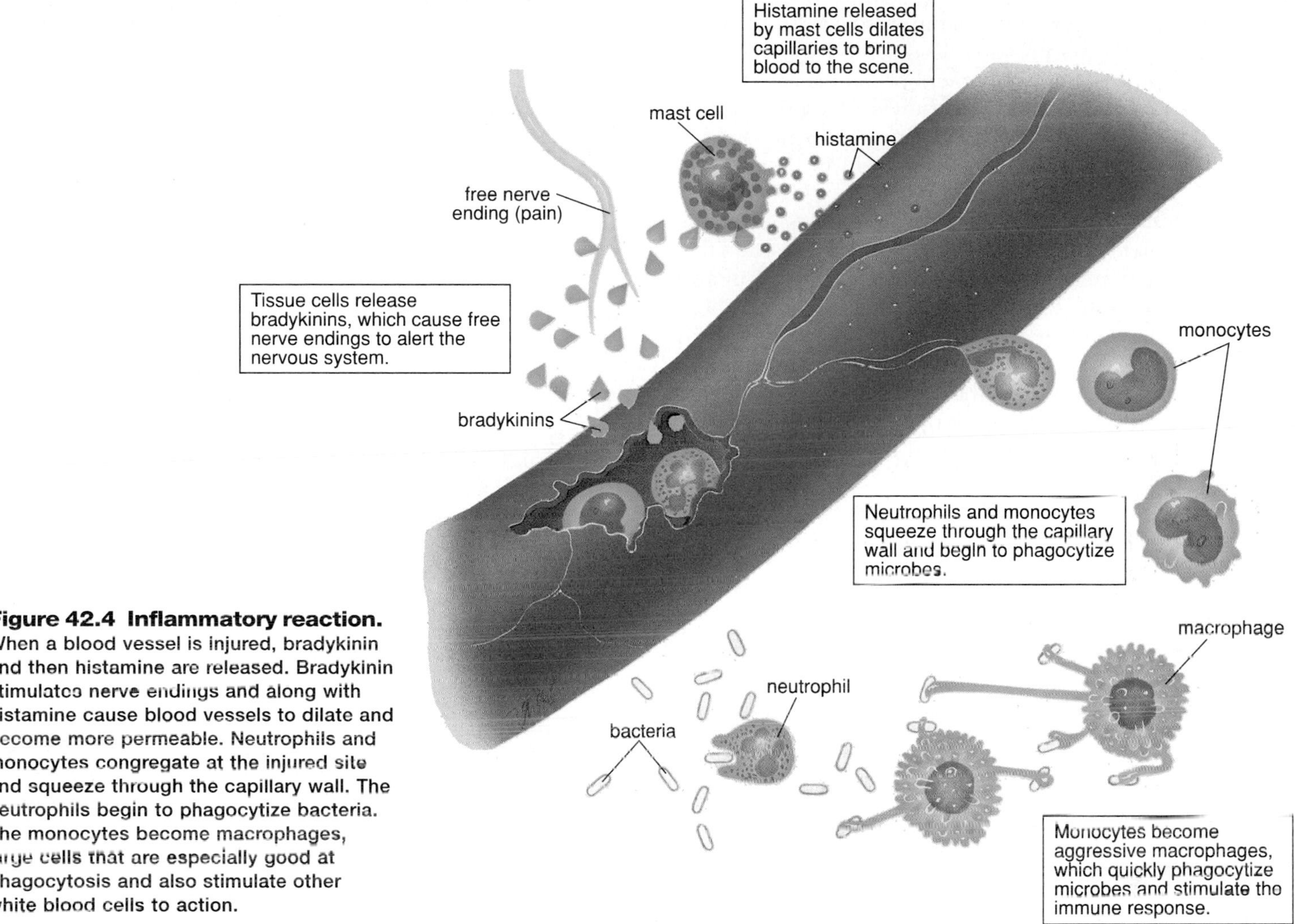
Histamine released by mast cells dilates capillaries to bring blood to the scene.
mast cell
histamine
free nerve ending (pain)
Tissue cells release bradykinins, which cause free nerve endings to alert the nervous system.
monocytes
bradykinins
Neutrophils and monocytes squeeze through the capillary wall and begin to phagocytize microbes.
macrophage
neutrophil
bacteria
Monocytes become aggressive macrophages, which quickly phagocytize microbes and stimulate the immune response.

Figure 42.4 Inflammatory reaction. When a blood vessel is injured, bradykinin and then histamine are released. Bradykinin stimulates nerve endings and along with histamine cause blood vessels to dilate and become more permeable. Neutrophils and monocytes congregate at the injured site and squeeze through the capillary wall. The neutrophils begin to phagocytize bacteria. The monocytes become macrophages, large cells that are especially good at phagocytosis and also stimulate other white blood cells to action.

When an injury occurs, a capillary and several tissue cells are apt to rupture and to release *bradykinin* [Gk. *bradys,* slow, and *kineo,* move]. This molecule *stimulates receptors that initiate* nerve impulses, resulting in the sensation of pain, and stimulates mast cells to release **histamine** [Gk. *histos,* tissue] which, together with bradykinin, causes the capillaries to dilate and become more permeable. The enlarged capillaries cause the skin to redden, and the increased permeability allows proteins and fluids to escape so swelling results. A rise in temperature reduces the number of invading microbes and increases phagocytosis by white blood cells.

Any break in the skin allows microbes to enter the body and triggers a migration of neutrophils and monocytes to the site of injury. Neutrophils and monocytes are amoeboid; they can change shape and squeeze through capillary walls to enter tissue fluid. Neutrophils phagocytize bacteria and when phagocytosis occurs, an endocytic vesicle forms. The engulfed bacteria are destroyed by hydrolytic enzymes when the endocytic vesicle combines with a lysosome, one of the cellular organelles.

Monocytes differentiate into **macrophages** [Gk. *makros,* long, and *phagein,* to eat], large phagocytic cells that are able to devour a hundred bacteria or viruses and still survive. Some tissues, particularly connective tissue, have resident macrophages, which routinely act as scavengers, devouring old blood cells, bits of dead tissue, and other debris. Macrophages can also bring about an explosive increase in the number of leukocytes by liberating a growth factor; this passes by way of blood to the red bone marrow, where it stimulates the production and the release of white blood cells, *primarily* neutrophils.

As the infection is being overcome, some neutrophils die. These—along with dead tissue, cells, bacteria, and living white blood cells—form *pus* [L. *pus,* corrupt matter from a sore], a whitish material. Pus indicates that the body is trying to overcome the infection.

The inflammatory reaction is a "call to arms"—it marshals phagocytic white blood cells to the site of bacterial invasion.

Protective Proteins

The **complement system,** often simply called complement, is a number of plasma proteins designated by the letter *C* and a subscript. Once a complement protein is activated, it activates another protein, and the result is a set series of reactions. A limited amount of activated protein is needed because a domino effect occurs: each protein in the series is capable of activating many other proteins.

Complement is activated when microbes enter the body. It "complements" certain immune responses, and this accounts for its name. For example, it is involved in and amplifies the inflammatory response because complement proteins attract phagocytes to the scene. Some complement proteins bind to the surface of microbes already coated with antibodies, which ensures that the microbes will be phagocytized by a neutrophil or macrophage. Facilitated phagocytosis by complement is referred to as *opsonization* [Gk. *opsonin*, to prepare food].

Another series of reactions is complete when complement proteins result in a *membrane attack complex* that produces holes in the bacterial cell walls and plasma membranes of bacteria. When potassium ions leave, fluids and salt enter the bacterial cell to the point that it bursts (Fig. 42.5).

Interferon is a protein produced by virus-infected cells. Interferon binds to receptors of noninfected cells, causing the cells to prepare for possible attack by producing substances that interfere with viral replication. Interferon is specific to the species; therefore, only human interferon can be used in humans. Although it once was a problem to collect enough interferon for clinical and research purposes, interferon is now made by recombinant DNA (deoxyribonucleic acid) technology.

Immunity includes these nonspecific defenses: barriers to entry, the inflammatory reaction, and protective proteins.

42.3 Other Defenses Are Specific

Sometimes, it is necessary to rely on a specific defense rather than a nonspecific defense against a particular antigen. An **antigen** is any foreign substance (often a protein or polysaccharide) that stimulates the immune system to react to it. Microbes have antigens, but antigens can also be part of a foreign cell or a cancer cell. Because we do not ordinarily become immune to our own cells, it is said that the immune system is able to tell self from nonself.

Immunity usually lasts for some time. For example, once we recover from the measles, we usually do not get the illness a second time. Immunity is primarily the result of the action of the **B lymphocytes** [L. *lympha,* clear water, and Gk. *kytos* cell] and the **T lymphocytes.** B lymphocytes[1] mature in the *b*one marrow, and T lymphocytes mature in the *t*hymus gland. B lymphocytes, also called B cells, give rise to plasma cells, which produce **antibodies** [Gk. *anti,* against], proteins that are capable of combining with and neutralizing antigens. These antibodies are secreted into the blood and the lymph. In contrast, T lymphocytes, also called T cells, do not produce antibodies. Instead, certain T cells directly attack cells that bear antigens. Other T cells regulate the immune response.

Lymphocytes are capable of recognizing an antigen because they have receptor molecules on their surface. The shape of the receptors on any particular lymphocyte is complementary to a specific antigen. It is often said that the receptor and the antigen fit together like *a lock and a key.* It is estimated that during our lifetime, we encounter a million different antigens, so we need the same number of different lymphocytes for protection against those antigens. It is remarkable that diversification occurs to such

[1] Historically, the B stands for *bursa of Fabricius,* an organ in the chicken where these cells were first identified. As it turns out, however, the B can conveniently be thought of as referring to bone marrow.

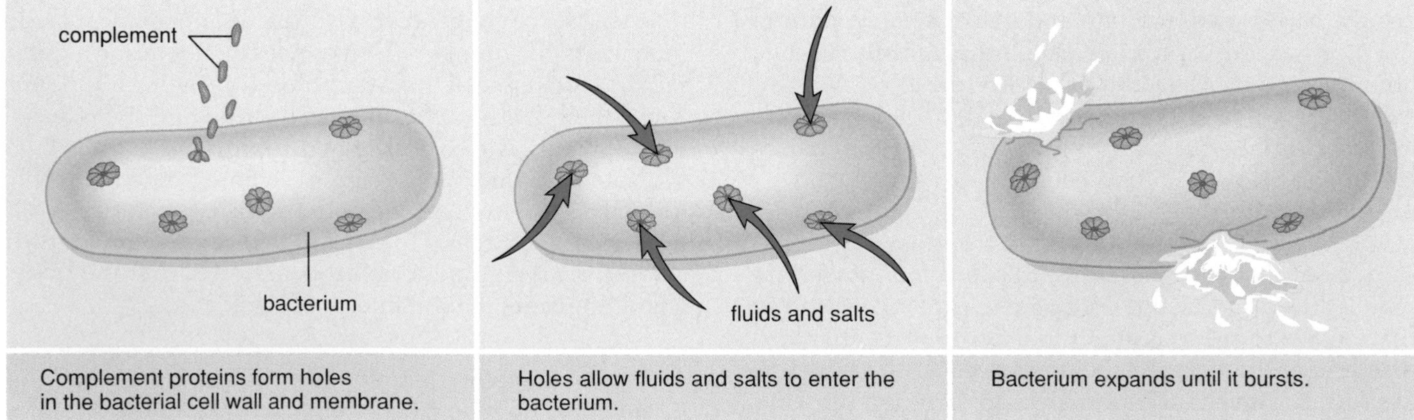

complement
bacterium
fluids and salts
Complement proteins form holes in the bacterial cell wall and membrane.
Holes allow fluids and salts to enter the bacterium.
Bacterium expands until it bursts.

Figure 42.5 Action of the complement system against a bacterium.
When complement proteins in the plasma are activated by an immune reaction, they form holes in bacterial cell walls and plasma membranes, allowing fluids and salts to enter until the cell eventually bursts.

an extent during the maturation process that there is a different lymphocyte type for each possible antigen. Just how this occurs is discussed in the reading on page 763. Despite this great diversity, none of the lymphocytes ordinarily attacks the body's own cells. It is believed that if by chance a lymphocyte arises that is equipped to respond to the body's own proteins, it is normally suppressed and develops no further.

B Cells Make Plasma Cells and Memory Cells

Each type of B cell carries its specific antibody, as a membrane-bound receptor, on its surface. When a B cell in a lymph node of the spleen encounters a bacterial cell or a toxin bearing an appropriate antigen, it becomes activated to divide many times. Most of the resulting cells are plasma cells, which secrete antibodies against this antigen. A **plasma cell** is a mature B cell that mass-produces antibodies in the blood or lymph.

The *clonal selection theory* states that the antigen selects which B cell will produce a clone of plasma cells (Fig. 42.6). Notice that a B cell does not clone until its antigen is present and binds to its receptors. B cells are then stimulated to clone by helper T cells secretions, as is discussed in the next section.

Once antibody production is sufficient to agglutinate and/or neutralize the amount of antigen present in the body, the development of plasma cells ceases. Those members of a clone that do not participate in antibody production remain in the bloodstream as memory B cells. *Memory B cells* are the means by which long-term immunity is possible. If the same antigen enters the system again, memory B cells quickly divide and give rise to new antibody-producing plasma cells.

Defense by B cells is called *antibody-mediated immunity* because the various types of B cells produce antibodies. It is also called *humoral immunity* because these antibodies are present in blood and lymph. A *humor* is any fluid normally occurring in the body.

Characteristics of B cells:

- Antibody-mediated immunity
- Produced and mature in bone marrow
- Undergo clonal selection in spleen and lymph nodes
- Directly recognize antigen
- Clonal expansion produces antibody-secreting plasma cells as well as memory B cells

---

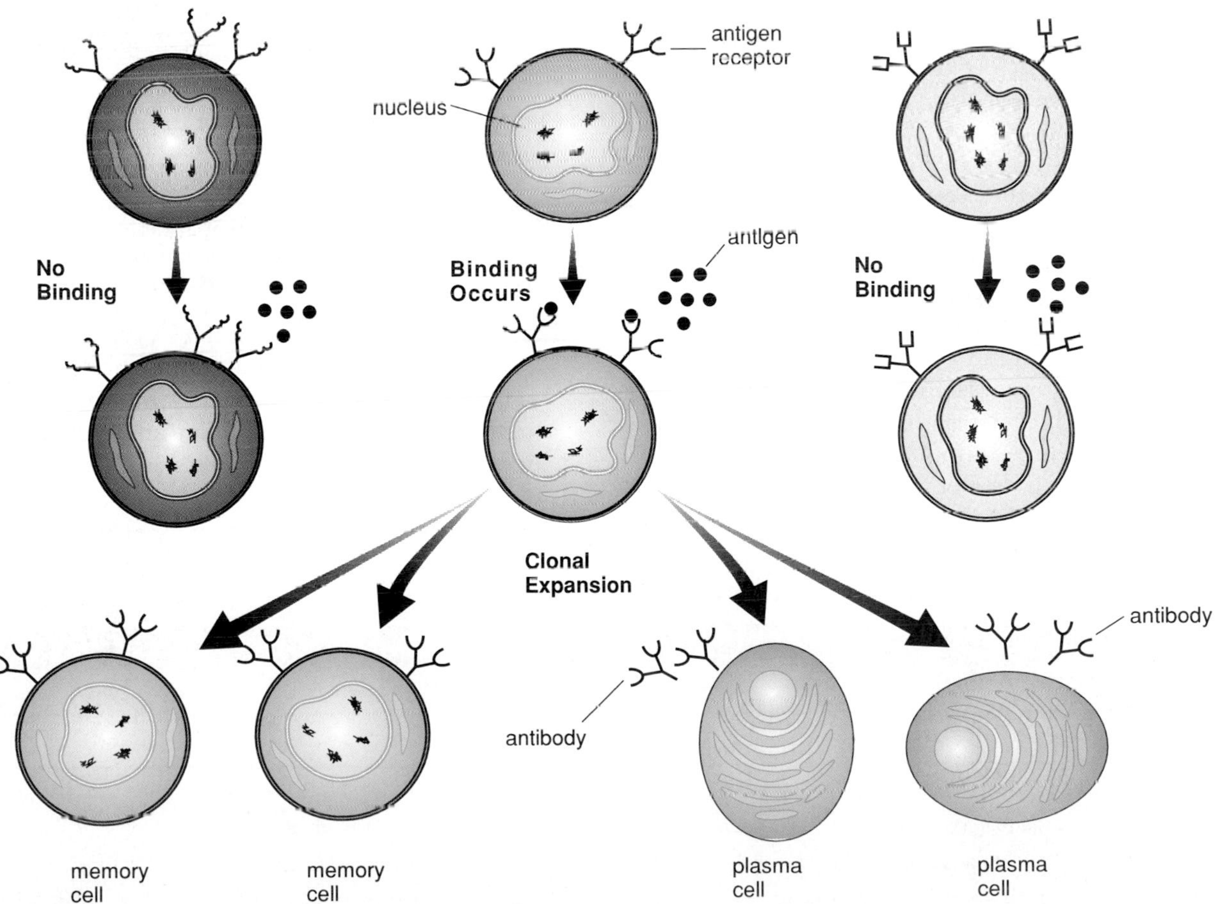

Figure 42.6 Clonal selection theory as it applies to B cells.
In this diagram, color is used to signify that the membrane-bound antibody differs on three B cells. An antigen activates only the B cell with the rounded receptors, then it undergoes clonal expansion. During the process many plasma cells, which produce specific antibodies against this antigen, are produced. Memory cells, which retain the ability to recognize this antigen are produced also.

How Antibodies Work

The most common type of antibody (IgG) is a Y-shaped protein molecule with two arms. Each arm has a "heavy" (long) polypeptide chain and a "light" (short) polypeptide chain. These chains have *constant regions,* where the sequence of amino acids is set, and *variable regions,* where the sequence of amino acids varies (Fig. 42.7). The constant regions are not identical among all the antibodies. Instead, they are the same within different classes of antibodies. The variable regions form an antigen-binding site, and their shape is specific to a particular antigen, as discussed in the reading on the next page. The antigen combines with the antibody at the antigen-binding site in a lock-and-key manner.

The antigen-antibody reaction can take several forms, but quite often the reaction produces complexes of antigens combined with antibodies. Such antigen-antibody complexes, sometimes called the immune complexes, mark the antigens for destruction by other forces. For example, an antigen-antibody complex may be engulfed by neutrophils or macrophages, or it may activate complement. Complement makes microbes more susceptible to phagocytosis, as discussed previously.

How Antibodies Differ

There are five different classes of circulating antibodies (Table 42.1). IgG antibodies are the major type in blood, and lesser amounts are also found in lymph and tissue fluid. IgG antibodies attack microbes and their toxins. A *toxin* is a specific chemical (produced by bacteria, for example) that is poisonous to other living things. IgM antibodies are pentamers, meaning that they contain five of the Y-shaped structures shown in Figure 42.7*a*. These antibodies appear in blood soon after an infection begins and disappear before it is over. They are good activators of the complement system. IgA antibodies are dimers and contain two Y-shaped structures. They are the main type of antibody found in bodily secretions. They attack microbes and their toxins before they reach the bloodstream. The role of IgD antibodies in immunity is uncertain, but a very limited number is present in blood. IgE antibodies, which are responsible for allergic reactions, are discussed on page 769.

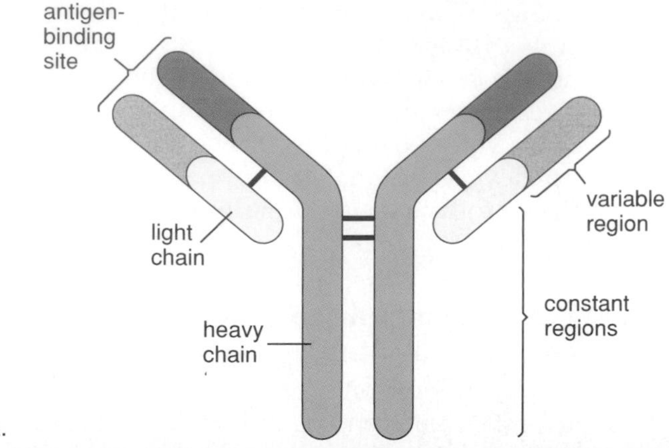

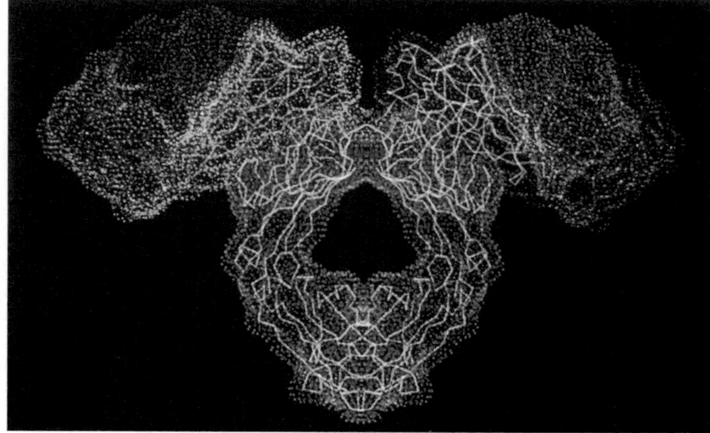

Figure 42.7 Structure of the most common antibody (IgG).
a. An IgG antibody contains two heavy (long) polypeptide chains and two light (short) chains arranged so there are two variable regions, where a particular antigen is capable of binding with the antibody. **b.** Computer model of an antibody molecule. In this model, an antigen combines with the two side branches.

An antigen combines with an antibody at the antigen-binding site in a lock-and-key manner. The reaction can produce antigen-antibody complexes, which contain several molecules of antibody and antigen.

Table 42.1
Antibodies

Classes	Presence	Function
IgG	Main antibody type in circulation	Attacks microbes* and bacterial toxins; enhances phagocytosis
IgM	Antibody type found in circulation; largest antibody	Activates complement; clumps cells
IgA	Main antibody type in secretions such as saliva and milk	Attacks microbes and bacterial toxins
IgD	Antibody type found in circulation in extremely low quantity	Presence signifies maturity of B cell
IgE	Antibody type found as membrane-bound receptor on basophils in blood and on mast cells in tissues	Responsible for allergic reactions

* *Viruses and bacteria*

Doing science

Susumu Tonegawa and Antibody Diversity

In 1987, Susumu Tonegawa became the first Japanese scientist to win the Nobel Prize in Physiology or Medicine. Described by his colleagues as a creative genius who intuitively knows how to design experiments, Dr. Tonegawa had dedicated himself to finding the solution to an engrossing puzzle. Immunologists and geneticists knew that each B cell makes an antibody especially equipped to recognize the specific shape of a particular antigen. But how could the human genome contain enough genetic information to permit the production of up to a billion different antibody types needed to combat all of the microbes we are likely to encounter during the course of our lives?

As you know, an antibody is composed of two light and two heavy polypeptide chains, which are divided into constant and variable regions. The constant region determines the antibody class and the variable region determines the specificity of the antibody, because this is where an antigen binds to a specific antibody (Fig. 42.7). Each B cell must make use of a unique gene to code for the variable region of the light chain and another for the variable region of the heavy chain.

Working with embryonic and mature mice, Tonegawa found that the DNA sequences coding for the variable and constant regions were scattered throughout the genome in embryonic B cells and that only certain DNA segments had come together in mature antibody-secreting B cells. He said that as a B cell matures, it randomly selects certain DNA segments, which collectively code for a specific variable region, and then these are joined to another segment coding for a constant region (Fig. 42A). As an analogy, consider that each person entering a supermarket chooses various items for purchase, and that the possible combination of items in any particular grocery bag is astronomical. Tonegawa also found that mutations occur as the variable segments are undergoing rearrangements. Such mutations are another source of antibody diversity.

Tonegawa received his B.S. in chemistry in 1963 at Kyoto University and earned his Ph.D. in biology from the University of California at San Diego (UCSD) in 1969. After that he worked as a research fellow at UCSD and the Salk Institute. In 1971, he moved to the Basel Institute for Immunology and began the experiments that eventually led to his Nobel Prize-winning discovery. Since the 1970s, Tonegawa has been doing experiments to isolate the receptors of T cells. This is an even more challenging area of research than the diversity of antibodies produced by B cells. Since 1981, he has been a full professor at the Massachusetts Institute of Technology (MIT), where he has a reputation for being an "aggressive, determined researcher" who often works late into the night.

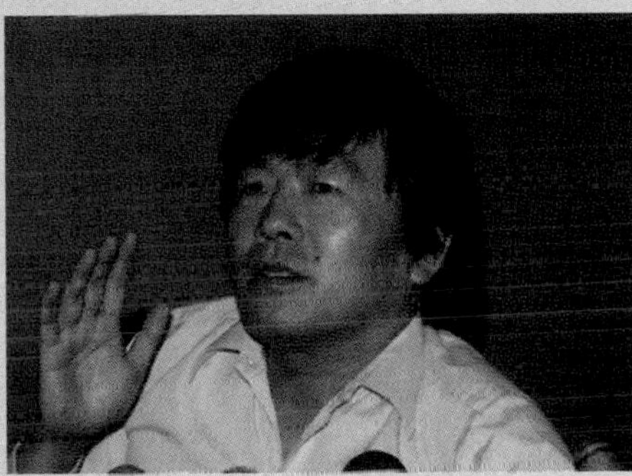

a.

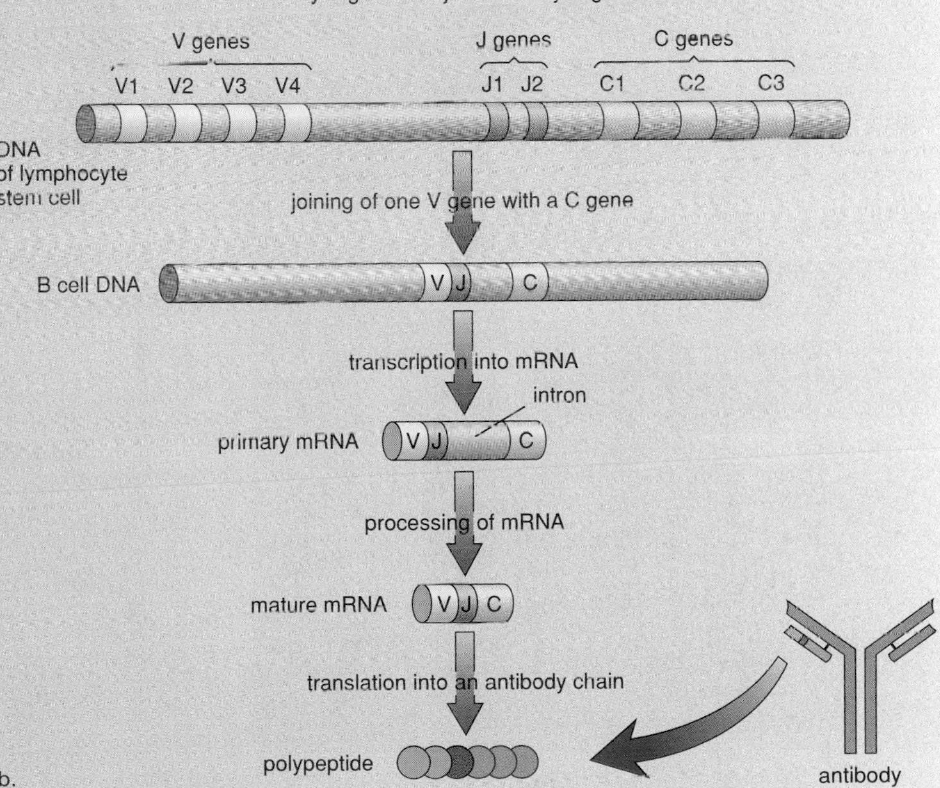

b.

Figure 42A Antibody diversity.
a. Susumu Tonegawa, who received a Nobel Prize for his findings. **b.** Immature B cells have V (for variable) genes, J (for junction) genes, and C (for constant) genes. During maturation, any V gene can join with any C gene to produce a sequence for an antibody chain. Each B cell has its own sequence and therefore produces its own unique antibody.

T Cells Become Cytotoxic, Helper, Memory, or Suppressor Cells

There are different types of T cells including cytotoxic T cells, helper T cells, memory T cells, and suppressor T cells. (Some doubt the existence of suppressor T cells and believe instead that suppression may depend upon the mixture of cytokines released by T cells.) Cytotoxic and helper T cells differentiate in the thymus; memory T cells and suppressor T cells arise at a later time. All types of T cells look alike but can be distinguished by their functions.

Cytotoxic T cells [Gk. *kytos*, cell, and *toxikon*, poison] sometimes are called killer T cells. They attack and destroy antigen-bearing cells, such as virus-infected or cancer cells. Cytotoxic T cells have storage vacuoles containing perforin molecules. Perforin molecules perforate a plasma membrane, forming a hole that allows water and salts to enter. The cell under attack then swells and eventually bursts (Fig. 42.8). It often is said that T cells are responsible for *cell-mediated immunity*, characterized by destruction of antigen-bearing cells. Of all the T cells, only cytotoxic T cells are involved in this type of immunity.

Helper T cells regulate immunity by enhancing the response of other immune cells. When exposed to an antigen, they enlarge and secrete *lymphokines* [L. *lympha*, clear water, and Gk. *kineo*, move], stimulatory molecules that cause helper T cells to clone and other immune cells to perform their functions. For example, lymphokines stimulate macrophages to phagocytize and stimulate B cells to become antibody-producing plasma cells. Because HIV that

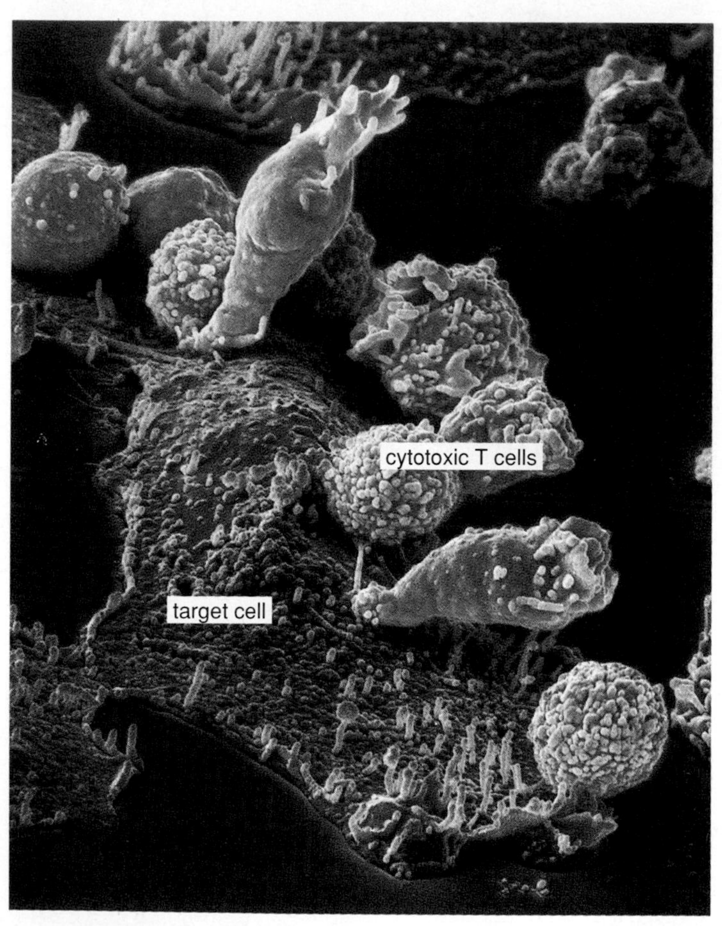

a.

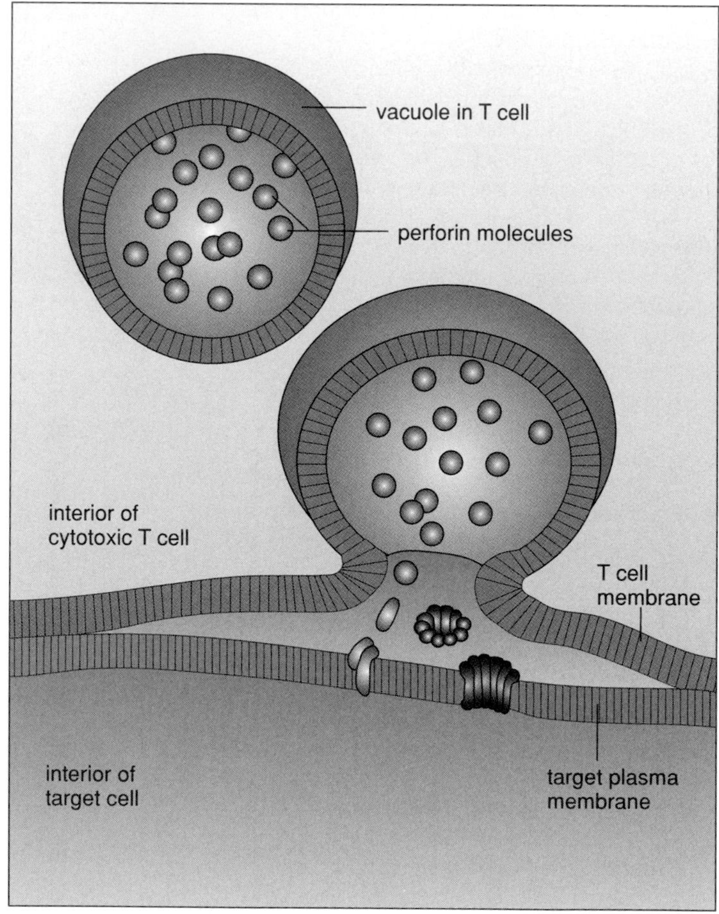

b.

Figure 42.8 Cell-mediated immunity.
a. The scanning electron microscope shows cytotoxic T cells attacking and destroying a cancer cell. **b.** During the killing process, the vacuoles in a cytotoxic T cell fuse with the cancer cell's plasma membrane and release units of the protein perforin. These units combine to form holes in the target plasma membrane. Thereafter, fluid and salts enter so that the target cell eventually bursts.

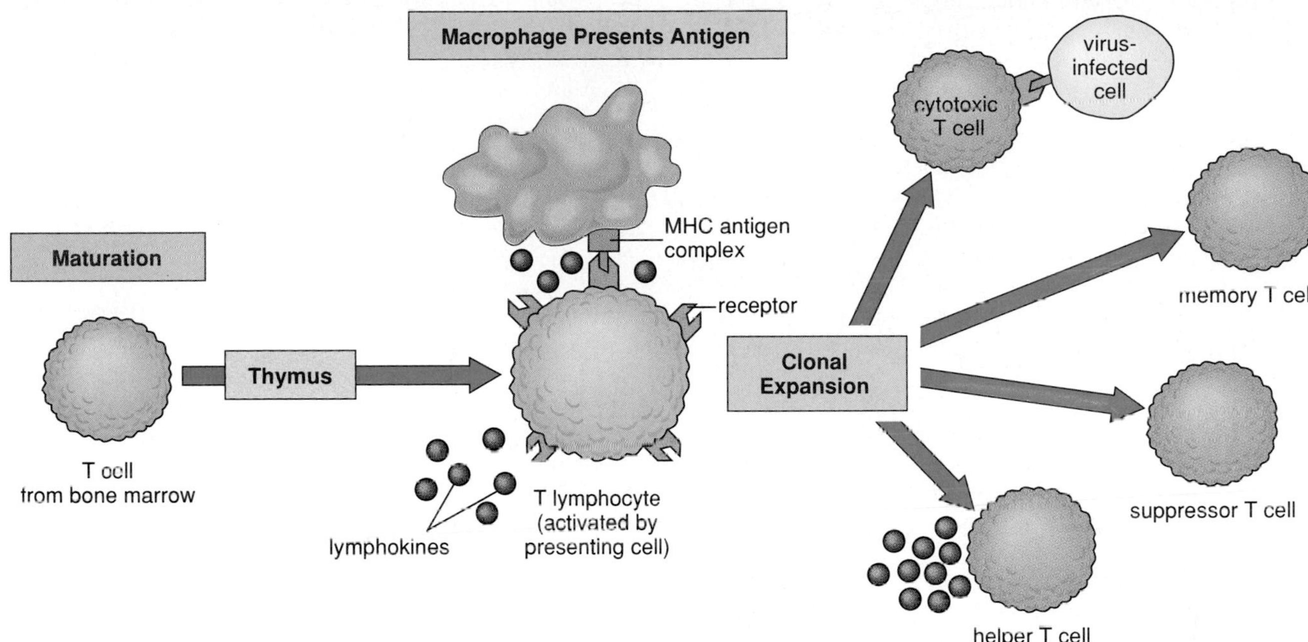

Figure 42.9 Activation and diversity of T cells.
T cells mature in the thymus and are activated when presented with an antigen by a macrophage. Clonal expansion follows with the production of the four types of T cells. Cytotoxic T cells are responsible for cell-mediated immunity.

causes AIDS attacks helper T cells and certain other cells of the immune system, it inactivates the immune response. AIDS is discussed in the reading on page 766.

Suppressor T cells also regulate the immune response by suppressing further development of helper T cells. Since helper T cells also stimulate antibody production by B cells, the suppressor T cells help to prevent both the B cell and T cell immune response from getting out of hand. Following suppression, a population of *memory T cells* persists, perhaps for life. These cells are able to secrete lymphokines and to stimulate macrophages and B cells whenever the same antigen reenters the body. In this way they contribute to active immunity.

Activating Cytotoxic and Helper T Cells

T cells have receptors just as B cells do. Unlike B cells, however, cytotoxic T cells and helper T cells are unable to recognize an antigen that simply is present in lymph or blood. Instead, the antigen must be presented to them by an *antigen-presenting cell (APC)*. When an APC, usually a macrophage, engulfs a microbe, the microbe is broken down to fragments within an endocytic vesicle. These fragments are antigenic; that is, they have the properties of an antigen. The fragments are linked to a **major histocompatibility complex (MHC) protein** in the plasma membrane, and then they can be presented to a T cell. MHC proteins mark the cell as belonging to a particular individual, and therefore, they are *self antigens*. In humans, MHC proteins are called HLA (human leukocyte associated) proteins.

The importance of MHC proteins was first recognized when it was discovered that they contribute to the specificity of tissues and make it difficult to transplant tissue from one animal (including humans) to another. In other words, when the donor and the recipient are histo-(tissue) compatible (the same or nearly so), a transplant is more likely to be successful.

Figure 42.9 shows a macrophage presenting an antigen to a helper T cell. Once a helper T cell recognizes an antigen, it undergoes clonal expansion, producing memory T cells, which can also recognize this same antigen. Once a cytotoxic T cell recognizes an antigen, it attacks and destroys any cell that is infected with the same virus. In this way T cells contribute to active immunity.

Characteristics of T cells:

- **Cell-mediated immunity**
- **Produced in bone marrow, mature in thymus**
- **Antigen must be presented, usually by macrophage**
- **Cytotoxic T cells search and destroy antigen-bearing cells**
- **Helper T cells secrete lymphokines and stimulate other immune cells**

Health Focus

The AIDS Epidemic

AIDS (acquired immunodeficiency syndrome) is caused by a group of related retroviruses known as HIV (human immunodeficiency viruses). Worldwide, as many as 20 million people may now be infected with HIV, and there could be as many as 100 million infected by the end of the century. A new infection is believed to occur every 15 seconds, the majority between heterosexuals.

HIV is transmitted by sexual contact with an infected person, including vaginal or rectal intercourse and oral/genital contact. Also, needle-sharing among intravenous drug users is high-risk behavior. Babies born to HIV-infected women may become infected before or during birth, or through breast-feeding after birth.

HIV first spread through the homosexual community, and male-to-male sexual contact still accounts for the largest percentage of new AIDS cases in the United States. But the largest increases of HIV infections are occurring through heterosexual contact or by intravenous drug use. Now, women account for 19% of all newly diagnosed cases of AIDS. The rise in the incidence of AIDS among women of reproductive age is paralleled by a rise in the incidence of AIDS in children younger than 13.

Stages of an HIV Infection

The primary target of HIV is a type of T cell known as a T4 lymphocyte (Fig. 42B). The Centers for Disease Control and Prevention recognize three stages of an HIV-1 infection called Category A, B, and C. During a Category A stage, the T4 lymphocyte count is 500 per mm^3 or greater. For a period of time after the initial infection with HIV, people don't usually have any symptoms at all. A few (1–2%) do have mononucleosis-like symptoms that may include fever, chills, aches, swollen lymph nodes, and an itchy rash. These symptoms disappear, however, and there are no other symptoms for quite some time. Although there are no symptoms, the person is highly infectious. Although there is a large numbers of viruses in the plasma, the HIV blood test is not yet positive because it tests for the presence of antibodies and not for the presence of HIV itself. This means that HIV can still be transmitted before the HIV blood test is positive.

Several months to several years after a non-treated infection, the infected individual will probably progress to category B in which the T lymphocyte count is 200 to 499 per mm^3. During this stage, there will be swollen lymph nodes in the neck, armpits, or groin that persist for three months or more. Other symptoms that indicate category B are severe fatigue not related to exercise or drug use; unexplained persistent or recurrent fevers, often with night sweats; persistent cough not associated with smoking, a cold, or the flu; and persistent diarrhea.

When the individual develops non-life-threatening but recurrent infections, it is a signal that full-blown AIDS will occur shortly. Previously, the majority of infected persons proceeded to category C, in which the T4 lymphocyte count is 200 per mm^3 and the lymph nodes degenerated. The patient, who is now suffering from "slim disease" (as AIDS is called in Africa)—characterized by severe weight loss and weakness due to persistent diarrhea and coughing—will most likely succumb to one of the opportunistic infections. An opportunistic infection is one that has the opportunity to occur only because the immune system is severely weakened.

Treatment for HIV Infection

There is no cure for AIDS, but there are two types of drugs presently available for treatment. The well-known drug called AZT is a nucleotide analog, which stops viral DNA replication. A proteinase inhibitor blocks the action of the viral enzyme called proteinase, which is required for viral assembly. Physicians now believe that drug therapy should begin as soon as an HIV infection is detected. Recently, multidrug therapy with the analogs AZT and 3TC and a proteinase inhibitor has met with encouraging success.

Many investigators are working on a vaccine for AIDS. Some are trying to develop a vaccine in the traditional way. Others are working on subunit vaccines that utilize just a single HIV protein as the vaccine. For example, the spike gp120 (Fig. 42B), which can be produced by genetic engineering of bacteria, is undergoing a clinical trial in Thailand.

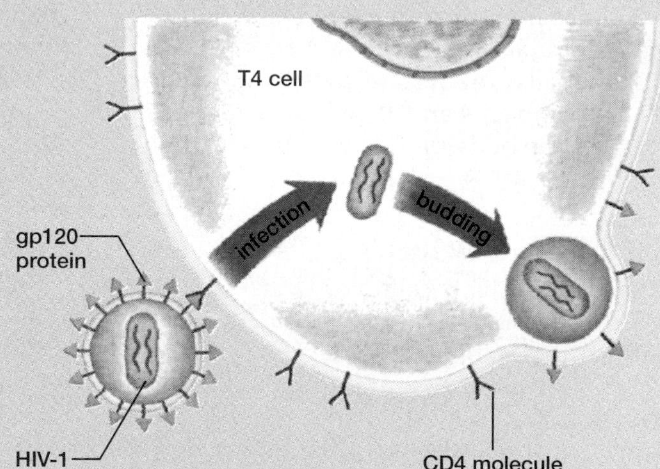

Figure 42B
HIV-1, the most common cause of AIDS in the United States, has an envelope molecule called gp120, which allows it to attach to CD4 receptors that project from a T4 cell. Infection of the T4 cell follows, and HIV eventually buds from the infected T4 cell. If the immune system can be trained by the use of a vaccine to attack and destroy all cells that bear gp120, a person could not be infected with HIV-1.

42.4 Immunity Can Be Induced

Immunity occurs naturally through infection or is brought about artificially by medical intervention. There are two types of induced immunity: active and passive. In active immunity, the individual alone produces antibodies against an antigen; in passive immunity, the individual is given prepared antibodies.

Active Immunity Is Long-Lived

Active immunity sometimes develops naturally after a person is infected with a microbe. However, active immunity is often induced when a person is well so that possible future infection will not take place. To prevent infections, people can be artificially immunized against them.

Immunization involves the use of **vaccines** [L. *vaccinus*, of cows], substances that contain an antigen to which the immune system responds. The use of vaccines dates back to the late eighteenth century when the English physician Edward Jenner used matter from cowpox pustules to protect individuals from smallpox. By the end of the nineteenth century, the renowned French researcher, Louis Pasteur, succeeded in producing vaccines for cholera, anthrax, and rabies, which are diseases caused by microbes. To prepare vaccines for these diseases, the microbes themselves were treated so they were no longer virulent (able to cause disease). Only recently has an entirely different method of producing vaccines been developed. Today, it is possible to genetically engineer bacteria to mass-produce a protein from microbes, and this protein can be used as vaccine. This method now has been used to produce a vaccine against hepatitis B, a viral disease, and is being used to prepare a vaccine against malaria, a protozoan disease.

After a vaccine is given, it is possible to follow an immune response by determining the amount of antibody present in a sample of serum—this is called the *antibody titer*. After the first exposure to a vaccine, a primary response occurs. For a period of several days, no antibodies are present; then, there is a slow rise in the titer, followed by first a plateau and then a gradual decline as the antibodies bind to the antigen or simply break down (Fig. 42.10). After a second exposure, a secondary response is expected. The titer rises rapidly to a plateau level much greater than before. The second exposure is called a "booster" because it boosts the antibody titer to a high level. The high antibody titer now is expected to help prevent disease symptoms even if the individual is exposed to the disease-causing antigen. Immunological memory causes an individual to be actively immune.

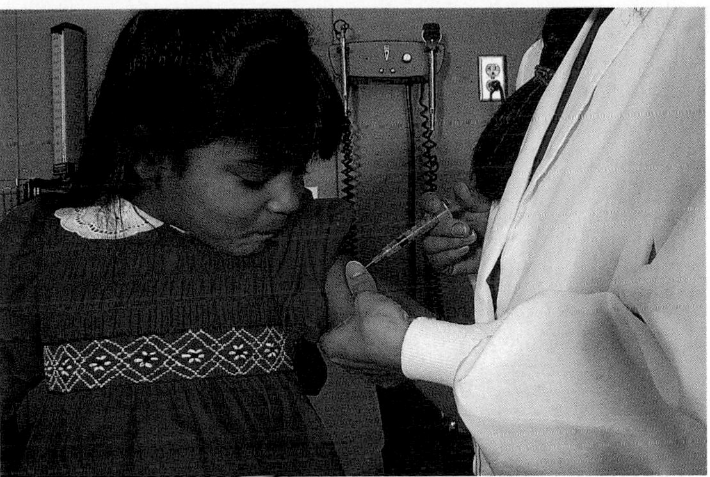

a.

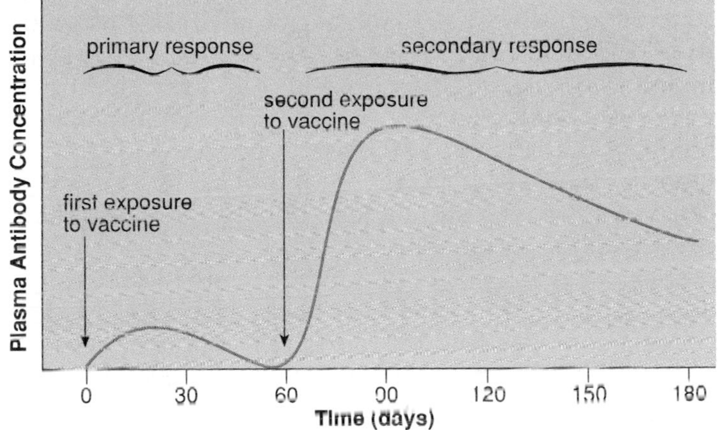

b.

Figure 42.10 Active immunity due to immunizations. The primary response, after the first exposure to a vaccine, is minimal, but the secondary response, which may occur after the second exposure, shows a dramatic rise in the amount of antibody present in serum.

Memory Cells Provide a State of Readiness

Immunological memory is dependent upon the number of memory B and memory T cells capable of responding to a particular antigen. The receptors of memory B cells usually have a higher affinity for the antigen due to the selection process that occurred during the first exposure to the antigen, and therefore they are prone to make IgG earlier. Both memory B cells and memory T cells can respond to lower doses of antigen. Good active immunity lasts as long as clones of memory B and memory T cells are present in blood. Active immunity is usually long-lived.

Active (long-lived) immunity can be induced by the use of vaccines when a person is well and in no immediate danger of contracting an infectious disease. Active immunity is dependent upon the presence of memory B cells and memory T cells in the body.

Passive Immunity Is Short-Lived

Passive immunity occurs when an individual is given prepared antibodies (immunoglobulins) to combat a disease. Since these antibodies are not produced by the individual's B cells, passive immunity is short-lived. For example, newborn infants are passively immune to some diseases because antibodies have crossed the placenta from the mother's blood. These antibodies soon disappear, however, so that within a few months, infants become more susceptible to infections. Breast-feeding prolongs the natural passive immunity an infant receives from the mother because antibodies are present in the mother's milk (Fig. 42.11).

Even though passive immunity does not last, it sometimes is used to prevent illness in a patient who has been unexpectedly exposed to an infectious disease. Usually, the patient receives a gamma globulin injection (serum that contains antibodies), perhaps taken from individuals who have recovered from the illness. In the past, horses were immunized, and serum was taken from them to provide the needed antibodies against such diseases as diphtheria, botulism, and tetanus. In the past, a patient who received these antibodies occasionally became ill because the serum contained proteins that the individual's immune system recognized as foreign. This was called serum sickness. But problems can still occur with products produced in other ways. An immunoglobulin intravenous product, called Gammagard, was withdrawn from the market because of possible implication in the transmission of hepatitis.

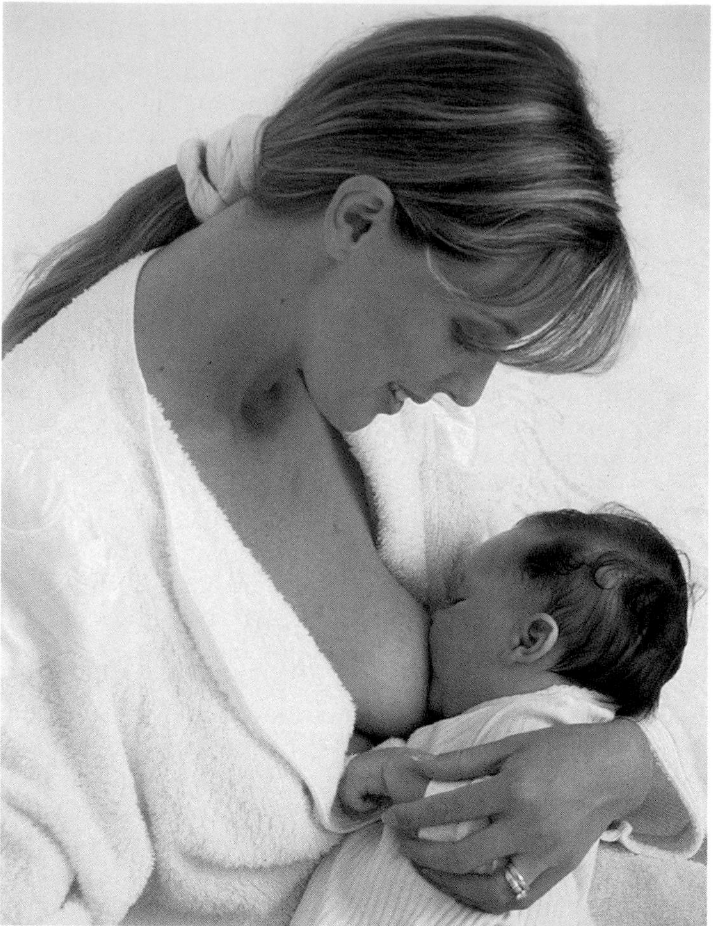

Figure 42.11 Passive immunity.
Breast-feeding is believed to prolong the passive immunity an infant receives from the mother because antibodies are present in the mother's milk.

Passive immunity is needed when an individual is in immediate danger of succumbing to an infectious disease. Passive immunity is short-lived because the antibodies are administered to and not made by the individual.

Cytokines Boost White Blood Cells

Cytokines are messenger molecules produced by either lymphocytes or monocytes. They are called *lymphokines* when produced by lymphocytes and *monokines* when produced by monocytes. Because cytokines stimulate white blood cell formation and/or function, they are being investigated as possible adjunct therapy for cancer and AIDS. Both interferon and various other types of cytokines called *interleukins* have been used as immunotherapeutic drugs, particularly to enhance the ability of the individual's own T cells (and possibly B cells) to fight cancer.

Interferon, discussed previously on page 760, is a substance produced by leukocytes, fibroblasts, and probably most cells in response to a viral infection. When it is produced by T cells, interferon is called a lymphokine. Interferon still is being investigated as a possible cancer drug, but so far it has proven to be effective only in certain patients, and the exact reasons for this as yet cannot be discerned.

When and if cancer cells carry an altered protein on their cell surface, they should be attacked and destroyed by cytotoxic T cells. Whenever cancer does develop, it is possible that the cytotoxic T cells have not been activated. In that case, lymphokines might awaken the immune system and lead to the destruction of the cancer. In one technique being investigated, researchers first withdraw T cells from the patient and activate the cells by culturing them in the presence of an interleukin. The cells then are reinjected into the patient, who is given doses of interleukin to maintain the killer activity of the T cells.

Those who are actively engaged in interleukin research believe that interleukins soon will be used as adjuncts for vaccines, for the treatment of chronic infectious diseases, and perhaps for the treatment of cancer. Interleukin antagonists also may prove helpful in preventing skin and organ rejection, autoimmune diseases, and allergies.

The interleukins and other lymphokines show some promise of potentiating the individual's own immune system.

Monoclonal Antibodies Have Same Specificity

As previously discussed, every plasma cell derived from the same B cell secretes antibodies against a specific antigen. These are **monoclonal antibodies** [Gk. *monos*, one, and *klonos*, clone, and *anti*, against] because all of them are the same type and because they are produced by plasma cells derived from the same B cell.

One method of producing monoclonal antibodies in vitro (outside the body in glassware) is depicted in Figure 42.12. B lymphocytes are removed from an animal (today, usually mice are used) and are exposed to a particular antigen. The activated B lymphocytes are fused with myeloma cells (malignant plasma cells that live and divide indefinitely). The fused cells are called hybridomas; *hybrid* because they result from the fusion of two different cells, and *oma* because one of the cells is a cancer cell.

At present, monoclonal antibodies are being used for quick and certain diagnosis of various conditions. For example, a particular hormone is present in the urine of a pregnant woman. A monoclonal antibody can be used to detect this hormone and if it is present, the woman is pregnant. Monoclonal antibodies also are used to identify infections. They are so accurate they can even sort out the different types of T cells in a blood sample. And because they can distinguish between cancer and normal tissue cells, they are used to carry radioactive isotopes or toxic drugs to tumors so that they can be selectively destroyed.

Monoclonal antibodies are considered to be a biotechnology product because a living system is used to mass-produce the product.

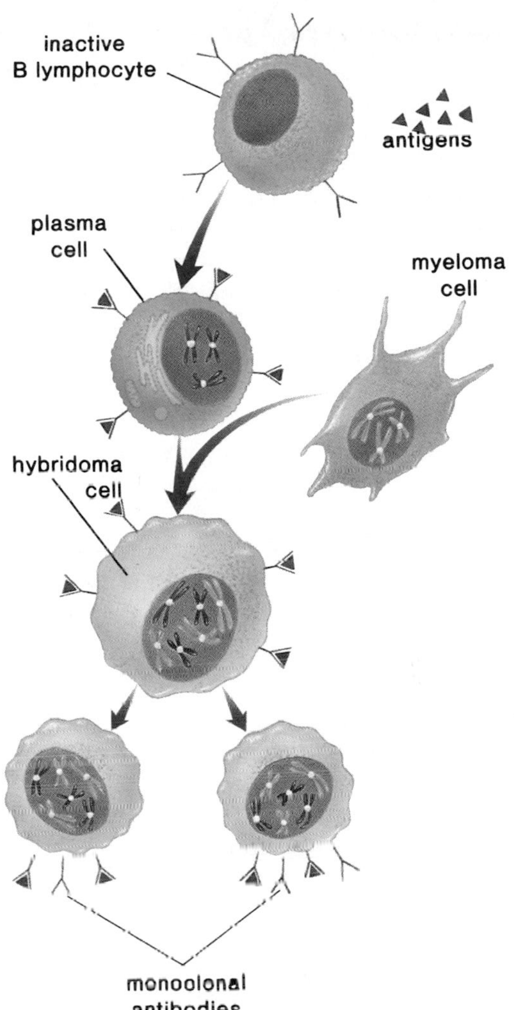

Figure 42.12 Production of monoclonal antibodies. Inactive lymphocytes are exposed to an antigen so that they become plasma cells that can produce antibodies. The plasma cells are fused with myeloma (cancer) cells, producing hybridoma cells that are "immortal." Hybridoma cells divide and continue to produce the same type of antibody, called monoclonal antibodies.

Monoclonal antibodies are produced in pure batches—they all react to just one type of molecule (antigen). Therefore, they can distinguish one cell, or even one molecule, from another.

42.5 Immunity Has Side Effects

The immune system protects us from disease because it can tell self from nonself. Sometimes, however, the immune system is underprotective, as when an individual develops cancer, or is overprotective, as when individuals have allergies.

Allergies: Overactive Immune System

Allergies are caused by an overactive immune system, which forms antibodies to substances that usually are not recognized as foreign substances. Unfortunately, allergies usually are accompanied by coldlike symptoms or, at times, by severe systemic reactions such as anaphylactic shock, which is a sudden drop in blood pressure.

Of the five varieties of antibodies—IgG, IgM, IgA, IgD, and IgE (see Table 42.1)—IgE antibodies are involved in allergic reactions. IgE antibodies are found in the bloodstream; but they, unlike the other types of antibodies, also reside in the membrane of mast cells found in the tissues. As mentioned, *mast cells* are basophils that have left the bloodstream and taken up residence in the tissues. In any case, when the *allergen*, an antigen that provokes an allergic reaction, attaches to the IgE antibodies on mast cells, these cells release histamine and other substances, which cause mucus secretion and airway constriction. This results in the characteristic allergy symptoms. On occasion, basophils and other white blood cells release these substances into the bloodstream. The increased capillary permeability that results from this can lead to fluid loss and shock.

Allergy shots sometimes prevent the onset of allergic symptoms. Injections of the allergen cause the body to build up high quantities of IgG antibodies, and these combine with allergens received from the environment before they have a chance to reach the IgE antibodies located in the membrane of mast cells.

Histamine and other substances released by mast cells cause allergic symptoms.

Autoimmune Diseases: The Body Attacks Itself

Certain human illnesses are referred to as **autoimmune diseases** [Gk. *aut,* self, and L. *immun,* safe, free] because they are due to an attack on tissues by the body's own antibodies and T cells. Exactly what causes autoimmune diseases is not known, but they seem to appear after the individual has recovered from an infection. Some bacteria have been observed to produce toxic products that can cause T cells to bind prematurely to macrophages. Perhaps at that time the T cell learns to recognize the body's own tissues. This might be the cause of at least some autoimmune diseases. In myasthenia gravis, neuromuscular junctions do not work properly and muscular weakness results. In multiple sclerosis (MS), the myelin sheath of nerve fibers is attacked, and this causes various neuromuscular disorders. A person with systemic lupus erythematosus (SLE) has various symptoms prior to death due to kidney damage. In rheumatoid arthritis, the joints are affected. It is suspected that heart damage following rheumatic fever and type I diabetes are also autoimmune illnesses. There are no cures for autoimmune diseases.

Autoimmune diseases seem to be preceded by an infection that results in cytotoxic T cells or antibodies attacking the body's own organs.

Tissue Rejection: Foreign HLA Proteins

Certain organs, such as skin, the heart, and the kidneys, could be transplanted easily from one person to another if the body did not attempt to *reject* them. Rejection occurs because cytotoxic T cells bring about destruction of foreign tissue in the body.

Organ rejection can be controlled by careful selection of the organ to be transplanted and the administration of immunosuppressive drugs. It is best if the transplanted organ has the same type of HLA proteins as those of the recipient, because cytotoxic T cells recognize foreign HLA proteins. The immunosuppressive drug cyclosporine has been used for many years. A new drug, tacrolimus (formerly known as FK-506), shows some promise, especially in liver transplant patients. However both drugs, which act by inhibiting the production of interleukin-2, are known to adversely affect the kidneys.

Transplantation of bone marrow between individuals that are only partially HLA-matched has now gained acceptance. To reduce rejection it is imperative to remove the T cells from the transferred bone marrow This can be done by using monoclonal anti-T antibodies and complement.

When an organ is rejected, the immune system is attacking cells that bear different MHC proteins from those of the individual.

connecting concepts

The abundance of invertebrates suggests that they must have some way to defend themselves against infectious bacteria and viruses. As far back as 1882, the Russian zoologist Elie Metchnikoff stuck a rose thorn into a starfish and noted that phagocytes gathered in an attempt to remove the foreign object. You might surmise that phagocytosis would be an ancient means of defense because protozoa carry out phagocytosis when they feed. In humans, as we have seen, white blood cells, notably the neutrophils and macrophages, phagocytize invaders, especially if they have been coated by complement. Chemo warfare has been found among invertebrates also. In 1979 Swedish Hans G. Boman discovered a class of silk moth antibacterial peptides, which he named cecropins. Like complement, they perforate bacteria, causing them to burst. (These peptides are currently being developed as antibacterial agents for use in humans.) Similarly, many invertebrates possess molecules related to vertebrate cytokines. In the sea star, cells called coelomocytes, which are the equivalent of macrophages, release a chemical that functions the same as an interleukin.

However, B and T cell equivalents have not been found among invertebrates. Apparently, specific defense mechanisms evolved only among the vertebrates. Gary W. Litman, who is associated with the University of South Florida, has studied specific immunity in the horned shark. He has found that diversity in antigen receptors has a genetic basis in this shark, just as it does in humans. But the horned shark relies more heavily on "inherited" diversity. This shark more quickly produces antibodies directed against microbes it is apt to encounter than antibodies against new and different antigens. This characteristic of antibody-mediated immunity may be an advantage if the environment remains relatively constant. Investigators have hypothesized that cell-mediated immunity based on the presence of T cells predates antibody-mediated immunity based on B cells. Therefore, Litman decided to test the horned shark for evidence of T cell activity. He used PCR to replicate human T cell receptor genes prior to sequencing them. Then he found evidence of corresponding proteins in the horned shark. The existence of many clusters of immune system genes in sharks as well as mammals offers the potential for mutation and diversity of the immune response among all the vertebrates.

Summary

42.1 Lymphatic System Helps Cardiovascular System

The lymphatic system consists of lymphatic vessels and lymphoid organs. The lymphatic vessels collect fat molecules at intestinal villi, collect excess tissue fluid, and carry these to the bloodstream.

Lymphocytes are produced and accumulate in the lymphoid organs (red bone marrow, lymph nodes, spleen, and thymus gland).

The lymph nodes are divided into sinus-containing nodules in which lymph is cleansed of infectious organisms and/or their toxins. The thymus is divided into lobules, where T lymphocytes mature. The spleen is divided into sinus-containing lobules in which blood is cleansed of infectious organisms and/or their toxins. Red bone marrow produces all types of blood cells. White blood cells are necessary for the development of immunity.

42.2 Some Defenses Are Nonspecific

Immunity involves nonspecific and specific defenses. Nonspecific defenses include barriers to entry, the inflammatory reaction, and protective proteins.

42.3 Other Defenses Are Specific

Specific defenses require lymphocytes, which are produced in the bone marrow. B cells mature in the bone marrow and undergo clonal selection in the lymph nodes and the spleen. T cells mature in the thymus.

B cells are responsible for antibody-mediated immunity. An antibody is a Y-shaped molecule that has two binding sites. Each antibody is specific for a particular antigen. Activated B cells become antibody-secreting plasma cells and memory B cells. Memory B cells respond if the same antigen enters the body at a later date.

There are various types of T cells. Cytotoxic T cells kill cells on contact; helper T cells stimulate other immune cells and produce lymphokines; suppressor T cells suppress the immune response; and memory T cells remain in the body to provide long-lasting immunity.

For the T cell to recognize an antigen, the antigen must be presented by an antigen-presenting cell (APC), usually a macrophage, along with an HLA (human lymphocyte-associated complex) protein. Cytotoxic T cells are responsible for cell-mediated immunity.

AIDS is an HIV infection of helper T cells, primarily, and therefore it destroys the cells that are needed to mount an immune response.

42.4 Immunity Can Be Induced

Immunity can be induced in various ways. Vaccines are available to promote long-lived active immunity, and antibodies sometimes are available to provide an individual with short-lived passive immunity.

Cytokines, notably interferon and interleukins, are used in an attempt to promote the body's ability to recover from cancer and to treat AIDS.

42.5 Immunity Has Side Effects

Allergies result when an overactive immune system forms antibodies to substances not normally recognized as foreign. Cytotoxic T cells attack transplanted organs as nonself; therefore immunosuppressive drugs must be administered. Autoimmune illnesses occur when antibodies and T cells attack the body's own tissues.

Reviewing the Chapter

1. What is the lymphatic system, and what are its three functions? 756
2. Describe the structure and the function of the lymph nodes, the spleen, the thymus, and bone marrow. 757–58
3. What are the body's nonspecific defense mechanisms? 758–59
4. Describe the inflammatory reaction, and give a role for each type of cell and molecule that participates in the reaction. 758–59
5. What is the clonal selection theory? B cells are responsible for which type of immunity? 761
6. Describe the structure of an antibody, and define the terms variable regions and constant regions. 762
7. Name the four types of T cells, and state their functions. 764–65
8. Explain the process by which a T cell is able to recognize an antigen. 765
9. How is active immunity achieved? How is passive immunity achieved? 767–68
10. What are lymphokines, and how are they used in immunotherapy? 768
11. How are monoclonal antibodies produced, and what are their applications? 769
12. Discuss allergies, tissue rejection, and autoimmune diseases as they relate to the immune system. 769–70

Testing Yourself

Choose the best answer for each question.

1. Complement
   a. is a general defense mechanism.
   b. is a series of proteins present in the plasma.
   c. plays a role in destroying bacteria.
   d. All of these are correct.
2. Which of these pertain(s) to T cells?
   a. have specific receptors
   b. have cell-mediated immunity
   c. stimulate antibody production by B cells
   d. All of these are correct.
3. Which one of these does not pertain to B cells?
   a. have passed through the thymus
   b. specific receptors
   c. antibody-mediated immunity
   d. synthesize and liberate antibodies
4. The clonal selection theory says that
   a. an antigen selects certain B cells and suppresses them.
   b. an antigen stimulates the multiplication of B cells that produce antibodies against it.
   c. T cells select those B cells that should produce antibodies, regardless of antigens.
   d. T cells suppress all B cells except the ones that should multiply and divide.
5. Plasma cells are
   a. the same as memory cells.
   b. formed from blood plasma.
   c. B cells that are actively secreting antibody.
   d. inactive T cells carried in the plasma.

6. For a T cell to recognize an antigen, it must interact with
   a. complement.
   b. a macrophage.
   c. a B cell.
   d. All of these are correct.
7. Antibodies combine with antigens
   a. at variable regions.
   b. at constant regions.
   c. only if macrophages are present.
   d. Both a and c are correct.
8. Which one of these is mismatched?
   a. helper T cells—help complement react
   b. cytotoxic T cells—active in tissue rejection
   c. suppressor T cells—shut down the immune response
   d. memory T cells—long-living line of T cells
9. Vaccines are
   a. the same as monoclonal antibodies.
   b. treated bacteria or viruses or one of their proteins.
   c. MHC proteins.
   d. All of these are correct.
10. The theory behind the use of lymphokines in cancer therapy is that
    a. if cancer develops, the immune system has been ineffective.
    b. lymphokines stimulate the immune system.
    c. cancer cells bear antigens that should be recognizable by cytotoxic T cells.
    d. All of these are correct.
11. Give the function of the four types of T cells shown.

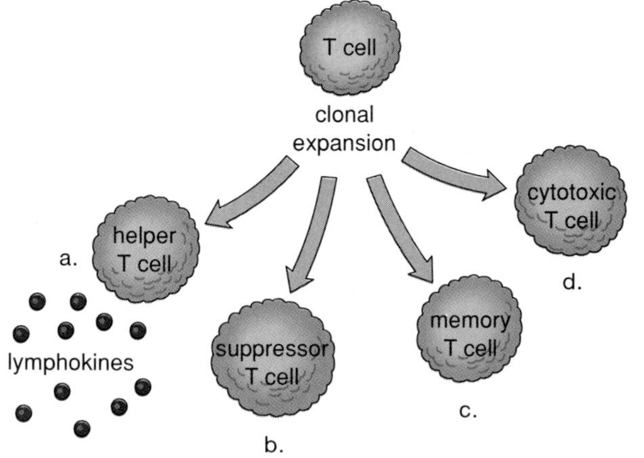

Applying the Concepts

1. *The body defends its integrity.*
   How does your study of immunity support this concept?
2. *Multitiered mechanisms maintain homeostasis.*
   How does your study of immunity support this concept?
3. *Organs belong to organ systems.*
   Argue that red bone marrow is a part of the skeletal, cardiovascular, and lymphatic systems.

Using Technology

Your study of lymph transport and immunity is supported by these available technologies:

Exploring the Internet
The Mader Home Page provides resources for and help with studying this chapter.

http://www.mhhe.com/sciencemath/biology/mader/
(Click on Biology.)

The Dynamic Human CD-ROM
Lymphatic System

Explorations in Human Biology CD-ROM
Immune Response (#12)
AIDS (#13)

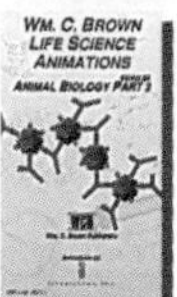

Life Science Animations Video
Video #4: Animal Biology II
B-Cell Immune Response (#41)
Structure and Function of Antibodies (#42)
Types of T-cells (#43)
Relationship of Helper T-cells and Killer T-cells (#44)

Understanding the Terms

antibody 760	lymph 756
antigen 760	lymphatic system 756
autoimmune disease 770	lymph nodes 757
B lymphocyte 760	macrophage 759
complement system 760	major histocompatibility complex (MHC) protein 765
cytokine 768	monoclonal antibody 769
edema 756	plasma cell 761
histamine 759	T lymphocyte 760
immunity 758	vaccine 767
inflammatory reaction 758	

Match the terms to these definitions:

a. _______ Ability of the body to protect itself from foreign substances and cells, including infectious microbes.
b. _______ Antigens prepared in such a way that they can promote active immunity without causing disease.
c. _______ Fluid, derived from tissue fluid, that is carried in lymphatic vessels.
d. _______ Foreign substance, usually a protein or a polysaccharide, that stimulates the immune system to react, such as to produce antibodies.
e. _______ Protein produced in response to the presence of an antigen; each type combines with a specific antigen.
f. _______ Lymphocyte that matures in the thymus and exists in four varieties, one of which kills antigen-bearing cells outright.

Digestion and Nutrition

CHAPTER

Chapter Concepts

African chameleon, *Chamaeleo*, catching a fly

Animals are heterotrophic organisms that must take in preformed food. The variety of diets found in the animal kingdom is astounding, from beetles which feed on rotting material to killer whales which go after live prey. The adaptations of animals to digest food are extremely varied and can be related to an animal's body plan and diet. Some animals have an incomplete digestive tract, and the same opening is used for entrance and exit. With this type of tract there is no specialization of parts. Most species, however, have a complete digestive tract with one opening (the mouth) that serves as entrance, and another that serves as an exit (the anus). With this type of tract, it is possible to have specialized parts. Among mammals, carnivores have teeth that shear off meat and a short, nonspecialized digestive tract. Herbivores have teeth that clip and crush grasses and a complex digestive tract with a part where bacteria digest cellulose.

In a study of mantled howling monkeys, it was concluded that howlers choose foods that give them the greatest nutrient return. In contrast, adult humans often must be taught good nutrition. There is increasing evidence to suggest that eating correctly may help prevent the major killer diseases in the U.S.—cardiovascular disease and cancer.

43.1 Comparing Digestive Tracts

Most animals have some sort of gut or digestive tract where food is digested to small nutrient molecules that can cross plasma membranes. Digestion contributes to homeostasis by providing the body with the nutrients needed to sustain the life of cells. A digestive tract (1) ingests food, (2) breaks food down into small molecules that can cross plasma membranes, (3) absorbs these nutrient molecules, and (4) eliminates nondigestible remains.

Gut Is Complete or Incomplete

An *incomplete gut* has a single opening, usually called a mouth. The planarian is an example of an animal with an incomplete gut (Fig. 43.1). It is carnivorous and feeds largely on smaller aquatic animals. Its digestive system contains only a mouth, a pharynx, and an intestine. When the worm is feeding, the pharynx actually extends beyond the mouth. It wraps its body about the prey and uses its muscular pharynx to suck up minute quantities at a time. Digestive enzymes present in the *gastrovascular cavity* allow some extracellular ("outside the cells") digestion to occur. Digestion is finished intracellularly by the cells that line the cavity, which branches throughout the body. No cell in the body is far from the intestine, and therefore diffusion alone is sufficient to distribute nutrient molecules.

The digestive system of a planarian is notable for its lack of specialized parts. It is saclike because the pharynx serves not only as an entrance for food but also as an exit for nondigestible material. Specialization of parts does not occur under these circumstances.

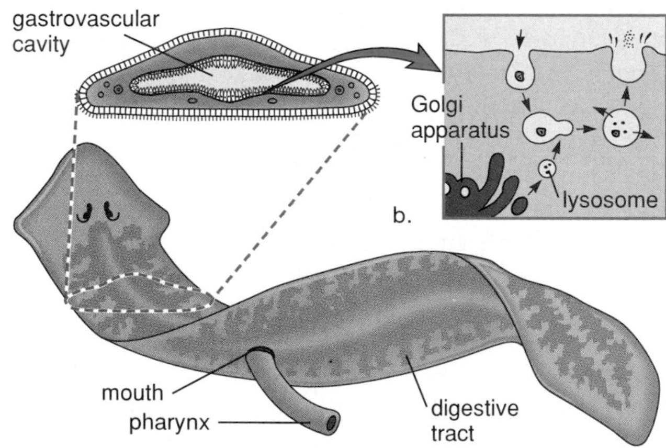

Figure 43.1 Incomplete digestive tract of a planarian. a. Planaria have a gastrovascular cavity with a single opening that acts as both an entrance and an exit. Planaria rely on intracellular digestion to complete the digestive process. b. Phagocytosis produces a vacuole, which joins with an enzyme-containing lysosome. The digested products pass from the vacuole into the cytoplasm before any nondigestible material is eliminated at the plasma membrane.

A planarian has some modified parasitic relatives. A tapeworm, for example, has no digestive system at all—it simply absorbs nutrient molecules from the intestinal juices that surround its body. The body wall is highly modified for this purpose: it has millions of microscopic fingerlike projections that increase the surface area for absorption.

In contrast to the planarian, an earthworm has a *complete gut,* meaning that it has a mouth and an anus (Fig. 43.2). Earthworms feed mainly on decayed organic matter in soil. The muscular *pharynx* draws in food with a sucking action. The *crop* is a storage area that has thin expansive walls. The *gizzard* has thick muscular walls for churning and grinding the food. Digestion is extracellular—in the intestine. The surface area of digestive tracts is often increased for absorption of nutrient molecules, and in the earthworm this is accomplished by an intestinal fold called the typhlosole. Undigested remains pass out of the body at the anus. Specialization of parts is obvious in the earthworm because the pharynx, the crop, the gizzard, and the intestine each has a particular function in the digestive process.

It is of interest to recall that some marine relatives of an earthworm are feather-duster worms, which live in a tube of various construction. The feathery tentacles of a feather-duster worm are specialized for gathering fine particles and microscopic plankton from the water. In these worms the digestive tract has the appearance of a simple uniform tube.

In contrast to the incomplete, saclike gut, the complete gut, with both a mouth and an anus, can have many specialized parts depending on the way of life of the animal.

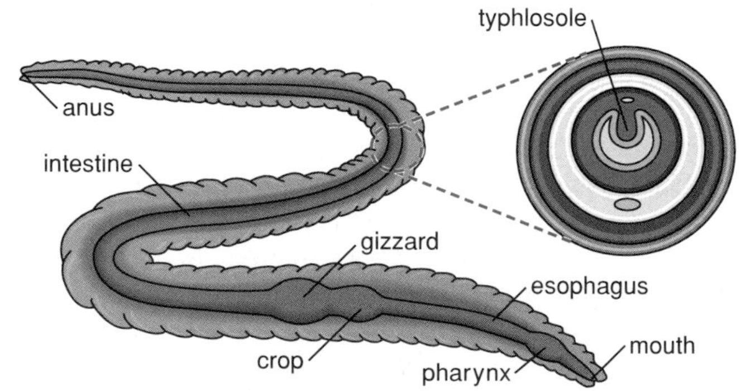

Figure 43.2 Complete digestive tract of an earthworm.
Complete digestive tracts have both a mouth and an anus and can have many specialized parts as those labeled in this drawing. Also in earthworms, as in many polychaetes, the absorptive surface of the intestine is increased by an internal fold called the typhlosole.

Figure 43.3 Nutritional mode of a clam compared to a squid.
A clam burrows in the sand or mud, where it filter feeds, whereas the squid swims freely in open waters and captures prey. **a.** A flow of water (see arrows) brings debris (food particles) into and carries waste out of a clam's mantle cavity. **b.** In contrast, a squid captures other aquatic animals with its tentacles and bites off pieces with its jaws. A strong contraction of the mantle forces water (see arrows), which has entered the mantle cavity, out the funnel, resulting in a type of "jet propulsion."

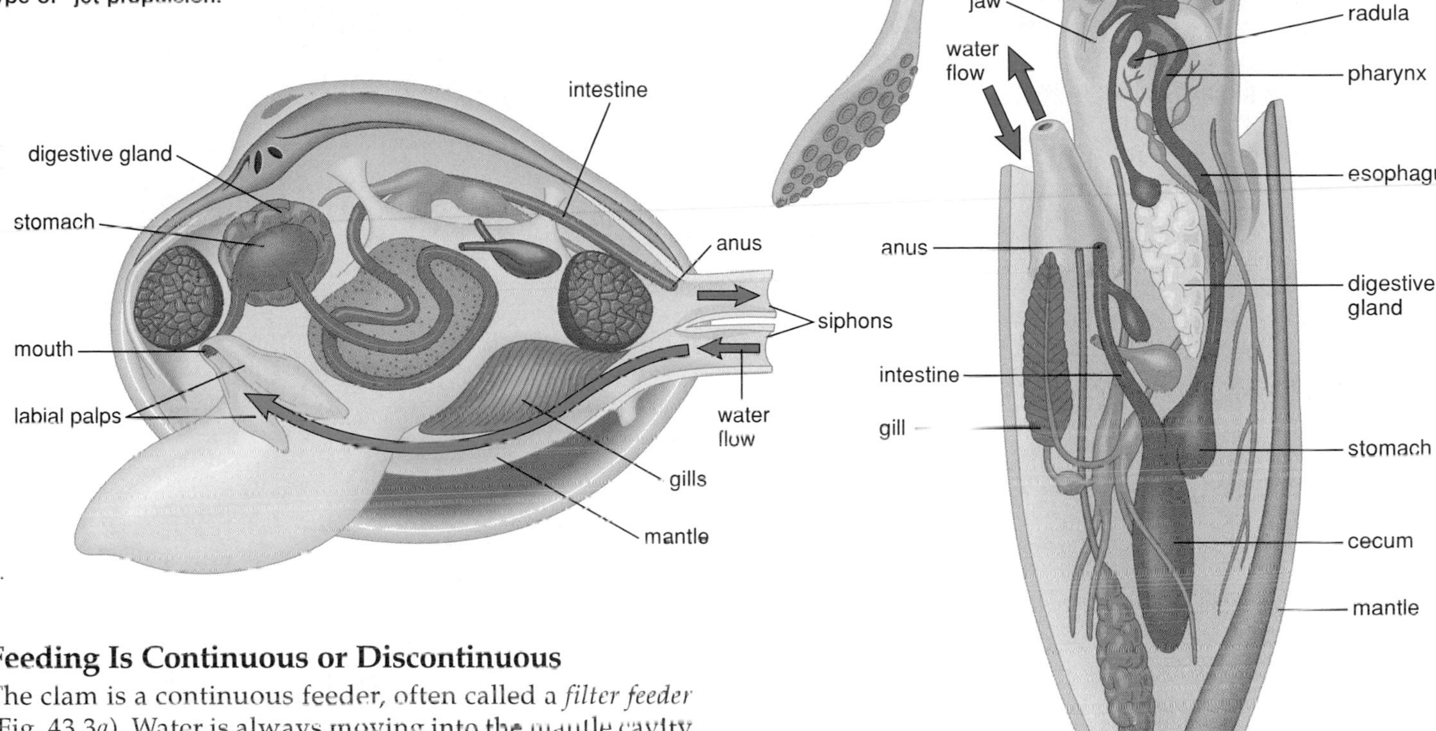

Feeding Is Continuous or Discontinuous

The clam is a continuous feeder, often called a *filter feeder* (Fig. 43.3*a*). Water is always moving into the mantle cavity by way of the incurrent siphon (slitlike opening) and depositing particles on the gills. The size of the incurrent siphon permits the entrance of only small particles, which adhere to the gills. Ciliary action moves suitably sized particles to the labial palps, which force them through the mouth into the stomach. Digestive enzymes are secreted by a large digestive gland, but amoeboid cells present throughout the tract are believed to complete the digestive process by intracellular digestion.

Marine feather-duster worms are termed filter feeders because they allow only small particles to enter their digestive tract. Larger particles are rejected. The baleen whale is an active filter feeder. The baleen, a curtainlike fringe, hangs from the roof of the mouth and filters small shrimp called krill from the water. The baleen whale filters up to a ton of krill every few minutes.

The squid is an example of a discontinuous feeder (Fig. 43.3*b*). The body of a squid is streamlined, and the animal moves rapidly through the water using jet propulsion (forceful expulsion of water from a tubular funnel). The head of a squid is surrounded by ten arms, two of which have developed into long, slender tentacles whose suckers have toothed, horny rings. These tentacles seize prey (fishes, shrimps, and worms) and bring it to the squid's beaklike jaws, which bite off pieces pulled into the mouth by the action of a radula, a toothy tongue. An esophagus leads to a stomach and a cecum (blind sac) where digestion occurs. The stomach, supplemented by the cecum, holds food until digestion is complete. Discontinuous feeders, whether they are carnivores or herbivores, require such a storage area.

Continuous feeders, such as sessile filter feeders, do not need a storage region for food; discontinuous feeders, such as herbivores and carnivores, do need a storage region for food.

Dentition and the Digestive Tract Are Suitable to Diet

Some animals are omnivores; they eat both plants and animals. Others are herbivores; they feed only on plants. Still others are carnivores; they eat only other animals. Among invertebrates, filter feeders like clams and tube worms are omnivores. Mollusks, such as land snails and some insects like grasshoppers and locusts, are herbivores. Spiders are carnivores, as are sea stars that feed on clams. A sea star positions itself above a clam and uses its tube feet to pull the valves of the shell apart. Then, it everts a part of its stomach to start the digestive process, even while the clam is trying to close its shell. Some invertebrates are cannibalistic. The female praying mantis, if starved, will feed upon her mate as the reproductive act is taking place!

Among mammals, the dentition differs according to mode of nutrition (Fig. 43.4). Humans, as well as raccoons, rats, and brown bears, are omnivores. Therefore, the dentition has a variety of specializations to accommodate both a vegetable diet and a meat diet. An adult human has thirty-two teeth. One-half of each jaw has teeth of four different types: two chisel-shaped *incisors* for shearing; one pointed *canine* for tearing; two fairly flat *premolars* for grinding; and three *molars,* well-flattened for crushing.

Among herbivores, the koala of Australia is famous for its diet of only eucalyptus leaves, and likewise many other mammals are browsers, feeding off bushes and trees. Grazers, like the horse, feed off grasses. The horse has sharp, even incisors for neatly clipping off blades of grass and large, flat premolars and molars for grinding and crushing the grass. Extensive grinding and crushing disrupts plant cell walls, allowing bacteria located in a part of the digestive tract called the cecum to get at and digest cellulose. Other mammalian grazers, like the cow and deer, are ruminants. In contrast to horses, they graze quickly and swallow partially chewed grasses into a special part of the stomach called a rumen. Here, microorganisms start the digestive process and the result, called cud, is regurgitated at a later time when the animal is no longer feeding. The cud is chewed again before being swallowed for complete digestion.

Many mammals, including dogs, toothed whales, and polar bears, are carnivores. A carnivore, like a lion, uses pointed incisors and enlarged canine teeth to shear off pieces small enough to be quickly swallowed. Meat is rich in protein and fat and is easier to digest than plant material. Therefore, the digestive system of carnivores is shorter and doesn't have the specialization seen in herbivores.

Omnivores have nonspecialized teeth. Herbivores have large, flat molars that grind food; carnivores have incisors and canines to tear off chunks of meat.

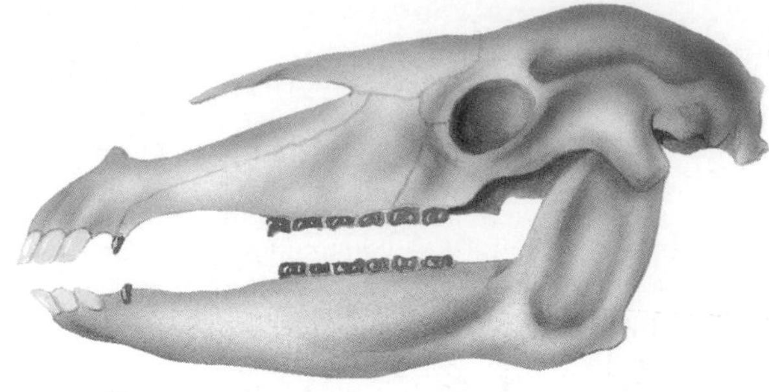

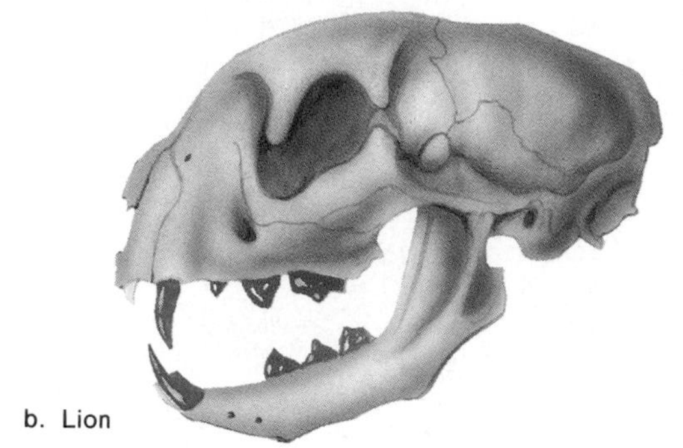

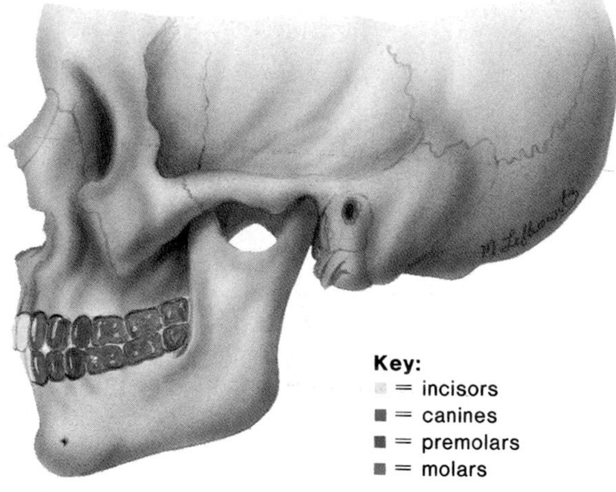

Figure 43.4 Dentition among vertebrates.
a. Horses are herbivores that graze on grasses. Note the sharp incisors, reduced canines, and large, flat premolars and molars.
b. Lions are carnivores that prey on other animals. Note the pointed incisors, enlarged canines, and jagged premolars and molars.
c. Humans are omnivores that have nonspecialized teeth.

43.2 Humans Have a Complete Tract

Humans have the tube-within-a-tube body plan, and therefore the human digestive tract is complete—there is both a mouth and an anus. Table 43.1 lists the parts and functions of the human digestive system, and Figure 43.5 depicts these parts.

Digestion of food in humans is an extracellular process, and digestive enzymes are secreted by the digestive tract or by glands that lie nearby. Food is never found within these accessory glands, only within the tract itself. Digestion requires a cooperative effort between different parts of the body. The production of hormones and the performance of the nervous system achieve the cooperation of body parts.

Mouth Receives Food

We have already mentioned that human dentition has a variety of specializations because humans are omnivores (see Fig. 43.4*c*). Food is chewed in the mouth, where it is mixed with saliva. There are three major pairs of **salivary glands** that send their juices by way of ducts to the mouth. Saliva contains the enzyme **salivary amylase**, which begins the process of starch digestion. The disaccharide maltose is a typical end product of salivary amylase digestion:

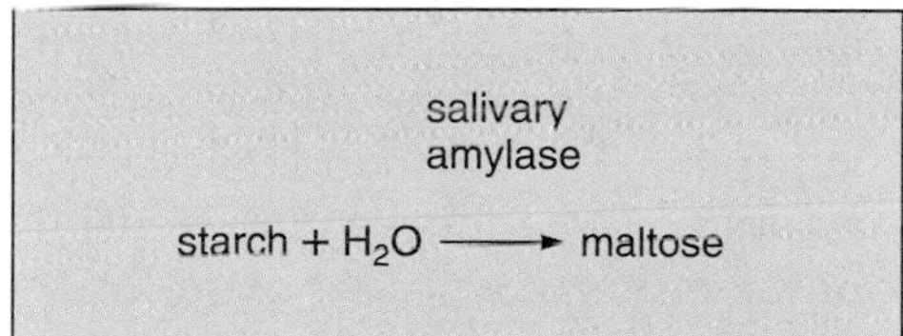

While in the mouth, food is manipulated by a muscular tongue, which has touch and pressure receptors similar to those in the skin. *Taste buds*, chemical receptors that are stimulated by the chemical composition of food, are also found primarily on the tongue as well as on the surface of the mouth and the wall of the pharynx. After the food has been thoroughly chewed and mixed with saliva, the tongue starts the process of swallowing by pushing the *bolus* (small mass of chewed food) back to the pharynx.

Table 43.1

Path of Food

Organ	Special Features	Function
Mouth	Teeth, tongue	Chewing of food; digestion of starch
Esophagus		Movement of food by peristalsis
Stomach	Gastric glands	Storage of food; acidity kills some bacteria; digestion of protein
Small intestine	Villi	Digestion of all foods; absorption of nutrients
Large intestine		Absorption of water; storage of nondigestible remains
Anus		Defecation

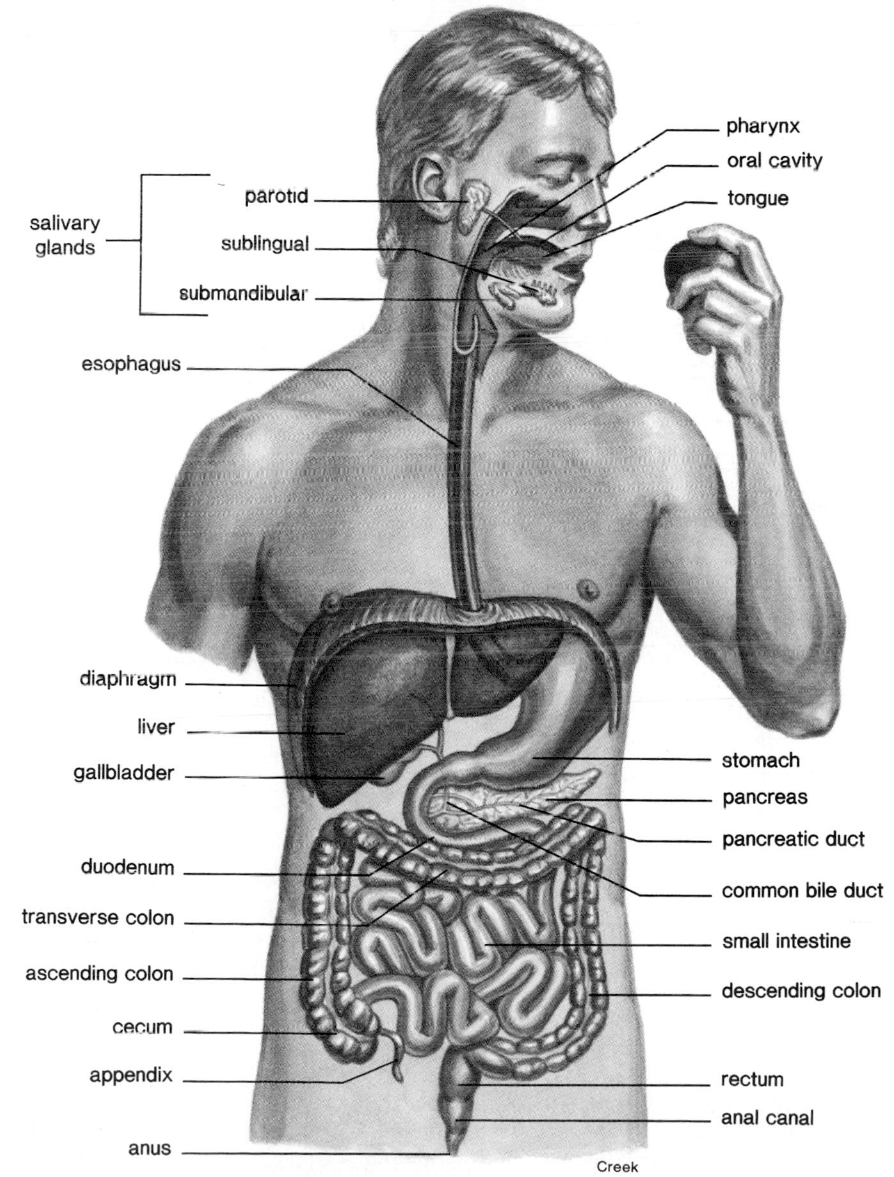

Figure 43.5 The human digestive tract.
Trace the path of food from the mouth to the anus. Note the placement of two accessory organs of digestion—the liver and the pancreas.

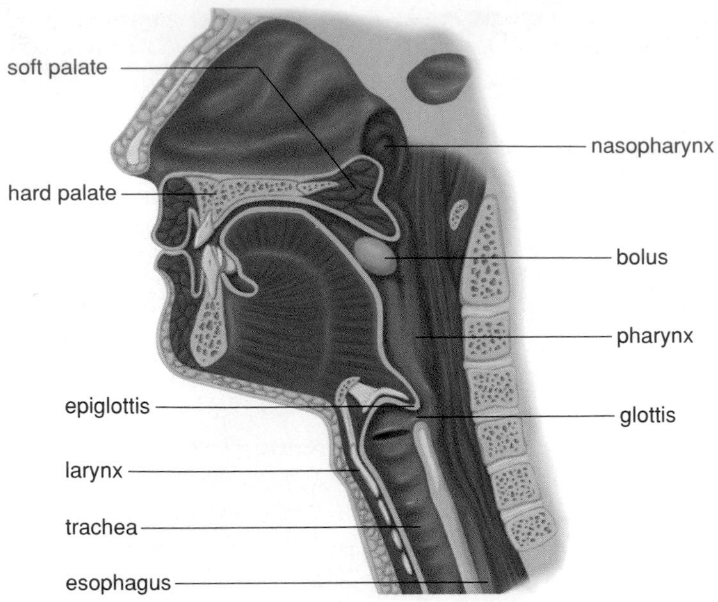

Figure 43.6 Swallowing.
Respiratory and digestive passages converge and diverge in the pharynx. During swallowing, the soft palate closes off the nasopharynx, and the epiglottis covers the glottis, forcing the bolus to pass down the esophagus. Therefore, you do not breathe when swallowing.

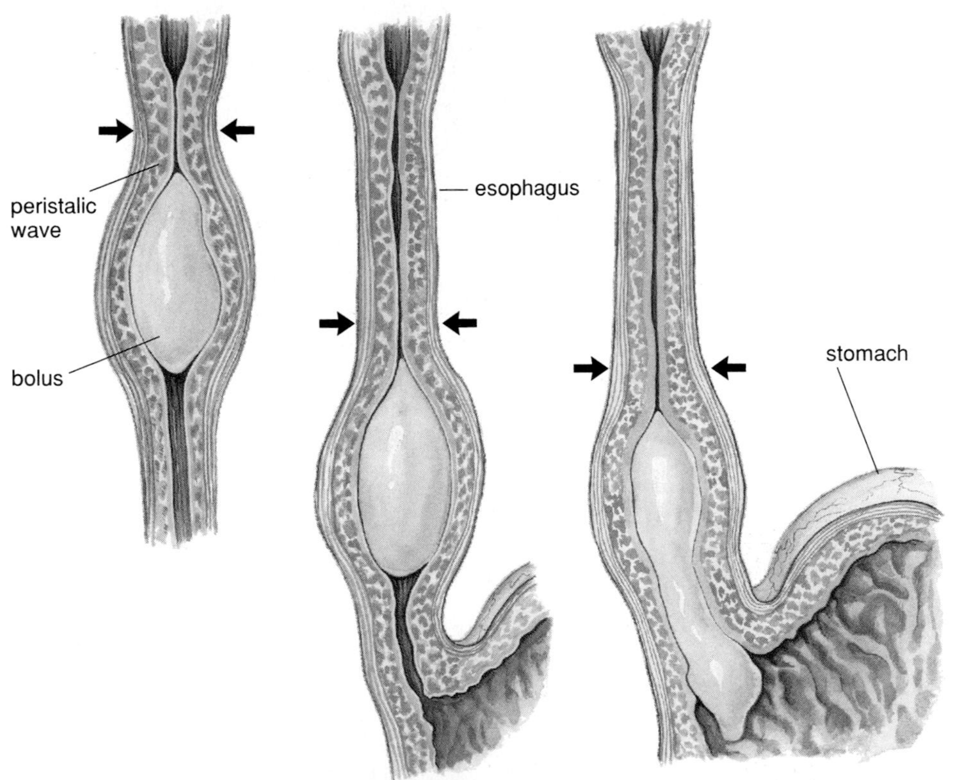

Figure 43.7 Peristalsis in the digestive tract.
Rhythmic waves of muscle contraction move material along the digestive tract. The three drawings show how a peristaltic wave moves through a single section of the esophagus over time.

Esophagus Conducts Food

The digestive and respiratory passages come together in the pharynx and then separate (Fig. 43.6). Therefore, during swallowing, the path of air to the lungs would be blocked if food entered the trachea (windpipe). Normally, however, a flap of tissue called the **epiglottis** [Gk. *epi*, over, and *glotta*, tongue] covers the opening into the **trachea** as muscles move the bolus through the pharynx into the esophagus. The **esophagus** [Gk. *eso*, within, and *phagein*, eat] is a tubular structure that takes food to the stomach. When food enters the esophagus, peristalsis begins (Fig. 43.7). **Peristalsis** [Gk. *peri*, around, and *stalsis*, compression] is a rhythmical contraction that serves to move the contents along in tubular organs, such as the digestive tract.

Stomach Stores Food

Usually, the stomach stores up to two liters of partially digested food. Therefore, humans can periodically eat relatively large meals and spend the rest of their time at other activities. But the stomach is much more than a mere storage organ, as was discovered by William Beaumont in the mid-nineteenth century. Beaumont, an American doctor, had a French Canadian patient, Alexis St. Martin. St. Martin had been shot in the stomach, and when the wound healed, he was left with a fistula, or opening, that allowed Beaumont to look inside the stomach and to collect gastric (stomach) juices produced by *gastric glands.* Beaumont was able to determine that the muscular walls of the stomach contract vigorously and mix food with juices that are secreted whenever food enters the stomach (Fig. 43.8). He found that *gastric juice* contains *hydrochloric acid* (HCl) and a substance active in digestion. (This substance was later identified as pepsin.) He also found that the gastric juices are produced independently of the protective mucus secretions of the stomach. Beaumont's work, which was very carefully and painstakingly done, pioneered the study of the physiology of digestion.

So much hydrochloric acid is secreted by the stomach that it routinely has a pH of about 2. Such a high acidity usually is sufficient to kill bacteria and other microorganisms that might be in food. This low pH also stops the activity of salivary amylase, which functions optimally at the near-neutral pH of saliva, but it promotes the activity of pepsin. **Pepsin** is a hydrolytic enzyme that acts on protein to produce peptides:

$$\text{protein} + H_2O \xrightarrow{\text{pepsin}} \text{peptides}$$

By now the stomach contents have a thick, soupy consistency and are called *chyme.* At the base of the stomach is a narrow opening controlled by a sphincter. A *sphincter* is a muscle that surrounds a tube and closes or opens the tube by contracting and relaxing. Whenever the sphincter relaxes, a small quantity of chyme passes through the opening into the **duodenum,** the first part of the **small intestine** (see Fig. 43.5). When chyme enters the duodenum, it sets off a neural reflex that causes the muscles of the sphincter to contract vigorously and to close the opening temporarily. Then the sphincter relaxes again and allows more chyme to enter. The slow manner in which chyme enters the small intestine allows for thorough digestion.

As with the rest of the digestive tract, a thick layer of mucus protects the wall of the stomach and the first part of the duodenum from enzymatic action. Still, an *ulcer,* which is an open sore in the wall caused by the gradual destruction of tissues, does occur in some individuals. It is now believed that ulcers are due to an infection by an acid-resistant bacterium, *Helicobacter pylori,* which is able to attach to the epithelial lining. Wherever the bacterium attaches, the lining stops producing mucus and the area becomes exposed to digestive action. Now an ulcer develops.

Prior to entering the small intestine, food passes through the mouth, pharynx, esophagus, and stomach. In the mouth, salivary amylase digests starch to maltose; in the stomach, pepsin digests protein to peptides.

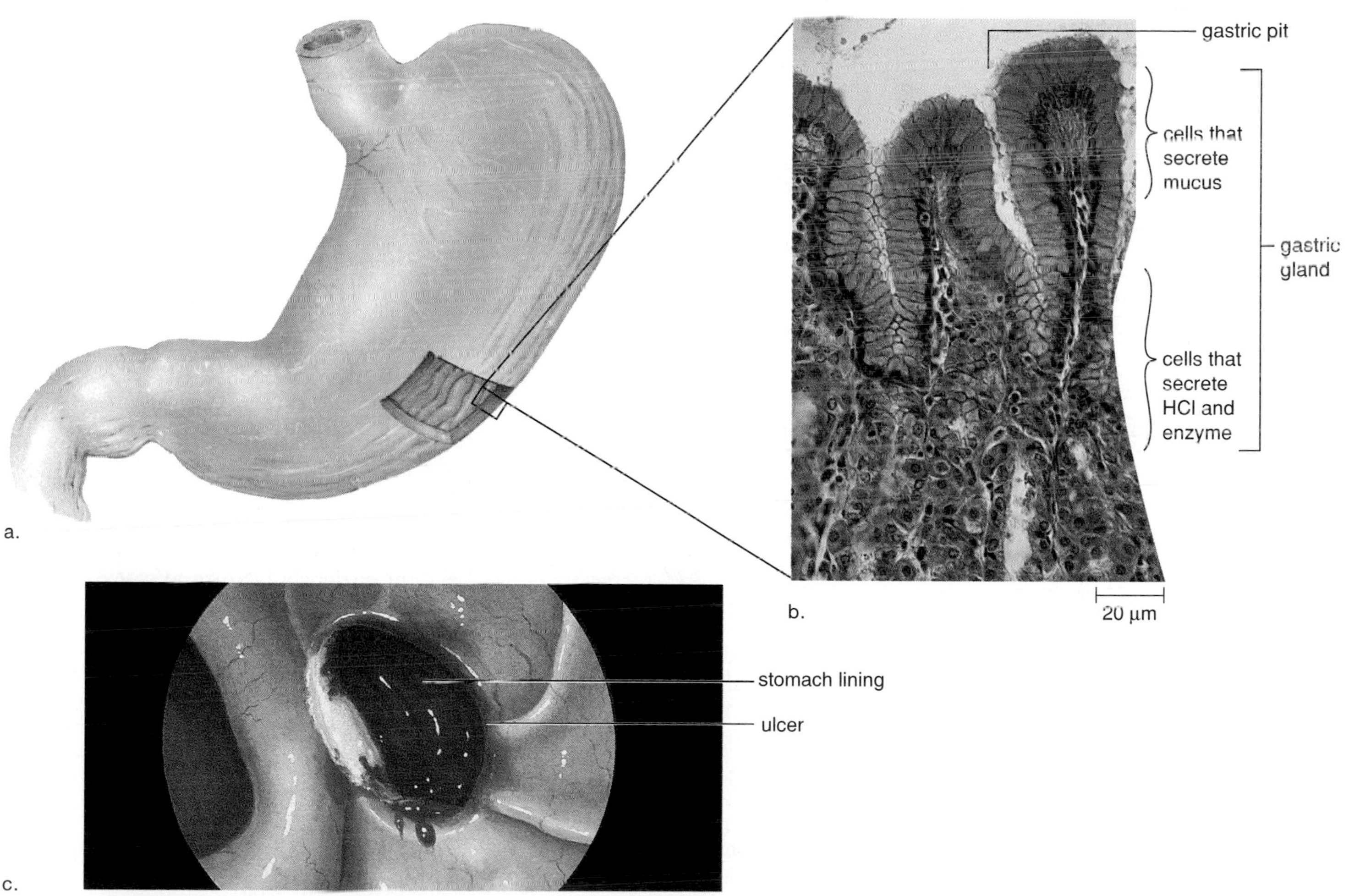

Figure 43.8 Anatomy of the stomach.
a. The stomach, which has thick walls, expands as it fills with food. **b.** The mucous membrane layer of its walls secretes mucus and contains gastric glands, which secrete a gastric juice active in the digestion of protein. **c.** View of a bleeding ulcer by using an endoscope (a tubular instrument bearing a tiny lens and a light source) that can be inserted into the abdominal cavity.

Table 43.2

Selected Digestive Enzymes

Reaction	Enzyme	Produced by	Site of Occurrence
Starch + $H_2O \rightarrow$ maltose	Salivary amylase Pancreatic amylase	Salivary glands Pancreas	Mouth Small intestine
Maltose + $H_2O \rightarrow$ glucose*	Maltase	Intestinal cells	Small intestine
Protein + $H_2O \rightarrow$ peptides	Pepsin Trypsin	Gastric glands Pancreas	Stomach Small intestine
Peptides + $H_2O \rightarrow$ amino acids*	Peptidases	Intestinal cells	Small intestine
Fat + $H_2O \rightarrow$ glycerol + fatty acids*	Lipase	Pancreas	Small intestine

** Absorbed by villi.*

Small Intestine Absorbs Nutrients

The human *small intestine* is a coiled tube about 3 meters long. (After death, the small intestine becomes as much as 6 meters long.) The mucous membrane layer of the small intestine has ridges and furrows that give it an almost corrugated appearance. On the surface of these ridges and furrows are small, fingerlike projections called **villi** [L. *villus*, shaggy hair]. Cells on the surfaces of the villi have minute projections called microvilli. The villi and microvilli greatly increase the effective surface area of the small intestine. If the small intestine were simply a smooth tube, it would have to be 500–600 meters long to have a comparable surface area.

When chyme enters the duodenum, proteins and carbohydrates are only partly digested, and fat digestion still needs to be carried out. Considerably more digestive activity is required before these nutrients can be absorbed through the intestinal wall. Two important accessory glands, the **liver,** the largest organ in the body and the **pancreas** located behind the stomach send secretions to the duodenum (see Fig. 43.5). The liver produces **bile,** which is stored in the **gallbladder,** an organ that acts on bile so that it is a thick, mucous-like material. Bile is sent to the duodenum by way of a duct. Bile looks green because it contains pigments that are products of hemoglobin breakdown. This green color is familiar to anyone who has observed how bruised tissue changes color. Hemoglobin within the bruise is breaking down into the same types of pigments found in bile. Bile also contains bile salts, which are *emulsifying* agents that break up fat into fat droplets.

$$\text{fat} \xrightarrow{\text{bile salts}} \text{fat droplets}$$

Fat droplets mix with the water and have more surface area for digestion by enzymes.

The pancreas produces pancreatic juice, which is conducted to the duodenum, also by way of a duct. Pancreatic juice contains sodium bicarbonate ($NaHCO_3$), which neutralizes the chyme and makes the pH of the small intestine slightly basic. A higher pH helps prevent autodigestion of the intestinal lining by pepsin and optimizes the pH for intestinal and pancreatic enzymes. Pancreatic juice also contains digestive enzymes that act on every major component of food (Table 43.2). **Pancreatic amylase** digests starch to maltose; **trypsin** and other enzymes digest protein to peptides; and **lipase** digests fat droplets to glycerol and fatty acids:

$$\text{starch} + H_2O \xrightarrow{\text{pancreatic amylase}} \text{maltose}$$

$$\text{protein} + H_2O \xrightarrow{\text{trypsin}} \text{peptides}$$

$$\text{fat droplets} + H_2O \xrightarrow{\text{lipase}} \text{glycerol} + \text{fatty acids}$$

The epithelial cells of the villi produce **intestinal enzymes,** which remain attached to the plasma membrane of microvilli. These enzymes complete the digestion of peptides and sugars. Peptides, which result from the first step in protein digestion, are digested by *peptidases* to amino acids. Maltose, which results from the first step in starch digestion, is digested by maltase to glucose:

$$\text{peptides} + H_2O \xrightarrow{\text{peptidases}} \text{amino acids}$$

$$\text{maltose} + H_2O \xrightarrow{\text{maltase}} \text{glucose}$$

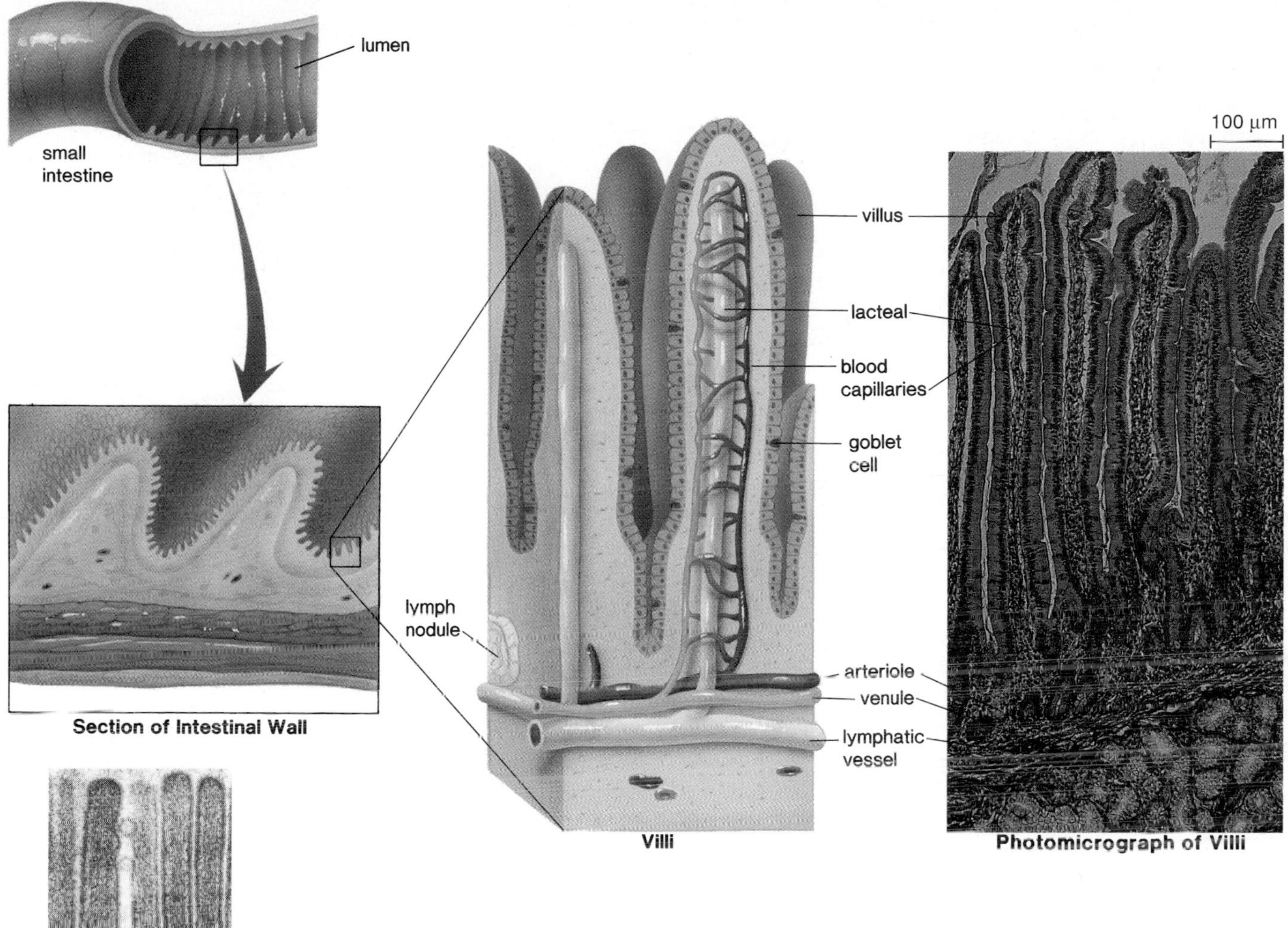

Figure 43.9 Anatomy of intestinal lining.
The products of digestion are absorbed by villi, fingerlike projections of the intestinal wall that contain blood vessels and a lacteal. Each villus has many microscopic extensions called microvilli.

Other disaccharides, each of which is acted upon by a specific enzyme, are digested in the small intestine. Table 43.2 reviews both the reactions required for the digestion of food and the various digestive enzymes discussed in this chapter.

Food is largely made up of carbohydrate (starch), protein, and fat. These very large macromolecules are broken down by digestive enzymes to small molecules that can be absorbed by intestinal villi.

Into the Villi

The small intestine is specialized for absorption. Molecules are absorbed by the huge number of villi that line the intestinal wall. Each villus has microscopic extensions called microvilli, which greatly increase its absorptive surface (Fig. 43.9). Each villus contains blood vessels and a lymphatic vessel called a **lacteal.** Sugars and amino acids enter villi cells and then are absorbed into the bloodstream. Because of their larger size, glycerol and fatty acids enter villi cells and are reassembled into fat molecules, which move into the lacteals.

Absorption continues until almost all products of digestion have been absorbed. Absorption involves both diffusion (passive transport) and active transport. Active transport requires an expenditure of cellular energy.

The wall of the small intestine is lined by villi. Sugars and amino acids enter blood vessels, and re-formed fats enter the lacteals of the villi.

A closer Look

Control of Digestive Juices

The study of the control of digestive gland secretion began in the late 1800s. At that time, Ivan Pavlov showed that dogs would begin to salivate at the ringing of a bell because they had learned to associate the sound of the bell with being fed. Pavlov's experiments demonstrated that even the thought of food can cause the nervous system to order the secretion of digestive juices. If food is present in the mouth, the stomach, and the small intestine, digestive secretion occurs because of simple reflex action. The presence of food sets off nerve impulses that travel to the brain. Thereafter, the brain stimulates the digestive glands to secrete.

In this century, investigators discovered that specific control of digestive secretions is achieved by hormones. A *hormone* is a substance produced by one set of cells that affects a different set of cells, the so-called target cells. Hormones are transported by the bloodstream. For example, when a person has eaten a meal particularly rich in protein, the gastric glands of the stomach wall produce the hormone gastrin (Fig. 43A). Gastrin enters the bloodstream, and soon stomach churning and the secretory activity of gastric glands increase.

Cells of the duodenal wall also produce hormones, two of which are of particular interest—secretin and CCK (cholecystokinin). Acid, especially hydrochloric acid (HCl) present in chyme, stimulates the release of secretin, while partially digested protein and fat stimulate the release of CCK. Soon after these hormones enter the bloodstream, the pancreas increases its output of pancreatic juice and the liver increases its output of bile. The gallbladder contracts to release bile.

Another hormone produced by the duodenal wall, GIP (gastric inhibitory peptide), works opposite to gastrin—it inhibits gastric gland secretion and stomach motility. This is not surprising, because the body's hormones often have opposite effects.

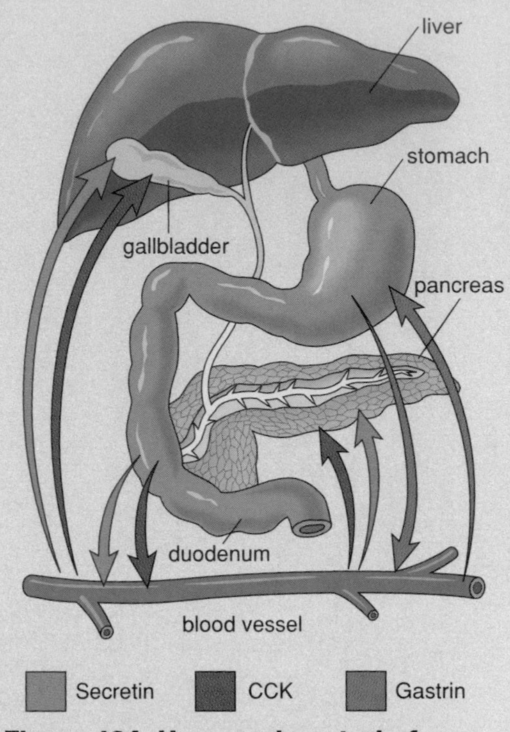

Figure 43A Hormonal control of digestive gland secretions.
The arrows to the blood vessel are secretion into the bloodstream, and the away arrows are hormone reception by target organs.

Two Accessory Organs Help Out

The pancreas and the liver are accessory organs of digestion along with the teeth, salivary glands, and gallbladder. Figure 43.5 shows how ducts conduct pancreatic juice from the pancreas, and bile from the liver, to the duodenum.

Pancreas Secretes Juices

The pancreas lies deep in the abdominal cavity, resting on the posterior abdominal wall. It is an elongated and somewhat flattened organ that is an endocrine gland when it produces and secretes insulin and glucagon into the bloodstream. It is an exocrine gland when it produces and secretes pancreatic juice into the duodenum. Just now we are interested in its exocrine function—most of its cells produce pancreatic juice, which contains enzymes for digestion of all types of food (see Table 43.2). The enzymes travel by way of ducts to the duodenum of the small intestine. Regulation of pancreatic secretion is discussed in the reading on this page.

Liver Secretes Bile

Blood vessels from the large and small intestines merge to form the hepatic portal vein, which leads to the liver (Fig. 43.10). The liver has numerous functions, including the following:

1. Detoxifies the blood by removing and metabolizing poisonous substances.
2. Makes the plasma proteins such as albumin and fibrinogen.
3. Destroys old red blood cells and converts hemoglobin to the breakdown products in bile (bilirubin and biliverdin).
4. Produces bile, which is stored in the gallbladder before entering the small intestine, where it emulsifies fats.
5. Stores glucose as glycogen and breaks down glycogen to glucose between meals to maintain a constant glucose concentration in the blood.
6. Produces urea from amino groups and ammonia.

Here we are interested in the last two functions listed. The liver helps maintain the glucose concentration in blood at about 0.1% by removing excess glucose from the hepatic portal vein and storing it as glycogen. Between meals, glycogen is broken down and glucose enters the hepatic vein. Glycogen is sometimes called *animal starch* because both

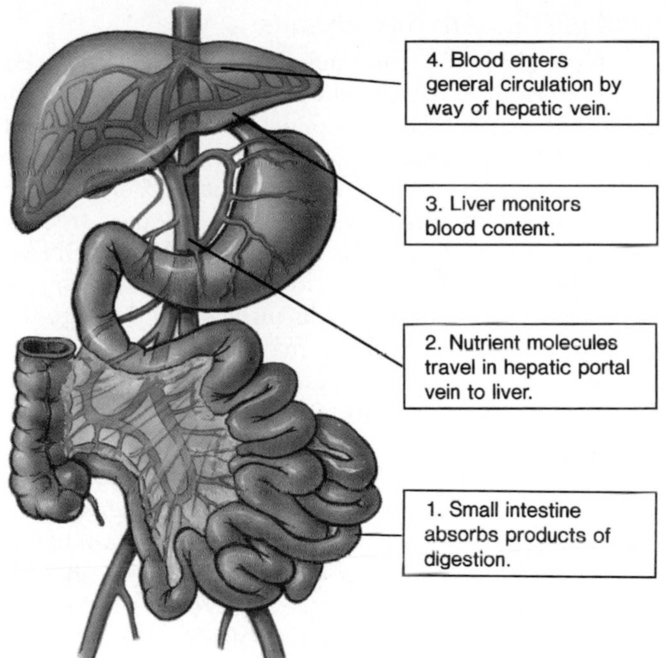
4. Blood enters general circulation by way of hepatic vein.
3. Liver monitors blood content.
2. Nutrient molecules travel in hepatic portal vein to liver.
1. Small intestine absorbs products of digestion.

Figure 43.10 Hepatic portal system.
The hepatic portal vein takes the products of digestion from the digestive system to the liver, where they are processed before entering the circulatory system proper.

starch and glycogen are made up of glucose molecules. If by chance the supply of glycogen and glucose runs short, the liver converts amino acids to glucose molecules. Amino acids contain nitrogen in the form of amino groups, whereas glucose contains only carbon, oxygen, and hydrogen. Therefore, before amino acids can be converted to glucose molecules, *deamination,* the removal of amino groups from amino acids must occur. By a complex metabolic pathway, the liver then converts the amino groups to urea, the most common nitrogenous waste product of humans. After urea is formed in the liver, it is transported by the bloodstream to the kidneys, and eventually it is excreted.

Blood from the small intestine enters the hepatic portal vein, which goes to the liver, a vital organ that has numerous important functions.

Liver Disorders Jaundice, hepatitis, and cirrhosis are three serious diseases that affect the entire liver and hinder its ability to repair itself. When a person is jaundiced, there is a yellowish tint to the skin due to an abnormally large amount of bilirubin in the blood. In *hemolytic jaundice,* red blood cells are broken down in abnormally large amounts; in *obstructive jaundice,* there is an obstruction of the bile duct or damage to the liver cells. Obstructive jaundice often occurs when crystals of cholesterol precipitate out of bile and form gallstones.

Jaundice can also result from *viral hepatitis,* a term that includes hepatitis A, hepatitis B, and hepatitis C. Hepatitis A is most often caused by eating contaminated food. Hepatitis B and C are commonly spread by blood transfusions, kidney dialysis, and injection with unsterilized needles. All three types of hepatitis can be spread by sexual contact.

Cirrhosis is a chronic liver disease in which the organ first becomes fatty. Liver tissue is then replaced by inactive fibrous scar tissue. In alcoholics, who often get cirrhosis, the condition most likely is caused at least in part by the excessive amounts of alcohol the liver is forced to break down.

Large Intestine Absorbs Water

The **large intestine** has four parts: the cecum, colon, rectum, and anal canal. The **colon** is subdivided into the ascending, transverse, and descending colon (see Fig. 43.5). About 1.5 liters of water enter the digestive tract daily as a result of eating and drinking. An additional 8.5 liters enter the digestive tract each day carrying the various substances secreted by the digestive glands. About 95% of this water is absorbed by the small intestine, and much of the remaining portion is absorbed into cells of the colon. If this water is not reabsorbed, then *diarrhea* will result, which can lead to serious dehydration and ion loss, especially in children.

The large intestine also functions in ion regulation. Notably, the large intestine absorbs some salts from the material passing through it. *Vitamin K,* produced by intestinal bacteria, is also absorbed in the colon.

Digestive wastes (feces) eventually leave the body through the anus, the opening of the anal canal. Feces are about 75% water and 25% solid matter. Almost one-third of this solid matter is made up of intestinal bacteria. The remainder is undigested plant material, fats, waste products (such as bile pigments), inorganic material, mucus, and dead cells from the intestinal lining.

Two serious medical conditions are associated with the large intestine. The small intestine joins the large intestine at the cecum, a blind sac. The cecum has a small projection about the size of the little finger, called the appendix. In the case of *appendicitis,* the appendix becomes infected and so filled with fluid that it may burst. If an infected appendix bursts before it can be removed, it can lead to a serious, generalized infection of the abdominal lining, called peritonitis.

The colon is subject to the development of polyps, which are small growths arising from the epithelial lining. Polyps, whether they are benign or cancerous, can be removed individually. A low-fat, high-fiber diet, which promotes regularity, is recommended as a protective measure against colon cancer because mutagenic agents, which might promote cancer, are quickly eliminated.

The large intestine does not produce digestive enzymes; it absorbs water and salts.

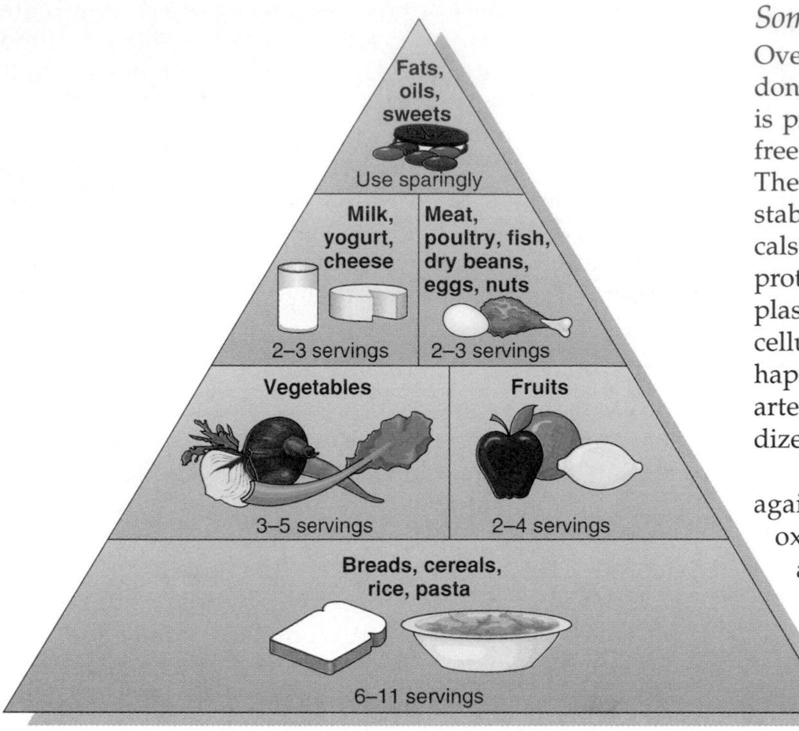

Figure 43.11 Ideal American diet.
The U.S. Department of Agriculture uses a pyramid to show the ideal American diet because it emphasizes the importance of grains in the diet and the undesirability of fats, oils, and sweets.
Source: U.S. Department of Agriculture.

43.3 Nutrition Affects Health

To be sure that the essential nutrients are included in the diet, it is necessary to eat a balanced diet. A balanced diet includes a variety of foods proportioned as shown in Figure 43.11.

Vitamins Help Metabolism

Vitamins are organic compounds (other than carbohydrate, fat, and protein) that the body is unable to produce but uses for metabolic purposes. Many vitamins are portions of coenzymes, which are enzyme helpers. For example, niacin is part of the coenzyme NAD^+ and riboflavin is part of FAD. Coenzymes are needed in only small amounts because each can be used over and over again. Not all vitamins are coenzymes; vitamin A, for example, is a precursor for the visual pigment that prevents night blindness. It has been known for several years that if vitamins are lacking in the diet, various symptoms develop. Altogether there are 13 vitamins, which are divided into those that are fat soluble and those that are water soluble (Table 43.3).

Some Vitamins Are Antioxidants

Over the past 20 years, numerous statistical studies have been done to determine whether a diet rich in fruits and vegetables is protective against cancer. Cellular metabolism generates free radicals, unstable molecules that carry an extra electron. The most common free radical in cells is oxygen in the unstable form, O_3^-. In order to stabilize themselves, free radicals donate an electron to another molecule like DNA, or proteins, including enzymes, and lipids, which are found in plasma membranes. Such attacks most likely damage these cellular molecules and thereby may lead to disorders, perhaps even cancer. Also, it appears that plaque formation in arteries may begin when arterial linings are injured by oxidized LDL cholesterol.

Vitamins C, E, and A are believed to defend the body against free radicals and, therefore, they are termed antioxidants. These vitamins are especially abundant in fruits and vegetables. The new dietary guidelines shown in Figure 43.11 suggest that we eat five servings of fruits and vegetables a day. It is not difficult to achieve the goal of having five servings of fruits and vegetables a day because a serving can be salad greens, raw or cooked vegetables, dried fruit, and fruit juice in addition to traditional apples and oranges and such.

Dietary supplements may provide a potential safeguard against cancer and cardiovascular disease, but it is important to realize that it is not appropriate to take supplements instead of improving intake of fruits and vegetables. There are many beneficial compounds in fruits that cannot be obtained from a vitamin pill. These substances enhance each other's absorption or action and also perform independent biological functions.

Vitamin D Makes Bones Strong

Skin cells contain a precursor cholesterol molecule that is converted to vitamin D after UV exposure. Since only a small amount of UV radiation is needed to change the precursor molecule to vitamin D, the skin should not be exposed unnecessarily. Vitamin D leaves the skin and is modified first in the kidneys and then in the liver until finally it is activated into a form called calcitriol. Calcitriol circulates throughout the body, regulating calcium uptake and metabolism. It promotes the absorption of calcium by the intestines, and the absence of vitamin D leads to rickets in children. Rickets, characterized by a bowing of the legs, is caused by defective mineralization of the skeleton. Most milk today is fortified with vitamin D, which helps prevent the occurrence of rickets.

Vitamins are essential to cellular metabolism; many are protective against identifiable illnesses and conditions.

Table 43.3

Vitamins: Their Role in the Body and Their Food Sources

Vitamin	Major Role in Body	Good Food Sources
Fat Soluble		
Vitamin A	Vision, health of skin, hair, bones, and sex organs	Deep green or yellow vegetables, dairy products
Vitamin D	Health of bones and teeth	Dairy products, tuna, eggs
Vitamin E	Strengthening of red blood cell membrane	Green leafy vegetables, whole grains
Vitamin K	Clotting of blood, bone metabolism	Green leafy vegetables, cabbage, cauliflower
Water Soluble		
Thiamine (B_1)	Carbohydrate metabolism	Pork, whole grains
Riboflavin (B_2)	Energy metabolism	Whole grains, milk, green vegetables
Niacin (B_3)	Energy metabolism	Organ meats, whole grains
Pyridoxine (B_6)	Amino acid metabolism	Meats, fish, whole grains
Vitamin B_{12}	Red blood cell formation	Meats, dairy foods
Biotin	Carbohydrate metabolism	Eggs, most foods
Folic acid	Formation of red blood cells, DNA and RNA	Green leafy vegetables, nuts, whole grains
Pantothenic acid	Energy metabolism	Most foods
Vitamin C	Collagen formation	Citrus fruits, tomatoes

Table 43.4

Minerals: Their Role in the Body and Their Food Sources

Mineral	Major Role in Body	Good Food Sources
Macrominerals		
Calcium (Ca)	Strong bones and teeth, nerve conduction, muscle contraction	Dairy products, green leafy vegetables
Phosphorus (P)	Strong bones and teeth	Meat, dairy products, whole grains
Potassium (K)	Nerve conduction, muscle contraction	Many fruits and vegetables
Sodium (Na)	Nerve conduction, pH balance	Table salt
Chlorine (Cl)	Water balance	Table salt
Magnesium (Mg)	Protein synthesis	Whole grains, green leafy vegetables
Microminerals		
Zinc (Zn)	Wound healing, tissue growth	Whole grains, legumes, meats
Iron (Fe)	Hemoglobin synthesis	Whole grains, legumes, eggs, green leafy vegetables
Fluorine (F)	Strong bones and teeth	Fluoridated drinking water, tea
Copper (Cu)	Hemoglobin synthesis	Seafood, whole grains, legumes
Iodine (I)	Thyroid hormone synthesis	Iodized table salt, seafood

Minerals Are for Structure and Function

In addition to vitamins, various minerals are also required by the body (Table 43.4). Some minerals (calcium, phosphorus, potassium, sulfur, sodium, chlorine, and magnesium) are recommended in amounts more than 100 mg per day. These macrominerals serve as constituents of cells and body fluids, and as structural components of tissues. Calcium is needed for the construction of bones and teeth and also for nerve conduction and muscle contraction.

Other minerals (iron, manganese, copper, iodine, cobalt, and zinc) are recommended in amounts of less than 20 mg per day. These microminerals are more likely to have very specific functions. Iron is needed for the production of hemoglobin, and iodine is used in the production of thyroxine, a hormone produced by the thyroid glands. As research continues, more and more elements have been added to the list of those considered essential. During the past three decades, very small amounts of molybdenum, selenium, chromium, nickel, vanadium, silicon, and even arsenic have been found to be essential to good health.

Some individuals do not receive enough iron (especially women), calcium, magnesium, or zinc in their diets.

Adult females need more iron in their diet than males (18 mg compared to 10 mg) because they lose hemoglobin each month during menstruation. Stress can bring on a magnesium deficiency, and a vegetarian diet may lack zinc, which is usually obtained from meat. A varied and complete diet, however, usually supplies the recommended daily allowances for minerals.

Calcium Builds Bones

There is much interest in calcium supplements to counteract the development of *osteoporosis* [Gk. *osteon,* bone, and *poros,* hole, passage], a degenerative bone disease that afflicts an estimated one-fourth of older men and one-half of older women in the United States. Osteoporosis develops because bone-eating cells, called osteoclasts, are more active than bone-forming cells, called osteoblasts. Therefore, the bones are porous, and they break easily because they lack sufficient calcium. Due to recent studies that show consuming more calcium does slow bone loss in elderly people, the guidelines have been revised. A calcium intake of 1,000–1,500 mg a day is recommended. To achieve this amount, supplemental calcium is most likely necessary.

Exercise is also effective in building bone mass. There are also medications that slow bone loss while increasing skeletal mass but these should be taken only when under a physician's care.

Sodium Causes Water Retention

The recommended amount of sodium intake per day is 400–3,300 mg, and the average American intake is 4,000–4,700 mg, mostly as salt (sodium chloride). In recent years, this imbalance has caused concern because high-sodium intake has been linked to hypertension in some people. About one-third of the sodium we consume occurs naturally in foods; another third is added during commercial processing; and we add the last third either during home cooking or at the table in the form of table salt.

Clearly, it is possible for us to cut down on the amount of sodium in the diet by reducing the amount of salt we eat.

Excess sodium in the diet can lead to hypertension; therefore, we should reduce sodium intake.

connecting concepts

The liver belongs to which organ system? Trying to answer this question illustrates that the body is arbitrarily divided into various systems. Granted, a good argument can be made for saying that the mouth, esophagus, stomach, and the intestines are in the digestive system. But actually, even these organs assist other body systems. The stomach and small intestine contribute to the endocrine system, for example, by producing hormones (see page 782). And how could muscles and nerves function without a supply of calcium from the digestive tract? Or for that matter, any other system in the body without a supply of nutrients absorbed by the digestive tract and distributed by the cardiovascular system?

The liver has so many functions it really belongs to many systems of the body. Doesn't it belong to the cardiovascular system because it produces plasma proteins and the urinary system because it produces urea, as well as the digestive system because it produces bile? The liver is a vital organ, meaning that we cannot live without it. Among its many functions, it detoxifies blood by removing and metabolizing poisonous substances. In alcoholics, cirrhosis of the liver is due to malnutrition and to the excessive amounts of alcohol the liver is forced to break down. First the organ becomes fatty, and then liver tissue is replaced by inactive fibrous scar tissue.

The liver has amazing regenerative powers and in some instances can recover if the rate of regeneration exceeds the rate of damage. If liver failure is present, however, there may not be enough time to let the liver heal itself. Liver transplantation is usually the preferred treatment for liver failure, but artificial livers have been developed and tried in a few cases. One type of artifical liver consists of a cartridge that contains liver cells. The patient's blood passes through a cellulose acetate tube of the cartridge and is serviced in the same manner as with a normal liver. In the meantime, the patient's liver has a chance to recover.

Summary

43.1 Comparing Digestive Tracts

Some animals (e.g., planaria) have an incomplete digestive tract, which shows little specialization of parts. Other animals (e.g., earthworms) have a complete digestive tract, which does show specialization of parts.

Some animals are continuous feeders (e.g., clams, which are sessile filter feeders); others are discontinuous feeders (e.g., squid). Discontinuous feeders need a storage area for food.

All mammals have teeth. Herbivores need teeth that can clip off plant material and grind it up. Also, the herbivore's stomach contains bacteria that can digest cellulose.

Carnivores need teeth that can tear and rip meat into pieces. Meat is easier to assimilate, so the digestive system of carnivores has less specialization of parts and a shorter intestine than that of herbivores.

43.2 Humans Have a Complete Tract

In the human digestive tract, food is chewed and manipulated in the mouth, where salivary glands secrete saliva. Saliva contains salivary amylase, which begins carbohydrate digestion.

Food then passes to the pharynx and down the esophagus by peristalsis to the stomach. The stomach stores and mixes food with mucus and gastric juice to produce chyme. Pepsin begins protein digestion here.

Chyme gradually enters the duodenum where bile, pancreatic juice, and the intestinal secretions are found. Enzymes in the small intestine hydrolyze all of the organic nutrients. Table 43.2 summarizes the enzymes involved in digesting food.

There are three hormones that regulate digestive-tract secretions: gastrin, which stimulates acid- and enzyme-secreting cells in the stomach; secretin, which stimulates the pancreas to release sodium bicarbonate; and CCK (cholecystokinin), which stimulates the gallbladder to release bile and the pancreas to release digestive enzymes.

The pancreas produces digestive enzymes for every major component of food.

The liver produces bile, which is stored in the gallbladder. The liver is also involved in the processing of absorbed nutrient molecules and in maintaining the blood concentration of nutrient molecules, such as glucose. The liver converts ammonia to urea and breaks down toxins.

Most nutrient absorption takes place in the small intestine, but some water and minerals are absorbed in the colon. Digestive wastes leave the colon by way of the anus.

43.3 Nutrition Affects Health

A balanced diet is required for good health. Food should provide us with all the necessary vitamins, minerals, amino acids, fatty acids, and an adequate amount of energy.

Reviewing the Chapter

1. Contrast the incomplete gut with the complete gut, using the planarian and earthworm as examples. 774
2. Contrast a continuous feeder with a discontinuous feeder, using the clam and squid as examples. 775
3. Contrast the dentition of the mammalian herbivore with that of the mammalian carnivore, using the horse and lion as examples. 776
4. List the parts of the human digestive tract, anatomically describe them, and state the contribution of each to the digestive process. 777–81
5. Assume that you have just eaten a ham sandwich. Discuss the digestion of the contents of the sandwich. 777–81
6. Discuss the absorption of the products of digestion into the cardiovascular system. 781
7. What are gastrin, secretin, and CCK (cholecystokinin)? Where are they produced, and what are their functions? 782
8. State the location and describe the functions of both the pancreas and the liver. 782–83
9. Explain why carbohydrates, fats, proteins, vitamins, and minerals are all necessary to good nutrition. 784–85

Testing Yourself

Choose the best answer for each question.

1. Animals that feed discontinuously
   a. must have digestive tracts that permit storage.
   b. are always filter feeders.
   c. exhibit extremely rapid digestion.
   d. have a nonspecialized digestive tract.
2. In which of the following types of animals would you expect the digestive tract to be more complex?
   a. those with a single opening for the entrance of food and exit of wastes
   b. those with two openings, one serving as an entrance and the other as an exit
   c. only those complex animals that also have a respiratory system
   d. Both b and c are correct.
3. The typhlosole within the gut of an earthworm compares best to which of these organs in humans?
   a. teeth in the mouth
   b. esophagus in the thoracic cavity
   c. folds in the stomach
   d. villi in the small intestine
4. Which of these animals is a continuous feeder with a complete gut?
   a. planarian
   b. clam
   c. squid
   d. lion
5. The products of digestion are
   a. large macromolecules needed by the body.
   b. enzymes needed to digest food.
   c. small nutrient molecules that can be absorbed.
   d. regulatory hormones of various kinds.
6. Which of these could be absorbed directly without need of digestion?
   a. glucose
   b. fat
   c. protein
   d. nucleic acid
7. Which association is incorrect?
   a. protein—trypsin
   b. fat—lipase
   c. maltose—pepsin
   d. starch—amylase
8. Most of the absorption of the products of digestion takes place in humans across the
   a. squamous epithelium of the esophagus.
   b. convoluted walls of the stomach.
   c. fingerlike villi of the small intestine.
   d. smooth wall of the large intestine.
9. The hepatic portal vein is located between the
   a. hepatic vein and the vena cava.
   b. mouth and the stomach.
   c. pancreas and the small intestine.
   d. small intestine and the liver.
10. Bile in humans
    a. is an important enzyme for the digestion of fats.
    b. is made by the gallbladder.
    c. emulsifies fat.
    d. All of these are correct.
11. Which of these is not a function of the liver in adults?
    a. produce bile
    b. store glucose
    c. produce urea
    d. make red blood cells
12. The large intestine in humans
    a. digests all types of food.
    b. is the longest part of the intestinal tract.
    c. absorbs water.
    d. is connected to the stomach.

13. Predict and explain the expected digestive results per test tube for this experiment.

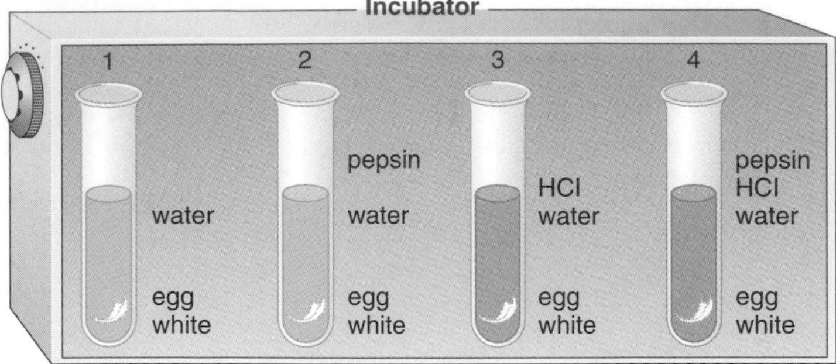

Applying the Concepts

1. *All systems of the animal's body contribute to homeostasis.*
   List several ways in which the human digestive system contributes to homeostasis.
2. *Organisms are adapted to obtaining a share of available resources.*
   Animals vary according to how they obtain and process food. Use the grasshopper, the earthworm, the clam, or the squid to support this concept.
3. *Forms of life depend on each other for materials and energy.*
   Explain how you are dependent on plants as a source of materials and energy.

Using Technology

Your study of digestion and nutrition is supported by these available technologies:

Exploring the Internet
The Mader Home Page provides resources for and help with studying this chapter.

http://www.mhhe.com/sciencemath/biology/mader/
(Click on Biology.)

The Dynamic Human CD-ROM
Digestive System

Explorations in Human Biology CD-ROM
Diet and Weight Loss (#7)

Life Science Animations Video
Video #4: Animal Biology II
Peristalsis (#33)

Understanding the Terms

Match the terms to these definitions:

a. __________ Essential requirement in the diet, needed in small amounts. They are often part of coenzymes.
b. __________ Fat-digesting enzyme secreted by the pancreas.
c. __________ Lymphatic vessel in an intestinal villus, it aids in the absorption of fats.
d. __________ Muscular tube for moving swallowed food from the pharynx to the stomach.
e. __________ Organ attached to the liver that serves to store and concentrate bile.
f. __________ Protein-digesting enzyme secreted by gastric glands.
g. __________ Protein-digesting enzyme secreted by the pancreas.
h. __________ Rhythmic, wavelike contraction that moves food through the digestive tract.
i. __________ Secretion of the liver that is temporarily stored and concentrated in the gallbladder before being released into the small intestine, where it emulsifies fat.
j. __________ Small, fingerlike projection of the inner small intestinal wall.

Respiration

CHAPTER

Chapter Concepts

The bottle-nosed dolphin, *Tursiops*, and boy, *Homo*, are mammals that breathe air.

In the last chapter, we saw that many animals have a storage area for food that allows them to eat first and digest later. When it comes to gases, animals have no such storage areas; oxygen and carbon dioxide are continually exchanged with the environment.

In many small aquatic organisms, gas exchange occurs directly between the animal's cells and the environment. Dissolved oxygen in water diffuses into cells, while carbon dioxide diffuses from the cells into the water. In these species, no specialized respiratory structures are required. Other aquatic species have gills, layers of very thin tissues filled with blood vessels. Water moving through the gills supplies oxygen and removes carbon dioxide. Among vertebrate animals, fishes and some amphibians rely on gills for respiration. Most adult amphibians have lungs, but also rely on a thin skin for gas exchange with the environment.

Reptiles, birds, and mammals have efficient vascularized lungs for respiration. Air tubes bring oxygen into the lungs where gas exchange occurs. Diffusion is the basic mechanism for gas exchange in all animals, whether the animal is a unicellular protozoan, aquatic invertebrate, fish, or mammal.

44.1 How Animals Exchange Gases

Breathing is only the first step of respiration, which can be said to include the following steps in terrestrial vertebrates:

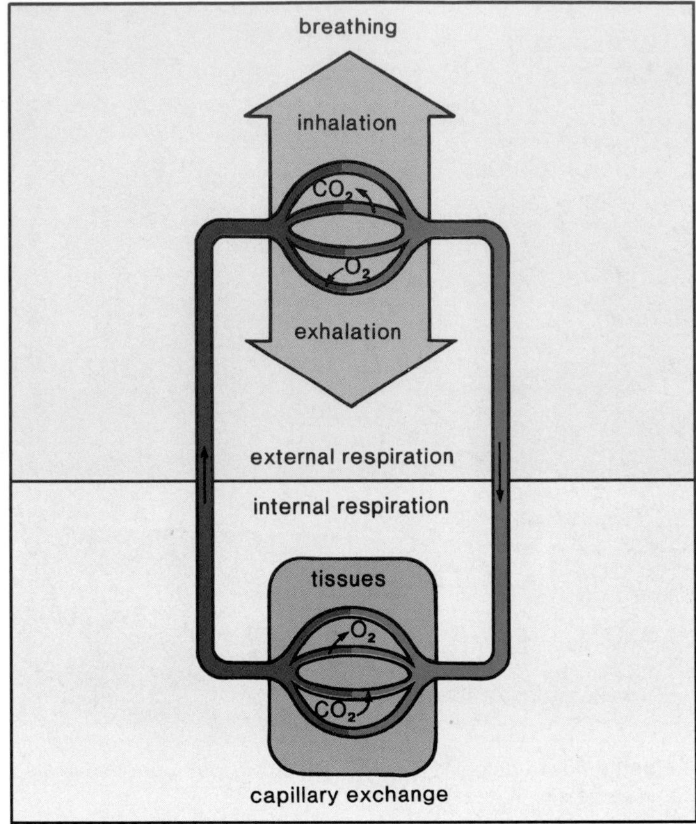

- Breathing: includes inspiration (entrance of air into lungs) and expiration (exit of air from the lungs).
- External respiration: gas exchange between air and blood.
- Internal respiration: gas exchange between blood and tissue fluid.
- Aerobic respiration: production of ATP (adenosine triphosphate) in cells.

Gas exchange takes place by the physical process of diffusion. For diffusion to be effective, the gas-exchange region must be (1) moist, (2) thin, and (3) large in relation to the size of the body. Some animals are small and shaped in a way that allows the surface of the animal to be the gas-exchange surface. Other animals are complex and have a specialized gas-exchange surface. The effectiveness of diffusion is enhanced by vascularization, and delivery to cells is promoted when the blood contains a respiratory pigment such as hemoglobin.

In the Water

It is more difficult for animals to obtain oxygen from water than from air. Water fully saturated with air contains only a fraction of the amount of oxygen in the same volume of air. Also, water is more dense than air. Therefore, aquatic animals expend more energy to breathe than do terrestrial animals. Fishes use up to 25% of their energy output to breathe, while terrestrial mammals use only 1–2% of their energy output.

Hydras and planaria have a large surface area in comparison to their size (Fig. 44.1). This makes it possible for most of their cells to exchange gases directly with the environment. In hydras, the outer layer of cells is in contact with the external environment, and the inner layer can exchange gases with the water in the gastrovascular cavity. In planaria, the flattened body permits cells to exchange gases with the external environment.

A tubular shape also provides a surface area adequate for gas-exchange purposes. In addition to a tubular shape, polychaete worms have extensions of the body wall called

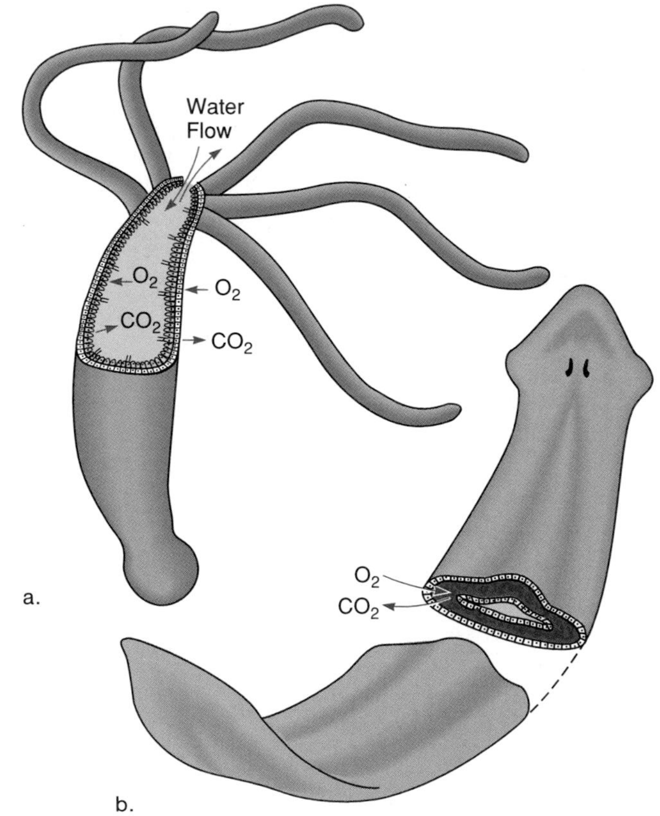

Figure 44.1 Animal shapes and gas exchange. Some small aquatic animals use the body surface for gas exchange. This works because the body surface is large compared to the size of the animal. **a.** In hydras, every cell is near a source of oxygen. **b.** The flattened body of planaria allows most cells to carry out gas exchange with the external environment.

parapodia, which are vascularized. Quite often, aquatic animals have **gills,** which are finely divided and vascularized outgrowths of either the outer or the inner body surface. Among mollusks, such as clams and many snails, water is drawn into the mantle cavity, where it passes through the gills. In decapod crustacea, the gills are located in brachial chambers covered by the exoskeleton. Water is kept moving by the action of specialized appendages located near the mouth.

Among vertebrates, the gills of bony fishes are outward extensions of the pharynx (Fig. 44.2). Ventilation is brought about by the combined action of the mouth and gill covers, or opercula (sing., operculum). When the mouth is open, the opercula are closed and water is drawn in. Then the mouth closes and the opercula open, drawing the water from the pharynx through the gill slits located between the gill arches. On the outside of the gill arches, the gills are composed of *filaments* that are folded into platelike *lamellae.*

In the capillaries of each lamella, the blood flows in a direction opposite to the movement of water across the gills. With a *countercurrent flow* as blood gains oxygen, it always encounters water having an even higher oxygen content and no equilibrium point is reached. If the flow were concurrent (oxygen rich water and oxygen poor blood flow in the same direction) an equilibrium point would occur and only about half the oxygen in the water would be captured. A countercurrent mechanism allows about 80–90% of the initial dissolved oxygen in water to be extracted.

Small aquatic animals sometimes use the body surface for gas exchange, but many larger ones have localized gas-exchange surfaces known as gills.

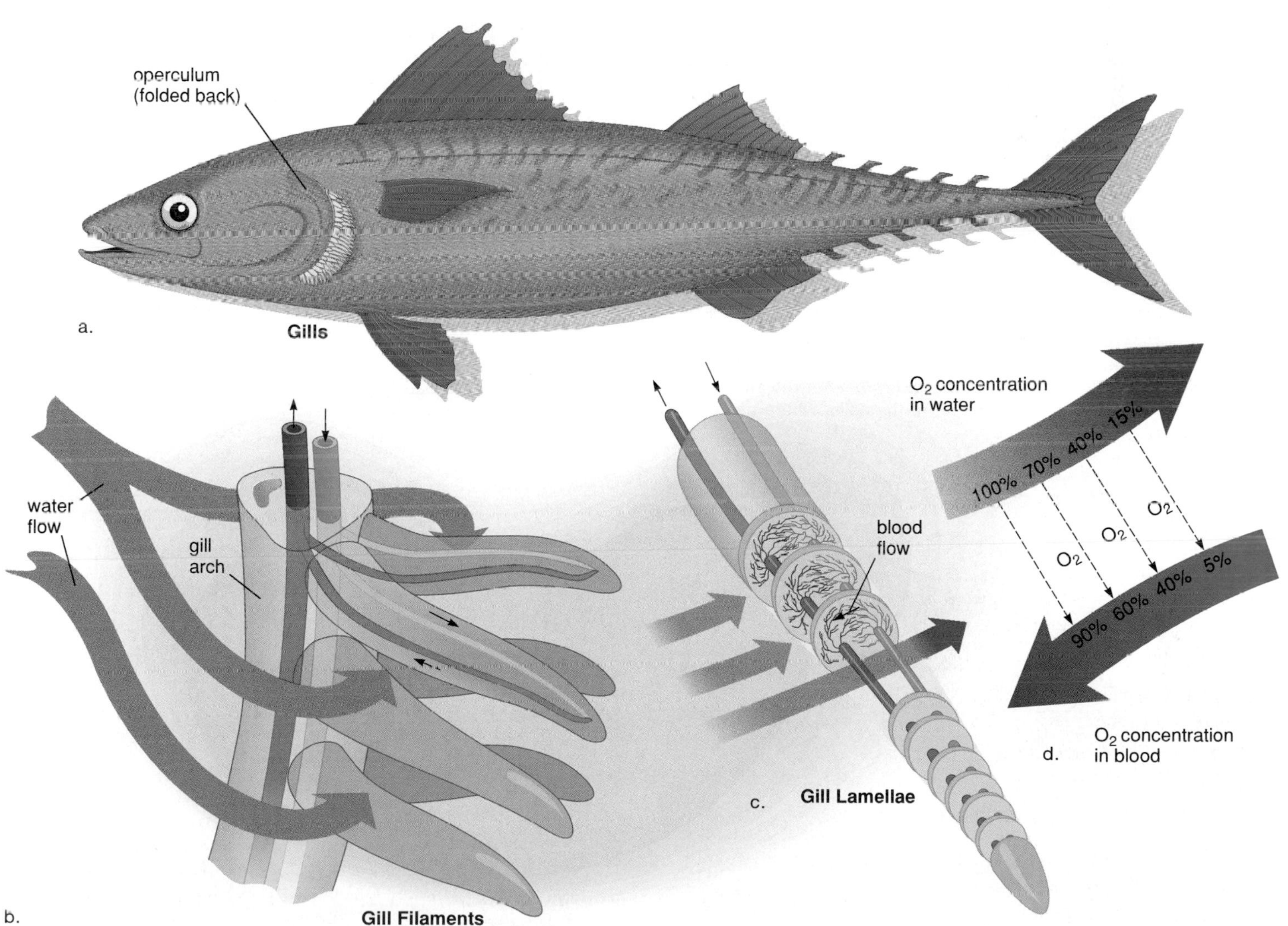

Figure 44.2 Anatomy of gills in bony fishes.
a. The operculum (folded back) covers and protects several layers of delicate gills. b. Each layer of gills has two rows of gill filaments. c. Each filament has many thin, platelike lamellae. Gases are exchanged between the capillaries inside the lamellae and the water that flows between the lamellae. d. Blood in the capillaries flows in the direction opposite to that of the water. Blood takes up almost all of the oxygen in the water as a result of this countercurrent flow.

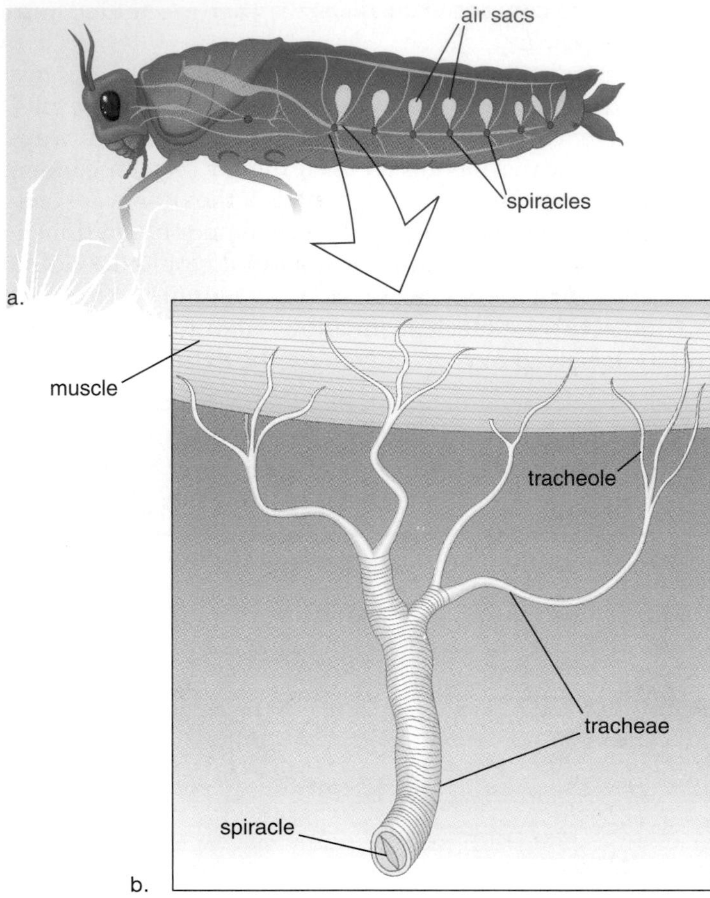

Figure 44.3 Tracheal system of insects.
a. A system of air tubes extends throughout the body of an insect and they, rather than blood, carry oxygen to the cells. **b.** Air enters the tracheae at openings called spiracles. From here, the air moves to the smaller tracheoles, which take it to the cells, where gas exchange takes place.

On the Land

Air is a rich source of oxygen compared to water; however, it does have a drying effect on respiratory surfaces. A human loses about 350 ml of water per day when the air has a relative humidity of only 50%.

The earthworm is an example of an invertebrate terrestrial animal that uses its body surface for respiration. An earthworm expends much energy to keep its body surface moist by secreting mucus and by releasing fluids from excretory pores. Further, the worm is behaviorally adapted to remain in damp soil during the day, when the air is driest.

Insects and certain other terrestrial arthropods have a respiratory system known as a tracheal system (Fig. 44.3). Oxygen enters the tracheae at spiracles, which are valvelike openings on each side of the body. The tracheae branch and then rebranch, ending in tiny channels, the tracheoles, that are in direct contact with the body cells. Larger insects have a ventilation system to keep the air moving in and out of the trachea. Many have air sacs located near major muscles; contraction of these muscles causes the air sacs to empty, and relaxation causes the air sacs to expand and draw air in. The tracheal system of insects is very effective in delivering oxygen to the cells so that the circulatory system plays no role in gas transport.

Terrestrial vertebrates, in particular, have evolved **lungs,** which are vascularized outgrowths from the lower pharyngeal region. The lungs of amphibians are simple, saclike structures (Fig. 44.4*a*). Many amphibians possess a short trachea, which divides into two bronchi that open

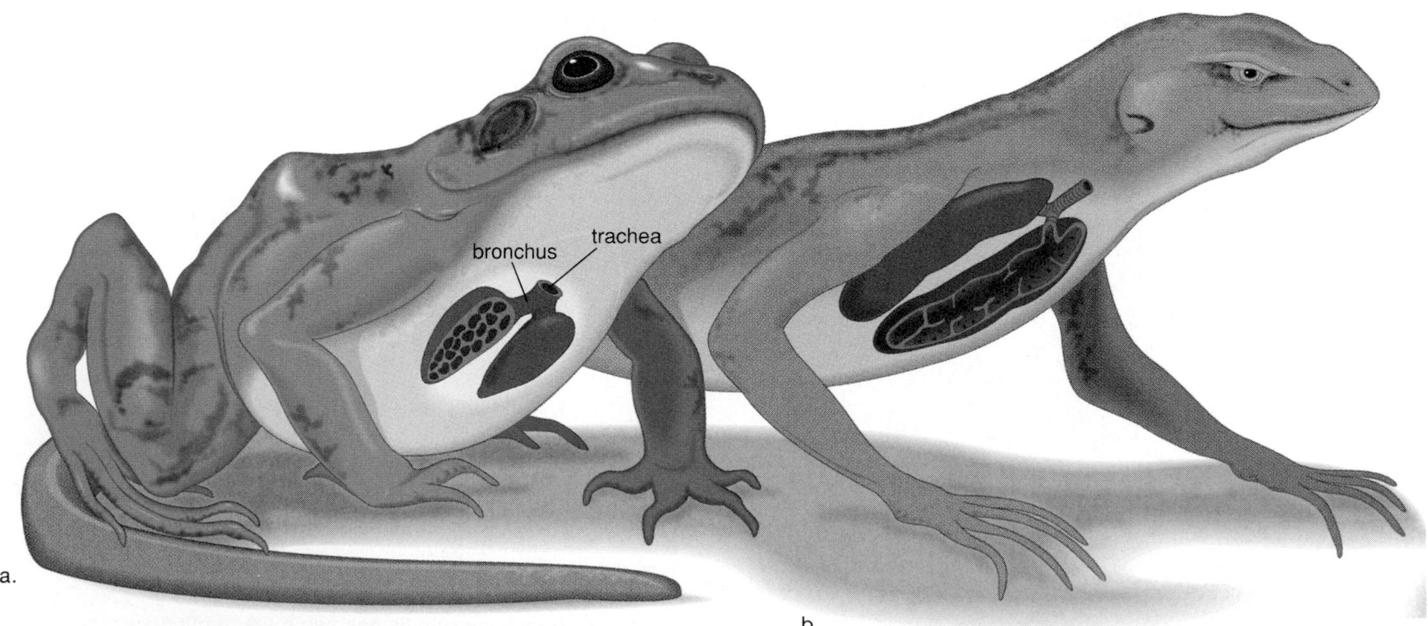

Figure 44.4 Respiration in amphibians compared to reptiles.
a. Amphibians use positive pressure to push air into small and saclike lungs. The moist, thin skin of amphibians is often used as an auxiliary gas-exchange surface. **b.** Reptiles use negative pressure to draw air into lungs, which are more convoluted. The tough, scaly skin protects the animal from drying out and is not used for respiration.

into the lungs. Most amphibians breathe to some extent through the skin, which is kept moist by the presence of mucus produced by numerous glands on the surface of the body. During the winter in temperate climates, amphibians burrow in the mud, and all gas exchange occurs by way of the skin.

The inner lining of the lungs is more finely divided in reptiles than in amphibians (Fig. 44.4*b*). The lungs of birds and mammals are elaborately subdivided into small passageways and spaces. It has been estimated that human lungs have a total surface area that is at least 50 times the skin's surface area. To keep the lungs from drying out, air is moistened as it moves through passageways leading to the lungs.

Terrestrial vertebrates ventilate the lungs by moving air into and out of the respiratory tract. Frogs use positive pressure to force air into the respiratory tract. With the nostrils firmly shut, the floor of the mouth rises and pushes the air into the lungs. Reptiles, birds, and mammals use negative pressure to move air into the lungs. Reptiles have jointed ribs that can be raised to expand the lungs. Mammals have a **rib cage** that is lifted up and out and a muscular **diaphragm** [Gk. *diaphragma*, partition wall] that is flattened. As the **thoracic cavity** (chest cavity) expands and lung volume increases, air will flow into the lungs due to differences in air pressure. Following **inhalation** (or inspiration), **exhalation** (or expiration) occurs. In reptiles, lowering the ribs exerts a pressure that forces air out. In mammals, when the rib cage is lowered and the diaphragm rises, the thoracic pressure increases, forcing air out of the lungs.

The lungs of amphibians, reptiles, and mammals are not completely emptied and refilled during each breathing cycle. Because of this *incomplete ventilation* method, the air entering mixes with used air remaining in the lungs. While this does help conserve water, it also decreases gas-exchange efficiency. The high oxygen requirement of flying birds, however, requires a method of *complete ventilation* (Fig. 44.5). Incoming air is carried past the lungs by a bronchus that takes it to a set of posterior air sacs. The air then passes forward through the lungs into a set of anterior air sacs. From here, it is finally expelled. Fresh, oxygen-rich air passes through the lungs in one direction only and does not mix with used air.

Most terrestrial animals have a specialized gas exchange region. Insects have a tracheal system, while terrestrial vertebrates depend on lungs, which are ventilated variously.

Figure 44.5 Respiratory system in birds.
Because of the presence of air sacs, there is a one-way flow of air through the lungs. Upon inhalation, air moves into the air sacs; upon exhalation, air moves through the lungs.

44.2 How Humans Exchange Gases

The human respiratory system includes all structures that conduct air to and from the lungs (Fig. 44.6 and Table 44.1). The lungs lie deep within the thoracic cavity, where they are protected from drying out. As air moves through the nose, the pharynx, the trachea, and the bronchi to the lungs, it is filtered so that it is free of debris, warmed, and humidified. By the time the air reaches the lungs, it is at body temperature and is saturated with water. In the nose, hairs and cilia act as a screening device. In the trachea and the bronchi, cilia beat upward, carrying mucus, dust, and occasional bits of food that "went down the wrong way" into the throat, where the accumulation may be swallowed or expectorated.

The hard and soft palates separate the nasal cavities from the mouth, but the air and food passages cross in the **pharynx** [Gk. *pharynx,* throat]. This may seem inefficient, and there is danger of choking if food accidentally enters the trachea, but this arrangement does have the advantage of letting you breathe through your mouth in case your nose is plugged up. In addition, it permits greater intake of air during heavy exercise, when greater gas exchange is required.

Air passes from the pharynx through the **glottis** [Gk. *glotti,* tongue], an opening into the **larynx** [Gk. *larynx,* gullet], or voice box. At the edges of the glottis, embedded in mucous membrane, are the **vocal cords.** These flexible and pliable bands of connective tissue vibrate and produce sound when air is expelled past them through the glottis from the larynx.

The larynx and the **trachea** [L. *trachia,* windpipe] are permanently held open to receive air. The larynx is held open by the complex of cartilages, among them is the cartilage of the Adam's apple. The trachea is held open by a series of C-shaped, cartilaginous rings that do not completely meet in the rear. When food is being swallowed, the larynx rises, and the glottis is closed by a flap of tissue called the **epiglottis** [Gk. *epi,* over, and *glotta,* tongue]. A backward movement of the soft palate covers the entrance of the nasal passages into the pharynx. The food then enters the esophagus, which lies behind the larynx.

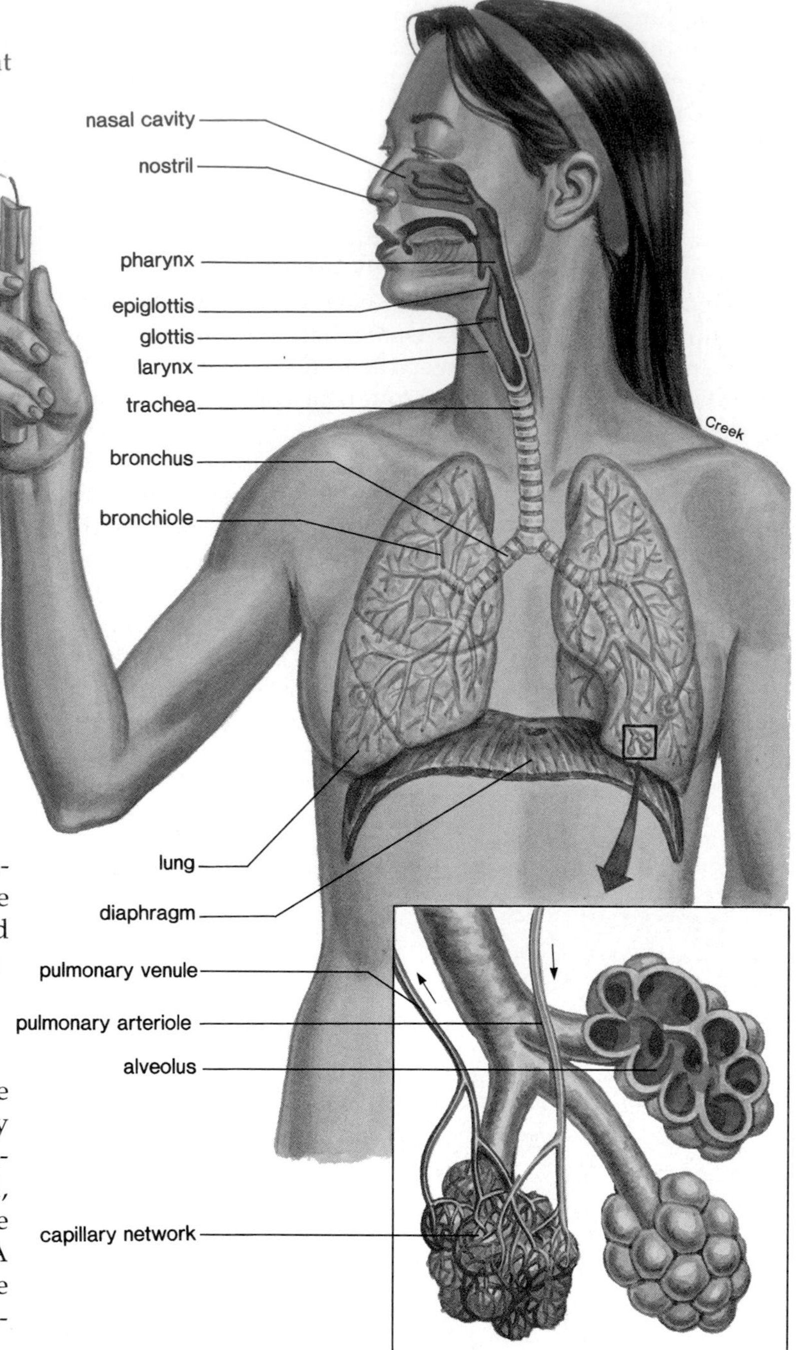

Figure 44.6 The human respiratory tract.
The respiratory tract extends from the nose to the lungs, which are composed of air sacs called alveoli. Gas exchange occurs between air in the alveoli and blood within a capillary network that surrounds the alveoli.

Table 44.1

Path of Air

Structure	Function
Nasal cavities	Filter, warm, and moisten
Pharynx (throat)	Connection to larynx
Glottis	Permits passage of air
Larynx (voice box)	Sound production
Trachea (windpipe)	Passage of air to bronchi
Bronchi	Passage of air to lung
Bronchioles	Passage of air to alveoli
Alveoli	Air sacs for gas exchange

The trachea divides into two primary **bronchi** [Gk. *bronchos,* windpipe], which enter the right and left lungs. Branching continues until there are a great number of smaller passages called **bronchioles** [Gk. dim. of *bronchos,* windpipe]. The two bronchi resemble the trachea in structure, but as the bronchial tubes divide and subdivide, their walls become thinner, and rings of cartilage are no longer present. Each bronchiole terminates in an elongated space enclosed by a multitude of air pockets, or sacs, called **alveoli** [L. *alveolus,* dim. of *alveus,* cavity], which make up the lungs.

Breathing In and Out

Humans breathe using the same mechanism employed by all other mammals. The volume of the thoracic cavity and lungs is increased by muscle contractions that lower the diaphragm and raise the ribs (Fig. 44.7). These movements create a negative pressure in the thoracic cavity and lungs, and air then flows into the lungs. When rib and diaphragm muscles relax, air is exhaled as a result of increased pressure in the thoracic cavity and lungs.

Increased carbon dioxide (CO_2) and hydrogen ion (H^+) concentrations in the blood are the primary stimuli that increase breathing rate. The chemical content of the blood is monitored by *chemoreceptors* called the *aortic* and *carotid* bodies, which are specialized structures located in the walls of the aorta and the carotid arteries. These receptors are very sensitive to changes in carbon dioxide and hydrogen ion concentrations, but they are only minimally sensitive to a lower oxygen (O_2) concentration. Information from the chemoreceptors goes to the respiratory center in the medulla oblongata of the brain, which then increases the breathing rate when carbon dioxide or hydrogen ion concentration increases. This respiratory center is itself sensitive to the chemical content of the blood reaching the brain.

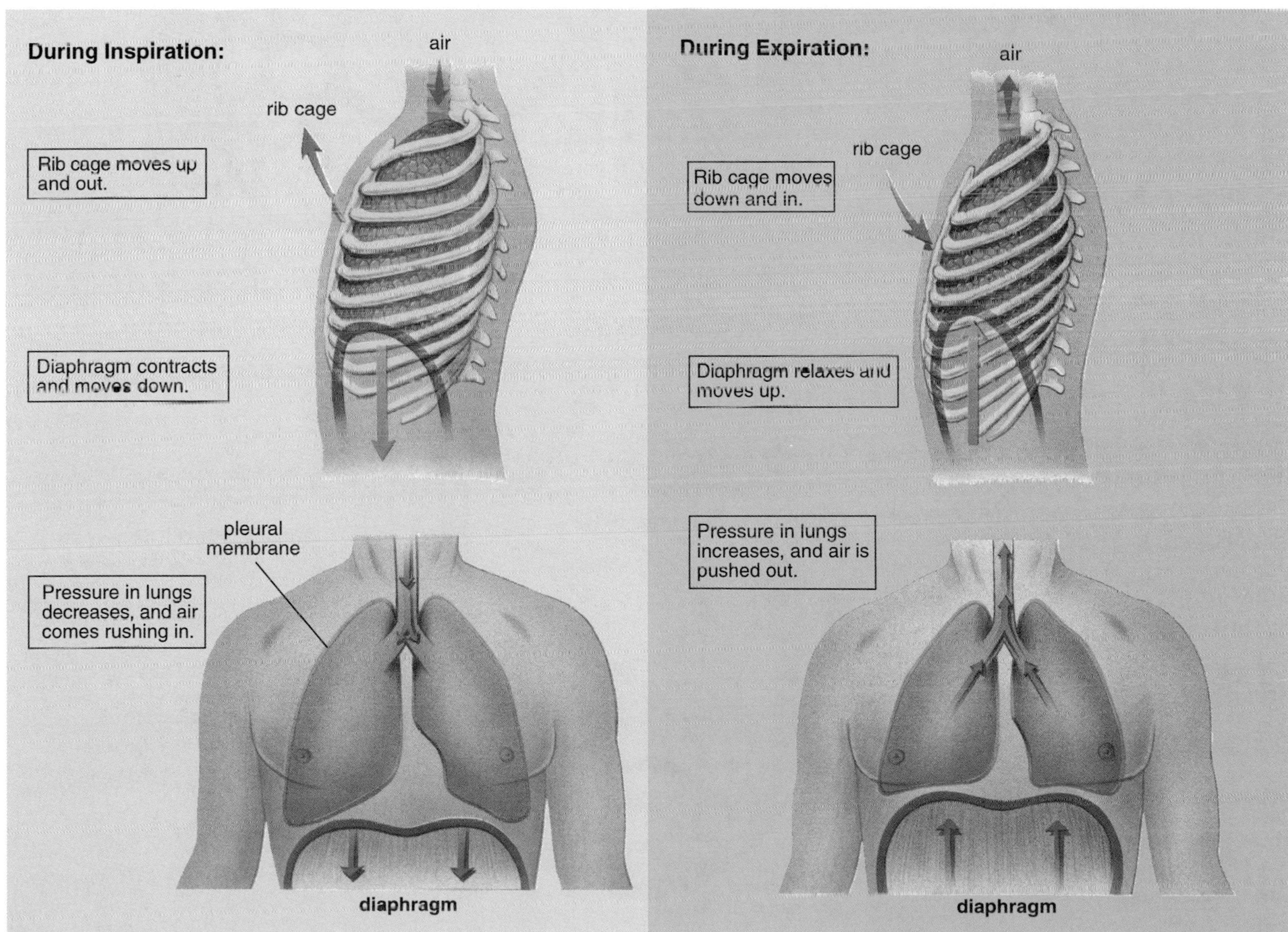

Figure 44.7 Inspiration versus expiration.
During inspiration, the thoracic cavity and lungs expand so that air is drawn in. During expiration, the thoracic cavity and lungs resume their original positions and pressures. Now, air is forced out.

Figure 44.8 External and internal respiration.
During external respiration in the lungs, carbon dioxide (CO_2) leaves blood and oxygen (O_2) enters blood. During internal respiration in the tissues, oxygen leaves blood and carbon dioxide enters blood. Steps necessary for gas exchange are shown for the lungs (*top*) and for the tissues (*bottom*).

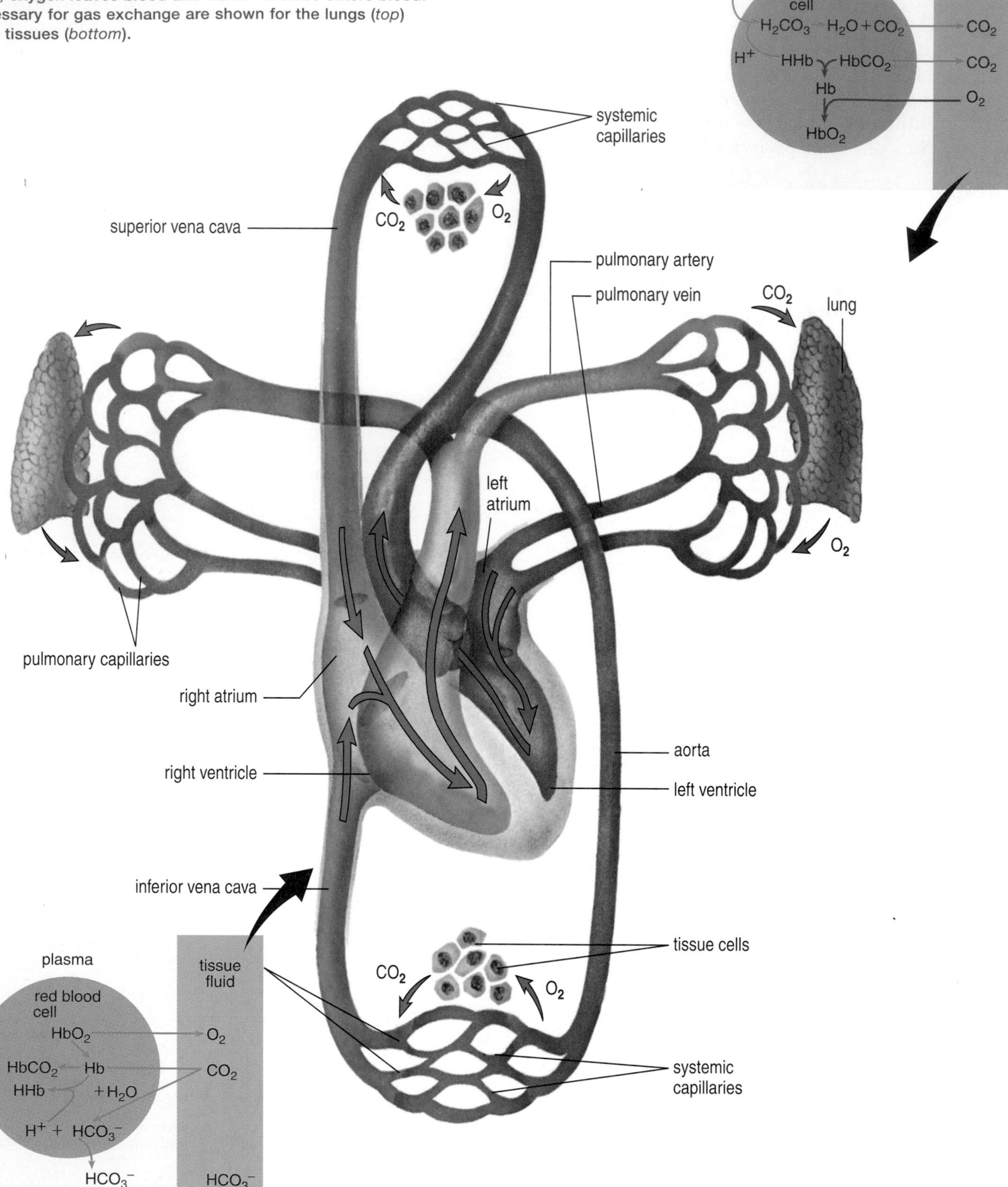

Exchanging and Transporting Gases

Diffusion primarily accounts for the exchange of gases between the air in the alveoli and the blood in the pulmonary capillaries (Fig. 44.8). Atmospheric air contains little CO_2, but blood flowing into the pulmonary capillaries is almost saturated with the gas. Therefore, CO_2 diffuses out of the blood and into the alveoli. The pattern is the reverse for O_2. Blood coming into the pulmonary capillaries is O_2 poor and the alveolar air is O_2 rich; therefore, O_2 diffuses into the capillaries.

Transporting O_2 and CO_2

Most O_2 entering the blood combines with **hemoglobin** (Hb) [Gk. *haima*, blood, and L. *globus*, ball] in red blood cells to form oxyhemoglobin (HbO_2):

$$Hb + O_2 \longrightarrow \underset{\text{oxyhemoglobin}}{HbO_2}$$

Each hemoglobin molecule contains four polypeptide chains, and each chain is folded around an iron-containing group called heme. It is actually the iron that forms a loose association with O_2. Since there are about 250 million hemoglobin molecules in each red blood cell, each cell is capable of carrying more than one billion molecules of oxygen.

Figure 44.9 shows the percent saturation of hemoglobin at various partial pressures, depending on the temperature and acidity. A partial pressure of a gas is simply the amount of pressure exerted by that gas among all the gases present. At the normal partial pressure of O_2 in the lungs, hemoglobin is practically saturated with O_2, but at the partial pressure in the tissues, oxyhemoglobin quickly gives up much of its O_2:

$$HbO_2 \longrightarrow Hb + O_2$$

The acid pH and warmer temperature of the tissues also promote the release of oxygen by hemoglobin.

In the tissues, some hemoglobin combines with CO_2 to form *carbaminohemoglobin*. Most of the CO_2, however, is transported in the form of the **bicarbonate ion** (HCO_3^-). First CO_2 combines with water, forming *carbonic acid*, and then this dissociates to a hydrogen ion (H^+) and HCO_3^-.

$$CO_2 + H_2O \longrightarrow \underset{\text{carbonic acid}}{H_2CO_3} \longrightarrow H^+ + \underset{\text{bicarbonate ion}}{HCO_3^-}$$

Carbonic anhydrase [Gk. *an*, without, and *hydrias*, water], an enzyme in red blood cells, speeds this reaction. The release of H^+, could drastically change the pH of the blood. However, the H^+ are absorbed by the globin portions of hemoglobin, and the HCO_3^- diffuse out of the red blood cells to be carried in the plasma. Hemoglobin that combines with a H^+ is called reduced hemoglobin and can be symbolized as HHb. HHb plays a vital role in maintaining the normal pH of the blood.

As blood enters the pulmonary capillaries, most of the CO_2 is present in plasma as HCO_3^-. But as free CO_2 begins to diffuse out, the following reaction occurs:

$$H^+ + HCO_3^- \longrightarrow H_2CO_3 \longrightarrow H_2O + CO_2$$

Carbonic anhydrase also speeds this reaction, and hemoglobin gives up the H^+ it has been carrying, and HHb becomes Hb.

Oxygen is transported in the blood by hemoglobin, and carbon dioxide is largely carried in plasma as the bicarbonate ion.

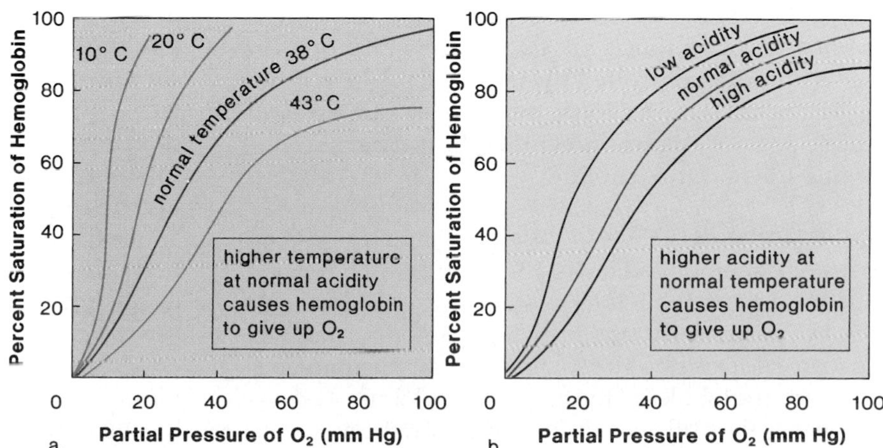

Figure 44.9 Environmental conditions and hemoglobin saturation.
These graphs show that the lower the partial pressure of O_2 and (a) the higher the temperature and (b) the higher the acidity of blood, the less saturated hemoglobin becomes.

44.3 Keeping the Respiratory Tract Healthy

We have seen that the entire respiratory tract has a warm, wet, mucous membrane lining, which is constantly exposed to environmental air. The quality of this air, determined by the pollutants and the microbes therein, can affect our health.

The Tract Gets Infected

Microbes frequently spread from one individual to another by way of the respiratory tract. Droplets from a single sneeze can carry billions of bacteria or viruses. The mucous membranes are protected by mucus and by the constant beating of the cilia, but if the number of infectious agents is large and/or our resistance is reduced, respiratory infections such as colds and influenza (flu) can result. Other more serious infections and disorders are discussed here.

Viral infections can spread from the nasal cavities to the sinuses (sinusitis), to the middle ears (otitis media), to the larynx (laryngitis), and to the bronchi (bronchitis). Acute bronchitis (Fig. 44.10) usually is caused by a secondary bacterial infection of the bronchi, resulting in a heavy mucus discharge with much coughing. Acute bronchitis usually responds to antibiotic therapy. Chronic bronchitis, on the other hand, is not necessarily due to infection. It is often caused by constant irritation of the lining of the bronchi, which consequently undergo degenerative changes, including the loss of cilia and their accompanying cleansing action. There is frequent coughing, and the individual is more susceptible to respiratory infections. Chronic bronchitis is most often seen in cigarette smokers or those exposed to secondhand smoke or other types of polluted air.

Strep throat is a very severe throat infection caused by the bacterium *Streptococcus pyogenes*. Swallowing may be difficult, and there is fever. Unlike a viral infection, strep throat should be treated with antibiotics. If not treated, it can lead to complications such as rheumatic fever, which can permanently damage the heart valves.

The Lungs Have Disorders

Pneumonia and tuberculosis are two serious infections of the lungs ordinarily controlled by antibiotics. Three other illnesses discussed, emphysema, pulmonary fibrosis, and lung cancer, are not due to infections; in most instances, they are due to cigarette smoking.

Pneumonia Can Kill

Most forms of pneumonia (Fig. 44.10) are caused by either a bacterium or a virus that has infected the lungs. AIDS patients are subject to a particularly rare form of pneumonia caused by the protozoan *Pneumocystis carinii.* Sometimes, pneumonia is localized in specific lobules of the lungs. These lobules become nonfunctional as they fill with fluid. Obviously, the more lobules involved, the more serious the infection.

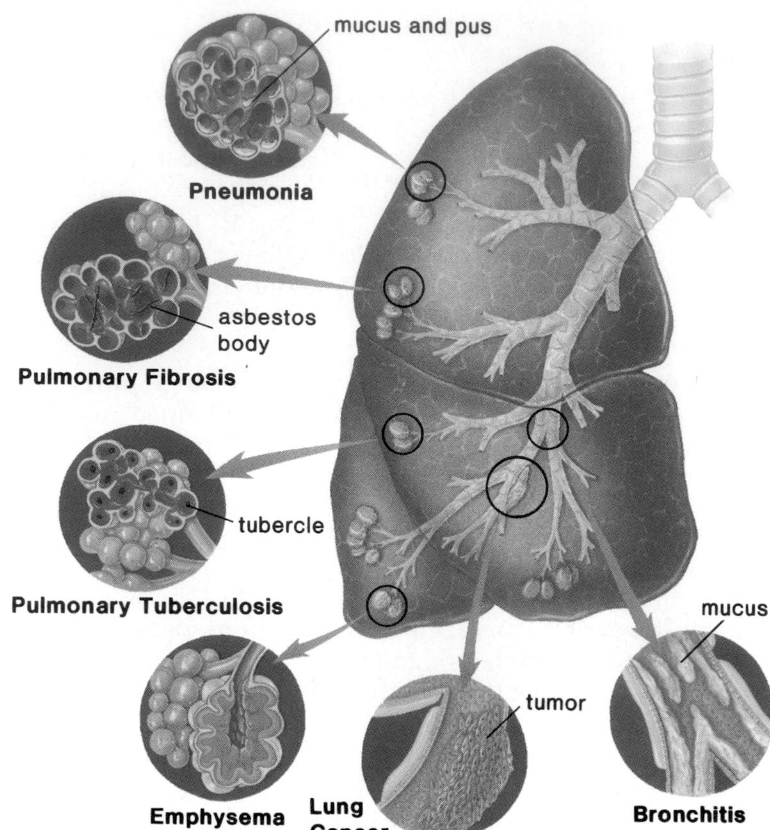

Figure 44.10 Common bronchial and pulmonary diseases.
Exposure to infectious microbes and/or polluted air, including tobacco smoke, causes the diseases and disorders shown here.

Tuberculosis Is Back

Pulmonary tuberculosis (Fig. 44.10) is caused by the tubercle bacillus, a type of bacterium. When a person has tuberculosis, the alveoli burst and are replaced by inelastic connective tissue. It is possible to tell if a person has ever been exposed to tuberculosis with a skin test, in which a highly diluted extract of the bacilli is injected into the skin of the patient. A person who has never been in contact with the bacillus shows no reaction, but one who has developed immunity to the organism shows an area of inflammation that peaks in about two days. If these bacilli invade the lung tissue, the cells build a protective capsule about the foreigners and isolate them from the rest of the body. This tiny capsule is called a *tubercle.* If the resistance of the body is high, the imprisoned organisms die, but if the resistance is low, the organisms eventually can be liberated. If a chest X ray detects active tubercles, the individual is put on appropriate drug therapy to ensure the localization of the disease and the eventual destruction of any live bacterial organisms.

Tuberculosis killed about 100,000 people in the United States annually before the middle of this century, when antibiotic therapy brought it largely under control. In recent years, however, the incidence of tuberculosis is on the rise, particularly among AIDS patients, the homeless, and the rural poor. Worse, the new strains are resistant to the usual antibiotic therapy. Therefore, some physicians would like to again use sanitoriums to quarantine patients needing treatment.

Emphysema Reduces Capacity

Emphysema (Fig. 44.10) refers to the destruction of lung tissue, with accompanying ballooning or inflation of the lungs due to trapped air. The trouble stems from the damage and the collapse of the bronchioles. When this occurs, the alveoli are cut off from renewed oxygen supply and the air within them is trapped. The trapped air very often causes the alveolar walls to rupture, and a loss of elasticity makes breathing difficult. The victim is breathless and may have a cough. Since the surface area for gas exchange is reduced, not enough oxygen reaches the heart and the brain. Even so, the heart works furiously to force more blood through the lungs, and this can lead to a heart condition. Lack of oxygen to the brain can make the person feel depressed, sluggish, and irritable.

Pulmonary Fibrosis Is Dangerous

Inhaling particles such as silica (sand), coal dust, and asbestos (Fig. 44.11) can lead to pulmonary fibrosis (Fig. 44.10), a condition in which fibrous connective tissue builds up in the lungs. Breathing capacity can be seriously impaired, and the development of cancer is common. Since asbestos has been used so widely as a fireproofing and insulating agent, unwarranted exposure has occurred. It is projected that 2 million deaths could be caused by asbestos exposure—mostly in the workplace—between 1990 and 2020.

Lung Cancer Due to Smoking

Lung cancer (Fig. 44.10) used to be more prevalent in men than in women, but recently it has surpassed breast cancer as a cause of death in women. This can be linked to an increase in the number of women who smoke today. Autopsies on smokers have revealed the progressive steps by which the most common form of lung cancer develops. The first event appears to be thickening and callusing of the cells lining the bronchi. (Callusing occurs whenever cells are exposed to irritants.) Then there is a loss of cilia so that it is impossible to prevent dust and dirt from settling in the lungs. Following this, cells with atypical nuclei appear in the callused lining. A tumor, consisting of disordered cells with atypical nuclei, is considered to be cancer in situ (at one location). When some of these cells finally break loose and penetrate other tissues—a process called metastasis—the cancer spreads.

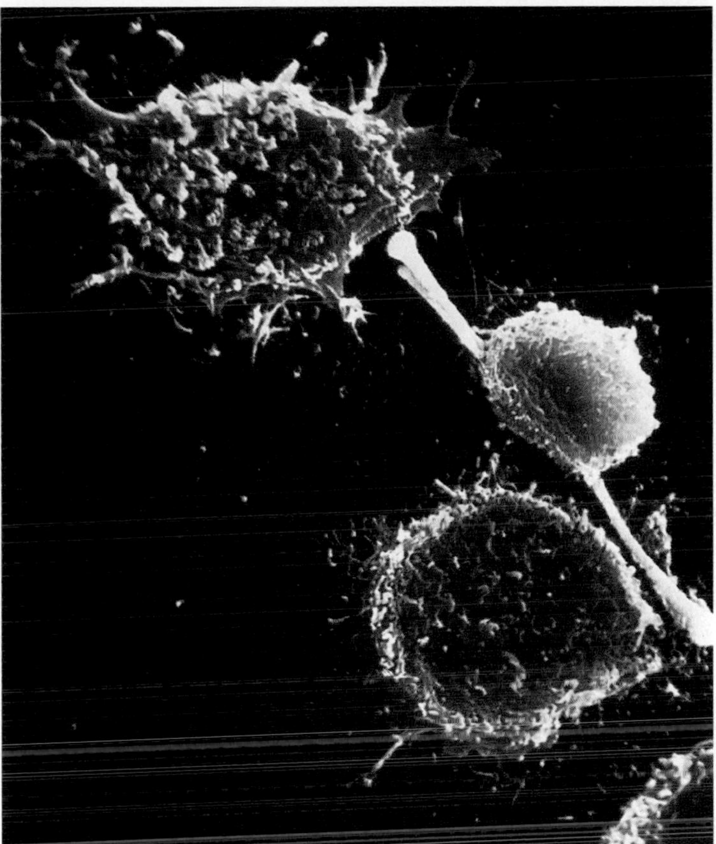

Figure 44.11 Asbestos fibers.
A scanning electron micrograph of macrophages reveals an "asbestos body" in the lung tissue of a person exposed to asbestos for some time. Asbestos bodies are fibers that have been coated with iron, plasma proteins, and other materials.

The tumor may grow until the bronchus is blocked, cutting off the supply of air to that lung. The entire lung then collapses, the secretions trapped in the lung spaces become infected, and pneumonia or a lung abscess (localized area of pus) results. The only treatment that offers a possibility of cure is to remove a lobe or the lung completely before secondary growths have had time to form. This operation is called *pneumonectomy.*

The incidence of lung cancer is over 20 times higher in individuals who smoke than in those who do not.

Current research indicates that *passive smoking*—simply breathing in air filled with cigarette smoke—can also cause lung cancer and a number of other illnesses associated with smoking. If a person stops both voluntary and involuntary smoking, and if the body tissues are not already cancerous, the tissues usually return to normal over time.

connecting concepts

Both the human intestine and lungs are exchange boundaries with the external environment. The lungs are composed of alveoli that number about 300 million, and each alveolus may have as many as 1,800 blood capillary contacts. The lungs are adapted to facilitate gas exchange, freely allowing oxygen to enter and carbon dioxide to exit the blood. The characteristics that enhance the function of the lungs for gas exchange also facilitate the entrance of substances and microbes that could be harmful to the body. Defense mechanisms are in place that limit possible microbe invasion of the body along the respiratory tract. Numerous lymphoid nodules located in the bronchial lining are a source of white blood cells that protect against infection.

The body, however, has no means by which to inhibit the entrance of dangerous chemicals across the alveolar wall. Breathing air pollutants can result in respiratory distress, headache, and exhaustion. On occasion, when smog gets trapped above a city by a blanket of warm air, the results can be disastrous. In 1966, about 168 people died in New York City due to an accumulation of air pollutants over several days. We know the manner in which one air pollutant causes death. Carbon monoxide (CO) is an air pollutant that comes from the incomplete combustion of natural gas and gasoline. Because carbon monoxide is a colorless, odorless gas, people can be unaware that it has entered their system by way of the lungs. But once CO is in the bloodstream, it combines with the iron of hemoglobin 200 times more tightly than oxygen. The result is that the delivery of oxygen to mitochondria is impaired and so is the functioning of mitochondria. The end result can be death.

Flushed red skin, especially on facial cheeks, is the first sign of carbon monoxide poisoning, because hemoglobin bound to carbon monoxide is a brighter red than oxygenated hemoglobin. Removing a person from the carbon monoxide source is not sufficient treatment because carbon monoxide, unlike oxygen, remains tightly bound to iron for many hours. A transfusion of red blood cells will help to increase the carrying capacity of the blood, and pure oxygen given under pressure will displace some carbon monoxide. Despite good medical care, some people still die each year from CO poisoning.

Carbon monoxide poisoning illustrates that we ordinarily take the steps of respiration for granted. Respiration involves breathing (gas exchange), external and internal respiration (delivery of oxygen to cells by hemoglobin in the blood), and aerobic respiration.

Summary

44.1 How Animals Exchange Gases

Some aquatic animals like hydras and planaria use their entire body surface for gas exchange.

Most animals have a special, localized gas-exchange area. Most large aquatic animals pass water over gills. On land, insects utilize tracheal systems and vertebrates have lungs.

Lungs are found inside the body, where water loss is reduced. To ventilate the lungs, some vertebrates use positive pressure, but most inhale, using muscular contraction to produce a negative pressure that causes air to rush into the lungs. When the breathing muscles relax, air is exhaled.

Birds have a series of air sacs that allow a one-way flow of air over the gas-exchange area. This is called a complete ventilation system.

44.2 How Humans Exchange Gases

Table 44.1 lists the structures found in the human respiratory system.

Humans breathe by negative pressure, as do other mammals. During inhalation, the rib cage goes up and out, and the diaphragm lowers. The lungs expand and air comes rushing in.

During exhalation, the rib cage goes down and in, and the diaphragm rises. Therefore, air rushes out.

The rate of breathing is dependent upon the amount of carbon dioxide in the blood, as detected by chemoreceptors such as the aortic and carotid bodies.

Gas exchange in the lungs and tissues is brought about by diffusion. Hemoglobin transports oxygen in the blood; carbon dioxide is mainly transported in plasma as the bicarbonate ion. The enzyme carbonic anhydrase found in red blood cells speeds the formation of the bicarbonate ion.

44.3 Keeping the Respiratory Tract Healthy

The respiratory tract is subject to infections. Two disorders of the lungs, emphysema and cancer, are usually due to cigarette smoking.

Reviewing the Chapter

1. Compare the respiratory organs of aquatic animals to those of terrestrial animals. 790–93
2. How does the countercurrent flow of blood within gill capillaries and water passing across the gills assist respiration in fishes? 791
3. Why don't insects require circulatory system involvement in air transport? Why is it beneficial for the body wall of earthworms to be moist? 792
4. Explain the phrase, "breathing by using negative pressure." 793
5. Contrast incomplete ventilation in humans with complete ventilation in birds. 793
6. Name the parts of the human respiratory system, and list a function for each part. 794
7. The concentration of what gas in the blood controls the breathing rate in humans? Explain. 795–97
8. Which conditions depicted in Figure 44.10 are due to infection? Which are due to behavioral or environmental factors? Explain. 798–99

Testing Yourself

Choose the best answer for each question.

1. One problem faced by terrestrial animals with lungs, but not by freshwater aquatic animals with gills, is that
   a. gas exchange involves water loss.
   b. breathing requires considerable energy.
   c. oxygen diffuses very slowly in air.
   d. All of these are correct.
2. In which animal is the circulatory system not involved in gas transport?
   a. mouse
   b. dragonfly
   c. trout
   d. sparrow
3. Birds have a more efficient lung than humans because the flow of air
   a. is the same during both inhalation and exhalation.
   b. travels in only one direction through the lungs.
   c. never backs up like in human lungs.
   d. is not hindered by a larynx.
4. Which animal breathes by positive pressure?
   a. fish
   b. human
   c. bird
   d. frog
5. Which of these is a true statement?
   a. In lung capillaries, carbon dioxide combines with water to produce carbonic acid.
   b. In tissue capillaries, carbonic acid breaks down to carbon dioxide and water.
   c. In lung capillaries, carbonic acid breaks down to carbon dioxide and water.
   d. In tissue capillaries, carbonic acid combines with hydrogen ions to form the carbonate ion.
6. Air enters the human lungs because
   a. atmospheric pressure is less than the pressure inside the lungs.
   b. atmospheric pressure is greater than the pressure inside the lungs.
   c. although the pressures are the same inside and outside, the partial pressure of oxygen is lower within the lungs.
   d. the residual air in the lungs causes the partial pressure of oxygen to be less than it is outside.
7. If the digestive and respiratory tracts were completely separate in humans, there would be no need for
   a. swallowing.
   b. a nose.
   c. an epiglottis.
   d. a diaphragm.
8. To trace the path of air in humans you would place the trachea
   a. directly after the nose.
   b. directly before the bronchi.
   c. before the pharynx.
   d. Both a and c are correct.
9. In humans, the respiratory center
   a. is stimulated by carbon dioxide.
   b. is located in the medulla oblongata.
   c. controls the rate of breathing.
   d. All of these are correct.
10. Carbon dioxide is carried in the plasma
    a. in combination with hemoglobin.
    b. as the bicarbonate ion.
    c. combined with carbonic anhydrase.
    d. All of these are correct.
11. Label this diagram of the human respiratory system:

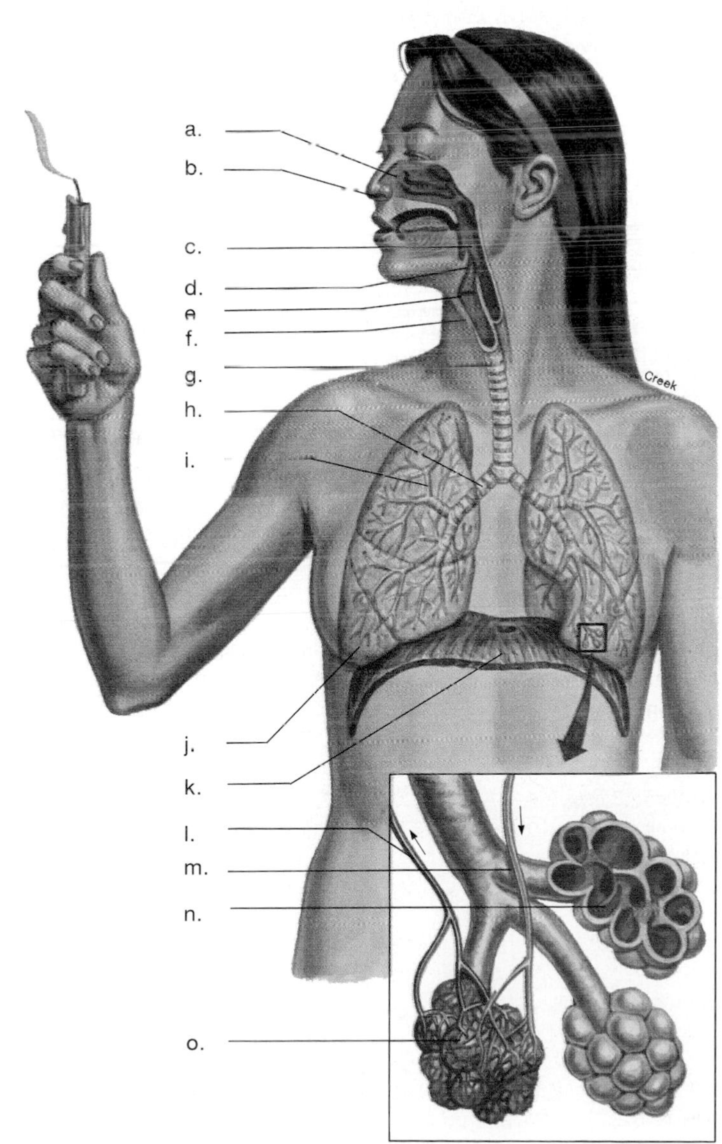

Applying the Concepts

1. *All systems of the animal's body contribute to homeostasis.*
   Describe two ways in which the human respiratory system contributes to homeostasis.
2. *Organisms have localized boundaries for exchanging materials with the environment.*
   Compare the similarities and differences between gills and lungs and the obvious localized boundaries with the environment.
3. *The structure of an organ suits its function.*
   How does the structure of the lungs suit their function?

Using Technology

Your study of respiration is supported by these available technologies:

Exploring the Internet
The Mader Home Page provides resources for and help with studying this chapter.

http://www.mhhe.com/sciencemath/mader/
(Click on Biology.)

The Dynamic Human CD-ROM
Respiratory System

Explorations in Human Biology CD-ROM
Life Span and Lifestyle (#3)
Smoking and Cancer (#6)

Understanding the Terms

Match the terms to these definitions:

a. __________ Common passageway for both food intake and air movement, located between the mouth and the esophagus.
b. __________ Dome-shaped muscularized sheet separating the thoracic cavity from the abdominal cavity in mammals.
c. __________ Fold of tissue within the larynx; creates vocal sounds when it vibrates.
d. __________ Form in which most of the carbon dioxide is transported in the bloodstream.
e. __________ Internal body space of some animals that contains the lungs, protecting them from desiccation.
f. __________ One of two main branches of the trachea in vertebrates that have lungs.
g. __________ Iron-containing respiratory pigment occurring in vertebrate red blood cells and in blood plasma of many invertebrates.
h. __________ Respiratory organ in most aquatic animals; in fish an outward extension of the pharynx.
i. __________ Stage during respiration when air is pushed out of the lungs.
j. __________ Terminal, microscopic, grapelike air sac found in vertebrate lungs.

Osmotic Regulation and Excretion

CHAPTER

Chapter Concepts

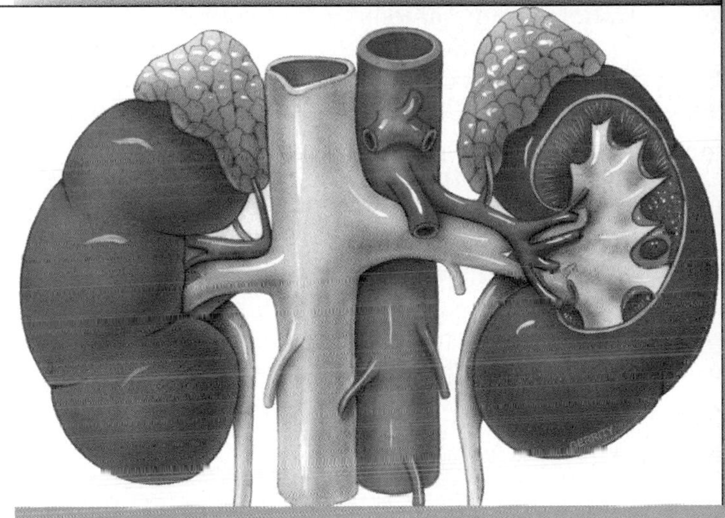

The human kidneys and their blood vessels

Like the digestive and respiratory organs, excretory organs play a major role in maintaining homeostasis. Excretory organs regulate the salt/water balance in body fluids—an abnormal osmolarity of internal fluids can cause an animal to lose or gain fluids to or from the external environment.

Maintenance of the normal salt/water balance of internal fluids is closely tied to the excretion of metabolic wastes in most animals. Cells release wastes into the internal environment and excretory organs discharge these wastes into the external environment. Urine is a general term for a fluid that contains nitrogenous wastes excreted by kidneys or any other kind of excretory organ.

Homeostasis not only includes the distribution of oxygen and nutrients to cells, it also includes maintaining the water and ion balance of the body and ridding the body of metabolic wastes. **Excretion** is the elimination of molecules that have taken part in metabolic reactions and should not be confused with defecation, which is the elimination of nondigested material from the digestive tract.

45.1 Osmotic Regulation

Osmotic regulation is the maintenance of the normal water and salt balance in body fluids. In many animals, water can enter the body through metabolism (e.g., aerobic respiration produces water); by eating foods that contain water; and by drinking water. Water is lost from the body through evaporation (e.g., from skin and lungs), feces formation, and excretion (Fig. 45.1). To be in fluid balance, the amount of water exiting the body must equal the amount that enters.

The osmolarity of body fluids is dependent upon the concentration of mineral ions such as sodium (Na^+), chloride (Cl^-), potassium (K^+), and the bicarbonate ion (HCO_3^-). Body fluids gain mineral ions as a result of eating foods and drinking fluids. Excretion is the primary way that the body loses ions.

When there are differences in the osmolarity between two regions, water tends to move into the region with the higher amount of solutes. A marine environment, which is high in salts, tends to promote the osmotic loss of water and the gain of ions by such means as drinking water. Fresh water tends to promote a gain of water by osmosis and a loss of ions as excess water is excreted. Terrestrial animals tend to lose both water and ions to the environment.

Aquatic Animals

Among aquatic animals, only marine invertebrates and cartilaginous fishes, such as sharks and rays, have body fluids that are nearly isotonic to seawater. These organisms have little difficulty maintaining their normal salt and water balance. It is surprising, though, that while they are isotonic, the body fluids of cartilaginous fishes do not contain the same amount of mineral ions as seawater. The answer to this paradox is that their blood contains a concentration of urea high enough to match the tonicity of the sea! For some unknown reason, this amount of urea is not toxic to them.

Bony Fishes

The body fluids of all bony fishes normally have only a moderate amount of salt. Apparently, their common ancestor evolved in fresh water, and only later did some groups invade the sea. Marine bony fishes (Fig. 45.2*a*) are therefore prone to water loss and could become dehydrated. To counteract this, they drink seawater almost constantly. On the average, marine bony fishes swallow an amount of water estimated to be equal to 1% of their body weight every hour. This is equivalent to a human drinking about 700 ml of water every hour around the clock. While they get water by drinking, this habit also causes these fishes to acquire salt. To rid the body of excess salt, they actively

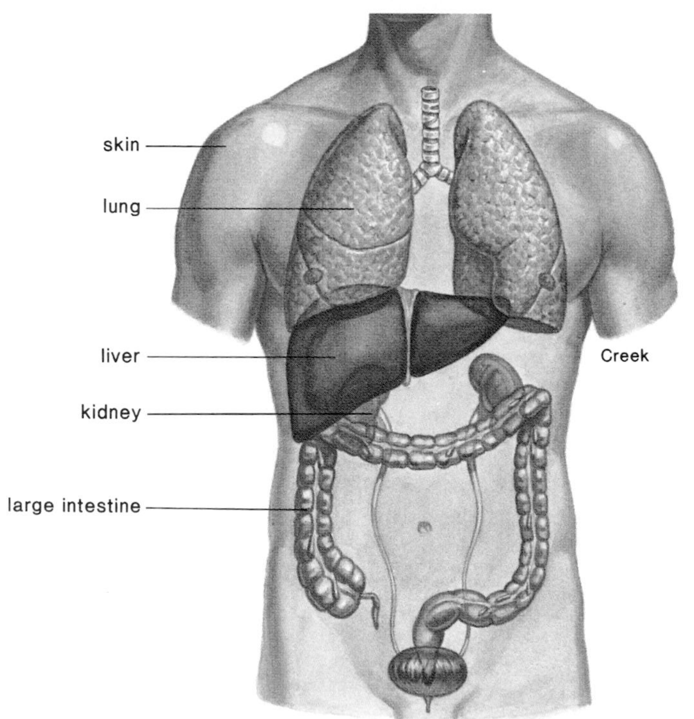

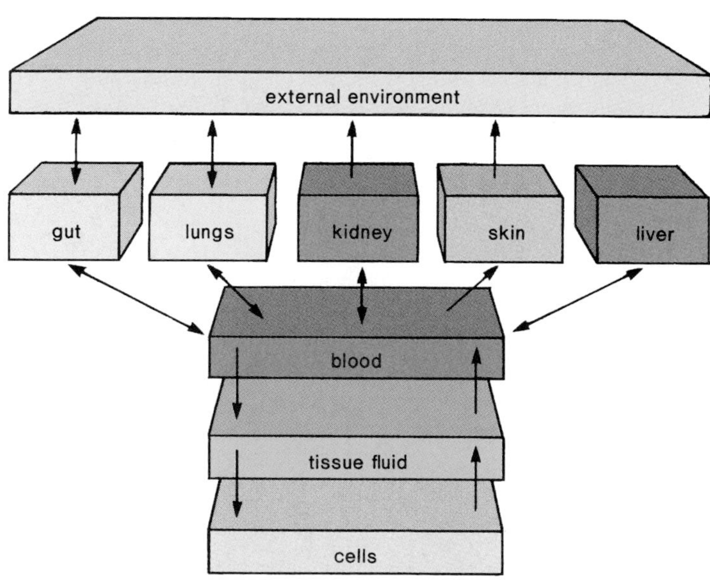

Figure 45.1 Osmotic regulation in humans.
a. The internal environment of cells (blood and tissue fluid) in humans and other animals stays relatively constant because the blood is continually exchanging substances with certain organs. **b.** Fluids and solutes enter blood and then tissue fluid by way of the gut; water and mineral ions are lost by way of the lungs, skin, and kidneys. Some of these organs are also involved in excretion: the lungs excrete carbon dioxide, the liver produces urea excreted by the kidneys. Urea is also present in sweat.

transport sodium (Na^+) and chloride (Cl^-) ions into the surrounding seawater at the gills. This causes a passive loss of water through gills.

The osmotic problems of freshwater bony fishes are exactly opposite to those of marine bony fishes (Fig. 45.2*b*). The body fluids of freshwater bony fishes are hypertonic to fresh water, and they are prone to passively gain water. These fishes never drink water, but instead eliminate excess water through production of large quantities of dilute (hypotonic) urine. They discharge a quantity of urine equal to one-third their body weight each day. Because this causes them to lose salts, they actively transport salts into the blood across the membranes of their gills.

The difference in adaptation between marine and freshwater bony fishes makes it remarkable that some fishes actually can move between the two environments during their life cycle. Salmon, for example, begin their lives in freshwater streams and rivers, move to the ocean for a period of time, and finally return to fresh water to breed. These fishes alter their behavior and their gill and kidney functions in response to the osmotic changes they encounter when moving from one environment to the other.

To regulate water and salt balance, marine bony fishes drink water constantly and excrete salt from the gills. Freshwater bony fishes never drink water; they excrete a dilute urine and actively transport salts into the body at the gills.

Terrestrial Animals

Most terrestrial animals need to drink water occasionally to make up for the water lost by excretion, respiration, in sweat and feces. Like marine bony fishes, birds and reptiles that live near the sea are able to drink seawater despite its high osmolarity because they have a nasal salt gland that can excrete large volumes of concentrated salt solution.

To prevent water loss, some animals excrete a nitrogenous waste that is rather insoluble. An impermeable outer covering also helps. Compare the moist, thin, permeable skin of a frog to the dry, horny, thick skin of a lizard and you know immediately which one is adapted to a dry terrestrial environment. Aside from these measures to prevent water loss, unique adaptations abound. To prevent loss of water during the process of breathing, certain desert animals, like the camel and kangaroo rat, have a nasal passage that has a highly convoluted mucous membrane surface. This surface allows them to capture moisture from exhaled air, which they use to humidify air that is being inhaled. In contrast, the air we exhale is full of moisture, which is why you can see it on cold winter mornings—the moisture in exhaled air is condensing. As we shall see, humans mainly conserve water by producing a hypertonic urine. The kangaroo rat also forms a very concentrated urine, and its fecal material is almost completely dry. This animal is so adapted to conserving water that it can survive using metabolic water derived from aerobic respiration.

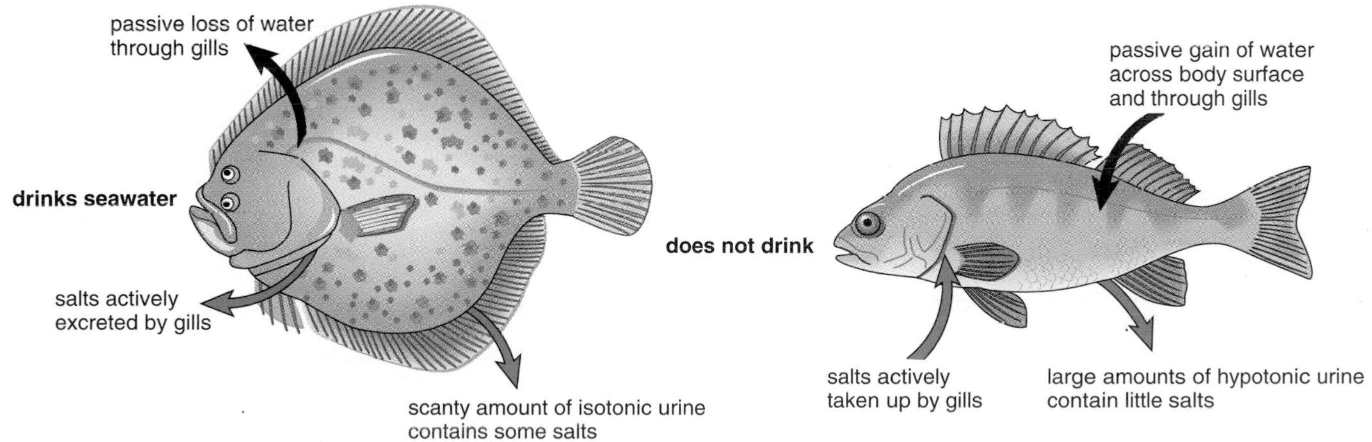

Figure 45.2 Osmotic regulation in bony fishes.
The black arrows represent passive transport from the environment, and the blue arrows represent active transport by the fishes to counteract environmental pressures. a. Marine bony fish. b. Freshwater bony fish.

45.2 Nitrogenous Waste Products

The breakdown of various molecules, including nucleic acids and amino acids, results in nitrogenous wastes. For simplicity's sake, however, we will limit our discussion to amino acid metabolism. Amino acids not used for synthesis by the body are oxidized to generate energy, or are converted to fats or carbohydrates that can be stored. In either case, the *amino groups* (—NH_2) must be removed because they are not needed when amino acids are converted to these forms of energy storage. Once the amino groups have been removed from amino acids, they may be excreted from the body in the form of ammonia, urea, or uric acid, depending on the species (Table 45.1). Removal of amino groups from amino acids requires a fairly constant amount of energy. The energy requirement for the conversion of amino groups to ammonia, urea, or uric acid, however, differs, as indicated in Figure 45.3.

Excreting Nitrogenous Wastes

Amino groups removed from amino acids immediately form **ammonia** (NH_3) by addition of a third hydrogen ion. Little or no energy is required to convert an amino group to ammonia by the addition of a hydrogen ion. Ammonia is quite toxic and can be used as a nitrogenous excretory product only if a good deal of water is available to wash it from the body. The high solubility of ammonia permits this means of excretion in bony fishes, aquatic invertebrates, and amphibians, whose gills and skin surfaces are in direct contact with the water of the environment.

Terrestrial amphibians and mammals usually excrete **urea** [Gk. *urina,* urine] as their main nitrogenous waste. Urea is much less toxic than ammonia and can be excreted in a moderately concentrated solution. This allows body water to be conserved, an important advantage for terrestrial animals with limited access to water. Production of urea, however, requires the expenditure of energy. Urea is produced in the liver by a set of energy-requiring enzymatic reactions known as the *urea cycle.* In the cycle, carrier molecules take up carbon dioxide and two molecules of ammonia, finally releasing urea.

Table 45.1

Nitrogenous Waste Excretion

Product	Habitat	Animals
Ammonia	Water	Aquatic invertebrates Bony fishes Amphibian larvae
Urea	Land	Adult amphibians Mammals
Uric acid	Land	Insects Birds Reptiles

Uric acid is routinely excreted by insects, reptiles, and birds. Uric acid is not very toxic, and is poorly soluble in water. Poor solubility is an advantage if water conservation is needed, because uric acid can be concentrated even more readily than can urea. In reptiles and birds, a dilute solution of uric acid passes from the kidneys to the *cloaca,* a common reservoir for the products of the digestive, urinary, and reproductive systems. The cloacal contents are refluxed into the large intestine, where water is reabsorbed. Embryos of reptiles and birds develop inside completely enclosed, shelled eggs. The production of insoluble, relatively nontoxic uric acid is advantageous for shelled embryos because all nitrogenous wastes are stored inside the shell until hatching takes place.

Uric acid is synthesized by a long, complex series of enzymatic reactions that requires expenditure of even more ATP than does urea synthesis. Here again, there seems to be a trade-off between the advantage of water conservation and the disadvantage of energy expenditure for synthesis of an excretory molecule.

Animals excrete nitrogenous wastes as ammonia, urea, or uric acid. Ammonia requires the most water to excrete; uric acid requires the most energy to produce.

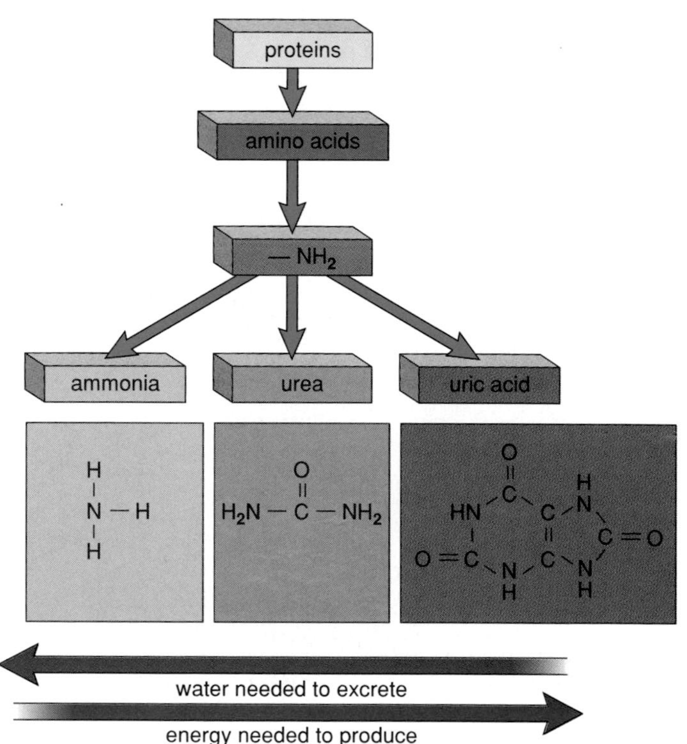

Figure 45.3 Nitrogenous wastes.
Proteins are hydrolyzed to amino acids, whose breakdown results in a carbon skeleton and amino groups. The carbon skeleton can be used as an energy source, but the amino groups must be excreted as ammonia, urea, or uric acid.

45.3 Organs of Excretion

Most animals have tubular organs that regulate water, salt, and waste concentration in the internal environment.

Planaria Have Flame Cells

Planaria, which live in fresh water, have two strands of branching excretory tubules that open to the outside of the body through excretory pores (Fig. 45.4*a*). Located along the tubules are bulblike **flame cells,** each of which contains a cluster of beating cilia that looks like a flickering flame under the microscope. The beating of flame-cell cilia propels hypotonic fluid through the excretory canals and out of the body. The system is believed to function in osmotic regulation and also in excreting wastes.

Earthworms Have Nephridia

The earthworm's body is divided into segments, and nearly every body segment has a pair of excretory structures called **nephridia** [Gk. *nephros,* kidney]. Each nephridium is a tubule with a ciliated opening (the nephridiostome) and an excretory pore (the nephridiopore) (Fig. 45.4*b*). As fluid from the coelom is propelled through the tubule by beating cilia, certain substances are reabsorbed and carried away by a network of capillaries surrounding the tubule. This process results in the formation of urine that contains only metabolic wastes, salts, and water.

Each day, an earthworm excretes a lot of water and may produce a volume of urine equal to 60% of its body weight. Its excretion of ammonia is consistent with this finding.

Insects Have Malpighian Tubules

Insects have a unique excretory system consisting of long, thin tubules, **Malpighian tubules,** attached to the gut. Uric acid simply flows from the surrounding hemolymph into these tubules, and water follows a salt gradient established by active transport of K^+. Water and other useful substances are reabsorbed at the rectum but the uric acid leaves the body at the anus. Insects that live in water or that eat large quantities of moist food reabsorb little water. But insects in dry environments reabsorb most of the water and excrete a dry, semisolid mass of uric acid.

Most animals have a primary excretory organ. Planaria have flame cells, earthworms have nephridia, and insects have Malpighian tubules.

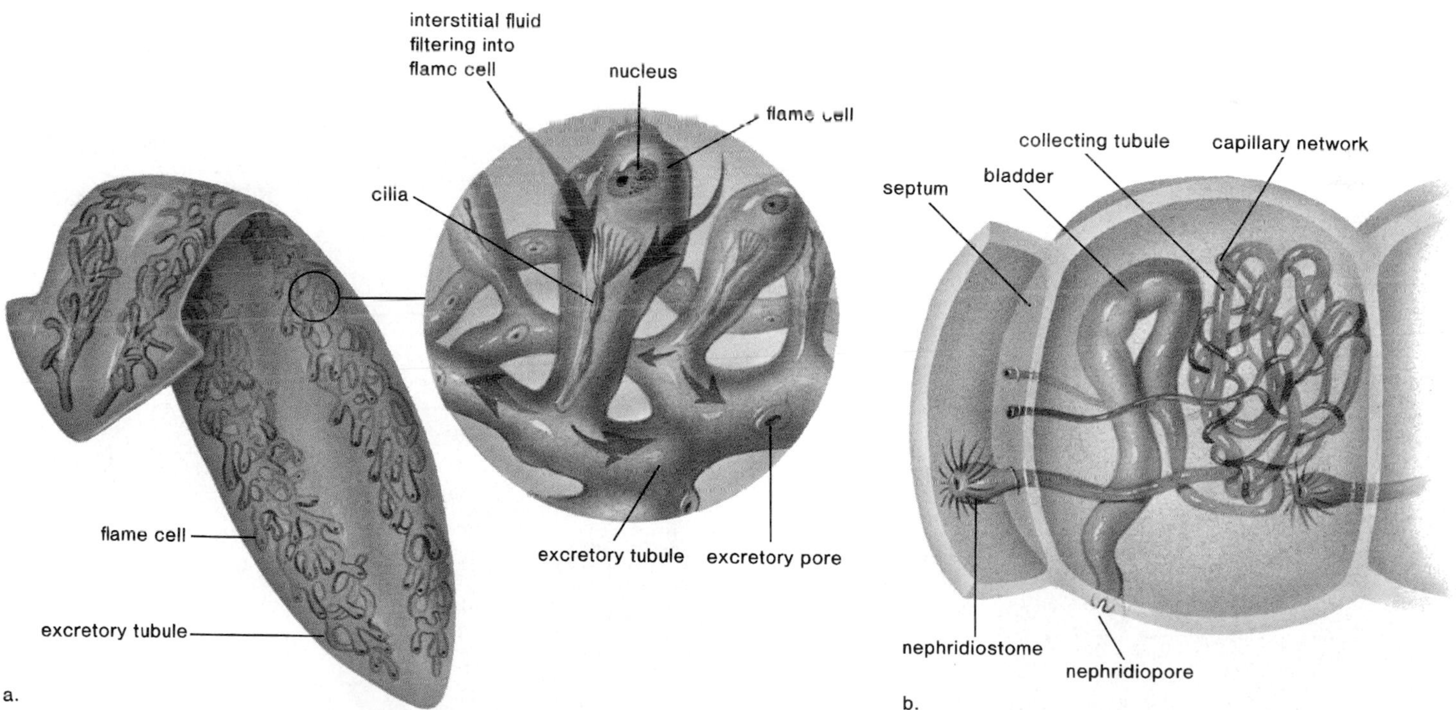

Figure 45.4 Excretory organs in animals.
a. The flame cell excretory system in planaria. Two or more tracts of branching tubules run the length of the body and open to the outside by pores. At the ends of side branches there are small flame cells, bulblike cells whose beating cilia cause fluid to enter the tubules that remove excess fluid from the body. **b.** The earthworm nephridium. The nephridium has a ciliated opening, the nephridiostome, that leads to a coiled tubule surrounded by a capillary network. Urine can be temporarily stored in the bladder before being released to the outside via a pore termed a nephridiopore. Most segments contain a pair of nephridia, one on each side.

45.4 Humans Have a Urinary System

The human urinary system consists of the organs shown in Figure 45.5. The human **kidneys** are bean-shaped, reddish brown organs, each about the size of a fist. They are located on either side of the vertebral column just below the diaphragm, where they are partially protected by the lower rib cage. **Urine** [Gk. *urina*, urine] made by the kidneys is conducted from the body by the other organs in the urinary system. Each kidney is connected to a **ureter,** a duct that conducts urine from the kidney to the **urinary bladder,** where it is stored until it is voided from the body through the single **urethra.** In males, the urethra passes through the penis, and in females, it opens ventral to the opening of the vagina. There is no connection between the genital (reproductive) and urinary systems in females, but there is a connection in males. In males, the urethra also carries sperm during ejaculation.

Kidneys Have Three Regions

If a kidney is sectioned longitudinally, three major parts can be distinguished (Fig. 45.6). The outer region is the **renal cortex,** which has a somewhat granular appearance. The **renal medulla** consists of striped, pyramid-shaped regions that lie on the inner side of the cortex. The innermost part of the kidney is a hollow chamber called the **renal pelvis.** Urine formed in the kidney collects in the renal pelvis before entering the ureter. Microscopically, each kidney is composed of about one million tiny tubules called **nephrons** [Gk. *nephros,* kidney]. Some nephrons are located primarily in the cortex, but others dip down into the medulla, as shown in Figure 45.6*b*.

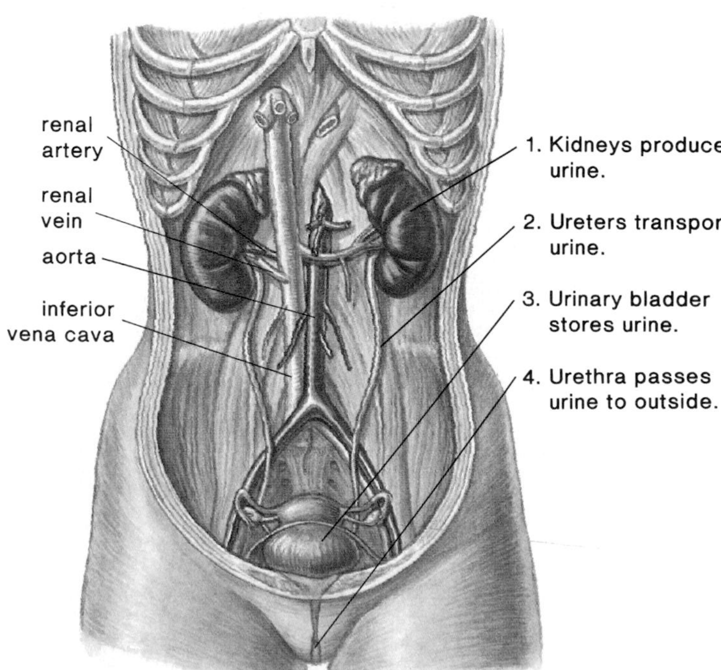

Figure 45.5 The human urinary system.
Urine is found only within the kidneys, the ureters, the urinary bladder, and the urethra.

Nephrons Are Numerous

Each nephron is made of several parts (Fig. 45.7). The blind end of a nephron is pushed in on itself to form a cuplike structure called the **glomerular capsule** [L. *glomeris,* ball] (formerly called Bowman's capsule). The outer layer of the glomerular capsule is composed of squamous epithelial cells; the inner layer is composed of specialized cells that allow easy passage of molecules. Leading from the glomerular capsule is a portion of the nephron known as the **proximal convoluted tubule** [L. *proximus,* nearest], which is lined by cells with many mitochondria and an inner brush border (tightly packed microvilli). Then simple squamous epithelium appears in a portion called the **loop of the nephron** (formerly called the loop of Henle). This is followed by the **distal convoluted tubule** [L. *distantia,* far]. Several distal convoluted tubules enter one **collecting duct.** The collecting duct transports urine down through the medulla and delivers it to the pelvis. The loop of the nephron and the collecting duct give the pyramids of the medulla their striped appearance (Fig. 45.6).

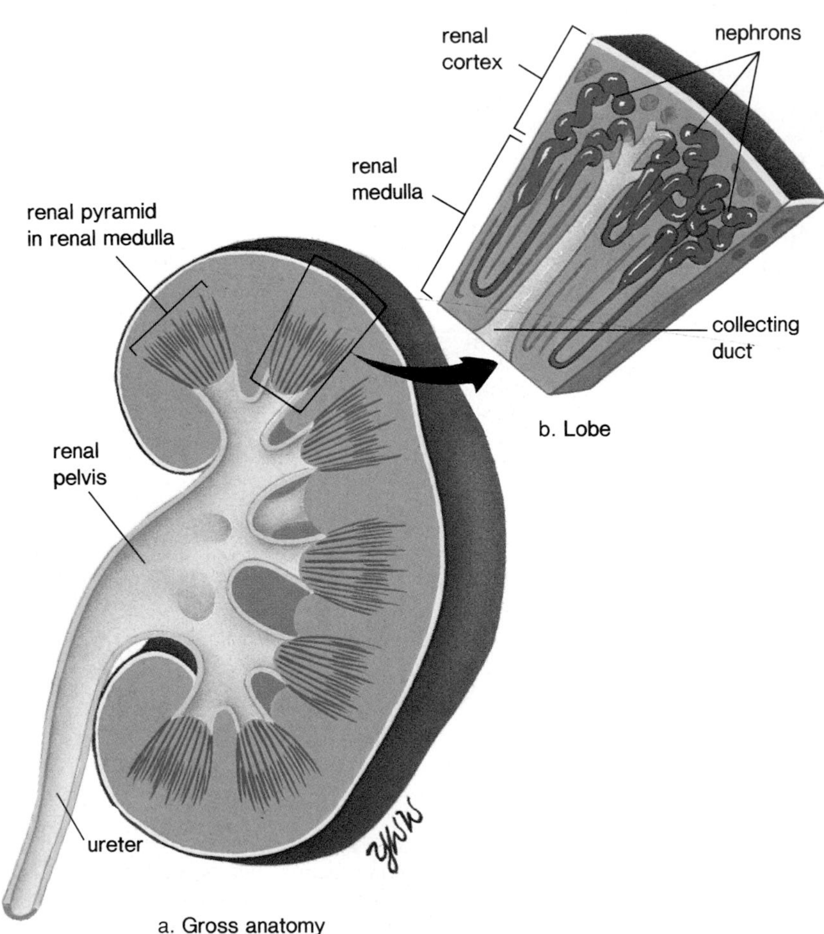

Figure 45.6 Macroscopic and microscopic anatomy of the kidney.
a. Longitudinal section of kidney showing the location of the renal cortex, the renal medulla, and the renal pelvis. **b.** An enlargement of one renal lobe showing the placement of nephrons.

Each nephron has its own blood supply (Fig. 45.7). The renal artery branches into numerous small arteries, which branch into arterioles, one for each nephron. Each arteriole, called an *afferent arteriole,* divides to form a capillary tuft, the **glomerulus** [L. *glomeris,* ball], which is surrounded by the glomerular capsule. The glomerular capillaries drain into an *efferent arteriole,* which subsequently branches into a second capillary network around the tubular parts of the nephron. These capillaries, called *peritubular capillaries,* lead to venules that join to form veins leading to the renal vein, a vessel that enters the inferior vena cava.

Microscopically, the human kidney is composed of nephrons, tubules with specific parts that are richly supplied with capillaries.

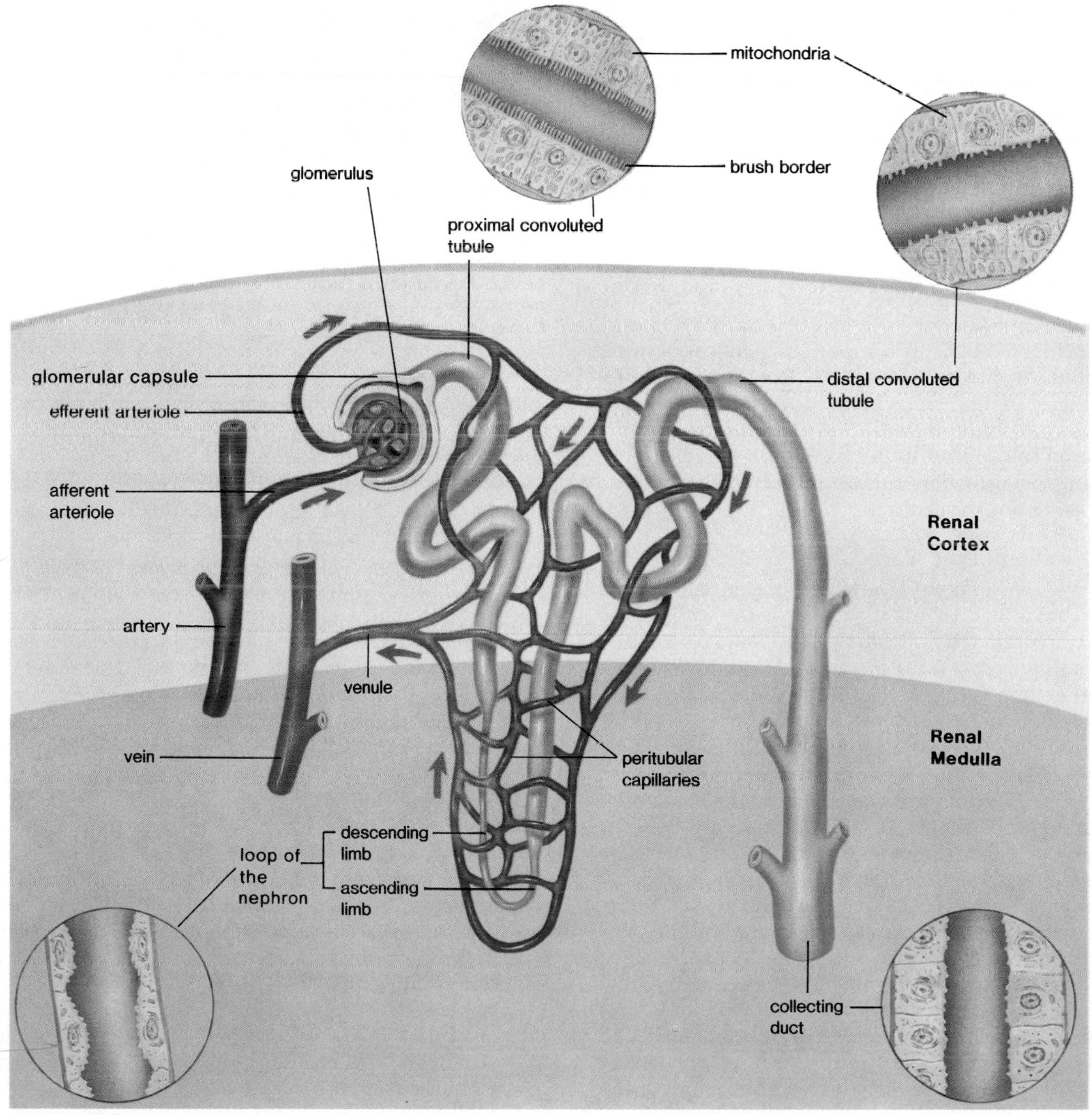

Figure 45.7 Nephron anatomy.
The enlargements show types of tissue at the locations noted. Trace the path of blood about the nephron by following the arrows.

How Urine Is Made

Urine production requires three distinct processes:

1. Glomerular filtration at the glomerular capsule;
2. Tubular reabsorption, including selective reabsorption, at the proximal convoluted tubule, in particular; and
3. Tubular secretion at the distal convoluted tubule in particular.

Glomerular Filtration

Glomerular filtration is the movement of small molecules across the glomerular wall as a result of blood pressure. When blood enters the glomerulus, blood pressure is sufficient to cause small molecules, such as water, nutrients, salts, and wastes, to move from the glomerulus to the inside of the glomerular capsule, especially since the glomerular walls are 100 times more permeable than the walls of most capillaries elsewhere in the body. The molecules that leave the blood and enter the glomerular capsule are called the *glomerular filtrate.* Blood proteins and blood cells are too large to be part of this filtrate, so they remain in the blood as it flows into the efferent arteriole.

Glomerular filtrate is essentially protein free but otherwise has the same composition as tissue fluid. If this composition were not altered in other parts of the nephron, death from loss of nutrients (starvation), and loss of water (dehydration) would quickly follow. The total blood volume averages about 5 liters and this amount of fluid is filtered every 40 minutes. Most of the filtered water is obviously quickly returned to the blood or a person would actually die from urination. Tubular reabsorption, however, prevents this from happening.

Tubular Reabsorption

Tubular reabsorption takes place when substances from the proximal convoluted tubule move across the walls of the proximal convoluted tubule into the associated peritubular capillaries (Fig. 45.8). The osmolarity of the blood is essentially the same as that of the filtrate within the glomerular capsule, and therefore osmosis of water from the filtrate into the blood cannot yet occur. However, sodium ions (Na^+) are actively pumped into the peritubular capillary and then chloride ions (Cl^-) follow passively. Now the osmolarity of the blood is such that water moves passively from the tubule into the blood. About 60–70% of salt and water are reabsorbed at the proximal convoluted tubule.

Nutrients, such as glucose and amino acids, also return to the blood at the proximal convoluted tubule. This is a selective process, because only molecules recognized by carrier proteins in plasma membranes are actively reabsorbed. The cells of the proximal convoluted tubule have numerous microvilli, which increase the surface area, and numerous mitochondria, which supply the energy needed for active transport (Fig. 45.8). Glucose is an example of a molecule that ordinarily is reabsorbed completely because there is a plentiful supply of carrier molecules for it. However, if there is more glucose in the filtrate than there are carriers to handle it, glucose will exceed its renal threshold, or transport maximum. When this happens, the excess glucose in the filtrate will appear in the urine. In diabetes mellitus, there is an abnormally large amount of glucose in the filtrate because the liver fails to store glucose as glycogen.

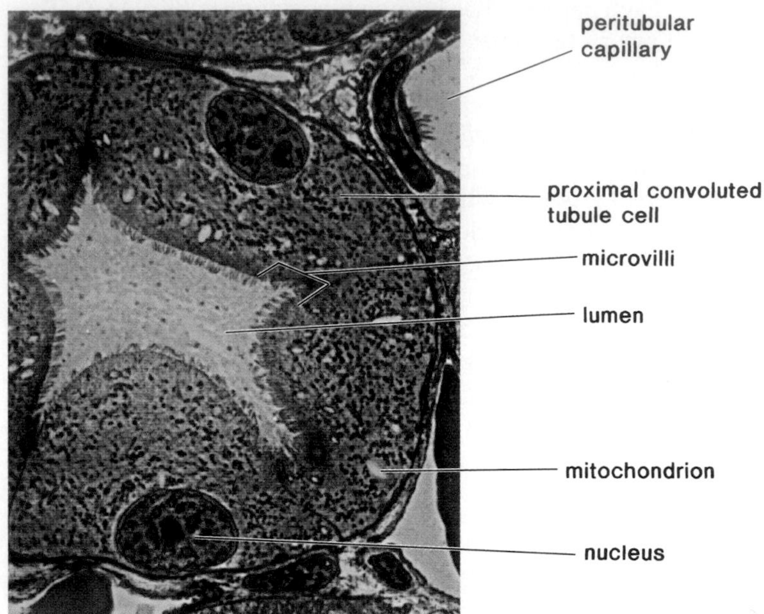

Figure 45.8 Proximal convoluted tubule. This photomicrograph shows that the cells lining the proximal convoluted tubule have a brush border composed of microvilli, which greatly increases the surface area exposed to the lumen. The peritubular capillary surrounds the cells. There are also numerous mitochondria, which provide the energy for active transport of nutrients.

Urea is an example of a substance that is passively reabsorbed from the filtrate. At first, the concentration of urea within the filtrate is the same as blood plasma. But after water is reabsorbed, the urea concentration is greater than that of peritubular plasma. This process causes reabsorption of about 50% of the filtered urea.

Tubular Secretion

Tubular secretion refers to the transport of substances into the tubular lumen by means other than glomerular filtration. For our purposes, tubular secretion may be particularly associated with the distal convoluted tubule (Fig. 45.9). Substances such as uric acid, hydrogen ions, ammonia, and penicillin are eliminated by tubular secretion. The process of secretion may be viewed as helping to rid the body of potentially harmful compounds that were not filtered into the glomerulus.

Following water reabsorption, which is discussed next, urine has the composition given in Table 45.2.

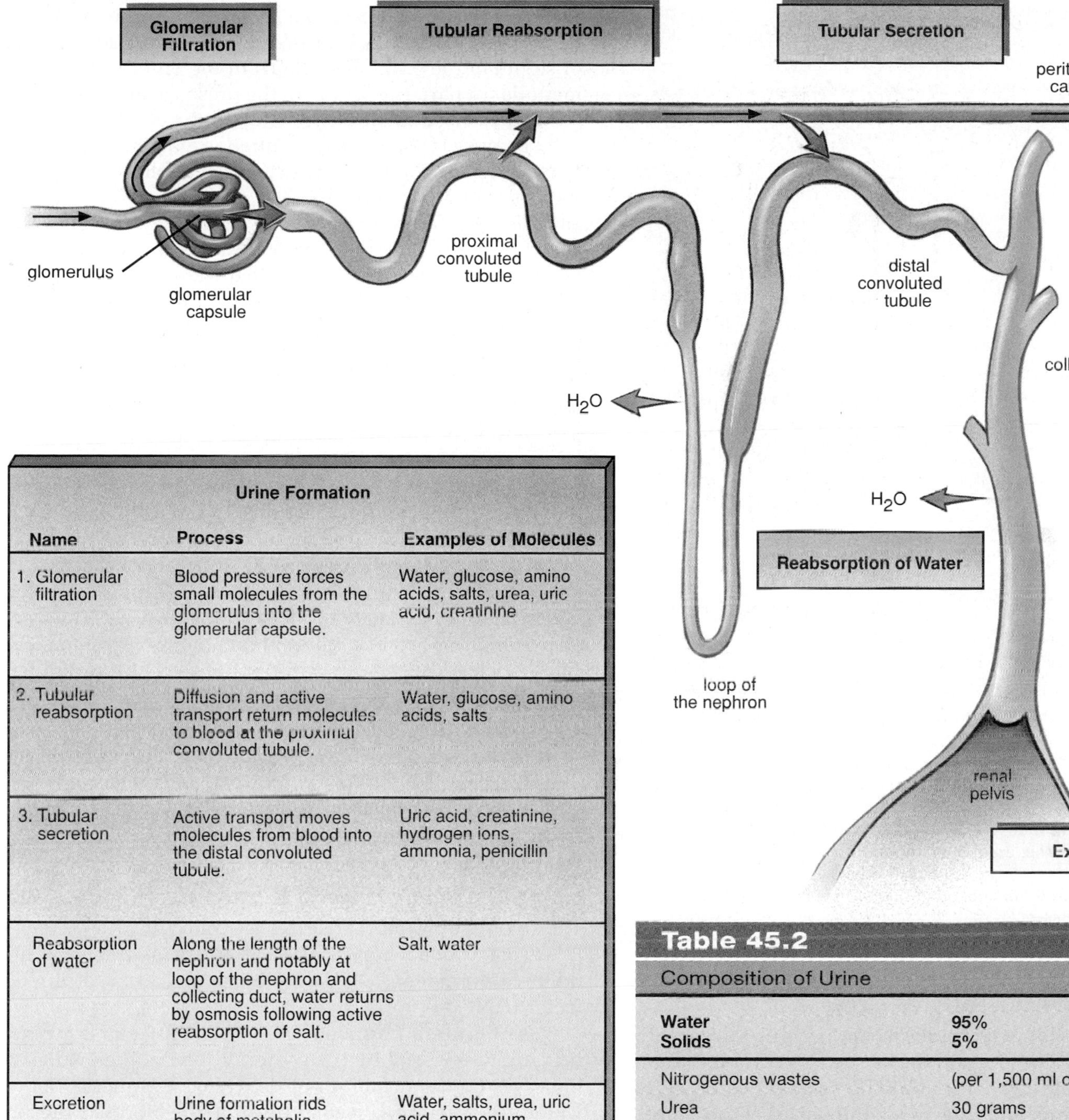

Urine Formation

Name	Process	Examples of Molecules
1. Glomerular filtration	Blood pressure forces small molecules from the glomerulus into the glomerular capsule.	Water, glucose, amino acids, salts, urea, uric acid, creatinine
2. Tubular reabsorption	Diffusion and active transport return molecules to blood at the proximal convoluted tubule.	Water, glucose, amino acids, salts
3. Tubular secretion	Active transport moves molecules from blood into the distal convoluted tubule.	Uric acid, creatinine, hydrogen ions, ammonia, penicillin
Reabsorption of water	Along the length of the nephron and notably at loop of the nephron and collecting duct, water returns by osmosis following active reabsorption of salt.	Salt, water
Excretion	Urine formation rids body of metabolic wastes.	Water, salts, urea, uric acid, ammonium, creatinine

Figure 45.9 Urine formation.
The three steps in urine formation are numbered. Reabsorption of water is not an individual step because it occurs along the length of the nephron and also at the loop of the nephron and collecting duct. Excretion is not a step because it is the end result.

Table 45.2

Composition of Urine

Water	**95%**
Solids	**5%**
Nitrogenous wastes	(per 1,500 ml of urine)
Urea	30 grams
Creatinine	1–2 grams
Ammonia	1–2 grams
Uric acid	1 gram
Ions (Salts)	**25 grams**
Positive Ions	*Negative Ions*
Sodium	Chlorides
Potassium	Sulfates
Magnesium	Phosphates
Ammonium	
Calcium	

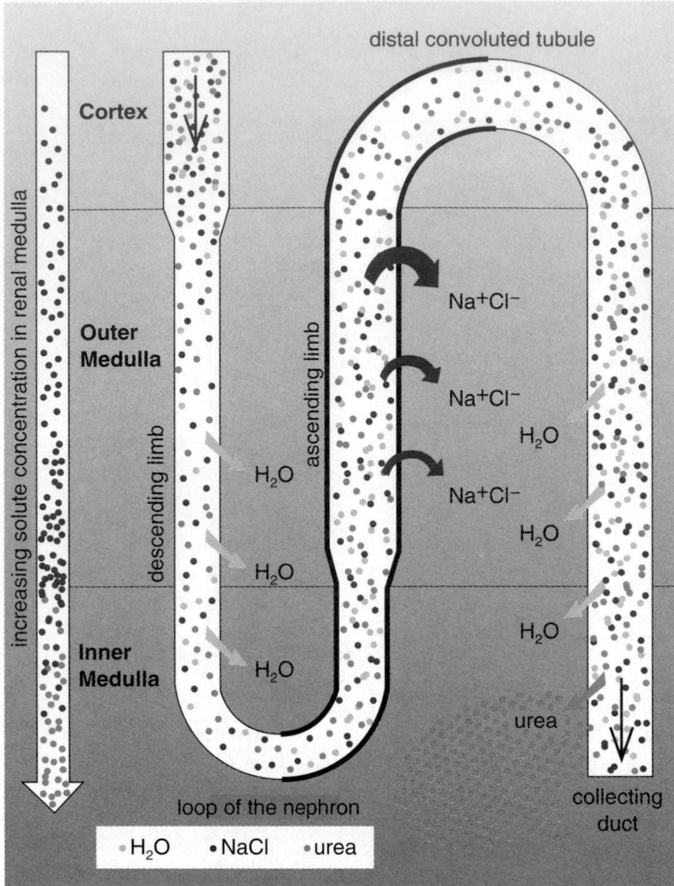

Figure 45.10 Countercurrent mechanism.
Salt (NaCl) diffuses and is actively transported out of the ascending limb of the loop of the nephron into the renal medulla. Urea is believed to leak from the collecting duct and to enter the tissues of the renal medulla. This creates an increasingly hypertonic environment (see far left arrow), which draws water out of the descending limb and the collecting duct. This water is returned to the circulatory system.

Production of Concentrated Urine

Reptiles and birds rely primarily on the gut to reabsorb water, but mammals rely on the kidneys. Nephrons have a loop of the nephron in which fluid flows downward within a descending limb and upward within an ascending limb. This countercurrent flow allows those nephrons that penetrated deeply into the renal medulla to produce a concentrated urine—a urine that is hypertonic to blood plasma.

First, we will consider the ascending limb of the loop of the nephron. Salt (NaCl) passively diffuses out of the lower portion of the ascending limb and the upper, thick portion of the limb actively transports salt out into the tissue of the outer renal medulla (Fig. 45.10). However, less and less salt is available for transport from the tubule as fluid moves up the thick portion of the ascending limb. (Water does not leave the ascending limb because it is impermeable to water.) Note the arrow on the far left of Figure 45.10 indicates that there is an increasing amount of solute in the renal medulla from the outer to the inner medulla. In part, this is due to the presence of urea, which has leaked from the lower portion of the collecting duct.

Because of the solute concentration gradient established by the ascending limb within the renal medulla, water leaves the descending limb of the loop of the nephron along its entire length. *This occurs because the decreasing number of water molecules in the descending limb encounters an increasing concentration of solute.* This ensures that water will continue to leave the descending limb as it penetrates into the inner medulla. As fluid rounds the bend within the loop of the nephron, it has a high concentration of salt—the very salt that leaves the ascending limb as it rises from the inner medulla to the outer medulla.

Fluid entering a collecting duct comes from the distal convoluted tubule. This fluid is isotonic to the cells of the cortex. But as the collecting duct passes through the renal medulla, water diffuses out of the collecting duct into the renal medulla due to the increasing solute concentration maintained by the descending limb. Finally, as stated previously, urea flows into the inner medulla from the collecting duct. This urea concentration provides the final step in making the urine hypertonic to blood plasma, because it draws still more water out of the collecting duct.

The ascending limb of the loop of the nephron establishes a concentration gradient in the medulla, which causes water to exit the descending limb of the loop of the nephron and the collecting duct along their entire lengths.

Hormones Maintain Water Balance

The amount of water that leaves the collecting duct is regulated by hormonal action. The hormones involved are antidiuretic hormone (ADH), aldosterone, and atrial natriuretic factor (ANF).

Antidiuretic hormone [Gk. *anti,* against, and L. *ouresis,* urination], released by the posterior lobe of the pituitary, increases the permeability of the collecting duct so more water leaves it and is reabsorbed into the blood. If the osmotic pressure of the blood increases, ADH is released, more water is reabsorbed, and consequently, there is less urine. On the other hand, if the osmotic pressure of the blood decreases, ADH is not released, more water is excreted, and consequently, more urine is formed. Drinking alcohol causes diuresis (increased urine flow) because it inhibits the secretion of ADH. Beer drinking causes diuresis primarily because of increased fluid intake. Drugs called diuretics are often prescribed for high blood pressure. These drugs increase urinary excretion and thus reduce blood volume and blood pressure.

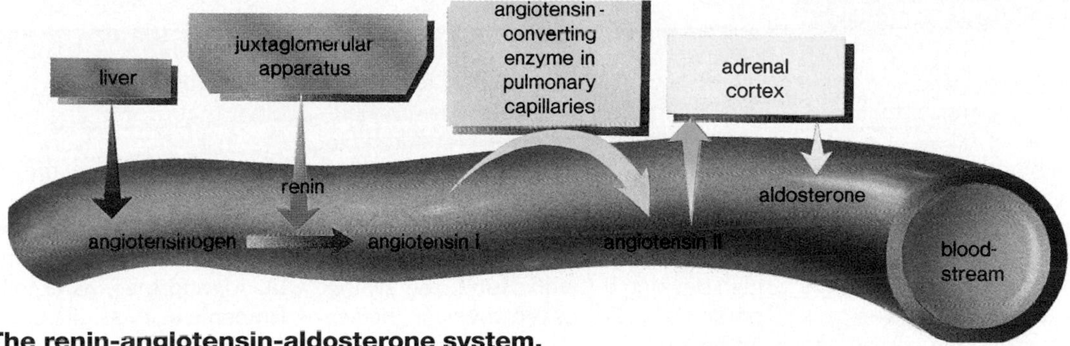

Figure 45.11 The renin-angiotensin-aldosterone system.
The organs named in the boxes above the blood vessel act on the substances listed in the bloodstream to bring about the release of aldosterone, which causes reabsorption of sodium ions and a subsequent rise in blood pressure.

The hormone **aldosterone,** which is secreted by the adrenal cortex, helps maintain the sodium (Na^+) and potassium (K^+) ion balance of the blood. It causes the distal convoluted tubule to reabsorb Na^+ and to excrete K^+. The increase of Na^+ in the blood causes water to be reabsorbed, leading to an increase in blood volume and blood pressure. Blood pressure is constantly monitored within the *juxtaglomerular apparatus,* a region of contact between the afferent arteriole and the distal convoluted tubule. When blood pressure is insufficient to promote glomerular filtration, afferent arteriole cells secrete renin. *Renin* is an enzyme that changes angiotensinogen (a large plasma protein produced by the liver) into angiotensin I. Later, angiotensin I is converted to angiotensin II in the lungs by angiotensin-converting enzyme. Angiotensin II, a powerful vasoconstrictor, stimulates the adrenal cortex to release *aldosterone,* which allows Na^+ to be reabsorbed and blood volume and blood pressure to rise. This sequence of events is called the renin-angiotensin-aldosterone system (Fig. 45.11).

Notice that both ADH and aldosterone act to increase blood volume and to raise the blood pressure. When blood pressure rises, the heart produces the peptide hormone called **atrial natriuretic hormone (ANH)** that inhibits the secretion of renin by the juxtaglomerular apparatus and the release of ADH by the posterior pituitary. This is an example of how hormones serve as checks and balances to one another.

Blood volume and blood pressure are raised when aldosterone and ADH are secreted. The actions of these hormones are opposed by ANH, and in this way, normal blood volume and pressure are maintained.

How the pH Is Adjusted

The kidneys help maintain the pH level of the blood within a narrow range, and the whole nephron takes part in this process. The excretion of hydrogen ions (H^+) and ammonia (NH_3), together with the reabsorption of sodium (Na^+) and bicarbonate ions (HCO_3^-), is adjusted to keep the pH within normal bounds. If the blood is acidic, hydrogen ions are excreted in combination with ammonia, while sodium and bicarbonate ions are reabsorbed. This will restore the pH because sodium ions promote the formation of hydroxide ions, while bicarbonate takes up hydrogen ions when carbonic acid is formed:

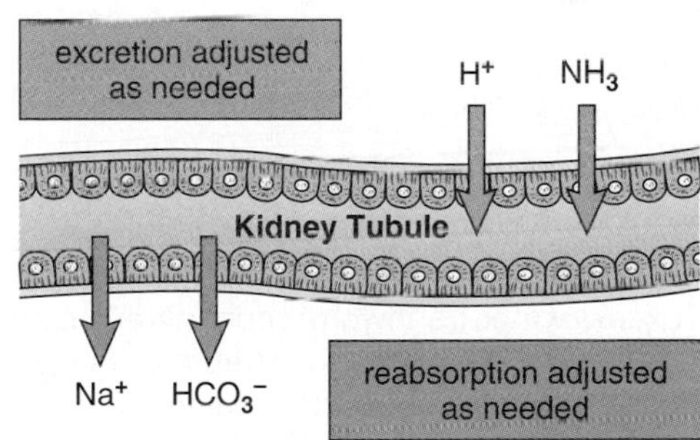

If the blood is basic, fewer hydrogen ions are excreted, and fewer sodium and bicarbonate ions are reabsorbed.

Reabsorption and/or excretion of ions by the kidneys illustrates their homeostatic ability: they maintain not only the pH of the blood but also its osmolarity.

The kidneys maintain blood pH level within a narrow range by making adjustments in the excretion of hydrogen ions and ammonium, and in the reabsorption of sodium and bicarbonate ions.

connecting concepts

An artificial kidney employs the principle of dialysis in order to cleanse the blood. Blood from a patient's artery is passed through a semimembranous tubing, which is surrounded by a dialysis solution that contains salt and various other small molecules. The concentration gradient between blood and the dialysis solution is such that waste molecules and excess salts diffuse through the membrane into the dialysis solution, while blood cells and plasma proteins stay behind. Because the volume of dialysis solution is large, wastes continually diffuse out of the blood. In the course of a 6-hour treatment, 50–250 grams of urea can be removed from a patient, an amount that greatly exceeds the urea clearance rate of normal kidneys. On the other hand, the dialysis contains salts at a level that maintains their proper concentration in the blood before it is returned to the patient.

How does a kidney machine function differently from an animal's excretory organ? In the human body, blood is also passing through a tube, namely the person's blood vessel; however, the solution that forms on the other side of this tube is free of proteins, but otherwise has the same composition as blood plasma. Why? Because blood pressure forces water and small molecules to cross a highly permeable membrane. There is no opposing solution that maintains the normal nutrient concentration and osmolarity of the blood, nor an extreme concentration gradient that draws urea out of the blood.

In the human kidney, energy must be expended to retrieve the water, salts, and nutrients that are on their way out of the body. The sodium-potassium pump is employed to actively return Na^+ to the blood, and then water follows passively. Nutrients such as glucose and amino acids are reabsorbed by carrier-mediated active transport processes. Therefore, we see that in contrast to an artificial kidney, an animal's kidney is selective about what small molecules are returned to the blood, rather than what small molecules leave the blood. Finally, hormones regulate the osmolarity of the blood and fine tune just how much water and salts are excreted.

Summary

45.1 Osmotic Regulation

Osmotic regulation is important to animals. Most have to balance their water and salt intake and their excretion to maintain normal solute and water concentration in body fluids. Marine fishes constantly drink water, excrete salts at the gills, and pass an isotonic urine. Freshwater fishes never drink water; they take in salts at the gills and pass a hypotonic urine.

45.2 Nitrogenous Waste Products

Animals excrete nitrogenous wastes, which differ as to the amount of water and energy required to excrete them. Aquatic animals usually excrete ammonia, and land animals excrete either urea or uric acid.

45.3 Organs of Excretion

Animals often have an excretory organ. The flame cells of planaria rid the body of excess water. Earthworm nephridia exchange molecules with the blood in a manner similar to vertebrate kidneys. Malpighian tubules in insects take up metabolic wastes and water from the hemolymph. Later, the water is absorbed by the gut.

45.4 Humans Have a Urinary System

Kidneys are a part of the human urinary system. Microscopically, each kidney is made up of nephrons, each of which has several parts and its own blood supply.

Urine formation requires three steps: pressure filtration, when nutrients, water, and wastes enter the glomerular capsule; selective reabsorption, when nutrients and most water is reabsorbed into the proximal convoluted tubule; and tubular secretion, when additional wastes are added to the tubule (e.g., the distal convoluted tubule).

Humans excrete a hypertonic urine. The ascending limb of the loop of the nephron actively extrudes salt so that the renal medulla is increasingly hypertonic relative to the contents of the descending limb and the collecting duct. Since urea leaks from the lower end of the collecting duct, the inner medulla has the highest concentration of solute. Therefore, a countercurrent mechanism assures that water will diffuse out of the descending limb and the collecting duct.

Three hormones are involved in maintaining the water content of the blood. The hormone ADH (antidiuretic hormone), which makes the collecting duct more permeable, is secreted by the posterior pituitary in response to an increase in the osmotic pressure of the blood.

The hormone aldosterone is secreted by the adrenal cortex after the low sodium ion (Na^+) content of the blood and the resultant low blood pressure have caused the kidneys to release renin. The presence of renin leads to the formation of angiotensin II, which causes the adrenal cortex to release aldosterone. Aldosterone causes the kidneys to retain Na^+; therefore, water is reabsorbed and blood pressure rises. The atrial natriuretic hormone prevents the secretion of ADH and aldosterone.

The kidneys adjust the pH of the blood by excreting or conserving hydrogen ions (H^+), ammonia (NH_3), and sodium (Na^+) and bicarbonate ions (HCO_3^-) as appropriate.

Reviewing the Chapter

1. Contrast the osmotic regulation of a marine bony fish with that of a freshwater bony fish. 804–5
2. Give examples of how other types of animals regulate their water and salt balance. 805
3. Relate the three primary nitrogenous wastes to the habitat of animals. 806
4. Describe how the excretory organs of the earthworm and the insect function. 807
5. Describe the path of urine, and give a function for each structure mentioned. 808
6. Describe the macroscopic anatomy of a human kidney, and relate it to the placement of nephrons. 808–9
7. List the parts of a nephron, and give a function for each structure mentioned. 808–9
8. Describe how urine is made by outlining what happens at each part of the nephron. 810–11
9. Explain the countercurrent mechanism by which water is reabsorbed and the urine is hypertonic. 812
10. What role do ADH (antidiuretic hormone), aldosterone, and ANH (atrial natriuretic hormone) play in regulating the tonicity of urine? How does this affect blood pressure? 812–13
11. How does the nephron regulate the pH of the blood? 813

Testing Yourself

Choose the best answer for each question.

1. Which of these is mismatched?
   a. insects—excrete uric acid
   b. humans—excrete urea
   c. fishes—excrete ammonia
   d. birds—excrete ammonia
2. One advantage of urea excretion over uric acid excretion is that urea
   a. requires less energy to form.
   b. can be concentrated to a greater extent.
   c. is not a toxic substance.
   d. requires no water to excrete.
3. Freshwater bony fishes maintain water balance by
   a. excreting salt across their gills.
   b. periodically drinking small amounts of water.
   c. excreting a hypotonic urine.
   d. excreting wastes in the form of uric acid.
4. Animals with which of these are most likely to excrete a semisolid nitrogenous waste?
   a. nephridia
   b. Malpighian tubules
   c. human kidneys
   d. All of these are correct.
5. In which of these human structures are you least apt to find urine?
   a. large intestine
   b. urethra
   c. ureter
   d. bladder
6. Excretion of a hypertonic urine in humans is associated best with the
   a. glomerular capsule.
   b. proximal convoluted tubule.
   c. loop of the nephron.
   d. distal convoluted tubule.
7. The presence of ADH (antidiuretic hormone) causes an individual to excrete
   a. sugars.
   b. less water.
   c. more water.
   d. Both a and c are correct.
8. In humans, water is
   a. found in the glomerular filtrate.
   b. reabsorbed from the nephron.
   c. in the urine.
   d. All of these are correct.
9. Pressure filtration is associated with the
   a. glomerular capsule.
   b. distal convoluted tubule.
   c. collecting duct.
   d. All of these are correct.
10. Normally in humans, glucose
    a. is always in the filtrate and urine.
    b. is always in the filtrate with little or none in urine.
    c. undergoes tubular secretion and is in urine.
    d. undergoes tubular secretion and is not in urine.
11. Label this diagram of a nephron, and give the steps for urine formation in the boxes:

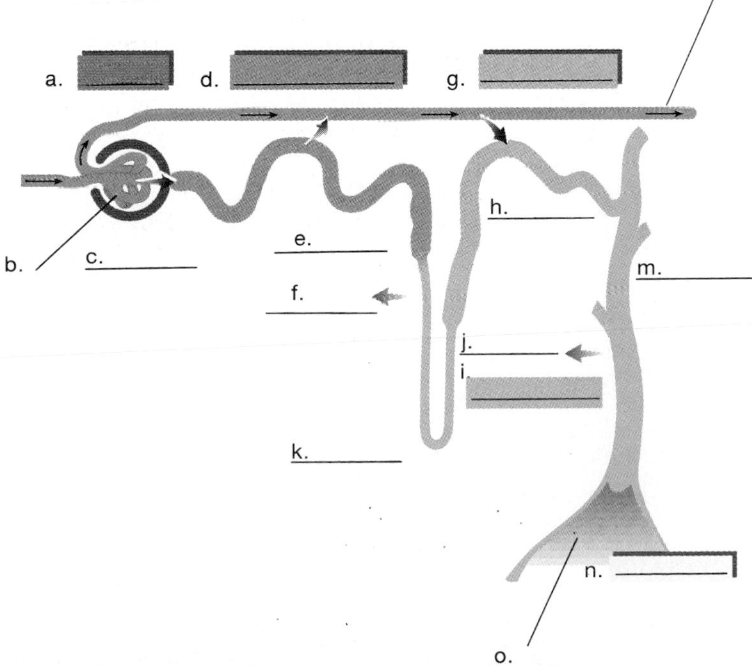

Applying the Concepts

1. *The excretory system plays a primary role in homeostasis.*
   Discuss several contributions of the human kidneys to homeostasis.
2. *Animals utilize countercurrent mechanisms to increase blood concentrations of substances.*
   Tell how the countercurrent mechanism in the gills of fishes and in the kidneys of mammals is an adaptation to their environments.
3. *Structure suits function.*
   Why would you expect the proximal convoluted tubule to be lined with cells that have many mitochondria and microvilli?

Using Technology

Your study of osmotic regulation and excretion is supported by these available technologies:

Exploring the Internet
The Mader Home Page provides resources for and help with studying this chapter.

http://www.mhhe.com/sciencemath/mader/
(Click on Biology.)

The Dynamic Human CD-ROM
Urinary System

Understanding the Terms

aldosterone 813
ammonia 806
antidiuretic hormone (ADH) 812
atrial natriuretic factor (ANF) 813
collecting duct 808
distal convoluted tubule 808
excretion 803
flame cell 807
glomerular capsule 808
glomerulus 809
kidney 808
loop of the nephron 808
Malpighian tubule 807
nephridium 807
nephron 808
proximal convoluted tubule 808
renal cortex 808
renal medulla 808
renal pelvis 808
urea 806
ureter 808
urethra 808
uric acid 806
urinary bladder 808
urine 808

Match the terms to these definitions:

a. _________ Blind, threadlike excretory tubule near the anterior end of an insect hindgut.
b. _________ Cuplike structure that is the initial portion of a nephron; where pressure filtration occurs.
c. _________ Final portion of a nephron that joins with a collecting duct; associated with tubular secretion.
d. _________ Hormone secreted by the adrenal cortex that regulates the sodium and potassium ion balance of the blood.
e. _________ Main nitrogenous waste of insects, reptiles, birds, and some dogs.
f. _________ Main nitrogenous waste of terrestrial amphibians and mammals.
g. _________ Microscopic kidney unit that regulates blood composition by glomerular filtration, tubular reabsorption, and tubular secretion; there are over a million of these per human kidney.
h. _________ Portion of a nephron between the proximal and distal convoluted tubules where water reabsorption occurs.
i. _________ Portion of a nephron following the glomerular capsule where selective reabsorption of filtrate occurs.
j. _________ Tubular structure that receives urine from the bladder and carries it to the outside of the body.

Neurons and Nervous Systems

CHAPTER 46

Chapter Concepts

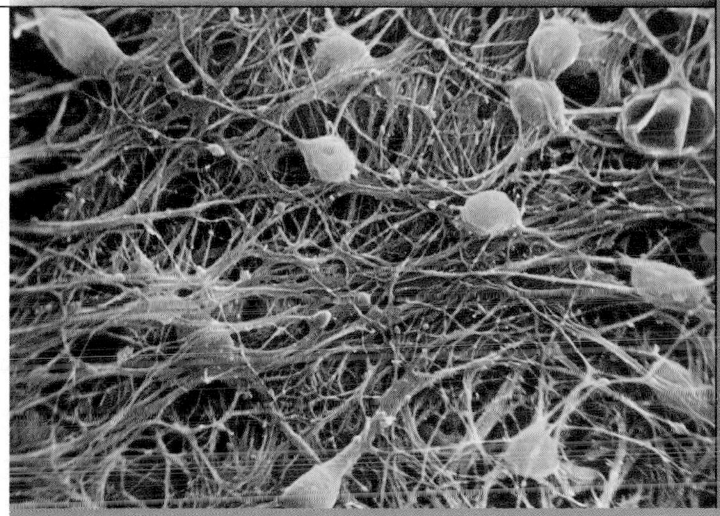

Neurons in nervous tissue of human cerebral cortex

The survival of all animals, from a minuscule rotifer to an enormous blue whale, depends on sensing the environment and responding to changes appropriately. The nervous system integrates incoming information and controls the musculoskeletal system, which allows animals to capture prey, avoid predators, and find a mate. Coordination of internal systems is especially important to achieve homeostasis. Digestion of food, breathing, and transport of nutrients are all regulated by the nervous system, with the help of the endocrine system.

Neurons and their supportive neuroglial cells make up the organs of a nervous system, such as the brain, spinal cord, and nerves of human beings. **Sensory receptors** detect changes in stimuli, and nerve impulses race through sensory neurons to the interneurons of the brain and spinal cord. The brain and spinal cord sum up the data before sending impulses via motor neurons to effectors (muscles and glands) so that a response to stimuli is possible. The principles of operation are simple, but the human nervous system is intricately complex.

46.1 Nervous Tissue

Although exceedingly complex, nervous tissue is made up of just two principal types of cells: (1) **neurons,** also called nerve cells, transmit nerve impulses; and (2) **neuroglial cells** support and nourish neurons.

In the human body, nervous tissue forms the central nervous system (CNS), which lies in the center or midline of the body, and the peripheral nervous system (PNS), which lies at the periphery—that is, to the sides—of the central nervous system.

Neurons

Neurons [Gk. *neuron,* nerve] are cells that vary in size and shape, but they all have three parts: the dendrites, an axon, and the cell body. In motor neurons (Fig. 46.1*a*), the **dendrites** [Gk. *dendron,* tree] are short processes that receive information from other neurons and conduct signals *toward* the cell body. The **axon** [Gk. *axon,* axis], on the other hand, is a process that conducts nerve impulses *away* from the cell body. The **cell body** contains the nucleus and other organelles typically found in cells. One of the main functions of the cell body is to manufacture neurotransmitters, which are chemicals stored in secretory vesicles at the ends of axons. When neurotransmitters are released, they influence the excitability of nearby neurons.

Any long axon is called a **nerve fiber.** Long axons are covered by a white **myelin sheath** [Gk. *myelos,* spinal cord] formed from the membranes of tightly spiraled neuroglial cells. In the PNS, neuroglial cells called **neurolemmocytes** (Schwann cells) perform this function, leaving gaps called neurofibril nodes (nodes of Ranvier). There is another type of neuroglial cell that performs a similar function in the CNS.

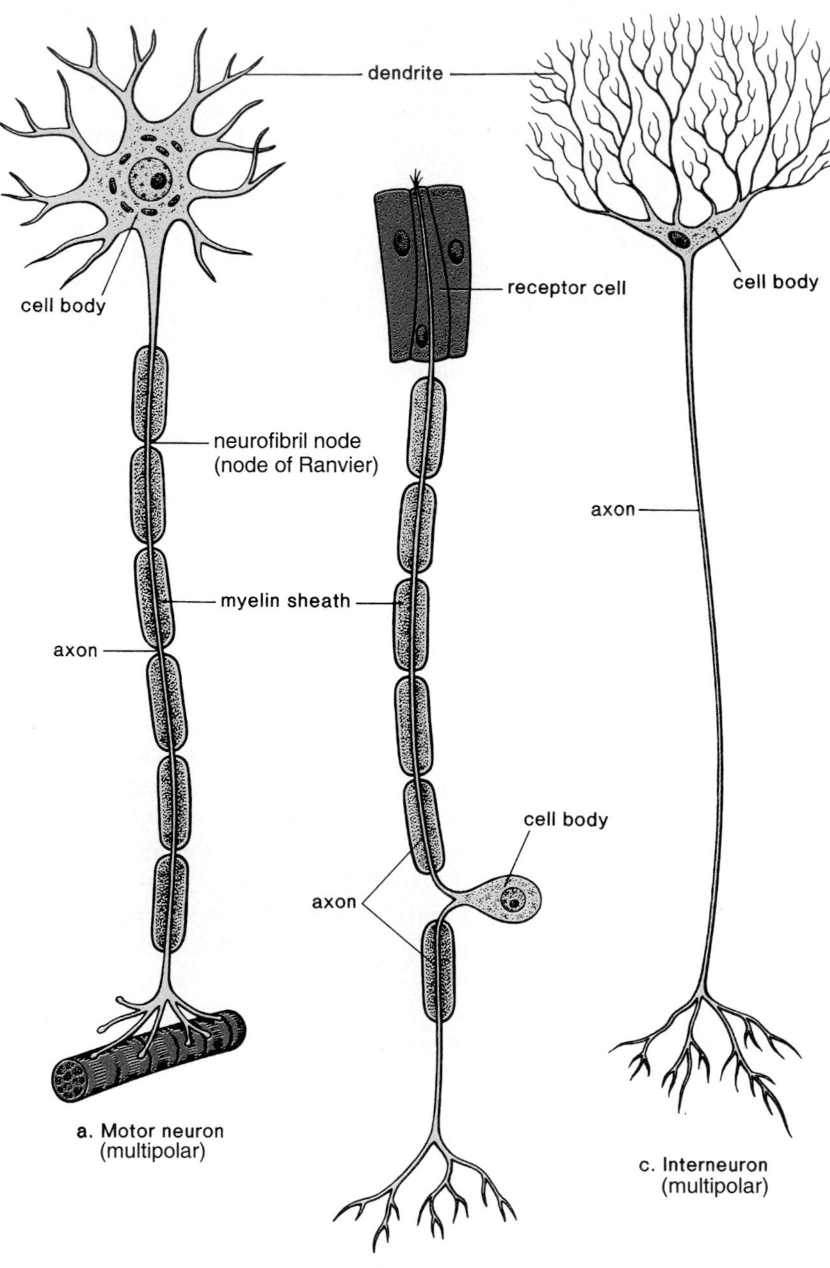

Types of Neurons

Neurons can be classified according to their function and shape. **Motor neurons** take messages from the CNS to muscle fibers or glands. Motor neurons are said to be *multipolar* because they have many dendrites and a single axon (Fig. 46.1*a*). Motor neurons cause muscle fibers or glands to react and therefore they are said to innervate these structures.

Sensory neurons take messages from sensory receptors to the CNS. Sometimes, the endings of sensory neurons are even modified as receptors (Fig. 46.1*b*). Almost all sensory neurons have a structure that is termed unipolar. In unipolar neurons, the process that extends from the cell body divides into a branch that extends to the periphery and another that extends to the CNS. Since both of these extensions are long and myelinated and transmit nerve impulses, it is now generally accepted to refer to them as axons.

Interneurons [L. *inter,* between, and Gk. *neuron,* nerve] occur within the CNS. Interneurons, which are typically multipolar (Fig. 46.1*c*), convey messages between various parts of the CNS. Some lie between sensory neurons and motor neurons, and some take messages from one side of the spinal cord to the other or from the brain to the cord, and vice versa. They also form complex pathways in the brain where processes accounting for critical thinking, memory, and language occur.

Figure 46.1 Neuron anatomy.
a. Motor neuron. Note the branched dendrites and the single, long axon, which branches only near its tip.
b. Sensory neuron with dendrite-like structures projecting from the peripheral end of the axon.
c. Interneuron (from the cortex of the cerebellum) with very highly branched dendrites.

Health Focus

A Cure for Alzheimer and Parkinson Disease

It appears that Alzheimer and Parkinson disease might someday be treatable, according to recent research findings. Alzheimer disease (AD) is a disorder characterized by a gradual loss of reason that begins with memory lapses and ends with an inability to perform any type of daily activity. Personality changes signal the onset of AD. A normal 50- to 60-year-old adult might forget the name of a friend not seen for years. However, people with AD forget the name of a neighbor who visits daily. With time, they have trouble finding their way and cannot perform simple errands. People afflicted with AD become confused and tend to repeat the same question over and over. Other signs of mental disturbances eventually appear, and patients gradually become bedridden and die of a complication such as pneumonia.

A neuron damaged by Alzheimer disease (AD) is shown in Figure 46A. The AD neuron has two abnormalities not seen in the normal neuron. Bundles of fibrous protein, called neurofibrillary tangles, surround the nucleus in the cell, and protein-rich accumulations, called amyloid plaques, envelop the axon branches. These abnormal neurons are especially seen in the portions of the brain that are involved in reason and memory (frontal lobe and limbic system). In order to see the abnormal brain neurons, brain tissue must be examined microscopically after the patient dies.

The likelihood of developing AD is higher if there is a family history of AD. Researchers have discovered that in some families whose members have a 50% chance of AD, a genetic defect is present on chromosome 21, the same chromosome associated with Down syndrome. Further, the genetic defect affects the normal production of amyloid precursor protein (APP), which may be the cause of the amyloid plaques.

Nevertheless, it appears that the neurotransmitter acetylcholine may be in short supply in the brains of patients with AD. Drugs that enhance acetylcholine production are now available in AD patients.

Parkinson is a disease that is usually not seen until after age 60. The three obvious signs of Parkinson disease are slowness of movement, tremor, and rigidity. The person walks with a small-stepped shuffle, and there is no swinging of the arms. An involuntary shaking of the hands occurs even when they are at rest. Rigidity of facial muscles gives the appearance of a masklike face although there is frequent eye blinking. All these symptoms are due to the degeneration of neurons that normally produce the neurotransmitter dopamine. Therefore, we know that dopamine is essential for normal coordinated movements.

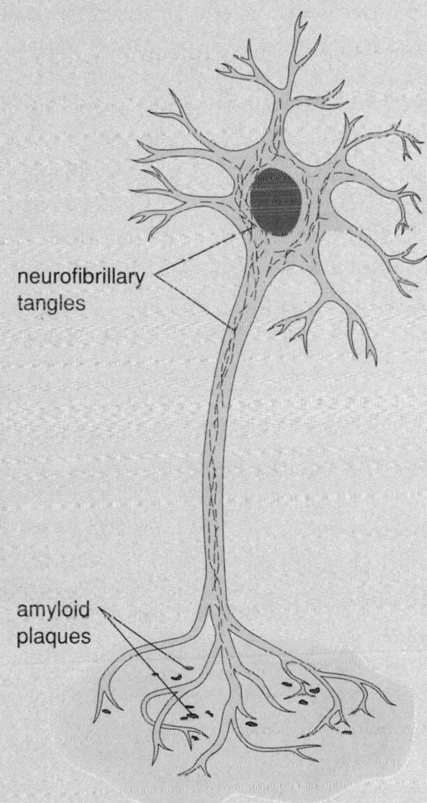

Figure 46A Alzheimer disease (AD).
An AD neuron has neurofibrillary tangles and amyloid plaques. AD neurons are particularly present in the frontal lobe and limbic system. This accounts for the development of symptoms of Alzheimer disease.

The cause of Parkinson disease is not identifiable in most cases. Some families have a high incidence of the disease, but Parkinson can strike one identical twin and not the other. Therefore, it appears that Parkinson is not inherited. Parkinson-like symptoms are known to develop after carbon monoxide poisoning, encephalitis (inflammation of the brain), or the intake of certain drugs.

Dopamine cannot be administered as a drug because it cannot cross the *blood-brain barrier*—the neuroglial cells associated with the capillaries in the brain prevent many substances from entering the brain. But L-dopa, which can be converted to dopamine by brain neurons, does cross the blood-brain barrier, and it is presently given to Parkinson patients. Unfortunately, L-dopa begins to lose its therapeutic effectiveness as more dopamine-secreting neurons die off. Presently, some Parkinson patients are receiving neuron transplants, but this requires a major operation in which cells are implanted in the brain. A significant number of individuals are opposed to the use of fetal brain cells for these operations. As an alternative, researchers are experimenting with cultured cells from a number of sources, even from pigs.

Medical treatment for Alzheimer and Parkinson disease is most desirable, and another may have been found. It's been discovered that neuroglial cells produce neurotropic proteins that can be mass produced through genetic engineering of bacteria. Nerve growth factor (NGF) is being considered for treatment of Alzheimer disease, and glial-derived growth factor (GDGF) is being considered for treatment of Parkinson disease. If the administration of these drugs can stop the deterioration of brain cells, additional medical treatment will be available.

Transmitting the Nerve Impulse

Italian investigator Luigi Galvani discovered in 1786 that a nerve can be stimulated by an electric current. But it was realized later that the speed of the nerve impulse is too slow to be due simply to the movement of electrons or current within an axon. In the early 1900s, Julius Bernstein, at the University of Halle, Germany, suggested that the nerve impulse is an electrochemical phenomenon involving the movement of unequally distributed ions on either side of an *axomembrane* (plasma membrane of an axon). It was not until later, however, that the investigators developed a technique that enabled them to support this hypothesis. A. L. Hodgkin and A. F. Huxley, English neurophysiologists, received the Nobel Prize in 1963 for their work in this field. They and a group of researchers, headed by K. S. Cole and J. J. Curtis, at Woods Hole, Massachusetts, managed to insert a very tiny electrode into the giant axon of the squid *Loligo*. This internal electrode was then connected to a voltmeter and an *oscilloscope*, an instrument with a screen that shows a trace or pattern indicating a change in voltage with time (Fig. 46.2). *Voltage* is a measure of the electrical potential difference between two points, which in this case is the difference between two electrodes—one placed inside and another placed outside the axon. (An electrical potential difference across a membrane is called the membrane potential.) When a potential difference exists, we can say that a plus pole and a minus pole exist; therefore, an oscilloscope indicates the existence of polarity and records polarity changes.

Resting Potential Is Baseline

When the axon is not conducting an impulse, the oscilloscope records a membrane potential equal to about –65 mV (millivolts), indicating that the inside of the neuron is more negative than the outside (Fig. 46.2*a*). This is called the **resting potential** because the axon is not conducting an impulse.

The existence of this polarity can be correlated with a difference in ion distribution on either side of the axomembrane. As Figure 46.2*a* shows, there is a higher concentration of sodium ions (Na^+) outside the axon and a higher concentration of potassium ions (K^+) inside the axon. The unequal distribution of these ions is in part due to the action of the sodium-potassium pump. This pump is an active transport system in the plasma membrane that pumps three sodium ions out of and two potassium ions into the axon. The pump is always working because the membrane is somewhat permeable to these ions and they tend to diffuse toward their lesser concentration. Since the membrane is more permeable to potassium ions than to sodium ions, there are always more positive ions outside the membrane than inside; this accounts for some of the polarity recorded by the oscilloscope. There are also large, negatively charged proteins in the cytoplasm of the axon; altogether, then, the oscilloscope records that the cytoplasm is –65 mV compared to tissue fluid. This is the resting potential.

Action Potential Brings Changes

The occurrence of an **action potential** is obvious when a rapidly moving pattern, indicating rapid changes in membrane potential, appears on the oscilloscope screen (Fig. 46.2*b*). An action potential requires two types of special protein-lined channels. There is a channel that allows sodium (Na^+) to pass through the membrane and another that allows potassium (K^+) to pass through the membrane. Each of these types of channels has a gate: the sodium channel has a gate called the sodium gate, and the potassium channel has a gate called the potassium gate.

The action potential is generated only after the occurrence of a threshold value. Threshold is the minimum change in polarity across the membrane that is required to generate an action potential. During a *depolarization*, the inside of a neuron at a particular location becomes negative because of the sudden entrance of sodium ions. If threshold depolarization occurs, many more sodium channels open, and the action potential begins. As sodium ions rapidly move across the membrane to the inside of the axon, the action potential swings up from –65 mV to +40 mV. This reversal in polarity causes the sodium channels to close and the potassium channels to open. Now potassium ions move from inside the axon to the outside of the axon. As potassium ions leave, the action potential swings down from +40 mV to –65 mV. In other words, a *repolarization* has occurred.

If an axon is unmyelinated, an action potential at one locale stimulates an adjacent part of the axomembrane to produce an action potential. In myelinated fibers, an action potential at one neurofibril node causes an action potential at the next node. This type of conduction, which is called **saltatory conduction** [L. *saltator*, hopper], is much faster than otherwise. In thin, unmyelinated axons, the action potential travels about 1.0 meter/second, and in thick, myelinated axons, the rate is more than 100 meters/second. In any case, an action potential is self-propagating; each action potential generates another along the length of an axon.

The conduction of a nerve impulse (action potential) is an *all-or-none* event; that is, either a fiber conducts a nerve impulse or it does not. Intensity of a message is determined by how many action potentials are generated within a given time span. A fiber can conduct a volley of nerve impulses because only a small number of ions are exchanged with each impulse. As soon as an impulse has passed by each successive portion of a fiber, it undergoes a short refractory period during which it is unable to conduct an impulse. This ensures a one-way direction of the impulse. During a refractory period, the sodium gates cannot yet open.

All neurons transmit the same type of nerve impulse—a change in potential across an axomembrane that is self-propagating.

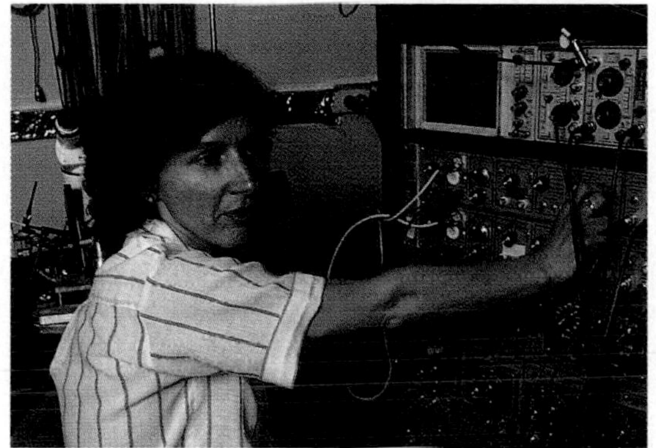

a. Resting Potential

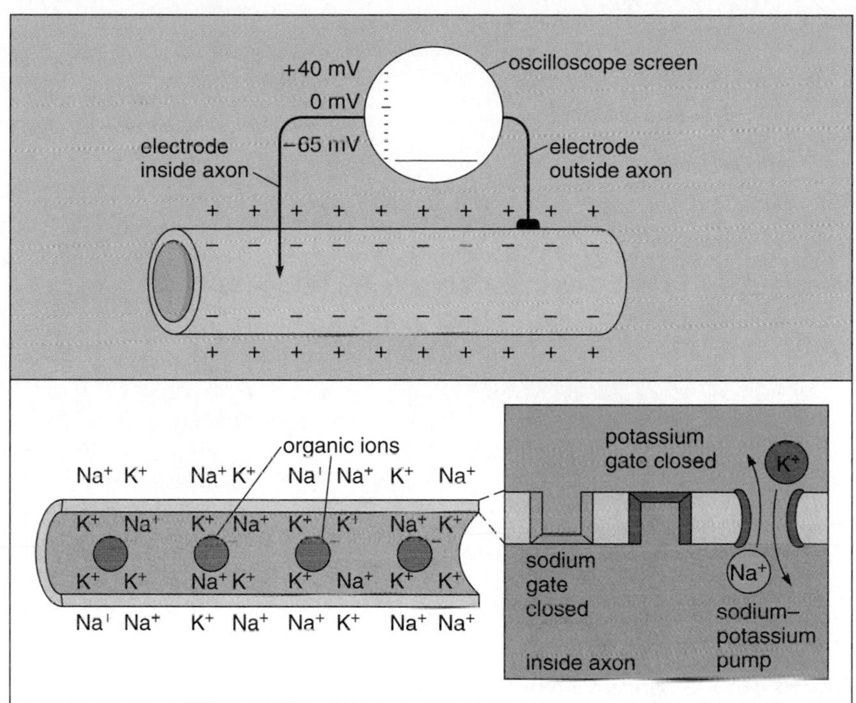

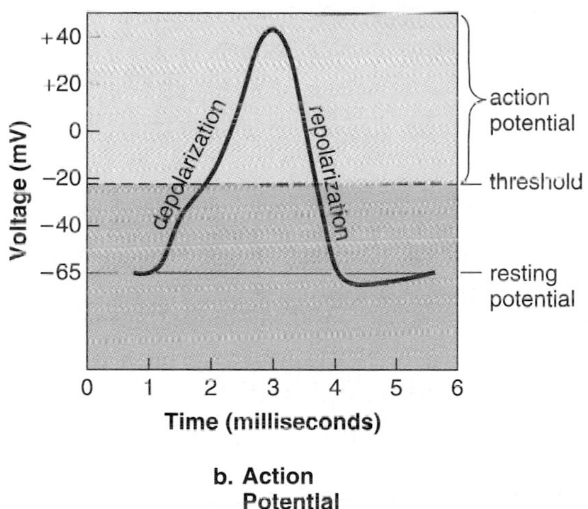

b. Action Potential

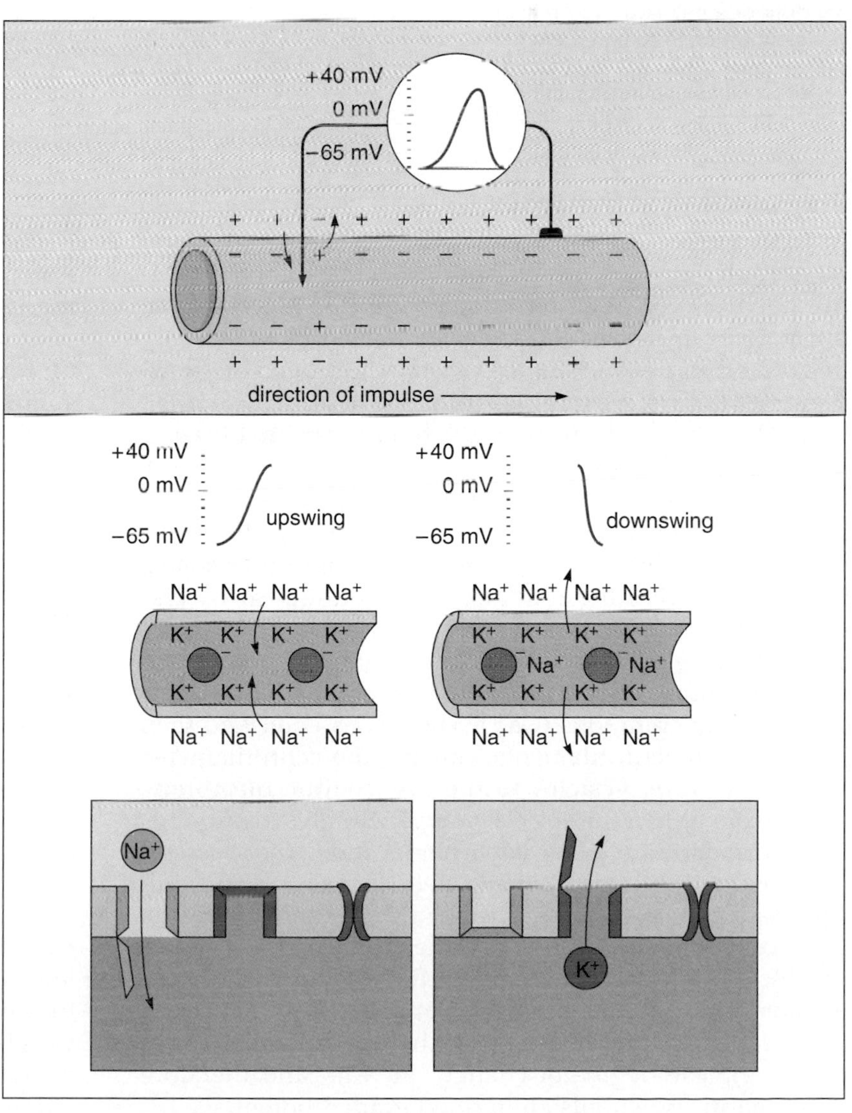

Figure 46.2 The resting and action potential. **a.** Resting potential. The oscilloscope (shown in photograph) records a resting potential of −65 millivolts due to the presence of large organic ions inside a fiber. Note also the unequal distribution of Na^+ and K^+ across the membrane due to the work of the sodium-potassium pump. **b.** Action potential (shown enlarged). A depolarization (upswing) is due to the movement of Na^+ ions to the inside of a fiber, and a repolarization (downswing) is due to the movement of K^+ ions to the outside of a fiber.

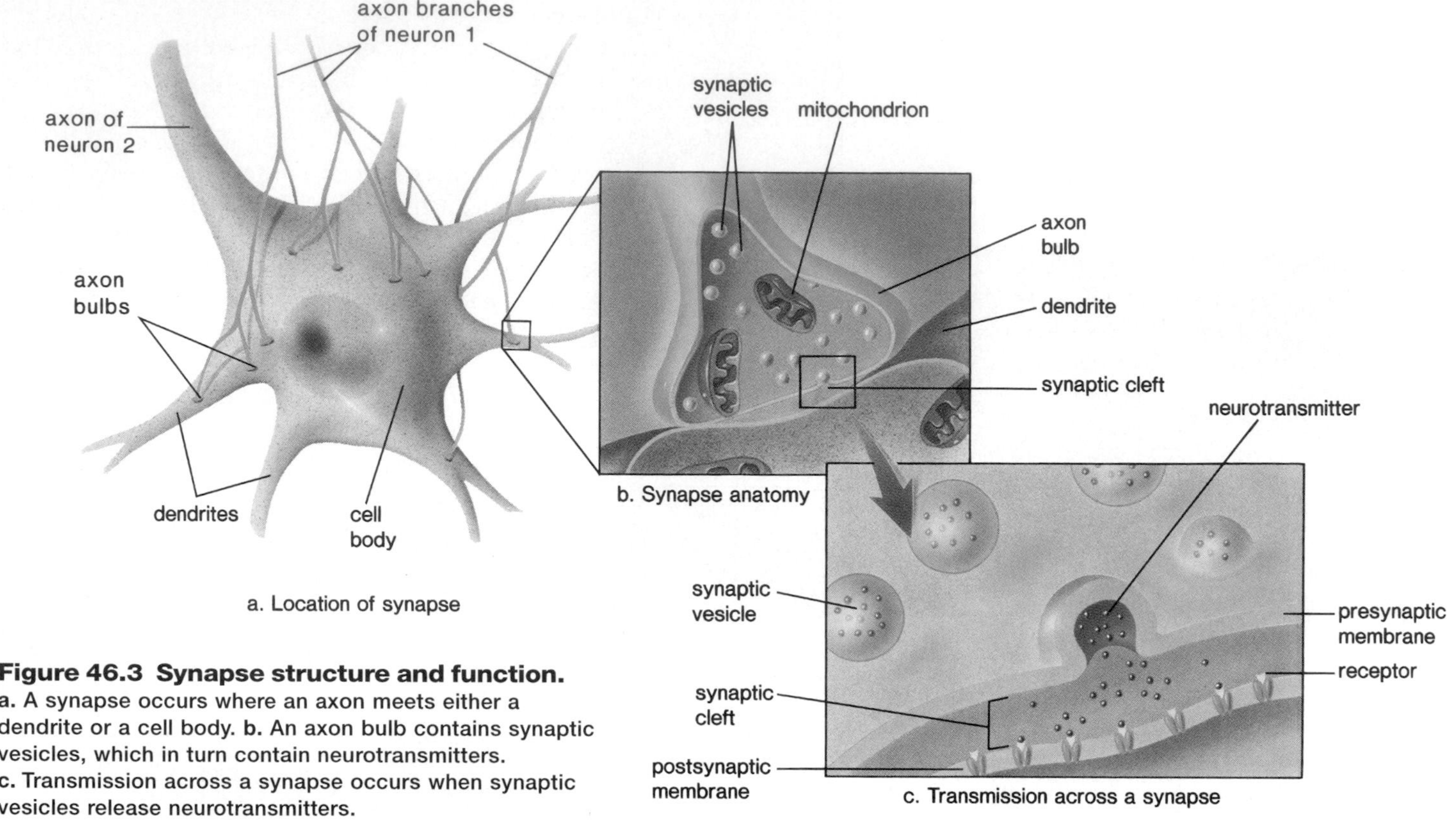

Figure 46.3 Synapse structure and function.
a. A synapse occurs where an axon meets either a dendrite or a cell body. **b.** An axon bulb contains synaptic vesicles, which in turn contain neurotransmitters. **c.** Transmission across a synapse occurs when synaptic vesicles release neurotransmitters.

Transmitting at Synapses

In 1897, the English scientist Sir Charles Sherrington and others noted two important aspects of nerve impulse transmission between neurons. First, an impulse passing from one vertebrate nerve cell to another always moves in only one direction. Second, there is a very short delay in transmission of the nerve impulse from one neuron to another. This latter observation led to the hypothesis that there is a minute space between neurons. Sherrington called the region where the impulse moves from one neuron to another a **synapse** [Gk. *synaptos*, united], meaning "to clasp." A synapse has three components: a presynaptic membrane, a gap now called the synaptic cleft, and a postsynaptic membrane (Fig. 46.3).

When nerve impulses traveling along an axon reach an axon ending called a bulb, the membrane becomes permeable to calcium ions (Ca^{2+}). These ions enter and then interact with actin filaments, causing the actin filaments to pull synaptic vesicles to the presynaptic membrane (Fig. 46.3*b*). When the vesicles merge with this membrane, a **neurotransmitter** [Gk. *neuron*, nerve, and L. *trans*, across] is discharged into the synaptic cleft (Fig. 46.3*c*). The neurotransmitter molecules diffuse across the cleft to the postsynaptic membrane, where they bind with a receptor in a lock-and-key manner. Depending on the type of neurotransmitter and the type of receptor, this binding creates a depolarization (less negative) or a hyperpolarization (more negative). The degree of change can vary and therefore postsynaptic potentials are termed graded potentials.

Graded Potentials Sum Up

A neuron is capable of comparing and integrating incoming information before passing on a message to another neuron. How does it do it, considering that nerve impulses do not vary in character? Integration is accomplished by **summation** or an adding up of all the graded potentials received. If the negative inputs—despite positive inputs—prevent threshold, then the postsynaptic neuron does not fire (generate an action potential), but if the positive inputs—despite the negative inputs—result in a threshold value, the neuron fires.

Neurotransmitters Act Quickly

Acetylcholine (ACh) and **norepinephrine (NE)** are well-known neurotransmitters, active in both the peripheral nervous system and the central nervous system. These are excitatory or inhibitory, according to the type of receptor at the postsynaptic membrane.

Once a neurotransmitter has been released into a synaptic cleft, it has only a short time to act. In some synapses, the cleft contains enzymes that rapidly inactivate the neurotransmitter. For example, the enzyme **acetylcholinesterase (AChE),** or simply cholinesterase, breaks down acetylcholine. In other synapses, the synaptic ending rapidly absorbs the neurotransmitter, possibly for repackaging in synaptic vesicles or for chemical breakdown. The short existence of neurotransmitters in the synapse prevents continuous stimulation (or inhibition) of postsynaptic membranes.

46.2 How the Nervous System Evolved

A comparative study of animal nervous organization indicates the evolutionary trends in nervous system organization that may have led to the nervous system of vertebrates.

Invertebrate Nervous Organization

Even sponges, which have the cellular level of organization, can respond to stimuli; the most common observable response is closure of the osculum (central opening). Hydras, which are cnidaria with the tissue level of organization, can contract and extend their bodies, move their tentacles to capture prey, and even turn somersaults. Cnidaria have a nerve net that is composed of neurons in contact with one another and with contractile epitheliomuscular cells in the gastrodermis and epidermis (Fig. 46.4*a*). Sea anemones and jellyfishes, which are also cnidaria, seem to have two nerve nets. A fast-acting one allows major responses, particularly in times of danger, and the other coordinates slower and more delicate movements.

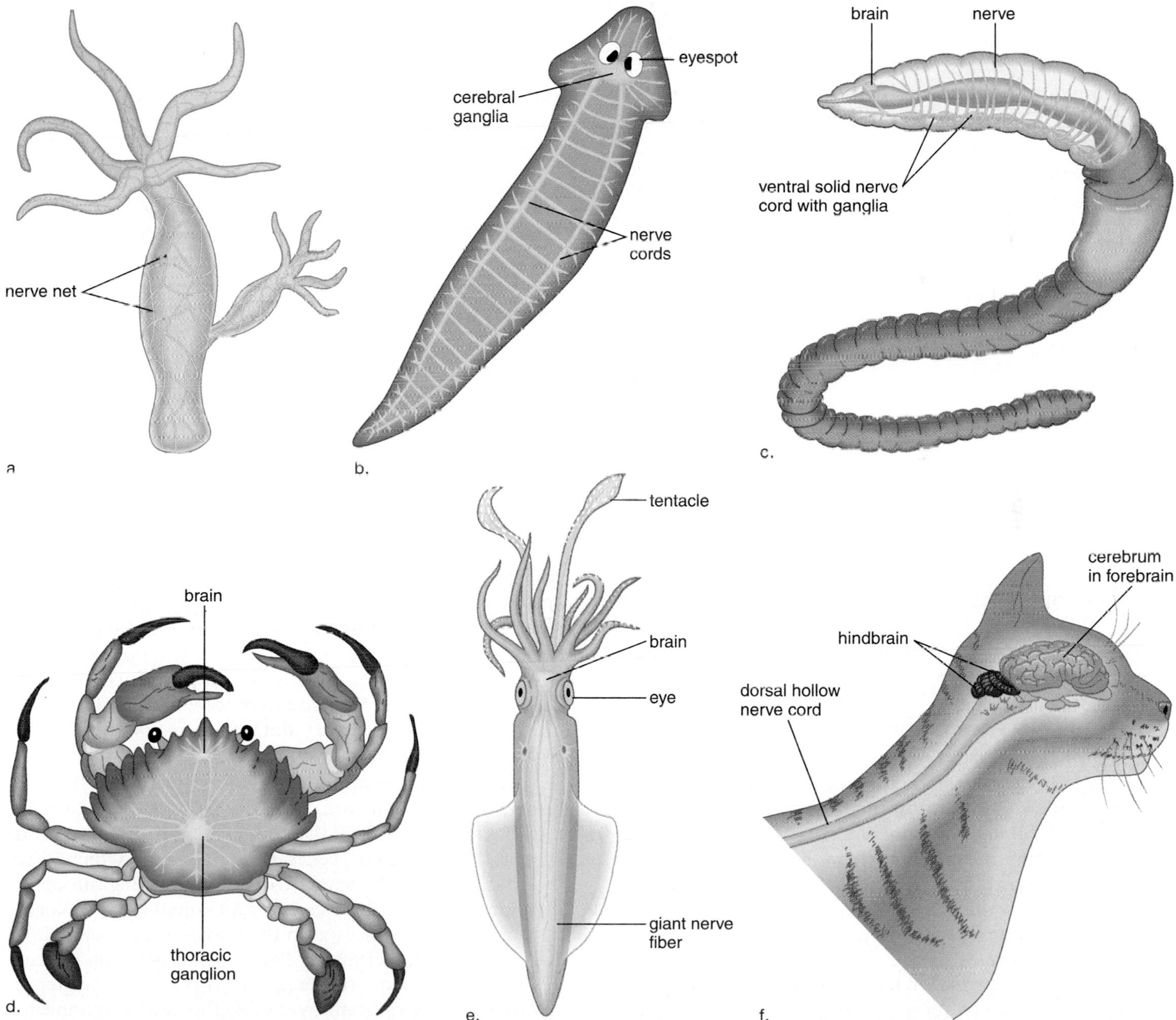

Figure 46.4 Evolution of the nervous system.
a. The nerve net of a hydra, a cnidarian. **b.** In planaria, a flatworm, the paired nerve cords with transverse nerves have the appearance of a ladder. **c.** The earthworm, an annelid, has a central nervous system consisting of a brain and a ventral solid nerve cord. It also has a peripheral nervous system consisting of nerves. **d.** The crab, an arthropod, has a nervous system that resembles that of annelids, but the ganglia are larger. **e.** The squid, a mollusk, has a definite brain with well-developed giant nerve fibers that produce rapid muscle contraction so the squid can move quickly. **f.** A cat, like other vertebrates, has a dorsal hollow nerve cord in the central nervous system.

Planaria, which are flatworms, have a nervous organization in keeping with their bilateral symmetry. They have two nerve cords (collections of neurons on either side of the body) and transverse nerves that join the nerve cords. The entire arrangement resembles a ladder. *Cephalization* is present because planaria have anterior cerebral ganglia that receive sensory information from photoreceptors in the eyespots and sensory cells in the auricles (Fig. 46.4*b*). The two longitudinal nerve cords allow a rapid transfer of information from the cerebral ganglia to the posterior end, and the transverse nerves between the nerve cords keep the movement of the two sides coordinated. Bilateral symmetry plus cephalization are two significant trends in the development of a nervous organization that is adaptive for an active way of life. Also, the nervous organization in planaria is a foreshadowing of a central nervous system (cerebral ganglia and nerve cords) and a peripheral nervous system (transverse nerves).

Annelids (e.g., earthworm, Fig. 46.4*c*), arthropods (e.g., crab, Fig. 46.4*d*), and mollusks (e.g., squid, Fig. 46.4*e*), are complex animals with true nervous systems. The annelids and arthropods have what is usually considered the typical invertebrate nervous system. There is a brain and a ventral solid nerve cord that has a ganglion in each segment. The brain, which normally receives sensory information, controls the activity of the ganglia and assorted nerves so that the muscle activity of the entire animal is coordinated. The crab and squid show marked cephalization—the anterior end has a well-defined brain, and there are well-developed sense organs. The presence of a brain and other ganglia in the body of all these animals indicate an increase in the number of neurons among invertebrates.

Bilateral symmetry, cephalization, and an increase in the number of neurons are trends that are observable among the invertebrates.

Vertebrate Nervous Organization

In vertebrates (e.g., cat, Fig. 46.4*f*) the central nervous system, consisting of a brain and spinal cord, develops from an embryonic dorsal hollow nerve cord. Cephalization, coupled with bilateral symmetry, results in several types of paired sensory receptors such as the paired eyes, ears, and olfactory structures that allow the animal to gather information from the environment. There are paired cranial and spinal nerves that contain the axons of numerous neurons. In vertebrates, there is a vast increase in the number of neurons. For example, an insect's entire nervous system may contain a total of about one million neurons, while a vertebrate nervous system may contain many thousand to several billion times that number.

The Vertebrate Brain

The vertebrate brain develops at the anterior end of the dorsal hollow nerve cord. It is customary to divide the brain into the hindbrain, midbrain, and forebrain. Nearly all vertebrates have a well-developed hindbrain that regulates organs below the level of consciousness. In humans, for example, the lungs and heart function even when we are sleeping. The hindbrain also functions in the coordination of motor activity associated with limb movement, posture, and balance.

The optic lobes are a part of the midbrain, which was originally a center for coordinating reflex responses to visual input. The forebrain receives sensory input from the midbrain and the hindbrain and regulates their output. The cerebrum, which is highly developed in mammals, integrates sensory and motor input and is particularly associated with conscious control of the body. In humans, the outermost part of the cerebrum, called the *cerebral cortex*, is particularly large and complex.

In vertebrates, the brain is organized into three areas: the hindbrain, the midbrain, and the forebrain. The forebrain is highly developed in mammals, particularly humans.

The Human Nervous System

A nervous system, including the human nervous system, has three specific functions: (1) it receives sensory input—sensory receptors present in skin and other organs respond to external and internal stimuli by generating nerve impulses that travel to the central nervous system (CNS); (2) it performs integration—the CNS sums up the input it receives from all over the body; and (3) it stimulates motor output—nerve impulses from the CNS go to the muscles and glands. Muscle contractions and gland secretions are therefore responses to stimuli received by sensory receptors. As an example, consider the events that occur as a person raises a glass to the lips. Continual sensory input to the CNS from the eyes and hand inform the CNS of the position of the glass, and the CNS continually sums up the incoming data before commanding the hand to proceed. At any time, integration with other sensory data might cause the CNS to command a different motion instead.

In humans, the **central nervous system (CNS)** includes the brain and spinal cord, which have a central location—they lie in the midline of the body. The **peripheral nervous system (PNS)** [Gk. *periphereia,* circumference]**,** which is further divided into the somatic division and the autonomic division, includes all the cranial and spinal nerves. Nerves have a peripheral location in the body—they project from the central nervous system. The division between the central nervous system and the peripheral nervous system is arbitrary; the two systems work together and are connected to one another. This comes about because the basic unit of the nervous system is the neuron. As discussed previously there are three types of neurons (see page 818). Interneurons (and neuroglial cells) make up the bulk of the CNS. Clusters of cell bodies in the CNS are called *nuclei*. Axons of motor neurons and sensory neurons make up the nerves. The cell bodies of these neurons are found in the CNS or in ganglia. **Ganglia** [Gk, *ganglion*, swelling under the skin] are collections of cell bodies found in the PNS.

The CNS and PNS of the human nervous system work together to perform the functions of the nervous system.

46.3 The Peripheral Nervous System Contains Nerves

The peripheral nervous system lies outside the central nervous system and contains both cranial nerves and spinal nerves. The paired **cranial nerves** connect to the brain and the paired **spinal nerves** lie on either side of the spinal cord. In the PNS, the **somatic system** controls the skeletal muscles and the autonomic system controls the smooth muscles, cardiac muscles, and glands. There are two parts to the autonomic system: the sympathetic system and the parasympathetic system (Fig. 46.5).

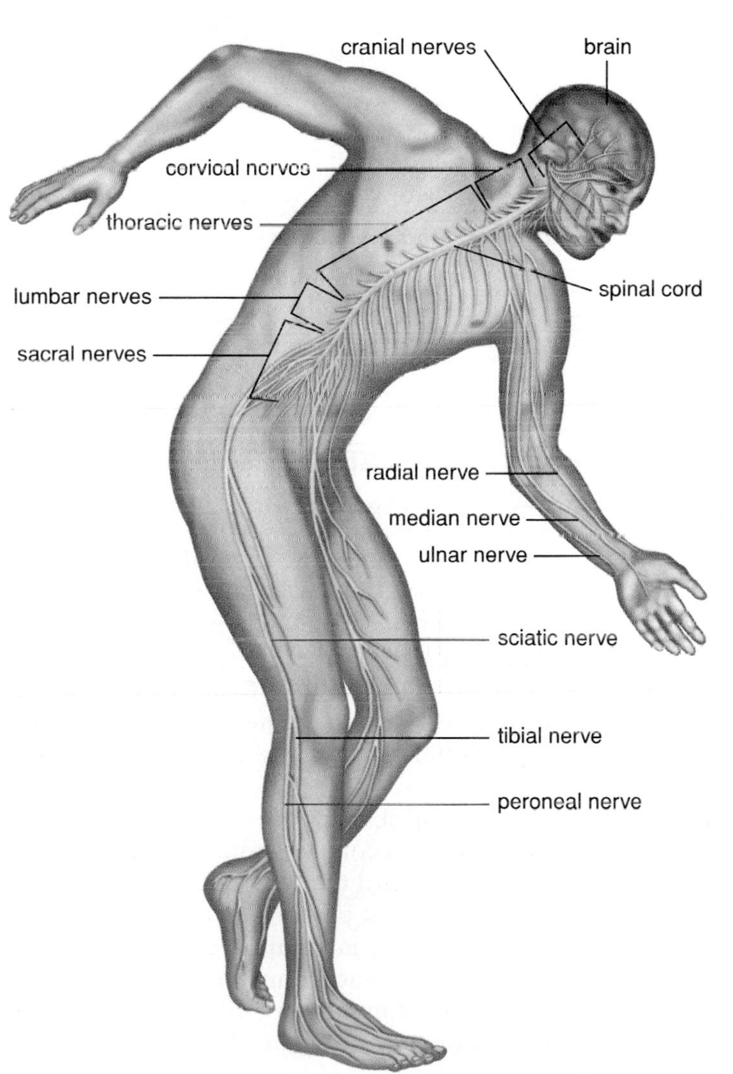

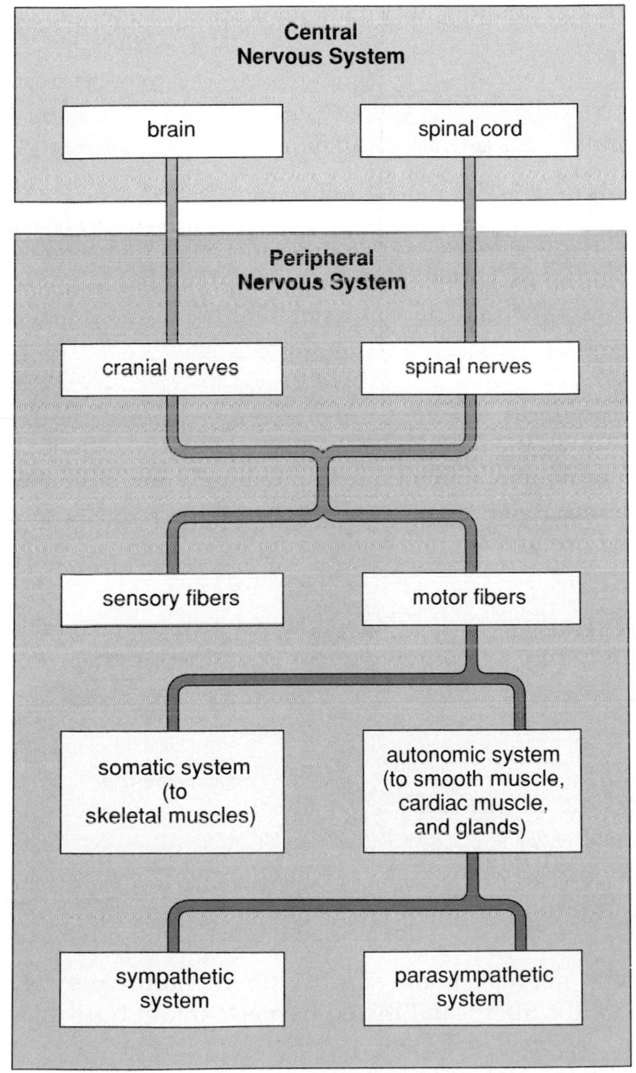

Figure 46.5 Organization of the nervous system in humans.
a. Pictorial representation. The central nervous system (CNS, composed of brain and spinal cord) and some of the nerves of the peripheral nervous system (PNS) are shown. **b.** The CNS communicates with the PNS, which contains nerves. In the somatic system, nerves send sensory impulses from receptors to the CNS and send motor impulses from the CNS to the skeletal muscles. In the autonomic system, consisting of the sympathetic and parasympathetic systems, motor impulses travel to smooth muscle, cardiac muscle, and the glands.

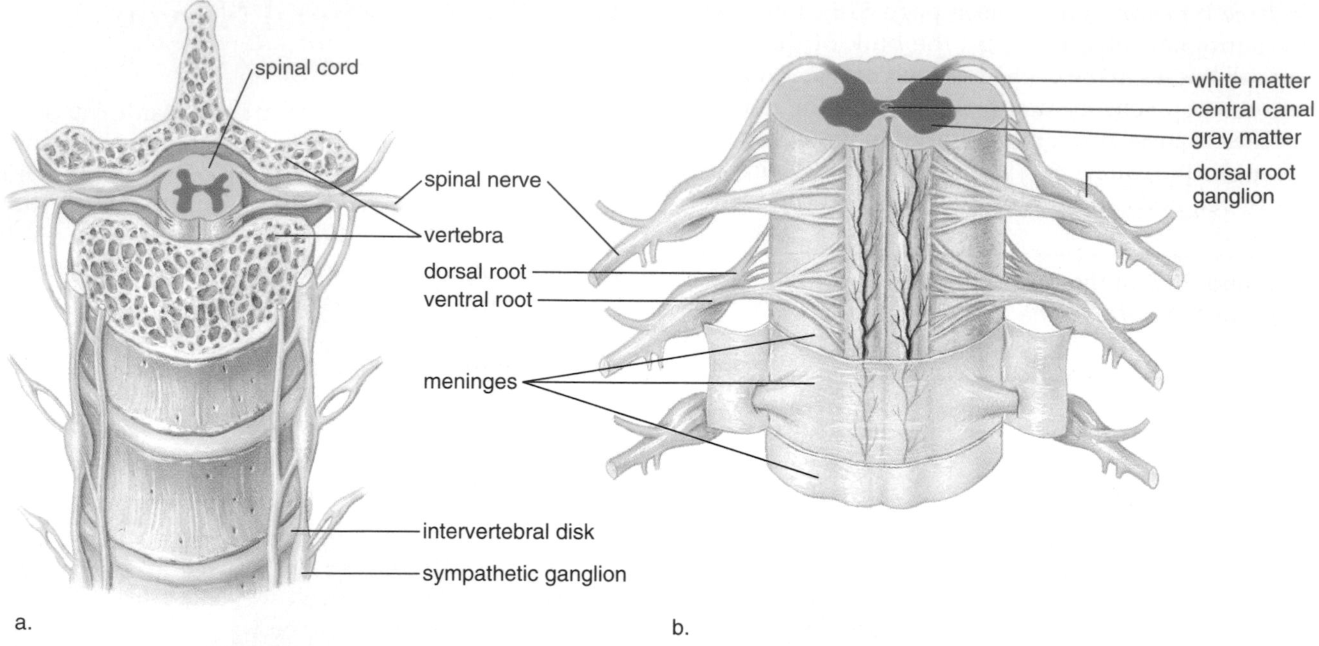

Figure 46.6 Spinal cord anatomy.
a. Cross section of the spine, showing spinal nerves. The human body has 31 pairs of spinal nerves. **b.** This cross section of the spinal cord shows that a spinal nerve has a dorsal root and a ventral root. Also, the cord is protected by three layers of tissue called the meninges. Spinal meningitis is an infection of these layers.

Cranial and Spinal Nerves

The peripheral nervous system is made up of nerves. **Nerves** are bundles of long neuron fibers held together by connective tissue.

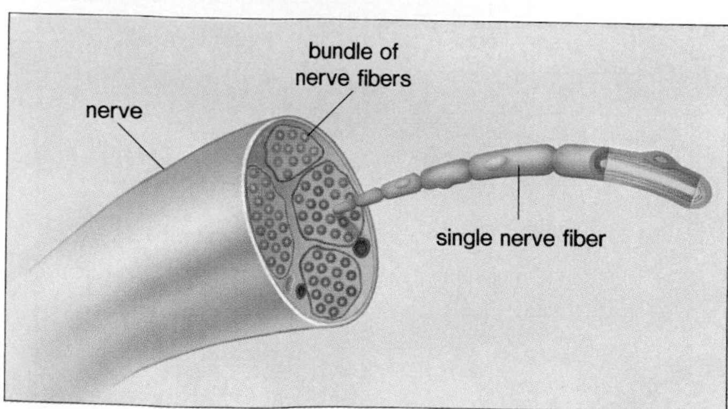

Sensory nerves contain sensory fibers, motor nerves contain motor fibers, and mixed nerves contain both types of fibers. The cell bodies of neurons are found in the central nervous system or in ganglia.

Humans have 12 pairs of *cranial nerves* attached to the brain. Some of these are sensory nerves and others are mixed nerves. Cranial nerves are largely concerned with the head, neck, and facial regions of the body; the *vagus nerve* has branches to the pharynx and larynx and also most of the internal organs.

Humans have 31 pairs of *spinal nerves*. Each spinal nerve emerges from the spinal cord (Fig. 46.6) by two short branches, or roots, which lie within the vertebral column. The dorsal root contains the axons of sensory neurons, which conduct impulses to the cord. The ventral root contains the axons of motor neurons, which conduct impulses away from the cord. These two roots join just before a spinal nerve leaves the vertebral column. Therefore, all spinal nerves are mixed nerves that take impulses to and from the **spinal cord.** The arrangement of spinal nerves shows that humans are segmented animals: there is a pair of spinal nerves for each segment.

Spinal nerves project from the spinal cord, which is a part of the central nervous system. The spinal cord is a thick, whitish nerve cord that extends longitudinally down the back, where it is protected by the vertebrae (sing., vertebra). The cord is composed of gray matter consisting of cell bodies and short unmyelinated fibers, and white matter consisting of myelinated fibers. At its center, the spinal cord contains a tiny **central canal** filled with cerebrospinal fluid.

In the peripheral nervous system, cranial nerves take impulses to and/or from the brain, and spinal nerves take impulses to and from the spinal cord.

n and Muscles

all nerves that serve the
he exterior sense organs, interior sense organs are *recep-*
mental stimuli and then initiate
ers and glands are *effectors,* which
the stimulus.

n back] are automatic, involuntary
curring inside or outside the body.
outside stimuli often initiate a re-
es, such as blinking the eye, involve
ich as withdrawing the hand from a
essarily involve the brain. Figure 46.7
the second type of reflex action in-
d and a spinal nerve, called a *spinal*

rp object, a sensory receptor in the
pulses, which move along the axon
vard the cell body and the central
ll body of a *sensory neuron* is lo-
ganglion, just outside the cord.
npulses continue along the axon of the sensory neuron. The impulses then pass to many interneurons, one of which connects with a motor neuron. The short dendrites and the cell body of the *motor neuron* lead to the axon, which leaves the cord by way of the ventral root of the spinal nerve. The nerve impulses travel along the axon to *muscle fibers,* which then contract so that you withdraw your hand from the sharp object. There are various other reactions—the person may look in the direction of the object, jump back, and cry out in pain. This whole series of responses occurs because the sensory neuron stimulates several interneurons. They take impulses to all parts of the central nervous system, including the cerebrum, which in turn makes the person conscious of the stimulus and his or her reaction to it. (In some reflex arcs, nerve impulses from sensory neurons do not pass to interneurons but pass directly to motor neurons.)

The reflex arc is a major functional unit of the nervous system. It allows us to react rapidly to internal and external stimuli.

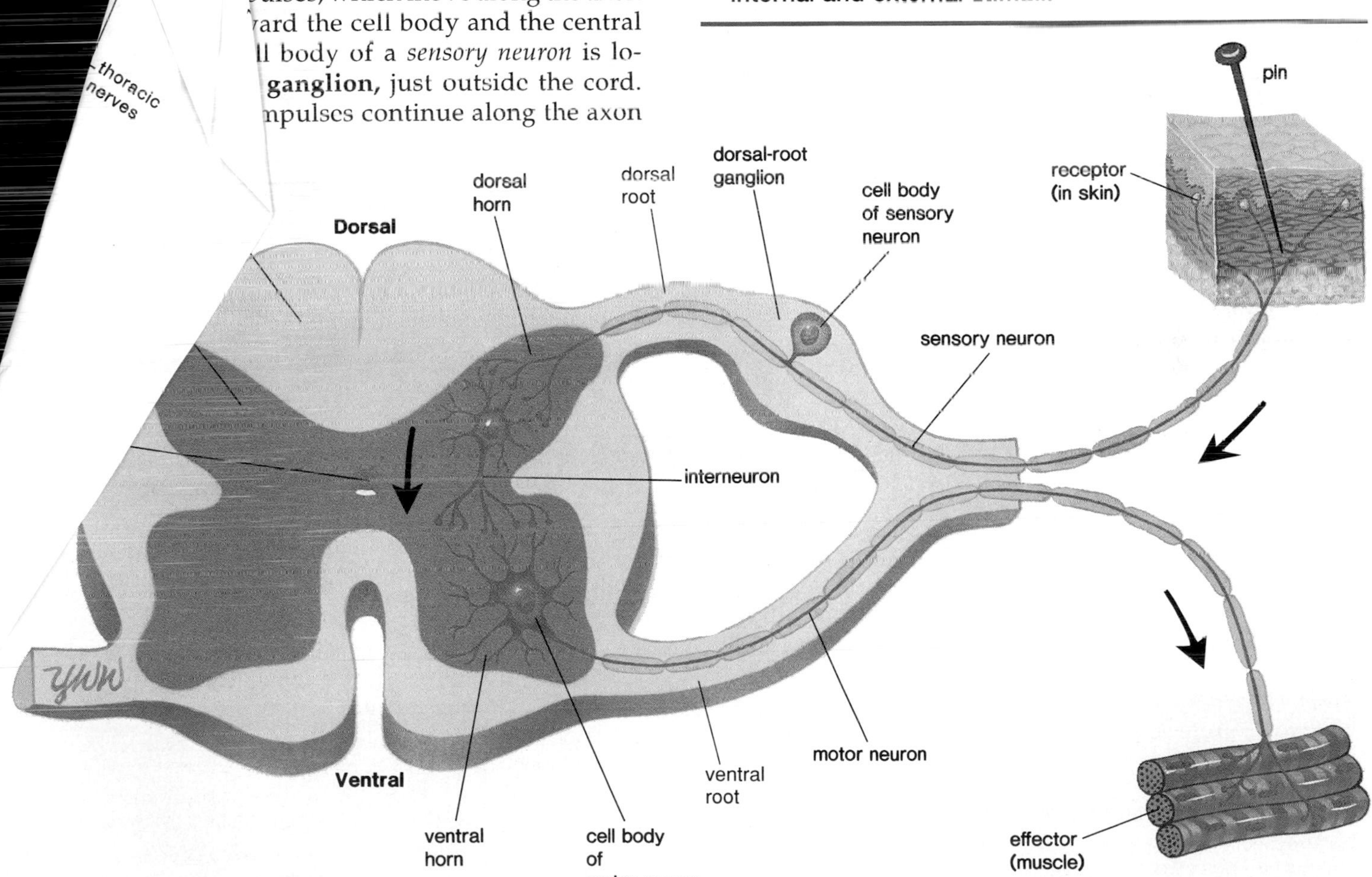

Figure 46.7 A reflex arc showing the path of a reflex.
When a receptor in the skin is stimulated, nerve impulses (see arrows) move along a sensory neuron to the spinal cord. (Note that the cell body of a sensory neuron is in a ganglion outside the cord.) The nerve impulses are picked up by an interneuron, which lies completely within the cord, and pass to the dendrites and the cell body of a motor neuron that lies ventrally within the cord. The nerve impulses then move along the motor neuron to an effector, such as a muscle, which contracts. The brain receives information concerning sensory stimuli by way of other interneurons, with long fibers in tracts that run up and down the cord within the white matter.

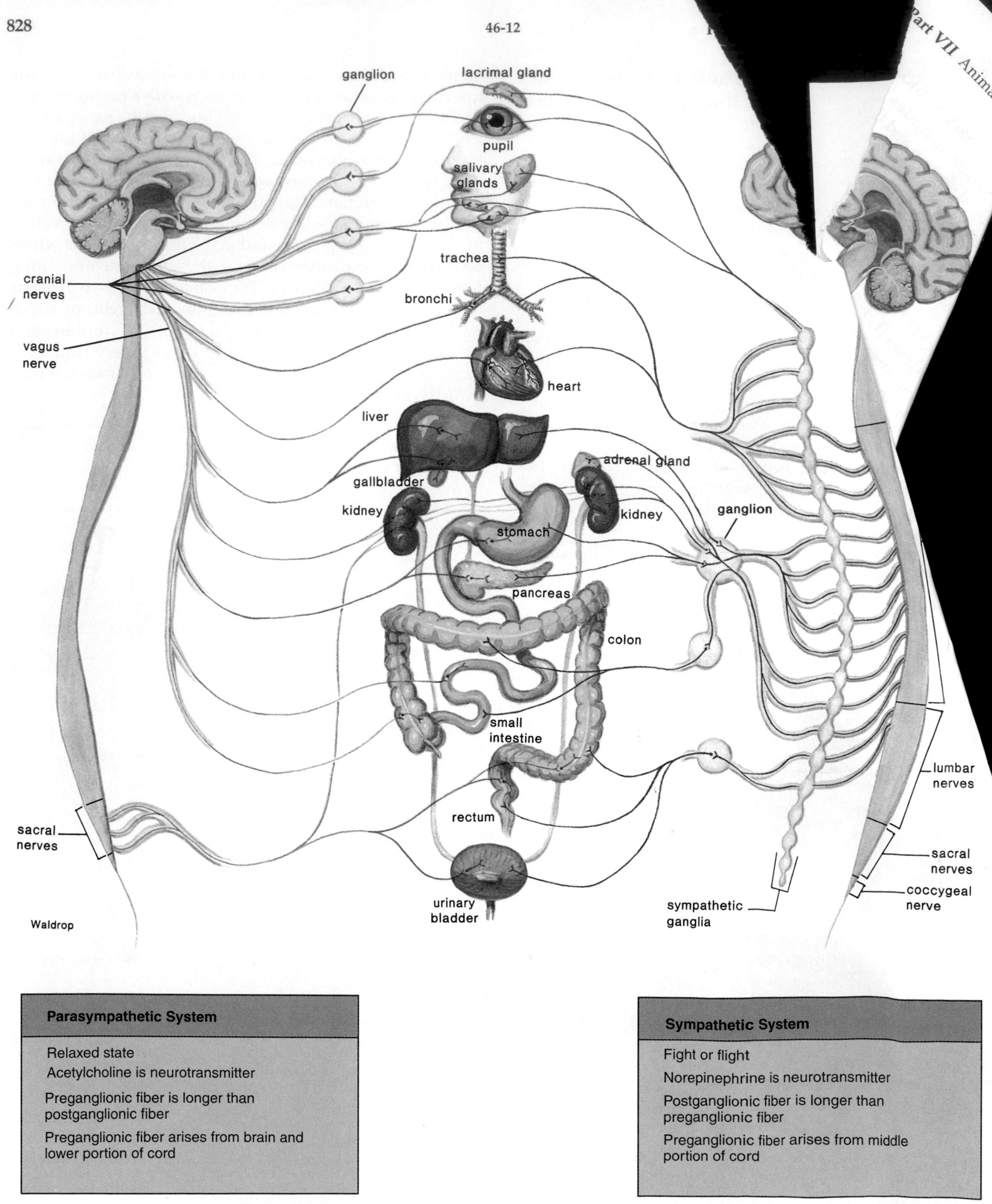

Parasympathetic System
Relaxed state
Acetylcholine is neurotransmitter
Preganglionic fiber is longer than postganglionic fiber
Preganglionic fiber arises from brain and lower portion of cord

Sympathetic System
Fight or flight
Norepinephrine is neurotransmitter
Postganglionic fiber is longer than preganglionic fiber
Preganglionic fiber arises from middle portion of cord

Figure 46.8 Autonomic system structure and function.
The sympathetic fibers arise from the thoracic and lumbar portions of the spinal cord; the parasympathetic fibers arise from the brain and the sacral portion of the spinal cord. Each system innervates the same organs but has contrary effects. For example, the sympathetic system increases the heart rate, while the parasympathetic system decreases it.

Autonomic System Serves Internal Organs

The **autonomic system** [Gk. *autos*, self, and *nomas*, roving] (Fig. 46.8), a part of the peripheral nervous system, is made up of neurons that serve the internal organs automatically and usually without need for conscious intervention. The sensory neurons that belong to the autonomic system and come from the internal organs allow us to perceive internal pain. The cell bodies for these sensory neurons are in dorsal-root ganglia, along with the cell bodies of somatic sensory neurons.

There are two divisions of the autonomic system: the sympathetic and parasympathetic systems. Both of these (1) function automatically and usually subconsciously in an involuntary manner; (2) innervate internal organs; and (3) utilize two motor neurons and one ganglion for each impulse. The first of these two neurons has a cell body within the central nervous system and a *preganglionic fiber*. The second neuron has a cell body within the ganglion and a *postganglionic fiber* (Table 46.1).

The autonomic system controls the function of internal organs, in the absence of conscious control.

Sympathetic System: Fight or Flight

Most preganglionic fibers of the **sympathetic system** (Fig. 46.8) arise from the middle, or *thoracic-lumbar*, portion of the spinal cord and almost immediately terminate in ganglia that lie near the cord. Therefore, this system is often referred to as the thoracolumbar portion of the autonomic system. In the sympathetic system, the preganglionic fiber is short, but the postganglionic fiber that makes contact with an organ is long.

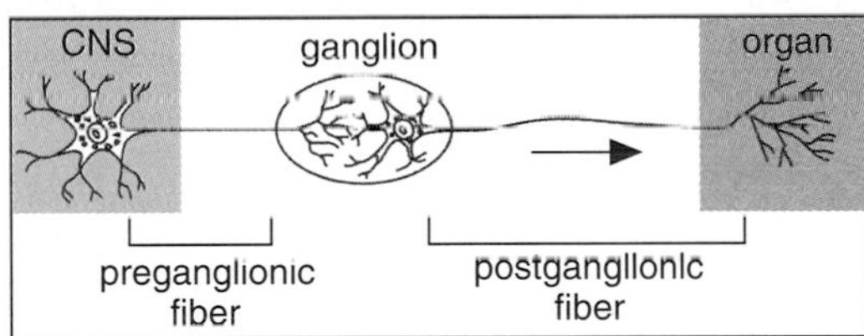

The sympathetic system is especially important during emergency situations and is associated with "fight or flight." If you need to fend off a foe or flee from danger, active muscles require a ready supply of glucose and oxygen. The sympathetic system accelerates the heartbeat, dilates the bronchi, and increases the breathing rate. On the other hand, the sympathetic system inhibits the digestive tract—digestion is not an immediate necessity if you are under attack. The neurotransmitter released by the postganglionic axon is primarily norepinephrine (NE), a chemical close in structure to epinephrine (adrenaline), a hormone used as a heart stimulant.

The sympathetic system brings about those responses we associate with "fight or flight."

Parasympathetic System: Relaxed State

A few cranial nerves, including the vagus nerve, together with fibers that arise from the sacral (bottom) portion of the spinal cord, form the **parasympathetic system** [Gk. *para*, alongside, and *sympathia*, fellow-feeling] (Fig. 46.8). Therefore, this system is often referred to as the *craniosacral portion* of the autonomic system. In the parasympathetic system, the preganglionic fiber is long and the postganglionic fiber is short because the ganglia lie near or within the organ.

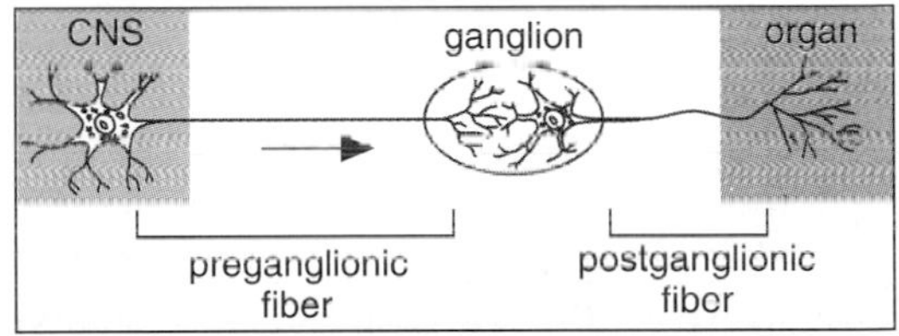

The parasympathetic system, sometimes called the "housekeeper system," promotes all the internal responses we associate with a relaxed state; for example, it causes the pupil of the eye to constrict, promotes digestion of food, and retards the heartbeat. The neurotransmitter utilized by the parasympathetic system is acetylcholine (ACh).

The parasympathetic system brings about the responses we associate with a relaxed state.

Table 46.1

Comparison of Somatic Motor and Autonomic Motor Pathways

Items	Somatic Motor Pathway	Autonomic Motor Pathways: *Sympathetic*	Autonomic Motor Pathways: *Parasympathetic*
Level of consciousness	Voluntary	Involuntary	Involuntary
Number of neurons per message	One	Two (preganglionic shorter than postganglionic)	Two (preganglionic longer than postganglionic)
Location of motor fiber and all spinal nerves	Most cranial nerves	Spinal nerves	Cranial (vagus) and sacral spinal nerves
Neurotransmitter	Acetylcholine	Norepinephrine	Acetylcholine
Effectors	Skeletal muscles	Smooth and cardiac muscle, glands	Smooth and cardiac muscle, glands

a.

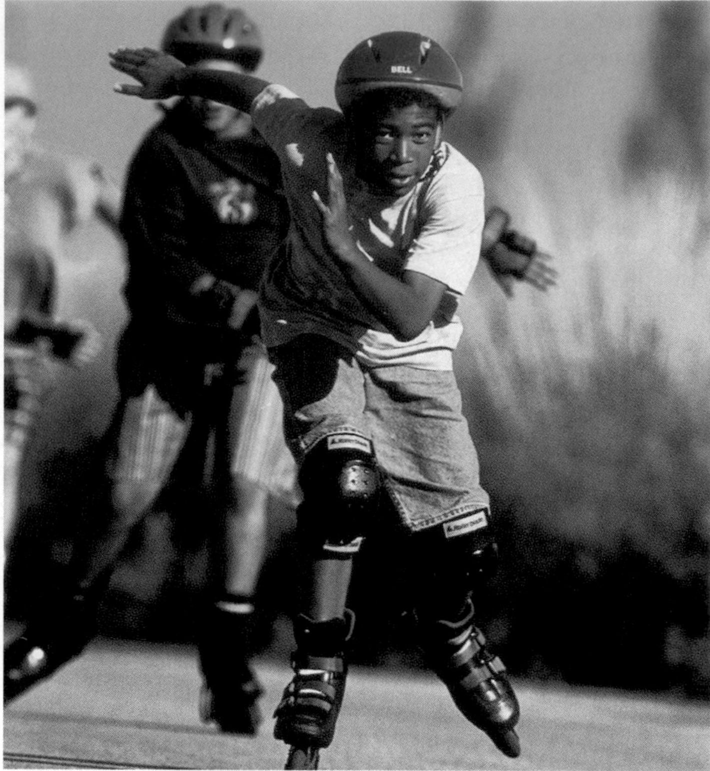

b.

Figure 46.9 Brain functions.
a. The medulla oblongata and hypothalamus function below the level of consciousness and maintain the vegetative functions of the body without input from the cerebrum. **b.** The cerebellum, which coordinates motor functions, is under the direct control of the cerebrum; necessary adjustments due to environmental stimuli are possible. The roller blader will stop rather than be hit by a car.

46.4 Central Nervous System: Brain and Spinal Cord

The central nervous system (CNS) consists of the spinal cord and the brain, where nerve impulses are coordinated and interpreted. The spinal cord is surrounded by vertebrae, and, like the brain, it is wrapped in three protective membranes known as **meninges** [Gk. *meninga,* membranes covering the brain] (see Fig. 46.6). The spaces between the meninges are filled with **cerebrospinal fluid** [L. *cerebrum,* brain, and *spina,* backbone], which cushions and protects the central nervous system. Cerebrospinal fluid is contained in the central canal of the spinal cord and within the *ventricles* of the brain, which are interconnecting spaces that produce and serve as reservoirs for cerebrospinal fluid.

Spinal Cord Communicates

The spinal cord has two main functions: (1) it is the center for many reflex actions, which are discussed on page 827; and (2) it provides a means of communication between the brain and the spinal nerves, which leave the spinal cord.

The spinal cord has white matter and gray matter (see Fig. 46.6). Cell bodies and short unmyelinated fibers give the *gray matter* its color. In cross section, the gray matter looks like a butterfly or the letter *H.* Portions of sensory neurons and motor neurons are found here, as are short interneurons that connect these two types of neurons.

Myelinated long fibers of interneurons that run together in bundles called *tracts* give *white matter* its color. These tracts connect the spinal cord to the brain. Dorsally, there are primarily ascending tracts taking information to the brain, and ventrally, there are primarily descending tracts carrying information from the brain. Because the tracts at one point cross over, the left side of the brain controls the right side of the body, and the right side of the brain controls the left side of the body.

The CNS, which lies in the midline of the body and consists of the brain and the spinal cord, receives sensory information and initiates motor control.

The Brain Commands

The human brain is divided into these parts: medulla oblongata, cerebellum, pons, midbrain, hypothalamus, thalamus, and cerebrum. The brain has four cavities, called **ventricles:** two lateral ventricles, the third ventricle, and the fourth ventricle.

Brain Stem

The medulla oblongata, the pons, and the midbrain lie in a portion of the brain known as the *brain stem* (Fig. 46.9*a*). The **medulla oblongata** [L. *medulla,* marrow, innermost part, and *oblongus,* longer than broad] lies between the spinal cord and the pons and is anterior to the cerebellum. It contains a number of *vital centers* for regulating heartbeat, breathing, and vasoconstriction (blood pressure). It also contains the reflex centers for vomiting, coughing, sneezing, hiccuping, and swallowing. The medulla contains tracts that ascend or descend between the spinal cord and the brain's higher centers.

The **pons** [L. *pontis,* bridge] contains bundles of axons traveling between the cerebellum and the rest of the central nervous system. In addition, the pons functions with

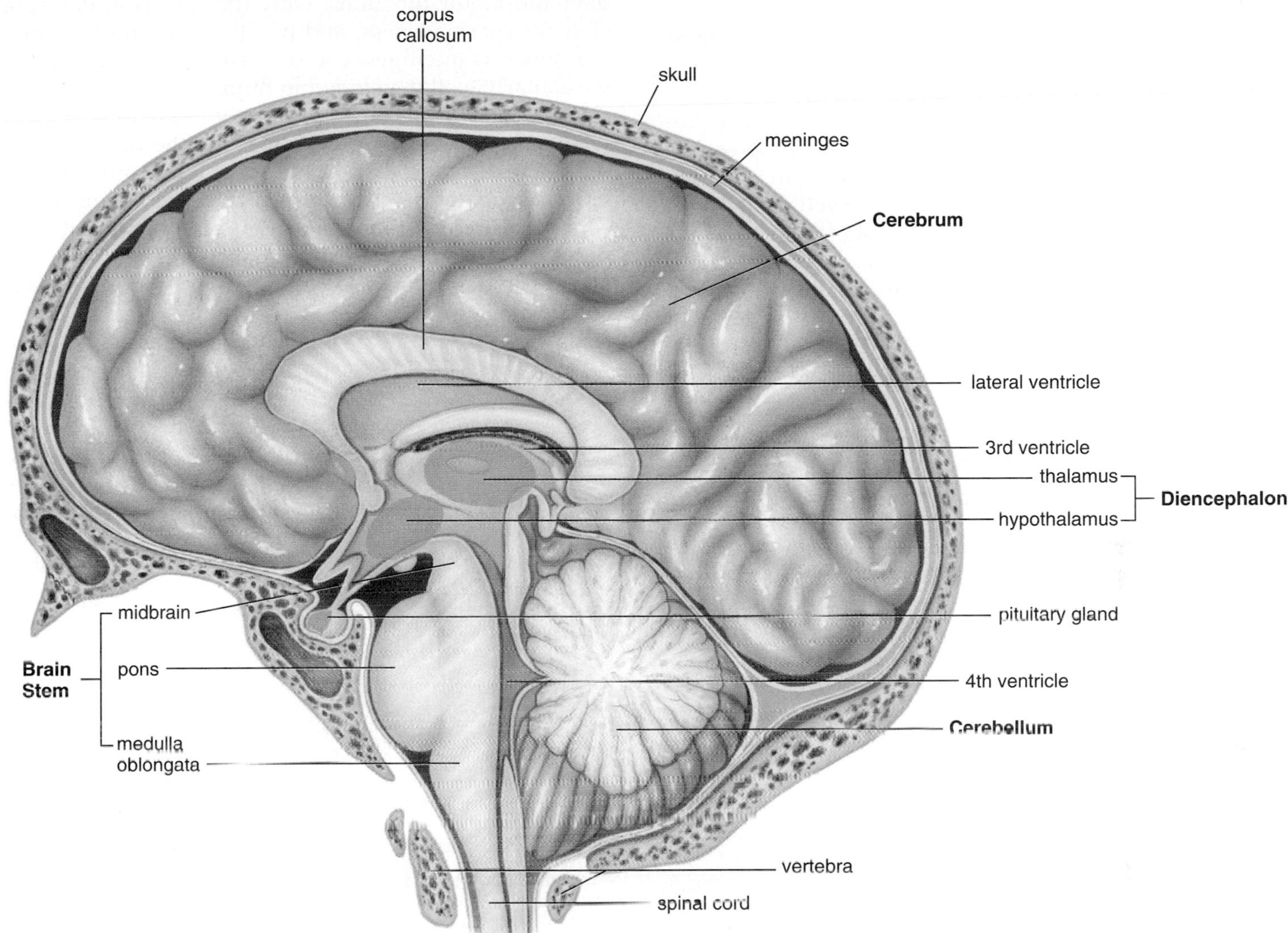

Figure 46.10 The human brain.
Note how large the cerebrum is, compared to the rest of the brain.

the medulla to regulate the breathing rate and has reflex centers concerned with head movements in response to visual and auditory stimuli. Aside from acting as a relay station for tracts passing between the cerebrum and the spinal cord or cerebellum, the **midbrain** has reflex centers for visual, auditory, and tactile responses.

Diencephalon

The hypothalamus and thalamus are in a portion of the brain known as the **diencephalon,** where the third ventricle is located. The hypothalamus forms the floor of the third ventricle. The **hypothalamus** [Gk. *hypo,* under] maintains homeostasis, or the constancy of the internal environment, and contains centers for regulating hunger, sleep, thirst, body temperature, water balance, and blood pressure. The hypothalamus controls the pituitary gland and thereby serves as a link between the nervous and endocrine systems.

The thalamus forms the sides and roof of the third ventricle. The **thalamus** is the last portion of the brain for sensory input before the cerebrum. It serves as a central relay station for sensory impulses traveling upward from other parts of the body and brain to the cerebrum. It receives all sensory impulses and channels them to appropriate regions of the cerebrum for interpretation.

Cerebellum

The **cerebellum,** which lies below the posterior portion of the cerebrum, is separated from the brain stem by the fourth ventricle. The cerebellum functions in muscle coordination, integrating impulses received from higher centers to ensure that all of the skeletal muscles work together to produce smooth and graceful motions (Fig. 46.9*b*). The cerebellum is also responsible for maintaining normal muscle tone and transmitting impulses to muscles that maintain posture. It receives information from the inner ear indicating position of the body and then sends impulses to the muscles, whose contraction maintains or restores balance.

Cerebrum

The **cerebrum,** the foremost part of the brain, is the largest part of the brain in humans (Fig. 46.10). It consists of two large masses called **cerebral hemispheres,** which are connected by a bridge of nerve fibers called the corpus callosum. The outer portion of the cerebral hemispheres, the *cerebral cortex,* is highly convoluted and gray in color because it contains cell bodies and short unmyelinated fibers.

The cerebral cortex of each hemisphere contains four surface lobes: frontal, parietal, temporal, and occipital. Different functions are associated with each lobe (Fig. 46.11). For example, the *frontal lobe* controls motor functions and permits us to control our muscles consciously. The *parietal lobe* receives information from receptors located in the skin, such as those for touch, pressure, and pain. The *occipital lobe* interprets visual input. The *temporal lobe* has sensory areas for hearing and smelling.

A comparative study of vertebrates indicates a progressive increase in the relative size of the cerebrum from fishes to humans, and the cerebral cortex is more convoluted in humans than other vertebrates. The function of the cerebrum has also changed. In fishes and amphibians, the cerebrum largely has an olfactory function, but in reptiles, birds, and mammals, the cerebrum receives information from other parts of the brain and coordinates sensory data and motor functions. Only the cerebrum is responsible for consciousness, and it is the portion of the brain that governs intelligence and reason. These qualities are particularly well developed in humans.

The cerebrum controls the activities of lower parts of the brain. The cerebrum can override the functioning of the brain stem and diencephalon, as when meditation or biofeedback helps control the heart rate. Acting on sensory input from the thalamus, the cerebrum initiates voluntary motor activities and controls the actions of the cerebellum. Certain areas of the cerebral cortex have been "mapped" in great detail (Fig. 46.11). We know which portions of the frontal lobe control various parts of the body and which portions of the parietal lobe receive sensory information from these same parts. Each of the four lobes of the cerebral cortex contains an association area, which receives information from the other lobes and integrates it into higher, more complex levels of consciousness. These areas are concerned with intellect, artistic and creative ability, learning, and memory.

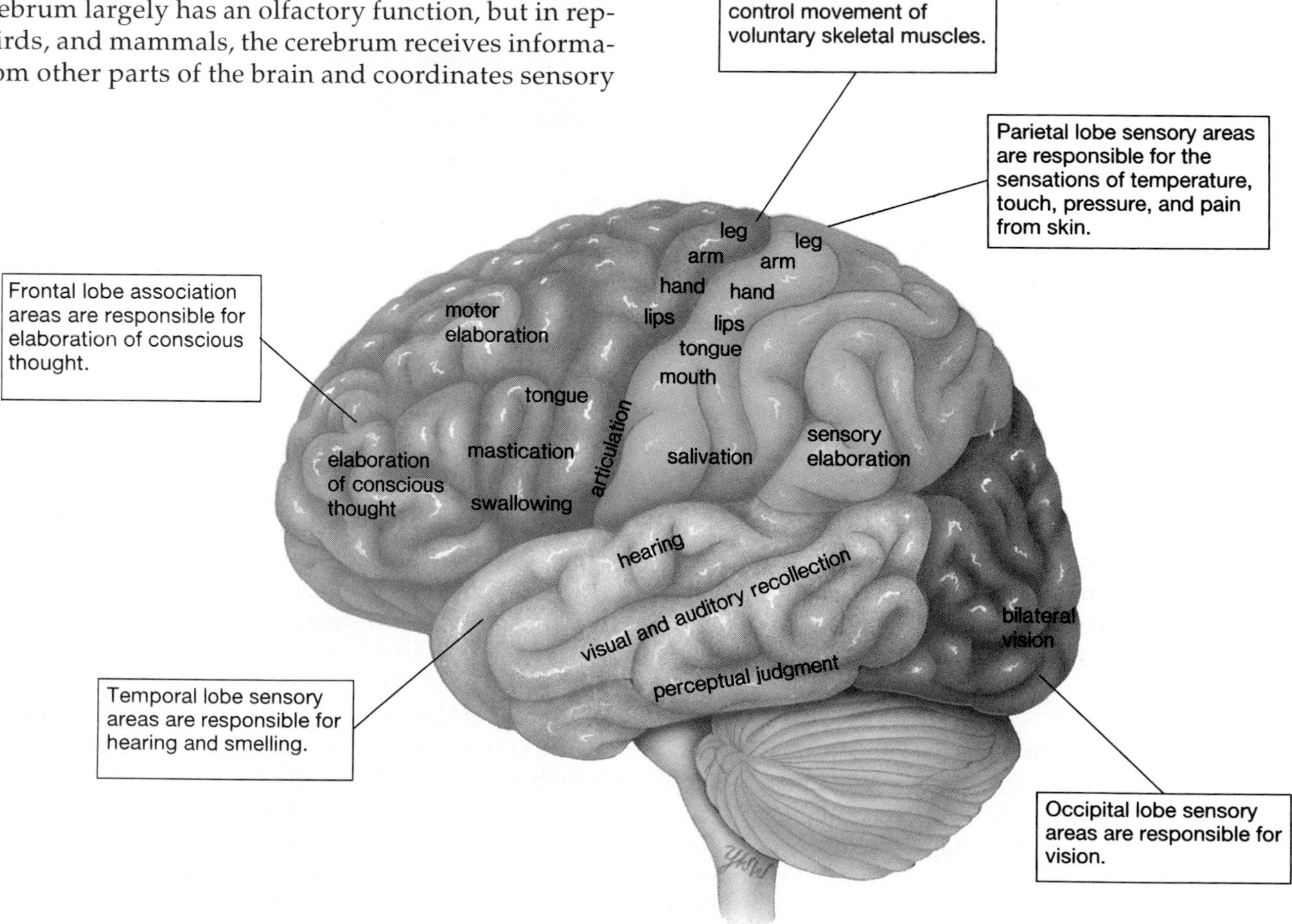

Figure 46.11 The cerebral cortex.
The convoluted cerebral cortex is divided into four surface lobes: frontal, parietal, temporal, and occipital. It is possible to map the cerebral cortex, since each area has a particular function.

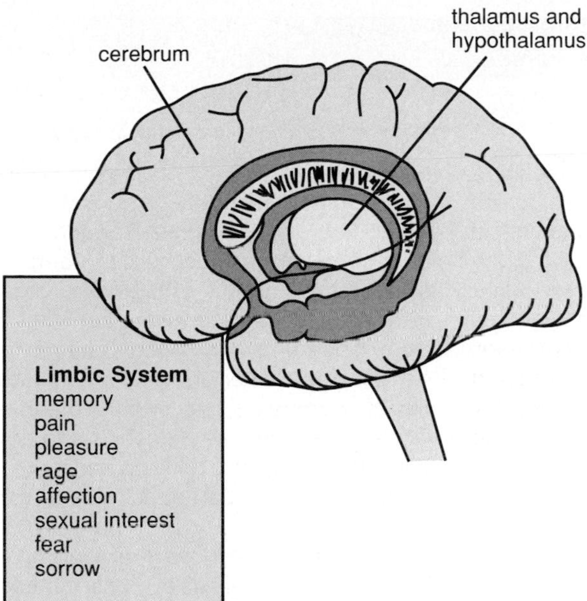

Figure 46.12 The limbic system.
The limbic system, which includes portions of the cerebrum, the thalamus, and the hypothalamus, is sometimes called the emotional brain because it seems to control the emotions listed.

Limbic System

The **limbic system** involves portions of both the unconscious and conscious brain. It lies just beneath the cerebral cortex and contains neural pathways that connect portions of the frontal lobes, the temporal lobes, the thalamus, and the hypothalamus (Fig. 46.12). Several masses of gray matter that lie deep within each hemisphere of the cerebrum, termed the *basal nuclei,* are also a part of the limbic system.

Stimulation of different areas of the limbic system causes the subject to experience rage, pain, pleasure, or sorrow. By causing pleasant or unpleasant feelings about experiences, the limbic system apparently guides the individual into behavior that is likely to increase the chance of survival.

Learning and Memory The limbic system is also involved in the processes of learning and memory. Learning requires memory, but just what permits memory development is not definitely known. Investigators have been working with invertebrates such as slugs and snails because their nervous system is very simple and yet they can be conditioned to perform a particular behavior. To study this simple type of learning, it has been possible to insert electrodes into individual cells and to alter or record the electrochemical responses of these cells (Fig. 46.13). This type of research has shown that learning is accompanied by an increase in the number of synapses, while forgetting involves a decrease in the number of synapses. In other words, the nerve-circuit patterns are constantly changing as learning, remembering, and forgetting occur. Within the individual neuron, learning involves a change in gene regulation and nerve protein synthesis and an increased ability to secrete transmitter substances.

Figure 46.13 Learning research.
Individual nerve cells in a snail, *Hermissenda,* are being stimulated by microelectrodes, which produce the signals scientists previously recorded when a snail learns to avoid light. (Normally, to teach snails to avoid light, they are placed on a table that rotates every time they venture toward light.) When this snail is freed, it automatically avoids the light and does not need to be taught like other snails.

At the other end of the spectrum, some investigators have been studying learning and memory in monkeys. This work has led to the conclusion that the limbic system is absolutely essential to both short-term and long-term memory. An example of short-term memory in humans is the ability to recall a telephone number long enough to dial the number; an example of long-term memory is the ability to recall the events of the day. After nerve impulses circulate within the limbic system, the basal nuclei stimulate the sensory areas where memories are stored. The involvement of the limbic system certainly explains why emotionally charged events result in our most vivid memories. The fact that the limbic system communicates with the sensory areas for touch, smell, vision, and so forth accounts for the ability of any particular sensory stimulus to awaken a complex memory.

The limbic system is particularly involved in the emotions and in memory and learning.

A closer Look

Five Drugs of Abuse

Alcohol

Drugs that people take to alter the mood and/or emotional state affect normal body functions, often by interfering with neurotransmitter release or uptake in the brain. It is possible to drink alcoholic beverages in moderation, but alcohol is often abused. Alcohol use becomes "abuse," or an illness, when alcohol ingestion impairs an individual's social relationships, health, job efficiency, or judgment. While it is general knowledge that alcoholics are prone to drink until they become intoxicated, there is much debate as to what causes alcoholism. Some believe that alcoholism is due to an underlying psychological disorder, while others maintain that the condition is due to an inherited physiological disorder.

Alcohol is primarily metabolized in the liver, where it disrupts the normal workings of glycolysis and the Krebs cycle. The liver contains dehydrogenase enzymes, which carry out these reactions, reducing NAD in the process.

$$\text{NAD} \longrightarrow \text{NADH}$$
$$\text{alcohol} \longrightarrow \longrightarrow \text{acetyl-CoA}$$

The supply of NAD in liver cells is used up by these reactions, and there is not enough free NAD left to keep glycolysis and the Krebs cycle running. The cell begins to ferment and lactic acid builds up. The pH of the blood decreases and becomes acidic.

Since the Krebs cycle is not working, excess active acetate cannot be broken down and it is converted to fat—the liver turns fatty. Fat accumulation, the first stage in liver deterioration, begins after only a single night of heavy drinking. If heavy drinking continues, fibrous scar tissue appears during a second stage of deterioration. If heavy drinking stops, the liver can still recover and become normal once again. If not, the final and irrevocable stage, cirrhosis of the liver, occurs: liver cells die, harden, and turn orange ("cirrhosis" means "orange").

The surgeon general recommends that pregnant women drink no alcohol at all. Alcohol crosses the placenta freely and can cause fetal alcohol syndrome, which is characterized by mental retardation and various physical defects.

Another problem is that heavy drinking interferes with good nutrition. Alcohol is energy intensive—the NADH molecules that result from its breakdown can be used to produce ATP molecules. However, these calories are empty because they do not supply any amino acids, vitamins, and minerals as other energy sources do. Without adequate vitamins, red and white blood cells cannot be formed in the bone marrow. The immune system becomes depressed, and the chances of stomach, liver, lung, pancreas, colon, and tongue cancer increase. Protein digestion and amino acid metabolism are so upset that even adequate protein intake will not prevent amino acid deficiencies. Muscles atrophy and weakness results. Fat deposits accumulate in the heart wall and hypertension develops. There is an increased risk of cardiac arrhythmias and stroke.

Nicotine

Nicotine, an alkaloid derived from tobacco, is a widely used neurological agent. When smoking a cigarette, nicotine is quickly distributed to all body organs including the central and peripheral nervous systems. In the central nervous system, nicotine causes neurons to release dopamine, a neurotransmitter associated with behavioral states. The excess of dopamine has a reinforcing effect that leads to dependence on the drug. In the peripheral nervous system, nicotine stimulates the same postsynaptic receptors as acetylcholine and leads to increased skeletal muscular activity. It also increases the heartbeat rate and blood pressure, and digestive tract mobility. Nicotine may even occasionally induce vomiting and/or diarrhea. It also causes water retention by the kidneys.

Many cigarette smokers find it difficult to give up the habit because nicotine induces both physiological and psychological dependence. Withdrawal symptoms include headache, stomach pain, irritability, and insomnia. Tobacco not only contains nicotine, it also contains many other harmful substances. Cigarette smoking contributes to early death from cancer, including not only lung cancer but also cancer of the larynx, mouth, throat, pancreas, and urinary bladder. Chronic diseases like bronchitis and emphysema are likely to develop and there is increased risk of heart attack due to cardiovascular disease.

Now that women are as apt to smoke as men, lung cancer has surpassed breast cancer as a cause of death. Cigarette smoking in young women who are sexually active is most unfortunate because nicotine, like other psychoactive drugs, adversely affects a developing embryo and fetus.

Marijuana

The dried flowering tops, leaves, and stems of the Indian hemp plant *Cannabis sativa* contain and are covered by a resin that is rich in THC (tetrahydrocannabinol). The names *Cannabis* and marijuana apply to either the plant or THC.

The effects of marijuana differ depending upon the strength and the amount consumed, the expertise of the user, and the setting in which it is taken. Usually, the user reports experiencing a mild euphoria along with alterations in vision and judgment, which result in distortions of space and time. Motor incoordination occurs, as well as the inability to concentrate and to speak coherently.

Intermittent use of low-potency marijuana generally is not associated with obvious symptoms of toxicity, but heavy use can produce chronic intoxication. Intoxication is recognized by the presence of hallucinations, anxiety, depression, rapid flow of ideas, body image distortions, paranoid reactions, and similar psychotic symptoms. The terms cannabis psychosis and cannabis delirium refer to such reactions.

Marijuana is classified as a hallucinogen. It is possible that, like LSD (lysergic acid diethylamide), it has an effect on the action of serotonin, an excitatory neurotransmitter in the brain.

Marijuana use does not seem to produce physical dependence, but a psychological dependence on the euphoric and sedative effects can develop. Craving can also occur as a part of regular heavy use.

Usually marijuana is smoked in a cigarette form called a joint. Since this allows toxic substances, including carcinogens, to enter the lungs, chronic respiratory disease and lung cancer are considered dangers of long-term, heavy use. Some researchers claim that marijuana use leads to long-term brain impairment as well. Others report that males and females suffer reproductive dysfunctions. Fetal cannabis syndrome, which resembles fetal alcohol syndrome, has also been reported. In addition, marijuana has been called a gateway drug because adolescents who have used marijuana also tend to try other drugs. For example, in a study of 100 cocaine abusers, 60% had smoked marijuana for more than ten years.

Some psychologists are very concerned about the use of marijuana among adolescents. Marijuana can be used to avoid dealing with the personal problems that often develop during this maturational phase.

Cocaine

Cocaine is an alkaloid derived from the shrub *Erythroxylon coca*. Cocaine is sold in powder form and as crack, a more potent extract. Users often describe the feeling of euphoria that follows intake of the drug as a rush. Snorting (inhaling) produces this effect in a few minutes, injection, within 30 seconds, and smoking, in less than 10 seconds. Persons dependent upon the drug are, therefore, most likely to smoke cocaine. The rush lasts only a few seconds and then is replaced by a state of arousal, which lasts from 5 to 30 minutes. Then the user begins to feel restless, irritable, and depressed. To overcome these symptoms, the user is apt to take more of the drug, repeating the cycle again and again. A binge of this sort can go on for days, after which the individual suffers a crash. During the binge period, the user is hyperactive and has little desire for food or sleep but has an increased sex drive. During the crash period, the user is fatigued, depressed, and irritable, has memory and concentration problems, and displays no interest in sex. Indeed, men are often impotent. Other drugs, such as marijuana, alcohol, or heroin, often are taken to ease the symptoms of the crash.

Cocaine affects the concentration of dopamine, a neurotransmitter associated with behavioral states. After release into a synapse, dopamine ordinarily is withdrawn into the presynaptic cell for recycling and reuse. Cocaine prevents the reuptake of dopamine by the presynaptic membrane; this causes an excess of dopamine in the synaptic cleft so that the user experiences the sensation of a rush. The epinephrine-like effects of dopamine account for the state of arousal that lasts for some minutes after the rush experience.

With continued cocaine use, the body begins to make less dopamine to compensate for a seemingly excess supply. The user then experiences tolerance, withdrawal symptoms, and an intense craving for the drug. Cocaine, then, is extremely addictive.

The number of deaths from cocaine and the number of emergency-room admissions for drug reactions involving cocaine have increased greatly. High doses can cause seizures and cardiac and respiratory arrest.

Individuals who snort the drug can suffer damage to the nasal tissues and even perforation of the septum between the nostrils. Whether or not long-term cocaine abuse causes brain damage is not yet known. It is known, however, that babies born to addicts suffer withdrawal symptoms and may suffer neurological and developmental problems.

Heroin

Heroin is derived from morphine, an alkaloid of opium. Heroin usually is injected. After intravenous injection, the onset of action is noticeable within one minute and reaches its peak in about five minutes. There is a feeling of euphoria along with relief of pain. Side effects can include nausea, vomiting, dysphoria, and respiratory and circulatory depression leading to death.

Heroin binds to receptors meant for the endorphins, the special neurotransmitters that kill pain and produce a feeling of tranquility. They are believed to alleviate pain by preventing the release of a neurotransmitter termed substance P from certain sensory neurons in the region of the spinal cord. When substance P is released, pain is felt, and when substance P is not released, pain is not felt. Endorphins and heroin also bind to receptors on neurons that travel from the spinal cord to the limbic system. Stimulation of these can cause a feeling of pleasure.

Individuals who inject heroin become physically dependent on the drug. With time, the body's production of endorphins decreases. Tolerance develops so that the user needs to take more of the drug just to prevent withdrawal symptoms. The euphoria originally experienced upon injection is no longer felt.

Heroin withdrawal symptoms include perspiration, dilation of pupils, tremors, restlessness, abdominal cramps, gooseflesh, defecation, vomiting, and increase in systolic pressure and respiratory rate. Those who are excessively dependent may experience convulsions, respiratory failure, and death. Infants born to women who are physically dependent also experience these withdrawal symptoms.

connecting concepts

Like the wiring of a modern office building, the peripheral nervous system of humans contains nerves that reach to all parts of the body. There is a division of labor among the nerves. The cranial nerves serve the face, teeth, and mouth, but below the head there is only one cranial nerve, the vagus nerve. All bodily movements are controlled by spinal nerves, and this is why paralysis may follow a spinal injury. Except for the vagus nerve, only spinal nerves make up the autonomic system, which controls the internal organs. Like most other animals, much of the work of the nervous system in humans is below the level of consciousness, and the same principles apply. The nervous system has just three functions: sensory input, integration, and motor output. Sensory input would be impossible without sensory receptors, which are sensitive to external and internal stimuli. You might even argue that sense organs like the eyes and ears should be considered a part of the nervous system, since there would be no nerve impulses without their ability to generate them. Nerve impulses are the same in all neurons, so how is it that stimulation of eyes causes us to see, and stimulation of ears causes us to hear? Essentially, the central nervous system carries out the function of integrating incoming data. The brain allows us to perceive our environment, reason, and remember. After sensory data have been processed by the CNS, motor output occurs. Muscles and glands are the effectors that allow us to respond to the original stimuli. Without the musculoskeletal system we would never be able to respond to a danger detected by our eyes and ears.

Summary

46.1 Nervous Tissue

The anatomical unit of the nervous system is the neuron, of which there are three types: sensory, motor, and interneuron. Each of these is made up of a cell body, an axon, and dendrites.

When an axon is not conducting a nerve impulse, the resting potential indicates that the inside of the fiber is negative compared to the outside. The sodium-potassium pump helps maintain a concentration of Na^+ ions outside the fiber and K^+ ions inside the fiber. When the axon is conducting a nerve impulse, an action potential (i.e., a change in membrane potential) travels along the fiber. Depolarization (inside becomes positive) due to the movement of Na^+ to the inside, and then repolarization (inside becomes negative again) due to the movement of K^+ to the outside of the fiber, occurs.

Transmission of the nerve impulse from one neuron to another takes place across a synapse. In humans, synaptic vesicles release a chemical, known as a neurotransmitter, into the synaptic cleft. The binding of the neurotransmitter to receptors in the postsynaptic membrane can either increase the chance of a nerve impulse (stimulation) or decrease the chance of a nerve impulse (inhibition) in the next neuron, depending on the type of neurotransmitter and/or the type of receptor.

46.2 How the Nervous System Evolved

A comparative study of the invertebrates shows a gradual increase in the complexity of the nervous system. The vertebrate nervous system, like that of the earthworm, is divided into the central and peripheral nervous systems.

46.3 The Peripheral Nervous System Contains Nerves

The peripheral nervous system contains the somatic system and the autonomic system. Reflexes are automatic, and some do not require involvement of the brain. A simple reflex requires the use of neurons that make up a reflex arc. In the somatic system, a sensory neuron conducts nerve impulses from a sensory receptor to an interneuron, which in turn transmits impulses to a motor neuron, which stimulates an effector to react.

While the motor portion of the somatic division of the PNS controls skeletal muscle, the motor portion of the autonomic division controls smooth muscle of the internal organs and glands. The sympathetic system, which is often associated with those reactions that occur during times of stress, and the parasympathetic system, which is often associated with those activities that occur during times of relaxation, are both parts of the autonomic system.

46.4 Central Nervous System: Brain and Spinal Cord

Most of the cells in the CNS are neuroglial cells, which recently have been shown to participate actively in the functioning of the system and to produce growth factors that can possibly be used to cure neurological diseases.

The CNS consists of the spinal cord and brain. The gray matter of the cord contains cell bodies; the white matter contains tracts that consist of the long axons of interneurons. These run from all parts of the cord, even up to the cerebrum.

The brain integrates all nervous system activity and commands all voluntary activities. In the brain stem, the medulla oblongata and pons have centers for visceral functions. The cerebellum coordinates muscle contractions. In the diencephalon, the hypothalamus in particular controls homeostasis, and the thalamus specializes in sense reception. The cerebrum, which is responsible for consciousness, can be mapped, and each lobe seems to have particular functions.

Research in invertebrates indicates that learning is accompanied by an increase in the number of synapses; research in monkeys indicates that the limbic system is involved. There are short-term and long-term memories; the involvement of the limbic system explains why emotionally charged events result in vivid long-term memories.

Reviewing the Chapter

1. Describe the structure of a neuron, and give a function for each part mentioned. 818
2. Name three types of neurons, and give a function for each. 818
3. What are the major events of an action potential, and what ion changes are associated with each event? 820–21
4. Describe the mode of action of a neurotransmitter at a synapse, including how it is stored and how it is destroyed. 822
5. Trace the evolution of the nervous system by contrasting the organization of the nervous system in hydras, planaria, earthworms, and humans. 823–24
6. Contrast the structure and function of the peripheral and central nervous systems. 824–25
7. Trace the path of a spinal reflex. 827
8. Contrast the sympathetic and parasympathetic divisions of the autonomic system. 829
9. Name the major parts of the human brain, and give a principal function for each part. 830–32
10. Describe the limbic system, and discuss its possible involvement in learning and memory. 833

Testing Yourself

Choose the best answer for each question.

1. Which is the most complete list of animals that have both a central nervous system (CNS) and a peripheral nervous system (PNS)?
   a. hydra, planarian, earthworm, rabbit, human
   b. planarian, earthworm, rabbit, human
   c. earthworm, rabbit, human
   d. rabbit, human
2. Which of these are the first and last elements in a spinal reflex?
   a. axon and dendrite
   b. sense organ and muscle effector
   c. ventral horn and dorsal horn
   d. motor neuron and sensory neuron
3. Which term does not belong with the others?
   a. cerebrum
   b. cerebral cortex
   c. cerebral hemispheres
   d. cerebellum
4. A spinal nerve takes nerve impulses
   a. to the CNS.
   b. away from the CNS.
   c. both to and away from the CNS.
   d. only inside the CNS.
5. Which of these correctly describes the distribution of ions on either side of an axon when it is not conducting a nerve impulse?
   a. more sodium ions (Na^+) outside and less potassium ions (K^+) inside
   b. K^+ outside and Na^+ inside
   c. charged protein outside; Na^+ and K^+ inside
   d. Na^+ and K^+ outside and water only inside
6. When the action potential begins, sodium gates open, allowing Na^+ to cross the membrane. Now the polarity changes to
   a. negative outside and positive inside.
   b. positive outside and negative inside.
   c. There is no difference in charge between outside and inside.
   d. Any one of these could be correct.
7. Transmission of the nerve impulse across a synapse is accomplished by the
   a. movement of Na^+ and K^+.
   b. release of neurotransmitters.
   c. Both of these are correct.
   d. Neither of these is correct.
8. The autonomic system has two divisions called the
   a. CNS and PNS.
   b. somatic and skeletal systems.
   c. efferent and afferent systems.
   d. sympathetic and parasympathetic systems.
9. Synaptic vesicles are
   a. at the ends of dendrites and axons.
   b. at the ends of axons only.
   c. along the length of all long fibers.
   d. All of these are correct.
10. Which of these is mismatched?
   a. cerebrum—consciousness
   b. thalamus—motor and sensory centers
   c. hypothalamus—internal environment regulator
   d. cerebellum—motor coordination

11. Label this diagram of a reflex arc.

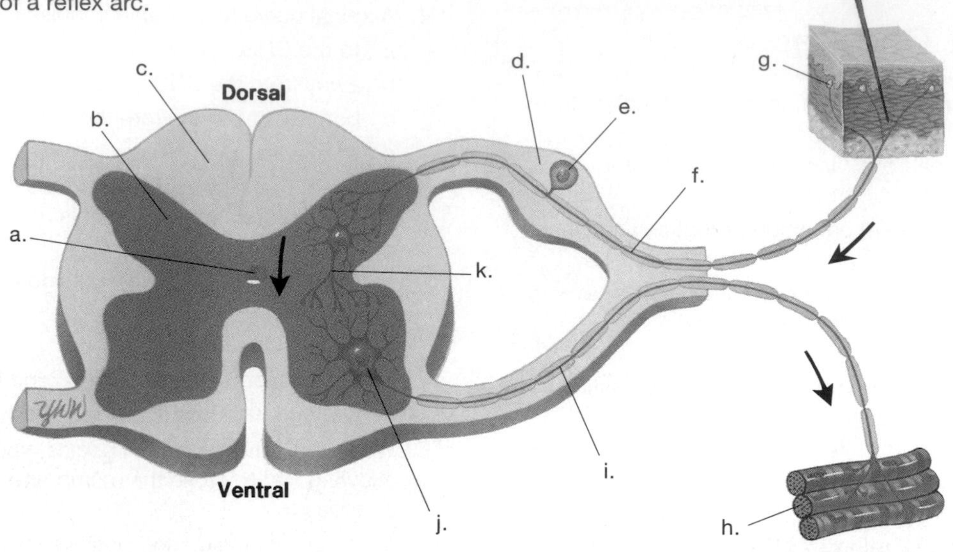

Applying the Concepts

1. *All systems of an animal's body contribute to homeostasis.*
   List several ways in which the nervous system contributes to homeostasis.
2. *Segmentation leads to specialization of parts.*
   How is segmentation, along with specialization of parts, reflected in the structure of the nervous system?
3. *Structure suits function.*
   How does the structure of a neuron suit its function?

Using Technology

Your study of neurons and nervous systems is supported by these available technologies:

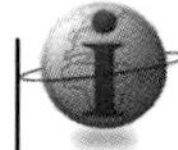

Exploring the Internet
The Mader Home Page provides resources for and help with studying this chapter.

http://www.mhhe.com/sciencemath/biology/mader/
(Click on Biology.)

The Dynamic Human CD-ROM
Nervous System

Explorations in Cell Biology & Genetics CD-ROM
Nerve Conduction (#8)
Synaptic Transmission (#9)

Explorations in Human Biology CD-ROM
Drug Addiction (#10)

Life Science Animations Video
Video #3: Animal Biology I
Formation of Myelin Sheath (#22)
Saltatory Nerve Conduction (#23)
Signal Integration (#24)
Reflex Arcs (#25)

Understanding the Terms

acetylcholine (ACh) 822
acetylcholinesterase (AChE) 822
action potential 820
autonomic system 829
axon 818
cell body 818
central canal 826
central nervous system (CNS) 824
cerebellum 831
cerebral hemisphere 832
cerebrospinal fluid 830
cerebrum 832
cranial nerve 825
dendrite 818
diencephalon 831
dorsal-root ganglion 827
ganglion 825
hypothalamus 831
interneuron 818
limbic system 833
medulla oblongata 830
meninges 830
midbrain 831
motor neuron 818
myelin sheath 818
nerve 826
nerve fiber 818
neuroglial cell 818
neurolemmocyte 818
neuron 818
neurotransmitter 822
norepinephrine (NE) 822
parasympathetic system 829
peripheral nervous system (PNS) 824
pons 830
reflex 827
resting potential 820
saltatory conduction 820
sensory neuron 818
sensory receptor 817
somatic system 825
spinal cord 826
spinal nerve 825
summation 822
sympathetic system 829
synapse 822
thalamus 831
ventricle 830

Match the terms to these definitions:

a. __________ Automatic, involuntary response of an organism to a stimulus.
b. __________ Chemical stored at the ends of axons that is responsible for transmission across a synapse.
c. __________ Division of the peripheral nervous system that regulates internal organs.
d. __________ Knot or bundle of neuron cell bodies usually outside the central nervous system.
e. __________ Neurotransmitter active in both the peripheral and central nervous systems.

Sense Organs

CHAPTER 47

Chapter Concepts

Young girl stoops to see and smell a tulip, *Tulipa*

When you stoop to smell a rose, molecules in the air bind to your olfactory cells, causing them to generate nerve impulses that travel to the brain. Interpretation of these impulses is the function of the brain, which has a specific region for receiving information from each of the sense organs. Impulses arriving at a particular sensory area of the brain can be interpreted in only one way; for example, those arriving at the olfactory area result in smell sensation, and those arriving at the visual area result in sight sensation. The brain integrates data from various receptors in order to perceive, for example, a flower that produced the sensations.

Our **sensory receptors** form an exchange area with the external environment, just as our digestive tract and lungs are exchange areas. They gather the information that allows the brain to make decisions about finding prey, escaping a predator, and any number of other adaptive behaviors. Therefore, receptors play a significant role in maintaining homeostasis. Information exchange is just as important to the maintenance of a relatively stable internal environment as are specific exchanges of materials with the environment.

47.1 Sensing Chemicals

The receptors responsible for taste and smell are termed **chemoreceptors** [Gk. *chemo,* pertaining to chemicals, and L. *receptor,* receiver] because they are sensitive to certain chemical substances in food, liquids, and air. Chemoreception is found almost universally in animals and is therefore believed to be the most primitive sense. Chemoreceptors are present all over the body of planaria, but they are concentrated on the auricles at the sides of the head. In insects, such as the housefly, chemoreceptors are found largely on the feet—flies taste with their feet instead of their mouth. Insects also detect airborne pheromones, which are chemical messages passed between individuals. In crustacea, (e.g., lobsters and crabs), chemoreceptors are widely distributed over all the appendages and antennae. In amphibians, chemoreceptors are located in the nose, in the mouth, and over the entire skin. They are used to locate mates, detect harmful chemicals, and find food. In mammals, the receptors for taste are located in the mouth, and the receptors for smell are located in the nose.

Tasting with Taste Buds

Taste buds are located primarily on the tongue (Fig. 47.1). Many lie along the walls of the papillae, the small elevations on the tongue that are visible to the naked eye. Isolated ones are also present on the hard palate, the pharynx, and the epiglottis.

Taste buds are pockets of cells that extend through the tongue epithelium and open at a taste pore. Taste buds have supporting cells and a number of elongated taste cells that end in *microvilli.* A plasma membrane receptor is a protein, or more likely a glycoprotein that projects from the surface of the cell and binds to a particular molecule. The microvilli bear plasma membrane receptors for certain chemicals. The binding of a molecule to the plasma membrane receptor causes electrochemical changes that lead to the generation of nerve impulses in associated sensory nerve fibers. These nerve impulses go to the parietal lobe of the cerebrum.

It is believed that there are four types of tastes (bitter, sour, salty, sweet) and that taste buds for each are concentrated on the tongue in particular regions (Fig. 47.1*a*). Sweet receptors are most plentiful near the tip of the tongue. Sour receptors occur primarily along the margins of the tongue. Salty receptors are most common on the tip and the upper front portion of the tongue. Bitter receptors are located toward the back of the tongue. Associated sensory fibers have graded, rather than all-or-nothing, sensitivities to the four taste types. The brain appears to survey the overall pattern of incoming sensory impulses and to take a "weighted average" of their taste messages as the perceived taste. Information goes directly to the cerebrum and also to the brain stem. This may have survival value because it suggests that babies will nurse without need of conscious control.

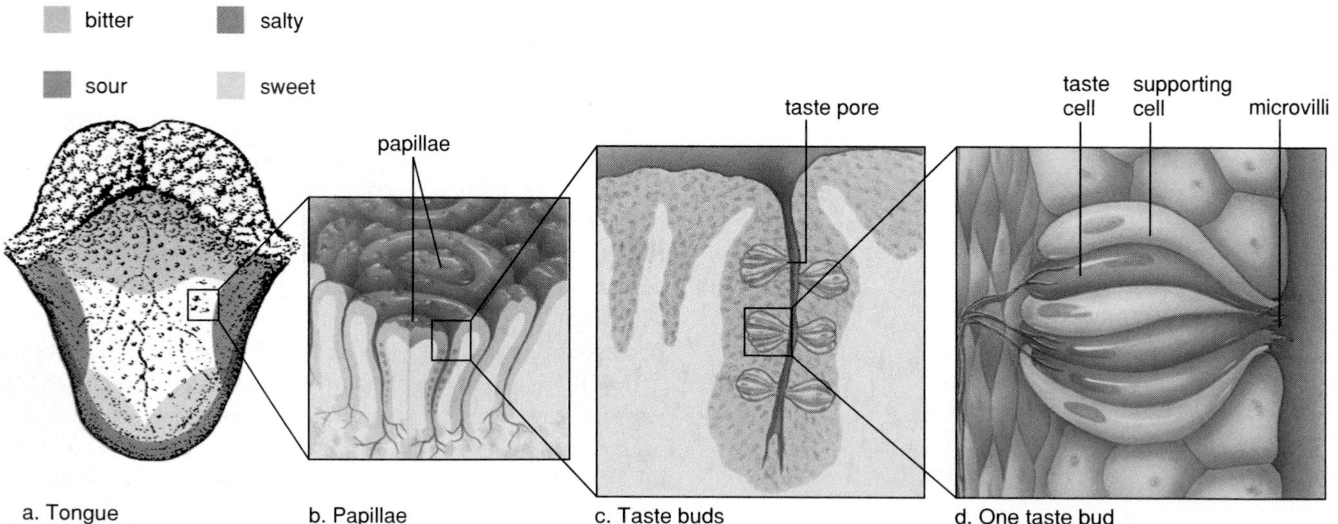

Figure 47.1 Taste buds in humans.
a. Elevations on the tongue are called papillae. The location of those containing taste buds responsive to sweet, sour, salt, and bitter is indicated. **b.** Enlargement of papillae. **c.** The taste buds are found along the walls of the papillae. **d.** The various cells that make up a taste bud. Taste cells in a bud end in microvilli that are sensitive to the chemicals exhibiting the tastes noted in (**a**). When the chemicals combine with plasma membrane receptors, nerve impulses are generated.

Smelling with the Nose

Our sense of smell is dependent on **olfactory cells** located high in the roof of the nasal cavity (Fig. 47.2). Olfactory cells are modified neurons. Each cell ends in a tuft of about five olfactory cilia, which bear plasma membrane receptors for various chemicals. When odorous molecules bind to the receptor, nerve impulses pass along nerve fibers to the olfactory bulb, which is located in the front of the brain. Some processing probably occurs here before olfactory information is sent primarily to the temporal lobe of the cerebrum, which produces the sensation of smell.

Recent work has determined that there are around one thousand different odor receptors. Many olfactory cells have the same specific type of receptor. A given odor activates a characteristic combination of cells, and this information is pooled in the olfactory bulb. The brain then determines the precise pattern of the types of receptors activated.

The olfactory receptors, like touch and temperature receptors, adapt to outside stimuli. In other words, after a while, the presence of a particular chemical no longer causes the olfactory cells to generate nerve impulses, and we are no longer aware of a particular smell.

The sense of taste and the sense of smell supplement each other, creating a combined effect when interpreted by the cerebrum. For example, when you have a cold, you think that food has lost its taste, but actually you have lost the ability to sense its smell. This method works in reverse also. When you smell something, some of the molecules move from the nose down into the mouth region and stimulate the taste buds there. Therefore, part of what we refer to as smell may actually be taste (Fig. 47.3).

The receptors for taste (taste cells) and the receptors for smell (olfactory cells) work together to give us our sense of taste and our sense of smell.

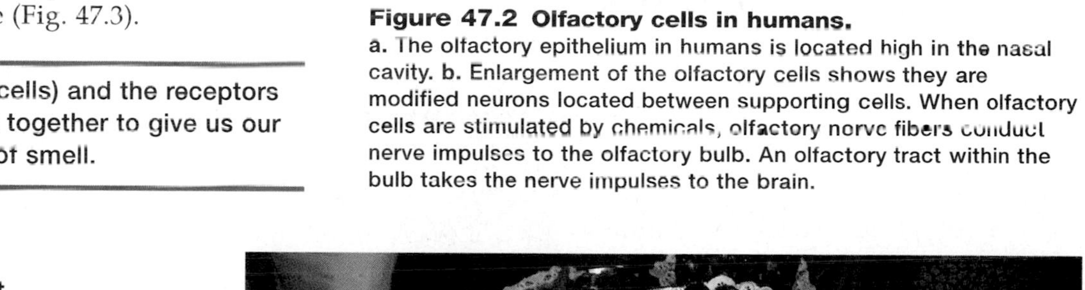

Figure 47.2 Olfactory cells in humans.
a. The olfactory epithelium in humans is located high in the nasal cavity. b. Enlargement of the olfactory cells shows they are modified neurons located between supporting cells. When olfactory cells are stimulated by chemicals, olfactory nerve fibers conduct nerve impulses to the olfactory bulb. An olfactory tract within the bulb takes the nerve impulses to the brain.

Figure 47.3 Sensation and diet.
If you are on a diet, it is safer to eat the same monotonous food each day instead of tempting the palate. A variety of interesting tastes and smells can cause you to eat more than is warranted.

47.2 Sensing Light

Animals that lack photoreceptors largely depend on the senses of hearing and smell rather than sight. Moles, which are mammals that live underground, utilize their sense of smell and touch rather than eyesight. In its simplest form, a **photoreceptor** [Gk. *photos*, light, and L. *receptor*, receiver] indicates only the presence of light and its intensity. The "eyespots" of planaria also allow these animals to determine the direction of light. Image-forming eyes are found among four invertebrate groups: cnidaria, annelids, mollusks, and arthropods. Arthropods have **compound eyes** composed of many independent visual units called *ommatidia* [Gk. *ommation*, dim. of *omma*, eye], each possessing all the elements needed for light reception (Fig. 47.4). Both the cornea and crystalline cone function as lenses to direct light rays toward the photoreceptors. The photoreceptors generate nerve impulses, which pass to the brain by way of optic nerve fibers. The outer pigment cells absorb stray light rays so that the rays do not pass from one visual unit to the other. The image, which results from all the stimulated visual units, is crude because the small size of compound eyes limits the number of visual units that still might number as many as 28,000. How arthropod brains integrate images from the compound eye to form perceptions is not known.

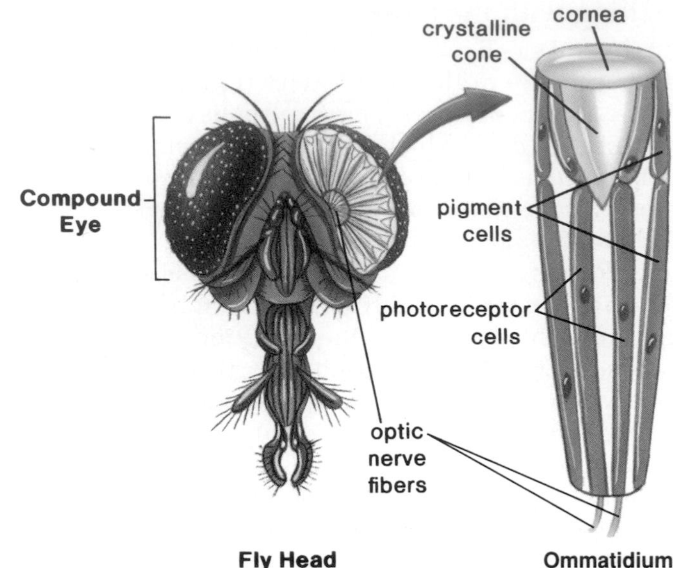

Figure 47.4 Compound eye.
Each visual unit of a compound eye has a cornea and a lens that focus light onto photoreceptors. The photoreceptors generate nerve impulses that are transmitted to the brain, where interpretation produces a mosaic image.

Insects have color vision, but they make use of a slightly shorter range of the electromagnetic spectrum compared to humans. They can see the longest of the ultraviolet rays, and this enables them to be especially sensitive to the reproductive parts of flowers, which reflect particular ultraviolet patterns (Fig. 47.5). Some fishes, reptiles, and most birds are believed to have color vision, but among mammals, only humans and other primates have color vision. It would seem, then, that this trait was adaptive for a diurnal habit (active during the day), which accounts for its retention in these few mammals.

Vertebrates and certain mollusks, like the squid and the octopus, have a *camera type of eye*. Since mollusks and vertebrates are not closely related, this similarity is an example of convergent evolution. A single lens focuses an image of the visual field on the photoreceptors, which are closely packed together. In vertebrates the lens changes shape to aid focusing, but in mollusks the lens moves back and forth. All of the photoreceptors taken together can be compared to a piece of film in a camera. The human eye is more complex than a camera, however, as we shall see.

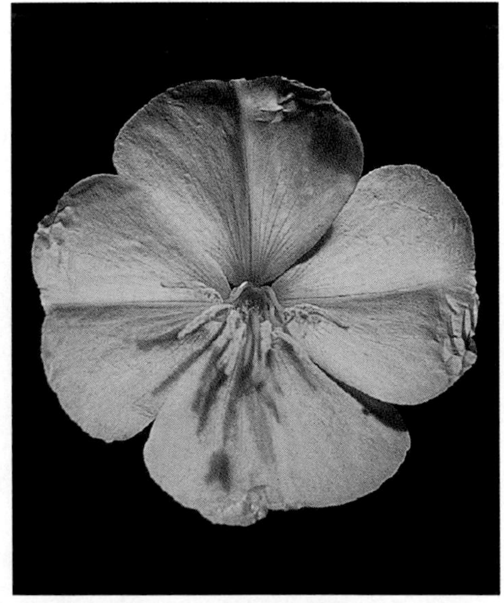

Figure 47.5 Nectar guides.
Evening primrose, *Oenothera*, as seen by humans *(left)* and insects *(right)*. Humans see no markings, but insects see distinct blotches because their eyes respond to ultraviolet rays. These types of markings, known as nectar guides, often highlight the reproductive parts of flowers where insects feed on nectar and pick up pollen at the same time.

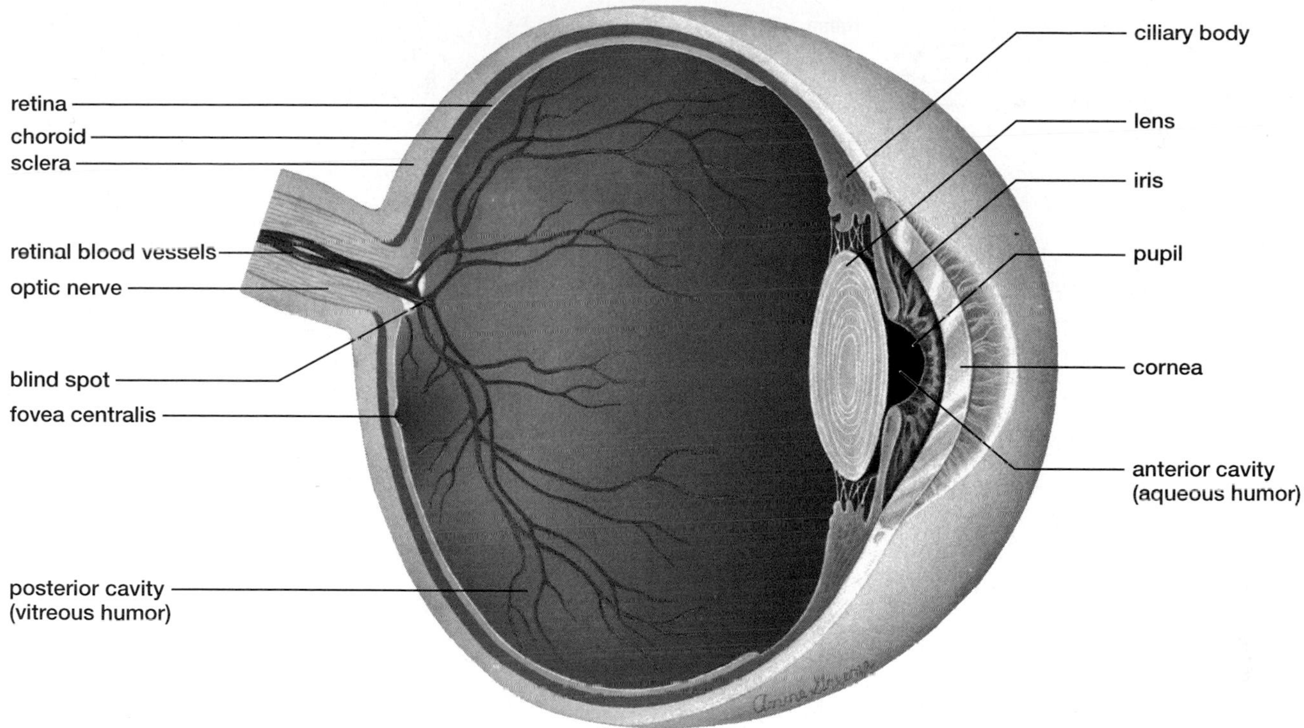

Figure 47.6 Anatomy of the human eye.
Notice that the sclera becomes the cornea and that the choroid is connected to the ciliary body and the iris. The retina contains the receptors for vision; the fovea centralis is the region where vision is most acute. A blind spot occurs where the optic nerve leaves the retina. There are no receptors for light in this location.

Table 47.1
Function of Parts of the Eye

Part	Function
Lens	Refracts and focuses light rays
Iris	Regulates light entrance
Pupil	Admits light
Choroid	Blood supply and absorbs stray light
Sclera	Protects and supports eyeball
Cornea	Refracts light rays
Humors	Refract light rays
Ciliary body	Holds lens in place, accommodation
Retina	Contains receptors for sight
Rod cells	Make black-and-white vision possible
Cone cells	Make color vision possible
Optic nerve	Transmits impulse to brain
Fovea centralis	Makes acute vision possible

Seeing with the Eye

The most important parts of the human eye and their functions are listed in Table 47.1. The human eye, which is an elongated sphere about 2.5 cm in diameter, has three layers, or coats: the sclera, the choroid, and the retina (Fig. 47.6). The outer layer, the **sclera** [Gk. *skleros,* hard], is an opaque, white, fibrous layer that covers most of the eye; in front of the eye the sclera becomes the transparent cornea, the window of the eye. The middle, thin, dark-brown layer, the **choroid** [Gk. *chorion,* membrane], contains many blood vessels and pigment that absorb stray light rays. Toward the front of the eye, the choroid thickens and forms the ring-shaped ciliary body and a thin, circular, muscular diaphragm, the *iris.* The iris regulates the size of an opening called the *pupil.* The *lens,* which is attached to the ciliary body by ligaments, divides the cavity of the eye into two portions. A basic, watery solution called *aqueous humor* fills the anterior cavity between the cornea and the lens. A viscous, gelatinous material, the *vitreous humor,* fills the large posterior cavity behind the lens.

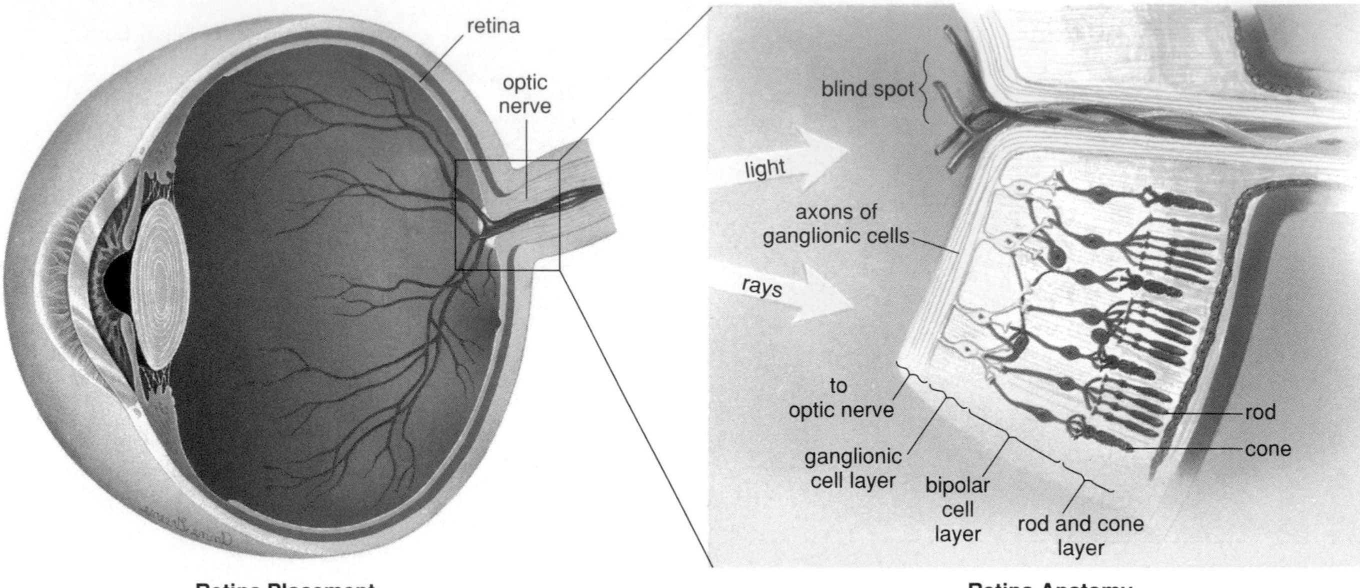

Figure 47.7 Anatomy of the retina.
The retina is the inner layer of the eye. Rods and cones are located at the back of the retina, followed by the bipolar cells and the ganglionic cells, whose fibers become the optic nerve. (Notice that rods share bipolar cells but cones do not. Cones, therefore, distinguish more detail.) The optic nerve carries impulses to the occipital lobe of the cerebrum via the optic chiasma.

How the Retina Is Structured

The inner layer of the eye, the **retina** [L. *retis,* net], contains the receptors for sight: **rod cells** and **cone cells** (Fig. 47.7). Nerve impulses initiated by the rods and the cones are passed to the bipolar cells, which in turn pass them to the ganglionic cells. The fibers of these cells pass in front of the retina, forming the **optic nerve** [Gk. *optikos,* related to sight], which carries the nerve impulses to the brain. Notice that there are many more rods and cones than nerve fibers leaving ganglionic cells. This means that there is considerable mixing of messages and a certain amount of integration before nerve impulses are sent to the brain. There are no rods or cones at the point where the optic nerve passes through the retina; therefore, this point is called the blind spot.

The center of the retina contains a special region called the **fovea centralis,** an oval, yellowish area with a depression where there are only cone cells (see Fig. 47.6). In the fovea centralis or fovea, color vision is most acute in daylight; at night, it is barely sensitive. At this time, the rods in the rest of the retina are active.

As discussed in the reading on page 847, blindness can have many causes, including glaucoma. Normally, aqueous humor produced by the ciliary body leaves the anterior cavity by way of tiny ducts. When a person has glaucoma, these drainage ducts are blocked, and aqueous humor builds up. The resulting pressure compresses the arteries that serve the nerve fibers of the retina. The nerve fibers begin to die due to lack of nutrients, and the person becomes partially blind. Over time, total blindness can result.

The human eye has three layers: the outer sclera, the middle choroid, and the inner retina. Only the retina contains receptors for sight.

Focusing Uses the Lens Light rays entering the eye are bent (refracted) as they pass through the cornea, the lens, and the humors and are brought to a focus on the retina. The lens is relatively flat when viewing distant objects but rounds up for near objects because light rays must be bent to a greater degree when viewing a near object. These changes of the lens shape are called accommodation (Fig. 47.8).

Because of refraction, the image on the retina is rotated 180° from the actual (Fig. 47.8*a*), but it is believed that this image is righted in the brain. In one experiment, scientists wore glasses that inverted and reversed the field. At first, they had difficulty adjusting to the placement of objects, but they soon became accustomed to their inverted world. Experiments such as this suggest that if the retina sees the world "upside down," the brain has learned to see it right side up.

The lens, assisted by the cornea and the humors, focuses images on the retina.

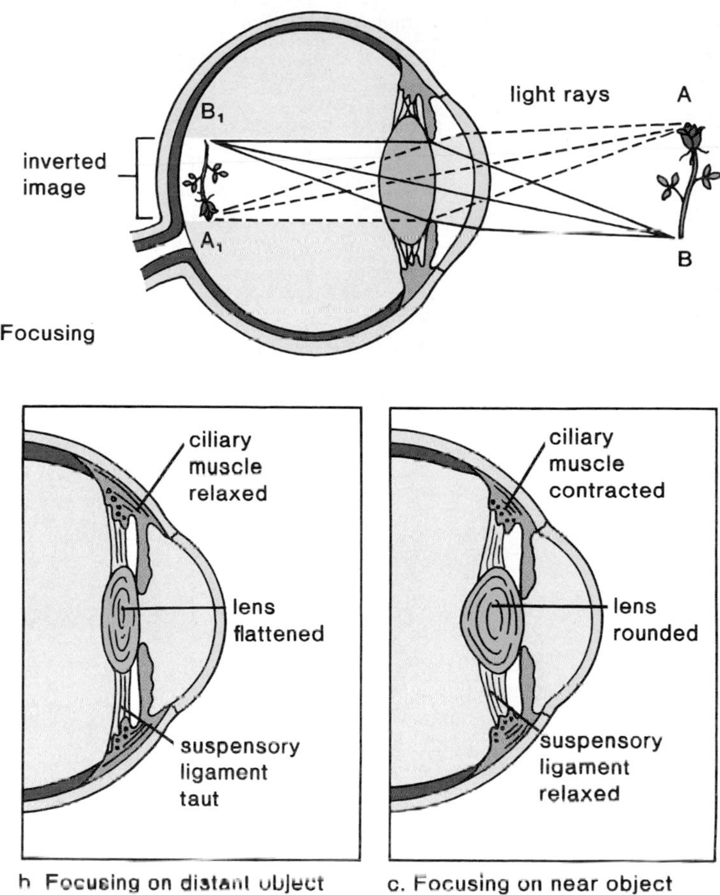

Figure 47.8 Focusing of the human eye.
a. Light rays from each point on an object are bent by the cornea and the lens in such a way that they are directed to a single point after emerging from the lens. By this process, an inverted image of the object forms on the retina. b. When focusing on a distant object, the lens is flat because the ciliary muscle is relaxed and the suspensory ligament is taut. c. When focusing on a near object, the lens accommodates: it becomes rounded because the ciliary muscle contracts, causing the suspensory ligament to relax.

Helping the Eye

With normal aging, the lens loses its ability to accommodate for near objects (Fig. 47.8*b*); therefore, persons frequently need reading glasses once they reach middle age. Aging, or possibly exposure to the sun, also makes the lens subject to cataracts; the lens can become opaque and therefore incapable of transmitting light rays. Currently surgery is the only viable treatment for cataracts. First, a surgeon opens the eye near the rim of the cornea. The enzyme zonulysin may be used to digest away the ligaments holding the lens in place. Most surgeons then use a cryoprobe, which freezes the lens for easy removal. An intraocular lens attached to the iris can then be implanted so that the patient does not need to wear thick glasses or contact lenses.

Persons who can see a near object but have trouble seeing what is designated as a size 20 letter 20 feet away are said to be nearsighted. These individuals often have an elongated eyeball, and when they attempt to look at a distant object, the image is brought to focus in front of the retina. Usually these people must wear concave lenses, which diverge the light rays so that the image can be focused on the retina. There is a new treatment for nearsightedness called radial keratotomy, or radial K. From four to eight cuts are made in the cornea so that they radiate out from the center like spokes on a wheel. When the cuts heal, the cornea is flattened. Although some patients are satisfied with the result, others complain of glare and varying visual acuity.

Persons who can easily see the optometrist's chart but cannot easily see near objects are farsighted. They often have a shortened eyeball, and when they try to see near objects, the image is focused behind the retina. These persons must wear a convex lens to increase the bending of light rays so that the image can be focused on the retina. When the cornea or lens is uneven, the image is fuzzy. This condition, called astigmatism, can be corrected by an unevenly ground lens to compensate for the uneven cornea.

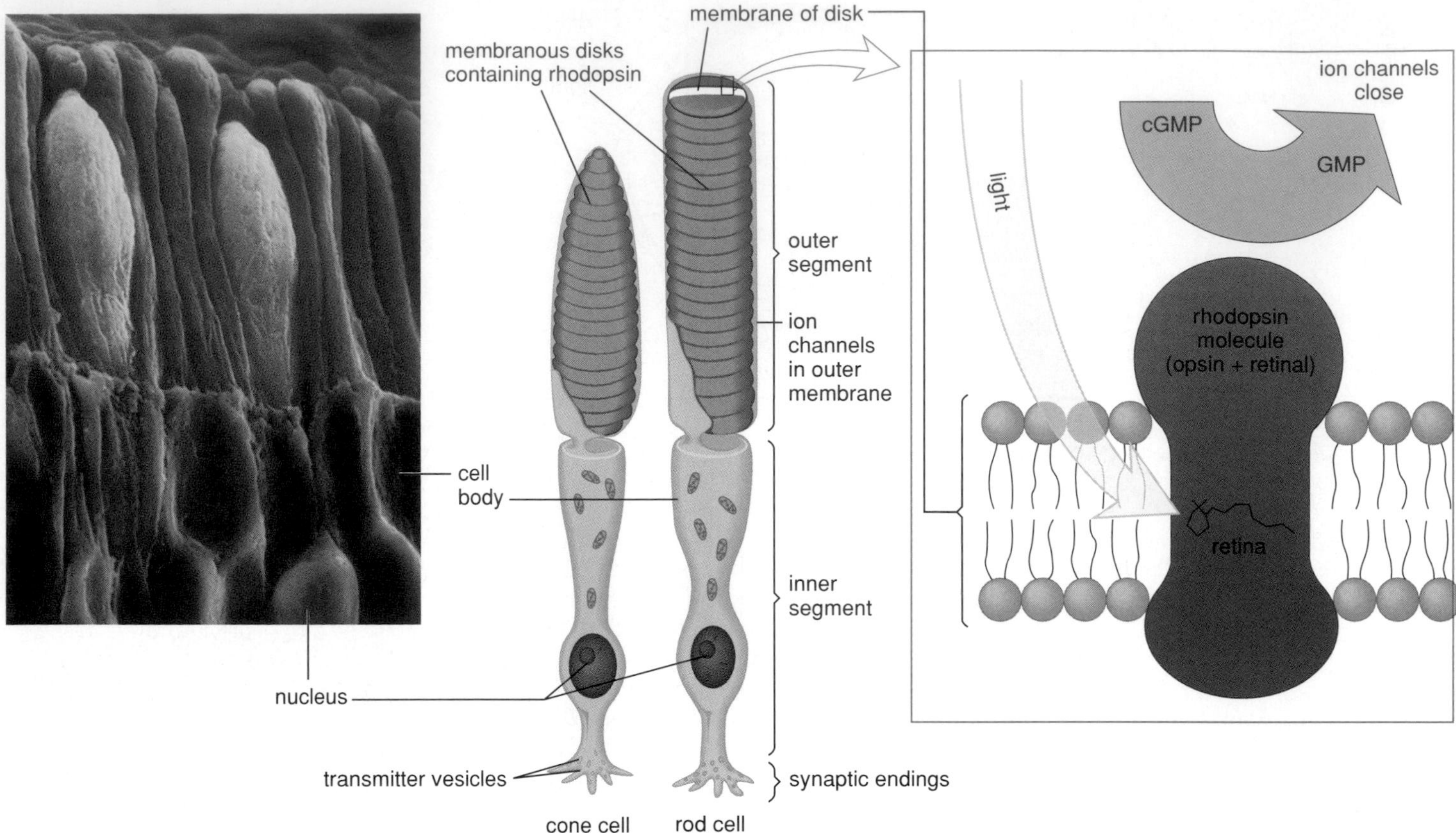

Figure 47.9 Structure and function of rods and cones.
The outer segment of rods and cones contains stacks of membranous disks, which contain visual pigments. In rods, the membrane of each disk contains rhodopsin, a complex molecule containing the protein opsin and the pigment retinal. When retinal absorbs light energy, it changes shape, activating rhodopsin to begin a series of reactions that end when cGMP (cyclic guanosine monophosphate) is converted to GMP. Thereafter, ion channels close, leading to nerve impulses.

Seeing Uses Chemistry Only dim light is required to stimulate rods; therefore, they are responsible for *night vision.* The numerous rods are also better at detecting motion than cones, but they cannot provide distinct and/or color vision. This causes objects to appear blurred and look gray in dim light. Many molecules of **rhodopsin** [Gk. *rodon,* rose, and *opsis,* sight] are located within the membrane of the disks (lamellae) found in the outer segment of the rods (Fig. 47.9). Rhodopsin is a complex molecule that contains protein (opsin) and a pigment molecule called *retinal,* which is a derivative of vitamin A. When light strikes retinal, it changes shape and rhodopsin is activated. The membrane of the outer segment of a rod cell has many ion channels that are held open in the dark by cyclic guanosine monophosphate (cGMP). Activation of rhodopsin leads to a reduction in the amount of cGMP and closure of some sodium ion channels. The resulting increased negativity of the rod interior leads to change in the frequency of nerve impulses in bipolar and then in the ganglionic cells that send messages to the brain. Rod cells are sensitive to dim light because one molecule of rhodopsin acts on many proteins that cause the cleavage of many cGMP molecules. There has been an *amplification* of the original stimulus!

The cones, located primarily in the fovea and activated by bright light, detect the fine detail and the color of an object. *Color vision* depends on three different kinds of cones, which contain pigments called the B (blue), G (green), and R (red) pigments. Each pigment is made up of retinal and opsin, but there is a slight difference in the opsin structure of each, which accounts for their individual absorption patterns. Various combinations of cones are believed to be stimulated by in-between shades of color, and the combined nerve impulses are interpreted in the brain as a particular color.

In the human eye, the receptors for sight are the rods and the cones. The rods are responsible for vision in dim light, and the cones are responsible for vision in bright light and for color vision.

Health Focus

Protecting Vision and Hearing

Age can be accompanied by a serious loss of vision and hearing. The time to start preventive measures for such problems, however, is when we are younger.

Preventing a Loss of Vision

The eye is subject to both injuries and disorders. Although flying objects sometimes penetrate the cornea and damage the iris, lens, or retina, careless use of contact lenses is the most common cause of injuries to the eye. Injuries cause only 4% of all cases of blindness; the most frequent causes are retinal disorders, glaucoma, and cataracts, in that order. Retinal disorders are varied. In diabetic retinopathy, which blinds many people between the ages of 20 and 74, capillaries to the retina burst and blood spills into the vitreous fluid. Careful regulation of blood glucose levels in these patients may be protective. In macular degeneration, the cones are destroyed because thickened choroid vessels no longer function as they should. Glaucoma occurs when the drainage system of the eyes fails, so that fluid builds up and destroys nerve fibers responsible for peripheral vision. Eye doctors always check for glaucoma, but it is advisable to be aware of the disorder in case it comes on quickly. Those who have experienced acute glaucoma report that the eyeball feels as heavy as a stone. In cataracts, cloudy spots on the lens of the eye eventually pervade the whole lens. The milky yellow-white lens scatters incoming light and blocks vision.

There are preventive measures that we can take to reduce the chance of defective vision as we age. Accumulating evidence suggests that both macular degeneration and cataracts, which tend to occur in the elderly, are caused by long-term exposure to the ultraviolet rays of the sun. It is recommended, therefore, that everyone, especially those who live in sunny climates or work outdoors, wear glass and not plastic sunglasses to absorb ultraviolet light. Large lenses worn close to the eyes offer further protection. Special purpose lenses that block at least 99% of UV-B and 60% UV-A, and 20–97% of visible light are good for bright sun combined with sand, snow, or water. Health care providers have found an increased incidence of cataracts in heavy cigarette smokers. In men, smoking 20 cigarettes or more a day, and in women, smoking more than 35 cigarettes a day doubles the risk of cataracts. It is possible that smoking reduces the delivery of blood and therefore nutrients to the lens.

Preventing a Loss of Hearing

Especially when we are young, the middle ear is subject to infections that can lead to hearing impairments if they are not treated promptly by a physician. The mobility of ossicles decreases with age, and in otosclerosis, new filamentous bone grows over the stirrup impeding its movement. Surgical treatment is the only remedy for this type of conduction deafness. However, age-associated nerve deafness due to stereocilia damage from exposure to loud noises is preventable. Hospitals are now aware that even the ears of the newborn need to be protected from noise and are taking steps to make sure neonatal intensive care units and nurseries are as quiet as possible.

In today's society, exposure to excessive noise is a possibility. Noise is measured in decibels, and any noise above a level of 80 decibels could result in damage to the hair cells of the spiral organ (organ of Corti). Eventually, the stereocilia and then the hair cells disappear completely (Fig. 47A). If listening to city traffic for extended periods can damage hearing, it stands to reason that frequent attendance at rock concerts, constantly playing a stereo loudly, or using earphones at high volume is also damaging to hearing. The first hint of danger could be temporary hearing loss, a "full" feeling in the ears, muffled hearing, or tinnitus (e.g., ringing in the ears). If you have any of these symptoms, modify your listening habits immediately to prevent further damage. If exposure to noise is unavoidable, specially designed noise-reduction earmuffs are available, and it is also possible to purchase ear plugs made from a compressible, spongelike material at the drugstore or sporting-goods store. These ear plugs are not the same as those worn for swimming, and they should not be used interchangeably.

Aside from loud music, noisy indoor or outdoor equipment, such as a rug-cleaning machine or a chain saw, is also troublesome. Even motorcycles and recreational vehicles such as snowmobiles and motocross bikes can contribute to a gradual loss of hearing. Exposure to intense sounds of short duration, such as a burst of gunfire, can result in an immediate hearing loss. Hunters may have a significant hearing reduction in the ear opposite to the shoulder where the gun is carried. The butt of the rifle offers some protection to the ear nearest the gun when it is shot.

Finally, people need to be aware that some medicines are ototoxic. Anticancer drugs, most notably cisplatin, and certain antibiotics (e.g., streptomycin, kanamycin, gentamicin) make ears especially susceptible to a hearing loss. Anyone taking such medications needs to be especially careful to protect the ears from any loud noises.

Figure 47A Hearing loss. Damaged hair cells in the spiral organ of a guinea pig. This damage occurred after 24-hour exposure to a noise level typical of rock concerts.

47.3 Sensing Mechanical Stimuli

Mechanoreceptors [Gk. *mechane,* machine, and L. *receptor,* receiver] are sensitive to mechanical stimuli, such as pressure, sound waves, and gravity. Human skin contains various types of mechanoreceptors, such as *touch receptors* and *pressure receptors.* A pressure receptor, called the Pacinian corpuscle, is shaped like an onion and consists of a series of concentric layers of connective tissue wrapped around the end of a sensory neuron. In contrast, *pain receptors* are only the unmyelinated ("naked") ends (dendrites) of the fibers of sensory neurons. Some pain receptors are especially sensitive to mechanical stimuli; others are most sensitive to temperature or chemicals.

Hair cells, which are named for the cilia they bear, are often mechanoreceptors in various specialized vertebrate sense organs. Fishes and amphibians have a series of such receptors called the lateral line system, which detect water currents and pressure waves from nearby objects. In primitive fishes and aquatic amphibians, the receptors are located on the body surface, but in advanced fishes, they are located within a canal that has openings to the outside (Fig. 47.10). A lateral line receptor is a collection of hair cells with cilia embedded in a mass of gelatinous material known as a cupula. When the cupula bends due to pressure waves, the hair cells initiate nerve impulses. The otic vesicles of fishes are derived from a portion of the lateral line system.

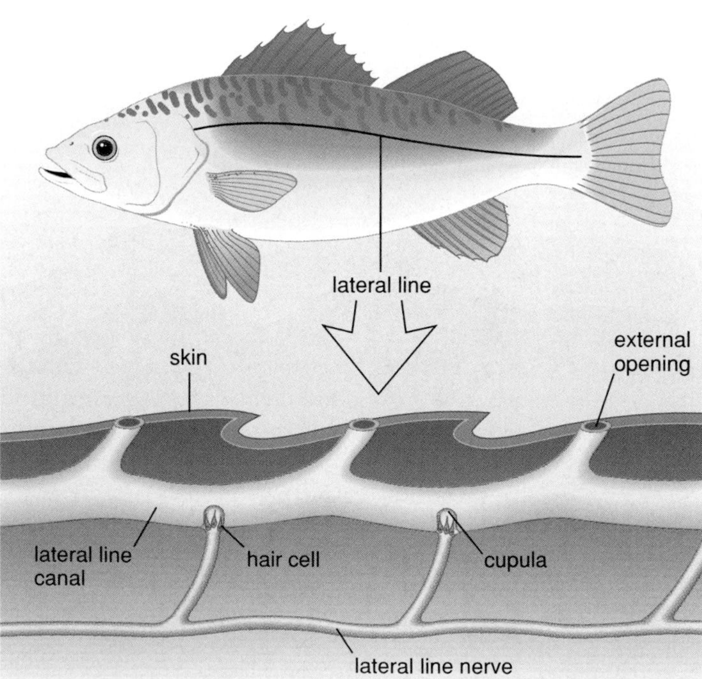

Figure 47.10 Lateral line system of fishes.
Location of the system (*upper*); longitudinal section of the system (*lower*). A main canal has openings to the exterior. Lining the canal are hair cells (embedded in cupulae) that act as sense receptors for pressure.

Structure of the Human Ear

Table 47.2 lists the most important parts of the human ear and their function. As Figure 47.11 shows, the human ear has an outer, a middle, and an inner portion. The evolution of the inner ear of humans can be traced back to the lateral line system of fishes. However, the inner ear of humans contains both equilibrium and sound receptors.

The *outer ear* consists of the pinna (external flap) and *auditory canal.* The auditory canal is lined with fine hairs, which filter the air. Modified sweat glands are located in the upper wall of the canal; these secrete earwax, which helps guard the ear against entrance of foreign materials such as air pollutants.

The *middle ear* begins at the **tympanic membrane** [Gk. *tympanum,* drum] and ends at a bony wall that has small openings—the *oval window* and the *round window*—covered by membranes. Three small bones are located between the tympanic membrane and the oval window. Collectively called *ossicles,* individually they are the malleus (hammer), the incus (anvil), and the stapes (stirrup), named for their structural resemblance to these objects. The auditory (eustachian) tube extends from the middle ear to the pharynx and permits the equalization of air pressure between the inside and the outside of the ear. Chewing gum, yawning, and swallowing help move air through the auditory tube during ascent and descent in airplanes or elevators. Atmospheric pressure pressing against the inner surface of the tympanic membrane then equals that pressing on its outer surface.

Table 47.2
Function of the Parts of the Ear

Part	Function
Outer Ear	
Pinna	Collects sound waves
Auditory canal	Filters air
Middle Ear	
Tympanic membrane and ossicles	Amplify sound waves
Auditory tube	Equalizes air pressure
Inner Ear	
Vestibule (contains utricle and saccule)	Balance (static equilibrium)
Semicircular canals	Balance (dynamic equilibrium)
Cochlea	Transmits pressure waves that cause the spiral organ to generate nerve impulses resulting in hearing

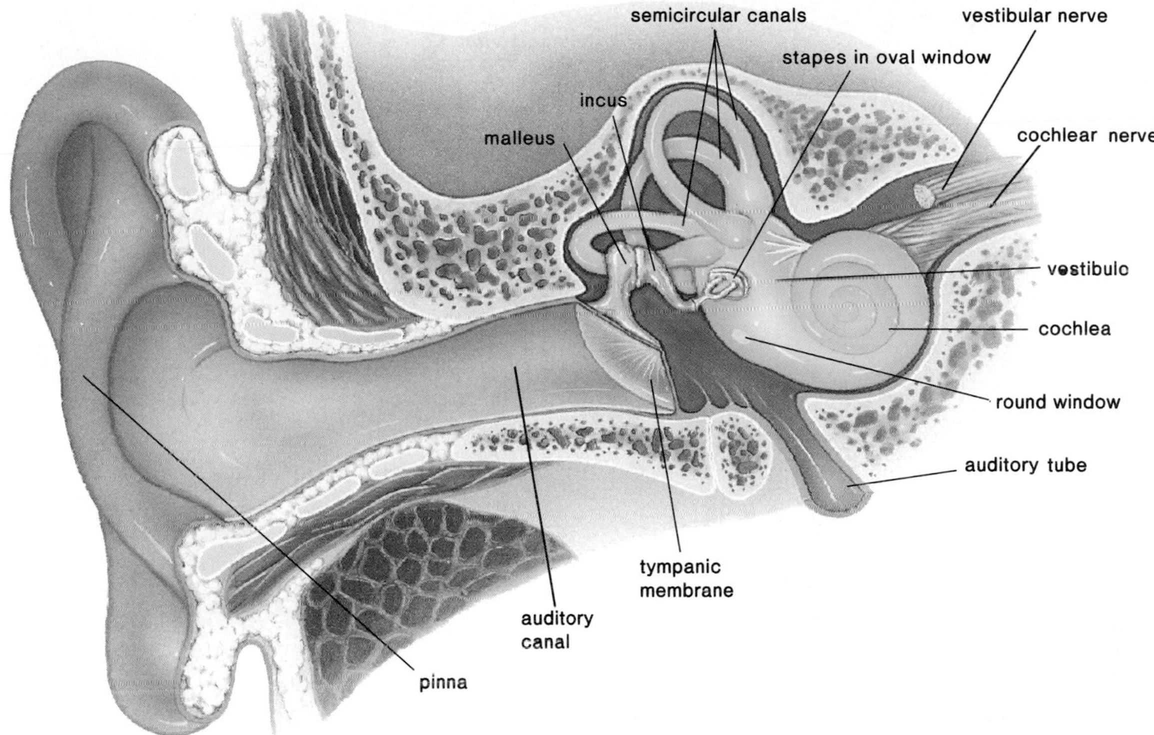

Figure 47.11 Anatomy of the human ear.
In the middle ear, the malleus (hammer), the incus (anvil), and the stapes (stirrup) amplify sound waves. Otosclerosis is a condition in which the stapes becomes attached to the inner ear and is unable to carry out its normal function. It can be replaced by a plastic piston, and thereafter the individual hears normally because sound waves are transmitted as usual to the cochlea, which contains the receptors for hearing.

The middle ear is continuous with mastoid air spaces in the temporal bone of the skull. Therefore, a middle ear infection can lead to mastoiditis, a serious medical condition that should be treated promptly.

The inner ear has three spaces that form a maze called the bony labyrinth. The spaces consist of the semicircular canals, the vestibule, and the cochlea. Membranous canals and sacs present in the spaces are filled with a clear fluid called endolymph.

The **semicircular canals** are arranged so that there is one in each dimension of space. The enlarged base of each semicircular canal is called an **ampulla**. In each ampulla, the stereocilia of hair cells are embedded within a gelatinous material called a cupula.

The vestibule, which lies between the semicircular canals and the cochlea, contains two small membranous sacs called the **utricle** and the **saccule.** Both sacs contain little hair cells with stereocilia embedded within a gelatinous material called an otolithic membrane. Calcum carbonate ($CaCO_3$) granules, or otoliths (Gk. *otos*, ear, and *litos*, stone), rest on the otolithic membrane.

The spiraling cochlea (L. *cochlea*, snail-shaped) resembles the shell of a snail. The cochlea contains three canals: the vestibular canal, the **cochlear canal**, and the tympanic canal. Stereocilia of hair cells, which rest on the basilar membrane (the lower wall of the cochlear canal), are embedded within a gelatinous material called the **tectorial membrane.** These hair cells, which are known as the **spiral organ** (organ of Corti) synapse with nerve fibers of the cochlear (auditory) nerve. The cochlear nerve carries nerve impulses to the brain.

The ear has three major divisions: outer ear, middle ear, and inner ear. The outer ear contains the auditory canal; the middle ear contains the ossicles; and the inner ear contains the semicircular canals, the vestibule, and the cochlea.

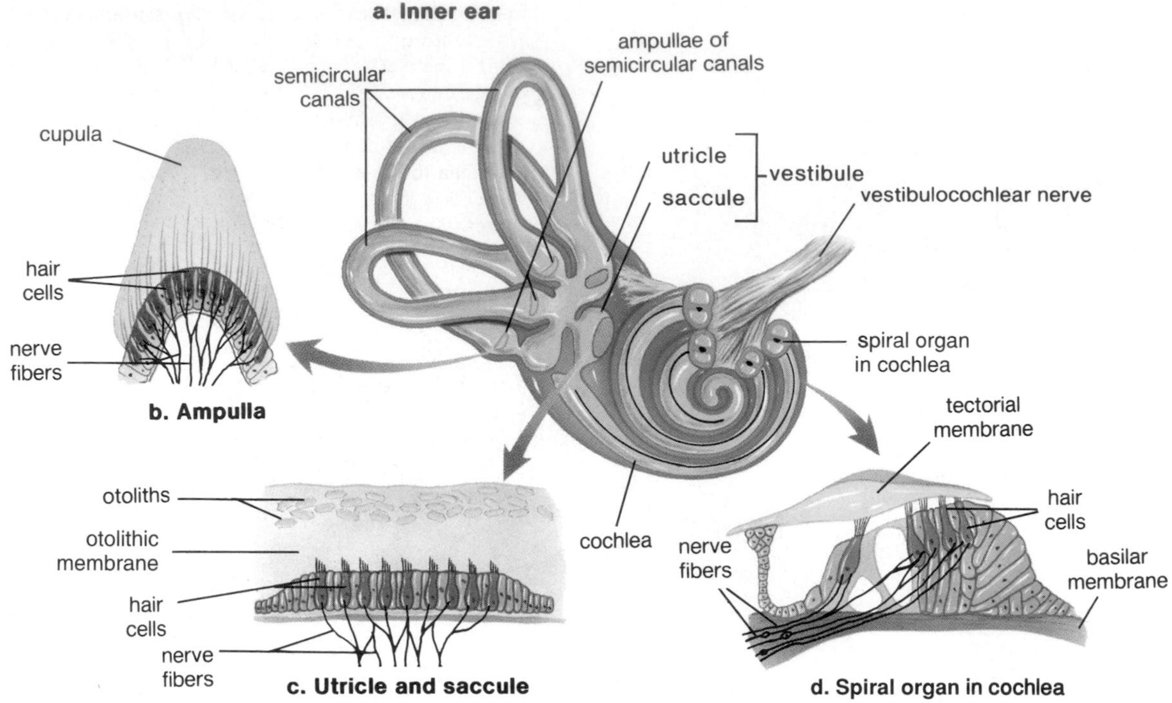

Figure 47.12 Anatomy of the inner ear.
a. The inner ear contains the semicircular canals, the utricle and the saccule within a vestibule, and the cochlea. The cochlea has been cut to show the location of the spiral organ. b. An ampulla at the base of each semicircular canal contains the receptors for dynamic equilibrium (hair cells). c. The utricle and the saccule are small sacs that contain the receptors for static equilibrium (hair cells). d. The receptors for hearing (hair cells) are in the spiral organ.

Hair Cells Keep Us Balanced

The sense of balance has been subdivided into two senses: *dynamic equilibrium,* involving angular and/or rotational movement of the head, and *static equilibrium,* involving movement of the head in one plane, either vertical or horizontal.

Because there are three semicircular canals, each ampulla responds to head rotation in a different plane of space. As fluid within a semicircular canal flows over and displaces a cupula, the stereocilia of the hair cells bend, and the pattern of impulses carried by the vestibular nerve to the brain changes (Fig. 47.12*a*). Continuous movement of fluid in the semicircular canals causes one form of motion sickness.

The utricle is especially sensitive to horizontal movements and the bending of the head, while the saccule responds best to vertical (up-down) movements. When the body is still, the otoliths in the utricle and the saccule rest on the otolithic membrane above the hair cells (Fig. 47.12*b*). When the head bends or the body moves in the horizontal and vertical planes, the otoliths are displaced and the otolithic membrane sags, bending the stereocilia of the hair cells beneath. The direction the stereocilia bend tells the brain the direction of the movement of the head. Similar types of organs, called statocysts, are also found in cnidaria, mollusks, and crustacea. These organs give information only about the position of the head; they are not involved in the sensation of movement (Fig. 47.13). They are therefore called *static equilibrium organs.*

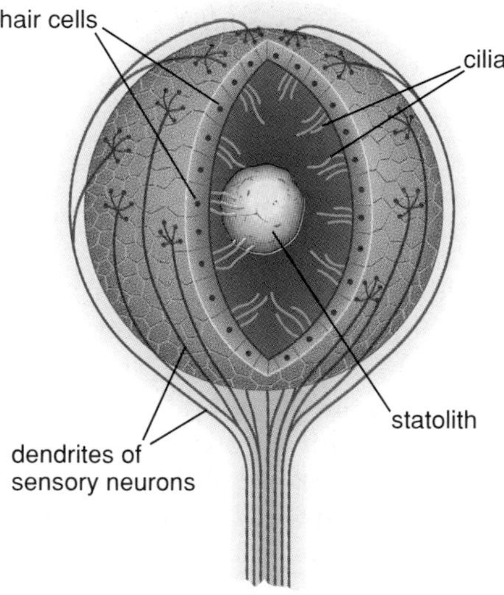

Figure 47.13 Statocysts.
In mollusks and crustacea, a small particle, the statolith, moves in response to a change in the animal's position. When the statolith stops moving, it stimulates the closest cilia of hair cells. These cilia generate impulses, indicating the position of the body.

Movement of cupula within the semicircular canals contributes to the sense of dynamic equilibrium. Movement of the otolithic membrane within the utricle and the saccule accounts for static equilibrium.

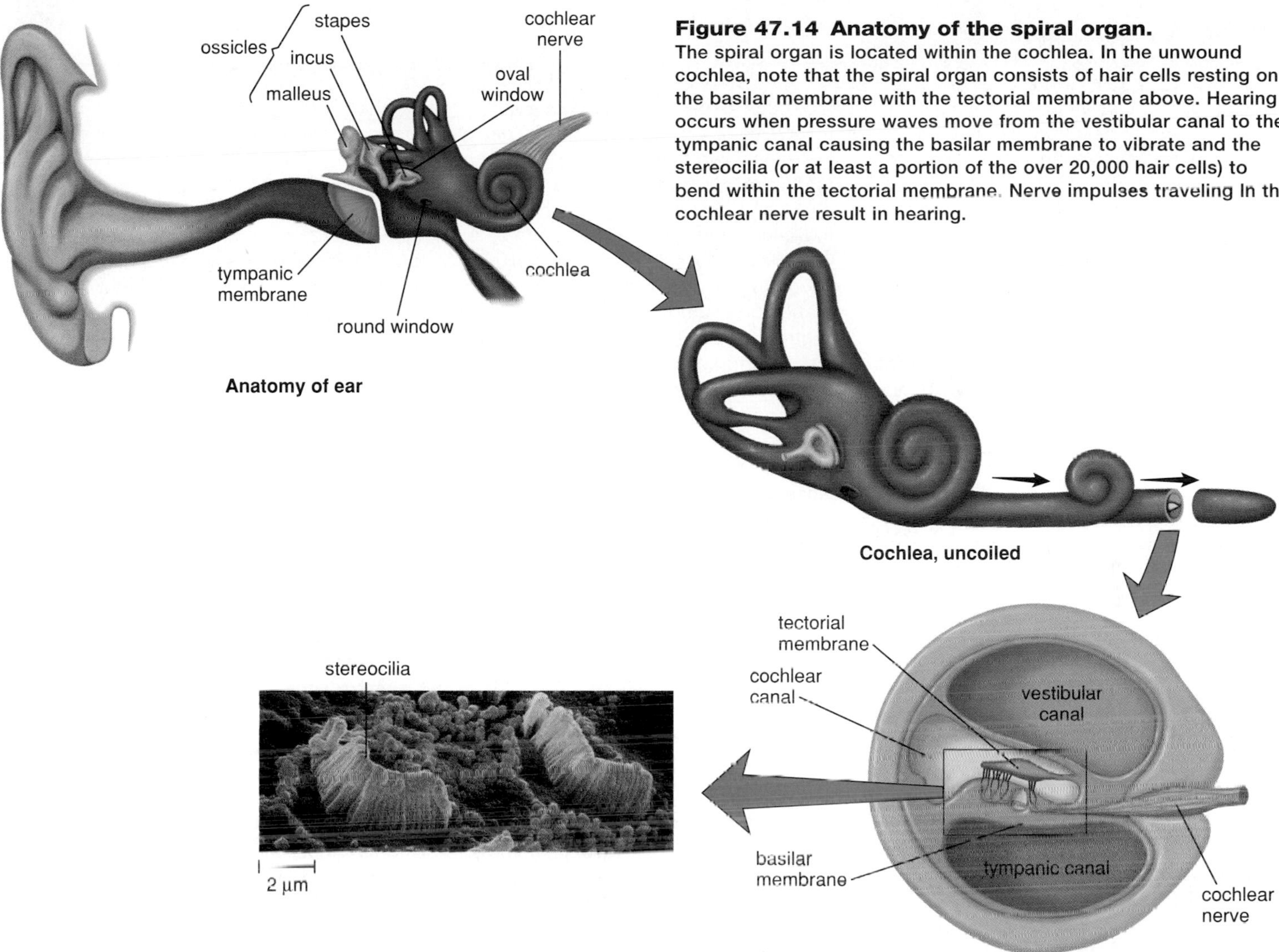

Figure 47.14 Anatomy of the spiral organ. The spiral organ is located within the cochlea. In the unwound cochlea, note that the spiral organ consists of hair cells resting on the basilar membrane with the tectorial membrane above. Hearing occurs when pressure waves move from the vestibular canal to the tympanic canal causing the basilar membrane to vibrate and the stereocilia (or at least a portion of the over 20,000 hair cells) to bend within the tectorial membrane. Nerve impulses traveling in the cochlear nerve result in hearing.

Hair Cells Account for Hearing

The process of hearing begins when sound waves enter the auditory canal, causing the ossicles to vibrate and the stapes to strike the membrane of the oval window. The sound is amplified because of a large size difference between the tympanic membrane and the oval window. The energy collected by the eardrum is concentrated onto the much smaller surface of the oval window.

When the stapes strikes the oval window, pressure waves occur in the inner ear. The vestibular canal connects with the tympanic canal, which leads to the round window membrane. When the stapes strikes the membrane of the oval window, pressure waves move from the vestibular canal to the tympanic canal and across the basilar membrane, and the round window bulges. As the basilar membrane moves up and down, the stereocilia of the hair cells embedded in the tectorial membrane bend (Fig. 47.14). Then, nerve impulses begin in the cochlear nerve and travel to the brain stem. Eventually they reach the temporal lobe of the cerebrum where they are interpreted as a sound.

The spiral organ is narrow at its base but widens as it approaches the tip of the cochlear canal. Each part of the organ is sensitive to different wave frequencies. Near the tip, the spiral organ responds to low frequencies, such as those of a tuba, and near the base, it responds to higher frequencies, such as those of a bell or a whistle. The nerve fibers from each region along the length of the spiral organ lead to slightly different areas in the brain. The pitch we perceive depends upon which region of the basilar membrane vibrates and which area of the brain is stimulated. It is believed that the tone of a sound is an interpretation of the brain based on the distribution of hair cells stimulated.

Volume is a function of the amplitude of sound waves. Loud noises cause the fluid of the cochlea to vibrate to a greater degree, and this, in turn, causes the basilar membrane to move up and down to a greater extent. The resulting increased stimulation is interpreted by the brain as volume. As discussed in the reading on page 847, extremely loud noises can cause deafness.

The sense receptors for sound are hair cells on the basilar membrane in the spiral organ. When the basilar membrane vibrates, the stereocilia of the hair cells bend, and nerve impulses are transmitted to the brain.

connecting concepts

An animal's information exchange with the internal and external environment is dependent upon just a few types of sensory receptors. We have examined three types: chemoreceptors, such as taste cells and olfactory cells; photoreceptors, such as eyes; and mechanoreceptors, such as those for balance and hearing. The senses are not equally developed in all animals. Male moths have chemoreceptors on the filaments of their antennae to detect minute amounts of an airborne sex attractant released by a female. This is certainly a more efficient method than searching for a mate by sight.

Birds that live in forested areas signal that a territory is occupied by singing—it is difficult to see a bird in a tree, as most birders know. On the other hand, hawks have such a keen sense of sight that they are able to locate a small mouse far below them. Insectivorous bats have an unusual adaptation for finding prey in the dark. They send out a series of sound pulses and listen for the echoes that come back. The time it takes for an echo to return indicates the location of an insect. A unique adaptation is found among the so-called electric fishes of Africa and Australia. They have electroreceptors that can detect disturbances in an electrical current they emit into the water. These disturbances indicate the location of obstacles and prey.

Animals that migrate use various senses to find their way. Salmon hatch in a freshwater stream, but drift to the ocean as larvae. By the end of the third or fourth year, they migrate back to where they were hatched. Like other migrating animals, they apparently can use the sun as a compass to find their way back to the vicinity of their home river, but then salmon switch to a sense of smell to find the exact location of their hatching. Perhaps the odor of the plants and soil in this stream was imprinted on the nervous system of the larval fish.

Through the evolutionary process, animals tend to rely on those stimuli and senses that are adaptive to their particular environment and way of life. In all cases, sensory receptors generate nerve impulses that travel to the brain where sensation occurs. In mammals, and particularly human beings, integration of the data received from various receptors results in a perception of events occurring in the external environment.

Summary

47.1 Sensing Chemicals

Chemoreception is found universally in animals and is, therefore, believed to be the most primitive sense.

Human olfactory cells and taste buds are chemoreceptors. They are sensitive to chemicals in water and air.

47.2 Sensing Light

Eyes are photoreceptors. The compound eye of arthropods is made up of many individual units, whereas the human eye is a camera-type eye with a single lens. Table 47.1 lists the parts of the eye and the function(s) of each part.

The receptors for sight in humans are the rod cells and cone cells, located in the retina. The rods work in minimum light and detect motion, but they do not detect color. The cones require bright light and do detect color.

Rhodopsin, the pigment found in rod cells, is a molecule composed of opsin and retinal. When light strikes rhodopsin, retinal changes shape and opsin is activated. Chemical changes that lead to nerve impulses follow. These impulses are eventually picked up by the optic nerve.

There are three kinds of cones, containing blue, green, or red pigment. Each pigment is also made up of retinal and opsin, but opsin structure varies among the three.

47.3 Sensing Mechanical Stimuli

Mechanoreceptors include the touch receptors and pressure receptors in the human skin. Many mechanoreceptors are hair cells with cilia, such as those found in the lateral line of fishes as well as the inner ear of humans.

Table 47.2 lists the parts of the ear and the function of each part of the human ear. The inner ear contains the sense organs for balance. Movement of fluid past hair cells in the semicircular canals gives us a sense of dynamic equilibrium. Just like the statocysts of invertebrates, portions of the human inner ear contain calcium carbonate granules (otoliths) resting on a gelatinous membrane. The movement of these granules gives us a sense of static equilibrium.

Hair cells on the basilar membrane (the spiral organ) are responsible for hearing. Pressure waves, which begin at the oval window, cause the basilar membrane to vibrate so that stereocilia embedded in the tectorial membrane bend. This causes the hair cells to initiate nerve impulses, which are carried by the auditory nerve to the brain.

Reviewing the Chapter

1. Discuss the structure and the function of human chemoreceptors. 840–41
2. In general, how does the eye in arthropods differ from that in humans? 842
3. What types of animals have eyes that are constructed similarly to the human eye? 842
4. Name the parts of the human eye, and give a function for each part. 843
5. Contrast the location and the function of rods to those of cones. 844, 846
6. What are the types of mechanoreceptors in human skin? 848
7. Describe how the lateral line system of fishes works and why it is considered to contain mechanoreceptors. 848
8. Describe the structure of the human ear. 848–49
9. Describe the role of the semicircular canals, utricle, and the saccule in balance. 850
10. Describe how we hear. 851

Testing Yourself

Choose the best answer for each question.

1. A receptor
   a. is the first portion of a reflex arc.
   b. initiates nerve impulses.
   c. responds to only one type of stimulus.
   d. All of these are correct.
2. Which of these gives the correct path for light rays entering the human eye?
   a. sclera, retina, choroid, lens, cornea
   b. fovea centralis, pupil, aqueous humor, lens
   c. cornea, pupil, lens, vitreous humor, retina
   d. optic nerve, sclera, choroid, retina, humors
3. Which gives an incorrect function for the structure?
   a. lens—focusing
   b. iris—regulation of amount of light
   c. choroid—location of cones
   d. sclera—protection
4. Which of these contain mechanoreceptors?
   a. human skin
   b. lateral line of fishes
   c. statocysts of arthropods
   d. All of these are correct.
5. Which association is incorrect?
   a. lateral line—fishes
   b. compound eye—arthropods
   c. camera-type eye—squid
   d. statocysts—sea stars
6. Which one of these wouldn't you mention if you were tracing the path of sound vibrations?
   a. auditory canal
   b. tympanic membrane
   c. semicircular canals
   d. cochlea
7. Which one of these correctly describes the location of the spiral organ?
   a. between the tympanic membrane and the oval window in the inner ear
   b. in the utricle and saccule within the vestibule
   c. between the tectorial membrane and the basilar membrane in the cochlear canal
   d. between the outer and inner ear within the semicircular canals
8. Which of these is mismatched?
   a. semicircular canals—inner ear
   b. utricle and saccule—outer ear
   c. auditory canal—outer ear
   d. ossicles—middle ear
9. Retinal is
   a. sensitive to light energy.
   b. a part of rhodopsin.
   c. found in both rods and cones.
   d. All of these are correct.
10. Both olfactory receptors and sound receptors have cilia, and they both
    a. are chemoreceptors.
    b. are mechanoreceptors.
    c. initiate nerve impulses.
    d. All of these are correct.
11. Label this diagram of the human eye. State a function for each structure labeled:

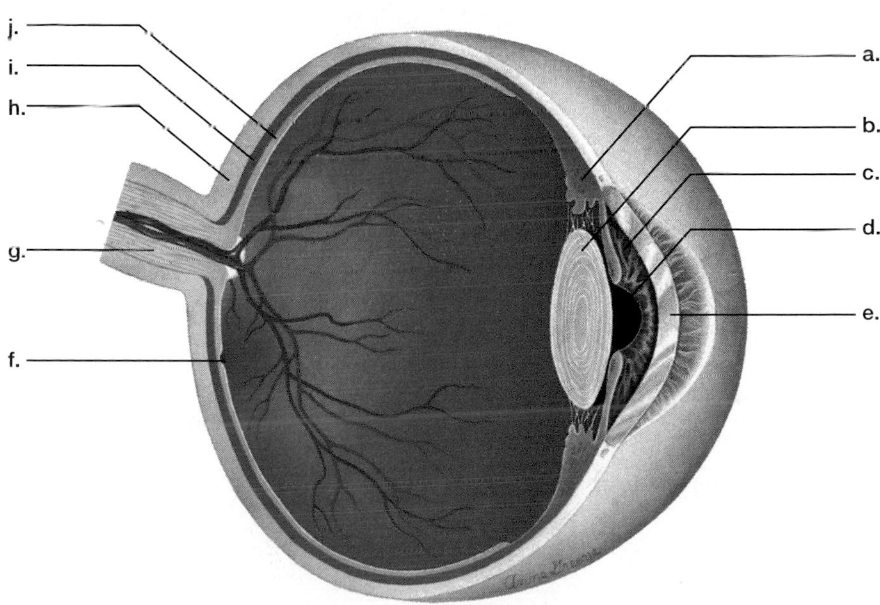

Applying the Concepts

1. *All organ systems contribute to homeostasis.*
   List several ways in which the sense organs contribute to homeostasis.
2. *In the whole animal, all systems work together and influence one another.*
   How do the nervous system and the sense organs work together?
3. *Animals are sensitive to only certain types of stimuli.*
   Why might animals be sensitive to only certain types of stimuli?

Using Technology

Your study of sense organs is supported by these available technologies:

Exploring the Internet

The Mader Home Page provides resources for and help with studying this chapter.

http://www.mhhe.com/sciencemath/biology/mader/
(Click on Biology.)

Life Science Animations Video

Video #3: Animal Biology I
Organ of Static Equilibrium (#26)
The Organ of Corti [i.e., spiral organ*] (#27)*

Understanding the Terms

Match the terms to these definitions:

a. __________ Structure that receives sensory stimuli and is a part of a sensory neuron or transmits signals to a sensory neuron.

b. __________ Innermost layer of the eyeball containing the photoreceptors—rods and cones.

c. __________ Membranous region that receives air vibrations in an auditory organ.

d. __________ Nerve that carries impulses from the retina of the eye to the brain.

e. __________ Outer, white, fibrous layer of the eye that surrounds the eye except for transparent cornea.

f. __________ Photoreceptor in the eyes that responds to bright light and allows color vision.

g __________ Receptor that is sensitive to chemical stimulation—for example, receptors for taste and smell.

h. __________ Receptor that is sensitive to mechanical stimulation, such as that from pressure, sound waves, and gravity.

i. __________ Specialized region of the cochlea containing the hair cells for sound detection and discrimination.

j. __________ Spiral-shaped structure of the inner ear containing the receptors for hearing.

Support Systems and Locomotion

CHAPTER 48

Chapter Concepts

48.1 Comparing Skeletons

- Animals have one of three types of skeletons: a hydrostatic skeleton, an exoskeleton, or an endoskeleton. 856
- All types of skeletons give support to the body and assist movement. An exoskeleton and endoskeleton also protect internal organs. 857

48.2 Humans Have an Endoskeleton

- Humans have a bony endoskeleton; bone is a rigid, but living, tissue. 859
- The human skeleton is divided into these portions: the axial skeleton consists of the skull, the ribs, the sternum, and the vertebrae, and the appendicular skeleton contains the girdles and the limbs. 860
- The human skeleton is jointed; the joints differ in movability. 863

48.3 How Muscles Function

- Macroscopically, human skeletal muscles work in antagonistic pairs and exhibit certain physiological characteristics. 865
- Microscopically, muscle fiber contraction is dependent on filaments of both actin and myosin, and a ready supply of calcium ions (Ca^{2+}) and ATP. 867

Crayola katydid, *Vestria* sp., has an exoskeleton.

Motility by contractile fibers is an animal characteristic. In order to create movement, muscle contraction must be directed against some sort of medium. This medium can be internal body fluids, a rigid exoskeleton, or a rigid endoskeleton. How do a planarian, hydra, earthworm, and other types of invertebrates move so well with no rigid skeleton at all? In these species, muscles push against body fluids inside a coelom or gastrovascular cavity. In vertebrates, the muscles are attached to a bony endoskeleton, which performs many other functions. It gives the body shape, protects internal organs, serves as a storage area for inorganic calcium and phosphorus salts, and produces blood cells. In terrestrial vertebrates, the skeleton also supports the body and its organs against the pull of gravity.

The term musculoskeletal system recognizes that the muscular system and the skeletal system work together to provide movement, whether it be a frog jumping, a fish swimming, or an eagle flying. Coordination is a necessity if movements are to be purposeful. Most animals have a nervous system, which integrates data received from sensory receptors and coordinates muscular activity so that animal behavior achieves goals like feeding, escaping enemies, reproducing, or simply playing.

48.1 Comparing Skeletons

Different types of skeletons occur in the animal kingdom. A hydrostatic skeleton is seen in cnidaria, flatworms, roundworms, and annelids. Some mollusks have a calcium carbonate exoskeleton, while arthropods have a chitinous one. Echinoderms and vertebrates have a bony endoskeleton (internal skeleton).

A Water-Filled Cavity

In animals that lack a hard skeleton, a fluid-filled gastrovascular cavity or coelom can act as a *hydrostatic skeleton*. A hydrostatic [Gk. *hydrias*, water, and *stasis*, a standing] skeleton also offers support and resistance to the contraction of muscles so that mobility results. As analogies, consider that a garden hose stiffens when filled with water, and that a water-filled balloon changes shape when squeezed at one end. Similarly, an animal with a hydrostatic skeleton can change shape and perform a variety of movements.

Cnidaria such as hydras, and flatworms such as planaria, use their fluid-filled gastrovascular cavity as a hydroskeleton that gives support. When muscle fibers at the base of epidermal cells in a hydra contract, the body or tentacles shorten rapidly. Planaria usually glide over a substrate with the help of muscular contractions that control the body wall and cilia. Nematodes (roundworms) have a fluid-filled pseudocoelom and move in a whiplike manner when their longitudinal muscles contract. Annelids, such as earthworms, are segmented and have septa that divide their coelom into compartments (Fig. 48.1). Each segment has its own set of longitudinal and circular muscles

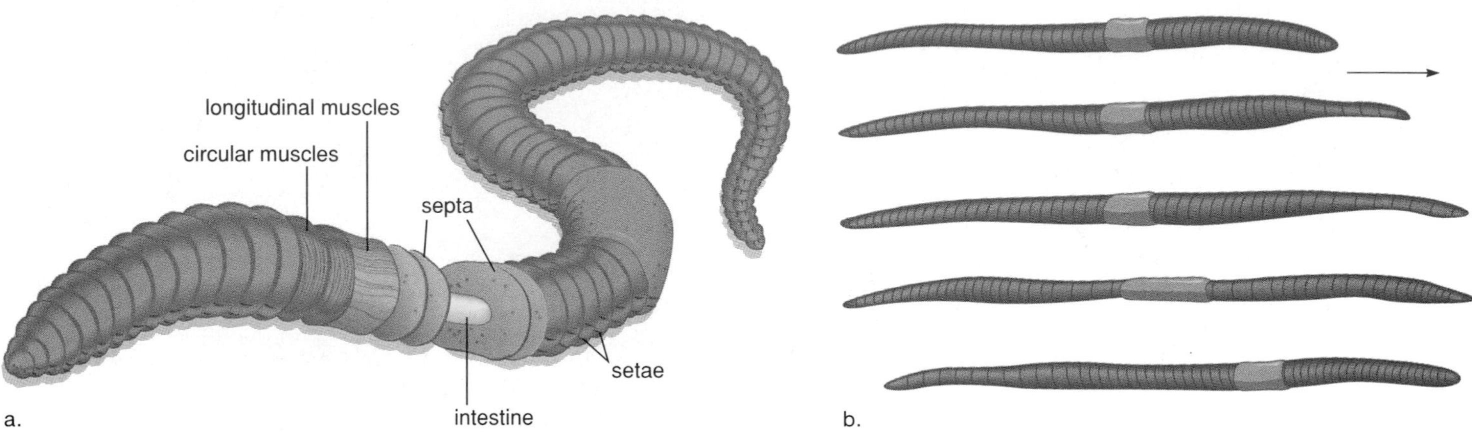

Figure 48.1 Locomotion in an earthworm.
a. The coelom is divided by septa, and each body segment is a separate locomotor unit. There are both circular and longitudinal muscles.
b. As circular muscles contract, a few segments extend. The worm is held in place by setae, needlelike chitinous structures on each segment of the body. Then, as longitudinal muscles contract, a portion of the body is brought forward. This series of events occurs down the length of the worm.

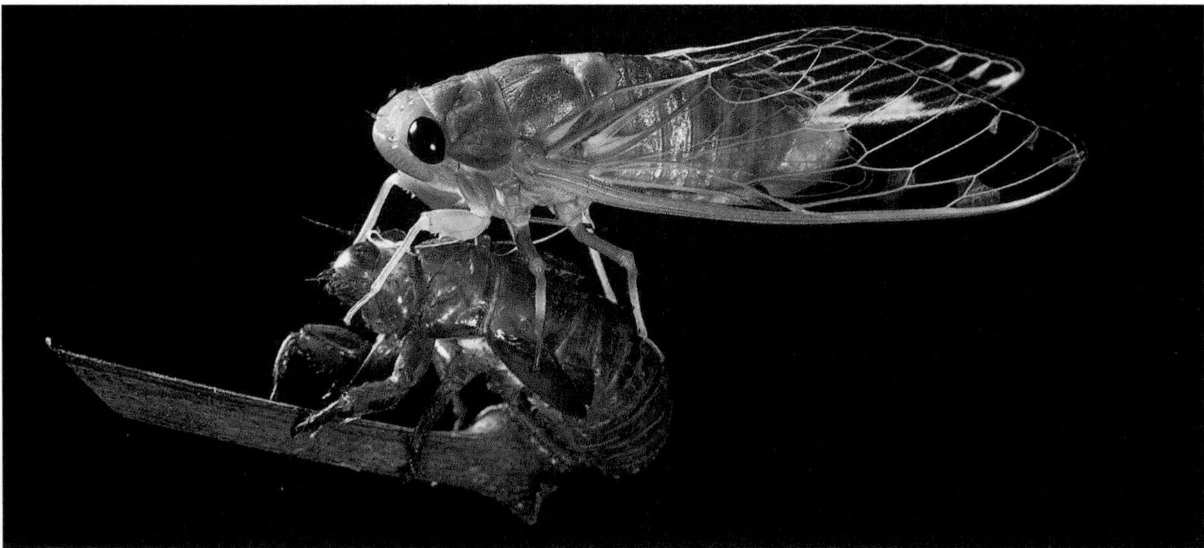

Figure 48.2 Exoskeleton.
Exoskeletons provide advantages by supporting muscle contraction and preventing drying out. The chitinous exoskeleton of arthropods is shed as the animal molts; until the new skeleton dries and hardens, the animal is vulnerable to predators, and muscle contractions may not translate into body movements. In this photo a dog-day cicada, *Tibicen*, has just finished molting.

and its own nerve supply, so each segment or group of segments may function independently. When circular muscles contract, the segments become thinner and elongate. When longitudinal muscles contract, the segments become thicker and shorten. By alternating circular muscle contraction and longitudinal muscle contraction, the animal moves forward.

Exoskeletons or Endoskeletons

An exoskeleton is an exterior skeleton. Calcium carbonate forms the exoskeleton of some animals such as corals and clams, but insects and crustacea have a chitinous exoskeleton. Chitin is a strong, flexible, nitrogenous polysaccharide. Besides providing protection against wear and tear and against enemies, an exoskeleton also prevents drying out. This is an important feature for animals that live on land. Although the stiffness of an exoskeleton provides superior support for muscle contractions, an exoskeleton is not as strong as an endoskeleton. Strength can be achieved by increasing the thickness of an exoskeleton, but this leaves less room for internal organs.

In mollusks, such as clams and snails, the exoskeleton grows as the animal grows. The thick and nonmobile calcium carbonate exoskeleton is largely for protection. The chitinous exoskeleton of arthropods, which is jointed and movable, is suitable for life on land. Arthropods, however, molt to rid themselves of an exoskeleton that has become too small, and molting makes an animal vulnerable to predators (Fig. 48.2).

Vertebrates have an endoskeleton, composed of bone and cartilage, which grows with the animal. Endoskeletons do not limit the space available for internal organs, and they support greater weight. The soft tissues that surround an endoskeleton protect it, and injuries to soft tissues are apt to be easier to repair than is a broken skeleton. However, endoskeletons usually do have elements that protect vital internal organs (Fig. 48.3).

Both arthropods (e.g., insects and crustacea) and vertebrates have a rigid and segmented skeleton. The exoskeleton of arthropods and the endoskeleton of vertebrates are jointed. A strong but flexible skeleton helped the arthropods and vertebrates successfully colonize the terrestrial environment.

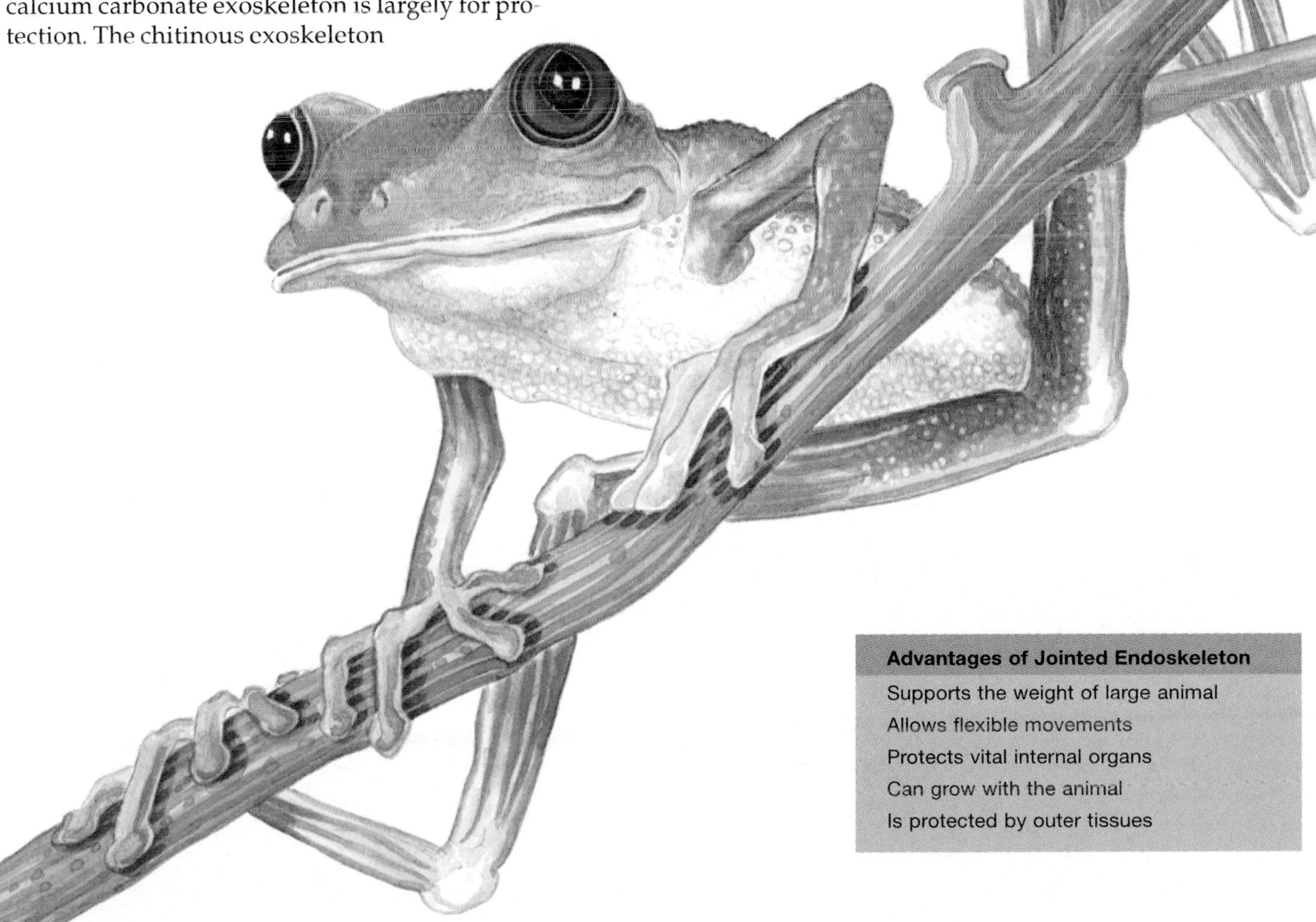

Advantages of Jointed Endoskeleton
Supports the weight of large animal
Allows flexible movements
Protects vital internal organs
Can grow with the animal
Is protected by outer tissues

Figure 48.3 The vertebrate endoskeleton.
The vertebrate jointed endoskeleton has the advantages listed. In addition, an endoskeleton lends itself to adaptation to the environment. Vertebrates move in various ways (e.g., jumping, flying, swimming, running).

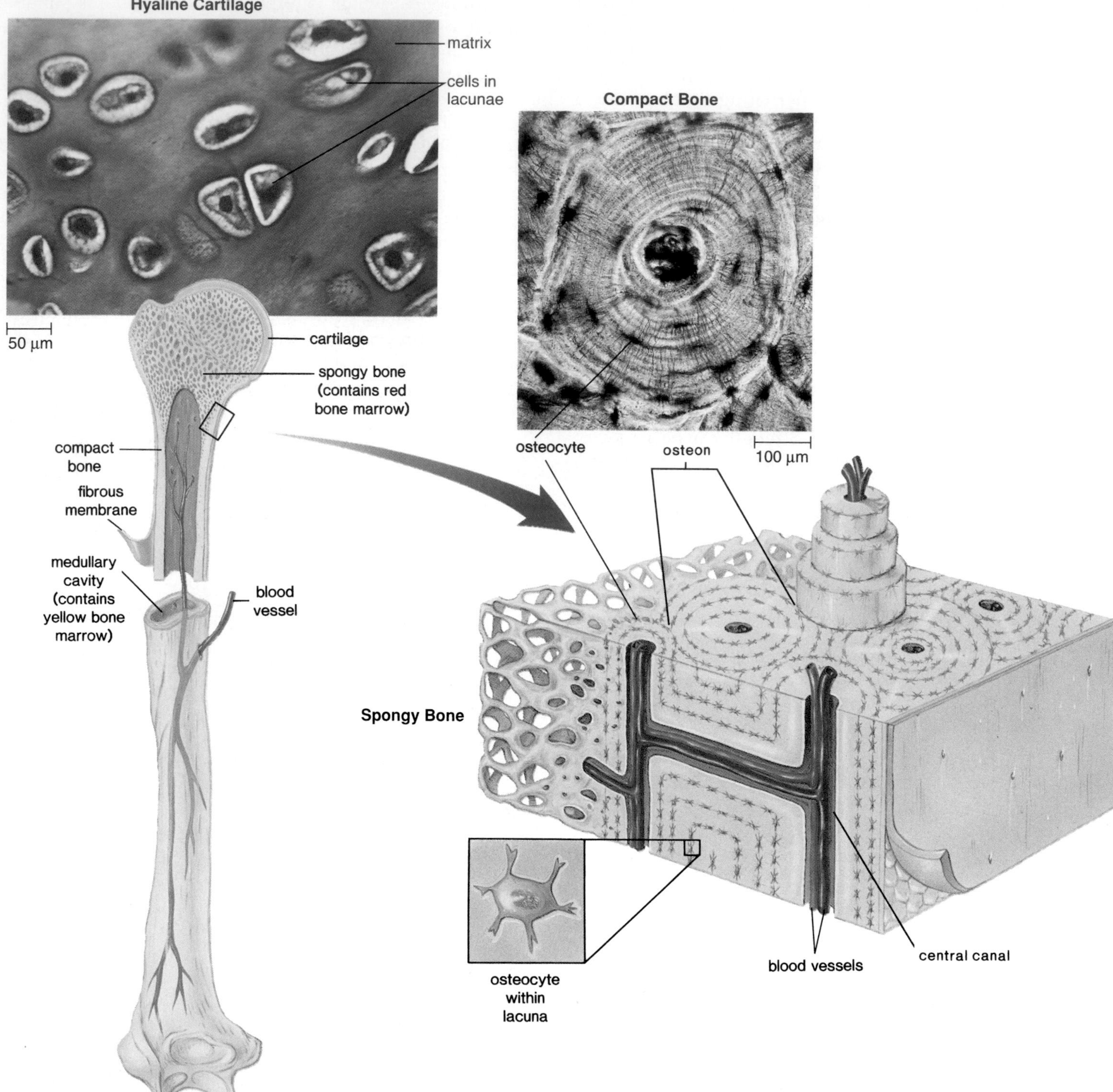

Figure 48.4 Anatomy of a long bone.
A long bone is encased by fibrous membrane except where it is covered by hyaline cartilage (see micrograph). Spongy bone located beneath the cartilage may contain red bone marrow. The compact bone of the central shaft, which is shown in the enlargement and micrograph, contains yellow bone marrow.

48.2 Humans Have an Endoskeleton

The human skeleton has many functions. The large, heavy bones of the legs support the body against the pull of gravity. The skeleton protects vital internal organs. For example, the skull forms a protective encasement for the brain, as does the rib cage for the heart and the lungs. Flat bones—such as those of the skull, the ribs, and the breastbone—produce red blood cells. All bones are storage areas for inorganic calcium and phosphorous salts. Bones also provide sites for muscle attachment. The long bones, particularly those of the legs and the arms, permit flexible body movement.

The skeleton not only permits flexible movement, it also supports and protects the body, produces red blood cells, and serves as a storehouse for certain inorganic salts.

How Bones Differ and Grow

A long bone, such as the humerus, illustrates principles of bone anatomy. When the bone is split open, as in Figure 48.4, the longitudinal section shows that it is not solid but has a cavity called the **medullary cavity** bounded at the sides by compact bone and at the ends by spongy bone. Beyond the spongy bone, there is a thin shell of compact bone and finally a layer of cartilage, a type of connective tissue with a flexible matrix.

Compact bone contains many osteons (Haversian systems) in which bone cells in tiny chambers called lacunae are arranged in concentric circles around central canals. The central canals contain blood vessels and nerves. The lacunae are separated by a matrix of collagen fibers and mineral deposits, primarily calcium and phosphorus salts.

Spongy bone contains numerous bony bars and plates separated by irregular spaces. Although lighter than compact bone, spongy bone is still designed for strength. Just as braces are used for support in buildings, the solid portions of spongy bone follow lines of stress. The spaces in spongy bone are often filled with **red bone marrow,** a specialized tissue that produces blood cells. The cavity of a long bone usually contains *yellow bone marrow,* which is a fat-storage tissue.

A long bone is designed for strength. It has a medullary cavity filled with yellow bone marrow and bounded by compact bone. The ends contain spongy bone and are covered by cartilage.

Bones Grow and Are Renewed

Most of the bones of the human skeleton are cartilaginous during prenatal development. Since the cartilaginous structures are shaped like the future bones, they provide "models" of these bones. The cartilaginous models are converted to bones when calcium salts are deposited in the matrix, first by certain of the cartilaginous cells and later by bone-forming cells called **osteoblasts** [Gk. *osteon,* bone, and *blastos,* bud]. The conversion of cartilaginous models to bones is called endochondral ossification.

There are also examples of ossification that have no previous cartilaginous model. Facial bones and certain other bones of the skull are formed in this way. During intramembranous ossification, bones develop between sheets of fibrous connective tissue. The bones of the skull are examples of intramembranous bones.

During endochondral ossification of a long bone, there is at first only a primary ossification center in the middle of the bone. Later, secondary ossification centers form at the ends of the bone. A cartilaginous disk remains between the primary ossification center and each secondary center. Growth can occur at a cartilaginous disk. The rate of growth is controlled by hormones, particularly the growth and the sex hormones. Eventually, the disks become ossified and the bone growth stops.

In the adult, bone is continually being broken down and built up again. Bone-absorbing cells, called **osteoclasts** [Gk. *osteon,* bone, and *klastos,* broken in pieces], break down bone, remove worn cells, and deposit calcium in the blood. Apparently after a period of about three weeks, they disappear. The destruction caused by the work of osteoclasts is repaired by osteoblasts. As they form bone, osteoblasts take calcium from the blood. Eventually some of these cells get caught in the matrix they secrete and are converted to **osteocytes** [Gk. *osteon,* bone, and *kytos,* cell], the cells found within the lacunae of osteons.

Thus, through a process of *remodeling,* old bone tissue is replaced by new bone tissue. Because of continual remodeling, the thickness of bones can change. Physical use and hormone balance affects the thickness of bones. Strange as it may seem, adults seem to require more calcium in the diet than do children in order to promote the work of osteoblasts.

The prenatal human skeleton is at first cartilaginous, but it is later replaced by a bony skeleton. During adult life, bone is constantly being broken down by osteoclasts and then rebuilt by osteoblasts that become osteocytes in osteons.

Bones Make Up the Skeleton

The human skeleton can be divided into two parts: the axial skeleton and the appendicular skeleton (Fig. 48.5).

Axial Skeleton Is in the Midline

The **axial skeleton** [L. *axis,* axis, hinge, and Gk. *skeleton,* dried body] lies in the midline of the body and consists of the skull, the vertebral column, the sternum, and the ribs.

The skull is formed by the cranium and the facial bones (Fig. 48.6). The cranium protects the brain and is composed of eight bones fitted tightly together in adults. In newborns, certain bones are not completely formed and instead are joined by membranous regions called **fontanels,** all of which usually close by the age of two years. The bones of the cranium contain the **sinuses** [L. *sinus,* hollow], air spaces lined by mucous membrane, which reduce the weight of the skull and give a resonant sound to the voice. Two sinuses called the mastoid sinuses drain into the middle ear. *Mastoiditis,* a condition that can lead to deafness, is an inflammation of these sinuses.

The major bones of the cranium have the same names as the lobes of the brain: frontal, parietal, temporal, and occipital. On the top of the cranium (Fig. 48.6*a*), the *frontal bone* forms the forehead, the *parietal bones* extend to the sides, and the *occipital bone* curves to form the base of the skull. Here there is a large opening, the **foramen magnum** [L. *foramen,* hole, and *magnus,* great, large] (Fig. 48.6*b*), through which the spinal cord passes and becomes the brain stem. Below the much larger parietal bones, each *temporal bone* has an opening that leads to the middle ear. The *sphenoid bone* not only completes the sides of the skull, it also contributes to the floors and walls of the eye sockets. Likewise, the *ethmoid bone,* which lies in front of the sphenoid, is a part of the orbital wall and, in addition, is a component of the nasal septum.

The cranium contains eight bones: the frontal, two parietal, the occipital, two temporal, the sphenoid, and the ethmoid.

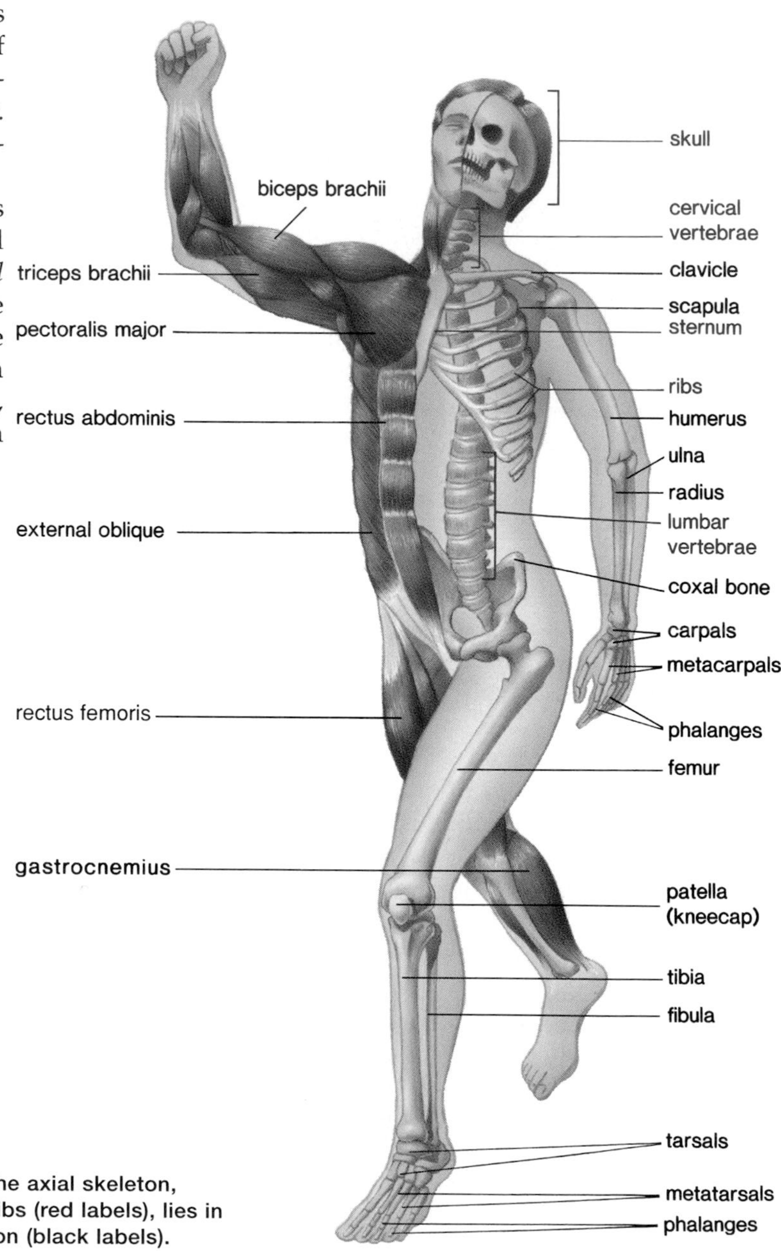

Axial Skeleton
Skull
Vertebral column
Sternum
Ribs
Appendicular Skeleton
Pectoral girdle: Clavicle, scapula
Arm: Humerus, ulna, radius
Hand: Carpals, metacarpals, phalanges
Pelvic girdle: Coxal bones
Leg: Femur, tibia, fibula, patella
Foot: Tarsals, metatarsals, phalanges

Figure 48.5 The human musculoskeletal system.
Major bones (*right*) and skeletal muscles (*left*) of the human body. The axial skeleton, composed of the skull, the vertebral column, the sternum, and the ribs (red labels), lies in the midline; the rest of the bones belong to the appendicular skeleton (black labels).

There are 14 facial bones. The **mandible** [L. *mandibula,* jaw], or lower jaw, is the only movable portion of the skull (Fig. 48.6*a*), and its action permits us to chew our food. Tooth sockets are located on this bone and on the **maxillae** [L. *maxilla,* jawbone] (maxillary bones), the upper jaw that also forms the anterior portion of the hard palate. The *palatine bones* make up the posterior portion of the hard palate and the floor of the nasal cavity. The *zygomatic bones* give us our cheekbone prominences, and the nasal bones form the bridge of the nose. Each thin, scalelike *lacrimal bone* lies between an ethmoid bone and a maxillary bone, and the thin, flat *vomer* [L. *vomer,* plowshare] joins with the perpendicular plate of the ethmoid to form the nasal septum.

Among the facial bones, the mandible is the lower jaw, the two maxillae are the upper jaw, the two zygomatic bones are the cheekbones, and the two nasal bones form the bridge of the nose.

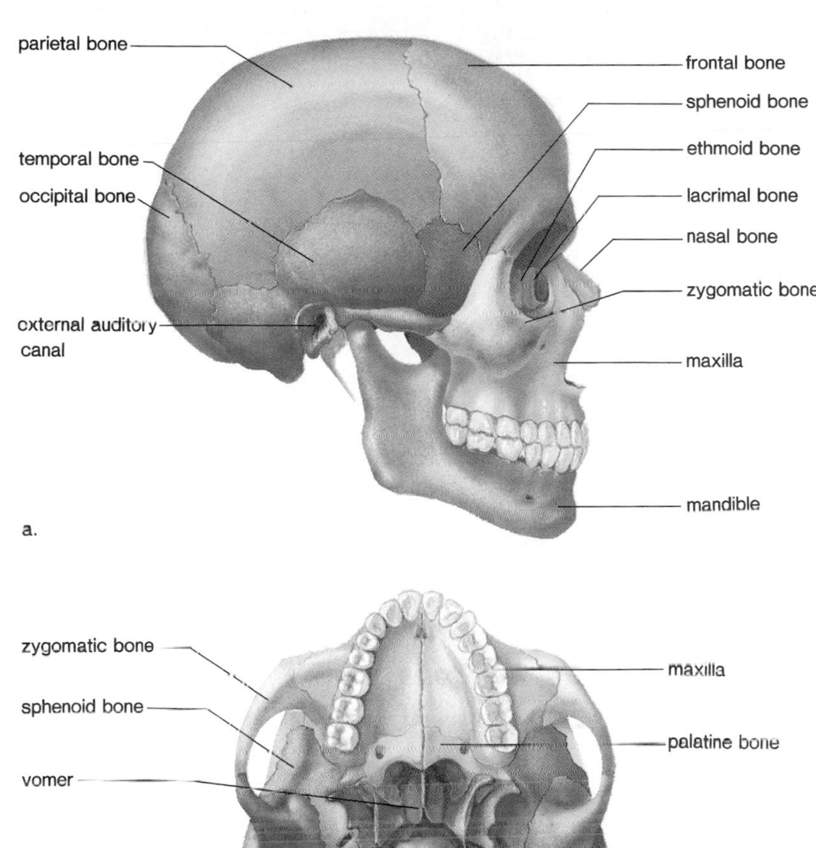

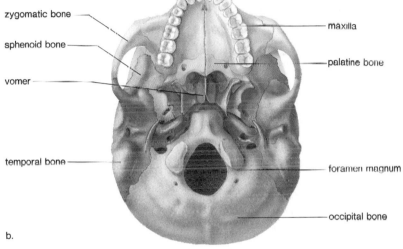

Figure 48.6 The human skull.
a. Lateral view. b. Inferior view.

Vertebral Column Supports The **vertebral column** [L. *vertebra,* bones of backbone] extends from the skull to the pelvis. Normally, the vertebral column has four curvatures that provide more resiliency and strength in an upright posture than a straight column could. The various vertebrae are named according to their location in the vertebral column (Fig. 48.5). When the vertebrae join, they form a canal through which the spinal cord passes. The *spinous processes* of the vertebrae can be felt as bony projections along the midline of the back.

There are intervertebral disks between the vertebrae that act as a kind of padding. They prevent the vertebrae from grinding against one another and absorb shock caused by movements such as running, jumping, and even walking. Unfortunately, these disks become weakened with age and can slip or even rupture. Pain results when the damaged disk presses against the spinal cord and/or spinal nerves. The body may heal itself, or the disk can be removed surgically. If the latter occurs, the vertebrae can be fused together, but this limits the flexibility of the body. The presence of the disks allows motion between the vertebrae so that we can bend forward, backward, and from side to side.

The vertebral column, directly or indirectly, serves as an anchor for all the other bones of the skeleton (see Fig. 48.5). All of the 12 pairs of **ribs** connect directly to the thoracic vertebrae in the back, and all but two pairs connect either directly or indirectly via shafts of cartilage to the **sternum** [Gk. *sternon,* breast] in the front. The lower two pairs of ribs are called "floating ribs" because they do not attach to the sternum.

The vertebral column contains the vertebrae and serves as the backbone for the body. Disks between the vertebrae provide padding and account for flexibility of the column.

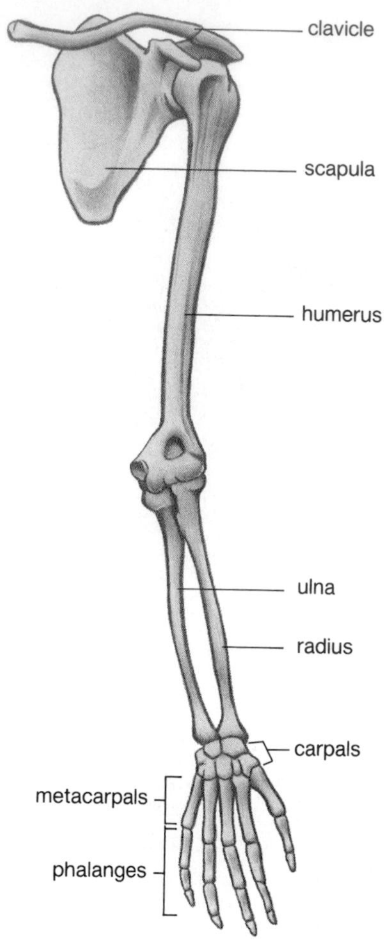

Figure 48.7 Bones of the pectoral girdle, the arm, and the hand.
The humerus is known as the "funny bone" of the elbow. The sensation upon bumping it is due to the activation of a nerve that passes across its end.

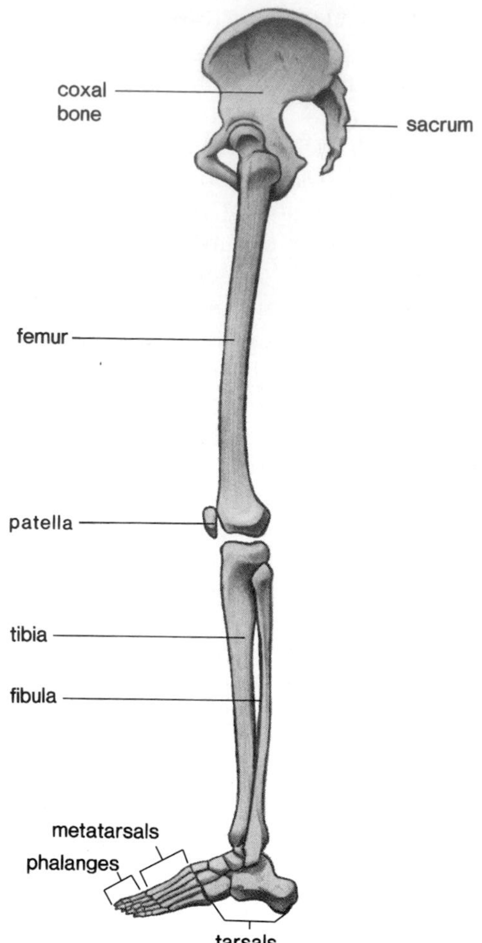

Figure 48.8 Bones of the pelvic girdle, the leg, and the foot.
The femur is our strongest bone—it withstands a pressure of 540 kilograms per 2.5 cm^3 when we walk.

Appendicular Skeleton Is Girdles and Limbs

The **appendicular skeleton** [L. *appendicula,* dim. of *appendix,* appendage] consists of the bones within the pectoral and pelvic girdles and the attached limbs (see Fig. 48.5). The pectoral (shoulder) girdle and upper limbs (arms) are specialized for flexibility, but the pelvic girdle (hipbones) and lower limbs (legs) are specialized for strength.

Pectoral Girdle and Arm The components of the **pectoral girdle** [Gk. *pechys,* forearm] are only loosely linked together by ligaments (Fig. 48.7). Each **clavicle** (collarbone) connects with the sternum in front and the **scapula** (shoulder blade) behind, but the scapula is freely movable and held in place only by muscles. This allows it to follow freely the movements of the arm. The single long bone in the upper arm, the **humerus,** has a smoothly rounded head that fits into a socket of the scapula. The socket, however, is very shallow and much smaller than the head. Although this means that the arm can move in almost any direction, there is little stability. Therefore, this is the joint that is most apt to dislocate. The opposite end of the humerus meets the two bones of the lower arm, the **ulna** and the **radius,** at the elbow. (The prominent bone in the elbow is the topmost part of the ulna.) When the arm is held so that the palm is turned frontward, the radius and ulna are about parallel to one another. When the arm is turned so that the palm is next to the body, the radius crosses in front of the ulna, a feature that contributes to the easy twisting motion of the forearm.

The many bones of the hand increase its flexibility. The wrist has eight **carpal** bones, which look like small pebbles. From these, five *metacarpal* bones fan out to form a framework for the palm. The metacarpal bone that leads to the thumb is placed in such a way that the thumb can reach out and touch the other digits. (**Digits** is a term that refers to either fingers or toes.) Beyond the metacarpals are the *phalanges,* the bones of the fingers and the thumb. The phalanges of the hand are long, slender, and lightweight.

Pelvic Girdle and Leg The **pelvic girdle** [L. *pelvis,* basin] (Fig. 48.8) consists of two heavy, large *coxal* bones (hipbones). The coxal bones are anchored to the sacrum, and together these bones form a hollow cavity that is wider in females (to accommodate childbearing) than in males. The weight of the body is transmitted through the pelvis to the legs and then onto the ground. The largest bone in the body is the femur, or thighbone.

In the lower leg, the larger of the two bones, the tibia, has a ridge we call the shin. Both of the bones of the lower leg have a prominence that contributes to the ankle—the *tibia* on the inside of the ankle and the *fibula* on the outside of the ankle. Although there are seven *tarsal bones* in the ankle, only one receives the weight and passes it on to the heel and the ball of the foot. If you wear high-heeled shoes, the weight is thrown even further anterior toward the front of your foot. The *metatarsal* bones participate in forming the arches of the foot. There is a longitudinal arch from the heel to the toes and a transverse arch across the foot. These provide a stable, springy base for the body. If the tissues that bind the metatarsals together become weakened, flat feet are apt to result. The bones of the toes are called *phalanges*, just like those of the fingers, but in the foot the phalanges are stout and extremely sturdy.

The appendicular skeleton contains the bones of the girdles and the limbs.

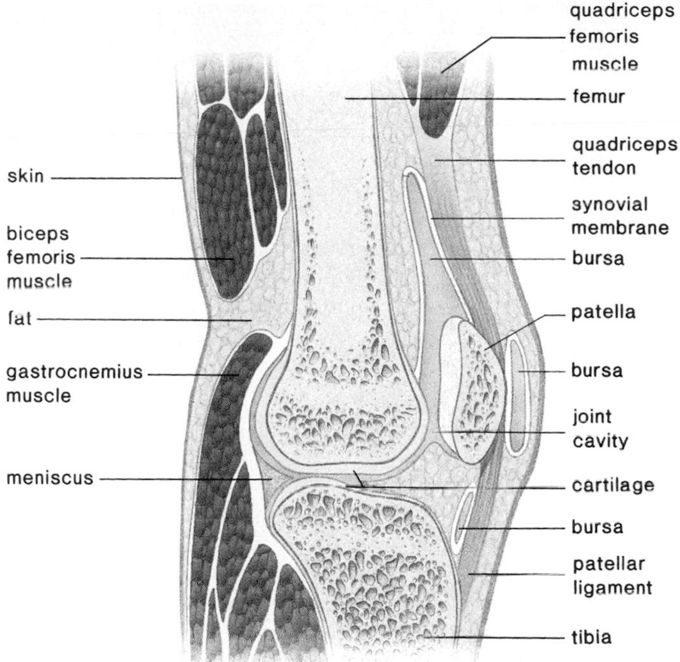

Figure 48.9 Knee joint.
The knee joint is an example of a synovial joint. Notice that there is a cavity between the bones that is encased by ligaments and lined by synovial membrane. The kneecap protects the joint.

Joints Join Bones

Bones are joined at the **joints,** which are classified as fibrous, cartilaginous, and synovial. Fibrous joints such as those between the cranial bones are immovable. Cartilaginous joints such as those between the vertebrae are slightly movable. The vertebrae are also separated by disks, which increase their flexibility. The two hipbones are slightly movable because they are ventrally joined by cartilage. Owing to hormonal changes, this joint becomes more flexible during late pregnancy, allowing the pelvis to expand during childbirth.

Synovial Joints Are Freely Movable

In *freely movable* **synovial joints,** the two bones are separated by a cavity. **Ligaments,** composed of fibrous connective tissue, bind the two bones to each other, holding them in place as they form a capsule. In a "double-jointed" individual, the ligaments are unusually loose. The joint capsule is lined by synovial membrane, which produces *synovial fluid,* a lubricant for the joint.

The knee is an example of a synovial joint (Fig. 48.9). In the knee, as in other freely movable joints, the bones are capped by cartilage, although there are also crescent-shaped pieces of cartilage between the bones called **menisci** [Gk. *meniscus,* crescent]. These give added stability, helping to support the weight placed on the knee joint. Unfortunately, athletes often suffer injury of the meniscus, known as torn cartilage. The knee joint also contains 13 fluid-filled sacs called bursae (sing., **bursa**), which ease friction between tendons and ligaments and between tendons and bones. Inflammation of the bursae is called bursitis. Tennis elbow is a form of bursitis.

There are different types of movable joints. The knee and elbow joints are *hinge joints* because, like a hinged door, they largely permit movement in one direction only. More movable are the ball-and-socket joints; for example, the ball of the femur fits into a socket on the hipbone. *Ball-and-socket joints* allow movement in all planes and even a rotational movement.

Synovial joints are subject to *arthritis*. In rheumatoid arthritis, the synovial membrane becomes inflamed and thickened. Degenerative changes take place that make the joint almost immovable and painful to use. There is evidence that these effects are brought on by an autoimmune reaction. In old-age arthritis, or osteoarthritis, the cartilage at the ends of the bones disintegrates so that the two bones become rough and irregular. This type of arthritis is apt to affect the joints that have received the greatest use over the years.

Bones are joined at joints. Synovial joins are freely movable and allow particular types of movements.

Health Focus

You Can Avoid Osteoporosis

Osteoporosis is a condition in which the bones are weakened due to a decrease in the bone mass that makes up the skeleton. Throughout life, bones are continuously remodeled. While a child is growing, the rate of bone formation is greater than the rate of bone breakdown. The skeletal mass continues to increase until ages 20 to 30. After that, there is an equal rate of formation and breakdown of bone mass until ages 40 to 50. Then, reabsorption begins to exceed formation, and the total bone mass slowly decreases.

Over time, men are apt to lose 25% and women lose 35% of their bone mass. But we have to consider that men tend to have denser bones than women anyway, and their testosterone (male sex hormone) level generally does not begin to decline significantly until after age 65. In contrast, the estrogen (female sex hormone) level in women begins to decline at about age 45. Since sex hormones play an important role in maintaining bone strength, this difference means that women are more likely than men to suffer a higher incidence of fractures, involving especially the hip vertebrae, long bones, and pelvis. Although osteoporosis may at times be the result of various disease processes, it is essentially a disease of aging.

There are measures that everyone can take to avoid osteoporosis when they get older. Adequate dietary calcium throughout life is an important protection against osteoporosis. The U.S. National Institutes of Health recommend a calcium intake of 1,200–1,500 mg per day during puberty. Males and females require 1,000 mg per day until the age of 65 and 1,500 mg per day after age 65. In older women not receiving estrogen replacement therapy, 1,500 milligrams per day is desirable.

A small daily amount of vitamin D is also necessary in order to use calcium correctly. Exposure to sunlight is required to allow skin to synthesize a precursor to vitamin D. If you reside on or north of a "line" drawn from Boston to Milwaukee, to Minneapolis, to Boise, chances are you're not getting enough vitamin D during the winter months. Therefore, you should avail yourself of vitamin D present in fortified foods such as low-fat milk and cereal.

Very inactive people, such as those confined to bed, lose bone mass 25 times faster than people who are moderately active. On the other hand, a regular, moderate, weight-bearing exercise like walking or jogging is another good way to maintain bone strength (Fig. 48A).

Postmenopausal women with any of these risk factors should have an evaluation of their bone density:

- white or Asian race
- thin body type
- family history of osteoporosis
- early menopause (before age 45)
- smoking
- a diet low in calcium, or excessive alcohol consumption and caffeine intake
- sedentary lifestyle

Presently bone density is measured by a method called dual energy X-ray absorptiometry (DEXA). This test measures bone density based on the absorption of photons generated by an X-ray tube. Soon there may be a blood and urine test to detect the biochemical markers of bone loss. Then it may be possible for physicians to screen all older women and at-risk men for osteoporosis.

If the bones are thin, it is worthwhile to take other measures to gain bone density because even a slight increase can significantly reduce fracture risk. Usually, experts believe that estrogen therapy in women is the treatment of choice, but there are other options. Evidence shows that estrogen replacement therapy will delay accelerated bone loss in postmenopausal women. Hormone replacement is most effective when begun at the start of menopause and continued over the long term. Some benefit, however, may still be obtained when treatment is begun later. A combination of hormone replacement and exercise apparently yields the best results. But there are other options, also. Calcitonin is a naturally occurring hormone whose main site of action is the skeleton where it inhibits the action of osteoclasts, the cells that break down bone. Also, alendronate is a drug that acts similarly to calcitonin. After three years of alendronate therapy, an increase in spinal density by about 8% and hip density by about 7% is obtained. Promising new drugs include slow-release fluoride therapy and certain growth hormones. These medications stimulate the formation of new bone.

Figure 48A Preventing osteoporosis.
A dietary intake of calcium and vitamin D can help prevent osteoporosis. Exercise is also important. When playing golf, you should carry your own clubs and walk instead of using a golf cart.

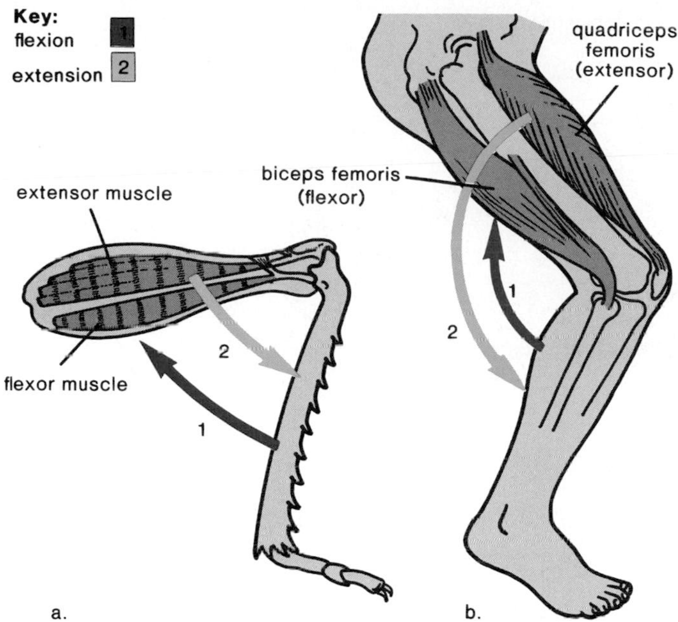

Figure 48.10 Antagonistic muscle pairs.
Muscles can exert force only by shortening. Movable joints are supplied with double sets of muscles, which work in opposite directions. As indicated by the arrows, flexor muscles decrease the angle at a joint, and extensors increase the angle at a joint. **a.** Muscles in an insect's leg as an example of antagonistic muscles attached to the inside of an arthropod exoskeleton. **b.** Muscles in the human leg as an example of antagonistic muscles attached to a vertebrate endoskeleton.

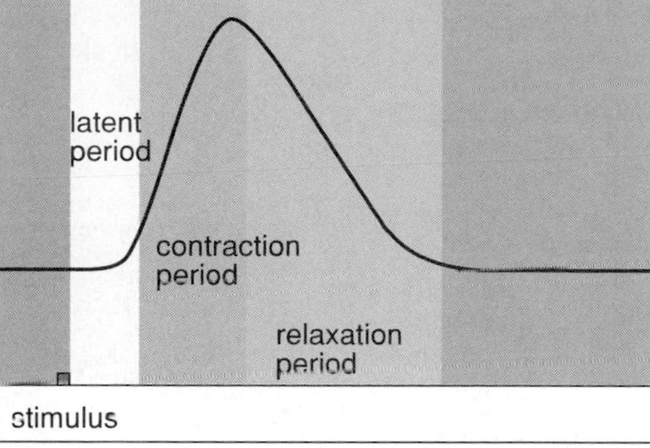

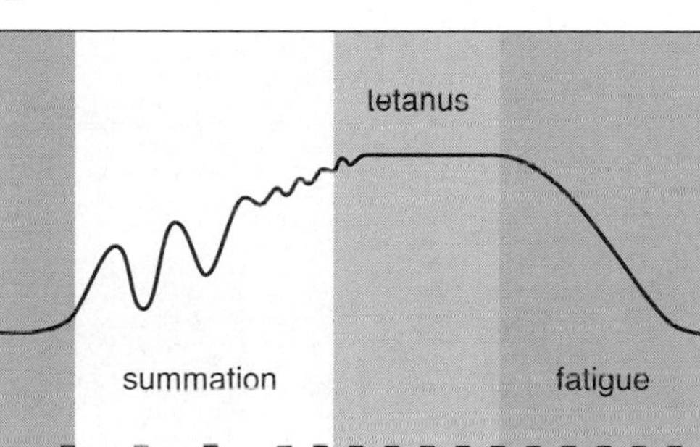

Figure 48.11 Physiology of muscle contraction.
a. When a muscle is electrically stimulated in the laboratory, it may contract. At first the stimulus may be so weak that no contraction occurs, but as soon as the strength of the stimulus reaches the threshold, the muscle contracts and then relaxes. If the strength of the stimulus is increased, the strength of the response increases. This is because more and more fibers are being stimulated to contract. **b.** If the strength of the stimulus is held constant, and the muscle is stimulated at an ever faster rate, relaxation disappears and the muscle contractions fuse into a smooth, sustained contraction called tetanus. Eventually, the muscle suffers fatigue and begins to relax. Fatigue in isolated muscles is associated with ATP depletion.

48.3 How Muscles Function

Skeletal muscle contraction is voluntary and allows an animal to move; smooth muscle contraction pushes food along in the digestive tract, and cardiac muscle allows the heart to pump the blood. Extensive experimental work has been done to understand vertebrate skeletal muscle contraction. It is assumed that the contraction of the other types of muscles is similar to this type of muscle.

On a Macroscopic Level

Skeletal muscles, comprising more than 40% of the body's weight, are attached to the skeleton by bands of fibrous connective tissue called **tendons** [L. *tendo,* stretch]. When muscles contract, they shorten. Therefore, muscles can only pull; they cannot push. Because of this, skeletal muscles must work in *antagonistic pairs.* If one muscle of an antagonistic pair bends the joint and brings the limb toward the body, the other one straightens the joint and extends the limb. Figure 48.10 illustrates this principle and compares the actions of muscles in the limb of an arthropod, in which the muscles are attached to an exoskeleton, with those of a human, in whom the muscles are attached to an endoskeleton.

It is possible to study the contraction of individual whole muscles in the laboratory. Customarily, the gastrocnemius muscle is removed from a frog and mounted so one end is fixed and the other is movable. The mechanical force of contraction is transduced into an electrical current recorded by an apparatus called a *Physiograph.* The resulting pattern is called a *myogram* (Fig. 48.11).

If a muscle is given a rapid series of threshold stimuli, it can respond to the next stimulus without relaxing completely. In this way, muscle contraction summates until maximal sustained contraction, called **tetanus,** is achieved. Tetanic contractions ordinarily occur in the body's muscles whenever skeletal muscles are actively used. Even when muscles appear to be at rest, they exhibit **tone,** in which some of their fibers are contracting. Muscle tone is particularly important in maintaining posture. If all the fibers within the muscles of the neck, trunk, and legs suddenly relax, the body collapses.

Muscles work in antagonistic pairs. A muscle at rest exhibits tone dependent on tetanic contractions.

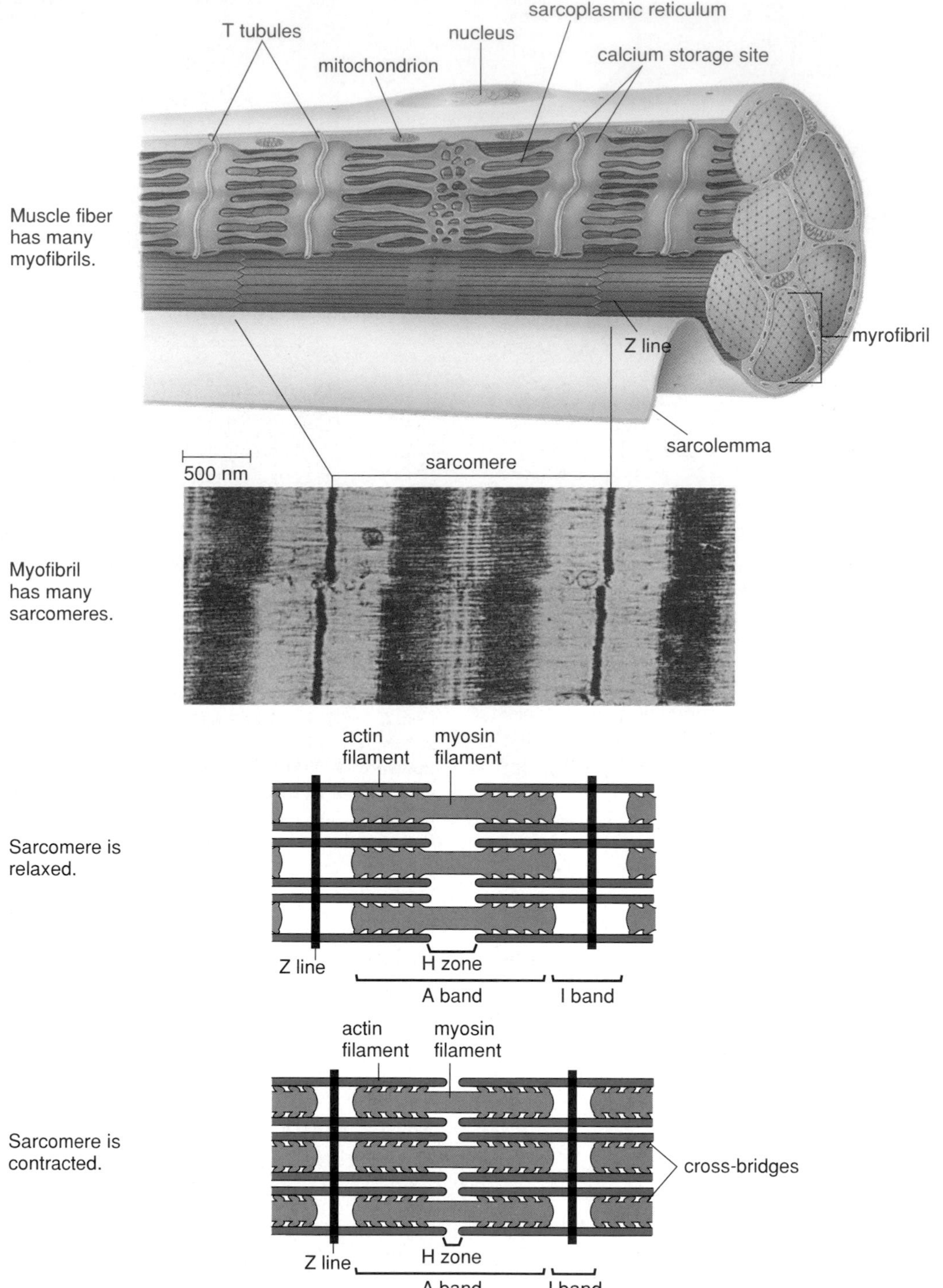

Figure 48.12 Skeletal muscle fiber structure and function.
A muscle fiber contains many myofibrils divided into sarcomeres, which are contractile. A sarcomere contains actin (thin) filaments and myosin (thick) filaments. When a muscle fiber contracts, the thin filaments move toward the center so that the H zone gets smaller, to the point of disappearing.

On a Microscopic Level

A whole skeletal muscle is composed of a number of muscle fibers in bundles. Each muscle fiber is a cell containing the usual cellular components but with special features (Fig. 48.12). The **sarcolemma** or plasma membrane [Gk. *sarkos,* flesh, and *lemma,* sheath], forms a *T* (for transverse) *system.* The *T tubules* penetrate, or dip down, into the cell so that they come into contact—but do not fuse—with expanded portions of modified endoplasmic reticulum called the **sarcoplasmic reticulum.** These expanded portions serve as storage sites for calcium ions (Ca^{2+}), which are essential for muscle contraction. The sarcoplasmic reticulum encases hundreds and sometimes even thousands of **myofibrils** [Gk. *myos,* muscle, and L. *fibra,* thread], which are the contractile portions of the fibers.

Myofibrils are cylindrical in shape and run the length of the muscle fiber. The light microscope shows that a myofibril has light and dark bands called *striations.* It is these bands that cause skeletal muscle to appear striated. The electron microscope shows that the striations of myofibrils are formed by the placement of protein filaments within contractile units called **sarcomeres** [Gk. *sarkos,* flesh, and *meros,* part]. A sarcomere extends between two dark lines called the Z lines. There are two types of protein filaments: the thick filaments are made up of **myosin,** and the thin filaments are made up of **actin.** The I band is light colored because it contains only actin filaments attached to a Z line. The dark regions of the A band contain overlapping actin and myosin filaments, and its H zone has only myosin filaments.

Filaments Slide

As a muscle fiber contracts, the sarcomeres within the myofibrils shorten. When a sarcomere shortens, the actin (thin) filaments slide past the myosin (thick) filaments and approach one another. This causes the I band to shorten and the H zone to nearly or completely disappear. The movement of actin filaments in relation to myosin filaments is called the **sliding filament theory** of muscle contraction. During the sliding process, the sarcomere shortens even though the filaments themselves remain the same length.

The participants in muscle contraction have the functions listed in Table 48.1. ATP supplies the energy for muscle contraction. Although the actin filaments slide past the myosin filaments, it is the myosin filaments that do the work. Myosin filaments break down ATP and form *cross-bridges* that attach to and pull the actin filaments toward the center of the sarcomere.

The sliding filament theory states that actin filaments slide past myosin filaments because myosin has cross-bridges, which pull the actin filaments inward.

Table 48.1

Muscle Contraction

Name	Function
Actin filaments	Slide past myosin, causing contraction
Ca^{2+}	Needed for myosin to bind to actin
Myosin filaments	Pull actin filaments by means of cross-bridges; are enzymatic and split ATP
ATP	Supplies energy for muscle contraction

ATP Is Needed ATP provides the energy for muscle contraction. Although muscle cells contain **myoglobin,** a molecule that stores oxygen, aerobic respiration does not immediately supply all the ATP that is needed. In the meantime, muscle fibers rely on **creatine phosphate** (phosphocreatine), a storage form of high-energy phosphate. Creatine phosphate cannot directly participate in muscle contraction. Instead, it is used to regenerate ATP by the following reaction:

$$\text{creatine} - \text{P} + \text{ADP} \longrightarrow \text{ATP} + \text{creatine}$$

This reaction occurs in the midst of sliding filaments, and therefore, this method of supplying ATP is the speediest energy source available to muscles.

When all of the creatine phosphate is depleted, fermentation is a second way muscles have of supplying ATP without consuming oxygen. Fermentation, which is apt to occur when strenuous exercise first begins, can supply ATP for only a short time because of lactate buildup. The buildup is noticeable when it produces muscle aches and fatigue upon exercising.

We all have had the experience of needing to continue deep breathing following strenuous exercise. This continued intake of oxygen required to complete the metabolism of lactate and restore cells to their original energy state represents an **oxygen debt.** The lactate is transported to the liver, where 20% of it is completely broken down to carbon dioxide (CO_2) and water (H_2O). The ATP gained by this respiration then is used to reconvert 80% of the lactate to glucose. In persons who train, the number of mitochondria increases, and there is a greater reliance on them rather than fermentation to produce ATP. Less lactate is produced, and there is less oxygen debt.

Working muscles require a supply of ATP. Anaerobic creatine phosphate breakdown and fermentation quickly generate ATP. Sustained exercise requires aerobic respiration.

At the Neuromuscular Junction

Nerves innervate muscles; that is, nerve impulses cause muscle to contract. The axon of a motor neuron branches to several muscle fibers, and each branch ends in several axon bulbs, with synaptic vesicles that contain the neurotransmitter acetylcholine (ACh).

The region where an axon bulb lies in close proximity to the sarcolemma of a muscle fiber is called a **neuromuscular junction.** A neuromuscular junction (Fig. 48.13) has the same components as a synapse: a presynaptic membrane, a synaptic cleft, and a postsynaptic membrane. Only in this case, the postsynaptic membrane is a portion of the sarcolemma of a muscle fiber. Nerve impulses travel down an axon and its branches until nerve impulses reach the axon bulbs. Then the synaptic vesicles merge with the presynaptic membrane and ACh is released into the synaptic cleft. When ACh diffuses to and binds with receptor sites on the sarcolemma, the sarcolemma is depolarized. **Muscle action potentials** then spread over the sarcolemma and down the T system to the region of calcium storage sites. Calcium ions are released and diffuse into the cytoplasm, where they participate in muscle contraction.

A neuromuscular junction functions like a synapse except muscle action potentials cause calcium ions (Ca^{2+}) to be released and thereafter muscle contraction occurs.

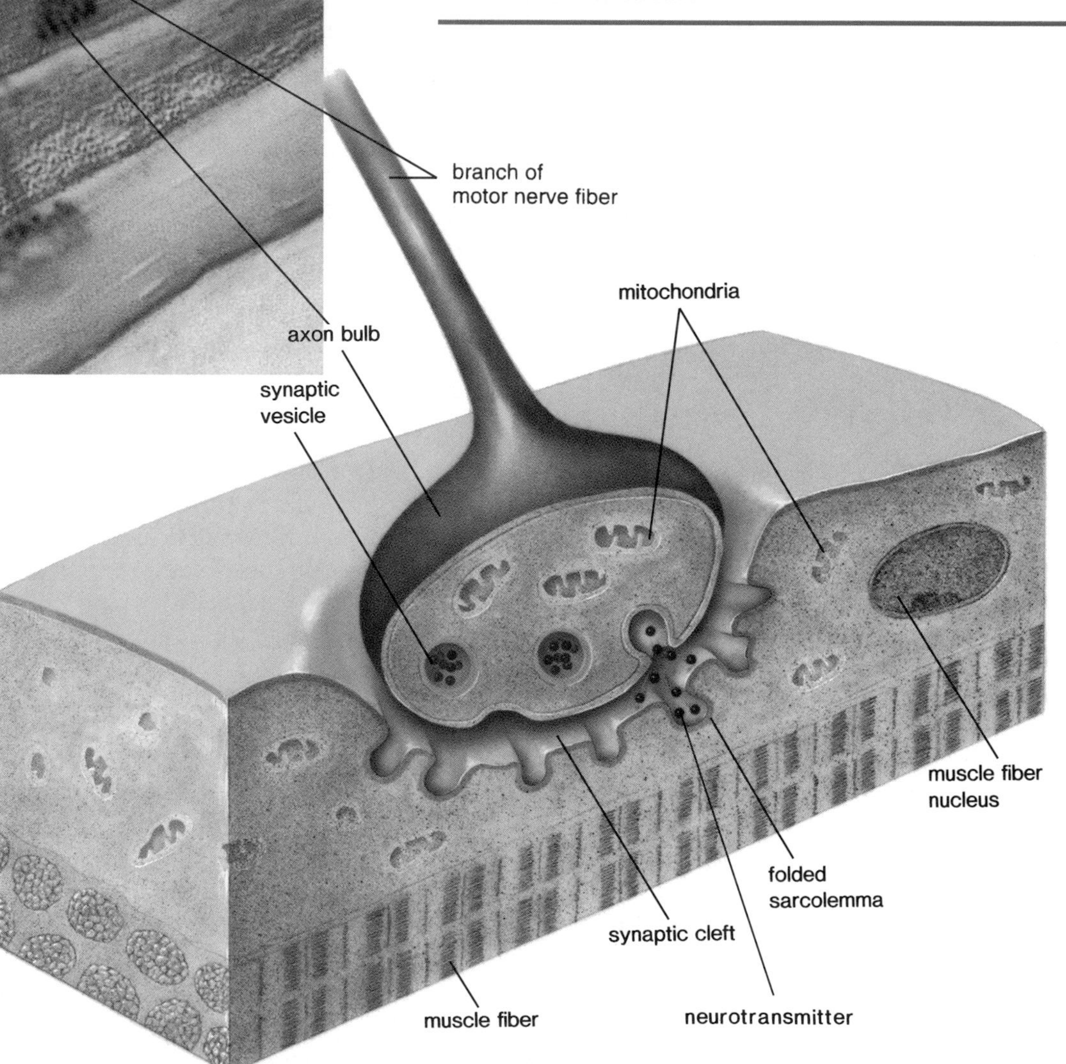

Figure 48.13 Neuromuscular junction.
A synaptic cleft separates an axon bulb from the sarcolemma of the muscle fiber. Nerve impulses traveling down a motor neuron cause synaptic vesicles to discharge a neurotransmitter, which diffuses across the synaptic cleft. When the neurotransmitter is received by the sarcolemma of a muscle fiber, a muscle action potential leads to contraction.

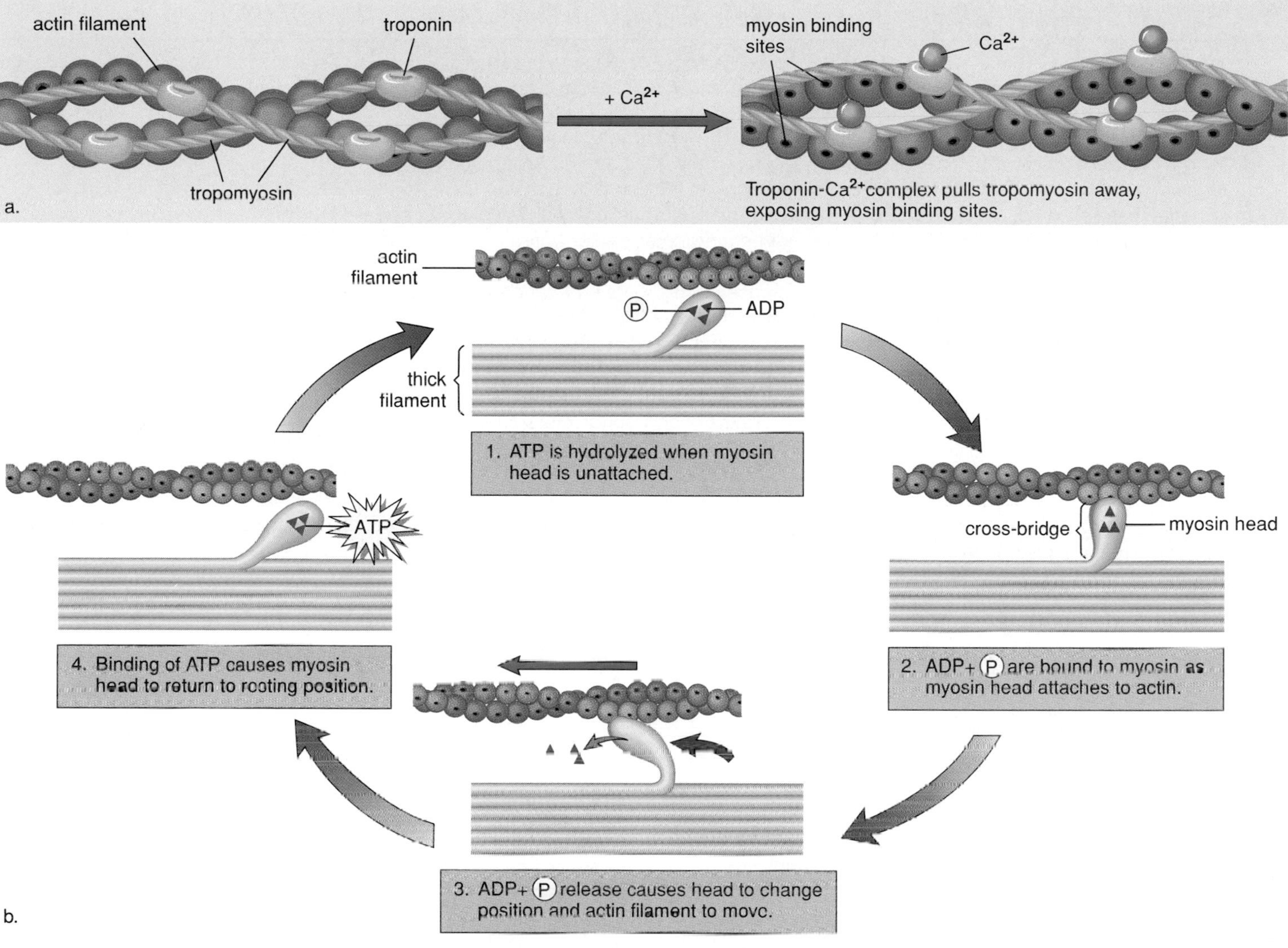

Figure 48.14 The role of calcium and myosin in muscle contraction.
a. Upon release, calcium binds to troponin, exposing myosin binding sites. b. After breaking down ATP (1), a myosin heads bind to an actin filament (2), and a power stroke causes the actin filament to move (3). When another ATP binds to myosin, the head detaches from actin (4), and the cycle begins again. Although only one myosin head is shown, many heads are active at the same time.

As Contraction Occurs

Figure 48.14 shows the placement of two other proteins associated with a thin filament, which is a double row of twisted actin molecules. Threads of *tropomyosin* wind about an actin filament, and *troponin* occurs at intervals along the threads. Calcium ions that have been released from their storage sac combine with troponin. After binding occurs, the tropomyosin threads shift their position, and myosin binding sites are exposed.

The thick filament is actually a bundle of myosin molecules, each having a double globular head with an ATP binding site. The heads function as ATPase enzymes, splitting ATP into ADP and Ⓟ. This reaction activates the heads so that they bind to actin. The ADP and Ⓟ remain on the myosin heads until the heads attach to actin, forming *cross-bridges.* Now, ADP and Ⓟ are released, and this causes the cross-bridges to change their positions. This is the *power stroke,* that pulls the thin filaments toward the middle of the sarcomere. When other ATP molecules bind to the myosin heads, the cross-bridges are broken as the heads detach from actin. The cycle begins again; the actin filaments move nearer the center of the sarcomere each time the cycle is repeated.

Contraction continues until nerve impulses cease and calcium ions are returned to their storage sites. The membranes of the sarcoplasmic reticulum contain active transport proteins that pump calcium ions back into the storage sites.

Myosin filament heads break down ATP and then attach to an actin filament, forming cross-bridges that pull the actin filament to the center of a sarcomere.

connecting concepts

The adage that structure fits function is nowhere more evident than when observing how different animals locomote. Among a group of widely diversified animals, such as mammals, various modes of locomotion have evolved. Humans are bipedal and walk on the soles of their feet formed by the tarsal bones. This form of locomotion allows the hands to be free and no doubt evolved from the habit of monkeys and apes to use only forelimbs as they swing through the branches of trees. Dexterity of hands and feet are actually the ancestral mammalian condition. In humans and apes, the bones of the hands and feet are not fused and the wrist and ankle can rotate in three dimensions.

In carnivores, like dogs and cats, the bones of the feet are fused and these animals walk on their toes. This is an obvious adaptation to running because we also raise the heel and engage the toes in order to run faster. Hoofed mammals, such as horses and deer, have greatly elongated legs and run on the tips of their digits. The hoof of a horse is its third digit only. A cheetah can sprint faster than a horse, but a horse will eventually outdistance the cheetah because it has more endurance.

Mammals that jump, like kangaroos and rabbits, have a squat shape and elongated hindlimbs that propel them forward. Several groups of arboreal mammals, like flying squirrels and sugar gliders, have membranes attached to their bodies that permit them to glide through the air from tree to tree. Among mammals, only bats truly fly. Their wings are membranes stretched between greatly elongated forelimbs and fingers. In both birds and bats, the wing is moved downward and forward in one motion, and then backward and upward in another.

In terrestrial animals, the skeleton gives the body its shape but also supports it against the pull of gravity. Because water is buoyant, gravity isn't much of a problem for aquatic animals, but a shape that reduces friction is quite helpful. Seals, sea lions, whales, and dolphins have a streamlined, torpedo shape that facilitates movement through water. A whale has few protruding parts; a male's penis is completely hidden within muscular folds, and the teats of the female lie behind slits on either side of the genital area. Thus, we see that while locomotion chiefly involves musculoskeletal adaptations, other body systems are involved as well.

Summary

48.1 Comparing Skeletons

Three types of skeletons are found in the animal kingdom: hydrostatic skeleton (cnidaria, flatworms, and segmented worms); exoskeleton (certain mollusks and arthropods); and endoskeleton (vertebrates). The rigid but jointed skeleton of arthropods and vertebrates helped them colonize the terrestrial environment.

48.2 Humans Have an Endoskeleton

The human skeleton gives support to the body, helps protect internal organs, provides sites for muscle attachment, and is a storage area for calcium and phosphorous salts, as well as a site for blood cell formation.

A long bone has a shaft of compact bone and two ends that contain spongy body. The shaft contains a medullary cavity with yellow marrow, and the ends contain red marrow. Bones are cartilaginous in the fetus but are converted to bone during development. A long bone undergoes endochondral ossification in which a cartilaginous disk remains between the primary ossification center in the middle and the secondary centers at the ends of the bones. Growth of the bone is possible as long as the cartilaginous disk is present, but eventually it, too, is converted to bone. Bone is constantly being renewed; osteoclasts break down bone, and osteoblasts build new bone. Osteocytes are in the lacunae of osteons.

The human skeleton is divided into two parts: (1) the axial skeleton, which is made up of the skull, the ribs, the sternum, and the vertebrae; and (2) the appendicular skeleton, which is composed of the girdles and their appendages.

Joints are classified as immovable, like those of the cranium; slightly movable, like those between the vertebrae; and freely movable or synovial joints, like those in the knee and hip. In synovial joints, ligaments bind the two bones together, forming a capsule in which there is synovial fluid.

48.3 How Muscles Function

Whole skeletal muscles can only shorten when they contract; therefore, they work in antagonistic pairs. For example, if one muscle bends the joint and brings the limb toward the body, the other one straightens the joint and extends the limb. A muscle at rest exhibits tone dependent on tetanic contractions.

A whole skeletal muscle is composed of muscle fibers. Each muscle fiber is a cell that contains myofibrils in addition to the usual cellular components. Longitudinally, myofibrils are divided into sarcomeres, which display the arrangement of actin and myosin filaments.

The sliding filament theory of muscle contraction says that myosin filaments have cross-bridges, which attach to and detach from actin filaments, causing actin filaments to slide and the sarcomere to shorten. (The H zone disappears as actin filaments approach one another.) Myosin is an ATPase—ATP is needed for muscle contraction. Anaerobic creatine phosphate breakdown and fermentation quickly generate ATP. Sustained exercise requires aerobic respiration for the generation of ATP.

Nerves innervate muscles. Nerve impulses traveling down motor neurons to neuromuscular junctions cause the release of ACh, which binds to receptor sites on the sarcolemma (plasma membrane of a muscle fiber). Muscle action potentials then begin and move down T tubules that approach the calcium storage sites of the sarcoplasmic reticulum (endoplasmic reticulum of muscle fibers.) Thereafter, calcium ions are released and bind to troponin. The troponin-Ca^{2+} complex causes tropomyosin threads winding around actin filaments to shift their position, revealing myosin binding sites. The myosin filament is composed of many myosin molecules, each containing a head with an ATP binding site. When a myosin head breaks down ATP, it is ready to attach to actin. The release of ATP + Ⓟ causes the head to change its position. This is the power stroke that causes the actin filament to slide toward the center of a sarcomere. When another ATP binds to myosin, the head detaches from actin and the cycle begins again.

Reviewing the Chapter

1. What are the three types of skeletons found in the animal kingdom and how do they differ. Cite some animals that have each type of skeleton? 856–57
2. Give several functions of the skeletal system in humans. 859
3. Describe the anatomy of a long bone; explain how bones grow and are renewed. 858–59
4. Distinguish between the axial and appendicular skeletons. 860, 862
5. List the bones that form the pectoral and pelvic girdles. 862–63
6. How are joints classified? Describe the anatomy of a freely movable joint. 863
7. Describe how muscles are attached to bones. What is accomplished by muscles acting in antagonistic pairs? 865
8. Discuss the microscopic structural features of a muscle fiber and a sarcomere. What is the sliding filament theory? 867
9. Discuss the availability and the specific role of ATP during muscle contraction. What is oxygen debt, and how is it repaid? 867
10. Describe the structure and function of a neuromuscular junction. 868
11. Describe the cyclical events as myosin pulls actin toward the center of a sarcomere. 869

Testing Yourself

Choose the best answer for each question.
For questions 1–4, match the following items with the correct locations given in the key.

Key:

a. upper arm
b. lower arm
c. pectoral girdle
d. pelvic girdle
e. upper leg
f. lower leg

1. ulna
2. tibia
3. clavicle
4. femur
5. Spongy bone
   a. contains osteons.
   b. contains red bone marrow where blood cells are formed.
   c. lends no strength to bones.
   d. All of these are correct.
6. Which of these is mismatched?
   a. slightly movable joint—vertebrae
   b. hinge joint—hip
   c. synovial joint—elbow
   d. immovable joint—sutures in cranium
7. In a muscle fiber
   a. the sarcolemma is connective tissue holding the myofibrils together.
   b. the T system contains calcium storage sites.
   c. both filaments have cross-bridges.
   d. All of these are correct.
8. When muscles contract
   a. sarcomeres increase in size.
   b. myosin slides past actin.
   c. the H zone disappears.
   d. calcium is taken up by calcium storage sites.
9. Which of these is a direct source of energy for muscle contraction?
   a. ATP
   b. creatine phosphate
   c. lactic acid
   d. Both a and b are correct.
10. Nervous stimulation of muscles
    a. occurs at a neuromuscular junction.
    b. results in an action potential that travels down the T system.
    c. causes calcium to be released from storage sites.
    d. All of these are correct.
11. Label this diagram of a sarcomere:

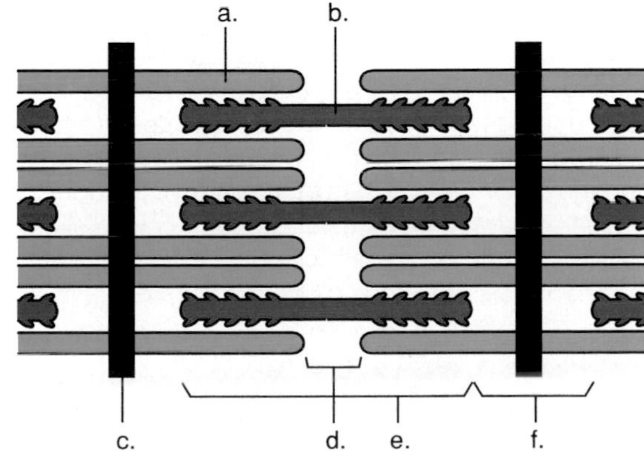

Applying the Concepts

1. *A bony skeleton determines the shape of an animal.*
   How does the skeleton aid adaptation to the environment?
2. *Movement is fundamental to the nature of animals.*
   Why is movement in animals more essential than in plants?
3. *Locomotion requires energy.*
   How do animals acquire the energy for locomotion, and specifically how is energy used to bring about locomotion?

Using Technology

Your study of support systems and locomotion is supported by these available technologies:

Exploring the Internet
The Mader Home Page provides resources for and help with studying this chapter.

http://www.mhhe.com/sciencemath/mader/
(Click on Biology.)

The Dynamic Human CD-ROM
Muscular System
Skeletal System

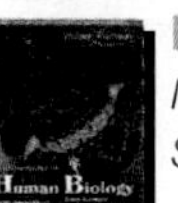

Explorations in Human Biology CD-ROM
Muscle Contraction (#4)
Synaptic Transmission (#9)

Life Science Animations Video
Video #3: Animal Biology I
Sliding Filament Model of Muscle Contraction (#30)
Regulation of Muscle Contraction (#31)

Understanding the Terms

actin 867
appendicular skeleton 862
axial skeleton 860
bursa 863
carpal 862
clavicle 862
compact bone 859
creatine phosphate 867
digit 862
fontanel 860
foramen magnum 860
humerus 862
joint 863
ligament 863
mandible 861
maxilla 861
medullary cavity 859
meniscus 863
muscle action potential 868
myofibril 867
myoglobin 867
myosin 867
neuromuscular junction 868
osteoblast 859
osteoclast 859
osteocyte 859
oxygen debt 867
pectoral girdle 862
pelvic girdle 862
radius 862
red bone marrow 859
rib 861
sarcolemma 867
sarcomere 867
sarcoplasmic reticulum 867
scapula 862
sinus 860
sliding filament theory 867
spongy bone 859
sternum 861
synovial joint 863
tendon 865
tetanus 865
tone 865
ulna 862
vertebral column 861

Match the terms to these definitions:

a. __________ Bone-forming cell.
b. __________ Electrochemical change due to increased sarcolemma permeability that is propagated down the T system and results in muscle contraction.
c. __________ Cell that causes erosion of bone.
d. __________ Movement of actin filaments in relation to myosin filaments, which accounts for muscle contraction.
e. __________ Muscle protein making up the thin filaments in a sarcomere; its movement shortens the sarcomere, yielding muscle contraction.
f. __________ Part of the skeleton that consists of the shoulder and pelvic girdles and the bones of the arms and legs.
g. __________ Part of the skeleton forming the vertical support or axis, including the skull, the rib cage, and the vertebral column.
h. __________ Portion of the skeleton that provides support and attachment for the arms.
i. __________ Portion of the skeleton to which the legs are attached.
j. __________ Region where an axon bulb approaches a muscle fiber; contains a presynaptic membrane, a synaptic cleft, and a postsynaptic membrane.

Hormones and Endocrine Systems

CHAPTER 49

Chapter Concepts

Blue-footed boobies, *Sula nebouxii*, courting

The nervous system integrates sensory information and controls the action of effectors by using chemical messengers called neurotransmitters. The **endocrine system** coordinates body parts by using chemical messengers called hormones. Hormones are usually secreted into the bloodstream and regulate whole body processes like growth, reproduction, and complex behaviors, including courtship and migration.

The nervous system can bring about immediate responses to environmental stimuli, such as the movement of a hand away from a flame or the strike of a snake to seize a mouse. The hormonal system is slower acting and regulates processes that occur over days or even months. Once hormones arrive at those cells with appropriate plasma membrane receptors, they influence the metabolism of the cell. It takes a while to metabolically change a cell and the effect is longer lasting.

Why is it beneficial for an animal to have two different systems for coordinating activities involving homeostasis? Having two systems allows an animal to respond to two different types of stimuli—those requiring immediate action and those that depend on a more long-term response.

49.1 Hormones Affect Cellular Metabolism

A **hormone** [Gk. *hormao*, instigate] is defined as an organic chemical produced by one set of cells that affects a different set. A hormone usually travels through the circulatory system to its target organ. It is important to stress that a hormone does not seek out a particular organ; to the contrary, the organ is awaiting the arrival of the hormone. Cells that can react to a hormone have specific receptors, which combine with the hormone in a lock-and-key manner. Therefore, certain cells respond to one hormone and not to another, depending on their receptors.

The distinction between the nervous system and the endocrine system is somewhat artificial since they both use chemical messengers—neurotransmitters versus hormones. And, as we shall see, the brain contains neurosecretory cells that produce hormones, some of which control the action of endocrine glands. A certain amount of overlap between the nervous and endocrine systems is to be expected since these two systems evolved simultaneously. We will examine several examples of associations between the two systems on the pages that follow.

Steroid Hormones Activate DNA

Some vertebrate hormones—produced by the adrenal cortex, the ovaries, and the testes—are steroids. **Steroid hormones** are derived from cholesterol by a series of metabolic reactions. These hormones are stored in fat droplets in the cell cytoplasm until their release at the plasma membrane. Steroid hormones do not bind to cell-surface receptors; they can enter the cell and the nucleus freely (Fig. 49.1). Once inside the nucleus, steroid hormones such as estrogen and progesterone bind to receptors. The hormone-receptor complex is believed to bind to transcription factors that in turn bind to DNA. Particular genes are activated; transcription of DNA and translation of messenger ribonucleic acid (mRNA) follow. In this manner, steroid hormones lead to protein synthesis.

Nonsteroid Hormones Activate Enzymes

Many vertebrate hormones are proteins or peptides that are coded for by genes and are synthesized within the cytoplasm at the ribosomes. There are other hormones, called catecholamines, that are derived from the amino acid tyrosine. Their production requires only a series of metabolic reactions within the cytoplasm. From the viewpoint of hor-

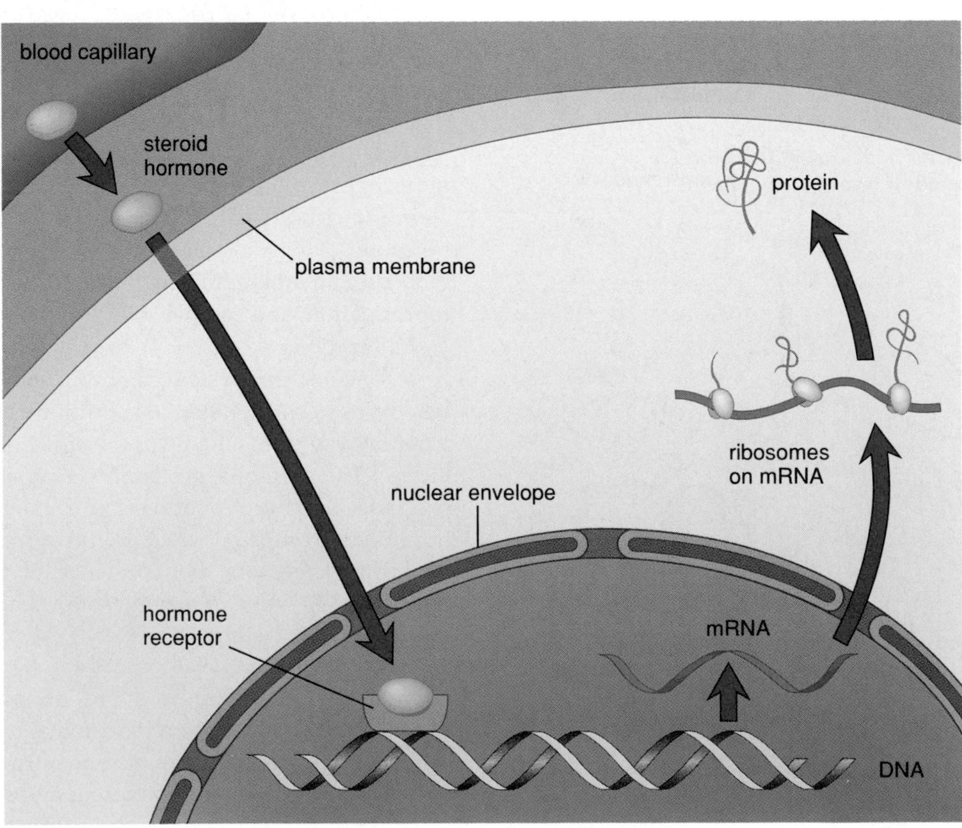

Figure 49.1 Cellular activity of steroid hormones.
After passing through the plasma membrane, a steroid hormone binds to a receptor inside the nucleus. Recent evidence indicates that the receptors reside in the nucleus rather than the cytoplasm. The hormone-receptor complex then binds to a transcription factor, and this leads to activation of certain genes and protein synthesis.

monal action, we can categorize the proteins, peptides, and catecholamines as **nonsteroid hormones.**

The mode of action of nonsteroid hormones was discovered in the 1950s by Earl W. Sutherland, Jr., who studied the effects of epinephrine on liver cells. Sutherland received a Nobel Prize for his hypothesis that when this hormone binds to a cell-surface receptor, the resulting complex leads to the activation of an enzyme that produces **cyclic AMP** (adenosine monophosphate) (Fig. 49.2). Cyclic AMP (cAMP) is a compound made from ATP, but it contains only one phosphate group, which is attached to adenosine at two locations.

Since nonsteroid hormones never enter the cell, they are sometimes called the *first messengers*, while cAMP, which sets the metabolic machinery in motion, is called the **second messenger.** Later research has shown that there is an intermediary between the first messenger and adenylate cyclase, the enzyme that converts ATP to cAMP. G proteins, named for their ability to bind to and break down GTP (guanosine triphosphate), an energy carrier similar to ATP, are directly stimulated by the hormone-receptor interaction. G proteins are located in the membrane and regulate the action of many other membrane proteins, particularly ion channels and enzymes. In this instance, the stimulated G protein activates adenylate cyclase and cAMP results.

cAMP is a second messenger that sets an *enzyme cascade* into motion. cAMP activates only one particular enzyme in the cell, but this enzyme in turn activates another, and so forth. The activated enzymes, of course, can be used repeatedly. Therefore, at every step in the enzyme cascade, more and more reactions take place—the binding of a single hormone molecule eventually results in a thousandfold response. Nonsteroid hormones tend to act relatively quickly but for a short period of time. cAMP is soon converted to an inactive product and the enzymes revert to an inactive state. Other second messengers have been discovered in cells. Inositol triphosphate (IP_3) is a second messenger that causes the release of calcium ions (Ca^{2+}) in muscle cells. Since calcium goes on to activate other proteins, it is called the third messenger.

It is apparent from our discussion that hormones are involved in a signal transduction pathway. Transduction is the sequence of steps that occur between signal reception and cell response. G proteins are intimately involved in transducing the signal so that the cell responds appropriately.

Sterioid hormones bring about the synthesis of new proteins, and nonsteroid hormones activate existing enzymes in cells.

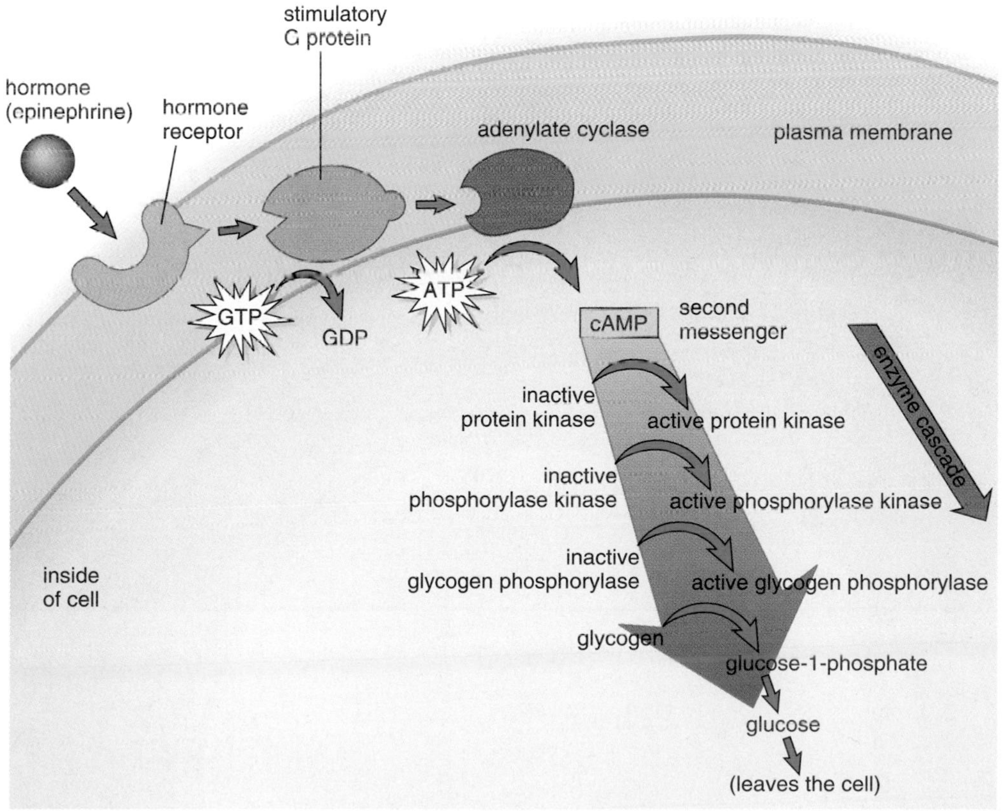

Figure 49.2 Cellular activity of nonsteroid hormones.
Nonsteroid hormones, which are called the first messengers, bind to a specific receptor protein in the plasma membrane. The hormone-receptor complex activates a G protein that binds to GTP; a portion of the complex splits off and activates adenylate cyclase. Adenylate cyclase acts on ATP to produce cAMP, the second messenger, which activates an enzyme cascade. There are also inhibitory G proteins.

Table 49.1

Principal Human Endocrine Glands and Their Hormones

Endocrine Gland	Hormone Released	Target Tissues/Organ	Chief Function(s) of Hormone
Hypothalamus	Hypothalamic-releasing and release-inhibiting hormones	Anterior pituitary	Regulate anterior pituitary hormones
Posterior pituitary (storage of hypothalamic hormones)	Antidiuretic hormone (ADH), also known as vasopressin	Kidneys	Stimulates water reabsorption by kidneys
	Oxytocin	Uterus, mammary glands	Stimulates uterine muscle contraction and release of milk by mammary glands
Anterior pituitary	Growth (GH), also known as somatotropin	Soft tissues, bones	Stimulates cell division, protein synthesis, and bone growth
	Prolactin (PRL)	Mammary glands	Stimulates milk production and secretion
	Thyroid-stimulating hormone (TSH)	Thyroid	Stimulates thyroid
	Adrenocorticotropic hormone (ACTH)	Adrenal cortex	Stimulates adrenal cortex
	Gonadotropic hormones	Gonads (testes and ovaries)	Control gamete and sex hormone production
	Melanocyte-stimulating hormone (MSH)	Skin	
Thyroid	Thyroxine and triiodothyronine	All tissues	Increases metabolic rate; helps to regulate growth and development
	Calcitonin	Bones, kidneys, intestine	Lowers blood calcium level
Parathyroids	Parathyroid hormone (PTH)	Bones, kidneys, intestine	Raises blood calcium level
Adrenal medulla	Epinephrine and norepinephrine	Cardiac and other muscles	Stimulate "fight-or-flight" reactions; raise blood glucose level
Adrenal cortex	Glucocorticoids* (e.g., cortisol)	All tissues	Raise blood glucose level
	Mineralocorticoids* (e.g., aldosterone)	Kidneys	Stimulate kidneys to reabsorb sodium and to excrete potassium
	Sex hormones*	Sex organs, skin, muscles, bones	Stimulate development of secondary sex characteristics
Pancreas	Insulin	Liver, muscles, adipose tissue	Lowers blood glucose level
	Glucagon	Liver, muscles, adipose tissue	Raises blood glucose level
Testes	Androgens (e.g., testosterone)*	Sex organs, skin, muscles	Stimulate spermatogenesis; develop and maintain primary and secondary male sex characteristics
Ovaries	Estrogen and progesterone*	Sex organs, skin, muscles, bones	Stimulate oogenesis; develop and maintain primary and secondary female sex characteristics
Thymus	Thymosins	T lymphocytes	Stimulate maturation of T lymphocytes
Pineal gland	Melatonin	Various tissues	Involved in daily rhythms; possibly involved in maturation of sex organs

* *Steroid hormones*

49.2 Human Endocrine System

Endocrine glands can be contrasted with exocrine glands. The latter have ducts and secrete their products into these ducts for transport into body cavities. For example, the salivary glands send saliva into the mouth by way of the salivary ducts. Endocrine glands are ductless; they secrete their hormones directly into the bloodstream for distribution throughout the body.

The human endocrine system can be used to exemplify the vertebrate endocrine system. Even so, there are marked differences in the vertebrate hormones. Prolactin in humans stimulates female breasts to secrete milk, but in pigeons it stimulates the secretion of crop milk, a product of the gut. Some hormones have entirely different functions in different vertebrates. Thyroxine in humans stimulates metabolism, but in frogs it induces metamorphosis (the transformation of tadpole to adult). Other hormones are species-specific; that is, they function only in one species.

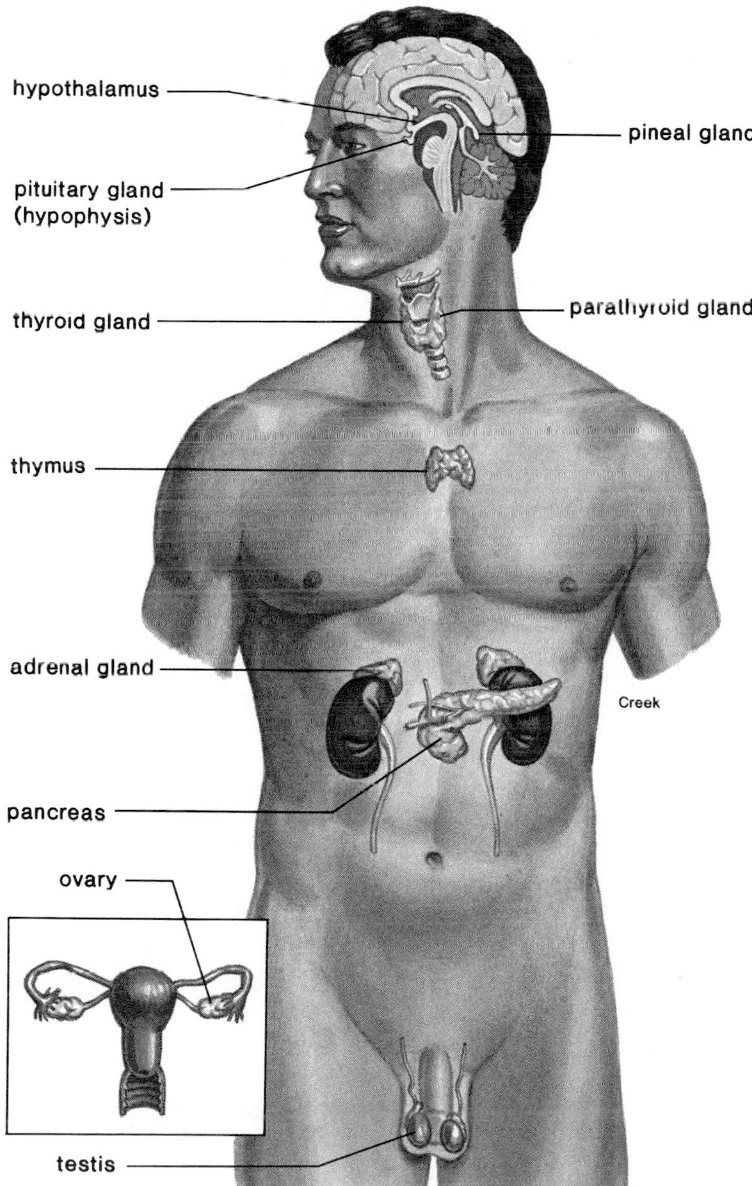

Figure 49.3 The human endocrine system.
Anatomical location of major endocrine glands in the body.

Table 49.1 lists the hormones released by the principal human endocrine glands, which are depicted in Figure 49.3. The hypothalamus and pituitary gland are located in the brain, the thyroid and parathyroids are located in the neck, while the adrenal glands and pancreas are located in the pelvic cavity. The **gonads** include the ovaries, located in the pelvic cavity, and the testes, located outside this cavity in the scrotum. Also listed in Table 49.1 are the pineal gland, located in the brain, and the thymus, which lies ventral to the thorax.

Like the nervous system, the endocrine system is especially involved in homeostasis; that is, the dynamic equilibrium of the internal environment. The internal environment is the blood and tissue fluid that surrounds the body's cells. Notice that several hormones directly affect the blood, water, glucose, calcium, and sodium levels. Other hormones are involved in the maturation and function of the reproductive organs. In fact, many people are most familiar with the effect of hormones on sexual functions.

There are two mechanisms that control the effect of endocrine glands. Quite often a negative feedback mechanism controls the secretion of hormones. An endocrine gland can be sensitive to either the condition it is regulating or to the blood level of the hormone it is producing. For example, when the blood glucose level rises, the pancreas produces insulin, which causes the liver to store glucose. The stimulus for the production of insulin has thereby been dampened, and therefore, the pancreas stops producing insulin. The effect of insulin, however, can also be offset by the production of glucagon by the pancreas. In other words, contrary actions of hormones is the second way their effects can be controlled.

Notice there are other examples of contrary hormonal actions in Table 49.1. The thyroid lowers the blood calcium level, but the parathyroids raise the blood calcium level. We will also have the opportunity to point out other instances in which hormones work opposite to one another and thereby bring about the regulation of a substance in the blood.

The effect of hormones is controlled in two ways: (1) negative feedback opposes their release, and (2) contrary hormones oppose each other's actions. The end result is regulation of a bodily substance or function within normal limits.

Hypothalamus Controls the Pituitary Gland

The **hypothalamus** is the portion of the brain that regulates the internal environment; it helps to control heart rate, body temperature, and water balance, as well as the glandular secretions of the **pituitary gland.** The pituitary, a small gland about 1 cm in diameter, lies just below the hypothalamus and is divided into two portions called the posterior pituitary and the anterior pituitary.

Posterior Pituitary Stores Two Hormones

The structural and functional relation between the hypothalamus and the posterior pituitary illustrates the overlap between the nervous and endocrine systems. The posterior pituitary is connected to the hypothalamus by means of a stalklike structure. There are neurons in the hypothalamus that are called neurosecretory cells because they both respond to neurotransmitters and produce the hormones that are stored in and released from the posterior pituitary. The hormones pass from the hypothalamus through axons that terminate in the posterior pituitary.

The axon endings in the posterior pituitary store **antidiuretic hormone (ADH)** [Gk. *anti,* against, and L. *ouresis,* urination]—sometimes called vasopressin—and oxytocin (Fig. 49.4). ADH promotes the reabsorption of water from collecting ducts, which receive filtrate produced by nephrons within the kidneys. The hypothalamus contains other nerve cells that act as a sensor because they are sensitive to the tonicity of the blood. When these cells determine that the blood is too concentrated, ADH is released into the bloodstream from the axon endings in the posterior pituitary, and upon reaching the kidneys, water is reabsorbed. As the blood becomes dilute, the hormone no longer is released. This is an example of control by negative feedback because the effect of the hormone (to dilute blood) acts to shut down the release of the hormone.

Oxytocin [Gk. *oxys,* quick, and *tokos,* birth] is the other hormone that is made in the hypothalamus and stored in the posterior pituitary. Oxytocin causes the uterus to contract and is used to artificially induce labor. It also stimulates the release of milk from the mammary glands for nursing.

The posterior pituitary stores two hormones, ADH and oxytocin, both of which are produced by and released from neurosecretory cells in the hypothalamus.

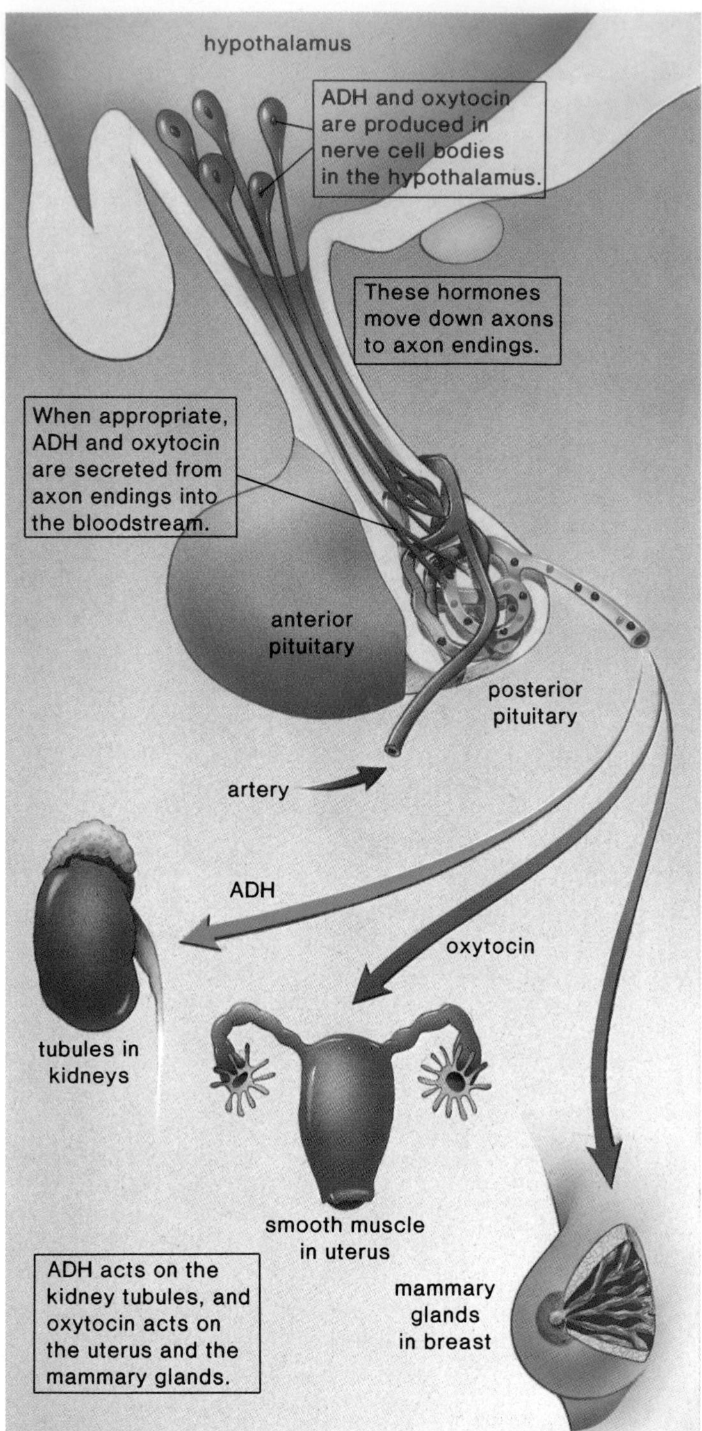
hypothalamus
ADH and oxytocin are produced in nerve cell bodies in the hypothalamus.
These hormones move down axons to axon endings.
When appropriate, ADH and oxytocin are secreted from axon endings into the bloodstream.
anterior pituitary
posterior pituitary
artery
ADH
oxytocin
tubules in kidneys
smooth muscle in uterus
mammary glands in breast
ADH acts on the kidney tubules, and oxytocin acts on the uterus and the mammary glands.

Figure 49.4 Hypothalamus.
The hypothalamus produces two hormones, ADH and oxytocin, which are stored and secreted by the posterior pituitary.

Anterior Pituitary Is the Master Gland

By the 1930s, biologists knew that the hypothalamus also controls the release of the anterior pituitary hormones. Electrical stimulation of the hypothalamus causes the release of anterior pituitary hormones, but *direct* stimulation of the anterior pituitary itself has no effect. Detailed anatomical studies show that there is a *portal* system consisting of blood vessels connecting a capillary bed in the hypothalamus with one in the anterior pituitary (Fig. 49.5). Some scientists believed that a different set of neurosecretory cells in the hypothalamus synthesizes a number of different releasing hormones that are sent to the anterior pituitary by way of the vascular portal system. There began a long and competitive struggle between two groups of investigators to identify one of these *hypothalamic-releasing hormones,* which cause the anterior pituitary to release hormones. The group headed by R. Guillemin processed nearly 2 million sheep hypothalami, and the group headed by A. V. Schally processed more than a million pig hypothalami until each group announced, in November of 1969, that it had isolated and determined the structure of one of these hormones, a peptide containing only three amino acids.

Later, it was found that some of the hypothalamic hormones inhibit the release of anterior pituitary hormones. Therefore, today, it is customary to speak of *hypothalamic-releasing hormones* and *hypothalamic-release-inhibiting hormones.* For example, there is a *gonadotropic-releasing hormone (GnRH)* and a *gonadotropic-release-inhibiting hormone (GnRIH).* The first hormone stimulates the anterior pituitary to release gonadotropic hormones, and the second inhibits the anterior pituitary from releasing the same hormones.

The anterior pituitary produces at least six different types of hormones, each by a distinct cell type (Fig. 49.5). Three of these hormones have a direct effect on the body. **Growth hormone (GH),** or somatotropic hormone, dramatically affects physical appearance since it determines the

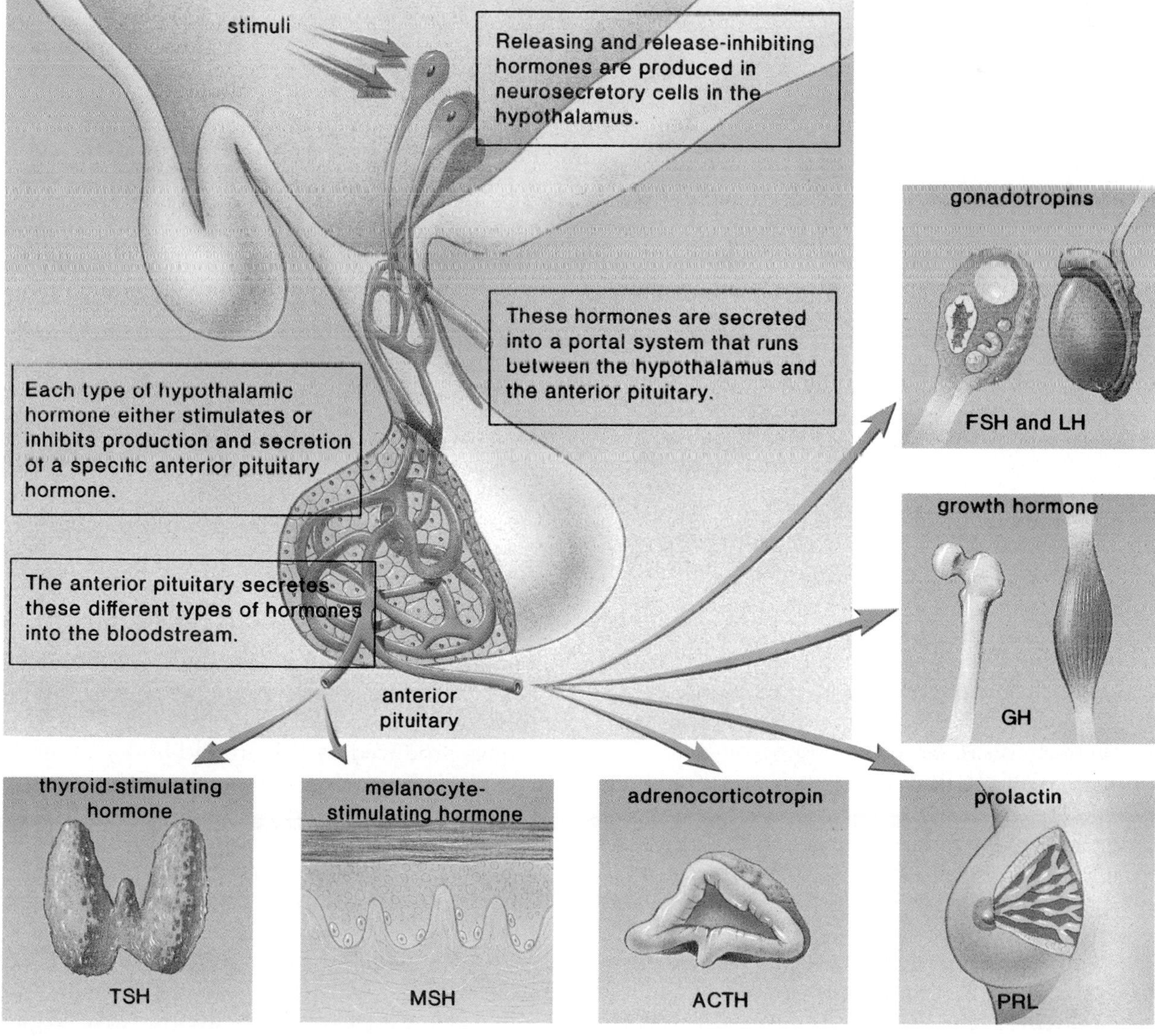

Figure 49.5 Hypothalamus and anterior pituitary.
The hypothalamus controls the secretions of the anterior pituitary. The thyroid, the adrenal cortex, and the gonads are also endocrine glands.

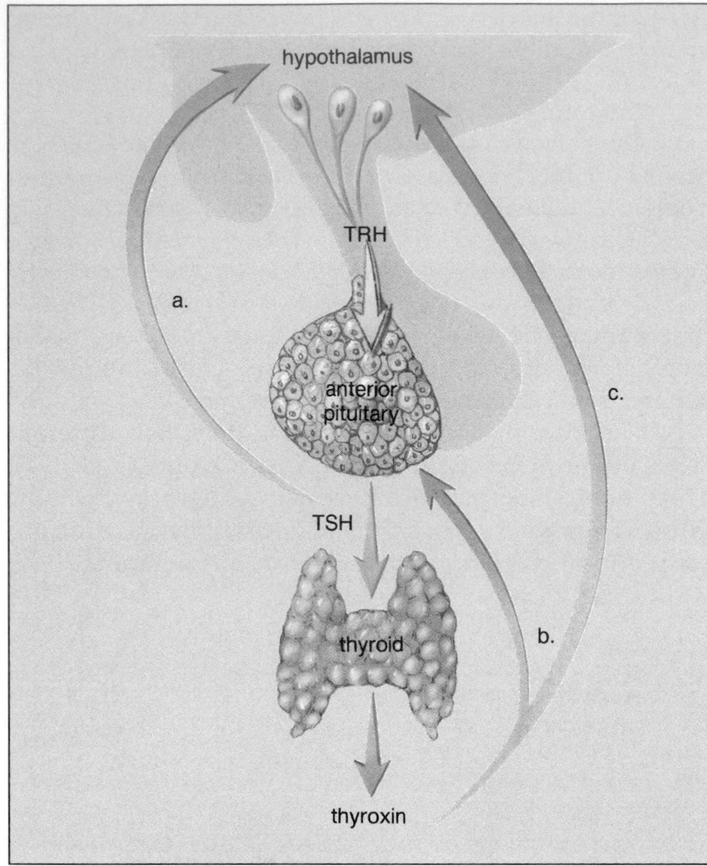

Figure 49.6 Hypothalamus-anterior pituitary-thyroid axis.
Negative feedback in the regulation of the hypothalamus, anterior pituitary, and thyroid. **a.** The level of TSH exerts feedback control over the hypothalamus. **b.** The level of thyroxine exerts feedback control over the anterior pituitary. **c.** The level of thyroxine exerts feedback control over the hypothalamus. In this way, thyroxine controls its own secretion. Cortisol and sex hormone levels are controlled in similar ways.

height of the individual. If too little GH is produced during childhood, the individual becomes a pituitary dwarf, and if too much is produced, the individual is a pituitary giant. In both instances, the individual has normal body proportions. On occasion, however, there is overproduction of growth hormone in the adult, and a condition called *acromegaly* results. Since only the feet, hands, and face (particularly chin, nose, and eyebrow ridges) can respond, these portions of the body become overly large.

GH promotes cell division, protein synthesis, and bone growth. It stimulates the transport of amino acids into cells and increases the activity of ribosomes, both of which are essential to protein synthesis. In bones, it promotes growth of the cartilaginous disks (p. 859) and causes osteoblasts to form bone. Evidence suggests that the effects on cartilage and bone may actually be due to hormones called *somatomedins,* which are released by the liver in response to GH.

Prolactin (PRL) [L. *pro,* before, and *lactis,* milk] is produced in quantity only after childbirth. It causes the mammary glands in the breasts to develop and produce milk. It also plays a role in carbohydrate and fat metabolism.

Melanocyte-stimulating hormone (MSH) [Gk. *melanos,* black, and *kytos,* cell] causes skin color changes in many fishes, amphibians, and reptiles who have melanophores, special skin cells that produce color variations. In humans, MSH stimulates melanocytes to increase their synthesis of melanin. It is derived from a molecule that is also the precursor for both ACTH and the anterior pituitary endorphins. These endorphins are structurally and functionally similar to the endorphins produced in brain nerve cells.

The anterior pituitary is sometimes called the *master gland* because it controls the secretion of some other endocrine glands (see Fig. 49.5). As indicated in Table 49.1, the anterior pituitary secretes the following hormones, which have an effect on other glands:

1. **Thyroid-stimulating hormone (TSH)**
2. **Adrenocorticotropic hormone (ACTH),** which stimulates the adrenal cortex
3. **Gonadotropic hormones (FSH and LH),** which stimulate the gonads—the testes in males and the ovaries in females

TSH causes the thyroid to produce thyroid hormones (e.g. thyroxine); ACTH causes the adrenal cortex to produce cortisol; and gonadotropic hormones cause the gonads to secrete sex hormones. A three-tiered relationship exists among the hypothalamus, the anterior pituitary, and the other endocrine glands; the hypothalamus produces releasing hormones, which control the anterior pituitary, and the anterior pituitary produces hormones that control the thyroid, the adrenal cortex, and the gonads. In turn these glands produce hormones that, through a negative feedback mechanism, regulate the secretion of the appropriate hypothalamic-releasing hormone. (Fig. 49.6)

The hypothalamus, the anterior pituitary, and the other endocrine glands controlled by the anterior pituitary are all involved in a self-regulating negative feedback mechanism.

Thyroid Gland Speeds Metabolism

The **thyroid gland** [Gk. *thyreos,* large, door-shaped shield] is a large gland located in the neck, where it is attached to the trachea just below the larynx (Fig. 49.7*a*). The two hormones produced by the thyroid both contain iodine. **Thyroxine,** or T_4, contains four atoms of iodine; it is secreted in greater amounts but is less potent than triiodothyronine, or T_3, which has only three atoms of iodine. Iodine is actively transported into the thyroid gland, and it may reach a concentration as much as 25 times greater than that of the blood.

Even before the structure of thyroxine was known, it was surmised that the hormone contained iodine because when iodine is lacking in the diet, the thyroid gland enlarges, producing a **goiter.** The cause of the enlargement becomes clear if we consider the relationship between the thyroid and the anterior pituitary.

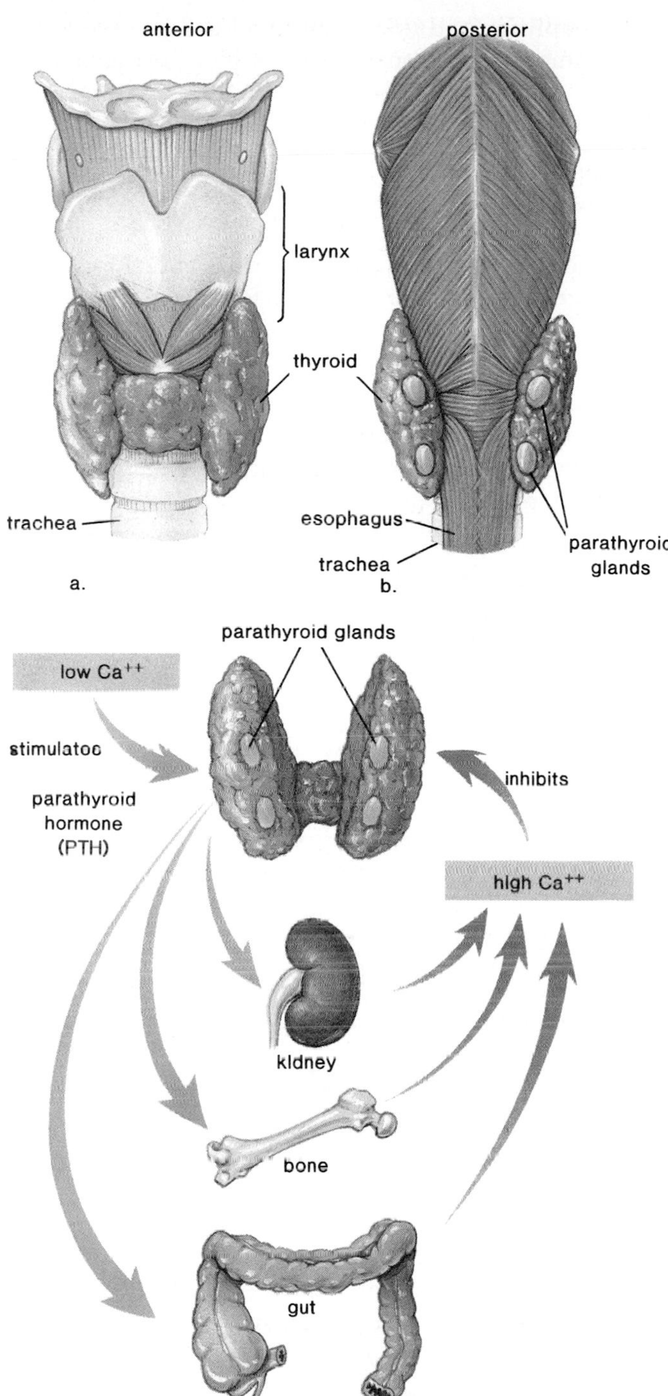

Figure 49.7 Thyroid and parathyroid glands.
a. The thyroid gland is located in the neck in front of the trachea. **b.** The four parathyroid glands are embedded in the posterior surface of the thyroid gland. Yet the parathyroid and thyroid glands have no anatomical or physiological connection with one another. **c.** Regulation of parathyroid hormone (PTH) secretion. A low blood calcium level causes the parathyroids to secrete PTH, which causes the kidneys and the intestine to retain calcium and osteoclasts to break down bone. The result is a higher blood calcium level. A high blood calcium level then inhibits the secretion of PTH.

The anterior pituitary produces TSH, which stimulates the thyroid to secrete thyroxine. The increased level of thyroxine exerts feedback control over the anterior pituitary, which ceases to produce TSH. When there is a low level of thyroxine in the blood, the anterior pituitary continues to produce TSH. The thyroid responds by increasing in size and producing a goiter, but this increase in size is ineffective because active thyroxine cannot be produced without iodine. Eventually scientists discovered that a goiter can be prevented by consumption of iodized salt.

In general, thyroxine increases the metabolic rate in all cells; the number of respiratory enzymes increases, as does oxygen uptake. Thyroxine is necessary in vertebrates for proper growth and development. For example, without thyroxine, frogs do not metamorphose properly, and humans do not mature properly. *Cretinism* occurs in individuals who have suffered from hypothyroidism (low thyroid function) since birth. They show reduced skeletal growth, sexual immaturity, and abnormal protein metabolism. The latter leads to mental retardation.

In addition to thyroxine, the thyroid gland also produces the hormone *calcitonin.* This hormone lowers the level of calcium in the blood and opposes the action of the parathyroid hormone, which we will discuss in the next section.

Thyroxine speeds up metabolism and helps regulate growth and development in immature animals.

Parathyroid Glands Regulate Calcium

The four **parathyroid glands** are embedded in the posterior surface of the thyroid gland (Fig. 49.7*b*). They produce a hormone called **parathyroid hormone (PTH).** Under the influence of PTH, the calcium (Ca^{2+}) level in the blood increases and the phosphate (PO_4) level decreases. PTH stimulates the *absorption* of calcium from the gut by activating vitamin D. It promotes the *retention* of calcium and the excretion of phosphate by the kidneys. PTH also brings about the *demineralization* of bone by promoting the activity of the osteoclasts, the bone-reabsorbing cells. When the blood calcium level reaches the appropriate level, the parathyroid glands no longer produce PTH (Fig. 49.7*c*).

If PTH is not produced in response to low blood calcium, *tetany* results because calcium plays an important role in both nervous conduction and muscle contraction. In tetany, the body shakes from continuous muscle contraction. The effect is brought about by increased excitability of the nerves, which fire spontaneously and without rest.

Parathyroid hormone (PTH) maintains a high blood calcium level. Its actions are opposed by calcitonin, which is produced by the thyroid.

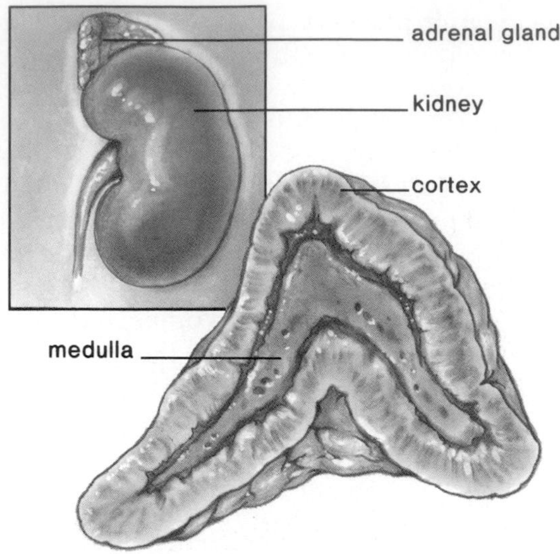

Figure 49.8 Adrenal glands.
Location and structure of adrenal glands.

Adrenal Glands Contain Two Parts

Each of the two **adrenal glands** [L. *ad,* toward, and *renis,* kidney] lies atop a kidney (Fig. 49.8). Each consists of an inner portion called the medulla and an outer portion called the cortex. These portions, like the anterior pituitary and the posterior pituitary, have no physiological connection with one another.

The adrenal hormones increase during times of stress. Those secreted by the adrenal medulla allow us to respond to emergency situations, and those from the adrenal cortex help us to recover from stress.

The hypothalamus exerts control over the activity of both portions of the adrenal glands. It can initiate nerve impulses that travel by way of the brain stem, spinal cord, and sympathetic nerve fibers to the adrenal medulla, which then secretes its hormones. The hypothalamus, by means of ACTH-releasing hormone, controls the anterior pituitary's secretion of ACTH, which, in turn, stimulates the adrenal cortex. Stress of all types, including both emotional and physical trauma, prompts the hypothalamus to stimulate the adrenal glands.

The adrenal glands have two parts, an outer cortex and an inner medulla. The adrenal medulla is under nervous control, and the cortex is under the control of ACTH, a hormone of the anterior pituitary.

Adrenal Medulla Responds to Stress

Epinephrine (adrenaline) and **norepinephrine** (noradrenaline) are produced by the **adrenal medulla.** The postganglionic fibers of the sympathetic system, which controls the adrenal medulla, also secrete norepinephrine.

Epinephrine and norepinephrine are involved in the body's immediate response to stress. They bring about all the bodily changes that occur when an individual reacts to an emergency:

- blood glucose level rises, and the metabolic rate increases
- bronchioles dilate, and breathing rate increases
- blood vessels to the digestive tract and skin constrict; those to skeletal muscles dilate
- cardiac muscle contracts more forcefully, and heart rate increases

The adrenal medulla releases epinephrine and norepinephrine into the bloodstream, helping us cope with emergency situations.

Adrenal Cortex Also Responds to Stress

The two major types of hormones produced by the **adrenal cortex** are the *glucocorticoids,* which help regulate the blood glucose level, and *mineralocorticoids,* which help regulate the level of minerals in the blood. It also secretes a small amount of male sex hormones and a small amount of female sex hormones in both sexes—that is, in the male, both male and female sex hormones are produced by the adrenal cortex and in the female, both male and female sex hormones are also produced by the adrenal cortex (Fig. 49.8).

Cortisol is the glucocorticoid responsible for the greatest amount of activity. Cortisol promotes the hydrolysis of muscle protein to amino acids, which enter the blood. This leads to a higher blood glucose level when the liver converts these amino acids to glucose. Cortisol also favors metabolism of fatty acids rather than carbohydrate. In opposition to insulin, therefore, cortisol raises the blood glucose level. Cortisol also counteracts the inflammatory response that leads to the pain and the swelling of joints in arthritis and bursitis. The administration of cortisol aids these conditions because it reduces inflammation.

Aldosterone is the most important of the mineralocorticoids. The primary target organ of aldosterone is the kidney, where it promotes renal absorption of sodium (Na^+) and renal excretion of potassium (K^+).

The secretion of mineralocorticoids is not under the control of the anterior pituitary. When the blood sodium level and therefore the blood volume are low, the kidneys secrete **renin** (Fig. 49.9). Renin is an enzyme that converts the plasma protein angiotensinogen to angiotensin I, which is changed to angiotensin II by a converting enzyme found in the lungs. Angiotensin II stimulates the adrenal cortex to release aldosterone. The effect of this system, called the renin-angiotensin-aldosterone system, is to raise the blood volume and pressure in two ways. First, angiotensin II constricts the arterioles directly, and secondly, aldosterone causes the kidneys to reabsorb sodium. When the blood sodium level rises, water is reabsorbed, and blood volume and pressure are maintained.

Two other hormones play a role in the homeostatic maintenance of blood volume. We have already discussed that the hormone ADH helps increase blood volume by causing the kidneys to reabsorb water. When the atria of the heart are stretched due to increased blood volume, cardiac cells release a hormone called *atrial natriuretic hormone (ANH),* which inhibits the secretion of renin by the kidneys and the secretion of aldosterone from the adrenal cortex. The effect of this hormone is, therefore, to cause the excretion of sodium, that is, *natriuresis.* When sodium is excreted, so is water, and therefore, blood volume and blood pressure decrease.

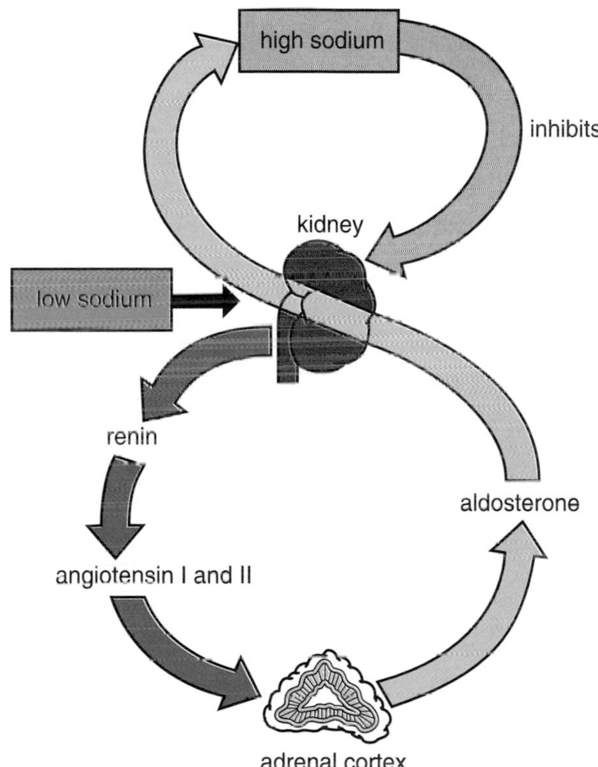

Figure 49.9 Renin-angiotensin-aldosterone homeostatic system.
If the blood level of sodium is low, the kidneys secrete renin. The increased renin acts via the increased production of angiotensin I and II to stimulate aldosterone secretion. Aldosterone promotes reabsorption of sodium by the kidneys; when the blood sodium level rises, the kidneys stop secreting renin.

Adrenal Cortex Can Malfunction

When there is a low level of adrenal cortex hormones due to hyposecretion, a person develops Addison disease. Typically, there is a peculiar bronzing of the skin in a person with **Addison disease** (Fig. 49.10). Because the lack of cortisol results in a drop in blood glucose level that can be severe, there is high susceptibility to any kind of stress due to an insufficient energy supply. Even a mild infection can cause death. Due to the lack of aldosterone, the blood sodium level is low, and the person experiences low blood pressure and possibly severe dehydration. Left untreated, Addison disease can be fatal.

When there is a high level of glucocorticoids in the body due to hypersecretion, a person develops **Cushing syndrome.** Excess cortisol causes a tendency toward diabetes mellitus, a decrease in muscular protein, and an increase in subcutaneous fat. Because of these effects, the person usually has an obese trunk, while the arms and legs remain normal. Due to the high blood sodium level, the blood is basic (pH greater than normal), hypertension occurs, and there is edema of the face, which gives it a moonlike shape. Masculinization may occur in women due to oversecretion of adrenal male sex hormone.

Addison disease is due to adrenal cortex hyposecretion, and Cushing syndrome is due to adrenal cortex hypersecretion.

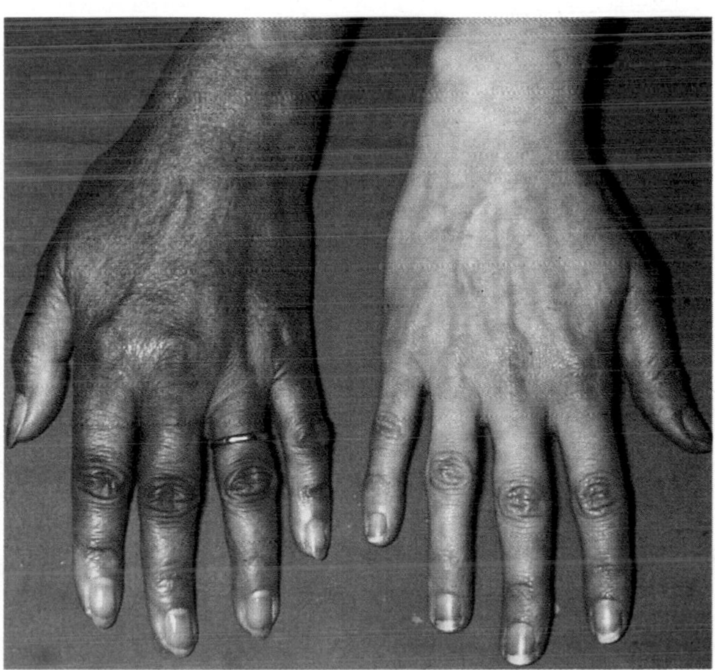

49.10 Addison disease.
This condition is characterized by a peculiar bronzing of the skin. Note the color of the hand on the left compared to the hand of an individual without the disease on the right.

Doing Science

Isolation of Insulin

The pancreas is both an exocrine gland and an endocrine gland. It sends digestive juices to the duodenum by way of the pancreatic duct, and it secretes the hormones insulin and glucagon into the bloodstream. In 1920, physician Frederick Banting decided to try to isolate insulin. Previous investigators had been unable to do this because the enzymes in the digestive juices destroyed insulin (a protein) during the isolation procedure. Banting hit upon the idea of tying off the pancreatic duct, which he knew from previous research would lead to the degeneration only of the cells that produce digestive juices and not of the pancreatic islets (of Langerhans), where insulin is made. J. J. Macleod made a laboratory available to him at the University of Toronto and also assigned a graduate student, Charles Best, to assist him. Banting and Best had limited funds and spent that summer working, sleeping, and eating in the lab. By the end of the summer, they had obtained pancreatic extracts that did lower the blood glucose level in diabetic dogs. Macleod then brought in biochemists, who purified the extract. Insulin therapy for the first human patient began in 1922, and large-scale production of purified insulin from pigs and cattle followed. Banting and Macleod received a Nobel Prize for their work in 1923. The amino acid sequence of insulin was determined in 1953. Insulin is presently synthesized using recombinant DNA technology. Banting and Best followed the required steps given in the chart to identify a chemical messenger.

Steps	Example
Identify the source of the chemical	Pancreatic islets are source
Identify the effect to be studied	Presence of pancreas in body lowers blood sugar
Isolate the chemical	Insulin isolated from pancreatic secretions
Show that the chemical alone has the effect	Insulin alone lowers blood sugar

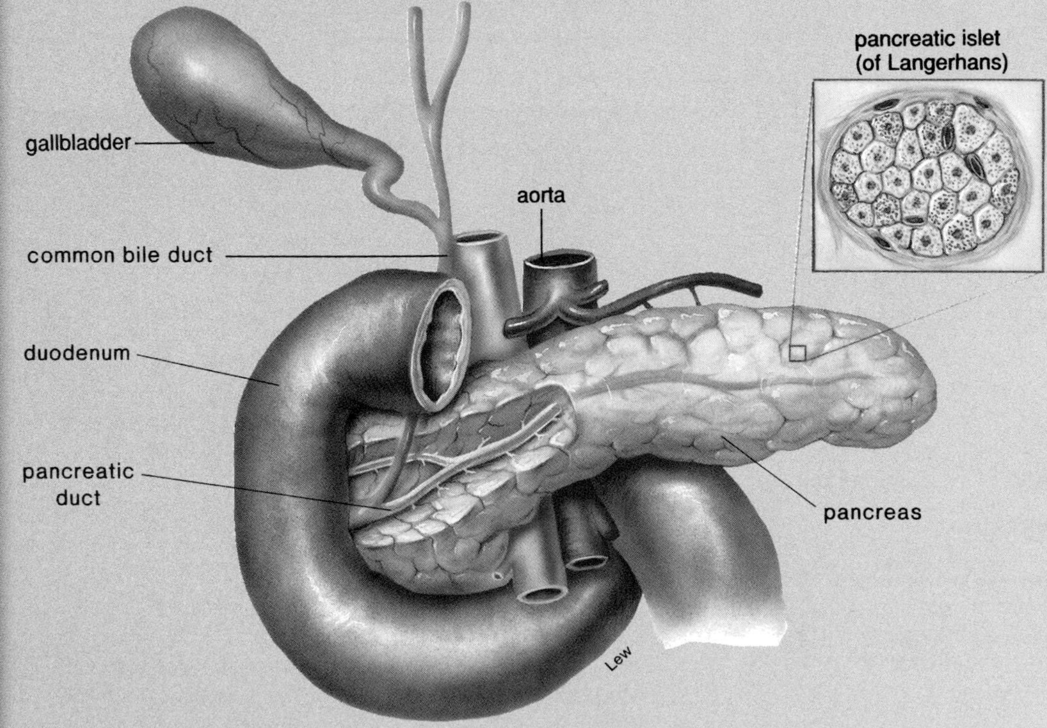

Figure 49A Anatomy of pancreas and associated ducts.

Pancreas Produces Two Hormones

The **pancreas** is composed of two types of tissue. Exocrine tissue produces and secretes *digestive juices* that go by way of ducts to the small intestine. Endocrine tissue, called the **pancreatic islets** (of Langerhans), produces and secretes the hormones **insulin** and **glucagon** directly into the blood.

All the cells of the body use glucose as an energy source; to preserve the health of the body, it is important that the glucose concentration remain within normal limits. Insulin is secreted after eating when there is ordinarily a high blood glucose level. Insulin has three primary actions: (1) it stimulates liver, fat, and muscle cells to take up and metabolize glucose; (2) it stimulates the liver and the muscles to store glucose as glycogen; and (3) it promotes the buildup of fats and proteins and inhibits their use as an energy source. It also causes all body cells to take up glucose. Therefore, insulin is a hormone that lowers the blood glucose level.

Glucagon is secreted from the pancreas in between eating, and its effects are to raise the blood glucose level. Glucagon stimulates the liver to break down glycogen and it also causes adipose tissue to break down fat to glycerol and fatty acids. The liver will then convert glycerol to glucose. The effects of glucagon are to raise the blood glucose level. (Fig. 49.11).

Diabetes Mellitus Is Deficient Insulin

Diabetes mellitus is a fairly common disease caused by an insulin deficit. Sugar in the urine, a common laboratory test for diabetes mellitus, means that the blood glucose level is high enough to cause the kidneys to excrete glucose. Therefore, the liver is not storing glucose as glycogen and the cells are not utilizing glucose as an energy source.

Since carbohydrate is not being metabolized, the body turns to the breakdown of protein and fat for energy, and this leads to the buildup of ketones in the blood. The resulting reduction in blood volume and acidosis (acid blood) can eventually lead to coma and death of the diabetic.

There are two types of diabetes. In *type I (insulin-dependent) diabetes,* the pancreas is not producing insulin. The condition is believed to be brought on by a viral infection that causes cytotoxic T cells to destroy the pancreatic islets (of Langerhans). The individual must have daily insulin injections—either an overdose or the absence of regular eating can bring on the symptoms of hypoglycemia (low blood sugar). Since the brain requires a constant supply of sugar, unconsciousness can result. The cure is quite simple: an immediate source of sugar, such as a sugar cube or fruit juice, can very quickly counteract hypoglycemia.

Of the 12 million people who now have diabetes in the United States, at least 10 million have *type II (non-insulin-dependent) diabetes.* This type of diabetes usually occurs in people of any age who are obese and inactive. The pancreas produces insulin, but the cells do not respond to it. If type II diabetes is untreated, the results can be as serious as those of type I diabetes. (Diabetics are prone to blindness, kidney disease, and circulatory disorders, including strokes. Pregnancy carries an increased risk of diabetic coma, and the child of a diabetic is somewhat more likely to be stillborn or to die shortly after birth.) It is important, therefore, to prevent or to at least control type II diabetes. The best defense is a low-fat diet and regular exercise. If this fails, there are oral drugs that make the cells more sensitive to the effects of insulin or that stimulate the pancreas to make more insulin.

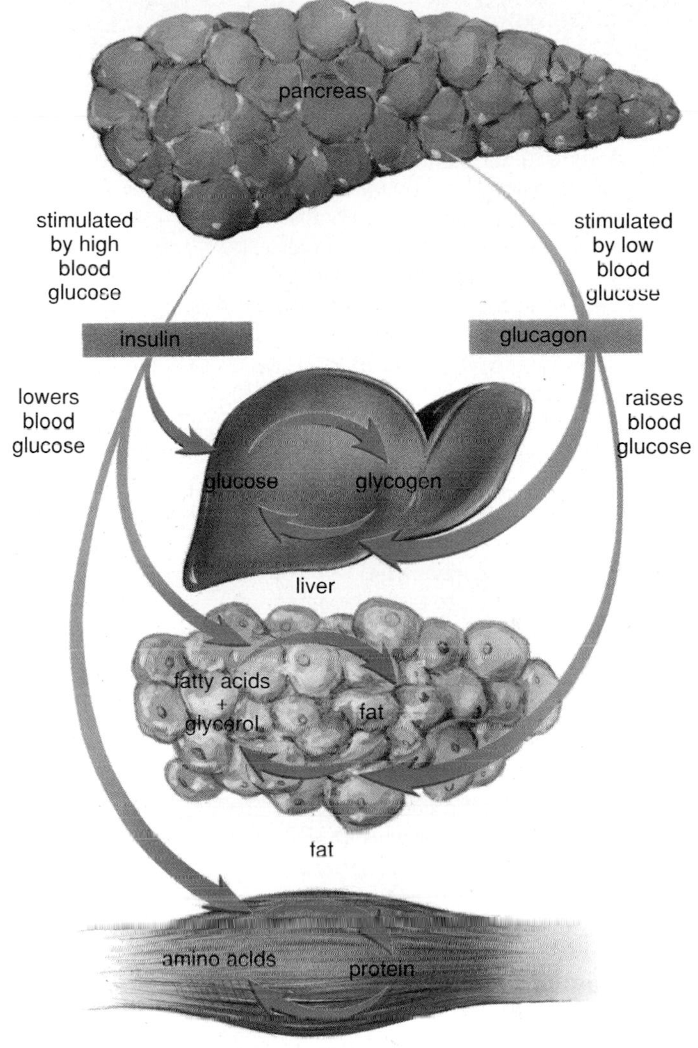

Figure 49.11 Insulin and glucagon homeostatic system.
When the blood glucose level is high, the pancreas secretes insulin. Insulin promotes the storage of glucose as glycogen and the synthesis of proteins and fats (as opposed to their use as energy sources). Therefore, insulin lowers the blood glucose level. When the blood glucose level is low, the pancreas secretes glucagon. Glucagon acts in opposition to insulin in all respects, particularly on the liver.

Diabetes mellitus is due to a lack of insulin or a lack of sensitivity to insulin. Insulin lowers the blood glucose level by promoting glucose uptake by cells and conversion of glucose to glycogen by the liver.

A Closer Look

Insect Growth Hormones

All animals use chemical messengers to regulate the activities of their cells. These messengers are now being identified in a wide variety of invertebrate animals. The work on insect growth hormones began in the 1930s, so they have been studied for some time.

Insects, like other arthropods, have an exoskeleton that must be shed periodically in order to accommodate growth. Typically, the larvae of insects molt (shed their skeleton) a number of times before they pupate and metamorphose into an adult (Fig. 49B). V. B. Wigglesworth showed in the 1930s that the insect brain was necessary for maturation to take place because it produced a hormone appropriately called *brain hormone*. Brain hormone stimulates the prothoracic gland, which lies in the region just behind the head, to secrete *ecdysone,* a steroid hormone. Ecdysone is also called the molt-and-maturation hormone because it promotes both molting and maturation. During larval stages, however, response to ecdysone is modified by the action of *juvenile hormone*. If juvenile hormone is present, the larva molts into another larval form; if it is minimally present, the larva pupates; and if it is absent, the larva undergoes metamorphosis into an adult.

Some plants produce compounds that are similar or identical to either ecdysone or juvenile hormone. These compounds apparently protect the plants by disrupting the development of the insects that feed upon them. Work is under way to extract these compounds for use as insecticides.

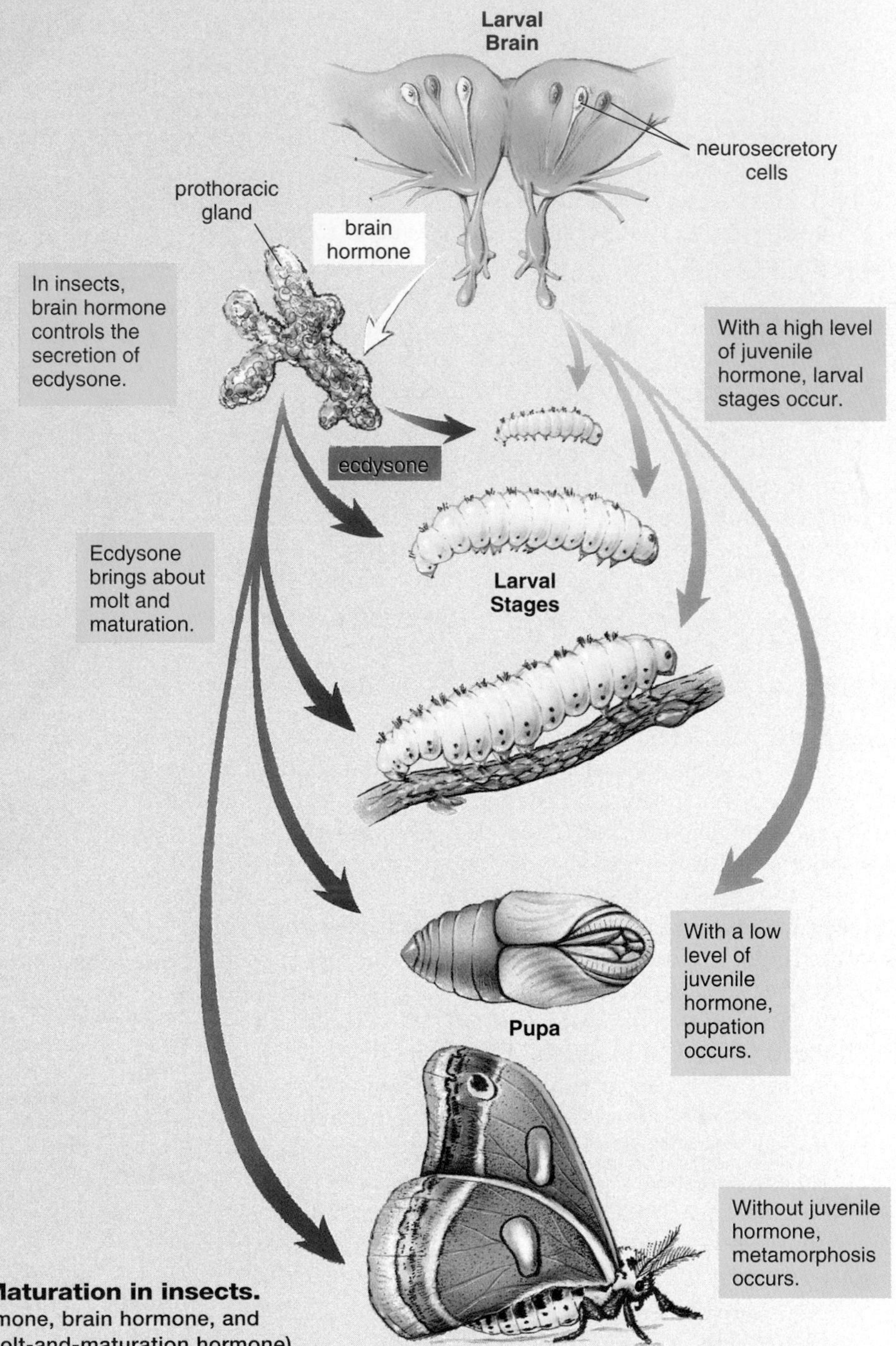

Figure 49B Maturation in insects. The juvenile hormone, brain hormone, and ecdysone (the molt-and-maturation hormone) interact to bring about maturation of insects.

Testes Are in Males and Ovaries Are in Females

The gonads are the testes in the male and the ovaries in the female. The *testes* are located in the scrotum, and the ovaries are located in the pelvic cavity. The testes produce the **androgens** [Gk. *andros*, male, and L. *genitus*, producing] (e.g., *testosterone*), which are the male sex hormones, and the *ovaries* produce **estrogens** and **progesterone,** the female sex hormones.

Puberty is the time of life in humans when the sexual organs mature and the secondary sexual organs appear. Testosterone secretion at the time of puberty stimulates maturation of the testes and other sexual organs. It also brings about the secondary sexual characteristics. Testosterone causes growth of a beard, axillary (underarm) hair, and pubic hair. It prompts the larynx and the vocal cords to enlarge, causing the voice to change. Testosterone is believed to be largely responsible for the sex drive and may even contribute to the supposed aggressiveness of males. It is responsible for the muscular strength of males, and this is the reason some athletes take supplemental amounts of **anabolic steroids,** which are either testosterone or related chemicals. The contraindications of taking anabolic steroids are listed in Figure 49.12. Testosterone is largely responsible for acne and body odor, and also baldness. Genes for baldness probably are inherited by both sexes, but baldness is seen more often in males because of the presence of testosterone.

The female sex hormones, *estrogens* and *progesterone,* have many effects on the body. In particular, estrogens secreted at the time of puberty stimulate the maturation of the ovaries and other sexual organs. Estrogen [Gk. *oistros*, sexual heat, and L. *genitus*, producing] is necessary for development of oocytes (egg production) and is largely responsible for the secondary sexual characteristics in females. It is responsible for female body hair and fat distribution. In general, females have a more rounded appearance than males because of a greater accumulation of fat beneath the skin. Also, the pelvic girdle enlarges in females, facilitating the process of giving birth. Both estrogen and progesterone are required for breast development and regulation of the uterine cycle, which includes monthly menstruation (discharge of blood and mucosal tissues from the uterus).

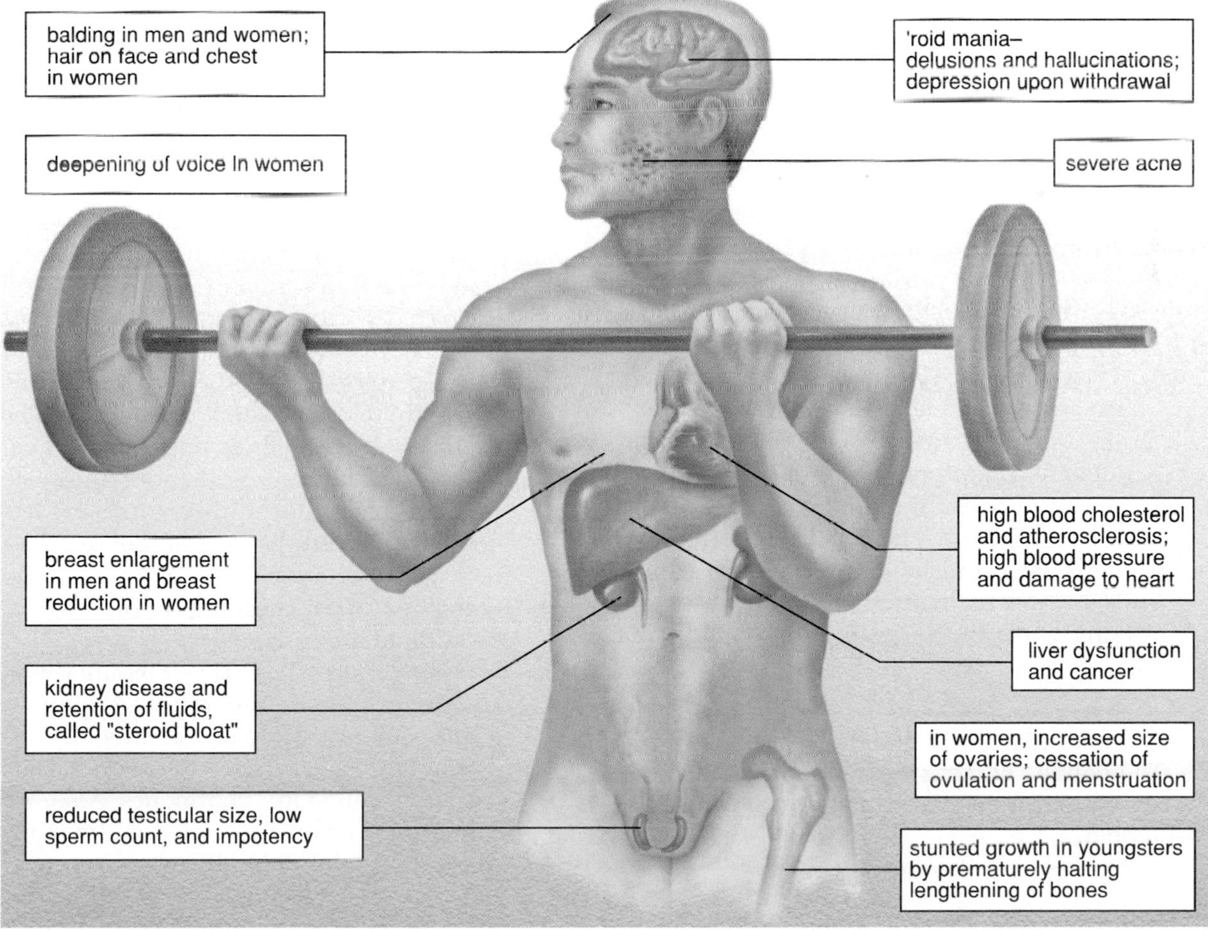

Figure 49.12 The effects of anabolic steroid use.

Thymus Is Most Active in Children

The **thymus** is a lobular gland that lies in the upper thoracic cavity (see Fig. 49.3). This organ reaches its largest size and is most active during childhood. With aging, the organ gets smaller and becomes fatty. Certain lymphocytes that originate in the bone marrow and then pass through the thymus are transformed into T cells. The thymus produces various hormones called *thymosins,* which aid the differentiation of T cells and may stimulate immune cells in general. There is hope that these hormones can be used in conjunction with lymphokine therapy to restore or to stimulate T cell function in patients suffering from AIDS or cancer.

Pineal Gland and Daily/Yearly Rhythms

The **pineal gland** produces the hormone called *melatonin;* this occurs primarily at night. In fishes and amphibians, the pineal gland is located near the surface of the body and is a "third eye," which receives light rays directly. In mammals, the pineal gland is located in the third ventricle of the brain and cannot receive direct light signals (see Fig. 49.3). However, it does receive nerve impulses from the eyes by way of the optic tract. Again, we see that the nervous and endocrine systems are connected.

The pineal gland and melatonin are involved in daily cycles called **circadian rhythms** [L. *circum,* around, and *dies,* day]. Normally we grow sleepy at night when melatonin levels are high and we awaken once daylight returns and melatonin levels are low. Shift work is usually troublesome because it upsets this normal daily rhythm. Similarly, international travel often results in jet lag because the body is still producing melatonin according to the old schedule.

Many animals go through a yearly cycle that includes enlargement of reproductive organs during the summer when melatonin levels are low. Mating occurs in the fall and young are born in the spring. It is of interest that children with a brain tumor that destroys the pineal gland experience early puberty. Therefore, it's possible that the pineal gland is also involved in human sexual development. Another disorder is seasonal affective disorder (SAD); sufferers become depressed and have an uncontrollable desire to sleep with the onset of winter. Giving melatonin makes their symptoms worse, but exposure to a bright light improves them.

The gonads, thymus, and pineal gland are also endocrines. The gonads secrete the sex hormones, the thymus secretes thymosins, and the pineal gland secretes melatonin.

49.3 Environmental Signals in Three Categories

In this chapter, we concentrated on describing the functions of the human endocrine glands and their hormonal secretions. We already know that hormones are only one type of chemical messenger or environmental signal between cells. In fact, the concept of the environmental signal now has been broadened to include at least the following three different categories of messengers (Fig. 49.13):

1. *Environmental signals that act at a distance between individuals.* Many organisms release chemical messengers, called *pheromones,* into the air or in externally deposited body fluids. These are intended to be messages for other members of the species. Ants lay down a pheromone trail to direct other ants to food, and the female silkworm moth releases bombykol, a sex attractant that is received by male moth antennae even several miles away. This chemical is so potent that it has been estimated that only 40 out of 40,000 receptors on the male antennae need to be activated in order for the male to respond. Mammals, too, release pheromones—the urine of dogs serves as a territorial marker. Studies are being conducted to determine if humans also have pheromones.
2. *Environmental signals that act at a distance between body parts.* This category includes the endocrine secretions, which traditionally have been called hormones. It also includes the secretions of the neurosecretory cells in the hypothalamus; the production and action of ADH and oxytocin illustrate the close relationship between the nervous system and the endocrine system. Neurosecretory cells produce these hormones, which are released when these cells receive nerve impulses. As another example of the overlap between the nervous and endocrine systems, consider that endorphins on occasion travel in the bloodstream, but they act on nerve cells to alter their membrane potential. Also, norepinephrine is both a neurotransmitter and hormone secreted by the adrenal medulla.
3. *Environmental signals that act locally between adjacent cells.* Neurotransmitters released by neurons belong in this category, as do substances like **prostaglandins** and growth factors, which are sometimes called local hormones. Also, when the skin is cut, histamine is released by mast cells and promotes the inflammatory response.

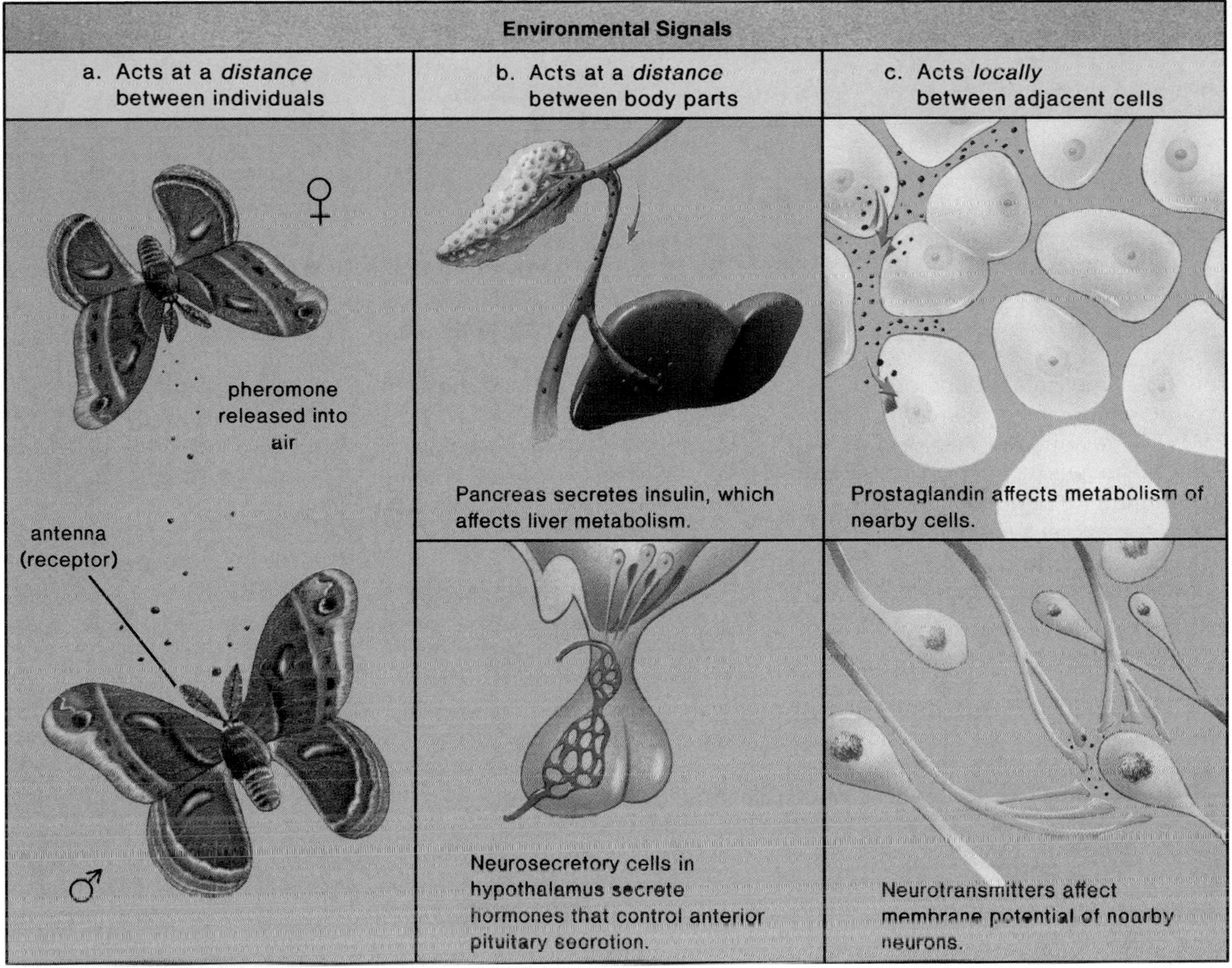
Environmental Signals
a. Acts at a *distance* between individuals
b. Acts at a *distance* between body parts
c. Acts *locally* between adjacent cells
♀
pheromone released into air
antenna (receptor)
♂
Pancreas secretes insulin, which affects liver metabolism.
Neurosecretory cells in hypothalamus secrete hormones that control anterior pituitary secretion.
Prostaglandin affects metabolism of nearby cells.
Neurotransmitters affect membrane potential of nearby neurons.

Figure 49.13 The three categories of environmental signals.
Pheromones are chemical messengers that act at a distance between individuals. Endocrine hormones and neurosecretions typically are carried in the bloodstream and act at a distance within the body of a single organism. Some chemical messengers have local effects only; they pass between cells that are adjacent to one another. This, of course, includes neurotransmitters.

Redefinition of a Hormone

Traditionally, a hormone was considered to be a secretion of an endocrine gland that was carried in the bloodstream to a target organ. In recent years, some scientists have broadened the definition of a hormone to include *all* types of chemical messengers. This change seemed necessary because those chemicals traditionally considered to be hormones now have been found in all sorts of tissues in the human body. It is impossible for insulin produced by the pancreas to enter the brain because of the blood-brain barrier—a tight fusion of endothelial cells of the capillary walls that prevents passage of larger molecules like peptides. Yet, insulin has been found in the brain. It now appears that the brain cells themselves can produce insulin, which is used locally to influence the metabolism of adjacent cells. Also, some chemicals identical to the hormones of the endocrine system have been found in lower organisms, even in bacteria! A moment's thought about the evolutionary process helps to explain this; these regulatory chemicals may have been present in the earliest cells and only became specialized as hormones as evolution proceeded.

There are chemical messengers that work at a distance between individuals or at a distance between body parts, or locally between adjacent cells. Perhaps all types of environmental signals should be considered hormones.

---

connecting concepts

Hormones are intimately involved in maintaining homeostasis, and several work in the same manner or in opposition to each other, thereby keeping the blood glucose level within the normal range of 90–100 mg glucose/100 ml. Insulin, released by the pancreas just after eating, acts on the liver and skeletal muscles to store glucose as glycogen and adipose tissue to convert glucose to fat. And it promotes the uptake and use of glucose, as opposed to fat, as an energy source by body cells. In these ways, insulin keeps the glucose level from rising too high.

In between eating, glucagon, also from the pancreas, keeps the glucose level from falling too low. In the presence of glucagon, adipose tissue releases fatty acids and glycerol into the blood. The liver breaks down glycogen and converts glycerol from the blood to glucose. Most tissues can use fatty acids instead of glucose as an energy source, but the brain must have a supply of glucose. The actions of glucagon assure a glucose supply for the brain. If a "fight-or-flight" emergency situation occurs at any time, the adrenal medulla releases epinephrine, which produces the same effects as glucagon. In addition, epinephrine promotes the breakdown of glycogen by muscles. Skeletal muscles store glycogen, but when they break it down, the glucose stays within their cells. Still, this action "spares" glucose in the sense that skeletal muscle cells will not be taking their glucose from the blood.

What effect does thyroxine ordinarily have on glucose levels? Consistent with thyroxine's effect to increase metabolism, it promotes the use of glucose and it also brings about the breakdown of fats. After all, once cells run out of glucose, most can use fatty acids as an energy source. It also causes cells to synthesize protein, which becomes useful if fasting occurs.

If you eat a high-protein meal, both insulin and glucagon are released. Why? Because if insulin were present alone, the blood glucose level would fall too low, and the presence of glucagon keeps that from happening. Amino acids can be used as an energy source, but first the liver has to convert their carbon chains to glucose. During a prolonged fast, the body turns to protein breakdown and the conversion of amino acids to glucose by the liver to keep the blood glucose level normal. Skeletal muscles are the first to be used as a source of animo acids, and the heart is spared until the skeletal muscles are wasted. Cortisol from the adrenal cortex promotes the use of fatty acids as an energy source and the breakdown of protein to amino acids, and this helps keep the blood glucose level normal.

Growth hormone balances out the effect of cortisol on protein breakdown. Consider that if a body was growing (which it is not when fasting), it would be building muscle (not breaking it down), and plenty of energy would be needed for synthesis. Growth hormone promotes the uptake of amino acids for protein synthesis and the use of fats as an energy source. By these effects, growth hormone also spares glucose and maintains its level so that the brain keeps functioning.

Several hormones are involved in maintaining glucose level within a normal range, even during a prolonged fast. This is only one example of how hormones serve to check and balance one another to keep the internal environment suitable to the operation of the body's organs and cells.

Summary

49.1 Hormones Affect Cellular Metabolism

There are two major types of compounds used as hormones; a few hormones are steroids; most hormones are classified as nonsteroid hormones.

The steroid hormones are lipid soluble and can pass through plasma membranes. Once inside the nucleus, they combine with a receptor molecule, and the complex attaches to transcription factors, which activate DNA. Transcription and translation lead to protein synthesis; in this way, steroid hormones alter the metabolism of the cell.

The nonsteroid hormones are usually received by a receptor located in the plasma membrane. Most often their reception leads to activation of an enzyme that changes ATP to cyclic AMP (cAMP). cAMP then activates another enzyme, which activates another, and so forth.

49.2 Human Endocrine System

The endocrine glands in humans, which are shown in Figure 49.3, produce hormones that are secreted into the bloodstream and circulate about the body until they are received by their target organs.

Neurosecretory cells in the hypothalamus produce antidiuretic hormone (ADH) and oxytocin, which are stored in axon endings in the posterior pituitary until they are released.

The hypothalamus produces hypothalamic-releasing and hypothalamic-release-inhibiting hormones, which pass to the anterior pituitary by way of a portal system. These hormones either stimulate or inhibit the release of a particular anterior pituitary hormone.

The anterior pituitary produces at least six types of hormones (see Figure 49.5). Because the anterior pituitary also stimulates certain other hormonal glands, it is sometimes called the master gland.

The thyroid gland produces thyroxine and triiodothyronine, hormones that play a role in growth and development of immature forms; in mature individuals they increase the metabolic rate. The thyroid gland also produces calcitonin, which helps lower the blood calcium level. The parathyroid glands raise the blood calcium and decrease the blood phosphate level.

The adrenal medulla secretes epinephrine and norepinephrine, which bring about responses we associate with the "fight-or-flight" reaction. The adrenal cortex primarily produces the glucocorticoids (cortisol) and the mineralocorticoids (aldosterone). Cortisol stimulates hydrolysis of proteins to amino acids that are converted to glucose; in this way, it raises the blood glucose level. It also counteracts the inflammatory reaction. Aldosterone causes the kidneys to reabsorb sodium ions (Na^+) and excrete potassium ions (K^+).

Pancreatic islets secrete insulin, which lowers the blood glucose level, and glucagon, which has the opposite effect.

The gonads produce the sex hormones; the thymus secretes thymosins, which stimulate T lymphocyte production and maturation; the pineal gland produces melatonin, whose function in mammals is uncertain—it may affect development of the reproductive organs.

49.3 Environmental Signals in Three Categories

There are three categories of chemical messengers: those that act at a distance between individuals (pheromones); those that act at a distance within the individual (traditional endocrine hormones and secretions of neurosecretory cells); and local messengers (such as neurotransmitters). Since there is great overlap between these categories, perhaps the definition of a hormone should be expanded to include all of them.

Reviewing the Chapter

1. Give a definition of endocrine hormones that includes how they are transported in the body, and how they are received. What does "target" organ mean? 874
2. Categorize endocrine hormones according to their chemical makeup. 874
3. Explain how the two major types of hormones influence the metabolism of the cell. 874–75
4. Give the location in the human body of all the major endocrine glands. Name the hormones secreted by each gland, and describe their chief functions. 876–77
5. Explain the concept of negative feedback and give an example involving antidiuretic hormone (ADH). 877–78
6. Explain the relationship of the hypothalamus to the posterior pituitary gland and to the anterior pituitary gland. 878–79
7. Give an example of the three-tiered relationship among the hypothalamus, the anterior pituitary, and other endocrine glands. Explain why the anterior pituitary can be called the master gland. 879
8. Draw a diagram to explain the contrary actions of insulin and glucagon. Use your diagram to explain the symptoms of type I diabetes mellitus. 885
9. Categorize chemical messengers into three groups, and give examples of each group. 888
10. Give examples to show that there is an overlap between the mode of operation of the nervous system and that of the endocrine system. Explain why the traditional definition of a hormone may need to be expanded. 889

Testing Yourself

Choose the best answer for each question.
Match the hormone in questions 1–5 to the correct gland in the key.

Key:

a. pancreas
b. anterior pituitary
c. posterior pituitary
d. thyroid
e. adrenal medulla
f. adrenal cortex

1. cortisol
2. growth hormone (GH)
3. oxytocin storage
4. insulin
5. epinephrine
6. The anterior pituitary controls the secretion(s) of
   a. both the adrenal medulla and the adrenal cortex.
   b. both cortisol and aldosterone.
   c. thyroxine.
   d. All of these are correct.
7. Nonsteroid hormones
   a. are received by a receptor located in the plasma membrane.
   b. are received by a receptor located in the cytoplasm.
   c. bring about the transcription of DNA.
   d. Both b and c are correct.
8. Aldosterone causes the
   a. kidneys to release renin.
   b. kidneys to reabsorb sodium.
   c. blood volume to increase.
   d. All of these are correct.
9. Diabetes mellitus is associated with
   a. too much insulin in the blood.
   b. too high a blood glucose level.
   c. blood that is too dilute.
   d. All of these are correct.
10. The blood cortisol level controls the secretion of
    a. a releasing hormone from the hypothalamus.
    b. adrenocorticotropic hormone (ACTH) from the anterior pituitary.
    c. cortisol from the adrenal cortex.
    d. All of these are correct.
11. One of the chief differences between pheromones and local hormones is
    a. the distance over which they act.
    b. that one is a chemical messenger and the other is not.
    c. that one is made by invertebrates and the other is made by vertebrates.
    d. All of these are correct.
12. Label this diagram and explain how a negative feedback mechanism keeps the blood level of aldosterone constant in the body:

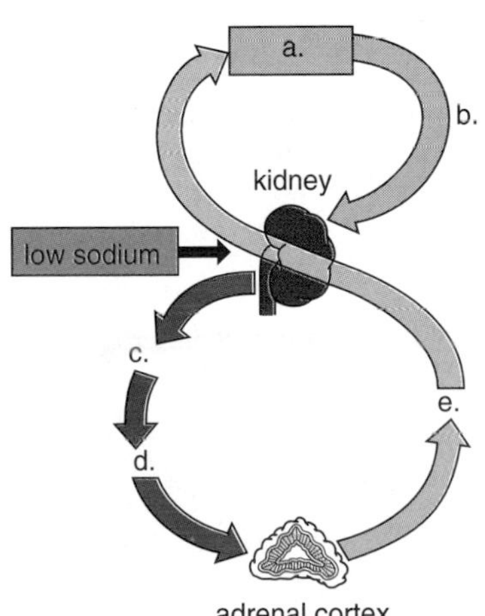

Applying the Concepts

1. *Hormone levels are maintained by feedback control.*
   Contrast control of neurotransmitter levels in the nervous system with control of hormone levels in the endocrine system.
2. *The nervous system is fast acting, and the endocrine system is fairly slow moving.*
   Contrast message delivery in the nervous system with that in the endocrine system.
3. *Hormone levels greatly affect the phenotype.*
   Use the effect of sex hormones to substantiate this concept.

Using Technology

Your study of hormones and endocrine systems is supported by these available technologies:

Exploring the Internet
The Mader Home Page provides resources for and help with studying this chapter.

http://www.mhhe.com/sciencemath/biology/mader/
(Click on Biology.)

The Dynamic Human CD-ROM
Endocrine System

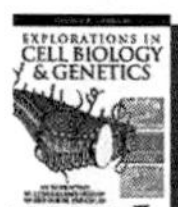

Explorations in Cell Biology & Genetics CD-ROM
Cell–Cell Interactions (#4)

Explorations in Human Biology CD-ROM
Hormone Action (#11)

Life Science Animations Video
Video #3: Animal Biology I
Peptide Hormone Action (cAMP) (#28)

Understanding the Terms

Addison disease 883
adrenal cortex 882
adrenal gland 882
adrenal medulla 882
adrenocorticotropic hormone (ACTH) 880
aldosterone 882
anabolic steroid 887
androgen 887
antidiuretic hormone (ADH) 878
circadian rhythm 888
cortisol 882
Cushing syndrome 883
cyclic AMP 875
diabetes mellitus 885
endocrine system 873
epinephrine 882
estrogen 887
glucagon 885
goiter 880
gonad 877
gonadotropic hormone 880
growth hormone (GH) 879
hormone 874
hypothalamus 878
insulin 885
melanocyte-stimulating hormone (MSH) 880
nonsteroid hormone 875
norepinephrine 882
oxytocin 878
pancreas 885
pancreatic islet 885
parathyroid gland 881
parathyroid hormone (PTH) 881
pineal gland 888
pituitary gland 878
progesterone 887
prolactin (PRL) 880
prostaglandin 888
renin 882
second messenger 875
steroid hormone 874
thymus 888
thyroid gland 880
thyroid-stimulating hormone (TSH) 880
thyroxine 880

Match the terms to these definitions:

a. __________ Hormone secreted by the adrenal cortex that regulates the sodium and potassium ion balance of the blood.
b. __________ Chemical messengers produced in one part of the body that controls the activity of other parts.
c. __________ Hormone secreted by the posterior pituitary that increases the permeability of the collecting ducts in a kidney.
d. __________ Gland—either at the skin surface (fish, amphibians) or in the third ventricle of the brain, (mammals)—that produces melatonin.
e. __________ Hormone secreted by the anterior lobe of the pituitary gland that stimulates activity in the adrenal cortex.
f. __________ Large gland in the neck that produces several important hormones, including thyroxine and calcitonin.
g. __________ Hormone, secreted by the pancreas, which causes the liver to break down glycogen and raises the blood glucose level.
h. __________ Substance secreted by the anterior pituitary; it promotes cell division, protein synthesis, and bone growth.

Reproduction

CHAPTER 50

Chapter Concepts

50.1 How Animals Reproduce

- Among animals, there are two patterns of reproduction: asexual reproduction and sexual reproduction. 894
- Sexually reproducing animals have gonads for the production of gametes, and many have accessory organs for the storage and passage of gametes into or from the body. 894
- Animals have various means of assuring fertilization of gametes and protecting immature stages. 895

50.2 Males Have Testes

- The human male reproductive system is designed for the continuous production of a large number of sperm that are transported within a fluid medium. 898
- Hormones control the production of sperm and maintain the primary and secondary sexual characteristics of males. 901

50.3 Females Have Ovaries

- The female reproductive system is designed for the monthly production of an egg and preparation of the uterus to house the developing fetus. 902
- Hormones control the monthly reproductive cycle in females and play a significant role in maintaining pregnancy, should it occur. 904

50.4 Humans Vary in Fertility

- There are alternative methods of reproduction today, including in vitro fertilization followed by introduction of the zygote to the uterus. 906
- Birth-control measures vary in effectiveness from those that are very effective to those that are minimally effective. 907

Ostrich, *Struthio camelus*, with hatchlings

The life processes we have discussed so far, like digesting food, exchanging gases, maintaining salt and water balance, and coordinating body systems, are necessary to the survival of the individual. Reproduction is different because it pertains to the survival of the species, not the individual. For the individual, reproduction can even be disadvantageous because much energy is usually spent on finding a mate, defending a territory, and raising young. For the species, reproduction is a necessity because it is the only way that the species will continue.

The many different ways that animals reproduce can be categorized as asexual or sexual. Asexual reproduction does not involve the use of sex cells like sperm and egg—the offspring have exactly the same traits as the single parent. The parent's adaptations for survival are passed on unchanged to the offspring, and this can be an advantage if the environment is not changing. Sexual reproduction requires two parents: the egg of one parent is fertilized by the sperm of another parent. Sexual reproduction has the advantage of producing offspring that are not exactly like either parent. The introduction of genetic variation among the offspring may very well help to ensure the survival of the species if environmental conditions are changing.

50.1 How Animals Reproduce

The two fundamental patterns of reproduction—asexual and sexual—can be diagrammed like this:

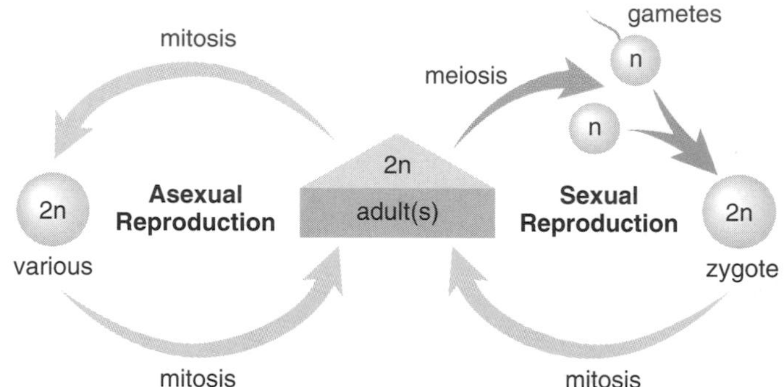

In **asexual reproduction,** there is only one parent, and in sexual reproduction there are two parents. It has long been thought that asexual reproduction could not occur among mammals, but the reading on page 897 describes how cloning of a mammal was carried out. Cloning requires laboratory manipulation and would not ordinarily occur.

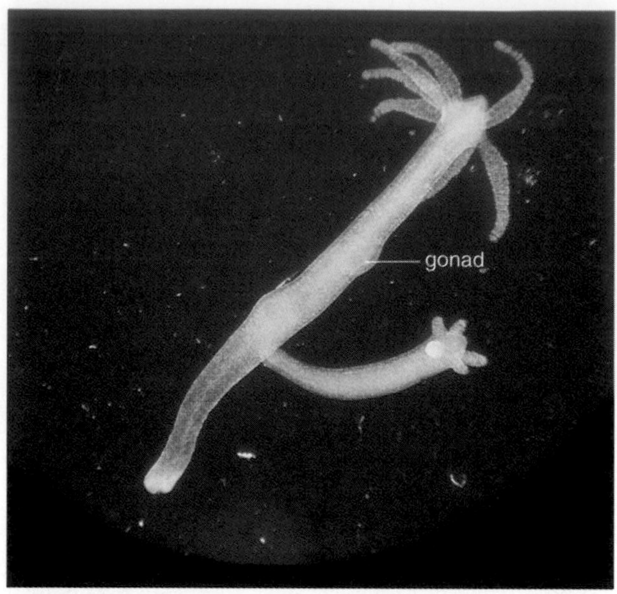

Figure 50.1 Reproduction in *Hydra.*
Hydras reproduce asexually and sexually. During asexual reproduction, a new polyp buds from the parental polyp. During sexual reproduction, temporary gonads develop in the body wall.

Asexual and Sexual Reproduction

There are various patterns of reproduction among animals, some of which are combinations of asexual and sexual reproduction. Some cnidaria, such as hydras, reproduce by *budding* (Fig. 50.1), during which the new individual arises as an outgrowth (bud) of the parent. In other cnidaria there are two diploid generations. The polypoid stage of *Obelia,* which is a colony made up of many hydralike polyps, is sessile and produces diploid medusae by budding. The medusa stage, which looks like a jellyfish, is motile and produces haploid eggs and sperm. The motile stage disperses the species. Many flatworms can constrict into two halves; each half regenerates to become a new individual. Fragmentation followed by regeneration is also seen among sponges and echinoderms. Chopping up a sea star does not kill it; instead, each fragment grows into another animal.

Several types of flatworms, roundworms, crustacea, annelids, and insects, fishes, and lizards have the ability to reproduce parthenogenetically. **Parthenogenesis** [Gk. *parthenos,* virgin, and *genitus,* producing] is a modification of sexual reproduction in which an unfertilized egg develops into a complete individual. In honeybees, the queen bee can fertilize eggs as she lays them or allow them to pass unfertilized. The fertilized eggs become diploid females called workers, and the unfertilized eggs become haploid males called drones.

Usually during **sexual reproduction,** the egg of one parent is fertilized by the sperm of another. Even among earthworms, which are hermaphroditic—each worm has both male and female sex organs—cross-fertilization occurs. Therefore, the offspring have a different combination of genes than either parent. Sequential hermaphroditism or sex reversal also occurs. In coral reef fishes called wrasses, a male has a harem of several females. If the male dies, the largest female becomes a male.

Reproductive Organs Are Primary or Accessory

Among animals that reproduce sexually, there are usually both primary sex organs for the production of gametes and accessory organs for the storage and transport of gametes. Animals usually produce gametes in specialized organs called **gonads** [Gk. *gone,* seed]. Sponges are an exception to this rule because in sponges the collar cells lining the central cavity give rise to sperm and eggs. Cnidaria, such as hydras, produce only temporary gonads in the fall when sexual reproduction occurs (Fig. 50.1). The animals in the other phyla have permanent reproductive organs. The gonads are **testes,** which produce sperm and **ovaries** [L. *ovaris,* egg-keeper], which produce eggs. Eggs or sperm are derived from germ cells, which become specialized for this purpose during early development. Other cells in a gonad support and nourish the developing gametes or produce hormones necessary to the reproductive process. There are also accessory organs—ducts and storage areas that aid in bringing the gametes together.

Copulation [L. *copulatus,* join] is sexual union to facilitate the reception of sperm by a partner, usually a female. Earthworms do not have a copulatory organ, but other animals do have a copulatory organ among their accessory organs of reproduction. In terrestrial animals, males typically have a penis for depositing sperm into the vagina of females. Aquatic animals also have other types of copulatory organs. Lobsters and crayfish have modified swimmerets; cuttlefish and octopuses use an arm; and sharks have a modified pelvic fin that passes packets of sperm to

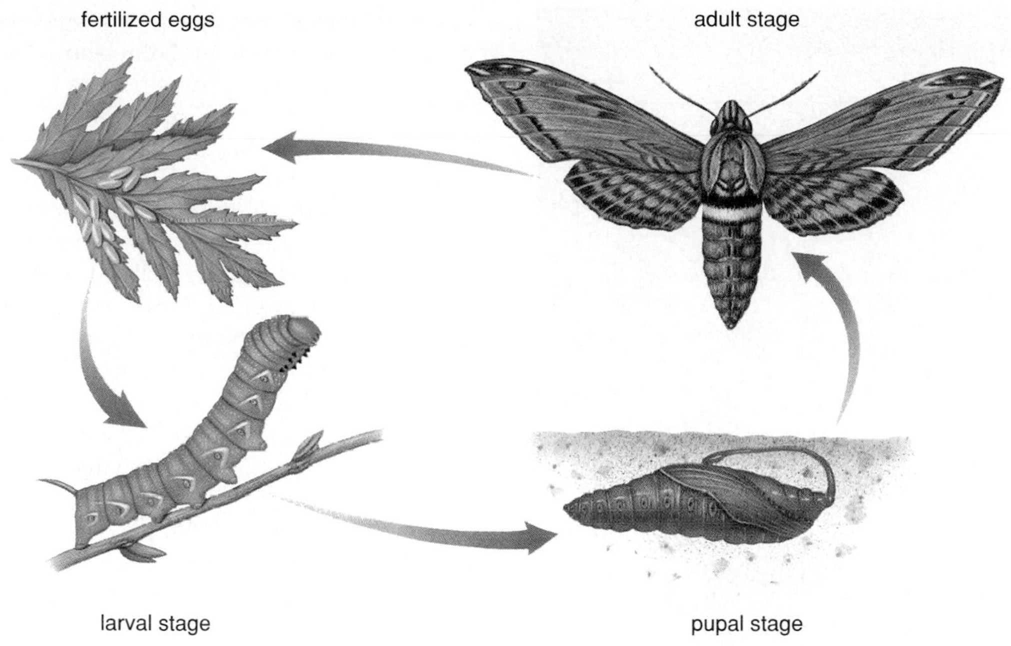

Figure 50.2 Life cycle of a moth.
Moths are oviparous and deposit fertilized eggs in a suitable location for their development. The eggs develop into larvae. There are several larval stages before pupation occurs and metamorphosis results in the adult.

the female. Among terrestrial animals, most birds lack a penis and vagina. They have a cloaca, a chamber that receives products from the digestive, urinary, and reproductive tracts. A male transfers sperm to a female after placing his cloacal opening adjacent to hers.

Protecting Zygote and Embryo

Many aquatic animals practice external fertilization; that is, eggs and sperm join outside the body in the water. Terrestrial animals tend to practice internal fertilization, meaning that egg and sperm join inside the female's body. Both types of animals are usually *oviparous,* meaning that they deposit eggs in the external environment. Consider the life cycle of insects (Fig. 50.2). Their eggs are produced in the ovaries, and as they mature they increase in size because yolk has been added to them. **Yolk** is stored food to be used by the developing embryo. Then, to prevent the eggs from drying out, they are covered by a shell consisting of several layers of protein- and wax-containing material. Small holes are left at one end of the egg for the entry of sperm. Some insects have a special internal organ for storing sperm for some time after copulation so that the eggs can be fertilized internally before they are deposited in the environment.

A **larva** is an independent form that is often quite different in appearance and way of life from the adult. A larva is able to seek its own food to sustain itself until it becomes an adult. In some terrestrial insects, there are several larval stages and then the animal pupates. A *pupa* is enclosed by a hardened cuticle, often within a cocoon. Here **metamorphosis** [Gk. *meta,* implying change, and *morphe,* shape, form], which is a dramatic change in shape, takes place. Then the adult insect emerges and flies off to find a mate and reproduce. Other insects (e.g., grasshoppers) undergo incomplete metamorphosis; pupation does not occur and there are a number of nymph stages, each one looking more like the adult.

Many aquatic forms also have a larval stage. Since the larva has a different lifestyle, it is able to make use of a different food source than the adult. In sea stars, the bilaterally symmetrical larva simply attaches itself to a substratum and undergoes metamorphosis to become a radially symmetrical juvenile. Among barnacles, the free-swimming larva metamorphoses into the sessile adult with calcareous plates. Crayfish, on the other hand, do not have a larval stage; the egg hatches into a tiny juvenile with the same form as the adult.

Reptiles and particularly birds provide their eggs with plentiful yolk; there is no larval stage. Complete development takes place within a shelled egg containing **extraembryonic membranes** [L. *extra,* on the outside] to serve the needs of the embryo. The outermost membrane, the *chorion,* lies next to the shell and functions in gas exchange. The *amnion* forms a water-filled sac around the embryo, ensuring that it will not dry out. The *yolk sac* holds the yolk, which nourishes the embryo, and the *allantois* holds nitrogen waste products (see Fig. 51.9). The shelled egg frees these animals from the need to reproduce in the water and is a significant adaptation to the terrestrial environment.

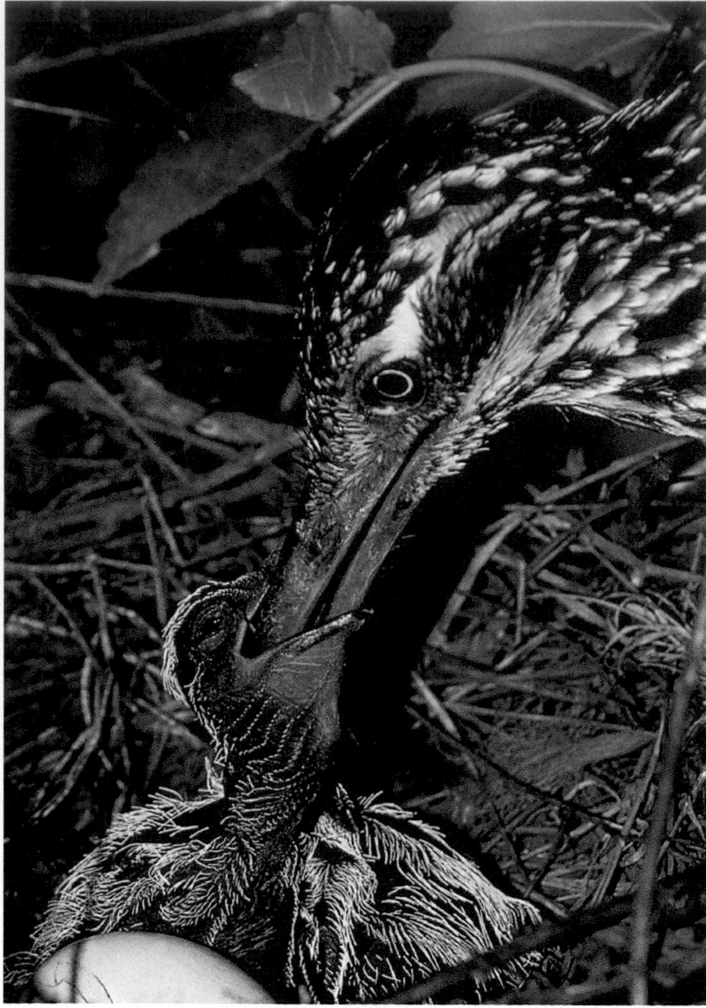

Figure 50.3 Parenting in birds.
Birds are oviparous and lay hard-shelled eggs. They are well-known for incubating their eggs and caring for their offspring after they hatch. This roadrunner, *Geococcyx*, feeds its young.

Birds in particular tend their eggs, and newly hatched birds usually have to be fed before they are able to fly away and seek food for themselves. Complex hormones and neural regulation are involved in reproductive behavior of parental birds (Fig. 50.3). Some animals take a different tactic. They do not deposit and tend their eggs; instead they are *ovoviviparous*, meaning that their eggs are retained in the body until they hatch. Then fully developed offspring, which have a way of life like the parent, are released. Oysters retain their eggs in the mantle cavity, and male sea horses have a special brood pouch in which the eggs develop. Garter snakes, water snakes, and pit vipers retain their eggs in their bodies until they hatch and give birth to living young.

Finally some animals, particularly mammals, practice internal fertilization and are *viviparous*. After offspring are born, the nutrients needed for development are supplied by the mother. Viviparity represents the ultimate in caring for the zygote and embryo. How did viviparity among certain mammals come about? Some mammals, such as the duckbill platypus and the spiny anteater, are egg-laying mammals. Marsupial offspring are born in a very immature state; they finish their development within a pouch, where they are nourished on milk. In marsupials, the embryos develop within a duplex uterus, having two chambers. The uterus of most placental mammals has two so-called horns where several embryos can attach and develop in sequence. Only among primates, including humans, is there a simplex uterus where usually a single embryo develops. The **placenta** is a complex structure derived in part from the chorion, which first appears in a shelled egg. Its evolution allowed the developing offspring to exchange materials with the mother and made the shelled egg unnecessary.

Finding a Mate

Sexually reproducing animals have all sorts of ways of making sure that the gametes find each other. Aquatic animals that practice external fertilization are programmed to release their eggs in the water only at certain times. One environmental signal that seems to work is the lunar cycle. Each month the moon moves closer to the earth and the tides become somewhat higher. Aquatic animals able to sense this change can release their gametes at the same time. Hundreds of thousands of palolo worms rise to the surface of the sea and release their eggs during a two- to four-hour period on two or three specific successive days of the year. Most likely they are under the control of a biological clock that can sense the passage of time so that their reproductive behavior is synchronized.

Among terrestrial animals, reproductive cycles are often tied to day-length changes. The photoperiod reliably indicates the proper time for reproductive behavior, including migration to distant places. Researchers have found that melatonin is produced by the pineal gland during the night, and that this hormone seems to be involved in reproductive cycles. It has been observed that fall-breeding animals can be induced to breed in the spring if they are subjected to artificially shortened day lengths.

Animals that copulate have courtship rituals—specific mating activities that bring the sexes together. These rituals ensure that male and female are of the same species and reduce any aggressive tendencies between the mating pair. They also promote hormonal responses that prepare the body for reproductive success. Reproductive success is evaluated in terms of how many fertile offspring an individual contributes to the next generation.

Sexually reproducing animals usually have primary and accessory sex organs, and they have various ways to protect developing offspring after finding a mate.

Doing Science

Cloning of Mammals

Vertebrates have no normal means of asexual reproduction; however, it is conceivable they could be cloned. Cloning—that is, the production of a copy of an individual—is a form of asexual reproduction because it requires only the genes of one cell. Each vertebrate cell is totipotent, meaning that the cell contains a copy of all the genes. But during development, certain genes are turned off as the cells become specialized. Muscle cells, for example, are specialized to contract; nerve cells to conduct nerve impulses; and glandular cells to secrete. Cloning of an adult vertebrate would require that all the genes of the chosen cell be turned on again. It has long been thought this would not be possible.

Despite formidable obstacles, investigators have never given up. Up to now, frogs and even monkeys have been cloned only under certain circumstances. For frogs, it was possible to take the nucleus from an intestinal cell of a tadpole and transplant it into a frog's egg whose own nucleus had been destroyed, and occasionally, normal development produced an adult frog. For monkeys, the nucleus had to be taken from an even earlier stage—an embryo consisting of several cells. Only if minimal differentiation of cells had occurred was it possible to have the nucleus "start over" to direct the development of a complete monkey. But it would be preferable to use the nuclei of adults for cloning, because only then is it possible to know what phenotypic characteristics you might be getting.

In March of 1997, a scientific breakthrough occurred. Ian Wilmut of the Roslin Institute in Edinburgh, Scotland, announced that he and his colleagues had cloned a sheep from a cell taken from an adult. They used the procedure depicted in Figure 50A to achieve their remarkable success. The donor cells were taken from an udder (mammary gland) of a Finn Dorset ewe (female sheep), and the egg cells were taken from a Blackface ewe. Although 29 clones were attempted, only one—named Dolly—resulted. How was this procedure different from all the others that have been attempted? Starving the donor cells caused them to stop dividing and go into a resting stage that made the nuclei amenable to cytoplasmic signals for initiation of development.

Scientists are watching to see if Dolly will age more rapidly than normal sheep—after all she is the product of a 2n nucleus that was already six years old. Or will she be more susceptible to diseases that will shorten her life? Assuming that all goes well, there are many advantages to the cloning of animals. Genetic similarity would do away with a variable that is often troublesome when doing experiments; it offers a possible means of saving endangered species from extinction; it might be coupled with genetic engineering for the production of transgenic animals, and so forth.

The public is fascinated with the idea that it might be possible to clone human beings. In fact, it took no time at all for President Clinton to announce an executive order that no federal funds were to be spent on experiments to clone human beings. And biologists wanted to be sure that the public understood that a human could never be an exact copy of the person cloned. The clone would start life as an infant in a different family situation and in a different social environment from its predecessor. Humans, after all, are especially the product not only of their genes, but also of their environment.

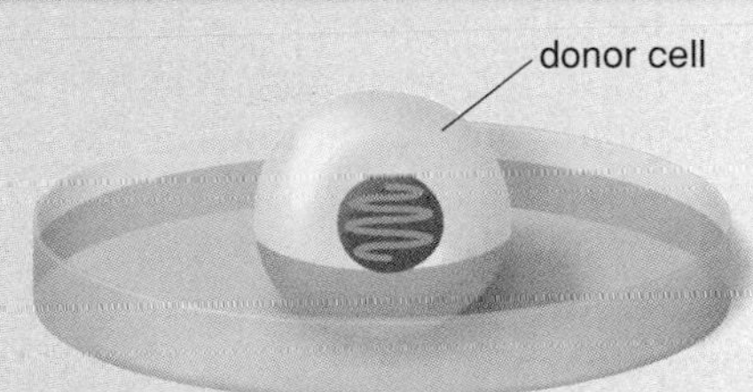

Step 1: Donor cells from udder of a Finn Dorset ewe were placed in a medium that caused all genes to become inactive.

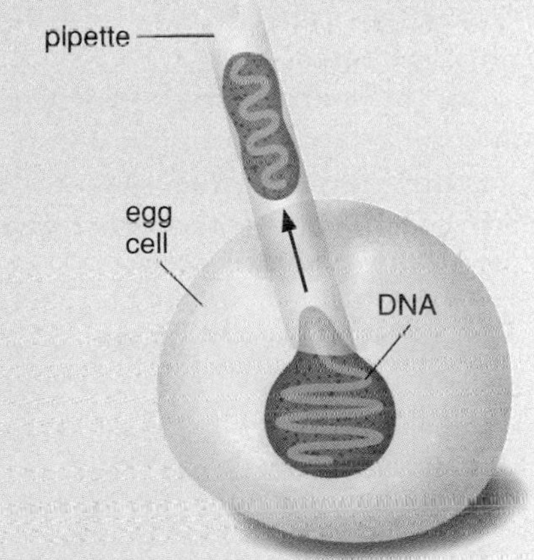

Step 2: Egg cells from the ovary of a Blackface ewe were enucleated.

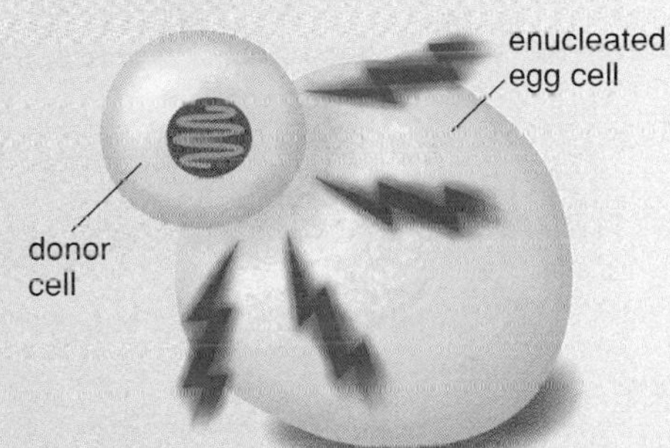

Step 3: Electric pulses were used to fuse the two types of cells and start development.

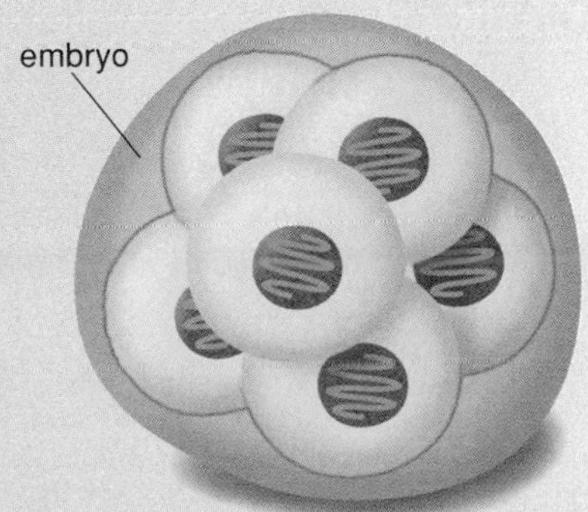

Step 4: Embryos were implanted in the uteri of Blackface ewes.

Figure 50A Cloning of a mammal.

Step 5: Finn Dorset ewe, named Dolly, is a clone of the ewe in step 1.

50.2 Males Have Testes

The human male reproductive system includes the organs pictured in Figure 50.4 and listed in Table 50.1. The male gonads are paired testes, which are suspended within the *scrotal sacs* of the **scrotum.** The testes begin their development inside the abdominal cavity, but they descend into the scrotal sacs as development proceeds. If the testes do not descend—and the male does not receive hormone therapy or undergo surgery to place the testes in the scrotum—sterility (the inability to produce offspring) results. *Sterility* occurs because normal sperm production is inhibited at body temperature; a cooler temperature of 1–1.5°C is required.

Sperm produced by the testes mature within the epididymides (sing., **epididymis**), which are tightly coiled tubules lying just outside the testes. Maturation seems to be required for the sperm to swim to the egg. Once the sperm have matured, they are propelled into the **vasa deferentia** (sing., vas deferens) by muscular contractions. Sperm are stored in both the epididymides and the vasa deferentia. When a male becomes sexually aroused, sperm enter the urethra, part of which is located within the penis.

The **penis** is a cylindrical organ that usually hangs in front of the scrotum. Three cylindrical columns of spongy, erectile tissue containing distensible blood spaces extends

Table 50.1

Male Reproductive System

Organ	Function
Testes	Produce sperm and sex hormones
Epididymides	Maturation and some storage of sperm
Vasa deferentia	Conduct and store sperm
Seminal vesicles	Contribute fluid to semen
Prostate gland	Contributes fluid to semen
Urethra	Conducts sperm (and urine)
Bulbourethral glands	Contribute fluid to semen
Penis	Organ of copulation

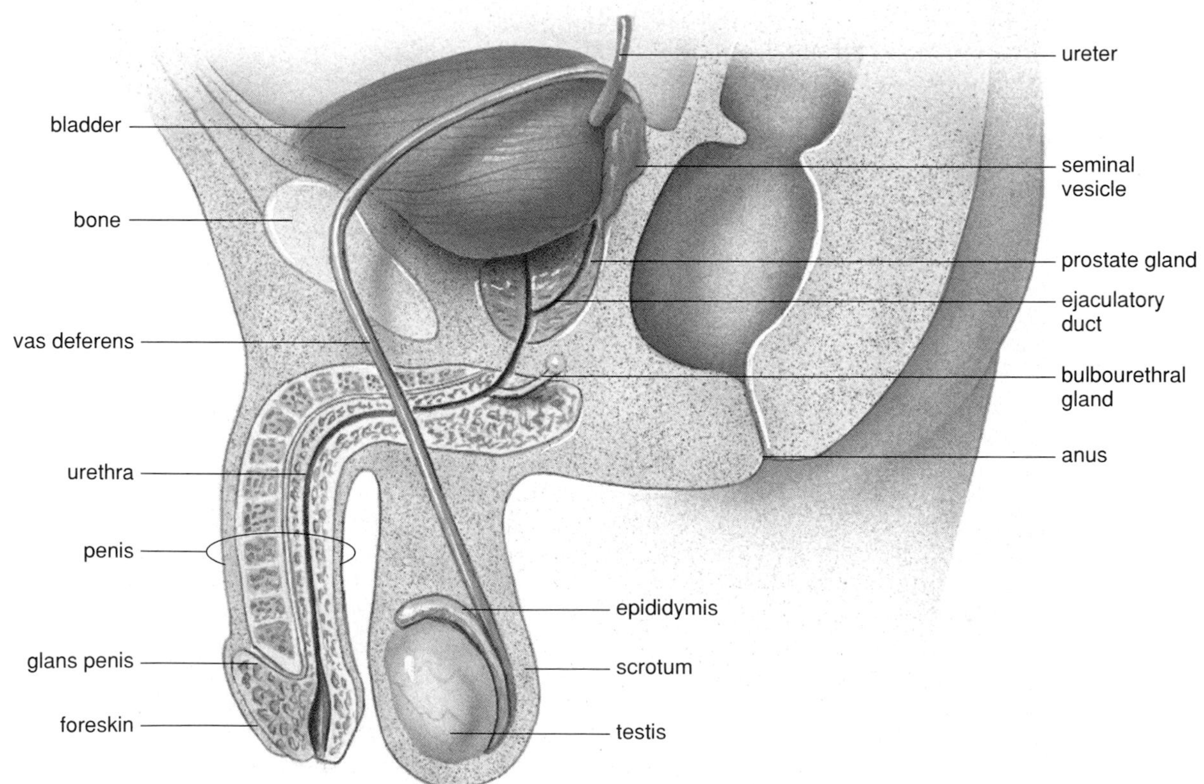

Figure 50.4 Side view of the male reproductive system.
The testes produce sperm. The seminal vesicles, the prostate gland, and the bulbourethral gland provide a fluid medium for the sperm. Circumcision is the removal of the foreskin. Notice that the penis in this drawing is not circumcised because the foreskin is present.

through the shaft of the penis (Fig. 50.5). During sexual arousal, nervous reflexes cause an increase in arterial blood flow to the penis. This increased blood flow fills the blood space in the erectile tissue, and the penis, which is normally limp (flaccid), stiffens and increases in size. These changes are called **erection.** If the penis fails to become erect, the condition is called *impotency.*

Semen (seminal fluid) [L. *semen*, seed] is a thick, whitish fluid that contains sperm and secretions from three glands (Table 50.1). The *seminal vesicles* lie at the base of the bladder. Each joins a vas deferens to form an ejaculatory duct that enters the urethra. As sperm pass from the vasa deferentia into the ejaculatory duct, these vesicles secrete a thick, viscous fluid containing nutrients for possible use by the sperm. Just below the bladder is the *prostate gland,* which secretes a milky alkaline fluid believed to activate or increase the motility of the sperm. In older men, the prostate gland may become enlarged, thereby constricting the urethra and making urination difficult. Also, prostate cancer is the most common form of cancer in men. Slightly below the prostate gland, on either side of the urethra, is a pair of small glands called *bulbourethral* glands, which have mucous secretions with a lubricating effect. Notice from Figure 50.5 that the urethra also carries urine from the bladder during urination.

Sperm produced by the testes mature in the epididymides and pass from the vasa deferentia to the urethra, where certain glands add fluid to semen.

Ejaculation

If sexual arousal reaches its peak, **ejaculation** follows an erection. The first phase of ejaculation is called *emission.* During emission, the spinal cord sends nerve impulses via appropriate nerve fibers to the epididymides and vasa deferentia. Their subsequent motility causes sperm to enter the ejaculatory duct, whereupon the seminal vesicles, prostate gland, and bulbourethral glands release their secretions. At this time, a small amount of secretion from the bulbourethral glands may leak from the end of the penis. Since this is a mucoid secretion, it is believed that this leakage may aid the process of intercourse by providing a certain amount of lubrication. During the second phase of ejaculation, called *expulsion,* rhythmical contractions of muscles at the base of the penis and within the urethral wall expel semen in spurts from the opening of the urethra. These rhythmical contractions are an example of release from *myotonia,* or muscle tenseness. Myotonia is another important sexual response.

An erection lasts for only a limited amount of time. The penis now returns to its normal flaccid state. Following ejaculation, a male may typically experience a period of time, called **refractory period,** during which stimulation does not bring about an erection. The contractions that expel semen from the penis are a part of male **orgasm** [Gk. *orgasmos,* sexual excitement], the physiological and psychological sensations that occur at the climax of sexual stimulation.

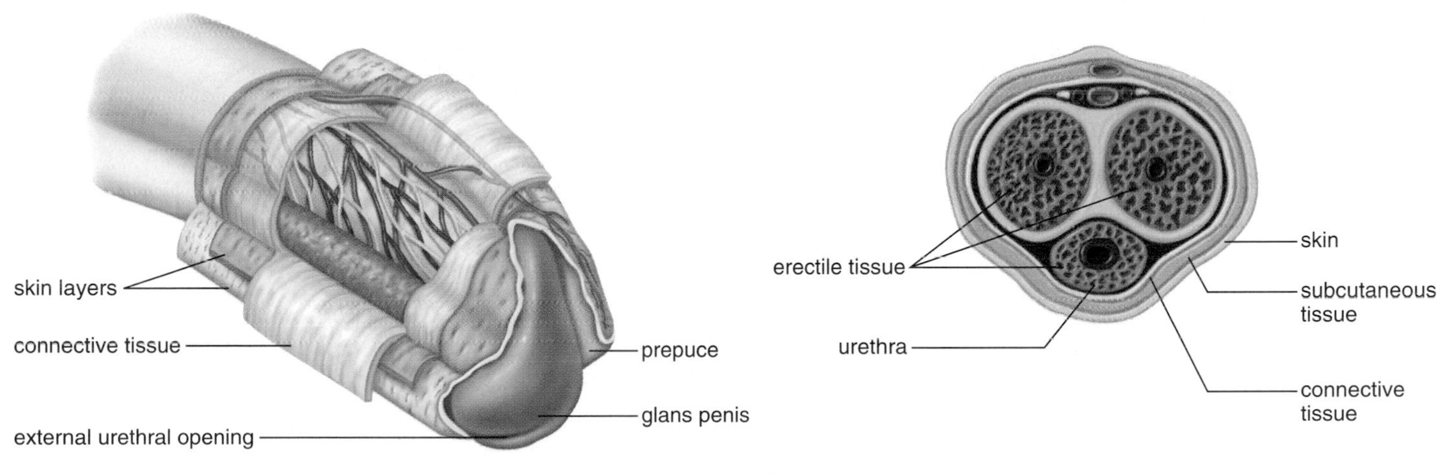

Figure 50.5 Penis anatomy.
a. Beneath the skin and the connective tissue lies the urethra, surrounded by erectile tissue. This tissue expands to form the glans penis, which in uncircumcised males is partially covered by the prepuce (foreskin). **b.** Two other columns of erectile tissue in the penis are dorsally located.

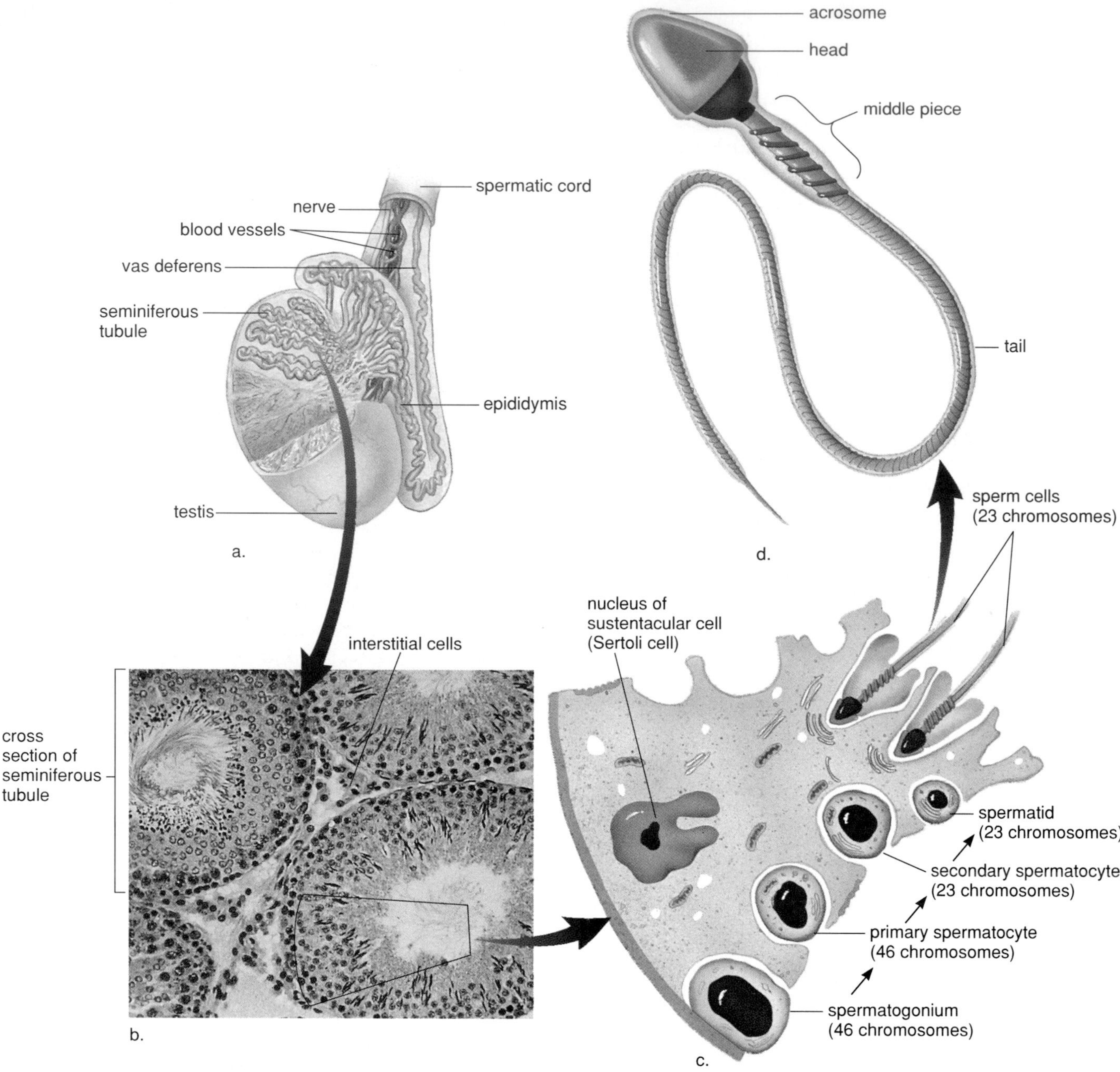

Figure 50.6 Anatomy of testis and sperm.
a. Longitudinal section showing lobules containing seminiferous tubules. **b.** Light micrograph of cross section of seminiferous tubules. **c.** Diagrammatic representation of spermatogenesis, which occurs in the wall of the tubules. **d.** Mature sperm consist of a head, a middle piece, and a tail. The nucleus is in the head, capped by an enzyme-containing acrosome.

Testes Produce Sperm and Hormones

A longitudinal section of a testis shows that it is composed of compartments called lobules, each of which contains one to three tightly coiled **seminiferous tubules** (Fig. 50.6*a*). Altogether, these tubules have a combined length of approximately 250 meters. A microscopic cross section of a seminiferous tubule shows that it is packed with cells undergoing **spermatogenesis** (Fig. 50.6*b*), a process that involves meiosis. Also present are sustentacular (Sertoli) cells, which support, nourish, and regulate the spermatogenic cells (Fig. 50.6*c*).

Mature **sperm,** or spermatozoa, have three distinct parts: a head, a middle piece, and a tail (Fig. 50.6*d*). The middle piece and the tail contain microtubules, in the characteristic 9 + 2 pattern of cilia and flagella. In the middle piece, mitochondria are wrapped around the microtubules and provide the energy for movement. The head contains a nucleus covered by a cap called the *acrosome* [Gk. *akros*, at the tip, and *soma*, body], which stores enzymes needed to penetrate the egg. The human egg is surrounded by several layers of cells and a thick membrane—the acrosomal enzymes play a role in allowing a sperm to reach the surface of the egg. The ejaculated semen of a normal human male contains several hundred million sperm, assuring an adequate number for fertilization to take place. Fewer than 100 ever reach the vicinity of the egg, however, and only one sperm normally enters an egg.

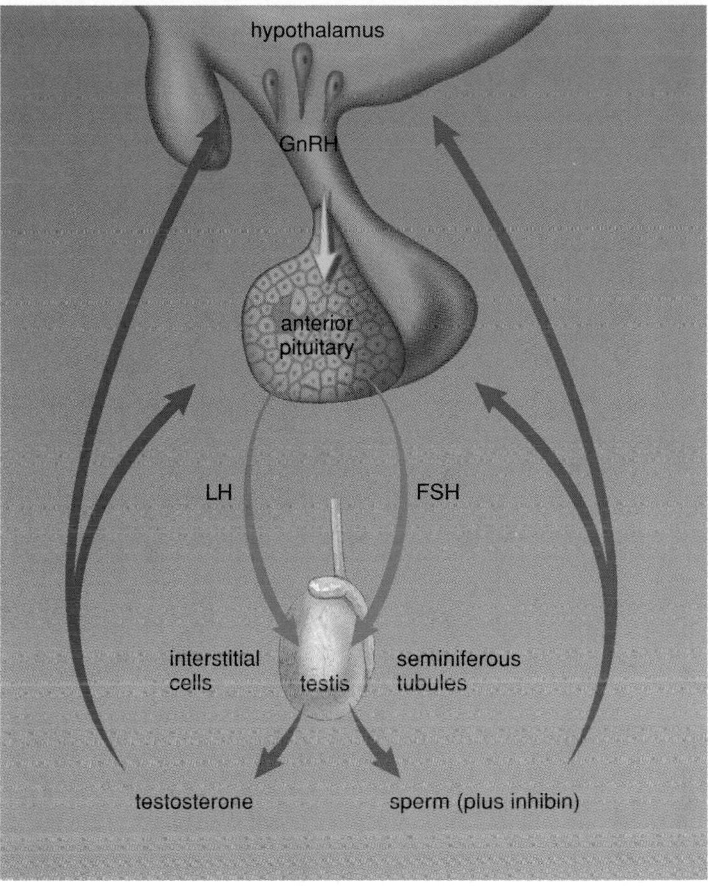

Figure 50.7 Hormonal control of testes.
GnRH (gonadotropin-releasing hormone) stimulates the anterior pituitary to secrete the gonadotropic hormones FSH and LH. FSH stimulates the testes to produce sperm, and LH stimulates the testes to produce testosterone. Testosterone and inhibin exert negative feedback control over the hypothalamus and the anterior pituitary, and this ultimately regulates the level of testosterone in the blood.

Hormonal Regulation in Males

The hypothalamus has ultimate control of the testes' sexual function because it secretes a hormone called gonadotropin-releasing hormone, or GnRH, that stimulates the anterior pituitary to produce the gonadotropic hormones. There are two gonadotropic hormones—follicle-stimulating hormone (FSH) and luteinizing hormone (LH)—in both males and females. In males, FSH promotes spermatogenesis in the seminiferous tubules, which also release the hormone inhibin.

LH in males is sometimes given the name *interstitial cell-stimulating hormone (ICSH)* because it controls the production of the androgen testosterone by the interstitial cells, scattered in the spaces between the seminiferous tubules (Fig. 50.6*b*). All these hormones are involved in a negative feedback relationship that maintains the fairly constant production of sperm and testosterone (Fig. 50.7).

Testosterone Is the Male Sex Hormone

Testosterone is the main sex hormone in males. It is essential for the normal development and functioning of the organs listed in Table 50.1. Testosterone is also necessary for the maturation of sperm.

Testosterone also brings about and maintains the male **secondary sexual characteristics** that develop at the time of **puberty.** Males are generally taller than females and have broader shoulders and longer legs relative to trunk length. The deeper voice of males compared to females is due to the fact they have a larger larynx with longer vocal cords. Since the so-called Adam's apple is a part of the larynx, it is usually more prominent in males than in females.

Testosterone is responsible for the greater muscular development in males. Knowing this, males and females sometimes take anabolic steroids (either the natural or synthetic form of testosterone) to build up their muscles (see page 887). Testosterone causes males to develop noticeable hair on the face, chest, and occasionally on other regions of the body such as the back. Testosterone also leads to the receding hairline and pattern baldness that occurs in males.

The gonads in males are the testes, which produce sperm and testosterone the most significant male sex hormone.

50.3 Females Have Ovaries

The human female reproductive system includes the ovaries, the oviducts, the uterus, and the vagina (Fig. 50.8 and Table 50.2). The ovaries, which produce a secondary **oocyte** each month, lie in shallow depressions, one on each side of the upper pelvic cavity. The **oviducts** [L. *ovum*, egg, and *duco*, lead out], also called uterine or fallopian tubes, extend from the ovaries to the uterus; however, the oviducts are not attached to the ovaries. Instead, they have fingerlike projections called fimbriae (sing., **fimbria**) that sweep over the ovaries. When an oocyte bursts from an ovary during ovulation, it usually is swept into an oviduct by the combined action of the fimbriae and the beating of cilia that line the oviducts. Fertilization, if it occurs, normally takes place in an oviduct, and the developing embryo is propelled slowly by ciliary movement and tubular muscle contraction to the uterus. The **uterus** [L. *uterus*, womb] is a thick-walled muscular organ about the size and shape of an inverted pear. The narrow end of the uterus is called the *cervix*. The embryo completes its development after embedding itself in the uterine lining, called the **endometrium** [Gk. *endon*, within, and *metra*, womb]. A small opening at the cervix leads to the vaginal canal. The **vagina** [L. *vagina*, sheath] is a tube at a 45° angle with the small of the back. The mucosal lining of the vagina lies in folds and can extend. This is especially important when the vagina serves as the birth canal, and it also can facilitate intercourse, when the vagina receives the penis during copulation.

The external genital organs of the female are known collectively as the **vulva** (Fig. 50.8*b*). The *mons pubis* and two folds of skin called *labia minora* and *labia majora* are on either side of the urethral and vaginal openings. At the juncture of the labia minora is the **clitoris,** which is homologous to the penis in males. The clitoris has a shaft of erectile tissue and is capped by a pea-shaped glans. The many sense receptors of the clitoris allow it to function as a sexually sensitive organ. Orgasm in the female is a release of neuromuscular tension in the muscles of the genital area, vagina, and uterus.

Table 50.2

Female Reproductive Organs

Organ	Function
Ovaries	Produce egg and sex hormones
Oviducts (fallopian tubes)	Conduct egg; location of fertilization
Uterus (womb)	Houses developing embryo and fetus
Cervix	Contains opening to uterus
Vagina	Receives penis during copulation and serves as birth canal

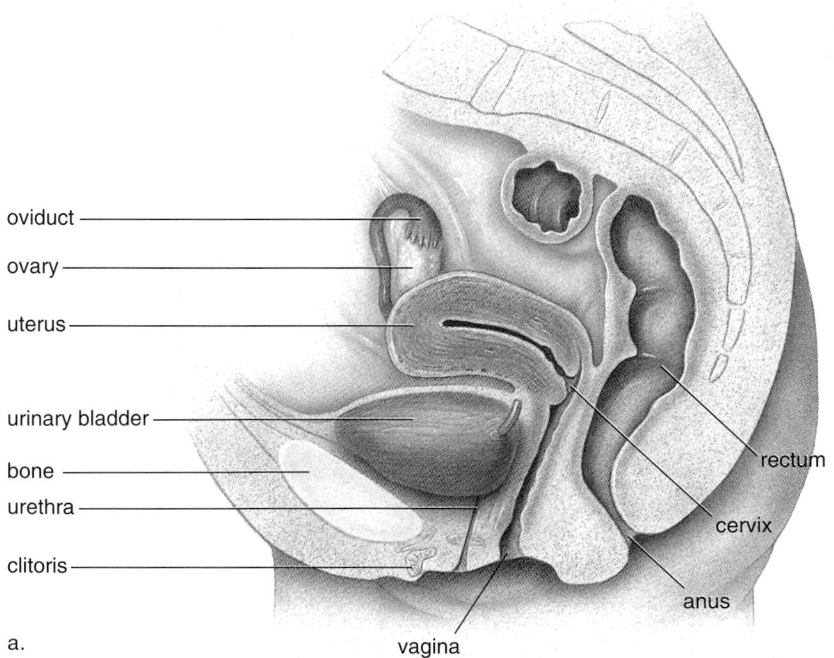

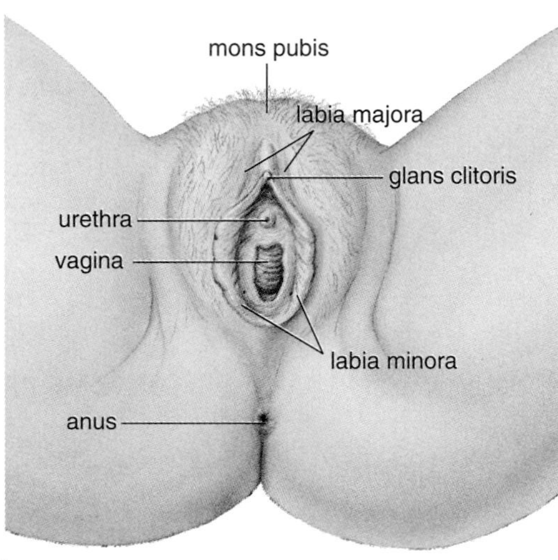

Figure 50.8 Female reproductive system.
a. Side view of organs. The ovaries produce one oocyte (egg) per month. Fertilization occurs in the oviduct, and development occurs in the uterus. The vagina is the birth canal and organ of copulation. **b.** Vulva. At birth, the opening of the vagina is partially occluded by a membrane called the hymen. Physical activities and sexual intercourse disrupt the hymen.

Ovaries Produce Oocytes and Hormones

The ovaries alternate in producing one oocyte each month. For the sake of convenience, the released oocyte is often called an **ovum**, or egg. The ovaries also produce the female sex hormones, **estrogens** and **progesterone,** during the ovarian cycle.

Ovarian Cycle

The **ovarian cycle** is described in Figure 50.9, a longitudinal section through an ovary. The ovary contains many cellular **follicles** [L. dim. of *folliculus,* bag], each containing an oocyte. At birth a female has as many as 2 million follicles, but the number is reduced to 300,000–400,000 by the time of puberty. Only a small number of follicles (about 400) ever mature, and produce an oocyte.

As the follicle undergoes maturation, it develops from a primary follicle to a secondary follicle to a *Graafian (vesticular)* follicle. **Oogenesis** has begun, and a secondary follicle contains a secondary oocyte with a reduced number of chromosomes. In a secondary follicle, the secondary oocyte is pushed to one side in a fluid-filled cavity. In a Graafian follicle, the fluid-filled cavity increases to the point that the follicle wall balloons out on the surface of the ovary and bursts, releasing the secondary oocyte surrounded by the zona pellucida (an amorphous glycoprotein layer) and follicular cells. The release of a secondary oocyte from a Graafian follicle is termed **ovulation**.

Oogenesis is completed when and if the secondary oocyte is fertilized by a sperm. In the meantime, the follicle is developing into the **corpus luteum** [L. *corpus,* body, and *luteus,* yellow]. If fertilization and pregnancy do not occur, the corpus luteum begins to degenerate after about ten days.

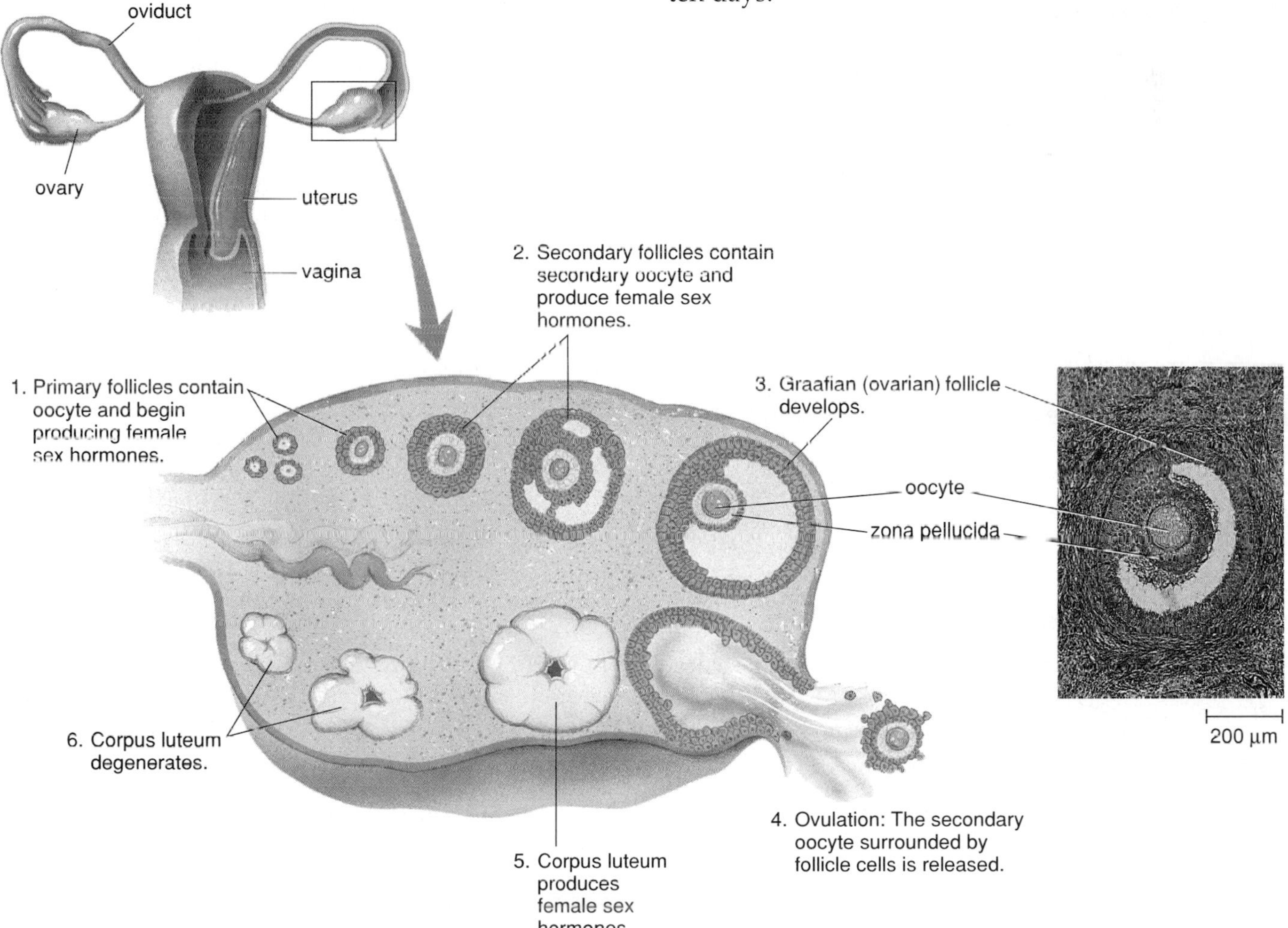
oviduct
ovary
uterus
vagina
2. Secondary follicles contain secondary oocyte and produce female sex hormones.
1. Primary follicles contain oocyte and begin producing female sex hormones.
3. Graafian (ovarian) follicle develops.
oocyte
zona pellucida
6. Corpus luteum degenerates.
5. Corpus luteum produces female sex hormones.
4. Ovulation: The secondary oocyte surrounded by follicle cells is released.
200 μm

Figure 50.9 Ovarian cycle.
As a follicle matures, the oocyte enlarges and is surrounded by layers of follicular cells and fluid. Eventually, ovulation occurs, the mature follicle ruptures, and an oocyte is released. A single follicle actually goes through all stages in one place within the ovary.

Table 50.3

Ovarian and Uterine Cycles (Simplified)

Ovarian Cycle	Events	Uterine Cycle	Events
Follicular phase—Days 1–13	FSH Follicle maturation Estrogens	Menstruation—Days 1–5 Proliferative phase—Days 6–13	Endometrium breaks down Endometrium rebuilds
Ovulation—Day 14*	LH spike		
Luteal phase—Days 15–28	LH Corpus luteum Progesterone	Secretory phase—Days 15–28	Endometrium thickens and glands are secretory

*Assuming a 28-day cycle

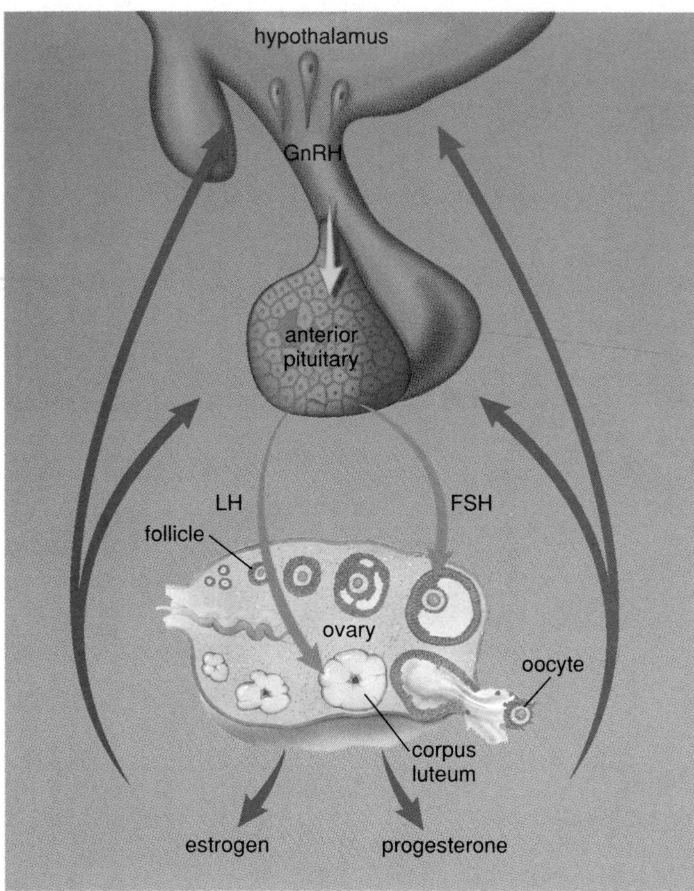

Figure 50.10 Hormonal control of ovaries.
GnRH is a hypothalamic-releasing hormone that stimulates the anterior pituitary to secrete FSH and LH. These gonadotropic hormones act on the ovaries. FSH promotes the development of the follicle that later, under the influence of LH, becomes the corpus luteum. Negative feedback controls the level of all hormones involved.

The ovarian cycle is under the control of the gonadotropic hormones, **follicle-stimulating hormone (FSH)** and **luteinizing hormone (LH)** (Fig. 50.10 and Table 50.3). The gonadotropic hormones are not present in constant amounts and instead are secreted at different rates during the cycle. For simplicity's sake, it is convenient to emphasize that during the first half, or **follicular phase,** of the cycle, FSH promotes the development of a follicle, which secretes estrogens. As the estrogen level in the blood rises, it exerts feedback control over the anterior pituitary secretion of FSH so that the follicular phase comes to an end.

Presumably, the high level of estrogens in the blood also causes the hypothalamus suddenly to secrete a large amount of GnRH. This leads to a surge of LH production by the anterior pituitary and to ovulation at about the fourteenth day of a 28-day cycle (Fig. 50.11).

During the second half, or **luteal phase,** of the ovarian cycle, it is convenient to emphasize that LH promotes the development of the corpus luteum, which secretes progesterone. Progesterone causes the uterine lining to build up. As the blood level of progesterone rises, it exerts feedback control over anterior pituitary secretion of LH so that the corpus luteum begins to degenerate. As the luteal phase comes to an end, menstruation occurs.

One ovarian follicle per month produces a secondary oocyte. Following ovulation, the follicle develops into the corpus luteum.

Uterine Cycle

The female sex hormones, estrogens and progesterone, have numerous functions. The effects of these hormones on the endometrium of the uterus causes the uterus to undergo a cyclical series of events known as the **uterine cycle** (Table 50.3 and Fig. 50.11). Twenty-eight-day cycles are divided as follows.

During *days 1–5,* there is a low level of female sex hormones in the body, causing the uterine lining to disintegrate and its blood vessels to rupture. A flow of blood, known as the **menses,** passes out of the vagina during **menstruation** [L. *menstrualis,* happening monthly], also known as the menstrual period.

During *days 6–13,* increased production of estrogens by an ovarian follicle causes the endometrium to thicken and to become vascular and glandular. This is called the proliferative phase of the uterine cycle.

Ovulation usually occurs on the fourteenth day of the 28-day cycle.

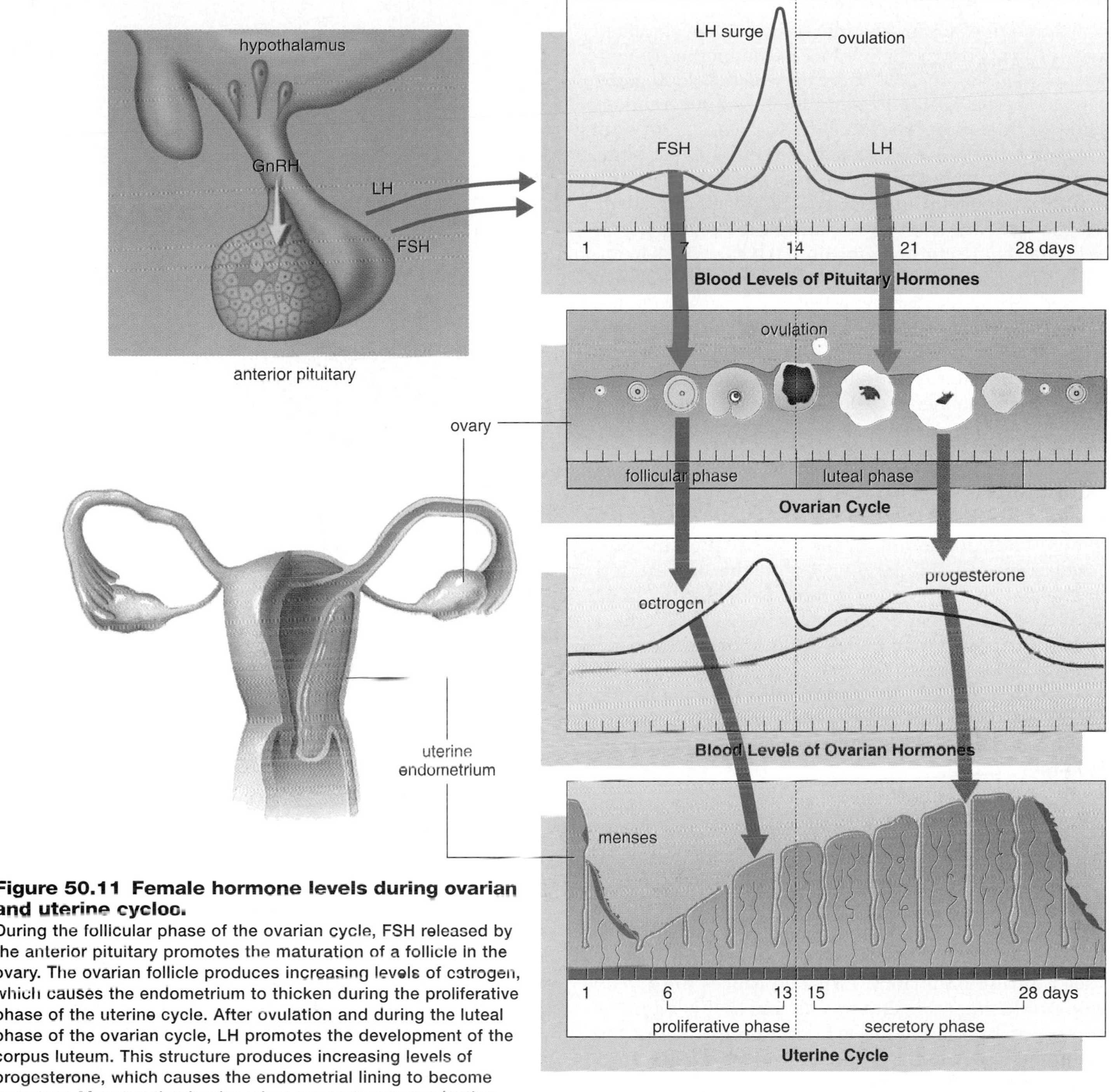

Figure 50.11 Female hormone levels during ovarian and uterine cycles.
During the follicular phase of the ovarian cycle, FSH released by the anterior pituitary promotes the maturation of a follicle in the ovary. The ovarian follicle produces increasing levels of estrogen, which causes the endometrium to thicken during the proliferative phase of the uterine cycle. After ovulation and during the luteal phase of the ovarian cycle, LH promotes the development of the corpus luteum. This structure produces increasing levels of progesterone, which causes the endometrial lining to become secretory. Menstruation begins when progesterone production declines to a low level.

During *days 15–28,* increased production of progesterone by the corpus luteum causes the endometrium to double in thickness and the uterine glands to mature, producing a thick mucoid secretion. This is called the secretory phase of the uterine cycle. The endometrium now is prepared to receive the developing embryo. If pregnancy does not occur, the corpus luteum degenerates and the low level of sex hormones in the female body causes the uterine lining to break down. The menstrual discharge begins at this time. Even while menstruation is occurring, the anterior pituitary begins to increase its production of FSH and a new follicle begins to mature. Table 50.3 indicates how the ovarian cycle controls the uterine cycle.

Events Following Fertilization

If fertilization does occur, an embryo begins development even as it travels down the oviduct to the uterus. The endometrium is now prepared to receive the developing embryo, which becomes embedded in the lining several days following fertilization. The *placenta* originates from both maternal and embryonic tissues. It is shaped like a large, thick pancake and is the site of exchange of gases and nutrients between fetal and maternal blood although there is rarely any mixing of the two. At first the placenta produces **human chorionic gonadotropin (HCG),** which maintains the corpus luteum until the placenta begins its own production of progesterone and estrogens. Progesterone and estrogens have two effects. They shut down the anterior pituitary so that no new follicles mature, and they maintain the lining of the uterus so that the corpus luteum is not needed. There is no menstruation during pregnancy.

Estrogens and Progesterone Are Female Sex Hormones

Estrogens in particular are essential for the normal development and functioning of the organs listed in Table 50.2. Estrogens are also largely responsible for the secondary sexual characteristics in females, including body hair and fat distribution. In general, females have a more rounded appearance than males because of a greater accumulation of fat beneath the skin. Also, the pelvic girdle enlarges so that females have wider hips than males and the thighs converge at a greater angle toward the knees. Both estrogen and progesterone are also required for breast development.

Breasts Produce Milk

A female breast contains between 15 and 24 lobules, each with its own mammary duct (Fig. 50.12). This duct begins at the nipple and divides into numerous other ducts, which end in blind sacs called *alveoli.* **Lactation** is the production of milk by the cells of the alveoli. Milk is not produced during pregnancy. Prolactin causes the alveoli to begin producing milk, and production of this hormone is suppressed by the feedback inhibition estrogens and progesterone have on the anterior pituitary during pregnancy. It takes a couple of days after delivery of a baby for milk production to begin. In the meantime, the breasts produce a watery, yellowish-white fluid called *colostrum,* which has a similar composition to milk but contains more protein and less fat. Colostrum (and later milk) is rich in IgA antibodies that may provide some degree of immunity to the newborn.

Breast cancer is the most common form of cancer in females. Women should regularly check their breasts for lumps and have mammograms (X-ray photographs) taken as recommended by their physician.

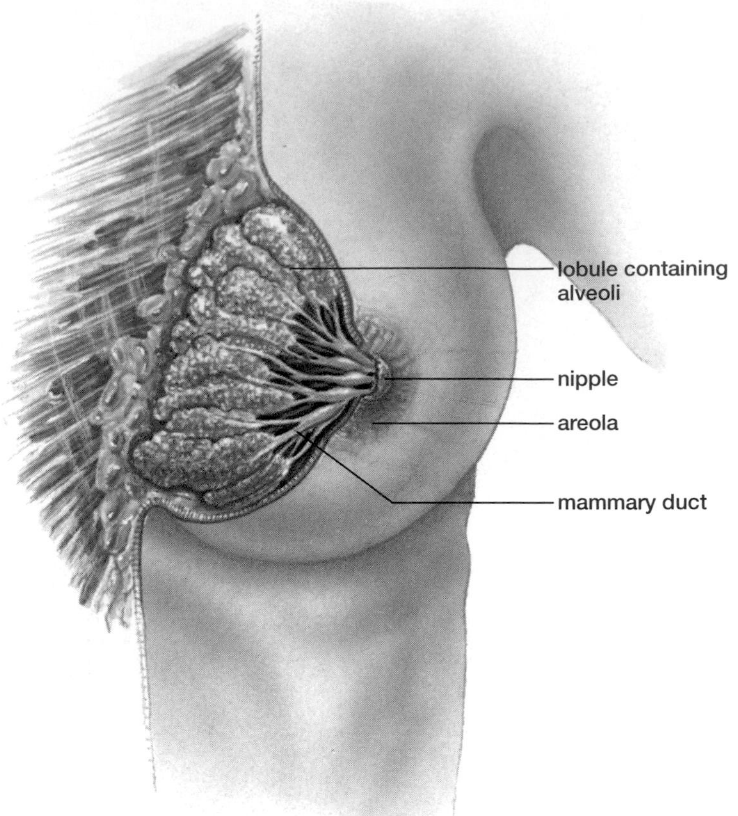

Figure 50.12 Anatomy of breast.
The female breast contains lobules consisting of ducts and alveoli. The alveoli are lined by milk-producing cells in the lactating (milk-producing) breast.

50.4 Humans Vary in Fertility

The two major causes of infertility in females are blocked oviducts, possibly due to a sexually transmitted disease, and failure to ovulate due to low body weight. *Endometriosis,* the spread of uterine tissue beyond the uterus, is also a cause. If no obstruction is apparent and body weight is normal, it is possible to give females HCG extracted from the urine of pregnant women, along with HMG (human menopausal gonadotropin) extracted from the urine of postmenopausal women. This treatment causes multiple ovulations and sometimes multiple pregnancies. The most frequent causes of sterility and infertility in males are low sperm count and/or a large proportion of abnormal sperm. Disease, radiation, chemical mutagens, too much heat near the testes, and the use of psychoactive drugs can contribute to this condition.

When reproduction does not occur in the usual manner, couples often seek alternative reproductive methods, which include artificial insemination (sperm are placed in the vagina by a physician), in vitro fertilization (fertilization takes place in laboratory glassware and the zygote is inserted into the uterus), and surrogate motherhood (a woman has another couple's child).

Table 50.4
Common Birth-Control Methods

Name	Procedure	Methodology	Effectiveness	Risk
Abstinence	Refrain from sexual intercourse	No sperm in vagina	100%	None
Vasectomy	Vasa deferentia cut and tied	No sperm in seminal fluid	Almost 100%	Irreversible sterility
Tubal ligation	Oviducts cut and tied	No eggs in oviduct	Almost 100%	Irreversible sterility
Oral contraception	Hormone medication is taken daily	Anterior pituitary does not release FSH and LH	Almost 100%	Thromboembolism, especially in smokers
Depo-Provera injection	Four injections of progesterone-like steroid given per year	Anterior pituitary does not release FSH and LH	About 99%	Breast cancer? Osteoporosis?
Contraceptive implants	Tubes of progestin (form of progesterone) implanted under skin	Anterior pituitary does not release FSH and LH	More than 90%	Presently none known
Intrauterine device (IUD)	Plastic coil inserted into uterus by physician	Prevents implantation	More than 90%	Infection (pelvic inflammatory disease, PID)
Diaphragm	Latex cup inserted into vagina to cover cervix before intercourse	Blocks entrance of sperm to uterus	With jelly, about 90%	Presently none known
Cervical cap over cervix	Latex cap held by suction	Delivers spermicide near cervix	Almost 85%	Cancer of cervix
Male condom	Latex sheath fitted over erect penis	Traps sperm and prevents STDs	About 85%	Presently none known
Female condom	Polyurethane liner fitted inside vagina	Blocks entrance of sperm to uterus and prevents STDs	About 85%	Presently none known
Coitus interruptus	Penis withdrawn before ejaculation	Prevents sperm from entering vagina	75%	Presently none known
Jellies, creams, foams	These spermicidal products inserted before intercourse	Kills a large number of sperm	About 75%	Presently none known
Natural family planning	Day of ovulation determined by record keeping; various methods of testing	Intercourse avoided on certain days of the month	About 70%	Presently none known
Douche	Vagina and uterus cleansed after intercourse	Washes out sperm	Less than 70%	Presently none known

Birth-Control Methods

Sometimes couples wish to prevent a possible pregnancy. The most reliable method of birth control is abstinence; that is, the absence of sexual intercourse. This form of birth control has the added advantage of preventing transmission of a sexually transmitted disease. Other, perhaps more common, means of birth control used in the United States, are listed in Table 50.4. The effectiveness of the method for all but abstinence refers to the number of women per year who will not get pregnant even though they are regularly engaging in sexual intercourse. The male and female condom also offer some protection against sexually transmitted diseases.

Investigators have long searched for a "male pill." Analogues of gonadotropic-releasing hormone prevent the hypothalamus from stimulating the anterior pituitary. Feminization due to low testosterone blood level has thus far made these possible birth-control methods undesirable. Inhibin has also been shown to inhibit spermatogenesis in males, but this hormone must be administered by injection.

Morning-After Pills

There are morning-after regimens available that, depending on when the woman begins medication, either prevent fertilization or stop the fertilized egg from ever implanting. These regimens involve taking pills containing synthetic progesterone and/or estrogen in a manner prescribed by a physician. Many women do not realize that this method of birth control is available, and yet the use of these regimens could greatly reduce the number of unintended pregnancies. Effective treatment sometimes causes nausea and vomiting, which can be severe.

Mifepristone, better known as RU-486, is a pill that causes the loss of an implanted embryo by blocking the progesterone receptors of the cells in the uterine lining. Without functioning receptors for progesterone, the uterine lining sloughs off, carrying the embryo with it. When taken in conjunction with a prostaglandin, a substance that induces uterine contractions, the drug is 95% effective. The pill can be taken by women who are experiencing delayed menstruation without knowing if they are actually pregnant.

Health focus

Sexually Transmitted Diseases

Sexually transmitted diseases (STDs) are caused by organisms ranging from viruses to arthropods; however, we will discuss only certain STDs caused by viruses and bacteria. Unfortunately, for unknown reasons, humans cannot develop good immunity to any STDs. Therefore, prompt medical treatment should be received when exposed to an STD. To prevent the spread of STDs, a latex condom can be used; the concomitant use of a spermicide containing nonoxynol 9 gives added protection.

It is difficult to cure the STDs caused by viruses (e.g., AIDS, genital herpes, and genital warts), but treatment is available for AIDS and genital herpes. Those STDs caused by bacteria (e.g., gonorrhea, chlamydia, and syphilis) are treatable with antibiotics.

AIDS Is Preventable

The organism that causes acquired immunodeficiency syndrome (AIDS) is a virus called **human immunodeficiency virus (HIV).** HIV attacks the type of lymphocyte known as helper T cells. Helper T cells, you will recall, stimulate the activities of B lymphocytes, which produce antibodies. After an HIV infection sets in, helper T cells begin to decline in number, and the person becomes more susceptible to other types of infections.

Symptoms

AIDS has three stages of infection called category A, B, and C. During a category A stage, which may last about a year, the individual is an asymptomatic carrier. There may be no symptoms, but the individual can pass on the infection. Immediately after infection, and before the blood test becomes positive, there is a large number of infectious viruses in the blood that could be passed on to another person. Even after the blood test becomes positive, the person remains well as long as the body produces sufficient helper T lymphocytes to keep the count higher than 500 mm^3. During the category B stage, which may last six to eight years, the lymph nodes swell and there may also be weight loss, night sweats, fatigue, fever, and diarrhea. Infections like thrush (white sores on the tongue and in the mouth) and herpes reoccur. Finally, the person may progress to category C, which is full-blown AIDS characterized by nervous disorders and by the development of an opportunistic disease such as an unusual type of pneumonia or skin cancer. Opportunistic diseases are ones that occur only in individuals who have little or no capability of fighting an infection. Without intensive medical treatment, the AIDS patient dies about seven to nine years after infection. Now, with a combination therapy of several drugs, AIDS patients are beginning to live longer in the United States.

Transmission

AIDS is transmitted by sexual contact with an infected person, including vaginal or rectal intercourse and oral/genital contact. Also, needle-sharing among intravenous drug users is high-risk behavior. A less common mode of transmission (and now rarely in countries where donated blood is screened for HIV) is through transfusions of infected blood or blood-clotting factors.

HIV first spread through the homosexual community, and male-to-male sexual contact still accounts for the largest percentage of new AIDS cases in the United States. But the largest increases of HIV infections are occurring through heterosexual contact or by intravenous drug use. Now, women account for 19% of all newly diagnosed cases of AIDS. The rise in the incidence of AIDS among women of reproductive age is paralleled by a rise in the incidence of AIDS in children younger than 13. Babies born to HIV-infected women may become infected before or during birth, or through breast-feeding after birth.

Genital Warts Can Cause Cancer

Genital warts are caused by the human papillomaviruses (HPVs). Many times, carriers do not have any sign of warts, or merely flat lesions may be present. When present, the warts commonly are seen on the penis and foreskin of men and near the vaginal opening in women. A newborn can become infected by passage through the birth canal.

Presently, there is no cure for an HPV infection, but it can be treated effectively by surgery, freezing, application of an acid, or laser burning, depending on severity. If visible warts are removed, they may recur. Genital warts are associated with cancer of the cervix, as well as tumors of the vulva, the vagina, the anus, and the penis. Some researchers believe that the viruses are involved in 90–95% of all cases of cancer of the cervix.

Genital Herpes Can Reoccur

Genital herpes is caused by herpes simplex virus—type 1 usually causes cold sores and fever blisters, while type 2 more often causes genital herpes.

Persons usually get infected with herpes simplex virus type 2 when they are adults. In some people there are no symptoms. Or there may be a tingling or itching sensation before blisters appear on the genitals. Once the blisters rupture, they leave painful ulcers that may take as long as three weeks or as little as five days to heal. The blisters may be accompanied by fever, pain on urination, swollen lymph nodes in the groin, and in women, a copious discharge. At this time, the individual has an increased risk of acquiring an AIDS infection.

After the ulcers heal, the disease is only latent, and blisters can recur although usually at less frequent intervals and with milder symptoms. Fever, stress, sunlight, and menstruation are associated with reoccurrence of symptoms. Exposure to herpes in the birth canal can cause an infection in the newborn, which leads to neurological disorders and even death. Birth by cesarean section prevents this possibility.

Hepatitis Infections

There are several types of hepatitis. The type of hepatitis and the virus that causes it is designated by the same letter. Hepatitis A is usually acquired from sewage-contaminated drinking water, but this infection can also be sexually transmitted through oral/anal contact. Hepatitis B, which is spread in the same manner as AIDS, is even more infectious. Fortunately, a vaccine is now available for hepatitis B. Hepatitis C is called post-transfusion hepatitis. Hepatitis infections cause infection of the liver and can lead to liver failure, liver cancer, and death.

Chlamydia Can Cause PID

Chlamydia is named for the tiny bacterium that causes it *(Chlamydia trachomatis).* The incidence of new chlamydial infections is higher than any other sexually transmitted disease (Fig. 50B).

Chlamydial infections of the lower reproductive tract usually are mild or asymptomatic, especially in women. About 8 to 21 days after infection, men may experience a mild burning sensation on urination and a mucoid discharge. Women may have a vaginal discharge along with the symptoms of a urinary tract infection. Chlamydia also causes cervical ulcerations, which increase the risk of acquiring AIDS.

If the infection is misdiagnosed, or the person does not seek medical help, there is a particular risk of the infection spreading from the cervix to the oviducts so that pelvic inflammatory disease (PID) results. This very painful condition can result in a blockage of the oviducts with the possibility of sterility and infertility. If a baby comes in contact with chlamydia during birth, inflammation of the eyes or pneumonia can result.

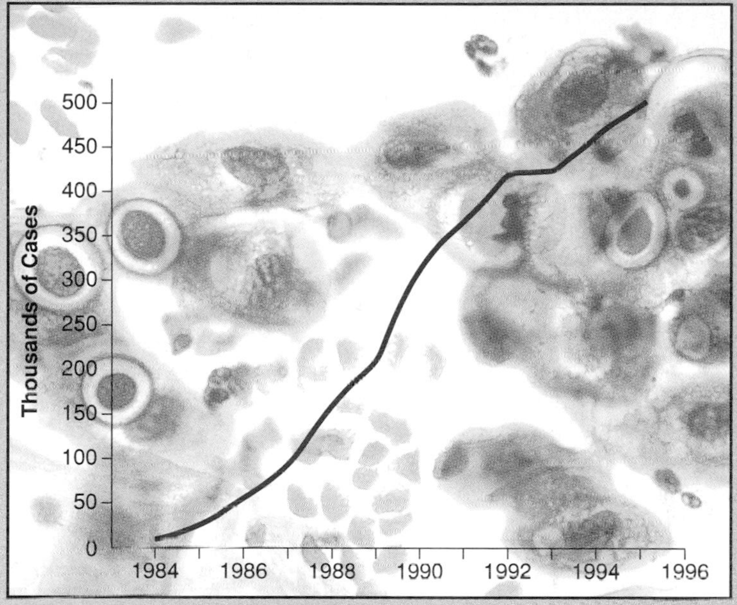

Figure 50B Chlamydial infection.
A graph depicting the incidence of new cases of chlamydia in the United States from 1984 to 1995 is superimposed on a micrograph of a cell containing different stages of the organism.

Source: Sexually Transmitted Disease Surveillance, 1994. Atlanta: Centers for Disease Control and Prevention, September 1995.

Gonorrhea Can Also Cause PID

Gonorrhea is caused by the bacterium *Neisseria gonorrhoeae.* Diagnosis in the male is not difficult, since typical symptoms are pain upon urination and a thick, greenish yellow urethral discharge. In males and females, a latent infection leads to pelvic inflammatory disease (PID), in which the vasa deferentia or oviducts are affected. As the inflamed tubes heal, they may become partially or completely blocked by scar tissue, resulting in sterility or infertility. If a baby is exposed during birth, an eye infection leading to blindness can result. All newborns are given eyedrops to prevent this possibility.

Gonorrhea proctitis, an infection of the anus, with symptoms including anal pain and blood or pus in the feces, also occurs in patients. Oral/genital contact can cause infection of the mouth, throat, and tonsils. Gonorrhea can spread to internal parts of the body, causing heart damage or arthritis. If, by chance, the person touches infected genitals and then touches his or her eyes, a severe eye infection can result. Up to now, gonorrhea was curable by antibiotic therapy, but resistance to antibiotic therapy is becoming more and more common, and 40% of all strains are now known to be resistant to therapy.

Syphilis Has Three Stages

Syphilis, which is caused by the bacterium *Treponema pallidum,* has three stages, which are typically separated by latent periods. In the primary stage, a hard chancre (ulcerated sore with hard edges) appears. In the secondary stage, a rash appears all over the body—even on the palms of the hands and the soles of the feet. During the tertiary stage, syphilis may affect the cardiovascular and/or nervous system. An infected person may become mentally retarded, become blind, walk with a shuffle, or show signs of insanity. Gummas, which are large destructive ulcers, may develop on the skin or within the internal organs. Syphilitic bacteria can cross the placenta, causing birth defects or a stillbirth. Unlike the other STDs discussed, there is a blood test to diagnose syphilis.

Syphilis is a very devastating disease. Control depends on prompt and adequate treatment of all new cases; therefore, it is very important for all sexual contacts to be traced so they can be treated with antibiotic therapy.

Vaginitis

Women are subject to vaginitis, an infection that is caused by the flagellated protozoa *Trichomonas vaginalis* or by the yeast *Candida albicans*. The protozoan infection causes a frothy white or yellow foul-smelling discharge accompanied by itching, and the yeast infection causes a thick, white, curdy discharge, also accompanied by itching. *Trichomonas* is most often acquired through sexual intercourse, and an asymptomatic partner is usually the reservoir of infection. *Candida albicans,* however, is a normal organism found in the vagina; its growth simply increases beyond normal under certain circumstances. Women taking the birth-control pill or who have been on antibiotic therapy are sometimes prone to yeast infections.

connecting concepts

The first chapter of this text stated that science does not decide ethical issues and that society at large had to make such decisions. There are many ethical decisions associated with reproduction in humans. First, there is the problem of overpopulation. The public is often under the impression that overpopulation is a worldwide problem. Actually, the population of most developed industrialized countries of Europe, parts of the Far East, and to a lesser extent North America, are growing slowly, or not at all, or even declining in size if the death rate exceeds the birthrate. Population growth is a problem in the developing countries of the world. We are told these countries will increase from 1 billion in 1900 to nearly 5 billion in 2000. According to a report issued by the Johns Hopkins University, one in every five women in the developing countries have had all the children they want but are not using birth control due to lack of information, poor access to quality services, and opposition from their husbands. Should the developed countries play a role in providing birth-control information and services?

The United States may not have an overpopulation problem, but there are still many ethical issues to be considered regarding regulation of reproduction. For example, should sex education be a part of a school's curriculum in order to curtail teenage pregnancies and the transmission of sexually transmitted diseases? If so, at what age should this education begin? Of all medical issues in the past decade, nothing has caused more discord than abortion. Abortion is the termination of pregnancy before the fetus is capable of surviving. Although induced abortions are not considered a preferred means of birth control, they are legally available in most states to women who can afford them. The debate about whether to use public funds to make abortion available to all women still continues. And there are some who believe that abortion should not be legally available to anyone.

About 15% of couples in this country are termed sterile because they can't have any children, and another 10% are termed infertile because they have fewer children than they wish to have. To what lengths should society go to help these couples have children? And is it moral to use methods like in vitro fertilization, accompanied perhaps by gene therapy to ensure the birth of a healthy child? And should society impose an age limit for having children? Today, in vitro fertilization followed by implantation into a hormonally primed uterus is a possibility for women who have been postmenopausal for some time.

Biologists had long thought that the cloning of mammals would be impossible, but now that Scottish researchers have cloned a sheep from a mature diploid cell, the possibility exists that one day cloning might be another means of reproducing humans. The procedure used by the Scottish researchers, which is described on page 897, would produce a baby with exactly the same genes as the individual who donated the cell. But as this baby grew to adulthood, it would have new and different environmental influences than the parent had. The same ethical considerations that surround in vitro fertilization also pertain to the cloning of humans. Under what circumstances should it be allowed, when and if it becomes possible?

Summary

50.1 How Animals Reproduce

Ordinarily, asexual reproduction may quickly produce a large number of offspring genetically identical to the parent. This is advantageous when environmental conditions are static.

Sexual reproduction involves gametes and produces offspring that are genetically slightly different from the parents. This may be advantageous if the environment is changing. The gonads are the primary sex organs, but there are also accessory organs. The accessory organs consist of storage areas for sperm and ducts that conduct the gametes.

Animals typically protect their eggs and embryos. Those that are oviparous provide them with yolk, and if the animal is terrestrial there is typically a shell to prevent drying out. The amount of yolk is dependent on whether there is a larval stage.

Reptiles and birds have extraembryonic membranes that allow vertebrates to develop on land; these same membranes are modified for internal development in mammals. Ovoviviparous animals retain their eggs until the offspring have hatched, and viviparous animals retain the embryo. Placental mammals exemplify viviparous animals.

50.2 Males Have Testes

In human males, sperm are produced in the testes, mature in the epididymides, and may be stored in the vasa deferentia before entering the urethra, along with seminal fluid (produced by seminal vesicles, prostate gland, and bulbourethral glands). Sperm are ejaculated during male orgasm, when the penis is erect.

Spermatogenesis occurs in the seminiferous tubules of the testes, which also produce testosterone in interstitial cells. Testosterone brings about the maturation of the primary sex organs during puberty and promotes the secondary sexual characteristics of males, such as low voice, facial hair, and increased muscle strength.

Follicle-stimulating hormone (FSH) from the anterior pituitary stimulates spermatogenesis, and luteinizing hormone (LH, also called ICSH) stimulates testosterone production. A hypothalamic-releasing hormone, gonadotropic-releasing hormone (GnRH), controls anterior pituitary production and FSH and LH release. The level of testosterone in the blood controls the secretion of GnRH and the anterior pituitary hormones by a negative feedback system.

50.3 Females Have Ovaries

In females, an oocyte produced by an ovary enters an oviduct, which leads to the uterus. The uterus opens into the vagina. The external genital area of women includes the vaginal opening, the clitoris, the labia minora, and the labia majora.

In either ovary, one follicle a month matures, produces a secondary oocyte, and becomes a corpus luteum. This is called the ovarian cycle. The follicle and the corpus luteum produce estrogens and progesterone, the female sex hormones.

The uterine cycle occurs concurrently with the ovarian cycle. In the first half of these cycles (days 1–13, before ovulation), the anterior pituitary produces FSH and the follicle produces estrogens. Estrogens cause the uterine lining to increase in thickness. In the second half of these cycles (days 15–28, after ovulation), the anterior pituitary produces LH and the follicle produces progesterone. Progesterone causes the uterine lining to become secretory. Feedback control of the hypothalamus and anterior pituitary causes the levels of estrogens and progesterone to fluctuate. When they are at a low level, menstruation begins.

If fertilization occurs, a zygote is formed and development begins. The resulting embryo travels down the oviduct and implants itself in the prepared uterine lining. A placenta, which is the region of exchange between the fetal blood and mother's blood, forms. At first, the placenta produces HCG, which maintains the corpus luteum; later, it produces progesterone and estrogens.

Estrogens and progesterone are the female sex hormones. Primarily estrogens bring about the maturation of the primary sex organs during puberty and promotes the secondary sexual characteristics of females, including less body hair than males, a wider pelvic girdle, a more rounded appearance, and development of breasts.

50.4 Humans Vary in Fertility

Infertile couples are increasingly resorting to alternative methods of reproduction. Numerous birth-control methods and devices are available for those who wish to prevent pregnancy.

Reviewing the Chapter

1. Give examples of asexual and sexual reproduction among animals. 894
2. What are the primary sex organs and the usual accessory sex organs? Give examples in reference to the earthworm. How does the reproductive system of earthworms differ from that of most animals? 894–95
3. Describe the life cycle of an insect utilizing these terms: internal fertilization, oviparous, larva, and metamorphosis. Would you expect an insect egg to contain much yolk? Why or why not? 895
4. Outline the path of sperm. What glands contribute fluids to semen? 898–99
5. Discuss the anatomy and physiology of the testes. Describe the structure of sperm. 900–1
6. Name the endocrine glands involved in maintaining the sexual characteristics of males and the hormones produced by each. 901
7. Name two functions of the vagina. 902
8. Discuss the anatomy and physiology of the ovaries. Describe ovulation. 903–4
9. Outline the path of the oocyte (egg). Where do fertilization and implantation occur? 903–4
10. Discuss hormonal regulation in the female by listing the events of the uterine cycle and relating these to the ovarian cycle. In what way is menstruation prevented if pregnancy occurs? 903–5
11. What means of birth control help prevent the spread of AIDS? What other measures can be taken to protect oneself from AIDS? 907–8

Testing Yourself

Choose the best answer for each question.

1. Which of these is a requirement for sexual reproduction?
   a. male and female parents
   b. production of gametes
   c. optimal environmental conditions
   d. aquatic habitat
2. Internal fertilization
   a. prevents the drying out of gametes and zygotes.
   b. must take place on land.
   c. is practiced by humans.
   d. Both a and c are correct.
3. Which of these is mismatched?
   a. interstitial cells—testosterone
   b. seminiferous tubules—sperm production
   c. vasa deferentia—seminal fluid production
   d. urethra—conducts sperm
4. Follicle-stimulating hormone (FSH)
   a. is secreted by females but not males.
   b. stimulates the seminiferous tubules to produce sperm.
   c. secretion is controlled by gonadotropic-releasing hormone (GnRH).
   d. Both b and c are correct.
5. Which of these combinations is most likely to be present before ovulation occurs?
   a. FSH, corpus luteum, estrogen, secretory uterine lining
   b. luteinizing hormone (LH), follicle, progesterone, thick uterine lining
   c. FSH, follicle, estrogen, uterine lining becoming thick
   d. LH, corpus luteum, progesterone, secretory uterine lining
6. In tracing the path of sperm, you would mention vasa deferentia before
   a. testes.
   b. epididymides.
   c. urethra.
   d. uterus.
7. An oocyte is fertilized in the
   a. vagina.
   b. uterus.
   c. oviduct.
   d. ovary.
8. During pregnancy
   a. the ovarian and uterine cycles occur more quickly than before.
   b. GnRH is produced at a higher level than before.
   c. the ovarian and uterine cycles do not occur.
   d. the female secondary sexual characteristics are not maintained.

9. Which of the following means of birth control is most effective in preventing AIDS?
   a. condom
   b. pill
   c. diaphragm
   d. spermicidal jelly
10. Label this diagram of the male reproductive system and trace the path of sperm:

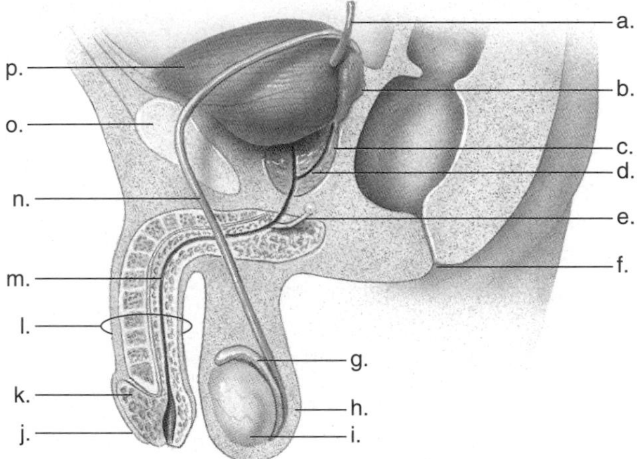

Using Technology

Your study of reproduction is supported by these available technologies:

Exploring the Internet
The Mader Home Page provides resources for and help with studying this chapter.

http://www.mhhe.com/sciencemath/biology/mader/
(Click on Biology.)

The Dynamic Human CD-ROM
Reproductive System

Explorations in Human Biology CD-ROM
AIDS (#13)

Applying the Concepts

1. *Successful reproduction on land requires certain adaptations.*
   Contrast the manner in which reptiles are adapted to reproduce on land with the manner in which humans are adapted to reproduce on land.
2. *Reproduction is under hormonal rather than nervous control.*
   Why would you have predicted hormonal rather than nervous control of reproduction?

Understanding the Terms

asexual reproduction 894
clitorius 902
copulation 894
corpus luteum 903
ejaculation 899
endometrium 902
epididymis 898
erection 899
estrogen 903
extraembryonic membrane 895
fimbria 902
follicle 903
follicle-stimulating hormone (FSH) 904
follicular phase 904
gonad 894
human chorionic gonadotropin (HCG) 906
human immunodeficiency virus (HIV) 908
lactation 906
larva 895
luteal phase 904
luteinizing hormone (LH) 904
menses 904
menstruation 904
metamorphosis 895
oocyte 902
oogenesis 903
orgasm 899
ovarian cycle 903
ovary 894
oviduct 902
ovulation 903
ovum 903
parthenogenesis 894
penis 898
placenta 896
progesterone 903
puberty 901
refractory period 899
scrotum 898
secondary sexual characteristic 901
semen (seminal fluid) 899
seminiferous tubule 901
sexual reproduction 894
sperm 901
spermatogenesis 901
testis 894
testosterone 901
uterine cycle 904
uterus 902
vagina 902
vas deferens 898
vulva 902
yolk 895

Match the terms to these definitions:

a. __________ Bursting of a follicle when an oocyte is released from the ovary.
b. __________ Change in shape and form that some animals, such as insects, undergo during development.
c. __________ Development of an egg cell into a whole organism without fertilization.
d. __________ Female sex hormone that causes the endometrium of the uterus to become secretory during the uterine cycle; along with estrogen, it maintains secondary sexual characteristics in females.
e. __________ Thick, whitish fluid consisting of sperm and secretions from several glands of the male reproductive tract.
f. __________ Organ that produces gametes; the ovary, which produces eggs, and the testis, which produces sperm.

Development

CHAPTER 51

Chapter Concepts

51.1 Development Has Stages

- Development begins when a sperm fertilizes an egg. 914
- The first stages of embryonic development in animals lead to the establishment of the embryonic germ layers. 915
- The presence of yolk affects the manner in which animal embryos go through the early developmental stages. 916
- In vertebrates, the nervous system develops above the notochord after formation of a neural tube. 917

51.2 Cells Become Specialized and Embryos Take Shape

- Cellular differentiation and morphogenesis are two processes that occur when specialized organs develop. 918
- Induction explains development of the nervous system and why development in general is an orderly process. 919
- Homeotic genes are involved in shaping the outward appearance of the developed animal. 921

51.3 Humans Are Embryos and Then Fetuses

- Humans, like chicks, are dependent upon extraembryonic membranes that perform various services and contribute to development. 922
- During the embryonic period of human development, all systems appear. 923
- Humans are placental mammals; the placenta is a unique organ where exchange between fetal blood and mother's blood takes place. 926
- During fetal development, the fetus grows large enough to live on its own. Birth is a multistage process that includes delivery of the child and the extraembryonic membranes. 927

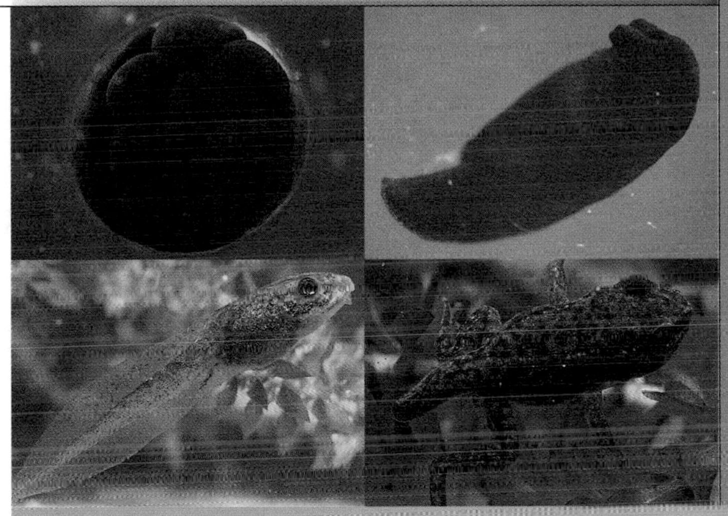

Development of a frog, *Rana* sp.

Every one of your billions of cells is totipotent; that is, each cell contains a copy of all your genes. Yet nerve cells have axons and dendrites, not myofibrils like muscle cells, and your bone cells deposit calcium salts, not skin pigment. The process whereby cells become specialized in structure and function, called cellular differentiation, is a significant part of development. In an early embryo, all the cells seem to be the same, yet as cell division proceeds, specialization becomes apparent. How does this occur, since all cells contain the same genes? Scientists do not have all the answers to this question, but they do know that signaling molecules play a role. Specific molecules induce specific genes to become active at different times. Also, as some cells differentiate, they migrate within the embryo, and as they move, they signal cells around them to migrate. Migration of cells contributes to the formation of embryonic organs.

The same processes observed during embryological development are also seen as the newly born or hatched organism matures, as lost parts regenerate, as a wound heals, and even as organisms age. Therefore, it has become increasingly clear that the study of development encompasses not only embryology (development of the embryo) but these other events as well.

51.1 Development Has Stages

Fertilization [L. *fertilis,* fruitful], which results in a zygote, requires that the sperm and egg interact. Figure 51.1 shows the manner in which an egg is fertilized by a sperm in sea stars. The sperm has three distinct parts: a head, a middle piece, and a tail. The tail is a flagellum, which allows the sperm to swim toward the egg, and the middle piece contains ATP-producing mitochondria. The head contains a haploid nucleus capped by a membrane-bounded acrosome containing enzymes that allow the sperm to penetrate the egg.

Several mechanisms have evolved to assure that fertilization takes place and in a species-specific manner (Fig. 51.1). A male releases so many sperm that the egg is literally covered by them. The sea star egg has a plasma membrane, a glycoprotein layer called the vitelline envelope, and a jelly coat. The acrosome enzymes digest away the jelly layer as the acrosome extrudes a filament that attaches to receptors located in the vitelline envelope. This is a lock-and-key reaction that is species-specific. Then, the egg plasma membrane and sperm nuclear membrane fuse, allowing the sperm nucleus to enter. Fusion takes place, and the zygote begins development.

As soon as the plasma membrane of sperm and egg fuse, the plasma membrane and the vitelline envelope undergo changes that prevent the entrance of any other sperm. The vitelline envelope is now called the fertilization envelope.

Early Developmental Stages

All chordate embryos go through the same early developmental stages of zygote, morula, blastula, early gastrula, and late gastrula. The presence of **yolk,** which is dense nutrient material, however, affects the manner in which embryonic cells complete the first three stages, and hence the appearance of the **embryo** at the end of each stage.

Following fertilization, the zygote undergoes **cleavage,** which is cell division without growth (Fig. 51.2). DNA replication and mitotic cell division occur repeatedly, and the cells get smaller with each division. Sea stars and lancelets, being deuterostomes, have a pattern of cleavage that is radial and indeterminate. The term *radial* means that any plane passing through the major axis will divide the em-

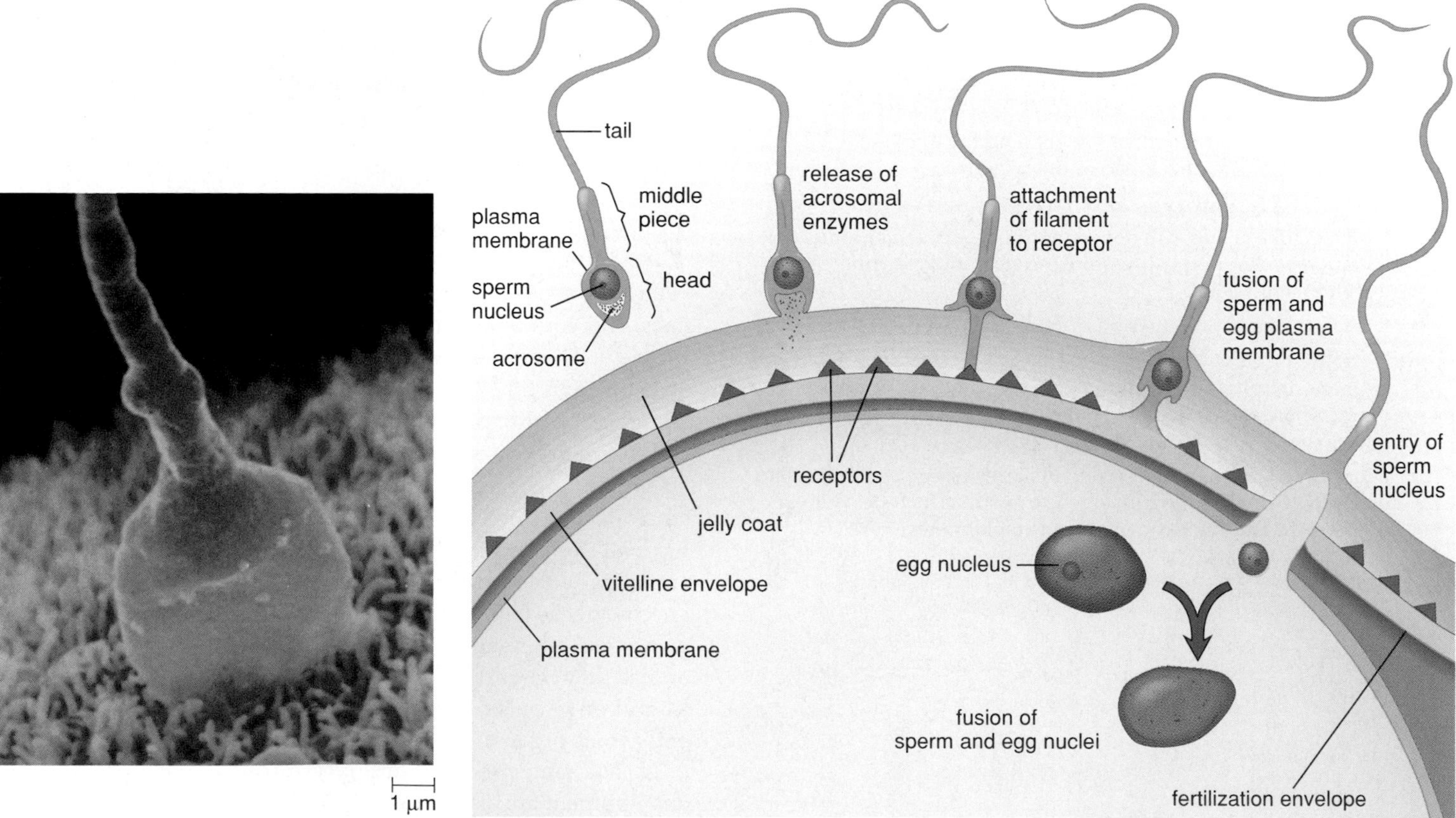

Figure 51.1 Fertilization of a sea star egg.
A head of a sperm has a membrane-bounded acrosome filled with enzymes. When released, these enzymes digest away the jelly coat around the egg, and the acrosome extrudes a filament that attaches to a receptor on the vitelline envelope. Now the sperm nucleus enters and fuses with the egg nucleus, and the resulting zygote begins to divide. The vitelline envelope becomes the fertilization envelope, which prohibits any more sperm from entering the egg.

bryo into two symmetrical halves. The term *indeterminate* means that the cleavage cells have not differentiated, and therefore their developmental fate is not yet set.

Because the lancelet has little yolk, the cell divisions are equal, and the cells are of uniform size in the resulting **morula** [L. dim. of *morus,* mulberry]. Then a cavity called the **blastocoel** [Gk. *blastos,* bud, and *koiloma,* cavity] develops and a hollow ball of cells known as the **blastula** [Gk. dim. of *blastos,* bud, and L. *ula,* little] forms.

The **gastrula** [Gk. dim. of *gastros,* stomach] stage is evident in a lancelet when certain cells begin to push, or invaginate, into the blastocoel, creating a double layer of cells. The outer layer is called the **ectoderm,** and the inner layer is called the **endoderm.** The space created by invagination will become the gut, but at this point it is termed either the primitive gut or the *archenteron.* The pore, or hole, created by invagination is the *blastopore,* and in a lancelet the blastopore eventually becomes the anus.

Gastrulation is not complete until three layers of cells are present. The third, or middle, layer of cells is called the **mesoderm.** In a lancelet, this layer begins as outpocketings from the primitive gut. These outpocketings grow in size until they meet and fuse. In effect then, two layers of mesoderm are formed, and the space between them is the coelom.

Ectoderm, mesoderm, and endoderm are called the embryonic **germ layers** of the embryo. No matter how gastrulation takes place, the end result is the same: three germ layers are formed. It is possible to relate the development of future organs to these germ layers:

Embryonic Germ Layer	Vertebrate Adult Structures
Ectoderm (outer layer)	Epidermis of skin; epithelial lining of mouth and rectum; nervous system
Mesoderm (middle layer)	Skeleton; muscular system; dermis of skin; circulatory system; excretory system; reproductive system—including most epithelial linings; outer layers of respiratory and digestive systems
Endoderm (inner layer)	Epithelial lining of digestive tract and respiratory tract; associated glands of these systems; epithelial lining of urinary bladder

Karl E. Von Baer, the nineteenth-century embryologist, first related development to the formation of germ layers. This is called the *germ layer theory.*

The three embryonic germ layers arise during gastrulation, when cells invaginate into the blastocoel. The development of organs can be related to the three germ layers: ectoderm, mesoderm, and endoderm.

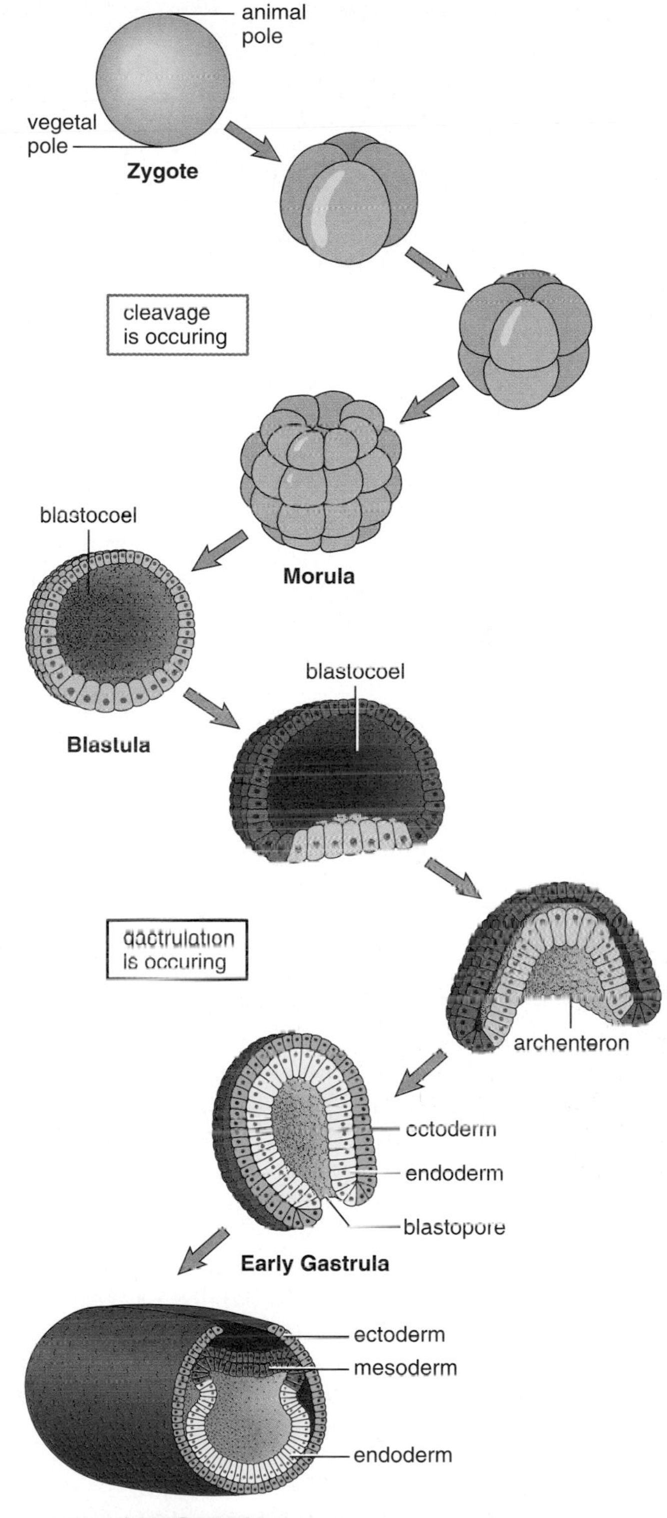

Figure 51.2 Lancelet early development.
A lancelet has little yolk as an embryo, and it can be used to exemplify the early stages of development in such animals. Cleavage produces a number of cells that form a cavity. Invagination during gastrulation produces the germ layers ectoderm and endoderm. Mesoderm arises from pouches that pinch off from the endoderm.

How Yolk Affects the Stages

Table 51.1 indicates the amount of yolk in four types of embryos and relates the amount of yolk to the environment in which the animal develops. The lancelet and frog develop in water, and they have less yolk than the chick because their development proceeds quickly to a swimming larval stage that can feed itself. The chick is representative of vertebrate animals that can develop on land because they lay a hard-shelled egg that contains plentiful yolk. Development continues in the shell until there is an offspring capable of land existence.

Early stages of human development resemble those of the chick embryo, yet this resemblance cannot be related to the amount of yolk because the human egg contains little yolk. But the evolutionary history of these two animals can provide an answer for this similarity. Both birds and mammals are related to reptiles. This explains why all three groups develop similarly, despite a difference in the amount of yolk in the eggs.

Figure 51.3 compares the appearance of early developmental stages in the lancelet, the frog, and the chick. In the frog embryo, cells at the animal pole have little yolk while those at the vegetal pole contain more yolk. The presence of yolk causes cells to cleave more slowly, and you can see that the cells of the animal pole are smaller than those of the vegetal pole. In the chick, cleavage is incomplete—only those cells lying on top of the yolk cleave. This means that although cleavage in the lancelet and the frog results in a morula, no such ball of cells is seen in the chick (Fig. 51.3*a*). Instead, during the morula stage the cells spread out on a portion of the yolk.

In the frog, the blastocoel is formed at the animal pole only. The heavily laden yolk cells of the vegetal pole do not participate in this step. In a chick, the blastocoel is created when the cells lift up from the yolk and leave a space between the cells and the yolk (Fig. 51.3*b*).

In the frog, the cells containing yolk do not participate in gastrulation and therefore they do not invaginate. Instead, a slitlike blastopore is formed when the animal pole cells begin to invaginate from above. Following this, other animal pole cells move down over the yolk, and the blastopore becomes rounded when these cells also

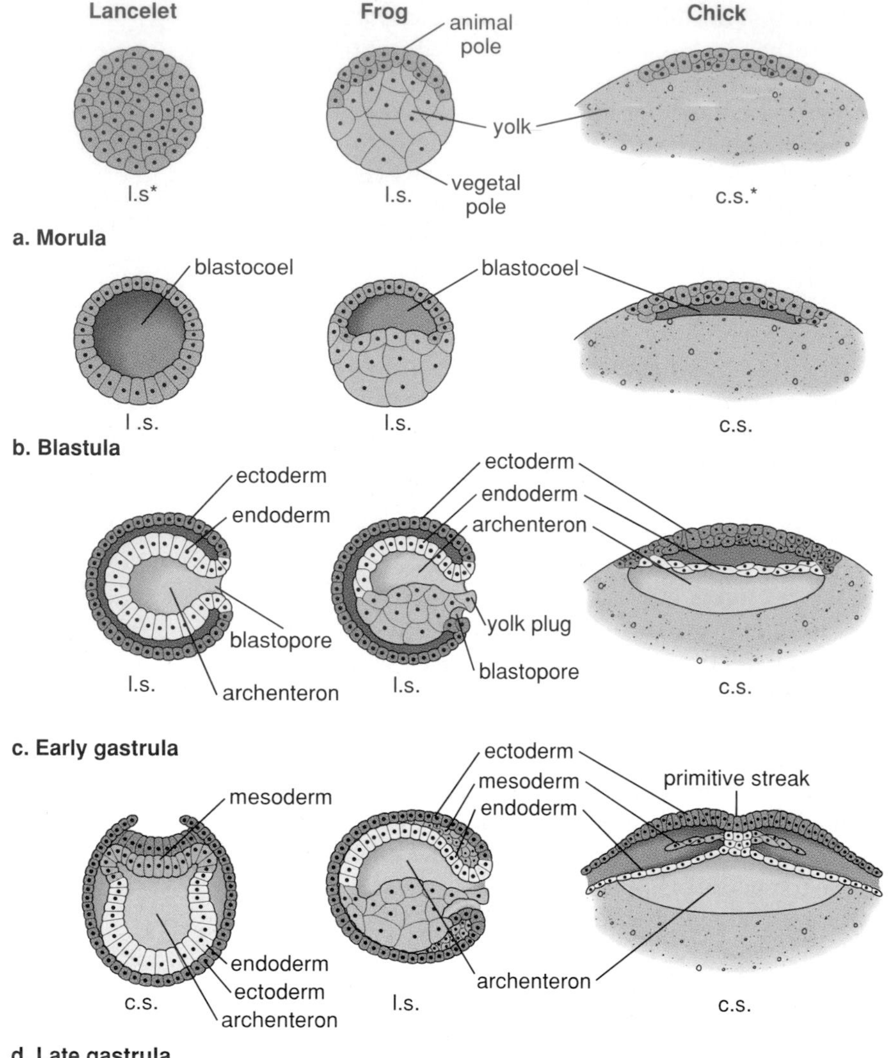

Figure 51.3 Comparative animal development.
a. Morula stages for a lancelet, a frog, and a chick. **b.** Blastula stages. **c.** Early gastrula stages. **d.** Late gastrula stages.

Table 51.1

Amount of Yolk in Eggs Versus Location of Development

Animal	Yolk	Location of Development
Lancelet	Little	External in water
Frog	Some	External in water
Chick	Much	Within hard shell
Human	Little	Inside mother

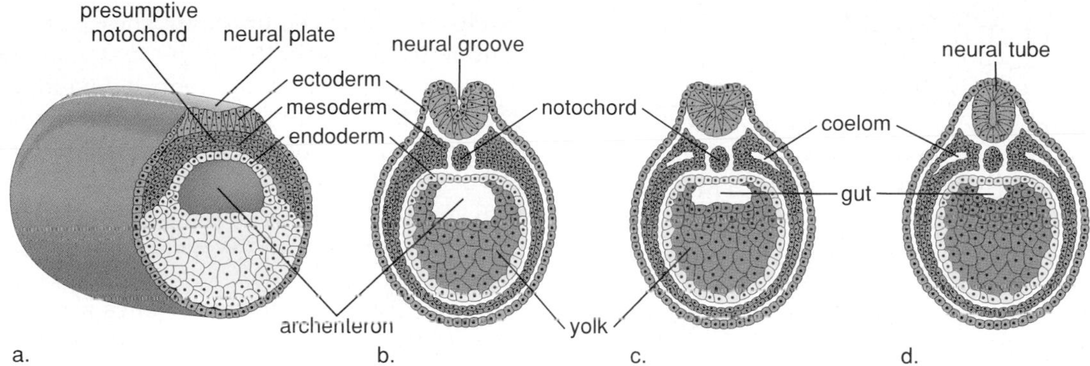

Figure 51.4 Development of neural tube and coelom in a frog embryo.
a. Ectoderm cells that lie above the future notochord (called presumptive notochord) thicken to form a neural plate. **b.** The neural groove and folds are noticeable as the neural tube begins to form. **c.** A splitting of the mesoderm produces a coelom, which is completely lined by mesoderm. **d.** A neural tube and a coelom have now developed.

invaginate from below. At this stage, there are some yolk cells temporarily left in the region of the pore; these are called the yolk plug. In the chick, there is so much yolk that endoderm formation does not occur by invagination. Instead, an upper layer of cells differentiates into ectoderm, and a lower layer differentiates into endoderm (Fig. 51.3*c*).

In the frog, cells from the dorsal lip of the blastopore migrate between the ectoderm and endoderm, forming the mesoderm. Later, a splitting of the mesoderm creates the coelom. In the chick, the mesoderm layer arises by an invagination of cells along the edges of a longitudinal furrow in the midline of the embryo. Because of its appearance, this furrow is called the *primitive streak* (Fig. 51.3*d*). Later, the newly formed mesoderm will split to produce a coelomic cavity.

The amount of yolk typically affects the manner in which animals complete the first three stages of development.

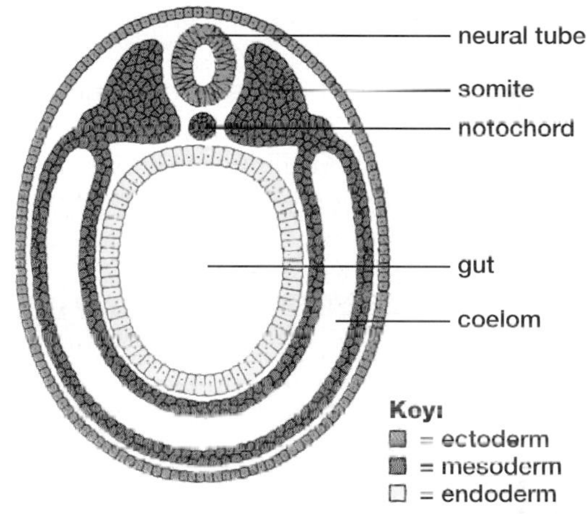

Figure 51.5 Chordate embryo, cross section.
At the neurula stage, each of the germ layers, indicated by color (see key), can be associated with the later development of particular parts. The somites give rise to the muscles of each segment and to the vertebrae, which replace the notochord.

Neurulation Produces the Nervous System

In chordate animals, newly formed mesoderm cells that lie along the main longitudinal axis of the animal coalesce to form a dorsal supporting rod called the **notochord** [Gk. *noto,* back, and *chord,* string]. The notochord persists in lancelets but in frogs, chicks, and humans, it is later replaced by the vertebral column.

The nervous system develops from midline ectoderm located just above the notochord. At first, a thickening of cells called the **neural plate** is seen along the dorsal surface of the embryo. Then, *neural folds* develop on either side of a neural groove, which becomes the **neural tube** when these folds fuse. Figure 51.4 shows cross sections of frog development to illustrate the formation of the neural tube. At this point, the embryo is called a *neurula.* Later, the anterior end of the neural tube develops into the brain.

Midline mesoderm cells that did not contribute to the formation of the notochord now become two longitudinal masses of tissue. These two masses become blocked off into the somites, which give rise to segmental muscles in all chordates. In vertebrates, the somites also produce the vertebral bones.

The chordate embryo in Figure 51.5 shows the location of various parts. This figure and the chart on page 915 will help you relate the formation of chordate structures and organs to the three embryonic layers of cells: the ectoderm, the mesoderm, and the endoderm.

During neurulation, the neural tube develops just above the notochord. At the neurula stage of development, a cross section of all chordate embryos is similar in appearance.

51.2 Cells Become Specialized and Embryos Take Shape

Development requires growth, cellular differentiation, and morphogenesis. **Cellular differentiation** occurs when cells become specialized in structure and function; a muscle cell looks different and acts differently than a nerve cell. **Morphogenesis** is a change in shape and form of a body part. The cellular slime mold, as discussed in the reading on page 920, is often the experimental material of choice for research into differentiation and morphogenesis.

How Cells Become Specialized

The process of differentiation most likely starts long before we can recognize different types of cells. Ectodermal, endodermal, and mesodermal cells in the gastrula look quite similar, but yet they must be different because they develop into different organs. What causes differentiation to occur, and when does it begin?

We know that differentiation cannot be due to a parceling out of genes into embryonic cells, because each cell in the body contains a full complement of chromosomes and, therefore, genes. However, we can note that the cytoplasm of a frog's egg is not uniform. It is polar and has both an anterior/posterior axis and a dorsal/ventral axis, which can be correlated with the **gray crescent,** a gray area that appears after the sperm fertilizes the egg (Fig. 51.6*a*). Hans Spemann, who received a Nobel Prize in 1935 for his extensive work in embryology, showed that if the gray crescent is divided equally by the first cleavage, each experimentally separated daughter cell develops into a complete embryo (Fig. 51.6*b*). If the egg divides so that only one daughter cell receives the gray crescent, however, only that cell becomes a complete embryo (Fig. 51.6*c*). We can therefore speculate that particular chemical signals within the gray crescent turn on the genes that control development in the frog.

It is hypothesized that genes are turned on/off due to *ooplasmic segregation,* which is the distribution of maternal cytoplasmic contents to the various cells of the morula:

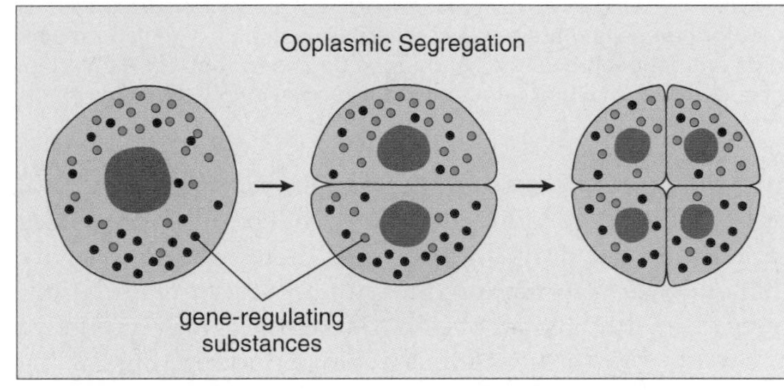

Cytoplasmic substances, parceled out during cleavage, initially influence which genes are activated and how cells differentiate.

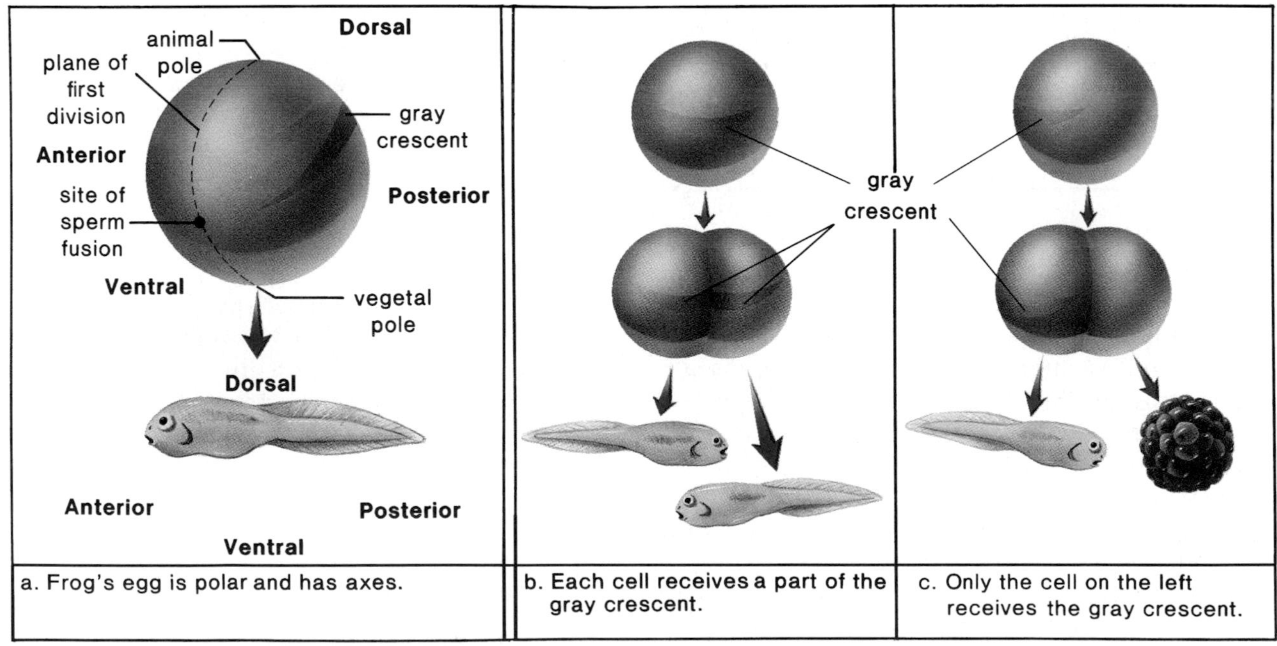

Figure 51.6 Cytoplasmic influence on development.
a. A frog's egg has anterior/posterior and dorsal/ventral axes that correlate with the position of the gray crescent. **b.** The first cleavage normally divides the gray crescent in half, and each daughter cell is capable of developing into a complete tadpole. **c.** But if only one daughter cell receives the gray crescent, then only that cell can become a complete embryo. This shows that chemical messengers are not uniformly distributed in the cytoplasm of frogs' eggs.

How Morphogenesis Occurs

As development proceeds, a cell's differentiation is influenced not only by its cytoplasmic content, but also by signals given off by neighboring cells. Migration of cells occurs during gastrulation, and there is evidence that one set of cells can influence the migratory path taken by another set of cells. Some cells produce an extracellular matrix that contains fibrils, and in the laboratory, it can be shown that the orientation of these fibrils influences migratory cells. The cytoskeletons of the migrating cells are oriented in the same direction as the fibrils. Although this may not be an exact mechanism at work during gastrulation, it suggests that formation of the germ layers is probably influenced by environmental factors.

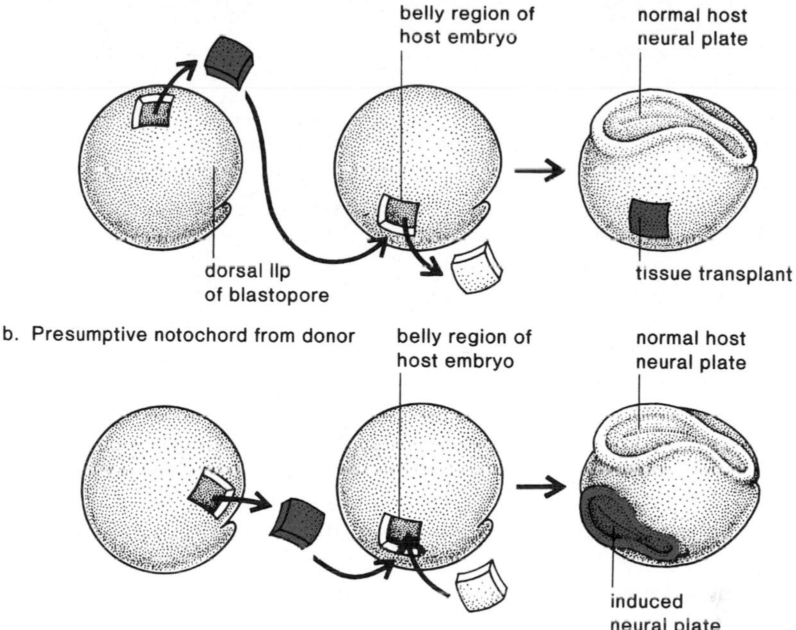

Figure 51.7 Control of nervous system development.
a. In this experiment, the presumptive nervous system (blue) does not develop into the neural plate if moved from its normal location. **b.** In this experiment, the presumptive notochord (red) can cause the belly ectoderm to develop into the neural plate (blue). This shows that the notochord induces ectoderm to become a neural plate, most likely by sending out chemical signals.

Spemann showed that a frog embryo's gray crescent becomes the dorsal lip of the blastopore, where gastrulation begins. Since this region is necessary for complete development, he called the dorsal lip of the blastopore the primary organizer. The cells closest to Spemann's primary organizer become endoderm, those farther away become mesoderm, and those farthest away become ectoderm. This suggests that there may be a molecular concentration gradient that acts as a signal to induce germ layer differentiation. **Induction** [L. *in,* into, and *duco,* lead] is the ability of a chemical or a tissue to influence the development of another tissue. The inducing chemical is called a *signal.*

Experiments performed by Jim Smith of the National Institute of Medical Research in London indicate that the peptide growth factor called activin may play a role in the signaling process. At low concentrations of activin, animal pole cells become epidermis, an ectoderm-derived tissue, and at higher concentrations, they become muscle and notochord, a mesoderm-derived tissue.

The gray crescent of a frog egg marks the dorsal side of the embryo where the notochord and nervous system develop. Spemann and his colleague Hilde Mangold showed that presumptive (potential) notochord tissue induces the formation of the nervous system (Fig. 51.7). If presumptive nervous system tissue, located just above the presumptive notochord, is cut out and transplanted to the belly region of the embryo, it does not form a neural tube. On the other hand, if presumptive notochord tissue is cut out and transplanted beneath what would be belly ectoderm, this ectoderm differentiates into neural tissue. Still other examples of induction are now known. In 1905, Warren Lewis studied the formation of the eye in frog embryos. He found that an optic vesicle, which is a lateral outgrowth of developing brain tissue, induces overlying ectoderm to thicken and become a lens. The developing lens in turn induces an optic vesicle to form an optic cup, where the retina develops.

Today investigators believe the process of induction goes on continuously—neighboring cells are always influencing one another. Either direct contact or the production of a chemical acts as a signal that activates certain genes and brings about protein synthesis. This diagram shows how morphogenesis can be a sequential process:

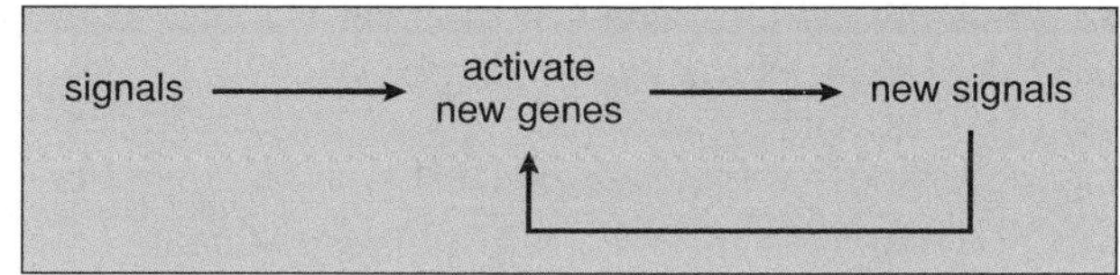

Doing Science

Movement of Cells Within a Cellular Slime Mold Slug

We do not often think how remarkable it is that most of our cells and tissues are in the correct places and in the correct proportion to other tissues. Further, we do not always realize that there are mechanisms controlling these events—until one of these mechanisms fails. Only then is it obvious that control mechanisms play such important roles in our development. Some of the most exciting questions in developmental biology involve these fundamental processes of positioning and proportioning of different cell types. In large complex organisms such as ourselves, the controls are likely to be more complex than in more simple species. Thus, research into these questions often centers on comparatively simple organisms such as the cellular slime mold, *Dictyostelium.*

Much of my research has involved study of the proportions and movements of cells in *Dictyostelium.* I first became interested in these questions as a graduate student in John Bonner's laboratory at Princeton University. Although I certainly found *Dictyostelium's* apparently simple development intriguing, I think what initially attracted me (a wide-eyed, eager graduate student) was my first visit to Dr. Bonner's lab. After introducing me around the lab, he spent a few minutes talking with his technician about a particular experiment. Dr. Bonner concluded that the experiment had not been fruitful, yet he was reluctant to drop it because it had been "fun with slime molds." I had known that Dr. Bonner was an excellent scientist, and now I saw that he found just doing experiments to be terrifically enjoyable. From that moment, I knew I wanted to join his lab; and since that time, I have also spent many pleasurable years working with slime molds.

Dictyostelium exhibits a fascinating life cycle that includes a unicellular stage and several multicellular stages. In the unicellular stage, individual amoeboid cells feed on bacteria in the soil and grow and divide. When the food supply is diminished, the cells aggregate into a tiny multicellular mound. The mound forms a slug about a millimeter long that subsequently develops into a fruiting body. The fruiting body stands erect on the substrate and consists of a stalk holding aloft a small mass of spores. Each spore can germinate to produce a single amoeba to begin the life cycle again.

Recently my research has focused on the movements of cells in the slug. A slug contains two kinds of anterior prestalk cells (PstA and PstB) in the anterior; prespore and anterior-like cells (they look like prestalk cells) in the posterior; and rearguard cells located at the very rear.

To study the movement of cells, I sometimes perform grafts using a microknife within a humid chamber built around the stage of a stereomicroscope. A typical grafting experiment is one in which the prestalk regions of a red-stained slug and blue-stained slug are exchanged, creating 2 two-colored slugs. When fewer cells need to be followed, I do transplantations by picking up a small group of cells with a small-bore pipette and injecting the cells into a differently stained slug. Slugs with differentially stained PstA and PstB cells, for instance, can be constructed. Whatever the experiment, I watch the movement of cells over time. Sometimes I make observations with time-lapse video photography, and sometimes I make sequential drawings using a drawing tube attachment for the microscope.

These experiments have revealed a unidirectional movement of cells within the slug. As the slug migrates, it leaves rearguard cells behind. These rearguard cells are replaced by PstB cells that move to the rear. The PstB cells are replaced, in turn, by the adjacent PstA cells. And the PstA cells are replaced by prespore cells. Clearly, repositioning of cells accompanied by conversion of one cell type into another type is occurring because the proportions of all cell types remain constant in the slug.

When we learn about events and mechanisms in simple organisms, such as *Dictyostelium,* we are able to turn to more complex organisms, including ourselves, and ask if similar processes occur during our development. The type of research I and the undergraduates in my lab do holds a fascination for me and many others in biology. For me personally, I get enjoyment in performing the experiments and in the discovery of new facts about *Dictyostelium.* In addition, an investigator knows that while basic research may not be directed toward solving some immediate problem of humankind, it may eventually contribute to the betterment of our species, either directly or indirectly.

John Sternfeld
State University of New York College at Cortland

Figure 51A Cellular slime mold slug.
A phase-contrast image of the rear of a slug (to the left) and the slug's slime track. The track contains rearguard cells left behind as the slug moved forward.

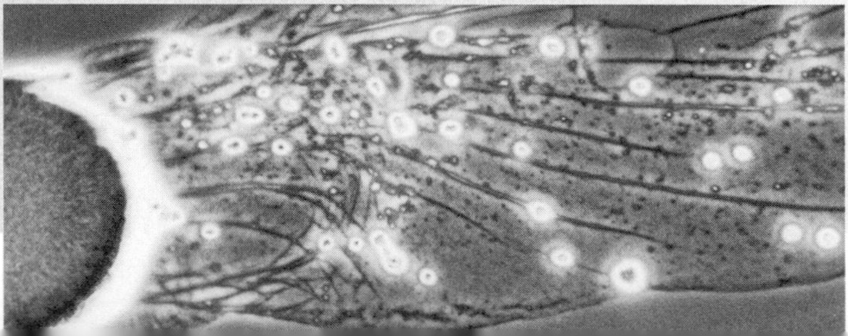

Genes That Control Pattern Formation

Investigators studying morphogenesis in *Drosophila* (fruit fly) have discovered that there are some genes that determine the animal's anterior/posterior and dorsal/ventral axes, others that determine the number and polarity of its segments, and still others, called *homeotic genes,* that determine how these segments develop. Homeotic genes have now been found in many other organisms and, surprisingly, they all contain the same particular sequence of nucleotides, called a **homeobox.**

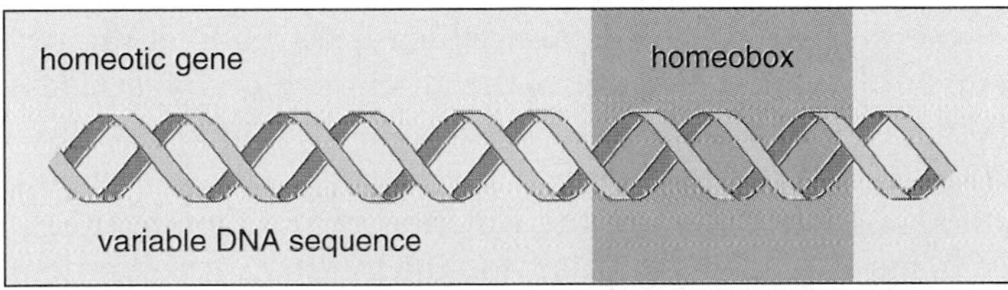

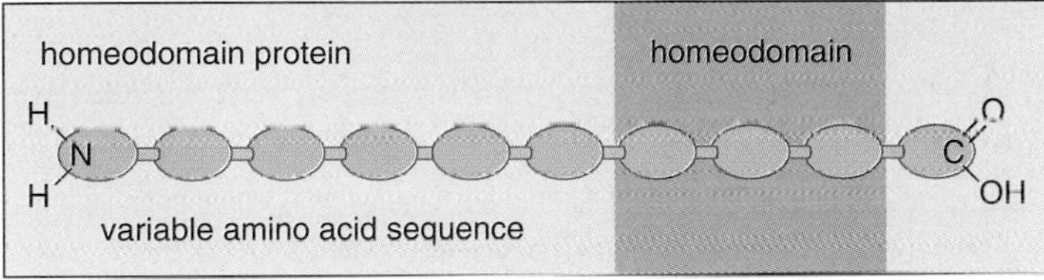

Since homeoboxes have been found in almost all eukaryotic organisms, it is believed that homeoboxes are derived from an original nucleotide sequence that has been largely conserved (maintained from generation to generation) because of its importance in the regulation of animal development.

In *Drosophila,* a homeotic mutation causes body parts to be misplaced—a homeotic mutant fly can have two pairs of wings or extra legs where antennae should be (Fig. 51.8). Similarly, in the frog *Xenopus,* if the expression levels of the homeotic genes are altered, headless and tailless embryos are produced.

Homeotic genes are clearly involved in **pattern formation**—that is, the shaping of an embryo so that the adult has a normal appearance. Homeotic genes are arranged in a definite order on a chromosome; those first in line determine the development of anterior segments of the animal, while those later in the sequence determine the development of posterior segments of the embryo. A homeotic gene codes for a homeodomain protein. Each of these proteins has a homeodomain, a sequence of 60 amino acids that is found in all other homeodomain proteins. Homeodomain proteins stay in the nucleus and regulate transcription of other genes during development. Researchers envision that a homeodomain protein produced by one homeotic gene binds to and turns on the next homeotic gene, and this orderly process determines the overall pattern of the embryo. It appears that homeotic genes also establish homeodomain protein gradients that affect the pattern development of specific parts, such as the limbs. One well-known gradient that determines wing formation in the chick involves retinoic acid, a chemical related to retinal, which is present in rods and cones of the eye.

a.

b.

Figure 51.8 Homeotic mutations in *Drosophila*.
Homeotic genes control pattern formation, an aspect of morphogenesis. If homeotic genes are activated at inappropriate times, abnormalities such as **(a)** a fly with four wings and **(b)** a fly with legs on its head can occur.

Many laboratories are engaged in discovering much more about homeotic genes and homeodomain proteins. We need to know, for example, how homeotic genes are turned on and how protein gradients are maintained in tissues. The roles of the cytoskeleton and extracellular matrix are being explored.

Morphogenesis is dependent upon signals (either contact or chemical) from neighboring cells. These signals are believed to activate particular genes, including homeotic genes.

51.3 Humans Are Embryos and Then Fetuses

In humans the length of the time from conception (fertilization followed by **implantation**) to birth (parturition) is approximately nine months. It is customary to calculate the time of birth by adding 280 days to the start of the last menstruation, because this date is usually known, whereas the day of fertilization is usually unknown. Because the time of birth is influenced by so many variables, only about 5% of babies actually arrive on the forecasted date.

Human development is often divided into embryonic development (months 1 and 2) and fetal development (months 3–9). The **embryonic period** consists of early formation of the major organs, and fetal development is the refinement of these structures.

Before we consider human development chronologically, we must understand the placement of **extraembryonic membranes** [L. *extra,* on the outside]. Extraembryonic membranes are best understood by considering their function in reptiles and birds. In reptiles, these membranes made development on land first possible. If an embryo develops in the water, the water supplies oxygen for the embryo and takes away waste products. The surrounding water prevents desiccation, or drying out, and provides a protective cushion. For an embryo that develops on land, all these functions are performed by the extraembryonic membranes.

In the chick, the extraembryonic membranes develop from extensions of the germ layers, which spread out over the yolk. Figure 51.9 shows the chick surrounded by the membranes. The **chorion** [Gk. *chorion,* membrane] lies next to the shell and carries on gas exchange. The **amnion** [Gk. *amnion,* membrane around fetus] contains the protective amniotic fluid, which bathes the developing embryo. The **allantois** [Gk. *allantos,* sausage] collects nitrogenous wastes, and the **yolk sac** surrounds the remaining yolk, which provides nourishment.

Humans (and other mammals) also have these extraembryonic membranes. The chorion develops into the fetal half of the placenta; the yolk sac, which lacks yolk, is the first site of blood cell formation; the allantoic blood vessels become the umbilical blood vessels; and the amnion contains fluid to cushion and protect the embryo, which develops into a fetus. Therefore, the function of the membranes in humans has been modified to suit internal development, but their very presence indicates our relationship to birds and to reptiles. It is interesting to note that all chordate animals develop in water, either in bodies of water or within amniotic fluid.

The presence of extraembryonic membranes in reptiles made development on land possible. Humans also have these membranes, but their function has been modified for internal development.

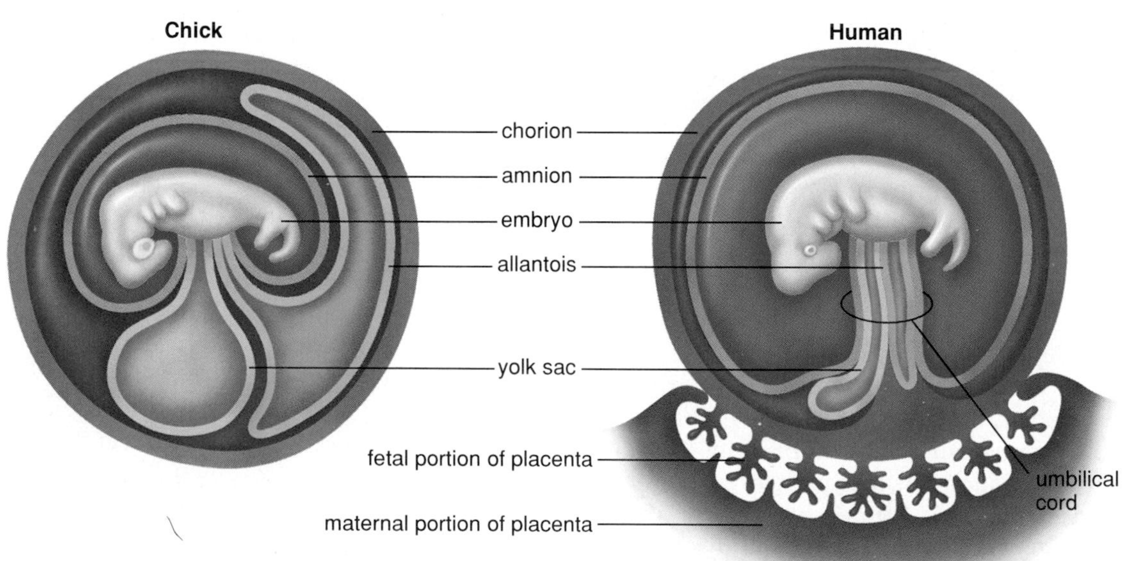

Figure 51.9 Extraembryonic membranes.
The membranes, which are not part of the embryo, are found during the development of chicks and humans, where each has a specific function.

Human Embryos Don't Look Human

Embryonic development includes the first two months of development.

First Week

Fertilization occurs in the upper third of an oviduct (Fig. 51.10), and cleavage begins even as the embryo passes down this tube to the uterus. By the time the embryo reaches the uterus on the third day, it is a *morula*. The morula is not much larger than the zygote because, even though multiple cell divisions have occurred, there has been no growth of these newly formed cells. By about the fifth day, the morula is transformed into the blastocyst. The **blastocyst** has a fluid-filled cavity, a single layer of outer cells called the **trophoblast** [Gk. *trophe*, food, and *blastos*, bud], and an inner cell mass. Later, the trophoblast, reinforced by a layer of mesoderm, gives rise to the *chorion*, one of the extraembryonic membranes (Fig. 51.9). The *inner cell* mass eventually becomes the embryo, which develops into a fetus.

Second Week

At the end of the first week, the embryo begins the process of *implanting* in the wall of the uterus. The trophoblast secretes enzymes to digest away some of the tissue and blood vessels of the uterine wall (Fig. 51.10). The embryo is now about the size of the period at the end of this sentence. The

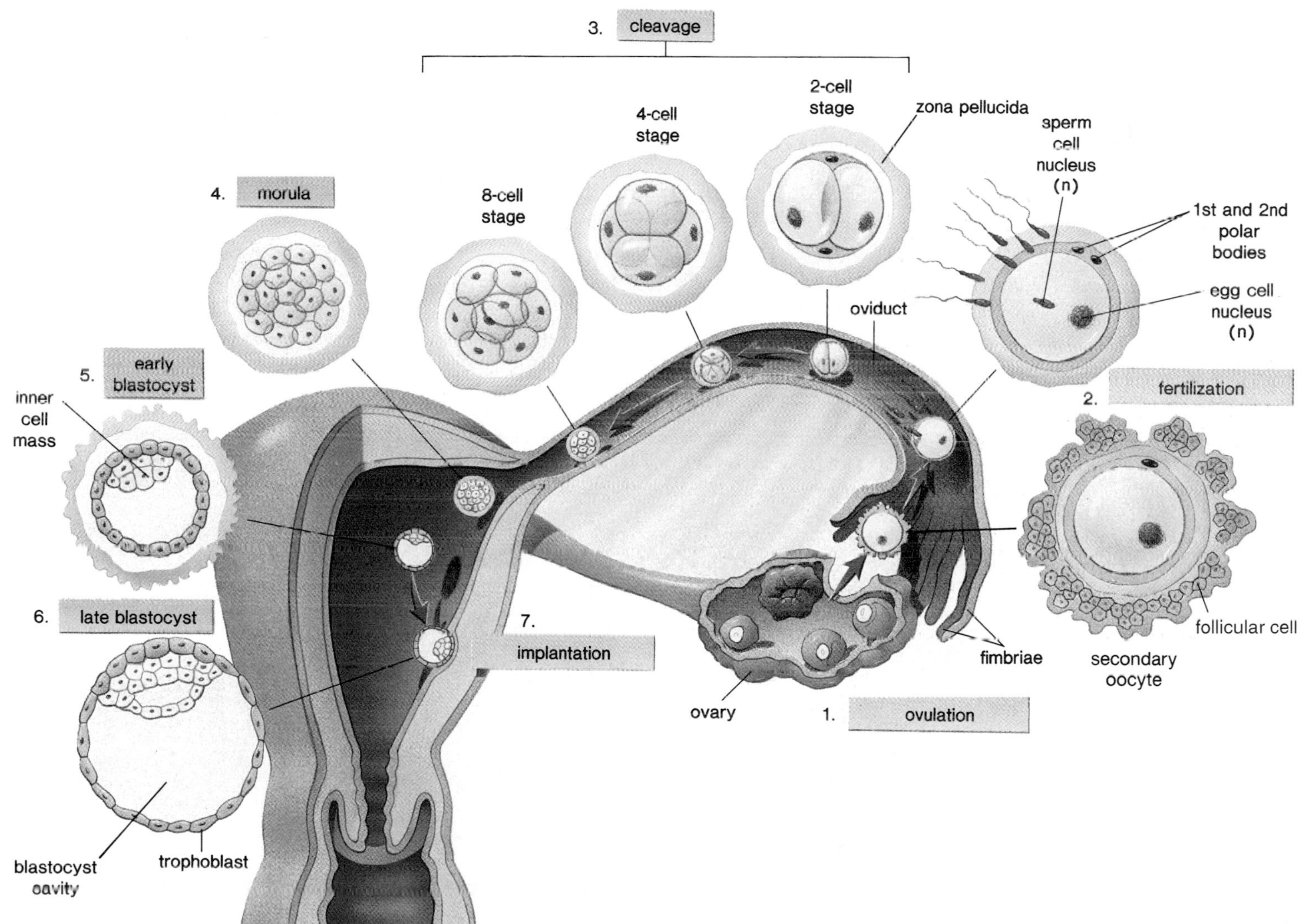

Figure 51.10 Human development before implantation.
Structures and events proceed counterclockwise. At ovulation (1), the secondary oocyte leaves the ovary. A single sperm penetrates the zona pellucida, and fertilization (2) occurs in the oviduct. As the zygote moves along the oviduct, it undergoes cleavage (3) to produce a morula (4). The blastocyst forms (5, 6) and implants itself in the uterine lining (7).

trophoblast begins to secrete **human chorionic gonadotropin (HCG),** the hormone that is the basis for the pregnancy test and that serves to maintain the corpus luteum past the time it normally disintegrates. Because of this, the endometrium is maintained and menstruation does not occur.

As the week progresses, the inner cell mass detaches itself from the trophoblast, and two more extraembryonic membranes form (Fig. 51.11*a*). The *yolk sac,* which forms below the embryonic disk, has no nutritive function as in chicks, but it is the first site of blood cell formation. However, the *amnion* and its cavity are where the embryo (and then the fetus) develops. In humans, amniotic fluid acts as an insulator against cold and heat and also absorbs shock, such as that caused by the mother exercising.

Gastrulation occurs during the second week. The inner cell mass now has flattened into the **embryonic disk,** composed of two layers of cells: *ectoderm* above and *endoderm* below. Once the embryonic disk elongates to form the *primitive streak,* similar to that found in birds, the third germ layer, mesoderm, forms by invagination of cells along the streak. The trophoblast is reinforced by mesoderm and becomes the chorion.

It is possible to relate the development of future organs to these germ layers (see page 915).

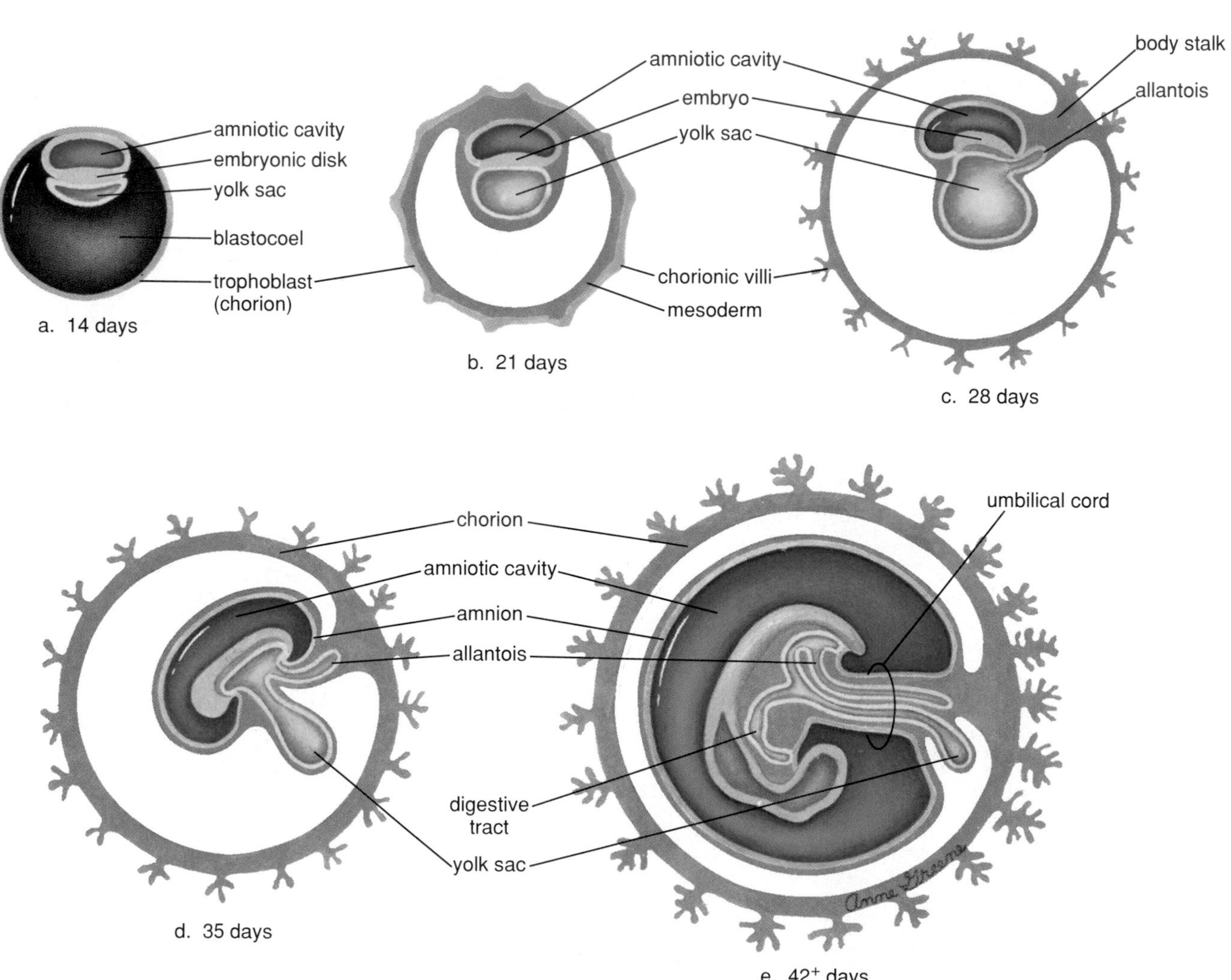

Figure 51.11 Human embryonic development.
a. At first there are only tissues present in the embryo. The amniotic cavity is above the embryo, and the yolk sac is below. **b.** The chorion is developing villi, so important to exchange between mother and child. **c.** The allantois and yolk sac are two more extraembryonic membranes. **d.** These extraembryonic membranes are positioned inside the body stalk as it becomes the umbilical cord. **e.** At 42 days, the embryo has a head region and a tail region. The umbilical cord takes blood vessels between the embryo and the chorion (placenta).

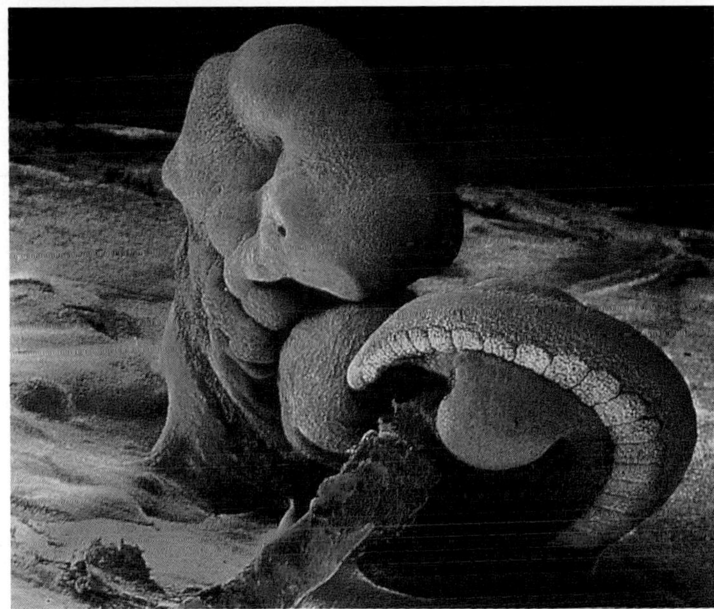

a.

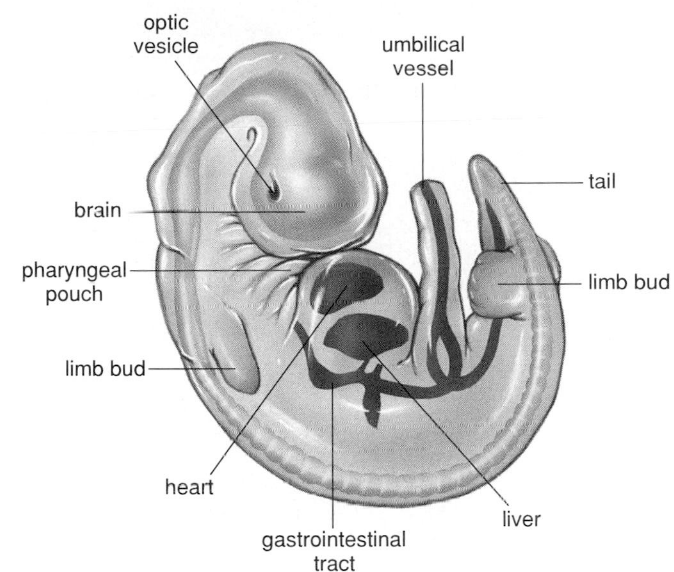

b.

Figure 51.12 Human embryo at beginning of fifth week.
a. Scanning electron micrograph. b. The embryo is curled so that the head touches the heart, the two organs whose development is further along than the rest of the body. The organs of the gastrointestinal tract are forming, and the arms and the legs develop from the bulges that are called limb buds. The tail is an evolutionary remnant; its bones regress and become those of the coccyx (tailbone). The pharyngeal arches become functioning gills only in fishes and amphibian larvae; in humans, the first pair of pharyngeal pouches becomes the auditory tubes. The second pair becomes the tonsils, while the third and fourth become the thymus gland and the parathyroids.

Third Week

Two important organ systems make their appearance during the third week. The nervous system is the first organ system to be visually evident. At first, a thickening appears along the entire dorsal length of the embryo, and then invagination occurs as neural folds appear. When the neural folds meet at the midline, the neural tube, which later develops into the brain and the nerve cord, is formed (see Fig. 51.4). After the notochord is replaced by the vertebral column, the nerve cord is called the spinal cord.

Development of the heart begins in the third week and continues into the fourth week. At first, there are right and left heart tubes; when these fuse, the heart begins pumping blood, even though the chambers of the heart are not fully formed. The veins enter posteriorly and the arteries exit anteriorly from this largely tubular heart, but later the heart twists so that all major blood vessels are located anteriorly.

Fourth and Fifth Weeks

At four weeks, the embryo is barely larger than the height of this print. A bridge of mesoderm called the body stalk connects the caudal (tail) end of the embryo with the chorion, which has projections called chorionic villi (Fig. 51.11*b*). The fourth extraembryonic membrane, the *allantois,* is contained within this stalk, and its blood vessels become the umbilical blood vessels. The head and the tail then lift up, and the body stalk moves anteriorly by constriction (Fig. 51.11*d*). Once this process is complete, the **umbilical cord** [L. *umbilicus,* navel], which connects the developing embryo to the placenta, is fully formed (Fig. 51.11*e*).

Little flippers called limb buds appear (Fig. 51.12); later, the arms and the legs develop from the limb buds, and even the hands and the feet become apparent. At the same time—during the fifth week—the head enlarges and the sense organs become more prominent. It is possible to make out the developing eyes, ears, and even nose.

Sixth Through Eighth Weeks

There is a remarkable change in external appearance during the sixth through eighth weeks of development—a form that is difficult to recognize as a human becomes easily recognized as human. Concurrent with brain development, the head achieves its normal relationship with the body as a neck region develops. The nervous system is developed well enough to permit reflex actions, such as a startle response to touch. At the end of this period, the embryo is about 38 mm (1.5 inches) long and weighs no more than an aspirin tablet, even though all organ systems are established.

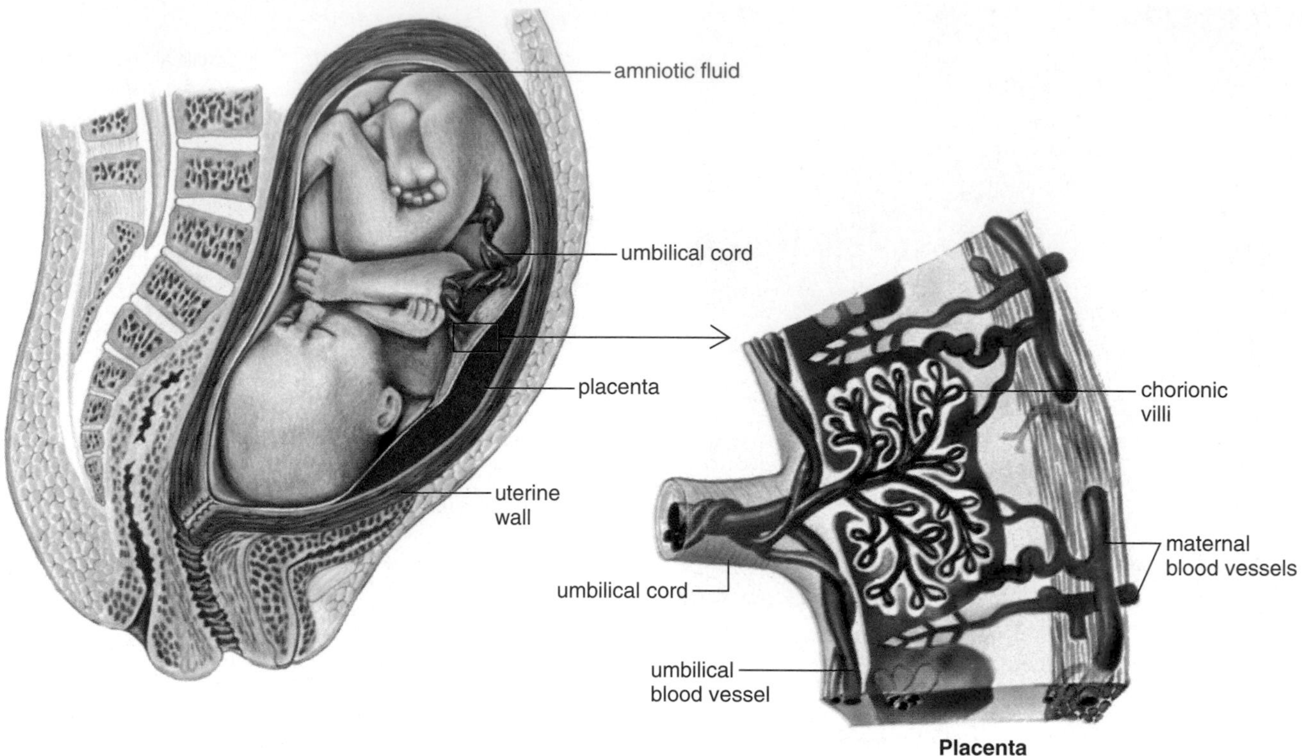

Figure 51.13 Anatomy of the placenta in a fetus at six to seven months.
The placenta is composed of both fetal and maternal tissues. Chorionic villi penetrate the uterine lining and are surrounded by maternal blood. Exchange of molecules between fetal and maternal blood takes place across the walls of the chorionic villi.

The Placenta Fulfills Needs

The **placenta** begins formation once the embryo is fully implanted. Treelike extensions of the chorion called **chorionic villi** [Gk. *chorion,* membrane, and L. *villus,* shaggy hair] project into the maternal tissues. Later, these disappear in all areas except where the placenta develops. By the tenth week, the placenta (Fig. 51.13) is fully formed and has already begun to produce progesterone and estrogen. These hormones have two effects: due to their negative feedback control of the hypothalamus and the anterior pituitary, they prevent any new follicles from maturing, and they maintain the lining of the uterus—now the corpus luteum is not needed. There is no menstruation during pregnancy.

The placenta has a fetal side contributed by the chorion and a maternal side consisting of uterine tissues. Notice in Figure 51.13 how the chorionic villi are surrounded by maternal blood sinuses; yet maternal and fetal blood never mix, since exchange always takes place across plasma membranes. Carbon dioxide and other wastes move from the fetal side to the maternal side, and nutrients and oxygen move from the maternal side to the fetal side of the placenta. The umbilical cord stretches between the placenta and the fetus. Although it may seem that the umbilical cord travels from the placenta to the intestine, actually the umbilical cord is simply taking fetal blood to and from the placenta. The umbilical cord is the lifeline of the fetus because it contains the umbilical arteries and vein, which transport waste molecules (carbon dioxide and urea) to the placenta for disposal and take oxygen and nutrient molecules from the placenta to the rest of the fetal circulatory system.

Harmful chemicals can also cross the placenta. This is of particular concern during the embryonic period, when various structures are first forming. Each organ or part seems to have a sensitive period during which a substance can alter its normal development. For example, if a woman takes the drug thalidomide, a tranquilizer, between days 27 and 40 of her pregnancy, the infant is likely to be born with deformed limbs. After day 40, however, the infant is born with normal limbs.

Fetuses Look Human

Fetal development (months 3–9) is marked by an extreme increase in size. Weight multiplies 600 times, going from less than 28 grams to 3 kilograms. In this time, too, the fetus grows to about 50 cm in length. The genitalia appear in the third month, so it is possible to tell if the fetus is male or female.

Soon, hair, eyebrows, and eyelashes add finishing touches to the face and head. In the same way, fingernails and toenails complete the hands and feet. A fine, downy hair **(lanugo)** covers the limbs and trunk, only to later disappear. The fetus looks very old because the skin is growing so fast that it wrinkles. A waxy, almost cheeselike substance **(vernix caseosa)** [L. *vernix*, varnish, and *caseus*, cheese] protects the wrinkly skin from the watery amniotic fluid.

The fetus at first only flexes its limbs and nods its head, but later it can move its limbs vigorously to avoid discomfort. The mother feels these movements from about the fourth month on. The other systems of the body also begin to function. After 16 weeks, the fetal heartbeat is heard through a stethoscope. A fetus born at 24 weeks has a chance of surviving, although the lungs are still immature and often cannot capture oxygen adequately. Weight gain during the last couple of months increases the likelihood of survival.

Birth Has Three Stages

The latest findings suggest that when the fetal brain is sufficiently mature, the hypothalamus causes the pituitary to stimulate the adrenal cortex so that androgens are released into the bloodstream. The placenta utilizes androgens as a precursor for estrogens, hormones that stimulate the production of prostaglandin (a molecule produced by many cells that acts as a local hormone) and oxytocin. All three of these molecules cause the uterus to contract and expel the fetus.

The process of birth (parturition) includes three stages: During the first stage, the cervix dilates to allow passage of the baby's head and body. The amnion usually bursts along about this time. During the second stage, the baby is born and the umbilical cord is cut. During the third stage, the placenta is delivered (Fig. 51.14).

During the embryonic period of fetal development, the extraembryonic membranes appear and serve important functions; the embryo acquires organ systems. During the fetal period there is a refinement of these systems. Finally birth occurs.

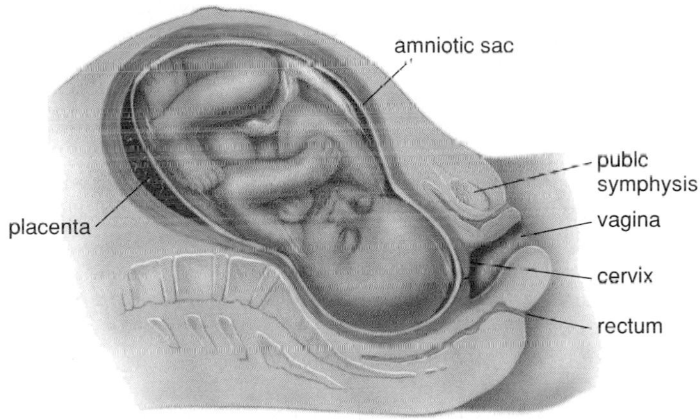

a. 9-month-old fetus

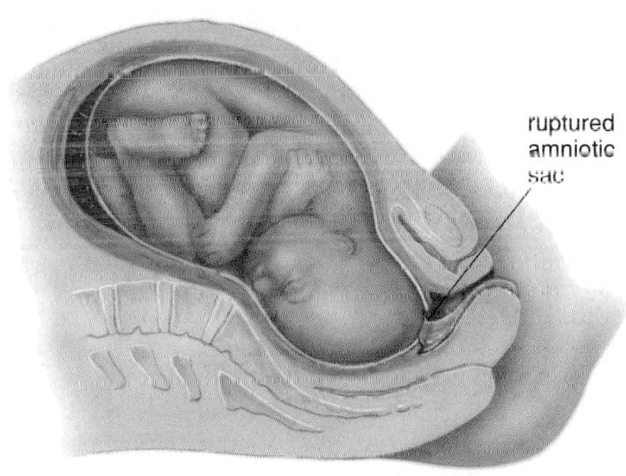

b. First stage of birth: cervix dilates

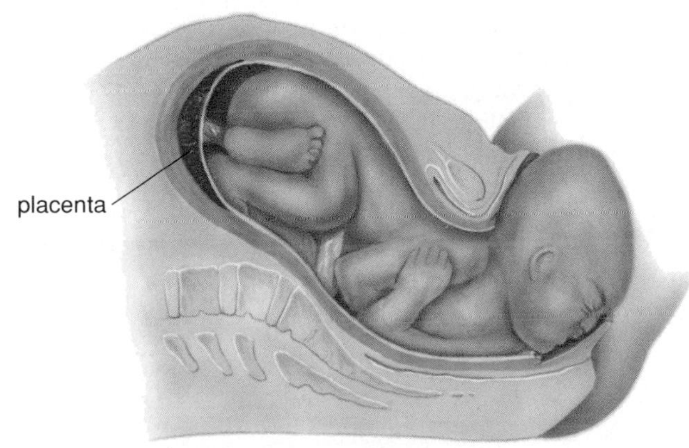

c. Second stage of birth: baby emerges

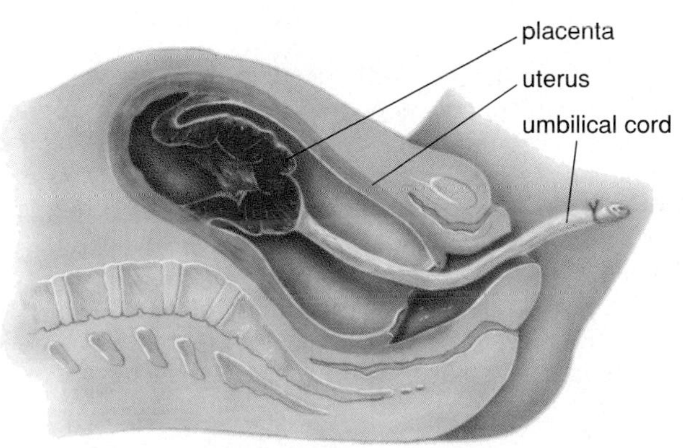

d. Third stage of birth: expelling afterbirth

Figure 51.14 Three stages of parturition.
a. Position of fetus just before birth begins. **b.** Dilation of cervix. **c.** Birth of baby. **d.** Expulsion of afterbirth.

connecting concepts

We have come full circle. We began our study of biology by considering the structure of the cell and its genetic machinery, including how the expression of genes is regulated. In this chapter, we have observed that animals go through the same early embryonic stages of morula, blastula, gastrula, and so forth. The set sequence of these stages is due to the expression of genes that bring about cellular changes. Therefore, once again, we are called upon to study organisms at the cellular level of organization.

We have seen that hormones are signals that affect cellular metabolism. Steroid hormones in animals, like gibberellins in plants, turn on the expression of genes. When a steroid hormone binds to a specific hormone receptor, a gene is transcribed, and then translation produces the corresponding protein. This same type of signal transduction pathway occurs during development. Transduction means that the signal has been transformed into an event that has an effect on the organism.

A set sequence of signaling molecules are produced as development occurs. Each new signal in the sequence turns on a specific gene, or more likely, a sequence of genes. Gene expression cascades are common during development. Homeotic genes are arranged in sets on the chromosomes, and a protein product from one set of genes acts as a transcription factor to turn on another set, and so forth. During the development of flies, it is possible to observe that first one region of a chromosome and then another puffs out, indicating that genes are transcribed in sequence as development occurs.

Developmental biology is now making a significant contribution to the field of evolution. Homeotic genes with the same homeoboxes (sequence of about 180 base pairs) have been discovered in many different types of organisms. This suggests that homeotic genes arose early in the history of life, and mutations in these genes could possibly account for macroevolution—the appearance of new species or even higher taxons.

Summary

51.1 Development Has Stages

Development occurs after fertilization. The acrosome of a sperm releases enzymes that digest away the jelly coat around the egg; the acrosome extrudes a filament that attaches to a receptor on the vitelline membrane. The sperm nucleus enters the egg and fuses with the egg nucleus. The resulting zygote begins to divide.

The early developmental stages in animals include these events. During cleavage, division occurs, but there is no overall growth. The result is a morula, which becomes the blastula when an internal cavity (the blastocoel) appears. During the gastrula stage, invagination of cells into the blastocoel results in formation of the germ layers: ectoderm, mesoderm, and endoderm. Later development of organs can be related to these layers.

The development of three types of animals (lancelet, frog, and chick) is compared. The first three stages (cleavage, blastulation, and gastrulation) differ according to the amount of yolk in the egg.

During neurulation, the nervous system develops from midline ectoderm, just above the notochord. At this point, it is possible to draw a typical cross section of a vertebrate embryo (see Fig. 51.4).

51.2 Cells Become Specialized and Embryos Take Shape

In many species, differentiation begins with cleavage, when the egg's cytoplasm is partitioned among the numerous cells. The cytoplasm is not uniform in content, and presumably each of the first few cells differ as to their cytoplasmic contents. Some probably contain substances that can influence gene activity—turning some genes on and others off.

After the first cleavage of a frog embryo, only a daughter cell that receives a portion of the gray crescent is able to develop into a complete embryo. This illustrates the importance of cytoplasmic inheritance to early development, especially in amphibians.

Morphogenesis involves the process of induction. The notochord induces the formation of the neural tube in frog embryos. The reciprocal induction that occurs between the lens and the optic vesicle is another good example of this phenomenon.

Today we envision induction as always present because cells are believed to constantly give off signals that influence the genetic activity of neighboring cells.

Studies of homeotic genes and pattern formation further indicate that morphogenesis involves genetic and environmental influences.

51.3 Humans Are Embryos and Then Fetuses

Human development can be divided into embryonic development (months 1 and 2) and fetal development (months 3–9). The early stages in human development resemble those of the chick. The similarities are probably due to their evolutionary relationship, not the amount of yolk the eggs contain, because the human egg has little yolk.

The extraembryonic membranes appear early in human development. The trophoblast of the blastocyst is the first sign of the chorion, which goes on to become the fetal part of the placenta. The placenta is where exchange occurs between fetal and maternal blood. The amnion contains the amniotic fluid, which cushions and protects the embryo. The yolk sac and allantois are also present.

Fertilization occurs in the oviduct, and cleavage occurs as the embryo moves toward the uterus. The morula becomes the blastocyst before implanting in the uterine lining. Organ development begins with neural tube and heart formation. There follows a steady progression of organ formation during embryonic development. During fetal development, refinement of features occurs, and the fetus adds weight. Birth occurs about 280 days after the start of the mother's last menstruation.

Reviewing the Chapter

1. Describe how fertilization of a sea star egg occurs. 914
2. State the germ layer theory, and tell which organs are derived from each of the germ layers. 915
3. Compare the process of cleavage and the formation of the blastula and gastrula in lancelets, frogs, and chicks. 916–17
4. Draw a cross section of a typical chordate embryo at the neurula stage, and label your drawing. 917
5. Explain how cellular differentiation and morphogenesis are dependent upon signals given off by neighboring cells. What effect do these signals have on the receiving cells? 918–19
6. Describe an experiment performed by Spemann suggesting that the notochord induces formation of the neural tube. Give another well-known example of induction between tissues. 919
7. What is the function of homeotic genes, and what is the significance of the homeobox within these genes? 921
8. List the human extraembryonic membranes, give a function for each, and compare their functions to those in the chick. 922
9. Tell where fertilization, cleavage, the morula stage, and the blastocyst stage occur in humans. What happens to the embryo in the uterus? 923
10. Describe the structure and the function of the placenta in humans. 926

Testing Yourself

Choose the best answer for each question.

1. Which of these stages is the first one out of sequence?
   a. cleavage
   b. blastula
   c. morula
   d. gastrula
2. Which of these stages is mismatched?
   a. cleavage—cell division
   b. blastula—gut formation
   c. gastrula—three germ layers
   d. neurula—nervous system
3. Which of the germ layers is best associated with development of the heart?
   a. ectoderm
   b. mesoderm
   c. endoderm
   d. All of these are correct.
4. In many embryos, differentiation begins at what stage?
   a. cleavage
   b. blastula
   c. gastrula
   d. neurula
5. Morphogenesis is best associated with
   a. overall growth.
   b. induction of one tissue by another.
   c. genetic mutations.
   d. oogenesis.
6. In humans, the placenta develops from the chorion. This indicates that human development
   a. resembles that of the chick.
   b. is associated with extraembryonic membranes.
   c. cannot be compared to lower animals.
   d. begins only upon implantation.
7. In humans, the fetus
   a. is surrounded by four extraembryonic membranes.
   b. has developed organs and is recognizably human.
   c. is dependent upon the placenta for excretion of wastes and acquisition of nutrients.
   d. Both b and c are correct.
8. Developmental changes
   a. require growth, differentiation, and morphogenesis.
   b. stop occurring when one is grown.
   c. are dependent upon a parceling out of genes into daughter cells.
   d. Both a and c are correct.
9. Mesoderm forms by out-pocketing from the primordial gut in the
   a. lancelet.
   b. frog.
   c. chick.
   d. All of these are correct.
10. Which of these is mismatched?
    a. brain—ectoderm
    b. gut—endoderm
    c. bone—mesoderm
    d. lens—endoderm
11. Label this diagram illustrating the placement of the extraembryonic membranes, and give a function for each membrane in humans:

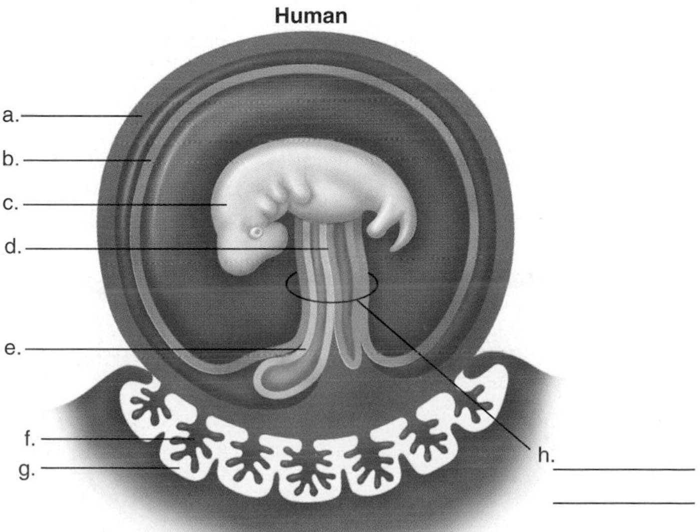

Applying the Concepts

1. *A complete organism develops from single cell.*
   In general, what processes are involved during development?
2. *Chemical signals are involved in development.*
   Should these chemical signals be considered hormones? Why or why not?
3. *Genes control development.*
   Reconcile this statement with the observation that development is sequential in nature.

Using Technology

Your study of development is supported by these available technologies:

Exploring the Internet

The Mader Home Page provides resources for and help with studying this chapter.

http://mhhe.com/sciencemath/biology/mader/
(Click on Biology.)

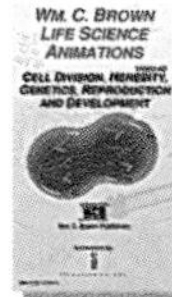

Life Science Animations Video

Video #2: Cell Division/Heredity/Genetics/Reproduction and Development
Human Embryonic Development (#21)

Understanding the Terms

allantois 922
amnion 922
blastocoel 915
blastocyst 923
blastula 915
cellular differentiation 918
chorion 922
chorionic villus 926
cleavage 914
ectoderm 915
embryo 914
embryonic disk 924
embryonic period 922
endoderm 915
extraembryonic membrane 922
fertilization 914
gastrula 915
gastrulation 915
germ layer 915
gray crescent 918
homeobox 921
human chorionic gonadotropin (HCG) 924
implantation 922
induction 919
lanugo 927
mesoderm 915
morphogenesis 918
morula 915
neural plate 917
neural tube 917
notochord 917
pattern formation 921
placenta 926
trophoblast 923
umbilical cord 925
vernix caseosa 927
yolk 914
yolk sac 922

Match the terms to these definitions:

a. __________ Ability of a chemical or a tissue to influence the development of another tissue.
b. __________ Cell division without cytoplasmic addition or enlargement; occurs during first stage of development.
c. __________ Primary tissue layer of a vertebrate embryo; namely ectoderm, mesoderm, or endoderm.
d. __________ Extraembryonic membrane of birds, reptiles, and mammals that forms an enclosing, fluid-filled sac.
e. __________ Hollow, fluid-filled ball of cells occurring during animal development prior to gastrula formation.
f. __________ A 180-nucleotide sequence located in all homeotic genes and serving to identify portions of the genome, in many different types of organisms, that are active in pattern formation.
g. __________ Movement of early embryonic cells to establish body outline and form.
h. __________ Specialization of early embryonic cells with regard to structure and function.
i. __________ Stage of animal development during which the germ layers form, at least in part, by invagination.
j. __________ Structure formed during the development of placental mammals from the chorion and the uterine wall that allows the embryo, and then the fetus, to acquire nutrients and rid itself of wastes.

Further Readings for Part vii

Alcamo, I. E. 1996. *AIDS: The biological basis*. Dubuque, Iowa: Wm. C. Brown Publishers. This easily understood book focuses on the biology of AIDS.

Alexander, N. J. March/April 1996. Barriers to sexually transmitted diseases. *Scientific American Science & Medicine* 3(2):32. Article discusses the effectiveness of certain contraceptives in protecting women against STDs.

Alexander, N. J. September 1995. Future contraceptives. *Scientific American* 273(3):136. Article discusses the possibility of contraceptive vaccines for both men and women.

Applegate, E. J. 1995. *The anatomy and physiology learning system*. Philadelphia: W. B. Saunders Publishing. Designed for a one-semester introductory course, this text provides fundamental information for students with minimal science background.

Axel, R. October 1995. The molecular logic of smell. *Scientific American* 273(4):154. Article discusses how the brain identifies a scent by neuron activation.

Beardsley, T. March 1996. Vital data. *Scientific American* 274(3):100. DNA tests for a wide array of conditions are becoming available.

Beardsley, T. August 1997. The machinery of thought. *Scientific American* 277(2):78. Researchers have identified the area of the brain responsible for memory.

Beck, G., and Habicht, G. S. November 1996. Immunity and the invertebrates. *Scientific American* 275(5):60. Nearly all aspects of the human immune system appear to have a cellular or chemical parallel among the invertebrates.

Benjamini, E., and Leskowitz, S. 1996. *Immunology: A short course*. 3d ed. New York: John Wiley & Sons. Presents the essential principles of immunology.

Bikle, D. D. March/April 1995. A bright future for the sunshine hormone. *Scientific American Science & Medicine* 2(2):58. Vitamin D receptors exist in many types of cells, suggesting therapeutic uses for the hormonally active metabolite of vitamin D.

Caldwell, J. C., and Caldwell, P. March 1996. The African AIDS epidemic. *Scientific American* 273(3):62. Article discusses a factor most likely responsible for causing the high rate of AIDS transmission in Africa.

Cox, F. D. 1996. *The AIDS booklet*. 4th ed. Dubuque, Iowa: Brown & Benchmark Publishers. This easy to read, informative booklet covers the transmission, prevention, and treatment of AIDS.

Davis, D. L., and Bradlow, H. L. April 1995. Can environmental estrogens cause breast cancer? *Scientific American* 273(4):166. Estrogenlike compounds found in the environment may contribute to breast cancer.

Fox, S. I. 1996. *Human physiology*. 5th ed. Dubuque, Iowa: Wm. C. Brown Publishers. This is an introductory physiology text.

Grillner, S. January 1996. Neural networks for vertebrate locomotion. *Scientific American* 274(1):64. Discoveries about how the brain coordinates muscle movement raise hopes for restoration of mobility for some accident victims.

Guyton, A. C., and Hall, J. E. 1996. *Textbook of medical physiology*. Presents physiological principles for those in the medical fields.

Hadley, M. E. 1996. *Endocrinology*, 4th ed. Upper Saddle River, NJ: Prentice Hall. This text discusses the role of chemical messengers in the control of neurological processes.

Julien, R. M. 1995. *A primer of drug action*. 7th ed. New York: W. H. Freeman and Company. A concise, nontechnical guide to the actions, uses, and side effects of psychoactive drugs.

MacDonald, P. C., and Casey, M. L. March/April 1996. Preterm birth. *Scientific American Science & Medicine* 3(2):42. Article discusses the role of oxytocin, prostaglandins, and infections in the initiation of human labor.

Mader, S. S. 1997. *Understanding anatomy and physiology*. 3d ed. Dubuque, Iowa: Wm. C. Brown Publishers. A text that emphasizes the basics for beginning allied health students.

Marieb, E. N. 1997. *Human anatomy and physiology*. 4th ed. Redwood City, Calif.: Benjamin/Cummings Publishing. A thorough anatomy and physiology text that can safely be used as a complete and accurate reference.

Marieb, E. N., and Mallatt, J. 1997. *Human anatomy*. 2d ed. Redwood City, Calif.: Benjamin/Cummings Publishing. An introductory anatomy text that covers gross, microscopic, developmental, and clinical anatomy.

Nolte, J. 1993. *The human brain*. 3d ed. St. Louis: Mosby-Year Book, Inc. Beginners are guided through the basic aspects of brain structure and function.

Nusslein-Volhard, C. August 1996. Gradients that organize embryo development. *Scientific American* 275(2):54. Nobel Prize winning researcher describes how chemical gradients of substances called morphogens give an evolving embryo its shape.

Scanlon, V. C., and Sanders, T. 1995. *Essentials of anatomy and physiology*. 2d ed. Philadelphia: F. A. Davis Company. This introductory text is designed for students with diverse educational backgrounds.

Schwartz, W. J. May/June 1996. Internal timekeeping. *Science & Medicine* 3(3):44. Article discusses circadian rhythm mechanisms.

Shier, D., et al. 1996. *Hole's human anatomy & physiology*. 7th ed. Dubuque, Iowa: Wm. C. Brown Publishers. An introductory anatomy and physiology text that has proved useful to beginning students.

Sussman, N. L., and Kelly, J. H. May/June 1995. The artificial liver. *Scientific American Science & Medicine* 2(3):68. An artificial liver might assist temporarily while the natural liver regenerates, restoring normal function.

Swerdlow, J. L. June 1995. The brain. *National Geographic* 187(6):2. New research leads to treatments for many age-old disorders.

Thomas, E. D. September/October 1995. Hematopoietic stem cell transplantation. *Scientific American Science & Medicine* 2(5):38. Article discusses reconstituting marrow from cultured stem cells for bone marrow transplants.

Tortora, G. J., and Grabowski, S. R. 1996. *Principles of anatomy and physiology*. 8th ed. New York: HarperCollins College Publishers. An introductory anatomy and physiology text that presents basic information in an easy to understand manner.

Valtin, H. 1995. *Renal function*. 3d ed. Boston: Little, Brown and Company. A good reference resource that discusses renal mechanisms for preserving fluid and solute balance.

Van De Graaff, K. M., and Fox, S. I. 1995. *Concepts of human anatomy & physiology*. 4th ed. Dubuque, Iowa: Wm. C. Brown Publishers. An introductory anatomy and physiology text that comprehensively presents basic principles.

West, J. B. 1995. *Respiratory physiology—the essentials*. 5th ed. Baltimore: Williams & Wilkins. A good reference resource that discusses all aspects of respiratory physiology including breathing, external, and internal respiration.

Youdim, M. B., and Riederer, P. January 1997. Understanding Parkinson's disease. *Scientific American* 276(1):52. The tremors and immobility of Parkinson's disease can be traced to damage in a part of the brain that regulates movement.

APPENDIX A

Answer Key

Chapter 1

Testing Yourself

1. d; **2.** c; **3.** b; **4.** e; **5.** c; **6.** b; **7.** c; **8.** b; **9.** c; **10.** b; **11. a.** Dye is spilled on culture plate. Investigator notices that bacteria live despite exposure to sunlight; **b.** Dye protects bacteria against death by UV light; **c.** Expose two culture plates to UV light. One plate contains bacteria and dye; the other plate contains only bacteria. The bacteria on both plates die; **d.** Dye does not protect bacteria against death by UV light.

Understanding the Terms

a. metabolism; **b.** evolution; **c.** experimental variable; **d.** photosynthesis; **e.** control group

Chapter 2

Testing Yourself

1. c; **2.** b; **3.** c; **4.** d; **5.** d; **6.** d; **7.** b; **8.** c; **9.** a; **10.** 7p and 7n in nucleus; two electrons in inner shell, and five electrons in outer shell. This means that nitrogen needs three more electrons in the outer shell to be stable; because each hydrogen contributes one electron, the formula for ammonia is NH_3.

Understanding the Terms

a. polar covalent bond; **b.** ion; **c.** acid; **d.** oxidation; **e.** pH scale; **f.** molecule; **g.** atom; **h.** compound; **i.** hydrophilic; **j.** hydrogen bond

Chapter 3

Testing Yourself

1. c; **2.** a; **3.** d; **4.** b; **5.** c; **6.** b; **7.** c; **8.** c; **9.** d; **10.** c; **11.** a; **12. a.** monomer; **b.** condensation; **c.** polymer; **d.** hydrolysis. The diagram shows the manner in which polymers are synthesized and degraded in cells; **13. a.** primary: sequence of amino acids; **b.** secondary: alpha (α) helix; **c.** tertiary: polypeptide folds and twists; **d.** quaternary: several polypeptides; **14.** c.

Understanding the Terms

a. carbohydrate; **b.** lipid; **c.** polymer; **d.** isomer; **e.** enzyme; **f.** amino acid; **g.** protein; **h.** nucleic acid; **i.** peptide

Chapter 4

Testing Yourself

1. c; **2.** c; **3.** d; **4.** a; **5.** c; **6.** c; **7.** a; **8.** d; **9.** b; **10.** d; **11.** c; **12. a.** example; **b.** Mitochondria and chloroplasts are a pair because they are both membranous structures involved in energy metabolism. **c.** Centrioles and flagella are a pair because they both contain microtubules; centrioles give rise to the basal bodies of flagella. **d.** ER and ribosomes are a pair because together they are rough ER, which produces proteins. **13. a.** Golgi apparatus processes and packages molecules; **b.** smooth ER synthesizes lipids; **c.** rough ER produces proteins; **d.** chromatin-DNA specifies the order of amino acids in proteins; **e.** nucleolus-forms ribosomal RNA, which participates in protein synthesis.

Understanding the Terms

a. nucleoid region; **b.** chromatin; **c.** nucleolus; **d.** cytoskeleton; **e.** lysosome; **f.** Golgi apparatus; **g.** cilium; **h.** flagellum; **i.** cell; **j.** endoplasmic reticulum

Chapter 5

Testing Yourself

1. b; **2.** b; **3.** a; **4.** c; **5.** c; **6.** d; **7.** d; **8.** b; **9.** b; **10. a.** hypertonic—Cell shrinks due to loss of water; **b.** hypotonic—Cell swells due to gain of water.

Understanding the Terms

a. differentially permeable; **b.** osmosis; **c.** hypertonic solution; **d.** osmotic pressure; **e.** cholesterol; **f.** turgor pressure; **g.** pinocytosis; **h.** carrier protein; **i.** isotonic solution; **j.** sodium-potassium pump

Chapter 6

Testing Yourself

1. a; **2.** d; **3.** c; **4.** a; **5.** d; **6.** c; **7.** d; **8.** c; **9. a.** active site; **b.** substrates; **c.** product; **d.** enzyme; **e.** enzyme-substrate complex; **f.** enzyme. The shape of an enzyme is important to its activity because it allows an enzyme-substrate complex to form. **10. a.** high H^+ concentration; **b.** low H^+ concentration; **c.** H^+ pump in electron transport system; **d.** H^+; **e.** ATP synthase complex; **f.** ADP; **g.** P; **h.** ATP. **i.** energy from electron transfers. See also Figure 6.9, page 108.

Understanding the Terms

a. metabolism; **b.** cofactor; **c.** kinetic energy; **d.** vitamin; **e.** denatured; **f.** entropy; **g.** feedback inhibition; **h.** coenzyme; **i.** ADP (adenosine diphosphate); **j.** enzyme

Chapter 7

Testing Yourself

1. d; **2.** d; **3.** a; **4.** d; **5.** c; **6.** d; **7.** c; **8.** d; **9. a.** outer membrane; **b.** inner membrane; **c.** stroma; **d.** thylakoid; **e.** thylakoid space; **f.** thylakoid; **g.** stroma; **10. a.** water; **b.** oxygen; **c.** carbon dioxide; **d.** carbohydrate; **e.** ADP + P $\rightarrow$ ATP; **f.** $NADP^+ \rightarrow$ NADPH

Understanding the Terms

a. photon; **b.** light-dependent reactions; **c.** chlorophyll; **d.** stroma; **e.** chloroplast; **f.** photosystem; **g.** electron transport system; **h.** photosynthesis; **i.** Calvin cycle; **j.** light-independent reactions

Chapter 8

Testing Yourself

1. b; **2.** c; **3.** a; **4.** c; **5.** c; **6.** c; **7.** a; **8.** b; **9.** d; **10.** c; **11.** a; **12.** b; **13.** d; **14.** c; **15.** b; **16.** a; **17.** d; **18.** a; **19.** c; **20.** b; **21. a.** cristae; **b.** matrix; **c.** outer membrane; **d.** intermembrane space; **e.** inner membrane

Understanding the Terms
a. aerobic respiration; **b.** fermentation; **c.** glycolysis; **d.** Krebs cycle; **e.** cellular respiration; **f.** metabolic pool; **g.** acetyl-CoA; **h.** oxygen debt; **i.** electron transport system; **j.** NAD^+

Chapter 9
Testing Yourself
1. c; **2.** b; **3.** d; **4.** d; **5.** a; **6.** c; **7.** b; **8.** c; **9.** b; **10.** b; **11. a.** chromatid of chromosome; **b.** centriole; **c.** spindle fiber or aster; **d.** nuclear envelope (fragment)

Understanding the Terms
a. diploid (2n) number; **b.** centrosome; **c.** centromere; **d.** cytokinesis; **e.** spindle; **f.** sister chromatid; **g.** mitosis

Chapter 10
Testing Yourself
1. b; **2.** c; **3.** c; **4.** b; **5.** a; **6.** c; **7.** d; **8.** d; **9.** c; **10.** The cell on the right represents metaphase I because bivalents are present at the metaphase plate.

Understanding the Terms
a. spermatogenesis; **b.** bivalent; **c.** polar body; **d.** secondary oocyte; **e.** homologue (homologous chromosome pair); **f.** synapsis; **g.** oogenesis; **h.** gamete; **i.** sexual reproduction

Chapter 11
Practice Problems 1
1. a. 100% *W*; **b.** 50% *W*, 50% *w*; **c.** 50% *T*, 50% *t*; **d.** 100% *T*; **2. a.** gamete; **b.** genotype, **c.** gamete

Practice Problems 2
1. *bb*; **2.** 75% yellow; 25% green; **3.** 3/4 or 75%; **4.** *Tt* × *tt*, *tt*

Practice Problems 3
1. Testcross; **2.** *Ll*, *ll*, 50% *Ll*, 50% *ll*; **3.** 200 have wrinkled seeds

Practice Problems 4
1. a. 50% *TC*, 50% *tG*; **b.** 25% *TG*, 25% *Tg*, 25% *tG*, 25% *tg*; **c.** 50% *TG*, 50% *Tg*; **2. a.** gamete; **b.** genotype; **c.** gamete

Practice Problems 5
1. *BbTt* only; **2. a.** *LlGg*, *llgg*; **b.** *LlGg*, *LlGg*; **3.** 9/16

Testing Yourself
1. b; **2.** a; **3.** c; **4.** d; **5.** c; **6.** d; **7.** a; **8.** b; **9.** b; **10.** b; **11.** c; **12.** d

Additional Genetics Problems
1. 100% chance for widow's peak and 0% chance for continuous hairline; **2.** *ee* (recessive); **3.** 50%; **4.** 210 gray bodies and 70 black bodies; 140 = heterozygous; cross fly with recessive (black body); **5.** *BBHH* = 2^0 = 1 possibility: *BH*; *BbHh* = 2^2 = 4 possibilities: *BH*, *Bh*, *bH*, *bh*; *BBHh* = 2^1 = 2 possibilities: *BH* and *Bh*; **6.** F_1 = all black with short hair; F_2 = 9 black, short: 3 black, long: 3 brown, short: 1 brown, long; offspring would be 1 brown long: 1 brown short : 1 black long : 1 black short; **7.** *Bbtt* × *bbTt* and *bbtt*; **8.** *GGLl*; **9.** 25%

Understanding the Terms
a. recessive allele; **b.** allele; **c.** dominant allele; **d.** testcross; **e.** genotype

Chapter 12
Practice Problems 1
1. Genotypes: *MM´*, *Mm*, *M´m*, *mm*; Phenotypes: *MM´*, *Mm* (melanic); *m´m* (Insularia); *mm* (typical); **2.** 1 pink, 1 white; or 50% pink, 50% white; **3.** 7; **4.** pleiotropy, epistasis

Practice Problems 2
1. 3 possible genotypes for females are: X^BX^B, X^BX^b (bar-eyed female), X^bX^b (normal-eyed female); 2 possible genotypes for males are: X^BY (bar-eyed male) and X^bY (normal-eyed male); gametes for males are: (X^B, Y) , (X^b, Y); **2.** b. ratio 1:1; **3.** 100%; none; 100%

Practice Problems 3
1. 9:3:3:1; linkage; **2.** 12 map units; **3.** bar eye, scalloped wings, garnet eye

Testing Yourself
1. b; **2.** c; **3.** c; **4.** a; **5.** b; **6.** b; **7.** d; **8.** b; **9.** c; **10.** d

Additional Genetics Problems
1. pleiotropy; **2.** codominance, F^BF^B, F^WF^W, F^BF^W; **3.** 50%; **4.** 6, yes; **5.** linkage, 10 map units; **6.** Both males and females are 1:1; **7.** $X^BX^B \times X^bY$ = all males and females are bar-eyed, $X^bX^b \times X^BY$ = females are bar-eyed, males are all normal; **8.** X^bX^b*Ww*; **9.** females , 1 short-haired tortoise shell : 1 short-haired yellow; males , 1 short-haired black : 1 short-haired yellow

Understanding the Terms
a. mutation; **b.** autosome; **c.** sex chromosome; **d.** polyploid; **e.** X-linked gene; **f.** polygenic inheritance; **g.** codominance; **h.** incomplete dominance; **i.** pleiotrophy; **j.** chromosomal theory of inheritance

Chapter 13
Testing Yourself
1. c; **2.** a; **3.** b; **4.** c; **5.** d; **6.** c; **7.** b; **8.** d; **9.** c; **10.** a; **11.** autosomal dominant condition

Additional Genetics Problems
1. 0%; 0%; 100%; **2.** 25%; **3.** AB; yes; A, B, AB, or O; **4.** X^cY*ww*; X^CX^c*Ww*, X^CY*Ww*; both sexes have normal vision with widow's peak; **5.** sex-linked recessive, X^AX^a

Understanding the Terms
a. autosome; **b.** sex-influenced trait; **c.** Punnett square; **d.** karyotype; **e.** nondisjunction; **f.** carrier; **g.** multiple allele

Chapter 14
Testing Yourself
1. b; **2.** d; **3.** a; **4.** c; **5.** d; **6.** c; **7.** a; **8.** d; **9.** d; **10.** The parental helix is heavy-heavy, and each daughter helix is heavy-light. This shows that each daughter helix is composed of one old strand and one new strand, which is consistent with semiconservative replication.

Understanding the Terms
a. mutation; **b.** complementary base pairing; **c.** DNA polymerase; **d.** purine; **e.** bacteriophage; **f.** thymine

Chapter 15
Testing Yourself
1. b; **2.** a; **3.** c; **4.** a; **5.** a; **6.** c; **7.** b; **8.** c; **9.** d; **10. a.** ACU CCU GAA UGC AAA; **b.** UGA GGA CUU ACG UUU; **c.** threonine-proline-glutamate-cysteine-lysine

Understanding the Terms
a. mutagen, **b.** RNA polymerase; **c.** exon; **d.** intron; **e.** transcription; **f.** translation; **g.** polyribsome; **h.** codon; **i.** anticodon; **j.** transfer RNA (tRNA)

Chapter 16
Testing Yourself
1. c; **2.** d; **3.** a; **4.** a; **5.** b; **6.** b; **7.** b; **8.** d; **9.** d; **10.** d; **11.** For the *trp* operon to be in the off position, a corepressor has to be attached to the repressor. **a.** DNA; **b.** regulator gene; **c.** promoter; **d.** operator; **e.** mRNA; **f.** active repressor protein. See also Figure 16.1*a*, page 253.

Understanding the Terms
a. oncogene; **b.** tumor; **c.** Barr body; **d.** euchromatin; **e.** carcinogen; **f.** tumor-suppressor gene; **g.** structural gene; **h.** operon; **i.** heterochromatin; **j.** promoter

Chapter 17
Testing Yourself
1. a; **2.** d; **3.** d; **4.** c; **5.** b; **6.** c; **7.** c; **8.** d; **9. a.** AATT; **b.** AATT; **10. a.** retrovirus; **b.** recombinant RNA; **c.** human genome; **d.** recombinant RNA; **e.** reverse transcription; **f.** recombinant DNA; **g.** defective gene. See also Figure 17.9, page 278.

Understanding the Terms
a. restriction enzyme; **b.** recombinant DNA (rDNA); **c.** transgenic organism; **d.** vector; **e.** probe; **f.** clone; **g.** plasmid; **h.** gene therapy; **i.** DNA ligase

Chapter 18

Testing Yourself
1. d; **2.** b; **3.** b; **4.** d; **5.** d; **6.** d; **7.** d; **8.** b; **9.** Life has a history, and it's possible to trace the history of individual organisms. **10.** Two different continents can have similar environments and, therefore, unrelated organisms that are similarly adapted. **11.** All vertebrates share a common ancestor, who had pharyngeal pouches during development. **12.** Genetic differences account for speciation and therefore evolution.

Understanding the Terms
a. biogeography; **b.** paleontology; **c.** vestigial structure; **d.** adaptation; **e.** inheritance of acquired characteristics; **f.** homologous structure; **g.** natural selection; **h.** evolution; **i.** catastrophism; **j.** fitness

Chapter 19

Testing Yourself
1. c; **2.** c; **3.** c; **4.** c; **5.** d; **6.** b; **7.** d; **8.** d; **9.** c; **10.** b; **11.** b; **12. a.** Label each circle "population" and each arrow "gene flow." **b.** Do away with five middle arrows so that two sets of circles remain. Only the populations of each species experience gene flow. **13. a.** See Figure 19.8, page 312. **b.** See Figure 19.7, page 312. **c.** See Figure 19.6, page 311.

Additional Genetics Problems
1. 16%; **2.** 81%; **3.** recessive allele = 0.2, dominant allele = 0.8; homozygous recessive = 0.04, homozygous dominant = 0.64, heterozygous = 0.32

Understanding the Terms
a. stabilizing selection; **b.** postmating isolating mechanism; **c.** genetic drift; **d.** adaptive radiation; **e.** population; **f.** species; **g.** Hardy-Weinberg Law; **h.** disruptive selection; **i.** speciation; **j.** gene flow

Chapter 20

Testing Yourself
1. d; **2.** c; **3.** b; **4.** c; **5.** b; **6.** b; **7.** b; **8.** a; **9.** a; **10.** d; **11.** c; **12. a.** Cambrian era begins; **b.** multicellularity; **c.** origin of eukaryotic cells; **d.** oxidizing atmosphere; **e.** oldest known fossils; **f.** formation of earth

Understanding the Terms
a. proteinoid; **b.** protocell; **c.** liposome; **d.** autotroph; **e.** ocean ridge; **f.** punctuated equilibrium; **g.** ozone shield; **h.** microsphere; **i.** chemical evolution; **j.** lineage

Chapter 21

Testing Yourself
1. a; **2.** b; **3.** d; **4.** d; **5.** d; **6.** c; **7.** b; **8.** c; **9.** b; **10. c; 11. a.** Chordata; **b.** Vertebrata; **c.** Class; **d.** Order; **e.** Anthropoidea; **f.** Hominoidea; **g.** Hominidae; **h.** Genus; **i.** *Homo sapiens*

Understanding the Terms
a. anthropoid; **b.** *Homo habilis*; **c.** Cro-Magnon; **d.** Neanderthal; **e.** *Homo erectus*; **f.** mammal; **g.** hominoid

Chapter 22

Testing Yourself
1. b; **2.** c; **3.** a; **4.** d; **5.** c; **6.** c; **7.** c; **8.** b; **9.** d; **10.** b

Understanding the Terms
a. sexual selection; **b.** pheromone; **c.** inclusive fitness; **d.** imprinting; **e.** operant conditioning; **f.** behavior; **g.** learning; **h.** communication; **i.** altruism; **j.** dominance hierarchy

Chapter 23

Testing Yourself
1. d; **2.** c; **3.** d; **4.** b; **5.** b; **6.** d; **7.** c; **8.** d; **9.** c; **10.** a

Understanding the Terms
a. demographic transition; **b.** population; **c.** exponential growth; **d.** carrying capacity; **e.** biotic potential; **f.** replacement reproduction; **g.** age structure diagram; **h.** *r*-selection; **i.** environmental resistance; **j.** sustainable world

Chapter 24

Testing Yourself
1. b; **2.** d; **3.** c; **4.** d; **5.** b; **6.** d; **7.** c; **8.** d; **9.** c; **10.** a; **11.** d; **12. a.** population densities; **b.** environmental gradient. A community contains species with overlapping tolerances to environmental factors (see also Fig. 24.2, p. 407).

Understanding the Terms
a. community; **b.** habitat; **c.** ecological succession; **d.** predation; **e.** coevolution; **f.** symbiosis; **g.** ecological niche; **h.** mutualism; **i.** parasitism; **j.** commensalism

Chapter 25

Testing Yourself
1. b; **2.** a; **3.** c; **4.** c; **5.** b; **6.** d; **7.** d; **8.** c; **9.** c; **10.** a; **11.** b; **12. a.** top carnivores; **b.** carnivores; **c.** herbivores; **d.** producers. The numbers are the dry biomass weights of the organisms at each level.

Understanding the Terms
a. ecosystem; **b.** biogeochemical cycle; **c.** food web; **d.** trophic level; **e.** producer; **f.** detritus; **g.** nitrogen fixation

Chapter 26

Testing Yourself
1. b; **2.** d; **3.** a; **4.** c; **5.** d; **6.** d; **7.** a; **8.** c; **9.** a; **10.** a; **11.** d; **12. a.** 4+ for both; **b.** 3+ for both; **c.** 2+ for both; **d.** 1+ for both

Understanding the Terms
a. estuary; **b.** plankton; **c.** biome; **d.** benthic division; **e.** pelagic division; **f.** taiga; **g.** epiphyte; **h.** savanna; **i.** biosphere; **j.** spring overturn

Chapter 27

Testing Yourself
1. a; **2.** d; **3.** c; **4.** b; **5.** b; **6.** c; **7.** d; **8.** c; **9.** a; **10.** c

Understanding the Terms
a. photochemical smog; **b.** ozone shield; **c.** pollutant; **d.** biological magnification; **e.** greenhouse effect

Chapter 28

Testing Yourself
1. a; **2.** c; **3.** d; **4.** a; **5.** c; **6.** a; **7.** d; **8.** b; **9.** b; **10.** d; **11. a.** three; **b.** vertebra; amniote egg and internal fertilization; **c.** the clades; they share derived characters. See also Figure 28A, page 503.

Understanding the Terms
a. taxonomy; **b.** phylogenetic tree; **c.** phylogeny; **d.** taxon; **e.** cladistics; **f.** classification; **g.** homology; **h.** parallel evolution; **i.** primitive character; **j.** systematics

Chapter 29

Testing Yourself
1. c; **2.** c; **3.** b; **4.** c; **5.** a and c; **6.** c; **7.** c; **8.** c; **9.** b; **10.** d; **11. a.** attachment; **b.** bacterial chromosome; **c.** penetration; **d.** integration; **e.** prophage; **f.** maturation; **g.** release; See also Figure 29.3, page 511.

Understanding the Terms
a. lysogenic cycle; **b.** virus; **c.** lytic cycle; **d.** saprotroph; **e.** cyanobacterium **f.** archaea

Chapter 30

Testing Yourself
1. b; **2.** b; **3.** d; **4.** b; **5.** b; **6.** d; **7.** c; **8.** d; **9.** b; **10. a.** sexual reproduction; **b.** isogametes pairing; **c.** zygote (2n); **d.** zygospore (2n); **e.** asexual reproduction; **f.** zoospores (n); **g.** nucleus; **h.** chloroplast; **i.** pyrenoid; **j.** starch granule; **k.** flagellum; **l.** eyespot; **m.** gamete formation. See also Figure 30.1, p. 524; **11.** Asexual: one parent, no gametes

produced, no genetic recombination. Sexual: two parents, gametes produced, genetic recombination.

Understanding the Terms
a. alga; **b.** spore; **c.** pseudopod; **d.** euglenoid; **e.** plankton; **f.** diatom; **g.** trypanosome; **h.** zooplankton; **i.** amoeboid; **j.** conjugation

Chapter 31

Testing Yourself
1. b; **2.** c; **3.** b; **4.** a; **5.** c; **6.** a; **7.** d; **8.** d; **9.** d; **10.** d; **11. a.** meiosis; **b.** basidiospores; **c.** dikaryotic mycelium; **d.** fruiting body (basidiocarp); **e.** stalk; **f.** gill; **g.** cap; **h.** dikaryotic (n + n); **i.** diploid (2n); **j.** zygote. See also Figure 31.6, page 546.

Understanding the Terms
a. sporangium; **b.** basidium; **c.** hypha; **d.** mycelium; **e.** fungus; **f.** conidiospore; **g.** fruiting body; **h.** lichen; **i.** mycorrhiza

Chapter 32

Testing Yourself
1. d; **2.** c; **3.** c; **4.** c; **5.** d; **6.** b; **7.** b; **8.** b; **9.** d; **10. a.** sporophyte (2n); **b.** meiosis; **c.** gametophyte (n); **d.** fertilization. See also Figure 32B, page 569.

Understanding the Terms
a. sporophyte; **b.** monocotyledon; **c.** dicotyledon; **d.** gametophyte; **e.** pollen grain; **f.** conifer; **g.** herbaceous plant; **h.** rhizoid; **i.** phloem; **j.** xylem

Chapter 33

Testing Yourself
1. d; **2.** c; **3.** d.; **4.** d; **5.** b; **6.** d; **7.** a; **8. a.** gastrovascular cavity; **b.** tentacle; **c.** mouth; **d.** mesoglea; **9. a.** Cnidaria; **b.** Nematoda; **c.** Cnidaria; **d.** Nematoda; **10. a.** tapeworm; **b.** planarian; **c.** sponge; **d.** hydra

Understanding the Terms
a. gastrovascular cavity; **b.** coelom; **c.** pseudocoelom; **d.** bilateral symmetry; **e.** tube-within-a-tube body plan; **f.** hermaphroditic

Chapter 34

Testing Yourself
1. b; **2.** d; **3.** b; **4.** c; **5.** c; **6.** a; **7.** d; **8.** c; **9.** a; **10. a.** all three; **b.** annelids, arthropods; **c.** all three; **d.** all three; **e.** all three; **f.** all three; **g.** arthropods; **h.** mollusks; **11. a.** earthworms; **b.** clams; **c.** clams; **d.** clams; **e.** earthworms; **f.** earthworms; **g.** clams; **h.** clams; **i.** earthworms; **j.** earthworms; **12. a.** head; **b.** antenna; **c.** simple eye; **d.** compound eye; **e.** thorax; **f.** tympanum; **g.** abdomen; **h.** forewing; **i.** hindwing; **j.** ovipositor; **k.** spiracles; **l.** air sac; **m.** spiracle; **n.** tracheae. See also Figure 34.14*a*, page 609.

Understanding the Terms
a. trachea; **b.** Malpighian tubule; **c.** cleavage; **d.** metamorphosis; **e.** jointed appendage; **f.** molt; **g.** exoskeleton; **h.** segmentation; **i.** nephridium; **j.** chitin

Chapter 35

Testing Yourself
1. b; **2.** b; **3.** d; **4.** a; **5.** b; **6.** d; **7.** a; **8.** c; **9.** d; **10.** b; **11. a.** pharyngeal pouches; **b.** dorsal hollow nerve cord; **c.** notochord; **d.** post-anal tail. See also illustration, page 616.

Understanding the Terms
a. amniote egg; **b.** homeothermic; **c.** monotreme; **d.** marsupial; **e.** ectothermic; **f.** amphibian; **g.** reptile; **h.** mammal; **i.** placental mammal; **j.** echinoderm

Chapter 36

Testing Yourself
1. c; **2.** b; **3.** c; **4.** c; **5.** b; **6.** b; **7.** b; **8.** c; **9.** c; **10.** b; **11. a.** epidermis; **b.** cortex; **c.** endodermis; **d.** phloem; **e.** xylem See also Figure 36.8, page 644. **12. a.** upper epidermis; **b.** palisade mesophyll; **c.** leaf vein; **d.** spongy mesophyll; **e.** lower epidermis. See also Figure 36.18, page 654.

Understanding the Terms
a. mesophyll; **b.** vascular cambium; **c.** parenchyma; **d.** cork; **e.** rhizome; **f.** cotyledon; **g.** stolon; **h.** phloem; **i.** xylem; **j.** cuticle

Chapter 37

Testing Yourself
1. d; **2.** a; **3.** c; **4.** b; **5.** c; **6.** d; **7.** c; **8.** d; **9.** The diagram shows that air pressure pushing down on mercury in the pan can raise a column of mercury only to 76 cm. When water above the column is transpired, it pulls on the mercury and raises it higher. This suggests that transpiration would be able to raise water to the top of trees.
10. See Figure 37.10, page 668. After K^+ enters guard cells, water follows by osmosis and the stomate opens.
11. There is more solute in the left bulb than in the right; therefore water enters the left bulb. This creates a positive pressure potential that causes water, along with solute, to flow toward the right bulb. See also illustration, page 671.

Understanding the Terms
a. root pressure; **b.** guttation; **c.** pressure-flow model; **d.** epiphyte; **e.** transpiration; **f.** water potential; **g.** girdling; **h.** root nodule; **i.** mycorrhiza; **j.** guard cell

Chapter 38

Testing Yourself
1. d; **2.** b; **3.** e; **4.** a; **5.** c; **6.** b; **7.** d; **8.** b; **9.** d; **10.** d; **11.** a; **12. a.** epidermal cell; **b.** guard cell. The hormone ABA in some unknown manner causes K^+ to leave the guard cell. Water follows the movement of K^+ and the guard cell closes.

Understanding the Terms
a. circadian rhythm; **b.** phototropism; **c.** gravitropism; **d.** abscission; **e.** tropism; **f.** gibberellin; **g.** auxin; **h.** ethylene; **i.** abscisic acid; **j.** photoperiodism

Chapter 39

Testing Yourself
1. c; **2.** a; **3.** b; **4.** a; **5.** a; **6.** b; **7.** a; **8.** d; **9.** c; **10.** d; **11. a.** sporophyte; **b.** meiosis; **c.** microspore; **d.** megaspore; **e.** microgametophyte (pollen grain); **f.** megagametophyte (embryo sac); **g.** egg and sperm; **h.** fertilization; **i.** zygote; **j.** seed. See also Figure 39.1, page 694.

Understanding the Terms
a. pistil; **b.** fruit; **c.** carpel; **d.** plumule; **e.** pollination; **f.** megagametophyte; **g.** seed; **h.** pollen grain; **i.** flower; **j.** cotyledon

Chapter 40

Testing Yourself
1. b; **2.** b; **3.** a; **4.** d; **5.** d; **6.** d; **7.** b; **8.** d; **9.** d; **10.** c; **11. a.** smooth muscular tissue; **b.** blood cells—connective tissue; **c.** nervous tissue; **d.** ciliated columnar epithelium

Understanding the Terms
a. adipose tissue; **b.** dermis; **c.** striated; **d.** homeostasis; **e.** neuron; **f.** epidermis; **g.** tendon; **h.** cardiac muscle; **i.** epithelial tissue; **j.** blood

Chapter 41

Testing Yourself
1. b; **2.** b; **3.** a; **4.** d; **5.** d; **6.** c; **7.** b; **8.** b; **9.** d; **10.** b; **11.** d; **12. a.** aorta; **b.** left pulmonary artery; **c.** pulmonary trunk; **d.** left pulmonary veins; **e.** left atrium; **f.** semilunar valves; **g.** atrioventricular (mitral) valve; **h.** left ventricle; **i.** septum; **j.** inferior vena cava; **k.** right ventricle; **l.** chordae tendineae; **m.** atrioventricular (tricuspid) valve; **n.** right atrium; **o.** right pulmonary veins; **p.** branches of right pulmonary artery; **q.** superior vena cava. See also Figure 41.6*a*, page 739.

Understanding the Terms
a. artery; **b.** agglutination; **c.** blood pressure; **d.** erythrocyte; **e.** atrioventricular; **f.** macrophage; **g.** plasma; **h.** aorta; **i.** hemoglobin; **j.** arteriole

Chapter 42

Testing Yourself

1. d; **2.** d; **3.** a; **4.** b; **5.** c; **6.** b; **7.** a; **8.** a; **9.** b; **10.** d; **11.** **a.** Helper T cells secrete lymphokines and stimulate other immune cells; **b.** suppressor T cells suppress further development of helper T cells, and prevent B and T cell immune responses from getting out of hand; **c.** memory T cells respond to particular antigens, resulting in active immunity; **d.** cytotoxic T cells search out and destroy antigen-bearing cells.

Understanding the Terms

a. immunity; **b.** vaccine; **c.** lymph; **d.** antigen; **e.** antibody; **f.** T lymphocyte

Chapter 43

Testing Yourself

1. a; **2.** b; **3.** d; **4.** b; **5.** c; **6.** a; **7.** c; **8.** c; **9.** d; **10.** c; **11.** d; **12.** c; **13.** *Test tube 1:* no digestion—no enzyme and no HCl; *Test tube 2:* some digestion—no HCl; *Test tube 3:* no digestion—no enzyme; *Test tube 4:* digestion—both enzyme and HCl are present

Understanding the Terms

a. vitamin; **b.** lipase; **c.** lacteal; **d.** esophagus; **e.** gallbladder; **f.** pepsin; **g.** trypsin; **h.** peristalsis; **i.** bile; **j.** villus

Chapter 44

Testing Yourself

1. a; **2.** b; **3.** b; **4.** d; **5.** c; **6.** b; **7.** c; **8.** b; **9.** d; **10.** b; **11.** **a.** nasal cavity; **b.** nostril; **c.** pharynx; **d.** epiglottis; **e.** glottis; **f.** larynx; **g.** trachea; **h.** bronchus; **i.** bronchiole; **j.** lung; **k.** diaphragm; **l.** pulmonary venule; **m.** pulmonary arteriole; **n.** alveolus; **o.** capillary network. See also Figure 44.6, page 794.

Understanding the Terms

a. pharynx; **b.** diaphragm; **c.** vocal cord; **d.** bicarbonate ion; **e.** thoracic cavity; **f.** bronchus; **g.** hemoglobin; **h.** gill; **i.** exhalation; **j.** alveolus

Chapter 45

Testing Yourself

1. d; **2.** a; **3.** c; **4.** b; **5.** a; **6.** c; **7.** b; **8.** d; **9.** a; **10.** b; **11.** **a.** pressure filtration; **b.** glomerulus; **c.** glomerular capsule; **d.** selective reabsorption; **e.** proximal convoluted tubule; **f.** water; **g.** tubular secretion; **h.** distal convoluted tubule; **i.** reabsorption of water; **j.** reabsorption of H_2O; **k.** loop of the nephron; **l.** peritubular capillary; **m.** collecting duct; **n.** excretion; **o.** renal pelvis. See also Figure 45.9, page 811.

Understanding the Terms

a. Malpighian tubule; **b.** glomerular capsule; **c.** distal convoluted tubule; **d.** aldosterone; **e.** uric acid; **f.** urea; **g.** nephron; **h.** loop of the nephron; **i.** proximal convoluted tubule; **j.** urethra

Chapter 46

Testing Yourself

1. c; **2.** b; **3.** d; **4.** c; **5.** a; **6.** a; **7.** b; **8.** d; **9.** b; **10.** b; **11.** **a.** central canal; **b.** gray matter; **c.** white matter; **d.** dorsal root ganglion; **e.** cell body of sensory neuron; **f.** sensory neuron; **g.** receptor; **h.** effector; **i.** motor neuron; **j.** cell body of motor neuron; **k.** interneuron

Understanding the Terms

a. reflex; **b.** neurotransmitter; **c.** autonomic system; **d.** ganglion; **e.** acetylcholine

Chapter 47

Testing Yourself

1. d; **2.** c; **3.** c; **4.** d; **5.** d; **6.** c; **7.** c; **8.** b; **9.** d; **10.** c; **11.** **a.** ciliary body (holds lens in place, accommodation); **b.** lens (refracts and focuses light rays); **c.** iris (regulates light entrance); **d.** pupil (admits light); **e.** cornea (refracts light rays); **f.** fovea centralis (makes acute vision possible); **g.** optic nerve (transmits impulse to brain); **h.** sclera (protects and supports eyeball); **i.** choroid (absorbs stray light); **j.** retina (contains receptors for sight). See also Figure 47.6, page 843.

Understanding the Terms

a. sensory receptor; **b.** retina; **c.** tympanic membrane; **d.** optic nerve; **e.** sclera; **f.** cone cell; **g.** chemoreceptor; **h.** mechanoreceptor; **i.** spiral organ; **j.** cochlea

Chapter 48

Testing Yourself

1. b; **2.** f; **3.** c; **4.** e; **5.** b; **6.** b; **7.** b; **8.** c; **9.** d; **10.** d; **11.** **a.** actin filament; **b.** myosin filament; **c.** Z line; **d.** H zone; **e.** A band; **f.** I band. See also Figure 48.12, page 866.

Understanding the Terms

a. osteoblast; **b.** muscle action potential; **c.** osteoclast; **d.** sliding filament theory; **e.** actin; **f.** appendicular skeleton; **g.** axial skeleton; **h.** pectoral girdle; **i.** pelvic girdle; **j.** neuromuscular junction

Chapter 49

Testing Yourself

1. f; **2.** b; **3.** c; **4.** a; **5.** e; **6.** c; **7.** a; **8.** d; **9.** b; **10.** d; **11.** a; **12.** **a.** high sodium; **b.** inhibits; **c.** renin; **d.** angiotensin I and II; **e.** aldosterone. When the blood sodium level is low, the kidneys secrete renin. Renin stimulates the production of angiotensin I and II, which stimulate the adrenal cortex to release aldosterone. The kidneys absorb sodium. See also Figure 49.9, page 883.

Understanding the Terms

a. aldosterone; **b.** hormone; **c.** antidiuretic hormone (ADH); **d.** pineal gland; **e.** adrenocorticotropic hormone; **f.** thyroid gland; **g.** glucagon; **h.** growth hormone (GH)

Chapter 50

Testing Yourself

1. b; **2.** d; **3.** c; **4.** d; **5.** c; **6.** c; **7.** c; **8.** c; **9.** a; **10.** **a.** ureter; **b.** seminal vesicle; **c.** prostate gland; **d.** ejaculatory duct; **e.** bulbourethral gland; **f.** anus; **g.** epididymis; **h.** scrotum; **i.** testis; **j.** foreskin; **k.** glans penis; **l.** penis; **m.** urethra; **n.** vas deferens; **o.** bone; **p.** urinary bladder. Path of sperm: testis, epididymis, vas deferens, urethra (in penis). See also Figure 50.4, page 898.

Understanding the Terms

a. ovulation; **b.** metamorphosis; **c.** parthenogenesis; **d.** progesterone; **e.** semen; **f.** gonad

Chapter 51

Testing Yourself

1. b; **2.** b; **3.** b; **4.** a; **5.** b; **6.** b; **7.** d; **8.** a; **9.** a; **10.** d; **11.** **a.** chorion (exchanges wastes for nutrients with the mother); **b.** amnion (protects and prevents desiccation); **c.** embryo; **d.** allantois (blood vessels become umbilical blood vessels); **e.** yolk sac (first site of blood cell formation); **f.** fetal portion of placenta; **g.** maternal portion of placenta; **h.** umbilical cord (connects developing embryo to the placenta). See also Figure 51.9, page 922.

Understanding the Terms

a. induction; **b.** cleavage; **c.** germ layer; **d.** amnion; **e.** blastula; **f.** homeobox; **g.** morphogenesis; **h.** cellular differentiation; **i.** gastrula; **j.** placenta

APPENDIX

Periodic Table of the Elements

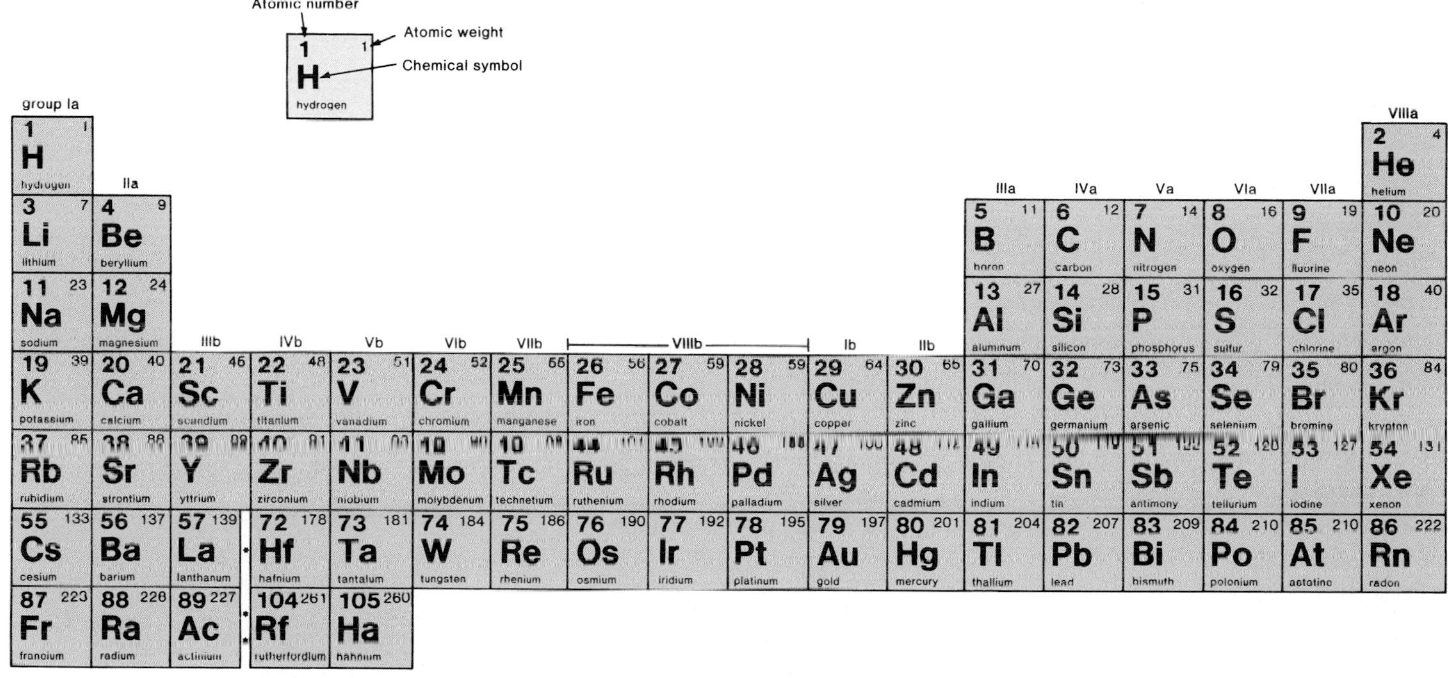

group Ia	IIa	IIIb	IVb	Vb	VIb	VIIb	VIIIb	VIIIb	VIIIb	Ib	IIb	IIIa	IVa	Va	VIa	VIIa	VIIIa
1 H hydrogen 1																	2 He helium 4
3 Li lithium 7	4 Be beryllium 9											5 B boron 11	6 C carbon 12	7 N nitrogen 14	8 O oxygen 16	9 F fluorine 19	10 Ne neon 20
11 Na sodium 23	12 Mg magnesium 24											13 Al aluminum 27	14 Si silicon 28	15 P phosphorus 31	16 S sulfur 32	17 Cl chlorine 35	18 Ar argon 40
19 K potassium 39	20 Ca calcium 40	21 Sc scandium 45	22 Ti titanium 48	23 V vanadium 51	24 Cr chromium 52	25 Mn manganese 55	26 Fe iron 56	27 Co cobalt 59	28 Ni nickel 59	29 Cu copper 64	30 Zn zinc 65	31 Ga gallium 70	32 Ge germanium 73	33 As arsenic 75	34 Se selenium 79	35 Br bromine 80	36 Kr krypton 84
37 Rb rubidium 85	38 Sr strontium 88	39 Y yttrium [illegible]	40 Zr zirconium 91	41 Nb niobium [illegible]	42 Mo molybdenum [illegible]	43 Tc technetium [illegible]	44 Ru ruthenium 101	45 Rh rhodium [illegible]	46 Pd palladium [illegible]	47 Ag silver [illegible]	48 Cd cadmium 112	49 In indium [illegible]	50 Sn tin 119	51 Sb antimony 122	52 Te tellurium 128	53 I iodine 127	54 Xe xenon 131
55 Cs cesium 133	56 Ba barium 137	57 La lanthanum 139 *	72 Hf hafnium 178	73 Ta tantalum 181	74 W tungsten 184	75 Re rhenium 186	76 Os osmium 190	77 Ir iridium 192	78 Pt platinum 195	79 Au gold 197	80 Hg mercury 201	81 Tl thallium 204	82 Pb lead 207	83 Bi bismuth 209	84 Po polonium 210	85 At astatine 210	86 Rn radon 222
87 Fr francium 223	88 Ra radium 226	89 Ac actinium 227 **	104 Rf rutherfordium 261	105 Ha hahnium 260													

*	58 Ce cerium 140	59 Pr praseodymium 141	60 Nd neodymium 144	61 Pm promethium 147	62 Sm samarium 150	63 Eu europium 152	64 Gd gadolinium 157	65 Tb terbium 159	66 Dy dysprosium 163	67 Ho holmium 165	68 Er erbium 167	69 Tm thulium 169	70 Yb ytterbium 173	71 Lu lutetium 175
**	90 Th thorium 232	91 Pa protactinium 231	92 U uranium 238	93 Np neptunium 237	94 Pu plutonium 242	95 Am americium 243	96 Cm curium 247	97 Bk berkelium 247	98 Cf californium 249	99 Es einsteinium 254	100 Fm fermium 253	101 Md mendelevium 256	102 No nobelium 254	103 Lr lawrencium 257

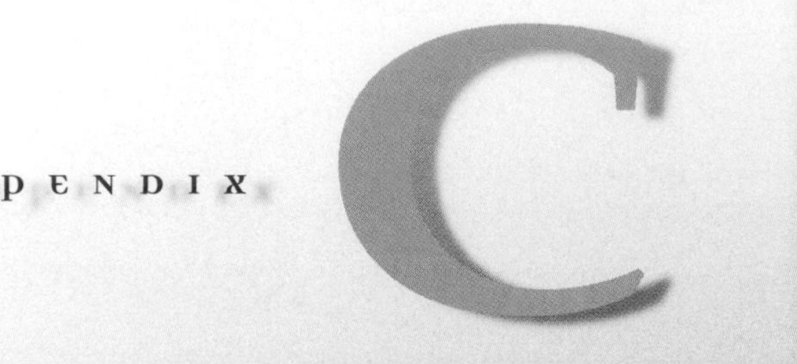

APPENDIX C METRIC SYSTEM

Unit and Abbreviation	Metric Equivalent	Approximate English-to-Metric Conversion Factor
Length		
nanometer (nm)	= 10^{-9} m	
micrometer (μm)	= 10^{-6} m	
millimeter (mm)	= 0.001 (10^{-3}) m	
centimeter (cm)	= 0.01 (10^{-2}) m	1 inch = 2.54 cm 1 foot = 30.5 cm
meter (m)	= 100 (10^{2}) cm = 1,000 mm	1 foot = 0.30 m 1 yard = 0.91 m
kilometer (km)	= 1,000 (10^{3}) m	1 mi = 1.6 km
Weight (mass)		
nanogram (ng)	= 10^{-9} g	
microgram (μg)	= 10^{-6} g	
milligram (mg)	= 10^{-3} g	
gram (g)	= 1,000 mg	1 ounce = 28.3 g 1 pound = 454 g
kilogram (kg)	= 1,000 (10^{3}) g	= 0.45 kg
metric ton (t)	= 1,000 kg	1 ton = 0.91 t
Volume		
microliter (μl)	= 10^{-6} l (10^{-3} ml)	
milliliter (ml)	= 10^{-3} liter = 1 cm^3 (cc) = 1,000 mm^3	1 tsp = 5 ml 1 fl oz = 30 ml
liter (l)	= 1,000 ml	1 pint = 0.47 liter 1 quart = 0.95 liter 1 gallon = 3.79 liters
kiloliter (kl)	= 1,000 liters	

°C	°F	
100	212	Water boils at standard temperature and pressure
71	160	Flash pasteurization of milk
57	134	Highest recorded temperature in the United States, Death Valley, July 10, 1913
41	105.8	Average body temperature of a marathon runner in hot weather
37	98.6	Human body temperature
20.3	68.6	Lowest recorded body temperature to be survived by a human
0	32.0	Water freezes at standard temperature and pressure

Units of Temperature

F° C°

212° — 210 | 100 — 100°
160° — 160 | 70 — 71°
134° | 57°
131° — 130
105.8° | 41°
98.6° — 100 | 37°
68.6° — 70 | 20 — 20.3°
32° — 30 | 0 — 0°

To convert temperature scales:

$$°C = \frac{5(°F-32)}{9}$$

$$°F = \frac{9°C}{5} + 32$$

The Five Kingdom System of Classification

APPENDIX

Kingdom Monera

Prokaryotic, unicellular organisms; heterotrophic by absorption, autotrophic by chemosynthesis or by photosynthesis; primarily asexual reproduction by binary fission but genetic exchange occurs by conjugation, transformation, transduction; motile forms move by flagella.

Group:

Aquificales Extremely thermophilic bacteria. *Aquifex, Hydrogenobacter.*

Thermotogales Extremely thermophilic bacteria; cells often surrounded by a toga. *Thermotoga, Thermosipho.*

Green nonsulfur bacteria Mostly phototrophic; some thermophilic. *Chloroflexus, Herpetosiphon, Thermomicrobium.*

Deinococci Atypical peptidoglycan cell wall; some Gram-positive; some Gram-negative thermophilic. *Deinococcus, Thermus.*

Proteobacteria Gram-negative bacteria; some phototrophic; some chemotrophic. *Rhodobacter, Nitrobacter, Escherichia, Salmonella, Shigella, Enterobacter, Pseudomonas, Rhizobium, Legionella.*

Gram-positive bacteria Includes endospore-forming bacteria, lactic acid bacteria, anaerobic and aerobic cocci, mycoplasmas; filamentous actinomycetes. *Staphylococcus, Clostridium, Bacillus, Mycoplasma, Mycobacterium, Streptomyces, Actinomyces.*

Cyanobacteria Phototrophic and producing oxygen. *Nostoc, Anabaena, Oscillatoria.*

Chlamydiae Lack peptidoglycan in cell wall; obligate intracellular parasites. *Chlamydia.*

Planctomycetes Gram-negative bacteria; divide by budding. *Planctomyces.*

Bacteroides and relatives Gram-negative rod-shaped bacteria; some fermentative anaerobes; some respiring aerobes. *Bacteroides, Fusobacterium, Cytophaga, Flavobacterium.*

Green sulfur bacteria Phototrophic. *Chlorobium.*

Spirochetes Gram-negative helical or spiral-shaped. *Spirochaeta, Borrelia, Treponema, Leptospira.*

Kingdom Protista

Eukaryotic, unicellular microorganisms and their immediate multicellular descendants; asexual and sexual reproduction; flagella and cilia with 9 + 2 microtubules. Nucleated algae, protozoa, slime molds, and water molds.

Phylum Chlorophyta Mostly freshwater, unicellular, colonial, or multicellular; chlorophylls *a* and *b*; starch accumulates within chloroplast. Cell wall; some with flagella. Green algae: *Chlamydomonas, Spirogyra, Oedogonim, Ulva.* 7,000 species.

Phylum Phaeophyta Almost all marine, often multicellular; common along rocky coasts; chlorophylls *a* and *c*. Brown algae: *Fucus, Laminaria, Nereocystis, Sargassum.* 1,500 species.

Phylum Chrysophyta Marine and freshwater, mostly unicellular; chlorophylls *a* and *c*. Cell wall in some contains silica. Diatoms and allies. 11,000 species.

Phylum Dinoflagella Mostly marine, unicellular; chlorophylls *a* and *c*. Cellulose plates for cell wall, two unequal flagella beat in grooves at right angles. Many are symbiotic and then known as zooxanthellae. Dinoflagellates. 1,000 species.

Phylum Euglenophyta Mostly freshwater, unicellular; often chlorophylls *a* and *b*. No cell wall; two flagella. Euglenoids: *Euglena.* 1,000 species.

Phylum Rhodophyta Mostly marine, multicellular; chlorophyll *a* and phycobilin like the cyanobacteria. Red algae. 4,000 species.

Phylum Sarcodina Freshwater and marine unicellular heterotrophs by ingestion; movement by pseudopods; no cell wall but marine forms may have shells; asexual reproduction. Amoebas and allies: *Amoeba proteus, Entamoeba histolytica.* 40,000 species.

Phylum Ciliophora Freshwater unicellular heterotrophs by ingestion; movement by cilia; no cell wall but some have a flexible pellicle; asexual reproduction with complex sexual exchange through conjugation. Ciliates: *Paramecium, Stentor, Vorticella.* 8,000 species.

Phylum Zoomastigophora Freshwater or parasitic unicellular heterotrophs by ingestion or absorption; movement by flagella. Asexual reproduction. Zooflagellates: *Giardia, Trichomonas, Trypanosoma* (African sleeping sickness).

Phylum Sporozoa Unicellular heterotrophs by absorption; nonmotile, spore-forming parasites of animals; complex life cycles. Sporozoa: *Plasmodium* (malaria), *Pneumocystis* (pneumonia), *Toxoplasma* (toxoplasmosis). 3,600 species.

Phylum Gymnomycota Individual amoeboid cells or multinucleated mass; heterotrophic by ingestion; nonmotile spores with rigid cell walls and motile cells with flexible walls are produced. Slime molds. 560 species.

Phylum Oomycota Filamentous mildews or molds with cellulose cell walls; heterotrophic by saprotrophism or parasitism; asexual reproduction by formation of zoospores; oogamy during sexual reproduction. Water molds. 580 species.

Kingdom Fungi

Multicellular eukaryotes; heterotrophic by absorption; lack flagella; nonmotile spores form during both asexual and sexual reproduction.

Division Zygomycota Hyphae nonseptate; asexual reproduction common by sporangiospores; sexual reproduction involves thick-walled zygospore. Zygospore fungi: soil and dung molds, black bread molds (*Rhizopus*). 600 species.

Division Ascomycota Hyphae septate; asexual reproduction common by conidiospores; sexual reproduction involves dikaryotic hyphae with formation of ascospores in asci. Sac fungi: many small wood-decaying fungi, yeasts (*Saccharomyces*), molds (*Neurospora*), morels, cup fungi, truffles; plant parasites: powdery mildews, ergots. 30,000 species.

Division Basidiomycota Hyphae septate; asexual reproduction by conidiospores; sexual reproduction common by long-lasting dikaryotic mycelium with formation of basidiospores on basidia. Club fungi: mushrooms, stinkhorns, puffballs, bracket and shelf fungi, coral fungi; plant parasites: rusts, smuts. 16,000 species.

Division Deuteromycota Hyphae septate; asexual reproduction common by conidiospores; sexual reproductive structures are not known. Imperfect fungi: athlete's foot, ringworm, candidiasis. 25,000 species.

Kingdom Plantae

Multicellular, primarily terrestrial eukaryotes with well-developed tissues; autotrophic by photosynthesis; alternation of generations life cycle. Like green algae, plants contain chlorophylls *a* and *b*, carotenoids; store starch in chloroplasts; cell wall contains cellulose.

Division Hepatophyta Strap-shaped (some leafy, some thallose), that creep along the substrate; dominant gametophyte and dependent sporophyte; mostly moist and shaded habitats. Liverworts. 10,000 species.

Division Bryophyta Stemlike, leaflike, and rootlike structures; dominant gametophyte and dependent sporophyte; mostly moist and shaded habitats. Mosses. 12,000 species.

Division Anthocerotophyta Thallose plants with green gametophytes; photosynthetic sporophytes at the upper surface of gametophyte; dominant gametophyte and dependent sporophyte; mostly moist and shaded habitats. Hornworts. 100 species.

Division Psilotophyta Rhizome with erect branched stems; no roots or leaves; terminal sporangia borne on stems; tropical and subtropical habitats. Whisk ferns. Several species.

Division Lycopodophyta In *Lycopodium,* rhizome with upright branches and roots; scalelike leaves (microphylls) cover stems and branches; sporangia mostly borne in terminal cones; mostly tropical and subtropical but also temperate to Arctic and dry habitats. Club mosses, spike mosses, quillworts. 1,000 species.

Division Equisetophyta In *Equisetum,* horizontal underground rhizome; jointed stems; whorled branches and scalelike leaves; sporangia borne in terminal cones; moist habitats. Horsetails. 15 species.

Division Pteridophyta Well-developed leaves (megaphylls) called fronds; sporangia on sporophylls; moist, shaded habitats common; sometimes dry areas. Ferns. 12,000 species.

Division Pinophyta Woody tree with thick trunk; needlelike or scalelike evergreen leaves (some shrubby); reproductive organs in pollen and seed cones; widespread distribution. Conifers: pines, firs, yews, redwoods, spruces. 550 species.

Division Cycadophyta Thick woody stem; large leaves resembling palm fronds; massive male and female cones; mostly tropical and subtropical. Cycads. 100 species.

Division Ginkgophyta Woody tree with long side branches; fan-shaped deciduous leaves; microsporangia in cones, naked ovules on stalks; gardens and parks. Maidenhair tree. One living species.

Division Gnetophyta In *Gnetum,* trees and climbing vines with large leathery leaves; moist tropical habitat. In *Ephedra,* a profusely branched shrub with small, scalelike leaves; desert habitat. In *Welwitschia,* massive woody, disk-shaped stem with straplike leaves; desert habitat. Gnetophytes. 70 species.

Division Magnoliophyta Trees, shrubs, herbs; most with broad leaves; herbs are perennials or annuals; reproductive organs in flower; pollen often carried by insects; double fertilization; seeds enclosed by fruits; widespread distribution; some aquatic. Flowering plants.

Class Liliopsida (monocot) Flower parts in threes or multiples of threes; parallel-veined leaves; one cotyledon. Grasses, lilies, pineapple. 65,000 species.

Class Magnoliopsida (dicot) Flower parts in fours, fives, or multiples of these; net-veined leaves; two cotyledons. Most flowering herbs, shrubs, trees. 170,000 species.

Kingdom Animalia

Multicellular organisms with well-developed tissues; usually motile; heterotrophic by ingestion, generally in a digestive cavity; diplontic life cycle.

Phylum Porifera Multicellular bodies perforated by many pores that admit water which exits through the only opening; internal skeleton of spicules; digestion within collar cells (choanocytes); regeneration possible. Mostly marine sponges. 5,000 species.

Phylum Cnidaria Two-layered radially symmetrical body; gastrovascular cavity has only one opening; characterized by presence of stinging cells (cnidocytes); regeneration possible; all aquatic and most marine. Hydras, jellyfishes. 9,000 species.

Class Cubozoa Medusa form square with toxic sting. Sea wasps, box jellyfish.

Class Anthozoa Polyp forms with no medusae. Sea anemones, corals.

Class Hydrozoa Polyp form dominant, often colonial, frequently with an alternation of generations. *Hydra,Obelia.*

Class Scyphozoa Medusa form dominant. *Aurelia.*

Phylum Ctenophora Free-swimming, often almost spherical animals; translucent, sometimes bioluminescent; plates of cilia and a thick layer of mesoglea; radially symmetrical with two very long, specialized tentacles. Comb jellies, sea walnuts. 90 species.

Phylum Platyhelminthes Three-layered bilaterally symmetrical body; gastrovascular cavity has only one opening; excretion by means of flame cells; complex hermaphroditic reproductive systems; regeneration possible. Flatworms. 13,000 species.

Class Turbellaria Ciliated, carnivorous with eyespots. Planaria.

Class Trematoda Parasitic with oral and ventral suckers; digestive cavity. Flukes.

Class Cestoda Parasitic with scolex having hooks and suckers and proglottids; no digestive cavity. Tapeworms.

Phylum Nemertea Usually marine worms with proboscis apparatus; acoelomate; a complete digestive system. Ribbon worms. 650 species.

Phylum Nematoda Nonsegmented worms; pseudocoelom and a tube-within-a-tube body plan; free living and parasitic. Roundworms: *Ascaris.* 50,000 species.

Phylum Rotifera Microscopic, pseudocoelomate wormlike or spherical animals with a corona (circle of cilia) on the head, the beating of which resembles a revolving wheel. Rotifers. 2,000 species.

Phylum Mollusca Body divided into foot, visceral mass, and mantle; usually also radula and shell; reduced coelom. Mollusks. 110,000 species.

Class Polyplacophora Body flattened dorsoventrally; shell consisting of eight dorsal plates; grazing marine herbivores. Chitons.

Class Bivalvia Body enclosed by a shell consisting of two valves; no head or radula; wedge-shaped foot; marine and freshwater sessile filter feeders. Clams, scallops, oysters, mussels.

Class Cephalopoda Tentacles about head; shell may be reduced or absent; head in line with elongated visceral mass, closed circulatory system; well-developed nervous system with cephalization, marine active predators. Squids, chambered nautilus, octopus.

Class Gastropoda Shell is coiled if present; body symmetry distorted by torsion; head with tentacles; marine, freshwater, and terrestrial grazing herbivores. Snails, slugs, nudibranchs.

Phylum Annelida Segmented worms with a long, cylindrical, soft body; protostome coelomates with internal septa; specialized digestive tract; definite central nervous system with brain and ventral solid nerve cord; closed circulatory system. Annelids. 12,000 species.

Class Polychaeta Marine with good cephalization; bundles of setae on parapodia. Clam worms (*Nereis*), tube worms.

Class Oligochaeta Fewer setae with no parapodia; no distinct head in terrestrial forms. Earthworms.

Class Hirudinea External parasites or scavengers, or with anterior and posterior suckers; parapodia and setae absent. Leeches.

Phylum Onychophora Iridescent green, blue-black, orange or whitish worms with 14 to 40 pairs of unjointed hollow legs, rigid due to hydrostatic pressure. Considered to have characteristics of both annelids and arthropods, the more or less cylindrical body shows no external segmentation except for the paired legs. The coelom is reduced to reproductive ducts and tiny sacs associated with paired nephridia. The open circulatory system does not distribute oxygen, and instead, there is a tracheal system somewhat different from that of insects. Onychophorans. 70 species.

Phylum Arthropoda Chitinous exoskeleton with jointed appendages specialized in structure and function; well-developed central nervous system with brain and ventral paired nerve cord; reduced coelom; hemocoel. Arthropods. Over 1 million species.

Subphylum Trilobitomorpha Three-lobed body with distinct head, thorax, and abdomen; serially repeated biramous appendages; extinct.

Subphylum Chelicerata Chelicerae, pedipalps, and four pairs of walking legs attached to a cephalothorax; no antennae, mandibles, or maxillae. Spiders, scorpions, horseshoe crabs.

Subphylum Crustacea Compound eyes and five pairs of walking appendages; antennae and antennules, mandibles and maxillae on head; biramous appendages on thorax and abdomen. Lobsters, crayfish, shrimps, crabs, many others.

Subphylum Uniramia Uniramous appendages; one pair of antennae, one pair of mandibles, and one or two pairs of maxillae; terrestrial with tracheae. Millipedes, centipedes, insects.

Phylum Echinodermata Radial symmetry; endoskeleton of spine-bearing plates; water vascular system with tube feet. Echinoderms. 6,000 species.

Class Crinoidea Filter feeders with feathery arms. Sea lilies, feather stars. 600 species.

Class Asteroidea Five arms project from central disk; movement by tube feet. Sea stars. 1,500 species.

Class Ophiuroidea Slender, long, often spiny, highly flexible arms from central disk. Brittle stars. 2,000 species.

Class Echinoidea No distinct arms; spines used for locomotion, defense, burrowing. Sea urchins, sand dollars. 950 species.

Class Holothuroidea Long, leathery body with tentacles about mouth. Sea cucumbers. 1,500 species.

Phylum Hemichordata Marine; wormlike body divided into proboscis, collar, trunk; pharyngeal gill slits and dorsal hollow nerve cord in proboscis. Hemichordates: acorn worms. 90 species.

Phylum Chordata Pharyngeal pouches; dorsal hollow nerve cord; notochord; post-anal tail. Chordates. 45,000 species.

Subphylum Urochordata Larvae free swimming with three chordate characteristics; adults sessile filter feeders with plentiful gill slits. Tunicates. 1,250 species.

Subphylum Cephalochordata Marine fishlike animals with three chordate characteristics as adults. Lancelets. 23 species.

Subphylum Vertebrata Notochord replaced by vertebrae that protect the nerve cord; skull that protects the brain; segmented with jointed appendages. Vertebrates. 43,700 species.

Superclass Agnatha Marine and freshwater fishes; lack jaws and paired appendages; cartilaginous plates added to notochordal skeleton; notochord. Lampreys, hagfishes. 63 species.

Superclass Gnathostomata Hinged jaws; paired appendages. Jawed fishes, tetrapods. 25,000 species.

Class Chondrichthyes Marine cartilaginous fishes; lack operculum and swim bladder; tail fin usually asymmetrical. Sharks, skates, rays. 850 species.

Class Osteichthyes Marine and freshwater bony fishes; operculum; swim bladder or lungs; tail fin usually symmetrical. Lungfishes, lobe-finned fishes, ray-finned fishes (herring, salmon, sturgeon, eels, sea horse). 20,000 species.

Class Amphibia Tetrapod with nonamniote egg; nonscaly skin; metamorphosis; three-chambered heart; ectothermic. Urodeles (salamanders, newts), anurans (frogs, toads). 3,900 species.

Class Reptilia Tetrapod with amniotic egg; scaly skin; ectothermic. Squamata (snakes, lizards), chelonians (turtles, tortoises); crocodilians (crocodiles and alligators). 6,000 species.

Class Aves Tetrapod with feathers; bipedal with wings; double circulation; endothermic. Sparrows, penguins, ostriches. 9,000 species.

Class Mammalia Tetrapods with hair, mammary glands; double circulation; endothermic; teeth differentiated. Monotremes (spiny anteater, duckbill platypus), marsupials (opossum, kangaroo), placental mammals (whales, rodents, dogs, cats, elephants, horses, bats, humans). 4,500 species.

Acronyms

APPENDIX

Acronym	Meaning
ACh	acetylcholine
AChE	acetylcholinesterase
ACTH	adrenocorticotropic hormone
ACV	acyclovir
AD	Alzheimer disease
ADH	antidiuretic hormone
ADP	adenosine diphosphate
AID	artificial insemination by donor
AIDS	acquired immunodeficiency syndrome
ANH	atrial natriuretic hormone
APC	antigen-presenting cell
ATP	adenosine triphosphate
AV node	atrioventricular node
AZT	azidothymidine
bGH	bovine growth hormone
BP	before present
CAM	crassulacean-acid metabolism
cAMP	cyclic adenosine monophosphate
CAPD	continuous ambulatory peritoneal dialysis
CCK	cholecystokinin
cDNA	complementary deoxyribonucleic acid
CFC	chlorofluorocarbon
cGMP	cyclic guanosine monophosphate
CHNOPS	carbon, hydrogen, nitrogen, oxygen, phosphorus, sulfur
CNS	central nervous system
CoA	coenzyme A
CVA	cardiovascular accident
CVD	cardiovascular disease
DES	diethylstilbestrol
DNA	deoxyribonucleic acid
DTP	diphtheria, tetanus, whopping cough
ECG (or EKG)	electrocardiogram
ELH	egg-laying hormone
ER	endoplasmic reticulum (rough ER, smooth ER)

Acronym	Meaning
EST	expressed sequence tag
FAD	flavin adenine dinucleotide
FAP	fixed action patterns
FAS	fetal alcohol syndrome
FSH	follicle-stimulating hormone
GABA	an inhibiting neurotransmitter
GDGF	glial-derived growth factor
GH	growth hormone
GIFT	gamete intrafallopian transfer
GIP	gastric inhibitory peptide
GMP	guanosine monophosphate
GnRH	gonadotropic-releasing hormone
GTP	guanosine triphosphate
Hb	deoxyhemoglobin
HbO_2	oxyhemoglobin
HC	hydrocarbon
HCG	human chorionic gonadotropin
HDL	high-density lipoprotein
HDN	hemolytic disease of the newborn
HepB	hepatitis B
HHb	hemoglobin
Hib	Haemophilus influenza, type b
HIV	human immunodeficiency virus
HPV	human papillomavirus
ICSH	interstitial cell-stimulating hormone
IUD	intrauterine device
IUI	intrauterine insemination
IVF	in vitro fertilization
LDC	less-developed country
LDL	low-density lipoprotein
LH	luteinizing hormone
LSD	lysergic acid diethylamide
MDC	more-developed country
MHC	major histocompatibility complex
MI	myocardial infarction

Acronym	Meaning
MMR	measles, mumps, rubella
MPPP	1-methyl-4-phenylprionoxy-piperidine
MSH	melanocyte-stimulating hormone
MYA	millions of years ago
NAD^+	nicotinamide adenine dinucleotide
$NADP^+$	nicotinamide adenine dinucleotide phosphate
NE	norepinephrine
NGF	nerve growth factor
NGU	nongonococcal urethritis
NO_X	nitrogen oxides
NPS	nail patella syndrome
OPV	oral polio vaccine
PAN	peroxyacetyl nitrate
PCB	polychlorinated biphenyl
P_{CO_2}	partial pressure for carbon dioxide
PCR	polymerase chain reaction
PG	prostaglandin
PGA	phosphoglycerate
PGAL	phosphoglyceraldehyde
PGAP	diphosphoglycerate
PHB	polyhydroxybutyrate
PID	pelvic inflammatory disease
PNS	peripheral nervous system
P_{O_2}	partial pressure for oxygen

Acronym	Meaning
PRL	prolactin
PS I	photosystem I
PS II	photosystem II
PTH	parathyroid hormone
rDNA	recombinant deoxyribonucleic acid
RFLP	restriction fragment length polymorphism
RNA	ribonucleic acid
rRNA	ribosomal ribonucleic acid
RuBP	ribulose bisphosphate
SA node	sinoatrial node
SAD	seasonal affective disorder
SCID	severe combined immunodeficiency syndrome
SEM	scanning electron microscope
SPF	sun protection factor
STD	sexually transmitted disease
Td	tetanus
TEM	transmission electron microscope
THC	tetrahydrocannabinol
tPA	tissue plasminogen activator
TRH	thyroid-releasing hormone
TRIH	thyroid-release-inhibiting hormone
TSH	thyroid stimulating hormone
UV	ultraviolet

Glossary

A

abscisic acid (ABA) (ab-SIZ-ik) Plant hormone that causes stomates to close and that initiates and maintains dormancy. 687

abscission (ab-SIZH-un) Dropping of leaves, fruits, or flowers from a plant. [L. *abscissus,* cut off] 687

acetylcholine (ACh) (uh-set-ul-KOH-leen) Neurotransmitter active in both the peripheral and central nervous systems. 822

acetylcholinesterase (AChE) (uh-set-ul-koh-luh-NES-tuh-rays) Enzyme that breaks down acetylcholine within a synapse. 822

acetyl-CoA Molecule made up of a two-carbon acetyl group attached to coenzyme A. During aerobic cellular respiration, the acetyl group enters the Krebs cycle for further breakdown. 134

acid Molecules tending to raise the hydrogen ion concentration in a solution and to lower its pH numerically. 30

acid deposition Return to earth as rain or snow of the sulfate or nitrate salts of acids produced by commercial and industrial activities. 476

actin Muscle protein making up the thin filaments in a sarcomere; its movement shortens the sarcomere, yielding muscle contraction. 867

action potential Nerve impulse; membrane potential changes in an active axon. 820

active site Region on the surface of an enzyme where the substrate binds and where the reaction occurs. 104

active transport Use of a plasma membrane carrier protein to move particles from a region of lower to higher concentration; it opposes equilibrium and requires energy. 90

adaptation Organism's modification in structure, function, or behavior suitable to the environment. [L. *ad,* toward, and *aptus,* fit, suitable] 4, 295

adaptive radiation Evolution of a large number of species from a common ancestor. 317

adenine (A) (AD-un-een) One of four nitrogen-containing bases in nucleotides composing the structure of DNA and RNA. 225

adipose tissue Connective tissue in which fat is stored. [L. *adipalis,* fatty] 720

ADP (adenosine diphosphate) (ah-den-ah-SEEN dy-FAHS-fayt) Nucleotide with two phosphate groups that can accept another phosphate group and become ATP. 51, 102

adrenal cortex Outer portion of the adrenal gland; secretes hormones such as mineralocorticoid aldosterone and glucocorticoid cortisol. 882

adrenal gland (uh-DREEN-ul) Gland that lies atop a kidney; the adrenal medulla produces the hormones epinephrine and norepinephrine and the adrenal cortex produces the corticoid hormones. [L. *ad,* toward, and *renis,* kidney] 882

adrenal medulla Inner portion of the adrenal gland; secretes the hormones epinephrine and norepinephrine. 882

adrenocorticotropic hormone (ACTH) (uh DREE-noh-kawrt-ih-koh-TROH-pik) Hormone secreted by the anterior lobe of the pituitary gland that stimulates activity in the adrenal cortex. 880

aerobic respiration Aerobic breakdown of glucose including glycolysis in cytosol and the Krebs cycle, and electron transport in mitochondria. Results in carbon dioxide and water, and typically 36 ATP. [Gk. *aeros,* air] 130

age structure diagram Representation of the number of individuals in each age group in a population. 391

agglutination (uh-gloot-un-AY-shun) Clumping of red blood cells due to a reaction between antigens on red blood cell plasma membranes and antibodies in the plasma. [L. *ad,* to, and *glutinis,* be sticky] 750

aldosterone (al-DAHS-tuh-rohn) Hormone secreted by the adrenal cortex that regulates the sodium and potassium ion balance of the blood. 813, 882

alga (pl., algae) Aquatic, plantlike organism carrying out photosynthesis and belonging to the kingdom Protista. [L. *alga,* seaweed] 524

allantois (uh-LANT-uh-wus) Extraembryonic membrane that accumulates nitrogenous wastes in birds and reptiles and contributes to the formation of umbilical blood vessels in mammals. [Gk. *allantos,* sausage] 922

allele (uh-LEEL) Alternative form of a gene—alleles occur at the same locus on homologous chromosomes. [Gk. *allelon,* reciprocal, parallel] 177

allopatric speciation (al-uh-PA-trik) Origin of new species in populations that are separated geographically. [Gk. *allo,* different, and *patri,* fatherland] 316

altruism Social interaction that has the potential to decrease the fitness of the member exhibiting the behavior. [L. *alter,* the other] 378

alveolus (pl., alveoli) (al-VEE-uh-lus) Terminal, microscopic, grapelike air sac found in vertebrate lungs. [L. *alveolus,* dim. of *alveus,* cavity] 795

amino acid Organic molecule having an amino group and an acid group, that covalently bonds to produce protein molecules. 46

amnion (AM-nee-ahn) Extraembryonic membrane of birds, reptiles, and mammals that forms an enclosing, fluid filled sac. [Gk. *amnion,* membrane around fetus] 922

amniote egg Egg that has an amnion, as seen during the development of reptiles, birds, and mammals. 624

amoeboid Protist that moves and engulfs prey with pseudopods; amoebalike in movement. 530

amphibian Member of a class of terrestrial vertebrates that includes frogs, toads, and salamanders; they are still tied to a watery environment for reproduction. [Gk. *amphibios,* living on both land and in water] 622

analogous structure Structures that have a similar function in separate lineages and differ in structure and origin. 497

androgen Male sex hormones; testosterone is an androgen. [Gk. *andros,* male, and L. *genitus,* producing] 887

angiosperm Flowering plant; the seeds are borne within a fruit. [Gk. *angion,* dim. of *angos,* vessel, and *sperma,* seed] 567, 637

Animalia Animal kingdom of organisms that includes eukaryotic heterotrophs that typically ingest food, are multicellular with specialized tissues, and are motile. 502

annelid Invertebrate of the phylum Annelida, which includes clam worms, tube worms, earthworms, and leeches; characterized by a segmented body. [L. *annelus,* dim. of *annulus,* ring] 600

anther In flowering plants, pollen-bearing portion of stamen. 568, 695

antheridium Reproductive organ found in nonvascular and some vascular plants which produces flagellated sperm. [Gk. *anthos,* flower, and *-idion,* small] 556

anthropoid (AN-thruh-poyd) Group of primates that includes only monkeys, apes, and humans. [Gk. *anthropos,* man, and *-eides,* like] 351

antibiotic Substance produced by a microorganism or semisynthetically that has the capacity to kill or inhibit growth of infectious microorganisms. 517

antibody Protein produced in response to the presence of an antigen; each antibody combines with a specific antigen [Gk. *anti,* against] 747, 760

anticodon Three nucleotides on a tRNA molecule attracted to a complementary codon on mRNA. 242

antidiuretic hormone (ant-ih-DY-yuu-RET-ik) (ADH) Hormone secreted by the posterior pituitary that increases the permeability of the collecting ducts in a kidney. [Gk. *anti,* against, and L. *ouresis,* urination] 812, 878

antigen Foreign substance, usually a protein or a polysaccharide, that stimulates the immune system to react, such as to produce antibodies. [Gk. *anti,* against, and L. *genitus,* forming, causing] 747, 760

aorta In humans, the major systemic artery that takes blood from the heart to the tissues. [L. *aorte,* great artery] 742

apoptosis (aPP-oh-TOE-sis) Process of programmed cell death involving a cascade of specific cellular events leading to the death and destruction of the cell. [Gk. *apo,* off, and *ptosis,* fall] 151, 263

aquifer (AHK-wuh-fur) Rock layers that contain water and will release it in appreciable quantities to wells or springs [L. *aqua,* water, and *fero,* to bear, carry] 435, 479

Archaea One of the three domains of life often found living in extreme habitats; prokaryotic cells that have unique genetic, biochemical, and physiological characteristics. 502, 519

arteriole Vessel that takes blood from an artery to capillaries. 736

artery Blood vessel that transports blood away from the heart. 736

arthropod (AR-throh-pahd) Invertebrate of the phylum Arthropoda, which includes lobsters, insects, and spiders; characterized by jointed appendages. [Gk. *arthron,* joint, and *podos,* foot] 603

ascus (pl., asci) Fingerlike sac in which nuclear fusion, meiosis, and ascospore formation occur during sexual reproduction of the sac fungi. [Gk. *askos,* bag, sac] 545

asexual reproduction Reproduction that requires only one parent and does not involve gametes. 148, 894

aster Short, radiating fibers produced by the centrosomes in animal cells. [Gk. *aster,* star] 153

asymmetry Body plan having no particular symmetry. 576

atom Smallest particle of an element that displays the properties of the element. [Gk. *atomos,* uncut, indivisible] 20

ATP (adenosine triphosphate) (ah-DEN-ah-seen try-FAHS-fayt) Nucleotide with three phosphate groups. The breakdown of ATP into ADP + P makes energy available for energy-requiring processes in cells. 51, 102

atrial natriuretic hormone (ANH) Hormone secreted by the heart that increases sodium excretion. 813

atrial ventricular valve Heart valve located between an atrium and a ventricle. 738

atrium (pl., atria) Chamber; particularly an upper chamber of the heart lying above a ventricle. 738

australopithecine (aw-stray-loh-PITH-uh-syn) One of several species of *Australopithecus,* a genus that contains the first generally recognized hominids. [L. *australis,* southern, *pithecus,* ape, and *ramus,* branch] 354

autonomic system (awt-uh-NAHM-ik) Division of the peripheral nervous system that regulates internal organs. [Gk. *autos,* self, and *nomas,* roving] 829

autotroph Organism that can make organic molecules from inorganic nutrients. [Gk. *autos,* alone, and *trophe,* food] 327

auxin (AHK-sun) Plant hormone regulating growth, particularly cell elongation; also called indoleacetic acid (IAA). [Gk. *auximos,* promoting growth] 572

axial skeleton (AK-see-ul) Part of the skeleton forming the vertical support or axis, including the skull, the rib cage, and the vertebral column. [L. *axi,* axis, hinge, and Gk. *skeleton,* dried body] 860

axon Elongated portion of a neuron that conducts nerve impulses typically from the cell body to the synapse. [Gk. *axon,* axis] 818

B

Bacteria One of the three domains of life; prokaryotic cells other than archaea with unique genetic, biochemical, and physiological characteristics. 502

bacteriophage Virus that infects bacteria. [Gk. *bacterion,* rod, and *phagein,* to eat] 223, 510

Barr body Dark-staining body (discovered by M. Barr) in the nuclei of female mammals that contains a condensed, inactive X chromosome. 256

base Molecules tending to lower the hydrogen ion concentration in a solution and raise its pH numerically. 30

basidium (pl., basidia) (buh-SID-ee-um) Clublike structure in which nuclear fusion, meiosis, and basidiospore production occur during sexual reproduction of club fungi. [Gk. *basis,* pedestal] 547

behavior Observable, coordinated responses to environmental stimuli. 368

benthic division (BEN-thik) Ocean floor, which supports a unique set of organisms in contrast to the pelagic division. [Gk. *benthos,* depths] 465

bicarbonate ion (HCO_3^-) Ion that participates in buffering the blood, and the form in which carbon dioxide is transported in the bloodstream. 797

bilateral symmetry Body plan having two corresponding or complementary halves. 576

bile Secretion of the liver that is temporarily stored and concentrated in the gallbladder before being released into the small intestine, where it emulsifies fat. 780

binary fission Splitting of a parent cell into two daughter cells; serves as an asexual form of reproduction in bacteria. [L. *binarius,* of two, and *fissura,* cleft, break] 148, 515

binomial system Assignment of two names to each organism, the first of which designates the genus and second of which is the specific epithet. 492

biogeochemical cycle (by-oh-jee-oh-KEM-i-kal) Circulating pathway of elements such as carbon and nitrogen from the environment, through biotic communities and also back to the environment. 434

biogeography Study of the geographical distribution of organisms. [Gk. *bios,* life, *geo,* earth, and *grapho,* writing] 290

biological clock Internal mechanism that maintains a biological rhythm in the absence of environmental stimuli. 679

biological magnification Process by which substances become more concentrated in organisms in the higher trophic levels of a food web. 480

biome (BY-ohm) Major terrestrial community characterized by certain climatic conditions and dominated by particular types of plants. 449

biosphere Zone of air, land, and water at the surface of the earth in which living organisms are found. [Gk. *bios,* life, and L. *sphaera,* ball] 6, 384, 428, 444

biotic potential Maximum population growth rate under ideal conditions. 387

bivalent (by-VAY-lent) Homologous chromosomes, each having sister chromatids that are joined by a nucleoprotein lattice during meiosis; also called tetrad. [L. *bis,* two, and *valens,* strength] 162

blastocoel Fluid-filled cavity of a blastula [Gk. *blastos,* bud, and *koiloma,* cavity] 915

blastocyst Early stage of human embryonic development that consists of a hollow fluid-filled ball of cells. 923

blastula (BLAST-chuh-luh) Hollow, fluid-filled ball of cells occurring during animal development prior to gastrula formation. [Gk. dim. of *blastos,* bud, and L. *ula,* little] 915

blood Type of connective tissue in which cells are separated by a liquid called plasma. 721

blood pressure Force of blood pushing against the inside wall of blood vessels. 743

B lymphocyte (LIM-fuh-syt) Lymphocyte that matures in the bone marrow and, when stimulated by the presence of a specific antigen, gives rise to antibody-producing plasma cells. [L. *lympha,* clear water, and Gk. *kytos,* cell] 760

bone Connective tissue in which the cells lie within lacunae embedded in a hard matrix of mineral salts deposited around protein fibers. 721

bronchiole (BRAHNG-kee ohl) Small tube that conducts air from a bronchus to the alveoli. [Gk. dim. of *bronchos,* windpipe] 795

bronchus (pl., bronchi) (BRAHNG-kus) Branch of the trachea in vertebrates that leads to lungs. [Gk. *bronchos,* windpipe] 795

bud In plants, undeveloped shoot, largely meristematic tissue covered by immature leaves and protected by bud scales; in animals, a protuberance from the body that develops into a new individual. 618

buffer Substance or group of substances that tend to resist pH changes in a solution, thus stabilizing its relative acidity and basidity. 31

C

C_3 photosynthesis Direct use of the Calvin cycle, in which CO_2 binds to ribulose bisphosphate (RuBP) to form two three-carbon phosphoglycerate (PGA) molecules; because CO_2 fixation produces a three-carbon product, this is called C_3 photosynthesis. 124

C_4 photosynthesis Involves CO_2 fixation resulting in the four-carbon oxaloacetate molecule; requires almost twice as much energy as C_3 photosynthesis, but it inhibits photorespiration and thus is advantageous in hot, dry climates. Used in numerous grasses and some other plant groups. 124

Calvin cycle Light-independent reactions of photosynthesis, carbon dioxide is fixed and reduced to yield a sugar-phosphate molecule that can be used to form glucose or regenerate RuBP, the first molecule of the cycle. The cycle uses ATP and NADPH from light-dependent reactions. 121

camouflage Method of hiding from predators in which the organism's behavior, form, and pattern of coloration allow it to blend into the background and prevent detection. 414

CAM plant Plant that fixes carbon dioxide at night to produce a C_4 molecule that releases carbon dioxide to the Calvin cycle during the day; CAM stands for crassulacean-acid metabolism. 124

cancer Malignant tumor whose nondifferentiated cells exhibit loss of contact inhibition, uncontrolled growth, and the ability to invade tissue and metastasize. 260

capillary Microscopic blood vessel; gas and other substances are exchanged across the walls of a capillary between blood and tissue fluid. [L. *capillus,* hair] 736

carbohydrate Class of organic compounds consisting of a carbon chain with hydrogen and oxygen atoms attached; includes monosaccharides, disaccharides, and polysaccharides. [L. *carbo,* charcoal, and Gk. *hydatos,* water] 39

carbon cycle Biogeochemical cycle in which carbon moves from the atmosphere through biotic communities and back to the atmosphere. 436

carbon dioxide (CO_2) fixation Binding of carbon dioxide to an organic molecule such as RuBP of the Calvin cycle. 122

carbonic anhydrase Enzyme in red blood cells that speeds the formation of carbonic acid from water and carbon dioxide. [Gk. *an,* without, and *hydrias,* water] 797

carcinogen (kar-SIN-uh-jen) Environmental agent that causes mutations leading to the development of cancer. [Gk. *carcino,* ulcer, and *gene,* origin] 264

cardiac muscle Striated, involuntary muscle tissue found only in the heart. [Gk. *kardia,* heart] 722

cardiovascular (or circulatory) system Organ system consisting of the blood, heart, and a series of blood vessels that distribute blood under the pumping action of the heart. [Gk. *kardia,* heart, and L. *vascular,* vessel] 734

carnivore Consumer in a food chain that eats other animals. 429

carpel (KAHR-pul) In flowering plants, a reproductive unit of a pistil; consisting of three parts—the stigma, the style, and the ovary. [Gk. *karpos,* fruit] 695

carrier Heterozygous individual who has no apparent abnormality but can pass on an allele for a recessively inherited genetic disorder. 215

carrier protein Protein that combines with and transports a molecule across the plasma membrane. 90

carrying capacity Largest number of organisms of a particular species that can be maintained indefinitely by a given environment. 389

cartilage Connective tissue in which the cells lie within lacunae embedded in a flexible proteinaceous matrix. [L. *cartilago,* gristle] 720

Casparian strip (kas-PAIR-ee-un) Layer of impermeable lignin and suberin bordering four sides of root endodermal cells; prevents water and solute transport between adjacent cells. 645, 662

catastrophism (ka-TAS-truh-fism) Belief espoused by Georges Cuvier that periods of catastrophic extinctions occurred, after which repopulation of surviving species took place, giving the appearance of change through time. [Gk. *katastrophe,* calamity, misfortune] 288

cell Smallest unit that displays the properties of life; composed of cytoplasm surrounded by a plasma membrane. 58

cell cycle Repeating sequence of cell growth and mitosis. 150

cell plate Structure across a dividing plant cell that signals the location of new plasma membranes and cell walls. 154

cell theory One of the major theories of biology which states that all organisms are made up of cells; cells are capable of self-reproduction and cells came only from pre-existing cells. 58

cellular respiration Metabolic reactions that use the energy from carbohydrate or fatty acid or amino acid breakdown to produce ATP molecules; includes fermentation, and aerobic respiration. 130

cellulose Polysaccharide that is the major complex carbohydrate in plant cell walls. 40

cell wall Structure that surrounds a plant, protistan, fungal or bacterial cell and maintains cell shape and rigidity. 63, 94

central nervous system (CNS) Brain and spinal cord of vertebrates. 824

centriole Cell organelle, existing in pairs, that occurs in the centrosome and may help organize a mitotic spindle for chromosome movement during animal cell division. [Gk. *centrum,* center] 74, 152

centromere (SEN-truh-mir) Constriction where sister chromatids of a chromosome are held together. [Gk. *centrum,* center, and *meros,* part] 149

centrosome (SEN-truh-sohm) Central microtubule organizing center of cells. In animal cells, it contains two centrioles. [Gk. *centrum,* center, and *soma,* body] 73, 152

cephalization (sef-uh-luh-ZAY-shun) Having a well-recognized anterior head with concentrated nerve masses and sensory receptors. [Gk. *kephale,* head] 584

cephalothorax Fused head and thorax regions of crustacea and some arachnids. [Gk. *kephale,* head, and *thorax,* breastplate] 604

cerebellum (ser-uh-BEL-um) Portion of the brain that coordinates skeletal muscles to produce smooth, graceful motions. 831

cerebral cortex Outer layer of cerebral hemispheres, receives sensory information and controls motor activities. 622

cerebral hemisphere Either of the two lobes of the cerebrum in vertebrates. 832

cerebrospinal fluid (suh-ree-broh-SPYN-ul) Fluid found in the ventricles of the brain, in the central canal of the spinal cord, and in association with the meninges. [L. *cerebrum,* brain, and *spina,* backbone] 830

cerebrum (suh-REE-brum) Main part of the brain consisting of two large masses, or cerebral hemispheres; the largest part of the brain in mammals. 832

chemical evolution Increase in the complexity of chemicals that could have led to the first cells. 325

chemiosmosis Ability of certain membranes to use a hydrogen ion gradient to drive ATP formation. 108

chemoreceptor Sensory receptor that is sensitive to chemical stimulation—for example, receptors for taste and smell. [Gk. *chemo,* pertaining to chemicals, and L. *receptor,* receiver] 840

chemosynthesis Process of making food by using atmospheric carbon dioxide and energy derived from the oxidation of inorganic compounds from the environment. [Gk. *chemo-,* pertaining to chemicals, *syn,* together, and *thesis,* an arranging] 516

chitin (KYT-un) Strong but flexible nitrogenous polysaccharide found in the exoskeleton of arthropods. [Gk. *chiton,* a tunic] 41, 603

chlorofluorocarbons (CFCs) Organic compounds containing carbon, chlorine, and fluorine atoms. CFCs like Freon can deplete the ozone shield by releasing chlorine atoms in the upper atmosphere. 478

chlorophyll Green pigment that absorbs solar energy and is important in algal and plant photosynthesis; occurs as chlorophyll *a* and chlorophyll *b*. 116

chloroplast Membrane-bounded organelle in algae and plants with chlorophyll-containing membranous thylakoids; where photosynthesis takes place. [Gk. *chloros,* green, and *plastos,* formed, molded] 70, 116

cholesterol One of the major lipids found in animal plasma membranes; makes the membrane impermeable to many molecules. 83

chordate Animals in the phylum Chordata that have a hollow dorsal nerve cord, a notochord, and pharyngeal gill pouches at some point in their life cycle; includes invertebrates such as lancelets, sea squirts, and vertebrates—the fishes, amphibians, reptiles, birds, and mammals. 616

chorion (KOR-ee-ahn) Extraembryonic membrane functioning for respiratory exchange in birds and reptiles; contributes to placenta formation in mammals. [Gk. *chorion,* membrane] 922

choroid (KOR-oyd) Vascular, pigmented middle layer of the eyeball. [Gk. *chorion,* membrane] 843

chromatin (KROH-mut-un) Network of fibrils consisting of DNA and associated proteins observed within a nucleus that is not dividing. [Gk. *chroma,* color, and *teino,* stretch] 66, 149

chromosomal theory of inheritance Theory that the genes are on the chromosomes, accounting for their similar behavior in the life cycle of organisms. 192

chromosome Structure consisting of DNA complexed with proteins that transmits genetic information from the previous generation of cells and organisms to the next generation. [Gk. *chroma,* color, and *soma,* body] 66, 148

cilium (pl., cilia) Short, hairlike projection from the plasma membrane, occurring usually in larger numbers. [L. *cilium,* eyelash, hair] 74

circadian rhythm Biological rhythm with a 24-hour cycle. [L. *circum,* about, and *dies,* day] 678, 888

clade In cladistics, a common ancestor and all the species descended from this common ancestor. [Gk. *klados,* branch, stem] 501

cladistics (kluh-DIS-tiks) School of systematics that determines the degree of relatedness by analyzing primitive and derived characters and constructing cladograms. 500

cladogram (KLAD-uh-gram) In cladistics, a branching diagram that shows the relationship among species in regard to their shared, derived characters. 500

classification Set of categories to which species are assigned on the basis of their evolutionary relationship to other species. 494

cleavage Cell division without cytoplasmic addition or enlargement; occurs during the first stage of animal development 594, 914

climate Weather condition of an area including especially prevailing temperature and average daily/yearly rainfall. 444

climax community In ecology, community that results when succession has come to an end. 421

clone Production of identical copies; in organisms, the production of organisms with same genes, in genetic engineering, the production of many identical copies of a gene. 271

cnidaria Invertebrate in the phylum Cnidaria existing as either a polyp or medusa with two tissue layers and radial symmetry. 580

cochlea (KOH-klee-uh) Spiral-shaped structure of the inner ear containing the sensory receptors for hearing. [L. *cochlea,* snail-shaped] 849

codominance Pattern of inheritance in which both alleles of a gene are equally expressed. 188

codon Three nucleotides of DNA or messenger RNA coding for a particular amino acid or termination of translation. 239

coelom Body cavity lying between the digestive tract and body wall that is completely lined by mesoderm. [Gk. *koiloma,* cavity] 576, 594

coenzyme Nonprotein organic molecule that aids the action of the enzyme to which it is loosely bound. 106

coevolution Joint evolution in which one species exerts selective pressure on the other species. 417, 698

cofactor Nonprotein adjunct required by an enzyme in order to function; many cofactors are metal ions, others are coenzymes. 106

cohesion-tension model Explanation for the transport of water to great heights in a plant due to a negative water potential (compared to the roots). This movement is brought about by transpiration and is dependent upon the ability of water molecules to cohere and to adhere to cell walls. 667

cohort Group of individuals having a statistical factor in common, such as year of birth, in a population study. 390

collecting duct Duct within the kidney that receives fluid from several nephrons; the reabsorption of water occurs here. 808

collenchyma Plant tissue composed of cells that fit closely together and with walls thickened at the angles of the cells; supports growth of stems and petioles. 642

colon Large intestine. 783

commensalism Symbiotic relationship in which one species is benefited, and the other is neither harmed nor benefited. 418, 517

common ancestor Ancestor held in common by at least two lines of descent. 494

communication Signal by a sender that influences the behavior of a receiver. 376

community Assemblage of populations interacting with one another within the same environment. 384, 406

companion cell Cell associated with sieve-tube cells in phloem of vascular plants. 664

competitive exclusion principle Theory that no two species can occupy the same niche. 410

complementary base pairing Hydrogen bonding between particular purines and pyrimidines in DNA. 51, 226

complement system Series of proteins in plasma that form a nonspecific defense mechanism against a microbe invasion; it complements the antigen-antibody reaction. 760

compound Substance having two or more different elements united chemically in fixed ratio. 24

compound eye Type of eye found in arthropods; it is composed of many independent visual units. 842

condensation Joining of monomers by covalent bonding with the accompanying loss of water molecules. 38

cone Structure comprised of scales which bear sporangia, pollen; cones bear microsporangia and seed cones bear megasporangia. 564

cone cell Photoreceptor in vertebrate eyes that responds to bright light and allows color vision. 844

conidiospore (kuh-NID-ee-uh-spohr) Spore produced by sac and club fungi during asexual reproduction. [Gk. *konis,* dust, and *spora,* seed] 545

conifer One of the four groups of gymnosperm plants; cone-bearing trees that include pine, cedar, and spruce. [L. *conus,* cone, and *fero,* carry] 564

conjugation Transfer of genetic material from one cell to another. [L. *conjugalis,* pertaining to marriage] 515, 526

connective tissue Type of animal tissue that binds structures together, provides support and protection, fills spaces, stores fat, and forms blood cells; adipose tissue, cartilage, bone, and blood are types of connective tissue. 720

consumer Organism that feeds on another organism in a food chain; primary consumers eat plants, and secondary consumers eat animals. 429

continental drift Movement of continents with respect to one another over the earth's surfaces. 338

control group Sample that goes through all the steps of an experiment except the one being tested; a standard against which results of an experiment are checked. 13

convergent evolution Similarity in structure in distantly related groups due to adaptation to the environment. 497

copulation Sexual union between a male and a female. [L. *copulatus,* join] 894

corepressor Molecule that binds to a repressor, allowing the repressor to bind to an operator in a repressible operon. 253

cork Outer covering of bark of trees; made of dead cells that may be sloughed off. 641

cork cambium Lateral meristem that produces cork. [L. *cambio,* exchange] 651

corpus luteum (KOR-pus LOOT-ee-um) Follicle that has released an oocyte and increases its secretion of progesterone. [L. *corpus,* body, and *luteus,* yellow] 903

cortex In animals, outer layer of an organ such as the cortex of the kidney or adrenal gland; in plants, ground tissue bounded by the epidermis and vascular tissue in stems and roots. 645

cotyledon (kaht-ul-EED-un) Seed leaf for embryo of a flowering plant; provides nutrient molecules for the developing plant before photosynthesis begins. [Gk. *cotyledon,* cup-shaped cavity] 640, 701

covalent bond (koh-VAY-lent) Chemical bond in which atoms share one pair of electrons. [L. *co,* together, with, and *valens,* strength] 25

Cro-Magnon Common name for the first fossils to be accepted as representative of modern humans. 360

crossing-over Exchange of segments between nonsister chromatids of a bivalent during meiosis. 163

crustacean Type of arthropod in the subphylum Crustacea named for their hard exoskeleton; head usually bears compound eyes and five pairs of appendages; all appendages on thorax and abdomen are biramous, having a gill branch and leg branch. 605

cuticle Waxy layer covering the epidermis of plants that protects the plant against water loss and disease-causing organisms. 554, 641, 667

cyanobacterium (sy-ah-noh-bak-TEE-ree-um) Photosynthetic bacterium that contains chlorophyll and releases oxygen; formerly called a blue-green alga. [Gk. *kyanos,* blue, and *bacterion,* rod] 517

cycad Type of gymnosperm with palmate leaves and massive cones that occurs in the tropics and subtropics. 566

cyclic AMP ATP-related compound that promotes chemical reactions in cells; as the second messenger in nonsteroid hormone transduction, it initiates an enzyme cascade. 875

cyclin Protein that cycles in quantity as the cell cycle progresses; combines with and activates the kinases that function to promote the events of the cycle. 150

cyst A sac that contains fluid; resting stage that contains reproductive bodies or embryos. 587

cytokinesis (syt-oh-kuh-NEE-sus) Division of the cytoplasm following mitosis and meiosis. 149

cytokinin (syt-uh-KY-nun) Plant hormone that promotes cell division; often works in combination with auxin during organ development in plant embryos. [Gk. *kytos,* cell, and *kineo,* move] 685

cytoplasm Contents of a cell between the nucleus (nucleoid) region of bacteria and the plasma membrane. 62

cytosine (C) (SYT-uh-seen) One of four nitrogen-containing bases in nucleotides composing the structure of DNA and RNA. 225

cytoskeleton Internal framework of the cell, consisting of microtubules, actin filaments, and intermediate filaments. [Gk. *kytos,* cell, and *skeleton,* dried body] 72

D

datum (pl., data) Fact or piece of information collected through observation and/or experimentation. 11

decomposer Organism, usually a bacterial or fungal species, that breaks down large organic molecules into elements that can be recycled in the environment. 429, 516

decomposition Process in which one or more substances break down into simpler molecular substances through the action of bacteria and fungi of decay. 429

deductive reasoning Process of logic and reasoning, using "if . . . then" statements. 12

deforestation Removal of trees from a forest in a way that ever reduces the size of the forest. 483

demographic transition Due to industrialization, a decline in the birthrate following a reduction in the death rate so that the population growth rate is lowered. 399

denatured Loss of an enzyme's normal shape so that it no longer functions; caused by a less than optimal pH and temperature. 46, 105

dendrite Part of a neuron that sends signals toward the cell body. [Gk. *dendron,* tree] 818

denitrification Conversion of nitrate or nitrite to nitrogen gas by bacteria in soil. 438

dependent variable Result or change that occurs when an experimental variable is utilized in an experiment. 13

derived character Structural, physiological, or behavioral trait that is present in a specific lineage and is not present in the common ancestor for several lineages. 496

dermis Deeper, thicker layer of the skin that consists of fibrous connective tissue and contains various structures such as sense organs. [Gk. *derma,* skin] 725

desertification (di-zurt-uh-fuh-KAY-shun) Transformation of marginal lands to desert conditions. 482

detrital food web Flow of nutrients and energy requiring decomposition of dead organic matter as a source food for other organisms. 433

detritivore Organism that feeds on dead organic matter; usually detritus feeding organisms other than bacteria and fungi. 429

detritus (dih-TRYT-us) Organic matter produced by the decay of a substance such as tissues and animal wastes. 429

deuterostome (DOOT-uh-ruh-stohm) Group of coelomate animals in which the second embryonic opening is associated with the mouth; the first embryonic opening, the blastopore, is associated with the anus. [Gk. *deuteros,* second, and L. *stoma,* mouth] 594, 614

diaphragm Dome-shaped muscularized sheet separating the thoracic cavity from the abdominal cavity in mammals. [Gk. *diaphragma,* partition, wall] 793

diastole (dy-AS-tuh-lee) Relaxation period of a heart during the cardiac cycle. [Gk. *diastole,* dilation, spreading] 740

dicotyledon (dy-KAHT-ul-eed-un) Flowering plant group; members have two embryonic leaves, (cotyledons) net-veined leaves, cylindrical arrangement of vascular bundles, flower parts in fours or fives, and other characteristics. [Gk. *dis,* two, and *cotyledon,* cup-shaped cavity] 568

differentially permeable Ability of plasma membranes to regulate the passage of substances into and out of the cell, allowing some to pass through and preventing the passage of others. 86

diffusion Movement of molecules from a region of higher to lower concentration; it requires no energy and tends to lead to an equal distribution. 87

dinoflagellate (dy-noh-FLAJ-uh-layt) Unicellular alga, with two flagella, one whiplash and the other located within a groove between protective cellulose plates, whose numbers periodically explode to cause a toxic "red tide" in ocean waters. [Gk. *dinos,* whirling, and L. *flagello,* whip] 528

diploid (2n) number Cell condition in which two of each type of chromosome is present. [Gk. *diplos,* twofold, and *-eides,* like] 149, 160

directional selection Outcome of natural selection in which an extreme phenotype is favored, usually in a changing environment. 310

disruptive selection Outcome of natural selection in which extreme phenotypes are favored over the average phenotype, leading to more than one distinct form. 312

distal convoluted tubule Final portion of a nephron that joins with a collecting duct; associated with tubular secretion. [L. *distantia,* far] 808

DNA (deoxyribonucleic acid) Nucleic acid polymer produced from covalent bonding of nucleotide monomers that contain the sugar deoxyribose; the genetic material of nearly all organisms. 50, 222

DNA fingerprinting Using DNA fragment lengths, resulting from restriction enzyme cleavage, to identify particular individuals. 273

DNA ligase (LY-gays) Enzyme that links DNA fragments; used during production of recombinant DNA to join foreign DNA to the vector DNA. [L. *ligo,* bind, tie] 270

DNA polymerase (PAHL-uh-muh-rays) During replication, an enzyme that joins the nucleotides complementary to a DNA template. 228

dominance hierarchy Social ranking within a group in which a higher ranking individual acquires more resources than a lower ranking individual. 373

dominant allele (uh-LEEL) Allele that exerts its phenotypic effect in the heterozygote; it masks the expression of the recessive allele. 177

dormancy In plants, a cessation of growth under conditions that seem appropriate for growth. 682

double fertilization In flowering plants, one sperm fuses with the egg and a second sperm fuses with the polar nucleus of an embryo sac. 700

doubling time Number of years it takes for a population to double in size. 399

duodenum (doo-uh-DEE-num) First part of the small intestine where chyme enters from the stomach. 779

E

echinoderm Member of a phylum of marine animals that includes sea stars, sea urchins, and sand dollars; characterized by radial symmetry and a water vascular system. [Gk. *echinos,* spiny, and *derma,* skin] 614

ecological niche Role an organism plays in its community, including its habitat and its interactions with other organisms. 409

ecological pyramid Pictorial graph representing biomass, organism number, or energy content of each trophic level in a food web—from the producer to the final consumer populations. 433

ecological succession Change following a disturbance in an ecosystem; primary succession occurs in regions with no soil and secondary succession occurs in regions that already have soil. 420

ecology Study of the interactions of organisms with other organisms and with the physical and chemical environment. [Gk. *oikos,* home, house, and *-logy,* "study of" from *logikos,* rational, sensible] 384

ecosystem Biological community together with the associated abiotic environment and characterized by a flow of energy and a cycling of inorganic nutrients. [Gk. *oikos,* home, house, and *systema,* ordered arrangement] 2, 384, 428

ectoderm In animal embryos, the outermost primary tissue layer that gives rise to the nervous system and the outer layer of the integument. 915

ectothermic Having a body temperature that varies according to the environmental temperature. [Gk. *ekto,* outer, and *therme,* heat] 622

electromagnetic spectrum Solar radiation divided on the basis of wavelength, with gamma rays having the shortest wavelength and radio waves having the longest wavelength. 114

electron Negative subatomic particle, moving about in an energy level around the nucleus of an atom. [Gk. *elektron,* amber, electricity] 20

electron transport system Passage of electrons along a series of membrane-bound carrier molecules from a higher to lower energy level; the energy released is used for the synthesis of ATP. 107, 118, 131

element Substance that cannot be broken down into substances with different properties; composed of only one type atom. 20

El Niño Warming of water in the Eastern Pacific equatorial region such that the Humboldt Current is displaced with possible negative results such as a reduction in marine life. 446

embryo Stage of a multicellular organism that develops from a zygote before it becomes free living; in seed plants the embryo is part of the seed. 700, 914

embryo sac Megagametophyte of flowering plants that contains an egg cell. 697

endocrine gland Ductless organ that secretes (a) hormone(s) into the bloodstream. 719

endocrine system One of the major systems that along with the nervous system coordinate body activities; uses chemical messengers called hormones, which are often secreted into the bloodstream. 873

endocytosis (en-doh-sy-TOH-sis) Process by which particles or debris are moved into the cell from the environment by phagocytosis (cellular eating) or pinocytosis (cellular drinking; includes receptor-mediated endocytosis). [Gk. *endon,* within, and *kytos,* cell] 92

endoderm In animal embryos, the inner most primary tissue layer that gives rise to the linings of the digestive tract and associated structures. 915

endodermis Internal plant root tissue forming a boundary between the cortex and the vascular cylinder. [Gk. *endon*, within, and *derma*, skin] 645, 666

endometrium (en-doh-MEE-tree-um) Mucous membrane lining the interior surface of the uterus. [Gk. *endon*, within, and *metra*, womb] 902

endoplasmic reticulum (en-doh-PLAZ-mik reh-TIK-yoo-lum) System of membranous saccules and channels in the cytoplasm, often with attached ribosomes. [Gk. *endon*, within, plasma, something molded, and L. *reticulum*, net] 67

endosperm In flowering plants, nutritive storage tissue that is derived from the fusion of a sperm cell and polar nuclei in the embryo sac. [Gk. *endon*, within, and *sperma*, seed] 700

endospore Bacterium that has shrunk its cell, rounded up within the former plasma membrane, and secreted a new and thicker cell wall in the face of unfavorable environmental conditions. [Gk. *endon*, within, and *spora*, seed] 515

energy Capacity to do work and bring about change; occurs in a variety of forms. 3, 22, 100

entropy Measure of disorder or randomness. 100

environmental resistance Sum total of factors in the environment that limit the numerical increase of a population in a particular region. 387

enzyme Organic catalyst, usually a protein, that speeds up a reaction in cells due to its particular shape. 46, 103

epidermis In plants, tissue that covers roots and leaves and stems of nonwoody organisms; in mammals, the outer, protective layer of the skin. [GK. *epi*, over, and *derma*, skin] 641, 725

epiglottis (ep-uh-GLAHT-us) Structure that covers the glottis, the air-tract opening, during the process of swallowing. [Gk. *epi*, over, and *glotta*, tongue] 778, 794

epiphyte (EP-uh-fyt) Plant that takes its nourishment from the air because its placement in other plants gives it an aerial position. [Gk. *epi*, over, and *phyton*, plant] 454, 663

epithelial tissue Type of animal tissue forming a continuous layer over most body surfaces (i.e., skin) and inner cavities; squamous, cuboidal, and columnar are the three types of epithelial tissue. [Gk. *epi*, over, and L. *theca*, case, container] 718

erythrocyte Red blood cell that contains hemoglobin and carries oxygen from the lungs or gills to the tissues in vertebrates. [Gk. *erythros*, red, and *kytos*, cell] 747

esophagus (i-SAHF-uh-gus) Muscular tube for moving swallowed food from the pharynx to the stomach in vertebrates. [Gk. *eso*, within, and *phagein*, eat] 778

estrogen In mammals, sex hormones that cause the endometrium of the uterus to proliferate during the uterine cycle; along with progesterone, estrogens maintain mammalian secondary sexual characteristics in females. 887, 903

estuary End of a river where fresh water and salt water mix as they meet. 462

ethylene (ETH-uh-leen) Plant hormone that causes ripening of fruit and is also involved in abscission. 687

eubacterium (yoo-bak-TEE-ree-um) Most common type of bacteria, including the cyanobacteria. [Gk. *eu*, good, and *bact*, rod] 518

euchromatin (yoo-KROH-mut-un) Diffuse chromatin, which is being transcribed. 256

euglenoid (yuu-GLEE-noyd) Flagellated and flexible freshwater unicellular organism that usually contains chloroplasts and is often characterized as having both animal-like and plantlike characteristics. 528

Eukarya One of the three domains of life consisting of organisms in the kingdoms Protista, Fungi, Plantae, and Animalia which have eukaryotic cells with unique genetic, biochemical, and physiological characteristics. 502

eukaryotic cell (yoo-kair-ee-AHT-ik) Type of cell that has a nucleus and membranous organelles; found in organisms of the eukaryal domain. [Gk. *eu*, true, and *karyon*, kernel, nucleus] 63

eutrophication In aquatic ecosystems, rapid nutrient cycling, high productivity by phytoplankton and relatively few species. 460

evolution Descent of organisms from common ancestors with the development of genetic and phenotypic changes over time that make them more suited to the environment. [L. *evolutio*, an unrolling] 4, 287

excretion Eliminaton of metabolic wastes by an organism at exchnage boundaries such as the plasma membrane of unicellular organisms and respiratory surfaces, excretory tubules and the integument of multicellular animals. 803

exhalation Stage during respiration when air is pushed out of the lungs; expiration. 793

exocrine gland Gland that secretes its product to an epithelial surface directly or through ducts. 719

exocytosis (ek-soh-sy-TOH-sis) Process by which particles or debris are moved out of the cell by vesicles that fuse with the plasma membrane. [Gk. *ex*, out of, and *kytos*, cell] 92

exon Portion of an interrupted gene that is expressed as the result of polypeptide formation. 241

exoskeleton (ek-soh-SKEL-ut-un) Protective external skeleton, as in arthropods. [Gk. *ex*, out of, and *skeleton*, dried body] 603

experimental variable Component that is tested in an experiment by manipulating it and observing the results. 13

exponential growth Growth, particularly of a population, in which the increase occurs in the same manner as compound interest. 387

extinction Total disappearance of a species or higher group. 340

extraembryonic membrane Membrane that is not a part of the embryo but is necessary to the continued existence and health of the embryo. [L. *extra*, on the outside] 895, 922

F

FAD Flavin adenine dinucleotide; a coenzyme that becomes $FADH_2$ and delivers electrons to the electron transport system in mitochondria during aerobic cellular respiration. 130

fall overturn Mixing process that occurs in fall in stratified lakes whereby the oxygen-rich top waters mix with nutrient-rich bottom waters. 460

fat Organic molecule that contains glycerol and fatty acids and is found in adipose tissue of vertebrates. 42

feedback inhibition Control mechanism whereby an increase in the concentration of some molecule inhibits the synthesis of that molecule or regulates the level of that molecule; important in the regulation of enzyme, hormone, and ion blood levels. 106

fermentation Anaerobic breakdown of glucose that results in a gain of two ATP and end products such as alcohol and lactate. 141

fern Seedless vascular plant characterized by large triangular fronds as the sporophyte and an independent heart-shaped prothallus as the gametophyte. 562

fertilization Fusion of sperm and egg nuclei producing a zygote which develops into a new individual. [L. *fertilis*, fruitful] 700, 914

fibroblast Cell type of loose and fibrous connective tissue with cells at some distance from one another and separated by a jellylike matrix containing collagen and elastin fibers. [L. *fibra*, thread, and Gk. *blastos*, bud] 720

fibrous root system In most monocots, a mass of similarly sized roots that cling to the soil. 646

fin In fish and other aquatic animals, membranous, winglike or paddlelike process used to propel, balance, or guide the body. 620

fish Type of vertebrate living in marine and fresh waters with gills, usually fins, and flattened body covered with scales; may be jawless, cartilaginous, or bony. 619

fitness Ability of an organism to survive and reproduce in its environment. 292, 310

flagellum (pl., flagella) Slender, long extension used for locomotion by some bacteria, protozoa, and sperm. [L. *flagello*, whip] 74, 514

flatworm Invertebrate in the phylum Platyhelminthes that has a flat body with three germ layers but no coelom or organ systems; planaria are free living but flukes and tapeworms are parasites. 584

flower Reproductive organ of a flowering plant, consisting of several kinds of modified leaves arranged in concentric rings and attached to a modified stem called the receptacle. 694

fluid-mosaic model Model for the plasma membrane based on the changing location and pattern of protein molecules in a fluid phospholipid bilayer. 82

follicle Structure in the ovary of animals that contains an oocyte; site of oocyte production. [L. dim. of *folliculus*, bag] 903

follicle-stimulating hormone (FSH) Gonadotropic hormone secreted by the anterior pituitary that promotes production of eggs in females and sperm in males. 904

food chain Succession of organisms in an ecosystem that are linked by an energy flow and by the order of who eats whom. 433

food web In ecosystems, complex pattern of interlocking and crisscrossing food chains. 433

foramen magnum (fuh-RAY-mun MAG-num) Opening in the occipital bone of the vertebrate skull through which the spinal cord passes. [L. *foramen*, hole, and *magnus*, great, large] 860

fossil Any past evidence of an organism that has been preserved in the earth's crust. [L. *fossilis*, dug up] 328

fossil fuel Fuels such as oil, coal, and natural gas that are the result of partial decomposition of plants and animals coupled with exposure to heat and pressure for millions of years. 473

founder effect Cause of genetic drift due to colonization by a limited number of individuals who, by chance, have different genotypic and allelic frequencies than the parent population. 309

free energy Useful energy in a system that is capable of performing work. 101

fruit Flowering plant structure consisting of one or more ripened ovaries that usually contain seeds. 568, 702

fruiting body Spore-producing and spore-disseminating structure found in sac and club fungi. 545

Fungi Kingdom of life that includes eukaryotic, multicellular saprotrophs, which form spores and lack flagella throughout their life cycle.

fungus (pl., fungi) Saprotrophic decomposer; the body is made up of filaments called hyphae that form a mass called a mycelium. 496, 540

G

gallbladder Organ attached to the liver that serves to store and concentrate bile. 780

gamete (GAM-eet) Haploid sex cell. 160

gametophyte (guh-MEET-uh-fyt) Haploid generation of the alternation of generations life cycle of a plant; it produces gametes that unite to form a diploid zygote. [Gk. *phyton*, plant] 555, 694

ganglion (pl., ganglia) (GANG-glee-un) Collection or bundle of neuron cell bodies usually outside the central nervous system. [Gk. *ganglion*, swelling under the skin] 825

gastrovascular cavity Blind digestive cavity that also serves a circulatory (transport) function in animals that lack a circulatory system. [Gk. *gastros*, stomach, and L. *vasculum*, dim. of *vas*, vessel] 580

gastrula (GAS-truh-luh) Stage of animal development during which the germ layers form, at least in part, by invagination. [Gk. dim. of *gastros*, stomach] 915

gene Unit of heredity existing as alleles on the chromosomes; in diploid organisms typically two alleles are inherited—one from each parent. 4

gene flow Sharing of genes between two populations through interbreeding. 308

gene locus Specific location of a particular gene on homologous chromosomes. 177

gene pool Total of all the genes of all the individuals in a population. 305

gene therapy Use of bioengineered cells or other biotechnology techniques to treat human genetic disorders. 278

genetic drift Change in the genotypic frequencies and allelic frequencies of a population due to chance (random) events; important in small populations or when only a few individuals mate. 308

genetic engineering Use of technology to alter the genome of a living cell for medical or industrial use. 270

genome Full set of genes in an individual whether a haploid or diploid organism or a virus. 271

genotype Genes of an organism for a particular trait or traits; for example, *BB* or *Aa*. [Gk. *genos*, birth, origin, race, and *typos*, image, shape] 177

genus Taxon above the species level, designated by the first word of a species' binomial name, containing species that are most closely related through evolution. 495

germination Beginning of growth of a seed, spore, or zygote, especially after a period of dormancy. 688

germ layer Primary tissue layer of a vertebrate embryo; namely, ectoderm, mesoderm, or endoderm. 915

gibberellin (jib-uh-REL-un) Plant hormone producing increased stem growth; also involved in flowering and seed germination. [L. dim. of *gibbus*, bent] 682

gill Respiratory organ in most aquatic animals; in fish, an outward extension of the pharynx. 616, 791

ginkgo Type of gymnosperm with long side branches, fan-shaped deciduous leaves with pollen cones, and fleshy-coated seeds on stalks. 566

girdling Removing a strip of bark from around a tree. 670

glomerular capsule (glu-MER-uh-lur) Cuplike structure that is the initial portion of a nephron. [L. *glomeris*, ball] 808

glomerulus (glu-MER-uh-lus) Capillary network within a glomerular capsule of a nephron. [L. *glomeris*, ball] 809

glottis (GLAHT-us) Opening for airflow in the larynx. [Gk. *glotti*, tongue] 794

glucagon Hormone secreted by the pancreas which causes the liver to break down glycogen, raising the blood glucose level. 885

glycolysis Anaerobic breakdown of glucose that results in a gain of two ATP and the end product pyruvate. 131

gnetophyte Type of gymnosperm occuring as a tree with climbing vines having large leathery leaves *(Gnetum)*, profusely branched shrub *(Ephedra)*, and massive woody, disk-shaped stem with straplike leaves *(Welwitschia)*. 567

Golgi apparatus Organelle consisting of saccules and vesicles that processes, packages, and distributes molecules about or from the cell. 68

gonad Organ that produces gametes; the ovary, which produces eggs, and the testis, which produces sperm. [Gk. *gone*, seed] 877, 894

gonadotropic hormone (goh-nad-uh-TRAHP-ik) Type of hormone that regulates the activity of the ovaries and testes; principally follicle-stimulating hormone (FSH) and luteinizing hormone (LH). 880

granum (pl., grana) (GRAY-num) Stack of chlorophyll-containing thylakoids in a chloroplast. 71, 116

gravitropism (grav-ih-TRUH-piz-um) Directional growth of plants in response to the earth's gravity; roots demonstrate positive gravitropism, and stems demonstrate negative gravitropism. [L. *gravis*, heavy, and Gk. *tropos*, turning] 677

grazing food web In ecosystems, interconnected food chains that begin with an autotroph (producer) and continue through a series of heterotrophs (consumers) to a top carnivore. 433

greenhouse effect Reradiation of solar heat toward the earth, caused by gases such as carbon dioxide, methane, nitrous oxide, water vapor, ozone, and nitrous oxide in the atmosphere. 474

ground tissue Tissue that constitutes most of the body of a plant; consists of parenchyma, collenchyma, and sclerenchyma cells which function in storage, basic metabolism, and support. 641

growth hormone (GH) Substance secreted by the anterior pituitary; it promotes cell division, protein synthesis, and bone growth. 879

guanine (G) (GWAHN-een) One of four nitrogen-containing bases in nucleotides composing the structure of DNA and RNA. 225

guard cell Type of plant cell that is found in pairs, with one on each side of a leaf stomate; changes in the turgor pressure of these cells regulate the size and passage of gases through the stomate. 668

guttation (guh-TAY-shun) Liberation of water droplets from the edges and tips of leaves. [L. *gutta*, drops, spots] 666

gymnosperm (JIM-nuh-sperm) Type of woody seed plant in which the seeds are not enclosed by fruit and are usually borne in cones such as those of the conifers. [Gk. *gymnos*, naked, and *sperma*, seed] 564

H

habitat Place where an organism lives and is able to survive and reproduce. 384, 409

haploid (n) number Cell condition in which only one of each type of chromosome is present. [Gk. *haplos*, single, and *-eides*, like] 149, 160

Hardy-Weinberg law Law stating that the allele frequency in a population remains stable under certain assumptions, such as random mating; therefore, no change or evolution occurs. 306

heart Muscular organ whose contraction causes blood to circulate in the body of an animal. 738

hemoglobin Iron-containing respiratory pigment occurring in vertebrate red blood cells and in blood plasma of many invertebrates. [Gk. *haima*, blood, and L. *globus*, ball] 747, 797

herbaceous plant Plant that lacks persistent woody tissue. [L. *herba*, vegetation, plant] 567

herbivore Primary consumer in a food chain; a plant eater. 429

hermaphroditic (hur-MAF-ruh-DIT-ik) Characterizes an animal having both male and female sex organs. 585

heterochromatin (het-uh-roh-KROH-mut-un) Highly compacted chromatin that is not being transcribed. 256

heterotroph Organism that cannot synthesize organic compounds from inorganic substances and therefore must take in preformed food. [Gk. *hetero*, different, and *trophe*, food] 327

heterozygous Possessing unlike alleles for a particular trait. [Gk. *hetero*, different, and *zygos*, balance, yoke] 177

homeobox 180 nucleotide sequence located in all homeotic genes and serving to identify portions of the genome, in many different types of organisms, that are active in pattern formation. 921

homeostasis Maintenance of normal internal conditions in a cell or an organism by means of a self-regulating mechanism. [Gk. *homoios*, like, resembling, and *stasis*, standing] 3, 728

hominid Member of the family hominid containing humans and their direct ancestors which are known only by the fossil record. 353

hominoid Member of a superfamily containing humans and the great apes. [L. *homo*, man, and Gk. *-eides*, like] 351

Homo erectus Hominid dated between 1.9 and 0.5 million years ago; first to have a posture and locomotion similar to modern humans. [L. *homo*, man, and *erectus*, upright] 358

Homo habilis Hominid of 2 million years ago; possibly a direct ancestor of modern humans. [L. *homo*, man, and *habilis*, suitable, handy] 357

homologous chromosome (hoh-MAHL-uh-gus) Member of a pair of chromosomes that carry genes for the same traits and synapse during prophase of the first meiotic division; a homologue. [Gk. *homologos*, agreeing, corresponding] 160

homologous structure In evolution, a structure that is similar in different organisms because these organisms are derived from a common ancestor. [Gk. *homologos*, agreeing, corresponding] 298, 497

homology Similarity in structure due to having a common ancestor. [Gk. *homologos*, agreeing, corresponding] 497

homozygous Possessing two identical alleles for a particular trait. [Gk. *homo*, same, and *zygos*, balance, yoke] 177

hormone Chemical messenger produced in one part of the body that controls the activity of other parts. [Gk. *hormao*, instigate] 680, 874

horsetail Common name for seedless vascular plants of the genus *Equisetum* which have hollow, jointed stems and an encircling sheath of small leaves at each joint. 561

host Organism that provides nourishment and/or shelter for a parasite. 416

hybridization Crossing of different varieties or species. [L. *hybrida*, mongrel] 708

hydrogen bond Weak bond that arises between a slightly positive hydrogen atom of one molecule and a slightly negative atom of another molecule or between parts of the same molecule. 26

hydrologic cycle Interdependent and continuous circulation of water from the ocean, to the atmosphere, to the land, and back to the ocean. 435

hydrolysis (hy-DRAH-lih-sis) Splitting of a compound by the addition of water, with the H^+ being incorporated in one fragment and the OH^- in the other. 38

hydrophilic (hy-druh-FIL-ik) Type of molecule that interacts with water by dissolving in water and/or by forming hydrogen bonds with water molecules. [Gk. *hydrias*, of water, and *phileo*, love] 28, 37

hydrophobic (hy-druh-FOH-bik) Type of molecule that does not interact with water because it is nonpolar. [Gk. *hydrias*, of water, and *phobos*, fear] 28, 37

hydroponics (hy-druh-PAHN-iks) Water culture method of growing plants that allows an experimenter to vary the nutrients and minerals provided so as to determine the essential nutrients. [Gk. *hydrias*, of water, and *ponos*, hard work] 660

hydrosphere Portion of the biosphere that is water whether liquid, ice, or water vapor. 446

hypertonic solution Higher solute concentration (less water) than the cytosol of a cell; causes cell to lose water by osmosis. 89

hypha (pl., hyphae) Filament of the vegetative body of a fungus. [Gk. *hyphe*, web] 540

hypothalamus (hy-poh-THAL-uh-mus) Part of the brain that helps regulate the internal environment of the body; involved in control of heart rate, body temperature, water balance, and glandular secretions of the stomach and pituitary gland. [Gk. *hypo*, under] 831, 878

hypothesis Supposition that is established by reasoning after consideration of available evidence; it can be tested by obtaining more data, often by experimentation. [Gk. *hypothesis*, assumption] 11

hypotonic solution Lower solute (more water) concentration than the cytosol of a cell; causes cell to gain water by osmosis. 89

I

immunity Ability of the body to protect itself from foreign substances and cells, including infectious microbes. [L. *immunis,* exempt, free] 758

implantation In placental mammals, the embedding of an embryo at the blastocyst stage into the endometrium of the uterus. 922

imprinting Form of learning that occurs early in the lives of animals; a close association is made that later influences sexual behavior. 371

inclusive fitness Fitness that results from direct selection and indirect selection. 378

incomplete dominance Pattern of inheritance in which the offspring shows characteristics intermediate between two extreme parental characteristics—for example, a red and a white flower producing pink offspring. 188

inducer Molecule that brings about activity of an operon by joining with a repressor and preventing it from binding to the operator. 252

induction Ability of a chemical or a tissue to influence the development of another tissue. [L. *in,* into, and *duco,* lead] 919

inductive reasoning Process of logic and reasoning, such as using specific observations to arrive at a hypothesis. 12

inflammatory reaction Tissue response to injury that is characterized by redness, swelling, pain, and heat. [L. *in,* into, and *flamma,* fire] 758

inhalation Stage during respiration when air is drawn into the lungs; inspiration. 793

inheritance of acquired characteristics Lamarckian belief that characteristics acquired during the lifetime of an organism can be passed on to offspring. 288

inorganic molecule Type of molecule that is not an organic molecule; not derived from a living organism. 36

insulin Hormone secreted by the pancreas that lowers the blood glucose level by promoting the uptake of glucose by cells and the conversion of glucose to glycogen by the liver. 885

interneuron Neuron, located within the central nervous system, conveying messages between parts of the central nervous system. [L. *inter,* between, and Gk. *neuron,* nerve] 818

internode In vascular plants, the region of a stem between two successive nodes. 639

interphase Stage of the cell cycle during which DNA synthesis occurs and the nucleus is not actively dividing. [L. *inter,* between, and Gk.*phasis*,appearance] 150

intertidal zone Part of a beach that lies between the high and low water lines. 463

intron Noncoding segments of DNA that are transcribed but removed before mRNA leaves the nucleus. [L. *intra,* within] 241

invertebrate Referring to an animal without a serial arrangement of vertebrae, or a backbone. [L. *in,* without, and *vertebra,* bones of backbone] 576

ion Charged particle that carries a negative or positive charge. 24

ionic bond Chemical bond in which ions are attracted to one another by opposite charges. 24

isomer (EYE-suh-mur) Molecules with the same molecular formula but different structure and, therefore, shape. [Gk. *isos,* equal, and *meros,* part, portion] 37

isotonic solution Solution that is equal in solute concentration to that of the cytosol of a cell; causes cell to neither lose nor gain water by osmosis. 88

isotopes Atoms having the same atomic number but a different atomic weight due to the number of neutrons. [Gk. *isos,* equal, and *topos,* place] 21

J

joint Articulation between two bones of a skeleton. 863

K

karyotype (KAR-ee-uh-typ) Chromosomes arranged by pairs according to their size, shape, and general appearance in mitotic metaphase. 202

keystone predator Predator whose activities have a significant role in determining community structure. 423

kidney One of the paired organs of the vertebrate urinary system that regulates the chemical composition of the blood and produces a waste product called urine. 808

kinase (KY-nays) Any one of several enzymes that phosphorylate their substrates. 150

kinetic energy Energy associated with motion. 100

Krebs cycle Cycle of reactions in mitochondria that begins with citric acid; it produces CO_2, ATP, NADH, and $FADH_2$; also called the citric acid cycle. 131

***K*-selection** The production of a few offspring with a high probability of survival resulting in a favorable life history strategy for increase in population size under certain conditions. 396

L

lactation Secretion of milk by mammary glands usually for the nourishment of an infant. 906

lacteal (LAK-tee-ul) Lymphatic vessel in an intestinal villus, it aids in the absorption of fats. 781

lancelet Invetebrate chordate with a body that resembles a lancet and has the three chordate characteristics as a adult: dorsal hollow nerve cord; notochord, and pharyngeal gill slits. 617

large intestine Portion of the digestive tract consisting of cecum, colon, rectum, and anal canal; absorbs water and prepares digestive remains for explusion from the body. 783

larva (pl., larvae) Immature form in the life cycle of some animals; the stage of development between the embryo and the adult form. It sometimes undergoes metamorphosis to become the adult form. 895

larynx (LAR-ingks) Cartilaginous organ located between the pharynx and the trachea in vertebrates which contains the vocal cords; voice box. [Gk. *larynx,* gullet] 794

lateral meristem In vascular plants, the type of meristem that accounts for increase in girth. 650

leaf Usually broad, flat structure of a plant shoot system, containing cells that carry out photosynthesis. 639

learning Relatively permanent change in an animal's behavior that results from practice and experience. 370

leukocyte White blood cell of which there are several types, each having a specific function in protecting the body from invasion by foreign substances and organisms. [Gk. *leukos,* white, and *kytos,* cell] 747

lichen (LY-kun) Symbiotic relationship between certain fungi and algae, in which the fungi possibly provide inorganic food or water and the algae provide organic food. 518, 549

ligament Tough cord or fibrous band of dense fibrous tissue that binds bone to bone at a joint. 720, 863

light-dependent reactions Energy capturing portion of photosynthesis that takes place in thylakoid membranes of chloroplasts and cannot proceed without solar energy; it produces ATP and NADPH. 116

light-independent reactions Synthesis portion of photosynthesis that takes place in the stroma of chloroplasts and does not directly require solar energy; it uses the products of the light-dependent reactions to reduce carbon dioxide to a carbohydrate. 116

limbic system Pathways linking the hypothalamus to some areas of the cerebral cortex; governs learning and memory and various emotions such as pleasure, fear, and happiness. 833

limiting factor Resource or environmental condition that restricts the abundance and distribution of an organism. 385

lineage Line of evolutionary descent from a common ancestor. 342

linkage group Alleles of different genes that are located on the same chromosome and tend to be inherited together. 194

lipase Fat-digesting enzyme secreted by the pancreas. 780

lipid (LIP-id) Class of organic compounds that tend to be soluble in nonpolar solvents such as alcohol; includes fats and oils. [Gk. *lipos,* fat] 42

liposome Droplet of phospholipid molecules formed in a liquid environment. [Gk. *lipos,* fat, and *soma,* body] 326

liver Large internal vertebrate organ that produces metabolic wastes and bile, detoxifies the blood, and regulates the amount of various substances, including blood glucose and plasma proteins. 780

logistic growth Population increase that results in an S-shaped curve; growth is slow at first, steepens, and then levels off due to environmental resistance. 388

loop of the nephron Portion of a nephron between the proximal and distal convoluted tubules where water reabsorption occurs. 808

lung Internal respiratory organ containing moist surfaces for gas exchange. 622, 792

luteinizing hormone (LH) (LOOT-ee-ny-zing) Gonadotropic hormone secreted by the anterior pituitary that stimulates the production of sex hormones in males and females. 904

lymph Fluid, derived from tissue fluid, that is carried in lymphatic vessels. [L. *lympha,* clear water] 749, 756

lymphatic system Mammalian organ system consisting of lymphatic vessels and lymphoid organs. [L. *lympha,* clear water] 756

lymph nodes Mass of lymphoid tissue located along the course of a lymphatic vessel. 757

lymphocyte Specialized white blood cell; occurs in two forms—T lymphocyte and B lymphocyte. [L. *lympha,* clear water, and Gk. *kytos,* cell] 747

lysogenic cycle Bacteriophage life cycle in which the virus incorporates its DNA into that of the bacterium; only later does it begin a lytic cycle, which ends with the destruction of the bacterium. [Gk. *lyo,* loose, and *genitus,* producing] 511

lysosome Membrane-bounded vesicle that contains hydrolytic enzymes for digesting macromolecules. [Gk. *lyo,* loose, and *soma,* body] 69

lytic cycle One of the bacteriophage life cycles in which the virus takes over the operation of the bacterium immediately upon entering it and subsequently destroys the bacterium. [Gk. *lyo,* loose] 510

M

macrophage In vertebrates, large phagocytic cell derived from a monocyte that ingests microbes and debris. [Gk. *makros,* long, and *phagein,* to eat] 747, 759

major histocompatibility complex (MHC) protein Membrane protein that marks the cell as belonging to the individual; called human lymphocyte-associated (HLA) protein in humans. 765

Malpighian tubule (mal-PIG-ee-un) Blind, threadlike excretory tubule near the anterior end of an insect hindgut. 609, 807

mammal Member of a class of vertebrates characterized especially by the presence of hair and mammary glands. [L. *mamma,* breast, teat] 348, 629

marsupial (mar- SOO-pee-ul) Mammal bearing immature young nursed in a marsupium, or pouch—for example, kangaroo and opossum. [Gk. *marsupium,* pouch] 630

mass extinction Episode of large-scale extinction in which large numbers of species disappear in a few million years or less. 340

mechanoreceptor Sensory receptor in animals that is sensitive to mechanical stimulation, such as that from pressure, sound waves, and gravity. [Gk. *mechane,* machine, and L. *receptor,* receiver] 848

medulla oblongata (muh-DUL-uh ahb-lawng-GAHT-uh) Part of the brain stem controlling heartbeat, blood pressure, breathing, and other vital functions. It also serves to connect the spinal cord and the cerebrum. [L. *medulla,* marrow, innermost part, and *oblongus,* longer than broad] 830

megagametophyte In seed plants, the gametophyte that produces an egg; in flowering plants, an embryo sac. [Gk. *megas,* great, large] 697

megaspore Spore produced by the megasporocyte of a seed plant; of the four produced, one develops into a megagametophyte (embryo sac). 564, 694

meiosis (my-OH-sus) Type of nuclear division that occurs as part of sexual reproduction in which the daughter cells receive the haploid number of chromosomes. [Gk. *mio,* less, and *-sis,* act or process of] 160

meninges (sing., meninx) (muh-NIN-jeez) Protective membranous coverings about the central nervous system. [Gk. *meninga,* membranes covering the brain] 830

menstruation Periodic shedding of tissue and blood from the inner lining of the uterus in primates. [L. *menstrualis,* happening monthly] 904

meristem Undifferentiated embryonic tissue in the active growth regions of plants. 641

mesoderm In animal embryos, the middle primary tissue layer that gives rise to muscle, several internal organs and connective tissue layers. 915

mesoglea (mez-uh-GLEE-uh) Jellylike layer between the epidermis and the gastrodermis of cnidaria. [Gk. *mesos,* middle, and *gloios,* glue] 580

mesophyll Inner, thickest layer of a leaf consisting of palisade and spongy mesophyll; the site of most of photosynthesis. [Gk. *mesos,* middle, and *phyllon,* leaf] 654

messenger RNA (mRNA) Type of RNA formed from a DNA template and bearing coded information that directs the amino acid sequence of a polypeptide. 238

metabolic pool Metabolites that are the products of and/or the substrates for key reactions in cells allowing one type of molecule to be changed into another type, such as the conversion of carbohydrates to fats. [Gk. *meta,* implying change] 139

metabolism All of the chemical reactions that occur in a cell during growth and repair. [Gk. *meta,* implying change] 3, 101

metamorphosis Change in shape and form that some animals, such as insects, undergo during development. [Gk. *meta,* implying change, and *morphe,* shape, form] 603, 895

metastasis (meh-TAS-tuh-sus) Spread of cancer from the place of origin throughout the body; caused by the ability of cancer cells to migrate and invade tissues. [Gk. *meta,* between, and L. *stasis,* standing, a position] 260

microgametophyte In seed plants, the gametophyte that produces sperm; a pollen grain. [Gk. *mikros,* small, little] 697

microsphere Formed from proteinoids exposed to water; has properties similar to today's cells. [Gk. *mikros,* small, little, and *sphaera,* ball] 326

microspore Spore produced by a microsporocyte of a seed plant; it develops into a microgametophyte (pollen grain). [Gk. *mikros,* small, little] 564, 694

microtubule Small cylindrical organelle composed of tubulin dimers about an empty central core present in the cytoplasm, centrioles, cilia and flagella. [Gk. *mikros,* small, little, and L. *tubus,* pipe] 73

midbrain Part of the brain located below the thalamus and above the pons. 831

mimicry Superficial resemblance of one organism to another organism of a different species; often used to avoid predation. 415

mitochondrion Membranous organelle in which aerobic respiration continues and produces ATP molecules. [Gk. *mito,* thread, and *chondro,* grain] 70, 134

mitosis (my-TOH-sus) Process in which a parent nucleus produces two daughter nuclei, each having the same number and kinds of chromosomes as the parent nucleus. 149

molecular clock Idea that the rate at which mutation changes accumulate in certain types of genes is constant over time and is not involved in adaptation to the environment. 353, 497

molecule Union of two or more atoms of the same element; also the smallest part of a compound that retains the properties of the compound. [L. *moles,* mass] 24

mollusk Member of the phylum Mollusca that includes squids, clams, snails, and chitons; characterized by a visceral mass, a mantle, and a foot. 596

molt Periodic shedding of the exoskeleton in arthropods. 603

Monera Kingdom of life that contains the bacteria including the cyanobacteria. 502

monoclonal antibody Antibody produced by a clone of hybridoma cells specific for a particular antigen. [Gk. *monos,* one, *klonos,* clone, and *anti,* against] 769

monocot Flowering plant group; members have one embryonic leaf (cotyledon), parallel-veined leaves, scattered arrangement of vascular bundles, flower parts in threes and other characteristics. 640

monocotyledon (mahn-uh-KAHT-ul-eed-un) Fowering plant group; members have one embryonic leaf, parallel-veined leaves, scattered vascular bundles, and other characteristics. 568

monotreme Egg-laying mammal—for example, duckbill platypus and spiny anteater. [Gk. *monos,* one, and *trema,* hole] 629

morphogenesis (mor-fuh-JEN-uh sus) Emergence of shape in tissues, organs, or entire embryo during development. 918

morula (MOR-yuh-luh) Spherical mass of cells resulting from cleavage during animal development prior to the blastula stage. [L. dim. of *morus,* mulberry] 915

motor neuron Nerve cell that conducts nerve impulses away from the central nervous system and innervates effectors (muscle and glands). 818

multiple allele Pattern of inheritance in which there are more than two alleles for a particular trait; each individual has only two of all possible alleles. 189, 213

muscle action potential Electrochemical change due to increased sarcolemma permeability that is propagated down the T system and results in muscle contraction. 868

muscular (contractile) tissue Type of animal tissue composed of fibers that shorten and lengthen to produce movements; skeletal, cardiac, and smooth (visceral) are the three types of vertebrate muscles. 722

mutagen (MYOOT-uh-jun) Agent, such as radiation or a chemical, that brings about a mutation. [L. *mutatas,* change, and *genitus,* producing] 247

mutation Alteration in chromosome structure or number and also an alteration in a gene due to a change in DNA composition. [L. *mutatus,* change] 196, 222

mutualism Symbiotic relationship in which both species benefit in terms of growth and reproduction. 418

mycelium (my-SEE-lee-um) Tangled mass of hyphal filaments composing the vegetative body of a fungus. [Gk. *mycelium,* fungus filaments] 540

mycorrhiza (my-kuh-RY-zuh) Symbiotic relationship between fungal hyphae and roots of vascular plants. The fungus allows the plant to absorb more mineral ions and obtains carbohydrates from the plant. [Gk. *mykes,* fungus, and *rhizon,* dim. for root] 550, 663

myelin sheath (MY-uh-lun) White, fatty material—derived from the membrane of neurolemmocytes—that forms a covering for nerve fibers. [Gk. *myelos,* spinal cord] 818

myocardium Muscle of the heart. [Gk. *myos,* muscle, and *kardia,* heart] 738

myofibril Specific muscle cell organelle containing a linear arrangement of sarcomeres, which shorten to produce muscle contraction. [Gk. *myos,* muscle, and L. *fibra,* thread] 867

myosin (MY-uh-sun) Muscle protein making up the thick filaments in a sarcomere; it pulls actin to shorten the sarcomere, yielding muscle contraction. 867

N

NAD$^+$ Nicotinamide adenine dinucleotide; coenzyme that becomes NADH + H$^+$ and carries electrons to the electron transport system in the mitochondria. 107, 130

NADP$^+$ Nicotinamide adenine dinucleotide phosphate; coenzyme that becomes NADPH + H$^+$ during the light-dependent reactions of photosynthesis and reduces participants in the Calvin cycle during the light-independent reactions. 107

natural selection Mechanism of evolution caused by environmental selection of organisms most fit to reproduce, resulting in adaptation. 292

Neanderthal Hominid with a sturdy build who lived during the last Ice Age in Europe and the Middle East; hunted large game and lived together in a kind of society. 360

negative feedback Mechanism of homeostatic response in which the output is counter to and cancels the input. 728

nematocyst (NEM-ut-uh-sist) In cnidaria, a capsule that contains a threadlike fiber whose release aids in the capture of prey. [Gk. *nema,* thread, and *kystis,* bladder] 580

nephridium (ni-FRID-ee-um) (pl., nephridia) Segmentally arranged, paired excretory tubules of many invertebrates, as in the earthworm, where the contents are released through a nephridiopore. [Gk. *nephros,* kidney] 600, 807

nephron (NEF-rahn) Microscopic kidney unit that regulates blood composition by glomerular filtration, selective reabsorption, and tubular reabsorption; there are over a million nephrons per human kidney. [Gk. *nephros,* kidney] 808

nerve Bundle of long nerve fibers (axons) and/or dendrites outside the central nervous system. 826

nerve cord In many complex animals, a centrally placed cord of nervous tissue that receives sensory information and exercises motor control. 616

nerve net Diffuse noncentralized arrangement of nerve cells in cnidaria. 580

nervous tissue Type of animal tissue; contains nerve cells (neurons), which conduct impulses, and neuroglial cells, which support, protect, and provide nutrients to neurons. 723

net reproductive rate In models of population growth, the combination of the birthrate and death rate to yield a constant symbolized as *r*. 386

neuroglial cell Nonconducting nerve cell that is intimately associated with neurons and functions in a supportive capacity. 818

neuromuscular junction Region where an axon bulb approaches a muscle fiber; contains a presynaptic membrane, a synaptic cleft, and a postsynaptic membrane. 868

neuron Nerve cell that characteristically has three parts: dendrites, cell body, and axon. [Gk. *neuron,* nerve] 723, 818

neurotransmitter Chemical stored at the ends of axons that is responsible for transmission across a synapse. [Gk. *neuron,* nerve, and L. *trans,* across] 822

neutron Neutral subatomic particle, located in the nucleus and having a weight of approximately one atomic mass unit. 20

neutrophil Granular leukocyte that is the most abundant of the white blood cells; first to respond to infection. [Gk. *neuter,* neither, and *phileo,* love] 747

nitrification Process by which nitrogen in ammonia and organic compounds is oxidized to nitrites and nitrates by soil bacteria. 438

nitrogen cycle Continuous process by which nitrogen circulates in the air, soil, water, and organisms of the earth. 437

nitrogen fixation Process whereby free atmospheric nitrogen is converted into compounds, such as ammonium and nitrates, usually by bacteria. 438, 516, 662

node In plants, the place where one or more leaves attach to a stem. 639

nondisjunction Failure of homologous chromosomes or daughter chromosomes to separate during meiosis I and meiosis II respectively. 203

nonrenewable energy source Resource that exists in a fixed amount and can only be renewed by physical and chemical processes that takes millions/billions of years. Examples are fossil fuels and minerals such as copper and aluminum. 473

nonsteroid hormone Type of hormone that is not a steroid such as a peptide hormone that is received by a plasma membrane receptor and brings about a change in metabolic reactions within a cell. 875

nonvascular plant Bryophytes such as mosses and liverworts that have no vascular tissue and either occur in moist locations or have special adaptations for living in dry locations. 555

norepinephrine Neurotransmitter of the post-ganglionic fibers in the sympathetic system of the autonomic nervous system; also produced in small quantity by the adrenal medulla. 822, 882

notochord Cartilaginous-like supportive dorsal rod in all chordates sometime in their life cycle; is replaced by vertebrae in vertebrates. [Gk. *notos,* back, and *chorde,* string] 616, 917

nuclear envelope Double membrane that surrounds the nucleus in eukaryotic cells and is connected to the endoplasmic reticulum; has pores that allow substances to pass between the nucleus and the cytoplasm. 66

nucleic acid Polymer of nucleotides; both DNA and RNA are nucleic acids. 50, 222

nucleoid region Portion of the bacterial cell that contains its genetic material. [L. *nucleus,* nucleus, kernel, and Gk. *-eides,* like] 62, 148, 514

nucleolus (pl., nucleoli) Dark-staining, spherical body in the cell nucleus that produces ribosomal subunits. [L. *nucleus,* nucleus, kernel] 66

nucleotide Monomer of DNA and RNA consisting of a five-carbon sugar bonded to a nitrogenous base and a phosphate group. 50

nucleus (NOO-klee-us) Region of a eukaryotic cell, containing chromosomes, that controls the structure and function of the cell. 66

O

ocean ridge Ridge on the ocean floor where oceanic crust forms and from which it moves laterally in each direction. 339

olfactory cell (ahl-FAK-tuh-ree) Modified neuron that is a sensory receptor for the sense of smell. 841

omnivore Organism in a food chain that feeds on both plants and animals. 429

oncogene (ONG-koh-jeen) Cancer-causing gene. [Gk. *onco,* a swelling, and L. *genitus,* producing] 262

oocyte Immature egg undergoing the process of mitosis. 902

oogenesis (oh-uh-JEN-uh-sus) Production of eggs in females by the process of meiosis and maturation. [Gk. *oon,* egg, and L. *genitus,* producing] 166, 903

operant conditioning Form of learning that results from rewarding or reinforcing a particular behavior. 370

operator In an operon the sequence of DNA to which the repressor protein binds. 252

operon (OP-er-on) Group of structural and regulating genes that functions as a single unit. 252

optic nerve Nerve that carries impulses from the retina of the eye to the brain. [Gk. *optikos,* related to sight] 844

orbital Volume of space around a nucleus where electrons can be found most of the time. 23

organ Combination of two or more different tissues performing a common function. 724

organelle Small, often membranous structure in the cytoplasm having a specific function. 63

organic molecule Type of molecule that contains carbon and hydrogen; it may also have oxygen attached to the carbon(s). 36

organ system Group of related organs working together. 724

osmosis Diffusion of water through a differentially permeable membrane. [Gk. *osmos,* a pushing] 88

osmotic pressure Measure of the tendency of water to move across a differentially permeable membrane; visible as an increase in liquid on the side of the membrane with higher solute concentration. 88

osteoblast Bone-forming cell. [Gk. *osteon,* bone, and *blastos,* bud] 859

osteoclast Cell that causes erosion of bone. [Gk. *osteon,* bone, and *klastos,* broken in pieces] 859

osteocyte Mature bone cell formed when an osteoblast secretes bony matrix outside of itself and then becomes trapped in its own matrix. [Gk. *osteon,* bone, and *kytos,* cell] 859

ovary In animals, female gonad in which eggs develop and which may produce the female sex hormones; in flowering plants, the enlarged ovule-bearing portion of the pistil which develops into a fruit. [L. *ovaris,* egg-keeper] 568, 695, 894

oviduct In animals, tube through which gametes pass from the ovary to the outside environment or to the uterus. [L. *ovum,* egg, and *duco,* lead out] 902

ovulation Bursting of a follicle when an oocyte is released from the ovary. 903

ovule In seed plants, a structure that contains the megasporangium, where meiosis occurs and the gametophyte is produced; develops into the seed. [L. *ovulum,* dim. of *ovum,* egg] 564, 695

ovum Haploid egg cell which is usually fertilized by a sperm to form a diploid zygote. 903

oxidation (ahk-sih-DAY-shun) Loss of one or more electrons from an atom or molecule; in biological systems, generally the loss of hydrogen atoms. 25

oxidative phosphorylation Process by which ATP production is tied to an electron transport system that uses oxygen as the final acceptor; occurs in mitochondria. 136

oxygen debt Use of oxygen to reconvert lactate, which builds up during anaerobic conditions to pyruvate. 141, 867

oxytocin Peptide hormone released by posterior pituitary that causes contraction of uterus and milk letdown. [Gk. *oxys,* quick, and *tokos,* birth] 878

ozone Gas with the chemical formula O_3 that occurs in the upper atmosphere and forms a protective layer against excess ultraviolet radiation; also a component of photochemical smog in the lower atmosphere. 478

ozone hole Seasonal thinning of the ozone shield in the lower stratosphere at the North and South Poles. 478

ozone shield Formed from oxygen in the upper atmosphere, it protects the earth from ultraviolet radiation. 331

P

paleontology (pay-lee-ahn-TAHL-uh-jee) Study of fossils that results in knowledge about the history of life. [Gk. *palaios,* ancient, old, and *ontos,* having existed; *-logy,* "study of" from *logikos,* rational, sensible] 288, 329

PAN (peroxyacyl nitrate) Type of chemical found in photochemical smog. 477

pancreas In animals, an internal organ that produces digestive enzymes and the hormones insulin and glucagon. 780, 885

pancreatic islet Masses of cells that constitute the endocrine portion of the pancreas. 885

parallel evolution Similarity in structure in related groups that cannot be traced to a common ancestor. 497

parasite Organism that lives off of and takes nourishment from an organism called the host. 416

parasitism Symbiotic relationship in which one species (parasite) benefits in terms of growth and reproduction to the harm of the other species (host). 416

parasympathetic system Division of the autonomic system that is active under normal conditions; uses acetylcholine as a neurotransmitter. [Gk. *para,* alongside, and *sympathia,* fellow-feeling] 829

parathyroid gland (par-uh-THY-royd) Gland embedded in the posterior surface of the thyroid gland; it produces parathyroid hormone. 881

parathyroid hormone (PTH) (par-uh-THY-royd) Hormone secreted by the four parathyoid glands that increases the blood calcium level and decreases the blood phosphate level. 881

parenchyma (puh-REN-kuh-muh) Least specialized of all plant cell or tissue types; found in all organs of a plant. [Gk. *para,* beside, and *enchyma,* infusion] 642

parthenogenesis (par-thuh-noh-JEN-uh-sus) Development of an egg cell into a whole organism without fertilization. [Gk. *parthenos,* virgin, and *genitus,* producing] 894

pattern formation Positioning of cells during development that determines the final shape of an organism. 921

pectoral girdle Portion of the skeleton that provides support and attachment for the arms. [Gk. *pechys,* forearm] 862

pelagic division (puh-LAJ-ik) Open portion of the sea. [Gk. *pelagos,* the sea] 464

pelvic girdle Portion of the skeleton to which the legs are attached. [L. *pelvis,* basin] 862

penis Male copulatory organ. 898

pepsin Protein-digesting enzyme secreted by gastric glands. 779

peptide Two or more amino acids joined together by covalent bonding. 46

perennial Flowering plant that lives more than one growing season because the underground vegetative parts regrow each season. 638

pericycle In plants, layer of cells surrounding the xylem and phloem of roots inside the vascular cylinder that produces branch roots. 645

peripheral nervous system (PNS) In animals, nerves which lie outside the central nervous system. [Gk. *periphereia,* circumference] 824

peristalsis In animals, rhythmic contractions of smooth muscle in the walls of tubular internal organs by which contents are forced onward as occurs in the digestive tract. [Gk. *peri,* around, and *stalsis,* compression] 778

permafrost Permanently frozen ground usually occuring in the tundra, a biome of arctic regions. 451

phagocytosis (fag-oh-suh-TOH-sis) Process by which amoeboid-type cells engulf large substances forming an intracellular vacuole. [Gk. *phagein,* to eat, and *kytos,* cell] 92

pharynx (FAR-ingks) Common passageway for both food intake and air movement, located between the mouth and the esophagus. [Gk. *pharynx,* throat] 794

phenotype Visible expression of a genotype—for example, brown eyes or attached earlobes. [Gk. *phaino,* appear, and *typos,* image, shape] 177

pheromone Chemical released by the body that causes a predictable reaction of another member of the same species. [Gk. *phero,* bear, carry, and *monos,* alone] 376

phloem (FLOH-um) Vascular tissue that conducts organic solutes in plants; contains sieve-tube cells and companion cells. [Gk. *phloios,* bark] 558, 642, 664

phospholipid (fahs-foh-LIP-id) Molecule having the same structure as a neutral fat except one bonded fatty acid is replaced by a group that contains phosphate; an important component of plasma membranes. [Gk. *photos,* light, and *lipos,* fat] 44, 83

phosphorus cycle Biogeochemical cycle in which phosphorus moves from the soil through the biotic community and back to the soil again. 438

photochemical smog Air pollution that contains nitrogen oxides and hydrocarbons which react to produce ozone and PAN (peroxylacetyl nitrate). [Gk. *photos,* light, and *chemo,* pertaining to chemicals] 477

photon Discrete packet of solar energy; the amount of energy in a packet is inversely related to the wavelength of the packet. [Gk. *photos,* light] 114

photoperiodism (foht-oh-PIR-ee-ud-iz-um) Relative lengths of daylight and darkness that affect the physiology and behavior of an organism. [Gk. *photos,* light, and *periodus,* completed course] 688

photoreceptor Light-sensitive sensory receptor. [Gk. *photos,* light, and L. *receptor,* receiver] 842

photosynthesis Process occurring usually within chloroplasts whereby chlorophyll-containing organelles trap solar energy to reduce carbon dioxide to carbohydrate. [Gk. *photos,* light, and *synthetos,* putting together] 3, 114

photosystem Photosynthetic unit where solar energy is absorbed and high energy electrons are generated; contains a pigment complex and an electron acceptor; occurs as PS (photosystem) I and PS II. 118

phototropism (foh-TAH-truh-piz-um) Directional growth of plants in response to light; stems demonstrate positive phototropism. [Gk. *photos,* light, and *tropos,* turning] 677

pH scale Measurement scale for the hydrogen ion concentration $[H^+]$ of a solution. 30

phyletic gradualism (fy-LET-ik) Evolutionary model that proposes evolutionary change resulting in a new species can occur gradually in an unbranched lineage. [Gk. *phyle,* tribe] 343

phylogenetic tree (fy-loh-jen-ET-ik) Diagram that indicates common ancestors and lines of descent. 496

phylogeny (fy-LAHJ-uh-nee) Evolutionary history of a group of organisms. 494

phytochrome (FYT-uh-krohm) Photoreversible plant pigment that is involved in photoperiodism and other responses of plants such as etiolation. [Gk. *phyton,* plant, and *chroma,* color] 688

phytoplankton (fyt-oh-PLANGK-tun) Part of plankton containing organisms that (1) photosynthesize and produce much of the oxygen in the atmosphere and (2) serve as food producers in aquatic ecosystems. [Gk. *phyton,* plant, and *planktos,* wandering] 461, 524

pineal gland (PY-nee-ul) Gland—either at the skin surface (fish, amphibians) or in the third ventricle of the brain, (mammals)—that produces melatonin. 888

pinocytosis (pin-oh-suh-TOH-sis) Process by which vesicles form around and bring macromolecules into the cell. [Gk. *pino,* drink, and *kytos,* cell] 92

pistil Flower structure consisting of one or more carpels. [L. *pistillum,* club-shaped pounder, pestle] 568, 695

pituitary gland Small gland that lies just inferior to the hypothalamus; the anterior pituitary produces several hormones, some of which control other endocrine glands; the posterior pituitary stores oxytocin and antidiuretic hormone produced by the posterior pituitary. 878

placenta Organ formed during the development of placental mammals from the chorion and the uterine wall that allows the embryo, and then the fetus, to acquire nutrients and rid itself of wastes; produces hormones that regulate pregnancy. 630, 896, 926

placental mammal Member of mammalian subclass characterized by the presence of a placenta during the development of an offspring. 30

plankton Freshwater and marine organisms that are suspended on or near the surface of the water. [Gk. *planktos,* wandering] 461, 524

Plantae Plant kingdom; includes eukaryotic multicellular organisms that protect the zygote internally and produce in their own food by photosynthesis. 502, 554, 761

plasma In vertebrates, the liquid portion of blood; contains nutrients, wastes, salts, and proteins. [Gk. *plasma,* something molded] 747

plasma membrane Membrane surrounding the cytoplasm that consists of a phospholipid bilayer with embedded proteins; functions to regulate the entrance and exit of molecules from cell. 62

plasmid Self-duplicating ring of accessory DNA in the cytoplasm of bacteria. 270

platelet Component of blood that is necessary to blood clotting. 748

plate tectonics (tek-TAHN-iks) Study of the behavior of the earth's crust in terms of moving plates that are responsible for continental drift. [Gk. *tektos,* fluid, molten, able to flow] 339

pleiotropy Condition in which one gene affects many characteristics of the individual. [Gk. *pleion,* more, and *tropos,* turning] 188

plumule (PLOO-myool) In flowering plants, the embryonic plant shoot that bears young leaves. [L. *plumulla,* dim. of *pluma,* feather] 705

polar body In oogenesis, a nonfunctional product: two to three meiotic products are of this type. 168

polar covalent bond Bond in which the sharing of electrons between atoms is unequal. 26

pollen grain Microgametophyte in seed plants. [L. *pollen,* fine dust] 564, 697

pollination In seed plants, the transfer of pollen from microsporangium to the ovule, which by this time contains a megagametophyte. [L. *pollen,* fine dust] 564, 700

pollutant Substance that is added to the environment and leads to undesirable effects for living organisms. 474

polygenic inheritance Pattern of inheritance in which a trait is controlled by several allelic pairs; each dominant allele contributes in an additive and like manner. [Gk. *polys,* many, and L. *genitus,* producing] 190

polymer Macromolecule consisting of covalently bonded monomers; for example, a protein is a polymer of monomers called amino acids. [Gk. *polys,* many, and *meros,* part, portion] 38

polyploid (polyploidy) Condition in which an organism has more than two complete sets of chromosomes. [Gk. *polys,* many, and *plo,* fold] 196

polyribosome String of ribosomes simultaneously translating regions of the same mRNA strand during protein synthesis. 243

pons Part of the brain stem above the medulla oblongata and below the midbrain. It also serves to connect the cerebellar hemispheres. [L. *pontis,* bridge] 830

population Group of organisms of the same species occupying a certain area and sharing a common gene pool. 2, 304, 384

population density Number of individuals of a species per unit area or volume. 385

population distribution Pattern of dispersal of the individuals in a species per unit area or volume. 385

population size Number of individuals that are contributing to the gene pool of a population. 386

portal system Pathway of blood flow that begins and ends in capillaries, such as the one found between the small intestine and liver. [L. *porto,* carry, transport] 742

positive feedback Mechanism of homeostatic response in which the output intensifies and increases the likelihood of response, instead of countering it and canceling it. 728

postmating isolating mechanism Anatomical or physiological difference between two species that prevents successful reproduction after mating has taken place. 315

potential energy Stored energy as a result of location or spatial arrangement. 100

predation Interaction in which one organism uses another, called the prey, as a food source. 412

predator Organism that practices predation. 412

premating isolating mechanism Anatomical or behavioral difference between two species that prevents the possibility of mating. 314

pressure-flow model Model explaining transport through sieve tubes of phloem by a positive pressure potential (compared to a sink) due to the active transport of sucrose and the passive transport of water. 670

prey Organism that provides nourishment for a predator. 412

primary productivity Rate at which autotrophs in an ecosystem produce food; gross productivity is the overall rate and net productivity is gross rate minus the rate of aerobic respiration. 430

primate Animal that belongs to the order Primates, the order of mammals that includes prosimians, monkeys, apes, and humans. [L. *primus,* first] 348

primitive character Structural, physiological, or behavioral trait that is present in a common ancestor and all members of a group. 496

probe Known sequences of DNA that are used to find complementary DNA strands; can be used diagnostically to determine the presence of particular genes. 271

Proconsul Possible hominoid ancestor; a forest-dwelling primate with some characteristics of living apes. 351

producer Organism at the start of a food chain that makes its own food (e.g., green plants on land and algae in water). 429

progesterone (proh-JES-tuh-rohn) Female sex hormone that causes the endometrium of the uterus to become secretory during the uterine cycle; along with estrogen, it maintains secondary sexual characteristics in females. 887, 903

prokaryotic cell (proh-kair-ee-AHT-ik) Lacking a membrane-bounded nucleus and membranous organelles, as in bacteria and archaea. [Gk. *pro,* before, and *karyon,* kernel, nucleus] 62

promoter In an operon, a sequence of DNA where RNA polymerase begins transcription. 240, 252

prosimian (proh-SIM-ee-un) Group of primates that includes lemurs and tarsiers and may resemble the first primates to have evolved. [L. *pro,* before, and *simia,* ape, monkey] 349

protein Polymer having, as its primary structure, a sequence of amino acids united through covalent bonding. 46

proteinoid (PROHT-en-oyd) Abiotically polymerized amino acids that are joined in a preferred manner; possible early step in cell evolution. 326

Protista Kingdom of life that contains protozoa, algae, slime molds, and water molds. 502

protocell Cell forerunner developed from cell-like microspheres. [Gk. *protos,* first] 326

proton Positive subatomic particle, located in the nucleus and having a weight of approximately one atomic mass unit. 20

proto-oncogene (PROH-toh-ONG-koh-jeen) Normal gene that can become an oncogene through mutation. 262

protoplast Plant cell from which the cell wall has been removed. [Gk. *protos,* first, and *plastos,* formed, molded] 710

protostome (PROH-toh-stohm) Group of coelomate animals in which the first embryonic opening (the blastopore) is associated with the mouth. [Gk. *protos,* first, and L. *stoma,* mouth] 594

protozoan (pl., protozoa) Animal-like, heterotrophic, unicellular organism. [Gk. *protos,* first, and *zoon,* animal] 530

proximal convoluted tubule Portion of a nephron following the glomerular capsule where selective reabsorption of filtrate occurs. [L. *proximus,* nearest] 808

pseudocoelom Body cavity lying between the digestive tract and body wall that is incompletely lined by mesoderm. [Gk. *pseudes,* false, and *koiloma,* cavity] 588

pseudopod Cytoplasmic extension of amoeboid protists; used for locomotion and engulfing food. [Gk. *pseudes,* false, and *podos,* foot] 530

pulmonary circuit Circulatory pathway between the lungs and the heart. [L. *pulmonarius,* of the lungs] 742

punctuated equilibrium Evolutionary model that proposes there are periods of rapid change dependent on speciation followed by long periods of stasis. 343

Punnett square Grid that enables one to calculate the results of simple genetic crosses by lining up alleles within the gametes of two parents on the outside margin and their recombination in boxes inside the grid. 179, 208

purine (PYUR-een) Type of nitrogen-containing base, such as adenine and guanine, having a double-ring structure. 225

pyrimidine (py-RIM-uh-deen) Type of nitrogen-containing base, such as cytosine, thymine, and uracil, having a single-ring structure. 225

pyruvate End product of glycolysis; its further fate, involving fermentation or entry into a mitochondrion, depends on oxygen availability. 131

R

radial symmetry Body plan in which similar parts are arranged around a central axis, like spokes of a wheel. 576

recessive allele (uh-LEEL) Allele that exerts its phenotypic effect only in the homozygote; its expression is masked by a dominant allele. 177

recombinant DNA (rDNA) DNA that contains genes from more than one source. 270

red bone marrow Substance that forms blood cells located in the spongy bone of certain bones. 859

reduction Gain of electrons by an atom or molecule with a concurrent storage of energy; in biological systems, generally the gain of hydrogen atoms. 25

reflex Automatic, involuntary response of an organism to a stimulus. [L. *reflexus,* turn back] 827

refractory period Time following an action potential when a neuron is unable to conduct another nerve impulse. 899

regulator gene In an operon, a gene that codes for a repressor. 252

renewable energy source Energy source that can potentially last indefinitely because it is being replaced through natural processes at a rate fast enough to keep up with demand. Examples are solar energy, falling water, and geothermal energy. 473

replacement reproduction Population in which each person is replaced by only one child. 400

repressible operon (OP-er-on) Operon that is normally active because the repressor must combine with a corepressor before the complex can bind to the operator. 253

repressor In an operon protein molecule that binds to an operator, preventing RNA polymerase from binding to the promoter site of an operon. 252

reproduce To produce a new individual of the same kind. 4

reptile Member of a class of terrestrial vertebrates with internal fertilization, scaly skin, and an egg with a leathery shell; includes snakes, lizards, turtles, and crocodiles. [L. *reptile,* snake] 624

resource partitioning Apportioning the supply of a resource such as food and living space between species as a means to increase the number of niches following competition between species. 410

resting potential Mmembrane potential of an inactive neuron. 820

restriction enzyme Bacterial enzyme that stops viral reproduction by cleaving viral DNA; used to cut DNA at specific points during production of recombinant DNA. 270

retina Innermost layer of the eyeball containing the photoreceptors—rod cells and cone cells. [L. *retis,* net] 844

retrovirus RNA virus containing the enzyme reverse transcriptase that carries out RNA/DNA transcription; retroviruses include oncogenes and the AIDS viruses. [L. *retro,* backward, and *virus,* poison] 512

rhizoid Rootlike hair that anchors a plant and absorbs minerals and water from the soil. [Gk. *rhizion,* dim. for root] 556

rhizome Rootlike, underground stem. [Gk. *rhiza,* root] 560, 652

rhodopsin (roh-DAHP-sun) Light-absorbing molecule in rods and cones that contains a pigment and the protein opsin. [Gk. *rodon,* rose, and *opsis,* sight] 846

rib One of 24 bones that, together with the sternum, form the rib cage. The first seven pairs are true ribs, the next three pairs are false ribs, and the last two pairs are floating ribs. 861

rib cage Top and side of the thoracic cavity in terrestrial vertebrates; contains ribs and intercostal muscles. 793

ribosomal RNA (rRNA) Type of RNA found in ribosomes that coordinates the coupling of anticodons with codons during polypeptide synthesis. 238

ribosome RNA and protein in two subunits; site of protein synthesis in the cytoplasm. 67

RNA (ribonucleic acid) Nucleic acid produced from covalent bonding of nucleotide monomers that contain the sugar ribose; occurs in three forms: messenger RNA, ribosomal RNA, and transfer RNA. 50, 222

RNA polymerase (PAHL-uh-muh-rays) Enzyme that speeds the formation of RNA from a DNA template. 240

rod cell Photoreceptor in vertebrate eyes that responds to dim light. 844

root In vascular plants, the underground organ that anchors the plant in the soil, absorbs water and minerals, and stores the products of photosynthesis. 558

root hair Extension of a root epidermal cell that collectively increases the surface area for the absorption of water and minerals. 641, 662

root pressure Force generated by an osmotic gradient that serves to elevate sap through xylem for a short distance. 666

roundworm Member of the phylum Nematoda with a cylindrical body that has a complete digestive tract and a pseudocoelom; some forms free-living in water, soil and many parasitic. 589

***r*-selection** Concept that a high reproductive rate is the most favorable life history strategy for increase in population size under certain conditions. 396

RuBP (ribulose bisphosphate) Five-carbon compound that combines with and fixes carbon dioxide during the Calvin cycle and is later regenerated by the same cycle. 122

S

sac body plan Body with a digestive cavity that has only one opening, as in cnidaria and flatworms. 584

salivary gland Gland associated with the mouth that secretes saliva. 777

saltatory conduction Movement of nerve impulses from one neurolemmal node to another along a myelinated axon. [L. *saltator*, hopper] 820

saprotroph (SAP-roh-trohf) Organism that secretes digestive enzymes and absorbs the resulting nutrients back across the plasma membrane. [Gk. *sapros*, rotten, and *trophe*, food] 516, 535

sarcolemma (sahr-kuh-LEM-uh) Plasma membrane of a muscle fiber that forms the tubules of the T system involved in muscular contraction. [Gk. *sarkos*, flesh, and *lemma*, sheath] 867

sarcomere (SAHR-kuh-mir) One of many units, arranged linearly within a myofibril, whose contraction produces muscle contraction. [Gk. *sarkos*, flesh, and *meros*, part] 867

sarcoplasmic reticulum Smooth endoplasmic reticulum of skeletal muscle cells; surrounds the myofibrils and stores calcium ions needed for myosin to bind to actin and therefore muscle contraction. 867

savanna Terrestrial biome that is a grassland in Africa, characterized by few trees and a severe dry season. 456

scientific method Step-by-step process for discovery and generation of knowledge—ranging from observation and hypothesis to theory and principle. 11

sclera (SKLER-uh) Outer, white, fibrous layer of the eye that surrounds the eye except for transparent cornea. [Gk. *skleros*, hard] 843

sclerenchyma In plants, a support tissue composed of cells with heavily lignified cells walls. 642

secondary oocyte (OH-uh-syt) In oogenesis the functional product of meiosis I; becomes the egg. [Gk. *oon*, egg, and *kytos*, cell] 168

secondary sexual characteristic Trait that is sometimes helpful but not absolutely necessary to reproduction and is maintained by the sex hormones in males and females. 901

second messenger Chemical signal such as cyclic AMP that causes the cell to respond to a hormone bound to plasma membrane receptor. 875

sedimentation Process by which particulate material accumulates and forms a stratum. [L. *sedimentum*, a settling] 328

seed Mature ovule that contains an embryo, with stored food enclosed in a protective coat. 564, 702

segmentation Repetition of body units as is seen in the earthworm. 600

semen Thick, whitish fluid consisting of sperm and secretions from several glands of the male reproductive tract. [L. *semen*, seed] 899

semicircular canal One of three half-circle-shaped canals of the inner ear that are fluid filled and register changes in motion. 849

semiconservative replication Duplication of DNA resulting in a double helix having one parental and one new strand. 228

sensory neuron Nerve cell that transmits nerve impulses to the central nervous sytem after a sensory receptor has been stimulated. 818

sensory receptor Structure that receives sensory stimuli and is a part of a sensory neuron or transmits signals to a sensory neuron. 817, 839

sessile Permanently fixed to a substratum so that movement from place to place does not occur. 576

sessile filter feeder Animal that stays in one place and filters its food from the water. 579

sewage treatment Any process by which the organic and bacterial content of sewage is reduced. 479

sex chromosome Chromosome that determines the sex of an individual; in humans, females have two X chromosomes and males have an X and Y chromosome. 192, 202

sex-influenced trait Autosomal phenotype controlled by an allele that is expressed differently in the two sexes; for example, the possibility of pattern baldness is increased by the presence of testosterone in human males. 217

sex-linked Allele that occurs on the sex chromosomes but may control a trait that has nothing to do with sexual characteristics of an animal. 215

sexual reproduction Reproduction involving meiosis, gamete formation, and fertilization; produces offspring with chromosomes inherited from each parent but a unique combination of genes. 160

sexual selection Changes in males and females, often due to male competition and female selectivity leading to reproductive success. 372

shoot system Aboveground portion of a plant consisting of the stem, leaves, and flowers. 638

sieve-tube cell Member that joins with others in the phloem tissue of plants as a means of transport for nutrient sap. 664

sinus Cavity, as with the sinuses in the human skull. [L. *sinus*, hollow] 860

sister chromatid (KROH-muh-tud) One of two genetically identical chromosomal units that are the result of DNA replication and are attached to each other at the centromere. 149

skeletal muscle Striated, voluntary muscle tissue that comprises skeletal muscles; also called striated muscle. 722

sliding filament theory Movement of actin filaments in relation to myosin filaments, which accounts for muscle contraction. 867

small intestine In vertebrates, the portion of the digestive tract that precedes the large intestine and where digestion of foods is completed and nutrient molecules are absorbed. 779

smooth (visceral) muscle Nonstriated, involuntary muscles found in the walls of internal organs. 722

society Group in which members of species are organized in a cooperative manner, extending beyond sexual and parental behavior. 376

sociobiology Application of evolutionary biological principles to the study of social behavior in animals. 378

sodium-potassium pump Carrier protein in the plasma membrane that moves sodium ions out of and potassium into animal cells, important in nerve and muscle cells. 90

solute Substance that is dissolved in a solvent, forming a solution. 87

solvent Liquid portion of a solution that serves to dissolve a solute. 87

somatic system Portion of the peripheral nervous system which controls the skeletal muscles. 825

speciation Origination of new species due to the evolutionary process of descent with modification. 422

species Taxonomic category whose members are characterized by anatomy and can only breed successfully with each other. [L. *species*, a kind] 4, 314, 495

sperm Male gamete having a haploid number of chromosomes and the ability to fertilize an egg produced by a female. 901

spermatogenesis (spur-mat-uh-JEN-uh-sus) Production of sperm in males by the process of meiosis and maturation. [Gk. *sperma*, seed, and L. *genitus*, producing] 166, 901

spicule (SPIK-yool) Skeletal structure of sponges composed of calcium carbonate or silicate. [L. *spicula*, dim. of *spika*, spear] 579

spinal cord Part of the central nervous system; the nerve cord that is continuous with the base of the brain and housed within the vertebral column. 826

spindle Microtubule structure that brings about chromosomal movement during nuclear division. 152

spiral organ Portion of inner ear that permits hearing and consists of hair cells located on the basilar membrane within the cochlea; also called organ of Corti. 849

sponge Invertebrate member of the phylum Porifera; pore-bearing filter feeders whose inner body wall is lined by collar cells. 578

spongy mesophyll In a plant leaf, the layer of mesophyll containing loosely packed, irregularly spaced cells that increase the amount of surface area for gas exchange. 654

sporangium (pl., sporangia) (spuh-RAN-jee-um) Capsule that produces sporangiospores. [Gk. *spora,* seed, and *angeion,* (dim. of *angos*), vessel] 535, 542, 556

spore Asexual reproductive structure that is resistant to unfavorable environmental conditions and develops into a haploid generation. 525, 541

sporophyte (SPOR-uh-fyt) Diploid generation of the alternation of generations life cycle of a plant; meiosis produces haploid spores that develop into the haploid generation. [Gk. *phyton,* plant] 555, 694

sporozoa Spore-forming protozoan that has no means of locomotion and is typically a parasite. 533

spring overturn Mixing process that occurs in spring in stratified lakes whereby the oxygen-rich top waters mix with nutrient-rich bottom waters. 461

stabilizing selection Outcome of natural selection in which extreme phenotypes are eliminated and the average phenotype is conserved. 312

stamen In flowering plants, the portion of the flower that consists of a filament and an anther containing pollen sacs where pollen is produced. 568, 695

stem Usually the upright, vertical portion of a plant, which transports substances to and from the leaves. 558, 639

steroid Type of lipid molecule having four interlocking rings; examples are cholesterol, progesterone, and testosterone. 44

steroid hormone Chemical messenger that is lipid soluble and therefore passes through the plasma and nuclear envelope to bind with a receptor inside the nucleus; the complex turns on specific genes leading to the production of particular proteins. 874

stigma Portion of the pistil where pollen grains adhere and germinate as a necessity to the process of fertilization in flowering plants. 568, 695

stimulus Change in the internal or external environment that a sensory receptor can detect leading to nerve impulses in sensory neurons. 723

stolon Stem that grows horizontally along the ground and establishes plantlets periodically where it contacts the soil (e.g., the runners of a strawberry plant). [L. *stolo,* shoot] 652

striated Having bands; in cardiac and skeletal muscle, alternating light and dark crossbands are produced by the distribution of contractile proteins. [L. *stria,* lined] 722

stroma (STROH-muh) Large, central compartment in a chloroplast that is fluid filled and contains enzymes used in photosynthesis. [Gk. *stroma,* bed, mattress] 71, 116

structural gene Gene that codes for an enzyme in a metabolic pathway. 252

substrate Reactant in a reaction controlled by an enzyme. 103

substrate-level phosphorylation Process in which ATP is formed by transferring a phosphate from a metabolic substrate to ADP. [Gk. *phos,* light, and *phoreus,* carrier] 132

summation Accumulation of effects as when the membrane potential of a neuron is determined by the total effect of excitatory and inhibitory neurotramitters bound to postsynaptic receptors. 822

survivorship Probability of newborn individuals of a cohort surviving to particular ages. 390

sustainable world Global way of life that can continue indefinitely, because the economic needs of all peoples are met while still protecting the environment. 401

symbiosis Relationship that occurs when two different species live together in a unique way; it may be beneficial, neutral, or detrimental to one and/or the other species. 416

sympathetic system Division of the autonomic system that is active when an organism is under stress; uses norepinephrine as a neurotransmitter. 829

sympatric speciation (sim-PA-trik) Origin of new species in populations that overlap geographically. [Gk. *sym,* together, and *patri,* fatherland] 316

synapse (SIN-aps) Junction between neurons consisting of the presynaptic (axon) membrane, the synaptic cleft, and the postsynaptic (usually dendrite) membrane. [Gk. *synaptos* united] 822

synapsis Pairing of homologous chromosomes during meiosis I. [Gk. *synaptos,* united, joined together] 162

systematics Study of the diversity of organisms to determine phylogenetic relationships and classify organisms. [Gk. *systema,* an orderly arrangement] 496

systemic circuit Circulatory pathway of blood flow between the tissues and the heart. 742

systole (SIS-tuh-lee) Contraction period of a heart during the cardiac cycle. [Gk. *systole,* contraction] 740

T

taiga (TY-guh) Terrestrial biome that is a coniferous forest extending in a broad belt across northern Eurasia and North America. 452

taproot Main axis of a root that penetrates deeply and is used by certain plants such as carrots for food storage. 646

taste bud Oral concentration of sensory nerve endings that function as taste receptors. 840

taxon (pl., taxa) Group of organisms that fills a particular classification category. [Gk. *tasso,* arrange, classify] 494

taxonomy Branch of biology concerned with identifying, describing, and naming organisms. [Gk. *tasso,* arrange, classify, and *nomos,* usage, law] 10, 492

tendon Strap of fibrous connective tissue that connects skeletal muscle to bone. [L. *tend,* stretch] 720, 865

territoriality Behavior related to the act of marking or defending a particular area against invasion by another species member; area often used for the purpose of feeding, mating, and caring for young. 374

testcross Cross between an individual with the dominant phenotype and an individual with the recessive phenotype to see if the individual with the dominant phenotype is homozygous or heterozygous. 180

testis In animals, male gonad which produces sperm and the male sex hormones. 894

testosterone (teh-STAHS-tuh-rohn) In mammals, major male sex hormone produced by interstitial cells in the testis; it stimulates development of primary sex organs and maintains secondary sexual characteristics in males. 901

tetrapod Four-footed vertebrates, including related animals that are limbless or bipedal; amphibians, reptiles, birds, and mammals. [Gk. *tetra,* four, and *podos,* foot] 622

thalamus (THAL-uh-mus) Part of the brain that serves as the integrating center for sensory input, it plays a role in arousing the cerebral cortex. 831

theory Conceptual scheme arrived at by the scientific method and supported by innumerable observations and experimentation. 11

thermal inversion Temperature inversion that traps cold air and its pollutants near the earth with the warm air above it. [Gk. *therme,* heat, and L. *in,* against or over, and *verto,* turn] 477

thigmotropism (thig-MAH-truh-piz-um) In plants, unequal growth due to contact with solid objects, as the coiling of tendrils around a pole. [Gk. *thigma,* touch, and *tropos,* turning] 678

thoracic cavity Internal body space of some tetrapods that contains the lungs, protecting them from desiccation; the chest. 793

thrombocyte Platelet; cell fragment in the blood that initiates the process of blood clotting. [Gk. *thrombos,* blood clot, and *kytos,* cell] 747

thylakoid (THY-luh-koyd) Fattened sac within a granum whose membrane contains chlorophyll and where the light-dependent reactions of photosynthesis occur. [Gk. *thylakos,* sack, and *-eides,* like, resembling] 71, 116

thymine (T) (THY-meen) One of four nitrogen-containing bases in nucleotides composing the structure of DNA. 225

thymus gland Lymphoid organ involved in the development and functioning of the immune system; T lymphocytes mature in the thymus gland. 888

thyroid gland Large gland in the neck that produces several important hormones, including thyroxine and calcitonin. [Gk. *thyreos,* large, door-shaped shield] 880

thyroxine (thy-RAHK-seen) Substance (also called T_4) secreted from the thyroid gland that promotes growth and development in vertebrates; in general, it increases the metabolic rate in cells. 880

tissue Group of similar cells combined to perform a common function. 718

tissue culture Process of growing tissue artificially in a usually liquid medium in laboratory glassware. 708

tissue fluid Filtrate, containing all the small molecules of blood plasma, that bathes all the cells of the body. 749

T lymphocyte (LIM-fuh-syt) Lymphocyte that matures in the thymus and exists in four varieties, one of which kills antigen-bearing cells outright. [L. *lympha,* clear water, and *kytos,* cell] 760

tonicity Osmolarity of a solution compared to that of a cell; if the solution is isotonic to the cell there is no net movement of water; if the solution is hypotonic, the cell gains water; and if the solution is hypertonic the cell loses water. 88

trachea (TRAY-kee-uh) Air tube (windpipe) in tetrapod vertebrates that runs between the larynx and the bronchi; also an air tube in insects that is located between the spiracles and the tracheoles. [L. *trachia,* windpipe] 603, 778, 794

tracheid In flowering plants, type of cell in xylem that has tapered ends and pits through which water and minerals flow. 664

transcription Process whereby a DNA strand serves as a template for the formation of mRNA. [L. *trans,* across, and *scriptio,* a writing] 238

transcription factor In eukaryotes protein required for the initiation of transcription by eukaryotic RNA polymerase. 258

transfer RNA (tRNA) Type of RNA that transfers a particular amino acid to a ribosome during protein synthesis; at one end it binds to the amino acid and at the other end it has an anticodon that binds to an mRNA codon. 238

transgenic organism Free-living organisms in the environment that have had a foreign gene inserted into them. 275

transition reaction Reaction that oxidizes pyruvate with the release of carbon dioxide; results in acetyl-CoA and connects glycolysis to the Krebs cycle. 131

translation Process whereby the sequence of codons in mRNA determines (is translated into) the sequence of amino acids in a polypeptide. [L. *trans,* across, and *latus,* carry, bear] 238

transpiration Plant's loss of water to the atmosphere, mainly through evaporation at leaf stomates. 667

trophic level Feeding level of one or more populations in a food web. 433

trophoblast (TROH-fuh-blast) Outer membrane surrounding the embryo in mammals, when thickened by a layer of mesoderm, it becomes the chorion, an extraembryonic membrane. [Gk. *trophe,* food, and *blastos,* bud] 923

tropism In plants, a growth response toward or away from a directional stimulus. [Gk. *tropos,* turning] 676

trypanosome (trip-AN-uh-sohm) Member of a genus of parasitic zooflagellates that cause severe disease in human beings and domestic animals, including a condition called sleeping sickness. 532

trypsin Protein-digesting enzyme secreted by the pancreas. 780

tube-within-a-tube body plan Body with a digestive tract that has both a mouth and an anus. 584

tumor Cells derived from a single mutated cell that has repeatedly undergone cell division; benign tumors remain at the site of origin, and malignant tumors metastasize. [L. *tumor,* swelling] 260

tumor-suppressor gene Gene that codes for a protein that ordinarily suppresses cell division. 263

turgor pressure Pressure of the cell contents against the cell wall, in plant cells, determined by the water content of the vacuole; gives internal support to the plant cell. [L. *turgor,* swelling] 89

tympanic membrane (tim-PAN-ik) Membranous region that receives air vibrations in an auditory organ. [Gk. *tympanum,* drum] 848

U

umbilical cord Cord connecting the fetus to the placenta through which blood vessels pass. [L. *umbilicus,* navel] 925

urea Main nitrogenous waste of terrestrial amphibians and mammals. [Gk. *urina,* urine] 806

ureter (YUUR-ut-ur) Tubular structure conducting urine from the kidney to the urinary bladder. 808

urethra (yuu-REE-thruh) Tubular structure that receives urine from the bladder and carries it to the outside of the body. 808

uric acid Main nitrogenous waste of insects, reptiles, birds, and some dogs. 806

urinary bladder Organ where urine is stored. 808

urine Liquid waste product made by the nephrons of the vertebrate kidney through the processes of glomerular filtration, tubular reabsorption, and tubular secretion. [Gk. *urina,* urine] 808

uterine cycle Cycle that runs concurrently with the ovarian cycle; it prepares the uterus to receive a developing zygote. 904

uterus In mammals, expanded portion of the female reproductive tract through which eggs pass to the environment or in which an embryo develops and is nourished before birth. [L. *uterus,* womb] 902

V

vaccine Antigens prepared in such a way that they can promote active immunity without causing disease. [L. *vaccinus,* of cows] 513, 767

vagina In mammals, distal portion of the female reproductive tract which receives the penis during copulation. 902

valve Membranous extension of a vessel or the heart wall that opens and closes, ensuring one-way flow; common to the systemic veins, the lymphatic veins, and the heart. 738

vascular bundle In plants, primary pholem and primary xylem enclosed by a bundle sheath. 640

vascular cambium In plants, lateral meristem that produces secondary phloem and secondary xylem. [L. *vasculum,* dim. of *vas,* vessel, and *cambio,* exchange] 648

vascular cylinder In plant roots, strand of conducting tissue consisting of primary phloem and primary xylem often surrounded by a sheath. 642

vascular plant Member of kingdom Plantae which contains vascular tissue, e.g. ferns, gymnosperms, and angiosperms [L. *vasculum,* dim. of *vas,* vessel] 558

vascular tissue Transport tissue in plants consisting of xylem and phloem. 641

vein Blood vessel that arises from venules and transports blood toward the heart. [L. *vena,* blood vessel] 640, 736

vena cava Large systemic vein that returns blood to the right atrium of the heart in tetrapods; either the superior or inferior vena cava. [L. *vena,* blood vessel, and *cavus,* hollow] 742

ventricle Cavity in an organ, such as a lower chamber of the heart; or the ventricles of the brain. 738, 830

venule Vessel that takes blood from capillaries to a vein. 736

vertebral column Backbone of vertebrates through which the spinal cord passes. [L. *vertebra,* bones of backbone] 861

vertebrate Referring to an animal with a backbone composed of vertebrae. 598, 619

vessel element Cell which joins with others to form a major conducting tube found in xylem 664

vestigial structure (ve-STIJ-(ee-)-ul) Remains of a structure that was functional in some ancestor but is no longer functional in the organism in question. [L. *vestigium,* trace, footprint] 298

villus (VIL-us) (pl., villi) Small, fingerlike projection of the inner small intestinal wall. [L. *villus,* shaggy hair] 780

viroid Infectious agent consisting of a small strand of RNA that is apparently replicated by host cell enzymes. 512

virus Nonliving, obligate, intracellular parasite consisting of an outer capsid and an inner core of nucleic acid. [L. *virus,* poison] 508

vitamin Essential requirement in the diet, needed in small amounts. They are often part of coenzymes. 106, 784

vocal cord In mammals, fold of tissue within the larynx; creates vocal sounds when it vibrates. 794

vulva In mammals, external genitalia of the female that lie near the opening of the vagina. 902

W

water potential Potential energy of water; it is a measure of the capability to release or take up water relative to another substance. 665

woody plant Plant that contains wood; usually trees such as evergreen trees (gymnosperms) and flowering trees (angiosperms). Alternative is a herbaceous plant. 567

X-linked In animals, gene located on the X chromosomes; does not necessarily control a sexual feature of the organism. 192, 215

xylem (ZY-lum) Vascular tissue that transports water and mineral solutes upward through the plant body; it contains vessel elements and tracheids. [Gk. *xylon,* wood, and L. *em,* in] 558, 642, 664

yolk sac One of the extraembryonic membranes which in shelled vertebrates contains yolk for the nourishment of the embryo and in placental mammals is the first site for blood cell formation. 922

Z

zero population growth No growth in population size. 400

zooflagellate (zoh-uh-FLAJ-uh-layt) Protozoan that moves by means of flagella. [Gk. *zoon,* animal, and L. *flagello,* whip] 532

zooplankton (zoh-uh-PLANGK-tun) Part of plankton containing protozoa and other types of microscopic animals. [Gk. *zoon,* animal, *planktos,* wandering] 461, 530

zygospore (ZY-guh-spohr) Thick-walled, resting cell formed during sexual reproduction of zygospore fungi. 542

Credits

Line Art and Text

Chapter 1

Closer Look: Courtesy of J. William Schopf, Director, UCLA Center for the Study of Evolution and the Origin of Life. **1.11 table:** Reprinted with permission from *Nature,* Vol. 279, May 17, 1979, page 233. Copyright © 1979 Macmillan Magazines Limited.

Chapter 8

Health Focus: Scott K. Powers and Edward T. Howley, *Exercise Physiology,* 2nd edition, 1994, McGraw-Hill Company, Inc., Dubuque, Iowa.

Chapter 9

9.7: From Peter H. Raven and George B. Johnson, *Biology,* 4th edition. Copyright © 1996 McGraw-Hill Company, Dubuque, Iowa. All Rights Reserved. Reprinted by permission.

Chapter 13

13.2B: From Robert F. Weaver and Philip W. Hedrick, *Genetics,* 2nd edition. Copyright © 1992 McGraw-Hill Company, Dubuque, Iowa. All Rights Reserved. Reprinted by permission.

Chapter 14

Doing Science: Courtesy of Joyce Haines.

Chapter 16

Doing Science: Courtesy of Thomas Gilmore, Boston University.

Chapter 18

18.17: From Margaret O. Dayhoff and Richard V. Eck, *Atlas of Protein Sequence and Structure, 1967–1968.* Reprinted by permission of National Biomedical Research Foundation, Washington, DC.

Chapter 19

Doing Science: Courtesy of Gerald D. Carr, University of Hawaii at Manoa.

Chapter 21

21.2A: Redrawn from *Vertebrates, Phylogeny, and Philosophy,* 1986, Contributions to Geology, University of Wyoming, Special Paper 3 (frontispiece and article by Rose and Bown; and Thomas M. Bown and Kenneth D. Rose, "Patterns of Dental Evolution in Early Eocene Anaptomorphine Primates (Omomyidae) from Bighorn Basin, Wyoming," in *Paleontological Society Memoir 23 (Journal of Paleontology,* Vol. 61, supplement to no. 5). Courtesy of Kenneth D. Rose of The Johns Hopkins University, Baltimore, MD. **21.4A:** Reprinted by permission from Alan Walker. **Doing Science:** Courtesy of Steven Stanley, The Johns Hopkins University.

Chapter 22

22.2: Data from S. J. Arnold, "The Microevolution of Feeding Behavior" in *Foraging Behavior: Ecology, Ethological, and Psychological Approaches,* edited by A. Kamil and T. Sargent, 1980, Garland Publishing Company, New York, NY. **22.7:** Data from G. Hausfater, "Dominance and Reproduction in Baboons (Papio cynocephalus): A Quantitative Analysis," *Contributions in Primatology,* 7:1-150, 1975.

Chapter 23

23.5A: From Raymond Pearl, *The Biology of Population Growth,* 1925. Copyright © 1925 McGraw-Hill Company, Inc.; **23.7:** Data from W.K. Purves, et al., *Life: The Science of Biology,* 4th edition, Sinauer and Associates; **23.9:** Data from Charles J. Krebs, *Ecology,* 3rd edition, 1984, Harper & Row; after Scheffer, 1951; **23.11B:** Data from Charles J. Krebs, *Ecology,* 3rd edition, 1984, Harper & Row; after Lack 1966 and J. Krebs, personal communication; **23.11C:** Data from Michael Begon, et al., *Population Ecology,* 3rd edition, 1996, Blackwell Science; **23B TOP:** Data from Karen Arms & Pamela Camp, *Biology,* 4th edition, 1995, Saunders College Publishing; **23B middle.** Data from C.J. Krebs, *Ecology,* 3rd edition, 1984, Harper & Row; **23.14:** Data from *Population Today* Vol. 24, #4, April, 1996, Population Reference Bureau, Inc., Washington, D.C.

Chapter 24

24.1C: Data from G. F. Gause, *The Struggle for Existence,* 1934, Williams & Wilkins Company, Baltimore, MD; **24.3**: Data from Charles J. Krebs, *Ecology,* 3rd edition, 1984, Harper & Row; **24.6:** Data from G.F. Gause, *The Struggle for Existence,* 1934, Williams & Wilkins Company, Baltimore, MD; **24.9:** From Peter H. Raven and George B. Johnson, *Biology,* 4th edition. Copyright © 1996 McGraw-Hill Company, Dubuque, Iowa. All Rights Reserved. Reprinted by permission; **24.10:** Data from G.F. Gause, *The Struggle for Existence,* 1934, Williams & Wilkins Company, Baltimore, MD; **24.11B:** Data from D. A. MacLulich, *Fluctuations in the Numbers of the Varying Hare (Lepus americanus),* University of Toronto Press, Toronto, 1937, reprinted 1974.

Chapter 26

26.1B, 26.2A 26.6B, 26.9A MAP, 26.10 MAP, 26.11 MAP, 26.13A MAP, 26.15 MAP, 26.16 MAP, 26.17A MAP: From Peter H. Raven and George B. Johnson, *Biology,* 4th edition. Copyright © 1996 McGraw-Hill Company, Dubuque, Iowa. All Rights Reserved. Reprinted by permission.

Chapter 27

27.9B: From Eldon D. Enger and Bradley F. Smith, *Environmental Science,* 5th edition. Copyright © 1995 McGraw-Hill Company, Dubuque, Iowa. All Rights Reserved. Reprinted by permission. **27.3B:** Data from R.J. Delmas, "Environmetal Information from Ice Cores," *Review of Geophysics,* 30, 1-21, 1992, American Geophysical Union. **27.3C:** Data from David M. Gates, *Climate Change and Its Biological Consequences,* 1993. Sinauer Associates, Inc., Sunderland, MA; **Ecology Focus:** From R.B. Primack, *Essentials of Conservation Biology,* 1993. Copyright © 1993 Sinauer Associates, Inc., Sunderland, MA.

Chapter 30

Doing Science: Courtesy of Susan Dutcher, University of Colorado at Boulder.

Chapter 33

33.7B: From Stephen Miller and John Harley, *Zoology,* 2d edition. Copyright © 1994 McGraw-Hill Company, Dubuque, Iowa. All Rights Reserved. Reprinted by permission.

Chapter 35

35.4: From Stephen Miller and John Harley, *Zoology,* 3rd edition. Copyright © 1996 McGraw-Hill Company, Dubuque, Iowa. All Rights Reserved. Reprinted by permission. **A Closer Look:** Courtesy of Gregory J. McConnell.

Chapter 37

Research Report: Courtesy of G. David Tilman, University of Minnesota.

Chapter 38

Research Report: Courtesy of Donald Briskin and Margaret Gawienowski, University of Illinois at Urbana-Champaign.

Chapter 39

39.3B: From T. Elliot Weier, et al., *Botany,* 6th edition. Copyright © 1982 John Wiley & Sons, Inc., New York, NY. Reprinted by permission of the authors; **39.8, 39.9:** From Kingsley R. Stern, *Introductory Plant Biology,* 6th edition. Copyright © 1994 McGraw-Hill Company, Inc., Dubuque, Iowa. All Rights Reserved. Reprinted by permission. **A Closer Look:** Courtesy of Charles Horn.

Chapter 40

40.4: From John W. Hole, Jr., *Human Anatomy & Physiology,* 6th edition. Copyright © 1993 McGraw-Hill Company, Inc., Dubuque, Iowa. All Rights Reserved. Reprinted by permission.

Chapter 41

41B: From Kent M. Van De Graaff and Stuart Ira Fox, *Concepts of Human Anatomy and Physiology,* 3rd edition. Copyright © 1992 McGraw-Hill Company, Dubuque, Iowa. All Rights Reserved. Reprinted by permission.

Chapter 43

43.7: From Kent M. Van De Graaff and Stuart Ira Fox, *Concepts of Human Anatomy and Physiology,* 4th edition. Copyright © 1995 McGraw-Hill Company, Dubuque, Iowa. All Rights Reserved. Reprinted by permission; **43.11:** From the U.S. Department of Agriculture, as seen in Ruth Bernstein and Stephen Bernstein, *Biology,* Copyright © 1996 McGraw-Hill Company, Inc., Dubuque, Iowa. All Rights Reserved. Reprinted by permission.

Chapter 44

44.8: From John W. Hole, Jr., *Human Anatomy and Physiology,* 5th edition. Copyright © 1990 McGraw-Hill Company, Dubuque, Iowa. All Rights Reserved. Reprinted by permission.

Chapter 45

45.5: From Kent M. Van De Graaff and Stuart Ira Fox, *Concepts of Human Anatomy and Physiology,* 4th edition. Copyright © 1995 McGraw-Hill Company, Dubuque, Iowa. All Rights Reserved. Reprinted by permission.

Chapter 47

47.8: From Kent M. Van De Graaff and Stuart Ira Fox, *Concepts of Human Anatomy and Physiology,* 4th edition. Copyright © 1995 McGraw-Hill Company, Inc., Dubuque, Iowa. All Rights Reserved. Reprinted by permission. **Doing Science:** Courtesy of Anita Zimmerman, Brown University, Providence, RI.

Chapter 48

48.6: From Kent M. Van De Graaff, *Human Anatomy,* 3d edition. Copyright © 1992 McGraw-Hill Company, Inc., Dubuque, Iowa. All Rights Reserved. Reprinted by permission.

Chapter 49

49.8: From John W. Hole, Jr., *Human Anatomy and Physiology,* 6th edition. Copyright © 1993 McGraw-Hill Company, Inc., Dubuque, Iowa. All Rights Reserved. Reprinted by permission; **49A:** From Stuart Ira Fox, *Human Physiology,* 4th edition. Copyright © 1993 McGraw-Hill Company, Inc., Dubuque, Iowa. All Rights Reserved. Reprinted by permission.

Chapter 50

50.4A, 50.6A, 50.8A: From Ruth Bernstein and Stephen Bernstein, *Biology,* Copyright © 1996 McGraw-Hill Company, Inc., Dubuque, Iowa. All Rights Reserved. Reprinted by permission; **50.8B:** From David Shier, et al., *Hole's Human Anatomy and Physiology,* 7th edition. Copyright © 1996 McGraw-Hill Company, Inc., Dubuque, Iowa. All Rights Reserved. Reprinted by permission; **50.12:** From Kent M. Van De Graaff and Stuart Ira Fox, *Concepts of Human Anatomy and Physiology,* 4th edition. Copyright © 1995 McGraw-Hill Company, Inc., Dubuque, Iowa. All Rights Reserved. Reprinted by permission; **TA 50.2:** From Ruth Bernstein and Stephen Bernstein, *Biology,* Copyright © 1996 McGraw-Hill Company, Inc., Dubuque, Iowa. All Rights Reserved. Reprinted by permission.

Chapter 51

51.14: Redrawn from Kent M. Van De Graaff and Stuart Ira Fox, *Concepts of Human Anatomy and Physiology,* 4th edition. Copyright © 1995 McGraw-Hill Company, Inc., Dubuque, Iowa. All Rights Reserved. Reprinted by permission. **Doing Science:** Courtesy of John Sternfeld.

Photographs

History of Biology

End Sheets: Darwin, Pasteur, Koch, Pavlov, Lorenz, Pauling, Leeuwenhoek: © The Bettmann Archive.

Chapter 1

Opener: © David C. Fritts/Animals Animals/Earth Scenes; **1.2 A:** ©Mitch Reardon/Photo Researchers, Inc.; **1.2 B:** © David C. Fritts/Animals Animals/Earth Scenes; **1.3:** © Francisco Erize/Bruce Coleman, Inc. **1 A:** © John D. Cunningham/ Visuals Unlimited; **1.5 A:** Red grouper © Carl Roessler/Animals Animals/Earth Scenes; **1.5 B:** Coral © F. Stuart Westmorland/Tom Stack & Assoc.; **1.5 C:** Palau coral © Ed Robinson/Tom Stack & Assoc.; **1.5 D:** Eel © David B. Fleetham/Tom Stack & Assoc.; **1.5 E:** Lionfish © Tom Stack/ Tom Stack & Assoc.; **1.6 A:** Rainforest © Barbara von Hoffman/Tom Stack & Assoc.; **1.6 B:** Toucan © Ed Reschke/Peter Arnold, Inc.; **1.6 C:** Butterfly © Kjell Sandved/Butterfly Alphabet; **1.6 D:** Jaguar © BIOS (Seitre)/Peter Arnold, Inc.; **1.6 E:** Orchid © Max and Bea Hunn/Visuals Unlimited; **1.6 F:** Frog © Kevin Schafer/ Marth Hill/Tom Stack & Assoc.; **1.7 A-1:** Nostoc © Eric Grave/Photo Researchers, Inc.; **1.7 A-2:** Euglena © John D. Cunningham/Visuals Unlimited; **1.7 A-3:** Mushroom © Rod Planck/Tom Stack & Assoc.; **1.7 A-4:** Rosa © Farell Grehan/Photo Researchers, Inc.; **1.7 A-5:** Lynx © Leonard L. Rue; **1.9 A:** © Breck Kent/Animals Animals/ Earth Scenes; **1.9 B:** © Michael Tweedie/ Photo Researchers, Inc.

Chapter 2

Opener: © Erwin & Peggy Bauer/Bruce Coleman, Inc.; **2.1 B:** © McGraw-Hill Higher Education/Carlyn Iverson, photographer; **2.5 B-1:** © Charles M. Falco/Photo Researchers, Inc.; **2.8 A:** © Holt Confer/ Grant Heilman Photography; **2.10 A:** © Di Maggio/Peter Arnold, Inc.; **2.9 B:** © Ken Edward/USDA/Photo Researchers, Inc.

Chapter 3

Opener: © Zig Leszczynsk/Animals Animals/ Earth Scenes; **3.1 A:** © John Gerlach/Tom Stack & Assoc.; **3.1 B:** © Leonard Lee Rue/Photo Researchers, Inc.; **3.1 C:** © H. Pol/CNRI/SPL/Photo Researchers, Inc.; **3.7 A:** © Jeremy Burgess/ Photo Researchers, Inc.; **3.7 B:** © Don W. Fawcett/Photo Researchers, Inc.; **3.8:** © BioPhoto Assoc./Photo Researchers, Inc.; **3.11 A:** ©Anthony Mercieca/Photo Researchers, Inc.; **3.11 B:** © W. Treat Davidson/Photo Researchers, Inc.; **3 A:** Courtesy Dr. Kirit Chapatwala.

Chapter 4

Opener: © CABISCO/Phototake; **4.1 A:** © Tony Stone Images; **4.1 B:** © CABISCO/ Phototake; **4.1 C:** © Barbara J. Miller/ Biological Photo Service; **4.1 D:** © Ed Reschke; **4 A A:** © Michael Abbey/Visuals Unlimited; **4 A B:** © M. Schliwa/Visuals Unlimited; **4 A C:** © Kessel/Shih/Peter Arnold, Inc.; **4 B A:** Brightfield © Ed Reschke; **4 B B:** Brightfield (stained) © Biophoto Assoc./Photo Researchers; **4 B C:** Differential © David M. Phillips/Visuals Unlimited; **4 B D:** Phase © David M. Phillips/Visuals Unlimited; **4 B E:** Darkfield © David M. Phillips/Visuals Unlimited; **4.3 A-2:** © David M. Phillips/Visuals Unlimited; **4.3 B-2:** © Biophoto Assoc./Photo Researchers, Inc.; **4.4 B:** ©Alfred Pasieka/ Photo Researchers, Inc.; **4.5 B:** © Newcomb/ Wergin/BPS/Tony Stone Images; **4.6 A:** © Don Fawcett/Photo Researchers, Inc.; **4.7A:** © Warren Rosenberg/Biological Photo Service; **4.8 A:** © Richard Rodewald/ Biological Photo Service; **4.8 B:** © David M. Phillips/Visuals Unlimited; **4.9:** © E.H. Newcomb & S.E. Frederick/Biological Photo Service; **4.10 A:** Courtesy Herbert W. Israel, Cornell University; **4.11 A:** Courtesy Dr. Keith Porter; **4.12 B-1:** © M. Schliwa/Visuals Unlimited; **4.12 C-1, D-1:** © K.G. Murti/ Visuals Unlimited; **4.13:** Courtesy Kent McDonald, University of Colorado Boulder; **4.14 A, B, C:** © William L. Dentler/Biological Photo Service.

Chapter 5

Opener: Courtesy Mrs. Doris Robertson, Photo by Dr. J. David Robertson, Duke Univ. Medical Center; **5.1 A:** © Warren Rosenberg/ Biological Photo Service; **5.1 D:** © Don Fawcett/Photo Researchers, Inc.; **5.12 B-1, B-2:** Courtesy Mark Bretscher; **5.15 A:** Courtesy Camillo Peracchia; **5.15 B:** © David M. Phillips/Visuals Unlimited; **5.15 C:** From Douglas E. Kelly, *J. Cell Biol.* 28 (1966): 51. Reproduced by copyright permission of The Rockefeller University Press.

Chapter 6

Opener: © Joe Distefano/Photo Researchers, Inc.; **6.1:** © Jim Shaffer **6.11 A:** © John D. Cunningham/Visuals Unlimited; **6.11 C:** © Sunset Moulu (20)/Peter Arnold, Inc.

Chapter 7

Opener: © David B. Fleetham/Visuals Unlimited; **7.2:** Courtesy Herbert W. Israel,

Cornell University; **7.6 A:** Courtesy Melvin P. Calvin; **7.7 B:** © The McGraw-Hill Companies, Inc./ Bob Coyle, photographer.

Chapter 8

Opener: © Don Smetzer/Tony Stone Images; **8.4 B:** Courtesy Dr. Keith Porter; **8 A:** © Tim Davis/Photo Researchers, Inc.

Chapter 9

Opener: © Prof. Motta/Photo Researchers, Inc.; **9.1 A-2, B-2, C-2:** © Stanley C. Holt/ Biological Photo Service; **9.2:** © Biophoto Assoc./Photo Researchers, Inc.; **9.5 A, B:** © Ed Reschke **9.5 C, D, E, F:** © Michael Abby/Photo Researchers, Inc.; **9.6 A, B, C, D:** © Ed Reschke; **9.7:** © B.A. Palevitz and E.H. Newcomb BPS/Tom Stack & Assoc.; **9.8 A-1, A-2:** © R.G. Kessel and C.Y. Shih (1974) *Scanning Electron Microscopy in Biology: A Student Atlas on Biological Organizaiton,* Springer-Verlag, New York.

Chapter 10

Opener: © David M. Phillips/Visuals Unlimited; **10.2:** Courtesy of Dr. D. Von Wettstein; **10.3:** Courtesy of Dr. Bernard John.

Chapter 11

Opener: © Tim Davis/Tony Stone Images; **11.1:** © The Bettmann Archive.

Chapter 12

Opener: © Biophoto Assoc./Photo Researchers, Inc.; **12.4:** © Jane Burton/Bruce Coleman, Inc.; **12.9 A, B, C:** © The McGraw-Hill Companies, Inc./Bob Coyle, photographer.

Chapter 13

Opener: © Mark E. Gibson/Visuals Unlimited; **13.3 A:** © Jill Cannefax/EKM-Nepenthe; **13.4 A, B:** From Kampheier, R., *Physical Examination of Health Diseases.* © 1958 F.A. Davis Company; **13.7:** © Steve Uzzell; **13.8:** Courtesy of Cystic Fibrosis Foundation; **13.11 B:** © Bill Longcore/Photo Researchers, Inc.

Chapter 14

Opener: © Ken Edwards/Photo Researchers, Inc.; **14.2, 14,2 (inset):** © Lee Simon/Photo Researchers, Inc.; **14.5B:** Courtesy of Biophysics Dept., Kings College, London; **14.6 A:** © Nelson Max/Peter Arnold, Inc.; **14 B:** Courtesy Cold Spring Harbor Archive; **14 C:** © John N.A. Lott/Biological Photo Service.

Chapter 15

Opener: © Dwight Kuhn; **15.2 A-1, A-2:** © Bill Longcore/Photo Researchers, Inc.; **15.6 A:** © Oscar L. Miller/Photo Researchers, Inc.; **15.8 B:** Courtesy University of California Lawrence Livermore National Library and the U.S. Department of Energy; **15.9 D:** Courtesy Alexander Rich.

Chapter 16

Opener: © Dr. Andrejs Liepins/Photo Researchers, Inc.; **16.4 A:** Courtesy NATURE 163:676 (1949); **16 B:** Courtesy Stephen Wolfe; **16.5 A:** From M.B. Roth and J.G. Gall *Journal of Cell Biology,* 105:1047-1054, 1978. The Rockefeller University Press; **16.6:** Courtesy Jose Mariano Amabis, University of Sao Paulo; **Page 261:** Courtesy Thomas Gilmore, Boston University; **16.12:** © Seth Joel/SPL/Photo Researchers, Inc.; **16 C:** © The McGraw-Hill Companies, Inc./Bob Coyle, photographer.

Chapter 17

Opener: © James Balog/Tony Stone Images; **17.6 A, B:** Courtesy General Electric Research & Development; **17.7:**Courtesy Monsanto; **17.8:** Courtesy Genzyme Corporation and Tufts University School of Medicine; **17 A:** Courtesy Seminis Vegetable Seeds, Inc.; **17.11:** Courtesy Bio-Rad Laboratory.

Chapter 18

Opener: © Kevin Schafer/Peter Arnold, Inc.; **18.1 B:** Rhea: © Tom Stack/Tom Stack & Associates; **18.1 C:** Desert: © C. Luiz Claudio Marigo/Peter Arnold, Inc.; **18.1 D:** Mountains: © Gary J. James/Biological Photo Service; **18.1 E:** Rain forest: © C. Luiz Claudio Marigo/Peter Arnold, Inc.; **18.1 F:** Iguanas: © Ken Lucas; **18.1 G:** Finch: © Miguel Castro/Photo Researchers, Inc.; **18.2:** © The Bettmann Archive; **18.3:** © John D. Cunningham/Visuals Unlimited; **18.4:** © Malcolm Boulton/Photo Researchers, Inc.; **18.5 B:** © J. & L. Weber/Peter Arnold, Inc.; **18.7:** © Juan & Carmecita Munoz/Photo Researchers, Inc.; **18.8 A:** © Walt Anderson/ Visuals Unlimited; **18.8 B:** © Michael Dick/ Animals Animals/Earth Scenes; **18.9 A:** © Adrienne T. Gibson/Animals Animals/ Earth Scenes; **18.9 B:** © Joe McDonald/ Animals Animals/Earth Scenes; **18.9 C:** © Leonard Lee Rue/Animals Animals/Earth Scenes; **18.10:** © Wendell Metzen/Bruce Coleman, Inc.; **18 A -1:** © Stock Montage/ Historical Picture Service; **18.11 A:** Timber wolf: © Gary Milburn/Tom Stack & Assoc.; **18.11 B:** Red chow: © Jeanne White/Photo Researchers; **18.11 C:** Dalmation: © Alexander Lowry/Photo Researchers, Inc.; **18.11 D:** Chihuahua: © Mary Eleanor Browning/Photo Researchers, Inc.; **18.11 E:** Bloodhound: © Mary Bloom/Peter Arnold, Inc.; **18.11 F:** French bulldog: © Salana/ Jacana/Photo Researchers, Inc.; **18.11 G:** Boston terrier: © Mary Bloom/Peter Arnold, Inc.; **18.11 H:** Shetland: © Carolyn McKeone/Photo Researchers, Inc.; **18.12 A, B, C:** Courtesy W. Atlee Burpee Company; **18.13 A:** © John D. Cunningham/Visuals Unlimited; **18.13 B:** Transp. #213 Courtesy Dept. of Library Services, American Museum of Natural History; **18.14-1:** Wombat: © Adrienne Gibson/Animals Animals/Earth Scenes; **18.14-2:** Sugar glider: © John Sundance/Janana/Photo Researchers; **18.14-3:** Kangaroo: © George Holton/Photo Researchers, Inc.; **18.14-4:** Dasyurus: © Tom McHugh/Photo Researchers, Inc.; **18.14-5:** Tasmanian wolf: © Tom McHugh/Photo Researchers, Inc.; **18.16 A:** © CABISCO/Phototake; **18.16 B:** © CABISCO/Visuals Unlimited.

Chapter 19

Opener: © Peter Fryer/Panos Pictures; **19.2 A:** © Breck Kent/Animals Animals/Earth Scenes; **19.2 B:** © Michael Tweedie/Photo Researchers, Inc.; **19.3 A:** © Visuals Unlimited; **19.3 B:** © William Weber/Visuals Unlimited; **19.3 C:** © Zig Leszczynski/ Animals Animals/Earth Scenes; **19.3 D:** © Dale Jackson/Visuals Unlimited; **19.3 E:** © Joseph Collins/Photo Researchers, Inc.; **19.3 F:** © Zig Leszczynski/Animals Animals/Earth Scenes; **19.5:** Courtesy Victor McKusick; **19.8 B:** © Bob Evans/Peter Arnold, Inc.; **19.10 A:** Least: © Stanley Maslowski/Visuals Unlimited, **19.10 B:** Acadian: © Karl Maslowski/Visuals Unlimited; **19.10 C:** Traill's: © Ralph Reinhold/Animals Animals/Earth Scenes; **Page 318:** Courtesy Gerald D. Carr; **19 A (bottom, top), 19B (bottom, top), 19C (bottom, top):** Courtesy Gerald D. Carr.

Chapter 20

Opener: © Jane Burton/Bruce Coleman, Inc.; **20.3 A:** © Science VU/Visuals Unlimited; **20.3 B** Courtesy Dr. David Deamer; **20.6:** © Sylvain Grandadam/Photo Researchers, Inc.; **20.7 A:** Courtesy J. William Schopf; **20.7 B:** © Francois Gohier/Photo Researchers, Inc.; **20.8 A:** Courtesy of The Field Museum; **20.8 B:** Courtesy Dr. Bruce N. Runnegar; **20.10, 20.11 A, B:** Courtesy of The Field Museum; **20 A A:** Courtesy Museum of the Rockies; **20 A B:** Courtesy Museum of the Rockies; **20.12, 20.13:** Courtesy of The Field Museum; **20.15 B:** © David Parker/ Photo Researchers, Inc.; **20.15 C:** © Matthew Shipp/Photo Researchers, Inc.; **20.17 A:** Courtesy Senckenberg Museum; **20.17 B:** © Ronald Seitre/Peter Arnold, Inc.

Chapter 21

Opener: © CABISCO/Visuals Unlimited; **21.2 B:** © Ron Austing/Photo Researchers, Inc.; **21.2 C:** © Howard Uible/Photo Researchers, Inc.; **21.4 C:** © National Museum of Kenya; **21.5 A:** Gibbon: © Martha Reeves/Photo Researchers, Inc.; **21.5 B:** Orangutan: © BIOS/Peter Arnold, Inc.; **21.5 C:** Gorilla: © George Holton/Photo Researchers, Inc.; **21.5 D:** Chimpanzee: © Tom McHugh/Photo Researchers, Inc.; **21.7 A:** © Dan Dryfus and Associates; **21.7 B:** © John Reader/Photo Researchers, Inc.; **21.8 A:** © Institute of Human Origins; **21 A:** © Margaret Miller/Photo Researchers, Inc.; **Page 356:** Courtesy Steven Stanley; **21.10:** © National Museum of Kenya; **21.12, 21.13:** Courtesy of The Field Museum.

Chapter 22

Opener: © Frank Lane Agency/Bruce Coleman, Inc.; **22.2 B, C:** © R. Andrew Odum/Peter Arnold, Inc.; **22.6:** © Frans Lanting/Minden Pictures; **22.8 A:** © Y. Arthus-Bertrand/Peter Arnold, Inc.; **22.8 B:** © FPG International; **22.9:** © Jonathan Scott/ Planet Earth; **22.10:** © Susan Kuklin/Photo Researchers, Inc.; **22.11 A:** © OSF/Animals Animals/Earth Scenes; **22.12:** © J & B Photo/ Animals Animals/Earth Scenes.

Chapter 23

Opener: © Nigel Dennis/Photo Researchers, Inc.; **23.1** © Mike Bacon/Tom Stack & Assoc.; **23.2 A-1:** © C. Palck/Animals Animals/ Earth Scenes; **23.2 B-1:** © S. J. Krasemann/

Peter Arnold, Inc.; **23.2 C-1:** © Peter Arnold/Peter Arnold, Inc.; **23.3 A:** © Breck P. Kent/Animals Animals/Earth Scenes; **23.3 B:** © John Shaw/Tom Stack & Assoc.; **23.6 A, B:** © The McGraw-Hill Companies, Inc./Bob Coyle, photographer; **23.9 B:** © Paul Janosi/Valan Photos; **23.10:** © Kent & Donna Dennen/Photo Researchers, Inc.; **23.11 A:** © Neil Bromhall/OSF/Animals Animals/Earth Scenes; **23.12 A:** © Strawberry arrow poison © Michael Fogden/Animals Animals/Earth Scenes; **23.12 B:** Wood frog pair: © Matt Meadows/Peter Arnold, Inc.; **23.12 C:** Surinam toad: © Tom McHugh/Photo Researchers, Inc.; **23.12 D:** Darwin's frog: © Michael Fogden/Animals Animals/Earth Scenes; **23.12 E:** Midwife toad: © Mike Linley/OSF/Animals Animals/Earth Scenes; **23.13 A:** © Ted Levin/Animals Animals/Earth Scenes; **23.13 B:** © Michio Hoshino/Minden Pictures; **23 C:** Courtesy Jeffrey Kassner.

Chapter 24

Opener: © Michael H. Francis; **24.1 A** © Charlie Ott/Photo Researchers, Inc.; **24.1 B:** © Michael Graybill and Jan Hodder/Biological Photo Service; **24.5-1, 24.5-2:** © Ken Wagner/Phototake; **24.9 A:** © Biophoto Assoc./Photo Researchers, Inc.; **24.10 A:** © Alan Carey/Photo Researchers, Inc.; **24.11 A, B:** © Herb Segars; **24.12 A:** © National Audubon Society/A. Cosmos Blank/Photo Researchers, Inc.; **24.12 B:** © Wyman Meinzer/Peter Arnold, Inc.; **24.12 C:** © Zig Leszczynski/Animals Animals/Earth Scenes; **24.13 A, B, C, D, E:** © Edward Ross; **24 A:** Courtesy Dr. Ian Wyllie; **24.15:** © Dave B. Fleetham/Visuals Unlimited; **24.16 A, B, C:** Courtesy Daniel Janzen; **24.17:** © Bill Wood/Bruce Coleman, Inc.; **24.18 A, B, C, D, E:** © Breck P. Kent/Animals Animals/Earth Scenes; **24.20 A:** © Jeff Foott/Bruce Coleman, Inc.; **24.21 A:** © William E. Townsend, Jr./Photo Researchers, Inc.

Chapter 25

Opener: © Larry Ulrich/Tony Stone Images; **25.1:** Courtesy NASA; **25.2 A-1:** © Hermann Eisenbeiss/Photo Researchers, Inc.; **25.2 A-2:** © Ed Reschke/Peter Arnold, Inc.; **25.2 B-1:** Giraffe © George W. Cox; **25.2 B-2:** Caterpillar © Muridsany et Perennoy/Photo Researchers, Inc.; **25.2 C-1:** Osprey © Joe McDonald/Visuals Unlimited; **25.2 C-2:** Mantis © Scott Camazine; **25.2 D-1:** Mushroom © Michael Beug; **25.2 D-2:** Acetobacter © David M. Phillips/Visuals Unlimited; **25.12:** © Hugh Spencer/Photo Researchers, Inc.

Chapter 26

Opener: © McGraw-Hill Higher Education/Carlyn Iverson, photographer; **26.9 A-2:** © John Shaw/Tom Stack & Assoc.; **26.9 B:** © John Eastcott/Animals Animals/Earth Scenes; **26.9 C:** © John Shaw/Bruce Coleman, Inc.; **26.10 A:** © Norman Owen Tomlin/Bruce Coleman, Inc.; **26.10 B:** © Bill Silliker, Jr./Animals Animals/Earth Scenes; **26.11 A:** Forest: © E.R. Degginger/Animals Animals/Earth Scenes; **26.11 B:** Chipmunk: © Zig Lesczynski/Animals Animals/Earth Scenes; **26.11 C:** Millipede: © OSF/Animals Animals/Earth Scenes; **26.11 D:** Bobcat: © Tom McHugh/Photo Researchers, Inc.; **26.11 E:** Marigolds: © Virginia Neefus/Animals Animals/Earth Scenes; **26.13 A:** Arboreal lizard: © Kjell Sandved/Butterfly Alphabet; **26.13 B:** Katydid: © James Castner; **26.13 C:** Dart poison frog: © James Castner; **26.13 D:** Butterfly: © Kjell Sandved/Butterfly Alphabet; **26.13 E:** Macaw: © Kjell Sandved/Butterfly Alphabet; **26.13 F:** Ocelot: © Martin Wendler/Peter Arnold, Inc.; **26.13 G:** Sifaka: © Erwin & Peggy Bauer/Bruce Coleman, Inc.; **26.14 A:** © Bruce Iverson; **26.14 B:** Inset: © Kathy Merrifield/Photo Researchers, Inc.; **26.15 B:** Prairie: © Jim Steinberg/Photo Researchers, Inc.; **26.15 C:** Bison: © Steven Fuller/Animals Animals/Earth Scenes; **26.16 A, B, D:** © Darla Cox; **26.16 C:** © George W. Cox; **26.17 B:** Desert: © John Shaw/Bruce Coleman, Inc.; **26.17 C:** Kangaroo rat: © Bob Calhoun/Bruce Coleman, Inc.; **26.17 D:** Roadrunner: © Jack Wilburn/Animals Animals/Earth Scenes; **26.18 B:** Stonefly: © Kim Taylor/Bruce Coleman, Inc.; **26.18 C:** Trout: © William H. Mullins/Photo Researchers, Inc.; **26.18 D:** Carp: © Robert Maier/Animals Animals/Earth Scenes; **26.19 A:** © Roger Evans/Photo Researchers, Inc.; **26.19 B:** © Michael Gadomski/Animals Animals/Earth Scenes; **26.21 B:** Pond skater: © G.I. Bernard/Animals Animals/Earth Scenes; **26.21 C:** Pike: © Robert Maier/Animals Animals/Earth Scenes; **26.22 B:** © Heather Angel; **26.23 A:** © John Eastcott/Yva Momatiuk/Animals Animals/Earth Scenes; **26.23 B:** © James Castner; **26.24 A, B:** © Anne Wertheim/Animals Animals/Earth Scenes; **26.24 C:** © Jeff Greenburg/Photo Researchers, Inc.; **26.27 A:** © Mike Bacon/Tom Stack & Assoc.

Chapter 27

Opener: © Tim Davis /Photo Researchers, Inc.; **27.4 A:** © John Shaw/Tom Stack & Assoc.; **27.4 B:** © Thomas Kitchin/Tom Stack & Assoc.; **27.5 C:** © Bill Aron/Photo Edit; **27.6:** Courtesy Arlin J. Krueger/Goddard Space Flight Center/NASA; **27 A:** © Tom McHugh/Steinhart Aquarium/Photo Researchers, Inc.; **27.9 A:** © John Eastcott/Yva Momatiuck/Image Works; **27.10 B:** © G.Prance/Visuals Unlimited; **27.11 A:** Shrimp trawler: © Ulrike Welsch/Photo Researchers, Inc.; **27.11 B:** Rhino: © James Hancock/Photo Researchers, Inc.; **27.11 C:** Mine: © Barbara Pfeffer/Peter Arnold, Inc.; **27.11 D:** Snake: © John Mitchell/Photo Researchers, Inc.; **27.11 E:** Tires: © Inga Spence/Tom Stack & Assoc.; **27.11 G:** Honey-bee: © L. West/Photo Researchers, Inc.

Chapter 28

Opener: © Michael Fogden/Bruce Coleman, Inc.; **28.2 A:** Courtesy Uppsala University Library, Sweden; **28.2 B:** Bubil lily: © Arthur Gurmankin/Visuals Unlimited; **28.2 C:** Canada lily: © Dick Poe/Visuals Unlimited; **28.3:** © Tim Davis/Photo Researchers, Inc.; **28.4:** © Jen & Des Bartlett/Bruce Coleman, Inc.; **28.7 A:** © John D. Cunningham/Visuals Unlimited; **28.7 B:** © John Shaw/Tom Stack & Assoc.; **28.12 A:** Wolf: © Art Wolfe/Tony Stone Images; **28.12 B:** Flower: © Ed Reschke/Peter Arnold, Inc.; **28.12 C:** Paramecium: © M. Abbey/Visuals Unlimited; **28.12 D:** E. Coli: © David M. Phillips/Visuals Unlimited; **28.12 E:** Mushroom: © Rod Planck/Tom Stack & Assoc.

Chapter 29

Opener: © Runk/Schoenberger/Grant Heilman; **29.1 A-2:** © Robert Caughey/Visuals Unlimited; **29.1 B-2:** © Michael Wurz/Biozentrum, Univ. of Basel/Photo Researchers, Inc.; **29.1 C-2:** © Science Source/Photo Researchers, Inc.; **29.1 D-2:** © K.G. Murti/Visuals Unlimited; **29.2:** © Ed Degginger/Color Pic Inc.; **29.7:** © Dr. Tony Brain/SPL/Photo Researchers, Inc.; **29.8:** Courtesy Nitragin Company, Inc.; **29.9 A:** Spirillum: © Science VU/Charles W. Stratton/Visuals Unlimited; **29.9 B:** Bacillus: © David M. Phillips/Visuals Unlimited; **29.9 C:** Coccus: © David M. Phillips/Visuals Unlimited; **29.10 A:** © R. Knauft/Biology Media/Photo Researchers, Inc.; **29.10 B:** © Eric Grave/Photo Researchers, Inc.

Chapter 30

Opener: © Andrew Syred/Photo Researchers, Inc.; **30.2 B:** © M.I. Walker/Science Source/Photo Researchers, Inc.; **30.3 A:** © William E. Ferguson; **30.4:** © R. Knauft/Photo Researchers, Inc.; **30.6 A:** © Dr. Ann Smith/Photo Researchers, Inc.; **30.6 B-1:** © Biophoto Assoc./Photo Researchers, Inc.; **30.8:** © Walter Hodge/Peter Arnold, Inc.; **30.9 B:** © Manfred Kage/Peter Arnold, Inc.; **30.9 C:** © CABISCO/Visuals Unlimited; **30.10 A, B:** © Eric Grave/Photo Researchers, Inc.; **30.10 C:** © CABISCO/Phototake; **30.11 A:** © M. Abbey/Visuals Unlimited; **30.11 B:** © Ed Reschke/Peter Arnold, Inc.; **30.13 B:** © V. Duran/Visuals Unlimited; **30.13 C:** © CABISCO/Visuals Unlimited.

Chapter 31

Opener: © Ray Coleman/Photo Researchers, Inc.; **31.1 A:** © Gary R. Robinson/Visuals Unlimited; **31.1 C:** © Gary T. Cole/Biological Photo Service; **31.2 A:** From C.Y. Shih and R.G. Kessel, *LIVING IMAGES*, Science Books International, Boston, 1982; **31.2 B:** © Jeffrey Lepore/Photo Researchers, Inc.; **31.3:** © David M. Phillips/Visuals Unlimited; **31.4 A-2:** © Walter H. Hodge/Peter Arnold, Inc.; **31.4 B-1:** © James Richardson/Visuals Unlimited; **31.4 B-2:** © Michael Viard/Peter Arnold, Inc.; **31.4 C-2:** © Kingsley Stern; **31.5:** © J. Forsdyke/Gene Cox/SPL/Photo Researchers, Inc.; **31.6 A-2:** © Biophoto Assoc./Photo Researchers, Inc.; **31.6 B:** © Glenn Oliver/Visuals Unlimited; **31.6 C-1:** © M. Eichelberger/Visuals Unlimited; **31.6 D-1:** © Dick Poe/Visuals Unlimited; **31.6 D-2:** © L. West/Photo Researchers, Inc.; **31.7 A:** © Leonard L. Rue/Photo Researchers, Inc.; **31.7 B:** © Arthur M. Siegelman/Visuals Unlimited; **31 A:** © G. Tomsich/Photo Researchers, Inc.; **31 B:** © R. Calentine/Visuals Unlimited; **31.8 A:** Courtesy G.L. Barron/University of Guelph;

31.9 B: © Stephen Krasemann/Peter Arnold, Inc.; **31.9 C:** © John Shaw/Tom Stack & Assoc.; **31.9 D:** © Kerry T. Givens/Tom Stack & Assoc.; **31.10:** © R. Roncadori/Visuals Unlimited.

Chapter 32

Opener: © Jeff Lepore/Photo Researchers, Inc.; **32.3 A-2:** © Ed Reschke/Peter Arnold, Inc.; **32.3 B-1:** © J.M. Conrarder/Nat'l Audubon Society/Photo Researchers; **32.3 B-2:** © R. Calentine/Visuals Unlimited; **32.4 B:** © John Gerlach/Visuals Unlimited; **32.5:** © Doug Sokell/Tom Stack & Assoc.; **32.6 A:** © CABISCO/Phototake; **32.7 A:** © Steve Solum/Bruce Coleman, Inc.; **32.8 A:** © Robert P. Carr/Bruce Coleman, Inc.; **32.9 A:** © John GerlachVisuals Unlimited; **32.9 B:** © Walter H. Hodge/Peter Arnold, Inc.; **32.9 C:** © Forest W. Buchanan/Visuals Unlimited; **32.10 B:** © Matt Meadows/Peter Arnold, Inc.; **32.11 B:** © CABISCO/Phototake; **32.12 A:** © David Hosking/Photo Researchers, Inc.; **32.12 B:** © Kingsley Stern; **32.12 C:** © Edward Ross.

Chapter 33

Opener: © James Castner; **33.1 A:** © Bio Media Associates; **33.1 B:** © Joe McDonald/Visuals Unlimited; **33.4 A:** © J. Mcullagh/Visuals Unlimited; **33.4 B:** © Jeff Rotman; **33.5 B:** © CABISCO/Phototake; **33.5 C:** © Ron Taylor/Bruce Coleman; **33.5 D:** © Runk/Schoenberger/Grant Heilman Photography; **33.5 E:** © Gregory Ochocki/Photo Researchers, Inc.; **33.6 B:** © CABISCO/Visuals Unlimited; **33.7 A:** © Runk/Schoenberger/Grant Heilman Photography; **33.8:** © Stan Elems/Visuals Unlimited; **33.9 A:** © CABISCO/Phototake; **33.11 B:** Courtesy Fred Whittaker; **33.12 A:** © Arthur Siegelman/Visuals Unlimited; **33.12 C:** © James Solliday/Biological Photo Service; **33.13:** From E.K. Markell and M. Voge *Medical Parasitology*, 1992 W.B. Saunders Co.

Chapter 34

Opener: © J.C. Carton/Bruce Coleman, Inc.; **34.2 A:** © Fred Bavendam/Peter Arnold, Inc.; **34.2 B:** © Gordon Leedale/Biophoto Associates; **34.2 C:** © Douglas Faulkner/Photo Researchers, Inc.; **34.2 D:** © Marty Snyderman/Visuals Unlimited; **34.4 A:** © Michael DiSpezio; **34.5 A:** © William E. Ferguson; **34.6 A:** © W.H. Hughes/Visuals Unlimited; **34.6 B:** © Paul Averbach/Visuals Unlimited; **34.7 B:** © Roger K. Burnard/Biological Photo Service; **34 A:** © St. Bartholomews Hospital/SPL/Photo Researchers, Inc.; **34.9 A:** © Tom McHugh/Photo Researchers, Inc.; **34.9 B:** © Zig Leszczynski/Animals Animals/Earth Scenes; **34.9 C-2:** © E.R. Degginger/Bruce Coleman, Inc.; **34.11 A-1:** © Dwight Kuhn; **34.11 B:** © John MacGregor/Peter Arnold, Inc.; **34 C-A:** © Edward Ross; **34 C-B:** Raymond A. Mendez/Animals Animals/Earth Scenes; **34.12 A:** Dragonfly: © John Gerlach/Tony Stone Images; **34.12 B:** Beetle: © Kjell Sandved/Bruce Coleman, Inc.; **34.12 C:** Grasshopper: © Alex Kerstitch/Visuals Unlimited; **34.12 D:** Lacewing: © Glenn Oliver/Visuals Unlimited; **34.12 E:** Walking stick: © Art Wolfe/Tony Stone Images; **34.12 F:** Scale: © Science VU/Visuals Unlimited; **34.13 B:** © Bill Beatty/Visuals Unlimited; **34.13 C:** © L. West/Bruce Coleman, Inc.

Chapter 35

Opener: © Tom McHugh/Photo Researchers, Inc.; **35.1 A:** © Randy Morse/Tom Stack & Assoc.; **35.1 B:** © Alex Kerstitch/Visuals Unlimited; **35.1 C:** © John D. Cunningham/Visuals Unlimited; **35.2 C:** © D.P. Wilson/FLPA; **35.3:** © Rick Harbo; **35.4 B, 35.6:** © Heather Angel; **35.7 A:** © Ron & Valarie Taylor/Bruce Coleman, Inc.; **35.8:** © Estate of Dr. Jerome Metzner/Peter Arnold, Inc.; **35.11 A, B, C, D:** © Jane Burton/Bruce Coleman, Inc.; **35.12 A:** © Bruce Davidson/Animals Animals/Earth Scenes; **35 A:** © William Weber/Visuals Unlimited; **35.14 A:** © R.F. Ashley/Visuals Unlimited; **35.15 B-1, B-2:** © Daniel J. Cox/Tony Stone Images; **35.16 A:** © Tom McHugh/Photo Researchers, Inc.; **35.16 B:** © Tony Stone Images; **35.16 C:** © Leonard Lee Rue.

Chapter 36

Opener: © Bob Gossington/Bruce Coleman, Inc.; **36.1 B:** © Dwight Kuhn; **36.2 A:** © Michael Gadomski/Photo Researchers, Inc.; **36.2 B:** © Norman Owen Tomalin/Bruce Coleman, Inc.; **36.2 C:** © Brian Stablyk/Tony Stone Images; **36.4 A:** © John D. Cunningham/Visuals Unlimited; **36.4 B:** © Ed Reschke/Peter Arnold, Inc.; **36.4 C:** © Biophoto Assoc./Photo Researchers, Inc.; **36.5 A:** © J. Robert Waaland/Biological Photo Service; **36.5 B, C:** © Biophoto Assoc./Photo Researchers, Inc.; **36.6 B:** © J. Robert Waaland/Biological Photo Service; **36.7 B:** © George Wilder/Visuals Unlimited; **36.8 B:** © CABISCO/Phototake; **36.9:** © Dwight Kuhn; **36.10 A:** © John D. Cunningham/Visuals Unlimited; **36.10 B:** Courtesy of George Ellmore, Tufts University; **35.11 A:** © G.R. Roberts; **36.11 B:** © Ed Degginger/Color Pic; **36.11 C:** © David Newman/Visuals Unlimited; **36.12 A:** © J. Robert Waaland/Biological Photo Service; **36.13 A, 36.14 A:** © CABISCO/Phototake; **36.14 B, C:** © Runk/Schoenberger/Grant Heilman Photography; **36.15 B:** © Ardea, London; **36 B:** © James Schnepf Photography, Inc.; **36.18 B:** © Jeremy Burgess/SPL/Photo Researchers, Inc.; **36.20 A:** Cactus: © Patti Murray Animals Animals/Earth Scenes; **36.20 B:** Cucumber: © Michael Gadomski/Photo Researchers, Inc.; **36.20 C:** Flytrap: © CABISCO/Phototake .

Chapter 37

Opener: © Runk/Schoenberger/Grant Heilman Photography; **37.1 A, B, C:** Courtesy Mary E. Doohan; **37.3 B:** © Dwight Kuhn; **37.4 A:** From *Plant Physiology, 4/e* by Salisbury and Ross, fig. 7.5 B, p. 139, Wadsworth Publishing Co.; **37.4 B:** Courtesy Dr. John Menge; **37.6 A, B:** © Dwight Kuhn; **37.8:** © Ed Reschke/Peter Arnold, Inc.; **37.10 A, B,:** © Jeremy Burgess/SPL/Photo Researchers, Inc.; **Page 669:** Courtesy G. David Tilman.

Chapter 38

Opener: © James Castner; **38.1:** © Kim Taylor/Bruce Coleman, Inc.; **38.2 A:** © Kingsley Stern; **38.2 B:** Courtesy Malcom Wilkins, Botany Department, Glascow University; **38.2 C:** © Biophot; **38.3:** © John D. Cunningham/Visuals Unlimited; **38.4 A, B:** © John Kaprielian/Photo Researchers, Inc.; **38.5 A, B:** © Tom McHugh/Photo Researchers, Inc.; **38.6:** © Kingsley Stern; **38.9:** © Robert E. Lyons/Visuals Unlimited; **Page 683:** Courtesy Donald Brisking; **38 A:** R.J. Weaver; **38.11 A, B, C, D:** Kiem Tran Thanh Van; **38 B, C, D:** Courtesy Elliot Meywerowitz; **38.13:** © Runk/Schoenberger/Grant Heilman Photography; **38.16:** Courtesy Frank Salisbury.

Chapter 39

Opener: © Stephen Dalton/Photo Researchers, Inc.; **39.2:** © Michael Viard/Peter Arnold, Inc.; **39 A-A:** © Comstock **39 A-B** © H. Eisenbeiss/Photo Researchers, Inc.; **39 A-C:** © Anthony Mercieca/Photo Researchers, Inc.; **39 A-D:** Donna Howell; **39.4 A:** © Biophoto Assoc./Photo Researchers, Inc.; **39.4 B:** © Jeremy Burgess/SPL/ Photo Researchers, Inc.; **39.7 A-1:** © Kingsley Stern; **39.7 A-2:** © Joe Munroe/Photo Researchers, Inc.; **39.7 B-1:** © Ralph Reinhold/Animals Animals/Earth Scenes; **39.7 B-2:** © W.Ormerod/Visuals Unlimited; **39.7 C-1, C-2:** © Dwight Kuhn; **39.7 D-1:** © Christopher Lobban/Biological Photo Service; **39.7 D-2:** © John N.A. Lott/Biological Photo Service; **39 B A-1:** © Adam Hart-Davis/SPL/Photo Researchers, Inc.; **39 B A-2:** © Mark S. Skalny/Visuals Unlimited; **39 B B-1:** © Phillip Hayson/Photo Researchers, Inc.; **39 B B-2:** © Cleveland P. Hickman/Visuals Unlimited; **39 B B-3:** © Scott Camazine; **39 B B-4:** © John Tiszler; **39 C A-1, A-2:** © Heather Angel; **39 C B-1:** © Will & Deni McIntyre/Photo Researchers, Inc.; **39 C B-2:** © Stephen King/Peter Arnold, Inc.; **39 C C-1:** © Bob Daemmrich/Image Works; **39 C C-2:** © Dale Jackson/Visuals Unlimited; **39.10:** © G.I. Bernard/Animals Animals/Earth Scenes; **39.11 A, B, C, D:** © Runk/Schoenberger/Grant Heilman Photography; **39.12 B:** © Biophoto Assoc./Photo Researchers, Inc.; **39.12 C:** Courtesy Carlsberg Laboratory; **39.13:** Courtesy Keith V. Wood; **39.14 A, B, C:** Courtesy Monsanto.

Chapter 40

Opener: © Stuart Westmorland/Tony Stone Images; **40.2 B, C, D, E:** © Ed Reschke; **40.2 F:** © Ed Reschke/Peter Arnold, Inc.; **40.3 A, B, C, D, 40.5 A, B, C:** © Ed Reschke; **40 A:** © Steve Bourgeois/Unicorn Stock Photo.

Chapter41

Opener: © Galen Rowell/Mountain Light Photography; **41.1 A:** © Eric Grave/Photo Researchers, Inc.; **41.1 B:** © CABISCO/Phototake; **41.1 C:** © Michael DiSpezio; **41 A:** © The Bettmann Archive; **41.12 B:** © SPL/Photo Researchers, Inc.; **41.14 A-2, 41.15 A-1:** Courtesy Stuart I. Fox.

Chapter 42

Opener: © NIBSC/SPL/Photo Researchers, Inc.; **42.3 A, B, C, D, E:** © Ed Reschke/Peter Arnold, Inc.; **42.7 B:** Courtesy Dr. Arthur J. Olson, Scripps Institute; **42 A A:** © AP/Wide World Photo; **42.8 A:** © Bohringer Ingelheim International, photo by Lennart Nilsson; **42.10 A:** © Matt Meadows/Peter Arnold, Inc.; **42.11:** © Chris Harvey/Tony Stone Images.

Chapter 43

Opener: © Stephen Dalton/Photo Researchers, Inc.; **43.8 B:** © Ed Reschke/ Peter Arnold, Inc.; **43.8 C:** © St. Bartholomew's Hospital/SPL/Photo Researchers, Inc.; **43.9 C:** © Manfred Kage/ Peter Arnold, Inc.; **43.9 D:** Photo by Susumu Ito, from Charles Flickinger *Medical Cellular Biology* W.B. Saunders, 1979.

Chapter 44

Opener: © Bates Littlehales/Animals Animals/Earth Scenes; **44.5:** © Paul R. Ehrlich/Biological Photo Service; **44.11:** © CNRI/SPL/Photo Researchers, Inc.; **45.8:** © J. Gennaro/Photo Researchers, Inc.

Chapter 46

Opener: © CNRI/SPL/Photo Researchers, Inc.; **46.2:** © Linda Bartlett; **46.9 A:** © Peter Miller/Photo Researchers, Inc.; **46.9 B:** © Lori Adamski Peek/Tony Stone Images; **46.13:** © Dan McCoy/Rainbow.

Chapter 47

Opener: © John Stern/Animals Animals/ Earth Scenes; **47.3:** © Kathy Husemann; **47.5 A, B:** © Heather Angel; **47.9:** © Lennart Nilsson, from *The Incredible Machine;* **47 A:** Robert S. Preston, courtesy Prof. J.E. Hawkins, Kresge Hearing Research Institute, Univ. of Michigan Medical School; **47.14 A:** © P. Motta/SPL/Photo Researchers, Inc.

Chapter 48

Opener: © James Castner; **48.2:** © Michael Fogden/OSF/Animals Animals/Earth Scenes; **48.4 A, B:** © Ed Reschke; **48 A:** © Royce Bair/Unicorn Stock Photo; **48.12 B:** Courtesy Hugh E. Huxley; **48.13 A:** © Victor B. Eichler.

Chapter 49

Opener: © David Frits/Animals Animals/ Earth Scenes; **49.10:** © NMSB/Custom Medical Stock.

Chapter 50

Opener: © John H. Hoffman/Bruce Coleman, Inc.; **50.1:** © Runk/Schoenberger/ Grant Heilman Photography; **50.3:** © Wyman Meinzer/Peter Arnold, Inc.; **50 A B:** Courtesy Mike Mckeen/The Roslin Institute; **50.6 B:** © Biophoto Assoc./Photo Researchers, Inc.; **50.9 B:** © Ed Reschke/ Peter Arnold, Inc.; **50 B:** © G.W. Willis/BPS/ Tony Stone Images.

Chapter 51

Opener A, B, C, D: © Dwight Kuhn; **51.1:** © David M. Phillips/Visuals Unlimited; **51 A:** Courtesy John Sternfield; **51.8 A, B:** Courtesy E.B. Lewis; **51.12 A:** © Lennart Nilsson *A Child Is Born,* Dell Publishing.

Illustrators

Observatory Group Inc.: 4.4a, 4.5a
Wilderness Graphics: 24A-b
Precision Graphics: 3.8, 3.14, 3.16, 3.19, 4.6, 4.7, 4.10, 4.11b, 4.13, TA4.2, 5.4, 5.6, 5.8a-f, 5.12a, 5.13b, 5.14, TA5.3, 6.11, 7.3, 7.4, 7.7, 8.2, 8.3(parts 1&2), 8.4, 8.5, 9.5a, 9.8, 12.4, 12.26, TA12.5, TA12.6, 13.6, 14.2, 14.6c, 14A, 15.9, 15.10a-c, 15.11, 15.12, 16.1, 16.2, TA16.1, TA16.2, 17.3, 17.4, 19.3, 19.7, 22.1, 23.1, 23.2, 23.4, 23.5, 23.7, 23.9a, 23.11b, 23.11c, 23.14, 23.16, 23A, 23B, TA23.1, 24.2, 24.3, 24.4, 24.5, 24.6, 24.9b, 24.19, 24.20b, 24.21b, TA24.1, TA24.2, 25.1, 25.3, 25.5, 25.6, 25.7, 25.9, 25.10, 25.11, 25.13, TA25.1, TA25.2, TA25.4, 26.1a, 26.2b,c, 26.3, 26.4, 26.5, 26.6a, 26.7, 26.8a, 26.8b, 26.18a, 26.27b, 27.1a, 27.1b, 27.2, 27.3a, 27.3b, 27.3c, 27.7, 27.10, TA27.1, 28A, 30.4b, 30.7, 31.3, 33.3, 33A, 34.3, 34.5, 34.9, 34.10a,b, 39.2, 39.3a, 46.4, 50.5, 50.7, 50A, 50.10, 51.3, 51.4, 51.9

Revisions prepared by Precision Graphics, Theis Graphics, and Wilderness Graphics.

Index

Note: Page numbers followed by f refer to figures; page numbers followed by t refer to tables.

A

C

D

E

F

G

I

J

K

L

M

N

O

P

Q

R

S

T

X

Y

Z